Percy Alexander MacMahon: Collected Papers

Volume 1

Mathematicians of our Time

Gian-Carlo Rota, series editor

Paul Erdös: The Art of Counting
edited by Joel Spencer

Einar Hille: Classical Analysis and Functional Analysis
Selected Papers of Einar Hille
edited by Robert R. Kallman

Charles Loewner: Theory of Continuous Groups
notes by Harley Flanders and Murray H. Protter

Percy Alexander MacMahon: Collected Papers
Volume I
Combinatorics
edited by George E. Andrews

George Pólya: Collected Papers
Volume I
Singularities of Analytic Functions
edited by R. P. Boas

George Pólya: Collected Papers
Volume II
Location of Zeros
edited by R. P. Boas

Collected Papers of Hans Rademacher
Volume I
edited by Emil Grosswald

Collected Papers of Hans Rademacher
Volume II
edited by Emil Grosswald

Stanislaw Ulam: Selected Works
Volume I
Sets, Numbers, and Universes
edited by W. A. Bayer, J. Mycielski, and G.-C. Rota

Norbert Wiener: Collected Works
Volume I
Mathematical Philosophy and Foundations; Potential Theory; Brownian Movement, Wiener Integrals, Ergodic and Chaos Theories; Turbulence and Statistical Mechanics
edited by P. Masani

Oscar Zariski: Collected Papers
Volume I
Foundations of Algebraic Geometry and Resolution of Singularities
edited by H. Hironaka and D. Mumford

Oscar Zariski: Collected Papers
Volume II
Holomorphic Functions and Linear Systems
edited by M. Artin and D. Mumford

Oscar Zariski: Collected Papers
Volume III
Topology of Curves and Surfaces, and Special Topics in the Theory of Algebraic Varieties
edited by M. Artin and B. Mazur

Percy Alexander MacMahon
Collected Papers

Volume I

Combinatorics

Edited by
George E. Andrews

Introduction by
Gian-Carlo Rota

The MIT Press
Cambridge, Massachusetts, and London, England

This book was set in Monophoto Baskerville by Asco Trade Typesetting Limited, Hong Kong, printed on R & E Book by Murray Printing Company and bound in Crown Linen in the United States of America

Library of Congress Cataloging in Publication Data

MacMahon, Percy Alexander, 1854–1929.
Percy Alexander MacMahon: collected papers.

(Mathematicians of our time)
Bibliography: p.
CONTENTS: v. 1. Combinatorics.
1. Mathematics—Collected works. I. Andrews, George E., 1938– II. Title.
III. Series.
QA3.M27 1978 510′.8 77-28962
ISBN 0-262-13121-8

Percy Alexander MacMahon

Contents

(Bracketed numbers refer to the order of listing in the bibliography.)

Chapter 5 Compositions and Simon Newcomb's problem

Chapter 6 Perfect Partitions

Chapter 7 Distributions Upon a Chess Board and Latin Squares

Chapter 8 Multipartite Numbers

Chapter 9 Partitions

Chapter 10 Partition Analysis

Chapter 11 Plane Partitions (Part 1)

Introduction

The history of mathematics, like all history generally, is replete with injustice. All publics, even the intellectual public of mathematicians, exhibit toward the past the forgetful, oversimplifying, hero-worshiping attitude that we have come to identify with mass behavior. Great advances in science are pinned on a few extraordinary white-maned individuals who alone, by the magic power of genius denied to ordinary mortals (thus safely getting us off the hook), are made responsible for Progress.

The public abhors detail and avidly adopts Bergson's dictum that every man is responsible for at most one idea. To reveal that behind every great man one can find a beehive of lesser known individuals who paved his way and obtained most of the results for which he is known is to commit a crime of lèse majesté. Whoever dares associate Apollonius with Euclid, Cavalieri with Leibniz, Saccheri with Lobachevskii, Student with Hilbert, MacMahon with Ramanujan, should stand ready for the scornful reaction of a disappointed majority.

One consequence of this sociological law is that, whenever a forgotten branch of mathematics comes back into use after a period of neglect, only the main outlines of the theory are remembered, those results you would find in the works of the Great Man. The bulk of the theory is likely to be rediscovered from scratch by innocent young mathematicians who have realized that their future careers depend on publishing research papers rather than on rummaging through dusty old journals. The result is wasted effort, unnecessary duplication, and eventually embarrassing rediscovery.

In all mathematics, it would be hard to find a more blatant instance of this regrettable state of affairs than the theory of symmetric functions. As G. Higman has written, each generation rediscovers them and presents them in the latest jargon as the dernier cri. Today it is K-theory, yesterday it was categories and functors, and the day before, group representations. Behind these and several other attractive theories stands one immutable source: the ordinary, crude definition of the symmetric functions and the identities they satisfy.

Thus anyone who pretends familiarity with symmetric functions in any guise cannot avoid wading through the papers of Major Percy A. MacMahon. With his moustache, his "British Empah" demeanor, and worst of all his military background, MacMahon was hardly the type to be chosen by Central Casting for the role of the Great Mathematician. Ramanujan, with his

Eastern aura, his frail physique, and his swarthy good looks qualified all the way. It would have been fascinating to be present at one of the battles of arithmetical wits at Trinity College, when MacMahon would regularly trounce Ramanujan by the display of superior ability for fast mental computation (as reported by D. C. Spencer, who heard it from G. H. Hardy). The written accounts of the lives of the characters, however, omit any mention of this episode, since it clashes against our prejudices.

The publication of the first volume of MacMahon's collected papers will accelerate our understanding of the catholic presence of symmetric functions in mathematics. Most of the material is not available elsewhere, and we hope that with its publication in one volume rediscovery will be, if not prevented, at least slowed down. But most of all, it is a source of satisfaction to know that historical justice has at last been done.

Gian-Carlo Rota

Preface

There are several compelling reasons for publishing the Collected Papers of Percy Alexander MacMahon (1854–1929):

First, MacMahon's researches in combinatorics were ahead of his time. In studying the literature of MacMahon's day, we find that while MacMahon was a prolific writer (as these Collected Papers confirm), his discoveries generated little work by others on combinatorics. Within the past twenty years, however, combinatorics has undergone a remarkable renaissance, and a random check through the Science Citation Index indicates clearly that MacMahon's work is no longer neglected.

Second, well over twenty-five per cent of MacMahon's papers appeared *after* the publication of his historic two volume work, *Combinatory Analysis* (Cambridge University Press 1915, 1916, Reprinted: Chelsea, New York, 1960). The later work includes his extensive researches on determinants, his book and papers on repeating patterns, and numerous contributions to combinatorics, notably his enumeration of the partitions of a multipartite number. Furthermore, less than twenty per cent of all his papers are referred to in *Combinatory Analysis*; the impact of *Combinatory Analysis* alone on the contemporary scientific community suggests the importance of publishing all MacMahon's papers.

Third, some of the expository papers by MacMahon (see chapter 19) are still a joy to read: for example, his Presidential Address to the London Mathematical Society, in which he gently chides J. J. Sylvester for not publishing his solution to "The Problem of the Virgins," and his Presidential Address before the British Association, which he concludes with a rousing description of the excitement he felt in the then emerging science of combinatorics.

In addition to the 121 papers of MacMahon that comprise the bulk of this project, we have included two obituaries (chapter 20, volume 2). The first, by H. F. Baker, describes the life and work of MacMahon from the viewpoint of a mathematician; the second, by H. H. Turner, was presented before the Royal Astronomical Society of which MacMahon was president in 1917–1918.

The following are some of the ways in which our presentation differs from the standard collected works:

INTRODUCTIONS. Substantial introductory material precedes most chapters. Since these Collected Papers are appearing long after MacMahon's death, the introductions serve both to place his work in proper historical context, and to relate it to contemporary mathematical developments. In each of five chapters (chapters 2, 7, 13, 14, and 18) the introduction includes an entire paper by another mathematician:

Chapter 2. P. Hall (1959) The algebra of partitions. Proc. 4th Canadian Math. Congr., Banff, 1957, 147–159.

Chapter 7. J. H. Redfield (1927) The theory of group reduced distributions, *American J. Math.*, *49*, 433–455.

Chapter 13. L. Solomon (1977) Partition identities and invariants of finite groups, *J. Combinatorial Th.*, *A-23*, 148–175.

Chapter 14. D. E. Littlewood and A. R. Richardson (1934) Group characters and algebra, *Phil. Trans.*, *A-233*, 99–141.

Chapter 18. A letter from A. C. Aitken to an Edinburgh colleague (1968) Proc. Edinburgh Math. Soc. (2), *16*, 166–172.

Each of the foregoing is quite readable by anyone with some mathematical maturity, and although each paper is significant, none has been reprinted in any recent collection. Furthermore, each of these papers provides a firm link between MacMahon's researches and modern combinatorics. I must express a special debt of gratitude to P. Hall, D. E. Littlewood, and L. Solomon for allowing me to include their papers.

CHAPTER ORGANIZATION. We have tried to make the Collected Papers parallel MacMahon's *Combinatory Analysis* and the following table shows their correspondence.

Collected Works Chapter	corresponds to	Combinatory Analysis Section
Chapter 1		Section I
Chapter 2		Section II
Chapter 3		Section III
Chapter 4		Section III
Chapter 5		Section IV
Chapter 6		Section V
Chapter 7		Section V
Chapter 8		Section VI

Chapter 9	corresponds to	Section VII
Chapter 10		Section VIII
Chapter 11		Section IX
Chapter 12		Section X

The first chapter of volume 2 (chapter 13) will correspond to section XI of *Combinatory Analysis*. There will be seven other chapters in volume 2 (chapters 14–20), which will treat determinants, repeating patterns, multiplicative number theory, miscellaneous topics, invariant theory, survey articles, and obituaries.

A word should be said about the problem of classifying the papers. Many times MacMahon would treat, simultaneously, problems in several areas such as compositions, partitions, permutations, multipartite numbers, etc. As a result a number of subjective classification judgments had to be made. However, MacMahon's guidance was followed whenever possible; each paper cited by MacMahon as a principal source of a particular section of *Combinatory Analysis* is placed in the corresponding chapter of these Collected Papers.

SUMMARIES. Every chapter contains a section with summaries of each MacMahon paper appearing therein. MacMahon seldom summarized a paper, nor did he often distinguish his main results with the title of "theorem." Thus it is difficult to "browse" in either *Combinatory Analysis* or in MacMahon's papers. It is hoped that the summary sections will make both the Collected Papers and *Combinatory Analysis* more readable.

I close this preface with the agreeable task of thanking those people (and institutions) who have substantially aided me in this undertaking: M. Abramson, M. Aissen, J. Arkin, R. Askey, D. E. Barton, E. A. Bender, J. H. Bennett, S. Blaha, N. G. deBruijn, L. Carlitz, P. Cartier, M. S. Cheema, E. Cobin, E. C. Cunningham, L. Comtet, J. F. Dillon, P. Doubilet, M. Farrell, D. C. Fielder, N. J. Fine, F. Fink, D. Foata, H. O. Foulkes, F. Freudenstein, M. Gardner, B. J. Gassner, J. Goldman, S. W. Golomb, I. J. Good, B. Gordon, H. W. Gould, R. C. Grimson, E. Grosswald, H. Gupta, P. Hall, M. Ismail, M. G. Kendall, D. Knuth, G. Kreweras, A. Lascoux, J. Levine, D. E. Littlewood, E. K. Lloyd, the London Mathematical Society, C. T. Long, A. O. Morris, W. O. J. Moser, R. E. Nather, H. L. Nelson, P. Nivat, T. H. O'Beirne, A. Oppenheim, T. D. Parsons, J. K. Percus, W. Philpott, J. Riordan, G. deB. Robinson, D. P. Roselle, the Royal Astronomical Society, the Royal Institution, the Royal Society, the Royal Society of Arts, St. John's College, Cambridge, M.-P. Schützenberger, D. Shanks, N. J. A. Sloane, D. A. Smith,

L. Solomon, B. Sorin, R. P. Stanley, P. R. Stein, J. R. N. Stone, M. V. Subbarao, P. V. Sukhatme, P. Thionet, G. Thomas, A. Unwin, P. Whittle, H. Wilf, S. G. Williamson, E. M. Wright, and F. Yates.

I would like to specially thank Gian-Carlo Rota for suggesting the inclusion of MacMahon's works in the series "Mathematicians of Our Time" and for inviting me to be the editor.

Finally I thank my wife, Joy, who has been an inspiration to me throughout this project, and who has played an important part in making the introduction and commentary in each chapter readable.

George E. Andrews
The Pennsylvania State University

Bibliography of the Mathematical Papers of Percy Alexander MacMahon

The following represents a complete chronological list of MacMahon's mathematical papers. We explain the identifying numbers with an example: [45:3] is the 45th paper in the chronological list and appears in chapter 3. For the most part we have not included in this list abstracts, announcements, translations, or papers unrelated to mathematics.

1881

[1:17] A property of pedal curves, *Messenger of Math.*, *10* (1881), 190–191.

[2:17] Sur un resultat de calcul obtenu par M. Allegret, *C. R. Acad. Sc. Paris*, *95* (1881), 831–832.

1882

[3:17] The cassinian, *Messenger of Math.*, *12* (1882), 118–120.

[4:17] An extension of Steiner's problem, *Messenger of Math.*, *12* (1882), 138–141.

[5:17] The three-cusped hypocycloid, *Messenger of Math.*, *12* (1882), 151–153.

1883

[6:17] A generalization of the nine-point properties of a triangle, *Proc. London Math. Soc.(1)*, *14* (1883), 129–132.

[7:1] Note on an algebraical identity, *Messenger of Math.*, *13* (1883), 142–144.

[8:18] On Professor Cayley's canonical form, *Quart. J. Math.*, *19* (1883), 337–341.

[9:17] On the differential equation $X^{-2/3}\,dx + Y^{-2/3}\,dy + Z^{-2/3}\,dz = 0$, *Quart. J. Math.*, *19* (1883), 158–162.

1884

[10:2] Algebraic identities arising out of an extension of Waring's formula, *Messenger of Math.*, *14* (1884), 8–11.

[11:1] On symmetric functions and in particular on certain inverse operators in connection therewith, *Proc. London Math. Soc.*, *15* (1884), 20–47.

[12:1] On the development of an algebraic fraction, *American J. Math.*, *6* (1884), 287–288.

[13:18] Seminvariants and symmetric functions, *American J. Math.*, *6* (1884), 131–163.

[14:18] Symmetric functions of the 13-ic, *American J. Math.*, *6* (1884), 289–301.

1885

[15:2] The multiplication of symmetric functions, *Messenger of Math.*, *14* (1885), 164–167.

[16:1] A new theorem in symmetric functions, *Quart. J. Math.*, *20* (1885), 365–369.

[17:1] Note on rationalisation, *Messenger of Math.*, *15* (1885), 65–67.

[18:18] On perpetuants, *American J. Math.*, *7* (1885), 26–46, 259–263.

[19:18] Operators in the theory of seminvariants, *Quart. J. Math.*, *20* (1885), 362–365.

1886

[20:6] Certain special partitions of numbers, *Quart. J. Math.*, *21* (1886), 367–373.

[21:18] Memoir on seminvariants, *American J. Math.*, *8* (1886), 1–18.

[22:18] On perpetuant reciprocants, *Proc. London Math. Soc.*, *17* (1886), 139–151.

1887

[23:18] The theory of a multilinear partial differential operator with applications to the theories of invariants and reciprocants, *Proc. London Math. Soc.*, *18* (1887), 61–88.

[24:18] The differential equation of the most general substitution of one variable, *Phil. Mag.(5)*, *23* (1887), 542–543.

[25:1] The law of symmetry and other theorems in symmetric functions, *Quart. J. Math.*, *22* (1887), 74–81.

[26:18] Observations on the generating functions of the theory of invariants, *American J. Math.*, *9* (1887), 189–192.

1888

[27:18] The expression of syzygies among perpetuants by means of partitions, *American J. Math.*, *10* (1888), 149–168.

[28:2] Properties of a complete table of symmetric functions, *American J. Math.*, *10* (1888), 42–46.

1889

[29:18] The algebra of multilinear partial differential operators, *Proc. London Math. Soc.*, *19* (1889), 112–128.

[30:18] The eliminant of two binary quantics, *Quart. J. Math.*, *23* (1889), 139–143.

[31:2] Memoir on a new theory of symmetric functions, *American J. Math.*, *11* (1889), 1–36.

[32:2] Symmetric functions and the theory of distributions, *Proc. London Math. Soc.*, *19* (1889), 220–256.

[33:10] On play "à outrance," *Proc. London Math. Soc.*, *20* (1889), 195–198.

1890

[34:13] Memoir on symmetric functions of the roots of systems of equations, *Phil. Trans.*, *181* (1890), 481–536.

[35:2] Second memoir on a new theory of symmetric functions, *American J. Math.*, *12* (1890), 61–102.

[36:18] A theorem in the calculus of linear partial differential operators, *Quart. J. Math.*, *24* (1890), 246–250.

[37:19] Weighing by a series of weights, *Nature*, *43* (1890), 113–114.

1891

[38:6] The theory of perfect partitions and the compositions of multipartite numbers, *Messenger of Math.*, *20* (1891), 103–119.

[39:2] Third memoir on a new theory of symmetric functions, *American J. Math.*, *13* (1891), 193–234.

[40:5] Yoke chains and "trees," *Proc. London Math. Soc.*, *22* (1891), 330–346.

1892

[41:4] Applications of a theory of permutations in circular procession to the theory of numbers, *Proc. London Math. Soc.*, *23* (1892), 305–313.

[42:5] The combinations of resistances, *The Electrician*, *28* (1892), 601–602.

[43:2] Fourth memoir on a new theory of symmetric functions, *American J. Math.*, *14* (1892), 15–38.

1893

[44:15] On the thirty cubes constructed with six colored squares, *Proc. London Math. Soc.*, *24* (1893), 145–155.

1894

[45:3] A certain class of generating functions in the theory of numbers, *Phil. Trans.*, *185* (1894), 111–160.

[46:5] Memoir on the theory of the compositions of numbers, *Phil. Trans.*, *184* (1894), 835–901.

1895

[47:18] The perpetuant invariants of binary quantics, *Proc. London Math. Soc.*, *26* (1895), 262–284.

[48:4] Self-conjugate permutations, *Messenger of Math.*, *24* (1895), 69–76.

1897

[49:19] Combinatory analysis: A review of the present state of knowledge (Presidential address), *Proc. London Math. Soc.*, *28* (1897), 5–32.

[50:19] James Joseph Sylvester, Nature, *59* (1897), 259–261; *Proc. Royal Soc.*, *63* (1898), ix–xxv.

[51:9] Memoir on the theory of the partition of numbers—Part I, *Phil. Trans.*, *187* (1897), 619–673.

1898

[52:7] A new method in combinatory analysis, Latin squares, *Trans. Cambridge Phil. Soc.*, *16* (1898), 262–290.

[53:4] Solution du problème de partition d'où résulte le dénombrement des genres distincts d'abaque relatifs aux équations à n variables, *Bull. Soc. Math. France, 26* (1898), 57–64.

1899

[54:10] Memoir on the theory of the partitions of numbers—Part II, *Phil. Trans.*, *192* (1899), 351–401.

[55:17] Mirage, *Nature*, *59* (1899), 259–261.

[56:12] Partitions of numbers whose graphs possess symmetry, *Trans. Cambridge Phil. Soc.*, *17* (1899), 149–170.

1900

[57:7] Combinatorial analysis. The foundations of a new theory, *Phil. Trans.*, *194* (1900), 361–386.

[58:10] Partition analysis and any system of consecutive integers, *Trans. Cambridge Phil. Soc.*, *18* (1900), 12–34.

1901

[59:19] Presidential address, Section A, British Association, Glasgow, British Assoc. Report (1901), 519–528.

1902

[60:19] Magic squares and other problems upon a chess board, *Nature*, *65* (1902), 447–452.

[61:3] The sums of powers of binomial coefficients, *Quart. J. Math.*, *33* (1902), 274–288.

1904

[62:10] The Diophantine inequality $\lambda x \geqq \mu y$, *Trans. Cambridge Phil. Soc.*, *19* (1904), 111–131.

[63:18] On the application of quaternions to the orthogonal transformation and invariant theory, *Proc. London Math. Soc.(2)*, *1* (1904), 210–229.

[64:18] Seminvariants of systems of binary quantics, *Trans. Cambridge Phil. Soc.*, *19* (1904), 234–248.

1905

[65:10] Note on the Diophantine inequality $\lambda x \geqq \mu y$, *Quart. J. Math.*, *36* (1905), 80–93.

[66:10] On a deficient multinomial expansion, *Proc. London Math. Soc.(2)*, *2* (1905), 478–485.

1906

[67:10] Memoir on the theory of the partitions of numbers—Part III, *Phil. Trans.*, *205* (1906), 37–58.

1907

[68:10] The Diophantine equation $x^n - Ny^n = z$, *Proc. London Math. Soc.(2)*, *5* (1907), 45–58.

1908

[69:18] Memoir on the orthogonal and other special systems of invariants, *Trans. Cambridge Phil. Soc.*, *20* (1908), 142–164.

[70:18] Preliminary note on the operational invariants of a binary quantic, *Proc. London Royal Soc.*, *A-80* (1908), 151–161.

[71:5] Second memoir on the composition of numbers, *Phil. Trans.*, *207* (1908), 65–134.

1909

[72:17] The determination of the apparent diameter of a fixed star, *London Astron. Soc. Monthly Notices*, *69* (1909), 126–127.

[73:10] Memoir on the theory of the partitions of numbers—Part IV, *Phil. Trans.*, *209* (1909), 153–175.

1910

[74:19] Algebraic forms, Encyclopedia Britannica, Edition 11, vol. 1, 620–641.

[75:19] Arthur Cayley, Encyclopedia Britannica, Edition 11, vol. 5, 589–590.

[76:19] Combinatory analysis, Encyclopedia Britannica, Edition 11, vol.6, 752–758.

1912

[77:11] Memoir on the theory of the partitions of numbers—Part V, *Phil. Trans.*, *211* (1912), 75–110.

[78:12] Memoir on the theory of the partitions of numbers—Part VI, *Phil. Trans.*, *211* (1912), 345–373.

[79:8] On compound denumeration, *Trans. Cambridge Phil. Soc.*, *22* (1912), 1–13.

[80:18] The operator reciprocants of Sylvester's theory of reciprocants, *Trans. Cambridge Phil. Soc.*, *21* (1912), 143–170.

[81:4] The problem of derangements in the theory of permutations, *Trans. Cambridge Phil. Soc.*, *21* (1912), 467–481.

1913

[82:4] The indices of permutations and the derivation therefrom of functions of a single variable associated with the permutations of any assemblage of objects, *American J. Math.*, *35* (1913), 281–322.

1914

[83:4] The superior and inferior indices of permutations, *Trans. Cambridge Phil. Soc.*, *29* (1914), 55–60.

1915

[84:18] The invariants of the Halphenian homographic substitution, *Trans. Cambridge Phil. Soc.*, *22* (1915), 101–131.

[85:18] On a modified form of pure reciprocants possessing the property that the algebraical sum of the coefficients is zero, *Proc. London Math. Soc. (2)*, *14* (1915), 67–70.

1916

[86:4] Two applications of general theorems in combinatory analysis: (1) to the theory of inversions of permutations; (2) to the ascertainment of the numbers of terms in the development of a determinant which has amongst its elements an arbitrary number of zeros, *Proc. London Math. Soc.(2)*, *15* (1916), 314–321.

1917

[87:8] Memoir on the theory of the partitions of numbers—Part VII, *Phil. Trans.*, *217* (1917), 81–113.

[88:2] Small contribution to combinatory analysis, *Proc. London Math. Soc.(2)*, *16* (1917), 352–354.

1918

[89:7] Combinations derived from m identical sets of n different letters, *Proc. London Math. Soc.(2)*, *17* (1918), 25–41.

[90:17] (with H. B. C. Darling) Reciprocal relations in the theory of integral equations, *Proc. Cambridge Phil. Soc.*, *19* (1918), 178–184.

1919

[91:17] (with H. B. C. Darling) Contribution to the theory of attraction when the force varies as any power of the distance, *Proc. London Royal Soc.*, *A-95* (1919), 390–399.

1920

[92:16] The divisors of numbers, *Proc. London Math. Soc.(2)*, *19* (1920), 305–340.

[93:17] Divisors of numbers and their continuations in the theory of partitions, *Proc. London Math. Soc.(2)*, *19* (1920), 75–113.

[94:9] On partitions into unequal and into uneven parts, *Quart. J. Math.*, *49* (1920), 40–45.

[95:4] Permutations, lattice permutations, and the hypergeometric series, *Proc. London Math. Soc.(2)*, *19* (1920), 216–227.

1921

[96:15] New Mathematical Pastimes, Cambridge University Press, Cambridge, 1921.

[97:9] Note on the parity of the number which enumerates the partitions of a number, *Proc. Cambridge Phil. Soc.*, *20* (1921), 281–283.

1922

[98:15] (with W. P. D. MacMahon) The design of repeating patterns, *Proc. London Royal Soc.*, *A-101* (1922), 80–94.

[99:15] The design of repeating patterns for decorative work, *J. Royal Soc. Arts*, *70* (1922), 567–582.

[100:15] Pythagoras's theorem as a repeating pattern, *Nature*, *109* (1922), 479.

1923

[101:2] The algebra of symmetric functions, *Proc. Cambridge Phil. Soc.*, *21* (1923), 376–390.

[102:2] An American tournament treated by the calculus of symmetric functions, *Quart. J. Math.*, *49* (1923), 1–36.

[103:2] Chess tournaments and the like treated by the calculus of symmetric functions, *Quart. J. Math.*, *49* (1923), 353–384.

[104:16] Congruences with respect to composite moduli, *Trans. Cambridge Phil. Soc.*, *22* (1923), 413–424.

[105:11] The connexion between the sum of the squares of the divisors and the number of partitions of a given number, *Messenger of Math.*, *52* (1923), 113–116.

[106:17] On a class of transcendants of which the Bessel functions are a particular case. *Proc. London Royal Soc.*, *A-104* (1923), 39–47.

[107:6] The partitions of infinity with some arithmetic and algebraic consequences, *Proc. Cambridge Phil. Soc.*, *21* (1923), 642–650.

[108:4] Prime lattice permutations, *Proc. Cambridge Phil. Soc.*, *21* (1923), 193–196.

[109:6] The prime numbers of measurement on a scale, *Proc. Cambridge Phil. Soc.*, *21* (1923), 651–654.

[110:9] The theory of modular partitions, *Proc. Cambridge Phil. Soc.*, *21* (1923), 197–204.

1924

[111:8] Dirichlet series and the theory of partitions, *Proc. London Math. Soc.(2)*, *22* (1924), 404–411.

[112:16] Properties of prime numbers deduced from the calculus of symmetric functions, *Proc. London Math. Soc.(2)*, *23* (1924), 290–316.

[113:14] Researches in the theory of determinants, *Trans. Cambridge Phil. Soc., 23* (1924), 89–135.

1925

[114:8] The enumeration of the partitions of multipartite numbers, *Proc. Cambridge Phil. Soc., 22* (1925), 951–963.

[115:14] On an X-determinant which includes as particular cases both "determinants" and "permanents," *Proc. Edinburgh Royal Soc., 44* (1925), 21–22.

[116:14] The symmetric functions of which the general determinant is a particular case, *Proc. Cambridge Phil. Soc., 22* (1925), 633–654.

1926

[117:8] Euler's ϕ-function and its connection with multipartite numbers, *Proc. London Math. Soc.(2), 25* (1926), 469–483.

[118:9] The parity of $p(n)$, the number of partitions of n, when $n \leqq 1000$, *J. London Math. Soc., 1* (1926), 225–226.

1927

[119:9] The elliptic products of Jacobi and the theory of linear congruences, *Proc. Cambridge Phil. Soc., 23* (1927), 337–355.

[120:14] The structure of a determinant (Rouse Ball Lecture) *J. London Math. Soc., 2* (1927), 273–286.

1928

[121:14] The expansion of determinants and permanents in terms of symmetric functions, Proc. International Congress (Toronto) 1928, 319–330.

In addition to the above works, MacMahon also published three books that are still in print:

Combinatory Analysis, vols. I and II, Cambridge University Press, Cambridge, 1915, 1916, Reprinted: Chelsea, New York, 1960.

An Introduction to Combinatory Analysis, Cambridge University Press, Cambridge, 1920, Reprinted: in Famous Problems and Other Essays, Chelsea, New York, 1955.

MacMahon also published a number of articles in the Minutes of the Proceedings of the Royal Artillery Institution. For the most part they concern mathematical analysis of some military problem, but we have not included any of them in these Collected Papers since their mathematical interest is secondary; however, we give a complete list.

1881 Trajectories of the 9-PR., 13-PR., and 16-PR. M.L.R. Guns, *11*, 421–426.

1884 Tables calculated by MacMahon for A. G. Greenhill's article, "Motion of a particle in a resisting medium," *12*, 26–31.

Note on Bashforth's method of working trajectories, *12*, 64.

On the motion of a projectile, *12*, 231–236.

On the motion of a projectile in a medium resisting as the cube of the velocity, *12*, 289–300.

1885 Practical rule for range finding, *13*, 194.

1894 Note on the correction of artillery fire, *21*, 55–60.

1895 Terrestrial refraction and mirage, *22*, 269–285 (discussion 285–288).

1897 Discussion of Capt. E. W. Wills's lecture "The total solar eclipse of January 1898," *24*, 257–258.

Discussion of Capt. A. G. Shoutt's article "The drift of service projectiles," *24*, 474.

Percy Alexander MacMahon: Collected Papers

Volume 1

Chapter 1
Symmetric Functions (Part 1)

1.1 Introduction and Commentary

MacMahon's work on symmetric functions was extensive—the great majority of his papers relate to them in some way—and his basic tools in studying them were the Hammond operators. Probably [32] best presents his development of the foundations while sections I and II of his *Combinatory Analysis* (volume 1) and his short book *An Introduction to Combinatory Analysis* present introductions to his work in this area prior to 1915.

Of the papers and books listed in section 1.3 we call special attention to the following:

Aissen (1970) extends the study of Ferrers graphs of partitions and applies his results to give a combinatorial interpretation of the standard proof of the Fundamental Theorem on Symmetric Functions (see the corollary to Theorem 2 in section 1.2 below).

The applications of symmetric functions are perhaps best introduced by the books of David and Barton (1962) and David, Barton, and Kendall (1966). The former devotes chapter 17 to "Symmetric Functions and Polykays," and discusses the relationship of symmetric functions to the statistical work of R. A. Fisher. The latter presents the most extensive and useful tables of symmetric functions in existence in addition to an examination, in the first 64 pages, of the various applications of symmetric functions that may be facilitated through their use.

Probably the most direct extensions of MacMahon's elementary work on symmetric functions are the papers of Levine (1955, 1956), Sukhatme (1938), Cartier (1971), and Doubilet (1972). In his very thorough exposition and extension of MacMahon's work, Levine fully develops the combinatorial aspects of symmetric functions, utilizing Hammond operators extensively; he also develops many algorithms for the actual representation of symmetric functions. Sukhatme (1938) proves a number of new theorems referred to by MacMahon as "reciprocity theorems" (see Theorem 4 below). (Levine (1955, chapter VI) presents Sukhatme's results and extends them.) Cartier (1971, chapter 3) presents the theory of symmetric functions with a view toward their application to MacMahon's Master Theorem and to the representation theory of the symmetric group; he also describes their connection with Baxter algebras (Rota, 1969).

The material in this chapter corresponds to section I of *Combinatory Analysis*.

1.2 The Combinatorial Development of Symmetric Functions

We have chosen to present an introductory exposition of the work of P. Doubilet (1972) on symmetric functions. There are two reasons for this choice: first, Doubilet's approach utilizes the theory of Möbius functions (Rota, 1964) and so ties symmetric functions in with some of the most recent developments in combinatorial theory; second, Doubilet's approach yields much insight concerning many of MacMahon's classical results.

We begin with some definitions of certain classes of symmetric functions. A partition λ of a nonnegative integer n is a finite nonincreasing sequence of integers $\lambda_1, \lambda_2, \ldots, \lambda_r$ such that $\lambda_1 + \lambda_2 + \ldots + \lambda_r = n$; we say that the partition has r parts, and we shall write $\lambda \vdash n$ (Philip Hall's notation; see section 2.2). Sometimes we shall write $\lambda = (\lambda_1, \lambda_2, \ldots)$ to denote the partition just described, and sometimes we shall write $\lambda = (1^{r_1}2^{r_2}\ldots)$ to indicate that the partition has exactly r_i parts equal to i.

Let A_n denote the set of all homogeneous symmetric functions of degree n in the infinitely many indeterminates $X = \{x_1, x_2, \ldots\}$ over the field of rational numbers Q. A_n is clearly a vector space over Q, and the set of all A_n can be made into a graded algebra

$$\overline{A} = A_0 \oplus A_1 \oplus A_2 \oplus \ldots$$

by using ordinary formal power series multiplication.

Define the monomial symmetric functions k_λ, where $\lambda \vdash n$, by

$$k_\lambda = \sum x_{i_1}^{\lambda_1} x_{i_2}^{\lambda_2} x_{i_3}^{\lambda_3} \ldots,$$

where the summation is over all distinct monomials wherein the exponents $\lambda_1 \geqq \lambda_2 \geqq \ldots$ form the partition λ. Clearly the k_λ form a basis for A_n, and so A_n is of dimension $p(n)$, the number of partitions of n.

Define the complete homogeneous symmetric functions (MacMahon calls these the homogeneous product sums) h_λ, where $\lambda \vdash n$, by

$$h_n = \sum_{\lambda \vdash n} k_\lambda, \qquad h_\lambda = h_{\lambda_1} h_{\lambda_2} h_{\lambda_3} \ldots.$$

The elementary symmetric functions a_λ, where $\lambda \vdash n$, are defined by

$$a_n = k_{(1^n)}, \qquad a_\lambda = a_{\lambda_1} a_{\lambda_2} a_{\lambda_3} \ldots.$$

Finally, the power sum symmetric functions s_λ, where $\lambda \vdash n$, are defined by

$$s_n = k_n, \qquad s_\lambda = s_{\lambda_1} s_{\lambda_2} s_{\lambda_3} \cdots .$$

We must now recall some of the elementary results concerning Möbius functions for partially ordered sets. Let P denote a locally finite (i.e., all intervals $\{z \in P | x \leqq z \leqq y\}$ are finite) partially ordered set. The incidence algebra A over P is the set of functions f from $P \times P$ to R, the real numbers, such that $f(x,y) = 0$ if $x \nleqq y$. Addition and scalar multiplication are done in the standard manner, while multiplication is defined by the convolution

$$fg(x,y) = \sum_z f(x,z) g(z,y).$$

We define the zeta-function of P by $\zeta(x,y) = 1$ if $x \leqq y$ and $\zeta(x,y) = 0$ otherwise. It is easy to prove that an $f \in P$ has an inverse if an only if $f(x, x) \neq 0$ for any $x \in P$. Hence ζ has an inverse which is called μ, the Möbius function on P. It is not difficult to show that left and right inverses are identical.

Next we treat a result of Weisner (see Wilson, 1971, pages 318–319) that will provide us with necessary combinatorial information.

Proposition 1. Let L be a finite lattice, and let $b \in L$, $b \neq 0$. Then for any $a \in L$, the following identity holds:

$$\sum_{x \vee b = a} \mu(0, x) = 0.$$

Proof. If $a = b$,

$$\sum_{x \vee b = b} \mu(0, x) = \sum_{0 \leqq x \leqq b} \mu(0, x) = \sum_{0 \leqq x \leqq b} \mu(0, x) \zeta(x, b) = \delta(0, b) = 0,$$

where $\delta(x,y)$ is the Kronecker δ-function, the identity of A.

Now let a be any one of the least elements of L for which the proposition is unverified. Then

$$\sum_{x \vee b \leqq a} \mu(0, x) = \sum_{x \vee b = a} \mu(0, x) + \sum_{x \vee b < a} \mu(0, x).$$

The final sum is 0 by the induction hypothesis while the left-hand side is 0 as above. Hence the result follows by mathematical induction.

We introduce the lattice Π_n (the partitions of a set D of n elements), which we shall concentrate on throughout the remainder of this section.

A partition of a set is a collection of mutually disjoint subsets whose union is the entire set. Each element of a partition is called a block. If $\delta, \pi \in \Pi_n$, we say that $\delta \leqq \pi$ if each block of δ is contained in some block of π. The partition containing only one block is denoted by 1, and the partition made up of n singleton blocks is denoted by 0.

Proposition 2. For the lattice Π_n,

$$\mu(0,1) \equiv \mu_n(0,1) = (-1)^{n-1}(n-1)!.$$

Proof. We begin by observing that reversing the order in the lattice L of Proposition 1 yields the following dual of Weisner's result: for $b \neq 1$,

$$\sum_{x \wedge b = a} \mu(x,1) = 0.$$

Let b be a co-atom in Π_n (say b is of the form $\{(n),(1,2,\ldots,n-1)\}$). Now there are precisely $n-1$ partitions that satisfy $x \wedge b = 0$, namely, those of the form $\{(i,n),(1,2,\ldots,i-1,i+1,\ldots,n-1)\}$. Hence by the dual of Weisner's result,

$$\mu_n(0,1) = -(n-1)\mu_{n-1}(0,1) = (-1)^{n-1}(n-1)!.$$

There are three notation conventions, concerning partitions of numbers, that will be useful to us in the following:

(i) $\lambda! \equiv \lambda_1!\lambda_2!\ldots = 1!^{r_1}2!^{r_2}\ldots$ if $\lambda = (\lambda_1,\lambda_2,\ldots) = (1^{r_1}2^{r_2}\ldots)$;

(ii) $|\lambda| = r_1!r_2!\ldots$ if $\lambda = (1^{r_1}2^{r_2}\ldots)$;

(iii) $\operatorname{sign}\lambda \equiv (-1)^{r_2+2r_3+3r_4+\cdots}$ if $\lambda = (1^{r_1}2^{r_2}3^{r_3}\ldots)$.

Next we catalog several observations and conventions concerning Π_n; much of this can be found in Rota (1964).

If $\sigma, \pi \in \Pi_n$, the segment $[\sigma,\pi]$ is isomorphic to the direct product of r_1 copies of Π_1, r_2 copies of Π_2, etc., where r_i denotes the number of blocks of π that are composed of i blocks of σ; furthermore, we let $\lambda(\sigma,\pi) = (1^{r_1}2^{r_2}3^{r_3}\ldots)$, and $\lambda(\pi) = \lambda(0,\pi)$. From Proposition 2 and the preceding comments, it is not difficult to show that

$$\mu(\sigma,\pi) = (-1)^{r_1+r_2+r_3+\cdots-n}\prod_i (i-1)!^{r_i}.$$

We define sign $(\sigma,\pi) = \text{sign}\,\lambda(\sigma,\pi)$, and sign $(\pi) = \text{sign}\,(0,\pi)$. It follows simply that $\text{sign}(\sigma,\pi) = (\text{sign}(\sigma))(\text{sign}(\pi))$, and that $\mu(\sigma,\pi) = \text{sign}(\sigma,\pi)\cdot|\mu(\sigma,\pi)|$.

Now we introduce certain generating functions that will be seen to be closely related to the various symmetric functions first described in this section. Let F denote the set of functions from D (our n element set) to X (our set of infinitely many indeterminates). Define the generating function $\gamma(f)$ for each $f \in F$ by

$$\gamma(f) = \prod_{d\in D} f(d) = \prod_i x_i^{|f^{-1}(x_i)|},$$

and, if T is a subset of F,

$$\gamma(T) = \sum_{f\in T} \gamma(f).$$

For each $f \in F$, we define a partition Ker f of D, called the kernel of f, by putting d_1 and d_2 in the same block of Ker f whenever $f(d_1) = f(d_2)$.

For each $\pi \in \Pi_n$, we define three special subsets of F:

(2.1) $$\mathscr{K}_\pi = \{f \in F \mid \text{Ker} f = \pi\},$$

(2.2) $$\mathscr{L}_\pi = \{f \in F \mid \text{Ker} f \geqq \pi\},$$

(2.3) $$\mathscr{A}_\pi = \{f \in F \mid \text{Ker} f \wedge \pi = 0\},$$

and we let $k_\pi = \gamma\mathscr{K}_\pi)$, $s_\pi = \gamma(\mathscr{L}_\pi)$, $a_\pi = \gamma(\mathscr{A}_\pi)$.

Our first theorem relates these three generating functions to symmetric functions. Throughout the remainder of this section we shall closely follow P. Doubilet (1972).

Theorem 1.

(2.4) $$k_\pi = |\lambda(\pi)|k_{\lambda(\pi)},$$

(2.5) $$s_\pi = s_{\lambda(\pi)},$$

(2.6) $$a_\pi = \lambda(\pi)!\,a_{\lambda(\pi)}.$$

Proof. We let $\lambda(\pi) = (1^{r_1}2^{r_2}3^{r_3}\ldots) = (\lambda_1,\lambda_2,\lambda_3,\ldots)$.

To prove (2.4) we consider an arbitrary $f \in \mathscr{K}_\pi$, i.e., Ker $f = \pi$. Then

$$\gamma(f) = x_{i_1}^{\lambda_1} x_{i_2}^{\lambda_2} x_{i_3}^{\lambda_3} \ldots$$

for some choice of $i_1, i_2, i_3, \ldots$ and each monomial arises from exactly $r_1!r_2!r_3! \ldots$ functions $f \in \mathscr{K}_\pi$. Hence

$$k_\pi = \gamma(\mathscr{K}_\pi) = r_1!r_2!r_3! \ldots k_\lambda = |\lambda(\pi)| k_{\lambda(\pi)}.$$

To prove (2.5) we consider

$$\mathscr{L}_\pi = \{f \in F | \operatorname{Ker} f \geqq \pi\} = \{f \in F | f \text{ is constant on blocks } B_i \text{ of } \pi\};$$

then

$$\begin{aligned}
\gamma(\mathscr{L}_\pi) &= \sum_{f \in \mathscr{S}_\pi} \gamma(f) = \sum_{f \in \mathscr{S}_\pi} \gamma(f|B_1)\gamma(f|B_2)\gamma(f|B_3) \ldots \\
&= \sum_{\substack{f: f \text{ constant} \\ \text{on each } B_i}} \gamma(f|B_1)\gamma(f|B_2)\gamma(f|B_3) \ldots \\
&= \prod_i \Bigg(\sum_{\substack{\text{constant} \\ \text{maps } f: B_i \to X}} \gamma(f) \Bigg) \\
&= \prod_i (x_1^{|B_i|} + x_2^{|B_i|} + x_3^{|B_i|} + \ldots) \\
&= s_{\lambda(\pi)}.
\end{aligned}$$

To prove (2.6) we consider

$$\mathscr{A}_\pi = \{f \in F | \operatorname{Ker} f \wedge \pi = 0\} = \{f \in F | f \text{ is 1–1 on blocks } B_i \text{ of } \pi\};$$

hence

$$\begin{aligned}
\gamma(\mathscr{A}_\pi) &= \sum_{f \in \mathscr{A}_\pi} \gamma(f) = \sum_{f \in \mathscr{A}_\pi} \gamma(f|B_1)\gamma(f|B_2)\gamma(f|B_3) \ldots \\
&= \sum_{\substack{f \in F: f \text{ is} \\ \text{1–1 on each } B_i}} \gamma(f|B_1)\gamma(f|B_2)\gamma(f|B_3) \ldots \\
&= \prod_i \Bigg(\sum_{\substack{f: B_i \to X \\ \text{where } f \text{ is 1–1}}} \gamma(f) \Bigg).
\end{aligned}$$

Now each multiplicand of this last product is the sum of monomials of $|B|$ distinct terms, and each such monomial can arise from exactly $|B|!$ functions $f: B \to X$; consequently, the multiplicand under consideration is just $|B|!$ $a_{|B|}$. Therefore

$$a_\pi = \gamma(\mathscr{A}_\pi) = |B_1|!|B_2|!|B_3|! \ldots a_{|B_1|}a_{|B_2|}a_{|B_3|} \ldots = \gamma(\pi)! a_{\lambda(\pi)}.$$

This concludes the proof of Theorem 1.

From (2.1), (2.2), and (2.3), we see that

(2.7)
$$s_\pi = \sum_{\sigma \geqq \pi} k_\sigma,$$

(2.8)
$$a_\pi = \sum_{\sigma : \sigma \wedge \pi = 0} k_\sigma.$$

Formulas (2.7) and (2.8) can be inverted by Möbius inversion (see Doubilet, 1972, for complete details), and this yields the following result:

Theorem 2.

(2.9)
$$k_\pi = \sum_{\sigma \geqq \pi} \mu(\pi, \sigma) s_\sigma,$$

(2.10)
$$k_\pi = \sum_{\sigma \geqq \pi} \frac{\mu(\pi, \sigma)}{\mu(0, \sigma)} \sum_{\tau \leqq \sigma} \mu(\tau, \sigma) a_\tau.$$

Corollary. The $\{s_\lambda\}$ and $\{a_\lambda\}$ are each a basis for A_n.

Proof. As we have already observed, the $\{k_\lambda\}$ form a basis for A_n. Theorem 2 shows that the $\{s_\lambda\}$ and $\{a_\lambda\}$ each generate A_n, and since each set has $p(n)$ elements just as $\{k_\lambda\}$ does, we see that each is a basis for A_n.

Next we prove a theorem which is one of the "Laws of Symmetry" that MacMahon was so fond of.

Theorem 3. Let

$$a_\lambda = \sum_\mu c_{\lambda\mu} k_\mu;$$

then

$$c_{\lambda\mu} = c_{\mu\lambda}.$$

Proof. It follows from (2.8) and Theorem 1 that the coefficient of $|\mu| k_\mu$ in $\lambda! a_\lambda$ is $\sum_{\sigma\in\mu} \delta(0, \sigma \wedge \pi)$, where π is some fixed partition of type λ. Thus the coefficient of k_μ in a_μ is

$$\frac{|\mu|}{\lambda!} \sum_{\sigma\in\mu} \delta(0, \sigma \wedge \pi) = \frac{|\mu|}{\lambda!} \frac{1}{\binom{n}{\lambda}} \sum_{\substack{\sigma\in\mu,\\ \pi\in\lambda}} \delta(0, \sigma \wedge \pi) = \frac{|\lambda|\,|\mu|}{n!} \sum_{\substack{\sigma\in\mu,\\ \pi\in\lambda}} \delta(0, \sigma \wedge \pi),$$

which is symmetric in λ and μ.

In order to introduce the homogeneous product sums (i.e., the complete homogeneous symmetric functions), we must introduce a further combinatorial construct, namely "placings."

To best understand the concept of placing, we define it in terms of the "boxes" and "balls" of elementary probability theory. From our picturesque description the reader can easily provide a purely set-theoretic definition. Our domain D will be referred to as a set of "balls," while $X = \{x_1, x_2, \ldots\}$ will be a set of "boxes." By a placing p we mean an arrangement of balls in boxes in some configuration (i.e., there is some ordering of the balls in each box). Clearly, we may associate a function in F with each placing by disregarding the order structure assigned in each box. Thus the terms "generating function" and "kernel" extend directly to placings p, and they are denoted by $\gamma(p)$ and Ker p, respectively.

We now define

$\mathscr{H}_\pi = \{$placings $p |$ within each box the balls from the same block of π are linearly ordered$\}$;

$h_\pi = \gamma(\mathscr{H}_\pi)$.

Theorem 4. $h_\pi = \lambda(\pi)!\, h_{\lambda(\pi)}$.

Proof. We dissect a placing $p \in \mathscr{H}_\pi$ by noting that, independently, (1) the balls in B_1 are placed in boxes and then linearly ordered in each box, (2) the balls in B_2 are placed in boxes and then linearly ordered in each box, and so on. Hence

$$h_\pi = \prod_{\substack{\text{blocks}\\ \text{of } B}} \gamma(\mathscr{H}_B),$$

where $\mathscr{H}_B$ is the set of placings of balls from B into X with linear ordering in each box. To evaluate $\gamma(\mathscr{H}_B)$ we note that each monomial $x_{i_1}^{u_1} x_{i_2}^{u_2} \ldots$ of degree $|B|$ appears $|B|!$ times in this generating function, since there is a one-to-one correspondence between the linear orderings of B and

$$\{p \in \mathscr{H}_B | \gamma(p) = x_{i_1}^{u_1} x_{i_2}^{u_2} \ldots\};$$

i.e., with each placing under consideration, we associate the linear ordering obtained by taking first the balls in x_{i_1} in their prescribed order, then those in x_{i_2} in their prescribed order, etc. Hence

$$\gamma(\mathscr{H}_B) = |B|!\, h_{|B|},$$

and therefore

$$h_\pi = \gamma(\mathscr{H}_\pi) = |B_1|!\,|B_2|! \ldots h_{|B_1|} h_{|B_2|} \ldots = \lambda(\pi)!\, h_{\lambda(\pi)}.$$

Theorem 5. $h_\pi = \sum_\sigma \lambda(\sigma \wedge \pi)!\, k_\sigma$.

Proof. For each $f \in F$, with $\operatorname{Ker} f = \sigma$, let us determine the number of associated placings. Now placings are obtained from functions by linearly ordering, within each box, the elements from the same block of π; this is precisely the number of ways of independently linearly ordering the elements within each block of $\sigma \wedge \pi$, i.e., $\lambda(\sigma \wedge \pi)!$. Since this depends only on σ (for fixed π) and not on f, Theorem 5 follows.

Corollary. If $h_\lambda = \sum_\mu d_{\lambda\mu} k_\mu$, then $d_{\lambda\mu} = d_{\mu\lambda}$.

Proof. Proceed exactly as in the proof of Theorem 3, replacing $\delta(0, \sigma \wedge \pi)$ by $\lambda(\sigma \wedge \pi)!$.

We now state a result (without proof) that gives the relationships between $\{h_\lambda\}$ and $\{k_\lambda\}$, $\{s_\lambda\}$ and $\{a_\lambda\}$. The proof (see Doubilet, 1972, for details) relies on Möbius inversion and manipulation of summations.

Theorem 6.

$$k_\pi = \sum_{\sigma \geqq \pi} \frac{\mu(\pi, \sigma)}{|\mu(0, \sigma)|} \sum_{\tau \leqq \sigma} \mu(\tau, \sigma)\, h_\tau, \tag{2.11}$$

$$h_\pi = \sum_{\sigma \leqq \pi} |\mu(0, \sigma)|\, s_\sigma, \tag{2.12}$$

$$s_\pi = \frac{1}{|\mu(0, \pi)|} \sum_{\sigma \leqq \pi} \mu(\sigma, \pi)\, h_\sigma, \tag{2.13}$$

(2.14) $$a_\pi = \sum_{\sigma \leqq \pi} (\text{sign } \sigma)\ \lambda(\sigma, \pi)!\ h_\sigma,$$

(2.15) $$h_\pi = \sum_{\sigma \leqq \pi} (\text{sign } \sigma)\ \lambda(\sigma, \pi)!\ a_\sigma,$$

(2.16) $$1 + \sum_{n \geqq 1} h_n t^n = \exp\left(\sum_{n \geqq 1} s_n t^n / n\right).$$

Corollary. $\{h_\lambda\}$ is a basis for A_n.

Proof. This follows immediately from (2.11).

Further discussion of Doubilet's work and its relationship to the theory of partitions appears in Andrews (1976).

1.3 References

M. I. Aissen (1970) Patterns, partitions, and polynomials, *Trans. New York Acad. Sci. II*, *32*, 535–544.

G. E. Andrews (1976) The Theory of Partitions, Encyclopedia of Mathematics and Its Applications, vol. 2, Addison-Wesley, Reading.

P. Cartier (1972) La serie géneratrice exponentielle applications probabilistes et algébriques, version préliminaire, Institut de Recherche Mathématique Avancée, Strasbourg.

L. Comtet (1974) Advanced Combinatorics, D. Reidel, Dordrecht.

F. N. David and D. E. Barton (1962) Combinatorial Chance, Griffin, London.

F. N. David, D. E. Barton and M. G. Kendall (1966) Symmetric Functions and Allied Tables, Cambridge University Press, Cambridge.

P. Doubilet (1972) On the foundations of combinatorial theory VII: Symmetric functions through the theory of distribution and occupancy, *Studies in Appl. Math.*, *51*, 377–396.

P. Doubilet, G.-C. Rota, and J. Stein (1974) On the foundations of combinatorial theory IX: Combinatorial methods in invariant theory, *Studies in Appl. Math.*, *53*, 185–216.

M. Eisen (1969) Elementary Combinatorial Analysis, Gordon and Breach, New York.

H. O. Foulkes (1948) Differential operators associated with S-functions, *J. London Math. Soc.*, *24*, 136–143.

H. O. Foulkes (1972) A survey of some combinatorial aspects of symmetric functions, Permutations Colloquium, Paris.

M. Hall (1958) Survey of combinatorial analysis, in Some Aspects of Analysis and Probability IV, J. Wiley, New York.

J. Hammond (1883) On the use of certain differential operators in the theory of equations, *Proc. London Math. Soc.*, *14*, 119–129.

J. Levine (1955) Combinatorial analysis: Part I. MacMahon's theory of distributions and symmetric functions. Section I. The algebraic theory of symmetric functions, O.N.R. Report, Contract N. 870(00).

J. Levine (1956) A theory of occupancy: Distribution in parcels (elementary theory), O.N.R. Report, Contract N. 870(00).

J. Levine (1959a) On the application of MacMahon diagrams to certain problems in the multiplication of monomial symmetric functions, *Duke Math. J.*, *26*, 419–436.

J. Levine (1959b) Monomial-monomial symmetric functions tables, *Biometrika*, *46*, 205–213.

J. Levine (1961) Coefficient identities derived from expansions of elementary symmetric function products in terms of power sums, *Duke Math. J.*, *28*, 89–106.

N. Metropolis and P. R. Stein (1970) An elementary solution to a problem in restricted partitions, *J. Combinatorial Th.*, *9*, 365–376.

E. Netto (1964) Lehrbuch der Kombinatorik, Reprinted: Chelsea, New York.

J. Riordan (1958) An Introduction to Combinatorial Analysis, J. Wiley, New York.

J. Riordan (1968) Combinatorial identities, J. Wiley, New York.

J. Riordan (1971) Symmetric function expansions of power sum products into monomials and their inverses, *Duke Math. J.*, *38*, 285–294.

G.-C. Rota (1964) On the foundations of combinatorial theory I: Theory of Möbius functions, *Z. Wahrschleinlichkeitstheorie und Verw. Gebiete*, *2*, 340–368.

G.-C. Rota (1969) Baxter algebras and combinatorial identities I and II, *Bull. A.M.S.*, *75*, 325–334.

G.-C. Rota (1971) Combinatorial Theory and Invariant Theory, notes by L. Guibas, Bowdoin College, Maine.

G.-C. Rota, P. Doubilet and R. P. Stanley (1971) On the foundations of combinatorial theory VI: The idea of generating function, Proc. Sixth Symposium on Mathematical Statistics and Probability, vol. 2, University of Calif. Press, Berkeley, pp. 267–318.

G.-C. Rota and D. A. Smith (1972) Fluctuation theory and Baxter algebras, *Inst. Naz. Alta Mat. Symp. Math.*, *9*, 179–201.

P. V. Sukhatme (1938) On bipartitional functions, *Phil. Trans.*, *A-237*, 375–409.

R. J. Wilson (1971) The Möbius function in combinatorial mathematics, from Combinatorial Mathematics and Its Applications, edited by D. J. A. Welsh, Academic Press, London.

1.4 Summaries of the Papers

[7] Note on an algebraic identity, *Messenger of Math.*, *13* (1883), 142–144.

MacMahon generalizes the following result due to C. H. Prior (published by J. W. L. Glaisher in *Messenger of Math.*, *10* (1880), 33, and *10* (1880), 54–60): if $a + b + c = 0$, then

$$\left(\frac{a}{b-c} + \frac{b}{c-a} + \frac{c}{a-b}\right)\left(\frac{b-c}{a} + \frac{c-a}{b} + \frac{a-b}{c}\right) = 9.$$

MacMahon proves the following n-parameter generalization of this result. If

$$\sum_{i=1}^{n} a_i^j = 0, \qquad 1 \leqq j \leqq n-2,$$

and if

$$\Delta(y_1, \ldots, y_{n-1}) = \prod_{1 \leqq i < j \leqq n-1} (y_i - y_j),$$

then

$$\left[\frac{a_1}{\Delta(a_2, a_3, \ldots, a_n)} - \frac{a_2}{\Delta(a_1, a_3, \ldots, a_n)} + \ldots\right]$$
$$\times \left[\frac{\Delta(a_2, a_3, \ldots, a_n)}{a_1} - \frac{\Delta(a_1, a_3, \ldots, a_n)}{a_2} + \ldots\right] = n^2.$$

[11] On symmetric functions, and in particular on certain inverse operators in connection therewith, *Proc. London Math. Soc.*, *15* (1884), 20–47.

MacMahon develops here many results useful for expanding certain symmetric functions in terms of other symmetric functions, especially monomial symmetric functions in terms of elementary symmetric functions.

Great use is made of the Hammond operators

$$D_\lambda = \frac{d}{da_\lambda} + a_1 \frac{d}{da_{\lambda+1}} + a_2 \frac{d}{da_{\lambda+2}} + \ldots,$$

where these operators are to be applied to symmetric functions written as polynomials in the elementary symmetric functions $a_1, a_2, a_3, \ldots$. He illustrates how the simple rules of operation of the Hammond operators may be applied to computation of symmetric function formulas and derives useful operator identities such as

$$\frac{d}{da_\lambda} = d_\lambda - H_{(1)} d_{\lambda+1} + H_{(2)} d_{\lambda+2} - H_{(3)} d_{\lambda+3} + \ldots,$$

where $H_{(n)}$ is the sum of all monomial symmetric functions of weight n (later called the homogeneous product sum h_n). He concludes section 3 with an illustration of his results by computing the representation of the monomial symmetric function $(5^2 3)$ in terms of elementary symmetric functions a_i.

He considers other new operators such as

$$V_{-r} = (r) \frac{d}{da_0} + (r, 1) \frac{d}{da_1} + (r, 1^2) \frac{d}{da_2} + \ldots$$

(where, for example, $(r, 1^3)$ is the monomial symmetric function $\sum \alpha^r \beta\gamma\delta$) and makes use of them in certain symmetric function computations.

After a study of generalizations of $H_{(n)}$, MacMahon presents a generalization of Newton's formula relating the sums of powers of roots of a polynomial: let S^r_ω denote the sum of all monomial symmetric functions related to partitions of ω into r parts; then

$$S^r_\omega - a_1 S^r_{\omega-1} + a_2 S^r_{\omega-2} - \dots + (-1)^{\omega+1} \frac{\omega!}{r!(\omega-r)!} a_\omega = 0.$$

Newton's formula corresponds to $r = 1$.

MacMahon concludes by presenting a table which represents all monomial symmetric functions related to nonunitary partitions of weight 13 in terms of elementary symmetric functions.

[12] Note on the development of an algebraic fraction, *American J. Math.*, *6* (1884), 287–288.

MacMahon gives a simple method for determining the power series expansion of

$$(\phi(x))^{-1} = (1 - a_1 x + a_2 x^2 - \dots + (-1)^n a_n x^n)^{-1} = \sum_{p=0}^{\infty} H_p x^p.$$

First, he notes that each H_p is the sum of the monomial symmetric functions of the roots of $x^n \phi(1/x)$; thus from well-known results on symmetric functions, he shows that

$$H_p = \begin{vmatrix} a_1 & a_2 & a_3 & \dots & a_p \\ 1 & a_1 & a_2 & \dots & a_{p-1} \\ 0 & 1 & a_1 & \dots & a_{p-2} \\ \cdot & \cdot & \cdot & \dots & \cdot \\ \cdot & \cdot & \cdot & \dots & \cdot \\ \cdot & \cdot & \cdot & \dots & \cdot \\ 0 & 0 & 0 & \dots & a_1 \end{vmatrix}$$

$$= \sum (-1)^{p+\alpha_1+\alpha_2+\alpha_3+\dots} \frac{(\alpha_1 + \alpha_2 + \alpha_3 + \dots)!}{\alpha_1! \alpha_2 \alpha_3! \dots} a_1^{\alpha_1} a_2^{\alpha_2} a_3^{\alpha_3} \dots,$$

where the last sum is extended over all partitions $\alpha_1 + 2\alpha_2 + 3\alpha_3 + \dots = p$

of p. In an appended note, Dr. Franklin neatly reverses MacMahon's approach to make the whole problem even simpler.

[16] A new theorem in symmetric functions, *Quart. J. Math.*, *20* (1885), 365–369.

In this paper MacMahon provides an algorithm whereby the Hammond operators D_i are made effective in the study of the representation of general symmetric functions by elementary symmetric functions.

The Hammond operator D_i is defined by

$$D_i = \frac{1}{i!}\left(a_0 \frac{d}{da_1} + a_1 \frac{d}{da_2} + \ldots\right)^i,$$

where the a_i denote the elementary symmetric functions. The basic rules of operation are

$$D_\kappa(\lambda^l \mu^m \nu^n \ldots) = 0, \qquad D_\lambda(\lambda^l \mu^m \nu^n \ldots) = (\lambda^{l-1} \mu^m \nu^n \ldots), \qquad D_\lambda(\lambda) = 1,$$

and more generally

$$\begin{aligned} &D_\kappa(\lambda_1 \mu_1 \nu_1 \ldots)(\lambda_2 \mu_2 \nu_2 \ldots)(\lambda_3 \mu_3 \nu_3 \ldots) \\ &= \sum D_\kappa(\mu_1 \nu_1 \ldots)(\lambda_2 \nu_2 \ldots)(\lambda_3 \mu_3 \ldots), \end{aligned}$$

where the summation is over all $\lambda_1 + \mu_2 + \nu_3 + \ldots = \kappa$; for example,

$$D_4(432)(21) = (32)(21) + (42)(2) + (43)(1).$$

Using the D_i and a related operator

$$\Delta_i = \frac{1}{i!}\left(a_1 \frac{d}{da_0} + a_2 \frac{d}{da_1} + a_3 \frac{d}{da_2} + \ldots\right)^i,$$

he produces an algorithm (equation (7)) for computing the T_j in

$$(\kappa\lambda\mu \ldots) = T_\kappa + T_{\kappa-1} + \ldots + T_1,$$

where T_j is the sum of all the terms of degree j in the unique representation of $(\kappa\lambda\mu \ldots)$ as a polynomial in the elementary symmetric functions.

[17] Note on rationalisation, *Messenger of Math.*, *15* (1885), 65–67.

In this note MacMahon seeks a polynomial in $a_1, \ldots, a_n$ that has a factor $a_1^{1/m} + \ldots + a_n^{1/m}$. Consider the polynomial

$$\prod_{\substack{1 \leqq j_i \leqq m \\ 2 \leqq i \leqq n}} (x - a_1^{1/m} - \omega^{j_2} a_2^{1/m} - \ldots - \omega^{j_n} a_n^{1/m})$$

$$= x^{m^{n-1}} - A_1 x^{m^{n-1}-1} + \ldots + (-1)^m A_{m^{n-1}},$$

where ω is an mth root of unity. It turns out that $A_{m^{n-1}}^{1/m}$ is a rationalisation of the type desired, and MacMahon deduces (by means of his general results on symmetric functions) a means for easily computing $A_{m^{n-1}}$, namely,

$$A_j = \sum \frac{\left(\frac{S_1}{1}\right)^{t_1}\left(\frac{S_2}{2}\right)^{t_2} \cdots \left(\frac{S_j}{j}\right)^{t_j}}{t_1!\, t_2! \ldots t_j!} (-1)^{j+\Sigma t},$$

where the summation is over all partitions of j of the form $j = t_1 \cdot 1 + t_2 \cdot 2 + \ldots + t_j \cdot j$. The S_i are given by

$$m^{1-n} S_i = \frac{\{m(l\lambda + m\mu + n\nu + \ldots)\}!}{\{(m\lambda)!\}\{(m\mu)!\}\{(m\nu)!\}\ldots} [\lambda^l \mu^m \nu^n \ldots],$$

and $[pqr \ldots]$ denotes the symmetric function $\sum a_1^p a_2^q a_3^r \ldots$.

Hence, if $m = n = 2$, we see that $S_1 = 2(a_1 + a_2)$, $S_2 = 2(a_1^2 + a_2^2 + 6a_1 a_2)$, and

$$A_2 = \frac{S_1^2}{2!} - \frac{S_2}{2} = a_1^2 + a_2^2 - 2a_1 a_2 = (a_1 - a_2)^2;$$

and indeed $a_1 - a_2$ is a rationalisation of $a_1^{1/2} + a_2^{1/2}$ since

$$(a_1^{1/2} + a_2^{1/2})(a_1^{1/2} - a_2^{1/2}) = a_1 - a_2.$$

[25] The law of symmetry and other theorems in symmetric functions, *Quart. J. Math.*, *22* (1887), 74–81.

Letting $a_i = (1^i)$ denote the ith elementary symmetric function of a set of numbers $\{\alpha_1, \alpha_2, \ldots, \alpha_n, \ldots\}$ and, in general, letting

$$(p_1 p_2 \ldots p_r) = \sum \alpha_{i_1}^{p_1} \alpha_{i_2}^{p_2} \ldots \alpha_{i_r}^{p_r},$$

where the sum runs over all possible r-tuples of αs, MacMahon establishes the following general symmetry law:

Theorem. If

$$\sum A_{\lambda^l \mu^m \nu^n \ldots} (\lambda^l \mu^m \nu^n \ldots) = \sum A'_{\lambda'^{l'} \mu'^{m'} \nu'^{n'} \ldots} a_{\lambda'}^{l'} a_{\mu'}^{m'} a_{\nu'}^{n'} \ldots$$

$$\sum B'_{\lambda^l \mu^m \nu^n \ldots} a_\lambda^l a_\mu^m a_\nu^n \ldots = \sum B_{\lambda'^{l'} \mu'^{m'} \nu'^{n'} \ldots} (\lambda'^{l'} \mu'^{m'} \nu'^{n'} \ldots),$$

then

$$\sum A_{\lambda^l \mu^m \nu^n \ldots} B'_{\lambda^l \mu^m \nu^n \ldots} = \sum A'_{\lambda'^{l'} \mu'^{m'} \nu'^{n'} \ldots} B_{\lambda'^{l'} \mu'^{m'} \nu'^{n'} \ldots}.$$

As an example, from the fact that

$$(31) - 2(2^2) = a_2 a_1^2 - 4a_2^2 - 3a_3 a_1$$

and

$$a_3 a_1 + 3a_2^2 = 7(21^2) + 3(2^2) + 18(1^4),$$

the law of symmetry established above implies that

$$1 \cdot 1 + (-2) \cdot 3 = 1 \cdot 7 + (-4) \cdot 3 + (-3) \cdot 0 + 0 \cdot 18$$

or

$$-5 = -5.$$

MacMahon then devotes several pages to the development of a theory of H-functions (the homogeneous product sums later to be denoted h_n); his object here is to show that the theory of H-functions is exactly parallel to that of the elementary symmetric functions developed by Hammond (see the reference in section 1.3 to J. Hammond, 1883). He concludes with a simple demonstration of how the Hammond operator acts on a composite symmetric function (see [16] for his original discussion of this theorem).

Reprints of the Papers

NOTE ON AN ALGEBRAIC IDENTITY.

By *Capt. P. A. MacMahon, R.A.*

THE present note relates to Mr. Prior's algebraic identity that has been considered by Mr. J. W. L. Glaisher in the *Messenger* for 1880. It is, that when $a+b+c=0$,

$$\left(\frac{a}{b-c}+\frac{b}{c-a}+\frac{c}{a-b}\right)\left(\frac{b-c}{a}+\frac{c-a}{b}+\frac{a-b}{c}\right)=9.$$

I proceed to shew that a similar theorem exists in the case of any number of quantities, subject to certain conditions. Writing a_1, a_2, a_3 for a, b, c, the identity may be written

$$\left(\frac{a_1}{a_2-a_3}-\frac{a_2}{a_1-a_3}+\frac{a_3}{a_1-a_2}\right)\left(\frac{a_2-a_3}{a_1}-\frac{a_1-a_3}{a_2}+\frac{a_1-a_2}{a_3}\right)=3^2.$$

Wherein now the suffixes, in the differences involved, are in numerical order.

In the case of two quantities, if $a_1+a_2=0$, it is easy to verify that

$$(a_1-a_2)\left(\frac{1}{a_1}-\frac{1}{a_2}\right)=2^2.$$

For four quantities I find that

$$\left\{\frac{a_1}{(a_2-a_3)(a_2-a_4)(a_3-a_4)}-\frac{a_2}{(a_1-a_3)(a_1-a_4)(a_3-a_4)}\right.$$
$$\left.+\frac{a_3}{(a_1-a_2)(a_1-a_4)(a_2-a_4)}-\frac{a_4}{(a_1-a_2)(a_1-a_3)(a_2-a_3)}\right\}$$
$$\times\left\{\frac{(a_2-a_3)(a_2-a_4)(a_3-a_4)}{a_1}-\frac{(a_1-a_3)(a_1-a_4)(a_3-a_4)}{a_2}\right.$$
$$\left.+\frac{(a_1-a_2)(a_1-a_4)(a_2-a_4)}{a_3}-\frac{(a_1-a_2)(a_1-a_3)(a_2-a_3)}{a_4}\right\}=4^2,$$

subject to the two conditions

$$a_1+a_2+a_3+a_4=0,$$
$$a_1^2+a_2^2+a_3^2+a_4^2=0.$$

This may be proved as follows:—

The left-hand side may be written

$$\frac{\Sigma a_1(a_1-a_2)(a_1-a_3)(a_1-a_4)}{(a_1-a_2)(a_1-a_3)(a_1-a_4)(a_2-a_3)(a_2-a_4)(a_3-a\ )}$$
$$\times\frac{a_2a_3a_4(a_2-a_3)(a_2-a_4)(a_3-a_4)-a_1a_3a_4(a_1-a_3))a_1-a_4)(a_3-a_4)+\ldots}{a_1a_2a_3a_4}.$$

Now $\quad\Sigma(a_1-a_2)(a_1-a_3)(a_1-a_4)$

$$=\Sigma\{a_1^4-(a_2+a_3+a_4)a_1^3+(a_3a_4+a_4a_2+a_2a_3)a_1^2-a_1a_2a_3a_4\}$$
$$=\Sigma(a_1^4+a_1^4+a_1^4-a_1a_2a_3a_4)$$
$$=3(a_1^4+a_2^4+a_3^4+a_4^4)-4a_1a_2a_3a_4,$$

but if a_1, a_2, a_3, a_4 be roots of the quartic

$$x^4-dx+e=0,$$
$$a_1^4+a_2^4+a_3^4+a_4^4=-4e=-12a_1a_2a_3a_4;$$

consequently

$$\Sigma a_1(a_1-a_2)(a_1-a_3)(a_1-a_4)=-16a_1a_2a_3a_4;$$

also since the expression

$$a_2a_3a_4(a_2-a_3)(a_2-a_4)(a_3-a_4)-a_1a_3a_4(a_1-a_3)(a_1-a_4)(a_3-a_4),$$
$$+a_1a_2a_4(a_1-a_2)(a_1-a_4)(a_2-a_4)-a_1a_2a_3(a_1-a_2)(a_1-a_3)(a_2-a_3)$$

vanishes for each of the substitutions

$$a_1=a_2,\quad a_1=a_3,\quad a_1=a_4,\quad a_2=a_3,\quad a_2=a_4,\quad a_3=a_4,$$

we may put it equal to

$$m(a_1-a_2)(a_1-a_3)(a_1-a_4)(a_2-a_3)(a_2-a_4)(a_3-a_4),$$

and m is evidently -1; thus the left-hand side is $+16$ as we wished to prove.

Suppose now we have n quantities

$$a_1,\ a_2,\ a_3,\ \dots a_n,$$

then if

$$\left.\begin{array}{l} a_1 + a_2 + a_3 + \dots + a_n = 0 \\ a_1^2 + a_2^2 + a_3^2 + \dots + a_n^2 = 0 \\ \dots\dots\dots\dots\dots\dots\dots\dots \\ a_1^{n-2} + a_2^{n-2} + a_3^{n-2} + \dots + a_n^{n-2} = 0 \end{array}\right\},$$

clearly

$$\begin{aligned} &\Sigma a_1(a_1-a_2)(a_1-a_3)\dots(a_1-a_n) \\ &= (n-1)(a_1^n + a_2^n + a_3^n + \dots + a_n^n) + (-)^{n+1} n a_1 a_2 a_3 \dots a_n \\ &= (-)^{n+1}\{n(n-1)+n\} a_1 a_2 a_3 \dots a_n \\ &= (-)^{n+1} n^2 a_1 a_2 a_3 \dots a_n; \end{aligned}$$

also, as before

$$\begin{aligned} &a_2 a_3 a_4 \dots a_n (a_2-a_3)(a_2-a_4)\dots(a_2-a_n)(a_3-a_n)\dots(a_3-a_n) \\ &\qquad \dots(a_{n-1}-a_n) - a_1 a_3 a_4 \dots a_n (a_1-a_3)(a_1-a_4)\dots(a_1-a_n), \\ &(a_3-a_4)(a_3-a_5)\dots(a_3-a_n)\dots(a_{n-1}-a_n) + \dots \\ &= (-)^{n+1}(a_1-a_2)(a_1-a_3)\dots(a_1-a_n)(a_2-a_3)\dots(a_2-a_n)\dots(a_{n-1}-a_n). \end{aligned}$$

Consequently we have proved the generalised identity :—

$$\begin{aligned} &\left[\frac{a_1}{(a_2-a_3)\dots(a_2-a_n)(a_3-a_4)\dots(a_3-a_n)\dots(a_{n-1}-a_n)}\right. \\ &\left.-\frac{a_2}{(a_1-a_3)\dots(a_1-a_n)(a_3-a_4)\dots(a_3-a_n)\dots(a_{n-1}-a_n)} + \dots\right] \\ &\times\left[\frac{(a_2-a_3)\dots(a_2-a_n)(a_3-a_4)\dots(a_3-a_n)\dots(a_{n-1}-a_n)}{a_1}\right. \\ &\left.-\frac{(a_1-a_3)\dots(a_1-a_n)(a_3-a_4)\dots(a_3-a_n)\dots(a_{n-1}-a_n)}{a_2} + \dots\right] = n^2. \end{aligned}$$

Royal Military Academy,
January, 1884.

On Symmetric Functions, and in particular on certain Inverse Operators in connection therewith. By Captain P. A. MacMahon, R.A.

[*Read Nov.* 8*th*, 1883.]

1. The present paper is more especially concerned with the non-unitary symmetric functions (that is, with those whose partitions contain no unit), their calculation and the development of their properties; for the reason that it has been recently shown (*vide* the author's paper in Vol. vi., No. 2, *American Journal of Mathematics*) that they are, in fact, seminvariants of an allied quantic, and the whole series contains potentially the complete solution of the syzygies which exist between the sources of covariants of binary quantics.

The equation considered is in every case

$$x^n - a_1 x^{n-1} + a_2 x^{n-2} - a_3 x^{n-3} + \ldots = 0 \ldots\ldots\ldots\ldots\ldots\ldots (1);$$

and, as much use will be made of Mr. Hammond's "Differential Operators," reference is made to his paper in the *Proceedings of the London Mathematical Society* for February, 1883, Vol. xiv., pp. 119—129; the operator most employed is

$$d_\lambda = \frac{d}{da_\lambda} + a_1 \frac{d}{da_{\lambda+1}} + a_2 \frac{d}{da_{\lambda+2}} + \ldots,$$

d_λ being defined by

$$d_\lambda = [\lambda],$$

where the partitions in [ ] refer to the symmetric functions of the roots of the equation

$$1 - y^{-1}D_1 + y^{-2}D_2 - y^{-3}D_3 + \ldots = 0;$$

and the law of operation of D on the partitions of the symmetric functions of the roots of (1) is, from Mr. Hammond's paper, as follows:—

$$D_\kappa(\lambda^l . \mu^m . \nu^n \ldots) = 0, \quad D_\lambda(\lambda^l . \mu^m . \nu^n \ldots) = (\lambda^{l-1} . \mu^m . \nu^n \ldots), \quad D_\lambda(\lambda) = 1.$$

If the symmetric functions be rendered homogeneous by multiplying each term by a_0 raised to the necessary power, then

$$d_\lambda = a_0 \frac{d}{da_\lambda} + a_1 \frac{d}{da_{\lambda+1}} + a_2 \frac{d}{da_{\lambda+2}} + \ldots \qquad \ldots\ldots\ldots (2).$$

Putting herein $a_0 = 0$, the character of the operator is not altered, and, in fact, the change is merely equivalent to a unit increase of suffixes throughout; this shows that the terms of highest degree in all symmetric functions of (1) are simply symmetric functions of the equation

$$x^n - a_2 x^{n-1} + a_3 x^{n-3} - a_4 x^{n-4} + \ldots = 0 \ldots\ldots\ldots\ldots\ldots\ldots (3),$$

a_1 being now the factor introduced for sake of homogeneity. Thus, every known symmetric function furnishes the terms of highest degree in an infinite number of higher symmetric functions by the simple process of increasing each suffix by unity, and then multiplying the whole by the necessary power of a_1; for example, if partitions in (( )) refer to equation (3), then the terms of highest degree in $(\lambda^l . \mu^m . \nu^n \ldots)$ are

$$a_1^{\lambda-\mu}((\mu^m . \nu^n \ldots)),$$

that is, from symmetric function $(\mu^m . \nu^n \ldots)$ we can, by giving λ different values > or = μ, derive the highest terms in an infinite number of others; *e.g.*, since

$$(3^2 2^2) = a_2 a_4^2 - 2a_2 a_3 a_5 - a_1 a_4 a_5 + 5a_5 + 2a_2^2 a_6 + 3a_1 a_3 a_6$$
$$-9a_4 a_6 - 7a_1 a_2 a_7 + 6a_3 a_7 + 7a_1^2 a_8 + a_2 a_8 - 15a_1 a_9 + 15a_{10},$$

we find

$$(3^3 2^2) = a_3 a_5^2 - 2a_3 a_4 a_6 - a_2 a_5 a_6 + 5a_1 a_6^2 + 2a_3^2 a_7 + 3a_2 a_4 a_7$$
$$-9a_1 a_5 a_7 - 7a_2 a_3 a_8 + 6a_1 a_4 a_8 + 7a_2^2 a_9 + a_1 a_3 a_9 - 15 a_1 a_2 a_{10}$$
$$+15 a_1^2 a_{11} + \text{terms of lower degree},$$

and so on in every case, the terms of highest degree in all symmetric functions of a given weight may be simply derived from those of lower weight.

It follows hence that the performance of the differential operation denoted by d_λ may be somewhat simplified; for, when $a_0 = 0$, let d_λ become d'_λ, so that

$$d_\lambda = \frac{d}{da_\lambda} + d'_\lambda,$$

and let the subject operated upon be

$$a_1^\kappa A + B,$$

wherein $a_1^\kappa A$ comprises the terms of highest degree then if, as is generally the case, the partition of the subject contain no partition of λ,

$$\left(\frac{d}{da_\lambda} + d'_\lambda\right)(a_1^\kappa A + B) = 0;$$

then $\lambda \neq 1, \quad a_1^\kappa \frac{dA}{d\lambda} + d_\lambda B = 0$

and $\lambda = 1, \quad a^{\kappa-1}\left(a_1 \frac{dA}{da_1} + \kappa A\right) + d_1 B = 0$ ....(4),

A being of course already known.

Suppose, for instance, the subject be $(\mu . \nu . \pi \dots)$; if the partition contain no partition of λ,

$$a_1^{\mu-\nu} \frac{d}{da_\lambda}((\nu . \pi \dots)) + d_\lambda B = 0,$$

or

$$a_1^{\mu-\nu}\left(\left(\frac{d}{da_{\lambda-1}}(\nu . \pi \dots)\right)\right) + d_\lambda B = 0,$$

2. A non-unitary symmetric function is completely given by its non-unitary portion; this follows at once from the fact that the complete expression of a seminvariant can be derived from its non-unitary portion, by substituting for the coefficients therein, the corresponding coefficients of Professor Cayley's generalized canonical form (*vide* Vol. xix., No. 76, *Quarterly Journal of Mathematics*); the corresponding theorem in symmetric functions is as follows, viz.:—forming the

equation $$x^n+A_2x^{n-2}-A_3x^{n-3}+A_4x^{n-4}-\ldots=0 \quad\ldots\ldots\ldots\ldots\ldots(5),$$
in which A_r is formed from the non-unitary symmetric functions of weight r, and is such that on putting therein $a_1=0$ it reduces to a_r: then we find

$$A_2=a_2-\frac{1}{2}a_1^2,$$

$$A_3=a_3-a_1a_2+\frac{2}{3!}a_1^3,$$

$$A_4=a_4-a_1a_3+\frac{1}{2!}a_1^2a_2-\frac{3}{4!}a_1^4,$$

$$A_5=a_5-a_1a_4+\frac{1}{2!}a_1^2a_3-\frac{1}{3!}a_1^3a_2+\frac{4}{5!}a_1^5,$$

$$A_6=a_6-a_1a_5+\frac{1}{2!}a_1^2a_4-\frac{1}{3!}a_1^3a_3+\frac{1}{4!}a_1^4a_2-\frac{5}{6!}a_1^6,$$

$$\ldots\quad\ldots\quad\ldots\quad\ldots\quad\ldots\quad\ldots\quad\ldots$$

$$A_r=a_r-a_1a_{r-1}+\frac{1}{2!}a_1^2a_{r-2}-\frac{1}{3!}a_1^3a_{r-3}+\frac{1}{4!}a_1^4a_{r-4}-\ldots$$
$$+(-)^r\frac{1}{(r-2)!}a_1^{r-2}a_2+(-)^{r+1}\frac{r-1}{r!}a_1^r.$$

Now the non-unitary symmetric functions of (5) are also non-unitary functions of (1), and the equations (1) and (5) become the same on putting therein $a_1=0$; consequently the non-unitary symmetric functions of (1) and (5) must be identical, and if

$$\phi(a_2, a_3, a_4 \ldots)$$

be the non-unitary portion of a function of (1),

$$\phi(A_2, A_3, A_4 \ldots)$$

must be its complete expression; now $\lambda>1$,

$$d_\lambda(a_1=0)=\frac{d}{da_\lambda}+a_2\frac{d}{da_{\lambda+2}}+a_3\frac{d}{da_{\lambda+3}}+\ldots,$$

and the operators included in this formula may be employed to calculate separately the non-unitary portions of the symmetric functions of (1).

*3. In this and the succeeding section, Mr. Hammond's operators (see *Proceedings*, Vol. xiv., p. 119), included in d_λ, are regarded from an inverse point of view; *i.e.*, instead of regarding d_λ as a function of certain differential coefficients $\frac{d}{da_\lambda}$, $\frac{d}{da_{\lambda+1}}$, ..., we look upon $\frac{d}{da^\lambda}$ as a

* The proofs of §§ 3 and 4 have in great measure been communicated to me by Mr. Hammond.

function of the operators d_λ, $d_{\lambda+1}$, $d_{\lambda+2}$, ..., which leads us at once to a system of inverse operators, enabling us to proceed by an operation from a lower to a higher symmetric function.

First, if $\alpha, \beta, \gamma, \ldots$ be the roots of

$$x^n - a_1 x^{n-1} + a_2 x^{n-2} - \ldots = 0,$$

$$\frac{1}{(1-\alpha x)(1-\beta x)(1-\gamma x)\ldots} = 1 + H_{(1)}x + H_{(2)}x^2 + H_{(3)}x^3 + \ldots = u \text{ suppose,}$$

$$= \frac{1}{1 - a_1 x + a_2 x^2 - a_3 x^3 + \ldots},$$

$H_{(r)}$ being the total symmetric function of weight r. Therefore

$$(1 - a_1 x + a_2 x^2 - a_3 x^3 + \ldots)(1 + H_{(1)}x + H_{(2)}x^2 + H_{(3)}x^3 + \ldots) = 1;$$

equating coefficients,

$$H_{(1)} - a_1 = 0,$$

$$H_{(2)} - a_1 H_{(1)} + a_2 = 0,$$

$$H_{(3)} - a_1 H_{(2)} + a_2 H_{(1)} - a_3 = 0,$$

$$\ldots \quad \ldots \quad \ldots \quad \ldots \quad \ldots \quad \ldots$$

generally $$H_{(r)} - a_1 H_{(r-1)} + a_2 H_{(r-2)} - \ldots + (-)^r a_r = 0 \quad \ldots\ldots\ldots\ldots(6).$$

Now, these equations may be solved either for a_r or H_r, giving either

$$a_r = \begin{vmatrix} H_{(1)}, & H_{(2)}, & H_{(3)}, & \ldots, & H_{(r-1)}, & H_{(r)} \\ 1, & H_{(1)}, & H_{(2)}, & \ldots, & H_{(r-2)}, & H_{(r-1)} \\ & 1, & H_{(1)}, & \ldots, & H_{(r-3)}, & H_{(r-2)} \\ & & 1, & \ldots, & H_{(r-4)}, & H_{(r-3)} \\ \ldots & \ldots & \ldots & \ldots & \ldots & \ldots \\ & & & & 1, & H_{(1)} \end{vmatrix},$$

or the corresponding formula with a and H interchanged. Consequently, if $$H_{(r)} = \phi(a_1, a_2, a_3, \ldots, a_r),$$

then $$a_r = \phi(H_{(1)}, H_{(2)}, H_{(3)}, \ldots, H_{(r)}).$$

Now, considering the system of equations

$$\frac{d}{da_1} + a_1 \frac{d}{da_2} + a_2 \frac{d}{da_3} + \ldots = d_1,$$

$$\frac{d}{da_2} + a_1 \frac{d}{da_3} + \ldots = d_2,$$

$$\frac{d}{da_3} + \ldots = d_3,$$

$$\ldots = \ldots,$$

and solving for $\frac{d}{da_\lambda}$, we obtain

$$\begin{vmatrix} 1, & a_1, & a_2, & \dots \\ & 1, & a_1, & \dots \\ & & 1, & \dots \\ \dots & \dots & & \dots \end{vmatrix} \frac{d}{da_\lambda} = \begin{vmatrix} d_\lambda, & a_1, & a_2, & a_3, & \dots \\ d_{\lambda+1}, & 1, & a_1, & a_2, & \dots \\ d_{\lambda+2}, & . & 1, & a_1, & \dots \\ d_{\lambda+3}, & . & . & 1, & \dots \\ \dots & \dots & \dots & \dots & \end{vmatrix};$$

whence $$\frac{d}{da_\lambda} = d_\lambda - H_1 d_{\lambda+1} + \begin{vmatrix} a_1, & a_2 \\ 1, & a_1 \end{vmatrix} d_{\lambda+2} - \begin{vmatrix} a_1, & a_2, & a_3 \\ 1, & a_1, & a_2 \\ & 1, & a_1 \end{vmatrix} d_{\lambda+3} + \dots,$$

that is, $$\frac{d}{da_\lambda} = d_\lambda - H_{(1)} d_{\lambda+1} + H_{(2)} d_{\lambda+2} - H_{(3)} d_{\lambda+3} + \dots \qquad \dots\dots\dots(7),$$

wherein if the subject be $(\lambda^l . \mu^m . \nu^n, \dots)$,

$$d_0 (\lambda^l . \mu^m . \nu^n \dots) = \lambda (\lambda^l . \mu^m . \nu^n \dots),$$

which is merely Euler's theorem of homogeneous functions.

It is necessary at this point to prove certain properties of the total symmetric function $H_{(r)}$.

Since $$u = \frac{1}{(1-\alpha x)(1-\beta x)\dots} = 1 + H_{(1)}x + H_{(2)}x^2 + \dots,$$

$$\frac{du}{dx} = H_{(1)} + 2H_{(2)}x + 3H_{(3)}x^2 + \dots,$$

and, differentiating u logarithmically,

$$\frac{1}{u}\frac{du}{dx} = \frac{\alpha}{1-\alpha x} + \frac{\beta}{1-\beta x} + \dots = S_1 + S_2 x + S_3 x^2 + \dots;$$

whence

$$(1 + H_{(1)}x + H_{(2)}x^2 + \dots)(S_1 + S_2 x + S_3 x^2 + \dots)$$
$$= H_{(1)} + 2H_{(2)}x + 3H_{(3)}x^2 + \dots,$$

and, equating coefficients, there follows the system

$$\left.\begin{aligned} H_{(1)} &= S_1 \\ 2H_{(2)} &= S_2 + H_{(1)} S_1, \\ 3H_{(3)} &= S_3 + H_{(1)} S_2 + H_{(2)} S_1 \\ \dots & \dots \dots \dots \dots \\ rH_{(r)} &= S_r + H_{(1)} S_{r-1} + H_{(2)} S_{r-2} + \dots + H_{(r-1)} S_1 \end{aligned}\right\} \dots\dots\dots(8).$$

Solving these equations for S_r and for $H_{(r)}$, we get

$$S_r = (-)^{r+1} \begin{vmatrix} H_{(1)}, & H_{(2)}, & H_{(3)}, & \dots, & H_{(r-1)}, & r\,.\,H_{(r)} \\ 1, & H_{(1)}, & H_{(2)}, & \dots, & H_{(r-2)}, & r-1\,.\,H_{(r-1)} \\ 0, & 1, & H_{(1)}, & \dots, & H_{(r-3)}, & r-2\,.\,H_{(r-2)} \\ \dots & \dots & \dots & \dots & \dots & \dots \\ . & . & . & \dots, & 1, & H_{(1)} \end{vmatrix} \quad \dots\dots(9),$$

$$r\,!\,H_{(r)} = \begin{vmatrix} S_1, & S_2, & S_3, & \dots, & S_r \\ -1, & S_1, & S_2, & \dots, & S_{r-1} \\ . & -2, & S_1, & \dots, & S_{r-2} \\ . & . & -3, & \dots, & S_{r-3} \\ \dots & \dots & \dots & \dots & \dots \\ & & & -(r-1), & S_1 \end{vmatrix} \quad \dots\dots\dots(10);$$

consequently, any expression of a symmetric function is unaltered by the substitution of H for a, if, when expressed in terms of the sums of powers of the roots, each term contains only sums of odd powers and an even number of sums of even powers. The symmetric functions, when expressed in terms of $H_{(1)}, H_{(2)}, H_{(3)}, \dots$, appear to obey the law of symmetry; that is, the coefficient of $H_{(\lambda)}^{l_1}\,.\,H_{(\mu)}^{m_1}\,\dots$ in symmetric function $(\lambda_1^{l_1}\,.\,\mu_1^{m_1}\,\dots)$ is equal to the coefficient of $H_{(\lambda_1)}^{l}\,.\,H_{(\mu_1)}^{m_1}\,\dots$ in symmetric function $(\lambda^{l}\,.\,\mu^{m}\,\dots)$; thus the table for weight 4 is

	$H_{(1)}^4$	$H_{(1)}^2 H_{(2)}$	$H_{(2)}^2$	$H_{(1)} H_{(3)}$	$H_{(4)}$
(4)	−1	+4	−2	−4	+4
(31)	+2	−7	+2	+7	−4
(2^2)	+1	−4	+3	+2	−2
(21^2)	−3	+10	−4	−7	+4
(1^4)	+1	−3	+1	+2	−1

...(11).

Now, referring to the set of equations (7), if S_m be the subject of operation, we find

$$\frac{d}{da_{m-r}} S_m = (-)^{m+r+1}\, m H_{(r)} \quad \dots\dots\dots\dots\dots\dots\dots(12),$$

and

$$H_{(m-1)} = \frac{1}{m}\,\frac{d}{da_1} S_m \quad \dots\dots\dots\dots\dots\dots\dots(13).$$

Now, since the general term in S_m is

$$\frac{(-)^{k+n}\,(k-1)!\,n}{a!\,b!\,c!\,\dots\,l!}\, a_1^a a_2^b a_3^c \dots a_n^l \quad \dots\dots\dots\dots\dots\dots(14),$$

in which k is the degree of the term, we deduce without difficulty that the general term in $H_{(n)}$ is

$$\frac{(-)^{k+n}\,k\,!}{a\,!\;b\,!\;c\,!\;\ldots\;l\,!}\;a_1^a a_2^b a_3^c \ldots a_n^l \quad \ldots\ldots\ldots\ldots\ldots\ldots(15).$$

This gives at once a curious theorem, viz. :—

"If in the expression of S_n in terms of the coefficients we multiply each term by the degree of the term, and then divide the whole expression by the degree of the expression, we obtain the sum of all the homogeneous symmetric functions of weight n; and conversely, if in the expression of $H_{(n)}$ in terms of the coefficients we multiply throughout by the degree of the expression, and then divide each term by the degree of the term, we obtain the expression of S_n."

Thus

$$\begin{aligned}
H_{(n)} = {} & a^n-(n-1)\,a_1^{n-2}a_2+(n-2)\,a_1^{n-3}a_3-(n-3)\,a_1^{n-4}\Big\{a_4-(n-2)\frac{a_2^2}{2!}\Big\} \\
& +(n-4)\,a_1^{n-5}\{a_5-(n-3)\,a_2a_3\}-(n-5)\,a_1^{n-6}\Big\{a_6-(n-4)\Big(a_2a_4+\frac{a_3^2}{2!}\Big) \\
& +(n-3)(n-4)\frac{a_2^3}{3!}\Big\}+(n-6)\,a_1^{n-7}\Big\{a_7-(n-5)(a_2a_5+a_3a_4) \\
& +(n-4)(n-5)\frac{a_2^2a_3}{2!}\Big\}-(n-7)\,a_1^{n-8}\Big\{a_8-(n-6)\Big(a_2a_6+a_3a_5+\frac{a_4^2}{2!}\Big) \\
& +(n-5)(n-6)\Big(\frac{a_2^2a_4}{2!}+\frac{a_2a_3^2}{2!}\Big)-(n-4)(n-5)(n-6)\frac{a_2^4}{4!}\Big\} \\
& +(n-8)\,a_1^{n-9}\Big\{a_9-(n-7)(a_2a_7+a_3a_6+a_4a_5) \\
& +(n-6)(n-7)\Big(\frac{a_2^2a_5}{2!}+a_2a_3a_4+\frac{a_3^3}{3!}\Big)-(n-5)(n-6)(n-7)\frac{a_2^3a_3}{3!}\Big\} \\
& -(n-9)\,a_1^{n-10}\Big\{a_{10}-(n-8)\Big(a_2a_8+a_3a_7+a_4a_6+\frac{a_5^2}{2!}\Big) \\
& +(n-7)(n-8)\Big(\frac{a_2^2a_6}{2!}+a_2a_3a_5+\frac{a_2a_4^2}{2!}+\frac{a_3^2a_4}{2!}\Big) \\
& -(n-6)(n-7)(n-8)\Big(\frac{a_2^3a_4}{3!}+\frac{a_2^2a_3^2}{2!2!}\Big) \\
& -(n-5)(n-6)(n-7)(n-8)\frac{a_2^5}{5!}\Big\}+\&c. \quad \ldots\ldots\ldots\ldots\ldots\ldots(16).
\end{aligned}$$

The process, indicated above, by which S_n is transformed into $H_{(n)}$, will, in what follows, be denoted by $\left(\frac{\delta}{\Delta}\right)$, and the reverse process by $\left(\frac{\Delta}{\delta}\right)$; thus

$$\left(\frac{\delta}{\Delta}\right) S_n = H_{(n)} \quad \text{.........................(17)},$$

$$\left(\frac{\Delta}{\delta}\right) H_{(n)} = S_n \quad \text{.............................(18)}.$$

It follows that a large portion of S_{n+1} can be obtained by the use of S_n alone; for from (17) we get $H_{(n)}$, and then from (13)

$$S_{n+1} = (n+1)\int_0^{a_1} H_{(n)}\,da_1 + \dots \quad \text{..............(19)},$$

e.g., $S_3 = a_1^3 - 3a_1a_2 + 3a_3$,

therefore $H_{(3)} = a_1^3 - 2a_1a_2 + a_3$,

therefore $S_4 = 4\left(\tfrac{1}{4}a_1^4 - a_1^2a_2 + a_1a_3\right) + \dots$

The operation denoted by $\left(\frac{\delta}{\Delta}\right)$ is equivalent to another, because, from the last equation of (7), if S_n be the subject, and we introduce a_0 for homogeneity, we find

$$\frac{d}{da_0} S_n = nS_n - nH_{(n)},$$

or

$$H_{(n)} = S_n - \frac{1}{n}\frac{d}{da_0} S_n \quad \text{....................(20)},$$

or

$$\left(\frac{\delta}{\Delta}\right) S_n = S_n - \frac{1}{n}\frac{d}{da_0} S_n \quad \text{....................(21)};$$

the operator $\left(\frac{\delta}{\Delta}\right)$ will be further considered presently.

Reverting now to the general theory, it is clear that the equations (7) enable us to calculate, separately, that portion of a symmetric function which contains any particular coefficient, and in every case by a simple process of integration, and the advantage of this method consists in the fact that the coefficient of a term containing p different coefficients will be subject to $p-1$ independent verifications; in terms whose degree is less than that of the function, a_0 of course takes its place as a coefficient.

The method is illustrated by the calculation of (5^23).

Putting $(5^23) = \phi$,

$$\frac{d\phi}{da_{13}} = d_{13}\phi = +13, \qquad \text{therefore } \phi = 13a_{13} + \dots,$$

$$\frac{d\phi}{da_{12}} = -H_{(1)}d_{13}\phi = -13a_1, \qquad \text{therefore } \phi = -13a_1a_{12} + \dots,$$

$$\frac{d\phi}{da_{11}} = +H_{(2)}d_{13}\phi = 13a_1^2 - 13a_2, \qquad \text{therefore } \phi = 13a_1^2a_{11} - 13a_2a_{11} + \dots,$$

$$\frac{d\phi}{da_{10}} = d_{10}\phi - H_{(3)}d_{13}\phi = +5\,(3) - 13H_{(3)} = -8a_1^3 + 11a_1a_2 + 2a_3,$$

therefore $\phi = -8a_1^3 a_{10} + 11a_1 a_2 a_{10} - 2a_3 a_{10} + \ldots,$

and so on;

$$\frac{d\phi}{da_5} = d_5\phi - H_{(3)} d_8\phi - H_{(5)} d_{10}\phi + H_{(8)} d_{13}\phi$$
$$= 5\,(53) - 8H_{(3)}\,(5) - 5H_{(5)}\,(3) + 13H_{(8)};$$

in reducing this, we reject at once all terms which either are, or give rise to, terms above the fourth degree, for ϕ, we know, is only of the fifth degree.

We find
$$\frac{d\phi}{da_5} = 3a_2^4 - 9a_1 a_2^2 a_3 + 8a_1^2 a_3^2 - 4a_2 a_3^2 + a_1^2 a_2 a_4$$
$$+ a_2^2 a_4 - 3a_1 a_3 a_4 - 7a_4^2 - 8a_1^3 a_5 + 22a_1 a_2 a_5$$
$$+ 6a_3 a_5 + a_1^2 a_6 - 14a_2 a_6 - 14a_1 a_7 + 27a_8,$$

and therefore

$$\phi = 3a_2^4 a_5 - 9a_1 a_2^2 a_3 a_5 + 8a_1^2 a_3^2 a_5 - 4a_2 a_3^2 a_5 + a_1^2 a_2 a_4 a_5 + a_2^2 a_4 a_5 - 3a_1 a_3 a_4 a_5$$
$$- 7a_4^2 a_5 - 4a_1^3 a_5^2 + 11a_1 a_2 a_5^2 + 3a_3 a_5^2 + a_1^2 a_5 a_6 - 14a_2 a_5 a_6 - 14a_1 a_5 a_7 + 27a_5 a_8$$
$$+ \ldots,$$

and so forth.

Finally, we obtain the amply verified result,

$$\begin{aligned}(5^2 3) = {} & a_2^2 a_3^3 - 2a_1 a_3^4 - 3a_2^3 a_3 a_4 + 6a_1 a_2 a_3^2 a_4 + 2a_3^3 a_4 + 3a_1 a_2^2 a_4^2 \\ & - 7a_1^2 a_3 a_4^2 - 4a_2 a_3 a_4^2 + 7a_1 a_4^3 + 3a_2^4 a_5 - 9a_1 a_2^2 a_3 a_5 + 8a_1^2 a_3^2 a_5 \\ & - 4a_2 a_3^2 a_5 + a_1^2 a_2 a_4 a_5 + a_2^2 a_4 a_5 - 3a_1 a_3 a_4 a_5 - 7a_4^2 a_5 - 4a_1^3 a_5^2 \\ & + 11a_1 a_2 a_5^2 + 3a_3 a_5^2 - 3a_1 a_2^3 a_6 + a_1^2 a_2 a_3 a_6 + 16a_2^2 a_3 a_6 - 9a_1 a_3^2 a_6 \\ & + 7a_1^3 a_4 a_6 - 23a_1 a_2 a_4 a_6 + 11a_3 a_4 a_6 + a_1^2 a_5 a_6 - 14a_3 a_5 a_6 + 13a_1 a_6^2 \\ & + 8a_1^2 a_2^2 a_7 - 13a_2 a_7 - 8a_1^3 a_3 a_7 + 7a_1 a_2 a_3 a_7 - 2a_3^2 a_7 + a_1^2 a_4 a_7 \\ & + 26a_2 a_4 a_7 - 14a_1 a_5 a_7 - 13a_6 a_7 - 8a_1^3 a_2 a_8 + 16a_1 a_2^2 a_8 + 16a_1^2 a_3 a_8 \\ & - 29a_2 a_3 a_8 - 14a_1 a_4 a_8 + 27a_5 a_8 + 8a_1^4 a_9 - 24a_1^2 a_2 a_9 + 13a_2^2 a_9 \\ & + 11a_1 a_3 a_9 - 13a_4 a_9 - 8a_1^3 a_{10} + 11a_1 a_2 a_{10} + 2a_3 a_{10} + 13a_1^2 a_{11} \\ & - 13a_2 a_{11} - 13a_1 a_{12} + 13a_{13} \quad \ldots\ldots (22).\end{aligned}$$

4. Since, from equation (13),

$$H_{(m)} = \frac{1}{m+1} \frac{d}{da_1} S_{m+1},$$

$$d_r H_{(m)} = \frac{1}{m+1} d_r \left(\frac{d}{da_1} S_{m+1}\right).$$

Now $\qquad d_r \dfrac{d}{da_1} = \dfrac{d^2}{da_r da_1} + a_1 \dfrac{d^2}{da_{r+1} da_1} + \ldots = P$ suppose.

But $\qquad \dfrac{d}{da_1} d_r = \dfrac{d}{da_1} \left(\dfrac{d}{da_r} + a_1 \dfrac{d}{da_{r+1}} + \ldots\right) = P + \dfrac{d}{da_{r+1}},$

and, r being $< m$, $d_r S_{m+1} = 0$, so that

$$PS_{m+1} + \frac{d}{da_{r+1}} S_{m+1} = 0.$$

Thus $$d_r H_{(m)} = -\frac{1}{m+1} \frac{d}{da_{r+1}} S_{m+1},$$

or, from (12),

$$d_r H_{(m)} = (-)^{r+1} H_{(m-r)} \quad \text{......................(23)};$$

this formula will be of use in the sequel.

Now, since

$$\frac{1}{u} = 1 - a_1 x + a_2 x^2 - a_3 x^3 + \ldots,$$

we have

$$\frac{1}{u}\frac{du}{dx} = (1 - a_1 x + a_2 x^2 - \ldots)(H_{(1)} + 2H_{(2)} x + 3H_{(3)} x^2 + \ldots)$$
$$= S_1 + S_2 x + S_3 x^2 + \ldots$$

Therefore, equating coefficients of like powers of x,

$$rH_{(r)} - (r-1)\, a_1 H_{(r-1)} + (r-2)\, a_2 H_{(r-2)} - \ldots = S_r,$$

or, changing r into $r+1$, and remembering that

$$H_{(r)} - a_1 H_{(r-1)} + \ldots = 0,$$

we find

$$a_1 H_{(r)} - 2a_2 H_{(r-1)} + 3a_3 H_{(r-2)} - \ldots = S_{r+1} \quad \text{...........(24)}.$$

Again, since $r > 1$,

$$(r \,.\, 1^s) = a_{s+1} S_{r-1} - a_{s+2} S_{r-2} + \ldots + (-)^{r+1} (r+s)\, a_{r+s}$$

(see *Proceedings*, Vol. xiv., p. 122); multiplying this by $(-)^s$ and adding Newton's series

$$S_{r+s} - a_1 S_{r+s-1} + a_2 S_{r+s-2} - \ldots + (-)^{r+s} (r+s)\, a_{r+s},$$

we obtain

$$(-)^s (r \,.\, 1^s) = S_{r+s} - a_1 S_{r+s-1} + a_2 S_{r+s-2} - \ldots + (-)^s a_s S_r;$$

consequently

$$H_{(t)}(r) - H_{(t-1)}(r \,.\, 1) + H_{(t-2)}(r \,.\, 1^2) - \ldots + (-)^t (r \,.\, 1^t)$$
$$= H_{(t)} S_r + H_{(t-1)} (S_{r+1} - a_1 S_r) + \ldots$$
$$+ [S_{r+t} - a_1 S_{r+t-1} + a_2 S_{r+t-2} - \ldots + (-)^t a_t S_r],$$
$$= S_{r+t} + S_{r+t-1}(H_{(1)} - a_1) + S_{r+t-2}(H_{(2)} - a_1 H_{(1)} + a_2) + \ldots$$
$$+ S_r [H_{(t)} - a_1 H_{(t-1)} + a_2 H_{(t-2)} - \ldots + (-)^t a_t];$$

therefore, by (6),

$$H_t(r) - H_{(t-1)}(r \,.\, 1) + H_{(t-2)}(r \,.\, 1^2) - \ldots + (-)^t (r \,.\, 1^t) = S_{r+t} \ldots (25).$$

Let now

$$V_{-1} = a_1 \frac{d}{da_0} + 2a_2 \frac{d}{da_1} + 3a_3 \frac{d}{da_2} + \ \ldots\ldots\ldots\ldots\ldots (26),$$

then, from equations (7),

$$V_{-1} = a_1 d_0 - (a_1 H_{(1)} - 2a_2)\, d_1 + (a_1 H_{(2)} - 2a_2 H_{(1)} + 3a_3)\, d_2 - \ldots,$$

and therefore, from equation (24),

$$V_{-1} = S_1 d_0 - S_2 d_1 + S_3 d_2 - S_4 d_3 + \ldots + (-)^r S_{r+1} d_r + \quad \ldots\ldots (27).$$

Also let

$$V_{-r} = (r) \frac{d}{da_0} + (r\,.\,1) \frac{d}{da_1} + (r\,.\,1^2) \frac{d}{da_2} + \ldots + (r\,.\,1^s) \frac{d}{da_s} + \ldots (28);$$

then, from equations (7),

$$V_{-r} = (r)\, d_0 - \{H_{(1)}(r) - (r\,.\,1)\}\, d_1 + \{H_{(2)}(r) - H_{(1)}(r\,.\,1) + (r\,.\,1^2)\} d_2$$
$$- \ldots \quad \ldots \quad \ldots \quad \ldots \quad \ldots \quad \ldots \quad \ldots \quad \ldots;$$

therefore, from equation (25),

$$V_{-r} = S_r d_0 - S_{r+1} d_1 + S_{r+2} d_2 - \ldots + (-)^t S_{r+t} d_t + \ldots\ldots\ldots\ldots (29).$$

Suppose now the subject of operation to be (λ^l), we find

$$V_{-r}(\lambda^l) = \lambda \{S_r(\lambda^l) - S_{r+\lambda}(\lambda^{l-1}) + S_{r+2\lambda}(\lambda^{l-2}) - \ldots + (-)^l S_{r+l\lambda}\};$$

and since

$$S_\mu(\lambda^l) = (\lambda^l.\mu) + (\lambda^{l-1}.\lambda+\mu),$$

unless $\lambda = \mu$, when

$$S_\lambda(\lambda^l) = (l+1)\,(\lambda^{l+1}) + (\lambda^{l-1}.\,2\lambda),$$

we obtain, by substituting in each term of the dexter of the foregoing,

$$V_{-r}(\lambda^l) = \lambda\,(\lambda^l r) \ \ldots\ldots\ldots\ldots\ldots\ldots\ldots\ldots\ldots\ldots (30),$$

$$V_{-\lambda}(\lambda^l) = \lambda\,(l+1)(\lambda^{l+1}) \ldots\ldots\ldots\ldots\ldots\ldots\ldots\ldots (31).$$

So that V_{-r} is an effective inverse operator by means of which $(\lambda^l r)$ can be calculated by means of a single operation performed upon (λ^l).

If $V^r_{-\lambda}$ denote the performance of $V_{-\lambda}$, r times, clearly

$$V^r_{-\lambda}(\lambda^l) = \lambda^r (l+1)(l+2) \ldots (l+r)\,(\lambda^{l+r}) \ \ldots\ldots\ldots\ldots (32).$$

It may further be shown that

$$V_{-\nu}(\lambda\,.\,\mu) = \lambda\,(\lambda\,.\,\mu\,.\,\nu) + \overline{\lambda-\mu}\,(\lambda\,.\,\mu+\nu) \ \ldots\ldots\ldots\ldots (33),$$

$$V_{-\lambda}(\lambda\,.\,\mu) = 2\lambda\,(\lambda^2\mu) + \overline{\lambda-\mu}\,(\lambda\,.\,\mu+\lambda) \ \ldots\ldots\ldots\ldots\ldots (34),$$

$$V_{-t}(\lambda\,.\,\mu\,.\,\nu) = \lambda\,(\lambda\,.\,\mu\,.\,\nu\,.\,t) + \overline{\lambda-\mu}\,(\lambda\,.\,\mu+t\,.\,\nu) + \overline{\lambda-\nu}\,(\lambda\,.\,\mu\,.\,\nu+t) \ldots (35),$$

$$V_{-\lambda}(\lambda\,.\,\mu\,.\,\nu) = 2\lambda\,(\lambda^2.\,\mu\,.\,\nu) + \overline{\lambda-\mu}\,(\lambda\,.\,\mu+\lambda\,.\,\nu) + \overline{\lambda-\nu}\,(\lambda\,.\,\mu\,.\,\nu+\lambda) \ldots (36);$$

for example

$$V_{-2}(5\,.\,4\,.\,3) = 5\,(5\,.\,4\,.\,3\,.\,2) + 2\,(5^2\,.\,4) + (6\,.\,5\,.\,3).$$

By means of the operator V_{-1} performed on (4^3), we find that

$$\begin{aligned}(4^3 1) = &-13a_{13}+9a_1a_{12}-13a_2a_{11}+13a_3a_{10}-23a_4a_9-7a_5a_8\\ &+13a_6a_7-9a_1^2a_{11}-22a_1a_2a_{10}+14a_1a_3a_9+5a_2^2a_9-6a_2a_3a_8\\ &-13a_3^2a_7+18a_1a_4a_8+10a_2a_4a_7+10a_3a_4a_6-7a_4^2a_5-2a_1a_5a_7\\ &-6a_2a_5a_6+7a_3a_5^2-11a_1a_6^2+9a_1^3a_{10}-5a_1^2a_2a_9+5a_2^2a_4a_5\\ &-9a_1^2a_3a_8+5a_1a_2^2a_8+14a_1a_2a_3a_7-a_1a_3^2a_6-5a_1a_2a_5^2-9a_1^2a_4a_7\\ &+11a_1^2a_5a_6-6a_1a_2a_4a_6-2a_1a_3a_4a_5+3a_1a_4^3+a_2^2a_3a_6-a_2a_3^2a_5\\ &-5a_2^3a_7-3a_2a_3a_4^2+a_3^3a_4.\end{aligned}$$

From §1, if in this we put $a_0 = 0$, and then decrease each suffix by unity, we obtain $(4^2 1)$, and if we put $a_0 = a_1 = 0$, and decrease each suffix by 2, we obtain (41): facts which afford a partial verification. The effect of any operator is easily examined; *ex. gr.*, consider the

operator $$\Gamma = a_1\frac{d}{da_0}+a_2\frac{d}{da_1}+a_3\frac{d}{da_2}+\ldots,$$

we have $$\Gamma = H_{(1)}d_0-H_{(2)}d_1+H_{(3)}d_2-\ldots$$

Hence, from the relation $d_\kappa H_{(\lambda)} = (-)^{\kappa+1}H_{(\lambda-\kappa)}$,

$$\Gamma H_{(\lambda)} = \lambda H_{(1)}H_{(\lambda)}-H_{(2)}H_{(\lambda-1)}-H_{(3)}H_{(\lambda-2)}-\ldots-H_{(\lambda+1)} \quad\text{.......(37)},$$

and $$\Gamma(\lambda) = \lambda\{S_1S_\lambda-H_{(\lambda+1)}\},$$

or, making use of the relation (20),

$$\Gamma(\lambda) = \lambda(\lambda 1)+\frac{\lambda}{\lambda+1}\frac{d}{da_0}S_{\lambda+1} \quad\text{......................(38)}.$$

It should be observed that, if a symmetric function be expressed in terms of the sums of the powers of the roots, we can employ a formula given by Brioschi to throw the operator V_{-r} into another form; for Brioschi's formula is

$$\frac{d}{ds_r} = -\frac{1}{r}\left(\frac{d}{da_r}+a_1\frac{d}{da_{r+1}}+a_2\frac{d}{da_{r+2}}+\ldots\right) = -\frac{1}{r}d_r.$$

Consequently,

$$V_{-r} = S_rd_0-S_{r+1}\frac{d}{ds_1}-2S_{r+2}\frac{d}{ds_2}-\ldots-\lambda S_{r+\lambda}\frac{d}{ds_\lambda} \quad\text{.........(39)}.$$

5. The expression $H_{(n)}$, for the sum of the homogeneous symmetric functions of weight n, will now be further examined, and it becomes necessary to amplify its definition, so that we may bring into view other like functions possessing closely analogous properties.

Definition.—Let $H_{(\lambda^l)}$ be a function such that

$$(-)^{l+1}\{H_{(\lambda^l)}-(\lambda^l)\}$$

represents the sum of all the homogeneous symmetric functions of weight $l\lambda$ and of degree $\lambda-1$ at most. Further, let $S\,\overline{w\,.\,j}$ be the sum of all the homogeneous symmetric functions of weight w, and of degree j at most; so that

$$S\,\overline{l\lambda\,.\,\lambda-1} = (-)^{l+1}\{H_{(\lambda^l)} - (\lambda^l)\}\ldots\ldots\ldots\ldots\ldots\ldots(40).$$

The meaning of $H_{(n)}$ is clearly unaltered by the change of definition; viz., it still represents the sum of all the homogeneous symmetric functions of weight n.

It may here be remarked with regard to $H_{(n)}$, that sum of all positive coefficients = sum of all negative coefficients = 2^{n-2}.

It has been proved (*ante*), and reflection shows it to be obvious, that the effect of the symbol $\frac{\delta}{\Delta}$ is equivalent to another operation; that is, if the subject be ϕ, of degree λ, then

$$\frac{\delta}{\Delta}\phi = \phi - \frac{1}{\lambda}\frac{d\phi}{da_0};$$

whence
$$\frac{\delta}{\Delta}(\lambda^2) = H_{(\lambda)}(\lambda) - H_{(2\lambda)},$$

$$\frac{\delta}{\Delta}(\lambda^3) = H_{(\lambda)}(\lambda^2) - H_{(2\lambda)}(\lambda) + H_{(3\lambda)};$$

and, generally,

$$\frac{\delta}{\Delta}(\lambda^l) = H_{(\lambda)}(\lambda^{l-1}) - H_{(2\lambda)}(\lambda^{l-2}) + H_{(3\lambda)}(\lambda^{l-3}) - \ldots(-)^{l+1}H_{l\lambda}\ldots(41).$$

Now, we find by trial that

$$\frac{\delta}{\Delta}(2^2) = (2^2)-(1^4),$$

$$\frac{\delta}{\Delta}(2^3) = (2^3)+(1^6),$$

$$\frac{\delta}{\Delta}(3^2) = (3^2)-(2^3)-(2^21^2)-(2\,1^4)-(1^6),$$

$$\frac{\delta}{\Delta}(3^3) = (3^3)+(2^41)+(2^31^3)+(2^21^5)+(2\,1^7)+(1^9),$$

$$\frac{\delta}{\Delta}(4^2) = (4^2)-(3^22)-(3^21^2)-(3\,2^21)-(2^4)-(3\,2\,1^3)-(3\,1^5)$$
$$-(2^31^2)-(2^21^4)-(2\,1^6)-(1^8),$$

and so on; so that apparently

$$(-)^{l+1}\left\{\frac{\delta}{\Delta}(\lambda^l)-(\lambda^l)\right\} = S\,\overline{l\lambda\,.\,\lambda-1},$$

or
$$H_{(\lambda^l)} = \frac{\delta}{\Delta}(\lambda^l)\ldots\ldots\ldots\ldots\ldots\ldots\ldots\ldots\ldots(42),$$

from which it appears that $H_{(\lambda^l)}$ is derived from (λ^l) in a precisely similar way to that in which $H_{(\lambda)}$ is obtained from (λ).

Also from (41) we obtain the formula

$$H_{(\lambda^l)} = H_{(\lambda)}(\lambda^{l-1}) - H_{(2\lambda)}(\lambda^{l-2}) + H_{(3\lambda)}(\lambda^{l-3}) - \dots (-)^{l+1} H_{(l\lambda)} \quad \dots (43).$$

Remembering the law of operation of d_λ, we have, starting from

$$S\,\overline{\underline{l\lambda\,.\,\lambda-1}} = (-)^{l+1}\{H_{(\lambda^l)} - (\lambda^l)\},$$

$$S\,\overline{\underline{l\lambda-1\,.\,\lambda-1}} = (-)^{l+1} d_1 H_{(\lambda^l)} = (-)^{l+1} d_1 \frac{\delta}{\Delta}(\lambda^l),$$

$$\dots \quad \dots \quad \dots \quad \dots \quad \dots \quad \dots \quad \dots \quad \dots \quad \dots \quad \dots \quad \dots$$

$$S\,\overline{\underline{l\lambda-k\,.\,\lambda-1}} = (-)^{l+k} d_k H_{(\lambda^l)} = (-)^{l+k} d_k \frac{\delta}{\Delta}(\lambda^l) \quad \dots\dots\dots\dots (44),$$

where $k \not> \lambda - 1$.

These equations, with those before obtained, suffice in every case to calculate $S\,\overline{\underline{w\,.\,j}}$.

Now, in equation (43), we see that, on the right-hand side, each term has a factor of the form $H_{(\mu)}$; therefore, on putting all the coefficients equal to unity, the right-hand side must vanish. Consequently, as regards the algebraic sum of the coefficients, the function $H_{(\lambda^l)}$ obeys the same law as $H_{(\lambda)}$; this law may be employed in conjunction with the operation $\frac{\delta}{\Delta}$ to verify the value of (λ^l).

The preceding results lead to the general expression of symmetric function (3^κ), for

$$S\,\overline{\underline{3\kappa\,.\,2}} = (-)^{\kappa+1}\left\{\frac{\delta}{\Delta}(3^\kappa) - (3^\kappa)\right\};$$

and, since
$$\frac{\delta}{\Delta}(3^\kappa) = (3^\kappa) - \frac{1}{3}\frac{d}{da_0}(3^\kappa),$$

we have
$$S\,\overline{\underline{3\kappa\,.\,2}} = (-)^\kappa \frac{1}{3}\frac{d}{da_0}(3^\kappa),$$

and
$$(3^\kappa) = 3(-)^\kappa \int^{a_0} S\,\overline{\underline{3\kappa\,.\,2}}\, da_0 + ((3^{\kappa-1}));$$

the double bracket indicating as before a unit increase of suffixes.

Now, the expression of all the symmetric functions $S\,\overline{\underline{3\kappa\,.\,2}}$ has been given by Serret (*vide Algèbre Supérieure*), whence we derive at once that the coefficient of $a_s a_{3\kappa-s}$ in $S\,\overline{\underline{3\kappa\,.\,2}}$ is

$$1 - \frac{3\kappa-2s}{1!} + \frac{(3\kappa-2s)(3\kappa-2s-3)}{2!} - \frac{(3\kappa-2s)(3\kappa-2s-4)(3\kappa-2s-5)}{3!}$$
$$+ \dots\dots,$$

which is equal to

$$1-(3\kappa-2s)\cdot\frac{1}{3\kappa-2s}\left\{1-2\cos\frac{(3\kappa-2s)\pi}{3}\right\}$$

(*vide* Todhunter's "Theory of Equations," 3rd ed., pp. 186, 187), and this is

$$2\cos\frac{3\kappa-2s}{3}\pi.$$

There are two obvious exceptions, viz., when

$$3\kappa-2s \text{ is either zero or } \pm 1,$$

and then the coefficient is unity.

Therefore, substituting and performing the integration, we arrive at finally the following result:—

$$(3^\kappa)=3!\,(-)^\kappa\left[\frac{1}{2!}\cos\kappa\pi\,.\,a_{3\kappa}+\sum_{s<\frac{1}{2}(3\kappa-1)}^{s=1}\cos\frac{(3\kappa-2s)\pi}{3}\,a_s a_{3\kappa-s}\right.$$
$$\left.+\frac{1}{2}\sum^{s\,\overline{\gtrless}\,\frac{1}{2}(3\kappa-1)}\{a_s a_{3\kappa-s}\}\right]+((3^{\kappa-1})).$$

Ex. gr., $(3^5)=3a_{15}-3a_1a_{14}-3a_2a_{13}+6a_3a_{12}-3a_4a_{11}-3a_5a_{10}$
$+6a_6a_9-3a_7a_8+((3^4)).$

This formula gives an important series of perpetuants.

6. It was proved in § 4 that

$$H_{(t)}(r)-H_{(t-1)}(r1)+H_{(t-2)}(r1^2)-\ldots+(-)^t(r1^t)=S_{r+t},$$

with a modification when $r=1$.

The object here is to examine in some cases the expression

$$H_{(t)}(\lambda\,.\,\mu\,.\,\nu\ldots)-H_{(t-1)}(\lambda\,.\,\mu\,.\,\nu\ldots 1)+H_{(t-2)}(\lambda\,.\,\mu\,.\,\nu\ldots 1^2)-\ldots$$
$$\ldots+(-)^s H_{(t-s)}(\lambda\,.\,\mu\,.\,\nu\ldots 1^s)+\ldots+(-)^t(\lambda\,.\,\mu\,.\,\nu\ldots 1^t),$$

which for brevity will be denoted by

$$F\{(t)\,.\,(\lambda\,.\,\mu\,.\,\nu\ldots)\}.$$

Firstly, F is in every case a non-unitary symmetric function, since the operation d_1 clearly makes it vanish.

Let $T(a, b, c, d)$ represent the sum of the non-unitary symmetric functions of weight a, of b parts, and having no part less than c or greater than d; then the following relations may be verified, viz.:—

$$F\{(0)\,.\,(2^2)\}=\quad(2^2)\quad=T(4\,.\,2\,.\,2\,.\,2+0),$$
$$F\{(1)\,.\,(2^2)\}=\quad(32)\quad=T(5\,.\,2\,.\,2\,.\,2+1),$$
$$F\{(2)\,.\,(2^2)\}=(42)+(3^2)=T(6\,.\,2\,.\,2\,.\,2+2),$$
$$\ldots\quad\ldots\quad\ldots\quad\ldots\quad\ldots\quad\ldots\quad\ldots$$
$$F\{(t)\,.\,(2^2)\}=T(t+4\,.\,2\,.\,2\,.\,2+t)\;\ldots\ldots\ldots\ldots(45),$$

and
$$F\{(t).(3^2)\} = T(t+6.2.3.3+t) \text{ ....(46)},$$
$$F\{(t).(4^2)\} = T(t+8.2.4.4+t) \text{ ....(47)},$$
$$\ldots \quad \ldots \quad \ldots \quad \ldots \quad \ldots$$

and, generally, $F\{(t).(\lambda^2)\} = T(t+2\lambda.2.\lambda.\lambda+t)$..............(48).

These, of course, lead to a series of inverse operators as before; thus

$$(\lambda^2)\frac{d}{da_0}+(\lambda^2 1)\frac{d}{da_1}+(\lambda^2 1^2)\frac{d}{da_2}+\ldots$$
$$=(\lambda^2)d_0 - F\{(1).(\lambda^2)\}d_1 + F\{(2).(\lambda^2)\}d_2 - \ldots + (-)^t F\{(t).(\lambda^2)\}\, d_t \ldots;$$

but, as the results do not appear to be very useful, no further mention will be made of them.

Proceeding, we find

$$F\{(t).(2^3)\} = T(t+6.3.2.2+t) \text{ ....(49)},$$
$$F\{(t).(3^3)\} = T(t+9.3.3.3+t) \text{ ....(50)},$$
$$\ldots \quad \ldots \quad \ldots \quad \ldots \quad \ldots$$
$$F\{(t).(\lambda^3)\} = T(t+3\lambda.3.\lambda.\lambda+t) \text{ ....(51)},$$

and, generally, $F\{(t).(\lambda^l)\} = T(t+l\lambda.l.\lambda.\lambda+t)$ ....(52).

An expression can now be obtained for the sum of all the non-unitary symmetric functions of a given weight; this in our notation is represented by (the weight being w)

$$T(w.1.2.w)+T(w.2.2.w-2)+T(w.3.2.w-4)$$
$$+T(w.4.2.w-6)+\ldots$$
$$= F\{(w-2).(2)\}+F\{(w-4).(2^2)\}+F\{(w-6).(2^3)\}$$
$$+F\{(w-8).(2^4)\}+\ldots$$
$$= H_{(w-2)}(2)-H_{(w-3)}(2\,1)+H_{(w-4)}\{(2\,1^2)+(2^2)\}$$
$$-H_{(w-5)}\{(2\,1^3)+(2^2 1)\}+H_{(w-6)}\{(2\,1^4)+(2^2\,1^2)+(2^3)\}$$
$$-\ldots+(-)^s H_{(w-s)}\{(2\,1^{s-2})+(2^2 1^{s-4})+(2^3 1^{s-6})+\ldots+(2^{\frac{1}{2}s}) \text{ or } (2^{\frac{1}{2}(s-1)}1)\}$$
$$+\ldots\ldots \qquad \text{..............................(53)},$$

the last term being that containing $H_0 = 1$ as a factor.

Ex. gr.—When $w = 6$,

$$H_{(4)}(2)-H_{(3)}(2\,1)+H_{(2)}\{(2\,1^2)+(2^2)\}-H_1\{(2\,1^3)+(2^2 1)\}$$
$$+(2\,1^4)+(2^2 1^2)+(2^3)$$
$$= a_1^6-6a_1^4a_2+10a_1^2a_2^2-3a_2^3+4a_1^3a_3-11a_1a_2a_3+4a_3^2-a_1^2a_4+3a_2a_4-a_1a_5+a_6$$
$$= (6)+(42)+(3^2)+(2^3).$$

An expression can also be deduced for the sum of the non-unitary symmetric functions of weight w, which contain no part less than j;

for, in the notation of this paper, this is

$$
\begin{aligned}
& T(w\,.\,1\,.\,j\,.\,w)+T(w\,.\,2\,.\,j\,.\,w-j)+T(w\,.\,3\,.\,j\,.\,w-2j) \\
& \qquad +T(w\,.\,4\,.\,j\,.\,w-3j)+\ldots \\
= \;& F\{(w-j)\,.\,(j)\}+F\{(w-2j)\,.\,(j^2)\}+F\{(w-3j)\,.\,(j^3)\}+\ldots \\
= \;& H_{(w-j)}(j)-H_{(w-j-1)}(j1)+H_{(w-j-2)}(j1^2)-\ldots \\
& +H_{(w-2j)}\{(j^2)+(-)^j(j1^j)\}-H_{(w-2j-1)}\{(j^21)+(-)^j(j1^{j+1})\} \\
& \qquad +H_{(w-2j-2)}\{(j^21^2)+(-)^j(j1^{j+2})\}-\ldots \\
& +H_{(w-3j)}\{(j^3)+(-)^{2j}(j1^{2j})\}-H_{(w-3j-1)}\{(j^31)+(-)^{2j}(j1^{2j+1})\} \\
& \qquad +H_{(w-3j-2)}\{(j^31^2)+(-)^{2j}(j1^{2j+2})\}-\ldots \\
& +\ldots\ldots \\
& +H_{(w-ij)}\{(j^i)+(-)^{ij-j}(j1^{ij-j})\}-H_{(w-ij-1)}\{(j^i1)+(-)^{ij-j}(j1^{ij-j+1})\} \\
& \qquad +H_{(w-ij-2)}\{(j^i1^2)+(-)^{ij-j}(j1^{ij-j+2})\}-\ldots \\
& +\ldots\ldots \qquad \ldots\ldots\ldots\ldots\ldots\ldots\ldots\ldots\ldots\ldots(54),
\end{aligned}
$$

the series being continued until the term containing H_0 is arrived at.

It is readily seen that any non-unitary symmetric function may be itself expressed in a similar form.

Considering $$F\{(t)\,.\,(\lambda\,.\,\mu\,.\,\nu\ldots)\},$$

$\lambda, \mu, \nu \ldots$ being all or some of them unequal, we obtain various interesting combinations of non-unitary symmetric functions, and the nature of the combination is found to depend upon the relation that t bears to the differences $\lambda-\mu$, $\mu-\nu$, ... ; generally, if the symmetric function $(\lambda_1\,.\,\lambda_2\,.\,\lambda_3\ldots\lambda_s)$ is of weight w, then $F\{(t)\,.\,(\lambda_1\,.\,\lambda_2\ldots\lambda_s)\}$ is composed of symmetric functions of weight $w+t$, of s parts, and having no part less than λ_s or greater than λ_1+t.

As particular cases,

$$
\begin{aligned}
F\{(1)\,.\,(3\,2)\} &= (4\,.\,2)+2\,(3^2), \\
F\{(2)\,.\,(3\,2)\} &= (5\,.\,2)+2\,(4\,.\,3), \\
F\{(1)\,.\,(4\,2)\} &= (5\,.\,2)+(4\,.\,3), \\
F\{(2)\,.\,(4\,2)\} &= (6\,.\,2)+(5\,.\,3)+2\,(4^2), \\
F\{(1)\,.\,(5\,2)\} &= (6\,.\,2)+(5\,.\,3), \\
F\{(2)\,.\,(5\,2)\} &= (7\,.\,2)+(6\,.\,3)+(5\,.\,4), \\
F\{(3)\,.\,(5\,2)\} &= (8\,.\,2)+(7\,.\,3)+(6\,.\,4)+2\,(5^2);
\end{aligned}
$$

the general law being

$$
\begin{aligned}
F\{(t)\,.\,(\mu+k\,.\,\mu)\} &= H_{(t)}(\mu+k\,.\,\mu)-H_{(t-1)}(\mu+k\,.\,\mu\,.\,1)+\ldots \\
&= (\mu+k+t\,.\,\mu)+(\mu+k+t-1\,.\,\mu+1)+\ldots+(\mu+t+1\,.\,\mu+k-1) \\
&\qquad +2\,\{(\mu+t\,.\,\mu+k)+(\mu+t-1\,.\,\mu+k+1)+\ldots\} \;\ldots\ldots\ldots(55),
\end{aligned}
$$

the terms being continued until the first part of the partition is about to become less than the second part.

Analogous results are obtained in the case of the symmetric functions of a greater number of parts, but it does not seem worth while presenting them.

7. Newton's formula for the sums of the powers of the roots,

$$S_w - a_1 S_{w-1} + a_2 S_{w-2} - \dots (-)^{w+1} w a_w = 0,$$

admits of generalisation.

For this purpose, S_w must not be regarded as essentially the sum of the wth powers of the roots, but simply as the sum of the one-part partition symmetric functions of weight w; the notation S^r_w is then adopted to signify the sum of the r-part partition symmetric functions of weight w, and what was before denoted by S_w becomes in the new notation S^1_w.

Let now the rth number in the series of polygonal numbers of the nth order be denoted by n_r, so that the series of numbers of the 1st, 2nd, 3rd, &c. orders are, written horizontally,

$$\begin{array}{lllll} 1_1, & 1_2, & 1_3, & 1_4 \dots\dots \\ 2_1, & 2_2, & 2_3, & 2_4 \dots\dots \\ 3_1, & 3_2, & 3_3, & 3_4 \dots\dots \\ \dots & \dots & \dots & \dots & \dots; \end{array}$$

then we have the following system of equations, viz.,—

$$\begin{aligned} S^0_w &= 1_1 a_0 H_{(w)} - 1_2 a_1 H_{(w-1)} + 1_3 a_2 H_{(w-2)} - \dots (-)^w 1_{w+1} a_w, \\ S^1_w &= 2_1 a_1 H_{(w-1)} - 2_2 a_2 H_{(w-2)} + 2_3 a_3 H_{(w-3)} - \dots (-)^{w+1} 2_w a_w, \\ S^2_w &= 3_1 a_2 H_{(w-2)} - 3_2 a_3 H_{(w-3)} + 3_3 a_4 H_{(w-4)} - \dots (-)^{w+2} 3_{w-1} a_w, \\ S^3_w &= 4_1 a_3 H_{(w-3)} - 4_2 a_4 H_{(w-4)} + 4_3 a_5 H_{(w-5)} - \dots (-)^{w+3} 4_{w-2} a_w, \\ & \dots \quad \dots \quad \dots \quad \dots \quad \dots \quad \dots \quad \dots \\ S^r_w &= \overline{r+1}_1 a_r H_{(w-r)} \overline{r+1}_2 a_{r+1} H_{(w-r-1)} + \overline{r+1}_3 a_{r+2} H_{(w-r-2)} - \dots \\ & \dots + (-)^{w+r} \overline{r+1}_{w-r+1} a_w \dots\dots\dots\dots \quad (56). \end{aligned}$$

These equations may be all readily verified by means of Mr. Hammond's operators; from the general term in $H_{(w)}$, S^r_w can therefore be easily calculated.

Also, starting with Newton's formula,

$$S^1_w - a_1 S^1_{w-1} + a_2 S^1_{w-2} - a_3 S^1_{w-3} + \dots + (-)^{w+1} 2_w a_w = 0,$$

we deduce

$$S^2_w - a_1 S^2_{w-1} + a_2 S^2_{w-2} - a_3 S^2_{w-3} + \dots + (-)^{w+1} 3_{w-1} a_w = 0,$$

$$S^3_w - a_1 S^3_{w-1} + a_2 S^3_{w-2} - a_3 S^3_{w-3} + \dots + (-)^{w+1} 4_{w-2} a_w = 0,$$

and generally

$$S^r_w - a_1 S^r_{w-1} + a_2 S^r_{w-2} - a_3 S^r_{w-3} + \dots + (-)^{w+1} \overline{r+1}_{w-r+1} a_w = 0;$$

or, putting for the polygonal number of the $\overline{r+1}^{\text{th}}$ order its value $\dfrac{w!}{r!\,(w-r)!}$, the formula is perhaps better written in the form

$$S_w^r - a_1 S_{w-1}^r + a_2 S_{w-2}^r - \ldots + (-)^{w+1} \frac{w!}{r!\,(w-r)!} a_w = 0 \ldots\ldots (57),$$

which is the generalisation of Newton's formula referred to at the commencement of this section. The function S_w^r further possesses other properties, analogous and very similar to those possessed by S_w^1, and in all cases the generalisation is effected by means of the polygonal numbers. To obtain the general term in S_w^r it is convenient to extend the notion of the weight of a term, as defined in Salmon's *Higher Algebra;* for this purpose consider any term

$$a_1^{a_1} a_2^{a_2} \ldots\ldots\ldots a_r^{a_r} a_{r+1}^{a_{r+1}} \ldots\ldots\ldots;$$

and define the r-gonal weight of this term to be

$$w_r = r_1 a_{r-1} + r_2 a_r + r_3 a_{r+1} + r_4 a_{r+2} + \ldots\ldots\ldots\ldots\ldots (58),$$

wherein $r_1, r_2, r_3 \ldots$ represent the successive r-gonal numbers; then what is ordinarily denoted by the weight will be the weight of the second order, viz.—

$$w_2 = 2_1 a_1 + 2_2 a_2 + 2_3 a_3 + 2_4 a_4 + \ldots\ldots;$$

then the theorem is that

$$S_w^r = \Sigma (-)^{w_2+k+r-1} \frac{(k-1)!\, w_{r+1}}{a_1!\, a_2!\, a_3! \ldots} a_1^{a_1} a_2^{a_2} a_3^{a_3} \ldots\ldots\ldots\ldots\ldots (59),$$

wherein $$a_1 + a_2 + a_3 + \ldots = k.$$

When $r = 1$, this becomes Waring's formula for the sums of the powers of the roots, and the general theorem is proved as follows:—

Since $$S_w^r = \overline{r+1}_1 a_r H_{(w-r)} - \overline{r+1}_2 a_{r+1} H_{(w-r-1)} + \ldots$$
$$+ (-)^{w+r}\, \overline{r+1}_{w-r+1} a_w,$$

and $$H_{(w)} = \Sigma (-)^{w+k} \frac{k!}{a_1!\, a_2!\, a_3! \ldots} a_1^{a_1} a_2^{a_2} a_3^{a_3} \ldots,$$

we obtain, without difficulty,

$$S_w^r = \Sigma (-)^{w_2+k+r-1} \frac{(k-1)!\, \{\overline{r+1}_1 a_r + \overline{r+1}_2 a_{r+1} + \ldots\}}{a_1!\, a_2!\, a_3!} a_1^{a_1} a_2^{a_2} a_3^{a_3}$$
$$= \Sigma (-)^{w_2+k+r-1} \frac{(k-1)!\, w_{r+1}}{a_1!\, a_2!\, a_3! \ldots} a_1^{a_1} a_2^{a_2} a_3^{a_3} \ldots$$

This theorem is extremely useful in the verification of symmetric functions, and Professor Cayley's law of symmetry enables it to answer a two-fold purpose, viz.,—firstly, as a summation theorem for the r-part partition symmetric functions of weight w; and, secondly, as a theorem for the algebraic sum of the numerical coefficients of the

r-part partition terms, in the expression in terms of the literal coefficients of all symmetric functions of weight w. For instance, the algebraic sum of the numerical coefficients of the r-part partition terms in symmetric function

$$(\lambda^l . \mu^m . \nu^n \ldots)$$

is
$$(-)^{w+l+m+n+\ldots+r-1}\frac{(l+m+n+\ldots-1)!\,w_{r+1}}{l!\,m!\,n!\ldots} \quad \ldots\ldots\ldots\ldots(60),$$

wherein w_{r+1} is the $r+1$-gonal weight of the partition; so that, in calculating a symmetric function independently, by means of the differential equation, we can write down at once relations that must be satisfied by the coefficients, equal in number to the highest index occurring in the expression in the roots.

The algebraic sum of all the coefficients is obviously

$$(-)^{w+l+m+n+\ldots}\frac{(l+m+n+\ldots)!}{l!\,m!\,n!\ldots},$$

so that, as regards a symmetric function of degree p in the coefficients, we have p independent verifications of the numerical coefficients.

The general expression of S^r_n may be thrown into the determinant form, for we have the system of equations

$$\begin{aligned}
&S^r_r \quad +(-)^{r+1}\overline{r+1}_1 a_r = 0,\\
&S^r_{r+1}-a_1 S^r_r+(-)^r\overline{r+1}_2 a_{r+1} = 0,\\
&S^r_{r+2}-a_1 S^r_{r+1}+a_2 S^r_r+(-)^{r+1}\overline{r+1}_3 a_{r+2} = 0,\\
&\ldots \quad \ldots \quad \ldots \quad \ldots \quad \ldots \quad \ldots \quad \ldots \quad \ldots \quad \ldots
\end{aligned}$$

whence, solving for S^r_n, we find if r be odd,

$$S^r_n = \left|\begin{array}{ccccccc}
a_1 & 1 & \ldots & & & & \\
a_2 & a_1 & 1 & \ldots & & & \\
a_3 & a_2 & a_1 & 1 & \ldots & & \\
\ldots & \ldots & \ldots & \ldots & \ldots & & \\
\ldots & \ldots & \ldots & \ldots & \ldots & \ldots & 1 \\
\overline{r+1}_{n-r+1}\,a_n, & \overline{r+1}_{n-r}\,a_{n-1}, & \ldots & \&c. & \ldots & & \overline{r+1}_1\,a_r
\end{array}\right| \quad \ldots\ldots\ldots\ldots(61);$$

if r be even,

$$S^r_n = \left|\begin{array}{ccccccc}
a_1 & -1 & & & & & \\
-a_2 & +a_1 & -1 & & & & \\
+a_3 & -a_2 & +a_1 & -1 & & & \\
\ldots & \ldots & \ldots & \ldots & & & \\
\ldots & \ldots & \ldots & \ldots & \ldots & \ldots & -1 \\
(-)^n\overline{r+1}_{n-r+1}\,a_n, & (-)^{n+1}\overline{r+1}_{n-r}, a_{n-1} & \ldots & \&c. & \ldots & & +\overline{r+1}_1\,a_r
\end{array}\right|$$

$$\ldots\ldots\ldots\ldots(62).$$

In the annexed table are given the expressions of the single partition seminvariants, otherwise non-unitary symmetric functions of weight 13; they are intended as a contribution to the study of the weight 13 sources of covariants of binary quantics; they were calculated as follows:—By the method of §2 the non-unitary portions of all the symmetric functions of weight 13 were calculated and tabulated vertically, headed by the partition of the function, the left-hand column representing the partitions of the terms; then of course, regarding the vertical partition column as pertaining to the functions and the horizontal partition row to the terms, the result is a table of the non-unitary symmetric functions.

As every possible verification afforded by the general expression for S^r_{13} was employed, it is scarcely possible that there should be an error; *ex. gr.*, consider the ten-part partition terms: we find

$$S^{10}_{13} = 11_1 a_{10} H_{(3)} - 11_2 a_{11} H_{(2)} + 11_3 a_{12} H_{(1)} - 11_4 a_{13},$$

$$= a_{10} \cdot a_3 - 11 a_{11}(-a_2) - \frac{13!}{10!\,3!} a_{13} + \text{terms containing } a_1,$$

$$= a_3 a_{10} + 11 a_2 a_{11} - 286 a_{13} + \ldots\ldots,$$

whilst the columns headed by partitions of ten parts are

	(4.1^9)	$(3.2.1^8)$	$(2^3.1^7)$		S^{10}_{13}
a_{13}	-13	-117	-156	=	-286
$a_2 a_{11}$	$+2$	$+18$	-9	=	$+11$
$a_3 a_{10}$	$+3$	-3	$+1$	=	$+1$

which is correct, and is given as a specimen of the verifications employed throughout.

The terms are arranged on Mr. Durfee's plan. The general values of S^2_n, S^3_n, ... for a few terms are:—

$$S^2_n = a_1^{n-2} a_2 - 3a_1^{n-3} a_3 + a_1^{n-4}\{6a_4 - (n-3)a_2^2\}$$

$$-a_1^{n-5}\{10a_5 - (1+3)(n-4)a_2 a_3\}$$

$$+a_1^{n-6}\left[15a_6 - (n-5)\{(1+6)a_2 a_4 + 3a_3^2\} + (n-4)(n-5)\frac{a_2^3}{2!}\right]$$

$$-a_1^{n-7}\left[21a_7 - (n-6)\{(1+10)a_2 a_5 + (3+6)a_3 a_4\} + (2.1+3)(n-5)(n-6)\frac{a_2^2 a_3}{2!}\right]$$

$$+a_1^{n-8}\left[28a_8 - (n-7)\{(1+15)a_2 a_6 + (3+10)a_3 a_5 + 6a_4^2\} + (n-6)(n-7)\left\{(2.1+6)\frac{a_2^2 a_4}{2!} + (1+2.3)\frac{a_2 a_3^2}{2!}\right\} - (n-5)(n-6)(n-7)\frac{a_2^4}{3!}\right]$$

$$-\ldots\ldots\ldots\ldots\ldots$$

$$S_n^3 = a_1^{n-3}a_3 - 4a_1^{n-4}a_4 + a_1^{n-5}\{10a_5 - (n-4)\,a_2a_3\}$$
$$-a_1^{n-6}[20a_6 - (n-5)\{4a_2a_4 + a_3^2\}]$$
$$+a_1^{n-7}\left[35a_7 - (n-6)\{10a_2a_5 + (1+4)a_3a_4\} + (n-5)(n-6)\frac{a_2^2a_3}{2\,!}\right]$$
$$-a_1^{n-8}\Big[56a_8 - (n-7)\{20a_2a_6 + (1+10)a_3a_5 + 4a_4^2\}$$
$$+(n-6)(n-7)\left\{a_2a_3^2 + 4\frac{a_2^2a_4}{2\,!}\right\}\Big]$$
$$+ \ldots\ldots\ldots\ldots\ldots\ldots$$

$$S_n^4 = a_1^{n-4}a_4 - 5a_1^{n-5}a_5 + a_1^{n-6}\{15a_6 - (n-5)\,a_2a_4\}$$
$$-a_1^{n-7}[35a_7 - (n-6)\{5a_2a_5 + a_3a_4\}]$$
$$+a_1^{n-8}\left[70a_8 - (n-7)\{15a_2a_6 + 5a_3a_5 + a_4^2\} + (n-6)(n-7)\frac{a_2^2a_4}{2\,!}\right]$$
$$-a_1^{n-9}\Big[126a_9 - (n-8)\{35a_2a_7 + 15a_3a_6 + (1+5)\,a_4a_5\}$$
$$+(n-7)(n-8)\left\{5\frac{a_2^2a_5}{2\,!} + a_2a_3a_4\right\}\Big]$$
$$+ \ldots\ldots\ldots\ldots\ldots\ldots,$$

and so on.

Another means of verifying any symmetric function may here be noted. Consider the equation

$$x^n + \frac{x^{n-1}}{1\,!} + \frac{x^{n-2}}{2\,!} + \frac{x^{n-3}}{3\,!} + \ldots + \frac{1}{n\,!} = 0,$$

or say $\phi(x) = 0$; we have

$$\frac{\phi(x)}{x^n} = e^{+\frac{1}{x}} - \frac{x^{-n-1}}{(n+1)\,!} - \frac{x^{-n-2}}{(n+2)\,!} - \ldots$$
$$= e^{+\frac{1}{x}}\left\{1 - \frac{e^{-\frac{1}{x}}x^{-n-1}}{(n+1)\,!} - \frac{e^{-\frac{1}{x}}x^{-n-2}}{(n+2)\,!} - \ldots\right\},$$

$$\therefore\ \log\frac{\phi(x)}{x^n} = +\frac{1}{x} + \log\left\{1 - \frac{e^{-\frac{1}{x}}x^{-n-1}}{(n+1)\,!} - \frac{e^{-\frac{1}{x}}x^{-n-2}}{(n+2)\,!} - \ldots\right\}$$
$$= +\frac{1}{x} + \text{terms involving } \frac{1}{x^{n+1}} \text{ and higher powers of } \frac{1}{x},$$

but (Todhunter's *Theory of Equations*, p. 181)

$$-\frac{1}{m}S_m^1 \text{ is the coefficient of } \frac{1}{x^m} \text{ in the dexter of this identity.}$$

Consequently the sums of all powers of the roots of the above equation from the 2nd to the nth inclusive vanish, and therefore also every symmetric function of the roots to the nth degree inclusive (with the exception of the coefficients of the equation) vanishes. Therefore the expression of a symmetric function vanishes on putting for each literal coefficient the reciprocal of the factorial of its suffix.

[I ought to make clear my acknowledgments to Mr. Hammond for many suggestions on this paper which I have almost invariably adopted.]

Single Partition Seminvariants, otherwise Non-unitary Symmetric Functions of Weight 13.

	(13)	(12 . 1)	(11 . 2)	(10 . 3)	(9 . 4)	(8 . 5)	(7 . 6)	$(11 . 1^2)$	(10 . 2 . 1)	(9 . 3 . 1)	$(9 . 2^2)$	(8 . 4 . 1)	(8 . 3 . 2)	(7 . 5 . 1)	(7 . 4 . 2)	$(7 . 3^2)$	$(6^2 . 1)$	(6 . 5 . 2)	(6 . 4 . 3)
(13)	+13	−13	−13	−13	−13	−13	−13	+13	+26	+26	+13	+26	+26	+26	+26	+13	+13	+26	+26
(11 . 2)	−13	+13	−9	+13	+13	+13	+13	−2	−4	−26	+9	−26	−4	−26	−4	−13	−13	−4	−26
(10 . 3)	−13	+13	+13	−17	+13	+13	+13	−13	+4	+4	−13	−26	+4	−26	−26	+17	−13	−26	+4
(9 . 4)	−13	+13	+13	+13	−23	+13	+13	−13	−26	+10	+5	+10	−26	−26	+10	−13	−13	−26	+10
(8 . 5)	−13	+13	+13	+13	+13	−27	+13	−13	−26	−26	−13	+14	+14	+14	−26	−13	−13	+14	−26
(7 . 6)	−13	+13	+13	+13	+13	+13	−29	−13	−26	−26	−13	−26	−26	+16	+16	+8	+29	+16	+16
$(9 . 2^2)$	+13	−13	+9	−13	+5	−13	−13	+2	+4	+8	0	+8	+4	+26	−14	+13	+13	+4	+8
(8 . 3 . 2)	+26	−26	−4	+4	−26	+14	−26	+15	0	+22	+4	+12	+8	+12	+30	−4	+26	−10	+22
(7 . 4 . 2)	+26	−26	−4	−26	+10	−26	+16	+15	+30	+16	−14	+16	+30	+10	+8	+5	−16	−12	−26
$(7 . 3^2)$	+13	−13	−13	+17	−13	−13	+8	+13	−4	−4	+13	+26	−4	+5	+5	+4	−8	+5	−25
(6 . 5 . 2)	+26	−26	−4	−26	−26	+14	+16	+15	+30	+52	+4	+12	−10	−30	−12	+5	−16	+8	+10
(6 . 4 . 3)	+26	−26	−26	+4	+10	−26	+16	+26	+22	−14	+8	+16	+22	+10	−26	−25	−16	+10	+16
$(5^2 . 3)$	+13	−13	−13	+2	−13	+27	−13	+13	+11	+11	+13	−14	−29	−14	+26	−2	+13	−14	+11
$(5 . 4^2)$	+13	−13	−13	−13	+23	+7	−13	+13	+26	−10	−5	−30	+6	+6	−10	+13	+13	+6	−10
$(7 . 2^3)$	−13	+13	−9	+13	−5	+13	−1	−2	−4	−8	0	−8	−4	−12	0	−6	+1	+10	+6
$(6 . 3 . 2^2)$	−39	+39	−5	+9	+21	−1	−3	−17	−4	−30	−4	−20	−12	+4	+26	+12	+3	−12	−24
$(5 . 4 . 2^2)$	−39	+39	−5	+39	−15	−1	−3	−17	−34	−24	+14	+16	+6	+4	+6	−18	+3	−12	+18
$(5 . 3^2 . 2)$	−39	+39	+17	−21	+39	−41	+18	−28	+4	−18	−17	+2	+36	+23	−35	0	−18	+15	+3
$(4^2 . 3 . 2)$	−39	+39	+17	+9	−33	+19	−3	−28	−26	+24	+19	+14	−30	−16	+2	+12	+3	+6	−6
$(5 . 2^4)$	+13	−13	+9	−13	+5	−3	+1	+2	+4	+8	0	−2	−6	+2	0	+6	−1	0	−6
$(4 . 3^3)$	−13	+13	+13	−17	+1	+13	−8	−13	+4	+16	−7	−14	+4	−5	+7	−4	+8	−5	+1
$(4 . 3 . 2^3)$	+52	−52	+14	−22	+20	−12	+4	+19	+8	+2	−14	−8	+16	+8	−6	−6	−4	+2	+6
$(3^3 . 2^2)$	+26	−26	−4	+19	−20	+14	−5	+15	−15	+1	+7	+6	−7	−9	+3	+2	+5	−1	−2
$(3 . 2^5)$	−13	+13	−9	+7	−5	+3	−1	−2	+2	−2	0	+2	0	−2	0	0	+1		

	$(5^2.3)$	(5.4^2)	(10.1^3)	$(9.2.1^2)$	$(8.3.1^2)$	$(8.2^2.1)$	$(7.4.1^2)$	$(7.3.2.1)$	(7.2^3)	$(6.5.1^2)$	$(6.4.2.1)$	$(6.3^2.1)$	$(6.3.2^2)$	$(5^2.2.1)$	$(5.4.3.1)$	$(5.4.2^2)$	$(5.3^2.2)$	$(4^3.1)$	(9.1^4)	$(8.2.1^3)$	$(7.2^2.1^2)$
(13)	+13	+13	−13	−39	−39	−39	−39	−78	−13	−39	−78	−39	−39	−39	−78	−39	−39	−13	+13	+52	+78
(11.2)	−13	−13	+2	+6	+28	−5	+28	+34	−9	+28	+34	+39	−5	+17	+78	−5	+17	+13	−2	−8	−1
(10.3)	+2	−13	+3	+9	+9	+9	+39	−12	+13	+39	+48	−21	+9	+24	+18	+39	−21	+13	−3	−12	−18
(9.4)	−13	+23	+13	+3	+3	+21	+3	+42	−5	+39	+6	+3	+21	+39	−30	−15	+39	−23	−4	−16	−24
(8.5)	+27	+7	+13	+39	−1	−1	−1	+38	+13	−1	+38	+39	−1	−41	−42	−1	−41	−7	−13	−12	−38
(7.6)	−13	−13	+13	+39	+39	+39	−3	−6	−1	−45	−48	−24	−3	−3	+36	−3	+18	+13	−13	−52	−15
(9.2^2)	+13	−5	−2	−6	−10	−4	−10	−16	0	−28	+2	−21	−4	−17	−24	+14	−17	+5	+2	+8	+10
$(8.3.2)$	−29	+6	−5	−15	−21	−12	−27	−30	−4	−27	−42	−18	−12	+39	+24	+6	+36	−6	+5	+20	+27
$(7.4.2)$	+26	−10	−15	−9	−31	−16	−17	−48	0	+17	+30	+21	+26	−14	−6	+6	−35	+10	+6	+24	+32
(7.3^2)	−2	+13	−3	−9	−9	−9	−18	−9	−6	+3	+15	+21	+12	−3	+3	−18	0	−13	+3	+12	+18
$(6.5.2)$	−14	+6	−15	−45	−27	+6	+15	+12	+10	+27	−6	−15	−12	+6	+6	−12	+15	−6	+15	+20	−24
$(6.4.3)$	+11	−10	−16	−12	−12	−30	0	+54	+6	+6	0	+9	−24	−21	−18	+18	+3	+10	+7	+28	−21
$(5^2.3)$	+3	−7	−8	−24	+16	+16	+1	+7	−13	+1	−23	−9	+16	+11	−3	+1	−4	+7	+8	−8	+8
(5.4^2)	−7	+7	−13	−3	+17	−1	+17	−22	+5	−19	+14	−3	−1	+1	+10	−5	+1	−7	+4	−4	+4
(7.2^3)	−13	+5	+2	+6	+10	+4	+10	+16	0	0	−16	0	−10	+3	+10	0	+10	−5	−2	−8	−10
$(6.3.2^2)$	+16	−1	+7	+21	+31	+16	−5	−38	−10	+1	+10	+12	+16	−4	−6	−2	−10	+1	−7	−28	+26
$(5.4.2^2)$	+1	−5	+17	+15	+1	−20	−13	+24	0	+1	−12	0	−2	+11	−10	0	+2	+5	−8	+8	−2
$(5.3^2.2)$	−4	+1	+8	+24	−10	−19	+5	−1	+10	−1	+17	−3	−10	−11	+3	+2	+4	−1	−8	+8	−5
$(4^2.3.2)$	−4	+3	+18	−18	−4	+11	0	+4	−5	+2	−2	−6	+5	−2	+8	−1	−2	−3	0	0	0
(5.2^4)	+3	0	−2	−6	0	+6	0	−6	0	0	+6	0	0	−3	0	0	0	0	+2	−2	0
(4.3^3)	+2	−1	+3	−3	−3	+3	+6	−3	0	−3	−3	+3	0	+3	−3	0	0	+1			
$(4.3.2^3)$	−3	0	−9	+9	−5	−2	+3	+2	0	−1	−2	0	0	+1							
$(3^3.2^2)$	+1																				

	(13)	(11.2)	(10.3)	(9.4)	(8.5)	(7.6)	(9.2^2)	(8.3.2)	(7.4.2)	(7.3^2)	(6.5.2)	(6.4.3)	$(5^2.3)$	(5.4^2)	(7.2^3)	$(6.3.2^2)$	$(5.4.2^2)$	$(5.3^2.2)$	$(4^2.3.2)$	(5.2^4)
$(5.2^2.1^4)$	+195	−19	−45	−60	+25	−6	+28	−28	+2	+3	0	0	0	0	0	0	0	0		
$(5.2^3.1^2)$	−130	−13	+40	+40	−30	+11	−14	+37	−6	−12	+2	+3	0	0	0	0	0	0		
$(5.3.2.1^3)$	−260	+106	+30	+80	−60	+8	−52	+48	−18	+12	−4	−2	0	0	+10	−2	0	0		
$(5.4.1^4)$	−65	+32	+45	+20	−15	−19	−14	+3	+4	−3	+7	−5	0	0	0	−1	0	0		
(5.2^4)	+13	+9	−13	+5	−3	+1	0	−6	0	+6	0	−6	+3	0	0	0	0	0		
$(5.3.2^2.1)$	+156	−24	−6	−84	+84	−30	+24	−66	+38	+6	−6	0	−9	+4	−10	+6	−2	0		
(6.1^7)	−13	+2	+3	+4	+5	0	−2	−5	0	0	0	0	0	0	0	0	0	0		
$(6.2.1^5)$	+78	−12	−18	−24	−30	+6	+12	+30	−6	−3	0	0	0	0	+2	0	0	0		
$(6.2^2.1^3)$	−130	+9	+30	+40	+50	−17	−18	−47	+28	+12	−2	−7	0	0	−10	+2	0	0		
(7.1^6)	+13	−2	−3	−4	−5	−6	+2	+5	+6	+3	0	0	0	0	−2	0	0	0		
$(5.3^2.1^2)$	+78	−67	+12	−6	+22	−15	+31	−21	+10	−12	−3	+3	+8	−4	−10	+2	+4	−2		
$(4^2.3.2)$	−39	+17	+9	−33	+19	−3	+19	−30	+2	+12	+6	−6	−4	+3	−5	+5	−1	−2	+1	
$(6.3.1^4)$	−65	+32	+15	+20	+25	−19	−14	−31	+4	+6	−3	+7	0	0	0	+3	0	0	0	
$(7.2.1^4)$	−65	+10	+15	+20	+25	+23	−10	−25	−30	−15	+7	+7	0	0	+10	−7	0	0	0	
(8.1^5)	−13	+2	+3	+4	+5	+13	−2	−5	−6	−3	−7	−7	0	0	+2	+7	0	0	0	
$(5.4.2.1^2)$	+156	−68	−96	−12	+44	+12	+14	−36	−4	+12	+6	+12	+1	−8	0	+4	0	−1	0	
$(5^2.1^3)$	+26	−15	−21	−26	+14	+16	+15	−4	−1	0	−11	+5	−4	+6	−1	+1	−4	+4	0	+1
$(6.2^3.1)$	+52	+14	−22	−16	−12	+4	+4	+16	−26	−6	+2	+18	−3	−4	+10	−6	+2	0	0	0
$(6.3.2\,1^2)$	+156	−68	−6	−48	−76	+54	+32	+66	−38	−36	+24	−12	+1	+8	+10	−14	−4	+5	0	0
$(6.4.1^3)$	+52	−30	−42	−16	−12	+32	+12	+32	−10	0	−12	−2	+7	−4	+2	−2	+8	−7	0	−2
$(7.3,1^3)$	+52	−30	−12	−16	−12	−10	+12	+26	+32	+12	0	−14	−8	−4	−12	+4	−4	+2	0	+2

	$(5.3.1^5)$	$(5.2.1^6)$	(5.1^8)	(4.3^3)	$(4^2.3.1^2)$	$(4^2.2^2.1)$	$(4.3^2.2.1)$	$(4.3.2^3)$	$(4^2.2.1^3)$	$(4.3^2.1^3)$	$(4.3.2^2.1^2)$	$(4.2^4.1)$	$(4^2.1^5)$	$(4.3.2.1^4)$	$(4.2^3.1^3)$	$(4.3.1^6)$	$(4.2^2.1^5)$	$(4.2.1^7)$	(4.1^9)	$(3^4.1)$
(13)	+78	−91	+13	−13	+78	+78	+156	+52	−130	−130	−390	−65	+39	+390	+260	−91	−273	+104	−13	+13
(11.2)	−34	+14	−2	+13	−67	−12	−90	+14	+53	+97	+93	−23	−17	−148	+4	+36	+31	−16	+2	−13
(10.3)	−18	+21	−3	−17	−33	−48	+54	−22	+75	0	+30	+35	−24	−60	−70	+21	+63	−24	+3	+17
(9.4)	−24	+28	−4	+1	+66	+48	+24	+20	−50	−50	−60	−25	+6	+60	+40	−8	−24	+4	0	−1
(8.5)	+10	−5	0	+13	+22	−18	−36	−12	−10	−10	+30	+15	+5	+10	−20	−5	+5	0	0	−13
(7.6)	+6	0	0	−8	−36	+6	+12	+4	+4	+25	−9	−5	+3	−12	+6	0	0	0	0	+8
(9.2^2)	+16	−14	+2	−7	−5	−33	0	−14	+19	−7	+42	+14	−1	−14	−28	0	+14	−2	0	+7
$(8.3.2)$	−4	+5	0	+4	+12	+24	+12	+16	−8	−5	−27	−10	+1	+8	+10	−1	−2	0	0	−4
$(7.4.2)$	+2	0	0	+7	−13	−8	+10	−6	+11	+2	−12	+6	−3	+4	−2	0	0	0	0	−7
(7.3^2)	−3	0	0	−4	+12	+6	−12	−6	−12	0	+12	0	+3	−3	0	0	0	0	0	+4
$(6.5.2)$	0	0	0	−5	+3	+6	−12	+2	−7	+3	+6	−2	0	0	0	0	0	0	0	+5
$(6.4.3)$	0	0	0	+1	−3	−12	+6	+6	+5	−1	−3	0	0	0	0	0	0	0	0	−1
$(5^2.3)$	0	0	0	+2	−7	+3	+6	−3	0	0	0	0	0	0	0	0	0	0	0	−2
(5.4^2)	0	0	0	−1	+4	+2	−4	0	0	0	0	0	0	0	0	0	0	0	0	+1
(7.2^3)	−2	0	0	0	+5	+5	0	0	−5	0	0	0	+1							
$(6.3.2^2)$	0	0	0	0	−1	−3	0	0	+1											
$(5.4.2^2)$	0	0	0	0	−2	+1														
$(5.3^2.2)$	0	0	0	0	+1															

	$(3^3.2^2)$	$(3^3.2.1^2)$	$(3^2.2^3.1)$	(3.2^5)	$(3^3.1^4)$	$(3^2.2^2.1^3)$	$(3.2^4.1^2)$	$(3^2.2.1^5)$	$(3.2^3.1^4)$	$(3^2.1^7)$	$(3.2^2.1^6)$	$(3.2.1^8)$	(3.1^{10})	$(2^6.1)$	$(2^5.1^3)$	$(2^4.1^5)$	$(2^3.1^7)$	$(2^2.1^9)$	(2.1^{11})	(1^{13})
(13)	+26	−130	−130	−13	+65	+390	+195	−273	−455	+52	+364	−117	+13	+13	−91	+182	−156	+65	−13	+1
(11.2)	−4	+75	−2	−9	−43	−104	+36	+97	+15	−19	−45	+18	−2	+9	−30	+27	−9	+1		
(10.3)	+19	−80	−35	+7	+20	+105	0	−42	−35	+3	+21	−3	0	−7	+14	−7	+1			
(9.4)	−20	+10	+40	−5	+10	−30	−15	−6	+20	+2	−4	0	0	+5	−5	+1				
(8.5)	+14	+30	−30	+3	−5	−10	+15	+5	−5	0	0	0	0	−3	+1					
(7.6)	−5	−17	+11	−1	−2	+9	−6	0	0	0	0	0	0	+1						
(9.2^2)	+7	−21	−7	0	+7	+14	0	−7	0	+1										
(8.3.2)	−7	+9	+5	0	−1	−5	0	+1												
(7.4.2)	+3	+6	−3	0	−2	+1														
(7.3^2)	+2	−4	0	0	+1															
(6.5.2)	−1	−3	+1																	
(6.4.3)	−2	+1																		
$(5^2.3)$	+1																			

Note on the Development of an Algebraic Fraction.

BY CAPT. P. A. MACMAHON, R. A.

In the *American Journal of Mathematics*, Vol. V, No. 3, M. Faà de Bruno has considered the development, in ascending powers of x, of the algebraic fraction

$$\phi(x) = \frac{1}{1 + a_1 x + a_2 x^2 + \ldots + a_n x^n},$$

and has obtained the coefficient of x^p in the form of a determinant.

His result may be simply obtained as follows.

For convenience I take the fraction to be

$$f(x) = \frac{1}{1 - a_1 x + a_2 x^2 - \ldots + (-)^n a_n x^n}.$$

Let

$$F(y) = y^n - a_1 y^{n-1} + a_2 y^{n-2} - \ldots + (-)^n a_n$$
$$= (y-\alpha)(y-\beta)(y-\gamma)\ldots$$

so that $\alpha, \beta, \gamma \ldots$ are the roots of the equation $F(y) = 0$;

then

$$\frac{1}{y-\alpha} = \frac{1}{y} + \frac{\alpha}{y^2} + \frac{\alpha^2}{y^3} + \ldots$$

and

$$\frac{1}{F(y)} = \frac{1}{y^n} + \frac{H_1}{y^{n+1}} + \frac{H_2}{y^{n+2}} + \ldots + \frac{H_p}{y^{n+p}} + \ldots$$

wherein H_p represents the sum of the homogeneous symmetric functions, of weight p, of the roots of the equation

$$F(y) = 0.$$

Write now $y = \frac{1}{x}$ and divide both sides of the resulting equation by x^n, thus obtaining

$$\frac{1}{1 - a_1 x + a_2 x^2 - \ldots + (-)^n a_n x^n} = 1 + H_1 x + H_2 x^2 + \ldots + H_p x^p + \ldots$$

It is well known that

$$H_p = \begin{vmatrix} a_1 & a_2 & a_3 & \ldots & a_p \\ 1 & a_1 & a_2 & \ldots & a_{p-1} \\ 0 & 1 & a_1 & \ldots & a_{p-2} \\ 0 & 0 & 1 & \ldots & a_{p-3} \\ \ldots & \ldots & \ldots & \ldots & \ldots \\ 0 & 0 & 0 & \ldots\ 1 & a_1 \end{vmatrix}$$

which is equivalent to M. Faà de Bruno's result.

Now $$H_p = \Sigma(-)^{p+\alpha_1+\alpha_2+\alpha_3+\dots}\frac{(\alpha_1+\alpha_2+\alpha_3+\dots)!}{\alpha_1!\,\alpha_2!\,\alpha_3!\dots}a_1^{\alpha_1}a_2^{\alpha_2}a_3^{\alpha_3}\dots$$
the summation extending to all integer, including zero, solutions of the equation $$\alpha_1+2\alpha_2+3\alpha_3+\dots=p;$$
consequently we have the result
$$\frac{1}{1-a_1x+a_2x^2-\dots+(-)^n a_n x^n}$$
$$=\sum_{p=\infty}^{p=0}\sum(-)^{p+k}\frac{k!}{\alpha_1!\,\alpha_2!\,\alpha_3!\dots}a_1^{\alpha_1}a_2^{\alpha_2}a_3^{\alpha_3}\dots x^p,$$
where
$$\alpha_1+\alpha_2+\alpha_3+\dots=k,$$
$$\alpha_1+2\alpha_2+3\alpha_3+\dots=p.$$

Royal Military Academy, Woolwich, *October* 2, 1883.

Note by Dr. Franklin.

The general coefficient in the expansion of
$$\frac{1}{a_0+a_1x+a_2x^2+\dots+a_nx^n}$$
is obviously given immediately in the form of a determinant by comparison of coefficients. If the required series is
$$b_0+b_1x+b_2x^2+\dots+b_nx^n,$$
we have
$$(a_0+a_1x+a_2x^2+\dots+a_nx^n)(b_0+b_1x+b_2x^2+\dots+b_nx^n)=1,$$
whence
$$\begin{aligned} a_0b_0 &= 1\\ a_1b_0+a_0b_1 &= 0\\ a_2b_0+a_1b_1+a_0b_2 &= 0, \text{ etc.,}\end{aligned}$$
whence, solving for the b's,
$$b_p=(-)^p\left(\frac{1}{a_0}\right)^{p+1}\begin{vmatrix} a_1 & a_0 & & & \\ a_2 & a_1 & a_0 & & \\ a_3 & a_2 & a_1 & a_0 & \\ \dots & \dots & \dots & \dots & \dots \\ \dots & \dots & \dots & \dots & \dots \\ a_p & a_{p-1} & \dots & \dots & a_0 \end{vmatrix}$$

The above is so obvious that I have been in the habit of regarding it as the natural method of obtaining the value of H_p, whereas, in the preceding note, Captain MacMahon has reversed the process.

A NEW THEOREM IN SYMMETRIC FUNCTIONS.

By Captain P. A. MacMahon, R.A.

REFERENCE is made to a paper on Symmetric Functions by Mr. Hammond (*Proc. Lond. Math. Soc.* vol. xiv. p. 119), and to a paper on the same subject, vol. xv. p. 21 of the same *Proceedings*.

The equation considered is invariably

$$a_0x^n - a_1x^{n-1} + a_2x^{n-2} - a_3x^{n-3} + \ldots = 0,$$

and the partition notation is adhered to throughout; that is to say, the symmetric function of the roots $\Sigma\alpha^\kappa\beta^\lambda\gamma^\mu\ldots$ is written symbolically $(\kappa\lambda\mu\ldots)$.

It is a well known theorem that, writing

$$(\kappa\lambda\mu\ldots) = T_\kappa + T_{\kappa-1} + T_{\kappa-2} + \ldots + T_2 + T_1 \ldots\ldots (1),$$

wherein T_κ represents the terms of the symmetric function of degree κ, and so on,

$$T_\kappa = a_1^{\kappa-\lambda}(\lambda\mu\ldots)_1,$$

the subscript $_1$ denoting that the function to which it is attached has been rendered homogeneous by the introduction of a_0 raised to the proper power, and that it has been then subjected to a unit increase of suffix throughout. The terms of highest degree in any function are thus very easily written down, and I purpose here to obtain the general theorem of which this is a particular case, and to shew how the terms of lower degree may be successively obtained on the same principle.

The truth of the theorem above-mentioned is at once seen from a consideration of Mr. Hammond's operator D_κ; in the paper to which reference has been made, the law of operation is proved to be

$$D_\kappa(\lambda^l\mu^m\nu^n\ldots) = 0, \quad D_\lambda(\lambda^l\mu^m\nu^n\ldots) = (\lambda^{l-1}\mu^m\nu^n\ldots), \quad D_\lambda(\lambda) = 1.$$

But it is further necessary to ascertain the law of operation upon a compound function, and a very slight extension of Mr. Hammond's proof suffices easily to prove that

$$D_\kappa(\lambda_1\mu_1\nu_1\ldots)(\lambda_2\mu_2\nu_2\ldots)(\lambda_3\mu_3\nu_3\ldots)\ldots$$
$$= \Sigma_\kappa(\mu_1\nu_1\ldots)(\lambda_2\nu_2\ldots)(\lambda_3\mu_3\ldots)\ldots,$$

where $\qquad \lambda_1 + \mu_2 + \nu_3 + \ldots = \kappa;$

that is to say, we have to take out in succession all possible partitions of κ that occur (not more than one part from each partition of the function), and then to add all the results together: for instance

$$D_4(432)(21) = (32)(21) + (42)(2) + (43)(1),$$

$$D_6(432)(21) = (32)(1).$$

I digress here to observe, that this last example shews that the process of decapitation of a symmetric function identity of degree κ, recently introduced into the theory of invariants by Professor Cayley, is merely the performance of the operator D_κ; we may similarly operate on the identity with $D_{\kappa-1}$, $D_{\kappa-2}$, ..., thus proceeding to other identities of weights lower by $\kappa - 1$, $\kappa - 2$, &c. Thus, since it may be proved that

$$(43^4)(2^5) - (3^5)(32^4) - (5^442) + 2(5^43^2) - (5^3432^2) + 3(5^33^32)$$
$$- (5^243^22^3) + 4(5^23^42^2) - (543^32^4) + 5(53^52^3) - (43^42^5) + 6(3^62^4) = 0;$$

operating with D_6, D_3, D_4, we get

$$(3^4)(2^4)-(3^4)(2^4)=0;$$

$$(43^3)(2^4)-(3^4)(32^3)-(5^342)+2(5^33^2)-(5^2432^2)+3(5^23^32)$$
$$-(543^22^3)+4(53^42^2)-(43^32^4)+5(3^52^3)=0;$$

$$(3^4)(2^5)-(5^42)-(5^332^2)-(5^23^22^3)-(53^32^4)-3^42^5=0.$$

Consider now equation (1), T_κ consist of terms of the form $a_1{}^{\alpha_1}a_2{}^{\alpha_2}a_3{}^{\alpha_3}a_4{}^{\alpha_4}\ldots$ to numerical coefficients *près*, which may be written

$$(1)^{\alpha_1}(1^2)^{\alpha_2}(1^3)^{\alpha_3}(1^4)^{\alpha_4}\ldots,$$

where $\qquad \alpha_1+\alpha_2+\alpha_3+\alpha_4+\ldots=\kappa;$

whence $\qquad D_\kappa a_1{}^{\alpha_1}a_2{}^{\alpha_2}a_3{}^{\alpha_3}a_4{}^{\alpha_4}\ldots=a_1{}^{\alpha_2}a_2{}^{\alpha_3}a_3{}^{\alpha_4}\ldots,$

and $\qquad D_\kappa T_\kappa=a_0{}^{\lambda-\kappa}(T_\kappa)_{-1},$

$(T_\kappa)_{-1}$ denoting T_κ subjected to a unit decrease of suffix; wherefore since $D_\kappa(\kappa\lambda\mu\ldots)=D_\kappa T_\kappa$,

$$T_\kappa=a_1{}^{\kappa-\lambda}(\lambda\mu\ldots)_1,$$

the theorem in question.

Write now $(\kappa\lambda\mu\ldots)=\phi$ for brevity, and suppose that the symmetric function is rendered homogeneous by the introduction of a_0 raised to the proper power, so that

$$\phi=T_\kappa+a_0T_{\kappa-1}+a_0^2T_{\kappa-2}+\ldots\ldots\ldots\ldots\ (1);$$

when D_i is written in the homogeneous form, viz:—

$$D_i=\frac{1}{i!}\left(a_0\frac{d}{da_1}+a_1\frac{d}{da_2}+\ldots\right)^i—$$

it is obvious that if D_i operating upon any homogeneous function of the coefficient does not make it vanish, it cannot in any case lower its degree in a_0, a_1, $a_2\ldots$.

From equation (1) we get

$$D_i\phi=D_iT_\kappa+a_0D_iT_{\kappa-1}+a_0^2D_iT_{\kappa-2}+\ldots+a^{\kappa-j}D_iT_j+\ldots,$$

It is to be shewn that

$$D_iT_j=\Delta_{j-i}(T_j)_{-1},$$

wherein

$$\Delta_{j-i}=\frac{1}{(j-i)!}\left(a_1\frac{d}{da_0}+a_2\frac{d}{da_1}+a_3\frac{d}{da_2}+\ldots\right)^{j-i},$$

and $(T_j)_{-1}$ denotes T_j when subjected to a unit decrease of suffix throughout.

Suppose $$T_j = f(a_1, a_2, a_3 \ldots),$$
so that $$(T_j)_{-1} = f(a_0, a_1, a_2 \ldots);$$
then $$f(a_1 + a_0\mu, a_2 + a_1\mu, a_3 + a_2\mu, \ldots)$$
$$= T_j + D_1 T_j \mu + D_2 T_j \mu^2 + \ldots + D_i T_j \mu^i + \ldots$$
$$= \sum_{i=j}^{i=0} D_i T_j \mu^i;$$
writing $\frac{1}{\mu}$ for μ and multiplying up by μ^j, we get
$$f(a_0 + a_1\mu, a_1 + a_2\mu, a_2 + a_3\mu, \ldots) = \sum_{i=j}^{i=0} D_i T_j \mu^{j-i};$$
but from Taylor's theorem
$$f(a_0 + a_1\mu, a_1 + a_2\mu, a_2 + a_3\lambda, \ldots) = \sum_{i=0}^{i=j} \Delta_{j-i} (T_j)_{-1} \mu^{j-i},$$
and consequently
$$D_i T_j = \Delta_{j-i} (T_j)_{-1}.$$

We now obtain the identity
$$D_i \phi = \Delta_{\kappa-i} (T_\kappa)_{-1} + a_0 \Delta_{\kappa-i-1} (T_{\kappa-1})_{-1} + a_0^2 \Delta_{\kappa-i-2} (T_{\kappa-2})_{-1} + \ldots$$
$$+ a_0^s \Delta_{\kappa-i-s} (T_{\kappa-s})_{-1} + \ldots \quad \ldots\ldots\ldots\ldots\ldots (2)$$
in which putting $i = \kappa$,
$$(T_\kappa)_{-1} = D_\kappa \phi = a_0^{\kappa-\lambda} (\lambda\mu \ldots)$$
or $$T_\kappa = a_1^{\kappa-\lambda} (\lambda\mu \ldots)_1 \ldots\ldots\ldots\ldots\ldots\ldots (3);$$
and putting $i = \kappa - 1$,
$$D_{\kappa-1} \phi = \Delta_1 (T_\kappa)_{-1} + a_0 (T_{\kappa-1})_{-1},$$
or $$a_0 (T_{\kappa-1})_{-1} = D_{\kappa-1} \phi - \Delta_1 (D_\kappa \phi),$$
that is
$$T_{\kappa-1} = \frac{1}{a_1} (D_{\kappa-1} \phi)_1 - \frac{1}{a_1} (\Delta_1)_1 (D_\kappa \phi)_1 \ldots\ldots\ldots (4),$$
which is the corresponding theorem for terms of highest-but-one degree.

Also
$$T_{\kappa-2} = \frac{1}{a_1^2} (D_{\kappa-2} \phi)_1 - \frac{1}{a_1} (\Delta_1)_1 T_{\kappa-1} - \frac{1}{a_1^2} (\Delta_2)_1 T_\kappa$$
$$= \frac{1}{a_1^2} (D_{\kappa-2} \phi)_1 - \frac{1}{a_1} (\Delta_1)_1 \frac{1}{a_1} (D_{\kappa-1} \phi)_1 - \left\{ \frac{1}{a_1^2} (\Delta_2)_1 - \frac{1}{a_1} \Delta_1 \frac{1}{a_1} \Delta_1 \right\} (D_\kappa \phi)_1,$$

or writing $$\frac{1}{a_1^u}(\Delta_u)_1 \frac{1}{a^v}(\Delta_v)_1 \ldots = \overline{uv\ldots},$$
this is
$$T_{\kappa-2} = \frac{1}{a_1^2}(D_{\kappa-2}\phi)_1 - \bar{1}\,(D_{\kappa-1}\phi)_1 - (\bar{2} - \overline{11})\,(D_\kappa\phi)_1 \ldots (5);$$
and $$T_{\kappa-3} = \frac{1}{a_1^3}(D_{\kappa-3}\phi)_1 - \bar{1}\frac{1}{a_1^2}(D_{\kappa-2}\phi)_1 - (\bar{2} - \overline{11})\frac{1}{a_1}(D_{\kappa-1}\phi)_1$$
$$- (\bar{3} - \overline{21} - \overline{12} + \overline{111})\,(D_\kappa\phi)_1 \ldots\ldots\ldots\ldots (6);$$
generally
$$T_{\kappa-s} = \frac{1}{a_1^s}(D_{\kappa-s}\phi)_1 - \bar{1}\frac{1}{a_1^{s-1}}(D_{\kappa-s+1}\phi)_1 - (\bar{2} - \overline{11})\frac{1}{a_1^{s-2}}(D_{\kappa-s+2}\phi)_1$$
$$- (\bar{3} - \overline{21} - \overline{12} + \overline{111})\frac{1}{a_1^{s-3}}(D_{\kappa-s+3}\phi)_1 - \ldots \quad \ldots (7)$$
the law of formation of the operators being obvious.

These formulæ are often very useful in calculating a symmetric function; as an example, to obtain the terms of the seventh degree in (8321) or say $T_7(8321)$, we have
$$T_7(8321) = -\frac{1}{a_1}(\Delta_1)_1\, a_1^5(321)_1\,;$$
hence, since
$$(321) = a_1a_2a_3 - 3a_0a_3^2 - 3a_1^2a_4 + 4a_0a_2a_4 + 7a_0a_1a_5 - 12a_0^2a_6,$$
$$T_7(8321) = -\frac{1}{a_1}\left(a_2\frac{d}{da_1} + a_3\frac{d}{da_2} + \ldots\right)(a_1^5a_2a_3a_4 - 3a_1^6a_4^2$$
$$- 3a_1^5a_2^2a_5 + 4a_1^6a_3a_5 + 7a_1^6a_2a_6 - 12a_1^7a_7)$$
$$= -\ 5a_1^3a_2^2a_3a\ - a_1^4a_3^2a_4 + 17a_1^4a_2a_4^2 + 15a_1^3a_2^3a_5 - 19a_1^4a_2a_3a$$
$$+ 2a_1^5a_4a_5 - 39a_1^4a_2^2a_6 - 11a_1^5a_3a_6 + 77a_1^5a_2a_7 + 12a_1^6a_8.$$

The algebraic sum of these coefficients should be (*vide Proc. Lon. Math. Soc.*, Vol. xv. p. 40)
$$(-)^{1+4+7-1}3\,!\ 8 = 48,$$
and thus a partial verification is afforded.

Again $$T_6 = -\frac{1}{a_1}(\Delta_1)_1\, T_7 - \frac{1}{a_1^2}(\Delta_2)_1\, T_8,$$
and with slightly more labour T_6 may be obtained; it appears then that these theorems are chiefly useful in the calculation of $T_{\kappa-1}$ and $T_{\kappa-2}$ when $\kappa - \lambda$ is large.

It may be added that if $\kappa - \lambda$ be not less that $2s$, $a_1^{\kappa-\lambda-2s}$ is a factor of $T_{\kappa-s}$.

Royal Military Academy, Woolwich,
October 21*st*, 1884.

NOTE ON RATIONALISATION.

By *Capt. P. A. MacMahon, R.A.*

To rationalise the equation

$$a_1^{\frac{1}{m}} + a_2^{\frac{1}{m}} + a_3^{\frac{1}{m}} + \ldots + a_n^{\frac{1}{m}} = 0,$$

is the same problem as that of finding the simplest rational algebraic function of the n quantities, which contains the sinister of the above equation as a factor. The case, $m = 2$, was considered by Meyer Hirsch at the beginning of this century, and it is given in his algebra; the general case may also have been worked out, but I have thought that its expression, by means of partitions, may not be without interest.

To rationalize the equation

$$y = a_1^{\frac{1}{m}} + a_2^{\frac{1}{m}} + \ldots + a_n^{\frac{1}{m}},$$

is merely the formation of the equation in y, of which this is one solution; this equation will be of degree m^n in y, since the dexter may assume m^n different values; and further, the powers of y will all be of the form $m^n - km$, k a positive integer, and we may thus put $y^m = x$, obtaining an equation in x wherein the indices decrease by unity, and which will be the rationalization of the equation

$$x^{\frac{1}{m}} = a_1^{\frac{1}{m}} + a_2^{\frac{1}{m}} + \ldots + a_n^{\frac{1}{m}}.$$

The equation will then be of degree m^{n-1}, its roots being the mth powers of the m^{n-1} internally different values of the expression

$$a_1^{\frac{1}{m}} + a_2^{\frac{1}{m}} + \ldots + a_n^{\frac{1}{m}}.$$

If ω_m be an mth root of unity, ω_m will not be the quotient of any two such values.

Let S_r be the sum of the rth powers of the roots of the resulting equation, and let the symmetric function $\Sigma a_1^p a_2^q a_3^r \ldots$ be denoted by the partition symbol $[pqr\ldots]$.

Then it may be shewn that

$$m^{1-n}S_1=[1],$$

$$m^{1-n}S_2=[2]+\frac{(2m)!}{(m!)^2}[1^2],$$

$$m^{1-n}S_3=[3]+\frac{(3m)!}{(2m)!\,m!}[21]+\frac{(3m)!}{(m!)^3}[1^3],$$

$$m^{1-n}S_4=[4]+\frac{(4m)!}{(3m)!\,m!}[31]+\frac{(4m)!}{(2m!)^2}[2^2],$$

$$+\frac{(4m)!}{(2m)!\,(m!)^2}[21^2]+\frac{(4m)!}{(m!)^4}[1^4],$$

....

This is, in fact, merely an interesting extension of the multinomial theorem of algebra which I have not before seen mentioned, though it may possibly have been long known.

The simple multinomial theorem may be expressed thus:—

$$(a_1+a_2+a_3+\ldots+a_n)^r=\Sigma\frac{(l\lambda+m\mu+n\nu+\ldots)!}{(\lambda!)^l\,(\mu!)^m\,(\nu!)^n\ldots}[\lambda^l\mu^m\nu^n\ldots].$$

The generalisation is

$$\Sigma\,(a_1^{\frac{1}{m}}+a_2^{\frac{1}{m}}+\ldots+a_n^{\frac{1}{m}})^{mr}$$

$$=m^{n-1}\Sigma\frac{\{m\,(l\lambda+m\mu+n\nu+\ldots)\}!}{\{(m\lambda)!\}^l\,\{(m\mu)!\}^m\,\{(m\nu)!\}^n\ldots}[\lambda^l\mu^m\nu^n\ldots]$$

where the summation on the sinister side has reference to the internally different values.

For $m=1$, the latter degrades to the former.

Whence $$m^{1-n}S_r$$

$$=\Sigma\frac{\{m\,(l\lambda+m\mu+n\nu+\ldots)\}!}{\{(m\lambda)!\}^l\,\{(m\mu)!\}^m\,\{(m\nu)!\}^n\ldots}[\lambda^l\mu^m\nu^n\ldots].$$

If the equation in x be

$$x^{m^{n-1}}-A_1x^{m^{n-1}-1}+A_2x^{m^{n-1}-2}-\ldots=0,$$

we know from the general theory of symmetric functions that

$$A_j=\Sigma\frac{\left(\frac{S_1}{1}\right)^{t_1}\left(\frac{S_2}{2}\right)^{t_2}\ldots\left(\frac{S_j}{j}\right)^{t_j}}{t_1!\,t_2!\,\ldots t_j!}(-)^{j+\Sigma t},$$

wherein $$t_1 + 2t_2 + 3t_3 + \ldots + jt_j = j.$$

We have thus the means of calculating the coefficients in succession.

The term not involving x when equated to zero will be the mth power of the rationalisation of

$$a_1^{\frac{1}{m}} + a_2^{\frac{1}{m}} + \ldots + a_n^{\frac{1}{m}} = 0,$$

which rationalisation consequently will consist of the continued product of the m^{n-1} internally different values of the sinister term, and its degree will be m^{n-2}.

The same result is arrived at much more easily by putting $x = \pm a_{n+1}$, and then putting $n-1$ for n.

The great advantage of this method is that it is only necessary to calculate about half the coefficients, the known symmetry of the result enabling one to write down the remaining terms.

It will suffice in this short note to give two results arrived at by this process,

$$a_1^{\frac{1}{5}} + a_2^{\frac{1}{5}} + a_3^{\frac{1}{5}} = 0,$$

leads to

$$(a_1+a_2+a_3)^5 - 5^4(a_1+a_2+a_3)^2 a_1a_2a_3 + 5^5(a_2a_3+a_3a_1+a_1a_2)a_1a_2a_3 = 0;$$

and $$a_1^{\frac{1}{7}} + a_2^{\frac{1}{7}} + a_3^{\frac{1}{7}} = 0,$$

leads to

$$(a_1+a_2+a_3)^7 - 5.7^4(a_1+a_2+a_3)^4 a_1a_2a_3$$
$$+ 2.7^6(a_1+a_2+a_3)^2(a_2a_3+a_3a_1+a_1a_2)\,a_1a_2a_3$$
$$- 7^7(a_2a_3+a_3a_1+a_1a_2)^2 a_1a_2a_3 + 7^8(a_1+a_2+a_3)\,a_1^2a_2^2a_3^2 = 0.$$

Royal Military Academy,
Woolwich.
August 24th, 1885.

THE LAW OF SYMMETRY AND OTHER THEOREMS IN SYMMETRIC FUNCTIONS.

By Captain P. A. MacMahon, R.A.

I GIVE what I conceive to be the complete statement of the law of symmetry which was first noticed by Cayley, and which has since been demonstrated by Betti and Hammond.

I consider the expression of a symmetric function in terms of the quantities H_0, H_1, H_2, ...; H_w denoting the weight-w homogeneous product sum of the roots of the equation; indicate a method of calculating such a function, and prove that a similar law of symmetry obtains.

It is shewn that if $f(a_0, a_1, a_2, a_3, \ldots)$ satisfy the partial differential equation

$$d_m \equiv a_0 \partial_{a_m} + a_1 \partial_{a_{m+1}} + a_2 \partial_{a_{m+2}} + \ldots = 0,$$

$f(H_0, H_1, H_2, H_3, \ldots)$ is also a solution.

As a particular result it is found that

$$H_0 \partial_{H_1} + H_1 \partial_{H_2} + H_2 \partial_{H_3} + \ldots$$

is a binariant operator.

I prove the law of operation of

$$D_t \equiv \frac{1}{t!} (a_0 \partial_{a_1} + a_1 \partial_{a_2} + a_2 \partial_{a_3} + \ldots)^t$$

upon a compound symmetric function.

---

* La méthode employée dans ce mémoire peut s'étendre à l'espace et sert à l'étude des systèmes de sphères; voir à ce sujet deux travaux que j'ai publiés entre les *Memorie* (Série II., Vol. XXVI.) et les *Atti* (Vol. XX.) de l'Académie de Sciences de Turin.

In this paper I take as my starting point a paper by J. Hammond (*Proc. Lond. Math. Soc.*, vol. XIV. pp. 119, &c.), "On the use of certain Differential Operators in the Theory of Equations."

Considering the equation ($n=\infty$),

$$a_0x^n - a_1x^{n-1} + a_2x^{n-2} - a_3x^{n-3} + \ldots = 0 \ldots\ldots\ldots(1),$$

he puts $$D_\kappa = \frac{1}{\kappa!}\overline{(a_0\partial_{a_1} + a_1\partial_{a_2} + a_2\partial_{a_3} + \ldots)^\kappa},$$

the symbols being combined according to the ordinary laws of quantity, and $(\lambda^l.\mu^m.\nu^n\ldots)$ being any symmetric function of (1), he deduces the law of operation of D_κ; his own statement being

$$D_\kappa(\lambda^l.\mu^m.\nu^n\ldots)=0,\ D_\lambda(\lambda^l.\mu^m.\nu^n\ldots)=(\lambda^{l-1}.\mu^m.\nu^n\ldots),\ D_\lambda(\lambda)=1;$$

he further shews that such operators follow the ordinary laws of quantity in their combinations with constants and with each other, and that denoting by partitions in [ ] the symmetric functions of the roots of the equation

$$1-y^{-1}D_1+y^{-2}D_2-y^{-3}D_3+\ldots=0 \ \ldots\ldots\ldots (2),$$

then $$[\lambda^l.\mu^m.\nu^n\ldots]=\overline{d_\lambda{}^l d_\mu{}^m d_\nu{}^n\ldots} \div l!\,m!\,n!\ldots,$$

wherein $$d_\lambda \equiv a_0\partial_{a_\lambda} + a_1\partial_{a_{\lambda+1}} + a_2\partial_{a_{\lambda+2}} + \ldots,$$

and the line placed over the dexter expression indicates that the symbols are to be combined according to the ordinary laws of quantity.

§1. Suppose any two results of the same weight in symmetric functions to be

$$\Sigma A_{\lambda^l,\mu^m,\nu^n\ldots}\ (\lambda^l.\mu^m.\nu^n\ldots) = \Sigma A'_{\lambda'^{l'},\mu'^{m'},\nu'^{n'}\ldots}\ a^{l'}_{\lambda'}a^{m'}_{\mu'}a^{n'}_{\nu'}\ldots \text{(I.)},$$

$$\Sigma B_{\lambda'^{l'},\mu'^{m'},\nu'^{n'}\ldots}\ (\lambda'^{l'}.\mu'^{m'}.\nu'^{n'}\ldots) = \Sigma B'_{\lambda^l,\mu^m,\nu^n\ldots}\ a^l_\lambda a^m_\mu a^n_\nu\ldots \text{(II.)},$$

wherein the A's and B's are numerical constants.

Translating (I.) into a relation between operators, we have

$$\Sigma A_{\lambda^l,\mu^m,\nu^n\ldots}\frac{\overline{d^l_\lambda d^m_\mu d^n_\nu\ldots}}{l!\,m!\,n!\ldots} = \Sigma A'_{\lambda'^{l'},\mu'^{m'},\nu'^{n'}\ldots} D^{l'}D^{m'}D^{n'}\ldots,$$

and operating with each side of this identity upon the opposite side of (II.), and remembering that in this case

$$\overline{d^l_\lambda d^m_\mu d^n_\nu\ldots} \equiv \left(\frac{d}{da_\lambda}\right)^l\left(\frac{d}{da_\mu}\right)^m\left(\frac{d}{da_\nu}\right)^n\ldots,$$

we have

$$\Sigma A_{\lambda^l.\mu^m.\nu^n\ldots} B'_{\lambda^l.\mu^m.\nu^n\ldots} = \Sigma B_{\lambda'^{l'}.\mu'^{m'}.\nu'^{n'}\ldots} A'_{\lambda'^{l'}.\mu'^{m'}.\nu'^{n'}\ldots} \ldots (3),$$

for, as remarked by Hammond, no other terms survive the operations.

This result (3) appears to be the most general statement of the law of symmetry, and it includes as particular cases the two portions of it that were noticed by Cayley.

A mutual verification is, as a general rule, afforded by any two results of like weight.

A particular case may be noticed:—If

$$(\lambda^l.\mu^m.\nu^n\ldots) = \Sigma A'_{\lambda'^{l'}.\mu'^{m'}.\nu'^{n'}\ldots} a^{l'}_{\lambda'} a^{m'}_{\mu'} a^{n'}_{\nu'} \ldots,$$

and $$a^l_\lambda a^m_\mu a^n_\nu \ldots = \Sigma B_{\lambda'^{l'}.\mu'^{m'}.\nu'^{n'}\ldots} (\lambda'^{l'}.\mu'^{m'}.\nu'^{n'}\ldots),$$

then $$\Sigma A'_{\lambda'^{l'}.\mu'^{m'}.\nu'^{n'}\ldots} B_{\lambda'^{l'}.\mu'^{m'}.\nu'^{n'}\ldots} = 1,$$

establishing a mutual verification between the two tables of symmetric functions of the same weight.

§ 2. The equation $x^n - a_1 x^{n-1} + a_2 x^{n-2} - a_3 x^{n-3} + \ldots = 0$ may be written in the form

$$x^n - H_1 x^{n-1} + (H_1^2 - H_2)\,x^{n-2} - (H_1^3 - 2H_1H_2 + H_3)\,x^{n-3} + \ldots = 0,$$

wherein H_w denotes the weight-w homogeneous product sum of the roots of the equation; multiplying this by the factor $x - \kappa$, so as to introduce a new root κ, it becomes converted into

$$x^{n+1} - (H_1 + \kappa)\,x^n + \{(H_1+\kappa)^2 - (H_2 + H_1\kappa + \kappa^2)\}\,x^{n-1}$$
$$- \{(H_1+\kappa)^3 - 2\,(H_1+\kappa)\,(H_2 + H_1\kappa + \kappa^2)$$
$$+ (H_3 + H_2\kappa + H_1\kappa^2 + \kappa^3)\}\,x^{n-2} - \ldots = 0,$$

since H_w becomes

$$H_w + H_{w-1}\kappa + H_{w-2}\kappa^2 + \ldots + \kappa^w;$$

if $\phi(H_1, H_2, H_3, \ldots)$ denotes the symmetric function $(\lambda^l.\mu^m.\nu^n\ldots)$, and the operator $\partial_{H_t} + H_1\partial_{H_{t+1}} + H_2\partial_{H_{t+2}} + \ldots$ be put $= \delta_t$, ϕ becomes transformed into

$$\phi\,(H_1 + \kappa,\ H_2 + H_1\kappa + \kappa^2,\ H_3 + H_2\kappa + H_1\kappa^2 + \kappa^3,\ \ldots)$$

$$= \phi + (\kappa\delta_1 + \kappa^2\delta_2 + \kappa^3\delta_3 + \ldots)\,\phi + \frac{1}{2!}\,(\kappa\delta_1 + \kappa^2\delta_2 + \kappa^3\delta_3 + \ldots)^2\phi + \ldots$$

$$=\phi+\kappa\delta_1\phi+\frac{\kappa^2}{2!}\begin{vmatrix}\delta_1, & 2\delta_2\\ -1, & \delta_1\end{vmatrix}\phi+\frac{\kappa^3}{3!}\begin{vmatrix}\delta_1, & 2\delta_2, & 3\delta_3\\ -2, & \delta_1, & 2\delta_2\\ 0, & -1, & \delta_1\end{vmatrix}\phi+\ldots$$

$$=\phi+\kappa\Delta_1\phi+\kappa^2\Delta_2\phi+\kappa^3\Delta_3\phi+\ldots,$$

where, generally,

$$s!\,\Delta_s=\begin{vmatrix}\delta_1 & , 2\delta_2 & , 3\delta_3, & \ldots, & s\delta_s\\ -(s-1), & \delta_1 & , 2\delta_2, & \ldots, & (s-1)\,\delta_{s-1}\\ 0 & , -(s-2), & \delta_1, & \ldots, & (s-2)\,\delta_{s-2}\\ \ldots & \ldots & \ldots & \ldots & \ldots\\ \ldots & \ldots & \ldots & \ldots & \ldots\\ 0 & , 0 & , 0, & \ldots -1, & \delta_1\end{vmatrix};$$

the introduction of the new root κ changes $(\lambda^l.\mu^m.\nu^n\ldots)$ into

$$(\lambda^l.\mu^m.\nu^n\ldots)+\kappa^\lambda(\lambda^{l-1}.\mu^m.\nu^n\ldots)+\kappa^\mu(\lambda^l.\mu^{m-1}.\nu^n\ldots)$$
$$+\kappa^\nu(\lambda^l.\mu^m.\nu^{n-1}\ldots)+\ldots,$$

so that Δ_s, when operating upon a partition, is precisely equivalent to D_s.

Also
$$\delta_{\lambda-1}\Delta_1=\overline{\delta_{\lambda-1}\Delta_1}+\delta_\lambda,$$
$$\delta_{\lambda-2}\Delta_2=\overline{\delta_{\lambda-2}\Delta_2}+\delta'^{*}_{\lambda-2}\,\Delta_2,\dagger$$
$$=\overline{\delta_{\lambda-2}\Delta_2}+\delta_{\lambda-1}\,\Delta_1,$$
$$=\overline{\delta_{\lambda-2}\Delta_2}+\overline{\delta_{\lambda-1}\Delta_1}+\delta_\lambda,$$

and generally

$$\delta_{\lambda-s}\Delta_s=\overline{\delta_{\lambda-s}\Delta_s}+\overline{\delta_{\lambda-s+1}\Delta_{s-1}}+\overline{\delta_{\lambda-s+2}\Delta_{s-2}}+\ldots+\delta_\lambda,$$

a theorem which is proved as follows:—

Put

$$J=1+(\delta_1\kappa+\delta_2\kappa^2+\ldots)+\frac{1}{2!}(\delta_1\kappa+\delta_2\kappa^2+\ldots)^2+\frac{1}{3!}(\delta_1\kappa+\delta_2\kappa^2+\ldots)^3+\ldots$$

$$=1+\Delta_1\kappa+\Delta_2\kappa^2+\Delta_3\kappa^3+\ldots;$$

therefore

$$\delta_{\lambda-s}J=\overline{\delta_{\lambda-s}J}+\overline{(\delta_{\lambda-s+1}\kappa+\delta_{\lambda-s+2}\kappa^2+\delta_{\lambda-s+3}\kappa^3+\ldots)\,J},$$

† The asterisk denoting the operation of $\delta_{\lambda-2}$ upon Δ_2 considered as an explicit function of the H's.

and equating coefficients of κ^s there results

$$\delta_{\lambda-s}\Delta_s = \overline{\delta_{\lambda-s}\Delta_s} + \overline{\delta_{\lambda-s+1}\Delta_{s-1}} + \overline{\delta_{\lambda-s+2}\Delta_{s-2}} + \ldots + \delta_\lambda,$$

the theorem in question.

It may be noticed also that

$$\delta_{\lambda-s} * \Delta_s = \delta_{\lambda-s+1}\Delta_{s-1},$$

and further, that

$$s!\,\Delta_s = \begin{vmatrix} \delta_1, & \delta_2, & \delta_3, & \ldots, & \delta_s \\ -(s-1), & \delta_1, & \delta_2, & \ldots, & \delta_{s-1} \\ 0, & -(s-2), & \delta_1, & \ldots, & \delta_{s-2} \\ 0, & 0, & -(s-3), & \ldots, & \delta_{s-3} \\ \cdots & \cdots & \cdots & \cdots & \cdots \\ \cdots & \cdots & \cdots & \cdots & \cdots \\ 0, & 0, & \ldots & -1, & \delta_1 \end{vmatrix},$$

wherein the symbols in the expanded form denote complete operations.

This determinant involves the law for the expression of the homogeneous product sums in terms of the sums of powers of the roots, whence

$$\delta_\lambda + \delta_{\lambda-1}\Delta_1 + \delta_{\lambda-2}\Delta_2 + \delta_{\lambda-3}\Delta_3 + \ldots + \delta_1\Delta_{\lambda-1}$$
$$= \overline{\delta_1\Delta_{\lambda-1}} + 2\,\overline{\delta_2\Delta_{\lambda-2}} + 3\overline{\delta_3\Delta_{\lambda-3}} + \ldots + \lambda\delta_\lambda = \lambda\Delta_\lambda,$$

which, in regard to the equation

$$1 - y^{-1}\Delta_1 + y^{-2}(\Delta_1^2 - \Delta_2) - y^{-3}(\Delta_1^3 - 2\Delta_1\Delta_2 + \Delta_3) + \ldots = 0 \ldots (4),$$

is Newton's rule for the sums of the powers of the roots, and gives

$$\delta_1 = \Delta_1,$$
$$\delta_2 = -(\Delta_1^2 - 2\Delta_2),$$
$$\delta_3 = \Delta_1^3 - 3\Delta_1\Delta_2 + 3\Delta_3,$$
$$\delta_4 = -(\Delta_1^4 - 4\Delta_1^2\Delta_2 + 2\Delta_2^2 + 4\Delta_1\Delta_3 - 4\Delta_4),$$
$$\vdots \qquad\qquad \vdots$$

and generally $\delta_\kappa = \{\kappa\}$, the partition having reference to the roots of equation (4).

The conclusion is that there exists in regard to the Π-functions a theory exactly parallel to that of Hammond in regard to the a-functions, and that the same methods of calculation may be applied.

We shall have

$$\delta^l_\lambda \delta^m_\mu \delta^n_\nu \ldots \div l!\, m!\, n! = \{\lambda^l . \mu^m . \nu^n \ldots\},$$

and if two results of the same weight be

$$\Sigma A_{\lambda^l.\mu^m.\nu^n\ldots} \quad (\lambda^l \mu^m \nu^n \ldots) = \Sigma A'_{\lambda'^{l'}.\mu'^{m'}.\nu'^{n'}\ldots} H^{l'}_{\lambda'} H^{m'}_{\mu'} H^{n'}_{\nu'} \ldots \text{(III)},$$

$$\Sigma B_{\lambda'^{l'}.\mu'^{m'}.\nu'^{n'}\ldots} (\lambda'^{l'}.\mu'^{m'}.\nu'^{n'}) = \Sigma B'_{\lambda^l.\mu^m.\nu^n\ldots} H^l_\lambda . H^m_\mu . H^n_\nu \ldots\ldots \text{(IV)},$$

we have the operator relation corresponding to (III), viz.

$$\Sigma A_{\lambda^l.\mu^m.\nu^n\ldots} \overline{\frac{\delta^l_\lambda \delta^m_\mu \delta^n_\nu}{l!\, m!\, n! \ldots}} = \Sigma A'_{\lambda'^{l'}.\mu'^{m'}.\nu'^{n'}\ldots} \Delta^{l'}_{\lambda'} \Delta^{m'}_{\mu'} \Delta^{n'}_{\nu'} \ldots,$$

operating with which upon (IV) we find

$$\Sigma A_{\lambda^l.\mu^m.\nu^n\ldots} B'_{\lambda^l.\mu^m.\nu^n\ldots} = \Sigma B_{\lambda'^{l'}.\mu'^{m'}.\nu'^{n'}\ldots} A'_{\lambda'^{l'}.\mu'^{m'}.\nu'^{n'}\ldots},$$

proving that a precisely similar law of symmetry obtains for symmetric functions expressed in terms of H-functions.

§ 3. From § 2 it appears that

$$\delta_1 \equiv \Delta_1 \equiv H_0 \partial_{H_1} + H_1 \partial_{H_2} + H_2 \partial_{H_3} + \ldots$$

has the same effect upon a partition as

$$d_1 \equiv D_1 \equiv a_0 \partial_{a_1} + a_1 \partial_{a_2} + a_2 \partial_{a_3} + \ldots,$$

from which it is easy to conclude that if $f(a_0, a_1, a_2, a_3, \ldots)$ be a non-unitary symmetric function of the equation

$$a_0 x_n - a_1 x^{n-1} + a_2 x^{n-2} - a_3 x^{n-3} + \ldots = 0,$$

or, which is the same thing, a seminvariant of the quantic

$$a_0 x^n - n a_1 x^{n-1} y + n(n-1) a_2 x^{n-2} y^2 - n(n-1)(n-2) a_3 x^{n-3} y^3 + \ldots;$$

then $f(H_0, H_1, H_2, H_3, \ldots)$ will be so also.

This is otherwise instantaneously evident from an observation of the effect of the operator

$$d_m \equiv a_0 \partial_{a_m} + a_1 \partial_{a_{m+1}} + a_2 \partial_{a_{m+2}} + \ldots$$

upon H_s; for, as is well known,

$$d_m H_s = (-)^{m+1} H_{s-m},$$

and a comparison of this result with the fact

$$d_m a_s = a_{s-m}$$

shews at once that if $f(a_0, a_1, a_2, a_3, \ldots)$ be a solution of the partial differential equation $d_m = 0$, $f(H_0, H_1, H_2, H_3, \ldots)$ must be a solution also.

Putting $m = 0$ indicates then that a seminvariant is annihilated by the operator

$$H_0 \partial_{H_1} + H_1 \partial_{H_2} + H_2 \partial_{H_3} + \ldots,$$

and generally that a solution of $d_m = 0$ is also a solution of $\delta_m = 0$.

§4. I have already given (*Quart. Math. Jour.*, vol. XX. p. 365 *et seq.*) the law of operation of

$$D_t = \frac{1}{t!}(a_0 \partial_{a_1} + a_1 \partial_{a_2} + \partial_{a_3} + \ldots)^t$$

upon a composite symmetric function

$$(\lambda^l.\mu^m, \nu^n \ldots) = (\lambda_1^{l_1}.\mu_1^{m_1}.\nu_1^{n_1}\ldots)(\lambda_2^{l_2}.\mu_2^{m_2}.\nu_2^{n_2})(\quad)\ldots(\lambda_s^{l_s}.\mu_s^{m_s}.\nu_s^{n_s}\ldots),$$

and I here append a simple proof.

Suppose $(\lambda^l.\mu^m.\nu^n\ldots) = \phi(a_0, a_1, a_2, a_3, \ldots)$

and $$\phi = \phi_1\phi_2\phi_3\ldots\phi_s.$$

Put $a_t + a_{t-1}\kappa$ for a_t, and expand each member of the identity in ascending powers of κ by Taylor's theorem; there results

$$\phi + \kappa D_1\phi + \kappa^2 D_2\phi + \kappa^3 D_3\phi + \ldots = \prod_{s=s}^{s=1}(\phi_s + \kappa D_1\phi_s + \kappa^2 D_2\phi_s + \kappa^3 D_3\phi^s + \ldots),$$

and equating the coefficients, on each side, of κ^t, we find

$$D_t\phi = \Sigma\Sigma D_{\pi_1}\phi_{\sigma_1} D_{\pi_2}\phi_{\sigma_2} D_{\pi_3}\phi_{\sigma_3}\ldots D_{\pi_t}\phi_{\sigma_t},$$

where $$t = \pi_1 + \pi_2 + \pi_3 + \ldots + \pi_t,$$

and the double summation is in regard to every partition of t and to every possible permutation of the ϕ's; whence

$$D_t(\lambda^l.\mu^m.\nu^n\ldots)$$

$$= \Sigma(\lambda_1^{l_1-1}.\mu_1^{m_1}.\nu_1^{n_1}\ldots)(\lambda_2^{l_2}.\mu_2^{m_2-1}.\nu_2^{n_2}\ldots)(\lambda_3^{l_3}.\mu_3^{m_3}.\nu_3^{n_3-1}\ldots)\ldots$$

$$\ldots(\lambda_s^{l_s}.\mu_s^{m_s}.\nu_s^{n_s}\ldots\zeta_s^{z_s-1}),$$

where $$\lambda_1 + \mu_2 + \nu_3 + \ldots + \zeta_s = t,$$

and the summation is in regard to every partition of t that can be picked out from the product of partitions, one part only being taken from each partition.

Ex. gr.
$$D_t a_1 a_2 a_3 \ldots a_t = D_t (1)(1^2)(1^3)\ldots(1^t)$$
$$= (1)(1^2)\ldots(1^{t-1})$$
$$= a_1 a_2 \ldots a_{t-1};$$

also
$$D_6 (43)(2)(1) = (3)(1) + (4),$$

the partitions picked out being
$$4 + 2 = 6 \text{ and } 3 + 2 + 1 = 6.$$

This theorem seems to be a fundamental one in the science of algebraic forms (its importance in the theory of invariants need not here be insisted upon), and to find its proper place in the differential calculus as a pendant to Leibnitz' theorem.

Royal Military Academy, Woolwich,
June 14th, 1886.

Chapter 2
Symmetric Functions (Part 2)

2.1 Introduction and Commentary

The papers of MacMahon included in this chapter are almost all devoted to applications of the Hammond operators to symmetric functions and combinatorics. J. G. van der Corput (1940, 1949, 1950) and H. O. Foulkes (1949, 1974) have greatly extended such operators; Foulkes (1974) presents a nice survey of these more general operators and other aspects of symmetric functions.

A paper by Philip Hall (1957) is extremely important in that it interrelates several areas of MacMahon's work. Much of the work by R. P. Stanley (1971) and others confirms that the program outlined by Hall is quite significant. Since Hall's paper is somewhat difficult to find, we reproduce it in its entirety in section 2.2. We refer the reader to chapters 1, 12, and 14 which include further comments and examples related to symmetric functions. In particular, the aspects of MacMahon's work related to Young tableaux and the representation theory of the symmetric group are covered in chapter 12 and 14.

The material in this chapter corresponds to section II of *Combinatory Analysis*.

2.2 The Algebra of Partitions, by Philip Hall

(Reprinted with the permission of the Canadian Math. Congress)

We may define a partition λ of the integer $n \geqslant 0$ to be a solution in non-negative integers $r_1, r_2, \ldots,$ of the equation

$$(1) \qquad r_1 + 2r_2 + 3r_3 + \ldots = n.$$

The number r_k is then the multiplicity of k as a part of λ and one denotes λ explicitly by the symbol

$$(2) \qquad (1^{r1}\, 2^{r2}\, 3^{r3} \ldots).$$

According to this definition, there is one and only one partition of zero and it has no parts. On the other hand, zero itself is never considered as a part of a partition.

The number of parts of the partition λ defined by (1) is, of course, $r = r_1 + r_2 + r_3 + \ldots$ and it is convenient to denote these parts in descending order of magnitude by $\lambda_1, \lambda_2, \ldots, \lambda_r$ so that

$$(3) \qquad \lambda_1 \geqslant \lambda_2 \geqslant \ldots \geqslant \lambda_r$$

and to define $\lambda_s = 0$ for $s = r + 1, r + 2, \ldots$. We can represent λ by a diagram of n squares arranged in r rows, the ith row containing just λ_i squares. Thus the diagram

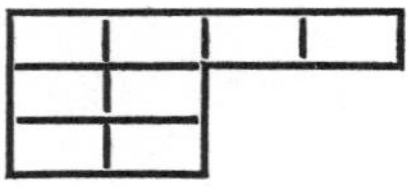

represents the partition $(2^2 4)$.

If we transpose the diagram of a partition λ by reflection in the main diagonal, we obtain the diagram of another partition λ', in general distinct from λ, which is called the *conjugate* of λ. Thus the conjugate of $(2^2 4)$ is $(1^2 3^2)$. We can express the multiplicities r_k of the parts of λ in terms of the parts of the conjugate partition λ' by the equation

$$(4) \qquad r_k = \lambda'_k - \lambda'_{k+1}.$$

For the number of parts of λ which are equal to k is just the number by which the length of the kth column of the λ-diagram exceeds the length of the $(k + 1)$th column. And the kth column of the λ-diagram becomes on transposition the kth row of the λ'-diagram, which has length λ_k'.

There are two quite distinct but equally natural ways of adding together two partitions λ and μ. If we add each row of the λ-diagram to the corresponding row of the μ-diagram, we obtain the diagram of a partition $\lambda + \mu$ whose kth largest part is $\lambda_k + \mu_k$. If on the other hand we add each column of the λ-diagram to the corresponding column of the μ-diagram, we get the diagram of a partition $\lambda \oplus \mu$. It is clear that these two kinds of addition are dual to one another in the sense that

$$\lambda \oplus \mu = (\lambda' + \mu')'. \tag{5}$$

And, in view of (4), we see also that the multiplicity of k as a part of $\lambda \oplus \mu$ is just the sum of its multiplicities as a part of λ and as a part of μ. Thus the parts of $\lambda \oplus \mu$ are the parts of λ together with those of μ.

It is the same with multiplication. We may define $\lambda\mu$ as the partition whose kth largest part is $\lambda_k\mu_k$, the product of the kth largest parts of λ and μ. The dual multiplication will then be defined by the equation

$$\lambda \otimes \mu = (\lambda'\mu')'. \tag{6}$$

And again, in view of (4), there is a simple way of describing the parts of $\lambda \otimes \mu$; they consist of all the numbers

$$\min(\lambda_i, \mu_j) \quad (i = 1, 2, \ldots, r; \;\; j = 1, 2, \ldots, s) \tag{7}$$

where r and s are the numbers of parts of λ and μ.

It is well known that any finite Abelian p-group G is expressible as the direct product of cyclic subgroups $G_1, G_2, \ldots, G_r$. If the order of G_i is p^{λ_i}, we may suppose that (3) holds, and thus obtain a partition λ of n, where p^n is the order of G. This partition λ determines G to within isomorphism, so that we may speak of G as being of *type* λ. If H is another Abelian p-group, of type μ say, then it is immediately obvious that the direct product $G \times H$ will be of type $\lambda \oplus \mu$. And in view of (7), it is equally clear that their tensor product $G \otimes H$, which is isomorphic with the group of homomorphisms of G into H, and equally of H into G, will be of type $\lambda \otimes \mu$.

These simple facts suggest that a suitably devised algebra of partitions may well throw light on some aspects of the theory of finite Abelian groups and of subjects, such as nilpotent matrices, where there is a similar natural relation with partitions.

The simplest way of forming an algebra of partitions is to associate each partition λ with a corresponding basis vector b_λ and to form the set B of all finite linear combinations of the b_λ, regarded as linearly independent, with coefficients in some suitable ring, say the field $\mathfrak{r}$ of rational numbers. Addition in B is defined in the obvious way; and multiplication by

$$b_\lambda b_\mu = b_{\lambda \oplus \mu} \tag{8}$$

and the distributive law. B then becomes a commutative and associative algebra over $\mathfrak{r}$, with b_0 as unit element, where 0 here stands for the partition of zero. (If instead of (8) we defined $b_\lambda b_\mu = b_{\lambda+\mu}$ we should get nothing essentially different, owing to (5).)

If $\lambda = (1^{r_1} 2^{r_2} 3^{r_3} \ldots)$, then (8) implies that

$$b_\lambda = b_1^{r_1} b_2^{r_2} b_3^{r_3} \ldots,$$

where for simplicity we have written b_k for $b_{(k)}$, the basis vector associated with the one-part partition (k). Thus B is simply the ring of all polynomials in the countably infinite set of variables $b_1, b_2, b_3, \ldots$. But in order to maintain a suitable relation with partitions, we must regard B as a *graded* algebra, the direct sum of the "homogeneous" subspaces B_n, $(n = 0, 1, 2, \ldots,)$, where B_n has as basis those b_λ for which λ is a partition of n. Thus the dimension of B_n is $p(n)$, the total number of partitions of n, defined by Euler's generating function

$$\prod_{k=1}^{\infty} \frac{1}{(1 - t^n)} = \sum_{n=0}^{\infty} p(n)t^n.$$

When considered as a graded algebra in this way, it is natural to identify B with the algebra A of all *symmetric* "polynomials" in a countably infinite set of variables $x_1, x_2, \ldots$, with coefficients in $\mathfrak{r}$. (The "polynomials" in A are, of course, to be understood in the obvious formal sense; except for the constants, they will always contain an infinite number of terms.) This relation between the algebra of partitions and the algebra of symmetric functions is, of course, very well known. And it is reasonable to expect that the highly developed formalism of symmetric functions will also prove useful in some of the other fields of mathematics where partitions play a part. I hope to illustrate this later on in one or two cases. But before doing so, I must review very briefly the main features of this formalism.

Let A_n be the space of all symmetric polynomials in the x's which are homogeneous of degree n. A basis of A_n will then be formed by the polynomials

(9) $$k_\lambda = \sum x_1^{\lambda_1} x_2^{\lambda_2} \ldots$$

where λ runs through all partitions of n and, by the usual convention, the sum is taken over all distinct monomials which are obtainable from the one actually written by arbitrary permutations of the variables. Thus the dimension of A_n is $p(n)$.

To express that λ is a partition of n, it is convenient to write

$$\lambda \vdash n.$$

Let us call any basis c_λ of A_n, where $\lambda \vdash n$, an *integral basis* if the matrix which relates it to the k_λ's is unimodular, that is, has integer coefficients and determinant ± 1. Thus the c_λ's will form an integral basis of A_n if and only if the elements of A_n which have integer coefficients when expressed as polynomials in the x's coincide with the elements of A_n which have integer coefficients when expressed as linear combinations of the c_λ's.

Besides the k_λ's themselves, there are several other important integral bases of A_n. For example, if we define the elementary symmetric functions a_n and the complete homogeneous functions h_n by

(10) $$a_n = k_{(1^n)},$$

and

(11) $$h_n = \sum_{\lambda \vdash n} k_\lambda;$$

and then, for any partition μ, we define $a_0 = h_0 = 1$ and

(12) $$a_\mu = a_{\mu_1} a_{\mu_2} a_{\mu_3} \ldots$$

and

(13) $$h_\mu = h_{\mu_1} h_{\mu_2} h_{\mu_3} \ldots,$$

we obtain two integral bases of A_n, viz. the a_λ with $\lambda \vdash n$ and the h_λ with $\lambda \vdash n$. Owing to (12) and (13), these may be called *multiplicative* bases. They exhibit the fact that A may be regarded also as the algebra of all polynomials in the sequence of "variables" $a_1, a_2, a_3, \ldots$; and equally well as the algebra of all polynomials in $h_1, h_2, h_3, \ldots$. (The identification of A with B may therefore be made by making a_λ correspond to b_λ.)

The elementary functions a_n and the complete functions h_n may also be defined by the generating functions

(14) $$\prod_{i=1}^{\infty} (1 + x_i t) = \sum_{n=0}^{\infty} a_n t^n$$

and

(15) $$\prod_{i=1}^{\infty} \frac{1}{(1 - x_i t)} = \sum_{n=0}^{\infty} h_n t^n,$$

which yield at once the fundamental recurrence relations

(16) $$\sum_{r=0}^{n} (-1)^r a_r h_{n-r} = 0 \qquad (n > 0).$$

The symmetry of these relations as between the a's and the h's shows that the automorphism of the algebra A which is uniquely determined by the mapping

(17) $$\theta : a_n \longrightarrow h_n \qquad (n = 1, 2, 3, \ldots,)$$

has period 2. It is an involution.

The next important step is to introduce a metric into each A_n by means of a suitable symmetric scalar product. That this can be done in a significant way results from a simple combinatorial fact: that the matrix relating the bases (13) and (9) of A_n, formed respectively by the h_λ's and the k_λ's with $\lambda \vdash n$, is a symmetric matrix. A very simple proof is as follows. Let $y_1, y_2, \ldots,$ be a second countably infinite set of variables and let $z_{ij} = x_i y_j$ $(i, j = 1, 2, 3, \ldots,)$. Then the z_{ij} form a third set of variables and we have

$$\sum_{n=0}^{\infty} h_n(z)t^n = \prod_{i,j=1}^{\infty} \frac{1}{(1 - x_i y_j t)} = \prod_{i=1}^{\infty} \sum_{m=0}^{\infty} h_m(y) t^m x_i^m$$
$$= \sum_{n=0}^{\infty} \Big(\sum_{\lambda \vdash n} h_\lambda(y) k_\lambda(x) \Big) t^n,$$

so that

(18) $$h_n(z) = \sum_{\lambda \vdash n} h_\lambda(y) k_\lambda(x) = \sum_{\lambda \vdash n} h_\lambda(x) k_\lambda(y)$$

by interchange of x's and y's, which does not affect $h_n(z)$.

Reverting once more to the algebra A of symmetric polynomials $u = u(x)$ in the x's alone, we now define a bilinear scalar product (u, v) for any two elements u and v of A by requiring that

(19) $$(h_\lambda, k_\mu) = \delta_{\lambda\mu}$$

for all λ and μ. Since the h_λ and the k_μ each form a basis of A, the value of (u, v) is determined uniquely by (19). And the identity (18) shows that this scalar product is symmetric:

(20) $$(u, v) = (v, u)$$

for all u and v of A.

With respect to the scalar product so defined, the homogeneous subspaces A_n are mutually orthogonal and we need only consider the geometry of the separate A_n's. Equation (19) expresses that the two integral bases (h_λ) and (k_λ) are dual to one another. And it is natural to ask the following question: does there exist an integral basis $(e_\lambda)_{\lambda \vdash n}$ of A_n which is self-dual, that is, which consists of mutually orthogonal unit vectors, satisfying

(21) $$(e_\lambda, e_\mu) = \delta_{\lambda\mu}$$

for all λ and μ?

What is immediately clear is that such a self-dual integral basis, if it exists, is uniquely determined apart from the signs and the labelling of its terms. For suppose $(\bar{e}_\lambda)_{\lambda \vdash n}$ is another such basis. Then the matrix which expresses the $\bar{e}_\lambda$ in terms of the e_λ must be at the same time orthogonal and unimodular. Hence it can only be a "signed" permutation matrix, with a single non-zero coefficient equal to ± 1 in each row and column.

The existence of this integral self-dual basis (e_λ) may be regarded as the central fact in the algebra of symmetric functions. These e_λ are the *Schur functions*, so-called because it was Schur who first discovered their central role in connection with the representation theory of the general linear groups. Here it will be possible to mention only a few of their many equivalent definitions.

(i) First, there is the combinatorial definition. Suppose $\lambda \vdash n$. If in each of the n squares of the λ-diagram we enter one of the variables x_i, subject to the condition that the suffixes i of the entries strictly increase as we pass down a column and never decrease as we pass along a row from left to right, we obtain a *standard* λ*-tableau* T in the sense of A. Young. Let $p(T)$ be the product of all the entries in T. Then

$$e_\lambda = \sum p(T), \tag{22}$$

the sum being taken over all standard λ-tableaux T.

This definition shows that the leading term of e_λ is precisely

$$x_1^{\lambda_1} x_2^{\lambda_2} \ldots,$$

with coefficient $+1$; and thus fixes the signs and the labelling of the e_λ's unambiguously.

(ii) Clearly every u of A_n, being symmetric in the x's and homogeneous of degree n, is uniquely determined by those of its terms which involve $x_1, x_2, \ldots, x_n$ alone. Let the sum of these terms be denoted by $u^{(n)}$. For any $\lambda \vdash n$, let

$$l_i = \lambda_i + n - i \qquad (i = 1, 2, \ldots, n)$$

and let

$$\Delta_\lambda = \begin{vmatrix} x_1^{l_1} x_1^{l_2} \ldots x_1^{l_n} \\ x_2^{l_1} x_2^{l_2} \ldots x_2^{l_n} \\ \ldots \\ x_n^{l_1} x_n^{l_2} \ldots x_n^{l_n} \end{vmatrix}.$$

Thus Δ_0 is simply the Vandermonde determinant of $x_1, \ldots, x_n$:

$$\Delta_0 = \prod_{\substack{i,j=1 \\ i<j}}^{n} (x_i - x_j),$$

and each of the quotients Δ_λ/Δ_0 is a symmetric polynomial of degree n in $x_1, \ldots, x_n$. In fact

$$\Delta_\lambda/\Delta_0 = e_\lambda^{(n)}. \tag{23}$$

(iii) Expressing the Schur functions in terms of the complete functions h_n we have the Jacobi-Trudi formulae where $h_{-1} = h_{-2} = \ldots = 0$ and r is any number $\geqslant \lambda'_1$, the number of parts of λ.

$$(24) \qquad e_\lambda = \begin{vmatrix} h_{\lambda_1} & h_{\lambda_1+1} & h_{\lambda_1+2} & \ldots & h_{\lambda_1+r-1} \\ h_{\lambda_2-1} & h_{\lambda_2} & h_{\lambda_2+1} & \ldots & h_{\lambda_2+r-2} \\ & & \ldots & & \\ h_{\lambda_r-r+1} & & \ldots & & h_{\lambda_r} \end{vmatrix}$$

There is a similar expression for the e_λ in terms of the elementary functions a_n. This is obtained by applying the involution θ to (24), for it turns out that θ merely interchanges e_λ with $e_{\lambda'}$:

$$(25) \qquad \theta\colon e_\lambda \leftrightarrow e_{\lambda'}.$$

This brief sketch of the algebra A would remain essentially incomplete if I failed to mention in conclusion a fifth basis which, like the (a_λ) and the (h_λ), is multiplicative, but unlike these and the (k_λ) and the (e_λ) is not an integral basis. This fifth basis consists of the power-sums

$$s_n = k_{(n)} = \sum_{i=1}^{\infty} x_i^n \qquad (n = 1, 2, \ldots,)$$

and their products

$$s_\lambda = s_{\lambda_1} s_{\lambda_2} s_{\lambda_3} \ldots,$$

where we take $s_0 = 1$. It has the merit that it consists of eigenfunctions of the involution. In fact

$$(26) \qquad \theta\colon s_n \rightarrow (-1)^{n-1} s_n.$$

It is also an orthogonal basis:

$$(27) \qquad (s_\lambda, s_\mu) = 0 \qquad \text{if } \lambda \neq \mu.$$

But it is not self-dual because, if

$$\lambda = (1^{r_1}\, 2^{r_2}\, 3^{r_3} \ldots),$$

then

$$(28) \qquad (s_\lambda, s_\lambda) = 1^{r_1} \,.\, r_1!\, 2^{r_2} \,.\, r_2!\, 3^{r_3} \,.\, r_3! \ldots.$$

These numbers arise naturally in the theory of the symmetric groups S_n of order $n!$ The classes of conjugate elements in S_n correspond one-one to the partitions λ of n, the class S_λ consisting of those permutations ξ in S_n which for each k have just r_k cycles of order k. The number of elements in the class S_λ is then equal to $n!/(s_\lambda, s_\lambda)$.

From the general theory of representations of finite groups over the complex field it then follows that the irreducible representations of S_n can also be put into one-one correspondence with the partitions λ of n. If χ^λ is the character of the irreducible representation corresponding to λ, and if $\chi_\mu{}^\lambda$ is its value for an element ξ of the class S_μ, we have the formulae

$$s_\mu = \sum_{\lambda \vdash n} \chi_\mu^\lambda e_\lambda \qquad (\mu \vdash n), \tag{29}$$

so that, by (21),

$$\chi_\mu^\lambda = (e_\lambda, s_\mu), \tag{30}$$

and, by (27),

$$e_\lambda = \sum_{\mu \vdash n} \chi_\mu^\lambda s_\mu/(s_\mu, s_\mu) \qquad (\lambda \vdash n). \tag{31}$$

If in (29) we take $\mu = (1^n)$, so that S_μ is the unit class of S_n, the coefficients $\chi_\mu{}^\lambda$ become the degrees

$$f^\lambda = \chi_{(1^n)}^\lambda$$

of the irreducible representations of S_n. These numbers have several interesting definitions of which I will only mention two. First

$$f^\lambda = n!/\prod_{i,j} (\lambda_i + \lambda'_j - i - j + 1), \tag{32}$$

the product being taken over all the n squares (i, j) of the λ-diagram, the square (i, j) being where the ith row meets the jth column. Secondly, from (30) we have $f^\lambda = (e_\lambda, s_1{}^n)$. But $s_1 = a_1 = h_1$ is just the sum of the variables x_i. Hence $f^\lambda = (e_\lambda, h_1{}^n)$ and, in view of (19), this means that f^λ is the coefficient of $k_{(1^n)}$ when e_λ is expressed in terms of the k's. In other words, f^λ is the coefficient of $x_1x_2 \ldots x_n$ when e_λ is expressed as a polynomial in the x's. By (22), it follows that f^λ is the number of standard λ-tableaux which contain one each of the first n variables.

I should now like to mention one or two cases, less well explored than the representation theory of the symmetric groups, in which the formalism of symmetric functions may prove useful.

Let us consider first the geometry, or perhaps one should say the lattice structure, of finite Abelian p-groups. If p is any prime and λ any partition, let $G_\lambda(p)$ be an Abelian p-group of type λ. Then we may construct an algebra $A(p)$ which will reflect the lattice structure of these groups in the following way. As basis of $A(p)$, over the rational field $\mathfrak{r}$, we take the $G_\lambda(p)$ themselves, and define in $A(p)$ a distributive multiplication by the rule

$$(33) \qquad G_\lambda(p)G_\mu(p) = \sum_\rho g_{\lambda\mu}{}^\rho(p)G_\rho(p),$$

where $g_{\lambda\mu}{}^\rho(p)$ is the number of subgroups H of $G_\rho(p)$ such that

$$(34) \qquad H \cong G_\lambda(p),\ G_\rho(p)/H \cong G_\mu(p).$$

From this it will follow at once that the coefficients $g_{\lambda\mu\dots\nu}{}^\rho(p)$ defined by

$$(35) \qquad G_\lambda(p)G_\mu(p)\dots G_\nu(p) = \sum_\rho g_{\lambda\mu\dots\nu}{}^\rho(p)G_\rho(p)$$

give the number of chains of subgroups

$$H_1 \leqslant H_2 \leqslant \dots \leqslant H_k$$

in $G = G_\rho(p)$, such that $H_1, H_2/H_1, \dots, G/H_k$ are respectively of types $\lambda, \mu, \dots, \nu$.

That this multiplication is associative is clear. That it is also commutative follows easily from the character theory of Abelian groups which implies that

$$(36) \qquad g_{\lambda\mu}{}^\rho(p) = g_{\lambda\mu}{}^\rho(p).$$

Thus we obtain, for each prime p, an algebra $A(p)$ which is in fact isomorphic with the algebra A of symmetric functions, the "homogeneous" subspace $A_n(p)$ which corresponds to A_n having as basis those $G_\lambda(p)$ for which $\lambda \vdash n$. This isomorphism may be established most conveniently by identifying the sums

$$(37) \qquad H_n(p) = \sum_{\lambda \vdash n} G_\lambda(p)$$

with the complete symmetric functions h_n.

For $\lambda = (1^n)$, $G_\lambda(p)$ becomes the elementary Abelian p-group

$A^*_n(p)$ of order p^n. And it is easy to establish that, with the definition (33), (34) for multiplication in $A(p)$, we have the identities

$$(38) \qquad \sum_{r=0}^{n} (-1)^r p^{\binom{r}{2}} A^*_r(p) H_{n-r}(p) = 0, \qquad n > 0,$$

which are closely analogous to the recurrence relations (16) connecting the h_n with the elementary symmetric functions a_n. The identification

$$(39) \qquad H_n(p) = h_n$$

now implies that

$$(40) \qquad p^{\frac{1}{2}r(r-1)} A^*_r(p) = a_n.$$

And it also implies, of course, that for any given partition λ of n, $G_\lambda(p)$ is to be identified with a certain uniquely determined symmetric function in A_n, though it is not so easy to specify which this is explicitly.

What is not difficult to show is that the multiplication constants $g_{\lambda\mu}{}^{\rho}(p)$, considered as functions of the prime p, are polynomials with integer coefficients. Moreover $g_{\lambda\mu}{}^{\rho}(p)$ vanishes identically in p if and only if the corresponding multiplication constant for the Schur functions, viz.

$$(41) \qquad (e_\lambda e_\mu, e_\rho)$$

is zero. And more generally, when (41) is different from zero, it is equal to the coefficient of the highest power of p in $g_{\lambda\mu}{}^{\rho}(p)$ and the degree of $g_{\lambda\mu}{}^{\rho}(p)$ in p is then equal to

$$(42) \qquad \sum_{k=1}^{\infty} (k-1)(\rho_k - \lambda_k - \mu_k).$$

Similar results hold for the coefficients $g_{\lambda\mu\ldots\nu}{}^{\rho}(p)$.

If therefore we define the numbers

$$(43) \qquad m_\lambda = \sum_{i=1}^{\infty} (i-1)\lambda_i = \sum_{j=1}^{\infty} \tfrac{1}{2}\lambda'_j(\lambda'_j - 1)$$

and write

$$(44) \qquad F_\lambda(p) = p^{m_\lambda} G_\lambda(p),$$

then the $F_\lambda(p)$ will be again a basis of $A(p)$ and their multiplication constants $f_{\lambda\mu}{}^{\rho}(p)$ will be polynomials in $1/p$ with $(e_\lambda e_\mu, e_\rho)$ as constant term. Thus

$$F_\lambda(p) \to e_\lambda \quad \text{as} \quad p \to \infty \tag{45}$$

It can also be shown that the numbers $g_{\lambda\mu}{}^{\rho}(1)$ are equal to the corresponding multiplication constants for the symmetric functions k_λ. If, therefore, we think of p as a real variable, we shall have

$$G_\lambda(p) \to k_\lambda \quad \text{as} \quad p \to 1. \tag{46}$$

As a corollary of these facts, we obtain a simple asymptotic formula for the number of composition series of the group $G_\lambda(p)$. As $p \to \infty$, this number is asymptotically equivalent to

$$f^\lambda p^{m_\lambda},$$

where f^λ is the degree of the irreducible representation of S_n associated with the partition λ of n to which we referred above.

Before leaving this topic, I should mention that the polynomials $g_{\lambda\mu}{}^{\rho}(p)$ play an interesting part in J. A. Green's important recent work on the representation theory of the general linear groups over a finite field. If $t_\lambda(p)$ denotes the order of the group of automorphisms of $G_\lambda(p)$ and if we define the functions $\zeta^\lambda{}_\mu(p)$ by the equations

$$s_\mu = \sum_{\lambda \vdash n} \zeta^\lambda_\mu(p) G_\lambda(p) \qquad (\mu \vdash n) \tag{47}$$

which express the basis (s_μ) in terms of the basis $(G_\lambda(p))$, Green obtains the elegant orthogonality relations

$$\sum_{\lambda \vdash n} \zeta^\lambda_\mu(p)\zeta^\lambda_\nu(p)/t_\lambda(p) = 0 \qquad \text{if } \mu \neq \nu \tag{48}$$

and

$$\sum_{\lambda \vdash n} \zeta^\lambda_\mu(p)\left\{\zeta^\lambda_\mu(p) - (s_\mu, s_\mu)\right\}/t_\lambda(p) = 0.$$

Here μ and ν are partitions of n.

Finally, I would like to mention the *Schubert manifolds*. Here also we have a class of objects in one-one correspondence with the partitions. These manifolds may be defined as follows.

Let λ be a partition of n with r parts. We choose in a suitable linear space L over any field k a fixed chain of subspaces

$$0 = L_0 < L_1 < L_2 < \ldots < L_r = L,$$

such that the dimension of L_i is $\lambda_{r-i+1} + i$. Thus the dimension of L itself is $\lambda_1 + r = \lambda_1 + \lambda_1'$. The Schubert manifold M_λ is then the representative manifold on the appropriate Grassmannian of the set of all r-dimensional subspaces R of L such that the dimension of the intersection $R \cap L_i$ is at least equal to i for each $i = 1, 2, \ldots, r$.

It is rather easy to see that, when λ is the "rectangular" partition (s^r) with r parts all equal to s, the conditions on R are automatically fulfilled and consequently $M_{(s^r)}$ is the Grassmannian itself, whose points are in one-one correspondence with the totality of r-dimensional subspaces of the $(r + s)$-dimensional space L.

It turns out that the dimension of M_λ is n, the number which λ is a partition of; and its order is the number f^λ which we have met in so many other connections. I can mention one further interesting fact. Suppose that the field k over which M_λ is defined is finite, with q elements. Then the number of points on M_λ, that is, the number of subspaces R satisfying the required conditions will also be finite. In fact it is a polynomial $\phi_\lambda(q)$, of degree n in q, with non-negative integer coefficients. And the coefficient of q^k in $\phi_\lambda(q)$ is equal to the number of partitions μ of k such that $\mu_i \leqslant \lambda_i$ for each i. This condition means simply that the λ-diagram completely covers the μ-diagram.

This fact is, of course, well known in the special case of the Grassmannians $M_{(s^r)}$. For here

$$\phi_{(s^r)}(q) = \frac{(q^{r+s} - 1)(q^{r+s-1} - 1) \ldots (q^{r+1} - 1)}{(q^s - 1)(q^{s-1} - 1) \ldots (q - 1)},$$

which is the polynomial which enumerates the partitions having at most r parts none exceeding s.

2.3 References

S. Blaha (1969) Character analysis of U(N) and SU(N), *J. Math. Physics*, *10*, 2156–2168.

S. Blaha (1974) The calculation of the coefficients relating the irreducible and compound characters of the symmetric groups (to appear).

P. Cartier (1972) La serie géneratrice exponentielle applications probabilistes et algébriques, version préliminaire, Institut de Recherche Mathématique Avancée, Strasbourg.

F. N. David, M. G. Kendall, and D. E. Barton (1966) Symmetric Functions and Allied Tables, Cambridge University Press, Cambridge.

P. Doubilet (1972) On the foundations of combinatorial theory VII: Symmetric functions through the theory of distribution and occupancy, *Studies in Appl. Math.*, *51*, 377–396.

P. Doubilet (1973) An inversion formula involving partitions *B.A.M.S.*, *79*, 177–179.

C. J. Everett and P. R. Stein (1971) The asymptotic number of integer stochastic matrices, *Discr. Math.*, *1*, 55–72.

H. O. Foulkes (1949) Differential operators associated with S-functions, *J. London Math.* Soc., *24*, 136–143.

H. O. Foulkes (1974) A survey of some combinatorial aspects of symmetric functions, Permutations Colloquium (held at Paris, July 10–13, 1972).

J. A. Green (1955) The characters of the finite general linear groups, *Trans. A.M.S.*, *80*, 402–447.

P. Hall (1957) The algebra of partitions, Proc. 4th Canadian Math. Cong., 147–159.

H. Jack (1970) A class of symmetric polynomials with a parameter, *Proc. Royal Soc. Edinburgh*, *A-69*, 1–18.

H. Jack (1972) A surface integral and symmetric functions, *Proc. Royal Soc. Edinburgh*, *A-69*, 347–364.

A. T. James (1961) Zonal polynomial coefficients of the real positive definite symmetric matrices, *Ann. Math.*, *74*, 456–469.

A. T. James (1968) Calculation of zonal polynomial coefficients by use of the Laplace-Beltrami operator, *Ann. Math. Stat.*, *39*, 1711–1718.

J. Levine (1955) Combinatorial analysis: Part I. MacMahon's theory of distributions and symmetric functions. Section I. The algebraic theory of symmetric functions, O.N.R. Report, Contract No. 870(00).

J. Levine (1956) A theory of occupancy: Distributions in parcels (elementary theory), O.N.R. Report, Contract No. 870(00).

J. Levine (1959) On the application of MacMahon diagrams to certain problems in the multiplication of monomial symmetric functions, *Duke Math. J.*, *26*, 419–436.

J. Levine (1961) Coefficient identities derived from expansions of elementary symmetric function products in terms of power sums, *Duke Math. J.*, *28*, 89–106.

D. E. Littlewood (1950) The Theory of Group Characters, Oxford University Press, Oxford.

D. E. Littlewood (1961) On certain symmetric functions, *Proc. London Math. Soc.(3)*, *11*, 485–498.

N. Metropolis and P. R. Stein (1970) An elementary solution to a problem in restricted partitions, *J. Combinatorial Th.*, *9*, 365–376.

A. O. Morris (1963a) The characters of the group $GL(n, q)$, *Math. Zeit.*, *81*, 112–123.

A. O. Morris (1963b) The multiplication of Hall functions, *Proc. London Math. Soc.(3)*, *13*, 733–742.

A. O. Morris (1964) A note on symmetric functions, *American Math. Monthly*, *71*, 50–53.

A. O. Morris (1965) On an algebra of symmetric functions, *Quart. J. Math.(2)*, *16*, 53–64.

A. O. Morris (1967) A note on lemmas of Green and Kondo, *Proc. Cambridge Phil. Soc.*, *63*, 83–85.

A. O. Morris (1971) Generalizations of the Cauchy and Schur identities, *J. Combinatorial Th.*, *11*, 163–169.

J. Riordan (1971) Symmetric function expansions of power sum products into monomials and their inverses, *Duke Math. J.*, *38*, 285–294.

J. Riordan and P. R. Stein (1972) Arrangements on chessboards, *J. Combinatorial Th.(A)*, *12*, 72–80.

G.-C. Rota (1964) On the foundations of combinatorial theory I: Theory of Möbius functions, *Z. Wahrscheinlichkeitstheorie und Verw. Gebiete*, *2*, 340–368.

G.-C. Rota (1969) Baxter algebras and combinatorial identities I and II, *Bull. A.M.S.*, *75*, 325–334.

G.-C. Rota (1971a) Combinatorial theory and invariant theory, notes by L. Guibas, Bowdoin College, Brunswick, Maine.

G.-C. Rota, P. Doubilet, and R. P. Stanley (1971) On the foundations of combinatorial theory VI: The idea of generating function, Proc. Sixth Symp. on Mathematical Statistics and Probability, vol. 2, 267–318, University of California Press, Berkeley.

R. P. Stanley (1971) Theory and application of plane partitions I and II, *Studies in Appl. Math.*, *50*, 167–188, 259–279.

M. L. Stein and P. R. Stein (1970) Enumeration of stochastic matrices with integer elements, Los Alamos Sci. Lab. Publ. No. LA-4434.

P. V. Sukhatme (1938) On bipartitional functions, *Phil. Trans.*, *A-237*, 375–409.

J. G. van der Corput (1940) Symmetric functions, *Chr. Huyghens*, *18*, 251–277.

J. G. van der Corput (1949) Sur les fonctions symétriques, Math. Centrum Amsterdam, Scriptum no. 3.

J. G. van der Corput (1950) Sur les fonctions symétriques, *Indagationes Math.*, *12*, 216–230.

2.4 Summaries of the papers

[10] Algebraic identities arising out of an extension of Waring's formula, *Messenger of Math.(2)*, *14* (1884), 8–11.

MacMahon discusses an identity of Cauchy, namely, the expression of

$$x^n + y^n + (-1)^n(x+y)^n$$

as a rational function of $x^2 + xy + y^2$ and $xy(x+y)$. Noting that the key to the identity lies in Waring's formula for the sums of the powers of the roots of an equation, he applies his generalization of Waring's formula to produce new algebraic identities of the Cauchy type.

[15] The multiplication of symmetric functions, *Messenger of Math.*, *14* (1885), 164–167.

MacMahon utilizes Hammond operators to derive formulas related to the expression of products of monomial symmetric functions by linear combinations of monomial symmetric functions. This work is a special case of a general theorem he eventually discovered and published in [101].

[28] Properties of a complete table of symmetric functions, *American J. Math.*, *10* (1888), 42–46.

MacMahon describes certain symmetries and appearances of 0 and ± 1 in tables of symmetric functions compiled by W. P. Durfee (*American J. Math.*, *5* (1882), 45–61). He proves that the phenomena observed in such tables are true in general by means of one of the "laws of symmetry" for symmetric functions (see [25]).

[31] Memoir on a new theory of symmetric functions, *American J. Math.*, *11* (1889), 1–36.

MacMahon begins by reviewing some of the notation and terminology of his paper [32]; in particular, he describes in detail his "Algebraic Reciprocity Theorem." He then proceeds to develop a general theorem of expressibility:

Theorem. Given a monomial symmetric function $(\lambda\mu\nu\ldots)$ and given $\sum\lambda_i$ a partition of λ, $\sum\mu_i$ a partition of μ, $\sum\nu_i$ a partition of ν, . . . , the monomial symmetric function $(\lambda\mu\nu\ldots)$ is expressible in terms (i.e., as a polynomial with integral coefficients) of the separations (herein called species partitions) of $(\lambda_1\lambda_2\lambda_3\ldots\mu_1\mu_2\mu_3\ldots\nu_1\nu_2\nu_3\ldots\ldots)$.

The well-known Fundamental Theorem of Symmetric Functions is merely the case $\lambda_1 = \lambda_2 = \ldots = \mu_1 = \mu_2 = \ldots = \nu_1 = \nu_2 = \ldots = 1$. From the above theorem a means of computing symmetric function tables is presented, and tables related to partitions of numbers not exceeding 6 are given.

As an example of the use of the tables, we consider Table 2.1.

Table 2.1
Weight 3. Partition (1^3)

	(3)	(21)	(1^3)
(1^3)			1
$(1^2)(1)$		1	3
$(1)^3$	1	3	6

	(1^3)	$(1^2)(1)$	$(1)^3$
(3)	3	−3	1
(21)	−3	1	
(1^3)	1		

We may now read off from the first table:

$$\begin{aligned}(1^3) &= (1^3),\\ (1^2)(1) &= (21) + 3(1^3),\\ (1)^3 &= (3) + 3(21) + 6(1^3);\end{aligned}$$

and from the second table:

$$\begin{aligned}(3) &= 3(1^3) - 3(1^2)(1) + (1)^3,\\ (21) &= -3(1^3) + (1^2)(1),\\ (1^3) &= (1^3).\end{aligned}$$

The existence of such tables is guaranteed by MacMahon's Theorem of Expressibility. He also points out that his previous generalization of Waring's theorem is merely an example of the type of result obtained from his Theorem of Expressibility.

Newton's theorem relating sums of powers (i.e., monomial symmetric functions of the form (n)) to elementary symmetric functions is greatly generalized, and this work is applied to further expressibility problems.

The final major section of this paper is devoted to two general theorems concerning "groups of separations"; the definition of this concept is somewhat lengthy but is clearly illustrated on page 29 of this paper. The first of these theorems follows:

Theorem. In the expression of the symmetric function $(p_1^{\pi_1} p_2^{\pi_2} \ldots)$ by means of separations of $(t_1^{\tau_1} t_2^{\tau_2} \ldots)$, the algebraic sum of all the coefficients will be zero if the partition $(p_1^{\pi_1} p_2^{\pi_2} \ldots)$ possesses no separations of specification $(\tau_1 t_1, \tau_2 t_2, \tau_3 t_3, \ldots)$.

The terms "separation" and "specification" (or "species partition") are illustrated by the following example: $(3^2 2)(2^3 1^2)(1)$ is a separation of $(3^2 2^4 1^3)$ of specification $(3 + 3 + 2, 2 + 2 + 2 + 1 + 1, 1) = (881)$.

[32] Symmetric functions and the theory of distributions, *Proc. London Math. Soc.*, *19* (1889), 220–256.

Herein MacMahon lays the foundations for the relationship of symmetric functions to combinatorial analysis. This work eventually forms the basis for much of the first two sections of *Combinatory Analysis* and *Introduction to Combinatory Analysis*.

In the first sixteen sections, he methodically derives theorems for enumerating the number of ways that sets of objects may be distributed into parcels. Various restrictions concerning the number and type of objects and parcels are considered; note that when the order of objects in a parcel is to be considered, the parcel is called a group. As an example of one of the more elementary theorems presented, we mention the following result.

Theorem. Suppose $p + q + r + \dots = p_1 + q_1 + r_1 + \dots = n$. The number of distribution of n objects, p of one color, q of a second color, r of a third color, etc., into n parcels, p_1 of one color, q_1 of a second color, r_1 of a third color, etc., is the coefficient of $(pqr \dots)$ in $h_{p_1}h_{q_1}h_{r_1} \dots$, where $(pqr \dots)$ is the symmetric function $\sum \alpha^p \beta^q \gamma^r \dots$, and $h_n = \sum (abc \dots)$, the latter sum running over all partitions $a + b + c + \dots = n$.

As a result of this study of distributions, MacMahon is able to present several interesting and important symmetric function theorems in section 17, which include the following.

Algebraic Reciprocity Theorem. Let

$$X_n = \sum (abc \dots) x_a x_b x_c \dots,$$

where the summation runs over all partitions $a + b + c + \dots$ of n. Suppose that

$$X_{p_1}^{\pi_1} X_{p_2}^{\pi_2} X_{p_3}^{\pi_3} \dots = \dots + \theta (\lambda_1^{l_1} \lambda_2^{l_2} \lambda_3^{l_3} \dots) x_{s_1}^{\sigma_1} x_{s_2}^{\sigma_2} x_{s_3}^{\sigma_3} \dots;$$

then

$$X_{\lambda_1}^{l_1} X_{\lambda_2}^{l_2} X_{\lambda_3}^{l_3} \dots = \dots + \theta (p_1^{\pi_1} p_2^{\pi_2} p_3^{\pi_3} \dots) x_{s_1}^{\sigma_1} x_{s_2}^{\sigma_2} x_{s_3}^{\sigma_3} \dots.$$

Generalized Waring Formula. Suppose $l\lambda + m\mu + n\nu + \dots = N$; then

$$(-1)^{l+m+n+\dots}(l + m + n + \dots - 1)!S(\lambda^l \mu^m \nu^n \dots)/l!m!n!\dots$$
$$= \sum (-1)^{j_1 + j_2 + j_3 + \dots}(j_1 + j_2 + j_3 + \dots - 1)!$$
$$(J_1)^{j_1}(J_2)^{j_2} \dots /j_1!j_2!j_3! \dots,$$

where the sum runs over all separations $(J_1)^{j_1}(J_2)^{j_2}(J_3)^{j_3}\ldots$ of $(\lambda^l\mu^m\nu^n\ldots)$, and where $S(\lambda^l\mu^m\nu^n\ldots)$ denotes the sum of the Nth powers of $\alpha_1, \alpha_2, \alpha_3, \ldots$, the variables from which the various symmetric functions are formed.

Waring's original formula corresponds to the case $(\lambda^l\mu^m v^n\ldots) = (1^N)$.

MacMahon concludes the paper with some general results concerning the expressibility of certain symmetric functions in terms of others. Included is a set of tables from which one may deduce formulas relating products of the h-functions and the monomial symmetric functions of the form $(pqr\ \ldots)$.

[35] Second memoir on a new theory of symmetric functions, *American J. Math.*, *12* (1890), 61–102.

This is the successor to [31]. MacMahon now extends his theory of symmetric functions to include negative powers of the variables. For example,

$$(2, 1, 0, -1) = \sum \alpha^2 \beta \gamma^0 \delta^{-1},$$

where α, β, γ, δ run over all quadruples of a given set of variables.

Section 1 is devoted to a short review of [32]. Section 2 is devoted to a simple analytic proof of MacMahon's generalization of Waring's theorem given initially in [32]. Section 3 is devoted to a simple proof of the main result on groups of separations that concluded [31].

In section 4 MacMahon proceeds with symmetric functions related to partitions with zero and negative parts. Putting

$$1 + \sum_{m=-\infty}^{\infty} X_m \mu^m = \prod_{i=1}^{n} \left(1 + \sum_{m=-\infty}^{\infty} \alpha_i^m x_m \mu^m\right),$$

he develops a theory of the X-functions similar to that obtained previously relative to partitions with positive parts (see [31]). In particular, he obtains an analytic formulation of a result that further generalizes the Waring formula:

$$\begin{aligned} &\sum (-1)^{l_1+l_2+\ldots-1} (l_1 + l_2 + \ldots - 1)!\, X_{\lambda_1}^{l_1} X_{\lambda_2}^{l_2} \ldots / l_1! l_2! \ldots \\ &= (m) \sum (-1)^{l_1+l_2+\ldots-1} (l_1 + l_2 + \ldots - 1)!\, x_{\lambda_1}^{l_1} x_{\lambda_2}^{l_2} \ldots / l_1! l_2! \ldots, \end{aligned}$$

where the summation now runs over all partitions of $m = l_1\lambda_1 + l_2\lambda_2 + l_3\lambda_3 + \ldots$, with no restrictions as to positivity of any of the integers λ_i or m.

Section 5 extends the Algebraic Reciprocity Theorem of [32] and [31] to the more general situation in which the partitions involved may have nonpositive parts. MacMahon's proof depends on the extension of the concept of "distribution" to account for partitions with nonpositive parts. He concludes the paper with a description of the formation of tables of symmetric functions as extended here, and he discusses extensions of his theorems on "groups of separations" and "expressibility" that were treated extensively in [31].

[39] Third memoir on a new theory of symmetric functions, *American J. Math.*, *13* (1891), 193–234.

In this memoir MacMahon extends his work on operators of the Hammond type in sections 8 and 9 (the sections of this paper are numbered consecutively with those of paper [35]).

Suppose we are considering symmetric functions of $\alpha_1, \alpha_2, \ldots, \alpha_n$ so that $(0) = \sum_{i=1}^{n} \alpha_i^0 = n$. MacMahon's first object is to prove

$$\begin{aligned} &\exp\{(0) + (1)x + (-1)x^{-1}\} \text{ symbolic} \\ &= \overline{\exp}\{(0) + (1)x + (-1)x^{-1}\} \\ &= \exp\left\{\sum_{n=1}^{\infty} (-1)^{n-1}(n)x^n/n + \sum_{n=1}^{\infty} (-1)^{n-1}(-n)x^{-n}/n\right\}, \end{aligned}$$

a purely formal identity in which the left side is interpreted as meaning that after the exponential function is expanded, $(1)^{\kappa}(0)^{\lambda}(-1)^{\mu}/\kappa!\lambda!\mu!$ is to be replaced by $(1^{\kappa}0^{\lambda}(-1)^{\mu})$. The main results from sections 8 and 9 are summarized in certain formal identities given in paragraphs 146–148 at the end of section 9.

In section 10 MacMahon presents analytic (he says "algebraical") proofs of various "laws of reciprocity" discussed in papers [31] and [35]. He obtains new, more general results by considering identities relating symmetric functions of three sets of variables $\{\alpha_1, \ldots, \alpha_n\}$, $\{\beta_1, \ldots, \beta_n\}$, and $\{\gamma_1, \ldots, \gamma_n\}$. From this work, he obtains parallel identities related to various partial differential operators studied in section 9.

In section 11 the above mentioned operators are studied to produce the following elegant law of symmetry:

Theorem. Let

$$(pqr\ldots)_\alpha = \sum \alpha_{i_1}^p \alpha_{i_2}^q \alpha_{i_3}^r \ldots$$

denote the monomial symmetric function of $\alpha_1, \ldots, \alpha_n$ related to the partition $p + q + r + \ldots$. Suppose

$$(p_1^{\pi_1} p_2^{\pi_2} \ldots)_\alpha = \ldots + A(\lambda_1^{l_1} \lambda_2^{l_2} \ldots) B_{s_1}^{\sigma_1} B_{s_2}^{\sigma_2} \ldots + \ldots ;$$

then

$$(s_1^{\sigma_1} s_2^{\sigma_2} \ldots) = \ldots + A(\lambda_1^{l_1} \lambda_2^{l_2} \ldots) B_{p_1}^{\pi_1} B_{p_2}^{\pi_2} \ldots + \ldots ,$$

where the B_i arise from the formal identity

$$1 + \sum_{m=-\infty}^{\infty} B_m x^m = e^{n'} \prod_{j=1}^{n'} (1 + \beta_j x)(1 + \beta_j^{-1} x^{-1}),$$

with $n' = (0)_\beta$.

The remainder of section 11 and all of section 12 are devoted to examples of the usefulness of this symmetry law and to the proof of two other symmetry laws (paragraphs 170 and 173).

Section 13 is devoted to an exposition of Cayley's algorithm (*American J. Math.*, *7* (1885), 2–9) for expressing the product of two monomial symmetric functions in terms of monomial symmetric functions. MacMahon then applies his method of operators to solve these and similar problems, remarking that while Cayley's algorithm is suitable for treating only a two-term product, the method of operators may be successfully used to treat n-term products.

The final section of this paper considers expansions of the g-functions that were introduced in section 10.

[43] Fourth memoir on a new theory of symmetric functions, *American J. Math.*, *14* (1892), 15–38.

This concludes the series of memoirs on symmetric functions. MacMahon extends his work on operators to facilitate further the representation of symmetric functions by other symmetric functions.

In section 19 MacMahon presents a general symmetry law (see paragraph 240) related to symmetric functions of n systems of variables; the case $n = 3$ was treated in section 10 of [39]. The final section is devoted to the extension of MacMahon's theory of distributions from the symmetric functions of a single system of variables to the general setting of n systems of variables. In the conclusion, MacMahon points out that paper [34] is the natural successor to this series of memoirs.

[88] Small contribution to combinatory analysis, *Proc. London Math. Soc.(2)*, *16* (1917), 352–354.

This note establishes the following result:

Theorem. The coefficient of x^n in the expansion of

$$(1 - x)^{-p_1}(1 - x^2)^{-p_2} \ldots (1 - x^t)^{-p_t}$$

(where $p_1 \geqq p_2 \geqq \ldots \geqq p_t$) is equal to the number of ways of distributing n indistinguishable objects into boxes of specification $(m_1 m_2 \ldots m_s)$, where $(p_1 p_2 \ldots p_t)$ is the partition conjugate to $(m_1 m_2 \ldots m_s)$.

[101] The algebra of symmetric functions, *Proc. Cambridge Phil. Soc., 21* (1923), 376–390.

In this paper MacMahon extends special results obtained by himself [15] and Cayley (*American J. Math., 7* (1885), 2–9) to a general theorem.

The following basic problem is solved: Find $C_{r_1 r_2 r_3} \ldots$ in

$$(p_1 p_2 p_3 \ldots)(q_1 q_2 q_3 \ldots) = \ldots + C_{r_1 r_2 r_3} \ldots (r_1 r_2 r_3 \ldots) + \ldots,$$

where $(pqr \ldots)$ is the monomial symmetric function $\sum \alpha_1^p \alpha_2^q \alpha_3^r \cdot \ldots$ An immediate solution is

$$D_{r_1} D_{r_2} D_{r_3} \ldots (p_1 p_2 p_3 \ldots)(q_1 q_2 q_3 \ldots) = C_{r_1 r_2 r_3} \ldots,$$

where D_r is the Hammond operator. Thus in order to solve our problem, we must have a means for calculating the effects of several Hammond operators operating on a product of monomial symmetric functions.

Theorem. Let U be any linear function of monomial symmetric functions of the same weight; then

$$D_{p_1}^{\pi_1} D_{p_2}^{\pi_2} D_{p_3}^{\pi_3} \ldots (\ldots 3^{r_3} 2^{r_2} 1^{r_1}) U$$

$$= [\text{ those terms not involving } a, b, c, \ldots \text{ in}$$

$$\{a^{r_1} b^{r_2} c^{r_3} \ldots (D_{p_1} + a^{-1} D_{p_1 - 1} + b^{-1} D_{p_1 - 2} + c^{-1} D_{p_1 - 3} + \ldots)^{\pi_1}$$
$$\times (D_{p_2} + a^{-1} D_{p_2 - 1} + b^{-1} D_{p_2 - 2} + c^{-1} D_{p_3 - 3} + \ldots)^{\pi_3}$$
$$\times (D_{p_3} + a^{-1} D_{p_3 - 1} + b^{-1} D_{p_3 - 2} + c^{-1} D_{p_3 - 3} + \ldots)^{\pi_3}$$
$$\times \ldots \}] U,$$

and this is precisely the coefficient of $(p_1^{\pi_1} p_2^{\pi_2} p_3^{\pi_3} \ldots)$ in the expansion of $(\ldots 3^{r_3} 2^{r_2} 1^{r_1}) U$.

MacMahon concludes by demonstrating the usefulness of this theorem in calculating a large number of examples.

[102] An American tournament treated by the calculus of symmetric functions, *Quart. J. Math.*, *49* (1923), 1–36.

Suppose we consider a tournament of two-person games (no draw games allowed) in which each player plays every other player. MacMahon analyzes the possible outcomes of such tournaments by means of the calculus of symmetric functions.

Simple combinatorial reasoning allows one to establish that if

$$\prod_{1 \leqq i < j \leqq n} (\alpha_i + \alpha_j) = \ldots + C\,(k_1 k_2 \ldots k_n) + \ldots,$$

where

$$(k_1 k_2 \ldots k_n) = \sum \alpha_{i_1}^{k_1} \alpha_{i_2}^{k_2} \alpha_{i_3}^{k_3} \ldots \alpha_{i_n}^{k_n},$$

then C is the number of tournaments in which player #1 wins k_1 games, player #2 wins k_2 games, etc. MacMahon utilizes Hammond operators to compute such coefficients and examines the outcomes of tournaments with at most 8 players.

Part 2 is devoted to a study of those partitions $(k_1 k_2 \ldots k_n)$ that arise and are related to nonzero terms of the above expansion of $\prod_{1 \leqq i < j \leqq n} (\alpha_i + \alpha_j)$.

Part 3 treats tournaments in which every possible pair of players plays every possible other pair. At the conclusion, the general case in which every possible combination of s players plays every other possible combination of s players is briefly mentioned.

[103] Chess tournaments and the like treated by the calculus of symmetric functions, *Quart. J. Math.*, *49* (1923), 353–384.

This paper is the successor to [102]. We are now considering tournaments of two-person games in which each player plays every other player, and a drawn game is now a possible outcome; a chess tournament provides a good example. The problem of analyzing the outcome of such tournaments is reduced by MacMahon to that of a problem in symmetric functions: If

$$\prod_{1 \leqq i < j \leqq n} (\alpha_i^2 + \alpha_i \alpha_j + \alpha_j^2) = \ldots + C(k_1 k_2 \ldots k_n) + \ldots,$$

where

$$(k_1 k_2 \ldots k_n) = \sum \alpha_{i_1}^{k_1} \alpha_{i_2}^{k_2} \alpha_{i_3}^{k_3} \ldots \alpha_{i_n}^{k_n},$$

then C is the number of tournaments in which player #1 scored k_1 (each win counts 2, draws count 1, losses count 0), player #2 scored k_2, etc. MacMahon utilizes Hammond operators to compute such coefficients, and examines the outcomes of tournaments with at most 6 players.

In part II MacMahon considers generalizations of such tournaments. He now assumes that in each game a total score of r is divided among the two players, and that each player plays every other player exactly t times. The symmetric function of interest in this case is

$$\prod_{1 \leqq i < j \leqq n} (\alpha_i^r + \alpha_i^{r-1} \alpha_j + \alpha_i^{r-2} \alpha_j^2 + \ldots + \alpha_j^r)^t.$$

Reprints of the Papers

ALGEBRAIC IDENTITIES ARISING OUT OF AN EXTENSION OF WARING'S FORMULA.

By *Capt. P. A. MacMahon, R.A.*

CAUCHY'S identity is the expression of

$$x^n + y^n + (-)^n (x + y)^n$$

as a rational function of the two expressions $x^2 + xy + y^2$, $xy (x + y)$; the development is given at length in Todhunter's *Theory of Equations*, page 189, and the theorem has been discussed and extended to any number of quantities x, y, z ... by Mr. Glaisher, Mr. Muir, and Prof. Cayley, in various numbers of the *Messenger* and the *Quarterly Journal*; the key to the identity was, in fact, found to lie in Waring's formula for the sums of the powers of the roots of an equation.

As I have, in vol. XV. No. 216 of the *Proceedings of the Lond. Math. Soc.*, generalised Waring's formula, so as to give the general term in the sum of all those symmetric functions which contain any given number of roots, I thought it would be interesting to examine some of the associated identities.

It may be remarked with reference to Cauchy's identity, that the extensions that have been made of it in the direction of increasing the number of quantities, assume very neat forms if certain relations exist between the quantities.

Thus if $$x^2+y^2+z^2+yz+zx+xy=0,$$
$$x^n+y^n+z^n+(-)^n(x+y+z)^n,$$
is always expressible as a rational function of
$$a_3=(x+y+z)(x^2+y^2+z^2)+xyz,$$
$$a_4=xyz(x+y+z);$$
for instance,
$$\frac{1}{n}\{x^n+y^n+z^n+(-)^n(x+y+z)^n\}=\Sigma(-)^{n+\alpha_3+2\alpha_4}\frac{(\alpha_1+\alpha_2-1)!}{\alpha_1!\,\alpha_2!}a_3^{\alpha_3}a_4^{\alpha_4},$$
wherein $$3\alpha_3+4\alpha_4=n,$$
which for $n=9$ gives
$$\tfrac{1}{3}\{x^9+y^9+z^9-(x+y+z)^9\}=\{(x+y+z)(x^2+y^2+z^2)+xyz\}^3.$$

2. What may be called the weight of the rth order of a term $a_\lambda^l a_\mu^m\ldots$ is formed on the same principle as the ordinary weight with the modification that previous to multiplying the suffixes by the indices, each suffix is to be decreased by $r-2$, and the numbers so decreased are to represent the different figurate numbers of the rth order; thus, in the above written term, to form what will be denoted by w_r, we have to change λ, μ into the $(\lambda-r+2)^{\text{th}}$ and $(\mu-r+2)^{\text{th}}$ r-agonal numbers respectively, and if λ', μ' represent the numbers so obtained,
$$w_r=l.\lambda'+m.\mu'+\ldots.$$
Then denoting by S_n^r the sum of all the symmetric functions of weight n which contain r and only r roots, the theorem is
$$S_n^r=\Sigma(-)^{n+k+r-1}\frac{(k-1)!\,w_{r+1}}{\alpha_1!\,\alpha_2!\,\alpha_3!\ldots}a_1^{\alpha_1}a_2^{\alpha_2}a_3^{\alpha_3}\ldots,$$
where $$k=\alpha_1+\alpha_2+\alpha_3+\ldots,$$
and the equation is $\xi^n-a_1\xi^{n-1}+a_2\xi^{n-2}-\ldots=0$.

Taking the simplest case

$$S_n^2 = \Sigma (-)^{n+k+1} \frac{(k-1)!\, w_3}{\alpha_1!\, \alpha_2!\, \alpha_3! \ldots} a_1^{\alpha_1} a_2^{\alpha_2} a_3^{\alpha_3} \ldots,$$

suppose x, y, z three roots of the above equation, and the remainder to vanish; then

$$S_n^2 = xy \frac{x^{n-1} - y^{n-1}}{x - y} + yz \frac{y^{n-1} - z^{n-1}}{y - z} + zx \frac{z^{n-1} - x^{n-1}}{z - x};$$

putting $$x + y + z = 0,$$
so that
$$a_2 = -(x^2 + xy + y^2),$$
$$a_3 = -xy(x + y),$$

$$S_n^2 = xy \frac{x^{n-1} - y^{n-1}}{x - y} - y(x + y) \frac{y^{n-1} + (-)^n (x+y)^{n-1}}{x + 2y}$$
$$- x(x + y) \frac{x^{n-1} + (-)^n (x+y)^{n-1}}{2x + y},$$

or $(x - y)(x + 2y)(y + 2x)\, S_n^2 = x(x + 2y)\{2y(x + y) - x^2\} x^{n-1}$
$- y(2x+y)\{2x(x+y) - y^2\} y^{n-1} + (-)^{n+1}(x^2 - y^2)(x^2 + 4xy + y^2)(x+y)^{n-1}$
$= \{(x + 2y) x^n + (-)^{n+1}(x - y)(x + y)^n\}\{2y(x + y) - x^2\}$
$- \{(2x + y) y^n + (-)^{n+1}(y - x)(x + y)^n\}\{2x(x + y) - y^2\},$
in virtue of the relation

$$2y(x + y) - x^2 + 2x(x + y) - y^2 = x^2 + 4xy + y^2,$$

thus

$$S_n^2 = \frac{(x + 2y) x^n + (-)^{n+1} (x - y)(x + y)^n}{(x - y)(x + 2y)(2x + y)} \{2y(x + y) - x^2\}$$
$$+ \frac{(y + 2x) y^n + (-)^{n+1} (y - x)(y + x)^n}{(y - x)(y + 2x)(2y + x)} \{2x(y + x) - y^2\},$$

the second term being derived from the first by the interchange of x and y.

And, finally, there results the identity

$$\frac{(x + 2y) x^n + (-)^{n+1} (x - y)(x + y)^n}{(x - y)(x + 2y)(2x + y)} \{2y(x + y) - x^2\}$$
$$+ \frac{(y + 2x) y^n + (-)^{n+1} (y - x)(y + x)^n}{(y - x)(y + 2x)(2y + x)} \{2x(x + y) - y^2\}$$
$$= \Sigma (-)^{\alpha_3+1} \frac{(\alpha_2 + \alpha_3 - 1)!\, (\alpha_2 + 3\alpha_3)}{\alpha_2!\, \alpha_3!} (x^2 + xy + y^2)^{\alpha_2} \{xy(x + y)\}^{\alpha_3}.$$

As a particular case when $n=7$, the dexter of the identity is $5xy(x+y)(x^2+xy+y^2)^2$, which may be readily verified.

In the case of three quantities, the identity is much less simple; the final form is

$$\frac{(x+y+z)(2xy-3yz+2zx-x^2)+x(xy+yz+zx)-3xyz}{(x-y)(x-z)(2x+y+z)}x^n$$

+ the same with x and y interchanged

+ the same with x and z interchanged

$$+\frac{(x+y+z)^3+2(x+y+z)(yz+zx+xy)+3xyz}{(2x+y+z)(x+2y+z)(x+y+2z)}(-)^{n+1}(x+y+z)^n$$

$$\equiv \Sigma(-)^{n+1}\frac{(\alpha_2+\alpha_3+\alpha_4-1)!\,(\alpha_2+3\alpha_3+6\alpha_4)}{\alpha_2!\,\alpha_3!\,\alpha_4!}b_2^{\alpha_2}b_3^{\alpha_3}b_4^{\alpha_4},$$

where
$$b_2=x^2+y^2+z^2+yz+zx+xy,$$
$$b_3=(x+y+z)(yz+zx+xy)-xyz,$$
$$b_4=xyz(x+y+z),$$

and
$$2\alpha_2+3\alpha_3+4\alpha_4=n.$$

Passing now to the case of the formula for symmetric functions containing three roots, one very simple result is obtained; namely, when two quantities are involved, then if

$$a_2=x^2+xy+y^2,\quad a_3=xy(x+y);$$

$$\frac{x^n}{(x-y)(2x+y)}+\frac{y^n}{(y-x)(2y+x)}+(-)^n\frac{(x+y)^n}{(2x+y)(2y+x)}$$

$$=\Sigma(-)^n\frac{(\alpha_2+\alpha_3-1)!\,(\alpha_3)}{\alpha_2!\,\alpha_3!}a_2^{\alpha_2}a_3^{\alpha_3-1},\quad(2\alpha_2+3\alpha_3=n+1),$$

for $n=3$, since then α_3-1 is negative,

$$\frac{x^3}{(x-y)(2x+y)}+\frac{y^3}{(y-x)(2y+x)}-\frac{(x+y)^3}{(2x+y)(2y+x)}=0,$$

when $n=4$, the dexter is x^2+xy+y^2 and so on. Hereafter the identities get very much more complicated, and it does not seem worth while to pursue the subject further.

Royal Military Academy,
March 23*rd*, 1884.

THE MULTIPLICATION OF SYMMETRIC FUNCTIONS.

By *Capt. P. A. MacMahon, R.A.*

IN vol VII., No. 1, of the *American Journal of Mathematics*, Professor Cayley has developed a theory for the multiplication of symmetric functions and given analytical formulæ for certain forms of products; in particular he finds that

$$3^{\alpha}2^{\beta}.3^{\gamma}2^{\delta} = \Sigma \Lambda 6^{A}5^{B}4^{C}3^{D}2^{E} \dots\dots\dots\dots\dots (1),$$

wherein, in the general case, Λ is a sum of terms, each of which is a product of three binomial coefficients; he does not give, however, the actual expression of the whole coefficient, but only its form. The object of this paper is to obtain the value of Λ in the form of a series and to shew a new method of treating such questions.

Suppose

$$D_{\lambda} = \left(a_0 \frac{d}{da_1} + a_1 \frac{d}{da_2} + a_2 \frac{d}{da_3} + \dots\right)^{\lambda} \frac{1}{\lambda!},$$

the development of the differential expression being symbolic as in the generalised form of Taylor's theorem.

Then $\quad D_6^A D_5^B D_4^C D_3^D D_2^E (6^A 5^B 4^C 3^D 2^E) = 1,$

(*vide Proc. Lond. Math. Soc.*, vol. XIV., No. 200) and, performing this operation on each side of the identity (1), there results

$$\Lambda = D_6^A D_5^B D_4^C D_3^D D_2^E (3^{\alpha}2^{\beta}.3^{\gamma}2^{\delta});$$

we have $\quad D_6^A (3^{\alpha}2^{\beta}.3^{\gamma}2^{\delta}) = 3^{\alpha-A}2^{\beta}.3^{\gamma-A}2^{\delta}.$

$$D_6^A D_5 P = 3^{\alpha-A-1}2^{\beta}.3^{\gamma-A}2^{\delta-1} + 3^{\alpha-A}2^{\beta-1}.3^{\gamma-A-1}2^{\delta};$$

putting $3^{\alpha}2^{\beta}.3^{\gamma}2^{\delta} = P$, for brevity;

$$D_6^A D_5^2 P = 3^{\alpha-A-2}2^{\beta}.3^{\gamma-A}2^{\delta-2} + 2(3^{\alpha-A-1}2^{\beta-1}.3^{\gamma-A-1}2^{\delta-1})$$
$$+ 3^{\alpha-A}2^{\beta-2}.3^{\gamma-A-2}2^{\delta},$$

and proceeding we find

$$D_6^A D_5^B P = 3^{\alpha-A-B} 2^{\beta} . 3^{\gamma-A} 2^{\delta-B} + B(3^{\alpha-A-B+1} 2^{\beta-1} . 3^{\gamma-A-1} 2^{\delta-B+1})$$
$$+ \frac{B(B-1)}{2\,!} (3^{\alpha-A-B+2} 2^{\beta-2} . 3^{\gamma-A-2} 2^{\delta-B+2}) + \ldots,$$

whence

$$D_6^A D_5^B D_4^C P = 3^{\alpha-A-B} 2^{\beta-C} . 3^{\gamma-A} 2^{\delta-\beta-C}$$
$$+ B(3^{\alpha-A-B+1} 2^{\beta-1-C} . 3^{\gamma-A-1} 2^{\delta-B-C+1})$$
$$+ \frac{B(B-1)}{2\,!} (3^{\alpha-A-B+2} 2^{\beta-2-C} \cdot 3^{\gamma-A-2} 2^{\delta-B-C+2}) + \ldots.$$

Consider now the product $3^x 2^y . 3^v 2^w$; we have

$$D_3\ (3^x 2^y . 3^v 2^w) = 3^{x-1} 2^y . 3^v 2^w + 3^x 2^y . 3^{v-1} 2^w,$$

$$D_3^x\ (3^x 2^y . 3^v 2^w) = 2^y . 3^v 2^w + x(32^y . 3^{v-1} 2^w) + \frac{x(x-1)}{2\,!} (3^x 2^y . 3^{v-2} 2^w) + \ldots,$$

and

$$D_3^{x+v} (3^x 2^y . 3^v 2^w) = 2^y . 2^w + x . v\,(2^y . 2^w) + \frac{x(x-1)}{2\,!} . \frac{v(v-1)}{2\,!} (2^y . 2^w) + \ldots$$
$$= \frac{(x+v)\,!}{x\,!\ v\,!} 2^y . 2^w \text{ by an obvious theorem,}$$

and further

$$D_3^{x+v} D_2^{y+w} (3^x 2^y . 3^v 2^w) = \frac{(x+v)\,!}{x\,!\ v\,!} . \frac{(y+w)\,!}{y\,!\ w\,!};$$

applying this result, we have

$$D_6^A D_5^B D_4^C D_3^D D_2^E P = \frac{D\,!\ E\,!}{(\alpha-A-B)\,!\ (\gamma-A)\,!\ (\beta-C)\,!\ (\delta-B-C)\,!}$$
$$+ B \frac{D\,!\ E\,!}{(\alpha-A-B+1)\,!\ (\gamma-A-1)\,!\ (\beta-C-1)\,!\ (\delta-B-C+1)\,!}$$
$$+ \frac{B(B-1)}{2\,!} \frac{D\,!\ E\,!}{(\alpha-A-B+2)\,!\ (\gamma-A-2)\,!\ (\beta-C-2)\,!\ (\delta-B-C+2)\,!}$$
$$+ \ldots\ldots\ldots\ldots\ldots\ldots\ldots\ldots\ldots\ldots\ldots\ldots\ldots\ldots\ldots\ldots\ldots\ldots\ldots\ldots$$
$$+ \frac{B\,!\ D\,!\ E\,!}{r\,!\ (B-r)\,!\ (\alpha-A-B+r)\,!\ (\gamma-A-r) . (\beta-C-r) . (\delta-B-C+r)\,!}$$
$$+ \ldots\ldots\ldots\ldots\ldots\ldots\ldots\ldots\ldots\ldots\ldots\ldots\ldots\ldots\ldots\ldots\ldots\ldots\ldots\ldots$$
$$= \Lambda.$$

This agrees with Professor Cayley's result.

§ 2. Consider next the expansion

$$4^{\phi}3^{\alpha}2^{\beta}.3^{\gamma}2^{\delta}=\Sigma M 7^{A}6^{B}5^{C}4^{D}3^{E}2^{F},$$

the product of a quartic and a cubic form; in this case M is found by operating on each side with $D_7^A D_6^B D_5^C D_4^D D_3^E D_2^F$ and is expressed by a series, each term of which is the product of five binomial coefficients.

Putting

$$t=3A+3B+2C+2D+E+F-\alpha-\beta-\gamma-\delta-2\phi,$$

the result I find to be

$$M=\frac{B!}{t!\,(B-t)!}\,\frac{C!}{0!\,C!}\,\frac{D}{(\phi-A-B+t)!\,(D-\phi+A+B-t)}$$

$$\times\frac{E!}{(\alpha-C-t)!\,(\gamma-A-t)!}\,\frac{F!}{(\phi+\beta-A-B-D+t)!\,(\phi+\delta-A-2B-C-D+2t)!}$$

$$+\frac{B!}{t!\,(B-t)!}\,\frac{C!}{1!\,(C-1)!}\,\frac{D!}{(\phi-A-B+t)!\,(D-\phi+A+B-t)!}$$

$$\times\frac{E!}{(\alpha-C-t+1)!\,(\gamma-A-t-1)!}$$

$$\times\frac{F!}{(\phi+\beta-A-B-D+t-1)!\,(\phi+\delta-A-2B-C-D+2t+1)!}$$

$$+\frac{B!}{t!\,(B-t)!}\,\frac{C!}{2!\,(C-2)!}\,\frac{D!}{(\phi-A-B+t)!\,(D-\phi+A+B-t)!}$$

$$\times\frac{E!}{(\alpha-C-t+2)!\,(\gamma-A-t-2)!}$$

$$\times\frac{F!}{(\phi+\beta-A-B-D+t-2)!\,(\phi+\delta-A-2B-C-D+2t+2)!}$$

$$+\ldots\ldots\ldots\ldots\ldots\ldots\ldots\ldots\ldots\ldots\ldots\ldots\ldots,$$

when $\phi=A=0$, this formula should reduce to the foregoing one for the product of two cubic forms, and that it actually does so, is obvious from the remark, that in that case we have also $t=B$.

To obtain similarly from this formula, that for the product of a quartic and quadric form, it should be observed that in that case $t=0$; the formula then at once reduces to Professor Cayley's form.

Unitary symmetric functions may of course be treated in a similar way.

§ 3. The multinomial theorem is part of the same theory, and may be arrived at in the same way; ex. gr. suppose

$$(1)^n = \Sigma M \lambda^l \mu^m \nu^n \pi^p \ldots\ldots,$$

wherein $\lambda^l \mu^m \nu^n \pi^p \ldots$ is any partition of n, then

$$M = D_\lambda^l D_\mu^m D_\nu D_\pi^p \ldots (1)^n,$$

we have

$$D_\lambda (1)^n = \frac{n!}{\lambda!\,(n-\lambda)!}(1)^{n-\lambda},$$

$$D_\lambda^2 (1)^n = \frac{n!}{\lambda!\,(n-\lambda)!} \cdot \frac{(n-\lambda)!}{(\lambda!)(n-2\lambda)!}(1)^{n-2\lambda} = \frac{n!}{(\lambda!)^2\,(n-2\lambda)!}(1)^{n-2\lambda},$$

$$D_\lambda^l \quad = \frac{n!}{(\lambda!)\,(n-l\lambda)!}(1)^{n-l\lambda},$$

and proceeding, we find

$$M = \frac{n!}{(\lambda!)^l\,(\mu!)^m\,(\nu!)^n\,(\pi!)^p \ldots},$$

that is $\quad (1)^n = \Sigma \dfrac{n!}{(\lambda!)^l\,(\mu!)^m\,(\nu!)^n\,(\pi!)^p\ldots}\lambda^l \mu^m \nu^n \pi^p \ldots,$

which is the partition expression of the multinomial theorem of algebra.

We may generalize this theorem by expanding $(1^m)^n$, but even in the case $m = 2$, the general coefficient is of such complexity that I have not thought it worth while to give it.

Royal Military Academy,
Woolwich.
Jan. 6th, 1885.

Properties of a Complete Table of Symmetric Functions.

By Captain P. A. MacMahon, R. A., *Royal Military Academy, Woolwich.*

§1. In Vol. V, *American Journal of Mathematics*, Mr. Durfee has set forth the only complete and perfectly arranged table of symmetric functions in existence.

I propose to establish some remarkable features of such a tabulation which, so far as my knowledge extends, have not as yet been noticed.

If we represent the symmetric functions by partitions in ( ) and the literal products by partitions in ( )′, Mr. Durfee's table exhibits each partition ( ) in terms of the partitions ( )′, and inversely each partition ( )′ in terms of the partitions ( ); say these constitute the first and second portions of the table; the secondary diagonal of the square is mainly composed of units, in the exceptional cases a zero replacing a unit. The first and second portions of the table lie, in the main, above and below the secondary diagonal; the fact that this is not invariably the case being entirely due to the peculiar properties of the self-conjugate partitions; in both the 12^{ic} and 13^{ic} tables there are three such, with the consequence that the corresponding portion of the secondary diagonal becomes twisted about its middle point into coincidence with the principal diagonal of the square.

The terminal units, whether lying in a principal or secondary diagonal of the square, are common to both portions of the table in such wise that any unit, together with the numbers in the same row or column lying left of it or above it, belong to the first portion, whilst the same unit, together with the numbers lying in the same row or column to the right of it or below it, belong to the second portion.

It is important to observe that terminal units must lie either in the secondary or principal diagonal, and those in the principal diagonal correspond invariably to self-conjugate partitions.

The number of partitions of the weight being uneven, the number of self-conjugate partitions is also uneven, and in this case one of the terminal units, corresponding to a self-conjugate partition, is at the point of intersection of the two diagonals, the remaining units being symmetrically distributed on the principal diagonal in adjacent squares; on the other hand, if the number of partitions of the weight be even, so also is the number of self-conjugate partitions; and then there is no place for a number at the intersection of the diagonals, the self-conjugate units being now none of them in the secondary diagonal, but symmetrically placed about it in adjacent squares.

§2. In the Quarterly Journal of Pure and Applied Mathematics, No. 85, 1886, I gave the complete statement of the Cayley-Betti law of symmetry; viz. if any two results of the same weight be

$$\sum A_{\lambda^l\mu^m}\ldots(\lambda^l\mu^m\ldots) = \sum A'_{\lambda'^{l'}\mu'^{m'}}\ldots(\lambda'^{l'}\mu'^{m'}\ldots)',$$

$$\sum B_{\lambda'^{l'}\mu'^{m'}}\ldots(\lambda'^{l'}\mu'\ \ldots) = \sum B'_{\lambda^l\mu^m}\ldots\lambda_l\mu^m\ldots)',$$

then

$$\sum A_{\lambda^l\mu^m}\ldots B'_{\lambda^l\mu^m}\ldots = \sum A'_{\lambda'^{l'}\mu'^{m'}}\ldots B_{\lambda'^{l'}\mu'^{m'}}\ldots$$

I proceed to the application of this theorem.

§3. Regarding the whole square as a matrix of order equal to the number of partitions of the weight, I consider a minor square matrix of any order whose secondary diagonal is coincident with that of the whole matrix; and for clearness I first suppose such a minor matrix to be situated so that its secondary diagonal contains only units, or, what is the same thing, so that it does not intersect or include the square matrix whose principal diagonal consists entirely of units.

Represent any such minor of order s by

$$\begin{array}{cccccccccc}
a_{11} & a_{12} & a_{13} & \ldots a_{1,s-5} & a_{1,s-4} & a_{1,s-3} & a_{1,s-2} & a_{1,s-1} & 1, \\
a_{21} & a_{22} & a_{23} & \ldots a_{2,s-5} & a_{2,s-4} & a_{2,s-3} & a_{2,s-2} & 1 & a_{2s}, \\
a_{31} & a_{32} & a_{33} & \ldots a_{3,s-5} & a_{3,s-4} & a_{3,s-3} & 1 & a_{3,s-1} & a_{3s}, \\
a_{41} & a_{42} & a_{43} & \ldots a_{4,s-5} & a_{4,s-4} & 1 & a_{4,s-2} & a_{4,s-1} & a_{4s}, \\
a_{51} & a_{52} & a_{53} & \ldots a_{5,s-5} & 1 & a_{5,s-3} & a_{5,s-2} & a_{5,s-1} & a_{5s}, \\
a_{61} & a_{62} & a_{63} & \ldots 1 & a_{6,s-4} & a_{6,s-3} & a_{6,s-2} & a_{6,s-1} & a_{6s}, \\
\ldots & \ldots & \ldots & \ldots & \ldots & \ldots & \ldots & \ldots & \ldots \\
\ldots & \ldots & \ldots & \ldots & \ldots & \ldots & \ldots & \ldots & \ldots \\
a_{s-1,1} & 1 & a_{s-1,3} & \ldots a_{s-1,s-5} & a_{s-1,s-4} & a_{s-1,s-3} & a_{s-1,s-2} & a_{s-1,s-1} & a_{s-1,s}, \\
1 & a_{s2} & a_{s3} & \ldots a_{s,s-5} & a_{s,s-4} & a_{s,s-3} & a_{s,s-2} & a_{s,s-1} & a_{ss},
\end{array}$$

wherein the α's and the a's belong respectively to the first and second portions of the table.

Applications of the law of symmetry enable us to write down the $s-1$ relations:

$$
\begin{aligned}
&\alpha_{1,s-1}+a_{2s}=0,\\
&\alpha_{1,s-2}+a_{3,s-1}\alpha_{1,s-1}+a_{3s}=0,\\
&\alpha_{1,s-3}+a_{4,s-2}\alpha_{1,s-2}+a_{4,s-1}\alpha_{1,s-1}+a_{4s}=0,\\
&\alpha_{1,s-4}+a_{5,s-3}\alpha_{1,s-3}+a_{5,s-2}\alpha_{1,s-2}+a_{5,s-1}\alpha_{1,s-1}+a_{5s}=0,\\
&\alpha_{1,s-5}+a_{6,s-4}\alpha_{1,s-4}+a_{6,s-3}\alpha_{1,s-3}+a_{6,s-2}\alpha_{1,s-2}+a_{6,s-1}\alpha_{1,s-1}+a_{6s}=0,\\
&\cdots\cdots\cdots\cdots\cdots\cdots\cdots\cdots\cdots\cdots\cdots\cdots\\
&\cdots\cdots\cdots\cdots\cdots\cdots\cdots\cdots\cdots\cdots\cdots\cdots\\
&\alpha_{12}+a_{s-1,3}\alpha_{13}+\ldots+a_{s-1,s-5}\alpha_{1,s-5}+a_{s-1,s-4}\alpha_{1,s-4}+a_{s-1,s-3}\alpha_{1,s-3}\\
&\qquad+a_{s-1,s-2}\alpha_{1,s-2}+a_{s-1,s-1}\alpha_{1,s-1}+a_{s-1,s}=0,\\
&\alpha_{11}+a_{s2}\alpha_{12}+a_{s3}\alpha_{13}+\ldots+a_{s,s-5}\alpha_{1,s-5}+a_{s,s-4}\alpha_{1,s-4}+a_{s,s-3}\alpha_{1,s-3}\\
&\qquad+a_{s,s-2}\alpha_{1,s-2}+a_{s,s-1}\alpha_{1,s-1}+a_{ss}=0,
\end{aligned}
$$

which are very convenient for purposes of verification. From them is deduced

$$
\alpha_{11}=\left|\begin{array}{ccccccccc}
-a_{2s} & 0 & 0 & \ldots 0 & 0 & 0 & 0 & 1\\
-a_{3s} & 0 & 0 & \ldots 0 & 0 & 0 & 1 & a_{3,s-1}\\
-a_{4s} & 0 & 0 & \ldots 0 & 0 & 1 & a_{4,s-2} & a_{4,s-1}\\
-a_{5s} & 0 & 0 & \ldots 0 & 1 & a_{5,s-3} & a_{5,s-2} & a_{5,s-1}\\
-a_{6s} & 0 & 0 & \ldots 1 & a_{6,s-4} & a_{6,s-3} & a_{6,s-2} & a_{6,s-1}\\
\cdots & \cdots & \cdots & \cdots & \cdots & \cdots & \cdots & \cdots\\
\cdots & \cdots & \cdots & \cdots & \cdots & \cdots & \cdots & \cdots\\
-a_{s-1,s} & 1 & a_{s-1,3} & \ldots a_{s-1,s-5} & a_{s-1,s-4} & a_{s-1,s-3} & a_{s-1,s-2} & a_{s-1,s-1}\\
-a_{ss} & a_{s2} & a_{s3} & \ldots a_{s,s-5} & a_{s,s-4} & a_{s,s-3} & a_{s,s-2} & a_{s,s-1}
\end{array}\right|\div\Delta,
$$

where Δ is a determinant of order $s-1$ whose secondary diagonal consists entirely of units and having nothing but zeros above this diagonal; so that

$$\Delta=(-)^{s+s-1+s-2+\ldots+2}=(-)^{\frac{1}{2}s(s+1)-1};$$

and in the numerator, if we perform on the columns a cyclical substitution, so as to make the first column the last and then change the sign of the last column, we shall obtain a determinant multiplied by $(-)^{s-1}$; whence the sign of the fraction becomes $\qquad (-)^{s-1+\frac{1}{2}(s)(s+1)-1}=(-)^{\frac{1}{2}s(s-1)}$,

and accordingly

$$
\alpha_{11}=(-)^{\frac{1}{2}s(s-1)}\left|\begin{array}{cccccccc}
0 & 0 & 0 & \ldots 0 & 0 & 1 & a_{2s}\\
0 & 0 & 0 & \ldots 0 & 1 & a_{3,s-1} & a_{3s}\\
0 & 0 & 0 & \ldots 1 & a_{4,s-2} & a_{4,s-1} & a_{4s}\\
0 & 0 & 0 & \ldots a_{5,s-3} & a_{5,s-2} & a_{5,s-1} & a_{5s}\\
\cdots & \cdots & \cdots & \cdots & \cdots & \cdots & \cdots\\
\cdots & \cdots & \cdots & \cdots & \cdots & \cdots & \cdots\\
1 & a_{s-1,3} & a_{s-1,4} & \ldots a_{s-1,s-3} & a_{s-1,s-2} & a_{s-1,s-1} & a_{s-1,s}\\
a_{s2} & a_{s3} & a_{s4} & \ldots a_{s,s-3} & a_{s,s-2} & a_{s,s-1} & a_{ss}
\end{array}\right|;
$$

but the matrix of this determinant is, as regards the original matrix, none other than the matrix complementary to α_{11}, in which all the a's are put equal to zero; further, we shall find α_{1j} equal to the product of $(-)^{\frac{1}{2}s(s-1)+1+j}$ and the determinant of its complementary matrix, with the above convention as regards zeros.

Again, consider any α element in the ρth row and κ^{th} column $\alpha_{\rho\kappa}$, and in conjunction therewith the matrix obtained from the original matrix by deletion of the first $\rho - 1$ rows and the last $\rho - 1$ columns.

The matrix now under view is of order $s - \rho + 1$, and, by the theorem already established, the element considered is the product of $(-)^{\frac{1}{2}(s-\rho+1)(s-\rho)+\kappa+1}$ and the determinant of the matrix which is its complementary with regard to the matrix of order $s - \rho + 1$.

If we now form the complementary matrix of $\alpha_{\rho\kappa}$ with regard to the matrix of order s, we find that its determinant has the same numerical value as before, but that its sign is by the added rows and columns multiplied by

$$(-)^{s+s-1+\ldots.+s-\rho+2} = (-)^{\frac{1}{2}(\rho-1)(2s-\rho+2)};$$

whence the element $\alpha_{\rho\kappa}$ is equal to the determinant of the matrix which is its complement with regard to the matrix of order s multiplied by

$$(-)^{\frac{1}{2}(s-\rho+1)(s-\rho)+\kappa+1+\frac{1}{2}(\rho-1)(2s-\rho+2)},$$

that is, by $(-)^{\frac{1}{2}s(s-1)+\rho+\kappa}$.

Similarly, the a's are expressed in terms of the α's, and we may enunciate the general theorem:

"Given a complete table of symmetric functions arranged on the Durfee system, and isolating any square matrix of order s whose secondary diagonal, being coincident with that of the whole square, consists solely of terminal units, the value of any element belonging to the first or second portions of the table and situated in the ρth row and κ^{th} column is equal to the product of

$$(-)^{\frac{1}{2}s(s-1)+\rho+\kappa}$$

and the determinant of its complementary matrix, when in such matrix all other elements belonging to the first or second portions respectively are replaced by zero."

The theorem may be otherwise in part exhibited by writing down the identity

$$\begin{vmatrix} a_{11} & a_{12} \cdots\cdots a_{1,s-3} & a_{1,s-2} & a_{1,s-1} & 1 \\ 0 & 0 \cdots\cdots 0 & 0 & 1 & a_{2s} \\ 0 & 0 \cdots\cdots 0 & 1 & a_{3,s-1} & a_{3s} \\ 0 & 0 \cdots\cdots 1 & a_{4,s-2} & a_{4,s-1} & a_{4s} \\ \cdots & \cdots & \cdots & \cdots & \cdots \\ \cdots & \cdots & \cdots & \cdots & \cdots \\ 0 & 1 \cdots\cdots a_{s-1,s-3} & a_{s-1,s-2} & a_{s-1,s-1} & a_{s-1,s} \\ 1 & a_{s2} \cdots\cdots a_{s,s-3} & a_{s,s-2} & a_{s,s-1} & a_{ss} \end{vmatrix}$$

$$= (-)^{\frac{1}{2}s(s-1)}\{a_{11}^2 + a_{12}^2 + \cdots\cdots + a_{1,s-3}^2 + a_{1,s-2}^2 + a_{1,s-1}^2 + 1\},$$

which follows at once from the foregoing results.

§4. We might now investigate a similar theorem for those cases in which the matrix intersects or includes the matrix whose principal diagonal consists entirely of units, and the requisite modification of the determinant rule is easily reached. But, instead of doing this, I will indicate a slight modification of Durfee's arrangement whereby the law above established becomes universally applicable.

It is well known, and moreover very easily proved, that a symmetric function whose partition is self-conjugate contains one, and only one, term whose partition is self-conjugate; the partition of this one term is identical with that of the symmetric function, and, as we know, its coefficient is a terminal unit; it hence follows that the matrix whose principal diagonal consists entirely of terminal units has every other element necessarily zero; as a consequence, if we arrange the self-conjugate partitions of symmetric functions in any order and then place the self-conjugate partitions of the terms in the reverse order, we necessarily confine each portion of the tabulation to half a square and bring all the terminal units into the secondary diagonal.

I therefore propose that Mr. Durfee's arrangement be modified in this manner, so that the self-conjugate partitions are arranged in only a quasi-symmetrical manner. The theorem of §3 will then be applicable to each of the secondary diagonal matrices of the whole matrix.

Royal Military Academy, Woolwich, England, July 8th.

Memoir on a New Theory of Symmetric Functions.

By Captain P. A. MacMahon, R. A., *Woolwich, England.*

In a communication recently made to the London Mathematical Society, I have sketched out an extension of the algebra of the theory of symmetrical functions and have established the bases of a wide development.

For the sake of unity, as well as for the convenience of the readers of this journal, I repeat some definitions and preliminary theorems which are of great moment to the due comprehension of what follows.

The main object of this memoir is to show clearly the proper place of the "Symmetric Function Tables" as studied by Hirsch, Cayley, Durfee and other mathematicians in Europe and America, in the algebra of such functions; to point out that the fact of their existence depends upon a wide theorem of algebraic reciprocity which leads to an equally wide theorem of algebraic expressibility, and that they are a particular case, and not the most important case from the point of view of application, of a system of such tables.

I indicate an application to the general theory of binary forms, which as regards ground forms and syzygies is of considerable promise.

It has been usual to discuss and develop the theory by reference to the weight of the involved symmetric functions; tables have thus been constructed of weights 1, 2, 3, ... , and laws and formulae appertaining thereto have been evolved; some of these laws and formulae are in regard to an arbitrary weight, but so far as my knowledge extends no attempt has hitherto been made to develop the theory from a more general standpoint.

In what follows, I regard the whole theory as arising from the discussion of an arbitrary partition of an arbitrary weight, and bring out the ordinary theory as the particular case corresponding to that partition of w (the arbitrary weight) which is composed of w units.

The relation of the ordinary to the present inclusive theory will become clear as the investigation proceeds, and it will in particular be interesting to note those theorems which are perfectly general for every partition, and those which are peculiar to single partitions.

Definitions.

A number is partitioned into parts by writing down a set of positive numbers which result in the number when added together; each of the constituent numbers is a part of the partition, and the parts are usually written in descending order from left to right and enclosed in a bracket ( ).

A partition is separated into separates by writing down a set of partitions, each separate partition in its own brackets, from left to right, so that when all the parts of these partitions are assembled in a single bracket, the partition which is separated is reproduced. Thus of a partition

$$(p_1p_2p_3p_4p_5),$$

separations are

$$(p_1p_2)(p_3p_4)(p_5),$$
$$(p_1p_2p_3)(p_4p_5).$$

Professor Sylvester has termed a number, quâ its partitions, the partible number; so here we may term a partition, quâ its separations, the separable partition. It is convenient to order the separates of a separation from left to right, in descending order as regards weight.

If the successive weights of the separates be

$$w_1, w_2, w_3, \ldots,$$

I speak of a separation of species partition

$$(w_1w_2w_3 \ldots).$$

The sum of the highest parts of the several separates I further call the "degree" of the separation. The characteristics of a separation are

(i) the weight,
(ii) the separable partition,
(iii) the species partition,
(iv) the degree;

to which may be added

(v) the multiplicity,

where, if the separation be

$$(p_1 \ldots .)^{j_1}(p_s \ldots .)^{j_2}(p_t \ldots .)^{j_3} \ldots .,$$

the multiplicity is defined by the succession of indices

$$j_1 j_2 j_3 \ldots .$$

Theorem of Reciprocity.

Partitions being symbolical representations of symmetric functions, let

$$X_1 = (1)\,x_1,$$
$$X_2 = (2)\,x_2 + (1^2)\,x_1^2,$$
$$X_3 = (3)\,x_3 + (21)\,x_2 x_1 + (1^3)\,x_1^3,$$
$$X_4 = (4)\,x_4 + (31)\,x_3 x_1 + (2^2)\,x_2^2 + (21^2)\,x_2 x_1^2 + (1^4)\,x_1^4,$$
$$\ldots\ldots\ldots\ldots\ldots\ldots\ldots\ldots\ldots\ldots\ldots$$
$$X_m = \sum (m_1 m_2 m_3 \ldots .)\, x_{m_1} x_{m_2} x_{m_3} \ldots .,$$

the summation being in regard to every partition

$$(m_1 m_2 m_3 \ldots .)$$

of the number m.

Form the product $\quad X_{p_1}^{\pi_1} X_{p_2}^{\pi_2} X_{p_3}^{\pi_3} \ldots .$

and observe that, on performing the multiplication, the coefficient of the x term

$$x_{s_1}^{\sigma_1} x_{s_2}^{\sigma_2} x_{s_3}^{\sigma_3} \ldots .$$

is a sum of compound symmetric functions, each of which, to a numerical factor près, is represented symbolically by a separation of the partition

$$(s_1^{\sigma_1} s_2^{\sigma_2} s_3^{\sigma_3} \ldots .),$$

and further, that each such separation has a species partition

$$(p_1^{\pi_1} p_2^{\pi_2} p_3^{\pi_3} \ldots .).$$

Write then $\quad X_{p_1}^{\pi_1} X_{p_2}^{\pi_2} X_{p_3}^{\pi_3} \ldots . = \sum P x_{s_1}^{\sigma_1} x_{s_2}^{\sigma_2} x_{s_3}^{\sigma_3} \ldots .;$

P, being a sum of compound symmetric functions, can be expanded in a sum of monomial symmetric functions. Thus, suppose

$$P = \sum \theta\,(\lambda_1^{l_1} \lambda_2^{l_2} \lambda_3^{l_3} \ldots .),$$

we may write

$$X_{p_1}^{\pi_1} X_{p_2}^{\pi_2} X_{p_3}^{\pi_3} \ldots . = \sum \sum \theta\,(\lambda_1^{l_1} \lambda_2^{l_2} \lambda_3^{l_3} \ldots .)\, x_{s_1}^{\sigma_1} x_{s_2}^{\sigma_2} x_{s_3}^{\sigma_3} \ldots .$$

I have established, in a practically instantaneous manner (loc. cit.), by considering a certain restricted distribution of objects amongst parcels, that in this result the coefficient θ remains unaltered if we interchange the species partition

$$(p_1^{\pi_1}p_2^{\pi_2}p_3^{\pi_3}\ldots.)$$

and the partition $(\lambda_1^{l_1}\lambda_2^{l_2}\lambda_3^{l_3}\ldots.)$;
so that we have also

$$X_{\lambda_1}^{l_1}X_{\lambda_2}^{l_2}X_{\lambda_3}^{l_3}\ldots. = \sum\sum\theta(p_1^{\pi_1}p_2^{\pi_2}p_3^{\pi_3}\ldots.)x_{s_1}^{\sigma_1}x_{s_2}^{\sigma_2}x_{s_3}^{\sigma_3}\ldots.;$$

in other words, the theorem of reciprocity states that the coefficient of

$$(\lambda_1^{l_1}\lambda_2^{l_2}\lambda_3^{l_3}\ldots.)x_{s_1}^{\sigma_1}x_{s_2}^{\sigma_2}x_{s_3}^{\sigma_3}\ldots.$$

in the development of $X_{p_1}^{\pi_1}X_{p_2}^{\pi_2}X_{p_3}^{\pi_3}\ldots.$
is equal to the coefficient of

$$(p_1^{\pi_1}p_2^{\pi_2}p_3^{\pi_3}\ldots.)x_{s_1}^{\sigma_1}x_{s_2}^{\sigma_2}x_{s_3}^{\sigma_3}\ldots.$$

in the development of $X_{\lambda_1}^{l_1}X_{\lambda_2}^{l_2}X_{\lambda_3}^{l_3}\ldots.$

Observe that the Cayley-Betti law of symmetry connected with ordinary tables of symmetric functions is obtained from this theorem by merely attending to the powers of x_1.

Formation of New Tables.

Writing as before

$$X_{p_1}^{\pi_1}X_{p_2}^{\pi_2}X_{p_3}^{\pi_3}\ldots. = \sum P x_{s_1}^{\sigma_1}x_{s_2}^{\sigma_2}x_{s_3}^{\sigma_3}\ldots.,$$

$$P = \sum\theta(\lambda_1^{l_1}\lambda_2^{l_2}\lambda_3^{l_3})\ldots.,$$

P is an aggregate of separations of

$$(s_1^{\sigma_1}s_2^{\sigma_2}s_3^{\sigma_3}\ldots.)$$

of species partition $(p_1^{\pi_1}p_2^{\pi_2}p_3^{\pi_3}\ldots.)$.

Of any separable partition $(s_1^{\sigma_1}s_2^{\sigma_2}s_3^{\sigma_3}\ldots.)$

there is a definite number of species partitions, which is in general less than the number of separations. Forming a product

$$X_{p_1}^{\pi_1}X_{p_2}^{\pi_2}X_{p_3}^{\pi_3}\ldots.$$

for each such species partition, we get an equal number of expressions P, each of which is an aggregate of separations of the corresponding species; on expanding all these expressions P in a series of monomials, we obtain a certain number of different symmetric functions

$$(\lambda_1^{l_1}\lambda_2^{l_2}\lambda_3^{l_3}\ldots.),$$

each of which, by the law of reciprocity, is a species partition of the separable partition $(s_1^{\sigma_1}s_2^{\sigma_2}s_3^{\sigma_3}\ldots\ldots)$;

hence if there be altogether k species partitions of the separable partition

$$(s_1^{\sigma_1}s_2^{\sigma_2}s_3^{\sigma_3}\ldots\ldots),$$

the development of the corresponding k products

$$X_{p_1}^{\pi_1}X_{p_2}^{\pi_2}X_{p_3}^{\pi_3}\ldots\ldots$$

will, through the k expressions P, lead to precisely k different monomial symmetric functions $(\lambda_1^{l_1}\lambda_2^{l_2}\lambda_3^{l_3}\ldots\ldots)$

symbolized by identically the same partitions as the species partitions; hence we see directly that, given a separable partition

$$(s_1^{\sigma_1}s_2^{\sigma_2}s_3^{\sigma_3}\ldots\ldots),$$

the species partitions $(p_1^{\pi_1}p_2^{\pi_2}p_3^{\pi_3}\ldots\ldots)$

are, in some order, the same as the partitions

$$(\lambda_1^{l_1}\lambda_2^{l_2}\lambda_3^{l_3}\ldots\ldots),$$

and that writing the expressions

$$P$$

in any order vertically and the corresponding species partitions in the same order from left to right, we are able to express the separations

$$P$$

in terms of monomial symmetric functions by means of a table possessing the same row and column symmetry as given by the Cayley-Betti law in the particular case of existing tables.

Theorem: Of a separable partition

$$(s_1^{\sigma_1}s_2^{\sigma_2}s_3^{\sigma_3}\ldots\ldots)$$

the separations P

are expressible by means of monomial symmetric functions, symbolized by partitions which are identical with the species partitions, and a symmetrical table may be thus formed.

By solving a set of simultaneous linear equations, we may now express the monomial symmetric functions which are symbolized by the species partitions of the separations of $(s_1^{\sigma_1}s_2^{\sigma_2}s_3^{\sigma_3}\ldots.)$
in terms of the separations P
which arise from the X-products.

We must also thus get a table possessing the same law of symmetry, as may be easily gathered from the elementary properties of determinants.

Theorem: The monomial symmetric functions, symbolized by the species partitions of the separations of

$$(s_1^{\sigma_1}s_2^{\sigma_2}s_3^{\sigma_3}\ldots.)$$

are expressible in terms of compound symmetric functions, symbolized by the separations P

which arise from the X-products corresponding to the several species partitions, and a table thus formed will possess row and column symmetry.

This last theorem is in fact a law of algebraic expressibility which may be further enunciated as follows:

Theorem of expressibility: If a symmetric function be symbolized by

$$(\lambda\mu\nu\ldots.)$$

and $(\lambda_1\lambda_2\lambda_3\ldots.)$ be any partition of λ,
$(\mu_1\mu_2\mu_3\ldots.)$ " " " μ,
$(\nu_1\nu_2\nu_3\ldots.)$ " " " ν,
. . . .

the symmetric function $(\lambda\mu\nu\ldots.)$

is expressible by means of separations of

$$(\lambda_1\lambda_2\lambda_3\ldots.\mu_1\mu_2\mu_3\ldots.\nu_1\nu_2\nu_3\ldots.).$$

In the ordinary theory, we have the particular case that every symmetric function whatever is expressible by means of the elementary symmetric functions (that is, by the coefficients of the equation, the symmetric functions of whose roots we are considering); observe that these coefficients have partitions composed wholly of units, and that the theorem necessarily arises because *every* member is expressible as a sum of units; the theorem in this case is also one of reducibility, but in general expressibility is not coincident with reducibility, as will appear later on.

I now proceed to show the practical method of constructing symmetrical symmetric function tables corresponding to any partition of any number. For clearness I will take a simple case, viz the partition

$$21^3$$

of the number 5.

We are concerned with symmetric functions of weight 5. Write down the separations of 21^3, viz:

(21^3)	of species	partition	(5),
$(21^2)(1)$	"	"	(41),
$(21)(1^2)$, $(1^3)(2)$	"	"	(32),
$(21)(1)^2$	"	"	(31^2),
$(2)(1^2)(1)$	"	"	(2^21),
$(2)(1)^3$	"	"	(21^3).

To each of these separations must be attached the numerical coefficient which arises from the X-product corresponding to the species partition. Thus to find the coefficient of the separation

$$(2)(1^2)(1)$$

we form the product

$$X_2^2X_1 \equiv \{(2)x_2 + (1^2)x_1^2\}^2\{(1)x_1\},$$

because (2^21) is the species partition of

$$(2)(1^2)(1),$$

and pick out the coefficient of

$$(2)(1^2)(1)\,x_2x_1^3$$

in its development. Since

$$X_2^2X_1 = \ldots . + 2\,(2)(1^2)(1)\,x_2x_1^3 + \ldots . ,$$

the required coefficient is 2.

In this manner the proper coefficient of each separation has to be determined, and they may then be written down in a vertical column according to the dictionary, alphabetical or Durfee-order of their species partitions as may seem appropriate to the purpose in hand.

The species partitions themselves are then written in the *same order* from left to right and the skeleton table then appears in the form :

	(5)	(41)	(32)	(31^2)	(2^21)	(21^3)
(21^3)						
$(21^2)(1)$						
$(21)(1^2)+(1^3)(2)$						
$(21)(1)^2$						
$2(2)(1^2)(1)$						
$(2)(1)^3$						

a table which, when filled in, will express the separations in terms of the monomial symmetric functions in a symmetrical manner.

The companion table will express inversely the monomial symmetric functions in terms of the separations, and will have the skeleton form :

	(21^3)	$(21^2)(1)$	$(21)(1^2)+(1^3)(2)$	$(21)(1)^2$	$2(2)(1^2)(1)$	$(2)(1)^3$
(5)						
(41)						
(32)						
(31^2)						
(2^21)						
(21^3)						

This example will, I think, make the method in general clear; we can readily obtain the coefficient to be used with any particular separation by considering the corresponding X-product. Thus, suppose the separation to be

$$(a_1)^{\alpha_1}(a_2)^{\alpha_2}\dots\dots(b_1)^{\beta_1}(b_2)^{\beta_2}\dots\dots(c_1)^{\gamma_1}(c_2)^{\gamma_2}\dots\dots,$$

wherein $a_1, a_2, \dots\dots$ are partitions each of weight a,

$b_1, b_2, \dots\dots$ " " " b,

$c_1, c_2, \dots\dots$ " " " c,

. . . .

then we have to put the numerical coefficient of

$$(a_1)^{\alpha_1}(a_2)^{\alpha_2}\ldots\ldots(b_1)^{\beta_1}(b_2)^{\beta_2}\ldots\ldots(c_1)^{\gamma_1}(c_2)^{\gamma_2}\ldots\ldots$$

in the development of the product

$$X_a^{\alpha_1+\alpha_2+\ldots\ldots}X_b^{\beta_1+\beta_2+\ldots\ldots}X_c^{\gamma_1+\gamma_2+\ldots\ldots}\ldots\ldots,$$

which is $$\frac{(\alpha_1+\alpha_2+\ldots\ldots)!\,(\beta_1+\beta_2+\ldots\ldots)!\,(\gamma_1+\gamma_2+\ldots\ldots)!\ldots\ldots}{\alpha_1!\,\alpha_2!\ldots\ldots\quad\beta_1!\,\beta_2!\ldots\ldots\quad\gamma_1!\,\gamma_2!\ldots\ldots}.$$

The process just laid down for the formation of symmetrical tables is that which would be adopted *a priori* as a result of the general law of reciprocity; there is, however, a practical convenience in modifying the process, in a manner which in no way interferes with the symmetrical character of the tables, so as either to lessen the magnitude of the table numbers or to get rid of fractions.

In the tables which express the separations in terms of the monomials, no fractions can possibly occur, but the numbers may be reduced in magnitude by a modification now to be explained.

In the table which expresses the separations of (21^3), the skeleton of which is given above, the symmetry will remain unchanged if we simultaneously write

$$(2)(1^2)(1) \text{ for } 2\,(2)(1^2)(1)$$

in the vertical column, and

$$2\,(2^21) \text{ for } (2^21)$$

in the horizontal row. The effect of this is to diminish the numbers without introducing fractions. To see the reason of this, observe that the coefficient of

$$(2^21)$$

in the development of every separation is necessarily of the form

$$0 \bmod 2$$

(vide Cayley, *Amer. Jour. of Math.*, Vol. VII, p. 59), so that in the original form of table, the vertical column headed (2^21) is necessarily divisible by 2; further, the horizontal row $2\,(2)(1^2)(1)$ is obviously simultaneously divisible by 2 also.

These divisions will be accomplished, as stated above, by simultaneously writing $$(2)(1^2)(1) \text{ for } 2\,(2)(1^2)(1)$$
and $$2\,(2^21) \text{ for } (2^21).$$

Generally, if in the vertical column any separation occur singly with a coefficient p, the coefficient of the corresponding species partition in the development of every separation of the separable partition is of the form

$$0 \bmod p$$

2

(vide Cayley loc. cit.); we can thus divide the separation row and the corresponding species partition column simultaneously by p, and this clearly does not interfere with the symmetry. Suppose now that we have in the vertical column of separations an aggregate of separations of a certain species partition, each separation having attached to it its proper coefficient as obtained from the X-product; of these separations, a certain one may be termed the leading separation. To make clear what I mean, I must explain what I wish to be understood by the "leading term" in the development of a separation. Taking any separation, say

$$(51)(432)(1^5),$$

write down the separates in a column, thus

$$\begin{array}{r} 51 \\ 432 \\ 11111 \\ \hline 10,5311 \end{array}$$

and add up the parts vertically, thus obtaining the monomial

$$(10,531^2).$$

I call this term $(10,531^2)$ the leading term in the development of the separation

$$(51)(432)(1^5),$$

and observe particularly that this term, in the development, of necessity occurs with a coefficient unity and precedes in dictionary order any other term arising in the development.

On this understanding I call the leading separation of an aggregate of separations of the same species partitions, that separation whose leading term precedes in dictionary order the leading terms of all other separations in the aggregate. (As to "dictionary order," vide Durfee, *Amer. Jour. of Math.*, Vol. V, p. 349). The rule for modifying the process is to divide the horizontal row and corresponding vertical column by the coefficient of the leading separation. Observe that we thus obtain each leading term with coefficient unity, in itself a considerable advantage.

The inverse tables are further modified by multiplication throughout by some number which will get rid of fractions.

It will be convenient at this point to have before us the complete tables for the first six weights.

Tables of Symmetric Functions.

WEIGHT 1. PARTITION (1).

	(1)
(1)	1

WEIGHT 2. PARTITION (1^2).

	(2)	(1^2)
(1^2)		1
$(1)^2$	1	2

	(1^2)	$(1)^2$
(2)	$\bar{2}$	1
(1^2)	1	

WEIGHT 2. PARTITION (2),

	(2)
(2)	1

WEIGHT 3. PARTITION (1^3).

	(3)	(21)	(1^3)
(1^3)			1
(1^2)(1)		1	3
$(1)^3$	1	3	6

	(1^3)	(1^2)(1)	$(1)^3$
(3)	3	$\bar{3}$	1
(21)	$\bar{3}$	1	
(1^3)	1		

WEIGHT 3. PARTITION (21).

	(3)	(21)
(21)		1
(2)(1)	1	1

	(21)	(2)(1)
(3)	$\bar{1}$	1
(21)	1	

WEIGHT 3. PARTITION (3).

	(3)
(3)	1

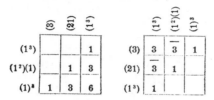

WEIGHT 4. PARTITION (1^4).

	(4)	(31)	(2^2)	(21^2)	(1^4)
(1^4)					1
$(1^3)(1)$				1	4
$(1^2)^2$			1	2	6
$(1^2)(1)^2$		1	2	5	12
$(1)^4$	1	4	6	12	24

	(1^4)	$(1^3)(1)$	$(1^2)^2$	$(1^2)(1)^2$	$(1)^4$
(4)	$\bar{4}$	4	2	$\bar{4}$	1
(31)	4	$\bar{1}$	$\bar{2}$	1	
(2^2)	2	$\bar{2}$	1		
(21^2)	$\bar{4}$	1			
(1^4)	1				

WEIGHT 4. PARTITION (21^2).

	(4)	(31)	$2(2^2)$	(21^2)
(21^2)				1
(21)(1)		1	1	2
$(2)(1^2)$		1		1
$(2)(1)^2$	1	2	1	2

	(21^2)	(21)(1)	$(2)(1^2)$	$(2)(1)^2$
(4)	1	$\bar{1}$	$\bar{1}$	1
(31)	$\bar{1}$		1	
$2(2^2)$	$\bar{1}$	1	$\bar{1}$	
(21^2)	1			

WEIGHT 4. PARTITION (2^2).

	(4)	(2^2)
(2^2)		1
$(2)^2$	1	2

	(2^2)	$(2)^2$
(4)	$\bar{2}$	1
(2^2)	1	

WEIGHT 4. PARTITION (31).

	(4)	(31)
(31)		1
(3)(1)	1	1

	(31)	(3)(1)
(4)	$\bar{1}$	1
(31)	1	

WEIGHT 4. PARTITION (4).

	(4)
(4)	1

WEIGHT 5. PARTITION (1^5).

	(5)	(41)	(32)	(31^2)	(2^21)	(21^3)	(1^5)
(1^5)							1
(1^4)(1)						1	5
(1^3)(1^2)					1	3	10
(1^3)$(1)^2$				1	2	7	20
$(1^2)^2$(1)			1	2	5	12	30
(1^2)$(1)^3$		1	3	7	12	27	60
$(1)^5$	1	5	10	20	30	60	120

	(1^5)	(1^4)(1)	(1^3)(1^2)	(1^3)$(1)^2$	$(1^2)^2$(1)	(1^2)$(1)^3$	$(1)^5$
(5)	5	$\bar{5}$	$\bar{5}$	5	5	$\bar{5}$	1
(41)	$\bar{5}$	1	5	$\bar{1}$	$\bar{3}$	1	
(32)	$\bar{5}$	5	$\bar{1}$	$\bar{2}$	1		
(31^2)	5	$\bar{1}$	$\bar{2}$	1			
(2^21)	5	$\bar{3}$	1				
(21^3)	$\bar{5}$	1					
·(1^5)	1						

WEIGHT 5. PARTITION (21^3).

	(5)	(41)	(32)	(31^2)	2(2^21)	(21^3)
(21^3)						1
(21^2)(1)				1	1	3
(21)(1^2) + (1^3)(2)			1	3	1	4
(21)$(1)^2$		1	3	4	3	6
(2)(1^2)(1)		1	1	3	1	3
(2)$(1)^3$	1	3	4	6	3	6

	(21^3)	(21^2)(1)	(21)(1^2) + (1^3)(2) $\div$ 5	(21)$(1)^2$	(2)(1^2)(1)	(2)$(1)^3$
(5)	$\bar{5}$	5	5	$\bar{5}$	$\overline{10}$	5
(41)	5		$\bar{5}$		5	
(32)	5	$\bar{5}$	1	2	$\bar{2}$	
(31^2)	$\bar{5}$		2	$\bar{1}$	1	
2(2^21)	$\overline{10}$	5	$\bar{2}$	1	$\bar{1}$	
(21^3)	5					

WEIGHT 5. PARTITION (2^21).

	(5)	(41)	(32)	(2^21)
(2^21)				1
(2^2)(1)			1	2
(21)(2)		1	1	2
$(2)^2$(1)	1	1	2	2

	(2^21)	(2^2)(1)	(21)(2)	$(2)^2$(1)
(5)	1	$\bar{1}$	$\bar{1}$	1
(41)	$\bar{1}$	$\bar{1}$	1	
(32)	$\bar{1}$	1		
(2^21)	1			

Weight 5. Partition (31^2).

	(5)	(41)	(32)	(31^2)
$(31)^2$				1
(31)(1)		1	1	2
$(3)(1^2)$		1		1
$(3)(1^2)$	1	2	1	2

	(31^2)	(31)(1)	$(3)(1^2)$	$(3)(1)^2$
(5)	1	$\bar{1}$	$\bar{1}$	1
(41)	$\bar{1}$		1	
(32)	$\bar{1}$	1	$\bar{1}$	
(31^2)	1			

Weight 5. Partition (32).

	(5)	(32)
(32)		1
(3)(2)	1	1

	(32)	(3)(2)
(5)	$\bar{1}$	1
(32)	1	

Weight 5. Partition (41).

	(5)	(41)
(41)		1
(4)(1)	1	1

	(41)	(4)(1)
(5)	$\bar{1}$	1
(41)	1	

Weight 5. Partition (5).

	(5)
(5)	1

WEIGHT 6. PARTITION (1^6).

	(6)	(51)	(42)	(3^2)	(41^2)	(321)	(31^3)	(2^3)	(2^21^2)	(21^4)	(1^6)
(1^6)											1
$(1^5)(1)$										1	6
$(1^4)(1^2)$									1	4	15
$(1^3)^2$								1	2	6	20
$(1^4)(1)^2$							1		2	9	30
$(1^3)(1^2)(1)$						1	3	3	8	22	60
$(1^3)(1)^3$					1	3	10	6	18	48	120
$(1^2)^3$				1		3	6	6	15	36	90
$(1^2)^2(1)^2$			1	2	2	8	18	15	34	78	180
$(1^2)(1)^4$		1	4	6	9	22	48	36	78	168	360
$(1)^6$	1	6	15	20	30	60	120	90	180	360	720

	(1^6)	$(1^5)(1)$	$(1^4)(1^2)$	$(1^3)^2$	$(1^4)(1)^2$	$(1^3)(1^2)(1)$	$(1^3)(1)^3$	$(1^2)^3$	$(1^2)^2(1)^2$	$(1^2)(1)^4$	$(1)^6$
(6)	$\bar{6}$	6	6	3	$\bar{6}$	$\overline{12}$	6	$\bar{2}$	9	$\bar{6}$	1
(51)	6	$\bar{1}$	$\bar{6}$	$\bar{3}$	1	7	$\bar{1}$	2	$\bar{4}$	1	
(42)	6	$\bar{6}$	2	$\bar{3}$	2	4	$\bar{2}$	$\bar{2}$	1		
(3^2)	3	$\bar{3}$	$\bar{3}$	3	3	$\bar{3}$		1			
(41^2)	$\bar{6}$	1	2	3	$\bar{1}$	$\bar{3}$	1				
(321)	$\overline{12}$	7	4	$\bar{3}$	$\bar{3}$	1					
(31^3)	6	$\bar{1}$	$\bar{2}$		1						
(2^3)	$\bar{2}$	2	$\bar{2}$	1							
(2^21^2)	9	$\bar{4}$	1								
(21^4)	$\bar{6}$	1									
(1^6)	1										

WEIGHT 6. PARTITION (21^4)

	(6)	(51)	(42)	$2(3^2)$	(41^2)	(321)	(31^3)	$3(2^3)$	$2(2^21^2)$	(21^4)
(21^4)										1
$(21^3)(1)$							1		1	4
$(21^2)(1^2)+(1^4)(2)$						1	4	1	2	7
$(21)(1^3)$						1	3		1	4
$(21^2)(1)^2$					1	3	6	2	5	12
$(21)(1^2)(1)+(1^3)(2)(1)$			1	1	3	6	13	2	6	16
$(21)(1)^3$		1	4	3	6	13	18	6	12	24
$(2)(1^2)^2$			1		2	2	6		2	6
$(2)(1^2)(1)^2$		1	2	1	5	6	12	2	5	12
$(2)(1)^4$	1	4	7	4	12	16	24	6	12	24

$\div 30.$

	(21^4)	$(21^3)(1)$	$(21^2)(1^2)+(1^4)(2)$	$(21)(1^3)$	$(21^2)(1)^2$	$(21)(1^2)(1)+(1^3)(2)(1)$	$(21)(1)^3$	$(2)(1^2)^2$	$(2)(1^2)(1)^2$	$(2)(1)^4$
(6)	30	$\overline{30}$	$\overline{30}$	$\overline{30}$	30	60	$\overline{30}$	30	$\overline{90}$	30
(51)	$\overline{30}$		30	30		$\overline{30}$		$\overline{30}$	30	
(42)	$\overline{30}$	30	$\overline{10}$	30	$\overline{10}$	$\overline{20}$	10	30	$\overline{10}$	
$2(3^2)$	$\overline{30}$	30	30	$\overline{60}$	$\overline{30}$	30		$\overline{30}$		
(41^2)	30		$\overline{10}$	$\overline{30}$	5	10	$\overline{5}$		5	
(321)	60	$\overline{30}$	$\overline{20}$	30	10	$\overline{4}$	2		$\overline{2}$	
(31^3)	$\overline{30}$		10		$\overline{5}$	2	$\overline{1}$		1	
$3(2^3)$	30	$\overline{30}$	30	$\overline{30}$						
$2(2^21^2)$	$\overline{90}$	30	$\overline{10}$		5	$\overline{2}$	1		$\overline{1}$	
(21^4)	30									

Weight 6. Partition (2^21^2). $\div 15$.

	(6)	(51)	(42)	(3^2)	(41^2)	(321)	$3(2^3)$	(2^21^2)
(2^21^2)								1
$(2^21)(1)$						1	1	2
$(21^2)(2)+(2^2)(1^2)$				1	1	2		3
$(21)^2$			1	2	2	2	2	4
$(2^2)(1)^2$			1	2		2	1	2
$(21)(2)(1)$		1	2	2	2	3	2	4
$(2)^2(1^2)$		1		2	1	2		2
$(2)^2(1)^2$	1	2	3	4	2	4	2	4

	(2^21^2)	$(2^21)(1)$	$(21^2)(2)+(2^2)(1^2)$	$(21)^2$	$(2^2)(1)^2$	$(21)(2)(1)$	$(2)^2(1^2)$	$(2)^2(1)^2$
(6)	$\bar{10}$	10	10	5	$\bar{10}$	$\bar{20}$	$\bar{10}$	15
(51)	10	5	$\bar{10}$	$\bar{5}$	$\bar{5}$	5	10	
(42)	10	$\bar{10}$	2	$\bar{5}$	4	8	$\bar{8}$	
(3^2)	5	$\bar{5}$	$\bar{5}$	5	5	$\bar{5}$	5	
(41^2)	$\bar{10}$	$\bar{5}$	4	5	$\bar{7}$	1	$\bar{1}$	
(321)	$\bar{20}$	5	8	$\bar{5}$	1	2	$\bar{2}$	
$3(2^3)$	$\bar{10}$	10	$\bar{8}$	5	$\bar{1}$	$\bar{2}$	2	
(2^21^2)	15							

Weight 6. Partition (2^3).

	(6)	(42)	(2^3)
(2^3)			1
$(2^2)(2)$		1	3
$(2)^3$	1	3	6

	(2^3)	$(2^2)(2)$	$(2)^3$
(6)	3	$\bar{3}$	1
(42)	$\bar{3}$	1	
(2^3)	1		

Weight 6. Partition (31^3).

	(6)	(51)	$2(3^2)$	(42)	(41^2)	(321)	(31^3)
(31^3)							1
$(31^2)(1)$					1	1	3
$(3)(1^3)$					1		1
$(31)(1^2)$				1	2	1	3
$(31)(1)^2$		1	1	2	4	3	6
$(3)(1^2)(1)$		1		1	3	1	3
$(3)(1)^3$	1	3	1	3	6	3	6

	(31^3)	$(31^2)(1)$	$(3)(1^3)$	$(31)(1^2)$	$(31)(1)^2$	$(3)(1^2)(1)$	$(3)(1)^3$
(6)	$\bar{1}$	1	1	1	$\bar{1}$	$\bar{2}$	1
(51)	1		$\bar{1}$	$\bar{1}$		1	
$2(3^2)$	1	$\bar{1}$	2	$\bar{1}$	1	$\bar{1}$	
(42)	1	$\bar{1}$	$\bar{1}$	1			
(41^2)	$\bar{1}$		1				
(321)	$\bar{2}$	1	$\bar{1}$				
(31^3)	1						

3

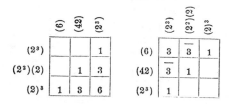

Weight 6. Partition (321). $\div$ 2.

	(6)	(51)	(42)	2(3²)	(321)
(321)					1
(32)(1)			1	1	1
(31)(2)		1		1	1
(3)(21)		1	1		1
(3)(2)(1)	1	1	1	1	1

	(321)	(32)(1)	(31)(2)	(3)(21)	(3)(2)(1)
(6)	1	$\bar{1}$	$\bar{1}$	$\bar{1}$	2
(51)	$\bar{1}$	$\bar{1}$	1	1	
(42)	$\bar{1}$	1	$\bar{1}$	1	
2(3²)	$\bar{1}$	1	1	$\bar{1}$	
(321)	1				

Weight 6. Partition (3²).

	(6)	(3²)
(3²)		1
(3)²	1	2

	(3²)	(3)²
(6)	$\bar{2}$	1
(3²)	1	

Weight 6. Partition (41²).

	(6)	(51)	(42)	(41²)
(41²)				1
(41)(1)		1	1	2
(4)(1²)		1		1
(4)(1)²	1	2	1	2

	(41²)	(41)(1)	(4)(1²)	(4)(1)²
(6)	1	$\bar{1}$	$\bar{1}$	1
(51)	$\bar{1}$		1	
(42)	$\bar{1}$	1	$\bar{1}$	
(41²)	1			

Weight 6. Partition (42).

	(6)	(42)
(42)		1
(4)(2)	1	1

	(42)	(4)(2)
(6)	$\bar{1}$	1
(42)	1	

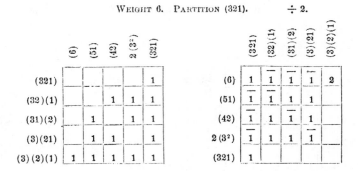

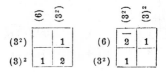

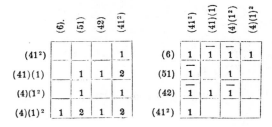

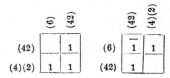

WEIGHT 6. PARTITION (51).

	(6)	(51)
(51)		1
(5)(1)	1	1

	(51)	(5)(1)
(6)	1	1
(51)	1	

WEIGHT 6. PARTITION (6).

	(6)
(6)	1

In these tables I have slightly varied the order of the partitions as I found it convenient in different cases; the proper order of the partitions for a table of given partitions remains, I think, to be discovered.

The sums of the powers of the roots may be at once written down in terms of the table separations by means of an extension of Waring's formula which I have previously given (loc. cit.), and it is convenient to repeat it here. If

$$(\lambda^l \mu^m \ldots .)$$

be any partition of n, and

$$(J_1)^{j_1}(J_2)^{j_2} \ldots .$$

one of its separations, then

$$(-)^{l+m+\ldots} \frac{(l+m+\ldots -1)!}{l!\, m! \ldots} S_n$$
$$= \sum (-)^{j_1+j_2+\ldots} \frac{(j_1+j_2+\ldots -1)!}{j_1!\, j_2! \ldots} (J_1)^{j_1}(J_2)^{j_2} \ldots ,$$

where it will be observed that any coefficient depends merely upon the multiplicity of the separation which it multiplies, and that the right-hand side is a function of the same assemblages of separations as are employed, by rule, in the tables. This very important result is a good example of the extent and scope of the new theory; the mere existence of so suggestive a formula, of which the simplicity vies with the generality, points with a clearness which cannot be mistaken, to the conclusion that up to the present the algebraic theory of symmetric functions has been regarded from a point of view so little elevated that its chief beauties remained obscured. It is convincing proof that much may be expected from a comprehensive survey from the new vantage point.

General Remarks upon Syzygies.

In general the expression of any symmetric function in terms of the table separations is unique, but where we are not restricted to these combinations this is not so. There are as many table separations as there are species partitions of those separations, and in general the whole number of separations exceeds the number of corresponding species partitions; this difference indicates the number of syzygies which exist between the separations. Thus, turning to the table of partition (21^4) we observe 10 species partitions and 12 separations pointing to $12-10=2$ syzygies between the separations; the actual forms of the syzygies are easily obtained, for consider the product

$$(21^2)(2),$$

we may *either* express (21^2) by means of separations of (1^4) *or* (2) by means of separations of (1^2); in either way we arrive at the expression of the product by means of separations of (21^4). Thus

$$\begin{aligned}(21^2)(2) &= \{(1^3)(1)-4(1^4)\}(2)\\ &= (21^2)\{(1)^2-2(1^2)\}\end{aligned}$$

leading to the syzygy

$$2(21^2)(1^2)-(21^2)(1)^2+(1^3)(2)(1)-4(2)(1^4)=0;$$

the other syzygy is obtained from the product

$$(21)(2)(1)$$

and is found to be

$$2(21)(1^2)(1)-3(1^3)(2)(1)-(21)(1)^3+(2)(1^2)(1)^2=0.$$

More conveniently we may regard the first of these syzygies as arising from the partition (42),
for taking the product $(4)(2)$,

we may simultaneously express (4) in terms of separations of (1^4) and (2) in terms of separations of (2), *or* simultaneously (4) in terms of separations of (21^2) and (2) in terms of separations of (1^2).

In a similar manner the second syzygy arises from the partition

$$(321),$$

where observe that the partitions

$$(42),$$
$$(321)$$

are the only ones containing two different parts besides unity. So, considering the separations of

$$(21^{n-2})$$

and

$$\lambda_1^{l_1}\lambda_2^{l_2}\ldots\ldots\lambda_s^{l_s}1^{t}),$$

any one of its species partitions, we see that corresponding thereto there must be

$$s-1$$

syzygies, which may be written down according to the method above explained. The number of species partitions is the whole number of partitions less one, the generating function being

$$\frac{1}{(1-x)(1-x^2)(1-x^3)\ldots\ldots}-\frac{1}{1-x}.$$

If we write down all the partitions of a number and find that p_2 partitions contain a part 2, p_3 partitions a part 3, and so on, the number

$$p_2+p_3+\ldots\ldots+p_n$$

indicates the number of separations of

$$(21^{n-2});$$

the generating function for this number is

$$\frac{x^2}{(1-x)^2(1-x^2)(1-x^3)\ldots\ldots},$$

and hence the generating function for the syzygies between the separations of

$$(21^{n-2})$$

is

$$\frac{x^2}{(1-x)^2(1-x^2)(1-x^3)\ldots\ldots}-\frac{1}{(1-x)(1-x^2)(1-x^3)\ldots\ldots}+\frac{1}{1-x},$$

which is

$$\frac{-1+x+x^2+(1-x)(1-x^2)(1-x^3)\ldots\ldots}{(1-x)^2(1-x^2)(1-x^3)\ldots\ldots},$$

or

$$\frac{\sum_{j=2}^{j=\infty}(-)^j(1+x^j)x^{\frac{1}{2}(3j^2-j)}}{(1-x)^2(1-x^2)(1-x^3)\ldots\ldots}.$$

The syzygies between the separations of other partitions may be worked out in a similar manner, though the calculations soon become laborious.

All these syzygies are linear relations between the separations of partitions, each syzygy involving only the separations of a single partition.

Beyond these there are syzygies connecting separations of different partitions which are at once seen by comparing the different tables of a given weight.

As regards any one table which expresses monomial symmetric functions in terms of separations of a given partition, any one row exhibits either such monomials as reducible quâ the separations, or else a congruence between the separations of highest degree.

By a comparison of the expressions of the same monomial symmetric function in different tables, syzygies are obtained between separations of different partitions; this circumstance I hope to discuss at some future time when applying the present theory to that of the covariants of binary forms.

Analogue of Newton's Theorem of the Sums of Powers.

The general expression of s_n in terms of the table separations of any partition of n was obtained by a comparison with a known particular case; for the proper discussion of the extended theory of symmetric functions brought forward here, it is necessary to see how the formula arises in another manner. We start with Newton's theorem for the expression of the sums of the powers in terms of the elementary coefficient, which is usually written as a series of relations, viz:

$$\begin{aligned}&s_1-(1)=0,\\&s_2-(1)s_1+2(1^2)=0,\\&s_3-(1)s_2+(1^2)s_1-3(1^3)=0,\\&s_4-(1)s_3+(1^2)s_2-(1^3)s_1+4(1^4)=0,\\&\cdots\cdots\cdots\cdots\cdots\cdots\end{aligned}$$

enabling the successive calculations of $s_2, s_3, s_4, \ldots.$ These relations are all exhibited by the single identity

$$\frac{(1)x-2(1^2)x^2+3(1^3)x^3-\ldots.}{1-(1)x+(1^2)x^2-(1^3)x^3+\ldots.}=s_1x+s_2x^2+s_3x^3+\ldots.,$$

for, on clearing of fractions and equating coefficients of like powers of x, the set of relations is produced.

Here, it is very important to observe, we can immediately obtain Waring's summation formula for the sums of the powers by expanding the denominator of the left-hand side by the multinomial theorem. This fact seems to have hitherto escaped the notice of writers upon the subject. This formula thus represents the expression of s_n in terms of separations of (1^n); the object in view is the corresponding formula for any other partition of n.

We can at once solve the problem for the partitions

$$(21^{n-2}),\ (31^{n-3}),\ (41^{n-4}),\ \ldots.\ (\lambda 1^{n-\lambda})\ \ldots.,$$

for, omitting the first of the series of relations, we may write

$$s_2 - (2) = 0,$$
$$s_3 - (1)\,s_2 + (21) = 0,$$
$$s_4 - (1)\,s_3 + (1^2)\,s_2 - (21^2) = 0,$$
$$\cdots\cdots\cdots\cdots$$

a series which leads to the identity

$$\frac{(2)\,x^2 - (21)\,x^3 + (21^2)\,x^4 - \ldots.}{1 - (1)\,x + (1^2)\,x^2 - (1^3)\,x^3 + \ldots.} = s_2x^2 + s_3x^3 + s_4x^4 + \ldots.,$$

and now expanding the sinister by the multinomial theorem we necessarily exhibit s_n in terms of separations of (21^{n-2}) and arrive at the corresponding summation formula.

Omitting the first of the last set of relations, we may write

$$s_3 - (3) = 0,$$
$$s_4 - (1)\,s_3 + (31) = 0,$$
$$s_5 - (1)\,s_4 + (1^2)\,s_3 - 31^2 = 0,$$
$$\cdots\cdots\cdots\cdots$$

and thence

$$\frac{(3)\,x^3 - (31)\,x^4 + (31^2)\,x^5 - \ldots.}{1 - (1)\,x + (1^2)\,x^2 - (1^3)\,x^3 + \ldots.} = s_3x^3 + s_4x^4 + s_5x^5 + \ldots.,$$

and proceeding in a similar manner we get finally

$$\frac{(\lambda)\,x^\lambda - (\lambda 1)\,x^{\lambda+1} + (\lambda 1^2)\,x^{\lambda+2} - \ldots.}{1 - (1)\,x + (1^2)\,x^2 - (1^3)\,x^3 + \ldots.} = s_\lambda x^\lambda + s_{\lambda+1}x^{\lambda+1} + s_{\lambda+2}x^{\lambda+2} + \ldots.,$$

a formula which enables us to exhibit s_n by means of separations of $(\lambda 1^{n-\lambda})$. Put now

$$1 - (1)\,x + (1^2)\,x^2 - (1^3)\,x^3 + \ldots. = (1 - \alpha x)(1 - \beta x)(1 - \gamma x)\ \ldots.$$

and

$$\sum_\lambda{}' = \sum \frac{\alpha^\lambda x^\lambda}{1 - \alpha x} = s_\lambda x^\lambda + s_{\lambda+1}x^{\lambda+1} + s_{\lambda+2}x^{\lambda+2} + \ldots.,$$

$$\sum_\lambda{}'' = \sum \frac{\alpha^\lambda x^\lambda}{(1 - \alpha x)^2} = s_\lambda x^\lambda + 2s_{\lambda+1}x^{\lambda+1} + 3s_{\lambda+2}x^{\lambda+2} + \ldots.,$$

$$\cdots\cdots\cdots\cdots$$

$$\sum_\lambda{}^{(m)} = \sum \frac{\alpha^\lambda x^\lambda}{(1 - \alpha x)^m} = s_\lambda x^\lambda + \frac{1}{2!}\,m\,(m+1)\,s_{\lambda+1}x^{\lambda+1}$$
$$+ \frac{1}{3!}\,m\,(m+1)(m+2)\,s_{\lambda+2}x^{\lambda+2} + \ldots.,$$

we have

$$\sum'_{\lambda}\sum'_{\mu}-\sum''_{\lambda+\mu}=\sum\frac{\alpha^{\lambda}x^{\lambda}}{1-\alpha x}\sum\frac{\alpha^{\mu}x^{\mu}}{1-\alpha x}-\sum\frac{\alpha^{\lambda+\mu}x^{\lambda+\mu}}{(1-\alpha x)^2}$$

$$=\frac{\sum\alpha^{\lambda}\beta^{\mu}x^{\lambda+\mu}(1-\gamma x)(1-\delta x)}{(1-\alpha x)(1-\beta x)(1-\gamma x)(1-\delta x)\ldots.}$$

$$=\frac{(\lambda\mu)x^{\lambda+\mu}-(\lambda\mu 1)x^{\lambda+\mu+1}+(\lambda\mu 1^2)x^{\lambda+\mu+2}-\ldots.}{1-(1)x+(1^2)x^2-(1^3)x^3+\ldots.},$$

or, taking previous results into account, this identity may be written

$$\frac{(\lambda)x^{\lambda}-(\lambda 1)x^{\lambda+1}+(\lambda 1^2)x^{\lambda+2}-\ldots.}{1-(1)x+(1^2)x^2-\ldots.}\cdot\frac{(\mu)x^{\mu}-(\mu 1)x^{\mu+1}+(\mu 1^2)x^{\mu+2}-\ldots.}{1-(1)x+(1^2)x^2-\ldots.}$$

$$-\frac{(\lambda\mu)x^{\lambda+\mu}-(\lambda\mu 1)x^{\lambda+\mu+1}+(\lambda\mu 1^2)x^{\lambda+\mu+2}-\ldots.}{1-(1)x+(1^2)x^2-\ldots.}$$

$$=s_{\lambda+\mu}x^{\lambda+\mu}+2s_{\lambda+\mu+1}x^{\lambda+\mu+1}+3s_{\lambda+\mu+2}x^{\lambda+\mu+2}+\ldots.,$$

a formula which leads to the expression of s_n in terms of separations of $(\lambda\mu 1^{n-\lambda-\mu})$. Proceeding, we find

$$\sum'_{\lambda}\sum'_{\mu}\sum'_{\nu}-\sum'_{\lambda}\sum''_{\mu+\nu}-\sum'_{\mu}\sum''_{\lambda+\nu}-\sum'_{\nu}\sum''_{\lambda+\mu}+2\sum'''_{\lambda+\mu+\nu}$$

$$=\frac{(\lambda\mu\nu)x^{\lambda+\mu+\nu}-(\lambda\mu\nu 1)x^{\lambda+\mu+\nu+1}+(\lambda\mu\nu 1^2)x^{\lambda+\mu+\nu+2}-\ldots.}{1-(1)x+(1^2)x^2-(1^3)x^3+\ldots.},$$

which should be compared and contrasted with the ordinary symmetric function formula

$$s_{\lambda}s_{\mu}s_{\nu}-s_{\lambda}s_{\mu+\nu}-s_{\mu}s_{\lambda+\nu}-s_{\nu}s_{\lambda+\mu}+2s_{\lambda+\mu+\nu}=(\lambda\mu\nu).$$

We are now led to the formula

$$\frac{(\lambda)x^{\lambda}-(\lambda 1)x^{\lambda+1}+\ldots.}{1-(1)x+(1^2)x^2-\ldots.}\cdot\frac{(\mu)x^{\mu}-(\mu 1)x^{\mu+1}+\ldots.}{1-(1)x+(1^2)x^2-\ldots.}\cdot\frac{(\nu)x^{\nu}-(\nu 1)x^{\nu+1}+\ldots.}{1-(1)x+(1^2)x^2-\ldots.}$$

$$-\frac{1}{2}\frac{(\lambda)x^{\lambda}-(\lambda 1)x^{\lambda+1}+\ldots.}{1-(1)x+(1^2)x^2-\ldots.}\cdot\frac{(\mu\nu)x^{\mu+\nu}-(\mu\nu 1)x^{\mu+\nu+1}+\ldots.}{1-(1)x+(1^2)x^2-\ldots.}$$

$$-\frac{1}{2}\frac{(\mu)x^{\mu}-(\mu 1)x^{\mu+1}+\ldots.}{1-(1)x+(1^2)x^2-\ldots.}\cdot\frac{(\lambda\nu)x^{\lambda+\nu}-(\lambda\nu 1)x^{\lambda+\nu+1}+\ldots.}{1-(1)x+(1^2)x^2-\ldots.}$$

$$-\frac{1}{2}\frac{(\nu)x^{\nu}-(\nu 1)x^{\nu+1}+\ldots.}{1-(1)x+(1^2)x^2-\ldots.}\cdot\frac{(\lambda\mu)x^{\lambda+\mu}-(\lambda\mu 1)x^{\lambda+\mu+1}+\ldots.}{1-(1)x+(1^2)x^2-\ldots.}$$

$$+\frac{1}{2}\frac{(\lambda\mu\nu)x^{\lambda+\mu+\nu}-(\lambda\mu\nu 1)x^{\lambda+\mu+\nu+1}+\ldots.}{1-(1)x+(1^2)x^2-\ldots.}$$

$$=s_{\lambda+\mu+\nu}x^{\lambda+\mu+\nu}+3s_{\lambda+\mu+\nu+1}x^{\lambda+\mu+\nu+1}+6s_{\lambda+\mu+\nu+2}x^{\lambda+\mu+\nu+2}+\ldots.,$$

enabling us to express s_n in terms of separations of

$$(\lambda\mu\nu 1^{n-\lambda-\mu-\nu}).$$

From the mode in which these results have been evolved, it is absolutely certain, without further demonstration, that the expression

$$\frac{(\lambda\mu\nu\ldots\ldots)x^{\lambda+\mu+\nu+\ldots\ldots}-(\lambda\mu\nu\ldots\ldots1)x^{\lambda+\mu+\nu+\ldots\ldots+1}+\ldots\ldots}{1-(1)x+(1^2)x^2-\ldots\ldots}$$

is given in terms of

$$\sum\nolimits'_{\lambda},\ \sum\nolimits'_{\mu},\ \sum\nolimits'_{\nu},\ldots\ldots\sum\nolimits''_{\lambda+\mu},\ \sum\nolimits''_{\lambda+\nu},\ \sum\nolimits''_{\mu+\nu},\ldots\ldots\sum\nolimits'''_{\lambda+\mu+\nu},\ldots\ldots$$

by precisely the same law as

$$(\lambda\mu\nu\ldots\ldots)$$

is given in terms of

$$s_\lambda,\ s_\mu,\ s_\nu,\ldots\ldots s_{\lambda+\mu},\ s_{\lambda+\nu},\ s_{\mu+\nu},\ldots\ldots s_{\lambda+\mu+\nu},\ldots\ldots$$

If any number of equalities exist between the numbers

$$\lambda,\ \mu,\ \nu,\ \ldots\ldots$$

the same modifications are requisite in both systems of formulae; accordingly both systems are inverted in a similar manner and following the same law; thus the formula above written is obtained by a comparison with the formula

$$(\lambda)(\mu)(\nu)-\frac{1}{2}(\lambda)(\mu\nu)-\frac{1}{2}(\mu)(\lambda\nu)-\frac{1}{2}(\nu)(\lambda\mu)+\frac{1}{2}(\lambda\mu\nu)=s_{\lambda+\mu+\nu};$$

in this manner is established the identity which for any separable partition is the direct analogue of the elementary formula

$$\frac{(1)x-2(1^2)x^2+3(1^3)x^3-\ldots\ldots}{1-(1)x+(1^2)x^2-(1^3)x^3+\ldots\ldots}=s_1x+s_2x^2+s_3x^3+\ldots\ldots,$$

and this leads by multinomial expansions to a proof of the summation formula which expresses s_n in terms of separations of any partition of n.

Derivation of Symmetric Functions from the Sums of Powers.

In the ordinary theory we derive the symmetric functions $(\lambda\mu)$, $(\lambda\mu\nu)$, ... from the formulae

$$(\lambda\mu)=s_\lambda s_\mu-s_{(\lambda+\mu)}\equiv s(1^\lambda)s(1^\mu)-s(1^{\lambda+\mu}),^*$$

$$\begin{aligned}(\lambda\mu\nu)&=s_\lambda s_\mu s_\nu-s_\lambda s_{\mu+\nu}-s_\mu s_{\nu+\lambda}-s_\nu s_{\lambda+\mu}+2s_{\lambda+\mu+\nu},\\&\equiv s(1^\lambda)s(1^\mu)s(1^\nu)-s(1^\lambda)s(1^{\mu+\nu})-s(1^\mu)s(1^{\nu+\lambda})-s(1^\nu)s(1^{\lambda+\mu})\\&\quad+2s(1^{\lambda+\mu+\nu}),\end{aligned}$$

* N. B. $s(\lambda\mu\ldots\ldots)$ denotes the expression of $s_{\lambda+\mu+\ldots\ldots}$ by means of separations of $(\lambda\mu\ldots\ldots)$.

4

and no difficulty is presented, because the products

$$s(1^\lambda)\,s(1^\mu),$$
$$s(1^\lambda)\,s(1^\mu)\,s(1^\nu),\ s(1^\lambda)\,s(1^{\mu+\nu}),\ s(1^\mu)\,s(1^{\nu+\lambda}),\ s(1^\nu)\,s(1^{\lambda+\mu}),$$

are, by direct multiplication, obtained in terms of separations of $(1^{\lambda+\mu})$, $(1^{\lambda+\mu+\nu})$ respectively. Had we been concerned with a separate partition composed of dissimilar parts, we could not thus have proceeded by direct multiplication; we could indeed have obtained the products in terms of separations, but the process would not have been a unique one and we would not have obtained an expression involving the tabular assemblages of separations. It is this expression that is required. Take the separable partion

$$(\lambda 1^{n-\lambda});$$

supposing it to be possible to express the product

$$s_p s_{n-p}$$

in terms of its tabular separations, we have, in the first place, two alternatives (in general). We may express s_p in terms of separations of (1^p), and s_{n-p} in terms of separations of $(\lambda 1^{n-p-\lambda})$, or *vice versa;* but neither of the products

$$s(1^p)\,s(\lambda 1^{n-p-\lambda}),$$
$$s(\lambda 1^{p-\lambda})\,s(1^{n-p})$$

are in terms of tabular separations of $(\lambda 1^{n-\lambda})$; we have in fact to take a linear function of these products. The investigation is facilitated by proceeding at once to the general case.

Let us then consider the problem of expressing the product

$$s_\lambda^l s_\mu^m \ldots.$$

by means of tabular separations of the partition

$$(t_1^{\tau_1} t_2^{\tau_2} t_3^{\tau_3} \ldots.).$$

Suppose $(\lambda_1^{l_1}\mu_1^{m_1}\ldots.)^{L_1}(\lambda_2^{l_2}\mu_2^{m_2}\ldots.)^{L_2}\ldots.$
to be any separation of $(t_1^{\tau_1} t_2^{\tau_2} t_3^{\tau_3} \ldots.)$,
having the species partition

$$(\lambda^l \mu^m \ldots.);$$

the general summation formula gives us

$$s(\lambda_1^{l_1}\mu_1^{m_1}\ldots.) = \ldots. + (-)^{l_1+m_1+\ldots+1} \frac{l_1!\,m_1!\ldots.}{(l_1+m_1+\ldots.-1)!}(\lambda_1^{l_1}\mu_1^{m_1}\ldots.);$$

hence, observing that

$$L_1(l_1+m_1+\ldots.)+L_2(l_2+m_2+\ldots.)+\ldots.=\tau_1+\tau_2+\tau_3+\ldots.$$

we find

$$\{s(\lambda_1^{l_1}\mu_1^{m_1}\ldots.)\}^{L_1}\{s(\lambda_2^{l_2}\mu_2^{m_2}\ldots.)\}^{L_2}\ldots.$$

$$=\ldots.+(-)^{\tau_1+\tau_2+\ldots.+L_1+L_2+\ldots.}\left\{\frac{l_1!\,m_1!\ldots.}{(l_1+m_1+\ldots.-1)!}\right\}^{L_1}\left\{\frac{l_2!\,m_2!\ldots.}{(l_2+m_2+\ldots.-1)!}\right\}^{L_2}\ldots.(\lambda_1^{l_1}\mu_1^{m_1}\ldots.)^{L_1}(\lambda_2^{l_2}\mu_2^{m_2}\ldots.)^{L_2}\ldots.$$

Taking the summation for every separation having the species partition

$$(\lambda^l\mu^m\ldots.),$$

we may write

$$\sum\left\{\frac{(l_1+m_1+\ldots.-1)!}{l_1!\,m_1!\ldots.}\right\}^{L_1}\left\{\frac{(l_2+m_2+\ldots.-1)!}{l_2!\,m_2!\ldots.}\right\}^{L_2}\ldots.\frac{(L_1+L_2+\ldots.-1)!}{L_1!\,L_2!\ldots.}\{s(\lambda_1^{l_1}\mu_1^{m_1}\ldots.)\}^{L_1}\{s(\lambda_2^{l_2}\mu_2^{m_2}\ldots.)\}^{L_2}\ldots.$$

$$=\ldots.+\sum(-)^{\tau_1+\tau_2+\ldots.+L_1+L_2+\ldots.}\frac{(L_1+L_2+\ldots.-1)!}{L_1!\,L_2!\ldots.}(\lambda_1^{l_1}\mu_1^{m_1}\ldots.)^{L_1}(\lambda_2^{l_2}\mu_2^{m_2}\ldots.)^{L_2}\ldots.$$

Observe that the dexter of this identity agrees with the corresponding portion of the expression of $s(t_1^{\tau_1}t_2^{\tau_2}\ldots.)$, for

$$s(t_1^{\tau_1}t_2^{\tau_2}\ldots.)=\ldots.$$

$$+\sum(-)^{\tau_1+\tau_2+\ldots+L_1+L_2+\ldots}\frac{(L_1+L_2+\ldots-1)!}{L_1!\,L_2!\ldots}(\lambda_1^{l_1}\mu_1^{m_1}\ldots)^{L_1}(\lambda_2^{l_2}\mu_2^{m_2}\ldots)^{L_2}\ldots+\ldots,$$

and that we cannot obtain this agreement unless we take the linear function of the s products which appears on the sinister side; but we know that the s product

$$s_\lambda^l s_\mu^m\ldots.$$

is necessarily expressible by means of tabular separations, because each of the monomial forms, which arise on performing the multiplication, is so; hence the above linear function of the s products must be a function of tabular separations of the partition $(t_1^{\tau_1}t_2^{\tau_2}\ldots.)$.

We thus arrive at the result:

$$s_\lambda^l s_\mu^m\ldots.=\frac{\displaystyle\sum\left\{\frac{(l_1+m_1+\ldots.-1)!}{l_1!\,m_1!\ldots.}\right\}^{L_1}\left\{\frac{(l_2+m_2+\ldots.-1)!}{l_2!\,m_2!\ldots.}\right\}^{L_2}\ldots.\frac{(L_1+L_2+\ldots.-1)!}{L_1!\,L_2!\ldots.}\{s(\lambda_1^{l_1}\mu_1^{m_1}\ldots.)\}^{L_1}\{s(\lambda_2^{l_2}\mu_2^{m_2}\ldots.)\}^{L_2}\ldots.}{\displaystyle\sum\left\{\frac{(l_1+m_1+\ldots.-1)!}{l_1!\,m_1!\ldots.}\right\}^{L_1}\left\{\frac{(l_2+m_2+\ldots.-1)!}{l_2!\,m_2!\ldots.}\right\}^{L_2}\ldots.\frac{(L_1+L_2+\ldots.-1)!}{L_1!\,L_2!\ldots.}}$$

where

(i) the separable partition is $(t_1^{\tau_1}t_2^{\tau_2})$,

(ii) $(\lambda_1^{l_1}\mu_1^{m_1}\ldots)^{L_1}(\lambda_2^{l_2}\mu_2^{m_2}\ldots)^{L_2}\ldots$ is any separation having the species partition $(\lambda^l\mu^m\ldots)$,

(iii) the summations are in regard to all such separations; the result being that the dexter side is a function of tabular separations of $(t_1^{\tau_1}t_2^{\tau_2}\ldots)$.

I illustrate the process by the calculation of (42) in terms of separations of (2^21^2). We have $(42) = s_4s_2 - s_6$,

and

$$s_4s_2 = \frac{\frac{1}{2}\cdot\frac{1}{2}s(2^2)s(1^2) + s(21^2)s(2)}{\frac{1}{2}\cdot\frac{1}{2}+1} = \frac{1}{5}s(2^2)s(1^2) + \frac{4}{5}s(21^2)s(2)$$

$$= \frac{1}{5}\{-2(2^2)+(2)^2\}\{-2(1^2)+(1)^2\} + \frac{4}{5}\{(21^2)-(21)(1)-(2)(1^2)+(2)(1)^2\}(2),$$

$$= \frac{4}{5}(21^2)(2) + \frac{4}{5}(2^2)(1^2) - \frac{2}{5}(2^2)(1)^2 - \frac{6}{5}(2)^2(1^2) + (2)^2(1)^2,$$

which, since the separations $(21^2)(2)$, $(2^2)(1^2)$ occur with the same coefficient $\frac{4}{5}$, is, as it should be, a function of tabular separations.

Hence, extracting the value of $s_6 = s(2^21^2)$ from the tables, we find

$$15(42) = 10(2^21^2) - 10(2^21)(1) + 2\{(21^2)(2) + (2^2)(1^2)\} - 5(21)^2$$
$$+ 4(2^2)(1)^2 + 8(21)(2)(1) - 8(2)^2(1^2).$$

In this general way we are enabled, as in the ordinary theory, to combine the expressions for the sums of the powers so as to obtain the expressions of other symmetric functions by means of separations of any selected partition.

It will be gathered that this process would be very laborious in the case of symmetric functions whose partitions contained many parts. This is also the case in the ordinary or unitary theory (where by "unitary" it is meant that separable partitions composed wholly of parts, unity, are alone considered); it will be seen hereafter that this process will, for such forms, be naturally rejected in favor of easier methods of calculation.

Properties of Coefficients and Groups of Separations.

The general expression for the sum of the powers in terms of separations of any partition has, in regard to the numerical coefficients, a certain interesting and important property.

If the separable partition be (1^n) or (λ^n), it is a known theorem that the sum of the coefficients is

$$(-)^{n+1}.$$

Now, whenever the separable partition contains *dissimilar* parts, the sum of all the coefficients is *invariably* zero. This will be established by proving another theorem of a much more refined character from which it is immediately deducible. The separations of a given partition may be grouped in a manner which is independent of their species partitions. For clearness I take as a particular case the separable partition

$$(\lambda^3\mu^2)$$

and write down any one of its separations, say

$$(\lambda^2)(\lambda\mu)(\mu).$$

This separation may be regarded as compounded of the two separations:

$$(\lambda^2)(\lambda) \text{ of } (\lambda^3)$$

and $$(\mu)^2 \text{ of } (\mu^2);$$

and moreover we will find that three other separations possess the same property, viz:

$$(\lambda^2)(\lambda)(\mu)^2,$$
$$(\lambda^2\mu)(\lambda)(\mu),$$
$$(\lambda^2\mu)(\lambda\mu);$$

for on suppressing the μ's in each we are left with

$$(\lambda^2)(\lambda),$$

and on suppressing the λ's there remains

$$(\mu)^2.$$

I say that these four separations

$$(\lambda^2)(\lambda\mu)(\mu),$$
$$(\lambda^2)(\lambda)(\mu)^2,$$
$$(\lambda^2\mu)(\lambda)(\mu),$$
$$(\lambda^2\mu)(\lambda\mu),$$

form a group, each member of which is compounded of the two separations

$$(\lambda^2)(\lambda),\ (\mu)^2.$$

I will call it the group $\{(\lambda^2)(\lambda),\ (\mu)^2\}$.

Consider now the separations of

$$(\lambda^3\mu^2)$$

as divided into six groups, viz:

The group $\{(\lambda)^3 \quad , (\mu)^2\}$,
" $\{(\lambda)^3 \quad , (\mu^2)\}$,
" $\{(\lambda^2)(\lambda), (\mu)^2\}$,
" $\{(\lambda^2)(\lambda), (\mu^2)\}$,
" $\{(\lambda^3) \quad , (\mu)^2\}$,
" $\{(\lambda^3) \quad , (\mu^2)\}$,

where, be it observed, there is a group corresponding to every compound of a separation of (λ^3) with a separation of (μ^2).

So in general, if the separable partition be

$$(\lambda^l\mu^m\nu^n \ldots .),$$

we may take any separations of (λ^l), (μ^m), $(\nu^n) \ldots .$ and form the group which is compounded of these separations. If (λ^l), (μ^m), $(\nu^n) \ldots .$ possess $L, M, N \ldots .$ separations* respectively, there will thus be

$$LMN \ldots . \text{ groups.}$$

It will be observed that the grouping depends merely upon the multiplicities $l, m, n, \ldots .$ and not at all upon the parts $\lambda, \mu, \nu, \ldots .$

I now enunciate the theorem:

Theorem: "In the expression of s_n in terms of separations of any partition of n, which contains dissimilar parts, the sum of the coefficients of the separations in each group is zero." I subjoin an example:

$$\begin{aligned} \tfrac{3}{2} s(\lambda^2\mu^2) = {} & \tfrac{3}{2}(\lambda)^2(\mu)^2 - (\lambda)^2(\mu^2) + (\lambda^2)(\mu)^2 + (\lambda^2)(\mu^2) \\ & - 2(\lambda\mu)(\lambda)(\mu) + (\lambda\mu^2)(\lambda) + (\lambda^2\mu)(\mu) - (\lambda^2\mu^2) \\ & + \tfrac{1}{2}(\lambda\mu)^2 \\ \hline & \pm 2 \qquad\qquad \pm 1 \qquad\qquad \pm 1 \qquad\qquad \pm 1 \end{aligned}$$

wherein the first, second, third, fourth columns of separations constitute the groups

$$\{(\lambda)^2, (\mu)^2\}, \ \{(\lambda)^2, (\mu^2)\}, \ \{(\lambda^2), (\mu)^2\}, \ \{(\lambda^2), (\mu^2)\}$$

respectively, and the sum of the coefficients of the separations in each group is seen to be zero.

* Or, what is the same thing, if the numbers $l, m, n \ldots .$ possess $L, M, N \ldots .$ partitions respectively.

To establish this theorem, consider in the first place the identity

$$\frac{(\lambda)x^\lambda - (\lambda 1)x^{\lambda+1} + (\lambda 1^2)x^{\lambda+2} - \ldots .}{1 - (1)x + (1^2)x^2 - (1^3)x^3 + \ldots .} = s_\lambda x^\lambda + s_{\lambda+1}x^{\lambda+1} + s_{\lambda+2}x^{\lambda+2} + \ldots . ;$$

write the left-hand side in the form

$$\frac{(\lambda)y_\lambda - (\lambda 1)y_\lambda x_1 + (\lambda 1^2)y_\lambda x_2 - (\lambda 1^3)y_\lambda x_3 + \ldots .}{1 - (1)x_1 + (1^2)x_2 - (1^3)x_3 + \ldots .},$$

wherein y_λ, x_t are symbolical representations of

x^λ and x^t respectively.

If this expression be developed by the multinomial theorem, we obtain a product

$$y_\lambda x_1^{l_1} x_2^{l_2} x_3^{l_3} \ldots .$$

multiplied by a certain symmetric function. This function consists of a number of symmetric functions, each of which is a separation of

$$(\lambda 1^{l_1+2l_2+3l_3+\ldots .})$$

and each of which is a member of the group of separations

$$\{(\lambda), (1)^{l_1}(1^2)^{l_2}(1^3)^{l_3} \ldots . \}.$$

The complete symmetric function which multiplies

$$y_\lambda x_1^{l_1} x_2^{l_2} x_3^{l_3} \ldots .$$

constitutes those terms in the expression of

$$s_{\lambda+l_1+2l_2+3l_3+\ldots .}$$

which belong to the group of separations

$$\{(\lambda), (1)^{l_1}(1^2)^{l_2}(1^3)^{l_3} \ldots . \}.$$

Putting
$$(\lambda) = (\lambda 1) = (\lambda 1^2) = \ldots . = 1,$$
$$(1) = (1^2) = \ldots . = 1,$$

the left-hand side of the identity reduces to

$$y_\lambda ;$$

hence, in the development by the multinomial theorem, all the other terms must vanish; but under these circumstances the aggregate of those separations which occur in any group of separations of a partition containing dissimilar parts, becomes the algebraic sum of the numerical coefficients which occur in such group; it follows that if s_n be expressed by means of separations of

$$(\lambda 1^{n-\lambda}) \qquad n > \lambda$$

the algebraic sum of the coefficients in each group must be zero.

Next consider the two identities

$$\frac{(\lambda)x^{\lambda}-(\lambda 1)x^{\lambda+1}+\ldots\ldots}{1-(1)x+(1^2)x^2-\ldots\ldots}\cdot\frac{(\mu)x^{\mu}-(\mu 1)x^{\mu+1}+\ldots\ldots}{1-(1)x+(1^2)x^2-\ldots\ldots}-\frac{(\lambda\mu)x^{\lambda+\mu}-(\lambda\mu 1)x^{\lambda+\mu+1}+\ldots\ldots}{1-(1)x+(1^2)x^2-\ldots\ldots}$$
$$=s_{\lambda+\mu}x^{\lambda+\mu}+2s_{\lambda+\mu+1}x^{\lambda+\mu+1}+3s_{\lambda+\mu+2}x^{\lambda+\mu+2}+\ldots\ldots,$$

$$\left\{\frac{(\lambda)x^{\lambda}-(\lambda 1)x^{\lambda+1}+\ldots\ldots}{1-(1)x+(1^2)x^2-\ldots\ldots}\right\}^2-2\frac{(\lambda^2)x^{2\lambda}-(\lambda^2 1)x^{2\lambda+1}+\ldots\ldots}{1-(1)x+(1^2)x^2-\ldots\ldots}$$
$$=s_{2\lambda}x^{2\lambda}+2s_{2\lambda+1}x^{2\lambda+1}+3s_{2\lambda+2}x^{2\lambda+2}+\ldots\ldots$$

These may, as regards their left-hand sides, be written symbolically

$$\frac{(\lambda)y_{\lambda}-(\lambda 1)y_{\lambda}x_1+(\lambda 1^2)y_{\lambda}x_2-\ldots\ldots}{1-(1)x_1+(1^2)x_2-(1^3)x_3+\ldots\ldots}\cdot\frac{(\mu)z_{\mu}-(\mu 1)z_{\mu}x_1+(\mu 1^2)z_{\mu}x_2-\ldots\ldots}{1-(1)x+(1^2)x_2-(1^3)x_3+\ldots\ldots}$$
$$-\frac{(\lambda\mu)y_{\lambda}z_{\mu}-(\lambda\mu 1)y_{\lambda}z_{\mu}x_1+\ldots\ldots}{1-(1)x+(1^2)x_2-\ldots\ldots}\left\{\frac{(\lambda)y_{\lambda}-(\lambda 1)y_{\lambda}x_1+(\lambda 1^2)y_{\lambda}x_2+\ldots\ldots}{1-(1)x_1+(1^2)x_2-(1^3)x_3+\ldots\ldots}\right\}^2$$
$$-2\frac{(\lambda^2)y_{2\lambda}-(\lambda^2 1)y_{2\lambda}x_1+(\lambda^2 1^2)y_{2\lambda}x_2-\ldots\ldots}{1-(1)x_1+(1^2)x_2-\ldots\ldots},$$

and herein putting all the symmetric functions equal to unity, we obtain respectively zero,

and $y_{\lambda}^2-2y_{2\lambda}$;

hence, when s_n is expressed by means of separations of

$$(\lambda\mu 1^{n-\lambda-\mu}),$$

the algebraic sum of the coefficients in each group of separations is zero, and also when s_n is expressed by means of separations of

$$(\lambda^2 1^{n-2\lambda}) \qquad n>2\lambda$$

the algebraic sum of the coefficients in each group is zero also.

This method of proof is perfectly general and leads to the important conclusion that whenever s_n is expressed by means of separations of a partition, which does not consist merely of repetitions of a single part, the algebraic sum of the coefficients in each group is zero.

In the proof it has been assumed that the algebraic sum of the coefficients on the dexter of such relation as

$$s(\lambda\mu\nu)=(\lambda)(\mu)(\nu)-\tfrac{1}{2}(\lambda\mu)(\nu)-\tfrac{1}{2}(\lambda\nu)(\mu)-\tfrac{1}{2}(\mu\nu)(\lambda)+\tfrac{1}{2}(\lambda\mu\nu),$$

is zero. It is easily seen that this is so, for writing down the relations

$$(\lambda) = s(\lambda),$$
$$(\lambda\mu) = s(\lambda)\,s(\mu) - s(\lambda\mu),$$
$$\text{etc.} = \text{etc.}$$

if we put $(\lambda) = (\lambda\mu) = \ldots\ldots = 1$, we have

$$s(\lambda) = s(\mu) = 1,$$

and hence from the second relation

$$s(\lambda\mu) = 0.$$

Also the third relation being

$$(\lambda\mu\nu) = s(\lambda)\,s(\mu)\,s(\nu) - s(\lambda\mu)\,s(\nu) - s(\lambda\nu)\,s(\mu) - s(\mu\nu)\,s(\lambda) + 2s(\lambda\mu\nu),$$

putting
$$(\lambda\mu\nu) = 1,\ s(\lambda) = s(\mu) = s(\nu) = 1,$$
$$s(\lambda\mu) = s(\lambda\nu) = s(\mu\nu) = 0,$$

we find $s(\lambda\mu\nu) = 0$, and so on in general, the expression for s_n becomes zero when, the separable partition containing dissimilar parts, the separates of the separations are each put equal to unity.

It is noteworthy that this property of the coefficients is not a generalization or extension of any known property, but is one which, by its nature, only comes into existence with the new theory; there is in fact no corresponding property in the ordinary theory, for in that theory the "group" does not exist *as a group*, but merely as an isolated term without interest. The theorem is of great importance as a verification of the coefficients of the separations, as well as for ascertaining that no separation has been accidentally omitted. It constitutes an example of departure from absolute generality.

This theory of the group may be easily extended to other symmetric functions.

Consider for a moment the expression of (3^2) by means of separations of (21^4). We have

$$2(3^2) = s_3^2 - s_6 = s(21)\,s(1^3) - s(21^4),$$

wherein we know that in both $s(21)$ and $s(21^4)$ the algebraic sum of the coefficients is zero; hence also in (3^2) the algebraic sum of all the coefficients must vanish; if, on the dexter side, there had been such a product as

$$s(1^4)\,s(2),$$

5

this could not have been the case, because the algebraic sum of the coefficients is not zero either in $s(1^4)$ or in $s(2)$. This s product does not occur simply because the partition (3^2) possesses no separation of species partition (42).

This reasoning is manifestly general, and we are thus led to the important theorem:

Theorem: "In the expression of symmetric function

$$(p_1^{\pi_1} p_2^{\pi_2} \ldots\ldots)$$

by means of separations of

$$(t_1^{\tau_1} t_2^{\tau_2} \ldots\ldots),$$

the algebraic sum of all the coefficients will be zero if the partition

$$(p_1^{\pi_1} p_2^{\pi_2} \ldots\ldots)$$

possesses no separations of species partition

$$(\tau_1 t_1, \tau_2 t_2)."$$

The theorem may be verified from the tables in the cases of

Separable partition.		Symmetric functions.
(21^4)	for	$(51), (3^2)$,
$(2^2 1^2)$	for	$(51), (3^2)$,
(31^3)	for	$(51), (42), (41^2)$,
(321)	for	$(51), (42), (3^2)$,
(41^2)	for	(51).

The group coefficient theory obtains also under the same circumstances; to establish this it will suffice to show that if we take any two sums of powers, say

$$s(\lambda^5 \mu^4) \text{ and } s(\lambda^3 \mu^2),$$

multiply out their product and arrange the resulting separations in groups, the algebraic sum of the coefficients in each group is zero. In $s(\lambda^5\mu^4)$ there is a group $\{(\lambda)^5, (\mu^2)(\mu)^2\}$, and in $s(\lambda^3\mu^2)$ a group $\{(\lambda^2)(\lambda), (\mu^2)\}$, and if we multiply the separations in these two groups together, we will, quâ the separable partitions $(\lambda^8\mu^6)$, obtain separations belonging to the group

$$\{(\lambda)^6(\lambda^2),\ (\mu^2)^2(\mu)^2\}.$$

In this batch of separations the algebraic sum of the coefficients is necessarily zero; we obtain another batch of separations comprised in the *same* group by multiplication of the groups

$$\{(\lambda)^3(\lambda^2),\ (\mu^2)^2\} \text{ and } \{(\lambda)^3,\ (\mu)^2\},$$

and again for this batch of separations the algebraic sum of the coefficients is zero. Observe that by thus multiplying certain *complete* groups of the one expression by certain *complete* groups of the other, we must obtain all the separations of $(\lambda^8\mu^6)$ belonging to the group

$$\{(\lambda)^6(\lambda^2),\ (\mu^2)^2(\mu)^2\}$$

which occur in the product $s(\lambda^5\mu^4)\,s(\lambda^3\mu^2)$. Hence, if in such product there occurs one factor at least of which the separable partition does not consist merely of repetitions of a single part, the algebraic sum of the coefficients in each group of the product will be zero. We are thus enabled to enunciate as follows:

Theorem: "In the expression of symmetric function

$$(p_1^{\pi_1}p_2^{\pi_2}\ldots.),$$

by means of separations of

$$(t_1^{\tau_1}t_2^{\tau_2}\ldots.),$$

the algebraic sum of the coefficients *in each group* will be zero, if the partition

$$(p_1^{\pi_1}p_2^{\pi_2}\ldots.)$$

possesses no separations of species partition

$$(\tau_1 t_1,\ \tau_2 t_2\ldots.)."$$

Concluding Remarks.

So far I have merely stood upon the threshold of the theory; the further development must be reserved for some future occasion; systems of partial differential equations exist analogous to those employed with such marked success by Brioschi, Hammond and others; and until this theory has been properly worked out, it does not seem advisable to enter upon the discussion of the calculation of the various tables.

The theory of those separable partitions which contain no part equal to unity is, from another point of view, a theory of the covariants of binary forms; in particular, in such cases the general theorem of algebraic reciprocity becomes a remarkable theorem of binary forms which promises to be chiefly of use in the discussion of perpetuants.

The "theorem of symmetric function expressibility" becomes a theorem, of a fundamental character, of covariant expressibility.

It is to be hoped that some of these facts may help to forward the algebraical (as distinct from the symbolical) treatment of the theory of invariants; as yet, however, a purely algebraical demonstration of Gordan's great theorem concerning the finality of the ground covariants seems as far distant as ever. This theorem has been given by Gordan, in a much simplified form, in a tract which appears to be little known; the reference was given to me by Mr. A. R. Forsyth of Trinity College, Cambridge, and it may be useful to give it here: "Ueber das Formensystem binärer Formen." Teubner (Leipzig, 1875).

Royal Military Academy, Woolwich, May 9th, 1888.

Symmetric Functions and the Theory of Distributions.

By Captain P. A. MacMahon, R.A.

[*Read March 8th,* 1888.]

The theory of distributions is discussed in an elementary manner in Whitworth's *Choice and Chance*, Third Edition, Ch. III. The subject is studied in France under the title *L'Analyse Combinatoire.* There have been very few researches during recent years, and none, so far as my knowledge extends, which proceed by the method employed in this paper. This method is essentially constructive in its nature.

The investigation has for its object the bringing forward of the theory as an analytical weapon of considerable power in algebraical research. The notation employed is new, and possesses the advantage of being the simplest that it is possible to use.

Among results of minor importance and interest, four important and very general purely algebraical theorems are established. These are—

(1) A comprehensive law of algebraic reciprocity.

(2) A cardinal theorem of symmetric function expressibility.

(3) A generalisation of Vandermonde's (or Waring's) formula in symmetric functions.

(4) The formation of symmetrical symmetric-function tables corresponding to every partition of every number.

The research is continued, from the point of view of the Algebra of Symmetric Functions, in a paper by the author ("Memoir on a New Theory of Symmetric Functions"), which will shortly appear in No. 4, Vol. x. of the *American Journal of Mathematics.*

The notation employed throughout is that of partitions.

The Theory of Partitions, from the point of view of the Theory of Numbers, has been studied chiefly by Cayley, Sylvester, Glaisher, Franklin, and Hammond. These researches have appeared principally in the *Philosophical Transactions of the Royal Society*, the *American Journal of Mathematics*, the *Quarterly Journal of Mathematics*, and the *Messenger of Mathematics.*

An important reference is "A constructive Theory of Partitions," by J. J. Sylvester, *American Journal of Mathematics*, Vol. v., p. 251.

The first mathematician who employed the notation of a partition in ordinary algebra was Meyer Hirsch in his Algebra published in 1812; since then the idea has been further developed by Cayley, Hammond, the author of this paper, and probably a few others. The following memoirs may be consulted :—

Cayley: "A Memoir on the Symmetric Functions," *Phil. Trans. R. S.*, 1857.

The Author: "Seminvariants and Symmetric Functions," *Amer. Jour. of Math.*, Vol. vi.; the Author: "On Perpetuants," *Amer. Jour. of Math.*, Vol. vii.; the Author: "Memoir on Seminvariants," *Amer. Jour. of Math.*, Vol. viii.; Hammond: "On Perpetuants," *Amer. Jour. of Math.*, Vol. viii.; the Author: "The Expression of Syzygies," &c., *Amer. Jour. of Math.*, Vol. x.

Preliminary.

As defined by Whitworth (*loc. cit.*), "Distribution" is the separation of a series of elements into a series of classes; in the general problem, the things to be distributed may be of any species, viz., there may be n things, of which p are of one kind, q of a second kind, r of a third, &c. ..., where $p+q+r+ \ldots = n$; it is then convenient to speak of things or objects $(pqr \ldots)$ where, in this particular connection, the partition $(pqr \ldots)$ is to be regarded as defining the objects in regard to species; again, the classes into which the objects are to be distributed may be of any species, and this leads us to speak of classes $(p_1 q_1 r_1 \ldots)$, where $p_1+q_1+r_1+ \ldots = n_1 =$ the number of classes; the partition $(p_1 q_1 r_1 \ldots)$ here defines the classes in regard to species, indicating p_1 classes of one description, q_1 of a second, r_1 of a third, and so forth.

It should be observed that, in the use of partitions, repetitions of the same part are indicated by an index; for instance

$$(pppqqr \ldots) \text{ is written } (p^3 q^2 r \ldots).$$

If no attention is paid to the order of the objects (whatever be their species) in a class, the distribution may be described as one into "parcels"; each parcel is a class of unarranged objects.

If, however, permutations are permissible amongst objects in the same class, the distribution is said to be one into "groups"; each group is a class of arranged objects.

Two chief problems may be enunciated as follows:—

"To determine the number of distributions of objects $(pqr \ldots)$ into parcels $(p_1 q_1 r_1 \ldots)$."

"To determine the number of distributions of objects $(pqr \ldots)$ into groups $(p_1 q_1 r_1 \ldots)$."

Further, we may discuss each of these problems when the distributions are subject to certain restrictions; it is from the consideration of restricted distributions that most of the analytical results of this paper are evolved.

Section 1.

The Distribution Function.

Let $\alpha, \beta, \gamma, \ldots$ be the roots of the equation

$$x^n - a_1 x^{n-1} + a_2 x^{n-2} - \ldots = 0.$$

The symmetric function $\Sigma \alpha^p \beta^q \gamma^r \ldots$, where $p+q+r+\ldots = n$, is, in the partition notation, written

$$(pqr \ldots).$$

Let
$$A_{(pqr \ldots),(p_1 q_1 r_1 \ldots)}$$

denote the number of ways of distributing objects, defined by the partition $(pqr \ldots)$, into parcels, defined by the partition $(p_1 q_1 r_1 \ldots)$. I suppose there to be m parcels, so that $p_1+q_1+r_1+\ldots = m$.

It will be convenient henceforward to speak simply of the distribution of objects $(pqr \ldots)$ into parcels $(p_1 q_1 r_1 \ldots)$.

I attach the number

$$A_{(pqr \ldots),(p_1 q_1 r_1 \ldots)}$$

to the symmetric function

$$(pqr \ldots),$$

and construct the expression

$$\Sigma A_{(pqr\ldots),(p_1q_1r_1\ldots)}(pqr\ldots),$$

by taking the summation over every partition $(pqr\ldots)$ of the number n.

Definition. The Distribution Function of n objects into parcels $(p_1q_1r_1\ldots)$ is the expression

$$\Sigma A_{(pqr\ldots),(p_1q_1r_1\ldots)}(pqr\ldots),$$

where $$p+q+r+\ldots = n.$$

I write also

$$\Sigma A_{(pqr\ldots),(p_1q_1r_1\ldots)}(pqr\ldots) = Dp\,(n),\ (p_1q_1r_1\ldots).$$

Let, also,

$$B_{(pqr\ldots),(p_1q_1r_1\ldots)}$$

denote the number of ways of distributing objects $(pqr\ldots)$ into groups $(p_1q_1r_1\ldots)$.

Definition. The Distribution Function of n objects into groups $(p_1q_1r_1\ldots)$ is the expression

$$\Sigma B_{(pqr\ldots),(p_1q_1r_1\ldots)}(pqr\ldots),$$

where $$p+q+r+\ldots = n.$$

In this case I write

$$\Sigma B_{(pqr\ldots),(p_1q_1r_1\ldots)}(pqr\ldots) = Dg\,(n),\ (p_1q_1r_1\ldots).$$

My present purpose is the study of these two Distribution Functions.

SECTION 2.

Parcels, m in number (i.e., m = n).

Let h_s be the homogeneous product-sum, of degree s, of the n quantities $\alpha, \beta, \gamma, \ldots$; so that

$$\begin{aligned} h_1 &= \Sigma\alpha = (1), \\ h_2 &= \Sigma\alpha^2+\Sigma\alpha\beta = (2)+(1^2), \\ h_3 &= \Sigma\alpha^3+\Sigma\alpha^2\beta+\Sigma\alpha\beta\gamma = (3)+(21)+(1^3), \\ \&c. &= \&c. \end{aligned}$$

Consider the product $\qquad h_{p_1}h_{q_1}h_{r_1}\ldots.$

The symmetric function $(pqr\ldots)$ will, on performing the multiplication, be produced a certain number of times. In the factor h_{p_1} every term is of degree p_1 in the quantities. Taking any particular term, write down the p_1 quantities occurring therein in any order with a dot between each pair of consecutive quantities. We may consider these p_1 quantities as distributed into p_1 similar parcels, one quantity into each parcel. In the same way, any q_1 quantities which occur in any term of h_{q_1} may be considered to be q_1 quantities distributed into q_1 parcels, similar to one another, but different from the former. Hence it is clear that the number of times that the symmetric function $(pqr\ldots)$ occurs in the development of the product $h_{p_1}h_{q_1}h_{r_1}\ldots$ is precisely the number of ways that it is possible to distribute objects $(pqr\ldots)$ into parcels $(p_1q_1r_1\ldots)$, one object in each parcel. Hence, when $m = n$, and no parcel is empty,

$$Df(n), (p_1q_1r_1\ldots) = \Sigma A_{(pqr\ldots), (p_1q_1r_1\ldots)}(pqr\ldots) = h_{p_1}h_{q_1}h_{r_1}\ldots .$$

Consider, for a moment, the distribution of objects (43) into parcels (52), and represent objects and parcels by small and capital letters respectively. One distribution is represented by the scheme

$$\begin{matrix} A & A & A & A & A & B & B \\ a & a & a & a & b & b & b \end{matrix}$$

wherein an object denoted by a small letter is placed in a parcel denoted by the capital letter directly above it. Corresponding to this distribution of objects (43) into parcels (52), we have a distribution of objects (52) into parcels (43), given by the scheme

$$\begin{matrix} A & A & A & A & B & B & B \\ a & a & a & a & a & b & b \end{matrix}$$

derived from the former by interchanging rows as well as small and capital letters. The process is clearly general and exhibits a one-to-one correspondence between the distributions of objects $(pqr\ldots)$ into parcels $(p_1q_1r_1\ldots)$, and the distributions of objects $(p_1q_1r_1\ldots)$ into parcels $(pqr\ldots)$. It is, in fact, an intuitive observation, that we may either consider an object placed in or attached to a parcel, *or* a parcel placed in or attached to an object.

Hence the very important theorem

$$A_{(pqr\ldots), (p_1q_1r_1\ldots)} = A_{(p_1q_1r_1\ldots), (pqr\ldots)}.$$

Analytically this result leads to a law of algebraic symmetry which I now enunciate.

Theorem.—" The coefficient of symmetric function $(pqr\ldots)$ in the development of the product $h_{p_1} h_{q_1} h_{r_1} \ldots$ is equal to the coefficient of symmetric function $(p_1 q_1 r_1 \ldots)$ in the development of the product $h_p h_q h_r \ldots$."

This law of symmetry I established in the *Quarterly Journal of Mathematics*.

The problem of the distribution of n objects into n parcels, one object into each parcel, is thus completely solved by means of a table of symmetric functions which expresses the h-products as linear functions of the single partition forms. (*Vide* the Tables at the end of the paper.)

Section 3.

Parcels of species (1^m), *where* $m < n$.

I now discuss the distributions of n objects into m parcels, no two of which are similar. Whitworth would describe the problem as a distribution into m DIFFERENT parcels.

Let $$(p_1^{\pi_1} p_2^{\pi_2} p_3^{\pi_3} \ldots), \qquad [\Sigma\pi = m, \ \Sigma\pi p = n],$$

be any partition of n into m parts.

Of the whole number of distributions, there will be a certain number such that π_s parcels each contain p_s objects,

$$(s = 1, 2, 3, \ldots).$$

The distribution function of this particular case of the distribution is

$$\frac{m!}{\pi_1!\, \pi_2!\, \pi_3! \ldots} h_{p_1}^{\pi_1} h_{p_2}^{\pi_2} h_{p_3}^{\pi_3} \ldots .$$

To see how this is, observe that the product $h_{p_1}^{\pi_1} h_{p_2}^{\pi_2} h_{p_3}^{\pi_3} \ldots$ is susceptible of $\dfrac{m!}{\pi_1!\, \pi_2!\, \pi_3! \ldots}$ permutations. The parcels are all different, and hence there are distributions corresponding to each of these permutations. By the last section, for each of these permutations there will be a distribution function

$$h_{p_1}^{\pi_1} h_{p_2}^{\pi_2} h_{p_3}^{\pi_3} \ldots,$$

and for the aggregate of permutations a distribution function

$$\frac{m!}{\pi_1!\, \pi_2!\, \pi_3! \ldots} h_{p_1}^{\pi_1} h_{p_2}^{\pi_2} h_{p_3}^{\pi_3} \ldots .$$

Hence the distribution function of n objects into parcels (1^m) is

$$\Sigma \frac{m!}{\pi_1!\,\pi_2!\,\pi_3!\,\ldots}\, h_{p_1}^{\pi_1} h_{p_2}^{\pi_2} h_{p_3}^{\pi_3} \ldots, \qquad [\Sigma\pi = m,\ \Sigma\pi p = n],$$

that is, it is the coefficient of x^n in the expansion of

$$(h_1x+h_2x^2+h_3x^3+\ldots)^m.$$

We may write this result

$$Dp\,(n),\,(1^m) = \Sigma A_{(pqr\ldots),\,(1^m)}\,(pqr\ldots) = \Sigma \frac{m!}{\pi_1!\,\pi_2!\,\pi_3!\,\ldots}\, h_{p_1}^{\pi_1} h_{p_2}^{\pi_2} h_{p_3}^{\pi_3} \ldots,$$

where $\qquad \Sigma\pi = m, \quad \Sigma p\pi = n.$

Section 4.

General value of $A_{(p_1^{\pi_1}p_2^{\pi_2}p_3^{\pi_3}\ldots),\,(1^m)}$.

We require the coefficient of x^n in the expansion of

$$(h_1x+h_2x^2+h_3x^3+\ldots)^m = u^m, \text{ suppose.}$$

Put $\qquad f(x) = 1+h_1x+h_2x^2+h_3x^3+\ldots = 1+u,$

then $\qquad (1+u)^m = (1-\alpha x)^{-m}(1-\beta x)^{-m}(1-\gamma x)^{-m}\ldots.$

and $\qquad u^m = (1+u-1)^m$

$$= (1+u)^m - m\,(1+u)^{m-1} + \frac{m\,(m-1)}{2!}\,(1+u)^{m-2} - \ldots + (-)^m\,1.$$

Now, the coefficient of $(p_1^{\pi_1}p_2^{\pi_2}p_3^{\pi_3}\ldots)\,x^n$ in

$$(1+u)^s = (1-\alpha x)^{-s}(1-\beta x)^{-s}(1-\gamma x)^{-s}\ldots$$

is

$$\left\{\frac{(s+p_1-1)!}{p_1!\,(s-1)!}\right\}^{\pi_1} \left\{\frac{(s+p_2-1)!}{p_2!\,(s-1)!}\right\}^{\pi_2} \left\{\frac{(s+p_3-1)!}{p_3!\,(s-1)!}\right\}^{\pi_3}\ldots.$$

Hence $\qquad A_{(p_1^{\pi_1}p_2^{\pi_2}p_3^{\pi_3}\ldots),\,(1^m)}$

$$= \left\{\frac{(m+p_1-1)!}{p_1!\,(m-1)!}\right\}^{\pi_1} \left\{\frac{(m+p_2-1)!}{p_2!\,(m-1)!}\right\}^{\pi_2} \left\{\frac{(m+p_3-1)!}{p_3!\,(m-1)!}\right\}^{\pi_3}\ldots$$

$$-m\left\{\frac{(m+p_1-2)!}{p_1!\,(m-2)!}\right\}^{\pi_1} \left\{\frac{(m+p_2-2)!}{p_2!\,(m-2)!}\right\}^{\pi_2} \left\{\frac{(m+p_3-2)!}{p_3!\,(m-2)!}\right\}^{\pi_3}\ldots$$

$$+\frac{m\,(m-1)}{2!}\left\{\frac{(m+p_1-3)!}{p_1!\,(m-3)!}\right\}^{\pi_1} \left\{\frac{(m+p_2-3)!}{p_2!\,(m-3)!}\right\}^{\pi_2} \left\{\frac{(m+p_3-3)!}{p_3!\,(m-3)!}\right\}^{\pi_3}\ldots$$

$-\ldots$ to $m+1$ terms.

Observe that, when

$$p_1 = p_2 = \ldots = \pi_1 = \pi_2 = \ldots = 1,$$

this expression reduces to the mth divided difference of 0^n.

Section 5.

Parcels of species (m).

We now discuss what is commonly known as the distribution of n objects into m indifferent parcels, but here the objects are of type

$$(p_1^{\pi_1} p_2^{\pi_2} p_3^{\pi_3} \ldots).$$

We may separate the distribution function into portions corresponding to every partition of the number n into exactly m parts. First, consider such a partition which consists wholly of *unrepeated* parts, say

$$(r_1 r_2 r_3 \ldots r_m), \qquad [\Sigma r = n],$$

the corresponding distribution function is necessarily

$$h_{r_1} h_{r_2} h_{r_3} \ldots h_{r_m},$$

but in any other case of distribution the function is much less simple.

For clearness first take $n = 4$, $m = 2$, and let us examine the distribution function corresponding to two objects in each parcel.

We have $h_2^2 = (\alpha^2 + \beta^2 + \gamma^2 + \ldots + \alpha\beta + \beta\gamma + \gamma\alpha + \ldots)^2$,

and here the distribution, $\alpha\alpha$ in one parcel, $\beta\beta$ in the other, occurs twice instead of once, as would have to be the case if this were really the distribution function.

Take the expression

$$\frac{1}{(1-\alpha^2x^2)(1-\beta^2x^2)(1-\gamma^2x^2)\ldots(1-\alpha\beta x^2)(1-\beta\gamma x^2)(1-\gamma\alpha x^2)\ldots},$$

and expand it in ascending powers of x; herein the coefficient of x^{2s} will be the sum of order s of the homogeneous products of the quantities

$$\alpha^2, \beta^2, \gamma^2, \ldots \alpha\beta, \beta\gamma, \gamma\alpha, \ldots,$$

which compose the function h_2. This homogeneous product sum consists of a number of terms each of which is obtained by multiplying together s of the quantities

$$\alpha^2, \beta^2, \gamma^2, \ldots \alpha\beta, \beta\gamma, \gamma\alpha, \ldots,$$

repeated or unrepeated; clearly then, in this homogeneous product sum, the symmetric function

$$(p_1 p_2 p_3 \ldots), \qquad [\Sigma p = 2s],$$

will occur just as many times as it is possible to distribute $2s$ objects $(p_1 p_2 p_3 \ldots)$ into parcels (s), two objects being in each parcel.

If then we write

$$\frac{1}{(1-\alpha^2x^2)(1-\beta^2x^2)(1-\gamma^2x^2)\ldots(1-\alpha\beta x^2)(1-\beta\gamma x^2)(1-\gamma\alpha x^2)\ldots}$$
$$= 1 + h_{2^1}x^2 + h_{2^2}x^4 + h_{2^3}x^6 + \ldots,$$

h_{2^s} will be the distribution function corresponding to the particular case of $2s$ objects in parcels (s), each parcel containing 2 objects.

Now take rs objects in parcels (s), each parcel containing r objects.

Form a fraction whose denominator contains a factor corresponding to each component member of h_r, and then suppose

$$\frac{1}{\left[\begin{array}{l}(1-\alpha^rx^r)(1-\beta^rx^r)\ldots(1-\alpha^{r-1}\beta x^r)(1-\alpha\beta^{r-1}x)\\ \qquad\ldots(1-\alpha^{r-2}\beta^2x^r)\ldots(1-\alpha\beta\gamma\ldots x^r)\end{array}\right]}$$
$$= 1 + h_{r^1}x^r + h_{r^2}x^{2r} + h_{r^3}x^{3r} + \ldots.$$

Previous reasoning shows that the distribution function is

$$h_{r^s}.$$

Reserving for the present the particular examination of this important symmetric function, I continue the general discussion.

We have already considered the distribution function corresponding to the particular case of the partition of n into unrepeated parts; we are now in a position to determine the function corresponding to the case of τ_1 parcels each containing t_1 objects, τ_2 parcels each containing t_2 objects, &c., or say, corresponding to the partition of n,

$$(t_1^{\tau_1} t_2^{\tau_2} t_3^{\tau_3} \ldots t_\nu^{\tau_\nu}) \qquad [\text{where } \Sigma\tau = m].$$

For, form the symmetric function

$$h_{t_1^{\tau_1}} h_{t_2^{\tau_2}} h_{t_3^{\tau_3}} \ldots h_{t_\nu^{\tau_\nu}},$$

and observe the meaning of the coefficient of the symmetric function

$$(p_1^{\pi_1} p_2^{\pi_2} p_3^{\pi_3} \ldots),$$

which will appear when the symmetric function product is developed.

The function $h_{\tau_1}^{t_1}$ contains terms corresponding to every selection of $t_1\tau_1$ objects of the total number n, and corresponding to every distribution of each of these selections into parcels (τ_1), each parcel containing exactly t_1 objects. Hence in the product

$$h_{t_1^{\tau_1}} h_{t_2^{\tau_2}} h_{t_3^{\tau_3}} \dots h_{t_\nu^{\tau_\nu}};$$

the symmetric function $\qquad (p_1^{\pi_1} p_2^{\pi_2} p_3^{\pi_3} \dots)$

will occur just so many times as it is possible to distribute objects $(p_1^{\pi_1} p_2^{\pi_2} p_3^{\pi_3} \dots)$ into m parcels, of which τ_1 contain exactly t_1 objects, τ_2 exactly t_2, τ_3 exactly t_3, &c., and τ_ν parcels exactly t_ν objects, &c.

Hence the particular distribution function sought for is

$$h_{t_1^{\tau_1}} \dots h_{t_\nu^{\tau_\nu}} \dots .$$

Finally, noticing that h_{r_1} and $h_{r'_1}$ are identical, we see that the distribution function of n objects into parcels (m) is

$$\Sigma\, h_{t_1^{\tau_1}} h_{t_2^{\tau_2}} h_{t_3^{\tau_3}} \dots,$$

the summation taking place over every partition

$$(t_1^{\tau_1} t_2^{\tau_2} t_3^{\tau_3} \dots)$$

of n which contains exactly $m\ [= \Sigma\tau]$ parts.

We may write this theorem in the form—

$$Dp\,(n), (m) = \Sigma A_{(p_1^{\pi_1} p_2^{\pi_2} p_3^{\pi_3} \dots),\ (m)} (p_1^{\pi_1} p_2^{\pi_2} p_3^{\pi_3} \dots)$$

$$= \Sigma h_{t_1^{\tau_1}} h_{t_2^{\tau_2}} h_{t_3^{\tau_3}} \dots,$$

where $\qquad \Sigma\tau = m, \quad \Sigma\tau t = n.$

It is now clear that $Df\,(n)\,(m)$ is the coefficient of $x^n a^m$ in the expression

$$(1 + h_{1^1} xa + h_{1^2} x^2 a^2 + \dots)(1 + h_{2^1} x^2 a + h_{2^2} x^4 a^2 + \dots)(1 + h_{3^1} x^3 a + h_{3^2} x^6 a^2 + \dots) \dots$$

$$\dots (1 + h_{s^1} x^s a + h_{s^2} x^{2s} a^2 + \dots) \dots$$

$$\equiv \prod_{s=1}^{s=\infty} (1 + h_{s^1} x^s a + h_{s^2} x^{2s} a^2 + \dots),$$

which is, therefore, its generating function.

Section 6.

Parcels of type $(m_1 m_2)$.

In this case, we are concerned with m_1 similar parcels of one kind and m_2 similar parcels of another kind. Of the n objects we may have m_1 objects only distributed amongst the m_1 similar parcels, and the remaining $n-m_1$ objects distributed amongst the m_2 similar parcels; or we may have m_1+s objects distributed amongst the m_1 similar parcels, and the remaining $n-m_1-s$ objects amongst the m_2 similar parcels, where $s \not> n-m_1-m_2$.

Hence

$$Dp\,(n), (m_1 m_2) = \sum_{s=0}^{n-m_1-m_2} Dp\,(m_1+s), (m_1) \,.\, Dp\,(n-m_1-s), (m_2),$$

and $Dp\,(n), (m_1 m_2)$ is the coefficient of $x^n a^{m_1} b^{m_2}$ in the product

$$\prod_{s=1}^{s=\infty} (1+h_{s^1} x^s a + h_{s^2} x^{2s} a^2 + \ldots)(1+h_{s^1} x^s b + h_{s^2} x^{2s} b^2 + \ldots),$$

which is its generating function.

Section 7.

Parcels of type $(m_1 m_2 m_3 \ldots)$.

By similar reasoning we find:—

$$Dp\,(n), (m_1 m_2 m_3) = \sum_{s=0}^{s=n-m_1-m_2-m_3} Dp(m_1+s), (m_1).Dp\,(n-m_1-s), (m_2 m_3)$$

$$= \sum_{s=0}^{s=n-m_1-m_2-m_3} \Big[Dp\,(m_1+s), (m_1) \sum_{t=0}^{t=n-m_1-m_2-m_3-s}$$

$$Dp\,(m_2+t), (m_2) \,.\, Dp\,(n-m_1-m_2-s-t), (m_3) \Big]$$

$$= \sum_{s=0}^{s=n-\Sigma m} \sum_{t=0}^{t=n-\Sigma m-s} Dp\,(m_1+s), (m_1) \,.\, Dp\,(m_2+t), (m_2)$$

$$\times Dp\,(n+m_3-\Sigma m-s-t), (m_3),$$

and generally,

$$Dp\,(n), (m_1 m_2 m_3 \ldots m_r) = \sum_{s_1=0}^{s_1=n-\Sigma m,} \sum_{s_2=0}^{s_2=n-\Sigma m-s_1} \sum_{s_3=0}^{s_3=n-\Sigma m-s_1-s_2} \ldots$$

$$Dp\,(m_1+s_1), (m_1) \,.\, Dp\,(m_2+s_2), (m_2) \ldots Dp\,(n+m_r-\Sigma m-\Sigma s), (m_r);$$

which is the coefficient of $x^n \mu_1^{m_1} \mu_2^{m_2} \ldots \mu_r^{m_r}$ in the product

$$\prod_{s=1}^{s=\infty} \prod_{t=1}^{t=\infty} (1+h_{s^1} x^s \mu_t + h_{s^2} x^{2s} \mu_t^2 + h_{s^3} x^{3s} \mu_t^3 + \ldots),$$

the generating function.

This determination completes analytically the solution of the problem of the distribution of objects $(p_1^{\pi_1} p_2^{\pi_2} \ldots)$ into parcels $(m_1 m_2 \ldots m_r)$.

Before proceeding to the subject of distributions, involving restrictions, I will draw up a list of some of the simpler results.

Section 8.

The simplest cases of Distribution into Parcels.

No. of Objects.	No. of Parcels.	Type of Parcels.	Distribution Function.
1	1	(1)	h_1,
2	1	(1)	h_2,
2	2	(2)	h_2,
2	2	(1^2)	h_1^2,
3	1	(1)	h_3,
3	2	(2)	$h_2 h_1$,
3	2	(1^2)	$2h_2 h_1$,
3	3	(3)	h_3,
3	3	(21)	$h_2 h_1$,
3	3	(1^3)	h_1^3,
4	1	(1)	h_4,
4	2	(2)	$h_4 + h_2^2$,
4	2	(1^2)	$2h_3 h_1 + h_2^2$,
4	3	(3)	h_2^2,
4	3	(21)	$h_2^2 + h_2 h_1^2$,
4	3	(1^3)	$3h_2 h_1^2$,
4	4	(4)	h_4,
4	4	(31)	$h_3 h_1$,
4	4	(2^2)	h_2^2,
4	4	(21^2)	$h_2 h_1^2$,
4	4	(1^4)	h_1^4,
5	1	(1)	h_5,
5	2	(2)	$h_4 h_1 + h_3 h_2$,
5	2	(1^2)	$2h_4 h_1 + 2h_3 h_2$,
5	3	(3)	$h_4 h_1 + h_3 h_2 - h_3 h_1^2 + h_2^2 h_1$,
5	3	(21)	$h_4 h_1 + h_3 h_2 + 2h_2^2 h_1$,
5	3	(1^3)	$3h_3 h_1^2 + 3h_2^2 h_1$,

No. of Objects.	No. of Parcels.	Type of Parcels.	Distribution Function.
5	4	(4)	h_3h_2,
5	4	(31)	$h_3h_2+h^2h_1$,
5	4	(2^2)	$2h_2^2h_1$,
5	4	(21^2)	$2h_2^2h_1+h_2h_1^3$,
5	4	(1^4)	$4h_2h_1^3$,
6	1	(1)	h_6,
6	2	(2)	$h_6+2h_4h_2$,
6	2	(1^2)	$2h_5h_1+2h_4h_2+h_3^2$,
6	3	(3)	$h_6-h_5h_1+h_4h_2+h_4h_1^2+h_3^2-h_3h_2h_1+h_2^3$,
6	3	(21)	$2h_4h_2+h_4h_1^2+2h_3h_2h_1+h_2^3$,
6	3	(1^3)	$3h_4h_1^2+6h_3h_2h_1+h_2^3$,
6	4	(4)	$h_4h_2+h_3^2-h_3h_2h_1+h_2^3$,
6	4	(31)	$h_4h_1^2+h_3^2+h_3h_2h_1-h_3h_1^3+h_2^3+h_2^2h_1^2$,
6	4	(2^2)	$2h_4h_2+2h_2^3+h_2^2h_1^2$,
6	4	(21^2)	$h_4h_1^2+2h_3h_2h_1+h_2^3+3h_2^2h_1^2$,
6	4	(1^4)	$4h_3h_1^3+6h_2^2h_1^2$,
6	5	(5)	h_4h_2,
6	5	(41)	$h_4h_2+h_3h_2h_1$,
6	5	(32)	$h_3h_2h_1+h_2^3$,
6	5	(31^2)	$2h_3h_2h_1+h_2^2h_1^2$,
6	5	(21^2)	$h_2^3+2h_2^2h_1^2$,
6	5	(21^3)	$3h_2^2h_1^2+h_2h_1^4$,
6	5	(1^5)	$5h_2h_1^4$.

This table may be continued with little labour, the distribution functions being derived from those corresponding to a lesser number of objects whenever the parcel is of such a type that its partition contains more than a single part. For instance, we may employ either of the two formulæ

$$Dp\,(6), (31^2) = Dp\,(3), (3)\; Dp\,(3), (1^2) + Dp\,(4), (3)\; Dp\,(2), (1^2),$$

$$Dp\,(6), (31^2) = Dp\,(4), (31)\; Dp\,(2), (1) + Dp\,(5), (31)\; Dp\,(1), (1),$$

for the calculation of $Dp\,(6), (31^2)$.

The Distribution Functions can then be evaluated in terms of single partition forms by means of the tables subsequently given.

I proceed now to show how to express the symmetric function

$$h_{r^s}$$

in terms of $h_1, h_2, h_3, \ldots$ so as to obtain the expression generally of $Df(n)(m)$.

Section 9.

The symmetric function h_{rs}.

This function is a homogeneous product sum, formed by taking s and s together the terms which compose the homogeneous product sum h_r. h_r is the homogeneous product sum of the roots of the equation

$$x^n - a_1 x^{n-1} + a_2 x^{n-2} - \ldots = 0 \quad \ldots\ldots\ldots\ldots\ldots\ldots\text{(i.)}.$$

Form the equation whose roots are the several terms of h_r, viz.,

$$x^p - j_1 x^{p-1} + j_2 x^{p-2} - \ldots = 0 \quad \ldots\ldots\ldots\ldots\ldots\ldots\text{(ii.)},$$

where $$p = \frac{(n+r-1)!}{(n-1)!\,r!}, \text{ and } j_1 = h_{r'} = h_r.$$

Form also the equation

$$x^n - h_1 x^{n-1} + h_2 x^{n-2} - \ldots = 0 \quad \ldots\ldots\ldots\ldots\ldots\ldots\text{(iii.)}.$$

Let partitions in ( ) and [ ] denote respectively the symmetric functions of the roots of (i.) and (iii.), and σ_κ the sum of the κth powers of the roots of (ii.).

We may easily establish the two results

$$(\kappa) = (-)^{\kappa+1}[\kappa],$$
$$[\kappa^r] = (-)^{r(\kappa+1)}\sigma_\kappa\,;*$$

---

* We have in fact

$$\frac{1}{1 - \frac{a_1}{x} + \frac{a_2}{x^2} - \ldots + (-)^n \frac{a_n}{x^n}} = 1 + \frac{h_1}{x} + \frac{h_2}{x^2} + \ldots + \frac{h_n}{x^n} + \ldots,$$

which may be written

$$\frac{1 - \frac{\phi}{x^{n+1}} - \frac{\psi}{x^{n+2}} - \ldots}{1 - \frac{a_1}{x} + \frac{a_2}{x^2} - \ldots + (-)^n \frac{a_n}{x^n}} = 1 + \frac{h_1}{x} + \frac{h_2}{x^2} + \ldots + \frac{h_n}{x^n},$$

or

$$\frac{1 - \frac{\phi}{x^{n+1}} - \frac{\psi}{x^{n+2}} - \ldots}{\left(1 - \frac{\alpha}{x}\right)\left(1 - \frac{\beta}{x}\right)\ldots} = \left(1 + \frac{\alpha'}{x}\right)\left(1 + \frac{\beta'}{x}\right)\ldots,$$

wherein $\alpha, \beta, \ldots$ are the roots of (i.), and $\alpha', \beta', \ldots$ the roots of (iii.).

whence $h_{r^1} = j_1 = \sigma_1 = [1^r]$,

$$h_{r^2} = j_1^2 - j_2 = \tfrac{1}{2}(\sigma + \sigma_2)$$

$$= \frac{\sigma_1^2}{2!} + \frac{\sigma_2}{2}$$

$$= \frac{[1^r]^2}{2!} + (-)^r \frac{[2^r]}{2};$$

Hence
$$\log\left(1 - \frac{\phi}{x^{n+1}} - \frac{\psi}{x^{n+2}} - \dots\right) + \frac{(1)}{x} + \frac{1}{2}\frac{(2)}{x^2} + \dots + \frac{1}{n}\frac{(n)}{x^n} + \dots$$
$$= \frac{[1]}{x} - \frac{1}{2}\frac{[2]}{x^2} + \frac{1}{3}\frac{[3]}{x^3} - \dots + (-)^{n+1}\frac{[n]}{x^n} + \dots,$$

leading to the result

$$(\kappa) = (-)^{\kappa+1}[\kappa], \quad \text{where } \kappa \not> n.$$

Next, consider the identity

$$\frac{1}{\left(1-\frac{\alpha}{x}\right)\left(1-\frac{\beta}{x}\right)\left(1-\frac{\gamma}{x}\right)\dots} = \left(1+\frac{u}{x}\right)\left(1+\frac{v}{x}\right)\left(1+\frac{w}{x}\right)\dots,$$

wherein u, v, w, ... are the roots of the equation

$$x^n - h_1 x^{n-1} + h_2 x^{n-2} - \dots = 0,$$

and n is supposed indefinitely grea.

Let the κ, κ^{th} roots of unity be denoted by

$$\epsilon_1, \epsilon_2, \epsilon_3, \dots \epsilon_\kappa;$$

then
$$\prod_{s=1}^{s=\kappa} \frac{1}{\left(1-\frac{\epsilon_s\alpha}{x}\right)\left(1-\frac{\epsilon_s\beta}{x}\right)\left(1-\frac{\epsilon_s\gamma}{x}\right)\dots} = \prod_{s=1}^{s=\kappa}\left(1+\frac{\epsilon_s u}{x}\right)\left(1+\frac{\epsilon_s v}{x}\right)\left(1+\frac{\epsilon_s w}{x}\right)\dots,$$

or
$$\frac{1}{\left(1-\frac{\alpha^\kappa}{x^\kappa}\right)\left(1-\frac{\beta^\kappa}{x^\kappa}\right)\left(1-\frac{\gamma^\kappa}{x^\kappa}\right)\dots}$$
$$= \left\{1+(-)^{\kappa+1}\frac{u^\kappa}{x^\kappa}\right\}\left\{1+(-)^{\kappa+1}\frac{v^\kappa}{x^\kappa}\right\}\left\{1+(-)^{\kappa+1}\frac{w^\kappa}{x^\kappa}\right\}\dots,$$

which is

$$\frac{1}{1-\frac{(\kappa)}{x^\kappa}+\frac{(\kappa^2)}{x^{2\kappa}}-\frac{(\kappa^3)}{x^{3\kappa}}+\dots} = 1+(-)^{\kappa+1}\frac{[\kappa]}{x^\kappa}+(-)^{2(\kappa+1)}\frac{[\kappa^2]}{x^{2\kappa}}+\dots+(-)^{r(\kappa+1)}\frac{[\kappa^r]}{x^{r\kappa}}+\dots.$$

The coefficient of $\frac{1}{x^{r\kappa}}$ in the development of the sinister of this identity, according to ascending powers of $\frac{1}{x^\kappa}$, is the homogeneous product sum of order r of the quantities α^κ, β^κ, γ^κ, ...; it is thus equal to σ_κ, the sum of the κth powers of the roots of (ii.).

Hence
$$\sigma_\kappa = (-)^{r(\kappa+1)}[\kappa^r],$$
which is equivalent to the second of the two results.

$$h_{r^3} = j_1^3 - 2j_1j_2 + j_3$$
$$= \frac{\sigma_1^3}{3!} + \frac{\sigma_1\sigma_2}{2} + \frac{\sigma_3}{3}$$
$$= \frac{[1^r]^3}{3!} + (-)^r \frac{[1^r][2^r]}{2} + \frac{[3^r]}{3};$$

$$h_{r^4} = j_1^4 - 3j_1^2j_2 + j_2^2 + 2j_1j_3 - j_4$$
$$= \frac{\sigma_1^4}{4!} + \frac{\sigma_1^2\sigma_2}{2!\,2} + \frac{\sigma_1\sigma_3}{3} + \frac{\sigma_2^2}{2^2.\,2!} + \frac{\sigma_4}{4}$$
$$= \frac{[1^r]^4}{4!} + (-)^r \frac{[1^r]^2[2^r]}{2!\,2} + \frac{[1^r][3^r]}{3} + \frac{[2^r]^2}{2^2.\,2!} + (-)^r \frac{[4^r]}{4},$$

and so forth.

The law is identical with that which obtains in the expression of the elementary symmetric functions in terms of the sums of powers, with the exception that the signs are all positive when r is even.

Hence we can express h_{r^s} in terms of $h_1, h_2, h_3, \ldots$.

In particular we thus find

$$h_{1^1} = h_1,$$

and generally

$$h_{1^r} = h_{r^1} = h_r,$$

$$h_{2^2} = \frac{1}{2!}\{h_2^2 + h_2^2 - 2h_1h_3 + 2h_4\} = h_2^2 - h_1h_3 + h_4,$$

$$h_{2^3} = \frac{1}{3!}\{h_2^3 + 3h_2(h_2^2 - 2h_1h_3 + 2h_4)$$
$$+ 2(h_2^3 - 3h_3h_2h_1 + 3h_3^2 + 3h_4h_1^2 - 3h_4h_2 - 3h_5h_1 + 3h_6)\}$$
$$= h_6 - h_5h_1 + h_4h_1^2 + h_3^2 - 2h_3h_2h_1 + h_2^3,$$

$$h_{3^2} = \frac{1}{2!}\{h_3^2 - h_3^2 + 2h_4h_2 - 2h_5h_1 + 2h_6\} = h_6 - h_5h_1 + h_4h_2,$$

$$h_{2^4} = \frac{1}{4!}\{h_2^4 + 6h_2^2(h_2^2 - 2h_1h_3 + 2h_4)$$
$$+ 8h_2(h_2^3 - 3h_3h_2h_1 + 3h_3^2 + 3h_4h_1^2 - 3h_4h_2 - 3h_5h_1 + 3h_6)$$
$$+ 3(h_2^4 + 4h_3^2h_1^2 + 4h_4^2 - 4h_3h_2^2h_1 + 4h_4h_2^2 - 8h_4h_3h_1)$$
$$+ 6(h_2^4 - 4h_3h_2^2h_1 + 2h_3^2h_1^2 + 4h_3^2h_2 + 4h_4h_2h_1^2 - 4h_4h_2^2$$
$$- 8h_4h_3h_1 + 6h_4^2 - 4h_5h_1^3 + 8h_5h_2h_1 - 4h_5h_3$$
$$+ 4h_6h_1^2 - 4h_6h_2 - 4h_7h_1 + 4h_8)\}$$
$$= h_2^4 - 3h_3h_2^2h_1 - h_4h_2^2 + 2h_3^2h_2 + 2h_4h_2h_1^2 + 2h_5h_2h_1 + h_3^2h_1^2 + 2h_4^2$$
$$- 3h_4h_3h_1 - h_5h_1^3 - h_5h_3 + h_6h_1^2 - h_6h_2 - h_7h_1 + h_8$$
$$= h_8 - h_7h_1 - h_6h_2 + h_6h_1^2 - h_5h_3 + 2h_5h_2h_1 - h_5h_1^3 + 2h_4^2 - 3h_4h_3h_1 - h_4h_2^2$$
$$+ 2h_4h_2h_1^2 + 2h_3^2h_2 + h_3^2h_1^2 - 3h_3h_2^2h_1 + h_2^4,$$

$$h_{4^2} = \frac{1}{2!}\{h_4^2 + h_4^2 - 2h_5 h_3 + 2h_6 h_2 - 2h_7 h_1 + 2h_8\}$$

$$= h_8 - h_7 h_1 + h_6 h_2 - h_5 h_3 + h_4^2,$$

and for present purposes we need calculate no further.

Section 10.

Groups of type (1^m).

Consider the expansion of

$$h_1^n = (\alpha + \beta + \gamma + \ldots)^n.$$

It consists of products of the quantities $\alpha, \beta, \gamma, \ldots$ of the nth degree taken in all possible ways, repetitions and permutations being alike allowable. On this understanding the expansion consists of a number of terms each with coefficient unity. Suppose any such term to be

$$\alpha_1 \beta_1 \beta_2 \alpha_2 \alpha_3 \gamma_1 \beta_3 \alpha_4 \ldots,$$

and place dots in any $m-1$ out of the $n-1$ intervals between the letters; this can be done in

$$\frac{(n-1)!}{(n-m)!\,(m-1)!} \text{ ways.}$$

A distribution (1^m) will correspond to each of these ways for every term of the expansion h_1^n.

Thus the distribution function of n objects into groups (1^m) is

$$Dg\,(n), (1^m) = \frac{(n-1)!}{(n-m)!\,(m-1)!} h_1^n,$$

and denoting by $\quad B_{(p_1^{\pi_1} p_2^{\pi_2} \ldots),\,(1^m)}$

the number of distributions of objects $(p_1^{\pi_1} p_2^{\pi_2} \ldots)$ into groups (1^m), we have

$$B_{(p_1^{\pi_1} p_2^{\pi_2} p_3^{\pi_3} \ldots),\,(1^m)} = \frac{n!}{(p_1!)^{\pi_1}(p_2!)^{\pi_2}(p_3!)^{\pi_3}\ldots} \cdot \frac{(n-1)!}{(n-m)!\,(m-1)!}.$$

The distribution function is the coefficient of x^n in

$$(h_1 x + h_1^2 x^2 + h_1^3 x^3 + h^4 x^4 + \ldots)^m.$$

Section 11.

Groups of type (m).

Consider the symmetric function sum

$$\Sigma \frac{n!}{(p_1!)^{\pi_1}(p_2!)^{\pi_2}\ldots} (p_1^{\pi_1} p_2^{\pi_2} \ldots) = h_1^n$$

arranged as a sum of products of letters α, β, γ, ..., each permutation of every product occurring as a term, so that only coefficients equal to unity present themselves.

We require the homogeneous product sum of all these terms, of any desired order.

Putting $n = 2, 3, \dots r$ successively, we may write as generating functions

$$\frac{1}{(1-\alpha^2ax^2)(1-\beta^2ax^2)\dots(1-\alpha\beta ax^2)^2(1-\alpha\gamma ax^2)^2\dots}$$
$$= 1+H_2ax^2+H_{2^2}a^2x^4+H_{2^3}a^3x^6+\dots$$

$$\frac{1}{(1-\alpha^3ax^3)(1-\beta^3ax^3)\dots(1-\alpha^2\beta ax^3)^3\dots(1-\alpha\beta\gamma ax^3)^6\dots}$$
$$= 1+H_3ax^3+H_{3^2}a^2x^6+\dots,$$

&c.,

and generally

$$\frac{1}{(1-\alpha^r ax^r)\dots(1-\alpha^{r-1}\beta ax^r)^r\dots(1-\alpha\beta\gamma\dots ax^r)^{r!}\dots}$$
$$= 1+H_rax^r+H_{r^2}a^2x^{2r}+H_{r^3}a^3x^{3r}+\dots,$$

wherein H_{r^s} represents the sth order homogeneous product sum of all the separate terms which arise when h_1^r is multiplied out *in extenso.*

By reasoning similar to that employed in the discussion of "Parcels," we see that

$$H_{r^s}$$

denotes the distribution function of sr objects in groups (s) in suchwise that each group shall consist of r objects.

Also that the distribution function of n objects into groups (m) is

$$\Sigma H_{t_1^{r_1}} H_{t_2^{r_2}} H_{t_3^{r_3}} \dots,$$

the summation taking place over every partition of n

$$(t_1^{r_1}\, t_2^{r_2}\, t_3^{r_3} \dots)$$

which contains exactly m $[=\Sigma r]$ parts.

Thus $$Dg\,(n),\,(m) = \Sigma H_{t_1^{r_1}} H_{t_2^{r_2}} H_{t_3^{r_3}} \dots,$$

and it is the coefficient of x^na^m in the expansion of

$$\prod_{s=1}^{s=\infty}(1+H_sx^sa+H_{s^2}x^{2s}a^2+\dots),$$

which is, therefore, the generating function.

Section 12.

Groups of type $(m_1 m_2 m_3 \ldots)$.

The law of derivation of the distribution functions of groups of many part partition types is precisely the same in the case of groups as in the case of parcels.

I, therefore, proceed at once to the examination of the new symmetric function

$$H_{r^s}.$$

Section 13.

The symmetric function H_{r^s}.

Let

$$x^w - k_1 x^{w-1} + k_2 x^{w-2} - \ldots = 0$$

be the equation, having for its roots the several quantities of which H_{r^s} is the homogeneous product sum of order s.

Then

$$k_1 = h_1^r = H_{r^1}.$$

Further, let σ_t denote the sum of the tth powers of the roots of this equation.

If partitions in ( ) refer to the symmetric functions of the equation

$$(x-\alpha)(x-\beta)(x-\gamma) \ldots = 0,$$

we have

$$\sigma_t = (t)^r;$$

also

$$k_2 = \tfrac{1}{2}(\sigma_1^2 - \sigma_2) = \frac{(1)^{2r}}{2!} - \frac{(2)^r}{2};$$

hence

$$H_{r^2} = k_1^2 - k_2 = \frac{\sigma_1^2}{2!} + \frac{\sigma_2}{2}$$

$$= \frac{(1)^{2r}}{2!} + \frac{(2)^r}{2}.$$

Also, since

$$k_3 = \frac{\sigma_1^3}{3!} - \frac{\sigma_1 \sigma_2}{2} + \frac{\sigma_3}{3},$$

we find

$$H_{r^3} = k_1^3 - 2k_1 k_2 + k_3$$

$$= \frac{\sigma_1^3}{3!} + \frac{\sigma_1 \sigma_2}{2} + \frac{\sigma_3}{3}$$

$$= \frac{(1)^{3r}}{3!} + \frac{(1)^r (2)^r}{2} + \frac{(3)^r}{3},$$

and so forth.

Hence, finally, transforming as before to symmetric functions of the roots of the equation

$$x^n - h_1 x^{n-1} + h_2 x^{n-2} - \dots = 0,$$

$$H_{r^1} = (1)^r = [1]^r,$$

$$H_{r^2} = \frac{[1]^{2r}}{2\,!} + (-)^r \frac{[2]^r}{2},$$

$$H_{r^3} = \frac{[1]^{3r}}{3\,!} + (-)^r \frac{[1]^r [2]^r}{2} + \frac{[3]^r}{3},$$

$$\dots \quad \dots \quad \dots \quad \dots \quad \dots \quad \dots$$

The symmetric function H_{r^s} can be thus expressed in terms of h_1, h_2, h_3.

These results should be compared with those obtained in section 9 for the case of distribution into parcels.

It will be noticed that H_{r^s} is derived from h_{r^s} by simply writing $[k]^r$ in place of $[k^r]$.

Section 14.

Restricted distributions into Parcels.

The distributions considered in the foregoing sections were not subject to any restriction. There was no limit to the number of similar objects that it was permissible to distribute either into a single parcel or into a set of similar parcels. This freedom from restriction led naturally to the invariable appearance of the symmetric functions, which express the sums of the homogeneous products of the quantities, in the distribution functions.

In order to find the distribution function of n objects in n parcels, one object in each parcel, subject to the restriction that no two objects of the same kind are to appear in parcels of the same kind, we have merely to employ the elementary symmetric functions

$$a_1, a_2, a_3, \dots$$

instead of the homogeneous product sums

$$h_1, h_2, h_3, \dots .$$

The product $$a_{p_1} a_{q_1} a_{r_1} \dots$$

is necessarily the distribution function of n objects into parcels $(p_1 q_1 r_1 \dots)$, where $p_1 + q_1 + r_1 + \dots = n$, subject to the restriction that no two similar objects are to appear in similar parcels. Thus, since

$$a_3 a_2 = (1^3)(1^2) = (2^2 1) + 3\,(21^3) + 10\,(1^5),$$

we discover that, subject to the restriction, objects (21^3) can be distributed into parcels (32) in three different ways. These three ways are apparent in the scheme:—

A	A	A	B	B
α	β	γ	δ	α
α	γ	δ	β	α
α	δ	β	γ	α

We wish now to impose the restriction that not more than t similar objects are to be distributed into similar parcels. For this purpose, form the symmetric functions

$$t_1, t_2, t_3, t_4, \ldots,$$

where t_s is defined to be that portion of the homogeneous product sum h_s in which no quantity occurs to a higher power than t.

In the product $$t_{p_1} t_{q_1} t_{r_1} \ldots,$$

we may suppose any term composing t_{p_1} to be written out with the letters in any order and a dot placed between each consecutive pair of letters. We consider the p_1 letters to denote p_1 objects distributed into p_1 similar parcels. Obviously, not more than t similar objects thus appear in similar parcels. By reasoning similar to that employed in section 1, it is established that, in the product

$$t_{p_1} t_{q_1} t_{r_1} \ldots,$$

when expanded, the symmetric function $(pqr\ldots)$ will appear with a coefficient which represents the number of ways that it is possible to distribute objects $(pqr\ldots)$ into parcels $(p_1q_1r_1\ldots)$, one object in each parcel, subject to the restriction that not more than t similar objects are to appear in similar parcels. This restriction does not alter the reciprocal nature of the distribution. It is immaterial whether we regard the objects distributed into the parcels or the parcels distributed into the objects. We may say that not more than t similar objects are to be contained in similar parcels, or we may say that not more than t similar parcels are to contain similar objects. The restriction does not affect the reciprocity.

Theorem.—The number of ways of distributing objects $(pqr\ldots)$ into parcels $(p_1q_1r_1\ldots)$ is equal to the number of ways of distributing objects $(p_1q_1r_1\ldots)$ into parcels $(pqr\ldots)$; the distributions being subject to the restriction that not more than t similar objects are to present themselves in similar parcels.

This theorem points to a general algebraic law of symmetry.

Theorem.—The coefficient of symmetric function $(pqr\ldots)$ in the development of

$$t_{p_1}\,t_{q_1}\,t_{r_1}\ldots$$

is equal to the coefficient of symmetric function $(p_1\,q_1\,r_1\ldots)$ in the development of

$$t_p\,t_q\,t_r\ldots.$$

This theorem includes all previous laws of symmetry.

The observation is made that, if any table of functions be found to possess symmetry of this nature, it follows, as a necessary and easily established result, that the "inverse table" also possesses the same symmetry.

The laws of symmetry, as apparent in ordinary tables of symmetric functions, are included in the above theorem. Still retaining the same restriction, it is easy to prove that the distribution function of n objects into parcels (1^m) is

$$\Sigma_t A_{(pqr\ldots),\,(1^m)}\,(pqr\ldots) = \Sigma\frac{m\,!}{\pi_1\,!\;\pi_2\,!\;\pi_3\,!\ldots}\,t_{p_1}^{\pi_1}\,t_{p_2}^{\pi_2}\,t_{p_3}^{\pi_3}\ldots,$$

wherein $\qquad \Sigma\pi = m;\quad \Sigma p\pi = n.$

Section 15.

General value of ${}_tA_{(p_1^{\pi_1}p_2^{\pi_2}p_3^{\pi_3}\ldots),\,(1^m)}$.

We require the coefficient of x^n in the expansion of

$$(t_1x+t_2x^2+t_3x^3+\ldots)^m,$$

and therein, the coefficient of the symmetric function

$$(p_1^{\pi_1}p_2^{\pi_2}p_3^{\pi_3}\ldots).$$

Put $\qquad t_1x+t_2x^2+t_3x^3+\ldots = u,$

so that $\qquad 1+u = \dfrac{1-\alpha^{t+1}x^{t+1}}{1-\alpha x}\cdot\dfrac{1-\beta^{t+1}x^{t+1}}{1-\beta x}\cdot\dfrac{1-\gamma^{t+1}x^{t+1}}{1-\gamma x}\ldots,$

and $\qquad (1+u)^m = \Pi\left\{1-m\alpha^tx^t+\dfrac{m(m-1)}{2!}\alpha^{2t}x^{2t}-\ldots\right\}$

$$\times\left\{1+m\alpha x+\frac{m(m+1)}{2!}\alpha^2x^2+\ldots\right\}.$$

In this product, the coefficient of the symmetric function

$$(p_1^{\pi_1}p_2^{\pi_2}p_3^{\pi_3}\ldots)$$

is
$$\left\{\frac{(m+p_1-1)!}{(m-1)!\,p_1!}-m\frac{(m+p_1-t-2)!}{(m-1)!\,(p_1-t-1)!}+\ldots\right\}^{\pi_1}$$
$$\times\left\{\frac{(m+p_2-1)!}{(m-1)!\,p_2!}-m\frac{(m+p_2-t-2)!}{(m-1)!\,(p_2-t-1)!}+\ldots\right\}^{\pi_2}\ldots,$$

and, since
$$u^m=(1+u-1)^m=(1+u)^m-m(1+u)^{m-1}+\frac{m(m-1)}{2!}(1+u)^{m-2}-\ldots,$$

we find
$${}_tA_{(p_1^{\pi_1}p_2^{\pi_2}p_3^{\pi_3}\ldots),\,(1^m)}$$
$$=\left\{\frac{(m+p_1-1)!}{(m-1)!\,p_1!}-m\frac{(m+p_1-t-2)!}{(m-1)!\,(p_1-t-1)!}+\ldots\right\}^{\pi_1}$$
$$\times\left\{\frac{(m+p_2-1)!}{(m-1)!\,p_2!}-m\frac{(m+p_2-t-2)!}{(m-1)!\,(p_2-t-1)!}+\ldots\right\}^{\pi_2}\ldots$$
$$-m\left\{\frac{(m+p_1-2)!}{(m-2)!\,p_1!}-(m-1)\frac{(m+p_1-t-3)!}{(m-2)!\,(p_1-t-1)!}+\ldots\right\}^{\pi_1}$$
$$\times\left\{\frac{(m+p_2-2)!}{(m-2)!\,p_2!}-(m-1)\frac{(m+p_2-t-3)!}{(m-2)!\,(p_2-t-1)!}+\ldots\right\}^{\pi_2}\ldots$$
$$+\frac{m(m-1)}{2!}\left\{\frac{(m+p_1-3)!}{(m-3)!\,p_1!}-(m-2)\frac{(m+p_1-t-4)!}{(m-3)!\,(p_1-t-1)!}+\ldots\right\}^{\pi_1}$$
$$\times\left\{\frac{(m+p_2-3)!}{(m-3)!\,p_2!}-(m-2)\frac{(m+p_2-t-4)!}{(m-3)!\,(p_2-t-1)!}+\ldots\right\}^{\pi_2}\ldots$$
$$-\ldots\ldots.$$

There is no difficulty in continuing the theory of this restriction. I have not thought it advantageous to proceed further with it in the case of distributions into parcels.

Section 16.

Restricted Distributions into Groups.

It is convenient to write
$$h_1^s=H_s.$$

The distribution function of the unrestricted distribution of n objects into groups
$$(1^m)$$

is then the coefficient of x^n in the expansion of
$$(H_1x+H_2x^2+H_3x^3+\ldots)^m,$$

where
$$H_1 = h_1 = (1),$$
$$H_2 = h_1^2 = (2) + 2\,(1^2),$$
$$H_3 = h_1^3 = (3) + 3\,(21) + 6\,(1^3),$$
$$\ldots \quad \ldots \quad \ldots \quad \ldots \quad \ldots$$

I further denote by $A_s, B_s, C_s, \ldots T_s, \ldots$

those portions of H_s which involve partitions containing no part greater than

$$1, 2, 3, \ldots t, \ldots \text{ respectively.}$$

It is easily seen that the distribution function of n objects into groups (1^m), subject to the restriction that not more than t objects of the same kind are to present themselves in groups of the same kind, is given by the coefficient of x^n in

$$(T_1 x + T_2 x^2 + T_3 x^3 + \ldots)^m.$$

Section 17.

Algebraic Theorems derived from the Theory of Distributions.

Definition.

Of a number n, take any partition

$$(\lambda_1 \lambda_2 \lambda_3 \ldots \lambda_s).$$

It becomes necessary to consider the separation of such a partition into component partitions. Such a separation may be represented by enclosing the component partitions in brackets; thus:

$$(\lambda_1 \lambda_2)\,(\lambda_3 \lambda_4 \lambda_5)\,(\lambda_6) \ldots .$$

It is convenient to arrange the components in descending order as regards their weight or content, and, if these successive weights are in order

$$p, q, r, \ldots,$$

to speak of a separation of species $(pqr\ldots)$.

Just as we speak of the degree of a partition, meaning the magnitude of the largest part in such partition, so we may speak of the degree of a separation, meaning the sum of the largest parts in its components.

We have thus, primarily, three characteristics of a separation, viz.,

(i.) the separable partition,

(ii.) the species,

(iii.) the degree.

R 2

General Theorem of Algebraic Reciprocity.

In § 1, I considered the distribution of n objects into n parcels, and showed that the distribution function of objects into parcels

$$(p_1 q_1 r_1 \ldots)$$

is

$$h_{p_1} h_{q_1} h_{r_1} \ldots.$$

We may analyse this result in the following manner:—

Write
$$\begin{aligned} X_1 &= (1)\, x_1, \\ X_2 &= (2)\, x_2 + (1^2)\, x_1^2, \\ X_3 &= (3)\, x_3 + (21)\, x_2 x_1 + (1^3)\, x_1^3, \\ X_4 &= (4)\, x_4 + (31)\, x_3 x_1 + (2^2)\, x_2^2 + (21^2)\, x_2 x_1^2 + (1^4)\, x_1^4, \\ &\ldots \quad \ldots \quad \ldots \quad \ldots \quad \ldots \quad \ldots \quad \ldots \end{aligned}$$

and generally
$$X_s = \Sigma\, (\lambda\mu\nu\ldots)\, x_\lambda x_\mu x_\nu \ldots,$$

the summation being in regard to every partition of s.

Consider the result of multiplication

$$X_{p_1} X_{q_1} X_{r_1} \ldots = \Sigma P x_{s_1}^{\sigma_1} x_{s_2}^{\sigma_2} x_{s_3}^{\sigma_3} \ldots.$$

P consists of an aggregate of terms, each of which, to a numerical factor *près*, is a separation of the partition

$$(s_1^{\sigma_1} s_2^{\sigma_2} s_3^{\sigma_3} \ldots)$$

of species
$$(p_1 q_1 r_1 \ldots).$$

P, further, is the distribution function of objects into parcels

$$(p_1 q_1 r_1 \ldots),$$

subject to certain restrictions.

If in any distribution of n objects into n parcels (one object into each parcel) we write down a number

$$\xi$$

whenever we observe ξ similar objects in similar parcels, we write down a succession of numbers

$$\xi_1,\ \xi_2,\ \xi_3,\ \ldots,$$

where
$$(\xi_1 \xi_2 \xi_3 \ldots)$$

is some partition of n.

We may be given these numbers, and say that the distribution is subject to a restriction of partition

$$(\xi_1 \xi_2 \xi_3 \ldots).$$

Subject to this restriction, there are a certain number of distributions. In the present case, if we put

$$x_1 = x_2 = x_3 = \dots = 1,$$

$$\Sigma P$$

is obviously the distribution function of n objects into n parcels without restriction.

P itself is manifestly the distribution function subject to the restriction of partition

$$(s_1^{\sigma_1} s_2^{\sigma_2} s_3^{\sigma_3} \dots).$$

Employing a more general notation, we may write

$$X_{p_1}^{\pi_1} X_{p_2}^{\pi_2} X_{p_3}^{\pi_3} \dots = \Sigma P x_{s_1}^{\sigma_1} x_{s_2}^{\sigma_2} x_{s_3}^{\sigma_3} \dots,$$

and then P is the distribution function of objects into parcels

$$(p_1^{\pi_1} p_2^{\pi_2} p_3^{\pi_3} \dots),$$

subject to the restriction of partition

$$(s_1^{\sigma_1} s_2^{\sigma_2} s_3^{\sigma_3} \dots).$$

Multiplying out P, we get the result

$$X_{p_1}^{\pi_1} X_{p_2}^{\pi_2} X_{p_3}^{\pi_3} \dots = \Sigma \theta \, (\lambda_1^{l_1} \lambda_2^{l_2} \lambda_3^{l_3} \dots) \, x_{s_1}^{\sigma_1} x_{s_2}^{\sigma_2} x_{s_3}^{\sigma_3} \dots,$$

indicating that, with a restriction of partition

$$(s_1^{\sigma_1} s_2^{\sigma_2} s_3^{\sigma_3} \dots),$$

there are precisely θ ways of distributing n objects

$$(\lambda_1^{l_1} \lambda_2^{l_2} \lambda_3^{l_3} \dots)$$

amongst n parcels $\quad (p_1^{\pi_1} p_2^{\pi_2} p_3^{\pi_3} \dots),$

one object into each parcel.

Now, it is seen intuitively that, since there is one object in every parcel, it is immaterial whether we regard an object attached to a parcel or a parcel attached to an object, and that making this exchange does not alter the partition of restriction.

Hence the number of distributions must be the same, and if

$$X_{p_1}^{\pi_1} X_{p_2}^{\pi_2} X_{p_3}^{\pi_3} \dots = \dots + \theta \, (\lambda_1^{l_1} \lambda_2^{l_2} \lambda_3^{l_3} \dots) \, x_{s_1}^{\sigma_1} x_{s_2}^{\sigma_2} x_{s_3}^{\sigma_3} \dots,$$

then also $\quad X_{\lambda_1}^{l_1} X_{\lambda_2}^{l_2} X_{\lambda_3}^{l_3} \dots = \dots + \theta \, (p_1^{\pi_1} p_2^{\pi_2} p_3^{\pi_3} \dots) \, x_{s_1}^{\sigma_1} x_{s_2}^{\sigma_2} x_{s_3}^{\sigma_3} \dots.$

This extensive theorem of algebraic reciprocity includes all known theorems of symmetry in symmetric functions.

Limiting attention to the powers of

$$x_1,$$

we immediately obtain Cayley's law of symmetry.

Putting, further, $x_1 = 0,$

we obtain a theorem of wide application in the multiplication of covariants of binary quantics.

We may enunciate it as follows:—

Theorem.—Selecting at pleasure any three partitions of n

$$(p_1^{\pi_1} p_2^{\pi_2} p_3^{\pi_3} \dots),$$
$$(\lambda_1^{l_1} \lambda_2^{l_2} \lambda_3^{l_3} \dots),$$
$$(s_1^{\sigma_1} s_2^{\sigma_2} s_3^{\sigma_3} \dots),$$

separate in any manner the numbers occurring in

$$(s_1^{\sigma_1} s_2^{\sigma_2} s_3^{\sigma_3} \dots)$$

into

π_1	portions of content		p_1,
π_2	,,	,,	p_2,
π_3	,,	,,	p_3,
$\vdots$	,,	,,	$\vdots$

Multiply the product of partitions thus formed by the number which expresses the number of ways of permuting the product, the only permutations allowable being those amongst partitions of the same content; take the sum of all such separations of the partition

$$(s_1^{\sigma_1} s_2^{\sigma_2} s_3^{\sigma_3} \dots),$$

each multiplied by the proper number determined as explained above. The coefficient of the symmetric function

$$(\lambda_1^{l_1} \lambda_2^{l_2} \lambda_3^{l_3} \dots),$$

in this sum of compound symmetric functions, will be precisely the same as if in the process we had interchanged the partitions

$$(p_1^{\pi_1} p_2^{\pi_2} p_3^{\pi_3} \dots), \quad (\lambda_1^{l_1} \lambda_2^{l_2} \lambda_3^{l_3} \dots).$$

Generalisation of Waring's Formula.

Waring's formula for the expression of the nth power sum of the roots of an equation

$$x^n - a_1 x^{n-1} + a_2 x^{n-2} - \dots = 0,$$

in terms of the coefficients, is usually written

$$S_m = \Sigma \frac{(-)^{m+\Sigma\lambda}(\Sigma\lambda-1)!\, m}{\lambda_1!\, \lambda_2! \dots \lambda_n!} a_1^{\lambda_1} a_2^{\lambda_2} \dots a_n^{\lambda_n}.$$

Write this in the following form, viz.,

$$\frac{(-)^m (m-1)!}{m!} S(1^m) = \Sigma \frac{(-)^{\Sigma\lambda}(\Sigma\lambda-1)!}{\lambda_1!\, \lambda_2! \dots \lambda_n!} (1)^{\lambda_1} (1^2)^{\lambda_2} \dots (1^n)^{\lambda_n}.$$

Observe that this formula expresses the sum of the mth powers of the roots in terms of separations of the partition

$$(1^m);$$

the typical separation $(1)^{\lambda_1} (1^2)^{\lambda_2} \dots (1^n)^{\lambda_n}$

is of species $(1^{\lambda_1} 2^{\lambda_2} \dots n^{\lambda_n})$,

and of degree $\Sigma\lambda$.

I proceed to demonstrate a formula for the expression of the mth power sum of the roots as a linear function of separations of any partition whatever of m.

The general formula to be established is

$$(-)^{l+m+\dots} \frac{(l+m+\dots-1)!}{l!\, m! \dots} S(\lambda^l \mu^m \dots)$$
$$= \Sigma (-)^{j_1+j_2+\dots} \frac{(j_1+j_2+\dots-1)!}{j_1!\, j_2! \dots} (J_1)^{j_1} (J_2)^{j_2} \dots,$$

wherein $(\lambda^l \mu^m \dots)$ is the separable partition, $(J_1)^{j_1} (J_2)^{j_2} \dots$ is a separation of $(\lambda^l \mu^m \dots)$, and the summation is in regard to every such separation.

In this formula, $S(\lambda^l \mu^m \dots)$

denotes the sum of the nth powers of the roots $(l\lambda + m\mu + \dots = n)$ in terms of separations of $(\lambda^l \mu^m \dots)$.

Write down the series of relations

$$(a_1) = S_{a_1},$$
$$(a_1 a_2) = S_{a_1} S_{a_2} - S_{a_1+a_2},$$
$$(a_1 a_2 a_3) = S_{a_1} S_{a_2} S_{a_3} - S_{a_1} S_{a_2+a_3} - S_{a_2} S_{a_1+a_3} - S_{a_3} S_{a_1+a_2} + 2S_{a_1+a_2+a_3},$$
$$\dots \quad \dots \quad \dots \quad \dots \quad \dots \quad \dots \quad \dots \quad \dots \quad \dots$$

from the first two of these we find

$$S_{a_1+a_2} = (a_1)(a_2) - (a_1 a_2),$$

or, as this may be written,

$$S(a_1a_2) = (a_1)(a_2) - (a_1a_2),$$

and from the third we get, after reduction,

$$2S(a_1a_2a_3) = 2(a_1)(a_2)(a_3) - (a_1)(a_2a_3) - (a_2)(a_1a_3) - (a_3)(a_1a_2) + (a_1a_2a_3).$$

It is obvious that we can continue this series indefinitely, and express $S(a_1a_2a_3a_4)$, $S(a_1a_2a_3a_4a_5)$, ... in terms of separations of the partitions

$$(a_1a_2a_3a_4),\quad (a_1a_2a_2a_4a_5), \ldots.$$

This holds also notwithstanding any equalities that may exist between the parts a_1, a_2, a_3, ... of the separable partition. The formulæ would, however, require modification in those cases.

First, suppose that no equalities exist between the parts of the separable partition; we require the expression of

$$S(a_1a_2 \ldots a_n)$$

in terms of separations of $\quad (a_1a_2 \ldots a_n)$.

One such separation is, for example,

$$(a_{11}a_{12} \ldots a_{1p})(a_{21}a_{22} \ldots a_{2q}) \ldots (a_{t1}a_{t2} \ldots a_{t\nu}),$$

where the successive component partitions have p, q, ... ν parts, respectively, and there are t components.

Of this type there are in all

$$\frac{n!}{p!\,q! \ldots \nu!} \text{ separations,}$$

and by symmetry we see that in the expression of

$$S(a_1a_2 \ldots a_n),$$

each such separation must be affected by the same coefficient.

Write, then,

$$S(a_1a_2 \ldots a_n) = \Sigma P \Sigma (a_{11}a_{12} \ldots a_{1p})(a_{21}a_{22} \ldots a_{2q}) \ldots (a_{t1}a_{t2} \ldots a_{t\nu}).$$

To determine P, observe that if

$$a_1 = a_2 = \ldots = a_n,$$

the formula should reduce to Waring's, viz.—

$$S(a_1^n) = \Sigma (-)^{n+t} (t-1)!\, n (a_1^p)(a_1^q) \ldots (a_1^\nu),$$

where for simplicity it is supposed that no equalities exist between the ntegers p, q, ... ν.

On this supposition of the equality of the parts of the separable partition, the assumed formula becomes

$$S(a_1^n) = \Sigma P \frac{n!}{p!\,q!\,\ldots\,\nu!}\, p!\,(a_1^p)\, q!\,(a_1^q)\,\ldots\,\nu!\,(a_1^\nu),$$

and, equating these two expressions for $S(a_1^n)$, we find

$$P = (-)^{n+t}\frac{(t-1)!}{(n-1)!},$$

and we thus reach the formula

$$(-)^n (n-1)!\, S(a_1 a_2 \ldots a_n)$$

$$= \Sigma (-)^t (t-1)!\, \Sigma (a_{11}a_{12}\ldots a_{1p})(a_{21}a_{22}\ldots a_{2q})\ldots(a_{t1}a_{t2}\ldots a_{t\nu}),$$

or, as this may be written,

$$(-)^n (n-1)!\, S(a_1 a_2 \ldots a_n)$$

$$= \Sigma (-)^t (t-1)!\, (a_{11}a_{12}\ldots a_{1p})(a_{21}a_{22}\ldots a_{2q})\ldots(a_{t1}a_{t2}\ldots a_{t\nu}).$$

The supposition of any number of equalities between the integers $p, q, \ldots \nu$ renders requisite an easy modification of the proof, and leads to the same final result.

I pass on to the general case

$$S(\lambda^l \mu^m \ldots),$$

and put

$$S(\lambda^l \mu^m \ldots) = \Sigma P (\lambda^{l_1}\mu^{m_1}\ldots)^{j_1}(\lambda^{l_2}\mu^{m_2}\ldots)^{j_2}\ldots(\lambda^{l_\nu}\mu^{m_\nu}\ldots)^{j_\nu}.$$

Starting with the formula

$$(-)^n (n-1)!\, S(a_1 a_2 \ldots a_n)$$

$$= \Sigma (-)^t (t-1)!\, (a_{11}a_{12}\ldots a_{1p})(a_{21}a_{22}\ldots a_{2q})\ldots(a_{t1}a_{t2}\ldots a_{tu}),$$

suppose that of the numbers

$$p, q, \ldots u,$$

j_1 have the value $l_1+m_1+\ldots$,

j_2 have the value $l_2+m_2+\ldots$,

... ...

j_ν have the value $l_\nu+m_\nu+\ldots$.

We may give such values to the quantities a, that certain of the separations under the summation sign shall become

$$(\lambda^{l_1}\mu^{m_1}\ldots)^{j_1}(\lambda^{l_2}\mu^{m_2}\ldots)^{j_2}\ldots(\lambda^{l_\nu}\mu^{m_\nu}\ldots)^{j_\nu};$$

viz.,—we must put l of them equal to λ, m of them equal to μ, and so on. The number of separations which thus become of the required form is easily found to be

$$\frac{l!\,m!}{(l_1!\,m_1!\,\dots)^{j_1}(l_2!\,m_2!\,\dots)^{j_2}\dots(l_\nu!\,m_\nu!\,\dots)^{j_\nu}\,j_1!\,j_2!\dots j_\nu!}.$$

Also, on replacing a component $(\alpha_{11}\alpha_{12}\dots\alpha_{1p})$ by $(\lambda^{l_1}\mu^{m_1}\dots)$, where

$$l_1+m_1+\dots=p,$$

we must multiply by $\quad l_1!\,m_1!\dots;$

we thus get a multiplier

$$(l_1!\,m_1!\dots)^{j_1}(l_2!\,m_2!\dots)^{j_2}\dots(l_\nu!\,m_\nu!\dots)^{j_\nu},$$

and, further, t is equivalent to Σj.

Thus,

$$P=(-)^{l+m+\dots+\Sigma j}\frac{(\Sigma j-1)!}{(l+m+\dots-1)!}$$

$$\times\frac{l!\,m!\dots}{(l_1!\,m_1!\dots)^{j_1}(l_2!\,m_2!\dots)^{j_2}\dots(l_\nu!\,m_\nu!\dots)^{j_\nu}j_1!\,j_2!\dots j_\nu!}$$

$$\times(l_1!\,m_1!\dots)^{j_1}(l_2!\,m_2!\dots)^{j_2}\dots(l_\nu!\,m_\nu!\dots)^{j_\nu}$$

$$=(-)^{l+m+\dots+\Sigma j}\frac{l!\,m!\dots}{(l+m+\dots-1)!}\cdot\frac{(\Sigma j-1)!}{j_1!\,j_2!\dots j_\nu!},$$

leading to the formula

$$\frac{(-)^{l+m+\dots}(l+m+\dots-1)!}{l!\,m!\dots}S(\lambda^l\mu^m\dots)$$

$$=\Sigma(-)^{\Sigma j}\frac{(\Sigma j-1)}{j_1!\,j_2!\dots j_\nu!}(\lambda^{l_1}\mu^{m_1}\dots)^{j_1}(\lambda^{l_2}\mu^{m_2}\dots)^{j_2}\dots(\lambda^{l_\nu}\mu^{m_\nu}\dots)^{j_\nu}.$$

Assuming the form

$$S(\lambda^l\mu^m\dots)=\Sigma P\Sigma(\lambda^{l_1}\mu^{m_1}\dots)^{j_1}(\lambda^{l_2}\mu^{m_2}\dots)^{j_2}\dots(\lambda^{l_\nu}\mu^{m_\nu}\dots)^{j_\nu},$$

another proof may be given.

In this form the sums—

$$l_1+m_1+\dots,$$
$$l_2+m_2+\dots,$$
$$\dots\quad\dots\quad\dots$$
$$l_\nu+m_\nu+\dots,$$

are each considered constant.

Putting each part equal to λ, we must multiply every resulting component $(\lambda^{l_1+m_1+\dots})$ by

$$\frac{(l_1+m_1+\dots)\,!}{l_1\,!\;m_1\,!\,\dots}.$$

Thus $$S\,(\lambda^{l+m+\dots})$$

$$= \Sigma P\left\{\Sigma\left(\frac{(l_1+m_1+\dots)\,!}{l_1\,!\;m_1\,!\,\dots}\right)^{j_1}\left(\frac{(l_2+m_2+\dots)\,!}{l_2\,!\;m_2\,!\,\dots}\right)^{j_2}\dots\left(\frac{(l_\nu+m_\nu+\dots)\,!}{l_\nu\,!\;m_\nu\,!\,\dots}\right)^{j_\nu}\right\}$$

$$\times\,(\lambda^{l_1}\mu^{m_1}\dots)^{j_1}\dots(\lambda^{l_\nu}\mu^{m_\nu}\dots)^{j_\nu}.$$

Now, $$\Sigma\,\frac{\{(l_1+m_1+\dots)\,!\}^{j_1}}{(l_1\,!\;m_1\,!\,\dots)^{j_1}}\,\frac{\{(l_2+m_2+\dots)\,!\}^{j_2}}{(l_2\,!\;m_2\,!\,\dots)^{j_2}}\dots\frac{\{(l_\nu+m_\nu+\dots)\,!\}^{j_\nu}}{(l_\nu\,!\;m_\nu\,!\,\dots)^{j_\nu}}$$

$$=\frac{(l+m+\dots)\,!}{l\,!\;m\,!\,\dots},$$

for each represents the total number of permutations of $l+m+\dots$ things, of which l are of one sort, m of a second, &c.

Hence

$$S\,(\lambda^{l+m+\dots})=\Sigma\,\frac{(l+m+\dots)\,!}{l\,!\;m\,!\,\dots}\,P\,(\lambda^{l_1+m_1+\dots})^{j_1}\,(\lambda^{l_2+m_2+\dots})^{j_2}\dots(\lambda^{l_\nu+m_\nu+\dots})^{j_\nu}.$$

Comparing this with the known formula

$$\frac{(-)^{l+m+\dots}}{l+m+\dots}\,S\,(\lambda^{l+m+\dots})$$

$$=\Sigma\,(-)^{\Sigma j}\,\frac{(\Sigma j-1)\,!}{j_1\,!\;j_2\,!\,\dots j_\nu\,!}\,(\lambda^{l_1+m_1+\dots})^{j_1}\,(\lambda^{l_2+m_2+\dots})^{j_2}\dots(\lambda^{l_\nu+m_\nu+\dots})^{j_\nu},$$

we find, as before,

$$P=(-)^{l+m+\dots+\Sigma j}\,\frac{l\,!\;m\,!\,\dots}{(l+m+\dots-1)\,!}\cdot\frac{(\Sigma j-1)\,!}{j_1\,!\;j_2\,!\,\dots}.$$

It will be noticed that the general result involves only the numbers $l, m, \dots j_1, j_2, \dots$; so that, merely attending to these multiplicities, we may write the result in the hypersymbolic and compact form—

$$(-)^{l+m+\dots}\frac{(l+m+\dots-1)\,!}{l\,!\;m\,!\,\dots}\,S\,|\,lm\,\dots|=\Sigma\,(-)^{\Sigma j}\,\frac{(\Sigma j-1)\,!}{j_1\,!\;j_2\,!\,\dots}\,|\,j_1 j_2 j_3\dots|,$$

where $|\,j_1 j_2 j_3\dots|$ denotes the sum of all the corresponding separations.

This theorem enables us at once to write down an expression for the sth power of the roots corresponding to every partition of s. Thus for $s = 6$, the series is

$$S(6) = (6),$$

$$S(51) = (5)(1) - (51),$$

$$S(42) = (4)(2) - (42),$$

$$S(41^2) = (4)(1)^2 - (41)(1) - (4)(1^2) + (41^2),$$

$$\tfrac{1}{2}S(3^2) = \tfrac{1}{2}(3)^2 - (3^2),$$

$$2S(321) = 2(3)(2)(1) - (32)(1) - (31)(2) - (21)(3) + (321),$$

$$S(31^3) = (3)(1)^3 - (31)(1)^2 - 2(3)(1^2)(1) + (31)(1^2) + (31^2)(1) + (3)(1^3) - (31^3),$$

$$\tfrac{1}{3}S(2^3) = \tfrac{1}{3}(2)^3 - (2^2)(2) + (2^3),$$

$$\tfrac{3}{2}S(2^2 1^2) = \tfrac{3}{2}(2)^2(1)^2 - 2(21)(2)(1) - (2^2)(1)^2 - (1^2)(2)^2 + (2^2)(1^2) + \tfrac{1}{2}(21)^2 + (21^2)(2) + (2^2 1)(1) - (2^2 1^2),$$

$$S(21^4) = (2)(1)^4 - (21)(1)^3 - 3(2)(1^2)(1)^2 + (21)(1)^2 + 2(1^3)(1)(2) + 2(21)(1^2)(1) + (2)(1^2)^2 - (2)(1^4) - (21)(1^3) - (21^2)(1^2) - (21^3)(1) + (21^4),$$

$$\tfrac{1}{6}S(1^6) = \tfrac{1}{6}(1)^6 - (1^2)(1)^4 + \tfrac{3}{2}(1^2)^2(1)^2 + (1^3)(1)^3 - \tfrac{1}{3}(1^2)^3 - 2(1^3)(1^2)(1) - (1^4)(1)^2 + \tfrac{1}{2}(1^3)^2 + (1^4)(1^2) + (1^5)(1) - (1^6).$$

New Tables of Symmetric Functions.

It may be gathered from the foregoing section that it is possible to form tables of symmetric functions, of a symmetrical character, corresponding to every partition of every number. We may select at pleasure any partition as the partition of restriction, and write down partitions representing every possible species of its separations; by the side of these partitions we may write down the compound symmetric functions represented by the corresponding separations. In the expansion of these compounds in a series of monomial symmetric functions, only those monomials will occur which have partitions identical with those representing the species of the separations; this follows naturally from the law of algebraic reciprocity. Thus a symmetrical table necessarily results. To make the method clear, I instance the partition

$$(21^3),$$

and exemplify, in full, the corresponding symmetric function table.

Form two columns—

(5)	(21^3)
(41)	$(21^2)(1)$
(32)	$(21)(1^2)+(1^3)(2)$
(31^2)	$(21)(1)^2$
(2^21)	$2(2)(1^2)(1)$
(21^3)	$(2)(1)^3$.

The left-hand column gives the species of possible separations of

$$(21^3).$$

The right-hand column gives the corresponding separations as derived from the X products (*vide* previous section).

Thus $(21)(1^2)+(1^3)(2)$ is coefficient of $x_2x_1^3$ in X_3X_2,

and $2(2)(1^2)(1)$ „ $x_2x_1^3$ in $X_3^2X_1$.

We may then set out in any convenient way the following table :—

	(5)	(41)	(32)	(31^2)	(2^21)	(21^3)
(21^3)						1
$(21^2)(1)$				1	2	3
$(21)(1^2)+(1^3)(2)$			1	3	2	4
$(21)(1)^2$		1	3	4	6	6
$2(2)(1^2)(1)$		2	2	6	4	6
$(2)(1)^3$	1	3	4	6	6	6

which reads the same by rows as by columns.

In this way we may treat every partition of every number.

We may invert these tables so as to exhibit the single partition symmetric functions in terms of compound symmetric functions symbolised by separations.

We reach then the cardinal and very important theorem of expressibility, which I now enunciate.

Theorem.—" Being given any symmetric function, of partition

$$(\lambda\mu\nu\ldots),$$

let $(\lambda_1\lambda_2\lambda_3 \ldots)$ be any partition of λ,

$(\mu_1\mu_2\mu_3 \ldots)$ " " μ,

$(\nu_1\nu_2\nu_3 \ldots)$ " " ν,

... ...

Then the symmetric function

$$(\lambda\mu\nu \ldots)$$

is expressible by means of compound symmetric functions which are symbolised by separations of the partition

$$(\lambda_1\lambda_2\lambda_3 \ldots \mu_1\mu_2\mu_3 \ldots \nu_1\nu_2\nu_3 \ldots).\text{"}$$

In the example above of the partition

$$(21^3),$$

it will be noticed that there are 7 separations and 6 species of separations; there is thus

$$7-6=1,$$

syzygy between the separations.

The syzygy in question is, in fact, derivable from the separation

$$(21)(2),$$

for we may either express (21) in terms of separations of (1^3), leaving (2) unchanged, or we may leave (21) unchanged and express (2) in terms of separations of (1^2); thus the syzygy is

$$(2)\{(1^2)(1)-3(1^3)\}-(21)\{(1)^2-2(1^2)\}=0,$$

or
$$(2)(1^2)(1)-3(2)(1^3)-(21)(1)^2+2(21)(1^2)=0.$$

In general, if there are θ separations of any partition and ϕ species of separation, there must be

$$\theta-\phi$$

syzygies between the θ separations.

The h Tables direct.

	(1)
h_1	1

	(2)	(1^2)
h_2	1	1
h_1^2	1	2

	(3)	(21)	(1^3)
h_3	1	1	1
h_2h_1	1	2	3
h_1^3	1	3	6

	(4)	(31)	(2^2)	(21^2)	(1^4)
h_4	1	1	1	1	1
h_3h_1	1	2	2	3	4
h_2^2	1	2	3	4	6
$h_2h_1^2$	1	3	4	7	12
h_1^4	1	4	6	12	24

	(5)	(41)	(32)	(31^2)	(2^21)	(21^3)	(1^5)
h_5	1	1	1	1	1	1	1
h_4h_1	1	2	2	3	3	4	5
h_3h_2	1	2	3	4	5	7	10
$h_3h_1^2$	1	3	4	7	8	13	20
$h_2^2h_1$	1	3	5	8	11	18	30
$h_2h_1^3$	1	4	7	13	18	33	60
h_1^5	1	5	10	20	30	60	120

	(6)	(51)	(42)	(41^2)	(3^2)	(321)	(31^3)	(2^3)	(2^21^2)	(21^4)	(1^6)
h_6	1	1	1	1	1	1	1	1	1	1	1
h_5h_1	1	2	2	3	2	3	4	3	4	5	6
h_4h_2	1	2	3	4	3	5	7	6	8	11	15
$h_4h_1^2$	1	3	4	7	4	8	13	9	14	21	30
h_3^2	1	2	3	4	4	6	8	7	10	14	20
$h_3h_2h_1$	1	3	5	8	6	12	19	15	24	38	60
$h_3h_1^3$	1	4	7	13	8	19	34	24	42	72	120
h_2^3	1	3	6	9	7	15	24	21	33	54	90
$h_2^2h_1^2$	1	4	8	14	10	24	42	33	58	102	180
$h_2h_1^4$	1	5	11	21	14	38	72	54	102	192	360
h_1^6	1	6	15	30	20	60	120	90	180	360	720

The h Tables—inverse.

	h_1
(1)	1

	h_2	h_1^2
(2)	2	−1
(1^2)	−1	1

	h_3	h_2h_1	h^3
(3)	3	−3	1
(21)	−3	4	−2
(1^3)	1	−2	1

	h_4	h_3h_1	h_2^2	$h_2h_1^2$	h_1^4
(4)	4	−4	−2	4	−1
(31)	−4	7	2	−7	2
(2^2)	−2	2	3	−4	1
(21^2)	4	−7	−4	10	−3
(1^4)	−1	2	1	−3	1

	h_5	h_4h_1	h_3h_2	$h_3h_1^2$	$h_2^2h_1$	$h_2h_1^3$	h_1^5
(5)	5	−5	−5	5	5	−5	1
(41)	−5	9	5	−9	−7	9	−2
(32)	−5	5	11	−8	−11	10	−2
(31^2)	5	−9	−8	12	10	−13	3
(2^21)	5	−7	−11	10	14	−14	3
(21^3)	−5	9	10	−13	−14	17	−4
(1^5)	1	−2	−2	3	3	−4	1

	h_6	h_5h_1	h_4h_2	$h_4h_1^2$	h_3^2	$h_3h_2h_1$	h_3h_1	h_2^3	$h_2^2h_1^2$	$h_2h_1^4$	h_1^6
(6)	6	−6	−6	6	−3	12	−6	2	−9	6	−1
(51)	−6	11	6	−11	3	−17	11	−2	14	−11	2
(42)	−6	6	14	−10	3	−20	10	−6	19	−12	2
(41^2)	6	−11	−10	15	−3	21	−15	4	−20	16	−3
(3^2)	−3	3	3	−3	6	−15	6	−1	9	−6	1
(321)	12	−17	−20	21	−15	61	−30	8	−48	34	−6
(31^3)	−6	11	10	−15	6	−30	19	−4	26	−21	4
(2^3)	2	−2	−6	4	−1	8	−4	4	−10	6	−1
(2^21^2)	−9	14	19	−20	9	−48	26	−10	46	−33	6
(21^4)	6	−11	−12	16	−6	34	−21	6	−33	26	−5
(1^6)	−1	2	2	−3	1	−6	4	−1	6	−5	1

Second Memoir on a New Theory of Symmetric Functions.

By Captain P. A. MacMahon, R. A.

In my first memoir on this subject (Vol. XI, No. 1) I introduced the notion of the "separation" of a partition, but restricted myself to the discussion of rational integral symmetric functions.

In the present memoir I am engaged with functions which are not necessarily integral, but require partitions, with positive, zero, and negative parts for their symbolical expression.

The chief results which I obtain are

(i). A simple proof of a generalized Vandermonde-Waring power law which presents itself in the guise of an invariantive property of a transcendental transformation.

(ii). The law of "Groups of Separations."

(iii). The fundamental law of algebraic reciprocity; the proof here given being purely arithmetical.

(iv). The fundamental law of algebraic expressibility which asserts that certain indicated symmetric functions can be exhibited as linear functions of the separations of any given partition.

(v). The existence is established of a pair of symmetrical tables in association with every partition into positive, zero, and negative parts, of every number positive, zero, or negative.

The results (iv) and (v) are immediate deductions from (iii), which I believe to be a theorem of great importance and a natural origin of research in symmetrical algebra.

Attention may be drawn to the free introduction of the zero part into the partitions; this forms a connecting link between arithmetic and algebra, and enables us to pass in a novel and natural manner from theorems of quantity to theorems of number. An illustration of this may be found at the conclusion of this memoir, where I have given symmetrical tables of binomial coefficients. By employing zero parts, any algebraic function of one quantity may be expressed by means of partitions, and further, every unsymmetrical algebraic function of the quantity x is expressible as a symmetric function of any arbitrary quantities x in number; this is in fact equivalent to the development of $\phi(x)$, a given rational and integral algebraic function of x, in a series of factorials, but it is interesting as showing that all algebra is in reality included in the algebra of symmetric functions; for this reason I think the theorems here given are entitled to rank as theorems in general algebra, and should not be regarded as appertaining exclusively to symmetrical algebra.

In one or two succeeding memoirs I hope to be permitted to further develop the theory of the $X - x$ transformation which possesses many properties of great elegance, and to exhibit, with some approach to completeness, the theory of the allied differential operations, a large and important part of the subject upon which I have not entered in these two memoirs, although I have it by me in manuscript.

Readers should consult "Symmetric Functions and the Theory of Distributions," Lond. Math. Soc., Vol. XIX, p. 220, and "Théorie des Formes Binaires," by Faà de Bruno.

Section 1.

1. The theory of symmetric functions is a part of the general theory of permutations, combinations and distributions. Formulae in the former are merely elegant analytical expressions of propositions in the latter theory; this fact I have dwelt upon at some length in a paper, "Symmetric Functions and the Theory of Distributions," Proceedings of the London Mathematical Society, Vol. XIX, p. 220 et seq.

As an illustration, I give the interpretations of two well known theorems in symmetric functions and refer readers to the paper above quoted for the necessary explanations and elucidations.

2. If

$$(1 - a_1x + a_2x^2 - a_3x^3 + \ldots)^{-1} = 1 + h_1x + h_2x^2 + h_3x^3 + \ldots,$$

then a_m and h_m are designated respectively "the elementary symmetric function of weight m," and "the homogeneous product sum of weight m" of the quantities

$$\alpha, \beta, \gamma, \delta, \ldots,$$

where

$$1 - a_1x + a_2x^2 - a_3x^3 + \ldots = (1-\alpha x)(1-\beta x)(1-\gamma x)(1-\delta x)\ldots.$$

We have the well known theorems

$$(-)^m a_m = \sum \frac{(-)^{\Sigma\lambda}(\Sigma\lambda)!}{\lambda_1!\,\lambda_2!\,\lambda_3!\ldots} h_1^{\lambda_1}h_2^{\lambda_2}h_3^{\lambda_3}\ldots, \qquad \text{(i)}$$

the summation being controlled by the relation $\Sigma s\lambda_s = m$ and

$$(-)^m h_m = \sum \frac{(-)^{\Sigma\lambda}(\Sigma\lambda)!}{\lambda_1!\,\lambda_2!\,\lambda_3!\ldots} a_1^{\lambda_1}a_2^{\lambda_2}a_3^{\lambda_3}\ldots, \qquad \text{(ii)}$$

with, as before, the relation $\Sigma s\lambda_s = m$.

3. It will be observed that (ii) is derivable from (i) by the interchange of a and h.

4. These formulae give rise respectively to

Theorem I. "Considering n objects of any species whatever, the number of distinct ways of distributing them into an even number of different parcels is precisely equal to the number of distributions into an uneven number of different parcels, except when the objects are all of different species; in this case, the former number is in excess or in defect of the latter number by unity, according as the number of objects is even or uneven."

5. *Theorem* II. "Considering n objects of any species whatever with the restriction that no parcel may contain two objects of the same species, the number of distributions into an even number of different parcels is in excess or in defect by unity of the number of distributions into an uneven number of different parcels according as n, the number of objects, is even or uneven."

6. In these theorems it is to be understood that the phrase "of any species whatever" means that the objects are not restricted to be all of the same kind or to be all of different kinds, but may be of any kinds whatever; the phrase "different parcels" means that no two parcels are of the same description.

7. As an example of the first theorem, suppose there are four objects, say three pears and an apple, we have the distributions:

Four parcels.	Three parcels.	Two parcels.	One parcel.
p, p, p, a	pp, p, a	pp, pa	$pppa$
p, p, a, p	pp, a, p	pa, pp	
p, a, p, p	p, pp, a	ppp, a	
a, p, p, p	a, pp, p	a, ppp	
	p, a, pp	ppa, p	
	a, p, pp	p, ppa	
	pa, p, p		
	p, pa, p		
	p, p, pa		
No. = 4	9	6	1

and

$$4 + 6 = 9 + 1,$$

as stated by the theorem.

8. Again take three different objects, say a pear, an apple, and an orange; the distributions are

Three parcels.	Two parcels.	One parcel.
p, a, o	pa, o	pao
p, o, a	o, pa	
a, p, o	ao, p	
a, o, p	p, ao	
o, p, a	op, a	
o, a, p	a, op	
No. = 6	6	1

and

$$6 + 1 - 6 = 1,$$

as stated by the theorem.

9. As an example of the second theorem, take two pears and two apples, and remember that now no two similar objects can appear in the same parcel; we have thus

Four parcels.	Three parcels.	Two parcels.	One parcel.
p, p, a, a	pa, p, a	pa, pa	no way.
p, a, p, a	pa, a, p		
a, p, p, a	p, pa, a		
p, a, a, p	a, pa, p		
a, p, a, p	p, a, pa		
a, a, p, p	a, p, pa		
No. = 6	6	1	0

and

$$6 + 1 - (6 + 0) = 1,$$

as should be the case.

SECTION 2.

The Vandermonde-Waring Law.

10. Referring readers to the "Definitions" given on page 2 of my former memoir, I pass on to a further consideration of the separation theorem given on page 19 (loc. cit.), viz.

$$(-)^{l+m+\cdots}\frac{(l+m+\cdots-1)!}{l!\,m!\cdots}S(\lambda^l\mu^m\cdots)$$
$$=\sum(-)^{j_1+j_2+\cdots}\frac{(j_1+j_2+\cdots-1)!}{j_1!\,j_2!\cdots}(J_1)^{j_1}(J_2)^{j_2}\cdots,$$

where $S(\lambda^l\mu^m\ldots)$ denotes the sum of the nth powers of the quantities expressed by means of separations of the partition $(\lambda^l\mu^m\ldots)$ of the number n; $(J_1)^{j_1}(J_2)^{j_2}\ldots$ is any one of these separations and the summation is in regard to all the separations.

11. I established this theorem in the Proceedings of the London Mathematical Society, Vol. XIX, p. 247 et seq., but having recently obtained a far simpler proof, I give it here as a preparation for a far more general result which will be established subsequently.

12. Write

$$\begin{aligned}
X_1 &= (1)\,x_1,\\
X_2 &= (2)\,x_2 + (1^2)\,x_1^2,\\
X_3 &= (3)\,x_3 + (21)\,x_2x_1 + (1^3)\,x_1^3,\\
X_4 &= (4)\,x_4 + (31)\,x_3x_1 + (2^2)\,x_2^2 + (21^2)\,x_2x_1^2 + (1^4)\,x_1^4,\\
&\cdots\cdots\cdots\cdots\cdots\cdots\cdots\cdots\\
X_m &= \Sigma\,(m_1m_2m_3\ldots)\,x_{m_1}x_{m_2}x_{m_3}\ldots,
\end{aligned}$$

the summation having reference to every partition $(m_1m_2m_3\ldots)$ of the number m.

13. We may regard the quantities $X_1, X_2, X_3, \ldots$ as transformed into the quantities $x_1, x_2, x_3, \ldots$ by means of these relations, and we may enquire whether there exists a system of invariants of this transformation; whether in fact we can form a system of relations between $X_1, X_2, X_3, \ldots$ which, to symmetric function multipliers *près*, are equal to the like functions of $x_1, x_2, x_3, \ldots$ A complete system of such invariants does exist, and they are of fundamental importance.

9

In the first place, X_1 is such an invariant; the complete system is found in the following manner:

14. I suppose that the symmetric functions on the dexter of the above relations refer to quantities

$$\alpha, \beta, \gamma, \ldots,$$

which I further consider to be infinite in number.

15. I observe that the expression

$$1 + X_1 + X_2 + X_3 + \ldots$$

may be broken up into factors of the form

$$1 + \alpha x_1 + \alpha^2 x_2 + \alpha^3 x_3 + \ldots,$$

so that there is the identity

$$1 + X_1 + X_2 + X_3 + \ldots = \Pi_\alpha (1 + \alpha x_1 + \alpha^2 x_2 + \alpha^3 x_3 + \ldots),$$

a factor appearing for each of the quantities

$$\alpha, \beta, \gamma, \ldots;$$

this relation indeed, from another point of view, serves to define the quantities $X_1, X_2, X_3, \ldots$ in a concise manner and *a posteriori* one is directly convinced of its truth.

16. It is convenient to introduce an arbitrary quantity μ and to write

$$1 + \mu X_1 + \mu^2 X_2 + \mu^3 X_3 + \ldots = \Pi_\alpha (1 + \mu\alpha x_1 + \mu^2\alpha^2 x_2 + \mu^3\alpha^3 x_3 + \ldots)$$

Taking logarithms, we find

$$\log(1 + \mu X_1 + \mu^2 X_2 + \mu^3 X_3 + \ldots) = \sum_\alpha \log(1 + \mu\alpha x_1 + \mu_2\alpha^2 x_2 + \mu^3\alpha^3 x_3 + \ldots);$$

the left-hand side of this identity is, when expanded,

$$\mu X_1 + \mu^2\left(X_2 - \frac{1}{2}X_1^2\right) + \mu^3\left(X_3 - X_2 X_1 + \frac{1}{3}X_1^3\right) + \ldots,$$

the general term being

$$\mu^l \sum (-)^{l_1 + l_2 + \ldots + 1} \frac{(l_1 + l_2 + \ldots - 1)!}{l_1!\, l_2! \ldots} X_{\lambda_1}^{l_1} X_{\lambda_2}^{l_2} \ldots,$$

where $$l = \sum l\lambda;$$

whereas the right-hand side has the general term

$$\left\{\sum a^l\right\}\mu^l\sum(-)^{l_1+l_2+\ldots\ldots+1}\frac{(l_1+l_2+\ldots\ldots-1)!}{l_1!\,l_2!\ldots\ldots}x_{\lambda_1}^{l_1}x_{\lambda_2}^{l_2}\ldots\ldots$$

17. Hence, equating coefficients of like powers of μ, we have a system of invariants shown by the relations

$$X_1=(1)x_1,$$

$$X_2-\frac{X_1^2}{2}=(2)\left\{x_2-\frac{x_1^2}{2}\right\},$$

$$X_3-X_2X_1+\frac{X_1^3}{3}=(3)\left\{x_3-x_2x_1+\frac{x_1^3}{3}\right\},$$

. . . .

. . . .

$$\sum(-)^{l_1+l_2+\ldots\ldots-1}\frac{(l_1+l_2+\ldots\ldots-1)!}{l_1!\,l_2!\ldots\ldots}X_{\lambda_1}^{l_1}X_{\lambda_2}^{l_2}\ldots\ldots$$

$$=(l)\sum(-)^{l_1+l_2+\ldots\ldots-1}\frac{(l_1+l_2+\ldots\ldots-1)!}{l_1!\,l_2!\ldots\ldots}x_{\lambda_1}^{l_1}x_{\lambda_2}^{l_2}\ldots\ldots$$

18. If we now multiply out the left-hand side in order to find the cofactor therein of

$$x_{\lambda_1}^{l_1}x_{\lambda_2}^{l_2}\ldots\ldots$$

we see that the cofactor consists of products of symmetric functions, and that each product is necessarily a separation of the symmetric function

$$(\lambda_1^{l_1}\lambda_2^{l_2}\ldots\ldots).$$

Moreover, the coefficient of

$$x_{\lambda_1}^{l_1}x_{\lambda_2}^{l_2}\ldots\ldots$$

in the product $X_{\mu_1}^{m_1}X_{\mu_2}^{m_2}\ldots\ldots$ is (vide first memoir loc. cit., p. 9)

$$\sum\frac{m_1!\,m_2!\ldots\ldots}{j_1!\,j_2!\ldots\ldots}(J_1)^{j_1}(J_2)^{j_2}\ldots\ldots,$$

wherein $(J_1)^{j_1}(J_2)^{j_2}\ldots$ is any separation of $(\lambda_1^{l_1}\lambda_2^{l_2}\ldots)$ of specification $(\mu_1^{m_1}\mu_2^{m_2}\ldots)$.* Hence

$$(-)^{m_1+m_2+\ldots\ldots-1}\frac{(m_1+m_2+\ldots\ldots-1)!}{m_1!\,m_2!\ldots\ldots}X_{\mu_1}^{m_1}X_{\mu_2}^{m_2}\ldots\ldots$$

$$=\ldots+(-)^{m_1+m_2+\ldots-1}\frac{(m_1+m_2+\ldots-1)!}{m_1!\,m_2!\ldots}m_1!\,m_1!\ldots\left\{\sum\frac{(J_1)^{j_1}(J_2)^{j_2}\ldots}{j_1!\,j_2!\ldots}\right\}x_{\lambda_1}^{l_1}x_{\lambda_2}^{l_2}\ldots+\ldots$$

$$=\ldots+\left\{\sum(-)^{j_1+j_2+\ldots-1}\frac{(j_1+j_2+\ldots-1)!}{j_1!\,j_2!\ldots}(J_1)^{j_1}(J_2)^{j_2}\ldots\right\}x_{\lambda_1}^{l_1}x_{\lambda_2}^{l_2}\ldots+\ldots,$$

* At Professor Cayley's suggestion, I abandon the expression "species partition" in favor of "specification," which is a far more appropriate word.

since $$m_1 + m_2 + \ldots\ldots = j_1 + j_2 + \ldots\ldots$$
Therefore

$$\sum_r (-)^{l_1+l_2+\ldots\ldots-1} \frac{(l_1+l_2+\ldots\ldots-1)!}{l_1!\, l_2! \ldots\ldots} X_{\lambda_1}^{l_1} X_{\lambda_2}^{l_2} \ldots\ldots$$
$$= \sum\sum (-)^{j_1+j_2+\ldots\ldots-1} \frac{(j_1+j_2+\ldots\ldots-1)!}{j_1!\, j_2! \ldots\ldots} (J_1)^{j_1} (J_2)^{j_2} \ldots\ldots x_{\lambda_1}^{l_1} x_{\lambda_2}^{l_2} \ldots\ldots,$$

wherein $(J_1)^{j_1}(J_2)^{j_2}\ldots\ldots$ is any separation of $(\lambda_1^{l_1}\lambda_2^{l_2}\ldots\ldots)$.

Hence, substituting

$$\sum\sum (-)^{j_1+j_2+\ldots\ldots-1} \frac{(j_1+j_2+\ldots\ldots-1)!}{j_1!\, j_2! \ldots\ldots} (J_1)^{j_1}(J_2)^{j_2} \ldots\ldots x_{\lambda_1}^{l_1} x_{\lambda_2}^{l_2} \ldots\ldots$$
$$= (l) \sum (-)^{l_1+l_2+\ldots\ldots-1} \frac{(l_1+l_2+\ldots\ldots-1)!}{l_1!\, l_2! \ldots\ldots} x_{\lambda_1}^{l_1} x_{\lambda_2}^{l_2} \ldots\ldots,$$

and equating coefficients of $x_{\lambda_1}^{l_1} x_{\lambda_2}^{l_2} \ldots\ldots$ we obtain

$$(-)^{l_1+l_2+\ldots\ldots} \frac{(l_1+l_2+\ldots\ldots-1)!}{l_1!\, l_2! \ldots} (l)$$
$$= \sum (-)^{j_1+j_2+\ldots\ldots-1} \frac{(j_1+j_2+\ldots\ldots-1)!}{j_1!\, j_2! \ldots\ldots} (J_1)^{j_1}(J_2)^{j_2} \ldots\ldots,$$

which is the theorem to be proved.

19. It will be observed that the theorem arises at once from the invariant property exhibited by the formula

$$\sum (-)^{l_1+l_2+\ldots\ldots} \frac{(l_1+l_2+\ldots\ldots-1)!}{l_1!\, l_2! \ldots\ldots} X_{\lambda_1}^{l_1} X_{\lambda_2}^{l_2} \ldots\ldots$$
$$= (l) \sum (-)^{l_1+l_2+\ldots\ldots} \frac{(l_1+l_2+\ldots\ldots-1)!}{l_1!\, l_2! \ldots\ldots} x_{\lambda_1}^{l_1} x_{\lambda_2}^{l_2} \ldots\ldots,$$

an application of the multinomial theorem in algebra being in reality all that is necessary; the formula in fact establishes the theorem at once for all partitions of all numbers, and is itself a condensed and exceedingly elegant analytical representation of it. It is very interesting to find an extensive proposition like the one under view appearing under the guise of an invariant of an algebraical transformation; I remark it particularly, as I have never met with a case at all similar to it before.

SECTION 3.

Property of the Coefficients of a Group.

20. On page 28 of my former memoir I defined a "Group" as applied to separations of a partition, and I recall here that the separation $(\lambda^2\mu)(\lambda)(\mu)$ belongs to the group

$$G\{(\lambda^2)(\lambda);\ (\mu)^2\}$$

because (λ^3) and (μ^2) occur in the separations $(\lambda^2)(\lambda)$ and $(\mu)^2$ respectively.

21. To put the group in evidence, it is expedient to substitute for the relations between $X_1, X_2, X_3, \ldots$ and $x_1, x_2, x_3, \ldots$ another set, as follows:

$$Y_1 = (1)\, y_1,$$
$$Y_2 = (2)\, y_2 + (1^2)\, y_{1^2},$$
$$Y_3 = (3)\, y_3 + (21)\, y_2 y_1 + (1^3)\, y_{1^3},$$
$$Y_4 = (4)\, y_4 + (31)\, y_3 y_1 + (2^2)\, y_{2^2} + (21^2)\, y_2 y_{1^2} + (1^4)\, y_{1^4},$$
$$\ldots\ldots\ldots\ldots\ldots\ldots\ldots\ldots\ldots\ldots\ldots\ldots$$
$$Y_\mu = \Sigma\, (\mu_1^{m_1}\mu_2^{m_2}\ldots)\, y_{\mu_1{}^{m_1}} y_{\mu_2{}^{m_2}} \ldots ;$$

we then find

$$Y_4 - Y_3 Y_1 - \frac{1}{2} Y_2^2 + Y_2 Y_1^2 - \frac{1}{4} Y_1^4$$
$$= (4)\, y_4 + \{(31) - (3)(1)\} y_3 y_1 + (2^2)\, y_{2^2} - \frac{1}{2}(2)^2 y_2^2 + \{(21^2) - (2)(1^2)\}\, y_2 y_{1^2}$$
$$+ \{(2)(1)^2 - (21)(1)\}\, y_2 y_1^2 + (1^4)\, y_{1^4} - (1^3)(1)\, y_{1^3} y_1 - \frac{1}{2}(1^2)^2 y_{1^2}^2$$
$$+ (1^2)(1)^2 y_1^2 y_{1^2} - \frac{1}{4}(1)^4 y_1^4.$$

22. Observe that the cofactor of $y_3 y_1$ is composed of members of $G\{(3);\ (1)\}$,
" " " $y_2 y_{1^2}$ " " " $G\{(2);\ (1^2)\}$,
" " " $y_2 y_1^2$ " " " $G\{(2);\ (1)^2\}$,

and that these are the only y products which are multiplied by separations of a partition composed of different parts. Generally, in the cofactor of a y product,

$$y^{m_1'}_{\mu_1 m_1} y^{m_2'}_{\mu_2 m_2} \ldots,$$

the equations must belong to the group

$$G\{(\mu_1^{m_1})^{m_1'};\ (\mu_2^{m_2})^{m_2'};\ldots\}.$$

23. I have before given the theorem that if the symmetric function (l) be expressed by means of separations of any partition of l which does not merely consist of repetitions of a single part, the algebraic sum of the coefficients of the separations of each group is zero. To establish this, it is merely necessary to

prove that, forming in succession

$$\begin{aligned}
&Y_1,\\
&Y_2 - \frac{1}{2} Y_1^2,\\
&Y_3 - Y_2 Y_1 + \frac{1}{3} Y_1^3,\\
&Y_4 - Y_3 Y_1 - \frac{1}{2} Y_2^2 + Y_2 Y_1^2 - \frac{1}{4} Y_1^4,\\
&\dots\dots\dots\dots\dots\dots
\end{aligned}$$

every product $$y^{m'_1}_{\mu_1^{m_1}} y^{m'_2}_{\mu_2^{m_2}}$$

vanishes on putting all the symmetric functions (1), (2), (1^2), (3), (21), (1^3), ... equal to unity, unless $\mu_1 = \mu_2 = \dots$.

24. For this purpose put

$$\begin{aligned}
'Y_1 &= y_1,\\
'Y_2 &= y_2 + y_{1^2},\\
'Y_3 &= y_3 + y_2 y_1 + y_{1^3},\\
'Y_4 &= y_4 + y_3 y_1 + y_{2^2} + y_2 y_{1^2} + y_{1^4},\\
&\dots\dots\dots\dots\dots\dots
\end{aligned}$$

so that

$$\begin{aligned}
&1 + 'Y_1 + 'Y_2 + 'Y_3 + \dots\\
&\quad = (1+y_1+y_{1^2}+y_{1^3}+\dots)(1+y_2+y_{2^2}+y_{2^3}+\dots)(1+y_3+y_{3^2}+y_{3^3}+\dots)\dots
\end{aligned}$$

and taking logarithms

$$\begin{aligned}
&\log(1 + 'Y_1 + 'Y_2 + 'Y_3 + \dots) = \log(1 + y_1 + y_{1^2} + y_{1^3} + \dots)\\
&\quad + \log(1 + y_2 + y_{2^2} + y_{2^3} + \dots) + \log(1 + y_3 + y_{3^2} + y_{3^3} + \dots) + \dots,
\end{aligned}$$

and on expansion

$$\begin{aligned}
&'Y_1 + \left('Y_2 - \frac{1}{2}\,'Y_1^2\right) + \left('Y_3 - 'Y_2\,'Y_1 + \frac{1}{3}\,'Y_1^3\right)\\
&\quad + \left('Y_4 - 'Y_3\,'Y_1 - \frac{1}{2}\,'Y_2^2 + 'Y_2\,'Y_1^2 - \frac{1}{4}\,'Y_1^4\right) + \dots\\
&= y_1 + \left(y_{1^2} - \frac{1}{2} y_1^2\right) + \left(y_{1^3} - y_{1^2} y_1 + \frac{1}{3} y_1^3\right)\\
&\quad + \left(y_{1^4} - y_{1^3} y_1 - \frac{1}{2} y_{1^2}^2 + y_{1^2} y_1^2 - \frac{1}{4} y_1^4\right) + \dots\\
&\quad + y_2 + \left(y_{2^2} - \frac{1}{2} y_2^2\right) + \left(y_{2^3} - y_{2^2} y_2 + \frac{1}{3} y_2^3\right)\\
&\quad + \left(y_{2^4} - y_{2^3} y_2 - \frac{1}{2} y_{2^2}^2 + y_{2^2} y_2^2 - \frac{1}{4} y_2^4\right) + \dots\\
&\quad + y_3 + \left(y_{3^2} - \frac{1}{2} y_3^2\right) + \left(y_{3^3} - y_{3^2} y_3 + \frac{1}{3} y_3^3\right)\\
&\quad + \left(y_{3^4} - y_{3^3} y_3 - \frac{1}{2} y_{3^2}^2 + y_{3^2} y_3^2 - \frac{1}{4} y_3^4\right) + \dots\\
&\quad + \dots\dots\dots\dots\dots\dots
\end{aligned}$$

which may be written

$$\sum (-)^{l_1+l_2+\ldots\ldots-1} \frac{(l_1+l_2+\ldots\ldots-1)!}{l_1!\,l_2!\ldots\ldots} Y^{l_1}_{\lambda_1} Y^{l_2}_{\lambda_2}\ldots\ldots$$

$$= \sum (-)^{l_1+l_2+\ldots\ldots-1} \frac{(l_1+l_2+\ldots\ldots-1)!}{l_1!\,l_2!\ldots\ldots} y^{l_1}_{1\lambda_1} y^{l_2}_{1\lambda_2}\ldots\ldots$$

$$+ \sum (-)^{l_1'+l_2'+\ldots\ldots-1} \frac{(l_1'+l_2'+\ldots\ldots-1)!}{l_1'!\,l_2'!\ldots\ldots} y^{l_1'}_{2\lambda_1} y^{l_2'}_{2\lambda_2}\ldots\ldots$$

$$+ \sum (-)^{l_1''+l_2''+\ldots\ldots-1} \frac{(l_1''+l_2''+\ldots\ldots-1)!}{l_1''!\,l_2''!\ldots\ldots} y^{l_1''}_{3\lambda_1} y^{l_2''}_{3\lambda_2}\ldots\ldots+\ldots\ldots$$

where the right-hand side, visibly, contains only products

$$y^{m_1'}_{\mu_1 m_1} y^{m_2'}_{\mu_2 m_2}\ldots\ldots,$$

in which $\mu_1 = \mu_2 = \ldots\ldots$

25. It is thus established that when we express the symmetric function (l) by means of separations of a partition of the number l, which does not merely consist of repetitions of a single part, the algebraic sum of the coefficients in each group of separations is zero.

This proof seems far preferable to the one given in the former memoir.

SECTION 4.

The Theory of Rational Symmetric Functions.

26. I propose to discuss symmetric functions which are rational, but are freed from the restriction of being integral.

Such an expression is $\quad \sum \frac{\alpha^p \beta^q}{\gamma^r} = \sum \alpha^r \beta^q \gamma^{-r};$

attending merely to the indices, this may be written

$$(pq, -r),$$

in which form it appears as a partition with negative as well as positive parts. As far as I have discovered, Meyer Hirsch was the first who employed partitions with negative parts, but neither he nor any subsequent writer appears to have developed this part of the theory (vide Hirsch's Collection of Examples, Formulae and Calculations on the Literal Calculus and Algebra, translated by Rev. J. A. Ross, London, 1827).

27. As a matter of convenience, I write the partition $(p, q, -r)$ in the form $(pq\bar{r})$, and writing the parts of such a partition in descending order of algebraical magnitude, thus:

$$(pq \dots \bar{r}\bar{s}).$$

28. I call p and s respectively the positive and negative degrees of the partition or of the symmetric function.

29. The sum $p + q + \dots - r - s$ is the weight of the partition or symmetric function, or quâ partitions it may be alluded to as the partible number.

30. Strictly speaking, the partition $(pq \dots \bar{r}\bar{s})$ may be spoken of as an algebraic partition of the partible number, but no confusion need arise in the comprehension of what follows if we speak merely of the partition instead of the algebraic partition.

31. For the sake of continuity, as well as for other weighty reasons which will appear, it is advisable to admit the zero as a possible part in such partitions. The general function to be studied then becomes

$$\sum \alpha^p\beta^q \dots \gamma^0\delta^0 \dots \epsilon^{-r}\theta^{-s},$$

which may be written

$$(pq \dots 00 \dots \bar{r}\bar{s}),$$

where $p, q, \dots r, s$ are integers.

32. Repetitions of the same part are as usual denoted by power indices, so that

$$(pp000\bar{r}\bar{r}\bar{r})$$

is written

$$(p^2 0^3 \bar{r}^3).$$

33. Regarding $p_1, p_2, p_3, \dots p_s$ as positive or negative integers excluding zero, we have evidently

$$(p_1p_2 \dots p_s0) = n - s . (p_1p_2 \dots p_s),$$

$$(p_1p_2 \dots p_s0^2) = \frac{n-s.n-s-1}{1.2}(p_1p_2 \dots p_s),$$

$$\dots\dots\dots\dots\dots\dots\dots$$

from which we obtain in succession

$$n(p_1p_2\ldots.p_s) = s(p_1p_2\ldots.p_s) + (p_1p_2\ldots.p_s0),$$
$$n^2(p_1p_2\ldots.p_s) = s^2(p_1p_2\ldots.p_s) + \overline{2s+1}.(p_1p_2\ldots.p_s0) + 2(p_1p_2\ldots.p_s0^2),$$
$$\ldots\ldots\ldots\ldots\ldots\ldots\ldots\ldots\ldots\ldots\ldots\ldots$$

so that the function $(p_1 p_2 \ldots. p_s)$ multiplied by any rational integral algebraical function of n is expressible as a linear function of the expressions

$$(p_1p_2\ldots.p_s),\quad (p_1p_2\ldots.p_s0),\quad (p_1p_2\ldots.p_s0^2),\ldots.,$$

in which the coefficients are independent of n.

34. Hence we are considering symmetric functions which are rational algebraic functions of the n quantities

$$\alpha,\ \beta,\ \gamma,\ldots.$$

and at the same time rational and integral algebraic functions of n.

35. Having in view a comprehensive study of the whole theory, I proceed as in the former case and put

$$1 + X_0\mu^0 + X_1\mu + X_2\mu^2 + \ldots.$$
$$+ X_{-1}\frac{1}{\mu} + X_{-2}\frac{1}{\mu^2} + \ldots.$$
$$= \left(\begin{array}{l} 1 + \alpha^0 x_0\mu^0 + \alpha x_1\mu + \alpha^2 x_2\mu^2 + \ldots. \\ \qquad + \frac{1}{\alpha}x_{-1}\frac{1}{\mu} + \frac{1}{\alpha^2}x_{-2}\frac{1}{\mu^2} + \ldots. \end{array}\right)$$
$$\times \left(\begin{array}{l} 1 + \beta^0 x_0\mu^0 + \beta x_1\mu + \beta^2 x_2\mu^2 + \ldots. \\ \qquad + \frac{1}{\beta}x_{-1}\frac{1}{\mu} + \frac{1}{\beta^2}x_{-2}\frac{1}{\mu^2} + \ldots. \end{array}\right)$$
$$\times \left(\begin{array}{l} 1 + \gamma^0 x_0\mu^0 + \gamma x_1\mu + \gamma^2 x_2\mu^2 + \ldots. \\ \qquad + \frac{1}{\gamma}x_{-1}\frac{1}{\mu} + \frac{1}{\gamma^2}x_{-2}\frac{1}{\mu^2} + \ldots. \end{array}\right)$$
$$\times \text{ etc.}$$
$$= \prod_{\alpha} \left(\begin{array}{l} 1 + \alpha^0 x_0\mu^0 + \alpha x_1\mu + \alpha^2 x_2\mu^2 + \ldots. \\ \qquad + \frac{1}{\alpha}x_{-1}\frac{1}{\mu} + \frac{1}{\alpha^2}x_{-2}\frac{1}{\mu^2} + \ldots. \end{array}\right)$$

10

36. On multiplying out the right-hand side of this equation, the cofactor of μ^s (s positive, zero, or negative) is found to contain symmetric functions which are symbolical by all partitions of s into positive, zero and negative integers, and moreover, each of these symmetric functions (infinite in number) is attached to the corresponding x product.

Equating coefficients of like powers of μ, we obtain

$$\begin{aligned}
X_0 = {} & (0)x_0 + (1\bar{1})x_1x_{-1} + (2\bar{2})x_2x_{-2} + (2\bar{1}^2)x_2x_{-1}^2 \\
& \qquad + (1^2\bar{2})x_1^2x_{-2} + (1^2\bar{1}^2)x_1^2x_{-1}^2 + \dots \\
& + (0^2)x_0^2 + (10\bar{1})x_1x_0x_{-1} + (20\bar{2})x_2x_0x_{-2} + (20\bar{1}^2)x_2x_0x_{-1}^2 \\
& \qquad + (1^20\bar{2})x_1^2x_0x_{-2} + (1^20\bar{1}^2)x_1^2x_0x_{-1}^2 + \dots \\
& + (0^3)x_0^3 + (10^2\bar{1})x_1x_0^2x_{-1} + (20^2\bar{2})x_2x_0^2x_{-2} + (20^2\bar{1}^2)x_2x_0^2x_{-1}^2 \\
& \qquad + (1^20^2\bar{2})x_1^2x_0^2x_{-2} + (1^20^2\bar{1}^2)x_1^2x_0^2x_{-1}^2 + \dots \\
& + \dots + \dots \quad + \dots \quad + \dots + \dots
\end{aligned}$$

$$\begin{aligned}
X_1 = {} & (1)x_1 + (2\bar{1})x_2x_{-1} + (1^2\bar{1})x_1^2x_{-1} + \dots \\
& + (10)x_1x_0 + (20\bar{1})x_2x_0x_{-1} + (1^20\bar{1})x_1^2x_0x_{-1} + \dots \\
& + (10^2)x_1x_0^2 + (20^2\bar{1})x_2x_0^2x_{-1} + (1^20^2\bar{1})x_1^2x_0^2x_{-1} + \dots \\
& + \dots \quad + \dots \quad + \dots
\end{aligned}$$

$$\begin{aligned}
X_{-1} = {} & (\bar{1})x_{-1} + (1\bar{2})x_1x_{-2} + (1\bar{1}^2)x_1x_{-1}^2 + \dots \\
& + (0\bar{1})x_0x_{-1} + (10\bar{2})x_1x_0x_{-2} + (10\bar{1}^2)x_1x_0x_{-1}^2 + \dots \\
& + (0^2\bar{1})x_0^2x_{-1} + (10^2\bar{2})x_1x_0^2x_{-2} + (10^2\bar{1}^2)x_1x_0^2x_{-1}^2 + \dots \\
& + \dots \quad + \dots \quad + \dots \\
& \text{etc.}
\end{aligned}$$

and generally in the expression of X_s, s being positive, zero, or negative, the summation is taken for every partition of s into positive, zero, and negative integers.

37. Observe that we may write these relations in the form

$$\begin{aligned}
1 + X_0 = {} & (1+x_0)^n + (1\bar{1})x_1(1+x_0)^{n-2}x_{-1} + (2\bar{2})x_2(1+x_0)^{n-2}x_{-2} \\
& + (2\bar{1}^2)x_2(1+x_0)^{n-3}x_{-1}^2 + (1^2\bar{2})x_1^2(1+x_0)^{n-3}x_{-2} + (1^2\bar{1}^2)x_1^2(1+x_0)^{n-4}x_{-1}^2 \\
& + \dots \\
X_1 = {} & (1)x_1(1+x_0)^{n-1} + (2\bar{1})x_2(1+x_0)^{n-2}x_{-1} + (1^2\bar{1})x_1^2(1+x_0)^{n-3}x_{-1} + \dots, \\
X_{-1} = {} & (\bar{1})(1+x_0)^{n-1}x_{-1} + (1\bar{2})x_1(1+x_0)^{n-2}x_{-2} + (1\bar{1}^2)x_1(1+x_0)^{n-3}x_{-1}^2 + \dots, \\
& \dots\dots\dots\dots\dots\dots\dots\dots\dots\dots
\end{aligned}$$

38. And also in the forms

$$\frac{1+X_0}{(1+x_0)^n} = 1 + (1\bar{1})\frac{x_1x_{-1}}{(1+x_0)^2} + (2\bar{2})\frac{x_2x_{-2}}{(1+x_0)^2} + (2\bar{1}^2)\frac{x_2x_{-1}^2}{(1+x_0)^3}$$
$$+ (1^2\bar{2})\frac{x_1^2x_{-2}}{(1+x_0)^{n-3}} + (1^2\bar{1}^2)\frac{x_1^2x_{-1}^2}{(1+x_0)^4} + \ldots,$$

$$\frac{X_1}{(1+x_0)^n} = (1)\frac{x_1}{1+x_0} + (2\bar{1})\frac{x_2x_{-1}}{(1+x_0)^2} + (1^2\bar{1})\frac{x_1^2x_{-1}}{(1+x_0)^3} + \ldots,$$

$$\frac{X_{-1}}{(1+x_0)^n} = (\bar{1})\frac{x_{-1}}{1+x_0} + (1\bar{2})\frac{x_1x_{-2}}{(1+x_0)^2} + (1\bar{1}^2)\frac{x_1x_{-1}^2}{(1+x_0)^3} + \ldots$$

. . . .

39. These relations may be regarded as defining a transformation of

$$\begin{array}{lll} X_0, X_1, X_2, \ldots & \text{into functions of} & x_0, x_1, x_2, \ldots, \\ X_{-1}, X_{-2}, \ldots & \text{" \quad " \quad "} & x_{-1}, x_{-2}, \ldots, \end{array}$$

and we may seek the invariants of the transformation.

40. Recalling the relation

$$\begin{aligned} &1 + X_0\mu^0 + X_1\mu + X_2\mu^2 + \ldots = \prod_\alpha \{1 + \alpha^0x_0\mu^0 + \alpha x_1\mu + \alpha^2x_2\mu^2 + \ldots\}, \\ &\qquad + X_{-1}\frac{1}{\mu} + X_{-2}\frac{1}{\mu^2} \qquad\qquad\qquad + \frac{1}{\alpha}x_{-1}\frac{1}{\mu} + \frac{1}{\alpha^2}x_{-2}\frac{1}{\mu^2}, \end{aligned}$$

and taking logarithms, we find

$$\begin{aligned} &\log(1 + X_0\mu^0 + X_1\mu + X_2\mu^2 + \ldots) = \sum_\alpha \log(1 + \alpha^0x_0\mu^0 + \alpha x_1\mu + \alpha^2x_2\mu^2 + \ldots), \\ &\qquad + X_{-1}\frac{1}{\mu} + X_{-2}\frac{1}{\mu^2} \qquad\qquad\qquad + \frac{1}{\alpha}x_{-1}\frac{1}{\mu} + \frac{1}{\alpha^2}x_{-2}\frac{1}{\mu^2}, \end{aligned}$$

which may be written

$$\begin{aligned} &\log(1+X_0) + \log\Big\{1 + \frac{X_1}{1+X_0}\mu + \frac{X_2}{1+X_0}\mu^2 + \ldots \\ &\qquad\qquad + \frac{X_{-1}}{1+X_0}\frac{1}{\mu} + \frac{X_{-2}}{1+X_0}\frac{1}{\mu^2} + \ldots\Big\} \\ &= \sum_\alpha \Big[\log(1+x_0) + \log\Big\{1 + \alpha\frac{x_1}{1+x_0}\mu + \alpha^2\frac{x_2}{1+x_0}\mu^2 + \ldots \\ &\qquad\qquad + \frac{1}{\alpha}\frac{x_{-1}}{1+x_0}\frac{1}{\mu} + \frac{1}{\alpha^2}\frac{x_{-2}}{1+x_0}\frac{1}{\mu^2} + \ldots\Big\}\Big] \end{aligned}$$

41. We may now expand each side by the multinomial theorem and equate coefficients of like powers of μ.

Taking first the zero power of μ, we have

$$
\begin{aligned}
&\log(1+X_0) - \frac{X_1X_{-1}}{(1+X_0)^2} \\
&\quad - \frac{X_2X_{-2}}{(1+X_0)^2} + \frac{X_1^2X_{-2}}{(1+X_0)^3} \\
&\quad + \frac{X_2X_{-1}^2}{(1+X_0)^3} - \frac{3}{2}\frac{X_1^2X_{-1}^2}{(1+X_0)^4} \\
&\quad - \frac{X_3X_{-3}}{(1+X_0)^2} + 2\frac{X_2X_1X_{-3}}{(1+X_0)^3} - \frac{X_1^3X_{-3}}{(1+X_0)^4} \\
&\quad + 2\frac{X_3X_{-2}X_{-1}}{(1+X_0)^3} - 6\frac{X_2X_1X_{-2}X_{-1}}{(1+X_0)^4} + 4\frac{X_1^3X_{-2}X_{-1}}{(1+X_0)^5} \\
&\quad - \frac{X_3X_{-1}^3}{(1+X_0)^4} + 4\frac{X_2X_1X_{-1}^3}{(1+X_0)^5} - \frac{10}{3}\frac{X_1^3X_{-1}^3}{(1+X_0)^6} \\
&\quad + \ldots\ldots
\end{aligned}
$$

$$
\begin{aligned}
= (0)\Big[&\log(1+x_0) - \frac{x_1x_{-1}}{(1+x_0)^2} \\
&\quad - \frac{x_2x_{-2}}{(1+x_0)^2} + \frac{x_1^2x_{-2}}{(1+x_0)^3} \\
&\quad + \frac{x_2x_{-1}^2}{(1+x_0)^3} - \frac{3}{2}\frac{x_1^2x_{-1}^2}{(1+x_0)^4} \\
&\quad - \frac{x_3x_{-3}}{(1+x_0)^2} + 2\frac{x_2x_1x_{-3}}{(1+x_0)^3} - \frac{x_1^3x_{-3}}{(1+x_0)^4} \\
&\quad + 2\frac{x_3x_{-2}x_{-1}}{(1+x_0)^3} + 6\frac{x_2x_1x_{-2}x_{-1}}{(1+x_0)^4} + 4\frac{x_1^3x_{-2}x_{-1}}{(1+x_0)^5} \\
&\quad - \frac{x_3x_{-1}^3}{(1+x_0)^4} + 4\frac{x_2x_1x_{-1}^3}{(1+x_0)^5} - \frac{10}{3}\frac{x_1^3x_{-1}^3}{(1+x_0)^6} \\
&\quad + \ldots\ldots\ldots\ldots\ldots\ldots\ldots\ldots\Big],
\end{aligned}
$$

from which it appears that the left-hand side of the identity is an invariant of the transformation.

42. Observe that this invariant consists of a logarithmic term, together with an infinite succession of square blocks of terms; each of these blocks possesses row and column symmetry, both as regards the numerical coefficients and as regards the forms of the X products.

43. An invariant is likewise obtained from every other power of μ positive and negative, thus:

$$
\begin{aligned}
&\frac{X_1}{1+X_0} \\
&-\frac{X_2X_{-1}}{(1+X_0)^2}+\frac{X_1^2X_{-1}}{(1+X_0)^3} \\
&-\frac{X_3X_{-2}}{(1+X_0)^2}+2\frac{X_2X_1X_{-2}}{(1+X_0)^3}-\frac{X_1^3X_{-2}}{(1+X_0)^4} \\
&+\frac{X_3X_{-1}^2}{(1+X_0)^3}-3\frac{X_2X_1X_{-1}^2}{(1+X_0)^4}+2\frac{X_1^3X_{-1}^2}{(1+X_0)^5} \\
&+\ldots\ldots\ldots\ldots\ldots\ldots \\
=(1)\Big[&\frac{x_1}{1+x_0} \\
&-\frac{x_2x_{-1}}{(1+x_0)^2}+\frac{x_1^2x_{-1}}{(1+x_0)^3} \\
&-\frac{x_3x_{-2}}{(1+x_0)^2}+2\frac{x_2x_1x_{-2}}{(1+x_0)^3}-\frac{x_1^3x_{-2}}{(1+x_0)^4} \\
&+\frac{x_3x_{-1}^2}{(1+x_0)^3}-3\frac{x_2x_1x_{-1}^2}{(1+x_0)^4}+2\frac{x_1^3x_{-1}^2}{(1+x_0)^5} \\
&+\ldots\ldots\ldots\ldots\ldots\ldots\Big],
\end{aligned}
$$

and

$$
\begin{aligned}
&\frac{X_{-1}}{1+X_0} \\
&-\frac{X_1X_{-2}}{(1+X_0)^2}+\frac{X_1X_{-1}^2}{(1+X_0)^3} \\
&-\frac{X_2X_{-3}}{(1+X_0)^2}+2\frac{X_2X_{-1}X_{-2}}{(1+X_0)^3}-\frac{X_2X_{-1}^3}{(1+X_0)^4} \\
&+\frac{X_1^2X_{-3}}{(1+X_0)^3}-3\frac{X_1^2X_{-1}X_{-2}}{(1+X_0)^4}+2\frac{X_1^2X_{-1}^3}{(1+X_0)^5} \\
&+\ldots\ldots\ldots\ldots\ldots\ldots \\
=(\bar{1})\Big[&\frac{x_{-1}}{1+x_0} \\
&-\frac{x_1x_{-2}}{(1+x_0)^2}+\frac{x_1x_{-1}^2}{(1+x_0)^3} \\
&-\frac{x_2x_{-3}}{(1+x_0)^2}+2\frac{x_2x_{-1}x_{-2}}{(1+x_0)^3}-\frac{x_2x_{-1}^3}{(1+x_0)^4} \\
&+\frac{x_1^2x_{-3}}{(1+x_0)^3}-3\frac{x_1^2x_{-1}x_{-2}}{(1+x_0)^4}+2\frac{x_1^2x_{-1}^3}{(1+x_0)^5} \\
&+\ldots\ldots\ldots\ldots\ldots\ldots\Big],
\end{aligned}
$$

and so forth.

44. These invariants may be written

$$\log(1+X_0)+\sum(-)^{l_1+l_2+\ldots-1}\frac{(l_1+l_2+\ldots-1)!}{l_1!\,l_2!\ldots}=\left(\frac{X_{\lambda_1}}{1+X_0}\right)^{l_1}\left(\frac{X_{\lambda_2}}{1+X_0}\right)^{l_2}\ldots,$$

the summation being for all solutions of the equation

$$l_1\lambda_1+l_2\lambda_2+\ldots=0,$$

in positive and negative integers, but excluding zero; and

$$\sum(-)^{l_1+l_2+\ldots-1}\frac{(l_1+l_2+\ldots-1)!}{l_1!\,l_2!\ldots}\left(\frac{X_{\lambda_1}}{1+X_0}\right)^{l_1}\left(\frac{X_{\lambda_2}}{1+X_0}\right)^{l_2}\ldots,$$

where the summation is for all integer and non-zero solutions of the equation

$$l_1\lambda_1+l_2\lambda_2+\ldots=m,$$

m being any positive or negative but not a zero integer.

45. We may expand the logarithm in the invariant of weight zero, and moreover in all the invariants we may expand the factors

$$\left(\frac{1}{1+X_0}\right)^{l_1},\ \left(\frac{1}{1+X_0}\right)^{l_2},\ldots,$$

and we see that we may write the whole system of invariants in the form

$$\sum(-)^{l_1+l_2+\ldots-1}\frac{(l_1+l_2+\ldots-1)!}{l_1!\,l_2!\ldots}X_{\lambda_1}^{l_1}X_{\lambda_2}^{l_2}\ldots,$$

where now $\qquad l_1\lambda_1+l_2\lambda_2+\ldots=m;$

m may be any integer, positive, zero or negative, and the summation is in regard to all solutions of the indeterminate equation

$$l_1\lambda_1+l_2\lambda_2+\ldots=m,$$

in positive, zero and negative integers.

46. We can now enunciate as follows:

Theorem. If

$$\begin{aligned}1+X_0+X_1+X_2+\ldots &= \prod_\alpha(1+\alpha^0x_0+\alpha x_1+\alpha^2x_2+\ldots),\\ +X_{-1}+X_{-2}\qquad &\qquad\qquad +\frac{1}{\alpha}x_{-1}+\frac{1}{\alpha^2}x_{-2}\end{aligned}$$

then

$$\sum(-)^{l_1+l_2+\ldots-1}\frac{(l_1+l_2+\ldots-1)!}{l_1!\,l_2!\ldots}X_{\lambda_1}^{l_1}X_{\lambda_2}^{l_2}\ldots$$

$$=(m)\sum(-)^{l_1+l_2+\ldots-1}\frac{(l_1+l_2+\ldots-1)!}{l_1!\,l_2!\ldots}x_{\lambda_1}^{l_1}x_{\lambda_2}^{l_2}\ldots,$$

where the summations are for all solutions of the indeterminate equation

$$l_1\lambda_1 + l_2\lambda_2 + \ldots\ldots = m,$$

in positive, zero, and negative integers, and m is any integer, positive, zero, or negative.

47. This invariant property that has just been established is fundamental and of very great importance.

We now proceed as in the previous more simple case, to multiply out the sinister of the identity, in order to find out therein the cofactor of $x_{\lambda_1}^{l_1}x_{\lambda_2}^{l_2}\ldots\ldots$; this cofactor is an assemblage of symmetric function products, each of which is symbolized by a separation of the partition $(\lambda_1^{l_1}\lambda_2^{l_2}\ldots\ldots)$, and we obtain the numerical coefficients by application of the ordinary multinomial theorem: the reasoning is the same as in the previous case, and we are thus led to the comprehensive theorem

$$(-)^{l_1+l_2+\ldots\ldots-1}\frac{(l_1+l_2+\ldots\ldots-1)!}{l_1!\,l_2!\ldots\ldots}(m)$$
$$=\sum(-)^{j_1+j_2+\ldots\ldots-1}\frac{(j_1+j_2+\ldots\ldots-1)!}{j_1!\,j_2!\ldots\ldots}(J_1)^{j_1}(J_2)^{j_2}\ldots\ldots,$$

or, as this may be written,

48. $$(-)^{l_1+l_2+\ldots\ldots-1}\frac{(l_1+l_2+\ldots\ldots-1)!}{l_1!\,l_2!\ldots\ldots}S(\lambda_1^{l_1}\lambda_2^{l_2}\ldots\ldots)$$
$$=\sum(-)^{j_1+j_2+\ldots\ldots-1}\frac{(j_1+j_2+\ldots\ldots-1)!}{j_1!\,j_2!\ldots\ldots}(J_1)^{j_1}(J_2)^{j_2}\ldots\ldots,$$

wherein $(\lambda_1^{l_1}\lambda_2^{l_2}\ldots\ldots)$ is any partition of $m\,(=\Sigma l\lambda)$ into positive, zero and negative integers; $S(\lambda_1^{l_1}\lambda_2^{l_2}\ldots\ldots)$ denotes the symmetric function (m) expressed by means of separations of the partition $(\lambda_1^{l_1}\lambda_2^{l_2}\ldots\ldots)$ of the number m; $(J_1)^{j_1}(J_2)^{j_2}\ldots\ldots$ is any separation of the partition

$$(\lambda_1^{l_1}\lambda_2^{l_2}\ldots\ldots);$$

and the summation is in regard to all such separations. Two examples of this theorem are subjoined.

Example I.

49. To express the symmetric function (2) by means of separations of the symmetric function

$$(210\bar{1}).$$

We form two columns, the first consisting of the different separations, and the second involving the coefficients given by the theorem. We thus have

Separable Partition	Coefficient
$(210\bar{1})$	-6

Separations	Coefficients
$(2)(1)(0)(\bar{1})$	-6
$(21)(0)(\bar{1})$	$+2$
$(20)(1)(\bar{1})$	$+2$
$(2\bar{1})(1)(0)$	$+2$
$(10)(2)(\bar{1})$	$+2$
$(1\bar{1})(2)(0)$	$+2$
$(0\bar{1})(2)(1)$	$+2$
$(21)(0\bar{1})$	-1
$(20)(1\bar{1})$	-1
$(2\bar{1})(10)$	-1
$(210)(\bar{1})$	-1
$(21\bar{1})(0)$	-1
$(20\bar{1})(1)$	-1
$(10\bar{1})(2)$	-1
$(210\bar{1})$	$+1.$

Hence

$$
\begin{aligned}
-6(2) = &-6(2)(1)(0)(\bar{1}) \\
&+2\{(21)(0)(\bar{1})+(20)(1)(\bar{1})+(2\bar{1})(1)(0)+(10)(2)(\bar{1}) \\
&\qquad\qquad\qquad\qquad +(1\bar{1})(2)(0)+(0\bar{1})(2)(1)\} \\
&-\ \{(21)(0\bar{1})+(20)(1\bar{1})+(2\bar{1})(10\} \\
&-\ \{(210)(\bar{1})+(21\bar{1})(0)+(20\bar{1})(1)+(10\bar{1})(2)\} \\
&+\ (210\bar{1}).
\end{aligned}
$$

50. To verify this identity, observe that

$$
\begin{aligned}
(0) &= n, \\
(\lambda 0) &= n-1.(\lambda), \\
(\lambda\mu 0) &= n-2.(\lambda\mu),
\end{aligned}
$$

so that the identity leads to

$$\begin{aligned}
-6(2) = & -6n(2)(1)(\bar{1}) \\
& + 2n\{(21)(\bar{1}) + (2\bar{1})(1) + (1\bar{1})(2)\} \\
& + 6(n-1)(2)(1)(\bar{1}) \\
& -(n-1)\{(21)(\bar{1}) + (2\bar{1})(1) + (1\bar{1})(2)\} \\
& -n(21\bar{1}) \\
& -(n-2)\{(21)(\bar{1}) + (2\bar{1})(1) + (1\bar{1})(2)\} \\
& +(n-3)(21\bar{1});
\end{aligned}$$

which reduces to

$$\begin{aligned}
+2(2) = & +2(2)(1)(\bar{1}) \\
& -\{(21)(\bar{1}) + (2\bar{1})(1) + (1\bar{1})(2)\} \\
& +(21\bar{1}),
\end{aligned}$$

a result which is precisely that given by the theorem for the expression of (2) by means of separations of

$$(21\bar{1}).$$

51. Written in the algebraic form, this last result is

$$2\Sigma\alpha^2 = 2\Sigma\alpha^2\Sigma\alpha'\Sigma\alpha^{-1} - \Sigma\alpha^2\beta'\Sigma\alpha^{-1} - \Sigma\alpha^2\beta^{-1}\Sigma\alpha' - \Sigma\alpha'\beta^{-1}\Sigma\alpha^2 + \Sigma\alpha^2\beta'\gamma^{-1}.$$

52. Example II.

To express S_3 by means of separations of $(3^3\bar{3}^2)$. The result arranged by groups is as follows:

$$\begin{array}{llllll}
2S_3 = \;\; 2(3)^3(\bar{3})^2 & -(3)^3(\bar{3}^2) & -3(3^2)(3)(\bar{3})^2 & +2(3^2)(3)(\bar{3}^2) & +(3^3)(\bar{3})^2 & -(3^3)(\bar{3}^2) \\
 & & & & & -(3^3)(\bar{3}^2) \\
-3(3)^2(3\bar{3})(\bar{3}) & +(3)^2(3\bar{3}^2) & +2(3^2\bar{3})(3)(\bar{3}) & -(3^2\bar{3}^2)(3) & -(3^3\bar{3})(\bar{3}) & +(3^3\bar{3}^2) \\
+(3)(3\bar{3})^2 & & +2(3^2)(3\bar{3})(\bar{3}) & -(3^2)(3\bar{3}^2) & & \\
 & & -(3^2\bar{3})(3\bar{3}) & & & \\
\hline
\pm 3 & \pm 1 & \pm 4 & \pm 2 & \pm 1 & \pm 1 \\
\hline
\end{array}$$

53. To establish the law that the algebraic sum of the coefficients in each group is constant, we proceed, as in paragraph 24, and put

$$\begin{aligned}
1 + {}'Y_0 + {}'Y_1 & + {}'Y_2 + \ldots . \\
= & (1 + y_0 + y_{0^2} + y_{0^3} + \ldots)(1 + y_1 + y_{1^2} + y_{1^3} + \ldots)(1 + y_2 + y_{2^2} + y_{2^3} + \ldots)\ldots \\
+ {}'Y_{-1} + {}'Y_{-2} & \quad \times (1 + y_{\bar{1}} + y_{\bar{1}^2} + y_{\bar{1}^3} + \ldots)(1 + y_{\bar{2}} + y_{\bar{2}^2} + y_{\bar{2}^3} + \ldots)\ldots
\end{aligned}$$

11

Now, taking logarithms, the demonstration proceeds *pari passu* with the former and simpler case.

SECTION 5.

The Law of Reciprocity.

54. I pass on to the generalization of the law of reciprocity which was established in the former memoir, p. 3 et seq.

55. The theorem to be proved is:

Theorem. "Writing

$$\begin{aligned}1 + X_0\mu^0 + X_1\mu \quad + X_2\mu^2 + \ldots\ldots &= \prod_\alpha (1 + \alpha^0 x_0\mu^0 + \alpha x_1\mu \quad + \alpha^2 x_2\mu^2 + \ldots\ldots),\\ + X_{-1}\frac{1}{\mu} + X_{-2}\frac{1}{\mu^2} \qquad\qquad & \qquad\qquad + \frac{1}{\alpha}x_{-1}\frac{1}{\mu} + \frac{1}{\alpha^2}x_{-2}\frac{1}{\mu^2},\end{aligned}$$

where the product extends to each of the n quantities

$$\alpha, \beta, \gamma, \ldots\ldots \qquad (n = \infty),$$

and forming and developing the product

$$X_{p_1}^{\pi_1} X_{p_2}^{\pi_2} X_{p_3}^{\pi_3} \ldots\ldots,$$

we obtain a result

$$X_{p_1}^{\pi_1} X_{p_2}^{\pi_2} X_{p_3}^{\pi_3} \ldots\ldots = \ldots\ldots + \theta\,(\lambda_1^{l_1}\lambda_2^{l_2}\lambda_3^{l_3}\ldots\ldots)\, x_{s_1}^{\sigma_1} x_{s_2}^{\sigma_2} x_{s_3}^{\sigma_3} \ldots\ldots + \ldots\ldots,$$

θ being the numerical coefficient of the term

$$(\lambda_1^{l_1}\lambda_2^{l_2}\lambda_3^{l_3}\ldots\ldots)\, x_{s_1}^{\sigma_1} x_{s_2}^{\sigma_2} x_{s_3}^{\sigma_3} \ldots\ldots,$$

in the development of the product

$$X_{p_1}^{\pi_1} X_{p_2}^{\pi_2} X_{p_3}^{\pi_3} \ldots\ldots;$$

then $\quad X_{\lambda_1}^{l_1} X_{\lambda_2}^{l_2} X_{\lambda_3}^{l_3} \ldots\ldots = \ldots\ldots + \theta\,(p_1^{\pi_1} p_2^{\pi_2} p_3^{\pi_3} \ldots\ldots)\, x_{s_1}^{\sigma_1} x_{s_2}^{\sigma_2} x_{s_3}^{\sigma_3} \ldots\ldots + \ldots\ldots,$

that is to say, the coefficient of the term

$$(p_1^{\pi_1} p_2^{\pi_2} p_3^{\pi_3} \ldots\ldots)\, x_{s_1}^{\sigma_1} x_{s_2}^{\sigma_2} x_{s_3}^{\sigma_3} \ldots\ldots$$

in the development of the product

$$X_{\lambda_1}^{l_1} X_{\lambda_2}^{l_2} X_{\lambda_3}^{l_3} \ldots\ldots$$

is the same number θ."

56. The proof here presented is, as was the one in the former memoir, purely arithmetical in its nature, and depends upon the consideration of a particular mode of distribution of a given number of objects into the same number

of parcels, no parcel being empty; we have invariably one object in each parcel. The distribution is of a more general character than the one previously considered and includes the latter as a particular case. It will be seen that when once the character of the distribution has been precisely defined and its connection, with the subject treated of, established by a close examination of a particular case, the actual proof is instantaneous; it arises in fact from a single observation which is of such an elementary character that it admits of no dispute.

It is necessary to make some definitions more extended than those given in Proc. Lond. Math. Soc., Vol. XIX, p. 243.

57. Suppose any number of objects, all of the same kind, to be separated into an upper group and a lower group, in such wise that the upper group consists of λ_1 more objects than the lower group; such an assemblage of similar objects, so separated, may be spoken of as "Objects of type (λ_1)"; the actual number of objects is immaterial; so long as the number of objects in the upper group exceeds the number in the lower group by λ_1, the objects are of type (λ_1).

58. The number λ_1 may be positive, zero, or negative.

Ex. gr. Objects of type (0) may be $\begin{matrix} a \\ a \end{matrix}$ or $\begin{matrix} aaa \\ aaa \end{matrix}$ or etc.

and objects of type $(\bar{2})$ may be $\begin{matrix} . \\ aa \end{matrix}$ or $\begin{matrix} a \\ aaa \end{matrix}$ or $\begin{matrix} aaa \\ aaaaa \end{matrix}$ or etc.

59. I make a distinction between "Objects of type (λ_1)" and "Objects (λ_1)."

I consider "Objects of *type* (λ_1)" to have reference to objects of any, the same, kind, so that objects

$$\begin{matrix} a \\ a \end{matrix} \text{ or } \begin{matrix} b \\ b \end{matrix}, \text{ etc.}$$

are alike of type (0); whereas, when the objects are restricted to be of a certain definite kind a, I speak of "Objects (λ_1)."

60. Again: "Objects of type $(\lambda_1\lambda_2\lambda_3 \ldots .)$" is defined to mean

(i). "Objects of type (λ_1) of one kind.
(ii). "Objects of type (λ_2) of a second kind.
(iii). "Objects of type (λ_3) of a third kind.
. . . . ;

thus "objects of type $(30\bar{1})$" may be such as

$$\begin{matrix} ccc & ddd & ee \\ & ddd & eee \end{matrix} \text{ or } \begin{matrix} eeee & f & g \\ e & f & gg \end{matrix},$$

where the species of object obtaining in each group is not specified; whereas, if it be stated or implied that the objects in the three groups are of given species, say a, b, c respectively, we would speak of "objects $(30\bar{1})$"; then "objects $(30\bar{1})$" might mean an assemblage such as

$$\begin{matrix} aaaaa & bb & \\ aa & bb & c \end{matrix},$$

the excesses of the objects in the upper group over those in the lower groups being respectively 3, 0 and -1.

The distinction made between "objects of type $(\lambda_1\lambda_2\lambda_3 \ldots)$" and "objects $(\lambda_1\lambda_2\lambda_3 \ldots)$" will be now understood.

61. Observe that "objects (0)" refers to a set of at least two objects, one in each group.

62. If no restriction be placed upon the number of objects, there is an infinite number of assemblages included in the phrase "objects $(\lambda_1\lambda_2\lambda_3 \ldots)$"; by fixing the number of objects we obtain a finite number of assemblages; fixing the number of objects at 8, "objects $(30\bar{1})$" will comprise the three assemblages:

$$\begin{matrix} aaa & b & c \\ & b & cc \end{matrix}; \quad \begin{matrix} aaa & bb & \\ & bb & c \end{matrix}; \quad \begin{matrix} aaaa & b & \\ a & b & c \end{matrix}.$$

63. We have now objects of various kinds, divided into upper and lower groups, and we may have boxes or parcels of various kinds, similarly divided into certain upper and lower groups, to contain these objects in such wise that one parcel of an upper group contains one object of an upper group, and one parcel of a lower group contains one object of a lower group, there being as many parcels as objects.

64. "Parcels of type $(\lambda_1\lambda_2\lambda_3 \ldots)$" and "Parcels $(\lambda_1\lambda_2\lambda_3 \ldots)$" are defined precisely as in the case of objects, capital letters being employed to exhibit them instead of small ones.

65. Thus 9 "Parcels $(10^2\bar{2})$" will comprise the four assemblages of parcels:

$$\begin{matrix} A & B & C & D \\ & B & C & DDD \end{matrix}; \quad \begin{matrix} A & BB & C & \\ & BB & C & DD \end{matrix};$$

$$\begin{matrix} A & B & CC & \\ & B & CC & DD \end{matrix}; \quad \begin{matrix} AA & B & C & \\ A & B & C & DD \end{matrix}.$$

66. Let us now take 8 "objects $(30\bar{1})$," viz. the three assemblages

$$\begin{array}{ccc} aaa & b & c \\ & b & cc \end{array}; \quad \begin{array}{ccc} aaa & bb & \\ & bb & c \end{array}; \quad \begin{array}{cccc} aaaa & b & \\ a & & b & c \end{array};$$

and also 8 "parcels $(4\bar{2})$," viz. the two assemblages

$$\begin{array}{cc} AAAAA & \\ A & BB \end{array}; \quad \begin{array}{cc} AAAA & B \\ & BBB \end{array}.$$

We make a distribution of 8 "objects $(30\bar{1})$" into 8 "parcels $(4\bar{2})$" by placing the objects which occur in any one of the assemblages of objects into the parcels which occur in any one of the assemblages of parcels, in such wise that objects of upper and lower groups appear only in parcels of upper and lower groups respectively, and one parcel contains one and only one object.

67. This distribution is practicable because the partitions

$$(30\bar{1}) \text{ and } (4\bar{2})$$

are each of the same weight, viz. 2. In this manner a definite number of distributions is obtained.

Let us place the second assemblage of objects in the first assemblage of parcels: thus, as one case, we have

$$\begin{array}{ll} AAAAA & \\ aaabb & \\ A & BB \\ b & bc \end{array}.$$

68. An examination of this distribution shows us that we can separate it into four portions, so that each portion consists of but one kind of parcel and of but one kind of object; the four portions are

I	II	III	IV
AAA	AA		
aaa	bb		
	A	B	B
	b	b	c

,

wherein portion I contains "objects of type (3)" placed in "parcels of type (3),"
II " " " (1)" " " " " (1),"
III " " " $(\bar{1})$" " " " " $(\bar{1})$,"
IV " " " $(\bar{1})$" " " " " $(\bar{1})$";

this particular case of distribution possesses a property which is indicated by the succession of numbers 3, 1, -1, -1; thus the property may be defined by the partition $(31\bar{1}^2)$ whose weight is 2, which is of necessity the same as that common to the partitions $(30\bar{1})$, $(4\bar{2})$ which define the assemblages of objects and parcels.

69. We may now restrict ourselves to those distributions of assemblages of objects into assemblages of parcels which possess the property defined by the partition $(31\bar{1}^2)$.

70. This partition $(31\bar{1}^2)$ will be spoken of as the "partition of restriction."

71. The whole number of distributions of assemblages of objects $(30\bar{1})$ into assemblages of parcels $(4\bar{2})$, subject to the restriction of partition $(31\bar{1}^2)$, are now given; they are four in number, viz.

$$\left\{\begin{array}{ll} AAAAA & \\ aaabb & \\ A & BB \\ b & bc \end{array}\right. ,$$

$$\left\{\begin{array}{ll} AAAAA & \\ aaaab & \\ A & BB \\ a & bc \end{array}\right. ,$$

$$\left\{\begin{array}{ll} AAAA & B \\ aaab & c \\ & BBB \\ & bcc \end{array}\right. ,$$

$$\left\{\begin{array}{ll} AAAA & B \\ aaab & b \\ & BBB \\ & bbc \end{array}\right. .$$

72. It is to be understood that the distributions now under examination are connected with three partitions of the same number; the partition of the objects, the partition of the parcels, and the partition of restriction.

73. The weight of the partitions may be any integer, positive, zero, or negative.

74. The number of objects may be any whatever, subject merely to a lower limit which is fixed by the partition of restriction; if a positive part λ occur in this partition, λ objects at least are thereby implied; a negative part $\bar{\lambda}$ also implies at least λ objects, whilst each part zero necessitates at least two objects; thus if p be the sum of the positive parts, if there be q zeros, and if r be the sum of the negative parts,

$$p + 2q + r$$

is a lower limit to the number of objects which can be taken, while in general we may take $p + 2q + r + 2m$ objects, where m is zero or any positive integer.

75. For present purposes it is necessary to consider a minimum number of objects as taking part in the distributions; this, as above mentioned, is known as soon as we decide upon the partition of restriction.

In the example already given, 8 objects were taken, but 6 objects may be taken, as is evident from the partition of restriction $(31\bar{1}^2)$. Reducing the number to 6, we find one assemblage of objects

$$\begin{array}{lll} aaa & b & \\ & b & c \end{array}$$

and also one assemblage of parcels

$$\begin{array}{ll} AAAA & \\ & BB \end{array}$$

and subject to the restriction, but one distribution, viz.

$$\left\{\begin{array}{ll} AAAA & \\ a\,a\,a\,b & \\ & BB \\ & b\,c \end{array}\right. .$$

76. In general, therefore, our distributions are precisely defined by three partitions of the same number, and in every case their number will be perfectly definite.

77. It is now necessary to make a minute examination of a particular case of the general theorem, in order that we may see the bearing of this theory of distribution upon the multiplication of symmetric functions.

78. Since

$$
\begin{aligned}
X_1 = {} & (1)\,x_1 + (10)\,x_1x_0 + (10^2)\,x_1x_0^2 + \dots \\
& + (2\bar{1})\,x_2x_{-1} + (20\bar{1})\,x_2x_0x_{-1} + (20^2\bar{1})\,x_2x_0^2x_{-1} + \dots \\
& + (1^2\bar{1})\,x_1^2x_{-1} + (1^20\bar{1})\,x_1^2x_0x_{-1} + (1^20^2\bar{1})\,x_1^2x_0^2x_{-1} + \dots \\
& + (3\bar{2})\,x_3x_{-2} + (30\bar{2})\,x_3x_0x_{-2} + (30^2\bar{2})\,x_3x_0^2x_{-2} + \dots \\
& + (3\bar{1}^2)\,x_3x_{-1}^2 + (30\bar{1}^2)\,x_3x_0x_{-1}^2 + (30^2\bar{1}^2)\,x_3x_0^2x_{-1}^2 + \dots \\
& + (21\bar{2})\,x_2x_1x_{-2} + (210\bar{2})\,x_2x_1x_0x_{-2} + (210^2\bar{2})\,x_2x_1x_0^2x_{-2} + \dots \\
& + (21\bar{1}^2)\,x_2x_1x_{-1}^2 + (210\bar{1}^2)\,x_2x_1x_0x_{-1}^2 + (210^2\bar{1}^2)\,x_2x_1x_0^2x_{-1}^2 + \dots \\
& + (1^3\bar{2})\,x_1^3x_{-2} + (1^30\bar{2})\,x_1^3x_0x_{-2} + (1^30^2\bar{2})\,x_1^3x_0^2x_{-2} + \dots \\
& + (1^3\bar{1}^2)\,x_1^3x_{-1}^2 + (1^30\bar{1}^2)\,x_1^3x_0x_{-1}^2 + (1^30^2\bar{1}^2)\,x_1^3x_0^2x_{-1}^2 + \dots \\
& + \dots\dots\dots\dots\dots\dots
\end{aligned}
$$

$$
\begin{aligned}
X_{-2} = {} & (\bar{2})\,x_{-2} + (0\bar{2})\,x_0x_{-2} + (0^2\bar{2})\,x_0^2x_{-2} + \dots \\
& + (1\bar{3})\,x_1x_{-3} + (10\bar{3})\,x_1x_0x_{-3} + (10^2\bar{3})\,x_1x_0^2x_{-3} + \dots \\
& + (\bar{1}^2)\,x_{-1}^2 + (0\bar{1}^2)\,x_0x_{-1}^2 + (0^2\bar{1}^2)\,x_0^2x_{-1}^2 + \dots \\
& + (1\bar{1}\bar{2})\,x_1x_{-1}x_{-2} + (10\bar{1}\bar{2})\,x_1x_0x_{-1}x_{-2} + (10^2\bar{1}\bar{2})\,x_1x_0^2x_{-1}x_{-2} + \dots \\
& + (1\bar{1}^3)\,x_1x_{-1}^3 + (10\bar{1}^3)\,x_1x_0x_{-1}^3 + (10^2\bar{1}^3)\,x_1x_0^2x_{-1}^3 + \dots \\
& + \dots\dots\dots\dots\dots\dots
\end{aligned}
$$

we have

$$X_1X_{-2} = \dots + \{(1^2\bar{1})(0\bar{2}) + (1^20\bar{1})(\bar{2}) + (1)(10\bar{1}\bar{2}) + (10)(1\bar{1}\bar{2})\}\,x_1^2x_0x_{-1}x_{-2} + \dots$$

79. The partition of the term $x_1^2x_0x_{-1}x_{-2}$ is $(1^20\bar{1}\bar{2})$; each of the products

$$(1^2\bar{1})(0\bar{2}),\quad (1^20\bar{1})(\bar{2}),\quad (1)(10\bar{1}\bar{2}),\quad (10)(1\bar{1}\bar{2}),$$

is a separation of the partition $(1^20\bar{1}\bar{2})$ of specification $(1\bar{2})$; this follows of necessity because $(1\bar{2})$ is the partition of the term X_1X_{-2}.

80. When the products which occur in the coefficient of the term $x_1^2x_0x_{-1}x_{-2}$ are multiplied, a monomial symmetric function $(1^2\bar{1}\bar{2})$ will be presented attached to a certain numerical coefficient; supposing the symmetric functions refer to quantities $a, b, c, \dots,$ we have

$$(1^2\bar{1})(0\bar{2}) = \sum \frac{a}{1}\cdot\frac{b}{1}\cdot\frac{1}{c}\sum\frac{a}{a}\cdot\frac{1}{b^2},$$

and also

$$(1^2\bar{1}\bar{2}) = \sum \frac{a}{1}\cdot\frac{b}{1}\cdot\frac{1}{c}\cdot\frac{1}{d^2}.$$

81. A term $\frac{a}{1}\cdot\frac{b}{1}\cdot\frac{1}{c}\cdot\frac{1}{d^2}$ of the symmetric function $(1^2\bar{1}\bar{2})$ arises from the multiplication $(1^2\bar{1})(0\bar{2})$ in the three ways:

$$\left(\frac{a}{1}\cdot\frac{b}{1}\cdot\frac{1}{c}\right)\left(\frac{a}{a}\cdot\frac{1}{d^2}\right),$$

$$\left(\frac{a}{1}\cdot\frac{b}{1}\cdot\frac{1}{c}\right)\left(\frac{b}{b}\cdot\frac{1}{d^2}\right),$$

$$\left(\frac{a}{1}\cdot\frac{b}{1}\cdot\frac{1}{c}\right)\left(\frac{c}{c}\cdot\frac{1}{d^2}\right),$$

for each of the terms $\frac{a^2}{a}\cdot\frac{b}{1}\cdot\frac{1}{c}\cdot\frac{1}{d^2}$, $\frac{a}{1}\cdot\frac{b^2}{b}\cdot\frac{1}{c}\cdot\frac{1}{d^2}$, $\frac{a}{1}\cdot\frac{b}{1}\cdot\frac{c}{c^2}\cdot\frac{1}{d^2}$, is the same term $\frac{a}{1}\cdot\frac{b}{1}\cdot\frac{1}{c}\cdot\frac{1}{d^2}$ of the function $(1^2\bar{1}\bar{2})$.

82. Observe that such a product as

$$\left(\frac{a}{1}\cdot\frac{b}{1}\cdot\frac{1}{c}\right)\left(\frac{e}{e}\cdot\frac{1}{d^2}\right)$$

gives rise to a term $\frac{a}{1}\cdot\frac{b}{1}\cdot\frac{e}{e}\cdot\frac{1}{c}\cdot\frac{1}{d^2}$ which belongs to the function $(1^20\bar{1}\bar{2})$ and not to $(1^2\bar{1}\bar{2})$; the coefficient of $(1^2\bar{1}\bar{2})$ in the product $(1^2\bar{1})(0\bar{2})$ is thus 3.

83. To connect this result with the preceding theory of distribution, observe that the terms

$$\frac{a^2}{a}\cdot\frac{b}{1}\cdot\frac{1}{c}\cdot\frac{1}{d^2},\quad \frac{a}{1}\cdot\frac{b^2}{b}\cdot\frac{1}{c}\cdot\frac{1}{d^2},\quad \frac{a}{1}\cdot\frac{b}{1}\cdot\frac{c}{c^2}\cdot\frac{1}{d^2},$$

may each be considered as representing an assemblage of 7 objects $(1^2\bar{1}\bar{2})$, the numerator and denominator letters denoting objects in the upper and lower groups respectively; the term $\frac{a^2}{a}\cdot\frac{b}{1}\cdot\frac{1}{c}\cdot\frac{1}{d^2}$ arose from the multiplication

$$\left(\frac{a}{1}\cdot\frac{b}{1}\cdot\frac{1}{c}\right)\left(\frac{a}{a}\cdot\frac{1}{d^2}\right),$$

and, conversely, we may regard it as decomposed in this manner; we may further consider this decomposition of the term $\frac{a^2}{a}\cdot\frac{b}{1}\cdot\frac{1}{c}\cdot\frac{1}{d^2}$ to denote a distribution of the assemblage of 7 objects represented by the term; this distribu-

12

tion will be into an assemblage of 7 parcels $(1\bar{2})$ and will be indicated by the scheme

$$\begin{matrix} AA & B \\ ab & a \\ A & BBB \\ c & add \end{matrix}$$

84. Drawing a vertical line between the A and the B parcels, the scheme breaks up into two portions; the left-hand portion denotes a distribution of objects $(1^2\bar{1})$ into parcels (1), whilst the right-hand shows a distribution of objects $(0\bar{2})$ into parcels $(\bar{2})$; the distribution is of objects $(1^2\bar{1}\bar{2})$ into parcels $(1\bar{2})$, and it is necessarily subject to a restriction whose partition is $(1^2 0\bar{1}\bar{2})$ because the term $x_1^2 x_0 x_{-1} x_{-2}$ has this partition; the two portions into which the distribution may be divided are respectively restricted by partitions $(1^2\bar{1})$ and $(0\bar{2})$ because these partitions are factors of the separation

$$(1^2\bar{1})(0\bar{2})$$

which is being discussed.

85. Two more distributions of precisely the same nature correspond to the two terms

$$\frac{a}{1}\cdot\frac{b^2}{b}\cdot\frac{1}{c}\cdot\frac{1}{d^2},\quad \frac{a}{1}\cdot\frac{b}{1}\cdot\frac{c}{c^2}\cdot\frac{1}{d^2};$$

these are

$$\begin{matrix} AA & B \\ ab & b \\ A & BBB \\ c & bdd \end{matrix}, \quad \begin{matrix} AA & B \\ ab & c \\ A & BBB \\ c & cdd \end{matrix};$$

each of the three distributions is of objects $(1^2\bar{1}\bar{2})$ into parcels $(1\bar{2})$, and is not only subject to the restriction whose partition is $(1^2 0\bar{1}\bar{2})$, but also more minutely to the compound restriction indicated by the separation $(1^2\bar{1})(0\bar{2})$.

86. It is thus clear that, corresponding to the algebraical result

$$(1^2\bar{1})(0\bar{2}) = \ldots\ldots + 3\,(1^2\bar{1}\bar{2}) + \ldots\ldots,$$

we have a distribution theorem, viz.

"There are 3 ways of distributing objects $(1^2\bar{1}\bar{2})$ into parcels $(1\bar{2})$, $(1\bar{2})$ being the specification of the separation $(1^2\bar{1})(0\bar{2})$, subject to the compound restriction, of separation $(1^2\bar{1})(0\bar{2})$."

87. Consider next the product

$$(1^2 0 \bar{1})(\bar{2}) = \sum \frac{a}{1} \cdot \frac{b}{1} \cdot \frac{c}{c} \cdot \frac{1}{d} \sum \frac{1}{a^2};$$

the term $\frac{a}{1} \cdot \frac{b}{1} \cdot \frac{1}{c} \cdot \frac{1}{d^2}$ can only arise from the product

$$\left(\frac{a}{1} \cdot \frac{b}{1} \cdot \frac{1}{c} \cdot \frac{d}{d}\right)\left(\frac{1}{d^2}\right);$$

the coefficient of $(1^2 \bar{1}\bar{2})$ in the product $(1^2 0 \bar{1})(\bar{2})$ is therefore unity; the corresponding distribution is seen to be

$$\begin{matrix} AAA & \\ abd & \\ AA & BB \\ cd & dd \end{matrix};$$

the restrictions in the A and B parcels are respectively $(1^2 0 \bar{1})$ and $(\bar{2})$; hence we have a distribution of objects $(1^2 \bar{1}\bar{2})$ into parcels $(1\bar{2})$ subject to the composite restriction $(1^2 0 \bar{1})(\bar{2})$.

88. Again the product

$$(1)(10\bar{1}\bar{2}) = \sum \frac{a}{1} \sum \frac{a}{1} \cdot \frac{b}{b} \cdot \frac{1}{c} \cdot \frac{1}{d^2};$$

the term $\frac{a}{1} \cdot \frac{b}{1} \cdot \frac{1}{c} \cdot \frac{1}{d^2}$ is obtained from 2 products

$$\left(\frac{b}{1}\right)\left(\frac{a}{1} \cdot \frac{b}{b} \cdot \frac{1}{c} \cdot \frac{1}{d^2}\right),$$
$$\left(\frac{a}{1}\right)\left(\frac{a}{a} \cdot \frac{b}{1} \cdot \frac{1}{c} \cdot \frac{1}{d^2}\right);$$

thus the coefficient of $(1^2 \bar{1}\bar{2})$ in the product $(1)(10\bar{1}\bar{2})$ is 2, and the corresponding distributions are

$$\begin{matrix} A & BB \\ b & ab \\ & BBBB \\ & bcdd \end{matrix}, \qquad \begin{matrix} A & BB \\ a & ab \\ & BBBB \\ & acdd \end{matrix};$$

which are distributions of objects $(1^2 \bar{1}\bar{2})$ into parcels $(1\bar{2})$, subject to the composite restriction $(1)(10\bar{1}\bar{2})$.

89. Finally we have the product

$$(10)(1\bar{1}\bar{2}) = \sum \frac{a}{1} \cdot \frac{b}{b} \sum \frac{a}{1} \cdot \frac{1}{b} \cdot \frac{1}{c^2};$$

the term $\frac{a}{1} \cdot \frac{b}{1} \cdot \frac{1}{c} \cdot \frac{1}{d^2}$ is obtained from 6 products

$$\left(\frac{a}{1} \cdot \frac{b}{b}\right)\left(\frac{b}{1} \cdot \frac{1}{c} \cdot \frac{1}{d^2}\right), \quad \left(\frac{a}{a} \cdot \frac{b}{1}\right)\left(\frac{a}{1} \cdot \frac{1}{c} \cdot \frac{1}{d^2}\right),$$

$$\left(\frac{a}{1} \cdot \frac{c}{c}\right)\left(\frac{b}{1} \cdot \frac{1}{c} \cdot \frac{1}{d^2}\right), \quad \left(\frac{b}{1} \cdot \frac{c}{c}\right)\left(\frac{a}{1} \cdot \frac{1}{c} \cdot \frac{1}{d^2}\right),$$

$$\left(\frac{a}{1} \cdot \frac{d}{d}\right)\left(\frac{b}{1} \cdot \frac{1}{c} \cdot \frac{1}{d^2}\right), \quad \left(\frac{b}{1} \cdot \frac{d}{d}\right)\left(\frac{a}{1} \cdot \frac{1}{c} \cdot \frac{1}{d^2}\right);$$

the coefficient of $(1^2\bar{1}\bar{2})$ in the product $(10)(1\bar{1}\bar{2})$ is thus 6, and the corresponding distributions are

$$\begin{array}{ll} AA & B \\ ab & b \\ A & BBB \\ b & cdd \end{array}, \quad \begin{array}{ll} AA & B \\ ab & a \\ A & BBB \\ a & cdd \end{array},$$

$$\begin{array}{ll} AA & B \\ ac & b \\ A & BBB \\ c & cdd \end{array}, \quad \begin{array}{ll} AA & B \\ bc & a \\ A & BBB \\ c & cdd \end{array},$$

$$\begin{array}{ll} AA & B \\ ad & b \\ A & BBB \\ d & cdd \end{array}, \quad \begin{array}{ll} AA & B \\ bd & a \\ A & BBB \\ d & cdd \end{array};$$

these are distributions of objects $(1^2\bar{1}\bar{2})$ into parcels $(1\bar{2})$, subject to the composite restriction $(10)(1\bar{1}\bar{2})$.

90. Altogether, in the product X_1X_{-2}, the coefficient of $(1^2\bar{1}\bar{2})\, x_1^2 x_0 x_{-1} x_{-2}$ is 12 $(= 3 + 1 + 2 + 6)$; the 12 corresponding distributions have been exhibited; each of these had reference to objects $(1^2\bar{1}\bar{2})$ and parcels $(1\bar{2})$; each, further, was associated with a composite restriction which was denoted by a separation of the partition $(1^2 0\bar{1}\bar{2})$ because the term $x_1^2 0 x_{-1} x_{-2}$ has this partition; each of these separations had the specification $(1\bar{2})$ because $(1\bar{2})$ is the partition of the term X_1X_{-2}; the 12 distributions were complete, that is, they included all those that

were possible under the given conditions; this must be so because there is a one-to-one correspondence between the distributions and term products, and care was taken to consider the whole of the latter. Amongst the separations which denoted composite restrictions were included all separations of $(1^2 0 \bar{1} \bar{2})$ which had the specification $(1\bar{2})$; this is a consequence of the forms of the functions X and X_{-2}. Hence if we consider the whole cofactor of $x_1^2 x_0 x_{-1} x_{-2}$, which arises from the product $X_1 X_{-2}$ and therein the coefficient of $(1^2 \bar{1}\bar{2})$, we find that this coefficient denotes the number of ways of distributing objects $(1^2\bar{1}\bar{2})$ into parcels $(1\bar{2})$ subject to the restriction whose partition is $(1^2 0\bar{1}\bar{2})$; this restriction does and must involve all the composite restrictions whose separations have a specification $(1\bar{2})$, and there is no need to specifically mention the circumstance in describing the distribution; we may simply state that the analytical result

$$X_1 X_{-2} = \ldots . + 12\,(1^2\bar{1}\bar{2})\,x_1^2 x_0 x_{-1} x_{-2} + \ldots .$$

is the analytical statement of the arithmetical theorem: "There are 12 ways of distributing objects $(1^2\bar{1}\bar{2})$ into parcels $(1\bar{2})$, subject to the restriction whose partition is $(1^2 0\bar{1}\bar{2})$."

91. In the case just considered there is a one-to-one correspondence between the literal products and the distributions; this, however, does not always obtain. Suppose that we take the product of symmetric functions $(1^2 0)(2)$, in which each factor is of the same weight 2, and seek the coefficient of (21^2) in its development; proceeding in the usual manner, we have

$$(1^2 0)(2) = \sum \frac{a}{1} \cdot \frac{b}{1} \cdot \frac{c}{c} \sum \frac{a^2}{1}$$

and

$$(21^2) = \sum \frac{a^2}{1} \cdot \frac{b}{1} \cdot \frac{c}{1};$$

the term $\frac{a^2}{1} \cdot \frac{b}{1} \cdot \frac{c}{1}$ arises only from the product

$$\left(\frac{b}{1} \cdot \frac{c}{1} \cdot \frac{a}{a}\right)\left(\frac{a^2}{1}\right),$$

but corresponding to this decomposition, there are two distributions of 6 objects (21^2) into 6 parcels (2^2), viz.

AAA	*BB*	*AA*	*BBB*
a b c	*a a* ,	*a a*	*a b c* ;
A			*B*
a			*a*

the fact is that the component partitions $(1^2 0)$ and (2) being of the *same* weight but *different*, we obtain an additional distribution by the interchange of A and B; but if we form the product X_2^2 we obtain a term $2(1^2 0)(2)$, the 2 appearing for the very reason that $(1^2 0)$ and (2) are of the same weight but different; we may therefore effect a one-to-one correspondence between the literal products in $2(1^2 0)(2)$ and the distributions thence arising. Similarly, if we form the product X_λ^l, and $(\daleth_1)$, $\daleth_2)$, ... denote different partitions of weight λ, we will on development obtain a term which involves

$$\frac{(l_1 + l_2 + \ldots.)!}{l_1!\, l_2! \ldots.} (\daleth_1)^{l_1} (\daleth_2)^{l_2} \ldots.;$$

and, moreover, corresponding to a literal product in $(\daleth_1)^{l_1}(\daleth_2)^{l_2} \ldots.$ there will be precisely

$$\frac{(l_1 + l_2 + \ldots.)!}{l_1!\, l_2! \ldots.}$$

distributions, since we may permute the capital letters in any one distribution in all possible ways; thus we may consider that there exists a one-to-one correspondence between the literal products in

$$\frac{(l_1 + l_2 + \ldots.)!}{l_1!\, l_2! \ldots.} (\daleth_1)^{l_1} (\daleth_2)^{l_2} \ldots.$$

and the distributions which arise from them.

92. In general, if partitions of the same weight p_s (where p_s is positive, zero, or negative) be denoted by (P_s'), (P_s''), (P_s'''), ..., the development of the product

$$X_{p_1}^{\pi_1} X_{p_2}^{\pi_2} X_{p_3}^{\pi_3} \ldots.$$

will produce a term involving

$$\frac{(\pi_1' + \pi_1'' + \ldots.)!}{\pi_1'!\, \pi_1''! \ldots.} \cdot \frac{(\pi_2' + \pi_2'' + \ldots.)!}{\pi_2'!\, \pi_2''! \ldots.} \ldots. (P_1)^{\pi_1'} (P_1')^{\pi_1''} \ldots. (P_2)^{\pi_2'} (P_2')^{\pi_2''} \ldots.,$$

and there will be a one-to-one correspondence between the literal terms occurring therein and the distributions arising therefrom.

93. Hence, from what has gone before, the result

$$X_{p_1}^{\pi_1} X_{p_2}^{\pi_2} X_{p_3}^{\pi_3} \ldots. = \ldots. + \theta (\lambda_1^{l_1} \lambda_2^{l_2} \lambda_3^{l_3} \ldots.) x_{s_1}^{\sigma_1} x_{s_2}^{\sigma_2} x_{s_3}^{\sigma_3} \ldots. + \ldots.$$

is the analytical statement of the arithmetical theorem: "There are θ ways of distributing objects $(\lambda_1^{l_1} \lambda_2^{l_2} \lambda_3^{l_3} \ldots.)$ into parcels $(p_1^{\pi_1} p_2^{\pi_2} p_3^{\pi_3} \ldots.)$, subject to the restriction whose partition is $(s_1^{\sigma_1} s_2^{\sigma_2} s_3^{\sigma_3} \ldots.)$."

94. Recalling our former result

$$X_1X_{-2} = \ldots . + 12\,(1^2\bar{1}\bar{2})\,x_1^2x_0x_{-1}x_{-2} + \ldots . ,$$

we can now establish, in an instantaneous manner, the reciprocal result

$$X_1^2X_{-1}X_{-2} = \ldots . + 12\,(1\bar{2})\,x_1^2x_0x_{-1}x_{-2} + \ldots . ;$$

for, take any one of the foregoing 12 distributions, viz.

$$\begin{array}{ll} AA & B \\ ab & a \\ A & BBB \\ a & cdd \end{array},$$

and change the small letters into capitals and vice versa, we get thus a distribution

$$\begin{array}{ll} aa & b \\ AB & A \\ a & bbb \\ A & CDD \end{array},$$

which may be put into the form

$$\begin{array}{llll} AA & B & & \\ ab & a & & \\ A & & C & DD \\ a & & b & bb \end{array},$$

and this denotes a distribution of objects $(1\bar{2})$ into parcels $(1^2\bar{1}\bar{2})$, subject to a restriction whose partition is $(1^20\bar{1}\bar{2})$.

95. We have thus passed from a distribution of objects $(1^2\bar{1}\bar{2})$ into parcels $(1\bar{2})$ to a distribution of objects $(1\bar{2})$ into parcels $(1^2\bar{1}\bar{2})$ without altering the restriction which still possesses the partition $(1^20\bar{1}\bar{2})$.

96. This interchange of small and capital letters (in reality an interchange of objects and parcels) cannot possibly alter the partition of restriction; this is manifest from the definition of the latter.

97. Further, the process is reversible; from every distribution of the second kind we are able to pass to a distribution of the first kind and vice versa.

98. There is thus a one-to-one correspondence between the two natures of distribution, and the numbers of the distributions of the two kinds must be identical. Hence

$$X_1^2X_{-1}X_{-2} = \ldots . + 12\,(1\bar{2})\,x_1^2x_0x_{-1}x_{-2} + \ldots . ,$$

for this is merely the analytical statement of the arithmetical fact that there are 12 distributions of objects $(1\bar{2})$ into parcels $(1^2\bar{1}\bar{2})$ subject to a restriction whose partition is $(1^2 0\bar{1}\bar{2})$.

99. The general theorem is now practically established, for if

$$X_{p_1}^{\pi_1} X_{p_2}^{\pi_2} X_{p_3}^{\pi_3} \ldots \ldots = \ldots \ldots + \theta\,(\lambda_1^{l_1}\lambda_2^{l_2}\lambda_3^{l_3}\ldots\ldots)\, x_{s_1}^{\sigma_1} x_{s_2}^{\sigma_2} x_{s_3}^{\sigma_3} \ldots\ldots + \ldots\ldots,$$

there are θ ways of distributing objects $(\lambda_1^{l_1}\lambda_2^{l_2}\lambda_3^{l_3}\ldots\ldots)$ into parcels $(p_1^{\pi_1}p_2^{\pi_2}p_3^{\pi_3}\ldots\ldots)$ subject to a restriction whose partition is $(s_1^{\sigma_1}s_2^{\sigma_2}s_3^{\sigma_3}\ldots\ldots)$, and the above reversible process proves that there must be also exactly θ ways of distributing objects $(p_1^{\pi_1}p_2^{\pi_2}p_3^{\pi_3}\ldots\ldots)$ into parcels $(\lambda_1^{l_1}\lambda_2^{l_2}\lambda_3^{l_3}\ldots\ldots)$, subject to the same restriction; hence

$$X_{\lambda_1}^{l_1} X_{\lambda_2}^{l_2} X_{\lambda_3}^{l_3} \ldots\ldots = \ldots\ldots + \theta\,(p_1^{\pi_1}p_2^{\pi_2}p_3^{\pi_3}\ldots\ldots)\, x_{s_1}^{\sigma_1} x_{s_2}^{\sigma_2} x_{s_3}^{\sigma_3} \ldots\ldots + \ldots\ldots,$$

the theorem to be demonstrated.

100. This proposition is cardinal in symmetrical algebra and of great importance; I hope, in a subsequent memoir in this Journal, to give another proof of it by means of differential operators.

Section 6.

The Formation of Symmetrical Tables.

101. One of the consequences of the theorem of reciprocity is the possibility of forming a pair of tables of symmetric functions, of a symmetrical character, in association with every partition, in positive, zero, and negative integers, of every number, positive, zero, or negative.

102. For, let the separations of the partition $(s_1^{\sigma_1}s_2^{\sigma_2}s_3^{\sigma_3}\ldots\ldots)$ possess in all r specifications which may be

$$\varkappa_1,\ \varkappa_2,\ \varkappa_3,\ \ldots\ldots\ \varkappa_r,$$

and let, moreover,

$$[X_{\varkappa_1}],\ [X_{\varkappa_2}],\ [X_{\varkappa_3}],\ \ldots\ldots\ [X_{\varkappa_r}]$$

denote the corresponding X-products, so that if

$$\varkappa_m = (\mu_1\mu_2\mu_3\ \ldots\ldots),$$
$$[X_{\varkappa_m}] = X_{\mu_1}X_{\mu_2}X_{\mu_3}\ \ldots\ldots$$

103. The law of reciprocity shows that if

$$X_{p_1}^{\pi_1} X_{p_2}^{\pi_2} X_{p_3}^{\pi_3} \ldots\ldots = \ldots\ldots + P x_{s_1}^{\sigma_1} x_{s_2}^{\sigma_2} x_{s_3}^{\sigma_3} \ldots\ldots + \ldots\ldots,$$

so that P consists of an assemblage of separations of the partition $(s_1^{\sigma_1}s_2^{\sigma_2}s_3^{\sigma_3}\ldots\ldots)$, each of which has the specification $(p_1^{\pi_1}p_2^{\pi_2}p_3^{\pi_3}\ldots\ldots)$ which is one of the series of specifications

$$\varkappa_1,\ \varkappa_2,\ \varkappa_3,\ \ldots\ldots\ \varkappa_r,$$

P on development will only give rise to symmetric functions which are symbolized by partitions included in the specification set $\varkappa_1, \varkappa_2, \varkappa_3, \ldots\ldots \varkappa_r$; for otherwise the law of reciprocity could not be true.

104. Now form X-products corresponding to all the specifications; let $P_{\varkappa_1}, P_{\varkappa_2}, P_{\varkappa_3}, \ldots\ldots P_{\varkappa_r}$ be the corresponding values of P, and further let $\theta_{t,m}$ be the numerical coefficient of $\varkappa_m x_{s_1}^{\sigma_1}x_{s_2}^{\sigma_2}x_{s_3}^{\sigma_3}\ldots$ in the development of $[X_{\varkappa_t}]$, or what is the same thing, the coefficient of $\varkappa_m$ in the development of $P_{\varkappa_t}$; thus,

$$X_{\varkappa_1} = \ldots\ldots + P_{\varkappa_1}x_{s_1}^{\sigma_1}x_{s_2}^{\sigma_2}x_{s_3}^{\sigma_3}\ldots\ldots + \ldots\ldots$$

$$= \ldots\ldots + (\theta_{1,1}\varkappa_1 + \theta_{1,2}\varkappa_2 + \theta_{1,3}\varkappa_3 + \ldots\ldots + \theta_{1,r}\varkappa_r)\,x_{s_1}^{\sigma_1}x_{s_2}^{\sigma_2}x_{s_3}^{\sigma_3}\ldots\ldots + \ldots\ldots,$$

$$X_{\varkappa_2} = \ldots\ldots + P_{\varkappa_2}x_{s_1}^{\sigma_1}x_{s_2}^{\sigma_2}x_{s_3}^{\sigma_3}\ldots\ldots + \ldots\ldots$$

$$= \ldots\ldots + (\theta_{21}\varkappa_1 + \theta_{22}\varkappa_2 + \theta_{23}\varkappa_3 + \ldots\ldots + \theta_{2r}\varkappa_r)\,x_{s_1}^{\sigma_1}x_{s_2}^{\sigma_2}x_{s_3}^{\sigma_3}\ldots\ldots + \ldots\ldots,$$

$$X_{\varkappa_3} = \ldots\ldots + P_{\varkappa_3}x_{s_1}^{\sigma_1}x_{s_2}^{\sigma_2}x_{s_3}^{\sigma_3}\ldots\ldots + \ldots\ldots$$

$$= \ldots\ldots + (\theta_{31}\varkappa_1 + \theta_{32}\varkappa_2 + \theta_{33}\varkappa_3 + \ldots\ldots + \theta_{3r}\varkappa_r)\,x_{s_1}^{\sigma_1}x_{s_2}^{\sigma_2}x_{s_3}^{\sigma_3}\ldots\ldots + \ldots\ldots,$$

. . . .

. . . .

. . . .

$$X_{\varkappa_r} = \ldots\ldots + P_{\varkappa_r}x_{s_1}^{\sigma_1}x_{s_2}^{\sigma_2}x_{s_3}^{\sigma_3}\ldots\ldots + \ldots\ldots$$

$$= \ldots\ldots + (\theta_{r1}\varkappa_1 + \theta_{r2}\varkappa_2 + \theta_{r3}\varkappa_3 + \ldots\ldots + \theta_{rr}\varkappa_r)\,x_{s_1}^{\sigma_1}x_{s_2}^{\sigma_2}x_{s_3}^{\sigma_3}\ldots\ldots + \ldots\ldots$$

105. This result shows that the assemblages of separations

$$P_{\varkappa_1},\ P_{\varkappa_2},\ P_{\varkappa_3},\ \ldots\ldots\ P_{\varkappa_r}$$

are linearly connected with the specifications

$$\varkappa_1,\ \varkappa_2,\ \varkappa_3,\ \ldots\ldots\ \varkappa_r,$$

and that we may form a table, viz.

	$\varkappa_1$	$\varkappa_\kappa$	$\varkappa_3$	....	$\varkappa_r$
$P_{\varkappa_1}$	θ_{11}	θ_{12}	θ_{13}	....	θ_{1r}
$P_{\varkappa_2}$	θ_{21}	θ_{22}	θ_{23}	....	θ_{2r}
$P_{\varkappa_3}$	θ_{31}	θ_{32}	θ_{33}	....	θ_{3r}
	...	...	...	....	...
$P_{\varkappa_r}$	θ_{r1}	θ_{r2}	θ_{r3}	....	θ_{rr}

13

and then by the law of reciprocity

$$\theta_{pq} = \theta_{qp},$$

and the table enjoys row and column (i. e. diagonal) symmetry.

106. We may similarly invert the table and express $x_1, x_2, x_3, \ldots x_r$ as linear functions of $P_{\kappa_1}, P_{\kappa_2}, P_{\kappa_3}, \ldots P_{\kappa_r}$ in a table enjoying the same symmetry.

107. To make the meaning clear, omit in the first instance all partitions which contain zero or negative parts, and write down a complete system of X-products for any given weight, as follows, e. g. weight $= 5$:

	5 X_5	41 X_4X_1	32 X_3X_2	31^2 $X_3X_1^2$	2^21 $X_2^2X_1$	21^3 $X_2X_1^3$	1^5 X_1^5
x_5	(5)						
x_4x_1	(41)	$(4)(1)$					
x_3x_2	(32)		$(3)(2)$				
$x_3x_1^2$	(31^2)	$(31)(1)$	$(3)(1^2)$	$(3)(1)^2$			
$x_2^2x_1$	(2^21)	$(2^2)(1)$	$(2)(21)$		$(2)^2(1)$		
$x_2x_1^3$	(21^3)	$(21^2)(1)$	$(2)(1^3)+(21)(1^2)$	$(21)(1)^2$	$2\,(2)(1^2)(1)$	$(2)(1)^3$	
x_1^5	(1^5)	$(1^4)(1)$	$(1^3)(1^2)$	$(1^3)(1)^2$	$(1^2)^2(1)$	$(1^2)(1)^3$	$(1)^5$

here each line is a set of "assemblages of separations," each assemblage having its own specification, as appearing by the top line. The assemblages and specifications represent symmetric functions, and the theorem is that these symmetric functions are linearly connected, the coefficients being symmetrical in regard to a diagonal. Thus, from the last line but one we have the assemblages (separations of (21^3))

$$(21^3),\ (21^2)(1),\ (2)(1^3)+(21)(1^2),\ (21)(1)^2,\ 2\,(2)(1^2)(1),\ (2)(1)^3$$

linearly connected with the specifications

$$(5),\quad (41),\quad (32),\quad (31^2),\quad (2^21),\quad (21^3).$$

108. Again, let us take the weight -2 and the separable partition $(0^2\bar{1}^2)$; the corresponding portion of the table of X-products is

	$(\bar{2})$	$(0\bar{2})$	$(0^2\bar{2})$	$(\bar{1}^2)$	$(0\bar{1}^2)$	$(0^2\bar{1}^2)$
	X_{-2}	X_0X_{-2}	$X_0^2X_{-2}$	X_{-1}^2	$X_0X_{-1}^2$	$X_0^2X_{-1}^2$
$x_0^2x_{-1}^2$	$(0^2\bar{1}^2)$	$(0)(0\bar{1}^2)+(0^2)(\bar{1}^2)$	$(0)^2(\bar{1}^2)$	$2(0^2\bar{1})(\bar{1})+(0\bar{1})^2$	$(0^2)(\bar{1})^2+2(0)(0\bar{1})(\bar{1})$	$(0)^2(\bar{1})^2$

,

showing that we have the assemblages (separations of $(0^2\bar{1}^2)$) indicated in the bottom line, linearly connected with the specifications shown in the top line.

109. Writing down the assemblages in a vertical column and the specifications in a horizontal row, we may then form a table which calculation shows to be

	$(\bar{2})$	$(0\bar{2})$	$(0^2\bar{2})$	$(\bar{1}^2)$	$(0\bar{1}^2)$	$(0^2\bar{1}^2)$
$(0^2\bar{1}^2)$						1
$(0)(0\bar{1}^2)+(0^2)(\bar{1}^2)$				1	5	3
$(0)^2(\bar{1}^2)$				4	5	2
$2(0^2\bar{1})(\bar{1})+(0\bar{1})^2$		1	4	2	10	8
$(0^2)(\bar{1})^2+2(0)(0\bar{1})(\bar{1})$		5	5	10	20	10
$(0)^2(\bar{1})^2$	1	3	2	8	10	4

where the third line is to be read

$$(0)^2(\bar{1}^2) = 4(\bar{1}^2) + 5(0\bar{1}^2) + 2(0^2\bar{1}^2),$$

or in an algebraic form

$$\left(\sum \alpha^0\right)^2 \sum \alpha^{-1}\beta^{-1} = 4\sum \alpha^{-1}\beta^{-1} + 5\sum \alpha^0\beta^{-1}\gamma^{-1} + 2\sum \alpha^0\beta^0\gamma^{-1}\delta^{-1},$$

verified through the medium of the identity

$$n^2 \qquad = 4 \qquad + 5(n-2) \qquad + 2.\tfrac{1}{2}(n-2)(n-3).$$

110. The table already given possesses diagonal symmetry as a direct consequence of the law of reciprocity; the inverse table, which expresses the specifi-

cations as linear functions of the assemblages of separations, necessarily enjoys the same symmetry. Its form is

	$(0^2\bar{1}^2)$	$(0)(0\bar{1}^2)+(0^2)(\bar{1}^2)$	$(0)^2(\bar{1}^2)$	$2(0^2\bar{1})(\bar{1})+(0\bar{1})^2$	$(0^2)(\bar{1})^2+2(0)(0\bar{1})(\bar{1})$	$(0)^2(\bar{1})^2$
$(\bar{2})$	$-\frac{2}{3}$	$\frac{2}{3}$	$-\frac{2}{3}$	$\frac{1}{3}$	$-\frac{2}{3}$	1
$(0\bar{2})$	$\frac{2}{3}$	$\frac{2}{15}$	$-\frac{8}{15}$	$-\frac{1}{3}$	$\frac{4}{15}$	
$(0^2\bar{2})$	$-\frac{2}{3}$	$-\frac{8}{15}$	$\frac{2}{15}$	$\frac{1}{3}$	$-\frac{1}{15}$	
$(\bar{1}^2)$	$\frac{1}{3}$	$-\frac{1}{3}$	$\frac{1}{3}$			
$(0\bar{1}^2)$	$-\frac{2}{3}$	$\frac{4}{15}$	$-\frac{1}{15}$			
$(0^2\bar{1}^2)$	1					

111. It has thus been demonstrated that a pair of symmetrical tables exist in the case of every partition into positive, zero, and negative integers of every number positive, zero, or negative.

112. The theorem in regard to the coefficients in a group, given on page 35 of the former memoir, is extended easily to this enlarged theory, and we may enunciate as follows:

113. *Theorem.* "In the expression of symmetric function

$$(p_1^{\pi_1}p_2^{\pi_2}p_3^{\pi_3}\ldots.)$$

by means of separations of

$$(s_1^{\sigma_1}s_2^{\sigma_2}s_3^{\sigma_3}\ldots.),$$

where the parts of the partitions are positive, zero, or negative, the algebraic sum of the coefficients in each group will be zero if the partition

$$(p_1^{\pi_1}p_2^{\pi_2}p_3^{\pi_3}\ldots.)$$

possesses no separation of specification

$$(\sigma_1 s_1,\ s_2\sigma_2,\ s_3\sigma_3,\ \ldots\ldots).$$"

114. This theorem may be verified in the second of the tables above given in the cases of $(\bar{2})$ and $(\bar{1}^2)$ only, as all the other symmetric functions in the left-hand vertical column possess separations of specification $(0\bar{2})$. Ex. gr.

$$\begin{aligned}(\bar{2}) = {} & (0)^2(\bar{1})^2 - \tfrac{2}{3}(0^2)(\bar{1})^2 - \tfrac{2}{3}(0)^2(\bar{1}^2) + \tfrac{2}{3}(0^2)(\bar{1}^2) \\ & - \tfrac{4}{3}(0)(0\bar{1})(\bar{1}) + \tfrac{2}{3}(0^2\bar{1})(\bar{1}) + \tfrac{2}{3}(0)(0\bar{1}^2) - \tfrac{2}{3}(0^2\bar{1}^2) \\ & + \tfrac{1}{3}(0\bar{1})^2 \\ & \pm\tfrac{4}{3} \qquad \pm\tfrac{2}{3} \qquad \pm\tfrac{2}{3} \qquad \pm\tfrac{2}{3}\end{aligned}$$

$$\begin{aligned}(\bar{1}^2) = {} & \tfrac{1}{3}(0)^2(\bar{1}^2) - \tfrac{1}{3}(0^2)(\bar{1}^2) \\ & - \tfrac{1}{3}(0)(0\bar{1}^2) + \tfrac{1}{3}(0^2\bar{1}^2) \\ & \pm\tfrac{1}{3} \qquad \pm\tfrac{1}{3}.\end{aligned}$$

SECTION 7.

The Law of Expressibility.

115. The law given on page 6 of former memoir may now be extended as follows:

116. *Theorem.* "If a symmetric function be symbolized by

$$(\lambda\mu\nu\ \ldots\ldots)$$

the parts λ, μ, ν, ... being positive, zero, or negative, and

$(\lambda_1\lambda_2\lambda_3\ \ldots\ldots)$ be any partition of λ,
$(\mu_1\mu_2\mu_3\ \ldots\ldots)$ " " " μ,
$(\nu_1\nu_2\nu_3\ \ldots\ldots)$ " " " ν,
. . . .

the symmetric function $(\lambda\mu\nu\ \ldots\ldots)$

is expressible as a linear function of separations of

$$\lambda_1\lambda_2\lambda_3\ \ldots\ldots\ \mu_1\mu_2\mu_3\ \ldots\ \nu_1\nu_2\nu_3\ \ldots\ldots).$$"

117. As an example of this, we may express the function (0^3) as a linear

function of separations of (0^4); it will be interesting to give, as well, the complete tables of separations of (0^4) which includes this result.

	(0)	(0^2)	(0^3)	(0^4)
$(0)^4$	1	14	36	24
$3(0^2)(0)^2$		12	45	36
$(0^2)^2 + 2(0^3)(0)$		1	12	14
(0^4)				1

	$(0)^4$	$3(0^2)(0)^2$	$(0^2)^2 + 2(0^3)(0)$	(0^4)
(0)	1	$-\frac{4}{3}$	2	-4
(0^2)		$\frac{4}{33}$	$-\frac{5}{11}$	2
(0^3)		$-\frac{1}{99}$	$\frac{4}{33}$	$-\frac{4}{3}$
(0^4)				1

from which

$$(0^2) = \tfrac{4}{33} \cdot 3(0^2)(0)^2 - \tfrac{5}{11}\{(0^2)^2 + 2(0^3)(0)\} + 2(0^4),$$

and this merely exhibits a relation connecting the second, third, fourth and fifth binomial coefficients in the expansion of $(1 + x)^n$; for

$$(1 + x)^n = 1 + (0)x + (0^2)x^2 + (0^3)x^3 + (0^4)x^4 + \ldots + (0^n)x^n.$$

118. The subject of "Expansion by factorials," which is usually discussed in works on Finite Differences, is thus clearly within the domain of this theory, and the two tables last given might have been expressed by the notation and symbols of the calculus of Finite Differences.

On this subject I hope to say more upon a future occasion.

119. It is, in conclusion, to be particularly observed that all algebra is expressible by means of factorials, and thus any algebraical expression whatever, of a finite nature, may be exhibited as a symmetric function of one or more sets of quantities.

ROYAL ARSENAL, WOOLWICH, ENGLAND, December 1st, 1888.

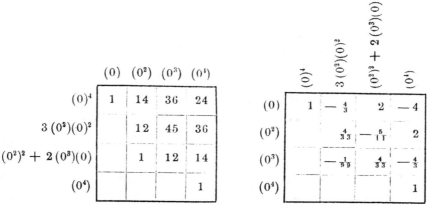

Third Memoir on a New Theory of Symmetric Functions.

BY MAJOR P. A. MACMAHON, R. A.

In this memoir I carry on the development of the Theory of Separations. It is divided into seven sections (8 to 14), and is numbered continuously with the second memoir. It brings the theory up to the point where modes of calculating tables of separations may be advantageously discussed. Sections 8 and 9 lay down the fundamental laws of operation which in sections 10, 11 and 12 are applied to the deduction of some comprehensive theorems of algebraic symmetry. Section 13 is concerned with the multiplication of symmetric functions. Section 14 commences the application of the operators to the functions which appear in a table of separations and establishes the theorem which is preliminary to further researches which may possibly appear in a future number of this journal.

I desire to draw attention to the fundamental theorem in operations of which a statement merely is given in Art. 140 of Section 9. It is a generalization of a theorem of Sylvester, to be found in the Philosophical Magazine, 1877, under the title, "A generalization of Taylor's theorem."

§8.

The Differential Operators.

120. I purpose to keep in view throughout the following investigation the analogy between quantity and operation which was pointed out by Hammond for the first time in Vol. XIII of the Proceedings of the London Mathematical Society.

I present the analogy in an extended form and from two distinct points of view.

Supposing n to be indefinitely great, I write down two relations, viz:

$$1 + a_1x + a_2x^2 + a_3x^3 + \dots + a_nx^n$$
$$= (1 + \alpha_1x)(1 + \alpha_2x)(1 + \alpha_3x)\dots(1 + \alpha_nx),$$
$$1 + a_{-1}\frac{1}{x} + a_{-2}\frac{1}{x^2} + a_{-3}\frac{1}{x^3} + \dots + a_{-n}\frac{1}{x^n}$$
$$= \left(1 + \frac{1}{\alpha_1x}\right)\left(1 + \frac{1}{\alpha_2x}\right)\left(1 + \frac{1}{\alpha_3x}\right)\dots\left(1 + \frac{1}{\alpha_nx}\right),$$

which lead to the algebraic theory of the expression of any rational function of the quantities

$$\alpha_1, \alpha_2, \alpha_3, \dots \alpha_n$$

as a rational *integral* function of the number n and of the quantities

$$a_1, a_{-1}, a_2, a_{-2}, a_3, a_{-3}, \dots a_n, a_{-n}.$$

121. By multiplying the two equations together and subsequently multiplying each side by exp n, we obtain

$$e^n\left(1 + a_1x + a_2x^2 + a_3x^3 + \dots\right)\left(1 + a_{-1}\frac{1}{x} + a_{-2}\frac{1}{x^2} + a_{-3}\frac{1}{x^3} + \dots\right)$$
$$= e^n\left(1 + \alpha_1x\right)\left(1 + \alpha_2x\right)\left(1 + \alpha_3x\right)\dots\left(1 + \frac{1}{\alpha_1x}\right)\left(1 + \frac{1}{\alpha_2x}\right)\left(1 + \frac{1}{\alpha_3x}\right)\dots$$

Observe that the right-hand side is

$$e^n\{1 + (1)x + (1^2)x^2 + (1^3)x^3 + \dots\}\left\{1 + (\bar{1})\frac{1}{x} + (\bar{1}^2)\frac{1}{x^2} + (\bar{1}^3)\frac{1}{x^3} + \dots\right\},$$

that is,

$$e^n\Big[1 + (1)x + (\bar{1})\frac{1}{x} + (1^2)x^2 + (1)(\bar{1}) + (\bar{1}^2)\frac{1}{x^2} + (1^3)x^3 + (1^2)(\bar{1})x$$
$$+ (1)(\bar{1}^2)\frac{1}{x} + (\bar{1}^3)\frac{1}{x^3} + \dots\Big],$$

and this is

$$e^n\Big\{1 + (1)x + (\bar{1})\frac{1}{x} + (1^2)x^2 + (0) + (1\bar{1}) + (\bar{1}^2)\frac{1}{x^2}$$
$$+ (1^3)x^3 + (10)x + (1^2\bar{1})x + (0\bar{1})\frac{1}{x} + (1\bar{1}^2)\frac{1}{x} + (\bar{1}^3)\frac{1}{x^3} + \dots\Big\}$$
$$= e^n\sum(1^\kappa 0^\lambda \bar{1}^\mu)x^{\kappa-\mu},$$

the summation being for all zero and positive integer values of κ, λ and μ.

But
$$\sum(1^\kappa 0^\lambda \bar{1}^\mu)x^{\kappa-\mu} = \exp\left\{(0) + (1)x + (\bar{1})\frac{1}{x}\right\}$$

symbolically, where

$$\frac{(1)^{\kappa}(0)^{\lambda}(\bar{1})^{\mu}}{\kappa!\ \lambda!\ \mu!} \text{ is the symbolic expression for } (1^{\kappa}0^{\lambda}\bar{1}^{\mu}).$$

Hence we may write

$$e^n(1 + a_1x + a_2x^2 + \dots)\left(1 + a_{-1}\frac{1}{x} + a_{-2}\frac{1}{x^2} + \dots\right)$$
$$= e^n \exp\left\{(0) + (1)x + (\bar{1})\frac{1}{x}\right\} \text{ symbolically.}$$

Returning to the former identity and taking logarithms, we find

$$n + \log(1 + a_1x + a_2x^2 + \dots) + \log\left(1 + a_{-1}\frac{1}{x} + a_{-2}\frac{1}{x^2} + \dots\right)$$
$$= (0) + (1)x - \tfrac{1}{2}(2)x^2 + \tfrac{1}{3}(3)x^3 - \dots$$
$$+ (\bar{1})\frac{1}{x} - \tfrac{1}{2}(\bar{2})\frac{1}{x^2} + \tfrac{1}{3}(\bar{3})\frac{1}{x^3} - \dots,$$

an identity which indicates that any symmetric function which can be exhibited by means of partitions composed of positive, zero and negative integers, is uniquely expressible by the use of the quantities

$$n,\ a_1,\ a_{-1},\ a_2,\ a_{-2},\ \dots;$$

it further gives the law of such expression of the sums of the powers of the quantities $\alpha_1, \alpha_2, \alpha_3, \dots$

122. We are now led to the result

$$e^n \exp\left\{(0) + (1)x + (\bar{1})\frac{1}{x}\right\} \text{ symbolic}$$
$$= \exp\left\{\begin{array}{l}(0) + (1)x - \tfrac{1}{2}(2)x^2 + \tfrac{1}{3}(3)x^3 - \dots \\ \quad + (\bar{1})\frac{1}{x} - \tfrac{1}{2}(\bar{2})\frac{1}{x^2} + \tfrac{1}{3}(\bar{3})\frac{1}{x^3} - \dots,\end{array}\right\}$$

which gives

$$\exp\left\{(0) + (1)x + (\bar{1})\frac{1}{x}\right\} \text{ symbolic}$$
$$= \exp\left\{\begin{array}{l}(1)x - \tfrac{1}{2}(2)x^2 + \tfrac{1}{3}(3)x^3 - \dots \\ + (\bar{1})\frac{1}{x} - \tfrac{1}{2}(\bar{2})\frac{1}{x^2} + \tfrac{1}{3}(\bar{3})\frac{1}{x^3} - \dots\end{array}\right\}$$

a formula of great importance, as will appear presently.

123. Write now

$$\begin{aligned}&1+A_0+A_1x \quad + A_2x^2 \quad + \ldots . \\ &\qquad + A_{-1}\frac{1}{x} + A_{-2}\frac{1}{x^2} + \ldots . \\ &= \exp\left\{ \begin{aligned}&(0)+(1)x - \tfrac{1}{2}(2)x^2 + \tfrac{1}{3}(3)x^3 - \ldots . \\ &\quad + (\bar{1})\frac{1}{x} - \tfrac{1}{2}(\bar{2})\frac{1}{x^2} + \tfrac{1}{3}(\bar{3})\frac{1}{x^3} - \ldots .\end{aligned}\right\}\end{aligned}$$

so that

$$\begin{aligned}1+A_0 &= e^n(1 \quad + a_1a_{-1} + a_2a_{-2} + a_3a_{-3} + \ldots .), \\ A_1 &= e^n(a_1 \quad + a_2a_{-1} + a_3a_{-2} + \ldots .), \\ A_{-1} &= e^n(a_{-1} + a_1a_{-2} + a_2a_{-3} + \ldots .), \\ A_2 &= e^n(a_2 \quad + a_3a_{-1} + a_4a_{-2} + \ldots .), \\ A_{-2} &= e^n(a_{-2} + a_1a_{-3} + a_2a_{-4} + \ldots .), \\ &\ldots\ldots\ldots\ldots\ldots\ldots\ldots\end{aligned}$$

124. The quantities

$$A_0,\ A_1,\ A_{-1},\ A_2,\ A_{-2},\ \ldots .,$$

now, I believe, introduced into analysis for the first time, are of great and fundamental importance in the theory; subsequently they will be freely adopted as arguments of the symmetric functions, and it will appear that, for the purposes of analysis, they are the proper functions to consider.

125. As I wish to show the complete correspondence which exists between the algebraic theory and the theory of the related partial differential operations, it is necessary to write down some obvious results which, however, are in correspondence with theorems in operations which are by no means obvious.

Thus the last written identity, through expansion of the exponential function and comparison of coefficients, leads to the manifest conclusions:

$$\begin{aligned}1+A_0 &= e^{(0)}\left\{1+(1)(\bar{1}) + \frac{(1)^2-(2)}{2!}\cdot\frac{(\bar{1})^2-(\bar{2})}{2!}\right. \\ &\qquad \left. + \frac{(1)^3-3(2)(1)+2(3)}{3!}\cdot\frac{(\bar{1})^3-3(\bar{2})(\bar{1})+2(\bar{3})}{3!} + \ldots .\right\}, \\ A_1 &= e^{(0)}\left\{(1) + \frac{(1)^2-(2)}{2!}(\bar{1}) + \frac{(1)^3-3(2)(1)+2(3)}{3!}\cdot\frac{(\bar{1})^2-(\bar{2})}{2!} + \ldots .\right\}, \\ A_{-1} &= e^{(0)}\left\{(\bar{1}) + \frac{(\bar{1})^2-(\bar{2})}{2!}(1) + \frac{(\bar{1})^3-3(\bar{2})(\bar{1})+2(\bar{3})}{3!}\cdot\frac{(1)^2-(2)}{2!} + \ldots .\right\}, \\ &\ldots\ldots\ldots\ldots\ldots\ldots\ldots\end{aligned}$$

relations which also are simply reached from the expressions of A_0, A_1, A_{-1}, ... in terms of n, a_1, a_{-1}, a_2, a_{-2} .... by expressing these last quantities in terms of the sums of the powers.

126. Again, from the identity,

$$\begin{aligned}1 + A_0 + A_1x \quad &+ A_2x^2 \quad + A_3x^3 \quad + \dots \\ + A_{-1}\frac{1}{x} &+ A_{-2}\frac{1}{x^2} + A_{-3}\frac{1}{x^3} + \dots \\ = \exp\Big\{(0) + (1)x &- \tfrac{1}{2}(2)x^2 + (3)\tfrac{1}{3}x^3 - \dots \\ + (\bar{1})\frac{1}{x} &- \tfrac{1}{2}(\bar{2})\frac{1}{x^2} + \tfrac{1}{3}(\bar{3})\frac{1}{x^3} - \dots\Big\}\end{aligned}$$

we obtain by taking logarithms,

$$\begin{aligned}(0) + (1)x \quad &- \tfrac{1}{2}(2)x^2 + \tfrac{1}{3}(3)x^3 - \dots \\ + (\bar{1})\frac{1}{x} &- \tfrac{1}{2}(\bar{2})\frac{1}{x^2} + \tfrac{1}{3}(\bar{3})\frac{1}{x^3} - \dots \\ = \log\Big(1 + A_0 + A_1x \quad &+ A_2x^2 \quad + A_3x^3 \quad + \dots \\ + A_{-1}\frac{1}{x} &+ A_{-2}\frac{1}{x^2} + A_{-3}\frac{1}{x^3} + \dots\Big)\end{aligned}$$

and on expanding and comparing coefficients of like powers of x, we have

$$\begin{aligned}(0) = \log(1 + A_0) - \frac{A_1A_{-1}}{(1+A_0)^2} &- \frac{A_2A_{-2}}{(1+A_0)^2} + \frac{A_1^2A_{-2}}{(1+A_0)^3} \\ &- \frac{A_3A_{-3}}{(1+A_0)^2} + 2\frac{A_2A_1A_{-3}}{(1+A_0)^3} \\ - \frac{A_1^3A_{-3}}{(1+A_0)^4} - \dots &+ \frac{A_2A_{-1}^2}{(1+A_0)^3} - \frac{3}{2}\frac{A_1^2A_{-1}^2}{(1+A_0)^4} \\ + 2\frac{A_3A_{-1}A_{-2}}{(1+A_0)^3} &- 6\frac{A_2A_1A_{-1}A_{-2}}{(1+A_0)^4} + 4\frac{A_1^3A_{-1}A_{-2}}{(1+A_0)^5} + \dots \\ - \frac{A_3A_{-1}^3}{(1+A_0)^4} &+ 4\frac{A_2A_1A_{-1}^3}{(1+A_0)^5} - \frac{10}{3}\frac{A_1^3A_{-1}^3}{(1+A_0)^6} - \dots \\ = \sum \frac{(-)^{\Sigma\pi-1}(\Sigma\pi - 1)!}{\dots \pi_3!\,\pi_2!\,\pi_1!\,\pi_0!\,\pi_{-1}!\,\pi_{-2}!\,\pi_{-3}!\dots} &\dots A_3^{\pi_3}A_2^{\pi_2}A_1^{\pi_1}A_0^{\pi_0}A_{-1}^{\pi_{-1}}A_{-2}^{\pi_{-2}}A_{-3}^{\pi_{-3}}\dots,\end{aligned}$$

the summation being in regard to every solution of the indeterminate equation

$$\Sigma t\pi_t = 0$$

in positive, zero and negative integers.

127. In general,

$$\frac{(-)^{m-1}}{m}(m) = \sum \frac{(-)^{\Sigma\pi-1}(\Sigma\pi-1)!}{\ldots\pi_2!\,\pi_1!\,\pi_0!\,\pi_{-1}!\,\pi_{-2}!\ldots} \ldots A_2^{\pi_2}A_1^{\pi_1}A_0^{\pi_0}A_{-1}^{\pi_{-1}}A_{-2}^{\pi_{-2}}\ldots,$$

$$\frac{(-)^{m-1}}{m}(\overline{m}) = \sum \frac{(-)^{\Sigma\pi-1}(\Sigma\pi-1)!}{\ldots\pi_2!\,\pi_1!\,\pi_0!\,\pi_{-1}!\,\pi_{-2}!\ldots} \ldots A_2^{\pi_2}A_1^{\pi_1}A_0^{\pi_0}A_{-1}^{\pi_{-1}}A_{-2}^{\pi_{-2}}\ldots,$$

the summations being in regard to the solutions of the indeterminate equations

$$\Sigma t\pi_t = m,$$
$$\Sigma t\pi_t = -m$$

respectively, in positive, zero and negative integers.

§9.

128. We may express symmetric functions, of the nature here considered, as functions either of the quantities

$$n,\ a_1,\ a_{-1},\ a_2,\ a_{-2},\ \ldots.$$

or of the quantities

$$A_0,\ A_1,\ A_{-1},\ A_2,\ A_{-2},\ \ldots,$$

where observe that the latter set of arguments does not involve the number n explicitly. Both sets of quantities are defined by the identities

$$e^n(1 + a_1x + a_2x^2 + a_3x^3 + \ldots.)\left(1 + a_{-1}\frac{1}{x} + a_{-2}\frac{1}{x^2} + a_{-3}\frac{1}{x^3} + \ldots.\right)$$
$$= 1 + A_0 + A_1x \quad + A_2x^2 \quad + A_3x^3 \quad + \ldots.$$
$$+ A_{-1}\frac{1}{x} + A_{-2}\frac{1}{x^2} + A_{-3}\frac{1}{x^3} + \ldots.$$
$$= e^n(1+\alpha_1x)(1+\alpha_2x)(1+\alpha_3x)\ldots.\left(1+\frac{1}{\alpha_1x}\right)\left(1+\frac{1}{\alpha_2x}\right)\left(1+\frac{1}{\alpha_3x}\right)\ldots.$$

129. Let us now introduce a quantity μ in addition to the n quantities

$$\alpha_1,\ \alpha_2,\ \alpha_3,\ \ldots.\ \alpha_n.$$

We thus obtain the identities

$$e^{n+1}(1+a_1x+a_2x^2+a_3x^3+\ldots)\left(1+a_{-1}\frac{1}{x}+a_{-2}\frac{1}{x^2}+a_{-3}\frac{1}{x^3}+\ldots\right)(1+\mu x)\left(1+\frac{1}{\mu x}\right)$$
$$=\begin{pmatrix}1 + A_0 + A_1x \quad + A_2x^2 \quad + \ldots. \\ + A_{-1}\frac{1}{x} + A_{-2}\frac{1}{x^2} + \ldots.\end{pmatrix} e(1+\mu x)\left(1+\frac{1}{\mu x}\right)$$
$$= e^{n+1}(1+\alpha_1x)(1+\alpha_2x)\ldots.(1+\mu x)\left(1+\frac{1}{\alpha_1x}\right)\left(1+\frac{1}{\alpha_2x}\right)\ldots.\left(1+\frac{1}{\mu x}\right),$$

for it is merely necessary to multiply throughout by the factor

$$e(1+\mu x)\left(1+\frac{1}{\mu x}\right).$$

130. The identities may be written

$$e^{n+1}\{1+(a_1+\mu)x+(a_2+a_1\mu)x^2+(a_3+a_2\mu)x^3+\ldots\}$$
$$\times\left\{1+\left(a_{-1}+\frac{1}{\mu}\right)\frac{1}{x}+\left(a_{-2}+a_{-1}\frac{1}{\mu}\right)\frac{1}{x^2}+\left(a_{-3}+a_{-2}\frac{1}{\mu}\right)\frac{1}{x^3}+\ldots\right\}$$
$$\begin{aligned}
={}&1+A_0+(2e-1)(1+A_0)+eA_{-1}\mu+eA_1\frac{1}{\mu}\\
&+A_1+(2e-1)A_1+e(1+A_0)\mu+eA_2\frac{1}{\mu}\\
&+A_{-1}+(2e-1)A_{-1}+eA_{-2}\mu+e(1+A_0)\frac{1}{\mu}\\
&+A_2+(2e-1)A_2+eA_1\mu+eA_3\frac{1}{\mu}\\
&+A_{-2}+(2e-1)A_{-2}+eA_{-3}\mu+eA_{-1}\frac{1}{\mu}\\
&+\ldots\ldots\ldots\ldots\ldots\ldots\ldots\ldots
\end{aligned}$$
$$=e^{n+1}(1+a_1x)(1+a_2x)\ldots(1+\mu x)\left(1+\frac{1}{a_1x}\right)\left(1+\frac{1}{a_2x}\right)\ldots\left(1+\frac{1}{\mu x}\right),$$

from which it appears that the introduction of the new quantity μ has the effect of changing the quantities

$$\ldots a_{-3},\quad a_{-2},\quad a_{-1},\quad n,\quad a_1,\quad a_2,\quad a_3,\ldots$$

into

$$a_{-3}+a_{-2}\frac{1}{\mu},\quad a_{-2}+a_{-1}\frac{1}{\mu},\quad a_{-1}+\frac{1}{\mu},\quad n+1,\quad a_1+\mu,\quad a_2+a_1\mu,\quad a_3+a_2\mu,\ldots$$

respectively, and moreover changes

$$\begin{aligned}
&A_0 \text{ into } A_0+(2e-1)(1+A_0)+eA_{-1}\mu+eA_1\frac{1}{\mu},\\
&A_1 \text{ into } A_1+(2e-1)A_1+e(1+A_0)\mu+eA_2\frac{1}{\mu},\\
&A_{-1} \text{ into } A_{-1}+(2e-1)A_{-1}+eA_{-2}\mu+e(1+A_0)\frac{1}{\mu},\\
&A_2 \text{ into } A_2+(2e-1)A_2+eA_1\mu+eA_3\frac{1}{\mu},\\
&A_{-2} \text{ into } A_{-2}+(2e-1)A_{-2}+eA_{-3}\mu+eA_{-1}\frac{1}{\mu},\\
&\ldots\ldots\ldots\ldots\ldots\ldots\ldots\ldots
\end{aligned}$$

131. Hence if any symmetric function be

$$\phi\{n, a_1, a_{-1}, a_2, a_{-2}, a_3, a_{-3}, \ldots\} = \psi\{A_0, A_1, A_{-1}, A_2, A_{-2}, \ldots\}$$
$$= \phi = \psi,$$

the altered value will be

$$\phi\left\{n+1, a_1+\mu, a_{-1}+\frac{1}{\mu}, a_2+a_1\mu, a_{-2}+a_{-1}\frac{1}{\mu}, \ldots\right\},$$

which by Taylor's theorem is symbolically

$$\exp\left(d_0+d_1\mu+d_{-1}\frac{1}{\mu}\right)\phi,$$

where

$$\begin{aligned} d_0 &= \partial_n, \\ d_1 &= \partial_{a_1} + a_1\partial_{a_2} + a_2\partial_{a_3} + \ldots, \\ d_{-1} &= \partial_{a_{-1}} + a_{-1}\partial_{a_{-2}} + a_{-2}\partial_{a_{-3}} + \ldots, \end{aligned}$$

and the multiplication of operators is symbolic, and will also be

$$\begin{aligned} \psi\Big\{&A_0 + (2e-1)(1+A_0) + eA_{-1}\mu + eA_1\frac{1}{\mu}, \\ &A_1 + (2e-1)A_1 + e(1+A_0)\mu + eA_2\frac{1}{\mu}, \\ &A_{-1} + (2e-1)A_{-1} + eA_{-2}\mu + e(1+A_0)\frac{1}{\mu}, \ldots\Big\} \end{aligned}$$

which by Taylor's theorem is symbolically

$$\exp\left\{e\left(g_0+g_1\mu+g_{-1}\frac{1}{\mu}\right)+(e-1)g_0\right\}.\psi,$$

where

$$\begin{aligned} g_0 &= (1+A_0)\partial_{A_0} + A_1\partial_{A_1} + A_2\partial_{A_2} + \ldots \\ &\quad + A_{-1}\partial_{A_{-1}} + A_{-2}\partial_{A_{-2}} + \ldots, \\ g_1 &= (1+A_0)\partial_{A_1} + A_1\partial_{A_2} + A_2\partial_{A_3} + \ldots \\ &\quad + A_{-1}\partial_{A_0} + A_{-2}\partial_{A_{-1}} + \ldots, \\ g_{-1} &= (1+A_0)\partial_{A_{-1}} + A_1\partial_{A_0} + A_2\partial_{A_1} + \ldots \\ &\quad + A_{-1}\partial_{A_{-2}} + A_{-2}\partial_{A_{-3}} + \ldots, \end{aligned}$$

and the multiplication of operators is symbolic.

132. If, moreover, the symmetric function be

$$(pq\ldots o^{\omega}\bar{r}\bar{s}\ldots) = \phi = \psi,$$

the introduction of the new quantity μ results in the addition to $(pq\ldots o^{\omega}\bar{r}\bar{s}\ldots)$ of the new terms

$$(q\ldots o^{\omega}\bar{r}\bar{s}\ldots)\mu^p + (p\ldots o^{\omega}\bar{r}\bar{s}\ldots)\mu^q + \ldots$$
$$+ (pq\ldots o^{\omega-1}\bar{r}\bar{s})\mu^o + (pq\ldots o^{\omega}\bar{s}\ldots)\frac{1}{\mu^r} + (pq\ldots o^{\omega}\bar{r}\ldots)\frac{1}{\mu^s} + \ldots,$$

so that we are face to face with the identities

$$(pq \ldots . o^{\omega}\overline{rs} \ldots .) + (pq \ldots . o^{\omega-1}\overline{rs})\,\mu^{o}$$
$$+ (q \ldots o^{\omega}\overline{rs} \ldots .)\,\mu^{p} \quad + (p \ldots . o^{\omega}\overline{rs} \ldots .)\,\mu^{q} \; + \ldots .$$
$$+ (pq \ldots . o^{\omega}\overline{s} \ldots .)\,\frac{1}{\mu^{r}} + (pq \ldots . o^{\omega}\overline{r} \ldots .)\,\frac{1}{\mu^{s}} + \ldots .$$
$$= \exp\left(d_0 + d_1\mu + d_{-1}\frac{1}{\mu}\right).\phi$$
$$= \exp\left\{e\left(g_0 + g_1\mu + g_{-1}\frac{1}{\mu}\right) + (e-1)\,g_0\right\}.\psi.$$

133. We now write

$$\exp\left(d_0 + d_1\mu + d_{-1}\frac{1}{\mu}\right) = 1 + D_0\mu^0 + D_1\mu \quad + D_2\mu^2 \quad + \ldots$$
$$+ D_{-1}\frac{1}{\mu} + D_{-2}\frac{1}{\mu^2} + \ldots ,$$
$$\exp\left\{e\left(g_0 + g_1\mu + g_{-1}\frac{1}{\mu}\right) + (e-1)\,g_0\right\} = 1 + G_0\mu^0 + G_1\mu \quad + G_2\mu^2 \quad + \ldots$$
$$+ G_{-1}\frac{1}{\mu} + G_{-2}\frac{1}{\mu^2} + \ldots ,$$

the expansions being of course symbolic, and we find on equating coefficients of like powers of μ,

$$D_0\,(pq \ldots . o^{\omega}\overline{rs} \ldots .) = G_o\,(pq \ldots . o^{\omega}\overline{rs} \ldots .) = (pq \ldots . o^{\omega-1}\overline{rs} \ldots .),$$
$$D_p\,(pq \ldots . o^{\omega}\overline{rs} \ldots .) = G_p\,(pq \ldots . o^{\omega}\overline{rs} \ldots .) = (q \quad \ldots . o^{\omega} \quad \overline{rs} \ldots .),$$
$$D_q\,(pq \ldots . o^{\omega}\overline{rs} \ldots .) = G_q\,(pq \ldots . o^{\omega}\overline{rs} \ldots .) = (p \quad \ldots . o^{\omega} \quad \overline{rs} \ldots .),$$
$$D_{-r}(pq \ldots . o^{\omega}\overline{rs} \ldots .) = G_{-r}(pq \ldots . o^{\omega}\overline{rs} \ldots .) = (pq \ldots . o^{\omega} \quad \overline{s} \ldots .),$$
$$D_{-s}(pq \ldots . o^{\omega}\overline{rs} \ldots .) = G_{-s}(pq \ldots . o^{\omega}\overline{rs} \ldots .) = (pq \ldots . o^{\omega} \quad \overline{r} \ldots .),$$
$$\ldots\ldots\ldots\ldots\ldots\ldots\ldots\ldots\ldots\ldots\ldots\ldots\ldots$$

or x, being positive, zero, or negative, D_x or G_x is an operating symbol which, performed upon a monomial symmetric function, has the effect of striking out one part x from its partition.

Also, if P denote a partition which contains no part x,

$$D_xP = G_xP \quad = \text{zero}.$$

Moreover, $$D_x\,(x) = G_x\,(x) = 1,$$

(Compare Hammond, Proceedings of the London Mathematical Society, Vol. XIII, p. 79.)

26

134. As remarked above, in the two results

$$1 + D_0 + D_1\mu + D_2\mu^2 + \ldots = \exp\left(d_0 + d_1\mu + d_{-1}\frac{1}{\mu}\right)$$
$$+ D_{-1}\frac{1}{\mu} + D_{-2}\frac{1}{\mu^2} + \ldots,$$
$$1 + G_0 + G_1\mu + G_2\mu^2 + \ldots = \exp\left\{e\left(g_0 + g_1\mu + g_{-1}\frac{1}{\mu}\right) + (e-1)\,g_0\right\}$$
$$+ G_{-1}\frac{1}{\mu} + G_{-2}\frac{1}{\mu^2} + \ldots.$$

the multiplications arising from the developments of the exponential functions are symbolic.

135. It is convenient to indicate this, as Hammond has done, by placing a horizontal line over the operators which are so multiplied.*

136. We have now the relations which follow, obtained at once by a comparison of coefficients on either side of the two identities

$$1 + D_0 = e^{d_0}\left\{1 + \frac{\overline{d_1 d_{-1}}}{1!\,1!} + \frac{\overline{d_1^2 d_{-1}^2}}{2!\,2!} + \frac{\overline{d_1^3 d_{-1}^3}}{3!\,3!} + \ldots\right\},$$
$$= e^{d_0}\sum_{s=0}^{s=\infty}\frac{\overline{d_1^s d_{-1}^s}}{s!\,s!};$$
$$D_1 = e^{d_0}\left\{d_1 + \frac{\overline{d_1^2 d_{-1}}}{2!\,1!} + \frac{\overline{d_1^3 d_{-1}^2}}{3!\,2!} + \frac{\overline{d_1^4 d_{-1}^3}}{4!\,3!} + \ldots\right\},$$
$$= e^{d_0}\sum_{s=0}^{s=\infty}\frac{\overline{d_1^{s+1} d_{-1}^s}}{(s+1)!\,s!};$$
$$D_{-1} = e^{d_0}\left\{d_{-1} + \frac{\overline{d_1 d_{-1}^2}}{1!\,2!} + \frac{\overline{d_1^2 d_{-1}^3}}{2!\,3!} + \frac{\overline{d_1^3 d_{-1}^4}}{3!\,4!} + \ldots\right\},$$
$$= e^{d_0}\sum_{s=0}^{s=\infty}\frac{\overline{d_1^s d_{-1}^{s+1}}}{s!\,(s+1)!};$$

---

* It will be convenient also in future to write $\overline{\exp}\, u$ when the multiplication of operators that occur in u is symbolic, and $\exp u$ in other cases.

and in general

$$D_\kappa \ = e^{d_0} \sum_{s=0}^{s=\infty} \frac{\overline{d_1^{s+\kappa} d_{-1}^s}}{(s+\varkappa)!\ s!}\,;$$

$$D_{-\kappa} = e^{d_0} \sum_{s=0}^{s=\infty} \frac{\overline{d_1^s d_{-1}^{s+\kappa}}}{s!\ (s+\varkappa)!}\,;$$

whatever be the value of $\varkappa$.

137. Also

$$1 + G_0 \ = e^{(2e-1)g_0} \sum_{s=0}^{s=\infty} e^{2s} \frac{\overline{g_1^s g_{-1}^s}}{s!\ s!}\,,$$

$$G_\kappa \ = e^{(2e-1)g_0} \sum_{s=0}^{s=\infty} e^{2s+\kappa} \frac{\overline{g_1^{s+\kappa} g_{-1}^s}}{(s+\varkappa)!\ s!}\,,$$

$$G_{-\kappa} = e^{(2e-1)g_0} \sum_{s=0}^{s=\infty} e^{2s+\kappa} \frac{\overline{g_1^s g_{-1}^{s+\kappa}}}{s!\ (s+\varkappa)!}\,.$$

138. In these relations the factors $\overline{\exp d_0}$, $\overline{\exp}\,(2e-1)\,g_0$ are to be multiplied symbolically into all that follows; this is of no importance in the case of the factor $\overline{\exp d_0}$, because neither of the operations d_1, d_{-1} involves the quantity n as a symbol of quantity; we may in this case, if we choose, perform the operations

$$\left\{1 + \frac{\overline{d_1 d_{-1}}}{1!\ 1!} + \ldots\ldots\right\}$$

and

$$\exp d_0 = 1 + \frac{\partial_n}{1!} + \frac{\partial_n^2}{2!} + \ldots\ldots$$

successively, for their successive operation is equivalent to their symbolic multiplication. Also the symbols which occur in d_1 and d_{-1} are altogether independent, so that the operation

$$\overline{d_1^{s+\kappa} d_{-1}^s}$$

may be regarded as the successive performance of the operations

$$\overline{d_1^{s+\kappa}}$$

and

$$\overline{d_{-1}^s},$$

and may likewise be written

$$\overline{d_1^{s+\kappa}}\ \overline{d_{-1}^s};$$

hence we may also write

$$D_\kappa = \sum_{t=0}^{t=\infty}\sum_{s=0}^{s=\infty} \frac{\overline{d_1^{s+\kappa} d_0^t d_{-1}^s}}{(s+\kappa)!\, t!\, s!},$$

$$D_{-\kappa} = \sum_{t=0}^{t=\infty}\sum_{s=0}^{s=\infty} \frac{\overline{d_1^{s} d_0^t d_{-1}^{s+\kappa}}}{s!\, t!\, (s+\kappa)!}.$$

With regard to the operations g_0, g_1, g_{-1} the case is different, since the same quantities occur as symbols of quantity in all three. Hence it is absolutely necessary to actually perform all the symbolic multiplications.

139. It is now necessary to find the expressions for

$$\overline{\exp}\left(d_0 + d_1\mu + d_{-1}\frac{1}{\mu}\right)$$

and

$$\overline{\exp}\left\{e\left(g_0 + g_1\mu + g_{-1}\frac{1}{\mu}\right) + (e-1)\,g_0\right\}$$

in series of products of linear operators in which the multiplication is not symbolic.

In the first place write

$$d_\lambda = \partial_{a_\lambda} + a_1\partial_{a_{\lambda+1}} + a_2\partial_{a_{\lambda+2}} + \ldots.$$
$$d_{-\lambda} = \partial_{a_{-\lambda}} + a_{-1}\partial_{a_{-\lambda-1}} + a_{-2}\partial_{a_{-\lambda-2}} + \ldots.$$

Now

$$\overline{\exp}\, d_1\mu = 1 + d_1\mu + \frac{\overline{d_1^2}}{2!}\mu^2 + \frac{\overline{d_1^3}}{3!}\mu^3 + \ldots,$$

and the correspondence between the algebra of the operators d_λ and that of symmetric functions was pointed out and established by Hammond (loc. cit. Proc. Lond. Math. Soc.).

Hence since, as is well known,

$$1 + (1)\mu + (1^2)\mu^2 + (1^3)\mu^3 + \ldots = \exp\{(1)\mu - \tfrac{1}{2}(2)\mu^2 + \tfrac{1}{3}(3)\mu^3 - \ldots\}$$

we derived at once

$$1 + d_1\mu + \frac{\overline{d_1^2}}{2!}\mu^2 + \frac{\overline{d_1^3}}{3!}\mu^3 + \ldots = \exp(d_1\mu - \tfrac{1}{2}d_2\mu^2 + \tfrac{1}{3}d_3\mu^3 - \ldots);$$

that is,

$$\overline{\exp}\, d_1\mu = \exp(d_1\mu - \tfrac{1}{2}d_2\mu^2 + \tfrac{1}{3}d_3\mu^3 - \ldots),$$

where the multiplications on the dexter are not symbolic.

Similarly

$$\overline{\exp}\, d_{-1}\frac{1}{\mu} = \exp\left(d_{-1}\frac{1}{\mu} - \tfrac{1}{2} d_{-2}\frac{1}{\mu^2} + \tfrac{1}{3} d_{-3}\frac{1}{\mu^3} - \ldots\right).$$

Hence, remarking what has gone before, the absolute identity

$$\overline{\exp}\left(d_0 + d_1\mu + d_{-1}\frac{1}{\mu}\right) = \exp\left\{\begin{array}{l} d_0 + d_1\mu - \tfrac{1}{2} d_2\mu^2 + \tfrac{1}{3} d_3\mu^3 - \ldots \\ \qquad + d_{-1}\frac{1}{\mu} - \tfrac{1}{2} d_{-2}\frac{1}{\mu^2} + \tfrac{1}{3} d_{-3}\frac{1}{\mu^3} - \ldots \end{array}\right\}.$$

140. In the second place, writing in general

$$\begin{aligned} g_\lambda = (1 + A_0)\,\partial_{A_\lambda} + A_1\partial_{A_{\lambda+1}} + A_2\partial_{A_{\lambda+2}} + \ldots \\ + A_{-1}\partial_{A_{\lambda-1}} + A_{-2}\partial_{A_{\lambda-2}} + \ldots, \end{aligned}$$

where λ is any positive, zero, or negative integer, it will be convenient to at once put on record a theorem of great generality and importance that I have been led to by the present investigation.

Let

$$\exp\left\{\begin{array}{l} f_0 + f_1 y + f_2 y^2 + \ldots \\ + f_{-1}\frac{1}{y} + f_{-2}\frac{1}{y^2} + \ldots \end{array}\right\} = \begin{array}{l} 1 + F_0 + F_1 y + F_2 y^2 + \ldots \\ \qquad + F_{-1}\frac{1}{y} + F_{-2}\frac{1}{y^2} + \ldots, \end{array}$$

where y is arbitrary, be an absolute identity. Then the theorem asserts the absolute identity

$$\exp\left(\begin{array}{l} f_0 g_0 + f_1 g_1 + f_2 g_2 + \ldots \\ + f_{-1} g_{-1} + f_{-2} g_{-2} + \ldots \end{array}\right) = \overline{\exp}\left(\begin{array}{l} F_0 g_0 + F_1 g_1 + F_2 g_2 + \ldots \\ \qquad + F_{-1} g_{-1} + F_{-2} g_{-2} + \ldots \end{array}\right).$$

Of this theorem I have two independent proofs which will be communicated elsewhere; it enables us from any linear function P of the operators to determine another linear function Q such that

$$\exp P = \overline{\exp}\, Q,$$

or conversely.

141. In the case before us we are given

$$\overline{\exp}\left\{(2e - 1)\, g_0 + e\mu g_1 + \frac{e}{\mu} g_{-1}\right\},$$

so that $F_0 = 2e - 1$, $F_1 = e\mu$, $F_{-1} = \dfrac{e}{\mu}$ and the remainder of the F's vanish.

Hence taking logarithms of the ruling identity

$$\begin{aligned} f_0 + f_1 y \quad + f_2 y^2 \quad + \ldots\ldots &= \log\left(2e + e\mu y + \frac{e}{\mu y}\right) \\ + f_{-1}\frac{1}{y} + f_{-2}\frac{1}{y^2} + \ldots\ldots & \\ &= 1 + \log(1+\mu y) + \log\left(1 + \frac{1}{\mu y}\right) \\ &= 1 + \mu y \quad - \tfrac{1}{2}\mu^2 y^2 + \tfrac{1}{3}\mu^3 y^3 - \ldots\ldots \\ &\quad + \frac{1}{\mu y} - \tfrac{1}{2}\frac{1}{\mu^2 y^2} + \tfrac{1}{3}\frac{1}{\mu^3 y^3} - \ldots\ldots, \end{aligned}$$

and thus $f_0, f_1, f_{-1}, f_2, f_{-2}, \ldots\ldots$ are determined, and we reach the relation

$$\exp\begin{pmatrix} g_0 + g_1\mu \quad - \tfrac{1}{2}g_2\mu^2 \quad + \tfrac{1}{3}g_3\mu^3 \quad - \ldots\ldots \\ + g_{-1}\frac{1}{\mu} - \tfrac{1}{2}g_{-2}\frac{1}{\mu^2} + \tfrac{1}{3}g_{-3}\frac{1}{\mu^3} - \ldots\ldots \end{pmatrix} = \overline{\exp}\left\{(2e-1)g_0 + eg_1\mu + eg_{-1}\frac{1}{\mu}\right\}.$$

142. For present purposes it is useful to mention another theorem in regard to these operations.

Let P denote any linear function of the operators and consider the combination

$$\exp g_0 \,.\, \overline{\exp}\, P,$$

where $\overline{\exp}\, P$ and $\exp g_0$ denote two successive operations.

It is easy to prove that

$$\exp g_0 \,.\, \overline{\exp}\, P = \overline{\exp}\,\{eP + (e-1)g_0\},$$

and writing $P = g_0 + g_1\mu + g_{-1}\dfrac{1}{\mu}$, we have

$$\exp g_0 \,.\, \overline{\exp}\left(g_0 + g_1\mu + g_{-1}\frac{1}{\mu}\right) = \overline{\exp}\left\{(2e-1)g_0 + eg_1\mu + eg_{-1}\frac{1}{\mu}\right\},$$

and this leads to

$$\exp g_0 \,.\, \overline{\exp}\left(g_0 + g_1\mu + g_{-1}\frac{1}{\mu}\right) = \exp\begin{pmatrix} g_0 + g_1\mu \quad - \tfrac{1}{2}g_2\mu^2 \quad + \ldots\ldots \\ + g_{-1}\frac{1}{\mu} - \tfrac{1}{2}g_{-2}\frac{1}{\mu^2} + \ldots\ldots \end{pmatrix}$$

and thence to

$$\overline{\exp}\left(g_0 + g_1\mu + g_{-1}\frac{1}{\mu}\right) = \exp\begin{pmatrix} g_1\mu \quad - \tfrac{1}{2}g_2\mu^2 \quad + \tfrac{1}{3}g_3\mu^3 \quad - \ldots\ldots \\ + g_{-1}\frac{1}{\mu} - \tfrac{1}{2}g_{-2}\frac{1}{\mu^2} + \tfrac{1}{3}g_{-3}\frac{1}{\mu^3} - \ldots\ldots \end{pmatrix}.$$

143. Recalling previous results, we now obtain

$$1 + D_0 + D_1\mu + D_2\mu^2 + \dots + D_{-1}\frac{1}{\mu} + D_{-2}\frac{1}{\mu^2} + \dots = \exp\left\{d_0 + d_1\mu - \tfrac{1}{2}d_2\mu^2 + \dots + d_{-1}\frac{1}{\mu} - \tfrac{1}{2}d_{-2}\frac{1}{\mu^2} + \dots\right\},$$

$$1 + G_0 + G_1\mu + G_2\mu^2 + \dots + G_{-1}\frac{1}{\mu} + G_{-2}\frac{1}{\mu^2} + \dots = \exp\left\{g_0 + g_1\mu - \tfrac{1}{2}g_2\mu^2 + \dots + g_{-1}\frac{1}{\mu} - \tfrac{1}{2}g_{-2}\frac{1}{\mu^2} + \dots\right\},$$

and from these

144.

$$(1+D_0) = e^{d_0}\left\{1 + d_1d_{-1} + \frac{d_1^2 - d_2}{2!}\cdot\frac{d_{-1}^2 - d_{-2}}{2!} + \frac{d_1^3 - 3d_2d_1 + 2d_3}{3!}\cdot\frac{d_{-1}^3 - 3d_{-2}d_{-1} + 2d_{-3}}{3!} + \dots\right\},$$

$$D_1 = e^{d_0}\left\{d_1 + \frac{d_1^2 - d_2}{2!}\cdot d_{-1} + \frac{d_1^3 - 3d_2d_1 + 2d_3}{3!}\cdot\frac{d_{-1}^2 - d_{-2}}{2!} + \dots\right\},$$

$$D_{-1} = e^{d_0}\left\{d_{-1} + d_1\frac{d_{-1}^2 - d_{-2}}{2!} + \frac{d_1^2 - d_2}{2!}\cdot\frac{d_{-1}^3 - 3d_{-2}d_{-1} + 2d_{-3}}{3!} + \dots\right\},$$

etc. = etc.

$$1 + G_0 = e^{g_0}\left\{1 + g_1g_{-1} + \frac{g_1^2 - g_2}{2!}\cdot\frac{g_{-1}^2 - g_{-2}}{2!} + \frac{g_1^3 - 3g_2g_1 + 2g_3}{3!}\cdot\frac{g_{-1}^3 - 3g_{-2}g_{-1} + 2g_{-3}}{3!} + \dots\right\},$$

$$G_1 = e^{g_0}\left\{g_1 + \frac{g_1^2 - g_2}{2!}g_{-1} + \frac{g_1^3 - 3g_2g_1 + 2g_3}{3!}\cdot\frac{g_{-1}^2 - g_{-2}}{2!} + \dots\right\},$$

$$G_{-1} = e^{g_0}\left\{g_{-1} + g_1\frac{g_{-1}^2 - g_{-2}}{2!} + \frac{g_1^2 - g_2}{2!}\cdot\frac{g_{-1}^3 - 3g_{-2}g_{-1} + 2g_{-3}}{3!} + \dots\right\},$$

etc. = etc.

145. And also by taking logarithms

$$d_0 + d_1\mu - \tfrac{1}{2}d_2\mu^2 + \tfrac{1}{3}d_3\mu^3 - \dots + d_{-1}\frac{1}{\mu} - \tfrac{1}{2}d_{-2}\frac{1}{\mu^2} + \tfrac{1}{3}d_{-3}\frac{1}{\mu^3} - \dots$$

$$= \log\left\{1 + D_0 + D_1\mu + D_2\mu^2 + \dots + D_{-1}\frac{1}{\mu} + D_{-2}\frac{1}{\mu^2} + \dots\right\},$$

$$g_0+g_1\mu\quad-\tfrac{1}{2}g_2\mu^2\quad+\tfrac{1}{3}g_3\mu^3\quad-\dots$$
$$+g_{-1}\frac{1}{\mu}-\tfrac{1}{2}g_{-2}\frac{1}{\mu^2}+\tfrac{1}{3}g_{-3}\frac{1}{\mu^3}-\dots$$
$$=\log\left\{1+G_0+G_1\mu\quad+G_2\mu^2\quad+\dots\right.$$
$$\left.+G_{-1}\frac{1}{\mu}+G_{-2}\frac{1}{\mu^2}+\dots\right\},$$

which lead to the relations

$$g_0=\sum\frac{(-)^{\Sigma\pi-1}(\Sigma\pi-1)!}{\dots\pi_2!\ \pi_1!\ \pi_0!\ \pi_{-1}!\ \pi_{-2}!\dots}\dots G_2^{\pi_2}G_1^{\pi_1}G_0^{\pi_0}G_{-1}^{\pi_{-1}}G_{-2}^{\pi_{-2}}\dots,$$
$$\frac{(-)^{m-1}}{m}g_m\quad=\text{ditto},$$
$$\frac{(-)^{m-1}}{m}g_{-m}=\text{ditto},$$

the summations having regard respectively to the solutions of the indeterminate equations

$$\Sigma t\pi_t=0;\ \Sigma t\pi_t=m;\ \Sigma t\pi_t=-m.$$

We have also similar relations between the d and D operations, or if

$$\theta_1,\theta_2,\theta_3,\dots;\ \phi_1,\phi_2,\phi_3,\dots$$

be two sets of fictitious quantities such that

$$\left(1+d_1x+\frac{d_1^2-d_2}{2}+\dots\right)\left(1+d_{-1}\frac{1}{x}+\frac{d_{-1}^2-d_{-2}}{2}\frac{1}{x^2}+\dots\right)$$
$$=(1+\theta_1x)(1+\theta_2x)\dots\left(1+\frac{1}{\theta_1x}\right)\left(1+\frac{1}{\theta_2x}\right)\dots,$$
$$\left(1+g_1x+\frac{g_1^2-g_2}{2}+\dots\right)\left(1+g_{-1}\frac{1}{x}+\frac{g_{-1}^2-g_{-2}}{2}\frac{1}{x^2}+\dots\right)$$
$$=(1+\phi_1x)(1+\phi_2x)\dots\left(1+\frac{1}{\phi_1x}\right)\left(1+\frac{1}{\phi_2x}\right)\dots,$$

and we represent symmetric functions of the two sets by partitions in brackets [ ] and [ ]′ respectively, we have the following correspondence between quantity and operations:

I.

Quantity.

146.

$$e^n(1+a_1x+a_2x^2+\dots)\left(1+a_{-1}\frac{1}{x}+a_{-2}\frac{1}{x^2}+\dots\right)$$
$$=e^n(1+\alpha_1x)(1+\alpha_2x)\dots\left(1+\frac{1}{\alpha_1x}\right)\left(1+\frac{1}{\alpha_2x}\right)\dots$$
$$=e^n\,\overline{\exp}\left\{(0)+(1)x+(\bar{1})\frac{1}{x}\right\},$$

where $\frac{(1)^\lambda (0)^\mu (\bar{1})^\nu}{\lambda!\ \mu!\ \nu!}$ is symbolic expression for $(1^\lambda 0^\mu \bar{1}^\nu)$

$$= 1 + A_0 + A_1 x + A_2 x^2 + \ldots. + A_{-1}\frac{1}{x} + A_{-2}\frac{1}{x^2} + \ldots.$$

$$= \exp\left\{(0) + (1)x - \tfrac{1}{2}(2)x^2 + \ldots. + (\bar{1})\frac{1}{x} - \tfrac{1}{2}(\bar{2})\frac{1}{x^2} + \ldots.\right\}$$

II.

d-Operations.

147.

$$e^{d_0}\left(1 + d_1 x + \frac{d_1^2 - d_2}{2} + \ldots.\right)\left(1 + d_{-1}\frac{1}{x} + \frac{d_{-1}^2 - d_{-2}}{2}\frac{1}{x^2} + \ldots.\right)$$

$$= e^{d_0}(1 + \theta_1 x)(1 + \theta_2 x)\ldots.\left(1 + \frac{1}{\theta_1 x}\right)\left(1 + \frac{1}{\theta_2 x}\right)\ldots.$$

$$= \overline{\exp}\left\{[0] + [1]x + [\bar{1}]\frac{1}{x}\right\},$$

where $\frac{[1]^\lambda [0]^\mu [\bar{1}]^\nu}{\lambda!\ \mu!\ \nu!}$ is symbolic expression for $[1^\lambda 0^\mu \bar{1}^\nu]$,

$$= 1 + D_0 + D_1 x + D_2 x^2 + \ldots. + D_{-1}\frac{1}{x} + D_{-2}\frac{1}{x^2} + \ldots.$$

$$= \exp\left\{[0] + [1]x - \tfrac{1}{2}[2]x^2 + \ldots. + [\bar{1}]\frac{1}{x} - \tfrac{1}{2}[\bar{2}]\frac{1}{x^2} + \ldots.\right\}$$

so that $$[m] = d_m,$$

and $$[p_1^{\pi_1} p_2^{\pi_2}\ldots.] = \frac{\overline{d_{p_1}^{\pi_1} d_{p_2}^{\pi_2}\ldots.}}{\pi_1!\ \pi_2!\ldots.}.$$

III.

g-Operations.

148.

$$e^{g_0}\left(1 + g_1 x + \frac{g_1^2 - g_2}{2} + \ldots.\right)\left(1 + g_{-1}\frac{1}{x} + \frac{g_{-1}^2 - g_{-2}}{2}\frac{1}{x^2} + \ldots.\right)$$

$$= e^{g_0}(1 + \phi_1 x)(1 + \phi_2 x)\ldots.\left(1 + \frac{1}{\phi_1 x}\right)\left(1 + \frac{1}{\phi_2 x}\right)\ldots.$$

$$= \exp[0]'\ \overline{\exp}\left\{[0]' + [1]'x + [\bar{1}]'\frac{1}{x}\right\},$$

27

where $\frac{[1]'^{\lambda}[0]'^{\mu}[\bar{1}]'^{\nu}}{\lambda!\,\mu!\,\nu!}$ is a symbolic expression for $[1^{\lambda}0^{\mu}\bar{1}^{\nu}]'$,

$$= 1 + G_0 + G_1x + G_2x^2 + \ldots .$$
$$+ G_{-1}\frac{1}{x} + G_{-2}\frac{1}{x^2} + \ldots .$$
$$= \exp\{[0]' + [1]'x - \tfrac{1}{2}[2]'x^2 + \ldots .$$
$$+ [\bar{1}]'\frac{1}{x} - \tfrac{1}{2}[\bar{2}]'\frac{1}{x^2} + \ldots .\},$$

so that

$$[m]' = g_m,$$

$$[p_1^{\pi_1}p_2^{\pi_2}]' = \frac{g_{p_1}^{\pi_1}g_{p_2}^{\pi_2}\ldots .}{\pi_1!\,\pi_2!\ldots .}.$$

§10.

149. I now apply the foregoing section to a new demonstration of the "Law of Reciprocity" in the theory of separations of which a purely arithmetical proof was given in the second memoir.

Consider three identities

$$1 + A_0 + A_1x + A_2x^2 + \ldots . + A_{-1}\frac{1}{x} + A_{-2}\frac{1}{x^2} + \ldots .,$$
$$= e^n(1+\alpha_1x)(1+\alpha_2x)\ldots .\left(1+\frac{1}{\alpha_1x}\right)\left(1+\frac{1}{\alpha_2x}\right)\ldots .$$
$$1 + B_0 + B_1x + B_2x^2 + \ldots . + B_{-1}\frac{1}{x} + B_{-2}\frac{1}{x^2} + \ldots .,$$
$$= e^{n'}(1+\beta_1x)(1+\beta_2x)\ldots .\left(1+\frac{1}{\beta_1x}\right)\left(1+\frac{1}{\beta_2x}\right)\ldots .,$$
$$1 + C_0 + C_1x + C_2x^2 + \ldots . + C_{-1}\frac{1}{x} + C_{-2}\frac{1}{x^2} + \ldots .,$$
$$= e^{n''}(1+\gamma_1x)(1+\gamma_2x)\ldots .\left(1+\frac{1}{\gamma_1x}\right)\left(1+\frac{1}{\gamma_2x}\right)^{*}\ldots .,$$

and let symmetric functions of the quantities

$\alpha_1, \alpha_2, \alpha_3, \ldots .$ be denoted by partitions in $(\;)_\alpha$,
$\beta_1, \beta_2, \beta_3, \ldots .$ " " " " " $(\;)_\beta$,
$\gamma_1, \gamma_2, \gamma_3, \ldots .$ " " " " " $(\;)_\gamma$,

---

* n, n', n'' are each to be supposed indefinitely great, and further $n = n'$.

then we have a triad of identities.

150.

$$\left.\begin{aligned}1 + A_0 + A_1 x + A_2 x^2 + \ldots &= \exp\{(0)_\alpha + (1)_\alpha x - \tfrac{1}{2}(2)_\alpha x^2 + \ldots \\ + A_{-1}\frac{1}{x} + A_{-2}\frac{1}{x^2} + \ldots &\qquad + (\bar{1})_\alpha \frac{1}{x} - \tfrac{1}{2}(\bar{2})_\alpha \frac{1}{x^2} + \ldots\end{aligned}\right\},$$

$$\left.\begin{aligned}1 + B_0 + B_1 x + B_2 x^2 + \ldots &= \exp\{(0)_\beta + (1)_\beta x - \tfrac{1}{2}(2)_\beta x^2 + \ldots \\ + B_{-1}\frac{1}{x} + B_{-2}\frac{1}{x^2} + \ldots &\qquad + (\bar{1})_\beta \frac{1}{x} - \tfrac{1}{2}(\bar{2})_\beta \frac{1}{x^2} + \ldots\end{aligned}\right\},$$

$$\left.\begin{aligned}1 + C_0 + C_1 x + C_2 x^2 + \ldots &= \exp\{(0)_\gamma + (1)_\gamma x - \tfrac{1}{2}(2)_\gamma x^2 + \ldots \\ + C_{-1}\frac{1}{x} + C_{-2}\frac{1}{x^2} + \ldots &\qquad + (\bar{1})_\gamma \frac{1}{x} - \tfrac{1}{2}(\bar{2})_\gamma \frac{1}{x^2} + \ldots\end{aligned}\right\}.$$

151. Now assume that, between the quantities herein involved, there exists the relation

$$\begin{aligned}&1 + C_0 + C_1 y + C_2 y^2 + \ldots . \\ &\quad + C_{-1}\frac{1}{y} + C_{-2}\frac{1}{y^2} + \ldots . \\ &\qquad = \prod_{s=1}^{s=n}\left(1 + B_0 + \alpha_s B_1 y + \alpha_s^2 B_2 y^2 + \ldots . \right. \\ &\qquad\qquad \left. + \frac{1}{\alpha_s} B_{-1}\frac{1}{y} + \frac{1}{\alpha_s^2} B_{-2}\frac{1}{y^2} + \ldots . \right),\end{aligned}$$

a relation which also implies the identity

$$1 + c_1 y + c_2 y^2 + c_3 y^3 + \ldots . = \prod_{s=1}^{s=n}(1 + \alpha_s b_1 y + \alpha_s^2 b_2 y^2 + \alpha_s^3 b_3 y^3 + \ldots .),$$

y being arbitrary, so that

$$\begin{aligned}&\log\left(1 + C_0 + C_1 y + C_2 y^2 + \ldots . \right. \\ &\qquad \left. + C_{-1}\frac{1}{y} + C_{-2}\frac{1}{y^2} + \ldots . \right) \\ &\qquad = \sum_{s=1}^{s=n}\log\left(1 + B_0 + \alpha_s B_1 y + \alpha_s^2 B_2 y^2 + \ldots . \right. \\ &\qquad\qquad \left. + \frac{1}{\alpha_s} B_{-1}\frac{1}{y} + \frac{1}{\alpha_s^2} B_{-2}\frac{1}{y^2} + \ldots . \right),\end{aligned}$$

leading to

$$\begin{aligned}(0)_\gamma + (1)_\gamma y - \tfrac{1}{2}(2)_\gamma y^2 + \ldots . &= (0)_\alpha(0)_\beta + (1)_\alpha(1)_\beta y - \tfrac{1}{2}(2)_\alpha(2)_\beta y^2 + \ldots . \\ + (\bar{1})_\gamma\frac{1}{y} - \tfrac{1}{2}(\bar{2})_\gamma\frac{1}{y^2} + \ldots . &\qquad + (\bar{1})_\alpha(\bar{1})_\beta\frac{1}{y} - \tfrac{1}{2}(\bar{2})_\alpha(\bar{2})_\beta\frac{1}{y^2} + \ldots .\end{aligned}$$

and thence to $\qquad (m)_\gamma = (m)_\alpha(m)_\beta,$

where m is any integer, positive, zero, or negative.

152. This result, which is of great importance, shows that the function of $C_0, C_1, C_{-1}, C_2, C_{-2}, \ldots$, denoted by $(m)_\gamma$, is unaltered when the n quantities $\alpha_1, \alpha_2, \alpha_3, \ldots$ and the several n' quantities $\beta_1, \beta_2, \beta_3, \ldots$ are interchanged; but every symmetric function is expressible in terms of sums of powers of the quantities, and it hence follows that every symmetric function of the n quantities

$$\gamma_1, \gamma_2, \gamma_3, \ldots$$

remains unaltered by the interchange of the n quantities

$$\alpha_1, \alpha_2, \alpha_3, \ldots$$

with the several n' quantities

$$\beta_1, \beta_2, \beta_3, \ldots;$$

we may say, in fact, that if any assemblage of partitions in brackets

$$(\quad)_\gamma$$

be expressed in terms of partitions in brackets

$$(\quad)_\alpha \text{ and } (\quad)_\beta,$$

it remains unaltered by the interchange of the brackets $(\quad)_\alpha$ and $(\quad)_\beta$.

153. For example, it is shown in this way that if we have a result

$$(s_1^{\sigma_1} s_2^{\sigma_2} \ldots)_\gamma = \ldots + J (p_1^{\pi_1} p_2^{\pi_2} \ldots)_\alpha (\lambda_1^{l_1} \lambda_2^{l_2} \ldots)_\beta + \ldots$$

we must also have

$$(s_1^{\sigma_1} s_2^{\sigma_2} \ldots)_\gamma = \ldots + J \{ (p_1^{\pi_1} p_2^{\pi_2} \ldots)_\alpha (\lambda_1^{l_1} \lambda_2^{l_2} \ldots)_\beta + (p_1^{\pi_1} p_2^{\pi_2} \ldots)_\beta (\lambda_1^{l_1} \lambda_2^{l_2} \ldots)_\alpha \} + \ldots,$$

and this fact will be shown to involve the "Law of Reciprocity" brought forward in the first and second memoirs.

154. It should be remarked that, as a consequence, the assumed relation may be also written in the form

$$\begin{aligned} 1 + C_0 + C_1 y &+ C_2 y^2 + \ldots \\ + C_{-1}\frac{1}{y} &+ C_{-2}\frac{1}{y^2} + \ldots \\ &= \prod_{s=1}^{s=n} \left(\begin{aligned} 1 + A_0 + \beta_s A_1 y &+ \beta_s^2 A_2 y^2 + \ldots \\ + \frac{1}{\beta_s} A_{-1} \frac{1}{y} &+ \frac{1}{\beta_s^2} A_{-2} \frac{1}{y^2} + \ldots \end{aligned} \right). \end{aligned}$$

155. Associated with the triad of identities above set forth, we have a triad of operator relations, which I will write as follows:

$$\begin{aligned} 1 + {}_{\alpha}G_0 + {}_{\alpha}G_1 y &+ {}_{\alpha}G_2 y^2 + \dots \\ + {}_{\alpha}G_{-1}\frac{1}{y} &+ {}_{\alpha}G_{-2}\frac{1}{y^2} + \dots \\ &= \exp\left\{\begin{array}{l}[0]'_{\alpha} + [1]'_{\alpha}y - \tfrac{1}{2}[2]'_{\alpha}y^2 + \dots \\ \qquad + [\bar{1}]'_{\alpha}\dfrac{1}{y} - \tfrac{1}{2}[\bar{2}]'_{\alpha}\dfrac{1}{y^2} + \dots\end{array}\right\}, \end{aligned}$$

where $[m]'_{\alpha} = {}_{\alpha}g_m,$

$$[p_1^{\pi_1}p_2^{\pi_2}\dots]'_{\alpha} = \frac{\overline{{}_{\alpha}g_{p_1}^{\pi_1}\,{}_{\alpha}g_{p_2}^{\pi_2}\dots}}{\pi_1!\,\pi_2!\dots},$$

$$\begin{aligned} 1 + {}_{\beta}G_0 + {}_{\beta}G_1 y &+ {}_{\beta}G_2 y^2 + \dots \\ + {}_{\beta}G_{-1}\frac{1}{y} &+ {}_{\beta}G_{-2}\frac{1}{y^2} + \dots \\ &= \exp\left\{\begin{array}{l}[0]'_{\beta} + [1]'_{\beta}y - \tfrac{1}{2}[2]'_{\beta}y^2 + \dots \\ \qquad + [\bar{1}]_{\beta}\dfrac{1}{y} - \tfrac{1}{2}[\bar{2}]'_{\beta}\dfrac{1}{y^2} + \dots\end{array}\right\}, \end{aligned}$$

where $[m]'_{\beta} = {}_{\beta}g_m,$

$$[p_1^{\pi_1}p_2^{\pi_2}\dots]'_{\beta} = \frac{\overline{{}_{\beta}g_{p_1}^{\pi_1}\,{}_{\beta}g_{p_2}^{\pi_2}\dots}}{\pi_1!\,\pi_2!\dots},$$

$$\begin{aligned} 1 + {}_{\gamma}G_0 + {}_{\gamma}G_1 y &+ {}_{\gamma}G_2 y^2 + \dots \\ + {}_{\gamma}G_{-1}\frac{1}{y} &+ {}_{\gamma}G_{-2}\frac{1}{y^2} + \dots \\ &= \exp\left\{\begin{array}{l}[0]'_{\gamma} + [1]'_{\gamma}y - \tfrac{1}{2}[2]'_{\gamma}y^2 + \dots \\ \qquad + [\bar{1}]'_{\gamma}\dfrac{1}{y} - \tfrac{1}{2}[\bar{2}]'_{\gamma}\dfrac{1}{y^2} + \dots\end{array}\right\}, \end{aligned}$$

where $[m]'_{\gamma} = {}_{\gamma}g_m,$

$$[p_1^{\pi_1}p_2^{\pi_2}\dots]'_{\gamma} = \frac{\overline{{}_{\gamma}g_{p_1}^{\pi_1}\,{}_{\gamma}g_{p_2}^{\pi_2}\dots}}{\pi_1!\,\pi_2!\dots}.$$

156. Now writing the assumed relation, viz.

$$\begin{aligned} 1 + C_0 + C_1 y &+ C_2 y^2 + \dots \\ + C_{-1}\frac{1}{y} &+ C_{-2}\frac{1}{y^2} + \dots \\ &= \prod_{s=1}^{s=n}\left(\begin{array}{l}1 + B_0 + \alpha_s B_1 y + \alpha_s^2 B_2 y^2 + \dots \\ \qquad + \dfrac{1}{\alpha_s}B_{-1}\dfrac{1}{y} + \dfrac{1}{\alpha_s^2}B_{-2}\dfrac{1}{y^2} + \dots\end{array}\right) \end{aligned}$$

in the abbreviated form

$$U = u_{\alpha_1} u_{\alpha_2} u_{\alpha_3} \ldots ,$$

we have

$$_\beta g_m U = ({}_\beta g_m u_{\alpha_1}) u_{\alpha_2} u_{\alpha_3} \ldots + u_{\alpha_1} ({}_\beta g_m u_{\alpha_2}) u_{\alpha_3} \ldots + u_{\alpha_1} u_{\alpha_2} ({}_\beta g_m u_{\alpha_3}) \ldots + \ldots$$

and

$$\begin{aligned} {}_\beta g_m u_{\alpha_1} &= \left\{ \begin{matrix} (1+B_0)\partial_{B_m} + B_1 \partial_{B_{m+1}} + \ldots \\ + B_{-1}\partial_{B_{m-1}} + \ldots \end{matrix} \right\} \left(\begin{matrix} 1 + B_0 + \alpha_1 B_1 y + \alpha_1^2 B_2 y^2 + \ldots \\ + \frac{1}{\alpha_1} B_{-1} \frac{1}{y} + \frac{1}{\alpha_1^2} B_{-2} \frac{1}{y^2} + \ldots \end{matrix} \right) \\ &= \alpha_1^m y^m u_{\alpha_1}; \end{aligned}$$

hence $\quad {}_\beta g_m U = (\alpha_1^m + \alpha_2^m + \alpha_3^m + \ldots) y^m U = (m)_\alpha y^m U,$

leading to

$$\begin{aligned} &{}_\beta g_m (1 + C_0) + y\, {}_\beta g_m C_1 + y^2\, {}_\beta g_m C_2 + \ldots \\ &+ \frac{1}{y}\, {}_\beta g_m C_{-1} + \frac{1}{y^2}\, {}_\beta g_m C_{-2} + \ldots \\ &= (m)_\alpha y^m \left\{ \begin{matrix} 1 + C_0 + C_1 y + C_2 y^2 + \ldots \\ + C_{-1} \frac{1}{y} + C_{-2} \frac{1}{y^2} + \ldots \end{matrix} \right\}, \end{aligned}$$

which gives, on equating coefficients of like powers of y,

$$\begin{aligned} {}_\beta g_m (1 + C_0) &= (m)_\alpha C_{-m} , \\ {}_\beta g_m C_1 &= (m)_\alpha C_{-m+1} , \\ {}_\beta g_m C_{-1} &= (m)_\alpha C_{-m-1} , \\ &\ldots\ldots\ldots\ldots \\ {}_\beta g_m C_m &= (m)_\alpha (1 + C_0); \end{aligned}$$

but regarding $B_0, B_1, B_{-1}, B_2, B_{-2}, \ldots$ as functions of $C_0, C_1, C_{-1}, C_2, C_{-2}, \ldots$ only, we have

$$\begin{aligned} {}_\beta g_m &\equiv {}_\beta g_m (1 + C_0) \partial_{C_0} + {}_\beta g_m C_1 \partial_{C_1} + {}_\beta g_m C_2 \partial_{C_2} + \ldots \\ &\qquad + {}_\beta g_m C_{-1} \partial_{C_{-1}} + {}_\beta g_m C_{-2} \partial_{C_{-2}} + \ldots \\ &= (m)_\alpha \left\{ \begin{matrix} C_{-m} \partial_{C_0} + C_{-m+1} \partial_{C_1} + C_{-m+2} \partial_{C_2} + \ldots \\ + C_{-m-1} \partial_{C_{-1}} + C_{-m-2} \partial_{C_{-2}} + \ldots \end{matrix} \right\}, \end{aligned}$$

or $\quad {}_\beta g_m = (m)_\alpha\, {}_\gamma g_m;$

we thus arrive at the conclusion that, regarding the assumed relation as defining a transformation of any function of $C_0, C_1, C_{-1}, C_2, C_{-2}, \ldots$ into a function of B_0, B_1, B_{-1}, B_2, B_{-2}, ..., α_1, α_2, α_3, ... being the constants of the transformation, the operation

$${}_{\gamma}g_m$$

is an invariant.

157. It may be remarked by the way that

$${}_{\gamma}g_m \log \left\{ \begin{array}{l} 1 + C_0 + C_1 x \quad + C_2 x^2 \quad + \ldots. \\ \quad + C_{-1}\dfrac{1}{x} + C_{-2}\dfrac{1}{x^2} + \ldots. \end{array} \right\} = x^m,$$

whence

$${}_{\gamma}g_m \left\{ \begin{array}{l} (0)_\gamma + (1)_\gamma x \quad - \tfrac{1}{2}(2)_\gamma x^2 \; + \ldots. \\ \quad + (\overline{1})_\gamma \dfrac{1}{x} - \tfrac{1}{2}(\overline{2})\dfrac{1}{x^2} + \ldots. \end{array} \right\} = x^m,$$

that is,

$${}_{\gamma}g_m \frac{(-)^{m+1}}{m}(m)_\gamma = 1,$$

$${}_{\gamma}g_{-m} \frac{(-)^{m+1}}{m}(\overline{m})_\gamma = 1,$$

where m differs from zero, and

$${}_{\gamma}g_0 (0)_\gamma = 1,$$

whilst

$${}_{\gamma}g_m (s)_\gamma = 0 \text{ if } s \neq m;$$

we have thus a set of transcendental solutions of the partial differential equation

$${}_{\gamma}g_m = 0,$$

and all of these solutions have otherwise been proved to be invariants of the supposed transformation (vide second memoir).

158. Returning to the relation

$${}_{\beta}g_m = (m)_\alpha \, {}_{\gamma}g_m,$$

we may write

$$\begin{array}{l} {}_{\beta}g_0 + {}_{\beta}g_1 y \quad - \tfrac{1}{2}\,{}_{\beta}g_2 y^2 \quad + \ldots. \\ \quad + {}_{\beta}g_{-1}\dfrac{1}{y} - \tfrac{1}{2}\,{}_{\beta}g_{-2}\dfrac{1}{y^2} + \ldots. \end{array} = \begin{array}{l} (0)_\alpha + (1)_\alpha \, {}_{\gamma}g_1 y \quad - \tfrac{1}{2}(2)_\alpha \, {}_{\gamma}g_2 y^2 \quad + \ldots. \\ \quad + (\overline{1})_\alpha \, {}_{\gamma}g_{-1}\dfrac{1}{y} - \tfrac{1}{2}(\overline{2})_\alpha \, {}_{\gamma}g_{-2}\dfrac{1}{y^2} + \ldots., \end{array}$$

y being arbitrary, and also in the form

$$\exp \left\{ \begin{array}{l} [0]'_\beta + [1]'_\beta y \quad - \tfrac{1}{2}[2]'_\beta y^2 \quad + \ldots. \\ \quad + [\overline{1}]'_\beta \dfrac{1}{y} - \tfrac{1}{2}[\overline{2}]'_\beta \dfrac{1}{y^2} + \ldots. \end{array} \right\},$$

$$= \exp \left\{ \begin{array}{l} (0)_\alpha [0]'_\gamma + (1)_\alpha [1]'_\gamma y \quad - \tfrac{1}{2}(2)_\alpha [2]'_\gamma y^2 \quad + \ldots. \\ \quad + (\overline{1})_\alpha [\overline{1}]'_\gamma \dfrac{1}{y} - \tfrac{1}{2}(\overline{2})_\alpha [\overline{2}]'_\gamma \dfrac{1}{y^2} + \ldots. \end{array} \right\},$$

and then from previous work we are led to the result

159.

$$1 + {}_\beta G_0 + {}_\beta G_1 y + {}_\beta G_2 y^2 + \dots + {}_\beta G_{-1}\frac{1}{y} + {}_\beta G_{-2}\frac{1}{y^2} + \dots$$

$$= \prod_{s=1}^{s=n}\left(1 + {}_\gamma G_0 + \alpha_s\, {}_\gamma G_1 y + \alpha_s^2\, {}_\gamma G_2 y^2 + \dots + \frac{1}{\alpha_s}\, {}_\gamma G_{-1}\frac{1}{y} + \frac{1}{\alpha_s^2}\, {}_\gamma G_{-2}\frac{1}{y^2} + \dots\right),$$

and a comparison with the assumed relation

$$1 + C_0 + C_1 y + C_2 y^2 + \dots + C_{-1}\frac{1}{y} + C_{-2}\frac{1}{y^2} + \dots$$

$$= \prod_{s=1}^{s=n}\left(1 + B_0 + \alpha_s B_1 y + \alpha_s^2 B_2 y^2 + \dots + \frac{1}{\alpha_s} B_{-1}\frac{1}{y} + \frac{1}{\alpha_s^2} B_{-2}\frac{1}{y^2} + \dots\right)$$

leads to the following theorem:

160. "In any relation connecting the quantities

$$C_0,\ C_1,\ C_{-1},\ C_2,\ C_{-2},\ \dots$$

with the quantities

$$B_0,\ B_1,\ B_{-1},\ B_2,\ B_{-2},\ \dots$$

we are at liberty to substitute

$${}_\beta G_\kappa \text{ for } C_\kappa$$

and

$${}_\gamma G_\kappa \text{ for } B_\kappa,$$

and we so obtain a relation between operators."

This very important theorem can be applied forthwith.

161. By means of the initial relation

$$1 + C_0 + C_1 x + C_2 x^2 + \dots + C_{-1}\frac{1}{x} + C_{-2}\frac{1}{x^2} + \dots$$

$$= \prod_{s=1}^{s=n}\left(1 + B_0 + \alpha_s B_1 x + \alpha_s^2 B_2 x^2 + \dots + \frac{1}{\alpha_s} B_{-1}\frac{1}{x} + \frac{1}{\alpha_s^2} B_{-2}\frac{1}{x^2} + \dots\right).$$

Where x is arbitrary,* we can express C_0, C_1, C_{-1}, C_2, C_{-2}, ... in terms of the quantities B_0, B_1, B_{-1}, B_2, B_{-2}, ... and monomial symmetric functions of the n quantities

$$\alpha_1,\ \alpha_2,\ \ldots.\ \alpha_n;$$

and, multiplying out, we obtain a result such as

$$C_{p_1}^{\pi_1} C_{p_2}^{\pi_2} \ldots. = \ldots. + L B_{s_1}^{\sigma_1} B_{s_2}^{\sigma_2} \ldots. + \ldots.,\qquad\text{(I)}$$

and also a result such as

$$C_{\lambda_1}^{l_1} C_{\lambda_2}^{l_2} \ldots. = \ldots. + M B_{s_1}^{\sigma_1} B_{s_2}^{\sigma_2} \ldots. + \ldots.\qquad\text{(II)}$$

Join to these two, a third, viz.

$$(s_1^{\sigma_1} s_2^{\sigma_2} \ldots.)_\gamma = \ldots. + A\,(p_1^{\pi_1} p_2^{\pi_2} \ldots.)_\beta + B\,(\lambda_1^{l_1} \lambda_2^{l_2} \ldots.)_\beta + \ldots.\qquad\text{(III)}$$

Now the equation (I) yields the equation of operators†

$${}_\beta G_{p_1}^{\pi_1}\,{}_\beta G_{p_2}^{\pi_2} \ldots. = \ldots. + L\,{}_\gamma G_{s_1}^{\sigma_1}\,{}_\gamma G_{s_2}^{\sigma_2} \ldots. + \ldots.,$$

and performing each side of this upon the opposite side of the relation (III), we obtain

$$L\,{}_\gamma G_{s_1}^{\sigma_1}\,{}_\gamma G_{s_2}^{\sigma_2} \ldots.\ (s_1^{\sigma_1} s_2^{\sigma_2} \ldots.)_\gamma = A\,{}_\beta G_{p_1}^{\pi_1}\,{}_\beta G_{p_2}^{\pi_2} \ldots.\ (p_1^{\pi_1} p_2^{\pi_2} \ldots.)_\beta,$$

no other terms surviving the operation; but

$${}_\gamma G_{s_1}^{\sigma_1}\,{}_\gamma G_{s_2}^{\sigma_2} \ldots.\ (s_1^{\sigma_1} s_2^{\sigma_2} \ldots.)_\gamma = {}_\beta G_{p_1}^{\pi_1}\,{}_\beta G_{p_2}^{\pi_2} \ldots.\ (p_1^{\pi_1} p_2^{\pi_2} \ldots.)_\beta = 1;$$

hence $$L = A,$$

and also from the identities (II) and (III) we obtain

$$M = B,$$

and this leads to

$$(s_1^{\sigma_1} s_2^{\sigma_2} \ldots.)_\gamma = \ldots. + L\,(p_1^{\pi_1} p_2^{\pi_2} \ldots.)_\beta + M\,(\lambda_1^{l_1} \lambda_2^{l_2} \ldots.)_\beta + \ldots.$$

Now, it has been shown previously that

$$(s_1^{\sigma_1} s_2^{\sigma_2} \ldots)_\gamma = \ldots + J\,\{(\lambda_1^{l_1} \lambda_2^{l_2} \ldots)_\alpha (p_1^{\pi_1} p_2^{\pi_2} \ldots)_\beta + (\lambda_1^{l_1} \lambda_2^{l_2} \ldots)_\beta (p_1^{\pi_1} p_2^{\pi_2} \ldots)_\alpha\} + \ldots;$$

* This relation takes the place of the relation in the second memoir, viz.

$$\begin{aligned}1 + X_0 + X_1 x + X_2 x^2 + \ldots. &= \prod_{s=1}^{s=n}\Big(1 + x_0 + a_s x_1 x + a_s^2 x_2 x^2 + \ldots.\\ + X_{-1}\frac{1}{x} + X_{-2}\frac{1}{x^2} + \ldots. &\qquad + \frac{1}{a_s} x_{-1}\frac{1}{x} + \frac{1}{a_s^2} x_{-2}\frac{1}{x^2} + \ldots.\Big),\end{aligned}$$

the notation alone being changed.

† Compare Hammond, loc. cit.

28

hence we must have

$$L = \ldots . + J(\lambda_1^{l_1}\lambda_2^{l_2} \ldots .)_\alpha + \ldots .,$$
$$M = \ldots . + J(p_1^{\pi_1}p_2^{\pi_2} \ldots .)_\alpha + \ldots .,$$

162. And then the relations (I) and (II) become

$$C_{p_1}^{\pi_1}C_{p_2}^{\pi_2} \ldots . = \ldots . + J(\lambda_1^{l_1}\lambda_2^{l_2} \ldots .)_\alpha B_{s_1}^{\sigma_1}B_{s_2}^{\sigma_2} \ldots . + \ldots .,$$
$$C_{\lambda_1}^{l_1}C_{\lambda_2}^{l_2} \ldots . = \ldots . + J(p_1^{\pi_1}p_2^{\pi_2} \ldots .)_\alpha B_{s_1}^{\sigma_1}B_{s_2}^{\sigma_2} \ldots . + \ldots .,$$

which is the law of reciprocity it was required to establish.

§11.

New Law of Symmetry.

163. Suppose that we find the relation

$$(p_1^{\pi_1}p_2^{\pi_2} \ldots .)_\gamma = \ldots . + PB_{s_1}^{\sigma_1}B_{s_2}^{\sigma_2} \ldots . + \ldots .,$$

leading to the operator relation

$$\frac{\overline{{}_Bg_{p_1}^{\pi_1}\,{}_Bg_{p_2}^{\pi_2} \ldots .}}{\pi_1!\ \pi_2! \ldots .} = \ldots . + P\,{}_\gamma G_{s_1}^{\sigma_1}\,{}_\gamma G_{s_2}^{\sigma_2} \ldots . + \ldots .,$$

where P is an aggregate of symmetric functions of the quantities $\alpha_1, \alpha_2, \ldots \alpha_n$.

Further, suppose that

$$(s_1^{\sigma_1}s_2^{\sigma_2} \ldots .)_\gamma = \ldots . + QB_{p_1}^{\pi_1}B_{p_2}^{\pi_2} \ldots . + \ldots .,$$

since

$$\begin{aligned} {}_Bg_m = \partial_{B_m} + B_0\partial_{B_m} &+ B_1\partial_{B_{m+1}} + \ldots . \\ &+ B_{-1}\partial_{B_{m-1}} + \ldots ., \end{aligned}$$

we have $\dfrac{\overline{{}_Bg_{p_1}^{\pi_1}\,{}_Bg_{p_2}^{\pi_2} \ldots .}}{\pi_1!\ \pi_2! \ldots .} = \dfrac{\partial_{Bp_1}^{\pi_1}\ \partial_{B_{p_2}}^{\pi_2} \ldots .}{\pi_1!\ \pi_2! \ldots .}$ + terms which, operating upon a function of $B_0, B_1, B_{-1}, \ldots .$, do not diminish its degree; hence, attending only to terms of like weight and degree,

$$P\,{}_\gamma G_{s_1}^{\sigma_1}\,{}_\gamma G_{s_2}^{\sigma_2} \ldots . (s_1^{\sigma_1}s_2^{\sigma_2} \ldots .)_\gamma = Q\frac{\partial_{B_{p_1}}^{\pi_1}\ \partial_{B_{p_2}}^{\pi_2} \ldots .}{\pi_1!\ \pi_2! \ldots .}B_{p_1}^{\pi_1}B_{p_2}^{\pi_2} \ldots .,$$

that is, $P = Q$; therefore we have the theorem

164. "If

$$(p_1^{\pi_1}p_2^{\pi_2} \ldots .)_\gamma = \ldots . + A(\lambda_1^{l_1}\lambda_2^{l_2} \ldots .)_\alpha B_{s_1}^{\sigma_1}B_{s_2}^{\sigma_2} \ldots . + \ldots .,$$

then

$$(s_1^{\sigma_1}s_2^{\sigma_2} \ldots .)_\gamma = \ldots . + A(\lambda_1^{l_1}\lambda_2^{l_2} \ldots .)_\alpha B_{p_1}^{\pi_1}B_{p_2}^{\pi_2} \ldots . + \ldots . ."$$

165. To make manifest the importance of this theorem I take a concrete case and, for simplicity, restrict myself to symmetric functions which are expressible by means of partitions composed merely of positive integers. We have now, as arguments, the quantities

$$b_1,\ b_2,\ b_3,\ \ldots\ ;\ c_1,\ c_2,\ c_3,\ \ldots$$

derived from the relations

$$1 + b_1x + b_2x^2 + \ldots = (1 + \beta_1x)(1 + \beta_2x)(1 + \beta_3x)\ldots,$$
$$1 + c_1x + c_2x^2 + \ldots = (1 + \gamma_1x)(1 + \gamma_2x)(1 + \gamma_3x)\ldots,$$

and then the fundamental relation leads to the set of identities

$$c_1 = (1)_a b_1,$$
$$c_2 = (2)_a b_2 + (1^2)_a b_1^2,$$
$$c_3 = (3)_a b_3 + (21)_a b_2 b_1 + (1^3)_a b_1^3,$$
$$c_4 = (4)_a b_4 + (31)_a b_3 b_1 + (2^2)_a b_2^2 + (21^2)_a b_2 b_1^2 + (1^4)_a b_1^4,$$
$$\ldots\ldots\ldots\ldots\ldots\ldots\ldots\ldots\ldots$$

and we can, for example, form a table of the fourth order by expressing the symmetric functions of $\gamma_1, \gamma_2, \gamma_3, \ldots$ of weight four in terms of the quantities $b_1, b_2, b_3, b_4, \ldots$

166. Such a table is now given:

	b_4	b_3b_1	b_2^2	$b_2b_1^2$	b_1^4
$(4)_\gamma$	$-4(4)$	$4(3)(1)-4(31)$	$2(2)^2-4(2^2)$	$-4(2)(1)^2+4(2)(1^2)$ $+4(21)(1)\ -4(21^2)$	$(1)^4-4(1^2)(1)^2+2(1^2)^2$ $+4(1^3)(1)\ -4(1^4)$
$(31)_\gamma$	$4(4)$	$-\ (3)(1)+4(31)$	$-2(2)^2+4(2^2)$	$(2)(1)^2-\ (21)(1)$ $-4(2)(1^2)+4(21^2)$	$(1)^2(1^2)-\ (1^3)(1)$ $-2(1^2)^2\ +4(1^4)$
$(2^2)_\gamma$	$2(4)$	$-2(3)(1)+2(31)$	$(2)^2+2(2^2)$	$2(2)(1^2)-2(21)(1)$ $+2(21^2)$	$(1^2)^2-2(1^3)(1)$ $+2(1^4)$
$(21^2)_\gamma$	$-4(4)$	$(3)(1)-4(31)$	$-4(2^2)$	$(21)(1)-4(21^2)$	$(1^3)(1)-4(1^4)$
$(1^4)_\gamma$	(4)	(31)	(2^2)	(21^2)	(1^4)

This is to be read from left to right; for instance, the last line is read

$$(1^4)_\gamma = (4)\,b_4 + (31)\,b_3b_1 + (2^2)\,b_2^2 + (21^2)\,b_2b_1^2 + (1^4)\,b_1^4.$$

167. In this table the suffix α has been for convenience omitted.

The theorem shows that the sth row and the sth column are identical; the table in fact possesses, what is termed, row and column symmetry.

In any column the terms are all separations of the partition of the B-product at the head of the column.

In any row the assemblages of separations are formed according to a law defined by the γ partition at the left of the row.

168. To explain this I form the assemblage of separations of (21^2) according to the law defined by the partition (31); the process consists in first writing down the expression of (31) in terms of separations of (1^4), thus:

$$(31) = (1^2)(1)^2 - 2(1^2)^2 - (1^3)(1) + 4(1^4),$$

the specification of each term is then written down and, beneath, the corresponding coefficient, thus:

Specifications, ...	(21^2)	(2^2)	(31)	(4)
Coefficients, ...	$+1$	-2	-1	$+4$

The two lines, last written, define the law which is applied to the case in hand as follows: The line of specifications is again written down, and underneath each specification those separations of (21^2) which are of that specification, care being taken to write down a separation in correspondence with each permutation amongst separates of the same weight.

We thus obtain two lines, viz.

Specifications.	(21^2)	(2^2)	(31)	(4)
Separations.	$(2)(1)^2$	$(2)(1^2)$ $(1^2)(2)$	$(21)(1)$	(21^2)

We finally attach the coefficients, which appear in the definition of the law, to the separations of corresponding specification; the result is the assemblage

$$(2)(1)^2 - 2(2)(1^2) - (21)(1) + 4(21^2)$$
$$- 2(1^2)(2),$$

or

$$(2)(1)^2 - 4(2)(1^2) - (21)(1) + 4(21^2),$$

which will be found in the second row and fourth column of the above table.

169. The process is conveniently placed in four rows as follows:

(31)	Specifications,	(21^2)	(2^2)	(31)	(4)
	Coefficients,	$+1$	-2	-1	$+4$
(21^2)	Separations,	$(2)(1)^2$	$(2)(1^2)$ $(1^2)(2)$	$(21)(1)$	(21^2)
	Assemblage,	$(2)(1)^2$	$-4(2)(1^2)$	$-(21)(1)$	$+4(21^2)$

Now, observe that the assemblage which occurs in the fourth row and second column is identical with that just found, and moreover is formed from separations of (31) according to the law defined by the partition (21^2), since

$$(21^2) = (1^3)(1) - 4(1^4).$$

The process is as under:

(21^2)	Specifications,	(31)	(4)
	Coefficients,	$+1$	-4
(31)	Separations,	$(3)(1)$	(31)
	Assemblage,	$(3)(1)$	$-4(31)$

and the assemblage thus found is necessarily in the fourth row and second column of the table.

170. I now enunciate a law of symmetry.

Theorem:

"If any symmetric function $(p_1^{\pi_1} p_2^{\pi_2} \ldots .)$ of weight n be expressed in terms of separations of the symmetric function (1^n), a term of specification $(\lambda_1^{l_1} \lambda_2^{l_2} \ldots .)$ is found attached to a certain numerical coefficient x; two rows of specifications and corresponding numerical coefficients respectively are said to define the 'law of the symmetric function $(p_1^{\pi_1} p_2^{\pi_2} \ldots .)$.' The assemblage of separations of a symmetric function $(s_1^{\sigma_1} s_2^{\sigma_2} \ldots .)$, formed according to the law of the symmetric function $(p_1^{\pi_1} p_2^{\pi_2} \ldots .)$ is identical with the assemblage, of separations of the symmetric function $(p_1^{\pi_1} p_2^{\pi_2} \ldots .)$, formed according to the law of the symmetric function $(s_1^{\sigma_1} s_2^{\sigma_2} \ldots .)$."

171. I subjoin an additional example of the theorem, indicating the reciprocity between the functions

$$(31^3) \text{ and } (2^2 1^2) \text{ of weight } 6.$$

$$(2^2 1^2) = (1^4)(1^2) \quad - 4(1^5)(1) \quad + 9(1^6),$$

$(2^2 1^2)$	Specifications,	(42)	(51)	(6)
	Coefficients,	$+1$	-4	$+9$
(31^3)	Separations,	$(31)(1^2)$	$(31^2)(1)$	(31^3)
	Assemblage,	$(31)(1^2)$	$-4(31^2)(1)$	$+9(31^3)$

$$(31^3) = (1^4)(1)^2 \quad - 2(1^4)(1^2) \quad - (1^5)(1) \quad + 6(1^6)$$

(31^3)	Specifications,	(41^2)	(42)	(51)	(6)
	Coefficients,	$+1$	-2	-1	$+6$
$(2^2 1^2)$	Separations,	$(2^2)(1)^2$	$(2^2)(1^2)$ $(21^2)(2)$	$(2^2 1)(1)$	$(2^2 1^2)$
	Assemblage,	$(2^2)(1)^2$	$-2(2^2)(1^2)$ $-2(21^2)(2)$	$-(2^2 1)(1)$	$+6(2^2 1^2)$

and the identity of the two assemblages, viz.

$$(31)(1^2) - 4(31^2)(1) + 9(31^3) = (2^2)(1)^2 - 2(2^2)(1^2) - 2(21^2)(2) - (2^2 1)(1) + 6(2^2 1^2)$$

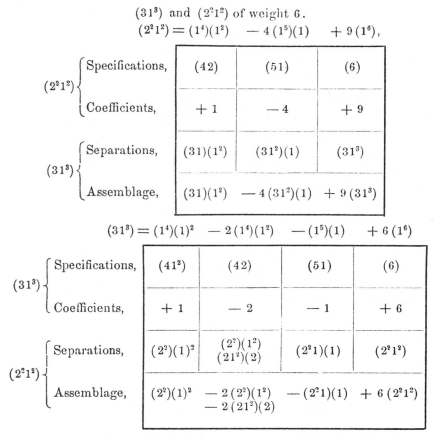

is easily established; each is, in fact, equal to

$$(42) - 2(41^2) - 3(321).$$

172. The theorem is not restricted to positive integers, but there seems to be no advantage to be gained practically in extending it to the most general case.

§12.

173. In the foregoing section another law of symmetry was incidentally met with. As it is of importance, I propose to further examine it.

From the relation

$$C_{p_1}^{\pi_1} C_{p_2}^{\pi_2} \ldots\ldots = \ldots\ldots + L B_{s_1}^{\sigma_1} B_{s_2}^{\sigma_2} \ldots\ldots + \ldots\ldots, \qquad \text{(I)}$$

is derived the operator relation

$$_{\beta}G_{p_1}^{\pi_1}\,{}_{\beta}G_{p_2}^{\pi_2} \ldots\ldots = \ldots\ldots + L\,{}_{\gamma}G_{s_1}^{\sigma_1}\,{}_{\gamma}G_{s_2}^{\sigma_2} \ldots\ldots + \ldots\ldots,$$

and operating with these two sides upon opposite sides of an *assumed* relation

$$(s_1^{\sigma_1} s_2^{\sigma_2} \ldots\ldots)_{\gamma} = \ldots\ldots + A\,(p_1^{\pi_1} p_2^{\pi_2} \ldots\ldots)_{\beta} + \ldots\ldots \qquad \text{(II)}$$

we obtain at once

$$L = A,$$

so that a law of symmetry is involved in the two results

$$\begin{cases} C_{p_1}^{\pi_1} C_{p_2}^{\pi_2} \ldots\ldots = \ldots\ldots + L B_{s_1}^{\sigma_1} B_{s_2}^{\sigma_2} \ldots\ldots + \ldots\ldots, \\ (s_1^{\sigma_1} s_2^{\sigma_2} \ldots\ldots)_{\gamma} = \ldots\ldots + L\,(p_1^{\pi_1} p_2^{\pi_2} \ldots\ldots)_{\beta} + \ldots\ldots \end{cases}$$

By direct multiplication of the product

$$C_{p_1}^{\pi_1} C_{p_2}^{\pi_2} \ldots\ldots,$$

we observe that L is composed of separations of the symmetric function

$$(s_1^{\sigma_1} s_2^{\sigma_2} \ldots\ldots)_{\alpha},$$

of specification $(p_1^{\pi_1} p_2^{\pi_2} \ldots\ldots)$;

L is in fact composed of the tabular assemblages of separations which appear in the tables of the first two memoirs.

174. Hence in the result

$$(s_1^{\sigma_1} s_2^{\sigma_2} \ldots\ldots)_{\gamma} = \ldots\ldots + L\,(p_1^{\pi_1} p_2^{\pi_2} \ldots\ldots)_{\beta} + \ldots\ldots$$

L is composed of tabular separations of the symmetric function

$$(s_1^{\sigma_1} s_2^{\sigma_2} \ldots\ldots)_{\alpha}$$

of specification $(p_1^{\pi_1} p_2^{\pi_2} \ldots\ldots)$.

175. It is now easy to form any function

$$(s_1^{\sigma_1} s_2^{\sigma_2} \ldots .)_\gamma .$$

Example I.

To form $(1^3 0)_\gamma$, first write down all the separations of $(1^3 0)$, and underneath them their respective specifications.

Separations:	$(1)^3(0)$,	$(10)(1)^2$,	$(1^2)(1)(0)$,	$(1^2)(10)$,	$(1^3)(0)$,	$(1^2 0)(1)$,	$(1^3 0)$,
Specifications:	$(1^3 0)$	(1^3)	(210)	(21)	(30)	(21)	(3).

We have now to form the C products corresponding to the several specifications and therein pick out the terms involving the B-product $B_1^3 B_0$.

$$\begin{aligned}
\text{Thus}\quad C_1^3 C_0 &= \ldots . + (1)^3(0)\, B_1^3 B_0 && + \ldots ., \\
C_1^3 &= \ldots . + 3\,(1)^2(10)\, B_1^3 B_0 && + \ldots ., \\
C_2 C_1 C_0 &= \ldots . + (1^2)(1)(0)\, B_1^3 B_0 && + \ldots ., \\
C_2 C_1 &= \ldots . + \{(1^2)(10) + (1^2 0)(1)\}\, B_1^3 B_0 && + \ldots ., \\
C_3 C_0 &= \ldots . + (1^3)(0)\, B_1^3 B_0 && + \ldots ., \\
C_3 &= \ldots . + (1^3 0)\, B_1^3 B_0 && + \ldots .
\end{aligned}$$

Hence

$$\begin{aligned}
(1^3 0)_\gamma = {} & (1)^3(0)(1^3 0)_\beta + 3\,(1)^2(10)(1^3)_\beta + (1^2)(1)(0)(210)_\beta \\
& + \{(1^2)(10) + (1^2 0)(1)\}\,(21)_\beta + (1^3)(0)(30)_\beta + (1^3 0)(3)_\beta .
\end{aligned}$$

Observe that from this point of view $(1^3 0)_\gamma$ is a generating function for the tabular separations of symmetric function $(1^3 0)$.

176. If we now take the operator relation

$$\begin{aligned}
1 + {}_\beta G_0 + {}_\beta G_1 y \quad + \ldots . &= \prod_{s=1}^{s=n} \left(1 + {}_\gamma G_0 + \alpha_s\, {}_\gamma G_1 y \quad + \ldots . \right. \\
+ {}_\beta G_{-1} \frac{1}{y} + \ldots . \qquad & \qquad\qquad \left. + \frac{1}{\alpha_s}\, {}_\gamma G_{-1} \frac{1}{y} + \ldots . \right)
\end{aligned}$$

we obtain

$$\begin{aligned}
{}_\beta G_0 &= (0)\, {}_\gamma G_0 + \ldots ., \\
{}_\beta G_1 &= (1)\, {}_\gamma G_1 + (10)\, {}_\gamma G_1\, {}_\gamma G_0 + \ldots .,
\end{aligned}$$

where only those terms which are significant for present purposes have been retained.

Operating with these relations on opposite sides of the expression for $(1^3 0)_\gamma$, we have

$$(0)(1^3)_\gamma = (1)^3(0)(1^3)_\beta + (1^2)(1)(0)(21)_\beta + (1^3)(0)(3)_\beta ,$$

which is

$$(1^3)_\gamma = (1)^3(1^3)_\beta + (1^2)(1)(21)_\beta + (1^3)(3)_\beta$$

and

$$(1)(1^20)_\gamma + (10)(1^2)_\gamma = (1)\{(1)^2(0)(1^20)_\beta + 2(1)(10)(1^2)_\beta + (1^2)(0)(20)_\beta + (1^20)(2)_\beta\}$$
$$+ (10)\{(1)^2(1^2)_\beta + (1^2)(2)_\beta\},$$

which is true, because the process explained gives

$$(1^20)_\gamma = (1)^2(0)(1^20)_\beta + 2(1)(10)(1^2)_\beta + (1^2)(0)(20)_\beta + (1^20)(2)_\beta$$

and $$(1^2)_\gamma = (1)^2(1^2)_\beta + (1^2)(2)_\beta.$$

This example will serve to show the method of utilizing the operators for the derivation of new results from those already obtained.

177. The most important cases are those which have reference merely to partitions composed of parts which are positive, non-zero, integers.

We may write down a table of weight 4.

$$(4)_\gamma = (4)(4)_\beta,$$
$$(31)_\gamma = (31)(4)_\beta + (3)(1)(31)_\beta,$$
$$(2^2)_\gamma = (2^2)(4)_\beta + (2)^2(2^2)_\beta,$$
$$(21^2)_\gamma = (21^2)(4)_\beta + (21)(1)(31)_\beta + 2(2)(1^2)(2^2)_\beta + (2)(1)^2(21^2)_\beta,$$
$$(1^4)_\gamma = (1^4)(4)_\beta + (1^3)(1)(31)_\beta + (1^2)^2(2^2)_\beta + (1^2)(1)^2(21^2)_\beta + (1)^4(1^4)_\beta,$$

and thence we may immediately write down the corresponding operator relations, viz. writing in general

$$\frac{\overline{g_\lambda^l g_\mu^m \ldots}}{l!\, m! \ldots} = \overline{g_\lambda^{(l)} g_\mu^{(m)}} \ldots,$$

we have

$${}_\beta g_4 = (4)\, {}_\gamma g_4,$$
$$\overline{{}_\beta g_3\, {}_\beta g_1} = (31)\, {}_\gamma g_4 + (3)(1)\, \overline{{}_\gamma g_3\, {}_\gamma g_1},$$
$$\overline{{}_\beta g_2^{(2)}} = (2^2)\, {}_\gamma g_4 \qquad + (2)^2\, \overline{{}_\gamma g_2^{(2)}},$$
$$\overline{{}_\beta g_2\, {}_\beta g_1^{(2)}} = (21^2)\, {}_\gamma g_4 + (21)(1)\, \overline{{}_\gamma g_3\, {}_\gamma g_1} + 2(2)(1^2)\, \overline{{}_\gamma g_2^{(2)}} + (2)(1)^2\, \overline{{}_\gamma g_2\, {}_\gamma g_1^{(2)}},$$
$$\overline{{}_\beta g_1^{(4)}} = (1^4)\, {}_\gamma g_4 + (1^3)(1)\, \overline{{}_\gamma g_3\, {}_\gamma g_1} + (1^2)^2\, \overline{{}_\gamma g_2^{(2)}} \qquad + (1^2)(1)^2\, \overline{{}_\gamma g_2\, {}_\gamma g_1^{(2)}} + (1)^4\, {}_\gamma g_1^{(4)},$$

and so in general it is clear that on the right-hand side we must obtain each tabular separation of the partition of the β operation on the left-hand side attached to a symbolic product of γ operations of the corresponding specification. These results may be utilized in a variety of ways, though at present I do not stop to further discuss them.

§13.

The Multiplication of Symmetric Functions.

178. Professor Cayley has, in this Journal (Vol. VII, p. 2), laid down an algorithm for the multiplication of two symmetric functions which can be sym-

29

bolically expressed by partitions composed of positive integers. Moreover, I have, in the Messenger of Mathematics (Vol. XIV, p. 164), shown how to apply differential operators to the same purpose.

The method of operators may be used when the product contains any number whatever of functions, the coefficient of each term in the result being obtained by a separate and direct process.

Professor Cayley's process on the other hand, while more simple in the case of a product of two functions, is not adapted when the number of functions exceeds two; it may be employed when the partitions contain, as well, zero and negative integer parts.

Of this I give two examples.

179. *Example* I. To multiply

$$\Sigma\alpha^3\beta\gamma^0\delta^{-1} \text{ and } \Sigma\alpha^{-1}\beta^{-1};$$

that is, to form the product

$$(310\bar{1})(\bar{1}^2),$$

3	1	0	$\bar{1}$	$\bar{1}\bar{1}$	F	$\div 2$	M	C
$\bar{1}$	$\bar{1}$				2	$(20^2\bar{1})$	2	2
$\bar{1}$		$\bar{1}$			2	$(21\bar{1}^2)$	2	2
$\bar{1}$			$\bar{1}$		2	$(210\bar{2})$	1	1
	$\bar{1}$	$\bar{1}$			2	$(30\bar{1}^2)$	2	2
	$\bar{1}$		$\bar{1}$		2	$(30^2\bar{2})$	2	2
		$\bar{1}$	$\bar{1}$		2	$(31\bar{1}\bar{2})$	1	1
$\bar{1}$				$\bar{1}$	2	$(210\bar{1}^2)$	2	2
	$\bar{1}$			$\bar{1}$	2	$(30^2\bar{1}^2)$	4	4
		$\bar{1}$		$\bar{1}$	2	$(31\bar{1}^3)$	6	6
			$\bar{1}$	$\bar{1}$	2	$(310\bar{1}\bar{2})$	1	1
				$\bar{1}\bar{1}$	1	$(310\bar{1}^3)$	6	3

where F means "frequency," M, multiplicity, and the coefficients in the column headed C are to be attached to the partitions in the same rows.

Reference should be made to Cayley (loc. cit.)

180. *Example* II. To multiply

$$\Sigma\alpha^0\beta^0\gamma^0 \text{ and } \Sigma\alpha^0\beta^0,$$

that is, to form the product

$$(0^3)(0^2),$$

we have

000	00	F	$\div 12$	M	C
00		6	(0^3)	6	3
0	0	6	(0^4)	24	12
	00	1	(0^5)	120	10

and the result is

$$(0^3)(0^2) = 3\,(0^3) + 12\,(0^4) + 10\,(0^5).$$

181. To establish the method of operators, suppose ϕ to be any rational integral function of

$$A_0,\ A_1,\ A_{-1},\ A_2,\ A_{-2},\ \ldots\ldots,$$

and let

$$\phi = \phi_1\phi_2\phi_3 \ldots\ldots \phi_s;$$

if now we make the transformation of Art. 129, we shall have

$$\begin{pmatrix} 1 + G_0 + G^1\mu & + G_2\mu^2 & + \ldots\ldots \\ + G_{-1}\dfrac{1}{\mu} & + G_{-2}\mu^2 & + \ldots\ldots \end{pmatrix}\phi$$

$$= \begin{pmatrix} 1 + G_0 + G_1\mu & + G_2\mu^2 & + \ldots\ldots \\ + G_{-1}\dfrac{1}{\mu} & + G_{-2}\dfrac{1}{\mu^2} & + \ldots\ldots \end{pmatrix}\phi_1$$

$$\times \begin{pmatrix} 1 + G_0 + G_1\mu & + G_2\mu^2 & + \ldots\ldots \\ + G_{-1}\dfrac{1}{\mu} & + G_{-2}\dfrac{1}{\mu^2} & + \ldots\ldots \end{pmatrix}\phi_2$$

$$\times \ldots\ldots\ldots\ldots\ldots\ldots\ldots\ldots\ldots$$

$$\times \begin{pmatrix} 1 + G_0 + G_1\mu & + G_2\mu^2 & + \ldots\ldots \\ + G_{-1}\dfrac{1}{\mu} & + G_{-2}\dfrac{1}{\mu^2} & + \ldots\ldots \end{pmatrix}\phi_s$$

or

$$\left(\begin{array}{l} 1 + G_0 + G_1\mu \quad + G_2\mu^2 \quad + \ldots \\ \quad + G_{-1}\frac{1}{\mu} + G_{-2}\frac{1}{\mu^2} + \ldots \end{array}\right)\phi$$

$$= \prod_{s=1}^{s=s}\left(\begin{array}{l} \phi_s + G_0\phi_s + G_1\phi_s\mu \quad + G_2\phi_s\mu^2 \quad + \ldots \\ \quad + G_{-1}\phi_s\frac{1}{\mu} + G_{-2}\phi_s\frac{1}{\mu^2} + \ldots \end{array}\right).$$

Whence, equating cofactors of like powers of μ, we obtain

$$G_\lambda\phi = \Sigma\Sigma\, G_{\lambda_1}\phi_{l_1} \cdot G_{\lambda_2}\phi_{l_2} \cdot G_{\lambda_3}\phi_{l_3} \ldots,$$

where

(1). $(\lambda_1\lambda_2\lambda_3 \ldots)$ is any partition whatever of λ into positive, zero and negative integers.

(2). $\phi_{l_1}, \phi_{l_2}, \phi_{l_3}, \ldots$ are different members of the set $\phi_1, \phi_2, \phi_3, \ldots \phi_s$.

(3). The summation is taken for every partition of λ into positive, zero and negative parts.

(4). The summation is further taken for all the expressions obtained by permuting the numbers $l_1, l_2, l_3, \ldots$ in all possible ways.

182. Hence if ϕ be a symmetric function expressed by a product of partitions, the operation G_λ is performed by abstracting every partition of λ in *all possible ways* from the product, one part at most being taken from each partition.

The process is in general simple enough and it must be made perfectly clear by a series of examples.

183. *Example* I. Let

$$\phi = (0^4)(0^3)(0^2).$$

Then

$$\begin{aligned} G_0\phi = \; & G_0(0^4).(0^3)(0^2) + (0^4).G_0(0^3).(0^2) + (0^4)(0^3).G_0(0^2) \\ & + G_0(0^4).G_0(0^3).(0^2) + G_0(0^4).(0^3).G_0(0^2) + (0^4).G_0(0^3).G_0(0^2) \\ & + G_0(0^4).G_0(0^3).G_0(0^2). \\ = \; & (0^3)(0^3)(0^2) + (0^4)(0^2)(0^2) + (0^4)(0^3)(0) \\ & + (0^3)(0^2)(0^2) + (0^3)(0^3)(0) + (0^4)(0^2)(0) \\ & + (0^3)(0^2)(0); \end{aligned}$$

where observe, in the first, second and third lines respectively, the partitions (0), (0^2), (0^3), each of which is a partition of zero, have been abstracted in all possible ways from the product in such wise that one part only is taken from each partition of the product.

184. *Example* II. Let

$$\phi = (0)^m;$$

then

$$G_0\phi = m\,(0)^{m-1} + \frac{1}{2!}\,m\,(m-1)(0)^{m-2} + \ldots . + \tfrac{1}{2}\,m\,(m-1)(0^2) + m\,(0) + 1,$$

the successive terms in the value of $G_0\phi$ corresponding to the abstraction of $(0), (0^2), \ldots . (0^{m-2}), (0^{m-1}), (0^m)$.

Thus $$G_0(0)^m = \{1 + (0)\}^m - (0)^m;$$

that is, $$G_0 n^m \;= (n+1)^m - n^m,$$
which is clearly right.

In general $$G_0 f(n) = f(n+1) - f(n)$$
$$= (E-1)f(n)$$

in the notation of the calculus of finite differences. Hence G_0 is precisely equivalent to the symbol Δ of the same calculus, or

$$G_0 f(n) = \Delta f(n).$$

185. *Example* III. Let

$$\phi = (2)(1)(3\bar{1})(0).$$

To operate with G_1 we have to consider the following partitions of unity, viz.

$$(2\bar{1}),\ (2\bar{1}0),\ (1),\ (10).$$

Hence $$G_1\phi = (1)(3)(0) + (1)(3) + (2)(3\bar{1})(0) + (2)(3\bar{1}).$$

186. *Example* IV. Coming now to multiplication, suppose we are required to find the coefficient of the term

$$(30^2\bar{2})$$

in the product $$(310\bar{1})(\bar{1}^2).$$

If $$(310\bar{1})(\bar{1}^2) = \ldots . + A\,(30^2\bar{2}) + \ldots . ,$$

then, operating throughout with $G_3G_0^2G_{-2}$, we find

$$G_3G_0^2G_{-2}\,(310\bar{1})(\bar{1}^2) = \ldots . + A + \ldots . ,$$

where on the right-hand side the only term which is a simple number is the sought coefficient A; we have then to operate upon the product with the operation $G_3G_0^2G_{-2}$ and the numerical term that issues will be the result we seek.

To carry this out, we find

$$G_{-2}(310\bar{1})(\bar{1}^2) = (310)(\bar{1}),$$
$$G_0 G_{-2}(310\bar{1})(\bar{1}^2) = (31)(\bar{1}) + (30),$$
$$G_0^2 G_{-2}(310\bar{1})(\bar{1}^2) = (3) + (3) = 2(3),$$
$$G_3 G_0^2 G_{-2}(310\bar{1})(\bar{1}^2) = 2.$$

Hence $$(310\bar{1})(\bar{1}^2) = \ldots\ldots + 2(30^2\bar{2}) + \ldots\ldots,$$

which agrees of course with the result obtained by Cayley's algorithm.

§14.

The Linear Partial Differential Operators of the Theory of Separations.

187. I propose to adapt the operations

$$g_0,\ g_1,\ g_{-1},\ g_2,\ g_{-2},\ \ldots\ldots,$$

so that they may be performed on the expression of any symmetric function in terms of separations of a given partition.

It will be remembered that a partition is separated into partitions termed "separates." Any combination whatever of the parts of a partition may present itself as a separate. If the separable partition be

$$(\lambda_1^{l_1}\lambda_2^{l_2}\lambda_3^{l_3}\ldots\ldots),$$

precisely $(l_1+1)(l_2+1)(l_3+1)\ldots\ldots-1$ distinct separates *may* occur.

It is necessary to consider all these separates as independent variables.

Let then $$(\ldots\ldots 2^{p_2+\pi_2}1^{p_1+\pi_1}0^{p_0+\pi_0}\bar{1}^{p_{-1}+\pi_{-1}}\bar{2}^{p_{-2}+\pi_{-2}}\ldots\ldots)$$

be any separate of a given separable partition

$$P,$$

then, by a known theorem,

$$g_s \equiv \Sigma g_s(\ldots 2^{p_2+\pi_2}1^{p_1+\pi_1}0^{p_0+\pi_0}\bar{1}^{p_{-1}+\pi_{-1}}\bar{2}^{p_{-2}+\pi_{-2}}\ldots)\,\partial_{(\ldots 2^{p_2+\pi_2}1^{p_1+\pi_1}0^{p_0+\pi_0}\bar{1}^{p_{-1}+\pi_{-1}}\bar{2}^{p_{-2}+\pi_{-2}}\ldots)},$$

the summation being in regard to every separate. Moreover, we have proved the relation

$$\frac{(-)^s}{s}g_s = \sum \frac{(-)^{\Sigma\pi}(\Sigma\pi-1)!}{\ldots\ldots\pi_2!\,\pi_1!\,\pi_0!\,\pi_{-1}!\,\pi_{-2}!\ldots\ldots}\ldots\ldots G_2^{\pi_2}G_1^{\pi_1}G_0^{\pi_0}G_{-1}^{\pi_{-1}}G_{-2}^{\pi_{-2}}\ldots\ldots,$$

when the summation is in regard to the solutions in positive, zero and negative integers of the indeterminate equation

$$\Sigma t\pi_t = s,$$

and also

$$\ldots . G_2^{\pi_2} G_1^{\pi_1} G_0^{\pi_0} G_{-1}^{\pi_{-1}} G_{-2}^{\pi_{-2}} \ldots . (\ldots . 2^{p_2+\pi_2} 1^{p_1+\pi_1} 0^{p_0+\pi_0} \bar{1}^{p_{-1}+\pi_{-1}} \bar{2}^{p_{-2}+\pi_{-2}} \ldots .)$$
$$= (\ldots . 2^{p_2} 1^{p_1} 0^{p_0} \bar{1}^{p_{-1}} \bar{2}^{p_{-2}} \ldots .).$$

188. Hence

$$\frac{(-)^s}{s} g_s$$
$$\equiv \sum \sum \frac{(-)^{\Sigma\pi} (\Sigma\pi - 1)!}{\ldots . \pi_2! \, \pi_1! \, \pi_0! \, \pi_{-1}! \, \pi_{-2}! \ldots .}$$
$$(\ldots . 2^{p_2} 1^{p_1} 0^{p_0} \bar{1}^{p_{-1}} \bar{2}^{p_{-2}} \ldots .) \, \partial_{(\ldots . 2^{p_2+\pi_2} 1^{p_1+\pi_1} 0^{p_0+\pi_0} \bar{1}^{p_{-1}+\pi_{-1}} \bar{2}^{p_{-2}+\pi_{-2}} \ldots .)},$$

the summation being in regard

(1) to every separate;

(2) to every solution of the indeterminate equation $\Sigma t\pi_t = s$ in positive, zero, and negative integers.

189. To take a concrete case, consider the separable partition

$$(21^{n-2});$$

we then have

$$g_1 \equiv \partial_{(1)} + (1)\,\partial_{(1^2)} + (1^2)\,\partial_{(1^3)} + \ldots . + (1^{n-3})\,\partial_{(1^{n-2})}$$
$$+ (2)\,\partial_{(21)} + (21)\,\partial_{(21^2)} + (21^2)\,\partial_{(21^3)} + \ldots . + (21^{n-3})\,\partial_{(21^{n-2})},$$
$$g_s \equiv \partial_{(1^s)} + (1)\,\partial_{(1^{s+1})} + (1^2)\,\partial_{(1^{s+2})} + \ldots . + (1^{n-s-2})\,\partial_{(1^{n-2})}$$
$$+ (2)\,\partial_{(21^s)} + (21)\,\partial_{(21^{s+1})} + \ldots . + (21^{n-s-2})\,\partial_{(21^{n-2})}$$
$$- s\,\{\partial_{(21^{s-2})} + (1)\,\partial_{(21^{s-1})} + \ldots . + (1^{n-s})\,\partial_{(21^{n-2})}\},$$

relations which are obtained at once from the general formula.

190. Again, consider the separable partition

$$(321^2);$$

the separates are

$$\begin{array}{llll} (1) & , (2) & , (3) & , \\ (1^2) & , (21) , & (31) & , (32), \\ (21^2) & , (31^2), & (321), & \\ (321^2); & & & \end{array}$$

then

$$g_1 \equiv \partial_{(1)} + (1)\,\partial_{(1^2)} + (2)\,\partial_{(21)} + (21)\,\partial_{(21^2)} + (3)\,\partial_{(31)} + (31)\,\partial_{(31^2)} + (32)\,\partial_{(321)} + (321)\partial_{(321^2)};$$

wherein only those separates which contain the part unity occur as independent variables. Also

$$g_2 \equiv \partial_{(1^2)} + (2)\,\partial_{(21^2)} + (3)\,\partial_{(31^2)} + (32)\,\partial_{(321^2)}$$
$$- 2\,\{\partial_{(2)} + (1)\,\partial_{(21)} + (1^2)\,\partial_{(21^2)} + (3)\,\partial_{(32)} + (31)\,\partial_{(321)} + (31^2)\,\partial_{(321^2)}\},$$

wherein the operation is comprised of two portions corresponding to the partitions (1^2) and (2) of the number 2.

In the first and second portions respectively the independent variables contain the partitions (1^2) and (2).

This should be compared with the known formula

$$s_2 = a_1^2 - 2a_2.$$

Similarly

$$g_3 \equiv -3\{\partial_{(21)} + (1)\partial_{(21^2)} + (3)\partial_{(321)} + (31)\partial_{(321^2)}\}$$
$$+ 3\{\partial_{(3)} + (1)\partial_{(31)} + (1^2)\partial_{(31^2)} + (2)\partial_{(32)} + (21)\partial_{(321)} + (21^2)\partial_{(321^2)}\}$$

in correspondence with

$$s_3 = \dots - 3a_2a_1 + 3a_3,$$

and

$$g_4 \equiv -4\partial_{(21^2)}$$
$$+ 4\{\partial_{(31)} + (1)\partial_{(31^2)} + (2)\partial_{(321)} + (21)\partial_{(321^2)}\}$$

in correspondence with

$$s_4 = -4a_2a_1^2 + 4a_3a_1 + \dots.$$

Also

$$g_5 \equiv 5\{\partial_{(31^2)} + (2)\partial_{(321^2)}\} - 5\{\partial_{(32)} + (1)\partial_{(321)} + (1^2)\partial_{(321^2)}\},$$
$$g_6 \equiv -12\{\partial_{(321)} + (1)\partial_{(321^2)}\},$$
$$g_7 \equiv -21\partial_{(321^2)},$$

in correspondence respectively with

$$s_5 = 5a_3a_1^2 + \dots,$$
$$s_6 = -12a_3a_2a_1 + \dots,$$
$$s_7 = -21a_3a_2a_1^2 + \dots.$$

Any of these results may be easily verified. Thus from the first memoir

$$(7) = \dots - \tfrac{1}{3}(321^2) + \dots,$$

and since

$$g_7(7) = +7,$$

it is obvious that g_7 and $-21\partial_{(321^2)}$ are equivalent operations.

191. The formation of these operators is seen to be particularly simple in the case of a separable partition composed merely of positive (non-zero) integers. They are written down at once from the expression of Vandermonde for the sums of the powers of the quantities in terms of the elementary (that is, unitary) symmetric functions.

Let us now consider a separable partition which contains as well zero and negative integers.

For example, the partition

$$(210^2\bar{1})$$

of weight 2.

We must now consider the expressions of the sums of the powers in terms of the arguments

$$A_0,\ A_1,\ A_{-1},\ A_2,\ A_{-2},\ \ldots\ldots,$$

merely retaining necessary terms of the infinite expressions

$$\begin{aligned}
s_{-1} &= \quad A_{-1} - A_0A_{-1} + A_0^2A_{-1} + \ldots\ldots, \\
s_0 &= \quad A_0 - \tfrac{1}{2}A_0^2 - A_1A_{-1} + 2A_1A_0A_{-1} - 3A_1A_0^2A_{-1} + \ldots\ldots, \\
s_1 &= \quad A_1 - A_1A_0 + A_1A_0^2 - A_2A_{-1} + 2A_2A_0A_{-1} - 3A_2A_0^2A_{-1} + \ldots\ldots, \\
s_2 &= -2A_2 + 2A_2A_0 - 2A_2A_0^2 - 4A_2A_1A_{-1} + 12A_2A_1A_0A_{-1} - 24A_2A_1A_0^2A_{-1} + \ldots, \\
s_3 &= -3A_2A_1 + 6A_2A_1A_0 - 9A_2A_1A_0^2 + \ldots.
\end{aligned}$$

192. Hence the following expressions for

$$g_{-1},\ g_0,\ g_1,\ g_2,\ g_3,$$

viz.

$$\begin{aligned}
g_{-1} \equiv \{\partial_{(\bar{1})} &+ (0)\partial_{(0\bar{1})} + (1)\partial_{(1\bar{1})} + (2)\partial_{(2\bar{1})} + (0^2)\partial_{(0^2\bar{1})} + (10)\partial_{(10\bar{1})} + (20)\partial_{(20\bar{1})} \\
&+ (21)\partial_{(21\bar{1})} + (210)\partial_{(210\bar{1})} + (20^2)\partial_{(20^2\bar{1})} + (10^2)\partial_{(10^2\bar{1})} + (210^2)\partial_{(210^2\bar{1})}\} \\
-\{\partial_{(0\bar{1})} &+ (0)\partial_{(0^2\bar{1})} + (1)\partial_{(10\bar{1})} + (2)\partial_{(20\bar{1})} + (10)\partial_{(10^2\bar{1})} + (20)\partial_{(20^2\bar{1})} \\
&+ (21)\partial_{(210\bar{1})} + (210)\partial_{(210^2\bar{1})}\} \\
+\{\partial_{(0^2\bar{1})} &+ (1)\partial_{(10^2\bar{1})} + (2)\partial_{(20^2\bar{1})} + (21)\partial_{(210^21)}\}; \\
g_0 \equiv \partial_{(0)} &+ (2)\partial_{(20)} + (1)\partial_{(10)} + (0)\partial_{(0^2)} + (\bar{1})\partial_{(0\bar{1})} + (21)\partial_{(210)} \\
&+ (0\bar{1})\partial_{(0^2\bar{1})} + (10)\partial_{(10^2)} + (20)\partial_{(20^2)} + (1\bar{1})\partial_{(10\bar{1})} + (2\bar{1})\partial_{(20\bar{1})} \\
&+ (210)\partial_{(210^2)} + (21\bar{1})\partial_{(210\bar{1})} + (20\bar{1})\partial_{(20^2\bar{1})} + (10\bar{1})\partial_{(10^2\bar{1})} + (210\bar{1})\partial_{(210^2\bar{1})} \\
-\tfrac{1}{2}\{\partial_{(0^2)} &+ (1)\partial_{(10^2)} + (2)\partial_{(20^2)} + (\bar{1})\partial_{(0^2\bar{1})} + (21)\partial_{(210^2)} \\
&+ (2\bar{1})\partial_{(20^2\bar{1})} + (1\bar{1})\partial_{(10^2\bar{1})} + (21\bar{1})\partial_{(210^2\bar{1})}\} \\
-\{\partial_{(1\bar{1})} &+ (0)\partial_{(10\bar{1})} + (0^2)\partial_{(10^2\bar{1})} + (2)\partial_{(21\bar{1})} + (20)\partial_{(210\bar{1})} + (20^2)\partial_{(210^2\bar{1})}\} \\
+2\{\partial_{(10\bar{1})} &+ (0)\partial_{(10^2\bar{1})} + (2)\partial_{(210\bar{1})} + (20)\partial_{(210^2\bar{1})}\} \\
-3\{\partial_{(10^2\bar{1})} &+ (2)\partial_{(210^2\bar{1})}\}; \\
g_1 \equiv \partial_{(1)} &+ (0)\partial_{(10)} + (\bar{1})\partial_{(1\bar{1})} + (2)\partial_{(21)} + (0^2)\partial_{(0^21)} + (0\bar{1})\partial_{(10\bar{1})} + (2\bar{1})\partial_{(21\bar{1})} \\
&+ (20)\partial_{(210)} + (0^2\bar{1})\partial_{(10^2\bar{1})} + (20\bar{1})\partial_{(210\bar{1})} + (20^2)\partial_{(210^2)} + (20^2\bar{1})\partial_{(210^2\bar{1})} \\
-\{\partial_{(10)} &+ (\bar{1})\partial_{(10\bar{1})} + (0)\partial_{(10^2)} + (2)\partial_{(210)} + (0\bar{1})\partial_{(10^2\bar{1})} + (2\bar{1})\partial_{(210\bar{1})} \\
&+ (20)\partial_{(210^2)} + (20\bar{1})\partial_{(210^2\bar{1})}\} \\
+\{\partial_{(10^2)} &+ (\bar{1})\partial_{(10^2\bar{1})} + (2)\partial_{(210^2)} + (2\bar{1})\partial_{(210^2\bar{1})}\} \\
-\{\partial_{(2\bar{1})} &+ (0)\partial_{(20\bar{1})} + (1)\partial_{(21\bar{1})} + (0^2)\partial_{(20^2\bar{1})} + (10)\partial_{(210\bar{1})} + (10^2)\partial_{(210^2\bar{1})}\} \\
+2\{\partial_{(20\bar{1})} &+ (0)\partial_{(20^2\bar{1})} + (1)\partial_{(210\bar{1})} + (10)\partial_{(210^2\bar{1})}\} \\
-3\{\partial_{(20^2\bar{1})} &+ (1)\partial_{(210^2\bar{1})}\};
\end{aligned}$$

30

$$
\begin{aligned}
g_2 \equiv -2\{ & \partial_{(2)} + (\bar{1})\partial_{(2\bar{1})} + (0)\partial_{(20)} + (1)\partial_{(21)} + (0\bar{1})\partial_{(20\bar{1})} + (0^2)\partial_{(20^2)} \\
& + (1\bar{1})\partial_{(21\bar{1})} + (10)\partial_{(210)} + (0^2\bar{1})\partial_{(20^2\bar{1})} + (10\bar{1})\partial_{(210\bar{1})} \\
& + (10^2)\partial_{(210^2)} + (10^2\bar{1})\partial_{(210^2\bar{1})}\} \\
+ 2\{ & \partial_{(20)} + (\bar{1})\partial_{(20\bar{1})} + (0)\partial_{(20^2)} + (1)\partial_{(210)} + (0\bar{1})\partial_{(20^2\bar{1})} \\
& + (1\bar{1})\partial_{(210\bar{1})} + (10)\partial_{(210^2)} + (10\bar{1})\partial_{(210^2\bar{1})}\} \\
- 2\{ & \partial_{(20^2)} + (\bar{1})\partial_{(20^2\bar{1})} + (1)\partial_{(210^2)} + (1\bar{1})\partial_{(210^2\bar{1})}\} \\
- 4\{ & \partial_{(21\bar{1})} + (0)\partial_{(210\bar{1})} + (0^2)\partial_{(210^2\bar{1})}\} \\
+ 12\{ & \partial_{(210\bar{1})} + (0)\partial_{(210^2\bar{1})}\} \\
- 24 & \partial_{(210^2\bar{1})},
\end{aligned}
$$

$$
\begin{aligned}
g_3 \equiv -3\{ & \partial_{(21)} + (\bar{1})\partial_{(21\bar{1})} + (0)\partial_{(210)} + (0\bar{1})\partial_{(210\bar{1})} + (0^2)\partial_{(210^2)} + (0^2\bar{1})\partial_{(210^2\bar{1})}\} \\
+ 6\{ & \partial_{(210)} + (\bar{1})\partial_{(210\bar{1})} + (0)\partial_{(210^2)} + (0\bar{1})\partial_{(210^2\bar{1})}\} \\
- 9\{ & \partial_{(210^2)} + (\bar{1})\partial_{(210^2\bar{1})}\}.
\end{aligned}
$$

193. These operations appear to be somewhat complicated, but it will be seen subsequently that for practical use they may in general be broken up into effective fragments. Their application to calculations in the separation theory must be reserved for a future occasion; I hope then also to bring forward a new theorem of distribution of an extended character and to apply the method of this memoir to its analytical solution.

The main result was communicated by me verbally to the London Mathematical Society at its February meeting.

WOOLWICH, ENGLAND, *April* 24, 1889.

Fourth Memoir on a New Theory of Symmetric Functions.

By Major P. A. MacMahon, R. A., F. R. S.

§15.

194. At the conclusion of the previous memoir I applied the linear operations

$$g_0,\ g_1,\ g_{-1},\ \ldots\ldots$$

directly to the theory of separations; I proceed to the further development of this part of the subject; it will be merely necessary in general to consider symmetric functions symbolized by positive non-zero integers.

A previous result, given in §14, may be written

$$\frac{(-)^s}{s}g_s = \sum_\pi \sum_p \frac{(-)^{\Sigma\pi}(\Sigma\pi - 1)!}{\ldots\ \pi_3!\ \pi_2!\ \pi_1!}(\ldots\ 3^{p_3}2^{p_2}1^{p_1})\,\partial_{(\ldots\ 3^{p_3+\pi_3}2^{p_2+\pi_2}1^{p_1+\pi_1})};$$

the right-hand side may be broken up into fragments, in each of which the numbers

$$\ldots\ \pi_3,\ \pi_2,\ \pi_1$$

are constant.

We may thus write

$$\frac{(-)^s g_s}{s} = \sum_\pi \frac{(-)^{\Sigma\pi}(\Sigma\pi - 1)!}{\ldots\ \pi_3!\ \pi_2!\ \pi_1!}\sum_p(\ldots\ 3^{p_3}2^{p_2}1^{p_1})\,\partial_{(\ldots\ 3^{p_3+\pi_3}2^{p_2+\pi_2}1^{p_1+\pi_1})},$$

where the linear operator

$$\sum_p(\ldots\ 3^{p_3}2^{p_2}1^{p_1})\,\partial_{(\ldots\ 3^{p_3+\pi_3}2^{p_2+\pi_2}1^{p_1+\pi_1})},$$

in which the numbers

$$\ldots\ \pi_3,\ \pi_2,\ \pi_1$$

are constant, and the summation is merely in regard to the numbers

$$\ldots\ p_3,\ p_2,\ p_1,$$

that is, to every separate

$$(\ldots\ldots 3^{p_3+\pi_3}2^{p_2+\pi_2}1^{p_1+\pi_1}),$$

is one of the fragments above mentioned.

This fragmentary operation has of course a weight s; but further it may be regarded as having a partition

$$(\ldots\ldots 3^{\pi_3}2^{\pi_2}1^{\pi_1})$$

of the number s; we may so define the fragment and may write, for brevity and convenience,

$$\sum_p (\ldots\ldots 3^{p_3}2^{p_2}1^{p_1})\,\partial_{(\ldots\ldots 3^{p_3+\pi_3}2^{p_2+\pi_2}1^{p_1+\pi_1})} = g_{(\ldots\ldots 3^{\pi_3}2^{\pi_2}1^{\pi_1})}\,.$$

The operator relation is now written

$$\frac{(-)^s g_s}{s} = \sum_\pi \frac{(-)^{\Sigma\pi}(\Sigma\pi - 1)!}{\ldots\ldots \pi_3!\,\pi_2!\,\pi_1!} g_{(\ldots\ldots 3^{\pi_3}2^{\pi_2}1^{\pi_1})},$$

the summation being in regard to every partition

$$(\ldots\ldots 3^{\pi_3}2^{\pi_2}1^{\pi_1}),$$

of the number s, which occurs in the given separable partition.

195. Considering a perfectly general separable partition, every partition of s may occur, and it is, in consequence, convenient to discuss the full result

$$\frac{(-)^s}{s} g_s = \sum \frac{(-)^{\Sigma\pi}(\Sigma\pi - 1)!}{\ldots\ldots \pi_3!\,\pi_2!\,\pi_1!} g_{(\ldots\ldots 3^{\pi_3}2^{\pi_2}1^{\pi_1})},$$

the summation having reference to every partition of the number s.

196. Just as in the ordinary theory we meet with a linear differential operation corresponding to every number, so here in the wider theory of separations we are brought face to face with a linear differential operation which is in correspondence with an arbitrary partition of an arbitrary number.

197. In the simplest cases we have the equivalences

$$g_1 = g_{(1)},$$
$$g_2 = g_{(1^2)} - 2g_{(2)},$$
$$g_3 = g_{(1^3)} - 3g_{(21)} + 3g_{(3)},$$
$$g_4 = g_{(1^4)} - 4g_{(21^2)} + 2g_{(2^2)} + 4g_{(31)} - 4g_{(4)},$$
$$\ldots\ldots\ldots\ldots\ldots\ldots$$

each *weight operator* $g_1, g_2, g_3, g_4, \ldots$ being expressible as a linear function of its fragmentary *partition operators* according to the same law as the sums of powers are represented by elementary (that is, unitary) symmetric functions.

198. Let $\qquad g_{(\ldots 3^{\pi_3}2^{\pi_2}1^{\pi_1})}, \; g_{(\ldots 3^{\rho_3}2^{\rho_2}1^{\rho_1})},$

be any two partition operators of the same or of different weights. We have the known theorem:

$$g_{(\ldots 3^{\pi_3}2^{\pi_2}1^{\pi_1})}\, g_{(\ldots 3^{\rho_3}2^{\rho_2}1^{\rho_1})} = \overline{g_{(\ldots 3^{\pi_3}2^{\pi_2}1^{\pi_1})}\, g_{(\ldots 3^{\rho_3}2^{\rho_2}1^{\rho_1})}} + g_{(\ldots 3^{\pi_3}2^{\pi_2}1^{\pi_1})} \dagger g_{(\ldots 3^{\rho_3}2^{\rho_2}1^{\rho_1})},$$

where the bar written over the product on the right denotes that the operators are to be multiplied together symbolically and the symbol $\dagger$ denotes the performance of the operation $g_{(\ldots 3^{\pi_3}2^{\pi_2}1^{\pi_1})}$ upon the operator $g_{(\ldots 3^{\rho_3}2^{\rho_2}1^{\rho_1})}$, where the latter is considered to be a function of symbols of quantity only.

Now since

$$g_{(\ldots 3^{\pi_3}2^{\pi_2}1^{\pi_1})} = \sum (\ldots\, 3^{j_3}2^{j_2}1^{j_1})\, \partial_{(\ldots 3^{j_3+\pi_3}2^{j_2+\pi_2}1^{j_1+\pi_1})},$$

$$g_{(\ldots 3^{\rho_3}2^{\rho_2}1^{\rho_1})} = \sum (\ldots\, 3^{j_3+\pi_3}2^{j_2+\pi_2}1^{j_1+\pi_1})\, \partial_{(\ldots 3^{j_3+\pi_3+\rho_3}2^{j_2+\pi_2+\rho_2}1^{j_1+\pi_1+\rho_1})},$$

there results

$$g_{(\ldots 3^{\pi_3}2^{\pi_2}1^{\pi_1})} \dagger g_{(\ldots 3^{\rho_3}2^{\rho_2}1^{\rho_1})} = g_{(\ldots 3^{\pi_3+\rho_3}2^{\pi_2+\rho_2}1^{\pi_1+\rho_1})},$$

and, thence, the multiplication theorem

$$g_{(\ldots 3^{\pi_3}2^{\pi_2}1^{\pi_1})}\, g_{(\ldots 3^{\rho_3}2^{\rho_2}1^{\rho_1})} = \overline{g_{(\ldots 3^{\pi_3}2^{\pi_2}1^{\pi_1})}\, g_{(\ldots 3^{\rho_3}2^{\rho_2}1^{\rho_1})}} + g_{(\ldots 3^{\pi_3+\rho_3}2^{\pi_2+\rho_2}1^{\pi_1+\rho_1})},$$

analogous to the known theorem

$$g_s g_t = \overline{g_s g_t} + g_{s+t},$$

with which it should be contrasted.

199. We are now led to the result:

$$g_{(\ldots 3^{\pi_3}2^{\pi_2}1^{\pi_1})}\, g_{(\ldots 3^{\rho_3}2^{\rho_2}1^{\rho_1})} - g_{(\ldots 3^{\rho_3}2^{\rho_2}1^{\rho_1})}\, g_{(\ldots 3^{\pi_3}2^{\pi_2}1^{\pi_1})} = 0.$$

The expression on the left has been termed by Sylvester (in his Lectures on Reciprocants and elsewhere) the alternant of the operators involved.

Theorem. The *alternant* of any two *partition operators* vanishes.

200. This theorem again leads us to two corollaries—

Corollary 1. The *alternant* of any *partition operator* and any weight operator vanishes.

3

Corollary 2. The *alternant* of any two weight operators vanishes.

This last corollary is already well known.

The theorem may be otherwise stated by enunciating a necessary consequence.

201. *Theorem.* Any partition operator and any weight operator is commutable with any other partition or weight operator.

202. Consider now the solutions of the linear partial differential equation

$$P = 0,$$

where P is any partition or weight operator.

If ϕ be one solution, so that identically

$$P\phi = 0,$$

it follows that $Q\phi$ must be another solution, where Q is any other partition or weight operator.

For $$PQ\phi - QP\phi = 0,$$
and since $$P\phi = 0,$$
therefore also $$P(Q\phi) = 0,$$
or $Q\phi$ is also a solution of

$$P = 0.$$

203. *Theorem.* If ϕ be a solution of $P = 0$, $Q\phi$ is also a solution when P and Q are any two partition or weight operations.

204. The partition operators are of most importance in the case of the differential equation

$$g_s = 0.$$

For we have seen that

$$\frac{(-)^s g_s}{s} = \sum_\pi \frac{(-)^{\Sigma\pi}(\Sigma\pi - 1)!}{\ldots\ldots \pi_3!\,\pi_2!\,\pi_1!} g_{(\ldots\ldots 3^{\pi_3} 2^{\pi_2} 1^{\pi_1})},$$

and if $g_s\phi = 0$, where ϕ is expressed in terms of separations of the separable partition

$$(\ldots\ldots 3^{P_3} 2^{P_2} 1^{P_1}),$$

the effect of the partition operator

$$g_{(\ldots\ldots 3^{\pi_3} 2^{\pi_2} 1^{\pi_1})}$$

will be the production of terms each of which is a separation of the partition

$$(\dots . \, 3^{P_3-\pi_3}2^{P_2-\pi_2}1^{P_1-\pi_1}).$$

No other partition operator can produce separations of this partition. Since therefore

$$g_s\Phi$$

is identically zero, we must have also

$$g_{(\dots . \, 3^{\pi_3}2^{\pi_2}1^{\pi_1})}\,\Phi = 0\,.$$

205. *Theorem.* If a function be annihilated by a *weight operator*, it must also be annihilated by each *partition operator* of that weight.

This important and comprehensive theorem renders the calculation of tables of separations a straightforward and comparatively easy matter.

206. As an example, suppose we have to calculate the function (3^2) in terms of separations of (31^3).

Assume

$$2(3^2) = A(31^3) + B(31^2)(1) + C(3)(1^3) + D(31)(1^2) + E(31)(1)^2 + F(3)(1^2)(1),$$

there being of necessity no term $(3)(1)^3$.

We have

$$\begin{aligned}
g_{(1)} &= \partial_{(1)} + (1)\partial_{(1^2)} + (1^2)\partial_{(1^3)} + (3)\partial_{(31)} + (31)\partial_{(31^2)} + (31^2)\partial_{(31^3)},\\
g_{(1^2)} &= \partial_{(1^2)} + (1)\partial_{(1^3)} + (3)\partial_{(31^2)} + (31)\partial_{(31^3)},\\
g_{(1^3)} &= \partial_{(1^3)} + (3)\partial_{(31^3)},\\
g_{(3)} &= \partial_{(3)} + (1)\partial_{(31)} + (1^2)\partial_{(31^2)} + (1^3)\partial_{(31^3)},\\
g_{(31)} &= \partial_{(31)} + (1)\partial_{(31^2)} + (1^2)\partial_{(31^3)},\\
g_{(31^2)} &= \partial_{(31^2)} + (1)\partial_{(31^3)},\\
g_{(31^3)} &= \partial_{(81^3)},
\end{aligned}$$

and these are the only operations we are concerned with; their number is $(1+1)(3+1)-1=7$, viz. one for each separate. The weight operators which annihilate the function are g_1, g_2, g_4 and g_5. Hence we have as annihilators the partition operators

$$g_{(1)},\ g_{(1^2)},\ g_{(31)},\ g_{(31^2)}.$$

Further,

$$\begin{aligned}
g_3 2(3^2) &= -6(3),\\
g_6 2(3^2) &= +6,
\end{aligned}$$

that is,

$$[\partial_{(1^6)} + (3)\,\partial_{(31^3)} + 3\{\partial_{(3)} + (1)\,\partial_{(31)} + (1^2)\,\partial_{(31^2)} + (1^3)\,\partial_{(31^3)}\}]\;2\,(3^2) = -6\,(3),$$
$$6\,\partial_{(31^3)}\,2\,(3^2) = +6\,.$$

These operations are more than sufficient to determine all the coefficients and to verify the result to be found in Vol. XI, p. 17 of this Journal.

§16.

207. Recalling the equivalences

$$\begin{aligned} g_1 &= g_{(1)}, \\ g_2 &= g_{(1^2)} - 2g_{(2)}, \\ g_3 &= g_{(1^3)} - 3g_{(21)} + 3g_{(3)}, \\ g_4 &= g_{(1^4)} - 4g_{(21^2)} + 2g_{(2^2)} + 4g_{(31)} - 4g_{(4)}, \\ &\cdots\cdots\cdots\cdots\cdots\cdots \end{aligned}$$

I find it convenient to write them in a new notation, so as to bring into better evidence their law of formation. I write

$$\begin{aligned} g_{[1]} &= g_{[1]}, \\ g_{[2]} &= g_{[1]^2} - 2g_{[1^2]}, \\ g_{[3]} &= g_{[1]^3} - 3g_{[1^2][1]} + 3g_{[1^3]}, \\ g_{[4]} &= g_{[1]^4} - 4g_{[1^2][1]^2} + 2g_{[1^2]^2} + 4g_{[1^3][1]} - 4g_{[1^4]}, \\ &\cdots\cdots\cdots\cdots\cdots\cdots \end{aligned}$$

where $g_{[1^\lambda][1^\mu]\ldots\ldots}$ takes the place of $g_{(\lambda\mu\ldots\ldots)}$ and $g_{[\lambda]}$ that of g_λ. This notation is quite consistent. In general $g_{[\lambda\ldots\ldots][\mu\ldots\ldots]\ldots\ldots}$ denotes that the operator is formed according to the same law as the symmetric function $(\lambda\ldots\ldots)(\mu\ldots\ldots)\ldots\ldots$ when expressed in terms of elementary symmetric functions.

The theorems

$$\begin{aligned} g_\lambda \dagger g_\mu &= g_{\lambda+\mu}, \\ g_{(\lambda\mu\ldots\ldots)} \dagger g_{(\pi\rho\ldots\ldots)} &= g_{(\lambda\mu\ldots\ldots\pi\rho\ldots\ldots)} \end{aligned}$$

now become in the new notation

$$\begin{aligned} g_{[\lambda]} \dagger g_{[\mu]} &= g_{[\lambda][\mu]}, \\ g_{[1^\lambda][1^\mu]\ldots\ldots} \dagger g_{[1^\pi][1^\rho]\ldots\ldots} &= g_{[1^\lambda][1^\mu]\ldots\ldots[1^\pi][1^\rho]\ldots\ldots}, \end{aligned}$$

that is, a partition suffix addition is transformed into a partition suffix multiplication.

208. We have now in succession

$$g_{[\lambda]}g_{[\mu]}=\overline{g_{[\lambda]}g_{[\mu]}}+g_{[\lambda][\mu]},$$
$$g_{[\kappa]}g_{[\lambda]}g_{[\mu]}=\overline{g_{[\kappa]}g_{[\lambda]}g_{[\mu]}}+\overline{g_{[\kappa]}g_{[\lambda][\mu]}}+\overline{g_{[\lambda]}g_{[\mu][\kappa]}}+\overline{g_{[\mu]}g_{[\kappa][\lambda]}}+g_{[\kappa][\lambda][\mu]},$$
$$\cdots\cdots\cdots\cdots\cdots\cdots$$

which are collateral with the symmetric function theorems:

$$(\lambda)(\mu)=(\lambda\mu)+(\lambda+\mu),$$
$$(\kappa)(\lambda)(\mu)=(\kappa\lambda\mu)+(\kappa,\ \lambda+\mu)+(\lambda,\ \mu+\kappa)+(\mu,\ \kappa+\lambda)+(\kappa+\lambda+\mu),$$
$$\cdots\cdots\cdots\cdots\cdots\cdots$$

and further,

$$\overline{g_{[\lambda]}g_{[\mu]}}=g_{[\lambda]}g_{[\mu]}-g_{[\lambda][\mu]},$$
$$\overline{g_{[\kappa]}g_{[\lambda]}g_{[\mu]}}=g_{[\kappa]}g_{[\lambda]}g_{[\mu]}-g_{[\kappa]}g_{[\lambda][\mu]}-g_{[\lambda]}g_{[\mu][\kappa]}-g_{[\mu]}g_{[\kappa][\lambda]}+2g_{[\kappa][\lambda][\mu]},$$
$$\cdots\cdots\cdots\cdots\cdots\cdots$$

in correspondence with the known results:

$$(\lambda\mu)=s_\lambda s_\mu-s_{\lambda+\mu},$$
$$(\kappa\lambda\mu)=s_\kappa s_\lambda s_\mu-s_\kappa s_{\lambda+\mu}-s_\lambda s_{\mu+\kappa}-s_\mu s_{\kappa+\lambda}+2s_{\kappa+\lambda+\mu},$$
$$\cdots\cdots\cdots\cdots\cdots\cdots$$

209. In both systems of formulae the same modifications, in particular cases, are necessary.

N. B.—Note the result,

$$g_{[\lambda][\mu]}=g_{[\lambda\mu]}+g_{[\lambda+\mu]},$$

and so on; similar developments proceed exactly as in the case of symmetric functions.

§17.

210. Reverting again to the previous notation,

$$g_1=g_{(1)},$$
$$g_2=g_{(1^2)}-2g_{(2)},$$
$$\text{etc.}=\text{etc.},$$

and recalling the well-known relations, viz.

$$g_1=G_1,$$
$$g_2=G_1^2-2G_2,$$
$$g_3=G_1^3-3G_2G_1+3G_3,$$
$$\text{etc.}=\text{etc.},$$

where G_1, G_2, G_3 .... are the obliterators before met with, we notice the similarity of the two laws of operation.

211. We deduce

$$\begin{aligned}
G_1 &= g_1 = g_{(1)},\\
2G_2 &= g_1^2 - g_2 = g_{(1)}^2 - g_{(1^2)} + 2g_{(2)},\\
6G_3 &= g_1^3 - 3g_2g_1 + 2g_3,\\
&= g_{(1)}^3 - 3g_{(1^2)}g_{(1)} + 2g_{(1^3)} + 6\{g_{(2)}g_{(1)} - g_{(21)}\} + 6g_{(3)},\\
24G_4 &= g_1^4 - 6g_2g_1^2 + 3g_2^2 + 8g_3g_1 - 6g_4,\\
&= g_{(1)}^4 - 6g_{(1^2)}g_{(1)}^2 + 3g_{(1^2)}^2 + 8g_{(1^3)}g_{(1)} - 6g_{(1^4)}\\
&\quad + 12\{g_{(2)}g_{(1)}^2 - g_{(2)}g_{(1^2)} - 2g_{(21)}g_{(1)} + 2g_{(21^2)}\} + 12\{g_{(2)}^2 - g_{(2^2)}\}\\
&\quad + 24\{g_{(3)}g_{(1)} - g_{(31)}\} + 24g_{(4)},
\end{aligned}$$

after some slight reduction.

212. We can thus always express the obliterators G_1, G_2, G_3, ... in terms of successive operations of the linear partition operators.

To see the law, on the right-hand side of the identity, last obtained, replace each partition by a partition containing a single part equal in magnitude to the weight of the partition and omit the literal symbols altogether. We thus obtain

$$\begin{aligned}
&\{(1)^4 - 6(2)(1)^2 + 3(2)^2 + 8(3)(1) - 6(4)\} + 12\{(2)(1)^2 - (2)^2 - 2(3)(1) + 2(4)\}\\
&\qquad + 12\{(2)^2 - (4)\} + 24\{(3)(1) - (4)\} + 24(4),
\end{aligned}$$

which is

$$\begin{aligned}
&24(1^4) + 12.2(21^2) + 12.2(2^2) + 24(31) + 24(4)\\
&\qquad = 24\{(1^4) + (21^2) + (2^2) + (31) + (4)\}.
\end{aligned}$$

It is easy to establish *à priori* that this must be so. The expression of $24G_4$ breaks up into 5 portions corresponding to the 5 partitions of the number 4. In general the expression for G_s breaks up into as many portions as there are partitions of s, and in general

$$G_s = \sum \frac{(-)^{L_1+L_2+\ldots+l+m+\ldots}}{L_1!\,L_2!\ldots} \left\{\frac{(l_1+m_1+\ldots-1)!}{l_1!\,m_1!\ldots}\right\}^{L_1} \left\{\frac{(l_3+m_2+\ldots-1)!}{l_2!\,m_2!\ldots}\right\}^{L_2} \ldots g^{L_1}_{(\lambda_1^{l_1}\mu_1^{m_1}\ldots)} g^{L_2}_{(\lambda_2^{l_2}\mu_2^{m_2}\ldots)} \ldots,$$

where $(\lambda_1^{l_1}\mu_1^{m_1}\ldots)^{L_1}(\lambda_2^{l_2}\mu_2^{m_2}\ldots)^{L_2}\ldots$

is any separation of any partition

$$(\lambda^l\mu^m\ldots)$$

of the number s.

213. Instead of expressing G_s in terms of successive operations of linear partition operators, we may express it in terms of operators formed by multiplying together the partition operators symbolically. The system of relations is

$$G_1 = g_{(1)},$$
$$2G_2 = \overline{g_{(1)}^2} + 2g_{(2)},$$
$$6G_3 = \overline{g_{(1)}^3} + 6\overline{g_{(2)}g_{(1)}} + 6g_{(3)},$$
$$24G_4 = \overline{g_{(1)}^4} + 12\overline{g_{(2)}g_{(1)}^2} + 12\overline{g_{(2)}^2} + 24\overline{g_{(3)}g_{(1)}} + 24g_{(4)},$$
$$\text{etc.} = \text{etc.},$$

and in general

$$G_s = \sum \frac{1}{l!\,m!\ldots\ldots}\overline{g_{(\lambda)}^l g_{(\mu)}^m \ldots\ldots}$$

§18.

214. In what has preceded in respect of the linear operations $g_1, g_2, g_3, \ldots\ldots$ a generalization has been made from a number to the partition of a number and weight operators were broken up into linear functions of partition operators.

A like generalization can be made in respect of the obliterating operators $G_1, G_2, G_3, \ldots\ldots$

215. Suppose a symmetric function

$$f(a_1, a_2, a_3, \ldots\ldots a_s, \ldots\ldots) = f$$

to be the product of m monomial functions, and write

$$f = f_1 f_2 f_3 \ldots\ldots f_m.$$

If a_s be changed into $a_s + \mu a_{s-1}$, we have from previous work

$$(1 + \mu G_1 + \mu^2 G_2 + \mu^3 G_3 + \ldots\ldots + \mu^s G_s + \ldots\ldots) f$$
$$= (1 + \mu G_1 + \mu^2 G_2 + \ldots\ldots + \mu^s G_s + \ldots\ldots) f_1$$
$$\times (1 + \mu G_1 + \mu^2 G_2 + \ldots\ldots + \mu^s G_s + \ldots\ldots) f_2$$
$$\times \ldots\ldots\ldots\ldots\ldots\ldots\ldots\ldots$$
$$\times (1 + \mu G_1 + \mu^2 G_2 + \ldots\ldots + \mu^s G_s + \ldots\ldots) f_m,$$

and now expanding and equating coefficients of like powers of μ, there result:

$$G_1 f = \Sigma (G_1 f_1) f_2 f_3 \ldots\ldots f_m,$$
$$G_2 f = \Sigma (G_1 f_1)(G_1 f_2) f_3 \ldots\ldots f_m + \Sigma (G_2 f_1) f_2 \ldots\ldots f_m,$$
$$G_3 f_1 = \Sigma (G_1 f_1)(G_1 f_2)(G_1 f_3) f_4 \ldots\ldots f_m$$
$$+ \Sigma (G_2 f_1)(G_1 f_2) f_3 \ldots\ldots f_m + \Sigma (G_3 f_1) f_2 f_3 \ldots\ldots f_m,$$

and so on, the summations being in regard to the different terms obtained by permutation of the m suffixes of the functions $f_1, f_2, \ldots . f_m$.

216. In general, in the expression of G_s there appears a summation corresponding to each partition of the number s.

The summation in correspondence with a partition $(p_1, p_2, \ldots . p_s)$ is

$$\Sigma (G_{p_1} f_1)(G_{p_2} f_2) \ldots . (G_{p_s} f_s) f_{s+1} \ldots . f_m.$$

Thus, when performed upon a product of functions, the operator G_s breaks up into as many distinct operations as the weight s possesses partitions.

I denote the operation indicated by the summation

$$\Sigma (G_{p_1} f_1)(G_{p_2} f_2) \ldots . (G_{p_s} f_s) f_{s+1} \ldots . f_m$$

by

$$G_{(p_1 p_2 \ldots . p_s)},$$

and speak of it as a partition obliterating operator.

217. We have now the equivalence

$$G_s = \Sigma G_{(p_1 p_2 \ldots . p_s)},$$

the summation being in regard to every partition of the weight s.

This theorem indicates the method of operating with G_s upon a product of symmetric functions.

In particular,

$$\begin{aligned} G_1 &= G_{(1)}, \\ G_2 &= G_{(1^2)} + G_{(2)}, \\ G_3 &= G_{(1^3)} + G_{(21)} + G_{(3)}, \\ &\ldots\ldots\ldots\ldots \end{aligned}$$

218. Interesting relations may be established between the partition g and the partition G operators.

From the relation

$$\sum_\pi \frac{(-)^{\Sigma\pi - 1}(\Sigma\pi - 1)!}{\pi_1!\, \pi_2! \ldots .} g_{(p_1^{\pi_1} p_2^{\pi_2} \ldots .)} = \sum_\pi \frac{(-)^{\Sigma\pi - 1}(\Sigma\pi - 1)!}{\pi_1!\, \pi_2! \ldots .} G_{p_1}^{\pi_1} G_{p_2}^{\pi_2} \ldots .$$

there arise the relations

$$\begin{aligned} g_{(1)} &= G_1 = G_{(1)}, \\ g_{(1^2)} - 2g_{(2)} &= G_1^2 - 2G_2 = G_{(1)}^2 - 2G_{(1^2)} - 2G_{(2)}, \\ g_{(1^3)} - 3g_{(21)} + 3g_{(3)} &= G_1^3 - 3G_2 G_1 + 3G_3 \\ &= G_{(1)}^3 - 3G_{(1^2)}G_{(1)} + 3G_{(1^3)} - 3\{G_{(2)}G_{(1)} - G_{(21)}\} + 3G_{(3)}, \end{aligned}$$

and so forth.

219. Considering particularly the relation last written, I say that it may be broken up into three relations, viz.

$$g_{(1^3)} = G^3_{(1)} - 3G_{(1^2)}G_{(1)} + 3G_{(1^3)},$$
$$g_{(21)} = G_{(2)}G_{(1)} - G_{(21)},$$
$$g_{(3)} = G_{(3)},$$

for it is easy to see that the two sides of the unfractured operator relation must produce upon any operand the same result identically. Hence after operation those functions which are separations of the same function must separately vanish, and hence we must have the equivalences of operations above set forth.

220. In general there exists the relation

$$\frac{(-)^{\Sigma\pi-1}(\Sigma\pi-1)!}{\pi_1!\,\pi_2!\ldots.}\,g_{(p_1^{\pi_1}p_2^{\pi_2}\ldots.)} = \sum_j \frac{(-)^{\Sigma j-1}(\Sigma j-1)!}{j_1!\,j_2!\ldots.}\,G^{j_1}_{(J_1)}G^{j_2}_{(J_2)}\ldots.,$$

the summation being for all separations

$$(J_1)^{j_1}(J_2)^{j_2}\ldots.$$

of the partition $(p_1^{\pi_1}p_2^{\pi_2}\ldots.)$.

221. This result gives the general relation between the partition g and the partition G operators, and should be compared with the formula which expresses a function symbolized by a single part in terms of separations of $(p_1^{\pi_1}p_2^{\pi_2}\ldots.)$.

222. The relation when reversed is

$$(-)^{\Sigma\pi-1}G_{(p_1^{\pi_1}p_2^{\pi_2}\ldots.)}$$
$$= \sum_j \frac{(-)^{\Sigma j-1}(\Sigma\pi_1-1)!\,(\Sigma\pi_2-1)!\ldots.}{j_1!\,j_2!\ldots.\,\pi_{11}!\,\pi_{12}!\ldots.\,\pi_{21}!\,\pi_{22}!\ldots.}\,g^{j_1}_{(p_1^{\pi_{11}}p^{\pi_{12}}\ldots.)}g^{j_2}_{(p_1^{\pi_{21}}p^{\pi_{22}}\ldots.)}\ldots.,$$

the summation being for every separation of the partition $(p_1^{\pi_1}p_2^{\pi_2}\ldots.)$.

223. The following results of operations should be remarked:

$$G_{(p_1^{\pi_1}p_2^{\pi_2}\ldots.)}s_{p_1}^{\pi_1}s_{p_2}^{\pi_2}\ldots. = 1,$$
$$\frac{1}{j_1!}\,\frac{1}{j_2!}\ldots.\,g^{j_1}_{(p_{11}^{\pi_{11}}p_{12}^{\pi_{12}}\ldots.)}g^{j_2}_{(p_{21}^{\pi_{21}}p_{22}^{\pi_{22}}\ldots.)}\ldots.\,(p_{11}^{\pi_{11}}p_{12}^{\pi_{12}}\ldots.)^{j_1}(p_{21}^{\pi_{21}}p_{22}^{\pi_{22}}\ldots.)^{j_2}\ldots. = 1.$$

§19.

224. There is a more extensive law of reciprocity than that established in §10 of the Third Memoir. The latter sprang from three identities of the form

$$\begin{aligned}1+A_0+A_1x\quad &+A_2x^2\quad +\ldots = e^n(1+\alpha_1x)(1+\alpha_2x)\ldots\left(1+\frac{1}{\alpha_1x}\right)\left(1+\frac{1}{\alpha_2x}\right)\ldots,\\ +A_{-1}\frac{1}{x}\ &+A_{-2}\frac{1}{x^2}+\ldots\end{aligned}$$

4

which for brevity may be written

$$F(A)=f(\alpha).$$

225. We may instead consider any number of such identities; but first of all, for the sake of simplicity, let us consider four identities

$$\begin{aligned} F(A)&=f(\alpha),\\ F(B)&=f(\beta),\\ F(C)&=f(\gamma),\\ F(D)&=f(\delta), \end{aligned}$$

and add the auxiliary identity

$$F(K)=f(\varkappa).$$

226. Assume the quantities herein involved to be connected by the two relations

$$\begin{aligned} &1+K_0+K_1y+K_2y^2+\ldots+K_{-1}\frac{1}{y}+K_{-2}\frac{1}{y^2}+\ldots=\Pi_s\left(1+B_0+\alpha_sB_1y+\alpha_s^2B_2y^2+\ldots+\frac{1}{\alpha_s}B_{-1}\frac{1}{y}+\frac{1}{\alpha_s^2}B_{-2}\frac{1}{y^2}+\ldots\right),\\ &1+D_0+D_1y+D_2y^2+\ldots+D_{-1}\frac{1}{y}+D_{-2}\frac{1}{y^2}+\ldots=\Pi_s\left(1+K_0+\gamma_sK_1y+\gamma_s^2K_2y^2+\ldots+\frac{1}{\gamma_s}K_{-1}\frac{1}{y}+\frac{1}{\gamma_s^2}K_{-2}\frac{1}{y^2}+\ldots\right). \end{aligned}$$

Denoting $\Sigma\alpha_1^m$ by $(m)_\alpha$, it has been shown in §10 that these relations lead to the identities

$$\begin{aligned} (m)_\varkappa&=(m)_\alpha(m)_\beta,\\ (m)_\delta&=(m)_\varkappa(m)_\gamma, \end{aligned}$$

so that eliminating $(m)_\varkappa$ we have

$$(m)_\delta=(m)_\alpha(m)_\beta(m)_\gamma;$$

this is equivalent to the result of the elimination of the quantities K between the two assumed relations; in fact we find immediately

$$1+D_0+D_1y+D_2y^2+\ldots+D_{-1}\frac{1}{y}+D_{-2}\frac{1}{y^3}+\ldots=\Pi_s\Pi_t\left(1+B_0+\alpha_s\gamma_tB_1y+\alpha_s^2\gamma_t^2B_2y^2+\ldots+\frac{1}{\alpha_s\gamma_t}B_{-1}\frac{1}{y}+\frac{1}{\alpha_s^2\gamma_t^2}B_{-2}\frac{1}{y^2}+\ldots\right),$$

and the result is then reached by taking logarithms and expanding in powers of y.

In the found relation

$$(m)_\delta = (m)_\alpha (m)_\beta (m)_\gamma,$$

m may be any integer, positive, zero or negative.

227. The expression $(m)_\delta$ remains unchanged for any permutation of the sets of quantities

$$\alpha_1, \alpha_2, \alpha_3, \ldots .$$
$$\beta_1, \beta_2, \beta_3, \ldots .$$
$$\gamma_1, \gamma_2, \gamma_3, \ldots .$$

and hence every symmetric function of the quantities

$$\delta_1, \delta_2, \delta_3, \ldots .$$

remains unchanged for all permutations amongst the three sets mentioned.

228. Thus if we have found

$$(s_1^{\sigma_1} s_2^{\sigma_2} s_3^{\sigma_3} \ldots .)_\delta = \ldots . + J(p_1^{\pi_1} p_2^{\pi_2} \ldots .)_\alpha (\lambda_1^{l_1} \lambda_2^{l_2} \ldots .)_\beta (\mu_1^{m_1} \mu_2^{m_2} \ldots .)_\gamma + \ldots .,$$

we necessarily have

$$(s_1^{\sigma_1} s_2^{\sigma_2} \ldots .)_\delta = \ldots . + J\{(p_1^{\pi_1} p_2^{\pi_2} \ldots .)_\alpha (\lambda_1^{l_1} \lambda_2^{l_2} \ldots .)_\beta (\mu_2^{m_1} \mu_1^{m_2} \ldots .)_\gamma + 5 \text{ similar expressions obtained by permuting } \alpha, \beta \text{ and } \gamma\} + \ldots .$$

229. The two assumed relations lead, as shown in §10, to the operator relations

$$_\beta g_m = (m)_\alpha \, _\kappa g_m,$$
$$_\kappa g_m = (m)_\gamma \, _\delta g_m,$$

and thence

$$(m)_\delta \, _\delta g_m = (m)_\alpha \, _\alpha g_m = (m)_\beta \, _\beta g_m = (m)_\gamma \, _\gamma g_m,$$

showing the invariant character of the operation

$$(m)_\delta \, _\delta g_m,$$

for any transformation of a function of the quantities D into a function of either of the sets of quantities A, B, C as given by the relations assumed.

230. Since

$$_\beta g_m = (m)_\alpha (m)_\gamma \, _\delta g_m,$$

we may write

$$\begin{aligned} &_\beta g_0 + {_\beta g_1} y - \tfrac{1}{2} {_\beta g_2} y^2 + \ldots . \\ &\quad + {_\beta g_{-1}} \frac{1}{y} - \tfrac{1}{2} {_\beta g_{-2}} \frac{1}{y^2} + \ldots . \\ &= (0)_\alpha (0)_\gamma \, _\delta g_0 + (1)_\alpha (1)_\gamma \, _\delta g_1 y - \tfrac{1}{2} (2)_\alpha (2)_\gamma \, _\delta g_2 y^2 + \ldots . \\ &\quad + (\overline{1})_\alpha (\overline{1})_\gamma \, _\delta g_{-1} \frac{1}{y} - \tfrac{1}{2} (\overline{2})_\alpha (\overline{2})_\gamma \, _\delta g_{-2} \frac{1}{y^2} + \ldots ., \end{aligned}$$

and by taking the exponential of each side, we reach by previous work the relation

$$1 + {}_\beta G_0 + {}_\beta G_1 y + {}_\beta G_2 y^2 + \dots + {}_\beta G_{-1}\frac{1}{y} + {}_\beta G_{-2}\frac{1}{y^2} + \dots$$

$$= \prod_s \prod_t \left(1 + {}_\delta G_0 + \alpha_s\gamma_t\, {}_\delta G_1 y + \alpha_s^2\gamma_t^2\, {}_\delta G_2 y^2 + \dots + \frac{1}{\alpha_s\gamma_t}\, {}_\delta G_{-1}\frac{1}{y} + \frac{1}{\alpha_s^2\gamma_t^2}\, {}_\delta G_{-2}\frac{1}{y^2} + \dots\right),$$

showing that in any relation, connecting the quantities D with the quantities B, we obtain an operator relation by writing

$${}_\beta G_m \text{ for } D_m,$$

and

$${}_\delta G_m \text{ for } B_m.$$

231. In the relation just established we may make any permutation of the letters α, β, γ, so that, regarding the quantities D as expressed in terms of the quantities A or C, we may write

$${}_\alpha G_m \text{ for } D_m,$$
$${}_\delta G_m \text{ for } A_m,$$

or

$${}_\gamma G_m \text{ for } D_m,$$
$${}_\delta G_m \text{ for } C_m$$

in either case.

232. Consider now the quantities D expressed in terms of the quantities C, so that by multiplication we obtain a relation such as

$$D_{p_1}^{\pi_1} D_{p_2}^{\pi_2} \dots = \dots + L(\lambda_1^{l_1}\lambda_2^{l_2}\dots)_\alpha(\mu_1^{m_1}\mu_2^{m_2}\dots)_\beta C_{s_1}^{\sigma_1} C_{s_2}^{\sigma_2}\dots + \dots \qquad \text{I}$$

and also two others such as

$$D_{\lambda_1}^{l_1} D_{\lambda_2}^{l_2} \dots = \dots + M(p_1^{\pi_1}p_2^{\pi_2}\dots)_\alpha(\mu_1^{m_1}\mu_2^{m_2}\dots)_\beta C_{s_1}^{\sigma_1} C_{s_2}^{\sigma_2}\dots + \dots \qquad \text{II}$$

$$D_{\mu_1}^{m_1} D_{\mu_2}^{m_2} \dots = \dots + N(\lambda_1^{l_1}\lambda_2^{l_2}\dots)_\alpha(p_1^{\pi_1}p_2^{\pi_2}\dots)_\beta C_{s_1}^{\sigma_1} C_{s_2}^{\sigma_2}\dots + \dots \qquad \text{III}$$

233. Assume, moreover,

$$(s_1^{\sigma_1}s_2^{\sigma_2}\dots)_\delta = \dots + J\{(\lambda_1^{l_1}\lambda_2^{l_2}\dots)_\alpha(\mu_1^{m_1}\mu_2^{m_2}\dots)_\beta(p_1^{\pi_1}p_2^{\pi_2}\dots)_\gamma + \dots\} + \dots \qquad \text{IV}$$

The relation I yields the operator relation

$${}_\gamma G_{p_1}^{\pi_1}\, {}_\gamma G_{p_2}^{\pi_2} \dots = \dots + L(\lambda_1^{l_1}\lambda_2^{l_2}\dots)_\alpha(\mu_1^{m_1}\mu_2^{m_2}\dots)_\beta\, {}_\delta G_{s_1}^{\sigma_1}\, {}_\delta G_{s_2}^{\sigma_2}\dots + \dots,$$

which, performed upon opposite sides of IV, gives

$$\dots + L(\lambda_1^{l_1}\lambda_2^{l_2}\dots)_\alpha(\mu_1^{m_1}\mu_2^{m_2}\dots)_\beta = \dots + J(\lambda_1^{l_1}\lambda_2^{l_2}\dots)_\alpha(\mu_1^{m_1}\mu_2^{m_2}\dots)_\beta + \dots.$$

whence $$L = J,$$
and similarly, $$M = N = J;$$
that is, $$L = M = N.$$

234. Hence in the relation

$$D_{p_1}^{\pi_1} D_{p_2}^{\pi_2} \ldots\ldots = \ldots\ldots + L(\lambda_1^{l_1}\lambda_2^{l_2}\ldots\ldots)_\alpha (\mu_1^{m_1}\mu_2^{m_2}\ldots\ldots)_\beta C_{s_1}^{\sigma_1} C_{s_2}^{\sigma_2} \ldots\ldots + \ldots\ldots,$$

if any permutation be impressed upon the three partitions

$$(\lambda_1^{l_1}\lambda_2^{l_2}\ldots\ldots),\ (\mu_1^{m_1}\mu_2^{m_2}\ldots\ldots),\ (p_1^{\pi_1}p_2^{\pi_2}\ldots\ldots),$$

the numerical coefficient L remains unchanged.

235. More generally consider n identities

$$F(A_1) = f(\alpha_1),$$
$$F(A_2) = f(\alpha_2),$$
$$\ldots\ldots\ldots\ldots$$
$$\ldots\ldots\ldots\ldots$$
$$F(A_n) = f(\alpha_n),$$

and therewith, $n - 3$ auxiliary identities

$$F(K_1) = f(x_1),$$
$$F(K_2) = f(x_2),$$
$$\ldots\ldots\ldots\ldots$$
$$\ldots\ldots\ldots\ldots$$
$$F(K_{n-3}) = f(x_{n-3}).$$

236. Assume to exist between the quantities involved $n - 2$ relations, viz.

$$\begin{aligned}1 + K_{1,0} + K_{1,1}y &+ K_{1,2}y^2 + \ldots = \Pi_s \left(1 + A_{1,0} + \alpha_{2,s}A_{1,1}y + \alpha_{2,s}^2 A_{1,2}y^2 + \ldots \right. \\ + K_{1,-1}\frac{1}{y} &+ K_{1,-2}\frac{1}{y^2} + \ldots \qquad \left. + \frac{1}{\alpha_{2,s}}A_{1,-1}\frac{1}{y} + \frac{1}{\alpha_{2,s}^2}A_{1,-2}\frac{1}{y^2} + \ldots \right)\end{aligned}$$

which for brevity write

$$\Phi(K_1) = \phi(\alpha_2, \alpha_1),$$

and also

$$\Phi(K_2) = \phi(\alpha_3, x_1),$$
$$\ldots\ldots\ldots\ldots\ldots\ldots$$
$$\ldots\ldots\ldots\ldots\ldots\ldots$$
$$\Phi(K_{n-3}) = \phi(\alpha_{n-2}, x_{n-4}),$$
$$\Phi(A_n) = \phi(\alpha_{n-1}, x_{n-3}).$$

237. Observe that the relation

$$\Phi(K_s) = \phi(\alpha_{s+1}, x_{s-1}),$$

is symmetrical with regard to the sets of quantities

$$\alpha_{s+1,1},\ \alpha_{s+1,2},\ \alpha_{s+1,3},\ \ldots\ \text{and}\ \kappa_{s-1,1},\ \kappa_{s-1,2},\ \kappa_{s-1,3},\ \ldots .$$

238. We can now eliminate all the auxiliary quantities.

We have

$$(m)_{\kappa_1} = (m)_{\alpha_2}(m)_{\alpha_1},$$
$$(m)_{\kappa_2} = (m)_{\alpha_3}(m)_{\kappa_1},$$
$$\cdots\cdots\cdots\cdots$$
$$\cdots\cdots\cdots\cdots$$
$$(m)_{\kappa_{n-3}} = (m)_{\alpha_{n-2}}(m)_{\kappa_{n-4}},$$
$$(m)_{\alpha_n} = (m)_{\alpha_{n-1}}(m)_{\kappa_{n-3}},$$

and thence

$$(m)_{\alpha_n} = (m)_{\alpha_1}(m)_{\alpha_2}\ldots(m)_{\alpha_{n-1}},$$

and thence, as before, we obtain the identity

$$1 + A_{n,0} + A_{n,1}y + A_{n,2}y^2 + \ldots$$
$$+ A_{n,-1}\frac{1}{y} + A_{n,-2}\frac{1}{y^2} + \ldots$$
$$= \prod_{s_2}\ldots\prod_{s_{n-1}}\left(\begin{array}{l}1 + A_{1,0} + \alpha_{2,s_2}\alpha_{3,s_3}\ldots\alpha_{n-1,s_{n-1}}A_{1,1}y + \alpha^2_{2,s_2}\alpha^2_{3,s_3}\ldots\alpha^2_{n-1,s_{n-1}}A_{1,2}y^2 + \ldots \\ + \dfrac{1}{\alpha_{2,s_2}\alpha_{3,s_3}\ldots\alpha_{n-1,s_{n-1}}}A_{1,-1}\dfrac{1}{y} + \dfrac{1}{\alpha^2_{2,s_2}\alpha^2_{3,s_3}\ldots\alpha^2_{n-1,s_{n-1}}}A_{1,-2}\dfrac{1}{y^2} + \ldots\end{array}\right)$$

239. This result is invariant for every substitution that may be impressed upon the sets of quantities

$$\alpha_1,\ \alpha_2,\ \alpha_3,\ \ldots\ \alpha_{n-1}.$$

240. Proceeding by the same method, as in the preceding particular case, it is easy to prove that if the identity above given lead to a result

$$A^{l_{n,1}}_{n,\lambda_{n,1}}A^{l_{n,2}}_{n,\lambda_{n,2}}\ldots = \ldots$$
$$+ L(\lambda_{21}^{l_{21}}\lambda_{22}^{l_{22}}\ldots)_{\alpha_2}(\lambda_{31}^{l_{31}}\lambda_{32}^{l_{32}}\ldots)_{\alpha_3}\ldots(\lambda_{n-1,1}^{l_{n-1,1}}\lambda_{n-1,2}^{l_{n-1,2}}\ldots)_{\alpha_{n-1}}A^{l_{11}}_{1,\lambda_{11}}A^{l_{12}}_{1,\lambda_{12}}\ldots + \ldots,$$

wherein L is a numerical coefficient, the number L remains unaltered when any substitution is impressed upon the $n-1$ partitions

$$(\lambda_{s_1}^{l_{s_1}}\lambda_{s_2}^{l_{s_2}}\ldots);$$
$$s = 2,\ 3,\ \ldots\ n.$$

241. This theorem expresses a very general law of symmetry. For its further examination I shall restrict myself to partitions which contain positive integers, zero excluded. This restriction is made with the sole idea of avoiding complexity in the expressions.

242. Consider these relations

$$1 + A_{s,1}x + A_{s,2}x^2 + \dots = (1 + \alpha_{s,1}x)(1 + \alpha_{s,2}x)\dots,$$
$$(s = 1, 2, \dots n),$$

and
$$1 + K_{s,1}x + K_{s,2}x^2 + \dots = (1 + \kappa_{s,1}x)(1 + \kappa_{s,2}x)\dots,$$
$$\{s = 1, 2, \dots (n-3)\}.$$

243. The relations, assumed to exist between the quantities involved, become

$$1+K_{1,1}y \quad +K_{1,2}y^2 \quad + \dots = \prod_{s_2}(1+\alpha_{2,s_2}A_{1,1}y \qquad + \alpha_{2,s_2}^2A_{1,2}y^2 \qquad + \dots),$$
$$1+K_{2,1}y \quad +K_{2,2}y^2 \quad + \dots = \prod_{s_3}(1+\alpha_{3,s_3}K_{1,1}y \qquad + \alpha_{3,s_3}^2K_{1,2}y^2 \qquad + \dots),$$
$$\dots\dots\dots\dots\dots\dots\dots\dots\dots\dots\dots\dots$$
$$1+K_{n-3,1}y + K_{n-3,2}y^2 + \dots = \prod_{s_{n-2}}(1+\alpha_{n-2,s_{n-2}}K_{n-4,1}y + \alpha_{n-2,s_{n-2}}^2K_{n-4,2}y^2 + \dots),$$
$$1+A_{n,1}y \quad +A_{n,2}y^2 \quad + \dots = \prod_{s_{n-1}}(1+\alpha_{n-1,s_{n-1}}K_{n-3,1}y + \alpha_{n-1,s_{n-1}}^2K_{n-3,2}y^2 + \dots),$$

y being an undetermined quantity.

244. Multiplying out the right-hand sides of these identities, equating coefficients of like powers of y in each and employing the partition notation, we obtain

$$K_{1,1} = (1)_{\alpha_2}A_{1,1},$$
$$K_{1,2} = (2)_{\alpha_2}A_{1,2} + (1^2)_{\alpha_2}A_{1,1}^2,$$
$$K_{1,3} = (3)_{\alpha_2}A_{1,3} + (21)_{\alpha_2}A_{1,2}A_{1,1} + (1^3)_{\alpha_2}A_{1,1}^3,$$
$$\dots\dots\dots\dots\dots\dots\dots\dots$$
$$K_{2,1} = (1)_{\alpha_3}K_{1,1},$$
$$K_{2,2} = (2)_{\alpha_3}K_{1,2} + (1^2)_{\alpha_3}K_{1,1}^2,$$
$$K_{2,3} = (3)_{\alpha_3}K_{1,3} + (21)_{\alpha_3}K_{1,2}K_{1,1} + (1^3)_{\alpha_3}K_{1,1}^3,$$
$$\dots\dots\dots\dots\dots\dots\dots\dots$$
$$\dots\dots\dots\dots\dots\dots\dots\dots$$
$$K_{n-3,1} = (1)_{\alpha_{n-2}}K_{n-4,1},$$
$$K_{n-3,2} = (2)_{\alpha_{n-2}}K_{n-4,2} + (1^2)_{\alpha_{n-2}}K_{n-4,1}^2,$$
$$K_{n-3,3} = (3)_{\alpha_{n-2}}K_{n-4,3} + (21)_{\alpha_{n-2}}K_{n-4,2}K_{n-4,1} + (1^3)_{\alpha_{n-2}}K_{n-4,1}^3,$$
$$\dots\dots\dots\dots\dots\dots\dots\dots$$
$$A_{n,1} = (1)_{\alpha_{n-1}}K_{n-3,1},$$
$$A_{n,2} = (2)_{\alpha_{n-1}}K_{n-3,2} + (1^2)_{\alpha_{n-1}}K_{n-3,1}^2,$$
$$A_{n,3} = (3)_{\alpha_{n-1}}K_{n-3,3} + (21)_{\alpha_{n-1}}K_{n-3,2}K_{n-3,1} + (1^3)_{\alpha_{n-1}}K_{n-3,1}^3,$$
$$\dots\dots\dots\dots\dots\dots\dots\dots$$

245. We can now, by direct and successive substitution, express the quantities

$$A_{n,\kappa}$$

in terms of the quantities

$$A_{1,s},$$

and the symmetric functions

$$(\ )_{a_2},\ (\ )_{a_3},\ \ldots\ldots (\ )_{a_{n-1}},$$

and thence form directly any product

$$A_{n,s_1}A_{n,s_2}A_{n,s_3},\ \ldots\ldots$$

expressed in the same manner.

246. We can also verify the relations

$$A_{n,1} = (1)_{a_2}(1)_{a_3}\ldots\ldots(1)_{a_{n-1}}A_{1,1},$$

$$A_{n,1}^2 - 2A_{n,2} = (2)_{a_2}(2)_{a_3}\ldots\ldots(2)_{a_{n-1}}(A_{1,1}^2 - 2A_{1,2}),$$

$$A_{n,1}^3 - 3A_{n,2}A_{n,1} + 3A_{n,3} = (3)_{a_2}(3)_{a_3}\ldots\ldots(3)_{a_{n-1}}(A_{1,1}^3 - 3A_{1,2}A_{1,1} + 3A_{1,3}),$$

$$\ldots\ldots\ldots\ldots\ldots\ldots\ldots\ldots\ldots\ldots\ldots\ldots\ldots\ldots\ldots\ldots$$

$$(m)_{a_n} = (m)_{a_2}(m)_{a_3}\ldots\ldots(m)_{a_{n-1}}(m)_{a_1},$$

which have been previously established.

247. Reverting to the law of symmetry, which it is convenient to express in the form

$$A_{n,\lambda_{n,1}}^{l_{n,1}}A_{n,\lambda_{n,2}}^{l_{n,2}}\ldots = \ldots + L(\lambda_{21}^{l_{21}}\lambda_{22}^{l_{22}}\ldots)_{a_2}\ldots(\lambda_{n-1,1}^{l_{n-1,1}}\lambda_{n-2,2}^{l_{n-2,2}}\ldots)_{a_{n-1}}A_{1\lambda_{11}}^{l_{11}}A_{1\lambda_{12}}^{l_{12}}\ldots+\ldots,$$

where L is unchanged for any substitution impressed upon the $n-1$ partitions

$$(\lambda_{s1}^{l_{s1}}\lambda_{s2}^{l_{s2}}\ldots\ldots)$$
$$(s = 2,\ 3,\ \ldots\ldots n);$$

I recall that in the present instance the partitions are restricted to contain as parts, positive, non-zerȯ integers.

248. Guided by the distribution theorems of the first and second memoirs in this Journal, we may enquire the meaning of the number L in the theory of distributions and of the law of symmetry connected with it which is here brought forward.

§20.

249. In the first and second memoir I considered the distribution of objects of a certain type into parcels of a certain type and obtained a law of symmetry

by observing that it was immaterial whether an object was supposed attached to a parcel or a parcel attached to an object.

250. The consideration of both objects and parcels is in point of fact unnecessary. The parcels may be considered to be objects also, but of a different nature, and the distribution to be of objects of the first set with objects of the second set so as to form a number of pairs of objects, each pair consisting of an object from each set.

251. The result of this distribution may be regarded as the formation of a new set of objects of a two-fold character. Thus if the two sets of objects be

$$a_1,\ a_1,\ a_2,\ a_2, \text{ the first set,}$$
$$b_1,\ b_1,\ b_1,\ b_2, \text{ the second set,}$$

we may make a distribution

$$a_1b_1,\ a_1b_1,\ a_2b_1,\ a_2b_2,$$

and look upon this as a new set of four 2-fold objects. We may then say that we have distributed objects of type (2^2) with objects of type (31) so as to form two-fold objects of type (21^2).*

252. The new method of statement arises naturally from the observed reciprocity between parcels and objects. Together with the new set of two-fold objects we may now consider another set of objects, say

$$c_1,\ c_1,\ c_1,\ c_2,$$

and making any distribution

$$\begin{matrix} a_1b_1, & a_1b_1, & a_2b_1, & a_2b_2, \\ c_1\ , & c_1\ , & c_1\ , & c_2\ , \end{matrix}$$

we arrive at a new set of three-fold objects

$$a_1b_1c_1,\ a_1b_1c_1,\ a_2b_1c_1,\ a_2b_2c_2,$$

which constitutes objects (3-fold) of type (21^2) obtained by distributing objects (2-fold) of type (21^2) with objects of type (31).

253. Thus from the three sets of objects

$$a_1,\ a_1,\ a_2,\ a_2 \text{ of type } (2^2),$$
$$b_1,\ b_1,\ b_1,\ b_2 \text{ of type } (31),$$
$$c_1,\ c_1,\ c_1,\ c_2 \text{ of type } (31),$$

* In the first memoir this was stated to be a distribution of objects of type (2^2) into parcels of type (31) with a partition of restriction (21^2).

5

we have obtained a distribution

$$a_1b_1c_1,\ a_1b_1c_1,\ a_2b_1c_1,\ a_2b_2c_2,$$

which constitutes objects (3-fold) of type (21^2).

254. Consider the problem of finding the number of distributions obtained from these three sets of objects, given by their types, so that a distribution may be constituted of 3-fold objects of given type (21^2).

255. The two sets of objects

$$a_1,\ a_1,\ a_2,\ a_2,$$
$$b_1,\ b_1,\ b_1,\ b_2,$$

may be distributed into sets of two-fold objects of a variety of types. Each of these sets may be then distributed with the objects

$$c_1,\ c_1,\ c_1,\ c_2,$$

and one or more sets of three-fold objects of type (21^2) may or may not be thus reached.

256. Forming the relations

$$A_{3,1} = (1)_2 K_{1,1},$$
$$A_{3,3} = (3)_2 K_{1,3} + (21)_2 K_{1,2} K_{1,1} + (1^3)_2 K_{1,1}^3,$$

we find

$$A_{3,3} A_{3,1} = \ldots . + 2(2^2)_2 K_{1,2} K_{1,1}^2 + \ldots . ,$$

which (see first memoir) shows that when objects of type (2^2) are distributed with objects of type (31), the set of two-fold objects formed is necessarily of type (21^2), and they are 2 in number.

257. We have now to find the number of distributions of objects of type (21^2) with objects of type (31) so that the distributions may be of type (21^2).

Writing
$$K_{1,1} = (1)_1 T_1,$$
$$K_{1,2} = (2)_1 T_2 + (1^2)_1 T_1^2$$

we find
$$K_{1,2} K_{1,1}^2 = \ldots . + 2(31)_1 T_2 T_1^2 + \ldots .$$

From this it appears that any set of objects of type (21^2) may be distributed in two different ways with any set of objects of type (31) in such wise that the distribution is of type (21^2).

258. Combining the two results, we have

$$A_{3,3} A_{3,1} = \ldots . + 4(2^2)_2 (31)_1 T_2 T_1^2 + \ldots . ,$$

showing that the whole number of distributions, of the given type, of the three sets of objects is 4.

These are in fact

$$\begin{array}{llll} a_1b_1c_1, & a_1b_1c_1, & a_2b_1c_1, & a_2b_2c_2; \\ a_1b_1c_1, & a_1b_1c_1, & a_2b_1c_2, & a_2b_2c_1; \\ a_2b_1c_1, & a_2b_1c_1, & a_1b_1c_1, & a_1b_2c_2; \\ a_2b_1c_1, & a_2b_1c_1, & a_1b_1c_2, & a_1b_2c_1. \end{array}$$

259. Consider n sets of objects of types

$$P_1^{(1)}, \; P_1^{(2)}, \; \ldots . \; P_1^{(r_1)} \qquad (r_1 = n)$$

and let it be demanded to find the number of distributions into n-fold objects of type $$T.$$

260. The two sets of objects whose types are $P_1^{(1)}$, $P_1^{(2)}$ may be distributed into a set of two-fold objects of r_2 different types, say

$$P_{12}^{(1)}, \; P_{12}^{(2)}, \; \ldots . \; P_{12}^{(r_2)}.$$

261. Selecting at pleasure any one of these sets of two-fold objects, say one of type P_{12}, we may distribute it with the set of objects of type $P_1^{(3)}$ and thus obtain a set of three-fold objects which may be of r_3 different types; say these are $$P_{123}^{(1)}, \; P_{123}^{(2)}, \; \ldots . \; P_{123}^{(r_3)}.$$

262. Again selecting one of these sets at pleasure, we may proceed successively until finally we arrive at a set of n-fold objects which may be of r_n different types which may be denoted by

$$P_{123\ldots}^{(1)}, \; P_{123\ldots n}^{(2)}, \; \ldots . \; P_{123\ldots n}^{(r_n)}.$$

One of these types may or may not be identical with the given type T.

263. Performing the process above indicated in all possible ways, we reach all the distributions into n-fold objects of the given type T.

264. The analytical process for arriving at the number of such distributions is simple and elegant. We have merely to combine the successive processes into a single process, as has been already done in the simple case considered which has gone before.

265. Recalling the previous notation and writing down the relation

$$A_{n,\lambda_{n,1}}^{l_{n,1}} A_{n,\lambda_{n,2}}^{l_{n,2}} \ldots = \ldots + L\left(\lambda_{21}^{l_{21}} \lambda_{22}^{l_{22}} \ldots\right)_{a_2} \ldots \left(\lambda_{n-1,1}^{l_{n-1,1}} \lambda_{n-2,2}^{l_{n-2,2}} \ldots\right)_{a_{n-1}} A_{1\lambda_{11}}^{l_{11}} A_{1\lambda_{12}}^{l_{12}} \ldots + \ldots$$

I now state definitely the meaning to be attached to the number L which may be gathered from the preceding.

266. Let there be $n-1$ sets of objects of types

$$\begin{aligned}&(\lambda_{21}^{l_{21}}\lambda_{22}^{l_{22}}\ldots\ldots),\\&(\lambda_{31}^{l_{31}}\lambda_{32}^{l_{32}}\ldots\ldots),\\&\ldots\ldots\ldots\\&\ldots\ldots\ldots\\&(\lambda_{n,1}^{l_{n,1}}\lambda_{n,2}^{l_{n,2}}\ldots\ldots).\end{aligned}$$

The number of distributions into $n-1$-fold sets of objects of type

$$(\lambda_{11}^{l_{11}}\lambda_{12}^{l_{12}}\ldots\ldots)$$

is equal to L.

267. From this statement it is immediately obvious that any substitution whatever may be impressed upon the $n-1$ partitions

$$(\lambda_{s1}^{l_{s1}}\lambda_{s2}^{l_{s2}}\ldots\ldots),$$
$$(s=2,3,\ldots\ldots n),$$

and hence in the above written identity the number L is unchanged when any substitution is impressed upon these $n-1$ partitions.

268. This distribution theorem thus involves an intuitive proof of the general law of symmetry.

269. It should be borne in mind that the general identity is constructed through the medium of the identities

$$\begin{aligned}&K_{1,1}=(1)_{a_2}A_{1,1},\\&K_{1,2}=(2)_{a_2}A_{1,2}+(1^2)_{a_2}A_{1,1}^2,\\&K_{1,3}=(3)_{a_2}A_{1,3}+(21)_{a_2}A_{1,2}A_{1,1}+(1^3)_{a_2}A_{1,1}^3,\\&\ldots\ldots\ldots\ldots\ldots\ldots\ldots\ldots\ldots\end{aligned}$$

$$s=2,3,\ldots,(n-3)\left\{\begin{aligned}&K_{s,1}=(1)_{a_s+1}K_{s-1,1},\\&K_{s,2}=(2)_{a_s+1}K_{s-1,2}+(1^2)_{a_s+1}K_{s-1,1}^2,\\&K_{s,3}=(3)_{a_s+1}K_{s-1,3}+(21)_{a_s+1}K_{s-1,2}K_{s-1,1}+(1^3)_{a_s+1}K_{s-1,1}^3,\\&\ldots\ldots\ldots\ldots\ldots\ldots\ldots\ldots\ldots\end{aligned}\right.$$

$$\begin{aligned}&A_{n,1}=(1)_{a_n-1}K_{n-3,1},\\&A_{n,2}=(2)_{a_n-1}K_{n-3,2}+(1^2)_{a_n-1}K_{n-3,1}^2,\\&A_{n,3}=(3)_{a_n-1}K_{n-3,3}+(21)_{a_n-1}K_{n-3,2}K_{n-3,1}+(1^3)_{a_n-1}K_{n-3,1}^3\\&\ldots\ldots\ldots\ldots\ldots\ldots\ldots\ldots\ldots\end{aligned}$$

270. I propose to obtain the general result for the case of $n=4$ and three objects in each set. There are then three sets of objects.

We have

$$K_{1,1}=(1)_{a_2}A_{1,1},$$
$$K_{1,2}=(2)_{a_2}A_{1,2}+(1^2)_{a_2}A^2_{1,1},$$
$$K_{1,3}=(3)_{a_2}A_{1,3}+(21)_{a_2}A_{1,2}A_{1,1}+(1^3)_{a_2}A^3_{1,1},$$
$$\cdots\cdots\cdots\cdots\cdots$$
$$A_{4,1}=(1)_{a_3}K_{1,1},$$
$$A_{4,2}=(2)_{a_3}K_{1,2}+(1^2)_{a_3}K^2_{1,1},$$
$$A_{4,3}=(3)_{a_3}K_{1,3}+(21)_{a_3}K_{1,2}K_{1,1}+(1^3)_{a_3}K^3_{1,1},$$
$$\cdots\cdots\cdots\cdots\cdots$$

in order to eliminate the quantities K.

271. The result is, writing $(3)_{a_s}=(3)_s$ for brevity, and so on,

$$A_{4,3}=(3)_3(3)_2A_{1,3}+\{(3)_3(21)_2+(21)_3(3)_2\}A_{1,2}A_{1,1}+\{(3)_3(1^3)_2+(1^3)_3(3)_2\}A^3_{1,1}+(21)_3(21)_2(A_{12}A_{11}+A^3_{11})+3\{(21)_3(1^3)_2+(1^3)_3(21)_2\}A^3_{1,1}+6(1^3)_3(1^3)_2A^3_{1,1},$$

$$A_{4,2}A_{4,1}=(3)_3(3)_2A_{12}A_{11}+\{(3)_3(21)_2+(21)_3(3)_2\}(A_{12}A_{11}+A^3_{11})+3\{(3)_3(1^3)_2+(1^3)_3(3)_2\}A^3_{11}+(21)_3(21)_2(A_{12}A_{11}+4A^3_{11})+9\{(21)_3(1^3)_2+(1^3)_3(21)_2\}A^3_{11}+18(1^3)_3(1^3)_2A^3_{11},$$

$$A^3_{4,1}=(3)_3(3)_2A^3_{11}+3\{(3)_3(21)_2+(21)_3(3)_2\}A^3_{11}+6\{(3)_3(1^3)_2+(1^3)_3(3)_2\}A^3_{11}+9(21)_3(21)_2A^3_{11}+18\{(21)_3(1^3)_2+(1^3)_3(21)_2\}A^3_{11}+36(1^3)_3(1^3)_2A^3_{11}.$$

272. The symmetry is manifest, and representing by

$$[(3)(21)(1^3)]$$

a distribution of three sets of objects of types (3), (21) and (1^3) and so forth, we obtain for the numbers of the different types

$$[(3),(3),(3)]=A_{1,3},$$
$$[(3),(3),(21)]=A_{1,2}A_{1,1},$$
$$[(3),(3),(1^3)]=A^3_{1,1},$$
$$[(3),(21),(21)]=A_{1,2}A_{1,1}+A^3_{1,1},$$
$$[(3),(21),(1^3)]=3A^3_{1,1},$$
$$[(3),(1^3),(1^3)]=6A^3_{1,1},$$
$$[(21),(21),(21)]=A_{1,2}A_{1,1}+4A^3_{1,1},$$
$$[(21),(21),(1^3)]=9A^3_{1,1},$$
$$[(21),(1^3),(1^3)]=18A^3_{1,1},$$
$$[(1^3),(1^3),(1^3)]=36A^3_{1,1},$$

where for example the seventh of these results is to be interpreted as indicating that of three sets of objects of types (21), (21) and (21) respectively, there is one distribution of type (21) and four of type (1^3).

273. Taking the objects to be

$$a_1,\ a_1,\ a_2 \text{ of type } (21),$$
$$b_1,\ b_1,\ b_2 \text{ of type } (21),$$
$$c_1,\ c_1,\ c_2 \text{ of type } (21),$$

the five distributions are in fact

$$a_1b_1c_1,\ a_1b_1c_1,\ a_2b_2c_2 \text{ of type } (21)$$

and

$$\left.\begin{array}{l} a_1b_1c_1,\ a_1b_1c_2,\ a_2b_2c_1 \\ a_1b_1c_1,\ a_1b_2c_2,\ a_2b_1c_1 \\ a_1b_2c_1,\ a_1b_1c_1,\ a_2b_1c_2 \\ a_1b_1c_2,\ a_1b_2c_1,\ a_2b_1c_1 \end{array}\right\} \text{of type } (1^3),$$

and observe that there are no others.

274. I have extended the subject of these four memoirs in a paper under the title "Memoir on Symmetric Functions of the Roots of Systems of Equations" in the Philosophical Transactions of the Royal Society of London, Vol. 181 (1890), A, pp. 481–536.

WOOLWICH, *January* 10, 1890.

SMALL CONTRIBUTION TO COMBINATORY ANALYSIS

By Major P. A. MACMAHON.

[Received January 23rd, 1917.—Read February 1st, 1917.]

THE enumerating generating function for the number of partitions of the number n into m or fewer parts is

$$\frac{1}{(1-x)(1-x^2)(1-x^3)\dots(1-x^m)}.$$

In the Theory of Distributions we consider the number of ways of distributing objects of given specification into boxes of given specification. If there be n objects, their specification may be given by any partition of n. Thus, if the specification be the partition

$$(n_1 n_2 n_3 \dots n_r)$$

of the number n, we mean that n_1 objects are of one kind, n_2 of a second, n_3 of a third, and so on. The objects might be denoted by the factors of the product

$$a_1^{n_1} a_2^{n_2} a_3^{n_3} \dots a_r^{n_r}.$$

Similarly the m boxes might have a specification

$$(m_1 m_2 m_3 \dots m_s),$$

any partition of m. The boxes might be denoted by the factors of the product

$$A_1^{m_1} A_2^{m_2} A_3^{m_3} \dots A_s^{m_s}.$$

The problem of enumerating the partitions of n into m or fewer parts is the same as that of enumerating the distribution of n objects of specification (n) into m or fewer boxes of specification (m).

From this point of view we may generalize the enumeration of ordinary partitions by considering the number of ways of distributing n objects of specification (n) into m or fewer boxes of specification

$$(m_1 m_2 m_3 \dots m_s).$$

We may, without loss of generality, take $m_1, m_2, m_3, \ldots, m_s$ to be integers in descending order of magnitude.

For brevity denote the algebraic fraction

$$\frac{1}{(1-x)(1-x^2)(1-x^3)\ldots(1-x^m}$$

by

$$X_m.$$

The slightest consideration suffices to establish that the distributions before us are given by the coefficient of x^n in the expansion of the fraction

$$X_{m_1}X_{m_2}X_{m_3}\ldots X_{m_s},$$

which, if

$$(p_1 p_2 p_3 \ldots p_t)$$

be the partition conjugate to

$$(m_1 m_2 m_3 \ldots m_s),$$

may be written

$$\frac{1}{(1-x)^{p_1}(1-x^2)^{p_2}(1-x^3)^{p_3}\ldots(1-x^t)^{p_t}},$$

wherein, of course, $p_1, p_2, p_3, \ldots, p_t$ are integers in descending order of magnitude.

Observe that, if $m_1 = m$, we have

$$(p_1, p_2, p_3, \ldots, p_t) = (1, 1, 1, \ldots, 1),$$

and $t = m$, the case of ordinary partition.

Moreover, if $\quad m_1 = m_2 = m_3 = \ldots = m_s = 1,$

we reach the case of distribution of objects of specification (n) into boxes of specification $(1, 1, 1, \ldots, 1)$, wherein there are s units.

Here there are no similarities between the boxes, and corresponding thereto we have the compositions of the number n into s or fewer parts, the zero part not being excluded.

The generating function is $\quad (1-x)^{-s},$

because (s) is the partition conjugate to (1^s).

As an example of the general theorem consider the distribution of objects of specification (n) into boxes of specification

$$(221).$$

Since (32) is the partition conjugate to (221), we derive the generating function

$$(1-x)^{-3}(1-x^2)^{-2}.$$

Expanded to a few terms this is

$$1+3x+8x^2+16x^3+30x^4+50x^5+\dots.$$

To verify the coefficient 30 of x^4, suppose the boxes to be denoted by A, A, B, B, C, and suppose m_A to denote m objects placed in a box A. We have the distributions

4_A	3_A1_A	3_B1_C	2_A2_A	$2_A1_A1_B$	$2_C1_A1_B$	$1_A1_A1_B1_B$
4_B	3_A1_B	3_B1_A	2_B2_B	$2_A1_A1_C$	$2_A1_B1_B$	$1_A1_A1_B1_C$
4_C	3_B1_B	3_C1_A	2_A2_B	$2_B1_B1_C$	$2_B1_A1_A$	$1_A1_B1_B1_C$
	3_A1_C	3_C1_B	2_A2_C	$2_A1_B1_C$	$2_C1_A1_A$	
			2_B2_C	$2_B1_B1_A$	$2_C1_B1_B$	
				$2_B1_A1_C$		

The theorem that has been established is as follows:—

"The coefficient of x^n in the ascending expansion of the algebraic fraction

$$\frac{1}{(1-x)^{p_1}(1-x^2)^{p_2}(1-x^3)^{p_3}\dots(1-x^t)^{p_t}},$$

where $p_1, p_2, p_3, \dots, p_t$ are integers in descending order of magnitude, is equal to the number of ways of distributing n objects of specification (n) into boxes of specification

$$(m_1m_2m_3\dots m_s),$$

where $(m_1m_2m_3\dots m_s)$ is the partition conjugate to the partition

$$(p_1p_2p_3\dots p_t)."$$

The Algebra of Symmetric Functions. By Major P. A. MACMAHON.

[*Read* 30 October 1922.]

1. The necessity for an algebra of symmetric functions arose in the first place in 1884 from the circumstance that, in the Theory of Invariants, the seminvariants of binary quantics were shewn to be transformations of non-unitary symmetric functions of the roots of an equation of infinite order*, where a non-unitary symmetric function is defined to be

$$\Sigma\alpha_1^{p_1}\alpha_2^{p_2}\alpha_3^{p_3}\dots\alpha_s^{p_s},$$

where the number of quantities α is infinite, s may be any integer and exponents p may be any integers $\nless 2$.

Cayley† was thus enabled to attack the problem of determining the seminvariants of the quantic of infinite order which, *qua* degree, are irreducible. In the paper (*loc. cit.*) he introduced an algorithm for the multiplication of two non-unitary symmetric functions and then for the first time definitely broached an algebra of these functions.

Cayley employs the partition notation of symmetric functions and the algorithm is the production of an abbreviated method of multiplication on combinatory principles. In fact, one part of the process consists, if the two functions to be multiplied be

$$(p_1p_2p_3\dots p_s),\quad (q_1q_2q_3\dots q_t),$$

in placing *all or any* of the numbers $q_1, q_2, \dots q_t$ underneath the numbers $p_1, p_2, \dots p_s$ in all the really distinct ways and in assigning a number to denote the frequency of each way. The rest of the algorithm is automatically completed and does not involve a possibility of errors in counting.

He was able to give a general formula for the multiplication

$$(3^a2^b)(3^c2^d),$$

a, b, c, d denoting repetitions of parts, and was thence able to advance the theory of the perpetuant seminvariants.

2. The next step in the algebra was taken by the present writer‡ who introduced the Hammond operators $D_1, D_2, D_3, \dots$ to the subject. The result of multiplication, being a linear function of monomial symmetric functions, we write

$$(p_1p_2p_3\dots)(q_1q_2q_3\dots) = \dots + C_{r_1r_2r_3\dots}(r_1r_2r_3\dots) + \dots.$$

* MacMahon, 'Seminvariants and Symmetric Functions,' *A.J.M.* Vol. VI, p. 131.

† 'A Memoir on Seminvariants,' *A.J.M.* Vol. VII, 1885.

‡ MacMahon, 'The Multiplication of Symmetric Functions,' *The Messenger of Mathematics*, New Series, No. 167, March 1885; *Combinatory Analysis*, Vol. I, p. 43.

Then we have by the laws appertaining to the operator

$$D_{r_1}D_{r_2}D_{r_3}\dots(p_1p_2p_3\dots)(q_1q_2q_3\dots)$$
$$= C_{r_1r_2r_3}\dots D_{r_1}D_{r_2}D_{r_3}\dots(r_1r_2r_3\dots)$$
$$= C_{r_1r_2r_3}\dots;$$

that is to say that the number $C_{r_1r_2r_3}\dots$ is found as the direct result of the performance of certain differential operations.

Moreover, this method is directly applicable to the multiplication of *three or more* symmetric functions. The method depends upon the law by which a Hammond operator is performed upon a product of two or more functions. The way in which D_r is performed upon a product of s functions depends upon the compositions of the number r into s parts (zero parts being taken into account). These compositions are enumerated by

$$\binom{r+s-1}{r}$$

because this number is the coefficient of x^r in $(1-x)^{-s}$.

The operation of D_r divides up, therefore, into

$$\binom{r+s-1}{r}$$

separate operations, one of which is

$$(D_{c_1}F_1)(D_{c_2}F_2)\dots(D_{c_s}F_s),$$

$c_1c_2\dots c_s$ being a composition of r into s parts

$$c_1+c_2+\dots+c_s=r.$$

Ex. gr.

$$D_2F_1F_2F_3$$
$$=(D_2F_1)(D_0F_2)(D_0F_3)+(D_0F_1)(D_2F_2)(D_0F_3)$$
$$+(D_0F_1)(D_0F_2)(D_2F_3)+(D_0F_1)(D_1F_2)(D_1F_3)$$
$$+(D_1F_1)(D_0F_2)(D_1F_3)+(D_1F_1)(D_1F_2)(D_0F_3)$$

because 2 has, into 3 parts, the six compositions

200, 020, 002

011, 101, 110.

Not all of these separate operations may be effective because $D_{c_k}F_k$ is zero unless F_k involves a part c_k.

The complete operation

$$D_{r_1}D_{r_2}D_{r_3}\dots F_1F_2F_3\dots F_s$$

is performed by associating single compositions of the numbers $r_1, r_2, r_3, \dots$, each into s parts, in all possible ways. If r_k has R_k compositions into s parts we may have to deal with

$$R_1R_2R_3\dots \text{ separate operations.}$$

Each of these may be denoted by a tableau.

For if a composition of r_k be

$$c_{1k},\ c_{2k},\ c_{3k},\ \dots\ c_{sk}$$

a tableau such as

c_{11}	c_{21}	c_{31}	.	.	c_{s1}
c_{12}	c_{22}	c_{32}	.	.	c_{s2}
c_{13}	c_{23}	c_{33}	.	.	c_{s3}
.	.	.	.	.	.
.	.	.	.	.	.

denotes one of the $R_1 R_2 R_3 \dots$ separate operations. But some of these may vanish as mentioned above. Those which do not vanish form a set of tableaux. These possess a common property, viz. the sums of the numbers in the successive rows are $r_1, r_2, r_3, \dots$ respectively and the collections of numbers in the successive columns are those which define the symmetric functions $F_1, F_2, F_3, \dots F_s$ respectively. If one is asked: What is the number of tableaux each of which possesses the property above-defined, the mathematician's answer is

$$D_{r_1} D_{r_2} D_{r_3} \dots F_1 F_2 \dots F_s,$$

which denotes a number which is readily evaluated. We have, in fact, a good example of the use of the Hammond operators in solving questions of enumeration of the Magic Square Class. The multiplication of symmetric functions invariably supplies an example of this kind of enumeration.

3. Let U be any linear function of symmetric functions of the same weight w.

Consider the multiplication

$$(1^{r_1})\, U,$$

and therein the term involving the function

$$(p_1^{\pi_1} p_2^{\pi_2} p_3^{\pi_3} \dots) \text{ of weight } w + r.$$

We have

$$D_{p_1} (1^{r_1})\, U = \{(1^{r_1})\, D_{p_1} + (1^{r_1 - 1})\, D_{p_1 - 1}\}\, U,$$

$$D_{p_1}^2 (1^{r_1})\, U = \{(1^{r_1})\, D_{p_1}^2 + 2\,(1^{r_1-1})\, D_{p_1} D_{p_1-1} + (1^{r_1-2})\, D_{p_1-1}^2\}\, U.$$

Now put $(1^{r_1}) = a^{r_1}$ symbolically, so that these results may be written

$$D_{p_1} (1^{r_1})\, U = a^{r_1} (D_{p_1} + a^{-1} D_{p_1 - 1})\, U,$$

$$D_{p_1}^2 (1^{r_1})\, U = a^{r_1} (D_{p_1} + a^{-1} D_{p_1 - 1})^2\, U,$$

and generally

$$D_{p_1}^{\pi_1} (1^{r_1})\, U = a^{r_1} (D_{p_1} + a^{-1} D_{p_1 - 1})^{\pi_1}\, U,$$

and more generally

$$D_{p_1}^{\pi_1} D_{p_2}^{\pi_2} D_{p_3}^{\pi_3} \ldots (1^{r_1})\, U$$

$$= a^{r_1} (D_{p_1} + a^{-1} D_{p_1-1})^{\pi_1} (D_{p_2} + a^{-1} D_{p_2-1})^{\pi_2} (D_{p_3} + a^{-1} D_{p_3-1})^{\pi_3} \ldots U.$$

When this operator is developed and applied to U it is obvious that every term which involves an operator product $D_{s_1}^{\sigma_1} D_{s_2}^{\sigma_2} \ldots$ of a weight $> w$ causes U to vanish. This happens whenever the associated power of a is positive.

When the power of a is zero the attached D product is of weight w and either does or does not cause U to vanish. If U involves a term

$$C_{s_1^{\sigma_1} s_2^{\sigma_2} s_3^{\sigma_3} \ldots} (s_1^{\sigma_1} s_2^{\sigma_2} s_3^{\sigma_3} \ldots)$$

the operation of $D_{s_1}^{\sigma_1} D_{s_2}^{\sigma_2} D_{s_3}^{\sigma_3} \ldots$ produces the number $C_{s_1^{\sigma_1} s_2^{\sigma_2} s_3^{\sigma_3} \ldots}$. In the contrary case the operator produces zero.

The negative powers of a in the development have no real existence symbolically so that all terms involving such must be put equal to zero.

It thus appears that the effective operator upon U consists entirely of the terms in the developments which are *free from the symbol a*.

If the process of picking out this portion of the whole operator be denoted by T_a, we may write our result

$$D_{p_1}^{\pi_1} D_{p_2}^{\pi_2} D_{p_3}^{\pi_3} \ldots (1^{r_1})\, U$$

$$= T_a \{a^{r_1} (D_{p_1} + a^{-1} D_{p_1-1})^{\pi_1} (D_{p_2} + a^{-1} D_{p_2-1})^{\pi_2}$$

$$(D_{p_3} + a^{-1} D_{p_3-1})^{\pi_3} \ldots\} \, U.$$

This theorem enables us, from the known value of U as a sum of monomial functions, to proceed to the similar expression of

$$(1^{r_1})\, U.$$

Ex. gr. Suppose

$$U = (1^2)^2 = (2^2) + 2\,(21^2) + 6\,(1^4),$$

and that we require the term of $(1^2)^3$ which involves

$$(321),$$

$$D_3 D_2 D_1 (1^2) . (1^2)^2$$

$$= T_a \{a^2 (D_3 + a^{-1} D_2)(D_2 + a^{-1} D_1)(D_1 + a^{-1})\} (1^2)^2$$

$$= T_a \{a D_2 (D_2 + a^{-1} D_1)(D_1 + a^{-1})\} (1^2)^2$$

(because, since U is here of degree 2, we may put $D_3 = 0$)

$$= (D_2^2 + D_2 D_1^2) \{(2^2) + 2\,(21^2) + 6\,(1^4)\} = 1 + 2 = 3.$$

Thence $\qquad (1^2)^3 = \ldots + 3\,(321) + \ldots.$

4. The expression of

$$D_{p_1}^{\pi_1} D_{p_2}^{\pi_2} D_{p_3}^{\pi_3} \dots (1^{r_1})\, U$$

may be presented in a more convenient and suggestive form by the employment of a further symbolism.

If we write

$$\pi_k(\pi_k - 1)(\pi_k - 2) \dots (\pi_k - s + 1) = \pi_k^s \text{ symbolically,}$$

we have

$$(D_{p_k} + a^{-1} D_{p_{k-1}})^{\pi_1} = D_{p_k}^{\pi_1} \exp \pi_k a^{-1} \frac{D_{p_{k-1}}}{D_{p_k}} \text{ effectively,}$$

and thence the expression

$$T_a \left\{ a^{r_1} D_{p_1}^{\pi_1} D_{p_2}^{\pi_2} D_{p_3}^{\pi_3} \dots \exp\left(\pi_1 \frac{D_{p_1-1}}{D_{p_1}} + \pi \frac{D_{p_2-1}}{D_{p_2}} + \pi_3 \frac{D_{p_3-1}}{D_{p_3}} + \dots\right) a^{-1} \right\} U$$

$$= \frac{1}{r_1!} D_{p_1}^{\pi_1} D_{p_2}^{\pi_2} D_{p_3}^{\pi_3} \dots \left(\pi_1 \frac{D_{p_1-1}}{D_{p_1}} + \pi_2 \frac{D_{p_2-1}}{D_{p_2}} + \pi_3 \frac{D_{p_3-1}}{D_{p_3}} + \dots\right)^{r_1} U,$$

a result which, as will be seen presently, is generalizable.

Applied to the particular case we find

$$\frac{1}{2!} D_3 D_2 D_1 \left(\frac{D_2}{D_3} + \frac{D_1}{D_2} + \frac{1}{D_1}\right)^2 (1^2)^2$$

$$= (D_2^2 + D_2 D_1^2 + D_3 D_1)\{(2)^2 + 2(21^2) + 6(1^4)\}$$

$$= 1 + 2 = 3 \text{ as above.}$$

5. Next consider the product

$$(2^{r_2} 1^{r_1})\, U.$$

We have

$$D_{p_1}(2^{r_2} 1^{r_1})\, U = \{(2^{r_2} 1^{r_1}) D_{p_1} + (2^{r_2} 1^{r_1-1}) D_{p_1-1} + (2^{r_2-1} 1^{r_1}) D_{p_1-2}\}\, U,$$

$$D_{p_1}^2 (2^{r_2} 1^{r_1})\, U$$

$$= \left\{\begin{aligned} &(2^{r_2} 1^{r_1}) D_{p_1}^2 + (2^{r_2} 1^{r_1-1}) D_{p_1} D_{p_1-1} \\ &\qquad + (2^{r_2-1} 1^{r_1}) D_{p_1} D_{p_1-2} \\ &+ (2^{r_2} 1^{r_1-1}) D_{p_1} D_{p_1-1} + (2^{r_2} 1^{r_1-2}) D_{p_1-1}^2 \\ &\qquad + (2^{r_2-1} 1^{r_1-1}) D_{p_1-1} D_{p_1-2} \\ &+ (2^{r_2-1} 1^{r_1}) D_{p_1} D_{p_1-2} + (2^{r_2-1} 1^{r_1-1}) D_{p_1-1} D_{p_1-2} \\ &\qquad + (2^{r_2-2} 1^{r_1}) D_{p_1-2} D_{p_2-2} \end{aligned}\right\} U.$$

Thence writing symbolically

$$(2^{r_2} 1^{r_1}) = a^{r_1} b^{r_2},$$

so that symbolically

$$D_{p_1}(2^{r_2} 1^{r_1})\, U = a^{r_1} b^{r_2} (D_{p_1} + a^{-1} D_{p_1-1} + b^{-1} D_{p_1-2})\, U,$$

$$D_{p_1}^2 (2^{r_2} 1^{r_1})\, U = a^{r_1} b^{r_2} (D_{p_1} + a^{-1} D_{p_1-1} + b^{-1} D_{p_1-2})^2\, U,$$

and generally

$$D_{p_1}^{\pi_1}(2^{r_2}1^{r_1})\,U = a^{r_1}b^{r_2}(D_{p_1} + a^{-1}D_{p_1-1} + b^{-1}D_{p_1-2})^{\pi_1}\,U,$$

and more generally

$$D_{p_1}^{\pi_1}D_{p_2}^{\pi_2}D_{p_3}^{\pi_3}\ldots(2^{r_2}1^{r_1})\,U$$

$$= a^{r_1}b^{r_2}(D_{p_1} + a^{-1}D_{p_1-1} + b^{-1}D_{p_1-2})^{\pi_1}(D_{p_2} + a^{-1}D_{p_2-1} + b^{-1}D_{p_2-2})^{\pi_2}$$

$$(D_{p_3} + a^{-1}D_{p_3-1} + b^{-1}D_{p_3-2})^{\pi_3}\ldots U.$$

By reasoning similar to that advanced above we have here the coefficient of a^0b^0 as the effective part in the development of the operator, and denoting this by T_{ab}, we find that

$$D_{p_1}^{\pi_1}D_{p_2}^{\pi_2}D_{p_3}^{\pi_3}\ldots(2^{r_2}1^{r_1})\,U$$

$$= T_{ab}\{a^{r_1}b^{r_2}(D_{p_1} + a^{-1}D_{p_1-1} + b^{-1}D_{p_1-2})^{\pi_1}$$

$$\times(D_{p_2} + a^{-1}D_{p_2-1} + b^{-1}D_{p_2-2})^{\pi_2}$$

$$\times(D_{p_3} + a^{-1}D_{p_3-1} + b^{-1}D_{p_3-2})^{\pi_3}\ldots\}\,U,$$

and introducing as above the symbolism

$$\pi_k(\pi_k - 1)\ldots(\pi_k - s + 1) = \pi_k^s,$$

we find finally

$$\frac{1}{r_1!\,r_2!}D_{p_1}^{\pi_1}D_{p_2}^{\pi_2}D_{p_3}^{\pi_3}\ldots\left(\pi_1\frac{D_{p_1-1}}{D_{p_1}} + \pi_2\frac{D_{p_2-1}}{D_{p_2}} + \ldots\right)^{r_1}$$

$$\left(\pi_1\frac{D_{p_1-2}}{D_{p_1}} + \pi_2\frac{D_{p_2-2}}{D_{p_2}} + \ldots\right)^{r_2}U$$

as the symbolic form of the operator which when performed upon U produces the coefficients of the term

$$(p_1^{\pi_1}p_2^{\pi_2}p_3^{\pi_3}\ldots)$$

in the development of

$$(2^{r_2}1^{r_1})\,U.$$

6. We now readily proceed to the general result

$$D_{p_1}^{\pi_1}D_{p_2}^{\pi_2}D_{p_3}^{\pi_3}\ldots(\ldots 3^{r_3}2^{r_2}1^{r_1})\,U$$

$$= T_{abc\ldots}\{a^{r_1}b^{r_2}c^{r_3}\ldots(D_{p_1} + a^{-1}D_{p_1-1} + b^{-1}D_{p_1-2} + c^{-1}D_{p_1-3} + \ldots)^{\pi_1}$$

$$\times(D_{p_2} + a^{-1}D_{p_2-1} + b^{-1}D_{p_2-2} + c^{-1}D_{p_2-3} + \ldots)^{\pi_2}$$

$$\times(D_{p_3} + a^{-1}D_{p_3-1} + b^{-1}D_{p_3-2} + c^{-1}D_{p_3-3} + \ldots)^{\pi_3}$$

$$\times\ \ldots\ldots\ldots\ldots\ldots\ldots\ldots\ldots\}\,U,$$

and this, adopting the π symbolism as above, is

$$\left\{\frac{1}{r_1!\,r_2!\,r_3!\,\dots} D_{p_1}^{\pi_1} D_{p_2}^{\pi_2} D_{p_3}^{\pi_3} \dots \right.$$
$$\times \left(\pi_1 \frac{D_{p_1-1}}{D_{p_1}} + \pi_2 \frac{D_{p_2-1}}{D_{p_2}} + \pi_3 \frac{D_{p_3-1}}{D_{p_3}} + \dots\right)^{r_1}$$
$$\times \left(\pi_1 \frac{D_{p_1-2}}{D_{p_1}} + \pi_2 \frac{D_{p_2-2}}{D_{p_2}} + \pi_3 \frac{D_{p_3-2}}{D_{p_3}} + \dots\right)^{r_2}$$
$$\times \left(\pi_1 \frac{D_{p_1-3}}{D_{p_1}} + \pi_2 \frac{D_{p_2-3}}{D_{p_2}} + \pi_3 \frac{D_{p_3-3}}{D_{p_3}} + \dots\right)^{r_3}$$
$$\left. \times \dots\dots\dots\dots\dots\dots \right\} U,$$

the value of the coefficients of the term

$$(p_1^{\pi_1} p_2^{\pi_2} p_3^{\pi_3} \dots),$$

in the development of $(\dots 3^{r_3} 2^{r_2} 1^{r_1})\, U$.

7. I will now apply this result to the cases dealt with by Cayley (*loc. cit.*).

The first multiplication he took was

$$(2^\alpha)\,(2^\beta),$$

where I think it best to retain his original notation. I take the precisely equivalent case

$$(1^\alpha)\,(1^\beta).$$

Here U is (1^β) and the general form of monomial term in the product is

$$(2^A\, 1^{\alpha+\beta-2A}).$$

To find the coefficients of this term we have, as above,

$$D_2^A D_1^{\alpha+\beta-2A}\,(1^\alpha).(1^\beta)$$
$$= T_a \{a^\alpha (D_2 + a^{-1} D_1)^A (D_1 + a^{-1})^{\alpha+\beta-2A}\}.(1^\beta)$$
$$= T_a \{a^{\alpha-A} D_1^A (D_1 + a^{-1})^{\alpha+\beta-2A}\}.(1^\beta)$$
$$= \binom{\alpha+\beta-2A}{\alpha-A} D_1^\beta\,(1^\beta) = \binom{\alpha+\beta-2A}{\alpha-A}.$$

So that $$(1^\alpha)\,(1^\beta) = \sum_A \binom{\alpha+\beta-2A}{\alpha-A} (2^A 1^{\alpha+\beta-2A})$$

(cf. *loc. cit.* and *Collected Papers*, Vol. XII, p. 244).

It will be observed that the present process is not concerned with 'frequencies' and 'multiplicities' but is altogether algebraical.

Cayley's second case is

$$(3^\alpha 2^\beta)\,(3^\gamma 2^\delta)$$

because he was only concerned with non-unitary forms.

Here $$U = (3^\gamma 2^\delta),$$

and the general form of monomial in the product is

$$(6^A 5^B 4^C 3^D 2^E),$$

where $D = \alpha + \gamma - 2A - B, \quad E = \beta + \delta - B - 2C;$

to find the coefficient of this general term we have

$$T_{bc}\{b^\beta c^\alpha (D_6 + b^{-1} D_4 + c^{-1} D_3)^A (D_5 + b^{-1} D_3 + c^{-1} D_2)^B$$
$$\times (D_4 + b^{-1} D_2 + c^{-1} D_1)^C (D_3 + b^{-1} D_1 + c^{-1})^D$$
$$\times (D_2 + b^{-1})^E\} . (3^\gamma 2^\delta).$$

Since the operand is annihilated by each of the four operators D_6, D_5, D_4, D_1 we put each of these equal to zero and find

$$T_{bc}\{b^{\beta-C} c^{\alpha-A} D_3^A (b^{-1} D_3 + c^{-1} D_2)^B D_2^C$$
$$(D_3 + c^{-1})^D (D_2 + b^{-1})^E\} . (3^\gamma 2^\delta),$$

whence we observe at once that the sought coefficient is expressible as a linear function of terms each of which is a product of these binomial coefficients (cf. *loc. cit.* p. 246, the final conclusion).

Before calculating we put

$$D = \alpha + \gamma - 2A - B, \quad E = \beta + \delta - B - 2C,$$

and if (with Cayley) we take

$$y, z, r \text{ to be integers}$$
$$\not< B, \not< D, \not< E \text{ respectively},$$

we find as the coefficient sought

$$\Sigma \binom{B}{y} \binom{D}{z} \binom{E}{r},$$

the summation being controlled by the relations

$$-y + r = \delta - B - C$$
$$y + z = \gamma - A.$$

8. Having thus verified the results of Cayley by pure algebra I now resume, from a general point of view, by considering the next case in order to $(1^\alpha)(1^\beta)$, viz.

$$(2^{\alpha_2} 1^{\alpha_1})(2^{\beta_2} 1^{\beta_1}).$$

Here $U = (2^{\beta_2} 1^{\beta_1})$ and

$$D_4^A D_3^B D_2^C D_1^D (2^{\alpha_2} 1^{\alpha_1}) . (2^{\beta_2} 1^{\beta_1})$$
$$= T_{ab}\{a^{\alpha_1} b^{\alpha_2} (D_4 + a^{-1} D_3 + b^{-1} D_2)^A \ (D_3 + a^{-1} D_2 + b^{-1} D_1)^B$$
$$(D_2 + a^{-1} D_1 + b^{-1} \quad)^C \ (D_1 + a^{-1} \qquad)^D\}$$
$$(2^{\beta_2} 1^{\beta_1}).$$

Since here $D_4 = D_3 = 0$ this is

$$T_{ab}\{a^{a_1}b^{a_2-A}D_2^A(a^{-1}D_2+b^{-1}D_1)^B$$
$$\times(D_2+a^{-1}D_1+b^{-1})^C(D_1+a^{-1})^D\}(2^{\beta_2}1^{\beta_1}),$$

and we observe that the coefficient sought is a linear function of terms each of which is a product of a binomial, a trinomial and a binomial coefficient, in the order named.

We say that the structure of the coefficient is

232.

Observe that in the case $(1^a)(1^\beta)$ the coefficient is a single binomial coefficient so that its structure is denoted by

2.

9. It is not necessary to carry this case further and I proceed to the product

$$(3^{a_3}2^{a_2}1^{a_1})(3^{\beta_3}2^{\beta_2}1^{\beta_1}).$$

We have to consider

$$D_6^A D_5^B D_4^C D_3^D D_2^E D_1^F(3^{a_3}2^{a_2}1^{a_1}).(3^{\beta_3}2^{\beta_2}1^{\beta_1}),$$

where U is $(3^{\beta_3}2^{\beta_2}1^{\beta_1})$.

Merely writing down the D suffixes which occur in the operator factors, viz.

(6543) (5432) (4321) (3210) (210) (10),

and striking out the numbers greater than 3 we obtain

(3) (32) (321) (3210) (210) (10),

and we observe that the character of the sought coefficient is

23432.

Again for the case

$$(4^{a_4}3^{a_3}2^{a_2}1^{a_1}).(4^{\beta_4}3^{\beta_3}2^{\beta_2}1^{\beta_1}),$$

we have first of all

(87654) (76543) (65432) (54321) (43210) (3210) (210) (10),

and thence

(4) (43) (432) (4321) (43210) (3210) (210) (10),

shewing that the character of the coefficient is

2345432.

10. It is now clear that in the case of the product

$$(k^{a_k}\dots 3^{a_3}2^{a_2}1^{a_1})(k^{\beta_k}\dots 3^{\beta_3}2^{\beta_2}1^{\beta_1}),$$

the character of the coefficient is

$$234\dots k,\ k+1,\ k\dots 432.$$

11. The same principle is available for the product of three functions through the composition of numbers into three parts.

Consider the simplest case

$$(1^\alpha)(1^\beta)U,$$

where U is any given function of weight w.

$$D_{p_1}(1^\alpha)(1^\beta)U = [(1^\alpha)(1^\beta)D_{p_1} + \{(1^{\alpha-1})(1^\beta) + (1^\alpha)(1^{\beta-1})\}D_{p_1-1} + (1^{\alpha-1})(1^{\beta-1})D_{p_1-2}]U,$$

wherein writing $(1^\alpha) = a^\alpha$, $(1^\beta) = b^\beta$ symbolically

$$D_{p_1}(1^\alpha)(1^\beta)U = a^\alpha b^\beta\left\{D_{p_1} + \left(\frac{1}{a}+\frac{1}{b}\right)D_{p_1-1} + \frac{1}{ab}D_{p_1-2}\right\}U,$$

leading to

$$D_{p_1}^{\pi_1}(1^\alpha)(1^\beta)U = a^\alpha b^\beta\left\{D_{p_1} + \left(\frac{1}{a}+\frac{1}{b}\right)D_{p_1-1} + \frac{1}{ab}D_{p_1-2}\right\}^{\pi_1}U,$$

and if

$$(p_1^{\pi_1}p_2^{\pi_2}p_3^{\pi_3}\ldots)$$

be a partition of $w+\alpha+\beta$, representative of a symmetric function which appears in the development of

$$(1^\alpha)(1^\beta)U,$$

$$D_{p_1}^{\pi_1}D_{p_2}^{\pi_2}D_{p_3}^{\pi_3}\ldots(1^\alpha)(1^\beta)U$$

$$= T_{ab}\left[a^\alpha b^\beta\left\{D_{p_1} + \left(\frac{1}{a}+\frac{1}{b}\right)D_{p_1-1} + \frac{1}{ab}D_{p_1-2}\right\}^{\pi_1}\right.$$
$$\times\left\{D_{p_2} + \left(\frac{1}{a}+\frac{1}{b}\right)D_{p_2-1} + \frac{1}{ab}D_{p_2-2}\right\}^{\pi_2}$$
$$\times\left\{D_{p_3} + \left(\frac{1}{a}+\frac{1}{b}\right)D_{p_3-1} + \frac{1}{ab}D_{p_3-2}\right\}^{\pi_3}$$
$$\left.\times\ \ldots\ldots\ldots\ldots\ldots\ldots\ldots\ldots\ldots\ldots\right]U.$$

The operator which survives the process T_{ab} is of weight w and when performed upon U produces the coefficient of the term

$$(p_1^{\pi_1}p_2^{\pi_2}p_3^{\pi_3}\ldots)$$

in the development of $(1^\alpha)(1^\beta)U$.

The reader will have no difficulty in proceeding to the result which follows from introducing the further symbolism in regard to $\pi_1, \pi_2, \pi_3, \ldots$.

12. Take as an example $U = (1^\gamma)$ and, without loss of generality, we may suppose

$$\alpha \geqslant \beta \geqslant \gamma.$$

Let the general term in the developed product involve

$$(3^A 2^B 1^{\alpha+\beta+\gamma-3A-2B}).$$

We have (writing $\alpha+\beta+\gamma-3A-2B=C$ for short)

$$D_3^A D_2^B D_1^C (1^\alpha)(1^\beta)(1^\gamma)$$

$$= T_{ab}\left[a^\alpha b^\beta \left\{D_3 + \left(\frac{1}{a}+\frac{1}{b}\right)D_2 + \frac{1}{ab}D_1\right\}^A \right.$$

$$\times\left\{D_2 + \left(\frac{1}{a}+\frac{1}{b}\right)D_1 + \frac{1}{ab}\right\}^B$$

$$\left.\times\left\{D_1 + \left(\frac{1}{a}+\frac{1}{b}\right)\right\}^C\right](1^\gamma)$$

$$= T_{ab}\left[a^{\alpha-A} b^{\beta-A} D_1^A \left\{\left(\frac{1}{a}+\frac{1}{b}\right)D_1 + \frac{1}{ab}\right\}^B \left\{D_1 + \left(\frac{1}{a}+\frac{1}{b}\right)\right\}^C\right](1^\gamma).$$

If y, z be integers $\leqslant B$, $\leqslant C$ respectively, we find

$$\Sigma T_{ab}\left[a^{\alpha-A} b^{\beta-A} D_1^A \binom{B}{y}\left(\frac{1}{a}+\frac{1}{b}\right)^y D_1^y \left(\frac{1}{ab}\right)^{B-y}\binom{C}{z} D_1^z \left(\frac{1}{a}+\frac{1}{b}\right)^{C-z}\right](1^\gamma)$$

or $$\Sigma\binom{B}{y}\binom{C}{z} D_1^{A+y+z} T_{ab}\left[a^{\alpha-A-B+y} b^{\beta-A-B+y}\left(\frac{1}{a}+\frac{1}{b}\right)^{y+C-z}\right](1^\gamma).$$

If x be an integer $\leqslant y+C-z$, we find

$$\Sigma\binom{B}{y}\binom{C}{z}\binom{y+C-z}{x} D_1^{A+y+z} T_{ab}\left(a^{\alpha-A-B+y-x} b^{\beta-A-B-C+z+x}\right)(1^\gamma),$$

or $$\Sigma\binom{B}{y}\binom{C}{z}\binom{y+C-z}{x},$$

where
$$A+y+z=\gamma,$$
$$\alpha-A-B+y-x=0,$$
$$\beta-A-B-C+z+x=0,$$

three relations to control the summation, reducing to the two

$$y+z=\gamma-A,$$
$$y-x=-\alpha+A+B.$$

The coefficient is thus a linear function of terms each of which is a product of three binomial coefficients.

In particular, the coefficient of $(1^{\alpha+\beta+\gamma})$ reduces to

$$\binom{\alpha+\beta+\gamma}{\gamma}\binom{\alpha+\beta}{\alpha},$$

or $$\frac{(\alpha+\beta+\gamma)!}{\alpha!\,\beta!\,\gamma!},$$

which is obviously correct.

We may similarly treat the product

$$(2^{\alpha_2}1^{\alpha_1})(2^{\beta_2}1^{\beta_1})\,U.$$

$$D_{p_1}(2^{\alpha_2}1^{\alpha_1})(2^{\beta_2}1^{\beta_1})\,U$$

is in fact dealt with by considering the composition of p_1

$$0, 0, p_1;\ 0, 1, p_1-1;\ 1, 0, p_1-1;\ 0, 2, p_1-2;\ 1, 1, p_1-2;$$
$$2, 0, p_1-2;\ 2, 1, p_1-3;\ 1, 2, p_1-3;\ 2, 2, p_1-4;$$

yielding, symbolically,

$$a_2^{\alpha_2}a_1^{\alpha_1}b_2^{\beta_2}b_1^{\beta_1}\left\{D_{p_1}+\left(\frac{1}{a_1}+\frac{1}{b_1}\right)D_{p_1-1}+\left(\frac{1}{a_2}+\frac{1}{a_1b_1}+\frac{1}{b_2}\right)D_{p_1-2}\right.$$
$$\left.+\left(\frac{1}{a_2b_1}+\frac{1}{a_1b_2}\right)D_{p_1-3}+\frac{1}{a_2b_2}D_{p_1-4}\right\}U,$$

and we proceed in the usual manner to express

$$D_{p_1}^{\pi_1}D_{p_2}^{\pi_2}D_{p_3}^{\pi_3}\ldots(2^{\alpha_2}1^{\alpha_1})(2^{\beta_2}1^{\beta_1})\,U.$$

Finally, the procedure is clear in the case of the product of any number of symmetric functions.

The process is valuable when we require the calculation of the successive powers of a symmetric function.

13. Suppose, for example, that we have calculated the development of $(1^2)^{m-1}$,

and we require that of $(1^2)^m$.

$$D_{p_1}^{\pi_1}(1^2).(1^2)^{m-1}$$
$$=a^2\left(D_{p_1}+\frac{1}{a}D_{p_1-1}\right)^{\pi_1}.(1^2)^{m-1}$$
$$=a^2\left\{D_{p_1}^{\pi_1}+\binom{\pi_1}{1}\frac{1}{a}D_{p_1}^{\pi_1-1}D_{p_1-1}+\binom{\pi_1}{2}\frac{1}{a^2}D_{p_1}^{\pi_1-2}D_{p_1-1}^{2}\right\}(1^2)^{m-1},$$

since powers of $\frac{1}{a}$ greater than the second cannot contribute to the final result.

Thence

$$D_{p_1}^{\pi_1}D_{p_2}^{\pi_2}D_{p_3}^{\pi_3}\ldots(1^2).(1^2)^{m-1}$$
$$=T_a\left[a^2\left\{D_{p_1}^{\pi_1}+\binom{\pi_1}{1}\frac{1}{a}D_{p_1}^{\pi_1-1}D_{p_1-1}+\binom{\pi_1}{2}\frac{1}{a^2}D_{p_1}^{\pi_1-2}D_{p_1-1}^{2}\right\}\right.$$
$$\times\left\{D_{p_2}^{\pi_2}+\binom{\pi_2}{1}\frac{1}{a}D_{p_2}^{\pi_2-1}D_{p_2-1}+\binom{\pi_2}{2}\frac{1}{a^2}D_{p_2}^{\pi_2-2}D_{p_2-1}^{2}\right\}$$
$$\times\left\{D_{p_3}^{\pi_3}+\binom{\pi_3}{1}\frac{1}{a}D_{p_3}^{\pi_3-1}D_{p_3-1}+\binom{\pi_3}{2}\frac{1}{a^2}D_{p_3}^{\pi_3-2}D_{p_3-1}^{2}\right\}$$
$$\left.\times\ldots\ldots\ldots\ldots\ldots\ldots\ldots\ldots\ldots\ldots\ldots\right](1^2)^{m-1}$$

$$= \left\{\Sigma \binom{\pi_1}{2} D_{p_1}^{\pi_1 - 2} D_{p_1 - 1}^2 D_{p_2}^{\pi_2} D_{p_3}^{\pi_3} \ldots\right\} (1^2)^{m-1}$$

$$+ \left\{\Sigma \binom{\pi_1}{1} \binom{\pi_2}{1} D_{p_1}^{\pi_1 - 1} D_{p_1 - 1} D_{p_2}^{\pi_2 - 1} D_{p_2 - 1} D_{p_3}^{\pi_3} \ldots\right\} (1^2)^{m-1},$$

so that if we multiply certain coefficients of $(1^2)^{m-1}$ by numbers $\binom{\pi_1}{2}$ and certain others by numbers $\binom{\pi_1}{1}\binom{\pi_2}{1}$ and all these numerical results together we obtain the coefficients of the term

$$(p_1^{\pi_1} p_2^{\pi_2} p_3^{\pi_3} \ldots)$$

in the development of $(1^2)^m$.

Ex. gr. We know that

$$(1^2)^3 = (3^2) + 3\,(321) + 6\,(31^3) + 6\,(2^3) + 15\,(2^2 1^2) + 36\,(21^4) + 90\,(1^6),$$

and we find for the coefficient of $(32^2 1)$ in $(1^2)^4$

$$T_a \left\{a^2 \left(D_3 + \frac{1}{a} D_2\right) \left(D_2^2 + \frac{2}{a} D_2 D_1 + \frac{1}{a^2} D_1^2\right) \left(D_1 + \frac{1}{a}\right)\right\} (1^2)^3$$

$$= (D_3 D_1^3 + 2 D_3 D_2 D_1 + D_2^3 + 2 D_2^2 D_1^2)\,(1^2)^3$$

$$= \quad 6 \quad + \quad 2.3 \quad + \; 6 \; + \; 2.15$$

$$= \; 48.$$

14. The general result

$$\left\{\Sigma \binom{\pi_1}{2} D_{p_1}^{\pi_1 - 2} D_{p_1 - 1}^2 D_{p_2}^{\pi_2} D_{p_3}^{\pi_3} \ldots\right\} (1^2)^{m-1}$$

$$+ \left\{\Sigma \binom{\pi_1}{1} \binom{\pi_2}{1} D_{p_1}^{\pi_1 - 1} D_{p_1 - 1} D_{p_2}^{\pi_2 - 1} D_{p_2 - 1} D_{p_3}^{\pi_3} \ldots\right\} (1^2)^{m-1}$$

can be put into an interesting and suggestive form by writing

$$\pi_k (\pi_k - 1)(\pi_k - 2) \ldots (\pi_k - s + 1) = \pi_k^s \text{ symbolically,}$$

for it then appears in the symbolic form

$$D_{p_1}^{\pi_1} D_{p_2}^{\pi_2} D_{p_3}^{\pi} \ldots \frac{1}{2} \left(\pi_1 \frac{D_{p_1 - 1}}{D_{p_1}} + \pi_2 \frac{D_{p_2 - 1}}{D_{p_2}} + \pi_3 \frac{D_{p_3 - 1}}{D_{p_3}} + \ldots\right)^2 (1^2)^{m-1}.$$

So that in the particular case exemplified we have

$$D_3 D_2^2 D_1 \frac{1}{2} \left(1 \frac{D_2}{D_3} + 2 \frac{D_1}{D_2} + 1 \frac{D_0}{D_1}\right)^2 (1^2)^3,$$

where, by the symbolism, the squared bracket is to be read

$$1.0 \frac{D_2^2}{D_3^2} + 2.1.2. \frac{D_1}{D_3} + 2.1 \frac{D_1^2}{D_2^2} + 2.1.1. \frac{D_2}{D_3 D_1} + 2.2.1. \frac{1}{D_2} + 1.0 \frac{1}{D_1^2},$$

or
$$4\frac{D_1}{D_3}+2\frac{D_1^2}{D_2^2}+2\frac{D_2}{D_3D_1}+4\frac{1}{D_2},$$

and thence $\quad (2D_2^2D_1^2+D_3D_1^3+D_2^3+2D_3D_2D_1)(1^2)^3.$

15. Consider next $\quad (1^n).(1^n)^{m-1},$

$$\begin{aligned}&D_{p_1}(1^n).(1^n)^{m-1}\\&\quad=\{(1^n)D_{p_1}+(1^{n-1})D_{p_1-1}\}(1^n)^{m-1}\\&\quad=a^n\left(D_{p_1}+\frac{1}{a}D_{p_1-1}\right)(1^n)^{m-1}\text{ symbolically,}\end{aligned}$$

and thence

$$\begin{aligned}&(D_{p_1}^{\pi_1}D_{p_2}^{\pi_2}D_{p_3}^{\pi_3}\ldots)(1^n).(1^n)^{m-1}\\
&=T_a\left[a^n\left(D_{p_1}+\frac{1}{a}D_{p_1-1}\right)^{\pi_1}\left(D_{p_2}+\frac{1}{a}D_{p_2-1}\right)^{\pi_2}\left(D_{p_3}+\frac{1}{a}D_{p_3-1}\right)^{\pi_3}\ldots\right](1^n)^{m-1}\\
&=T_a\left[a^n\left\{D_{p_1}^{\pi_1}+\binom{\pi_1}{1}\frac{1}{a}D_{p_1}^{\pi_1-1}D_{p_1-1}+\binom{\pi_1}{2}\frac{1}{a^2}D_{p_1}^{\pi_1-2}D_{p_1-1}^2+\ldots\right.\right.\\
&\qquad\qquad\left.+\binom{\pi_1}{n}\frac{1}{a^n}D_{p_1}^{\pi_1-n}D_{p_1-1}^{n}\right\}\\
&\quad\times\left\{D_{p_2}^{\pi_2}+\binom{\pi_2}{1}\frac{1}{a}D_{p_2}^{\pi_2-1}D_{p_2-1}+\binom{\pi_2}{2}\frac{1}{a^2}D_{p_2}^{\pi_2-2}D_{p_2-1}^2+\ldots\right.\\
&\qquad\qquad\left.+\binom{\pi_2}{n}\frac{1}{a^n}D_{p_2}^{\pi_2-n}D_{p_2-1}^{n}\right\}\\
&\quad\times\left\{D_{p_3}^{\pi_3}+\binom{\pi_3}{1}\frac{1}{a}D_{p_3}^{\pi_3-1}D_{p_3-1}+\binom{\pi_3}{2}\frac{1}{a^2}D_{p_3}^{\pi_3-3}D_{p_3-1}^2+\ldots\right.\\
&\qquad\qquad\left.+\binom{\pi_3}{n}\frac{1}{a^n}D_{p_3}^{\pi_3-n}D_{p_3-1}^{n}\right\}\\
&\quad\left.\times\ \ldots\ldots\ldots\ldots\ldots\ldots\ldots\ldots\ldots\ldots\ldots\ldots\right](1^n)^{m-1}.\end{aligned}$$

Introducing the symbolism

$$\pi_k(\pi_k-1)(\pi_k-2)\ldots(\pi_k-s+1)=\pi_k^s$$

this is

$$\begin{aligned}&T_a\left[a^nD_{p_1}^{\pi_1}D_{p_2}^{\pi_2}D_{p_3}^{\pi_3}\ldots\exp\left(\frac{\pi_1}{a}\frac{D_{p_1-1}}{D_{p_1}}+\frac{\pi_2}{a}\frac{D_{p_2-1}}{D_{p_2}}+\frac{\pi_3}{a}\frac{D_{p_3-1}}{D_{p_3}}+\ldots\right)\right](1^n)^{m-1}\\
&\quad=\frac{1}{n!}D_{p_1}^{\pi_1}D_{p_2}^{\pi_2}D_{p_3}^{\pi_3}\ldots\left(\pi_1\frac{D_{p_1-1}}{D_{p_1}}+\pi_2\frac{D_{p_2-1}}{D_{p_2}}+\pi_3\frac{D_{p_3-1}}{D_{p_3}}+\ldots\right)^n(1^n)^{m-1},\end{aligned}$$

a noteworthy generalization of the ordinary multinomial theorem.

In fact we have established the formula

$$(1^n)^m$$
$$= \Sigma \frac{1}{n!} D_{p_1}^{\pi_1} D_{p_2}^{\pi_2} D_{p_3}^{\pi_3} \dots \left(\pi_1 \frac{D_{p_1-1}}{D_{p_1}} + \pi_2 \frac{D_{p_2-1}}{D_{p_2}} + \pi_3 \frac{D_{p_3-1}}{D_{p_3}} + \dots\right)^n (1^n)^{m-1},$$

and the ordinary multinomial emerges as

$$(1)^m = \Sigma D_{p_1}^{\pi_1} D_{p_2}^{\pi_2} D_{p_3}^{\pi_3} \dots \left(\pi_1 \frac{D_{p_1-1}}{D_{p_1}} + \pi_2 \frac{D_{p_2-1}}{D_{p_2}} + \pi_3 \frac{D_{p_3-1}}{D_{p_3}} + \dots\right) (1)^{m-1},$$

wherein the summations are in respect of all symmetric functions

$$(p_1^{\pi_1} p_2^{\pi_2} p_3^{\pi_3} \dots),$$

which present themselves in the developments.

The formula above for $n = 1$ is readily seen to be equivalent to the formula

$$(1)^m = \Sigma \frac{m!}{(p_1!)^{\pi_1} (p_2!)^{\pi_2} (p_3!)^{\pi_3} \dots} (p_1^{\pi_1} p_2^{\pi_2} p_3^{\pi_3} \dots).$$

16. In the discussion of American Tournaments which has appeared recently* the present author has dealt with the symmetric function algebra involved in an analogous manner—but in that case the applications were of a particular and not of a general character.

However, the study of those Tournaments is responsible for bringing to light the advance in symmetrical algebra which is brought forward in this paper.

* 'An American Tournament treated by the Calculus of Symmetric Functions,' *The Quarterly Journal of Pure and Applied Mathematics*, Vol. XLIX, No. 193, 1920.

AN AMERICAN TOURNAMENT TREATED BY THE CALCULUS OF SYMMETRIC FUNCTIONS.

By Major P. A. MacMahon.

Part I.

1. In a tournament of n players, where each player plays every other player, there are $\frac{1}{2}n(n-1)$ games. Since each game may be won or lost there are $2^{\frac{1}{2}n(n-1)}$ events and I propose to analyse them by means of the powerful calculus of symmetric functions. The final result of the play is that the players are arranged in a definite order, each with a certain number of games to his credit. These numbers constitute a partition of the number $\frac{1}{2}n(n-1)$, and we may ask how many of the $2^{\frac{1}{2}n(n-1)}$ events will yield a given partition of $\frac{1}{2}n(n-1)$ when the players are or are not in an assigned order.

2. Consider the symmetric function

$$(\alpha_1+\alpha_2)(\alpha_1+\alpha_3)(\alpha_2+\alpha_3)\ldots(\alpha_{n-1}+\alpha_n)$$

of the n quantities $\alpha_1, \alpha_2, \alpha_3, \ldots, \alpha_n$.

It involves $\frac{1}{2}n(n-1)$ factors, and the terms, after carrying out the multiplication, are grouped together in monomial symmetric functions.

Consider a factor of the product

$$\alpha_p+\alpha_q, \quad p<q,$$

and let us agree to form a factor of a term in the development by selecting α_p or α_q, according as a player, denoted by p, defeats or is defeated by a player, denoted by q.

If a term in the development be

$$C\alpha_1^{k_1}\alpha_2^{k_2}\ldots\alpha_n^{k_n},$$

it is clear that in C of the possible $2^{\frac{1}{2}n(n-1)}$ events the players denoted by 1, 2, 3, ..., n will be successful in

$$k_1,\ k_2,\ k_3,\ \ldots,\ k_n$$

games respectively out of the $n-1$ games engaged in by each.

Write now

$$(\alpha_1+\alpha_2)(\alpha_1+\alpha_3)(\alpha_2+\alpha_3)\ldots(\alpha_{n-1}+\alpha_n)=\Sigma C(k_1k_2k_3\ldots k_n),$$

where $$(k_1k_2k_3\ldots k_n)=\Sigma\alpha_1^{k_1}\alpha_2^{k_2}\alpha_3^{k_3}\ldots\alpha_n^{k_n},$$

and we conclude that the players in any assigned order are successful in $k_1, k_2, k_3, \ldots, k_n$ games in C out of the possible $2^{\frac{1}{2}n(n-1)}$ events.

As we shall have to consider equalities between the numbers $k_1, k_2, k_3, \ldots, k_n$, we now write

$$(k_1k_2k_3\ldots k_n)=(p_1^{\pi_1}p_2^{\pi_2}p_3^{\pi_3}\ldots p_s^{\pi_s}),$$

and writing the literal product, briefly, Π_n we have

$$\Pi_n=\Sigma C(p_1^{\pi_1}p_2^{\pi_2}p_3^{\pi_3}\ldots p_s^{\pi_s}).$$

When the symmetric function

$$(p_1^{\pi_1}p_2^{\pi_2}p_3^{\pi_3}\ldots p_s^{\pi_s})$$

is written out at length it involves

$$\frac{n!}{\pi_1!\,\pi_2!\,\pi_3!\ldots\pi_s!}\ \text{terms},$$

so that, disregarding the final order of the players, the numbers of games to their credit at the conclusion of the tournament will constitute the partition

$$(p_1^{\pi_1}p_2^{\pi_2}p_3^{\pi_3}\ldots p_s^{\pi_s})\ \text{of}\ \tfrac{1}{2}n(n-1)$$

in $$C\times\frac{n!}{\pi_1!\,\pi_2!\,\pi_3!\ldots\pi_s!}$$

events out of the possible $2^{\frac{1}{2}n(n-1)}$, and since the number of terms in Π_n is $2^{\frac{1}{2}n(n-1)}$ we must have

$$\Sigma C\,.\,\frac{n!}{\pi_1!\,\pi_2!\,\pi_3!\ldots\pi_s!}=2^{\frac{1}{2}n(n-1)},$$

where the summation is for every partition

$$(p_1^{\pi_1} p_2^{\pi_2} p_3^{\pi_3} \ldots p_s^{\pi_s})$$

that occurs in the development of Π_n.

This relation supplies a ready means of verifying our results.

3. We have therefore, for study, the development of Π_n as a linear function of monomial symmetric functions, since such linear function completely analyses the events of the tournament.

We commence by noting simple properties of the partition

$$(p_1^{\pi_1} p_2^{\pi_2} p_3^{\pi_3} \ldots p_s^{\pi_s})$$

that is before us. We have

$$\Sigma \pi p = \binom{n}{2}, \quad \Sigma \pi = n.$$

Since no player can win more than $n-1$ games, and, moreover, two players cannot both win $n-1$ games,

$$p_1 \not> n-1, \quad \pi_1 \not> 1, \text{ if } p_1 = n-1.$$

Since one player, but not more than one, can lose $n-1$ games, p_s may be zero but $\pi_s \not> 1$ if $p_s = 0$.

Including a single zero as a possible part, every partition has n parts.

If n be uneven and equal to $2m+1$, no partition can have a greatest part $< m$.

If n be even and equal to $2m$, no partition can have a greatest part $< m$.

In order to make other properties clear it is convenient to have before us the early calculations

$$\Pi_2 = (10),$$

$$\Pi_3 = (210) + 2(1^3),$$

$$\Pi_4 = (3210) + 2(31^3) + 2(2^30) + 4(2^21^2),$$

$$\Pi_5 = (43210) + 2(431^3) + 2(3^310) + 2(42^30) + 4(42^21^2)$$
$$+ 4(3^22^20) + 8(3^221^2) + 14(32^31) + 24(2^5).$$

Consider the structure of Π_n.

In performing the multiplication we obtain a term by selecting the first or second term of each factor. We obtain a corresponding term by selecting the second or first term of each factor. The consequence is that the symmetric functions, of which these two selected terms are types (or members), occur with the same coefficient in the development.

Thus, *e.g.*, in Π_5 we have the corresponding terms

$$\alpha_1\alpha_1\alpha_1\alpha_1\alpha_2\alpha_2\alpha_2\alpha_3\alpha_5\alpha_4 \equiv \alpha_1^4\alpha_2^3\alpha_3\alpha_4\alpha_5,$$

$$\alpha_2\alpha_3\alpha_4\alpha_5\alpha_3\alpha_4\alpha_5\alpha_4\alpha_3\alpha_5 \equiv \alpha_3^3\alpha_4^3\alpha_5^3\alpha_2,$$

and we find that the functions

$$(431^3),\ \ (3^310)$$

occur with the same coefficient in Π_5.

Note that $$(431^3)$$

is $$\{(4-0),\ 4-1,\ (4-3)^3\},$$

and $$(3^310)$$

is $$\{(4-1)^3,\ 4-3,\ 4-4\};$$

and, in general, to derive, from a partition, the complementary, we have merely to substitute, for the part p, the part $n-p-1$. Hence every partition has either a complementary with the same coefficient or it is self-complementary.

This circumstance may be regarded either as nearly halving the labour of calculation or as supplying a very important check upon the accuracy of the calculations.

The two complementary partitions involve the same repetitional exponents, and consequently involve the same number of terms.

Hence the product of the coefficient derived from the expansion of Π_n and the number of terms in the function is the same for both symmetric functions. We thus see that the number

$$C \times \frac{n!}{\pi_1!\,\pi_2!\ldots\pi_s!}$$

which arises in the tournament problem is the same for each.

Again, an examination of the product Π_n shews that all terms of Π_n, which involve one zero part, are derivable from the terms of Π_{n-1} by increasing each part (including zero when it occurs) by unity. *E.g.*, from

$$\Pi_4 = (3210) + 2\,(31^3) + 2\,(2^30) + 4\,(2^21^2)$$

we derive

$$\Pi_5 = (43210) + 2\,(42^30) + 2\,(3^310) + 4\,(3^22^20)$$

+ other terms which do not involve a zero part.

4. I now write

$$(\alpha_1+\alpha_2)(\alpha_1+\alpha_3)\dots(\alpha_1+\alpha_n)=\alpha_1^{n-1}+(1)'\alpha_1^{n-2}+(1^2)'\alpha_1^{n-3}+\dots+(1^{n-1})',$$

where the dashed partitions denote symmetric functions of

$$\alpha_2,\ \alpha_3,\ \dots,\ \alpha_n.$$

In Π_n the factor of

$$(\alpha_1+\alpha_2)(\alpha_1+\alpha_3)\dots(\alpha_1+\alpha_n),$$

when, in it we write α_{s-1} for α_s, is Π_{n-1}.

A portion of the result of multiplication is therefore

$$(1^m)'\alpha_1^{n-m-1}\,\Pi'_{n-1},$$

where Π'_{n-1} is the same function of $\alpha_2, \alpha_3, \dots, \alpha_n$ that Π_{n-1} is of $\alpha_1, \alpha_2, \dots, \alpha_{n-1}$.

Hence if D_k be Hammond's well-known differential operator of order k

$$D_{n-m}\Pi_n=(1^{m-1})\,\Pi_{n-1},$$

and in particular $\qquad D_{n-1}\Pi_n=\Pi_{n-1}.$

D_k as an operator performed upon a symmetric function, represented by a partition, deletes the number k from such partition and causes it to vanish when the number k is not present. Further, as is well known,

$$D_{p_1}^{\pi_1}D_{p_2}^{\pi_2}\dots D_{p_s}^{\pi_s}\,(p_1^{\pi_1}p_2^{\pi_2}\dots p_s^{\pi_s})=1.$$

Now, $n-1$ is the highest part in any partition which occurs in Π_n. This leading number is thus struck out by D_{n-1} from all partitions in which it presents itself. This process was named by Cayley 'Decapitation'. Thus

$$D_4\Pi_5=(3210)+2\,(31^3)+2\,(2^30)+4\,(2^21^2)=\Pi_4.$$

The converse process of 'Capitation' may thus be employed, and from

$$\Pi_4=(3210)+2\,(31^3)+2\,(2^30)+4\,(2^21^2)$$

we at once proceed to

$$\Pi_5=(43210)+2\,(431^3)+2\,(42^30)+4\,(42^21^2)$$

+ other terms, no one of which contains the highest possible part 4.

Thus from Π_{n-1} we find directly all terms of Π_n (i) which involve the highest part $n-1$; (ii) which involve the part zero; (iii) the complementary terms of (i) and (ii).

B 2

The remaining terms can be derived from the theorem

$$D_{n-m}\Pi_n = (1^{m-1})\,\Pi_{n-1},$$

which readily leads to a convenient calculus.

By the well-known properties of D_k we have

$$D_k(1^m)\,\Pi_n = \{(1^{m-1})\,D_{k-1} + (1^m)\,D_k\}\,\Pi_{n-1},$$

and it is convenient to denote (1^m) by a^m symbolically, so that

$$D_k a^m \Pi_n = (a^{m-1}D_{k-1} + a^m D_k)\,\Pi_{n-1} = a^{m-1}(aD_k + D_{k-1})\,\Pi_{n-1},$$

leading to

$$D_{k_1}D_{k_2}a^m\Pi_n = a^{m-2}(aD_{k_1} + D_{k_1-1})(aD_{k_2} + D_{k_2-1})\,\Pi_{n-1},$$

and generally to

$$D_{p_1}^{\pi_1}D_{p_2}^{\pi_2}\ldots D_{p_s}^{\pi_s}a^m\Pi_n$$

$$= a^{m-\Sigma\pi}(aD_{p_1}+D_{p_1-1})^{\pi_1}(aD_{p_2}+D_{p_2-1})^{\pi_2}\ldots(aD_{p_s}+D_{p_s-1})^{\pi_s}\,\Pi_{n-1}.$$

Now $\quad D_{p_1}\Pi_n = a^{n-p_1-1}\Pi_{n-1},$

$$D_{p_1}^{\pi_1}\Pi_n = a^{n-p_1-1}(D_{p_1} + a^{-1}D_{p_1-1})^{\pi_1-1}\,\Pi_{n-1},$$

and in general $\quad D_{p_1}^{\pi_1}D_{p_2}^{\pi_2}D_{p_3}^{\pi_3}\ldots D_{p_s}^{\pi_s}\Pi_n$

$$= a^{n-p_1-1}(D_{p_1} + a^{-1}D_{p_1-1})^{\pi_1-1}(D_{p_2} + a^{-1}D_{p_2-1})^{\pi_2}$$

$$\times(D_{p_3} + a^{-1}D_{p_3-1})^{\pi_3}\ldots(D_{p_s} + a^{-1}D_{p_s-1})^{\pi_s}\,\Pi_{n-1},$$

a result which can be put into other forms by varying the order in which the operations

$$D_{p_1}^{\pi_1},\ D_{p_2}^{\pi_2},\ D_{p_3}^{\pi_3},\ \ldots,\ D_{p_s}^{\pi_s}$$

are taken.

If $\quad (p_1^{\pi_1}p_2^{\pi_2}p_3^{\pi_3}\ldots p_s^{\pi_s})$

be the partition of a symmetrical function which occurs in Π_n, the result of the operation

$$D_{p_1}^{\pi_1}D_{p_2}^{\pi_2}D_{p_3}^{\pi_3}\ldots D_{p_s}^{\pi_s}\Pi_n$$

must be that part of

$$a^{n-p_1-1}(D_{p_1} + a^{-1}D_{p_1-1})^{\pi_1-1}(D_{p_2} + a^{-1}D_{p_2-1})^{\pi_2}\ldots$$

$$\times(D_{p_s} + a^{-1}D_{p_s-1})^{\pi_s}\,\Pi_{n-1},$$

which on development of the *operator* is free from a. This part is a linear function of products of D operators, which when performed upon Π_{n-1} gives the value of

$$D_{p_1}^{\pi_1} D_{p_2}^{\pi_2} D_{p_3}^{\pi_3} \ldots D_{p_s}^{\pi_s} \Pi_n ,$$

or in other words it gives the coefficient of symmetric function

$$(p_1^{\pi_1} p_2^{\pi_2} p_3^{\pi_3} \ldots p_s^{\pi_s})$$

in the expression of Π_n as a linear function of monomial symmetric functions.

5. A few examples will make the method clear to the reader. The expression Π_5 has been given in Art. 3. Suppose that we desire to derive therefrom the coefficient of symmetric function $(3^4 21)$ in the expression of Π_6. We have

$$D_3^4 D_2 D_1 \Pi_6 = a^2 (D_3 + a^{-1} D_2)^3 (D_2 + a^{-1} D_1)(D_1 + a^{-1}) \Pi_5$$

$$= (a^2 D_3^3 + 3a D_3^2 D_2 + 3 D_3 D_2^2 + \ldots)(D_2 + a^{-1} D_1)(D_1 + a^{-1}) \Pi_5 ,$$

where the fourth term in the first bracket is omitted because it has no bearing upon the result.

The part free from a upon the right-hand side is

$$(D_3^3 D_1 + 3 D_3^2 D_2^2 + 3 D_3^2 D_2 D_1^2 + 3 D_3 D_2^3 D_1) \Pi_5 ,$$

and, on referring to the expression of Π_5, this is

$$(2 + 3 \times 4 + 3 \times 8 + 3 \times 14) = 2 + 12 + 24 + 42 = 80 ;$$

so that a portion of Π_6 is $80 (3^4 21)$.

To verify this we will find the coefficient of (432^4) by the same method; we know that this is also 80 because (432^4) and $(3^4 21)$ are complementary partitions. We have

$$D_4 D_3 D_2^4 \Pi_6 = a (D_3 + a^{-1} D_2)(D_2 + a^{-1} D_1)^4 \Pi_5$$

$$= (a D_3 + D_2)(D_2^4 + 4a^{-1} D_2^3 D_1 + \ldots) \Pi_5$$

$$= (+ 4 D_3 D_2^3 D_1 + D_2^5 + \ldots) \Pi_5$$

$$= 56 + 24 = 80.$$

The method is very convenient because it is allied to a means of verifying results, which provides a constant check of the work.

6. The application to the American Tournament is obtained by writing

$$\Pi_n = \Sigma C \frac{(p_1^{\pi_1} p_2^{\pi_2} \dots p_s^{\pi_s})}{n!/(\pi_1!\,\pi_2!\dots\pi_s!)},$$

and then C denotes the number of events out of the $2^{\binom{n}{2}}$ possible events, which give, without specification of the players, the final score

$$p_1, p_1 \dots \pi_1 \text{ times}, \; p_2, p_2 \dots \pi_2 \text{ times}, \; \dots, \; p_s, p_s \dots \pi_s \text{ times}.$$

Thus we write, when $n = 3$,

$$\Pi_3 = 6\,\frac{(210)}{6} + 2\,(1^3),$$

indicating that the final score will be

2, 1, 0 in 6 events,

1, 1, 1 in 2 events,

2^3

If we write instead $\Pi_3 = (210) + 2\,(1^3)$,

we learn that only 1 event gives the score 2, 1, 0 when the order of the players is specified and that as before 2 events give the score 1, 1, 1 because the order of the players is of no moment.

For $n = 4$, we write

$$\Pi_4 = 24\,\frac{(3210)}{24} + 8\,\frac{(31^3)}{4} + 8\,\frac{(2^3 0)}{4} + 24\,\frac{(2^2 1^2)}{6},$$

indicating that the final score will be

3, 2, 1, 0 in 24 events,

3, 1, 1, 1 in 8 events,

2, 2, 2, 0 in 8 events,

2, 2, 1, 1 in 24 events,

2^6

while if we write

$$\Pi_4 = (3210) + 2\,(31^3) + 2\,(2^3 0) + 4\,(2^2 1^2),$$

we see that if the order of the players be specified the events are in numbers 1, 2, 2, 4 for the scores

$$(3, 2, 1, 0), \quad (3, 1, 1, 1), \quad (2, 2, 2, 0), \quad (2, 2, 1, 1),$$

respectively.

In the above case $n=4$ we see that an absolute winner emerges in exactly half of the events. It is an even wager. A tie of two players comes out in 24 events and a tie of three players in 8 events.

The case of five players is better arranged as under

43210	1	120	120
431^3	2	20	40
$\{3^3 10$	2	20	40
$\{42^3 0$	2	20	40
$\{42^2 1^2$	4	30	120
$\{3^2 2^2 0$	4	30	120
$3^2 21^2$	8	30	240
$32^3 1$	14	20	280
2^5	24	1	24
			$1024 = 2^{10}$

The left-hand column gives the partitions of the monomial symmetric functions due to Π_5, *or* the partitions which give the final scores in the tournament. The second column gives the coefficients of the symmetric functions in the first column as they arise in Π_5. The third column gives the numbers of terms in the parallel symmetric functions, and the fourth column, being the product of the two preceding columns, gives the numbers of events which appertain to the scores in the first column.

We gather that the most probable score is

$$3, 2, 2, 2, 1,$$

and that it emerges from 280 events.

An absolute winner emerges in

$120 + 40 + 40 + 120 + 280$ or 600 events.

A tie of two winners emerges in $120 + 240$ or 360 events.

A tie of three winners emerges in or 40 events.

A tie of five winners emerges in or 24 events.

1024

7. The attached tables give the complete results as far as the tournament of eight players inclusive.

The first column gives in partition notation the symmetric functions that arise from the function Π; the second column the coefficients of these functions; the third column the numbers of terms in the functions when they are written out *in extenso*; the fourth column numbers are the products of the numbers in the second and third columns.

In the table for eight players the sixth row is interpretable as follows.

The final score of the tournament is

$$7, 6, 5, 3, 3, 2, 2, 0$$

in 40320 out of 2^{28} events.

The same final score arises for the players in a definite order in 4 out of 2^{28} events.

The development of Π_8 is possibly the most elaborate symmetric function that has ever been computed.

Sym. Function or Score.	Coeff. of Sym. Func.	No. of terms in Sym. Func.	Events in Tournament.

Three Players. $\Pi_3 = (\alpha_1 + \alpha_2)(\alpha_1 + \alpha_3)(\alpha_2 + \alpha_3)$.

210	1	6	6
1^3	2	1	2

Four Players. $\Pi_4 = (\alpha_1 + \alpha_2)(\alpha_1 + \alpha_3)(\alpha_2 + \alpha_3)\ldots(\alpha_3 + \alpha_4)$.

3210	1	24	24
31^3	2	4	8
2^30	2	4	8
2^21^2	4	6	24

Sum of numbers in last column $= 2^6$.

Five Players. $\Pi_5 = (\alpha_1 + \alpha_2)(\alpha_1 + \alpha_3)(\alpha_2 + \alpha_3)\ldots(\alpha_4 + \alpha_5)$.

43210	1	120	120
431^3	2	20	40
42^30	2	20	40
42^21^2	4	30	120
3^310	2	20	40
3^22^20	4	30	120
3^221^2	8	30	240
32^31	14	20	280
2^5	24	1	24

Sum of numbers in the last column $= 2^{10}$.

Six Players. $\Pi_6 = (\alpha_1 + \alpha_2)(\alpha_1 + \alpha_3)(\alpha_2 + \alpha_3)\ldots(\alpha_5 + \alpha_6)$.

543210	1	720	720
5431^3	2	120	240
542^30	2	120	240
542^21^2	4	180	720
53^310	2	120	240
53^22^20	4	180	720
53^221^2	8	180	1440
532^31	14	120	1680
52^5	24	6	144
4^3210	2	120	240
4^31^3	4	20	80
4^23^210	4	180	720
4^232^20	8	180	1440
4^2321^2	16	180	2880
4^22^31	28	60	1680
43^320	14	120	1680
43^31^2	28	60	1680
43^22^21	48	180	8640
432^4	80	30	2400
3^50	24	6	144
3^421	80	30	2400
3^32^3	132	20	2640

Sum of numbers in last column $= 2^{15}$.

The most probable score is 4, 3, 3, 2, 2, 1, which arises in 8640 out of 2^{15} events.

Seven players. $\Pi_7 = (\alpha_1 + \alpha_2)(\alpha_1 + \alpha_3)(\alpha_2 + \alpha_3)\ldots(\alpha_6 + \alpha_7)$.

6543210	1	5040	5040
65431^3	2	840	1680
6542^30	2	840	1680
6542^21^2	4	1260	5040
653^310	2	840	1680
653^22^20	4	1260	5040
653^221^2	8	1260	10080
6532^31	14	840	11760
652^5	24	42	1008
64^3210	2	840	1680
64^31^3	4	140	560
64^23^210	4	1260	5040
64^232^20	8	1260	10080
64^2321^2	16	1260	20160
64^22^31	28	420	11760

$643^3 20$	14	840	11760
$643^3 1^2$	28	420	11760
$643^2 2^2 1$	48	1260	60480
6432^4	80	210	16800
$63^5 0$	24	42	1008
$63^4 21$	80	210	16800
$63^3 2^3$	132	140	18480
$5^3 3210$	2	840	1680
$5^3 31^3$	4	140	560
$5^3 2^3 0$	4	140	560
$5^3 2^2 1^2$	8	210	1680
$5^2 4^2 210$	4	1260	5040
$5^2 4^2 1^3$	8	210	1680
$5^2 43^2 10$	8	1260	10080
$5^2 432^2 0$	16	1260	20160
$5^2 4321^2$	32	1260	40320
$5^2 42^3 1$	56	420	23520
$5^2 3^3 20$	28	420	11760
$5^2 3^3 1^2$	56	210	11760
$5^2 3^2 2^2 1$	96	630	60480
$5^2 32^4$	160	105	16800
$54^3 310$	14	840	11760
$54^3 2^2 0$	28	420	11760
$54^3 21^2$	56	420	23520
$54^2 3^2 20$	48	1260	60480
$54^2 3^2 1^2$	96	630	60480
$54^2 32^2 1$	164	1260	206640
$54^2 2^4$	272	105	28560
$543^4 0$	80	210	16800
$543^3 21$	266	840	223440
$543^2 2^3$	436	420	183120
$53^5 1$	424	42	17808
$53^4 2^2$	688	105	72240
$4^5 10$	24	42	1008
$4^4 320$	80	210	16800
$4^4 31^2$	160	105	16800
$4^4 2^2 1$	272	105	28560
$4^3 3^3 0$	132	140	18480
$4^3 3^2 21$	436	420	183120
$4^3 32^3$	712	140	99680
$4^2 3^4 1$	688	105	72240
$4^2 3^3 2^2$	1112	210	233520
$43^5 2$	1720	42	72240
3^7	2640	1	2640

Sum of numbers in last column $= 2^{21}$.

The most probable score is 4, 4, 3, 3, 3, 2, 2, which arises in 233520 out of 2^{21} events.

Eight Players. $\Pi_8 = (\alpha_1 + \alpha_2)(\alpha_1 + \alpha_3)(\alpha_2 + \alpha_3)\ldots(\alpha_7 + \alpha_8)$.

76543210	1	40320	40320
$7654 31^3$	2	6720	13440
$76542^3 0$	2	6720	13440
$76542^2 1^2$	4	10080	40320
$7653^3 10$	2	6720	13440
$7653^2 2^2 0$	4	10080	40320
$7653^2 21^2$	8	10080	80640
$76532^3 1$	14	6720	94080
7652^5	24	336	8064
$764^3 210$	2	6720	13440
$764^3 1^3$	4	1120	4480
$764^2 3^2 10$	4	10080	40320
$764^2 32^2 0$	8	10080	80640
$764^2 321^2$	16	10080	161280
$764^2 2^3 1$	28	3360	94080
$7643^3 20$	14	6720	94080
$7643^3 1^2$	28	3360	94080
$7643^2 2^2 1$	48	10080	483840
76432^4	80	1680	134400
$763^5 0$	24	336	8064
$763^4 21$	80	1680	134400
$763^3 2^3$	132	1120	147840
$75^3 3210$	2	6720	13440
$75^3 31^3$	4	1120	4480
$75^3 2^3 0$	4	1120	4480
$75^3 2^2 1^2$	8	1680	13440
$75^2 4^2 210$	4	10080	40320
$75^2 4^2 1^3$	8	1680	13440
$75^2 43^2 10$	8	10080	80640
$75^2 432^2 0$	16	10080	161280
$75^2 4321^2$	32	10080	322560
$75^2 42^3 1$	56	3360	188160
$75^2 3^3 20$	28	3360	94080
$75^2 3^3 1^2$	56	1680	94080
$75^2 3^2 2^2 1$	96	5040	483840
$75^2 32^4$	160	840	134400
$754^3 310$	14	6720	94080
$754^3 2^2 0$	28	3360	94080
$754^3 21^2$	56	3360	188160
$754^2 3^2 20$	48	10080	483840
$754^2 3^2 1^2$	96	5040	483840

$754^2 32^2 1$	164	10080	1653120
$754^2 2^4$	272	840	228480
$7543^4 0$	80	1680	134400
$7543^3 21$	266	6720	1787520
$7543^2 2^3$	436	3360	1464960
$753^5 1$	424	336	142464
$753^4 2^2$	688	840	577920
$74^5 10$	24	336	8064
$74^4 320$	80	1680	134400
$74^4 31^2$	160	840	134400
$74^4 2^2 1$	272	840	228480
$74^3 3^3 0$	132	1120	147840
$74^3 3^2 21$	436	3360	1464960
$74^3 32^3$	712	1120	797440
$74^2 3^4 1$	688	840	577920
$74^2 3^3 2^2$	1112	1680	1868160
$743^5 2$	1720	336	577920
73^7	2640	8	21120
$6^3 43210$	2	6720	13440
$6^3 431^3$	4	1120	4480
$6^3 42^3 0$	4	1120	4480
$6^3 42^2 1^2$	8	1680	13440
$6^3 3^3 10$	4	1120	4480
$6^3 3^2 2^2 0$	8	1680	13440
$6^3 3^2 21^2$	16	1680	26880
$6^3 32^3 1$	28	1120	31360
$6^3 2^5$	48	56	2688
$6^2 5^2 3210$	4	10080	40320
$6^2 5^2 31^3$	8	1680	13440
$6^2 5^2 2^3 0$	8	1680	13440
$6^2 5^2 2^2 1^2$	16	2520	40320
$6^2 54^2 210$	8	10080	80640
$6^2 54^2 1^3$	16	1680	26880
$6^2 543^2 10$	16	10080	161280
$6^2 5432^2 0$	32	10080	322560
$6^2 54321^2$	64	10080	645120
$6^2 542^3 1$	112	3360	376320
$6^2 53^3 20$	56	3360	188160
$6^2 53^3 1^2$	112	1680	188160
$6^2 53^2 2^2 1$	192	5040	967680
$6^2 532^4$	320	840	268800
$6^2 4^3 310$	28	3360	94080
$6^2 4^3 2^2 0$	56	1680	94080
$6^2 4^3 21^2$	112	1680	188160
$6^2 4^2 3^2 20$	96	5040	483840

$6^2 4^2 3^2 1^2$	192	2520	483840
$6^2 4^2 3 2^2 1$	328	5040	1653120
$6^2 4^2 2^4$	544	420	228480
$6^2 4 3^4 0$	160	840	134400
$6^2 4 3^3 2 1$	532	3360	1787520
$6^2 4 3^2 2^3$	872	1680	1464960
$6^2 3^5 1$	848	168	142464
$6^2 3^4 2^2$	1376	420	577920
$6 5^3 4 2 1 0$	14	6720	94080
$6 5^3 4 1^3$	28	1120	31360
$6 5^3 3^2 1 0$	28	3360	94080
$6 5^3 3 2^2 0$	56	3360	188160
$6 5^3 3 2 1^2$	112	3360	376320
$6 5^3 2^3 1$	196	1120	219520
$6 5^2 4^2 3 1 0$	48	10080	483840
$6 5^2 4^2 2^2 0$	96	5040	483840
$6 5^2 4^2 2 1^2$	192	5040	967680
$6 5^2 4 3^2 2 0$	164	10080	1653120
$6 5^2 4 3^2 1^2$	328	5040	1653120
$6 5^2 4 3 2^2 1$	560	10080	5644800
$6 5^2 4 2^4$	928	840	779520
$6 5^2 3^4 0$	272	840	228480
$6 5^2 3^3 2 1$	904	3360	3037440
$6 5^2 3^2 2^3$	1480	1680	2486400
$6 5 4^4 1 0$	80	1680	134400
$6 5 4^3 3 2 0$	266	6720	1787520
$6 5 4^3 3 1^2$	532	3360	1787520
$6 5 4^3 2^2 1$	904	3360	3037440
$6 5 4^2 3^3 0$	436	3360	1464960
$6 5 4^2 3^2 2 1$	1440	10080	14515200
$6 5 4^2 3 2^3$	2348	3360	7889280
$6 5 4 3^4 1$	2256	1680	3790080
$6 5 4 3^3 2^2$	3640	3360	12230400
$6 5 3^5 2$	5584	336	1876224
$6 4^5 2 0$	424	336	142464
$6 4^5 1^2$	848	168	142464
$6 4^4 3^2 0$	688	840	577920
$6 4^4 3 2 1$	2256	1680	3790080
$6 4^4 2^3$	3664	280	1025920
$6 4^3 3^3 1$	3504	1120	3924480
$6 4^3 3^2 2^2$	5632	1680	9461760
$6 4^2 3^4 2$	8576	840	7203840
$6 4 3^6$	12960	56	725760
$5^5 2 1 0$	24	336	8064
$5^5 1^3$	48	56	2688

$5^4 4310$	80	1680	134400
$5^4 42^2 0$	160	840	134400
$5^4 421^2$	320	840	268800
$5^4 3^2 20$	272	840	228480
$5^4 3^2 1^2$	544	420	228480
$5^4 32^2 1$	928	840	779520
$5^4 2^4$	1536	70	107520
$5^3 4^3 10$	132	1120	147840
$5^3 4^2 320$	436	3360	1464960
$5^3 4^2 31^2$	872	1680	1464960
$5^3 4^2 2^2 1$	1480	1680	2486400
$5^3 43^3 0$	712	1120	797440
$5^3 43^2 21$	2348	3360	7889280
$5^3 432^3$	3824	1120	4282880
$5^3 3^4 1$	3664	280	1025920
$5^3 3^3 2^2$	5904	560	3306240
$5^2 4^4 20$	688	840	577920
$5^2 4^4 1^2$	1376	420	577920
$5^2 4^3 3^2 0$	1112	1680	1868160
$5^2 4^3 321$	3640	3360	12230400
$5^2 4^3 2^3$	5904	560	3306240
$5^2 4^2 3^3 1$	5632	1680	9461760
$5^2 4^2 3^2 2^2$	9040	2520	22780800
$5^2 43^4 2$	13712	840	11518080
$5^2 3^6$	20640	28	577920
$54^5 30$	1720	336	577920
$54^5 21$	5584	336	1876224
$54^4 3^2 1$	8576	840	7203840
$54^4 32^2$	13712	840	11518080
$54^3 3^3 2$	20676	1120	23157120
$54^2 3^5$	30960	168	5201280
$4^7 0$	2640	8	21120
$4^6 31$	12960	56	725760
$4^6 2^2$	20640	28	577920
$4^5 3^2 2$	30960	168	5201280
$4^4 3^4$	46144	70	3230080

Sum of numbers in last column $= 2^{28}$.

The most probable score is 5, 4, 4, 4, 3, 3, 3, 2, which arises in 23157120 out of 2^{28} events.

Part II.

The application of Symmetric Functions to this question naturally arises because the players enter the tournament in a symmetrical manner. The fortunes of the players do not depend upon a random fixing of pairs of opponents, the pairing is fixed *à priori* by the nature of the contest. The analysis of the $2^{\frac{1}{2}n(n-1)}$ events depends upon the expression of the symmetric function

$$\Pi_n = (\alpha_1 + \alpha_2)(\alpha_1 + \alpha_3)(\alpha_2 + \alpha_3)\ldots(\alpha_{n-1} + \alpha_n)$$

as a linear function of monomial symmetric functions. The partitions of these monomials may be roughly specified.

Each of them has the content $\frac{1}{2}n(n-1)$ and either n or $n-1$ parts. Admitting zero as part, we may say that there are always n parts, one of which may be zero. The highest part cannot exceed $n-1$ and only one part can be so large.

If a certain partition presents itself so also does the partition obtained by substituting, for every part p, the part $n-p-1$. We call these complementary partitions. Every partition having the above specifications does not present itself.

Several questions come up for consideration:—

(1) How many partitions occur for a given value of n?

(2) What is the most convenient and expeditious way of calculating the coefficients?

(3) How can the results be best verified?

I recall that in Part I. it was shown that the whole linear function appertaining to Π_n becomes by 'Capitation' by n a portion of the linear function appertaining to Π_{n+1}; also that if the same linear function be subjected to a unit increase of part and the part zero added the result is a portion of the linear function appertaining to Π_{n+1}; further the complementary symmetric functions, corresponding to complementary partitions, are affected with the same coefficients.

The theorems are guides to what follows.

It is convenient to divide the partitions appertaining to Π_n into four sets.

Set I. includes all partitions which involve the part $n-1$, but do not involve the part zero.

Set II. includes those which involve both the parts $n-1$ and zero.

Set III. includes those which involve neither $n-1$ nor zero.

Set IV. includes those which do not involve $n-1$, but do involve zero.

In fact, briefly:—

Set I. gives $n-1$ yes, zero no.
Set II. gives $n-1$ yes, zero yes.
Set III. gives $n-1$ no, zero no.
Set IV. gives $n-1$ no, zero yes.

The parts $n-1$ zero are complementary in a partition and neither can occur more than once in any partition.

The partitions in Set IV. are complementary to those in Set I. These sets consequently comprise equal numbers of partitions, and the corresponding linear functions involve the same coefficients.

Set II. and Set III. are both self-complementary because every partition

$$\begin{pmatrix} n-1 \text{ yes, zero no} \\ n-1 \text{ no, zero yes} \end{pmatrix}$$

converts into a partition

$$\begin{pmatrix} n-1 \text{ yes, zero no} \\ n-1 \text{ no, zero yes} \end{pmatrix}.$$

Let $N_1(n),\ N_2(n),\ N_3(n),\ N_4(n)$

denote the numbers of partitions in

Set I., Set II., Set III., Set IV.,

respectively, and put $N(n)$ for the whole numbers of partitions appertaining to Π_n. Then

$$N_1(n) + N_2(n) + N_3(n) + N_4(n) = N(n).$$

From the 'Capitation' theorem

$$N_1(n) + N_2(n) = N(n-1).$$

From the 'Unit Increase' theorem, which it should be observed arises directly from the theorem of Part I.,

$$D_k\Pi_n = (1^{n-k-1})\,\Pi_{n-1},$$

by putting $k=0$, we obtain

$$N_2(n) + N_4(n) = N(n-1)$$

equivalent to the relation derived from the 'Capitation' theorem.

Finally, if we take any partition appertaining to Π_{n-2}, viz.,

$$p_1 p_2 p_3 \dots p_{n-2}.$$

When p_1 may be as large as $n-3$ and p_{n-2} may be zero, we give a unit increase of part and add zero, obtaining

$$p_1+1,\ p_2+1,\ p_3+1,\ \dots,\ p_{n-2}+1,\ 0,$$

and now capitation with $n-1$ yields

$$n-1,\ p_1+1,\ p_2+1,\ \dots,\ p_{n-2}+1,\ 0,$$

which is a partition, appertaining to Π_n, of Set II., since

$$n-2+n-1+\binom{n-2}{2}=\binom{n}{2}.$$

Hence $\qquad N_2(n)=N(n-2),$

so that from (3)

$$N_1(n)=N(n-1)-N(n-2),$$

and from (2)

$$N_3(n)=N(n)-2N(n-1)+N(n-2),$$

shewing that $N_1(n)$ and $N_3(n)$ are first and second differences of the numbers $N(n)$.

It may be gathered that if we know the numbers

$$N(2),\ N(3),\ \dots,\ N(n-1),$$

we know $\qquad N_1(n),\ N_2(n),$ and $N_4(n),$

and the value of $N(n)$ only depends upon our finding the value of $N_3(n)$. In fact, $N(n)$ is expressible entirely in terms of the numbers $N_3(n)$. We readily find the relation

$$N(n)-1=(n-2)N_3(3)+(n-3)N_3(4)+\dots+N_3(n).$$

Consider next the linear functions of monomials corresponding to the sets of partitions, and write

$$L_1(n)+L_2(n)+L_3(n)+L_4(n)=L(n),$$

where $\qquad \Pi_n \equiv L(n).$

Denote by K_{n-1}, when applied to the partition of a symmetric function, capitation with $n-1$ and by I. a unit increase of part and the addition of the part zero. Then

$$L_1(n)+L_2(n)=K_{n-1}L(n-1),$$

$$L_2(n)+L_4(n)=IL(n-1),$$

$$L_2(n)=K_{n-1}IL(n-2),$$

so that
$$L_1(n) = K_{n-1}L(n-1) - K_{n-1}IL(n-2),$$
$$L_4(n) = IL(n-1) - K_{n-1}IL(n-2),$$
and
$$L_3(n) = L(n) - K_{n-1}L(n-1) - IL(n-1) + K_{n-1}IL(n-2).$$
If we know the expressions for
$$L(2),\ L(3),\ \ldots,\ L(n-1),$$
we only require to calculate $L_3(n)$ in order to obtain the expression of $L(n)$. It thus suffices to calculate the expressions $L_3(n)$ in order to obtain the expressions $L(n)$. Moreover, $L_3(n)$ is a self-complementary function, so that it is only necessary to calculate the coefficients of symmetric monomial functions associated with self-complementary partitions and one half of the remainder. Applying this to the calculation of $L(9)$ we find it to be a linear function of 490 monomial functions; $L_3(9)$ is a linear function of 215 such functions, 33 of which are associated with self-complementary partitions, and therefore the calculation is only necessary of
$$33 + \tfrac{1}{2}(215 - 33) \text{ or } 124$$
coefficients.

The calculation of the coefficients.

In Part I. it was established that the coefficient of the function, whose partition is
$$(p_1^{\pi_1}p_2^{\pi_2}\ldots p_s^{\pi_s}),$$
in the development of Π_n, is obtained from the operator expression
$$a^{n-p_1-1}(D_{p_1} + a^{-1}D_{p_1-1})^{\pi_1-1}(D_{p_2} + a^{-1}D_{p_2-1})^{\pi_2}\ldots(D_{p_s} + a^{-1}D_{p_s-1})^{\pi_s},$$
where $(1^s) = a^s$ symbolically. We seek therein the portion of the operator product which, on expansion, is free from a, and apply this operator which emerges to Π_{n-1}, but a better way of determining the useful portion of the operator is as follows:—

If we write $\pi_1 + 1$ for π_1 and put
$$\pi_m(\pi_m - 1)(\pi_m - 2)\ldots(\pi_m - k + 1) = \pi_m^k$$
symbolically the operator product may be effectively written
$$a^{n-p-1}.\exp\frac{1}{a}(\pi_p D_p^{-1}D_{p-1} + \pi_{p-1}D^{-1}_{p-1}D_{p-2} + \ldots$$
$$+ \pi_2 D_2^{-1}D_1 + \pi_1 D_1^{-1}).D_p^{\pi_p}D_{p-1}^{\pi_{p-1}}\ldots D_1^{\pi_1},$$

and we are immediately led to the result

$$D_p^{\pi_p+1} D_{p-1}^{\pi_{p-1}} \dots D_2^{\pi_2} D_1^{\pi_1} \Pi_n$$

$$= \frac{1}{(n-p-1)!} (\pi_p D_p^{-1} D_{p-1} + \pi_{p-1} D^{-1}_{p-1} D_{p-2}$$

$$+ \dots \pi_2 D_2^{-1} D_1 + \pi_1 D_1^{-1})^{n-p-1} D_p^{\pi_p} D_{p-1}^{\pi_{p-1}} \dots D_1^{\pi_1} \Pi_{n-1}.$$

This formula enables us to express the coefficients in Π_n as linear functions of the coefficients in Π_{n-1} and is very convenient to work with. I give some examples from the calculation of Π_9.

To find the coefficient of

$$\Sigma \alpha_1^7 \alpha_2^6 \alpha_3^6 \alpha_4^6 \alpha_5^3 \alpha_6^3 \alpha_7^2 \alpha_8^2 \alpha_9,$$

we have $$D_7 D_6^3 D_3^2 D_2^2 D_1 \Pi_9$$

$$= (3 D_6^{-1} D_5 + 2 D_3^{-1} D_2 + 2 D_2^{-1} D_1 + D_1^{-1}) D_6^3 D_3^2 D_2^2 D_1 \Pi_8$$

$$= (3 D_6^2 D_5 D_3^2 D_2^2 D_1 + 2 D_6^3 D_3 D_2^3 D_1$$

$$+ 2 D_6^3 D_3^2 D_2 D_1^2 + D_6^3 D_3^2 D_2^2) \Pi_8,$$

and looking to the Table for Π_8 in Part I. and ascertaining the coefficients associated with the partitions

$$(6^2 5 3^2 2^2 1), \quad (6^3 3 2^3 1), \quad (6^3 3^2 2 1^2), \quad (6^3 3^2 2^2),$$

we find $$3 \times 192 + 2 \times 28 + 2 \times 16 + 8 = 672.$$

Again, to find the coefficient of

$$\Sigma \alpha_1^6 \alpha_2^4 \alpha_3^6 \alpha_4^5 \alpha_5^5 \alpha_6^2 \alpha_7^2 \alpha_8^2 \alpha_9^2,$$

we have $$D_6^3 D_5^2 D_2^4 \Pi_9$$

$$= \frac{1}{2!} (2 D_6^{-1} D_5 + 2 D_5^{-1} D_4 + 4 D_2^{-1} D_1)^2 D_6^2 D_5^2 D_2^4 \Pi_8$$

$$= \frac{1}{2!} (2 D_6^{-2} D_5^2 + 2 D_5^{-2} D_4^2 + 12 D_2^{-2} D_1^2 + 8 D_6^{-1} D_4$$

$$+ 16 D_6^{-1} D_5 D_2^{-1} D_1 + 16 D_5^{-1} D_4 D_2^{-1} D_1) D_6^2 D_5^2 D_2^4 \Pi_8$$

$$= (D_5^4 D_2^4 + D_6^2 D_4^2 D_2^2 + 6 D_6^2 D_5^2 D_2^2 D_1^2 + 4 D_6 D_5^2 D_4 D_2^7$$

$$+ 8 D_6 D_5^3 D_2^3 D_1 + 8 D_6^2 D_5 D_4 D_2^3 D_1).$$

Whence, operation upon Π_8, and reference to the Π_8 Table in Part I., gives

$$1536 + 544 + 96 + 3712 + 1568 + 896 = 8.352$$

for the coefficient sought.

C 2

As a verification of the linear function

$$L_3(n),$$

we recall that when the symmetric function in $L(n)$ are written out *in extenso* the number of terms is

$$2^{\frac{1}{2}n(n-1)}.$$

To find the number of terms in $L_3(n)$ we observe that

$$L_3(n) = L(n) - K_{n-1}L(n-1) - IL(n-1) + K_{n-1}IL(n-2).$$

The number of terms in a symmetric function included in $L(n-1)$ becomes multiplied by n when it is capitated by K_{n-1} and the unit increase due to I has a similar effect. Also the unit increase and capitation performed upon a function included in $L(n-2)$ has the effect of multiplying the number of terms by $n(n-1)$. Hence the number of terms in $L_3(n)$ is

$$2^{\frac{1}{2}n(n-1)} - 2n.2^{\frac{1}{2}(n-1)(n-2)} + n(n-1).2^{(\frac{1}{2}n-2)(n-3)}$$

or

$$2^{\frac{1}{2}(n-2)(n-3)}\{2^{2n-3} - n.2^{n-1} + n(n-1)\}.$$

For $n = 9$, we find

number of terms in $L_3(9) = 64, 038, 633, 472.$

I append the expression of $L_3(9)$.

The first column of numbers gives the coefficients of the symmetric function whose partitions are in the left-hand column. The second column of numbers gives the number of terms in the functions. The right-hand column is the product of the numbers in the first and second columns of numbers. I have verified that the sum of the numbers in the right-hand column is

64, 038, 633, 472,

and have thus the satisfaction of knowing that there is not a single numerical error in the calculation.

Π_9. The Portion $L_3(9)$.

Sym. Function or Score.	Coeff. of Sym. Func.	No. of terms in Sym. Func.	Events in Tournament.
$7^3 5431^3$	4	10080	40320
$7^3 542^2 1^2$	8	15120	120960
$7^3 53^2 21^2$	16	15120	241920
$7^3 532^3 1$	28	10080	282240
$7^3 52^5$	48	504	24192

$7^3 4^3 1^3$	8	1680	13440
$7^3 4^2 3 2 1^2$	32	15120	483840
$7^3 4^2 2^3 1$	56	5040	282240
$7^3 4 3^3 1^2$	56	5040	282240
$7^3 4 3^2 2^2 1$	96	15120	1451520
$7^3 4 3 2^4$	160	2520	403200
$7^3 3^4 2 1$	160	2520	403200
$7^3 3^3 2^3$	264	1680	443520
$7^2 6^2 4 3 1^3$	8	15120	120960
$7^2 6^2 4 2^2 1^2$	16	22680	362880
$7^2 6^2 3^2 2 1^2$	32	22680	725760
$7^2 6^2 3 2^3 1$	56	15120	846720
$7^2 6^2 2^5$	96	756	72576
$7^2 6 5^2 3 1^3$	16	15120	241920
$7^2 6 5^2 2^2 1^2$	32	22680	725760
$7^2 6 5 4^2 1^3$	32	15120	483840
$7^2 6 5 4 3 2 1^2$	128	90720	11612160
$7^2 6 5 4 2^3 1$	224	30240	6773760
$7^2 6 5 3^3 1^2$	224	15120	3386880
$7^2 6 5 3^2 2^2 1$	384	45360	17418240
$7^2 6 5 3 2^4$	640	7560	4838400
$7^2 6 4^3 2 1^2$	224	15120	3386880
$7^2 6 4^2 3^2 1^2$	384	22680	8709120
$7^2 6 4^2 3 2^2 1$	656	45360	29756160
$7^2 6 4^2 2^4$	1088	3780	4112640
$7^2 6 4 3^3 2 1$	1064	30240	32175360
$7^2 6 4 3^2 2^3$	1744	15120	26369280
$7^2 6 3^5 1$	1696	1512	2564352
$7^2 6 3^4 2^2$	2752	3780	10402560
$7^2 5^3 4 1^3$	56	5040	282240
$7^2 5^3 3 2 1^2$	224	15120	3386880
$7^2 5^3 2^3 1$	392	5040	1975680
$7^2 5^2 4^2 2 1^2$	384	22680	8709120
$7^2 5^2 4 3^2 1^2$	656	22680	14878080
$7^2 5^2 4 3 2^2 1$	1120	45360	50803200
$7^2 5^2 4 2^4$	1856	3780	7015680
$7^2 5^2 3^3 2 1$	1808	15120	27336960
$7^2 5^2 3^2 2^3$	2960	7560	22377600
$7^2 5 4^3 3 1^2$	1064	15120	16087680
$7^2 5 4^3 2^2 1$	1808	15120	27336960
$7^2 5 4^2 3^2 2 1$	2880	45360	130636800
$7^2 5 4^2 3 2^3$	4696	15120	71003520
$7^2 5 4 3^4 1$	4512	7560	34110720
$7^2 5 4 3^3 2^2$	7280	15120	110073600
$7^2 5 3^5 2$	11168	1512	16886016

$7^2 4^5 1^2$	1696	756	1282176
$7^2 4^4 321$	4512	7560	34110720
$7^2 4^4 2^3$	7328	1260	9233280
$7^2 4^3 3^3 1$	7008	5040	35320320
$7^2 4^3 3^2 2^2$	11264	7560	85155840
$7^2 4^2 3^4 2$	17152	3780	64834560
$7^2 4 3^6$	25920	252	6531840
$7 6^3 5 3 1^3$	28	10080	282240
$7 6^3 5 2^2 1^2$	56	15120	846720
$7 6^3 4^2 1^3$	56	5040	282240
$7 6^3 4 3 2 1^2$	224	30240	6773760
$7 6^3 4 2^3 1$	392	10080	3951360
$7 6^3 3^3 1^2$	392	5040	1975680
$7 6^3 3^2 2^2 1$	672	15120	10160640
$7 6^3 3 2^4$	1120	2520	2822400
$7 6^2 5^2 4 1^3$	96	15120	1451520
$7 6^2 5^2 3 2 1^2$	384	45360	17418240
$7 6^2 5^2 2^3 1$	672	15120	10160640
$7 6^2 5 4^2 2 1^2$	656	45360	29756160
$7 6^2 5 4 3^2 1^2$	1120	45360	50803200
$7 6^2 5 4 3 2^2 1$	1912	90720	173456640
$7 6^2 5 4 2^4$	3168	7560	23950080
$7 6^2 5 3^3 2 1$	3084	30240	93260160
$7 6^2 5 3^2 2^3$	5048	15120	76325760
$7 6^2 4^3 3 1^2$	1808	15120	27336960
$7 6^2 4^3 2^2 1$	3072	15120	46448640
$7 6^2 4^2 3^2 2 1$	4888	45360	221719680
$7 6^2 4^2 3 2^3$	7968	15120	120476160
$7 6^2 4 3^4 1$	7648	7560	57818880
$7 6^2 4 3^3 2^2$	12336	15120	186520320
$7 6^2 3^5 2$	18896	1512	28570752
$7 6 5^4 1^3$	160	2520	403200
$7 6 5^3 4 2 1^2$	1064	30240	32175360
$7 6 5^3 3^2 1^2$	1808	15120	27336960
$7 6 5^3 3 2^2 1$	3084	30240	93260160
$7 6 5^3 2^4$	5104	2520	12862080
$7 6 5^2 4^2 3 1^2$	2880	45360	130636800
$7 6 5^2 4^2 2^2 1$	4888	45360	221719680
$7 6 5^2 4 3^2 2 1$	7744	90720	702535680
$7 6 5^2 4 3 2^3$	12608	30240	381265920
$7 6 5^2 3^4 1$	12064	7560	91203840
$7 6 5^2 3^3 2^2$	19432	15120	293811840
$7 6 5 4^4 1^2$	4512	7560	34110720
$7 6 5 4^3 3 2 1$	11918	60480	720800640
$7 6 5 4^3 2^3$	19324	10080	194785920

$7654^2 3^3 1$	18404	30240	556536960
$7654^2 3^2 2^2$	29528	45360	1339390080
$76543^4 2$	44688	15120	675682560
7653^6	67104	504	33820416
$764^5 21$	18136	3024	54843264
$764^4 3^2 1$	27792	7560	210107520
$764^4 32^2$	44416	7560	335784960
$764^3 3^3 2$	66804	10080	673384320
$764^2 3^5$	99760	1512	150837120
$75^5 21^2$	1696	1512	2564352
$75^4 431^2$	4512	7560	34110720
$75^4 42^2 1$	7648	7560	57818880
$75^4 3^2 21$	12064	7560	91203840
$75^4 32^3$	19616	2520	49432320
$75^3 4^3 1^2$	7008	5040	35320320
$75^3 4^2 321$	18404	30240	556536960
$75^3 4^2 2^3$	29800	5040	150192000
$75^3 3^4 2$	68416	2520	172408320
$75^3 43^3 1$	28316	10080	285425280
$75^3 43^2 2^2$	45368	15120	685964160
$75^2 4^4 21$	27792	7560	210107520
$75^2 4^3 3^2 1$	42440	15120	641692800
$75^2 4^3 32^2$	67728	15120	1024047360
$75^2 4^2 3^3 2$	101528	15120	1535103360
$75^2 43^5$	151120	1512	228493440
$754^5 31$	63144	3024	190947456
$754^5 2^2$	100368	1512	151756416
$754^4 3^2 2$	149664	7560	1131459840
$754^3 3^4$	221728	2520	558754560
$74^7 1$	93360	72	6721920
$74^6 32$	219360	504	110557440
$74^5 3^3$	323600	504	163094400
$6^5 31^3$	48	504	24192
$6^5 2^2 1^2$	96	756	72576
$6^4 541^3$	160	2520	403200
$6^4 5321^2$	640	7560	4838400
$6^4 52^3 1$	1120	2520	2822400
$6^4 4^2 21^2$	1088	3780	4112640
$6^4 43^2 1^2$	1856	3780	7015680
$6^4 432^2 1$	3168	7560	23950080
$6^4 42^4$	5248	630	3306240
$6^4 3^3 21$	5104	2520	12862080
$6^4 3^2 2^3$	8352	1260	10523520
$6^3 5^3 1^3$	264	1680	443520
$6^3 5^2 421^2$	1744	15120	26369280

$6^3 5^2 3^2 1^2$	2960	7560	22377600
$6^3 5^2 3 2^2 1$	5048	15120	76325760
$6^3 5^2 2^4$	8352	1260	10523520
$6^3 5 4^2 3 1^2$	4696	15120	71003520
$6^3 5 4^2 2^2 1$	7968	15120	120476160
$6^3 5 4 3^2 2 1$	12608	30240	381265920
$6^3 5 4 3 2^3$	20520	10080	206841600
$6^3 5 3^4 1$	19616	2520	49432320
$6^3 5 3^3 2^2$	31584	5040	159183360
$6^3 4^4 1^2$	7328	1260	9233280
$6^3 4^3 3 2 1$	19324	10080	194785920
$6^3 4^3 2^3$	31320	1680	52617600
$6^3 4^2 3^3 1$	29800	5040	150192000
$6^3 4^2 3^2 2^2$	47792	7560	361297520
$6^3 4 3^4 2$	72224	2520	182004480
$6^3 3^6$	108288	84	9096192
$6^2 5^4 2 1^2$	2752	3780	10402560
$6^2 5^3 4 3 1^2$	7280	15120	110073600
$6^2 5^3 4 2^2 1$	12336	15120	186520320
$6^2 5^3 3^2 2 1$	19432	15120	293811840
$6^2 5^3 3 2^3$	31584	5040	159183360
$6^2 5^2 4^3 1^2$	11264	7560	85155840
$6^2 5^2 4^2 3 2 1$	29528	45360	1339390080
$6^2 5^2 4^2 2^3$	47792	7560	361297520
$6^2 5^2 4 3^3 1$	45368	15120	685964160
$6^2 5^2 4 3^2 2^2$	72656	22680	1647838080
$6^2 5^2 3^4 2$	109408	3780	413562240
$6^2 5 4^4 2 1$	44416	7560	335784960
$6^2 5 4^3 3^2 1$	67728	15120	1024047360
$6^2 5 4^3 3 2^2$	108032	15120	1633443840
$6^2 5 4^2 3^3 2$	161704	15120	2444964480
$6^2 5 4 3^5$	240320	1512	363363840
$6^2 4^5 3 1$	100368	1512	151756416
$6^2 4^5 2^2$	159456	756	120548736
$6^2 4^4 3^2 2$	237408	3780	897402240
$6^2 4^3 3^4$	351168	1260	442471680
$6 5^5 3 1^2$	11168	1512	16886016
$6 5^5 2^2 1$	18896	1512	28570752
$6 5^4 4^2 1^2$	17152	3780	64834560
$6 5^4 4 3 2 1$	44688	15120	675682560
$6 5^4 4 2^3$	72224	2520	182004480
$6 5^4 3^3 1$	68416	2520	172408320
$6 5^4 3^2 2^2$	109408	3780	413562240
$6 5^3 4^3 2 1$	66804	10080	673384320
$6 5^3 4^2 3^2 1$	101528	15120	1535103360

$65^3 4^2 3 2^2$	161704	15120	2444964480
$65^3 4 3^3 2$	241312	10080	2432424960
$65^3 3^5$	357600	504	180230400
$65^2 4^4 31$	149664	7560	1131459840
$65^2 4^4 2^2$	237408	3780	897402240
$65^2 4^3 3^2 2$	352472	15120	5329376640
$65^2 4^2 3^4$	519968	3780	1965479040
$654^6 1$	219360	504	110557440
$654^5 32$	512184	3024	1548844416
$654^4 3^3$	752464	2520	1896209280
$64^7 2$	740880	72	53343360
$64^6 3^2$	1084320	252	273248640
$5^6 41^2$	25920	252	6531840
$5^6 321$	67104	504	33820416
$5^6 2^3$	108288	84	9096192
$5^5 4^2 21$	99760	1512	150837120
$5^5 43^2 1$	151120	1512	228493440
$5^5 432^2$	240320	1512	363363840
$5^5 3^3 2$	357600	504	180230400
$5^4 4^3 31$	221728	2520	558754560
$5^4 4^3 2^2$	351168	1260	442471680
$5^4 4^2 3^2 3$	519968	3780	1965479040
$5^4 43^4$	765120	630	482025600
$5^3 4^5 1$	323600	504	163094400
$5^3 4^4 32$	752464	2520	1896209280
$5^3 4^3 3^3$	1102824	1680	1852744320
$5^2 4^6 2$	1084320	252	273248640
$5^2 4^5 3^2$	1583360	756	1197020160
$54^7 3$	2265200	72	163094400
4^9	3230080	1	3230080

Sum of numbers in last column is

$$2^{\frac{1}{2}.7.6}(2^{15} - 9.2^8 + 9.8) \text{ or } 64038633472;$$

the most probable score is

6, 5, 5, 4, 4, 4, 3, 3, 2,

which emerges in 5329376640 of the events.

PART III.

I now consider a tournament in which every pair of players plays every other pair. As the players enter the tournament in a perfectly symmetrical manner, the calculus of symmetric functions is peculiarly fitted to the discussion.

If we take four players, denoted by α, β, γ, δ, we have three games, α, β v. γ, δ; α, γ v. β, δ; α, δ v. β, γ, and we form the symmetric product

$$(\alpha\beta + \gamma\delta)(\alpha\gamma + \beta\delta)(\alpha\delta + \beta\gamma),$$

and perform the multiplication. To obtain a term of the product we take the first or second term of a factor according as we suppose that the corresponding pair of players wins or loses the game denoted by the factor.

Thus the term formed by the selected product

$$\gamma\delta . \alpha\gamma . \beta\gamma$$

denotes that the pairs

$$\gamma, \delta; \ \alpha, \gamma; \ \beta, \gamma$$

have been successful in the games

$$\alpha, \beta \text{ v. } \gamma, \delta; \ \alpha, \gamma \text{ v. } \beta, \delta; \ \alpha, \delta \text{ v. } \beta, \gamma$$

respectively.

The selection gives the term, of the product,

$$\alpha\beta\gamma^3\delta,$$

shewing that the players α, β, γ, δ have been successful in 1, 1, 3, 1 games respectively.

Since the product

$$(\alpha\beta + \gamma\delta)(\alpha\gamma + \beta\delta)(\alpha\delta + \beta\gamma)$$

is symmetrical the product will be, on multiplication, a sum of monomial symmetric functions. It is, in fact,

$$\Sigma\alpha^2\beta^2\gamma^2 + \Sigma\alpha^3\beta\gamma\delta,$$

and we observe that of the $2^3 = 8$ events, four result in the final scores 2, 2, 2, 0 with the players in undefined order and four result in the score 3, 1, 1, 1 with the players in undefined order. These scores are equally probable if it is an even wager on each game. There is only one chance in eight that the final score will be named with the players in a given order of merit. If n players take part in the tournament, four players can be selected in $\binom{n}{4}$ ways, and each of these selections gives rise to three games; hence the total number of games to be played is $3\binom{n}{4}$, and there are $2^{3\binom{n}{4}}$ events to be analysed.

Before proceeding to general principles I give the complete discussion for the case of five players.

Here there are 15 games and $2^{15} = 32768$ events to be examined.

We form the symmetric product

$$(\alpha_1\alpha_2+\alpha_3\alpha_4)(\alpha_1\alpha_2+\alpha_3\alpha_5)(\alpha_1\alpha_2+\alpha_4\alpha_5)(\alpha_1\alpha_3+\alpha_2\alpha_4)(\alpha_1\alpha_3+\alpha_2\alpha_5)(\alpha_1\alpha_3+\alpha_4\alpha_5)$$
$$(\alpha_1\alpha_4+\alpha_2\alpha_3)(\alpha_1\alpha_4+\alpha_2\alpha_5)(\alpha_1\alpha_4+\alpha_3\alpha_5)(\alpha_1\alpha_5+\alpha_2\alpha_3)(\alpha_1\alpha_5+\alpha_2\alpha_4)(\alpha_1\alpha_5+\alpha_3\alpha_4)$$
$$(\alpha_2\alpha_3+\alpha_4\alpha_5)(\alpha_2\alpha_4+\alpha_3\alpha_5)(\alpha_2\alpha_5+\alpha_3\alpha_4),$$

a factor corresponding to each game.

We must expand this into a sum of monomial symmetric functions.

Denoting the product, for the case of n quantities, by $\Pi_{n,2}$ we have

$$\Pi_{5,2} = (\alpha_1\alpha_2+\alpha_3\alpha_4)(\alpha_1\alpha_2+\alpha_3\alpha_5)(\alpha_1\alpha_2+\alpha_4\alpha_5)(\alpha_1\alpha_3+\alpha_2\alpha_4)(\alpha_1\alpha_3+\alpha_2\alpha_5)$$
$$\times(\alpha_1\alpha_3+\alpha_4\alpha_5)(\alpha_1\alpha_4+\alpha_2\alpha_3)(\alpha_1\alpha_4+\alpha_2\alpha_5)(\alpha_1\alpha_4+\alpha_3\alpha_5)(\alpha_1\alpha_5+\alpha_2\alpha_3)$$
$$\times(\alpha_1\alpha_5+\alpha_2\alpha_4)(\alpha_1\alpha_5+\alpha_3\alpha_4)\times\Pi_{4,2},$$

and we have seen that, in the partition notation,

$$\Pi_{4,2} = (2220) + (3111).$$

The product which multiplies $\Pi_{4,2}$ is of degree 12 in α_1, and we will denote it by $F_{5,2}$, so that

$$\Pi_{5,2} = F_{5,2}\Pi_{4,2}.$$

In $\Pi_{5,2}$ the coefficient of α_1^{12} is $(\alpha_2\alpha_3\alpha_4\alpha_5)^3 \equiv (3333)$,

and that of α_1^{0} is $(\alpha_2\alpha_3\alpha_4\alpha_5)^6 \equiv (6666)$.

We write

$$F_{5,2} = (3333)\left\{\alpha_1^{12} + \frac{N_1}{(1111)}\alpha_1^{11} + \frac{N_2}{(2222)}\alpha_1^{10} + \frac{N_3}{(3333)}\alpha_1^{9} + \ldots + \frac{N_{12}}{(12, 12, 12, 12)}\right\},$$

where it is clear that $N_1 = (2210)$ and N_s is equal to the sum of the products, s at a time and unrepeated, of the terms of $\Sigma\alpha_2^2\alpha_3^2\alpha_4$ in respect of the four quantities α_2, α_3, α_4, α_5.

We observe also that

$$N_{12}=(\alpha_2\alpha_3\alpha_4\alpha_5)^{15}=(15,\ 15,\ 15,\ 15).$$

We have to find the elementary symmetric functions of the twelve expressions

$$\alpha_2^2\alpha_3^2\alpha_4,\quad \alpha_2^2\alpha_3\alpha_4^2,\quad \alpha_2\alpha_3^2\alpha_4^2,\quad \alpha_2^2\alpha_3^2\alpha_5,\quad \alpha_2^2\alpha_3\alpha_5^2,\quad \alpha_2\alpha_3^2\alpha_5^2,$$
$$\alpha_2^2\alpha_4^2\alpha_5,\quad \alpha_2^2\alpha_4\alpha_5^2,\quad \alpha_2\alpha_4^2\alpha_5^2,\quad \alpha_3^2\alpha_4^2\alpha_5,\quad \alpha_3^2\alpha_4\alpha_5^2,\quad \alpha_3\alpha_4^2\alpha_5^2,$$

and denoting these by

$$\{1\},\quad \{1^2\},\quad \{1^3\},\quad \&c.,$$

we have

$$1!\,N_1=1!\,\{1\}=(221),$$
$$2!\,N_2=2!\,\{1^2\}=(221)^2-(442),$$
$$3!\,N_3=3!\,\{1^3\}=(221)^3-3\,(221)(442)+2\,(663),$$
$$4!\,N_4=4!\,\{1^4\}=(221)^4-6\,(221)^2(442)+3\,(442)^2$$
$$+\,8\,(221)(663)-6\,(884),$$
$$5!\,N_2=5!\,\{1^5\}=(221)^5-10\,(221)^3(442)+15\,(221)(442)^2$$
$$+20\,(221)^2(663)-20\,(442)(663)-30\,(221)(884)+24\,(10,\ 10,\ 5),$$

etc.

We might proceed to develop the right-hand sides of the relations by means of Hammond's differential operators, and in carrying this out there is more than one mode of procedure.

Having obtained the expressions we can multiply out by $\Pi_{4,2}$, and thus obtain the expression of $\Pi_{5,2}$ in descending powers of α_1. This, as will be seen, leads at once to the desired expression of $\Pi_{5,2}$ as a sum of monomial symmetric functions. Newton's theorem connecting the elementary symmetric functions of a number of quantities with the sums of powers of the same leads to the relations

$$2N_2=(221)\,N_1-(442),$$
$$3N_3=(221)\,N_2-(442)\,N_1+(663),$$
$$4N_4=(221)\,N_3-(442)\,N_2+(663)\,N_1-(884),$$
$$5N_5=(221)\,N_4-(442)\,N_3+(663)\,N_2-(884)\,N_1+(10,\ 10,\ 5),$$

etc.,

relations which are convenient for the successive calculations of N_2, N_3, N_4, N_5, &c.

The convenience arises from the circumstance that the Hammond operators are performed upon products of not more than two monomials. This is easy to carry out with small risk of a numerical error, and a check upon the coefficients is provided by the counting of the terms in the development of the monomials.

Ex. gr. consider the product

$$P = (8444)(221),$$
$$D_{10} D_6 D_5 D_4 P = 1,$$
$$D_9 D_6^2 D_4 \qquad = 1,$$
$$D_8 D_6^2 D_5 \qquad = 1,$$

so that
$$P = (10{,}654) + (9664) + (8665),$$

and counting the terms in the development

$$4 \times 12 = 24 + 12 + 12$$

a verification, as it shows that no terms have been omitted.

It will be seen too that when sN_s has been calculated, a certain verification is secured by the fact that each of the coefficients in the resulting linear function of monomials must be divisible by s.

The expressions obtained are

$N_1 = (2210),$

$N_2 = (4411) + (4330) + (4321) + 3\,(4222) + 2\,(3322),$

$N_3 = (6531) + 2\,(6432) + 2\,(6333) + (5550) + (5541)$
$\quad + 2\,(5532) + 3\,(5442) + 3\,(5433) + 7\,(4443),$

$N_4 = (8633) + (8552) + (8543) + 3\,(8444) + (7751) + 2\,(7733)$
$\quad + 2\,(7652) + 2\,(7643) + 4\,(7553) + 4\,(7544) + 5\,(6653)$
$\quad + 7\,(6644) + 9\,(6554) + 9\,(5555),$

$N_5 = (10{,}654) + (9853) + (9770) + (9763) + 2\,(9754) + 3\,(9664)$
$\quad + 3\,(9655) + 3\,(8854) + 3\,(8773) + 4\,(8764) + 7\,(8755)$
$\quad + 9\,(8665) + 9\,(7774) + 11\,(7765) + 15\,(7666),$

$N_6 = (12{,}666) + (11{,}874) + (11{,}865) + 2\,(11{,}766) + (10{,}1055)$
$\quad + (10{,}974) + (10{,}965) + 3\,(10{,}875) + 5\,(10{,}866) + 5\,(10{,}776)$
$\quad + (9993) + 2\,(9984) + 5\,(9975) + 4\,(9966) + 5\,(9885)$
$\quad + 10\,(9876) + 11\,(9777) + 11\,(8886) + 17\,(8877),$

etc.,

whence we obtain

$$F_{5,3}$$

$$= (3333)\,\alpha_1^{12}$$

$$+ (4432)\,\alpha_1^{11}$$

$$+ \{(5522) + (5441) + (5432) + 3\,(5333) + 2\,(4433)\}\,\alpha_1^{10}$$

$$+ \{(6531) + 2\,(6432) + 2\,(6333) + (5550) + (5541)$$
$$+ 2\,(5532) + 3\,(5442) + 3\,(5433) + 7\,(4443)\}\,\alpha_1^{9}$$

$$+ \{(7522) + (7441) + (7432) + 3\,(7333) + (6640)$$
$$+ 2\,(6622) + 2\,(6541) + 2\,(6532) + 4\,(6442) + 4\,(6433)$$
$$+ 5\,(5542) + 7\,(5533) + 9\,(5443) + 9\,(4444)\}\,\alpha_1^{8}$$

$$+ \{(8432) + (7631) + (7550) + (7541) + 2\,(7532)$$
$$+ 3\,(7442) + 3\,(7433) + 3\,(6632) + 3\,(6551) + 4\,(6542)$$
$$+ 7\,(6533) + 9\,(6443) + 9\,(5552) + 11\,(5543) + 15\,(5444)\}\,\alpha_1^{7}$$

$$+ \{(9333) + (8541) + (8532) + 2\,(8433) + (7722)$$
$$+ (7641) + (7632) + 3\,(7542) + 5\,(7533) + 5\,(7443)$$
$$+ (6660) + 2\,(6651) + 5\,(6642) + 4\,(6633) + 5\,(6552)$$
$$+ 10\,(6543) + 11\,(6444) + 11\,(5553) + 17\,(5544)\}\,\alpha_1^{6}$$

$$+ \&c.$$

It is not necessary for my present purpose to calculate the expression of N with a higher suffix than 6 or to complete the expression of $F_{5,3}$, but it should be remarked that the expression of N_{12-s} can be at once written down from that of N_s by merely substituting for every part p in the latter the part $15-p$. For if

$$N_s = \Sigma\,(a, b, c, d),$$

$$N_{12-s} = N_{15}\,\Sigma\,(-a, -b, -c, -d)$$
$$= (15, 15, 15, 15)\,\Sigma\,(-a, -b, -c, -d)$$
$$= \Sigma\,(15-d, 15-c, 15-b, 15-a).$$

In agreement with this law it will be noticed that N_6 is unaltered by the substitution referred to. As a direct consequence of this fact the coefficient of α_1^{12-s} in $F_{5,3}$ is obtainable from that of α_1^{s} by the substitution of $9-p$ for p. It will be noticed that the coefficient of α_1^{6} is unaltered by this substitution. A verification of the expression for N_s is afforded by the circumstance that when it is written out *in extenso* the number of its terms is

$$\binom{12}{s}.$$

Ex. gr. for $s = 2$,

$$6 + 12 + 24 + 3 \times 4 + 2 \times 6 = 66 = \binom{12}{2}.$$

Obviously also in $F_{5,2}$ the number of terms in the coefficients of α_1^s is

$$\binom{12}{s}.$$

We have now to multiply $F_{5,2}$ by Π $\{= (2220) + (3111)\}$ to obtain the expression of $\Pi_{5,2}$ in descending powers of α_1. If we then insert the part s in each of the partitions which occurs in the coefficient of α_1^s, we obtain all that part of $\Pi_{5,2}$ which involves partitions containing the part s. We have thus, obviously, an ample verification of our results. The multiplication of $F_{5,2}$ by $(2220) + (3111)$ is performed with little labour, and the final result is

$\Pi_{5,2} =$

Partition of Func.	Coeff.	No. of Terms.	Partition of Func.	Coeff.	No. of Terms.
12,6444	1	20	97752	3	180
12,5553	1	20	97743	7	420
11,7543	1	120	97662	6	360
11,6652	1	60	97653	10	1200
11,6643	1	60	97644	13	780
11,6553	1	60	97554	18	1080
11,6544	2	120	96663	14	280
11,5554	3	60	96654	23	1380
10,8633	1	60	96555	27	540
10,8552	1	60	88860	1	20
10,8543	1	120	88842	3	60
10,8444	3	60	88833	2	20
10,7742	1	60	88761	2	120
10,7661	1	60	88752	5	300
10,7652	2	240	88743	7	420
10,7643	3	360	88662	6	180
10,7553	3	180	88653	13	780
10,7544	5	300	88644	26	600
10,6653	6	360	88554	20	600

10,6644	6	180	87771	3	60
10,6554	9	540	87762	9	540
10,5555	12	60	87753	18	1080
99642	1	60	87744	20	600
99543	2	120	87663	23	1380
99444	2	20	87654	34	4080
98751	1	120	87555	45	900
98742	1	120	86664	42	840
98733	2	120	86655	54	1620
98661	1	60	77772	12	60
98652	3	360	77763	27	540
98643	4	480	77754	45	900
98553	7	420	77664	54	1620
98544	7	420	77655	64	1920
97770	1	20	76665	84	1680
97761	1	60	66666	88	88

The sum of the numbers in the column headed 'No. of Terms' is 2^{15}—a verification.

If a partition in the left-hand column be

$$p_1^{\pi_1} p_2^{\pi_2} \dots p_t^{\pi_t} \dots,$$

containing a given part p_t, π_t times, where $\pi_t > 0$, and C be its coefficient

$$\Sigma C \frac{(\pi_1 + \pi_2 + \dots + \pi_t - 1 + \dots)!}{\pi_1 ! \, \pi_2 ! \dots (\pi_t - 1)!} = 8 \binom{12}{p_t},$$

where the summation is in regard to those partitions which contain the part p_t. Thus for $s = 11$, the sum is

$$24 + 12 + 12 + 12 + 24 + 12 = 96 = 8 \binom{12}{11}.$$

In performing the multiplication of the fifteen factors, which compose $\Pi_{5,2}$, a selection of terms implies the winning of games by certain pairs and the losses by other pairs. If we take the complementary selection so that the pairs who were winners are now losers we see intuitively that the coefficient in $\Pi_{5,3}$ of a particular symmetric function is the same as that of the complementary function obtained by substituting for all parts p the parts $12 - p$.

Ex. gr. the coefficients of the functions (98553), (97743) are the same. The functions are thus associated in pairs; certain functions, such as (76665) which are self-complementary, being outstanding.

It will be gathered that there are so many verifications possible throughout the work that a numerical error in the final result may be regarded as out of the question.

We may notice the particular result, extracted from the Table,

87654 34 4080

which gives the highest number in the 'No. of Terms' column. It shows that the most probable score is

8, 7, 6, 5, 4,

and that this occurs in 4080 out of the 2^{15} events. If this is to be the score with the players in a named order we see that it occurs in 34 out of 2^{15} events. The result

66666 88 88

shews that the five players will end up 'all square' in 88 out of 2^{15} events.

The symmetric functions which arise in these American Tournaments between n players, s players engaging s players in all possible ways, have similar properties whatever values are given to n and s. The actual calculation becomes impracticable very soon with the increase of n. The particular case $n=6$, $s=3$ may be set forth because the calculation is not at all laborious. Treated as in the previous cases for $s=2$, we find the result

$$F_{6,3}$$
$$=(\alpha_1\alpha_2\alpha_3+\alpha_4\alpha_5\alpha_6)(\alpha_1\alpha_2\alpha_4+\alpha_3\alpha_5\alpha_6)(\alpha_1\alpha_2\alpha_5+\alpha_3\alpha_4\alpha_6)(\alpha_1\alpha_2\alpha_6+\alpha_3\alpha_4\alpha_5)$$
$$\times(\alpha_1\alpha_3\alpha_4+\alpha_2\alpha_5\alpha_6)(\alpha_1\alpha_3\alpha_5+\alpha_2\alpha_4\alpha_6)(\alpha_1\alpha_3\alpha_6+\alpha_2\alpha_4\alpha_5)$$
$$\times(\alpha_1\alpha_4\alpha_5+\alpha_2\alpha_3\alpha_6)(\alpha_1\alpha_4\alpha_6+\alpha_2\alpha_3\alpha_5)(\alpha_1\alpha_5\alpha_6+\alpha_2\alpha_3\alpha_4),$$

as shown in the following Table:

Sym. Function.	Coeff.	No. of Terms.
10,44444	1	6
955533	1	60
866442	1	180
864444	3	90
777333	1	20
775551	1	60
775533	2	180
755553	6	180
666660	1	6
666642	3	90
666444	7	140
555555	12	12
		$1024=2^{10}$.

It will be noticed that the whole function is unchanged when, for each part p in the partitions, $10-p$ is written for p, any particular function being complementary either to itself or to one of the other functions.

If a partition be

$$p_1^{\pi_1} p_2^{\pi_2} \ldots p_t^{\pi_t} \ldots,$$

containing a given part p_t, π_t times, where $\pi_t > 0$, and C be its coefficient

$$\Sigma C \frac{(\pi_1 + \pi_2 + \ldots + \pi_t - 1 + \ldots)!}{\pi_1!\, \pi_2! \ldots (\pi_t - 1)! \ldots} = \binom{10}{p_t},$$

where the summation is in regard to those partitions which have a part p_t. For instance, taking $p_t = 6$, we are concerned with the third, fourth, ninth, tenth, and eleventh of the functions above appearing. We then have

$$1.60 + 3.5 + 1.5 + 3.20 + 7.10 = 210 = \binom{10}{6}.$$

The theorem is universally applicable, and provides a good check upon the accuracy of the numbers.

CHESS TOURNAMENTS AND THE LIKE TREATED BY THE CALCULUS OF SYMMETRIC FUNCTIONS.

By Major P. A. MacMahon.

Part I.

1. IN a recent paper* I discussed the American Tournament for such contests that a player must either win or lose, and was led to consider the symmetric function

$$(\alpha_1+\alpha_2)(\alpha_1+\alpha_3)(\alpha_2+\alpha_3)\ldots(\alpha_{n-1}+\alpha_n)$$

of the n quantities $\alpha_1, \alpha_2, \alpha_3, \ldots, \alpha_n$, which involves a complete solution of the problem.

In certain other games there is a third possibility—viz. the games may be drawn. Of these, the Chess Tournament is a type and the best example.

My attention was called to this question by the Masters' Chess Tournament which has been held this year (1922) in London. There were sixteen Masters engaged. Eight games were played on each of fifteen days, and thus $\binom{16}{2}=120$ games took place, the players being paired in all possible ways.

The method of pairing the players is not immediately obvious, and it may not be out of place to present one to the readers of this paper. This is given in Table II., with the necessary explanations.

2. In a Chess Tournament between Masters the players are not handicapped. A player scores one point for a game won and half a point for a game drawn.

Consider the symmetric function

$$(\alpha_1+\alpha_1^{\frac12}\alpha_2^{\frac12}+\alpha_2)(\alpha_1+\alpha_1^{\frac12}\alpha_3^{\frac12}+\alpha_3)(\alpha_2+\alpha_2^{\frac12}\alpha_3^{\frac12}+\alpha_3)\ldots(\alpha_{n-1}+\alpha_{n-1}^{\frac12}\alpha_n^{\frac12}+\alpha_n)$$

of the n quantities $\alpha_1, \alpha_2, \alpha_3, \ldots, \alpha_n$ in connection with a Tournament between n players.

* "An American Tournament treated by the Calculus of Symmetric Functions", *Quarterly Journal of Pure and Applied Mathematics*, vol. xlix., No. 193 (1920).

Denote the players by the numbers 1, 2, 3, ..., n, and consider the factor

$$\alpha_s + \alpha_s^{\frac{1}{2}}\alpha_t^{\frac{1}{2}} + \alpha_t \quad (s < t)$$

in connection with the game between the players denoted by the numbers s, t.

The symmetric function involves $\binom{n}{2}$ factors, and in multiplying it out we arrive at $3^{\binom{n}{2}}$ terms. This is the case because, in forming a term of the product, we may select a term from each factor in three different ways. In the game between the players s, t we see that the player s either wins, draws or loses the game with the player t, and we associate with these possibilities the selection of the terms

$$\alpha_s,\ \alpha_s^{\frac{1}{2}}\alpha_t^{\frac{1}{2}},\ \alpha_t$$

from the factor
$$\alpha_s + \alpha_s^{\frac{1}{2}}\alpha_t^{\frac{1}{2}} + \alpha_t$$

in order to form a term of the product of the $\binom{n}{2}$ factors.

If a term in the product be

$$C\alpha_1^{k_1}\alpha_2^{k_2}\alpha_3^{k_3}\ldots\alpha_n^{k_n},$$

it is clear that in C of the possible $3^{\binom{n}{2}}$ events the players, denoted by 1, 2, 3, ..., n, will emerge with the *scores* k_1, k_2, k_3, ..., k_n respectively.

If we write

$$\prod_{s,t} (\alpha_s^2 + \alpha_s^{\frac{1}{2}}\alpha_t^{\frac{1}{2}} + \alpha_t) = \Sigma C (k_1k_2k_3\ldots k_n) \quad (s < t),$$

where
$$(k_1k_2k_3\ldots k_n) = \Sigma\alpha_1^{k_1}\alpha_2^{k_2}\alpha_3^{k_3}\ldots\alpha_n^{k_n};$$

we may conclude that the players in *any assigned order* have the *scores* k_1, k_2, k_3, ..., k_n in C out of the possible $3^{\binom{n}{2}}$ events. With the object of avoiding fractional exponents, I suppose that a player scores 2 for a game won and 1 for a game drawn. Thus we have to consider the symmetric function

$$\prod_{s,t} (\alpha_s^2 + \alpha_s\alpha_t + \alpha_t^2) = \Sigma C (k_1k_2k_3\ldots k_n),$$

and, as we have to consider equalities between the *integers* k_1, k_2, k_3, ..., k_n, I employ *repetitional* exponents π_1, π_2, ..., π_n and write

$$(k_1k_2k_3\ldots k_n) = (p_1^{\pi_1}p_2^{\pi_2}p_3^{\pi_3}\ldots p_n^{\pi_n}),$$

and, putting the symmetric function product in notation χ_n, we have

$$\chi_n = \Sigma C (p_1^{\pi_1}p_2^{\pi_2}\ldots p_n^{\pi_n}).$$

When the symmetric function with coefficient C is written out at length it involves

$$\frac{(\Sigma\pi)!}{\pi_1!\,\pi_2!\ldots\pi_n!}\text{ terms,}$$

and, disregarding the final order of the players, the assemblage of *scores* will constitute the partition

$$(p_1^{\pi_1}p_2^{\pi_2}\ldots p_n^{\pi_n})\text{ of } 2\binom{n}{2}$$

in

$$C\frac{(\Sigma\pi)!}{\pi_1!\,\pi_2!\ldots\pi_n!}\text{ events.}$$

Thence the numerical relation

$$\Sigma C\frac{(\Sigma\pi)!}{\pi_1!\,\pi_2!\ldots\pi_n!}=3^{\binom{n}{2}},$$

where the summation is for every partition that occurs in the development of the function χ_n.

We have now to study the development of χ_n as a linear function of monomial symmetric functions, so as to determine the number of events C which yield the score denoted by the partition

$$(p_1^{\pi_1}p_2^{\pi_2}\ldots p_n^{\pi_n}).$$

We note that

$$\Sigma\pi p=2\binom{n}{2},\quad \Sigma\pi=n.$$

One of the numbers $p_1, p_2, \ldots, p_n$ may be zero, and its repetitional exponent cannot exceed unity because not more than one player can be defeated in every game that he plays. Also the greatest of the numbers $p_1, p_2, \ldots, p_n$ cannot exceed $2n-2$ because no player can win more than $n-1$ games. Including a single zero as a possible part every partition has exactly n parts, each part denoting the score of one of the players.

The early calculations yield the results

$$\chi_2=(20)+(1^2),$$

$$\chi_3=(420)+(41^2)+(3^20)+2(321)+3(2^3).$$

The term $2(321)$ of χ_3 indicates that three players, in an assigned order, will emerge with the scores 3, 2, 1 in precisely 2 out of the 27 $(=3^3)$ events. If the order be not assigned the scores 3, 2, 1 will arise in $2\times 6=12$ events, because, here,

$$\frac{(\Sigma\pi)!}{\pi_1!\,\pi_2!\ldots}=6.$$

Similarly the scores 3, 3, 0 arise in one or three ways according as the order of the players is or is not assigned.

It must be remembered that in the above a player scores 2, 1 for a game won and drawn respectively. If the scores be 1, $\frac{1}{2}$, as usual in Tournaments, the scores in the present case will be

$\frac{3}{2}$, 1, $\frac{1}{2}$ instead of 3, 2, 1,

$\frac{3}{2}$, $\frac{3}{2}$, 0 „ 3, 3, 0.

The players 1, 2, 3 as shewn above emerge with the scores 3, 2, 1 respectively in two ways—viz. these are

1 defeats 2	1 draws with 2
1 draws with 3	1 defeats 3
2 defeats 3	2 draws with 3.
Scores—3, 2, 1.	Scores—3, 2, 1.

Counting 2 for a game won, 1 for a game drawn.

3. An important property of the symmetric function product is easily established. Write

$$\prod_{s,t} (\alpha_s^2 + \alpha_s\alpha_t + \alpha_t^2) = \Sigma C_{k_1k_2k_3k_n} \alpha_1^{k_1}\alpha_2^{k_2}\alpha_3^{k_3}\ldots\alpha_n^{k_n},$$

so that

$$\prod_{s,t} \frac{\alpha_s^2 + \alpha_s\alpha_t + \alpha_t^2}{\alpha_s^2\alpha_t^2} = \Sigma \frac{C_{k_1k_2k_3\ldots}\alpha_1^{k_1}\alpha_2^{k_2}\alpha_3^{k_3}\ldots\alpha_n^{k_n}}{(\alpha_1\alpha_2\alpha_3\ldots\alpha_n)^{2n-2}},$$

or

$$\prod_{s,t} \left(\frac{1}{\alpha_s^2} + \frac{1}{\alpha_s\alpha_t} + \frac{1}{\alpha_t^2}\right) = \Sigma C_{k_1k_2k_3\ldots}\alpha_1^{k_1-2n+2}\alpha_2^{k_2-2n+2}\alpha_3^{k_3-2n+2}\ldots.$$

Now put $\alpha_s = \frac{1}{\alpha_s}$, $\alpha_t = \frac{1}{\alpha_t}$

for all values of s, t, and we obtain

$$\prod_{s,t} (\alpha_s^2 + \alpha_s\alpha_t + \alpha_t^2) = \Sigma C_{k_1k_2k_3\ldots}\alpha_1^{2n-2-k_1}\alpha_2^{2n-2-k_2}\alpha_3^{2n-2-k_3}\ldots$$
$$= \Sigma C_{2n-2-k_1,\, 2n-2-k_2,\, 2n-2-k_3,\ldots}\alpha_1^{2n-2-k_1}\alpha_2^{2n-2-k_2}\alpha_3^{2n-2-k_3}\ldots .$$

Thence the relation

$$C_{k_1k_2k_3\ldots} = C_{2n-2-k_1,\, 2n-2-k_2,\, 2n-2-k_3,\ldots},$$

establishing that the score

$$k_1,\ k_2,\ k_3,\ \ldots,\ k_n$$

occurs in as many ways as the score

$$2n-2-k_1,\ 2n-2-k_2,\ 2n-2-k_3,\ \ldots,\ 2n-2-k_n.$$

This theorem much lessens the labour of calculating the development of the symmetric functions.

4. The development of χ_n. Write

$$(\alpha_1^2+\alpha_1\alpha_2+\alpha_2^2)(\alpha_1^2+\alpha_1\alpha_3+\alpha_3^2)\dots(\alpha_1^2+\alpha_1\alpha_n+\alpha_n^2)$$
$$=\alpha_1^{2n-2}+B_1'\alpha_1^{2n-3}+B_2'\alpha_1^{2n-4}+\dots+B'_{n-2},$$

and we note that

$$B_{2m} = (2^m)'+(2^{m-1}1^2)'+(2^{m-2}1^4)'+\dots+(1^{2m})',$$
$$B_{2m+1}=(2^m1)'+(2^{m-1}1^3)'+(2^{m-2}1^5)'+\dots+(1^{2m+1})',$$

where the symmetric functions, which are denoted by partitions in dashed brackets, refer to the elements

$$\alpha_2,\ \alpha_3,\ \alpha_4,\ \dots\ .$$

Certain terms in these two series vanish automatically by reason of the circumstance that the number of elements is limited by the number $n-1$. Thence

$$\chi_n=(\alpha_1^{2n-2}+B_1'\alpha_1^{2n-3}+\dots+B'_{2n-2})\,\chi'_{n-1},$$

where

$$\chi'_{n-1}=(\alpha_2^2+\alpha_2\alpha_3+\alpha_3^2)(\alpha_2^2+\alpha_2\alpha_4+\alpha_4^2)\dots(\alpha_2^2+\alpha_2\alpha_n+\alpha_n^2).$$

Let D_k be Hammond's differential operator of order k, and refer to the infinite series of elements

$$\alpha_1,\ \alpha_2,\ \alpha_3,\ \dots,$$

and further let D_k' refer to the infinite series

$$\alpha_2,\ \alpha_3,\ \alpha_4,\ \dots\ .$$

In order to obtain the laws of operation of D_k it is necessary to consider the elements to be unrestricted in number. In the application to the particular number of elements, certain terms will then vanish automatically. We have evidently

$$D_1'B_r'=B'_{r-1},\quad D_2'B_r'=B'_{r-2},\quad D_k'B_r'=0 \quad \text{when } k>2.$$

A portion of χ_n is $B'_{2n-k-2}\alpha_1^k\chi'_{n-1}$, so that, by a well-known property of D_k,

$$D_k\chi_n=B_{2n-k-2}\chi_{n-1}.$$

The operation of D_k upon a product of symmetric functions has been explained by me in various papers and particularly in *Combinatory Analysis.**

Repeating the operation of D_k, we find

$$D_k^2\chi_n=B_{2n-k-2}D_k\chi_{n-1}+D_1B_{2n-k-2}D_{k-1}\chi_{n-1}+D_2B_{2n-k-2}D_{k-2}\chi_{n-1},$$

* Cambridge University Press, 1916.

AA 2

the partitions of k into two parts, one at least of which does not exceed 2, being the guide to the operation. These partitions are $(0, k)$, $(1, k-1)$, $(2, k-2)$.

The part first written in these partitions determines the D operator to be performed upon B_{2n-k-2}. There are no others because

$$D_s B_{2n-k-2} = 0 \quad \text{when } s > 2.$$

Thus we are led to

$$D_k^2 \chi_n = (B_{2n-k-2} D_k + B_{2n-k-3} D_{k-1} + B_{2n-k-4} D_{k-2}) \chi_{n-1},$$

wherein, putting symbolically $B_s = b^s$, we are brought to

$$D_k^2 \chi_n = b^{2n-k-2} (D_k + b^{-1} D_{k-1} + b^{-2} D_{k-2}) \chi_{n-1}.$$

Supposing
$$(p_1^{\pi_1} p_2^{\pi_2} \dots p_n^{\pi_n})$$
to be the partition representation of a symmetric function which occurs in χ_m, we derive in succession

$$D_{p_1} \chi_n = b^{2n-p_1-2} \chi_{n-1},$$

$$D_{p_1}^2 \chi_n = b^{2n-p_1-2} (D_{p_1} + b^{-1} D_{p_1-1} + b^{-2} D_{p_1-2}) \chi_{n-1},$$

$$D_{p_1}^{\pi_1} \chi_n = b^{2n-p_1-2} (D_{p_1} + b^{-1} D_{p_1-1} + b^{-2} D_{p_1-2})^{\pi_1-1} \chi_{n-1};$$

and finally
$$\begin{aligned}
&(D_{p_1}^{\pi_1} D_{p_2}^{\pi_2} \dots D_{p_n}^{\pi_n}) \chi_n \\
&= b^{2n-p_1-2} (D_{p_1} + b^{-1} D_{p_1-1} + b^{-2} D_{p_1-2})^{\pi_1-1} \\
&\times (D_{p_2} + b^{-1} D_{p_2-1} + b^{-2} D_{p_2-2})^{\pi_2} \\
&\times (D_{p_3} + b^{-1} D_{p_3-1} + b^{-2} D_{p_3-2})^{\pi_3} \\
&\times \dots\dots\dots\dots\dots\dots\dots\dots \\
&\times (D_{p_n} + b^{-1} D_{p_n-1} + b^{-2} D_{p_n-2})^{\pi_n} \chi_{n-1}.
\end{aligned}$$

On the left-hand side we arrive, as the result of the operations, at the coefficient of symmetric function

$$(p_1^{\pi_1} p_2^{\pi_2} \dots p_n^{\pi_n})$$

in the development of χ_n because

$$D_{p_1}^{\pi_1} D_{p_2}^{\pi_2} \dots D_{p_n}^{\pi_n} (p_1^{\pi_1} p_2^{\pi_2} \dots p_n^{\pi_n}) = 1.$$

On the right-hand side the effective portion of the operator is that part which, on carrying out the multiplications, is free from the symbol b. This is a linear function of operator products

$$D_{\mu_1}^{m_1} D_{\mu_2}^{m_2} \dots D_{\mu_{n-1}}^{m_{n-1}},$$

which correspond to symmetric functions which occur in the development of χ_{n-1} and are denoted by partitions

$$(m_1^{\mu_1} m_2^{\mu_2} \ldots m_{n-1}^{\mu_{n-1}}).$$

If then the surviving operation be

$$\Sigma K_{m_1}^{\mu_1} \ldots D_{m_1}^{\mu_1} \ldots$$

and
$$\chi_{n-1} = \Sigma C_m^{\mu_1} \ldots (m_1^{\mu_1} \ldots),$$

the right-hand side becomes

$$\Sigma K_m^{\mu_1} \ldots C_m^{\mu_1} \ldots,$$

and the acquired coefficients in the development of χ_n is determined from a knowledge of the coefficients appertaining to χ_{n-1}.

The formula enables us to find the complete development of χ_n from a knowledge of that of χ_{n-1} (cf. *loc. cit.* p. 6). For the purpose of calculation the formula can be given the convenient form

$$\begin{aligned}
&(D_p^{\rho_p} D_{p-1}^{\rho_{p-1}} D_{p-2}^{\rho_{p-2}} \ldots D_2^{\rho_2} D_1^{\rho_1}) \chi_n \\
&= b^{2n-p-2} (D_p + b^{-1} D_{p-1} + b^{-2} D_{p-2})^{\rho_p - 1} \\
&\times (D_{p-1} + b^{-1} D_{p-2} + b^{-2} D_{p-3})^{\rho_{p-1}} \\
&\times (D_{p-2} + b^{-1} D_{p-3} + b^{-2} D_{p-4})^{\rho_{p-2}} \\
&\times \ldots\ldots\ldots\ldots\ldots\ldots\ldots\ldots \\
&\times (D_2 + b^{-1} D_1 + b^{-2})^{\rho_2} \\
&\times (D_1 + b^{-1})^{\rho_1} . \chi_{n-1}.
\end{aligned}$$

The coefficients of symmetric functions in χ_n, whose partitions involve the maximum part $2n-2$ are obtained at once from the development of χ_{n-1} because the formula

$$D_k \chi_n = b^{2n-k-2} \chi_{n-1}$$

gives
$$D_{2n-2} \chi_n = \chi_{n-1}.$$

This establishes that the whole of χ_{n-1} is obtainable by considering the part of the partition of χ_n, which involves partitions which have a part $2n-2$ and striking out the part $2n-2$ from each of those partitions. This is the process named by Cayley 'Decapitation'. Conversely, if we add the part $2n-2$ to each of the partitions which present themselves in χ_{n-1}, we obtain that portion of χ_n which involves partitions having $2n-2$ as highest part. This is Cayley's process of 'Capitation'.

Again, the partitions whose coefficients are thus determined have, some of them, a single zero part. The remainder have not a zero part, but if we substitute for each part in such another, by subtraction from $2n-2$, we obtain a partition which has one zero part and a highest part $< 2n-2$ which connotes the complementary symmetric function which we know, by the complementary theorem, has the same coefficient.

Hence the complementary theorem yields, without calculation, the coefficients of all symmetric functions whose partitions contain a zero part.

5. It is convenient in this place to give the results for 2, 3, 4, 5, 6 players in order to elucidate the remarks above, and to make what follows about the calculations intelligible.

TABLE I.

Two players $\chi_2 = \alpha_1^2 + \alpha_1\alpha_2 + \alpha_2^2$.

Sym. Function or Score	Coeff. of Sym. Funct.	No. of terms in Sym. Funct.	Events in Tournament
20	1	2	2
1^2	1	1	1

Sum of numbers in last column $= 3^1$.

Three players $\psi_3 = (\alpha_1^2 + \alpha\ \alpha_2 + \alpha_2^2)(\alpha_1^2 + \alpha_1\alpha_3 + \alpha_3^2)(\alpha_2^2 + \alpha_2\alpha_3 + \alpha_3^2)$.

Sym. Function or Score	Coeff. of Sym. Funct.	No. of terms in Sym. Funct.	Events in Tournament
420	1	6	6
41^2	1	3	3
3^20	1	3	3
321	2	6	12
2^3	3	1	3

Sum of numbers in last column $= 3^3$.

Four players $\chi_4 = (\alpha_1^2 + \alpha_1\alpha_2 + \alpha_2^2)(\alpha_1^2 + \alpha_1\alpha_3 + \alpha_3^2)\ldots(\alpha_3^2 + \alpha_3\alpha_4 + \alpha_4^2)$.

Sym. Function or Score	Coeff. of Sym. Funct.	No. of terms in Sym. Funct.	Events in Tournament
6420	1	24	24
641^2	1	12	12
63^20	1	12	12
6321	2	24	48
62^3	3	4	12
5^220	1	12	12
5^21^2	1	6	6
5430	2	24	48
5421	4	24	96
53^21	5	12	60
532^2	7	12	84
4^30	3	4	12
4^231	7	12	84
4^22^2	10	6	60
43^22	12	12	144
3^4	15	1	15

Sum of numbers in last column $= 3^6$.

Five players $\chi_5 = (\alpha_1{}^2 + \alpha_1\alpha_2 + \alpha_2{}^2)(\alpha_1{}^2 + \alpha_1\alpha_3 + \alpha_3{}^2)\ldots(\alpha_4{}^2 + \alpha_4\alpha_5 + \alpha_5{}^2)$.

86420	1	120	120
8641^2	1	60	60
863^20	1	60	60
86321	2	120	240
862^3	3	20	60
85^220	1	60	60
85^21^2	1	30	30
85430	2	120	240
85421	4	120	480
853^21	5	60	300
8532^2	7	60	420
84^30	3	20	60
84^231	7	60	420
84^22^2	10	30	300
843^22	12	60	720
83^4	15	5	75
7^2420	1	60	60
7^241^2	1	30	30
7^23^20	1	30	30
7^2321	2	60	120
7^22^3	3	10	30
76520	2	120	240
7651^2	2	60	120
76430	4	120	480
76421	8	120	960
763^21	10	60	600
7632^2	14	60	840
75^230	5	60	300
75^221	10	60	600
754^20	7	60	420
75431	18	120	2160
7542^2	25	60	1500
753^22	31	60	1860
74^31	24	20	480
74^232	41	60	2460
743^3	51	20	1020
6^320	3	20	60
6^31^2	3	10	30
6^2530	7	60	420
6^2521	14	60	840
6^24^20	10	30	300
6^2431	25	60	1500
6^242^2	35	30	1050
6^23^22	43	30	1290

$65^2 40$	12	60	720
$65^2 31$	31	60	1860
$65^2 2^2$	43	30	1290
$654^2 1$	41	60	2460
65432	70	120	8400
653^3	87	20	1740
$64^3 2$	90	20	1800
$64^2 3^2$	111	30	3330
$5^4 0$	15	5	75
$5^3 41$	51	20	1020
$5^3 32$	87	20	1740
$5^2 4^2 2$	111	30	3330
$5^2 43^2$	138	30	4140
$54^3 3$	174	20	3480
4^5	219	1	219

Sum of numbers in last column $= 3^{10}$.

Six players $\chi_6 = (\alpha_1^2 + \alpha_1\alpha_2 + \alpha_2^2)(\alpha_1^2 + \alpha_1\alpha_3 + \alpha_3^2)\ldots(\alpha_5^2 + \alpha_5\alpha_6 + \alpha_6^2)$.

1086420	1	720	720
1086411	1	360	360
1086330	1	360	360
1086321	2	720	1440
1086222	3	120	360
1085520	1	360	360
1085511	1	180	180
1085430	2	720	1440
1085421	4	720	2880
1085331	5	360	1800
1085322	7	360	2520
1084440	3	120	360
1084431	7	360	2520
1084422	10	180	1800
1014332	12	360	4320
1083333	15	30	450
1077420	1	360	360
1077411	1	180	180
1077330	1	180	180
1077321	2	360	720
1077222	3	60	180
1076520	2	720	1440
1076511	2	360	720
1076430	4	720	2880
1076421	8	720	5760
1026331	10	360	3600
1026322	14	360	5040

1075530	5	360	1800
1075521	10	360	3600
1075440	7	360	2520
1075431	18	720	12960
1075422	25	360	9000
1075332	31	360	11160
1074441	24	120	2880
1074432	41	360	14760
1074333	51	120	6120
1066620	3	120	360
1066611	3	60	180
1066530	7	360	2520
1066511	14	360	5040
1066440	10	180	1800
1066431	25	360	9000
1066422	35	180	6300
1066332	43	183	7740
1065540	12	360	4320
1065531	31	360	11160
1065522	43	180	7740
1065441	41	360	14760
1065432	70	720	50400
1065333	87	120	10440
1064442	90	120	10800
1064433	111	180	19980
1055550	15	30	450
1055541	51	120	6120
1055532	87	120	10440
1055442	111	180	19980
1055433	138	180	24840
1054443	174	120	20880
1044444	219	6	1314
996411	1	180	180
996321	2	360	720
996222	3	60	180
995511	1	90	90
995421	4	360	1440
995331	5	180	900
995322	7	180	1260
994431	7	180	1260
994422	10	90	900
994332	12	180	2160
993333	15	15	225
987411	2	360	720
986511	4	360	1440

987321	4	720	2880
987222	6	120	720
986421	16	720	11520
986331	20	360	7200
986322	28	360	10080
985521	20	360	7200
985431	36	720	25920
985422	50	360	18000
985332	62	360	22320
984441	48	120	5760
984432	82	360	29520
984333	202	120	12240
977511	5	180	900
976611	7	180	1260
977421	20	360	7200
976521	36	720	25920
977331	25	180	4500
977322	35	180	6300
976431	65	720	46800
976422	90	360	32406
976332	112	360	40320
975531	82	360	29520
975522	113	180	20340
975441	108	360	38880
975432	185	720	133200
975333	231	120	27720
974442	237	120	28440
974433	295	180	53100
966621	48	120	5760
966531	108	360	38880
966522	149	180	26820
966441	142	180	25560
966432	243	360	87480
966333	303	60	18180
965541	176	360	63360
965532	301	360	108360
965442	382	360	137520
965433	476	360	171360
964443	597	120	71640
955551	219	30	6570
955542	471	120	56520
955533	588	60	35280
955443	735	180	132300
954444	915	30	27450
996420	1	360	360

995520	1	180	180
996330	1	180	180
995430	2	360	720
994440	3	60	180
987420	2	720	1440
986520	4	720	2880
987330	2	360	720
986430	8	720	5760
985530	10	360	3600
985440	14	360	5040
977520	5	360	1800
976620	7	360	2520
977430	10	360	3600
976530	18	720	12960
976440	25	360	9000
975540	31	360	11160
966630	24	120	2880
966540	41	360	14760
965550	51	120	6120
888420	3	120	360
888411	3	60	180
188330	3	60	180
888321	6	120	720
888222	9	20	180
887520	7	360	2520
887511	7	180	1260
887430	14	360	5040
887421	28	360	10080
887331	35	180	6300
887322	49	180	8820
886620	10	180	1800
886611	10	90	900
886530	25	360	9000
886521	50	360	18000
886440	35	180	6300
886431	90	360	32400
886422	125	180	22500
886332	155	180	27900
885540	43	180	7740
885531	113	180	20340
885522	156	90	14040
885441	149	180	26820
885432	255	360	91800
885333	318	60	19080
884442	327	60	19620

884433	406	90	36540
877620	12	360	4320
877611	12	180	2160
877530	31	360	11160
877521	62	360	22320
877440	43	180	7740
877431	112	360	40320
877422	155	180	27900
877332	193	180	34740
876630	41	360	14760
876621	82	360	29520
876540	70	720	50400
876531	185	720	133200
876522	255	360	91800
876441	243	360	87480
876432	416	720	299520
876333	519	120	62280
875550	87	120	10440
875541	301	360	108360
875532	515	360	185400
875442	653	360	235080
875433	814	360	293040
874443	1020	120	122400
866640	90	120	10800
866631	237	120	28440
866622	327	60	19620
866550	111	180	19980
866541	382	360	137520
866532	653	360	235080
866442	826	180	148680
866433	1028	180	185040
865551	471	120	56520
865542	1009	360	363240
865533	1258	180	226440
865443	1566	360	563760
864444	1941	30	58230
855552	1233	30	36990
855543	1908	120	228960
855444	2358	60	141480
777720	15	30	450
777711	15	15	225
777630	51	120	6120
777621	102	120	12240
777540	87	120	10440
777531	231	120	27720

777522	318	60	19080
777441	303	60	18180
777432	519	120	62280
777333	648	20	12960
776640	111	180	19980
776631	295	i80	53100
776622	406	90	36540
776550	138	180	24840
776541	476	360	171360
776532	814	360	293040
776442	1028	180	185040
776433	1282	180	230760
775551	588	60	35280
775542	1258	180	226440
775533	1570	90	141300
775443	1953	180	351540
774444	2418	15	36270
766650	174	120	20880
766641	597	120	71640
766632	1020	120	122400
766551	735	180	132300
766542	1566	360	563760
766533	1953	180	351540
766443	2424	180	436320
765552	1908	120	228960
765543	2943	360	1059480
865444	3627	120	435240
755553	3570	40	107100
755544	4389	60	263340
666660	219	6	1314
666651	915	30	27450
666642	1941	30	58230
666633	2418	15	36270
666552	2358	60	141480
666543	3627	120	435240
666444	4464	20	89280
665553	4389	60	263340
665544	5388	90	484920
655554	6501	30	195030
555555	7839	1	7839

Sum of numbers in last column is 3^{15}.

6. The Tables exhibit the expression of χ_n as a linear function of the monomial symmetric function which are denoted

by partitions in the first column. The second column gives the coefficients which appear in the linear function. The third column gives the number of terms which appear when the symmetric function in the same row is written out *in extenso*. The numbers in the fourth column are obtained by multiplying the numbers in the same row and second and third columns. Since

$$\chi_n = \prod_{s,t} (\alpha_s^2 + \alpha_s\alpha_t + \alpha_t^2),$$

we see, by putting $\alpha_1 = \alpha_2 = \ldots = \alpha_n = 1$, that the sum of the numbers in the fourth column is

$$3^{\binom{n}{2}}.$$

This fact furnishes an important check upon the accuracy of the calculations. It is readily seen that the fourth column numbers, due to symmetric functions whose highest part is $2n-2$ (the maximum part), sum up to

$$n.3^{\binom{n-1}{2}},$$

because this portion of the Table is obtained from the Table of χ_{n-1} by simply multiplying the third and fourth column numbers by n. This is in fact the first of a series of auxiliary summation theorems which will be now considered. The theorem which has just been stated yields, with the aid of the complementary theorem (*ante*), a complementary summation theorem—viz. the sum of the fourth column numbers due to symmetric functions whose partitions possess a *zero part* is

$$n.3^{\binom{n-1}{2}}.$$

The parts $2n-2$ and zero can be discussed together, and in general parts s and $2n-2-s$ give complementary summation theorems.

We have for consideration the theorems of summations connected with the symmetric functions in the developments whose partitions contain one or more parts equal to s, where s may have any value from $2n-2$ to zero. Take that portion of χ_n which is of the form

$$\Sigma C_{s^\sigma\ldots}(s^\sigma\ldots) \quad \text{where } s \nless 1.$$

If we operate with D_s upon χ_n, this is the only part of χ_n which it does not cause to vanish. We may write

$$D_s \sum_\sigma C_{s^\sigma\ldots}(s^\sigma\ldots) = B_{2n-s-2}\chi_{n-1}$$

or $$\sum_\sigma C_{s^\sigma\ldots}(s^{\sigma-1}\ldots) = B_{2n-s-2}\chi_{n-1}.$$

The left-hand side is a sum of linear functions; one linear function corresponding to each value of σ. Consider the linear function

$$\Sigma C_{s^\sigma\ldots}(s^{\sigma-1}\ldots),$$

in which the value of σ is *fixed.*

What is the result of putting herein $\alpha_1 = \alpha_2 = \ldots = \alpha_n = 1$? The fourth column associated with the single term

$$C_{s^\sigma\ldots}(s^\sigma\ldots)$$

being $$M_{s^\sigma\ldots},$$

that associated with the *single* term

$$C_{s^\sigma\ldots}(s^{\sigma-1}\ldots)$$

is $$\frac{\sigma}{n} M_{s^\sigma\ldots},$$

Hence the sum of the fourth column numbers associated with the linear function

$$\Sigma C_{s^\sigma\ldots}(s^{\sigma-1}\ldots),$$

in which σ has a *fixed* value, is

$$\frac{\sigma}{n} \Sigma M_{s^\sigma\ldots},$$

the summation being for all functions whose partitions involve σ parts equal to s. Thence we find that the sum of the fourth column numbers associated with the linear function

$$\sum_\sigma C_{s^\sigma\ldots}(s^{\sigma-1}\ldots)$$

is $$\sum_\sigma \left(\frac{\sigma}{n} \Sigma M_{s^\sigma\ldots}\right).$$

On the right-hand side we have to find the number which results from putting $\alpha_1 = \alpha_2 = \ldots = \alpha_n = 1$. In $B_{2n-s-2}\chi_{n-1}$, χ_{n-1} becomes $3^{\binom{n-1}{2}}$.

When

$s = 2n-3$ we have B_1 or (1) becoming $\binom{n-1}{1}$,

$2n-4$ „ B_2 or $(2)+(1^2)$ „ $\binom{n-1}{1} + \binom{n-1}{2}$,

$2n-5$ „ B_3 or $(4)+(1^3)$ „ $\binom{2}{1}\binom{n-1}{2} + \binom{n-1}{3}$, and in general when $s = 2n - 2s' - 1$, we have

$$B_{2s'-1} \text{ or } (2^{s'-1}1) + (2^{s'-2}1^3) + (2^{s'-3}1^5) + \ldots$$

becoming

$$\binom{s'}{1}\binom{n-1}{s'} + \binom{s'+1}{3}\binom{n-1}{s'+1} + \binom{s'+2}{5}\binom{n-1}{s'+2} + \ldots,$$

and when $s = 2n - 2s'$,

$$B_{2s'-2} \text{ or } (2^{s'-1}) + (2^{s'-2}1^2) + (2^{s'-3}1^4) + \ldots$$

becoming

$$\binom{s'-1}{0}\binom{n-1}{s'-1} + \binom{s'}{2}\binom{n-1}{s'} + \binom{s'+1}{4}\binom{n-1}{s'+1} + \ldots.$$

We have thus the convenient formulæ

$$\frac{1}{n}\Sigma M_{2n-2,\ldots} = 3^{\binom{n-1}{2}},$$

$$\frac{1}{n}\Sigma M_{2n-3,\ldots} + \frac{2}{n}\Sigma M_{2n-3,2n-3,\ldots} = \binom{n-1}{1} 3^{\binom{n-1}{2}},$$

$$\frac{1}{n}\Sigma M_{2n-4,\ldots} + \frac{2}{n}\Sigma M_{2n-4,2n-4,\ldots} + \ldots = \left\{\binom{n-1}{1} + \binom{n-1}{2}\right\} 3^{\binom{n-1}{2}},$$

$$\frac{1}{n}\Sigma M_{2n-5,\ldots} + \frac{2}{n}\Sigma M_{2n-5,2n-5,\ldots} + \ldots = \left\{2\binom{n-1}{2} + \binom{n-1}{3}\right\} 3^{\binom{n-1}{2}},$$

and it must be remembered that owing to the complementary law the right-hand sides of these formulæ are unchanged when on the left-hand sides we substitute

$$M_{s\ldots} \text{ for } M_{2n-s-2\ldots}.$$

The application to χ_n is as follows (as can be verified by the Tables):

$$\Sigma M_{6\ldots} = 24 + 12 + 12 + 48 + 12 = 108 = \Sigma M_{0\ldots}.$$

The formula gives

$$\tfrac{1}{4}\Sigma M_{6\ldots} = \tfrac{1}{4}M_{0\ldots} = 3^3,$$

$$\Sigma M_{5\ldots} = 48 + 96 + 60 + 84 = 288 = \Sigma M_{1\ldots},$$

$$\Sigma M_{55\ldots} = 12 + 6 \qquad = 18 \quad = \Sigma M_{11\ldots}.$$

The formula gives

$$\tfrac{1}{4}\Sigma M_{5\ldots} + \tfrac{2}{4}\Sigma M_{55\ldots} = \binom{3}{1} 3^3 = \tfrac{1}{4}\Sigma M_{1\ldots} + \tfrac{2}{4}\Sigma M_{11\ldots},$$

$$\Sigma M_{3\ldots} = 48 + 48 + 84 + 84 = 264,$$

$$\Sigma M_{33\ldots} = 12 + 60 + 144 = 216,$$

$$\Sigma M_{3333} = 15 \qquad = 15.$$

The formula gives

$$\tfrac{1}{4}\Sigma M_{3\ldots} + \tfrac{2}{4}\Sigma M_{33\ldots} + \tfrac{4}{4}\Sigma M_{3333\ldots} = 7.3^3,$$

a verification.

$$\Sigma M_{4\ldots} = 24 + 12 + 48 + 96 + 144 = 324 = \Sigma M_{2\ldots},$$

$$\Sigma M_{44\ldots} = 84 + 60 \qquad = 144 = \Sigma M_{22\ldots},$$

$$\Sigma M_{444\ldots} = 12 \qquad = 12 = \Sigma M_{222\ldots}.$$

The formula gives

$$\tfrac{1}{4}\Sigma M_{4\ldots} + \tfrac{2}{4}\Sigma M_{44\ldots} + \tfrac{3}{4}\Sigma M_{444\ldots} = \left\{\binom{3}{1} + \binom{3}{2}\right\} 3^3 = 162,$$

$$\tfrac{1}{4}\Sigma M_{2\ldots} + \tfrac{2}{4}\Sigma M_{22\ldots} + \tfrac{3}{4}\Sigma M_{222\ldots} = \left\{\binom{3}{2} + 3\binom{3}{3}\right\} 3^3 = 162,$$

verifying.

We have thus a very satisfactory way of verifying the numbers that are separately calculated. In this manner errors have been detected and eliminated, and the Tables may be regarded as free from errors due to faulty calculation.

The process of calculating the coefficients in the development of χ_n is shewn by the following determination of the coefficients of the symmetric function (766443) in the development of χ_6. This is

$$D_7 D_6^2 D_4^2 D_3 \chi_6$$

$$= b^3 \{D_6 + b^{-1} D_5 + b^{-2} D_4\}^2 \{D_4 + b^{-1} D_3 + b^{-2} D_2\}^2 \{D_3 + b^{-1} D_2 + b^{-2} D_1\} \chi_5.$$

The part on the right-hand side which is free from b is found to be

$$\left\{\begin{array}{l} 3D_3^2 D_3^2 D_2 + 8 D_6 D_5 D_4 D_3 D_2 + 2 D_6 D_5 D_3^3 + 4 D_6 D_4^2 D_3^2 \\ + 2 D_5^2 D_4 D_3^2 + 2 D_5 D_4^3 D_3 + 2 D_6^2 D_4 D_2^2 + 2 D_6 D_4^3 D_2 \\ + D_5^2 D_4^2 D_2 + 3 D_6^2 D_4 D_3 D_1 + 2 D_6 D_5 D_4^2 D_1 \end{array}\right\} \chi_5.$$

A reference to the Table of χ_5 now leads to the sum

$$\begin{array}{l} 129 + 560 + 174 + 444 \\ + 276 + 348 + 70 + 180 \\ + 111 + 50 + 82 \\ = 2424. \end{array}$$

7. A Chess Tournament of 16 players. The players are denoted by the numbers which appear along the top and down the side of the triangle. The other numbers refer to the simultaneous pairings in the Tournament. A player denoted by a top number plays with one denoted by a side number in the pairing denoted by the number found at the intersection of the column through the former with the row through the latter.

TABLE II.

	1	2	3	4	5	6	7	8	9	10	11	12	13	14	15
16	15	2	4	6	8	10	12	14	1	3	5	7	9	11	13
15	14	15	1	2	3	4	5	6	7	8	9	10	11	12	
14	13	14	15	1	2	3	4	5	6	7	8	9	10		
13	12	13	14	15	1	2	3	4	5	6	7	8			
12	11	12	13	14	15	1	2	3	4	5	6				
11	10	11	12	13	14	15	1	2	3	4					
10	9	10	11	12	13	14	15	1	2						
9	8	9	10	11	12	13	14	15							
8	7	8	9	10	11	12	13								
7	6	7	8	9	10	11									
6	5	6	7	8	9										
5	4	5	6	7											
4	3	4	5												
3	2	3													
2	1														

For example, the pairings No. 7 are formed by the pairs 4 and 5, 3 and 6, 2 and 7, 1 and 8, 11 and 13, 10 and 14, 9 and 15, 12 and 16. The construction of the triangle for any even number of players may be gathered from that, above, for 16 players.

At the moment of writing (16 September 1922), the Hastings Chess Tournament between six players has been played through once with the result: Alekhine, 4; Rubenstein, $3\frac{1}{2}$; Tanesch, Thomas and Bogoljubow, 2; Yates, $3\frac{1}{2}$. Multiplying these scores by 2 we obtain the partition of 30,

874443.

Reference to χ_6 in Table I. shews that this exact result was obtainable in 1020 different ways; while if the order of the players be not specified the number of different ways of arriving at such a score is 122400.

PART II.

8. From a mathematical point of view the theory of Part I. is capable of generalization. The American Tournament in which games cannot be drawn was examined by means of the symmetric function

$$\Pi_n (\alpha_1 + \alpha_2),$$

and it will be noticed that one mark is awarded to each game, and that mark goes to the winner. We have in fact the two compositions of unity into two parts, zero counting as a part. These are 10, 01, and yield the two possible ways of awarding the mark. We call this a Tournament of Class I.

The Tournament considered in Part I of this paper allowed a game to be drawn, and assigned two marks to each game, the ways of awarding the marks depending upon an examination of the symmetric function

$$\Pi_n (\alpha_1^2 + \alpha_1\alpha_2 + \alpha_2^2)$$

and upon the compositions of the number 2 into two parts

$$20,\ 11,\ 02.$$

We call this a Tournament of Class 2. Similarly we may have a Tournament of Class r depending upon the development of the symmetric function

$$\Pi_n (\alpha_1^r + \alpha_1^{r-1}\alpha_2 + \ldots + \alpha_1\alpha_2^{r-1} + \alpha_2^r)$$

and upon the compositions of the number r into two parts

$$r, 0;\ r-1, 1;\ r-2, 2;\ \ldots\ 2, r-2;\ 1, r-1;\ 0, r,$$

where for a game there are r marks at disposal and the players may be assigned $r-s$ and s marks respectively according to the skill displayed.

The question then is as to the number of different ways in which a given final score may be reached in the Tournament.

Consider the case $r=3$. Here is a Tournament game between two players, one player may be awarded 3, 2, 1, 0 marks, and the second 0, 1, 2, 3 marks. No provision is made for awarding equal marks to the players. This is the case whenever r is an uneven number, but there is no difference in the mathematical theory. Writing

$$\Xi_n = \Pi_n (\alpha_1^3 + \alpha_1^2\alpha_2 + \alpha_1\alpha_2^2 + \alpha_2^3),$$

BB 2

the simplest results are

Table III.

Two players $\underset{2}{\Xi} = (\alpha_1^3 + \alpha_1^2\alpha_2 + \alpha_1\alpha_2^2 + \alpha_2^3)$.

Sym. Func. or Score.	Coeff. of Sym. Func.	No. of Terms in Sym. Func.	Events in Tournament.
(30)	1	2	2
(21)	1	2	2

Sum of numbers in fourth column $= 4'$.

Three players $\underset{3}{\Xi} = (\alpha_1^3+\alpha_1^2\alpha_2+\alpha_1\alpha_2^2+\alpha_2^3)(\alpha_1^3+\alpha_1^2\alpha_3+\alpha_1\alpha_3^2+\alpha_3^3)(\alpha_2^3+\alpha_2^2\alpha_3+\alpha_2\alpha_3^2+\alpha_3^3)$.

630	1	6	6
621	1	6	6
540	1	6	6
531	2	6	12
522	2	3	6
441	2	3	6
432	3	6	18
333	4	1	4

Sum of numbers in fourth column $= 4^3$.

Four players $\underset{4}{\Xi} = (\alpha_1^3 + \alpha_1^2\alpha_2 + \alpha_1\alpha_2^2 + \alpha_2^3)(\quad)(\quad)(\quad)(\alpha_3^3 + \alpha_3^2\alpha_4 + \alpha_3\alpha_1^3 + \alpha_4^3)$.

9630	1	24	24
9621	1	24	24
9540	1	24	24
9531	2	24	48
9522	2	12	24
9441	2	12	24
9432	3	24	72
9333	4	4	16
8730	1	24	24
8721	1	24	24
8640	2	24	48
8631	4	24	96
8622	4	12	48
8550	2	12	24
8541	5	24	120
8532	7	24	168
8442	8	12	96
8433	10	12	120
7740	2	12	24
7731	4	12	48
7722	4	6	24

7650	3	24	72
7641	7	24	168
7632	10	24	240
7551	8	12	96
7542	13	24	312
7533	16	12	192
7443	18	12	216
6660	4	4	16
6651	10	12	120
6642	16	12	192
6633	20	6	120
6552	18	12	216
6543	25	24	600
6444	28	4	112
5553	28	4	112
5544	32	6	192

Sum of numbers in fourth column $= 4^6$.

9. We write

$$\underset{n}{\Xi} = (\alpha_1^3 + \alpha_1^2\alpha_2 + \alpha_1\alpha_2^2 + \alpha_2^3)\ldots(\alpha_1^3 + \alpha_1^2\alpha_n + \alpha_1\alpha_n^2 + \alpha_n^3)\ \underset{n-1}{\Xi'}$$

$$= (\alpha_1^{3n-3} + C_1'\alpha_1^{3n-4} + C_2'\alpha_1^{3n-5} + \ldots + C'_{3n-3})\ \underset{n-1}{\Xi},$$

where C'_s is the sum of all symmetric functions of weight 5 of the elements $\alpha_2, \alpha_3, \ldots, \alpha_n$, which are such that no element occurs with a higher exponent than 3.

As before, we find that

$$D_s \underset{n}{\Xi} = C_{3n-s-3}\ \underset{n-1}{\Xi},$$

where C_{3n-s-3} has the above definition, but with the elements $\alpha_1, \alpha_2, \ldots, \alpha_n$.

To find the law of operation of D_s upon $c_1, c_2, c_3, \ldots$ let the number of elements be unrestricted in magnitude. Then it is clear that

$$D_s C_k = C_{k-s} \quad \text{for } s = 1, 2, 3,$$
$$D_s C_k = 0 \quad \text{for } s > 3.$$

Hence

$$D_p \underset{n}{\Xi} = C_{3n-p-3}\ \underset{n-1}{\Xi},$$

$$D_p^2 \underset{n}{\Xi} = (C_{2n-p-3}D_p + C_{3n-p-4}D_{p-1} + C_{3n-p-5}D_{p-2} + C_{3n-p-6}D_{p-3})\ \underset{n-1}{\Xi},$$

or, if $C_k = c^k$ symbolically,

$$D_p^2 = c^{3n-p-3}(D_p + c^{-1}D_{p-1} + c^{-2}D_{p-2} + c^{-3}D_{p-3})\ \underset{n-1}{\Xi},$$

leading to the formula

$$\begin{aligned}&(D_p^{\rho_p} D_{p-1}^{\rho_{p-1}} \ldots D_2^{\rho_2} D_1^{\rho_1}) \underset{n}{\Xi} \\ &= c^{3n-p-3}(D_p + c^{-1}D_{p-1} + c^{-2}D_{p-2} + c^{-3}D_{p-3})^{\rho_p - 1} \\ &\quad \times (D_{p-1} + c^{-1}D_{p-2} + c^{-2}D_{p-3} + c^{-3}D_{p-4})^{\rho_{p-1}} \\ &\quad \times \ldots\ldots\ldots\ldots\ldots\ldots\ldots\ldots \\ &\quad \times (D_3 + c^{-1}D_2 + c^{-2}D_1 + c^{-3})^{\rho_3} \\ &\quad \times (D_2 + c^{-1}D_1 + c^{-2})^{\rho_2} \\ &\quad \times (D_1 + c^{-1})^{\rho_1} \underset{n-1}{\Xi},\end{aligned}$$

which, as in the previous instances, enables us to calculate the coefficients in the development of $\underset{n}{\Xi}$ as soon as we know those which appertain to $\underset{n-1}{\Xi}$.

It is now clear that if

$$\underset{n,\,r}{\Xi} = \underset{n}{\Pi}\,(\alpha_1^{\,r} + \alpha_1^{\,r-1}\alpha_2 + \ldots + \alpha_1\alpha_2^{\,r-1} + \alpha_2^{\,r}),$$

we have the formula

$$\begin{aligned}&(D_p^{\rho_p} D_{p-1}^{\rho_{p-1}} \ldots D_2^{\rho_2} D_1^{\rho_1}) \underset{n,\,r}{\Xi} \\ &= c^{rn-p-r}(D_p + c^{-1}D_{p-1} + c^{-2}D_{p-2} + \ldots + c^{-r}D_{p-r})^{\rho_p - 1} \\ &\quad \times (D_{p-1} + c^{-1}D_{p-2} + c^{-2}D_{p-3} + \ldots + c^{-r}D_{p-r-1})^{\rho_{p-1}} \\ &\quad \times \ldots\ldots\ldots\ldots\ldots\ldots\ldots\ldots \\ &\quad \times (D_r + c^{-1}D_{r-1} + c^{-2}D_{r-2} + \ldots + c^{-r})^{\rho_r} \\ &\quad \times \ldots\ldots\ldots\ldots\ldots\ldots\ldots\ldots \\ &\quad \times (D_2 + c^{-1}D_1 + c^{-2})^{\rho_2} \\ &\quad \times (D_1 + c^{-1})^{\rho_1} \underset{n-1,\,r}{\Xi},\end{aligned}$$

which enables the calculation of the coefficients in the developments of $\underset{n,\,r}{\Xi}$ when we know those which appertain to $\underset{n-1,\,r}{\Xi}$.

Auxiliary summation theorems may be obtained, and they are very similar to those which have been investigated for the cases $r = 1$, $r = 2$.

10. In some cases the n players in an American Tournament play more than one round. Each player plays every

other player t times. We have then to consider the development of the symmetric function product

$$\prod_n (\alpha_1^r + \alpha_1^{r-1}\alpha_2 + \ldots + \alpha_1\alpha_2^{r-1} + \alpha_2^r)^t$$

for various values of n, r and t.

Let us first consider the cases for which $r = 1$ and put

$$\chi'_{n,1} = \Pi\,(\alpha_1 + \alpha_2)^t = (\alpha_1^{n-1} + A_1'\alpha_1^{n-2} + \ldots + A'_{n-1})^t\,\chi'^t_{n-1,1}.$$

Put

$$(\alpha_1^{n-1} + A_1'\alpha_1^{n-2} + A_2'\alpha_1^{n-3} + \ldots + A'_{n-1})^t = \alpha_1^{tn-t} + A'_{1,t}\alpha_1^{tn-t-1} + \ldots + A_{tn-t,t}.$$

Let $$A_1,\ A_2,\ \ldots,\ A_{1,t},\ A_{2,t}\ \ldots$$

have reference to the series of elements $\alpha_1, \alpha_2, \alpha_3, \ldots$ infinite in number so that we may study the effort of D_s upon $A_{1,t}, A_{2,t}, \ldots$.

Write

$$(1 + A_1x + A_2x^2 + \ldots)^t = 1 + A_{1,t}x + A_{2,t}x^2 + \ldots .$$

We know that $D_1A_s = A_{s-1}$, $D_kA_s = 0$ if $k > 1$. Since

$$D_1(1 + A_1x + A_2x^2 + \ldots)^t = tx(1 + A_1x + A_2x^2 + \ldots)^t,$$

$$D_1(1 + A_{1,t}x + A_{2,t}x^2 + \ldots) = tx(1 + A_{1,t}x + A_{2,t}x^2 + \ldots),$$

whence comparison of the coefficients of powers of x gives

$$D_1A_{s,t} = tA_{s-1,t}.$$

To operate with D_2 upon

$$(1 + A_1x + A_2x^2 + \ldots)^t,$$

we consider the compositions of 2 into t parts, no part > 1 and zero parts admitted.

Ex. gr. $t = 4$, such compositions are

1100, 1010, 1001, 0110, 0101, 0011,

and putting $1 + A_1x + A_2x^2 + \ldots = X$, D_2 operates upon X^4 in the manner

$$D_1XD_1X.X.X + D_1X.X.D_1X.X + D_1X.X.X.D_1X$$
$$+ X.D_1X.D_1X.X + X.D_1X.X.D_1X + X.X.D_1X.D_1X,$$

producing $$\binom{4}{2} X^2 (D_1X)^2.$$

So, in general, the operation of D_2 produces

$$\binom{t}{2} X^{t-2} (D_1X)^2,$$

and, since $D_1 X = xX$, this is

$$\binom{t}{2} x^2 X^t.$$

Hence $$D_2 X^t = \binom{t}{2} x^2 X^t,$$

yielding $$D_2 A_{s,t} = \binom{t}{2} A_{s-2,t},$$

while generally $$D_k A_{s,t} = \binom{t}{k} A_{s-k,t}.$$

It is important to observe here that $\binom{t}{k}$ is the coefficient β_k in $(1+\beta)^t$.

We obtain, as on a previous page,

$$D_p \chi^t_{n,1} = A_{tn-p-t,t} \chi^t_{n-1,1},$$

$$D_p^2 \chi^t_{n,1} = \Big\{ A_{tn-p-t,t} D_p + \binom{t}{1} A_{tn-p-t-1,t} D_{p-1} + \binom{t}{2} A_{tn-p-t-2,t} D_{p-2} + \ldots\ldots + \binom{t}{t} A_{tn-p-2t,t} D_{p-t} \Big\} \chi^t_{n-1,1}.$$

This result is not further developed for the moment, as it will be included in a more general formula further on.

11. Next consider the $r=2$.

$$\chi^t_{n,2} = \prod_n (\alpha_1^2 + \alpha_1\alpha_2 + \alpha_2^2)^t = (\alpha_1^{2n-2} + \beta_1'\alpha_1^{2n-3} + \ldots + B'_{2n-2})^t \chi^t_{n-1,2}$$
$$= (\alpha_1^{2tn-2t} + \beta'_{1,t}\alpha_1^{2tn-2t-1} + \ldots + B_{2tn-2t,t}) \chi^t_{n-1,2}$$

$$D_p \chi^t_{n,2} = B_{2tn-p-2t,t} \chi^t_{n-1,2}.$$

We make the same conditions as in the case of $r=1$, and consider the identity

$$(1 + B_1 x + B_2 x^2 + \ldots)^t = 1 + B_{1,t} x + B_{2,t} x^2 + \ldots .$$

B_s is expressed as a linear function of partitions which have no part > 2. To operate upon the left-hand side with D_k we work through the compositions of the number k by means of parts 0, 1, 2, the whole number of such parts being t.

Ex. gr., writing $1 + B_1 x + B_2 x^2 + \ldots = X$, and taking $t=3$,

$$D_4 X^3 = D_2 X . D_2 X . X + D_2 X . X . D_2 X + X . D_2 X . D_2 X$$
$$+ D_2 X . D_1 X . D_1 X + D_1 X . D_2 X . D_1 X + D_1 X . D_1 X . D_2 X$$
$$= 3X (D_2 X)^2 + 3 (D_2 X)(D_1 X)^2,$$

the associated compositions of the number 4 being

220, 202, 002 211, 121, 112.

Thence $$D_4 X^3 = 6x^4 X^3$$

and $$D_4 B_{s,t} = 6 B_{s-4,t},$$

where, observe, 6 is the coefficient of β^4 in the expansion of

$$(1 + \beta + \beta^2)^3.$$

In general, $D_k B_{s,t} = B_{s-k,t} \times$ the coefficient of β^k in the expansion of $(1 + \beta + \beta^2)^t$, because this is the function which enumerates, by the coefficient of β^k, the compositions of k into t parts limited to the magnitudes 0, 1, 2. The numbers $f_{r,t;k}$, upon which the whole investigation depends, are easily determined in simple cases by forming a series of tableaux.

$r = 1$, we write down a row of two units

1	1			
1	2	1		
1	3	3	1	
1	4	6	4	1

....

and underneath each write the sum of two numbers—viz. the the number directly over it and the number directly on the left of the latter. Proceed to 3rd and 4th, &c., rows on the same rule. Thus, the third number in the 4th row is the sum of the second and third numbers in the 3rd row. The sth number in the tth row is the sum of the $s-1^{th}$ and s^{th} numbers in the $t-1^{th}$ row.

$f_{1,t;k}$ is the $k+1^{th}$ number in the tth row.

$r = 2$, we write down a row of three units

1	1	1						
1	2	3	2	1				
1	3	6	7	6	3	1		
1	4	10	16	19	16	10	4	1

....

the sth number in the tth row is the sum of the $s-2^{th}$, $s-1^{th}$ and s^{th} numbers in the $t-1^{th}$ row.

$r=3$, we write down a row of four units

1	1	1	1									
1	2	3	4	3	2	1						
1	3	6	10	12	12	10	6	3	1			
1	4	10	20	31	40	44	40	31	20	10	4	1

....

the sth number in the tth row is the sum of the $s-3^{th}$, $s-2^{th}$, $s-1^{th}$, s^{th} numbers in the $t-1^{th}$ row.

In general, we write down $r+1$ units in a row and form the tableau by making the sth number in the tth row the sum of the $s-r^{th}$, $s-r+1^{th}$, ..., s^{th} numbers in the $t-1^{th}$ row; $f_{r,t:k}$ is the $k+1^{th}$ number in the tth row.

We write

$$(1+\beta+\beta^2)^t=\sum_k f_{2,t:k}\beta_k,$$

and then $$D_k B_{s,t}=f_{2,t:k}B_{s-k,t}.$$

Hence from $$D_p \chi^t_{n,2}=B_{2tn-p-2t,t}\chi^t_{n-1,2},$$

we proceed to

$$D_p^2\chi^t_{n,2}=(\sum_{s=0} f_{2,t:s}B_{2tn-p-2t-s,t}D_{p-s})\chi^t_{n-1,2},$$

and, putting $B_{m,t}=b_t^m$ symbolically,

$$D_p^2\chi^t_{n,2}=b_t^{2tn-p-2t}(D_p+f_{2,t:1}b^{-1}D_{p-1}+f_{2,t:2}b^{-2}D_{p-2}+\ldots)\chi^t_{n,2}.$$

In general, if

$$(1+\beta+\beta^2+\ldots+\beta^r)^t=\sum_k f_{r,t:k}\beta^k,$$

$$D_p\chi^t_{n,r}=B_{rtn-p-rt}\chi^t_{n-1,r},$$

$$D_p^2\chi^t_{n,r}=b_t^{rtn-p-rt}(D_p+f_{r,t:1}b^{-1}D_{p-1}+f_{r,t:2}b^{-2}D_{p-2}+\ldots)\chi^t_{n-1,r},$$

$$\begin{aligned}(D_p^{\rho_p}D_{p-1}^{\rho_{p-1}}\ldots D_2^{\rho_2}D_1^{\rho_1})\chi^t_{n,r} &= b_t^{rtn-p-rt}(D_p+f_{r,t:1}b^{-1}D_{p-1}+f_{r,t:2}b^{-2}D_{p-2}+\ldots)^{\rho_p-1}\\ &\times(D_{p-1}+f_{r,t:1}b^{-1}D_{p-2}+f_{r,t:2}b^{-2}D_{p-3}+\ldots)^{\rho_{p-1}}\\ &\times\ldots\ldots\ldots\ldots\ldots\ldots\\ &\times(D_2+f_{r,t:1}b^{-1}D_1+f_{r,t:2}b^{-2})^{\rho_2}\\ &\times(D_1+f_{r,t:1}b^{-1})^{\rho_1}\chi^t_{n-1,r},\end{aligned}$$

the general formula for use with all values of n, r and t.

Some further results are given in Table IV.

TABLE IV.

$(n, r, t) = (3, 1, 2)$

420	1	6	6
411	2	3	6
330	2	3	6
321	6	6	36
222	10	1	10
			64

$(n, r, t) = (4, 1, 2)$

6420	1	24	24
6411	2	12	24
6330	2	12	24
6321	6	24	144
6222	10	4	40
5520	2	12	24
5511	4	6	24
5430	6	24	144
5421	18	24	432
5331	28	12	336
5322	44	12	528
4440	10	4	40
4431	44	12	528
4422	68	6	408
4332	102	12	1224
3333	152	1	152
			4096

$(n, r, t) = (2, 2, 2)$

40	1	2	2
31	2	2	4
22	3	1	3
			9

$(n, r, t) = (3, 2, 2)$

840	1	6	6
831	2	6	12
822	3	3	9
750	2	6	12
741	6	6	36
732	10	6	60
660	3	3	9
651	10	6	60
642	20	6	120
633	24	3	72
552	24	3	72
543	36	6	216
444	45	1	45
			729

$(n, r, t) = (2, 2, 3)$

60	1	2	2
51	3	2	6
42	6	2	12
33	7	1	7
			27

$(n, r, t) = (3, 2, 3)$

1260	1	6	6
1251	3	6	18
1242	6	6	36
1233	7	3	21
1170	3	6	18
1161	12	6	72
1152	27	6	162
1143	39	6	234
1080	6	6	36
1071	27	6	162
1062	69	6	414
1053	114	6	684
1044	135	3	405
990	7	3	21
981	39	6	234
972	114	6	684
963	218	6	1308
954	297	6	1782
882	135	3	405

873	297	6	1782
864	468	6	2808
855	540	3	1620
774	540	3	1620
765	720	6	4320
666	831	1	831

Sum of numbers in fourth column $= 3^9$.

$$(n, r, t) = (3, 1, 3)$$

630	1	6	6
621	3	6	18
540	3	6	18
531	12	6	72
522	18	3	54
441	18	3	54
432	39	6	234
333	56	1	56

Sum of numbers in fourth column $= 2^9$.

$$(n, r, t) = (2, 3, 2)$$

60	1	2	2
51	2	2	4
42	3	2	6
33	4	1	4

Sum of numbers in fourth column $= 4^2$.

$$(n, r, t) = (2, 3, 3)$$

90	1	2	2
81	3	2	6
72	6	2	12
63	10	2	20
54	12	2	24

Sum of numbers in fourth column $= 4^3$.

$(n, r, t) = (3, 3, 2)$

1260	1	6	6
1251	2	6	12
1242	3	6	18
1233	4	3	12
1170	2	6	12
1161	6	6	36
1152	10	6	60
1143	14	6	84
1080	3	6	18
1071	10	6	60
1062	20	6	120
1053	30	6	180
1044	34	3	102
990	4	3	12
981	14	6	84
972	30	6	180
963	50	6	300
954	62	6	372
882	34	3	102
873	62	6	372
864	87	6	522
855	96	3	288
774	96	3	288
765	120	6	720
666	136	1	136

Sum of numbers in fourth column $= 4^6$.

Chapter 3
The Master Theorem

3.1 Introduction and Commentary

One of MacMahon's most surprising and valuable contributions to combinatorics was his discovery of the following.

Master Theorem. The coefficient of $X_1^{p_1} X_2^{p_2} \ldots X_n^{p_n}$ in the product of linear forms

$$(A_{\binom{1}{1}} X_1 + \ldots + A_{\binom{1}{n}} X_n)^{p_1} (A_{\binom{2}{1}} X_1 + \ldots + A_{\binom{2}{n}} X_n)^{p_2}$$
$$\ldots (A_{\binom{n}{1}} X_1 + \ldots + A_{\binom{n}{n}} X_n)^{p_n}$$

is identical with the coefficient of $X_1^{p_1} X_2^{p_2} \ldots X_n^{p_n}$ in

$$\begin{vmatrix} 1 - A_{\binom{1}{1}} X_1 & - A_{\binom{1}{2}} X_2 \ldots & - A_{\binom{1}{n}} X_n \\ - A_{\binom{2}{1}} X_1 & 1 - A_{\binom{2}{2}} X_2 \ldots & - A_{\binom{2}{n}} X_n \\ \cdot & \cdot & \cdot \\ \cdot & \cdot & \cdot \\ \cdot & \cdot & \cdot \\ - A_{\binom{n}{1}} X_1 & - A_{\binom{n}{2}} X_2 \ldots & 1 - A_{\binom{n}{n}} X_n \end{vmatrix}^{-1}$$

$$= (\det(\delta_{ij} - A_{\binom{i}{j}} X_j))^{-1},$$

where $A_{\binom{i}{j}}$ is our special notation for the doubly subscripted coefficient A_{ij}.

This theorem is studied in great detail in [45], and MacMahon's motivation to do so may be clearly seen in his first memoir on compositions [42]. In fact, we find the most extensive examples of the power and usefulness of the Master Theorem in these two papers, where MacMahon applies his result to numerous permutation problems.

A number of proofs of the Master Theorem have been given since MacMahon's original proof. Probably the most important contribution was made by D. Foata (1965) who explained, in his own proof, the real combinatorial significance of the Master Theorem. Subsequently, Foata (in

---

The material in this chapter corresponds to section III, chapters II and IV in *Combinatory Analysis.*

collaboration with P. Cartier) presented a simplified approach to his proof (Cartier and Foata, 1969). Cartier (1972) provided three striking proofs, one using symmetric functions, one using the symmetric group, and one using homological algebra (Lefschetz's Lemma). I. J. Good (1962a) and J. Percus (1971) each provided a proof using multiple integral complex variable techniques.

While the Master Theorem is historically the first really important theorem of this type, other mathematicians (back to Jacobi) have studied related problems. An excellent account of such researches was given by T. Muir (1903), who placed the Master Theorem in a setting which also anticipates recent work of P. Whittle (1956).

Applications of the Master Theorem have been discussed extensively by Percus (1971), and also by L. Carlitz (1974, 1977), D. Foata (1965), P. Cartier and D. Foata (1969), I. J. Good (1962b), M. Hall (1958), and H. S. Wilf (1968). Askey, Ismail, and Rashed (1975) and Askey and Ismail (1976) have presented a particularly interesting relationship between the Master Theorem and positivity problems in analysis.

To give some of the flavor of recent work on the Master Theorem, we shall present a modified account of the proof of Foata (1965). (See also Cartier and Foata, 1969) in section 4.2. In section 4.3 we shall discuss a conjecture of F. J. Dyson (1962) and one of the three subsequent proofs (I. J. Good, 1970); D. Foata has remarked that Dyson's conjecture is closely related to the Master Theorem and has suggested that many further results of combinatorial significance in this area still await discovery. Conjectures of F. J. Dyson and M. L. Mehta (see Mehta, 1974, and 1967, chapter 4) provide added evidence for this view.

3.2 The Master Theorem

We begin with a small lemma that reduces the proof of the Master Theorem to a special case. By doing this, we facilitate the proof of the full result.

Lemma. If the Master Theorem is true for

$$A_{\binom{1}{1}} = A_{\binom{2}{2}} = \ldots = A_{\binom{n}{n}} = 0,$$

then it is true in general.

Proof. We begin by observing that the actual coefficient arising in the product of the linear forms in the Master Theorem may be obtained by use of the multinomial theorem; thus the coefficient is

$$\sum{}^{*} p_1!p_2!\dots p_n! \prod_{i,j=1}^{n} \frac{A_{\binom{j}{i}}^{h_{ji}}}{h_{ji}!},$$

where $\sum^*$ is summed over all nonegative integers h_{ji} subject to

$$\sum_{i=1}^{n} h_{ji} = p_j, \qquad \sum_{j=1}^{n} h_{ji} = p_i.$$

Hence the Master Theorem is equivalent to

$$(1) \qquad \sum_{p_1 \geqq 0, \dots, p_n \geqq 0} p_1!p_2!\dots p_n! \sum{}^{*} \prod_{i,j=1}^{n} \frac{A_{\binom{j}{i}}^{h_{ji}}}{h_{ji}} X_1^{p_1} X_2^{p_2} \dots X_n^{p_n}$$

$$= (\det(\delta_{ij} - A_{\binom{i}{j}} X_j))^{-1}.$$

Our object is to prove that the truth of (1) when

$$A_{\binom{1}{1}} = A_{\binom{2}{2}} = A_{\binom{3}{3}} = \dots A_{\binom{n}{n}} = 0$$

implies the truth of (1) in general.

We proceed to treat the sum in (1) by replacing each p_j by $\sum_{i=1}^{n} h_{ji}$, and then we sum on each h_{jj}, using the binomial series. Therefore,

$$\sum_{p_1 \geqq 0, \dots, p_n \geqq 0} p_1!p_2!\dots p_n! \sum{}^{*} \prod_{i,j=1}^{n} \frac{A_{\binom{j}{i}}^{h_{ji}}}{h_{ji}!} X_1^{p_1} X_2^{p_2} \dots X_n^{p_n}$$

$$= \sum{}^{\dagger} \frac{(h_{11} + h_{12} + \dots + h_{1n})! \dots (h_{n1} + h_{n2} + \dots + h_{nn})!}{\prod_{i,j=1}^{n} h_{ji}!}$$

$$\times \prod_{i,j=1}^{n} A_{\binom{j}{i}}^{h_{ji}} X_1^{h_{11} + \dots + h_{1n}} \dots X_n^{h_{n1} + \dots + h_{nn}}$$

$$= \sum{}^{\#} \frac{(h_{12} + \dots + h_{1n})!\,(h_{21} + h_{23} + \dots + h_{2n})! \dots (h_{n1} + \dots + h_{n\,n-1})!}{\prod_{\substack{i,j=1 \\ i \neq j}}^{n} h_{ji}!}$$

$$\times \left\{ \prod_{\substack{i,j=1 \\ i \neq j}}^{n} A_{\binom{j}{i}}^{h_{ji}} \right\} X_1^{h_{12} + \dots + h_{1n}} \dots X_n^{h_{n1} + \dots + h_{nn-1}}$$

$$\times\ (1 - A_{\binom{1}{1}} X_1)^{-h_{12}-\dots-h_{1n}-1} \dots (1 - A_{\binom{n}{n}} X_n)^{-h_{n1}-\dots-h_{nn-1}-1}$$

$$= (1 - A_{\binom{1}{1}} X_1)^{-1}(1 - A_{\binom{2}{2}} X_2)^{-1} \dots (1 - A_{\binom{n}{n}} X_n)^{-1}$$

$$\times \begin{vmatrix} 1 & -A_{\binom{1}{2}}\left(\frac{X_2}{1 - A_{\binom{2}{2}} X_2}\right) & \dots & -A_{\binom{1}{n}}\left(\frac{X_n}{1 - A_{\binom{n}{n}} X_n}\right) \\ -A_{\binom{2}{1}}\left(\frac{X_1}{1 - A_{\binom{1}{1}} X_1}\right) & 1 & \dots & -A_{\binom{2}{n}}\left(\frac{X_n}{1 - A_{\binom{n}{n}} X_n}\right) \\ \vdots & & & \vdots \\ -A_{\binom{n}{1}}\left(\frac{X_1}{1 - A_{\binom{1}{1}} X_1}\right) & -A_{\binom{n}{2}}\left(\frac{X_2}{1 - A_{22} X_2}\right) & \dots & 1 \end{vmatrix}^{-1}$$

$$= (\det(\delta_{ij} - A_{\binom{i}{j}} X_j))^{-1}.$$

In the above, $\sum^{\dagger}$ is summed over all nonnegative integers h_{ji} subject to $\sum_{i=1}^{n} h_{ji} = \sum_{i=1}^{n} h_{ij}$; $\sum^{\#}$ is the same as $\sum^{\dagger}$ with the added proviso that each $h_{jj} = 0$. The penultimate equation above follows from (1) in the special case in which all the diagonal coefficients are zero. Hence our lemma is established.

Proof of the Master Theorem. Let us consider the set P of permutations without fixed points of the multiset $1^{a_1} 2^{a_2} \dots r^{a_r}$ (a multiset is a set in which repetitions may occur; e.g., in the case under consideration, 1 appears a_1 times, 2 appears a_2 times, etc.). We shall denote such a permutation by using the standard notation

$$\begin{pmatrix} 1\,1\,1 & \dots & 1 & 2 & \dots & 2 & \dots & r & \dots & r \\ i_1 i_2 i_3 & \dots & i_{a_1} & i_{a_1+1} & \dots & i_{a_1+a_2} & \dots & i_{a_1+\dots+a_{r-1}+1} & \dots & i_{a_1+\dots+a_r} \end{pmatrix},$$

where each entry in the lower line is the image of the corresponding entry in the upper line under the permutation. When we say that the permutation is without fixed points, we mean that the upper entry is never equal to the corresponding lower entry.

We now define a semi-group structure on P where the operation involved is the juxtaposition (or intercalation) of permutations. For example,

$$\begin{pmatrix} 1 & 1 & 2 & 2 & 2 & 3 & 4 \\ 2 & 2 & 1 & 3 & 4 & 1 & 2 \end{pmatrix} = \begin{pmatrix} 1 & 2 \\ 2 & 1 \end{pmatrix}\begin{pmatrix} 1 & 2 & 3 \\ 2 & 3 & 1 \end{pmatrix}\begin{pmatrix} 2 & 4 \\ 4 & 2 \end{pmatrix}.$$

Note that the above procedure is exactly like that used in the treatment of permutations of sets, except that columns with the same upper elements, such as $\begin{smallmatrix}2&2&2\\1&3&4\end{smallmatrix}$, must retain their left-to-right order in the factorization. We observe that primes in this semi-group are precisely the cycles of length at least 2 (since there are no fixed points), and each permutation factors uniquely (up to permisseble commutations) into primes.

With these facts in mind, we replace each $A_{\binom{i}{j}}$ by the symbol $\binom{i}{j}$. Recalling the lemma, we assume that $\binom{1}{1} = \ldots = \binom{n}{n} = 0$. Then the desired coefficient in the product of linear forms is nothing but

$$\begin{pmatrix} i_1 i_2 \ldots i_N \\ j_1 j_2 \ldots j_N \end{pmatrix},$$

where the sum is extended over all permutations without fixed points of the multiset $1^{p_1} 2^{p_2} \ldots n^{p_n}$. If we multiply this sum by $X_1^{p_1} X_2^{p_2} \ldots X_n^{p_n}$ and sum over all $p_i \geqq 0$, we obtain the generating function for all multiset permutations without fixed points on the first n positive integers:

$$G = \sum \begin{pmatrix} i_1 i_2 \ldots i_N \\ j_1 j_2 \ldots j_N \end{pmatrix} X_{j_1} X_{j_2} \ldots X_{j_N}.$$

On the other hand,

$$D = \begin{vmatrix} 1 & -\binom{1}{2} X_2 \ldots & -\binom{1}{n} X_n \\ -\binom{2}{1} X_1 & 1 \quad\ldots & -\binom{2}{n} X_n \\ \cdot & \cdot & \cdot \\ \cdot & \cdot & \cdot \\ \cdot & \cdot & \cdot \\ -\binom{n}{1} X_1 & -\binom{n}{2} X_2 \ldots & 1 \end{vmatrix},$$

when expanded, equals

$$\sum \begin{pmatrix} i_1 i_2 \ldots i_M \\ j_1 j_2 \ldots j_M \end{pmatrix} (-1)^c X_{j_1} X_{j_2} \ldots X_{j_M},$$

where $\begin{pmatrix} i_1 i_2 \dots i_M \\ j_1 j_2 \dots j_M \end{pmatrix}$ is a permutation without fixed points of some subset of 1, 2, 3, . . . , n, and c is the number of cycles in this permutation.

To conclude our proof of the Master Theorem, we need only show that $D \cdot G = 1$. Now

$$(2) \qquad D \cdot G = \sum \Biggl(\sum_{\pi_1 \left| \begin{pmatrix} i_1 i_2 \dots i_N \\ j_1 j_2 \dots j_N \end{pmatrix} \right.} (-1)^{c(\pi_1)} \Biggr) \begin{pmatrix} i_1 \dots i_N \\ j_1 \dots j_N \end{pmatrix} X_{j_1} X_{j_2} \dots X_{j_N},$$

where the outer sum runs over all multiset permutations $\begin{pmatrix} i_1 i_2 \dots i_N \\ j_1 j_2 \dots j_N \end{pmatrix}$ without fixed points on the first n positive integers, and the inner sum runs over all left factors π_1 of $\begin{pmatrix} i_1 i_2 \dots i_N \\ j_1 j_2 \dots j_N \end{pmatrix}$ made up of disjoint cycles. If the left prime factors of $\begin{pmatrix} i_1 i_2 \dots i_N \\ j_1 j_2 \dots j_N \end{pmatrix}$ are denoted by $\sigma_1, \sigma_2, \dots, \sigma_r$, and if we define

$$\mu(\pi) = \begin{cases} (-1)^{c(\pi)} \text{ if } \pi \text{ factors into } c(\pi) \text{ disjoint cycles,} \\ 0 \text{ otherwise,} \end{cases}$$

then

$$\sum_{\pi_1 \left| \begin{pmatrix} i_1 \dots i_N \\ j_1 \dots j_N \end{pmatrix} \right.} (-1)^{c(\pi_1)} = \prod_{j=1}^{r} (1 + \mu(\sigma_j)) = \begin{cases} 1 \text{ if } r = 0, \\ 0 \text{ otherwise.} \end{cases}$$

Thus the only term on the right side of (2) that does not vanish corresponds to the empty permutation of the empty set. Therefore $D \cdot G = 1$, and the Master Theorem is proved.

3.3 Good's Proof of Dyson's Conjecture

Let us examine for a moment a special case of the Master Theorem, namely,

$$A_{\binom{1}{1}} = A_{\binom{2}{2}} = A_{\binom{3}{3}} = 0, \qquad A_{\binom{1}{2}} = A_{\binom{2}{3}} = A_{\binom{3}{1}} = 1,$$
$$A_{\binom{1}{3}} = A_{\binom{2}{1}} = A_{\binom{3}{2}} = -1.$$

Thus the coefficient of

$$X_1^{a_2+a_3} X_2^{a_1+a_3} X_3^{a_1+a_2}$$

in

$$(X_2 - X_3)^{a_2+a_3}(X_3 - X_1)^{a_1+a_3}(X_1 - X_2)^{a_1+a_2}$$

is identical with the same coefficient in

$$\begin{vmatrix} 1 & -X_2 & X_3 \\ X_1 & 1 & -X_3 \\ -X_1 & X_2 & 1 \end{vmatrix}^{-1} = (1 + X_2X_3 + X_1X_2 + X_1X_3)^{-1}$$

$$= \sum_{m \geqq 0} (-1)^m \sum_{j_1+j_2+j_3=m} \frac{m!}{j_1!j_2!j_3!} X_1^{j_2+j_3} X_2^{j_1+j_3} X_3^{j_1+j_2}.$$

Hence the desired coefficient arises from the lone term $j_1 = a_1$, $j_2 = a_2$, $j_3 = a_3$, and is therefore

$$(-1)^{a_1+a_2+a_3}(a_1 + a_2 + a_3)!/a_1!a_2!a_3!.$$

D. Foata points out that this result may be stated in the following, slightly altered form as a corollary of the Master Theorem (see [61]).

Corollary. The constant term in the expansion of $(1 - X_3/X_2)^{a_3}(1 - X_2/X_3)^{a_2}(1 - X_1/X_3)^{a_1}(1 - X_3/X_1)^{a_3}(1 - X_2/X_1)^{a_2}(1 - X_1/X_2)^{a_1}$ is $(a_1 + a_2 + a_3)!/a_1!a_2!a_3!$.

This result is actually Dixon's theorem on the summability of the well-poised ${}_3F_2$ (see W. N. Bailey, Generalized Hypergeometric Series, Cambridge University Press, 1935, p.13), and in the case $a_1 = a_2 = a_3 = p$, it reduces to Dixon's theorem concerning the sum of the cubes of the binomial coefficients.

Interestingly enough, this corollary has a generalization which seems to be related to the Master Theorem and suggests the existence of more general results in this area:

Theorem (Dyson's Conjecture). The constant term in the expansion of

$$\prod_{1 \leqq i \neq j \leqq n} (1 - X_j/X_i)^{a_j}$$

is $(a_1 + a_2 + \ldots + a_n)!/a_1!a_2!\ldots a_n!$.

Remark. F. J. Dyson (1962) conjectured this result in an extensive work on statistical mechanics. Shortly thereafter, J. Gunson (1962) and K. Wilson (1962) proved this result. D. Foata pointed out the intersection of Dyson's conjecture and MacMahon's Master Theorem. I. J. Good (1970) gave a very short and elegant proof which we now present.

Proof. We begin by noting that

$$1 = \sum_{k=1}^{n} \prod_{\substack{j=1 \\ j \neq k}}^{n} \frac{(x - x_j)}{(x_k - x_j)}, \tag{3}$$

since the right side of (3) is a polynomial of degree at most $n - 1$ that assumes the value 1 for n different values of x, and is therefore identically 1. Setting $x = 0$ in (3), we obtain

$$1 = \sum_{k=1}^{n} \prod_{\substack{j=1 \\ j \neq k}}^{n} \frac{1}{(1 - x_k/x_j)}. \tag{4}$$

Let us write

$$F(x_1, \ldots, x_n; a_1, \ldots, a_n) = \prod_{1 \leqq i \neq j \leqq n} (1 - x_j/x_i)^{a_j}, \tag{5}$$

and observe that if we multiply equation (4) by this function, then when no $a_j = 0$,

$$F(x_1, \ldots, x_n; a_1, \ldots, a_n) = \sum_{j=1}^{n} F(x_1, \ldots, x_n; a_1, \ldots, a_{j-1}, a_j - 1, a_{j+1}, \ldots, a_n).$$

Consequently, if $G(a_1, \ldots, a_n)$ denotes the constant term in (5), then when no $a_j = 0$,

$$G(a_1, \ldots, a_n) = \sum_{j=1}^{n} G(a_1, \ldots, a_{j-1}, a_j - 1, a_{j+1}, \ldots, a_n). \tag{6}$$

If a particular $a_j = 0$, then x_j occurs to nonpositive powers only in (5), so that $G(a_1, \ldots, a_n)$ is equal to the constant term in

$$F(x_1, \ldots, x_{j-1}, x_{j+1}, \ldots, x_n; a_1, \ldots, a_{j-1}, a_{j+1}, \ldots, a_n),$$

i.e.,

$$G(a_1, \ldots, a_n) = G(a_1, \ldots, a_{j-1}, a_{j+1}, \ldots, a_n) \text{ if } a_j = 0. \tag{7}$$

Finally,

$$G(0, \ldots, 0) = 1. \tag{8}$$

Now equations (6), (7), and (8) uniquely define $G(a_1, \ldots, a_n)$, and since $(a_1 + a_2 + \ldots + a_n)!/a_1!a_2! \ldots a_n!$ also satisfies (6), (7), and (8), we must have

$$G(a_1, \ldots, a_n) = (a_1 + a_2 + \ldots + a_n)!/a_1!a_2! \ldots a_n!.$$

This concludes our proof.

3.4 References

R. Askey, M. Ismail, and T. Rashed (1975) A derangement problem, Mathematics Research Center, University of Wisconsin, Madison, Technical Report Summary No. 1522.

R. Askey and M. Ismail (1976) Permutation problems and special functions, *Canadian J. Math.*, *28*, 853–874.

L. Carlitz (1965) Some multiple sums and binomial identities, *J. S.I.A.M.*, *13*, 469–486.

L. Carlitz (1974) An application of MacMahon's Master Theorem, *S.I.A.M. J. Appl. Math.*, *26*, 431–436.

L. Carlitz (1977) Some expansion and convolution formulas related to MacMahon's master theorem, *S.I.A.M. J. Math. Anal.*, *8*, 320–336.

P. Cartier and D. Foata (1969) Problèmes combinatoires de commutation et réarrangements, Lecture Notes in Math. No. 85, Springer, Berlin.

P. Cartier (1972) La serie géneratrice exponentielle applications probabilistes et algébriques, version préliminaire, Institut de Recherche Mathématique Avancée, Strasbourg.

L. Comtet (1974) Advanced Combinatorics, Reidel, Dordrecht.

F. J. Dyson (1962) Statistical theory of energy levels of complex systems (I), *J. Math. Physics*, *3*, 140–156.

D. Foata (1965) Etude algébrique de certains problémes d'analyse combinatoire et du calcul des probabilités, *Publ. Inst. Stat. Univ. Paris*, *14*, 81–241.

D. Foata (1975) Studies in enumeration, Inst. Stat. Mimeo Series, No. 974, University of North Carolina, Chapel Hill.

I. J. Good (1960) Generalization to several variables of Lagrange's expansion, with applications to stochastic processes, *Proc. Cambridge Phil. Soc.*, *56*, 367–380.

I. J. Good (1962a) A short proof of MacMahon's "master theorem," *Proc. Cambridge Phil. Soc.*, *58*, 160.

I. J. Good (1962b) Proofs of some "binomial" identities by means of MacMahon's "master theorem," *Proc. Cambridge Phil. Soc.*, *58*, 161–162.

I. J. Good (1970) Short proof of a conjecture of Dyson, *J. Math. Physics*, *11*, 1884.

I. J. Good (1976) The relationship of a formula of Carlitz to the generalized Lagrange expansion, *S.I.A.M. J. Appl. Math.*, *30*, 103.

J. Gunson (1962) Proof of a conjecture by Dyson in the statistical theory of energy levels, *J. Math. Physics*, *3*, 752–753.

M. Hall (1958) Survey of Combinatorial Analysis, in Some Aspects of Analysis and Probability IV, J. Wiley, New York.

M. L. Mehta (1967) Random Matrices and the Statistical Theory of Energy Levels, Academic Press, New York.

M. L. Mehta (1974) Problem 74–2, *S.I.A.M. Review*, *16*, 92–93.

T. Muir (1903) The generating function of the reciprocal of a determinant, *Trans. Royal Soc. Edinburgh*, *40*, 615–629.

J. K. Percus (1971) Combinatorial Methods, Springer, New York.

P. Whittle (1955) Some distribution and moment formulae for the Markov chain, *J. Royal Stat. Soc.*, *17*, 235–242.

P. Whittle (1956) Some combinatorial results for matrix powers, *Quart. J. Math. (2)*, *7*, 316–320.

H. S. Wilf (1968) A mechanical counting method and combinatorial applications, *J. Combinatorial Th.*, *4*, 246–258.

K. G. Wilson (1962) Proof of a conjecture by Dyson, *J. Math. Physics*, *3*, 1040–1043.

3.5 Summaries of the Papers

[45] A certain class of generating functions in the theory of numbers, *Phil. Trans.*, *185* (1894), 111–160.

This is the extremely important paper in which MacMahon introduces and proves his celebrated "Master Theorem." (See section 3.1. for a statement of the theorem.)

In section 1 MacMahon proves the Master Theorem. In section 2 he applies it to several problems in the theory of permutations, including two problems that appeared earlier (in [46]); the latter presumably led to his discovery of the Master Theorem. In sections 3–6 he considers when a form V_n linear in $x_1, \ldots, x_n$ can be written in the form $\det(\delta_{ij} - a_{ij}x_j)$. The paper concludes with a further exploration of permutation problems.

[61] The sums of powers of the binomial coefficients, *Quart. J. Math.*, *33* (1902), 274–288.

This paper applies MacMahon's Master Theorem to summations of binomial coefficients. MacMahon proves (among many other results) that

$$\sum_{i=0}^{p} \binom{p}{i}\binom{q}{i} = \binom{p+q}{p} \qquad \text{(the Chu-Vandermonde sum),}$$

and,

$$\sum_{i=0}^{p}(-1)^i\binom{2p}{i}^3 = (-1)^p(2p)!/(p!)^3 \qquad \text{(Dixon's summation).}$$

Use of the Master Theorem makes the derivation of these and many other results extremely elegant (see section 3.3).

In this way MacMahon treats the sum

$$\sum_{i=0}^{p}\binom{p}{i}^\alpha$$

for $\alpha = 2, 3,$ or 4.

Reprints of the Papers

IV. *A Certain Class of Generating Functions in the Theory of Numbers.*

By Major P. A. MacMahon, *R.A., F.R.S.*

Received November 3.—Read November 24, 1893.

Introductory Abstract.

The present investigation arose from my "Memoir on the Compositions of Numbers," recently read before the Royal Society and now in course of publication in the 'Philosophical Transactions.' The main theorem may be stated as follows :—

If $X_1, X_2, \ldots, X_n$ be linear functions of quantities $x_1, x_2, \ldots, x_n$ given by the matricular relation

$$(X_1, X_2, \ldots\ldots X_n) = \left(\begin{array}{cccc} a_{11} & a_{12} & \cdots & a_{1n} \\ a_n & a_{22} & \cdots & a_{2n} \\ \cdot & \cdot & \ddots & \cdot \\ \cdot & \cdot & & \cdot \\ a_{n1} & a_{n2} & {}_{n3} & a_{nn} \end{array}\right)(x_1, x_2, \ldots\ldots, x_n),$$

that portion of the algebraic fraction

$$\frac{1}{(1 - s_1X_1)(1 - s_2X_2)\ldots\ldots(1 - s_nX_n)}$$

which is a function of the products

$$s_1x_1, s_2x_2, \ldots\ldots, s_nx_n,$$

only, is $1/V_n$, where (putting $s_1 = s_2 = \ldots\ldots = s_n = 1$)

$$V_n = (-)^n x_1x_2\ldots\ldots x_n \left|\begin{array}{cccc} a_{11} - 1/x_1, & a_{12}, & \cdots & a_{1n} \\ a_{21}, & a_{22} - 1/x_2 & \cdots & a_{2n} \\ \cdot & \cdot & \ddots & \cdot \\ \cdot & \cdot & & \\ a_{n1} & a_{n2} & \cdots & a_{nn} - 1/x_n \end{array}\right|$$

The proof of this theorem rests upon an identity which, for order 3, is

$$\begin{vmatrix} a_{11}s_1x_1 - 1, & a_{12}s_1x_1, & a_{13}s_1x_1, \\ a_{21}s_2x_2, & a_{22}s_2x_2 - 1, & a_{23}s_2x_2, \\ a_{31}s_3x_3, & a_{32}s_3x_3, & a_{33}s_3x_3 - 1, \end{vmatrix}$$

$$= \begin{vmatrix} 1 - s_1X_1, & 0, & 0, \\ 0, & 1 - s_2X_2, & 0, \\ 0, & 0, & 1 - s_3X_3, \end{vmatrix}$$

$$\times \begin{vmatrix} \frac{s_1(a_{11}x_1 - X_1)}{1 - s_1X_1} - 1, & \frac{a_{12}s_1x_1}{1 - s_1X_1}, & \frac{a_{13}s_1x_1}{1 - s_1X_1}, \\ \frac{a_{21}s_2x_2}{1 - s_2X_2}, & \frac{s_2(a_{22}x_2 - X)}{1 - s_2X_2} - 1, & \frac{a_{23}s_2x_2}{1 - s_2X_2}, \\ \frac{a_{31}s_3x_3}{1 - s_3X_3}, & \frac{a_{32}s_3x_3}{1 - s_3X_3}, & \frac{s_3(a_{33}x_3 - X_3)}{1 - s_3X_3} - 1, \end{vmatrix}$$

and is very easily established.

An instantaneous deduction of the general theorem is the result that the generating function for the coefficients of $x_1^{\xi_1}x_2^{\xi_2} \dots x_n^{\xi_n}$ in the product

$$X_1^{\xi_1}X_2^{\xi_2} \dots X_n^{\xi_n}$$

is

$$1/V_n.$$

The expression V_n involves the several coaxial minors of the determinant of the linear functions. Thus

$$V_3 = 1 - a_{11}x_1 - a_{22}x_2 - a_{33}x_3 + | a_{11}a_{22} | x_1x_2 + | a_{11}a_{33} | x_1x_3 - | a_{22}a_{33} | x_2x_3 \\ - | a_{11}a_{22}a_{33} | x_1x_2x_3.$$

The theorem is of considerable arithmetical importance and is also of interest in the algebraical theories of determinants and matrices.

The product

$$X_1^{\xi_1}X_2^{\xi_2} \dots X_n^{\xi_n},$$

often appears in arithmetic as a redundant form of generating function. The theorem above supplies a condensed or exact form of generating function.

Ex. gr. It is clear that the number of permutations of the $\Sigma\xi$ symbols in the product

$$x_1^{\xi_1}x_2^{\xi_2} \dots x_n^{\xi_n}$$

which are such that every symbol is displaced, is obviously the coefficient of

$$x_1^{\xi_1}x_2^{\xi_2} \dots x_n^{\xi_n}$$

in the product

$$(x_2 + \dots + x_n)^{\xi_1}(x_1 + x_3 + \dots + x_n)^{\xi_2} \dots (x_1 + x_2 + \dots + x_{n-1})^{\xi_n},$$

and thence we easily pass to the true generating function

$$\frac{1}{1 - \Sigma x_1 x_2 - 2\,\Sigma x_1 x_2 x_3 - 3\,\Sigma x_1 x_2 x_3 x_4 - \ldots - (n-1)\,x_1 x_2 \ldots x_n}.$$

In the paper many examples are given.

Frequently the redundant and condensed generating functions are differently interpretable; we then obtain an arithmetical correspondence, two cases of which presented themselves in the "Memoir on the Compositions of Numbers."

A more important method of obtaining arithmetical correspondences is developed in the researches which follow the statement and proof of the theorem.

The general form of V_n is such that the equation

$$V_n = 0$$

gives each quantity x_s as a homographic function of the remaining $n - 1$ quantities, and it is interesting to enquire whether, assuming the coefficients of V_n arbitrarily, it is possible to pass to a corresponding redundant generating function.

I find that the coefficients of V_n must satisfy

$$2^n - n^2 + n - 2$$

conditions, and, assuming the satisfaction of these conditions, a redundant form can be constructed which involves

$$n - 1$$

undetermined quantities. In fact, when a redundant form exists at all, it is necessarily of a $(n - 1)$-tuply infinite character.

We are now able to pass from any particular redundant generating function to an equivalent generating function which involves $n - 1$ undetermined quantities. Assuming these quantities at pleasure, we obtain a number of different algebraic products, each of which may have its own meaning in arithmetic, and thus the number of arithmetical correspondences obtainable is subject to no finite limit.

This portion of the theory is given at length in the paper, with illustrative examples.

Incidentally interesting results are obtained in the fields of special and general determinant theory. The special determinant, which presents itself for examination, provisionally termed "inversely symmetric," is such that the constituents symmetrically placed in respect to the principal axis have, each pair, a product unity, whilst the constituents on the principal axis itself are all of them equal to unity. The determinant possesses many elegant properties which are of importance to the principal investigation of the paper. The theorems concerning the general determinant are connected entirely with the co-axial minors.

I find that the general determinant of even order, greater than two, is expressible in precisely two ways as an irrational function of its co-axial minors, whilst no determinant of uneven order is so expressible at all.

Of order superior to 3, it is not possible to assume arbitrary values for the determinant itself and all of its co-axial minors. In fact of order n the values assumed must satisfy

$$2^n - n^2 + n - 2$$

conditions, but, these conditions being satisfied, the determinant can be constructed so as to involve $n - 1$ undetermined quantities.

§ 1.

ART. 1. In a Memoir on "The Theory of the Composition of Numbers," recently communicated to the Royal Society (as above-mentioned), there occurred certain generating functions which admitted important transformations to redundant forms.

I proceed to the general theory of these transformations, and subsequently discuss the algebraical and arithmetical consequences. The main theorem is, in reality, a theorem in determinants, of considerable interest, as will appear.

Art. 2. Consider the algebraic fraction

$$\frac{1}{(1 - s_1X_1)(1 - s_2X_2)\dots(1 - s_nX_n)},$$

wherein $X_1, X_2, \dots X_n$ are linear functions, of n quantities $x_1, x_2, \dots x_n$, as given by the matricular relation

$$(X_1, X_2, \dots X_n) = \left(\begin{array}{l} a_1, a_2, \dots a_n \\ b_1, b_2, \dots b_n \\ \dots\dots \\ n_1, n_2, \dots n_n \end{array}\right)(x_1, x_2, \dots x_n).$$

I assume the quantities involved to have such values that the fraction is capable of expansion in ascending powers, and products of $x_1, x_2, \dots x_n$ by a convergent series.

Art. 3. A certain portion of this expansion is a function of $s_1x_1, s_2x_2, \dots s_nx_n$, and of the coefficients of the linear functions $X_1, X_2, \dots X_n$ only. One object of this investigation is the isolation of this portion of the expansion which, for some purposes, in the Theory of Numbers is the only portion of importance.*

* It will occur to mathematicians, who are familiar with the Theory of Invariants, that generating functions not unfrequently present themselves in a redundant form. In particular, it is frequently necessary to isolate that portion of a generating function which includes the whole of the positive terms of the expansion, the negative terms, though admitting of interpretation, being of little moment.

Without specifying at present the arithmetical meaning of the generating function, I will call the portion above-written the "redundant form," and the essential portion, to which reference has been made, the "condensed form."

Art. 4. As typical of the general case, put $n = 3$.

It will be shown that the condensed form is $1/\mathrm{N}$, where

$$\mathrm{N} = 1 - a_1s_1x_1 - b_2s_2x_2 - c_3s_3x_3 \\ + |a_1b_2|\, s_1s_2x_1x_2 + |a_1c_3|\, s_1s_3x_1x_3 + |b_2c_3|\, s_2s_3x_2x_3 - |a_1b_2c_3|\, s_1s_2s_3x_1x_2x_3.$$

The notation is that in use in the Theory of Determinants, the coefficients of N being the several co-axial minors of the determinant $|a_1b_2c_3|$; this determinant is the content of the matrix which occurs in the definition of the linear quantics X_1, X_2, X_3.

Art. 5. In determinant form N may be written

$$\begin{vmatrix} 1 - a_1s_1x_1, & -a_2s_1x_1, & -a_3s_1x_1 \\ -b_1s_2x_2, & 1 - b_2s_2x_2, & -b_3s_3x_3 \\ -c_1s_3x_3, & -c_2s_3x_3, & 1 - c_3s_3x_3 \end{vmatrix}$$

and also in the important symbolic form

$$|(1 - a_1s_1x_1)(1 - b_2s_2x_2)(1 - c_3s_3x_3)|,$$

wherein, after multiplication, the a, b, c products are to be written in determinant brackets. Such symbolic multiplication will be denoted by external determinant brackets as shown.

Art. 6. We have now

$$\frac{\mathrm{N}}{(1 - s_1\mathrm{X}_1)(1 - s_2\mathrm{X}_2)(1 - s_3\mathrm{X}_3)}$$

$$= \frac{|(1 - a_1s_1x_1)(1 - b_2s_2x_2)(1 - c_3s_3x_3)|}{(1 - s_1\mathrm{X}_1)(1 - s_2\mathrm{X}_2)(1 - s_3\mathrm{X}_3)}$$

$$= \frac{|(1 - s_1\mathrm{X}_1 + s_1\mathrm{X}_1 - a_1s_1x_1)(1 - s_2\mathrm{X}_2 + s_2\mathrm{X}_2 - b_2s_2x_2)(1 - s_3\mathrm{X}_3 + s_3\mathrm{X}_3 - c_3s_3x_3)|}{(1 - s_1\mathrm{X}_1)(1 - s_2\mathrm{X}_2)(1 - s_3\mathrm{X}_3)}$$

$$= 1 + \frac{s_1(\mathrm{X}_1 - a_1x_1)}{1 - s_1\mathrm{X}_1} + \frac{s_2(\mathrm{X}_2 - b_2x_2)}{1 - s_2\mathrm{X}_2} + \frac{s_3(\mathrm{X}_3 - c_3x_3)}{1 - s_3\mathrm{X}_3} + \frac{s_2s_3|(\mathrm{X}_2 - b_2x_2)(\mathrm{X}_3 - c_3x_3)|}{(1 - s_2\mathrm{X}_2)(1 - s_3\mathrm{X}_3)}$$

$$+ \frac{s_3s_1|(\mathrm{X}_3 - c_3x_3)(\mathrm{X}_1 - a_1x_1)|}{(1 - s_3\mathrm{X}_3)(1 - s_1\mathrm{X}_1)} + \frac{s_1s_2|(\mathrm{X}_1 - a_1x_1)(\mathrm{X}_2 - b_2x_2)|}{(1 - s_1\mathrm{X}_1)(1 - s_2\mathrm{X}_2)},$$

Q 2

since, as will be seen presently, the determinant

$$| (X_1 - a_1x_1)(X_2 - b_2x_2)(X_3 - c_3x_3) |$$

vanishes identically.

The right-hand side of their identity does not, on expansion, contain any terms which are functions of s_1x_1, s_2x_2, s_3x_3 and of the coefficients a, b, c only.

Art. 7. Before proceeding to establish this, it may be remarked that the above identity may be written in the determinant form :—

$$\frac{\begin{vmatrix} a_1s_1x_1 - 1, & a_2s_1x_1, & a_3s_1x_1 \\ b_1s_2x_2, & b_2s_2x_2 - 1, & b_3s_2x_2 \\ c_1s_3x_3, & c_2s_3x_3, & c_3s_3x_3 - 1 \end{vmatrix}}{\begin{vmatrix} 1 - s_1X_1, & 0, & 0 \\ 0, & 1 - s_2X_2, & 0 \\ 0, & 0, & 1 - s_3X_3 \end{vmatrix}}$$

$$= \begin{vmatrix} \frac{s_1(a_1x_1 - X_1)}{1 - s_1X_1} - 1, & \frac{a_2s_1x_1}{1 - s_1X_1}, & \frac{a_3s_1x_1}{1 - s_1X_1} \\ \frac{b_1s_2x_2}{1 - s_2X_2}, & \frac{s_2(b_2x_2 - X_2)}{1 - s_2X_2} - 1, & \frac{b_3s_2x_2}{1 - s_2X_2} \\ \frac{c_1s_3x_3}{- s_3X_3}, & \frac{c_2s_3x_3}{1 - s_3X_3}, & \frac{s_3(c_3x_3 - X_3)}{1 - s_3X_3} - 1 \end{vmatrix},$$

and, in this form, is very easily established.

Art. 8. Consider, in regard to the order n, the algebraic fraction

$$\frac{s_1s_2 \ldots s_t \,| (X_1 - a_1x_1)(X_2 - b_2x_2) \ldots (X_t - t_tx_t) |}{(1 - s_1X_1)(1 - s_2X_2) \ldots (1 - s_tX_t)},$$

wherein t has an integer value not superior to n. This fraction is specified by the first t natural numbers, but this is merely for convenience, as what follows can be readily modified to meet the case of a fraction specified by any selection of t natural numbers, which are unequal and not superior to n.

To show that this fraction contains, on expansion, no terms which are functions of s_1x_1, s_2x_2, ... s_nx_n only, it is merely necessary to show that every term in the development of the determinant

$$| (X_1 - a_1x_1)(X_2 - b_2x_2) \ldots (X_t - t_tx_t) |,$$

contains either $x_{t+1}, x_{t+2}, \ldots x_n$; viz., that every term contains an x with a suffix that does not occur in the s-product

$$s_1 s_2 \ldots s_t ;$$

for visibly the fraction contains neither

$$s_{t+1},\ s_{t+2},\ \ldots \text{ nor } s_n ;$$

or, the same thing, the quantities s, occurring in the product

$$s_1 s_2 \ldots s_t,$$

are the only ones that are found in the fraction, the determinant should therefore vanish by putting

$$x_{t+1} = x_{t+2} = \ldots = x_n = 0.$$

The determinant is

$$\begin{vmatrix} X_1 - a_1x_1, & -a_2x_1, & \ldots & -a_tx_1 \\ -b_1x_2, & X_2 - b_2x_2, & \ldots & -b_tx_2 \\ \cdot & \cdot & \ldots & \cdot \\ \cdot & \cdot & \ldots & \cdot \\ -t_1x_t, & -t_2x_t, & \ldots & X_t - t_tx_t \end{vmatrix};$$

putting

$$x_{t+1} = x_{t+2} = \ldots = x_n = 0,$$

the first row is

$$a_2x_2 + a_3x_3 + \ldots + a_tx_t,\ \ -a_2x_1,\ \ -a_3x_1,\ \ldots\ -a_tx_1,$$

and adding together, x_1 times the first element, x_2 times the second, . . . , &c., x_t times the tth element, we obtain zero.

A similar operation, performed on the elements of all the other rows, likewise results in zero.

Hence the determinant vanishes on the supposition

$$x_{t+1} = x_{t+2}, = \ldots = x_n = 0,$$

and accordingly every term, in its development, contains as factor one at least of the quantities

$$x_{t+1},\ x_{t+2},\ \ldots\ x_n.$$

This proves the proposition and also shows that the determinant

$$|(X_1 - a_1x_1)(X_2 - b_2x_2)\ldots(X_n - n_nx_n)|,$$

of the nth order, vanishes identically.

Art. 9. Hence, of order 3, we have the identity

$$\frac{1}{(1 - s_1X_1)(1 - s_2X_2)(1 - s_3X_3)} = \frac{1}{|(1 - a_1s_1x_1)(1 - b_2s_2x_2)(1 - c_3s_3x_3)|},$$

multiplied by

$$1 + \frac{s_1(X_1 - a_1x_1)}{1 - s_1X_1} + \frac{s_2(X_2 - b_2x_2)}{1 - s_2X_2} + \frac{s_3(X_3 - c_3x_3)}{1 - s_3X_3} + \frac{s_2s_3|(X_2 - b_2x_2)(X_3 - c_3x_3)|}{(1 - s_2X_2)(1 - s_3X_3)}$$
$$+ \frac{s_3s_1|(X_3 - c_3x_3)(X_1 - a_1x_1)|}{(1 - s_3X_3)(1 - s_1X_1)} + \frac{s_1s_2|(X_1 - a_1x_1)(X_2 - b_2x_2)|}{(1 - s_1X_1)(1 - s_2X_2)};$$

and, of order n, the identity

$$\frac{1}{(1 - s_1X_1)(1 - s_2X_2)\ldots(1 - s_nX_n)} = \frac{1}{|(1 - a_1s_1x_1)(1 - b_2s_2x_2)\ldots(1 - n_ns_nx_n)|},$$

multiplied by

$$1 + \Sigma\frac{s_1(X_1 - a_1x_1)}{1 - s_1X_1} + \Sigma\frac{s_1s_2\,|(X_1\,a_1x_1)(X_2 - b_2x_2)|}{(1 - s_1X_1)(1 - s_2X_2)}$$
$$+ \ldots + \Sigma\frac{s_1s_2\ldots s_t\,|(X_1 - a_1x_1)(X_2 - b_2x_2)\ldots(X_t - t_tx_t)|}{(1 - s_1X_1)(1 - s_2X_2)\ldots(1 - s_tX_t)} + \ldots,$$

the last batch of fractions involving, each, $n - 1$ denominator factors, and the numbers of fractions, under the summation signs, being in order

$$\binom{n}{1}, \binom{n}{2}, \ldots \binom{n}{t}, \ldots \binom{n}{n-1}.$$

Moreover, it has been shown that the fraction

$$\frac{1}{|(1 - a_1s_1x_1)(1 - b_2s_2x_2)\ldots(1 - n_ns_nx_n)|}$$

is the condensed form of the fraction

$$\frac{1}{(1 - s_1X_1)(1 - s_2X_2)\ldots(1 - s_nX_n)},$$

or we may regard the latter as a redundant form of the former.

Art. 10. The coefficients of the terms

$$(s_1x_1)^{\xi_1}(s_2x_2)^{\xi_2}\ldots(s_nx_n)^{\xi_n},$$

in the expansions of both fractions, are the same.

Hence, the coefficient of the product

$$x_1^{\xi_1}x_2^{\xi_2}\ldots x_n^{\xi_n},$$

in the expansion of algebraic fraction

$$\frac{1}{|(1-a_1x_1)(1-b_2x_2)\ldots(1-n_nx_n)|},$$

is equal to the same coefficient in the product

$$(a_1x_1+\ldots+a_nx_n)^{\xi_1}(b_1x_1+\ldots+b_nx_n)^{\xi_2}\ldots(n_1x_1+\ldots+n_nx_n)^{\xi_n},$$

where this product is a "particular redundant generating function," the use of which renders the quantities $s_1, s_2, \ldots s_n$ unnecessary to the statement of the theorem.

Art. 11. The theorem regarded as a proposition concerning the coaxial minors of a general determinant is very remarkable; for it will be observed that we are able to exhibit the coefficient of

$$x_1^{\xi_1}x_2^{\xi_2}\ldots x_n^{\xi_n}$$

in the "particular redundant generating function" as a function of the coaxial minors of the determinant of the n quantities.

§ 2. *Arithmetical Interpretations.*

Art. 12. Most of the arithmetical results that can be deduced arise from duality of interpretation from algebra to arithmetic in particular cases. In the memoir to which reference has been made two particular cases presented themselves.

Art. 13. The first one was connected with the matricular relation

$$(X_1, X_2, X_3 \ldots X_n) = \begin{pmatrix} k, & 1, & 1, & \ldots & 1 \\ k, & k, & 1, & \ldots & 1 \\ k, & k, & k, & \ldots & 1 \\ \cdot & \cdot & \cdot & \ldots & 1 \\ \cdot & \cdot & \cdot & \ldots & 1 \\ \cdot & \cdot & \cdot & \ldots & 1 \\ k, & k, & k, & \ldots & k \end{pmatrix} x_1, x_2, x_3 \ldots x_n).$$

and the condensed form, thence derivable, which has the form

$$\frac{1}{1 - k\Sigma x_1 + k(k-1)\Sigma x_1 x_2 - k(k-1)^2 \Sigma x_1 x_2 x_3 + \ldots + (-)^n k(k-1)^{n-1} x_1 x_2 \ldots x_n}.$$

The latter generating function occurs in the Theory of the Composition of Numbers. The corresponding redundant form is not unique (this will appear in the sequel, but that given above is one of the most useful.

Art. 14. The second one was founded on the relation

$$(X_1, X_2, X_3 \ldots X_n) = \left(\begin{array}{|cccccc|} 1, & \lambda_{21}, & \lambda_{31}, & . & . & \lambda_{n1} \\ 1, & 1, & \lambda_{32}, & . & . & \lambda_{n2} \\ 1, & 1, & 1, & . & . & \lambda_{n3} \\ . & . & . & . & . & . \\ . & . & . & . & . & . \\ 1, & 1, & 1, & 1, & . & 1 \end{array}\right)(x_1, x_2, x_3 \ldots x_n)$$

leading to the condensed form

$$\frac{1}{\left[\begin{array}{l} 1 - \Sigma x_1 - \Sigma(\lambda_{\beta\alpha} - 1) x_\alpha x_\beta - \Sigma(\lambda_{\beta\alpha} - 1)(\lambda_{\gamma\beta} - 1) x_\alpha x_\beta x_\gamma \\ \quad - \ldots - (\lambda_{21} - 1)(\lambda_{32} - 1)(\lambda_{43} - 1) \ldots (\lambda_{n, n-1} - 1) x_1 x_2 x_3 x_4 \ldots x_{n-1} x_n \end{array}\right]}$$

wherein the numbers $\alpha, \beta, \gamma, \ldots$ are in ascending order of magnitude.

These particular cases gave rise to dual interpretations in arithmetic.

Art. 15. The general theorem, as so far developed, apparently only admits of a single interpretation.

Regarding the product

$$(a_1 x_1 + a_2 x_2 + \ldots + a_n x_n)^{\xi_1} (b_1 x_1 + b_2 x_2 + \ldots + b_n x_n)^{\xi_2} \ldots (n_1 x_1 + n_2 x_2 + \ldots + n_n x_n)^{\xi_n},$$

the coefficient of

$$a_1^{\alpha_1} b_1^{\beta_1} \ldots n_1^{\nu_1} a_2^{\alpha_2} b_2^{\beta_2} \ldots n_2^{\nu_2} \ldots a_n^{\alpha_n} b_n^{\beta_n} \ldots n_n^{\nu_n} x_1^{\xi_1} x_2^{\xi_2} \ldots x_n^{\xi_n}$$

may be interpreted in the theory of permutations.

Considering the permutations of the $\Sigma\xi$ quantities which form the product

$$x_1^{\xi_1} x_2^{\xi_2} \ldots x_n^{\xi_n},$$

the coefficient indicates the number of permutations which possess the property that

x_1	occurs	α_1	times in places	originally occupied	by an	x_1
,,	,,	β_1	,,	,,	,,	x_2
.	.	.	.	.	.	.
,,	,,	ν_1	,,	,,	,,	x_n
x_2	,,	α_2	,,	,,	,,	x_1
,,	,,	β_2	,,	,,	,,	x_2
.	.	.	.	.	.	.
,,	,,	ν_2	,,	,,	,,	x_n
		:	:	:	:	:
x_n	,,	α_n	,,	,,	,,	x_1
,,	,,	β_n	,,	,,	,,	x_2
.	.	.	.	.	.	.
,,	,,	ν_n	,,	,,	,,	x_n.

Accordingly the proper generating function for the enumeration of the permutations possessing this property is

$$\frac{1}{|(1-a_1x_1)(1-b_2x_2)\ldots(1-n_nx_n)|}.$$

Art. 16. As an interesting particular case we can find the generating function for the enumeration of those permutations of the quantities in

$$x_1^{\xi_1}x_2^{\xi_2}\ldots x_n^{\xi_n}$$

which possess the property that no quantity is in the place originally occupied; that is, in the permutation, no x_s is to occupy a position formerly occupied by an x_s, s having all values from 1 to n.

Clearly we have merely to put

$$a_1 = b_2 = c_3 = \ldots = n_n = 0,$$

and the remaining letters, $a, b, c, \ldots n$ equal to unity. The generating function involves the coaxial minors of the determinant of the nth order

$$\begin{vmatrix} 0, & 1, & 1, & .. & 1 \\ 1, & 0, & 1, & .. & 1 \\ 1, & 1, & 0, & .. & 1 \\ . & . & . & . & . \\ 1, & 1, & 1, & .. & 0 \end{vmatrix}$$

This determinant has the value

$$(-)^n(n-1),$$

while its first coaxial minors have each the value

$$(-)^{n-1}(n-2),$$

and its sth coaxial minors each the value

$$(-)^{n-s}(n-s-1).$$

Hence the generating function is

$$\frac{1}{\{1-\Sigma x_1x_2-2\Sigma x_1x_2x_3-3\Sigma x_1x_2x_3x_4-\ldots-s\Sigma x_1x_2\ldots x_{s+1}-\ldots-(n-1)\,x_1x_2\ldots x_n\}},$$

or writing

$$(x-x_1)(x-x_2)\ldots(x-x_n)=x^n-a_1x^{n-1}+a_2x^{n-2}-\ldots,$$

this is

$$\frac{1}{1-a_2-2a_3-3a_4-\ldots-(n-1)\,a_n}.$$

Art. 17. As another example, again consider the permutations of the quantities in

$$x_1^{\xi_1}x_2^{\xi_2}\ldots x_n^{\xi_n}.$$

Divide the places occupied by the quantities into compartments

$$A_1\,A_2\ldots A_n,$$

such that the first ξ_1 places are in compartment A_1
next ξ_2 ,, ,, A_2
⋮
last ξ_n ,, ,, A_n,

and let us find the number of the permutations which have the property that no quantity with an uneven suffix is in a compartment with an uneven suffix, and no quantity with an even suffix is in a compartment with an even suffix.

In the "particular redundant generating function" we have merely to put

$$\begin{gathered}a_1=a_3=a_5=\ldots=0,\\ b_2=b_4=b_6=\ldots=0,\\ c_1=c_3=c_5=\ldots=0,\\ \&c.,\qquad \&c.,\end{gathered}$$

and the remaining $a, b, c, \ldots$ letters equal to unity.

For the true general (or condensed) generating function we have thus to evaluate the coaxial minors of the chess-board pattern determinant of the nth order,

$$\begin{vmatrix} 0, & 1, & 0, & 1, & 0 & \ldots \\ 1, & 0, & 1, & 0, & 1 & \ldots \\ 0, & 1, & 0, & 1, & 0 & \ldots \\ 1, & 0, & 1, & 0, & 1 & \ldots \\ 0, & 1, & 0, & 1, & 0 & \ldots \\ \vdots & \vdots & \vdots & \vdots & \vdots & \end{vmatrix}$$

Here, all the minors of Order 1 are zero.

A minor (coaxial) of Order 2 has either the value zero or negative unity. If the minor be formed by deletion of all rows except the pth and q^{th} and all columns except the pth and q^{th} $(q > p)$ the value will be zero, if $q - p \equiv 0 \bmod 2$, and will be negative unity in all other cases.

Coaxial minors of Order > 2 as well as the whole determinant vanish, because in every case two rows are found to be identical.

Hence the true generating function is

$$\frac{1}{1 - x_1(x_2 + x_4 + \ldots) - x_2(x_3 + x_5 + \ldots) - x_3(x_4 + x_6 + \ldots) - \ldots - x_{n-1}x_n},$$

which may be written

$$\frac{1}{1 - \sum^{\alpha} \sum^{m} x_\alpha x_{\alpha + 2m + 1}}.$$

Art. 18. Again for the enumeration of the permutations which are such that no quantity with an uneven suffix is in a compartment with an even suffix, and also no quantity with an even suffix is in a compartment with an uneven suffix, we are led to the complementary chess-board pattern determinant:—

$$\begin{vmatrix} 1, & 0, & 1, & 0, & 1 & \ldots \\ 0, & 1, & 0, & 1, & 0 & \ldots \\ 1, & 0, & 1, & 0, & 1 & \ldots \\ 0, & 1, & 0, & 1, & 0 & \ldots \\ 1, & 0, & 1, & 0, & 1 & \ldots \\ \vdots & \vdots & \vdots & \vdots & \vdots & \end{vmatrix}$$

and thence to the true generating function

R 2

$$\frac{1}{[1-x_1-x_2-x_3-\ldots-x_n+x_1(x_2+x_4+\ldots)+x_2(x_3+x_5+\ldots)+x_3(x_4+x_6+\ldots)+\ldots+x_{n-1}x_n]}$$

which may be written

$$\frac{1}{1-\Sigma x_1+\overset{a}{\Sigma}\overset{m}{\Sigma} x_a x_{a+2m+1}}.$$

Art. 19. Again, if it be necessary to enumerate the permutations of

$$x_1^{\xi_1}x_2^{\xi_2}\ldots x_n^{\xi_n},$$

in which x_1 occurs α_1 times in the compartment A_1,
,, β_1 ,, ,, ,, A_2,
,, γ_1 ,, ,, ,, A_3,
,, ⋮

we are led to the true generating function

$$\frac{1}{1-a_1x_1-x_2-x_3-\ldots-x_n+(a_1-b_1)x_1x_2+(a_1-c_1)x_1x_3+\ldots+(a_1-n_1)x_1x_n},$$

in which we have to seek the coefficient of

$$a_1^{\alpha_1}b_1^{\beta_1}c_1^{\gamma_1}\ldots n_1^{\nu_1}x_1^{\xi_1}x_2^{\xi_2}\ldots x_n^{\xi_n}.$$

Art. 20. Again consider the general problem of "Derangements in the Theory of Permutations."

In regard to the permutations of

$$x_1^{\xi_1}x_2^{\xi_2}\ldots x_n^{\xi_n}$$

it is necessary to determine the number of permutations such that exactly m of the symbols are in the places they originally occupied.

We have the particular redundant product

$$(ax_1+x_2+\ldots+x_n)^{\xi_1}(x_1+ax_2+\ldots+x_n)^{\xi_2}\ldots(x_1+x_2+\ldots+ax_n)^{\xi_n},$$

in which the number sought is the coefficient of

$$a^m x_1^{\xi_1}x_2^{\xi_2}\ldots x_n^{\xi_n}.$$

The true generating function (*i.e.*, condensed form) is derived from the coaxial minors of the determinant of order n :—

$$\begin{vmatrix} a & 1 & 1 & 1 & \dots \\ 1 & a & 1 & 1 & \dots \\ 1 & 1 & a & 1 & \dots \\ 1 & 1 & 1 & a & \dots \\ \vdots & \vdots & \vdots & \vdots & \end{vmatrix} \begin{aligned} &= (a-1)^n + n(a-1)^{n-1} \\ &= (a-1)^{n-1}(a+n-1). \end{aligned}$$

Thence the true generating function

$$\frac{1}{\{1 - a\Sigma x_1 + (a-1)(a+1)\Sigma x_1x_2 - (a-1)^2(a+2)\Sigma x_1x_2x_3 + \dots + (-)^n(a-1)^{n-1}(a+n-1)x_1x_2\dots x_n\}},$$

which constitutes a *perfect* solution of the problem of "derangement."

§ 3. *The General Theory Resumed.*

Art. 21. The denominator of a perfect generating function, of the type under consideration, is the most general function linear in each of n variables $x_1, x_2, \dots x_n$.

Let V_n be the most general linear function of the n quantities, involving $2^n - 1$ independent coefficients.

Art. 22. I enquire, irrespective of arithmetical interpretation or correspondence, into the possibility of expressing the fraction

$$V_n^{-1}$$

in a factorized redundant form.

Art. 23. The coefficients of V_n must be the several coaxial minors of some determinant, and the question arises: Can a determinant be constructed such that its coaxial minors assume given values?

The redundant form of order n involves n^2 coefficients. In general, in order that the fraction

$$V_n^{-1}$$

may be expressible in a redundant form, its coefficients must satisfy

$$\sigma_n$$

conditions, and, assuming the satisfaction of these conditions, a redundant form involving

$$n^2 - (2^n - 1 - \sigma_n)$$

arbitrary coefficients can be constructed.

Art. 24. The relation

$$n^2 - (2^n - 1 - \sigma_n) = n - 1$$

will be established, and this leads to the conclusion that the redundant form, when possible, is always of a

$$(n - 1)^{\text{tuply}}$$

infinite character.

Art. 25. The fact, subject to the above-mentioned conditions, that there is an infinite flexibility in the redundant forms is of great importance in the Theory of Numbers, because the potentiality of arithmetical interpretation would appear to have no finite limit.

Art. 26. Observe that

$$\sigma_n$$

denotes the number of identical relations or syzygies connecting the coaxial minors of a general determinant of order n.

Art. 27. The discussion of the theory of the first few orders forms a convenient method of approaching the general theory.

I take the general form of V_n as

$$1 - p_1 s_1 x_1 - p_2 s_2 x_2 - \ldots + p_{12} s_1 s_2 x_1 x_2 + \ldots + (-)^n p_{12\ldots n} s_1 s_2 \ldots s_n x_1 x_2 \ldots x_n.$$

Art. 28. *The case* $n = 1$.

This case is trivial because the perfect form

$$V_1^{-1} = \frac{1}{1 - p_1 s_1 x_1}$$

coincides with the redundant form

$$\sigma_1 = 0;$$

$$n^2 - (2^n - 1 - \sigma_1) = 0.$$

Art. 29. *The case* $n = 2$.

In order that

$$\frac{1}{\{1 - s_1(a_{11}x_1 + a_{12}x_2)\}\{1 - s_2(a_{21}x_1 + a_{22}x_2)\}}$$

may be a redundant form of

$$V_2^{-1} = \frac{1}{1 - p_1 s_1 x_1 - p_2 s_2 x_2 + p_{12} s_1 s_2 x_1 x_2},$$

we have

$$a_{11} = p_1, \qquad a_{22} = p_2,$$
$$| a_{11}, a_{22} | = p_{12},$$

and thence $a_{12}a_{21} = p_1p_2 - p_{12} = q_{12}$ (suppose); introducing an undetermined quantity α_{12}, we may put :—

$$a_{12} = \alpha_{12}q_{12},$$
$$a_{21} = 1/\alpha_{12},$$

where α_{12} may be a *certain* function of the quantities

$$p_1, p_2, p_{12}, x_1, x_2;$$

but, numerically, may not be either zero or infinity.

The matricular relation is

$$(X_1, X_2) = \begin{pmatrix} a_{11}, & a_{12} \\ a_{21}, & a_{22} \end{pmatrix} (x_1, x_2) = \begin{pmatrix} p_1, & \alpha_{12}q_{12} \\ 1/\alpha_{12}, & p_2 \end{pmatrix} (x_1, x_2)$$

and the redundant form

$$\frac{1}{\{1 - s_1(p_1x_1 + \alpha_{12}q_{12}x_2)\}\{1 - s_2(1/\alpha_{12}x_1 + p_2x_2)\}}$$

of a singly infinite character.

$$\sigma_2 = 0;$$
$$n^2 - (2^n - 1 - \sigma_2) = 1.$$

Art. 30. *The case* $n = 3$.

The matrix being that connected with the determinant

$$| a_{13} |,$$

we have the following relations

$$a_{11} = p_1, \qquad a_{22} = p_2, \qquad a_{33} = p_3,$$
$$| a_{11}, a_{22} | = p_{12}, \qquad | a_{11}, a_{33} | = p_{13}, \qquad | a_{22}, a_{33} | = p_{23},$$
$$| a_{11}, a_{22}, a_{33} | = p_{123};$$

and thence

$$a_{12}a_{21} = q_{12}, \qquad a_{13}a_{31} = q_{13}, \qquad a_{23}a_{32} = q_{23},$$

where

$$(q_{12}, q_{13}, q_{23}) = (p_1p_2 - p_{12}, p_1p_3 - p_{13}, p_2p_3 - p_{23});$$

introducing the undetermined quantities

$$\alpha_{12},\quad \alpha_{13},\quad \alpha_{23},$$

write

$$a_{12} = \alpha_{12}q_{12},\qquad a_{13} = \alpha_{13}q_{13},\qquad a_{23} = \alpha_{23}q_{23},$$

$$a_{21} = \frac{1}{\alpha_{12}},\qquad a_{31} = \frac{1}{\alpha_{13}},\qquad a_{32} = \frac{1}{\alpha_{23}},$$

and thence by substitution

$$\begin{vmatrix} p_1 & \alpha_{12}q_{12} & \alpha_{13}q_{13} \\ \frac{1}{\alpha_{12}} & p_2 & \alpha_{23}q_{23} \\ \frac{1}{\alpha_{13}} & \frac{1}{\alpha_{23}} & p_3 \end{vmatrix} = p_{123},$$

which may be written

$$\begin{vmatrix} p_1 & q_{12} & \frac{\alpha_{13}}{\alpha_{12}\alpha_{23}}q_{13} \\ 1 & p_2 & q_{23} \\ \frac{\alpha_{12}\alpha_{23}}{\alpha_{13}} & 1 & p_3 \end{vmatrix} = p_{123};$$

this is a quadratic equation for the evaluation of $\alpha_{13}/\alpha_{12}\alpha_{23}$, which may be written

$$\left(\frac{\alpha_{13}}{\alpha_{12}\alpha_{23}} - \frac{1}{c_{13}}\right)\left(\frac{\alpha_{13}}{\alpha_{12}\alpha_{23}} - \frac{1}{c_{31}}\right) = 0.$$

Thus two of the three quantities α_{12}, α_{13}, α_{23} remain undetermined, and the coefficients of V_3 are not subject to any condition.

The matricular relation is either

$$(X_1, X_2, X_3) = \begin{pmatrix} p_1 & \alpha_{12}q_{12} & \frac{\alpha_{12}\alpha_{23}}{c_{13}}q_{13} \\ \frac{1}{\alpha_{12}} & p_2 & \alpha_{23}q_{23} \\ \frac{c_{13}}{\alpha_{12}\alpha_{23}} & \frac{1}{\alpha_{23}} & p_3 \end{pmatrix}(x_1, x_2, x_3)$$

or the one involving the matrix similar to the above with c_{31} written for c_{13}.

α_{12}, α_{23} are undetermined quantities, and $c_{13}{}^{-1}$, $c_{31}{}^{-1}$ are the roots of the above-given quadratic equation, which are expressible as irrational functions of the coefficients of V_3. The redundant form is

$$\frac{1}{(1-s_1X_1)(1-s_2X_2)(1-s_3X_3)},$$

of a doubly infinite character.

Also

$$\sigma_3 = 0,$$
$$n^2 - (2^n - 1 - \sigma_n) = 2, \text{ for } n = 3.$$

Art. 31. *The case* $n = 4$.

The matrix being that connected with the determinant $| a_{14} |$ we have the relations :—

$$a_{11} = p_1, \quad a_{22} = p_2, \quad a_{33} = p_3, \quad a_{44} = p_4,$$

$$| a_{11}a_{22} | = p_{12}, \quad | a_{11}a_{33} | = p_{13}, \quad | a_{22}a_{33} | = p_{23},$$
$$| a_{11}a_{44} | = p_{14}, \quad | a_{22}a_{44} | = p_{24}, \quad | a_{33}a_{44} | = p_{34},$$

$$| a_{11}a_{22}a_{33} | = p_{123}, \quad | a_{11}a_{22}a_{44} | = p_{124},$$
$$| a_{11}a_{33}a_{44} | = p_{134}, \quad | a_{22}a_{33}a_{44} | = p_{234},$$

$$| a_{11}a_{22}a_{33}a_{44} | = p_{1234};$$

and thence

$$a_{12}a_{21} = q_{12}, \quad a_{13}a_{31} = q_{13}, \quad a_{23}a_{32} = q_{23},$$
$$a_{14}a_{41} = q_{14}, \quad a_{24}a_{42} = q_{24}, \quad a_{34}a_{43} = q_{34};$$

and introducing six undetermined quantities,

$$a_{12} = \alpha_{12}q_{12}, \quad a_{13} = \alpha_{13}q_{13}, \quad a_{14} = \alpha_{14}q_{14}, \quad a_{23} = \alpha_{23}q_{23}, \quad a_{24} = \alpha_{24}q_{24}, \quad a_{34} = \alpha_{34}q_{34},$$
$$a_{21} = \frac{1}{\alpha_{12}}, \quad a_{31} = \frac{1}{\alpha_{13}}, \quad a_{41} = \frac{1}{\alpha_{14}}, \quad a_{32} = \frac{1}{\alpha_{23}}, \quad a_{42} = \frac{1}{\alpha_{24}}, \quad a_{43} = \frac{1}{\alpha_{34}},$$

and thence by substitution in the remaining relations,

$$\begin{vmatrix} p_1, & \alpha_{12}q_{12}, & \alpha_{13}q_{13} \\ \frac{1}{\alpha_{12}}, & p_2, & \alpha_{23}q_{23} \\ \frac{1}{\alpha_{13}}, & \frac{1}{\alpha_{23}}, & p_3 \end{vmatrix} = p_{123}, \qquad \begin{vmatrix} p_1, & \alpha_{12}q_{12}, & \alpha_{14}q_{14} \\ \frac{1}{\alpha_{12}}, & p_2, & \alpha_{24}q_{24} \\ \frac{1}{\alpha_{14}}, & \frac{1}{\alpha_{24}}, & p_4 \end{vmatrix} = p_{124},$$

$$\begin{vmatrix} p_1, & \alpha_{13}q_{13}, & \alpha_{14}q_{14} \\ \frac{1}{\alpha_{13}}, & p_3, & \alpha_{34}q_{34} \\ \frac{1}{\alpha_{14}}, & \frac{1}{\alpha_{34}}, & p_4 \end{vmatrix} = p_{134}, \qquad \begin{vmatrix} p_2, & \alpha_{23}q_{23}, & \alpha_{24}q_{24} \\ \frac{1}{\alpha_{23}}, & p_3, & \alpha_{34}q_{34} \\ \frac{1}{\alpha_{24}}, & \frac{1}{\alpha_{34}}, & p_4 \end{vmatrix} = p_{234},$$

$$\begin{vmatrix} p_1, & \alpha_{12}q_{12}, & \alpha_{13}q_{13}, & \alpha_{14}q_{14} \\ \frac{1}{\alpha_{12}}, & p_2, & \alpha_{23}q_{23}, & \alpha_{24}q_{24} \\ \frac{1}{\alpha_{13}}, & \frac{1}{\alpha_{23}}, & p_3, & \alpha_{34}q_{34} \\ \frac{1}{\alpha_{14}}, & \frac{1}{\alpha_{24}}, & \frac{1}{\alpha_{34}} & p_4 \end{vmatrix} = p_{1234}.$$

The six undetermined quantities that have been introduced must satisfy these five equations. However, the six quantities only enter the equations in three combinations; for, writing

$$\gamma_{13} = \frac{\alpha_{13}}{\alpha_{12}\alpha_{23}}, \qquad \gamma_{14} = \frac{\alpha_{14}}{\alpha_{12}\alpha_{23}\alpha_{34}}, \qquad \gamma_{24} = \frac{\alpha_{24}}{\alpha_{23}\alpha_{34}},$$

the five equations are easily transformed into the following five—

$$\begin{vmatrix} p_1, & q_{12}, & \gamma_{13}q_{13} \\ 1, & p_2, & q_{23} \\ \frac{1}{\gamma_{13}}, & 1, & p_3 \end{vmatrix} = p_{123}, \qquad \begin{vmatrix} p_1, & q_{12}, & \frac{\gamma_{14}}{\gamma_{24}}q_{14} \\ 1, & p_2, & q_{24} \\ \frac{\gamma_{24}}{\gamma_{14}}, & 1, & p_4 \end{vmatrix} = p_{124},$$

$$\begin{vmatrix} p_1, & q_{13}, & \frac{\gamma_{14}}{\gamma_{13}}q_{14} \\ 1, & p_3, & q_{34} \\ \frac{\gamma_{13}}{\gamma_{14}}, & 1, & p_4 \end{vmatrix} = p_{134}, \qquad \begin{vmatrix} p_2, & q_{23}, & \gamma_{24}q_{24} \\ 1, & p_3, & q_{34} \\ \frac{1}{\gamma_{24}}, & 1, & p_4 \end{vmatrix} = p_{234},$$

$$\begin{vmatrix} p_1, & q_{12}, & \gamma_{13}q_{13} & \gamma_{14}q_{14} \\ 1, & p_2, & q_{23}, & \gamma_{24}q_{24} \\ \frac{1}{\gamma_{13}}, & 1, & p_3, & q_{34} \\ \frac{1}{\gamma_{14}}, & \frac{1}{\gamma_{24}}, & 1, & p_4 \end{vmatrix} = p_{1234},$$

which involve only the three undetermined quantities

$$\gamma_{13},\ \gamma_{14},\ \gamma_{24}.$$

From these five equations we can eliminate the three quantities

$$\gamma_{13}, \qquad \gamma_{14}, \qquad \gamma_{24},$$

and thus obtain two independent relations between the coefficients of V_4. These are the two conditions that the coefficients must satisfy in order that a redundant form may be possible.

Since also these coefficients are the several co-axial minors of the determinant

$$|\ \alpha_{14}\ |$$

we establish the fact that these co-axial minors are connected by two relations or syzygies. Thus

$$\sigma_4 = 2\,;$$

and assuming the satisfaction of these two conditions we can solve the equations so as to express

$$\gamma_{13}, \qquad \gamma_{14}, \qquad \gamma_{24}$$

as functions of the coefficients of V_4.

Solving these equations and writing

$$P_{123} = p_{123} - p_1 p_{23} - p_2 p_{13} - p_3 p_{12} + 2 p_1 p_2 p_3,$$

we find

$$\gamma_{13} = \frac{1}{2q_{13}} \{P_{123} \pm \surd (P^2_{123} - 4 q_{12} q_{13} q_{23})\}\,,$$

$$\gamma_{24} = \frac{1}{2q_{24}} \{P_{234} \pm \surd (P^2_{234} - 4 q_{23} q_{24} q_{34})\}\,,$$

$$\frac{\gamma_{14}}{\gamma_{13}} = \frac{1}{2q_{14}} \{P_{134} \pm \surd (P^2_{134} - 4 q_{13} q_{34} q_{14})\}\,,$$

$$\frac{\gamma_{14}}{\gamma_{24}} = \frac{1}{2q_{14}} \{P_{124} \pm \surd (P^2_{124} - 4 q_{12} q_{24} q_{14})\}\,;$$

and assuming these four equations, as well as the fifth equation, consistent, there are just two systems of values of

$$\gamma_{13}, \qquad \gamma_{14}, \qquad \gamma_{24},$$

which satisfy all the equations.

Let the two values of γ_{13} be

$$1/c_{13} \quad \text{and} \quad 1/c_{31},$$

S 2

corresponding to the positive and negative signs respectively, and further taking the signs *all* positive, let γ_{xy} have the value

$$1/c_{xy},$$

and taking all the signs negative, let the value be

$$1/c_{yx}.$$

We have the solutions

$$(\gamma_{13}, \gamma_{14}, \gamma_{24}) = \left(\frac{1}{c_{13}}, \frac{1}{c_{14}}, \frac{1}{c_{24}}\right)$$

$$(\gamma_{13}, \gamma_{14}, \gamma_{24}) = \left(\frac{1}{c_{31}}, \frac{1}{c_{41}}, \frac{1}{c_{42}}\right)$$

and we may write either

$$(\alpha_{13}, \alpha_{14}, \alpha_{24}) = \left(\frac{\alpha_{12}\alpha_{23}}{c_{13}}, \frac{\alpha_{12}\alpha_{23}\alpha_{34}}{c_{14}} \frac{\alpha_{23}\alpha_{34}}{c_{24}}\right),$$

or

$$(\alpha_{13}, \alpha_{14}, \alpha_{24}) = \left(\frac{\alpha_{12}\alpha_{23}}{c_{31}}, \frac{\alpha_{12}\alpha_{23}\alpha_{34}}{c_{41}} \frac{\alpha_{23}\alpha_{34}}{c_{42}}\right).$$

The undetermined quantities are thus reduced to the three

$$\alpha_{12}, \alpha_{23}, \alpha_{34}.$$

Writing for brevity,

$$(\alpha_{12}\alpha_{23}, \alpha_{23}\alpha_{34}, \alpha_{12}\alpha_{23}\alpha_{34}) = (\beta_{13}, \beta_{24}, \beta_{14}),$$

and also

$$\alpha_{x,\, x+1} = \beta_{x,\, x+1},$$

the matrix that defines X_1, X_2, X_3, X_4 is either

$$\begin{pmatrix} p_1 & \beta_{12}q_{12} & \frac{\beta_{13}}{c_{13}}q_{13} & \frac{\beta_{14}}{c_{14}}q_{14} \\ \frac{1}{\beta_{12}} & p_2 & \beta_{23}q_{23} & \frac{\beta_{24}}{c_{24}}q_{24} \\ \frac{c_{13}}{\beta_{13}} & \frac{1}{\beta_{23}} & p_3 & \beta_{34}q_{34} \\ \frac{c_{14}}{\beta_{14}} & \frac{c_{24}}{\beta_{24}} & \frac{1}{\beta_{34}} & p_4 \end{pmatrix}$$

or the same matrix with the substitution of c_{yx} for c_{xy}.

The redundant form is

$$\frac{1}{(1 - s_1X_1)(1 - s_2X_2)(1 - s_3X_3)(1 - s_4X_4)}$$

of a triply infinite character and of two forms.

Also for $n = 4$,

$$n^2 - (2^n - 1 - \sigma_n) = 3.$$

Art. 32. In order to proceed to the general case it is necessary to make a digression for the purpose of establishing certain properties of a determinant of special form.

§ 4. *Digression on the Theory of Inversely Symmetrical Determinants.*

Art. 33. The determinant of special form which I have provisionally termed "inversely symmetrical" is

$$\begin{vmatrix} 1, & \alpha_{12}, & \alpha_{13} & \cdot & \cdot & \cdot & \alpha_{1n} \\ \frac{1}{\alpha_{12}}, & 1, & \alpha_{23} & \cdot & \cdot & \cdot & \alpha_{2n} \\ \frac{1}{\alpha_{13}}, & \frac{1}{\alpha_{23}}, & 1 & \cdot & \cdot & \cdot & \alpha_{3n} \\ \cdot & \cdot & \cdot & \cdot & \cdot & \cdot & \cdot \\ \cdot & \cdot & \cdot & \cdot & \cdot & \cdot & \cdot \\ \cdot & \cdot & \cdot & \cdot & \cdot & \cdot & \cdot \\ \frac{1}{\alpha_{1n}}, & \frac{1}{\alpha_{2n}}, & \frac{1}{\alpha_{3n}} & \cdot & \cdot & \cdot & 1 \end{vmatrix},$$

which involves $\binom{n}{2}$ different quantities α, and is such that the elements on the principal axis are all unity, and is inversely axi-symmetric in the sense that elements, symmetrically placed in regard to the principal axis, have a product equal to unity.

Art. 34. The property of this determinant, which is of vital import to the present investigation, may be stated as follows :—

"The determinant, as well as all of its co-axial minors, may be exhibited as functions of $\binom{n-1}{2}$ combinations of the $\binom{n}{2}$ quantities α_{xy}."

To establish this, first, consider the determinant itself, and put

$$\beta_{xy} = \alpha_{x, x+1}\, \alpha_{x+1, x+2} \ldots \alpha_{y-1, y}, \quad (x < y),$$

$$\gamma_{xy} = \alpha_{xy}/\beta_{xy},$$

so that

$$\beta_{x, x+1} = \alpha_{x, x+1},$$

$$\gamma_{x, x+1} = 1.$$

Observe that the combinations

$$\gamma_{x, y} \quad (x < y - 1)$$

are $\binom{n-1}{2}$ in number; it will be shown that the quantities $\gamma_{x,y}$ are those to which reference has been made in the above statement of theorem.

Art. 35. With the new symbols the determinant may be written :—

$$\begin{vmatrix}
1 & \beta_{12} & \beta_{13}\gamma_{13} & \beta_{14}\gamma_{14} & \cdot & \cdot & \beta_{1,n-1}\gamma_{1,n-1} & \beta_{1n}\gamma_{1n} \\
\frac{1}{\beta_{12}} & 1 & \beta_{23} & \beta_{24}\gamma_{24} & \cdot & \cdot & \beta_{2,n-1}\gamma_{2,n-1} & \beta_{2n}\gamma_{2n} \\
\frac{1}{\beta_{13}\gamma_{13}} & \frac{1}{\beta_{23}} & 1 & \beta_{34} & \cdot & \cdot & \beta_{3,n-1}\gamma_{3,n-1} & \beta_{3n}\gamma_{3n} \\
\frac{1}{\beta_{14}\gamma_{14}} & \frac{1}{\beta_{24}\gamma_{24}} & \frac{1}{\beta_{34}} & 1 & \cdot & \cdot & \beta_{4,n-1}\gamma_{4,n-1} & \beta_{4n}\gamma_{4n} \\
\cdot & \cdot & \cdot & \cdot & \cdot & \cdot & \cdot & \cdot \\
\cdot & \cdot & \cdot & \cdot & \cdot & \cdot & \cdot & \cdot \\
\frac{1}{\beta_{1,n-1}\gamma_{1,n-1}} & \frac{1}{\beta_{2,n-1}\gamma_{2,n-1}} & \frac{1}{\beta_{3,n-1}\gamma_{3,n-1}} & \frac{1}{\beta_{4,n-1}\gamma_{4,n-1}} & \cdot & \cdot & 1 & \beta_{n-1,n} \\
\frac{1}{\beta_{1n}\gamma_{1n}} & \frac{1}{\beta_{2n}\gamma_{2n}} & \frac{1}{\beta_{3n}\gamma_{3n}} & \frac{1}{\beta_{4n}\gamma_{4n}} & \cdot & \cdot & \frac{1}{\beta_{n-1,n}} & 1
\end{vmatrix}$$

and may be transformed, without alteration of value, by the following operations performed successively.

Multiply

1st column by β_{12}

,, row $\frac{1}{\beta_{12}}$

3rd column $\frac{1}{\beta_{23}}$

,, row β_{23}

4th column $\frac{1}{\beta_{24}}$

,, row β_{24}

,, ,, ,,

sth column $\frac{1}{\beta_{23}}$

,, row β_{23}

,, ,, ,,

nth column $\frac{1}{\beta_{2n}}$

,, row β_{2n}

it then assumes the form—

$$\left|\begin{array}{ccccccccc}
1 & 1 & \gamma_{13} & \gamma_{14} & \cdot & \cdot & \gamma_{1,n-2} & \gamma_{1,n-1} & \gamma_{1n} \\
1 & 1 & 1 & \gamma_{24} & \cdot & \cdot & \gamma_{2,n-2} & \gamma_{2,n-1} & \gamma_{2n} \\
\frac{1}{\gamma_{13}} & 1 & 1 & 1 & \cdot & \cdot & \gamma_{3,n-2} & \gamma_{3,n-1} & \gamma_{3n} \\
\frac{1}{\gamma_{14}} & \frac{1}{\gamma_{24}} & 1 & 1 & \cdot & \cdot & \gamma_{4,n-2} & \gamma_{4,n-1} & \gamma_{4n} \\
\cdot & \cdot & \cdot & \cdot & \cdot & \cdot & \cdot & \cdot & \cdot \\
\cdot & \cdot & \cdot & \cdot & \cdot & \cdot & \cdot & \cdot & \cdot \\
\frac{1}{\gamma_{1,n-2}} & \frac{1}{\gamma_{2,n-2}} & \frac{1}{\gamma_{3,n-2}} & \frac{1}{\gamma_{4,n-2}} & \cdot & \cdot & 1 & 1 & \gamma_{n-2,n} \\
\frac{1}{\gamma_{1,n-1}} & \frac{1}{\gamma_{2,n-1}} & \frac{1}{\gamma_{3,n-1}} & \frac{1}{\gamma_{4,n-1}} & \cdot & \cdot & 1 & 1 & 1 \\
\frac{1}{\gamma_{1,n}} & \frac{1}{\gamma_{2,n}} & \frac{1}{\gamma_{3,n}} & \frac{1}{\gamma_{4,n}} & \cdot & \cdot & \frac{1}{\gamma_{n-2,n}} & 1 & 1
\end{array}\right|$$

which involves only the $\binom{n-1}{2}$ combinations $\gamma_{x,y}$ of the $\binom{n}{2}$ quantities $\alpha_{x,y}$.

Art. 36. The determinant is also inversely symmetrical, and not only the principal diagonals, but also the adjacent minor diagonals consist wholly of units. In regard to the occurrence of three diagonals of units, we have here the normal form of inversely symmetrical determinant.

Art. 37. We have next to consider the coaxial minor of order $n-1$ obtained by deletion of the sth row and s^{th} column.

The following successive operations, which do not alter the value, have then to be performed—

Multiply

1st column by	β_{12}	and	1st row by	$\frac{1}{\beta_{12}}$.
3rd ,,	$\frac{1}{\beta_{23}}$	,,	3rd ,,	β_{23}.
.	.		.	.
$(s-1)^{\text{th}}$,,	$\frac{1}{\beta_{2,s-1}}$	,,	$(s-1)^{\text{th}}$,,	$\beta_{2,s-1}$.
$(s+1)^{\text{th}}$,,	$\frac{1}{\gamma_{s-1,s+1}\beta_{2,s+1}}$	,,	$(s+1)^{\text{th}}$,,	$\gamma_{s-1,s+1}\beta_{2,s+1}$.
$(s+2)^{\text{th}}$,,	$\frac{1}{\gamma_{s-1,s+1}\beta_{2,s+2}}$	,,	$(s+2)^{\text{th}}$,,	$\gamma_{s-1,s+1}\beta_{2,s+2}$.
.	.	.		.
n^{th} ,,	$\frac{1}{\gamma_{s-1,s+1}\beta_{2,n}}$	,,	n^{th} ,,	$\gamma_{s-1,s+1}\beta_{2,n}$.

Art. 38. To represent the result conveniently, suppose the determinant divided into four compartments by the lines of deletion, thus—

I.	II.
III.	IV.

We then obtain—

$$
\begin{array}{c|ccccccc|ccccccc|c}
 & 1 & 1 & \gamma_{13} & \gamma_{14} & \cdot\ \cdot & \gamma_{1,s-2} & \gamma_{1,s-1} & \frac{\gamma_{1,s+1}}{\gamma_{s-1,s+1}} & \frac{\gamma_{1,s+2}}{\gamma_{s-1,s+1}} & \cdot\ \cdot & \frac{\gamma_{1,n-3}}{\gamma_{s-1,s+1}} & \frac{\gamma_{1,n-2}}{\gamma_{s-1,s+1}} & \frac{\gamma_{1,n-1}}{\gamma_{s-1,s+1}} & \frac{\gamma_{1,n}}{\gamma_{s-1,s+1}} & \\
 & 1 & 1 & 1 & \gamma_{24} & \cdot\ \cdot & \gamma_{2,s-2} & \gamma_{2,s-1} & \frac{\gamma_{2,s+1}}{\gamma_{s-1,s+1}} & \frac{\gamma_{2,s+2}}{\gamma_{s-1,s+1}} & \cdot\ \cdot & \frac{\gamma_{2,n-3}}{\gamma_{s-1,s+1}} & \frac{\gamma_{2,n-2}}{\gamma_{s-1,s+1}} & \frac{\gamma_{2,n-1}}{\gamma_{s-1,s+1}} & \frac{\gamma_{2,n}}{\gamma_{s-1,s+1}} & \\
 & \frac{1}{\gamma_{13}} & 1 & 1 & 1 & \cdot\ \cdot & \gamma_{3,s-2} & \gamma_{3,s-1} & \frac{\gamma_{3,s+1}}{\gamma_{s-1,s+1}} & \frac{\gamma_{3,s+2}}{\gamma_{s-1,s+1}} & \cdot\ \cdot & \frac{\gamma_{3,n-3}}{\gamma_{s-1,s+1}} & \frac{\gamma_{3,n-2}}{\gamma_{s-1,s+1}} & \frac{\gamma_{3,n-1}}{\gamma_{s-1,s+1}} & \frac{\gamma_{3,n}}{\gamma_{s-1,s+1}} & \\
\text{I.} = & \frac{1}{\gamma_{14}} & \frac{1}{\gamma_{24}} & 1 & 1 & \cdot\ \cdot & \gamma_{4,s-2} & \gamma_{4,s-1} & \frac{\gamma_{4,s+1}}{\gamma_{s-1,s+1}} & \frac{\gamma_{4,s+2}}{\gamma_{s-1,s+1}} & \cdot\ \cdot & \frac{\gamma_{4,n-3}}{\gamma_{s-1,s+1}} & \frac{\gamma_{4,n-2}}{\gamma_{s-1,s+1}} & \frac{\gamma_{4,n-1}}{\gamma_{s-1,s+1}} & \frac{\gamma_{4,n}}{\gamma_{s-1,s+1}} & = \text{II.} \\
 & \vdots & \vdots & \vdots & \vdots & \therefore & \vdots & \vdots & \vdots & \vdots & \therefore & \vdots & \vdots & \vdots & \vdots & \\
 & \frac{1}{\gamma_{1,s-2}} & \frac{1}{\gamma_{2,s-2}} & \frac{1}{\gamma_{3,s-2}} & \frac{1}{\gamma_{4,s-2}} & \cdot\ \cdot & 1 & 1 & \frac{\gamma_{s-2,s+1}}{\gamma_{s-1,s+1}} & \frac{\gamma_{s-2,s+2}}{\gamma_{s-1,s+1}} & \cdot\ \cdot & \frac{\gamma_{s-2,n-3}}{\gamma_{s-1,s+1}} & \frac{\gamma_{s-2,n-2}}{\gamma_{s-1,s+1}} & \frac{\gamma_{s-2,n-1}}{\gamma_{s-1,s+1}} & \frac{\gamma_{s-2,n}}{\gamma_{s-1,s+1}} & \\
 & \frac{1}{\gamma_{1,s-1}} & \frac{1}{\gamma_{2,s-1}} & \frac{1}{\gamma_{3,s-1}} & \frac{1}{\gamma_{4,s-1}} & \cdot\ \cdot & 1 & 1 & 1 & \frac{\gamma_{s-1,s+2}}{\gamma_{s-1,s+1}} & \cdot\ \cdot & \frac{\gamma_{s-1,n-3}}{\gamma_{s-1,s+1}} & \frac{\gamma_{s-1,n-2}}{\gamma_{s-1,s+1}} & \frac{\gamma_{s-1,n-1}}{\gamma_{s-1,s+1}} & \frac{\gamma_{s-1,n}}{\gamma_{s-1,s+1}} & \\
\hline
 & \frac{\gamma_{s-1,s+1}}{\gamma_{1,s+1}} & \frac{\gamma_{s-1,s+1}}{\gamma_{2,s+1}} & \frac{\gamma_{s-1,s+1}}{\gamma_{3,s+1}} & \frac{\gamma_{s-1,s+1}}{\gamma_{4,s+1}} & \cdot\ \cdot & \frac{\gamma_{s-1,s+1}}{\gamma_{s-2,s+1}} & 1 & 1 & 1 & \cdot\ \cdot & \gamma_{s+1,n-3} & \gamma_{s+1,n-2} & \gamma_{s+1,n-1} & \gamma_{s+1,n} & \\
 & \frac{\gamma_{s-1,s+1}}{\gamma_{1,s+2}} & \frac{\gamma_{s-1,s+1}}{\gamma_{2,s+2}} & \frac{\gamma_{s-1,s+1}}{\gamma_{3,s+2}} & \frac{\gamma_{s-1,s+1}}{\gamma_{4,s+2}} & \cdot\ \cdot & \frac{\gamma_{s-1,s+1}}{\gamma_{s-2,s+2}} & \frac{\gamma_{s-1,s+1}}{\gamma_{s-1,s+2}} & 1 & 1 & \cdot\ \cdot & \gamma_{s+2,n-3} & \gamma_{s+2,n-2} & \gamma_{s+2,n-1} & \gamma_{s+2,n} & \\
 & \vdots & \vdots & \vdots & \vdots & \therefore & \vdots & \vdots & \vdots & \vdots & \therefore & \vdots & \vdots & \vdots & \vdots & \\
\text{III.} = & \frac{\gamma_{s-1,s+1}}{\gamma_{1,n-3}} & \frac{\gamma_{s-1,s+1}}{\gamma_{2,n-3}} & \frac{\gamma_{s-1,s+1}}{\gamma_{3,n-3}} & \frac{\gamma_{s-1,s+1}}{\gamma_{4,n-3}} & \cdot\ \cdot & \frac{\gamma_{s-1,s+1}}{\gamma_{s-2,n-3}} & \frac{\gamma_{s-1,s+1}}{\gamma_{s-1,n-3}} & \frac{1}{\gamma_{s+1,n-3}} & \frac{1}{\gamma_{s+2,n-3}} & \cdot\ \cdot & 1 & 1 & \gamma_{n-3,n-1} & \gamma_{n-3,n} & = \text{IV.} \\
 & \frac{\gamma_{s-1,s+1}}{\gamma_{1,n-2}} & \frac{\gamma_{s-1,s+1}}{\gamma_{2,n-2}} & \frac{\gamma_{s-1,s+1}}{\gamma_{3,n-2}} & \frac{\gamma_{s-1,s+1}}{\gamma_{4,n-2}} & \cdot\ \cdot & \frac{\gamma_{s-1,s+1}}{\gamma_{s-2,n-2}} & \frac{\gamma_{s-1,s+1}}{\gamma_{s-1,n-2}} & \frac{1}{\gamma_{s+1,n-2}} & \frac{1}{\gamma_{s+2,n-2}} & \cdot\ \cdot & 1 & 1 & 1 & \gamma_{n-2,n} & \\
 & \frac{\gamma_{s-1,s+1}}{\gamma_{1,n-1}} & \frac{\gamma_{s-1,s+1}}{\gamma_{2,n-1}} & \frac{\gamma_{s-1,s+1}}{\gamma_{3,n-1}} & \frac{\gamma_{s-1,s+1}}{\gamma_{4,n-1}} & \cdot\ \cdot & \frac{\gamma_{s-1,s+1}}{\gamma_{s-2,n-1}} & \frac{\gamma_{s-1,s+1}}{\gamma_{s-1,n-1}} & \frac{1}{\gamma_{s+1,n-1}} & \frac{1}{\gamma_{s+2,n-1}} & \cdot\ \cdot & \frac{1}{\gamma_{n-3,n-1}} & 1 & 1 & 1 & \\
 & \frac{\gamma_{s-1,s+1}}{\gamma_{1,n}} & \frac{\gamma_{s-1,s+1}}{\gamma_{2,n}} & \frac{\gamma_{s-1,s+1}}{\gamma_{3,n}} & \frac{\gamma_{s-1,s+1}}{\gamma_{4,n}} & \cdot\ \cdot & \frac{\gamma_{s-1,s+1}}{\gamma_{s-2,n}} & \frac{\gamma_{s-1,s+1}}{\gamma_{s-1,n}} & \frac{1}{\gamma_{s+1,n}} & \frac{1}{\gamma_{s+2,n}} & \cdot\ \cdot & \frac{1}{\gamma_{n-3,n}} & \frac{1}{\gamma_{n-2,n}} & 1 & 1 & \\
\end{array}
$$

Art. 39. This is an inversely symmetrical determinant of normal form involving the $\binom{n-1}{2}$ quantities γ_{xy}. In the compartment II, the elements, other than the units, have the denominator $\gamma_{s-1,\,s+1}$. The transformed of the minor is derived from the transformed complete determinant by deletion of the sth row and s^{th} column, and the subsequent division of each γ element in the compartment II by $\gamma_{s-1,\,s+1}$ and multiplication of each γ element in compartment III by $\gamma_{s-1,\,s+1}$.

It is now obvious that if a minor be formed from the untransformed determinant by deletion of the

$$s^{\text{th}}\ (s+1)^{\text{th}}\ .\ .\ .\ (s+\sigma)^{\text{th}} \text{ rows}$$

and the

$$s^{\text{th}}\ (s+1)^{\text{th}}\ .\ .\ .\ (s+\sigma)^{\text{th}} \text{ columns},$$

the transformed minor will be obtained from the transformed complete determinant by deletion of the aforesaid rows and columns, and subsequent division of all γ elements which are at once above the sth row and to the right of the $(s+\sigma)^{\text{th}}$ column by $\gamma_{s-1,\,s+\sigma+1}$ and corresponding multiplication of the inversely symmetrical elements by the same quantity. Or, as before, we may suppose the minor divided into four compartments and state the rule with reference to them. It will be convenient to allude to these compartments as I_s, II_s, III_s, IV_s.

In addition to the aforesaid rows and columns, suppose the $t^{\text{th}}\ (t+1)^{\text{th}}\ .\ .\ .\ (t+\tau)^{\text{th}}$ rows and columns deleted.

In correspondence we have other four compartments, I_t, II_t, III_t, IV_t; and there will be a certain extent of overlapping of compartments.

Art. 40. The rule is (after deletion from transformed complete determinant) :—

Divide γ elements in II_s by $\gamma_{s-1,\,s+\sigma+1}$,

" " II_t " $\gamma_{t-1,\,t+\tau+1}$,

with corresponding multiplication of the inversely symmetrical elements.

If this be carried out it will be found that those γ elements which are in both II_s and II_t will be divided by $\gamma_{s-1,\,s+\sigma+1}\,\gamma_{t-1,\,t+\tau+1}$.

The general rule guiding the formation of the minor when there are any number of sets of compartments arising from the deletions will be now perfectly clear.

Art. 41. We are thus enabled to exhibit all the co-axial minors of the determinant as functions of the $\binom{n-1}{2}$ quantities γ.

So much of the theory of these interesting determinants suffices for present purposes.

§ 5.

Art. 42. *The general case.*

The matrix being that connected with the determinant

$$| a_{1n} |,$$

we have the relations

$$a_{xx} = p_x,$$

$$| a_{xx} a_{yy} | = p_{xy},$$

as well as

$$2^n - 1 - n - \binom{n}{2}$$

other relations

$$| a_{xx} a_{yy} a_{zz} \dots | = p_{xyz\dots},$$

connected with the co-axial minors of order greater than 2.

From the relation

$$| a_{xx} a_{yy} | = p_{xy}$$

is derived

$$a_{xy} a_{yx} = p_x p_y - p_{xy} = q_{xy} \text{ (suppose).}$$

We now introduce $\binom{n}{2}$ undetermined quantities α_{xy} such that

$$a_{xy} = \alpha_{xy} q_{xy},$$
$$a_{yx} = 1/\alpha_{xy},$$

and substitute in the remaining

$$2^n - 1 - n - \binom{n}{2}$$

relations.

The typical relation

$$| a_{xx} a_{yy} a_{zz} \dots | = p_{xyz\dots}$$

then becomes

T 2

$$\begin{vmatrix} p_x, & \alpha_{xy}q_{xy}, & \alpha_{xz}q_{xz} & \cdot & \cdot & \cdot \\ \dfrac{1}{\alpha_{xy}}, & p_y, & \alpha_{yz}q_{yz} & \cdot & \cdot & \cdot \\ \dfrac{1}{\alpha_{xz}}, & \dfrac{1}{\alpha_{yz}}, & p_z & \cdot & \cdot & \cdot \\ \cdot & \cdot & \cdot & \cdot & \cdot & \cdot \\ \cdot & \cdot & \cdot & \cdot & \cdot & \cdot \\ \cdot & \cdot & \cdot & \cdot & \cdot & \cdot \end{vmatrix} = p_{xyz} \ldots$$

In the determinant the quantities α occur in an inversely symmetrical manner, and the determinant becomes inversely symmetrical on putting the quantities p and q equal to unity.

Art. 43. The determinant is transformable in the same manner as the corresponding inversely symmetrical form, and the foregoing "Digression" establishes the fact that the quantities α will then occur in only some or all of $\binom{n-1}{2}$ combinations $\gamma_{x,y}$, where

$$\gamma_{x,y} = \frac{\alpha_{xy}}{\alpha_{x,x+1}\alpha_{x+1,x+2}\ldots\alpha_{y-1,y}} = \frac{\alpha_{xy}}{\beta_{xy}}.$$

Hence we are presented with

$$2^n - 1 - n - \binom{n}{2} \text{ equations}$$

involving $\binom{n-1}{2}$ quantities $\gamma_{x,y}$.

Art. 44. Eliminating these $\binom{n-1}{2}$ quantities, we find

$$2^n - 1 - n - \binom{n}{2} - \binom{n-1}{2} = 2^n - n^2 + n - 2$$

relations or syzygies between the coaxial minors

$$p_{xyz} \ldots$$

of the determinant

$$| \alpha_{1n} |.$$

Art. 45. This shows that the coefficients of V_n must satisfy

$$2^n - n^2 + n - 2$$

independent conditions.

Art. 46. Assuming the satisfaction of these conditions we can solve the equations so as to express the $\binom{n-1}{2}$ quantities $\gamma_{x,y}$ in terms of the coefficients of V_n.

Hence we can express the $\frac{1}{2}(n-1)(n-4)$ quantities

$$\alpha_{x,y} \qquad (y > x+1),$$

in terms of the $n-1$ quantities

$$\alpha_{x,x+1},$$

thus reducing the number of undetermined quantities to

$$n-1.$$

Art. 47. We have

$$\sigma_n = 2^n - n^2 + n - 2,$$

while the matrix, which defines

$$X_1, X_2, \ldots X_n$$

of the redundant form, is :—

$$\begin{pmatrix} p_1 & \beta_{12}q_{12} & \frac{\beta_{13}}{c_{13}}q_{13} & \frac{\beta_{14}}{c_{14}}q_{14} & \cdot\ \cdot & \frac{\beta_{1n}}{c_{1n}}q_{1n} \\ \frac{1}{\beta_{12}} & p_2 & \beta_{23}q_{23} & \frac{\beta_{24}}{c_{24}}q_{24} & \cdot\ \cdot & \frac{\beta_{2n}}{c_{2n}}q_{2n} \\ \frac{c_{13}}{\beta_{13}} & \frac{1}{\beta_{23}} & p_3 & \beta_{34}q_{34} & \cdot\ \cdot & \frac{\beta_{3n}}{c_{3n}}q_{3n} \\ \frac{c_{14}}{\beta_{14}} & \frac{c_{24}}{\beta_{24}} & \frac{1}{\beta_{34}} & p_4 & \cdot\ \cdot & \frac{\beta_{4n}}{c_{4n}}q_{4n} \\ \cdot & \cdot & \cdot & \cdot & \cdot\ \cdot & \cdot \\ \cdot & \cdot & \cdot & \cdot & \cdot\ \cdot & \cdot \\ \frac{c_{1n}}{\beta_{1n}} & \frac{c_{2n}}{\beta_{2n}} & \frac{c_{3n}}{\beta_{3n}} & \frac{c_{4n}}{\beta_{4n}} & \cdot\ \cdot & p_n \end{pmatrix}$$

or the matrix similar to this with c_{yx} written for c_{xy}.

Postponing particular explanation in regard to the quantities c_{xy} I merely remark that $c_{xy}{}^{-1}$ is a value of $\gamma_{x,y}$ deduced from the equations.

The quantity β_{xy} has been defined to be

$$\alpha_{x,x+1}\alpha_{x+1,x+2}\ldots\alpha_{y-1,y}.$$

The matrix involves $n-1$ undetermined quantities

$$\alpha_{12}, \alpha_{23}, \ldots \alpha_{n-1,n},$$

or since

$$\beta_{xy} = \beta_{1y}/\beta_{1x},$$

we may take the undetermined quantities to be

$$\beta_{12}, \beta_{13}, \ldots \beta_{1,n}.$$

Each redundant form is thus of the nature

$$\infty^{n-1}$$

as was to be shown.

Art. 48. The equations for the determination of the $\binom{n-1}{2}$ quantities $\gamma_{x,y}$ can be taken from amongst the $\binom{n}{3}$ equations connected with the co-axial minors of Order 3.

One such equation is

$$| a_{xx}a_{yy}a_{zz} | = p_{x,y,z},$$

which may be written

$$\begin{vmatrix} p_x & q_{xy} & \dfrac{\gamma_{xz}}{\gamma_{xy}\gamma_{yz}} q_{xz} \\ 1 & p_y & q_{yz} \\ \dfrac{\gamma_{xy}\gamma_{yz}}{\gamma_{xz}} & 1 & p_z \end{vmatrix} = p_{xyz},$$

and this is a quadratic equation for $\gamma_{xz}/\gamma_{xy}\gamma_{yz}$.

If x, y, z be consecutive integers, this is simply a quadratic equation for γ_{xz}. Hence, the $n - 2$ quantities $\gamma_{x,x+2}$ are at once determined. The $n - 3$ quantities $\gamma_{x,x+3}$ are found by the aid of $\gamma_{x,x+1}$, which is unity, and $\gamma_{x+1,x+3}$. Thence, $\gamma_{x,x+s}$ is found in terms of $\gamma_{x+1,x+s}$, and all the quantities γ_{xy} are easily found.

Assuming the coefficients of V_n to satisfy the above-mentioned

$$2^n - n^2 + n - 2$$

conditions, we have to find systems of values of the quantities γ_{xy} which satisfy the

$$2^n - 1 - n - \binom{n}{2} \text{ equations}$$

in which they appear.

I find that there are only two such systems, obtained respectively by taking the positive and the negative signs in the solutions of the quadratic equations. In the one solution the signs are all taken positive and in the other all negative.

Let c_{xy}^{-1} be the value of γ_{xy} obtained by always taking positive signs and c_{yx}^{-1} that value obtained by always taking negative signs.

We have the system c_{xy}^{-1} and the system c_{yx}^{-1}. There are thus two representations of the redundant form, each involving $n - 1$ undetermined quantities.

Art. 49. Given a redundant form of order n, involving the matrix

$$\left(\begin{matrix} a_{1n} \end{matrix} \right),$$

we may exhibit its two representations, each involving $n - 1$ undetermined quantities.

The coefficients of the condensed form now necessarily satisfy the proper conditions, and passing through the condensed form we must, in the matrix of Art. 48, write

$$p_x = a_{xx}$$
$$q_{xy} = a_{xx}a_{yy} - | a_{xx}a_{yy} | = a_{xy}a_{yx},$$

and then it only remains to find the values of c_{xy} and c_{yx} in terms of the elements of the determinant

$$| a_{1n} | .$$

Solving the quadratic equation

$$\begin{vmatrix} a_{xx} & a_{xy}a_{yx} & \frac{\gamma_{xz}}{\gamma_{xy}\,\gamma_{yz}}\,a_{xz}a_{zx} \\ 1 & a_{yy} & a_{yz}a_{zy} \\ \frac{\gamma_{xy}\,\gamma_{yz}}{\gamma_{xz}} & 1 & a_{zz} \end{vmatrix} = | a_{xx}a_{yy}a_{zz} | ,$$

transformed from Art. 48, we find

$$\frac{\gamma_{xz}}{\gamma_{xy}\,\gamma_{yz}} = \frac{(a_{xy}a_{yz}a_{zx} + a_{yx}a_{zy}a_{xz}) \pm (a_{xy}a_{yz}a_{zx} - a_{yx}a_{zy}a_{xz})}{2a_{xz}a_{zx}},$$

or taking the positive sign

$$\frac{\gamma_{xz}}{\gamma_{xy}\,\gamma_{yz}} = \frac{a_{xy}\,a_{yz}}{a_{xz}},$$

and taking the negative sign

$$\frac{\gamma_{xz}}{\gamma_{xy}\,\gamma_{yz}} = \frac{a_{yx}\,a_{zy}}{a_{zx}}.$$

Hence, if c_{xy}^{-1}, be the value of γ_{xy} deduced by always taking positive signs and c_{yx}^{-1} that value arising from the negative signs, we find

$$c_{xy} = \frac{a_{xy}}{a_{x,\,x+1}\,a_{x+1,\,x+2}\,\ldots\,a_{y-1,\,y}} = \frac{a_{xy}}{b_{xy}},$$

$$c_{yx} = \frac{a_{yx}}{a_{y,\,y-1}\,a_{y-1,\,y-2}\,\ldots\,a_{x+1,\,x}} = \frac{a_{yx}}{b_{yx}},$$

where the symbols b_{xy} have been introduced, so that now

$$a_{xy},\ b_{xy},\ c_{xy}$$

in regard to the elements of the matrix of the fundamental form are analogous to

$$\alpha_{xy},\ \beta_{xy},\ \gamma_{xy}$$

in regard to the undetermined quantities.

It is easy to verify that the two systems of values

$$c_{xy}^{-1},\ c_{yx}^{-1},$$

of the quantities γ_{xy}, satisfy the whole of the $2^n - 1 - n - \binom{n}{2}$ equations, but I do not stop to prove that these are the only systems of values of γ_{xy}.

Substituting in the matrix of Art. 47 we obtain the two representations

$$\begin{pmatrix} a_{11} & \beta_{12}a_{21}b_{12} & \beta_{13}a_{31}b_{13} & \beta_{14}a_{41}b_{14} & \cdot\ \cdot & \beta_{1n}a_{n1}b_{1n} \\ \dfrac{1}{\beta_{12}} & a_{22} & \beta_{23}a_{32}b_{23} & \beta_{24}a_{42}b_{24} & \cdot\ \cdot & \beta_{2n}a_{n2}b_{2n} \\ \dfrac{a_{13}}{\beta_{13}b_{13}} & \dfrac{1}{\beta_{23}} & a_{33} & \beta_{34}a_{43}b_{34} & \cdot\ \cdot & \beta_{3n}a_{n3}b_{3n} \\ \dfrac{a_{14}}{\beta_{14}b_{14}} & \dfrac{a_{24}}{\beta_{24}b_{24}} & \dfrac{1}{\beta_{34}} & a_{44} & \cdot\ \cdot & \beta_{4n}a_{n4}b_{4n} \\ \cdot & \cdot & \cdot & \cdot & \cdot\ \cdot & \cdot \\ \dfrac{a_{1n}}{\beta_{1n}b_{1n}} & \dfrac{a_{2n}}{\beta_{2n}b_{2n}} & \dfrac{a_{3n}}{\beta_{3n}b_{3n}} & \dfrac{a_{4n}}{\beta_{4n}b_{4n}} & \cdot\ \cdot & a_{nn} \end{pmatrix}$$

$$\begin{pmatrix} a_{11} & \beta_{12}a_{12}b_{21} & \beta_{13}a_{13}b_{31} & \beta_{14}a_{14}b_{41} & \cdot\ \cdot & \beta_{1n}a_{1n}b_{n1} \\ \dfrac{1}{\beta_{12}} & a_{22} & \beta_{23}a_{23}b_{32} & \beta_{24}a_{24}b_{42} & \cdot\ \cdot & \beta_{2n}a_{2n}b_{n2} \\ \dfrac{a_{31}}{\beta_{13}b_{31}} & \dfrac{1}{\beta_{23}} & a_{33} & \beta_{34}a_{34}b_{43} & \cdot\ \cdot & \beta_{3n}a_{3n}b_{n3} \\ \dfrac{a_{41}}{\beta_{14}b_{41}} & \dfrac{a_{42}}{\beta_{24}b_{42}} & \dfrac{1}{\beta_{34}} & a_{44} & \cdot\ \cdot & \beta_{4n}a_{4n}b_{n4} \\ \cdot & \cdot & \cdot & \cdot & \cdot\ \cdot & \cdot \\ \dfrac{a_{n1}}{\beta_{1n}b_{n1}} & \dfrac{a_{n2}}{\beta_{2n}b_{n2}} & \dfrac{a_{n3}}{\beta_{3n}b_{n3}} & \dfrac{a_{n4}}{\beta_{4n}b_{n4}} & \cdot\ \cdot & a_{nn} \end{pmatrix}$$

and the second is obtainable from the first by writing

$$(a_{xy}, b_{xy}) = (a_{yx}, b_{yx}).$$

These redundant forms all lead to the same condensed form, viz.:—that derivable from the matrix

$$\left(\begin{array}{c} a_{1n} \end{array} \right).$$

Further we have here the most general forms of determinants such that their co-axial minors coincide with those of the determinant

$$| a_{1n} | .$$

The matrix reverts to its primary form on putting

$$\beta_{xy} = a_{xy}/a_{yx}b_{xy}$$

in the first representation, or, on putting

$$\beta_{xy} = 1/b_{yx}$$

in the second representation.

The transverse matrix is obtained, from the first representation, by putting

$$\beta_{xy} = 1/b_{xy}.$$

Art. 50. The function V which has entered in such a fundamentally important manner into the foregoing analysis appears to have a place in the general theory of matrices. Confining ourselves, for simplicity, to the third order, it may be recalled that SYLVESTER terms the function

$$\begin{vmatrix} a_{11} - x & a_{12} & a_{13} \\ a_{21} & a_{22} - x & a_{23} \\ a_{31} & a_{32} & a_{33} - x \end{vmatrix}$$

the latent function of the matrix

$$\left(\begin{array}{ccc} a_{11} & a_{12} & a_{13} \\ a_{21} & a_{22} & a_{23} \\ a_{31} & a_{32} & a_{33} \end{array} \right)$$

This function appears very frequently in pure mathematics, and also in applications to physics. From it can be derived a function of three variables, viz.:—

$$\begin{vmatrix} a_{11} - x_1 & a_{12} & a_{13} \\ a_{21} & a_{22} - x_2 & a_{23} \\ a_{31} & a_{32} & a_{33} - x_3 \end{vmatrix}$$

and herein writing $1/x_1$ for x_1, &c., and multiplying by $x_1x_2x_3$ and by -1 when the order is uneven, we get

$$V = \begin{vmatrix} 1 - a_{11}x_1 & - a_{12}x_1 & - a_{31}x_1 \\ - a_{21}x_2 & 1 - a_{22}x_2 & - a_{32}x_2 \\ - a_{31}x_3 & - a_{32}x_3 & 1 - a_{33}x_3 \end{vmatrix}$$

Thus the latent function is a particular case of the function V.

In the discussion of the roots of the latent function we are concerned with the order of vacuity of the matrix which may be any integer of the series 0, 1, 2, . . . n. In the case of the function V, which may be called the homographic function of the matrix, it is evident that a more refined nature of vacuity is pertinent to the discussion. We have to consider not merely the vanishing of the sum of all the co-axial minors whose order exceeds a given integer, but rather the vanishing of each separate co-axial minor.

It may be remarked that the homographic function V vanishes for the system of values of x_1, x_2, x_3, which satisfies the equations

$$X_1 = X_2 = X_3 = 1.$$

§ 6. *Digression on the General Theory of Determinants.*

Art. 51. The foregoing investigation has established the fact that the co-axial minors, of a general determinant of Order n, are connected by $2^n - n^2 + n - 2$ relations, or in other words, that but $n^2 - n + 1$ of them can assume given values.

Of these relations a certain number are connected in a special manner with the determinant of Order n, in that they are not relations merely between the coaxial minors of one of the principal coaxial minors of the determinant.

Let this number be

$$\psi(n),$$

and put

$$2^n - n^2 + n - 2 = \phi(n).$$

Then

$$\phi(n) = \psi(n) + \binom{n}{1}\psi(n-1) + \binom{n}{2}\psi(n-2) + \ldots + \binom{n}{n-4}\psi(4);$$

whence

$$\psi(4) = \phi(4) = 2,$$

and

$$\psi(n) = \phi(n) - \binom{n}{1}\phi(n-1) + \binom{n}{2}\phi(n-2) - \ldots (-)^{n-4}\binom{n}{n-4}\phi(4);$$

and, by summation, we obtain the result

$$\psi(n) = 1 + (-1)^n; \quad (n \neq 2)$$

shewing that

$$\psi(2m) = 2. \quad (m > 1)$$
$$\psi(2m+1) = 0.$$

Hence, when the determinant is of even order greater than two, there are two special relations between the coaxial minors and these two relations can each be thrown into a form which exhibits the determinant as an irrational function of its coaxial minors.

In the case of a determinant of uneven order no *special* relations exist between the coaxial minors, and it is not possible to express the determinant as a function of its coaxial minors.*

Art. 52. In the investigation we met with $\binom{n}{3}$ equations

$$\begin{vmatrix} p_x & q_{xy} & \dfrac{\gamma_{xz}}{\gamma_{xy}\gamma_{yz}} q_{xz} \\ 1 & p_y & q_{yz} \\ \dfrac{\gamma_{xy}\gamma_{yz}}{\gamma_{xz}} & 1 & p_z \end{vmatrix} = p_{xyz},$$

involving the $\binom{n-1}{2}$ quantities γ_{xy} and the coaxial minors of the first three orders of the determinant $|\alpha_{1n}|$. Hence, by elimination, we find $\binom{n-1}{3}$ identical relations between such coaxial minors.

Also we found

$$\binom{n}{3} + \binom{n}{4} + \ldots + \binom{n}{s}$$

* It is evident that these relations must occur in pairs in accordance with the 'Law of Complementaries' which is so important in the general theory of determinants.

equations involving the $\binom{n-1}{2}$ quantities γ_{xy} and the co-axial minors of the first s orders of the determinant $| a_{1n} |$. Hence, by elimination, we find

$$\binom{n-1}{3} + \binom{n}{4} + \ldots + \binom{n}{s}$$

relations between such coaxial minors.

Special to the coaxial minors of order s, we thus find $\binom{n}{s}$ relations if n be greater than 3. The one relation, special (from this standpoint) to the determinant of even order (greater than two), is obtained by eliminating the determinant itself from the two special identical relations above referred to.

Art. 53. I take this opportunity of verifying the statements made in Art. 49 in regard to the systems of values of the quantities

$$\gamma_{xy}$$

which satisfy the

$$2^n - 1 - n - \binom{n}{2} \text{ equations.}$$

It is, in reality, a question concerning the properties of determinants.

To ensure that the coefficients of the condensed form satisfy the requisite conditions, assume them to be derived from the determinant

$$| a_{1n} | .$$

We have $\binom{n}{2}$ equations of the type

$$\begin{vmatrix} a_{xx} & a_{xy}a_{yz} & \dfrac{\gamma_{xz}}{\gamma_{xy}\gamma_{yz}}\, a_{xz}a_{zx} \\ 1 & a_{yy} & a_{yz}a_{zy} \\ \dfrac{\gamma_{xy}\gamma_{yz}}{\gamma_{xz}} & 1 & a_{zz} \end{vmatrix} = | a_{xx}a_{yy}a_{zz} |$$

This equation, being a quadratic for $\gamma_{xz} / \gamma_{xy}\gamma_{yz}$, has only two roots, and it is easy to verify that the equation is satisfied by the values

$$\frac{a_{xy}a_{yz}}{a_{xz}}, \qquad \frac{a_{yx}a_{zy}}{a_{zx}}$$

In Art. 49, these values have been obtained by solving the quadratic, and it was found that the values corresponded to the positive and negative sign respectively.

Taking always the positive sign, let $c_{xy}{}^{-1}$ be the value deduced for γ_{xy}.
Then

$$c_{xy} = a_{xy}/b_{xy},$$

and

$$\gamma_{xz}/\gamma_{xy}\gamma_{yz} = c_{xy}c_{yz}/c_{xz}.$$

Hence, the $\binom{n}{2}$ equations are all satisfied by the system

$$\gamma_{xy} = c_{xy}{}^{-1}.$$

Similarly, they are all satisfied by the system

$$\gamma_{xy} = c_{yx}{}^{-1},$$

where

$$c_{yx} = a_{yx}/b_{yx}.$$

Art. 54. To show that each of these systems satisfies the remaining equations, it suffices to consider the typical determinant equation of the fourth order.

We have—

$$\begin{vmatrix} a_{xx} & a_{xy}a_{yx} & \dfrac{\gamma_{xz}}{\gamma_{xy}\gamma_{yz}}a_{xz}a_{zx} & \dfrac{\gamma_{xw}}{\gamma_{xy}\gamma_{yz}\gamma_{zw}}a_{xw}a_{wx} \\ 1 & a_{yy} & a_{yz}a_{zy} & \dfrac{\gamma_{yw}}{\gamma_{yz}\gamma_{zw}}a_{yw}a_{wy} \\ \dfrac{\gamma_{xy}\gamma_{yz}}{\gamma_{xz}} & 1 & a_{zz} & a_{zw}a_{wz} \\ \dfrac{\gamma_{xy}\gamma_{yz}\gamma_{zw}}{\gamma_{xw}} & \dfrac{\gamma_{yz}\gamma_{zw}}{\gamma_{yw}} & 1 & a_{ww} \end{vmatrix}$$

$$= \mid a_{xx}a_{yy}a_{zz}a_{ww} \mid .$$

On the left-hand side put

$$\gamma_{xy} = c_{xy}{}^{-1} = b_{xy}/a_{xy},$$

and the determinant becomes

$$\begin{vmatrix} a_{xx} & a_{xy}a_{yx} & a_{xy}a_{yz}a_{zx} & a_{xy}a_{yz}a_{zw}a_{wx} \\ 1 & a_{yy} & a_{yz}a_{zy} & a_{yz}a_{zw}a_{wy} \\ \dfrac{a_{xz}}{a_{xy}a_{yz}} & 1 & a_{zz} & a_{zw}a_{wz} \\ \dfrac{a_{xw}}{a_{xy}a_{yz}a_{zw}} & \dfrac{a_{yw}}{a_{yz}a_{zw}} & 1 & a_{ww} \end{vmatrix}$$

In succession, multiply the first column by a_{xy}, divide the first row by a_{xy}; multiply the third row by a_{yz}, divide the third column by a_{yz}; multiply the fourth row by a_{zw}, divide the fourth column by a_{zw}; the determinant is then $| a_{xx}a_{yy}a_{zz}a_{ww} |$.

Similarly it is shown that the equation is satisfied by the system

$$\gamma_{xy} = c_{yx}^{-1} = b_{yx}/a_{yx}.$$

The equations, involving determinants of higher order, can similarly be shown to be satisfied by both systems of values, and since the $\binom{n}{3}$ quadratic equations have each but two roots, it follows at once that

$$c_{xy}^{-1},\ c_{yx}^{-1}$$

are the only systems.

§ 7. *Arithmetical Interpretations resumed.*

Art. 55. The arithmetical interpretations drawn from the theory have been so far of two kinds. In the examples taken from the "Memoir on the Compositions of Numbers" we had a redundant form of generating function and an exact or condensed form; the redundant form and the exact form could be differently interpreted, and this led to an arithmetical correspondence which was duly noted in the memoir quoted. The interpretations, subsequently considered in this paper, were single, and there was no arithmetical correspondence; the condensed forms did not admit of easy and useful interpretations, but only the redundant forms. The redundant forms were not considered in the most general form which, as we have seen, involves $n - 1$ undetermined quantities, but each of these quantities was given a special numerical value; this process led to simple and useful arithmetical results but it will be obvious that the possibility of interpretation does not stop here.

Art. 56. In proceeding from the condensed form to the redundant form we met with $n - 1$ undetermined quantities

$$\alpha_{12},\ \alpha_{23},\ \ldots\ \alpha_{n-1,n}.$$

As before remarked, we may, if we please, put these quantities equal to certain functions of the quantities

$$x_1,\ x_2,\ \ldots\ x_n.$$

We are not at liberty to choose *any* functions. The functions must satisfy certain conditions, otherwise the coefficient of

$$x_1^{\xi_1}x_2^{\xi_2}\ \ldots\ x_n^{\xi_n}$$

in the particular redundant product will not remain unchanged.

I propose to examine this question.

Art. 57. Of order 2 we have the product

$$(p_1x_1 + \alpha_{12}q_{12}x_2)^{\xi_1}\left(\frac{1}{\alpha_{12}}x_1 + p_2x_2\right)^{\xi_2},$$

and in performing the multiplication we find a term involving

$$(p_1x_1)^m(\alpha_{12}q_{12}x_2)^{\xi_1-m}\left(\frac{x_1}{\alpha_{12}}\right)^{\xi_1-m}(p_2x_2)^{\xi_2-\xi_1+m} = p_1{}^m p_2{}^{\xi_2-\xi_1+m}q_{12}{}^{\xi_1-m}x_1{}^{\xi_1}x_2{}^{\xi_2},$$

and if α_{12} be not a function of x_1 and x_2 the terms involving $x_1{}^{\xi_1}x_2{}^{\xi_2}$ can only arise in a manner similar to this.

If, however, α_{12} be such that $\alpha_{12}x_2$ is a multiple of x_1, and consequently x_1/α_{12} a multiple of x_2, we at once get an addition to the coefficient of $x_1{}^{\xi_1}x_2{}^{\xi_2}$. In the present case the coefficient becomes

$$(p_1 + cq_{12})^{\xi_1}\left(\frac{1}{c} + p_2\right)^{\xi_2}$$

Hence, considering monomial values of α_{12} only, the inequality

$$\frac{\alpha_{12}x_2}{cx_1} \neq 1$$

must be satisfied in assigning to α_{12} a function of x_1 and x_2.

We may put α_{12}, subject to the above condition, equal to any monomial integral or fractional function of x_1 and x_2.

We may *not* put

$$\alpha_{12} = c\frac{x_1}{x_2},$$

where c is any function of p_1, p_2.

We may not, in fact, realize a portion of the coefficient of $x_1{}^{\xi_1}x_2{}^{\xi_2}$ as

$$(p_1x_1)^m(\alpha_{12}q_{12}x_2)^{\xi_1-m}\left(\frac{x_1}{\alpha_{12}}\right)^{\xi_1-n}(p_2x_2)^{\xi_2-\xi_1+n},$$

wherein n differs from m.

Art. 58. Of Order 3, the particular redundant product is

$$(p_1x_1 + \alpha_{12}q_{12}x_2 + c_{13}\alpha_{12}\alpha_{23}q_{13}x_3)^{\xi_1}\left(\frac{x_1}{\alpha_{12}} + p_2x_2 + \alpha_{23}q_{23}\right)^{\xi_2}\left(\frac{c_{13}x_1}{\alpha_{12}\alpha_{23}} + \frac{x_2}{\alpha_{23}} + p_3x_3\right)^{\xi_3},$$

and we must realize the coefficient of

$$x_1{}^{\xi_1}x_2{}^{\xi_2}x_3{}^{\xi_3}$$

in the manner

$$(p_1x_1)^{m_1}\left(\frac{x_1}{\alpha_{12}}\right)^{n_1}\left(\frac{x_1}{\gamma_{13}\alpha_{12}\alpha_{23}}\right)^{\xi_1-m-n} \times (\alpha_{12}q_{12}x_2)^{m_2}(p_2x_2)^{n_2}\left(\frac{x_2}{\alpha_{23}}\right)^{\xi_2-m_2-n_2}$$
$$\times \text{ a multiple of } x_3^{\xi_3},$$

where, of the three portions, the first accounts wholly for $x_1^{\xi_1}$, the second wholly for $x_2^{\xi_2}$, and so on; and not in any other manner.

Put

$$(\alpha_{12}, \alpha_{23}) = (\phi_1, \phi_2),$$

where ϕ_1, ϕ_2 are fractions of x_1, x_2, x_3, and consider the simplified matrix,

$$\left(\begin{array}{ccc} x_1, & \phi_1 x_2, & \phi_1\phi_2 x_3 \\ x_1, & x_2, & \phi_2 x_3 \\ \frac{x_1}{\phi_1\phi_2}, & \frac{x_2}{\phi_2}, & x_3 \end{array}\right),$$

in which unnecessary quantities are omitted.

Further, omitting a multiplier, independent of x_1, x_2, x_3, on the right-hand sides, the following six inequalities must be satisfied,

$$\phi_1^2\phi_2 \neq \frac{x_1^2}{x_2x_3}, \qquad \frac{\phi_2}{\phi_1} \neq \frac{x_2^2}{x_1x_3}, \qquad \frac{1}{\phi_1\phi_2^2} \neq \frac{x_3^2}{x_1x_2},$$

$$\phi_2 \neq \frac{x_2}{x_3}, \qquad \frac{1}{\phi_1\phi_2} \neq \frac{x_3}{x_1}, \qquad \phi_1 \neq \frac{x_1}{x_2},$$

putting

$$\Phi_1 = \phi_1\frac{x_2}{x_1}, \qquad \Phi_2 = \phi_2\frac{x_3}{x_2};$$

these conditions are representable by the single inequality

$$\Phi_1^3\Phi_2 + \frac{1}{\Phi_1^3\Phi_2} + \frac{\Phi_2^2}{\Phi_1} + \frac{\Phi_1}{\Phi_2^2} + \Phi_1^2\Phi_2^3 + \frac{1}{\Phi_1^2\Phi_2^3}$$
$$\neq \Phi_2^3\Phi_1 + \frac{1}{\Phi_2^3\Phi_1} + \frac{\Phi_1^2}{\Phi_2} + \frac{\Phi_2}{\Phi_1^2} + \Phi_2^2\Phi_1^3 + \frac{1}{\Phi_2^2\Phi_1^3}.$$

As regards functions of x_1, x_2, x_3, this inequality being satisfied, ϕ_1 and ϕ_2 may be put equal to any functions that may be desired. Like inequalities may be obtained in respect of the fourth and higher orders.

Art. 59. The important point to notice is that it is legitimate to put the undetermined quantities equal to any *integral* functions of $x_1, x_2, \ldots x_n$—a fact, for the general order, that becomes obvious on examination of the above processes.

As subsequently appears, it is such integral functions that usually present themselves in arithmetical applications.

Art. 60. As an example of the applications to arithmetic which swarm about the theory, consider the important condensed form (*vide* Art. 14):—

$$\frac{1}{\left[\begin{array}{l}1 - \Sigma x_1 - \Sigma (\lambda_{\beta\alpha} - 1) x_\alpha x_\beta - \Sigma (\lambda_{\beta\alpha} - 1)(\lambda_{\gamma\beta} - 1) x_\alpha x_\beta x_\gamma \\ \quad - \ldots - (\lambda_{21} - 1)(\lambda_{32} - 1)(\lambda_{23} - 1)\ldots(\lambda_{n, n-1} - 1) x_1 x_2 x_3 \ldots x_n\end{array}\right]}$$

and, at first, consider the form of Order 3.

The matrix of the redundant form is easily found to be either

$$\left|\begin{array}{ccc} 1 & \alpha_{12}\lambda_{21} & \dfrac{\beta_{13}\lambda_{31}}{c_{13}} \\ \dfrac{1}{\alpha_{12}} & 1 & \alpha_{23}\lambda_{32} \\ \dfrac{c_{13}}{\beta_{13}} & \dfrac{1}{\alpha_{23}} & 1 \end{array}\right|$$

or the similar matrix with c_{31} written for c_{13}. Since

$$c_{13} = \frac{\lambda_{31}}{\lambda_{21}\lambda_{32}}, \qquad c_{31} = 1,$$

we have, taking c_{31} and putting $(\alpha_{12}, \alpha_{23}) = (1, 1)$ a particular redundant product

$$(x_1 + \lambda_{21}x_2 + \lambda_{31}x_3)^{\xi_1} (x_1 + x_2 + \lambda_{32}x_3)^{\xi_2} (x_1 + x_2 + x_3)^{\xi_3}.$$

In this, the coefficient of $x_1^{\xi_1}x_2^{\xi_2}x_3^{\xi_3}$ (which is equal to the coefficient of the same term in the condensed form) is arithmetically interpretable as in Art. 15.

Art. 61. If, however, we put (*vide* Art. 59)

$$(\alpha_{12}, \alpha_{23}; c_{13}^{-1}) = \left(x_1, x_2; \frac{\lambda_{32}\lambda_{21}}{\lambda_{31}}\right),$$

we obtain a form which may be written :—

$$(x_1 + \lambda_{21}x_2x_1 + \lambda_{32}\lambda_{21}x_3x_2x_1)^{\xi_1} (1 + x_2 + \lambda_{32}x_3x_2)^{\xi_2} \left(\frac{\lambda_{31}x_3x_1}{\lambda_{32}\lambda_{21}x_3x_2x_1} + 1 + x_3\right)^{\xi_3},$$

and herein we see that the coefficient of

$$\lambda_{21}^{s_{21}}\lambda_{31}^{s_{31}}\lambda_{32}^{s_{32}}x_1^{\xi_1}x_2^{\xi_2}x_3^{\xi_3}$$

represents the number of permutations of the symbols in

$$x_1^{\xi_1}x_2^{\xi_2}x_3^{\xi_3},$$

which possess exactly

$$\begin{array}{lll} s_{21}, & x_2x_1 & \text{contacts} \\ s_{31}, & x_3x_1 & ,, \\ s_{32}, & x_3x_2 & ,, \end{array}$$

Here is an entirely new interpretation and we see that the true generating function for the enumeration of the indicated permutations is

$$\frac{1}{1-x_1-x_2-x_3-(\lambda_{21}-1)x_1x_2-(\lambda_{31}-1)x_1x_3-(\lambda_{32}-1)x_2x_3-(\lambda_{21}-1)(\lambda_{32}-1)x_1x_2x_3},$$

a result which does not lie by any means on the surface.

The arithmetical correspondence should also be noted.

Art. 62. For the order n we have the matrix

$$\left(\begin{array}{cccccc} 1 & \alpha_{12}\lambda_{21} & \frac{\beta_{13}\lambda_{31}}{c_{13}} & \frac{\beta_{14}\lambda_{41}}{c_{14}} & \cdot\ \cdot & \frac{\beta_{1n}\lambda_{n1}}{c_{1n}} \\ \frac{1}{\alpha_{12}} & 1 & \alpha_{23}\lambda_{32} & \frac{\beta_{24}\lambda_{42}}{c_{24}} & \cdot\ \cdot & \frac{\beta_{2n}\lambda_{n2}}{c_{2n}} \\ \frac{c_{13}}{\beta_{13}} & \frac{1}{\alpha_{23}} & 1 & \alpha_{34}\lambda_{43} & \cdot\ \cdot & \frac{\beta_{3n}\lambda_{n3}}{c_{3n}} \\ \frac{c_{14}}{\beta_{14}} & \frac{c_{24}}{\beta_{24}} & \frac{1}{\alpha_{34}} & 1 & \cdot\ \cdot & \frac{\beta_{4n}\lambda_{n4}}{c_{4n}} \\ \cdot & \cdot & \cdot & \cdot & \cdot\ \cdot & \cdot \\ \cdot & \cdot & \cdot & \cdot & \cdot\ \cdot & \cdot \\ \frac{c_{1n}}{\beta_{1n}} & \frac{c_{2n}}{\beta_{2n}} & \frac{c_{3n}}{\beta_{3n}} & \frac{c_{4n}}{\beta_{4n}} & \cdot\ \cdot & 1 \end{array}\right)$$

and we obtain another form by writing c_{yx} for c_{xy}.

Moreover $(y > x)$ we have

$$c_{yx} = 1, \quad c_{xy} = \lambda_{yx}/\mu_{yx},$$

where

$$\mu_{yx} = \lambda_{y,\,y-1}\lambda_{y-1,\,y-2}\ldots\lambda_{x+1,\,x};$$

whence writing

$$(\alpha_{xy}, c_{yx}) = (1, 1)$$

we obtain the matrix

$$\begin{pmatrix} 1 & \lambda_{21} & \lambda_{31} & . & . & \lambda_{n1} \\ 1 & 1 & \lambda_{32} & . & . & \lambda_{n2} \\ 1 & 1 & 1 & . & . & \lambda_{n3} \\ . & . & . & . & . & . \\ 1 & 1 & 1 & . & . & 1 \end{pmatrix}$$

and we can interpret the coefficient of $x_1^{\xi_1}x_2^{\xi_2} \ldots x^{\xi_n}$ in the corresponding particular redundant product as in Art. 15.

Again, writing

$$(\alpha_{p,p+1}, c_{xy}) = \left(x_p, \frac{\lambda_{yx}}{\mu_{yx}}\right),$$

which, as far as $\alpha_{p,p+1}$ is concerned, Art. 59 shows to be legitimate, we have

$$\beta_{p,q} = x_p x_{p+1} \ldots x_{q-1} = \mathrm{X}_{p,q-1} = \mathrm{X}_{q-1,p} \text{ suppose,}$$

and the matrix

$$\begin{pmatrix} 1 & \lambda_{21}x_1 & \mu_{31}\mathrm{X}_{12} & \mu_{41}\mathrm{X}_{13} & . & . & \mu_{n1}\mathrm{X}_{1,n-1} \\ \frac{1}{x_1} & 1 & \lambda_{32}x_2 & \mu_{42}\mathrm{X}_{23} & . & . & \mu_{n2}\mathrm{X}_{2,n-1} \\ \frac{\lambda_{31}}{\mu_{31}\mathrm{X}_{12}} & \frac{1}{x_2} & 1 & \lambda_{43}x_3 & . & . & \mu_{n3}\mathrm{X}_{3,n-1} \\ \frac{\lambda_{41}}{\mu_{41}\mathrm{X}_{13}} & \frac{\lambda_{42}}{\mu_{42}\mathrm{X}_{23}} & \frac{1}{x_3} & 1 & . & . & \mu_{n4}\mathrm{X}_{4,n-1} \\ . & . & . & . & . & . & . \\ \frac{\lambda_{n1}}{\mu_{n1}\mathrm{X}_{1,n-1}} & \frac{\lambda_{n2}}{\mu_{n2}\mathrm{X}_{2,n-1}} & \frac{\lambda_{n3}}{\mu_{n3}\mathrm{X}_{3,n-1}} & \frac{\lambda_{n4}}{\mu_{n4}\mathrm{X}_{4,n-1}} & . & . & 1 \end{pmatrix}$$

and the new particular redundant product is :—

$$\begin{aligned} &\left(x_1 + \lambda_{21}x_2x_1 + \mu_{31}\mathrm{X}_{31} + \mu_{41}\mathrm{X}_{41} + \ldots + \mu_{n1}\mathrm{X}_{n1}\right)^{\xi_1} \\ &\left(1 + x_2 + \lambda_{32}x_3x_2 + \mu_{42}\mathrm{X}_{42} + \ldots + \mu_{n2}\mathrm{X}_{n2}\right)^{\xi_2} \\ &\left(\frac{\lambda_{31}}{\mu_{31}x_2} + 1 + x_3 + \lambda_{43}x_4x_3 + \ldots + \mu_{n3}\mathrm{X}_{n3}\right)^{\xi_3} \\ &\left(\frac{\lambda_{41}}{\mu_{41}\mathrm{X}_{32}} + \frac{\lambda_{42}}{\mu_{42}x_3} + 1 + x_4 + \ldots + \mu_{n4}\mathrm{X}_{n4}\right)^{\xi_4} \\ &\left(. \quad . \quad . \quad . \quad \ldots \quad . \right) \\ &\left(\frac{\lambda_{n1}}{\mu_{n1}\mathrm{X}_{n-1,2}} + \frac{\lambda_{n2}}{\mu_{n2}\mathrm{X}_{n-1,3}} + \frac{\lambda_{n3}}{\mu_{n3}\mathrm{X}_{n-1,4}} + \frac{\lambda_{n4}}{\mu_{n4}\mathrm{X}_{n-1,5}} + \ldots + x_n\right)^{\xi_n} \end{aligned} .$$

Art. 63. In this product we may interpret the coefficient of

$$x_1^{\xi_1} x_2^{\xi_2} \dots x_n^{\xi_n}.$$

From the nature of the condensed form we know that this coefficient is an integral function of the quantities λ_{xy}. We may prove that if a portion of the expansion be

$$c\lambda_{21}^{s_{21}}\lambda_{32}^{s_{32}} \dots \lambda_{qp}^{s_{qp}}\, x_1^{\xi_1} x_2^{\xi_2} \dots x_n^{\xi_n},$$

the number c indicates the number of permutations of the $\Sigma\xi$ quantities in

$$x_1^{\xi_1} x_2^{\xi_2} \dots x_n^{\xi_n},$$

which possess exactly s_{21} contacts $x_2 x_1$

$$\begin{array}{ccc} s_{32} & ,, & x_3 x_2 \\ \vdots & ,, & \vdots \\ s_{qp} & ,, & x_q x_p \end{array}$$

Regard the above product, as written, as being a square form of n rows and n columns involving n^2 elements.

Observe that if $s \nless t$ the element common to the sth row and t^{th} column is

$$\frac{\lambda_{st}}{\mu_{st}} \cdot \frac{x_t}{X_{t,s-1}}$$

while the element common to the tth row and s^{th} column is

$$\mu_{st} X_{st},$$

and that the product of these two elements is

$$\lambda_{st} x_s x_t.$$

Now, take a particular permutation of the $\Sigma\xi$ quantities and observe how it may be considered to arise in the multiplication. Let a portion of the permutation be

$$x_2 \mid x_4 x_2 x_1 \mid x_8 x_5 x_3 x_2 x_1 \mid x_5 \mid x_5 \dots$$

divided off by bars into compartments in such wise that in any compartment the suffixes are in descending order.

The portion is a permutation of

$$\ldots x_1^2 x_2^3 x_3 x_4 x_5^3 x_8 \ldots$$

and we can obtain this portion by selecting for multiplication

2 elements from the row appertaining to the exponent	ξ_1
3 ,, ,, ,, ,, ,, ,,	ξ_2
1 ,, ,, ,, ,, ,, ,,	ξ_3
1 ,, ,, ,, ,, ,, ,,	ξ_4
3 ,, ,, ,, ,, ,, ,,	ξ_5
1 ,, ,, ,, ,, ,, ,,	ξ_8

The permutation is divided into five compartments as shown.

In the first compartment we have simply x_2 which is to be taken from the 2nd row 2nd column. In the second compartment we have

$$x_4 x_2 x_1$$

which is obtainable by multiplication of elements taken from the 4th, 2nd, and 1st rows, as follows :—

In row 4, column 2, we take $\frac{\lambda_{42}}{\mu_{42}} \cdot \frac{1}{x_3}$

,, 2, ,, 1, ,, 1

,, 1, ,, 4, ,, $\mu_{41} x_4 x_3 x_2 x_1$;

multiplication gives

$$\lambda_{42}\lambda_{21} x_4 x_2 x_1.$$

In the third compartment we find

$$x_8 x_5 x_3 x_2 x_1$$

From row 8, column 5, we take $\frac{\lambda_{85}}{\mu_{85}} \frac{1}{x_7 x_6}$.

,, 5, ,, 3, ,, $\frac{\lambda_{53}}{\mu_{53}} \cdot \frac{1}{x_4}$.

,, 3, ,, 2, ,, 1.

,, 2, ,, 1, ,, 1.

,, 1, ,, 8, ,, $\mu_{81} x_8 x_7 x_6 x_5 x_4 x_3 x_2 x_1$.

Multiplication of these five elements yields

$$\lambda_{85}\lambda_{53}\lambda_{32}\lambda_{21} x_8 x_5 x_3 x_2 x_1.$$

In the fourth and fifth compartments we have simply x_5, and in each case the element selected is that in the 5th row and 5th column. Altogether we have obtained the product

$$\lambda_{85}\lambda_{53}\lambda_{42}\lambda_{32}\lambda^2_{21}x_2x_4x_2x_1x_8x_5x_3x_2x_1x_5x_5,$$

and we observe that the contacts

$$x_qx_p \quad (q > p)$$

are correctly indicated by the quantities

$$\lambda_{qp}.$$

Art. 64. The process is obviously a general one, and the rule of element selection to demonstrate the desired result may be set forth as follows :—

If a compartment of the permutation be

$$x_ax_bx_cx_dx_e,$$

a, b, c, d, e being in descending order of magnitude, we take elements in

row	a,	column	b,
,,	b,	,,	c,
,,	c,	,,	d,
,,	d,	,,	e,
,,	e,	,,	a,

and thus obtain the product,

$$\lambda_{ab}\lambda_{bc}\lambda_{cd}\lambda_{de}x_ax_bx_cx_dx_e,$$

wherein the contacts are correctly represented by the quantities λ.

If a compartment contain the single quantity x_s, we take the element in the sth row and sth column.

By the above process

ξ_1	elements are taken from row			1,
ξ_2	,,	,,	,,	2,
$\vdots$				$\vdots$
ξ_n	,,	,,	,,	n,

to form the product

$$x_1^{\xi_1}x_2^{\xi_2}\ldots x_n^{\xi_n}.$$

Art. 65. Hence it has been established that the coefficient of the term

$$\lambda_{21}^{s_{21}} \lambda_{32}^{s_{32}} \dots \lambda_{qp}^{s_{qp}} x_1^{\xi_1} x_2^{\xi_2} \dots x_n^{\xi_n},$$

in the product, enumerates the permutations of the $\Sigma\xi$ quantities in

$$x_1^{\xi_1} x_2^{\xi_2} \dots x_n^{\xi_n},$$

which possess exactly

$$\begin{array}{lcl} s_{21} & \text{contacts} & x_2 x_1, \\ s_{32} & ,, & x_3 x_2, \\ \cdot & \cdot & \cdot \\ s_{qp} & ,, & x_q x_p ; \end{array}$$

and since the redundant product can assume the appearance derived from the matrix

$$\left(\begin{array}{cccc} 1 & \lambda_{21} & \dots & \lambda_{n1} \\ 1 & 1 & \dots & \lambda_{n2} \\ \cdot & \cdot & \dots & \cdot \\ 1 & 1 & \dots & 1 \end{array}\right)$$

we find that the enumeration is identical with that of the permutations which are such that the quantity x_q occurs s_{qp} times in places originally occupied by the quantity x_p, when $q > p$, and, as before, we take the coefficient of

$$\lambda_{21}^{s_{21}} \lambda_{32}^{s_{32}} \dots \lambda_{qp}^{s_{qp}} x_1^{\xi_1} x_2^{\xi_2} \dots x_n^{\xi_n}.$$

Hence, an arithmetical correspondence, and, also, the fact that the true generating function for the enumeration of these permutations is

$$\frac{1}{\left[\begin{array}{l} 1 - \Sigma x_1 - \Sigma(\lambda_{\beta\alpha} - 1) x_\alpha x_\beta - \Sigma(\lambda_{\beta\alpha} - 1)(\lambda_{\gamma\beta} - 1) x_\alpha x_\beta x_\gamma \\ - \dots - (\lambda_{21} - 1)(\lambda_{32} - 1) \dots (\lambda_{n\,n-1} - 1) x_1 x_2 x_3 \dots x_{n-1} x_n \end{array}\right]}$$

The above example is only a solitary one of a large number that might be furnished. An advantageous method for procedure appears to be to take some simple interpretable redundant product, and to then pass through the condensed form to the general redundant product, involving $n - 1$ undetermined quantities as well as quantities c_{xy}, which admit of a choice of values. The assignment of these quantities then leads to a variety of arithmetical correspondences which, as before remarked, is absolutely limitless.

The theory, moreover, includes an exhaustive Theory of Permutations, and gives in every case the true condensed Generating Functions. Its importance in the General Theory of Determinants has been touched upon.

In conclusion, the paper will have achieved its object if it is successful in indicating the arithmetical and algebraical power of the main theorem considered.

THE SUMS OF POWERS OF THE BINOMIAL COEFFICIENTS.

By Major P. A. MacMahon, D.Sc., F.R.S.

Introduction.

I HAVE shown elsewhere* that the coefficient of $x_1^{\xi_1} x_2^{\xi_2} \ldots x_n^{\xi_n}$ in the product

$$(a_1 x_1 + a_2 x_2 + \ldots + a_n x_n)^{\xi_1} (b_1 x_1 + b_2 x_2 + \ldots + b_n x_n)^{\xi_2} \ldots$$

$$(n_1 x_1 + n_2 x_2 + \ldots + n_n x_n)^{\xi_n}$$

is equal to the coefficient of the same term in the expansion of the algebraic fraction

$$\frac{1}{|(1 - a_1 x_1)(1 - b_2 x_2)(1 - c_3 x_3) \ldots (1 - n_n x_n)|},$$

wherein the denominator denotes symbolically the expression

$$1 - a_1 x_1 - b_2 x_2 - c_3 x_3 - \ldots - n_n x_n$$
$$+ |a_1 b_2| x_1 x_2 + |a_1 c_3| x_1 x_3 + |b_2 c_3| x_2 x_3 + \ldots$$
$$- |a_1 b_2 c_3| x_1 x_2 x_3 - \ldots$$
$$\cdots\cdots\cdots\cdots\cdots\cdots\cdots$$
$$+ (-)^n |a_1 b_2 c_3 \ldots n_n| x_1 x_2 x_3 \ldots x_n,$$

where the coefficients are the several coaxial minors of the determinant

$$|a_1 b_2 c_3 \ldots n_n|.$$

I propose to apply this theorem to the coordination of some known theorems concerning binomial coefficients and to the discovery of some new ones; further I wish to show a general method by which such questions can be discussed.

Adopting a convenient notation, I write

$$\frac{(p+q\ !}{p!\,q!} = \binom{p+q}{p} = \binom{p+q}{q}.$$

I make the remark that the binomial theorem for a positive integral index may be regarded as an expression of the sum of

---

* "A certain class of generating functions in the Theory of Numbers," *Phil. Trans. Roy. Soc. London*, A. 1894, p. 111.

a certain linear function of the binomial coefficients; we may, in fact, regard the theorem as being concerned with a summation instead of with an expansion. From this point of view we write

$$\Sigma\alpha^p\beta^q\binom{p+q}{p}=(\alpha+\beta)^{p+q},$$

wherein we have the numerical magnitudes α, β at disposal in order to determine the values of various linear functions of the coefficients $\binom{p+q}{p}$.

One object of the paper will be to show how to obtain analogous expressions for the sum

$$\Sigma\alpha^p\beta^q\binom{p+q}{p}^m$$

m a positive integer, and to give the actual expression when m is 2, 3, or 4.

I commence with some trivial cases concerning well-known results in order that the research may proceed in a clear and orderly manner.

§ 1.

Combinations of the second order.

The sum

$$1+\binom{p}{1}\binom{q}{1}+\binom{p}{2}\binom{q}{2}+\ldots$$

is the coefficient of x^py^q in $(x+y)^p(x+y)^q$, and the general theorem shows that it is also the coefficient of x^py^q in

$$\frac{1}{1-x-y},$$

for the determinant that presents itself is

$$\begin{vmatrix} x, & y \\ x, & y \end{vmatrix},$$

and thence denominator of the generating function is seen to be $1-x-y$.

Accordingly the sum is $\binom{p+q}{p}$; or, when $q=p$, $\binom{2p}{p}$.

In the next place consider the sum

$$1-\binom{p}{1}\binom{q}{1}+\binom{p}{2}\binom{q}{2}-\ldots,$$

which is the coefficient of $x^p y^q$ in $(x+y)^p(-x+y)^q$.

The determinant is here

$$\begin{vmatrix} x, & y \\ -x, & y \end{vmatrix},$$

and the derived generating function

$$\frac{1}{1-x-y+2xy}$$

To obtain the coefficient of $x^p y^q$ in the expansion of the fraction we may proceed in a variety of ways; if we write it

$$\frac{1}{1-x.1-y}\cdot\frac{1}{1+\dfrac{xy}{1-x.1-y}}=\Sigma\frac{(xy)^\sigma}{(1-x)^{\sigma+1}(1-y)^{\sigma+1}},$$

we find that the coefficient of $x^p y^q$ is

$$\sum_\sigma\binom{p}{\sigma}\binom{q}{\sigma}x^p y^q,$$

and we merely get a verification of the generating fraction, but if we write it

$$\sum_\sigma(x+y-2xy)^\sigma$$

$$=\sum_{\sigma_1,\sigma_2}\sum_\sigma(-)^{\sigma-\sigma_1-\sigma_2}2^{\sigma-\sigma_1-\sigma_2}\frac{\sigma!}{\sigma_1!\,\sigma_2!\,(\sigma-\sigma_1-\sigma_2)!}x^{\sigma-\sigma_1}y^{\sigma-\sigma_2},$$

and put $\sigma-\sigma_2=p$, $\sigma-\sigma_1=q$, we find, assuming p, q to be in descending order, that the sum is

$$\sum_{\sigma=p}^{\sigma=p+q}(-)^{p+q-\sigma}2^{p+q-\sigma}\frac{\sigma!}{(\sigma-p)!\,(\sigma-q)!\,(p+q-\sigma)!},$$

or

$$\frac{(p+q)!}{p!\,q!\,0!}-2\frac{(p+q-1)!}{(p-1)!\,(q-1)!\,1!}+2^2\frac{(p+q-2)!}{(p-2)!\,(q-2)!\,2!}-\ldots$$

$$+(-)^{q-1}2^{q-1}\frac{(p+1)!}{1!\,(p-q+1)!\,(q-1)!}+(-)^q 2^q\frac{p!}{0!\,(p-q)!\,q!}.$$

As a verification, taking $p=3$, $q=2$,

$$1.1-3.2+3.1$$

$$=\frac{5!}{3!\,2!\,0!}-2\frac{4!}{2!\,1!\,1!}+2^2\frac{3!}{1!\,0!\,2!}=10-24+12=-2.$$

If we put $q=p$, we find

$$1-\binom{p}{1}^2+\binom{p}{2}^2-\dots$$

$$=\frac{(2p)!}{(p!)^2\,0!}-2\frac{(2p-1)!}{\{(p-1)!\}^2\,1!}+2^2\frac{(2p-2)!}{\{(p-2)!\}^2\,2!}-\dots.$$

Now this sum is also the coefficient of $(xy)^p$ in $(y^2-x^2)^p$, and has therefore the expression

$$(-)^{\frac{1}{2}p}\frac{p!}{\left(\frac{p!}{2}\right)^2},\ \text{or zero,}$$

according as p is even or uneven.

To obtain a more general result consider the coefficient of x^py^q in $(ax+by)^p(bx+ay)^q$.

This is

$$a^{p+q}+\binom{p}{1}\binom{q}{1}a^{p+q-2}b^2+\binom{p}{2}\binom{q}{2}a^{p+q-4}b^4+\dots,$$

while the generating function, derived from the determinant

$$\begin{vmatrix} ax, & by \\ bx, & ay \end{vmatrix},$$

is

$$\frac{1}{1-a(x+y)-(b^2-a^2)xy}.$$

Herein the coefficient of x^py^q is readily found to be

$$\sum_{\sigma=p}^{\sigma=p+q}\frac{\sigma!}{(p+q-\sigma)!\,(\sigma-q)!\,(\sigma-p)!}a^{2\sigma-p-q}(b^2-a^2)^{p+q-\sigma},$$

or

$$\frac{p!}{q!\,(p-q)!\,0!}a^{p-q}(b^2-a^2)^q$$

$$+\frac{(p+1)!}{(q-1)!\,(p+q+1)!\,1!}a^{p-q+2}(b^2-a^2)^{q-1}+\dots+\frac{(p+q)!}{0!\,p!\,q!}a^{p+q},$$

a result which includes all of those obtained above.

As a particular case put $q=p$, $a^2=\alpha$, $b^2=\beta$, we find

$$\alpha^p+\binom{p}{1}^2\alpha^{p-1}\beta+\binom{p}{2}^2\alpha^{p-2}\beta^2+\dots$$

$$=(\beta-\alpha)^p+\binom{p+1}{2}\binom{2}{1}\alpha(\beta-\alpha)^{p-1}$$

$$+\binom{p+2}{4}\binom{4}{2}\alpha^2(\beta-\alpha)^{p-2}+\dots+\binom{2p}{p}\alpha^p$$

$$=(\alpha-\beta)^p+\binom{p+1}{2}\binom{2}{1}\beta(\alpha-\beta)^{p-1}$$

$$+\binom{p+2}{4}\binom{4}{2}\beta^2(\alpha-\beta)^{p-2}+\dots+\binom{2p}{p}\beta^p,$$

since α and β may be interchanged.

This result is expressible in the notation of hypergeometric series in the form

$$\alpha^pF\left(-p,\ -p,\ 1,\ \frac{\beta}{\alpha}\right)=(\alpha-\beta)^pF\left(p+1,\ -p,\ 1,\ -\frac{\beta}{\alpha-\beta}\right),$$

and will be recognised as a particular case of the known transformation

$$F(\alpha,\beta,\gamma,x)=(1-x)^{-\alpha}F\left(\alpha,\gamma-\beta,\gamma,\frac{x}{x-1}\right)$$

derivable from a known property of the linear differential equation of the second orders satisfied by the hypergeometric series.*

This curious result gives an expression for the same linear function of the squares of the binomial coefficients, as the Binomial Theorem does in respect of the first powers of the coefficients.

§ 2.

Combinations of the Third Order.

In § 1 we have been on the familiar and well beaten ground of combinations of the second order, and yet the novel method brought into use has been of service in extending and coordinating known results.

In this section we find ourselves at the outset upon untrodden ground. The only investigation that a diligent

---

* Cf. Forsyth's, *Differential Equations*, Chapter VI.

search has brought to light is that by Dixon* which deals with a result which will be presently referred to.

First consider the sum

$$1-\binom{p}{1}^3+\binom{p}{2}^3-\binom{p}{3}^3+\ldots,$$

considered by Dixon (*loc. cit.*).

I say that it is the coefficient of $(xyz)^p$ in

$$(y-z)^p(z-x)^p(x-y)^p,$$

and that the determinant

$$\begin{vmatrix} 0, & y, & -z \\ -x, & 0, & z \\ x, & -y, & 0 \end{vmatrix}$$

gives us the generating function

$$\frac{1}{1+yz+yx+xy},$$

a striking result which gives immediately much information concerning the sum under consideration.

The term $(xyz)^p$ can only occur in the term

$$(-)^{\frac{3p}{2}}(yz+zx+xy)^{\frac{3p}{2}},$$

and hence the sum vanishes unless p be even, and when p is even is positive or negative according as $\frac{1}{2}p$ is even or uneven.

To evaluate the coefficient put

$$yz=u^2,\ zx=v^2,\ xy=w^2,$$

so that

$$xyz=uvw,$$

and then the coefficients of $(uvw)^p$ in

$$\frac{1}{1+u^2+v^2+w^2}$$

is

$$(-)^{\frac{3p}{2}}\frac{\left(\frac{3p}{2}\right)!}{\left\{\left(\frac{p}{2}\right)!\right\}^3},$$

* *Messenger of Mathematics*, Vol. xx., p. 79.

and we have the result

$$1-\binom{2p}{1}^3+\binom{2p}{2}^3-\ldots=(-)^p\frac{(3p)!}{(p!)^3}. \qquad \text{(Cf. Dixon, } l.c.\text{).}$$

An observation of considerable algebraic importance can now be made. Since, in the product

$$(y-z)^p\,(z-x)^p\,(x-y)^p,$$

the factors $y-z$, $z-x$, $x-y$ are raised *to the same power*, we may interchange them in any manner and thus we are presented with six different guiding determinants, and these, in the present instance, yield six different but equivalent generating functions.

These are

Determinant.	Generating Function.
$\begin{vmatrix} 0, & y, & -z \\ -x, & 0, & z \\ x, & -y, & 0 \end{vmatrix}$,	$\dfrac{1}{1+yz+zx+xy}$,
$\begin{vmatrix} 0, & y, & -z \\ x, & -y, & 0 \\ -x, & 0, & z \end{vmatrix}$,	$\dfrac{1}{1+y-z-yz-zx-xy}$,
$\begin{vmatrix} -x, & 0, & z \\ 0, & y, & -z \\ x, & -y, & 0 \end{vmatrix}$,	$\dfrac{1}{1+x-y-yz-zx-xy}$,
$\begin{vmatrix} -x, & 0, & z \\ x, & -y, & 0 \\ 0, & y, & -z \end{vmatrix}$,	$\dfrac{1}{1+x+y+z+yz+zx+xy}$,
$\begin{vmatrix} x, & -y, & 0 \\ 0, & y, & -z \\ -x, & 0, & z \end{vmatrix}$,	$\dfrac{1}{1-x-y-z+yz+zx+xy}$,
$\begin{vmatrix} x, & -y, & 0 \\ -x, & 0, & z \\ 0, & y, & -z \end{vmatrix}$,	$\dfrac{1}{1-x+z-yz-zx-xy}$.

Since we may interchange x, y, z by any substitution and may also change their signs at pleasure, the essentially different generating functions may be taken to be the first, second, and fifth of those above written.

In each of these algebraic fractions the part which involves powers of xyz is the same.

In general, if we seek the coefficient of $(xyz)^p$ in the product

$$(a_1x + a_2y + a_3z)^p (b_1x + b_2y + b_3z)^p (c_1x + c_2y + c_3z)^p,$$

we obtain $3!$ different generating functions obtained by permutation of the rows of the determinant

$$| a_1b_2c_3 |.$$

The sum of the cubes of the binomial coefficients

$$1 + \binom{p}{1}^3 + \binom{p}{2}^3 + \ldots$$

is the coefficient of $(xyz)^p$ in

$$(y + z)^p (z + x)^p (x + y)^p;$$

and thence, from the determinant

$$\begin{vmatrix} 0, & y, & z \\ x, & 0, & z \\ x, & y, & 0 \end{vmatrix},$$

and the five other determinants obtained by permutation of rows, we derive six generating functions, but of these only five are different and only three essentially distinct; these are

$$\frac{1}{1 - yz - zx - xy - 2xyz},$$

$$\frac{1}{1 - y - z + yz - zx - xy + 2xyz},$$

$$\frac{1}{1 - x - y - z + yz + zx + xy - 2xyz}.$$

From the first of these it is easy to find an expression for the series, giving us the identity

$$1 + \binom{p}{1}^3 + \binom{p}{2}^3 + \ldots$$

$$= 2^p + \binom{p+1}{3}\frac{3!}{(1!)^3} + \binom{p+2}{6}\frac{6!}{(2!)^3} + \ldots + \text{a final term},$$

which is $\dfrac{(\frac{3}{2}v)!}{\{(\frac{1}{2}p)!\}^3}$ or $\dbinom{\frac{3p-1}{2}}{\frac{3p-3}{2}} \dfrac{\left(\frac{3p-3}{2}\right)!}{\left\{\left(\frac{p-1}{2}\right)!\right\}^3}$,

according as p is even or uneven.

These results are thrown together by considering the coefficient of $(xyz)^p$ in the product

$$(ay+bz)^p\,(az+bx)^p\,(ax+by)^p.$$

We are then led to the three essentially different generating functions

$$\frac{1}{1-ab\,(yz+zx+xy)-(a^3+b^3)\,xyz},$$

$$\frac{1}{1-by-az+abyz-b^2zx-a^2xy+(a^3+b^3)\,xyz},$$

$$\frac{1}{1-b\,(x+y+z)+b^2\,(yz+zx+xy)-(a^3+b^3)\,xyz},$$

and selecting the first of these we obtain the identity

$$a^{3p}+\binom{p}{1}^3 a^{3p-3}b^3+\binom{p}{2}^3 a^{3p-6}b^6+\dots$$

$$=\frac{p!}{(0!)^3\,p!}(a^3+b^3)^p+\frac{(p+1)!}{(1!)^3\,(p-2)!}(ab)^3(a^3+b^3)^{p-2}$$

$$+\frac{(p+2)!}{(2!)^3\,(p-4)!}(ab)^6(a^3+b^3)^{p-4}+\dots;$$

or, writing $a^3=\alpha$, $b^3=\beta$,

$$\alpha^p+\binom{p}{1}^3\alpha^{p-1}\beta+\binom{p}{2}^3\alpha^{p-2}\beta^2+\dots$$

$$=\frac{p!}{(0!)^3\,p!}(\alpha+\beta)^p+\frac{(p+1)!}{(1!)^3\,(p-2)!}\alpha\beta\,(\alpha+\beta)^{p-2}$$

$$+\frac{(p+2)!}{(2!)^3\,(p-4)!}(\alpha\beta)^6(\alpha+\beta)^{p-4}+\dots,$$

a result obviously including the two results obtained above.

Writing the hypergeometric series

$$1+\frac{\alpha\beta\gamma}{\theta\epsilon}x+\frac{\alpha(\alpha+1)\,\beta(\beta+1)\,\gamma(\gamma+1)}{1.2.\theta(\theta+1)\,\epsilon(\epsilon+1)}x^2+\dots$$

in the notation

$$F\left\{\begin{pmatrix} \alpha, & \beta, & \gamma \\ \theta, & \epsilon \end{pmatrix}, x\right\},$$

we may write the above result in the form

$$\alpha^p F\left\{\begin{pmatrix} -p, & -p, & -p \\ 1, & 1 \end{pmatrix}, -\frac{\beta}{\alpha}\right\}$$

$$= (\alpha + \beta)^p F\left\{\begin{pmatrix} p+1, & -\frac{p}{2}, & -\frac{p-1}{2} \\ 1, & 1, \end{pmatrix}, \frac{4\alpha\beta}{(\alpha+\beta)^2}\right\},$$

which doubtless involves a known theorem of transformation of hypergeometric series.

If now we take the coefficient of $x^p y^q z^r$ in the product

$$(ay + bx)^p (az + bx)^q (ax + by)^r,$$

we have the series

$$\begin{pmatrix} p \\ p-q \end{pmatrix} \begin{pmatrix} q \\ p-r \end{pmatrix} \begin{pmatrix} r \\ 0 \end{pmatrix} a^{-p+2q+2r} b^{2p-q-r}$$

$$+ \begin{pmatrix} p \\ p-q+1 \end{pmatrix} \begin{pmatrix} q \\ p-r+1 \end{pmatrix} \begin{pmatrix} r \\ 1 \end{pmatrix} a^{-p+2q+2r-3} b^{2p-q-r+3} + \dots,$$

and the generating function as before, viz.

$$\frac{1}{1 - ab\,(yz + zx + xy) - (a^3 + b^3)\,xyz}.$$

In this for the coefficient of $x^p y^q z^r$ it is easy to obtain the series

$$\frac{\dfrac{p+q+r!}{2}}{0!\ \dfrac{-p+q+r!}{2}\ \dfrac{p-q+r!}{2}\ \dfrac{p+q-r!}{2}}\,(ab)^{\frac{1}{2}(p+q+r)}$$

$$+ \frac{\dfrac{p+q+r-2!}{2}}{2!\ \dfrac{-p+q+r-2!}{2}\ \dfrac{p-q+r-2!}{2}\ \dfrac{p+q-r-2!}{2}}\,(ab)^{\frac{1}{2}(p+q+r\ 6)}(a^3+b^3)^2$$

$$+ \frac{\dfrac{p+q+r-4!}{2}}{4!\ \dfrac{-p+q+r-4!}{2}\ \dfrac{p-q+r-4!}{2}\ \dfrac{p+q-r-4!}{2}}\,(ab)^{\frac{1}{2}(p+q+r-12)}(a^3+b^3)^4$$

$$+\dots,$$

when $p+q+r$ is even; and the series

$$\frac{\dfrac{p+q+r-1}{2}!}{1!\,\dfrac{-p+q+r-1}{2}!\,\dfrac{p-q+r-1}{2}!\,\dfrac{p+q-r-1}{2}!}(ab)^{\frac{1}{2}(p+q+r-3)}(a^3+b^3)$$

$$+\frac{\dfrac{p+q+r-3}{2}!}{3!\,\dfrac{-p+q+r-3}{2}!\,\dfrac{p-q+r-3}{2}!\,\dfrac{p+q-r-3}{2}!}(ab)^{\frac{1}{2}(p+q+r-9)}(a^3+b^3)^3$$

$$+\frac{\dfrac{p+q+r-5}{2}!}{5!\,\dfrac{-p+q+r-5}{2}!\,\dfrac{p-q+r-5}{2}!\,\dfrac{p+q-r-5}{2}!}(ab)^{\frac{1}{2}(p+p+r-15)}(a^3+b^3)^5$$

$+\ldots,$

when $p+q+r$ is uneven.

Consider next the series

$$a^{p+q+r}\left\{1+\binom{p}{1}\binom{q}{1}\binom{r}{1}\frac{b^3}{a^3}+\binom{p}{2}\binom{q}{2}\binom{r}{2}\frac{b^6}{a^6}+\ldots\right\},$$

which is the coefficient of $x^p y^q z^r$ in the product

$$(ax+by)^p\,(ay+bz)^q\,(az+bx)^r.$$

The generating function derived from the matrix

$$\begin{vmatrix} ax, & by, & 0 \\ 0\,, & ay, & bz \\ bx, & 0\,, & az \end{vmatrix}$$

assumes the form

$$\frac{1}{1-a(x+y+z)+a^2(yz+zx+xy)-(a^3+b^3)\,xyz},$$

or

$$\sum_\sigma \frac{b^{3\sigma}(xyz)^\sigma}{(1-ax)^{\sigma+1}(1-ay)^{\sigma+1}(1-az)^{\sigma+1}},$$

and herein the coefficient of $x^p y^q z^r$ is

$$\Sigma\binom{p}{\sigma}\binom{q}{\sigma}\binom{r}{\sigma}a^{p+q+r-3\sigma}b^{3\sigma},$$

which merely verifies the generating function.

Another mode of summation, however, gives the result in an interesting form, but it is too complicated to be worth preserving.

Combinations of order 4.

The series

$$1+\binom{p}{1}^4+\binom{p}{2}^4+\ldots$$

is the coefficient of $(xyzu)^p$ in

$$(y+z)^p\,(z+u)^p\,(u+x)^p\,(x+y)^p.$$

We have 4! matrices at disposal. One of these is

$$\begin{vmatrix} 0, & y, & z, & 0 \\ 0, & 0, & z, & u \\ x, & 0, & 0, & u \\ x, & y, & 0, & 0 \end{vmatrix},$$

which yields the generating function

$$\frac{1}{1-xz-yu-xyz-xyu-xzu-yzu}.$$

If we put $xz=a$, $yu=b$, we have to find the coefficient of $(ab)^p$ in

$$\frac{1}{1-a-b-a(y+u)-b(x+z)},$$

and find the general term

$$\frac{(2\sigma_1+2\sigma_2+\sigma_3)!}{(\sigma_1!)^2\,(\sigma_2!)^2\,\sigma_4!\,(\sigma_3-\sigma_4)!}\,a^{2\sigma_1+\sigma_2+\sigma_4}b^{\sigma_1+2\sigma_2+\sigma_3-\sigma_4},$$

and putting

$$2\sigma_1+\sigma_2+\sigma_4=\sigma_1+2\sigma_2+\sigma_3-\sigma_4=p,$$

we arrive at the identity

$$1+\binom{p}{1}^4+\binom{p}{2}^4+\ldots$$

$$=\sum_{\sigma_1}\sum_{\sigma_2}\frac{(2p-\sigma_1-\sigma_2)!}{(p-2\sigma_1-\sigma_2)!\,(p-\sigma_1-2\sigma_2)!\,(\sigma\,!)^2\,(\sigma_2!)^2},$$

for all values of σ_1, σ_2 for which the expression has a meaning. Of the remaining 23 generating functions 21 are different, but only 4 are essentially distinct: these are

$$\frac{1}{1-x-yz-yu+xyz+xyu-xzu-yzu},$$

$$\frac{1}{1-x-z-xy-yu-zu+xz+xyz+xyu+xzu+yzu},$$

$$\frac{1}{1-y-z-u-xy+yz+yu+zu+xyz+xyu-xzu-yzu},$$

$$\frac{1}{1-x-y-z-u+xy+xz+xu+yz+yu+zu-xyz-xyu-xzu-yzu}.$$

The last written generating function is of a type which is common to all orders.

Let ξ_1, ξ_2, ξ_3, ... be the elementary symmetric functions of the quantities x, y, z, u, ...; then the value of the series

$$1+\binom{p}{1}^n+\binom{p}{2}^n+\ldots$$

is the coefficient of ξ_n^p in the development of

$$\frac{1}{1-\xi_1+\xi_2-\xi_3+\ldots+(-)^{n-1}\xi_{n-1}},$$

wherein, after expansion, the symmetric functions are to be multiplied out. Thence, by known processes in symmetric functions, the series can be evaluated.

The coefficient of $(xyzu)^p$ in

$$(ay+bz)^p(az+bu)^p(au+bx)^p(ax+by)^p$$

is equal to the coefficients of the same term in

$$\frac{1}{1-A(x_1+y_1)-B\{x_1(y+u)+y_1(x+z)-Cx_1y_1\}},$$

where
$$x_1=xz,\quad y_1=yu,$$

$$A=b^2,\quad B=a^2b,\quad C=(a^4-b^4).$$

In this the general term is

$$\frac{\sigma!}{(\sigma_2!)^2(\sigma_3!)^2\sigma_4!\sigma_5!\sigma_6!}A^{\sigma_1}B^{2\sigma_2+2\sigma_3}C^{\sigma_4}x_1^{2\sigma_2+\sigma_3+\sigma_4+\sigma_5}y_1^{\sigma_2+2\sigma_3+\sigma_4+\sigma_6},$$

whence, writing

$$2\sigma_2+\sigma_3+\sigma_4+\sigma_5=p=\sigma_2+2\sigma_3+\sigma_4+\sigma_6,$$

the coefficient of $(xyzu)^p$ becomes

$$\Sigma\Sigma\Sigma\frac{(2p-\sigma_2-\sigma_3-\sigma_4)!}{(\sigma_2!)^2(\sigma_3!)^2\sigma_4!(p-2\sigma_2-\sigma_3-\sigma_4)!(p-\sigma_2-2\sigma_3-\sigma_4)!}$$
$$\times a^{4\sigma_2+4\sigma_3}b^{4p-4\sigma_2-4\sigma_3-4\sigma_4}(a^4-b^4)^{\sigma_4},$$

and, writing $a^4=\alpha$, $b^4=\beta$, we obtain the identity

$$\alpha^p+\binom{p}{1}^4\alpha^{p-1}\beta+\binom{p}{2}^4\alpha^{p-2}\beta^2+\ldots+\beta^p$$

$$=\sum_{\sigma_2}\sum_{\sigma_3}\sum_{\sigma_4}\frac{(2p-\sigma_2-\sigma_3-\sigma_4)}{(\sigma_2!)^2(\sigma_3!)^2\sigma_4!(p-2\sigma_2-\sigma_3-\sigma_4)(p-\sigma_2-2\sigma_3-\sigma_4)!}$$
$$\times\alpha^{\sigma_2+\sigma_3}\beta^{p-\sigma_2-\sigma_3-\sigma_4}(\alpha-\beta)^{\sigma_4}.$$

If herein we put $\alpha=\beta$, σ_4 becomes zero, and we obtain the expression obtained above for the sum of the fourth powers of the binomial coefficients.

In any product such as

$$(a_1x+a_2y)^p(b_1x+b_2y)^q$$

we may regard a_1, a_2, b_1, b_2 as real coefficients or as umbræ in the symbolic notation of Aronhold. In either case the generating function for the coefficient of x^py^q is

$$\frac{1}{1-a_1x-b_2y+(ab)xy},$$

where $(ab)=a_1b_2-a_2b_1$.

Expanding this fraction, we find for the coefficient in question

$$\sum_\sigma(-)^{p+q-\sigma}\frac{\sigma!}{(p+q-\sigma)!(\sigma-p)!(\sigma-q)!}a_1^{\sigma-q}b_2^{\sigma-p}(ab)^{p+q-\sigma}.$$

Hence we reach the identity for $p \geq q$,

$$a_1^p b_2^q + \binom{p}{1}\binom{q}{1} a_1^{p-1} a_2 b_1^{q-1} b_2$$

$$+ \binom{p}{2}\binom{q}{2} a_1^{p-2} a_2^2 b_1^{r-2} b_2^2 + \ldots + \binom{p}{q}\binom{q}{q} a_1^{p-q} a_2^q b_2^q$$

$$= \frac{(p+q)!}{0!\,p!\,q!} a_1^p b_2^q - \frac{(p+q-1)!}{1!\,(p-1)!\,(q-1)!} a_1^{p-1} b_2^{q-1} (ab)$$

$$+ \frac{(p+q-2)!}{2!\,(p-2)!\,(q-2)!} a_1^{p-2} b_2^{q-2} (ab)^2 - \ldots$$

$$+ (-)^q \frac{p!}{q!\,(p-q)!\,0!} a_1^{p-q} (ab)^q$$

$$= \kappa_0 a_1^p b_2^q - \kappa_1 a_1^{p-1} b_2^{q-1} (ab) + \ldots + (-)^q \kappa_q a_1^{p-q} (ab)^q,$$

where
$$\kappa_s = \frac{(p+q-s)!}{s!\,(p-s)!\,(q-s)!}.$$

Hence, if

$$(a_1 x + a_2 y)^p = A_0 x^p + \binom{p}{1} A_1 x^{p-1} y + \binom{p}{2} A_2 x^{p-2} y^2 + \ldots,$$

$$(b_1 x + b_2 y)^q = B_0 x^q + \binom{q}{1} B_1 x^{q-1} y + \binom{q}{2} B_2 x^{q-2} y^2 + \ldots,$$

we have

$$A_0 B_0 + \binom{p}{1}\binom{q}{1} A_1 B_1 + \binom{p}{2}\binom{q}{2} A_2 B_2 + \ldots$$

$$= \kappa_0 A_0 B_0 - \kappa_1 (A_0 B_0 - A_1 B_1) + \kappa_2 (A_0 B_0 - 2A_1 B_1 + A_2 B_2) - \ldots,$$

where the law is obvious.

Comparing the coefficients of $A_0 B_0$, $A_1 B_1$, ..., we have the identities

$$\kappa_0 - \kappa_1 + \kappa_2 - \ldots + (-)^q \kappa_q = 1,$$

$$\kappa_1 - 2\kappa_2 + \ldots + (-)^{q-1} a\kappa_q = \binom{p}{1}\binom{q}{1},$$

$$\kappa_2 - 3\kappa_3 + \ldots (-)^q \binom{q}{2} \kappa_q = \binom{p}{2}\binom{q}{2},$$

....

$$\kappa_q = \binom{p}{q}\binom{q}{q}.$$

Chapter 4
Permutations

4.1 Introduction and Commentary

In MacMahon's *Combinatory Analysis*, volume 1, it was quite logical to include the Master Theorem in the work on permutations. However, among modern workers, only J. K. Percus (1971) has utilized and extended the applications of the Master Theorem to permutation problems. For example, Percus (1971, pages 21–26) uses the Master Theorem to give a new solution to the Ménage Problem.

Apart from the Master Theorem, much of MacMahon's work on permutations and indices of permutations was done in his investigations of plane partitions. In regard to this, paper [82] is probably the most important one in this chapter. Paper [86] presents an interesting relationship between indices and inversions that has been the subject of more recent investigations; it is not mentioned in volume 1 of *Combinatory Analysis* and is only briefly cited on page viii of volume 2. Consequently, some modern authors (Abramson, 1971, and Carlitz, 1970) have rediscovered MacMahon's result, independently.

The work of Foata (1965) (see also Cartier and Foata, 1969) provides an algebraic approach to many of MacMahon's permutation problems. We have already seen this approach used to prove the Master Theorem in chapter 3; in the next section we shall present a related proof of one of MacMahon's permutation theorems. We also draw attention to the work of G. Kreweras (see also chapter 5), and to that of Carlitz and his colleagues. Carlitz has utilized generating functions extensively in his work on permutation problems.

4.2 A Factorization Theorem and its Consequences

MacMahon was interested in permutations of "assemblages of objects." Modern combinatorialists refer to "permutations of multisets," a subject introduced in section 3.2 in the proof of the Master Theorem. In this section we shall present the proof, due to Foata (1965), of a theorem that appears in Paragraph 63 of [46]. While MacMahon discovered this theorem as a result of his work on compositions (see chapter 5), our proof will nonetheless be very much in the spirit of this chapter. We shall utilize the basic notation and results concerning permutations of multisets that appear in section 3.2. In addition, we shall denote cycles by

---

The material in this chapter corresponds to section III, chapters I, III, V, and VI in *Combinatory Analysis*.

$$\begin{pmatrix} a\ b\ c \dots e \\ b\ c\ d \dots a \end{pmatrix} = (a, b, c, d, \dots, e) = (b, c, d, \dots, e, a).$$

Theorem 1. (Foata, 1965). Each multiset permutation π may be factored as:

$$\pi = (x_{11}, x_{12}, \dots, x_{1n}, y_1)(x_{21}, x_{22}, \dots, x_{2m}, y_2) \dots,$$

where $y_1 \leqq y_2 \leqq \dots, y_i < x_{ij}$ for all j.

Proof. We first factor π into primes (i.e., cycles) and proceed by mathematical induction on r, the length of the factorization. Clearly we may find a shortest left factor of π, say ρ, that contains y_1, the minimal element of the multiset M. If $r = 1$, then we are done. If $r < 1$, then we may cancel ρ from π and proceed by the induction hypothesis. Hence Theorem 1 is true.

In the following two definitions and in Theorem 2, we shall be considering multiset permutations π of the form

$$\begin{pmatrix} 1 & 1 & \dots & 1 & 2 & 2 & \dots & 2 & 3 & 3 & \dots \\ i_1 & i_2 & \dots & i_j & i_{j+1} & i_{j+2} & \dots & i_{j+k} & i_{j+k+1} & \dots & \end{pmatrix}.$$

Definition 1. Let $\phi(\pi)$ denote the number of pairs (i_h, i_{h+1}) with $i_h > i_{h+1}$.

Definition 2. Let $r(\pi)$ denote the number of columns $\begin{pmatrix} a \\ \cdot \\ i_a \end{pmatrix}$ in π with $a > i_a$.

Theorem 2 (from [46], paragraph 63). Let M be a multiset whose elements are taken from the first n positive integers. Then for each t, the number of permutations π of M with $\phi(\pi) = t$ equals the number of permutations π' of M with $r(\pi') = t$.

Proof (Foata, 1965; see also Cartier and Foata, 1969). We shall establish a one-to-one correspondence between the two types of permutations as follows: Let π be a permutation of M with $r(\pi) = t$; factor according to Theorem 1; now define a new permutation π' whose lower line is

$$x_{11}, x_{12}, \dots, x_{1n}, y_1, x_{21}, x_{22}, \dots, x_{2m}, y_2, \dots.$$

Then we see that each column $\begin{pmatrix} a \\ \cdot \\ i_a \end{pmatrix}$ in the original permutation π is transformed into exactly one juxtaposed a, i_a in the lower line of π' whenever $a > i_a$. Consequently the mapping $\pi \to \pi'$ yields $r(\pi) = \phi(\pi')$. Finally we note that this mapping must be a bijection, since the y_i in the lower line of π' are distinguished as those entries that do not exceed all succeeding entries. Thus π is the only preimage of π', and each π' has a preimage.

4.3 References

M. Abramson (1971) On the generating function for permutations with repetitions and inversions, *Canadian Math. Bull.*, *14* (1), 101–102.

M. Abramson (1975a) Sequences by number of w-rises, *Canadian Math. Bull.*, *18*, 317–319.

M. Abramson (1975b) Enumeration of sequences by levels and rises, *Discr. Math.*, *12*, 101–112.

M. Abramson (1975c) A note on permutations with fixed pattern, *J. Combinatorial Th.*, *A-19*, 237–239.

M. Abramson and W. O. J. Moser (1966) A note on combinations, *Canadian Math. Bull.*, *9*, 675–677.

M. Abramson and W. O. J. Moser (1969a) Enumeration of combinations with restricted differences and cospan, *J. Combinatorial Th.*, *7*, 162–170.

M. Abramson and W. O. J. Moser (1969b) Generalizations of Terquem's problem, *J. Combinatorial Th.*, *7*, 171–180.

G. E. Andrews (1975) A theorem on reciprocal polynomials with applications to permutations and compositions, *American Math. Monthly*, *82*, 830–833.

G. E. Andrews (1976) The Theory of Partitions, Encyclopedia of Mathematics and Its Applications, vol. 2, Addison-Wesley, Reading.

D. E. Barton and F. N. David (1962) Combinatorial Chance, Griffin, London.

D. E. Barton and C. L. Mallows (1965) Some aspects of the random sequence, *Ann. Math. Stat.*, *36*, 236–260.

C. Berge (1971) Principles of Combinatorics, Academic Press, New York.

L. Carlitz (1969) Solution of certain recurrences, *S.I.A.M. J. Appl. Math.*, *17*, 251–259.

L. Carlitz (1970) Sequences and inversions, *Duke Math. J.*, *37*, 193–198.

L. Carlitz (1972a) Enumeration of sequences by rises and falls: a refinement of the Simon Newcomb problem, *Duke Math. J.*, *39*, 267–280.

L. Carlitz (1972b) Rectangular arrays of zeros and ones, *Duke Math. J.*, *39*, 153–164.

L. Carlitz (1972c) Sequences, paths, ballot numbers, *Fibonacci Quart.*, *10*, 531–549.

L. Carlitz (1973) Enumeration of permutations by rises and cycle structure, *J. reine und angew. Math.*, *262/263*, 220–237.

L. Carlitz (1974a) Permutations and sequences, *Advances in Math.*, *14*, 92–120.

L. Carlitz (1974b) Up-down and down-up partitions, Proc. of Eulerian Series and Applications Conference, Pennsylvania State University.

L. Carlitz (1975a) Generating functions for a special class of permutations, *Proc. A.M.S.*, *47*, 251–256.

L. Carlitz (1975b) Permutations, sequences, and special functions, *S.I.A.M. Rev.*, *17*, 298–322.

L. Carlitz and J. Riordan (1955) The number of labeled two-terminal series-parallel networks, *Duke Math. J.*, *23*, 435–446.

L. Carlitz and J. Riordan (1964) Two element lattice permutation numbers and their q-generalizations, *Duke Math. J.*, *31*, 371–388.

L. Carlitz, D. P. Roselle, and R. A. Scoville (1966) Permutations and sequences with repetitions by number of increases, *J. Combinatorial Th.*, *1*, 350–374.

L. Carlitz, D. P. Roselle, and R. A. Scoville (1971) Some remarks on ballot type sequences of positive integers, *J. Combinatorial Th.*, *11*, 258–271.

L. Carlitz and R. A. Scoville (1974a) Generalized Eulerian numbers: combinatorial applications, *J. reine und angew. Math.*, *265*, 110–137.

L. Carlitz and R. Scoville (1974b) Enumeration of permutations by rises, falls, rising maxima and falling minima, *Acta Math. Acad. Sci. Hungar.*, *25*, 269–277.

L. Carlitz and R. Scoville (1975a) Generating functions for certain types of permutations, *J. Combinatorial Th.*, *A-18*, 262–275.

L. Carlitz and R. A. Scoville (1975b) Enumeration of up-down permutations by upper records, *Monatsh. Math.*, *79*, 3–12.

L. Carlitz, R. Scoville, and T. Vaughan (1976a) Enumeration of pairs of sequences by rises, falls and levels, *Manuscripta Math.*, *19*, 211–243.

L. Carlitz, R. Scoville, and T. Vaughan (1976b) Enumeration of pairs of permutations, *Discr. Math.*, *14*, 215–239.

L. Carlitz and T. Vaughan (1974) Enumeration of sequences of given specification according to rises, falls and maxima, *Discr. Math.*, *8*, 147–167.

P. Cartier and D. Foata (1969) Problèmes combinatoires de commutation et réarrangements, Lecture Notes in Math. No. 85, Springer, Berlin.

J. F. Dillon and D. P. Roselle (1968) Eulerian numbers of higher order, *Duke Math. J.*, *35*, 247–256.

J. F. Dillon and D. P. Roselle (1969) Simon Newcomb's problem, *S.I.A.M. J. Appl. Math.*, *17*, 1086–1093.

D. Dumont (1974) Interprétations combinatoires des nombres Genocchi, *Duke Math. J.*, *41*, 305–318.

D. Dumont and D. Foata (1975) A symmetry property of the Genocchi numbers, Inst. of Statist. Mimeo Series No. 981, University of North Carolina, Chapel Hill.

M. Eisen (1969) Elementary Combinatorial Analysis, Gordon and Breach, New York.

R. C. Entringer (1966) A combinatorial interpretation of the Euler and Bernoulli numbers, *Nieuw Archief der Wisk. (3)*, *14*, 241–246.

D. Foata (1965) Étude algébrique de certains problèmes d'analyse combinatoire et du calcul des probabilités, *Publ. Inst. Stat. Univ. Paris*, *14*, 81–241.

D. Foata (1968) On the Netto inversion number of a sequence, *Proc. A.M.S.*, *19*, 236–240.

D. Foata (1972) Groupes de réarrangements et nombres d'Euler, *C.R. Acad Sc. Paris*, *A-275*, 1147–1150.

D. Foata (1974) La série génératrice exponentielle dans les problèmes d'enumération, Séminaire de Mathematique Supérieure, Les Presses de l'Université de Montréal.

D. Foata (1975) Studies in Enumeration, Inst. of Statist. Mimeo Series No. 974, University of North Carolina, Chapel Hill.

D. Foata (1977) Distributions Eulériennes et Mahoniennes sur le groupe des permutations, in Proceedings of the N.A.T.O. Conference on Advanced Combinatorics, Berlin, September 1–10, 1976, D. Reidel, Dordrecht.

D. Foata and J. Riordan (1974) Mappings of acyclic and parking functions, *Aequationes Math.*, *10*, 10–22.

D. Foata and M.-P. Schützenberger (1970) Théorie géométrique des polynômes euleriens, Lecture Notes in Math. No. 138, Springer, Berlin.

D. Foata and M.-P. Schützenberger (1973) Nombres d'Euler et Permutations Alternantes, A Survey of Combinatorial Theory (J. N. Srivastava et al., eds.), North-Holland, 173–187.

D. Foata and M.-P. Schützenberger (1977a) Major index and inversion number of permutations, *Math. Nach.* (to appear).

D. Foata and M.-P. Schützenberger (1977b) La troisième transformation fondamentale du groupe des permutations (to appear).

D. Foata and V. Strehl (1974) Rearrangements of the symmetric group and enumerative properties of the tangent and secant numbers, *Math. Zeit.*, *137*, 257–264.

H. O. Foulkes (1974) Paths in ordered structures of partitions, *Discr. Math.*, *9*, 365–374.

J. Françon (1974) Preuves combinatoires des identités d'Abel, *Discr. Math.*, *8*, 331–343.

R. D. Fray and D. P. Roselle (1971) Weighted lattice paths, *Pacific J. Math.*, *37*, 85–96.

I. Gessel (1977) Counting permutations by descents, inversions, and greater index, *Notices A.M.S.*, *24*, A-36.

J. R. Goldman (1975) An identity for fixed points of permutations, *Aequationes Math.*, *13*, 155–156.

I. J. Good (1960) Contribution to the discussion of S. W. Golomb, "A Mathematical theory of discrete classification" in Information Theory: Fourth London Symposium, Colin Cherry, ed., Butterworths, London, p. 423.

I. J. Good (1971) The factorization of a sum of matrices and the multivariate cumulants of a set of quadratic expressions, *J. Combinatorial Th.*, *11*, 27–37.

H. Gupta (1971) A problem in permutations and Stirling's numbers, *Math. Student*, *39*, 341–345.

D. Jackson and J. Reilly (1976) Permutations with a prescribed number of p-runs, *Ars Combinatoria*, *1*, 297–305.

I. Kaplansky and J. Riordan (1945) Multiple matching and runs by the symbolic method, *Ann. Math. Stat.*, *16*, 272–277.

D. Knuth (1970) A note on solid partitions, *Math. Comp.*, *24*, 955–961.

D. Knuth (1973) The Art of Computer Programming, vol. 3, Searching and Sorting, Addison-Wesley, Reading.

G. Kreweras (1965) Sur une classe de problèmes de dènombrement liés au treillis des partitions des entiers, Cahiers du B.U.R.O., no. 6, Paris.

G. Kreweras (1966) Dénombrements de chemins minimaux à sauts imposès, *C.R. Acad. Sc. Paris*, *263*, 1–3.

G. Kreweras (1967) Traitement simultané du "Problème de Young" et du "Problème de Simon Newcomb," Cahiers du B.U.R.O., no. 10, Paris.

G. Kreweras (1969a) Inversion des polynômes de Bell bidimensionnels et application au dénombrement des relations binaires connexes, *C.R. Acad. Sc. Paris*, *268*, 577–579.

G. Kreweras (1969b) Dénombrement systematiques de relations binaires externes, *Math. et Sc. Humaines*, *7*, 5–15.

G. Kreweras (1970) Sur les éventails de segments, Cahiers du B.U.R.O., no. 15, Paris.

G. Kreweras (1972) Classification des permutations suivant certaines propriétés ordinales de leur representation plane, Actes du colloque sur les permutations, Juillet 1972, Gauthier-Villars, Montreuil, 97–115.

E. Netto (1901) Lehrbuch der Kombinatorik, Reprinted: Chelsea, New York, 1964.

P. Nivat (1972a) Sorting of permutations, to appear in Actes du colloque sur les permutations, Gauthiers-Villars, Montreuil

P. Nivat (1972b) Calcul du nombre moyen de comparaisons pour la methode d'insertion dichotomique, No. 1040, Laboratoire de Statistique, Université Paris.

J. K. Percus (1969) One more technique for the dimer problem, *J. Math. Physics*, *10*, 1881–1884.

J. K. Percus (1971) Combinatorial Methods, Springer, New York.

M. A. Rashid (1974) The number of permutations on symbols which contain at least one cycle of length i ≧ 1, *Punjab Univ. J. Math. (Lahore)*, *7*, 35–37.

J. Riordan (1958) An Introduction to Combinatorial Analysis, J. Wiley, New York.

J. Riordan (1968a) Combinatorial Identities, J. Wiley, New York.

J. Riordan (1968b) A note on a q-extension of ballot numbers, *J. Combinatorial Th.*, *4*, 191–192.

J. Riordan and C. E. Shannon (1942) The number of two-terminal series-parallel networks, *J. Math. and Phys.*, *21*, 83–93.

D. P. Roselle (1969) Permutations by number of rises and successions, *Proc. A.M.S.*, *19*, 8–16.

R. P. Stanley (1972) Ordered structures and partitions, Memoirs of the A.M.S., No. 119.

R. P. Stanley (1976) Binomial posets, Mobius inversion, and permutation enumeration, *J. Combinatorial Th.*, *A-20*, 336–356.

S. M. Tanny (1976) Permutations and successions, *J. Combinatorial Th.*, *A-21*, 196–202.

H. G. Williams (1971) An extension of Simon Newcomb's problem, Ph.D. Thesis, Duke University.

4.4 Summaries of the Papers

[41] Applications of a theory of permutations in circular procession to the theory of numbers, *Proc. London Math. Soc.*, *23* (1892), 305–313.

MacMahon states a general theorem that relates linear permutations to circular permutations (attributed to Jablonski and Moreau independently). From this result he derives a number of nice results about the Euler quotient function. For example:

Theorem. The number of circular permutations of p different objects taken n at a time, allowing repetitions, is

$$\frac{1}{n}\sum_{d/n}\phi(d)\,p^{n/d}.$$

This result has often been attributed to Burnside who published it at a later date. MacMahon also derives the generalization of Fermat's "little theorem" due to Euler:

$$a^{\phi(n)} \equiv 1 \pmod{n}, \text{ if } (a, n) = 1.$$

The paper concludes with another generalization of Fermat's little theorem.

[48] Self-conjugate permutations, *Messenger of Math.*, *24* (1895), 69–76.

A permutation σ of 1, 2, 3, . . . , n is called self-conjugate if σ^2 is the identity permutation, i.e., $\sigma(i) = j$ if and only if $\sigma(j) = i$ for $1 \leqq i \leqq n$. Let U_n denote the number of self-conjugate permutations of 1, 2, . . . , n. Then MacMahon notes H. A. Rothe's result:

$$U_1 = 1, \qquad U_2 = 2, \qquad U_n = U_{n-1} + (n-1)\,U_{n-2}.$$

MacMahon considers generalized self-conjugate permutations as follows: Consider $x_1^{\xi_1} x_2^{\xi_2} \ldots x_n^{\xi_n}$, a product of $\sum \xi_i$ factors. In any product of $\sum \xi_i$ terms we shall say that the first ξ_1 terms make up the first compartment, the next ξ_2 terms the second compartment, etc. With these conventions, we say that a permutation of the product $x_1^{\xi_1} x_2^{\xi_2} \ldots x_n^{\xi_n}$ is self-conjugate, whenever x_s occurs j times in the tth compartment, if and only if x_t occurs j times in the sth compartment.

MacMahon applies his Master Theorem to obtain concise expressions for the related generating functions. For example, when $n = 3$, the coefficient of $x_1^{\xi_1} x_2^{\xi_2} x_3^{\xi_3}$ in

$$(1 - x_1 - x_2 - x_3)^{-1/2} (1 - x_1 - x_2 - x_3 + 4x_1x_2x_3)^{-1/2}$$

is the number of self-conjugate permutations of $x_1^{\xi_1} x_2^{\xi_2} x_3^{\xi_3}$.

[53] Solution du problème de partition d'où résulte le dénombrement des genres distincts d'abaque relatifs aux équations à n variables, *Bull. Soc. Math. France*, *26* (1898), 57–64.

MacMahon treats the following partition problem due to d'Ocagne: In how many ways can n be written as the sum of 8 nonnegative numbers

$$\begin{matrix} n_1 & n_2 & n_3 & n_4 \\ n_1' & n_2' & n_3' & n_4', \end{matrix}$$

where two solutions are considered identical if one can be transformed into another by permutations of rows and columns? MacMahon directly reduces the problem to one of partitioning bipartite numbers. In particular, he shows that the number of such partitions of n (for n odd) with s columns is the coefficient of $a^s x^n$ in

$$\frac{1}{2} \prod_{j=0}^{\infty} (1 - ax^j)^{-j-1}.$$

The case n even provides a few more difficulties but is handled similarly. The paper concludes with a consideration of the same problem in which there are limitations on the size of parts.

[81] The problem of 'derangement' in the theory of permutations, *Trans. Cambridge Phil. Soc.*, *21* (1912), 467–481.

In this paper, MacMahon obtains further results for the generalization of the game "Rencontre" that he introduced in [45]. We consider the number $\{m; \xi_1 \xi_2 \ldots \xi_n\}$ of permutations of the multiset (assemblage) $x_1^{\xi_1} x_2^{\xi_2} \ldots x_n^{\xi_n}$ such that exactly m letters are in their original positions. The problem of "Rencontre" is to show that

$$\{0; 1^n\} = n! \sum_{j=0}^{n} (-1)^j / j!,$$

and this is the first case that MacMahon mentions. Various identities are derived for $\{m; \xi_1 \xi_2 \ldots \xi_n\}$ throughout the remainder of the paper and the methods of proof rely either on the Master Theorem or Hammond operators.

[82] The indices of permutations and the derivation therefrom of functions of a single variable associated with the permutations of any assemblage of of objects, *American J. Math.*, *35* (1913), 281–322.

This paper arose from the work on permutations in [77]. In the latter, MacMahon introduces certain generating functions related to lattice permutations in order to solve problems on plane partitions. Here MacMahon considers general permutation questions related to sets of letters with possible repetitions.

In particular, we are to consider permutations of the multiset (assemblage) $\alpha_1^{i_1} \alpha_2^{i_2} \alpha_3^{i_3} \ldots$. A contact in a permutation of these letters, say $\alpha_s \alpha_t$, is called a major contact, an equal contact, or a minor contact, according as $s >, =,$ or $<$ t. Subsequently, major contacts have become known as "falls" and minor contacts as "rises." MacMahon now defines greater, equal, and lesser indices of permutations as follows: If in a given permutation, the p_1st, p_2nd, p_3rd, ... letters are the lefthand members of major (resp. equal, minor) contacts, then the greater (resp. equal, lesser) index of the permutation is given by $p_1 + p_2 + p_3 + \dots$. (In general, q_i are used for equal contacts and r_i for lesser contacts.)

MacMahon obtains an immense number of truly elegant formulas for the generating functions related to these and other indices. For example,

$$\sum x^p = \frac{(1-x)(1-x^2)\dots\dots\dots\dots\dots\dots\dots\dots(1-x^{i+j+k+\dots})}{(1-x)(1-x^2)\dots(1-x^i)(1-x)(1-x^2)\dots(1-x^j)\dots},$$

where the sum is over all permutations of $\alpha^i\beta^j\gamma^k\dots$ and p denotes the greater index. We should point out that further surprising relationships of this type are found in [86].

[83] The superior and inferior indices of permutations, *Trans. Cambridge Phil. Soc.*, *22* (1914), 55–60.

This little paper is really a footnote to [82]. MacMahon begins by establishing that the greater index of a permutation (see [82] for definitions) equals the superior index (see [86] wherein the superior index is called the inversion number). As a biproduct of this investigation, MacMahon obtains a purely combinatorial proof of the q-analog of the Vandermonde convolution. In MacMahon's notation the result reads

$$F_x(i,j) = \sum_{l=0}^{\sigma} x^{(\sigma-l)(j-l)} F_x(\sigma-l,l)\,F_x(i-\sigma+l,j-l),$$

where the $F_x(i,j)$ are the Gaussian polynomials defined by

$$F_x(i,j) = \prod_{m=1}^{j} \frac{(1-x^{i+m})}{(1-x^m)}.$$

MacMahon concludes the paper by refining his definition of the superior index. For a permutation of the assemblage $\alpha_1^{i_1}\alpha_2^{i_2}\dots\alpha_s^{i_s}$, we denote by $p'_{uv,\sigma}$ the number of letters α_v which precede exactly σ letters α_u, and we write

$$p'_{vu} = p'_{vu,1} + 2p'_{vu,2} + 3p'_{vu,3} + \dots + i_u p'_{vu,i_u}.$$

Formulas for refined indices of this nature are obtained.

[86] Two applications of general theorems in combinatory analysis: (1) to the theory of inversion of permutations; (2) to the ascertainment of the numbers of terms in the development of a determinant which has amongst its elements an arbitrary number of zeros, *Proc. London Math. Soc.(2), 15* (1916), 314–321.

(1) MacMahon has previously discussed the greater index p of a permutation (see [82] for definitions). He has shown (see [82]) that, relative to the assemblage $1^i2^j3^k \ldots$, the number of permutations with greater index p is the coefficient of x^p in the expansion of

$$(*) \qquad \frac{(x)_{i+j+k+\ldots}}{(x)_i(x)_j(x)_k\ldots},$$

where $(x)_i = (1-x)(1-x^2)(1-x^3)\ldots(1-x^i)$.

MacMahon now defines the inversion number v of a permutation. For example, if we consider the permutation 4121533 of the multiset (assemblage) $1^2 2 3^2 4 5$, then we see that 4 precedes five smaller numbers, 1 none, 2 one, 1 none, 5 two, 3 none, and 3 none; the inversion number of this permutation is therefore $5+0+1+0+2+0+0=8$.

Theorem. The number of permutations of $1^i2^j3^k \ldots$ with greater index p equals the number of permutations with inversion number p.

The proof relies on the establishment of $(*)$ as the generating function for inversion numbers.

(2) MacMahon quotes the following observation of Laisant: The number of nonzero terms in the expansion of $\det(a_{ij})$ $1 \leqq i, j \leqq n$, is clearly the coefficient of $x_1x_2\ldots x_n$ in

$$\prod_{i=1}^{n} (\varepsilon_{i1}x_1 + \varepsilon_{i2}x_2 + \ldots + \varepsilon_{in}x_n),$$

where $\varepsilon_{ij} = 0$ if $a_{ij} = 0$, and $\varepsilon_{ij} = 1$ otherwise.
MacMahon notes that his Master Theorem (see [45]) implies that this number is also the coefficient of $x_1x_2\ldots x_n$ in

$$[\det(\delta_{ij} - \varepsilon_{ij}x_j)]^{-1},$$

where $\delta_{ij} = 1$ if $i = j$, and $\delta_{ij} = 0$ otherwise.

[95] Permutations, lattice permutations, and the hypergeometric series, *Proc. London Math. Soc.(2)*, *19* (1920), 216–227.

An assemblage of letters

$$x_1^{m_1+m_2+\ldots+m_n} x_2^{m_2+m_3+\ldots+m_n} \ldots x_{n-1}^{m_{n-1}+m_n} x_n^{m_n},$$

in which $m_i \geqq 0$, is termed a lattice assemblage with respect to the order $x_1 \geqq x_2 \geqq \ldots \geqq x_n$. A permutation of this lattice assemblage is said to be a lattice permutation if on drawing a line between any two letters of the permutation, the assemblage to the left of the line is a lattice assemblage. We let

$$P(m_1 + m_2 + \ldots + m_n, m_2 + m_3 + \ldots + m_n, \ldots, m_{n-1} + m_n, m_n)$$

and

$$LP(m_1 + m_2 + \ldots + m_n, m_2 + m_3 + \ldots + m_n, \ldots, m_{n-1} + m_n, m_n)$$

denote the number of permutations and the number of lattice permutations of the assemblage, respectively.

MacMahon deduces a number of results for generating functions related to these permutation functions. For example, if

$$m_{in} = \sum_{j=i}^{n} m_j, \qquad M_i = \sum_{j=1}^{i} jm_j,$$

then

$$\sum_{m_{nn}=0}^{\infty} P(m_{1n}, m_{2n}, \ldots, m_{nn})\, x_1^{m_{1n}} x_2^{m_{2n}} \ldots x_n^{m_{nn}}$$

$$= P(m_{1,n-1}, m_{2,n-1}, \ldots, m_{n-1,n-1})\, x_1^{m_{1,n-1}} x_2^{m_{2,n-1}} \ldots x_{n-1}^{m_{n-1,n-1}}$$

$$\times\ {}_nF_{n-1}\begin{bmatrix}(M_{n-1}+1)/n, (M_{n-1}+2)/n, \ldots, (M_{n-1}+n)/n; n^n x_1 x_2 \ldots x_n \\ m_{n-1,n-1}+1, \ldots, m_{1,n-1}+1\end{bmatrix}$$

where

$${}_aF_b\begin{bmatrix} r_1, r_2, \ldots, r_a; t \\ s_1, s_2, \ldots, s_b\end{bmatrix}$$

$$= \sum_{n\geqq 0} \frac{t^n}{n!}\left\{\prod_{i=j}^{a}\prod_{j=0}^{n-1}(r_i + j)\right\}\left\{\prod_{i=1}^{b}\prod_{j=0}^{n-1}(s_i + j)\right\}^{-1}.$$

The paper concludes with an investigation of that portion of $(1 - x_1 - x_2 - \dots - x_n)^{-1}$ that involves lattice assemblages.

[108] Prime lattice permutations, *Proc. Cambridge Phil. Soc.*, *21* (1923), 193–196.

The assemblage of letters $\alpha_1^{k_1}\alpha_2^{k_2}\dots\alpha_n^{k_n}$ (with respect to the linear ordering $\alpha_1 > \alpha_2 > \dots > \alpha_n$) is said to be a lattice assemblage if $k_1 \geqq k_2 \geqq \dots \geqq k_n$. A permutation of a lattice assemblage is called a lattice permutation if, when a line is drawn between any two letters or after the last letter, the assemblage to the left of such line is a lattice assemblage. A composite lattice permutation is one for which it is possible to draw a line between some two letters such that the assemblages to both the left and the right are lattice assemblages. A prime lattice permutation is one that is not composite.

Let $L(k_1, k_2, \dots, k_n)$ denote the number of lattice permutations of the type described above, and let $P(k_1, k_2, \dots, k_n)$ denote the number of these that are prime. MacMahon interrelates these functions with the following formulas:

$$L(k_1, k_2, \dots, k_n) =$$

$$\sum \frac{(\Sigma\kappa)!}{\kappa_1!\kappa_2!\dots} \{P(k'_1, k'_2, \dots)\}^{\kappa_1} \{P(k''_1, k''_2, \dots)\}^{\kappa_2} \dots ;$$

$$P(k_1, k_2, \dots, k_n) =$$

$$\sum \frac{(-1)^{\Sigma\kappa+1}(\Sigma\kappa)!}{\kappa_1!\kappa_2!\dots} \{L(k'_1, k'_2, \dots)\}^{\kappa_1} \{L(k''_1, k''_2, \dots)\}^{\kappa_2} \dots .$$

The summations are over every partition

$$(k'_1, k'_2, \dots)^{\kappa_1} (k''_1, k''_2, \dots)^{\kappa_2} \dots$$

of the multipartite number $(k_1, k_2, k_3, \dots, k_n)$ satisfying

$$k'_1 \geqq k'_2 \geqq k'_3 \geqq \dots ; k''_1 \geqq k''_2 \geqq k''_3 \geqq \dots ; \dots .$$

Reprints of the Papers

Applications of a Theory of Permutations in Circular Procession to the Theory of Numbers. *By* Major P. A. MacMahon.

Received, and read, Thursday, May 12th, 1892.

1. In the *Comptes Rendus* of the French Academy, 11th April, 1892, there appeared a note by M. E. Jablonshi, presented by M. C. Jordan, on the subject of permutations in circular procession.

Also, in the Appendix to his "Théorie des Nombres," t. 1, M. E. Lucas gives a short investigation, which he attributes to M. Moreau.

Both investigators reach the same result, which, in a slightly different notation, may be stated as follows:—

Let there be n things, of which α are of one kind, β of a second, γ of a third, and so forth,

$$\alpha+\beta+\gamma+\ldots=n;$$

let N be the greatest common factor of the numbers $\alpha, \beta, \gamma, \ldots$, and write

$$N=\frac{\alpha}{\alpha'}=\frac{\beta}{\beta'}=\frac{\gamma}{\gamma'}=\ldots;$$

further, let $$LP(\alpha, \beta, \gamma, \ldots)$$

and $$CP(\alpha, \beta, \gamma, \ldots)$$

denote respectively the numbers of permutations of the n things in linear and in circular procession, so that

$$LP(\alpha, \beta, \gamma, \ldots)=\frac{n!}{\alpha!\,\beta!\,\gamma!\ldots},$$

and $$CP(\alpha, \beta, \gamma \ldots)$$

is the number whose expression is sought.

If d be any divisor of N, including N itself and unity, and $\phi(d)$ the *totient* of d (Sylvester's nomenclature for the numbers of integers prime to and not superior to d), the result is

$$nCP(N\alpha', N\beta', N\gamma', \ldots)=\Sigma\phi(d)\,LP\left(\frac{N}{d}\alpha', \frac{N}{d}\beta', \frac{N}{d}\gamma', \ldots\right),$$

otherwise written

$$nCP(N\alpha', N\beta', N\gamma', \ldots) = \Sigma\phi(d)\frac{\left(\frac{n}{d}\right)!}{\left(\frac{N}{d}\alpha'\right)!\left(\frac{N}{d}\beta'\right)!\left(\frac{N}{d}\gamma'\right)!\ldots}.$$

This theorem involves results of interest and importance in the pure theory of numbers.

If n be prime, N is unity, and

$$nCP(\alpha, \beta, \gamma, \ldots) = \frac{n!}{\alpha!\,\beta!\,\gamma!\ldots},$$

and this is also true whenever, n being composite, $\alpha, \beta, \gamma, \ldots$ possess no common factor greater than unity. Hence

Theorem.—The multinomial coefficient

$$\frac{n!}{\alpha!\,\beta!\,\gamma!\ldots}$$

is divisible by n, provided that the numbers $\alpha, \beta, \gamma, \ldots$ constitute a prime assemblage.

The usual statement and proof in treatises is in reference to the divisibility by n, when n is prime, of every coefficient of the multinomial expansion except those attached to powers of single letters. The theorem above is more general and shows the nature of the quotient.

Next suppose $\quad N = n = \alpha = N\alpha'$;

we find $\quad nCP(n) = \Sigma\phi(d)$,

and manifestly $\quad CP(n) = 1$;

therefore $\quad \Sigma\phi(d) = n$,

where d is any divisor of n, including n itself and unity; the well-known theorem due to Gauss.

The complete theorem may be viewed as a generalization of that of Gauss. The expression

$$\Sigma\phi(d)\frac{\left(\frac{n}{d}\right)!}{\left(\frac{N}{d}\alpha'\right)!\left(\frac{N}{d}\beta'\right)!\left(\frac{N}{d}\gamma'\right)!\ldots}$$

gives, for every partition of n, a linear function of the totients of the divisors of n which is divisible by n. Moreover the quotient is

$$CP\,(N\alpha',\ N\beta',\ N\gamma',\ \ldots).$$

It may be noted that, when N is a prime,

$$nCP\,(N\alpha', N\beta', N\gamma' \ldots) = \frac{n!}{(N\alpha')!\,(N\beta')!\,(N\gamma')!\ldots} + (N-1)\frac{\left(\frac{n}{N}\right)!}{\alpha'!\,\beta'!\,\gamma'!\ldots},$$

and then $(\alpha'\beta'\gamma'\ldots)$ is any partition of $\frac{n}{N}$.

2. The next problem is with reference to the permutations of p different objects, n at a time, when repetitions of objects are permissible.

The number of line permutations may be denoted by $RLP\,(p, n)$, and the number of circular permutations by $RCP\,(p, n)$.

The value of $RLP\,(p, n)$ is known to be p^n.

In a circular permutation there are n objects which may be all similar, or they may be of one, two, &c. ... p different kinds.

A permutation which involves a_1 objects of one kind, a_2 of a second, a_3 of a third, &c., may be said to be of type

$$(a_1 a_2 \ldots a_p),$$

where $$a_1 + a_2 + \ldots + a_p = n,$$

and 1, 2, ... or $p-1$ of the quantities a may be zero.

In order to place in evidence the existence of equalities amongst the quantities a, it is convenient to consider the general type

$$(a_1^{\kappa_1}\, a_2^{\kappa_2}\, a_3^{\kappa_3} \ldots),$$

where $$\kappa_1 + \kappa_2 + \kappa_3 + \ldots = p,$$

and $$\kappa_1 a_1 + \kappa_2 a_2 + \kappa_3 a_3 + \ldots = n.$$

We have to find the number of different combinations of objects which come under a given type.

Were the type $(a_1, a_2, \ldots a_p)$, with the quantities a all different, the number of different combinations would be clearly $p!$.

Hence, for the general type, the number of different combinations of the objects must be

$$\frac{p!}{\kappa_1!\,\kappa_2!\,\kappa_3!\ldots}.$$

X 2

Ex. gr., if a permutation involves $n-2$ objects of one kind, one of a second, and one of a third, the type is

$$(n-2, 1^2, 0^{p-3}),$$

and the number of different combinations of this type is

$$\frac{p!}{1!\,2!\,(p-3)!}.$$

Denote by $CP(a_1^{\kappa_1}, a_2^{\kappa_2}, a_3^{\kappa_3}, \ldots)$ the number of circular permutations of a set of objects of the type

$$(a_1^{\kappa_1} a_2^{\kappa_2} a_3^{\kappa_3} \ldots);$$

then $$RCP(p, n) = \Sigma \frac{p!}{\kappa_1!\,\kappa_2!\,\kappa_3!\ldots} CP(a_1^{\kappa_1} a_2^{\kappa_2} a_3^{\kappa_3} \ldots).$$

But $$nCP(a_1^{\kappa_1} a_2^{\kappa_2} a_3^{\kappa_3} \ldots) = \Sigma\phi(d)\, LP\left\{\left(\frac{a_1}{d}\right)^{\kappa_1}\left(\frac{a_2}{d}\right)^{\kappa_2}\left(\frac{a_3}{d}\right)^{\kappa_3}\ldots\right\},$$

by the previous theorem, d denoting a divisor of the highest common factor of the numbers a.

We can now express $RCP(p, n)$ as a linear function of the expressions LP. We find that the coefficient of $\frac{1}{n}\phi(d)$ is

$$\Sigma \frac{p!}{\kappa_1!\,\kappa_2!\,\kappa_3!\ldots} LP\left\{\left(\frac{a_1}{d}\right)^{\kappa_1}\left(\frac{a_2}{d}\right)^{\kappa_2}\left(\frac{a_3}{d}\right)^{\kappa_3}\ldots\right\},$$

and this series is manifestly the value of

$$RLP\left(p, \frac{n}{d}\right),$$

arrived at by summation in regard to types. Hence

$$RCP(p, n) = \frac{1}{n}\Sigma\phi(d)\, RLP\left(p, \frac{n}{d}\right),$$

or $$RCP(p, n) = \frac{1}{n}\Sigma\phi(d)\, p^{n/d},$$

the summation having reference to every divisor d of the number n.

Theorem.—The number of circular permutations of p different objects n together, repetitions permissible, is equal to

$$\frac{1}{n}\Sigma\phi(d)\, p^{n/d},$$

where $\phi(d)$ represents the totient of d, and the summation is for all divisors d of the number n.

We have here another extension of Gauss's theorem concerning totients. Corresponding to every integer p, we have a linear function of the number $\phi(d)$, d a divisor of n, which is divisible by n, and the theorem shows the nature of the quotient.

Theorem.—If n and p be any positive integers, and d a divisor of n, the sum

$$\Sigma\,\phi(d)\,p^{n/d} \equiv 0 \mod n,$$

and the quotient is equal to the number of circular permutations of p different things n together, repetitions permitted.

Gauss's theorem is given by p equal to unity.

The theorem is also a generalization of Fermat's theorem concerning the divisibility of $p^{n-1}-1$ by n, when n is a prime and p prime to n.

For, since $$\Sigma\,\phi(d)\,p^{n/d} \equiv 0 \mod n,$$

if n be prime, $$p^n + (n-1)\,p \equiv 0 \mod n,$$

or $$p^{n-1} - 1 \equiv 0 \mod n,$$

if p be prime to n, and

$$\frac{1}{n}(p^{n-1}-1) = \frac{1}{p}\,RCP\,(p, n) - 1,$$

showing the connexion between the quotient in Fermat's theorem and the number $RCP\,(p, n)$.

The usual extension of Fermat's theorem asserts the congruence

$$p^{\phi(N)} - 1 \equiv 0 \mod N,$$

when p is prime to N.

To deduce this from the general formula, suppose N to involve the prime m to the power μ, and put

$$N = N'm^{\mu}.$$

Consider the permutations in circular procession of $p^{\phi(N')}$ objects m^{μ} together, repetitions allowed.

By the formula established above

$$m^{\mu}RCP\,\{p^{\phi(N')}, m^{\mu}\}$$

$$= p^{\phi(N')m^{\mu}} + (m-1)\,p^{\phi(N')m^{\mu-1}} + \ldots + (m^{\mu}-m^{\mu-1})\,p^{\phi(N')}$$

$$= p^{\phi(N')m^{\mu-1}}\{p^{\phi(N'm^{\mu})}-1\} + mp^{\phi(N')m^{\mu-2}}\{p^{\phi(N'm^{\mu-1})}-1\} + \ldots + m^{\mu}p^{\phi(N')};$$

or
$$RCP\left\{p^{\phi(N')}, m^{\mu}\right\}$$
$$= p^{\phi(N')} + \frac{p^{\phi(N')}}{m}\left\{p^{\phi(N'm)}-1\right\} + \frac{p^{\phi(N')m}}{m^2}\left\{p^{\phi(N'm^2)}-1\right\}$$
$$+ \ldots + \frac{p^{\phi(N')m^{\mu-1}}}{m^{\mu}}\left\{p^{\phi(N)}-1\right\},$$
where the number of terms on the dexter is $\mu+1$.

Giving μ successive integer values, we establish the congruence
$$p^{\phi(N)}-1 \equiv 0 \mod m^{\mu},$$
and thence
$$p^{\phi(N)}-1 \equiv 0 \mod N.$$
To find the quotient of $p^{\phi(N)}-1$ by m^{μ}, write
$$RCP\left\{p^{\phi(N')}, m^{\mu}\right\} = R_{\mu},$$
$$\frac{1}{m^{\mu}}\left\{p^{\phi(N'm^{\mu})}-1\right\} = \vartheta_{\mu},$$
$$p^{\phi(N')} = P,$$
so that
$$R_{\mu} = P + P\vartheta_1 + P^m\vartheta_2 + \ldots + P^{m^{\mu-1}}\vartheta_{\mu},$$
and hence
$$\vartheta_{\mu} = P^{-m^{\mu-1}}(R_{\mu}-R_{\mu-1}),$$
or
$$\frac{1}{m^{\mu}}\left\{p^{\phi(N)}-1\right\}$$
$$= p^{-\phi(N/m^{\mu})m^{\mu-1}}\left[RCP\left\{p^{\phi(N/m^{\mu})}, m^{\mu}\right\} - RCP\left\{p^{\phi(N/m^{\mu})}, m^{\mu-1}\right\}\right]$$
$$= \vartheta_{\mu}.$$

Let now
$$N = m_1^{\mu_1} m_2^{\mu_2} \ldots m_s^{\mu_s},$$
then
$$\frac{\left\{p^{\phi(N)}-1\right\}^s}{N} = \vartheta_{\mu_1}\vartheta_{\mu_2}\ldots\vartheta_{\mu_s},$$
leading to
$$\frac{p^{\phi(N)}-1}{N} = (\vartheta_{\mu_1}\vartheta_{\mu_2}\ldots\vartheta_{\mu_s})^{1/s} N^{-(s-1)/s},$$
giving the complete quotient.

Ex. gr., $N = 15$, $m_1 = 3$, $m_2 = 5$, $\mu_1 = \mu_2 = 1$, $s = 2$,
$$\vartheta_{\mu_1} = \tfrac{1}{4}RCP\,(4, 5)-1 = \tfrac{1}{20}(4^5+4^2)-1 = 51,$$
$$\vartheta_{\mu_2} = \tfrac{1}{16}RCP\,(16, 3)-1 = \tfrac{1}{48}(16^3+2\,.\,16)-1 = 85;$$

therefore $$(\vartheta_{\mu_1}\vartheta_{\mu_2})^{\frac{1}{2}}N^{-\frac{1}{2}} = \sqrt{51}\,\sqrt{85}\,\frac{1}{\sqrt{15}} = 17,$$

which is right.

3. To extend Fermat's theorem in another direction, expand the right-hand side of the formula

$$nRCP(p, n) = \Sigma\,\phi(d)\,p^{n/d},$$

according to powers of the prime factors of n.

It is convenient to adopt a symbolic notation by which

$$p^{[a]} = p^{a-1},$$

$$p^{[a]}p^{[b]}p^{[c]}\ldots = p^{[abc\ldots]} = p^{abc\ldots-1}.$$

As a simple case, put $n = m_1m_2m_3$, and we have

$$\begin{aligned} &p^{m_1m_2m_3} + (m_1-1)\,p^{m_2m_3} + (m_2-1)\,p^{m_1m_3} + (m_3-1)\,p^{m_1m_2} \\ &+ (m_1-1)(m_2-1)\,p^{m_3} + (m_2-1)(m_3-1)\,p^{m_1} + (m_3-1)(m_1-1)\,p^{m_2} \\ &+ (m_1-1)(m_2-1)(m_3-1)\,p \equiv 0 \bmod m_1m_2m_3; \end{aligned}$$

p being prime to n, a rearrangement gives

$$\begin{aligned} &(p^{[m_1]}-1)(p^{[m_2]}-1)(p^{[m_3]}-1) \\ &+ m_1(p^{[m_2]}-1)(p^{[m_3]}-1) + m_2(p^{[m_3]}-1)(p^{[m_1]}-1) \\ &+ m_3(p^{[m_1]}-1)(p^{[m_2]}-1) \\ &+ m_2m_3(p^{[m_1]}-1) + m_3m_1(p^{[m_2]}-1) + m_1m_2(p^{[m_3]}-1) \\ &+ m_1m_2m_3 \equiv 0 \bmod m_1m_2m_3. \end{aligned}$$

Now, by Fermat's theorem,

$$p^{[m_1]}-1 \equiv 0 \bmod m_1,$$

and, taking n the product of two primes, a formula similar to the above shows the congruence

$$(p^{[m_2]}-1)(p^{[m_3]}-1) \equiv 0 \bmod m_2m_3.$$

Hence the congruence

$$(p^{[m_1]}-1)(p^{[m_2]}-1)(p^{[m_3]}-1) \equiv 0 \bmod m_1m_2m_3,$$

which is equivalent to

$$\begin{aligned} &p^{m_1m_2m_3-1} - p^{m_2m_3-1} - p^{m_3m_1-1} - p^{m_1m_2-1} \\ &\qquad + p^{m_1-1} + p^{m_2-1} + p^{m_3-1} - 1 \equiv 0 \bmod m_1m_2m_3. \end{aligned}$$

To generalize this congruence write

$$\Phi_p(m_1^{\mu_1} m_2^{\mu_2} \dots m_s^{\mu_s}) = p^{[m_1^{\mu_1-1}]}(p^{[m_1]}-1) \dots p^{[m_s^{\mu_s-1}]}(p^{[m_s]}-1),$$

according to the law of the totient formula

$$\phi(m_1^{\mu_1} m_2^{\mu_2} \dots m_s^{\mu_s}) = m_1^{\mu_1-1}(m_1-1)\, m_2^{\mu_2-1}(m_2-1) \dots m_s^{\mu_s-1}(m_s-1).$$

The above-established congruence is then written

$$\Phi_p(m_1 m_2 m_3) \equiv 0 \bmod m_1 m_2 m_3.$$

We can establish the congruence

$$\Phi_p(m_1^{\mu_1} m_2^{\mu_2} \dots m_s^{\mu_s}) \equiv 0 \bmod m_1^{\mu_1} m_2^{\mu_2} \dots m_s^{\mu_s},$$

p being prime to $m_1^{\mu_1} m_2^{\mu_2} \dots m_s^{\mu_s}$.

If $m_1^{a_1} m_2^{a_2} \dots m_s^{a_s}$ and $m_1^{b_1} m_2^{b_2} \dots m_s^{b_s}$ be conjugate divisors of $m_1^{\mu_1} m_2^{\mu_2} \dots m_s^{\mu_s}$, we may write

$$n\Sigma\phi(d)\, p^{n/d} = \Sigma\phi(m_1^{a_1} m_2^{a_2} \dots m_s^{a_s})\, p^{m_1^{b_1} m_2^{b_2} \dots m_s^{b_s}},$$

where

$$\phi(m_1^{a_1} m_2^{a_2} \dots m_s^{a_s}) = (m_1^{a_1} - m_1^{a_1-1})(m_2^{a_2} - m_2^{a_2-1}) \dots (m_s^{a_s} - m_s^{a_s-1});$$

therefore $\Sigma\phi(m_1^{a_1} m_2^{a_2} \dots m_s^{a_s})\, p^{[m_1^{b_1} m_2^{b_2} \dots m_s^{b_s}]} \equiv 0 \bmod n.$

On the left-hand side, expansion of

$$\phi(m_1^{a_1} m_2^{a_2} \dots m_s^{a_s})$$

shows that the whole coefficient of $m_1^{a_1} m_2^{a_2} \dots m_s^{a_s}$ is

$$p^{[m_1^{b_1} m_2^{b_2} \dots m_s^{b_s}]} - p^{[m_1^{b_1-1} m_2^{b_2} \dots m_s^{b_s}]} - p^{[m_1^{b_1} m_2^{b_2-1} \dots m_s^{b_s}]} - \dots$$
$$+ p^{[m_1^{b_1-1} m_2^{b_2-1} \dots m_s^{b_s}]} + p^{[m_1^{b_1} m_2^{b_2-1} m_3^{b_3-1} \dots m_s^{b_s}]} + \dots$$
$$- \dots,$$

which is

$$(p^{[m_1^{b_1}]} - p^{[m_1^{b_1-1}]})(p^{[m_2^{b_2}]} - p^{[m_1^{b_2-1}]}) \dots (p^{[m_s^{b_s}]} - p^{[m_s^{b_s-1}]}),$$

that is, $\Phi_p(m_1^{b_1} m_2^{b_2} \dots m_s^{b_s}).$

Hence $\Sigma m_1^{a_1} m_2^{a_2} \ldots m_s^{a_s} \Phi_p (m_1^{b_1} m_2^{b_2} \ldots m_s^{b_s}) \equiv 0 \text{ mod } n$;

or, more briefly,

$$\Sigma\, d\, \Phi_p \left(\frac{n}{d}\right) \equiv 0 \text{ mod } n,$$

which is simply another form of the congruence

$$\Sigma\, \phi(d)\, p^{n/d} \equiv 0 \text{ mod } n.$$

The congruence $$\Sigma\, d\, \Phi_p \left(\frac{n}{d}\right) \equiv 0 \text{ mod } n$$

may be written $$\Phi_p(n) + \Sigma' d \Phi_p \left(\frac{n}{d}\right) \equiv 0 \text{ mod } n,$$

wherein the summation sign Σ' has reference to every divisor of *excluding* unity.

Assuming the truth of the congruence

$$\Phi_p \left(\frac{n}{d}\right) \equiv 0 \text{ mod } \frac{n}{d},$$

for every divisor d of n which exceeds unity, manifestly

$$\Sigma' d \Phi_p \left(\frac{n}{d}\right) \equiv 0 \text{ mod } n,$$

and hence $$\Phi_p(n) \equiv 0 \text{ mod } n;$$

therefore by induction universally

$$\Phi_p(n) \equiv 0 \text{ mod } n;$$

or, otherwise, $\Phi_p (m_1^{\mu_1} m_2^{\mu_2} \ldots m_s^{\mu_s}) \equiv 0 \text{ mod } m_1^{\mu_1} m_2^{\mu_2} \ldots m_s^{\mu_s}$;

and, since $\Phi_p (m_1^{\mu_1} m_2^{\mu_2} \ldots m_s^{\mu_s})$ contains a factor $p^{N/(m_1 m_2 \ldots m_s) - 1}$, we have finally

$$p^{1 - N/(m_1 m_2 \ldots m_s)}\, \Phi_p (m_1^{\mu_1} m_2^{\mu_2} \ldots m_s^{\mu_s}) \equiv 0 \text{ mod } m_1^{\mu_1} m_2^{\mu_2} \ldots m_s^{\mu_s}.$$

Ex. gr., Take $$n = m_1^3 m_2,$$

$$p^{1-m_1^2} (p^{[m_1^3]} - p^{[m_1^2]}) (p^{[m_2]} - 1) \equiv 0 \text{ mod } m_1^3 m_2,$$

or $$p^{1-m_1^2} (p^{m_1^3 m_2 - 1} - p^{m_1^3 - 1} - p^{m_1^2 m_2 - 1} + p^{m_1^2 - 1}) \equiv 0 \text{ mod } m_1^3 m_2,$$

or $$p^{m_1^3 m_2 - m_1^2} - p^{m_1^3 - m_1^2} - p^{m_1^2 m_2 - m_1^2} + 1 \equiv 0 \text{ mod } m_1^3 m_2.$$

SELF-CONJUGATE PERMUTATIONS.

By *Major P. A. MacMahon, R.A., F.R.S.*

CONJUGATE Permutations (verwandte permutationen) were apparently first studied by H. A. Rothe,* and a reference may be made to "The Theory of Determinants in historical order of its development," part I., by Thomas Muir, M.A., LL.D., F.R.S.E., p. 59. The definition is:—

Two permutations of the numbers 1, 2, 3 ..., n are called conjugate when each number and the number of the place which it occupies in the one permutation are interchanged in the case of the other permutation.

Rothe obtained the result

$$U_n = U_{n-1} + (n-1)\ U_{n-2},$$

* See also Théorie des Nombres' par Edward Lucas p. 215 et seq.

where U_n is the number of self-conjugate permutations of the first n integers, and

$$U_1 = 1,\ U_2 = 2.$$

The above definition may be extended as follows.

Symbols composing the product

$$x_1^{\xi_1} x_2^{\xi_2} x_3^{\xi_3} \ldots x_n^{\xi_n}$$

occupy $\Sigma\xi$ places and n compartments, where the first ξ_1 places form the first compartment

next ξ_2	"	second	"
&c.	"	&c.	"
last ξ_n	"	n^{th}	"

and two permutations of the symbols are called conjugate when corresponding to a symbol x_s in compartment t, in the one permutation, we have a symbol x_t, in compartment s, in the other permutation.

We may consider the problem of the enumeration of the self-conjugate permutations.

Take the algebraic product

$$\begin{aligned}
&(x_1 + \alpha_{12}x_2 + \alpha_{13}x_3 + \ldots + \alpha_{1n}x_n)^{\xi_1}\\
&\times\left(\frac{x_1}{\alpha_{12}} + x_2 + \alpha_{23}x_3 + \ldots + \alpha_{2n}x_n\right)^{\xi_2}\\
&\times\left(\frac{x_1}{\alpha_{13}} + \frac{x_2}{\alpha_{23}} + x_3 + \ldots + \alpha_{3n}x_n\right)^{\xi_3}\\
&\ldots\ldots\ldots\ldots\ldots\ldots\ldots\ldots\ldots\ldots\\
&\times\left(\frac{x_1}{\alpha_{1n}} + \frac{x_2}{\alpha_{2n}} + \frac{x_3}{\alpha_{3n}} + \ldots + x_n\right)^{\xi_n}.
\end{aligned}$$

Performing the multiplication we find that the coefficient of

$$x_1^{\xi_1} x_2^{\xi_2} x_3^{\xi_3} \ldots x_n^{\xi_n}$$

is a certain function of

$$\alpha_{12}, \alpha_{13}, \ldots, \alpha_{23}, \ldots, \alpha_{n-1,n},$$

which, on putting each of these quantities equal to unity, becomes

$$\frac{(\Sigma\xi)!}{\xi_1!\,\xi_2!\,\xi_3!\ldots\xi_n!},$$

the complete number of permutations.

That portion of the function, which does not contain α_{st}, corresponds to those permutations in which x_t occurs as often in the compartment s as x_s does in the compartment t; and that portion which is independent of the coefficients, altogether, clearly corresponds to the self-conjugate permutations.

In general the coefficient of

$$\alpha_{12}^{\lambda_{12}}\alpha_{13}^{\lambda_{13}} \ldots \alpha_{23}^{\lambda_{23}} \ldots \alpha_{st}^{\lambda_{st}} \ldots x_1^{\xi_1}x_2^{\xi_2}x_3^{\xi_3} \ldots x_n^{\xi_n}$$

gives the number of permutations in which x_1 occurs λ_{12} times oftener in compartment two than x_2 does in compartment one, and in general x_s occurs λ_{st} times oftener in compartment t than x_t does in compartment s.

The enumeration of the self-conjugate permutations therefore is given by that portion of the coefficient of

$$x_1^{\xi_1}x_2^{\xi_2}x_3^{\xi_3} \ldots x_n^{\xi_n},$$

which is independent of the quantities α. As shewn in the paper "A certain class of generating functions in the Theory of Numbers,' *Phil. Trans.*, R.S. of London, Vol. CLXXXV. (1894) A. § 4, the above-written product may be replaced by another which involves $\binom{n-1}{2}$ combinations of the $\binom{n}{2}$ quantities α without affecting the coefficient therein of

$$x_1^{\xi_1}x_2^{\xi_2}x_3^{\xi_3} \ldots x_n^{\xi_n}.$$

A reference to the paper quoted (§ 4) will shew that this product is

$$(x_1 + x_2 + \gamma_{13}x_3 + \gamma_{14}x_4 + \ldots + \gamma_{1n}x_n)^{\xi_1}$$
$$\times\, (x_1 + x_2 + x_3 + \gamma_{24}x_4 + \ldots + \gamma_{2n}x_n)^{\xi_2}$$
$$\times \left(\frac{x_1}{\gamma_{13}} + x_2 + x_3 + x_4 + \ldots + \gamma_{3n}x_n\right)^{\xi_3}$$
$$\ldots\ldots\ldots\ldots\ldots\ldots\ldots\ldots\ldots\ldots\ldots\ldots$$
$$\times \left(\frac{x_1}{\gamma_{1,n-1}} + \frac{x_2}{\gamma_{2,n-1}} + \frac{x_3}{\gamma_{3,n-1}} + \frac{x_4}{\gamma_{4,n-1}} + \ldots + x_n\right)^{\xi_{n-1}}$$
$$\times \left(\frac{x_1}{\gamma_{1,n}} + \frac{x_2}{\gamma_{2,n}} + \frac{x_3}{\gamma_{3,n}} + \frac{x_4}{\gamma_{4,n}} + \ldots + x_n\right)^{\xi_n}$$

wherein $$\gamma_{xy} = \frac{\alpha_{xy}}{\alpha_{x,x+1}\alpha_{x+1,x+2} \ldots \alpha_{y-1,y}}.$$

This shews that self-conjugate permutations are subject to really only $\binom{n-1}{2}$ conditions. The original definition implies

$$\binom{n}{2} - \binom{n-1}{2} = n - 1$$

redundant conditions, the exact definition being:—
A permutation of the symbols in the product

$$x_1^{\xi_1} x_2^{\xi_2} x_3^{\xi_3} \dots x_n^{\xi_n}$$

is termed self-conjugate when, corresponding to a symbol x_s in compartment t, there is a symbol x_t in compartment s where $t \sim s > 1$.

The coefficient of $x_1^{\xi_1} x_2^{\xi_2} x_3^{\xi_3} \dots x_n^{\xi_n}$ in the product depends, as shewn in the paper quoted, on the co-axial minors of the determinant

$$\begin{vmatrix} 1, & 1, & \gamma_{13}, & \gamma_{14}, & \dots, & \gamma_{1,n-2}, & \gamma_{1,n-1}, & \gamma_{1n} \\ 1, & 1, & 1, & \gamma_{24}, & \dots, & \gamma_{2,n-2}, & \gamma_{2,n-1}, & \gamma_{2n} \\ \frac{1}{\gamma_{13}}, & 1, & 1, & 1, & \dots, & \gamma_{3,n-2}, & \gamma_{3,n-1}, & \gamma_{3,n} \\ \frac{1}{\gamma_{14}}, & \frac{1}{\gamma_{24}}, & 1, & 1, & \dots, & \gamma_{4,n-2}, & \gamma_{4,n-1}, & \gamma_{4,n} \\ \dots & & & & & & & \\ \frac{1}{\gamma_{1,n-2}}, & \frac{1}{\gamma_{2,n-2}}, & \frac{1}{\gamma_{3,n-2}}, & \frac{1}{\gamma_{4,n-2}}, & \dots, & 1, & 1, & \gamma_{n-2,n} \\ \frac{1}{\gamma_{1,n-1}}, & \frac{1}{\gamma_{2,n-1}}, & \frac{1}{\gamma_{3,n-1}}, & \frac{1}{\gamma_{4,n-1}}, & \dots, & 1, & 1, & 1 \\ \frac{1}{\gamma_{1,n}}, & \frac{1}{\gamma_{2,n}}, & \frac{1}{\gamma_{3,n}}, & \frac{1}{\gamma_{4,n}}, & \dots, & \frac{1}{\gamma_{n-2,n}}, & 1, & 1 \end{vmatrix}$$

I recall that it was shewn that the coefficient of

$$x_1^{\xi_1} x_2^{\xi_2} \dots x_n^{\xi_n}$$

in the product

$$X_1^{\xi_1} X_2^{\xi_2} \dots X_n^{\xi_n},$$

where in the matricular notation

$$(X_1, X_2, \dots, X_n) = \left(\begin{array}{cccc} a_{11}, & a_{12}, & \dots, & a_{1n} \\ a_{21}, & a_{22}, & \dots, & a_{2n} \\ \dots & & & \\ a_{n1}, & a_{n2}, & \dots, & a_{nn} \end{array} \right)(x_1, x_2, \dots, x_n)$$

is equal to the coefficient of the same term in the expansion of V_n^{-1},

where

$$V_n = 1 - \Sigma \mid a_{ss} \mid x_s + \Sigma \mid a_{ss} a_{tt} \mid x_s x_t - \Sigma \mid a_{ss} a_{tt} a_{uu} \mid x_s x_t x_u$$
$$+ \ldots + (-)^n \mid a_{11} a_{22} \ldots a_{nn} \mid x_1 x_2 \ldots x_n$$

the co-axial minors being in the usual notation.

To apply this result to the particular case which has presented itself we have to calculate the co-axial minors of the determinant in the quantities γ.

We have at once

$$\mid a_{ss} \mid = 1,$$

$$\mid a_{ss} a_{tt} \mid = 0;$$

but thereafter the minors are increasingly complicated in form.

To commence with the trivial case of two quantities x_1 and x_2, we find the reduced generating function to be

$$\frac{1}{1 - x_1 - x_2},$$

which does not involve any quantities γ, an indication that every permutation of the symbols in

$$x_1^{\xi_1} x_2^{\xi_2}$$

is self-conjugate.

In the case of three quantities

$$x_1, x_2, x_3,$$

observing that the determinant

$$\begin{vmatrix} 1 & , 1, & \gamma_{13} \\ 1 & , 1, & 1 \\ \frac{1}{\gamma_{13}} & , 1, & 1 \end{vmatrix}$$

has the value

$$\frac{(\gamma_{13} - 1)^2}{\gamma_{13}},$$

the reduced generating function is

$$\frac{1}{1 - x_1 - x_2 - x_3 - \frac{(\gamma_{13} - 1)^2}{\gamma_{13}} x_1 x_2 x} .$$

We have to determine the portion of the development of this fraction which is independent of γ_{13}. For this purpose we will expand it in the form

$$U_0 + \sum_1^\infty U_n (\gamma_{13}{}^n + \gamma_{13}{}^{-n}),$$

and then U_0 will be the generating function of the self-conjugate permutations of the symbols in

$$x_1^{\xi_1} x_2^{\xi_2} x_3^{\xi_3},$$

while U_n will enumerate those in which x_3 occurs n times oftener or less often in the first ξ_1 places than x_1 does in the last ξ_3 places.

Observe that the full statement of the problem needs no reference to the second compartment of ξ_2 places.

Putting $x_1 + x_2 + x_3 = b$, $x_1 x_2 x_3 = d$, the reduced generating function may be written

$$\frac{1}{1 - b - d(\gamma_{13}{}^{\frac{1}{2}} - \gamma_{13}{}^{-\frac{1}{2}})^2}$$

$$= -\frac{1}{d} \frac{\gamma_{13}}{\gamma_{13}{}^2 - \dfrac{1 - b + 2d}{d} \gamma_{13} + 1}.$$

If this be regarded as a function of a complex variable γ_{13}, it may be described as a rational algebraical meromorphic function with one zero and two poles in the finite part of the plane.

The poles are

$$c_1 = \frac{1}{2d} \{1 - b + 2d + (1 - b)^{\frac{1}{2}} (1 - b + 4d)^{\frac{1}{2}}\},$$

$$c_2 = \frac{1}{2d} \{1 - b + 2d - (1 - b)^{\frac{1}{2}} (1 - b + 4d)^{\frac{1}{2}}\};$$

and observe that we may also write

$$c_1 = \left\{\frac{(1 - b + 4d)^{\frac{1}{2}} + (1 - b)^{\frac{1}{2}}}{2d^{\frac{1}{2}}}\right\}^2,$$

$$c_2 = \left\{\frac{(1 - b + 4d)^{\frac{1}{2}} - (1 - b)^{\frac{1}{2}}}{2d^{\frac{1}{2}}}\right\}^2,$$

and that

$$c_1 - c_2 = \frac{1}{d}(1-b)^{\frac{1}{2}}(1-b+4d)^{\frac{1}{2}},$$

$$c_1 c_2 = 1.$$

The reduced generating function is therefore written

$$-\frac{1}{d}\frac{\gamma_{13}}{(\gamma_{13}-c_1)\left(\gamma_{13}-\frac{1}{c_1}\right)},$$

and expressing it by partial fractions

$$-\frac{1}{d}\frac{c_1}{c_1-\frac{1}{c_1}}\frac{1}{\gamma_{13}-c_1}+\frac{1}{d}\frac{\frac{1}{c_1}}{c_1-\frac{1}{c_1}}\frac{1}{\gamma_{13}-\frac{1}{c_1}},$$

which is

$$\frac{1}{d}\frac{1}{c_1-\frac{1}{c_1}}\left\{1+\frac{\frac{\gamma_{13}}{c_1}}{1-\frac{\gamma_{13}}{c_1}}+\frac{\frac{1}{\gamma_{13}c_1}}{1-\frac{1}{\gamma_{13}c_1}}\right\},$$

or $$\frac{1}{d}\frac{1}{c_1-\frac{1}{c_1}}+\frac{1}{d}\frac{1}{c_1-\frac{1}{c_1}}\sum_1^\infty\frac{1}{c_1^n}(\gamma_{13}^n+\gamma_{13}^{-n}).$$

This is obviously the expansion required and gives us

$$U_0=\frac{1}{d}\frac{1}{c_1-\frac{1}{c_1}}=(1-b)^{-\frac{1}{2}}(1-b+4d)^{-\frac{1}{2}},$$

$$U_n=\frac{1}{d}\frac{1}{c_1-\frac{1}{c_1}}\frac{1}{c_1^n}$$

$$=(1-b)^{-\frac{1}{2}}(1-b+4d)^{-\frac{1}{2}}\left\{\frac{(1-b+4d)^{\frac{1}{2}}-(1-b)^{\frac{1}{2}}}{2d^{\frac{1}{2}}}\right\}^{2n}.$$

Written at length, we finally obtain the generating functions

$$U_0 = (1 - x_1 - x_2 - x_3)^{-\frac{1}{2}} (1 - x_1 - x_2 - x_3 + 4x_1x_2x_3)^{-\frac{1}{2}},$$

$$U_n = (1 - x_1 - x_2 - x_3)^{-\frac{1}{2}} (1 - x_1 - x_2 - x_3 + 4x_1x_2x_3)^{-\frac{1}{2}}$$
$$\times \left\{ \frac{(1 - x_1 - x_2 - x_3 + 4x_1x_2x_3)^{\frac{1}{2}} - (1 - x_1 - x_2 - x_3)^{\frac{1}{2}}}{2x_1^{\frac{1}{2}}x_2^{\frac{1}{2}}x_3^{\frac{1}{2}}} \right\}^{2n},$$

the enumeration being given by the coefficient of

$$x_1^{\xi_1} x_2^{\xi_2} x_3^{\xi_3}$$

for all values of ξ_1, ξ_2, ξ_3.

These generating functions are very remarkable.

U_0 it is seen is the square root of the discriminant of the denominator of the reduced generating function.

U_0 is also obtained by evaluating a definite integral; for in

$$\frac{1}{1 - b + 2d - \left(a + \frac{1}{a}\right) d},$$

putting $a = \exp. i\theta$, we have by Lament's theorem

$$U_0 = \frac{1}{2\pi} \int_0^{2\pi} \frac{d\theta}{1 - b + 2d - 2d \cos \theta}$$
$$= \{(1 - b)(1 - b + 4d)\}^{-\frac{1}{2}}.$$

In the case of four quantities it is necessary to evaluate a definite triple integral, and in the general case a $\left(\frac{n-1}{2}\right)^{\text{ple}}$ integral. At an early stage the integrations involve the introduction of elliptic and higher transcendents, and it does not appear to be worth while to pursue the subject further.

SOLUTION DU PROBLÈME DE PARTITION D'OÙ RÉSULTE LE DÉNOMBREMENT DES GENRES DISTINCTS D'ABAQUE RELATIFS AUX ÉQUATIONS A n VARIABLES;

Par M. le Major P.-A. Mac-Mahon, R. A., F. R. S.

Un problème très intéressant de partition a été proposé par M. Maurice d'Ocagne sous l'énoncé que voici :

« De combien de manières peut-on composer le nombre n comme somme de 8 nombres

$$\begin{matrix} n_1 & n_2 & n_3 & n_4 \\ n'_1 & n'_2 & n'_3 & n'_4 \end{matrix}$$

(pouvant être nuls), deux solutions n'étant considérées comme distinctes que si l'on ne peut passer de l'une à l'autre par permutation soit des deux lignes, soit des quatre colonnes ci-dessus (¹)? »

Il a, en outre, fait la remarque : « Pour $n = 3$ et $n = 4$, j'ai trouvé ce nombre égal à 7 et à 19. Je voudrais une solution générale. »

Le problème posé ne s'applique qu'à deux lignes et à quatre colonnes; mais j'observerai dès le début que le problème relatif à deux lignes et à un nombre arbitraire s de colonnes n'est exactement pas plus difficile.

Prenant pour l'instant $s = 4$, tout arrangement

$$\begin{matrix} n_1 & n_2 & n_3 & n_4 \\ n'_1 & n'_2 & n'_3 & n'_4 \end{matrix}$$

où

$$n_1 + n_2 + n_3 + n_4 = \alpha, \qquad n'_1 + n'_2 + n'_3 + n'_4 = \alpha'$$

et

$$\alpha + \alpha' = n,$$

nous donne la partition bipartitive

$$(\overline{n_1 n'_1} \quad \overline{n_2 n'_2} \quad \overline{n_3 n'_3} \quad \overline{n_4 n'_4}$$

du nombre bipartitif $(\alpha\alpha')$.

Tout arrangement obtenu en partant de là par permutation des colonnes donne la même partition bipartitive de $(\alpha\alpha')$. Si nous prenons en considération les divers arrangements des colonnes, nous avons en correspondance les différentes compositions associées à la partition. Comme les arrangements ne sont pas tenus pour distincts s'ils ne diffèrent que par une permutation des colonnes, nous n'avons évidemment à nous occuper que des partitions.

D'autre part, puisque deux solutions sont regardées comme identiques si l'on peut passer de l'une à l'autre par interéchange des lignes, nous pourrons toujours, dans la suite, considérer α comme supérieur ou égal à α', et nous n'avons à envisager que les partitions de n en deux parties. La partition $(\alpha\alpha')$ peut prendre

(¹) Si l'on se reporte aux pages 21 et 22 de ce Volume, on voit que ce problème équivaut à celui-ci : Combien y a-t-il de genres distincts d'abaques applicables à des équations à n variables? (M. O.)

l'une quelconque des formes

$$(n, 0), \quad (n-1, 1), \quad (n-2, 2), \quad \ldots.$$

Si n est impair, α doit être pris supérieur à α'.

Il nous faut dénombrer les partitions de chacun des nombres bipartitifs

$$(\alpha\alpha') \qquad [\alpha + \alpha' = n, \ \alpha \geqq \alpha'],$$

en s ou un nombre moindre de nombres bipartitifs.

La fonction génératrice peut immédiatement s'écrire. C'est

$$\frac{1}{\begin{array}{c} 1-a \\ (1-ax)(1-ay) \\ (1-ax^2)(1-axy)(1-ay^2) \\ (1-ax^3)(1-ax^2y)(1-axy^2)(1-ay^3) \\ (1-ax^4)(1-ax^3y)(1-ax^2y^2)(1-axy^3)(1-ay^4) \\ \ldots\ldots\ldots\ldots\ldots\ldots\ldots\ldots \end{array}}.$$

Les partitions de $(\alpha\alpha')$ en s ou un nombre moindre de parties sont dénombrées par les coefficients de

$$a^s x^\alpha y^{\alpha'},$$

dans le développement suivant les puissances croissantes de x et y.

Prenant la somme de ces coefficients pour

$$\alpha + \alpha' = n, \qquad \alpha \geqq \alpha',$$

nous obtenons le dénombrement demandé par les termes du problème proposé, si toutefois n est impair.

Si n est impair, α ne peut pas être égal à α' et le coefficient de $a^s x^\alpha y^{\alpha'}$ est la demi-somme des coefficients de $a^s x^\alpha y^{\alpha'}$ et $a^s x^{\alpha'} y^\alpha$ dans chaque cas. Par conséquent, le nombre cherché est la demi-somme des coefficients de tous les termes $a^s x^\alpha y^{\alpha'}$ où $\alpha + \alpha' = n$, et α, α' ne sont sujets à aucune autre condition. Donc, lorsque n est impair, nous pouvons faire $y = x$, ce qui nous donne la fonction génératrice

$$\frac{1}{2}\,\frac{1}{(1-a)(1-ax)^2(1-ax^2)^3(1-ax^3)^4\ldots},$$

et le nombre cherché est donné par le coefficient de $a^s x^n$ dans le développement suivant les puissances croissantes de x.

Dans le cas de n impair, on a ainsi la solution complète du problème.

Quand n est pair, la sommation des coefficients ne se fait pas aussi aisément. Cela provient en partie du fait que le terme en $x^{\frac{1}{2}n} y^{\frac{1}{2}n}$ se présente seul et ne s'associe pas avec un autre terme.

Soit $n = 2\alpha$.

Dans l'arrangement des deux lignes supposons $s = 4$:

1° Nous avons une solution

$$\begin{matrix} n_1 & n_2 & n_3 & n_4 \\ n_1 & n_2 & n_3 & n_4 \end{matrix}$$

où $n_1 + n_2 + n_3 + n_4 = \alpha$, qui est unique et ne peut être doublée par un interéchange des lignes;

2° Nous avons une solution telle que

$$\begin{matrix} n_1 & n_2 & n_3 & 0 \\ n_1 & n_2 & 0 & n_3 \end{matrix}$$

qui possède la propriété de ne pas se modifier par interéchange des deux lignes complété par un interéchange subséquent de colonnes.

En général, si

$$n_1 + n_2 + n_3 + n_4 = n'_1 + n'_2 + n'_3 + n'_4 = \alpha,$$

la solution

$$\begin{matrix} n_1 & n_2 & n_3 & n_4 \\ n'_1 & n'_2 & n'_3 & n'_4 \end{matrix}$$

est unique si

$$\begin{matrix} n'_1 & n'_2 & n'_3 & n'_4 \\ n_1 & n_2 & n_3 & n_4 \end{matrix}$$

peut lui être rendue identique par une permutation de colonnes.

Dans ce cas, la partition bipartitive

$$(\overline{n_1 n'_1} \quad \overline{n_2 n'_2} \quad \overline{n_3 n'_3} \quad \overline{n_4 n'_4})$$

est la même que

$$(\overline{n'_1 n_1} \quad \overline{n'_2 n_2} \quad \overline{n'_3 n_3} \quad \overline{n'_4 n_4}).$$

Il est clair par conséquent que, quand n est pair et égal à 2α, nous n'avons pas à introduire dans le dénombrement toutes les partitions de $(\alpha\alpha)$, mais seulement celles qui restent inaltérées par la substitution

$$\begin{pmatrix} n_1 & n_2 & n_3 & n_4 \\ n'_1 & n'_2 & n'_3 & n'_4 \end{pmatrix},$$

suivant la notation habituelle de la théorie des substitutions, ajoutées à la moitié de celles qui ne restent pas inaltérées.

Si A est le nombre de celles qui restent inaltérées, B le nombre de celles qui sont altérées, nous avons le nombre

$$A + \tfrac{1}{2}B.$$

Par la fonction génératrice déjà construite, nous avons obtenu

$$\tfrac{1}{2}A + \tfrac{1}{2}B.$$

Il nous reste conséquemment à ajouter

$$\tfrac{1}{2}A.$$

Pour dénombrer les formes A, on peut observer que si

$$(\overline{n_1 n'_1} \quad \overline{n_2 n'_2} \quad \overline{n_3 n'_3} \quad \overline{n_4 n'_4})$$

est une des bipartitions en question, on doit pouvoir la décomposer en parties qui soient de la forme $(n_1 n_1)$ ou de la forme $(\overline{n_1 n_2} \ \overline{n_2 n_1})$.

Un terme $x^\gamma y^\gamma$ peut être retenu dans la fonction génératrice pour une puissance quelconque et un terme $x^\gamma y^\delta$ doit être associé à un terme $x^\delta y^\gamma$.

Nous prendrons donc comme fonction génératrice pour les formes A :

$$\frac{1}{\begin{array}{c}(1-a)\\ (1-pax)\left(1-\frac{1}{p}ay\right)\\ (1-qax^2)(1-axy)\left(1-\frac{1}{q}ay^2\right)\\ (1-rax^3)(1-sax^2y)\left(1-\frac{1}{s}axy^2\right)\left(1-\frac{1}{r}ay^3\right)\\ (1-tax^4)(1-uax^3y)(1-ax^2y^2)\left(1-\frac{1}{u}axy^3\right)\left(1-\frac{1}{t}ay^4\right)\\ \ldots\ldots\ldots\ldots\ldots\ldots\end{array}}.$$

dans laquelle nous devons prendre le coefficient de

$$a^s x^\alpha y^\alpha,$$

en excluant tous les termes qui renferment les puissances positives ou négatives de $p, q, r, s, t, u, \ldots$

Cela peut être beaucoup simplifié parce que

$$\frac{1}{(1-jaX)\left(1-\frac{1}{j}aY\right)} = \frac{1}{1-a^2XY} + \text{des termes qui dépendent de } j.$$

Conséquemment, la fonction génératrice devient

$$\frac{1}{\begin{matrix}(1-a)\\(1-a^2xy)\\(1-axy)(1-a^2x^2y^2)\\(1-a^2x^3y^3)^2\\(1-ax^2y^2)(1-a^2x^4y^4)^2\\\ldots\ldots\ldots\ldots\ldots\ldots\end{matrix}},$$

ou, mieux,

$$\frac{1}{\begin{matrix}(1-a)(1-axy)(1-ax^2y^2)(1-ax^3y^3)(1-ax^4y^4)\ldots\\(1-a^2xy)(1-a^2x^2y^2)(1-a^2x^3y^3)^2(1-a^2x^4y^4)^2\ldots\end{matrix}}.$$

Pour déduire de là $\frac{1}{2}$A, nous devons prendre la moitié du coefficient de $a^s x^\alpha y^\alpha$. Posant $y = x$, nous avons la fonction génératrice

$$\frac{1}{2}\cdot\frac{1}{\begin{matrix}(1-a)(1-ax^2)(1-ax^4)(1-ax^6)\ldots(1-ax^{4t-2})(1-ax^{4t})\ldots\\(1-a^2x^2)(1-a^2x^4)(1-a^2x^6)^2\ldots(1-a^2x^{4t-2})^t(1-a^2x^{4t})^t\ldots\end{matrix}},$$

dans laquelle nous devons prendre le coefficient de $a^s x^n$.

Si l'on remarque que le développement de cette fonction ne contient que des puissances paires de x, on peut dire que, *dans tous les cas,* le nombre des partitions demandées pour le nombre n admet la fonction génératrice

$$\frac{1}{2}\cdot\frac{1}{(1-a)(1-ax)^2(1-ax^2)^3\ldots(1-ax^{t-1})^t\ldots}$$
$$+\frac{1}{2}\cdot\frac{1}{\begin{matrix}(1-a)(1-ax^2)(1-ax^4)\ldots(1-ax^{4t-2})(1-ax^{4t})\\(1-a^2x^2)(1-a^2x^4)\ldots(1-a^2x^{4t-2})^t(1-a^2x^{4t})^t\end{matrix}},$$

dans le développement de laquelle on doit prendre le coefficient de $a^s x^n$.

Dans une lettre subséquente, M. d'Ocagne me signalait l'intérêt, en vue des recherches qu'il poursuit, de dénombrer les solutions dans lesquelles les nombres n_i et n'_i ne doivent pas excéder les valeurs 1 et 2 (¹).

Il est facile d'imposer une limite quelconque à la grandeur de ces nombres.

Supposons cette limite d'abord égale à 1.

Dans la fonction génératrice en x et y nous devons rejeter tous les termes qui renferment x et y à des puissances supérieures à 1.

Par conséquent, nous obtenons

$$\frac{1}{2}\cdot\frac{1}{(1-a)(1-ax)(1-ay)(1-axy)} + \frac{1}{2}\cdot\frac{1}{(1-a)(1-axy)(1-a^2xy)}$$

et, en posant $x=y$,

$$\frac{1}{2}\cdot\frac{1}{(1-a)(1-ax)^2(1-ax^2)} + \frac{1}{2}\cdot\frac{1}{(1-a)(1-ax^2)(1-a^2x^2)},$$

pour la fonction demandée.

De même, si la grandeur des nombres n_i et n'_i est limitée à 2, nous obtenons

$$\frac{1}{2}\cdot\frac{1}{(1-a)(1-ax)^2(1-x^2)^3(1-ax^3)^2(1-ax^4)} + \frac{1}{2}\cdot\frac{1}{(1-a)(1-ax^2)(1-ax^4)(1-a^2x^2)(1-a^2x^4)(1-a^2x^6)},$$

et, en général, si la grandeur est limitée à k, nous obtenons faci-

(¹) Cela fait connaître, pour les équations à n variables, le nombre des genres d'abaques constitués au moyen d'éléments à deux cotes ou à une cote au maximum. (M. O.)

lement la fonction génératrice

$$\frac{1}{2}\cdot\frac{1}{(1-a)(1-ax)^2(1-ax^2)^3\ldots(1-ax^k)^{k+1}(1-ax^{k+1})^k\ldots(1-ax^{2k})}$$

$$+\frac{1}{2}\cdot\frac{1}{\begin{array}{c}(1-a)(1-ax^2)(1-ax^4)\ldots(1-ax^{2k})\\ (1-a^2x^2)(1-a^2x^4)(1-a^2x^6)^2(1-a^2x^8)^2\ldots\\ (1-a^2x^{4k-2})(1-a^2x^{4k-4})(1-a^2x^{4k-6})^2(1-a^2x^{4k-8})^2\ldots\end{array}},$$

où, dans le dénominateur de la seconde fraction, la deuxième et la troisième ligne procèdent suivant les indices

$$\begin{array}{cc|cc|cc|c|cc|c|} 1 & 1 & 2 & 2 & 3 & 3 & & \tfrac{1}{2}(k-1) & \tfrac{1}{2}(k-1) & \tfrac{1}{2}(k-1) \\ 1 & 1 & 2 & 2 & 3 & 3 & \ldots & \tfrac{1}{2}(k-1) & \tfrac{1}{2}(k-1) & \end{array}$$

si k est impair; suivant les indices

$$\begin{array}{cc|cc|cc|c|cc|cc|} 1 & 1 & 2 & 2 & 3 & 3 & & \tfrac{1}{2}(k-2) & \tfrac{1}{2}(k-2) & \tfrac{1}{2}k & \tfrac{1}{2}k \\ 1 & 1 & 2 & 2 & 3 & 3 & \ldots & \tfrac{1}{2}(k-2) & \tfrac{1}{2}(k-2) & \tfrac{1}{2}k & \end{array}$$

si k est pair.

J'espère, dans une autre occasion, traiter le problème plus général pour lequel le nombre des lignes est trois ou plus, problème qui ne me semble pas présenter de difficultés insurmontables.

SUR LA PROBABILITÉ DES ÉVÉNEMENTS COMPOSÉS;

Par M. H. Delannoy.

Dans les *Comptes rendus de l'Association française pour l'avancement des Sciences* (Congrès de Carthage, p. 1; 1896), un mathématicien distingué, M. le Rév. T.-C. Simmons, examine la probabilité des événements composés, dans le cas où les événements sont indépendants. Il critique la définition de Moivre (1) et le troisième principe de Laplace (2).

(1) Deux événements sont indépendants quand ils n'influent pas l'un sur l'autre et que l'arrivée de l'un n'avance ni ne retarde l'arrivée de l'autre. Deux événements sont dépendants quand ils se rattachent tellement que la probabilité de l'arrivée de l'un est changée par l'arrivée de l'autre.

(2) Si les événements sont indépendants les uns des autres, la probabilité de leur ensemble est le produit de leurs probabilités particulières.

XVIII. *The Problem of 'Derangement' in the Theory of Permutations.*

By Major P. A. MacMahon, R.A., Sc.D., LL.D., F.R.S.
Honorary Member Cambridge Philosophical Society.

[*Received* May 1, 1912. *Read* May 6, 1912.]

Section I.

1. In a paper entitled 'A Certain Class of Generating Functions in the Theory of Numbers,' *Phil. Trans. Roy. Soc.*, 1894, A. pp. 111—160, I have given a solution of the general problem of 'derangement' in the form of a symmetric function generating function. It was therein established that the number of permutations of the assemblage of letters

$$x_1^{\xi_1} x_2^{\xi_2} \dots x_n^{\xi_n},$$

which are such that exactly m of the letters are in the places they originally occupied, is equal to the coefficient of the term

$$a^m x_1^{\xi_1} x_2^{\xi_2} \dots x_n^{\xi_n},$$

in the development of the product

$$(ax_1 + x_2 + \dots + x_n)^{\xi_1} (x_1 + ax_2 + x_3 + \dots + x_n)^{\xi_2} \dots (x_1 + x_2 + \dots + x_{n-1} + ax_n)^{\xi_n}.$$

This is the redundant generating function.

It was also shewn that the same number is the coefficient of the term $a^m x_1^{\xi_1} x_2^{\xi_2} \dots x_n^{\xi_n}$ or of $a^m (\xi_1 \xi_2 \dots \xi_n)$, in the notation of symmetric functions, in the development of the algebraic fraction

$$\frac{1}{1 - a\Sigma x_1 + (a-1)(a+1)\Sigma x_1 x_2 - (a-1)^2 (a+2) \Sigma x_1 x_2 x_3 + \dots + (-)^n (a-1)^{n-1} (a+n-1) x_1 x_2 \dots x_n}.$$

This does not involve the numbers $\xi_1, \xi_2, \dots \xi_n$ and is the condensed generating function. If the number in question be denoted by

$$\{m;\ \xi_1 \xi_2 \dots \xi_n\}$$

the condensed generating function is clearly

$$\Sigma \{m;\ \xi_1 \xi_2 \dots \xi_n\} (\xi_1 \xi_2 \dots \xi_n) a^m.$$

I propose, in this short paper, to submit the generating function to examination so as to determine the properties of the number

$$\{m;\ \xi_1 \xi_2 \dots \xi_n\}.$$

2. I write $(x - x_1)(x - x_2) \dots (x - x_n) = x^n - p_1 x^{n-1} + p_2 x^{n-2} - \dots,$

so that the generating function is

$$\frac{1}{1 - ap_1 + (a-1)(a+1)p_2 - (a-1)^2(a+2)p_3 + \dots + (-)^n (a-1)^{n-1}(a+n-1)p_n}.$$

When, in the generating functions, we put $a = 0$ the numbers generated are

$$\{0;\ \xi_1 \xi_2 \dots \xi_n\};$$

and the functions become

$$(x_2 + x_3 + \dots + x_n)^{\xi_1} (x_1 + x_3 + \dots + x_n)^{\xi_2} \dots (x_1 + x_2 + \dots + x_{n-1})^{\xi_n},$$

and
$$\frac{1}{1 - p_2 - 2p_3 - \dots - (n-1)p_n}.$$

If $\xi_1 = \xi_2 = \dots = \xi_n = 1$ some properties of the numbers are obtained in a very simple manner from the redundant function

$$(x_2 + x_3 + \dots + x_n)(x_1 + x_3 + \dots + x_n) \dots (x_1 + x_2 + \dots + x_{n-1}),$$

for this may be written

$$(p_1 - x_1)(p_1 - x_2) \dots (p_1 - x_n),$$

or
$$p_1^{n-2} p_2 - p_1^{n-3} p_3 + \dots + (-)^n p_n,$$

and observing that

$$p_1^{n-s} p_s = (1)^{n-s}(1^s) = \dots + \frac{n!}{s!} p_n + \dots,$$

when $(1)^{n-s}(1^s)$ is multiplied out so as to be expressed as a linear function of monomial symmetric functions, the coefficient of $x_1 x_2 \dots x_n$ or of p_n is seen to be

$$\frac{n!}{2!} - \frac{n!}{3!} + \dots + (-)^n \frac{n!}{n!},$$

which may be written

$$n!\left\{1 - \frac{1}{1!} + \frac{1}{2!} - \frac{1}{3!} + \dots + (-)^n \frac{1}{n!}\right\},$$

the well known value of the number

$$\{0;\ 1^n\}$$

which is met with in the 'Problème des Rencontres.'

3. Similarly it is easy to find an expression for

$$\{m;\ 1^n\};$$

for, retaining a in the redundant generating function and putting

$$1 - a = b,$$

it becomes
$$(p_1 - bx_1)(p_1 - bx_2) \dots (p_1 - bx_n);$$

or
$$p_1^n - bp_1^n + b^2 p_1^{n-2} p_2 - \dots + (-)^n b^n p_n.$$

Developing this expression, so as to obtain the coefficient of p_n, we find

$$n!\left\{1 - b + \frac{b^2}{2!} - \dots + (-)^n \frac{b^n}{n!}\right\};$$

and expressing this in terms of a we readily find

$$\{0;\ 1^n\} + \binom{n}{1} a \{0;\ 1^{n-1}\} + \binom{n}{2} a^2 \{0;\ 1^{n-2}\} + \ldots + \binom{n}{n} a^n;$$

and thence
$$\{m;\ 1^n\} = \binom{n}{m} \{0;\ 1^{n-m}\}.$$

It is readily seen *à priori* that this result is correct; for we can select the m letters, which are to remain undisplaced, in $\binom{n}{m}$ ways; and, for each of these selections, the remaining letters can be permuted so as to be all displaced in $\{0;\ 1^{n-m}\}$ ways.

The whole of the permutations can be found by summing $\{m;\ 1^n\}$ from $m=0$ to $m=n$. Hence the well-known formula

$$\{0;\ 1^n\} + \binom{n}{1} \{0;\ 1^{n-1}\} + \binom{n}{2} \{0;\ 1^{n-2}\} + \ldots + \binom{n}{n} = n!$$

The reader is reminded that the early values of $\{0;\ 1^n\}$ are

1, 0, 1, 2, 9, 44, 265, 1854, 14833, ...

From the above results it is easy to shew that

$$\{0;\ 1^n\} = (n-1)\left[\{0;\ 1^{n-1}\} + \{0;\ 1^{n-2}\}\right],$$
$$\{0;\ 1^n\} = n\{0;\ 1^{n-1}\} + (-)^n,$$

both of them well known relations.

4. We will now consider the condensed generating function

$$\frac{1}{1 - p_2 - 2p_3 - 3p_4 - \ldots - (n-1)p_n},$$

wherein $\{0;\ \xi_1\xi_2\ldots\xi_n\}$ is the coefficient of the symmetric function $(\xi_1\xi_2\ldots\xi_n)$.

We may regard $\xi_1, \xi_2, \ldots \xi_n$ as numbers in descending order of magnitude and n as indefinitely great. The numbers $\xi_1, \xi_2, \ldots \xi_n$ may be any integers, zero not excluded, but there are certain symmetric functions that, from *à priori* considerations, must be absent. For clearly

$$\{0;\ \xi_1\xi_2\ldots\xi_n\} = 0,$$

if $\xi_1 > \xi_2 + \xi_3 + \ldots + \xi_n$. For example such functions as (21), (421), ... do not present themselves in the development.

In the first instance we will restrict ourselves to the numbers $\{0;\ 1^s\}$ which we will write P_s for convenience.

Write
$$\frac{1}{1 - p_2 - 2p_3 - 3p_4 - \ldots} = 1 + P_1p_1 + P_2p_2 + \ldots + P_np_n + \text{other terms};$$

or
$$1 = (1 - p_2 - 2p_3 - 3p_4 - \ldots)(1 + P_1p_1 + P_2p_2 + \ldots + P_np_n) + \text{other terms}.$$

On the right-hand side to find the coefficient of (1^n) or p_n the relevant terms are

$$P_np_n - P_{n-2}p_2p_{n-2} - 2P_{n-3}p_3p_{n-3} - \ldots - (n-3)P_2p_{n-2}p_2 - (n-2)P_1p_{n-1}p_1 - (n-1)p_n,$$

62—2

and since, if $n > 0$, the coefficient of p_n must be zero we have

$$P_n = \binom{n}{2} P_{n-2} + 2\binom{n}{3} P_{n-3} + 3\binom{n}{4} P_{n-4} + \ldots + (n-1)\binom{n}{n},$$

or

$$\{0;\ 1^n\} = \binom{n}{2}\{0;\ 1^{n-2}\} + 2\binom{n}{3}\{0;\ 1^{n-3}\} + \ldots + (n-1)\binom{n}{n},$$

a new relation, and the verification for $n=6$ is

$$265 = 1.15.9 + 2.20.2 + 3.15.1 + 4.6.0 + 5.1.1.$$

The law that has been established can be exhibited by putting

$$(n-s-1)P_s = Q_{s,n};$$

so that

$$Q_{n,n} = -P_n \text{ and } Q_{n-1,n} = 0;$$

then

$$0 = Q_{n,n} + \binom{n}{1} Q_{n-1,n} + \binom{n}{2} Q_{n-2,n} + \ldots + \binom{n}{n} Q_{0,n},$$

and now writing symbolically

$$Q_{s,n} = Q^s_{0,n},$$

$$(Q_{0,n} + 1)^n = 0.$$

5. Next consider the expansion of $(1 - p_2 - 2p_3 - 3p_4 - \ldots)^{-1}$ in ascending powers of $(p_2 + 2p_3 + 3p_4 + \ldots)$; we have

$$(p_2 + 2p_3 + 3p_4 + \ldots)^s = \Sigma \frac{s!}{s_2!\, s_3!\, s_4! \ldots} 1^{s_2} . 2^{s_3} . 3^{s_4} \ldots p_2^{s_2} p_3^{s_3} p_4^{s_4} \ldots,$$

where

$$s_2 + s_3 + s_4 + \ldots = s.$$

The coefficient herein of p_n or (1^n) is readily obtained because the coefficient of (1^n) in the development of the product

$$(1^2)^{s_2} (1^3)^{s_3} (1^4)^{s_4} \ldots$$

is, by a well known theorem of symmetry, equal to the coefficient of symmetric function $(2^{s_2} 3^{s_3} 4^{s_4} \ldots)$ in the expansion of (1^n); this, by the multinomial theorem, is

$$\frac{n!}{(2!)^{s_2} (3!)^{s_3} (4!)^{s_4} \ldots}.$$

Hence the portion of the right-hand side that we require is

$$\Sigma \frac{s!\, n!}{s_2!\, s_3!\, s_4! \ldots (2.0!)^{s_2} (3.1!)^{s_3} (4.2!)^{s_4} \ldots} (1^n),$$

the summation being in respect of all values of s_2, s_3, s_4, ... such that

$$2s_2 + 3s_3 + 4s_4 + \ldots = n,$$

$$s_2 + s_3 + s_4 + \ldots = s.$$

Thence the coefficient of (1^n) in the expansion of $(1 - p_2 - 2p_3 - 3p_4 - \ldots)^{-1}$ is

$$\{0;\ 1^n\} = \sum_s \Sigma \frac{s!\, n!}{s_2!\, s_3!\, s_4! \ldots (2.0!)^{s_2} (3.1!)^{s_3} (4.2!)^{s_4} \ldots},$$

where we have the additional summation in regard to all integer values of s.

This solution depends upon the non-unitary partitions of n; for $n=8$ the calculation is

Partition	gives	Number
2^4	,,	2520
$3^2 2$	,,	6720
42^2	,,	3780
4^2	,,	630
53	,,	896
62	,,	280
8	,,	7
	Total	14833 the value of $\{0;\ 1^8\}$.

6. The above method is only appropriate for obtaining results over a limited range of the expanded function. We require theorems of a more general character and the symmetric function operators are competent to produce them. The operators available are

$$d_s = \frac{d}{dp_s} + p_1 \frac{d}{dp_{s+1}} + p_2 \frac{d}{dp_{s+2}} + \dots,$$

$$D_s = \frac{1}{s!}(d_1{}^s).$$

Writing $p_2 + 2p_3 + 3p_4 + \dots = B$ it will be found that for the particular operand $(1-B)^{-1}$ these operators are connected by special relations and that every such relation is of significance in the theory of the generating function. A special object of the following investigation is the discovery of operators which have the effect of leaving the special operand unaltered.

Since $$d_s B = (s-1) + s(p_1 + p_2 + p_3 + \dots) + B, \text{ when } s > 0,$$
we find $$(d_s - d_t) B = (s-t)(1 + p_1 + p_2 + p_3 + \dots), \text{ when } t > 0;$$
whence if u, v be integers also, greater than zero,

$$(u-v)(d_s - d_t) B = (s-t)(d_u - d_v) B;$$

shewing that, for the particular operand B,

$$(u-v)(d_s - d_t) \text{ and } (s-t)(d_u - d_v)$$

are equivalent operators. Moreover since these operators are linear the equivalence obtains for any power of B and for the operand $(1-B)^{-1}$, u, v, s, t being any positive integers, zero excluded. This result has been reached by the elimination of $p_1 + p_2 + p_3 + \dots$ and B; if we only eliminate $p_1 + p_2 + p_3 + \dots$ we find

$$(td_s - sd_t) B = (s-t)(1-B);$$

leading to the important result

$$\frac{td_s - sd_t}{s-t} \frac{1}{1-B} = \frac{1}{1-B};$$

shewing that the operation

$$\frac{td_s - sd_t}{s-t} \text{ or } \frac{1}{s-t}\begin{vmatrix} d_s & d_t \\ s & t \end{vmatrix}$$

leaves the operand $(1-B)^{-1}$ unaltered.

7. The two results that have now been established lead to a large number of relations between the operators which may be applied forthwith to the study of the properties of the coefficients which arise in the development of the generating function $(1-B)^{-1}$. The difficulty lies in selecting the relations so as to best exhibit those properties. Generally, in applying the operator to the expanded form of the generating function, we first express the operators d_s in terms of the operators D_s in order to take advantage of the facility with which the latter operators are performed upon symmetric functions which are denoted by partitions.

It was shewn by Hammond* that the linear operators d_s have the expressions

$$d_1 = D_1,$$
$$d_2 = D_1^2 - 2D_2,$$
$$d_3 = D_1^3 - 3D_2D_1 + 3D_3,$$
$$\ldots\ldots\ldots\ldots\ldots\ldots$$

the law being the same as that which expresses the sums of powers of quantities in terms of their elementary symmetric functions. Moreover Hammond also shewed (*loc. cit.*) that the operators D_s have the effect

$$D_s(\lambda\mu\nu\ldots) = 0,$$

if s is not included among the integers $\lambda, \mu, \nu, \ldots$

$$D_\lambda(\lambda^l\mu\nu\ldots) = (\lambda^{l-1}\mu\nu\ldots),$$
$$D_\lambda^l D_\mu^m D_\nu^n \ldots (\lambda^l\mu^m\nu^n) = 1.$$

We will in the first place consider the equivalence

$$(u-v)(d_s-d_t) \equiv (s-t)(d_u-d_v),$$

in the simple particular case obtained by putting $(u, v, s, t) = (3, 2, 2, 1)$, viz.

$$d_3 \equiv 2d_2 - d_1.$$

Transforming to the operators D_s, we see that the operation

$$(D_1^3 - 3D_2D_1 + 3D_3) - 2(D_1^2 - 2D_2) + D_1$$

reduces the function $(1-B)^{-1}$ to zero; writing the function in the form

$$\Sigma\{0;\ 1^{\pi_1}2^{\pi_2}3^{\pi_3}\ldots\}.(1^{\pi_1}2^{\pi_2}3^{\pi_3}\ldots),$$

and, after operating, equating the coefficient of $(1^{\pi_1}2^{\pi_2}3^{\pi_3}\ldots)$ to zero, we find

$$\{0;\ 1^{\pi_1+3}2^{\pi_2}3^{\pi_3}\ldots\} - 3\{0;\ 1^{\pi_1+1}2^{\pi_2+1}3^{\pi_3}\ldots\} + 3\{0;\ 1^{\pi_1}2^{\pi_2}3^{\pi_3+1}\ldots\}$$
$$-2[\{0;\ 1^{\pi_1+2}2^{\pi_2}3^{\pi_3}\ldots\} - 2\{0;\ 1^{\pi_1}2^{\pi_2+1}3^{\pi_3}\ldots\}] + \{0;\ 1^{\pi_1+1}2^{\pi_2}3^{\pi_3}\ldots\} = 0.$$

We thus obtain a linear relation between certain groups of numbers which are found throughout the whole extent of the expanded generating function; for the numbers $\pi_1, \pi_2, \pi_3, \ldots$ are entirely at our disposal. The way in which the specification of the numbers is connected with the formula which expresses sums of powers in terms of elementary symmetric functions will be noted. In mathematical shorthand we may denote the above relation by

$$(3)\{\ \} - 2(2)\{\ \} + (1)\{\ \} = 0.$$

* *Proc. Lond. Math. Soc.*

As a simple example put $\pi_1 = 1$, $\pi_2 = \pi_3 = \ldots = 0$; then

$$\{0;\ 1^4\} - 3\{0;\ 21^2\} + 3\{0;\ 31\} - 2\{0;\ 1^3\} + 4\{0;\ 21\} + \{0;\ 1^2\} = 0,$$

and this relation, since $\{0;\ 31\} = \{0;\ 21\} = 0$, yields

$$3\{0;\ 21^2\} = \{0;\ 1^4\} - 2\{0;\ 1^3\} + \{0;\ 1^2\} = 9 - 4 + 1 = 6;$$

so that $\{0;\ 21^2\} = 2$, which is obviously correct.

8. In general from the relation

$$(u - v)(d_s - d_t) = (s - t)(d_u - d_v),$$

we proceed to the relation

$$(u - v)[(s)\{\ \} - (t)\{\ \}] = (s - t)[(u)\{\ \} - (v)\{\ \}],$$

a valuable property of the numbers under examination.

9. Next we see that the result

$$\frac{td_s - sd_t}{s - t}\frac{1}{1 - B} = \frac{1}{1 - B}$$

gives rise to the equivalence $td_s - sd_t \equiv s - t$. It will be found that there are several ways of dealing with this.

We will first consider the particular case

$$td_1 - d_t = -(t - 1);$$

putting $t = 2$, we deduce

$$2!\,D_2 = D_1^2 - 2D_1 - 1;$$

putting $t = 3$ and reducing by means of the relation just found there results

$$3!\,D_3 = D_1^3 - 6D_1^2 + 3D_1 + 4,$$

and, thence similarly

$$4!\,D_4 = D_1^4 - 12D_1^3 + 30D_1^2 + 4D_1 - 15,$$

$$5!\,D_5 = D_1^5 - 20D_1^4 + 110D_1^3 - 140D_1^2 - 95D_1 + 56,$$

&c.

and it is clear that we can express D_s in terms of D_1.

To calculate these relations we remark that the algebraic equivalent of the relation $td_1 - d_t = -(t - 1)$ is

$$tp_1 - (t) = -(t - 1),$$

and since p_s corresponds to D_s we have to express p_s in terms of p_1 being given that

$$tp_1 - s_t = -(t - 1),$$

where (t) has been replaced by s_t; thus since

$$2!\,p_2 = s_1^2 - s_2,$$

$$3!\,p_3 = s_1^3 - 3s_2s_1 + 2s_3,$$

$$4!\,p_4 = s_1^4 - 6s_2s_1^2 + 3s_2^2 + 8s_3s_1 - 6s_4,$$

&c.

we find
$$2!\,p_2 = p_1^2 - (2p_1 + 1),$$
$$3!\,p_3 = p_1^3 - 3(2p_1 + 1)\,p_1 + 2(3p_1 + 2),$$
$$4!\,p_4 = p_1^4 - 6(2p_1 + 1)\,p_1^2 + 3(2p_1 + 1)^2 + 8(3p_1 + 2)\,p_1 - 6(4p_1 + 3),$$
&c.

and now we have merely to write D_s for p_s to arrive at the relations under consideration. We can of course write down the general formula for D_s expressed as a function of the elements
$$D_1,\ 2D_1 + 1,\ 3D_1 + 2,\ 4D_1 + 3,\ \ldots.$$

10. Applying these relations to the expanded generating function we first obtain
$$2!\,\{0;\ 1^{\pi_1}2^{\pi_2+1}3^{\pi_3}\ldots\} = \{0;\ 1^{\pi_1+2}2^{\pi_2}3^{\pi_3}\ldots\} - 2\{0;\ 1^{\pi_1+1}2^{\pi_2}3^{\pi_3}\ldots\} - \{0;\ 1^{\pi_1}2^{\pi_2}3^{\pi_3}\ldots\}.$$

A particular case, putting $\{0;\ 1^s\} = P_s$, is
$$2!\,\{0;\ 21^s\} = P_{s+2} - 2P_{s+1} - P_s;$$
a convenient formula for $\{0;\ 21^s\}$ which may be given another form by utilizing known properties of the numbers P_s, viz.:—
$$2\{0;\ 21^s\} = (s^2 + s - 1)\,P_s + (-)^{s-1}(s - 1).$$

Thence we obtain values of $\{0;\ 21^s\}$

for
$$s = 0,\ 1,\ 2,\ 3,\ 4,\ 5,\ 6,\ 7,\ 8,\ \ldots$$
$$\{0;\ 21^s\} = 0,\ 0,\ 2,\ 12,\ 84,\ 640,\ 5430,\ 50988,\ 526568,\ \ldots.$$

We next obtain
$$3!\,\{0;\ 1^{\pi_1}2^{\pi_2}3^{\pi_3+1}\ldots\} = \{0;\ 1^{\pi_1+3}2^{\pi_2}3^{\pi_3}\} - 6\{0;\ 1^{\pi_1+2}2^{\pi_2}3^{\pi_3}\ldots\}$$
$$+ 3\{0;\ 1^{\pi_1+1}2^{\pi_2}3^{\pi_3}\ldots\} + 4\{0;\ 1^{\pi_1}2^{\pi_2}3^{\pi_3}\ldots\};$$

and thence
$$6\{0;\ 31^s\} = P_{s+3} - 6P_{s+2} + 3P_{s+1} + 4P_s.$$

We derive values of $\{0;\ 31^s\}$, viz.:—for
$$s = 0,\ 1,\ 2,\ 3,\ 4,\ 5,\ \ldots$$
$$\{0;\ 31^s\} = 0,\ 0,\ 0,\ 6,\ 72,\ 780,\ \ldots.$$

Similarly
$$24\{0;\ 41^s\} = P_{s+4} - 12P_{s+3} + 30P_{s+2} + 4P_{s+1} - 15P_s,$$
$$120\{0;\ 51^s\} = P_{s+5} - 20P_{s+4} + 110P_{s+3} - 140P_{s+2} - 95P_{s+1} + 56P_s;$$
and generally, in the relation which expresses D_t in terms of powers of D_1, we are at liberty to substitute
$$\{0;\ t1^s\} \text{ for } D_t \text{ and } P_{s+\kappa} \text{ for } D_1^\kappa.$$

Also we may, more generally, substitute
$$\{0;\ 1^{\pi_1}2^{\pi_2}\ldots t^{\pi_t+1}\ldots\} \text{ for } D_t \text{ and } \{0;\ 1^{\pi_1+\kappa}2^{\pi_2}3^{\pi_3}\ldots\} \text{ for } D_1^\kappa.$$

The reader may also proceed to the relation
$$24\{0;\ 41^s\} = \left\{\frac{(s+4)!}{s!} - 12\frac{(s+3)!}{s!} + 30\frac{(s+2)!}{s!} + 4(s+1) - 15\right\}P_s + (-)^{s+1}(s-1)(s^2 - 3s - 1);$$
and will find no difficulty in reaching the general formula.

Again, from the relation

$$2D_2 = D_1^2 - 2D_1 - 1,$$

is derived the formula of reduction

$$\{0;\ 2^r 1^s\} = \tfrac{1}{2}\{0;\ 2^{r-1}1^{s+2}\} - \{0;\ 2^{r-1}1^{s+1}\} - \tfrac{1}{2}\{0;\ 2^{r-1}1^s\};$$

and a multitude of similar results are obtainable.

Section II.

11. The generating function

$$\frac{1}{1 - ap_1 + (a-1)(a+1)p_2 - (a-1)^2(a+2)p_3 + \ldots},$$

which when expanded is

$$\Sigma\{m;\ 1^{\pi_1}2^{\pi_2}3^{\pi_3}\ldots\}\, a^m (1^{\pi_1}2^{\pi_2}3^{\pi_3}\ldots),$$

may be similarly dealt with. For write it $(1-C)^{-1}$ where

$$C = ap_1 - (a-1)(a+1)p_2 + (a-1)^2(a+2)p_3 - \ldots;$$

then

$$d_s C = (-)^{s+1}(a-1)^s(1-C) + (-)^{s+1}s(a-1)^{s-1}E,$$

where

$$E = 1 - (a-1)p_1 + (a-1)^2 p_2 - \ldots.$$

From this relation we obtain

$$\{d_s + (-)^{s-t+1}(a-1)^{s-t}d_t\}\,C = (-)^{s-1}(a-1)^{s-1}(s-t)\,E;$$

and, herein putting $1 - a = b$,

$$(b^{1-s}d_s - b^{1-t}d_t)\,C = (s-t)\,E;$$

shewing that, for an operand C,

$$(u-v)(b^{1-s}d_s - b^{1-t}d_t) \equiv (s-t)(b^{1-u}d_u - b^{1-v}d_v)$$

are equivalent operations. Since the operations are linear the equivalence persists when the operand is $(1-C)^{-1}$.

If, from the original expression of $d_s C$, we eliminate E we find

$$(tb^t d_s - sb^s d_t)\,C = (s-t)\,b^{s+t}(1-C);$$

whence

$$\frac{tb^{-s}d_s - sb^{-t}d_t}{s-t}\cdot\frac{1}{1-C} = \frac{1}{1-C};$$

establishing that the operator

$$\frac{tb^{-s}d_s - sb^{-t}d_t}{s-t}$$

leaves the generating function unchanged. In other words the operator

$$tb^{-s}d_s - sb^{-t}d_t - s + t$$

causes the function to vanish.

In regard to the two operators

$$(u-v)(b^{1-s}d_s - b^{1-t}d_t) - (s-t)(b^{1-u}d_u - b^{1-v}d_v),$$

$$tb^{-s}d_s - sb^{-t}d_t - s + t,$$

which cause the generating function to vanish, it is to be remarked that, regarding b as being of weight unity and d_s as of weight s, the first operator is of weight unity and the second of weight zero. We are thus able to proceed to these operators from those obtained for the case $b = 1$, viz.:—

$$(u-v)(d_s-d_t)-(s-t)(d_u-d_v),$$
$$td_s - sd_t - s + t,$$

by introducing, in each term, such a power of b as will make every term of the *same* weight.

12. From the first operator taking the simple case

$$d_3 - 2bd_2 + b^2 d_1 = 0,$$

we apply it to the expanded form

$$\Sigma\,\{m;\ 1^{\pi_1} 2^{\pi_2} 3^{\pi_3} \ldots\}\, a^m (1^{\pi_1} 2^{\pi_2} 3^{\pi_3} \ldots)$$

of the generating function. We thus obtain

$$\sum_{m,\,\pi} \{m;\ 1^{\pi_1} 2^{\pi_2} 3^{\pi_3}\}\, a^m \{(1^{\pi_1-3} 2^{\pi_2} 3^{\pi_3} \ldots) - 3\,(1^{\pi_1-1} 2^{\pi_2-1} 3^{\pi_3} \ldots) + 3\,(1^{\pi_1} 2^{\pi_2} 3^{\pi_3-1} \ldots)\}$$
$$-2 \sum_{m,\,\pi} \{m;\ 1^{\pi_1} 2^{\pi_2} 3^{\pi_3} \ldots\}\,(a^m - a^{m+1})\,\{(1^{\pi_1-2} 2^{\pi_2} 3^{\pi_3} \ldots) - 2\,(1^{\pi_1} 2^{\pi_2-1} 3^{\pi_3} \ldots)\}$$
$$+ \Sigma\,\{m;\ 1^{\pi_1} 2^{\pi_2} 3^{\pi_3} \ldots\}\,(a^m - 2a^{m+1} + a^{m+2})\,(1^{\pi_1-1} 2^{\pi_2} 3^{\pi_3} \ldots) = 0.$$

Herein selecting the coefficient of $(1^{\pi_1} 2^{\pi_2} 3^{\pi_3} \ldots)$ we have

$$\sum_m [\{m;\ 1^{\pi_1+3} 2^{\pi_2} 3^{\pi_3} \ldots\} - 3\,\{m;\ 1^{\pi_1+1} 2^{\pi_2+1} 3^{\pi_3} \ldots\} + 3\,\{m;\ 1^{\pi_1} 2^{\pi_2} 3^{\pi_3+1} \ldots\}]\, a^m$$
$$-2 \sum_m [\{m;\ 1^{\pi_1+2} 2^{\pi_2} 3^{\pi_3} \ldots\} - 2\,\{m;\ 1^{\pi_1} 2^{\pi_2+1} 3^{\pi_3} \ldots\}]\,(a^m - a^{m+1})$$
$$+ \sum_m \{m;\ 1^{\pi_1+1} 2^{\pi_2} 3^{\pi_3} \ldots\}\,(a^m - 2a^{m+1} + a^{m+2}) = 0,$$

and, herein selecting the coefficient of a^m, we find that

$$\{m;\ 1^{\pi_1+3} 2^{\pi_2} 3^{\pi_3} \ldots\} - 3\,\{m;\ 1^{\pi_1+1} 2^{\pi_2+1} 3^{\pi_3} \ldots\} + 3\,\{m;\ 1^{\pi_1} 2^{\pi_2} 3^{\pi_3+1} \ldots\}$$
$$-2\,[\{m;\ 1^{\pi_1+2} 2^{\pi_2} 3^{\pi_3} \ldots\} - 2\,\{m;\ 1^{\pi_1} 2^{\pi_2+1} 3^{\pi_3} \ldots\}]$$
$$+2\,[\{m-1;\ 1^{\pi_1+2} 2^{\pi_2} 3^{\pi_3} \ldots\} - 2\,\{m-1;\ 1^{\pi_1} 2^{\pi_2-1} 3^{\pi_3} \ldots\}]$$
$$+\{m;\ 1^{\pi_1+1} 2^{\pi_2} 3^{\pi_3} \ldots\} - 2\,\{m-1;\ 1^{\pi_1+1} 2^{\pi_2} 3^{\pi_3} \ldots\} + \{m-2;\ 1^{\pi_1+1} 2^{\pi_2} 3^{\pi_3} \ldots\} = 0,$$

a relation connecting ten of the coefficients.

In applying this formula it must be noticed that

$$\{m;\ 1^{\pi_1} 2^{\pi_2} 3^{\pi_3} \ldots\},$$

denoting as it does the number of permutations in which exactly m of the letters are not displaced, must be zero,

(i) when m is negative,

(ii) when $m > \Sigma s\pi_s$, i.e. greater than the number of letters in the permutation.

Also it is manifest that

$$\{\Sigma s\pi_s;\ 1^{\pi_1} 2^{\pi_2} 3^{\pi_3} \ldots\} = \{\Sigma s\pi_s - 1;\ 1^{\pi_1} 2^{\pi_2} 3^{\pi_3} \ldots\},$$

since if all but one of the letters are undisplaced then all must be so.

Bearing these facts in mind there is no difficulty in verifying the formula in some cases.

13. Generally we may obtain the relation between the coefficients corresponding to the relation

$$(u-v)(b^{1-s}d_s - b^{1-t}d_t) - (s-t)(b^{1-u}d_u - b^{1-v}d_v) = 0,$$

for writing this

$$(u-v)\,d_s - (u-v)\,b^{s-t}\,d_t - (s-t)\,b^{s-u}\,d_u + (s-t)\,b^{s-v}d_v = 0,$$

where s, t, u, v are in descending order of magnitude, and denoting the expressions

$$\{m;\ 1^{\pi_1+1}2^{\pi_2}3^{\pi_3}\ldots\},$$
$$\{m;\ 1^{\pi_1+2}2^{\pi_2}3^{\pi_3}\ldots\} - 2\{m;\ 1^{\pi_1}2^{\pi_2+1}3^{\pi_3}\ldots\},$$
$$\{m;\ 1^{\pi_1+3}2^{\pi_2}3^{\pi_3}\ldots\} - 3\{m;\ 1^{\pi_1+1}2^{\pi_2+1}3^{\pi_3}\ldots\} + 3\{m;\ 1^{\pi_1}2^{\pi_2}3^{\pi_3+1}\ldots\},$$

by $(1)\{\,\}_m$, $(2)\{\,\}_m$, $(3)\{\,\}_m$, ... respectively, we find

$$\begin{aligned}&(u-v)(s)\{\,\}_m\\ &-(u-v)\left[(t)\{\,\}_m - \binom{s-t}{1}(t)\{\,\}_{m-1} + \binom{s-t}{2}(t)\{\,\}_{m-2} - \ldots\right]\\ &-(s-t)\left[(u)\{\,\}_m - \binom{s-u}{1}(u)\{\,\}_{m-1} + \binom{s-u}{2}(u)\{\,\}_{m-2} - \ldots\right]\\ &+(s-t)\left[(v)\{\,\}_m - \binom{s-v}{1}(v)\{\,\}_{m-1} + \binom{s-v}{2}(v)\{\,\}_{m-2} - \ldots\right] = 0,\end{aligned}$$

a relation which, if N_s denotes the number of partitions of s and s, t, u, v are all *different*, involves

$$(s+1)(N_s + N_t + N_u + N_v) - (sN_s + tN_t + uN_u + vN_v) \text{ coefficients.}$$

14. Passing now to the relation

$$tb^{-s}d_s - sb^{-t}d_t - s + t = 0,$$

and putting $s=1$, $t=2$, we find

$$2!\,D_2 = D_1^2 - 2bD_1 - b^2;$$

and without difficulty we reach the further results

$$3!\,D_3 = D_1^3 - 6bD_1^2 + 3b^2D_1 + 4b^3,$$
$$4!\,D_4 = D_1^4 - 12bD_1^3 + 30b^2D_1^2 + 4b^3D_1 - 15b^4,$$
$$5!\,D_5 = D_1^5 - 20bD_1^4 + 110b^2D_1^3 - 140b^3D_1^2 - 95b^4D_1 + 56b^5,$$

&c.,

which can be written down, from those given by the case $b=1$, by simply introducing the proper power of b in each term.

Application of the first of these

$$2D_2 = D_1^2 - 2bD_1 - b^2,$$

yields the relation

$$\begin{aligned}2\{m;\ 1^{\pi_1}2^{\pi_2+1}\ldots\} = \{m;\ 1^{\pi_1+2}2^{\pi_2}\ldots\} - 2\{m;\ 1^{\pi_1+1}2^{\pi_2}\ldots\} - \{m;\ 1^{\pi_1}2^{\pi_2}\ldots\}\\ + 2\{m-1;\ 1^{\pi_1+1}2^{\pi_2}\ldots\} + 2\{m-1;\ 1^{\pi_1}2^{\pi_2}\ldots\}\\ - \{m-2;\ 1^{\pi_1}2^{\pi_2}\ldots\},\end{aligned}$$

63—2

of which a particular case is

$$2\{m;\ 21^s\} = \{m;\ 1^{s+2}\} - 2\{m;\ 1^{s+1}\} - \{m;\ 1^s\}$$
$$+ 2\{m-1;\ 1^{s+1}\} + 2\{m-1;\ 1^s\}$$
$$- \{m-2;\ 1^s\};$$

and since

$$\{m;\ 1^s\} = \binom{s}{m}\{0;\ 1^{s-m}\},$$

we find

$$2\{m;\ 21^s\} = \left\{\binom{s+2}{m} + 2\binom{s+1}{m-1} - \binom{s}{m-2}\right\}\{0;\ 1^{s-m+2}\}$$
$$-2\left\{\binom{s+1}{m} - \binom{s}{m-1}\right\}\{0;\ 1^{s-m+1}\} - \binom{s}{m}\{0;\ 1^{s-m}\}.$$

The number $\{m;\ 1^{\pi_1}2^{\pi_2}3^{\pi_3}\ldots\}$ is ultimately expressible as a linear function of the numbers $\{0;\ 1^s\}$.

15. It is worth while remarking that the numbers $\{0;\ 1^s\}$ may be studied by means of the elementary notions of the Theory of Substitutions.

Every substitution which displaces the whole of the letters may be represented by a product of circular substitutions of order not less than 2. Such a substitution, displacing the whole of the letters, may be termed a non-unitary substitution and there is a one-to-one correspondence between the arrangement in which every letter is displaced and the non-unitary substitutions. We have therefore merely to enumerate the non-unitary substitutions. A certain number of such substitutions correspond to a particular non-unitary partition

$(2^{\pi_2}3^{\pi_3}\ldots)$ of the number n.

If we distribute the n letters in any manner into $\pi_2 + \pi_3 + \ldots$ parcels so that π_s parcels each contain s letters, where s has the values 2, 3, ..., we obtain a definite circular substitution corresponding to any assigned order of the letters in the parcels. Now we observe that a parcel which contains s letters may have the letters permuted in $(s-1)!$ different ways so as to give $(s-1)!$ different circular substitutions because $(s-1)!$ is the number of permutations of s different letters which are arranged in circular order; so that if N be the number of ways of distributing n different letters into $\pi_2 + \pi_3 + \ldots$ parcels so that $\pi_2, \pi_3, \ldots$ parcels contain 2, 3, ... letters respectively, the number of substitutions thence derivable is

$$N(1!)^{\pi_2}(2!)^{\pi_3}(3!)^{\pi_4}\ldots.$$

We can find N because it is known to be the coefficient of (1^n) in the development which arises when the product

$$\frac{p_2^{\pi_2}p_3^{\pi_3}p_4^{\pi_4}\ldots}{\pi_2!\,\pi_3!\,\pi_4!\ldots}$$

is multiplied out. The coefficient is

$$\frac{n!}{(2!)^{\pi_2}(3!)^{\pi_3}(4!)^{\pi_4}\ldots\pi_2!\,\pi_3!\,\pi_4!\ldots},$$

and

$$N(1!)^{\pi_2}(2!)^{\pi_2}(3!)^{\pi_3}\ldots = \frac{n!}{2^{\pi_2}.3^{\pi_3}.4^{\pi_4}\ldots\pi_2!\,\pi_3!\,\pi_4!\ldots}.$$

Hence the total number of non-unitary substitutions or of permutations which displace every letter is

$$\{0;\ 1^n\} = \sum_{\pi} \frac{n!}{2^{\pi_2}.3^{\pi_3}.4^{\pi_4}\ldots\pi_2!\,\pi_3!\,\pi_4!\ldots},$$

the summation being for every non-unitary partition

$$(2^{\pi_2}\,3^{\pi_3}\,4^{\pi_4}\ldots),$$

of the number n.

16. The interest of the above solution lies in the comparison with the result previously reached in Section I. Art. 5, viz. :—

$$\{0;\ 1^n\} = \sum_{\pi} \left\{\frac{n!}{2^{\pi_2}.3^{\pi_3}.4^{\pi_4}\ldots.\pi_2!\,\pi_3!\,\pi_4!\ldots} \cdot \frac{(\pi_2+\pi_3+\pi_4+\ldots)!}{(2!)^{\pi_4}(3!)^{\pi_5}\ldots}\right\}.$$

This old expression gives for $n=6$

$$120.\frac{1}{24}+90.\frac{2}{2}+40\frac{2}{1}+15.6=265.$$

The new expression gives

$$\frac{6!}{6}+\frac{6!}{2.4}+\frac{6!}{9.2}+\frac{6!}{8.6}=265,$$

the four terms corresponding in each case to the partitions

(6), (42), (3^2), (2^3) respectively.

The identity which presents itself, of the form

$$\sum_{\pi} A_{\pi_1,\pi_2,\pi_3,\ldots} = \sum A_{\pi_1,\pi_2,\pi_3,\ldots}\, B_{\pi_1,\pi_2,\pi_3,\ldots},$$

is remarkable.

Section III.

Laplace's Problem.

17. In the *Théorie des Probabilités* Laplace discusses and solves a problem of a somewhat similar kind. He supposes an urn to contain nr tickets which are in n sets each set involving r tickets. The tickets in a first set are each numbered one, in a second set two and so on till those in an nth set are each numbered n. He supposes the tickets to be well mixed and then n tickets to be drawn in succession. If the mth ticket that is drawn happens to be numbered m he calls this a coincidence and he inquires into the probability of there being *at least s* coincidences. Observing that s cannot be superior to n the method of this paper leads quickly to the solution.

We have to determine the number of permutations of

$$x_1{}^r x_2{}^r \ldots x_n{}^r,$$

which are such that an x_m occurs in the mth place from the left on *at least s* occasions.

Consider a redundant generating function

$$(ax_1+x_2+\dots+x_n)(x_1+ax_2+\dots+x_n)\dots(x_1+x_2+\dots+ax_n)(x_1+x_2+\dots+x_n)^{rn-m};$$

it is clear that the coefficient C_s of

$$a^s x_1^r x_2^r \dots x_n^r$$

in this product denotes the number of permutations in which an x_m occurs in the mth place *exactly* s times.

Hence the number of permutations we seek is the sum of the coefficients of the terms

$$a^s x_1^r x_2^r \dots x_n^r,\ a^{s+1} x_1^r x_2^r \dots x_n^r,\ \dots\ a^n x_1^r x_2^r \dots x_n^r,$$

in the product. Denoting the product by u we have to find the coefficient of a^s in

$$\left(1+\frac{1}{a}+\frac{1}{a^2}+\dots\frac{1}{a^{n-s}}\right)u,$$

that is in $\dfrac{u}{1-\dfrac{1}{a}}$ or in $\dfrac{au}{a-1}$, since a does not occur to a higher power than n.

Now, writing $$p_1=\Sigma x_1,\ p_2=\Sigma x_1x_2,\ \dots,$$

$$\begin{aligned} n &= \{p_1+(a-1)x_1\}\{p_1+(a-1)x_2\}\dots\{p_1+(a-1)x_n\}p_1^{rn-n},\\ &= \{p_1^n+(a-1)p_1^n+(a-1)^2p_1^{n-2}p_2+\dots+(a-1)^np_n\}p_1^{rn-n},\\ &= ap_1^{rn}+(a-1)^2p_1^{rn-2}p_2+(a-1)^3p_1^{rn-3}p_3+\dots+(a-1)^np_1^{rn-n}p_n \end{aligned}$$

and $$\frac{au}{a-1}=\frac{a^2p_1^{rn}}{a-1}+a(a-1)p_1^{rn-2}p_2+a(a-1)^2p_1^{rn-3}p_3+\dots+a(a-1)^{n-1}p_1^{rn-n}p_n.$$

Herein the coefficient of a^s is

$$p_1^{rn-s}p_s-\binom{s}{1}p_1^{rn-s-1}p_{s+1}+\binom{s+1}{2}p_1^{rn-s-2}p_{s+2}-\dots+(-)^{n-s}\binom{n-1}{n-s}p_1^{rn-n}p_n.$$

The coefficient herein of $x_1^r x_2^r \dots x_n^r$ is obtained as the result of operating upon it with D_r^n.

The reader will have no difficulty in proving that

$$D_r^n p_1^{nr-m}p_m=\binom{n}{m}\frac{(nr-m)!}{(r!)^{n-m}\{(r-1)!\}^m}.$$

Thence the number we require is

$$\binom{n}{s}\frac{(nr-s)!}{(r!)^{n-s}\{(r-1)!\}^s}-\binom{s}{1}\binom{n}{s+1}\frac{(nr-s-1)!}{(r!)^{n-s-1}\{(r-1)!\}^{s+1}}$$
$$+\binom{s+1}{2}\binom{n}{s+2}\frac{(nr-s-2)!}{(r!)^{n-s-2}\{(r-1)!\}^{s+2}}-\dots+(-)^{n-s}\binom{n-1}{n-s}\binom{n}{n}\frac{(nr-n)!}{\{(r-1)!\}^n};$$

and dividing this by the whole number of permutations, viz. :—

$$\frac{(nr)!}{(r!)^n},$$

we find a result which is readily identified with that of Laplace.

18. If we put $s=1$ we obtain the whole number of permutations which exhibit coincidences in the first n places; this number is therefore

$$\frac{(nr)!}{(r!)^n}-\binom{n}{2}\frac{(nr-2)!}{(r!)^{n-2}\{(r-1)!\}^2}+\binom{n}{3}\frac{(nr-3)!}{(r!)^{n-3}\{(r-1)!\}^3}-\ldots+(-)^{n-1}\frac{(nr-n)!}{\{(r-1)!\}^n}.$$

If we now subtract this number from

$$\frac{(nr)!}{(r!)^n},$$

we must obtain the whole number of permutations which do *not* exhibit coincidences in the first n places. This number is

$$\binom{n}{2}\frac{(nr-2)!}{(r!)^{n-2}\{(r-1)!\}^2}-\binom{n}{3}\frac{(nr-3)!}{(r!)^{n-3}\{(r-1)!\}^3}+\ldots+(-)^n\frac{(nr-n)!}{\{(r-1)!\}^n}.$$

If herein we put $r=1$, the first n places are in fact the whole of the places so that the expression becomes the value of the number we have denoted by $\{0;\ 1^n\}$. We thus find again

$$\{0;\ 1^n\}=n!\left\{\frac{1}{2!}-\frac{1}{3!}+\ldots+(-)^n\frac{1}{n!}\right\},$$

a verification.

The Indices of Permutations and the Derivation therefrom of Functions of a Single Variable Associated with the Permutations of any Assemblage of Objects.

By Major P. A. MacMahon.

Table of Contents.

§ 2.

Introduction.

This paper arises out of the postscript to the "Memoir on the Theory of the Partitions of Numbers," Part V. *Phil. Trans. R. S.*, Vol. CCXI, A 473, January, 1911. In that paper I derived algebraic functions of a single variable from *certain* permutations of any assemblage of objects. The permutations were restricted to be such as fulfilled certain conditions, and by reason of their intimate association with lattices or gratings of nodes complete or incomplete they were termed lattice-permutations. The discovery of the functions led to the solution of the problem of two-demensional partition at the points of a Ferrers-Sylvester graph such as is figured below,

```
• • • • •
• • •
•
•
```

where we have a system of nodes constituting the graph of a partition of a number. In two-dimensional partition we have a *certain* generalization of ordinary or one-dimensional partition. In the latter the partition is constituted by numbers placed at the nodes of a single row in descending order of magnitude, the sum of the numbers being equal to the number which is partitioned. In the former numbers are placed at the nodes of the two-dimensional system in such wise that the numbers are in descending order of magnitude alike in *each row and in each column* of the system. The system of nodes is supposed to be given, and when the sum of the numbers placed at the nodes is equal to the partitioned number, the system of numbers in two dimensions is a two-dimensional partition of the given number *quâ* the given system of nodes. It will be observed that the system of nodes is itself defined by a one-dimensional partition of some number and the enumerative problem that was solved may be stated as the determination of the number of partitions of any number *quâ* a system of nodes defined by any one-dimensional partition of any number. The generalization that was made was brought to a conclusion in a subsequent paper, "Memoir on the Theory of the Partitions of Numbers," Part VI. *Phil. Trans. R. S.*, Vol. CCXI, A 479. It depends upon the relative position of the parts and may be made to depend upon linear Diophantine analysis. The problem is certainly very simply stated in Diophantine terms, but the actual solution reached was obtained from generating functions derived from *permutations* in association with the given system of nodes. The permutations were made to lead to algebraic functions through the agency of numbers derived from the permutations. In the postscript alluded to there was an adumbration of a similar theory connected with the entire set of permutations of any assemblage of objects. The present paper opens the discussion of this question and formally introduces various kinds of indices of permutations. So much of the matter that the postscript discussed as involved a limit to the magnitude of the parts is not broached here except in one place in an incidental manner. The theory of the greater and lesser indices is complete. That of the equal index is incomplete so far as the algebraic functions are concerned, but the theory of the average values is complete in this case also. As regards the theory of the major and minor indices, the discussion is complete when the assemblage only contains three *kinds* of objects and the *crude* form of generating function has been obtained for the most general assemblage. The theory of average values is complete in this case also. I hope, if permitted, to continue the discussion on a future occasion.

A communication on the subject-matter of this paper was made to the Mathematical Subsection of Section A of the British Association for the Advancement of Science at the Dundee meeting, September, 1912. P. A. M.

§ 1. *Indices of the First Genus.*

1. The assemblage of objects I take to be

$$\alpha^i\beta^j\gamma^k\ldots\quad\text{or}\quad\alpha_1^{i_1}\alpha_2^{i_2}\alpha_3^{i_3}\ldots$$

indifferently.

A contact in a permutation of these letters, which is (say) $\alpha_s\alpha_t$, is called a major contact, an equal contact or a minor contact, according as $s >, =,$ or $< t$.

The first definitions have reference to the major contacts of a permutation.

2. *The Greater Index of a Permutation.* Of $\alpha^4\beta^2\gamma^3$ let any permutation be (say)

$$\beta\alpha\alpha\alpha\gamma\gamma\beta\alpha\gamma.$$

Whenever a letter is the left-hand member of a major contact, I write under it a number which shows how many places it is from the left of the permutation. Thus if the s-th letter stands before a letter prior to it in alphabetical (or numerical) order, I place under it the number s. In this way we obtain

$$\begin{array}{ccccccccc}\beta&\alpha&\alpha&\alpha&\gamma&\gamma&\beta&\alpha&\gamma.\\ 1&&&&&6&7&&\end{array}$$

We add up the numbers so placed and obtain $1+6+7=14$, a number which I call the "greater index" of the permutation. Similarly, if, in any permutation of any assemblage of letters or of ordered objects, the p_1-th, p_2-th, p_3-th, ... letters are the left-hand members of major contacts, we have the definition

$$\text{Greater index} = p_1+p_2+p_3+\ldots\ldots=p.$$

If, in so forming the number p, we have to add m numbers or, in other words, if the permutation possesses m major contacts, I speak of the permutation as being of "Class m" *quâ* major contacts.

3. *The Equal Index of a Permutation.* If, in any permutation, the q_1-th, q_2-th, q_3-th, ... letters immediately precede letters identical with themselves, so that such letters are the left-hand members of equal contacts, I make the definition

$$\text{Equal index} = q_1+q_2+q_3+\ldots\ldots=q.$$

Thus from

$$\begin{array}{ccccccccc}\beta&\alpha&\alpha&\alpha&\gamma&\gamma&\beta&\alpha&\gamma\\ &2&3&&5&&&&\end{array}$$

we obtain the equal index $2+3+5=10$.

If, in so forming the number q, we have to add m numbers, or if, in other words, the permutation possesses m equal contacts, I speak of the permutation as being of "Class m" *quâ* equal contacts.

4. *The Lesser Index of a Permutation.* If, in any permutation, the r_1-th, r_2-th, r_3-th, ... letters immediately precede letters which are later in alphabetical order, so that such letters are the left-hand members of minor contacts, I make the definition

$$\text{Lesser index} = r_1+r_2+r_3+\ldots\ldots=r.$$

If in so forming the number r we have to add m numbers, or, in other words, if the permutation possesses m minor contacts, I speak of the permutation as being of "Class m" *quâ* minor contacts.

5. *The Greater and Equal Index.* This refers to letters which immediately precede letters which are *not later* in alphabetical order and is equal to the sum of the greater and equal indices, or to $p+q$.

In the same manner we may have to consider other combinations of the indices p, q, r.

6. *Evaluation of Σx^p for Permutations of Any Given Assemblage.* In the case of the assemblage $\alpha^i\beta^j\gamma^k\ldots.$ I wish to give a formal proof of the result

$$\Sigma x^p = \frac{(1-x)(1-x^2)\ldots(1-x^{i+j+k+\ldots})}{(1-x)(1-x^2)\ldots(1-x^i)\cdot(1-x)(1-x^2)\ldots(1-x^j)\cdot(1-x)(1-x^2)\ldots(1-x^k)\cdot\ldots}$$

of which an indication only was given in the postscript to the paper, to which reference has been made in the preamble above.

In Cayley's notation I write

$$1-x^s=(\mathbf{s}),$$

so that the result to be established is written

$$\Sigma x^p = \frac{(\mathbf{1})(\mathbf{2})\ldots.(\mathbf{i}+\mathbf{j}+\mathbf{k}+\ldots.)}{(\mathbf{1})(\mathbf{2})\ldots.(\mathbf{i})\cdot(\mathbf{1})(\mathbf{2})\ldots.(\mathbf{j})\cdot(\mathbf{1})(\mathbf{2})\ldots.(\mathbf{k})\cdot\ldots.}.$$

Consider a system of rows of nodes, containing i, j, k, ... nodes respectively and at these nodes numbers to be placed in such wise that there is a descending order of magnitude in each row. The number zero is not excluded from being placed at a node and there is no condition whatever in regard to the numbers which appear in the columns of nodes. If it be asked,—In how many ways can numbers be so placed that their sum is w?—it is readily seen that the answer is supplied by the coefficient of x^w in the ascending expansion of the algebraic fraction

$$\frac{1}{(\mathbf{1})(\mathbf{2})\ldots.(\mathbf{i})\cdot(\mathbf{1})(\mathbf{2})\ldots.(\mathbf{j})\cdot(\mathbf{1})(\mathbf{2})\ldots.(\mathbf{k})\cdot\ldots.};$$

because if w_i, w_j, w_k, ... be any numbers such that $w_i+w_j+w_k+\ldots.=w$, the members w_i, w_j, w_k, ... can be partitioned in the 1st, 2d, 3d, ... rows in numbers of ways indicated by the generating functions

$$\frac{1}{(\mathbf{1})(\mathbf{2})\ldots.(\mathbf{i})},\quad \frac{1}{(\mathbf{1})(\mathbf{2})\ldots.(\mathbf{j})},\quad \frac{1}{(\mathbf{1})(\mathbf{2})\ldots.(\mathbf{k})},\quad\ldots.$$

respectively.

The present proof consists in showing that the same generating function

36

can be exhibited as the quotient of Σx^p, p being the greater index of a permutation of $\alpha^i\beta^j\gamma^k\ldots\ldots$, by the function

$$(1)\,(2)\ldots\ldots(i+j+k+\ldots\ldots).$$

Consider a simple particular case, $i=3$, $j=2$, to which belongs the system of nodes

$$\begin{matrix}\bullet & \bullet & \bullet\\ \bullet & \bullet & \end{matrix}$$

and suppose numbers to be placed at the nodes, viz.,

$$\begin{matrix}a_1 & a_2 & a_3\\ b_1 & b_2 & \end{matrix}$$

in such wise that $a_1 \geq a_2 \geq a_3$, $b_1 \geq b_2$. Place also the quantities α, β of the assemblage $\alpha^3\beta^2$ at the nodes in the manner

$$\begin{matrix}\alpha & \alpha & \alpha\\ \beta & \beta & \end{matrix}$$

We are about to enquire into the number of ways of placing the numbers at the nodes by arranging them in all possible descending orders, and then enumerating each order separately. We must determine the lines of route through the system of nodes which may give a descending order of magnitude of the numbers a_1, a_2, a_3, b_1, b_2 collectively. These numbers may be in any permutation so long as the two orders a_1, a_2, a_3; b_1, b_2 are in evidence. Thus through the system of nodes we may have the lines of route

$\alpha\alpha\alpha\beta\beta$ $\qquad$ $\alpha\alpha\beta\alpha\beta$ $\qquad$ $\alpha\beta\alpha\alpha\beta$ $\qquad$ $\beta\alpha\alpha\alpha\beta$ $\qquad$ $\alpha\alpha\beta\beta\alpha$

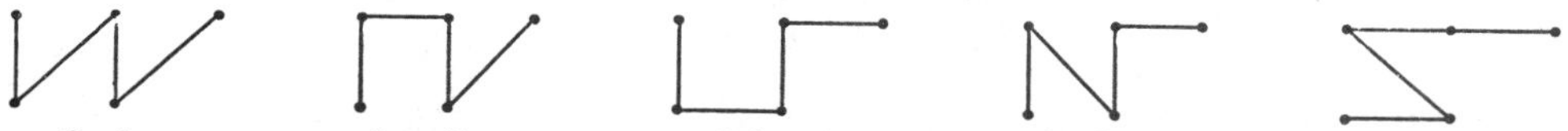

$\alpha\beta\alpha\beta\alpha$ $\qquad$ $\beta\alpha\alpha\beta\alpha$ $\qquad$ $\alpha\beta\beta\alpha\alpha$ $\qquad$ $\beta\alpha\beta\alpha\alpha$ $\qquad$ $\beta\beta\alpha\alpha\alpha$

Beneath each line of route is placed the corresponding permutation of $\alpha^3\beta^2$. There is obviously one line of route for each permutation.

If we take any one of these lines of route, we may lay out five numbers in descending order of magnitude along it. In the general case each line of route is $i+j+k+\ldots\ldots$ nodes long, and we may lay out $i+j+k+\ldots\ldots$ numbers in descending order along the one selected. The number of ways in which this can be done is given by the coefficient of x^w in

$$\{(1)\,(2)\ldots\ldots(i+j+k+\ldots\ldots)\}^{-1};$$

and if we so proceeded, in the case of each permutation of $\alpha^i\beta^j\gamma^k\ldots$, we would obtain this coefficient multiplied by the number of permutations of $\alpha^i\beta^j\gamma^k\ldots$. We would not, however, thus obtain the required generating function, because, in general, many systems of numbers may be laid out along more than one line of route. For example, the arrangement

4 2 2
3 2

occurs along each of the lines of route

For the enumeration before us we must arrange that each line of route is utilized without overlapping. This is accomplished by imposing certain conditions upon the successions of numbers which appertain to the various lines of route.

Observe the subjoined scheme of descending orders for the particular case under examination:

1.	$\alpha\alpha\alpha\beta\beta$,	$a_1 \geq a_2 \geq a_3 \geq b_1 \geq b_2$,
2.	$\alpha\alpha\beta\alpha\beta$,	$a_1 \geq a_2 \geq b_1 > a_3 \geq b_2$,
3.	$\alpha\beta\alpha\alpha\beta$,	$a_1 \geq b_1 > a_2 \geq a_3 \geq b_2$,
4.	$\beta\alpha\alpha\alpha\beta$,	$b_1 > a_1 \geq a_2 \geq a_3 \geq b_2$,
5.	$\alpha\alpha\beta\beta\alpha$,	$a_1 \geq a_2 \geq b_1 \geq b_2 > a_3$,
6.	$\alpha\beta\alpha\beta\alpha$,	$a_1 \geq b_1 > a_2 \geq b_2 > a_3$,
7.	$\beta\alpha\alpha\beta\alpha$,	$b_1 > a_1 \geq a_2 \geq b_2 > a_3$,
8.	$\alpha\beta\beta\alpha\alpha$,	$a_1 \geq b_1 \geq b_2 > a_2 \geq a_3$,
9.	$\beta\alpha\beta\alpha\alpha$,	$b_1 > a_1 \geq b_2 > a_2 \geq a_3$,
10.	$\beta\beta\alpha\alpha\alpha$,	$b_1 \geq b_2 > a_1 \geq a_2 \geq a_3$,

where the descending order on the right appertains to the permutation on the left and therefore to the line of route which the permutation denotes. Wherever in the permutation there is either an equal contact or a minor contact, the symbol $\geq$ is placed in the corresponding place in the descending order; but wherever there is a major contact, the symbol employed is $>$. The successions of numbers thus defined are all distinct. We obtain ten sets of numbers in descending order without overlapping. It will be noticed that the symbol $>$ invariably occurs in correspondence with one of the numbers which, when added together, become the greater index of the permutation. We have now to form the ten generating functions which enumerate the ten sets of numbers.

To enumerate set No. 6, put

$$b_2 = a_3+A+1,$$
$$a_2 = a_3+A+B+1,$$
$$b_1 = a_3+A+B+C+1+1,$$
$$a_1 = a_3+A+B+C+D+1+1,$$

where A, B, C, D are arbitrary positive (including zero) integers; then

$$\Sigma x^{a_1+a_2+a_3+b_1+b_2} = \Sigma x^{5a_3+4A+3B+2C+D+2+4},$$

where the two numbers 2, 4, which present themselves in the exponent of x, are necessarily the two numbers which are added together to form the greater index of the associated permutation $\alpha\beta\alpha\beta\alpha$. This generating function is therefore

$$\frac{x^6}{(1)(2)(3)(4)(5)};$$

and it is readily seen that the sum of the ten generating functions must be

$$\frac{\Sigma x^p}{(1)(2)(3)(4)(5)},$$

the sum being in respect of the greater indices of the ten permutations. Hence

$$\frac{\Sigma x^p}{(1)(2)(3)(4)(5)} = \frac{1}{(1)(2)(3)\cdot(1)(2)},$$

or

$$\Sigma x^p = \frac{(1)(2)(3)(4)(5)}{(1)(2)(3)\cdot(1)(2)}.$$

The above reasoning is of general application. When the system of nodes has rows containing $i, j, k, \ldots.$ nodes and we consider the number of ways of placing numbers in descending orders in each row so that the sum of all the numbers may be w, we have to consider

$$\frac{(i+j+k+\ldots.)!}{i!\,j!\,k!\ldots.}$$

conditioned descending orders corresponding to the permutations of $\alpha^i\beta^j\gamma^k\ldots.$; whenever the contact is major we employ the symbol $>$, and $\geq$ in other cases, as conditions of the descending order, and we find in each case that x^p is the numerator of the generating function, where p is the greater index of the permutation. The reader will note that when the p_1-th interval between the letters $a_1, a_2, a_3, \ldots.; b_1, b_2, \ldots.; c_1, c_2, c_3, \ldots.; \ldots.$, in appropriate order, is occupied by the symbol $>$, one outstanding number, in the exponent of x which first arises, is p_1; and that the whole outstanding number must be p.

Hence in general

$$\Sigma x^p = \frac{(1)(2)\ldots\ldots(\mathbf{i}+\mathbf{j}+\mathbf{k}+\ldots\ldots)}{(1)(2)\ldots\ldots(\mathbf{i})\cdot(1)(2)\ldots\ldots(\mathbf{j})\cdot(1)(2)\ldots\ldots(\mathbf{k})\cdot\ldots\ldots}.$$

7. When x is put equal to unity the generating function naturally reduces to the number which enumerates the permutations of the assemblage.

8. *Important Property of* Σx^p. From the symmetry of the expression just obtained we gather the information that the expression of Σx^p is unaltered by any substitution impressed upon the numbers $i, j, k, \ldots\ldots$

Ex. gr., Σx^p is the same for $\alpha^3\beta^2$ as for $\alpha^2\beta^3$.

9. *Proof that* $\Sigma x^r = \Sigma x^p$. The last remark enables us to establish that in general $\Sigma x^r = \Sigma x^p$. Consider a permutation of any assemblage, viz.

$$\beta\alpha\alpha\alpha\gamma\gamma\beta\alpha\gamma,$$

and transfrom it by the substitution

$$\begin{pmatrix}\alpha\beta\gamma\\ \gamma\beta\alpha\end{pmatrix}$$

(equivalent to changing the assemblage $\alpha^4\beta^2\gamma^3$ into $\alpha^3\beta^2\gamma^4$). The permutation becomes

$$\beta\gamma\gamma\gamma\alpha\alpha\beta\gamma\alpha,$$

and its lesser index is necessarily equal to the greater index of the original permutation. Hence Σx^r for the transformed is equal to Σx^p for the original assemblage; and since, as we have seen above, Σx^p remains unchanged by the transformation, it follows at once that for every assemblage

$$\Sigma x^r = \Sigma x^p;$$

for we have merely to make the substitution

$$\begin{pmatrix}\alpha_1\alpha_2 & \alpha_3 & \ldots\ldots\alpha_s\\ \alpha_s\alpha_{s-1} & \alpha_{s-2} & \ldots\ldots\alpha_1\end{pmatrix},$$

and employ the same reasoning.

10. *The Expression of* Σx^{p+q}. Of course p, q are the greater and equal indices of the same permutation. We have

$$p+q+r = 1+2+\ldots\ldots+(\Sigma i-1) = \binom{\Sigma i}{2},$$

p, q, r referring to the same permutation, and $\binom{\Sigma i}{2}$ denoting $\frac{1}{2}\Sigma i(\Sigma i-1)$. Hence

$$\Sigma x^{p+q} = x^{p+q+r}\Sigma x^{-r}$$

$$= x^{\binom{\Sigma i}{2}} \frac{\left(1-\frac{1}{x}\right)\left(1-\frac{1}{x^2}\right)\ldots\left(1-\frac{1}{x^{i_1+i_2+\ldots\ldots}}\right)}{\left(1-\frac{1}{x}\right)\left(1-\frac{1}{x^2}\right)\ldots\left(1-\frac{1}{x^{i_1}}\right)\left(1-\frac{1}{x}\right)\left(1-\frac{1}{x^2}\right)\ldots\left(1-\frac{1}{x^{i_2}}\right)\ldots}$$

$$= x^{\binom{\Sigma i}{2}+\binom{i_1+1}{2}+\binom{i_2+1}{2}+\ldots\ldots-\binom{\Sigma i+1}{2}} \frac{(1)(2)\ldots\ldots(\mathbf{i}_1+\mathbf{i}_2+\ldots\ldots)}{(1)(2)\ldots\ldots(\mathbf{i}_1)\cdot(1)(2)\ldots\ldots(\mathbf{i}_2)\cdot\ldots\ldots};$$

and thence, on reduction,

$$\Sigma x^{p+q} = x^{\binom{i}{2}+\binom{j}{2}+\binom{k}{2}+\dots}\Sigma x^p,$$

a noteworthy result, the truth of which can also be seen *à priori*.

We have merely to recur to the method by which the expression of Σx^p was investigated. In the descending orders of numbers associated with the permutations we must clearly have the symbol $>$ *both* when there are major and when there are equal contacts. This will happen if we impress a condition upon the numbers in each row. The condition is that in any row of numbers, placed at the nodes of a row of nodes, there must be no two numbers equal. This is secured by adding to every row of numbers

$$a_1 a_2 a_3 \dots a_i$$

the numbers $i-1, i-2, i-3, \dots, 1, 0$ respectively to each number in succession from left to right, so that the row becomes

$$a_1+i-1,\ a_2+i-2,\ a_3+i-3,\ \dots,\ a_{i-1}+1,\ a_i.$$

The result of this is to multiply Σx^p by $x^{1+2+\dots+i-1}$ or by $x^{\binom{i}{2}}$ in respect of each row. Thence at once

$$\Sigma x^{p+q} = x^{\binom{i}{2}+\binom{j}{2}+\binom{k}{2}+\dots}\Sigma x^p.$$

11. *Difference Equation for* Σx^p. Taking the assemblage to be $\alpha_1^{i_1}\alpha_2^{i_2}\dots\alpha_s^{i_s}$ write

$$\Sigma x^p = u(i_1, i_2, \dots, i_s);$$

and if Σx^p is restricted to those permutations which terminate with a letter α_m, write

$$\underset{m}{\Sigma} x^p = \underset{m}{u}(i_1, i_2, \dots, i_s), \quad \text{and} \quad \Sigma i = S;$$

so that

$$\Sigma x^p = \sum_{1}^{s} \underset{m}{u}(i_1, i_2, \dots, i_s).$$

Obviously

$$\underset{s}{u}(i_1 i_2 \dots i_s) = u(i_1, i_2, \dots, i_{s-1}, i_s-1).$$

The function $\underset{s-1}{u}(i_1 i_2 \dots i_s)$ refers to permutations that may end with $\alpha_s\alpha_{s-1}$. These give rise to the function $x^{S-1}\underset{s}{u}(i_1, i_2, \dots, i_{s-1}-1, i_s)$. The remaining permutations yield the function

$$u(i_1, i_2, \dots, i_{s-1}-1, i_s) - \underset{s}{u}(i_1, i_2, \dots, i_{s-1}-1, i_s).$$

Hence

$$\underset{s-1}{u}(i_1, i_2, \dots, i_s)$$
$$= u(i_1, i_2, \dots, i_{s-1}-1, i_s) - (1-x^{S-1})u(i_1, i_2, \dots, i_{s-1}-1, i_s-1).$$

Similarly it will be found that

$$\begin{aligned}\underset{s-2}{u}(i_1, i_2, \ldots, i_s) = & u(i_1, i_2, \ldots, i_{s-2}-1, i_{s-1}, i_s) \\ & -(1-x^{S-1})u(i_1, i_2, \ldots, i_{s-2}-1, i_{s-1}-1, i_s) \\ & -(1-x^{S-1})u(i_1, i_2, \ldots, i_{s-2}-1, i_{s-1}, i_s-1) \\ & +(1-x^{S-1})(1-x^{S-2})u(i_1, i_2, \ldots, i_{s-2}-1, i_{s-1}-1, i_s-1),\end{aligned}$$

and so forth. Thence

$$\begin{aligned}u(i_1, i_2, \ldots, i_s) = & u(i_1, i_2, \ldots, i_{s-1}, i_s-1) + \{u(i_1, i_2, \ldots, i_{s-1}-1, i_s) \\ & -(\mathbf{S}-1)u(i_1, i_2, \ldots, i_{s-1}-1, i_s-1)\} \\ & +\{u(i_1, i_2, \ldots, i_{s-2}-1, i_{s-1}, i_s) \\ & -(\mathbf{S}-1)u(i_1, i_2, \ldots, i_{s-2}-1, i_{s-1}-1, i_s) \\ & -(\mathbf{S}-1)u(i_1, i_2, \ldots, i_{s-2}-1, i_{s-1}, i_s-1) \\ & +(\mathbf{S}-1)(\mathbf{S}-2)u(i_1, i_2, \ldots, i_{s-2}-1, i_{s-1}-1, i_s-1)\} + \ldots,\end{aligned}$$

and we obtain without difficulty the general equation of differences

$$\begin{aligned}u(i_1, i_2, \ldots, i_s) = & \sum_{l} u(i_1, \ldots, i_l-1, \ldots, i_s) \\ & -(\mathbf{S}-1)\sum_{l,m} u(i_1, \ldots, i_l-1, \ldots, i_m-1, \ldots, i_s) \\ & +(\mathbf{S}-1)(\mathbf{S}-2)\sum_{l,m,n} u(i_1, \ldots, i_l-1, \ldots, i_m-1, \ldots, i_n-1, \ldots, i_s) \\ & -\ldots\ldots\end{aligned}$$

This relation will be of service in the sequel.

12. *The Particular Case of the Assemblage* $\alpha^i\beta^j$. Merely considering the main result for the permutations of $\alpha^i\beta^j$, viz.

$$\Sigma x^p = \frac{(1)(2)\ldots\ldots(\mathbf{i}+\mathbf{j})}{(1)(2)\ldots\ldots(\mathbf{i})\cdot(1)(2)\ldots\ldots(\mathbf{j})},$$

we notice that Σx^p is equal to the generating function which enumerates the partitions of all numbers into parts limited in magnitude by i and in number by j. This indicates that the number of permutations of $\alpha^i\beta^j$ which have p for greater index is equal to the number of partitions of p, the parts being restricted as above stated by the numbers i, j.

If the permutation is of class m (see Art. 2) the greater index is obtained as the sum of m numbers. If we restrict ourselves to such permutations, I showed, in the postscript of the paper above referred to, that

$$\overset{m}{\Sigma} x^p = x^{m^2}\frac{(\mathbf{i}-\mathbf{m}+1)(\mathbf{i}-\mathbf{m}+2)\ldots\ldots(\mathbf{i})}{(1)(2)\ldots\ldots(\mathbf{m})}\cdot\frac{(\mathbf{j}-\mathbf{m}+1)(\mathbf{j}-\mathbf{m}+2)\ldots\ldots(\mathbf{j})}{(1)(2)\ldots\ldots(\mathbf{m})},$$

a result which yields the well-known expansion

$$\frac{(1)(2)\ldots\ldots(\mathbf{i}+\mathbf{j})}{(1)(2)\ldots\ldots(\mathbf{i})\cdot(1)(2)\ldots\ldots(\mathbf{j})} = 1 + x\frac{(\mathbf{i})(\mathbf{j})}{(1)\cdot(1)} + x^4\frac{(\mathbf{i}-1)(\mathbf{i})\cdot(\mathbf{j}-1)(\mathbf{j})}{(1)(2)\cdot(1)(2)} + \ldots\ldots$$

This fact is referred to here in order that I may make the observation that the expansion is identical with that which Sylvester obtained from Durfee's dissection of the graph of a partition into a square of nodes with lateral and subjacent appendages. (Compare *Collected Mathematical Papers*, Vol. IV, pp. 1 *et seq.*).

13. For the assemblage $\alpha^i\beta^j$ we can find the equal index function Σx^q. It has first to be shown that the number $p-r$ is equal to $-i$ or $+j$ according as the permutation terminates with β or α.

Assume the law to obtain for $\alpha^i\beta^j$ and denote by $(\alpha^i\beta^j)_\alpha$, $(\alpha^i\beta^j)_\beta$ any permutations terminating with α, β respectively.

Then for $(\alpha^i\beta^j)_\alpha\alpha$, $\quad p-r=+j$;
for $(\alpha^i\beta^j)_\alpha\beta$, $\quad p-r=+j-(i+j)=-i$;
for $(\alpha^i\beta^j)_\beta\alpha$, $\quad p-r=-i+i+j=+j$;
for $(\alpha^i\beta^j)_\beta\beta$, $\quad p-r=+j$;
for $(\alpha^i\beta^j)_\alpha\alpha\beta$, $\quad p-r=+j-(i+j+1)=-i-1$;
for $(\alpha^i\beta^j)_\alpha\beta\alpha$, $\quad p-r=+j-(i+j)+(i+j+1)=+j+1$;
for $(\alpha^i\beta^j)_\beta\alpha\beta$, $\quad p-r=-i+(i+j)-(i+j+1)=-i-1$;
for $(\alpha^i\beta^j)_\beta\beta\alpha$, $\quad p-r=-i+(i+j+1)=+j+1$.

These results establish the theorem.

Hence, for the permutations ending with β,

$$\Sigma x^{p-r}=\binom{i+j-1}{i}x^{-i};$$

and, for those ending with α,

$$\Sigma x^{p-r}=\binom{i+j-1}{j}x^{j}.$$

Since, for all permutations,

$$\Sigma x^{p+q+r}=\binom{i+j}{i}x^{\binom{i+j}{2}},$$

it follows that

$$\Sigma x^{2p+q}=\binom{i+j-1}{i}x^{\binom{i+j}{2}-i}+\binom{i+j-1}{j}x^{\binom{i+j}{2}+j}.$$

Writing Σx^p, for the whole of the permutation, equal to $F_{i,j}(x)$, we see that, for those ending with β, α,

$$\Sigma x^p=F_{i,j-1}(x) \quad\text{and}\quad x^jF_{i-1,j}(x)$$

respectively. (Compare Art. 11.) Hence

$$\Sigma x^q=\Sigma x^{2p+q-2p}=x^{\binom{i+j}{2}-i}F_{i,j-1}\left(\frac{1}{x^2}\right)+x^{\binom{i+j}{2}+j-2j}F_{i-1,j}\left(\frac{1}{x^2}\right);$$

easily reducing to the formula

$$\Sigma x^q = x^{\binom{i-j+1}{2}} F_{i,j-1}(x^2) + x^{\binom{j-i+1}{2}} F_{i-1,j}(x^2).$$

It will be noted that the index q is *always* even when $i-j \equiv 0 \bmod 4$ and *always* uneven when $i-j \equiv 2 \bmod 4$.

14. We may obtain the expression of Σx^{p+r}; for this is

$$\Sigma x^{p-r+2r};$$

and, since Σx^r is $x^i F_{i,j-1}(x)$ or $F_{i-1,j}(x)$ according as the permutation ends with β or α,

$$\Sigma x^{p+r} = x^{-i} \cdot x^{2i} F_{i,j-1}(x^2) + x^j F_{i-1,j}(x^2),$$
$$= x^i F_{i,j-1}(x^2) + x^j F_{i-1,j}(x^2).$$

The reader will have no difficulty in obtaining the expression of $\Sigma x^{Ap+Bq+Cr}$ for this assemblage.

15. Resuming the consideration of the assemblage $\alpha^i \beta^j \gamma^k \ldots$, it is to be remarked that $p-r$ has no longer any simple relation to the permutations when the number of Greek letters exceeds 2. It is not possible to obtain expressions for Σx^q and Σx^{p+r} in the foregoing manner. Similar theorems, however, will be reached when major and minor indices come under discussion later in the paper. Before proceeding to those matters I give some interesting properties of the indices that have been already under discussion.

16. *The Parity of the Greater Index.* The permutations of $\alpha^i \beta^j \gamma^k \ldots$ are such that the number with an even index is equal to the number with an uneven index whenever more than one of the numbers $i, j, k, \ldots$ are uneven. For consider the function

$$\frac{(1)(2)\ldots(\mathbf{i}+\mathbf{j}+\mathbf{k}+\ldots)}{(1)(2)\ldots(\mathbf{i}) \cdot (1)(2)\ldots(\mathbf{j}) \cdot (1)(2)\ldots(\mathbf{k}) \cdot \ldots};$$

if we put therein $x=-1$ and the function happens to vanish it must be because the permutations with even index are equal in number to those with uneven index. The factor (m) becomes 2 for $x=-1$ whenever m is uneven. Eliminating these factors we are left with a function which may be an indeterminate form and may be evaluated when $x=-1$. This happens when numerator and denominator contain the same number of factors. If this be not the case, the numerator must contain *more* factors than the denominator and the function must therefore vanish for $x=-1$. The numerator cannot contain less factors than the denominator, for in that case the function would become infinite for $x=-1$ and this result, from the nature of the function, is absurd. We have then to find out the circumstances under which the numerator has more factors than the denominator after the elimination of factors referred to above. If $E\sigma$ denotes the greatest integer in σ, the difference in number between the numerator and and denominator factors is

$$E\frac{1}{2}(i+j+k+\ldots.)-E\frac{i}{2}-E\frac{j}{2}-E\frac{k}{2}-\ldots.;$$

and this is greater than zero whenever more than one of the numbers $i, j, k, \ldots.$ are uneven. This establishes the theorem.

17. There is an analogous result in the case of every *prime* modulus μ. For, putting x equal to ωx, where ω is an imaginary μ-th root of unity, we set apart all factors which involve ω^t, where t is not equal to μ or $\equiv 0 \bmod \mu$. We are left with a number of factors free from ω, and the function involving them is seen to vanish when $x=\omega$, whenever

$$\Sigma\frac{1}{\mu}(i+j+k+\ldots.)-E\frac{i}{\mu}-E\frac{j}{\mu}-E\frac{k}{\mu}-\ldots.>0.$$

This happens if $i\equiv i' \bmod \mu$, $j\equiv j' \bmod \mu$, $k\equiv k' \bmod \mu$, etc., whenever

$$i'+j'+k'+\ldots.>\mu-1.$$

Under these circumstances the number of permutations, which have indices $\equiv \sigma \bmod \mu$, does not vary by variation of σ.

Ex. gr., take the assemblage $\alpha^2\beta\gamma$:

	$p=$		$p=$
$\alpha\alpha\beta\gamma$	0	$\beta\alpha\alpha\gamma$	1
$\alpha\alpha\gamma\beta$	3	$\gamma\alpha\alpha\beta$	1
$\alpha\beta\alpha\gamma$	2	$\beta\alpha\gamma\alpha$	4
$\alpha\gamma\alpha\beta$	2	$\gamma\alpha\beta\alpha$	4
$\alpha\beta\gamma\alpha$	3	$\beta\gamma\alpha\alpha$	2
$\alpha\gamma\beta\alpha$	5	$\gamma\beta\alpha\alpha$	3

The reader will verify that the even and uneven indices are equal in number; also, since 2, 1, 1 are $\equiv$ 2, 1, 1 mod 3 respectively and $2+1+1>3$, that the indices $\equiv$ 0, 1, 2 mod 3 are equal in number.

18. *Average Values of the Indices.* The expression obtained for Σx^p enables us to find the average values of the different indices of a permutation.

Write $\Sigma x^p=\Sigma C_p x^p$, so that C_p permutations have the greater index equal to p. The average value of p is the quotient of $\Sigma C_p p$ by

$$\frac{(i+j+k+\ldots.)!}{i!\,j!\,k!\ldots.}.$$

To find $\Sigma C_p p$ we may differentiate Σx^p with regard to x and then put x

equal to unity. This is not the only, or the most obvious, way of finding $\Sigma C_p p$, but it is the best way for present purposes, as will appear later. To evaluate

$$\left\{\partial_x \frac{(1)(2)\ldots\ldots(i+j)}{(1)(2)\ldots\ldots(i)\cdot(1)(2)\ldots\ldots(j)}\right\}_{x=1},$$

we require the value of

$$\left\{\partial_x \frac{(i+\sigma)}{(\sigma)}\right\}_{x=1}.$$

This is readily found to be

$$\frac{i(i+\sigma)}{2\sigma},$$

so that

$$\left\{\partial_x \frac{(i+1)\ldots\ldots(i+j)}{(1)\ldots\ldots(j)}\right\}_{x=1} = \frac{(i+j)\,!}{i\,!\,j\,!}\sum_{\sigma=1}^{\sigma=j}\frac{\dfrac{i(i+\sigma)}{2\sigma}}{\dfrac{i+\sigma}{\sigma}} = \frac{(i+j)\,!}{i\,!\,j\,!}\cdot\frac{1}{2}ij;$$

establishing that the average value of the greater (or lesser) index of the permutation of $\alpha^i\beta^j$ is

$$\frac{1}{2}ij.$$

Thence

$$\left\{\partial_x \frac{(1)(2)\ldots(i+j+k)}{(1)(2)\ldots(i)\cdot(1)(2)\ldots(j)\cdot(1)(2)\ldots(k)\cdot\ldots}\right\}_{x=1}$$

$$=\left\{\frac{(1)(2)\ldots(i+j+k)}{(1)(2)\ldots(i)\cdot(1)(2)\ldots(j+k)}\right\}_{x=1}\left\{\partial_x \frac{(1)(2)\ldots(j+k)}{(1)(2)\ldots(j)\cdot(1)(2)\ldots(k)}\right\}_{x=1}$$

$$+\left\{\frac{(1)(2)\ldots(j+k)}{(1)(2)\ldots(j)\cdot(1)(2)\ldots(k)}\right\}_{x=1}\left\{\partial_x \frac{(1)(2)\ldots(i+j+k)}{(1)(2)\ldots(i)\cdot(1)(2)\ldots(j+k)}\right\}_{x=1}$$

$$=\frac{(i+j+k)\,!}{i\,!\,j\,!\,k\,!}\left\{\frac{1}{2}jk+\frac{1}{2}i(j+k)\right\};$$

establishing that the average value of the greater (or lesser) index of the permutations of $\alpha^i\beta^j\gamma^k$ is

$$\frac{1}{2}(ij+ik+jk).$$

From the above it is quite clear that, for the permutations of $\alpha^i\beta^j\gamma^k\ldots\ldots$, the average value of the greater (or lesser) index is

$\frac{1}{2}$ (sum of the products two and two together of the numbers $i, j, k, \ldots\ldots$).

19. *Average Value of the Equal Index.* From the general formula

$$\Sigma x^{p+q} = x^{\binom{i}{2}+\binom{j}{2}+\binom{k}{2}+\cdots}\Sigma x^p,$$

it is seen that the average value of the equal index is

$$\binom{i}{2}+\binom{j}{2}+\binom{k}{2}+\ldots.;$$

or from the circumstance that the average value of $p+r$ is

$$ij+ik+jk+\ldots.,$$

and that the sum $p+q+r$ is invariably

$$\binom{i+j+k+\ldots.}{2},$$

the average value of q is

$$\binom{i+j+k+\ldots.}{2}-(ij+ik+jk+\ldots.),$$

which is

$$\binom{i}{2}+\binom{j}{2}+\binom{k}{2}+\ldots..$$

20. *Average Value of the Square of the Greater Index.* The value in question is derived from the average value of $p(p-1)$, which is clearly the quotient of

$$\left\{\partial_x^2 \frac{(1)(2)\ldots.(\mathbf{i}+\mathbf{j}+\mathbf{k}+\ldots.)}{(1)(2)\ldots.(\mathbf{i})\cdot(1)(2)\ldots.(\mathbf{j})\cdot(1)(2)\ldots.(\mathbf{k})\cdot\ldots.}\right\}_{x=1}$$

by

$$\frac{(i+j+k+\ldots.)!}{i!\,j!\,k!\,\ldots.}.$$

To evaluate this we require the value of

$$\left\{\partial_x^2\frac{1-x^\lambda}{1-x^\mu}\right\}_{x=1}$$

when $\lambda>\mu$. This I find to be

$$\frac{1}{6}\frac{\lambda}{\mu}(\lambda-\mu)(2\lambda-\mu-3).$$

There is theoretically no difficulty in determining the average value of any given positive integral power of the greater index. It depends upon finding the value of

$$\left\{\partial_x^s\frac{1-x^\lambda}{1-x^\mu}\right\}_{x=1}$$

when $\lambda>\mu$.

The simplest cases, corresponding to $s=1$ and $s=2$, may be written

$$\left\{\partial_x\frac{1-x^\lambda}{1-x^\mu}\right\}_{x=1}=\mu\left\{\frac{1}{2}\frac{\lambda}{\mu}\left(\frac{\lambda}{\mu}-1\right)\right\},$$

$$\left\{\partial_x^2\frac{1-x^\lambda}{1-x^\mu}\right\}_{x=1}=\mu^2\left\{\frac{1}{6}\frac{\lambda}{\mu}\left(\frac{\lambda}{\mu}-1\right)\left(2\frac{\lambda}{\mu}-1\right)\right\}-\mu\left\{\frac{1}{2}\frac{\lambda}{\mu}\left(\frac{\lambda}{\mu}-1\right)\right\},$$

and the general result, at which I have arrived, may be stated in the following manner: Let

$$\mu(\mu-1)(\mu-2)\ldots\ldots(\mu-s+1)=\mu^s+A_{s-1}\mu^{s-1}+A_{s-2}\mu^{s-2}+\ldots\ldots+A_1\mu,$$

and

$$1^\nu+2^\nu+\ldots\ldots+n^\nu={}_nS_\nu\,,$$

then

$$\left\{\partial_x^s\frac{1-x^\lambda}{1-x^\mu}\right\}_{x=1}=\{\mu^s{}_nS_s+A_{s-1}\mu^{s-1}{}_nS_{s-1}+A_{s-2}\mu^{s-2}{}_nS_{s-2}+\ldots\ldots+A_1\mu\,{}_nS_1\}_{n=\frac{\lambda}{\mu}-1}.$$

I do not interrupt the present investigation by giving a proof of this elegant theorem. I write at length the two cases which follow those given above.

$$\left\{\partial_x^3\frac{1-x^\lambda}{1-x^\mu}\right\}_{x=1}=\mu^3\left\{\frac{1}{4}\frac{\lambda^2}{\mu^2}\left(\frac{\lambda}{\mu}-1\right)^2\right\}-3\mu^2\left\{\frac{1}{6}\frac{\lambda}{\mu}\left(\frac{\lambda}{\mu}-1\right)\left(2\frac{\lambda}{\mu}-1\right)\right\}+2\mu\left\{\frac{1}{2}\frac{\lambda}{\mu}\left(\frac{\lambda}{\mu}-1\right)\right\},$$

$$\left\{\partial_x^4\frac{1-x^\lambda}{1-x^\mu}\right\}_{x=1}=\mu^4\left\{\frac{1}{30}\frac{\lambda}{\mu}\left(\frac{\lambda}{\mu}-1\right)\left(2\frac{\lambda}{\mu}-1\right)\left(3\frac{\lambda^2}{\mu^2}-3\frac{\lambda}{\mu}-1\right)\right\}-6\mu^3\left\{\frac{1}{4}\frac{\lambda^2}{\mu^2}\left(\frac{\lambda}{\mu}-1\right)^2\right\}$$
$$+11\mu^2\left\{\frac{1}{6}\frac{\lambda}{\mu}\left(\frac{\lambda}{\mu}-1\right)\left(2\frac{\lambda}{\mu}-1\right)\right\}-6\mu\left\{\frac{1}{2}\frac{\lambda}{\mu}\left(\frac{\lambda}{\mu}-1\right)\right\}.$$

These formulæ may be often simplified. Thus I find that

$$\left\{\partial_x^3\frac{1-x^\lambda}{1-x^\mu}\right\}_{x=1}=\frac{1}{4}\frac{\lambda}{\mu}(\lambda-\mu)(\lambda-\mu-2)(\lambda-2).$$

The result depends upon the sums of the powers of the natural numbers and therefore ultimately upon the numbers of Bernouilli. I notice that

$${}_{-n}S_\nu=(-)^{\nu+1}{}_{n-1}S_\nu,$$

a useful relation. Writing

$$\left\{\partial_x^s\frac{1-x^\lambda}{1-x^\mu}\right\}_{x=1}=K_s,\quad\text{and}\quad({}_nS_\nu)_{n=\frac{\lambda}{\mu}-1}=S'_\nu,$$

we have

$$\begin{aligned}
&K_1=\mu S'_1, && \mu S'_1=K_1,\\
&K_2=\mu^2S'_2-\mu S'_1, && \mu^2S'_2=K_1+K_2,\\
&K_3=\mu^3S'_3-3\mu^2S'_2+2\mu S'_1, && \mu^3S'_3=K_1+3K_2+K_3,\\
&K_4=\mu^4S'_4-6\mu^3S'_3+11\mu^2S'_2-6\mu S'_1, && \mu^4S'_4=K_1+7K_2+6K_3+K_4,\\
&\ldots\ldots\ldots\ldots\ldots\ldots, && \ldots\ldots\ldots\ldots\ldots\ldots,
\end{aligned}$$

where the coefficients in the expression of K_s are derived from the product $x(x-1)\ldots\ldots(x-s+1)$, and those in the expression of $\mu^sS'_s$ from the expression of x^s as a linear function of x, $x(x-1)$, $x(x-1)(x-2)$, ...

To apply these results I write

$$(X-i)(X-j)(X-k)\ldots\ldots=X^s-\{1\}X^{s-1}+\{11\}X^{s-2}-\ldots\ldots$$

I find that the average value of $p(p-1)$ is

$$\frac{1}{12}[3\{22\}+6\{211\}+18\{1111\}+\{21\}+2\{111\}-5\{11\}],$$

where, in the notation of symmetric functions, $\{22\}$ denotes $\Sigma\, i^2 j^2$, etc. Since we have established that the average value of p is $\frac{1}{2}\{11\}$, we see that the average value of p^2 is

$$\frac{1}{12}[3\{22\}+6\{211\}+18\{1111\}+\{21\}+2\{111\}+\{11\}],$$

which may be written

$$\frac{1}{4}\{11\}^2+\frac{1}{12}[\{21\}+2\{111\}+\{11\}],$$

wherein it will be noticed that the first term is the square of the average value of p.

As a verification observe that, for the permutations of $\alpha\beta\gamma\delta$,

$$\Sigma x^p = 1+3x+5x^2+6x^3+5x^4+3x^5+x^6;$$

so that the sum of the squares of the indices is

$$3\cdot 1^2+5\cdot 2^2+6\cdot 3^2+5\cdot 4^2+3\cdot 5^2+1\cdot 6^2=268,$$

and the average value of the square $=\frac{268}{24}=\frac{67}{6}$.

The formula gives $\frac{1}{12}(3\cdot 6+6\cdot 12+18\cdot 1+12+2\cdot 4+6)=\frac{76}{6}$. 67.

For the case $\alpha^i\beta^j$ the average value of the square of p is

$$\frac{1}{12}ij(3ij+i+j+1).$$

21. *Average Value of the Cube of the Greater Index.* Similarly the reader will find with little difficulty that the average value of $p(p-1)(p-2)$ for the assemblage $\alpha^i\beta^j$ is

$$\frac{1}{8}ij(i^2j^2+i^2j+ij^2-5ij-2i-2j+6);$$

and, since $p^3=p+3p(p-1)+p(p-1)(p-2)$, we find the average value of p^3 by adding to the above

$$\frac{1}{2}ij+\frac{1}{4}ij(3ij+i+j-5);$$

thence the average value of p^3 is found to be

$$\frac{1}{8}i^2j^2(i+1)(j+1).$$

22. For the assemblage $\alpha^i\beta^j\gamma^k$ if we write

$$\frac{(1)(2)\ldots.(\mathrm{i}+\mathrm{j}+\mathrm{k})}{(1)(2)\ldots.(\mathrm{i})\cdot(1)(2)\ldots.(\mathrm{j})\cdot(1)(2)\ldots.(\mathrm{k})}$$

$$=\frac{(1)(2)\ldots.(\mathrm{i}+\mathrm{j}+\mathrm{k})}{(1)(2)\ldots.(\mathrm{i})\cdot(1)(2)\ldots.(\mathrm{j}+\mathrm{k})}\cdot\frac{(1)(2)\ldots.(\mathrm{j}+\mathrm{k})}{(1)(2)\ldots.(\mathrm{j})\cdot(1)(2)\ldots.(\mathrm{k})}$$

$$=F_{i,\,j+k}\cdot F_{j,\,k},$$

since

$$\partial_x^3 F_{i,j,k}=F_{i,j+k}\,\partial_x^3 F_{j,k}+3\,\partial_x F_{i,j+k}\,\partial_x^2 F_{j,k}+3\,\partial_x^2 F_{i,j+k}\,\partial_x F_{j,k}+\partial_x^3 F_{i,j+k}\cdot F_{j,k},$$

the average value of $p(p-1)(p-2)$ is seen to be

$$\frac{1}{8}jk(j^2k^2+j^2k+jk^2-5jk-2j-2k+6)+\frac{3}{2}i(j+k)\cdot\frac{1}{12}jk(3jk+j+k-5)$$

$$+3\cdot\frac{1}{12}i(j+k)\{3i(j+k)+i+j+k-5\}\cdot\frac{1}{2}jk$$

$$+\frac{1}{8}i(j+k)\{i^2(j+k)^2+i^2(j+k)+i(j+k)^2-5i(j+k)-2i-2(j+k)+6\},$$

which is

$$\frac{1}{8}\{\Sigma i^3j^3+3\Sigma i^3j^2k+6\Sigma i^2j^2k^2+\Sigma i^3j^2+2\Sigma i^3jk+4\Sigma i^2j^2k$$
$$-5\Sigma i^2j^2-10\Sigma i^2jk-2\Sigma i^2j-4ijk+6\Sigma ij\}.$$

Recalling that, for the same assemblage,

Average value of p is $\frac{1}{2}\Sigma ij$,

and of $p(p-1)$ is $\frac{1}{12}(3\Sigma i^2j^2+6\Sigma i^2jk+\Sigma i^2j+2ijk-5\Sigma ij)$,

we find that the average value of p^3 is

$$\frac{1}{8}\{\Sigma i^3j^3+3\Sigma i^3j^2k+6i^2j^2k^2+\Sigma i^3j^2+2\Sigma i^3jk+4\Sigma i^2j^2k+2\Sigma i^2jk+\Sigma i^2j^2\}.$$

23. With small labor I find that for the general assemblage the average value of p^3 is

$$\frac{1}{8}[6\{3111\}+15\{2211\}+\{33\}+3\{321\}+6\{222\}+9\{2111\}$$
$$+\{32\}+2\{311\}+4\{221\}+\{22\}+2\{211\}+6\{1111\}].$$

24. *Average Values from Another Standpoint.* For the assemblage $\alpha_1^{i_1}\alpha_2^{i_2}\ldots.\alpha_s^{i_s}$ let us inquire into the effect which the particular major contact $\alpha_\mu\alpha_\lambda$ $(\mu>\lambda)$ has upon the sum of the greater indices Σp. When the contact has σ letters to the left of it, it adds $\sigma+1$ to the index of the permutation. This happens in the case of

$$\frac{(S-2)!}{i_1!\,i_2!\ldots.i_s!}\,i_\lambda i_\mu \text{ permutations, where } S=\Sigma i,$$

so that, altogether, the effect of the contact $\alpha_\mu\alpha_\lambda$ is to add to Σp the number

$$\sum_{\sigma=0}^{\sigma=S-2}\frac{(S-2)!}{i_1!\,i_2!\ldots.i_s!}\,i_\lambda i_\mu\,(\sigma+1),$$

or

$$\frac{S!}{i_1!\,i_2!\ldots.i_s!}\,\frac{1}{2}\,i_\lambda i_\mu.$$

Hence the contact $\alpha_\mu\alpha_\lambda$, on the average, adds the number $\frac{1}{2}\,i_\lambda i_\mu$ to the index of a permutation, and thence we find that the average value of p is

$$\frac{1}{2}\,\Sigma\, i_\lambda i_\mu,$$

verifying the conclusion of Art. 18.

25. The number p is obtained as the sum of m numbers $p_1, p_2, \ldots., p_m$,* where m is the class of the permutation *quâ* the greater index. If we had under consideration the sum

$$\Sigma(p_1^\nu+p_2^\nu+\ldots.+p_m^\nu),$$

we might inquire into the effect which the particular contact $\alpha_\mu\alpha_\lambda$ has upon such sum. The answer to this question is that the effect is to add to the sum the number

$$\sum_{\sigma=0}^{\sigma=S-2}\frac{(S-2)!}{i_1!\,i_2!\ldots.i_s!}\,i_\lambda i_\mu\,(\sigma+1)^\nu,$$

which is

$$\frac{S!}{i_1!\,i_2!\ldots.i_s!}\;\frac{1^\nu+2^\nu+\ldots.+(S-1)^\nu}{S(S-1)}\,i_\lambda i_\mu.$$

Hence the contact $\alpha_\mu\alpha_\lambda$, on the average, adds to the sum

$$p_1^\nu+p_2^\nu+\ldots.+p_m^\nu,$$

with regard to a permutation, the number

$$\frac{1^\nu+2^\nu+\ldots.+(S-1)^\nu}{S(S-1)}\,i_\lambda i_\mu.$$

From this result it follows that the average value of $p_1^\nu+p_2^\nu+\ldots.+p_m^\nu$ is

$$\frac{1^\nu+2^\nu+\ldots.+(S-1)^\nu}{S(S-1)}\,\Sigma\, i_\lambda i_\mu.$$

26. The average value of $p_1^2+p_2^2+\ldots.+p_m^2$ is therefore

$$\frac{1}{6}\,(2\Sigma i_1-1)\,\Sigma i_1 i_2;$$

* These numbers it is convenient to term the "components" of the index p.

and of $p_1^3+p_2^3+p_3^3+\ldots\ldots+p_m^3$,

$$\frac{1}{4}\Sigma i_1(\Sigma i_1-1)\Sigma i_1 i_2;$$

and so forth.

To verify these results, take the assemblage $\alpha^2\beta\gamma$:

	p_1	p_2	$p_1^2+p_2^2$	$p_1^3+p_2^3$		p_1	p_2	$p_1^2+p_2^2$	$p_1^3+p_2^3$
$\alpha\alpha\beta\gamma$	0	0	0	0	$\beta\alpha\alpha\gamma$	1	0	1	1
$\alpha\alpha\gamma\beta$	3	0	9	27	$\gamma\alpha\alpha\beta$	1	0	1	1
$\alpha\beta\alpha\gamma$	2	0	4	8	$\beta\alpha\gamma\alpha$	1	3	10	28
$\alpha\gamma\alpha\beta$	2	0	4	8	$\gamma\alpha\beta\alpha$	1	3	10	28
$\alpha\beta\gamma\alpha$	3	0	9	27	$\beta\gamma\alpha\alpha$	2	0	4	8
$\alpha\gamma\beta\alpha$	2	3	13	35	$\gamma\beta\alpha\alpha$	1	2	5	9
			39	105				31	75

giving the average of $p_1^2+p_2^2$ and $p_1^3+p_2^3$ equal to $\frac{35}{6}$ and 15 respectively.

The formulæ give, since $i_1=2$, $i_2=1$, $i_3=1$, $\Sigma i_1=4$, $\Sigma i_1 i_2=5$,

$$\text{Average of } p_1^2+p_2^2=\frac{1}{6}(2\cdot 4-1)\cdot 5=\frac{35}{6},$$

$$\text{Average of } p_1^3+p_2^3=\frac{1}{4}\cdot 4\cdot 3\cdot 5=15,$$

a verification.

27. *Average Value of the Class of a Permutation quâ the Greater Index.* If in the foregoing results we put $\nu=0$, then

$$p_1^0+p_2^0+\ldots\ldots+p_m^0=m,$$

the class of the permutation. Hence the average value of the class is

$$\frac{S-1}{S(S-1)}\Sigma i_1 i_2 \text{ or } \frac{\Sigma i_1 i_2}{\Sigma i_1}.^*$$

In the above example

$$\frac{\Sigma i_1 i_2}{\Sigma i_1}=\frac{5}{4},$$

and we observe that one permutation is of class 0, seven of class 1 and four of class 2, so that the average class is

$$\frac{0+7+8}{12}=\frac{5}{4},$$

* Another corollary, paying attention to Art. 25, is that the average value of $p_s^\nu\,(p_s\neq 0)$ is

$$\frac{1}{\Sigma i-1}\{1^\nu+2^\nu+\ldots\ldots+(\Sigma i-1)^\nu\}.$$

a verification. In other words we may say that the average number of major contacts possessed by a permutation is

$$\frac{\Sigma i_1 i_2}{\Sigma i_1}.$$

28. An easy corollary is that a permutation possesses on the average

$$\frac{2\Sigma i_1 i_2}{\Sigma i_1}$$

contacts which are either major or minor.

Since, moreover, the whole number of contacts is $\Sigma i_1 - 1$, it follows that, on the average, a permutation possesses a number of equal contacts equal to

$$\Sigma i_1 - 1 - 2\frac{\Sigma i_1 i_2}{\Sigma i_1} \quad \text{or} \quad 2\frac{\Sigma \binom{i_1}{2}}{\Sigma i_1},$$

an elegant result.

This last result may be obtained independently.

29. *Average Values Connected with the Equal Index.* The particular contact $\alpha_\lambda \alpha_\lambda$ has an effect upon the sum of the equal indices $\Sigma q = q_1 + q_2 + \ldots\ldots + q_m$, where m is the class of the permutation *quâ* equal contacts. When such a contact has σ letters to the left of it, it adds $\sigma+1$ to the equal index of the permutation. This happens in the case of

$$\frac{(S-2)\,!}{i_1\,!\,i_2\,!\ldots\ldots i_s\,!}\cdot 2\binom{i_\lambda}{2}$$

permutations; so that, altogether, the contact adds to Σq the number

$$\sum_{\sigma=0}^{\sigma=S-2} \frac{(S-2)\,!}{i_1\,!\,i_2\,!\ldots\ldots i_s\,!}\cdot 2\binom{i_\lambda}{2}(\sigma+1),$$

or

$$\frac{S\,!}{i_1\,!\,i_2\,!\ldots\ldots i_s\,!}\binom{i_\lambda}{2}.$$

Hence the contact $\alpha_\lambda \alpha_\lambda$, on the average, adds the number

$$\binom{i_\lambda}{2}$$

to the equal index of a permutation. Thence it appears that the average value of the equal index is

$$\Sigma\binom{i_1}{2}$$

(see Art. 19).

Moreover, the contact $\alpha_\lambda \alpha_\lambda$, on the average, adds to the sum

$$q_1^\nu + q_2^\nu + \ldots\ldots + q_m^\nu$$

(where $q=q_1+q_2+\ldots\ldots+q_m$) the number

$$2\frac{1^\nu+2^\nu+\ldots\ldots+(S-1)^\nu}{S(S-1)}\binom{i_\lambda}{2};$$

and thence the average value of $q_1^\nu+q_2^\nu+\ldots\ldots+q_m^\nu$ is

$$2\cdot\frac{1^\nu+2^\nu+\ldots\ldots+(S-1)^\nu}{S(S-1)}\Sigma\binom{i_1}{2}.$$

The average values of $q_1^2+q_2^2+\ldots\ldots+q_m^2$ and $q_1^3+q_2^3+\ldots\ldots+q_m^3$ respectively are

$$\frac{1}{3}(2\Sigma i_1-1)\Sigma\binom{i_1}{2}$$

and

$$\frac{1}{2}\Sigma i(\Sigma i-1)\Sigma\binom{i_1}{2}.$$

Moreover, putting $\nu=0$, we find for the average class, *quâ* the equal index,

$$\frac{2}{\Sigma i_1}\Sigma\binom{i_1}{2},$$

agreeing with the result of Art. 28.

An easy corollary is that the average value of q_g^ν $(q_g\neq 0)$ is

$$\frac{1}{\Sigma i-1}\{1^\nu+2^\nu+\ldots\ldots+(\Sigma i-1)^\nu\}.$$

In fact q_g^ν and p_g^ν have the same average values in respect of those permutations for which they exist as positive non-zero numbers.

§ 2. *The Indices of the Second Genus.*

30. When we come to consider the permutations of $\alpha^i\beta^j\gamma^k$, we find that the value of $p-r$ does not follow any simple law, and moreover the only linear function of p, q and r that does so is the sum $p+q+r$. I now import a new idea into the subject by defining the major and minor indices of a permutation.

The Major Index of a Permutation. Consider two adjacent letters $\alpha_\mu\alpha_\lambda$ $(\mu>\lambda)$ of a permutation of the assemblage

$$\alpha_1^{i_1}\alpha_2^{i_2}\ldots\ldots\alpha_s^{i_s}.$$

If $\alpha_\mu\alpha_\lambda$ be the g-th major contact of the permutation, counting from the beginning, and α_μ be the p_g-th letter of the permutation, put

$$(\mu-\lambda)p_g=P_g.$$

Then $\sum_{g=1}^{g=m}(\mu-\lambda)p_g$ or $\sum_{g=1}^{g=m}P_g$ is the major index of the permutation and m is the class of the permutation *quâ* the major index. We put the major index equal to P, so that

$$P_1+P_2+\ldots\ldots+P_m=P.$$

31. *The Minor Index of a Permutation.* Consider two adjacent letters $\alpha_\lambda\alpha_\mu$ $(\mu>\lambda)$ of a permutation.

If $\alpha_\lambda\alpha_\mu$ be the g-th minor contact of the permutation, counting from the beginning, and α_λ be the r_g-th letter of the permutation, put

$$(\mu-\lambda)r_g=R_g.$$

Then $\sum_{g=1}^{g=m}(\mu-\lambda)r_g$ or $\sum_{g=1}^{g=m}R_g$ is the minor index of the permutation and m is the class of the permutation *quâ* the minor index. We put the minor index equal to R, so that

$$R_1+R_2+\ldots\ldots+R_m=R.$$

In this system of indices the equal index is zero since $(\mu-\lambda)$ vanishes for $\lambda=\mu$.

Ex. gr., for the permutation $\alpha_4\alpha_1\alpha_3\alpha_1\alpha_5$

$$\text{Major index}=3\cdot1+2\cdot3=9,$$
$$\text{Minor index}=2\cdot2+4\cdot4=20.$$

It will be at once noticed that for the particular assemblage $\alpha^i\beta^j$ the major and minor indices are the same as the greater and lesser indices respectively.

32. *Theorem in regard to P—R. We have now the important circumstance that, for the most general assemblage, the number P—R has only s values corresponding to the s different letters that may terminate the permutations.*

This will now be established. To bring something definite under the reader's eye I give the major and minor indices of the permutations of $\alpha\beta\gamma\delta$.

	P	R	$P-R$		P	R	$P-R$
$\alpha\beta\gamma\delta$	0	6	−6	$\alpha\beta\delta\gamma$	3	5	−2
$\alpha\gamma\beta\delta$	2	8	−6	$\alpha\delta\beta\gamma$	4	6	−2
$\beta\alpha\gamma\delta$	1	7	−6	$\beta\alpha\delta\gamma$	4	6	−2
$\beta\gamma\alpha\delta$	4	10	−6	$\beta\delta\alpha\gamma$	6	8	−2
$\gamma\alpha\beta\delta$	2	8	−6	$\delta\alpha\beta\gamma$	3	5	−2
$\gamma\beta\alpha\delta$	3	9	−6	$\delta\beta\alpha\gamma$	4	6	−2

	P	R	$P-R$		P	R	$P-R$
$\alpha\gamma\delta\beta$	6	4	2	$\beta\gamma\delta\alpha$	9	3	6
$\alpha\delta\gamma\beta$	5	3	2	$\beta\delta\gamma\alpha$	8	2	6
$\gamma\alpha\delta\beta$	8	6	2	$\gamma\beta\delta\alpha$	10	4	6
$\gamma\delta\alpha\beta$	6	4	2	$\gamma\delta\beta\alpha$	7	1	6
$\delta\alpha\gamma\beta$	6	4	2	$\delta\beta\gamma\alpha$	8	2	6
$\delta\gamma\alpha\beta$	5	3	2	$\delta\gamma\beta\alpha$	6	0	6

The symmetry of these numbers is striking. It gives a notion of the theorems to be established. Let

$$(\alpha_1^{i_1}\alpha_2^{i_2}\ldots\ldots\alpha_s^{i_s})$$

denote any permutation of the assemblage $\alpha_1^{i_1}\alpha_2^{i_2}\ldots\ldots\alpha_s^{i_s}$, and let

$$(\alpha_1^{i_1}\alpha_2^{i_2}\ldots\ldots\alpha_s^{i_s})_l$$

denote any permutation of the same assemblage which terminates with α_l. Assume that, for every permutation

$$(\alpha_1^{i_1}\alpha_2^{i_2}\ldots\ldots\alpha_s^{i_s})_l,$$

$P-R$ has the value $-l\Sigma i+i_1+2i_2+3i_3+\ldots\ldots+si_s$. Now consider the value of $P-R$ when α_k is placed as terminal letter to the above permutation. We have three cases to consider.

(i) If $k<l$, P for $(\alpha_1^{i_1}\alpha_2^{i_2}\ldots\ldots\alpha_s^{i_s})_l\,\alpha_k$ becomes $P+(l-k)\Sigma i$ and R remains unaltered, so that $P-R$ becomes

$$-k(\Sigma i+1)+i_1+2i_2+\ldots\ldots+k(i_k+1)+\ldots\ldots+si_s,$$

and, since $\Sigma i+1$ and i_k+1 are the new values of Σi and i_k, we see that the value of $P-R$ is independent of l and thus obeys the law assumed.

(ii) If $k=l$, P and R are both unaltered and we may write the value of $P-R$

$$-l(\Sigma i+1)+i_1+2i_2+\ldots\ldots+l(i_l+1)+\ldots\ldots+si_s,$$

which is in accord with the assumed law.

(iii) If $k>l$, P is unchanged but R is increased by $(k-l)\Sigma i$, so that $P-R$ is diminished by $(k-l)\Sigma i$ and thus becomes

$$-k(\Sigma i+1)+i_1+2i_2+\ldots\ldots+k(i_k+1)+\ldots\ldots+si_s,$$

and again the assumed law is verified.

Therefore by induction from simple cases the general law that for the permutations of $(\alpha_1^{i_1}\alpha_2^{i_2}\ldots\ldots\alpha_s^{i_s})_l$ we have

$$P-R=-l\Sigma i+i_1+2i_2+\ldots\ldots+si_s$$

is established.

33. The law is, however, more extensive than this, for, instead of having letters $\alpha_1, \alpha_2, \ldots\ldots, \alpha_s$ with subscripts forming the series of natural numbers, the subscripts may be any numbers whatever. Thus, take the assemblage to be

$$\alpha_{m_1}^{i_{m_1}}\alpha_{m_2}^{i_{m_2}}\ldots\ldots\alpha_{m_s}^{i_{m_s}},$$

where $m_1, m_2, \ldots\ldots, m_s$ are *any* integers, of which one is l. Assume the law

$$P-R \text{ for } \left(\alpha_{m_1}^{i_{m_1}}\alpha_{m_2}^{i_{m_2}}\ldots\ldots\alpha_{m_s}^{i_{m_s}}\right)_l=-l\Sigma i+m_1i_{m_1}+m_2i_{m_2}+\ldots\ldots+li_l+\ldots\ldots+m_si_{m_s}.$$

Now adding α_k to the permutation as before we get for the new value of $P-R$

$$-k(\Sigma i+1)+m_1 i_{m_1}+m_2 i_{m_2}+\ldots\ldots+k(i_k+1)+\ldots\ldots+m_s i_{m_s},$$

whether k be $>$, $=$ or $<$ l and whether k is one of the numbers $m_1, m_2, \ldots\ldots, m_s$ or not. This is independent of l and obeys the assumed law, so that the general law is established by induction. Setting forth the values of $P-R$ for the assemblages $\alpha_1^{i_1}\alpha_2^{i_2}$, $\alpha_1^{i_1}\alpha_2^{i_2}\alpha_3^{i_3}$, ... we obtain the schemes

$$\begin{array}{c|ll} \alpha_2 & -i_1 & \\ \alpha_1 & & +i_2, \end{array}\qquad \begin{array}{c|lll} \alpha_3 & -2i_1-i_2 & & \\ \alpha_2 & -i_1 & +i_3 & \\ \alpha_1 & & +i_2+2i_3, & \end{array}\qquad \begin{array}{c|llll} \alpha_4 & -3i_1-2i_2-i_3 & & & \\ \alpha_3 & -2i_1-i_2 & & +i_4 & \\ \alpha_2 & -i_1 & +i_3 & +2i_4 & \\ \alpha_1 & & i_2\ +2i_3+3i_4, & & \end{array}\quad \text{etc.,}$$

the letters α denoting the terminal letters of the permutations.

The values for the terminal letters α_m, α_{s-m+1} are symmetrically related. Each line of a scheme is derived from the preceding line by adding Σi.

34. *General Relation between Σx^P and Σx^R.* It is not true in general that $\Sigma x^P=\Sigma x^R$, although this equality happens to obtain for the particular case of the assemblage $\alpha_1^{i_1}\alpha_2^{i_2}\alpha_3^{i_3}$, as will be seen presently. The above investigation shows that if $\underset{l}{\Sigma} x^P$ refers to the permutations which terminate with α_l,

$$\underset{l}{\Sigma} x^P=x^{-l\Sigma i+i_1+2i_2+\ldots\ldots+si_s}\underset{l}{\Sigma} x^R.$$

This appears to be the most general relation obtainable, but, if the assemblage be unaltered by the substitution

$$\begin{pmatrix}\alpha_1\alpha_2 & \alpha_3 & \ldots\ldots\alpha_s\\ \alpha_s\alpha_{s-1} & \alpha_{s-2} & \ldots\ldots\alpha_1\end{pmatrix},$$

it is clear that the P and R of any permutation are the same as the R and P respectively of the permutation obtained by impressing the substitution. Thus in this case

$$\underset{l}{\Sigma} x^P=\underset{s-l+1}{\Sigma} x^R=x^{l\Sigma i-i_1-2i_2-\ldots\ldots-si_s}\underset{s-l+1}{\Sigma} x^P$$

and

$$\Sigma x^P=\Sigma x^R.$$

35. *Difference Equation for Σx^P.* Let

$$\Sigma x^P=v(i_1, i_2, \ldots\ldots, i_s),$$

and

$$\underset{m}{\Sigma} x^P=\underset{m}{v}(i_1, i_2, \ldots\ldots, i_s).$$

It is clear that $\underset{s}{v}(i_1, i_2, \ldots\ldots, i_s)=v(i_1, i_2, \ldots\ldots, i_{s-1}, i_s-1)$. The function $\underset{s-1}{v}(i_1, i_2, \ldots\ldots, i_s)$ refers to permutations that may end with $\alpha_s\alpha_{s-1}$; these give rise to the function

$$x^{\Sigma i-1}\underset{s}{v}(i_1, i_2, \ldots\ldots, i_{s-1}-1, i_s);$$

the remaining permutations give rise to the function

$$v(i_1, i_2, \ldots, i_{s-1}-1, i_s) - v(i_1, i_2, \ldots, i_{s-1}-1, i_s).$$

Hence

$$\underset{s-1}{v}(i_1 i_2 \ldots i_s) = v(i_1, i_2, \ldots, i_{s-1}-1, i_s) - (\Sigma\mathbf{i}-1)\, v(i_1, i_2, \ldots, i_{s-1}-1, i_s-1),$$

where $(\Sigma\mathbf{i}-1)$ means $1-x^{\Sigma i-1}$.

By similar reasoning we obtain

$$\begin{aligned}\underset{s-2}{v}(i_1 i_2 \ldots i_s) = v(i_1, i_2, \ldots, i_{s-2}-1, i_{s-1}, i_s)& \\ -(2\Sigma\mathbf{i}-2)\, v(i_1, i_2, \ldots, i_{s-2}-1, i_{s-1}, i_s-1)& \\ -(\Sigma\mathbf{i}-1)\, v(i_1, i_2, \ldots, i_{s-2}-1, i_{s-1}-1, i_s)& \\ +(\Sigma\mathbf{i}-1)(\Sigma\mathbf{i}-2)\, v(i_1, i_2, \ldots, i_{s-2}-1, i_{s-1}-1, i_s-1)&,\end{aligned}$$

and the general formula is easy to obtain, but we do not need it at present. An application will be made at once to a particular case.

36. *The Assemblage* $\alpha^i\beta^j\gamma^k$. Since $v(i,j,k) = \underset{3}{v}(i,j,k) + \underset{2}{v}(i,j,k) + \underset{1}{v}(i,j,k)$, we find

$$\begin{aligned}\Sigma x^P = v(i, j, k) &= v(i, j, k-1) + v(i, j-1, k) + v(i-1, j, k) \\ &-(\Sigma\mathbf{i}-1)\, v(i, j-1, k-1,) - (\Sigma\mathbf{i}-1)\, v(i-1, j-1, k) \\ &-(2\Sigma\mathbf{i}-2)\, v(i-1, j, k-1) + (\Sigma\mathbf{i}-1)(\Sigma\mathbf{i}-2)\, v(i-1, j-1, k-1).\end{aligned}$$

Also interchanging i and k,

$$\begin{aligned}v(k, j, i) &= v(k, j, i-1,) + v(k, j-1, i) + v(k-1, j, i) \\ &-(\Sigma\mathbf{i}-1)\, v(k, j-1, i-1) - (\Sigma\mathbf{i}-1)\, v(k-1, j-1, i) \\ &-(2\Sigma\mathbf{i}-2)\, v(k-1, j, i-1) + (\Sigma\mathbf{i}-1)(\Sigma\mathbf{i}-2)\, v(k-1, j-1, i-1).\end{aligned}$$

These results enable us to establish that

$$v(i, j, k) = v(k, j, i);$$

for if this law obtains in all cases when the sum of the three numbers i, j, k is less than w, it is seen that the right-hand sides of the expressions obtained for $v(i, j, k)$ and $v(k, j, i)$ are identical. If then we can show that the law obtains when $i+j+k$ is equal to any number w in all cases, the law must be universal.

We may note in the first place that

$$v(0, j, k) = \frac{(\mathbf{1})(\mathbf{2})\ldots(\mathbf{j}+\mathbf{k})}{(\mathbf{1})(\mathbf{2})\ldots(\mathbf{j})\cdot(\mathbf{1})(\mathbf{2})\ldots(\mathbf{k})} = F_{j,k}(x),$$

$$v(i, j, 0) = F_{i,j}(x), \quad v(i, 0, k) = F_{i,k}(x^2);$$

and that for these forms the law obtains. It follows that the law obtains when $w=2$, for then the forms are

$$v(2, 0, 0), \quad v(0, 2, 0), \quad v(0, 0, 2), \quad v(1, 1, 0), \quad v(1, 0, 1), \quad v(0, 1, 1).$$

Hence universally

$$v(i, j, k) = v(k, j, i).$$

The function is therefore symmetrical in i and k.

37. *Proof that for the Assemblage* $\alpha^i\beta^j\gamma^k$, $\Sigma x^P = \Sigma x^R$. Since the value of P for any permutation of the assemblage is equal to the value of R for the permutation obtained from the former by the substitution $(\alpha\gamma)$, it follows that Σx^P for the assemblage $\alpha^i\beta^j\gamma^k$ is equal to Σx^R for the assemblage $(\alpha^k\beta^j\gamma^i)$. Since, moreover, the substitution has been shown above not to alter the expression Σx^P, it follows that universally $\Sigma x^P = \Sigma x^R$.

38. *The Algebraic Expression of* Σx^P *for the Assemblage* $\alpha^i\beta^j\gamma^k$. Since it has been shown that the functions $u(0, j, k)$, $u(i, 0, k)$, $u(i, j, 0)$ are equal to $F_{j,k}(x)$, $F_{i,k}(x^2)$, $F_{i,j}(x)$ respectively, we are led to consider the function

$$\frac{(1)(2)\ldots.(\mathrm{i}+\mathrm{j}+\mathrm{k})}{(1)(2)\ldots.(\mathrm{j})\cdot(1)(2)\ldots.(\mathrm{i}+\mathrm{k})}\cdot\frac{(2)(4)\ldots.(2\mathrm{i}+2\mathrm{k})}{(2)(4)\ldots.(2\mathrm{i})\cdot(2)(4)\ldots.(2\mathrm{k})},$$

which possesses these properties and also becomes equal to the number of permutations of $\alpha^i\beta^j\gamma^k$ when $x=1$. It is, moreover, symmetrical in i and k, and is obviously correct in simple particular cases. *Ex. gr.*, it is $\frac{(3)(4)}{(1)(2)}$ when $i=j=k=1$.

In fact, it will be shown subsequently that

$$\Sigma x^P = F_{j,i+k}(x)\cdot F_{i,k}(x^2);$$

and it has been verified that this expression satisfies the difference equation for $v(i, j, k)$ obtained above.

Another verification of our results is obtained by showing algebraically that Σx^R has the same expression as Σx^P.

39. *Derivation of* Σx^R *from* Σx^P. Since $\Sigma x^R = \Sigma x^{R-P+P} = x^{R-P}\Sigma x^P$ for all sets of permutations for which $R-P$ has a constant value, we must have

$$\Sigma x^R = x^{2i+j}\underset{3}{v}(i, j, k) + x^{i-k}\underset{2}{v}(i, j, k) + x^{-j-2k}\underset{1}{v}(i, j, k);$$

and this will be found to be equal to the product of

$$\frac{(1)(2)\ldots.(\mathrm{i}+\mathrm{j}+\mathrm{k}-1)}{(1)(2)\ldots.(\mathrm{j})\cdot(1)(2)\ldots.(\mathrm{i}+\mathrm{k})}\cdot\frac{(2)(4)\ldots.(2\mathrm{i}+2\mathrm{k}-2)}{(2)(4)\ldots.(2\mathrm{i})\cdot(2)(4)\ldots.(2\mathrm{k})}$$

and

$$x^{2i+j}(\mathrm{i}+\mathrm{k})(2\mathrm{k}) + x^{i-k}(\mathrm{j})(2\mathrm{i}+2\mathrm{k}) - x^{i-k}(\mathrm{j})(\mathrm{i}+\mathrm{k})(2\mathrm{k})$$
$$+ x^{-j-2k}(\mathrm{i}+\mathrm{k})(2\mathrm{i}) - x^{j-2k}(\mathrm{j})(\mathrm{i}+\mathrm{k})(2\mathrm{i}) - x^{-2k}(\mathrm{i}+\mathrm{k})(2\mathrm{i})(2\mathrm{k}).$$

The expression last written is found to be

$$(\mathrm{i}+\mathrm{j}+\mathrm{k})(2\mathrm{i}+2\mathrm{k}),$$

establishing that

$$\Sigma x^R = \frac{(1)(2)\ldots\ldots(\mathrm{i}+\mathrm{j}+\mathrm{k})}{(1)(2)\ldots\ldots(\mathrm{j})\cdot(1)(2)\ldots\ldots(\mathrm{i}+\mathrm{k})}\cdot\frac{(2)(4)\ldots\ldots(2\mathrm{i}+2\mathrm{k})}{(2)(4)\ldots\ldots(2\mathrm{i})\cdot(2)(4)\ldots\ldots(2\mathrm{k})},$$

the expression already found for Σx^P.

40. *Evaluation of* Σx^{P+R}. We could also find the expression of $\Sigma x^{\lambda P+\mu R}$, where λ, μ are arbitrary positive or negative integers. To find Σx^{P+R} we consider

$$x^{R-P}\Sigma x^{2P}$$

in three fragments and if $\Sigma x^P = G_{i,j,k}(x)$, we find

$$\Sigma x^{P+R} = x^{2i+j}\underset{3}{G}_{i,j,k}(x^2) + x^{i-k}\underset{2}{G}_{i,j,k}(x^2) + x^{-j-2k}\underset{1}{G}_{i,j,k}(x^2),$$

where $\underset{t}{u}(i,j,k)$ is written $\underset{t}{G}_{i,j,k}(x)$.

I thence find

$$\Sigma x^{P+R} = \frac{(2)(4)\ldots\ldots(2\mathrm{i}+2\mathrm{j}+2\mathrm{k}-2)(+\mathrm{i}+\mathrm{j}+\mathrm{k})}{(2)(4)\ldots\ldots(2\mathrm{j})\cdot(2)(4)\ldots\ldots(2\mathrm{i}+2\mathrm{k}-2)}$$

$$\cdot\frac{(4)(8)\ldots\ldots(4\mathrm{i}+4\mathrm{k}-4)}{(4)(8)\ldots\ldots(4\mathrm{i})\cdot(4)(8)\ldots\ldots(4\mathrm{k})}\cdot(x^{2i}+x^{2k})(x^j+x^{i+k}).$$

Ex. gr., for $i=2$, $j=1$, $k=1$, I thus find

$$\Sigma x^{P+R} = x^3 + 3x^5 + 4x^7 + 3x^9 + x^{11},$$

which the reader can readily verify from the permutations of $\alpha^2\beta\gamma$.

41. *Average Values of the Major and Minor Indices.* Following the method of Art. 18, we have only to differentiate the major or minor index function of the assemblage $\alpha^i\beta^j\gamma^k$, viz.

$$F_{j,i+k}(x)\,F_{i,k}(x^2),$$

with regard to x, and then to put $x=1$, to obtain the value of ΣP or ΣR, and thence the mean value of P or R.

The mean value is in fact

$$\frac{\{\partial_x F_{j,i+k}(x)\}_{x=1}}{\binom{i+j+k}{j}} + \frac{\{\partial_x F_{i,k}(x^2)\}_{x=1}}{\binom{i+k}{i}},$$

which, making use of results obtained above, is found to be

$$\frac{1}{2}j(i+k)+ik$$

or

$$\frac{1}{2}(ij+2ik+jk).$$

39

42. The mean value of $P(P-1)$ or $R(R-1)$ is similarly found to be

$$\frac{\{\partial_x^2 F_{j,i+k}(x)\}_{x=1}}{\binom{i+j+k}{j}}+2\frac{\{\partial_x F_{j,i+k}(x)\}_{x=1}}{\binom{i+j+k}{j}}\cdot\frac{\{\partial_x F_{i,k}(x^2)\}_{x=1}}{\binom{i+k}{i}}+\frac{\{\partial_x^2 F_{i,k}(x^2)\}_{x=1}}{\binom{i+k}{i}};$$

and, making use of previous results, this is found to be

$$\frac{1}{12}\{12i^2k^2+4i^2k+4ik^2-8ik+j(12i^2k+12ik^2+i^2+k^2+2ik-5i-5k)$$
$$+j^2(3i^2+3k^2+6ik+i+k)\},$$

symmetrical in i, k, as should be the case.

Adding to the above the mean value of P, viz. $\frac{1}{2}(ij+2ik+jk)$, we find for the mean value of P^2 the expression

$$\frac{1}{12}\{4ik(3ik+i+k+1)+j(i+k)(12ik+i+k+1)+j^2(i+k)(3i+3k+1)\}.$$

There is of course no theoretical difficulty in obtaining the mean values of higher powers of P.

43. *Mean Values Discussed in the Second Manner.* Following the procedure taken with regard to the greater, equal and lesser indices, we can obtain some results which appertain to the major and minor indices of the general assemblage $\alpha_1^{i_1}\alpha_2^{i_2}\ldots\alpha_s^{i_s}$.

I suppose that P is formed by addition in the manner

$$P=P_1+P_2+\ldots+P_m,$$

where m is the class of the permutation *quâ* the major index.

We inquire into the effect which the particular major contact $\alpha_{\mu+\nu}\alpha_\mu$ has upon the sum of the indices ΣP. When the contact has σ letters to its left and $\Sigma i-\sigma-2$ letters to its right, it adds $\nu(\sigma+1)$ to the index of the permutation. This happens in the case of

$$\frac{(\Sigma i-2)!\,i_\mu i_{\mu+\nu}}{i_1!\,i_2!\ldots i_s!}$$

permutations; so that, altogether, the contact adds to ΣP the number

$$\sum_{\sigma=0}^{\sigma=\Sigma i-2}\frac{(\Sigma i-2)!\,i_\mu i_{\mu+\nu}}{i_1!\,i_2!\ldots i_s!}\nu(\sigma+1),$$

which is

$$\frac{1}{2}\nu\frac{(\Sigma i)!}{i_1!\,i_2!\ldots i_s!}i_\mu i_{\mu+\nu}.$$

Hence, on the average, it adds the number

$$\frac{1}{2}\nu\, i_\mu i_{\mu+\nu}$$

to the major index of a permutation.

Also the $s-\nu$ contacts $\alpha_{\nu+1}\alpha_1, \alpha_{\nu+2}\alpha_2, \ldots, \alpha_s\alpha_{s-\nu}$ add, on the average, the number

$$\frac{1}{2}\nu\,(i_1 i_{\nu+1}+i_2 i_{\nu+2}+\ldots+i_{s-\nu} i_s)$$

to the index.

Now making ν vary from 1 to $s-1$, we find that the average value of the major index is

$$\frac{1}{2}\sum_{\mu=2}^{\mu=s}\sum_{\lambda=1}^{\lambda=\mu-1}(\mu-\lambda)\,i_\lambda i_\mu,$$

where $\mu>\lambda$, a result which clearly embraces and generalizes that already found for the assemblage $\alpha^i\beta^j\gamma^k$ from other principles.

44. The result is essentially the average value of $P_1+P_2+\ldots+P_m$. To find the average value of $P_1^t+P_2^t+\ldots+P_m^t$, we have merely to note that, under circumstances such as the foregoing, the contact $\alpha_{\mu+\nu}\alpha_\nu$ adds the number

$$\sum_{\sigma=0}^{\sigma=\Sigma i-2}\frac{(\Sigma i-2)!\,i_\mu i_{\mu+\nu}}{i_1!\,i_2!\ldots i_s!}\nu^t(\sigma+1)^t$$

to the sum ΣP_1^t. Hence we readily find that

$$\text{Average value of } \Sigma P_1^2=\frac{1}{6}(2\Sigma i-1)\sum_{\mu=2}^{\mu=s}\sum_{\lambda=1}^{\lambda=\mu-1}(\mu-\lambda)^2 i_\lambda i_\mu, \text{ where } \mu>\lambda;$$

$$\text{Average value of } \Sigma P_1^3=\frac{1}{4}(\Sigma i)(\Sigma i-1)\sum_{\mu=2}^{\mu=s}\sum_{\lambda=1}^{\lambda=\mu-1}(\mu-\lambda)^3 i_\lambda i_\mu, \text{ where } \mu>\lambda.$$

There is no difficulty in continuing the series, which ultimately, of course, involve the numbers of Bernouilli.

It will be noted that the above results enable us to find the average value of

$$\Sigma P_1 P_2.$$

Applied to the assemblage $\alpha^i\beta^j\gamma^k$, we find

$$\text{Average value of } \Sigma P_1^2=\frac{1}{6}(2i+2j+2k-1)(4ik+ij+jk),$$

$$\text{Average value of } \Sigma P_1^3=\frac{1}{4}(i+j+k)(i+j+k-1)(8ik+ij+jk),$$

. . . .

45. The same course of reasoning applies to the minor index

$$R=R_1+R_2+\ldots+R_m,$$

and we thus find that, for the most general assemblage,

$$\Sigma R=\Sigma P;$$

and that the average values of ΣR_1^t and ΣP_1^t are identical. This is somewhat remarkable, because the minor index function Σx^R is not equal in general to the major index function Σx^P; nor is there in general any one-to-one equality between the values of R and the values of P for the same assemblage. The reader may test this statement by looking into the indices of the assemblage $\alpha_1^{i_1}\alpha_2^{i_2}\alpha_3^{i_3}\alpha_4^{i_4}$ when $i_1=i_2=i_4=1$ and $i_3=0$. However, there is a one-to-one equality between the numbers P_g and the numbers R_g for the whole of the permutations of the assemblage. Indeed, if this were not the case, we could not have

$$\Sigma P_g^t = \Sigma R_g^t,$$

the summation being for all the permutations, and our results absolutely necessitate this.

46. *General Connection between Σx^P and Partitions.* With the object of investigating the functions Σx^P, Σx^R, I next place them in general connection with the theory of partitions. Consider any permutation

$$\beta\alpha\gamma\gamma\alpha\gamma\beta\alpha\gamma$$

of any assemblage $\alpha^3\beta^2\gamma^4$. Associate with it three rows of numbers

$$\begin{array}{l} a_1\,a_2\,a_3 \\ b_1\,b_2 \\ c_1\,c_2\,c_3\,c_4 \end{array}$$

the numbers in each row being in descending order of magnitude from left to right. I suppose the permutation to correspond to a descending order of magnitude, subject to conditions, of these nine numbers. I regard the permutation as yielding the Diophantine relations

$$\begin{array}{c} \beta\quad\alpha\quad\gamma\quad\gamma\quad\overbrace{\alpha\qquad\quad\alpha}\quad\gamma\quad\beta\quad\alpha\quad\gamma \\ b_1 > a_1 \geq c_1 \geq c_2 > a_2+1;\; a_2 \geq c_3 > b_2 > a_3 \geq c_4 \end{array},$$

where it will be noted that in correspondence with the letters α, β, γ, we have the letters a, the letters b and the letters c, each in the *given* descending order. The contacts $\beta\alpha$, $\gamma\beta$ are associated with the symbol $>$; the contact $\gamma\alpha$ also with the symbol $>$; but the letter a, corresponding to α, in this case is increased by unity. So, in general, when the contact is $\alpha_{\mu+\nu}\alpha_\mu$, ν being >0, the symbol is $>$, but the letter corresponding to α_μ is increased by $\nu-1$. When the contact is $\alpha_\mu\alpha_{\mu+\nu}$, where ν is zero or positive, the symbol is $\geq$, and the letter corresponding to $\alpha_{\mu+\nu}$ is not increased or diminished.

Under these circumstances the enumerating generating function of the numbers a, b, c, which satisfy the conditions, is

$$\frac{x^{22}}{(1)(2)\ldots.(9)},$$

where the number which is the exponent of x in the numerator is necessarily the major index of the permutation (compare Art. 6). If we go through a like process for every permutation of the given assemblage and add the generating functions, we find

$$\frac{\Sigma x^P}{(1)(2)\ldots(9)},$$

Σx^P being the major index function of the given assemblage $\alpha^3\beta^2\gamma^4$.

47. The question now arises: What have we enumerated by this generating function? We clearly have not counted the number of ways of arranging the numbers a, b, c in the three rows so that there may be descending order in each row and also a sum equal to a given number w, for that is what we succeeded in doing by means of the greater index function Σx^p. We have in fact eliminated certain arrangements from those under enumeration. It is easy to see that with reference to the contacts $\alpha\gamma$, $\gamma\alpha$, the former gives

$$\text{Either } a_1 \text{ or } a_2 \text{ or } a_3 \geq \text{either } c_1, c_2, c_3, \text{ or } c_4,$$

while the latter gives

$$\text{Either } c_1 \text{ or } c_2 \text{ or } c_3 \text{ or } c_4 > \text{either } a_1+1 \text{ or } a_2+1 \text{ or } a_3+1,$$

so that in no circumstances can any number in the third row be *one* greater than any number in the first row. There is no other condition, so that we have enumerated all of the arrangements that were connected with the greater index function *except* those which involve any third row number greater by one than any first row number. Having obtained the precise condition of partition corresponding to the major index function of the assemblage $\alpha^i\beta^j\gamma^k$, we are now in a position to investigate the major index function. Before doing this it is convenient to establish the like conditions that are associated with the major index function of the general assemblage $\alpha_1^{i_1}\alpha_2^{i_2}\ldots\alpha_s^{i_s}$.

48. In correspondence with the contact $\alpha_\mu\alpha_{\mu+\nu}$, we have as a condition of summation

$$\text{Some } \mu\text{-th row number} \geq \text{some } (\mu+\nu)\text{-th row number},$$

and for the contact $\alpha_{\mu+\nu}\alpha_\mu$

$$\text{Some } (\mu+\nu)\text{-th row number} > \text{some } \mu\text{-th row number} + \nu-1;$$

hence under no circumstances can a $(\mu+\nu)$-th row number be greater than a μ-th row number by any of the numbers

$$1, 2, 3, \ldots, \nu-1.$$

In regard to the rows under examination, there are thus $(\nu-1)i_\mu i_{\mu+\nu}$ conditions. Between adjacent rows there are no conditions. Altogether there are

$$\Sigma(\nu-1)i_\mu i_{\mu+\nu}$$

conditions, the summation extending to every pair of the numbers $i_1, i_2, \ldots, i_s$ whose subscripts differ, at least, by the number 2.

If we now find the function which enumerates the ways in which numbers, of given sum, can be arranged in s rows (containing respectively $i_1, i_2, \ldots, i$ numbers, zero being permissible as a number) in descending order in each row so that the above-mentioned $\Sigma(\nu-1)i_\mu i_{\mu+\nu}$ conditions are satisfied, we shall evidently obtain an expression equivalent to

$$\frac{\Sigma x^P}{(1)(2)\ldots(\Sigma i)},$$

and thus we can arrive at the form of the major index function Σx^P.

The minor index function Σx^R presents no further difficulty, since it was shown in Art. 34 that Σx^R for the assemblage $\alpha_1^{i_1}\alpha_2^{i_2}\ldots\alpha_s^{i_s}$ is identical with Σx^P for the assemblage $\alpha_1^{i_s}\alpha_2^{i_{s-1}}\ldots\alpha_s^{i_1}$.

49. *Determination of Σx^P for the Assemblage $\alpha^i\beta^j\gamma^k$.* The associated problem in partitions is merely concerned with the first and third rows of the three rows of numbers. We have thus two rows

$$a_1a_2\ldots a_i,$$
$$c_1c_2\ldots c_k,$$

and we have to find $\Sigma x^{a_1+a_2+\ldots+a_i+c_1+c_2+\ldots+c_k}$ subject to the conditions

$$a_1 \geq a_2 \geq a_3 \geq \ldots \geq a_i,$$
$$c_1 \geq c_2 \geq c_3 \geq \ldots \geq c_k,$$

with the added set of conditions that no c is to be one greater than any a. Let this function be $\phi_{i,k}(x)$. Then we shall have

$$\frac{\Sigma x^P}{(1)(2)\ldots(i+j+k)} = \frac{\phi_{i,k}(x)}{(1)(2)\ldots(j)}.$$

We will first determine $\phi_{1,1}(x)$. We have to find Σx^{a+c} subject to the single condition that c is not to be equal to $a+1$. This is very simple, because we have merely to exclude the terms for which $c=a+1$ from the complete function $\frac{1}{(1)^2}$. Thus

$$\phi_{1,1}(x) = \frac{1}{(1)_2} - \Sigma x^{2a+1} = \frac{1}{(1)^2} - \frac{1}{(2)} = \frac{(4)}{(1)(2)^2}.$$

It is, however, more advantageous to proceed by the method that is of general application. For this purpose I construct the function

$$\chi\frac{\lambda}{(1-\lambda x)\left(1-\frac{x}{\lambda}\right)},$$

where the symbol χ denotes that the function is to be expanded in ascending powers of the variable; that, from such expansion, all terms are to be rejected which do not involve either a positive or a negative power of λ; and that then λ is to be put equal to unity. *Ex. gr.*, the general term is

$$\lambda(\lambda x)^a\left(\frac{x}{\lambda}\right)^c \quad \text{or} \quad \lambda^{a+1-c}x^{a+c};$$

so that the symbol χ rejects terms for which $c=a+1$, and thus the final function is Σx^{a+c}, the sum being for all positive integers a, c except those for which $c=a+1$. This is the required condition, so that

$$\phi_{1,1}(x)=\chi\frac{\lambda}{(1-\lambda x)\left(1-\frac{x}{\lambda}\right)}.$$

Observe that the χ function is *necessarily* unaltered when we write $\frac{1}{\lambda}$ for λ, although its form is changed. We have the identity

$$\frac{\lambda}{(1-\lambda x)\left(1-\frac{x}{\lambda}\right)}=\frac{\lambda}{(2)(1-\lambda x)}+\frac{\frac{x^2}{\lambda}}{(2)\left(1-\frac{x}{\lambda}\right)}+\frac{x}{(2)};$$

so that

$$\chi\frac{\lambda}{(1-\lambda x)\left(1-\frac{x}{\lambda}\right)}=\frac{1}{(1)(2)}+\frac{x^2}{(1)(2)}=\frac{(4)}{(1)(2)^2};$$

leading of course to the result that, for the assemblage $\alpha\beta^j\gamma$, Σx^p has the expression

$$\frac{(1)(2)\ldots\ldots(j+2)}{(1)(2)\ldots\ldots(j)}\frac{(4)}{(1)(2)^2} \quad \text{or} \quad \frac{(j+1)(i+2)\cdot(4)}{(1)(2)\cdot(2)}.$$

50. To determine $\phi_{2,1}(x)$ I form a function

$$\chi\cdot\Omega\cdot\frac{\lambda\mu}{(1-A\lambda x)\left(1-\frac{\mu x}{A}\right)\left(1-\frac{x}{\lambda\mu}\right)}.$$

The general term in the expansion of the function is

$$\lambda\mu(A\lambda x)^{a_1}\left(\frac{\mu x}{A}\right)^{a_2}\left(\frac{x}{\lambda\mu}\right)^{c_1} \quad \text{or} \quad A^{a_1-a_2}\lambda^{a_1+1-c_1}\mu^{a_2+1-c_1}x^{a_1+a_2+c_1}.$$

If, therefore, in the expansion, the exponent of A is zero or positive, the condition $a_1\geq a_2$ is satisfied. If also the exponents of λ and μ are neither of them zero, the conditions $a_1+1\neq c_1$, $a_2+1\neq c_1$ are satisfied. The symbol Ω is associated with the auxiliary coefficient A in such wise that its effect upon the expanded function is to abolish all terms which involve A to a negative power. The symbol χ is associated with the auxiliaries λ and μ in such wise that it abolishes from the expansion all terms which do not involve λ and μ (each of them) positively or negatively. Hence, from our definitions

$$\phi_{2,1}(x)=\chi\cdot\Omega\cdot\frac{\lambda\mu}{(1-A\lambda x)\left(1-\frac{\mu x}{A}\right)\left(1-\frac{x}{\lambda\mu}\right)}.$$

The effect of Ω can be seen with little trouble, so that we may write

$$\phi_{2,1}(x)=\chi\cdot\frac{\lambda\mu}{(1-\lambda x)(1-\lambda\mu x^2)\left(1-\frac{x}{\lambda\mu}\right)}.$$

The reader will have little difficulty in proving that, if σ be any auxiliary or combination of auxiliaries,

$$\chi\frac{\sigma}{(1-\sigma x^s)\left(1-\frac{x^t}{\sigma}\right)}=\frac{\sigma}{(\mathbf{s}+\mathbf{t})(1-\sigma x^s)}+\frac{\frac{1}{\sigma}x^{2t}}{(\mathbf{s}+\mathbf{t})\left(1-\frac{x^t}{\sigma}\right)},\quad\sigma=1;$$

$$\chi\frac{\frac{1}{\sigma}}{(1-\sigma x^s)\left(1-\frac{x^t}{\sigma}\right)}=\frac{\sigma x^{2s}}{(\mathbf{s}+\mathbf{t})(1-\sigma x^s)}+\frac{\frac{1}{\sigma}}{(\mathbf{s}+\mathbf{t})\left(1-\frac{x^t}{\sigma}\right)},\quad\sigma=1.$$

Applying the first of these results we can eliminate the auxiliary μ. We have

$$\phi_{2,1}(x)=\chi\frac{1}{1-\lambda x}\left\{\frac{\lambda}{(\mathbf{3})(1-\lambda x^2)}+\frac{\frac{1}{\lambda}x^2}{(\mathbf{3})\left(1-\frac{x}{\lambda}\right)}\right\}$$

$$=\frac{1}{(\mathbf{1})(\mathbf{2})(\mathbf{3})}+\frac{x^2}{(\mathbf{3})}\chi\frac{\frac{1}{\lambda}}{(1-\lambda x)\left(1-\frac{x}{\lambda}\right)};$$

and, since in the χ function we can write $\frac{1}{\lambda}$ for λ and $\phi_{2,0}=\frac{1}{(\mathbf{1})(\mathbf{2})}$, we find

$$\phi_{2,1}(x)=\frac{1}{(\mathbf{3})}\{\phi_{2,0}(x)+x^2\phi_{1,1}(x)\},$$

a suggestive formula of reduction.

51. In a precisely similar manner we find

$$\phi_{3,1}(x)=\chi\cdot\Omega\cdot\frac{\lambda\mu\nu}{(1-A_1\lambda x)\left(1-\frac{A_2\mu x}{A_1}\right)\left(1-\frac{\nu x}{A_2}\right)\cdot\left(1-\frac{x}{\lambda\mu\nu}\right)},$$

where Ω abolishes the terms which involve negative powers of A_1 and A_2, and χ acts in the usual manner on the auxiliaries λ, μ, ν.

We find at once

$$\phi_{3,1}(x)=\chi\frac{\lambda\mu\nu}{(1-\lambda x)(1-\lambda\mu x^2)(1-\lambda\mu\nu x^3)\cdot\left(1-\frac{x}{\lambda\mu\nu}\right)};$$

and now, eliminating ν, we find

$$\phi_{3,1}(x)=\chi\frac{1}{(1-\lambda x)(1-\lambda\mu x^2)}\cdot\left\{\frac{\lambda\mu}{(4)(1-\lambda\mu x^3)}+\frac{\frac{1}{\lambda\mu}x^2}{(4)\left(1-\frac{x}{\lambda\mu}\right)}\right\},$$

$$=\frac{1}{(4)}\phi_{3,0}(x)+\frac{x^2}{(4)}\chi\frac{\frac{1}{\lambda\mu}}{(1-\lambda x)(1-\lambda\mu x^2)\left(1-\frac{x}{\lambda\mu}\right)}.$$

Now

$$\chi\frac{\frac{1}{\lambda\mu}}{(1-\lambda x)(1-\lambda\mu x^2)\left(1-\frac{x}{\lambda\mu}\right)}$$

is not of the same form as the expression for $\phi_{2,1}(x)$; but it is equivalent to it, because, from the fact that $\phi_{i,k}(x)=\phi_{k,i}(x)$, established above, if we proceed to form $\phi_{1,2}(x)$ we find

$$\phi_{1,2}(x)=\chi\frac{\lambda\mu}{(1-\lambda\mu x)\left(1-\frac{x}{\lambda}\right)\left(1-\frac{x^2}{\lambda\mu}\right)},$$

and herein we may substitute $\frac{1}{\lambda}$ for λ and $\frac{1}{\mu}$ for μ.

Hence

$$\phi_{3,1}(x)=\frac{1}{(4)}\phi_{3,0}(x)+\frac{x^2}{(4)}\phi_{2,1}(x).$$

52. Examination of the foregoing shows that

$$\phi_{s,1}(x)=\frac{1}{(s+1)}\phi_{s,0}(x)+\frac{x^2}{(s+1)}\phi_{s-1,1}(x),$$

and assuming that

$$\phi_{s-1,1}(x)=\frac{(2s)}{(1)(2)^2(3)\ldots\ldots(s)}$$

we find

40

$$\phi_{s,1}(x)=\frac{1}{(1)\,(2)\ldots\ldots(s+1)}+\frac{x^2(1-x^{2s})}{(1)\,(2)^2(3)\ldots\ldots(s+1)}=\frac{(2s+2)}{(1)\,(2)^2(3)\ldots\ldots(s+1)},$$

verifying the assumption.

Hence Σx^P for the assemblage $\alpha^i\beta^j\gamma$ is

$$\frac{(1)\,(2)\ldots\ldots(i+j+1)}{(1)\,(2)\ldots\ldots(j)\cdot(1)\,(2)\ldots\ldots(i+1)}\cdot\frac{(2)\,(4)\ldots\ldots(2i+2)}{(2)\,(4)\ldots\ldots(2i)\cdot(2)}.$$

53. We now examine the series $\phi_{2,2}(x)$, $\phi_{3,2}(x)$, ... It will be agreed that

$$\phi_{2,2}(x)=\chi\cdot\Omega\frac{\lambda_{11}\lambda_{12}\lambda_{21}\lambda_{22}}{(1-A\lambda_{11}\lambda_{12}x)\left(1-\frac{\lambda_{21}\lambda_{22}}{A}x\right)\cdot\left(1-\frac{B}{\lambda_{11}\lambda_{21}}x\right)\left(1-\frac{x}{B\lambda_{12}\lambda_{22}}\right)},$$

where Ω has reference to A and B, and χ to the auxiliaries λ_{11}, λ_{12}, λ_{21}, λ_{22}. We immediately eliminate A and B and find

$$\phi_{2,2}(x)=\chi\cdot\frac{\lambda_{11}\lambda_{12}\lambda_{21}\lambda_{22}}{(1-\lambda_{11}\lambda_{12}x)\,(1-\lambda_{11}\lambda_{12}\lambda_{21}\lambda_{22}x^2)\cdot\left(1-\frac{x}{\lambda_{11}\lambda_{21}}\right)\left(1-\frac{x^2}{\lambda_{11}\lambda_{12}\lambda_{21}\lambda_{22}}\right)}.$$

Eliminating λ_{22} we find

$$\phi_{2,2}(x)=\chi\cdot\frac{1}{(1-\lambda_{11}\lambda_{12}x)\left(1-\frac{x}{\lambda_{11}\lambda_{21}}\right)(4)}\left\{\frac{\lambda_{11}\lambda_{12}\lambda_{21}}{1-\lambda_{11}\lambda_{12}\lambda_{21}x^2}+\frac{\frac{1}{\lambda_{11}\lambda_{12}\lambda_{21}}x^4}{1-\frac{x^2}{\lambda_{11}\lambda_{12}\lambda_{21}}}\right\}.$$

$\phi_{2,2}(x)$ is now the sum of two χ functions; in the first of these we may put $\lambda_{12}=1$, since λ_{12} occurs to a positive power in every term of the expansion, and similarly we may put $\lambda_{21}=1$ in the second function, for it occurs to a negative power in every term of the expansion. Hence

$$\phi_{2,2}(x)=\frac{1}{(4)}\chi\cdot\frac{\lambda_{11}\lambda_{21}}{(1-\lambda_{11}x)\,(1-\lambda_{11}\lambda_{21}x^2)\left(1-\frac{x}{\lambda_{11}\lambda_{21}}\right)}$$

$$+\frac{x^4}{(4)}\chi\cdot\frac{\frac{1}{\lambda_{11}\lambda_{12}}}{(1-\lambda_{11}\lambda_{12}x)\left(1-\frac{x}{\lambda_{11}}\right)\left(1-\frac{x^2}{\lambda_{11}\lambda_{12}}\right)},$$

and from previous results we may write

$$\phi_{2,2}(x)=\frac{1}{(4)}\phi_{2,1}(x)+\frac{x^4}{(4)}\phi_{1,2}(x).$$

54. Similarly we find

$$\phi_{3,2}(x) = \frac{1}{(5)}\phi_{3,1}(x) + \frac{x^4}{(5)}\phi_{2,2}(x);$$

and in general

$$\phi_{i,2}(x) = \frac{1}{(i+2)}\phi_{i,1}(x) + \frac{x^4}{(i+2)}\phi_{i-1,2}(x).$$

We know the value of $\phi_{1,2}(x)$, and assuming that

$$\phi_{i-1,2}(x) = \frac{1}{(1)(2)\ldots\ldots(i+1)}\cdot\frac{(2)(4)\ldots\ldots(2i+2)}{(2)(4)\ldots\ldots(2i-2)\cdot(2)(4)},$$

we find

$$\phi_{i,2}(x) = \frac{(2i+2)}{(1)(2)^2(3)(4)\ldots\ldots(i+2)} + \frac{x^4}{(1)(2)\ldots\ldots(i+2)}\cdot\frac{(2i)(2i+2)}{(2)(4)},$$

$$= \frac{(2i+2)}{(1)(2)^2(3)(4)^2(5)\ldots\ldots(i+2)}\{1-x^4+x^4-x^{2i+4}\},$$

$$= \frac{(2i+2)(2i+4)}{(1)(2)^2(3)(4)^2(5)\ldots\ldots(i+2)},$$

$$= \frac{1}{(1)(2)\ldots\ldots(i+2)}\cdot\frac{(2)(4)\ldots\ldots(2i+4)}{(2)(4)\ldots\ldots(2i)\cdot(2)(4)};$$

establishing the assumed law.

Hence Σx^P for the assemblage $\alpha^i\beta^j\gamma^2$ is

$$\frac{(1)(2)\ldots\ldots(i+j+2)}{(1)(2)\ldots\ldots(j)\cdot(1)(2)\ldots\ldots(i+2)}\cdot\frac{(2)(4)\ldots\ldots(2i+4)}{(2)(4)\ldots\ldots(2i)\cdot(2)(4)}.$$

55. On precisely the same reasoning we find

$$\phi_{i,k}(x) = \frac{1}{(i+k)}\phi_{i,k-1}(x) + \frac{x^{2k}}{(i+k)}\phi_{i-1,k}(x);$$

and assuming that

$$\phi_{i,k-1}(x) = \frac{1}{(1)(2)\ldots\ldots(i+k-1)}\cdot\frac{(2)(4)\ldots\ldots(2i+2k-2)}{(2)(4)\ldots\ldots(2i)\cdot(2)(4)\ldots\ldots(2k-2)},$$

$$\phi_{i-1,k}(x) = \frac{1}{(1)(2)\ldots\ldots(i+k-1)}\cdot\frac{(2)(4)\ldots\ldots(2i+2k-2)}{(2)(4)\ldots\ldots(2i-2)\cdot(2)(4)\ldots\ldots(2k)},$$

we find

$$\phi_{i,k}(x) = \frac{1}{(1)(2)\ldots\ldots(i+k)}\cdot\frac{(2)(4)\ldots\ldots(2i+2k-2)}{(2)(4)\ldots\ldots(2i)\cdot(2)(4)\ldots\ldots(2k)}\{1-x^{2k} + x^{2k}(1-x^{2i})\},$$

or

$$\phi_{i,k}(x) = \frac{1}{(1)(2)\ldots\ldots(i+k)}\cdot\frac{(2)(4)\ldots\ldots(2i+2k)}{(2)(4)\ldots\ldots(2i)\cdot(2)(4)\ldots\ldots(2k)},$$

the final result.

Hence Σx^P for the assemblage $\alpha^i\beta^j\gamma^k$ is

$$\frac{(1)(2)\ldots.(i+j+k)}{(1)(2)\ldots.(j)\cdot(1)(2)\ldots.(i+k)}\cdot\frac{(2)(4)\ldots.(2i+2k)}{(2)(4)\ldots.(2i)\cdot(2)(4)\ldots.(2k)},$$

or

$$F_{j,i+k}(x)\cdot F_{i,k}(x^2),$$

as previously conjectured.

56. *The Expression of Σx^P for the Assemblage $\alpha^i\beta^j\gamma^k\delta^l$.* The actual determination of this expression is reserved for a future occasion, but a few remarks upon the problem will not be out of place. For the assemblage $\alpha^i\beta^j\gamma^k$ the second row of numbers was not involved in the partition question, so that we had only two rows to deal with. In the present instance we have all four rows essentially involved in the partition question. We may write

$$\frac{\Sigma x^P}{(1)(2)\ldots.(i+j+k+l)}=\phi_{i,j,k,l}(x),$$

and the question in partition is the determination of $\phi_{i,j,k,l}(x)$ by considering the four rows of numbers

$$\begin{array}{l} a_1\,a_2\,a_3\ldots.\,a_i, \\ b_1\,b_2\,b_3\ldots.\,b_j, \\ c_1\,c_2\,c_3\ldots.\,c_k, \\ d_1\,d_2\,d_3\ldots.\,d_l. \end{array}$$

We have to find $\Sigma x^{\Sigma a+\Sigma b+\Sigma c+\Sigma d}$ from the conditions

$$\begin{array}{l} a_1\geq a_2\geq a_3\geq\ldots.\geq a_i, \\ b_1\geq b_2\geq b_3\geq\ldots.\geq b_j, \\ c_1\geq c_2\geq c_3\geq\ldots.\geq c_k, \\ d_1\geq d_2\geq d_3\geq\ldots.\geq d_l, \end{array}$$

with the added conditions:

(i) No c is to be one greater than any a.
(ii) No d is to be one greater than any b.
(iii) No d is to be one greater or two greater than any a.

There is no difficulty in writing down the crude generating function involving the symbols χ and Ω and the associated auxiliary coefficients. There is, however, considerable difficulty in handling the function in question so as to eliminate all the auxiliaries and arrive at the desired formula.

The function Σx^P or $(1)(2)\ldots.(i+j+k+l)\,\phi_{i,j,k,l}(x)$ has the following properties:

(i) For $x=1$ it must reduce to the number which enumerates the permutation of the assemblage $\alpha^i\beta^j\gamma^k\delta^l$.

(ii) If x be put equal to unity after differentiation with regard to x, we must obtain the value of ΣP that has been set forth in Art. 43.

(iii) For $j=k=0$ it must reduce to

$$F_{i,l}(x^3).$$

(iv) For $l=0$ and $i=0$ it must reduce to

$$F_{j,i+k}(x)\cdot F_{i,k}(x^2) \quad \text{and} \quad F_{k,j+l}(x)\cdot F_{j,l}(x^2)$$

respectively.

If we put $k=0$ we must get the expression of Σx^P for the assemblage $\alpha^i\beta^j\delta^l$; if we put $j=0$ we must get the expression of Σx^P for the assemblage $\alpha^i\gamma^k\delta^l$; and it would seem to be imperative to investigate these expressions in the first instance. In fact, when we come to the general assemblage $\alpha_1^{i_1}\alpha_2^{i_2}\ldots\alpha_s^{i_s}$, we may put any $s-3$ of the repetitional exponents equal to zero, and we would thus realize the expression of Σx^P for the assemblage

$$\alpha_{s_1}^{i_{s_1}}\alpha_{s_2}^{i_{s_2}}\alpha_{s_3}^{i_{s_3}},$$

where s_1, s_2, s_3 are any three different numbers in ascending order of magnitude. It would thus appear to be imperative ultimately to investigate this case also.

It is of course obvious that Σx^P for the assemblage $\alpha_{s_1}^{i_{s_1}}\alpha_{s_2}^{i_{s_2}}$ is

$$F_{i_{s_1}},\ F_{i_{s_2}}(x^{s_2-s_1}),$$

and we thus obtain three indications to help us in the investigation of the assemblage

$$\alpha_{s_1}^{i_{s_1}}\alpha_{s_2}^{i_{s_2}}\alpha_{s_3}^{i_s}.$$

To these we may add the known results of putting $x=1$ before and after differentiation.

57. *Crude Expression for Σx^P for the Assemblage $\alpha\beta\gamma\delta$.* The actual expression of Σx^P can be obtained from the permutation set forth in Art. 32. It is

$$1+x+2x^2+3x^3+4x^4+2x^5+5x^6+x^7+3x^8+x^9+x^{10}.$$

Its unsymmetrical character will be noted. It shows that we can not expect the expression of Σx^P for the assemblage $\alpha^i\beta^j\gamma^k\delta^l$ to be in the form of a single algebraic fraction involving factors all of the form (s).

For the assemblage $\alpha\beta\gamma\delta$ I construct the function

$$\chi\cdot\frac{\lambda_{13}\lambda_{14}\lambda_{24}\mu_{14}^2}{(1-\lambda_{13}\lambda_{14}\mu_{14}x)(1-\lambda_{24}x)\left(1-\frac{x}{\lambda_{13}}\right)\left(1-\frac{x}{\lambda_{14}\lambda_{24}\mu_{14}}\right)}$$

and note that the general term in the expansion is

$$\lambda_{13}\lambda_{14}\lambda_{24}\mu_{14}^2(\lambda_{13}\lambda_{14}\mu_{14}x)^a(\lambda_{24}x)^b\left(\frac{x}{\lambda_{13}}\right)^c\left(\frac{x}{\lambda_{14}\lambda_{24}\mu_{14}}\right)^d$$

or

$$\lambda_{13}^{a+1-c}\lambda_{14}^{a+1-d}\lambda_{24}^{b+1-d}\mu_{14}^{a+2-d}x^{a+b+c+d}.$$

The operation of χ excludes zero powers of each of the four auxiliaries, so that the required conditions are satisfied, and the constructed function is therefore the crude expression of $\phi_{1,1,1,1}(x)$. Then

$$\Sigma x^P = (1)(2)(3)(4)\phi_{1,1,1,1}(x).$$

I have actually verified that

$$\phi_{1,1,1,1}(x) = \frac{1+x+2x^2+3x^3+4x^4+2x^5+5x^6+x^7+3x^8+x^9+x^{10}}{(1)(2)(3)(4)},$$

but as I have not yet succeeded in carrying out the reduction in a satisfactory manner, so as to show how the general case should be handled, I withhold the algebra.

London, England, *October* 21, 1912.

IV. *The Superior and Inferior Indices of Permutations.*

By Major P. A. MacMahon, F.R.S., Hon. Member Camb. Phil. Soc.

[*Received* 12 January 1914.]

Reference is made to the paper on Indices of Permutations*.

I here define indices of a new kind and subject them to investigation.

Let any assemblage of letters be

$$\alpha_1^{i_1}\alpha_2^{i_2}\ldots\alpha_s^{i_s}$$

and consider any permutation of them.

If any letter precedes p_1' letters which have a smaller subscript we obtain the component p_1' of the Superior Index.

The Superior Index of the permutation is defined to be

$$\Sigma p_1' = p',$$

the summation being in respect to every letter of the permutation.

On the other hand if any letter precedes r_1' letters which have a larger subscript we obtain the component r_1' of the Inferior Index.

The Inferior Index is defined to be

$$\Sigma r_1' = r',$$

the summation being in respect to every letter of the permutation.

Ex. gr. Consider the assemblage $\alpha^4\beta^3\gamma^2\delta$ and the permutation

$\beta\alpha\alpha\delta\gamma\alpha\alpha\beta\beta\gamma$

4 64 , $4+6+4=14$ the Superior Index

355 3311 , $3+5+5+3+3+1+1=21$ the Inferior Index.

If the permutation be reversed

$\gamma\beta\beta\alpha\alpha\gamma\delta\alpha\alpha\beta$

744 33 , $7+4+4+3+3=21$ the Superior Index

122331 11 , $1+2+2+3+3+1+1+1=14$ the Inferior Index,

and we see that the Superior and Inferior Indices of the permutation are respectively equal to

* "The Indices of Permutations and the Derivation therefrom of Functions of a Single Variable associated with the Permutations of any Assemblage of Objects." *American Journal of Mathematics*, Vol. xxxv. No. 3, 1913.

the Inferior and Superior Indices of the reversed permutation. This is obviously true in general so that we can assert in regard to the assemblage

$$\alpha_1^{i_1}\alpha_2^{i_2}\dots\alpha_s^{i_s}$$

that the collection of numbers which specifies the superior indices of the permutations is identical with the collection which specifies the inferior indices.

Hence, in regard to the permutations of any assemblage,

$$\Sigma x^{p'} = \Sigma x^{r'}.$$

It is also readily established that, for every permutation,

$$p' + r' = \Sigma i_1 i_2;$$

for consider that part of a permutation which involves two letters α_k, α_m.

Suppose it to be

$$\dots \alpha_k^{i_k'} \dots \alpha_m^{i_m'} \dots \alpha_m^{i_m''} \dots \alpha_k^{i_k''} \dots \alpha_k^{i_k'''} \dots \alpha_m^{i_m'''} \dots .$$

The portion of the superior index due to these two letters is, if $k > m$,

$$i_k' (i_m' + i_m'' + i_m''' + \dots) + (i_k'' + i_k''') (i_m''' + \dots)$$

and the portion of the inferior index is

$$(i_m' + i_m'') (i_k'' + i_k''' + \dots) + i_m''' (\dots).$$

Adding these together we find that the two letters contribute to the sums of the two indices the number

$$i_m i_k.$$

Thence obviously $$p' + r' = \Sigma i_1 i_2,$$

leading to the relation $$\Sigma x^{p'} = x^{\Sigma i_1 i_2} \Sigma x^{-p'}.$$

The maximum value of p' is clearly $\Sigma i_1 i_2$ and its average value $\frac{1}{2}\Sigma i_1 i_2$.

The function $\Sigma x^{p'}$ is of degree $\Sigma i_1 i_2$ in x and if it be divided by $x^{\frac{1}{2}\Sigma i_1 i_2}$ it is unaltered by the substitution of $\frac{1}{x}$ for x because

$$\Sigma x^{p' - \frac{1}{2}\Sigma i_1 i_2} = \Sigma x^{\frac{1}{2}\Sigma i_1 i_2 - p'}.$$

A function of x which satisfies these conditions is

$$\frac{(1-x)(1-x^2)\dots\dots(1-x^{i_1+i_2+\dots+i_s})}{(1-x)(1-x^2)\dots(1-x^{i_1}).(1-x)(1-x^2)\dots(1-x^{i_2})\dots\dots(1-x)(1-x^2)\dots(1-x^{i_s})}$$

and it will be shewn that this is in fact equal to $\Sigma x^{p'}$.

In the first place consider the assemblage $\alpha^i\beta^j$, and write $\Sigma x^{p'} = F_x(i, j)$.

All permutations which terminate with β, contribute

$$F_x(i, j-1)$$

to $F_x(i, j)$, and those which terminate with α, contribute

$$x^j F(i-1, j).$$

Hence the difference equation

$$F_x(i, j) = x^j F_x(i-1, j) + F_x(i, j-1),$$

the solution of which, satisfying the above conditions, is

$$F_x(i, j) = \frac{(\mathbf{1})(\mathbf{2})\ldots(\mathbf{i}+\mathbf{j})}{(\mathbf{1})(\mathbf{2})\ldots(\mathbf{i}).(\mathbf{1})(\mathbf{2})\ldots(\mathbf{j})}$$

where $(\mathbf{m})$ has been written to denote (after Cayley) $1-x^m$.

Similarly for the assemblage $\alpha^i\beta^j\gamma^k$, write $\Sigma x^{p'} = F_x(i, j, k)$. The permutations terminating with γ, β, α respectively contribute

$$F_x(i, j, k-1),\ x^k F_x(i, j-1, k) \text{ and } x^{j+k} F_x(i-1, j, k-1) \text{ to } F_x(i, j, k),$$

leading us to the difference equation

$$F_x(i, j, k) = x^{j+k} F_x(i-1, j, k) + x^k F_x(i, j-1, k) + F_x(i, j, k-1),$$

the solution of which, satisfying the conditions, is

$$F_x(i, j, k) = \frac{(\mathbf{1})\ldots(\mathbf{i}+\mathbf{j}+\mathbf{k})}{(\mathbf{1})\ldots(\mathbf{i}).(\mathbf{1})\ldots(\mathbf{j}).(\mathbf{1})\ldots(\mathbf{k})}.$$

Similarly we reach the difference equation

$$F_x(i_1, i_2, i_3 \ldots i_s) = x^{i_2+i_3+\ldots+i_s} F_x(i_1-1, i_2, \ldots i_s)$$
$$+ x^{i_3+\ldots+i_s} F_x(i_1, i_2-1, \ldots i_s) + \ldots + F_x(i_1, i_2, \ldots i_s-1),$$

the solution of which, satisfying the conditions, is

$$F_x(i_1, i_2, \ldots i_s) = \frac{(\mathbf{1})\ldots(\mathbf{i}_1+\mathbf{i}_2+\ldots+\mathbf{i}_s)}{(\mathbf{1})\ldots(\mathbf{i}_1).(\mathbf{1})\ldots(\mathbf{i}_2)\ldots\ldots(\mathbf{1})\ldots(\mathbf{i}_s)}.$$

This result is remarkable because it establishes that

$$\Sigma x^{p'} = \Sigma x^p,$$

where p is the Greater Index of a permutation (vide *American Journ. Math.* Vol. XXXV. No. 3, 1913). In fact the whole collection of Superior Indices coincides with the whole collection of Greater Indices, but it is not easy to establish this by a one-to-one correspondence.

Observe the permutations of $\alpha\alpha\beta\gamma$.

Permutation	Greater Index	Superior Index
$\alpha\alpha\beta\gamma$	0	0
$\alpha\alpha\gamma\beta$	3	1
$\alpha\beta\alpha\gamma$	2	1
$\alpha\gamma\alpha\beta$	2	2
$\alpha\beta\gamma\alpha$	3	2
$\alpha\gamma\beta\alpha$	5	3
$\beta\alpha\alpha\gamma$	1	2
$\gamma\alpha\alpha\beta$	1	3
$\beta\alpha\gamma\alpha$	4	3
$\gamma\alpha\beta\alpha$	4	4
$\beta\gamma\alpha\alpha$	2	4
$\gamma\beta\alpha\alpha$	3	5

$$\Sigma x^p = \Sigma x^{p'} = \frac{(\mathbf{1})(\mathbf{2})(\mathbf{3})(\mathbf{4})}{(\mathbf{1})(\mathbf{2}).(\mathbf{1}).(\mathbf{1})} = \frac{(\mathbf{3})(\mathbf{4})}{(\mathbf{1})^2}.$$

8—2

The particular result

$$\Sigma x^{p'} = \frac{(1)(2)\ldots(i+j)}{(1)\ldots(i).(1)\ldots(j)}$$

establishes that the permutations of the assemblage $\alpha^i\beta^j$ which have a superior (or greater) index equal to p' are equinumerous with the partitions of the number p' into parts, not exceeding i in magnitude and not exceeding j in number.

The property of $\Sigma x^{p'}$ that is before us leads to interesting relations between the functions $F_x(i, j)$.

Write the assemblage $\alpha^i\beta^j$ in the form

$$\alpha^{i-a}\beta^{j-b}.\alpha^a\beta^b$$

wherein a, b are any two numbers, such that $a \not> i$, $b \not> j$.

It is to be shewn that

$$F_x(i, j) = \Sigma x^{(j-b)a} F_x(i-a, j-b) F_x(a, b)$$

wherein the summation is in respect of every composition a, b of the constant number $a+b$. The number zero is not excluded so that if for instance $a+b=4$, the summation will be in respect of the compositions 40, 31, 22, 13, 04, it being understood, as above stated, that $a \not> i$, $b \not> j$.

It will be admitted that when the permutations admit of representation in the form

Some permutation of $\alpha^{i-a}\beta^{j-b}$ followed by some permutation of $\alpha^a\beta^b$,

the expression $x^{(j-b)a} F_x(i-a, j-b) F_x(a, b)$ denotes $\Sigma x^{p'}$ for the permutations in question. If we sum this expression for all values of a and b which give permutations involving $i+j-a-b$ letters followed by permutations involving $a+b$ letters we must arrive at the expression of $\Sigma x^{p'}$ for the whole of the permutations of $\alpha^i\beta^j$.

Hence $$F_x(i, j) = \Sigma x^{(j-b)a} F_x(i-a, j-b) F_x(a, b)$$

where $a+b=$ any constant number.

This interesting relation between the functions x has a very interesting particular case.

If $a+b=\sigma$ a constant, we have

$$F_x(i, j) = x^{\sigma j} F_x(\sigma, 0) F_x(i-\sigma, j) + x^{(\sigma-1)(j-1)} F_x(\sigma-1, 1) F_x(i-\sigma+1, j-1) + \ldots$$
$$+ F_x(0, \sigma) F_x(i, j-\sigma).$$

Putting $\sigma = j$ we obtain

$$F_x(i, j) = x^{j^2} F_x(j, 0) F_x(i-j, j) + x^{(j-1)^2} F_x(j-1, 1) F_x(i-j+1, j-1) + \ldots + F_x(0, j) F_x(i, 0)$$

and if we now put $i = j$

$$F_x(j, j) = x^{j^2}\{F_x(j, 0)\}^2 + x^{(j-1)^2}\{F_x(j-1, 1)\}^2 + x^{(j-2)^2}\{F_x(j-2, 2)\}^2 + \ldots + \{F_x(0, j)\}^2.$$

This result is a generalization of the theorem in regard to the sum of the squares of the binomial coefficients, for putting $x=1$, it becomes

$$\binom{2j}{j} = \binom{j}{0}^2 + \binom{j}{1}^2 + \binom{j}{2}^2 + \ldots + \binom{j}{j}^2.$$

In general the reader will see that we have the relation

$$F_x(i_1 i_2 \dots i_s)$$
$$= \Sigma x^{(i_s - a_s)(a_1 + a_2 + \dots + a_{s-1}) + (i_{s-1} - a_{s-1})(a_1 + a_2 + \dots + a_{s-2}) + \dots + (i_2 - a_2) a_1}$$
$$. F_x(i_1 - a_1,\ i_2 - a_2, \dots i_s - a_s) . F_x(a_1 a_2 \dots a_s),$$

the summation being for every composition of a given number $a_1 + a_2 + \dots + a_s$ into s or fewer parts $a_1, a_2, \dots a_s$ such that $a_s \not> i_s$ for all values of s.

The Superior Index as defined is obtained by adding several numbers together. This is the simplest way of obtaining the index, but the numbers so added are not the most interesting that come up for consideration. If $v > u$ the letter α_v adds a number to the index if it precedes one or more letters α_u. Denote by p'_{vu} the number added to the index due to the positions of the letters α_u, α_v. Moreover α_v may precede 1, 2, ... or i_u letters α_u. Denote by $p'_{vu,\sigma}$ the number of letters α_v which precede exactly σ letters α_u. Every time an α_v precedes exactly σ letters α_u the number σ is added to the index. Hence

$$p'_{vu} = p'_{vu,1} + 2p'_{vu,2} + 3p'_{vu,3} + \dots + i_u p'_{vu,i_u}.$$

Also if p''_{vu} denotes, in regard to the whole of the permutations, the sum of the numbers added to the indices by reason of the relative positions of the letters α_u, α_v and $p''_{vu,\sigma}$ the number of times in the whole of the permutations that a letter α_v precedes exactly σ letters α_u,

$$p''_{vu} = p''_{vu,1} + 2p''_{vu,2} + 3p''_{vu,3} + \dots + i_u p''_{vu,i_u}.$$

Now we know the value of p''_{vu} from the following consideration. In any permutation consider merely the letters α_u, α_v. If r'_{uv} denotes the number added to the Inferior Index by the relative positions of these letters we see that

$$p'_{vu} + r'_{uv} = i_u i_v,$$

for any one letter α_v contributes to the sum of the Superior and Inferior indices the number i_u and therefore the total of i_v letters α_v contributes the number $i_u i_v$. Hence the *average* value of p'_{vu} in a permutation is $\frac{1}{2} i_u i_v$ and thence the number contributed to the Superior Indices of all of the permutations by the relative positions of α_u and α_v is

$$p''_{vu} = \tfrac{1}{2} i_u i_v \frac{(\Sigma i)!}{i_1!\ i_2! \dots i_s!}.$$

It will now be proved that $$p''_{vu,\sigma}$$
has a value which is independent of the number σ.

Consider the permutations of the assemblage

$$\alpha_1^{i_1} \alpha_2^{i_2} \dots \alpha_{u-1}^{i_{u-1}} \alpha_u^{i_u+1} \alpha_v^{i_v-1} \alpha_{v+1}^{i_{v+1}} \dots \alpha_s^{i_s},$$

which is derived from the original assemblage by adding an α_u and subtracting an α_v.

In any permutation fix the attention upon the $i_u + 1$ letters α_u. Call the one on the extreme right the last α_u, the one nearest to it the last but one α_u, the next one again the last but two α_u and so on. Now delete the last α_u but σ from the permutation and substitute for it the letter α_v. We have thus an α_v followed by σ letters α_u and the assemblage is the original assemblage of letters. We thus construct a case of an α_v followed by exactly σ letters α_u from every one of the

$$\frac{(\Sigma i)!}{i_1!\, i_2! \dots i_{u-1}!\, (i_u + 1)!\, (i_v - 1)!\, i_{v+1}! \dots i_s!}$$

permutations of the assemblage

$$\frac{\alpha_u}{\alpha_v}\alpha_1^{i_1}\alpha_2^{i_2}\ldots\alpha_s^{i_s}.$$

Thence $$p''_{vu,\sigma}=\frac{i_v}{i_u+1}\frac{(\Sigma i)!}{i_1!\,i_2!\ldots i_s!},$$

a value which is independent of σ.

Therefore $$p''_{vu,\,0}=p''_{vu,\,1}=\ldots=p''_{vu,\,i_u},$$

leading to $$p''_{vu}=\binom{i_u+1}{2}p''_{vu,\sigma}=\tfrac{1}{2}i_u i_v\frac{(\Sigma i)!}{i_1!\,i_2!\ldots i_s!},$$

and $$p''_{vu,\sigma}=\frac{i_v}{i_u+1}\frac{(\Sigma i)!}{i_1!\,i_2!\ldots i_s!}.$$

We deduce that the average value of $p''_{vu,\sigma}$ in a permutation is

$$\frac{i_v}{i_u+1}.$$

To illustrate these results take the assemblage $\alpha\alpha\beta\gamma$ wherein $i_1=2,\ i_2=1,\ i_3=1.$

	$\beta\alpha$	$\gamma\alpha$	$\gamma\beta$
$\alpha\alpha\beta\gamma$	0	0	0
$\alpha\alpha\gamma\beta$	0	0	1
$\alpha\beta\alpha\gamma$	1	0	0
$\alpha\gamma\alpha\beta$	0	1	1
$\alpha\beta\gamma\alpha$	1	1	0
$\alpha\gamma\beta\alpha$	1	1	1
$\beta\alpha\alpha\gamma$	2	0	0
$\gamma\alpha\alpha\beta$	0	2	1
$\beta\alpha\gamma\alpha$	2	1	0
$\gamma\alpha\beta\alpha$	1	2	1
$\beta\gamma\alpha\alpha$	2	2	0
$\gamma\beta\alpha\alpha$	2	2	1

Here $$p''_{21,\,0}=p''_{21,\,1}=p''_{21,\,2}=\frac{1}{3}\frac{4!}{2!\,1!\,1!}=4,$$

and we verify that in the first column the numbers 0, 1 and 2 each occur 4 times.

Also $$p''_{21}=\binom{3}{2}.4=12,$$

and we verify that the sum of the numbers in the first column is 12.

Again $$p''_{31,\,0}=p''_{31,\,1}=p''_{31,\,2}=\frac{1}{3}.12=4,$$

$$p''_{31}=\binom{3}{2}.4=12,$$

and we verify that in the second column the numbers 0, 1 and 2 each occur 4 times and that the sum of the numbers is 12.

Again $$p''_{32,\,0}=p''_{32,\,1}=\frac{1}{2}.12=6,$$

$$p''_{32}=\binom{2}{2}6=6,$$

and we verify that in the third column the numbers 0 and 1 each occur 6 times and that the sum of the numbers is 6.

TWO APPLICATIONS OF GENERAL THEOREMS IN COMBINATORY ANALYSIS: (1) TO THE THEORY OF INVERSIONS OF PERMUTATIONS; (2) TO THE ASCERTAINMENT OF THE NUMBERS OF TERMS IN THE DEVELOPMENT OF A DETERMINANT WHICH HAS AMONGST ITS ELEMENTS AN ARBITRARY NUMBER OF ZEROS

By Major P. A. MACMAHON.

[Received and Read March 9th, 1916.]

1. THE THEORY OF INVERSIONS OF PERMUTATIONS.

In a recent paper* I defined the greater index of a permutation as depending upon the situation of the numbers or letters which *immediately* precede a number or letter which is prior in numerical or alphabetical order. Thus, if the assemblage, say of numbers, be 1, 1, 2, 3, 3, 4, 5, and any permutation

$$\begin{matrix} 4 & 1 & 2 & 1 & 5 & 3 & 3 \\ 1 & & 3 & & 5 & & \end{matrix}$$

we find that the first, third, and fifth numbers immediately precede numbers which are prior in numerical order; the sum of the numbers 1, 3, 5 is 9, which is called the greater index of the permutation. If p be the greater index of a permutation of the numbers in the assemblages

$$1^i\, 2^j\, 3^k \ldots,$$

the exponents, of course, denoting repetitions, I shewed that, the summation having regard to every such permutation, we have

$$\Sigma x^p = \frac{(1-x)(1-x^2)\ldots(1-x^{i+j+k+\ldots})}{(1-x)(1-x^2)\ldots(1-x^i).(1-x)(1-x^2)\ldots(1-x^j).(1-x)(1-x^2)\ldots(1-x^k)\ldots},$$

* *American Journal of Mathematics*, Vol. xxxv, No. 3, 1913, "The Indices of Permutations and the Derivation therefrom of Functions of a Single Variable associated with the Permutations of any Assemblage of Objects."

or in Cayley's notation, writing $1-x^s = (s)$,

$$\Sigma x^p = \frac{(1)(2)\ldots(i+j+k+\ldots)}{(1)(2)\ldots(i)\,.\,(1)(2)\ldots(j)\,.\,(1)(2)\ldots(k)\ldots}.$$

This notation is exceedingly illuminating, and is a striking example of the power mathematics has gained by an appropriate notation.*

The similarity of the expression to the number which enumerates the permutations, obtained by putting x equal to unity, is obvious. In my writings I have termed the expression "The Permutation Function." I wish to shew the application of this theorem to the study of inversions of permutations.† A permutation of an assemblage of numbers involves an inversion whenever a number precedes (not necessarily immediately) a number prior to it in numerical order. A permutation therefore involves a definite number of inversions, and the main question is the determination of the number of the permutations, each of which possesses a given number of inversions. The researches set forth in the *Lehrbuch* of Netto are concerned with permutations of assemblages of numbers between which there are no similarities. He indicates by $I_\kappa^{(n)}$ the number of the permutations of n *different* elements, each of which possesses exactly κ inversions, and by purely arithmetical reasoning he derives various relations between the numbers $I_\kappa^{(n)}$.

In the present note we are concerned with the inversions which are possessed by permutations of the general assemblage of numbers

$$1^i\, 2^j\, 3^k \ldots.$$

Consider first the permutation

$$4\ 1\ 2\ 1\ 5\ 3\ 3$$

which we found above to have a greater index equal to 9.

Here 4	precedes	1, 2, 1, 3, 3	giving	5	inversions,
2	„	1	„	1	„
5	„	3, 3	„	2	„

Hence the permutation possesses $5+1+2 = 8$ inversions, or, as we may say, has an inversion number equal to 8.

Next take the whole of the permutations of the assemblage

$$1,\ 1,\ 2,\ 3\,;$$

---

* See in this connexion *Napier's Tercentenary Memorial Volume*, Longmans, Green & Co., 1915; J. W. L. Glaisher, on *Logarithms and Computation*, p. 75.

† See Netto's *Combinatorik*, p. 92

there are twelve of them, and we will write down for each permutation the greater index and the inversion number.

Permutation.	Greater Index.	Inversion Number.
1123	0	0
1132	3	1
1213	2	1
1231	3	2
1312	2	2
1321	5	3
2113	1	2
2131	4	3
2311	2	4
3112	1	3
3121	4	4
3211	3	5

It will be observed that in this case the greater indices constitute the same *assemblage* of numbers as the inversion numbers. It will now be established that if p and v be the greater index and inversion number of a permutation of any assemblage of integers

$$\Sigma x^p = \Sigma x^v.$$

We have to shew that Σx^v has the same algebraic expression as has been already found for Σx^p.

Σx^v is a finite and integral function of x, in which the highest exponent of x is $ij+ik+jk+\ldots$, for clearly v in this case is derived from the permutation which is the assemblage written in descending order of magnitude. This is also the highest exponent of x in the algebraic expansion of the expression found for Σx^p; for this is

$$\binom{i+j+k+\ldots+1}{2}-\binom{i}{2}-\binom{j}{2}-\binom{k}{2}-\ldots$$

or

$$ij+ik+jk+\ldots.$$

For all permutations which commence with the integer s, denote Σx^v by $(\Sigma x^v)_s$. Restricting ourselves, for convenience, to the assemblage

$$1^i\, 2^j\, 3^k,$$

let
$$\Sigma x^{v} = F_{i,j,k}(x),$$
then we see that $(\Sigma x^{v})_1$ for the assemblage $1^i 2^j 3^k$ is equal to Σx^{v} for the assemblage $1^{i-1} 2^j 3^k$, since the leading integer 1 can add nothing to the inversion number v.

Thus
$$(\Sigma x^{v})_1 = F_{i-1,j,k}(x),$$
also
$$(\Sigma x^{v})_2 = x^{i} F_{i,j-1,k}(x),$$
since the leading integer 2 is prior to i ones, and therefore adds i to every inversion number of permutations of the assemblage $1^i 2^{j-1} 3^k$.

So also
$$(\Sigma x^{v})_3 = x^{i+j} F_{i,j,k-1}(x),$$
and, finally,
$$\Sigma x^{v} = F_{i,j,k}(x) = F_{i-1,j,k}(x) + x^{i} F_{i,j-1,k}(x) + x^{i+j} F_{i,j,k-1}(x).$$

Assuming the law to hold for the three functions
$$F_{i-1,j,k}(x), \quad F_{i,j-1,k}(x), \quad F_{i,j,k-1}(x),$$
we find that
$$F_{i,j,k}(x) = \frac{(1)(2)\ldots(i+j+k-1)}{(1)(2)\ldots(i-1).(1)(2)\ldots(j).(1)(2)\ldots(k)}$$
$$+x^{i}\frac{(1)(2)\ldots(i+j+k-1)}{(1)(2)\ldots(i).(1)(2)\ldots(j-1).(1)(2)\ldots(k)}$$
$$+x^{i+j}\frac{(1)(2)\ldots(i+j+k-1)}{(1)(2)\ldots(i).(1)(2)\ldots(j).(1)(2)\ldots(k-1)}$$
$$= \frac{(1)(2)\ldots(i+j+k)}{(1)(2)\ldots(i).(1)(2)\ldots(j).(1)(2)\ldots(k)}.$$

Since, by trial, we find the law to obtain for the smaller values of the sum $i+j+k$, we have here an inductive proof of the theorem, and since the process is perfectly general, we find that for the general assemblage
$$1^i 2^j 3^k \ldots,$$
$$\Sigma x^{v} = \Sigma x^{p} = \frac{(1)(2)\ldots(i+j+k+\ldots)}{(1)(2)\ldots(i).(1)(2)\ldots(j).(1)(2)\ldots(k)\ldots}.$$

The theory of the inversion number is identical with that of the greater index, and thus all of the results obtained in the *American Journal* paper, which has been quoted, are equally true of inversion numbers.

For the particular case discussed in Netto's *Combinatorik*,

$$\Sigma x^r = \frac{(1)(2)\dots(n)}{(1)^n},$$

and the results he gives are readily deduced from this expression.

The Number of Terms in a Determinant which has some Zero Elements.

In a determinant of order n, the number of terms which arises in the development is in general $n!$; but if some of the elements are zeros, we obtain a lesser number which it is the object of this note to determine. If the elements in the principal diagonal are zeros, and there are no other zeros, the determinant has been named by Sylvester "invertebrate," and it is well known that the number of terms is equal to the number of permutations of n different letters, which are such that no letter is in its original place. The number is therefore

$$n! - \binom{n}{1}(n-1)! + \binom{n}{2}(n-2)! - \dots \pm \binom{n}{n} 0!.$$

G. de Longchamps* also established that if the first q elements in the diagonal are zeros, the number is

$$n! - \binom{q}{1}(n-1)! + \binom{q}{2}(n-2)! - \dots \pm \binom{q}{q}(n-q)!.$$

The similar question in which the zero elements are arbitrarily chosen has also been considered by de Longchamps and by C. A. Laisant,† but their solutions fell short of what is desirable.‡ The matter may be handled in the following manner :—

In *Phil. Trans. Roy. Soc.*, 1894, A, "A certain Class of Generating Functions in the Theory of Numbers," I shewed that if $X_1, X_2, \dots, X_n$ be linear functions given by the matricular relation

$$(X_1, X_2, \dots, X_n) = \left(\begin{array}{cccc} a_1, & a_2, & \dots, & a_n \\ b_1, & b_2, & \dots, & b_n \\ \cdot & \cdot & \dots, & \cdot \\ n_1, & n_2, & \dots, & n_n \end{array}\right)(x_1, x_2, \dots, x_n),$$

* *Jour. de Math. Spéc.* (1891).

† C. R. Paris (1891); *Soc. Math. de France Bull.* (1891).

‡ Netto, *Lehrbuch der Combinatorik* (1901), p. 73.

the coefficient of the term

$$x_1^{\xi_1} x_2^{\xi_2} \dots x_n^{\xi_n},$$

in the development of the product

$$X_1^{\xi_1} X_2^{\xi_2} \dots X_n^{\xi_n},$$

is equal to the coefficient of the term

$$x_1^{\xi_1} x_2^{\xi_2} \dots x_n^{\xi_n},$$

in the expansion of the algebraic fraction

$$\frac{1}{|(1-a_1x_1)(1-b_2x_2)\dots(1-n_nx_n)|},$$

wherein the denominator is in symbolic form in such wise that on multiplication the products a_1b_2, $a_1b_2c_3$, ... are to be placed in determinant brackets, like $|a_1b_2|$, $|a_1b_2c_3|$, ..., and denote the coaxial minors of the determinant $|a_1b_2c_3\dots n_n|$, which appertains to the matricular relation. This result I have termed a master theorem in the theory of permutations in the work *Combinatory Analysis*, published 1915, by the Cambridge University Press.

Now Laisant has shewn, in the notation of this note, that considering the determinant $|a_1b_2c_3\dots n_n|$, if we form the product

$$(a_1x_1+a_2x_2+\dots+a_nx_n)(b_1x_1+b_2x_2+\dots+b_nx_n)\dots(n_1x_1+n_2x_2+\dots+n_nx_n),$$

and therein write $a_\kappa = b_\lambda = c_\mu = \dots = 1$ whenever the element is not zero, and $a_\kappa = b_\lambda = c_\mu = \dots = 0$ whenever the element is zero, the number of terms in the determinant is equal to the coefficient of $x_1x_2\dots x_n$ in the development of the product.

Hence by the master theorem the number of terms in the determinant is equal to the coefficient of

$$x_1x_2\dots x_n,$$

in the development of the fraction

$$\frac{1}{|(1-a_1x_1)(1-b_2x_2)\dots(1-n_nx_n)|},$$

when therein the elements are given the values unity or zero in the manner specified above.

Some examples are now given.

Suppose all the elements of the determinant to be zero except those in the principal diagonal. We put

$$a_1 = b_2 = c_3 = \dots = n_n = 1,$$

and the remaining elements equal to zero; the fraction is then

$$\frac{1}{1-\Sigma x_1+\Sigma x_1x_2-\dots+(-)^n x_1x_2\dots x_n}$$

or
$$\frac{1}{(1-x_1)(1-x_2)\dots(1-x_n)},$$

in which the coefficient of $x_1x_2 \dots x_n$ is unity.

This is correct because clearly the determinant has only one term formed from the elements of the principal diagonal.

Again, suppose that the only zero elements are those in the principal diagonal. We put

$$a_1 = b_2 = c_3 = \dots = n_n = 0,$$

and the remaining elements equal to unity; the fraction is then

$$\frac{1}{1-\Sigma x_1x_2-2\Sigma x_1x_2x_3-3\Sigma x_1x_2x_3x_4-\dots-(n-1)x_1x_2\dots x^n}$$

(see *Combinatory Analysis*, Vol. I, p. 100).

The coefficient of $x_1x_2 \dots x_n$ in the expansion denotes also the number of permutations of $x_1x_2 \dots x_n$ which are such that every element is displaced. The expansion gives various properties of these numbers.

When there are no zero elements the fraction becomes

$$\frac{1}{1-x_1-x_2-x_3-\dots-x_n},$$

in which the coefficient of $x_1x_2 \dots x_n$ is $n!$.

Next take the determinant

$$\begin{vmatrix} 0 & 0 & a_3 & a_4 \\ b_1 & b_2 & 0 & b_4 \\ c_1 & c_2 & c_3 & c_4 \\ d_1 & d_2 & d_3 & d_4 \end{vmatrix},$$

which, on development, is found to have ten terms.

The fraction, derived from the coaxial minors of the determinant

$$\begin{vmatrix} 0 & 0 & x_3 & x_4 \\ x_1 & x_2 & 0 & x_4 \\ x_1 & x_2 & x_3 & x_4 \\ x_1 & x_2 & x_3 & x_4 \end{vmatrix},$$

is

$$\frac{1}{1-x_2-x_3-x_4-x_1x_3-x_1x_4+x_2x_3},$$

and on development we find the term

$$10x_1x_2x_3x_4$$

a verification.

PERMUTATIONS, LATTICE PERMUTATIONS, AND THE HYPERGEOMETRIC SERIES

By Major P. A. MacMahon.

[Received December 1st, 1919.—Read December 11th, 1919.]

1. An assemblage of letters

$$x_1^{m_1+m_2+\ldots+m_n} x_2^{m_2+m_3+\ldots+m_n} \ldots x_{n-1}^{m_{n-1}+m_n} x_n^{m_n}$$

in which the quantities $m_1, m_2, \ldots, m_n$ may each be zero or any positive integer is termed a Lattice Assemblage with respect to the ordered system of letters $x_1, x_2, \ldots, x_n$. The reason for this nomenclature is that the assemblage may be denoted graphically by a regular lattice of nodes.

A permutation of this lattice assemblage is defined to be a Lattice Permutation if on drawing a line between *any two* letters of the permutation the assemblage to the left of the line is a lattice assemblage.

The essential characteristic of a lattice assemblage is that the repetitional exponents are in descending order of magnitude. Lattice permutations are of great importance in certain theories in Combinatory Analysis and have been studied by the writer.*

It is convenient to denote by

$$P(m_1+m_2+\ldots+m_n,\ m_2+m_3+\ldots+m_n,\ \ldots,\ m_{n-1}+m_n,\ m_n),$$

$$LP(m_1+m_2+\ldots+m_n,\ m_2+m_3+\ldots+m_n,\ \ldots,\ m_{n-1}+m_n,\ m_n),$$

the number of permutations and the number of lattice permutations of the assemblage, respectively.

The general hypergeometric series

$$1+\frac{a_1 a_2 \ldots a_k}{1.b_2 \ldots b_k}x+\frac{a_1(a_1+1)\,a_2(a_2+1)\ldots a_k(a_k+1)}{1.2.b_2(b_2+1)\ldots b_k(b_k+1)}x^2+\ldots$$

* *Phil. Trans. Roy. Soc.*, "Combinatory Analysis."

has been usually written

$$F\begin{pmatrix} a_1\, a_2\, a_3 \,\dots\, a_k & \\ b_2\, b_3 \,\dots\, b_k & x \end{pmatrix},$$

but from considerations of symmetry I prefer to write it in the present communication in the notation

$$F\begin{pmatrix} a_1\, a_2\, a_3 \,\dots\, a_k & \\ 1\, b_2\, b_3 \,\dots\, b_k & x \end{pmatrix},$$

which, in any case, appears to be in no need of justification.

Permutations and the Hypergeometric Series.

2. To establish a connexion between the permutations of

$$x_1^{m_1+m_2+\dots+m_n}\, x_2^{m_2+m_3+\dots+m_n} \dots x_{n-1}^{m_{n-1}+m_n}\, x_n^{m_n},$$

and the hypergeometric series, we may first note the easily verifiable relations

$$1+\binom{2}{1}x_1x_2+\binom{4}{2}x_1^2x_2^2+\dots+\binom{2s}{s}x_1^sx_2^s+\dots$$

$$= F\left\{\begin{matrix} \frac{1}{2}, & \frac{2}{2}, & \\ 1, & 1, & \end{matrix} \; 2^2x_1x_2\right\},$$

$$1+\frac{3!}{(1!)^3}x_1x_2x_3+\frac{6!}{(2!)^3}(x_1x_2x_3)^2+\dots+\binom{3s}{s}(x_1x_2x_3)^s+\dots$$

$$= F\left\{\begin{matrix} \frac{1}{3}, & \frac{2}{3}, & \frac{3}{3}, & \\ 1, & 1, & 1, & \end{matrix} \; 3^3x_1x_2x_3\right\},$$

and, in general, the relation

$$1+\frac{n!}{(1!)^n}x_1x_2\dots x_n+\frac{(2n)!}{(2!)^3}(x_1x_2\dots x_n)^2+\dots+\frac{(sn)!}{(s!)^n}(x_1x_2\dots x_n)^s+\dots$$

$$= F\left\{\begin{matrix} \frac{1}{n}, & \frac{2}{n}, & \frac{3}{n}, & \dots, & \frac{n-1}{n}, & \frac{n}{n}, & \\ 1, & 1, & 1, & \dots, & 1, & 1, & \end{matrix} \; n^nx_1x_2x_3\dots x_n\right\}.$$

which yields the hypergeometric series which generates the numbers which enumerate the permutations of the assemblage

$$(x_1 x_2 \dots x_n)^s$$

for all values of s.

It may be written

$$\sum_{s=0}^{\infty} P(s, s, \dots, s)(x_1 x_2 \dots x_n)^s = F\left\{\begin{matrix} \frac{1}{n}, & \frac{2}{n}, & \dots, & \frac{n}{n}, & n^n x_1 x_2 \dots x_n \\ 1, & 1, & \dots, & 1, & \end{matrix}\right\}.$$

To generalise by finding an expression for

$$\Sigma P(m_1+m_2+\dots+m_n,\ m_2+m_3+\dots+m_n,\ \dots,\ m_{n-1}+m_n,\ m_n)\, x_1^{m_1+\dots+m_n} \dots x_n^{m_n},$$

we note the relations

$$\sum_{m_2=0}^{\infty} P(m_1+m_2,\ m_2)\, x_1^{m_1+m_2} x_2^{m_2} = x_1^{m_1} F\left\{\begin{matrix} \frac{m_1+1}{2}, & \frac{m_1+2}{2}, & 2^2 x_1 x_2 \\ 1, & m_1+1, & \end{matrix}\right\}$$

$$\sum_{m_3=0}^{\infty} P(m_1+m_2,\ m_2+m_3,\ m_3)\, x_1^{m_1+m_2+m_3} x_2^{m_2+m_3} x_3^{m_3}$$

$$= P(m_1+m_2,\ m_2)\, x_1^{m_1+m_2} x_2^{m_2}$$

$$\times F\left\{\begin{matrix} \frac{m_1+2m_2+1}{3}, & \frac{m_1+2m_2+2}{3}, & \frac{m_1+2m_2+3}{3}, & 3^3 x_1 x_2 x_3 \\ 1, & m_2+1, & m_1+m_2+1, & \end{matrix}\right\},$$

which are verifiable without difficulty.

In fact the coefficient of $x_1^{m_1+m_2+m_3} x_2^{m_2+m_3} x_3^{m_3}$ on the right-hand side is

$$\frac{(m_1+2m_2)!}{(m_1+m_2)!\, m_2!} \frac{(m_1+2m_2+3m_3)!}{(m_1+2m_2)!} \div \frac{m_3!\,(m_2+m_3)!\,(m_1+m_2+m_3)!}{m_2!\,(m_1+m_2)!},$$

which is $\qquad P(m_1+m_2+m_3,\ m_2+m_3,\ m_3),$

in agreement with the left-hand side.

Guided by these relations we may conjecture a general formula and proceed to its verification.

Write
$$\begin{aligned} m_1+m_2+\ldots+m_n &= m_{1,n}, \\ m_2+\ldots+m_n &= m_{2,n}, \\ \ldots \quad \ldots \quad &\ldots \\ m_{n-1}+m_n &= m_{n-1,n}, \\ m_n &= m_{n,n}, \end{aligned}$$
$$m_1+2m_2+3m_3+\ldots+(n-1)\,m_{n-1} = M_{n-1},$$
and

$$\sum_{m_{n,n}=0}^{\infty} P(m_{1,n}, m_{2,n}, \ldots, m_{n-1,n}, m_{n,n})\, x_1^{m_{1,n}} x_2^{m_{2,n}} \ldots x_{n-1}^{m_{n-1,n}} x_n^{m_{n,n}}$$
$$= P(m_{1,n-1}, m_{2,n-1}, \ldots, m_{n-2,n-1}, m_{n-1,n-1})\, x_1^{m_{1,n-1}} x_2^{m_{2,n-1}} \ldots x_{n-2}^{m_{n-2,n-1}} x_{n-1}^{m_{n-1,n-1}}$$
$$F\left\{\begin{matrix} \frac{1}{n}(M_{n-1}+1), & \frac{1}{n}(M_{n-1}+2), & \frac{1}{n}(M_{n-1}+3), & \ldots, & \frac{1}{n}(M_{n-1}+n), \\ 1, & m_{n-1,n-1}+1, & m_{n-2,n-1}+1, & \ldots, & m_{1,n-1}+1 \end{matrix}\; n^n x_1 x_2 \ldots x_n\right\}$$

wherein $m_1, m_2, \ldots, m_{n-1}$ may be each zero or any positive integer, and the summation on the left is in regard to $m_n\ (\equiv m_{n,n})$ from zero to infinity. It suffices to show that the coefficients of

$$x_1^{m_{1,n}} x_2^{m_{2,n}} \ldots x_{n-1}^{m_{n-1,n}} x_n^{m_{n,n}}$$

agree on the two sides.

On the right-hand side we have to pick out the coefficients of

$$(x_1 x_2 \ldots x_n)^{m_{n,n}}$$

in the hypergeometric series. This is

$$\frac{M_n!}{M_{n-1}!}\;\frac{m_{1,n-1}!\, m_{2,n-1}! \ldots m_{n-1,n-1}!}{m_{1,n}!\, m_{2,n}! \ldots m_{n,n}!},$$

and now multiplication by

$$P(m_{1,n-1}, m_{2,n-1}, \ldots, m_{n-1,n-1}) \equiv \frac{M_{n-1}!}{m_{1,n-1}!\, m_{2,n-1}! \ldots m_{n-1,n-1}!},$$

gives
$$\frac{M_n!}{m_{1,n}!\, m_{2,n}! \ldots m_{n,n}!} \equiv P(m_{1,n}, m_{2,n}, \ldots, m_{n,n}).$$

Hence the conjectured theorem is established.

Lattice Permutations and the Hypergeometric Series.

3. It has been established (*loc. cit.*) that the number of lattice permutations of the assemblage

$$x_1^{m_1+m_2+\ldots+m_n} x_2^{m_2+m_3+\ldots+m_n} \ldots x_{n-1}^{m_{n-1}+m_n} x_n^{m_n},$$

which, in the above notation, may be written

$$x_1^{m_{1,n}} x_2^{m_{2,n}} \ldots x_{n-1}^{m_{n-1,n}} x_n^{m_{n,n}},$$

is $$LP(m_{1,n}, m_{2,n}, \ldots, m_{n-1,n}, m_{n,n})$$

$$= \frac{M_n! \, \Pi_{a,b}(m_a+m_{a+1}+\ldots+m_{b-1}+b-a)}{(m_{1,n}+n-1)! \, (m_{2,n}+n-2)! \ldots (m_{n-1,n})! \, m_{n,n}!},$$

where the product Π is for every pair of integers drawn from $1, 2, 3, \ldots, n$ such that $b > a$.

From this formula we find by putting

$$m_1 = m_2 = \ldots = m_{n-1} = 0 \quad \text{and} \quad m_n = s,$$

$$LP(s) = 1,$$

$$LP(s, s) = \frac{(2s)! \; 1!}{(s+1)! \, s!},$$

$$LP(s, s, s) = \frac{(3s)! \quad 2! \quad 1!}{(s+2)! \, (s+1)! \, s!},$$

... ...

$$LP(s, s, \ldots, s) = \frac{(ns)! \quad (n-1)! \quad (n-2)! \; \ldots \; 2! \quad 1!}{(s+n-1)! \, (s+n-2)! \, (s+n-3)! \ldots (s+1)! \, s!},$$

and at once verify the relations

$$\sum_{s=0}^{\infty} LP(s, s)(x_1 x_2)^s = F\left\{\begin{matrix} \frac{1}{2}, & \frac{2}{2}, & 2^2 x_1 x_2 \\ 1, & 2, & \end{matrix}\right\},$$

$$\sum_{s=0}^{\infty} LP(s, s, s)(x_1 x_2 x_3)^s = F\left\{\begin{matrix} \frac{1}{3}, & \frac{2}{3}, & \frac{3}{3}, & 3^3 x_1 x_2 x_3 \\ 1, & 2, & 3, & \end{matrix}\right\},$$

$$\sum_{s=0}^{\infty} LP(s, s, \ldots, s)(x_1 x_2 \ldots x_n)^s = F\left\{\begin{matrix} \frac{1}{n}, & \frac{2}{n}, & \ldots, & \frac{n}{n}, & n^n x_1 x_2 \ldots x_n \\ 1, & 2, & \ldots, & n, & \end{matrix}\right\},$$

which are for comparison with the analogous relations of § 2.

We further verify the relations

$$\sum_{m_2=0}^{\infty} LP(m_1+m_2,\ m_2)\,x_1^{m_1+m_2}x_2^{m_2}$$

$$= x^{m_1} F\left\{\begin{matrix}\tfrac{1}{2}(m_1+1), & \tfrac{1}{2}(m_1+2), \\ 1, & m_1+2,\end{matrix}\ 2^2 x_1 x_2\right\}$$

$$\sum_{m_3=0}^{\infty} LP(m_1+m_2+m_3,\ m_2+m_3,\ m_3)\,x_1^{m_1+m_2-m_3}x_2^{m_2+m_3}x_3^{m_3}$$

$$= LP(m_1+m_2,\ m_3)\,x_1^{m_1+m_2}x_2^{m_2}$$

$$\times F\left\{\begin{matrix}\tfrac{1}{3}(m_1+2m_2+1), & \tfrac{1}{3}(m_1+2m_2+2), & \tfrac{1}{3}(m_1+2m_2+3), \\ 1, & m_2+2, & m_1+m_2+3,\end{matrix}\ 3^3 x_1 x_2 x_3\right\},$$

suggesting the general relation

$$\sum_{m_n=0}^{\infty} LP(m_{1,n},\ m_{2,n},\ \ldots,\ m_{n,n})\,x_1^{m_{1,n}}x_2^{m_{2,n}}\ldots x_n^{m_{n,n}}$$

$$= LP(m_{1,n-1},\ m_{2,n-1},\ \ldots,\ m_{n-1,n-1})\,x_1^{m_{1,n-1}}x_2^{m_{2,n-1}}\ldots x_{n-1}^{m_{n-1,n-1}}$$

$$\times F\left\{\begin{matrix}\frac{1}{n}(M_{n-1}+1), & \frac{1}{n}(M_{n-1}+2), & \ldots, & \frac{1}{n}(M_{n-1}+n), \\ 1, & m_{n-1,n-1}+2, & \ldots, & m_{1,n-1}+n,\end{matrix}\ n^n x_1 x_2 \ldots x_n\right\},$$

where the lower row of terms in the series is

$$1,\ m_{n-1,n-1}+2,\ m_{n-2,n-1}+3,\ m_{n-3,n-1}+4,\ \ldots,\ m_{2,n-1}+n-1,\ m_{1,n-1}+n.$$

To prove this we find that the coefficient of $(x_1 x_2 \ldots x_n)^{m_n}$ in the hypergeometric series is

$$\frac{M_n!}{M_{n-1}!}\ \frac{(m_{n-1,n-1}+1)!\,(m_{n-2,n-1}+2)!\ldots(m_{1,n-1}+n-1)!}{m_{n,n}!\,(m_{n-1,n}+1)!\,(m_{n-2,n}+2)!\ldots(m_{1,n}+n-1)!},$$

and
$$\frac{LP(m_{1,n},\ m_{2,n},\ \ldots,\ m_{n,n})}{LP(m_{1,n-1},\ m_{2,n-1},\ \ldots,\ m_{n-1,n-1})}$$

$$= \frac{M_n!\,\Pi_{a,b}(m_a+\ldots+m_{b-1}+b-a)}{(m_{1,n}+n-1)!\,(m_{2,n}+n-2)!\ldots m_{n,n}!}$$

$$\times \frac{(m_{1,n-1}+n-2)!\,(m_{2,n-1}+n-2)!\ldots m_{n-1,n-1}!}{M_{n-1}!\,\Pi'_{a,b}(m_a+\ldots+m_{b-1}+b-a)},$$

where in the product Π' in the denominator the numbers a, b are selected from the series 1, 2, 3, ..., $n-1$, and not from 1, 2, 3, ..., n.

Hence

$$\frac{\Pi_{a,b}(m_a+\ldots+m_{b-1}+b-a)}{\Pi'_{a,b}(m_a+\ldots+m_{b-1}+b-a)}$$

$$= (m_{1,n-1}+n-1)(m_{2,n-1}+n-2)\ldots(m_{n-1,n-1}+1),$$

leading to

$$\frac{LP(m_{1,n}, m_{2,n}, \ldots, m_{n,n})}{LP(m_{1,n-1}, m_{2,n-1}, \ldots, m_{n-1,n-1})}$$

$$= \frac{M_n!}{M_{n-1}!}\,\frac{(m_{n-1,n-1}+1)!\,(m_{n-2,n-1}+2)!\ldots(m_{1,n-1}+n-1)!}{m_{n,n}!\,(m_{n-1,n}+1)!\,(m_{n-2,n}+2)!\ldots(m_{1,n}+n-1)!}$$

$$= \text{coefficient of } (x_1x_2\ldots x_n)^{m_{n,n}}$$

in the hypergeometric series.

This establishes the relation.

The result of the investigation is the two theorems :—

$$\sum_{m_{n,n}=0}^{\infty} P(m_{1,n}, m_{2,n}, \ldots, m_{n,n})\, x_1^{m_{1,n}} x_2^{m_{2,n}} \ldots x_n^{m_{n,n}}$$

$$= P(m_{1,n-1}, m_{2,n-1}, \ldots, m_{n-1,n-1})\, x_1^{m_{1,n-1}} x_2^{m_{2,n-1}} \ldots x_{n-1}^{m_{n-1,n-1}}$$

$$\times F\left\{\begin{matrix} \frac{1}{n}(M_{n-1}+1), & \frac{1}{n}(M_{n-1}+2), & \frac{1}{n}(M_{n-1}+3), & \ldots, & \frac{1}{n}(M_{n-1}+n), \\ 1, & m_{n-1,n-1}+1, & m_{n-2,n-1}+1, & \ldots, & m_{1,n-1}+1, \end{matrix}\quad n^n x_1x_2\ldots x_n\right\}$$

$$\sum_{m_{n,n}=0}^{\infty} LP(m_{1,n}, m_{2,n}, \ldots, m_{n,n})\, x_1^{m_{1,n}} x_2^{m_{2,n}} \ldots x_n^{m_{n,n}}$$

$$= LP(m_{1,n-1}, m_{2,n-1}, \ldots, m_{n-1,n-1})\, x_1^{m_{1,n-1}} x_2^{m_{2,n-1}} \ldots x_{n-1}^{m_{n-1,n-1}}$$

$$\times F\left\{\begin{matrix} \frac{1}{n}(M_{n-1}+1), & \frac{1}{n}(M_{n-1}+2), & \frac{1}{n}(M_{n-1}+3), & \ldots, & \frac{1}{n}(M_{n-1}+n), \\ 1, & m_{n-1,n-1}+2, & m_{n-2,n-1}+3, & \ldots, & m_{1,n-1}+n, \end{matrix}\quad n^n x_1x_2\ldots x_n\right\}$$

where

$$m_{1,n} = m_1+m_2+\ldots+m_n,$$

$$m_{2,n} = m_2+m_3+\ldots+m_n,$$

$$\ldots \quad \ldots \quad \ldots \quad \ldots$$

$$m_{n,n} = m_n,$$

$$M_{n-1} = m_1+2m_2+3m_3+\ldots+(n-1)\,m_{n-1},$$

and $$P(m_{1,n}, m_{2,n}, \ldots, m_{n,n}), \quad LP(m_{1,n}, m_{2,n}, \ldots, m_{n,n}),$$

denote the numbers of permutations and lattice permutations, respectively, of the assemblage of letters

$$x_1^{m_{1,n}} x_2^{m_{2,n}} \ldots x_n^{m_{n,n}}.$$

The two relations only differ in the lower rows of parameters of the hypergeometric series, and then only in a simple and interesting manner. In both $m_1, m_2, \ldots, m_{n-1}$ may be each zero or any positive integer.

4. If $$(1-x_1-x_2-\ldots-x_n)^{-1}$$

be expanded a certain portion will involve lattice assemblages of the letters *quâ* the ordered letters

$$x_1, x_2, \ldots, x_n.$$

This portion we will denote by

$$L(1-x_1-x_2-\ldots-x_n)^{-1}.$$

The connexion between this lattice portion of the function and the hypergeometric series is manifest from the investigation of § 2.

Thus for $n = 2$,

$$L(1-x_1-x_2)^{-1}$$

$$= F\left\{\begin{matrix} \frac{1}{2}, & \frac{2}{2}, & \\ & & 2^2x_1x_2 \\ 1, & 1, & \end{matrix}\right\}$$

$$+x_1F\left\{\begin{matrix} \frac{2}{2}, & \frac{3}{2}, & \\ & & 2^2x_1x_2 \\ 1, & 2, & \end{matrix}\right\}$$

$$+\ldots$$

$$+x_1^{m_1}F\left\{\begin{matrix} \frac{m_1+1}{2}, & \frac{m_1+2}{2}, & \\ & & 2^2x_1x_2 \\ 1, & m_1+1, & \end{matrix}\right\}$$

$$+\ldots,$$

and since $(1-x_1-x_2)^{-1}$ is a symmetric function we may further deduce that

$$(1-x_1-x_2)^{-1}$$

$$= F\left\{\begin{matrix}\frac{1}{2}, & \frac{2}{2}, & 2^2x_1x_2\\ 1, & 1, & \end{matrix}\right\}$$

$$+(x_1+x_2)F\left\{\begin{matrix}\frac{2}{2}, & \frac{3}{2}, & 2^2x_1x_2\\ 1, & 2, & \end{matrix}\right\}$$

$$+\ldots$$

$$+(x_1^{m_1}+x_2^{m_2})F\left\{\begin{matrix}\frac{m_1+1}{2}, & \frac{m_1+2}{2}. & 2^2x_1x_2\\ 1, & m_1+1, & \end{matrix}\right\}$$

$$+\ldots.$$

Again, for $n=3$,

$$L(1-x_1-x_2-x_3)^{-1}$$

$$= \sum_{m_1=0}^{\infty}\sum_{m_2=0}^{\infty} P(m_1+m_2,\, m_2)\, x_1^{m_1+m_2} x_2^{m_2}$$

$$\times F\left\{\begin{matrix}\tfrac{1}{3}(m_1+2m_2+1), & \tfrac{1}{3}(m_1+2m_2+2), & \tfrac{1}{3}(m_1+2m_2+3, & 3^3x_1x_2x_3\\ 1, & m_2+1, & m_1+m_2+1, & \end{matrix}\right\},$$

$$(1-x_1-x_2-x_3)^{-1}$$

$$= \sum_{m_1=0}^{\infty}\sum_{m_2=0}^{\infty}\Big[P(m_1+m_2,\, m_2)\,(\Sigma x_1^{m_1+m_2} x_2^{m_2})$$

$$\times F\left\{\begin{matrix}\tfrac{1}{3}(m_1+2m_2+1), & \tfrac{1}{3}(m_1+2m_2+2,\tfrac{1}{3} & (m_1+2m_2+3), & 3^3x_1x_2x_3\\ 1, & m_2+1, & m_1+m_2+1, & \end{matrix}\right\}\Big].$$

The general formulæ are

$$L(1-x_1-x_2-\dots-x_n)^{-1}$$

$$= \sum_{m_1=0}^{\infty} \sum_{m_2=0}^{\infty} \dots \sum_{m_{n-1}=0}^{\infty} P(m_{1,n-1}, m_{2,n-1}, \dots, m_{n-1,n-1}) x_1^{m_{1,n-1}} x_2^{m_{2,n-1}} \dots x_{n-1}^{m_{n-1,n-1}}$$

$$F\left\{\begin{matrix} \frac{1}{n}(M_{n-1}+1), \frac{1}{n}(M_{n-1}+2), \frac{1}{n}(M_{n-1}+3), \dots, \frac{1}{n}(M_{n-1}+n), \\ 1, \quad m_{n-1,n-1}+1, m_{n-2,n-1}+1, \dots, m_{1,n-1}+1, \end{matrix} n^n x_1 x_2 \dots x_n \right\},$$

$$(1-x_1-x_2-\dots-x_n)^{-1}$$

$$= \sum_{m_1=0}^{\infty} \sum_{m_2=0}^{\infty} \dots \sum_{m_{n-1}=0}^{\infty} \Big[P(m_{1,n-1}, m_{2,n-1}, \dots, m_{n-1,n-1}) (\Sigma x_1^{m_{1,n-1}} x_2^{m_{2,n-1}} \dots x_{n-1}^{m_{n-1,n-1}})$$

$$\times F\left\{\begin{matrix} \frac{1}{n}(M_{n-1}+1), \frac{1}{n}(M_{n-1}+2), \frac{1}{n}(M_{n-1}+3), \dots, \frac{1}{n}(M_{n-1}+n), \\ 1, \quad m_{n-1,n-1}+1, m_{n-2,n-1}+1, \dots, m_{1,n-1}+1, \end{matrix} n^n x_1 x_2 \dots x_n \right\}\Big].$$

5. Similarly if we expand

$$(1-x_1-x_2-\dots-x_n)^{-1} \Pi_{a,b}\left(1-\frac{x_b}{x_a}\right),$$

where the product has reference to every pair of letters x_a, x_b subject to the condition $b > a$, the coefficient of a lattice assemblage of letters enumerates the lattice permutations of such assemblage (*loc. cit.*).

The portion of the expansion which involves lattice assemblages *quâ* the ordered letters $x_1, x_2, \dots, x_n$ is here denoted by

$$L(1-x_1-x_2-\dots-x_n)^{-1} \Pi_{a,b}\left(1-\frac{x_b}{x_a}\right).$$

The investigation of § 3 shows that

$$L\frac{1-\frac{x_2}{x_1}}{1-x_1-x_2}$$

$$= F\left\{\begin{matrix}\frac{1}{2}, & \frac{2}{2}, \\ 1, & 2,\end{matrix}\; 2^2x_1x_2\right\}$$

$$+x_1F\left\{\begin{matrix}\frac{2}{2}, & \frac{3}{2}, \\ 1, & 3,\end{matrix}\; 2^2x_1x_2\right\}$$

$$+x_1^2F\left\{\begin{matrix}\frac{3}{2}, & \frac{4}{2}, \\ 1, & 4,\end{matrix}\; 2^2x_1x_2\right\}$$

$$+\ldots$$

$$+x_1^{m_1}F\left\{\begin{matrix}\frac{m_1+1}{2}, & \frac{m_1+2}{2}, \\ 1, & m_1+2,\end{matrix}\; 2^2x_1x_2\right\}$$

$$+\ldots;$$

but we cannot obtain an expression for

$$(1-x_1-x_2)^{-1}\left(1-\frac{x_2}{x_1}\right),$$

because it is neither symmetrical nor integral.

Similarly $$L\frac{\left(1-\frac{x_2}{x_1}\right)\left(1-\frac{x_3}{x_1}\right)\left(1-\frac{x_3}{x_2}\right)}{1-x_1-x_2-x_3}$$

$$= \sum_{m_1=0}^{\infty}\sum_{m_2=0}^{\infty} LP(m_1+m_2,\, m_2)\, x_1^{m_1+m_2}x_2^{m_2}$$

$$\times F\left\{\begin{matrix}\frac{1}{3}(m_1+2m_2+1), & \frac{1}{3}(m_1+2m_2+2), & \frac{1}{3}(m_1+2m_2+3), \\ 1, & m_2+2, & m_1+m_2+3,\end{matrix}\; 3^3x_1x_2x_3\right\}.$$

The general theorem is

$$L\,\frac{\Pi_{a,\,b}\left(1-\dfrac{x_b}{x_a}\right)}{1-x_1-x_2-\ldots-x_n}$$

$$= \Sigma\Sigma\ldots\Sigma\, LP(m_{1,\,n-1},\, m_{2,\,n-1},\, \ldots,\, m_{n-1,\,n-1})\, x_1^{m_{1,\,n-1}}\, x_2^{m_{2,\,n-1}} \ldots x_{n-1}^{m_{n-1,\,n-1}}$$

$$F\left\{\begin{matrix} \frac{1}{n}(M_{n-1}+1),\ \frac{1}{n}(M_{n-1}+2),\ \frac{1}{n}(M_{n-1}+3),\ \ldots,\ \frac{1}{n}(M_{n-1}+n), \\ 1,\quad m_{n-1,\,n-1}+2,\ m_{n-2,\,n-1}+3,\ \ldots,\ m_{1,\,n-1}+n, \end{matrix}\quad n^n x_1 x_2 \ldots x_n\right\}.$$

Prime Lattice Permutations. By Major P. A. MACMAHON, Sc.D., F.R.S.

[*Received* 6 June, 1922.]

1. An assemblage of letters $\alpha_1^{k_1}\alpha_2^{k_2}\ldots\alpha_n^{k_n}$ in respect of the letters in the *definite order*

$$\alpha_1,\ \alpha_2,\ \ldots,\ \alpha_n$$

is said to be a lattice assemblage if

$$k_1 \geqslant k_2 \geqslant \ldots \geqslant k_n.$$

Such an assemblage may be denoted by

$$|\, k_1 k_2 \ldots k_n \,| \text{ or } |\, \alpha_1^{k_1}\alpha_2^{k_2}\ldots\alpha_n^{k_n} \,|,$$

where in the usual notation $(k_1 k_2 \ldots k_n)$ is a partition of the number Σk. For each assemblage there is a corresponding partition.

If we take any permutation of the lattice assemblage we call it a lattice permutation if, a line being drawn between any two letters or after the last letter, the assemblage to the left of such line is a lattice assemblage. The line in question may be drawn in Σk positions so that a lattice permutation implies Σk lattice assemblages and the like number of partitions.

A lattice permutation is either prime or composite. It is composite if it be possible to draw a line between some two letters in such wise that on each side of the line a lattice permutation of some assemblage is in evidence. Thus $\alpha\beta\alpha\gamma\,|\,\alpha\beta$ is a composite lattice permutation of the assemblage $|\,\alpha^3\beta^2\gamma\,|$ because $\alpha\beta\alpha\gamma$, $\alpha\beta$ are lattice permutations of the assemblages $|\,\alpha^2\beta\gamma\,|$, $|\,\alpha\beta\,|$ respectively. On the other hand $\alpha\beta\alpha\alpha\gamma\beta$ is a prime lattice permutation of the assemblage $|\,\alpha^3\beta^2\gamma\,|$*.

2. We consider the lattice permutations of the assemblage $|\,k_1k_2\ldots k_n\,|$ and, further, the subdivision of these into prime and composite lattice permutations.

If a lattice permutation be composite, we must be able to draw one or more lines between letters dividing it into prime permutations. In correspondence therewith we can separate the whole assemblage into one or more component assemblages.

Denote by $l\,|\,k_1k_2\ldots k_n\,|$ any lattice permutation of $|\,k_1k_2\ldots k_n\,|$,

„ $L\,|\,k_1k_2\ldots k_n\,|$ the number of such.

„ $p\,|\,k_1k_2\ldots k_n\,|$ any prime lattice permutation of $|\,k_1k_2\ldots k_n\,|$,

„ $P\,|\,k_1k_2\ldots k_n\,|$ the number of such.

* MacMahon, *Combinatory Analysis*, vol. I, Cambridge, 1915, ch. V, p. 124.

If we can separate the lattice assemblage into components in the manner

$$|k_1k_2\ldots k_n| = |k_1'k_2'\ldots k_n'|\,.\,|k_1''k_2''\ldots k_n''|\ldots$$

we can proceed to a composite lattice permutation

$$p\,|k_1'k_2'\ldots k_n'|\,.\,p\,|k_1''k_2''\ldots k_n''|\ldots$$

in a number of ways denoted by

$$P\,|k_1'k_2'\ldots k_n'|\,.\,P\,|k_1''k_2''\ldots k_n''|\ldots.$$

Every lattice permutation

$$l\,|k_1k_2\ldots k_n|$$

can be dealt with and we can account for the whole of them. We may therefore write

$$\begin{aligned} L\,|k_1k_2\ldots k_n| = {} & P\,|k_1k_2\ldots k_n| \\ & + \Sigma P\,|k_1'k_2'\ldots k_n'|\,.\,P\,|k_1''k_2''\ldots k_n''| \\ & + \Sigma P\,| \qquad |\,.\,P\,| \qquad |\,.\,P\,| \qquad | \\ & + \ldots \end{aligned}$$

until we obtain a numerical identity.

The succession of numbers $k_1k_2\ldots k_n$ where $k_1 \geqslant k_2 \geqslant \ldots \geqslant k_n$ denotes on the one hand a lattice assemblage of letters and on the other a partition of a number. It also denotes a multipartite number which specifies a lattice assemblage without specification of the particular letters $\alpha_1\alpha_2\ldots\alpha_n$. There is a theory of the compositions (viz. partitions in which the order of the parts is of moment) of multipartite numbers which is involved in the present question. In general the part of a partition (or composition) of a multipartite number is or is not of the form

$$k_1'k_2'\ldots k_n' \text{ when } k_1' \geqslant k_2' \geqslant \ldots \geqslant k_n'.$$

If the former is the case we call it a *lattice part*, and we note that we are only concerned with compositions into lattice parts. We consider the lattice-part compositions of the multipartite number $k_1k_2\ldots k_n$.

3. A simplification arises when

$$k_1 = k_2 = \ldots = k_n = k,$$

for then the only lattice parts that present themselves are

$$(11\ldots),\ (22\ldots),\ (33\ldots),\ \ldots\ (kk\ldots),$$

there being n numbers in each bracket.

Ex. gr. If the multipartite be (333) the compositions are

$$(333),\ (222, 111),\ (111, 222),\ (111, 111, 111).$$

In fact in this particular case we are concerned with the compositions of the unipartite number k.

Thus when $k = 3$, we really have to deal with the compositions (3), (21), (12), (111) of the number 3.

We are led to the relations

$$L|111| = P|111|$$
$$L|222| = P|222| + \{P|111|\}^2$$
$$L|333| = P|333| + 2P|222|P|111| + \{P|111|\}^3$$
$$\dots\dots\dots\dots\dots\dots$$

yielding

$$P|111| = L|111|$$
$$P|222| = L|222| - \{L|111|\}^2$$
$$P|333| = L|333| - 2L|222|L|111| + \{L|111|\}^3$$
$$\dots\dots\dots\dots\dots\dots$$

Since we know (*loc. cit.*) that

$$L|kkk| = \frac{(3k)!\,2}{(k+2)!\,(k+1)!\,k!},$$

giving $\qquad L|111| = 1,\ L|222| = 5,\ L|333| = 42,$

we find $\qquad P|111| = 1,\ P|222| = 4,\ P|333| = 33.$

In general we have the two formulæ

$$L|kkk\dots|_n = \Sigma \frac{(\Sigma\mu)!}{\mu_1!\,\mu_2!\dots\mu_k!}\{P|111\dots|_n\}^{\mu_1}\{P|222\dots|_n\}^{\mu_2}\dots\{P|kkk\dots|_n\}^{\mu_k},$$

$$P|kkk\dots|_n = \Sigma \frac{(-)^{\Sigma\mu+1}(\Sigma\mu)!}{\mu_1!\,\mu_2!\dots\mu_k!}\{L|111\dots|_n\}^{\mu_1}\{L|222\dots|_n\}^{\mu_2}\dots\{L|kkk\dots|_n\}^{\mu_k},$$

where $\qquad L|kkk\dots|_n = \dfrac{(nk)!\,(n-1)!}{(k+n-1)!\,(k+n-2)!\dots k!},$

and the summations are for all partitions

$$1^{\mu_1}2^{\mu_2}\dots k^{\mu_k} \text{ of the number } k.$$

This is the solution of the problem of the enumeration of the prime lattice permutations of the assemblage

$$\alpha_1^k\alpha_2^k\dots\alpha_n^k.$$

Otherwise we may assert that in regard to the equation

$$x^k - L|111\dots|_n x^{k-1} - L|222\dots|_n x^{k-2} - \dots - L|kkk\dots|_n = 0,$$

the homogeneous product sum, of order $s\ (\not> k)$, of the roots is

$$P|sss\dots|_n.$$

In particular when $n = 2$,

$$x^k - x^{k-1} - 2x^{k-2} - 5x^{k-3} - 14x^{k-4} - \dots - \frac{(2k)!}{(k+1)!\,k!} = 0,$$

and we find that $\qquad P|s+1, s+1| = L|s, s|$*.

* Cf. *Netto Combinatorik*, Leipzig, 1901, § 122, p. 192 *et seq.*

4. The passage to the general case of the assemblage is now clear. Taking the assemblage

$$|\alpha_1^{k_1}\alpha_2^{k_2}\dots\alpha_n^{k_n}| \quad k_1 \geqslant k_2 \geqslant \dots \geqslant k_n,$$

we find the relation

$$\frac{1}{\begin{array}{llll} 1-P|1|x_1-P|11|x_1x_2-P|111|x_1x_2x_3-P|1111|x_1x_2x_3x_4-\dots \\ \quad -P|2|x_1^2 \quad -P|21|x_1^2x_2 \quad -P|211|x_1^2x_2x_3 \\ \quad\quad -P|3|x_1^3 \quad -P|22|x_1^2x_2^2 \\ \quad\quad\quad -P|31|x_1^3x_2 \\ \quad\quad\quad -P|4|x_1^4 \end{array}}$$

$$\begin{array}{l} =1+L|1|x_1+L|11|x_1x_2+L|111|x_1x_2x_3+L|1111|x_1x_2x_3x_4+\dots \\ \quad +L|2|x_1^2 \quad +L|21|x_1^2x_2 \quad +L|211|x_1^2x_2x_3 \\ \quad\quad +L|3|x_1^3 \quad +L|22|x_1^2x_2^2 \\ \quad\quad\quad +L|31|x_1^3x_2 \\ \quad\quad\quad +L|4|x_1^4 \end{array}$$

leading to the two formulæ

$$L|k_1k_2k_3\dots|=\Sigma\frac{(\Sigma\kappa)!}{\kappa_1!\,\kappa_2!\dots}\{P|k_1'k_2'k_3'\dots|\}^{\kappa_1}\{P|k_1''k_2''k_3''\dots|\}^{\kappa_2}\dots$$

$$P|k_1k_2k_3\dots|=\Sigma(-)^{\Sigma\kappa+1}\frac{(\Sigma\kappa)!}{\kappa_1!\,\kappa_2!\dots}\{L|k_1'k_2'k_3'\dots|\}^{\kappa_1}\{L|k_1''k_2''k_3''\dots|\}^{\kappa_2}\dots$$

the summations being in respect of every lattice-part partition

$$(k_1'k_2'k_3'\dots)^{\kappa_1}(k_1''k_2''k_3''\dots)^{\kappa_2}\dots$$

of the multipartite number

$$(k_1k_2k_3\dots).$$

Since the value of $L|k_1k_2k_3\dots|$ is known (*loc. cit.*) we thus obtain an expression for $P|k_1k_2k_3\dots|$.

The best way of calculating these numbers is to take the relation from which the formulæ are obtained, clear the fraction and equate coefficients of like powers of $x_1^{k_1}x_2^{k_2}x_3^{k_3}\dots$.

The coefficient of $x_1^2x_2$ gives

$$P|21|+P|11|L|1|+P|1|L|11|-L|21|=0,$$

and so forth.

The calculation is simplified by noting that $P|k_1k_2|=0$ if $k_1>k_2$ and that

$$L|k_1k_2k_3\dots|=L|j_1j_2j_3\dots|$$

if $(k_1k_2k_3\dots)$, $(j_1j_2j_3\dots)$ be conjugate partitions.

Chapter 5
Compositions and Simon Newcomb's Problem

5.1 Introduction and Commentary

Compositions of an integer are merely ordered partitions. Thus there are four compositions of 3, namely 3, 2 + 1, 1 + 2, 1 + 1 + 1. MacMahon's interest in this topic lies in problems concerning compositions of multipartite numbers (i.e., nonzero vectors with nonnegative integral coordinates). The study of such compositions yields solutions to various permutation problems, especially those of the Simon Newcomb type. In fact, MacMahon's second memoir on compositions [71] is entirely devoted to the original Simon Newcomb problem. Such permutation problems have subsequently had many applications; see for example Barton and David (1962), Foata and Schützenberger (1970), Gassner (1967), Knuth (1973), Kreweras (1965), Sorin (1971), and Stanley (1972).

Studies of permutation problems of this type and their relationship to Eulerian numbers date back to D. André (1879, 1881), and extensive advances have recently been made by L. Carlitz, D. Foata, H. Foulkes, M. Schützenberger, and others. An elegant solution of Simon Newcomb's original problem, contributed by Dillon and Roselle (1969), has a close relationship to the work of Carlitz (1964) on Eulerian numbers; this work will be presented in Section 9.2. We shall then give the solution (Andrews, 1976a) to a conjecture on compositions by Long (1970), and we shall use the solution to obtain further information about Simon Newcomb's problem (Andrews, 1975).

5.2 The Dillon-Roselle Solution of Simon Newcomb's Problem

Let $Z_n = \{1, 2, \ldots, n\}$. Let α be a permutation of a multiset with elements in Z_n (a multiset is a set with possible repetitions; see sections 3.2 and 4.2 for further discussions of multisets). We say that α is of specification $[s_1, s_2, \ldots, s_n]$ if j appears in α exactly s_j times. We call $N = \sum s_i$ the content of α, and we say that α has r rises if $\alpha = \{a_1, a_2, \ldots, a_N\}$ and $a_i < a_{i+1}$ occurs for exactly r indices i. Let $S = S([s])$ denote the set of all multiset permutations with elements in Z_n and specification $[s_1, s_2, \ldots, s_n]$. We shall say that S is the deck of specification $[s]$ and content N.

Simon Newcomb's Problem. Obtain a formula for $A([s], r)$, the number of multiset permutations in $S([s])$ that have exactly r rises.

The material in this chapter corresponds to section IV of *Combinatory Analysis.*

Lemma 1. Let $S([s]) = S$ be a deck with specification $[s] = [s_1, s_2, \ldots, s_n]$, and let σ be a permutation of $\mathcal{Z}_n$. Let $\sigma(S)$ be the deck obtained when S is transformed by σ so that $\sigma(S)$ has specification $[\tau(s)] = [s_{\tau(1)}, s_{\tau(2)}, \ldots, s_{\tau(n)}]$, where $\tau = \sigma^{-1}$. Then $A([\tau(s)], r) = A([s], r)$ for each r.

Proof. Since all permutations of $\mathcal{Z}_n$ may be obtained by repeated transpositions, we see that we need only prove the lemma in the case where σ interchanges u and $u + 1$. For each ρ in S, we define $\sigma(\rho)$ by interchanging appearances of u and $u + 1$ in ρ. We observe that if there are r rises in ρ, a occurrences of $(u, u + 1)$, and b occurrences of $(u + 1, u)$, then $\sigma(\rho)$ has $r + b - a$ rises, b occurrences of $(u, u + 1)$, and a occurrences of $(u + 1, u)$.

Now we define a mapping f of S onto itself by

$$f(U_1, V_1, \ldots, U_t, V_t) = (\overline{U}_1, V_1, \ldots, \overline{U}_t, V_t),$$

where $(U_1, V_1, \ldots, U_t, V_t)$ is the unique representation of an element of S in block form, the blocks U_i contain only u and $u + 1$, the blocks V_i do not contain u or $u + 1$, and $\overline{U}_i$ is the block obtained by reversing U_i. Note that either U_1 or V_t can be empty. Furthermore, if $\delta \in S$, and δ has r rises, a occurrences of $(u, u + 1)$, and b occurrences of $(u + 1, u)$, then $f(\delta)$ has $r + b - a$ rises, b occurrences of $(u, u + 1)$, and a occurrences of $(u + 1, u)$. Consequently, the mapping $g = \sigma \cdot f$ is a bijection, between S and $\sigma(S)$, that preserves rises.

Lemma 2. Let S be a deck with specification $[s_1, s_2, \ldots, s_n]$ and content $\mathcal{N}$. The maximum number $\mathcal{N}^*$ of rises in any of the multiset permutations in S is $\mathcal{N}^* = \mathcal{N} - s^*$, where $s^* = \max\{s_1, s_2, \ldots, s_n\}$. Furthermore,

$$A([s], \mathcal{N}^*) = \prod_{i=1}^{n} \binom{s^*}{s_i}.$$

Proof. By Lemma 1 we may assume that $s_1 \geqq s_2 \geqq \ldots \geqq s_n$; hence $s_1 = s^*$. Therefore $\mathcal{N}^* \leqq \mathcal{N} - s^*$ since 1 can never be the second element in a rise. We can, however, create a permutation with $\mathcal{N} - s^*$ rises by forming an n-column array (i.e., Young tableau or plane partition) with s_i is in the ith column, and then juxtaposing the rows. Therefore $\mathcal{N}^* = \mathcal{N} - s^*$.

Next we observe that if δ is in S and δ has $\mathcal{N}^*$ rises, then 1 must be the first element of δ, and elements which lie between successive 1's must occur in strictly increasing order. We obtain all such permutations as follows: Take $s^* = s_1$ containers $C_1, C_2, \ldots, C_{s^*}$; distribute the s_2 twos among the containers so that no container contains more than one 2 (this can be done in

$\binom{s^*}{s_2}$ ways); distribute the s_3 threes among the containers so that no container contains more than one 3 (this can be done in $\binom{s^*}{s_3}$ ways); etc. After all the numbers 2, 3, . . . , n have been distributed in this way, place one 1 in each container, order the elements in each container in increasing order, and juxtapose the containers. This procedure produces all the multiset permutations lying in S with $\mathcal{N}^*$ rises. Hence

$$A([s], \mathcal{N}^*) = \prod_{i=1}^{n} \binom{s^*}{s_i}.$$

Theorem 1.

$$A([s], r) = \sum_{j=0}^{r} (-1)^j \binom{\mathcal{N}+1}{j} \prod_{i=1}^{n} \binom{s_i + r - j}{s_i}.$$

Proof. We let S be a deck with specification $[s] = [s_1, s_2, \ldots, s_n]$ and content $\mathcal{N}$, while T denotes a deck with specification $[t] = [t, s_1, s_2, \ldots, s_n]$ (note that the multiset permutations in T are related to $\mathcal{Z}_{n+1}$) and content $t + \mathcal{N}$.

Our first goal is to prove that

$$A([t], r) = \sum_{j=0}^{r} \binom{\mathcal{N}-j}{r-j} \binom{t+j}{r} A([s], j). \tag{2.1}$$

We begin by noting that for each permutation enumerated by $A([t], r)$, we may determine one enumerated by $A([s], j)$ (where $j \leqq r$) by deleting all appearances of 1 and then subtracting 1 from all the remaining entries. However, it may well be that two permutations enumerated by $A([t], r)$ are transformed into the same permutation enumerated by $A([s], j)$. In fact, for each permutation counted by $A([s], j)$, let us first add 1 to each number appearing. Then we insert $r - j$ ones (without repetition) between the $\mathcal{N} - j - 1$ nonrise pairs or in the first position, which can be done in $\binom{(\mathcal{N}-j-1)+1}{r-j} = \binom{\mathcal{N}-j}{r-j}$ ways. Next we insert the remaining $t - r + j$ ones (with repetition allowed) between the current r rise pairs or in the last position, which can be done in $\binom{r+1+t-r+j-1}{t-r+j} = \binom{t+j}{r}$ ways.

Consequently, exactly $\binom{N-j}{r-j}\binom{t+j}{r}$ permutations from among those counted by $A([t], r)$ are associated with each permutation enumerated by $A([s], j)$. Summing over all j, we obtain (2.1).

We now put $t \geqq s^* = \max(s_1, s_2, \ldots, s_n)$. Hence by Lemma 2, N is the maximum number of rises in any permutation in T, since T has content $N + t$. Therefore by Lemma 2 and equation (2.1), we see that

$$\prod_{i=1}^{n} \binom{t}{s_i} = A([t], N) = \sum_{j=0}^{N^*} \binom{t+j}{N} A([s], j). \tag{2.2}$$

In (2.2) replace t by $s^* + r$ and reverse the order of summation:

$$\prod_{i=1}^{n} \binom{s^* + r}{s_i} = \sum_{j=0}^{r} \binom{N + r - j}{N} A([s], N^* - j), \tag{2.3}$$

where we use the fact that $N = N^* + s^*$. Applying binomial inversion (J. Riordan, 1968, chapter 2), we see that

$$A([s], N^* - r) = \sum_{j=0}^{r} (-1)^j \binom{N+1}{j} \prod_{i=1}^{n} \binom{s^* + r - j}{s_i}, \tag{2.4}$$

or equivalently,

$$A([s], r) = \sum_{j=0}^{N^*-r} (-1)^j \binom{N+1}{j} \prod_{i=1}^{n} \binom{N - r - j}{s_i}. \tag{2.5}$$

Next we observe the $\prod_{i=1}^{n} \binom{N-r}{s_i}$ is a polynomial in r of degree $s_1 + s_2 + \ldots + s_n = N$; therefore its Nth-order difference is zero:

$$\begin{aligned} 0 &= \sum_{j=0}^{N+1} (-1)^j \binom{N+1}{j} \prod_{i=1}^{n} \binom{N - r - j}{s_i} \\ &= \sum_{j=N+1-r}^{N} (-1)^{j+1} \binom{N+1}{j} \prod_{i=1}^{n} \binom{N - r - j}{s_i} \\ &= \sum_{j=0}^{r} (-1)^{N+j} \binom{N+1}{j} \prod_{i=1}^{n} \binom{-(r + 1 - j)}{s_i}. \end{aligned}$$

In light of the above, our theorem now follows from (2.5).

5.3 C. Long's Conjecture

A multipartite number (or n-partite number) is an ordered n-tuple of nonnegative integers, not all zero. A composition of a multipartite number is its representation as a sum of multipartite numbers, where two representations are considered distinct if either the summands or their order differs. Let $f(\alpha_1, \alpha_2, \ldots, \alpha_n)$ denote the number of compositions of the multipartite number $(\alpha_1, \alpha_2, \ldots, \alpha_n)$. The following theorem was conjectured by C. Long (1970), and the proof we give follows Andrews (1976a).

Theorem 2. If the polynomial

$$2^{\alpha_1 - 1} \prod_{i=2}^{r} \{x_1 x_2 \ldots x_{i-1} + (1 + x_1)(1 + x_2) \ldots (1 + x_{i-1})\}^{\alpha_i}$$

is fully expanded and each x_i^k is replaced by $\binom{\alpha_i}{\alpha_{i+1} + \ldots + \alpha_r - k}$, then the resulting number is $f(\alpha_1, \alpha_2, \ldots, \alpha_r)$.

Proof. We begin by letting $F_r(h_1, \ldots, h_r; \alpha_1, \alpha_2, \ldots, \alpha_r)$ denote the number of (monotone) lattice paths that start at the origin in r-dimensional space and end somewhere in the parallelotope $0 \leqq x_i \leqq \alpha_i$ $(1 \leqq i \leqq r)$, wherein exactly h_j edges are parallel to the hyperplane $x_{j+1} = x_{j+2} = \ldots = x_r = 0$, but not parallel to the hyperplane $x_j = x_{j+1} = \ldots = x_r = 0$.

Our first goal is to prove that

$$(3.1) \qquad F_r(h_1, \ldots, h_r; \alpha_1, \ldots, \alpha_r) =$$

$$\binom{\alpha_1}{h_1}\binom{\alpha_2}{h_2}\cdots\binom{\alpha_r}{h_r}\binom{\alpha_1 + h_2 + h_3 + \ldots + h_r}{h_2 + h_3 + \ldots + h_r}$$
$$\binom{\alpha_2 + h_3 + \ldots + h_r}{h_3 + \ldots + h_r}\cdots\binom{\alpha_{r-1} + h_r}{h_r}.$$

There are two kinds of edges in the lattice paths enumerated by $F_r(h_1, \ldots, h_r; \alpha_1, \ldots, \alpha_r)$: (i) the h_1 edges parallel to the x_1-axis, and (ii) the other $h_2 + \ldots + h_r$ edges. Suppose that for a path P the type (i) edges terminate on the hyperplanes $x_1 = a_i$ $(1 \leqq i \leqq h_1)$ and the type (ii) edges terminate on the hyperplanes $x_1 = b_j$ $(1 \leqq j \leqq h_2 + \ldots + h_r)$; then P is uniquely determined by its projection on the hyperplane $x_1 = 0$ together with the numbers a_i and b_j. Since the a_i are chosen from $\{1, 2, \ldots, \alpha_1\}$ without repetitions, while the b_j are chosen from $\{0, 1, \ldots, \alpha_1\}$ with repetitions, we see that

$$(3.2)\quad F_r(h_1, \ldots, h_r; \alpha_1, \ldots, \alpha_r) = \binom{\alpha_1}{h_1}\binom{\alpha_1 + h_2 + \ldots + h_r}{h_2 + \ldots + h_r} F_{r-1}(h_2, \ldots, h_r; \alpha_2, \ldots, \alpha_r).$$

Iteration of (3.2) yields (3.1).

Now to each composition of the multipartite number $(\alpha_1, \alpha_2, \ldots, \alpha_r)$, say $(A_1, \ldots, A_r) + (B_1 - A_1, \ldots, B_r - A_r) + (C_1 - B_1, \ldots, C_r - B_r) + \ldots + (\alpha_1 - J_1, \ldots, \alpha_r - J_r)$, there correspond two lattice paths of the type under consideration:

$$\{(A_1, \ldots, A_r), (B_1, \ldots, B_r), \ldots, (J_1, \ldots, J_r)\}$$

and

$$\{(A_1, \ldots, A_r), (B_1, \ldots, B_r), \ldots, (J_1, \ldots, J_r), (\alpha_1, \ldots, \alpha_r)\};$$

hence

$$(3.3)\quad f(\alpha_1, \ldots, \alpha_r) = \frac{1}{2} \sum_{h_1 \geqq 0, \ldots, h_r \geqq 0} F_r(h_1, \ldots, h_r; \alpha_1, \ldots, \alpha_r).$$

We formalize the substitutions called for in Long's Conjecture by defining linear operators L_i such that

$$L_i(x_i^k) = \binom{\alpha_i}{\alpha_{i+1} + \ldots + \alpha_r - k}.$$

We note that

$$\begin{aligned} L_i(x_i^C(1 + x_i)^D) &= L_i\left(\sum_{j=0}^{D}\binom{D}{j}x_i^{C+j}\right) \\ &= \sum_{j=0}^{D}\binom{D}{j}\binom{\alpha_i}{\alpha_{i+1} + \ldots + \alpha_r - C - j} \\ &= \binom{\alpha_i + D}{\alpha_{i+1} + \ldots + \alpha_r - C}, \end{aligned}$$

where the final equation follows from the Vandermonde convolution (J. Riordan, 1958, page 9).

We are now prepared to perform the substitutions called for in Long's conjecture:

$$L_1L_2 \ldots L_{r-1}\left\{2^{\alpha_1-1}\prod_{i=2}^{r}(x_1x_2\ldots x_{i-1}+(1+x_1)(1+x_2)\ldots (1+x_{i-1}))^{\alpha_i}\right\}$$

$$= L_1L_2\ldots L_{r-1}\left\{2^{\alpha_1-1}\sum_{h_2\geqq 0,\ldots,h_r\geqq 0}\binom{\alpha_2}{h_2}\cdots\binom{\alpha_r}{h_r}\prod_{j=2}^{r}\{x_{j-1}^{(\alpha j-hj)+\ldots+(\alpha_r-h_r)}(1+x_{j-1})^{hj+\ldots+h_r}\}\right\}$$

$$= 2^{\alpha_1-1}\sum_{h_2\geqq 0,\ldots,h_r\geqq 0}\binom{\alpha_2}{h_2}\cdots\binom{\alpha_r}{h_r}\prod_{j=1}^{r}L_{j-1}\{x_{j-1}^{(\alpha j-hj)+\ldots+(\alpha_r-h_r)}(1+x_{j-1})^{hj+\ldots+h_r}\}$$

$$= 2^{\alpha_1-1}\sum_{h_2\geqq 0,\ldots,h_r\geqq 0}\binom{\alpha_2}{h_2}\cdots\binom{\alpha_r}{h_r}\binom{\alpha_1+h_2+\ldots+h_r}{h_2+\ldots+h_r}\binom{\alpha_2+h_3+\ldots+h_r}{h_3+\ldots+h_r}\cdots\binom{\alpha_{r-1}+h_r}{h_r}$$

$$= \frac{1}{2}\sum_{h_1\geqq 0,\ldots,h_r\geqq 0}\binom{\alpha_1}{h_1}\cdots\binom{\alpha_r}{h_r}\binom{\alpha_1+h_2+\ldots+h_r}{h_2+\ldots+h_r}\binom{\alpha_2+h_3+\ldots+h_r}{h_3+\ldots+h_r}\cdots\binom{\alpha_{r-1}+h_r}{h_r}$$

$$= \frac{1}{2}\sum_{h_1\geqq 0,\ldots,h_r\geqq 0}F_r(h_1,\ldots,h_r;\alpha_1,\ldots,\alpha_r)$$

$$= f(\alpha_1,\ldots,\alpha_r),$$

and Long's Conjecture is established.

Let us now turn to a refinement of the Simon Newcomb Problem that may be solved utilizing our approach to Long's Conjecture. We shall only sketch the results—a complete account appears in Andrews (1975).

Definition 1. Let $\phi_r(h_1, \ldots, h_r; \alpha_1, \ldots, \alpha_r)$ denote the number of compositions of the multipartite number $(\alpha_1, \ldots, \alpha_r)$ in which exactly h_i of the parts have i as the largest coordinate with a nonzero entry.

Definition 2. Let $N_r(h_1, \ldots, h_r; \alpha_1, \ldots, \alpha_r)$ denote the number of permutations of the multiset on $\{1, 2, \ldots, n\}$ with specification $[\alpha_1, \alpha_2, \ldots, \alpha_r]$ in which exactly h_i runs end with i.

Theorem 3. $\phi_r(H_1, \ldots, H_r; \alpha_1, \ldots, \alpha_r) =$

$$\frac{H_1 + H_2 + \ldots + H_r}{\alpha_r} \binom{\alpha_1}{H_1} \cdots \binom{\alpha_r}{H_r} \binom{\alpha_1 + H_2 + \ldots + H_r - 1}{H_2 + \ldots + H_r} \cdots \binom{\alpha_{r-1} + H_r - 1}{H_r - 1}.$$

Proof. First we observe that

$$F_r(H_1, \ldots, H_r; \alpha_1, \ldots, \alpha_r) = \sum_{\substack{0 \leqq a_i \leqq \alpha_i \\ 1 \leqq i \leqq r}} \phi_r(H_1, \ldots, H_r; a_1, \ldots, a_r).$$

This relation follows immediately from the definition of $F_r(H_1, \ldots, H_r; \alpha_1, \ldots, \alpha_r)$, and the correspondence between compositions and lattice paths that was described immediately following equation (3.2). Theorem 3 now follows by taking the first difference with respect to each α_i in the above equation and then applying formula (3.1).

Theorem 4. $\mathcal{N}_r(H_1, \ldots, H_r; \alpha_1, \ldots, \alpha_r) =$

$$\sum_{\substack{0 \leqq h_i \leqq \alpha_i \\ 1 \leqq i \leqq r}} (-1)^{H_1 + \ldots + H_r + h_1 + \ldots + h_r} \binom{\alpha_1 - h_1}{\alpha_1 - H_1} \cdots \binom{\alpha_r - h_r}{\alpha_r - H_r} \phi_r(h_1, \ldots, h_r; \alpha_1, \ldots, \alpha_r).$$

Proof. We shall prove that

$$\phi_r(H_1, \ldots, H_r; \alpha_1, \ldots, \alpha_r) = \sum_{\substack{0 \leqq h_i \leqq H_i \\ 1 \leqq i \leqq r}} \binom{\alpha_1 - h_1}{H_1 - h_1} \cdots \binom{\alpha_r - h_r}{H_r - h_r} \mathcal{N}_r(h_1, \ldots, h_r; \alpha_1, \ldots, \alpha_r). \tag{3.4}$$

Our theorem follows from this identity by applying a simple inversion theorem (Riordan, 1968, page 49, Table 2.1, entry 3) to each of the r variables H_i.

To prove the latter identity we define a mapping f, from the compositions enumerated by $\phi_r(H_1, \ldots, H_r; \alpha_1, \ldots, \alpha_r)$ to the various multiset permutations enumerated by the $\mathcal{N}_r(h_1, \ldots, h_r; \alpha_1, \ldots, \alpha_r)$, for which $0 \leqq h_i \leqq H_i$, by merely dropping parentheses in the summation that represents the composition and letting each part be the specification of a run in the permutation. For example, the image of the composition $(1, 2, 0) + (2, 0, 0) + (0, 0, 1)$ is the permutation 122113.

Several compositions may have the same image under f; to obtain the cardinality of the preimage in each instance we observe that each run in the permutation may be split up into portions, each corresponding to a separate part of a composition. Initially we have i as the largest number in h_i runs. We choose $H_i - h_i$ of the remaining $\alpha_i - h_i$ is, and we place ") (" between each chosen i and the following entry; a composition of the desired form is now obtained by replacing each resulting portion of the permutation by its specification. Thus the total number of preimages in each case is

$$\binom{\alpha_1 - h_1}{H_1 - h_1} \cdots \binom{\alpha_r - h_r}{H_r - h_r}.$$

Summing over all possible h_i, we obtain (3.4); this establishes Theorem 4.

Theorems 3 and 4 may be combined to obtain an explicit formula for $\mathcal{N}_r(h_1, \ldots, h_r; \alpha_1, \ldots, \alpha_r)$. As a result we have a new formula for $A([\alpha], s)$, since

$$A([\alpha], s) = \sum_{h_1 + h_2 + \ldots + h_r = s} \mathcal{N}_r(h_1, \ldots, h_r; \alpha_1, \ldots, \alpha_r).$$

5.4 References

M. Abramson (1971) Combinations, compositions, and occupancy problems, *Fibonacci Quart.*, *9*, 225–236, 244.

M. Abramson (1975) A simple solution of Simon Newcomb's problem, *J. Combinatorial Th.*, *A-18*, 223–225.

M. Abramson (1976) Restricted combinations and compositions, *Fibonacci Quart.*, *14*, 439–452.

M. Abramson and W. O. J. Moser (1967) Permutations without rising or falling ω-sequences, *Ann. Math. Stat.*, *38*, 1245–1254.

D. André (1879) Développements de sec x et de tang x, *C.R. Acad. Sc.* Paris, *88*, 965–967.

D. André (1881) Mémoire sur le nombre des permutations alternées, *Journ. de Math.*, *7*, 167.

G. E. Andrews (1975) The theory of compositions II: Simon Newcomb's problem, *Utilitas Math.*, *7*, 33–54.

G. E. Andrews (1976a) The theory of compositions I: The ordered factorizations of n and a conjecture of C. Long, *Canadian Math. Bull.*, *18*, 479–484.

G. E. Andrews (1976b) The theory of compositions III: The MacMahon formula and the Stanton-Cowan numbers, *Utilitas Math.*, *9*, 283–190.

G. E. Andrews (1976c) The theory of Partitions, Encyclopedia of Mathematics and Its Applications, vol. 2, Addison-Wesley, Reading.

D. E. Barton and F. N. David (1962) Combinatorial Chance, Griffin, London.

D. E. Barton and C. L. Mallows (1965) Some aspects of the random sequence, *Ann. Math. Stat.*, *36*, 236–260.

L. Carlitz (1959) Eulerian numbers and polynomials, *Math. Mag.*, *33*, 247–260.

L. Carlitz (1960) Eulerian numbers and polynomials of higher order, *Duke Math. J.*, *27*, 401–424.

L. Carlitz (1964) Extended Bernoulli and Eulerian numbers, *Duke Math. J.*, *31*, 667–690.

L. Carlitz (1972a) Enumeration of sequences by rises and falls: A refinement of the Simon Newcomb problem, *Duke Math. J.*, *39*, 267–280.

L. Carlitz (1972b) Sequences, paths, and ballot numbers, *Fibonacci Quart.*, *10*, 531–549.

L. Carlitz (1972c) Eulerian numbers and operators, from The Theory of Arithmetic Functions, edited by A. A. Gioia and D. L. Goldsmith, Lecture Notes in Mathematics No. 251, Springer, New York.

L. Carlitz (1973a) Enumeration of a special class of permutations by rises, *Publ. Elek. Fak. Univ. Beogradu*, No. 451, 189–196.

L. Carlitz (1973b) Enumeration of up-down permutations by number of rises, *Pacific J. Math.*, *45*, 49–58.

L. Carlitz (1973c) Enumeration of up-down sequences, *Discr. Math.*, *4*, 273–286.

L. Carlitz (1973d) Permutations with prescribed pattern, *Math. Nach.*, *58*, 31–53.

L. Carlitz (1974a) Permutations and sequences, *Advances in Math.*, *14*, 92–120.

L. Carlitz (1974b) Up-down and down-up partitions, Proc. of Eulerian Series and Applications Conference, Pennsylvania State University.

L. Carlitz (1976) Polynomial representations and compositions II. Q-analogs, *Houston J. Math.*, *2*, 345–372.

L. Carlitz and J. Riordan (1955) The number of labeled two-terminal series-parallel networks, *Duke Math. J.*, *23*, 435–446.

L. Carlitz and J. Riordan (1971) Enumeration of some two-line arrays by extent, *J. Combinatorial Th.*, *10*, 271–283.

L. Carlitz, D. Roselle, and R. Scoville (1966) Permutations and sequences with repetitions by number of increases, *J. Combinatorial Th.*, *1*, 350–374.

L. Carlitz, D. Roselle, and R. Scoville (1971) Some remarks on ballot type sequences of positive integers, *J. Combinatorial Th.*, *11*, 258–271.

L. Carlitz and R. Scoville (1972) Up-down sequences, *Duke Math. J.*, *39*, 583–598.

L. Carlitz and R. Scoville (1973) Enumeration of rises and falls by position, *Discr. Math.*, *5*, 45–59.

L. Carlitz and R. Scoville (1974) Generalized Eulerian numbers: combinatorial applications, *J. reine und angew. Math.*, *265*, 110–137.

L. Carlitz and R. Scoville (1975) Enumeration of up-down permutations by upper records, *Monatsh. Math.*, *79*, 3–12.

J. F. Dillon and D. Roselle (1968) Eulerian numbers of higher order, *Duke Math. J.*, *35*, 247–256.

J. F. Dillon and D. Roselle (1969) Simon Newcomb's problem, S.I.A.M. J. Appl. Math., *17*, 1086–1093.

R. C. Entringer (1966) A combinatorial interpretation of the Euler and Bernoulli numbers, *Nieuw Archief der Wisk.* (3), *14*, 241–246.

R. Evans (1974) An asymptotic formula for the extended Eulerian numbers, *Duke Math. J.*, *41*, 161–175.

D. Foata (1965) Etude algébrique de certains problèmes d'analyse combinatoire et du calcul des probabilités, *Publ. Inst. Stat. Univ. Paris*, *14*, 81–241.

D. Foata (1977) Distributions Eulériennes et Mahoniennes sur le groupe des permutations, in Proceedings of the N.A.T.O. Conference on Advanced Combinatorics, Berlin, September 1–10, 1976, D. Reidel, Dordrecht.

D. Foata and J. Riordan (1974) Mappings of acyclic and parking functions, *Aequationes Math.*, *10*, 10–22.

D. Foata and M.-P. Schützenberger (1970) Théorie géométrique des polynômes eulériens, Lecture Notes in Math., No. 138, Springer, New York.

D. Foata and M.-P. Schützenberger (1973) Nombres d'Euler et permutations alternantes, A Survey of Combinatorial Theory, J. N. Srivastava et al., eds., North-Holland, Amsterdam, 173–187.

H. O. Foulkes (1975) Enumeration of permutations with prescribed up-down and inversion sequence, *Discr. Math.*, *15*, 235–252.

H. O. Foulkes (1977) A nonrecursive combinatorial rule for Eulerian numbers, *J. Combinatorial Th.*, *A-22*, 246–248.

R. D. Fray and D. Roselle (1971) Weighted lattice paths, *Pacific J. Math.*, *37*, 85–96.

B. J. Gassner (1967) Sorting by replacement selecting, *Comm. A.C.M.*, *10*, 89–93.

J. R. Goldman, J. T. Joichi, and D. E. White (1975) Rook theory I. Rook equivalence of Ferrers boards, *Proc. A.M.S.*, *52*, 485–492.

J. R. Goldman, J. T. Joichi, and D. E. White (1976a) Rook theory II. Boards of binomial type, *S.I.A.M. J. Appl. Math.*, *31*, 618–633.

J. R. Goldman, J. T. Joichi, and D. E. White (1976b) Rook theory V. Rook polynomial, Möbius inversion and the umbral calculus, *J. Combinatorial Th.*, *A-21*, 230–239.

I. J. Good (1960) Contribution to the discussion of S. W. Golomb's "A mathematical theory of discrete classification" in Information Theory: Fourth London Symposium, Colin Cherry, ed., Butterworths, London, p. 423.

I. J. Good (1971) The factorization of a sum of matrices and the multivariate cumulants of a set of quadratic expressions, *J. Combinatorial Th.*, *11*, 27–37.

I. Kaplansky and J. Riordan (1945) Multiple matching and runs by the symbolic method, *Ann. Math. Stat.*, *16*, 272–277.

D. A. Klarner (1966) Enumeration involving sums over compositions, Ph.D. thesis, University of Alberta, Edmonton.

D. Knuth (1973) The Art of Computer Programming, vol. 3, Searching and Sorting, Addison-Wesley, Reading.

G. Kreweras (1965) Sur une classe de problèmes de dénombrement liés au treillis des partitions des entiers, Cahiers du B.U.R.O., No. 6, Paris.

G. Kreweras (1966) Dénombrements de chemins minimaux à sauts imposés, *C.R. Acad. Sc. Paris*, *263*, 1–3.

G. Kreweras (1967) Traitement simultané du "Problème de Young" et du "Problème de Simon Newcomb," Cahiers du B.U.R.O., No. 10, Paris.

G. Kreweras (1969a) Inversion des polynômes de Bell bidimensionnels et application au dénombrement des relations binaires connexes, *C.R. Acad. Sc. Paris*, *268*, 577–579.

G. Kreweras (1969b) Dénombrement systématique de relations binaires externes, *Math. et Sc. Humaines*, *7*, 5–15.

G. Kreweras (1970) Sur les éventails de segments, Cahiers du B.U.R.O., No. 15, Paris.

C. Long (1967) Addition theorems for sets of integers, *Pacific J. Math.*, *23*, 107–112.

C. Long (1970) On a problem in partial difference equations, *Canadian Math. Bull.*, *13*, 333–335.

E. Netto (1964) Lehrbuch der Kombinatorik, Chelsea, New York.

I. Niven (1968) A combinatorial problem of finite sequences, *Nieuw Archief voor Wisk.*, *16*, 116–123.

F. Poussin (1968) Sur une propriété arithmétique de certains polynômes associés aux nombres d'Euler, *C.R. Acad. Sc. Paris*, *266*, 392–393.

J. Riordan (1958) An Introduction to Combinatorial Analysis, J. Wiley, New York.

J. Riordan (1968) Combinatorial Identities, J. Wiley, New York.

J. Riordan and C. E. Shannon (1942) The number of two-terminal series-parallel networks, *J. Math. and Phys.*, *21*, 83–93.

D. Roselle (1969) Permutations by number of rises and successions, *Proc. A.M.S.*, *19*, 8–16.

E. B. Shanks (1951) Iterated sums of powers of binomial coefficients, *American Math. Monthly*, *58*, 404–407.

B. Sorin (1971) La distribution combinatoire du nombre de differences premières positives, *Revue de l'Institut International de Statistique*, *39*, 9–20.

R. P. Stanley (1972) Ordered structures and partitions, Memoirs of the A.M.S., No. 119.

Z. Star (1975) An asymptotic formula in the theory of compositions, *Aequationes Math.*, *13*, 279–284.

P. Thionet (1966) Emploi d'une suite Markovienne (Processus Markovien nonhomogène à temps discret) pour retrouver quelques tests non paramétriques connus, 36th Session of the I.S.I., Sydney, 1967.

P. Thionet (1968) On a distribution of MacMahon and Fisher, European Meeting on Statistics, Econometrics and Management Science, Amsterdam, Sept. 2–7.

H. G. Williams (1971) An extension of the Simon Newcomb Problem, Ph.D. thesis, Duke University, Durham.

S. G. Williamson (1972) The combinatorial analysis of patterns and the principle of inclusion-exclusion, *Discr. Math.*, *1*, 357–388.

5.5 Summaries of the Papers

[40] Yoke-chains and multipartite compositions in connexion with the analytical forms called "trees," *Proc. London Math. Soc.*, *22* (1891), 330–346.

This paper begins with a study of what MacMahon terms "yoke-chains"; these combinatorial objects are precisely the ones studied in [42] where they were related to combinations of electrical resistances. Having presented a one-to-one correspondence between the yoke-chains and certain types of trees, he proceeds to enumerate the latter in terms of compositions of multipartite numbers:

Theorem. The number of compositions of the $(m-1)$-partite number $(1, 1, \ldots, 1)$ into n parts, zeros not excluded, is n^{m-1}, and there is a natural one-to-one correspondence between these compositions and the trees of altitude n with m terminal nodes.

[42] The combinations of resistances, *The Electrician, 28* (1892), 601–602. This paper is devoted to the following theorem:

Theorem. Let $2A_n$ denote the number of two-terminal electrical networks that can be composed from n resistances in various series and parallel combinations. Then

$$1 + x + 2 \sum_{n=2}^{\infty} A_n x^n = (1 - x)^{-1} \prod_{n=2}^{\infty} (1 - x^n)^{-A_n}.$$

[46] Memoir on the theory of compositions of numbers, *Phil. Trans., 185* (1894), 835–901.

Compositions are partitions in which the order of the parts is taken into account. For example, there are three partitions of 3, namely 3, 2 + 1, 1 + 1 + 1; however, there are four compositions, namely 3, 2 + 1, 1 + 2, 1 + 1 + 1. When we consider a multipartite (or n-partite) number (i.e., an n-dimensional, non-zero vector $(a_1, a_2, \ldots, a_n)$ with nonnegative integral coordinates), the same distinction between compositions and partitions is made. For example, consider the bipartite number (2, 1): the four partitions are (2, 1), (2, 0) + (0, 1), (1, 1) + (1, 0), (1, 0) + (1, 0) + (0, 1), while the eight compositions are (2, 1), (2, 0) + (0, 1), (0, 1) + (2, 0), (1, 0) + (1, 1), (1, 1) + (1, 0), (1, 0) + (1, 0) + (0, 1), (1, 0) + (0, 1) + (1, 0), (0, 1) + (1, 0) + (1, 0).

In this paper, graphical representations of compositions are studied extensively. MacMahon begins with generating functions and graphical representations; for example, the generating function for the number of compositions of n into p parts is

$$(x + x^2 + x^3 + \ldots)^p = x^p (1 - x)^{-p} = \sum_{m=p}^{\infty} \binom{m-1}{p-1} x^m;$$

hence the number of such compositions is $\binom{m-1}{p-1}$. MacMahon presents a graphical representation from which the above observation follows immediately: given the composition $n = c_1 + c_2 + \ldots + c_p$, we divide the interval $[0, n]$ into subsegments such that the ith segment is of length c_i. Thus to produce a composition of n into p parts, we need only choose $p - 1$ of the integers from among $\{1, 2, \ldots, n-1\}$ to form the end points of the first $p-1$ segments (n is the end point of the pth segment). According to this construction, the graphical representations of the compositions of 3 are shown in the following diagram.

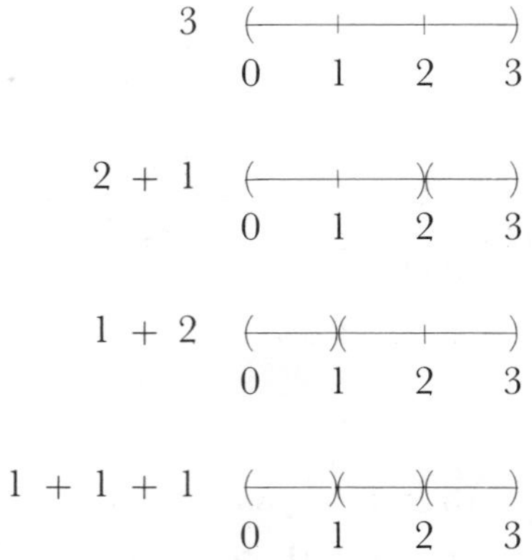

Since there are $\binom{n-1}{p-1}$ ways of choosing $p-1$ things from among $n-1$ things, we see directly from the graphical representation that there are $\binom{n-1}{p-1}$ compositions of n into p parts.

The above represents the foundation of section 1 in which MacMahon studies the compositions of unipartite numbers. In section 2 he uses the theory of symmetric functions from [32] to treat the generating functions of multipartite numbers. In section 3 he presents a graphical representation of bipartite numbers. The graphical representation is best described by an example: consider $(6,6) = (2,1) + (2,3) + (1,2) + (1,0)$; the graphical representation of this composition is the path indicated in Figure 5.1.

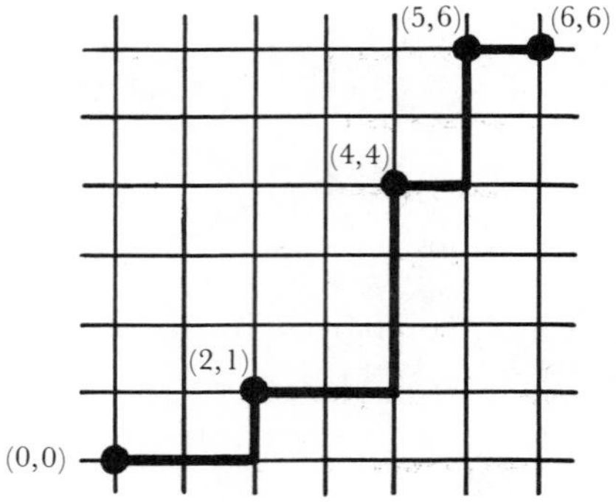

Such a representation may obviously be generalized to n dimensions; MacMahon does this in section 4 and shows that permutation problems arise quite naturally in the graphical representation of multipartite numbers. Problems of this kind are treated extensively in [45] and [82]. In solving these permutation problems, he is forced to prove special cases of his "Master Theorem," and he mentions at the end of this paper that it was this work that led him to the Master Theorem. In section 5 he notes a relationship between rooted trees and compositions and concludes by proving another permutation theorem that is again a special case of the Master Theorem.

[71] Second memoir on the compositions of numbers, *Phil. Trans.*, *207* (1908), 65–134.

This paper is the natural successor to [46]. MacMahon now develops the theory of compositions further in order to solve what has since become famous as the Simon Newcomb Problem:

A pack of cards of any specification is taken: p cards are marked 1, q cards 2, r cards 3, etc. After shuffling, one deals out the deck in the following manner: so long as the cards that appear have numbers that are in nonincreasing order, they are placed in one pack together, but as soon as the nonincreasing order is broken a fresh pack is started, and so on until all the cards have been dealt. The result of the deal will be m packs containing, in order, a, b, c, ... cards, respectively; n is the number of cards in the whole pack and $(abc \ldots)$ is some composition of the number n, the number of parts in the composition being m.

Problem 1. How many arrangements of the cards yield a given composition $(abc \ldots)$?

Problem 2. How many arrangements lead to a distribution into exactly m packs?

Part I of this paper is concerned with an elementary theory of the case in which the cards are numbered differently; part II treats the general case.

MacMahon begins by introducing a Ferrers graph type representation of a composition. For example, the representation of $(41^2 23)$ is

The above is appropriately termed a zig-zag graph; note that there are 4 nodes in the first row, 1 in the second, 1 in the third, 2 in the fourth, and 3 in the fifth. Reading the graph vertically, we obtain the conjugate composition: $(1^3 4 2 1^2)$. This representation of a composition is quite different from the one used in [46] but is much more important for the purposes of this paper.

In part I we consider a permutation of the first five natural numbers, say 21435. We separate the permutation into a minimal number of compartments such that the members of each compartment are in nonincreasing order: 21|43|5. We shall say that the permutation 21435 has *descending specification* (221) = $(2^2 1)$ since there are two members of the first compartment, two of the second, and one of the third. Let $\mathcal{N}(abc \ldots)$ denote the number of permutations of the first n integers that have a descending specification denoted by the composition $(abc \ldots)$ of the number n.

MacMahon easily obtains a formula for $\mathcal{N}(abc \ldots)$ in terms of a sum of multinomial coefficients and obtains some simple summations of these $\mathcal{N}$-functions. He also derives the following simple multiplication theorem:

$$(1) \qquad \binom{n}{a_1 + a_2 + \ldots + a_s} \mathcal{N}(a_1 a_2 \ldots a_s) \mathcal{N}(a_{s+1} a_{s+2} \ldots a_{s+t})$$

$$= \mathcal{N}(a_1 a_2 \ldots a_{s+t}) + \mathcal{N}(a_1 a_2 \ldots a_{s-1}, a_s + a_{s+1}, a_{s+2} \ldots a_{s+t}).$$

and proceeds to apply (1) to derive further summations of the $\mathcal{N}$-functions.

Section 2 of part I is devoted to $\mathcal{N}_m$, the number of permutations associated with compositions containing exactly m parts (see Problem 2 above). MacMahon applies some results from [32] to show that

$$\mathcal{N}_m = \sum_{j=0}^{m} (-1)^j \binom{n+1}{j} (m-j)^n.$$

Part I concludes with several other formulas for the $\mathcal{N}_m$ and sets the stage for Part II.

In part II, repetitions in our numbered pack of cards are now permitted. Suppose $\alpha > \beta > \gamma > \ldots$ and that p cards are numbered α, q cards β, r cards γ, etc. Then we speak of the card deck being given by the assemblage (i.e., multiset) $\alpha^p \beta^q \gamma^r \ldots$, and we say the deck is of specification $(pqr \ldots)$, a composition of the number of cards in the deck.

Let us suppose that the deck is of specification $(p_1^{\pi_1} p_2^{\pi_2} p_3^{\pi_3} \ldots)$, and let

$$h_{p_1}^{\pi_1} h_{p_2}^{\pi_2} h_{p_3}^{\pi_3} \ldots = \ldots + C(abc\ldots) \times (abc\ldots),$$

where h_n is the "homogeneous product sum" of weight n (see [32], section 2 for a full definition and discussion of the h_n) and where $(abc\ldots)$ denotes the monomial symmetric function $\sum \alpha_{i_1}^a \alpha_{i_2}^b \alpha_{i_3}^c \ldots$. MacMahon, in analogy with part I, now derives the $N(abc\ldots)$ in terms of the $C(abc\ldots)$. Important in these considerations are the two functions

$$\begin{aligned}\Theta_N\{(a)(b)(c)\ldots\} = {} & N(abc\ldots) + N(a+b, c\ldots) + N(a, b+c, \ldots) \\ & + \ldots + N(a+b+c, \ldots) + \ldots,\end{aligned}$$

and

$$\begin{aligned}\phi_C\{(a)(b)(c)\ldots\} = {} & C(abc\ldots) - C(a+b, c\ldots) - C(a, b+c, \ldots) \\ & - \ldots + C(a+b+c, \ldots) + \ldots.\end{aligned}$$

The principal laws of formation may be expressed by

$$\begin{aligned}\Theta_N\{(a)(b)(c)\ldots\} &= C(abc\ldots), \\ \phi_C\{(a)(b)(c)\ldots\} &= N(abc\ldots).\end{aligned}$$

N.B. When it is necessary to make the specification of the deck $(pqr\ldots)$ explicit, MacMahon writes $N(abc\ldots)_{(pqr\ldots)}$, etc.

Section 4 is devoted to the derivation of various relationships between the Θ_N and ϕ_C functions.

In section 5 MacMahon applies the familiar Hammond operator

$$D_s = \frac{1}{s!}(\partial_{\alpha_1} + a_1\partial_{a_2} + a_2\partial_{a_3} + \ldots)^s$$

to the Simon Newcomb problem. He now (in analogy with the above definition of ϕ_C) writes

$$\begin{aligned}\phi_h\{(a)(b)(c)\ldots\} &= h_a h_b h_c \ldots - h_{a+b} h_c \ldots - h_a h_{b+c} \ldots + h_{a+b+c} \ldots \\ &\quad + \ldots \\ &\equiv h_{abc\ldots},\end{aligned}$$

and proves the fundamental result:

$$h_{abc\ldots} = \sum \mathcal{N}(abc\ldots)_{(pqr\ldots)}(pqr\ldots).$$

The new symmetric functions $h_{abc\ldots}$ are thus seen to be extremely important in this study and are extensively examined in section 6; in fact they are linear combinations of the Schur functions described in section 14.2. Also in section 6, MacMahon points out that in the notation of part I,

$$\mathcal{N}(abc\ldots) = \mathcal{N}(abc\ldots)',$$

where $(abc\ldots)'$ is the conjugate composition to $(abc\ldots)$, and he investigates possible generalizations of this to the general result considered here in part II.

In section 7 a generalization of equation (1) is given. Section 8 parallels section 2 and provides the complete answer to the second part of the Simon Newcomb problem (Problem 2).

The two concluding sections treat results related to generating functions that arose in [46]. For example, there is a detailed treatment of the identity

$$\begin{aligned}&(1 - a_1 + (1-\lambda)a_2 - (1-\lambda)^2 a_3 + \ldots)^{-1}\\ &= \sum \mathcal{N}_{m,\,p_1p_2p_3\ldots}\lambda^{m-1}(p_1p_2\ldots p_k),\end{aligned}$$

where a_i is the ith elementary symmetric function (1^i), and $\mathcal{N}_{m,\,p_1p_2p_3\ldots}$ is the number of times that m packs result in the Simon Newcomb problem when the specification of the deck is $(p_1p_2p_3\ldots)$.

Reprints of the Papers

Yoke-Chains and Multipartite Compositions in connexion with the Analytical Forms called "Trees."

By Major P. A. MacMahon, R.A., F.R.S.

[*Read April* 9*th*, 1891.]

A yoke-chain, an expression adopted at the suggestion of Professor Cayley, is a geometrical configuration composed of line branches.

The simple line branch —— is indifferently a yoke or a chain.

The combinations  &c. are yokes;

whilst  &c. are chains;

and generally we form chains by combining chain-wise any number of yokes, and also generally we form yokes by combining yoke-wise any number of chains.

E.g.,  is a chain,

 is a yoke.

Yoke-chains may be viewed as diagrammatic representations of the combinations of resistances of linear electrical conductors, or of the capacities of electrical condensers. The yoke and the chain represent parallel and series combinations of resistances, but series and parallel combinations of capacities.

The theory of yoke-chains does not include every combination of resistances, but only those which are made up of two or more combinations, either in parallel or in series. In the case of five resistances, for example, we find a combination such as the Wheatstone net, which is not decomposable into other combinations either in parallel or in series. Such networks, considered by Kirchhoff and Maxwell, do not come into view here.

Consider in the first place the different yoke-chains that can be formed from a given number of branches.

From a single branch we can merely form the yoke or chain.

From two branches we form the chain of Fig. 1 or the yoke of Fig. 2,

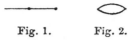

Fig. 1.

Fig. 2.

I define the yoke of Fig. 2 to be the yoke-chain conjugate to the chain of Fig. 1, and I further regard the yoke or chain —— as being self-conjugate.

Every chain is a chain of yokes, and every yoke is a yoke of chains.

I make the following definitions :—

Definition.—The conjugate of any chain is formed by placing the conjugates of the component yokes in a yoke.

Definition.—The conjugate of any yoke is formed by placing the conjugates of the component chains in a chain.

The process of conjugation is necessarily reversible, as is obvious from the definitions.

The second of the above definitions is derived from the first by interchanging the words chain and yoke.

We have thus the notion of conjugate yoke-chains, and it is clear that the yoke-chains of a given order (*i.e.*, of a given number of branches) may be arranged in conjugate pairs, each pair comprising a chain and a yoke.

Further, the whole of the yoke-chains may be arranged in a chain set and a yoke set, and either set is derivable by conjugation from the other.

Passing now to the case of three branches, we form the chain set by placing each of the forms of order 2 in chain with the form of order 1. We thus obtain the left-hand column below :

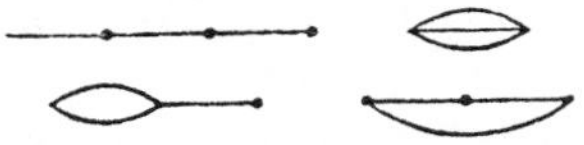

To form the right-hand column, we first place the three self-conjugate components of the form in yoke, and then place the conjugates of the components  —— of the form

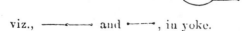

viz., and  , in yoke.

We thus obtain the left-hand chain set, and the right-hand yoke set.

In forming the forms of order 4, we place each of the four forms of order 3 in chain with the form of order 1, and further place the yoke form of order 2 in chain with itself. We have thus $5\,(=4+1)$ forms, constituting the chain-column of order 4. These are given below, and also the conjugate yoke-column.

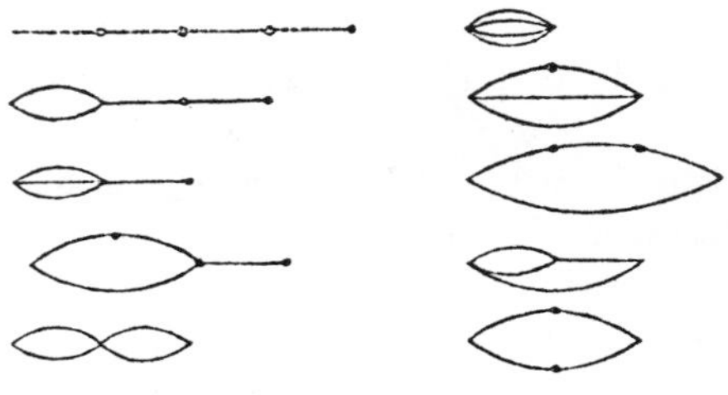

To form the chain-column of order 5, each of the ten forms of order 4 is placed in chain with the form of order 1, and further each of the yoke forms of order 3 is placed in chain with the yoke form of order 2. There are thus, in all, $2\,(10+2) = 24$ forms. In general, the forms of order n are formed from the forms of lower orders. To form the chain-column, as many processes must be performed as there are non-unitary partitions of n, or as there are partitions of n composed of the integers

$$2,\ 3,\ 4,\ \ldots\ n.$$

We first place each form of order $n-1$ in chain with the form of order 1. This is the first process, and thereafter we have a process associated with every non-unitary partition of n with the exception of the partition consisting merely of the number n itself.

For the partition $(n-2, 2)$ we take every yoke form of order $n-2$ in chain with the yoke form of order 2. For the partition $(n-p, p)$, $p>1$, we take every yoke form of order $n-p$ in chain with every yoke form of order p.

Denote by B_p the number of chain or of yoke forms of order p. Then, if $p=1$, the complete number of forms is $B_1 = 1$; but if $p>1$ the complete number of forms is $2B_p$.

In forming synthetically the number B_n we have above found a portion $2B_{n-1}$, and corresponding to the partition $(n-p, p)$, $p>1$, a portion $B_{n-p}\,B_p$, for this represents the number of ways of combining one of $n-p$ things with one of p things of a different sort.

If, however, $n-p=p$, the corresponding portion is not B_p^2, but $\frac{B_p(B_p+1)}{2}$, viz., the number of homogeneous products of B_p things, two and two together. So also, corresponding to the partition $(l^\lambda m^\mu \ldots)$, we have the number

$$\frac{B_l(B_l+1)\ldots(B_l+\lambda-1)}{\lambda!}\cdot\frac{B_m(B_m+1)\ldots(B_m+\mu-1)}{\mu!}\ldots,$$

and finally we have the relation

$$B_n = 2B_{n-1}+B_{n-2}B_2+B_{n-3}B_3+\ldots$$

$$\ldots+\Sigma\frac{B_l(B_l+1)\ldots(B_l+\lambda-1)}{\lambda!}\cdot\frac{B_m(B_m+1)\ldots(B_m+\mu-1)}{\mu!}\ldots,$$

the numbers l, m, &c., ... being >1.

This relation leads at once to the expression of the law of the numbers B in the form

$$(1-x^2)^{-B_2}(1-x^3)^{-B_3}(1-x^4)^{-B_4}\ldots$$

$$=1+(B_1-1)x+(2B_2-B_1)x^2+2(B_3-B_2)x^3+\ldots+2(B_p-B_{p-1})x^p+\ldots,$$

or, multiplying up by $1-x$,

$$(1-x)^{-B_1}(1-x^2)^{-B_2}(1-x^3)^{-B_3}(1-x^4)^{-B_4}\ldots$$

$$=1+B_1x+2(B_2x^2+B_3x^3+B_4x^4+\ldots).$$

This formula, wherein $B_1=B_2=1$, is convenient for calculating the numbers $2B_p$.

We find

B_1	$2B_2$	$2B_3$	$2B_4$	$2B_5$	$2B_6$	$2B_7$	$2B_8$	$2B_9$	$2B_{10}$ ...
1	2	4	10	24	66	180	522	1532	4984 ...

There is a paper by Cayley in the *Philosophical Magazine*, Vol. XIII. (1857), also *Collected Papers*, Vol. III., No. 203, "On the Theory of the Analytical Forms called Trees." The following passage occurs (*Collected Papers, loc. cit.*, pp. 245, 246)* :—

* Other references are—

"On the Analytical Forms called Trees," Cayley, *Phil. Mag.*, Vol. XX. (1860), pp. 337–341.

"On the Mathematical Theory of Isomers," Cayley, *Phil. Mag.*, Vol. XLVII. (1874), p. 444.

"On the Analytical Forms called Trees, with Application to the Theory of Chemical Combinations," Cayley, *British Association Report*, 1875, pp. 257–305.

"On the Analytical Forms called Trees," Cayley, *American Journal of Mathematics*, Vol. IV., pp. 266–268.

In the latter paper the notion of the centre or bicentre of number is stated to be due to M. Camille Jordan, but I cannot find the reference.

"I have had occasion for another purpose to consider the question of finding the number of trees with a given number of free branches, bifurcations at least. Thus, when the number of free branches is three, the trees of the form in question are those in the annexed figure, and the number is therefore two. It is not difficult to see that

we have in this case (B_r being the number of such trees with r free branches)

$$(1-x)^{-1}(1-x^2)^{-B_2}(1-x^3)^{-B_3}(1-x^4)^{-B_4}\ldots$$
$$= 1+x+2B_2x^2+2B_3x^3+2B_4x^4+\&c."$$

In view of this interesting identity of enumeration, though in the absence of information in regard to the purpose for which Cayley investigated the subject, I propose to examine in detail the correspondence between yoke-chains and the trees with free branches.

A single tree represents either member of a certain conjugate pair of yoke-chains, according to the interpretation placed upon the combination of knots and branches which constitutes the tree.

In the first place, we may restrict attention to the chain combinations of the several orders.

One, two, three, &c. branches in chain may be denoted by the trees

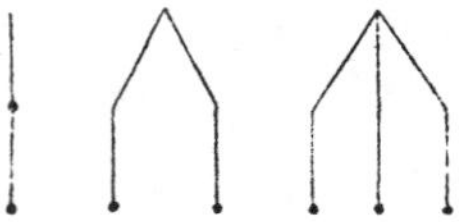

and this representation would be in some respects the most consistent with what follows. The idea is that a branch

denotes a chain of order 1, and the upper knots are then joined by branches to a single knot to denote that the several chains are to be joined in chain. The trees, however, become simplified if we agree to represent a single linear resistance, not by a branch but by a terminal knot. The three trees above then become

Consider next the tree 

which may be taken to denote the resistances

placed in series. In order that this tree may represent the combination

it is necessary to suppose that the tree

represents not the combination but rather its conjugate yoke form

The ultimate step in interpreting a tree will be to take a number of forms in chain. The trees which represent these forms will be found pendent to the second row of knots in the tree. We must therefore agree always to interpret the trees which originate from the second row of knots as yoke combinations. On this convention the tree

denotes the chain combination of the yoke combinations denoted by the trees

viz., the yokes 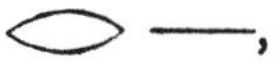, placed in chain, or

The above principle enables us to form in succession the free-branch trees of the various orders.

The five trees of order 4 are

Each corresponds to some partition of four other than the number 4 itself.

The tree I. [partition (1^4)] clearly denotes

 ;

the tree II., partition (21^2), denotes

 ;

the tree III., partition (2^2), gives

 ;

trees IV. and V. both belong to the partition (31); this follows from the fact that there are two trees of the order 3, and either may be placed as a pendent to a knot of the tree

in order to form a tree of order 4.

According to the rule, we must give parallel interpretations to the trees

prior to taking them in series, in each case, with the tree denoted by a single knot.

We have thus—

Tree IV. = 

Tree V. = 

and the five trees have been placed in correspondence with the five series combinations of order 4.

As another example I take the tree

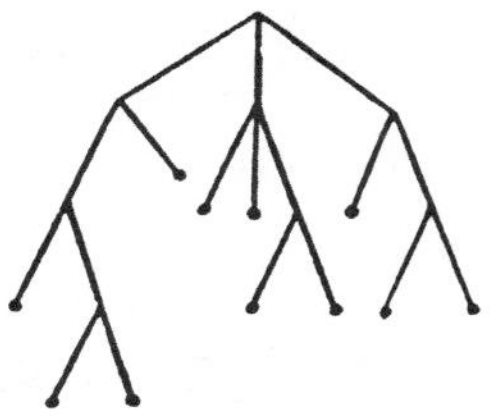

The series equivalences of the trees

are

To form the complete combination we have merely to take the conjugates of these in series.

The conjugates are

so that the tree denotes the combination

In this way any tree may be interpreted so as to denote a chain.

If in the foregoing rules the words chain and yoke be interchanged, the result will be the conjugate yoke combination.

A tree therefore may be taken to be a representation of either combination of a conjugate pair at pleasure, or if we please of both combinations of such a pair.

I give below a Table showing the correspondence of trees with yoke-chains of the first six orders.

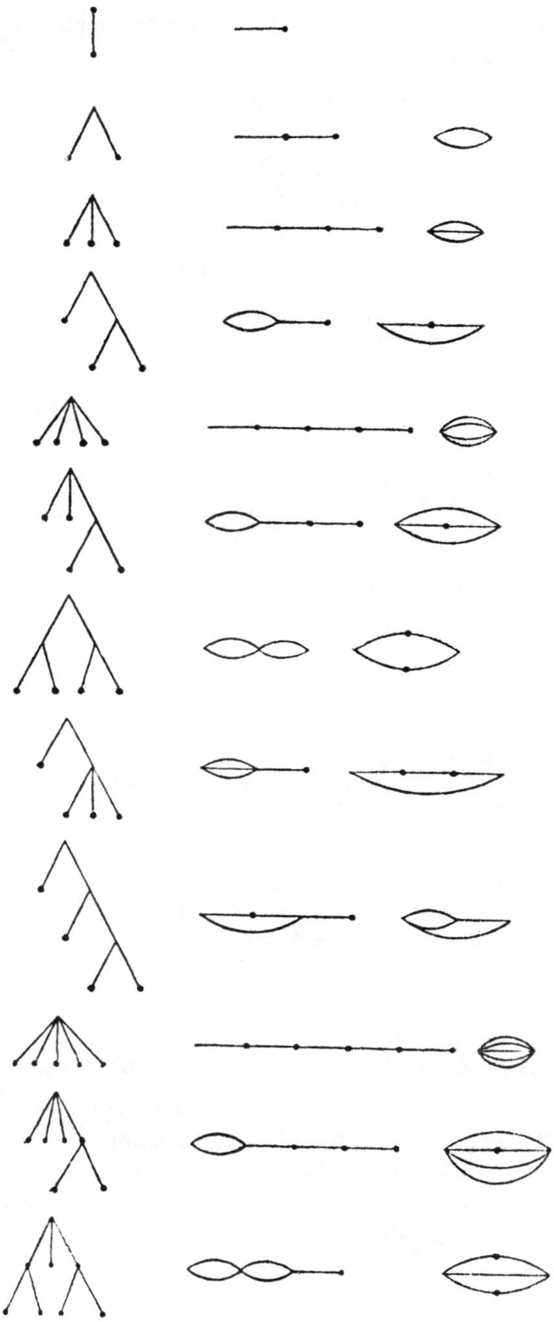

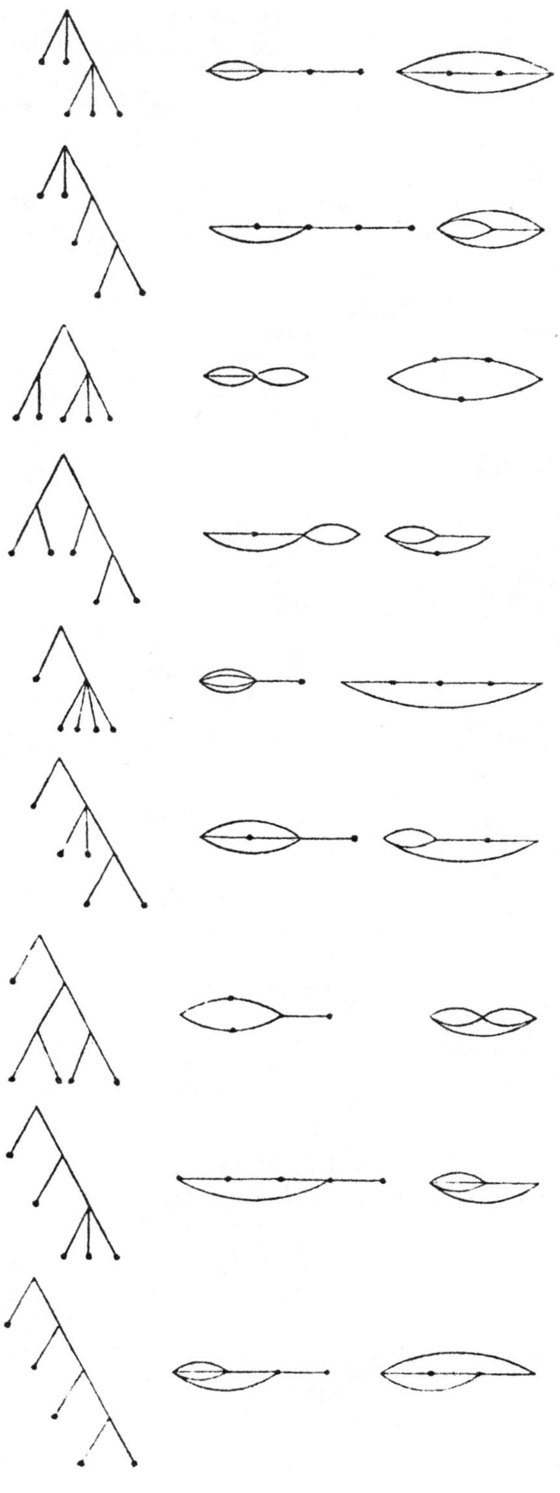

The capacity of condensers placed in parallel is formed according to the same law as the resistance of linear conductors placed in series.

In general the capacity of any combination of condensers is formed according to the same law as the resistance of the conjugate combination of linear conductors.

There is a very simple result connected with the conjugate combinations of equal linear conductors which may possibly have escaped notice.

If any combination of linear conductors, each of r ohms resistance, be formed so that the combined resistance is

$$\frac{m}{n} r \text{ ohms};$$

then the resistance of the conjugate combination is

$$\frac{n}{m} r \text{ ohms}.$$

The inductive proof is easy, for suppose two series combinations to have resistances

$$\frac{a}{b} r \quad \text{and} \quad \frac{a'}{b'} r \text{ ohms},$$

and to be such that the resistances of the conjugate combinations are

$$\frac{b}{a} r \quad \text{and} \quad \frac{b'}{a'} r \text{ ohms}.$$

Placing the two combinations in parallel, the resistance is

$$\frac{\dfrac{aa'}{bb'}}{\dfrac{a}{b} + \dfrac{a'}{b'}} r \text{ ohms},$$

which is

$$\frac{1}{\dfrac{b}{a} + \dfrac{b'}{a'}} r \text{ ohms},$$

and the reciprocal of the multiplier of r is

$$\frac{b}{a} + \frac{b'}{a'},$$

which, multiplied into r, represents the resistance of the combination formed by placing the conjugates of the component combinations in series.

Hence, since the law evidently holds in the simplest cases, it must be true in general.

2. In the *Philosophical Magazine*, 1860, Professor Cayley has enumerated the trees with a given number of terminal knots. He remarks :—

"We have here

$$\phi m = 1\,.\,2\,.\,3\,\ldots(m-1) \text{ coefficient } x^{m-1} \text{ in } \frac{1}{2-\exp x},$$

giving the values

$$\phi m = 1,\ 1,\ 3,\ 13,\ 75,\ 541,\ 4683,\ 47293,\ \ldots,$$

for $$m = 1,\ 2,\ 3,\ 4,\ 5,\ 6,\ 7,\ 8,\ \ldots."$$

This enumeration is identical with that of the compositions of certain multipartite numbers.*

The correspondence is between the trees with m terminal knots, and the compositions of the multipartite number

$$\overline{1^{m-1}},$$

where 1^{m-1} denotes 1 repeated $m-1$ times, and the bar —— means that the partition does not denote the partition (1^{m-1}) of a single number $m-1$, but rather the multipartite number having the multipartite weight 1, 1, 1, 1, ..., $m-1$ times.

To identify each tree with a composition, consider, for example, the 13 trees with 4 terminal knots.

---

* H. J. S. Smith and J. W. L. Glaisher have termed partitions in which the order of the parts is essential "compositions." For multipartite numbers see the author, "Memoir on Symmetric Functions of the Roots of Systems of Equations," *Phil. Trans. R. S. of London*, Vol. CLXXXI. (1890), A., pp. 481–536; and for compositions of multipartite numbers, the author, the *Messenger of Mathematics*, Vol. XX. (1890).

These are

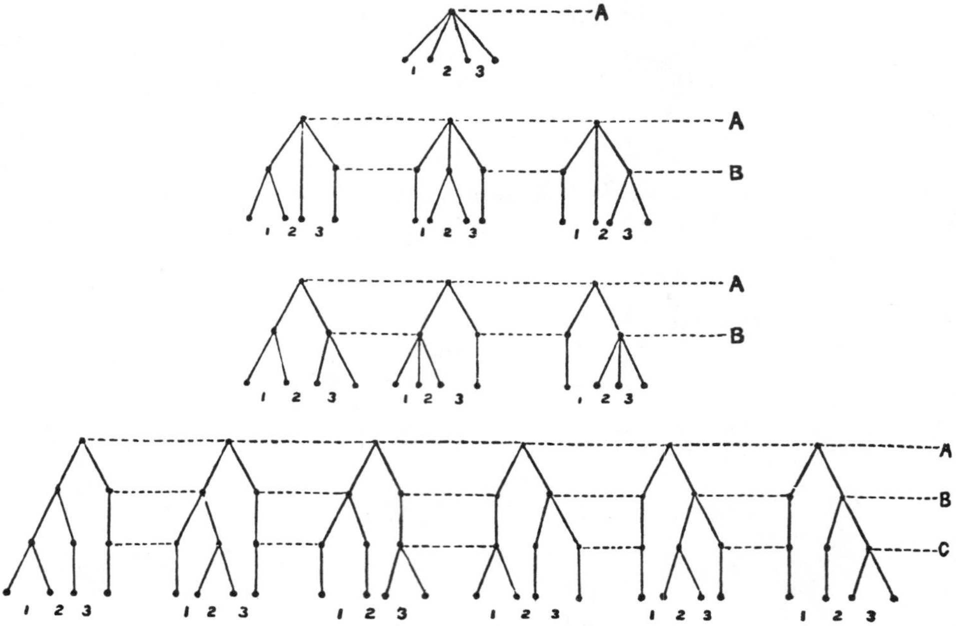

We have to identify each of these trees with a composition of the tripartite number $\overline{111}$. The number being tripartite, whilst the terminal knots are four in number, the idea is presented of considering not the terminal knots, but the spaces between them, which are but three in number. These spaces are numbered 1, 2, 3 from left to right in each tree. Consider, moreover, the knots in the first, second and third rows, omitting the terminal row of knots. These are marked *A*, *B*, *C*; in each row we have knots from which descend two or more branches. Of these branches any two that are adjacent bound an area which is in direct communication with one of the numbered spaces between the terminal knots. In any row of knots we must observe the connexion between each angle, formed by the

descent of two adjacent branches from a knot, and the terminal spaces.

In the second of the last line of trees, beginning with row *A*, we see an angle in communication with space 3, but not with spaces 1 or 2; this connexion between angles and spaces may be denoted by

$$\overline{001};$$

similarly for row B, there is one angle in connexion with space 1, so that we have $\overline{100}$;

in row C, one angle leads to space 2, so that we have

$$\overline{010};$$

the whole connexion between angles and spaces is represented by the composition $(\overline{001},\ \overline{100},\ \overline{010})$

of the multipartite number $\overline{111}$.

Interpreting each tree in succession, we get the whole of the 13 compositions of $\overline{111}$, viz.:—in order

$$(\overline{111}),$$

$$(\overline{011},\ \overline{100}),\quad (\overline{101},\ \overline{010}),\quad (\overline{110},\ \overline{001}),$$

$$(\overline{010},\ \overline{101}),\quad (\overline{001},\ \overline{110}),\quad (\overline{100},\ \overline{011}),$$

$$(\overline{001},\ \overline{010},\ \overline{100}),\quad (\overline{001},\ \overline{100},\ \overline{010}),\quad (\overline{010},\ \overline{100},\ \overline{001}),$$

$$(\overline{010},\ \overline{001},\ \overline{100}),\quad (\overline{100},\ \overline{001},\ \overline{010}),\quad (\overline{100},\ \overline{010},\ \overline{001}).$$

This principle of interpretation is perfectly general. The number of angles in a tree must be less by one than the number of terminal knots, and each row from the first (or top) to the penultimate must contain at least one angle.

If any composition of a multipartite number of the form which presents itself in this theory be given, it is extremely easy to form the corresponding tree. To form the tree, it is best to commence with the right-hand part of the composition, and to then proceed regularly towards the left.

Suppose given the composition

$$(\overline{0101},\ \overline{1000},\ \overline{0010}),$$

the successive operations are

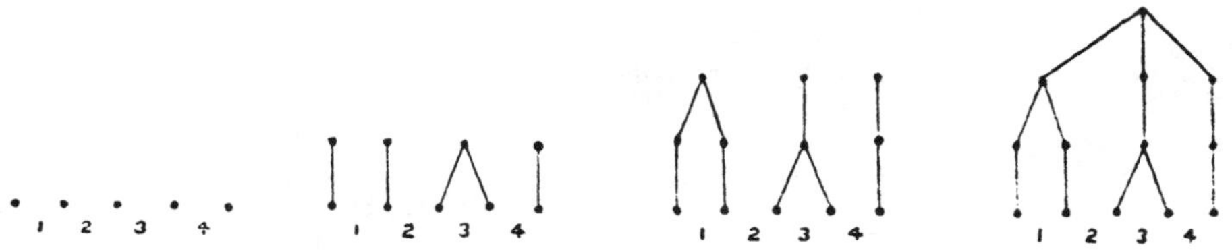

This correspondence is valuable, as putting us in possession of a graphical method through which the compositions may be studied.

3. The compositions of the multipartite numbers

$$\overline{1},\quad \overline{11},\quad \overline{111},\quad \overline{1111},\ \&c.,$$

into a definite number of parts, zero parts not excluded, may be graphically represented by trees with a definite number of terminal knots, and also a definite altitude (or number of rows of knots).

The trees of altitude 1 are trivial, as merely denoting the succession of integer numbers 1, 2, 3, &c.

Passing to those of altitude 2, we have, for 1, 2, 3, &c., terminal knots,

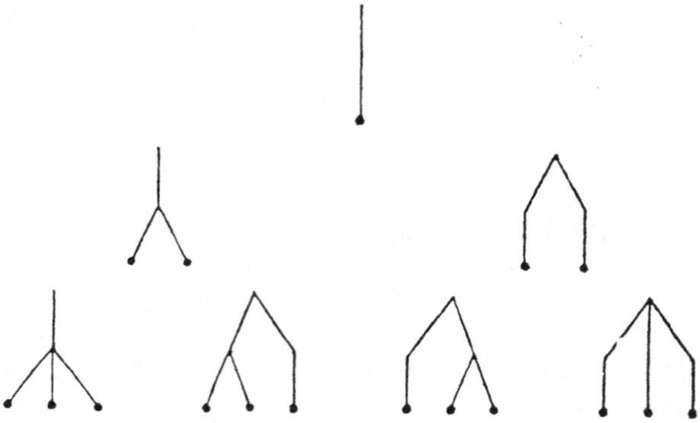

&c.;

and in the first place it is seen that these trees may be considered to be graphical representations of the compositions of the integer numbers 1, 2, 3, &c.; the last, for example, denoting the compositions of 3, viz.,

$$3,\quad 21,\quad 12,\quad 111.$$

Consequently the number of trees of altitude 2 with m terminal knots is 2^{m-1}.

We may also interpret these trees as before by paying attention to the inter-terminal knot spaces.

We thus obtain the compositions

$$(01),\ (10),$$

$$(\overline{00}\ \overline{11}),\quad (\overline{01}\ \overline{10}),\quad (\overline{10}\ \overline{01}),\quad (\overline{11}\ \overline{00}),$$

&c.,

in which we now have the compositions of the multipartites

$$1, \quad \overline{11}, \quad \overline{111}, \quad \&c.,$$

into two parts, zero not excluded.

In particular we have just obtained a graphical proof of the theorem:—

"The whole number of compositions of the number $m+1$ is equal to the number of compositions of the multipartite $\overline{1^m}$ into two parts, zero parts not excluded."

In general, the trees of altitude n and having m terminal knots are representations of the compositions of the multipartite

$$\overline{1^{m-1}},$$

into n parts, zeros not excluded.

It is to be shown that the trees of altitude n and having m terminal knots are

$$n^{m-1}$$

in number.

All the trees of a given altitude n are derivable from the trees of the next lower altitude.

We may start with a tree of altitude 1, having p terminal knots, $p \ngtr m$, and append any p trees of altitude $n-1$ of which the sum of the terminal knots is m. We must give p all integer values from 1 to m in succession, and append the appropriate trees.

Let then $T_{m,n}$ denote the number of trees in question.

It is easy to see that the above considerations lead to the relation

$$T_{m,n} = \Sigma \frac{(\pi_1+\pi_2+\pi_3+\ldots)!}{\pi_1!\,\pi_2!\,\pi_3!\ldots} T^{\pi_1}_{p_1,n-1}\, T^{\pi_2}_{p_2,n-1}\, T^{\pi_3}_{p_3,n-1}\ldots,$$

the summation being for all partitions

$$(p_1^{\pi_1} p_2^{\pi_2} p_3^{\pi_3} \ldots)$$

of the number m.

This relation is expressed algebraically by the formula

$$T_{1,n}x+T_{2,n}x^2+T_{3,n}x^3+\ldots = \frac{T_{1,n-1}x+T_{2,n-1}x^2+T_{3,n-1}x^3+\ldots}{1-(T_{1,n-1}x+T_{2,n-1}x^2+T_{3,n-1}x^3+\ldots)},$$

and since obviously $$T_{m,1} = 1,$$

we reach without difficulty the formula

$$T_{1,n}x+T_{2,n}x^2+T_{3,n}x^3+\ldots = \frac{x}{1-nx}.$$

Hence, as above stated, $T_{m,n} = n^{m-1}$.

We have established the theorem :—

" The number of compositions of the multipartite number

$$\overline{1^{m-1}}$$

into n parts, zeros not excluded, is

$$n^{m-1},$$

and there is a one-to-one correspondence between these compositions and the trees of altitude n which have m terminal knots."

THE COMBINATIONS OF RESISTANCES.

BY MAJOR P. A. MACMAHON, R.A., F.R.S.

Resistances may be placed in circuit either in series or in parallel, or in various combinations of these methods. There are also networks such as the Wheatstone net, which are neither series nor parallel arrangements, and these are excluded from what follows. Every combination considered here consists of other combinations either in series or in parallel. Each must either consist of parallel combinations arranged in series, or of series combinations arranged in parallel. For the study of the subject it is convenient to make a few definitions. Two linear conductors may be placed either in series or in parallel. These arrangements may be defined as being conjugate the one of the other. It will be observed that the second of the arrangements merely differs in description from the first by the substitution of the word "parallel" for the word "series." To generalise the notion we may define two combinations to be conjugate when their descriptions merely differ by the interchange of the words "series" and "parallel."

A single linear conductor should, for the sake of unity, be regarded as being either in series or in parallel, and, further, as being a combination which is self conjugate. A combination consisting of parallel combinations in series will be called a series combination, and one consisting of series combinations in parallel a parallel combination. Thus, in the usual diagrammatic representation,

is a series combination, consisting of the three parallel forms

in series; and

is a parallel combination, consisting of the three series forms

in parallel.

The combinations into which series and parallel combinations are decomposable will be termed component parallel combinations and component series combinations respectively. The conjugate of any series combination is formed by placing the conjugates of the component parallel combinations in parallel. Similarly, the conjugate of any parallel combination is formed by placing the conjugates of the component series combinations in series. That this is so is evident, because by taking the conjugates of the component combinations, we merely interchange the words "series" and "parallel" in their descriptions.

The form  might be described by saying that two single forms are placed in parallel, and the combination then placed in series with two single forms. So also the form  consists of two single forms placed in series, the combination being placed in parallel with two single forms. These two forms are thus conjugate, and may be said to constitute a conjugate pair.

It may be gathered that all combinations naturally arrange themselves in conjugate pairs.

For one, two, three, four resistances, we have

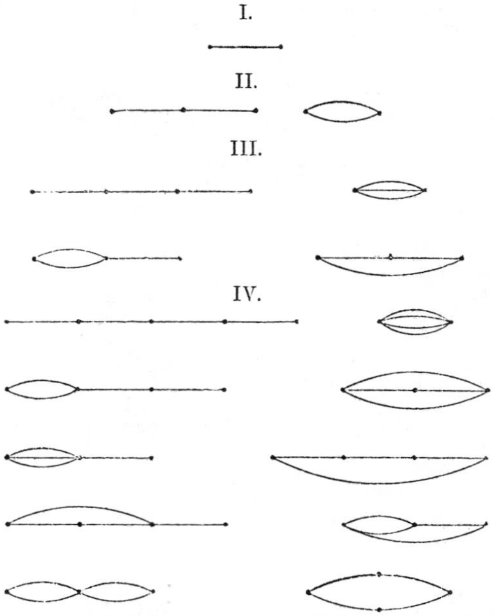

where the combinations are arranged in a left-hand series column and a right-hand parallel column. The columns are conjugate, the one of the other. Conjugate pairs are found in successive horizontal rows.

In a similar manner the combinations of any number of resistances may be arranged in conjugate pairs.

It will be remarked that no attention is paid to the actual values of the resistances, but only to the *forms* in which they can be combined. The enumeration of the forms of combinations of a given number of resistances is of considerable interest. It can be shown that if $2A_n$ represents the number of combinations in the case of n resistances (n greater than one)

$$(1-x)^{-1}(1-x^2)^{-A_2}(1-x^3)^{-A_3}(1-x^4)^{-A_4}\ldots\ldots$$
$$= 1+x+2(A_2x^2+A_3x^3+A_4x^4+\ldots\ldots\ldots$$

If the left-hand side of this identity be expanded by the binomial theorem, and comparison made with the right-hand side, the numbers in question are easily calculated. The results are

$2A_2$	$2A_3$	$2A_4$	$2A_5$	$2A_6$	$2A_7$	$2A_8$	$2A_9$	$2A_{10}\ldots\ldots$
2	4	10	24	66	180	522	1532	4984......

There is in regard to the combinations of equal resistances a very pretty and simple theorem which appears to have hitherto escaped notice.

If any number of resistances, each of r ohms, be combined so as to have a resistance of

$$\frac{p}{q}\,r \text{ ohms,}$$

then the conjugate combination of resistances will have a resistance of $\frac{q}{p} r$ ohms.

In fact, conjugation merely inverts the fraction which presents itself as the multiplier of r. As an example, the conjugates,

each single resistance being r ohms, have resistances $\frac{10}{3} r$ ohms and $\frac{3}{10} r$ ohms respectively.

The inductive proof of the theorem is very simple. For suppose the law to hold in the case of two combinations whose resistances are

$$\frac{a}{b} r \text{ ohms and } \frac{c}{d} r \text{ ohms.}$$

The resistance obtained by placing these two combinations in series is $\left(\frac{a}{b}+\frac{c}{d}\right) r$ ohms.

The conjugate of this new combination is obtained by placing the conjugates of the component combinations in parallel, and since the law is assumed to hold for the component combinations, the resistance obtained will be

$$\frac{\frac{b}{a} \cdot \frac{d}{c}}{\frac{b}{a}+\frac{d}{c}} r \text{ ohms,}$$

which is $\dfrac{1}{\frac{a}{b}+\frac{c}{d}} r$ ohms.

This proves that the law holds for the new combination.

Similarly assuming the law to hold in the case of two combinations of resistance, $\frac{a}{b} r$ ohms and $\frac{c}{d} r$ ohms, we obtain, by placing them in parallel, a new resistance

$$\frac{\frac{a}{b} \cdot \frac{c}{d}}{\frac{a}{b}+\frac{c}{d}} r \text{ ohms.}$$

And, moreover, the resistance obtained by placing their conjugates in series is $\left(\frac{b}{a}+\frac{d}{c}\right) r$ ohms,

and $$\frac{b}{a}+\frac{d}{c}=\frac{1}{\frac{\frac{a}{b} \cdot \frac{c}{d}}{\frac{a}{b}+\frac{c}{d}}}.$$

Hence, since every combination is decomposable into two combinations, either in series or in parallel, and the law is obviously true for the simpler cases, it must hold universally.

According to the definition here made of the word "conjugate," it will be remarked that capacities of condensers combine according to the law of the conjugate combinations of resistances.

XVII. *Memoir on the Theory of the Compositions of Numbers.*

By P. A. MacMahon, *Major R.A., F.R.S.*

Received November 17—Read November 24, 1892.

§ 1. *Unipartite Numbers.*

1. Compositions are merely partitions in which the order of occurrence of the parts is essential; thus, while the partitions of the number 3 are (3), (21), (111), the compositions are (3), (21), (12), (111).

The enumerations of the compositions of a number n into p parts, zeros excluded, is given by the coefficient of x^n in the expansion of

$$(x + x^2 + x^3 + \ldots)^p;$$

this expression may be written

$$\left(\frac{x}{1-x}\right)^p,$$

and the coefficient of x^n is seen to be

$$\binom{n-1}{p-1}.^*$$

The generating function of the total number of compositions of n is

$$\sum_1^\infty (x + x^2 + x^3 + \ldots)^p = \frac{x}{1-2x},$$

hence the number in question is

$$2^{n-1}.$$

2. If the parts of the compositions are limited not to exceed s in magnitude, the generating function of the number into p parts is

$$(x + x^2 + x^3 + \ldots + x^s)^p = x^p\left(\frac{1-x^s}{1-x}\right)^p,$$

* In the continental notation I write $\binom{n}{p}$ for $\frac{n!}{p!(n-p)!}$.

and herein the coefficient of x^n is

$$\binom{n-1}{p-1} - \binom{p}{1}\binom{n-s-1}{n-s-p} + \binom{p}{2}\binom{n-2s-1}{n-2s-p} - \dots$$

The number of parts being unrestricted the generating function is

$$\sum_p x^p \left(\frac{1-x^s}{1-x}\right)^p = \frac{x(1-x^s)}{1-2x+x^{s+1}}.$$

The expression $\binom{n-1}{p-1}$ is unchanged by the substitution of $n-p+1$ for p; hence the numbers of compositions of n into p parts and into $n-p+1$ parts are identical.

3. The graph of a number n is taken to be a straight line divided at $n-1$ points into n equal segments.

The graph of a composition of the number n is obtained by placing nodes at certain of these $n-1$ points of division.

AB being the graph of the number 7, for the representation of the composition (214), nodes are placed at the points P, Q, so that in moving from A to B by steps proceeding from node to node, 2, 1, and 4 segments of the line are passed over in succession. Although strictly speaking the initial and final points A, B are nodes on the graphs of all the compositions, it is only the inter-terminal nodes that will be considered in what follows, as appertaining to the graph.

The number of parts in the composition exceeds by unity the number of nodes on the graph.

For a composition of n into p parts we can place nodes at any $p-1$ out of the $n-1$ points of the graph of the number. The number of such composition graphs is at once seen to be

$$\binom{n-1}{p-1},$$

and further, since each of the $n-1$ points of the number graph is or is not the position of a node, the total number of composition graphs is

$$2^{n-1}.$$

4. Associated with any one graph, there is another graph obtained by obliterating the nodes and placing nodes at the points not previously occupied.

These graphs are said to be conjugate.

If a graph denotes a composition of n into p parts, the conjugate graph denotes a composition of n into $n-p+1$ parts.

This notion supplies, in consequence, a graphical proof of the theorem of Art. 2.

Compositions of a number are conjugate when their graphs are conjugate.

E.g. The conjugate graphs

yield the conjugate compositions

$$(214) \qquad\qquad (13111)$$

The composition conjugate to a given composition may be written down, without constructing the graph, by the rules about to be explained.

The composition must first be prepared—

(i.) By writing successions of units in the power symbolism; for example, s successive units must be written 1^s;

(ii.) By intercalating 1^0 between each successive pair of non-unitary parts; thus, when aa or ab occur (a and b being superior to unity) we have to write $a1^0a$, $a1^0b$ respectively.

For the moment, call the non-unitary parts and the symbolic powers of unity the "elements" of the composition.

When an element does not occur at either end of the composition it is called "non-terminal," when at one end only "terminal," and when at both ends (thus constituting the entire composition), "doubly terminal."

The rules for procession to the conjugate are :—

I. If m or 1^m be doubly terminal, substitute 1^m for m or m for 1^m.

II. If m or 1^m be terminal, substitute 1^{m-1} for m or $m+1$ for 1^m.

III. If m or 1^m be non-terminal, substitute 1^{m-2} for m or $m+2$ for 1^m.

The composition thus obtained is in the "prepared" form and can be transformed to the ordinary form.

E.g. To find the conjugate of (231141), take the "prepared" form

$$(21^031^241),$$

and, beginning from the left, by

Rule	II.	For	2	substitute	1,
,,	III.	,,	1^0	,,	2,
,,	III.	,,	3	,,	1,
,,	III.	,,	1^2	,,	4,
,,	III.	,,	4	,,	1^2,
,,	II.	,,	1	,,	2,

resulting in the conjugate composition,

$$(12141^22), \text{ or non-symbolically } (1214112)$$

An examination of the rules shows that they are reversible, and that the process gives a one-to-one correspondence between compositions of n into p and $n-p+1$ parts.

A composition in general has, when prepared for conjugation, four different forms, viz. :—

$$(1)\ a_1 1^{\alpha_1} a_2 1^{\alpha_2} \ldots a_{s-1} 1^{\alpha_{s-1}} a_s.$$

$$(2)\ a_1 1^{\alpha_1} a_2 1^{\alpha_2} \ldots a_{s-1} 1^{\alpha_{s-1}} a_s 1^{\alpha_s}$$

$$(3)\ 1^{\alpha_1} a_2 1^{\alpha_2} \ldots a_{s-1} 1^{\alpha_{s-1}} a_s$$

$$(4)\ 1^{\alpha_1} a_2 1^{\alpha_2} \ldots a_{s-1} 1^{\alpha_{s-1}} a_s 1^{\alpha_s};$$

in all the four forms $a_1, a_2, a_3, \ldots a_s$ may have any positive integral values superior to unity. The numbers, $\alpha_1, \alpha_2, \ldots \alpha_s$, may have any positive integral values, including zero, with the exceptions,

In form (2) α_s cannot be zero.

(3) α_1 ,,

(4) α_1 and α_s ,,

The conjugates of the forms are

$$(1)\ 1^{a_1-1}\,.\,\alpha_1+2\,.\,1^{a_2-2}\,.\,\alpha_2+2\ldots 1^{a_{s-1}-2}\,.\,\alpha_{s-1}+2\,.\,1^{a_s-1},$$

$$(2)\ 1^{a_1-1}\,.\,\alpha_1+2\,.\,1^{a_2-2}\,.\,\alpha_2+2\ldots 1^{a_{s-1}-2}\,.\,\alpha_{s-1}+2\,.\,1^{a_s-2}\,.\,\alpha_s+1,$$

$$(3)\ \alpha_1+1\,.\,1^{a_2-2}\,.\,\alpha_2+2\ldots 1^{a_{s-1}-2}\,.\,\alpha_{s-1}+2\,.\,1^{a_s-1},$$

$$(4)\ \alpha_1+1\,.\,1^{a_2-2}\,.\,\alpha_2+2\ldots 1^{a_{s-1}-2}\,.\,\alpha_{s-1}+2\,.\,1^{a_s-2}\,.\,\alpha_3+1.$$

5. Two compositions are said to be inverse (the one of the other) when the parts of the one, read from left to right, are identical with those of the other when read from right to left.

A composition may therefore be self-inverse.

In the graph of a self-inverse composition, the nodes must be symmetrically placed with respect to the extremities of the graph. If the number be even, the number of segments of the graph is even, and the two central nodes (nodes nearest to the centre of the graph) may be coincident, or they may include 2, 4, or any even number of segments. A self-inverse composition of an even number, say $2m$, into an even number, say $2p$, of parts, can only occur when the two central nodes of the graph are coincident and, attending to one side only of this node, we find that the number of self-inverse compositions of the number $2m$, composed of $2p$ parts, is equal to the number of compositions of m composed of p parts. In a notation, which is self-explanatory, we may write

$$\mathrm{SIC}\,(2m, 2p) = \mathrm{C}\,(m, p) = \binom{m-1}{p-1}.$$

Next consider the self-inverse compositions of $2m$ into an uneven number, $2p - 1$, of parts. The two central nodes must be distinct, and may include any even number of segments. If this even number be 2κ the corresponding number of self-inverse compositions is equal to the number of compositions of $m - \kappa$ into $p - 1$ parts.

Hence

$$\mathrm{SIC}\,(2m, 2p - 1) = \mathrm{C}\,(m - 1, p - 1) + \mathrm{C}\,(m - 2, p - 2) + \ldots + \mathrm{C}\,(p - 1, p - 1),$$

or

$$\mathrm{SIC}\,(2m, 2p - 1) = \mathrm{C}\,(m, p) = \binom{m-1}{p-1}.$$

Self-inverse compositions of uneven numbers occur only when the number of parts is uneven, and it is easy to prove that

$$\mathrm{SIC}\,(2m - 1, 2p - 1) = \mathrm{C}\,(m, p) = \binom{m-1}{p-1}.$$

Hence, without restriction of the number of parts,

$$\mathrm{SIC}\,(2m) = \mathrm{SIC}\,(2m + 1) = \mathrm{C}\,(m + 1) = 2^m.$$

This completes the enumeration of the self-inverse compositions.

6. Two compositions which are at once conjugate and inverse, may be termed "inverse conjugates."

A composition whose conjugate is its own inverse is said to be "inversely conjugate."

Inversely conjugate compositions of a number n which have p parts, can occur only when $p = n - p + 1$, or $n = 2p - 1$, an uneven number.

The inversely conjugate compositions of $2m + 1$ are composed of $m + 1$ parts.

Consider a graph in which white and black nodes have reference to the two inverse conjugates respectively.

The black nodes are placed to the right and left of the centre of the graph in a manner similar to the white nodes to the left and right. If on the right there are

s black nodes and $m - s$ white nodes; to the left there are similarly placed s white nodes and $m - s$ black nodes.

Of the graph of the number there are m points to the right of the centre at which white and black nodes can be placed in 2^m distinct ways. Hence it at once follows that the number $2m + 1$ possesses in all 2^m inversely conjugate compositions. Otherwise, we may say that the number of inversely conjugate compositions of $2m + 1$ is equal to the number of compositions of $m + 1$.

It will be observed that the number $2m + 1$ has precisely the same number, 2^m, of self-inverse compositions.

There is, in fact, a one-to-one correspondence between the compositions of $2m + 1$, which are inversely conjugate, and those which are self-inverse.

To explain this, take the graph last represented. Read according to the black nodes we obtain the inversely conjugate composition

(23121) of the number 9.

To proceed to the corresponding self-inverse composition obliterate the black nodes to the right of the centre, and also the white nodes to the left of the centre. Substituting white nodes for the black nodes then remaining we have the graph

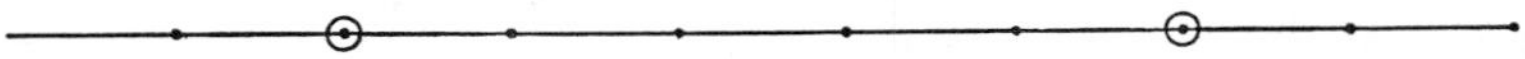

of the self-inverse composition

(252).

Again, reading the original graph according to white nodes, we have the inversely-conjugate composition

(12132),

and proceeding, as before, with the exception that black and white nodes are obliterated on the left and right of the centre respectively, we obtain the graph

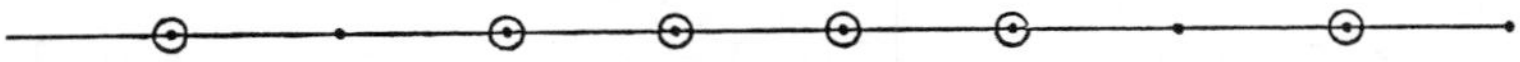

of the self-inverse composition

(1211121).

The process is a general one, and shows that we can always pass from a composition which is inversely conjugate to one which is self-inverse.

E.g., of the number 7 we have the correspondence

Inversely conjugate,	Self inverse.
(4111)	(7)
(3211)	(313)
(2221)	(232)
(2131)	(21112)
(1312)	(151)
(1222)	(12121)
(1123)	(11311)
(1114)	(1111111)

The general form of an inversely conjugate composition is

$$\alpha_1 1^{\alpha_1} \alpha_2 1^{\alpha_2} \alpha_3 1^{\alpha_3} \ldots \alpha_3 + 2 \;.\; 1^{\alpha_3 - 2} \;.\; \alpha_2 + 2 \;.\; 1^{\alpha_2 - 2} \;.\; \alpha_1 + 2 \;.\; 1^{\alpha_1 - 1}$$

in its form prepared for conjugation.

7. The compositions of a number, m, give rise to the compositions of $m + 1$ by rules somewhat similar to those in ARBOGAST's method of derivations. The rules are obvious as soon as stated.

Each composition of the number m gives rise to two compositions of the number $m + 1$.

I. By prefixing the part unity.

II. By increasing the magnitude of the first part by unity.

All the compositions thus obtained are necessarily distinct. As an example see the subjoined scheme for passing from the compositions of 3 to those of 4.

111		12		21		3.	
1111	211	112	22	121	31	13	4.

If the conjugates of these two lines be taken, the result is the same as the two lines inverted. We have the theorem, easily proved: "The conjugate of the $\begin{matrix}\text{first}\\\text{second}\end{matrix}$ derivative of a composition is the $\begin{matrix}\text{second}\\\text{first}\end{matrix}$ derivative of the conjugate composition."

8. The theory of the compositions of numbers is closely connected with the theory of the *perfect partitions* of numbers.* The connection is between the compositions of all multipartite numbers and the perfect partitions of unipartite numbers. The enumeration of the compositions of a single multipartite number enumerates also the perfect partitions of an infinite number of unipartite numbers. There is, moreover,

* "The Theory of Perfect Partitions of Numbers and the Compositions of Multipartite Numbers," The Author, 'Messenger of Mathematics.' New Series, No. 235. November, 1890.

as was shown, *loc. cit.*, a one-to-one correspondence between the compositions of the unipartite number m and the perfect partitions, comprising m parts, of the whole assemblage of unipartite numbers.

Defining a perfect partition of a number to be one which contains one, and only one, partition of every lower number, it was shown that if

$$(\alpha, \beta, \gamma, \delta, \ldots)$$

be any composition of the number

$$\alpha + \beta + \gamma + \delta + \ldots,$$

the partition,

$$\{1^\alpha \,.\, (1 + \alpha)^\beta \,.\, (1 + \alpha \,.\, 1 + \beta)^\gamma \,.\, (1 + \alpha \,.\, 1 + \beta \,.\, 1 + \gamma)^\delta \ldots\},$$

the exponents being symbolic, denoting repetitions of parts, is a perfect partition of the number

$$(1 + \alpha)(1 + \beta)(1 + \gamma)(1 + \delta) \ldots - 1.$$

§ 2. *Multipartite Numbers.*

9. The multipartite number $\overline{\alpha\beta\gamma \ldots}$ may be regarded as specifying $\alpha + \beta + \gamma + \ldots$ things, α of one sort, β of a second, γ of a third, and so forth.

To illustrate partitions and compositions of such numbers, I write down those appertaining to the bipartite number $\overline{21}$.

Partitions.	Compositions.
$(\overline{21})$	$(\overline{21})$
$(\overline{20}\,\overline{01})$	$(\overline{20}\,\overline{01})$, $(\overline{01}\,\overline{20})$
$(\overline{11}\,\overline{10})$	$(\overline{11}\,\overline{10})$, $(\overline{10}\,\overline{11})$
$(\overline{10}^2\,\overline{01})$	$(\overline{10}^2\,\overline{01})$, $(\overline{10}\,\overline{01}\,\overline{10})$, $(\overline{01}\,\overline{10}^2)$

I speak of the parts of the partition or composition, and observe that each part is a multipartite number of the same nature, or order of multiplicity, as the number partitioned.

There is (*see* Art. 8, *loc. cit.*) a one-to-one correspondence between the compositions of the multipartite

$$\overline{\alpha\beta\gamma \ldots}$$

and the perfect partitions of the unipartite number

$$a^\alpha b^\beta c^\gamma \ldots - 1,$$

where $a, b, c, \ldots$ are any different prime numbers.

This correspondence is entirely distinct from that alluded to in Art. 12.

10. Taking for the present the general multipartite number to be

$$\overline{p_1 p_2 p_3 \ldots}$$

it is easily seen that the enumeration of the compositions into r parts is the same problem as the enumeration of the distributions of $p_1 + p_2 + p_3 + \ldots$ things, of which p_1 are of one sort, p_2 of a second, p_3 of a third, and so forth into r different parcels.

This number* is the coefficient of $\alpha_1^{p_1}\alpha_2^{p_2}\alpha_3^{p_3} \ldots$ in the expansion of

$$(h_1 + h_2 + h_3 + \ldots)^r$$

wherein h_s denotes the sum of the homogeneous products, of degree s, of the quantities

$$\alpha_1, \alpha_2, \alpha_3, \ldots$$

Hence, the generating function of the total number of compositions is

$$\frac{h_1 + h_2 + h_3 + \ldots}{1 - h_1 - h_2 - h_3 - \ldots}.$$

The coefficient of $\alpha_1^{p_1}\alpha_2^{p_2}\alpha_3^{p_3} \ldots$ in the expansion of $(h_1 + h_2 + h_3 + \ldots)^r$ is readily found to be

$$\binom{p_1 + r - 1}{p_1}\binom{p_2 + r - 1}{p_2}\binom{p_3 + r - 1}{p_3} \ldots$$
$$- \binom{r}{1}\binom{p_1 + r - 2}{p_1}\binom{p_2 + r - 2}{p_2}\binom{p_3 + r - 2}{p_3} \ldots$$
$$+ \binom{r}{2}\binom{p_1 + r - 3}{p_1}\binom{p_2 + r - 3}{p_2}\binom{p_3 + r - 3}{p_3} \ldots$$
$$- \ldots \text{ to } r \text{ terms.}$$

and the enumeration is analytically complete.

11. This method is laborious in the case of high multipartite numbers since r may have all values from unity to $p_1 + p_2 + p_3 + \ldots$; but fortunately the series, above written, possesses some remarkable properties which can be utilized so as greatly to abridge the necessary labour.

Let $F(p_1 p_2 p_3 \ldots)$ and $f(p_1 p_2 p_3 \ldots, r)$ denote, respectively, the total number of

* See the Author, "Symmetric Functions and the Theory of Distributions," 'Proceedings of the London Mathematical Society,' vol. 19, Nos. 318–320.

compositions and the number comprised of r parts of the multipartite number $\overline{p_1p_2p_3 \ldots}$; so that,

$$F(p_1p_2p_3\ldots) = \sum_1^{\Sigma p} f(p_1p_2p_0 \ldots, r).$$

Writing

$$(1-\alpha_1)(1-\alpha_2)(1-\alpha_3)\ldots = 1 - a_1 + a_2 - a_3 + \ldots$$

$$\frac{h_1 + h_2 + h_3 + \ldots}{1 - h_1 - h_2 - h_3 - \ldots} = \frac{a_1 - a_2 + a_3 - \ldots}{1 - 2(a_1 - a_2 + a_3 - \ldots)}.$$

This new form of the generating function gives a relation connecting the number of compositions of any multipartite with numbers related to lower multipartites.

For unipartite numbers

$$\frac{\alpha_1}{1-2\alpha_1} = \Sigma\, F(p_1)\,\alpha_1^{p_1}$$

yielding

$$F(1) = 1$$

and

$$F(p_1) = 2F(p_1 - 1) \text{ when } p_1 > 1.$$

For bipartite numbers

$$\frac{\alpha_1 + \alpha_2 - \alpha_1\alpha_2}{1 - 2(\alpha_1 + \alpha_2 - \alpha_1\alpha_2)} = \Sigma\, F(p_1p_2)\,\alpha_1^{p_1}\alpha_2^{p_2}$$

giving

$$F(p_1p_2) = 2F(p_1 - 1, p_2) + 2F(p_1, p_2 - 1) - 2F(p_1 - 1, p_2 - 1),$$

and similarly for multipartite numbers

$$\begin{aligned} F(p_1p_2p_3\ldots) = {} & 2\{F(p_1 - 1, p_2, p_3, \ldots) + \ldots\} \\ & - 2\{F(p_1 - 1, p_2 - 1, p_3, \ldots) + \ldots\} \\ & + 2\{F(p_1 - 1, p_2 - 1\ p_3 - 1, \ldots) + \ldots\} \\ & - \ldots \end{aligned}$$

a formula absolutely true for all multipartites superior to $\overline{111\ldots}$, and universally true if F (000 . . .) when it occurs be interpreted to mean $\frac{1}{2}$.*

12. A simple expression is obtainable for $F(p_1p_2)$. We have to find the coefficient of $\alpha_1^{p_1}\alpha_2^{p_2}$ in

$$\frac{\alpha_1 + \alpha_2 - \alpha_1\alpha_2}{1 - 2(\alpha_1 + \alpha_2 - \alpha_1\alpha_2)}$$

or, this is the coefficient of $\alpha_1^{p_1}$ in

$$2^{p_2 - 1}\frac{(1-\alpha_1)}{(1-2\alpha_1)^{p_2+1}},$$

* See *post*, Art. 39.

and thence

$$F(p_1p_2) = 2^{p_1+p_2-1}\frac{(p_1+p_2)!}{p_1!\,p_2!} - 2^{p_1+p_2-2}\frac{(p_1+p_2-1)!}{1!(p_1-1)!(p_2-1)!}$$
$$+ 2^{p_1+p_2-3}\frac{(p_1+p_2-2)!}{2!(p_1-2)!(p_2-2)!} - \ldots$$

until one of the denominator factorials becomes zero.*

13. There is another method by which the value of $F(p_1p_2p_3\ldots\mu)$ can be easily obtained from the numbers $f(p_1p_2p_3\ldots, r)$ which compose the value of $F(p_1p_2p_3\ldots)$. Writing $h_1 + h_2 + h_3 + \ldots = H$, we may write the generating function

$$\frac{H}{1-H} = \Sigma\, F(p_1p_2p_3\ldots).(p_1p_2p_3\ldots)$$

where $(p_1p_2p_3\ldots)$ denotes the symmetric function

$$\Sigma\,\alpha_1^{p_1}\alpha_2^{p_2}\alpha_3^{p_3}$$

and

$$H^r = \Sigma f(p_1p_2p_3\ldots, r).(p_1p_2p_3\ldots).$$

Let

$$d_r = \partial_{\alpha_r} + \alpha_1\partial_{\alpha_{r+1}} + \alpha_2\partial_{\alpha_{r+2}} + \ldots$$
$$D_r = \frac{1}{r!}(\partial_{\alpha_1} + \alpha_1\,\partial_{\alpha_2} + \alpha_2\,\partial_{\alpha_3} + \ldots)^r,$$

D_r being an operator of the rth order obtained by symbolical multinomial expansion, as in TAYLOR'S theorem of the Differential Calculus, and not denoting r successive performances of linear operations. The effect of the operation of D_r upon a symmetric function of the quantities $\alpha_1, \alpha_2, \alpha_3 \ldots$ expressed in the notation of partitions, is well known; it obliterates one part r from every symmetric function partition which possesses such a part, and causes every other symmetric function to vanish.

Also $D_r(r) = 1$.

Hence

$$D_\mu H^r = \overset{p}{\Sigma} f(p_1p_2p_3\ldots\mu, r).(p_1p_2p_3\ldots).$$

To evaluate $D_\mu H^r$ we require the well-known formulæ

$$d_\mu h_s = (-)^{\mu+1}h_{s-\mu},$$

leading to

$$d_\mu H = (-)^{\mu+1}(1 + H).$$

When operating upon a function of H, d_μ is equivalent to d_1, when μ is uneven, and equivalent to $-d_1$ when μ is even. Hence, with such an operand, we have the equivalences

* Tables for the verification of formulæ will be found *post* pp. 898–900.

$$D_1 \equiv d_1,$$

$$D_2 \equiv \frac{1}{2!} d_1 (d_1 + 1),$$

$$D_3 \equiv \frac{1}{3!} d_1 (d_1 + 1)(d_1 + 2),$$

$$\cdot \quad \cdot \quad \cdot \quad \cdot \quad \cdot \quad \cdot \quad \cdot \quad \cdot \quad \cdot$$

$$D_\mu \equiv \frac{1}{\mu!} d_1 (d_1 + 1) \ldots (d_1 + \mu - 1),$$

the products on the right denoting successive operations.*

To evaluate $D_\mu H^r$ as an algebraic function of H, it is necessary to operate successively with $d_1, d_1 + 1, \ldots d_1 + \mu - 1$ and to divide the result by $\mu!$

We have

$$D_1 H^r = rH^{r-1} + rH^r,$$

$$D_2 H^r = \binom{r}{2} H^{r-2} + \binom{r}{1}^2 H^{r-1} + \binom{r+1}{2} H^r.$$

Suppose that generally

$$D_\mu H^r = \sum_{s=0}^{s=\mu} \binom{r+s-1}{s} \binom{r}{\mu - s} H^{r-\mu+s}.$$

Operation with

$$\frac{1}{\mu + 1} (d_1 + \mu)$$

gives

$$D_{\mu+1} H^r = \sum_{s=0}^{s=\mu+1} \binom{r+s-1}{s} \binom{r}{\mu + 1 - s} H^{r-(\mu+1)+s}.$$

so that the assumed law is established inductively.

* The truth of this law is seen by comparison with the corresponding algebraic relations. Denoting by $s_1, s_2, \ldots$ the sums of powers of $\alpha_1, \alpha_2, \alpha_3, \ldots$

$$1 - a_1 x + a_2 x^2 - \ldots = \exp. - (s_1 x + \tfrac{1}{2} s_2 x^2 + \tfrac{1}{3} s_3 x^3 + \ldots),$$

and putting

$$s_{2\kappa+1} = s_1, \quad s_{2\kappa} = - s_1,$$

$$1 - a_1 x + a_2 x^2 - \ldots = (1 + x)^{-s_1}.$$

Hence

$$a_\mu = \frac{1}{\mu!} s_1 (s_1 + 1) \ldots (s_1 + \mu - 1),$$

from which we pass to the operator relation

$$D_\mu = \frac{1}{\mu!} d_1 (d_1 + 1) \ldots (d_1 + \mu - 1).$$

Observe also the relation

$$h_\mu = \frac{1}{\mu!} s_1 (s_1 - 1) \ldots (s_1 - \mu + 1).$$

Substituting in a previous identity

$$\sum_{s=0}^{s=\mu} \binom{r+s-1}{s}\binom{r}{\mu-s} \mathrm{H}^{r-\mu+s} = \sum^{p} f(p_1p_2p_3 \ldots \mu, r) \, . \, (p_1p_2p_3 \ldots),$$

therefore,

$$\sum_{s=0}^{s=\mu} \binom{r+s-1}{s}\binom{r}{\mu-s} \sum^{p} f(p_1p_2p_3 \ldots, r-\mu+s) \, . \, (p_1p_2p_3 \ldots)$$

$$= \sum^{p} f(p_1p_2p_3 \ldots \mu, r) \, . \, (p_1p_2p_3 \ldots).$$

This is an absolute identity and, equating coefficients after writing

$$\binom{r+s-1}{s}\binom{r}{\mu-s} = \phi(r, s),$$

we obtain

$$\phi(r, \mu) f(p_1p_2p_3 \ldots, r) + \phi(r, \mu-1) f(p_1p_2p_3 \ldots, r-1) + \ldots$$
$$+ \phi(r, \mu-r+1) f(p_1p_2p_3 \ldots, 1) = f(p_1p_2p_3 \ldots \mu, r).$$

This formula enables the calculation of the number $f(p_1p_2p_3 \ldots \mu, r)$ from the successive numbers

$$f(p_1p_2p_3 \ldots, r), \quad f(p_1p_2p_3 \ldots, r-1), \ldots f(p_1p_2p_3 \ldots, 1).$$

14. A more useful result is obtained by summing each side of this identity for a values of r from 1 to $\Sigma p + \mu$. This result is

$$\sum^{r} \{\phi(r, \mu) + \phi(r+1, \mu-1) + \ldots + \phi(r+\mu, 0)\} f(p_1p_2p_3 \ldots r) = \mathrm{F}(p_1p_2p_3 \ldots \mu).$$

It can be shown that the expression in brackets { } to the left has the values

$$2^{\mu-1} \binom{r+\mu-1}{r} \frac{2r+\mu}{\mu};$$

therefore,

$$\mathrm{F}(p_1p_2p_3 \ldots \mu) = \sum^{r} 2^{\mu-1} \binom{r+\mu-1}{r} \frac{2r+\mu}{\mu} f(p_1p_2p_3 \ldots, r)$$

$$= 2^{\mu-1}(\mu+2) f(p_1p_2p_3 \ldots, 1) + 2^{\mu-1} \frac{(\mu+1)(\mu+4)}{2!} f(p_1p_2p_3 \ldots, 2)$$

$$+ 2^{\mu-1} \frac{(\mu+1)(\mu+2)(\mu+6)}{3!} f(p_1p_2p_3 \ldots, 3) + \ldots,$$

a formula of great service when μ is large, as the number of arithmetical operations is comparatively small.

As an example we can find another series for $F(p_1p_2)$.

Since

$$F(p_1) = 1 + \binom{p_1 - 1}{1} + \binom{p_1 - 1}{2} + \ldots + 1$$

wherein

$$\binom{p_1 - 1}{r} \text{ is the value of } f(p_1, r),$$

we find

$$F(p_1p_2) = 2^{p_2-1}(p_2 + 2) + 2^{p_2-1}\frac{(p_2+1)(p_2+4)}{2!}(p_1 - 1)$$
$$+ 2^{p_2-1}\frac{(p_2+1)(p_2+2)(p_2+6)}{3!}\binom{p_1-1}{2} + \ldots,$$

a series which it is easy to identify with that previously given (Art. 12).

15. Useful formulæ of verification are obtainable.

It has been shown in Art. 13 that the operation $(-)^{\mu+1} d_\mu$ is equivalent to d_1 when the operand is a function of H only.

Now

$$\frac{(-)^{\mu+1}}{\mu} d_\mu = \Sigma (-)^{\alpha+\beta+\ldots-1}\frac{(\alpha+\beta+\ldots-1)!}{\alpha!\,\beta!\ldots} D_1^{\alpha} D_2^{\beta}\ldots,$$

the summation having reference to all positive integer solutions of the equation,

$$\alpha + 2\beta + 3\gamma + \ldots = \mu.$$

Operating on the expression

$$\Sigma f(p_1p_2p_3\ldots, r).(p_1p_2p_3\ldots)$$

with $(-)^{\mu+1} d_\mu$ for successive positive integral values of μ, and equating the coefficients of the symmetric function $(p_1p_2p_3\ldots)$ we find the relations—

$$\begin{aligned} & f(p_1p_2p_3\ldots 1, r) \\ &= -\{f(p_1p_2p_3\ldots 1^2, r) - 2f(p_1p_2p_3\ldots 2, r)\} \\ &= +\{f(p_1p_2p_3\ldots 1^3, r) - 3f(p_1p_2p_3\ldots 21, r) + 3f(p_1p_2p_3\ldots 3, r)\} \\ &= -\ldots \\ &= (-)^{s+1}\Sigma(-)^{\alpha+\beta+\ldots-1}\frac{(\alpha+\beta+\ldots-1)!\,s}{\alpha!\,\beta!\ldots} f(p_1p_2p_3\ldots 2^{\beta}1^{\alpha}, r), \end{aligned}$$

the condition of summation being

$$\alpha + 2\beta + 3\gamma + \ldots = s.$$

Either summing these identities with respect to r, or operating as before upon the expression

$$\Sigma\, F(p_1p_2p_3\ldots).(p_1p_2p_3\ldots),$$

we obtain the identities

$$\begin{aligned} &F(p_1p_2p_3\ldots 1)\\ &\quad = -\{F(p_1p_2p_3\ldots 1^2) - 2F(p_1p_2p_3\ldots 2)\}\\ &\quad = +\{F(p_1p_2p_3\ldots 1^3) - 3F(p_1p_2p_3\ldots 21) + 3F(p_1p_2p_3\ldots 3)\}\\ &\quad = \ldots\\ &\quad = (-)^{s+1}\,\Sigma\,(-)^{\alpha+\beta+\ldots-1}\frac{(\alpha+\beta+\ldots-1)!\,s}{\alpha!\,\beta!\ldots}F(p_1p_2p_3\ldots 2^\beta 1^\alpha).\end{aligned}$$

These relations are readily verified in the particular case

$$(p_1p_2p_3\ldots 1) = (1).$$

16. Another very useful result is derived from the algebraical result noticed in the foot-note to Art. 13.

Since the supposition $(-)^{\mu-1}s_\mu = s_1$ leads to the formula

$$\frac{1}{\mu!}s_1(s_1-1)\ldots(s_1-\mu+1) = h_\mu$$

and

$$h_\mu = \Sigma\,(-)^{\mu+v_1+v_2+\ldots}\frac{(v_1+v_2+\ldots)!}{v_1!\,v_2!\ldots}a_1^{v_1}a_2^{v_2}\ldots$$

we reach the operator relation

$$\frac{1}{\mu!}d_1(d_1-1)\ldots(d_1-\mu+1) = \Sigma\,(-)^{\mu+v_1+v_2+\ldots}\frac{(v_1+v_2+\ldots)!}{v_1!\,v_2!\ldots}D_1^{v_1}D_2^{v_2}\ldots$$

whenever, as in the present case, the operand is such that $(-)^{\mu-1}d_\mu = d_1$.

Assuming

$$\frac{1}{\mu!}d_1(d_1-1)\ldots(d_1-\mu+1)H^r = \binom{r}{\mu}H^{r-\mu}(1+H)^\mu$$

and operating with $\frac{1}{\mu+1}(d_1-\mu)$ we find

$$\frac{1}{(\mu+1)!}d_1(d_1-1)\ldots(d_1-\mu)H^r = \binom{r}{\mu+1}H^{r-\mu-1}(1+H)^{\mu+1}$$

verifying the assumption.

Hence operation on the relation

$$\mathrm{H}^r = \Sigma f(p_1 p_2 p_3 \ldots, r) . (p_1 p_2 p_3 \ldots)$$

yields the relation

$$\binom{r}{\mu} \mathrm{H}^{r-\mu} (1 + \mathrm{H})^{\mu}$$
$$= \Sigma (-)^{\mu+v_1+v_2+\ldots} \frac{(v_1 + v_2 + \ldots)!}{v_1!\, v_2! \ldots} \mathrm{D}_1^{v_1} \mathrm{D}_2^{v_2} \ldots \{\Sigma f(p_1 p_2 p_3 \quad ., r) . (p_1 p_2 p_3 \ldots)\},$$

or

$$\binom{r}{\mu} \Sigma \left\{ f(p_1 p_2 p_3 \ldots, r - \mu) + \binom{\mu}{1} f(p_1 p_2 p_3 \ldots, r - \mu + 1) \right.$$
$$\left. + \binom{\mu}{2} f(p_1 p_2 p_3 \ldots, r - \mu + 2) + \ldots \right\} . (p_1 p_2 p_3 \ldots)$$
$$= \Sigma \left\{ f(p_1 p_2 p_3 \ldots 1^{\mu}, r) - (\mu - 1) f(p_1 p_2 p_3 \ldots 21^{\mu-2}, r) \right.$$
$$+ \binom{\mu - 2}{1} f(p_1 p_2 p_3 \ldots 31^{\mu-3}, r) + \binom{\mu - 2}{2} f(p_1 p_2 p_3 \ldots 2^2 1^{\mu-4}, r)$$
$$\left. + \ldots \right\} . (p_1 p_2 p_3 \ldots),$$

or, equating coefficients and inverting

$$f(p_1 p_2 p_3 \ldots 1^{\mu}, r) - (\mu - 1) f(p_1 p_2 p_3 \ldots 21^{\mu-2}, r)$$
$$+ (\mu - 2) f(p_1 p_2 p_3 \ldots 31^{\mu-3}, r) + \ldots$$
$$= \binom{r}{\mu} \left\{ f(p_1 p_2 p_3 \ldots, r - \mu) + \binom{\mu}{1} f(p_1 p_2 p_3 \ldots, r - \mu + 1) \right.$$
$$\left. + \binom{\mu}{2} f(p_1 p_2 p_3 \ldots, r - \mu + 2) + \ldots \right\}.$$

If $r < \mu$ the dexter vanishes and

$$f(p_1 p_2 p_3 \ldots 1^{\mu}, r) - (\mu - 1) f(p_1 p_2 p_3 \ldots 21^{\mu-2}, r)$$
$$+ (\mu - 2) f(p_1 p_2 p_3 \ldots 31^{\mu-3}, r) + \ldots = 0.$$

If $r = \mu$ the dexter is

$$f(p_1 p_2 p_3 \ldots, 0) + \binom{\mu}{1} f(p_1 p_2 p_3 \ldots, 1) + \binom{\mu}{2} f(p_1 p_2 p_3 \ldots, 2) + \ldots$$

where $f(p_1 p_2 p_3 \ldots, 0)$ in general vanishes, but

$$f(0, 0) = 1.$$

Performing the summation in regard to r,

$$F(p_1p_2p_3\ldots 1^\mu) - (\mu-1)F(p_1p_2p_3\ldots 21^{\mu-2}) + \ldots$$
$$+ (-)^{\mu+v_1+v_2+\ldots}\frac{(v_1+v_2+\ldots)!}{v_1!\,v_2!\ldots}F(p_1p_2p_3\ldots 2^{v_2}1^{v_1}) + \ldots$$
$$= \binom{\mu}{0}f(p_1p_2p_3\ldots, 0) + \left\{\binom{\mu}{1} + \binom{\mu+1}{1}\binom{\mu}{0}\right\}f(p_1p_2p_3\ldots, 1) + \ldots$$
$$+ \theta_{\mu,s}f(p_1p_2p_3\ldots, s) + \ldots,$$

where

$$\theta_{\mu,s} = \binom{\mu}{s} + \binom{\mu+1}{1}\binom{\mu}{s-1} + \binom{\mu+2}{2}\binom{\mu}{s-2} + \ldots + \binom{\mu+s}{s}$$

that is to say

$$(1+x)^\mu(1-x)^{-\mu-1} = \Sigma\,\theta_{\mu,s}\,x^s.$$

E.g., $r < \mu$,

$$f(21^3, 2) - 2f(2^21, 2) + f(32, 2) = 0,$$

verified by

$$22 - 2\times 16 + 10 = 0.$$ (See Tables, pp. 898–900.)

$r = \mu$,

$$f(21^3, 3) - 2f(2^21, 3) + f(32, 3) = f(2, 0) + 3f(2, 1) + 3f(2, 2)$$

verified by

$$93 - 2\times 57 + 27 = 0 + 3\times 1 + 3\times 1.$$

For the summation formula

$$\begin{aligned} &F(1),\\ &= F(1^2) - F(2),\\ &= F(1^3) - 2F(21) + F(3),\\ &= F(1^4) - 3F(21^2) + F(2^2) + 2F(31) - F(4),\\ &= \ldots\ldots\\ &= 1,\end{aligned}$$

verified by

$$\begin{aligned} &1,\\ &= 3 - 2,\\ &= 13 - 2\times 8 + 4,\\ &= 75 - 3\times 44 + 26 + 2\times 20 - 8,\\ &= \ldots\ldots\\ &= 1;\end{aligned}$$

also

$$F(41^3) - 2F(421) + F(43)$$
$$= f(4, 0) + (3 + 4 \times 1) f(4, 1) + (3 + 4 \times 3 + 10 \times 1) f(4, 2)$$
$$+ (1 + 4 \times 3 + 10 \times 3 + 20 \times 1) f(4, 3)$$
$$+ (4 \times 1 + 10 \times 3 + 20 \times 3 + 35 \times 1) f(4, 4),$$

verified by

$$3408 - 3776 + 768$$
$$= 0 + 7 \times 1 + 25 \times 3 + 63 \times 3 + 129 \times 1,$$
$$= 400.$$

§ 3. *The Graphical Representation of the Compositions of Bipartite Numbers.*

17. The graphical method, that has been employed in the case of unipartite compositions, can be extended so as to meet the cases of bipartite, tripartite, and multipartite numbers in general. For the present the bipartite case alone is under consideration. The graph of a bipartite number $(\overline{pq})$ is derived directly from the graphs of the unipartite numbers (p), (q).

Take $q + 1$ exactly similar graphs of the number p and place them parallel to one another, at equal distances apart, and so that their left hand extremities lie on a right line; corresponding points of the $q + 1$ graphs can then be joined by right lines and a reticulation will be formed which is the graph of the bipartite number $\overline{pq}$.

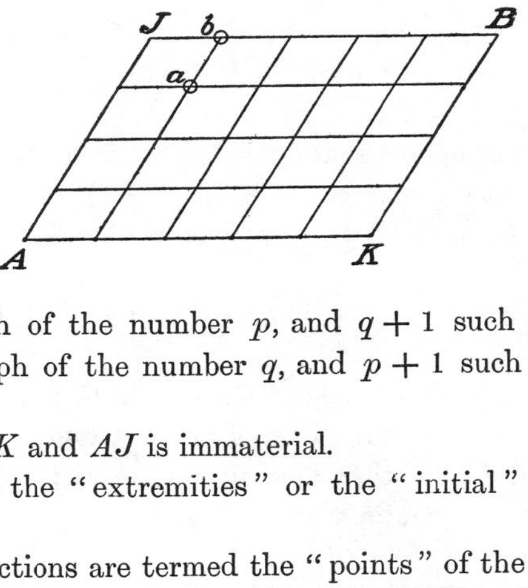

We have AK a graph of the number p, and $q + 1$ such graphs parallel to one another; and AJ a graph of the number q, and $p + 1$ such graphs parallel to one another.

The angle between AK and AJ is immaterial.

The points A, B, are the "extremities" or the "initial" and "final" points of the graph.

The remaining intersections are termed the "points" of the graph.

The lines of the graph have either the "direction" AK or the direction AJ. These will be called the α and β directions respectively. Through each point of the graph pass lines in each of these directions. Each line is made up of segments, and we speak of α segments and of β segments indicating that the corresponding lines (on which lie the segments) are in the α and β directions.

Suppose a traveller to proceed from A to B by successive steps. A step is performed by moving over a certain number of α segments and subsequently moving over a certain number of β segments. A step is thus made up of two figures—say an α figure and a β figure. The number of segments moved over may be zero in either, but not in both of these two figures of the step.

A step may be taken from A to any point a of the graph; a second step may be taken from a to any point of the graph aB which has a and B for its initial and final points; subsequent steps are taken on a similar principle, and the last step terminates at the point B and completes the procession from the point A to the point B. A step which takes x, α segments followed by y, β segments, is taken to be the representation of a bipartite part, $\overline{xy}$. A procession from the initial to the final point (from A to B) of the graph thus represents a sequence of bipartite parts which constitutes a composition of the bipartite number $\overline{pq}$. To every procession from A to B corresponds a composition of the bipartite $\overline{pq}$, and the enumeration of the different processions is identical with the enumeration of the total number of different compositions.

The steps of a procession are marked out by nodes placed at the points of the reticulation, which terminate the first, second, third, &c., and penultimate steps When nodes are thus placed, we have the graph of a composition. If nodes be placed at the points a, b of the graph, we obtain the graph of the composition

$$(\overline{13}\ \overline{01}\ \overline{40})$$

of the bipartite $\overline{54}$.

The number of parts in a composition is always one greater than the number of nodes in its graph.

Considering the number of compositions of $\overline{pq}$, of two parts, it is clear that we may place the defining node at any one of $(p+1)(q+1)-2$ points of the reticulation. Hence (and this may be verified by previous work) the number of two-part compositions is

$$(p+1)(q+1)-2.$$

18. The graph of a composition, traced from A to B, passes over certain segments, and may be said to follow a certain line of route through the reticulation. Other compositions in general follow the same line of route; they all have their defining nodes upon the line of route, and a certain number of defining nodes will, in general, be common to all of them.

Consider the path $AcbB$ in the graph given below. All compositions whose graphs follow this line of route must have the node b; b, in fact, is an essential node along this line of route.

Essential nodes occur on a line of route at all points where the course changes from the β to the α direction.

We may regard a line of route as defined by these essential nodes since the line of route is completely given by these nodes.

Every composition-graph involves nodes of two kinds—

(1) Those which are essential to its line of route;

(2) Those which, in this respect, are not essential.

There are $p + q - 1$ "points" along every line of route.

We have now to discuss the compositions whose graphs follow a given line of route, and we find that they naturally arrange themselves in pairs.

19. Associated with any one graph is another obtained by obliterating those nodes which are not essential to its line of route, and by then placing nodes at those points on the line of route not previously occupied by nodes.

These two graphs are said to be conjugate.

The compositions, represented by these graphs, are likewise said to be conjugate.

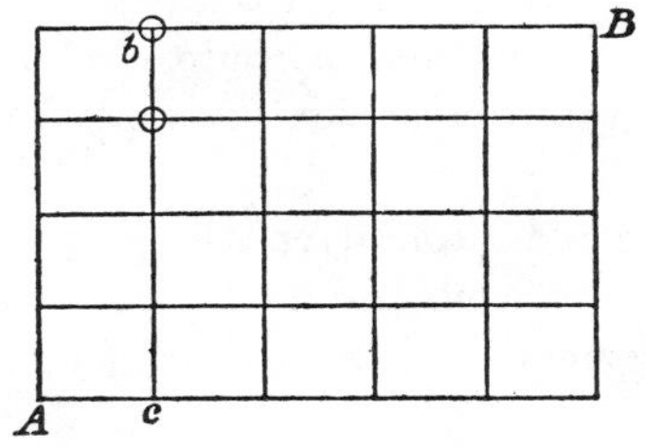

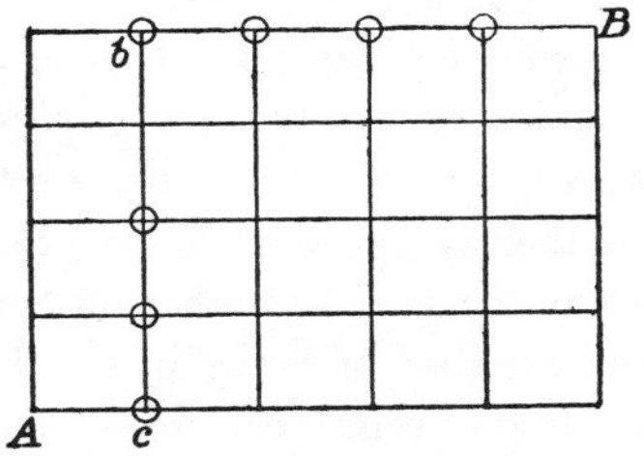

Conjugate graphs are shown above; $AcbB$ is the line of route, and b is the essential node.

The corresponding conjugate compositions of the bipartite number $\overline{54}$ are

$$(\overline{13}\ \overline{01}\ \overline{40}) \text{ and } (\overline{10}\ \overline{01}\ \overline{01}\ \overline{02}\ \overline{10}\ \overline{10}\ \overline{10}\ \overline{10}).$$

If the graph of a composition of $\overline{pq}$, of r parts, possesses s essential nodes, it is clear that the conjugate composition has $p + q - r + s + 1$ parts.

Of compositions, of the bipartite $\overline{pq}$, whose graphs possess s essential nodes, there is a one-to-one correspondence between those of r parts and those of $p + q - r + s + 1$ parts.

Corresponding to an essential node in the graph of a composition there exist in the composition itself adjacent parts ... $\overline{p_1q_1}\ \overline{p_2q_2}$ ... possessing the property of q_1 and p_2, being both superior to zero. Thus, from inspection of a composition, we are enabled to determine the number of essential nodes in its graph.

It is useful to recognise four species of contact between adjacent parts of a composition

$$\text{in} \ldots \overline{p_1q_1}\ \overline{p_2q_2} \ldots$$

if q_1 is zero and	p_2 zero	we have a	zero-zero	contact
q_1 ,,	p_2 positive	,,	zero-positive	,,
q_1 positive	p_2 zero	,,	positive-zero	,,
q_1 ,,	p_2 positive	,,	positive-positive	,,

In this nomenclature we may say that the graph of a composition possesses as many essential nodes as the composition itself possesses positive-positive contacts.

20. The theorem arrived at may be stated as follows :—

"Of compositions of $\overline{pq}$ possessing s positive-positive contacts, there is a one-to-one correspondence between those of r parts and those of $p + q - r + s + 1$ parts."

An essential node corresponding to a positive-positive contact in the composition occurs at a point of the graph where there is a change from a β direction to an α direction; we may say that this is a $\beta\alpha$ point on the line of route. Similarly to positive-zero, zero-positive, zero-zero contacts in the composition correspond $\beta\beta$, $\alpha\alpha$, $\alpha\beta$ points respectively on the line of route.

The number of different lines of route that can be traced on the graph of the bipartite number is the number of permutations of p symbols α, and q symbols β, for this is the number of ways in which the p α-segments and the q β-segments, which make up a line of route, can form a succession.

Hence the number of lines of route through the reticulation is

$$\binom{p+q}{p}.$$

The whole of the compositions of $\overline{pq}$ can be arranged in conjugate pairs.

E.g., The correspondence in regard to the compositions of the bipartite $\overline{22}$ is shown in parallel columns.

$$(\overline{22}) \quad s = 0, \quad r = 1 \qquad\qquad (\overline{10}\ \overline{10}\ \overline{01}\ \overline{01}) \quad s = 0, \quad r = 4$$

$$\left.\begin{matrix}(\overline{20}\ \overline{02})\\(\overline{21}\ \overline{01})\\(\overline{10}\ \overline{12})\end{matrix}\right\} s = 0, \quad r = 2 \qquad\qquad \left.\begin{matrix}(\overline{10}\ \overline{11}\ \overline{01})\\(\overline{10}\ \overline{10}\ \overline{02})\\(\overline{20}\ \overline{01}\ \overline{01})\end{matrix}\right\} s = 0, \quad r = 3$$

$$\left.\begin{array}{l}(\overline{02}\ \overline{20})\\(\overline{11}\ \overline{11})\\(\overline{01}\ \overline{21})\\(\overline{12}\ \overline{10})\end{array}\right\}\ s=1,\quad r=2 \qquad \left.\begin{array}{l}(\overline{01}\ \overline{01}\ \overline{10}\ \overline{10})\\(\overline{10}\ \overline{01}\ \overline{10}\ \overline{01})\\(\overline{01}\ \overline{10}\ \overline{10}\ \overline{01})\\(\overline{10}\ \overline{01}\ \overline{01}\ \overline{10})\end{array}\right\}\ s=1,\quad r=4$$

$$\left.\begin{array}{l}(\overline{01}\ \overline{20}\ \overline{01})\\(\overline{01}\ \overline{01}\ \overline{20})\\(\overline{10}\ \overline{02}\ \overline{10})\\(\overline{11}\ \overline{10}\ \overline{01})\end{array}\right\}\ s=1,\quad r=3 \qquad \left.\begin{array}{l}(\overline{01}\ \overline{10}\ \overline{11})\\(\overline{02}\ \overline{10}\ \overline{10})\\(\overline{11}\ \overline{01}\ \overline{10})\\(\overline{10}\ \overline{01}\ \overline{11})\end{array}\right\}\ s=1,\quad r=1$$

$$(\overline{01}\ \overline{11}\ \overline{10})\quad s=2,\quad r=3 \qquad (\overline{01}\ \overline{10}\ \overline{01}\ \overline{10})\quad s=2,\quad r=4.$$

The compositions present themselves in pairs of conjugate groups according to the several values of s and r. In the above example for $s=1$, $r=3$, there is a self-conjugate group. This happens when $2r=p+q+s+1$.

A self-conjugate group exists for even or uneven values of s, according as $p+q$ is uneven or even.

21. The enquiry now is in regard to the number of lines of route through the reticulation which possess exactly s essential nodes.

It may be observed, in passing, that from any line of route may be derived a composition whose graph exhibits only essential nodes and no others. This may be called the principal composition along the line of route; it will have $s+1$ parts, and each of the s contacts of its parts will be positive-positive.

There is a one-to-one correspondence between the lines of route having s essential nodes and the compositions of $s+1$ parts, all of whose contacts are positive-positive.

Also the number of compositions, number of parts unrestricted, all of whose contacts are positive-positive is equal to the number of different lines of route.

E.g., for the number $\overline{22}$,

$s=0$ $\qquad s=1$ $\qquad s=2$

Lines of route

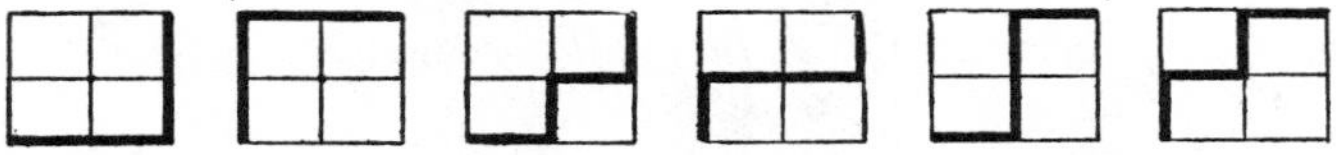

Compositions $(\overline{22})$ $\quad(\overline{02}\ \overline{20})$ $\quad(\overline{11}\ \overline{11})$ $\quad(\overline{01}\ \overline{21})$ $\quad(\overline{12}\ \overline{10})$ $\quad(\overline{01}\ \overline{11}\ \overline{10})$

and there are no other compositions having all contacts positive-positive.

22. The number of different lines of route with exactly s essential nodes is

$$\binom{p}{s}\binom{q}{s}.$$

Of this theorem I give three proofs, because its thorough examination is necessary for the purpose of leading up to the more difficult theories connected with tripartite and higher multipartite numbers.

First Proof.

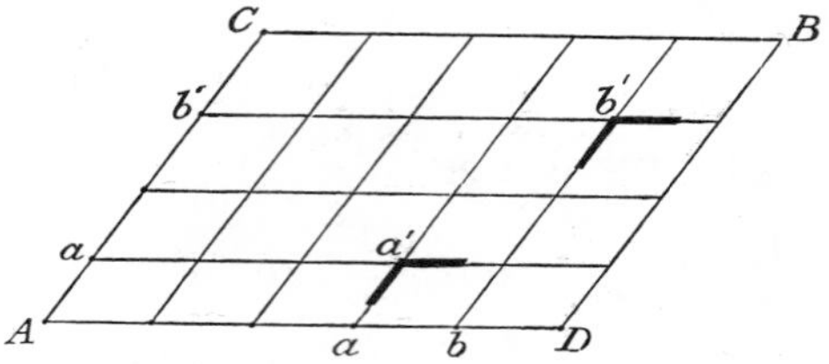

In each of the adjacent sides AD, AC of the graph of $\overline{pq}$, select any s "points" (see definition of "point") a, b, c, . . . in order from the point A. The two points a, a are seen to determine an essential node a'; the two points b, b an essential node b', &c., and a line of route necessarily exists which possesses these essential nodes and no others. The points along AD and AC, from which s points may be selected, are in number p and q respectively; along AD s points can be selected in $\binom{p}{s}$ ways, and along AC in $\binom{q}{s}$ ways; any selection on AD can be taken with any selection on AC. Hence the number of lines of route having exactly s essential nodes is

$$\binom{p}{s}\binom{q}{s}.$$

23. *Second Proof.*

We determined the total number of lines of route by considering the permutations of p, α-segments, and q, β-segments. Whenever in any such permutation there is a sequence $\beta\alpha$ there must be an essential node upon the corresponding line route. We have simply to find the number of permutations of the $p + q$ symbols in $\alpha^p\beta^q$ which possess exactly s, $\beta\alpha$-contacts.

Write down the s, $\beta\alpha$-sequences

$$\ldots\beta\alpha\ldots\beta\alpha\ldots\beta\alpha\ldots\beta\alpha\ldots$$

and the $s + 1$ intervals between them. In these intervals we have to distribute the letters in

$$\alpha^{p-s}\ \beta^{q-s}$$

in such manner as to introduce no fresh $\beta\alpha$-contacts. For each of these $p + q - 2s$ letters there is a choice of $s + 1$ intervals. The $p - s$ letters α may thus be distributed in $\binom{p}{s}$ ways, and the $q - s$ letters β in $\binom{q}{s}$, and each distribution of the letters α may occur with each distribution of the letters β. Hence the total number of permutations is

$$\binom{p}{s}\binom{q}{s}.$$

24. *Third Proof.**

I now come to a method which is valuable in the theory of multipartite numbers in general, and also intrinsically of great interest.

I consider, as in the second proof, the permutations of the $p+q$ letters $\alpha^p\beta^q$, which exhibit exactly s, $\beta\alpha$-contacts. The proof depends upon showing there exists a one-to-one correspondence between these permutations, and those in which the letter β occurs exactly s times in the first p places counted from the left.

Suppose the permutation with s, $\beta\alpha$-contacts to be

$$\alpha^{x_1}\beta^{y_1}\quad \beta\alpha \quad \alpha^{x_2}\beta^{y_2}\quad \beta\alpha \quad \alpha^{x_3}\beta^{y_3}\quad \beta\alpha \quad \alpha^{x_4}\beta^{y_4}\quad \beta\alpha \quad \alpha^{x_5}\beta^{y_5}\quad \beta\alpha \quad \alpha^{x_6}\beta^{y_6}$$

where, for convenience, s has the special value 5.

Any of the indices x, y may be zero.

Observe that

$$x_1 + x_2 + x_3 + x_4 + x_5 + x_6 + 5 = p,$$
$$y_1 + y_2 + y_3 + y_4 + y_5 + y_6 + 5 = q.$$

Obliterate the letters β, which do not occur in $\beta\alpha$-contacts, and the letters α, which do occur in $\beta\alpha$-contacts, there remains a succession

$$\alpha^{x_1}\beta\alpha^{x_2}\beta\alpha^{x_3}\beta\alpha^{x_4}\beta\alpha^{x_5}\beta\alpha^{x_6}$$

of p letters, β occurring s times.

Next obliterate in the original permutation the letters α, which do not occur in $\beta\alpha$-contacts, and the letters β, which do so occur; there remains a succession

$$\beta^{y_1}\alpha\beta^{y_2}\alpha\beta^{y_3}\alpha\beta^{y_4}\alpha\beta^{y_5}\alpha\beta^{y_6}$$

of q letters.

Take these two successions for the left and right portions of a new permutation, viz.:—

$$\alpha^{x_1}\beta\alpha^{x_2}\beta\alpha^{x_3}\beta\alpha^{x_4}\beta\alpha^{x_5}\beta\alpha^{x_6}\ .\ \beta^{y_1}\alpha\beta^{y_2}\alpha\beta^{y_3}\alpha\beta^{y_4}\alpha\beta^{y_5}\alpha\beta^{y_6}\alpha,$$

and we have made a perfectly definite transformation of a permutation involving exactly s $\beta\alpha$-contacts into another possessing the property that the letter β occurs s times in the first p places.

E.g., to transform

$$\beta^4\alpha^2\beta^3\alpha,$$

write it

$$\alpha^0\beta^3\quad \beta\alpha\quad \alpha\beta^2\quad \beta\alpha\,;$$

* This proof is of fundamental importance in the succeeding part of this investigation.

thence the successions

$$\alpha^0\beta\alpha\beta,$$

$$\beta^3\alpha\beta^2\alpha,$$

and the permutation

$$\beta\alpha\beta^4\alpha\beta^2\alpha,$$

in which β occurs twice in the first three places.

Now, it happens that these transformed permutations are very easily enumerated. The number of permutations of the letters in $\alpha^p\beta^q$, which possess the property that the letter β occurs exactly s times in the first p places is the coefficient of $\mu^s\alpha^p\beta^q$ in the development of

$$(\alpha + \mu\beta)^p (\alpha + \beta)^q.$$

This is evident from a consideration of the actual multiplication process.

Hence, in regard to the bipartite $\overline{pq}$,

$$(\alpha + \mu\beta)^p (\alpha + \beta)^q$$

is the generating function for the number of lines of route possessing a given number of essential nodes. The complete coefficient of $\alpha^p\beta^q$ is found to be

$$1 + \binom{p}{1}\binom{q}{1}\mu + \binom{p}{2}\binom{q}{2}\mu^2 + \ldots + \binom{p}{s}\binom{q}{s}\mu^s + \ldots + \binom{p}{q}\binom{q}{q}\mu^q,$$

generality not being lost by the supposition $p \nless q$. Hence the number of lines of route possessing s essential nodes is

$$\binom{p}{s}\binom{q}{s}.$$

Observe that the known formula

$$\binom{p}{0}\binom{q}{0} + \binom{p}{1}\binom{q}{1} + \ldots + \binom{p}{s}\binom{q}{s} + \ldots + \binom{p}{q}\binom{q}{q} = \binom{p+q}{p}$$

supplies a verification of the result.

It has also been proved that the number of compositions of $s + 1$ parts having all contacts positive-positive is $\binom{p}{s}\binom{q}{s}$.

25. On each of the $\binom{p}{s}\binom{q}{s}$ lines of route, which have s essential nodes, may be represented

$$2^{p+q-s-1}$$

5 R 2

compositions, because the $p+q-s-1$ non-essential nodes on each line of route may be selected as composition-nodes in this number of ways.

Hence, the total number of compositions is

$$F(pq) = \overset{s}{\Sigma} \binom{p}{s} \binom{q}{s} 2^{p+q-s-1}$$

(*cf.* the expressions for this number obtained in Arts. 12 and 14).

Analytically, this result is equivalent to the algebraic expansion

$$\frac{1}{1-2(x+y-xy)} = \frac{1}{(1-2x)(1-2y)} + \frac{2xy}{(1-2x)^2(1-2y)^2} + \frac{2^2x^2y^2}{(1-2x)^3(1-2y)^3} + \ldots$$

26. The important transformation of permutations established above is interesting when viewed graphically.

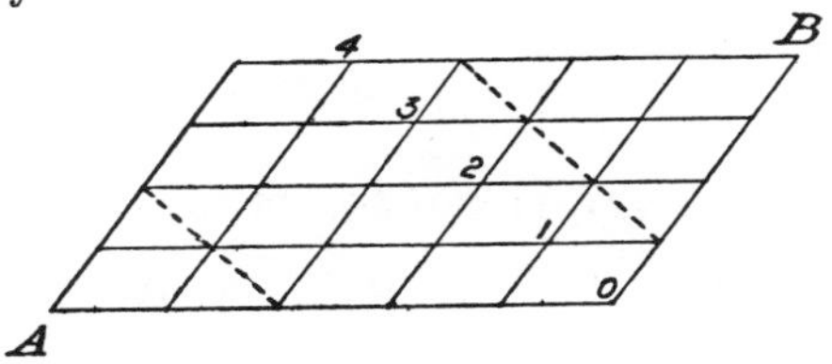

A line of route is traced by a certain succession of α and β segments. That portion of a line of route traced by the initial p segments terminates in one of the points 0, 1, 2, 3, 4 in the diagram. All these points lie on a straight line through the right-hand lower corner. If one of these points be marked s, s of the p segments will be β segments, and every line of route passing through the point s has the property that s symbols β occur in the first p places of the corresponding permutation of the symbols α and β. Hence, there is a one-to-one correspondence between the lines of route possessing 0, 1, 2, 3, 4, . . . s, . . . essential nodes and the lines of route passing through the points marked 0, 1, 2, 3, 4, . . . s, . . .

Observe that the number of lines of route in the graph, of which A and s are the initial and final points, is $\binom{p}{s}$, and in the graph of which s and B are the initial and final points, is $\binom{q}{s}$.

Hence of the graph AB the number of lines of route passing through the point s is

$$\binom{p}{s} \binom{q}{s},$$

and since every line of route must pass through one of the points 0, 1, 2, 3, 4, . . . s, . . . we have

$$\sum^{s} \binom{p}{s} \binom{q}{s} = \binom{p+q}{p} .*$$

Inverse Bipartite Compositions.

27. A line of route being marked out on a reticulation from A to B, the inverse line of route is obtained by rotating the reticulation through two right angles and interchanging the letters A and B.

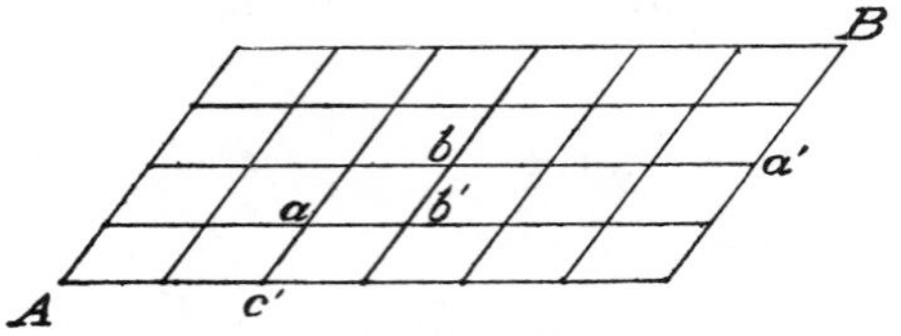

Consider a line of route from A to B having essential nodes a, b. On the inverse line of route a', b', c' are the essential nodes.

The principal compositions along these lines of route are—

$$(\overline{21}\ \overline{11}\ \overline{32}) \text{ from } A \text{ to } B.$$

$$(\overline{02}\ \overline{31}\ \overline{11}\ \overline{20}) \text{ from } B \text{ to } A.$$

* Consider also the points of the graph distant t segments from A. The number of such points is the coefficient of x^t in the product

$$(1 + x + x^2 + \ldots + x^p)\,(1 + x + x^2 + \ldots + x^q),$$

and if

$$\begin{array}{lll} t < q + 1 & \text{is equal to} & t + 1, \\ > q < p + 1 & \quad ,, & q + 1, \\ > p & \quad ,, & p + q - t + 1. \end{array}$$

These points lie on lines parallel to the line 01234 ... s ... and we obtain a graphical proof of the identities,

$$\binom{t}{0}\binom{p+q-t}{q} + \binom{t}{1}\binom{p+q-t}{q-1} + \ldots + \binom{t}{t}\binom{p+q-t}{q-t} = \binom{p+q}{p} \text{ for } t < q + 1;$$

$$\binom{t}{0}\binom{p+q-t}{q} + \binom{t}{1}\binom{p+q-t}{q-1} + \ldots + \binom{t}{q}\binom{p+q-t}{0} = \binom{p+q}{p} \text{ for } p + 1 > t > q;$$

$$\binom{t}{p}\binom{p+q-t}{0} + \binom{t}{p-1}\binom{p+q-t}{1} + \ldots + \binom{t}{t-q}\binom{p+q-t}{p+q-t} = \binom{p+q}{p} \text{ for } t > p$$

where the total number of terms in all these identities is

$$\sum_{0}^{q} (t+1) + (p-q)(q+1) + \sum_{t=p+1}^{t=p+q} (p+q-t+1) = (p+1)(q+1),$$

which is the total number of points in the graph (including A and B), and is therefore right.

All the contacts in these compositions are necessarily positive-positive; hence the leading and ending parts are the only ones that can involve a zero element; in the leading part the leading element may be zero, and in the ending part the ending element.

Every part, of a principal composition, which does not possess a zero element necessitates an essential node on the graph of the principal composition on the inverse line of route. Thus, if the one principal composition has s nodes, $s+1$ parts and t parts without a zero element, the other principal composition has t nodes, $t+1$ parts and s parts without a zero element.

These may be called "inverse principal compositions."

The number of pairs of inverse principal compositions is equal to half the number of distinct lines of route through the reticulation. The number of pairs is thus—

$$\tfrac{1}{2}\binom{p+q}{p}.$$

Otherwise, we may say that in regard to compositions, all of whose contacts are positive-positive, there is a one-to-one correspondence between those having $s+1$ parts and t parts without a zero element and those having $t+1$ parts and s parts without a zero element.

The number $s+1-t$ is either 0, 1, or 2.

E.g., for the bipartite $\overline{22}$ the correspondence is

$$s=0,\ t=1 \qquad\qquad s=1,\ t=0$$
$$(\overline{22}) \qquad\qquad (\overline{02}\ \overline{20})$$

$$s=1,\ t=1 \qquad\qquad s=1,\ t=1$$
$$(01\ 2\bar{1}) \qquad\qquad (\overline{12}\ \overline{10})$$

$$s=1,\ t=2 \qquad\qquad s=2,\ t=1$$
$$(\overline{11}\ \overline{11}) \qquad\qquad (\overline{01}\ \overline{11}\ \overline{10}).$$

28. This particular case of inversion leads easily to the general idea. Suppose a line route traced on a reticulation, and suppose marked the s essential nodes on the line drawn from A to B, and also the t essential nodes of the inverse line of route. These $s+t$ nodes are distinct. Along the line we may place an additional node at an unoccupied point and interpret the new compositions as read before and after rotation of the reticulation. The new compositions have each acquired an additional part; they each have the same number of parts without zero elements as before; the added node has either introduced a zero-positive contact into each or a positive-zero contact into each, according as the node is on an α or on a β line between adjacent nodes.

Altogether there are $p+q-1-s-t$ points at disposal, which may be selected as positions for nodes in $2^{p+q-1-s-t}$ ways. Suppose j_1+j_2 additional nodes taken, such that j_1 zero-positive and j_2 positive-zero contacts are introduced into each composition; the number of parts in each composition will be increased by j_1+j_2.

We have a one-to-one correspondence between the compositions, having

$s+1+j_1+j_2$ parts,
t parts without a zero element,
j_1 zero-positive contacts,
j_2 positive-zero contacts;

and those having

$t+1+j_1+j_2$ parts,
s parts without a zero element,
j_1 zero-positive contacts,
j_2 positive-zero contacts.

We have thus pairs of inverse compositions; the correspondence for the bipartite $\overline{22}$ is

$$(s, t, j_1, j_2) = (0, 1, 1, 0), \quad (1, 0, 1, 0), \quad (1, 1, 0, 1),$$
$$(\overline{10}\ \overline{12}) \text{ and } (\overline{02}\ \overline{10}\ \overline{10}); \quad (\overline{11}\ \overline{01}\ \overline{10});$$

$$(0, 1, 0, 1), \quad (1, 0, 0, 1), \quad (1, 1, 1, 0),$$
$$(\overline{21}\ \overline{01}) \text{ and } (\overline{01}\ \overline{01}\ \overline{20}); \quad (\overline{01}\ \overline{10}\ \overline{11}).$$

$$(0, 1, 1, 1), \quad (1, 0, 1, 1),$$
$$(\overline{10}\ \overline{11}\ \overline{01}) \text{ and } (\overline{01}\ \overline{01}\ \overline{10}\ \overline{10});$$

Of the above, two compositions $(\overline{11}\ \overline{01}\ \overline{10})$, $(\overline{01}\ \overline{10}\ \overline{11})$ may be termed self-inverse.

29. There are twelve compositions, viz., those with zero-zero contacts, which do not appear in the theory.

To enumerate these compositions, consider the lines of route with s essential nodes or bends ┌; every such line must have either $s-1$, s, or $s+1$ bends ┘. If the allied $\alpha\beta$ permutation neither commences with α nor ends with β it is $s-1$; if either α commences or β ends it is s; if both α commences and β ends it is $s+1$.

The enumeration of the lines of route gives

$$\binom{p-1}{s-1}\binom{q-1}{s-1} \text{ for } s-1 \text{ bends ┘};$$

$$\binom{p-1}{s}\binom{q-1}{s-1} + \binom{p-1}{s-1}\binom{q-1}{s} \text{ for } s \text{ bends ┘};$$

$$\binom{p-1}{s}\binom{q-1}{s} \text{ for } s+1 \text{ bends ┘}.$$

To obtain compositions along these lines of route which have not zero-zero contacts, we have at disposal

$$p + q - 2s,$$
$$p + q - 2s - 1,$$
$$p + q - 2s - 2$$

points respectively at which non-essential nodes may be placed. Hence the number of compositions which have s positive-positive contacts (see Art. 19) and no zero-zero contacts is—

$$2^{p+q-2s}\binom{p-1}{s-1}\binom{q-1}{s-1}$$
$$+\ 2^{p+q-2s-1}\left\{\binom{p-1}{s}\binom{q-1}{s-1} + \binom{p-1}{s-1}\binom{q-1}{s}\right\}$$
$$+\ 2^{p+q-2s-2}\binom{p-1}{s}\binom{q-1}{s},$$

which is

$$\frac{(p+s)(q+s)}{pq}\,2^{p+q-2s-2}\binom{p}{s}\binom{q}{s}.$$

Hence the total number of compositions which have no zero-zero contacts is

$$2^{p+q-2} + \frac{(p+1)(q+1)}{pq}\,2^{p+q-4}\binom{p}{1}\binom{q}{1} + \frac{(p+2)(q+2)}{pq}\,2^{p+q-6}\binom{p}{2}\binom{q}{2} + \ldots,$$

and, since the total number of compositions which have s positive-positive contacts is

$$2^{p+q-1-s}\binom{p}{s}\binom{q}{s},$$

we find that the number of compositions having s positive-positive contacts and also zero-zero contacts is

$$2^{p+q-2s-2}\left\{2^{s+1} - \frac{(p+s)(q+s)}{pq}\right\}\binom{p}{s}\binom{q}{s},$$

and the total number of compositions having zero-zero contacts is

$$2^{p+q-2} + 2^{p+q-4}\left(2^2 - \frac{(p+1)(q+1)}{pq}\right)\binom{p}{1}\binom{q}{1} + 2^{p+q-6}\left\{2^3 - \frac{(p+2)(q+2)}{pq}\right\}\binom{p}{2}\binom{q}{2}$$
$$+ \ldots$$

E.g., putting $\overline{pq} = \overline{22}$, we thus verify

$$4 + (4 - \tfrac{9}{4})\,4 + \tfrac{1}{4}(8 - 4) = 12$$

that the number is 12.

30. Self-inverse compositions can only occur upon self-inverse lines of route. For the existence of such a line, one at least of the elements of the bipartite number must be even. If both elements be even there is a central point in the reticulation, and the number of self-inverse lines of route for the bipartite $\overline{2p', 2q'}$ is equal to the number of lines of route of the bipartite $\overline{p'q'}$; that is to say, the number is

$$\binom{p' + q'}{p'}.$$

If the bipartite be $\overline{2p' + 1, 2q'}$ it is easy to see that the number is also

$$\binom{p' + q'}{p'}.$$

The number of self-inverse compositions in both cases is evidently equal to the total number of compositions of the bipartite $\overline{p'q'}$.

§ 4. *The Graphical Representation of the Compositions of Tripartite and Multipartite Numbers.*

31. The graph of a tripartite number may be in either two or three dimensions. It may be derived from a bipartite graph in a manner similar to that in which the bipartite has been derived from the unipartite graph. In the tripartite number $\overline{pqr}$ we take $r + 1$, exactly similar graphs of the bipartite $\overline{pq}$, and place them similarly with corresponding lines parallel, and like points lying on straight lines; when these straight lines are drawn the graph is complete. The $r + 1$ bipartite graphs may be in the same plane or in parallel planes according as the tripartite graph is required to be in two or three dimensions.

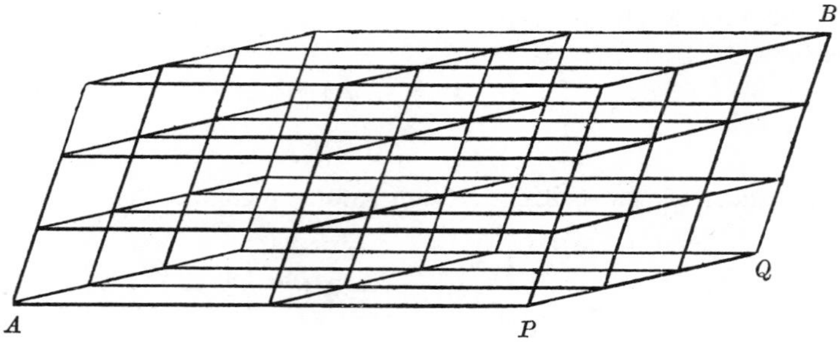

The figure depicts the graph of the tripartite $\overline{233}$. The points of the reticulation are identical with the points of the $r+1$ reticulations of the bipartites $\overline{pq}$. Observe that in two dimensions there are intersections of lines which are not points of the reticulation. On the other hand in three dimensions all intersections are also points.

Other than the initial and final points A and B there are $(p+1)(q+1)(r+1)-2$ points.

The graph involves lines in three different directions; say an α, a β, and a γ direction. These are parallel to AP, PQ, and QB respectively.

Through each point of the graph pass lines in all three directions, and a segment joining two adjacent points is called an α segment when it is in the α direction.

A line of route proceeds from A to B, from point to point of the reticulation. The number of lines of route is the number of permutations of the symbols in $\alpha^p\beta^q\gamma^r$, and is, therefore,

$$\binom{p+q+r}{p,\ q,\ }.$$

A step along a line of route traverses in succession any number of α, β, and γ segments, and in any one step the segments must be taken in the order, α, β, γ. The number of segments traversed may be zero in one or two, but not in three of these directions.

A step is represented by a tripartite number $\overline{p_1q_1r_1}$ and a succession of steps, the first starting at A, and the last terminating at B, is represented by a succession of tripartite numbers constituting a composition of the tripartite $\overline{pqr}$.*

The graph of a composition is obtained by placing nodes at the points which terminate the first, second, &c., and penultimate steps.

Essential nodes occur upon lines of route whenever the direction at a point changes from β to α, from γ to α, or from γ to β. These will be alluded to briefly as $\beta\alpha$, $\gamma\alpha$, or $\gamma\beta$ essential nodes. To these essential nodes on the graph of a composition correspond respectively zero-positive, positive-positive, and positive-zero contacts in the composition itself. It is convenient to call these collectively essential contacts. The theory of conjugate composition exists as in the bipartite theory. On every line of route there are $p+q+r-1$ points, and if there be s essential nodes, $2^{p+q+r-1-s}$ distinct composition graphs can be delineated along the line of route. Each of these has s essential contacts, and, as in the bipartite case, we establish a one-to-one correspondence between the compositions having s essential contacts, and $s+t+1$ parts, and those having s essential contacts, and $p+q+r-t$ parts. The graph of a quadripartite number is derived from the tripartite graph, and generally the graphs of the multipartite numbers of order n from those of the multipartite numbers of order $n-1$, on the same principle as the tripartite from the bipartite; and, moreover, in two dimensions.

32. For the multipartite number

$$\overline{p_1p_2\ldots p_{n-1}p_n}$$

* Observe that the thirteen compositions of the tripartite $\overline{111}$ are elegantly represented on the edges of a single cube, the six lines of route lying between opposite corners.

we take $p_n + 1$ exactly similar graphs of the multipartite number

$$\overline{p_1 p_2 \ldots p_{n-1}}$$

and place them similarly with corresponding lines parallel and like points lying on straight lines. When these straight lines are drawn the graph is complete. Other than the initial and final points A, B there are $(p_1 + 1)(p_2 + 1) \ldots (p_n + 1) - 2$ *points* in the graph.

There are lines in n different directions passing through or meeting at each point (including A and B). Call these directions the α_1 direction, the α_2, &c., the α_n direction. A segment joining two adjacent points is called an α segment when it is in the α direction. A step through the reticulation traverses in succession any number of $\alpha_1, \alpha_2, \ldots \alpha_n$ segments; but *in any one step* the segments must be taken in the order $\alpha_1, \alpha_2, \ldots \alpha_n$. In a step the number of steps traversed may be zero in any one, any two, &c., any $n - 1$ of the n different directions. If a step involve p'_1 segments in the α_1 direction, p'_2 segments in the α_2 direction and so on, it may be represented by the multipartite number

$$\overline{p'_1 p'_2 \ldots p'_n}.$$

A succession of steps, the first starting at A and the last terminating at B, is represented by a succession of multipartite numbers, constituting a composition of the multipartite

$$\overline{p_1 p_2 p_3 \ldots p_n}.$$

Any composition follows a certain line of route through the reticulation. The number of distinct lines of route is the number of permutations of the letters in

$$\alpha_1^{p_1} \alpha_2^{p_2} \ldots \alpha_n^{p_n},$$

and is therefore

$$\binom{p_1 + p_2 + \ldots + p_n}{p_1, p_2, \ldots p_{n-1}},$$

employing an obvious extension of notation.

The graph of a composition is obtained by placing nodes at the points which terminate the first, second, &c., and penultimate steps. Essential nodes occur upon a line of route whenever the direction at a point changes from α_u to α_t where $u > t$. At this point the contact between the adjacent parts of the composition is such that a part terminating with $n - u$ zero elements precedes a part commencing with $t - 1$ zero elements. We may speak of this as a contact of $n - u$ zeros with $t - 1$ zeros. The number of zeros in contact, being

$$n - 1 - (u - t),$$

may be any number from 0 to $n - 2$, according to the magnitude of $u - t$; and there are $j + 1$ different contacts for which the number of zeros in contact is j.

5 s 2

33. I now enquire, in respect of the tripartite reticulation, into the number of lines of route which possess exactly s_{21}, $\beta\alpha$ essential nodes, s_{32}, $\gamma\beta$ and s_{31}, $\gamma\alpha$ essential nodes. The second method of investigation adopted in the bipartite case will be employed.

Consider the permutations of the symbols in $\alpha^p\beta^q\gamma^r$ which possess s_{21}, $\beta\alpha$; s_{32}, $\gamma\beta$; and s_{31}, $\gamma\alpha$ contacts, and suppose

$$s_{21} + s_{32} + s_{31} = s.$$

It will be shown that the number of such permutations is equal to the number of permutations in which

β occurs s_{21} times in the first p places,
γ ,, s_{31} ,, ,, ,, ,,
γ ,, s_{32} ,, in the q places succeeding the first p.

A permutation of the former kind involves successions of letters

$$\beta\alpha,\ \gamma\beta,\ \gamma\alpha,\ \gamma\beta\alpha\,;$$

letters between consecutive successions being in alphabetical order.

As regards these successions, and attending to them alone, the permutation exhibits some permutation of the terms in

$$(\beta\alpha)^{s_{21}-\sigma}\ (\gamma\beta)^{s_{32}-\sigma}\ (\gamma\beta\alpha)^{\sigma}\ (\gamma\alpha)^{s_{31}}.$$

These permutations are

$$\frac{s_{21} + s_{32} + s_{31} - \sigma)\,!}{(s_{21} - \sigma)\,!\,(s_{32} - \sigma)\,!\,\sigma\,!\,s_{31}\,!}\ \text{in number.}$$

Selecting any one of these there are, with reference to the terms,

$$s_{21} + s_{32} + s_{31} - \sigma + 1$$

different positions in which other letters may be placed. It is clear that α cannot be placed after a term $(\gamma\beta)$, or γ before a term $(\beta\alpha)$, for these placings would lead to additional $(\beta\alpha)$ and $(\gamma\beta)$ contacts. The letter α may be placed before any of the terms $(\beta\alpha)$, $(\gamma\beta)$, $(\gamma\beta\alpha)$, $(\gamma\alpha)$, and after any of the terms $(\beta\alpha)$, $(\gamma\beta\alpha)$, $(\gamma\alpha)$. Hence out of the $s_{21} + s_{32} + s_{31} - \sigma + 1$ different positions (relative to the terms) $s_{21} + s_{31} + 1$ positions may be occupied by a letter α. The letter β may be placed in all the $s_{21} + s_{32} + s_{31} - \sigma + 1$ positions without the introduction of additional $\beta\alpha$, $\gamma\beta$, $\gamma\alpha$ terms. The letter γ must not be in a place preceding $(\beta\alpha)$, and hence it may occupy $s_{32} + s_{31} + 1$ different positions.

Besides the letters occurring in the terms

$$(\beta\alpha)^{s_{21}-\sigma}\ (\gamma\beta)^{s_{32}-\sigma}\ (\gamma\beta\alpha)^{\sigma}\ (\gamma\alpha)^{s_{31}},$$

α occurs $p - s_{21} - s_{31}$ times,
β ,, $q - s_{21} - s_{32} + \sigma$ times,
γ ,, $r - s_{32} - s_{31}$ times.

The $p - s_{21} - s_{31}$ letters α may be distributed in $s_{21} + s_{,1} + 1$ positions in

$$\binom{p}{s_{21} + s_{31}} \text{ ways.}$$

The $q - s_{21} - s_{32} + \sigma$ letters β may be distributed in $s_{21} + s_{32} + s_{31} - \sigma + 1$ positions in

$$\binom{q + s_{31}}{s_{21} + s_{32} + s_{31} - \sigma} \text{ ways.}$$

The $r - s_{32} - s_{31}$ letters γ may be distributed in $s_{32} + s_{31} + 1$ positions in

$$\binom{r}{s_{32} + s_{31}} \text{ ways.}$$

Hence, for any given permutation of the terms in

$$(\beta\alpha)^{s_{21}-\sigma} \, (\gamma\beta)^{s_{32}-\sigma} \, (\gamma\beta\alpha)^{\sigma} \, (\gamma\alpha)^{s_{31}}$$

there are

$$\binom{p}{s_{21} + s_{31}} \binom{q + s_{31}}{s_{21} + s_{32} + s_{31} - \sigma} \binom{r}{s_{32} + s_{31}}$$

arrangements of the remaining letters which do not introduce additional $\beta\alpha$, $\gamma\beta$ or $\gamma\alpha$ contacts.

Hence for a given value of σ there are

$$\binom{p}{s_{21} + s_{31}} \binom{q + s_{31}}{s_{21} + s_{32} + s_{31} - \sigma} \binom{r}{s_{32} + s_{31}} \frac{(s_{21} + s_{32} + s_{31} - \sigma)!}{(s_{21} - \sigma)! \, (s_{32} - \sigma)! \, \sigma! \, s_{31}!}$$

permutations.

To complete the enumeration we have to sum this expression in regard to σ.
We have

$$\Sigma \binom{q + s_{31}}{s_{21} + s_{32} + s_{31} - \sigma} \frac{(s_{21} + s_{32} + s_{31} - \sigma)!}{(s_{21} - \sigma)! \, (s_{32} - \sigma)! \, \sigma! \, s_{31}!}$$

$$= \Sigma \frac{(q + s_{31})!}{(q - s_{21} - s_{32} + \sigma)! \, (s_{21} - \sigma)! \, (s_{32} - \sigma)! \, \sigma! \, s_{31}!}$$

$$= \frac{(q + s_{31})!}{s_{31}! \, s_{32}! \, (q - s_{32})!} \, \Sigma \, \frac{s_{32}!}{\sigma! \, (s_{32} - \sigma)!} \cdot \frac{(q - s_{32})!}{(s_{21} - \sigma)! \, (q - s_{21} - s_{32} + \sigma)!}$$

$$= \frac{(q + s_{31})!}{s_{31}! \, s_{32}! \, (q - s_{32})!} \, \Sigma \binom{s_{32}}{\sigma} \binom{q - s_{32}}{s_{21} - \sigma}$$

$$= \frac{(q + s_{31})!}{s_{31}! \, s_{32}! \, (q - s_{32})!} \binom{q}{s_{21}}$$

$$= \binom{s_{21} + s_{31}}{s_{21}} \binom{q}{s_{32}} \binom{q + s_{31}}{s_{21} + s_{31}}.$$

Hence the number of permutations sought is

$$\binom{s_{21}+s_{31}}{s_{21}}\binom{p}{s_{21}+s_{31}}\binom{q}{s_{32}}\binom{q+s_{31}\cdot}{s_{21}+s_{31}}\binom{r}{s_{32}+s_{31}}.$$

34. It can be shown by direct expansion that this number is the coefficient of $\lambda_{21}{}^{s_{21}}\lambda_{31}{}^{s_{31}}\lambda_{32}{}^{s_{32}}\alpha^p\beta^q\gamma^r$ in the development of

$$(\alpha+\lambda_{21}\beta+\lambda_{31}\gamma)^p\,(\alpha+\beta+\lambda_{32}\gamma)^q\,(\alpha+\beta+\gamma)^r,$$

which expresses the number of permutations in which

β occurs s_{21} times in the first p places.
γ ,, s_{31} ,, ,,
γ ,, s_{32} times in the q places succeeding the first p.

This remarkable identity of numbers suggests a one-to-one correspondence between the permutations of the two kinds, but I have not as yet been able to determine the law of the transformation. It would appear to be a very difficult problem. The number of lines of route in the tripartite reticulation which possess

s_{21}, $\beta\alpha$ essential nodes,
s_{32}, $\gamma\beta$,, ,,
s_{31}, $\gamma\alpha$,, ,,

is thus

$$\binom{s_{21}+s_{31}}{s_{21}}\binom{p}{s_{21}+s_{31}}\binom{q}{s_{32}}\binom{q+s_{31}}{s_{21}+31}\binom{r}{s_{32}+s_{31}}.$$

35. Hence the identity

$$\frac{p+q+r)!}{p!\,q!\,r!}=\Sigma\binom{s_{21}+s_{31}}{s_{21}}\binom{p}{s_{21}+s_{31}}\binom{q}{s_{32}}\binom{q+s_{31}}{s_{21}+s_{31}}\binom{r}{s_{32}+s_{31}},$$

the summation being for all positive integral, including zero, values of s_{21}, s_{31}, s_{32}, which yield positive terms.

36. The generating function for the number of lines of route having s essential nodes; that is $s_{21}+s_{31}+s_{32}=s$; is

$$(\alpha+\lambda\beta+\lambda\gamma)^p\,(\alpha+\beta+\lambda\gamma)^q\,(\alpha+\beta+\gamma)^r,$$

the number in question being the coefficient of $\lambda^s\alpha^p\beta^q\gamma^r$.

37. The whole coefficient of $\alpha^p\beta^q\gamma^r$ being

$$C_0 + C_1\lambda + C_2\lambda^2 + \ldots + C_s\lambda^s + \ldots$$

$$C_s = \Sigma \binom{s_{21} + s_{31}}{s_{21}} \binom{p}{s_{21} + s_{31}} \binom{q}{s_{32}} \binom{q + s_{31}}{s_{21} + s_{31}} \binom{r}{s_{32} + s_{31}}$$

the summation being subject to the condition

$$s_{21} + s_{31} + s_{32} = s.$$

Moreover, denoting the whole number of compositions of the tripartite $\overline{pqr}$ by $F(pqr)$, we have

$$F(pqr) = \overset{s}{\Sigma}\, C_s\, 2^{p+q+r-s-1}$$
$$= 2^{p+q+r-1} \overset{s}{\Sigma}\, C_s\, (\tfrac{1}{2})^s.$$

38. Hence $F(pqr)$ is the coefficient of $\alpha^p\beta^q\gamma^r$ in the development of the product

$$2^{p+q+r-1} (\alpha + \tfrac{1}{2}\beta + \tfrac{1}{2}\gamma)^p (\alpha + \beta + \tfrac{1}{2}\gamma)^q (\alpha + \beta + \gamma)^r,$$

which is more conveniently written

$$\tfrac{1}{2} (2\alpha + \beta + \gamma)^p (2\alpha + 2\beta + \gamma)^q (2\alpha + 2\beta + 2\gamma)^r.$$

39. This product is a generating function which enumerates the compositions of the single multipartite number $\overline{pqr}$, but the generating function of all tripartite numbers can be at once derived from it.

It is

$$\tfrac{1}{2} \frac{1}{\{1 - s(2\alpha + \beta + \gamma)\} \{1 - t(2\alpha + 2\beta + \gamma)\} \{1 - u(2\alpha + 2\beta + 2\gamma)\}},$$

in which, when expanded, the coefficient of $(s\alpha)^p(t\beta)^q(u\gamma)^r$ is the number of compositions of the tripartite number $\overline{pqr}$.

The generating function previously obtained for tripartite numbers from the analytical theory was

$$\frac{\alpha + \beta + \gamma - \beta\gamma - \gamma\alpha - \alpha\beta + \alpha\beta\gamma}{1 - 2(\alpha + \beta + \gamma - \beta\gamma - \gamma\alpha - \alpha\beta + \alpha\beta\gamma)},$$

the number of compositions being given by the coefficient of $\alpha^p\beta^q\gamma^r$. The addition of $\tfrac{1}{2}$ to this fraction brings it to the better form

$$\tfrac{1}{2} \frac{1}{1 - 2(\alpha + \beta + \gamma - \beta\gamma - \gamma\alpha - \alpha\beta + \alpha\beta\gamma)},$$

which is consistent also with the circumstance that it was found convenient analytically to regard the number of compositions of multipartite zero as being the fraction $\frac{1}{2}$.

40. The number may be stated as the coefficient of $(s\alpha)^p(t\beta)^q(u\gamma)^r$ in the expansion of

$$\frac{1}{2}\frac{1}{1-2(s\alpha+t\beta+u\gamma-tu\beta\gamma-us\gamma\alpha-st\alpha\beta+stu\alpha\beta\gamma)},$$

and we have shown that this fraction is equivalent to that portion of the expansion of the fraction

$$\frac{1}{2}\frac{1}{\{1-s(2\alpha+\beta+\gamma)\}\{1-t(2\alpha+2\beta+\gamma)\}\{1-u(2\alpha+2\beta+2\gamma)\}}$$

which is a function only of $s\alpha$, $t\beta$, and $u\gamma$.

41. It will now be shown that the fraction

$$\frac{1}{2}\frac{1}{\{1-s_1(2\alpha_1+\alpha_2+\ldots+\alpha_n)\}\{1-s_2(2\alpha_1+2\alpha_2+\ldots+\alpha_n)\}\ldots\{1-s_n(2\alpha_1+2\alpha_2+\ldots+2\alpha_n)\}}$$

is, in fact, a generating function which enumerates the compositions of multipartite numbers of order n.

This important theorem will be demonstrated by showing that the aggregate of terms in the expansion of the last written fraction, which is composed entirely of powers of $s_1\alpha_1$, $s_2\alpha_2$, . . . $s_n\alpha_n$, is correctly represented by the fraction

$$\frac{1}{2}\frac{1}{1-2(\Sigma s_1\alpha_1-\Sigma s_1s_2\alpha_1\alpha_2+\ldots(-)^{n+1}s_1s_2\ldots s_n\alpha_1\alpha_2\ldots\alpha_n)},$$

which has been already shown to be a true generating function.

42. For brevity put

$$S_\kappa = s_\kappa(2\alpha_1+2\alpha_2+\ldots+2\alpha_\kappa+\alpha_{\kappa+1}+\ldots+\alpha_n) = s_\kappa(A_\kappa+2\alpha_\kappa)$$
$$b_1 = \Sigma s_1\alpha_1,\ b_2 = \Sigma s_1s_2\alpha_1\alpha_2,\ \&c.\ldots$$
$$M = (1-2s_1\alpha_1)(1-2s_2\alpha_2)\ldots(1-2s_n\alpha_n).$$

The two fractions under comparison are

$$\frac{1}{2}\frac{1}{-1+2(1-s_1\alpha_1)(1-s_2\alpha_2)\ldots(1-s_n\alpha_n)} = \frac{1}{2N},$$

and

$$\frac{1}{2}\frac{1}{(1-S_1)(1-S_2)\ldots(1-S_n)} = \frac{1}{2D}.$$

Since $M - N = (2^2 - 2) b_2 - (2^3 - 2) b_3 + . \; . + (-)^n (2^n - 2) b_n$,

$$\frac{N}{D} = \frac{M}{D} - \frac{M-N}{D} = \prod_1^n \frac{1 - 2s_\kappa \alpha_\kappa}{1 - S_\kappa} - \frac{(2^2-2) b_2 - (2^3-2) b_3 + \ldots + (-)^n (2^n - 2) b_n}{(1-S_1)(1-S_2)\ldots(1-S_n)}$$

$$= \prod_1^n \left(1 + \frac{s_\kappa A_\kappa}{1 - S_\kappa}\right) - \frac{(2^2-2) b_2 - \ldots + (-)^n (2^n - 2) b_n}{(1-S_1)(1-S_2)\ldots(1-S_n)};$$

the second member of the right-hand side of this identity has now to be transformed into a series of partial fractions of the same form as those which arise from the product

$$\prod_1^n \left(1 + \frac{s_\kappa A_\kappa}{1 - S_\kappa}\right).$$

43. Let $\kappa_1, \kappa_2, \kappa_3, \ldots, \kappa_t$ be any t different numbers selected from the first n integers $1, 2, \ldots n$, arranged in ascending order of magnitude.

Let

$$B^{(1)}_{\kappa_1\kappa_2} = 2s_{\kappa_1}\alpha_{\kappa_1}s_{\kappa_2}\alpha_{\kappa_2}$$

$$B^{(1)}_{\kappa_1\kappa_2\kappa_3} = B^{(1)}_{\kappa_2\kappa_3} S_{\kappa_1} + B^{(1)}_{\kappa_3\kappa_1} S_{\kappa_2} + B^{(1)}_{\kappa_1\kappa_2} S_{\kappa_3}$$

$$\cdot \quad \cdot \quad \cdot \quad \cdot \quad \cdot \quad \cdot \quad \cdot \quad \cdot \quad \cdot \quad \cdot \quad \cdot \quad \cdot \quad \cdot$$

$$B^{(1)}_{\kappa_1\kappa_2\ldots\kappa_t} = \Sigma B^{(1)}_{\kappa_1\kappa_2} S_{\kappa_3} \ldots S_{\kappa_t},$$

the summation being for the $\binom{t}{2}$ terms obtained from the $\binom{t}{2}$ expressions $B_{\kappa_1\kappa_2}{}^{(1)}$ that it is possible to construct from the t integers $\kappa_1, \kappa_2, \ldots \kappa_t$.

Further, let

$$B^{(j)}_{\kappa_1\kappa_2\ldots\kappa_{j+1}} = (2^{j+1} - 2) s_{\kappa_1}\alpha_{\kappa_1}s_{\kappa_2}\alpha_{\kappa_2} \ldots s_{\kappa_{j+1}}\alpha_{\kappa_{j+1}}$$

$$B^{(j)}_{\kappa_1\kappa_2\ldots\kappa_t} = \Sigma B^{(j)}_{\kappa_1\kappa_2\ldots\kappa_{j+1}} S_{j+2} \ldots S_t$$

the summation being for the $\binom{t}{j+1}$ terms obtained from the $\binom{t}{j+1}$ expressions $B_{\kappa_1\kappa_2\ldots\kappa_{j+1}}{}^{(j)}$ that it is possible to construct from the t integers $\kappa_1, \kappa_2, \ldots \kappa_t$.

44. The following lemma is required :—

"*Lemma.*

$$\Sigma B^{(j)}_{\kappa_1\kappa_2\ldots\kappa_{t-1}} S_{\kappa_t} = \binom{t-j-1}{1} B^{(j)}_{\kappa_1\kappa_2\ldots\kappa_t},$$

$$\Sigma B^{(j)}_{\kappa_1\kappa_2\ldots\kappa_{t-2}} S_{\kappa_{t-1}} S_{\kappa_t} = \binom{t-j-1}{2} B^{(j)}_{\kappa_1\kappa_2\ldots\kappa_t},$$

$$\cdot \quad \cdot \quad \cdot \quad \cdot \quad \cdot \quad \cdot \quad \cdot \quad \cdot \quad \cdot \quad \cdot \quad \cdot \quad \cdot \quad \cdot$$

$$\Sigma B^{(j)}_{\kappa_1\kappa_2\ldots\kappa_{t-s}} S_{\kappa_{t-s+1}} \ldots S_{\kappa_t} = \binom{t-j-1}{s} B^{(j)}_{\kappa_1\kappa_2\ldots\kappa_t}$$

the summations being subject to the same conditions as before. The truth of these relations is obvious from the elementary theory of permutations."

45. Consider now the series of functions—

$$P_{\kappa_1\kappa_2} = B^{(1)}_{\kappa_1\kappa_2},$$
$$P_{\kappa_1\kappa_2\kappa_3} = B^{(1)}_{\kappa_1\kappa_2\kappa_3} - B^{(2)}_{\kappa_1\kappa_2\kappa_3},$$
$$P_{\kappa_1\kappa_2\kappa_3\kappa_4} = B^{(1)}_{\kappa_1\kappa_2\kappa_3\kappa_4} - B^{(2)}_{\kappa_1\kappa_2\kappa_3\kappa_4} + B^{(3)}_{\kappa_1\kappa_2\kappa_3\kappa_4},$$
$$\cdots\cdots\cdots\cdots$$
$$P_{\kappa_1\kappa_2\ldots\kappa_t} = B^{(1)}_{\kappa_1\kappa_2\ldots\kappa_t} - B^{(2)}_{\kappa_1\kappa_2\ldots\kappa_t} + \ldots + (-)^t B^{(t-1)}_{\kappa_1\kappa_2\ldots\kappa_t}$$

From the foregoing lemma are derived the relations—

$$\Sigma P_{\kappa_1\kappa_2} S_{\kappa_3} \ldots S_{\kappa_t} = B^{(1)}_{\kappa_1\kappa_2\ldots\kappa_t},$$
$$\Sigma P_{\kappa_1\kappa_2\kappa_3} S_{\kappa_4} \ldots S_{\kappa_t} = \binom{t-2}{1} B^{(1)}_{\kappa_1\kappa_2\ldots\kappa_t} - B^{(2)}_{\kappa_1\kappa_2\ldots\kappa_t},$$
$$\Sigma P_{\kappa_1\kappa_2\kappa_3\kappa_4} S_{\kappa_5} \ldots S_{\kappa_t} = \binom{t-2}{2} B^{(1)}_{\kappa_1\kappa_2\ldots\kappa_t} - \binom{t-3}{1} B^{(2)}_{\kappa_1\kappa_2\ldots\kappa_t} + B^{(3)}_{\kappa_1\kappa_2\ldots\kappa_t},$$
$$\cdots\cdots\cdots\cdots$$
$$\Sigma P_{\kappa_1\kappa_2\ldots\kappa_s} S_{\kappa_s+1} \ldots S_{\kappa_t} = \binom{t-2}{s-2} B^{(1)}_{\kappa_1\kappa_2\ldots\kappa_t} - \binom{t-3}{s-3} B^{(2)}_{\kappa_1\kappa_2\ldots\kappa_t} + \ldots + (-)^s B^{(s-1)}_{\kappa_1\kappa_2\ldots\kappa_t}.$$
$$\cdots\cdots\cdots\cdots$$

Hence, by addition with alternate signs,

$$(-)^t B_{\kappa_1\kappa_2\ldots\kappa_t}{}^{(t-1)} = P_{\kappa_1\kappa_2\ldots\kappa_t} - \Sigma P_{\kappa_1\kappa_2\ldots\kappa_{t-1}} S_{\kappa_t} + \Sigma P_{\kappa_1\kappa_2\ldots\kappa_{t-2}} S_{\kappa_{t-1}} S_{\kappa_t} - \ldots$$
$$+ (-)^t \Sigma P_{\kappa_1\kappa_2} S_{\kappa_3} S_{\kappa_4} \ldots S_{\kappa_t}.$$

Here the summations are in respect of the t numbers $\kappa_1, \kappa_2, \ldots \kappa_t$.

46. A similar identity may be established for every selection of t integers out of the numbers 1, 2, 3, . . . n.

Summing all these identities, and remembering that

$$\overset{n}{\Sigma} B_{\kappa_1\kappa_2\ldots\kappa_t}{}^{(t-1)} = (2^t - 2)\, b_t,$$

we obtain

$$(-)^t (2^t - 2)\, b_t = \overset{n}{\Sigma} P_{\kappa_1\kappa_2\ldots\kappa_t} - \overset{n}{\Sigma} P_{\kappa_1\kappa_2\ldots\kappa_{t-1}} S_{\kappa_t} + \ldots + (-)^t \overset{n}{\Sigma} P_{\kappa_1\kappa_2} S_{\kappa_3} \ldots S_{\kappa_t},$$

or, attributing to t all integer values from 2 to n inclusive, we have the series of relations :—

$$(2^2-2)\,b_2=\sum^n \mathrm{P}_{\kappa_1\kappa_2}$$

$$-(2^3-2)\,b_3=\sum^n \mathrm{P}_{\kappa_1\kappa_2\kappa_3}-\Sigma\,\mathrm{P}_{\kappa_1\kappa_2}\mathrm{S}_{\kappa_3}$$

$$+(2^4-2)\,b_4=\sum^n \mathrm{P}_{\kappa_1\kappa_2\kappa_3\kappa_4}-\sum^n \mathrm{P}_{\kappa_1\kappa_2\kappa_3}\mathrm{S}_{\kappa_4}+\sum^n \mathrm{P}_{\kappa_1\kappa_2}\mathrm{S}_{\kappa_3}\mathrm{S}_{\kappa_4}$$

. . . .

$$(-)^n(2^n-2)\,b_n=\sum^n \mathrm{P}_{\kappa_1\kappa_2\ldots\kappa_n}-\sum^n \mathrm{P}_{\kappa_1\kappa_2\ldots\kappa_{n-1}}\mathrm{S}_{\kappa_n}+\ldots+(-)^n\sum^n \mathrm{P}_{\kappa_1\kappa_2}\mathrm{S}_{\kappa_3}\ldots\mathrm{S}_{\kappa_n}.$$

Now

$$\sum^n \mathrm{P}_{\kappa_1\kappa_2}\mathrm{S}_{\kappa_3}=\sum^n \mathrm{P}_{\kappa_1\kappa_2}(\mathrm{S}_{\kappa_3}+\mathrm{S}_{\kappa_4}+\ldots+\mathrm{S}_{\kappa_n})$$

$$\sum^n \mathrm{P}_{\kappa_1\kappa_2}\mathrm{S}_{\kappa_3}\mathrm{S}_{\kappa_4}=\sum^n \mathrm{P}_{\kappa_1\kappa_2}(\mathrm{S}_{\kappa_3}\mathrm{S}_{\kappa_4}+\mathrm{S}_{\kappa_3}\mathrm{S}_{\kappa_5}+\ldots+\mathrm{S}_{\kappa_{n-1}}\mathrm{S}_{\kappa_n})$$

with similar relations in the case of the other symmetric functions.

47. Hence, by addition,

$$\begin{aligned}&(2^2-2)\,b_2-(2^3-2)\,b_3+\ldots+(-)^n(2^n-2)\,b_n\\ &=\sum^n \mathrm{P}_{\kappa_1\kappa_2}(1-\mathrm{S}_{\kappa_3})(1-\mathrm{S}_{\kappa_4})\ldots(1-\mathrm{S}_{\kappa_n})\\ &\quad+\sum^n \mathrm{P}_{\kappa_1\kappa_2\kappa_3}(1-\mathrm{S}_{\kappa_4})(1-\mathrm{S}_{\kappa_5})\ldots(1-\mathrm{S}_{\kappa_n})\\ &\quad+\ldots\\ &\quad+\sum^n \mathrm{P}_{\kappa_1\kappa_1\ldots\kappa_{n-1}}(1-\mathrm{S}_{\kappa_n})\\ &\quad+\sum^n \mathrm{P}_{\kappa_1\kappa_2\ldots\kappa_n},\end{aligned}$$

and thence,

$$\begin{aligned}&\frac{(2^2-2)\,b_2-(2^3-2)\,b_3+\ldots+(-)^n(2^n-2)\,b_n}{(1-\mathrm{S}_1)(1-\mathrm{S}_2)\ldots(1-\mathrm{S}_n)}\\ &=\sum^n\frac{\mathrm{P}_{\kappa_1\kappa_2}}{(1-\mathrm{S}_{\kappa_1})(1-\mathrm{S}_{\kappa_2})}\\ &\quad+\sum^n\frac{\mathrm{P}_{\kappa_1\kappa_2\kappa_3}}{(1-\mathrm{S}_{\kappa_1})(1-\mathrm{S}_{\kappa_2})(1-\mathrm{S}_{\kappa_3})}\\ &\quad+\ldots\\ &\quad+\sum^n\frac{\mathrm{P}_{\kappa_1\kappa_2\ldots\kappa_t}}{(1-\mathrm{S}_{\kappa_1})(1-\mathrm{S}_{\kappa_2})\ldots(1-\mathrm{S}_{\kappa_t})}\\ &\quad+\ldots\\ &\quad+\frac{\mathrm{P}_{123\ldots n}}{(1-\mathrm{S}_1)(1-\mathrm{S}_2)\ldots(1-\mathrm{S}_n)}.\end{aligned}$$

5 T 2

The relation

$$\frac{N}{D} = \prod_1^n \left(1 + \frac{s_\kappa A_\kappa}{1 - S_\kappa}\right) - \frac{(2^2-2)b_2 - (2^3-2)b_3 + \ldots + (-)^n(2^n-2)b_n}{(1-S_1)(1-S_2)\ldots(1-S_n)}$$

now becomes

$$\frac{N}{D} = 1 + \sum^n \frac{s_{\kappa_1}A_{\kappa_1}}{1-S_{\kappa_1}} + \sum^n \frac{s_{\kappa_1}s_{\kappa_2}A_{\kappa_1}A_{\kappa_2} - P_{\kappa_1\kappa_2}}{(1-S_{\kappa_1})(1-S_{\kappa_2})} + \ldots + \sum^n \frac{s_{\kappa_1}\ldots s_{\kappa_{n-1}}A_{\kappa_1}\ldots A_{\kappa_{n-1}} - P_{\kappa_1\kappa_2\ldots\kappa_{n-1}}}{(1-S_{\kappa_1})(1-S_{\kappa_1})\ldots(1-S_{\kappa_{n-1}})},$$

the final fraction

$$\frac{s_{\kappa_1}s_{\kappa_2}\ldots s_{\kappa_n}A_{\kappa_1}A_{\kappa_2}\ldots A_{\kappa_{n-1}} - P_{\kappa_1\kappa_2\ldots\kappa_n}}{(1-S_1)(1-S_2)\ldots(1-S_n)}$$

vanishing as will be seen presently.

48. This form of the fraction $\frac{N}{D}$ will be employed in order to show that the expansions of $\frac{1}{D}$ and $\frac{1}{N}$ are effectively identical. This arises from the circumstance that the right-hand side of the above identity when expanded contains no single term expressible as a function of $s_1\alpha_1, s_2\alpha_2, s_3\alpha_3, \ldots s_n\alpha_n$.

49. Of the fractions $\Sigma \frac{s_{\kappa_1}A_{\kappa_1}}{1-S_{\kappa_1}}$ consider the typical fraction

$$\frac{s_{\kappa_1}A_{\kappa_1}}{1-S_{\kappa_1}} = \frac{s_{\kappa_1}(2\alpha_1 + \ldots + 2\alpha_{\kappa_1-1} + \alpha_{\kappa_1+1} + \ldots + \alpha_n)}{1 - s_{\kappa_1}(2\alpha_1 + \ldots + 2\alpha_{\kappa_1} + \alpha_{\kappa_1+1} + \ldots + \alpha_n)};$$

since the numerator contains a factor s_{κ_1} and no quantity α_{κ_1} the fraction when expanded can contain no term which is a function of $s_1\alpha_1, s_2\alpha_2, \ldots s_n\alpha_n$ alone.

Consider next the typical fraction

$$\frac{s_{\kappa_1}s_{\kappa_2}A_{\kappa_1}A_{\kappa_2} - P_{\kappa_1\kappa_2}}{(1-S_{\kappa_1})(1-S_{\kappa_2})};$$

the numerator is

$$s_{\kappa_1}s_{\kappa_2}(A_{\kappa_1}A_{\kappa_2} - 2\alpha_{\kappa_1}\alpha_{\kappa_2})$$

and

$$A_{\kappa_1}A_{\kappa_2} = (2\alpha_1 + \ldots + 2\alpha_{\kappa_1-1} + \ldots + \alpha_{\kappa_2} + \ldots + \alpha_n)(2\alpha_1 + \ldots + 2\alpha_{\kappa_1} + \ldots + 2\alpha_{\kappa_2-1} + \ldots + \alpha_n),$$

and, therefore, contains a term

$$2\alpha_{\kappa_1}\alpha_{\kappa_2};$$

hence the numerator is free from such a term and every term in it contains one of the quantities

$$\alpha_{\kappa_3}, \alpha_{\kappa_4}, \ldots \alpha_{\kappa_n};$$

but the whole fraction contains no quantities

$$s_{\kappa_3}, \alpha_{\kappa_4}, \ldots s_{\kappa_n}$$

and thus it is manifest that the fraction can contribute no term which is a function of

$$s_1\alpha_1, \ldots s_n\alpha_n.$$

50. To simplify the discussion of the remaining fractions it is necessary to consider some particular properties of the typical numerator.

The function

$$P_{\kappa_1\kappa_2 \ldots \kappa_t}$$

may be expressed as

$$B^{(1)}_{\kappa_1\kappa_2\ldots\kappa_t} - B^{(2)}_{\kappa_1\kappa_2\ldots\kappa_t} + \ldots + (-)^t B^{(t-1)}_{\kappa_1\kappa_2\ldots\kappa_t},$$

or as

$$\begin{aligned}\{&\overset{t}{\Sigma}(2^2-2)\,\alpha_{\kappa_1}\alpha_{\kappa_2}(A_{\kappa_3}+2\alpha_{\kappa_3})\ldots(A_{\kappa_t}+2\alpha_{\kappa_t})\\ &-\overset{t}{\Sigma}(2^3-2)\,\alpha_{\kappa_1}\alpha_{\kappa_2}\alpha_{\kappa_3}(A_{\kappa_4}+2\alpha_{\kappa_4})\ldots(A_{\kappa_t}+2\alpha_{\kappa_t})\\ &+\ldots\\ &+(-)^t(2^t-2)\,\alpha_{\kappa_1}\alpha_{\kappa_2}\ldots\alpha_{\kappa_t}\}s_{\kappa_1}s_{\kappa_2}\ldots s_{\kappa_t}\end{aligned}$$

the summations being in respect of the t numbers

$$\kappa_1, \kappa_2, \ldots \kappa_t.$$

Writing

$$P_{\kappa_1\kappa_2\ldots\kappa_t} = P'_{\kappa_1\kappa_2\ldots\kappa}\ s_{\kappa_1}s_{\kappa_2}\ldots s_{\kappa_t},$$

we have now the identity

$$\begin{aligned}P'_{\kappa_1\kappa_2\ldots\kappa_t} = \overset{t}{\Sigma}(2^2-2)\,\alpha_{\kappa_1}\alpha_{\kappa_2}A_{\kappa_3}A_{\kappa_4}\ldots A_{\kappa_t} + \Sigma(2^3-2)\,\alpha_{\kappa_1}\alpha_{\kappa_2}\alpha_{\kappa_3}A_{\kappa_4}\ldots A_{\kappa_t}\\ +\ldots+(2^t-2)\,\alpha_{\kappa_1}\alpha_{\kappa_2}\ldots\alpha_{\kappa}.\end{aligned}$$

51. To establish this it is sufficient to observe that the coefficient of

$$\alpha_{\kappa_1}\alpha_{\kappa_2}\ \ldots\alpha_{\kappa_s}A_{\kappa_{s+1}}\ldots A_{\kappa_t}$$

in the former series is

$$\binom{s}{2} 2^{s-2}(2^2-2) - \binom{s}{3} 2^{s-3}(2^3-2) + \ldots + (-)^{s+1}\binom{s}{s}(2^s-2),$$

which has the value $2^s - 2$.

52. Hence the numerator of the typical fraction may, to a factor $s_{\kappa_1}s_{\kappa_2}\ldots s_{\kappa_t}$ *près* be thrown into the form :—

$$2(A_{\kappa_1}+\alpha_{\kappa_1})(A_{\kappa_2}+\alpha_{\kappa_2})\ldots(A_{\kappa}+\alpha_{\kappa_t}) - (A_{\kappa_1}+2\alpha_{\kappa_1})(A_{\kappa_2}+2\alpha_{\kappa_2})\ldots(A_{\kappa_t}+2\alpha_{\kappa_t}),$$

which is of great service.

Of the series of quantities

$$s_1, s_2, \ldots s_n,$$

the typical fraction contains only the t quantities

$$s_{\kappa_1}, s_{\kappa_2}, \ldots s_{\kappa_t}.$$

If it can be shown that the numerator consists of terms each of which contains quantities of the series

$$\alpha_1, \alpha_2, \ldots \alpha_n,$$

which are not included in the set

$$\alpha_{\kappa_1}, \alpha_{\kappa_2}, \ldots \alpha_{\kappa_t},$$

it will follow at once that the fraction can, when expanded, contain no term which is a function of

$$s_1\alpha_1, s_2\alpha_2, \ldots s_n\alpha_n.$$

It suffices to show that the numerator vanishes when all the quantities of the set

$$\alpha_1, \alpha_2, \ldots \alpha_n,$$

not included in the set

$$\alpha_{\kappa_1}, \alpha_{\kappa_2}, \ldots \alpha_{\kappa_t},$$

are put equal to zero.

Under these circumstances, remembering that $\kappa_1\ \kappa_2 \ldots \kappa_n$ are in ascending order of magnitude, we have

$$A_{\kappa_1} = \alpha_{\kappa_2} + \alpha_{\kappa_3} + \ldots + \alpha_{\kappa_t}$$
$$A_{\kappa_2} = 2\alpha_{\kappa_1} + \alpha_{\kappa_3} + \ldots + \alpha_{\kappa_t}$$
$$A_{\kappa_3} = 2\alpha_{\kappa_1} + 2\alpha_{\kappa_2} + \alpha_{\kappa_4} + \ldots + \alpha_{\kappa_t}.$$
$$\cdots\cdots\cdots\cdots$$
$$A_{\kappa_{t-1}} = 2(\alpha_{\kappa_1} + \alpha_{\kappa_2} + \ldots + \alpha_{\kappa_{t-2}}) + \alpha_{\kappa_t}$$
$$A_{\kappa_t} = 2(\alpha_{\kappa_1} + \alpha_{\kappa_2} + \ldots + \alpha_{\kappa_{t-1}})$$

and the relations

$$\begin{aligned} A_{\kappa_1} + 2\alpha_{\kappa_1} &= A_{\kappa_2} + \alpha_{\kappa_2} \\ A_{\kappa_2} + 2\alpha_{\kappa_2} &= A_{\kappa_3} + \alpha_{\kappa_3} \\ &\cdots\cdots \\ A_{\kappa_{t-1}} + 2\alpha_{\kappa_{t-1}} &= A_{\kappa_t} + \alpha_{\kappa_t} \\ A_{\kappa_t} + 2\alpha_{\kappa_t} &= 2(A_{\kappa_1} + \alpha_{\kappa_1}) \end{aligned}$$

hence the numerator factor becomes

$$2(A_{\kappa_1} + \alpha_{\kappa_1})(A_{\kappa_2} + \alpha_{\kappa_2})\ldots(A_{\kappa_t} + \alpha_{\kappa_t}) - (A_{\kappa_2} + \alpha_{\kappa_2})\ldots(A_{\kappa_t} + \alpha_{\kappa_t})\,2(A_{\kappa_1} + \alpha_{\kappa_1}) = 0.$$

53. This result proves incidentally, as stated above, that the final fraction

$$\frac{s_{\kappa_1} s_{\kappa_2} \ldots s_{\kappa_n} A_{\kappa_1} A_{\kappa_2} \ldots A_{\kappa_n} - P_{\kappa_1 \kappa_2 \ldots \kappa_n}}{(1 - S_1)(1 - S_2)\ldots(1 - S_n)}$$

vanishes identically.

It also terminates the proof which it has been the object of this portion of the investigation to set forth.

54. The analytical result may be stated as follows :—

The fraction

$$\tfrac{1}{2}\frac{1}{\{1 - s_1(2\alpha_1 + \alpha_2 + \ldots + \alpha_n)\}\{1 - s_2(2\alpha_1 + 2\alpha_2 + \ldots + \alpha_n)\}\ldots\{1 - s_n(2\alpha_1 + 2\alpha_2 + \ldots + 2\alpha_\kappa)\}}$$

is equal to the product of the fraction

$$\tfrac{1}{2}\frac{1}{1 - 2(\Sigma s_1\alpha_1 - \Sigma s_1 s_2 \alpha_1 \alpha_2 + \ldots + (-)^{n+1} s_1 s_2 \ldots s_n \alpha_1 \alpha_2 \ldots \alpha_n)}$$

and the series

$$1 + \Sigma\frac{2(A_{\kappa_1} + \alpha_{\kappa_1})(A_{\kappa_2} + \alpha_{\kappa_2})\ldots(A_{\kappa_t} + \alpha_{\kappa_t}) - (A_{\kappa_1} + 2\alpha_{\kappa_1})(A_{\kappa_2} + 2\alpha_{\kappa_2})\ldots(A_{\kappa_t} + 2\alpha_{\kappa_t})}{(1 - S_{\kappa_1})(1 - S_{\kappa_2})\ldots(1 - S_{\kappa_t})}$$

where

$$S_\kappa = s_\kappa(2\alpha_1 + \ldots + 2\alpha_\kappa + \alpha_{\kappa+1} + \ldots + \alpha_n) = s_\kappa(A_\kappa + 2\alpha_\kappa)$$

and the summation is in regard to every selection of t integers from the series $1, 2, 3, \ldots n$, and t takes all values from 1 to $n - 1$.

55. It may be interesting to give the simplest cases at length.

Order 2.

$$\tfrac{1}{2}\frac{1}{\{1 - s_1(2\alpha_1 + \alpha_2)\}\{1 - s_2(2\alpha_1 + 2\alpha_2)\}}$$

$$= \tfrac{1}{2}\frac{1}{1 - 2(s_1\alpha_1 + s_2\alpha_2 - s_1 s_2 \alpha_1 \alpha_2)} \times \left[1 + \frac{s_1\alpha_2}{1 - s_1(2\alpha_1 + \alpha_2)} + \frac{2 s_2 \alpha_1}{1 - s_2(2\alpha_1 + 2\alpha_2)}\right].$$

56. *Order* 3.

$$\frac{1}{2}\frac{1}{\{1-s_1(2\alpha_1+\alpha_2+\alpha_3)\}\{1-s_2(2\alpha_1+2\alpha_2+\alpha_3)\}\{1-s_3(2\alpha_1+2\alpha_2+2\alpha_3)\}}$$

$$=\frac{1}{2}\frac{1}{1-2(s_1\alpha_1+s_2\alpha_2+s_3\alpha_3-s_2s_3\alpha_2\alpha_3-s_3s_1\alpha_3\alpha_1-s_1s_2\alpha_1\alpha_2+s_1s_2s_3\alpha_1\alpha_2\alpha_3)}$$

$$\times\left[1+\frac{s_1(\alpha_2+\alpha_3)}{1-s_1(2\alpha_1+\alpha_2+\alpha_3)}+\frac{s_2(2\alpha_1+\alpha_3)}{1-s_2(2\alpha_1+2\alpha_2+\alpha_3)}+\frac{s_3(2\alpha_1+2\alpha_2)}{1-s_3(2\alpha_1+2\alpha_2+2\alpha_3)}\right.$$

$$+\frac{s_1s_2\alpha_3(2\alpha_1+\alpha_2+\alpha_3)}{\{1-s_1(2\alpha_1+\alpha_2+\alpha_3)\}\{1-s_2(2\alpha_1+2\alpha_2+\alpha_3)\}}$$

$$+\frac{2s_1s_3\alpha_2(\alpha_1+\alpha_2+\alpha_3)}{\{1-s_1(2\alpha_1+\alpha_2+\alpha_3)\}\{1-s_3(2\alpha_1+2\alpha_2+2\alpha_3)\}}$$

$$\left.+\frac{2s_2s_3\alpha_1(2\alpha_1+2\alpha_2+\alpha_3)}{\{1-s_2(2\alpha_1+2\alpha_2+\alpha_3)\}\{1-s_3(2\alpha_1+2\alpha_2+2\alpha_3)\}}\right].$$

57. The general algebraical result, interpreted arithmetically, shows that

$$\frac{1}{2}\frac{1}{\{1-s_1(2\alpha_1+\alpha_2+\ldots+\alpha_n)\}\{1-s_2(2\alpha_1+2\alpha_2+\ldots+\alpha_n)\}\ldots\{1-s_n(2\alpha_1+2\alpha_2+\ldots+2\alpha_n)\}}$$

is a generating function for the enumeration of the compositions of multipartite numbers.

58. The original generating function of the earlier sections has by the addition of ineffective terms become factorized, and can thus be dissected for detailed examination.

The process seems to be analogous to the chemical operation by which the addition of a flux causes an element to be the more easily melted.

As a direct consequence of the geometrical representation of the compositions of multipartite numbers on a reticulation it is of great interest.

59. To resume. The number of compositions of the multipartite number $\overline{p_1p_2\ldots p_n}$ is the coefficient of

$$\alpha_1^{p_1}\alpha_2^{p_2}\ldots\alpha_n^{p_n}$$

in the expanded product

$$\tfrac{1}{2}(2\alpha_1+\alpha_2+\ldots+\alpha_n)^{p_1}(2\alpha_1+2\alpha_2+\ldots+\alpha_n)^{p_2}\ldots(2\alpha_1+2\alpha_2+\ldots+2\alpha_n)^{p_n},$$

or we may say that it is the coefficient of the symmetric function

$$\Sigma\,\alpha_1^{p_1}\alpha_2^{p_2}\ldots\alpha_n^{p_n}$$

in the development of the symmetric function

$$\tfrac{1}{2}\Sigma(2\alpha_1+\alpha_2+\ldots+\alpha_n)^{p_1}(2\alpha_1+2\alpha_2+\ldots+\alpha_n)^{p_2}\ldots(2\alpha_1+2\alpha_2+\ldots+2\alpha_n)^{p_n}$$

in a sequence of monomial symmetric functions.

60. The unsymmetrical form gives a direct connection between the numbers of the compositions and of the permutations of the letters in the product

$$\alpha_1^{p_1}\alpha_2^{p_2}\ldots\alpha_n^{p_n}.$$

In the unsymmetrical product the term $\alpha_1^{p_1}\alpha_2^{p_2}\ldots\alpha_n^{p_n}$, attached to some numerical coefficient, arises in connection with every permutation of the letters in the term. One factor of such coefficient is manifestly

$$2^{p_1-1};$$

if the letter α_2, in a particular permutation, occur q_2 times in the last $p_2+p_3+\ldots+p_n$ places of the permutation there will be a factor

$$2^{q_2};$$

further, if the letter α_s, in the same permutation, occur q_s times in the last $p_s+p_{s+1}+\ldots+p_n$ places of the permutation there will be a factor

$$2^{q_s}.$$

Hence for the permutation considered there arises a term

$$2^{p_1-1+q_2+q_3+\ldots+q_n},$$

and the number of compositions must therefore be

$$2^{p_1-1}\Sigma\, 2^{q_2+q_3+\ldots+q_n},$$

the summation being taken in regard to every permutation of the letters in the product

$$\alpha_1^{p_1}\alpha_2^{p_2}\ldots\alpha_n^{p_n}.$$

E.g., to find in this way the number of compositions of the bipartite $\overline{22}$, we have the following scheme :—

			q_2
$p_1=2$	$\alpha_1\alpha_1$	$\alpha_2\alpha_2$	2
	$\alpha_1\alpha_2$	$\alpha_1\alpha_2$	1
	$\alpha_1\alpha_2$	$\alpha_2\alpha_1$	1
	$\alpha_2\alpha_1$	$\alpha_1\alpha_2$	1
	$\alpha_2\alpha_1$	$\alpha_2\alpha_1$	1
	$\alpha_2\alpha_2$	$\alpha_1\alpha_1$	0

Therefore

$$F(22) = 2^{2-1}(2^2+2^1+2^1+2^1+2^1+2^0) = 26,$$

61. The unsymmetrical product may be written in the form

$$2^{p_1+p_2+\ldots+p_n-1}\{\alpha_1+\tfrac{1}{2}(\alpha_2+\ldots+\alpha_n)\}^{p_1}\{\alpha_1+\alpha_2+\tfrac{1}{2}(\alpha_3+\ldots+\alpha_n)\}^{p_2}\ldots$$
$$\ldots\{\alpha_1+\alpha_2+\ldots+\alpha_n\}^{p_n}$$

and herein the coefficient of $\alpha_1^{p_1}\alpha_2^{p_2}\ldots\alpha_n^{p_n}$ may be written

$$2^{\Sigma p-1}\overset{s}{\Sigma}\, C_s(\tfrac{1}{2})^s=\overset{s}{\Sigma}C_s2^{\Sigma p-s-1},$$

where C_0, C_1, C_2, ... are certain integers.

Observe that C_s is the coefficient of $\lambda^s\alpha_1^{p_1}\alpha_2^{p_2}\ldots\alpha_n^{p_n}$ in the product

$$\{\alpha_1+\lambda(\alpha_2+\ldots+\alpha_n)\}^{p_1}\{\alpha_1+\alpha_2+\lambda(\alpha_3+\ldots+\alpha_n)\}^{p_2}\ldots\{\alpha_1+\alpha_2+\ldots+\alpha_n\}^{p_n}$$

Considering the reticulation of the multipartite number $\overline{p_1p_2\ldots p_n}$, suppose there to be D_s lines of route which possess exactly s essential nodes. Upon these lines of route are represented

$$D_s2^{\Sigma p-s-1}$$

distinct compositions, and the whole number of compositions is manifestly

$$\Sigma\, D_s2^{\Sigma p-s-1}.$$

Hence

$$\Sigma\, D_s2^{\Sigma p-s-1}=\Sigma C_s2^{\Sigma p-s-1},$$

or $D_s=C_s$ presumably (see *post*, Sec. 5, Art. 71), that is to say, the number of distinct lines of route through the reticulation which possess exactly s essential nodes is the coefficient of

$$\lambda^s\alpha_1^{p_1}\alpha_2^{p_2}\ldots\alpha_n^{p_n}$$

in the product

$$\{\alpha_1+\lambda(\alpha_2+\ldots+\alpha_n)\}^{p_1}\{\alpha_1+\alpha_2+\lambda(\alpha_3+\ldots+\alpha_n)\}^{p_2}\ldots\{\alpha_1+\alpha_2+\ldots+\alpha_n\}^{p_n}.$$

62. To interpret the product arithmetically we find that for a particular permutation of the letters in

$$\alpha_1^{p_1}\alpha_2^{p_2}\ldots\alpha_n^{p_n}$$

the factor λ^s will occur if

α_2 occurring	r_2 times	in the first	p_1	places of the permutation
α_3 ,,	r_3	,,	p_1+p_2	,, ,,
. . .	. . .	. . .	. . .	. . .
α_n ,,	r_n	,,	$p_1+p_2+\ldots+p_{n-1}$	,, ,,

the sum $r_2+r_3+\ldots+r_n=s$.

63. Hence the number of lines of route through the reticulation which possess exactly s essential nodes is equal to the number of permutations of the letters in

$$\alpha_1^{p_1}\alpha_2^{p_2}\ldots\alpha_n^{p_n},$$

for which

$$r_2 + r_3 + \ldots + r_n = s;$$

r_t denoting the number of times that the letter α_t occurs in the first

$$p_1 + p_2 + \ldots + p_{t-1}$$

places of permutation.

64. We are at once led to an important identity of enumeration in the pure theory of permutations. Following the nomenclature of a previous section of the investigation, a line of route through the reticulation is traced out by a succession of steps, each step being an α_1 step or an α_2, &c., or an α_n step.

The whole length of the line of route there are altogether p_1, α_1 steps, p_2, α_2 steps, &c., . . . p_n, α_n steps.

Without regard to the characteristic of lines of route in respect of essential nodes, the whole number of lines of route is equal to the number of different orders in which these steps can be taken, viz., equal to the whole number of permutations of the letters in

$$\alpha_1^{p_1}\alpha_2^{p_2}\ldots\alpha_n^{p_n}.$$

An essential node occurs whenever a step α_u *immediately* precedes a step α_t, where $u > t$.

In the corresponding permutation there is a contact

$$\alpha_u\alpha_t, \quad u > t,$$

which may be called a *major* contact.

Hence a line of route with s essential nodes is represented by a permutation with s major contacts. We have thus the theorem :—

"In the reticulation of the multipartite number $\overline{p_1p_2\ldots p_n}$ the number of lines of route which possess exactly s essential nodes is equal to the number of permutations of the letters in the product

$$\alpha_1^{p_1}\alpha_2^{p_2}\ldots\alpha_n^{p_n},$$

which possess exactly s major contacts."

"Calling a contact $\alpha_u\alpha_t$ a major contact when $u > t$ the number of permutations of the letters in the product

$$\alpha_1^{p_1}\alpha_2^{p_2}\ldots\alpha_n^{p_n},$$

5 U 2

which possess exactly s major contacts, is given by the coefficient of

$$\lambda^s \alpha_1^{p_1} \alpha_2^{p_2} \ldots \alpha_n^{p_n}$$

in the product

$$\{\alpha_1 + \lambda(\alpha_2 + \ldots + \alpha_n)\}^{p_1} \{\alpha_1 + \alpha_2 + \lambda(\alpha_3 + \ldots + \alpha_n)\}^{p_2} \ldots \{\alpha_1 + \alpha_2 + \ldots + \alpha_n\}^{p_n}."$$

I am not aware if this problem has been previously solved in this form, or, indeed, if it has ever been attacked before.*

65. "The number of permutations of the letters in the product

$$\alpha_1^{p_1} \alpha_2^{p_2} \ldots \alpha_n^{p_n},$$

which possess exactly s major contacts, is equal to the number of permutations for which

$$r_2 + r_3 + \ldots + r_n = s,$$

r_t denoting the number of times that the latter α_t occurs in the first

$$p_1 + p_2 + \ldots + p_{t-1}$$

places of the permutation."

66. It is easy to obtain a refinement of these theorems which has been foreshadowed by the detailed bipartite and tripartite cases which have preceded. Major contacts in a permutation, and, consequently, also the essential nodes along a line of route, are of $\binom{n}{2}$ different kinds.

Consider the product

$$(\alpha_1 + \lambda_{21}\alpha_2 + \lambda_{31}\alpha_3 + \ldots + \lambda_{n1}\alpha_n)^{p_1} (\alpha_1 + \alpha_2 + \lambda_{32}\alpha_3 + \lambda_{42}\alpha_4 + \ldots + \lambda_{n2}\alpha_n)^{p_2} \ldots (\alpha_1 + \alpha_2 + \ldots + \lambda_{n,n-1}\alpha_n)^{p_{n-1}} (\alpha_1 + \alpha_2 + \ldots + \alpha_n)^{p_n}.$$

The coefficient of $\alpha_1^{p_1}\alpha_2^{p_2} \ldots \alpha_n^{p_n}$ consists of a number of monomial products of the quantities λ, each attached to a numerical coefficient. Of these products a certain number are of a definite degree s. The sum of the coefficients of these products gives the number of permutations of the letters in

$$\alpha_1^{p_1}\alpha_2^{p_2} \ldots \alpha_n^{p_n}$$

which possess exactly s major contacts.

* See a paper by H. FORTY, M.A., "On Contact and Isolation, a Problem in Permutations," 'Proceedings London Mathematical Society,' vol. 15.

Let one such product with its attached numerical coefficient be

$$\kappa\lambda_{x_1 1}{}^{\xi_1}\lambda_{x_2 1}{}^{\xi_2}\ldots\lambda_{y_1 2}{}^{\eta_1}\lambda_{y_2}{}^{\eta_2}\ldots\lambda_{z_1 3}{}^{\zeta_1}\lambda_{z_2 3}{}^{\zeta_2}\ldots$$

The number κ enumerates a certain number of the permutations which have s major contacts, and $\Sigma\kappa$ enumerates the entire number.

The problem is to determine the particular permutations enumerated by the number κ.

If p_1 and p_2 be both greater than zero, the contact $\alpha_2\alpha_1$ must occur in one or more of the enumerated permutations.

If p_1 be zero, λ_{21} is absent, while if p_2 be zero, α_2 has no existence, and also λ_{21} does not appear.

Hence, λ_{21} only appears when both p_1 and p_2 are greater than zero. It follows that the products which involve λ_{21} only enumerate by their coefficients those permutations which possess $\alpha_2\alpha_1$ contacts. In fact, the absence of either α_1 or α_2, by causing λ_{21} to vanish, diminishes the total number of permutations which have a definite number of major contacts; this would not be the case if the terms comprising λ_{21} as a factor did not enumerate $\alpha_2\alpha_1$ contacts; if $\alpha_2\alpha_1$ contacts were not enumerated, the vanishing of α_1 or α_2 would not alter the enumeration of the permutations possessing a given number of contacts, for this given number would be complete without $\alpha_2\alpha_1$ contacts at all.

It is next to be shown that a product involving $\lambda_{21}{}^{s_{21}}$ comprises exactly s_{21}, $\alpha_2\alpha_1$ contacts.

Suppose $p_1 \nless s_{21}$, $p_2 < s_{21}$.

Then α_2 occurs less than s_{21} times.

And, therefore, $\lambda_{21}\alpha_2$ cannot be raised to the power s_{21}.

And, therefore, λ_{21} cannot occur in any term to so high a power as s_{21}.

Similarly if $p_1 < s_{21}$ it is evident that λ_{21} cannot occur to so high a power as s_{21}. Therefore, unless both p_1 and p_2 are at least as great as s_{21}, λ_{21} cannot occur in any term to so high a power as s_{21}. It follows that the products which involve $\lambda_{21}{}^{s_{21}}$ cannot enumerate permutations possessing fewer than s_{21}, $\alpha_2\alpha_1$ contacts. For suppose that the permutations enumerated possessed only σ_{21}, $\alpha_2\alpha_1$ contacts where $\sigma_{21} < s_{21}$. If p_1 or p_2 be diminished so as to be less than s_{21} the permutations possessing σ_{21}, $\alpha_2\alpha_1$ contacts would not *necessarily* be diminished in number, whilst those possessing s_{21} such contacts as well as the products involving $\lambda_{21}{}^{s_{21}}$ would certainly vanish. Hence the products involving $\lambda_{21}{}^{s_{21}}$ cannot, through their coefficients, enumerate permutations possessing fewer than s_{21}, $\alpha_2\alpha_1$ contacts. Hence the number κ enumerates permutations having at least

$$\begin{array}{lll} \xi_1, & \alpha_{x_1}\alpha_1 & \text{contacts} \\ \xi_2, & \alpha_{x_2}\alpha_1 & ,, \\ \cdot & \cdot\ \cdot\ \cdot & \cdot \\ \eta_1, & \alpha_{y_1}\alpha_2 & ,, \\ \eta_2, & \alpha_{y_2}\alpha_2 & ,, \\ \cdot & \cdot\ \cdot\ \cdot & \cdot \\ \zeta_1, & \alpha_{z_1}\alpha_3 & ,, \\ \zeta_2, & \alpha_{z_2}\alpha_3 & ,, \\ \cdot & \cdot\ \cdot\ \cdot & \cdot \end{array}$$

and these numbers must represent also the exact numbers of the contacts in question because the sum

$$\Sigma\,\xi + \Sigma\,\eta + \Sigma\,\zeta + \ldots$$

gives the whole number of contacts under consideration.

The number κ obviously denotes the number of lines of route through the reticulation of the multipartite

$$\overline{p_1 p_2 \ldots p_n}$$

which possess

$$\begin{array}{lllll} \xi_1 & \text{essential nodes of the kinds} & & & \alpha_{x_1}\alpha_1 \\ \xi_2 & ,, & ,, & ,, & \alpha_{x_2}\alpha_1 \end{array}$$

and so forth.

67. By an easy arithmetical interpretation the number κ also denotes the number of permutations in which

α_{x_1} occurs ξ_1 times in the first p_1 places.
α_{x_2} ,, ξ_2 ,, ,, ,, ,,
. . . .
α_{y_1} ,, η_1 ,, between the p_1^{th} and p_2^{th} places.
α_{y_2} ,, η_2 ,, ,, ,, ,, ,,
. . . .
α_{z_1} ,, ζ_1 ,, ,, ,, p_2^{th} ,, p_3^{th} ,,
α_{z_2} ,, ζ_2 ,, ,, ,, ,, ,, ,, ,,
. . . .

§ 5. *Extension of the idea of Composition.*

68. The idea of composition is capable of enlargement from a particular point of view.

In regard to unipartite numbers, consider p units placed in a row

$$1\ 1\ 1\ 1\ 1\ 1\ldots,$$

there are $p-1$ spaces between them which may be occupied by algebraic symbols at pleasure. We may select a definite number of different symbols, and choose any one to occupy any one of the $p-1$ spaces. If this definite number be k, we can, by filling each space at pleasure, arrive at k^{p-1}, different expressions involving the p units.

Clearly we may take as one of these symbols the simple unoccupied blank space,* since such space left between any two numbers, quantities, or expressions of any kinds has, in every case, a well-understood signification, not always, however, the same, in mathematical work.

If we restrict ourselves to a single symbol, only one expression involving the units is possible; choosing this symbol to be that which indicates addition we merely get

$$1+1+1+1+1+1+\ldots,$$

which is p, or the number which enumerates the units. Had the chosen symbol been that denoting subtraction, the expression would have denoted $-(p-2)$; a blank space would have yielded a succession of p units; the symbol of multiplication, unity, and so forth.

All the modes of obtaining expressions from the p units may be called combinations of the first order in respect of the p units.

Passing to the case of two different symbols we may choose to employ the sign of addition and the blank space. We thus obtain 2^{p-1} different expressions which are the several compositions of the number p.

In general, expressions obtained by choosing from any k different symbols may be called combinations of order k in respect of the p fundamental units.

69. There is a one-to-one correspondence between these combinations and the "Trees" which have an altitude k and p terminal knots.

As an example take k and p each equal to 3, and, further, take as symbols, the sign of addition, the blank space and the symbol | of unspecified signification.

The correspondence between the nine trees and the nine combinations of order three is shown below:—

* The unoccupied blank space might be represented by a definite symbol such as O, as suggested to me by one of the referees.

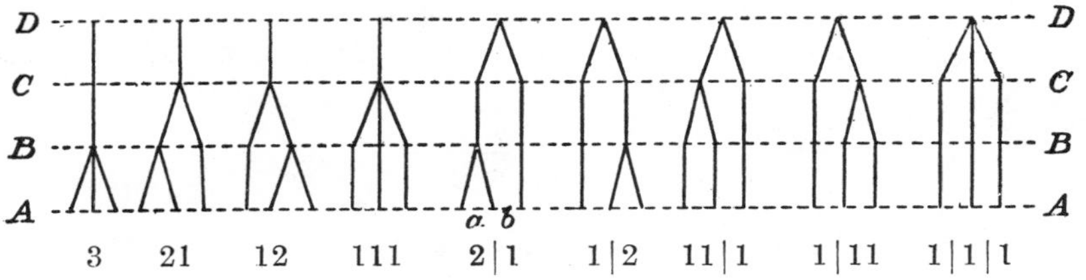

The process consists in writing down a unit for each terminal knot. The $p-1$ intervals between the units correspond to the $p-1$ inter-terminal-knot spaces. If a space leads to a bifurcation in row B the symbol is $+$; if in row C it is a blank space; if in row D it is $|$.

Thus, take the fifth tree above, we have

$$1+1|1$$

for the space a leads to a bifurcation in row B, giving the symbol $+$ and the space b leads to a bifurcation in row D giving the symbol $|$. Hence the corresponding combination

$$2|1.$$

On the same principle a tree can always be drawn to represent any combination of order k in respect of p units.

70. Passing to the case of multipartite numbers, consider tripartite numbers as representative of the general case. Arrange a row of multipartite units of the three kinds, viz.:—

$$100 \quad 100 \quad 100 \ldots 010 \quad 010 \quad 010 \ldots 001 \quad 001 \quad 001 \ldots$$

and employing two different symbols, viz., the sign of addition and the blank space, we arrive at a certain composition of the tripartite $\overline{p_1p_2p_3}$, supposing the numbers of the units 100, 010, 001 that appear in the row to be p_1, p_2, and p_3 respectively. Without altering the positions of the symbols introduced we may change the order of the units 100, 010, 001, and thus obtain other compositions. Permutations of these units between contiguous blank spaces are not permissible, so that for fixed positions of the pluses and blank spaces we do not obtain in general

$$\frac{(p_1+p_2+p_3)!}{p_1!\,p_2!\,p_3!}$$

compositions, but some lesser number.

This arises from the commutative nature of the symbol $+$.

Had neither of the symbols employed been such as obey the commutative law of algebra, the whole number of combinations of the second order would have been simply

$$2^{p_1+p_2+p_3-1}\frac{(p_1+p_2+p_3)!}{p_1!\,p_2!\,p_3!}.$$

Thus, in this case, the symbol + introduces a complexity into the theory of compositions.

Choosing similarly from k different symbols, none of which are commutative, there are

$$k^{p_1+p_2+p_3-1}\frac{(p_1+p_2+p_3)!}{p_1!\,p_2!\,p_3!},$$

combinations of order k, and the general generating function may clearly be written

$$\frac{1}{k}\cdot\frac{1}{1-k(\alpha_1+\alpha_2+\ldots+\alpha_n)};$$

but there is a lesser number of combinations if one or more of the symbols be commutative.

71. For the purpose of this investigation the most interesting case to consider is that in which one symbol, viz., the blank space, is non-commutative, and the remaining $k-1$ symbols commutative. This species of combination is not brought under view merely for the purpose of adding and discussing new complexity. Its introduction is absolutely vital to the investigation as clinching and confirming a conclusion which was momentarily assumed during the consideration of the compositions of multipartite numbers. (See *ante*, Art. 61.) Consider the reticulation of a multipartite number. That of a bipartite number will suffice as indicative of the general case.

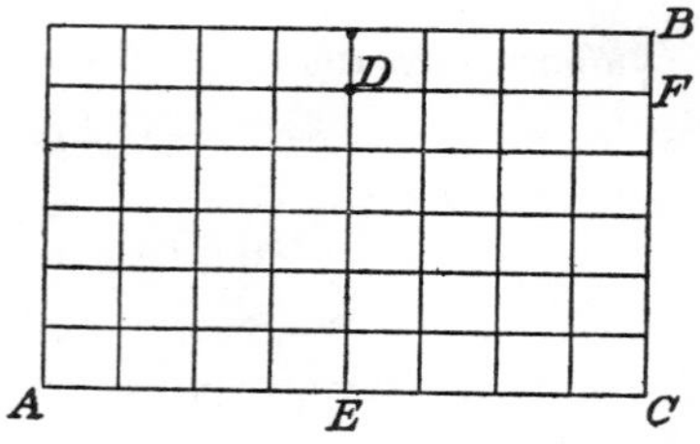

A combination of order k (of the nature under view) is regarded as having *m parts* when $m-1$ blank space symbols occur in the combination.

Let us first enquire how many combinations possess only a single part. We require nodes of k different kinds, viz., a blank space node and $k-1$ other different nodes. A blank space node may be either essential or non-essential, but an essential node must be also a blank space node. The only line of route through the reticulation which does not involve an essential (that is a blank space) node is ACB. Consequently the graphs of all the combinations having but one part must be along the line of route ACB. Similarly in the reticulation of the multipartite number

$$\overline{p_1 p_2 \ldots p_n},$$

the graphs of all the one-part combinations will be along the line of route traced out by p_1, α_1 segments, p_2, α_2 segments, and so on in succession. The $k-1$ different symbols at disposal may be placed at pleasure at $p_1 + p_2 + \ldots + p_n - 1$ points along this line of route. Hence

$$(k-1)^{p_1 + p_2 + \ldots + p_n - 1}$$

different one-part combinations are obtainable.

In other words the generating function for such combinations is

$$h_1 + (k-1)\,h_2 + (k-1)^2 h_3 + \ldots + (k-1)^{n'-1} h_{n'} \quad (n' = \infty,$$

where in a previous notation h_n is the homogeneous product sum, degree n, of the quantities

$$\alpha_1,\ \alpha_2,\ \ldots\ \alpha_n.$$

72. Next as to the combinations which have two parts. At any point D of the reticulation place a blank space node. All two-part combinations whose graphs pass through D must follow the line of route AEDFB, for otherwise an additional essential (and blank space) node would be introduced. The whole combination may be split up into a one-part combination along the line of route AED, followed first by a blank space, and then by a one-part combination along the line of route DFB. All the two part combinations whose graphs pass through the point D are obtained by associating every one-part combination in the reticulation AD with every one-part combination in the reticulation DB.

Hence the whole number of combinations, having two parts, of the multipartite number

$$\overline{p_1 p_2 \ldots p_n}$$

is

$$\Sigma\,(k-1)^{p'_1 + p'_2 + \ldots + p'_n - 1}(k-1)^{p''_1 + p''_2 + \ldots + p''_n - 1}$$

where

$$p'_1 + p''_1 = p_1;\ \ p'_2 + p''_2 = p_2,\ \ldots\ p'_n + p''_n = p_n.$$

Hence, the generating function for two-part combinations is

$$\{h_1 + (k-1)\,h_2 + (k-1)^2 h_3 + \ldots + (k-1)^{n'-1} h_{n'}\}^2. \qquad n' = \infty.$$

73. Pursuing this chain of reasoning it is completely manifest that the generating function of m-part combinations is

$$\{h_1 + (k-1)h_2 + (k-1)^2 h_3 + \ldots + (k-1)^{n'-1} h_{n'}\}^m \qquad (n' = \infty,$$

and hence the complete generating function of combinations of the multipartite number

$$\overline{p_1 p_2 \cdot \quad \cdot p_n}$$

is

$$\frac{h_1 + (k-1)h_2 + \ldots + (k-1)^{n'-1}h_{n'}}{1 - h_1 - (k-1)h_2 - \ldots - (k-1)^{n'-1}h_{n'}} \qquad (n' = \infty,$$

which is effectively the same as

$$\frac{1}{k} \cdot \frac{1}{1 - k\Sigma\alpha_1 + k(k-1)\Sigma\alpha_1\alpha_2 - k(k-1)^2\Sigma\alpha_1\alpha_2\alpha_3 + \ldots + (-)^n k(k-1)^{n-1}\alpha_1\alpha_2\ldots\alpha_n};$$

for the latter is obtained by adding the fraction $1/k$ to the former and then transforming from homogeneous product sums to elementary symmetric functions.

Just as in the case of $k = 2$, corresponding to compositions, this generating function admits of an important transformation to a factorized redundant form.

74. In the above fraction put $s_1\alpha_1, s_2\alpha_2, \ldots s_n\alpha_n$ for $\alpha_1, \alpha_2, \ldots \alpha_n$ respectively; it may then be written

$$\frac{k-1}{k} \cdot \frac{1}{-1 + k\{1-(k-1)s_1\alpha_1\}\{1-(k-1)s_2\alpha_2\}\ldots\{1-(k-1)s_n\alpha_n\}} = \frac{1}{k\mathrm{N}}.$$

For brevity put

$$\mathrm{S}_t = s_t(k\alpha_1 + k\alpha_2 + \ldots + k\alpha_t + \alpha_{t+1} + \ldots + \alpha_n) = s_t(\mathrm{A}_t + k\alpha_t)$$

$$b_1 = \Sigma s_1\alpha_1, \quad b_2 = \Sigma s_1 s_2 \alpha_1\alpha_2, \quad \&\text{c.},$$

$$\mathrm{M} = (1 - ks_1\alpha_1)(1 - ks_2\alpha_2)\ldots(1 - ks_n\alpha_n),$$

and

$$\frac{1}{k} \cdot \frac{1}{(1-\mathrm{S}_1)(1-\mathrm{S}_2)\ldots(1-\mathrm{S}_n)} = \frac{1}{k\mathrm{D}}.$$

It will be shown that $\frac{1}{k\mathrm{D}}$ is a generating function equivalent to the former in regard to terms which are products of powers of

$$s_1\alpha_1, s_2\alpha_2, \ldots s_n\alpha_n;$$

5 X 2

for

$$M-N=\{k^2-k(k-1)\}b_2-\{k^3-k(k-1)^2\}b_3+\ldots+(-)^n\{k^n-k(k-1)^{n-1}\}b_n,$$

and thence

$$\frac{N}{D}=\frac{M}{D}-\frac{(M-N)}{D}=\prod_1^n\left(1+\frac{s_tA_t}{1-S_t}\right)-\frac{\{k^2-k(k-1)\}b_2-\ldots+(-)^n\{k^n-k(k-1)^{n-1}\}b_n}{(1-S_1)(1-S_2)\ldots(1-S_n)}.$$

75. The investigation proceeds precisely as in the case of $k=2$ with the following result. The fraction

$$\frac{1}{k}\cdot\frac{1}{\{1-s_1(k\alpha_1+\alpha_2+\ldots+\alpha_n)\}\{1-s_2(k\alpha_1+k\alpha_2+\ldots+\alpha_n)\}\ldots\{1-s_n(k\alpha_1+k\alpha_2+\ldots+k\alpha_n)\}}$$

is equal to the product of the fraction

$$\frac{1}{k}\cdot\frac{1}{1-k\Sigma s_1\alpha_1+k(k-1)\Sigma s_1s_2\alpha_1\alpha_2-\ldots+(-)^nk(k-1)^{n-1}s_1s_2\ldots s_n\alpha_1\alpha_2\ldots\alpha_n},$$

and the series

$$1+\Sigma\frac{k(A_{t_1}+\alpha_{t_1})(A_{t_2}+\alpha_{t_2})\ldots(A_{t_u}+\alpha_{t_u})-(A_{t_1}+k\alpha_{t_1})(A_{t_2}+k\alpha_{t_2})\ldots(A_{t_u}+k\alpha_{t_u})}{(k-1)(1-S_{t_1})(1-S_{t_2})\ldots(1-S_{t_u})},$$

the summation having regard to every selection of u integers from the series $1, 2, 3, \ldots n$, and u takes all values from 1 to $n-1$.

As in the former case the relation

$$A_t+k\alpha_t=A_{t+1}+\alpha_{t+1},$$

which becomes

$$A_{t_u}+k\alpha_{t_u}=k(A_{t_1}+\alpha_{t_1})$$

where t_u and t_1 are the highest and lowest suffixes present, shows that the terms under the summation sign do not involve any products of $s_1\alpha_1, s_2\alpha_2, \ldots s_n\alpha_n$ only, and therefore as far as concerns the generating function may be put equal to zero.

76. Hence the number of combinations of order k of the multipartite number

$$\overline{p_1p_2\ldots p_n}$$

is the coefficient of $(s_1\alpha_1)^{p_1}(s_2\alpha_2)^{p_2}\ldots(s_n\alpha_n)^{p_n}$ in the expansion of the generating function

$$\frac{1}{k}\cdot\frac{1}{\{1-s_1(k\alpha_1+\alpha_2+\ldots+\alpha_n)\}\{1-s_2(k\alpha_1+k\alpha_2+\ldots+\alpha_n)\}\ldots\{1-s_n(k\alpha_1+\ldots+k\alpha_n)\}}$$

that is to say it is the coefficient of $\alpha_1^{p_1}\alpha_2^{p_2}\ldots\alpha_n^{p_n}$ in the product

$$\frac{1}{k}(k\alpha_1+\alpha_2+\ldots+\alpha_n)^{p_1}(k\alpha_1+k\alpha_2+\alpha_3+\ldots+\alpha_n)^{p_2}\ldots(k\alpha_1+k\alpha_2+\ldots+k\alpha_n)^{p_n},$$

which may be written

$$k^{p_1+p_2+\ldots+p_n-1}\{\alpha_1+\frac{1}{k}(\alpha_2+\ldots+\alpha_n)\}^{p_1}\{\alpha_1+\alpha_2+\frac{1}{k}(\alpha_3+\ldots+\alpha_n)\}^{p_2}$$
$$\ldots\{\alpha_1+\alpha_2+\ldots+\alpha_n\}^{p_n}.$$

The coefficient of $\alpha_1^{p_1}\alpha_2^{p_2}\ldots\alpha_n^{p_n}$ is

$$\overset{s}{\Sigma}\, C_s\, k^{\Sigma p-s-1},$$

where C_s is the coefficient of

$$\lambda^s\,\alpha_1^{p_1}\alpha_2^{p_2}\ldots\alpha_n^{p_n}$$

in the product

$$\{\alpha_1+\lambda(\alpha_2+\ldots+\alpha_n)\}^{p_1}\{\alpha_1+\alpha_2+\lambda(\alpha_3+\ldots+\alpha_n)\}^{p_2}\ldots\{\alpha_1+\alpha_2+\ldots+\alpha_n\}^{p_n}.$$

If in the reticulation of the multipartite number there be D_s lines of route which possess exactly s essential nodes,

$$D_s\, k^{\Sigma p-s-1}$$

combinations of order k may be represented upon these lines. Hence the whole number of combinations is

$$\overset{s}{\Sigma}\, D_s\, k^{\Sigma p-s-1}.$$

Hence

$$\overset{s}{\Sigma}\, D_s\, k^{\Sigma p-s-1}=\overset{s}{\Sigma}\, C_s\, k^{\Sigma p-s-1}$$

a relation which is true for *all positive integral values of k.*

77. Hence

$$D_s=C_s$$

the important relation temporarily assumed in the investigation concerning compositions. (Art. 61.)

78. The theorem thus established, viz., that the number of distinct lines of route through the reticulation of the multipartite

$$\overline{p_1p_2\ldots p_n},$$

which possess exactly s essential nodes, is given by the coefficient of

$$\lambda^s\alpha_1^{p_1}\alpha_2^{p_2}\ldots\alpha_n^{p_n}$$

in the product

$$\{\alpha_1 + \lambda(\alpha_2 + \ldots + \alpha_n)\}^{p_1} \{\alpha_1 + \alpha_2 + \lambda(\alpha_3 + \ldots + \alpha_n)\}^{p_2} \ldots \{\alpha_1 + \alpha_2 + \ldots + \alpha_n\}^{p_n}$$

gives the theorem in compositions :—

"The number of compositions of the multipartite

$$\overline{p_1 p_2 \ldots p_n}$$

which possess exactly s contacts of

$$n - u \text{ zeros with } \quad - 1 \text{ zeros,}$$

subject to the condition $u > t$, is

$$C_s 2^{p_1+p_2+ \ldots +p_n-s-1}$$

where C_s is the coefficient of

$$\lambda^s \alpha_1^{p_1} \alpha_2^{p_2} \ldots \alpha_n^{p_n}$$

in the above-mentioned product."

79. Also the further theorem :—

"The number of compositions of the multipartite

$$\overline{p_1 p_2 \ldots p_n}$$

which possess exactly

$s_{u_1 t_1}$ contacts of $n - u_1$ zeros with $t_1 - 1$ zeros.

$s_{u_2 t_2}$,, $n - u_2$,, $t_2 - 1$,,

. . . .

is the product of

$$2^{p_1+p_2+ \ldots +p_n-s-1} \qquad [s = \Sigma s_{ut}]$$

and the coefficient of

$$\lambda_{u_1 t_1}^{s_{u_1 t_1}} \lambda_{u_2 t_2}^{s_{u_2 t_2}} \ldots \alpha_1^{p_1} \alpha_2^{p_2} \ldots \alpha_n^{p_n}$$

in the product

$$(\alpha_1 + \lambda_{21}\alpha_2 + \ldots + \lambda_{n1}\alpha_n)^{p_1} (\alpha_1 + \alpha_2 + \lambda_{32}\alpha_3 + \ldots + \lambda_{n2}\alpha_n)^{p_2} \ldots (\alpha_1 + \alpha_2 + \ldots + \alpha_n)^{p_n}.$$

80. The generating function which enumerates the combinations of order k of multipartite numbers may be written

$$\frac{1}{k} + \frac{H'}{k-1} + \frac{H'^2}{(k-1)^2} + \frac{H'^3}{(k-1)^3} + \ldots + \frac{H'^m}{(k-1)^m} + \ldots$$

where

$$\alpha'_1 = (k-1)\,\alpha_1, \qquad \alpha'_2 = (k-1)\,\alpha_2, \text{ \&c.}$$

h'_m is the homogeneous product sum of degree m of the quantities

$$\alpha'_1, \alpha'_2, \ldots \alpha'_n,$$

and

$$H' = h'_1 + h'_2 + h'_3 + \ldots$$

The coefficient of

$$\alpha_1^{p_1}\alpha_2^{p_2}\ldots\alpha_n^{p_n}$$

in

$$\frac{H'^m}{(k-1)^m}$$

enumerates, in respect of the multipartite $\overline{p_1p_2\ldots p_n}$, the number of combinations having m parts.

From previous work this coefficient is

$$(k-1)^{p_1+p_2+\ldots+p_n-m} f(p_1p_2\ldots p_n, m),$$

where

$$f(p_1p_2\ldots p_n, m) = \binom{p_1+m-1}{p_1}\binom{p_2+m-1}{p_2}\ldots\binom{p_n+m-1}{p_n}$$
$$-\binom{m}{1}\binom{p_1+m-2}{p_1}\binom{p_2+m-2}{p_2}\ldots\binom{p_2+m-2}{p_n}$$
$$+ \text{\&c.},$$

to m terms.

81. Hence the whole number of combinations is

$$(k-1)^{\Sigma p-1} f(p_1p_2\ldots p_n, 1) + (k-1)^{\Sigma p-2} f(p_1p_2\ldots p_n, 2) + .$$

a result which is immediately obtainable from the reticulation.

82. There exists a very interesting correspondence between the compositions of the multipartite

$$\overline{1^{n-1}}$$

into k parts, zeros not excluded, and the combinations of order k of the unipartite number

$$n,$$

zeros excluded.

The generating function for the compositions of multipartite numbers into k parts zeros not excluded, is

$$(1 + h_1 + h_2 + \ldots)^k = (1-\alpha_1)^{-k}(1-\alpha_2)^{-k}\ldots(1-\alpha_n)^{-k},$$

hence the number in the case of $\overline{p_1p_2\ldots p_n}$ is

$$\binom{k+p_1-1}{p_1}\binom{k+p_2-1}{p_2}\binom{k+p_3-1}{p_3}\ldots\binom{k+p_n-1}{p_n},$$

which for the multipartite $\overline{1^{n-1}}$ is

$$k^{n-1}.$$

This expression also gives (*ante*) the number of combinations of order k of the unipartite number

$$n,$$

zeros excluded.

83. The correspondence may be shown by reference to the "Theory of Trees."

The trees to be considered are of altitude k and have n terminal knots.

For simplicity take $n = k = 3$.

The trees are

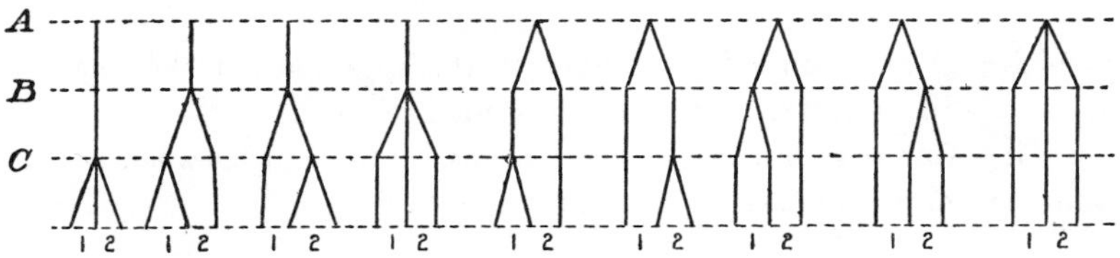

which, as shown above, represent combinations of order 3 of the unipartite number 3. These are

$$3 \quad 21 \quad 12 \quad 111 \quad 2|1 \quad 1|2 \quad 11|1 \quad 1|11 \quad 1|1|1,$$

the number of parts (observe that blank space symbols are between adjacent parts) being equal to the number of bifurcations in the second row from the top, increased by unity.

Now these trees may be interpreted so as to represent the compositions of the multipartite $(\overline{1^2})$ into three parts, zero parts not excluded, by attending to the connection between the two bifurcations of each tree and the two inter-terminal-knot spaces.

In any row of knots in a tree we have or we have not bifurcations, and each bifurcation communicates either with the first or with the second inter-terminal-knot space.

Beginning with the row marked A, if we find no bifurcation we write $\overline{00}$; if we find a bifurcation communicating with the first space but not with the second we write $\overline{10}$; if with the second space and not with the first $\overline{01}$; if there be two bifurcations, which necessarily in the present case communicate with both spaces, we write $\overline{11}$; and proceeding in the same way with the rows B and C in succession we will have finally written down three bipartite parts constituting a composition of the multipartite $\overline{11}$.

We thus obtain, beginning with the left hand tree,

$$(\overline{00}\,\overline{00}\,\overline{11}),\ (\overline{00}\,\overline{01}\,\overline{10}),\ (\overline{00}\,\overline{10}\,\overline{01}),\ (\overline{00}\,\overline{11}\,\overline{00}),\ (\overline{01}\,\overline{00}\,\overline{10}),\ (\overline{10}\,\overline{00}\,\overline{01}),$$
$$(\overline{01}\,\overline{10}\,\overline{00}),\ (\overline{10}\,\overline{01}\,\overline{00}),\ (\overline{11}\,\overline{00}\,\overline{00}),$$

in order of correspondence with the above written combinations, of order 3, of the unipartite number 3.

84. The process is perfectly general. Every tree of altitude k is representative alike of a combination of order k of a unipartite number, zeros excluded, and of a composition of a unitary multipartite, zeros not excluded, and each combination or composition of the nature considered is uniquely represented by a tree.

The number of compositions of the multipartite $\overline{1^{n-1}}$, zeros not excluded, into k or fewer parts is

$$1^{n-1} + 2^{n-1} + \ldots + k^{n-1}$$

a number which also represents the number of the aggregate of the combinations of the number n, of orders 1, 2, . . . k, zero parts not excluded.

The interesting fact here brought to light is the connection between the unipartite numbers and the unitary multipartite numbers.

85. We have seen that k^{m-1} expresses the number of combinations of order k possessed by the unipartite number m. Each combination involves a certain number of the k different symbols and we may inquire the number of combinations which involve exactly p out of the k symbols. It is clear that one combination can be formed which involves any one symbol and none of the others; hence, k combinations involve but a single symbol. Two out of k symbols may be selected in $\binom{k}{2}$ different ways; for each such selection we must take the number of combinations of the number m, of order 2, and subtract the number of these in which but a single symbol appears. Hence, the number of combinations involving exactly two symbols is

$$\binom{k}{2}(2^{m-1} - 2).$$

Similarly three out of k symbols may be selected in $\binom{k}{3}$ ways, and for each selection we take the whole number of combinations of order 3 and subtract those of them which involve exactly two symbols or exactly one.

Hence we arrive at the number

$$\binom{k}{3}\{3^{m-1} - 3(2^{m-1} - 2) - 3\}$$
$$= \binom{k}{3}\{3^{m-1} - 3 \,.\, 2^{m-1} + 3 \,.\, 1^{m-1}\}.$$

In this way it is easy to see that the number of combinations involving exactly p symbols is

$$\binom{k}{p}\{p^{m-1} - p(p-1)^{m-1} + \binom{p}{2}(p-2)^{m-1} - \ldots + (-)^{p+1}p\}$$
$$= \binom{k}{p}\Delta^p(0^{m-1})$$

in the notation of finite differences.

We have now the well known identity

$$k^{m-1} = \binom{k}{1}\Delta(0^{m-1}) + \binom{k}{2}\Delta^2(0^{m-1}) + \ldots + \binom{k}{p}\Delta^p(0^{m-1}) + \ldots + \binom{k}{k}\Delta^k(0^{m-1}),$$

and we have seen its interpretation in the theory of the combinations of order k of a given unipartite number m.

TABLES of Compositions of Multipartite Numbers.

No.	1	part	Total
1	1	. . .	1

No.	1	2	parts	Total
2	1	1	. . .	2
$\overline{1^2}$	1	2	. . .	3

No.	1	2	3	parts	Total
3	1	2	1	. . .	4
$\overline{21}$	1	4	3	. . .	8
$\overline{1^3}$	1	6	6	. . .	13

No.	1	2	3	4	parts	Total
4	1	3	3	1	. . .	8
$\overline{31}$	1	6	9	4	. . .	20
$\overline{2^2}$	1	7	12	6	. . .	26
$\overline{21^2}$	1	10	21	12	. . .	44
$\overline{1^4}$	1	14	36	24	. . .	75

No.	1	2	3	4	5	parts	Total
5	1	4	6	4	1	. .	16
$\overline{41}$	1	8	18	16	5	. .	48
$\overline{32}$	1	10	27	28	10	. .	76
$\overline{31^2}$	1	14	45	52	20	. .	132
$\overline{2^21}$	1	16	57	72	30	. .	176
$\overline{21^3}$	1	22	93	132	60	. .	308
$\overline{1^4}$	1	30	150	240	120	. .	541

No.	1	2	3	4	5	6	parts	Total
6	1	5	10	10	5	1	. .	32
$\overline{51}$	1	10	30	40	25	6	. .	112
$\overline{42}$	1	13	48	76	55	15	. .	208
$\overline{41^2}$	1	18	78	136	105	30	. .	368
3^2	1	14	55	92	70	20	. .	252
$\overline{321}$	1	22	111	220	190	60	. .	604
$\overline{31^3}$	1	30	177	388	360	120	. .	1076
2^3	1	25	138	294	270	90	. .	818
$\overline{2^21^2}$	1	34	219	516	510	180	. .	1460
$\overline{21^4}$	1	46	345	900	960	360	. .	2612
$\overline{1^6}$	1	62	540	1560	1800	720	.	4683

No.	1	2	3	4	5	6	7	parts	Total
7	1	6	15	20	15	6	1	. .	64
$\overline{61}$	1	12	45	80	75	36	7	. .	256
$\overline{52}$	1	16	75	160	175	96	21	. .	544
$\overline{51^2}$	1	22	120	280	325	186	42	. .	976
$\overline{43}$	1	18	93	216	255	150	35	. .	768
$\overline{421}$	1	28	183	496	655	420	105	. .	1888
$\overline{41^3}$	1	38	288	856	1205	810	210	. .	3408
$\overline{3^21}$	1	30	207	588	810	540	140	. .	2316
$\overline{32^2}$	1	34	255	772	1120	780	210	. .	3172
$\overline{321^2}$	1	46	399	1324	2050	1500	420	. .	5740
$\overline{31^4}$	1	62	621	2260	3740	2880	840	. .	10404
$\overline{2^31}$	1	52	489	1728	2820	2160	630	. .	7880
$\overline{2^21^3}$	1	70	759	2940	5130	4140	1260	. .	14300
$\overline{21^5}$	1	94	1173	4980	9300	7920	2520	. .	25988
$\overline{1^7}$	1	126	1806	8400	16800	15120	5040	. .	47293

[During the considerable time that has elapsed since this paper was read I have discovered the general theory of the transformations of Arts. 41 and 75.

Let X_1, X_2, X_3 be general linear functions of x_1, x_2, x_3 as exhibited in the Notation of the Theory of Matrices

$$(X_1, X_2, X_3) = \left(\begin{array}{ccc} a_1, & a_2, & a_3 \\ b_1, & b_2, & b_3 \\ c_1, & c_2, & c_3 \end{array} \right| (x_1, x_2, x_3),$$

and consider the algebraic fraction

$$\frac{1}{(1 - s_1 X_1)(1 - s_2 X_2)(1 - s_3 X_3)}.$$

I have established that that portion of the expansion of this fraction, which is a function of products of powers of s_1x_1, s_2x_2, s_3x_3 only, is represented by the fraction

$$\frac{1}{N},$$

where

$$\begin{aligned} N = 1 &- a_1s_1x_1 - b_2s_2x_2 - c_3s_3x_3 \\ &+ | a_1b_2 | s_1s_2x_1x_2 + | a_1c_3 | s_1s_3x_1x_3 + | b_2c_3 | s_2s_3x_2x_3 \\ &- | a_1b_2c_3 | s_1s_2s_3x_1x_2x_3 ; \end{aligned}$$

the notation being that in use in the Theory of Determinants.

The coefficients of N are the several co-axial minors of the determinant of the matrix defining X_1, X_2, and X_3, viz.:—

$$\begin{vmatrix} a_1 & a_2 & a_3 \\ b_1 & b_2 & b_3 \\ c_1 & c_2 & c_3 \end{vmatrix}.$$

The result is immediately deducible from the identity

$$\begin{vmatrix} \frac{s_1(a_1x_1 - X_1)}{1 - s_1X_1} - 1, & \frac{a_2s_1x_1}{1 - s_1X_1}, & \frac{a_3s_1x_1}{1 - s_1X_1} \\ \frac{b_1s_2x_2}{1 - s_2X_2}, & \frac{s_2(b_2x_2 - X_2)}{1 - s_2X_2} - 1, & \frac{b_3s_2x_2}{1 - s_2X_2} \\ \frac{c_1s_3x_3}{1 - s_3X_3}, & \frac{c_2s_3x_3}{1 - s_3X_3}, & \frac{s_3(c_3x_3 - X_3)}{1 - s_3X_3} - 1 \end{vmatrix}$$

$$\times \begin{vmatrix} 1 - s_1X_1, & 0, & 0 \\ 0, & 1 - s_2X_2, & 0 \\ 0, & 0, & 1 - s_3X_3 \end{vmatrix}$$

$$= \begin{vmatrix} a_1s_1x_1 - 1, & a_2s_1x_1, & a_3s_1x_1 \\ b_1s_2x_2, & b_2s_2x_2 - 1, & b_3s_2x_2 \\ c_1s_3x_3, & c_2s_3x_3, & c_3s_3x_3 - 1 \end{vmatrix}.$$

the determinant last written being, with changed sign, the value of N.

The theorem for the case of n variables $x_1, x_2, \ldots x_n$ will be completely manifest from the above.

It appears to be one of considerable importance with regard to the generating functions which present themselves in this domain of the Theory of Numbers.

The results of its further investigation I hope to bring before the Royal Society in the near future.—Added August 25, 1893. P. A. M.]

II. *Second Memoir on the Compositions of Numbers.*

By Major P. A. MacMahon, *R.A., D.Sc., F.R.S.*

Received August 23,—Read December 6, 1906.

Preamble.

In a Memoir on the Theory of the Compositions of Numbers, read before the Royal Society, November 24, 1892, and published in the 'Philosophical Transactions' for 1893, I discussed the compositions of multipartite numbers by a graphical method. The generating function produced by the method was of the form

$$\frac{1}{1-\Sigma\alpha_1+(1-\lambda)\,\Sigma\alpha_1\alpha_2-(1-\lambda)^2\,\Sigma\alpha_1\alpha_2\alpha_3+\ldots.},$$

a symmetrical function of the quantities α.

The investigation of the present paper leads, in part, to the same generating function which is subjected to a close examination. Moreover, the whole research has to do with the compositions of numbers, and appropriately follows the Memoir of 1893.

The problem under investigation, which was brought to my notice by Professor Simon Newcomb, may be stated as follows :—

A pack of cards of any specification is taken—say that there are p cards marked 1, q cards 2, r cards 3, and so on—and, being shuffled, is dealt out on a table ; so long as the cards that appear have numbers that are in descending order of magnitude, they are placed in one pack together—equality of number counting as descending order—but directly the descending order is broken a fresh pack is commenced, and so on until all the cards have been dealt. The result of the deal will be m packs containing, in order, $a, b, c, \ldots$ cards respectively, where, n being the number of cards in the whole pack,

$$(abc\ldots)$$

is some composition of the number n, the numbers of parts in the composition being m.

We have, then, for discussion—

(1) The number of ways of arranging the cards so as to yield a given composition

$$(abc\ldots);$$

(2) The number of arrangements which lead to a distribution into exactly m packs.

These problems, and many others of a like nature, are solved in this paper.

The first of the two questions has given rise to two new symmetric functions,

$$h_{abc...},\ a_{abc...},$$

of great interest, which supply the complete solution. The second gives rise to the same generating function that presented itself in the first Memoir. It is here attacked by the calculus of symmetric function differential operators, and a number of new results obtained.

If the whole pack be specified by the partition

$$(pqr...),$$

there is a one-to-one correspondence between the arrangements which lead to a distribution into m packs and the principal compositions, involving $m-1$ essential nodes, of the multipartite number

$$(\overline{pqr...}).$$

Part I. is concerned with an elementary theory of the case in which the cards are all numbered differently.

The general case, which is more difficult, is dealt with in Part II.

To make what follows clear to the reader, I commence with some elementary notions concerning the connection between the partitions and compositions of numbers on the one hand, and permutations and combinations of things on the other hand, and I also specify and describe the nomenclature and notation that I have found it convenient to adopt. A suitable notation is, indeed, of the first importance in this subject, as I hope to make evident as the investigation proceeds.

Introductory.

Art. 1. Any succession of numbers, written down from left to right at random, such as

142771,

is termed a "composition" of the number which is the sum of the numbers.

If the numbers be arranged in descending order from left to right,

774211,

the succession is termed a "descending partition," or simply a "partition" of the number which is the sum of the numbers.

Or, if we arrange in ascending order of magnitude,

112477,

the succession may be termed an "ascending partition."

Generally, in speaking of partitions, we understand that the descending order is meant; but it is convenient sometimes to consider them as being defined by an ascending order.

There is no other method of ordering a collection of numbers which is of general application.

We see that the same collection of numbers gives rise to only one partition, but, by permutation, to more than one composition.

Art. 2. Both partitions and compositions have an appropriate graphical representation. That of a partition was first given by FERRERS, and the notion was elaborated by SYLVESTER during the time he was at the Johns Hopkins University in Baltimore, U.S.A. It consisted merely in writing a row of nodes, or units, corresponding to each number (or part) of the partition, the left-hand nodes of the rows being placed in a vertical line. Thus

774211

is denoted by

```
● ● ● ● ● ● ●
● ● ● ● ● ● ●
● ● ● ●
● ●
●
●
```

Art. 3. A trial will show that this method is not suited to compositions. One method, effective for certain purposes, was given by the author.* To indicate it, consider the composition

142

of the number 7.

·—*—·—·—·—*—·—·

We take seven segments on a line, and place nodes, *, so that the line is divided off into 1, 4 and 2 segments respectively in order. The conjugate composition is reached from this by suppressing the existing nodes and placing nodes at the points of division which are free from nodes.

Thus

·—·—*—*—*—·—*—·

denotes the composition

21121.

Art. 4. There is a more illuminating mode of representation which is here given, it is believed, for the first time; it is akin to the method of FERRERS, and enables methods of research which SYLVESTER'S exertions have made familiar.

It consists in taking rows of nodes in order and placing the left-hand node of any row vertically beneath the right-hand node of the previous row.

Thus

142

is denoted by

```
●
● ● ● ●
      ● ●
```

* "Memoir on the Theory of the Compositions of Numbers," 'Phil. Trans. Roy. Soc.,' 1893.

K 2

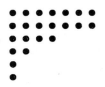

and

142771

by

This graph is read horizontally; the conjugate is obtained by reading vertically, giving

211221111121111112,

or, in brief notation,

$21^2 2^2 1^5 2 1^5 2$.

We may also read the graph horizontally from bottom to top and vertically from right to left, obtaining generally four compositions from the graph.

The graph is a zig-zag one and will be, without doubt, an important instrument of research.

Part I.—Section 1.

Art. 5. Consider the permutations of the first n integers, and for simplicity take $n = 9$.

Writing down a permutation at random,

31 | 4 | 5 | 92 | 76 | 8,

it is clear that lines can be drawn separating the numbers into compartments in such wise that in each compartment the numbers are in descending order of magnitude. We can then write down a succession of numbers which describe the size of the compartments, proceeding from left to right, and thus arrive at a composition

211221

of the number 9.

I say that the permutation under examination has a descending specification

(211221) or $(21^2 2^2 1)$.

Similarly, from the ascending character

3 | 1459 | 27 | 68

of the same permutation, I say that the ascending specification is

(1422) or (142^2),

where it is to be noticed that 142^2 is the composition of 9 which is conjugate to $21^2 2^2 1$, the composition which specifies the descending character. This is shown by the zig-zag graph

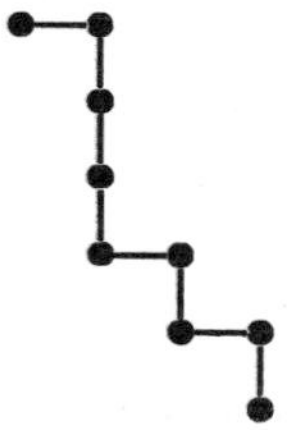

Art. 6. We can now formulate the question: Of the permutations of the first n numbers, how many have a descending specification denoted by a given composition of the number n? Whatever the answer, it is clear that the same answer must, in general, be given for three other compositions, viz., the three others associated with the zig-zag graph. In fact, from

314592768 of specification 211221,

we derive

867295413 ,, 2241;

and from these two by changing the number m into $n-m+1$,

796518342 of specification 1422,

243815679 ,, 122112

and so forth.

In two cases there are two associated compositions instead of four, viz.:—

(i) When the composition reads the same as its inverse (that is the same from left to right as from right to left),

(ii) When the conjugate and the inverse are identical, as in 221, whose conjugate is 122.

*The number of self-inverse compositions of an even number $2m$ and of an uneven number $2m+1$ is

$$2^m.$$

The number of inverse-conjugate compositions of an uneven number $2m+1$ is

$$2^m.$$

Hence, in the present theory, the number of different numbers that appear in the case of an even number $2m$ is, since the whole number of compositions is 2^{2m-1},

$$\tfrac{1}{2}.2^m+\tfrac{1}{4}(2^{2m-1}-2^m),$$
$$= 2^{m-2}(2^{m-1}+1);$$

and, in the case of an uneven number $2m+1$,

$$\tfrac{1}{2}2^m+\tfrac{1}{2}2^m+\tfrac{1}{4}(2^{2m}-2^{m+1}),$$
$$= 2^{m-1}(2^{m-1}+1);$$

viz., it is

$$2^{n-3}+2^{1/2(n-4)},$$

or

$$2^{n-3}+2^{1/2(n-3)},$$

according as n is even or uneven.

* See "Memoir on the Theory of the Compositions of Numbers," 'Phil. Trans. Roy. Soc.,' 1893.

Art. 7. Let N $(abc...)$ denote the number of permutations of the first n integers which have a descending specification denoted by the composition

$$(abc...)$$

of the number n.

Obviously
$$N(a) = 1, \quad a = n.$$

To determine N (ab), $a+b = n$, separate the n integers into two groups, a left-hand group of a numbers chosen at random and a right-hand group of the remaining b numbers. This can be done in

$$\binom{n}{a} \text{ different ways.}$$

$\left[\text{I write } \frac{n!}{a!\,(n-a)!} = \binom{n}{a} \text{ in a common notation}\right]$; now arrange each group of numbers in descending order of magnitude for each of the $\binom{n}{a}$ separations; we thus obtain each of the permutations enumerated by N (a, b) and the one permutation enumerated by N $(a+b)$.

Hence
$$N(ab) + N(a+b) = \binom{n}{a},$$

or
$$N(ab) = \binom{n}{a} - \binom{n}{a+b} = \binom{n}{a} - 1.$$

Again, to find N (abc), we separate the n integers into three groups containing a, b, and c integers respectively; this can be done in

$$\frac{n!}{a!\,b!\,c!}$$

different ways; placing the numbers in each group in descending order, we obtain all the permutations enumerated by

$$N(abc), \quad N(a+b, c), \quad N(a, b+c), \quad N(a+b+c).$$

Hence
$$N(abc) + N(a+b, c) + N(a, b+c) + N(a+b+c) = \frac{n!}{a!\,b!\,c!};$$

leading to
$$N(abc) = \frac{n!}{a!\,b!\,c!} - \frac{n!}{(a+b)!\,c!} - \frac{n!}{a!\,(b+c)!} + \frac{n!}{(a+b+c)!},$$

where $a+b+c = n$.

Similarly we find

$$N(abcd) = \frac{n!}{a!\,b!\,c!\,d!} - \frac{n!}{(a+b)!\,c!\,d!} - \frac{n!}{a!\,(b+c)!\,d!} - \frac{n!}{a!\,b!\,(c+d)!}$$
$$+ \frac{n!}{(a+b)!\,(c+d)!} + \frac{n!}{(a+b+c)!\,d!} + \frac{n!}{a!\,(b+c+d)!} - \frac{n!}{(a+b+c+d)!},$$

where $a+b+c+d = n$.

The general law is clear; the letters a, b, c, d are always in order in the denominators and the sign of a fraction depends upon the number of factors in its denominator.

We can thus calculate the number of permutations appertaining to each of the 2^{n-1} compositions of n.

It has been established independently, by the aid of the zig-zag graph, that these numbers

$$N(\ldots)$$

are equal in four's or in two's.

Art. 8. The sum of the numbers $N(\ldots)$ is of course $n!$

The details of the above results for

$$n = 2, 3, 4, 5, 6$$

are given for easy reference.

$$\begin{array}{rl} N(2) = 1 & 1 \\ N(1^2) = 1 & 1 \\ \hline & 2 = 2! \end{array}$$

$$\begin{array}{rl} N(3) = N(1^3) = 1 & 2 \\ N(21) = N(12) = 2 & 4 \\ \hline & 6 = 3! \end{array}$$

$$\begin{array}{rl} N(4) = N(1^4) = 1 & 2 \\ N(31) = N(13) = N(21^2) = N(1^22) = 3 & 12 \\ N(22) = N(121) = 5 & 10 \\ \hline & 24 = 4! \end{array}$$

$$\begin{array}{rl} N(5) = N(1^5) = 1 & 2 \\ N(41) = N(14) = N(21^3) = N(1^32) = 4 & 16 \\ N(32) = N(23) = N(121^2) = N(1^221) = 9 & 36 \\ N(31^2) = N(1^23) = 6 & 12 \\ N(2^21) = N(12^2) = 16 & 32 \\ N(131) = N(212) = 11 & 22 \\ \hline & 120 = 5! \end{array}$$

		$N(6)$	$= N(1^6)$	$= 1$	2
$N(51)$	$= N(15)$	$= N(21^4)$	$= N(1^4 2)$	$= 5$	20
$N(42)$	$= N(24)$	$= N(1^3 21)$	$= N(121^3)$	$= 14$	56
		$N(3^2)$	$= N(1^2 21^2)$	$= 19$	38
$N(41^2)$	$= N(1^2 4)$	$= N(31^3)$	$= N(1^3 3)$	$= 10$	40
		$N(141)$	$= N(21^2 2)$	$= 19$	38
$N(321)$	$= N(123)$	$= N(2^2 1^2)$	$= N(1^2 2^2)$	$= 35$	140
$N(312)$	$= N(213)$	$= N(1^2 31)$	$= N(131^2)$	$= 26$	104
$N(132)$	$= N(231)$	$= N(2121)$	$= N(1212)$	$= 40$	160
		$N(2^3)$	$= N(12^2 1)$	$= 61$	122
					$720 = 6!$

Art. 9. Some simple summations are obtainable from elementary considerations.

In regard to the permutations of the first n integers, let

$$\Sigma N(s\ldots),$$

where $s < n$, denote the sum of all numbers $N(\ldots)$, such that s is the first number in the specifying composition. Take any $s+1$ of the numbers

$$1, 2, 3, \ldots n,$$

and arrange them from left to right in such wise that the first s numbers are in descending order and the $s+1^{\text{th}}$ number greater than the sth; this can be done in

$$s\binom{n}{s+1} \text{ ways};$$

the remaining $n-s-1$ numbers can be arranged in $(n-s-1)!$ ways, so that, placing them to the right of the former, we arrive at the result

$$\Sigma N(s\ldots) = s\frac{n!}{(s+1)!}.$$

Art. 10. Again, denoting by

$$\Sigma N(1^{s-1}\ldots)$$

the sum of all numbers $N(\ldots)$ of which the specifying compositions commence with exactly $s-1$ units, the consideration of the properties of conjugate zig-zag graphs establishes that

$$\Sigma N(1^{s-1}\ldots) = \Sigma N(s\ldots),$$

with a single exception where $s = n$; *e.g.*,

$$\Sigma N(1^0\ldots) = {}^*\Sigma N(1\ldots) = \frac{n!}{2!},$$

$${}^\dagger\Sigma N(1\ldots) = \Sigma N(2\ldots) = 2\frac{n!}{3!},$$

and so on.

* No restriction is placed upon the number next to the unit in this case.

† Here the number following the unit must be >1.

Art. 11. Again, for the summation

$$\Sigma N(1^s\ldots),$$

where the composition begins with *at least* s units, we easily obtain the value

$$\frac{n!}{(s+1)!}.$$

The Multiplication Theorem.

Art. 12. A fundamental property of the numbers N (...) will be established from elementary considerations; it will, later on in the paper, be generalised.

Let

$$N(a_1a_2\ldots a_s)$$

be derived from the permutations of p different integers, and

$$N(a_{s+1}a_{s+2}\ldots a_{s+t})$$

from the permutations of $n-p$ different integers; it is to be shown that

$$\binom{n}{a_1+a_2+\ldots+a_s}N(a_1a_2\ldots a_s)\,N(a_{s+1}a_{s+2}\ldots a_{s+t})$$
$$= N(a_1a_2\ldots a_{s+t})+N(a_1a_2\ldots a_{s-1},\ a_s+a_{s+1},\ a_{s+2}\ldots a_{s+t}),$$

where on the right the reference is to the permutations of n different integers.

Out of the n numbers

$$1,\ 2,\ 3,\ldots n,$$

we can select

$$a_1+a_2+\ldots+a_s$$

numbers in

$$\binom{n}{a_1+a_2+\ldots+a_s}$$

ways, and arrange each selection, so as to have a descending specification

$$(a_1a_2\ldots a_s),$$

in

$$N(a_1a_2\ldots a_s)\text{ ways};$$

the remaining numbers can be arranged, to have a descending specification

$$(a_{s+1}a_{s+2}\ldots a_{s+t}),$$

in

$$N(a_{s+1}a_{s+2}\ldots a_{s+t})\text{ ways};$$

placing the latter to the right of the former there appears

$$\binom{n}{a_1+a_2+\ldots+a_s}N(a_1a_2\ldots a_s)\,N(a_{s+1}a_{s+2}\ldots a_{s+t})$$

arrangements.

Now, combining the two sets of numbers, we find that either there is or there is not a break in the descending order between

$$a_s \text{ and } a_{s+1};$$

hence the number of arrangements is also

$$\mathrm{N}(a_1a_2\ldots a_{s+t}) + \mathrm{N}(a_1a_2\ldots a_{s-1},\ a_s+a_{s+1},\ a_{s+2}\ldots a_{s+t}). \quad \text{Q.E.D.}$$

Art. 13. Regarded as a numerical theorem, the multiplication is commutative, but in regard to form it is not commutative; thus, by considering the multiplication

$$\mathrm{N}(a_{s+1}a_{s+2}\ldots a_{s+t})\,\mathrm{N}(a_1a_2\ldots a_s),$$

we obtain the linear relation

$$\mathrm{N}(a_1a_2\ldots a_{s+t}) + \mathrm{N}(a_1a_2\ldots a_{s-1},\ a_s+a_{s+1},\ a_{s+2}\ldots a_{s+t})$$
$$= \mathrm{N}(a_{s+1}a_{s+2}\ldots a_{s+t}a_1\ldots a_s) + \mathrm{N}(a_{s+1}\ldots a_{s+t-1},\ a_{s+t}+a_1,\ a_2\ldots a_s).$$

Observe also that the order of the numbers in brackets in any number N (...) can be reversed at pleasure and thus new forms of results obtained.

As a verification: from the tables

$$\binom{5}{3}\mathrm{N}(12)\,\mathrm{N}(11) = \mathrm{N}(121^2) + \mathrm{N}(131) = \mathrm{N}(1^32) + \mathrm{N}(12^2);$$
$$10 \;.\; 2 \;.\; 1 \;=\; 9 \;+\; 11 \;=\; 4 \;+\; 16$$

$$\mathrm{N}(123) + \mathrm{N}(15) = \mathrm{N}(312) + \mathrm{N}(42).$$
$$35 \;+\; 5 \;=\; 26 \;+\; 14$$

The fact that the multiplication is not commutative formally is of great importance in the theory of these numbers.

Art. 14. Extending the theorem to the product of three numbers

$$\mathrm{N}(a_1a_2\ldots a_s),\quad \mathrm{N}(b_1b_2\ldots b_t),\quad \mathrm{N}(c_1c_2\ldots c_u),$$

we find

$$\frac{n!}{(\Sigma a)!\,(\Sigma b)!\,(\Sigma c)!}\mathrm{N}(a_1a_2\ldots a_s)\,\mathrm{N}(b_1b_2\ldots b_t)\,\mathrm{N}(c_1c_2\ldots c_u)$$
$$= \mathrm{N}(a_1\ldots a_sb_1\ldots b_tc_1\ldots c_u) + \mathrm{N}(a_1\ldots a_{s-1},\ a_s+b_1,\ b_2\ldots b_tc_1\ldots c_u)$$
$$+ \mathrm{N}(a_1\ldots a_sb_1\ldots b_{t-1},\ b_t+c_1,\ c_2\ldots c_u) + \mathrm{N}(a_1\ldots a_{s-1},\ a_s+b_1,\ b_2\ldots b_{t-1},\ b_t+c_1,\ c_2\ldots c_u).$$

We may, in general, give the right-hand side 3! different forms corresponding to the 3! permutations of the numbers N (...) on the sinister.

If we take the product of m numbers N (...), to form the dexter, we combine the last integer of a number N (...) with the first integer of the next following number N (...),

0	times in	1	way ,
1	,,	$\binom{m-1}{1}$	ways,
2	,,	$\binom{m-1}{2}$	,, ,
.	,,	.	,, ,
$m-1$	,,	$\binom{m-1}{m-1}$	,, ;

hence 2^{m-1} numbers N (...) present themselves on the dexter.

Not counting reversals of order, the dexter can, in general, be given as many different forms as there are permutations of the numbers N (...) on the sinister. Counting reversals, the number of different forms is further multiplied by 2^m, subject to a diminution when one or more of the numbers N (...) is self-inverse.

Applications of the Theorem.

Art. 15. The theorems, already arrived at above, are particular cases of multiplication. Thus the formulæ, of which

$$\mathrm{N}\,(abc) + \mathrm{N}\,(a+b,\, c) + \mathrm{N}\,(a,\, b+c) + \mathrm{N}\,(a+b+c) = \frac{n!}{a!\,b!\,c!}$$

is a type, are equivalent to results, of which

$$\frac{n!}{a!\,b!\,c!}\mathrm{N}\,(a)\,\mathrm{N}\,(b)\,\mathrm{N}\,(c) = \mathrm{N}\,(abc) + \mathrm{N}\,(a+b,\, c) + \mathrm{N}\,(a,\, b+c) + \mathrm{N}\,(a+b+c)$$

is representative, since $\mathrm{N}\,(a) = \mathrm{N}\,(b) = \mathrm{N}\,(c) = 1.$

That the sum of all numbers N (...), of given weight n, is $n!$ is shown by the formula

$$n!\{\mathrm{N}\,(1)\}^n = \underset{\cdots}{\Sigma}\mathrm{N}\,(\ldots)\,;$$

since on the dexter occurs an N (...) corresponding to every composition of n.

Art. 16. Suppose that it is required to find the sum of all numbers N (...), of given weight, which are such that each associated composition commences with a given series of numbers

$$a_1a_2\ldots a_m,$$

or, in other words, suppose we wish to make the summation indicated by

$$\underset{\cdots}{\Sigma}\mathrm{N}\,(a_1a_2\ldots a_m\ldots)\,;$$

the solution is given at once by

$$\frac{n!}{(\Sigma a+1)!}\mathrm{N}\,(a_1a_2\ldots a_m1)\,\{\mathrm{N}\,(1)\}^{n-\Sigma a-1} = \underset{\cdots}{\Sigma}\mathrm{N}\,(a_1a_2\ldots a_m\ldots)\,;$$

for, by the multiplication process, the unit which terminates $\mathrm{N}\,(a_1a_2\ldots a_m1)$, combined with

$$\{\mathrm{N}\,(1)\}^{n-\Sigma a-1},$$

L 2

gives every composition of the number

$$n-\Sigma a.$$

Hence, since N (1) = 1,

$$\sum_{\dots} \mathrm{N}(a_1a_2\dots a_m\dots) = \frac{n!}{(\Sigma a+1)!}\mathrm{N}(a_1a_2\dots a_m 1).$$

Art. 17. By varying the order of the factors, on the sinister of the multiplication formula, a variety of interesting results present themselves; thus

$$\frac{n!}{(\Sigma a+1)!}\{\mathrm{N}(1)\}^p\,\mathrm{N}(a_1a_2\dots a_m 1)\{\mathrm{N}(1)\}^{n-\Sigma a-p-1} = \Sigma\Sigma\mathrm{N}(\dots a'_1a_2a_3\dots a_m\dots);$$

where after a_m, on the dexter, occurs every composition of

$$n-\Sigma a-p;$$

and the portion

$$\dots a'_1$$

includes every composition of

$$p+a_1$$

which terminates with a number not less than a_1.

Hence, for such a summation,

$$\Sigma\Sigma\mathrm{N}(\dots a'_1a_2\dots a_m\dots) = \frac{n!}{(\Sigma a+1)!}\mathrm{N}(a_1a_2\dots a_m 1);$$

a formula which is independent of p.

Art. 18. In particular from

$$\{\mathrm{N}(1)\}^{n-\Sigma a-1}\,\mathrm{N}(a_1a_2\dots a_m 1)$$

we obtain

$$\Sigma\mathrm{N}(\dots a'_1a_2\dots a_m 1) = \frac{n!}{(\Sigma a+1)!}\mathrm{N}(a_1a_2\dots a_m 1);$$

wherein the summation is for every composition of

$$n-a_2-\dots-a_m-1$$

which terminates with a number not less than a_1.

E.g., for $n = 6$, $a_1 = 1$, $a_2 = 1$,

$$\mathrm{N}(41^2)+\mathrm{N}(131^2)+\mathrm{N}(2^21^2)+\mathrm{N}(1^221^2) = \frac{6!}{4!}\mathrm{N}(211).$$

$$10 \;+\; 26 \;+\; 35 \;+\; 19 \;= 6.5.3$$

Art. 19. As another example of the power of the theorem, let

$$\Sigma\mathrm{N}(a_1a_2\dots a_{m_1}\dots b_1b_2\dots b_{m_2})$$

(the numbers a_1, $a_2\dots a_{m_1}$, b_1, $b_2\dots b_{m_2}$ being given) denote a summation in regard to compositions of

$$n-\Sigma a-\Sigma b$$

placed between a_{m_1} and b_1; we obtain

$$\sum_{\cdots} N(a_1 a_2 \ldots a_{m_1} \ldots b_1 b_2 \ldots b_{m_2})$$

$$= \frac{n!}{(\Sigma a+1)!\,(\Sigma b+1)!} N(a_1 a_2 \ldots a_{m_1} 1)\{N(1)\}^{n-\Sigma a-\Sigma b-2} N(1 b_1 b_2 \ldots b_{m_2}),$$

$$= \frac{n!}{(\Sigma a+1)!\,(\Sigma b+1)!} N(a_1 a_2 \ldots a_{m_1} 1)\, N(1 b_1 b_2 \ldots b_{m_2}),$$

$$= \frac{n!}{(\Sigma a+\Sigma b+2)!}\{N(a_1 a_2 \ldots a_{m_1} 1^2 b_1 b_2 \ldots b_{m_2}) + N(a_1 a_2 \ldots a_{m_1} 2 b_1 b_2 \ldots b_{m_2})\}.$$

By varying the order of the factors, other summations, leading to the same numerical result, can be effected.

Art. 20. Consider next the multiplication

$$\frac{n!}{(s_1+2)!\,(s_2+2)!\,(s_3+2)!}$$
$$\times \{N(1)\}^{w_1-1} N(1^{s_1+2}) \{N(1)\}^{w_2-2} N(1^{s_2+2}) \{N(1)\}^{w_3-2} N(1^{s_3+2}) \{N(1)\}^{w_4-1};$$

wherein, $\Sigma w+\Sigma s = n$,

w_1, w_2, w_3, w_4 are numbers not less than unity,

s_1, s_2, s_3 are any numbers, zero not excluded.

The result of the multiplication consists of numbers N (...), such that there is

(i) A composition of w_1 followed by s_1 units, succeeded by
(ii) A composition of w_2 followed by s_2 units, succeeded by
(iii) A composition of w_3 followed by s_3 units, succeeded by
(iv) A composition of w_4;

and the dexter is the sum of all such numbers N (...).

Denoting this sum by

$$\Sigma N(w_1 1^{s_1} w_2 1^{s_2} w_3 1^{s_3} w_4),$$

we find that its value is

$$\frac{n!}{(s_1+2)!\,(s_2+2)!\,(s_3+2)!},$$

since each number N (...) occurring in the product on the sinister has unity for its value.

Hence, in general, the remarkable theorem,

$$\Sigma N(w_1 1^{s_1} w_2 1^{s_2} w_3 1^{s_3} \ldots) = \frac{n!}{(s_1+2)!\,(s_2+2)!\,(s_3+2)!\ldots};$$

showing that the sum depends merely upon the numbers

$$s_1, s_2, s_3, \ldots,$$

and not at all upon the numbers

$$w_1, w_2, w_3, \ldots.$$

Observe that w_1 and the final number of the composition may or may not be unity, and that every composition of n may be written in the form

$$w_1 1^{s_1} w_2 1^{s_2} w_3 1^{s_3} \ldots.$$

If

$$s_1 = s_2 = s_3 = \ldots = 0,$$

$$\Sigma \mathrm{N}(w_1 w_2 w_3 w_4 \ldots) = \frac{n!}{2!2!2!\ldots};$$

and, in particular,

$$\Sigma \mathrm{N}(w_1 w_2 \ldots w_m) = \frac{n!}{(2!)^{m-1}},$$

wherein $w_2, w_3, \ldots w_{m-1}$ are non-unitary, but w_1, w_m may or may not be unitary.

As a simple example take

$$w_1 = 1, \quad s_1 = 4, \quad w_2 = 1,$$

so that

$$\mathrm{N}(1^6) = \Sigma \mathrm{N}(1, 1^4, 1) = \frac{6!}{(4+2)!} = 1,$$

a verification.

Art. 21. A more general theorem is yielded by

$$\frac{n!}{(\Sigma p+2)!(\Sigma q+2)!(\Sigma r+2)!}$$

$$\times \{\mathrm{N}(1)\}^{w_1-1}\mathrm{N}(1p_1\ldots p_{m_1}1)\,\{\mathrm{N}(1)\}^{w_2-2}\mathrm{N}(1q_1\ldots q_{m_2}1)\,\{\mathrm{N}(1)\}^{w_3-2}\mathrm{N}(1r_1\ldots r_{m_3}1)\,\{\mathrm{N}(1)\}^{w_4-1},$$

$$= \Sigma \mathrm{N}(w_1 p_1 \ldots p_{m_1} w_2 q_1 \ldots q_{m_2} w_3 r_1 \ldots r_{m_3} w_4),$$

wherein

$$p_1 \ldots p_{m_1},$$
$$q_1 \ldots q_{m_2},$$
$$r_1 \ldots r_{m_3},$$

are given integers and the summation indicated on the dexter is in respect of the whole of the compositions of the numbers

$$w_1, \quad w_2, \quad w_3, \quad w_4,$$

where

$$0 \eqslantless w_1-1, \quad w_2-2, \quad w_3-2 \quad \text{and} \quad w_4-1.$$

The value of the sum is thus

$$\frac{n!}{(\Sigma p+2)!(\Sigma q+2)!(\Sigma r+2)!}\mathrm{N}(1p_1\ldots p_{m_1}1)\mathrm{N}(1q_1\ldots q_{m_2}1)\mathrm{N}(1r_1\ldots r_{m_3}1),$$

which, by the multiplication theorem, may be given the form

$$\frac{n!}{(\Sigma p+\Sigma q+\Sigma r+6)!}\times[\ \mathrm{N}\,(1p_1\ldots p_{m_1}1^2q_1\ldots q_{m_2}1^2r_1\ldots r_{m_3}1)$$
$$+\mathrm{N}\,(1p_1\ldots p_{m_1}2q_1\ldots q_{m_2}1^2r_1\ldots r_{m_3}1)$$
$$+\mathrm{N}\,(1p_1\ldots p_{m_1}1^2q_1\ldots q_{m_2}2r_1\ldots r_{m_3}1)$$
$$+\mathrm{N}\,(1p_1\ldots p_{m_1}2q_1\ldots q_{m_2}2r_1\ldots r_{m_3}1)].$$

Evidently, from the above, comprehensive results can be obtained from the multiplication theorem.

SECTION 2.

Art. 22. The next problem I propose to solve is that of determining the number of the permutations of the first n integers, whose descending specifications contain a given number of integers, or, in other words, whose associated compositions involve a given number of parts. The solution is implicitly contained in a paper I wrote in the year 1888.*

Let N_m denote the number of permutations associated with compositions containing exactly m parts.

In the paper quoted, I had under view a collection of objects of any species—say p of one sort, q of a second sort, r of a third, and so on—and defined the objects as to species by these numbers placed in brackets. I thus formed a partition

$$(pqr\ldots)$$

of the number n, such partition being the species definition of the objects.

As equalities may occur between the numbers $p, q, r, \ldots$, I took, as a more general definition, the partition

$$(p_1^{\pi_1}p_2^{\pi_2}p_3^{\pi_3}\ldots),$$

where $\Sigma\pi p = n$.

In the case under consideration, where the integers (or objects) are all different, the species definition is the partition

$$(1^n).$$

I proved, in the general case, that the number of ways of distributing the objects, into m different parcels, is given by the series

$$\begin{aligned}\mathrm{F}_m = \quad & \binom{m+p_1-1}{p_1}^{\pi_1}\binom{m+p_2-1}{p_2}^{\pi_2}\binom{m+p_3-1}{p_3}^{\pi_3}\ldots\\ & -\binom{m}{1}\binom{m+p_1-2}{p_1}^{\pi_1}\binom{m+p_2-2}{p_2}^{\pi_2}\binom{m+p_3-2}{p_3}^{\pi_3}\ldots\\ & +\binom{m}{2}\binom{m+p_1-3}{p_1}^{\pi_1}\binom{m+p_2-3}{p_2}^{\pi_2}\binom{m+p_3-3}{p_3}^{\pi_3}\ldots\\ & -\ldots.\end{aligned}$$

* "Symmetric Functions and the Theory of Distributions," 'Proc. L. M. S.,' vol. xix., p. 226.

For the case in hand, $p_1 = 1$, $\pi_1 = n$,

$$F_m = m^n - \binom{m}{1}(m-1)^n + \binom{m}{2}(m-2)^n - \binom{m}{3}(m-3)^n + \dots.$$

Art. 23. I shall prove that

$$N_m = m^n - \binom{n+1}{1}(m-1)^n + \binom{n+1}{2}(m-2)^n - \binom{n+1}{3}(m-3)^n + \dots.$$

For consider the arrangements enumerated by F_m. Place the compartments (or parcels) in order, from left to right, in any one such arrangement, and, in each compartment, place the integers in descending order of magnitude. The arrangement is obviously one of those enumerated by

$$N_m,\ N_{m-1},\ N_{m-2},\ \dots \text{ or } N_1.$$

In the whole of the arrangements, enumerated by F_m, thus treated, each arrangement enumerated by N_m will occur once only.

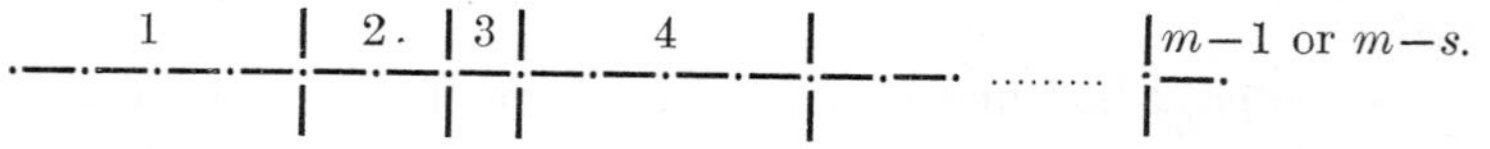

Let the illustration denote an arrangement enumerated by N_{m-1}. Each segment denotes an integer, and the $m-2$ vertical lines separate the integers into compartments.

By placing an extra vertical line at one of the unoccupied points of division, we obtain an arrangement enumerated by F_m. This can be done in $(n-1)-(m-2)$ different ways, showing that the particular arrangement, enumerated by N_{m-1}, is derivable by obliteration of a vertical line from

$$n-m+1$$

different arrangements enumerated by F_m.

Hence, the forms F_m include the forms N_{m-1} each $n-m+1$ times.

Again, let the illustration denote an arrangement enumerated by N_{m-s}. By placing s extra vertical lines, at unoccupied points of division, we obtain an arrangement enumerated by F_m. This can be done in

$$\binom{n-m+s}{s}$$

different ways; showing that the particular arrangement, enumerated by N_{m-s}, is derivable, by obliteration of s vertical lines, from

$$\binom{n-m+s}{s}$$

different arrangements enumerated by F_m.

Hence the forms F_m include the forms N_{m-s} each

$$\binom{n-m+s}{s} \text{ times.}$$

Hence

$$F_m = N_m + \binom{n-m+1}{1} N_{m-1} + \binom{n-m+2}{2} N_{m-2} + \ldots + \binom{n-1}{m-1} N_1.$$

Thence it is easy to show that

$$N_m = F_m - \binom{n-m+1}{1} F_{m-1} + \binom{n-m+2}{2} F_{m-2} - \ldots + (-)^{m+1} \binom{n-1}{m-1} F_1 ;$$

and also

$$N_m = m^n - \binom{n+1}{1}(m-1)^n + \binom{n+1}{2}(m-2)^n - \ldots + (-)^{m+1} \binom{n+1}{m-1} 1^n.$$

The relation $\sum_1^n N_m = n!$ may be verified.

Art. 24. It follows at once, from the zig-zag graphs, that

$$N_m = N_{n-m+1}.$$

Some of the simplest results are

$n =$	N_1.	N_2.	N_3.	N_4.	N_5.	N_6.
1	1					
2	1	1				
3	1	4	1			
4	1	11	11	1		
5	1	26	66	26	1	
6	1	57	302	302	57	1

Art. 25. There is another interesting series for N_m.

Let

$$(p+s)^n{}_{-1,0}$$

denote the expansion of

$$(p+s)^n$$

when deprived of the term which is linear in p and of the term independent of p; and put

$$P_s = (1+s)^n{}_{-1,0} ;$$

then

$$N_m = P_{m-1} - \frac{m-2}{m-1}\binom{n}{1} P_{m-2} + \frac{m-3}{m-1}\binom{n}{2} P_{m-3} - \ldots + (-)^m \frac{1}{m-1}\binom{n}{m-2} P_1.$$

I prove a general theorem, of which this is a particular case, later on in the paper.

Art. 26. Considering next p different numbers, defined by the partition

$$(1^p),$$

we have, by a previous definition,

$$\underset{a}{\Sigma} N(a_1 a_2 \ldots a_m) = N_{m, 1^p};$$

where $a_1, a_2, a_3, \ldots$ are each $\not< 1$ and such that

$$\Sigma a = p.$$

I have written

$$N_{m, 1^p}$$

instead of N_m, in order to specify the number of objects (or numbers) subjected to permutation.

Art. 27. I shall now prove that

$$\underset{a \cdots}{\Sigma\Sigma} N(a_1 a_2 \ldots a_m \ldots) = \frac{n!}{(p+1)!}(p-m+1) N_{m, 1^p},$$

where in

$$N(a_1 a_2 \ldots a_m \ldots)$$

the number of objects subjected to permutation is n, and the summation is in respect of all permutations such that the sum of the first m numbers in the descending specification is equal to p.

For, by Art. 16,

$$\underset{\cdots}{\Sigma} N(a_1 a_2 \ldots a_m \ldots) = \frac{n!}{(p+1)!} N(a_1 a_2 \ldots a_m 1);$$

hence

$$\underset{a \cdots}{\Sigma\Sigma} N(a_1 a_2 \ldots a_m \ldots) = \frac{n!}{(p+1)!} \underset{a}{\Sigma} N(a_1 a_2 \ldots a_m 1);$$

and, by the multiplication theorem,

$$(p+1) N(a_1 a_2 \ldots a_m) N(1) = N(a_1 a_2 \ldots a_m 1) + N(a_1 a_2 \ldots a_m + 1);$$

so that

$$\underset{a}{\Sigma} N(a_1 a_2 \ldots a_m 1) = (p+1) N_{m, 1^p} - \underset{a}{\Sigma} N(a_1 a_2 \ldots a_m + 1);$$

and since

$$\underset{a}{\Sigma} N(a_1 a_2 \ldots a_m + 1) = N_{m, 1^{p+1}} - \underset{a}{\Sigma} N(a_1 a_2 \ldots a_{m-1} 1),$$

$$\underset{a}{\Sigma} N(a_1 a_2 \ldots a_m 1) - \underset{a}{\Sigma} N(a_1 a_2 \ldots a_{m-1} 1) = (p+1) N_{m, 1^p} - N_{m, 1^{p+1}};$$

whence, by summation,

$$\underset{a}{\Sigma} N(a_1 a_2 \ldots a_m 1) = (p+1) \sum_1^m N_{m, 1^p} - \sum_1^m N_{m, 1^{p+1}};$$

but since

$$N_{m, 1^p} = m^p - \binom{p+1}{1}(m-1)^p + \binom{p+1}{2}(m-2)^p - \ldots$$

$$\sum_1^m N_{m, 1^p} = m^p - \binom{p}{1}(m-1)^p + \binom{p}{2}(m-2)^p - \ldots,$$

so that, substituting,

$$\sum_a \mathrm{N}\,(a_1 a_2 \ldots a_m 1) = (p-m+1)\,\mathrm{N}_{m,1^p}\,;$$

hence

$$\sum_a \sum_{\cdots} \mathrm{N}\,(a_1 a_2 \ldots a_m \ldots) = \frac{n\,!}{(p+1)\,!}(p-m+1)\,\mathrm{N}_{m,1^p}.$$

Art. 28. Further, summing each side with respect to m,

$$\sum_m \sum_a \sum_{\cdots} \mathrm{N}\,(a_1 a_2 \ldots a_m \ldots)$$

$$= \frac{n\,!}{(p+1)\,!}\{p\mathrm{N}_{1,1^p} + (p-1)\,\mathrm{N}_{2,1^p} + \ldots + \mathrm{N}_{p,1^p}\}$$

$$= \frac{n\,!}{(p+1)\,!}\{\mathrm{N}_{1,1^p} + 2\mathrm{N}_{2,1^p} + \ldots + p\mathrm{N}_{p,1^p}\}\,;$$

but the sinister is of the form

$$\Sigma \mathrm{N}\,(w_1 w_2) \quad \text{(see Art. 20)}$$

and thus has the value $\frac{1}{2}n\,!$; hence

$$\mathrm{N}_{1,1^p} + 2\mathrm{N}_{2,1^p} + \ldots + p\mathrm{N}_{p,1^p} = \tfrac{1}{2}(p+1)\,!,$$

an interesting result.

Art. 29. From a previous result

$$\sum_a \mathrm{N}\,(a_1 a_2 \ldots a_m + 1) = (p+1)\,\mathrm{N}_{m,1^p} - \sum_a \mathrm{N}\,(a_1 a_2 \ldots a_m 1) = m\mathrm{N}_{m,1^p}\,;$$

hence

$$\sum_a \mathrm{N}\,(a_1 a_2 \ldots a_m + 1) = \sum_a \mathrm{N}\,(a_1 a_2 \ldots a_{p-m+1} 1) = m\mathrm{N}_{m,1^p}\,;$$

and it may be observed that the numbers, included in

$$\sum_a \mathrm{N}\,(a_1 a_2 \ldots a_m + 1),$$

are the conjugates of those included in

$$\sum_a \mathrm{N}\,(a_1 a_2 \ldots a_{p-m+1} 1).$$

Art. 30. Also since

$$\sum_a \mathrm{N}\,(a_1 a_2 \ldots a_m + 1) = \mathrm{N}_{m,1^{p+1}} - \sum_a \mathrm{N}\,(a_1 a_2 \ldots a_{m-1} 1),$$

$$\sum_a \mathrm{N}\,(a_1 a_2 \ldots a_{m-1} 1) + \sum_a \mathrm{N}\,(a_1 a_2 \ldots a_{p-m+1} 1) = \mathrm{N}_{m,1^{p+1}}\,;$$

and this leads to the relation

$$(p-m+2)\,\mathrm{N}_{p-m+2,1^p} + m\mathrm{N}_{m,1^p} = \mathrm{N}_{m,1^{p+1}}.$$

E.g., for $p = 3$, $m = 2$, $p-m+2 = 3$,

$$3\mathrm{N}_{3,1^3} + 2\mathrm{N}_{2,1^3} = \mathrm{N}_{2,1^4}\,;$$

verified by

$$3.1 \;+\; 2.4 \;=\; 11.$$

The result is convenient for the calculation of the numbers $N_{m,1^{p+1}}$ from the numbers $N_{m,1^p}$.

We have also the remarkable result that the probability of obtaining a permutation, such that the sum of the first m numbers of the descending specification is p, is independent of n, and has the value

$$\frac{(p-m+1)}{(p+1)!}N_{m,1^p};$$

whenever p is $n-1$ or less.

Art. 31. From the definition we have in respect of the permutations of n numbers

$$N_1+N_2+N_3+\ldots = n!$$

I shall now show that

$$N_{n-\theta+1}+\binom{n-\theta+1}{1}N_{n-\theta+2}+\binom{n-\theta+2}{2}N_{n-\theta+3}+\ldots$$

$$= \sum_{\nu}\frac{n!}{(2!)^{\nu_2}(3!)^{\nu_3}\ldots}\cdot\frac{\theta!}{\nu_1!\,\nu_2!\ldots};$$

the summation being for all values of

$$\nu_1,\ \nu_2,\ \ldots$$

such that

$$\Sigma\nu = \theta,$$

$$\Sigma s\nu_s = n.$$

The theorem is the outcome of the multiplication theorem of Art. 12.

Observing that, for all values of s,

$$N(1^s) = 1,$$

we have

$$\frac{(s_1+s_2)!}{s_1!\,s_2!}N(1^{s_1})N(1^{s_2}) = N(1^{s_1+s_2})+N(1^{s_1-1}21^{s_2-1}),$$

$$\frac{(s_1+s_2+s_3)!}{s_1!\,s_2!\,s_3!}N(1^{s_1})N(1^{s_2})N(1^{s_3}) = N(1^{s_1+s_2+s_3})+N(1^{s_1+s_2-1}21^{s_3-1})$$
$$+N(1^{s_1-1}21^{s_2+s_3-1})+N(1^{s_1-1}21^{s_2-2}21^{s_3-1});$$

and generally, for the product

$$\{N(1)\}^{\nu_1}\{N(1^2)\}^{\nu_2}\ldots\{N(1^p)\}^{\nu_p},$$

since $\Sigma s\nu_s = n$,

$$\frac{n!}{(1!)^{\nu_1}(2!)^{\nu_2}\ldots(p!)^{\nu_p}} = \text{a linear function of numbers } N(\ldots).$$

We may write down a similar result for every permutation of the factors of

$$\{N(1)\}^{\nu_1}\{N(1^2)\}^{\nu_2}\ldots\{N(1^p)\}^{\nu_p}$$

and, by addition, obtain

$$\frac{n!}{(1!)^{\nu_1}(2!)^{\nu_2}\ldots(p!)^{\nu_p}}\cdot\frac{\theta!}{\nu_1!\,\nu_2!\ldots\nu_p!} = \text{linear function of numbers } N,$$

where $\Sigma\nu = \theta$.

Further, we obtain a result of this nature for all values of

$$\nu_1, \nu_2, \ldots \nu_p,$$

such that $\Sigma s\nu_s = n$, $\Sigma\nu = \theta$; and, by addition, we obtain

$$\Sigma \frac{n!}{(1!)^{\nu_1}(2!)^{\nu_2}\ldots(p!)^{\nu_p}} \cdot \frac{\theta!}{\nu_1!\,\nu_2!\ldots\nu_p!} = \text{linear function of numbers N},$$

where $\Sigma s\nu_s = n$, $\Sigma\nu = \theta$.

We have now to determine the linear function of numbers N which appears on the dexter.

If one such number be

$$\mathrm{N}\,(abc\ldots),$$

it is evident that

$$(abc\ldots)$$

is some composition of the number n.

Consider the product of θ factors

$$\mathrm{N}\,(1^{s_1})\,\mathrm{N}\,(1^{s_2})\ldots\mathrm{N}\,(1^{s_\theta}),$$

where $\Sigma s = n$.

The process of multiplication produces N numbers of θ different kinds.

In the first place we throw all the units together,

$$\mathrm{N}\,(1^{s_1+s_2+\ldots+s_\theta}),$$

one N number containing n parts.

In the second place we combine a consecutive pair of factors and throw the remainder of the units together, thus producing $\theta-1$ N numbers each containing $n-1$ parts, viz.,

$$\mathrm{N}\,(1^{s_1-1}21^{s_2-1+s_3+\ldots+s_\theta}),$$
$$\mathrm{N}\,(1^{s_1+s_2-1}21^{s_3-1+s_4+\ldots+s_\theta}),$$
$$\vdots$$
$$\mathrm{N}\,(1^{s_1+s_2+\ldots+s_{\theta-1}-1}21^{s_\theta-1}).$$

In the third place we combine two consecutive pairs (including, of course, a consecutive three) of factors and throw the rêmainder of the units together, thus producing

$$\binom{\theta-1}{2}$$

N numbers each containing $n-2$ parts, viz., the series of which one is

$$\mathrm{N}\,(1^{s_1-1}21^{s_2-2}21^{s_3-1+s_4+\ldots+s_\theta}).$$

Notice that, if $s_2 = 1$, this becomes

$$\mathrm{N}\,(1^{s_1-1}31^{s_3-1+s_4+\ldots+s_\theta}).$$

We proceed in this manner until finally we combine $\theta-1$ consecutive pairs and throw the remainder of the units together, thus producing

$$\binom{\theta-1}{\theta-1}$$

N numbers, each containing $n-\theta+1$ parts.

Hence the compositions that present themselves are included in those enumerated by

$$N_n,\ N_{n-1},\ \ldots,\ N_{n-\theta+1}.$$

We have to consider the product

$$N(1^{s_1})N(1^{s_2})\ldots N(1^{s_\theta})$$

in all of its permutations and for every system of values of

$$s_1,\ s_2,\ \ldots,\ s_\theta,$$

such that

$$s_1+s_2+\ldots+s_\theta = n.$$

Hence, from considerations of symmetry, and attending to the *modus operandi* of the multiplication theorem, we find that the whole of the compositions enumerated by

$$N_n,\ N_{n-1},\ \ldots,\ N_{n-\theta+1}$$

present themselves.

Hence the linear function we seek is a linear function of

$$N_{n-\theta+1},\ N_{n-\theta+2},\ \ldots,\ N_{n-1},\ N_n,$$

and it remains to determine the coefficients.

The number of products, including permutations,

$$N(1^{s_1})N(1^{s_2})\ldots N(1^{s_\theta}),$$

which we have to consider, is equal to the numbers of compositions of n into θ parts, viz., it is

$$\binom{n-1}{n-\theta};$$

each of these produces

$$\binom{\theta-1}{m}$$

N numbers, each containing $n-m$ parts.

There are thus

$$\binom{n-1}{n-\theta}\binom{\theta-1}{m}$$

N numbers, each containing $n-m$ parts.

But there are only

$$\binom{n-1}{m}$$

different N numbers, each containing $n-m$ parts, because

$$\binom{n-1}{m}$$

is equal to the number of compositions of n into $n-m$ parts.

Hence, each N number, comprised in

$$N_{n-m},$$

will occur

$$\frac{\binom{n-1}{n-\theta}\binom{\theta-1}{m}}{\binom{n-1}{m}} = \binom{n-m-1}{n-\theta} \text{ times.}$$

Hence the required linear function is

$$\Sigma\binom{n-m-1}{n-\theta}N_{n-m},$$

or

$$N_{n-\theta+1}+\binom{n-\theta+1}{1}N_{n-\theta+2}+\binom{n-\theta+2}{2}N_{n-\theta+3}+\ldots+\binom{n-1}{\theta-1}N_n,$$

and the final result is

$$\Sigma\frac{n!}{(1!)^{\nu_1}(2!)^{\nu_2}\ldots(p!)^{\nu_p}}\cdot\frac{\theta!}{\nu_1!\nu_2!\ldots\nu_p!}$$
$$= N_{n-\theta+1}+\binom{n-\theta+1}{1}N_{n-\theta+2}+\binom{n-\theta+2}{2}N_{n-\theta+3}+\ldots+\binom{n-1}{\theta-1}N_n,$$

where

$$\Sigma s\nu_s = n,\ \Sigma\nu = \theta.$$

PART II.—SECTION 3.

Art. 32. In the preceding pages we have had under view the permutations of n different numbers. As I am now taking in hand the general case of numbers which possess any number of similarities, I find it convenient to slightly alter the point of view.

Let

$$\alpha, \beta, \gamma, \ldots$$

denote numbers in descending order of magnitude, and suppose there are

p number equal to α,

q ,, ,, β,

r ,, ,, γ.

so that, placed in descending order, the assemblages may be written

$$\alpha^p\beta^q\gamma^r\ldots.$$

I say that the assemblage is specified by the composition

$$(pqr\ldots).$$

As equalities may occur between the numbers p, q, r, ..., I take, for greater generality, the specifying composition

$$(p_1^{\pi_1}p_2^{\pi_2}\ldots).$$

It will be seen later that the order of occurrence of the parts of this composition is immaterial, so that we may consider the parts p_1, p_2, ... to be in descending order of magnitude and the specification to be denoted by a partition

$$(p_1^{\pi_1}p_2^{\pi_2}\ldots).$$

E.g., we obtain the same results for each of the six assemblages,

$$\alpha\alpha\alpha\beta\beta\gamma,\quad \alpha\alpha\alpha\beta\gamma\gamma,\quad \alpha\alpha\beta\beta\beta\gamma,$$
$$\alpha\alpha\beta\gamma\gamma\gamma,\quad \alpha\beta\beta\beta\gamma\gamma,\quad \alpha\beta\beta\gamma\gamma\gamma,$$

the specification of each assemblage being

$$(321).$$

Every permutation has a descending specification.

E.g.,

$$\alpha\beta\alpha\alpha\gamma\beta$$

has the descending specification

$$(231).$$

In the case considered in Part I. the assemblage of numbers had the specification

$$(1^n)$$

since there were no similarities, and the numbers N (...) were expressed in terms of the coefficients obtained by the multinomial expansion

$$(\alpha_1+\alpha_2+\alpha_3+\ldots)^n.$$

E.g., we found

$$\mathrm{N}(a) = \text{coefficient of symmetric function } (a) \text{ in the expansion,}$$
$$\mathrm{N}(ab)+\mathrm{N}(a+b) = \quad ,, \quad ,, \quad (ab) \quad ,, \quad ,,$$

where, in the first case, $a = n$, and in the second, $a+b = n$.

In a usual notation let

$$h_1,\ h_2,\ h_3,\ \ldots$$

denote the homogeneous product sums, of the successive orders, of the roots of the equation

$$x^n - a_1x^{n-1} + a_2x^{n-2} - a_3x^{n-3} + \ldots = 0\,;$$

we may say that, in Part I., the auxiliary generating function was

$$(\alpha_1+\alpha_2+\alpha_3+\ldots)^n = h_1^n,$$

α_1, α_2, α_3, ... being the roots of the equation.

Art. 33. In the present case the auxiliary generating function is

$$h_{p_1}^{\pi_1}h_{p_2}^{\pi_2}h_{p_3}^{\pi_3}\ldots,$$

as will appear.

For it was shown, *loc. cit.*, that the number of ways of distributing the objects, as specified, into different parcels containing a, b, c... objects respectively is the coefficient of the symmetric function

$$(abc...)$$

in the development of the symmetric function

$$h_{p_1}^{\pi_1} h_{p_2}^{\pi_2} h_{p_3}^{\pi_3} \ldots$$

as a sum of monomial symmetric functions.

Let this coefficient be denoted by

$$\mathrm{C}\,(abc...),$$

and let the number of arrangements of the objects, which have a descending specification

$$(abc...),$$

be denoted by

$$\mathrm{N}\,(abc...).$$

Let the whole number of objects be

$$\Sigma \pi p = n.$$

Then, when $a = n$, clearly

$$\mathrm{N}\,(a) = \mathrm{C}\,(a) = 1,$$

and when $a+b = n$, $\mathrm{C}\,(ab)$ is the number of arrangements into two different parcels containing a, b objects respectively, and by previous reasoning

$$\mathrm{N}\,(ab) + \mathrm{N}\,(a+b) = \mathrm{C}\,(ab)\,;$$

and, when $a+b+c = n$,

$$\mathrm{N}\,(abc) + \mathrm{N}\,(a+b,\, c) + \mathrm{N}\,(a,\, b+c) + \mathrm{N}\,(a+b+c) = \mathrm{C}\,(abc),$$

and so forth as in the simple case already considered.

Hence

$$\begin{aligned} \mathrm{N}\,(ab) &= \mathrm{C}\,(ab) - \mathrm{C}\,(a+b), \\ \mathrm{N}\,(abc) &= \mathrm{C}\,(abc) - \mathrm{C}\,(a+b,\, c) - \mathrm{C}\,(a,\, b+c) + \mathrm{C}\,(a+b+c), \\ \mathrm{N}\,(abcd) &= \mathrm{C}\,(abcd) - \mathrm{C}\,(a+b,\, c,\, d) - \mathrm{C}\,(a,\, b+c,\, d) - \mathrm{C}\,(a,\, b,\, c+d) \\ &\quad + \mathrm{C}\,(a+b,\, c+d) + \mathrm{C}\,(a,\, b+c+d) + \mathrm{C}\,(a+b+c,\, d) \\ &\quad - \mathrm{C}\,(a+b+c+d), \end{aligned}$$

&c.,

the numbers N being all expressible in terms of coefficients of the auxiliary generating function.

Art. 34. *E.g.* Take objects $\alpha\alpha\alpha\beta\beta\gamma$, where α, β, γ are in descending order of magnitude.

Since

$$\begin{aligned} h_3 h_2 h_1 = {} & (6) + 3\,(51) + 5\,(42) + 8\,(41^2) + 6\,(3^2) + 12\,(321) \\ & + 19\,(31^3) + 15\,(2^3) + 24\,(2^2 1^2) + 38\,(21^4) + 60\,(1^6), \end{aligned}$$

we calculate, from the above formulæ,

$$N(6) = 1,$$

$$N(51) = 3-1 = 2,$$

$$N(3^2) = 6-1 = 5,$$

$$N(321) = 12-5-2-1 = 4$$

and so on.

The five arrangements, enumerated by $N(3^2)$, are

$$\alpha\alpha\beta\alpha\beta\gamma$$
$$\alpha\alpha\gamma\alpha\beta\beta$$
$$\alpha\beta\gamma\alpha\alpha\beta$$
$$\alpha\beta\beta\alpha\alpha\gamma$$
$$\beta\beta\gamma\alpha\alpha\alpha,$$

each having the descending specification (3^2).

The four arrangements, enumerated by $N(321)$, are

$$\alpha\alpha\beta\alpha\gamma\beta$$
$$\alpha\alpha\gamma\beta\beta\alpha$$
$$\alpha\beta\gamma\alpha\beta\alpha$$
$$\alpha\beta\beta\alpha\gamma\alpha,$$

each having the descending specification (321).

The complete results *quâ* numbers specified by (321) are

$N(31^3)$, $N(1^33)$, $N(21^4)$, $N(1^42)$, $N(2112)$, $N(121^3)$, $N(1^321)$, $N(1^221^2)$, $N(1^6)$	each	0 . .	0
$N(6)$, $N(41^2)$, $N(1^24)$, $N(131^2)$, $N(1^231)$, $N(2^21^2)$, $N(1^22^2)$, $N(2121)$, $N(1212)$	,,	1 . .	9
$N(51)$, $N(15)$, $N(312)$, $N(213)$	,,	2 . .	8
$N(141)$, $N(1221)$	,,	3 . .	6
$N(42)$, $N(24)$, $N(321)$, $N(123)$	,,	4 . .	16
$N(3^2)$, $N(231)$, $N(132)$	,,	5 . .	15
$N(2^3)$	,,	6 . .	6
			60

60 being, of course, the total number of permutations of the objects.

Art. 35. The method of calculation establishes that the number N (...) is unaltered by reversal of the order of the numbers in the bracket.

Also that the results are only dependent upon the magnitudes of the parts in the specification of the assemblage and not upon the order of their occurrence.

General Investigation of a Generating Function.

Art. 36. I have shown above that, for numbers specified by

$$(p_1^{\pi_1} p_2^{\pi_2} \ldots),$$

an auxiliary generating function is

$$h_{p_1}^{\pi_1} h_{p_2}^{\pi_2} \ldots,$$

for, from its expansion in terms of monomial symmetric functions, the numbers

$$\text{N}\,(abc\ldots)$$

can be successively calculated.

For present convenience I take the above generating function to be

$$h_p h_q h_r \ldots$$

and recall that

$$\text{N}\,(abc\ldots) + \text{N}\,(a+b,\ c,\ \ldots) + \text{N}\,(a,\ b+c,\ \ldots) + \ldots$$

is equal to the coefficient of symmetric function

$$(abc\ldots)$$

in the expansion of

$$h_p h_q h_r \ldots.$$

The above linear function of the numbers

$$\text{N}\,(\ldots)$$

is formed by adding *adjacent* numbers

$$0,\ 1,\ 2,\ 3,\ \ldots,\ k \text{ at a time,}$$

where the numbers a, b, c, ... are k in number.

It thus comprises 2^{k-1} terms in general.

Art. 37. Let this linear function be denoted by

$$\theta_{\text{N}}\{(a)(b)(c)\ldots\},$$

so that if we write

$$h_p h_q h_r \ldots = \Sigma\text{C}\,(abc\ldots).(abc\ldots),$$

$$\theta_{\text{N}}\{(a)(b)(c)\ldots\} = \text{C}\,(abc\ldots).$$

From this system of linear relations is determined the set

$\text{N}\,(a) = \text{C}\,(a)$, where $a = n$,

$\text{N}\,(ab) = \text{C}\,(ab) - \text{C}\,(a+b)$, where $a+b = n$,

$\text{N}\,(abc) = \text{C}\,(abc) - \text{C}\,(a+b,\ c) - \text{C}\,(a,\ b+c) + \text{C}\,(a+b+c)$, where $a+b+c = n$, and so on;

N 2

the law of formation of the linear functions of the numbers

$$C(\ldots)$$

being similar to that which occurs in

$$\theta_N\{(a)(b)(c)\ldots\},$$

with the exception that the signs are alternately positive and negative, depending upon the numbers of integers in the brackets.

Art. 38. Denote this linear function of the numbers $C(\ldots)$ by

$$\phi_C\{(a)(b)(c)\ldots\},$$

so that

$$N(abc\ldots) = \phi_C\{(a)(b)(c)\ldots\}.$$

When it is necessary to put in evidence the numbers whose permutations are under examination we may write the two formulæ

$$\theta_N\{(a)(b)(c)\ldots\}_{(pqr\ldots)} = C(abc\ldots)_{(pqr\ldots)};$$

$$N(abc\ldots)_{(pqr\ldots)} = \phi_C\{(a)(b)(c)\ldots\}_{(ppr\ldots)}.$$

SECTION 4.

Digression on the Forms θ_N, ϕ_C.

Art. 39. Define in general, so that

$$\theta_N\{(a_1\ldots a_{s-1}a_s)(b_1b_2\ldots b_{t-1}b_t)(c_1c_2\ldots c_{u-1}c_u)(d_1d_2\ldots d_v)\ldots(k_1k_2\ldots k_z)\},$$

where there are k symbols $a, b, c, d, \ldots, k$, denotes the 2^{k-1} terms forming the series

$$\begin{aligned}&N(a_1\ldots k_z)\\&+N(a_1\ldots a_{s-1}, a_s+b_1, b_2\ldots k_z)\\&+N(a_1\ldots b_{t-1}, b_t+c_1, c_2\ldots k_z)\\&+\ldots\\&+N(a_1\ldots a_{s-1}, a_s+b_1, b_2\ldots b_{t-1}, b_t+c_1, c_2\ldots k_z)\\&+\ldots,\end{aligned}$$

where additions take place,

$0, 1, 2, \ldots, k-1$ at a time between the pairs a_s, b_1; b_t, c_1; c_u, d_1; ....

Art. 40. Similarly define

$$\phi_C\{(a_1\ldots a_{s-1}a_s)(b_1b_2\ldots b_{t-1}b_t)(c_1c_2\ldots c_{u-1}c_u)(d_1d_2\ldots d_v)\ldots(k_1k_2\ldots k_z)\}$$

to denote the 2^{k-1} terms forming the series

$$\begin{aligned}&C(a_1\ldots k_z)\\&-C(a_1\ldots a_{s-1}, a_s+b_1, b_2\ldots k_z)\\&-C(a_1\ldots b_{t-1}, b_t+c_1, c_2\ldots k_z)\\&-\ldots\\&+C(a_1\ldots a_{s-1}, a_s+b_1, b_2\ldots b_{t-1}, b_t+c_1, c_2\ldots k_z)\\&+\ldots,\end{aligned}$$

formed according to the same law, but the successive blocks of terms having alternately positive and negative signs.

Art. 41. I proceed to generalise the two results

$$\theta_N\{(a)(b)(c)\ldots\} = \phi_C(abc\ldots),$$

$$\theta_N(abc\ldots) = \phi_C\{(a)(b)(c)\ldots\}.$$

By definition

$$\theta_N\{(a_1\ldots a_{s-1}a_s)(b_1b_2\ldots b_t)\}$$

$$= N(a_1\ldots b_t) + N(a_1\ldots a_{s-1},\ a_s+b_1,\ b_2\ldots b_t);$$

and since

$$N(abc\ldots) = \theta_N(abc\ldots) = \phi_C\{(a)(b)(c)\ldots\},$$

this

$$= \phi_C\{(a_1)(a_2)\ldots(b_t)\} + \phi_C\{(a_1)\ldots(a_{s-1})(a_s+b_1)(b_2)\ldots(b_t)\}.$$

Now the sum of these two terms is precisely

$$\phi_C\{(a_1)(a_2)\ldots(a_{s-1})(a_sb_1)(b_2)\ldots(b_t)\},$$

because the terms involving

$$a_s+b_1$$

in

$$\phi_C\{(a_1)(a_2)\ldots(b_t)\}$$

are the same, with opposite sign, as those involved in

$$\phi_C\{(a_1)\ldots(a_{s-1})(a_s+b_1)(b_2)\ldots(b_t)\},$$

and therefore cancel them.

Hence the result

$$\theta_N\{(a_1\ldots a_{s-1}a_s)(b_1b_2\ldots b_t)\} = \phi_C\{(a_1)(a_2)\ldots(a_{s-1})(a_sb_1)(b_2)\ldots(b_t)\}.$$

Art. 42. Again

$$\theta_N\{(a_1\ldots a_s)(b_1\ldots b_t)(c_1\ldots c_u)\}$$

$$= N(a_1\ldots c_u) + N(a_1\ldots a_{s-1},\ a_s+b_1,\ b_2\ldots c_u)$$
$$+ N(a_1\ldots b_{t-1},\ b_t+c_1,\ c_2\ldots c_u)$$
$$+ N(a_1\ldots a_{s-1},\ a_s+b_1,\ b_2\ldots b_{t-1},\ b_t+c_1,\ c_2\ldots c_u),$$

$$= \theta_N(a_1\ldots c_u) + \theta_N(a_1\ldots a_{s-1},\ a_s+b_1,\ b_2\ldots c_u)$$
$$+ \theta_N(a_1\ldots b_{t-1},\ b_t+c_1,\ c_2\ldots c_u) + \theta_N(a_1\ldots a_{s-1},\ a_s+b_1,\ b_2\ldots b_{t-1},\ b_t+c_1,\ c_2\ldots c_u),$$

$$= \phi_C\{(a_1)\ldots(c_u)\} + \phi_C\{(a_1)\ldots(a_{s-1})(a_s+b_1)(b_2)\ldots(c_u)\}$$
$$+ \phi_C\{(a_1)\ldots(b_{t-1})(b_t+c_1)(c_2)\ldots(c_u)\} + \phi_C\{(a_1)\ldots(a_s+b_1)\ldots(b_t+c_1)\ldots(c_u)\},$$

$$= \phi_C\{(a_1)\ldots(a_{s-1})(a_sb_1)(b_2)\ldots(c_u)\}$$
$$+ \phi_C\{(a_1)\ldots(a_{s-1})(a_sb_1)(b_2)\ldots(b_t+c_1)(c_2)\ldots(c_u)\},$$

$$= \phi_C\{(a_1)\ldots(a_{s-1})(a_sb_1)(b_2)\ldots(b_{t-1})(b_tc_1)(c_2)\ldots(c_u)\},$$

by successive use of the formula Art. 41 above.

Also, clearly, if $t = 1$

$$\theta_N\{(a_1 \ldots a_s)(b_1)(c_1 \ldots c_u)\}$$
$$= \phi_C\{(a_1)\ldots(a_{s-1})(a_s b_1 c_1)(c_2)\ldots(c_u)\}.$$

Art. 43. Therefore, by induction, we can express any form

$$\theta_N\{\quad\}$$

as a form

$$\phi_C\{\quad\}.$$

The law is well seen by a particular case, viz.,

$$\begin{aligned}
\theta_N\{(a)(b)(c)(d)\} &= \phi_C(abcd),\\
\theta_N\{(ab)(c)(d)\} &= \phi_C\{(a)(bcd)\},\\
\theta_N\{(a)(bc)(d)\} &= \phi_C\{(ab)(cd)\},\\
\theta_N\{(a)(b)(cd)\} &= \phi_C\{(abc)(d)\},\\
\theta_N\{(a)(bcd)\} &= \phi_C\{(ab)(c)(d)\},\\
\theta_N\{(ab)(cd)\} &= \phi_C\{(a)(bc)(d)\},\\
\theta_N\{(abc)(d)\} &= \phi_C\{(a)(b)(cd)\},\\
\theta_N\{(abcd)\} &= \phi_C\{(a)(b)(c)(d)\}.
\end{aligned}$$

We have, in respect of the four letters, $8 = 2^3$ relations; the letters always occur in the order

$$a, b, c, d,$$

and to obtain the form $\phi_C\{\quad\}$, which is equated to a form $\theta_N\{\quad\}$, we may make use of the zig-zag conjugate law; *e.g.*, connect with

$$(ab)(cd)$$

the composition 22; take the zig-zag conjugate of this, viz., 121, and then write

$$\theta_N\{(ab)(cd)\} = \phi_C\{(a)(bc)(d)\},$$

and

$$\theta_N\{(a)(bc)(d)\} = \phi_C\{(ab)(cd)\};$$

and so in every case.

Art. 44. In the general case of p letters we obtain 2^{p-1} relations corresponding to the 2^{p-1} compositions of p; the relations are obtainable from zig-zag conjugation of such compositions and, in any relation

$$\theta_N\{\quad\} = \phi_C\{\quad\},$$

we may interchange the form-symbols

$$\theta_N, \phi_C.$$

Art. 45. In the above investigation we obtained incidentally certain linear relations between the forms

$$\theta_N,$$

and also between the forms

$$\phi_{\mathrm{C}},$$

which must now be set forth in a regular manner.

The former relations are of the type

$$\begin{aligned}
&\theta_{\mathrm{N}}\{(a_1\ldots a_s)(b_1\ldots b_t)(c_1\ldots c_u)\ldots\}\\
=\;&\theta_{\mathrm{N}}(a_1\ldots a_s b_1\ldots b_t c_1\ldots c_u\ldots)\\
&+\theta_{\mathrm{N}}(a_1\ldots a_{s-1},\ a_s+b_1,\ b_2\ldots b_t c_1\ldots c_u\ldots)\\
&+\theta_{\mathrm{N}}(a_1\ldots a_s b_1\ldots b_{t-1},\ b_t+c_1,\ c_2\ldots c_u\ldots)\\
&+\ldots\\
&+\theta_{\mathrm{N}}(a_1\ldots a_{s-1},\ a_s+b_1,\ b_2\ldots b_{t-1},\ b_t+c_1,\ c_2\ldots c_u\ldots)\\
&+\ldots;
\end{aligned}$$

this follows directly from the definition of the form $\theta_{\mathrm{N}}\{\quad\}$, since

$$\theta_{\mathrm{N}}(abc\ldots) = \mathrm{N}(abc\ldots).$$

Art. 46. The latter relations are of the type

$$\begin{aligned}
&\phi_{\mathrm{C}}\{(a_1\ldots a_3)(b_1\ldots b_t)(c_1\ldots c_u)\ldots\}\\
=\;&\phi_{\mathrm{C}}(a_1\ldots a_s b_1\ldots b_t c_1\ldots c_u\ldots)\\
&-\phi_{\mathrm{C}}(a_1\ldots a_{s-1},\ a_s+b_1,\ b_2\ldots b_t c_1\ldots c_u\ldots)\\
&-\phi_{\mathrm{C}}(a_1\ldots a_s b_1\ldots b_{t-1},\ b_t+c_1,\ c_2\ldots c_u\ldots)\\
&-\ldots\\
&+\phi_{\mathrm{C}}(a_1\ldots a_{s-1},\ a_s+b_1,\ b_2\ldots b_{t-1},\ b_t+c_1,\ c_2\ldots c_u\ldots)\\
&+\ldots;
\end{aligned}$$

which also follows directly from the definition of the form $\phi_{\mathrm{C}}\{\quad\}$, since

$$\phi_{\mathrm{C}}(abc\ldots) = \mathrm{C}(abc\ldots).$$

Art. 47. We have other linear relations of the type

$$\begin{aligned}
&\theta_{\mathrm{N}}\{(a_1\ldots a_s)(b_1\ldots b_t)(c_1\ldots c_u)\}\\
=\;&\theta_{\mathrm{N}}\{(a_1\ldots b_t)(c_1\ldots c_u)\}\\
&+\theta_{\mathrm{N}}\{(a_1\ldots a_{s-1},\ a_s+b_1,\ b_2\ldots b_t)(c_1\ldots c_u)\};
\end{aligned}$$

$$\begin{aligned}
&\phi_{\mathrm{C}}\{(a_1\ldots a_s)(b_1\ldots b_t)(c_1\ldots c_u)\}\\
=\;&\phi_{\mathrm{C}}\{(a_1\ldots b_t)(c_1\ldots c_u)\}\\
&-\phi_{\mathrm{C}}\{(a_1\ldots a_{s-1},\ a_s+b_1,\ b_2\ldots b_t)(c_1\ldots c_u)\}.
\end{aligned}$$

In fact, the law may be taken to operate as between any sets of consecutive factors in

$$\phi_{\mathrm{N}}\{\quad\}\quad\text{and}\quad\phi_{\mathrm{C}}\{\quad\}\ \text{respectively,}$$

leaving the remaining factors untouched.

Thus it is easy to verify the three relations

$$\begin{aligned} & \theta_N\{(ab)(cd)(ef)(gh)\} \\ = & \ \theta_N\{(abcd)(ef)(gh)\} \\ & +\theta_N\{(a,\ b+c,\ d)(ef)(gh)\}, \\ = & \ \theta_N\{(ab)(cd)(efgh)\} \\ & +\theta_N\{(ab)(cd)(e,\ f+g,\ h)\}, \\ = & \ \theta_N\{(abcd)(efgh)\} \\ & +\theta_N\{(abcd)(e,\ f+g,\ h)\} \\ & +\theta_N\{(a,\ b+c,\ d)(efgh)\} \\ & +\theta_N\{(a,\ b+c,\ d)(e,\ f+g,\ h)\}\ ; \end{aligned}$$

and the further three

$$\begin{aligned} & \phi_C\{(ab)(cd)(ef)(gh)\} \\ = & \ \phi_C\{(abcd)(ef)(gh)\} \\ & -\phi_C\{(a,\ b+c,\ d)(ef)(gh)\}, \\ = & \ \phi_C\{(ab)(cd)(efgh)\} \\ & -\phi_C\{(ab)(cd)(e,\ f+g,\ h)\}, \\ = & \ \phi_C\{(abcd)(efgh)\} \\ & -\phi_C\{(a,\ b+c,\ d)(efgh)\} \\ & -\phi_C\{(abcd)(e,\ f+g,\ h)\} \\ & +\phi_C\{(a,\ b+c,\ d)(e,\ f+g,\ h)\}. \end{aligned}$$

Art. 48. From these relations we may obtain new relations by transforming from θ_N to ϕ_C, or *vice versâ*.

Thus from relations of type

$$\theta_N\{(a)(b)(c)\} = \theta_N(abc)+\theta_N(a+b,\ c)+\theta_N(a,\ b+c)+\theta_N(a+b+c),$$

we obtain those of type

$$\phi_C(abc) = \phi_C\{(a)(b)(c)\}+\phi_C\{(a+b)(c)\}+\phi_C\{(a)(b+c)\}+\phi_C(a+b+c)\ ;$$

and from those of type

$$\phi_C\{(a)(b)(c)\} = \phi_C(abc)-\phi_C(a+b,\ c)-\phi_C(a,\ b+c)+\phi_C(a+b+c),$$

we obtain others of type

$$\theta_N(abc) = \theta_N\{(a)(b)(c)\}-\theta_N\{(a+b)(c)\}-\theta_N\{(a)(b+c)\}+\theta_N(a+b+c).$$

These new expressions for

$$\theta_N(abc\ldots) \quad \text{and} \quad \phi_C(abc\ldots),$$

with an obviously analogous law to that we have frequently met with, are of great importance.

From the relation

$$\phi_C\{(a)(b)(c)(d)\} = \phi_C\{(ab)(cd)\} - \phi_C\{(a+b)(cd)\} - \phi_C\{(ab)(c+d)\} + \phi_C\{(a+b)(c+d)\},$$

we obtain

$$\theta_N(abcd) = \theta_N\{(a)(bc)(d)\} - \theta_N\{(a+b,\ c)(d)\} - \theta_N\{(a)(b,\ c+d)\} + \theta_N\{(a+b,\ c+d)\};$$

and there is no necessity to give further examples.

SECTION 5.

Art. 49. The differential operator, of order s, that is so frequently of use in the theory of symmetric functions, viz. :—

$$\frac{1}{s!}(\partial_{a_1} + a_1\partial_{a_2} + a_2\partial_{a_3} + \ldots)^s = D_s,$$

can now be employed.

Remembering that operating upon monomial symmetric functions,

$$D_a(a) = 1,$$
$$D_a(b) = 0 \text{ unless } b = a,$$
$$D_aD_bD_c\ldots(abc\ldots) = 1;$$

and generally that D_a obliterates a number a from the partition of a function and causes it to vanish if no such number presents itself, it is clear that

$$D_aD_bD_c\ldots h_ph_qh_r\ldots = C(abc\ldots)_{(pqr\ldots)};$$

and thence if we write

$$\phi_D\{(a)(b)(c)\ldots\} = D_aD_bC_c\ldots - D_{a+b}D_c\ldots - D_aD_{b+c}\ldots - \ldots$$

according to a law *derivable* from that which defines

$$\phi_C\{(a)(b)(c)\ldots\} \text{ (see Art. 38)},$$

we find

$$N(abc\ldots)_{(pqr\ldots)} = \phi_D\{(a)(b)(c)\ldots\}h_ph_qh_r\ldots.$$

Art. 50. Observe that in the paper to which reference has been made it was shown that

$$C(abc\ldots)_{pqr\ldots} = C(pqr\ldots)_{abc\ldots}.$$

Two consequences flow from this fact.

Firstly

$$\theta_N\{(a)(b)(c)\ldots\}_{(pqr\ldots)} = \theta_N\{(p)(q)(r)\ldots\}_{(abc\ldots)},$$

which is a theorem of reciprocity for the numbers

$$N(\ldots).$$

Secondly, since

$$D_aD_bD_c\ldots h_ph_qh_r\ldots = D_pD_qD_r\ldots h_ah_bh_c\ldots,$$

$$N(abc\ldots)_{(pqr\ldots)} = D_pD_qD_r\ldots(h_ah_bh_c\ldots - h_{a+b}h_c\ldots - h_ah_{b+c}\ldots - \ldots);$$

where, on the dexter, the operand is a function formed from the functions $h_1, h_2, h_3, \ldots$ in the same manner as

$$\phi_D\{(a)(b)(c)\ldots\}$$

is formed from the operators

$$D_1, D_2, D_3, \ldots.$$

Hence

$$N(abc\ldots)_{(pqr\ldots)} = D_pD_qD_r\ldots\phi_h\{(a)(b)(c)\ldots\},$$

where

$$\phi_h\{(a)(b)(c)\ldots\} = h_ah_bh_c\ldots - h_{a+b}h_c\ldots - h_ah_{b+c}\ldots - \ldots.$$

Art. 51. I now write

$$\phi_h\{(a)(b)(c)\ldots\} = h_{abc\ldots};$$

so that

$$N(abc\ldots)_{(pqr\ldots)} = D_pD_qD_r\ldots h_{abc\ldots};$$

and it appears that

$$h_{abc\ldots}$$

is the true generating function of the numbers

$$N(abc\ldots)$$

for the permutations of assemblages of numbers of *all specifications.*

In fact,

$$h_{abc\ldots} = \Sigma N(abc\ldots)_{(pqr\ldots)} \cdot (pqr\ldots);$$

and the expansion of

$$h_{abc\ldots}$$

as a linear function of monomial symmetric functions gives a complete account of numbers

$$N(abc\ldots).$$

Art. 52. Before proceeding to a rapid examination of this new and most important symmetric function

$$h_{abc\ldots},$$

never before I believe introduced into algebraic analysis, I give complete tables of the numbers N (...) as far as $n = 6$.

$n = 2.$

	(2)	(1^2)
N (2)	1	1
N (1^2)		1

= specification.

$n = 3.$

	(3)	(21)	(1^3)	= specification.
N (3)	1	1	1	
N (21)		1	2	
N (1^3)			1	

$n = 4.$

	(4)	(31)	(2^2)	(21^2)	(1^4)	= specification.
N (4)	1	1	1	1	1	
N (31)		1	1	2	3	
N (2^2)		1	2	3	5	
N (121)			1	2	5	
N (21^2)				1	3	
N (1^4)					1	

$n = 5.$

	(5)	(41)	(32)	(31^2)	(2^21)	(21^3)	(1^5)	= specification.
N (5)	1	1	1	1	1	1	1	
N (41)		1	1	2	2	3	4	
N (32)		1	2	3	4	6	9	
N (131)			1	2	3	6	11	
N (2^21)			1	2	4	8	16	
N (31^2)				1	1	3	6	
N (212)				1	2	5	11	
N (121^2)					1	3	9	
N (21^3)						1	4	
N (1^5)							1	

$n = 6.$

	(6)	(51)	(42)	(3^2)	(41^2)	(321)	(2^3)	(31^3)	(2^21^2)	(21^4)	(1^6)	= specification.
N (6)	1	1	1	1	1	1	1	1	1	1	1	
N (51)		1	1	1	2	2	2	3	3	4	5	
N (42)		1	2	2	3	4	5	6	7	10	14	
N (3^2)		1	2	3	3	5	6	7	9	13	19	
N (141)			1	1	2	3	4	6	7	12	19	
N (231)			1	2	2	5	7	9	13	23	40	
N (312)					1	2	3	5	7	14	26	
N (321)			1	1	2	4	6	8	11	20	35	
N (2^3)			1	2	2	6	10	11	18	33	61	
N (41^2)					1	1	1	3	3	6	10	
N (31^3)								1	1	4	10	
N (12^21)				1		3	6	6	13	28	61	
N (2^21^2)						1	2	3	6	15	35	
N (131^2)						1	2	3	5	12	26	
N (2121)						1	3	3	7	17	40	
N (21^22)								1	2	7	19	
N (1^221^2)							1		2	6	19	
N (121^3)									1	4	14	
N (21^4)										1	5	
N (1^6)											1	

To explain—it will be found that

$$h_{131} = h_3h_1^2 - 2h_4h_1 + h_5,$$
$$= (32) + 2(31^2) + 3(2^21) + 6(21^3) + 11(1^5),$$

corresponding to row 4 of the table for $n = 5$.

Art. 53. Another symmetric function

$$a_{p_1p_2p_3\ldots}$$

is formed from the elements

$$a_1, a_2, a_3, \ldots$$

in the same manner as the symmetric function

$$h_{p_1p_2p_3\ldots}$$

from the elements

$$h_1, h_2, h_3, \ldots.$$

SECTION 6.

The Symmetric Functions $h_{p_1p_2p_3...}$, $a_{p_1p_2p_3...}$.

Art. 54. These two new functions are of fundamental importance, not only in this investigation, but in the theory of symmetric functions generally.

In regard to the algebraic equation

$$x^n - a_1x^{n-1} + a_2x^{n-2} - ... = 0,$$

$h_1, h_2, h_3, ...$ are the homogeneous product sums of the roots and the two sets of elements

$$a_1, a_2, a_3, ...,$$
$$h_1, h_2, h_3, ...$$

have reciprocal properties which it is useful to briefly glance at.

We have

$$h_1 = a_1 = (1),$$
$$h_2 = a_1^2 - a_2 = (2) + (1^2),$$
$$h_3 = a_1^3 - 2a_1a_2 + a_3 = (3) + (21) + (1^3),$$

and, in general,

$$h_n = \Sigma(-)^{n+\Sigma k} \frac{(\Sigma k)!}{k_1!k_2!...k_s!} a_1^{k_1}a_2^{k_2}...a_s^{k_s}.$$

The two series of elements are connected in such wise that, in any relation between the elements, the symbols a, h may be interchanged. Thus, from

$$a_1^2 - 3a_2 = -2h_1^2 + 3h_2$$

is derived

$$h_1^2 - 3h_2 = -2a_1^2 + 3a_2.$$

As a particular case it is found that a_s is the same function of the elements $h_1, h_2, h_3, ...$ that h_s is of the elements $a_1, a_2, a_3, ...$

If functions of the elements $h_1, h_2, h_3, ...$ be denoted by

$$f(h), \quad \phi(h),$$

we see that, if

$$f(h) = \phi(a),$$

so that

$$\phi(h) = f(a),$$

then

$$f(h)\phi(h) = f(a)\phi(a);$$

showing that

$$f(h)\phi(h)$$

is an absolute invariant *quâ* the transformation which replaces the elements

$$h_1, h_2, h_3, ...$$

by the elements

$$a_1, a_2, a_3, ...$$

Art. 55. With these necessary preliminary remarks I define a new function of weight n, viz. :—

$$h_{p_1p_2p_3\ldots},$$

where $p_1p_2p_3\ldots$ is any composition of the number n; of the given weight there are

$$2^{n-1}$$

such functions, one of which is clearly

$$h_n.$$

The complete definition is given by the multiplication law

$$h_{p_1p_2\ldots p_s}h_{q_1q_2\ldots q_t}$$
$$= h_{p_1p_2\ldots p_sq_1q_2\ldots q_t}+h_{p_1\ldots p_{s-1},\, p_s+q_1,\, q_2\ldots q_t},$$

where the functions

$$h_{p_1p_2\ldots p_s},\quad h_{q_1q_2\ldots q_t}$$

are, or are not, of the same weight.

Art. 56. A second new function

$$a_{p_1p_2p_3\ldots}$$

is similarly defined by the same law; viz.,

$$a_{p_1p_2\ldots p_s}a_{q_1q_2\ldots q_t}$$
$$= a_{p_1\ldots p_sq_1\ldots q_t}+a_{p_1\ldots p_{s-1},\, p_s+q_1,\, q_2\ldots q_t}.$$

What follows applies generally to both of the new functions.

Art. 57. Since the multiplication is commutative, we have the first important property, viz.,

$$h_{p_1p_2\ldots p_sq_1q_2\ldots q_t}+h_{p_1\ldots p_{s-1},\, p_s+q_1,\, q_2\ldots q_t}$$
$$= h_{q_1\ldots q_tp_1\ldots p_s}+h_{q_1\ldots q_{t-1},\, q_t+p_1,\, p_2\ldots p_s}.$$

Art. 58. Every product of elementary functions is expressible in terms of the new functions, *e.g.*,

$$h_ph_q = h_{pq}+h_{p+q},$$
$$h_ph_qh_r = h_{pqr}+h_{p+q,\,r}+h_{p,\,q+r}+h_{p+q+r};$$

and in general

$$h_{p_1}h_{p_2}h_{p_3}\ldots h_{p_s} = \theta_h\{(p_1)(p_2)(p_3)\ldots(p_s)\},$$

where, in $\theta_h\{\quad\}$, the sum of the coefficients is

$$2^{s-1}.$$

These relations show that

$$h_{p_1p_2\ldots p_s} = h_{p_s\ldots p_2p_1}.$$

Art. 59. Similarly

$$h_{pq} = h_ph_q-h_{p+q},$$
$$h_{pqr} = h_ph_qh_r-h_{p+q}h_r-h_ph_{q+r}+h_{p+q+r};$$

and in general

$$h_{p_1p_2\ldots p_s} = \phi_h\{(p_1)(p_2)\ldots(p_s)\}.$$

If, moreover, we define

$$\theta_h\{(p_1p_2\ldots p_s)(q_1q_2\ldots q_t)(r_1r_2\ldots r_u)\}$$

as denoting

$$h_{p_1\ldots p_s q_1\ldots q_t r_1\ldots r_u}+h_{p_1\ldots p_{s-1},\, p_s+q_1,\, q_2\ldots q_t r_1\ldots r_u}$$
$$+h_{p_1\ldots p_s q_1\ldots q_{t-1},\, q_t+r_1,\, r_2\ldots r_u}+h_{p_1\ldots p_{s-1},\, p_s+q_1,\, q_2\ldots q_{t-1},\, q_t+r_1,\, r_2\ldots r_u},$$

and

$$\phi_h\{(p_1\ldots p_{s-1}p_s)\,(q_1q_2\ldots q_{t-1}q_t)\,(r_1r_2\ldots r_u)\}$$

as denoting

$$h_{p_1\ldots p_s}h_{q_1\ldots q_t}h_{r_1\ldots r_u}-h_{p_1\ldots p_{s-1},\, p_s+q_1,\, q_2\ldots q_t}h_{r_1\ldots r_u}$$
$$-h_{p_1\ldots p_s}h_{q_1\ldots q_{t-1},\, q_t+r_1,\, r_2\ldots r_u}+h_{p_1\ldots p_{s-1},\, p_s+q_1,\, q_2\ldots q_{t-1},\, q_t+r_1,\, r_2\ldots r_u},$$

according to the law usual in this subject; we find

$$\theta_h\{(p_1\ldots p_{s-1}p_s)\,(q_1q_2\ldots q_{t-1}q_t)\,(r_1r_2\ldots r_u)\}$$
$$= h_{p_1\ldots p_{s-1}p_s}h_{q_1q_2\ldots q_{t-1}q_t}h_{r_1r_2\ldots r_u};$$

and

$$\phi_h\{(p_1\ldots p_{s-1}p_s)\,(q_1q_2\ldots q_{t-1}q_t)\,(r_1r_2\ldots r_u)\}$$
$$= \phi_h\{(p_1\ldots p_{s-1}\,p_sq_1\ldots q_t)\,(r_1\ldots r_u)\},$$
$$= \phi_h\{(p_1\ldots p_s)\,(q_1\ldots q_tr_1\ldots r_u)\},$$
$$= \phi_h\{(p_1\ldots p_sq_1\ldots q_tr_1\ldots r_u)\},$$
$$= h_{p_1\ldots p_sq_1\ldots q_tr_1\ldots r_u}.$$

The reader may verify that

$$\phi_h\{(p)\,(q)\,(r)\},\ \phi_h\{(pq)\,(r)\},\ \phi_h\{(p)\,(qr)\},\ \phi_h\{(pqr)\}$$

have each the same value

$$h_ph_qh_r-h_{p+q}h_r-h_ph_{q+r}+h_{p+q+r},$$

which we have denoted by

$$h_{pqr}.$$

Art. 60. I pass on to drag into the light some important relations connecting

$$h_{p_1\ldots p_s} \text{ and } a_{p_1\ldots p_s}.$$

When the relation

$$h_n = \Sigma(-)^{n+\Sigma k}\frac{(\Sigma k)!}{k_1!\ldots k_s!}a_1^{k_1}\ldots a_s^{k_s}$$

was under observation just now, it will not have escaped notice that this is precisely the expansion of

$$a_{1^n}$$

for

$$a_{11} = a_1^2-a_2,$$
$$a_{111} = a_1^3-2a_1a_2+a_3;\ \&c.,$$

and by the law of formation we see that

$$a_{1^n} = h_n,$$

and thence

$$a_n = h_{1^n}.$$

The known value of h_n is thus given by a law identical with the multiplication law of this paper, and the expression of h_n in terms of

$$a_1,\ a_2,\ a_3,\ \ldots$$

is completely given by

$$h_n = a_{1^n}.$$

This new statement, of a well-known law, immediately suggests the generalization to which I proceed.

Observe that

$$n \text{ and } 1^n$$

are zig-zag conjugate compositions.

From the relation

$$a_{pq} = a_p a_q - a_{p+q}$$

is now deduced

$$a_{pq} = h_{1^p} h_{1^q} - h_{1^{p+q}};$$

and, since

$$h_{1^p} h_{1^q} = h_{1^{p+q}} + h_{1^{p-1}21^{q-1}},$$

$$a_{pq} = h_{1^{p-1}21^{q-1}};$$

and we again observe that

$$pq \text{ and } 1^{p-1}21^{q-1}$$

are zig-zag conjugate compositions.

Hence writing

$$(1^{p-1}21^{q-1}) = (pq)',$$

$$a_{(pq)} = h_{(pq)'};$$

and, in general, I have established (but reserve the proof for another occasion) that

$$a_{(p_1p_2\ldots)} = h_{(p_1p_2\ldots)'},$$

where

$$(p_1p_2\ldots),\ (p_1p_2\ldots)'$$

are zig-zag conjugate compositions.

Art. 61. The theorem has an interest of its own, but it is also of vital importance in this investigation. This importance consists partly in the circumstance that the functions

$$h_{pqr\ldots}$$

are those which naturally arise in the present theory of permutations. The present theorem enables the immediate expression of them in terms of the elementary symmetric functions

$$a_1,\ a_2,\ a_3, \ldots$$

and thus they may be more easily dealt with by symmetric functions differential operators. In fact, the homogeneous product sums

$$h_1,\ h_2,\ h_3, \ldots$$

can be made to disappear from the investigation; but, as will be seen, it is sometimes advantageous to retain them wholly or in part.

Art. 62. To gain familiarity with the new functions I give without proof some of their elementary properties.

$$s_n = a_{1^n} - a_{21^{n-2}} + a_{31^{n-3}} - \ldots (-)^{n+1} a_n$$
$$= (-)^{n+1} \{h_{1^n} - h_{21^{n-2}} + h_{31^{n-3}} - \ldots (-)^{n+1} h_n\},$$

where s_n is the sum of the nth power of the roots.

The following expression for $a_{s1^{n-s}}$

$$a_{s1^{n-s}} = a_{s-1} h_{n-s+1} - a_{s-2} h_{n-s+2} + \ldots (-)^{s+1} h_n.$$

The result of operations with D_p, viz.,

$$D_p a_{s1^{n-s}} = a_{s-1} h_{n-s-p+1}.$$

If $s_{n,t}$ denote the sum of the symmetric functions whose partitions contain exactly t parts, we have the companion tables, in which the law is obvious.

	$s_{n,1}$.	$s_{n,2}$.	$s_{n,3}$.	$s_{n,4}$.	$s_{n,5}$.	$s_{n,6}$.
a_{1^n}	1	+1	+1	+1	+1	+ 1
$a_{21^{n-2}}$		1	+2	+3	+4	+ 5
$a_{31^{n-3}}$			1	+3	+6	+10
$a_{41^{n-4}}$				1	+4	+10
$a_{51^{n-5}}$					1	+ 5
$a_{61^{n-6}}$						1

	a_{1^n}.	$a_{21^{n-2}}$.	$a_{31^{n-3}}$.	$a_{41^{n-4}}$.	$a_{51^{n-5}}$.	$a_{61^{n-6}}$.
$s_{n,1}$	1	- 1	+1	- 1	+1	- 1
$s_{n,2}$		1	- 2	+3	- 4	+ 5
$s_{n,3}$			1	- 3	+6	- 10
$s_{n,4}$				1	- 4	+10
$s_{n,5}$					1	- 5
$s_{n,6}$						1

The fundamental properties of these new symmetric functions were communicated by me to Section A of the British Association for the Advancement of Science, at the York meeting, 1906, August 1–8.

Art. 63. The generating function of N $(abc\ldots)$ is either

$$h_{abc\ldots} \quad \text{or} \quad a_{(abc\ldots)'}.$$

We can now determine the highest symmetric function, in dictionary order of the parts, which occurs in the development of $h_{abc\ldots}$. This, by the known theory of symmetric functions, is obtained from the form

$$a_{(abc\ldots)'}$$

by expressing $(abc\ldots)'$ as a partition and taking the Ferrers conjugate $(abc\ldots)''$; then we see that no symmetric function, prior in dictionary order to

$$(abc\ldots)'',$$

can appear.

Also the highest integer in
$$(abc\ldots)'$$
is the lower limit of the number of parts, occurring in the partition of a symmetric function, arising from the development of

$$h_{abc\ldots}.$$

E.g., since
$$h_{141} = a_{21^22},$$

we arrange 21^22 as a partition, obtaining 2^21^2, and taking the Ferrers conjugate from the graph

we reach (42) as the highest symmetric function in dictionary order that occurs in the development of h_{141}.

Hence
$$\mathrm{N}(141)_6 = \mathrm{N}(141)_{51} = 0.$$
(See the table of weight 6.)

Numerous relations such as
$$h_{141}+h_{51} = h_{41^2}+h_{42}$$
can be verified by the same table.

Art. 64. Before proceeding to establish the multiplication theorem, the generalization of that in Part I., it is necessary to examine the mode of operation of the differential operator
$$\mathrm{D}_a$$
upon a product
$$h_{p_1}^{\pi_1}h_{p_2}^{\pi_2}\ldots,$$
or
$$a_{p_1}^{\pi_1}a_{p_2}^{\pi_2}\ldots.$$

It is clear that
$$\mathrm{D}_a h_p = h_{p-a}.$$

In the paper it was shown that

$$\mathrm{D}_a h_p h_q = \Sigma\,(\mathrm{D}_{a'}h_p)(\mathrm{D}_{a''}h_q),$$

where $a'a''$ denotes a composition of a into two parts, zero not excluded, and the summation is for every such composition.

Hence $$D_a h_p h_q = \Sigma h_{p-a'} h_{q-a''}.$$

E.g., $$D_4 h_4 h_3 = h_3 + h_1 h_2 + h_3 + h_2 h_1 = 2h_3 + 2h_2 h_1,$$

where the compositions of 4 have been taken in the order

$$40,\ 31,\ 13,\ 22.$$

In general $$D_a h_{p_1} h_{p_2} \ldots h_{p_s} = \Sigma h_{p_1 - a'} h_{p_2 - a''} \ldots h_{p_s - a^{(s)}},$$

where $$a'a'' \ldots a^{(s)}$$

is a composition of a into s or fewer parts.

It is to be noted that in forming the compositions zeros are parts, so that, for instance, $$400,\ 040,\ 004$$

count as different compositions.

If the operand be $$a_{p_1} a_{p_2} \ldots a_{p_s},$$

since $$D_a (a_{p_1}) = 0 \text{ unless } a = 1,$$

we need only attend to the compositions composed of units and zeros.

Thus
$$D_2 a_4 a_5 a_6 a_7 = a_3 a_4 a_6 a_7 + a_3 a_5^2 a_7 + a_3 a_5 a_6^2 + a_4^2 a_5 a_7 + a_4^2 a_6^2 + a_4 a_5^2 a_6.$$

It is easy to show that $$D_1 h_{ab} = h_{a-1,b} + h_{a,b-1} + h_{a+b-1},$$

from which
$$N(ab)_{pqr\ldots 1} = N(a-1,\ b)_{pqr\ldots} + N(a,\ b-1)_{pqr\ldots} + N(a+b-1)_{pqr\ldots};$$

and, particularly,
$$N(42)_{2^2 1^2} = N(32)_{2^2 1} + N(41)_{2^2 1} + N(5)_{2^2 1}$$
$$7 \quad = \quad 4 \quad + \quad 2 \quad + \quad 1$$

from the table.

Similar formulæ can be established at pleasure.

The Conjugate Law.

Art. 65. It has been seen (Art. 6) that, when the numbers permuted are specified by

$$1^n,$$
$$N(pq\ldots) = N(pq\ldots)',$$

where $$(pq\ldots),\quad (pq\ldots)'$$

denote conjugate compositions.

We write the theorem $$N(pq\ldots)_{(1^n)} = N(pq\ldots)'_{(1^n)};$$

and we may inquire into the existence of an analogous theorem when the numbers permuted have any other specification.

Consider the expression
$$h_{(pq\ldots)}-h_{(pq\ldots)'},$$
which is the generating function for the difference between
$$\mathrm{N}\,(pq\ldots) \quad\text{and}\quad \mathrm{N}\,(pq\ldots)',$$
for all specifications of the numbers permuted.

The generating function may be written
$$h_{pq\ldots}-a_{pq\ldots},$$
according to the theorem proved above.

The differential operation
$$\mathrm{D}_1$$
has the equivalent forms
$$\partial_{a_1}+a_1\partial_{a_2}+a_2\partial_{a_3}+\ldots$$
$$\partial_{h_1}+h_1\partial_{h_2}+h_2\partial_{h_3}+\ldots;$$
hence
$$\mathrm{D}_1^{\mu}h_{pq\ldots}$$
is the same function of
$$h_1,\ h_2,\ h_3,\ \ldots$$
that
$$\mathrm{D}_1^{\mu}a_{pq\ldots}$$
is of
$$a_1,\ a_2,\ a_3,\ \ldots.$$

It follows at once that
$$\mathrm{D}_1^{n}\,(h_{pq\ldots}-a_{pq\ldots}) = 0,$$
equivalent to the known result
$$\mathrm{N}\,(pq\ldots)_{(1^n)} = \mathrm{N}\,(pq\ldots)'_{(1^n)}$$
already found.

Art. 66. Now, considering the generating functions
$$h_{(pq\ldots)}+h_{(pq\ldots)'}$$
or
$$h_{(pq\ldots)}+a_{(pq\ldots)},$$
$$\mathrm{D}_1^{n-2}\{h_{pq\ldots}+a_{pq\ldots}\}$$
must be of the form
$$\mathrm{A}h_2+\mathrm{B}h_1^2+\mathrm{A}a_2+\mathrm{B}a_1^2,$$
or
$$(\mathrm{A}+2\mathrm{B})\{(2)+2(1^2)\}.$$

Hence
$$\mathrm{D}_1^{n-2}\mathrm{D}_2\,(h_{pq\ldots}+a_{pq\ldots})$$
$$= \tfrac{1}{2}\mathrm{D}_1^{n}\,(h_{pq\ldots}+a_{pq\ldots})$$
$$= \mathrm{D}_1^{n}h_{pq\ldots},$$
equivalent to
$$\mathrm{N}\,(pq\ldots)_{(21^{n-2})}+\mathrm{N}\,(pq\ldots)'_{(21^{n-2})} = \mathrm{N}\,(pq\ldots)_{(1^n)}.$$

Thus, from the table $n = 6$,
$$\mathrm{N}\,(33)_{(21^4)}+\mathrm{N}\,(1^2 21^2)_{(21^4)} = \mathrm{N}\,(33)_{(1^6)}$$
$$13 \quad + \quad 6 \quad = \quad 19.$$

Art. 67. Again, operating with

$$D_1^{n-3},$$

upon

$$h_{pq\ldots}-a_{pq\ldots},$$

we obtain a result of the form

$$A(h_3-a_3)+B(h_2h_1-a_2a_1)$$

or

$$(A+B)\{(3)+(21)\};$$

hence

$$D_1^{n-3}D_3(h_{pq\ldots}-a_{pq\ldots}) = D_1^{n-2}D_2(h_{pq\ldots}-a_{pq\ldots}),$$

equivalent to

$$N(pq\ldots)_{(31^{n-3})}-N(pq\ldots)'_{(31^{n-3})} = N(pq\ldots)_{(21^{n-2})}-N(pq\ldots)'_{(21^{n-2})};$$

and, particularly, from the table

$$\begin{array}{ccccccc} N(321)_{(31^3)} & - & N(2^21^2)_{(31^3)} & = & N(321)_{(21^4)} & - & N(2^21^2)_{(21^4)} \\ 8 & - & 3 & = & 20 & - & 15. \end{array}$$

No new result is obtained by taking

$$h_{pq\ldots}+a_{pq\ldots}$$

as the operand.

Art. 68. Further,

$$D_1^{n-4}(h_{pq\ldots}-a_{pq\ldots})$$

has the form

$$A(h_4-a_4)+B(h_3h_1-a_3a_1)+C(h_2^2-a_2^2)+D(h_2h_1^2-a_2a_1^2),$$

reducing to

$$(A+B+C+D)(4)+(A+2B+2C+2D)\{(31)+(2^2)+(21^2)\},$$

equivalent to the new result

$$N(pq\ldots)_{(2^21^{n-4})}-N(pq\ldots)'_{(2^21^{n-4})} = N(pq\ldots)_{(31^{n-3})}-N(pq\ldots)'_{(31^{n-3})};$$

and, particularly, from the table

$$\begin{array}{ccccccc} N(2^3)_{(2^21^2)} & - & N(12^21)_{(2^21^2)} & = & N(2^3)_{(31^3)} & - & N(12^21)_{(31^3)} \\ 18 & - & 13 & = & 11 & - & 6. \end{array}$$

Art. 69. If we take here the operand to be

$$h_{pq\ldots}+a_{pq\ldots},$$

a new result is obtained, viz.,

$$2N(pq\ldots)_{(41^{n-4})}+2N(pq\ldots)'_{(41^{n-4})}+N(pq\ldots)_{(21^{n-2})}+N(pq\ldots)'_{(21^{n-2})}$$
$$= 2N(pq\ldots)_{(31^{n-3})}+2N(pq\ldots)'_{(31^{n-3})}+N(pq\ldots)_{2^21^{n-4}}+N(pq\ldots)'_{(2^21^{n-4})}.$$

The above is sufficient to indicate the nature of the results which present themselves; I have not attempted to generalise them. The question appears to be a difficult one.

SECTION 7.

Generalisation of the Multiplication Theorem.

Art. 70. I will establish the result

$$\Sigma\Sigma\Sigma\ldots\{\mathrm{N}(a_1a_2a_3\ldots)_{(p_1q_1r_1\ldots)}\mathrm{N}(b_1b_2b_3\ldots)_{(p_2q_2r_2\ldots)}\mathrm{N}(c_1c_2c_3\ldots)_{(p_3q_3r_3\ldots)}\ldots\}$$
$$= \theta_{\mathrm{N}}\{(a_1a_2a_3\ldots)(b_1b_2b_3\ldots)(c_1c_2c_3\ldots)\ldots\}_{(pqr\ldots)},$$

where the summation is for all solutions of the diophantine equations

$$p_1+q_1+r_1+\ldots = \Sigma a,$$
$$p_2+q_2+r_2+\ldots = \Sigma b,$$
$$p_3+q_3+r_3+\ldots = \Sigma c,$$
$$\cdot\quad\cdot\quad\cdot\quad\cdot\quad\cdot\quad\cdot\quad\cdot$$
$$p_1+p_2+p_3+\ldots = p,$$
$$q_1+q_2+q_3+\ldots = q,$$
$$r_1+r_2+r_3+\ldots = r,$$
$$\cdot\quad\cdot\quad\cdot\quad\cdot\quad\cdot\quad\cdot$$

For consider

$$\theta_{\mathrm{N}}\{(a_1a_2)(b_1b_2)\}$$
$$= \phi_{\mathrm{C}}\{(a_1)(a_2b_1)(b_2)\},$$
$$= \mathrm{C}(a_1a_2b_1b_2)_{(pqr\ldots)}-\mathrm{C}(a_1+a_2,\, b_1b_2)_{(pqr\ldots)}-\mathrm{C}(a_1a_2,\, b_1+b_2)_{(pqr\ldots)}+\mathrm{C}(a_1+a_2,\, b_1+b_2)_{(pqr\ldots)}$$
$$= (\mathrm{D}_{a_1}\mathrm{D}_{a_2}\mathrm{D}_{b_1}\mathrm{D}_{b_2}-\mathrm{D}_{a_1+a_2}\mathrm{D}_{b_1}\mathrm{D}_{b_2}-\mathrm{D}_{a_1}\mathrm{D}_{a_2}\mathrm{D}_{b_1+b_2}+\mathrm{D}_{a_1+a_2}\mathrm{D}_{b_1+b_2})h_ph_qh_r\ldots$$
$$= (\mathrm{D}_{a_1}\mathrm{D}_{a_2}-\mathrm{D}_{a_1+a_2})(\mathrm{D}_{b_1}\mathrm{D}_{b_2}-\mathrm{D}_{b_1+b_2})h_ph_qh_r\ldots$$
$$= \mathrm{D}_{a_1a_2}\mathrm{D}_{b_1b_2}h_ph_qh_r\ldots = \mathrm{D}_p\mathrm{D}_q\mathrm{D}_r\ldots h_{a_1a_2}h_{b_1b_2}.$$

Now

$$\mathrm{D}_p\mathrm{D}_q\mathrm{D}_r\ldots h_{a_1a_2}h_{b_1b_2}$$
$$= \Sigma\Sigma\Sigma\ldots(\mathrm{D}_{p_1}\mathrm{D}_{q_1}\mathrm{D}_{r_1}\ldots h_{a_1a_2})(\mathrm{D}_{p_2}\mathrm{D}_{q_2}\mathrm{D}_{r_2}\ldots h_{b_1b_2}),$$

the summation being for all solutions of the diophantine equations

$$p_1+q_1+r_1+\ldots = a_1+a_2,$$
$$p_2+q_2+r_2+\ldots = b_1+b_2,$$
$$p_1+p_2 = p,$$
$$q_1+q_2 = q,$$
$$r_1+r_2 = r,$$
$$\cdot\quad\cdot\quad\cdot\quad\cdot$$

Moreover,

$$\mathrm{D}_{p_1}\mathrm{D}_{q_1}\mathrm{D}_{r_1}\ldots h_{a_1a_2} = \mathrm{N}(a_1a_2)_{p_1q_1r_1\ldots}.$$

Hence

$$\theta_{\mathrm{N}}\{(a_1a_2)(b_1b_2)\}_{(pqr\ldots)}$$
$$= \Sigma\,\mathrm{N}(a_1a_2)_{p_1q_1r_1\ldots}\mathrm{N}(b_1b_2)_{p_2q_2r_2\ldots}\,;$$

and, by like reasoning, the theorem as enunciated follows.

As examples, $$N(321)_{(222)}+N(33)_{(222)} = 3N(32)_{(221)},$$

derived from $$\theta_N\{(32)(1)\}_{(222)};$$

and $$N(24)_{(222)}+N(231)_{(222)}+N(213)_{(222)}+N(222)_{(222)}$$
$$+N(2211)_{(222)}+N(2121)_{(222)}+N(2112)_{(222)}+N(21111)_{(222)}$$
$$= 6N(21)_{(111)}+18N(21)_{(21)},$$

derived from $$\theta_N\{(21)(1)^3\}_{(222)},$$

Art. 71. The enumeration of the permutations, whose specifications contain a given largest integer, will now be investigated.

Let $$I_m,\ J_m,\ K_m$$

denote respectively $$\Sigma N(abc\ldots)$$

in which

(i.) the highest of the integers $a, b, c, \ldots$ is m or less ; .
(ii.) ,, ,, ,, ,, or greater ;
(iii.) ,, ,, ,, ,, exactly ;

so that, when $a+b+c+\ldots = n$,

$$I_m = K_1+K_2+\ldots+K_m,$$
$$J_m = K_m+K_{m+1}+\ldots+K_m,$$
$$I_m = J_1 = I_m+J_m-K_m,$$
$$I_m-I_{m-1} = J_m-J_{m+1} = K_m,$$

$$\theta_N(n) = N(n) = J_n = K_n = I_n-I_{n-1};$$

and, since

$$\theta_N\{(n-1)(1)\} = N(n-1, 1)+N(n),$$
$$\theta_N\{(1)(n-1)\} = N(1, n-1)+N(n),$$

we find

$$2\theta_N\{(n-1)(1)\}$$
$$= K_{n-1}+2K_n = J_{n-1}+J_n = -I_{n-1}-I_{n-2}+2I_n;$$

also

$$\theta_N\{(n-2)(1)^2\} = N(n-2, 1^2)+N(n-1, 1)+N(n-2, 2)+N(n),$$
$$\theta_N\{(1)(n-2)(1)\} = N(1, n-2, 1)+N(n-1, 1)+N(1, n-1)+N(n),$$
$$\theta_N\{(1)^2(n-2)\} = N(1^2, n-2)+N(1, n-1)+N(2n-2)+N(n),$$

and by addition $$3\theta_N\{(n-2)(1)^2\} = K_{n-2}+2K_{n-1}+3K_n,$$
$$= J_{n-2}+J_{n-1}+J_n,$$
$$= -I_{n-3}-I_{n-2}-I_{n-1}+3I_n,$$

the law apparent here obtains so long as a number $n-\nu$ appearing in

$$N(\quad),$$

on the right-hand side, is not equal to any other number in the same bracket ; so that, *when* $s<\frac{1}{2}n$,

$$(s+1)\,\theta_{\mathrm{N}}\{(n-s)(1)^{s}\}$$
$$= \mathrm{K}_{n-s}+2\mathrm{K}_{n-s+1}+3\mathrm{K}_{n-s+2}+\ldots+(s+1)\,\mathrm{K}_{n},$$
$$= \mathrm{J}_{n-s}+\mathrm{J}_{n-s+1}+\mathrm{J}_{n-s+2}+\ldots+\mathrm{J}_{n},$$
$$= -\mathrm{I}_{n-s-1}-\mathrm{I}_{n-s}-\mathrm{I}_{n-s+1}-\ldots+(s+1)\,\mathrm{I}_{n}.$$

Hence

$$\mathrm{J}_{m} = (n-m+1)\,\theta_{\mathrm{N}}\{(m)(1)^{n-m}\}-(n-m)\,\theta_{\mathrm{N}}\{(m+1)(1)^{n-m-1}\},$$
$$\mathrm{K}_{m} = (n-m+1)\,\theta_{\mathrm{N}}\{(m)(1)^{n-m}\}-2(n-m)\,\theta_{\mathrm{N}}\{(m+1)(1)^{n-m-1}\}$$
$$+(n-m-1)\,\theta_{\mathrm{N}}\{(m+2)(1)^{n-m-2}\},$$

and, the specification of the numbers permuted being

$$(pqr\ldots),$$
$$\mathrm{I}_{m} = \frac{n!}{p!q!r!\ldots}-(n-m)\,\theta_{\mathrm{N}}\{(m+1)(1)^{n-m-1}\}+(n-m-1)\,\theta_{\mathrm{N}}\{(m+2)(1)^{n-m-2}\}.$$

Now

$$\theta_{\mathrm{N}}\{(a)(b)(c)\ldots\} = \mathrm{C}(abc\ldots) = \mathrm{D}_a\mathrm{D}_b\mathrm{D}_c\ldots h_p h_q h_r\ldots ;$$

thence

$$\mathrm{J}_{m}$$
$$= \{(n-m+1)\,\mathrm{D}_{m}\mathrm{D}_{1}^{n-m}-(n-m)\,\mathrm{D}_{m+1}\mathrm{D}_{1}^{n-m-1}\}\,h_p h_q h_r\ldots,$$
$$= \mathrm{D}_p\mathrm{D}_q\mathrm{D}_r\ldots\{(n-m+1)\,h_m h_1^{n-m}-(n-m)\,h_{m+1}h_1^{n-m-1}\} ;$$

or, m not being less than the greatest integer in $\frac{1}{2}(n+1)$,

$$(n-m+1)\,h_m h_1^{n-m}-(n-m)\,h_{m+1}h_1^{n-m-1}$$

is the generating function of the number J_m.

Similarly

$$\mathrm{K}_{m}$$
$$= \mathrm{D}_p\mathrm{D}_q\mathrm{D}_r\ldots\{(n-m+1)\,h_m h_1^{n-m}-2(n-m)\,h_{m+1}h_1^{n-m-1}+(n-m-1)\,h_{m+2}h_1^{n-m-2}\} ;$$

and, m not being less than the greatest integer in $\frac{1}{2}(n+1)$,

$$(n-m+1)\,h_m h_1^{n-m}-2(n-m)\,h_{m+1}h_1^{n-m-1}+(n-m-1)\,h_{m+2}h_1^{n-m-2}$$

is the generating function of the number K_m.

Similarly, but subject now to the condition that m must not be less than the greatest integer in $\frac{1}{2}(n-1)$,

$$(n-m-1)\,h_{m+2}h_1^{n-m-2}-(n-m)\,h_{m+1}h_1^{n-m-1}$$

is the function which generates the number

$$\mathrm{I}_m-\frac{n!}{p!q!r!\ldots}.$$

Subject to the conditions mentioned, we have a complete solution of the problem, but when m has other values, the solution is less simple and I see no way of effecting it.

SECTION 8.

Art. 72. I recall that the number of ways of distributing numbers (or objects) specified by

$$(p_1^{\pi_1}p_2^{\pi_2}p_3^{\pi_3}\ldots),$$

into m different parcels, is given by the series

$$\begin{aligned}F_m = &\binom{m+p_1-1}{p_1}^{\pi_1}\binom{m+p_2-1}{p_2}^{\pi_2}\binom{m+p_3-1}{p_3}^{\pi_3}\ldots\\ &-\binom{m}{1}\binom{m+p_1-2}{p_1}^{\pi_1}\binom{m+p_2-2}{p_2}^{\pi_2}\binom{m+p_3-2}{p_3}^{\pi_3}\ldots\\ &+\binom{m}{2}\binom{m+p_1-3}{p_1}^{\pi_1}\binom{m+p_2-3}{p_2}^{\pi_2}\binom{m+p_3-3}{p_3}^{\pi_3}\ldots\\ &-\ldots;\end{aligned}$$

and this, for brevity, I write

$$F_m = G_m - \binom{m}{1}G_{m-1} + \binom{m}{2}G_{m-2} - \ldots.$$

Let

$$N_{m, p_1^{\pi_1}p_2^{\pi_2}p_3^{\pi_3}\ldots}$$

denote the number of distributions, associated with a descending specification containing exactly m parts, and write this

$$N_m,$$

when there is no risk of misunderstanding.

Following the proof of Art. 23, it may be proved that

$$F_m = N_m + \binom{n-m+1}{1}N_{m-1} + \binom{n-m+2}{2}N_{m-2} + \ldots + \binom{n-1}{m-1}N_1;$$

and also

$$N_m = F_m - \binom{n-m+1}{1}F_{m-1} + \binom{n-m+2}{2}F_{m-2} + \ldots + (-)^{m+1}\binom{n-1}{m-1}F_1;$$

and thence

$$N_m = G_m - \binom{n+1}{1}G_{m-1} + \binom{n+1}{2}G_{m-2} - \ldots + (-)^{m+1}\binom{n+1}{m-1}G_1.$$

Art. 73. From this relation the following results are obtained :—

$n = 3.$

	(3).	(21).	(1^3).
N_1	1	1	1
N_2		2	4
N_3			1

$n = 4.$

	(4).	(31).	(2^2).	(21^2).	(1^4).
N_1	1	1	1	1	1
N_2		3	4	7	11
N_3			1	4	11
N_4					1

$n = 5.$

	(5).	(41).	(32).	(31^2).	(2^21).	(21^3).	(1^5).
N_1	1	1	1	1	1	1	1
N_2		4	6	10	12	18	26
N_3			3	9	15	33	66
N_4					2	8	26
N_5							1

$n = 6.$

	(6).	(51).	(42).	(41^2).	(3^2).	(321).	(31^3).	(2^3).	(2^21^2).	(21^4).	(1^6).
N_1	1	1	1	1	1	1	1	1	1	1	1
N_2		5	8	13	9	17	25	20	29	41	57
N_3			6	16	9	33	67	48	93	171	302
N_4					1	9	27	20	53	131	302
N_5								1	4	16	57
N_6											1

To explain, observe that the number at the intersection of the row N_3 and the column (2^21^2) shows that

$$N_{3,\,2^21^2} = 93.$$

These tables will be of constant service in verifying results to be obtained.

Art. 74. From the relation

$$N_m = G_m - \binom{n+1}{1} G_{m-1} + \binom{n+1}{2} G_{m-2} - \ldots,$$

we can obtain a system, for, summing each side from $m = 1$ to $m = m$,

$$N_m + N_{m-1} + \ldots + N_1 = G_m - \binom{n}{1} G_{m-1} + \binom{n}{2} G_{m-2} - \ldots,$$

and, repeating the summation θ times,

$$N_m+\binom{\theta+1}{1}N_{m-1}+\binom{\theta+2}{2}N_{m-2}+\ldots+\binom{\theta+m-1}{m-1}N_1 = G_m-\binom{n-\theta}{1}G_{m-1}+\binom{n-\theta}{2}G_{m-2}-\ldots;$$

so that, when $\theta = n$,

$$N_m+\binom{n+1}{1}N_{m-1}+\ldots+\binom{n+m-1}{m-1}N_1 = G_m.$$

Again, taking differences instead of summing, we get the series

$$N_m-N_{m-1} = G_m-\binom{n+2}{1}G_{m-1}+\binom{n+2}{2}G_{m-2}-\ldots,$$

$$N_m-2N_{m-1}+N_{m-2} = G_m-\binom{n+3}{1}G_{m-1}+\binom{n+3}{2}G_{m-2}-\ldots,$$

and in general

$$N_m-\binom{p}{1}N_{m-1}+\ldots\pm N_{m-p}$$

$$= G_m-\binom{n+p+1}{1}G_{m-1}+\binom{n+p+1}{2}G_{m-2}-\ldots.$$

These results are all given by the two formulæ

$$N_m+\binom{p}{1}N_{m-1}+\binom{p+1}{2}N_{m-2}+\ldots$$

$$= G_m-\binom{n+1-p}{1}G_{m-1}+\binom{n+1-p}{2}G_{m-2}-\ldots;$$

$$N_m-\binom{p}{1}N_{m-1}+\binom{p}{2}N_{m-2}-\ldots$$

$$= G_m-\binom{n+1+p}{1}G_{m-1}+\binom{n+1+p}{2}G_{m-2}-\ldots;$$

which become the same when $p = 0$.

Curious Expression for N_m.

Art. 75. I shall now prove that

$$N_m = P_{m-1}-\frac{m-2}{m-1}\binom{n}{1}P_{m-2}+\frac{m-3}{m-1}\binom{n}{2}P_{m-3}+\ldots+(-)^m\frac{1}{m-1}\binom{n}{m-2}P_1,$$

where

$$P_s = \frac{\binom{p_1+s-1}{p_1}^{\pi_1}\binom{p_2+s-1}{p_2}^{\pi_2}\ldots}{s^{\pi_1+\pi_2+\ldots}}\{(p_1+s)^{\pi_1}(p_2+s)^{\pi_2}\ldots\}_{-1,0},$$

where

$$\{(p_1+s)^{\pi_1}(p_2+s)^{\pi_2}\ldots\}_{-1,0}$$

denotes the expansion of

$$(p_1+s)^{\pi_1}(p_2+s)^{\pi_2}\ldots,$$

when deprived of the terms linear in p_1, p_2, ..., and of the term independent of $p_1, p_2, \ldots$.

For it is easy to show that two consecutive terms

$$(-)^t \frac{m-t-1}{m-1} \binom{n}{t} P_{m-t-1} + (-)^{t+1} \frac{m-t-2}{m-1} \binom{n}{t+1} P_{m-t-2}$$

may be given the form

$$(-)^t \frac{m-t-1}{m-1} \binom{n}{t} \binom{p_1+m-t-1}{p_1}^{\pi_1} \binom{p_2+m-t-1}{p_2}^{\pi_2} \ldots$$

$$+(-)^{t+1} \binom{n+1}{t+1} \binom{p_1+m-t-2}{p_1}^{\pi_1} \binom{p_2+m-t-2}{p_2}^{\pi_2} \ldots$$

$$+(-)^t \frac{n+m-t-2}{m-1} \binom{n}{t+1} \binom{p_1+m-t-3}{p_1}^{\pi_1} \binom{p_2+m-t-3}{p_2}^{\pi_2} \ldots,$$

and, giving t the values 0, 2, 4, ..., and summing and simplifying, we obtain

$$\binom{p_1+m-1}{p_1}^{\pi_1} \binom{p_2+m-1}{p_2}^{\pi_2} \ldots$$

$$-\binom{n+1}{1} \binom{p_1+m-2}{p_1}^{\pi_1} \binom{p_2+m-2}{p_2}^{\pi_2} \ldots$$

$$+\binom{n+1}{2} \binom{p_1+m-3}{p_1}^{\pi_1} \binom{p_2+m-3}{p_2}^{\pi_2} \ldots$$

$$-\ldots,$$

which we know to be the value of N_m.

Art. 76. The symmetry of the numbers N_{m,p^π} will not escape the notice of the reader.

SECTION 9.

Art. 77. My purpose now is to connect the preceding pages with my Memoir on the Compositions of Numbers, to which attention has already been directed. In the course of that investigation I had occasion to consider the permutations of the letters in

$$\alpha^p \beta^q \gamma^r,$$

with the object of determining the number of permutations containing given numbers of

$\beta\alpha$ contacts,

$\gamma\alpha$,,

$\gamma\beta$,,

If we take any permutation

$$\ldots\beta\alpha\ldots\gamma\alpha\ldots\gamma\beta\ldots\gamma\beta\alpha\ldots$$

and particularly notice all of such contacts, it is clear that the numbers of parts in the descending specification α, β, γ, ..., being numbers in descending order of magnitude, is necessarily one greater than the number of such contacts; in the present instance there are 6 parts in the descending specification and 5 contacts. The problem of the determination of the permutations having descending specifications containing m parts is identical with that which is concerned with those having $m-1$ contacts of the nature specified.

Art. 78. I established in the Memoir that the letters in

$$\alpha^p\beta^q\gamma^r$$

can be permuted in

$$\binom{s_{21}+s_{31}}{s_{21}}\binom{p}{s_{21}+s_{31}}\binom{q}{s_{32}}\binom{q+s_{31}}{s_{21}+s_{31}}\binom{r}{s_{31}+s_{32}}$$

ways so as to have exactly

s_{21} $\beta\alpha$ contacts,

s_{31} $\gamma\alpha$,,

s_{32} $\gamma\beta$,,

and I further discovered that this number is the coefficient of

$$\lambda_{21}{}^{s_{21}}\lambda_{31}{}^{s_{31}}\lambda_{32}{}^{s_{32}}\alpha^p\beta^q\gamma^r$$

in the development of the function

$$(\alpha+\lambda_{21}\beta+\lambda_{31}\gamma)^p\,(\alpha+\beta+\lambda_{32}\gamma)^q\,(\alpha+\beta+\gamma)^r.$$

Art. 79. In the same paper I showed that for this function may be substituted the function

$$\frac{1}{1-(\alpha+\beta+\gamma)+(1-\lambda_{21})\,\alpha\beta+(1-\lambda_{31})\,\alpha\gamma+(1-\lambda_{32})\,\beta\gamma-(1-\lambda_{21})\,(1-\lambda_{32})\,\alpha\beta\gamma},$$

which does not involve p, q, r, and may therefore be regarded as the general generating function of the numbers.

Art. 80. Reserving for the present the generalizations, which were also given in the papers referred to, it is clear that the application to the present question is obtained by putting

$$\lambda_{21}=\lambda_{31}=\lambda_{32}=\lambda,$$

when we find that the number of permutations of

$$\alpha^p\beta^q\gamma^r,$$

which have descending specifications containing m parts, is the coefficient of

$$\lambda^{m-1}\alpha^p\beta^q\gamma^r$$

in the development of

$$(\alpha+\lambda\beta+\lambda\gamma)^p\,(\alpha+\beta+\lambda\gamma)^q\,(\alpha+\beta+\gamma)^r,$$

or of

$$\frac{1}{1-(\alpha+\beta+\gamma)+(1-\lambda)\,(\alpha\beta+\alpha\gamma+\beta\gamma)-(1-\lambda)^2\alpha\beta\gamma}.$$

This, therefore, is the true generating function of the numbers N_m.

It may be verified, for example, that the complete coefficient of

$$\alpha^2\beta^2\gamma^2$$

is

$$(1+20\lambda+48\lambda^2+20\lambda^3+\lambda^4),$$

which agrees with a previous result.

From a previous result also the coefficient of

$$\lambda^{m-1}\alpha^p\beta^q\gamma^r$$

is

$$\binom{m+p-1}{p}\binom{m+q-1}{q}\binom{m+r-1}{r}$$

$$-\binom{n+1}{1}\binom{m+p-2}{p}\binom{m+q-2}{q}\binom{m+r-2}{r}$$

$$+\binom{n+1}{2}\binom{m+p-3}{p}\binom{m+q-3}{q}\binom{m+r-3}{r}$$

$$-\ldots,$$

where $n=p+q+r$.

Art. 81. Observe that the generating function is a symmetric function of α, β, γ, verifying a previous conclusion that an N_m number is not altered by any interchange of the letters α, β, γ.

When the numbers p, q, r are equal, that is when the objects are specified by the partition

$$(p^3),$$

we can establish a symmetrical property of the numbers N.

For coefficient

$$\lambda^{m-1}(\alpha\beta\gamma)^p \text{ in } (\alpha+\lambda\beta+\lambda\gamma)^p\,(\alpha+\beta+\lambda\gamma)^p\,(\alpha+\beta+\gamma)^p$$

is, by writing

$$\frac{1}{\lambda} \text{ for } \lambda \text{ and } \lambda\alpha,\ \lambda\beta,\ \lambda\gamma \text{ for } \alpha,\ \beta,\ \gamma,$$

equal to coefficient of

$$\lambda^{3p-m+1}(\alpha\beta\gamma)^p \text{ in } (\lambda\alpha+\beta+\gamma)^p\,(\lambda\alpha+\lambda\beta+\gamma)^p\,(\lambda\alpha+\lambda\beta+\lambda\gamma)^p,$$

equal to coefficient of

$$\lambda^{2p-m+1}(\alpha\beta\gamma)^p \text{ in } (\lambda\alpha+\beta+\gamma)^p\,(\lambda\alpha+\lambda\beta+\gamma)^p\,(\alpha+\beta+\gamma)^p,$$

equal to coefficient of

$$\lambda^{2p-m+1}(\alpha\beta\gamma)^p \text{ in } (\alpha+\lambda\beta+\lambda\gamma)^p\,(\alpha+\beta+\lambda\gamma)^p\,(\alpha+\beta+\gamma)^p.$$

Art. 82. Hence

$$N_m = N_{2p-m+2},$$

and the numbers N range from

$$N_1 \text{ to } N_{2p+1},$$

showing that $2p+1$ is the maximum number of parts in the descending specification, when the objects are specified by the partition

$$(p^3).$$

Art. 83. In general, when there are k *different* letters,

$$\alpha_1, \alpha_2, \ldots, \alpha_k,$$

the number of permutations of

$$\alpha_1^{p_1}\alpha_2^{p_2}\ldots\alpha_k^{p_k},$$

which have descending specifications containing m parts, is the coefficient of

$$\lambda^{m-1}\alpha_1^{p_1}\alpha_2^{p_2}\ldots\alpha_k^{p_k}$$

in the development of

$$\{\alpha_1+\lambda(\alpha_2+\ldots+\alpha_k)\}^{p_1}\{\alpha_1+\alpha_2+\lambda(\alpha_3+\ldots+\alpha_k)\}^{p_2}\ldots\{\alpha_1+\alpha_2+\ldots+\lambda\alpha_k\}^{p_{k-1}}\{\alpha_1+\alpha_2+\ldots+\alpha_k\}^{p_k};$$

or of

$$\frac{1}{1-\Sigma\alpha_1+(1-\lambda)\Sigma\alpha_1\alpha_2-(1-\lambda)^2\Sigma\alpha_1\alpha_2\alpha_3+\ldots+(-)^k(1-\lambda)^{k-1}\alpha_1\alpha_2\alpha_3\ldots\alpha_k}.$$

This is the general generating function of the numbers

$$\mathrm{N}_m.$$

Art. 84. Since it is symmetrical in regard to

$$\alpha_1, \alpha_2, \ldots, \alpha_k,$$

the value of N_m is not affected by permutation of the letters

$$\alpha_1, \alpha_2, \ldots, \alpha_k.$$

Art. 85. It can be shown also, as in the simpler case, that when

$$p_1 = p_2 = \ldots = p_k = p,$$

the coefficient of

$$\lambda^{m-1}(\alpha_1\alpha_2\ldots\alpha_k)^p$$

is equal to the coefficient of

$$\lambda^{(k-1)p-m+1}(\alpha_1\alpha_2\ldots\alpha_k)^p;$$

so that

$$\mathrm{N}_m = \mathrm{N}_{(k-1)p-m+2};$$

the numbers N range from

$$\mathrm{N}_1 \text{ to } \mathrm{N}_{(k-1)p+1};$$

and $(k-1)p+1$ is the maximum number of parts in a descending specification.

SECTION 10.

Art. 86. The generating function

$$\frac{1}{1-\Sigma\alpha_1+(1-\lambda)\Sigma\alpha_1\alpha_2-(1-\lambda)^2\Sigma\alpha_1\alpha_2\alpha_3+\ldots+(-)^k(1-\lambda)^{k-1}\alpha_1\alpha_2\ldots\alpha_k}$$

now presents itself for examination.

Introducing the elementary functions

$$a_1, a_2, \ldots,$$

and writing $1-\lambda = b$, it is written

$$\frac{1}{1-a_1+ba_2-b^2a_3+\ldots+(-)^kb^{k-1}a_k};$$

or

$$\frac{1}{1-A},$$

where

$$A = a_1-ba_2+b^2a_3-\ldots+(-)^{k+1}b^{k-1}a_k.$$

For the present purpose we may consider k to be infinite, and write

$$A = a_1-ba_2+b^2a_3-\ldots.$$

Art. 87. Taking the symmetric function operators

$$d_s = \partial_{a_s}+a_1\partial_{a_{s+1}}+a_2\partial_{a_{s+2}}+\ldots,$$

$$D_s = \frac{1}{s!}(\partial_{a_1}+a_1\partial_{a_2}+a_2\partial_{a_3}+\ldots)^s = \frac{1}{s!}(d_1^s),$$

and an auxiliary fictitious equation

$$x^r-D_1x^{r-1}+D_2x^{r-2}-D_3x^{r-3}+\ldots = 0,$$

r being an infinite number, it is necessary to remind the reader of the relations existing between the operators.

Successive linear operations of

$$d_\lambda,\ d_\mu,\ d_\nu,\ \ldots$$

are denoted by placing them in separate brackets, thus,

$$(d_\lambda)(d_\mu)(d_\nu)\ldots,$$

but when they are multiplied, as in TAYLOR'S theorem, so as to produce a single operator of higher order, they will be placed in one bracket, thus,

$$(d_\lambda d_\mu d_\nu\ldots).$$

Art. 88. Let monomial symmetric functions of the fictitious relation

$$x^r-D_1x^{r-1}+D_2x^{r-2}-\ldots = 0$$

be denoted by a partition in brackets with subscript D, thus,

$$(\quad)_D.$$

Then I have shown, in a previous paper,

$$d_1 = D_1 = (1)_D,$$

$$d_2 = D_1^2-2D_2 = (2)_D,$$

$$\cdot\quad\cdot\quad\cdot\quad\cdot\quad\cdot\quad\cdot\quad\cdot\quad\cdot$$

$$d_s = \quad .\quad .\quad .\quad = (s)_D,$$

and, in general,

$$\frac{1}{\pi_1!\,\pi_2!\ldots}(d_{p_1}^{\pi_1}d_{p_2}^{\pi_2}\ldots) = (p_1^{\pi_1}p_2^{\pi_2}\ldots)_D.$$

Art. 89. Every symmetric function identity has corresponding to it a relation between the operators; thus corresponding to the set

$$(1)^2 = (2)+2(1^2),$$
$$(1)^3 = (3)+3(21)+6(1^3),$$
$$(1)^4 = (4)+4(31)+6(2^2)+12(21^2)+24(1^4),$$
$$\cdots\cdots\cdots$$

we have the set

$$(d_1)^2 = (d_1^2)+d_2 = 2(1^2)_D+(2)_D,$$
$$(d_1)^3 = (d_1^3)+3(d_1d_2)+d_3 = 6(1^3)_D+3(21)_D+(3)_D,$$
$$(d_1)^4 = (d_1^4)+6(d_1^2d_2)+3(d_2^2)+4(d_3d_1)+(d_4),$$
$$= (24)(1^4)_D+12(21^2)_D+6(2^2)_D+4(31)_D+(4)_D,$$

and so on.

Art. 90. Also, corresponding to the set

$$2a_2 = s_1^2-s_2,$$
$$6a_3 = s_1^3-3s_1s_2+2s_3,$$
$$24a_4 = s_1^4-6s_1^2s_2+3s_2^2+8s_1s_3-6s_4,$$

&c., we have the set

$$2D_2 = (d_1^2) = (d_1)^2-d_2,$$
$$6D_3 = (d_1^3) = (d_1)^3-3(d_1)(d_2)+2d_3,$$
$$24D_4 = (d_1^4) = (d_1)^4-6(d_1)^2(d_2)+3(d_2)^2+8(d_1)(d_3)-6(d_4),$$

and so on.

Art. 91. For the special operand

$$\frac{1}{1-A}$$

these operator relations assume a special simple form which is of great importance in the theory of the generating function.

For $$d_sA = (-)^{s-1}b^{s-1}(1-bA) = (-)^{s-1}b^{s-1}d_1A;$$

or, *quâ* the above operand, $$d_s \equiv (-)^{s-1}b^{s-1}d_1;$$

and thence, from a set of relations given above,

$$2!\,D_2 \equiv D_1(D_1+b),$$
$$3!\,D_3 \equiv D_1(D_1+b)(D_1+2b),$$
$$\cdots\cdots\cdots$$
$$s!\,D_s \equiv D_1(D_1+b)\ldots\{D_1+(s-1)b\},$$
$$s!\,D_s \equiv t!\,D_t(D_1+tb)\{D_1+(t+1)b\}\ldots\{D_1+(s-1)b\}.$$

Art. 92. By means of these we can now arrive at a most important series of relations.

For

$$(p)_D = d_p = (-b)^{p-1}D_1\,;$$

$$(pq)_D = (d_pd_q) = (d_p)(d_q) - d_{p+q},$$

$$\equiv (-)^{p+q-2}b^{p+q-2}D_1^2 + (-)^{p+q-2}b^{p+q-1}D_1,$$

$$\equiv (-b)^{p+q-2}2\,!\,D_2\,;$$

$$(pqr)_D = (d_pd_qd_r)$$

$$= (d_p)(d_q)(d_r) - (d_{p+q})(d_r) - (d_{p+r})(d_q) - (d_{q+r})(d_p) + 2\,d_{p+q+r},$$

$$\equiv (-b)^{p+q+r-3}D_1^3, \; -3(-b)^{p+q+r-2}D_1^2, \; +2(-b)^{p+q+r-1}D_1,$$

$$\equiv (-b)^{p+q+r-3}3\,!\,D_3\,;$$

and generally

$$(p_1p_2\ldots p_s)_D \equiv (-b)^{\Sigma p-s}s\,!\,D_s\,;$$

and more generally

$$(p_1^{\pi_1}p_2^{\pi_2}\ldots)_D \equiv (-b)^{\Sigma\pi p-\Sigma\pi}\frac{(\Sigma\pi)!}{\pi_1!\,\pi_2!\ldots}D_{\Sigma\pi}\,;$$

or, if $\Sigma\pi p = n$, $\Sigma\pi = i$,

$$(p_1^{\pi_1}p_2^{\pi_2}\ldots)_D \equiv (-b)^{n-i}\frac{i!}{\pi_1!\,\pi_2!\ldots}D_i.$$

Art. 93. From the relations

$$s\,!\,D_s = D_1(D_1+b)\ldots\{D_1+(s-1)b\}\,;$$

we find the set

$$D_1^2 = 2D_2 - bD_1,$$

$$D_1^3 = 6D_3 - 6bD_2 + b^2D_1,$$

$$D_1^4 = 24D_4 - 36bD_3 + 14b^2D_2 - b^3D_1,$$

$$\cdots\cdots\cdots$$

Art. 94. And also the set

$$\binom{s}{1}D_s = D_{s-1}\left\{D_1 + \binom{s-1}{1}b\right\},$$

$$\binom{s}{2}D_s = D_{s-2}\left\{D_2 + \binom{s-2}{1}bD_1 + \binom{s-1}{2}b^2\right\},$$

$$\binom{s}{3}D_s = D_{s-3}\left\{D_3 + \binom{s-3}{1}bD_2 + \binom{s-2}{2}b^2D_1 + \binom{s-1}{3}b^3\right\},$$

$$\cdots\cdots\cdots$$

$$\binom{s}{t}D_s = D_{s-t}\left\{D_t + \binom{s-t}{1}bD_{t-1} + \ldots + \binom{s-1}{t}b^t\right\},$$

$$sD_s = D_1\{D_{s-1} + bD_{s-2} + b^2D_{s-3} + \ldots + b^{s-1}\}.$$

The Expressibility of D_s.

Art. 95. The fundamental relation

$$s!D_s = D_1(D_1+b)(D_1+2b)\ldots\{D_1+(s-1)b\},$$

exhibits D_s in terms of powers of D_1.

It is clear, *à priori*, that D_s is expressible in terms of D_1 and powers of D_2, *e.g.*,

$$\binom{3}{1}D_3 = D_2(D_1+2b),$$

$$\binom{4}{2}D_4 = D_2(D_2+2bD_1+3b^2),$$

$$\binom{3}{1}\binom{5}{3}D_5 = D_2(D_2D_1+8bD_2+9b^2D_1+12b^3),$$

$$\binom{4}{2}\binom{6}{4}D_6 = D_2(D_2^2+6bD_2D_1+29b^2D_2+24b^3D_1+30b^4), \text{ and so on,}$$

where notice, as a verification, that the sum of the numerical coefficients is the same on the two sides.

In every case D_2 appears as a factor.

In general the operator products, which appear on the right, are factors of

$$D_2^kD_1,$$

which contain the factor D_2, every weight of operator product being represented *once, and once only*, from the weight 2 up to the weight of the single operator on the left-hand side.

It is important to remark that

$$(2^k1)$$

is a perfect partition* of the number

$$2k+1;$$

because every lower number can be composed in exactly one way by the parts of the partition.

Art. 96. It will now appear that there exists an expression for

$$D_s$$

corresponding to every perfect partition that can be constructed.

The general expression of a perfect partition is

$$\ldots\{(1+\alpha)(1+\beta)(1+\gamma)\}^\delta\{(1+\alpha)(1+\beta)\}^\gamma(1+\alpha)^\beta1^\alpha;$$

where α, β, γ, δ, ... are any positive integers, zero excluded.

The perfect partition

$$(2^k1)$$

is the particular case

$$\alpha = 1, \beta = k, \gamma = \delta = \ldots = 0.$$

* 'Messenger,' 1890, p. 103.

R 2

In every case, if σ be the highest figure in the perfect partition, D_σ is a factor of the expression for D_s, *e.g.*, taking the perfect partition

$$3^k 1^2,$$

we have

$$40D_6 = 2D_3^2 + 3bD_3D_1^2 + 15b^2D_3D_1 + 20b^3D_3.$$

I do not interrupt the investigation by stopping to prove the theory of expressibility depending upon perfect partitions; its truth is intuitive.

Art. 97. It is necessary to labour the subject of the operator relations, *quâ* the special operands, because the whole theory of the numbers N_m is involved.

Art. 98. Perhaps the most interesting of the operator relations are those which do not involve b (or λ).

Recalling the relation of Art. 92, viz.,

$$(p_1^{\pi_1}p_2^{\pi_2}\ldots)_D = (-b)^{n-i}\frac{i!}{\pi_1!\pi_2!\ldots}D_i,$$

where

$$\Sigma\pi p = n, \ \Sigma\pi = i,$$

we may also write

$$(q_1^{\chi_1}q_2^{\chi_2}\ldots)_D = (-b)^{\nu-j}\frac{j!}{\chi_1!\chi_2!\ldots}D_j,$$

where

$$\Sigma\chi q = \nu, \ \Sigma\chi = j,$$

and, if

$$n-i = \nu-j,$$

we may eliminate b, obtaining

$$\frac{j!}{\chi_1!\chi_2!\ldots}D_j(p_1^{\pi_1}p_2^{\pi_2}\ldots)_D = \frac{i!}{\pi_1!\pi_2!\ldots}D_i(q_1^{\chi_1}q_2^{\chi_2}\ldots)_D.$$

Art. 99. The simplest formula thence obtained is found by putting

$$(p_1^{\pi_1}p_2^{\pi_2}\ldots) = (2^2),$$
$$(q_1^{\chi_1}q_2^{\chi_2}\ldots) = (31);$$

and this leads at once to

$$d_1d_3 - d_2^2 = 0,$$

or

$$D_2D_1^2 - 4D_2^2 + 3D_3D_1 = 0,$$

which also results by elimination of b from

$$2D_2 = D_1(D_1+b),$$
$$6D_3 = D_1(D_1+b)(D_1+2b).$$

Art. 100. To obtain some more relations in a simple manner, I write

$$(s)_D = (-b)^{s-1}D_1,$$
$$(t)_D = (-b)^{t-1}D_1,$$
$$(u)_D = (-b)^{u-1}D_1,$$
$$(t+u)_D = (-b)^{t+u-1}D_1,$$
$$(u+s)_D = (-b)^{u+s-1}D_1,$$
$$(s+t)_D = (-b)^{s+t-1}D_1;$$

and then
$$(s+t)_D\,(u)_D = (s)_D\,(t+u)_D = (t)_D\,(u+s)_D\,;$$

or, as these relations may be written,
$$d_{s+t}\,d_u = d_s\,d_{t+u} = d_t\,d_{u+s}\,;$$
or
$$(s+t,\ u)_D = (s,\ t+u)_D = (t,\ u+s)_D,$$

with the usual multiplier (*viz.*, 2), if either
$$s+t = u, \quad \text{or} \quad t+u = s, \quad \text{or} \quad u+s = t.$$

Art. 101. We are led to the series
$$(31)_D = 2\,(2^2)_D,$$
$$(41)_D = (32)_D,$$
$$(51)_D = (42)_D = 2\,(3^2)_D,$$
$$(61)_D = (52)_D = (43)_D,$$
$$(71)_D = (62)_D = (53)_D = 2\,(4^2)_D, \text{ \&c.},$$
and generally if
$$(p_1^{\pi_1}p_2^{\pi_2}\ldots), \quad (q_1^{\chi_1}q_2^{\chi_2}\ldots)$$
be functions of the same weight and degree, viz.,
$$\Sigma\pi p = \Sigma\chi q\,;\ \Sigma\pi = \Sigma\chi = i,$$
$$\pi_1!\,\pi_2!\ldots(p_1^{\pi_1}p_2^{\pi_2}\ldots)_D = \chi_1!\,\chi_2!\ldots(q_1^{\chi_1}q_2^{\chi_2}\ldots)_D.$$

Application of the Foregoing to the Generating Function.

Art. 102. It has been established that
$$\frac{1}{1-a_1+(1-\lambda)\,a_2-(1-\lambda)^2a_3+\ldots}$$
$$= \Sigma N_{m,\,p_1p_2\ldots p_k}\lambda^{m-1}\,(p_1p_2\ldots p_k).$$

We will first of all examine the result of the equivalence of operators
$$2!D_2 = D_1^{\,2}+(1-\lambda)\,D_1$$
(see Art. 93 *quâ* the operand on the right-hand side). Write the operand
$$\Sigma N_{m,\,1^{c_1}2^{c_2}3^{c_3}\ldots}\lambda^{m-1}\,(1^{c_1}2^{c_2}3^{c_3}\ldots)\,;$$
then
$$2!N_{m,\,1^{c_1}2^{c_2+1}3^{c_3}\ldots}$$
$$= N_{m,\,1^{c_1+2}2^{c_2}3^{c_3}\ldots}+(N_m-N_{m-1})_{1^{c_1+1}2^{c_2}3^{c_3}\ldots},$$
e.g., put
$$c_1 = 0, \quad c_2 = 2, \quad c_3 = c_4 = \ldots = 0,$$
$$m = 3,$$
$$2N_{3,\,2^3} = N_{3,\,1^22^2}+(N_3-N_2)_{12^2}\,;$$
verified (from the tables) by
$$2.48 = 93+(15-12).$$

Art. 103. Again, in the same formula, put

$$c_1 = n-2, \quad c_2 = 0, \quad c_3 = c_4 = \ldots = 0,$$

we find

$$2N_{m, 21^{n-2}} = N_{m, 1^n} + (N_m - N_{m-1})_{1^{n-1}}.$$

We obtain, from this, a useful result by writing

$$n-m+1 \text{ for } m,$$

for then

$$2N_{n-m+1, 21^{n-2}} = N_{n-m+1, 1^n} + (N_{n-m+1} - N_{n-m})_{1^{n-1}}.$$

Art. 104. Observe that

$$N_{m, 1^n} = N_{n-m+1, 1^n},$$
$$N_{m, 1^{n-1}} = N_{n-m, 1^{n-1}},$$
$$N_{m-1, 1^{n-1}} = N_{n-m+1, 1^{n-1}};$$

so that by addition and subtraction we obtain

$$N_{m, 21^{n-2}} + N_{n-m+1, 21^{n-2}} = N_{m, 1^n};$$
$$N_{m, 21^{n-2}} - N_{n-m+1, 21^{n-2}} = N_{m, 1^{n-1}} - N_{n-m+1, 1^{n-1}};$$

or, as we may conveniently write these relations,

$$(N_m + N_{n-m+1})_{21^{n-2}} = N_{m, 1^n};$$
$$(N_m - N_{n-m+1})_{21^{n-2}} = (N_m - N_{n-m+1})_{1^{n-1}};$$
$$= (N_m - N_{m-1})_{1^{n-1}}.$$

These are the relations connecting N_m and N_{n-m+1} *quâ* the subscript 21^{n-2} analogous to those connecting the same symbols *quâ* the subscript 1^n.

Art. 105. From any operator relation we can immediately derive a relation between the numbers N_m by substituting for

$$b^\sigma D_s D_t \ldots$$

the expressions

$$\left\{N_m - \binom{\sigma}{1} N_{m-1} + \binom{\sigma}{2} N_{m-2} - \ldots\right\}_{1^{c_1}2^{c_2}\ldots s^{c_s+1} t^{c_t+1}\ldots};$$

and this it is convenient to denote by

$$N^{(\sigma)}{}_{m, 1^{c_1}2^{c_2}\ldots s^{c_s+1} t^{c_t+1}\ldots}.$$

Art. 106. Thus, corresponding to the operator relation

$$6D_3 = D_1(D_1+b)(D_1+2b) = D_1^3 + 3bD_1^2 + 2b^2 D_1,$$

we obtain

$$6N_{m, 1^{c_1}2^{c_2}3^{c_3+1}\ldots} = N_{m, 1^{c_1+3}2^{c_2}3^{c_3}\ldots} + 3N^{(1)}{}_{m, 1^{c_1+2}2^{c_2}3^{c_3}\ldots} + 2N^{(2)}{}_{m, 1^{c_1+1}2^{c_2}3^{c_3}\ldots}.$$

As a particular case put

$$c_1 = n-3, \quad c_2 = c_3 = c_4 = \ldots = 0,$$

so that

$$6N_{m, 31^{n-3}} = N_{m, 1^n} + 3(N_m - N_{m-1})_{1^{n-1}} + 2(N_m - 2N_{m-1} + N_{m-2})_{1^{n-2}};$$

and thence

$$3\,(N_m+N_{n-m+1})_{31^{n-3}} = N_{m,1^n}+2\,(N_m-2N_{m-1}+N_{m-2})_{1^{n-2}}\,;$$

$$(N_m-N_{n-m+1})_{31^{n-3}} = (N_m-N_{m-1})_{1^{n-1}}.$$

For $n = 6$, these relations can be verified by the tables for all values of m.

Art. 107. Similarly the theorems derived from

$$s!\,D_s = D_1\,(D_1+b)\,(D_1+2b)\ldots\{D_1+(s-1)\,b\}$$

can be at once written down.

[It is worth noting that this operator relation can, by putting $D_1 = b\Delta_1$, be written

$$D_s = b^s\binom{\Delta_1+s-1}{s}\Big].$$

If

$$s_1!\,s_2!\ldots D_{s_1}D_{s_2}\ldots = D_1^{\Sigma s}+g_1D_1^{\Sigma s-1}+g_2D_1^{\Sigma s-2}+\ldots,$$

we can write down the corresponding relation between the numbers N_m.

It will be found that

$$(N_m+N_{n-m+1})_{s1^{n-s}}$$

is a linear function of

$$N_{m,1^n},\ N^{(2)}{}_{m,1^{n-2}},\ N^{(4)}{}_{m,1^{n-4}}\ldots\,;$$

and

$$(N_m-N_{n-m+1})_{s1^{n-s}}$$

a linear function of

$$N^{(1)}{}_{m,1^{n-1}},\ N^{(3)}{}_{m,1^{n-3}},\ N^{(5)}{}_{m,1^{n-5}}\ldots\,;$$

and that the same obtains when instead of

$$s1^{n-s}$$

we take

$$s_1s_2\ldots1^{n-\Sigma s}.$$

Art. 108. From the operator relation

$$(p_1!)^{\pi_1}\,(p_2!)^{\pi_2}\ldots D_{p_1}^{\pi_1}D_{p_2}^{\pi_2}\ldots = \prod_{p,\pi}[D_1\,(D_1+b)\ldots\{D_1+(p_1-1)\,b\}]^{\pi_1}$$

we find

$$D_{p_1}^{\pi_1}D_{p_2}^{\pi_2}\ldots = b^n\binom{\Delta_1+p_1-1}{p_1}^{\pi_1}\binom{\Delta_1+p_2-1}{p_2}^{\pi_2}\ldots,$$

$$= b^n\,(u_0\Delta_1^{\,n}+u_1\Delta_1^{\,n-1}+u_2\Delta_1^{\,n-2}+\ldots)\,;$$

where u_0, u_1, u_2, ... are numerical coefficients that may be determined.

Thence is derived the relation

$$N_{m,p_1^{\pi_1}p_2^{\pi_2}\ldots}$$

$$= u_0N_{m,1^n}+u_1N^{(1)}{}_{m,1^{n-1}}+u_2N^{(2)}{}_{m,1^{n-2}}+\ldots\,;$$

giving a hint to put

$$N^{(t)}{}_{m,1^{n-t}} = N_m^{\,n-t} \text{ symbolically,}$$

and then

$$N_{m,p_1^{\pi_1}p_2^{\pi_2}\ldots} = \binom{N_m+p_1-1}{p_1}^{\pi_1}\binom{N_m+p_2-1}{p_2}^{\pi_2}\ldots \text{ symbolically.}$$

Art. 109. It must now be remarked that, since

$$N_m^{\,n-t} = \left\{N_m-\binom{t}{1}N_{m-1}+\binom{t}{2}N_{m-2}\ldots\right\}_{1^{n-t}},$$

we obtain

$$N^{n-t}_{n-m+1} = \left\{N_{n-m+1}-\binom{t}{1}N_{n-m}+\binom{t}{2}N_{n-m-1}-\dots\right\}_{1^{n-t}};$$

and since

$$N_{s,1^{n-t}} = N_{n-s-t+1,1^{n-t}}, \quad N^{n-t}_{n-m+1} = (-)^t N_m^{\ n-t};$$

and

$$N_{n-m+1},\ p_1^{\pi_1}p_2^{\pi_2}\dots$$

$$= u_0 N_{m,1^n} - u_1 N^{(1)}_{m,1^{n-1}} + u_2 N^{(2)}_{m,1^{n-2}} - \dots$$

$$= u_0 N_m^{\ n} - u_1 N_m^{\ n-1} + u_2 N_m^{\ n-2} - \dots$$

$$= \binom{N_m}{p_1}^{\pi_1}\binom{N_m}{p_2}^{\pi_2}\dots;$$

we obtain

$$N_{m,\,p_1^{\pi_1}p_2^{\pi_2}\dots}$$

$$= \binom{N_{n-m+1}}{p_1}^{\pi_1}\binom{N_{n-m+1}}{p_2}^{\pi_2}\dots.$$

Art. 110. We have two alternative expressions for

$$N_{m,\,p_1^{\pi_1}p_2^{\pi_2}\dots}$$

in terms of numbers

$$N_{\mu,1^\sigma}.$$

I verify them in the case

$$p_1^{\pi_1}p_2^{\pi_2}\dots = 2^3;$$

(i.)

$$N_{m,2^3} = \tfrac{1}{8}(N_m^{\ 2}+N_m)^3,$$

or

$$8N_{m,2^3} = N_m^{\ 6}+3N_m^{\ 5}+3N_m^{\ 4}+N_m^{\ 3},$$

$$= N_{m,1^6}$$
$$+3(N_m-N_{m-1})_{1^5}$$
$$+3(N_m-2N_{m-1}+N_{m-2})_{1^4}$$
$$+\ (N_m-3N_{m-1}+3N_{m-2}-N_{m-3})_{1^3},$$

agreeing with, for $m = 3$,

$$8.48 = 302+3(66-26)+3(11-2.11+1)+(1-3.4+3.1).$$

(ii.)

$$N_{m,2^3} = \tfrac{1}{8}(N^2_{7-m}-N_{7-m})^3,$$

or

$$8N_{m,2^3} = N^6_{7-m}-3N^5_{7-m}+3N^4_{7-m}-N^3_{7-m},$$

and, for $m = 3$,

$$8N_{3,2^3} = N_4^{\ 6}-3N_4^{\ 5}+3N_4^{\ 4}-N_4^{\ 3},$$

$$= N_{4,1^6}$$
$$-3(N_4-N_3)_{1^5}$$
$$+3(N_4-2N_3+N_2)_{1^4}$$
$$-\ (N_4-3N_3+3N_2-N_1)_{1^3};$$

agreeing with

$$8.48 = 302-3(26-66)+3(1-2.11+11)-(-3.1+3.4-1).$$

Art. 111. We have seen that, in general, we have two expressions for

$$N_{m,\, p_1^{\pi_1} p_2^{\pi_2} \ldots},$$

but, since

$$N_{m,\, p^\pi} = N_{n-m-p+2,\, p^\pi},$$

we have four expressions for

$$N_{m,\, p^\pi},$$

viz.,

$$\binom{N_m + p - 1}{p}^\pi,$$

$$\binom{N_{n-m+1}}{p}^\pi,$$

$$\binom{N_{n-m-p+2} + p - 1}{p}^\pi,$$

$$\binom{N_{m+p-1}}{p}^\pi.$$

Art. 112. It is clear that the operator relations afford unlimited scope for obtaining theorems connecting the numbers

$$N_{m,\, p_1^{\pi_1} p_2^{\pi_2} \ldots}.$$

Relations, so far utilised, have involved the operator

$$D_1,$$

but it is easy to construct them so as not to contain D_1 and generally so as not to contain D_s, where s is less than a given integer.

E.g., from the symmetric function relation

$$(1^2)(1^2) = (2^2) + 2(21^2) + 6(1^4),$$

we find

$$D_2^2 = b^2 D_2 - 6bD_3 + 6D_4;$$

and generally the relation

$$(p^2)(q^2) = (\overline{p+q})^2 + (p+q,\, pq) + (p^2 q^2)$$

leads to

$$(-b)^{2p+2q-4} D_2^2 = (-b)^{2p+2q-2} D_2 + 6(-b)^{2p+2q-3} D_3 + 6(-b)^{2p+2q-4} D_4;$$

or, throwing out the factor

$$b^{2p+2q-4},$$

$$D_2^2 = b^2 D_2 - 6bD_3 + 6D_4,$$

the same relation as before.

Moreover, the relation

$$(pq)(rs) = (p+r,\, q+s) + (p+s,\, q+r) + (p+r,\, qs) + (q+r,\, ps) + (p+s,\, qr) + (q+s,\, pr) + (pqrs)$$

leads, after throwing out a power of b, to precisely the same relation.

Art. 113. This remarkable circumstance greatly limits the number of operator relations obtainable. It should be observed that *any* operator relation may be multiplied throughout by any power of b and may be then used to obtain relations between the numbers

$$N_{m,\, p_1^{\pi_1} p_2^{\pi_2} \ldots},$$

but no essentially new relations are thus obtainable; for take a simple case

$$2D_2 = D_1^2 + bD_1,$$

leading to

$$2N_{m, 1^{c_1}2^{c_2+1}\ldots} = N_{m, 1^{c_1+2}2^{c_2}\ldots} + (N_m - N_{m-1})_{1^{c_1+1}2^{c_2}\ldots},$$

true for all values of m.

If we take

$$2bD_2 = bD_1^2 + b^2D_1,$$

we are led to

$$2(N_m - N_{m-1})_{1^{c_1}2^{c_2+1}\ldots} = (N_m - N_{m-1})_{1^{c_1+2}2^{c_2}\ldots} + (N_m - 2N_{m-1} + N_{m-2})_{1^{c_1+1}2^{c_2}\ldots},$$

and if the former relation be written

$$f(m) = 0,$$

the latter is merely

$$f(m) - f(m-1) = 0;$$

and further multiplication by b leads to the series of which the general term is

$$f(m) - \binom{\sigma}{1} f(m-1) + \binom{\sigma}{2} f(m-2) \ldots;$$

so that no new information is obtained.

Art. 114. The operator relation of the form D_sD_t = a linear function of

$$b^tD_s,\ b^{t-1}D_{s+1},\ b^{t-2}D_{s+2}, \ldots$$

is not difficult to obtain.

I find that

$$(-)^tD_sD_t = \binom{s}{t}\binom{t}{0} b^tD_s - \binom{s+1}{t}\binom{1}{t} b^{t-1}D_{s+1} + \binom{s+2}{t}\binom{t}{2} b^{t-2}D_{s+2} - \ldots;$$

and thence the formula for

$$D_sD_tD_u$$

follows by taking D_u as the operand on each side and then reducing the products

$$D_sD_u,\ D_{s+1}D_u,\ D_{s+2}D_u, \ldots$$

by the formula for D_sD_t.

I find that

$$\begin{aligned}
&(-)^{t+u}D_sD_tD_u \\
= &\binom{s}{u}\binom{t}{0}\binom{s}{t}\binom{u}{0} b^{t+u}D_s \\
&- \binom{s+1}{u}\left\{\binom{t}{0}\binom{s}{t}\binom{u}{1} + \binom{t}{1}\binom{s+1}{t}\binom{u}{0}\right\} b^{t+u-1}D_{s+1} \\
&+ \binom{s+2}{u}\left\{\binom{t}{0}\binom{s}{t}\binom{u}{2} + \binom{t}{1}\binom{s+1}{t}\binom{u}{1} + \binom{t}{2}\binom{s+2}{t}\binom{u}{0}\right\} b^{t+u-2}D_{s+2} \\
&- \ldots;
\end{aligned}$$

and, generally, the product of any number of operators is expressible in the required linear form.

Art. 115. With the object of connecting this theory of the numbers N_m with that of the numbers $N(abc\ldots)$, the generating function

$$\frac{1}{1-a_1+(1-\lambda)a_2-(1-\lambda)^2a_3+\ldots}$$

will now be expanded in ascending powers of λ, the coefficients of λ being functions of the homogeneous product sums

$$h_1,\ h_2,\ h_3,\ \ldots.$$

The point of departure is the elementary formula

$$\frac{1}{1-a_1+a_2-a_3+\ldots} = 1+h_1+h_2+h_3+\ldots.$$

Remarking that

$$1-a_1+a_2-a_3+\ldots = (1-\alpha_1)(1-\alpha_2)(1-\alpha_3)\ldots,$$

I write

$$(1-\lambda)\alpha_s \text{ for } \alpha_s,$$

equivalent to writing

$$(1-\lambda)^s a_s \text{ for } a_s,$$

and

$$(1-\lambda)^s h_s \text{ for } h_s;$$

then

$$\frac{1}{1-(1-\lambda)a_1+(1-\lambda)^2a_2-(1-\lambda)^3a_3+\ldots} = 1+(1-\lambda)h_1+(1-\lambda)^2h_2+(1-\lambda)^3h_3+\ldots$$

$$= u \text{ suppose};$$

and, as before, write

$$a_1-(1-\lambda)a_2+(1-\lambda)^2a_3-\ldots = \mathrm{A};$$

so that

$$u = \frac{1}{1-(1-\lambda)\mathrm{A}};$$

whence, solving for A,

$$\mathrm{A} = \frac{1}{1-\lambda}\,\frac{u-1}{u},$$

and

$$\frac{1}{1-\mathrm{A}} = (1-\lambda)\frac{u}{1-\lambda u};$$

where $\frac{1}{1-\mathrm{A}}$ is the generating function under consideration.

Write

$$\mathrm{H}_1 = h_1+h_2+h_3+h_4+\ldots$$
$$\mathrm{H}_2 = h_2+2h_3+3h_4+4h_5+\ldots$$
$$\mathrm{H}_3 = h_3+3h_4+6h_5+10h_6+\ldots$$
$$\mathrm{H}_4 = h_4+4h_5+10h_6+20h_7+\ldots$$
$$\cdots\cdots\cdots$$

s 2

so that
$$u = 1+(1-\lambda)h_1+(1-\lambda)^2 h_2+(1-\lambda)^3 h_3+\dots$$
$$= 1+H_1-\lambda(H_1+H_2)+\lambda^2(H_2+H_3)+\dots+(-)^s(H_s+H_{s+1})+\dots$$
$$= 1+(1-\lambda)H_1-\lambda(1-\lambda)H_2+\lambda^2(1-\lambda)H_3-\dots;$$
therefore
$$\frac{1-\lambda u}{1-\lambda} = 1-\lambda H_1+\lambda^2 H_2-\lambda^3 H_3+\dots$$
and thence
$$\frac{1}{1-A} = 1+\frac{H_1-\lambda H_2+\lambda^2 H_3-\dots}{1-\lambda H_1+\lambda^2 H_2-\lambda^3 H_3+\dots}.$$
Now let functions
$$A_1, A_2, A_3, A_4, \dots$$
be connected with
$$H_1, H_2, H_3, H_4, \dots$$
in the same way that
$$a_1, a_2, a_3, a_4, \dots$$
are connected with
$$h_1, h_2, h_3, h_4, \dots$$
so that
$$A_1 = H_1,$$
$$A_2 = H_1^2-H_2,$$
$$A_3 = H_1^3-2H_1H_2+H_3,$$
. . . .

then
$$\frac{1}{1-\lambda H_1+\lambda^2 H_2-\lambda^3 H_3+\dots} = 1+\lambda A_1+\lambda^2 A_2+\lambda^3 A_3+\dots,$$
and
$$\frac{1}{1-A} = 1+(H_1-\lambda H_2+\lambda^2 H_3-\dots)(1+\lambda A_1+\lambda^2 A_2+\lambda^3 A_3+\dots).$$
On the dexter the co-factor of λ^s is
$$H_1A_s-H_2A_{s-1}+H_3A_{s-2}-\dots+(-)^s H_{s+1},$$
which has the value
$$A_{s+1},$$
since
$$a_{s+1}-h_1a_s+h_2a_{s-1}-\dots+(-)^{s+1}h_{s+1} = 0$$
is a well-known identity in the elementary theory of symmetric functions. Hence
$$\frac{1}{1-A} = 1+A_1+\lambda A_2+\lambda^2 A_3+\lambda^3 A_4+\dots;$$
or, as we may write it,
$$\frac{1}{1-a_1+(1-\lambda)a_2-(1-\lambda)^2 a_3+\dots}$$
$$= 1+A_1+\lambda A_2+\lambda^2 A_3+\lambda^3 A_4+\dots,$$
$$= 1+H_1+\lambda H_{11}+\lambda^2 H_{111}+\lambda^3 H_{1111}+\dots,$$
$$= 1+H_1+\lambda(H_1^2-H_2)+\lambda^2(H_1^3-2H_1H_2+H_3)$$
$$+\lambda^3(H_1^4-3H_1^2H_2+H_2^2+2H_1H_3-H_4)+\dots$$

Art. 116. The preceding pages show that the coefficient of

$$\lambda^{m-1}$$

in the expansion of

$$\frac{1}{1-a_1+(1-\lambda)a_2-(1-\lambda)^2a_3+\ldots}$$

is equal to

$$\Sigma N_{m,(pqr\ldots)},$$

the summation being for every partition

$$(pqr\ldots);$$

and this, from the theory of the numbers

$$N(p_1p_2\ldots p_m),$$

is equal to

$$\underset{p_1}{\Sigma}\underset{p_2}{\Sigma}\ldots\underset{p_m}{\Sigma}h_{p_1p_2\ldots p_m},$$

the summation being for all integer values of

$$p_1,\ p_2,\ldots p_m;$$

or, the same thing, for the compositions of all numbers into exactly m parts.

Hence

$$\Sigma h_{p_1} = A_1 = H_1,$$

$$\Sigma\Sigma h_{p_1p_2} = A_2 = H_{11} = H_1^2-H_2,$$

$$\Sigma\Sigma\Sigma h_{p_1p_2p_3} = A_3 = H_{111} = H_1^3-2H_1H_2+H_3,$$

. . . .

$$\Sigma\Sigma\ldots\Sigma h_{p_1p_2\ldots p_m} = A_m = H_1^m,$$

a remarkable result.

Art. 117. Since

$$\frac{1}{1-a_1+(1-\lambda)a_2-(1-\lambda)^2a_3+\ldots}$$

$$= 1+\Sigma h_{p_1}+\lambda\Sigma h_{p_1p_2}+\lambda^2\Sigma h_{p_1p_2p_3}+\ldots,$$

we find, putting $\lambda = 1$,

$$\frac{1}{1-a_1} = 1+\Sigma h_{p_1}+\Sigma h_{p_1p_2}+\Sigma h_{p_1p_2p_3}+\ldots,$$

and, since

$$D_1 = d_1 \equiv \partial_{a_1},$$

$$D_1^s(1+\Sigma h_{p_1}+\Sigma h_{p_1p_2}+\Sigma h_{p_1p_2p_3}+\ldots)$$

$$= s!(1+\Sigma h_{p_1}+\Sigma h_{p_1p_2}+\Sigma h_{p_1p_2p_3}+\ldots)^{s+1};$$

and thence, as an easy deduction,

$$D_s(1+\Sigma h_{p_1}+\Sigma h_{p_1p_2}+\Sigma h_{p_1p_2p_3}+\ldots)$$
$$=(1+\Sigma h_{p_1}+\Sigma h_{p_1p_2}+\Sigma h_{p_1p_2p_3})^{s+1}$$

(for observe that, for the operand $\frac{1}{1-\alpha_1}$, $D_1^s \equiv s!D_s$), and thence, by an easy step,

$$D_{p_1}^{\pi_1}D_{p_2}^{\pi_2}\ldots(1+\Sigma h_{p_1}+\Sigma h_{p_1p_2}+\Sigma h_{p_1p_2p_3}+\ldots)$$
$$=\frac{(\Sigma\pi p)!}{(p_1!)^{\pi_1}(p_2!)^{\pi_2}\ldots}(1+\Sigma h_{p_1}+\Sigma h_{p_1p_2}+\Sigma h_{p_1p_2p_3}+\ldots)^{\Sigma\pi p+1}.$$

Chapter 6
Perfect Partitions

6.1 Introduction and Commentary

A perfect partition π of n is a partition of n such that any positive integer less than n may be partitioned uniquely by a partition π' made up of parts taken from π. For example, $4 + 2 + 1$ is a perfect partition of 7 since $6 = 4 + 2$, $5 = 4 + 1$, $4 = 4$, $3 = 2 + 1$, $2 = 2$, $1 = 1$. MacMahon proved in [38] that the number of ordered factorizations of n equals the number of perfect partitions of $n - 1$. Since there are four ordered factorizations 8: 8, $4 \cdot 2$, $2 \cdot 4$, $2 \cdot 2 \cdot 2$, there must be four perfect partitions of 7: $4 + 2 + 1$, $4 + 1 + 1 + 1$, $2 + 2 + 2 + 1$, $1 + 1 + 1 + 1 + 1 + 1 + 1$.

The "prime numbers of measurement on a scale" are defined in [109] and are closely related to the study of perfect partitions in [107]. Several interesting questions are raised in the former paper (see also Andrews, 1976), but no answers are presented. N. J. A. Sloane (1973) lists the "prime numbers of measurement" (and the related difference sequence) in his Handbook of Integer Sequences (Sequences #363, #1044).

Perfect partitions have cropped up in various circumstances in number theory (Long, 1967, 1970; Andrews, 1976), and combinatorics (Riordan, 1958, chapter 6, section 5). We refer the reader to section 5.3 for a discussion of the relationship between perfect partitions and the compositions of a multipartite number.

6.2 Complementing Sets of Integers

Let C be a set of integers. Two subsets A and B of C are said to be complementing subsets of C if each $c \in C$ is uniquely representable by $a + b$ for some $a \in A$ and $b \in B$, and each $a + b$ is in C. C. Long (1967) has proved that the number of complementing subsets of $\{0, 1, \ldots, n\}$ is equal to the number of perfect partitions of n, and we shall present his proof here, including some necessary results taken from deBruijn (1956).

Let $N_n = \{0, 1, 2, \ldots, n - 1\}$ for each positive integer n. If A and B are complementing subsets of N_n, we say that they are of order n, and write $(A, B) \sim N_n$. We first note that the only complementing pair related to N_1 is $(\{0\}, \{0\})$. If $n > 1$ and $(A, B) \sim N_n$, we may draw a number of immediate conclusions about A and B: (i) $A \cap B = \{0\}$; (ii) since $1 \in A \cup B$, we may

The material in this chapter corresponds to section V, chapter I of *Combinatory Analysis*.

assume, without loss of generality, that $1 \in A$ and m is the least positive element of B (if $B \neq \{0\}$); (iii) from (ii) we see that $\mathcal{N}_m \subset A$ and $\{m + 1, m + 2, \ldots, 2m - 1\} \notin A \cup B$; (iv) finally, if $B = \{0\}$, then A must be $\mathcal{N}_n$. From now on we assume that $n > 1$ and we use the notation mS to denote $\{j \mid j/m \in S\}$.

Lemma 1. Let A, B, C and D be subsets of $\mathcal{N}_n$ such that, for a fixed integer $m \geqq 2$,

$$A = mC + \mathcal{N}_m \quad \text{and} \quad B = mD.$$

Then $(A, B) \sim \mathcal{N}_{mp}$ if and only if $(C, D) \sim \mathcal{N}_p$, where $p \geqq 1$.

Proof. First we assume that $(C, D) \sim \mathcal{N}_p$. Consequently, for each $s \in \mathcal{N}_{mp}$ there correspond $q \in \mathcal{N}_p$ and $r \in \mathcal{N}_m$ such that $s = mq + r$. Now $q = c + d$ for some $c \in C$ and $d \in D$ since $(C, D) \sim \mathcal{N}_p$. Therefore

$$s = m(c + d) + r = (mc + r) + md = a + b$$

where $a = mc + r \in A$ and $b = md \in B$. To see uniqueness, we assume that

$$s = a' + b' = (mc' + r') + md',$$

where $a' = mc' + r' \in A$ and $b' = md' \in B$. By the Division Lemma of the Euclidean Algorithm, $r = r'$; hence $c + d = q = c' + d'$, and therefore $c = c'$ and $d = d'$ since C and D are complementing sets.

On the other hand suppose that $(A, B) \sim \mathcal{N}_{mp}$. Then for each $s \in \mathcal{N}_p$,

$$sm = a + b = (mc + r) + md,$$

where $a \in A$, $b \in B$, $c \in C$, $d \in D$, and $r \in \mathcal{N}_m$. The Division Lemma of the Euclidean Algorithm now implies that $r = 0$ and therefore that $s = c + d$. This representation of s in $C + D$ is unique, for if $s = c' + d'$ for $c' \in C$ and $d' \in D$, then

$$sm = cm + dm = c'm + d'm;$$

this implies that $c = c'$ and $d = d'$ since sm must be uniquely represented in $A + B$.

Lemma 2. If $(A, B) \sim \mathcal{N}_n$, then there exists an integer $m \geqq 2$ such that $m|n$, and a complementing pair A', B' of order n/m, with $1 \in A'$ if $B \neq \{0\}$, such that

$$(*) \qquad A = mB' + N_m \quad \text{and} \quad B = mA'.$$

Proof. If $B = \{0\}$, then $A = N_n$, $A' = B' = \{0\}$, and $m = n$. If $B \neq \{0\}$, we let m denote the least positive integer in B, and since we are assuming (without loss of generality) that $1 \in A$, we must have $m > 1$. We determine the unique integer h such that

$$hm \leqq n < (h + 1)m,$$

and proceed by a finite mathematical induction on all nonnegative integers k less than h to show that

(i) if $R_k = \{km + 1, km + 2, \ldots, km + m - 1\}$, then $R_k \cap B = \phi$;

(ii) if $S_k = \{km, km + 1, \ldots, km + m - 1\}$, then either $S_k \subset A$ or $S_k \cap A = \phi$.

Statements (i) and (ii) are clearly true for $k = 0$ by the definition of m and the fact that A and B are complementing sets. Now assume that (i) and (ii) hold for all integers k smaller than p ($\leqq h - 1$). We must consider two cases:

Case 1. $pm \notin A$. Then in the unique representation $pm = a + b$, we must have $a < pm$; furthermore, by the induction hypothesis b must be a multiple of m, and so $a = lm\,(l < p)$. Also by the induction hypothesis, (ii) implies that S_l is contained in A; hence $(a + e) + b$ is the unique representation of $pm + e$. Consequently, $pm + e \notin A \cup B$, for otherwise there would be a second representation.

Case 2. $pm \in A$. Consider now $pm + m$; since $m \in B$, $pm + m$ is the unique representation of itself, with $pm \in A$ and $m \in B$. It follows that $pm + t \notin B$ $(0 < t < m)$, for otherwise $(m - t) + (pm + t)$ would constitute a second representation of $pm + m$. Thus (i) is established. Assuming (ii) false, we see that $pm + t \notin A$ for some t $(0 < t < m)$. Let $pm + t = a' + b'$ be the representation; we know that $m|b'$ by hypothesis, and therefore $a' \equiv t \pmod{m}$, $a' < pm + t$. By the induction hypothesis $a' - t \in A$, and hence

$$pm = pm + 0 = (a' - t) + b',$$

which is impossible since pm must be uniquely representable in $A + B$.

Therefore (i) and (ii) are established for all $k < h$.

Let us now show that $hm + r \notin A \cup B$ for each $r \geqq 0$. If $hm + r \in A$, then $hm + r + m \in A + B = \mathcal{N}_n$, which is impossible since $hm + r + m \geqq hm + m > n$. If $hm + r \in B$, then $(m - 1) + hm + r \in A + B = \mathcal{N}_n$, which is again impossible since $(m - 1) + hm + r \geqq hm + m - 1 > n - 1$.

The preceding paragraph taken together with (i) and (ii) provide the construction of A' and B' directly. Since $n = |C| = |A| \cdot |B| = |B'| \cdot |\mathcal{N}_m| \cdot |B| = |B'| \cdot m \cdot |B|$, we see that $m|n$. Finally by Lemma 1, $(A', B') \sim \mathcal{N}_{n/m}$.

Theorem. The number of complementing subsets of $\{0, 1, 2, \ldots, n\}$ equals the number of perfect partitions of n.

Proof. From Lemma 1 we see that for each ordered factorization of n, say $m_1 m_2 \ldots m_r$, we may form a unique complementing pair (A, B) of $\mathcal{N}_n$. By Lemma 2 we see that each complementing pair is formed in this way. Hence the number of complementing pairs equals the number of ordered factorizations of n, which in turn equals the number of perfect partitions of $n - 1$ (see [20]).

6.3 References

G. E. Andrews (1975) MacMahon's prime numbers of measurement, *American Math. Monthly*, *82*, 922–923.

G. E. Andrews (1976a) The theory of compositions (I): On the ordered factorizations of n and a conjecture of C. Long, *Canadian Math. Bull.*, *18*, 479–484.

G. E. Andrews (1976b) The Theory of Partitions, Encyclopedia of Mathematics and Its Applications, vol. 2, Addison-Wesley, Reading.

N. G. deBruijn (1956) On number systems, *Nieuw Archief Wisk.*, *4*, 15–17.

C. Long (1967) Addition theorems for sets of integers, *Pacific J. Math.*, *23*, 107–112.

C. Long (1970) On a problem in partial difference equations, *Canadian Math. Bull.*, *13*, 333–335.

S. Porubský (1977) On MacMahon's segmented numbers and related sequences, *Nieuw Archief Voor Wisk.* (to appear).

J. Riordan (1958) An Introduction to Combinatorial Analysis, J. Wiley, New York.

N. J. A. Sloane (1973) A Handbook of Integer Sequences, Academic Press, New York.

6.4 Summaries of the Papers

[20] Certain special partitions of numbers, *Quart. J. Math.*, *21* (1886), 367–373.

A perfect partition of a natural number is a partition that contains one and only one partition of each smaller natural number. A subperfect partition of a natural number is a partition for which, taking its parts positively or negatively (if a part is taken positively, no other part of the same magnitude may be taken negatively, and conversely), it is possible to compose each lower natural number in only one way.

MacMahon shows that the number of perfect partitions of n depends only on $a_1, a_2, \ldots, a_r$, where $n + 1 = p_1^{a_1} \ldots p_r^{a_r}$ is the prime factorization of $n + 1$; the number of perfect partitions of n is denoted $[a_1, a_2, \ldots, a_r]$. In general, $[a_1, a_2, \ldots, a_r]$ is difficult to determine; however, in special cases it is possible to derive nice recurrences. For example,

$$[1^0] = 1, \text{ and } [1^s] = \sum_{j=0}^{s-1} \binom{s}{j} [1^j] \text{ for } s \geqq 1,$$

a recurrence quite similar to that of the Bernoulli numbers.

The paper concludes with a consideration of the number of subperfect partitions of n which MacMahon shows to be $[a_1', a_2', \ldots, a_r']$, where $p_1^{a_1'} p_2^{a_2'} \ldots p_s^{a_s'} = 2n + 1$ is the prime factorization of $2n + 1$.

[38] The theory of perfect partitions of numbers and the compositions of multipartite numbers, *Messenger of Math.*, *20* (1891), 103–119.

This paper begins with an exposition of the theory of perfect partitions as presented in [20]. MacMahon extends the results in the previous work by relating perfect partitions to compositions as follows:

Theorem. If $a, b, c, \ldots$ are distinct primes, then the number $a^\alpha b^\beta c^\gamma \ldots - 1$ possesses as many perfect partitions as the multipartite number $(\overline{\alpha\beta\gamma \ldots})$ possesses compositions.

Various aspects of this correspondence are studied in detail.

[107] The partitions of infinity with some arithmetic and algebraic consequences, *Proc. Cambridge Phil. Soc.*, *21* (1923), 642–650.

This paper continues the work on perfect partitions begun in [20] and [38]. Using his standard notation for partitions, MacMahon notes that all perfect partitions are of the form

$$(1^{\alpha_1}\{1 + \alpha_1\}^{\alpha_2}\{(1 + \alpha_1)(1 + \alpha_2)\}^{\alpha_3} \ldots \{(1 + \alpha_1)(1 + \alpha_2) \ldots$$
$$(1 + \alpha_{r-1})\}^{\alpha_r}).$$

He goes on to consider a "partition of infinity" which is merely an infinite vector of which the first $\alpha_1 + \alpha_2 + \ldots + \alpha_r$ components are given in the above perfect partition: to each sequence of positive integers $\alpha_1, \alpha_2, \alpha_3, \ldots$ there is a corresponding partition of infinity. He then remarks that each partition of infinity corresponds to a scale of numeration and has an algebraic formula connected with it. For example, if all the $\alpha_i = 1$, the scale of numeration is binary representation and the algebraic formula is

$$(1 - q)^{-1} = \prod_{n=0}^{\infty} (1 + q^{2^n}).$$

MacMahon proceeds to apply the above observations to a variety of partition generating functions (including the generating functions related to the Rogers-Ramanujan identities). The final two pages of this paper are devoted to applications of this work to symmetric function formulas.

[109] The prime numbers of measurement on a scale, *Proc. Cambridge Math. Soc.*, *21* (1923), 651–654.

MacMahon considers the following problem: What is the sequence of integers $\{a_i\}_{i=1}^{\infty}$ such that $a_0 = 0$, every positive integer is uniquely representable as $a_i - a_j$, and $a_{i+1} - a_i$ strictly increases? He remarks that if just these integers are marked on the real line, then one has a measuring scale by which every positive integral distance may be uniquely measured, and gives a table of the a_i and $a_{i+1} - a_i$ for $1 \leqq i \leqq 42$.

Reprints of the Papers

[20] Certain special partitions of numbers

[38] The theory of perfect partitions of numbers and the compositions of multipartite numbers

[107] The partitions of infinity with some arithmetic and algebraic consequences

[109] The prime numbers of measurement on a scale

CERTAIN SPECIAL PARTITIONS OF NUMBERS.

By Captain P. A. MacMahon, R.A.

THE separation of positive integral numbers into primes and composites is dependent solely upon the divisors possessed by those numbers. Analogous separation may be made which depends upon the partitions of a certain kind possessed by numbers, and it will be seen that the two separations are very closely connected; in what follows I in fact discuss the relationship between the theories of prime numbers and of partitions, and in so doing supply the solutions of two questions of practical interest connected with the theory of weights and measures; viz. these two questions are:

(1) to partition a weight of u lbs. so as to be able to weigh any weight of an integral number of lbs. from 1 to u inclusive, it being only permissible to place the weights in one scale pan.

(2) to solve the same problem as (1) when we are allowed to place the weight in both scale pans.

It is necessary to explain what I mean by a perfect and a sub-perfect partition of a number.

Definition. A perfect partition of a number is one which contains one, and only one, partition of every lower number.

Definition. A sub-perfect partition of a number is such that, taking its parts positively or negatively (if a part be positively, no other part of the same magnitude must be taken negatively, and *vice versâ*), it is possible to compose in only one manner every lower number.

For example 311 is a perfect partition of 5, and 31 a sub-perfect partition of 4, since

$$1 = 1,$$
$$3 - 1 = 2,$$
$$3 = 3,$$
$$3 + 1 = 4.$$

Firstly, as regards perfect partitions it is obvious that every number possesses at least one, viz:—that composed entirely of units; a number which has no other perfect partition I call a partition prime number; similarly, in the theory of sub-perfect partitions we have partition sub-prime numbers, and any given number may be either at once a partition prime and a partition sub-prime, or it may be either of these singly, or it may be neither.

§1. Take u any positive integer and put

$$\phi_{p.q} = 1 + x^q + x^{2q} + \ldots + x^{pq},$$

so that

$$\phi_{u.1} = \sum_{k=u}^{k=0} x^k = \frac{1 - x^{u+1}}{1 - x}.$$

If $u + 1$ be not a prime number, $\phi_{u.1}$ may be broken up into factors of the form $\phi_{p.q}$; if then

$$\phi_{u.1} = \phi_{l.\lambda}\phi_{m.\mu}\phi_{n.\nu}\ldots\ldots \qquad \ldots\ldots(\text{I}),$$

$(\lambda^l\mu^m\nu^n\ldots)$ must be a perfect partition of u; for from the forms of $\phi_{p.q}$ and $\phi_{u.1}$ it is clear that all integers from 1 to u inclusive can be composed, and in one way only, by the parts of this partition taken positively.

u is a partition prime if $u + 1$ be a prime, and in all other cases the equation (I) *admits of more than one solution.* For, if $u = 14$ suppose, the irreducible factors of $1 - x^{15}$ are

$$1 + x + x^2,\ 1 + x + x^2 + x^3 + x^4,\ 1 - x + x^3 - x^4 + x^5 - x^7 + x^8,$$

and $\dfrac{1 - x^{15}}{1 - x}$ is either

$$(1 + x + x^2)(1 + x^3 + x^6 + x^9 + x^{12}),$$

or

$$(1 + x + x^2 + x^3 + x^4)(1 + x^5 + x^{10}).$$

We may at this point predict certain properties of a perfect partition $(\lambda^l\mu^m\nu^n\ldots)$.

Since every integer lower than u is uniquely composed by the parts of this partition, or, what is the same thing, this partition contains only one partition of every integer below u, it follows that the whole number of different combinations that can be formed of its parts must be precisely u. But this is, obviously, also equal to the number of divisors, excluding unity, of the composite number $a^lb^mc^n\ldots$, $a, b, c, \ldots$ being prime numbers, and is therefore

$$(l + 1)(m + 1)(n + 1)\ldots\ldots - 1,$$

wherefore

$$(l + 1)(m + 1)(n + 1)\ldots\ldots = u + 1 \ldots\ldots\ldots(\text{II}).$$

Having regard to equation (I) we see that the numbers

$$l+1,\ m+1,\ n+1,\ \ldots,\ \lambda,\ \mu,\ \nu,\ \ldots$$

must all be divisors of $u+1$; also

$$\left.\begin{aligned} u+1 &= (l+1)\lambda \\ \lambda &= (m+1)\mu \\ \mu &= (n+1)\nu \\ &\vdots \end{aligned}\right\} \quad \ldots\ldots\ldots\ldots\ldots\ldots\text{(III)},$$

$$\left.\begin{aligned} (m+1)(n+1)\ldots &= \lambda \\ (n+1)(p+1)\ldots &= \mu \\ &\vdots \end{aligned}\right\} \quad \ldots\ldots\ldots\ldots\ldots\text{(IV)}.$$

Equation (I) shews that at least one part is unity.

Equation (II) indicates that if u be odd, some part must be repeated an odd number of times, and that if u be even every part must be repeated an even number of times. The number of these perfect partitions of a given number u depends upon the form of $u+1$ when expressed as a product of prime numbers, and not at all upon the prime numbers themselves which compose it.

Put $$u+1 = a^{\alpha} b^{\beta} c^{\gamma} \ldots$$

$a, b, c, \ldots$ being primes other than unity; and represent the number of perfect partitions of u by

$$[\alpha\beta\gamma\ldots].$$

Consider $$\phi_{u.1} = \frac{1-x^{u+1}}{1-x}.$$

If t be any divisor of $u+1$, excluding $u+1$ and unity, a perfect partition corresponds to the identity

$$\phi_{u.1} = \frac{1-x^{t}}{1-x} \cdot \frac{1-x^{u+1}}{1-x^{t}},$$

the partition being t repeated $\dfrac{u+1-t}{t}$ times, and 1 repeated $t-1$ times. But if t_1 be a divisor of t, we have also

$$\phi_{u.1} = \frac{1-x^{t_1}}{1-x} \cdot \frac{1-x^{t}}{1-x^{t_1}} \cdot \frac{1-x^{u+1}}{1-x^{t}},$$

giving a perfect partition, viz.:—

$$t \text{ repeated } \frac{u+1-t}{t} \text{ times},$$

$$t_1 \;\ldots\ldots\ldots\; \frac{t-t_1}{t_1} \;\ldots\ldots,$$

$$1 \;\ldots\ldots\ldots\; t_1-1 \;\ldots\ldots.$$

Proceeding in this way we obtain the whole of the perfect partitions.

Thus if $u=44$, 8 perfect partitions are found, viz.:—

$$15^2 . 5^2 . 1^4,$$
$$15^2 . 3^4 . 1^2,$$
$$15^2 . 1^{14},$$
$$9^4 . 3^2 . 1^2,$$
$$9^4 . 1^8,$$
$$5^8 . 1^4,$$
$$3^{14} . 1^2,$$
$$1^{44}.$$

Excluding unity, $u+1$ has

$$(\alpha+1)(\beta+1)(\gamma+1)\ldots-1 \text{ divisors.}$$

If $a^{\alpha_1}b^{\beta_1}c^{\gamma_1}\ldots$ be any one of these divisors excluding $u+1$, $a^{\alpha_1}b^{\beta_1}c^{\gamma_1}\ldots$ has $(\alpha_1+1)(\beta_1+1)(\gamma_1+1)\ldots-2$ divisors excluding itself and unity; similarly if $a^{\alpha_2}b^{\beta_2}c^{\gamma_2}\ldots$ be any one of these latter divisors, it has

$$(\alpha_2+1)(\beta_2+1)(\gamma_2+1)\ldots-2$$

divisors excluding itself and unity; and so on; then

$$\begin{aligned}[\alpha\beta\gamma\ldots] = {} & (\alpha+1)(\beta+1)(\gamma+1)\ldots-1 \\ & + \Sigma\{(\alpha_1+1)(\beta_1+1)(\gamma_1+1)\ldots-2\} \\ & + \Sigma\Sigma\{(\alpha_2+1)(\beta_2+1)(\gamma_2+1)\ldots-2\} \\ & + \Sigma\Sigma\Sigma\{(\alpha_3+1)(\beta_3+1)(\gamma_3+1)\ldots-2\} \\ & + \ldots\ldots\end{aligned}$$

This formula may be used for the calculation of simple types; thus:—

$$[1\,]=1,$$

$$[1^2]=2^2-1=3,$$

$$[1^3]=2^3-1+3\,(2^2-2)=13,$$

$$[1^4]=2^4-1+4\,(2^3-2)+6\,(2^2-2)+4.3\,(2^2-2)$$
$$=75,$$

$$[1^5]=2^5-1+5\,(2^4-2)+10\,(2^3-2)+10\,(2^2-2)+5.4\,(2^3-2)+5.6\,(2^2-2)+10.3\,(2^2-2)+5.4.3\,(2^2-2)$$
$$=541;$$

but it is preferable to make use of elementary properties of the numbers $[\alpha\beta\gamma\ldots]$ to obtain their values. Thus, reflection convinces one of the truth of the following formula:—

$$[1^s]=1+s\,[1]+\frac{s.s-1}{2\,!}\,[1^2]+\frac{s.s-1.s-2}{3\,!}\,[1^3]+\ldots+s\,[1^{s-1}],$$

in which, giving s different values, the numbers of the above series are readily obtained, and the general value of $[1^s]$ is at once expressed as a determinant.

Numerous formulæ arise from this one; *ex. gr.*

$$[21^s]=2\left\{[1^{s+1}]+s\,[1]\,[1^s]+\frac{s.s-1}{2\,!}\,[1^2]\,[1^{s-1}]+\ldots+\frac{s.s-1}{2\,!}\,[1^{s-2}]\,[1^3]+s\,[1^{s-1}]\,[1^2]+[1^s]\,[1]\right\},$$

$$[31^s]=4\left\{[1^{s+1}]+s\,[2]\,[1^s]+\frac{s.s-1}{2\,!}\,[21]\,[1^{s-1}]+\ldots+\frac{s.s-1}{2\,!}\,[21^{s-3}]\,[1^3]+s\,[21^{s-2}]\,[1^2]+[21^{s-1}]\,[1]\right\};$$

and many others of a similar character which assist the calculation in particular cases.

Formulæ for a few of the simpler cases may be worthy of record.

Since clearly

$$[s] \;= 2\,[s-1]$$

we have $\quad [s] \;= 2^{s-1},$

and $\quad [s1] = 2\,[s-1.1] + [s]$

$$= 2^{s-1}(s+2),$$

also $\quad [s2] = 2\,[s-1.2] - 2\,[s-1.1] + 2\,[s1]$

$$= 2^{s-1}.\frac{1}{2!}(s^2 + 7s + 8).$$

A general formula of reduction is

$$[st] = 2\,\{[s-1.t] - [s-1.t-1] + [s.t-1]\},$$

from which are obtained

$$[s3] = 2^{s-1}\frac{1}{3!}(s^3 + 15s^2 + 56s + 6.4.2),$$

$$[s4] = 2^{s-1}\frac{1}{4!}(s^4 + 26s^3 + 203s^2 + 538s + 8.6.4\cdot 2),$$

$$\vdots$$

[s3] might have been calculated as follows, knowing *a posteriori* its form as a function of s. Put

$$[s3] = 2^{s-1}\frac{1}{3!}(s^3 + Bs^2 + Cs + D),$$

and giving s successively the vales 0, 1, 2 and using previous results, we get three equations for determining B, C and D, and so in other cases.

Other results are

$$[s1^2] \;= 2^{s-1}(s^2 + 6s + 6),$$

$$[s1^3] \;= 2^{s-1}(s^3 + 12s^2 + 36s + 26),$$

$$[s21] = 2^{s-2}(s^3 + 13s^2 + 42s + 32).$$

§ 3. The theory of the sub-perfect partitions is derived at once from what has gone before; for putting

$$\psi_{u.1} = \sum_{k=-u}^{k=u} x^k = \frac{1 - x^{2u+1}}{x^u(1-x)},$$

we have to consider the divisors of $2u+1$ instead of those of $u+1$.

We have

$$\psi_{p.q} = x^{-pq} + x^{-(p-1)q} + \ldots + x^{-q} + 1 + x^{q} + \ldots + x^{(p-1)q} + x^{pq},$$

and if
$$\psi_{u.1} = \psi_{l.\lambda}\psi_{m.\mu}\psi_{n.\nu}\ldots,$$

then $\lambda^l\mu^m\nu^n\ldots$ is a sub-perfect partition of u.

Also
$$(2l+1)(2m+1)(2n+1)\ldots = 2u+1,$$

and the numbers

$$2l+1,\ 2m+1,\ 2n+1,\ \ldots \lambda,\ \mu,\ \nu \ldots$$

must be divisors of $2u+1$.

$$\begin{aligned} 2u+1 &= (2l+1)\lambda, \\ \lambda &= (2m+1)\mu, \\ \mu &= (2n+1)\nu, \\ &\vdots \end{aligned}$$

u is a partition sub-prime if $2u+1$ be a prime; clearly also if $\lambda^l\mu^m\nu^n\ldots$ be a perfect partition of $2u$, then will $\lambda^{\frac{1}{2}l}\mu^{\frac{1}{2}m}\nu^{\frac{1}{2}n}\ldots$ be a sub-perfect partition of u. The number of sub-perfect partition of u is given by $[\alpha\beta\gamma\ldots]$, where

$$2u+1 = a^{\alpha}b^{\beta}c^{\gamma}\ldots$$

Royal Military Academy, Woolwich.
Oct. 12*th*, 1885.

THE THEORY OF PERFECT PARTITIONS OF NUMBERS AND THE COMPOSITIONS OF MULTIPARTITE NUMBERS.

By *Major P. A. MacMahon, R.A., F.R.S.*

1. In the *Quarterly Journal of Mathematics* for 1886 I brought to notice certain partitions of numbers which possess a peculiar and interesting property with respect to numbers less than the number partitioned. I called these at the time 'perfect partitions' and the nomenclature seems to be sufficiently appropriate. The definition is as follows:—

'A perfect partition of a number is one which contains one and only one partition of every lesser number.'

The word number is for the present taken to mean 'positive and integral number.' In some cases hereafter the restriction may be removed so as to bring under view all positive, zero, and negative integers.

Other partitions possess a very similar property. These may be termed provisionally subperfect, and may be defined as follows:—

'A subperfect partition of a number is one which contains one and only one partition of every lesser number if it be permissible to regard the several parts as affected with *either* the positive or negative sign.

Having recently discovered the law by which these partitions may in all cases be enumerated I here give an account of the investigation.

Every assemblage of the parts of a partition, or say every collection of the parts involving one or more parts, is a partition of some number. If the partition be 'perfect' no two of these collections can be partitions of the same number, and, moreover, the number of such collections must be equal to the partitioned number. Consider one of the perfect partitions of 7, viz.

$$(4111);$$

that it is perfect is at once obvious.

2. Before proceeding to the general case of enumeration it is convenient to first solve the problem for certain special numbers included in the formula

$$a^{\alpha} - 1,$$

where a is a prime number; these numbers when increased by unity are powers of primes. It will be shewn that this problem is identical with that of the enumeration of the compositions of numbers. Partitions in which account is also taken of the order of occurrence of the parts are termed by Glaisher 'compositions.'

Not only these numbers but all numbers have clearly one perfect partition composed wholly of unit parts. Represent this solution by the identity

$$\frac{1 - x^{a^{\alpha}}}{1 - x} = 1 + x + x^2 + \ldots + x^{a^{\alpha} - 1}.$$

If the right-hand side be factorized into expressions of the form

$$1 + x^q + x^{2q} + \ldots + x^{pq}$$

the arithmetical interpretation of the factorization proves the existence of a perfect partition of the number $a^{\alpha} - 1$ in correspondence therewith.

This factorization is most conveniently studied by means of the left-hand side

$$\frac{1 - x^{a^{\alpha}}}{1 - x},$$

for any exhibition of this expression in the form

$$\frac{1 - x^{a^{\alpha}}}{1 - x^p} \cdot \frac{1 - x^p}{1 - x^q} \cdot \frac{1 - x^q}{1 - x^r} \cdots \frac{1 - x^u}{1 - x},$$

wherein $p, q, r, \ldots, 1$ are factors of

$$a^{\alpha}, p, q, \ldots u \text{ respectively,}$$

yields a factorization of the form required. Hence we are led to the consideration of the factors of the number a^{α}. The factorization of the number a^{α} is the same problem as the partitioning of the number α. In the first place take a particular value of α, viz. 4. We have for one case of factorization

$$\frac{1-x^{a^4}}{1-x^a} \cdot \frac{1-x^a}{1-x}.$$

Let us agree for brevity to denote such a fraction as $\frac{1-x^{a^p}}{1-x^{a^q}}$ by $p-q$, which, observe, is the difference of the power indices of a in the numerator and denominator.

Then we may write

$$\frac{1-x^{a^4}}{1-x^a} \cdot \frac{1-x^a}{1-x} = 3,\ 1,$$

3, 1 being of necessity a partition of 4.

There is also the factorization represented by

$$\frac{1-x^{a^4}}{1-x^{a^3}} \cdot \frac{1-x^{a^3}}{1-x} = 1,\ 3,$$

where the same numbers 1, 3 occur, but in a different order. Whatever case of factorization be exhibited we must always obtain on the right a succession of numbers constituting a partition of 4. Since, moreover, the same numbers may occur in various orders it is more appropriate to speak of 'compositions' in lieu of partitions. The whole of the factorizations are

$$\frac{1-x^{a^4}}{1-x} = 4,$$

$$\frac{1-x^{a^4}}{1-x^a} \cdot \frac{1-x^a}{1-x} = 3,\ 1,$$

$$\frac{1-x^{a^4}}{1-x^{a^3}} \cdot \frac{1-x^{a^3}}{1-x} = 1,\ 3,$$

$$\frac{1-x^{a^4}}{1-x^{a^2}} \cdot \frac{1-x^{a^2}}{1-x} = 2,\ 2,$$

$$\frac{1-x^{a^4}}{1-x^{a^2}} \cdot \frac{1-x^{a^2}}{1-x^a} \cdot \frac{1-x^a}{1-x} = 2,\ 1,\ 1,$$

$$\frac{1-x^{a^4}}{1-x^{a^3}} \cdot \frac{1-x^{a^3}}{1-x^a} \cdot \frac{1-x^a}{1-x} = 1,\ 2,\ 1,$$

$$\frac{1-x^{a^4}}{1-x^{a^3}} \cdot \frac{1-x^{a^3}}{1-x^{a^2}} \cdot \frac{1-x^{a^2}}{1-x} = 1,\ 1,\ 2,$$

$$\frac{1-x^{a^4}}{1-x^{a^3}} \cdot \frac{1-x^{a^3}}{1-x^{a^2}} \cdot \frac{1-x^{a^2}}{1-x^a} \cdot \frac{1-x^a}{1-x} = 1,\ 1,\ 1,\ 1.$$

We have an instance of factorization corresponding to every composition of the number 4.

Every case of factorization is of necessity represented by a different composition of 4, and every composition of 4 is necessarily in correspondence with a separate factorization. Hence the number of factorizations of the nature considered must be equal to the number of compositions of 4. This means that the number of perfect partitions of the number $a^4 - 1$ is equal to the number of compositions of the number 4.

Returning to the more general case of the factorization of

$$\frac{1 - x^{a^\alpha}}{1 - x}$$

it is clear that a factorization

$$\frac{1 - x^{a^\alpha}}{1 - x^{a^{\alpha_1}}} \cdot \frac{1 - x^{a^{\alpha_1}}}{1 - x^{a^{\alpha_2}}} \cdot \frac{1 - x^{a^{\alpha_2}}}{1 - x^{a^{\alpha_3}}} \cdots$$

will yield the composition $\alpha - \alpha_1$, $\alpha_1 - \alpha_2$, $\alpha_2 - \alpha_3$, ... of the number α. Moreover, any composition A, B, C, ... of the number α can be put into the form

$$\alpha - (\alpha - A), \quad \alpha - A - (\alpha - A - B),$$
$$\alpha - A - B - (\alpha - A - B - C), \ldots,$$

and will thus correspond to a factorization

$$\frac{1 - x^{a^\alpha}}{1 - x^{a^{\alpha - A}}} \cdot \frac{1 - x^{a^{\alpha - A}}}{1 - x^{a^{\alpha - A - B}}} \cdot \frac{1 - x^{a^{\alpha - A - B}}}{1 - x^{a^{\alpha - A - B - C}}} \cdots,$$

Hence there is a one-to-one correspondence between the factorizations of $\frac{1 - x^{a^\alpha}}{1 - x}$ and the compositions of the number α, *i.e.* between the perfect partitions of the number $a^\alpha - 1$ and the compositions of the number α.

Hence this special problem is solved in the form:

A number of the form $a^\alpha - 1$, where a is a prime, possesses the same number of perfect partitions as the number α possesses compositions.

The number of compositions of α is, as is well known, equal to $2^{\alpha - 1}$.

3. For the comprehension of the general case it is necessary to consider the partitions and compositions of multipartite numbers.

An ordinary or unipartite number may be regarded as representing a succession of things, experiences, intervals or operations of any kinds, but, in any one number, of the same kind. It is an entity, and when partitioned its parts are entities of the same kind. The succession, being of similar things, is appropriately designated by a single number. Should however the succession be of things not all similar, the succession demands a succession of numbers for its adequate expression. Such succession of numbers may be viewed as a single entity, and provisionally be termed a multipartite number. When such a succession of things is distributed or (the same thing) when such a multipartite number is partitioned, the things in any portion of the distribution or any part of the partition may be regarded as an entity of the same character as the original entity. The parts into which the multipartite number is partitioned should be regarded as being themselves multipartite numbers. An entity such as 3 hours 43 minutes 37 seconds would be represented by a tripartite number 3,43,37, and would be partitioned into parts each tripartite. If one part were 1 hour, which might apparently be denoted by a unipartite number 1, still the part would be regarded as tripartite and would be represented by 1, 0, 0. It is thus clear that certain entities may be adequately represented by multipartite numbers, and that such numbers may be partitioned into multipartite parts (to the same order of multiplicity) which also are representative of entities of the same nature. It is in some respects convenient to denote a multipartite number involving a succession of unipartite numbers α, β, γ, ..., in the notation*

$$(\alpha\beta\gamma\ldots).$$

Observe that regard must be paid to the order of the succession of these unipartite numbers which go to make up the multipartite. The numbers $(\alpha\beta\gamma\ldots)$, $(\beta\alpha\gamma\ldots)$ must be viewed as distinct multipartites. A multipartite number may contain some zeros amongst the succession of numbers of which it is formed, but in the present theory must contain at least one non-zero number. In fact a zero part or number implies a part or number composed altogether of zeros, and as such is here excluded. Thus a part in a partition may contain any number of zeros less than the full number possible. In forming a partition of the number $(\alpha\beta\gamma\ldots)$ any part of α may be taken with any parts of β, γ, &c., ..., zeros not excluded but subject to the condition given above, to form a multipartite part.

* See *Phil. Trans. of the Royal Society*, Vol. 181 (1890), A, pp. 481—536.

Ex. gr. The partitions of the multipartite number $(\overline{111})$ are $(\overline{111})$, $(\overline{110}, \overline{001})$, $(\overline{101}, \overline{010})$, $(\overline{011}, \overline{100})$, $(\overline{100}, \overline{010}, \overline{001})$, they are five in number, and these, by attending to the order of the parts, give rise to thirteen compositions.

Leaving this short digression I return to the main subject.

4. The number of perfect partitions of the number $a^{\alpha}b^{\beta}c^{\gamma}\ldots - 1$ (where $a, b, c, \ldots$ are primes) depends upon the factorization of the expression

$$\frac{1 - x^{a^{\alpha}b^{\beta}c^{\gamma}\ldots}}{1 - x}$$

into factors of the form $\frac{1 - x^q}{1 - x^p}$, where p is a divisor of q.

Suppose such a factorization to be

$$\frac{1 - x^{a^{\alpha}b^{\beta}c^{\gamma}\ldots}}{1 - x^{a^{\alpha_1}b^{\beta_1}c^{\gamma_1}\ldots}} \cdot \frac{1 - x^{a^{\alpha_1}b^{\beta_1}c^{\gamma_1}\ldots}}{1 - x^{a^{\alpha_2}b^{\beta_2}c^{\gamma_2}\ldots}} \cdots \frac{1 - x^{a^{\alpha_s}b^{\beta_s}c^{\gamma_s}\ldots}}{1 - x} .$$

This, in analogy with what has gone before, may be denoted by the symbolism

$$\{(\alpha - \alpha_1, \beta - \beta_1, \gamma - \gamma_1, \ldots), (\alpha_1 - \alpha_2, \beta_1 - \beta_2, \gamma_1 - \gamma_2, \ldots), \ldots (\alpha_s \beta_s \gamma_s \ldots)\},$$

which may be viewed as a composition of the multipartite number

$$\{(\alpha\beta\gamma\ldots)\}.$$

Every factorization of the required form is necessarily symbolized by one composition of this multipartite number, and, further, every composition must be the symbol of some one factorization. Hence there is a one-to-one correspondence between the required factorizations of

$$\frac{1 - x^{a^{\alpha}b^{\beta}c^{\gamma}\ldots}}{1 - x}$$

and the composition of the multipartite number $\{(\alpha\beta\gamma\ldots)\}$. Hence their number must be equal, and we have the theorem: "The number $a^{\alpha}b^{\beta}c^{\gamma}\ldots - 1$ possesses as many perfect partitions as the multipartite number $\{(\alpha\beta\gamma\ldots)\}$ possesses compositions."

5. It is interesting to identify the various perfect partitions with the particular compositions to which they correspond. The perfect partition of the number $a^a - 1$ which arises from the factorization

$$\frac{1-x^{a^a}}{1-x^{a^{a_1}}} \cdot \frac{1-x^{a^{a_1}}}{1-x^{a^{a_2}}} \cdots \frac{1-x^{a^{a_{p-1}}}}{1-x}$$

consists of $a^{a-a_1}-1$ parts each equal to a^{a_1},

$a^{a_1-a_2}-1$ „ „ a^{a_2},

....

$a^{a_{p-1}}-1$ „ „ 1,

while the corresponding composition of the number α is as above shewn

$$(\alpha-\alpha_1,\ \alpha_1-\alpha_2,\ \ldots,\ \alpha_{p-1}).$$

We can therefore derive at once the composition from the partition or the partition from the composition.

Understanding by *frequencies* the numbers

$$a^{a-a_1}-1,\ a^{a_1-a_2}-1,\ \ldots$$

appertaining to the various different parts of the perfect partition, we see that the frequencies are, in some order, the same for all compositions derived by permutation from the same partition of the number α.

The number of perfect partitions which possess in some order the same frequencies is the same as the number of such compositions. All perfect partitions which possess, in some order, the same frequencies form a group in correspondence with some partition of the number α. Perfect partitions having the same frequencies have also the same number of parts. It is convenient to consider the perfect partitions as divided into groups. As an illustration I give now the classification of the perfect partitions of the number 15. Since 15 is 2^4-1 they will be associated with the compositions of the number 4.

Composition.	Frequency.	Perfect Partition.	Group.
(4)	15,	(1^{15})	(4),
(31)	7, 1,	$(2^7 1)$	(31),
(13)	1, 7,	(81^7)	
(22)	3, 3,	$(4^3 1^3)$	(22),
(211)	3, 1, 1,	$(4^3 21)$	(211).
(121)	1, 3, 1,	$(82^3 1)$	
(112)	1, 1, 3,	(841^3)	
(1111)	1, 1, 1, 1,	(8421)	(1111).

A perfect partition of the number $a^{\alpha}-1$ has in general

$$a^{\alpha-\alpha_1}+a^{\alpha_1-\alpha_2}+\ldots+a^{\alpha_{p-1}}-p \text{ parts},$$

and the associated composition p parts.

6. The subject of the minimum value of this number is now presented for enquiry.

The expression

$$a^{\alpha-\alpha_1}+a^{\alpha_1-\alpha_2}+\ldots+a^{\alpha_{p-1}},$$

is a sum of terms of which the continued product, being a^{α}, is constant. Suppose in the first place that the number p is constant so that the composition has a fixed number p of parts. The number of parts in all perfect partitions of the same group is as we have seen constant. How then does this number vary as we change the group while keeping the number of parts in the composition constant. The expression

$$a^{\alpha-\alpha_1}+a^{\alpha_1-\alpha_2}+\ldots+a^{\alpha_{p-1}},$$

where p is constant, has a certain minimum value for a certain partition of the number which defines the group to which the perfect partition belongs. Within this group the number denoted by the expression has a constant value.

Suppose any two unequal terms of the expression

$$a^{\alpha-\alpha_1},\ a^{\alpha_1-\alpha_2},$$

to be replaced by the two terms

$$(a^{\alpha-\alpha_1}\,.\,a^{\alpha_1-\alpha_2})^{\frac{1}{2}},\ (a^{\alpha-\alpha_1}\,.\,a^{\alpha_1-\alpha_2})^{\frac{1}{2}},$$

viz., by $a^{\frac{1}{2}(\alpha-\alpha_2)},\ a^{\frac{1}{2}(\alpha-\alpha_2)}$,

then the sum of the terms is diminished while their *product* remains constant.

Hence, so long as the sum contains two unequal terms the sum may be diminished without altering the constant product. Hence, the sum will be least when all the terms are equal. If then the number α possesses a partition containing p equal parts, this partition will denote the group which yields perfect partitions having the fewest parts for the given value of p.

The fewest number of parts is in this case seen to be

$$pa^{\frac{\alpha}{p}}-p,$$

and the perfect partition will consist of

$$a^{\frac{\alpha}{p}} - 1 \text{ parts each equal to } a^{\alpha - \frac{\alpha}{p}},$$

$$a^{\frac{\alpha}{p}} - 1 \quad \text{,,} \quad \text{,,} \quad \text{,,} \quad a^{\alpha - \frac{2\alpha}{p}},$$

....

$$a^{\frac{\alpha}{p}} - 1 \quad \text{,,} \quad \text{,,} \quad \text{,,} \quad 1,$$

and the associated composition of α is $\frac{\alpha}{p}$ repeated p times.

A number however does not in general possess a partition of this nature, so that a further examination is necessary.

Since we cannot always partition a number α into p equal parts where p is a fixed number, we must so partition the number so that the parts shall be as nearly equal as possible. It is always possible to partition a number α into p parts in such wise that only two different parts occur. We can further take the two parts to be respectively equal to the integers next above and next below the fraction $\alpha \div p$. It is postulated here that α is not of the form 0 mod. p. The partition is thus

$$\frac{\alpha+\xi}{p} \text{ repeated } \eta \text{ times and } \frac{\alpha-\eta}{p} \text{ repeated } \xi \text{ times,}$$

where $\frac{\alpha+\xi}{p}$ and $\frac{\alpha-\eta}{p}$ are the integers next above and next below $\alpha \div p$, and of necessity $\xi + \eta = p$.

This is the partition of α for a fixed value of p which yields perfect partitions with the fewest number of parts.

This fewest number is thus

$$\eta a^{\frac{\alpha-\xi}{p}} + \xi a^{\frac{\alpha-\eta}{p}} - p,$$

and all perfect partitions associated with compositions of the partition

$$\frac{\alpha+\xi}{p} \text{ repeated } \eta \text{ times and } \frac{\alpha-\eta}{p} \text{ repeated } \xi \text{ times}$$

possess this number of parts, which number is the least possible for a given value of p.

This group of perfect partitions is composed of $\frac{(\xi+\eta)!}{\xi!\,\eta!}$ members, since the partition of the group possesses this number of compositions. The perfect partitions under discussion may be all written down. The first is

$$\begin{array}{llllll}
a^{\frac{a+\xi}{p}}-1 & \text{parts each equal to} & & & a^{a-\frac{a+\xi}{p}}, \\
a^{\frac{a+\xi}{p}}-1 & ,, & ,, & ,, & a^{a-2\frac{a+\xi}{p}}, \\
\multicolumn{5}{c}{\dots\dots\dots\dots\dots\dots} \\
a^{\frac{a+\xi}{p}}-1 & ,, & ,, & ,, & a^{a-\eta\frac{a+\xi}{p}}, \\
a^{\frac{a-\eta}{p}}-1 & ,, & ,, & ,, & a^{a-\eta\frac{a+\xi}{p}-\frac{a-\eta}{p}}, \\
a^{\frac{a-\eta}{p}}-1 & ,, & ,, & ,, & a^{a-\eta\frac{a+\xi}{p}-2\frac{a-\eta}{p}}, \\
\multicolumn{5}{c}{\dots\dots\dots\dots\dots\dots} \\
a^{\frac{a-\eta}{p}}-1 & ,, & ,, & ,, & a^{a-\eta\frac{a+\xi}{p}-(\xi-1)\frac{a-\eta}{p}}, \\
a^{\frac{a-\eta}{p}}-1 & ,, & ,, & ,, & 1,
\end{array}$$

and the remainder are written down by taking the frequencies in the other orders.

The next point is to consider the minimum values of the expressions obtained as p changes from 1 to α.

When $\alpha \equiv 0 \bmod. p$, the expression is

$$p\,(a^{\frac{a}{p}}-1),$$

which obviously diminishes as p increases.

Hence, the minimum value is obtained by giving p its maximum value α.

It then becomes

$$\alpha\,(a-1).$$

The composition is now unity repeated α times and the group contains but a single partition. The perfect partition composed of the fewest parts is

$$\begin{array}{llllll}
a-1 & \text{parts each equal to} & & & a^{\alpha-1}, \\
a-1 & ,, & ,, & ,, & a^{\alpha-2}, \\
\multicolumn{5}{c}{\dots\dots\dots\dots\dots\dots} \\
a-1 & ,, & ,, & ,, & a, \\
a-1 & ,, & ,, & ,, & 1.
\end{array}$$

Next when α is not $\equiv 0$, mod. p. Consider the expression arrived at, viz:

$$\eta a^{\frac{\alpha+\xi}{p}} + \xi a^{\frac{\alpha+\eta}{p}} - p.$$

Where $\alpha \equiv -\xi$ mod. $p. \equiv \eta$, mod. p,

and $$\xi + \eta = p.$$

Put $\alpha + \xi = mp$, $\alpha - \eta = (m-1)p$,

and $$u_p = \eta a^{\frac{\alpha+\xi}{p}} + \xi a^{\frac{\alpha-\eta}{p}} - p,$$

$$= (\alpha - mp + p) a^m + (mp - \alpha) a^{m-1} - p.$$

Then

$$u_p - u_{p+1} = \{(m-1)a - m\} a^{m-1} + 1.$$

Since the least value of m is 2 and a is at least 2, the expression on the right is always a positive integer.

Hence u_{p+1} is always less than u_p, and the greater the number of parts in a composition the less the number of parts in the associated perfect partition.

7. Turning to the general case, the perfect partition of the number $a^\alpha b^\beta c^\gamma \ldots - 1$ which corresponds to the factorization

$$\frac{1 - x^{a^\alpha b^\beta c^\gamma \ldots}}{1 - x^{a^{\alpha_1} b^{\beta_1} c^{\gamma_1} \ldots}} \cdot \frac{1 - x^{a^{\alpha_1} b^{\beta_1} c^{\gamma_1} \ldots}}{1 - x^{a^{\alpha_2} b^{\beta_2} c^{\gamma_2} \ldots}} \cdots \frac{1 - x^{a^{\alpha_{p-1}} b^{\beta_{p-1}} c^{\gamma_{p-1}} \ldots}}{1 - x},$$

is

$a^{\alpha-\alpha_1} b^{\beta-\beta_1} c^{\gamma-\gamma_1} \ldots - 1$ parts each equal to $a^{\alpha_1} b^{\beta_1} c^{\gamma_1} \ldots$,

$a^{\alpha_1-\alpha_2} b^{\beta_1-\beta_2} c^{\gamma_1-\gamma_2} \ldots - 1$ „ „ „ $a^{\alpha_2} b^{\beta_2} c^{\gamma_2} \ldots$,

....

$a^{\alpha_{p-1}} b^{\beta_{p-1}} c^{\gamma_{p-1}} - 1$ „ „ „ 1,

and the associated composition of the multipartite number $\{(\alpha\beta\gamma)\} \ldots$ is

$$\{(\alpha - \alpha_1, \beta - \beta_1, \gamma - \gamma_1, \ldots \alpha_1 - \alpha_2, \beta_1 - \beta_2, \gamma_1 - \gamma_2, \ldots\ldots \alpha_{p-1}, \beta_{p-1}, \gamma_{p-1}, \ldots)\}.$$

There is a one-to-one correspondence between the perfect partitions and the compositions, and the former are derivable from the latter and *vice versâ*.

For compositions of the same partition the *frequencies* of the perfect partitions are, neglecting order, unaltered. We may generally therefore divide the perfect partitions with groups having reference to the partitions of the ruling multipartite number.

To take a concrete example, write down all the perfect partitions of 244, which is $7^2.5-1$, as follows:—

Multipartite Number.

$(\overline{21})$.

Partition	Composition	Perfect Partition.
$(\overline{21})$	$(\overline{21})$	(1^{244})
$(\overline{20}\ \overline{01})$	$(\overline{20}\ \overline{01})$	$(5^{48}.1^4)$
	$(\overline{01}\ \overline{20})$	$(49^4.1^{48})$,
$(\overline{11}\ \overline{10})$	$(\overline{11}\ \overline{10})$	$(7^{34}.1^6)$
	$(\overline{10}\ \overline{11})$	$(35^6.1^{34})$,
$(\overline{10}\ \overline{10}\ \overline{01})$	$(\overline{10}\ \overline{10}\ \overline{01})$	$(35^6.5^6.1^4)$
	$(\overline{10}\ \overline{01}\ \overline{10})$	$(35^6.7^4.1^6)$
	$(\overline{01}\ \overline{10}\ \overline{10})$	$(49^4.7^6.1^6)$.

Observe that corresponding to a part (xy) in a composition, the perfect partition involves a repetitional number $7^x.5^y-1$.

8. The general expression for the number of parts in a perfect partition is

$$a^{\alpha-\alpha_1}b^{\beta-\beta_1}c^{\gamma-\gamma_1}\ldots+a^{\alpha_1-\alpha_2}b^{\beta_1-\beta_2}c^{\gamma_1-\gamma_2}\ldots+\ldots$$
$$+a^{\alpha_{p-1}}b^{\beta_{p-1}}c^{\gamma_{p-1}}\ldots-p.$$

Consider a component portion of this, viz: the expression

$$a^{\alpha-\alpha_1}b^{\beta-\beta_1}c^{\gamma-\gamma_1}\ldots-1,$$

which may be put into the form

$$(1+a-1)^{\alpha-\alpha_1}(1+b-1)^{\beta-\beta_1}(1+c-1)^{\gamma-\gamma_1}\ldots-1,$$

and is thus seen to be

$$>(\alpha-\alpha_1)\,(a-1)+(\beta-\beta_1)\,(b-1)+(\gamma-\gamma_1)\,(c-1)+\ldots$$

(for a, b, c, ... are necessarily each >1), unless all but one of the numbers $\alpha-\alpha_1$, $\beta-\beta_1$, $\gamma-\gamma_1$, ... are equal to zero and the remaining number equal to unity, when the relation is one of equality.

Hence, from a summation it is clear that in general the expression

$$a^{\alpha-\alpha_1}b^{\beta-\beta_1}c^{\gamma-\gamma_1}\ldots+a^{\alpha_1-\alpha_2}b^{\beta_1-\beta_2}c^{\gamma_1-\gamma_2}\ldots+\ldots+a^{\alpha_{p-1}}b^{\beta_{p-1}}c^{\gamma_{p-1}}\ldots-p,$$

$$>\alpha\,(a-1)+\beta\,(b-1)+\gamma\,(c-1)+\ldots.$$

When however in *each* product

$$a^{\alpha_s-\alpha_{s+1}}b^{\beta_s-\beta_{s+1}}c^{\gamma_s-\gamma_{s+1}}\ldots,$$

all but one of the numbers $\alpha_s-\alpha_{s+1}$, $\beta_s-\beta_{s+1}$, $\gamma_s-\gamma_{s+1}$, ... are equal to zero, and the remaining number equal to unity, then the relation becomes one of equality.

Hence the least value of the expression for the number of parts in a perfect partition is

$$\alpha\,(a-1)+\beta\,(b-1)+\gamma\,(c-1)+\ldots.$$

Then also $p=\alpha+\beta+\gamma+\ldots$, and the composition is

$$\left(\overline{1000\ldots}^{\,\alpha}\;\overline{0100\ldots}^{\,\beta}\;\overline{0010\ldots}^{\,\gamma}\right),$$

or else some other composition obtained by permutation of its parts.

There thus exists a group of perfect partitions the members of which possess the same number of parts, which number is less than the number of parts in any perfect partition not comprised in the group.

The number of partitions in the group is obviously

$$\frac{(\alpha+\beta+\gamma+\ldots)!}{\alpha!\,\beta!\,\gamma!\ldots},$$

and one such perfect partition will consist of

$$
\begin{array}{llll}
a-1 & \text{parts each equal to} & & a^{\alpha-1}b^{\beta}c^{\gamma}\ldots, \\
a-1 & ,, \quad ,, & ,, & a^{\alpha-2}b^{\beta}c^{\gamma}\ldots, \\
\ldots\ldots & & & \\
a-1 & ,, \quad ,, & ,, & a^{\alpha-\alpha}b^{\beta}c^{\gamma}\ldots. \\
b-1 & ,, \quad ,, & ,, & a^{\alpha}b^{\beta-1}c^{\gamma}\ldots, \\
\ldots\ldots & & & \\
b-1 & ,, \quad ,, & ,, & a^{\alpha}b^{\beta-2}c^{\gamma}\ldots, \\
\ldots\ldots & & & \\
b-1 & ,, \quad ,, & ,, & a^{\alpha}b^{\beta-\beta}c^{\gamma}\ldots, \\
c-1 & ,, \quad ,, & ,, & a^{\alpha}b^{\beta}c^{\gamma-1}\ldots, \\
c-1 & ,, \quad ,, & ,, & a^{\alpha}b^{\beta}c^{\gamma-\gamma}\ldots, \\
\ldots\ldots & & & \\
c-1 & ,, \quad ,, & ,, & a^{\alpha}b^{\beta}c^{\gamma-\gamma}\ldots. \\
\&c. & & & \&c.
\end{array}
$$

The remaining perfect partitions of the group being obtained by taking the frequencies in the left-hand column in the other orders. The determination of the least number of parts in a perfect partition which corresponds to a composition with a given number p of parts is a more difficult matter into which I do not at present enter.

9. The generating function for the enumeration of the compositions of the multipartite number $(\alpha\beta\gamma\ldots)$ is easily obtained.

If X_r be the sum of the homogeneous products of order r of the quantities $x_1, x_2, x_3, \ldots$ the number in question is the coefficient of $x_1^{\alpha}x_2^{\beta}x_3^{\gamma}\ldots$ in the ascending expansion of the algebraic fraction

$$\frac{X_1 + X_2 + X_3 + \ldots}{1 - X_1 - X_2 - X_3 - \ldots},$$

which is

$$\frac{(1-x_1 . 1-x_2 . 1-x_3 \ldots)^{-1} - 1}{1 - \{(1-x_1 . 1-x_2 . 1-x_3 \ldots)^{-1} - 1\}},$$

or $$\frac{\Sigma x_1 - \Sigma x_1x_2 + \Sigma x_1x_2x_3 - \dots}{(1-x_1 . 1-x_2 . 1-x_3 \dots) - \Sigma x_1 + \Sigma x_1x_2 - \Sigma x_1x_2x_3 + \dots},$$

or $$\frac{\Sigma x_1 - \Sigma x_1x_2 + \Sigma x_1x_2x_3 - \dots}{1 - 2(\Sigma x_1 - \Sigma x_1x_2 + \Sigma x_1x_2x_3 - \dots)}.$$

As a particular case the compositions of the number α are given by the fraction

$$\frac{x_1}{1-2x_1},$$

which is a well-known result, shewing that the number of compositions is $2^{\alpha-1}$.

For the bipartite numbers $(\alpha\beta)$ the generating function is

$$\frac{x_1 + x_2 - x_1x_2}{1 - 2(x_1 + x_2 - x_1x_2)}.$$

The coefficient of $x_1^{\alpha}x_2^{\beta}$ in the expansion is

$$2^{\alpha+\beta-1}\frac{(\alpha+\beta)!}{\alpha!\,\beta!} - 2^{\alpha+\beta-2}\frac{(\alpha+\beta-1)!}{(\alpha-1)!(\beta-1)!} + 2^{\alpha+\beta-3}\frac{(\alpha+\beta-2)!}{(\alpha-2)!(\beta-2)!} - \dots,$$

which therefore represents alike the number of compositions of $(\alpha\beta)$ and the number of perfect partitions of the number $a^{\alpha}b^{\beta} - 1$.

10. It may be added that any perfect partition of any number gives rise directly to an unlimited number of perfect partitions of higher numbers.

For this purpose the repetitional index of the leading or highest part may be increased to any extent, or a part equal to the sum of *all* the parts with unity added may be prefixed. Regarded from this point of view, let A_p denote the number of perfect partitions of the number p so that a generating function when expanded is

$$A_1x + A_2x^2 + A_3x^3 + \dots + A_px^p + \dots.$$

The perfect partitions having a highest part equal to $p+1$ are derivable from the perfect partitions of the number p by prefixing to each of the latter the part $p+1$ any number of times repeated. Hence, the perfect partitions, of all numbers, which have a highest part equal to $p+1$ are given in number by the coefficients of x in the expansion of the fraction

$$\frac{A_px^{2p+1}}{1-x^{p+1}}.$$

Hence the algebraical identity

$$\frac{A_0x}{1-x}+\frac{A_1x^3}{1-x^2}+\frac{A_2x^5}{1-x^3}+\ldots+\frac{A_px^{2p+1}}{1-x^{p+1}}+\ldots,$$

$$=A_1x+A_2x^2+A_3x^3+\ldots+A_px^p+\ldots+A_{2p+1}x^{2p+1}+\ldots.$$

The arithmetical interpretation is the relation

$$A_{2p+1}=\Sigma A_s,$$

the summation being for all values of s which make $s+1$ a divisor of $p+1$.

For example,

$$A_{11}=A_0+A_1+A_2+A_5,$$

since $\frac{1}{2}(11+1)=6$ has the divisons 1, 2, 3, and 6.

The theorem may be written

$$A_{2p-1}=\Sigma A_{s-1},$$

where s is a divisor of p and the summation is for all such.

11. Instead of viewing the aggregate of perfect partitions of all numbers in respect of the highest part involved, it is interesting to regard the equation from the point of view of the number of parts involved, for herein lies another connexion with the theory of compositions.

It can be shewn that there is a one-to-one correspondence between the perfect partitions containing s parts and the compositions of the number s.

Write down the perfect partitions of less than five parts as follows:—

$$(1)$$

$$(1^2)\quad,\quad(21)$$

$$(1^3)\quad,\quad(31^2)\quad,\quad(2^21)\quad,\quad(421)$$

$$(1^4)\ ,\ (41^3)\ ,\ (3^21^2)\ ,\ (631^2)\ ,\ (2^31)\ ,\ (62^21)\ ,\ (4^221)\ ,\ (8421).$$

It will be observed that in any line each perfect partition gives birth to two perfect partitions in the line below

(*a*) by prefixing a part equal to the existing highest part.

(*b*) by prefixing a part which is greater by unity than the sum of all the existing parts.

Write down now, instead of the perfect partitions themselves, the repetitional numbers of the parts of those partitions.

We thus obtain a scheme:—

1

2 , 11

3 , 12 , 21 , 111

4 , 13 , 22 , 112 , 31 , 121 , 211 , 1111;

and the 1st, 2nd, 3rd and 4th lines are seen to represent the compositions of the numbers 1, 2, 3 and 4.

Each composition in a line gives rise to two compositions in the line below by rules derived from (*a*) and (*b*). These are:—

(*a'*) increase the first part by unity,

(*b'*) prefix a part unity.

These are in fact rules for deriving the compositions of a number from those of the next preceding number. That the rules are correct is clear, because no two compositions thus obtained can be identical, and, further, the number of compositions is right, viz. 2^{s-1} in the case of a number s.

Hence the theorem

"There is a one-to-one correspondence between the perfect partitions containing s parts and the compositions of the number S."

The number, as seen, is 2^{s-1}, and the correspondence is in reference to the repetitional numbers occurring in the perfect partitions.*

It is remarkable that the correspondence in this case should be with compositions of simple or unipartite numbers.

The main theorem of the paper exhibits a correspondence with the compositions of multipartite numbers in general, and I believe marks the first appearance of such compositions in the theory of Numbers. This, I think, shews that the subject of perfect partitions is one worthy of the attention of mathematicians.

* It will be readily seen that the general form of a perfect partition is

$$1^{\alpha}\,(1+\alpha)^{\beta}\,(1+\alpha.1+\beta)^{\gamma}\,(1+\alpha.1+\beta.1+\gamma)^{\delta}\,\ldots,$$

where $\alpha, \beta, \gamma, \delta, \ldots$, are any positive, non-zero integers taken at pleasure. There is thus one, and only one, perfect partition in correspondence with every succession of positive integers

$$\alpha, \beta, \gamma, \delta, \ldots .$$

The aggregate of these successions constitutes the aggregate of the compositions for all (unipartite) numbers.

The Partitions of Infinity with some Arithmetic and Algebraic consequences. By Major P. A. MACMAHON.

[*Received* 29 March 1923.]

1. The representation of numbers by means of a systematic scale of notation depends upon the well-known theorem:

"Let $r_1, r_2, r_3, \ldots, r_n, r_{n+1}, \ldots$ denote an infinite series of integers, restricted in no way except that each is to be greater than 1; then any integer N may be expressed in the finite form

$$N = p_0 + p_1 r_1 + p_2 r_1 r_2 + p_3 r_1 r_2 r_3 + \ldots + p_n r_1 r_2 \ldots r_n,$$

where $p_s < r_{s+1}$. When $r_1, r_2, r_3, \ldots$ are given, this can be done in one way only."

The place of this theorem in the theory of the Partition of Numbers has not, I believe, hitherto been explicitly stated.

In the year 1886* I discussed certain special partitions of numbers which I provisionally termed "Perfect Partitions" with the definition:

"A perfect partition of a number is one which contains one, and only one, partition of every lower number."

I recurred again to the subject† in "The Theory of the Perfect Partitions of Numbers and the Compositions of Multipartite Numbers" and subsequently‡ I made applications to various branches of Combinatory Analysis. The theory was an important auxiliary in the final solution of Euler's problem of the "Latin Square" and of its wide generalisation.

2. In the paper of 1890 (*l.c.*), p. 119, *foot-note*, I gave what is virtually the most general expression of the perfect partition of an infinite number.

This was

$$1^{\alpha_1}(1+\alpha_1)^{\alpha_2}\{(1+\alpha_1)(1+\alpha_2)\}^{\alpha_3}\{(1+\alpha_1)(1+\alpha_2)(1+\alpha_3)\}^{\alpha_4}\ldots \textit{ad inf.},$$

where (i) $\alpha_1, \alpha_2, \alpha_3, \alpha_4, \ldots$ are positive integers.

(ii) $\alpha_1, \alpha_2, \ldots$, when used as exponents, denote repetitions of the numbers which they affect, so that the expression denotes an infinite succession of integers in such wise that

1^{α_1} denotes a succession of α_1 units,

$(1+\alpha_1)^{\alpha_2}$ denotes a succession of α_2 integers each equal to $1+\alpha_1$, etc.

* *Quarterly Journal of Pure and Applied Mathematics*, No. 84.

† *The Messenger of Mathematics*, New Series, No. 235, Nov. 1890.

‡ *Combinatory Analysis*, vols. I, II, Camb. Univ. Press, 1915–16.

(iii) $\alpha_1, \alpha_2, \alpha_3, \ldots$ may each be given any one of the values 1, 2, 3, ... subject to

(iv) one number α_s may be an infinite integer and this implies that α with a suffix, $> s$, is zero.

Every number, from 1 to ∞, may be composed in one way (and one only) by means of integers selected from the infinite series whatever values be assigned to $\alpha_1, \alpha_2, \alpha_3, \ldots$ in agreement with the specified conditions.

This perfect partition of infinity is derived from the general theorem given at the commencement of the paper by writing

$$p_s = \alpha_{s+1}, \quad r_s = 1 + \alpha_s.$$

It is important to notice that the partition expression, which is in accord with the universally recognised notation, involves a *single* system of integers

$$\alpha_1, \alpha_2, \alpha_3, \ldots,$$

because the notation causes the system

$$p_0, p_1, p_2, \ldots$$

to disappear automatically.

3. The most interesting particular cases in the partition notation are:

I. $\alpha_1 = \infty$, $\alpha_s \ (s > 1) = 0$, an infinite succession of units

$$(1111\ldots);$$

II. $\alpha_s = r - 1$ for all values of s

$$\{1^{r-1} . r^{r-1} . (r^2)^{r-1} . (r^3)^{r-1} \ldots\},$$

involving the scale of numeration for radix r;

III. $\alpha_s = s$ for all values of s

$$(1! \, 2!^2 \, 3!^3 \, 4!^4 \ldots);$$

IV. $\alpha_1, \alpha_2, \alpha_3, \ldots$ in succession are, each of them, one less than the uneven primes 3, 5, 7, ...

$$\{1^2 . 3^4 . (3.5)^6 . (3.5.7)^{10} . (3.5.7.11)^{12} \ldots\}.$$

Every partition of infinity corresponds to a scale of numeration, and has an algebraic formula connected with it.

Thus I above give

$$\frac{1}{1-q} = 1 + q + q^2 + q^3 + \ldots.$$

II with $r = 2$,

$$\frac{1}{1-q} = (1+q)(1+q^2)(1+q^{2^2})(1+q^{2^3}) \ldots \textit{ ad inf.}$$

II in general,

$$\frac{1}{1-q}=(1+q+\ldots+q^{r-1})(1+q^r+q^{2r}+\ldots+q^{(r-1).r})(1+q^{r^2}+q^{2r^2}+\ldots+q^{(r-1).r^2})\ldots \textit{ad inf.}$$

III,

$$\frac{1}{1-q}=(1+q^{1!})(1+q^{2!}+q^{2.2!})(1+q^{3!}+q^{2.3!}+q^{3.3!})\ldots \textit{ad inf.}$$

etc.

Application to the Generating Function which enumerates the Partitions of Numbers.

4. The function is $\prod_1^\infty (1-q^s)^{-1}$,

if the partitions be unrestricted.

In the first place transform each factor by application of the binary scale formula

$$\frac{1}{1-q}=(1+q)(1+q^2)(1+q^4)(1+q^8)\ldots.$$

We find

$$\frac{1}{(1-q)(1-q^2)(1-q^3)\ldots}=(1+q)(1+q^2)^2(1+q^3)(1+q^4)^3\ldots(1+q^n)^{n_2}\ldots,$$

where $n_2=1+\nu$, and ν is the highest power of 2 which is a factor of n.

Otherwise, and preferably, we may define n_2 to be

"The number of representations of n in the form—a power of 2 multiplied by *any number*."

Observe that the phrase "*any number*" occurs in the definition because the exponents of q in the denominator of the function, which we are transforming, involve the whole of the integers 1, 2, 3, ...

If we had been transforming the function

$$\prod_0^\infty (1-q^{2s+1})^{-1}$$

the exponent of $(1+q^n)$ would have been unity because any number has only *one* representation in the form

"a power of 2 multiplied by an *uneven* number."

We now consider the factor

$(1+q^s)^{s_2}$, of the transformed function,

and find $$q\frac{d}{dq}\log(1+q^s)^{s_2}=s_2\left\{\frac{sq^s}{1-q^{2s}}-\frac{sq^{2s}}{1-q^{2s}}\right\}.$$

The general term in the expansion of $s_2 \frac{sq^s}{1-q^{2s}}$ is

$$s_2 s q^{s(2k+1)},$$

indicating that the coefficient of q^n in

$$\sum_1^\infty{}^s s_2 \frac{sq^s}{1-q^{2s}}$$

is

$$\Sigma d_2 d,$$

where d is a divisor of n whose conjugate is uneven.

I write $$\Sigma d_2 d = \Delta^{(1)}(n_2, n)\,*.$$

Similarly the coefficient of q^n in

$$\sum_1^\infty{}^s s_2 \frac{sq^{2s}}{1-q^{2s}}$$

is

$$\Sigma d_2 d,$$

where d ranges over the divisors of n which have even conjugates.

I write $$\Sigma d_2 d = D^{(1)}(n_2, n).$$

Hence

$$q\frac{d}{dq}\log \prod_1^\infty (1+q^s)^{s_2} = \sum_1^\infty \{\Delta^{(1)}(n_2, n) - D^{(1)}(n_2, n)\}\, q^n$$

$$= \sum_1^\infty \zeta^{(1)}(n_2, n)\, q^n\dagger.$$

Moreover, after Euler,

$$q\frac{d}{dq}\log \prod_1^\infty (1-q^s)^{-1} = \Sigma\sigma(n)\, q^n,$$

where $\sigma(n)$ is the sum of the divisors of n.

Thus we have the arithmetical identity

$$\zeta^{(1)}(n_2, n) = \sigma(n).$$

As a verification for $n = 20$:

Divisors	1,	2,	4,	5,	10,	20,
d_2	1,	2,	3,	1,	2,	3,
Conjugates	20,	10,	5,	4,	2,	1;

$$\Delta^{(1)}(n_2, n) = 3.4 + 3.20 = 72,$$

$$D^{(1)}(n_2, n) = 1.1 + 2.2 + 1.5 + 1.20 = 30,$$

$$\zeta^{(1)}(n_2, n) = 72 - 30 = 42 = 1 + 2 + 4 + 5 + 10 + 20 = \sigma(20).$$

* Glaisher denotes by $\Delta'(n)$ the sum of the divisors whose conjugates are uneven.
,, $D'(n)$,, ,, ,, even.

† Glaisher denotes $\Delta'(n) - D'(n)$ by $\zeta'(n)$.

5. I next apply the ternary scale formula

$$\frac{1}{1-q} = (1+q+q^2)(1+q^3+q^6)(1+q^9+q^{18})\dots(1+q^{3^k}+q^{2.3^k})\dots$$

and find that $\prod_1^\infty (1-q^n)^{-1} = \prod_1^\infty (1+q^s+q^{2s})^{s_3}$,

where s_3 is the number of representations of s in the form

"a power of 3 multiplied by *any number*."

We have

$$q\frac{d}{dq}\log(1+q^s+q^{2s})^{s_3} = \frac{s_3}{1-q^{3s}}(sq^s+sq^{2s}-2sq^{3s}).$$

The coefficient of q^n in

$$\sum_1^\infty{}_s \frac{s_3}{1-q^{3s}}(sq^s+sq^{2s})$$

is

$$\Sigma d_3 d,$$

where d ranges over those divisors of n whose conjugates are *not* of the form 0 mod 3.

I write $\Sigma d_3 d = \Delta^{(2)}(n_3, n)$.

And the coefficient of q^n in

$$\sum_1^\infty{}_s \frac{s_3}{1-q^{3s}}\cdot sq^{3s}$$

is

$$\Sigma d_3 d,$$

where d ranges over those divisors of n whose conjugates *are* of the form 0 mod 3.

I write $\Sigma d_3 d = D^{(2)}(n_3, n)$,

and thence $\zeta^{(2)}(n_3, n) = \Delta^{(2)}(n_3, n) - 2D^{(2)}(n_3, n) = \sigma(n)$.

6. In the general case of order r

$$\frac{1}{(1-q)(1-q^2)(1-q^3)\dots} = \prod_1^\infty (1+q^s+q^{2s}+\dots+q^{(r-1)s})^{s_r},$$

where s_r is the number of representations of s in the form

"a power of r multiplied by any number."

We have

$$q\frac{d}{dq}\log(1+q^s+q^{2s}+\dots+q^{(r-1)s})$$

$$= \frac{s_r}{1-q^{rs}}(sq^s+sq^{2s}+\dots+s^{(r-1)s}-(r-1)sq^{rs}).$$

The coefficient of q^n in

$$\sum_1^\infty{}_s \frac{s_r}{1-q^{rs}}(sq^s + sq^{2s} + \ldots + s^{(r-1)s})$$

is

$$\Sigma d_r d,$$

where d ranges over the divisors of n whose conjugates are *not* of the form 0 mod r.

I write $$\Sigma d_r d = \Delta^{(r-1)}(n_r, n),$$

and the coefficient of q^n in

$$\sum_1^\infty{}_s \frac{s_r s q^{rs}}{1-q^{rs}}$$

is

$$\Sigma d_r d,$$

where d ranges over the divisors whose conjugates *are* of the form 0 mod r.

I write $$\Sigma d_r d = D^{(r-1)}(n_r, n),$$

and thence

$$\zeta^{(r-1)}(n_r, n) = \Delta^{(r-1)}(n_r, n) - (r-1)\,D^{(r-1)}(n_r, n) = \sigma(n).$$

The verification for $r = 5$, $n = 20$ is

Divisors	1,	2,	4,	5,	10,	20,
d_5	1,	1,	1,	2,	2,	2,
Conjugates	20,	10,	5,	4,	2,	1;

$$\Delta^{(4)}(n_5, n) = 2.5 + 2.10 + 2.20 = 70,$$

$$D^{(4)}(n_5, n) = 1.1 + 1.2 + 1.4 = 7,$$

$$\zeta^{(4)}(n_5, n) = 70 - 4.7 = 42 = \sigma(20).$$

The arithmetical functions

$$\Delta^{(r-1)}(n_r, n),\ D^{(r-1)}(n_r, n),\ \zeta'(n_r, n) = \Delta^{(r-1)}(n_r, n) - (r-1)\,D^{(r-1)}(n_r, n),$$

which present themselves, appear to be new to the subject.

7. Other forms of arithmetical functions appear when other generating functions connected with the enumeration of partitions are considered. To give a general idea of their nature I will consider a generating function which, in consequence of the researches of Rogers and Ramanujan, has been much in evidence within the last few years. It is

$$\prod_0^\infty (1 - q^{5m \pm 1})^{-1} = \frac{1}{(1-q)(1-q^4)(1-q^6)(1-q^9)(1-q^{11})\ldots}.$$

Applying the formula for the r^{ary} scale of enumeration, we find

$$\prod_1^\infty (1 + q^s + q^{2s} + \ldots + q^{(r-1)s})^{f(s)},$$

where $f(s)$ denotes the number of representations of s in the form "a power of r multiplied by a number of the form ± 1 mod 5."

When we operate with $q \dfrac{d}{dq} \log$ we meet with two numerical functions

$$\text{(i)} \quad \Sigma f(d) \,.\, d = f_1(n),$$

where d ranges over the divisors of n whose conjugates are *not* of the form 0 mod r, and

$$\text{(ii)} \quad \Sigma f(d) \,.\, d = f_2(n),$$

where the range is over those divisors whose conjugates *are* of the form 0 mod r;

$$f_1(n) - (r-1) f_2(n) = g(n),$$

where $g(n)$ is the number of partitions of n into parts of the form $5m \pm 1$.

8. The numerical function n_r is such that

$$\Sigma n_r x^n$$

$$= \frac{x + x^2 + \ldots + x^{r-1}}{1 - x^r} + 2\,\frac{x^r + x^{2r} + \ldots + x^{(r-1)r}}{1 - x^{r^2}} + 3\,\frac{x^{r^2} + x^{2r^2} + \ldots + x^{(r-1)r^2}}{1 - x^{r^3}} + \ldots,$$

the general term being

$$s\,\frac{x^{r^{s-1}} + x^{2.r^{s-1}} + \ldots + x^{(r-1)r^{s-1}}}{1 - x^{r^s}};$$

and it can be seen at once that

$$\Sigma n_r x^n - 2\Sigma n_r x^{rn} + \Sigma n_r x^{r^2 n} = \frac{x + x^2 + \ldots + x^{r-1}}{1 - x^r}.$$

9. I pass on to the factorial scale of notation for which

$$\frac{1}{1-q} = (1 + q^{1!})(1 + q^{2!} + q^{2.2!})(1 + q^{3!} + q^{2.3!} + q^{3.3!}) \ldots,$$

and remark that, when application is made to

$$\prod_1^\infty (1 - q^n)^{-1},$$

it becomes

$$\prod_1^\infty \prod_1^\infty \frac{1 - q^{t(s+1)!}}{1 - q^{t.s!}}.$$

If we denote by f_n the number of representations of n in the form

"a factorial multiplied by *any number*,"

this may be written $$\prod_1^\infty \frac{(1-q^n)^{f_{n-1}}}{(1-q^n)^{f_n}}.$$

Operating as usual with $q\frac{d}{dq}\log$ we merely obtain the trivial result

$$\Sigma d f_d - \Sigma d\,(f_d - 1) = \Sigma d = \sigma\,(n).$$

The numerical function $\Sigma d f_d$ ranges over all of the divisors of n, and I write

$$\Sigma d f_d = \epsilon\,(n).$$

We have $$\Sigma f_m q^n = \frac{q^{1!}}{1-q^{1!}} + \frac{q^{2!}}{1-q^{2!}} + \frac{q^{3!}}{1-q^{3!}} + \ldots;$$

$$q\frac{d}{dq}\log \prod_1^\infty (1-q^n)^{-f_n} = \frac{f_1 q}{1-q} + \frac{2f_2 q^2}{1-q^2} + \frac{3f_3 q^3}{1-q^3} + \ldots + \frac{s f_s q^s}{1-q^s} + \ldots$$

$$= \sum_1^\infty \epsilon\,(n)\,q^n,$$

and by integration

$$\prod_1^\infty (1-q^n)^{-f_n} = \exp.\sum_1^\infty \frac{1}{n}\epsilon\,(n)\,q^n.$$

Application to the Reciprocal of a Polynomial.

10. Consider

$$\frac{1}{(1-\alpha_1 q)(1-\alpha_2 q^2)(1-\alpha_3 q^3)\ldots(1-\alpha_m q^m)} = \frac{1}{1-a_1 q + a_2 q^2 + \ldots + (-)^m a_m q^m},$$

and apply the formula

$$\frac{1}{1-q} = (1+q)\,(1+q^2)\,(1+q^4)\,(1+q^8)\ldots$$

to each factor. We obtain

$$\prod_0^\infty (1+\alpha_1^{2^s} q^{2^s})\,(1+\alpha_2^{2^s} q^{2^s})\ldots(1+\alpha_m^{2^s} q^{2^s}).$$

We express the symmetric functions of $\alpha_1, \alpha_2, \alpha_3, \ldots, \alpha_m$ in the partition notation and find

$$\begin{aligned}
&\{1 + (1)\,q + (1^2)\,q^2 + (1^3)\,q^3 + \ldots + (1^m)\,q^m\} \\
\times\,&\{1 + (2)\,q^2 + (2^2)\,q^4 + (2^3)\,q^6 + \ldots + (2^m)\,q^{2m}\} \\
\times\,&\{1 + (4)\,q^4 + (4^2)\,q^8 + (4^3)\,q^{12} + \ldots + (4^m)\,q^{4m}\} \\
\times\,&\{1 + (8)\,q^8 + (8^2)\,q^{16} + (8^3)\,q^{24} + \ldots + (8^m)\,q^{8m}\} \\
\times\,&\ldots\ \textit{ad inf.}
\end{aligned}$$

This infinite product is equal to

$$1 + h_1 q + h_2 q^2 + h_3 q^3 + \dots \textit{ ad inf.},$$

where h_s is the homogeneous product sum of degree s of the quantities

$$\alpha_1, \alpha_2, \alpha_3, \dots, \alpha_m.$$

Equating coefficients,

$$h_1 = (1),$$
$$h_2 = (1^2) + (2),$$
$$h_3 = (1^3) + (1)(2),$$
$$h_4 = (1^4) + (1^2)(2) + (2^2) + (4),$$
$$h_5 = (1^5) + (1^3)(2) + (1)(2^2) + (1)(4),$$
$$h_6 = (1^6) + (1^4)(2) + (1^2)(2^2) + (1^2)(4) + (2^3) + (2)(4),$$
$$h_7 = (1^7) + (1^5)(2) + (1^3)(2^2) + (1^3)(4) + (1)(2^3) + (1)(2)(4),$$
$$h_8 = (1^8) + (1^6)(2) + (1^4)(2^2) + (1^4)(4) + (1^2)(2^3) + (1^2)(2)(4) + (2^4) + (2^2)(4) + (4^2) + (8),$$

etc.

h_n is expressible as a sum of products of symmetric functions, where each symmetric function involves *one* of the numbers 1, 2, 2^2, 2^3, ... and no other number. This is an algebraic analogue of the unique expression of all integers by means of the addition of powers of 2.

The number of terms in the expression of h_n is equal to the number of solutions in integers of the equation

$$x_0 + 2x_1 + 2^2 x_2 + \dots + 2^s x_s + \dots = n.$$

The fact is that every perfect partition has an algebraic analogue of this character.

If we take the perfect partition corresponding to

$$(1 + q + q^2 + \dots + q^{k-1})(1 + q^k + q^{2k} + q^{3k} + \dots),$$

we find that we reach a result

$$h_n = \Sigma P \,.\, Q,$$

where P is a symmetric function whose partition expression contains no number greater than $k-1$; and Q is one whose partition expression involves only numbers which are multiples of k. The expression of h_n in this form is unique.

The Prime Numbers of measurement on a scale. By Major P. A. MacMahon.

[*Received* 7 April 1923.]

If we take a straight line of finite length l and divide it into n equal segments we obtain a scale or measuring rod which enables the measurement of any number of segments s where $0 \leqslant s \leqslant l$. If such measurements be the object in view it is clear that the rod is redundantly divided. There are more scale divisions than are necessary. Certain of the scale divisions may be obliterated. Thus a yard rod divided into three feet may have one scale division wiped out without interference with the measurements of one, two and three feet.

The problem which is here presented, in the case of a scale of finite length, has been discussed chiefly in connection with mathematical puzzles without leading to much of mathematical interest connected with the classical theory of numbers. The questions which arise are difficult but can be to some extent elucidated by tentative processes.

The maximum number of scale divisions that may be erased and the specification of the resulting scale of segments have not been determined as yet and it is clear that other questions quickly present themselves for solution.

The recent *History of the Theory of Numbers** in two volumes does not supply any references to scientific papers upon the subject.

The present communication deals with a scale of infinite length. This is, from one point of view, a simplification of the problem because we are practically freed from a boundary condition. In the scale of finite length if we take one end as origin, the other end presents a boundary condition which leads to difficulties immediately. In the case of the infinite rod these initial troubles are absent so that progress can be made up to a certain point, and it will be found, by a perusal of what follows, that an interesting system of numbers, of infinite extent, presents itself which appears to exhibit a certain analogy with the infinite series of prime numbers.

It is essential to specify precisely the problem of the infinite measuring rod. Beginning from the finite end, our zero point, we

* L. E. Dickson of Chicago, published by the Carnegie Institute of Washington, D.C.

are going to insert certain dividing lines, in correspondence with a segment of unit length, so as to enable the measurement on the rod of a length of any integral number of segments where such number may be any integer 1, 2, 3, ... ∞.

This measurement is to be made by *one* operation with two dividing lines of the rod. If we do not specify this condition, but allow a number of different measurements to be made and added together we are in face of a question which has already been completely solved. We have, for instance, only to take dividing lines which exhibit successive segments of lengths:

$$1, 2, 2^2, 2^3, \dots \textit{ ad inf.}$$

to be enabled, by *one or more* measurements, to measure any length which is equal to any integral number of segments.

In fact the above written succession of numbers constitutes a "perfect" partition of the number infinity in that every number can be composed by selecting members of the series, once only, in only one way.

This is the simplest solution of a particular case of the general theorem which states that if

$$\alpha_1, \alpha_2, \alpha_3, \dots$$

be *any integers at pleasure*, each $\geqslant 1$, a "perfect" partition of infinity is

α_1 numbers each equal to unity
α_2 ,, ,, $1 + \alpha_1$
α_3 ,, ,, $(1 + \alpha_1)(1 + \alpha_2)$
α_4 ,, ,, $(1 + \alpha_1)(1 + \alpha_2)(1 + \alpha_3)$
etc.*

The simplest solution is obtained by putting

$$\alpha_1 = \alpha_2 = \alpha_3 = \dots = 1.$$

In regard to the problem now before us we are to measure any integral number of units of length by a *single* operation:

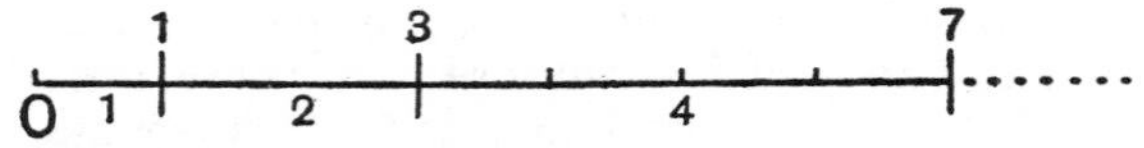

In order to arrive at a minimum number of dividing lines we further specify that, starting from the origin O, the lengths of the successive segments are to *increase* in length. The lengths are to be in ascending order of magnitude, but no two such lengths are to be equal.

Under these conditions it is clear that we require scale divisions

* *Combinatory Analysis*, Camb. Univ. Press, 1915, vol. I, p. 220.

at the points 1 and 3 in order to obtain segments of lengths 1, 2. We can now measure three units of length between the origin and the division at the point 3. We next require a division at the point 7 in order to measure a segment of length 4.

Proceeding in this way we obtain segments of successive lengths

$$1, 2, 4, 5, 8, 10, 14, \ldots$$

with divisions at the points

$$0, 1, 3, 7, 12, 20, 30, 44, \ldots$$

the segmental and divisional series of numbers respectively.

Observe that *every* number can be obtained

(i) by adding successive numbers of the segmental series;

(ii) by taking the difference of *two* properly selected numbers of the divisional series.

The first 347 numbers of the segmental series are now given*. This is the complete number $\not> 1000$. The corresponding divisional numbers are also given.

In the segmental series the longest sequence that appears is one of five from 629 to 633; the longest sequence that is absent is one of ten from 448 to 457. In the first hundred 42 numbers appear; in the tenth hundred there are 31. The density of the numbers appears to diminish very slowly from the results for the ten hundreds:

1	2	3	4	5	6	7	8	9	10
42	33	32	39	37	33	35	32	33	31

Every integer is the result of the addition of a consecutive set of numbers of the series, but the formation is not in every case unique. The earliest example occurs in regard to the number 29 which appears as

$$2 + 4 + 5 + 8 + 10 \text{ and as } 14 + 15.$$

In the classical theory of numbers we have on the one hand the set of primes which serve to express every integer uniquely, and on the other the set of square numbers with the theorem that every integer is expressible, but not uniquely, as the sum of four or fewer numbers of the set. The system of numbers before us seems to present analogies with both of these theories. Since we have already in the system of numbers $1, 2, 2^2, 2^3, \ldots$ what may be termed the prime numbers of addition which involve the unique construction by addition of every integer, it is not to be expected that in the present question there will be the same unique character. What is required is some method of dealing with the series of numbers analytically. The divisional numbers are also available.

* This number has been reduced to 42 on account of the cost of printing.

It would be interesting to determine how either set of numbers behaves at great distances from the origin of measurement.

Prime Numbers of Measurement.

Scale numbers.					
Ordinal number	Segmental number	Divisional number	Ordinal number	Segmental number	Divisional number
1	1	1	22	46	501
2	2	3	23	48	549
3	4	7	24	49	598
4	5	12	25	50	648
5	8	20	26	53	701
6	10	30	27	57	758
7	14	44	28	60	818
8	15	59	29	62	880
9	16	75	30	64	944
10	21	96	31	65	1009
11	22	118	32	70	1079
12	25	143	33	77	1156
13	26	169	34	80	1236
14	28	197	35	81	1317
15	33	230	36	83	1400
16	34	264	37	85	1485
17	35	299	38	86	1571
18	36	335	39	90	1661
19	38	373	40	91	1752
20	40	413	41	92	1844
21	42	455	42	100	1944

Chapter 7
Distributions upon a Chess Board and Latin Squares

7.1 Introduction and Commentary

MacMahon's work in this area inspired J. H. Redfield (1927) to develop extensively the applications of group theory to combinatorics and enumeration problems. The whole area has subsequently been titled "Polya-Redfield Enumeration Theory." Redfield's paper went unnoticed for many years; however, within the last 15 years F. Harary and E. Palmer (1967, 1973), H. O. Foulkes (1963), and R. C. Read (1959, 1960) have provided substantial evidence for the importance of Redfield's work.

Since Redfield's paper presents clearly the relationship between MacMahon's work and modern enumeration theory, we have decided to reprint Redfield's paper in its entirety. Of course, Redfield's paper did not enter the mainstream of combinatorics until 30 years after its publication; consequently Redfield's notation is out of date. We therefore include in section 2 a short description of the relevant terminology adapted from Harary and Palmer (1967).

Most of the references included in Section 7.4 contain accounts of the recent applications of the work of Polya and Redfield. We draw special attention to the thesis of D. A. Holton (1970) wherein a leisurely moderniza- of Redfield's work is presented, and we mention the two monographs of A. Kerber (1971), (1975) which discuss the relationship of group representation theory to combinatorial problems of the type considered by Redfield.

7.2 Redfield's Notation

In this section we shall follow closely the remarks by Harary and Palmer (1967) concerning Redfield's notation.

Let A and B be permutation groups acting on a set X, and let B be a normal subgroup of A. Then it is easy to see that the elements of A permute the orbits of B. Thus the factor group A/B induces a permutation group that acts on the orbits of B. No confusion should arise if we also denote this induced permutation group by A/B.

Let Y^{*X} denote the one-to-one mappings of Y into X. Let A and C denote

The material in this chapter corresponds to section V, chapters II and III in *Combinatory Analysis*.

two permutation groups acting on X and Y, respectively. The restricted power group C^{*A} is a permutation group acting on Y^{*X} that consists of pairs $(\alpha;\beta) \in C^{*A}$ such that for each f in Y^{*X} and $x \in X$, $(\alpha;\beta)f(x) = \beta f(\alpha x)$.

We now define Redfield's fundamental operations: cap $\Cap$ and cup $\Cup$. We consider polynomials in $a_1, a_2, \ldots, a_d$ with rational coefficients. The cap operation $\Cap$ is defined on a finite sequence of monomials by

$$(a_1^{j_1} a_2^{j_2} \ldots a_d^{j_d}) \Cap \ldots \Cap (a_1^{i_1} a_2^{i_2} \ldots a_d^{i_d}) = \left(\prod_{k=1}^{d} k^{j_k} j_k!\right)^{m-1},$$

if $j_k = \ldots = i_k$ for all k, and is zero otherwise (note the convention $0^0 = 1$). The cap operation is then extended by linearity. The cup operation $\Cup$ is defined on a finite sequence of monomials by

$$\begin{aligned}&(a_1^{j_1} a_2^{j_2} \ldots a_d^{j_d}) \Cup \ldots \Cup (a_1^{i_1} a_2^{i_2} \ldots a_d^{i_d})\\ &= [(a_1^{j_1} a_2^{j_2} \ldots a_d^{j_d}) \Cap \ldots \Cap (a_1^{i_1} a_2^{i_2} \ldots a_d^{i_d})]\, a_1^{j_1} a_2^{j_2} \ldots a_d^{j_d},\end{aligned}$$

and this operation is also extended by linearity.

As examples, we consider the cyclic and dihedral groups of degree 4:

$$\begin{aligned}\mathcal{Z}(C_4) &= (a_1^4 + a_2^2 + 2a_4)/4,\\ \mathcal{Z}(D_4) &= (a_1^4 + 2a_1^2 a_2 + 3a_2^2 + 2a_4)/8,\\ \mathcal{Z}(C_4) \Cap \mathcal{Z}(D_4) &= [(a_1^4 \Cap a_1^4) + 3(a_2^2 \Cap a_2^2) + 4(a_4 \Cap a_4)]/32\\ &= (24 + 24 + 16)/32 = 2,\\ \mathcal{Z}(C_4) \Cup \mathcal{Z}(D_4) &= (24a_1^4 + 24a_2^2 + 16a_4)/32,\end{aligned}$$

where $\mathcal{Z}(A)$ is the cycle index of the group A (Redfield refers to the cycle index as the group reduction function: $\mathrm{Grf}(A) = \mathcal{Z}(A)$).

We now define the matrix column group which Redfield referred to as the group of range correspondences. Let $B_1, B_2, \ldots, B_m$ denote permutation groups each acting on Y, a finite set of cardinality n. Let W be the set of $m \times n$ matrices in which the entries in each row are all the elements of Y. Two members of W are termed column-equivalent if one can be transformed into the other by a permutation of the columns; Redfield calls the resulting equivalence classes "range-correspondences." We denote the matrix column group related to the B_i by $[B_1, B_2, \ldots, B_m]$. For each $[\beta_1, \beta_2, \ldots, \beta_m]$ in $[B_1, B_2, \ldots, B_m]$ we define its action on W by

$$[\beta_1, \beta_2, \ldots, \beta_m][w_{ij}] = [\beta_i w_{ij}],$$

and we observe that this is well-defined on column equivalence classes.

Using the above notation, we may present an alternative definition of the matrix column group. We now introduce S_n, the symmetric group on n elements, say $X = \{a_1, a_2, \ldots, a_n\}$. Let C denote the permutation group product $(B_1^{*S_n}) \times \ldots \times (B_n^{*S_n})$ which acts on $(Y^{*X})^m$, the m-dimensional Cartesian product of sets of bijections of X onto Y. We let A denote the subgroup of C consisting of elements of the form $((\alpha;\beta_1), (\alpha;\beta_2), \ldots, (\alpha;\beta_m))$, and we let B denote the subgroup of A wherein each β_i is the identity permutation. Since B is a normal subgroup of A, we may form the factor permutation group A/B, and this group is identical with $[B_1, \ldots, B_m]$ as a permutation group.

Using Burnside's Lemma and the above concepts, Harary and Palmer (1967) present a very short proof of Redfield's first theorem.

Burnside's Lemma. Let $\mathcal{N}(A)$ denote the number of orbits determined by the permutation group A. Then

$$\mathcal{N}(A) = \frac{1}{|A|} \sum_{\alpha \in A} j_1(\alpha),$$

where $j_k(\alpha)$ is the number of cycles of length k in the disjoint cycle decomposition of α.

Redfield's Enumeration Theorem. The number $\mathcal{N}$ of orbits determined by the matrix column group $[B_1, B_2, \ldots, B_m]$ is

$$\mathcal{N} = \mathcal{Z}(B_1) \Omega \mathcal{Z}(B_2) \Omega \ldots \Omega \mathcal{Z}(B_m).$$

Proof (Harary and Palmer, 1967). To obtain the desired formula, apply Burnside's Lemma to the permutation group A defined above. This produces the appropriate $\mathcal{N}$ since the number of orbits determined by A is the same as the number determined by A/B, and A/B and $[B_1, B_2, \ldots, B_m]$ are identical.

We conclude this section by mentioning Redfield's "derived groups." Let A and B be permutation groups acting on X and Y, respectively, let $|X| = |Y| = n$, and let E_n denote the identity group that acts on Y. Then the restricted power group E_n^{*A} is a normal permutation subgroup of B^{*A}, and so we can form the factor group B^{*A}/E_n^{*A} which acts on the orbits of E_n^{*A}. Redfield refers to this factor permutation group as a "derived group B' of B"; he denotes its cycle index by $\mathcal{Z}(A)\,\mathfrak{z}\,\mathcal{Z}(B)$.

7.3 The Theory of Group-Reduced Distributions by J. H. Redfield

In view of the similarity which will be admitted to hold between the subject matters of the Theory of Finite Groups and of Combinatory Analysis, it is somewhat surprising to find that in their literatures the two branches have proceeded on their separate ways without developing their interrelationship, and with scarcely any reference to one another beyond the use by each of certain very elementary results of the other.

In the present paper the connection established between the two branches takes the form of associating every permutation group with a certain symmetric function. By means of these functions and certain differential operations closely allied with those of Hammond and MacMahon, a theory is developed which is in a sense an extension of MacMahon's theory of Chess-Board Diagrams (*Combinatory Analysis*, vol. I, p. 224 ff.). From the point of view here adopted, this theory of MacMahon may be regarded as a special one in which the groups involved are exclusively such as are direct products of symmetric groups on distinct sets of elements, and in which we deal with collections of objects divided into kinds and freely permutable within each kind. In the extended theory the groups may be of any type whatever. We are thus enabled to treat a much wider variety of combinatorial problems; some of them have previously been solved by special devices, but even as to these we gain generality of outlook. One class of such problems, striking because having a convenient geometrical interpretation, though otherwise not more important than some others, has to do with configurations based on the regular polygons and polyhedra; the groups of rotations involved clearly lie outside the domain of the chess-board diagram type of group.

At the same time nearly every problem which we are able to solve suggests others which do not yield to the methods here developed, but which seemingly should nevertheless be amenable to treatment by methods further perfected along the same lines. Belief is therefore warranted that this borderland region would repay much more extensive investigation. While for the moment Combinatory Analysis will be seen to be the chief beneficiary, some of the results obtained below are not without interest from the point of view of Group Theory, and both branches may be expected to profit by future work in this field.

1. *Preliminary Definitions.* We consider an *assemblage* $A(n, q)$ *of* nq *elements,* divided into q *ranges* $S_1, S_2, \cdots, S_q$ of n elements each. We form therewith correspondences, here termed *range-correspondences,* of type defined by a representation wherein the ranges are written in q lines so as to form a rectangular array of n columns. Range-correspondences will be regarded as identical if their arrays consist of identical columns, regardless of the order of the columns. Thus the number of possible range-correspondences will be $(n!)^{q-1}$.

With each range S_r $(r = 1, 2, \cdots, q)$ we associate a specified permutation group (its *range-group*) G_r, of degree n and order m_r, operating on the elements of S_r. The combined operations of these q groups will determine a group Γ of degree $(n!)^{q-1}$ and order $m_1 m_2 \cdots m_q$, whose elements will be the $(n!)^{q-1}$ range-correspondences, which Γ will interchange among themselves. In general, Γ will not be transitive, that is, the combined groups $G_1, G_2, \cdots, G_q$ will not transform every range-correspondence into every other. Such transformation will however be possible within each of a number θ of closed classes, here termed *group-reduced distributions* (being the transitive constituents of Γ), into which the $(n!)^{q-1}$ range-correspondences are parcelled by the given groups $G_1, G_2, \cdots, G_q$.

By the symbol $N(G_r; p_1^{\pi_1} p_2^{\pi_2} \cdots)$ we denote the number of operations, contained in the group G_r, which exhibit, when written in the usual cycle notation, π_1 cycles of order p_1, π_2 cycles of order p_2, etc.

A partition such as $p_1^{\pi_1} p_2^{\pi_2} \cdots$ which specifies the orders of the different cycles in the symbol of a group operation, will be called the *cycle-partition* of the operation; thus the operation $abc \cdot de \cdot fg \cdot h$ is said to be of cycle-partition $32^2 1$.

THEOREM: *The number* θ *of group-reduced distributions of an assemblage* $A(n, q)$ *determined by the range-groups* $G_1, G_2, \cdots, G_q$ *whose respective orders are* $m_1, m_2, \cdots, m_q$, *has the expression*

$$\theta = \frac{\sum\{(p_1^{\pi_1} p_2^{\pi_2} \cdots \pi_1!\, \pi_2! \cdots)^{q-1} \cdot N(G_1; p_1^{\pi_1} p_2^{\pi_2} \cdots) \cdot N(G_2; p_1^{\pi_1} p_2^{\pi_2} \cdots) \cdots N(G_q; p_1^{\pi_1} p_2^{\pi_2} \cdots)\}}{m_1 m_2 \cdots m_q}$$

in which under Σ *there is a term for every partition* $p_1^{\pi_1} p_2^{\pi_2} \cdots$ *of the degree* n *common to all the range-groups.*

Proof: Since the range-groups operate on distinct sets (ranges) of elements, operations of different range-groups will be commutative. The direct product of the range-groups is a group of degree qn, simply isomorphic with the group Γ defined above; Γ may be called the *group of range-correspondences.*

Let γ be an operation of Γ, and let $J(\gamma)$ denote the number of range-correspondences (i. e. elements of Γ) which are not altered by γ. The sum $\Sigma[J(\gamma)]$, taken for all the $m_1 m_2 \cdots m_q$ operations of Γ, gives an enumeration in which every range-correspondence β_1 is counted as many times as there are operations in Γ which do not alter β_1. Let $K(\beta_1)$ denote the number of operations of Γ which thus leave β_1 unaltered.

If β_1 is unaltered by γ_1 and γ_2, it is unaltered by γ_1^{-1}, γ_2^{-1}, $\gamma_1\gamma_2$ and $\gamma_2\gamma_1$. Hence those operations of Γ which do not alter β_1 form a sub-group U_1, whose order $K(\beta_1)$ is a divisor of $m_1 m_2 \cdots m_q$.

Let $1, u_2, u_3, \cdots, u_{K(\beta_1)}$ be the operations of U_1, and let v_2 be any operation of Γ which is not in U_1. Then v_2 will change β_1 into a range-correspondence β_2, different from β_1 but belonging to the same group-reduced distribution. It is clear that β_1 will be changed into β_2 by every operation of the set $v_2, u_2v_2, u_3v_2, \cdots, u_{K(\beta_1)}v_2$, and that β_2 will be unaltered by every operation of the set $1, v_2^{-1}u_2v_2, v_2^{-1}u_3v_2, \cdots, v_2^{-1}u_{K(\beta_1)}v_2^2$, which latter form a sub-group U_2 of Γ, isomorphic with U_1. Using other operations v_3, v_4, etc. we can in like manner reach every range-correspondence of the group-reduced distribution to which β_1 belongs, at the same time accounting for all the operations of Γ. The reasoning is identical with that employed in proving that the order of a group is a multiple of the order of each of its subgroups (e. g Miller, Blichfeldt, and Dickson, *Finite Groups,* p. 22), and establishes that the group-reduced distribution δ of which β_1 is a member contains altogether $m_1 m_2 \cdots m_q/K(\beta_1)$ range-correspondences. But every range-correspondence in δ is counted $K(\beta_1)$ times in the sum $\Sigma[J(\gamma)]$, and therefore δ itself is counted $K(\beta_1) \times [m_1 m_2 \cdots m_q/K(\beta_1)] = m_1 m_2 \cdots m_q$ times. Since this number $m_1 m_2 \cdots m_q$ is independent of $K(\beta_1)$ and the same for every one of the θ group-reduced distributions, it follows that $\theta = \Sigma[J(\gamma)]/m_1 m_2 \cdots m_q$.

It remains to evaluate $\Sigma[J(\gamma)]$, and we observe in the first place that $J(\gamma) = 0$ for every γ which is determined by a set of range-group operations which are not all of the same cycle-partition. For if to any range-correspondence β_1 we apply an operation (not the identity) of G_1 (say), the range-correspondence will be altered, and, to restore it, we must evidently apply to each of the ranges $S_2, S_3, \cdots, S_q$, an operation of the same cycle-partition with the operation first performed on S_1; while if the first operation is the identity of G_1, all the others must also be the identities of their range-groups.

Supposing, then, that γ is determined by the q operations $g_1, g_2, \cdots, g_q$, all of cycle-partition $p_1^{r_1} p_2^{r_2} \cdots$; let these operations be written in the cycle notation, in q lines, so that they form a rectangular array with cycles of equal order over one another. The elements forming columns will then determine a range-correspondence β_1 which is not altered by γ. Attending now to the first line of the array, the operation g_1 exhibits π_1 cycles of order p_1; each of these can be rewritten so as to begin with any one of its p_1 symbols, and the π_1 cycles can be permuted as wholes in $\pi_1!$ ways. We thus get $p_1^{\pi_1}\pi_1!$ variations, and when the same is done with the π_2 cycles of order p_2, and the rest, we have $p_1^{\pi_1}p_2^{\pi_2}\cdots\pi_1!\,\pi_2!\cdots$ equivalent notations for the same operation g_1. The matching of the orders of cycles in the q rows of the array is not disturbed, but instead of the single range-correspondence β_1 we now have $p_1^{\pi_1}p_2^{\pi_2}\cdots\pi_1!\,\pi_2!\cdots$ range-correspondences (including β_1) which are unaltered by γ. Similar variations can be effected on each of the remaining $q-1$ rows of the array. But if *all* the rows are varied, every range-correspondence occurring will be repeated $p_1^{\pi_1}p_2^{\pi_2}\cdots\pi_1!\,\pi_2!\cdots$ times, since permutations of whole columns of an array do not alter the range-correspondence which it represents. Evidently we shall obtain all the distinct range-correspondences each once only if we keep one row fixed as a reference base and vary independently the remaining $q-1$ rows. We thus find that the total number $J(\gamma)$ of distinct range correspondences which are unaltered by γ, is equal to $(p_1^{\pi_1}p_2^{\pi_2}\cdots\pi_1!\,\pi_2!\cdots)^{q-1}$. Hence

$$\theta = \frac{\Sigma[J(\gamma)]}{m_1m_2\cdots m_q} = \frac{\Sigma[(p_1^{\pi_1}p_2^{\pi_2}\cdots\pi_1!\,\pi_2!\cdots)^{q-1}]}{m_1m_2\cdots m_q},$$

the summation Σ covering every operation γ of Γ which is determined by a set of operations of $G_1, G_2, \cdots, G_q$ which are all of the same cycle-partition. The theorem as stated in terms of $N(G_r; p_1^{\pi_1}p_2^{\pi_2}\cdots)$ immediately follows.

2. *The Differential Operations* Ω *and* $\mho$. In the foregoing developments, for greater generality, no use has been made of the theory of Symmetric Functions, which at this point it becomes convenient to introduce. We shall have occasion to consider only such symmetric functions as are rational and integral, and the term is to be understood throughout with this restriction.

For the details of the theory reference is made to MacMahon's *Combinatory Analysis* (vol. I, sec. I, chap. I), of which the notation will in general be here followed. For convenience however we may here recall that we deal with the symmetric functions of a set of ν indeterminates or symbolic quantities $\alpha_1, \alpha_2, \cdots, \alpha_\nu$. The number ν is not specified, since ordinarily we make no use of expressions which are dependent on its value.

A partition (of a number w) enclosed in round brackets, as $(p_1p_2p_3\cdots p_k)$, the k parts being not necessarily unequal, denotes the sum $\sum[\alpha_{i_1}{}^{p_1}\alpha_{i_2}{}^{p_2}\cdots\alpha_{i_k}{}^{p_k}]$ of all the *different* terms which can be formed by taking for $i_1, i_2, \cdots, i_k$ every possible ordered set of k different suffixes selected from the integers $1, 2, 3, \cdots, \nu$. Specially, s_w is written for (w), a_w for (1^w), and h_w for the sum $\sum[(l_1^{\lambda_1}l_2^{\lambda_2}\cdots)]$ taken for every partition $l_1^{\lambda_1}l_2^{\lambda_2}\cdots$ of w; thus

$$h_4 = (4) + (31) + (2^2) + (21^2) + (1^4).$$

MacMahon (loc. cit., Vol. I, sec. I, chap. I) gives formulae for expressing the s-, a-, and h-functions in terms of one another. These can be derived, by the method of undetermined coefficients or otherwise, from the following generating function identities:

$$\text{(A)}\qquad \begin{cases} (1-\alpha_1x)(1-\alpha_2x)\cdots(1-\alpha_\nu x) \\ = 1 - a_1x + a_2x^2 - a_3x^3 + \cdots \\ = 1/(1+h_1x+h_2x^2+h_3x^3+\cdots) \\ = \exp(-s_1x - \tfrac{1}{2}s_2x^2 - \tfrac{1}{3}s_3x^3 - \cdots) \end{cases}$$

Of the Hammond differential operators, MacMahon makes d_w correspond to s_w, and D_w to a_w. Here however we shall use the equally valid correspondence of δ_w to s_w, and D_w to h_w, since by so doing we can replace certain alternating + and — signs in our expressions by + signs throughout.

It is convenient to take as fundamental the operator δ_w, which MacMahon (loc. cit., Vol. I, p. 36) defines thus:

$$\text{(B)}\qquad \delta_w = \frac{\partial}{\partial h_w} + h_1\frac{\partial}{\partial h_{w+1}} + h_2\frac{\partial}{\partial h_{w+2}} + \cdots$$

We proceed to show that δ_w is equivalent to $w(\partial/\partial s_w)$, a very simple relation of which MacMahon makes no use.

Equating logarithms of reciprocals of the third and fourth expressions of (A), we have

$$\text{(C)}\qquad \log(1+h_1x+h_2x^2+\cdots) = s_1x + \tfrac{1}{2}s_2x^2 + \tfrac{1}{3}s_3x^3 + \cdots .$$

Operating with δ_w on the right-hand member of (C), and with the equivalent

$$\frac{\partial}{\partial h_w} + h_1\frac{\partial}{\partial h_{w+1}} + h_2\frac{\partial}{\partial h_{w+2}} + \cdots$$

on the left-hand member, we obtain

$$\text{(D)} \qquad \frac{x^w + h_1 x^{w+1} + h_2 x^{w+2} + \cdots}{1 + h_1 x + h_2 x^2 + \cdots}$$
$$= \delta_w s_1 \cdot x + \tfrac{1}{2}\delta_w s_2 \cdot x^2 + \cdots + (1/w)\ \delta_w s_w \cdot x^w + \cdots.$$

The left-hand member of (D) reduces to x^w, and by equating terms containing like powers of x we find that $\delta_w s_v = 0$ if $w \neq v$, while $\delta_w s_w = w$, so that

$$\delta_w = w \ \frac{\partial}{\partial s_w}$$

It follows then that $\delta_{\lambda_1}{}^{l_1}\delta_{\lambda_2}{}^{l_2}\cdots s_{\mu_1}{}^{m_1}s_{\mu_2}{}^{m_2}\cdots = 0$ whenever $\lambda_1{}^{l_1}\lambda_2{}^{l_2}\cdots$ and $\mu_1{}^{m_1}\mu_2{}^{m_2}\cdots$ are different partitions of the same number, while $\delta_{\lambda_1}{}^{l_1}\delta_{\lambda_2}{}^{l_2}\cdots s_{\lambda_1}{}^{l_1}s_{\lambda_2}{}^{l_2}\cdots = \lambda_1{}^{l_1}\lambda_2{}^{l_2}\cdots l_1!\, l_2!\cdots$.

We introduce a new connective symbol of operation $\Cap$, and write by definition $s_{\lambda_1}{}^{l_1}s_{\lambda_2}{}^{l_2}\cdots \Cap s_{\mu_1}{}^{m_1}s_{\mu_2}{}^{m_2}\cdots$ for $\delta_{\lambda_1}{}^{l_1}\delta_{\lambda_2}{}^{l_2}\cdots s_{\mu_1}{}^{m_1}s_{\mu_2}{}^{m_2}\cdots$, with the further defining relation $(A + B) \Cap C = (A \Cap C) + (B \Cap C)$, expressing the distributive law with respect to addition. MacMahon's $D_{\lambda_1}{}^{l_1}D_{\lambda_2}{}^{l_2}\cdots h_{\mu_1}{}^{m_1}h_{\mu_2}{}^{m_2}\cdots$ or $D_{\mu_1}{}^{m_1}D_{\mu_2}{}^{m_2}\cdots h_{\lambda_1}{}^{l_1}h_{\lambda_2}{}^{l_2}\cdots$ is then readily seen to be equivalent to $h_{\lambda_1}{}^{l_1}h_{\lambda_2}{}^{l_2}\cdots \Cap h_{\mu_1}{}^{m_1}h_{\mu_2}{}^{m_2}\cdots$.

We extend the definition of $\Cap$ so as to write $s_{\lambda_1}{}^{l_1}s_{\lambda_2}{}^{l_2}\cdots \Cap s_{\lambda_1}{}^{l_1}s_{\lambda_2}{}^{l_2}\cdots \Cap \cdots \cdots \Cap s_{\lambda_1}{}^{l_1}s_{\lambda_2}{}^{l_2}\cdots$ (to q operands) $= (\lambda_1{}^{l_1}\lambda_2{}^{l_2}\cdots l_1!\, l_2!\cdots)^{q-1}$, while the result is zero if not all the operands exhibit the same partition $\lambda_1{}^{l_1}\lambda_2{}^{l_2}\cdots$. Thus extended, the operator $\Cap$ is commutative and distributive as to addition, but the result cannot be decomposed into simpler $\Cap$-operations, and the notion of associativeness has no relevance.

Another new symbol of operation $\Cup$ is defined by $s_{\lambda_1}{}^{l_1}s_{\lambda_2}{}^{l_2}\cdots \Cup s_{\lambda_1}{}^{l_1}s_{\lambda_2}{}^{l_2}\cdots = (s_{\lambda_1}{}^{l_1}s_{\lambda_2}{}^{l_2}\cdots \Cap s_{\lambda_1}{}^{l_1}s_{\lambda_2}{}^{l_2}\cdots)\, s_{\lambda_1}{}^{l_1}s_{\lambda_2}{}^{l_2}\cdots$, with $(A+B) \Cup C = (A \Cup C) + (B \Cup C)$; $s_{\lambda_1}{}^{l_1}s_{\lambda_2}{}^{l_2}\cdots \Cup s_{\mu_1}{}^{m_1}s_{\mu_2}{}^{m_2}\cdots$ vanishing as in the case of $\Cap$ when $\lambda_1{}^{l_1}\lambda_2{}^{l_2}\cdots$ and $\mu_1{}^{m_1}\mu_2{}^{m_2}\cdots$ are different partitions. Both $\Cap$ and $\Cup$ connect only symmetric functions of equal weights, but whereas $\Cap$ yields a pure number, $\Cup$ yields a symmetric function of the same weight with the operands. $\Cup$ is evidently commutative, distributive as to addition, and associative, so that we can write unambiguously without parentheses $A_1 \Cup A_2 \Cup A_3 \Cup \cdots \Cup A_q$. It is also readily verified that $(A \Cup B) \Cap C = A \Cap (B \Cup C) = A \Cap B \Cap C$.

3. *The Group-Reduction Function* of any permutation group G of degree n and order m, is defined to be the symmetric function

$$\mathrm{Grf}(G) = (1/m)\sum[N(G; p_1{}^{r_1}p_2{}^{r_2}\cdots)\ s_{p_1}{}^{r_1}s_{p_2}{}^{r_2}\cdots],$$

in which N has the same meaning as in §·1 and Σ covers every partition $p_1^{\pi_1}p_2^{\pi_2}\cdots$ of the weight n. It does not completely characterize its group, since the same G. R. F. may be shared by two or more distinct permutation groups (isomorphic or conformal).

The G. R. F.'s of the symmetric and alternating groups of degree n are, in MacMahon's notation, respectively h_n and $h_n + a_n$. For, the symmetric group of degree n contains all the possible operations of all cycle-partitions which can be formed with n elements. Considering the cycle-partition $p_1^{\pi_1}p_2^{\pi_2}\cdots$, we first write in the cycle notation a typical operation, putting blanks — in the places to be occupied by the n elements; thus for the cycle-partition 32^2 of 7 we should write (———)(——)(——). The blanks can now be filled by the n elements $\alpha_1, \alpha_2, \cdots, \alpha_n$ in any permutation; this gives $n!$ operation symbols, which are however not all distinct. For each cycle of p_r elements can be made to begin with any on^ of its elements, and if there are π_r such cycles, these can be permuted in $\pi_r!$ ways, and this can be done independently for each value of r; all these varieties are different notations for a single operation. Therefore the number of distinct operations of cycle-partition $p_1^{\pi_1}p_2^{\pi_2}\cdots$ in the symmetric group of degree n is

$$n!/(p_1^{\pi_1}p_2^{\pi_2}\cdots\pi_1!\,\pi_2!\cdots), \qquad \text{and the G. R. F. is}$$

$$(1/n!)\,\Sigma\left(\frac{n!}{p_1^{\pi_1}p_2^{\pi_2}\cdots\pi_1!\,\pi_2!\cdots}\,s_{p_1}^{\pi_1}s_{p_2}^{\pi_2}\cdots\right)$$

$$= \Sigma\left(\frac{s_{p_1}^{\pi_1}s_{p_2}^{\pi_2}\cdots}{p_1^{\pi_1}p_2^{\pi_2}\cdots\pi_1!\,\pi_2!\cdots}\right).$$

But this is the expression of h_n in terms of s-functions given by MacMahon (loc. cit., Vol. I, p. 7).—The alternating group of degree n contains the even permutations of the corresponding symmetric group, and its G. R. F. is therefore

$$(2/n!)\,\Sigma\left(\frac{n!}{p_1^{\pi_1}p_2^{\pi_2}\cdots\pi_1!\,\pi_2!\cdots}\,s_{p_1}^{\pi_1}s_{p_2}^{\pi_2}\cdots\right),$$

in which the summation includes only those cycle-partitions which correspond to even permutations, that is, those having an even number of even parts. Now the expression of a_n in terms of s-functions (MacMahon, loc. cit., Vol. I, p. 6) differs from the expression for h_n only in that the terms exhibiting partitions with an odd number of even parts, have the negative sign. In the sum $h_n + a_n$ these terms cancel out, while the terms exhibiting partitions with an even number of even parts are doubled. If the factor 2 thus arising is carried outside the Σ sign, the non-vanishing coefficients under the Σ remain as in the expression of h_n, and we find that the G. R. F. of the alternating group, as given above, is equal to $h_n + a_n$.

The G. R. F.'s of the cyclic groups are of considerable theoretical as well as practical importance. By $\mathrm{Cyc}(p_1^{\pi_1} p_2^{\pi_2} \cdots)$ we shall denote the G. R. F. of a cyclic group generated by an operation whose cycle-partition is $p_1^{\pi_1} p_2^{\pi_2} \cdots$, a partition of the degree w of the group; its order is of course the least common multiple of $p_1, p_2, \cdots$. Such a group is transitive if and only if it is generated by an operation whose cycle-partition is the one-part partition w, and in that case it is easily verified that

$$\mathrm{Cyc}(w) = (1/w)[s_1^w + \phi(\alpha) s_\alpha^{w/\alpha} + \phi(\beta) s_\beta^{w/\beta} + \cdots + \phi(w) s_w],$$

in which $1, \alpha, \beta, \cdots, w$ are all the divisors of w, and $\phi(\zeta)$ denotes the number of positive integers, including 1, less than and prime to ζ. Thus $\mathrm{Cyc}(6) = (1/6)(s_1^6 + s_2 \ + 2s_3^2 + 2s_6)$. In the G. R. F. of an intransitive cyclic group, the coefficients are the same as in the G. R. F. of the transitive cyclic group of the same order, but the literal parts of the terms are s-function products exhibiting the cycle-partitions of the generating operation and its powers. Thus $\mathrm{Cyc}(32^2) = (1/6)(s_1^7 + s_2^2 s_1^3 + 2s_3 s_1^4 + 2s_3 s_2^2)$.

The G. R. F. of a dihedral group derived from a transitive cyclic group of degree w will be denoted by $\mathrm{Dih}(w)$. For odd and even degrees we have respectively

$$\mathrm{Dih}(2k+1) = \tfrac{1}{2}\mathrm{Cyc}\,(2k+1) + \tfrac{1}{2} s_2^k s_1;$$
$$\mathrm{Dih}(2k) = \tfrac{1}{2}\mathrm{Cyc}(2k) + \tfrac{1}{4} s_2^k + \tfrac{1}{4} s_2^{k-1} s_1^2.$$

It is easily seen that the algebraic product of two or more G. R. F.'s is the G. R. F. of the direct product of the groups belonging to the factors, written on distinct sets of symbols.

The enumerating expression for group-reduced distributions can now be rewritten in the form

$$\theta = \mathrm{Grf}(G_1) \circledast \mathrm{Grf}(G_2) \circledast \cdots \circledast \mathrm{Grf}(G_q).$$

This results immediately from the Theorem of § 1 and the definition of $\circledast$ in § 2.

For every group G of degree n, $h_n \circledast \mathrm{Grf}(G) = 1$, and $h_n \circledast Grf(G) = \mathrm{Grf}(G)$. For in $\mathrm{Grf}(G)$, let the coefficient of $s_{p_1}^{\pi_1} s_{p_2}^{\pi_2} \cdots$ *be* $A(p_1^{\pi_1} p_2^{\pi_2} \cdots)$. In h_n the coefficient of $s_{p_1}^{\pi_1} s_{p_2}^{\pi_2} \cdots$ is $1/(p_1^{\pi_1} p_2^{\pi_2} \cdots \pi_1!\,\pi_2! \cdots)$. (MacMahon, *loc. cit.*, vol. I, p. 7). Then $A(p_1^{\pi_1} p_2^{\pi_2} \cdots) s_{p_1}^{\pi_1} s_{p_2}^{\pi_2} \cdots \circledast s_{p_1}^{\pi_1} s_{p_2}^{\pi_2} \cdots / p_1^{\pi_1} p_2^{\pi_2} \cdots \pi_1!\,\pi_2! \cdots = A(p_1^{\pi_1} p_2^{\pi_2} \cdots)$, by § 2. Thus $h_n \circledast \mathrm{Grf}(G) = \Sigma[A(p_1^{\pi_1} p_2^{\pi_2} \cdots)]$ taken for all partitions $p_1^{\pi_1} p_2^{\pi_2} \cdots$ of n. But this is the sum of the coefficients of $\operatorname{Grf}(G)$, and we know from the definition of a G. R. F. (§ 3) that this sum is equal to unity.—Also, since

$$A(p_1^{\pi_1}p_2^{\pi_2}\cdots)s_{p_1}^{\pi_1}s_{p_2}^{\pi_2}\cdots \between s_{p_1}^{\pi_1}s_{p_2}^{\pi_2}\cdots / p_1^{\pi_1}p_2^{\pi_2}\cdots\pi_1!\,\pi_2!\cdots \\ = A(p_1^{\pi_1}p_2^{\pi_2}\cdots)s_{p_1}^{\pi_1}s_{p_2}^{\pi_2}\cdots,$$

it follows immediately that the $\between$-product of h_n and any symmetric function (including any G. R. F.) of degree n, merely reproduces the original symmetric function.

If $h_{n-k}h_1^k \between \operatorname{Grf}(G) = 1$, then G is at least k-ply transitive, and conversely. For let the two ranges of the assemblage be S_1 and S_2, of which let S_1 be associated with the group G, and S_2 with a group H whose G. R. F. is $h_{n-k}h_1^k$; then H is the direct product of the symmetric group on $n-k$ symbols, by the group of identity on k other symbols. As elements of S_2 we may take $n-k$ similar symbols α, permutable among themselves in all possible ways, together with k distinct symbols $\beta_1, \beta_2, \cdots, \beta_k$, none of which are displaced by any operation of H. Let the elements of S_1 be $\gamma_1, \gamma_2, \cdots, \gamma_n$. If now G is at least k-ply transitive, the k elements of S_1 which are paired with $\beta_1, \beta_2, \cdots, \beta_k$ in any range-correspondence, can by some operation contained in G be replaced by any specified set of k elements of S_1, and this moreover in any one of the $k!$ possible correspondences. Thereafter the $n-k$ elements α can be permuted in any specified way by a suitable operation contained in H. Thus from any range correspondence we can pass to any other by operations of G and H, so that there is only one group-reduced distribution, and $h_{n-k}h_1^k \between \operatorname{Grf}(G) = 1$. But if G is not k-ply transitive, then not every set of k elements of S_1 can be brought into correspondence with $\beta_1, \beta_2, \cdots, \beta_k$, and there will be more than one group-reduced distribution; $h_{n-k}h_1^k \between \operatorname{Grf}(G) > 1$. Thus the proposition and its converse are established.

If $h_{n-1}h_1 \between \operatorname{Grf}(G) = t$, then G has t transitive constituents. For let S_1 as before be associated with the group G, and S_2 with a group H whose G. R. F. is $h_{n-1}h_1$; then we may take, as elements of S_2, $n-1$ similar symbols α, with a distinct symbol β. Starting with any range-correspondence in which an element γ of S_1 corresponds to β, the various operations of G will replace γ by those elements of S_1, and those alone, which belong to the same transitive constituent of G as does γ; while the operations of H, which affect only the α's, will unite into one group-reduced distribution all the range-correspondences in which a particular element of S_1 is paired with β. Thus there will be a group-reduced distribution for each of the t transitive constituents of G, and $h_{n-1}h_1 \between \operatorname{Grf}(G) = t$.

It is convenient, so far as it can be done significantly and unambiguously, to speak of a G. R. F. as transitive, cyclic, of degree n, etc., when the corresponding group has the property in question.

Examples of Application: (i) Required the number of distinct configurations which can be obtained by placing a solid node • at each of four vertices of a cube, and a hollow node ° at each of the four remaining vertices, configurations differing only in orientation not being regarded as distinct.

The group of rotations of the cube is simply isomorphic with the symmetric group on 4 symbols, and when written on 8 symbols representing the vertices, has the G. R. F.

$$V = (1/24)(s_1^8 + 9s_2^4 + 8s_3^2s_1^2 + 6s_4^2).$$

The group of all node-permutations which do not alter colors of nodes is the direct product of two symmetric groups of degree 4 on distinct sets of symbols, and has as G. R. F.

$$h_4^2 = [(1/24)(s_1^4 + 6s_2s_1^2 + 3s_2^2 + 8s_3s_1 + 6s_4)]^2$$

$$= (1/576)(s_1^8 + 12s_2s_1^6 + 42s_2^2s_1^4 + 36s_2^3s_1^2 + 9s_2^4 + 16s_3s_1^5 + 96s_3s_2s_1^3$$
$$+ 48s_3s_2^2s_1 + 64s_3^2s_1^2 + 12s_4s_1^4 + 72s_4s_2s_1^2 + 36s_4s_2^2 + 96s_4s_3s_1$$
$$36s_4^2).$$

The computation of h_4^2 ☊ V is as follows:

$1 \times 1 \times 1^8 \times 8!$	$=$	40320
$9 \times 9 \times 2^4 \times 4!$	$=$	31104
$64 \times 8 \times 3^2 1^2 \times (2!)^2$	$=$	18432
$36 \times 6 \times 4^2 \times 2!$	$=$	6912
Product of orders: $576 \times 24 = 13824$	)	96768
Number of configurations: h_4^2 ☊ $V =$		7

Be it noted that only those cycle-partitions need be considered which are represented in *all* the G. R. F.'s entering as operands. This "orthogonality" of the s-function products with respect to ☊ and ☋ greatly simplifies the computations. In the present case, however, and usually when one of the operands is a monomial product of h-functions, it would be simpler to use the D-operators and to compute D_4^2V by the rules which MacMahon gives (*loc. cit.*, p. 43).

The actual configurations, shown below, cannot be determined by the methods of the present theory, but must be found, as in all other cases, by detailed consideration of the groups involved, and this may of course be very laborious, except in simple cases, or where special devices are available.

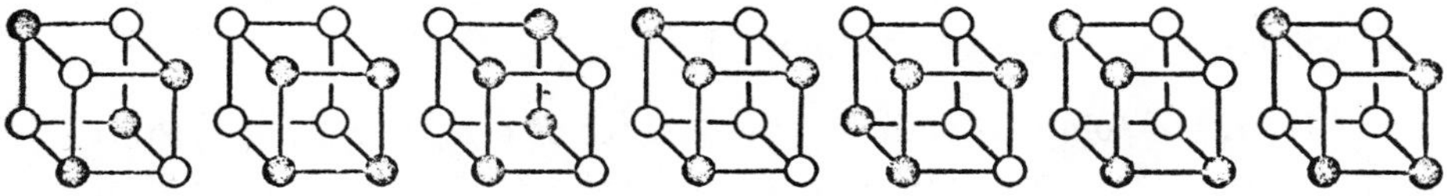

(1) (2) (3) (4) (5) (6) (7)

In connection with the present example we may note without proof certain other simple results obtainable.

Thus if in V we substitute $x^r + y^r$ for every s_r, we obtain the polynomial

$$x^8 + x^7y + 3x^6y^2 + 3x^5y^3 + 7x^4y^4 + 3x^3y^5 + 3x^2y^6 + xy^7 + y^8,$$

in which the coefficient of x^ty^{8-t} enumerates the distinct configurations possible with t nodes • and $8 - t$ nodes °.

The sum of the coefficients in the above expression is 23, which is the total number of configurations when the numbers of nodes of the two colors are not specified. This enumeration is also effected by substituting 2 for every s_r in V. Similarly if k colors are available we substitute k for every s_r; thus with 3 colors there are $(1/24)(3^8 + 9.3^4 + 8.3^4 + 6.3^2) = 333$ possible configurations.

If in V we put $1/(1 - x^r)$ for every s_r, we obtain the infinite series

$$1 + x + 4x^2 + 7x^3 + 21x^4 + 37x^5 + \cdots,$$

in which the coefficient of x^t enumerates the distinct configurations obtained by placing a zero or a positive integer at every vertex of the cube, subject to the condition that the sum of the 8 numbers is always t. For $t = 2$, the 4 configurations are

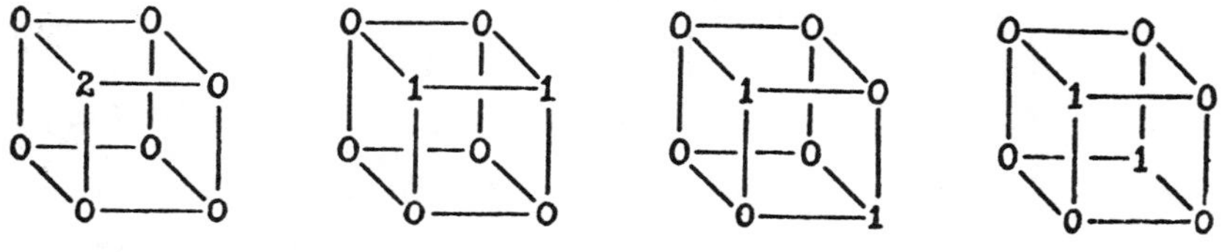

If in V we put 2 for every s_{2k} and 0 for every s_{2k+1}, we enumerate the configurations in which it is possible to change the color of every node into

the opposite color by a suitable rotation of the cube. The number is found to be $(1/24)(0+9.2^4+0+6.2^2)=7$, and it will be seen that in fact all the 7 configurations shown above have this property. In other analogous cases the like is not always true; thus for the 12 vertices of the icosahedron there are 24 distinct configurations with 6 nodes of each color, only 16 of which interchange colors by rotation.

(ii) Let n points be placed at the vertices of a regular n-gon. Starting from any one of these points, a closed route is traced, consisting of straight segments running from each point to the next in any order. Required the number of distinct configurations obtained, when no account is taken of differences of orientation or of any particular starting point of a route.

If we distinguish (a) between routes traversed in opposite senses and (b) between a configuration and its reflection, we shall have as range-groups two cyclic groups of degree and order n, and the number of configurations will be $\mathrm{Cyc}(n) \text{☊} \mathrm{Cyc}(n)$. Thus if $n=6$ we shall have

$$(1/6)(s_1^6+s_2^3+2s_3^2+2s_6) \text{☊} (1/6)(s_1^6+s_2^3+2s_3^2+2s_6)$$
$$=(1/36)(6!+2^3.3!+2^2.3^2.2!+2^2.6)=24.$$

If we disregard one only of the distinctions (a) and (b) above, the number of configurations will be $\mathrm{Cyc}(n) \text{☊} \mathrm{Dih}(n)$. For $n=6$ we shall have

$$(1/6)(s_1^6+s_2^3+2s_3^2+2s_6) \text{☊} (1/12)(s_1^6+3s_2^2s_1^2+4s_2^3+2s_3^2+2s_6)$$
$$=(1/72)(6!+4.2^3.3!+2^2.3^2.2!+2^2.6)=14.$$

In the diagrams shown below, distinction (a) is disregarded.

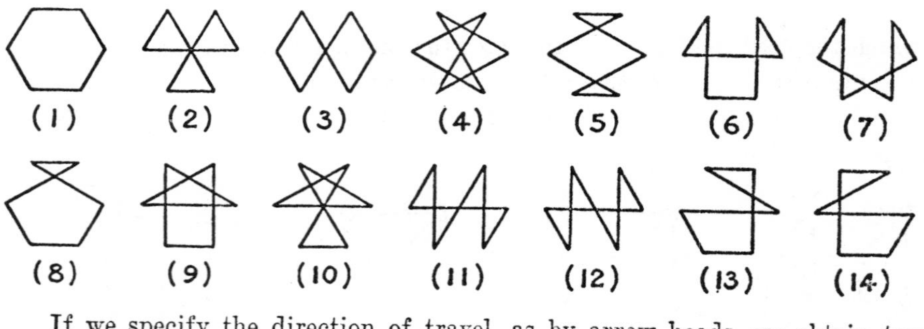

If we specify the direction of travel, as by arrow heads, we obtain two diagrams from each of the above; but these are not distinct in the cases of (3), (4), (11), and (12), since a reversal of direction is there equivalent to a rotation through 180° about an axis through the center of the figure and perpendicular to its plane. The remaining 10 pairs are distinct and increase the total to $14 + 10 = 24$, as computed.

If we abolish *both* distinctions (a) and (b), the enumeration is given by $\mathrm{Dih}(n) \,\&\, \mathrm{Dih}(n)$; for $n = 6$ this is

$$(1/12)(s_1^6 + 3s_2^2s_1^2 + 4s_2^3 + 2s_3^2 + 2s_6) \,\&\, (1/12)(s_1^6 + 3s_2^2s_1^2 + 4s_2^3 + 2s_3^2 + 2s_6)$$
$$= (1/144)(6! + 3^2.2^2.1^2.2!^2 + 4^2.2^3.3! + 2^2.3^2.2! + 2^2.6) = 12.$$

And in fact diagram (12) is the reflection of (11), and (14) of (13); thus the number of distinct configurations is now reduced to $14 - 2 = 12$.

§ 4. *Decomposition of* ⅋*-Products.*

THEOREM: *If* $\mathrm{Grf}(G_1) \,\&\, \mathrm{Grf}(G_2) \,\&\, \cdots \,\&\, \mathrm{Grf}(G_q) = \theta$, *then* $\mathrm{Grf}(G_1) \,⅋\, \mathrm{Grf}(G_2) \,⅋\, \cdots \,⅋\, \mathrm{Grf}(G_q)$ *can be expressed as the sum* $\Phi_1 + \Phi_2 + \cdots + \Phi_\theta$ *of* θ *G. R. F.'s, which can be associated in one-one correspondence with the* θ *group-reduced distributions* $\delta_1, \delta_2, \cdots, \delta_\theta$ *determined by* $G_1, G_2, \cdots, G_q$, *in such wise that if* Φ_σ *is the correspondent of* δ_σ, *and* β *is any range-correspondence of* δ_σ, *then those operations of any range-group* G_r *which leave* β *unaltered when they are performed in combination with suitable similar operations of the remaining G's, form a sub-group whose G. R. F. is* Φ_σ. For different β's belonging to the same δ these sub-groups are similar, but not necessarily identical.

Proof: In the Proof of the Theorem of § 1, we saw that

$$\theta = \sum [J(\gamma)]/m_1m_2 \cdots m_q,$$

and in § 3 we saw that this expression could be written in the form $\mathrm{Grf}(G_1) \,\&\, \mathrm{Grf}(G_2) \,\&\, \cdots \,\&\, \mathrm{Grf}(G_q)$. We also saw (§ 1, Proof) that whenever $J(\gamma)$ does not vanish, γ is determined by a set of operations of $G_1, G_2, \cdots, G_q$ which are all of the same cycle-partition, say $p_1^{\pi_1}p_2^{\pi_2}\cdots$ If now we attach to each $J(\gamma)$ a literal multiplier $s_{p_1}^{\pi_1}s_{p_2}^{\pi_2}\cdots$ embodying the cycle-partition $p_1^{\pi_1}p_2^{\pi_2}\cdots$ characterizing γ, it is evident that the modified expression $\sum[J(\gamma)\cdot s_{p_1}^{\pi_1}s_{p_2}^{\pi_2}\cdots]/m_1m_2\cdots m_q$ will be equal to $\mathrm{Grf}(G_1) \,⅋\, \mathrm{Grf}(G_2) \,⅋\, \cdots \,⅋\, \mathrm{Grf}(G_q)$.

But $\sum[J(\gamma)]$ contains a unit, and therefore $\sum[J(\gamma)\cdot s_{p_1}^{\pi_1}s_{p_2}^{\pi_2}\cdots]$ contains a summand $s_{p_1}^{\pi_1}s_{p_2}^{\pi_2}\cdots$, for every instance in which a range-correspondence β_1 is associated with an operation γ, of cycle-partition $p_1^{\pi_1}p_2^{\pi_2}\cdots$, which leaves β_1 unaltered. If we attend to the summands $s_{p_1}^{\pi_1}s_{p_2}^{\pi_2}\cdots$ associated with a single range-correspondence β_1, we see, since the corresponding operations form a group U_1 of order $K(\beta_1)$, that all these $s_{p_1}^{\pi_1}s_{p_2}^{\pi_2}\cdots$ taken together give an expression equal to $K(\beta_1)$ times the G. R. F. of U_1. The groups U_2, U_3, $\cdots$, etc., belonging to the range-correspondences β_2, β_3, $\cdots$ which are in the same group-reduced distribution with β_1, are all similar to U_1 and so have the same G. R. F. If we extend our summation to include this whole group-reduced distribution, which contains $m_1 m_2 \cdots m_q / K(\beta_1)$ range-correspondences, the result is

$$\frac{m_1 m_2 \cdots m_q}{K(\beta_1)} \cdot K(\beta_1) \Phi_\sigma = m_1 m_2 \cdots m_q \, \Phi_\sigma,$$

in which Φ_σ is the G. R. F. belonging indifferently to U_1, U_2, U_3, $\cdots$, etc. Finally, if we include all the θ group-reduced distributions, we obtain

$$m_1 m_2 \cdots m_q (\Phi_1 + \Phi_2 + \cdots + \Phi_\theta),$$

in which Φ_1, Φ_2, $\cdots$, Φ_θ is the complete set of G. R. F.'s of which Φ_σ is a typical one. But since we have now used all the summands $s_{p_1}^{\tau_1} s_{p_2}^{\tau_2} \cdots$, the above expression must be equal to $\sum[J(\gamma) \cdot s_{p_1}^{\tau_1} s_{p_2}^{\tau_2} \cdots]$, whence:

$$\begin{aligned} \Phi_1 + \Phi_2 + \cdots + \Phi_\theta &= \sum[J(\gamma) \cdot s_{p_1}^{\tau_1} s_{p_2}^{\tau_2} \cdots] / m_1 m_2 \cdots m_q \\ &= \mathrm{Grf}(G_1) \, ☋ \, \mathrm{Grf}(G_2) \, ☋ \cdots ☋ \, \mathrm{Grf}(G_q). \end{aligned}$$

The decomposition is thus proved to be theoretically possible in all cases. But for actually effecting it for the ☋-product of given G. R. F.'s, no general method has yet been found. Moreover, if the ☋-product involves operands which can belong to two or more distinct groups, the decomposition may not be unique.

Thus, the G. R. F. $(1/4)(s_1^6 + 3s_2^2 s_1^2)$ is shared by the two isomorphic but distinct groups

$$\begin{aligned} Q_1 &= [1;\ ab{\cdot}cd{\cdot}e{\cdot}f;\ ab{\cdot}c{\cdot}d{\cdot}ef;\ a{\cdot}b{\cdot}cd{\cdot}ef]; \\ Q_2 &= [1;\ ab{\cdot}cd{\cdot}e{\cdot}f;\ ac{\cdot}bd{\cdot}e{\cdot}f;\ ad{\cdot}bc{\cdot}e{\cdot}f]. \end{aligned}$$

Both are intransitive, with 3 transitive constituents, as shown by

$$h_5 h_1 \, ☋ \, (1/4)(s_1^6 + 3s_2^2 s_1^2) = 3;$$

but the transitive constituents are of degrees 2 2 2 and 4 1 1 respectively. $h_3 h_1 ☋ (1/4)(s_1^6 + 3s_2^2 s_1^2) = (1/4)(6s_1^6 + 6s_2^2 s_1^2)$, which has the two decompositions $3[(1/2)(s_1^6 + s_2^2 s_1^2)]$ and $2[(1/4)(s_1^6 + 3s_2^2 s_1^2)] + s_1^6$, the first corresponding to Q_1 and the second to Q_2.

Reverting to the example on cube configurations (§ 3, i), we have

$$\begin{aligned} h_4^2 \, ☋ \, (1/24)(s_1^8 + 9s_2^4 + 8s_3^2 s_1^2 + 6s_4^2) \\ = (1/24)(70s_1^8 + 54s_2^4 + 32s_3^2 s_1^2 + 12s_4^2), \end{aligned}$$

which decomposes into

(1)	$(1/12)(s_1^8 + 3s_2^4 + 8s_3^2 s_1^2)$
(2)	$+ (1/4)(s_1^8 + s_2^4 + 2s_4^2)$
(3)	$+ (1/4)(s_1^8 + 3s_2^4)$
(4)	$+ (1/3)(s_1^8 + 2s_3^2 s_1^2)$
(5) and (6)	$+ 2[(1/2)(s_1^8 + s_2^4)]$
(7)	$+ s_1^8.$

where the numbers to the left indicate the correspondence of the component G. R. F.'s with the seven diagrams shown in § 3, i. The reader will easily verify that each of them is the G. R. F. of the group of rotations which does not alter colors of nodes in the corresponding cube configuration. The decomposition

$$(1/4)(s_1^8 + s_2^4 + 2s_4^2) + 2[(1/3)(s_1^8 + 2s_3^2 s_1^2)] + 4[(1/2)(s_1^8 + s_2^4)]$$

is algebraically possible and involves none but admissible common sub-groups, yet it is not valid; this will give some notion of the difficulty of the general problem of decomposition.

5. *Special Cases of Decomposition.* The decomposition of $\between$-products into G. R. F.'s can be accomplished uniquely in two important special classes of cases:

CASE 1: THEOREM: *When the operands $F_1, F_2, \cdots, F_q$ are* all *of them G. R. F.'s of the form $h_{\lambda_1}^{l_1} h_{\lambda_2}^{l_2} \cdots$, then $\Phi = F_1 \between F_2 \between \cdots \between F_q$ is a sum of G. R. F.'s each of which is likewise a monomial product of h-functions, and the decomposition is found by transforming Φ, when it presents itself in terms of s-functions, into an expression in h-functions, using the formulae given by MacMahon* (loc. cit., Vol. I, p. 6).

Proof: For let the q ranges be $S_1, S_2, \cdots, S_q$, associated with groups $G_1, G_2, \cdots, G_q$, each G_r having a G. R. F. such as $h_{p_1}^{\pi_1} h_{p_2}^{\pi_2} \cdots$. We take, as elements of each S_r, p_1 symbols of each of π_1 kinds, p_2 symbols of each of π_2 kinds, etc., so that G_r admits all possible permutations among symbols of the same kind, but admits no interchange of symbols of different kinds. Let the ranges be written in a rectangular array of q rows and n columns, the elements of each range being in any arbitrary order; thus for 3 ranges of 7 elements each, the G. R. F.'s being $h_3 h_2^2$, $h_3^2 h_1$, $h_4 h_2 h_1$, we might have

$$
\begin{array}{ccccccc}
A & A & B & A & C & B & C \\
a & b & b & a & c & b & a \\
\alpha & \gamma & \gamma & \alpha & \beta & \alpha & \beta
\end{array}
$$

Such an array symbolizes a range-correspondence, which is not altered by permutations of entire columns. But such permutations of entire columns are admitted by $G_1, G_2, \cdots, G_q$ if and only if the elements interchanged in each row are alike, which requires that the entire columns interchanged be alike; and any set of like columns admits under $G_1, G_2, \cdots, G_q$ all its possible permutations. Thus the group of admissible column-permutations for any range-correspondence, is a direct product of symmetric groups, having a G. R. F. of the form $h_{\lambda_1}{}^{l_1} h_{\lambda_2}{}^{l_2} \cdots$. In the above example, the columns $Aa\alpha$ and $Bb\alpha$ occur twice each, the others once each; so that the range-correspondence represented admits a group whose G. R. F. is $h_2{}^2 h_1{}^3$. The same being true of every range-correspondence and of every group-reduced distribution involved in the problem, it follows that the $\cup$-product of all the G. R. F.'s can be decomposed into a sum of G. R. F.'s each of which is a product of h-functions.

Since the $\cup$-product is a symmetric function, and every symmetric function has a unique expression as a linear function of h-function products, it follows that this expression, when found by the formulae mentioned above, will give the required decomposition.

Chess Board Diagrams. MacMahon's theory of Chess Board Diagrams comes under this case (loc. cit., Vol. I, sec. V, chap. II). The coefficient of $h_{\lambda_1}{}^{l_1} h_{\lambda_2}{}^{l_2} \cdots$ in the decomposition of a $\cup$-product of monomial h-function products enumerates the diagrams in which the assemblage of numbers appearing in the various diagram compartments forms the partition $\lambda_1{}^{l_1} \lambda^{l_2} \cdots$. Using three or more operands we can enumerate analogous diagrams in three or more dimensions; the theory of these, though not explicitly given as such, is contained in Vol. I, sec. II, chap. V of *Combinatory Analysis*. The example which MacMahon gives at pp. 85-86 can be looked upon as an enumeration of three-dimensional block diagrams, having 2 1 1, 2 1 1, and 2 2 units in respective rows, columns, and layers. The enumerating expression for all the diagrams is $h_2 h_1{}^2 \cap h_2 h_1{}^2 \cap h_2{}^2 = 38$. The corresponding $\cup$-product is $h_2 h_1{}^2 \cup h_2 h_1{}^2 \cup h_2{}^2$

$$
\begin{aligned}
&= (1/2)(s_1{}^4 + s_2 s_1{}^2) \cup (1/2)(s_1{}^4 + s_2 s_1{}^2) \cup (1/4)(s_1{}^4 + 2s_2 s_1{}^2 + s_2{}^2) \\
&= \tfrac{1}{2}\cdot\tfrac{1}{2}\cdot\tfrac{1}{4}\cdot\{[1 \times 1 \times 1 \times (1^4 \cdot 4!)^2] s_1{}^4 + [1 \times 1 \times 2 \times (2.1^2.2!)^2] s_2 s_1{}^2\} \\
&= 36 s_1{}^4 + 2 s_2 s_1{}^2 = 34 s_1{}^4 + 4[(1/2)(s_1{}^2 + s_2)] s_1{}^2 = 34 h_1{}^4 + 4 h_2 h_1{}^2;
\end{aligned}
$$

so that there are 34 "unitary" diagrams with one unit in each of 4 compartments, and 4 diagrams with 2 units in one compartment and one unit in each of 2 others. If in MacMahon's expressions involving the magnitudes b_μ (loc. cit.) we put h_μ for every b_μ, we obtain the expression of a ♉-product in terms of h-functions..

CASE 2: THEOREM: *If of the operands $F_1, F_2, \cdots, F_q$ at least one is the G. R. F. of a cyclic group, say* $\mathrm{Cyc}(\lambda_1^{l_1}\lambda_2^{l_2}\cdots)$, *then* $\Phi = F_1 \text{♉} F_2 \text{♉} \cdots \text{♉} F_q$ *can be expressed as a sum of G. R. F.'s of cyclic groups, each of which is either* $\mathrm{Cyc}(\lambda_1^{l_1}\lambda_2^{l_2}\cdots)$ *itself, or the G. R. F. of a cyclic group generated by some power of an operation of cycle-partition $\lambda_1^{l_1}\lambda_2^{l_2}\cdots$. The decomposition is made as follows: From Φ obliterate the term $Cs_{\mu_1}^{m_1}s_{\mu_2}^{m_2}\cdots$ which corresponds to the operations of highest order involved, by subtracting a suitable multiple k* $\mathrm{Cyc}(\mu_1^{m_1}\mu_2^{m_2}\cdots)$ *of the G. R. F. of the cyclic group generated by a similar operation; treat the remainder* $\Phi - k\,\mathrm{Cyc}(\mu_1^{m_1}\mu_2^{m_2}\cdots)$ *in the same way; and continue the process until all the terms of Φ are obliterated; the G. R. F.'s thus subtracted will constitute the decomposition required.*

Proof: For every summand composing Φ is (by § 4, Theorem) the G. R. F. of a sub-group of each of the groups belonging to the F's, of which groups one, being by hypothesis cyclic, can have as sub-groups only itself and the cyclic groups generated by the powers of its own generator.

If Φ contains no terms which represent operations of higher order than does $Cs_{\mu_1}^{m_1}s_{\mu_2}^{m_2}\cdots$, then this term must be contributed by a component $\mathrm{Cyc}(\mu_1^{m_1}\mu_2^{m_2}\cdots)$ of Φ, repeated a number k of times sufficient to make up the coefficient C; and the same reasoning applies to the successive remainders left after subtraction of $k\,\mathrm{Cyc}(\mu_1^{m_1}\mu_2^{m_2}\cdots)$ and its analogues.

Thus $h_3h_2 \text{♉} \mathrm{Cyc}(32)$

$$= (1/12)(s_1^5 + 4s_2s_1^3 + 3s_2^2s_1 + 2s_3s_1^2 + 2s_3s_2)$$
$$\text{♉}\ (1/6)(s_1^5 + s_2s_1^3 + 2s_3s_1^2 + 2s_3s_2)$$
$$= (1/72)(120s_1^5 + 48s_2s_1^3 + 24s_3s_1^2 + 24s_3s_2)$$
$$= (1/6)(10s_1^5 + 4s_2s_1^3 + 2s_3s_1^2 + 2s_3s_2).$$

Subtraction of $\mathrm{Cyc}(32)$ or $(1/6)(s_1^5 + s_2s_1^3 + 2s_3s_1^2 + 2s_3s_2)$ leaves

$$(1/6)(9s_1^5 + 3s_2s_1^3);$$

from this subtracting $\mathrm{Cyc}(21^3)$ or $(1/2)(s_1^5 + s_2s_1^3)$ leaves s_1^5; from this subtracting $\mathrm{Cyc}(1^5)$ or s_1^5 leaves 0. Therefore

$$h_3h_2 \text{♉} \mathrm{Cyc}(32) = \mathrm{Cyc}(32) + \mathrm{Cyc}(21^3) + \mathrm{Cyc}(1^5).$$

6. *Derivation of G. R. F.'s of Isomorphs.* If $\phi(\alpha_1, \alpha_2, \cdots, \alpha_n)$, a function of n arguments, preserves the same value when these arguments are permuted in its symbol by any operation of a specified group H of degree n and order μ, but changes its value when its arguments are permuted by any operation not contained in H, then $\phi(\alpha_1, \alpha_2, \cdots, \alpha_n)$ assumes $n!/\mu$ different values in all when the same set of n arguments is inserted in its symbol in all possible permutations. This is a well-known result (comp. e. g. C. Jordan, *Traité des Substitutions,* pp. 25, 26), but we may here point out that it is readily established by a simple application of the theory of Group-Reduced Distributions, by taking as one range the n arguments $\alpha_1, \alpha_2, \cdots, \alpha_n$, with G. R. F. Grf$(H)$, and as the second range the n fixed positions in the symbol of ϕ, with the group of identity on n elements as range-group, G. R. F. $s_1{}^n$. Now Grf(H) is always of the form $(1/\mu)(s_1{}^n + \cdots)$, the omitted terms of which are irrelevant in the Ω-product $s_1{}^n \Omega (1/\mu)(s_1{}^n + \cdots) = n!/\mu$, which is the number of distinct values which ϕ can assume. It is unnecessary to think of ϕ as a particular algebraic or other function; it is merely a symbol which we agree to regard as equivalent to any similar symbol obtainable from it by any operation of H performed on its arguments.

Any permutation of the n α's determines a corresponding permutation of the $n!/\mu$ ϕ's, and since all the α's are on the same footing as to the positions which they may occupy in the symbol $\phi(—, —, \cdots, —)$, all permutations on the α's which are of a given cycle-partition P (a partition of n) will be in correspondence with permutations of the ϕ's which are of one and the same cycle-partition P' (a partition of $n!/\mu$).

The G. R. F. of any group G of degree n can be transformed, according to the correspondence so established, into the G. R. F. of a group G' of degree $n!/\mu$, isomorphic with G. In this transformation all the coefficients of Grf(G) are unaltered, and only the literal symbols $s_{\lambda_1}{}^{t_1} s_{\lambda_2}{}^{t_2} \cdots$ are changed in accordance with the correspondence of cycle-partitions. The isomorphism is not necessarily simple, and the G. R. F.'s determined by various groups H are not always distinct.

We adopt for the G. R. F. of the derived group G' the notation Grf$(H) \,\mathfrak{J}\,$ Grf(G). We also introduce the concept of the "powers" of a cycle-partition as follows: If P denotes any partition, P^k will denote that partition which is the cycle-partition of the k-th power u^k of any permutation-operation u which is itself of cycle-partition P. Thus if $P = 43$, we shall have

$$P = P^5 = P^7 = P^{11} = 43; \; P^2 = P^{10} = 32^2; \quad P^3 = P^9 = 41^3;$$
$$P^4 = P^8 = 31^4; \quad P^6 = 2^2 1^3; \quad P^{12} = 1^7.$$

THEOREM: *To every operation of order* ξ *and cycle-partition* P, *in a group* G *of degree* n, *there corresponds, in the derived group* G' *whose G. R. F. is* $\mathrm{Grf}(H) \between \mathrm{Grf}(G)$, *an operation of cycle-partition* $1^{A_1}\alpha^{A_\alpha}\beta^{A_\beta}\cdots\xi^{A_\xi}$, *where* $1, \alpha, \beta, \cdots, \xi$ *are the divisors of* ξ, *and the* A's *are the coefficients in the expression* $A_1\mathrm{Cyc}(P) + A_\alpha\mathrm{Cyc}(P^\alpha) + A_\beta\mathrm{Cyc}(P^\beta) + \cdots + A_\xi s_1^n$, *obtained by the decomposition of* $\mathrm{Grf}(H) \between \mathrm{Cyc}(P)$ *by Case* 2 *of* § 5 *above.*

Proof: We introduce for ϕ a new notation

$$\phi\begin{pmatrix}\alpha_f & \alpha_g & \cdots & \alpha_l \\ \beta_p & \beta_q & \cdots & \beta_z\end{pmatrix},$$

in which $f, g, \cdots, l$ and $p, q, \cdots, z$ are permutations of the numbers $1, 2, \cdots, n$, and a column such as $\begin{matrix}\alpha_g\\ \beta_q\end{matrix}$ indicates that α_g occupies the qth position in the old notation $\phi(\text{—}, \text{—}, \cdots, \text{—})$. The meaning of the new symbol is evidently not changed by permutations of whole columns; this purely notational invariance is to be distinguished from the special invariance determined by the group H operating on the upper row alone.

The α's and the β's may be regarded as two ranges S_α, S_β, with H as the group associated with S_α, while with S_β we associate a cyclic group C, generated by an operation u of order ξ and cycle-partition P, so that $\mathrm{Grf}(C) = \mathrm{Cyc}(P)$. Let η be the smallest integer such that the operation u^η, performed on the β's of a given ϕ, yields a symbol ϕ' equivalent to ϕ in virtue of the invariance under H. Then η must be either ξ or a divisor of ξ, and the operations $u, u^2, \cdots, u^\eta$ will yield successively η symbols $\phi_1, \phi_2, \cdots, \phi_{\eta-1}, \phi$ all distinct under the invariance determined by H. In the cyclic group C', into which H transforms C, the operation u will thus be represented by an operation u' of degree $n!/\mu$ and cycle-partition P', P' being a partition of $n!/\mu$ containing a part η, since one of the cycles of u' will be $(\phi\phi_1\phi_2\cdots\phi_{\eta-1})$. On the other hand the two-rowed symbol of ϕ can be read as a range-correspondence which is unaltered only by such operations as combine u^η or one of its powers with a suitable similar operation of H. But u^η generates a cyclic group whose G. R. F. is $\mathrm{Cyc}(P^\eta)$, and so $\mathrm{Grf}(H) \between \mathrm{Grf}(C) = \mathrm{Grf}(H) \between \mathrm{Cyc}(P)$ will have (by the Theorems of § 4 and of § 5 Case 2) a component $\mathrm{Cyc}(P^\eta)$, in correspondence with a part η in P'. Extension to the complete $\between$-product then establishes the rule stated for finding P'.

Example: Take $\mathrm{Grf}(H) = h_3h_2$; this being of order $3!2! = 12$ will transform any G. R. F. of degree 5 into an isomorphic G. R. F. of degree

$5!/12 = 10$. We found above (§ 5 Case 2) $h_3h_2 \between \mathrm{Cyc}(32) = \mathrm{Cyc}(32) + \mathrm{Cyc}(21^3) + \mathrm{Cyc}(1^5)$. Cyc(32) belongs to a group whose order 6 has the divisors 1, 2, 3, 6. We have $P = 32$, $P^2 = 31^2$, $P^3 = 21^3$, $P^6 = 1^5$; also $A_1 = A_3 = A_6 = 1$, $A_2 = 0$; whence $1^{A_1} 2^{A_2} 3^{A_3} 6^{A_6} = 631$. Thus we obtain the cycle-partition correspondence 32 : 631. Moreover, since a cycle-partition correspondence between two group operations implies that the same correspondence holds between like powers of those operations, we have at once the correspondences $31^2 : 3^31$, $21^3 : 2^31^4$, and $1^5 : 1^{10}$. Further correspondences are given by $h_3h_2 \between \mathrm{Cyc}(41) = \mathrm{Cyc}(2^21) + 2\mathrm{Cyc}(1^5)$, namely $41 : 4^22$, and $2^21 : 2^41^2$, and by $h_3h_2 \between \mathrm{Cyc}(5) = 2\mathrm{Cyc}(1^5)$, namely $5 : 5^2$. Then we have for the complete set of 7 partitions of 5:

Partitions of 5:	5	41	32	31^2	2^21	21^3	1^5
Partitions of 10:	5^2	4^22	631	3^31	2^41^2	2^31^4	1^{10}.

This set of correspondences serves to derive an isomorph of degree 10 from any group of degree 5. Thus from $(1/5)(s_1^5 + 4s_5) = \mathrm{Cyc}(5)$ comes $(1/5)(s_1^{10} + 4s_5^2) = \mathrm{Cyc}(5^2)$. Or taking for G the symmetric group of degree 5, we have from

$$h_5 = (1/120)(s_1^5 + 10s_2s_1^3 + 15s_2^2s_1 + 20s_3s_1^2 + 20s_3s_2 + 30s_4s_1 + 24s_5)$$

the isomorphic G. R. F.

$$h_3h_2 \mars h_5 = (1/120)(s_1^{10} + 10s_2^3s_1^4 + 15s_2^4s_1^2 + 20s_3^3s_1 + 30s_4^2s_2 + 24s_5^2 + 20s_6s_3s_1).$$

The function last written is typical of a class of G. R. F.'s $h_{n-2}h_2 \mars h_n$, of degree $n(n-1)/2$ and order $n!$, connected with the enumeration of the symmetrical aliorelative dyadic "relation-numbers" (Whitehead and Russell, *Principia Mathematica*, Vol. II, p. 301) on a field of n elements. If we represent field members by nodes o, and the holding of a typical relation by connecting lines, then there are 10 possible connecting lines for 5 nodes. If exactly 4 of these 10 are drawn, we get the configurations shown below, which are enumerated by $h_6h_4 \between (h_3h_2 \mars h_5) = 6$.

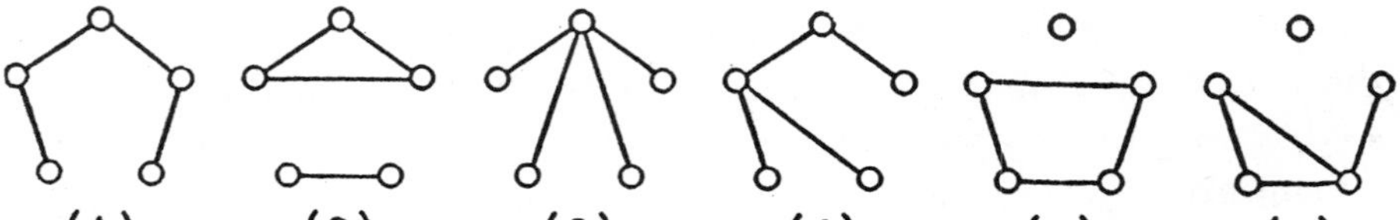

(1) (2) (3) (4) (5) (6)

Configurations (5) and (6), which contain isolated nodes, do not, according to the accepted definitions, represent relation-numbers on a field of *five* elements. What we have enumerated are in fact the relation-numbers of a certain type on fields of *five or fewer* elements, and to remove the redundancy we must subtract the corresponding expression for fields of four or fewer elements, thus: $h_6h_4 \mathbin{\text{♋}} (h_3h_2 \mathbin{\&} h_5) - h_2h_4 \mathbin{\text{♋}} (h_2{}^2 \mathbin{\&} h_4) = 6 - 2 = 4$.

By putting every $s_r = 2$ (see § 3 ex. i) in the expressions $h_{n-2}h_2 \mathbin{\&} h_n$ for various values of n, we find that the total number of symmetrical aliorelative dyadic relation-numbers on fields of n elements is as follows:

Elements in Field:	1	2	3	4	5	6	7	. . .
Relation-Numbers:	0	1	2	7	23	122	888	. . .

Other species of relation-numbers can be enumerated in a similar manner. Thus for asymmetrical aliorelative dyadic relation-numbers we use the functions $h_{n-2}h_1{}^2 \mathbin{\&} h_n$: for symmetrical aliorelative triadic relation numbers the functions $h_{n-3}h_3 \mathbin{\&} h_n$; for aliorelative triadic relation-numbers symmetrical in two only of their arguments, the functions $h_{n-3}h_2h_1 \mathbin{\&} h_n$.

Sometimes the isomorphism is not simple but j-fold; in such case every derived operation will be repeated j tim , and this will give a leading term $js_1{}^n$ inside the parentheses of the G. R. F. When this factor j is carried outside the parentheses, the G. R. F. will appear as of the correct order, so that no exception arises to the rule for finding the expression for $\mathrm{Grf}(H) \mathbin{\&} \mathrm{Grf}(G)$.

Other isomorphs can be obtained by using as transforming function, instead of a G. R. F. $\mathrm{Grf}(H)$, any symmetric function which is the sum of two or more G. R. F.'s of the proper degree. Thus for enumerating the symmetrical dyadic relation numbers which are not restricted to be aliorelative, we use functions $(h_{n-2}h_2 + h_{n-1}h_1) \mathbin{\&} h_n$. Such isomorphs are necessarily intransitive. On the other hand, when a single G. R. F. is used as transforming function, the derived isomorph *may* be intransitive, and if it is, still other isomorphs can be got by omitting some of the transitive constituents. The general problem of actually separating an intransitive G. R. F. into transitive constituents has however not been solved, and seems to be closely connected in nature and difficulty with the decomposition of ♋-products into sums of G. R. F.'s.

Some important general relations are (for G and H of degree n, H of order μ): $h_n \mathbin{\&} \mathrm{Grf}(G) = s_1$ for every G; $(h_n + a_n) \mathbin{\&} \mathrm{Grf}(G) = (1/2)(s_1{}^2 + s_2)$ or $s_1{}^2$ according as G has or has not odd permutations$\cdot$ $h_{n-1}h_1 \mathbin{\&} \mathrm{Grf}(G) = \mathrm{Grf}(G)$;

$$h_{(n!/\mu)-1}\, h_1 \mathbin{\text{♋}} [\mathrm{Grf}(H) \mathbin{\&} \mathrm{Grf}(G)] = \mathrm{Grf}(H) \mathbin{\text{♋}} \mathrm{Grf}(G).$$

7. *Determination of the G. R. F. from Given Generators.* If two symbols α and β occur in the same cycle of one of the given generators of a group G whose G. R. F. is to be determined, they belong to the same transitive constituent of G. By examination of the given generators from this point of view, the number θ of transitive constituents can be found. Writing Grf(G) in the general form $(1/m)\sum[N(G;p_1^{\pi_1}p_2^{\pi_2}\cdots)s_{p_1}^{\pi_1}s_{p_2}^{\pi_2}\cdots]$, the relation $h_{n-1}h_1 \cap (1/m)\sum[N(G;p_1^{\pi_1}p_2^{\pi_2}\cdots)s_{p_1}^{\pi_1}s_{p_2}^{\pi_2}\cdots]=\theta$ gives an equation with m and the N's as unknowns. Again, taking any known group H of degree n and order μ, we can form a function ϕ, of the original n symbols, which is invariant under H, and then determine the forms of the new generators, each containing now $n!/\mu$ symbols, and corresponding to the original generators, which generate the isomorph whose G. R. F. is Grf$(H) \cup$ Grf(G); and another equation is given by

$$h_{(n!/\mu)-1}\, h_1 \cap [\mathrm{Grf}(H) \cup \mathrm{Grf}(G)] = \mathrm{Grf}(H) \cap \mathrm{Grf}(G) = \theta',$$

θ' being found by inspection of the new generators just as θ was found previously. Continuing in the same way, using various other groups for H, a sufficient number of equations can in theory be found to determine m and all the N's. In practice some of the isomorphs will be of inconveniently high degree, a difficulty which can be mitigated to some extent by choosing the groups H of as high order as possible.

It is not necessary, however, to set up as many equations as there are unknowns. For without actually evaluating Grf$(H) \cap$ Grf(G), we know that it is a positive integer. Also $s_{\lambda_1}^{l_1}s_{\lambda_2}^{l_2}\cdots \cap$ Grf(G) is a positive integer or 0 for every partition $\lambda_1^{l_1}\lambda_2^{l_2}\cdots$ of n; while if H' is a sub-group of H, it is easily shown that Grf$(H') \cap$ Grf$(G) \geqq$ Grf$(H) \cap$ Grf(G). These and similar relations furnish a variety of congruences and diophantine inequalities which, in connection with a partial set of definite equations, will frequently suffice for the determination of Grf(G).

Even when so simplified, it may be doubted whether the method offers any substantial advantage over the alternative, frequently impracticable of course, of actually deriving the whole set of operations of the group from the given generators and then forming the G. R. F. by counting the operations of various cycle-partitions. It is here given rather as illustrating a principle on which it may be possible to base an effective method. For the present we must in practice rely on special devices when dealing with groups of high orders.

8. *Possible Algebraic Forms of G. R. F.'s.* The congruences and diophantine inequalities referred to in § 7 constitute a set of *necessary* conditions that a given symmetric function be the G. R. F. of an existent group. By means of them we can determine the possible forms of G. R. F.'s having specified properties, as for instance those belonging to groups of given degree and order. We can often thus demonstrate the non-existence of certain classes of groups, or narrow the possibilities to one or a few forms. We cannot however so demonstrate the existence of any proposed group, and a completely *sufficient* set of purely algebraic conditions has not been discovered, though nothing indicates that this would be impossible. Such methods might be useful in checking for omissions the lists of groups compiled by Cayley and others (*Quarterly Journal of Mathematics,* Vol. XXV, p. 71, etc.).

WAYNE, PA.
SEPTEMBER, 1926.

7.4 References

M. Abramson and W. O. J. Moser (1966) Combinations, successions and the n-kings problem, *Math. Mag.*, *39*, 269–273.

M. Abramson and W. O. J. Moser (1972) The problem of the second seating and generalizations, *Studia Scient. Math. Hungarica*, *7*, 211–224.

J. Arkin (1972a) The first solution of the classical Eulerian magic cube problem of order 10, *Fibonacci Quart.*, *11*, 174–178.

J. Arkin (1972b) A solution to the classical problem of finding systems of three mutually orthogonal numbers in a cube formed by three superposed 10 × 10 × 10 cubes, *Fibonacci Quart.*, *11*, 485–489.

C. Berge (1971) Principles of Combinatorics, Academic Press, New York.

F. Buchsbaum and F. Freudenstein (1970) Synthesis of kinematic structure of geared kinematic chains and other mechanisms, *Journal Mechanisms*, *5*, 357–392.

L. Comtet (1974) Advanced Combinatorics, Reidel, Dordrecht.

N. G. deBruijn (1959) Generalization of Polya's fundamental theorem in enumerative combinatorial analysis, *Indagationes Math.*, *21*, 59–69.

N. G. deBruijn (1963) Enumerative combinatorial problems concerning structures, *Nieuw Archief Wisk.(3)*, *11*, 142–161.

N. G. deBruijn (1964) Polya's Theory of Counting, in Applied Combinatorial Mathematics, E. F. Beckenbach ed., J. Wiley, New York, 144–184.

N. G. deBruijn (1971) A survey of generalizations of Polya's enumeration theorem, *Nieuw Archief Wisk.(2)*, *19*, 89–112.

M. Eisen (1969) Elementary Combinatorial Analysis, Gordon and Breach, New York.

R. A. Fisher and F. Yates (1934) The 6 × 6 Latin squares, *Proc. Cambridge Phil. Soc.*, *30*, 492–507.

H. O. Foulkes (1963) On Redfield's group reduction functions, *Canadian J. Math.*, *15*, 272–284.

M. Hall (1958) Survey of Combinatorial Analysis, in Some Aspects of Analysis and Probability IV, J. Wiley, New York.

F. Harary and E. Palmer (1967) The enumeration methods of Redfield, *American J. Math.*, *89*, 373–384.

F. Harary and E. Palmer (1973) Graphical Enumeration, Academic Press, New York.

D. A. Holton (1970) On Redfield's enumeration methods—application of group theory to combinatorics, Ph.D. thesis, McGill University.

A. Kerber (1971) Representations of Permutation Groups I, Lecture Notes in Math., No. 240, Springer-Verlag, Berlin-Heidelberg-New York.

A. Kerber (1975) Representations of Permutation Groups II, Lecture Notes in Math., No. 495, Springer-Verlag, Berlin-Heidelberg-New York.

O. Leue (1972) Methoden zur Lösung dreidimensionaler Zuordnungsprobleme, *Angewandte Informatik*, *4*, 154–162.

W. O. J. Moser (1967) The number of very reduced 4 × n Latin rectangles, *Canadian J. Math.*, *19*, 1011–1017.

J. K. Percus (1971) Combinatorial Methods, Springer, New York.

G. Polya (1937) Kombinatorische Anzahlbestimmungen für Gruppen, Graphen und chemische Verbindungen, *Acta Math.*, *68*, 145–254.

R. C. Read (1959) The enumeration of locally restricted graphs (I), *J. London Math. Soc.*, *34*, 417–436.

R. C. Read (1960) The enumeration of locally restricted graphs (II), *J. London Math. Soc.*, *35*, 344–351.

R. C. Read (1968) The use of S-functions in combinatorial analysis, *Canadian J. Math.*, *20*, 808–841.

J. H. Redfield (1927) The theory of group reduced distributions, *American J. Math.*, *49*, 433–455.

J. Riordan (1957) The combinatorial significance of a theorem of Polya, *J. Soc. Ind. Appl. Math.*, *5*, 225–237.

J. Riordan (1958) An Introduction to Combinatorial Analysis, J. Wiley, New York.

J. Riordan and P. R. Stein (1972) Arrangements on chessboards, *J. Combinatorial Th.*, *12*, 72–80.

G.-C. Rota (1969) Baxter algebras and combinatorial identities (I) and (II), *Bull. A.M.S.*, *75*, 325–334.

G.-C. Rota and D. A. Smith (1972) Fluctuation theory and Baxter algebras, *Inst. Naz. di Alta Mat. Symp. Math.*, *9*, 179–201.

J. Touchard (1934) Sur un problème de permutations, *C. R. Acad. Sc. Paris*, *198*, 631–633.

D. E. White (1975a) Redfield's theorems and multilinear algebra, *Canadian J. Math.*, *27*, 704–714.

D. E. White (1975b) Multilinear enumerative techniques, *Linear and multilinear algebra*, *2*, 341–352.

D. E. White and S. G. Williamson (1976) Combinatorial structures and group invariant partitions, *Proc. A.M.S.*, *55*, 233–236.

S. G. Williamson (1970) Operation theoretic invariants and the enumeration theory of Polya and deBruijn, *J. Combinatorial Th.*, *8*, 162–169.

S. G. Williamson (1971a) Polya's counting theorem and a class of tensor identities, *J. London Math. Soc.(2)*, *3*, 411–421.

S. G. Williamson (1971b) Symmetry operators of Kranz products, *J. Combinatorial Th.*, *11*, 122–138.

7.5 Summaries of the Papers

[52] A new method in combinatory analysis with application to Latin squares and associated questions, *Trans. Cambridge Phil. Soc.*, *16* (1898), 262–290.

In this important paper, MacMahon laid the foundation for what in subsequent years became Polya-Redfield enumeration theory. In paper [57] which follows he considers the general types of results that his techniques will produce.

The Graeco-Latin square problem is the following: Arrange n^2 pairs $(a_i \alpha_i)$, $1 \leqq i \leqq n$, in an n by n square so that each a_i and α_j appear once and only once in each row and column. The Latin square problem is to arrange n

copies each of $a_1, a_2, \ldots, a_n$ in an n by n square so that each a_i appears once and only once in each row and column.

In section 1 MacMahon applies the operators that are of importance in the theory of symmetric functions (see [29]) to the Latin square problem. His principal result is the following:

Theorem. In the identity

$$(\sum \alpha_{i_1}^{2^{n-1}} \alpha_{i_2}^{2^{n-2}} \ldots \alpha_{i_{n-1}}^2 \alpha_{i_n})^n = \ldots + K\sum (\alpha_{i_1}\alpha_{i_2} \ldots \alpha_{i_n})^{2^n - 1} + \ldots,$$

K is the number of n by n Latin squares. ($K/n!(n-1)!$ is the number of reduced Latin squares.)

Similar problems are treated in sections 2–4. In section 2, the Menage problem is treated; in section 3, Latin rectangles, and in section 4, restricted types of Latin squares.

Section 5 concludes the paper with an introduction to the ideas of [57]. An amusing result arising in this section is the following:

Theorem. n^2 different towns form a square and there are n^2 inspectors, n of each of n different nationalities. Then there are $(n!)^n$ arrangements of the inspectors in the towns subject to the condition that one inspector of each nationality must be in each row or column of towns, with no restriction on the number of inspectors that may be stationed in a particular town.

The above is, of course, merely the Latin square problem with the removal of the requirement that each town contain exactly one inspector.

[57] Combinatorial analysis. The foundations of a new theory, *Phil. Trans.*, *194* (1900), 12–34.

This paper is the natural successor to [52]. An expository introduction to the subject matter of this paper is presented in [60].

The object here is to construct both an operator and a function such that when the operator is applied to the function, the result is a number that enumerates a given set of combinatorial structures. The simplest example of this idea is given by the operator $(d/dx)^n$ and the function x^n. In this case $(d/dx)^n x^n = n!$, the number of permutations of n letters. MacMahon devotes most of section 1 to examples of this technique.

A more complex example, but one that is suggestive of Latin square type problems, is the following: How many m by l matrices of zeros and ones and there that have μ_i ones in the ith row ($1 \leqq i \leqq m$) and λ_j ones in the jth column ($1 \leqq j \leqq l$)? In this case the answer is obtained by applying the operator $D_{\mu_1} D_{\mu_2} \ldots D_{\mu_m}$ to the symmetric function

$$(1^{\lambda_1})\,(1^{\lambda_2})\,\ldots\,(1^{\lambda_l}),$$

where D_j is the Hammond operator and (1^j) is the jth elementary symmetric function of an infinite set of variables.

"Magic square" problems and a general symmetry theorem for symmetric functions conclude section 1.

In section 2 MacMahon extends the operators under consideration to include those from the calculus of finite differences; this allows our consideration of symmetric functions to include monomial symmetric functions related to partitions, in which the parts are merely nonnegative.

Finally, in section 3 MacMahon introduces the symmetric functions of several systems of quantities that are introduced in [34]; the operators are the appropriate generalizations of the Hammond operators and the formal solution of the Latin Square problem now follows from these considerations.

[89] Combinations derived from m identical sets of n different letters and their connexion with general magic squares, *Proc. London Math. Soc.(2), 17* (1918), 25–41.

MacMahon shows that there is a one-to-one correspondence between general magic squares (i.e., N by N square matrices of nonnegative integers (each $\leqq N - 1$) whose row sums and column sums are identical) and the number of ways of choosing n m-tuples from among m copies of n different letters. For example if $n = 3$, $m = 2$, there are five choices:

$$(aa)\,(bb)\,(cc),$$
$$(aa)\,(bc)\,(bc),$$
$$(ac)\,(bb)\,(ac),$$
$$(ab)\,(ab)\,(cc),$$
$$(ab)\,(ac)\,(bc).$$

Writing these choices in partition notation, we have

$$(200)\,(020)\,(002),$$
$$(200)\,(011)\,(011),$$
$$(101)\,(020)\,(101),$$
$$(110)\,(110)\,(002),$$
$$(110)\,(101)\,(011).$$

These in turn produce the five corresponding general magic squares with row and column sum equal to 2:

$$\begin{matrix} 2&0&0 \\ 0&2&0 \\ 0&0&2 \end{matrix} \qquad \begin{matrix} 2&0&0 \\ 0&1&1 \\ 0&1&1 \end{matrix} \qquad \begin{matrix} 1&0&1 \\ 0&2&0 \\ 1&0&1 \end{matrix} \qquad \begin{matrix} 1&1&0 \\ 1&1&0 \\ 0&0&2 \end{matrix} \qquad \begin{matrix} 1&1&0 \\ 1&0&1 \\ 0&1&1. \end{matrix}$$

MacMahon investigates the combinations described above and provides several means for computing them. For example, let p_n denote the number of n by n general magic squares of row and column sum equal to 2; then

$$(n - 3)p_n - (n^2 - 2n - 1)p_{n-1} + (n - 1)p_{n-2}$$
$$+ \binom{n-1}{2}(3n - 7)p_{n-3} - 3\binom{n-1}{3}p_{n-4} - 12\binom{n-1}{4}p_{n-5} = 0.$$

Other less manageable formulas are considered in more general cases. For example, if $f(m, n)$ denotes the number of n by n general magic squares of row and column sum m, then $f(m, n)$ is the coefficient of the symmetric function $\sum \alpha_1^m \alpha_2^m \ldots \alpha_n^m$ in the expansion of

$$\begin{aligned} &(1 - \alpha_1^m)(1 - \alpha_2^m) \ldots \\ &(1 - \alpha_1^{m-1}\alpha_2) \ldots \\ &(1 - \alpha_1^{m-2}\alpha_2) \ldots \\ &(1 - \alpha_1^{m-2}\alpha_2\alpha_3) \ldots \\ &\ldots \\ &\ldots \end{aligned}$$

Reprints of the Papers

XI. *A New Method in Combinatory Analysis, with application to Latin Squares and associated questions.* By Major P. A. MacMahon, R.A., Sc.D., F.R.S., Hon. Mem. Camb. Phil. Soc.

[*Received* January 11, 1898. *Read* January 24, 1898.]

Introduction.

Euler in the *Verhandelingen uitgegeven door het Zeeuwsch Genootschap der Wetenschappen te Vlissingen,* vol. 9, 1782, has a paper entitled "Recherches sur une nouvelle espèce de Quarrés Magiques."

He commences as follows:—

"Une question fort curieuse, qui a exercé pendant quelque temps la sagacité de bien du monde, m'a engagé à faire les recherches suivantes, qui semblent ouvrir une nouvelle carrière dans l'Analyse, et en particulier dans la doctrine des combinaisons. Cette question rouloit sur une assemblée de 36 officiers de six différens grades et tirés de six Régimens différens, qu'il s'agissoit de ranger dans un quarré, de manière, que sur chaque ligne tant horizontale que verticale il se trouva six officiers tant de différens caractères que de Régimens différens. Or après toutes les peines qu'on s'est donné pour résoudre ce Problème, on a été obligé de reconnoître qu'un tel arrangement est absolument impossible, quoiqu'on ne puisse pas en donner de démonstration rigoureuse."

He denotes the six regiments by the Latin letters a, b, c, d, e, f and the six ranks or grades by the Greek letters α, β, γ, δ, ϵ, ζ, and remarks that the 'character' of an officer is determined by two letters, the one Latin and the other Greek, and that the problem consists in arranging the 36 combinations

$$\begin{array}{cccccc} a\alpha & a\beta & a\gamma & a\delta & a\epsilon & a\zeta \\ b\alpha & b\beta & b\gamma & b\delta & b\epsilon & b\zeta \\ c\alpha & c\beta & c\gamma & c\delta & c\epsilon & c\zeta \\ d\alpha & d\beta & d\gamma & d\delta & d\epsilon & d\zeta \\ e\alpha & e\beta & e\gamma & e\delta & e\epsilon & e\zeta \\ f\alpha & f\beta & f\gamma & f\delta & f\epsilon & f\zeta \end{array}$$

in a square in such a manner that every row and column contains the six Latin and the six Greek letters.

He finds no solution of this particular problem and gives his opinion that none can be obtained whenever the order of the square is of the form 2 mod 4.

In other cases as far as the order 9 he obtains solutions.

The first step is to arrange the Latin letters in a square so that no letter is missing either from any row or any column. He calls this a Latin Square, and in regard to their enumeration for a given order observes, § 148, p. 230:

"J'observe encore à cette occasion que le parfait dénombrement de tous les cas possibles de variations semblables seroit un objet digne de l'attention des Géomètres, d'autant plus que tous les principes connus dans la doctrine des combinaisons n'y sçauroient prêter le moindre secours."

And again, § 152, p. 234:

"J'avois observé ci-dessus, qu'un parfait dénombrement de toutes les variations possibles des quarrés latins seroit une question très importante, mais qui me paroissoit extrèmement difficile et presque impossible dès que le nombre n surpassoit 5. Pour approcher de cette énumération il faudroit commencer par cette question:

En combien de manières différentes, la première bande horizontale étant donnée, peut-on varier la seconde bande horizontale pour chaque nombre proposé n?"

He in fact gives a solution of the last-mentioned question. Many different ones are now in existence and, incidentally, a new one is given in this paper.

It is the well-known 'Problème des rencontres' which was first proposed by Montmort.

For the rest the paper is entirely concerned with the actual construction of what may be termed Graeco-Latin Squares, to which one is led by considering the problem of the officers above mentioned, and with their transformation so as to enable one to obtain many solutions from any one that has been arrived at.

Euler himself admits the unsatisfactory nature of his investigation. He remarks § 11, p. 94 "La formation des formules directrices est donc le premier et le principal objet dans ces recherches; mais je dois avoüer, que jusqu'ici je n'avois aucune méthode sûre qui puisse conduire à cette investigation. Il semble même qu'on doit se contenter d'une espèce de simple tâtonnement que je vais expliquer pour le quarré latin de 49 cases rapporté ci-dessus."

To explain the meaning of the phrase 'formules directrices' which will be referred to in the sequel, I take Euler's Graeco-Latin Square of order 7, writing with him the natural numbers instead of Latin and Greek letters and writing the latter as exponents to the former:

1^1	2^6	3^4	4^3	5^7	6^5	7^2
2^2	3^7	1^5	5^4	4^1	7^6	6^3
3^3	6^1	5^6	7^5	1^2	4^7	2^4
4^4	5^2	6^7	1^6	7^3	2^1	3^5
5^5	1^3	7^1	2^7	6^4	3^2	4^6
6^6	7^4	4^2	3^1	2^5	5^3	1^7
7^7	4^5	2^3	6^2	3^6	1^4	5^1

Consider the Latin Square as given by erasing the exponent numbers.

In order to find possible places for the exponent 1 we must select 7 different numbers, from the square, one being in each column and one in each row.

Thus we may select the set

$$1 \quad 6 \quad 7 \quad 3 \quad 4 \quad 2 \quad 5$$

and give each the exponent 1 as above.

This is called 'une formule directrice' for the exponent 1.

We may similarly find a formula for the exponent 2, viz.:—

$$2 \quad 5 \quad 4 \quad 6 \quad 1 \quad 3 \quad 7$$

and if we are successful for each of the 7 numbers and can fill in all the exponents clearly the Graeco-Latin Square will have been constructed. In this case if 'une formule directrice' be written for each number, so that they form a square, it is clear that it must be a Latin Square.

I pass now to the paper by Cayley, *Messenger of Mathematics*, vol. XIX. (1890), pp. 135—137. This is little more than a statement of the problem involved in the enumeration of Latin Squares. With Euler he reduces the number of squares by taking the top row and left-hand column in the same determinate order; say for the order 5,

$$\begin{matrix} a & b & c & d & e \\ b & d & e & a & c \\ c & e & d & b & a \\ d & a & e & c & b \\ e & c & b & a & d \end{matrix}$$

and he remarks that "if the number of such squares be $= N$, then obviously the whole number of squares that can be formed with the same n arrangements is $= N[n]^n$." This however is not so; the number is $N . n!(n-1)!$ the factor $n!$ appearing from permutation of columns and $(n-1)!$ from permutation of the lower $n-1$ rows. He speaks of the possible arrangements of the second line, the 'Problème des rencontres' and of the difficulty of proceeding a step further to the enumeration of the arrangements of the second and third lines. This problem, previous to the present paper, had never been solved, and to quote from Cayley (*loc. cit.*) "the difficulty of course increases for the next following lines." He further makes the valuable, if obvious, observation "when all the lines are filled up except the bottom line, the bottom line is completely determined." In the above quotation I have substituted 'bottom' for 'top,' as I read with Euler from top to bottom instead of with Cayley from bottom to top.

The 'Problème des ménages' involves to some extent the consideration of the second and third lines of the square and a fairly satisfactory solution has been obtained. (See post.)

For the rest nothing of value has been accomplished and the whole question awaits elucidation.

SECTION 1.

Art. 1. It occurred to me to attack the problem by the powerful methods of the calculus of symmetric functions, and I have been led to the complete analytical solution of the main and many allied questions. The enumerations are given in the form of coefficients in the developments of certain generating functions whose inner structures are seen to implicitly involve the solutions.

In particular I have obtained the following simple and elegant theorem.

If the symmetric function

$$\Sigma \alpha_1^{2^{n-1}} \alpha_2^{2^{n-2}} \ldots\ldots \alpha_{n-1}^2 \alpha_n$$

be raised to the power n, so that

$$(\Sigma \alpha_1^{2^{n-1}} \alpha_2^{2^{n-2}} \ldots\ldots \alpha_{n-1}^2 \alpha_n)^n = \ldots\ldots + K\Sigma \alpha_1^{2^n-1} \alpha_2^{2^n-1} \ldots\ldots \alpha_{n-1}^{2^n-1} \alpha_n^{2^n-1} + \ldots\ldots,$$

then K is the number of Latin Squares of order n and division by $n!(n-1)!$ is merely necessary to obtain the number of reduced Latin Squares.

I make the following references to papers by myself where the principles employed are set forth and employed in various interesting questions of combinatory analysis.

"Symmetric Functions and the Theory of Distributions" (*Proc. L. M. S.*, vol. XIX. pp. 220—256).

"A theorem in the calculus of linear partial differential equations" (*Q. M. J.*, vol. XXIV. pp. 246—250).

"Memoir on a new Theory of Symmetric Functions" (*Amer. Journ. Math.*, vol. XI. 1889, pp. 1—36).

"Second Memoir......" (*ibid.* vol. XII. 1890, pp. 61—102).

"Third Memoir......" (*ibid.* vol. XIII. 1891, pp. 193—234).

"Fourth Memoir......" (*ibid.* vol. XIV. 1892, pp. 15—32).

"A certain class of Generating Functions in the Theory of Numbers" (*Phil. Trans. R. S.*, vol. 185 A, 1894, pp. 111—160).

"The Algebra of Multi-Linear Partial Differential Operators" (*Proc. L. M. S.* vol. XIX.).

"The Multiplication of Symmetric Functions" (*Mess. of Math.*, New Series, No. 167, March 1885).

Art. 2. I reproduce, with slightly altered notation, the master theorem, given in the last but one quoted paper, which has special reference to the present investigation.

Let $u_1, u_2, \ldots u_s$ be any linear operators, whatever, in regard to the elements

$$p_1, p_2, p_3, \ldots,$$

and put

$$\Theta = \lambda_1 u_1 + \lambda_2 u_2 + \ldots + \lambda_s u_s$$
$$= \Theta_1 \partial_{p_1} + \Theta_2 \partial_{p_2} + \Theta_3 \partial_{p_3} + \ldots.$$

Further let

$$\phi_1,\ \phi_2, \ldots \phi_m$$

be any m functions of $p_1,\ p_2,\ p_3, \ldots$

and $$\phi = \phi_1\phi_2\phi_3 \ldots\ldots \phi_m.$$

Then

$$\phi_t(p_1 + \Theta_1,\ p_2 + \Theta_2,\ p_3 + \Theta_3, \ldots)$$

$$= \phi_t + \Theta\phi_t + \frac{(\Theta^2)}{2!}\phi_t + \frac{(\Theta^3)}{3!}\phi_t + \ldots,$$

and $$\phi + \Theta\phi + \frac{(\Theta^2)}{2!}\phi_t + \frac{\Theta^3}{3!}\phi_t + \ldots$$

$$= \prod_{t=1}^{t=m}\left\{\phi_t + \Theta\phi_t + \frac{(\Theta^2)}{2!}\phi_t + \frac{(\Theta^3)^3}{3!}\phi_t + \ldots\ldots\right\},$$

that is

$$\phi + (\lambda_1 u_1 + \lambda_2 u_2 + \ldots\ldots + \lambda_s u_s)\phi + \frac{1}{2!}(\lambda_1 u_1 + \lambda_2 u_2 + \ldots\ldots + \lambda_s u_s)^2\phi + \ldots\ldots$$

$$= \prod_{t=1}^{t=m}\left\{\phi_t + (\lambda_1 u_1 + \lambda_2 u_2 + \ldots\ldots + \lambda_s u_s)\phi_t + \frac{1}{2!}(\lambda_1 u_1 + \lambda_2 u_2 + \ldots\ldots + \lambda_s u_s)^2\phi_t + \ldots\ldots\right\}.$$

We now compare the coefficients of

$$\lambda_1^{\chi_1}\lambda_2^{\chi_2}\ldots\ldots\lambda_s^{\chi_s},$$

on the two sides of the identity, and obtain the result

$$\frac{(u_1^{\chi_1}u_2^{\chi_2}\ldots u_s^{\chi_s})\,\phi}{\chi_1!\,\chi_2!\ldots\chi_s!} = \Sigma\Sigma\,\frac{(u_1^{\alpha_1}u_2^{\alpha_2}\ldots u_s^{\alpha_s})\,\phi_1}{\alpha_1!\,\alpha_2!\ldots\alpha_s!}\,\frac{(u_1^{\beta_1}u_2^{\beta_2}\ldots u_s^{\beta_s})\,\phi_2}{\beta_1!\,\beta_2!\ldots\beta_s!}\ldots\frac{(u_1^{\mu_1}u_2^{\mu_2}\ldots u_s^{\mu_s})\,\phi_m}{\mu_1!\,\mu_2!\ldots\mu_s!},$$

where $$\alpha_t + \beta_t + \ldots + \mu_t = \chi_t,\ (t = 1,\ 2,\ 3, \ldots s),$$

and the double summation is in regard to every positive integral solution of these s equations and to every permutation of

$$\phi_1,\ \phi_2,\ \phi_3 \ldots \phi_m.$$

This important result is similar to the well-known theorem of Leibnitz in form, but in form only, for here such an operator product as

$$u_1^{\delta_1}u_2^{\delta_2}\ldots u_s^{\delta_s}$$

does not mean $\Sigma\delta$ successive operations but the single operation of an operator of order $\Sigma\delta$ obtained by symbolic multiplication.

Art. 3. Putting s equal to unity we find

$$\frac{(u^\chi)\,\phi}{\chi!} = \Sigma\Sigma\,\frac{(u^\alpha)\,\phi_1}{\alpha!}\,\frac{(u^\beta)\,\phi_2}{\beta!}\ldots\frac{(u^\mu)\,\phi_m}{\mu!},$$

and now putting $$u = \partial_{p_1} + p_1\partial_{p_2} + p_2\partial_{p_3} + \ldots$$

$$\frac{(u^\delta)}{\delta!} = D_\delta,$$

$$D_\chi\phi = \Sigma\Sigma\, D_\alpha\phi_1 D_\beta\phi_2 \ldots D_\mu\phi_m,$$

where $$\alpha + \beta + \ldots + \mu = \chi,$$

and the double summation has reference to every solution of this equation in positive integers (including zero) and to every permutation of $\phi_1, \phi_2, \ldots \phi_m$.

Art. 4. As a further particular case,

$$D_\chi \phi^m = \Sigma\Sigma D_\alpha \phi D_\beta \phi \ldots D_\mu \phi,$$

and we have a summation in regard to every partition of χ into μ parts, zero being counted as a part, and to every permutation of the parts of each partition.

The operation $$\Sigma D_\alpha \phi_1 D_\beta \phi_2 \ldots D_\mu \phi_m,$$

the summation being for every permutation of the parts $\alpha, \beta, \ldots \mu$, may be denoted by

$$D_{(\alpha\beta\ldots\mu)},$$

and thence we have the equivalence

$$D_\chi \equiv \Sigma D_{(\alpha\beta\ldots\mu)},$$

where the summation is for all partitions of χ.

It may happen that $(\alpha\beta \ldots \mu)$ is the only partition of χ, for which $D_{(\alpha\beta\ldots\mu)}\phi$ does not vanish, so that

$$D_\chi \equiv D_{(\alpha\beta\ldots\mu)}.$$

As will be seen the assignment of ϕ, so that this may be the case, is the key to the solution of the problem of the Latin Square.

Art. 5. Take the polynomial

$$x^n - p_1 x^{n-1} + p_2 x^{n-2} - \ldots = (x - \alpha_1)(x - \alpha_2) \ldots (x - \alpha_n),$$

and, with the object of investigating Latin Squares with the n elements

$$a_1, a_2, \ldots a_n,$$

consider the symmetric function

$$\Sigma \alpha_1^{a_1} \alpha_2^{a_2} \ldots \alpha_n^{a_n},$$

wherein $a_1, a_2, \ldots a_n$ are to be regarded as unspecified different integers.

Let $$a_1 + a_2 + \ldots + a_n = w,$$

and form the partial differential operator

$$D_w = \frac{1}{w!} (\partial_{p_1} + p_1 \partial_{p_2} + p_2 \partial_{p_3} + \ldots)^w,$$

the linear operator being raised to the power w by symbolic multiplication, so that D_w is an operator of order w.

The symmetric function $$\Sigma \alpha_1^{a_1} \alpha_2^{a_2} \ldots \alpha_n^{a_n}$$

may be conveniently symbolised, in the usual manner, by the partition notation

$$(a_1 a_2 \ldots a_n).$$

As is well known the operation of D is shewn by

$$D_{a_1}(a_1 a_2 \ldots a_n) = (a_2 \ldots a_n),$$

$$D_{a_s}(a_1 a_2 \ldots a_{s-1} a_s \ldots a_n) = (a_1 a_2 \ldots a_{s-1} a_{s+1} \ldots a_n),$$

$$D_{a_1}(a_1) = 1,$$

$$D_{a_1} D_{a_2} D_{a_3} \ldots D_{a_n}(a_1 a_2 a_3 \ldots a_n) = 1.$$

I now construct the symmetric function

$$(a_1a_2 \dots a_n)^n,$$

and suppose it multiplied out until its equivalent representation as a sum of monomial symmetric functions is arrived at.

We have $$(a_1a_2 \dots a_n)^n = \Sigma K(b_1b_2b_3\dots),$$

and operating on both sides with

$$D_{b_1}D_{b_2}D_{b_3}\dots,$$

we find $$D_{b_1}D_{b_2}D_{b_3}\dots(a_1a_2\dots a_n)^n = K,$$

and we can calculate K by performing a definite series of operations.

If we operate with D_w only those symmetric functions, on the right-hand side, survive which include a part w in their partitions, and, in each of these, the number of times w occurs as a part is diminished by unity. We must consider the operation of D_w upon the left-hand side

$$(a_1a_2 \dots a_n)^n.$$

First take for simplicity $n = 4$, and write

$$(a_1, a_2, a_3, a_4) = (a, b, c, d),$$

and $$w = a + b + c + d.$$

Let a, b, c, d be so assigned that

$$(abcd)$$

is the *only partition* of $w = a + b + c + d$ into four or fewer parts, repeated or not, drawn from the parts a, b, c, d.

Ex. gr. we may take $(a, b, c, d) = (8, 4, 2, 1)$,

and it will be noticed that 15 has the single partition (8421) into four or fewer parts, repeated or not, drawn from the parts 8, 4, 2, 1.

Then $$D_w(abcd)^4 = D_{(abcd)}(abcd)^4 \text{ (see \textit{ante} Art. 4)},$$

and $$D_w(abcd)^4 = \Sigma D_{v_1}(abcd)\,D_{v_2}(abcd)\,D'_{v_3}(abcd)\,D_{v_4}(abcd),$$

wherein $v_1v_2v_3v_4$ is a permutation of the letters a, b, c, d, and the summation is for every such permutation.

(Cf. *Q. M. J.* No. 85, 1886, "The Law of Symmetry and other Theorems in Symmetric Functions," § 4.)

Art. 6. It is this valuable property of the operation D, when performed upon a product, that is the essential feature of this investigation. The right-hand side consists of 24 (= 4 !) terms, each of which is

$$(bcd)(acd)(abd)(abc),$$

and we have the identity

$$D_w(abcd)^4 = 4\,!\,(bcd)(acd)(abd)(abc),$$

and clearly also

$$D_w(a_1a_2a_3 \dots a_{n-1}a_n)^n = n\,!\,(a_2a_3 \dots a_n)(a_1a_3 \dots a_n) \dots (a_1a_2 \dots a_{n-1}).$$

As remarked above we may consider, in general, Latin Rectangles. If we consider a Rectangle of n columns and 1 row we obtain a trivial result, which however it is proper, for the orderly development of the subject, to notice. We are subject here to no condition. The whole number of rectangles is $n!$; of reduced rectangles

$$\frac{n!}{n!}=1.$$

This we may take to be indicated by the above operation of D_w upon

$$(a_1a_2 \ldots a_n)^n.$$

To give it analytical expression, in the identity

$$D_w(a_1a_2 \ldots a_n)^n = n!(a_2a_3 \ldots a_n) \ldots (a_1a_2 \ldots a_{n-1})$$

put
$$\alpha_1=\alpha_2=\ldots=\alpha_n=1,$$
so that
$$[D_w(a_1a_2 \ldots a_n)^n]_{\alpha_1=\alpha_2=\ldots=\alpha_n=1}$$
$$=n!\left(\frac{n!}{1!}\right)^n n_1,$$

where n_1 is the number of reduced Rectangles of n columns and 1 row.

I call to mind that the case is trivial, and $n_1=1$.

Thence
$$n_1=\frac{[D_w(a_1a_2 \ldots a_n)^n]_{\alpha_1=\alpha_2=\ldots=\alpha_n=1}}{n!\left(\frac{n!}{1!}\right)^n}.$$

Art. 7. I pass on to consider $D_w^2(abcd)^4$

$$=4!D_w(bcd)(acd)(abd)(abc)$$
$$=4!\Sigma D_{v_1}(bcd)D_{v_2}(acd)D_{v_3}(abd)D_{v_4}(abc),$$

where $v_1v_2v_3v_4$ is any permutation of the letters a, b, c, d, and the summation is for all such permutations.

Now observe that certain terms out of the 24 vanish. $D_{v_1}(bcd)$ vanishes when $v_1=a$, $D_{v_2}(acd)$ when $v_2=b$, $D_{v_3}(abd)$ when $v_3=c$, $D_{v_4}(abc)$ when $v_4=d$. We are *only* concerned with those permutations

$$v_1v_2v_3v_4$$

of a, b, c, d which are competent to form a Latin Rectangle of two rows, the top row being a, b, c, d.

We have in fact the 'Problème des rencontres.'

$v_1v_2v_3v_4$ can only assume the 9 permutations

$$\begin{array}{cccc} b & c & d & a \\ b & d & a & c \\ b & a & d & c \\ c & d & a & b \\ c & d & b & a \\ c & a & d & b \\ d & a & b & c \\ d & c & a & b \\ d & c & b & a, \end{array}$$

36—2

9 being the solution of the corresponding 'Problème des rencontres,' and representing the number of Latin Rectangles in which the top row is *abcd* but in which the first letter in the second row is not necessarily *b*.

To obtain the number of reduced rectangles we must divide the number 9 by 3 $(= n - 1)$, and so reach the number 3.

On the right-hand side of the identity

$$D_w^2 (abcd)^4$$
$$= 4 ! \Sigma D_{v_1} (bcd) D_{v_2} (acd) D_{v_3} (abd) D_{v_4} (abc)$$

we obtain, after operation, 9 terms; 3 for each reduced Latin Rectangle of two rows.

Let n_2 denote the number of reduced Latin Rectangles of n columns and two rows. On the right-hand side of the identity

$$D_w^2 (a_1 a_2 \ldots a_n)^n$$
$$= n ! D_{v_1} (a_2 a_3 \ldots a_n) D_{v_2} (a_1 a_3 \ldots a_n) \ldots D_{v_n} (a_1 a_2 \ldots a_{n-1})$$

we obtain after operation

$$(n - 1) n_2 \text{ terms,}$$

corresponding to the Latin Rectangles, reduced as regards columns but unreduced as regards rows,—or

$$n ! (n - 1) n_2 \text{ terms,}$$

corresponding to the totality of unreduced Latin Rectangles.

We have

$$n_2 = \frac{\text{sum of coefficients of terms in the development of } D_w^2 (a_1 a_2 \ldots a_n)^n}{n ! (n - 1)},$$

the development being that reached by the performance of D_w^2 in the manner indicated, and the numerator of the fraction necessarily representing the number of unreduced Latin Rectangles of n columns and two rows.

As before we can give n_2 an analytical expression.

Each of the $n ! (n - 1) n_2$ terms in $D_w^2 (a_1 a_2 \ldots a_n)^n$ assumes the value

$$\left(\frac{n !}{2 !}\right)^n$$

for $\alpha_1 = \alpha_2 = \ldots = \alpha_n = 1$.

Hence
$$n_2 = \frac{[D_w^2 (a_1 a_2 \ldots a_n)^n]_{\alpha_1 = \alpha_2 = \ldots = \alpha_n = 1}}{n ! (n - 1) \left(\frac{n !}{2 !}\right)^n},$$

which is a new solution of the much considered 'Problème des rencontres*.'

Art. 8. Observe that in the above process we have a series of operations corresponding to every unreduced Latin Rectangle. I pass now to the Latin Rectangle of three rows, a problem hitherto regarded as unassailable.

* Another solution has been given by the author, *Phil. Trans. R. S.* vol. 185 A, 1894.

Suppose

$$v_1\ v_2\ \ldots\ v_n,$$
$$v_1'\ v_2'\ \ldots\ v_n',$$
$$v_1''v_2''\ldots\ v_n'',$$

to be such a rectangle.

In operating with D_w upon $(a_1a_2\ldots a_n)^n$, we get one term corresponding to

$$D_{v_1} * D_{v_2} * \ldots\ D_{v_n} *.$$

Operating upon *this term* with D_w we get one term corresponding to the double operation

$$D_{v_1} * D_{v_2} * \ldots D_{v_n} *,$$
$$D_{v_1'} * D_{v_2'} * \ldots D_{v_n'} *,$$

and again operating upon *this last term* with D_w we get one term corresponding to the treble operation

$$D_{v_1} * D_{v_2} * \ldots D_{v_n} *,$$
$$D_{v_1'} * D_{v_2'} * \ldots D_{v_n'} *,$$
$$D_{v_1''} * D_{v_2''} * \ldots D_{v_n''} *;$$

that is to say, in performing D_w^3, we have a one-to-one correspondence between the terms involved and the unreduced Latin Rectangles of n columns and 3 rows. To obtain the reduced Latin Rectangles we have merely to divide the number of terms by

$$n\,!\,(n-1)(n-2).$$

Hence calling n_3 the number of reduced Rectangles, we have, by previous reasoning,

$$n_3 = \frac{[D_w^3\,(a_1a_2\ldots a_n)^n]_{a_1=a_2=\ldots=a_n=1}}{n\,!\,(n-1)(n-2)\left(\frac{n\,!}{3\,!}\right)^n}.$$

Art. 9. It is now easy to pass to the general case and to demonstrate the result

$$n_s = \frac{[D_w^s\,(a_1a_2\ldots a_n)^n]_{a_1=a_2=\ldots=a_n=1}}{n\,!\,\frac{(n-1)\,!}{(n-s)\,!}\left(\frac{n\,!}{s\,!}\right)^n}.$$

In the case of the Latin Square

$$s = n,$$

and we should be able to shew, after Cayley's remark, that

$$n_{n-1} = n_n.$$

Now

$$n_{n-1} = \frac{[D_w^{n-1}\,(a_1a_2\ldots a_n)^n]_{a_1=a_2=\ldots=a_n=1}}{n\,!\,(n-1)\,!\,n^n},$$

$$n_n = \frac{[D_w^n\,(a_1a_2\ldots a_n)^n]_{a_1=a_2=\ldots=a_n=1}}{n\,!\,(n-1)\,!},$$

and it will be shewn that these expressions have the same value.

After performance of D_w^{n-1} we obtain a number of terms of the nature

$$(c_1)(c_2)\dots(c_n),$$

and putting herein $\alpha_1=\alpha_2=\dots=\alpha_n=1$, we obtain a factor n^n.

On proceeding to perform D_w again, before putting $\alpha_1=\alpha_2=\dots=\alpha_n=1$, we obtain simply unity and the quantities $\alpha_1, \alpha_2, \dots \alpha_n$ have disappeared.

Hence

$$[D_w^{n-1}(a_1a_2\dots a_n)^n]_{\alpha_1=\alpha_2=\dots=\alpha_n=1}$$

$$=n^n[D_w^n\,(a_1a_2\dots a_n)^n]_{\alpha_1=\alpha_2=\dots=\alpha_n=1},$$

and thence

$$n_{n-1}=n_n.$$

The expression for n_n may be simplified because

$$D_w^n\,(a_1a_2\dots a_n)^n$$

is an integer and does not involve the roots $\alpha_1, \alpha_2, \dots \alpha_n$.

$$\therefore\ n_n=\frac{D_w^n\,(a_1a_2\dots a_n)^n}{n!\,(n-1)!}.$$

The numerator of the fraction is the coefficient of the symmetric function

$$(w^n)$$

in the development of the power

$$(a_1a_2\dots a_n)^n,$$

w being equal to Σa.

Art. 10. In the symmetric functions the letters $a_1, a_2, \dots a_n$ are taken to represent different integers and so far their values have been unspecified. They must be appropriately chosen or the analysis will fail.

It is essential that the number

$$a_1+a_2+\dots+a_n$$

shall possess but a single partition, into n or fewer parts, drawn from the numbers $a_1, a_2, a_3, \dots a_n$, repetitions of parts permissible.

Thus of order 4 the simplest system is

$$(a_1, a_2, a_3, a_4)=(8, 4, 2, 1),$$

the number 15 possesses the single partition 8421, of four or fewer parts, drawn from the integers 8, 4, 2, 1, repetitions permitted.

The system 7421 would not do because 14 possesses the partition 7, 7.

In general, of order n, the simplest system is

$$(a_1, a_2, \dots a_{n-1}, a_n)=(2^{n-1}, 2^{n-2}, \dots 2, 1).$$

To perform the operations indicated we have to express the function

$$(a_1a_2\dots a_n)^n$$

in terms of the coefficients

$$p_1, p_2, \dots p_n,$$

and operate with

$$D_w = \frac{1}{w!}(\partial_{p_1} + p_1\partial_{p_2} + \dots + p_{n-1}\partial_{p_n})^n,$$

as many times successively as may be necessary. We then write $\binom{n}{s}$ for p_s and substitute in the formulæ.

The calculations will, no doubt, be laborious but that is here not to the point, as an enumeration problem may be considered to be solved when definite algebraical processes are set forth which lead to the solution.

In the case of the Latin Square we may write the result

$$(2^{n-1}2^{n-2}\dots 21)^n = \dots + n!\,(n-1)!\,n_n\,\overline{(2^n-1)}^{\,n} + \dots,$$

or in the form

$$(\Sigma\alpha_1^{2^{n-1}}\alpha_2^{2^{n-2}}\dots\alpha^2_{n-1}\alpha_n)^n = \dots + n!\,(n-1)!\,n_n\Sigma\alpha_1^{2^n-1}\alpha_2^{2^n-1}\dots\alpha_n^{2^n-1} + \dots,$$

$n!\,(n-1)!\,n_n$ and n_n enumerating respectively the unreduced and reduced Latin Squares of order n.

It will be noticed that $(2^{n-1}2^{n-2}\dots 21)$ is what I have elsewhere* called a perfect partition of the number $2^n - 1$; that is, from its parts can be composed, in one way only, the number $2^n - 1$ and every lower number.

SECTION 2.

Art. 11. I proceed to discuss the 'Problème des ménages' by the same method.

Lucas in his *Théorie des Nombres* thus enunciates the question :—

"Des femmes, en nombre n, sont rangées autour d'une table, dans un ordre determiné ; on demande quel est le nombre des manières de placer leurs maris respectifs, de telle sorte qu'un homme soit placé entre deux femmes, sans se trouver à côté de la sienne ?"

He then remarks that it is necessary to determine the number of 'permutations discordantes' with the two permutations

$$\begin{matrix} 1 & 2 & 3 & 4 \dots & n-1, & n \\ 2 & 3 & 4 & 5 \dots & n, & 1. \end{matrix}$$

He remarks as follows :—

"Nous ne connaissons aucune solution simple de cette question, dont l'énoncé donne lieu à l'étude du nombre des permutations discordantes de deux permutations déjà discordantes et plus généralement, du nombre des permutations discordantes de deux permutations quelconques."

* "The theory of perfect partitions of numbers and the compositions of multipartite numbers." *Messenger of Mathematics*, Vol. 20, 1891, pp. 103—119.

He gives solutions due to M. Laisant and M. C. Moreau of which the most convenient is represented by the difference equation

$$(n-1)\lambda_{n+1} = (n^2-1)\lambda_n + (n+1)\lambda_{n-1} + 4(-1)^n$$

with the initial values

$$(\lambda_3, \lambda_4) = (1, 2).$$

The reader, who has mastered the preceding solution of the problem of the Latin Rectangle, will have no difficulty in applying the same method here.

Construct the symmetric function

$$(a_3a_4 \dots a_n)(a_1a_4 \dots a_n)(a_1a_2a_5 \dots a_n) \dots (a_2a_3 \dots a_{n-1}),$$

where the sth factor from the left is deprived of the symbols a_s, a_{s+1} (a suffix when $>n$ being taken to the modulus n) by the operation of $D_{a_s}D_{a_{s+1}}$.

We now operate with D_w and obtain

$$\Sigma D_{v_1}(a_3a_4 \dots a_n) D_{v_2}(a_1a_4 \dots a_n) \dots D_{v_n}(a_2a_3 \dots a_{n-1}),$$

the summation being for every permutation

$$v_1v_2 \dots v_n$$

of the letters $a_1a_2 \dots a_n$.

The number of products that survive is precisely the number of *ménages* denoted by Lucas by the symbol λ_n. Each factor of each product contains $n-3$ symbols a in brackets and for

$$\alpha_1 = \alpha_2 = \dots = \alpha_n = 1$$

has the value

$$\frac{n!}{3!}.$$

Hence $$[D_w(a_3a_4 \dots a_n)(a_1a_4 \dots a_n) \dots (a_2a_3 \dots a_{n-1})]_{\alpha_1=\alpha_2=\dots=\alpha_n=1} = \lambda_n \left(\frac{n!}{3!}\right)^n,$$

or $$\lambda_n = \left(\frac{n!}{3!}\right)^{-n} [D_w(a_3a_4 \dots a_n)(a_1a_4 \dots a_n) \dots (a_2a_3 \dots a_{n-1})]_{\alpha_1=\alpha_2=\dots=\alpha_n=1}.$$

As before we may take in the calculation

$$(a_1, a_2, \dots a_n) = (2^{n-1}, 2^{n-2}, \dots 1).$$

Art. 12. Similarly we can find the number of permutations discordant with each of any two permutations whatever that are mutually discordant.

If these two permutations be

$$v_1 v_2 \dots v_n,$$
$$v_1'v_2' \dots v_n',$$

we have merely to form a product in which the sth factor from the left is deprived of the symbols v_s, v_s' and proceed as before. We thus arrive at a similar but, of course, not an identical result.

Art. 13. It is equally easy to find the number of permutations discordant with each of two permutations which are not mutually discordant.

Let these permutations be

$$v_1 v_2 \ldots v_s v_{s+1} \ldots v_n$$

$$v_1 v_2 \ldots v_s v'_{s+1} \ldots v_n'.$$

We take a product in which the left-hand factor is without v_1, the next without v_2, &c....the sth without v_s, but beyond this the tth factor is without the two symbols v_t, v_t'.

Denoting this product by P we, by the usual method, reach the solution

$$\mu_n = \left(\frac{n\,!}{2\,!}\right)^{-s} \left(\frac{n\,!}{3\,!}\right)^{-(n-s)} [D_w P]_{\alpha_1=\alpha_2=\ldots=\alpha_n=1}.$$

Art. 14. On a similar principle we can enumerate the number of permutations discordant with *any number* of given permutations whatever.

Let

$$\begin{array}{l} v_1 \;\; v_2 \;\; v_3 \;\ldots v_n, \\ v_1' \;\; v_2' \;\; v_3' \ldots v_n', \\ v_1'' \;\; v_2'' \;\; v_3'' \ldots v_n'', \\ \ldots\ldots\ldots\ldots\ldots\ldots \end{array}$$

be the m permutations and of the m letters v_s, v_s', v_s'', ...; let the different ones be u_s, u_s', u_s'', ... k_s in number.

Take for the sth factor $(\alpha_1\alpha_2 \ldots \alpha_n)$,

deprived of the letters u_s, u_s', u_s'', ... ,

and form the resulting product P.

Proceeding as before we obtain the result

$$\nu_n = \left\{\frac{n\,!}{(k_1+1)\,!}\right\}^{-j_1} \left\{\frac{n\,!}{(k_2+1)\,!}\right\}^{-j_2} \ldots \left\{\frac{n\,!}{(k_s+1)\,!}\right\}^{-j_s} \ldots \times [D_w P]_{\alpha_1=\alpha_2=\ldots=\alpha_n=1},$$

where $j_1, j_2, \ldots j_s, \ldots$ are numbers, at once ascertainable, and $\Sigma j = n$.

Art. 15. A more direct generalisation of the 'Problème des ménages' is obtained by imposing the condition that no husband is to have less than $2m$ persons between himself and his wife.

In the problem above considered $m=1$. If $m=2$ we must enumerate the permutations discordant with

$$\begin{array}{cccccccc} a_1 & a_2 & a_3 & a_4 \ldots a_{n-3} & a_{n-2} & a_{n-1} & a_n \\ a_2 & a_3 & a_4 & a_5 \ldots a_{n-2} & a_{n-1} & a_n & a_1 \\ a_3 & a_4 & a_5 & a_6 \ldots a_{n-1} & a_n & a_1 & a_2 \\ a_4 & a_5 & a_6 & a_7 \ldots a_n & a_1 & a_2 & a_3 \end{array}$$

and we form the product

$$P_4 = (\alpha_5 \ldots \alpha_n)(\alpha_1\alpha_6 \ldots \alpha_n)(\alpha_1\alpha_2\alpha_7 \ldots \alpha_n) \ldots (\alpha_4 \ldots \alpha_{n-1}),$$

the solution being given by

$$\text{number} = \left(\frac{n\,!}{5\,!}\right)^{-n} [D_w P_4]_{a_1=a_2=\ldots=a_n=1},$$

and in general

$$\text{number} = \left(\frac{n\,!}{(2m+1)\,!}\right)^{-n} [D_w P_{2m}]_{a_1=a_2=\ldots=a_n=1}.$$

SECTION 3.

Art. 16. The notion of a Latin Rectangle may be generalised. Instead of n different letters we may have s_1 of one kind, s_2 of a second, s_3 of a third, and so on. The letters may be, ex. gr.,

$$a_1^{s_1} a_2^{s_2} a_3^{s_3} \ldots\ldots a_k^{s_k},$$

where $$\Sigma s = n.$$

To obtain a Latin Rectangle of t rows we take t permutations of the letters such that in no column does a_1 occur more than s_1 times, a_2 more than s_2, a_3 more than s_3, and so on. The reduced rectangles have the top row and left-hand column in the same assigned order and evidently we can obtain their number by dividing the number of unreduced rectangles by

$$\frac{n\,!}{s_1!\,s_2!\,s_3!\,\ldots\,s_k!} \times \frac{(n-1)\,!}{(s_1-1)\,!\,s_2!\,s_3!\,\ldots\,s_k!}$$

in the case when the rectangle is a square, and by factors of similar forms in the other particular cases.

Examples of such quasi-Latin Squares are

$$\begin{matrix} a & a & b & c \\ a & b & c & a \\ b & c & a & a \\ c & a & a & b \end{matrix} \qquad \begin{matrix} a & a & b & b \\ a & b & a & b \\ b & b & a & a \\ b & a & b & a \end{matrix} \qquad \begin{matrix} a & a & a & b \\ a & a & b & a \\ a & b & a & a \\ b & a & a & a. \end{matrix}$$

Art. 17. We have Latin Squares and Rectangles associated with every partition of every number. The three, given above, correspond to the partitions 21^2, 2^2, 31; we have already, in the first part of the paper, considered the case *abcd* corresponding to 1^4 and there remains the case *aaaa*, of partition 4, which is trivial. Part of this theory is intimately connected with certain chessboard problems that might be proposed.

Take the unreduced Latin Squares on the letters

$$a\ a\ a\ a\ a\ a\ a\ b.$$

The enumeration gives the number of ways of placing 8 rooks on the board so that no one can take any of the others.

Similarly the enumeration connected with

$$a\ a\ a\ a\ a\ a\ b\ c$$

gives the number of ways of placing 16 rooks, 8 white and 8 black, on the board so that no rook can be taken by another of the same colour.

Like problems can be connected with other cases.

Art. 18. Let us consider the general question of the enumeration.

First take the simple case

$$a_1^{n-1} a_2,$$

where

$$(n-1) a_1 + a_2 = w.$$

Assuming a_1 and a_2 to be undetermined integers and remembering the law of operation of D_w, we have

$$D_w (a_1^{n-1} a_2)^n = n (a_1^{n-2} a_2)^{n-1} (a_1^{n-1}).$$

The coefficient n indicates that for unreduced rectangles there are n possible first rows, viz.:—the n permutations of

$$a_1^{n-1} a_2,$$

and

$$D_w^2 (a_1^{n-1} a_2)^n = n (n-1) (a_1^{n-3} a_2)^{n-2} (a_1^{n-2})^2,$$

the coefficient $n(n-1)$ shewing that there are $n(n-1)$ possible pairs of two first rows in unreduced rectangles.

Also

$$D_w^s (a_1^{n-1} a_2)^n = \frac{n!}{(n-s)!} (a_1^{n-s-1} a_2)^{n-s} (a_1^{n-s})^s,$$

giving $\frac{n!}{(n-s)!}$ unreduced rectangles of s rows.

Hence

$$D_w^{n-1} (a_1^{n-1} a_2)^n = n! (a_2)(a_1)^{n-1}$$

$$D_w^n (a_1^{n-1} a_2)^n = n!$$

intimating (as is otherwise immediately evident) that the number of unreduced rectangles of $n-1$ rows or of squares is $n!$.

To enumerate the reduced rectangles observe that in Row 1 we have one place for a_2 instead of n places; in Row 2, $n-2$ places instead of $n-1$; in Row 3, $n-3$ instead of $n-2$, &c.

Therefore for the rectangle of s rows we have a divisor

$$\frac{n}{1} \cdot \frac{n-1}{n-2} \cdot \frac{n-2}{n-3} \cdots\cdots \frac{n-s+1}{n-s};$$

which is

$$\frac{n!}{(n-s)!} \cdot \frac{(n-s-1)!}{(n-2)!}.$$

Therefore the reduced rectangles, of s rows, are in number

$$\frac{(n-2)!}{(n-s-1)!} \qquad (s < n).$$

37—2

If $s = n$, the case of the square, the last fraction factor of the divisor must be omitted and we find the number

$$(n-2)\,!.$$

Suppose the symmetric functions to appertain to the quantities

$$\alpha_1,\ \alpha_2,\ \alpha_3,\ \ldots\ldots\ \alpha_n,$$

and in the identity

$$D_w^s (a_1^{n-1} a_2)^n = \frac{n\,!}{(n-s)\,!} (a_1^{n-s-1} a_2)^{n-s} (a_1^{n-s})^s,$$

suppose

$$\alpha_1 = \alpha_2 = \ldots\ldots = \alpha_n = 1.$$

Then

$$\frac{n!}{(n-s)!} = \left\{\frac{n!}{(n-s-1)!\,s!}\right\}^{-(n-s)} \left\{\frac{n\,!}{(n-s)!\,s!}\right\}^{-s} \times [D_w^s (a_1^{n-2} a_2)^n]_{\alpha_1=\alpha_2=\ldots=\alpha_n=1},$$

for

$$s \not> n-1.$$

When $s = n-1$ we have also the case of the Square as before remarked.

The right-hand side is therefore an analytical expression for the number of unreduced rectangles and we have merely to divide by $\frac{n\,!}{(n-s)!} \cdot \frac{(n-s-1)!}{(n-2)!}$ to obtain that of the reduced rectangles.

Art. 19. This simple case has been worked out to shew that the desired number can be obtained as the result of definite algebraical processes performed upon a certain symmetric function. In the actual working it is essential to select a_1 and a_2 in such wise that

$$a_1^{n-1} a_2$$

is the *only* partition of

$$w = (n-1)\,a_1 + a_2$$

into n or fewer parts drawn from the symbols a_1 and a_2 each any number of times repeated.

It will be found that the simplest system is

$$a_1 = 1,\ a_2 = n,$$

necessitating the consideration of the symmetric function

$$(n1^{n-1})^n.$$

Art. 20. If we next proceed to enumerate the Latin Rectangles of

$$a_1^{n-2} a_2^2$$

we find that the Square enumeration is most easily expressed, those connected with Rectangles having complicated expressions.

The reason for this will be obvious from the results

$$D_w(a_1^{n-2}a_2^2)^n = \binom{n}{2}(a_1^{n-3}a_2^2)^{n-2}(a_1^{n-2}a_2)^2,$$

$$D_w^2(a_1^{n-2}a_2^2)^n = \binom{n}{2}\Big\{(a_1^{n-4}a_2^2)^{n-2}(a_1^{n-2}a_2^2)$$
$$+ 2(n-2)(a_1^{n-4}a_2^2)^{n-3}(a_1^{n-3}a_2)^2(a_1^{n-2})$$
$$+ \binom{n-2}{2}(a_1^{n-4}a_2^2)^{n-4}(a_1^{n-3}a_2)^4\Big\},$$

for it will be seen that the right-hand side of the identity just obtained, as containing terms of three different types, is not simply evaluated for unit values of the quantities

$$\alpha_1, \alpha_2, \alpha_3, \ldots \alpha_n,$$

and the terms will not have a single type until we reach D_w^{n-1} which is the case of the square.

Then every term is some permutation of

$$(a_1)^{n-2}(a_2)^2.$$

If K be the *whole* number of these terms, K is the number of unreduced squares, and then putting

$$\alpha_1 = \alpha_2 = \ldots = \alpha_n = 1,$$

we obtain
$$K = n^{-n}[D_w^{n-1}(a_1^{n-2}a_2^2)^n]_{\alpha_1=\alpha_2=\ldots=\alpha_n=1}$$
$$= D_w^n(a_1^{n-2}a_2^2)^n,$$

and dividing K by
$$\frac{n!}{(n-2)!\,2!} \times \frac{(n-1)!}{(n-3)!\,2!}$$

we obtain the number of reduced squares.

Art. 21. We get a precisely similar result for the enumeration of the squares derived from

$$a_1^{s_1}a_2^{s_2}\ldots a_k^{s_k} \qquad \begin{pmatrix}\Sigma s = k\\ \Sigma sa = w\end{pmatrix},$$

viz. for the unreduced squares

$$K = n^{-n}[D_w^{n-1}(a_1^{s_1}a_2^{s_2}\ldots a_k^{s_k})^n],$$

and then division by

$$\frac{n!}{s_1!\,s_2!\ldots s_k!} \times \frac{(n-1)!}{(s_1-1)!\,s_2!\ldots s_k!}$$

gives the number of reduced squares.

The choice of $a_1, a_2, \ldots a_k$ is determined by the circumstance that, for the validity of the process, $w = \Sigma sa$ must possess no partition of n into n or fewer parts drawn from the set

$$a_1, a_2, \ldots a_k,$$

each repeatable as many as n times ($n = \Sigma s$), except

$$(a_1^{s_1}a_2^{s_2}\ldots a_k^{s_k}).$$

This condition is satisfied if

$$a_2 > s_1 a_1,$$
$$a_3 > s_1 a_1 + s_2 a_2,$$
$$\dots\dots\dots\dots\dots\dots$$
$$a_k > s_1 a_1 + s_2 a_2 + \dots + s_{k-1} a_{k-1}.$$

Putting therefore $a_1 = 1$, we can take

$$a_1 = 1,$$
$$a_2 = s_1 + 1,$$
$$a_3 = (s_1+1)(s_2+1),$$
$$a_4 = (s_1+1)(s_2+1)(s_3+1),$$
$$\dots\dots\dots\dots\dots\dots\dots\dots\dots\dots$$
$$a_k = (s_1+1)(s_2+1)\dots(s_{k-1}+1),$$

and the symmetric function to be considered is

$$\{1^{s_1}\,\overline{s_1+1}^{\,s_2}\,\overline{(s_1+1)(s_2+1)}^{\,s_3}\;\dots\dots\;\overline{(s_1+1)(s_2+1)\dots(s_{k-1}+1)}^{\,s_k}\}.$$

Observe that the partition, which here presents itself, is of necessity a perfect partition.

Ex. gr. To determine the Latin Squares on the base *aabb*, we take the function

$$(a^2b^2)^4,$$

or

$$(3^2 1^2)^4,$$

$$(3^2 1^2) = p_2^2 p_4 - 2p_1 p_3 p_4 + 2p_4^2 - p_1 p_2 p_5 + 3p_3 p_5 + 5p_1^2 p_6 - 9p_2 p_6 - 5p_1 p_7 + 12 p_8,$$

$$D_8 = \frac{1}{8!}\left(\frac{d}{dp_1} + p_1\frac{d}{dp_2} + p_2\frac{d}{dp_3} + p_3\frac{d}{dp_4} + p_4\frac{d}{dp_5} + p_5\frac{d}{dp_6} + p_6\frac{d}{dp_7} + p_7\frac{d}{dp_8}\right)^8,$$

and I find

$$D_8\,(3^2 1^2)^4 = 6\,(3^2 1)^2 (31^2)^2,$$
$$D_8^2\,(3^2 1^2)^4 = 6\,\{(3^2)^2 (1^2)^2 + 4\,(3^2)(31)^2(1^2) + (31)^4\},$$
$$D_8^3\,(3^2 1^2)^4 = 90\,(3)^2 (1)^2,$$
$$D_8^4\,(3^2 1^2)^4 = 90.$$

Hence the number of unreduced Latin Squares is 90, and to obtain the reduced forms we divide by $\binom{4}{2}\binom{3}{1} = 18$ and obtain 5 for the number of reduced squares.

These are

$$\begin{array}{cccc} a & a & b & b \\ a & a & b & b \\ b & b & a & a \\ b & b & a & a \end{array} \qquad \begin{array}{cccc} a & a & b & b \\ a & b & a & b \\ b & a & b & a \\ b & b & a & a \end{array} \qquad \begin{array}{cccc} a & a & b & b \\ a & b & b & a \\ b & b & a & a \\ b & a & a & b \end{array}$$

$$\begin{array}{cccc} a & a & b & b \\ a & b & a & b \\ b & b & a & a \\ b & a & b & a \end{array} \qquad \begin{array}{cccc} a & a & b & b \\ a & b & b & a \\ b & a & a & b \\ b & b & a & a \end{array}$$

SECTION 4.

Art. 22. Let us take into consideration the Græco-Latin Square of Euler (see Introduction).

Instead of Greek letters I find it more convenient to use accented Latin letters, so that for instance a Græco-Latin Square is

$$\begin{matrix} a^{a'} & b^{c'} & c^{b'}, \\ b^{b'} & c^{a'} & a^{c'}, \\ c^{c'} & a^{b'} & b^{a'}, \end{matrix} \quad \text{or more conveniently} \quad \begin{matrix} aa' & bc' & cb' \\ bb' & ca' & ac' \\ cc' & ab' & ba'. \end{matrix}$$

I remark that the Latin and accented Latin letters form, separately, Latin Squares, and that two other Latin Squares are obtainable,

(1) by taking the bases to the exponents a', b', c' in succession,

(2) by taking the exponents to the bases a, b, c in succession.

In order to apply to the question the method of this paper it is necessary to construct suitable operators and operands for use in the master operator theorem of § 1.

It is necessary to form symmetric functions of two systems of quantities

$$\alpha_1 \ \alpha_2 \ \ldots \ \alpha_n,$$
$$\alpha_1' \ \alpha_2' \ \ldots \ \alpha_n'.$$

Write
$$(1 + \alpha_1 x + \alpha_1' y)(1 + \alpha_2 x + \alpha_2' y) \ldots (1 + \alpha_n x + \alpha_n' y)$$
$$= 1 + p_{10} x + p_{01} y + \ldots + p_{ww'} x^w y^{w'} + \ldots,$$
$$g_{10} = \Sigma p_{w-1,\, w'} \partial p_{ww'}; \quad g_{01} = \Sigma p_{w,\, w'-1} \partial p_{ww'},$$
$$G_{ww'} = \frac{1}{w!\, w'!} \overline{g_{10}{}^w g_{01}{}^{w'}},$$

where $\overline{g_{10}{}^w g_{01}{}^{w'}}$ denotes that the multiplication of operators is symbolic, or non-operational, as in the symbolic form of Taylor's Theorem.

The reader should refer to the author's paper "Memoir on Symmetric Functions of the Roots of Systems of Equations," *Phil. Trans. R. S.*, 181 A, 1890, § 3, p. 488 et seq.

Denote the symmetric function

$$\Sigma \alpha_1^{a_1} \alpha_1'^{a_1'} \alpha_2^{a_2} \alpha_2'^{a_2'} \alpha_3^{a_3} \alpha_3'^{a_3'} \ldots$$

by
$$(\overline{a_1 a_1'}\ \overline{a_2 a_2'}\ \overline{a_3 a_3'} \ldots),$$

and observe the results given (*loc. cit.*, p. 490),

$$G_{a_1 a_1'} (\overline{a_1 a_1'}\ \overline{a_2 a_2'}\ \overline{a_3 a_3'} \ldots) = (\overline{a_2 a_2'}\ \overline{a_3 a_3'} \ldots),$$
$$G_{a_1 a_1'} (\overline{a_1 a_1'}) = 1,$$
$$G_{a_1 a_1'} G_{a_2 a_2'} \ldots G_{a_n a_n'} (\overline{a_1 a_1'}\ \overline{a_2 a_2'} \ldots \overline{a_n a_n'}) = 1.$$

Also (see § 10, p. 516 et seq.) if

$$f_1, \ f_2, \ f_3, \ \dots f_m$$

denote, each, any symmetric functions and

$$f = f_1 f_2 f_3 \ \dots \ f_m,$$

$$G_{ww'} = \Sigma\Sigma \left(G_{a_1 a_1'} f_1\right)\left(G_{a_2 a_2'} f_2\right) \dots \left(G_{a_s a_s'} f_s\right) f_{s+1} \dots f_m.$$

where the double summation is for every partition

$$\left(\overline{a_1 a_1'} \ \overline{a_2 a_2'} \ \dots \ \overline{a_s a_s'}\right)$$

of the bipartite number $\overline{ww'},$

and for every permutation of the m suffixes of the functions $f_1, \ f_2, \ f_3, \ \dots f_m.$

We may denote the operation indicated by the single summation

$$\Sigma \left(G_{a_1 a_1'} f_1\right)\left(G_{a_2 a_2'} f_2\right) \dots \left(G_{a_s a_s'} f_s\right) f_{s+1} \dots f_m$$

by $G_{\left(\overline{a_1 a_1'} \ \overline{a_2 a_2'} \ \dots \ \overline{a_s a_s'}\right)},$

so that there is the operator equivalence

$$G_{ww'} = \Sigma G_{\left(\overline{a_1 a_1'} \ \overline{a_2 a_2'} \ \dots \ \overline{a_s a_s'}\right)}$$

the summation having regard to every partition of the bipartite $\overline{ww'}.$

Let

$$a_1 + a_2 + \dots + a_n = w$$
$$a_1' + a_2' + \dots + a_n' = w',$$

and suppose the integers

$$a_1, \quad a_2, \ \dots \ a_n$$
$$a_1', \quad a_2', \ \dots \ a_n'$$

so chosen that on the one hand w possesses the single partition $(a_1 a_2 \dots a_n)$ composed of n or fewer parts drawn from the parts $a_1, \ a_2, \ \dots \ a_n$ repetitions permissible, and on the other hand w' possesses the single partition $(a_1' a_2' \dots a_n')$ composed of n or fewer parts drawn from the parts $a_1', \ a_2', \ \dots \ a_n'$ repetitions permissible. Then we have

$$G_{ww'} = \Sigma G_{\left(\overline{a_{s_1} a_{t_1}'} \ \overline{a_{s_2} a_{t_2}'} \ \dots \ \overline{a_{s_n} a_{t_n}'}\right)}$$

where $s_1 s_2 \dots s_n, \quad t_1 t_2 \dots t_n$

are some permutations of $1 . 2 . 3 \dots n$ respectively, and the summation is in regard to every association of a permutation

$$s_1 s_2 \dots s_n$$

with a permutation $t_1 t_2 \dots t_n.$

Art. 23.   First take $n = 3$

$$w = a_1 + a_2 + a_3$$
$$w' = a_1' + a_2' + a_3',$$

and, as operand, the product

$$\left(\overline{a_1 a_1'} \ \overline{a_2 a_2'} \ \overline{a_3 a_3'}\right)\left(\overline{a_1 a_2'} \ \overline{a_2 a_3'} \ \overline{a_3 a_1'}\right)\left(\overline{a_1 a_3'} \ \overline{a_2 a_1'} \ \overline{a_3 a_2'}\right)$$

where a_1, a_2, a_3 are in the same order in each factor, and the dashed letters in successive factors, being written in successive lines, a Latin Square is formed, viz. :—

$$\begin{matrix} a_1' & a_2' & a_3' \\ a_2' & a_3' & a_1' \\ a_3' & a_1' & a_2'. \end{matrix}$$

We find

$$G_{ww'}\left(\overline{a_1a_1'}\ \overline{a_2a_2'}\ \overline{a_3a_3'}\right)\left(\overline{a_1a_2'}\ \overline{a_2a_3'}\ \overline{a_3a_1'}\right)\left(\overline{a_1a_3'}\ \overline{a_2a_1'}\ \overline{a_3a_2'}\right)$$

$$= \left(\overline{a_2a_2'}\ \overline{a_3a_3'}\right)\left(\overline{a_1a_2'}\ \overline{a_3a_1'}\right)\left(\overline{a_1a_3'}\ \overline{a_2a_1'}\right)$$

$$+ \left(\overline{a_1a_1'}\ \overline{a_3a_3'}\right)\left(\overline{a_1a_2'}\ \overline{a_2a_3'}\right)\left(\overline{a_2a_1'}\ \overline{a_3a_2'}\right)$$

$$+ \left(\overline{a_1a_1'}\ \overline{a_2a_2'}\right)\left(\overline{a_2a_3'}\ \overline{a_3a_1'}\right)\left(\overline{a_1a_3'}\ \overline{a_3a_2'}\right)$$

three terms respectively derived from the partition operators

$$G_{(\overline{a_1a_1'}\ \overline{a_2a_3'}\ \overline{a_3a_2'})}$$
$$G_{(\overline{a_2a_2'}\ \overline{a_3a_1'}\ \overline{a_1a_3'})}$$
$$G_{(\overline{a_3a_3'}\ \overline{a_1a_2'}\ \overline{a_2a_1'})}.$$

Operating again with $G_{ww'}$ we obtain

$$\left(\overline{a_3a_3'}\right)\left(\overline{a_1a_2'}\right)\left(\overline{a_2a_1'}\right) + \left(\overline{a_2a_2'}\right)\left(\overline{a_3a_1'}\right)\left(\overline{a_1a_3'}\right)$$

$$+ \left(\overline{a_1a_1'}\right)\left(\overline{a_2a_3'}\right)\left(\overline{a_3a_2'}\right) + \left(\overline{a_3a_3'}\right)\left(\overline{a_1a_2'}\right)\left(\overline{a_2a_1'}\right)$$

$$+ \left(\overline{a_2a_2'}\right)\left(\overline{a_3a_1'}\right)\left(\overline{a_1a_3'}\right) + \left(\overline{a_1a_1'}\right)\left(\overline{a_2a_3'}\right)\left(\overline{a_3a_2'}\right)$$

terms respectively derived from the operators

$$G_{(\overline{a_2a_2'}\ \overline{a_3a_1'}\ \overline{a_1a_3'})},\quad G_{(\overline{a_3a_3'}\ \overline{a_1a_2'}\ \overline{a_2a_1'})}$$
$$G_{(\overline{a_3a_3'}\ \overline{a_1a_2'}\ \overline{a_2a_1'})},\quad G_{(\overline{a_1a_1'}\ \overline{a_2a_3'}\ \overline{a_3a_2'})}$$
$$G_{(\overline{a_1a_1'}\ \overline{a_2a_3'}\ \overline{a_3a_2'})},\quad G_{(\overline{a_2a_2'}\ \overline{a_3a_1'}\ \overline{a_1a_3'})}.$$

Operating again, on each term with the corresponding partition operator, we obtain the number 6.

We have obtained this number from the six combinations of operators

$$\left\{\begin{matrix} G_{(\overline{a_1a_1'}\ \overline{a_2a_3'}\ \overline{a_3a_2'})} \\ G_{(\overline{a_3a_3'}\ \overline{a_1a_2'}\ \overline{a_2a_1'})} \\ G_{(\overline{a_2a_2'}\ \overline{a_3a_1'}\ \overline{a_1a_3'})} \end{matrix}\right\}\quad \left\{\begin{matrix} G_{(\overline{a_2a_2'}\ \overline{a_3a_1'}\ \overline{a_1a_3'})} \\ G_{(\overline{a_1a_1'}\ \overline{a_2a_3'}\ \overline{a_3a_2'})} \\ G_{(\overline{a_3a_3'}\ \overline{a_1a_2'}\ \overline{a_2a_1'})} \end{matrix}\right\}\quad \left\{\begin{matrix} G_{(\overline{a_3a_3'}\ \overline{a_1a_2'}\ \overline{a_2a_1'})} \\ G_{(\overline{a_2a_2'}\ \overline{a_3a_1'}\ \overline{a_1a_3'})} \\ G_{(\overline{a_1a_1'}\ \overline{a_2a_3'}\ \overline{a_3a_1'})} \end{matrix}\right\}$$

$$\left\{\begin{matrix} G_{(\overline{a_1a_1'}\ \overline{a_2a_3'}\ \overline{a_3a_2'})} \\ G_{(\overline{a_2a_2'}\ \overline{a_3a_1'}\ \overline{a_1a_3'})} \\ G_{(\overline{a_3a_3'}\ \overline{a_1a_2'}\ \overline{a_2a_1'})} \end{matrix}\right\}\quad \left\{\begin{matrix} G_{(\overline{a_2a_2'}\ \overline{a_3a_1'}\ \overline{a_1a_3'})} \\ G_{(\overline{a_3a_3'}\ \overline{a_1a_2'}\ \overline{a_2a_1'})} \\ G_{(\overline{a_1a_1'}\ \overline{a_2a_3'}\ \overline{a_3a_2'})} \end{matrix}\right\}\quad \left\{\begin{matrix} G_{(\overline{a_3a_3'}\ \overline{a_1a_2'}\ \overline{a_2a_1'})} \\ G_{(\overline{a_1a_1'}\ \overline{a_2a_3'}\ \overline{a_3a_2'})} \\ G_{\overline{a_2a_2'}\ \overline{a_3a_1'}\ \overline{a_1a_3'})} \end{matrix}\right\}$$

and to these correspond the six Græco-Latin Squares

$$\begin{matrix} a_1a_1' & a_2a_3' & a_3a_2' \\ a_3a_3' & a_1a_2' & a_2a_1' \\ a_2a_2' & a_3a_1' & a_1a_3' \end{matrix} \qquad \begin{matrix} a_2a_2' & a_3a_1' & a_1a_3' \\ a_1a_1' & a_2a_3' & a_3a_2' \\ a_3a_3' & a_1a_2' & a_2a_1' \end{matrix} \qquad \begin{matrix} a_3a_3' & a_1a_2' & a_2a_1' \\ a_2a_2' & a_3a_1' & a_1a_3' \\ a_1a_1' & a_2a_3' & a_3a_2' \end{matrix}$$

$$\begin{matrix} a_1a_1' & a_2a_3' & a_3a_2' \\ a_2a_2' & a_3a_1' & a_1a_3' \\ a_3a_3' & a_1a_2' & a_2a_1' \end{matrix} \qquad \begin{matrix} a_2a_2' & a_3a_1' & a_1a_3' \\ a_3a_3' & a_1a_2' & a_2a_1' \\ a_1a_1' & a_2a_3' & a_3a_2' \end{matrix} \qquad \begin{matrix} a_3a_3' & a_1a_2' & a_2a_1' \\ a_1a_1' & a_2a_3' & a_3a_2' \\ a_2a_2' & a_3a_1' & a_1a_3' \end{matrix}$$

By forming the operand, as above, we have insisted upon the left-hand column of the square involving only the three products

$$a_1a_1', \qquad a_2a_2', \qquad a_3a_3',$$

but by permuting a_1', a_2', a_3' we get an additional factor 6 and by permuting the 2nd and 3rd columns a further factor 2! we find that the unreduced number of Græco-Latin Squares of order 3 is

$$6 \times 6 \times 2 = 72.$$

If we insist upon the suffixes appearing in numerical order in the left-hand column for both undashed and dashed letters and also in the top row in the case of undashed letters, we obtain the reduced squares. In this instance there is but one, viz.

$$\begin{matrix} a_1a_1' & a_2a_3' & a_3a_2' \\ a_2a_2' & a_3a_1' & a_1a_3' \\ a_3a_3' & a_1a_2' & a_2a_1'. \end{matrix}$$

In general the enumeration of the reduced squares is obtained by dividing the number of unreduced squares by

$$(n\,!)^2\,(n-1)\,!\,.$$

Above an operand was formed corresponding to the single reduced Latin Square,

$$\begin{matrix} a_1' & a_2' & a_3' \\ a_2' & a_3' & a_1' \\ a_3' & a_1' & a_2' \end{matrix}$$

of order 3.

Operating with $G^3_{ww'}$, viz.:—three times successively with $G_{ww'}$, we obtained the number 6 and this has been shewn to give 3! times the number of reduced Græco-Latin Squares for, as remarked above, 3! is the number of ways in which the products a_1a_1', a_2a_2', a_3a_3' can be permuted.

Hence the number of reduced Græco-Latin Squares is

$$\frac{6}{3\,!} = 1.$$

In general we shall find that we must form an operand corresponding with each reduced Latin Square in the dashed letters, operate upon each, with $G_{ww'}$ n times successively, take the sum of the resulting numbers, and divide by $n!$.

The result will be the number of reduced Græco-Latin Squares of order n.

Art. 24. To elucidate the matter I will work out (not quite in full) the case of order 4 and deduce the reduced Græco-Latin Squares. There are four operands, since there are four reduced Latin Squares of order 4, viz. :—

$$\begin{array}{cccc|cccc|cccc|cccc}
a_1' & a_2' & a_3' & a_4' & a_1' & a_2' & a_3' & a_4' & a_1' & a_2' & a_3' & a_4' & a_1' & a_2' & a_3' & a_4' \\
a_2' & a_3' & a_4' & a_1' & a_2' & a_1' & a_4' & a_3' & a_2' & a_1' & a_4' & a_3' & a_2' & a_4' & a_1' & a_3' \\
a_3' & a_4' & a_1' & a_2' & a_3' & a_4' & a_1' & a_2' & a_3' & a_4' & a_2' & a_1' & a_3' & a_1' & a_4' & a_2' \\
a_4' & a_1' & a_2' & a_3' & a_4' & a_3' & a_2' & a_1' & a_4' & a_3' & a_1' & a_2' & a_4' & a_3' & a_2' & a_1'
\end{array}$$

These are

$$(A) \quad (\overline{a_1a_1'}\ \overline{a_2a_2'}\ \overline{a_3a_3'}\ \overline{a_4a_4'})\ (\overline{a_1a_2'}\ \overline{a_2a_3'}\ \overline{a_3a_4'}\ \overline{a_4a_1'})\ (\overline{a_1a_3'}\ \overline{a_2a_4'}\ \overline{a_3a_1'}\ \overline{a_4a_2'})\ (\overline{a_1a_4'}\ \overline{a_2a_1'}\ \overline{a_3a_2'}\ \overline{a_4a_3'}),$$

$$(B) \quad (\overline{a_1a_1'}\ \overline{a_2a_2'}\ \overline{a_3a_3'}\ \overline{a_4a_4'})\ (\overline{a_1a_2'}\ \overline{a_2a_1'}\ \overline{a_3a_4'}\ \overline{a_4a_3'})\ (\overline{a_1a_3'}\ \overline{a_2a_4'}\ \overline{a_3a_1'}\ \overline{a_4a_2'})\ (\overline{a_1a_4'}\ \overline{a_2a_3'}\ \overline{a_3a_2'}\ \overline{a_4a_1'}),$$

$$(C) \quad (\overline{a_1a_1'}\ \overline{a_2a_2'}\ \overline{a_3a_3'}\ \overline{a_4a_4'})\ (\overline{a_1a_2'}\ \overline{a_2a_1'}\ \overline{a_3a_4'}\ \overline{a_4a_3'})\ (\overline{a_1a_3'}\ \overline{a_2a_4'}\ \overline{a_3a_2'}\ \overline{a_4a_1'})\ (\overline{a_1a_4'}\ \overline{a_2a_3'}\ \overline{a_3a_1'}\ \overline{a_4a_2'}),$$

$$(D) \quad (\overline{a_1a_1'}\ \overline{a_2a_2'}\ \overline{a_3a_3'}\ \overline{a_4a_4'})\ (\overline{a_1a_2'}\ \overline{a_2a_4'}\ \overline{a_3a_1'}\ \overline{a_4a_3'})\ (\overline{a_1a_3'}\ \overline{a_2a_1'}\ \overline{a_3a_4'}\ \overline{a_4a_2'})\ (\overline{a_1a_4'}\ \overline{a_2a_3'}\ \overline{a_3a_2'}\ \overline{a_4a_1'}).$$

The operation of $D^4_{ww'}$ upon (A), (C) and (D) causes them to vanish and on (B) the result is 48; hence the number of reduced Græco-Latin Squares is

$$\frac{48}{4!} = 2.$$

These are

$$\begin{array}{cccc|cccc}
a_1a_1' & a_2a_4' & a_3a_2' & a_4a_3' & a_1a_1' & a_2a_3' & a_3a_4' & a_4a_2' \\
a_2a_2' & a_1a_3' & a_4a_1' & a_3a_4' & a_2a_2' & a_1a_4' & a_4a_3' & a_3a_1' \\
a_3a_3' & a_4a_2' & a_1a_4' & a_2a_1' & a_3a_3' & a_4a_1' & a_1a_2' & a_2a_4' \\
a_4a_4' & a_3a_1' & a_2a_3' & a_1a_2' & a_4a_4' & a_3a_2' & a_2a_1' & a_1a_3'
\end{array}$$

Observe that the undashed Latin Square is the same in both cases and that the second square is obtainable from the first by a cyclical interchange of the 2nd, 3rd and 4th columns of dashed letters. By the method we can, by regular process, determine the number of Græco-Latin Squares appertaining to any given Latin Square. The enumeration, however, in all but the simplest cases is so laborious as to be impracticable. And I do not see the way to prove that no Græco-Latin of the order 6, and generally of order $\equiv 2 \bmod 4$, exists.

38—2

SECTION 5.

Art. 25. It naturally occurs to one to seek other systems of operators and operands which lead to the solution of interesting problems. In the master theorem

$$\frac{(u_1^{\chi_1}u_2^{\chi_2}\ldots u_s^{\chi_s})\,\phi}{\chi_1!\,\chi_2!\ldots\chi_s!}=\Sigma\Sigma\frac{(u_1^{\alpha_1}u_2^{\alpha_2}\ldots u_s^{\alpha_s})\,\phi_1}{\alpha_1!\,\alpha_2!\ldots\alpha_s!}\frac{(u_1^{\beta_1}u_2^{\beta_2}\ldots u_s^{\beta_s})\,\phi_2}{\beta_1!\,\beta_2!\ldots\beta_s!}\cdots\frac{(u_1^{\mu_1}u_2^{\mu_2}\ldots u_s^{\mu_s})\,\phi_m}{\mu_1!\,\mu_2!\ldots\mu_s!},$$

where $\phi=\phi_1\phi_2\ldots\phi_m$,

put $\phi=\theta^m,\quad u_t=\partial_{x_t},$

$\chi_1=\chi_2=\ldots=\chi_s=1,$

and $s=n.$

Then

$$(\partial_{x_1}\partial_{x_2}\ldots\partial_{x_n})\,\theta^n=\Sigma\Sigma\,(X_1)\,\theta\,(X_2)\,\theta\ldots(X_k)\,\theta\,.\,\theta^{n-k},$$

where $X_1X_2\ldots X_k=\partial_{x_1}\partial_{x_2}\ldots\partial_{x_n},$

and as usual we must take every factorization of the operator and then distribute the operations upon the right-hand side in all possible ways.

Take $\theta=x_1x_2\ldots x_n$ and putting $n=3$, we have

$$\begin{aligned}(\partial_{x_1}\partial_{x_2}\partial_{x_3})\,(x_1x_2x_3)^3=&\,(\partial_{x_1}\partial_{x_2}\partial_{x_3})\,(x_1x_2x_3)\,.\,(x_1x_2x_3)\,.\,(x_1x_2x_3)\\&+(x_1x_2x_3)\,.\,(\partial_{x_1}\partial_{x_2}\partial_{x_3})\,(x_1x_2x_3)\,.\,(x_1x_2x_3)\\&+(x_1x_2x_3)\,.\,(x_1x_2x_3)\,.\,(\partial_{x_1}\partial_{x_2}\partial_{x_3})\,(x_1x_2x_3)\\&+(\partial_{x_1}\partial_{x_2})\,(x_1x_2x_3)\,.\,\partial_{x_3}\,(x_1x_2x_3)\,.\,(x_1x_2x_3)\\&+5\text{ similar terms}\\&+(\partial_{x_2}\partial_{x_3})\,(x_1x_2x_3)\,.\,\partial_{x_1}\,(x_1x_2x_3)\,.\,(x_1x_2x_3)\\&+5\text{ similar terms}\\&+\partial_{x_3}\partial_{x_1}\,(x_1x_2x_3)\,.\,\partial_{x_2}\,(x_1x_2x_3)\,.\,(x_1x_2x_3)\\&+5\text{ similar terms}\\&+\partial_{x_1}\,(x_1x_2x_3)\,.\,\partial_{x_2}\,(x_1x_2x_3)\,.\,\partial_{x_3}\,(x_1x_2x_3)\\&+5\text{ similar terms.}\end{aligned}$$

In all $27\;(=3^3)$ terms corresponding to the 27 permuted partitions of $x_1x_2x_3$ into exactly 3 parts, zero being reckoned as a part.

Selecting any term on the right-hand side, say

$$\partial_{x_1}\partial_{x_2}\,(x_1x_2x_3)\,.\,\partial_{x_3}\,(x_1x_2x_3)\,.\,(x_1x_2x_3)$$

we obtain $(x_3)\,.\,(x_1x_2)\,.\,(x_1x_2x_3),$

and if we were to proceed to perform the operation $\partial_{x_1}\partial_{x_2}\partial_{x_3}$ a second time, of the whole number of 27 operations, into which the operator is seen to break up, only a certain number will be effective in producing a non-zero term.

We are subject to the conditions

(1) 1st operator factor must not contain ∂_{x_1} or ∂_{x_2},

(2) 2nd factor must not contain ∂_{x_3}.

As one operation we can take

$$(x_3)\,.\,\partial_{x_1}\partial_{x_2}(x_1x_2)\,.\,\partial_{x_3}(x_1x_2x_3),$$

resulting in

$$(x_3)\,.\,(\cdot)\,(x_1x_2).$$

Again operating with $\partial_{x_1}\partial_{x_2}\partial_{x_3}$, we find that only one of the 27 operations can be performed, and we have

$$\partial_{x_3}(x_3)\,.\,(\cdot)\,.\,\partial_{x_1}\partial_{x_2}(x_1x_2),$$

resulting in

$$(\cdot)\,.\,(\cdot)\,.\,(\cdot).$$

Forming a square table of these operations we find

$\partial_{x_1}\partial_{x_2}$	∂_{x_3}	
	$\partial_{x_1}\partial_{x_2}$	∂_{x_3}
∂_{x_3}		$\partial_{x_1}\partial_{x_2}$

and it will be seen that each of the three operators ∂_{x_1}, ∂_{x_2}, ∂_{x_3} occurs exactly once in each row and in each column.

This feature is a necessary result of the process.

Art. 26. We may symbolise the above taken successive differential operations by the scheme

12	3	
	12	3
3		12

and selecting the operations in any manner possible, so that an annihilating effect is not produced, we will obtain a Square of order 3, having the property that each of the three numbers 1, 2, 3 appears exactly once in each row and column without restriction in regard to the number of them that may appear in each compartment.

We have in fact the Latin Square freed from the condition that one letter must appear in each compartment. Hence it is seen that these squares are enumerated by the number of terms which survive the operations performed on the right-hand side of the identity after three successive operations of $\partial_{x_1}\partial_{x_2}\partial_{x_3}$.

Therefore the enumeration is given by

$$(\partial_{x_1}\partial_{x_2}\partial_{x_3})^3 (x_1^3 x_2^3 x_3^3) = (3!)^3.$$

In general for the order n the enumeration of these squares is given by

$$(\partial_{x_1}\partial_{x_2} \dots \partial_{x_n})^n (x_1^n x_2^n \dots x_n^n) = (n!)^n.$$

We may state the problem in the following way:—

"n^2 different towns form a square and there are n^2 inspectors, n of each of n different nationalities. Find the number of arrangements of the inspectors in the towns subject to the condition that one inspector of each nationality must be in each row or column of towns combined with the circumstance that no restriction is placed upon the number of inspectors that may be stationed in a particular town."

The result is, as shewn, $(n!)^n$.

Art. 27. We may also consider the operator

$$(\partial_{x_1}\partial_{x_2} \dots \partial_{x_n})^m,$$

in conjunction with the operand

$$(x_1 x_2 \dots x_n)^m,$$

where

$$m \gtreqless n.$$

Thus, in particular, taking $(\partial_{x_1}\partial_{x_2})^3 (x_1 x_2)^3$, we find arrangements such as

1		2
	12	
2		1

which is a square of 3^2 compartments, and the numbers 1, 2 are arranged, in such manner, that each is contained once in each row and in each column.

The enumeration is given by

$$(\partial_{x_1}\partial_{x_2})^3 (x_1 x_2)^3 = (3!)^2,$$

and in general the result

$$(\partial_{x_1}\partial_{x_2} \dots \partial_{x_n})^m (x_1 x_2 \dots x_n)^m = (m!)^n$$

shews that, in a square of m^2 compartments, the n numbers 1, 2, 3, ... n can be arranged, in such a manner, that each is contained once in each row and in each column in exactly $(m!)^n$ ways.

Art. 28. It is very interesting to see that these results can also be obtained by means of the symmetric function operators employed in the body of the paper.

For take as operand $\{(a_1)(a_2)(a_3)\dots(a_n)\}^n$,

where of course $(a_1)=\Sigma\alpha^{a_1}$, &c.;

and as operator $D^n_{a_1+a_2+a_3+\dots+a_n}=D^n_w$,

where as usual w possesses the single partition $(a_1a_2\dots a_n)$ into n or fewer parts drawn from $a_1, a_2, \dots a_n$, repetitions permissible.

For simplicity take $n=3$, and write

$$(a_1,\ a_2,\ a_3)=(a,\ b,\ c).$$

Then $\{(a)(b)(c)\}^3=\{(a+b+c)+(a+b,\ c)+(a+c,\ b)+(b+c,\ a)+(abc)\}^3$,

and $D_{a+b+c}\equiv D_{(a+b+c)}+D_{(a+b,\ c)}+D_{(a+c,\ b)}+D_{(b+c,\ a)}+D_{(abc)}$,

according to the notation explained above (Art. 4).

Now

$$D^3_{a+b+c}\{(a)(b)(c)\}^3=D^3_{a+b+c}(a)^3(b)^3(c)^3$$
$$=D^3_{(abc)}(a)^3(b)^3(c)^3=(3!)^3,$$

and, performing the developed operator upon

$$\{(a+b+c)+(a+b,\ c)+(a+c,\ b)+(b+c,\ a)+(abc)\}^3,$$

we have to consider the 125 terms of which the expanded power is composed.

One of these is $(a+b+c)(a+c,\ b)(abc)$,

and, performing D_{a+b+c}, we obtain two terms,

$$\begin{array}{ccc} \cdot & (a+c,\ b) & (abc) \\ +(a+b+c) & (b) & (ac), \end{array}$$

and now it is easy to see that

$$D^3_{a+b+c}(a+b+c)(a+c,\ b)(abc)=0.$$

But selecting out of the 125 the term

$$(a+b,\ c)(a+c,\ b)(abc),$$

the operation of D_{a+b+c} produces

$$\begin{array}{ccc} (c) & (a+c,\ b) & (ab) \\ +(a+b,\ c) & (b) & (ac) \\ +(a+b) & (a+c) & (bc); \end{array}$$

the terms corresponding respectively to

$$\begin{array}{ccc} D_{a+b}* & \cdot & *\cdot D_c* \\ \cdot & *D_{a+c} & *\cdot D_b* \\ D_c & *D_b & *\ \ D_a*, \end{array}$$

and, selecting the first of the terms produced,

$$D_c*D_b*D_a*$$

yields $(a+c)(b)$;

and now operating again, $*D_{a+c}*D_b*$

yields unity.

Hence we obtain one resulting term corresponding to the operator scheme

$a+b$		c
c	b	a
	$a+c$	b

or say

12		3
3	2	1
	13	2

which is a square having the desired property, viz.:—each row, as well as each column, contains each of the three numbers, without restriction in regard to the number of numbers appearing in a compartment.

And when we carry out the whole process we must arrive at

$$(3!)^3$$

such squares, each square typifying a succession of operations.

Art. 29. Hence we establish the general theorem, above enunciated, as the enumeration is given by

$$D^n_{a_1+a_2+\ldots+a_n}(a_1)^n(a_2)^n\ldots(a_n)^n,$$

which is

$$D^n_{(a_1a_2\ldots a_n)}(a_1^n)(a_2^n)\ldots(a_n^n)\times(n!)^n,$$

or

$$(n!)^n.$$

Clearly, the enlarged theorem corresponds to

$$D^m_{a_1+a_2+\ldots+a_n}(a_1)^m(a_2)^m\ldots(a_n)^m,$$

which gives

$$(m!)^n,$$

as before.

Art. 30. This interesting result shews that we may expect to meet with many pairs of operators and operands differing widely in character which conduct to the same theorem in combinatory analysis. I believe that the method of research, above set forth, is of considerable promise and worthy of the attention of mathematicians. It is probable that known theorems in combinatory analysis will lead conversely to theorems connected with operations which will prove both interesting and valuable.

IX. *Combinatorial Analysis. The Foundations of a New Theory.*

By Major P. A. MACMAHON, *D.Sc.*, *F.R.S.*

Received March 19,—Read April 5, 1900.

INTRODUCTION.

IN the 'Transactions of the Cambridge Philosophical Society' (vol. 16, Part IV., p. 262), I brought forward a new instrument of research in Combinatorial Analysis, and applied it to the complete solution of the great problem of the "Latin Square," which had proved a stumbling block to mathematicians since the time of Euler. The method was equally successful in dealing with a general problem of which the Latin Square was but a particular case, and also with many other questions of a similar character. I propose now to submit the method to a close examination, to attempt to establish it firmly, and to ascertain the nature of the questions to which it may be successfully applied. We shall find that it is not merely an enumerating instrument but a powerful reciprocating instrument, from which a host of theorems of algebraical reciprocity can be obtained with facility.

We will suppose that combinations defined by certain laws of combination have to be enumerated; the method consists in designing, on the one hand, an operation and, on the other hand, a function in such manner that when the operation is performed upon the function a number results which enumerates the combinations. If this can be carried out we, in general, obtain far more than a single enumeration; we arrive at the point of actually representing graphically all the combinations under enumeration, and solve by the way many other problems which may be regarded as leading up to the problem under consideration. In the case of the Latin Square it was necessary to design the operation and the function the combination of which was competent to yield the solution of the problem. It is a much easier process, and from my present standpoint more scientific, to start by designing the operation and the function, and then to ascertain the questions which the combination is able to deal with.

§ 1.

Art. 1.—I will commence by taking the simplest possible question to which the method is applicable. Let us inquire into the number of permutations of n different letters. A knowledge of the result would at once lead us to design

An operation.	A function.
$(d/dx)^n$	x^n

since $(d/dx)^n x^n = n!$; but once we observe the way in which d/dx operates upon x^n we require no previous knowledge of the result to aid us in the design. Conceive x^n written as a product

$$\overset{xx}{xxxxxnn} \ldots$$

the operation of d/dx consists in substituting unity for x in all possible ways, and summing the results obtained.

$$\frac{d}{dx} x^n = 1 . xxx \ldots + x1xx \ldots + xx1xx \ldots + \ldots = nx^{n-1}.$$

We have, in fact, to perform n operations of substitution; let us select one of these, say—

$$xx1xx \ldots$$

and denote the minor operation, by which it has been obtained, by the scheme

		1_a			---

the suffix a denoting that the first operation of d/dx has resulted in the appearance of the unit.

To obtain $\frac{d}{dx}(xx1xx \ldots)$ we have $n - 1$ minor operations by which x is replaced by unity in all possible ways. If one term obtained be

$$1x1x\overset{x}{n} \ldots$$

the operations by which this has been reached may be denoted by the scheme

		1_a			---
1_b					---

and by proceeding in this manner we finally reach a lattice, square and of n^2 compartments, which is the diagrammatic representation of one of the $n!$ combinations of minor operations which results from the operation of $(d/dx)^n$ upon x^n. If we transfer the 1_b, 1_c, . . . to the top row we see that to each diagram corresponds a permutation of the n different letters $a, b, c, \ldots$ Moreover suppressing the letters $a, b, c, \ldots$ we see that we have solved the following problem, viz. :—To place n units in the compartments of the square of order n, so that each row and each column contains one and only one unit. In general we find that the problems that can be solved have some simple definition upon a lattice, as in the present instance. Writing $a, b, c, \ldots$ as $a_1, a_2, a_3, \ldots$ the suffix of the letter is given by the row and the place in the permutation by the column, so that to a_s standing t^{th} in a permutation would correspond a unit in the sth row and t^{th} column of the lattice.

It may be remarked, and will afterwards appear, that in general many different designs of operation and function are appropriate to a particular problem.

Art. 2.—With the immediate object of applying the method to the general case of permutation, there being any number of identities of letters, we must first obtain another solution of the foregoing problem.

Let $\alpha_1, \alpha_2, \alpha_3 \ldots$ be a number of quantities and $a_1, a_2, a_3, \ldots$ their elementary symmetric functions. Further, let

$$d_1 = D_1 = \frac{d}{da_1} + a_1 \frac{d}{da_2} + a_2 \frac{d}{da_3} \ldots$$

and $a_s = \Sigma\alpha_1\alpha_2 \ldots \alpha_3 = (1^s)$ in partition notation.

We may take as operation and function

$$D_1^n \text{ and } (1)^n,$$

equivalent to $(d/da_1)^n$ and a_1^n, which we had before, but more convenient as being readily generalisable.

Let $D_s = \frac{1}{s!} d_1^s$, d_1^s denoting an operator of order s, obtained by symbolical multiplication as in Taylor's theorem. Suppose the question be the enumeration of the permutations of the quantities in $\alpha_1^{\pi_1}\alpha_2^{\pi_2} \ldots \alpha_n^{\pi_n}$, where $\Sigma\pi = n$. I say that the operation and function are respectively

$$D_{\pi_1}D_{\pi_2} \ldots D_{\pi_n} \text{ and } (1)^n.$$

Observe that this is merely the multinomial theorem for

$$(1)^n = \ldots + \frac{n!}{\pi_1!\pi_2! \ldots \pi_n!} \Sigma\alpha_1^{\pi_1}\alpha_2^{\pi_2} \ldots \alpha_n^{\pi_n} + \ldots$$

$$= \ldots + \frac{n!}{\pi_1!\pi_2! \ldots \pi_n!} (\pi_1\pi_2 \ldots \pi_n) + \ldots$$

in partition notation; and

$$D_{\pi_1}D_{\pi_2}D_{\pi_3} \ldots D_{\pi_n}(\pi_1\pi_2 \ldots \pi_n) = 1.*$$

Hence

$$D_{\pi_1}D_{\pi_2}D_{\pi_3} \ldots D_{\pi_n}(1)^n = \frac{n!}{\pi_1!\pi_2!\pi_3! \ldots \pi_n!}$$

the result we require.

The important operator D_π has been discussed by the author.† Its effect upon a monomial symmetric function is to erase a part π from the partition expression of the function.

Thus

$$D_\pi \Sigma\alpha^\pi\beta^\rho\gamma^\sigma \ldots = D_\pi(\pi\rho\sigma \ldots) = (\rho\sigma \ldots) = \Sigma\alpha^\rho\beta^\sigma \ldots.$$

* See HAMMOND, 'Proc. Lond. Math. Soc.,' vol. 13, p. 79; also 'Trans. Camb. Phil. Soc.,' *loc. cit.*

† 'Messenger of Mathematics,' vol. 14, p. 164. 'American Journal of Mathematics,' "Third Memoir on a New Theory of Symmetric Functions," vol. 13, p. 8 *et seq.*, p. 34 *et seq.* 'Trans. Camb. Phil. Soc.,' vol. 16, part IV., p. 262.

If no part π presents itself in the operand, D_π causes the monomial function to vanish. Thus

$$D_\pi(\rho\sigma\tau\ldots) = 0.$$

The compound operation $D_{\pi_1}D_{\pi_2}D_{\pi_3}\ldots$ denotes the *successive* performance of the operations

$$D_{\pi_1},\ D_{\pi_2},\ D_{\pi_3},\ \ldots \text{ of orders } \pi_1,\ \pi_2,\ \pi_3,\ \ldots \text{ respectively.}$$

The law of operation of D_π establishes that the component operations $D_{\pi_1}, D_{\pi_2}, D_{\pi_3}, \ldots$ may be performed *in any order*. Thus

$$D_{\pi_1}D_{\pi_2}(\pi_1\pi_2\rho\sigma\ldots) = D_{\pi_1}(\pi_1\rho\sigma\ldots) = D_{\pi_2}(\pi_2\rho\sigma\ldots) = (\rho\sigma\ldots).$$

As the order of operation is immaterial, it is found convenient in most cases to operate with $D_{\pi_1}D_{\pi_2}D_{\pi_3}\ldots$ in the order $D_{\pi_1}, D_{\pi_1}, D_{\pi_3}, \ldots$; this may seem at first sight at variance with the ordinary usage in the Differential Calculus, but there is a convenience in ordering the operator from left to right in agreement with the practice of ordering a partition from left to right. If, further, we note the result—

$$D_\pi(\pi) = 1$$

we have a complete account of the operator so far as it is concerned with an operand, which is merely a monomial symmetric function. The operation of D_π upon a symmetric function product is of even greater importance in the present theory. It has the effect of erasing a partition of π from the product, one part from each factor, in all possible ways; the result of the operation being a sum of products, one product arising from each such erasure of a partition. This has been set forth at length in the papers to which reference has been given, but in deference to the suggestion of one of the Referees appointed by the Royal Society to report upon the present paper, a number of examples are given to familiarise readers with the processes which are so much employed in what follows.

Example 1.—Consider

$$D_\pi(1)^n.$$

The operand consists of n factors, each of which is (1); the operator D_π is performed through the partition of the number π which involves π units; this partition must be erased from $(1)^n = (1)(1)(1)\ldots$ to n factors in each of the $\binom{n}{\pi}$ possible ways, and the results added. Thus

$$D_\pi(1)^n = \binom{n}{\pi}(1)^{n-\pi}$$

As a particular case

$$D_2(1)^4 = (\not{1})(\not{1})(1)(1) + (\not{1})(1)(\not{1})(1) + (\not{1})(1)(1)(\not{1}) + (1)(\not{1})(\not{1})(1)$$
$$+ (1)(\not{1})(1)(\not{1}) + (1)(1)(\not{1})(\not{1}) = 6(1)^2 = \binom{4}{2}(1)^2.$$

This, the simplest example that could be taken, shows clearly the great value of the operator D_π as an instrument in combinatorial analysis.

Example 2.—It follow from the first example that if $\Sigma\pi = n$,

$$D_{\pi_1} D_{\pi_2} D_{\pi_3} \ldots D_{\pi_n} (1)^n = \frac{n!}{\pi_1!\,\pi_2!\,\pi_3!\ldots\pi_n!}.$$

Example 3.—Consider

$$D_4 (2)^2 (1)^2.$$

We are concerned with the partitions of the number 4 into 2,2 and 2,1,1, and

$$D_4 (2)(2)(1)(1) = (\not{2})(\not{2})(1)(1) + (\not{2})(2)(\not{1})(\not{1}) + (2)(\not{2})(\not{1})(\not{1}) = (1)^2 + 2(2).$$

If now we operate with D_2 we have to take account of the partitions 1,1 and 2 of the number 2, and we find

$$D_4 D_2 (2)^2 (1)^2 = (\not{1})(\not{1}) + 2(\not{2}) = 3,$$

and we have the result

$$(2)^2 (1)^2 = \ldots + 3(42) + \ldots$$

as a consequence.

Similarly, reversing the order of the operations

$$D_2 (2)(2)(1)(1) = 2(2)(1)(1) + (2)(2) \quad \text{and} \quad D_4 (2)(1)(1) = D_4 (2)(2) = 1,$$

verifying the previous result.

If no partition of π can be picked out in this way from the partitions of the functions forming the product, the result of the operation is zero.

Example 4.—

$$D_2 (1^2)^2 = D_2 (1^2)(1^2) = (1)(1) = (1)^2.$$

It is important to notice here that a unit is erased from $(1^2) = (11)$ in *only one way*, and that for present purposes a number of similar figures enclosed in a bracket are to be considered as the same, and not different; we have already seen that when the figures are similar, but in different brackets, they are, for the purpose of selection, to be considered as different figures.

Observe that since

$$D_2 (1^2)^2 = (1)^2 = (2) + 2(1^2),$$

we may say that

$$(1^2)^2 = (2^2) + 2(21^2) + \ldots,$$

the terms to be added on the right for the full expression being such as do not contain a figure 2.

To obtain the lattice representations, suppose $\pi_1, \pi_2, \pi_3, \ldots \pi_n$ to be in descending order, and thus to be an ordered partition of the number n.

Since $D_{\pi_1}(1)^n = \binom{n}{\pi_1}(1)^{n-\pi_1}$, D_{π_1} may be regarded as breaking up into $\binom{n}{\pi_1}$ minor operations, each of which consists in erasing π_1 of the n factors

$$(1)(1)(1)\ .\ .\ .\ (1),$$

and replacing them by units. On the diagram we denote one such minor operation by (say),

1	1		1		1	1				----n° of units = π_1

Keeping to this term,

$$1\ .\ 1\ .\ (1)\ .\ 1\ .\ (1)\ .\ 1\ .\ 1\ .\ (.\ 1)(1)\ .\ .\ .$$

and operating with D_{π_2}, we find the operation breaking up into $\binom{n-\pi_1}{\pi_2}$ minor operations, each of which consists in erasing π_2 of the factors, and replacing them by units. Selecting one of these minor operations, we find that the corresponding term has been obtained by operations conforming to the diagram of two rows

1	1		1		1	1				----n° of units = π_1
		1					1	1		----n° of units = π_2

Proceeding in this manner, we finally arrive at a lattice of n rows and n columns, such that there is one and only one unit in each column, while the numbers of units in the 1st, 2nd . . . nth rows are $\pi_1, \pi_2, \ldots \pi_n$ respectively.

We have thus the associated problem on the lattice, and we obtain $\dfrac{n!}{\pi_1!\,\pi_2!\ldots\pi_n!}$ representations on the lattice corresponding to the permutations of the quantities in $\alpha_1^{\pi_1}\alpha_2^{\pi_2}\ldots\alpha_n^{\pi_n}$.

The simple introductory examples lead one to expect that the method will be found capable of dealing with questions either of a chess board character or which are concerned with rectangular lattices. Further, the idea of lattice rotation gives promise of leading to theorems of algebraic reciprocity and of reciprocity in the theory of numbers, features almost inseparable from any lattice theory.

Art. 3.—*The graph of a partition.*—It is convenient to have before us the connexion between this theory and the SYLVESTER-FERRERS graph of a unipartite partition.

Consider the operation $D_3D_2^2D_1$ and the symmetric function $(1^4)(1^3)(1)$, (32^21) being the partition conjugate to (431).

$D_3D_2^2D_1\ (1^4)(1^3)(1) = D_2^2D_1\ (1^3)(1^2)$, the operation D_3 erasing one part, viz.: unity from each factor; this we denote as usual by

1	1	1

;

again $D_2^2D_1\,(1^3)(1^2) = D_2D_1\,(1^2)(1)$ and the two operations together give us

1	1	1
1	1	

and $D_2D_1\,(1^2)(1) = D_1\,(1) = 1$ and the complete lattice representation is

1	1	1
1	1	
1	1	
1		

which is none other than the graph of the partition (32^21) or of (431) according as it is read by rows or by columns. We might also have operated with $D_4D_3D_1$ upon $(1^3)(1^2)^2(1)$ and, in general, if $(\pi_1\pi_2\pi_3\dots)$, $(\rho_1\rho_2\rho_3\dots)$ be conjugate partitions, we obtain their graphs either by operating with $D_{\pi_1}D_{\pi_2}D_{\pi_3}\dots$ upon $(1^{\rho_1})(1^{\rho_2})(1^{\rho_3})\dots$ *or* with $D_{\rho_1}D_{\rho_2}D_{\rho_3}\dots$ upon $(1^{\pi_1})(1^{\pi_2})(1^{\pi_3})\dots$

Art. 4.—I proceed to consider some less obvious but equally interesting examples of the method. The diagrams obtained depend upon the law by which the operation is performed upon the function which is the operand. The operator D_π in connexion with symmetric function operands is of commanding importance. It would be difficult to imagine an operation better adapted to research in combinatorial analysis. We shall find later that an analogous operation exists which can be employed when symmetric functions of several systems of quantities are taken as operands. As an example of diagram formation, take as operator $D_4D_3^2$ and as function $(3)(21)(2)(1)(1)$ the weight of operator and of function being the same.

We have

$$
\begin{aligned}
&D_4(3)(21)(2)(1)(1)\\
&\quad=(.)(2.)(2)(1)(1)+(.)(21)(2)(.)(1)+(.)(21)(2)(1)(.)\\
&\qquad\qquad+(3)(.1)(.)(1)(1)\\
&\quad+(3)(.1)(2)(.)(.)+(3)(2.)(.)(1)(.)+(3)(2.)(.)(.)(1)\\
&\qquad\qquad+(3)(21)(.)(.)(.),
\end{aligned}
$$

the eight terms arising from the partitions 31, 22, 211 of the number 4. The dots take the place of the picked out partitions.

Hence
$$
\begin{aligned}
&D_4(3)(21)(2)(1)(1)\\
&=(2)^2(1)^2+2(21)(2)(1)+(3)(1)^3+3(3)(2)(1)+(3)(21).
\end{aligned}
$$

The operation D_4 breaks up here into eight minor operations; taking any one of these—say that one which consists in taking 3 from the factor (3), and 1 from the factor (21), we form the first row of our diagram, viz.:—

3	1			

The term resulting from the selected minor operation is

$$(.)(2)(2)(1)(1);$$

the operation of D_3 results in four minor operations corresponding to the four ways of picking out a 2 and a 1 from different factors; we may select the particular minor operation which results in

$$(.)(.)(2)(1)(.),$$

and now we add on the second row which denotes this minor operation, and obtain the diagram of two rows.

3	1			
	2			1

We can now only operate in one way with D_3 upon (.) (.) (2) (1) (.) and we finally obtain the diagram of three rows :—

3	1			
	2			1
		2	1	

which possesses the property that the sums of the numbers in the successive rows are 4, 3, 3, respectively, while the successive columns involve the partitions (3), (21), (2), (1), (1) respectively.

The number of such diagrams is A where

$$(3)(21)(2)(1)^2 = \ldots + A(43^2) + \ldots,$$

and A has the analytical expression

$$D_4 D_3^2 (3)(21)(2)(1)^2.$$

Let us now consider the problem of placing units in the compartments of a lattice of m rows and l columns, not more than one unit in each compartment, in such wise that we can count $\mu_1, \mu_2, \ldots \mu_m$ units in the successive rows, and $\lambda_1, \lambda_2, \ldots \lambda_l$ units in the successive columns. Take

As operation.		As function.
$D_{\mu_1} D_{\mu_2} \ldots D_{\mu_m}$	and	$(1^{\lambda_1})(1^{\lambda_2}) \ldots (1^{\lambda_l})$.

If

$$(1^{\lambda_1})(1^{\lambda_2}) \ldots (1^{\lambda_l}) = \ldots + A(\mu_1 \mu_2 \ldots \mu_m) + \ldots$$

$$D_{\mu_1} D_{\mu_2} \ldots D_{\mu_m} (1^{\lambda_1})(1^{\lambda_2}) \ldots (1^{\lambda_l}) = A,$$

and we can show that the number A enumerates the lattices under investigation. The operation D_{μ_1} makes selections of every μ_1 of the l factors and erases a part unity from each; one minor operation of D_{μ_1} therefore is denoted by μ_1 units placed in μ_1 compartments of the first row of a lattice of l rows; the operation D_{μ_2} adds on a second row, in which units appear in μ_2 of the compartments, and so on we finally arrive at a lattice possessing the desired property as regards rows, and as obviously the column property obtains, the problem is solved.

Ex. gr. Take $\lambda_1 = 3, \lambda_2 = 2, \lambda_3 = 1, \mu_1 = 2, \mu_2 = 2, \mu_3 = 1, \mu_4 = 1$

$$(1^3)(1^2)(1) = a_3 a_2 a_1 = \ldots + 8(2211) + \ldots$$

The eight diagrams are

1	1	
1	1	
1		
		1

1	1	
1	1	
		1
1		

	1	1
1	1	
1		
1		

1	1	
	1	1
1		
1		

1		1
1	1	
1		
	1	

1		1
1	1	
	1	
1		

1	1	
1		1
1		
	1	

1	1	
1		1
	1	
1		

and no others possess the desired property.

We can now apply the method so as to be an instrument of reciprocation in algebra. If we transpose the diagrams so as to read by rows as they formerly did by columns, the effect is to interchange the set of numbers $\lambda_1, \lambda_2, \ldots \lambda_l$ with the set $\mu_1, \mu_2, \ldots \mu_m$, and the number of diagrams is not altered. Hence the reciprocal theorem.

If $\qquad (1^{\lambda_1})(1^{\lambda_2}) \ldots (1^{\lambda_l}) = \ldots + A(\mu_1\mu_2 \ldots \mu_m) + \ldots$

then $\qquad (1^{\mu_1}))1^{\mu_2}) \ldots (1^{\mu_m}) = \ldots + A(\lambda_1\lambda_2 \ldots \lambda_l) + \ldots$

a theorem known to algebraists as the Cayley-Betti Law of Symmetry in Symmetric Functions.

The easy intuitive nature of this proof of the theorem is very remarkable.

Art. 5.—In the above the magnitude of the numbers, appearing in the compartments of the lattice, has been restricted so as not to exceed unity. This restriction may be removed in the following manner. Consider the symmetric functions known as the homogeneous product sums of the quantities $\alpha_1, \alpha_2, \alpha_3, \ldots$ viz.—

$$h_1 = (1),$$
$$h_2 = (2) + (1^2),$$
$$h_3 = (3) + (21) + (1^3),$$
$$\ldots\ldots\ldots\ldots$$

and note the result

$$D_s h_\lambda = h_{\lambda-s},$$

and also $\quad D_s h_{\lambda_1} h_{\lambda_2} \ldots h_{\lambda_l} = \Sigma h_{\lambda_1-\sigma_1} h_{\lambda_2-\sigma_2} h_{\lambda_3-\sigma_3} \ldots h_{\lambda_l-\sigma_l}$

where $(\sigma_1\, \sigma_2\, \sigma_3 \ldots \sigma_l)$ is a partition of s and the sum is for all such partitions, and for a particular partition is for all ways of operating upon the suffixes with the parts of the partition. Thus

$$\begin{aligned} D_3 h_\lambda h_\mu h_\nu = & \; h_{\lambda-3} h_\mu h_\nu + h_\lambda h_{\mu-3} h_\nu + h_\lambda h_\mu h_{\nu-3} \\ & + h_{\lambda-2} h_{\mu-1} h_\nu + h_\lambda h_{\mu-2} h_{\nu-1} + h_{\lambda-1} h_\mu h_{\nu-2} \\ & + h_{\lambda-1} h_{\mu-2} h_\nu + h_\lambda h_{\mu-1} h_{\nu-2} + h_{\lambda-2} h_\mu h_{\nu-1} \\ & + h_{\lambda-1} h_{\mu-1} h_{\nu-1}. \end{aligned}$$

If from the result of $D_s h_{\lambda_1}, h_{\lambda_2} \ldots h_{\lambda_l}$ we select the term product $h_{\lambda_1-\sigma_1}$, $h_{\lambda_2-\sigma_2} \ldots h_{\lambda_l-\sigma_l}$, the corresponding lattice will have as first row

σ_1	σ_2	σ_3	σ_4	...	σ_l

the sum of the numbers being s, and if in the selected product we now operate with D_t we can select a term product from the result, and the two minor operations may be indicated by the two-row lattice,

σ_1	σ_2	σ_3	σ_4	...	σ_l
τ_1	τ_2	τ_3	τ_4	...	τ_l

the sum of the numbers τ being t.

Hence if we take as operation

$$D_{\mu_1} D_{\mu_2} \ldots D_{\mu_m},$$

and as function $\quad h_{\lambda_1} h_{\lambda_2} \ldots h_{\lambda_l},$

we will obtain a number of lattices of m rows and l columns, which possess the property that the sums of the numbers in the successive rows are $\mu_1, \mu_2, \ldots \mu_m$, and in the successive columns $\lambda_1, \lambda_2, \ldots \lambda_l$, no restriction being placed upon the magnitude of the numbers. The number of such lattices is A, where

$$h_{\lambda_1} h_{\lambda_2} \ldots h_{\lambda_l} = \ldots + A(\mu_1 \mu_2 \ldots \mu_m) + \ldots$$

and now transposition of lattices shows that

$$h_{\mu_1} h_{\mu_2} \ldots h_{\mu_m} = \ldots + A(\lambda_1 \lambda_2 \ldots \lambda_l) + \ldots$$

yielding a proof of a law of symmetry discovered by the present author many years ago. The process involves the actual formation of the things enumerated by the number A. The secret of its success in this instance lies in the result $D_\lambda h_s = h_{s-\lambda}$.

Ex. gr. We have

$$D_2^2 D_1^2 \, h_3 h_2 h_1 = 18$$

i.e. $$h_3 h_2 h_1 = \ldots + 18(2211) + \ldots$$
$$h_2^2 h_1^2 = \ldots + 18(321) + \ldots$$

and we must have 18 lattices; now eight of these, in which the compartment numbers do not exceed unity, have been depicted above; the remaining 10 are

2		
	2	
1		
		1

2		
	2	
		1
1		

2		
1	1	
	1	
		1

2		
1	1	
		1
	1	

2		
	1	1
1		
	1	

2		
	1	1
	1	
1		

2		
1		1
	1	
	1	

	2	
2		
1		
		1

	2	
2		
		1
1		

	2	
1		1
1		
1		

Art. 6.—The next problem I will consider is that in which the magnitude of the compartment numbers has a superior limit k.

Let k_s denote the homogeneous product sum of order s in which none of the quantities $\alpha_1, \alpha_2, \alpha_3, \ldots$ is raised to a higher power than k. *Ex. gr.* If $k = 2$, k_3 will be $(21) + (1^3)$, and not $(3) + (21) + (1^3)$.

We have $$D_\lambda k_s = k_{s-\lambda} \quad \text{where } \lambda \leqq k,$$

and $$D_\lambda k_s = 0 \quad \text{if } \lambda > k.$$

Take as operation

$$D_{\mu_1} D_{\mu_2} \ldots D_{\mu_m}$$

and as function $k_{\lambda_1}, k_{\lambda_2}, \ldots k_{\lambda_l}$ and we will obtain a number of lattices of m rows and l columns which possess the property that the sums of the numbers in the successive rows are $\mu_1, \mu_2, \ldots \mu_m$, and in the successive columns $\lambda_1, \lambda_2, \ldots \lambda_l$, the magnitude of the compartment numbers being restricted not to exceed k.

The number of such lattices is

$$D_{\mu_1} D_{\mu_2} \ldots D_{\mu_m} \, k_{\lambda_1} k_{\lambda_2} \ldots k_{\lambda_l} = A,$$

where $$k_{\lambda_1} k_{\lambda_2} \ldots k_{\lambda_l} = \ldots + A(\mu_1 \mu_2 \ldots \mu_m) + \ldots,$$

and by transposing the lattices

$$k_{\mu_1} k_{\mu_2} \ldots k_{\mu_m} = \ldots + A(\lambda_1 \lambda_2 \ldots \lambda_l) + \ldots;$$

establishing a law of symmetry in symmetrical algebra.

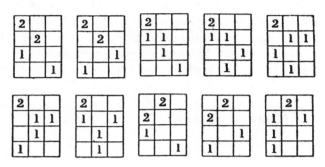

I observe that if $k = 9$, the lattices associated with and enumerated by

$$D^3_{15} k^3_{15}$$

include all the row and column-magic squares connected with the natural series of numbers 1, 2, 3, 4, 5, 6, 7, 8, 9. In general if $N = \frac{1}{2}n(n^2 + 1)$, the lattices enumerated by $D^n_N\, k^n_N$, where $k = n^2$, include all the magic squares of order n connected with the first n^2 numbers. If we could further impress the condition that no compartment number is to be twice repeated, we would be successful in enumerating the magic squares divorced from the diagonal property. This seems to be a matter of difficulty, which is increased if an attempt be made to introduce diagonal and other conditions to which certain classes of magic squares are subject.

It may be gathered from what has been said, that every case of symmetric function multiplication is connected with a theory of lattice combinations. For if we take as function

$$(\lambda_1\mu_1\nu_1 \ldots)(\lambda_2\mu_2\nu_2 \ldots) \ldots (\lambda_s\mu_s\nu_s \ldots),$$

and as operation

$$D_pD_qD_r \ldots ,$$

where

$$p + q + r + \ldots = \Sigma\lambda + \Sigma\mu + \Sigma\nu + \ldots ,$$

we have

$$D_pD_qD_r \ldots (\lambda_1\mu_1\nu_1 \ldots ((\lambda_2\mu_2\nu_2 \ldots)(\ldots) \ldots (\lambda_s\mu_s\nu_s \ldots) = A,$$

where

$$(\lambda_1\mu_1\nu_1 \ldots)(\lambda_2\mu_2\nu_2 \ldots) \ldots (\lambda_s\mu_s\nu_s \ldots) = \ldots + A(pqr \ldots) + \ldots ;$$

that is to say, we multiply together a number of monomial symmetric functions so as to exhibit it as a sum of monomial functions; in this sum we find a particular monomial function affected with a numerical coefficient A which, as shown by the present theory, is the number which enumerates lattices of a certain class easily definable. Thus, in the present instance, if the partition $(pqr \ldots)$ involve t parts, the lattices have s columns and t rows; the operation D_p acts, through its various partitions, upon the product of monomials, and any mode of picking out a partition of p from the factors of the product, one part from each factor, constitutes a minor operation which yields the first row of a lattice; the operation D_q is similarly responsible for all the second rows of the lattices, and finally every resulting lattice possesses a property which may be defined as under:—

The numbers in the successive rows are partitions of the numbers $p, q, r, \ldots$ respectively, and in the successive columns are the partitions $(\lambda_1\mu_1\nu_1 \ldots)$, $(\lambda_2\eta_2\nu_2 \ldots)$, $\ldots$ $(\lambda_s\mu_s\nu_s \ldots)$ respectively. Such are the lattices enumerated by the number A. One is reminded somewhat of CAYLEY's well-known algorithm for symmetric function multiplication (invented by him for use in his researches in the theory of Invariants), but here the determination is representative as well as enumerative, and has moreover analytical expression.

Ex. gr. Take as function (987)(654)(321), and as operation $D_{13}D_{15}D_{17}$; then

$$D_{13}D_{15}D_{17}(987)(654(321) = A,$$

where $$(987)(654)(321) = \ldots + A\,(13.15.17) + \ldots .$$

One of the associated lattices is

7	5	1
8	4	3
9	6	2

where observe that the numbers in the successive rows constitute partitions of the numbers 13, 15, 17 respectively, whilst in the successive columns the numbers constitute the partitions (987), (654), (321) respectively. The number of lattices possessing this property, is A, and A is readily found to have the value 6. If we had to find an expression for the number of row and column-magic squares of order 3, it would be necessary to write down the sum of all products

$$(762)(951)(843)$$

formed from the first 9 ($= n^2$) numbers in such wise that the content of each partition factor is $15 = \frac{1}{2}n(n^2+1)$, attention being paid to the order of the partitions, and to take as operation D_{15}^3 or in general $D_{\frac{1}{2}n(n^2+1)}^{n}$. The resulting lattices will all be magic squares in which the diagonal property is not essential, and the result of the operation upon the function will give the enumerating number.

Art. 7. To resume; in the lattice compartments we find invariably the numbers $\lambda_1, \mu_1, \nu_1, \ldots \lambda_2, \mu_2, \nu_2, \ldots \lambda_s, \mu_s, \nu_s, \ldots$ such numbers being subject to certain conditions for each row and each column. The assemblages of numbers in the successive columns do not vary from lattice to lattice, but those in the successive rows do vary from lattice to lattice.

Let $$\lambda_1 + \mu_1 + \nu_1 + \ldots = \lambda;\ \lambda_2 + \mu_2 + \nu_2 + \ldots = \mu;$$
$$\lambda_3 + \mu_3 + \nu_3 + \ldots = \nu, \&c. \ldots$$

Then we have the following facts :—

(i.) The whole assemblage of numbers, $\lambda_1, \mu_1, \nu_1, \ldots \lambda_2, \mu_2, \nu_2, \ldots \lambda_3\, \mu_3\, \nu_3, \ldots$ is unaltered from lattice to lattice.

(ii.) The numbers $\lambda, \mu, \nu, \ldots$ appertaining to the columns, and the numbers $p_1\, q_1\, r, \ldots$ appertaining to the rows, are unaltered from lattice to lattice.

These conditions do not define the lattices in question, because other lattices comply with them, viz., those in which, the whole assemblages of compartment numbers remaining unchanged, the column partitions, while satisfying the condition (ii.), are other than

$$(\lambda_1\mu_1\nu_1 \ldots), (\lambda_2\mu_2\nu_2 \ldots), (\lambda_3\mu_3\nu_3 \ldots), \ldots$$

successively.

Let
$$\lambda_1' + \mu_1' + \nu_1' + \ldots = \lambda,$$
$$\lambda_2' + \mu_2' + \nu_2' + \ldots = \mu,$$
$$\lambda_3' + \mu_3' + \nu_3' + \ldots = \nu,$$
$$\ldots\ldots\ldots\ldots\ldots$$

the assemblage of dashed numbers being in some order identical with the assemblages of undashed numbers. The new conditions include lattices enumerated by

$$D_{pqr} \ldots (\lambda_1'\mu_1'\nu_1' \ldots)(\lambda_2'\mu_2'\nu_2' \ldots)(\lambda_3'\mu_3'\nu_3' \ldots) \ldots$$

and the totality of lattices, implied by them, is enumerated by

$$D_{pqr} \ldots \Sigma(\lambda_1'\mu_1'\nu_1' \ldots)(\lambda_2'\mu_2'\nu_2' \ldots)(\lambda_3'\mu_3'\nu_3' \ldots) \ldots$$

the summation being for every separation of the assemblage of numbers

$$\lambda_1, \mu_1, \nu_1, \ldots \lambda_2, \mu_2, \nu_2, \ldots \lambda_3, \mu_3, \nu_3 \ldots$$

into partitions

$$(\lambda_1'\mu_1'\nu_1' \ldots), (\lambda_2'\mu_2'\nu_2' \ldots), (\lambda_3'\mu_3'\nu_3' \ldots), \ldots$$

such that

$$\lambda_1' + \mu_1' + \nu_1' + \ldots = \lambda,$$
$$\lambda_2' + \mu_2' + \nu_2' + \ldots = \mu,$$
$$\lambda_3' + \mu_3' + \nu_3' + \ldots = \nu;$$

or, as it is convenient to say, for every separation of the given assemblage of numbers which has the *specification* $\lambda, \mu, \nu \ldots$ With this nomenclature we may say that the successive row partitions have a specification $p, q, r \ldots$ and we may assert that the lattices under enumeration are associated with a definite assemblage of numbers and with two specifications, all three of which denote partitions of the same number, $\Sigma\lambda' + \Sigma\mu' + \Sigma\nu' + \ldots = \lambda + \mu + \nu + \ldots = p + q + r + \ldots$ We thus associate the lattice with three partitions of one number.

There is a law of symmetry connected with these lattices the true nature of which is not at once manifest; it is *not* obtained by simple transposition of the above lattices, and we are *not* permitted to simply exchange the partitions $(\lambda\mu\nu \ldots)$, $(pqr \ldots)$ preserving the assemblage of compartment numbers with the object of obtaining identity of enumeration. The difficulty presents itself whenever two or more partitions, $(\lambda_1'\mu_1'\nu_1' \ldots)$, $(\lambda_2'\mu_2'\nu_2' \ldots)$, &c. ... are *different* but have the *same specification*. I will obtain the true theorem by the examination of a particular case. Let the assemblage of numbers be 2, 2, 1, 1, and consider the two results

$$(2)^2(1)^2 = \ldots + 6(2^3) + \ldots$$

connected with $D_2{}^3(2)^2(1)^2 = 6$,

and $$(2)^2(1^2) = \ldots + 2(2^21^2) + \ldots$$

connected with $D_2{}^2D_1{}^2(2)^2(1^2) = 2$.

In the first case the row and column specifications are 2, 2, 2, and 2, 2, 1, 1, respectively; and in the second case 2, 2, 1, 1, and 2, 2, 2, respectively.

The first case yields the six lattices

2			
	2		
		1	1

	2		
2			
		1	1

2			
		1	1
	2		

	2		
		1	1
2			

		1	1
2			
	2		

		1	1
	2		
2			

the second case the two lattices

2		
	2	
		1
		1

	2	
2		
		1
		1

If we transpose the six lattices we obtain four lattices in addition to these two, viz.:—

2		
		2
	1	
	1	

		2
2		
	1	
	1	

;

	2	
		2
1		
1		

		2
	2	
1		
1		

The first pair of these would be derived from $D_2{}^2D_1{}^2(2)(1^2)(2)$, and the second pair from $D_2{}^2D_1{}^2(1^2)(2)^2$. Hence it is clear that to obtain identity of enumeration we must multiply $(2)^2(1^2)$ by a number equal to the number of ways of permuting the factors which have the same specification, viz., by $\frac{(2+1)!}{2!\,1!} = 3$.

The corresponding multiplier of $(2)^2(1^2)$ is $\frac{2!}{2!}\cdot\frac{2!}{2!} = 1$.

Let then an operand be

$$(L_1)^{l_1}(L_2)^{l_2}(L_3)^{l_3}\ldots(M_1)^{m_1}(M_2)^{m_2}(M_3)^{m_3}\ldots$$

$L_1, L_2, L_3, \ldots$ denoting different partitions of the same weight,
$M_1, M_2, M_3 \ldots$ " " "
&c., &c., &c.

we attach a coefficient

$$\frac{(l_1+l_2+l_3+\ldots)!}{l_1!\,l_2!\,l_3!\ldots}\cdot\frac{(m_1+m_2+m_3\ldots)!}{m_1!\,m_2!\,m_3!\ldots}\ldots$$

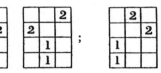

Let any operand

$$(\lambda_1'\mu_1'\nu_1' \ldots)(\lambda_2'\mu_2'\nu_2' \ldots)(\lambda_3'\mu_3'\nu_3' \ldots) \ldots$$

so multiplied be denoted by

$$\text{Co}\,(\lambda_1'\mu_1'\nu_1' \ldots)(\lambda_2'\mu_2'\nu_2' \ldots)(\lambda_3'\mu_3'\nu_3' \ldots) \ldots$$

then we have the following law of symmetry:

From a given finite assemblage of numbers

$$\lambda_1, \mu_1, \nu_1, \ldots, \quad \lambda_2, \mu_2, \nu_2, \ldots, \quad \lambda_3, \mu_3, \nu_3, \ldots \quad \ldots,$$

construct all the products

$$(\lambda_1'\mu_1'\nu_1' \ldots)(\lambda_2'\mu_2'\nu_2' \ldots)(\lambda_3'\mu_3'\nu_3' \ldots) \ldots$$

which have a given specification $(\lambda\mu\nu \ldots)$ and all the products

$$(p_1q_1r_1 \ldots)(p_2q_2r_2 \ldots)(p_3q_3r_3 \ldots) \ldots$$

which have a given specification $(pqr \ldots)$.

If

$$\Sigma\text{Co}\,(\lambda_1'\mu_1'\nu_1' \ldots)(\lambda_2'\mu_2'\nu_2' \ldots)(\lambda_3'\mu_3'\nu_3' \ldots) \ldots = \ldots + \text{A}\,(pqr \ldots) + \ldots$$

then

$$\Sigma\text{Co}\,(p_1q_1r_1 \ldots)(p_2q_2r_2 \ldots)(p_3q_3r_3 \ldots) \ldots = \ldots + \text{A}\,(\lambda\mu\nu \ldots) + \ldots$$

the lattices being derived from

$$\text{D}_{pqr\ldots}\,\Sigma\text{Co}\,(\lambda_1'\mu_1'\nu_1' \ldots)(\lambda_2'\mu_2'\nu_2' \ldots)(\lambda_3'\mu_3'\nu_3' \ldots) \ldots = \text{A}$$

$$\text{D}_{\lambda\mu\nu\ldots}\,\Sigma\text{Co}\,(p_1q_1r_1 \ldots)(p_2q_2r_2 \ldots)(p_3q_3r_3 \ldots) \ldots = \text{A}.$$

This is the most refined law of symmetry that has yet come to light in the algebra of a single system of quantities (*cf.* "Memoirs on Symmetric Functions," 'Amer. J.,' *loc. cit.*). The actual representation of the things enumerated by the number A is obtained with ease by this theory of the lattice.

§ 2.

Art. 8.—So far the operations have been those of the infinitesimal calculus, and the numbers involved in the partitions of the functions have been positive integers excluding zero. If we admit zero as a part in the partitions, we find that we have to do with the operations of the calculus of finite differences. At the commencement of the paper d/dx was shown to be a combinatorial symbol, in that when operating upon a power of x, the said power being positive and integral, it had the effect of summing the results obtained by substituting unity for x in all possible ways in the product of x's. Now the corresponding operator of the calculus of finite differences, viz., Δ operates upon a power of x by striking out one x, two x's, three x's, &c., in all possible ways and summing the results. Thus

$$\Delta x^3 = \not{x}xx + x\not{x}x + xx\not{x} + x\not{x}\not{x} + \not{x}x\not{x} + \not{x}\not{x}x + \not{x}\not{x}\not{x} = 3x^2 + 3x + 1.$$

This simple fact shows that we may expect a corresponding theory of lattices, and that this is, in fact, the case is seen immediately one introduces the part zero into the partitions of the functions. I have introduced zero parts into partitions in the Memoirs on Symmetric Functions above alluded to, and have imported into the theory the corresponding operators d_0 and D_0.* It was there shown that, if n be the number of quantities of which the symmetric functions are formed,

$$d_0 = \frac{d}{dn}, \quad D_0 = e^{\frac{d}{dn}} - 1;$$

and thence it appears that we have operations D_0, d_0, $1 + D_0$ corresponding to the operations Δ, d/dn, E of the calculus of finite differences.

Considering partitions which only involve zero parts, we have only finite difference operations; if we have other integers, we have mixed operations drawn both from the finite and infinitesimal calculus.

The partition (0^p) is derived from

$$\Sigma\alpha_1^q\alpha_2^q\alpha_3^q \ldots \alpha_p^q$$

by putting $q = 0$, and obviously has the value $\binom{n}{p}$ and, in the paper referred to, it has been shown that D_0 operates upon a monomial by erasing one zero part from its partition, so that

$$D_0(0^p) = (0^{p-1}),$$

which is to be compared with the operation of Δ, viz.:—

$$\Delta x^{(m)} = mx^{(m-1)}$$

where
$$x^{(m)} = x(x-1)(x-2)\ldots(x-m+1)$$

in the notation of the finite calculus.

Further, it has been shown that D_0 operates upon a product of monomials through its partitions 0, 00, 000, 0000, . . . which are infinite in number, viz.:—we are to strike out one zero, two zeros, three zeros, &c., in all possible ways; but in any one such operation not striking out more than one zero from any monomial factor.

Ex. gr.
$$\begin{aligned} D_0(0^3)(0^2)(0) = {} & (0^2)(0^2)(0) + (0^3)(0)(0) + (0^3)(0^2) \\ & + (0^2)(0)(0) + (0^2)(0^2) + (0^3)(0) \\ & + (0^2)(0) \end{aligned}$$

the successive lines being due to the partitions 0, 00, 000 respectively.

* 'American Journal of Mathematics,' vol. 12, second memoir, "On a New Theory of Symmetric Functions," p. 71 *et seq.*; and vol. 13, third memoir, &c., pp. 8 *et seq.*

Compare the difference formula

$$\begin{aligned}\Delta u_x v_x w_x &= (\mathrm{EE'E''} - 1)\, u_x v_x w_x \\ &= (\Delta + \Delta' + \Delta'' + \Delta'\Delta'' + \Delta''\Delta + \Delta\Delta' + \Delta\Delta'\Delta'')\, u_x v_x w_x,\end{aligned}$$

where u_x is only operated upon by E and Δ, v_x by E′ and Δ', w_x by E″ and Δ''.

Art. 9.—Consider the lattice theory connected with the operation D_0 and zero-part partition functions.

Take the function

$$(0^\lambda)(0^\mu)(0^\nu)\ldots$$

$\lambda, \mu, \nu, \ldots$ being in descending order; if it be multiplied out, it will appear as a linear function of (0^λ), $(0^{\lambda+1})$, ... $(0^{\lambda+\mu+\nu+\ldots})$, the coefficients being positive (*cf.* Second Memoir, *loc. cit.*, p. 102).

To find therein the coefficient of the term (0^s) we must operate with D_0^s, and the sought coefficient is the resulting *numerical* term. If the factors $(0^\lambda)(0^\mu)(0^\nu)\ldots$ be t in number, we are concerned with lattices of t columns and s rows. The first operation of D_0 results in a first row whose compartments contain t or fewer zeros placed in any manner so that not more than one zero is in each compartment; similarly, for the successive rows and the final lattice is subject to the single condition that the numbers of zeros in the successive columns are $\lambda, \mu, \nu, \ldots$ respectively. The number of such lattices is

$$\{D_0^s(0^\lambda)(0^\mu)(0^\nu)\ldots\}_{n=0} = \{(e^{\frac{d}{dn}} - 1)^s \tbinom{n}{\lambda}\tbinom{n}{\mu}\tbinom{n}{\nu}\ldots\}_{n=0}$$

or, symbolically, $$(e^{\frac{d}{do}} - 1)^s \tbinom{o}{\lambda}\tbinom{o}{\mu}\tbinom{o}{\nu}\ldots$$

We have thus the analytical solution of a distribution problem upon a lattice.

It may be convenient to give the lattice a literal form by writing a for zero in the compartments.

Art. 10.—Contrast the result obtained with that which arises from

$$D_1^s(1^\lambda)(1^\mu)(1^\nu)\ldots$$

The lattices are similar to those above, with the additional condition that each row is to contain but one letter a. Again, from

$$D_{p_1}D_{p_2}\ldots D_{p_s}(1^\lambda)(1^\mu)(1^\nu)\ldots$$

arise lattices of t columns and s rows with the same condition as the zero lattices, but with the additional conditions that the numbers of letters a in the successive rows are to be $p_1, p_2, \ldots p_s$ respectively. This remark leads to a relationship between the coefficients in the developments of $(1^\lambda)(1^\mu)(1^\nu)\ldots$ and $(0^\lambda)(0^\mu)(0^\nu)\ldots$ respectively. For let

$$\begin{aligned}(1^\lambda)(1^\mu)(1^\nu)\ldots = \ldots &+ A_{p_1p_2\ldots p_s}(p_1p_2\ldots p_s) \\ &+ A_{p_1'p_2'\ldots p_s'}(p_1'p_2'\ldots p_s') + \ldots,\end{aligned}$$

the terms written comprising all monomial functions whose partitions contain exactly s parts; and

$$(0^\lambda)(0^\mu)(0^\nu)\ldots = \ldots + B_s(0^s) + \ldots$$

If $P_{p_1 p_2 \ldots p_s}$ denote the number of permutations of the numbers $p_1, p_2, \ldots p_s$, I say that the above lattice theory establishes the relation

$$B_s = P_{p_1 p_2 \ldots p_s} A_{p_1 p_2 \ldots p_s} + P_{p_1' p_2' \ldots p_s'} A_{p_1' p_2' \ldots p_s'} + \ldots$$

Ex. gr. Observe the two results

$$(1^2)(1)(1) = (31) + 2(2^2) + 5(21^2) + 12(1^4),$$
$$(0^2)(0)(0) = 4(0^2) + 15(0^3) + 12(0^4),$$

and verify that the relation given obtains between the coefficients.

Art. 11.—This zero theory is really nothing more than a calculus of binomial coefficients, which enables the study of their properties by means of the powerful instruments appertaining to the Theory of Symmetric Functions. The Law of Symmetry established by the author in the Second Memoir (*loc. cit.*) is easily established by means of the lattice; it may be stated in a simple form, because all the functions $(0), (0^2), (0^3), \ldots$ are to be regarded as having the same specification, viz. (0); further, the specification of $(0^\lambda)(0^\mu)(0^\nu)\ldots$ to s factors is (0^s)). Select all the products formed from a *given number* of zeros which have the same specification (0^i), and attach to each a coefficient equal to the number of permutations of which it is susceptible. Denote the sum of such products by $\Sigma\text{Co}\,(0^\lambda)(0^\mu)(0^\nu)\ldots$. Similarly, for a specification (0^j) denote the sum of such products by $\Sigma\text{Co}\,(0^p)(0^q)(0^r)\ldots$. Then

$$\Sigma\text{Co}(0^\lambda)(0^\mu)(0^\nu)\ldots = \ldots + A(0^j) + \ldots$$
$$\Sigma\text{Co}(0^p)(0^q)(0^r)\ldots = \ldots + A(0^i) + \ldots$$

the coefficient A being the same in both cases.

Ex. gr. Verify that

$$2(0^3)(0) + (0^2)^2 = \ldots + 12(0^3) + \ldots$$
$$3(0^2)(0)^2 = \ldots + 12(0^2) + \ldots$$

§ 3.

Art. 12.—I continue the general plan of this paper, viz.:—I do not attempt the solution of any particular problems, unless they are suggested by the general course of the investigation, but rather start with definite operations and functions, and seek to discover the problems of which they furnish the solution. This is the reverse process to that employed in the 'Trans. Camb. Phil. Soc.' (*loc. cit.*), where I particularly investigated a number of questions more or less directly associated with the famous Problem of the Latin Square. I anticipate what follows to the extent of observing that the Latin Square again presents itself without special effort on the part of the investigator, and that a new and very simple solution of that and associated problems is obtained.

I seek to obtain theorems which flow from a consideration of symmetric functions of several systems of quantities, taken in conjunction with appropriate operations.

I make the reference MACMAHON, "Memoir on the Roots of Systems of Equations," 'Phil. Trans.,' A, 1890.

Consider the systems of quantities

$$\begin{array}{l} \alpha_1, \alpha_2, \alpha_3, \ldots \\ \beta_1, \beta_2, \beta_3, \ldots \\ \gamma_1, \gamma_2, \gamma_3, \ldots \\ \ldots\ldots\ldots \end{array}$$

and write

$$(1 + \alpha_1 x_1 + \beta_1 x_2 + \gamma_1 x_3 + \ldots)(1 + \alpha_2 x_1 + \beta_2 x_2 + \gamma_2 x_3) \ldots$$
$$= 1 + a_{100\ldots} x_1 + a_{010\ldots} x_2 + \ldots + a_{pqr\ldots} x_1^p x_2^q x_3^r \ldots + \ldots$$

Denote the symmetric function

$$\Sigma \alpha_1^{p_1} \beta_1^{q_1} \gamma_1^{r_1} \ldots \alpha_2^{p_2} \beta_2^{q_2} \gamma_2^{r_2} \ldots \alpha_3^{p_3} \beta_3^{q_3} \gamma_3^{r_3} \ldots \ldots$$

by $(\overline{p_1 q_1 r_1 \ldots}\ \overline{p_2 q_2 r_2 \ldots}\ \overline{p_3 q_3 r_3 \ldots} \ldots)$

so that $a_{pqr\ldots} = (\overline{100\ldots}^p\ \overline{010\ldots}^q\ \overline{001\ldots}^r \ldots)$

Ex. gr. $a_{1111} = \Sigma \alpha_1 \beta_2 \gamma_3 \delta_4 = (\overline{1000}\ \overline{0100}\ \overline{0010}\ \overline{0001})$

The quantities $a_{pqr\ldots}$ are the elementary symmetric functions.

The linear operator $d_{\pi\kappa\rho\ldots}$ is defined by

$$d_{\pi\kappa\rho\ldots} = \sum^{p,q,r,\ldots} a_{p-\pi,\, q-\kappa,\, r-\rho,\ldots} \frac{d}{da_{pqr\ldots}},$$

so that

$$d_{100\ldots} = \Sigma\ a_{p-1,q,r\ldots} \frac{d}{da_{pqr\ldots}},$$

$$d_{010\ldots} = \Sigma\ a_{p,q-1,r\ldots} \frac{d}{da_{pqr\ldots}},$$

$$d_{001\ldots} = \Sigma\ a_{p,q,r-1,\ldots} \frac{d}{da_{pqr\ldots}},$$

and then $$D_{pqr\ldots} = \frac{1}{p!\,q!\,r!\ldots} d_{100\ldots}^p\, d_{010\ldots}^q\, d_{001\ldots}^r \ldots,$$

the multiplication of operators being symbolic as in TAYLOR's theorem, so that $D_{pqr\ldots}$ is an operator of the order $p + q + r + \ldots$ and does *not* denote

$p+q+r+\ldots$ successive linear operations. The operation of $D_{pqr\ldots}$ upon a monomial symmetric function has been explained (*loc. cit.*). It has the effect of obliterating a part $\overline{pqr\ldots}$ from the partition of the function when such a part is present, and an annihilating effect is every other case. The operation upon a product has the effect of erasing a partition of $\overline{pqr\ldots}$ from the product, one part from each factor in all possible ways, the result of the operation being a sum of products, one product arising from each erasure of a partition.

Ex. gr.
$$D_{43}(\overline{43}\ \overline{22}) = (\overline{22}).$$

If we have to operate with D_{43} upon

$$(\overline{32}\ \overline{22})\ (\overline{21}\ \overline{11})$$

we have to erase the two partitions $(\overline{32}\ \overline{11})$, $(\overline{22}\ \overline{21})$, and arrive at

$$D_{43}(\overline{32}\ \overline{22})\ (\overline{21}\ \overline{11}) = (\overline{22})\ (\overline{21}) + (\overline{32}\ \overline{11}).$$

Art. 13.—It will suffice to consider three systems of quantities as typical of the general case.

Take the function

$$a_{\lambda_1\mu_1\nu_1}a_{\lambda_2\mu_2\nu_2}\ldots a_{\lambda_s\mu_s\nu_s} = (\overline{100}^{\lambda_1}\overline{010}^{\mu_1}\overline{001}^{\nu_1})(\overline{100}^{\lambda_2}\overline{010}^{\mu_2}\overline{001}^{\nu_2})\ldots(\overline{100}^{\lambda_s}\overline{010}^{\mu_s}\overline{001}^{\nu_s})$$

and the operation
$$D_{p_1q_1r_1}D_{p_2q_2r_2}\ldots D_{p_tq_tr_t}$$

$$(\overline{\lambda_1\mu_1\nu_1}\ \overline{\lambda_2\mu_2\nu_2}\ \ldots\ \overline{\lambda_s\mu_s\nu_s}) \text{ and } (\overline{p_1q_1r_1}\ \overline{p_2q_2r_2}\ \ldots\ \overline{p_tq_tr_t})$$

being each partitions of the same tripartite number.

If
$$a_{\lambda_1\mu_1\nu_1}a_{\lambda_2\mu_2\nu_2}\ldots a_{\lambda_s\mu_s\nu_s} = \ldots + A(\overline{p_1q_1r_1}\ \overline{p_2q_2r_2}\ \ldots\ \overline{p_tq_tr_t}) + \ldots,$$
$$D_{p_1q_1r_1}D_{p_2q_2r_2}\ldots D_{p_tq_tr_t}\ a_{\lambda_1\mu_1\nu_1}a_{\lambda_2\mu_2\nu_2}\ldots a_{\lambda_s\mu_s\nu_s} = A\ ;$$

and we have to determine the nature of the lattices enumerated by the number A. The tripartite number $(\overline{p_1q_1r_1})$ has a partition of $p_1+q_1+r_1$ parts, viz.:— $\overline{100}^{p_1}\ \overline{010}^{q_1}\ \overline{001}^{r_1}$ so that, in operating with $D_{p_1q_1r_1}$ upon the operand, we have to select this partition from the product, one part from each factor, in all possible ways; the operation breaks up into minor operations as usual, and the first row of the lattice of s columns and t rows will contain in $p_1+q_1+r_1$ of its compartments the tripartite numbers 100, 010, 001 (p_1 of the first, q_1 of the second, and r_1 of the third) in some order; the assemblage of numbers in this row is the partition of the elementary function $a_{p_1q_1r_1}$. Similarly a minor operation of $D_{p_2q_2r_2}$ produces a second row containing tripartite numbers, the assemblage of which constitutes the elementary function $a_{p_2q_2r_2}$. We finally arrive at a lattice such that the tripartites in the successive rows constitute the elementary functions $a_{p_1q_1r_1}, a_{p_2q_2r_2}, \ldots a_{p_tq_tr_t}$ respectively, and in the successive columns the elementary functions $a_{\lambda_1\mu_1\nu_1}, a_{\lambda_2\mu_2\nu_2}, \ldots a_{\lambda_s\mu_s\nu_s}$ respectively, and the number A enumerates the lattices possessing this property.

We may give this case a purely literal form by writing $100 = a$, $010 = b$, $001 = c$, and then we have a lattice of s columns and t rows, such that the products of letters in the successive rows are $a^{p_1}b^{q_1}c^{r_1}, a^{p_2}b^{q_2}c^{r_2}, \ldots a^{p_t}b^{q_t}c^{r_t}$ respectively, and in the successive columns $a^{\lambda_1}b^{\mu_1}c^{\nu_1}, a^{\lambda_2}b^{\mu_2}c^{\nu_2}, \ldots a^{\lambda_s}b^{\mu_s}c^{\nu_s}$ respectively.

Art. 14.—Stated in this form the problem appears to have a close relationship to the problem of the Latin Square. It is in fact a new generalization of that problem; for put $s = t = 3$ and

$$p_1 = q_1 = r_1 = p_2 = q_2 = r_2 = p_3 = q_3 = r_3 = 1$$
$$\lambda_1 = \mu_1 = \nu_1 = \lambda_2 = \mu_2 = \nu_2 = \lambda_3 = \mu_3 = \nu_3 = 1$$

so that the operation is D^3_{111} and the function a^3_{111}. One lattice is then

010	100	001
001	010	100
100	001	010

or in literal form

b	a	c
c	b	a
a	c	b

which is a Latin Square. Hence the numbers of Latin Squares of order 3 is

$$D^3_{111}\, a^3_{111},$$

and in general of order n

$$D^n_{111\ldots1}\, a^n_{111\ldots1},$$

a very simple solution of the problem. If reference be made to the solution arrived at (*loc. cit.*) by considerations relating to a single system of quantities, it will be noticed that the peculiar difficulties intrinsically present in that solution disappear at once when n systems of quantities are brought in as auxiliaries. The Latin Square appears at the outset of this investigation, and in a perfectly natural manner.

Art. 15.—Now put

$$p_1 = p_2 = \ldots = p_{\lambda+\mu+\nu} = \lambda = \lambda_1 = \lambda_2 = \ldots = \lambda_{\lambda+\mu+\nu}$$
$$q_1 = q_2 = \ldots = q_{\lambda+\mu+\nu} = \mu = \mu_1 = \mu_2 = \ldots = \mu_{\lambda+\mu+\nu}$$
$$r_1 = r_2 = \ldots = r_{\lambda+\mu+\nu} = \nu = \nu_1 = \nu_2 = \ldots = \nu_{\lambda+\mu+\nu},$$

so that $s = t = \lambda + \mu + \nu$. We have then lattices enumerated by

$$D^{\lambda+\mu+\nu}_{\lambda\mu\nu}\, a^{\lambda+\mu+\nu}_{\lambda\mu\nu},$$

and, in the literal form, they are such that the product of letters in each of the $\lambda + \mu + \nu$ rows and $\lambda + \mu + \nu$ columns is $a^\lambda b^\mu c^\nu$, one letter appearing in each compartment of the lattice. This is the extension of the idea of the Latin Square which was successfully considered in the former paper (*loc. cit.*), but now the enumeration is given in a simpler form and from simpler considerations.

In general, it has been established above that the Latin Squares based upon the product

$$a^\lambda b^\mu c^\nu \ldots$$

are enumerated by the expression

$$D_{\lambda\mu\nu\ldots}^{\lambda+\mu+\nu+\ldots} \quad a_{\lambda\mu\nu\ldots}^{\lambda+\mu+\nu+\ldots}$$

the simplicity of which leaves nothing to be desired.

Art. 16.—Consider the particular case of the general theorem which is such that no compartment is empty; the lattice has n columns and m rows.

$$p_1 + q_1 + r_1 = p_2 + q_2 + r_2 = \ldots = p_m + q_m + r_m = n$$
$$\lambda_1 + \mu_1 + \nu_1 = \lambda_2 + \mu_2 + \nu_2 = \ldots = \lambda_n + \mu_n + \nu_n = m,$$

the corresponding lattices being enumerated by

$$D_{p_1q_1r_1} D_{p_2q_2r_2} \ldots D_{p_mq_mr_m} \quad a_{\lambda_1\mu_1\nu_1} a_{\lambda_2\mu_2\nu_2} \ldots a_{\lambda_n\mu_n\nu_n},$$

These, when given the literal form, possess the property that the products of letters in the successive rows are $a^{p_1}b^{q_1}c^{r_1}$, $a^{p_2}b^{q_2}c^{r_2}$, $\ldots$ $a^{p_m}b^{q_m}c^{r_m}$ respectively, and in the successive columns $a^{\lambda_1}b^{\mu_1}c^{\nu_1}$, $a^{\lambda_2}b^{\mu_2}c^{\nu_2}$, $\ldots$ $a^{\lambda_n}b^{\mu_n}c^{\nu_n}$ respectively.

Ex. gr. Suppose the row products to be a^3b, a^2b^2, a^2b^2, a^4, and the column products a^4, a^3b, a^3b, ab^3

$$D_{31}D_{22}^2 D_{40} a_{40} a_{31}^2 a_{13}$$
$$= D_{31}D_{22}^2 D_{40}(\overline{10}^4)(\overline{10}^3\,\overline{01})^2(\overline{10}\,\overline{01}^3)$$
$$= D_{31}D_{22}^2(\overline{10}^3)(\overline{10}^2\,\overline{01})^2(\overline{01}^3)$$
$$= D_{22}^2(\overline{10}^2)(\overline{10}\,\overline{01})^2(\overline{01}^2)$$
$$= D_{22}\, 2(\overline{10})^2(\overline{01})^2$$
$$= 2.$$

and the two lattices are

a	*a*	*a*	*b*
a	*a*	*b*	*b*
a	*b*	*a*	*b*
a	*a*	*a*	*a*

a	*a*	*a*	*b*
a	*b*	*a*	*b*
a	*a*	*b*	*b*
a	*a*	*a*	*a*

.

Again, suppose the row products to be a^3b, a^2b^2, a^2b^2, and the column products a^3, a^2b, a^2b, b^3

$$
\begin{aligned}
&D_{31}D_{22}^2\, a_{30}\, a_{21}^2\, a_{03} \\
&= D_{31}D_{22}^2 (\overline{10}^3)(\overline{10}^2\,\overline{01})^2(\overline{01}^2) \\
&= D_{22}^2 (\overline{10}^2)(\overline{10}\,\overline{01})^2(\overline{01}^2) \\
&= 2D_{22}(\overline{10})^2(\overline{01})^2 = 2,
\end{aligned}
$$

and the lattices are

a	*a*	*a*	*b*
a	*a*	*b*	*b*
a	*b*	*a*	*b*

a	*a*	*a*	*b*
a	*b*	*a*	*b*
a	*a*	*b*	*b*

Art. 17.—In general, we may state that the lettered lattices, which are such that the row products are in order

$$a^{p_1}b^{q_1}c^{r_1} \ldots k^{z_1},\ a^{p_2}b^{q_2}c^{r_2} \ldots k^{z_2},\ \ldots\ a^{p_m}b^{q_m}c^{r_m} \ldots k^{z_m},$$

and the column products in order

$$a^{\lambda_1}b^{\mu_1}c^{\nu_1} \ldots k^{\zeta_1},\ a^{\lambda_2}b^{\mu_2}c^{\nu_2} \ldots k^{\zeta_2},\ \ldots\ a^{\lambda_n}b^{\mu_n}c^{\nu_n} \ldots k^{\zeta_n},$$

one letter being in each compartment, are enumerated by

$$D_{p_1q_1r_1\ldots z_1}\, D_{p_2q_2r_2\ldots z_2} \ldots D_{p_mq_mr_m\ldots z_m}\, a_{\lambda_1\mu_1\nu_1\ldots\zeta_1}\, a_{\lambda_2\mu_2\nu_2\ldots\zeta_2} \ldots a_{\lambda_n\mu_n\nu_n\ldots\zeta_n};$$

a very interesting development of the Latin Square problem.

We have found above the nature of the lattices enumerated by the number

$$D_{p_1q_1r_1\ldots}\, D_{p_2q_2r_2\ldots} \ldots D_{p_tq_tr_t\ldots}\, a_{\lambda_1\mu_1\nu_1\ldots}\, a_{\lambda_2\mu_2\nu_2\ldots} \ldots a_{\lambda_s\mu_s\nu_s\ldots}$$

any number of systems of quantities being involved, and the mere fact of the existence of the lattices indicates a law of symmetry which may be stated as follows :—

If

$$a_{\lambda_1\mu_1\nu_1\ldots}\, a_{\lambda_2\mu_2\nu_2\ldots} \ldots a_{\lambda_s\mu_s\nu_s\ldots} = \ldots + A(\overline{p_1q_1r_1\ldots}\ \overline{p_2q_2r_2\ldots}\ \ldots\ \overline{p_tq_tr_t\ldots}) + \ldots$$

then

$$a_{p_1q_1r_1\ldots}\, a_{p_2q_2r_2\ldots} \ldots a_{p_tq_tr_t\ldots} = \ldots + A(\overline{\lambda_1\mu_1\nu_1\ldots}\ \overline{\lambda_2\mu_2\nu_2\ldots}\ \ldots\ \overline{\lambda_s\mu_s\nu_s\ldots}) + \ldots$$

Art. 18.—The next case that comes forward for examination is that connected with the homogeneous product sums $h_{\lambda\mu\nu}\ldots$ We require the theorem

$$D_{pqr\ldots}\, h_{\lambda\mu\nu\ldots} = h_{\lambda-p,\,\mu-q,\,\nu-r\ldots}$$

and also

$$D_{pqr\ldots}h_{\lambda_1\mu_1\nu_1\ldots}h_{\lambda_2\mu_2\nu_2\ldots}\ldots\ldots h_{\lambda_s\mu_s\nu_s\ldots}$$
$$= \Sigma h_{\lambda_1-p_1,\,\mu_1-q_1,\,\nu_1-r_1,\ldots}h_{\lambda_2-p_2,\,\mu_2-q_2,\,\nu_2-r_2,\ldots}\ldots\ldots h_{\lambda_s-p_s,\,\mu_s-q_s,\,\nu_s-r_s,\ldots}$$

where $(\overline{p_1q_1r_1\ldots}\ \overline{p_2q_2r_2\ldots}\ \ldots\ \overline{p_sq_sr_s\ldots})$ is a partition of $(\overline{pqr\ldots})$, and the sum is for all such partitions and for a particular partition is for all ways of operating upon the suffixes with the parts of the partition. *Ex. gr.*

$$D_{11}h_{11}h_{22} = h_{22} + h^2_{11} + h_{01}h_{21} + h_{10}h_{12}$$

Taking only tripartite functions for convenience, consider the function

$$h_{\lambda_1\mu_1\nu_1}h_{\lambda_2\mu_2\nu_2}\ldots h_{\lambda_s\mu_s\nu_s}$$

and the operation $D_{p_1q_1r_1}D_{p_2q_2r_2}\ldots D_{p_tq_tr_t}$;

we have $D_{p_1q_1r_1}D_{p_2q_2r_2}\ldots D_{p_tq_tr_t}h_{\lambda_1\mu_1\nu_1}h_{\lambda_2\mu_2\nu_2}\ldots h_{\lambda_s\mu_s\nu_s} = A$;

where $(\overline{p_1q_1r_1}\ \overline{p_2q_2r_2}\ \ldots\ \overline{p_tq_tr_t})$ and $(\overline{\lambda_1\mu_1\nu_1}\ \overline{\lambda_2\mu_2\nu_2}\ \ldots\ \overline{\lambda_s\mu_s\nu_s})$

being partitions of the same tripartite number,

$$h_{\lambda_1\mu_1\nu_1}h_{\lambda_2\mu_2\nu_2}\ldots h_{\lambda_s\mu_s\nu_s} = \ldots + A(\overline{p_1q_1r_1}\ \overline{p_2q_2r_2}\ \ldots\ \overline{p_tq_tr_t}) + \ldots$$

The operation of $D_{p_1q_1r_1}$ upon the product splits up as usual into a number of minor operations, one of which, as shown above, is connected with one of its partitions operating in a definite manner upon the suffixes $\lambda_1\mu_1\nu_1$, $\lambda_2\mu_2\nu_2$, ... $\lambda_s\mu_s\nu_s$. Hence the first row of the lattice has in certain of its s compartments the tripartite parts ot some partition of $\overline{p_1q_1r_1}$; the second row also will have in certain of its compartments the tripartite parts of some partition of $\overline{p_2q_2r_2}$; and finally we must arrive at a lattice whose rows are associated with partitions of $\overline{p_1q_1r_1}$, $\overline{p_2q_2r_2}$, ... $\overline{p_tq_tr_t}$ respectively, and whose t columns are associated with partitions of $\overline{\lambda_1\mu_1\nu_1}$, $\overline{\lambda_2\mu_2\nu_2}$, ... $\overline{\lambda_s\mu_s\nu_s}$ respectively. There is no restriction on the magnitude of the constituents of the various tripartite numbers which appear in the compartments. The lattices thus defined are enumerated by the number A. We may give the lattice a literal form by writing $a^\rho b^\sigma c^\tau$ for $(\rho\sigma\tau)$ in a compartment. We then have a theorem which may be stated as follows:—

Monomial products of letters a, b, c, ... may be placed in the compartments of a lattice of t rows and s columns in such wise that the multiplication of products in successive rows produces $a^{p_1}b^{q_1}c^{r_1}$ ..., $a^{p_2}b^{q_2}c^{r_2}$ ..., ... $a^{p_t}b^{q_t}c^{r_t}$ ... respectively, and in successive columns produces $a^{\lambda_1}b^{\mu_1}c^{\nu_1}$ ..., $a^{\lambda_2}b^{\mu_2}c^{\nu_2}$ ..., ... $a^{\lambda_s}b^{\mu_s}c^{\nu_s}$ ... respectively in a number of ways enumerated by the number A above defined.

It is scarcely necessary to observe that

$$\Sigma p - \Sigma\lambda = \Sigma q - \Sigma\mu = \Sigma r - \Sigma\nu = \ldots = 0,$$

and that only $s + t - 1$ of the $s + t$ literal products are independent.

Art. 19.—In conclusion, it may be remarked that there is no difficulty in evolving a mixed theory which involves the operation both of the infinitesimal calculus and of the finite calculus. Operations and functions may be designed which lead to lattices which are not rectangular. The theory may be connected with complete or incomplete lattices in three or more dimensions; and finally one of the most promising paths of research appears to be connected with a multipartite zero. These matters may be the subjects of future investigation. For the present enough has been said to indicate the apparent scope of the new method.

COMBINATIONS DERIVED FROM m IDENTICAL SETS OF n DIFFERENT LETTERS AND THEIR CONNEXION WITH GENERAL MAGIC SQUARES

By Major P. A. MACMAHON.

[Received January 23rd, 1917.—Read February 1st, 1917.—Revised February 20th, 1917.]

IN this paper I consider combinations which are derived from m identical sets of n different letters, and shew that there is a one-to-one correspondence between them and squares of order n which are such that the numbers, placed in the cells, have for each row and column a sum m. I generalize this theory and give various theorems and numerical results. The squares in question have been named by me "General Magic Squares," and have been the subject of investigation,* but, in the present paper, squares are considered to be identical if they can be made so by interchange of rows.

The combinations have not, I believe, been treated by any previous investigator.

The problem involved is the enumeration of particular sets of partitions of certain multipartite numbers.

1. If a set of objects be such that p are of one kind, q of a second, r of a third, *et cetera*, the set is said to have a specification

$$(pqr\ldots),$$

which, since $p, q, r, \ldots$, may be taken to be numbers in descending order of magnitude with loss of generality, is a partition of the number

$$p+q+r+\ldots.$$

When these objects are distributed into an equal number of boxes of specification

$$(p'q'r'\ldots),$$

one object into each box, the number of the distributions is well known.†

* "Combinatory Analysis," *Camb. Univ. Press*, 1916, Vol. 2, §VIII, Ch. VII.

† *loc. cit.*, Vol. 1, § I, Ch. II.

This number is not altered when the specifications of the objects and of the boxes are interchanged. In fact, the boxes may be also regarded as objects, and the distribution to be a process of forming a set of pairs of objects, one object from each set being a constituent of a pair. Observe that there are no similarities between the objects in the one set and those in the other. If there be any similarities the enumeration of the sets of pairs of objects is a different matter.

Ex. gr.—Denoting the boxes and objects by capital and small letters, respectively, suppose that the boxes and objects both have the specification (111). The distributions are six in number

$$\begin{matrix} ABC & ABC & ABC & ABC & ABC & ABC \\ a\,b\,c & a\,c\,b & b\,a\,c & b\,c\,a & c\,a\,b & c\,b\,a \end{matrix}$$

Six sets of pairs of objects can be formed when we regard the boxes as objects.

Suppose now that instead of boxes we have objects of specification (111) precisely similar to the objects a, b, c. We then find the sets of pairs

$$\begin{matrix} abc & abc & abc & abc & abc & abc \\ abc & acb & bac & bca & cab & cba \end{matrix}$$

and we note that the sets of pairs shewn by

$$\begin{matrix} abc & abc \\ bca & cab \end{matrix}$$

are identical, each yielding ab, bc, ca.

The number of different sets of pairs is five, not six.

I now discuss the enumeration of the complete pairing of the members of the two identical sets

$$a_1 a_2 a_3 \ldots a_n, \qquad a_1 a_2 a_3 \ldots a_n,$$

each of which has the specification (1^n).

When it is convenient the letters a, b, c, ... will be used.

When $n = 1$, we have one pairing aa.

When $n = 2$, we have the two aa, bb; ab, ab.

A pairing which involves no pair, such as aa or bb, it is convenient to term a "linear pairing."

We thus have, for $n = 2$, two pairings one of which is linear.

When $n = 3$, it has been shewn that there are five pairings, one of which, viz., ab, ac, bc, is linear.

When $n = 4$, denote the number of pairings by

$$\begin{vmatrix} abcd \\ abcd \end{vmatrix} \quad \text{or} \quad p_4.$$

If l_4 denotes the number of linear pairings, we reach the relation

$$p_4 = \begin{vmatrix} a \\ a \end{vmatrix} \begin{vmatrix} b \\ b \end{vmatrix} \begin{vmatrix} c \\ c \end{vmatrix} \begin{vmatrix} d \\ d \end{vmatrix} + 6 \begin{vmatrix} a \\ a \end{vmatrix} \begin{vmatrix} b \\ b \end{vmatrix} l_2 + 4 \begin{vmatrix} a \\ a \end{vmatrix} l_3 + l_4,$$

or

$$p_4 = 1 + 6l_2 + 4l_3 + l_4,$$

and, generally,

$$p_n = 1 + \binom{n}{2} l_2 + \binom{n}{3} l_3 + \ldots + \binom{n}{n} l_n,$$

which, remembering that

$$l_1 = 0,$$

may be written symbolically

$$p^n = (1+l)^n,$$

wherein p^s, l^s are symbols for p_s, l_s.

Thence symbolically

$$p = 1 + l,$$

and

$$l^n = (p-1)^n,$$

leading to the non-symbolic relation

$$l_n = p_n - \binom{n}{1} p_{n-1} + \binom{n}{2} p_{n-2} - \ldots + (-1)^n.$$

We can thus derive the numbers p_n from the numbers l_n and *vice versa.*

Returning to the case $n = 4$, we note that the two objects a may occur in the single pair aa or in the pairs

$$ab,\ ab\ ; \quad ac,\ ac\ ; \quad ad,\ ad\ ;$$

$$ab,\ ac\ ; \quad ab,\ ad\ ; \quad ac,\ ad\ ;$$

and we derive the formula

$$p_4 = p_3 + 3p_2 + 3 \begin{vmatrix} cd \\ bd \end{vmatrix},$$

whence

$$p_4 = 5 + 6 + 6 = 17.$$

The above principle leads to a difference equation for p_n.

For consider the general case

$$\left.\begin{matrix} a_1\, a_2\, a_3\, \dots\, a_n \\ a_1\, a_2\, a_3\, \dots\, a_n \end{matrix}\right| = p_n.$$

We find that

$$p_n = p_{n-1} + \binom{n-1}{1} p_{n-2} + \binom{n-1}{2} \left|\begin{matrix} a_2\, a_4\, a_5\, \dots\, a_n \\ a_3\, a_4\, a_5\, \dots\, a_n \end{matrix}\right|.$$

Moreover, since

$$\left|\begin{matrix} a_2\, a_4\, a_5\, \dots\, a_n \\ a_3\, a_4\, a_5\, \dots\, a_n \end{matrix}\right| = p_{n-3} + (n-3) \left|\begin{matrix} a_4\, a_5\, \dots\, a_n \\ a_3\, a_5\, \dots\, a_n \end{matrix}\right|,$$

we reach the formula

$$p_n = p_{n-1} + \binom{n-1}{1} p_{n-2}$$
$$+ \binom{n-1}{2} \{p_{n-3} + (n-3)\, p_{n-4} + (n-3)(n-4)\, p_{n-5} + \dots\},$$

a difference equation which involves $p_0\,(=1)$, $p_1, p_2, \dots, p_n$.

Now, writing

$$q_n = p_{n-3} + (n-3)\, p_{n-4} + (n-3)(n-4)\, p_{n-5} + \dots,$$

we find $$q_n = (n-2)\, q_{n-1} - \binom{n-4}{2} q_{n-3},$$

and thus $$p_n = p_{n-1} + \binom{n-1}{1} p_{n-2} + \binom{n-1}{2} q_n,$$

and finally, a difference equation for p_n which involves six consecutive terms, viz.,

$$(n-3)\, p_n - (n^2-2n-1)\, p_{n-1} + (n-1)\, p_{n-2} + \binom{n-1}{2} (3n-7)\, p_{n-3}$$
$$-3 \binom{n-1}{3} p_{n-4} - 12 \binom{n-1}{4} p_{n-5} = 0.$$

From these formulæ the following values of p_n, q_n, l_n have been calculated.

n	p	q	l
1	1		0
2	2		1
3	5	1	1
4	17	2	6
5	73	6	22
6	388	23	130
7	2461	109	822
8	18155	618	6202
9	152531	4096	52552
10	1436714	31133	499194
11	14986879	267219	5238370
12	171453343	2557502	60222844

2. These numbers may be studied from another point of view. We may make use of the notation of the Theory of Substitutions. When we denote the pairing by placing the constituents of a pair in the same column, in the manner

$$\begin{matrix} a & b & c \\ c & a & b \end{matrix}$$

to denote the pairing *ac*, *ab*, *bc*, we observe that

$$\begin{pmatrix} a & b & c \\ c & a & b \end{pmatrix}$$

denotes a substitution. It is a linear pairing and also a circular substitution of the maximum order 3. In the Theory of Substitutions it may be denoted by

$$(a\,c\,b).$$

Hence we may consider that the substitution $(a\,c\,b)$ also denotes the pairing *ac*, *ab*, *bc*.

The same pairing also corresponds to the substitution $(a\,b\,c)$, because $(a\,c\,b)$, $(a\,b\,c)$ are inverse substitutions.

In general there is a one-to-two correspondence between the related pairing and the two substitutions

$$(a_1 a_2 a_3 \dots a_n), \quad (a_1 a_n a_{n-1} \dots a_2 a_1),$$

because these are inverse substitutions.

It is thus evident that, regarding the substitutions in the circular notaon, we can derive the theory of the pairing.

Ex. gr., of the Order 6 there are substitutions which in circular form correspond to every partition of 6. When the substitution is expressed by a circular substitution of six elements, we find 5! substitutions, giving rise to $\frac{1}{2}.5!$ or 60 distinct pairings. When the substitution involves λ circular substitutions, each of which involves more than two letters, we divide the number of substitutions of the given form by 2^λ to obtain the corresponding number of pairings. The following scheme shews the complete calculation for the Order 6.

Order 6.

Partition.	Type of Substitution.	No. of Substitutions.	Divisor.	No. of Pairings.	
6	(123456)	120	2	60	130 (sum of first four)
24	(12)(3456)	90	2	45	
33	(123)(456)	40	4	10	
222	(12)(34)(56)	15	1	15	
15	(1)(23456)	144	2	72	
123	(1)(23)(456)	120	2	60	
1^24	(1)(2)(3456)	90	2	45	
1^22^2	(1)(2)(34)(56)	45	1	45	
1^33	(1)(2)(3)(456)	40	2	20	
1^42	(1)(2)(3)(4)(56)	15	1	15	
1^6	(1)(2)(3)(4)(5)(6)	1	1	1	
		720		388	

verifying the number 388.

The first four partitions of six which do not contain the part unity correspond to the substitutions which displace every letter and yield the linear pairings, shewn above to be 130 in number.

For the order n we have, for the partition

$$1^\alpha 2^\beta 3^\gamma 4^\delta \ldots,$$

a number of substitutions

$$\frac{(\alpha+2\beta+3\gamma+4\delta+\ldots)!}{\alpha!\,\beta!\,\gamma!\,\delta!\ldots(2!)^{\beta-\gamma}(3!)^{\gamma-\delta}\ldots}.$$

As divisor we have $2^{\gamma+\delta+\dots}$, so that the number of pairings is

$$\frac{(\alpha+2\beta+3\gamma+4\delta+\dots)!}{\alpha!\,\beta!\,\gamma!\,\delta!\dots(2!)^{\beta-\gamma}(3!)^{\gamma-\delta}\dots 2^{\gamma+\delta+\dots}}.$$

The total number of pairings is therefore

$$P_n = \Sigma\frac{(\alpha+2\beta+3\gamma+4\delta+\dots)!}{\alpha!\,\beta!\,\gamma!\,\delta!\dots(2!)^{\beta-\gamma}(3!)^{\gamma-\delta}\dots 2^{\gamma+\delta+\dots}},$$

the summation being for every partition

$$1^\alpha 2^\beta 3^\gamma 4^\delta \dots$$

of n. Also

$$l_u = \Sigma\frac{(2\beta+3\gamma+4\delta+\dots)!}{\beta!\,\gamma!\,\delta!\dots(2!)^{\beta-\gamma}(3!)^{\gamma-\delta}\dots 2^{\gamma+\delta+\dots}},$$

the summation being for every partition

$$2^\beta 3^\gamma 4^\delta \dots$$

of n.

6. There is a third point of view from which the theory of the pairings can be regarded, and it has a larger outlook than those previously examined. We may consider m identical sets of n different letters, and one letter to be taken from each set so as to form an m-ad of letters; this process may be repeated until the sets are exhausted, and we are left with a set of n m-ads of letters. We may enquire as to how many different sets of n m-ads can thus be formed. It is the combination of m-ads which is in question and not permutation between them.

If we look at the elementary case already dealt with for which $n = 3$, $m = 2$, we find the five sets of duads

$$\begin{matrix} a^2, & b^2, & c^2 \\ a^2, & bc, & bc \\ ac, & b^2, & ac \\ ab, & ab, & c^2 \\ ab, & ac, & bc \end{matrix}$$

which are in co-relation with five partitions of the multipartite (here tripartite because $n = 3$) number viz.,

(200, 020, 002)
(200, 011, 011)
(101, 020, 101)
(110, 110, 002)
(110, 101, 011)

—where observe that 2 occurs in the partitioned multipartite, because here $m = 2$.

The partitions may be defined as being those into three parts for which the weight of each part is 2.

We have, in fact, to find the partitions

$$(a_1 a_2 a_3, \quad a_4 a_5 a_6, \quad a_7 a_8 a_9),$$

such that the following six Diophantine relations are satisfied, viz.,

$$a_1+a_4+a_7 = 2,$$
$$a_2+a_5+a_8 = 2,$$
$$a_3+a_6+a_9 = 2,$$
$$a_1+a_2+a_3 = 2,$$
$$a_4+a_5+a_6 = 2,$$
$$a_7+a_8+a_9 = 2,$$

and we thus see that there is a one-to-one correspondence between the partitions before us and Magic Squares of a certain nature. If in a square of order n the row and column properties are in evidence but not the diagonal properties, we may call it a General Magic Square of order n.

In the case before us $n = 3$, and the sum of the numbers in any row or column must be $m = 2$. The squares are

$$\begin{matrix} 2 & 0 & 0 \\ 0 & 2 & 0 \\ 0 & 0 & 2 \end{matrix} \qquad \begin{matrix} 2 & 0 & 0 \\ 0 & 1 & 1 \\ 0 & 1 & 1 \end{matrix} \qquad \begin{matrix} 1 & 0 & 1 \\ 0 & 2 & 0 \\ 1 & 0 & 1 \end{matrix}$$

$$\begin{matrix} 1 & 1 & 0 \\ 1 & 1 & 0 \\ 0 & 0 & 2 \end{matrix} \qquad \begin{matrix} 1 & 1 & 0 \\ 0 & 1 & 1 \\ 1 & 0 & 1 \end{matrix}$$

If we consider such squares to furnish new squares by interchange of rows, we could obtain altogether twenty-one squares. We may call the above written squares reduced general magic squares, from which the complete number can be obtained by interchanging rows. We have thus in this case a one-to-one correspondence between the possible pairings of the two sets

$$a\ b\ c$$
$$a\ b\ c,$$

the partitions of the multipartite (222)

into three parts, such that each part is of weight 2, and the reduced general magic squares of order 3, such that the sum of the numbers in any row or column is 2.

There is also a one-to-one correspondence between the sets of pairings, where permutation amongst the pairs is held to give distinct pairings, of the two sets

$$a\ b\ c$$
$$a\ b\ c,$$

the compositions of the partitions of the multipartite

$$(222)$$

into three parts, such that each part is of weight 2, and the general (*i.e.*, unreduced) magic squares of order 3, such that the sum of the numbers in any row or column is 2.

Similarly, in the general case, we find a one-to-one correspondence between the sets of m-ads that can be formed from m sets identical with

$$a\ b\ c\ \ldots\ n,$$

the partitions of the n-partite number

$$(m\ m\ m\ \ldots\ m)$$

into n parts, such that each part is of weight m, and the reduced general magic squares of order n, such that the sum of the numbers in any row or column is m.

Also a like correspondence when we take compositions in lieu of partitions.

To make the matter clear it is proper to give other examples of the correspondence.

When $n = 4$, $m = 2$, we have the two sets

$$a\ b\ c\ d$$
$$a\ b\ c\ d$$

giving rise to the 17 pairings

$a^2,\ b^2,\ c^2,\ d^2$	$ab,\ ab,\ c^2,\ d^2$	$ab,\ ac,\ bc,\ d^2$
$a^2,\ b^2,\ cd,\ cd$	$ab,\ ab,\ cd,\ cd$	$ab,\ ac,\ bd,\ cd$
$a^2,\ bd,\ c^2,\ bd$	$ac,\ ac,\ b^2,\ d^2$	$ab,\ ad,\ bd,\ c^2$
$a^2,\ bc,\ bc,\ d^2$	$ac,\ ac,\ bd,\ bd$	$ab,\ ad,\ bc,\ cd$
$a^2,\ bc,\ cd,\ bd$	$ad,\ ad,\ b^2,\ c^2$	$ac,\ ad,\ cd,\ b^2$
	$ad,\ ad,\ bc,\ bc$	$ac,\ ad,\ bc,\ bd$

in correspondence with the 17 partitions of (2222)

(2000, 0200, 0020, 0002)
(2000, 0200, 0011, 0011)
(2000, 0101, 0020, 0101)
(2000, 0110, 0110, 0002)
(2000, 0110, 0011, 0101)

(1100, 1100, 0020, 0002)	(1100, 1010, 0110, 0002)
(1100, 1100, 0011, 0011)	(1100, 1010, 0101, 0011)
(1010, 1010, 0200, 0002)	(1100, 1001, 0101, 0020)
(1010, 1010, 0101, 0101)	(1100, 1001, 0110, 0011)
(1001, 1001, 0200, 0020)	(1010, 1001, 0011, 0200)
(1001, 1001, 0110, 0110)	(1010, 1001, 0110, 0101)

and the 17 reduced general magic squares

2000	2000	2000	2000	2000
0200	0200	0101	0110	0110
0020	0011	0020	0110	0011
0002	0011	0101	0002	0101

1100	1100	1010	1010	1001	1001
1100	1100	0200	0101	0200	0110
0020	0011	1010	1010	0020	0110
0002	0011	0002	0101	1001	1001

1100	1100	1100	1100	1010	1010
0110	0101	0101	0110	0200	0101
1010	1010	0020	0011	0011	0110
0002	0011	1001	1001	1001	1001

where it must be borne in mind that squares are to be considered as identical if they can be made the same by interchange of rows. If this condition be not present it will be found that there are 282 different squares. The 17 representative squares have been chosen so that each is vertebrate; that is to say, there is no zero in the principal diagonal. When $n = 3$, $m = 3$, we have the three sets

$$\begin{matrix} a\ b\ c \\ a\ b\ c \\ a\ b\ c \end{matrix}$$

giving rise to the ten sets of 3-ads

a^3, b^3, c^3	a^2b, ab^2, c^3	ab^2, abc, ac^2
a^3, b^2c, bc^2	a^2b, abc, bc^2	abc, abc, abc
	a^2b, ac^2, b^2c	
	a^2c, ac^2, b^3	
	a^2c, abc, b^2c	
	a^2c, ab^2, bc^2	

in correspondence with the ten partitions of (333),

(300, 030, 003)	(210, 120, 003)	(120, 111, 102)
(300, 021, 012)	(210, 111, 012)	(111, 111, 111)
	(210, 102, 021)	
	(201, 102, 030)	
	(201, 111, 021)	
	(201, 120, 012)	

D 2

and the ten reduced general magic squares

300	210	201	120
030	120	030	111
003	003	102	102

300	210	201	111
021	111	111	111
012	012	021	111

210	201
021	120
102	012

where again squares are taken to be identical if they can be made so by interchange of rows. Without this condition the total number of squares is seen to be 55. The ten representative squares are vertebrate. The enumeration of general magic squares of order n with an associated number m has been discussed by the author in *Phil. Trans. Roy. Soc.*, A, Vol. 205, 1905, pp. 37–59.

Looking to the 17 magic squares above delineated it will be noticed that those which involve units only correspond to the linear pairings of the two sets. They are six in number in agreement with previous results. In the squares of order 3 last written there is but one square which involves units only in correspondence with the single linear set of triads.

Let the number which enumerates the sets of m-ads that can be formed from m sets of quantities identical with

$$a\, b\, c \,\ldots\, n,$$

as well as the partitions and magic squares associated with them, be denoted by

$$f(m, n).$$

In Arts. 1 and 2 we have discussed the numbers $f(2, n)$.

If we consider the correspondence with partitions, and suppose that the partitions denote symmetric functions, we can construct a function which determines the number $f(m, n)$.

From this point of view the number

$$f(2, n)$$

is equal to the coefficient of the symmetric function

$$\Sigma a_1^2 a_2^2 \ldots a_n^2 \equiv (2^n),$$

in the development of the algebraic fraction

$$\frac{1}{(1-a_1^2)(1-a_2^2)(1-a_3^2) \ldots (1-a_1 a_2)(1-a_1 a_3)(1-a_2 a_3) \ldots}.$$

Moreover, if we denote by $\quad g(2, n),$

the number of linear sets of pairings,

$$g(2, n)$$

is equal to the coefficient of

$$\Sigma a_1^2 a_2^2 \ldots a_n^2 \equiv (2^n)$$

in the development of

$$\frac{1}{(1-a_1 a_2)(1-a_1 a_3)(1-a_2 a_3) \ldots}.$$

To bring these notions into agreement with previous results we observe that

$$\frac{1}{(1-a_1^2)(1-a_2^2)(1-a_3^2) \ldots} = 1+(2)+(2^2)+(2^3)+\ldots+\text{irrelevant terms},$$

$$\frac{1}{(1-a_1 a_2)(1-a_1 a_3)(1-a_2 a_3) \ldots} = 1+(2^2)+(2^3)+6(2^4)+22(2^5)+130(2^6)$$

$$+822(2^7)+\ldots+\text{irrelevant terms}.$$

This last result has been taken from the results of Arts. 1 and 2. The direct expansion will be considered presently.

Hence

$$\frac{1}{(1-a_1^2)(1-a_2^2)(1-a_3^2) \ldots (1-a_1 a_2)(1-a_1 a_3)(1-a_2 a_3) \ldots}$$

$$= \{1+(2)+(2^2)+(2^3)+\ldots\} \{1+(2^2)+(2^3)+6(2^4)+22(2^5)+\ldots\}+\ldots,$$

and observing that

$$(2^s)(2^t) = \ldots + \binom{s+t}{s}(2^{s+t})+\ldots,$$

we readily obtain the expansion

$$1+(2)+2(2^2)+5(2^3)+17(2^4)+73(2^5)+388(2^6)+\ldots,$$

in agreement with Arts. 1 and 2.

When $m=3$, we have to consider the development of the function

$$\frac{1}{(1-a_1^3)(1-a_2^3)(1-a_3^3)\ldots(1-a_1^2a_2)(1-a_1a_2^2)(1-a_1a_3^2)\ldots(1-a_1a_2a_3)(1-a_1a_2a_4)\ldots},$$

and therein the coefficient of the symmetric function

$$\Sigma a_1^3a_2^3a_3^3\ldots a_n^3\equiv(3^n).$$

The number of linear sets of triads $g(3, n)$ is given by the coefficient of the same symmetric function in the expansion of

$$\frac{1}{(1-a_1a_2a_3)(1-a_1a_2a_4)(1-a_2a_3a_4)\ldots}.$$

The number of sets of triads in which each triad is of the form

$$a_s^2a_t$$

is given similarly by

$$\frac{1}{(1-a_1^2a_2)(1-a_1a_2^2)(1-a_1a_3^2)\ldots},$$

and, finally, when each triad is of the form

$$a_s^3,$$

the function to be considered is

$$\frac{1}{(1-a_1^3)(1-a_2^3)(1-a_3^3)\ldots}.$$

Since the last function is *effectively*

$$1+(3)+(3^2)+(3^3)+\ldots,$$

it is clear that we may consider separately the function

$$1/(1-a_1^2a_2)(1-a_1a_2^2)\ldots(1-a_1a_2a_3)(1-a_1a_2a_4)\ldots,$$

and then multiply together the two expansions so as to find the desired coefficients.

In general, to find the number $f(m, n)$, we must seek the coefficient of the symmetric function

$$(m^n)$$

in the expansion of a fraction whose numerator is unity and whose denominator is

$$(1-a_1^m)(1-a_2^m)\dots$$
$$\times(1-a_1^{m-1}a_2)\dots$$
$$\times(1-a_1^{m-2}a_2^2)\dots$$
$$\times(1-a_1^{m-2}a_2a_3)\dots$$
$$\times\dots,$$

there being a factor for every term in the product sum h_m homogeneous in respect of the letters

$$a_1,\ a_2,\ a_3,\ \dots,\ a_n.$$

To find the numbers of sets of m-ads which are linear, viz., $g(m, n)$, we take factors corresponding to every term of the elementary symmetric function

$$\Sigma a_1 a_2 \dots a_m$$

of the same n letters.

The numbers $g(m, n)$ enjoy the property

$$g(m, n) = g(n-m, n).$$

For consider the linear pairing enumerated by $g(2, 5)$, viz.,

$$ab,\ ab,\ cd,\ ce,\ de.$$

We can derive, by dividing each combination of two letters into $abcde$, the linear set of 3-ads

$$cde,\ cde,\ abc,\ abd,\ abc,$$

which is one of the linear sets of 3-ads, enumerated by $g(3, 5)$, derived from the three sets

$$abcde$$
$$abcde$$
$$abcde$$

So that $$g(2, 5) = g(3, 5),$$

and generally $$g(m, n) = g(n-m, n).$$

The writer has calculated some of the numbers as shewn in the two tables annexed.

$f(m, n)$

m \ $n =$	1	2	3	4	5
1	1	1	1	1	1
2	1	2	5	17	73
3	1	2	10	93	1317
4	1	3	23	465	
5	1	3	40		
6	1	4	73		
7	1	4	114		

$g(m, n)$

m \ $n =$	1	2	3	4	5	6	7
1	1	1	1	1	1	1	1
2		1	1	6	22	130	822
3			1	1	22	550	16700
4				1	1	130	16700
5					1	1	822
6							1

These numbers may be obtained from the associated generating functions.

The expansion of such a function consists of the homogeneous product sums of the combinations which appear in the denominator factors. These sums are expressible in terms of the sums of the powers of those combinations. An application of Hammond's differential operator

$$D_m^n$$

is then all that is necessary.

This soon becomes laborious. The alternative, failing a difference equation when $m > 2$, is a systematic enumeration of the sets of m-ads. All of the numbers that appear in the tables above have been calculated, for verification, by both methods.

Chapter 8
Multipartite Numbers

8.1 Introduction and Commentary

The title of this chapter might also be "Partitions of Multipartite Numbers" or "Vector Partitions" or "Multipartitions." We have already discussed compositions of multipartite numbers in chapter 5. Since many of the problems encountered concerning multipartite partitions are most effectively handled by the techniques of R. P. Stanley, this chapter also overlaps with chapter 10.

Nearly half of the references (see 8.3.) for this chapter are by L. Carlitz, H. Gupta, and E. M. Wright. Most of their research has focused on generating functions, asymptotics, and enumeration methods related to multipartite numbers. Especially noteworthy is the article by Cheema and Motzkin (1971) containing numerous asymptotic and combinatorial results together with an extensive bibliography (see also Andrews, 1976, chapter 12).

MacMahon's most important paper on multipartite numbers [87] is devoted to the discovery of (2.2) below. E. M. Wright (1956) asked whether the polynomials $\lambda_r(x_1, \ldots, x_s)$ in (2.3) below all have nonnegative coefficients, a question apparently overlooked by MacMahon. We shall follow B. Gordon (1963) in his elegant proof that these polynomials do indeed have positive coefficients. L. Solomon (1977) and I. Gessel have both obtained new proofs of Gordon's theorem.

8.2 Multipartite Partitions and Gordon's Theorem

Let $v = (n_1, \ldots, n_s)$ be a multipartite number, i.e., a vector with nonnegative integer coordinates. Let $q_r(v) = q_r(n_1, \ldots, n_s)$ denote the number of partitions of v into r parts, i.e., $q_r(v)$ is the number of solutions of

$$v = v_1 + v_2 + \ldots + v_r,$$

where two solutions $\{v_1, \ldots, v_r\}$ and $\{v_1', \ldots, v_r'\}$ are considered the same if one is a reordering of the other. Next we define

$$\phi_r(\boldsymbol{x}) = \phi_r(x_1, \ldots, x_s) = \sum_{n_1 \geqq 0, \ldots, n_s \geqq 0} q_r(n_1, \ldots, n_s) x_1^{n_1} \ldots x_s^{n_s}.$$

The material in this chapter corresponds to section VI of *Combinatory Analysis*.

It is then clear that

$$(2.1)\qquad F(z) = \sum_{r\geqq 0} \phi_r(\boldsymbol{x}) z^r = \prod_{k_1\geqq 0,\ldots k_s\geqq 0} (1 - x_1^{k_1} x_2^{k_2} \ldots x_s^{k_s} z)^{-1}.$$

To see (2.1), just expand each term in the infinite product as a geometric series and collect terms.

Now

$$\begin{aligned}\log F(z) &= - \sum_{k_1\geqq 0,\ldots,k_s\geqq 0} \log(1 - x_1^{k_1} x_2^{k_2} \ldots x_s^{k_s} z)\\ &= \sum_{m=1}^{\infty} m^{-1}\Bigg(\sum_{k_1\geqq 0,\ldots,k_s\geqq 0} x_1^{mk_1} x_2^{mk_2} \ldots x_s^{mk_s}\Bigg) z^m\\ &= \sum_{m=1}^{\infty} \frac{z^m}{m(1 - x_1^m)(1 - x_2^m)\ldots(1 - x_s^m)}.\end{aligned}$$

Hence

$$\begin{aligned}\sum_{r\geqq 0} \phi_r(\boldsymbol{x}) z^r &= \exp\left\{\sum_{m=1}^{\infty} \frac{z^m}{m(1 - x_1^m)(1 - x_2^m)\ldots(1 - x_s^m)}\right\}\\ &= \prod_{m=1}^{\infty} \exp\left\{\frac{z^m}{m(1 - x_1^m)(1 - x_2^m)\ldots(1 - x_s^m)}\right\}\\ &= \sum_{r\geqq 0} z^r \sum \prod_{m=1}^{r} \frac{1}{h_m! m^{h_m} (1 - x_1^m)^{h_m} (1 - x_2^m)^{h_m} \ldots (1 - x_s^m)^{h_m}},\end{aligned}$$

where the inner summation extends over all partitions $h_1 \cdot 1 + h_2 \cdot 2 + \ldots + h_r \cdot r$ of r. For any such partition of r, we see that

$$\frac{(1 - x)(1 - x^2)\ldots(1 - x^r)}{\prod_{m=1}^{r} (1 - x^m)^{h_m}}$$

is a polynomial in x since it is a rational function of x and, at the possible poles $exp(2\pi ij/k)$, $0 \leqq j < k \leqq r$, g.c.d. $(j, k) = 1$, the numerator has a zero of order $\left[\frac{r}{k}\right]$, while in the denominator the zero is of order

$$\sum_{l \leqq [\frac{r}{k}]} h_{lk} \leqq k^{-1} \sum_{l \leqq [\frac{r}{k}]} h_{lk} \cdot lk \leqq k^{-1} \sum_{l \geqq 1} h_l \cdot l = \frac{r}{k};$$

hence our rational function has no poles and is therefore a polynomial. Consequently, there exists a polynomial $\lambda_r(\boldsymbol{x})$ such that

$$\lambda_r(\boldsymbol{x}) = \phi_r(\boldsymbol{x}) \prod_{k=1}^{r} \prod_{j=1}^{s} (1 - x_j^k) \tag{2.3}$$

$$= \sum_{R \geqq m_1 \geqq 0, \ldots, R \geqq m_s \geqq 0} \lambda_r(m_1, \ldots, m_s) x_1^{m_1} \ldots x_s^{m_s}.$$

Our object now is to prove B. Gordon's elegant theorem concerning the positivity of the coefficients of $\lambda_r(\boldsymbol{x})$. We shall closely follow Gordon's proof.

Gordon's Theorem. $\lambda_r(m_1, \ldots, m_s) \geqq 0$.

Proof. As Gordon remarks, the general case is somewhat complicated. We shall therefore begin by doing the simple case $s = 3$, $r = 2$, i.e., we examine the partitions of (a, b, c) into two parts $(a_1, b_1, c_1) + (a_2, b_2, c_2)$. Since partitions do not consider the order of summands, let us always write the summands in increasing order lexicographically (we shall say that $(a_1, b_1, c_1) \leqq (a_2, b_2, c_2)$ if $a_1 \leqq a_2$; or $a_1 = a_2$ and $b_1 \leqq b_2$; or $a_1 = a_2$, $b_1 = b_2$, and $c_1 \leqq c_2$). Assuming increasing order lexicographically, we see that there are exactly four possibilities for inequalities among the corresponding parts:

(i) $a_1 \leqq a_2, b_1 \leqq b_2, c_1 \leqq c_2$,

(ii) $a_1 < a_2, b_2 < b_1, c_1 \leqq c_2$,

(iii) $a_1 \leqq a_2, b_1 < b_2, c_2 < c_1$,

(iv) $a_1 < a_2, b_2 \leqq b_1, c_2 < c_1$.

To see that these exhaust all possibilities, we begin from the right and note that either $c_1 \leqq c_2$ or $c_2 < c_1$; furthermore, if $c_1 \leqq c_2$, then either $b_1 \leqq b_2$ or $b_2 < b_1$, etc. In each case the inequality for the a_i is prescribed in order to insure lexicographic ordering.

Let $q_2(n)$ denote the number of (ordinary) partitions of n into two nonnegative parts. Then the number of partitions of (a, b, c) of type (i) is clearly $q_2(a) q_2(b) q_2(c)$ (one merely dissects each partition coordinate-wise). For partitions of type (ii) no result is immediately apparent because of the strict inequalities. Now we note that with each partition $a = a_1 + a_2$ we may associate a partition $a - 1 = a_1 + (a_2 - 1)$, and the condition $a_1 < a_2$ for the first partition is equivalent to $a_1 \leqq a_2 - 1$ for the second partition; we now see that the number of partitions related to case (ii) is precisely $q_2(a-1)q_2(b-1)q_2(c)$. For case (iii) we find $q_2(a)q_2(b-1)q_2(c-1)$, and for case (iv) we find $q_2(a-1)q_2(b)q_2(c-1)$. Consequently,

$$q_2(a, b, c) = q_2(a)q_2(b)q_2(c) + q_2(a-1)q_2(b-1)q_2(c) + q_2(a)q_2(b-1)q_2(c-1) + q_2(a-1)q_2(b)q_2(c-1).$$

Since $\sum_{n=0}^{\infty} q_2(n)x^n = (1-x)^{-1}(1-x^2)^{-1}$, we see that

$$\phi(x_1, x_2, x_3) = (1 + x_1x_2 + x_2x_3 + x_1x_3)\prod_{i=1}^{3}(1-x_i)^{-1}(1-x_i^2)^{-1}$$

is the corresponding generating function identity.

The proof of the general case is clearly motivated by the above example. We shall divide the partitions of the multipartite number $(n_1, \ldots, n_s)$ into a number of disjoint classes, and shall prove an identity of the following nature:

$$q_r(n_1, \ldots, n_s) = \sum_{m_i \geqq 0} \mu_r(m_1, \ldots, m_s)\, q_r(n_1 - m_1, \ldots, n_s - m_s), \tag{2.4}$$

where the $\mu_r(m_1, \ldots, m_s)$ are nonnegative integers. Identity (2.4) directly implies the validity of (2.3) wherein $\lambda_r(m_1, \ldots, m_s)$ has been replaced by $\mu_r(m_1, \ldots, m_s)$. By the uniqueness of polynomials, this means that $\lambda_r(m_1, \ldots, m_s) = \mu_r(m_1, \ldots, m_s)$, and the positivity of the $\mu_r(m_1, \ldots, m_s)$ implies Gordon's theorem. *Hence we need only establish* (2.4) *with nonnegative* $\mu_r(m_1, \ldots, m_s)$.

To proceed let us consider the partition

$$(n_1, \ldots, n_s) = (n_1^{(1)}, \ldots, n_s^{(1)}) + \ldots + (n_1^{(r)}, \ldots, n_s^{(r)}).$$

We assume that the parts are ordered lexicographically so that if j is the least integer for which $n_j^{(i)} \neq n_j^{(i+1)}$, then $n_j^{(i)} < n_j^{(i+1)}$. Next we arrange the numbers $n_s^{(1)}, \ldots, n_s^{(r)}$ in nondecreasing order by stipulating that if two are equal, the one with the smaller subscript comes first. We thus have $n_s^{(i_1)} \leqq \ldots \leqq n_s^{(i_r)}$ where strict inequality holds whenever $i_\alpha > i_{\alpha+1}$. This procedure allows us to split the partitions of $(n_1, \ldots, n_s)$ into $r!$ disjoint classes (one for each permutation of the first r positive integers). In our initial example, the preceding procedure distinguishes the cases $c_1 \leqq c_2$ and $c_2 < c_1$.

Having produced $r!$ classes by examining the rightmost coordinate, we divide each class into $r!$ subclasses by examining the $(s - 1)$st coordinate. We arrange $n_{s-1}^{(1)}, \ldots, n_{s-1}^{(r)}$ in nondecreasing order by stipulating that if two of them are equal, the one whose superscript appears first in the sequence $i_1, i_2, \ldots, i_r$ (see the previous paragraph for the definition of the i_j) shall be placed first. Thus in our initial example, the case $c_1 \leqq c_2$ was broken into two subcases $b_1 \leqq b_2$ and $b_2 < b_1$, since $i_1 = 1$ and $i_2 = 2$; however, the case $c_2 < c_1$ was broken into two subcases $b_2 \leqq b_1$ and $b_1 < b_2$, since the natural order is now $i_1 = 2$, $i_2 = 1$.

We proceed next to the $(s - 2)$nd coordinates, then to the $(s - 3)$rd, etc., finally stopping with the second coordinates. The first coordinates are treated as follows: We must always have $n_1^{(1)} \leqq \ldots \leqq n_1^{(r)}$ for consistency with lexicographic ordering. We must require strict inequality $n_1^{(i)} < n_1^{(i+1)}$ if and only if $n_2^{(i+1)}$ precedes $n_2^{(i)}$ in the above arrangement. We need only inspect the above construction to see that our requirements concerning the first coordinates are necessary and sufficient to guarantee lexicographic ordering.

We conclude by enumerating the number of partitions in each of the $(r!)^{s-1}$ classes defined above; the treatment of our initial example shows that this is a simple matter. We proceed coordinate-wise: For each j, the object is to count the number of solutions of $n_j = t_1 + \ldots + t_r$, where the t_i are certain nonnegative integers subject to $t_1 \leqq \ldots \leqq t_r$, with these inequalities being strict in certain specific places. We assume that the strict inequalities occur for $t_{i_\alpha} < t_{i_\alpha+1}$, where $S = \{i_\alpha\} \subset \{1, \ldots, r - 1\}$. With each such partition of n_j we may associate the partition

$$n_j - \sum_{i_\alpha \in S} (r - i_\alpha) = t_1 + (t_2 - u_2) + \ldots + (t_r - u_r), \tag{2.5}$$

where u_k is the number of i smaller than k. The new summands now satisfy

$$t_1 \leqq t_2 - u_2 \leqq \ldots \leqq t_r - u_r.$$

Consequently, with $m_j = \sum_{i_\alpha \in S} (r - i_\alpha)$, we see that (2.5) constitutes an ordinary partition of $n_j - m_j$ into r parts. Since the above process is clearly reversible, we see that the number of partitions in the class under consideration is precisely $q_r(n_1 - m_1)\, q_r(n_2 - m_2) \ldots q_r(n_s - m_s)$. Therefore, summing over all classes, we obtain (2.4) where the $\mu_r(m_1, \ldots, m_s)$ are nonnegative integers, and so by the remarks immediately following (2.4), we see that Gordon's Theorem is established.

8.3 References

M. Abramson and W. O. J. Moser (1973) Arrays with fixed row and column sum, *Discr. Math.*, *6*, 1–14.

H. Anand, V. C. Dumir and H. Gupta (1966) A combinatorial distribution problem, *Duke J.*, *33*, 757–770.

G. E. Andrews (1976) The Theory of Partitions, Encyclopedia of Mathematics and Its Applications, vol. 2, Addison-Wesley, Reading.

G. E. Andrews (1977) An extension of Carlitz's bipartition identity, *Proc. A.M.S.*, *63*, 180–184.

F. C. Auluck (1953) On partitions of bipartite numbers, *Proc. Cambridge Phil. Soc.*, *49*, 72–83.

R. Bellman (1955) Research Problem 3, *Bull. A.M.S.*, *61*, 92.

R. Bellman (1957) The expansions of some infinite products, *Duke Math. J.*, *24*, 353–356.

G. R. Blakeley (1964a) Formal solution of nonlinear simultaneous equations: reversion of series in several variables, *Duke Math. J.*, *31*, 347–358.

G. R. Blakley (1964b) Combinatorial remarks on partitions of a multipartite number, *Duke Math. J.*, *31*, 335–340.

G. R. Blakley (1964c) Algebra of formal power series, *Duke Math. J.*, *31*, 341–346.

L. Carlitz (1956) The expansion of certain products, *Proc. A.M.S.*, *7*, 558–564.

L. Carlitz (1962) The coefficients in the expansion of certain products, *Proc. A.M.S.*, *13*, 944–949.

L. Carlitz (1963) A problem in partitions, *Duke Math. J.*, *30*, 203–214.

L. Carlitz (1965) Generating functions and partition problems, *Proc. Symp. in Pure Math.*, *8*, 144–169.

L. Carlitz (1966) Enumeration of symmetric arrays, *Duke Math. J.*, *33*, 771–782.

L. Carlitz (1971) Enumeration of symmetric arrays II, *Duke Math. J.*, *38*, 717–731.

L. Carlitz and D. P. Roselle (1966) Restricted bipartite partitions, *Pacific J. Math.*, *19*, 221–228.

M. S. Cheema (1964) Vector partitions and combinatorial identities, *Math. Comp.*, *18*, 414–420.

M.S. Cheema (1971) Computers in the theory of partitions, from Computers in Number Theory, edited by A.O.L. Atkin and B. J. Birch, Academic Press, London.

M. S. Cheema and H. Gupta (1968) The maxima of $P_r(n_1, n_2)$, *Math. Comp.*, *22*, 199–200.

M. S. Cheema and T. S. Motzkin (1971) Multipartitions and multipermutations, *Proc. Symp. in Pure Math.*, *19*, 39–70.

F. N. David, M. G. Kendall, and D. E. Barton (1966) Symmetric Functions and Allied Tables, Cambridge University Press, Cambridge.

R. Evans (1974) An asymptotic formula for extended Eulerian numbers, *Duke Math. J.*, *41*, 161–175.

I. Gessel (1977) Counting permutations by descents, inversions, and greater index, *Notices A.M.S.*, *24*, *A*-36.

B. Gordon (1963) Two theorems on multipartite numbers, *J. London Math. Soc.*, *38*, 459–464.

R.C. Grimson (1969) Enumeration of rectangular arrays, Ph.D. thesis, Duke University.

R. C. Grimson (1971) Some results on the enumeration of symmetric arrays, *Duke Math. J.*, *38*, 711–715.

R. C. Grimson (1974) The evaluation of certain arithmetic sums, *Fibonacci Quart.*, *12*, 373–380.

H. Gupta (1958a) Partitions of *j*-partite numbers, *Res. Bull. Panjab University*, No. 146, 119–121.

H. Gupta (1958b) Partitions of *j*-partite numbers into *k* summands, *J. London Math. Soc.*, *33*, 403–405.

H. Gupta (1959) Graphic representation of a j-partite number, *Res. Bull. (N.S.) Panjab University*, *10*, 189–196.

H. Gupta (1963) Partition of *j*-partite numbers, *Math. Student*, *31*, 179–186.

H. Gupta (1968) Enumeration of symmetric matrices, *Duke Math. J.*, *35*, 653–660.

H. Gupta (1970) Partitions—a survey, *J. Res. Nat. Bur. Standards (B), Math. Soci.*, *74B*, 1–29.

A. Oppenheim (1926a) On an arithmetic function, *J. London Math. Soc.*, *1*, 205–211.

A. Oppenheim (1926b) On an arithmetic function II, *J. London Math. Soc.*, *2*, 123–130.

M. M. Robertson (1977) Partitioning multipartite numbers into a fixed number of parts which satisfy congruence conditions, *J. reine und angew. Math.*, *293/294*, 428–434.

M. M. Robertson and D. Spencer (1976) Partitions of large multipartites with congruence conditions I, *Trans. A.M.S.*, *219*, 299–322.

D. P. Roselle (1966a) Generalized Eulerian functions and a problem in partitions, *Duke Math. J.*, *33*, 293–304.

D. P. Roselle (1966b) Restricted *k*-partite partitions, *Math. Nach.*, *32*, 139–148.

D. P. Roselle (1974) Coefficients associated with the expansion of certain products, *Proc. A.M.S.*, *45*, 144–150.

L. Solomon (1977) Partition identities related to symmetric polynomials in several sets of variables, J. Combinatorial Th., Ser. A (to appear).

E. M. Wright (1956) Partitions of multipartite numbers, *Proc. A.M.S.*, *7*, 880–890.

E. M. Wright (1957) The number of partitions of a large bipartite number, *Proc. London Math. Soc.(3)*, *7*, 150–160.

E. M. Wright (1958) Partitions of large bipartites, *American J. Math.*, *80*, 643–658.

E. M. Wright (1959) The asymptotic behavior of the generating functions of partitions of multipartites, *Quart. J. Math.*, *10*, 60–69.

E. M. Wright (1961) Partition of multipartite numbers into a fixed number of parts, *Proc. London Math. Soc.(3)*, *11*, 499–510.

E. M. Wright (1964a) Partition of multipartite numbers into *k* parts, *J. reine und angew. Math.*, *216*, 101–112.

E. M. Wright (1964b) Direct proof of the basic theorem on multipartite partitions, *Proc. A.M.S.*, *15*, 469–472.

E. M. Wright (1965) An extension of a theorem of Gordon, *Proc. Glassgow Math. Assoc.*, *7*, 39–41.

8.4 Summaries of the Papers

[79] On compound denumeration, *Trans. Cambridge Phil. Soc.*, *22* (1912), 1–13.

In this paper, MacMahon presents a method (called "compound

denumeration") for enumerating the number of partitions of a multipartite number. He presents formulas for this enumeration that depend on his work on Hammond operators and the theory of distributions [32]. For example, the number of partitions of the bipartite number (pq) is given by

$$\sum_{\pi} \sum_{\chi} D_{\pi_1} D_{\pi_2} \ldots D_{\Sigma\chi} h_{\chi_1} h_{\chi_2} \ldots h_{\Sigma\pi},$$

where the sum is extended over every partition $(p_1^{\pi_1} p_2^{\pi_2} \ldots)$ of p and $(q_1^{\chi_1} q_2^{\chi_2} \ldots)$ of q, with

$$\sum \pi = \sum_{i \geqq 1} \pi_i, \quad \sum \chi = \sum_{i \geqq 1} \chi_i,$$

and D_s denotes the sth Hammond operator.

For tripartite (and multipartite numbers in general) MacMahon reduces the enumeration to a problem in distributions that was solved in [43]. This approach introduces the symmetric functions of several systems of variables. It should be mentioned that this entire topic is handled in a more complete manner in [87].

[87] Seventh memoir on the partition of numbers. A detailed study of the enumeration of the partitions of multipartite numbers, *Phil. Trans.*, *217* (1917), 81–113.

In this paper, MacMahon extends his work on Hammond operators to effectively enumerate the partitions of multipartite numbers. He had treated this topic previously in papers [32] and [79]; however, the previous results only produced formulas in which several-fold products of Hammond operators were to be applied to appropriate products of homogeneous product sums.

Letting $(pqr \ldots)$ denote the monomial symmetric function $\sum \alpha^p \beta^q \gamma^r \ldots$, we define

$$F(\alpha, \beta, \gamma, \ldots) = \mathbf{Q}_1 = \sum (pqr \ldots),$$

where the sum is extended over all partitions with positive parts. We then define

$$\mathbf{Q}_i = F(\alpha^i, \beta^i, \gamma^i, \ldots).$$

The result that yields the proposed enumeration is the following:

Theorem. Let $F_q(m; 1^{k_1} 2^{k_2} \ldots i^{k_i})$ denote the coefficient of x^m in the expansion of $(1 - x)^{-k_1}(1 - x^2)^{-k_2} \ldots (1 - x^i)^{-k_i}$. Then

$$D_m \mathbf{Q}_1^{k_1}\mathbf{Q}_2^{k_2}\ldots\mathbf{Q}_i^{k_i} = F_q(m;\, 1^{k_1}2^{k_2}\ldots i^{k_i})\,\mathbf{Q}_1^{k_1}\mathbf{Q}_2^{k_2}\ldots\mathbf{Q}_i^{k_i}.$$

We remark in passing that $F_q(m;\, 1^{k_1}2^{k_2}\ldots i^{k_i})$ is the solution to the distribution problem considered in [88].

MacMahon's principal result in this memoir is that there are

$$\sum_{k_1\cdot 1+\ldots+k_i\cdot i=k} \frac{F_q(m_1;\, 1^{k_1}2^{k_2}\ldots i^{k_i})\,F_q(m_2;\, 1^{k_1}2^{k_2}\ldots i^{k_i})\ldots F_q(m_s;\, 1^{k_1}2^{k_2}\ldots i^{k_i})}{1^{k_1}2^{k_2}\ldots i^{k_i}\,k_1!k_2!\ldots k_i!}$$

partitions of the multipartite number $(m_1, m_2, \ldots, m_s)$ into k or fewer parts. It should be mentioned that this result appears as identity (2.2.16) (disguised) in the paper by Cheema and Motzkin (1971) listed in section 8.3. MacMahon treats explicitly partitions of multipartite numbers into 2, 3, and 4 parts, and he concludes with a discussion of how to extend his procedures to the cases where various restrictions are put upon the consitiuent parts of the partitions.

[111] Dirichlet series and the theory of partitions, *Proc. London Math. Soc.(2)*, *22* (1924), 404–411.

MacMahon considers the potency of n, $pt(n)$, defined by

$$\prod_p (1 - b^p p^{-s})^{-1} = \sum_{n=1}^{\infty} b^{pt(n)} n^{-s},$$

where $\prod_p$ runs over all the primes, i.e., $pt(n) = \sum_{i=1}^{\infty} \pi_i p_i$, where $n = p_1^{\pi_1}p_2^{\pi_2}\ldots$. He notes several results on potency, including

$$\prod_p (1 - b^p)^{-1} = \sum_{n=1}^{\infty} N_n b^n,$$

where N_n is the number of integers with potency n.

He concludes by considering

$$\prod_{n=2}^{\infty} (1 - n^{-s})^{-1} = \sum_{n=1}^{\infty} a_n n^{-s},$$

and he observes that a_n is the number of ways of expressing n as a product of positive integers. Furthermore, if $n = p_1^{\pi_1}p_2^{\pi_2}\ldots p_r^{\pi_r}$, then a_n is easily seen to be the number of partitions of the multipartite number $(\pi_1, \pi_2, \ldots,$

π_r). The papers of A. Oppenheim listed in Section 8.3 obtain asymptotic formulas for the a_n.

[114] The enumeration of the partitions of multipartite numbers, *Proc. Cambridge Phil. Soc.*, *22* (1925), 951–963.

In the first section of this paper, MacMahon considers multipartite partitions utilizing Dirichlet series generating functions as opposed to power series generating functions (this extends the work begun in [111]): if

$$\eta(s) = \Pi_{n \geqq 2}(1 - n^{-s})^{-1},$$

then

$$\eta(s) = \sum_{n=1}^{\infty} fc(n)\, n^{-s},$$

where $fc(n)$ denotes the number of factorizations of n; a moment's reflection convinces one that if

$$n = p_1^{\pi_1} p_2^{\pi_2} \ldots p_r^{\pi_r},$$

then $fc(n)$ also denotes the number of partitions of the multipartite number $(\pi_1, \pi_2, \ldots, \pi_r)$.

If we set $\zeta(s) = \sum_{n \geqq 1} n^{-s}$, then it is well-known that

$$(\zeta(s))^r = \sum_{n=1}^{\infty} d_r(n)\, n^{-s},$$

where $d_r(n)$ (MacMahon writes $\underset{(r-1 \text{ times})}{\sigma_{000\ldots0}}(n)$ for $d_r(n)$) denotes the number of r-tuples $(x_1, x_2, \ldots, x_r)$ of positive integers whose product is n. In analogy with this, MacMahon studies

$$(\eta(s))^r = \sum_{n=1}^{\infty} d_r^{(f)}(n)\, n^{-s},$$

(here he writes $\underset{(r-1 \text{ times})}{\sigma^{(f)}_{000\ldots0}}(n)$ for $d_r^{(f)}(n)$).

MacMahon treats power series generating functions in the next three sections culminating with the following result:

Theorem. Let $(n_2 n_3 \ldots n_s)$ be any multipartite number and let

$$(a_2a_3 \ldots a_s)^{\pi_1} (b_2b_3 \ldots b_s)^{\pi_2} \ldots$$

be one of its partitions. Then

$$\left\{\prod_{j=1}^{\infty} (1 - x^j)^{-1}\right\} \times \sum{}^* \left\{\prod_{r=2}^{s}\prod_{n_r=1}^{\pi_r} (1 - x^{n_r})^{-1}\right\} = \sum_{n_1=0}^{\infty} p_{n_1 n_2 \ldots n_s} x^{n_1},$$

where $\sum^*$ is extended over each partition of $(n_2n_3 \ldots n_s)$ and where $p_{n_1 n_2 \ldots n_k}$ is the number of partitions of the multipartite number $(n_1n_2 \ldots n_k)$.

In the final section of this paper, MacMahon devotes a few brief comments to the function $j(n)$ defined by

$$d_2^{(f)}(n) \equiv \sigma_0^{(f)}(n) = \sum_{d|n} j(d).$$

The last page of the paper presents tables enumerating bipartite and multipartite partitions.

[117] Euler's ϕ-function and its connexion with multipartite numbers, *Proc. London Math. Soc. (2)*, *25* (1926), 469–483.

This is an account of possible generalizations of Euler's totient function $\phi(n)$, the number of integers not exceeding n that are relatively prime to n.

Definition 1. $\mu_k(N)$ denotes the number of $(k+1)$-partite numbers $(a_1a_2 \ldots a_k a_{k+1})$ such that $\sum a_i = N$, each $a_i > 0$, and $g.c.d.(a_1, a_2, \ldots a_k, a_{k+1}) = 1$.

Definition 2. $\phi((N_1N_2 \ldots N_s))$ denotes the number of multipartite numbers $(n_1n_2 \ldots n_s)$ such that $g.c.d.(g.c.d.(N_1, N_2, \ldots, N_s), g.c.d.(n_1, n_2, \ldots, n_s)) = 1$, where $0 < n_i \leqq N_i$ for $1 \leqq i \leqq s$.

MacMahon derives a number of results that are generalizations of well-known results for Euler's function (note $\mu_1(N) = \phi(N)$ from Definition 1 and $\phi((N_1)) = \phi(N_1)$ from Definition 2). For example,

$$\sum_{\substack{d|N \\ d\neq 1}} \mu_k(d) = \binom{N-1}{k},$$

$$\sum \phi((N_1/d, N_2/d, \ldots, N_s/d)) = N_1N_2 \ldots N_s,$$

where the latter sum runs over all d that divide $g.c.d.(N_1, N_2, \ldots, N_s)$. The paper concludes with an extension of $\mu_k(N)$ to $\mu_k((N_1N_2 \ldots N_s))$, and the final page is devoted to the construction of a table of values of the $\mu_k(N)$.

Reprints of the Papers

I. *On Compound Denumeration.*

By Major P. A. MacMahon, R.A., Sc.D., LL.D., F.R.S.
Honorary Member Cambridge Philosophical Society.

[*Received* May 1, 1912. *Read* May 6, 1912.]

Art. 1. I propose to examine the subject of compound denumeration, otherwise the partitions of multipartite numbers, by a direct application of the Theory of Distributions which was developed by me in the *Proceedings of the London Mathematical Society* *. It will be shewn that the actual denumeration may be made to depend upon the theory of the symmetric functions of a single system of quantities. Such a system is

$$\alpha_1, \alpha_2, \alpha_3, \ldots,$$

and I write in the usual notation

$$(1-\alpha_1 x)(1-\alpha_2 x)(1-\alpha_3 x)\ldots = 1 - a_1 x + a_2 x^2 - a_3 x^3 + \ldots$$
$$= \frac{1}{1 + h_1 x + h_2 x^2 + h_3 x^3 + \ldots},$$

so that the quantities a are the elementary symmetric functions and the quantities h the homogeneous product sums of the quantities α of the system respectively. With these functions are associated the differential operators

$$d_s = \partial_{a_s} + a_1 \partial_{a_{s+1}} + a_2 \partial_{a_{s+2}} + \ldots,$$

$$D_s = \frac{(d_1^s)}{s!},$$

where (d_1^s) denotes that the linear operator d_1 is raised to the sth power in symbolic manner so that it denotes *not* the performance of d_1, s times in succession, but rather an operator of the order s.

I first consider the partitions of a bipartite number (pq) and note, as observed long ago by me †, that the partitions are separable into groups which depend upon the partitions of the unipartite numbers (p), (q) respectively. Thus the partitions of the bipartite number (22) nine in number are separated into four groups:

Gr $(2, 2)$,	Gr $(2, 1^2)$,	Gr $(1^2, 2)$,	Gr $(1^2, 1^2)$,
(22)	(21 01)	(12 10)	(11 11)
(20 02)	(20 01 01)	(10 10 02)	(11 10 01)
			(10 10 01 01),

* *Proc. Lond. Math. Soc.* vol. XIX. 1887, "Symmetric Functions and the Theory of Distributions."

† *American Jour. of Math.* vol. XI. 1888, p. 29, "Memoir on a New Theory of Symmetric Functions."

where it is to be observed that supposing the partible number to be (pq) (here $p=q=2$) the p number is in partition (2) in the two first groups and in partition (1^2) in the last two; while the q number is in partition (2) in the first and third groups and in partition (1^2) in the second and fourth. In fact if the numbers p, q have P, Q partitions respectively the partitions of (pq) are separable into PQ groups for every partition of p may be associated with every partition of q.

We will now study the enumeration of the partitions appertaining to a given group. Consider the group

$$\text{Gr}\,\{(p_1^{\pi_1}p_2^{\pi_2}\ldots),\ (q_1^{\chi_1}q_2^{\chi_2}\ldots)\},$$

where $(p_1^{\pi_1}p_2^{\pi_2}\ldots)$, $(q_1^{\chi_1}q_2^{\chi_2}\ldots)$ are given partitions of p and q.

The most extended partition of the group contains $\Sigma\pi+\Sigma\chi$ parts, while that which is least extended contains a number of parts equal to the greatest of the integers $\Sigma\pi$, $\Sigma\chi$.

No generality is lost by the supposition $\Sigma\pi \geqslant \Sigma\chi$.

In a partition of the group the biparts may be ordered so that, as regards the partition of p, the first $\Sigma\pi$ biparts are

$$(p_1\,.)\ \pi_1 \text{ times } (p_2\,.)\ \pi_2 \text{ times, \&c.,}$$

and this will be the case for every partition of the group.

The second element of any bipart may be either zero or one of the parts $q_1, q_2, \ldots$ of the partition $(q_1^{\chi_1}q_2^{\chi_2}\ldots)$ of q. There may be also biparts of the form $(0q_1)$, $(0q_2)$, ... The biparts are therefore of one of the three forms (p_sq_t), (p_s0), $(0q_t)$, and their number has lower and upper limits $\Sigma\pi$ and $\Sigma\pi+\Sigma\chi$.

We now suppose there to be

π_1 parcels of one kind
π_2 „ a second kind
&c.
$\Sigma\chi$ parcels of another kind differing from those above.

Altogether $\Sigma\pi+\Sigma\chi$ parcels of a specification which may be denoted by the partition $(\pi_1\pi_2\ldots\Sigma\chi)$ of the number $\Sigma\pi+\Sigma\chi$.

We also suppose there to be

χ_1 objects of one kind
χ_2 „ a second kind
&c.
$\Sigma\pi$ objects of another kind differing from those above.

Altogether $\Sigma\pi+\Sigma\chi$ objects of a specification which may be denoted by the partition $(\chi_1\chi_2\ldots\Sigma\pi)$ of the number $\Sigma\pi+\Sigma\chi$.

The number of objects is equal to the number of parcels and we may consider the number of ways of distributing the objects in the parcels so that each parcel contains one object. When, as in the present case, the number of objects is equal to the number of parcels and one object goes into each parcel the notion of the parcel is not essential and we may consider two sets of objects of specifications

$$(\pi_1\pi_2\ldots\Sigma\chi),\quad (\chi_1\chi_2\ldots\Sigma\pi) \text{ respectively;}$$

and the problem is the enumeration of the sets of two-fold objects that can be formed by making $\Sigma\pi + \Sigma\chi$ pairs of objects, each pair consisting of an object from each set of objects.

This problem is precisely the same as that of determining the number of partitions of the bipartite number (pq) which appertain to the group

$$\text{Gr}\,\{(p_1^{\pi_1} p_2^{\pi_2} \ldots),\quad (q_1^{\chi_1} q_2^{\chi_2} \ldots)\}.$$

To explain this consider the partitions of the bipart (33) which appertain to the group

$$\text{Gr}\,\{(21),\ (1^3)\}.$$

Here $\pi_1 = 1,\ \pi_2 = 1,\ \Sigma\pi = 2,\ \chi_1 = 3,\ \Sigma\chi = 3.$

We consider objects specified by (113) as being the *first* elements of biparts; these are

2, 1, 0, 0, 0;

or if we want to exhibit the fact that they are first elements we may write them

2*, 1*, 0*, 0*, 0*.

With these consider objects specified by (32) as being the second elements of biparts; these are

1, 1, 1, 0, 0;

or as they may be written

*1, *1, *1, *0, *0.

Combining the two sets of objects in all possible ways so as to form a single set of two-fold objects we obtain the four sets

21, 11, 01, 00, 00,

21, 10, 01, 01, 00,

20, 11, 01, 01, 00,

20, 10, 01, 01, 01,

corresponding to the four partitions

(21, 11, 01),

(21, 10, 01, 01),

(20, 11, 01, 01),

(20, 10, 01, 01, 01),

of the bipartite number (33) appertaining to the group

$$\text{Gr}\,\{(21),\ (1^3)\}.$$

It is clear that there is in every case a one-to-one correspondence between the distributions as defined and the partitions under examination.

The number of the distributions was shewn (*loc. cit.*) to have either of the two expressions

$$D_{\pi_1} D_{\pi_2} \ldots D_{\Sigma\chi}\, h_{\chi_1} h_{\chi_2} \ldots h_{\Sigma\pi},$$

$$D_{\chi_1} D_{\chi_2} \ldots D_{\Sigma\pi}\, h_{\pi_1} h_{\pi_2} \ldots h_{\Sigma\chi}.$$

1—2

The expressions are equivalent and may be evaluated by means of theorems given (*loc. cit.*).

The whole number of partitions of the bipart (pq) is in consequence

$$\sum_{\pi}\sum_{\chi} D_{\pi_1} D_{\pi_2} \ldots D_{\Sigma\chi}\, h_{\chi_1} h_{\chi_2} \ldots h_{\Sigma\pi},$$

the double summation being for every partition $(p_1^{\pi_1} p_2^{\pi_2} \ldots)$ of (p) and for every partition $(q_1^{\chi_1} q_2^{\chi_2} \ldots)$ of (q).

We apply the method to find the number of partitions of the bipartite (33). We have

Group	π_1	π_2	$\Sigma\chi$	χ_1	χ_2	$\Sigma\pi$		Partitions in Group
$\{(3), (3)\}$	1	0	1	1	0	1	$D_1^2 h_1^2 =$	2
$\{(3), (21)\}$	1	0	2	1	1	1	$D_2 D_1 h_1^3 =$	3
$\{(3), (1^3)\}$	1	0	3	3	0	1	$D_3 D_1 h_3 h_1 =$	2
$\{(21), (3)\}$	1	1	1	1	0	2	$D_1^3 h_2 h_1 =$	3
$\{(21), (21)\}$	1	1	2	1	1	2	$D_2 D_1^2 h_2 h_1^2 =$	7
$\{(21), (1^3)\}$	1	1	3	3	0	2	$D_3 D_1^2 h_3 h_2 =$	4
$\{(1^3), (3)\}$	3	0	1	1	0	3	$D_3 D_1 h_3 h_1 =$	2
$\{(1^3), (21)\}$	3	0	2	1	1	3	$D_3 D_2 h_3 h_1^2 =$	4
$\{(1^3), (1^3)\}$	3	0	3	3	0	3	$D_3^2 h_3^2 =$	4
							Total	31

No calculation is required if we are given Tables which express the h products in terms of monomial symmetric functions.

Thus since a Table shews

$$h_2 h_1^2 = \ldots + 7\,(21^2) + \ldots$$
$$D_2 D_1^2 h_2 h_1^2 = 7.$$

The above resulting numbers are all shewn in the Tables which proceed as far as the weight 6.

In the above case where $p = q$, it is not necessary to consider all the groups because the two partitions that define the Group may be interchanged. Thus the two Groups $\{(3), (21)\}$, $\{(21), (3)\}$ are identical and the whole numbers of partitions might have been written

$$D_1^2 h_1^2 + 2D_2 D_1 h_1^3 + 2D_3 D_1 h_3 h_1 + D_2 D_1^2 h_2 h_1^2 + 2D_3 D_1^2 h_3 h_2 + D_3^2 h_3^2.$$

We now remark that, formally and algebraically but not operationally, this expression may be written in the factorized form

$$(D_1 h_1 + D_1^2 h_2 + D_3 h_3)(D_1 h_1 + D_2 h_1^2 + D_3 h_3);$$

for the multiplication gives

$$D_1^2 h_1^2 + D_1 D_2 h_1^3 + D_1 D_3 h_1 h_3 + D_1^3 h_2 h_1 + D_1^2 D_2 h_2 h_1^2 + D_1^2 D_3 h_2 h_3 + D_3 D_1 h_3 h_1 + D_3 D_2 h_3 h_1^2 + D_3^2 h_3^2,$$

and, observing that, by the well-known theorem of reciprocity

$$D_2 D_1 h_1^3 = D_1^3 h_2 h_1$$
$$D_3 D_1^2 h_3 h_2 = D_3 D_2 h_3 h_1^2,$$

the truth of the statement is verified.

In fact, formally and algebraically but not operationally, the double sum

$$\sum_\pi \sum_\chi D_{\pi_1} D_{\pi_2} \ldots D_{\Sigma\chi} h_{\chi_1} h_{\chi_2} \ldots h_{\Sigma\pi},$$

may be written

$$\{\sum_\pi D_{\pi_1} D_{\pi_2} \ldots h_{\Sigma\pi}\} \,.\, \{\sum_\chi D_{\Sigma\chi} h_{\chi_1} h_{\chi_2} \ldots\}.$$

The factorized form may be regarded as a symbolical expression.

By the above method the following numbers have been calculated

	1	2	3	4	5	$\ldots = q$
1	2	—	—	—	—	
2	4	9	—	—	—	
3	7	16	31	—	—	
4	12	29	57	109	—	
5	19	47	97	189	336	
$\vdots$ $\parallel$ p						

Thus, from the table, to find the number of partitions of the bipartite (43) we take the row commencing 4 and the column headed 3 and find at the intersection the number 57.

The numbers agree with those obtained by expansion of the generating function

$$\frac{1}{(1-x)(1-y)(1-x^2)(1-xy)(1-y^2)(1-x^3)(1-x^2y)(1-xy^2)(1-y^3)\ldots}.$$

Art. 2. The distribution of $\Sigma\pi + \Sigma\chi$ objects of type $(\pi_1\pi_2 \ldots \Sigma\chi)$ into $\Sigma\pi + \Sigma\chi$ parcels of type $(\chi_1\chi_2 \ldots \Sigma\pi)$ one object in each parcel has necessarily resulted in our obtaining the whole of the partitions of the group under view. Remembering that $\Sigma\pi \geqslant \Sigma\chi$ we may if we please make a distribution of $\Sigma\pi + s$ objects of type $(\pi_1\pi_2 \ldots s)$ into $\Sigma\pi + s$ parcels of type $(\chi_1\chi_2 \ldots \Sigma\pi - \Sigma\chi + s)$ where s is any number included in the series

$$0,\ 1,\ 2, \ldots \Sigma\chi.$$

We will thus obtain a number for the enumeration which is

$$D_{\pi_1} D_{\pi_2} \ldots D_s h_{\chi_1} h_{\chi_2} \ldots h_{\Sigma\pi - \Sigma\chi + s};$$

and this number also gives the number of partitions of (pq) which contain $\Sigma\pi + s$ or fewer parts and also appertain to the given group.

Hence also the number of such partitions which contain exactly $\Sigma\pi + s$ parts is

$$D_{\pi_1} D_{\pi_2} \ldots D_s h_{\chi_1} h_{\chi_2} \ldots h_{\Sigma\pi - \Sigma\chi + s} - D_{\pi_1} D_{\pi_2} \ldots D_{s-1} h_{\chi_1} h_{\chi_2} \ldots h_{\Sigma\pi - \Sigma\chi + s - 1};$$

or as it may be written

$$D_{\pi_1} D_{\pi_2} \ldots \{D_s h_{\chi_1} h_{\chi_2} \ldots h_{\Sigma\pi - \Sigma\chi + s} - D_{s-1} h_{\chi_1} h_{\chi_2} \ldots h_{\Sigma\pi - \Sigma\chi + s - 1}\}.$$

The whole number of partitions which contain $\Sigma\pi + s$ or fewer parts is

$$\sum_\pi \sum_\chi D_{\pi_1} D_{\pi_2} \ldots D_s h_{\chi_1} h_{\chi_2} \ldots h_{\Sigma\pi - \Sigma\chi + s},$$

the double summation being for all partitions

$$(p_1^{\pi_1} p_2^{\pi_2} \ldots),\ (q_1^{\chi_1} q_2^{\chi_2} \ldots) \text{ of the numbers } (p) \text{ and } (q);$$

and a similar summation gives the whole number of partitions which contain exactly $\Sigma\pi + s$ parts. The expression for the number of partitions which contain $\Sigma\pi + s$ or fewer parts can be given the factorized symbolic form

$$(\sum_\pi D_{\pi_1} D_{\pi_2} \ldots h_{\Sigma\pi - \Sigma\chi + s}) \cdot (\sum_\chi D_s h_{\chi_1} h_{\chi_2} \ldots).$$

As an example let us consider the partitions of the bipartite (44) which appertain to the group $\{(211), (211)\}$.

Here s may have the values 0, 1, 2, 3.

$$\pi_1 = 1,\ \pi_2 = 2,\ \Sigma\chi = 3,\ \chi_1 = 1,\ \chi_2 = 2,\ \Sigma\pi = 3.$$

For $s = 0$, we have $D_2 D_1 h_2 h_1 = 2$, and the two partitions into 3 parts are

(22, 11, 11),
(21, 12, 11).

For $s = 1$, we have $D_2 D_1^2 h_2 h_1^2 = 7$, shewing that there are 7 partitions into 4 or fewer parts; in addition to the 2 which have exactly 3 parts already written down we have 5 which contain exactly 4 parts; these are

(22, 11, 10, 01),
(21, 12, 10, 01),
(21, 11, 10, 02),
(20, 12, 11, 01),
(20, 11, 11, 02).

For $s = 2$, we have $D_2^2 D_1 h_2^2 h_1 = 11$, and we find that in addition to the 7 forms already written we have 4 which contain exactly 5 parts; these are

(22 10 10 01 01),
(21 10 10 02 01),
(20 12 10 01 01),
(20 11 10 02 01).

Finally for $s = 3$ we have $D_3 D_2 D_1 h_3 h_2 h_1 = 12$, and we have $12 - 11 = 1$ partition which contains exactly 6 parts; this is

(20 10 10 02 01 01).

*To explain the general method of calculation it is to be noted that

$$D_s h_m = h_{m-s},$$

and that when operating upon a product, D_s acts through each of the partitions of s. Thus

$$D_2 h_l h_m = h_{l-2} h_m + h_l h_{m-2} + h_{l-1} h_{m-1}$$

$$\begin{aligned} D_3 h_l h_m h_n = {} & h_{l-3} h_m h_n + h_l h_{m-3} h_n + h_l h_m h_{n-3} \\ & + h_{l-2} h_{m-1} h_n + h_{l-2} h_m h_{n-1} + h_l h_{m-2} h_{n-1} \\ & + h_{l-1} h_{m-2} h_n + h_{l-1} h_m h_{n-2} + h_l h_{m-1} h_{n-2} \\ & + h_{l-1} h_{m-1} h_{n-1}. \end{aligned}$$

* *Vide Proc. Lond. Math. Soc.* vol. xix. 1887, pp. 127—128, "The Algebra of Multi-linear partial differential operators."

The calculation of $D_3D_2D_1 . h_3h_2h_1$ therefore proceeds as follows:—

$$D_3D_2D_1 . h_3h_2h_1 = D_3D_2(h_2^2h_1 + h_3h_1^2 + h_3h_2)$$
$$= D_3(2h_2h_1 + h_1^3 + 2h_2h_1 + h_1^3 + h_3 + 2h_2h_1 + h_2h_1 + h_3 + h_2h_1)$$
$$= D_3(2h_1^3 + 8h_2h_1 + 2h_3) = 2 + 8 + 2 = 12.$$

Art. 3. In the next place we examine the effect of employing the elementary functions $a_1, a_2, a_3, \ldots$ instead of the homogeneous product sums $h_1, h_2, h_3, \ldots$. The distributions enumerated by the number

$$D_{\pi_1}D_{\pi_2}\ldots D_s a_{\chi_1}a_{\chi_2}\ldots a_{\Sigma\pi-\Sigma\chi+s},$$

are those of objects of type $(\pi_1\pi_2\ldots s)$ into parcels of type $(\chi_1\chi_2\ldots\Sigma\pi-\Sigma\chi+s)$ one object being placed in each parcel subject to the restriction that no two similar objects are to be placed in similar parcels.

The corresponding partitions of the bipartite number (pq) are those which appertain to the group $\{(p_1^{\pi_1}p_2^{\pi_2}\ldots), (q_1^{\chi_1}q_2^{\chi_2}\ldots)\}$, which contain exactly $\Sigma\pi+s$ parts, the zero bipart 00 not being excluded as a permissible bipart, and in which no particular part (including the bipart 00) occurs more than once.

Of course the double sum

$$\sum_\pi\sum_\chi D_{\pi_1}D_{\pi_2}\ldots D_s a_{\chi_1}a_{\chi_2}\ldots a_{\Sigma\pi-\Sigma\chi+s},$$

enumerates such partitions for the totality of the groups.

To see the meaning of this result consider again the partitions of the bipartite number (44) which appertain to the group $\{(211), (211)\}$.

For $s=0$, we have $D_2D_1a_2a_1=1$; since $\Sigma\pi=3$, this means that of all the partitions of the group which contain exactly 3 parts, the zero part 00 being admissible, there is but one in which there are no similarities of parts. This partition is in fact

(21 12 11).

For $s=1$, we have $D_2D_1^2a_2a_1^2=5$; for a Table which expresses a products in terms of monomial symmetric functions gives

$$a_2a_1^2 = \ldots + 5(21^2) + \ldots,$$

giving $$D_2D_1^2a_2a_1^2 = 5.$$

Thence we conclude that there are just 5 partitions which contain 4 parts involving no similarities. These are

(21 12 11 00),
(22 11 10 01),
(21 12 10 01),
(21 11 10 02),
(20 12 11 01),

the set including the one previously found with the part 00 added.

For $s=2$, we have $D_2^2 D_1 a_2^2 a_1 = 5$, since

$$a_2^2 a_1 = \ldots + 5\,(2^2 1) + \ldots;$$

and the 5 forms indicated are found to be

$$\begin{array}{l}(22\;\;11\;\;10\;\;01\;\;00),\\(21\;\;12\;\;10\;\;01\;\;00),\\(21\;\;11\;\;10\;\;02\;\;00),\\(20\;\;12\;\;11\;\;01\;\;00),\\(20\;\;11\;\;10\;\;02\;\;01),\end{array}$$

the parts in each involving no similarities.

Finally for $s=3$, we have $D_3 D_2 D_1 a_3 a_2 a_1 = 1$, since

$$a_3 a_2 a_1 = \ldots + (321) + \ldots$$

and the form indicated is

$$(20\;\;11\;\;10\;\;02\;\;01\;\;00)$$

containing no similarities of parts.

We obtain information concerning the partitions of the group which contain different parts when 00 is excluded as a part; for denote by Q_s the numbers of partitions of the group which contain exactly s different parts, the zero part being excluded, we have

$$Q_3 = 1,\;\; Q_3 + Q_4 = 5,\;\; Q_4 + Q_5 = 5,\;\; Q_5 = 1,$$

whence $Q_4 = 4$ and $Q_3 + Q_4 + Q_5 = 6$.

This number 6 which enumerates the partitions of the group which possess different parts is either

$$D_3 D_2 D_1 a_3 a_2 a_1 + D_2 D_1^2 a_2 a_1^2,$$

or

$$D_2^2 D_1 a_2^2 a_1 + D_2 D_1 a_2 a_1.$$

In general we have the relations

$$\begin{aligned} Q_{\Sigma\pi} &= D_{\pi_1} D_{\pi_2} \ldots a_{\chi_1} a_{\chi_2} \ldots a_{\Sigma\pi - \Sigma\chi},\\ Q_{\Sigma\pi} + Q_{\Sigma\pi+1} &= D_{\pi_1} D_{\pi_2} \ldots D_1 a_{\chi_1} a_{\chi_2} \ldots a_{\Sigma\pi - \Sigma\chi + 1},\\ &\ldots\ldots\ldots\ldots\ldots\ldots\ldots\ldots\ldots\ldots\ldots\ldots\\ Q_{\Sigma\pi+\Sigma\chi-2} + Q_{\Sigma\pi+\Sigma\chi-1} &= D_{\pi_1} D_{\pi_2} \ldots D_{\Sigma\chi-1} a_{\chi_1} a_{\chi_2} \ldots a_{\Sigma\pi-1},\\ Q_{\Sigma\pi+\Sigma\chi-1} &= D_{\pi_1} D_{\pi_2} \ldots D_{\Sigma\chi} a_{\chi_1} a_{\chi_2} \ldots a_{\Sigma\pi}.\end{aligned}$$

Hence the number which enumerates those partitions of the group which have different parts, the zero part being excluded, has two expressions; for

$$\begin{aligned} &Q_{\Sigma\pi} + Q_{\Sigma\pi+1} + \ldots + Q_{\Sigma\pi+\Sigma\chi-1}\\ &= D_{\pi_1} D_{\pi_2} \ldots a_{\chi_1} a_{\chi_2} \ldots a_{\Sigma\pi-\Sigma\chi} + D_{\pi_1} D_{\pi_2} \ldots D_2 a_{\chi_1} a_{\chi_2} \ldots a_{\Sigma\pi-\Sigma\chi+2}\\ &\qquad + D_{\pi_1} D_{\pi_2} \ldots D_4 a_{\chi_1} a_{\chi_2} \ldots a_{\Sigma\pi-\Sigma\chi+4} + \ldots\\ &= D_{\pi_1} D_{\pi_2} \ldots D_1 a_{\chi_1} a_{\chi_2} \ldots a_{\Sigma\pi-\Sigma\chi+1} + D_{\pi_1} D_{\pi_2} \ldots D_3 a_{\chi_1} a_{\chi_2} \ldots a_{\Sigma\pi-\Sigma\chi+3}\\ &\qquad + D_{\pi_1} D_{\pi_2} \ldots D_5 a_{\chi_1} a_{\chi_2} \ldots a_{\Sigma\pi-\Sigma\chi+5} + \ldots,\end{aligned}$$

where if $\Sigma\chi$ be uneven both series extend to $\frac{1}{2}(\Sigma\chi + 1)$ terms, whilst if $\Sigma\chi$ be even the first and second series extend to $\frac{1}{2}\Sigma\chi$ and $\frac{1}{2}\Sigma\chi - 1$ terms respectively.

That these two series are equivalent may be shewn algebraically as follows.

For brevity put $\Sigma\pi - \Sigma\chi = \theta$ and note that

$$D_1 a_{\chi_1} a_{\chi_2} \ldots a_{\theta+1} = a_{\chi_1} a_{\chi_2} \ldots a_\theta + \Sigma a_{\chi_1 - 1} a_{\chi_2} \ldots a_{\theta+1},$$

$$D_2 a_{\chi_1} a_{\chi_2} \ldots a_{\theta+2} = \Sigma a_{\chi_1 - 1} a_{\chi_2} \ldots a_{\theta+1} + \Sigma a_{\chi_1 - 1} a_{\chi_2 - 1} a_{\chi_3} \ldots a_{\theta+2},$$

$$D_3 a_{\chi_1} a_{\chi_2} \ldots a_{\theta+3} = \Sigma a_{\chi_1 - 1} a_{\chi_2 - 1} a_{\chi_3} \ldots a_{\theta+2} + \Sigma a_{\chi_1 - 1} a_{\chi_2 - 1} a_{\chi_3 - 1} a_{\chi_4} \ldots a_{\theta+3},$$

....

&c.

Directly we operate upon the relations with $D_{\pi_1} D_{\pi_2} \ldots$ the equivalence is obvious.

Art. 4. There is no difficulty in extending this theory by filling up the gap between the elementary functions and the homogeneous product sums. For suppose $k_1, k_2, k_3, \ldots$ be functions derived from the homogeneous product sums by deleting therefrom all terms which involve quantities of the system (from which the symmetric functions are derived) to a higher power than k. Then the distributions enumerated by the number

$$D_{\pi_1} D_{\pi_2} \ldots D_s k_{\chi_1} k_{\chi_2} \ldots k_{\Sigma\pi - \Sigma\chi + s},$$

are those of objects of type $(\pi_1 \pi_2 \ldots s)$ into parcels of type $(\chi_1 \chi_2 \ldots \Sigma\pi - \Sigma\chi + s)$ one object being placed in each parcel subject to the restriction that more than k similar objects are not to be placed in similar parcels. The corresponding partitions of the bipartite number (pq) are those which appertain to the group $\{(p_1^{\pi_1} p_2^{\pi_2} \ldots), (q_1^{\chi_1} q_2^{\chi_2} \ldots)\}$, which contain exactly $\Sigma\pi + s$ parts, the zero part 00 not being excluded from being an admissible part, and in which no particular part (including the part 00) occurs more than k times.

Art. 5. I pass on to consider the similar theory of tripartite partitions and it will be found to shew what the theory is for multipartite partitions in general. Consider the tripartite number (pqr) and the partitions appertaining to the group

$$\{(p_1^{\pi_1} p_2^{\pi_2} \ldots), (q_1^{\chi_1} q_2^{\chi_2} \ldots), (r_1^{\rho_1} r_2^{\rho_2} \ldots)\},$$

wherein we will suppose $\Sigma\pi \geqslant \Sigma\chi \geqslant \Sigma\rho$.

The partitions involve at least $\Sigma\pi$ and at most $\Sigma\pi + \Sigma\chi + \Sigma\rho$ parts. Reasoning as in the bipartite case we find that for partitions into $\Sigma\pi + s$ parts, where $0 \leqslant s \leqslant \Sigma\chi + \Sigma\rho$, we have to do with three assemblages of objects of types

$$(\pi_1 \pi_2 \ldots s), \ (\chi_1 \chi_2 \ldots t), \ (\rho_1 \rho_2 \ldots u) \text{ respectively,}$$

where $$\Sigma\pi + s = \Sigma\chi + t = \Sigma\rho + u.$$

We have to consider the number of ways of forming $\Sigma\pi + s$ triads of objects by taking one object from each assemblage to form a triad.

This number is also the number of partitions, appertaining to the group, which involve $\Sigma\pi + s$ or fewer parts.

This problem in 'Distributions' was solved by me in *American Journal of Mathematics*, vol. XIV. 1892, pp. 33 *et seq.*, "Fourth Memoir on a New Theory of Symmetric Functions."

Let there be two systems of quantities

$$\alpha_1, \quad \alpha_2, \quad \alpha_3, \ \ldots$$
$$\beta_1, \quad \beta_2, \quad \beta_3, \ \ldots$$

and let their symmetric functions be denoted by partitions with suffixes 1 and 2 respectively. Then write

$$A_1 = (1)_1$$
$$A_2 = (2)_1 + (1^2)_1$$
$$A_3 = (3)_1 + (21)_1 + (1^3)_1$$
$$\ldots\ldots\ldots\ldots\ldots\ldots$$
$$B_1 = (1)_2 A_1$$
$$B_2 = (2)_2 A_2 + (1^2)_2 A_1^2$$
$$B_3 = (3)_2 A_3 + (21)_2 A_2 A_1 + (1^3)_2 A_1^3$$
$$\ldots\ldots\ldots\ldots\ldots\ldots\ldots\ldots$$

where it will be noted that $A_1, A_2, A_3, \ldots$ are the successive homogeneous product sums of the quantities $\alpha_1, \alpha_2, \alpha_3, \ldots$*.

We now form the product

$$B_{\pi_1} B_{\pi_2} \ldots B_s,$$

and eliminate the quantities $A_1, A_2, A_3, \ldots$ so as to express it as a linear function of terms each of which is a product of two symmetric functions denoted by partitions with suffixes 1 and 2 respectively.

One of these terms will be

$$M (\chi_1 \chi_2 \ldots t)_1 (\rho_1 \rho_2 \ldots u)_2,$$

where M is an integer which is equal to the number of distributions in question.

We have therefore to find the coefficient of

$$(\chi_1 \chi_2 \ldots t)_1 (\rho_1 \rho_2 \ldots u)_2,$$

in the development of the product

$$B_{\pi_1} B_{\pi_2} \ldots B_s.$$

Let $D_1^{(1)}, D_2^{(1)}, D_3^{(1)} \ldots; D_1^{(2)}, D_2^{(2)}, D_3^{(2)}, \ldots,$ be obliterating operators associated with symmetric functions of the quantities $\alpha_1, \alpha_2, \alpha_3, \ldots; \beta_1, \beta_2, \beta_3, \ldots$ respectively. Then the operators $D^{(1)}, D^{(2)}$, act upon functions which are denoted by partitions in brackets $(\ )_1$, $(\ )_2$, respectively.

From the well-known properties of these operators we know that

$$D_{\chi_1}^{(1)} D_{\chi_2}^{(1)} \ldots D_t^{(1)} . D_{\rho_1}^{(2)} D_{\rho_2}^{(2)} \ldots D_u^{(2)} . B_{\pi_1} B_{\pi_2} \ldots B_s = M.$$

The reader will have no difficulty in establishing that

$$D_s^{(1)} A_m = A_{m-s},$$
$$D_s^{(2)} B_m = A_s B_{m-s},$$

* These quantities were denoted by $h_1, h_2, h_3, \ldots$ in Art. 2.

relations which much facilitate the calculation of the number M. To take a simple example, consider the partitions of the tripartite number (333) which appertain to the group

$$\{(21),\ (21),\ (21)\}.$$

Here
$$\pi_1 = 1,\ \pi_2 = 1,\ \chi_1 = 1,\ \chi_2 = 1,\ \rho_1 = 1,\ \rho_2 = 1,$$
$$t = u = s,$$

and s may have the values 0, 1, 2, 3, 4, for

$$s = 0, \quad (D_1^{(1)})^2 (D_1^{(2)})^2 B_1^2 = 4,$$
$$s = 1, \quad (D_1^{(1)})^3 (D_1^{(2)})^3 B_1^3 = 36,$$
$$s = 2, \quad (D_1^{(1)})^2 (D_2^{(1)}) (D_1^{(2)})^2 (D_2^{(2)}) B_1^2 B_2 = 74,$$
$$s = 3, \quad (D_1^{(1)})^2 (D_3^{(1)}) (D_1^{(2)})^2 (D_3^{(2)}) B_1^2 B_3 = 86,$$
$$s = 4, \quad (D_1^{(1)})^2 (D_4^{(1)}) (D_1^{(2)})^2, (D_4^{(2)}) B_1^2 B_4 = 87\ ;$$

shewing that the number of partitions of the group which contain

exactly	2	parts is	4,
,,	3	,,	32,
,,	4	,,	38,
,,	5	,,	12,
,,	6	,,	1;

and of course the total number of the partitions of the group is 87.

To explain the above calculations the reader is reminded that $D_s^{(1)}$ and $D_s^{(2)}$ operate through the whole of the partitions of s upon a B product. Thus for example

$$\begin{aligned} D_3^{(2)} B_s B_t B_u &= (D_3^{(2)} B_s) B_t B_u + B_s (D_3^{(2)} B_t) B_u + B_s B_t (D_3^{(2)} B_u) \\ &+ (D_2^{(2)} B_s)(D_1^{(2)} B_t) B_u + (D_2^{(2)} B_s) B_t (D_1^{(2)} B_u) + B_s (D_2^{(2)} B_t)(D_1^{(2)} B_u) \\ &+ (D_1^{(2)} B_s)(D_2^{(2)} B_t) B_u + (D_1^{(2)} B_s) B_t (D_2^{(2)} B_u) + B_s (D_1^{(2)} B_t)(D_2^{(2)} B_u) \\ &+ (D_1^{(2)} B_s)(D_1^{(2)} B_t)(D_1^{(2)} B_u) \\ &= A_3 (B_{s-3} B_t B_u + B_s B_{t-3} B_u + B_s B_t B_{u-3}) \\ &+ A_2 A_1 (B_{s-2} B_{t-1} B_u + B_{s-2} B_t B_{u-1} + B_s B_{t-2} B_{u-1} \\ &+ B_{s-1} B_{t-2} B_u + B_{s-1} B_t B_{u-2} + B_s B_{t-1} B_{u-2}) \\ &+ A_1^3 B_{s-1} B_{t-1} B_{u-1}, \end{aligned}$$

where the partitions of 3 being (3), (21), (1^3) the first line, the next two lines, and the fourth line are given by the three partitions respectively. The result

$$(D_1^{(1)})^2 (D_4^{(1)}) . (D_1^{(2)})^2 (D_4^{(2)}) . B_1^2 B_4 = 87,$$

is obtained as follows :—

$$\begin{aligned} (D_1^{(1)})^2 (D_4^{(2)}) . B_1^2 B_4 &= (D_1^{(2)}) (D_4^{(2)}) (2A_1 B_1 B_4 + A_1 B_1^2 B_3) \\ &= A_1 D_4^{(2)} (2A_1 B_4 + 2A_1 B_1 B_3 + 2A_1 B_1 B_3 + A_1 B_1^2 B_2) \\ &= A_1^2 (2A_4 + 4A_1 A_3 + A_1^2 A_2)\ ; \end{aligned}$$

therefore $(D_1^{(1)})^2 (D_4^{(1)}) . (D_1^{(2)})^2 (D_4^{(2)}) . B_1^2 B_4$

$$\begin{aligned} &= (D_1^{(1)}) (D_4^{(1)}) (4A_1 A_4 + 2A_1^2 A_3 + 12 A_1^2 A_3 + 4A_1^3 A_2 + 4A_1^3 A_2 + A_1^5) \\ &= (D_1^{(1)}) (D_4^{(1)}) (4A_1 A_4 + 14 A_1^2 A_3 + 8A_1^3 A_2 + A_1^5) \\ &= (D_4^{(1)}) (4A_4 + 4A_1 A_3 + 28 A_1 A_3 + 14\ A_1^2 A_2 + 24 A_1^2 A_2 + 8A_1^4 + 5A_1^4) \\ &= (D_4^{(1)}) (4A_4 + 32 A_1 A_3 + 38 A_1^2 A_2 + 13 A_1^4) \\ &= 4 + 32 + 38 + 13 = 87. \end{aligned}$$

We have found that the number of partitions of the group which have $\Sigma\pi + s$ or fewer parts is equal to the coefficient of

$$(\chi_1\chi_2 \dots t)_1 (\rho_1\rho_2 \dots u)_2,$$

in the development of the product

$$B_{\pi_1} B_{\pi_2} \dots B_s.$$

By a well known theorem of symmetry we may in this theorem interchange in any manner the partitions

$$(\pi_1\pi_2 \dots s),\ (\chi_1\chi_2 \dots t),\ (\rho_1\rho_2 \dots u).$$

We may therefore carry out the calculation in 3! different ways; a circumstance that is convenient for the purpose of verification.

The total number of partitions of the tripartite (pqr) is

$$\sum_\chi \sum_\rho \sum_\pi D_{\chi_1}^{(1)} D_{\chi_2}^{(1)} \dots D_t^{(1)} . D_{\rho_1}^{(2)} D_{\rho_2}^{(2)} \dots D_u^{(2)} . B_{\pi_1} B_{\pi_2} \dots B_s,$$

the summation being for the whole of the partitions of p, q and r.

Art. 6. We have also the theory of the partitions of the group which are composed of different parts; it is merely necessary in the above to substitute for the homogeneous product sums $A_1, A_2, A_3, \dots$ the elementary functions $a_1, a_2, a_3, \dots$.

Thus in the above particular case

$$(D_1^{(1)})^2 . (D_1^{(2)})^2 . B_1^2$$
$$= (D_1^{(1)})^2 . (2a_1^2) = D_1^{(1)} . 4a_1 = 4,$$

$$(D_1^{(1)})^3 . (D_1^{(2)})^3 . B_1^3$$
$$= (D_1^{(1)})^3 . (6a_1^3) = (D_1^{(1)})^2 . 18a_1^2 = 36,$$

$$(D_1^{(1)})^2 (D_2^{(1)}) . (D_1^{(2)})^2 (D_2^{(2)}) . B_1^2 B_2$$
$$= (D_1^{(1)})^2 (D_2^{(1)}) . (5a_1^4 + 2a_1^2 a_2) = (D_1^{(1)})^2 (34a_1^2 + 2a_2) = 70,$$

$$(D_1^{(1)})^2 (D_3^{(1)}) . (D_1^{(2)})^2 (D_3^{(2)}) . B_1^2 B_3$$
$$= (D_1^{(1)})^2 (D_3^{(1)}) . (a_1^5 + 4a_1^3 a_2 + 2a_1^2 a_3) = (D_1^{(1)})^2 (22a_1^2 + 6a_2) = 50,$$

$$(D_1^{(1)})^2 (D_4^{(1)}) . (D_1^{(2)})^2 (D_4^{(2)}) . B_1^2 B_4$$
$$= (D_1^{(1)})^2 (D_4^{(1)}) . (a_1^4 a_2 + 4a_1^3 a_3 + 2a_1^2 a_4) = (D_1^{(1)})^2 (4a_1^2 + 5a_2) = 13.$$

For a given number of parts we take all the corresponding partitions of the group zero parts 000 being admissible as parts; then the numbers found indicate the number of partitions which have no repeated parts.

Thus of 2 parts there are 4 partitions in which no part is repeated

3	„	36	„	„
4	„	70	„	„
5	„	50	„	„
6	„	13	„	„

Moreover if Q_s denote the number of partitions of the groups which contain exactly s different parts, zero parts excluded,

$$Q_2 = 4,$$
$$Q_2 + Q_3 = 36,$$
$$Q_3 + Q_4 = 70,$$
$$Q_4 + Q_5 = 50,$$
$$Q_5 + Q_6 = 13;$$

whence $$Q_2 = 4,\ Q_3 = 32,\ Q_4 = 38,\ Q_5 = 12,\ Q_6 = 1.$$

Comparing these numbers with those found in Art. 6 we see that there are in fact no partitions, of the group, which involve repeated parts.

Art. 7. The theory in respect of multipartite numbers in general is now clear. We take the multipartite number $(pqrs\ldots)$ and the partitions appertaining to the group

$$\{(p_1^{\pi_1} p_2^{\pi_2}\ldots),\ (q_1^{\chi_1} q_2^{\chi_2}\ldots),\ (r_1^{\rho_1} r_2^{\rho_2}\ldots),\ (s_1^{\sigma_1} s_2^{\sigma_2}\ldots)\ldots\}.$$

We continue the series of relations

$$A_1 = (1)_1 \qquad B_1 = (1)_2 A_1 \qquad C_1 = (1)_3 B_1 \qquad \&c.\ldots \qquad M_1 = (1)_{n-1} L_1$$
$$A_2 = (2)_1 + (1^2)_1 \qquad B_2 = (2)_2 A_2 + (1^2)_2 A_1^2, \qquad C_2 = (2)_3 B_2 + (1^2)_3 B_1^2 \qquad M_2 = (2)_{n-1} L_2 + (1^2)_{n-1} L_1^2$$
$$\&c. \qquad \&c. \qquad \&c.$$

where if the multipartite number be n-partite, L, M are the $n-2$th and $n-1$th letters of the alphabet. We have then to find the coefficient of

$$(\chi_1\chi_2\ldots t_2)_1 (\rho_1\rho_2\ldots t_3)_2 (\sigma_1\sigma_2\ldots t_4)_3\ldots$$

in the development of $M_{\pi_1} M_{\pi_2}\ldots M_{t_1}$. We have the sought number equal to

$$D_{\chi_1}{}^{(1)} D_{\chi_2}{}^{(1)}\ldots D_{t_2}{}^{(1)} . D_{\rho_1}{}^{(2)} D_{\rho_2}{}^{(2)}\ldots D_{t_3}{}^{(2)} . D_{\sigma_1}{}^{(3)} D_{\sigma_2}{}^{(3)}\ldots D_{t_4}{}^{(3)}\ldots\ldots M_{\pi_1} M_{\pi_2}\ldots M_{t_1};$$

and observe that

$$D_s{}^{(3)} . C_m = B_s C_{m-s},$$
$$D_s{}^{(4)} . D_m = C_s D_{m-s},$$
$$\vdots$$
$$D_s{}^{(n-1)} M_m = L_s M_{m-s},$$

relations which enable the regular and progressive calculations of the sought number.

In all cases the theory of the partitions into dissimilar parts is reached by substituting the elementary functions $a_1, a_2, a_3, \ldots$ for the homogeneous product sums $A_1, A_2, A_3, \ldots$.

The totality of the partitions into $\Sigma\pi + t_1$ or fewer parts is given by the expression

$$\sum_\chi \sum_\rho \sum_\sigma \ldots D_{\chi_1}{}^{(1)} D_{\chi_2}{}^{(1)}\ldots D_{t_2}{}^{(1)} . D_{\rho_1}{}^{(2)} D_{\rho_2}{}^{(2)}\ldots D_{t_3}{}^{(2)} . D_{\sigma_1}{}^{(3)} D_{\sigma_2}{}^{(3)}\ldots D_{t_3}{}^{(3)}\ldots\ldots M_{\pi_1} M_{\pi_2} M_{\pi_3}\ldots M_{t_1},$$

the summation being in regard to all the partitions

$$(p_1^{\pi_1} p_2^{\pi_2}\ldots),\ (q_1^{\chi_1} q_2^{\chi_2}\ldots),\ (r_1^{\rho_1} r_2^{\rho_2}\ldots),\ (s_1^{\sigma_1} s_2^{\sigma_2}\ldots),\ \ldots$$

of the numbers p, q, r, s, ... respectively.

IV. *Seventh Memoir on the Partition of Numbers.*

A Detailed Study of the Enumeration of the Partitions of Multipartite Numbers.

By Major P. A. MacMahon, *R.A., D.Sc., Sc.D., F.R.S.*

Received April 13,—Read May 11, 1916.

Introduction.

In this paper I give the complete solution of the problem of the partition of multipartite numbers. This is the same subject as that named by Sylvester, "Compound Denumeration." Twenty-nine years have elapsed since I announced that the algebra of symmetric functions is co-extensive with the grand problems of the combinatory analysis. The theory of symmetric function supplies generating functions which enumerate all the combinations, while the operators of Hammond* are the instruments which are effective in actually evaluating the coefficients of the terms of the expanded generating functions. When these operators fell from the hands of Hammond, they were already of much service as mining tools in extracting the ore from the mine field of symmetric functions; but they were only partially adequate. They required sharpening and general adaptation to the work in hand. The first step was to decompose an operator of given order into the sum of a number of operators in correspondence with every partition of the number which defines the order. Since there is a Hammond operator corresponding to every positive integer, this process resulted in there being an operator in correspondence with every partition of every integer. The outcome of this decomposition was that the operators were able to deal with the symmetric operands in a much more effective manner. The surface material of the mine could not only be removed, but the strata to a considerable depth could be dealt with. But this was not sufficient. It became necessary to effect a further decomposition by showing that every partition operator could be represented by a sum of composition operators. There emerged a composition operator in correspondence with every permutation of the parts of the partition of the operator. The operators at once became effective in dealing with the material in the lower strata of the mine field. The operators had, in fact, been handled with particular reference to the operands with which they were to be associated. It was now necessary to deal with the material of the mine with particular reference to the tools which had

* 'Proc. Lond. Math. Soc.,' 1883, vol. xiv., pp. 119–129.

VOL. CCXVII.—A 552. N [Published June 23, 1917.

been forged. To evaluate the coefficients we have to operate repeatedly with the appropriate operators until a numerical result is reached. In order to accomplish this with facility and to establish laws we have to put the generating functions in such a form that these operations are carried out in a regular and simple manner. To make my meaning clear, I will instance the case of the simple operation of differentiation ∂_x and the exponential function e^{ax}. We have

$$\partial_x e^{ax} = ae^{ax},$$

the effect of the operation being, to merely multiply the operand by the numerical magnitude a.

Thence

$$\partial_x^n e^{ax} = a^n e^{ax}$$

and we arrive at the conclusion, that if a given operand, a function of x, could be expressed as a linear function of exponential functions of x, the r times repeated operation of ∂_x could take place with facility upon each term of the linear function, and a general law for the repeated operation of ∂_x upon the operand would be obtainable. This reflection suggests the possibility of finding symmetric function operands in a form which will enable the repeated performance of HAMMOND'S operators in a practically effective manner. It is quite certain that any such operand must possess at least two properties in common with the exponential function : (i) its first term must be unity ; (ii) it must contain an infinite number of terms. The first step was to find a symmetric function Q_1 of the elements $\alpha, \beta, \gamma, \ldots$ such that the effect of every Hammond operator upon it is to leave it unchanged ; or, as I prefer to say, to multiply it by unity. Q_1 is, in fact, the sum of unity and the whole of the monomial symmetric functions $\Sigma\alpha^p\beta^q\gamma^k \ldots \equiv (pqr \ldots)$. It is in the partition notation

$$Q_1 = 1+(1)+(2)+(1^2)+(3)+(21)+(1^3)+\ldots \text{ ad inf.}$$

It was then found that the effect of any Hammond operator upon any power of Q_1 is merely to multiply it by a positive integer. It then appeared that, denoting Q_1 by $F(\alpha, \beta, \gamma, \ldots)$, the function

$$Q_i = F(\alpha^i, \beta^i, \gamma^i, \ldots)$$

possesses properties of a character similar to those appertaining to Q_1. The fact is that any Hammond operator when performed upon any power of Q_i, say $Q_i^{k_i}$ has the effect of merely multiplying it by an integer, which may exceptionally be zero. Finally the important fact emerged that the performance of any Hammond operator upon the product

$$Q_1^{k_1} Q_2^{k_2} \ldots Q_i^{k_i} \ldots,$$

where $k_1, k_2, \ldots k_i, \ldots$ may, each of them, be zero or any positive integer, is merely to multiply it by a positive integer, which may, exceptionally, be zero

This discovery involves the complete enumerative solution of the unrestricted partition of multipartite numbers into a given number of parts. The reason of this is that the enumerating symmetric function generating function can be expanded in ascending powers of the functions $Q_1, Q_2, Q_3, \ldots$. On every term of this expansion the repeated performance of Hammond operators is practically effective and is successful in forcing out the sought numerical coefficients. When the magnitudes of the integer constituents of the multipartite parts are restricted in any manner there exists similarly an appropriate series of symmetric functions,

$$U_1, U_2, \ldots U_i \ldots,$$

the formation of which is explained in the paper, which in their properties are analogous to the series $Q_1, Q_2, \ldots Q_i, \ldots$. This circumstance involves the complete enumerative solution when the magnitudes of the constituents of the parts are restricted in any manner whatever.

SECTION I.

The Partition of Multipartite Numbers.

Art. 1. One of the problems which has engaged the attention of writers on the combinatory analysis is that of enumerating the different modes of exhibiting a given composite integer as the product of a given number of factors. For instance, the number 30, which is the product of three unrepeated primes, can be given as the product of two factors in the three ways,

$$2 \times 15, \qquad 3 \times 10, \qquad 5 \times 6.$$

When the given composite number is a product of different primes the question is very easy and is completely solved by means of the generating function

$$1/(1-x)(1-2x)\ldots(1-kx).$$

In the ascending expansions the coefficient of x^{q-k} is the number of ways of factorizing a number, which is the product of q different primes, into exactly k factors.*

Generating functions of the same character have also been obtained for some other simple forms of the composite number such as $p_1^2 p_2 \ldots p_q$, $p_1^2 p_2^2 p_3 \ldots p_q$; $p_1, p_2, \ldots$ denoting primes.

It is, of course, obvious that the absolute magnitudes of the prime factors have nothing whatever to do with the question, which necessarily appertains to the exponents of the primes and to nothing else.

* Compare 'NETTO's Combinatorik,' 1901, pp. 168 *et seq.*

N 2

Art. 2. Writers upon the problem have not usually observed that the general question is identical with the partition of a multipartite number into a given number of parts. Thus the problem discussed by NETTO and others is simply the enumeration of the partitions, into exactly k parts, of the multipartite number

$$1111 \ldots q \text{ times repeated.}$$

Ex. gr. when $q = 3$, $k = 2$ (see the two-factor factorization of $2 \times 3 \times 5$ above), we have three partitions of the multipartite 111 into exactly two parts. These partitions are

$$(110, 001), \quad (101, 010), \quad (011, 100).$$

In general the enumeration of the factorizations, involving k factors, of the composite number

$$p_1^{m_1} p_2^{m_2} \ldots p_s^{m_s},$$

yields the same number as the enumeration of the partitions of the multipartite number

$$m_1 m_2 \ldots m_s$$

into exactly k (multipartite) parts.

Art. 3. It is the same problem also to enumerate the separations of a given (unipartite) partition. Thus in relation to the partition (321) of the number 6, there is a one-to-one correspondence between the separations which involve two separates and the partitions of the multipartite number 111, which involve exactly two parts.

The separations are in fact

$$(32)(1), \quad (31)(2), \quad (21)(3).$$

In general there is a one-to-one correspondence between the separations of the partition

$$(q_1^{m_1} q_2^{m_2} \ldots q_s^{m_s}),$$

which involve k separates, and the partitions of the multipartite number

$$m_1 m_2 \ldots m_s,$$

which involve exactly k parts.

Art. 4. The general question of multipartite partition I have already discussed* by a method of grouping the partitions and a particular theory of distribution. The present investigation which depends upon other principles leads to results of a different and more general character. I showed many years ago† that in regard to the system of infinitely numerous quantities

$$\alpha, \beta, \gamma, \ldots,$$

* 'Phil. Trans. Camb. Phil. Soc.,' vol. xxi., No. xviii., pp. 467–481, 1912.

† 'Proc. Lond. Math. Soc.,' vol. xix., 1887, pp. 220 *et seq.*

the enumerating generating function is the symmetric function

$$\frac{1}{(1-a)} \times \frac{1}{(1-\alpha a)(1-\beta a)(1-\gamma a)\ldots}$$
$$\times \frac{1}{(1-\alpha^2 a)(1-\beta^2 a)(1-\gamma^2 a)\ldots(1-\alpha\beta a)(1-\alpha\gamma a)(1-\beta\gamma a)\ldots}$$
$$\times \frac{1}{(1-\alpha^3 a)\ldots(1-\alpha^2\beta a)\ldots(1-\alpha\beta\gamma a)\ldots}$$
$$\times \quad . \quad . \quad . \quad . \quad . \quad . \quad . \quad . \quad . \quad . \quad . \quad . \quad .$$

wherein if h_s denote the sum of homogeneous products of weight s of the quantities $\alpha, \beta, \gamma, \ldots$, the $s-1^{\text{th}}$ fractional factor of the generating function possesses a denominator factor corresponding to every separate term of h_s. The function is to be developed in ascending powers of a and, replacing for the moment the series $\alpha, \beta, \gamma, \ldots$ by $\alpha_1, \alpha_2, \alpha_3, \ldots$, we seek the coefficient of

$$a^k \left(\Sigma\alpha_1^{m_1}\alpha_2^{m_2}\ldots\alpha_s^{m_s}\right).$$

We write this, usually, in the notation

$$a^k (m_1 m_2 \ldots m_s).$$

The coefficient mentioned enumerates the partitions of the multipartite number

$$(m_1 m_2 \ldots m_s),$$

into *k or fewer* parts. If the first fractional factor $1/1-a$ had been omitted the coefficient would have denoted the number of the partitions into *exactly* k parts. The inclusion of the factor $1/1-a$ is of great importance to the investigation and equally yields the enumerations into exactly k parts, because from the coefficients of $a^k (m_1 m_2 \ldots m_s)$ we have merely to subtract the coefficients of $a^{k-1}(m_1 m_2 \ldots m_s)$. The importance is due to the circumstance that the symmetric functions which present themselves in the expansion are in the best possible form for the performance of the Hammond operators. This is not the case when the factor $1/1-a$ is excluded, as then a transformation, the necessity for which is not at once clear, is needed to obtain the proper form.

I will remind the reader that, writing

$$(1-\alpha x)(1-\beta x)(1-\gamma x)\ldots = 1-a_1 x+a_2 x^2-a_3 x^3+\ldots,$$

HAMMOND's differential operator of order m is

$$D_m = \frac{1}{m!}(\partial_{a_1}+a_1\partial_{a_2}+a_2\partial_{a_3}+\ldots)^m;$$

and its cardinal property is

$$D_{m_1}D_{m_2}\ldots D_{m_s}(m_1m_2\ldots m_s) = 1;$$

and this operation does not result in unity when it is performed upon any other symmetric function.

In order to obtain the coefficient of

$$a^k(m_1m_2\ldots m_s),$$

in the expanded function, we first of all find the complete coefficient of a^k and then operate upon it with the Hammond combination of operators

$$D_{m_1}D_{m_2}\ldots D_{m_s}.$$

The result is an *integer* followed by the sum of an infinite series of symmetric functions. The integer mentioned is the number we seek.

Art. 5. We now expand the generating function. On well-known principles we can assert that the coefficient of a^k in the expansion is the *homogeneous product-sum of order k* of unity, and of the whole of the $\alpha, \beta, \gamma, \ldots$ products which occur in the denominator factors of the generating function. The *elements*, of which we must form homogeneous product sums are, in fact,

$$1$$

$$\alpha, \beta, \gamma, \ldots,$$

$$\alpha^2, \beta^2, \gamma^2, \ldots \alpha\beta, \alpha\gamma, \beta\gamma, \ldots,$$

$$\alpha^3, \beta^3, \gamma^3, \ldots \alpha^2\beta, \alpha^2\gamma, \ldots\alpha\beta\gamma, \alpha\beta\delta, \ldots.$$

. . . .

We can form these product-sums from the sums of the powers of these elements, because we have before us the well-known symmetric function formula

$$h_k = \Sigma_\sigma \frac{s_1^{k_1}s_2^{k_2}\ldots s_i^{k_i}}{1^{k_1}2^{k_2}\ldots i^{k_i}.k_1!\,k_2!\ldots k_i!}.$$

The sum of the powers are readily formed; for, calling them

$$Q_1, Q_2, \ldots Q_i, \ldots,$$

it is clear that Q_i is the sum of unity and the whole of the monomial (that is to say merely involving in the partition notation a single partition), symmetric functions of weights one to infinity. Hence

$$Q_1 = 1+(1)+(2)+(1^2)+(3)+(21)+(1^3)+\ldots \text{ ad inf.};$$

and, regarding Q_1 as $F(\alpha, \beta, \gamma, \ldots)$, it is obvious that

$$Q_i = F(\alpha^i, \beta^i, \gamma^i, \ldots);$$

showing us that

$$Q_2 = 1+(2)+(4)+(2^2)+(6)+(42)+(2^3)+\ldots,$$
$$Q_i = 1+(i)+(2i)+(i^2)+(3i)+(2i, i)+(i^3)+\ldots.$$

Thence the expansion

$$\begin{aligned}&1+aQ_1\\&+\frac{a^2}{2!}(Q_1^2+Q_2)\\&+\frac{a^3}{3!}(Q_1^3+3Q_1Q_2+2Q_3)\\&+\frac{a^4}{4!}(Q_1^4+6Q_1^2Q_2+3Q_2^2+8Q_1Q_3+6Q_4)\\&+\ldots\\&+a^kF_k(Q)\\&+\ldots\end{aligned}$$

where

$$F_k(Q) = \Sigma\frac{Q_1^{k_1}Q_2^{k_2}\ldots Q_i^{k_i}}{1^{k_1}2^{k_2}\ldots i^{k_i}k_1!\,k_2!\ldots k_i!}.$$

Art. 6. The importance of this expansion lies in the fact that the infinite series of Hammond operators on the one hand and the infinite series of Q functions on the other hand have very remarkable properties in relation to one another. The first property we notice is that from the well-known law of operation,

$$D_mQ_1 = Q_1,$$

for all values of m. Also

$$D_mQ_2 = Q_2, \text{ or zero},$$

ascending as m is, or is not, a multiple of *two*. And generally

$$D_mQ_i = Q_i, \text{ or zero},$$

according as m is, or is not, a multiple of i.

When D_m is performed upon a product of k separate functions, it operates through the medium of a number of operators associated with the compositions of m into k parts, zero being regarded as a part and D_0 being regarded as a symbol for unity. Thus the compositions of the number 4 into three parts being 400, 040, 004; 310,

301, 130, 103, 031, 013 ; 220, 202, 022 ; 211, 121, 112, the law of operation is as follows :—

$$D_4 Q_a Q_b Q_c = (D_4 Q_a)(D_0 Q_b)(D_0 Q_c) + (D_0 Q_a)(D_4 Q_b)(D_0 Q_c) + (D_0 Q_a)(D_0 Q_b)(D_4 Q_c)$$
$$+ (D_3 Q_a)(D_1 Q_b)(D_0 Q_c) + (D_3 Q_a)(D_0 Q_b)(D_1 Q_c) + (D_1 Q_a)(D_3 Q_b)(D_0 Q_c)$$
$$+ (D_1 Q_a)(D_0 Q_b)(D_3 Q_c) + (D_0 Q_a)(D_3 Q_b)(D_1 Q_c) + (D_0 Q_a)(D_1 Q_b)(D_3 Q_c)$$
$$+ (D_2 Q_a)(D_2 Q_b)(D_0 Q_c) + (D_2 Q_a)(D_0 Q_b)(D_2 Q_c) + (D_0 Q_a)(D_2 Q_b)(D_2 Q_c)$$
$$+ (D_2 Q_a)(D_1 Q_b)(D_1 Q_c) + (D_1 Q_a)(D_2 Q_b)(D_1 Q_c) + (D_1 Q_a)(D_1 Q_b)(D_2 Q_c)$$

D_0, being unity, may be omitted but has been retained above to make the connexion with the compositions quite clear. This method of performing D_m upon a product I have explained and used in previous papers during the last five and twenty years. Upon this example some observations can be made. In the first place the operation breaks up into 15 portions because the number 4 has 15 three-part compositions. The result of each portion must be moreover either $Q_a Q_b Q_c$ or zero, because $D_i Q_t$ is either Q_t or zero. Hence the result of the whole operation must be merely to multiply $Q_a Q_b Q_c$ by some integer $\eqslantless 15$. In general the result of the operation

$$D_m Q_1{}^{k_1} Q_2{}^{k_2} \ldots Q_i{}^{k_i}$$

must be merely to multiply the product

$$Q_1{}^{k_1} Q_2{}^{k_2} \ldots Q_i{}^{k_i}$$

by some integer equal to, or less, than the number of compositions of m into $k_1 + k_2 + \ldots + k_i$ parts, zero always counting as a part.

Hence also the result of the operation

$$D_{m_1} D_{m_2} \ldots D_{m_s} Q_1{}^{k_1} Q_2{}^{k_2} \ldots Q_i{}^{k_i}$$

must be merely to multiply the product

$$Q_1{}^{k_1} Q_2{}^{k_2} \ldots Q_i{}^{k_i}$$

by an integer.

This valuable result shows that the Hammond operators may be performed with facility upon the function

$$F_k (Q)$$

which is before us.

Art 7. The determination of the result of the operation

$$D_m Q_1{}^{k_1} Q_2{}^{k_2} \ldots Q_i{}^{k_i}$$

is now entered upon. We have to find the value of the multiplying integer

The value is clearly equal to the number of compositions of m which do not have the effect of multiplication by zero. Suppose that we write out the operand at length

$$Q_1 Q_1 Q_1 \dots Q_2 Q_2 Q_2 \dots\dots Q_i Q_i Q_i \dots,$$

in i blocks, containing $k_1, k_2, \dots k_i$ factors respectively.

Underneath the k factors we will suppose written any composition of m into k parts, zero being included as a part.

In order that the corresponding operation may result in unity and not in zero we have the conditions :—

(i.) Any number, including zero, may occur underneath any of the k_1 factors Q_1 ;
(ii.) Zero, or any multiple of 2, may occur underneath any of the k_2 factors Q_2 ;
(iii.) Zero, or any multiple of 3, may occur underneath any of the k_3 factors Q_3 ;
(iii. ...) And lastly, zero, or any multiple of i, may occur underneath any of the k_i factors Q_i.

How many such compositions exist ?

We have merely to find the coefficient of x^m in the expansion of

$$(1+x+x^2+\dots)^{k_1}(1+x^2+x^4+\dots)^{k_2}(1+x^3+x^6+\dots)^{k_3}\dots\dots(1+x^i+x^{2i}+\dots)^{k_i};$$

for this is the function which enumerates the compositions which possess these properties. In fact to form the composition we take a power of x from each of the first k_1 factors; then a power of x from each of the next k_2 factors, observing that the exponents of x are all zero or multiples of two; then a power of x from the next k_3 factors, observing that the exponents are all zero or multiples of three; and so on, until finally in the k_i factors we find that the exponents are all zero or multiples of i.

Hence it follows that the operation

$$D_m Q_1^{k_1} Q_2^{k_2} \dots Q_i^{k_i}$$

has the effect of multiplying $Q_1^{k_1} Q_2^{k_2} \dots Q_i^{k_i}$ by a number which is given by the coefficient of x^m in the expansion of

$$(1-x)^{-k_1}(1-x^2)^{-k_2}(1-x^3)^{-k_3}\dots(1-x^i)^{-k_i};$$

an elegant theorem.

Let this coëfficient be denoted by

$$F_q(m;\ 1^{k_1}2^{k_2}\dots i^{k_i});$$

so that

$$D_m Q_1^{k_1} Q_2^{k_2} \dots Q_i^{k_i} = F_q(m;\ 1^{k_1}2^{k_2}\dots i^{k_i}) \,.\, Q_1^{k_1} Q_2^{k_2} \dots Q_i^{k_i}$$

Art. 8. Looking to the symmetric function expressions of $Q_1, Q_2, \dots Q_i$ it will be noted that the only portion of the product

$$Q_1^{k_1} Q_2^{k_2} \dots Q_i^{k_i}$$

which does not involve the elements α, β, γ, ... is unity.

Hence the portion of

$$D_m \mathbf{Q}_1^{k_1} \mathbf{Q}_2^{k_2} \ldots \mathbf{Q}_i^{k_i}$$

that is free from the elements is

$$F_q(m;\ 1^{k_1} 2^{k_2} \ldots i^{k_i}),$$

which is obtained directly from the result of the operation by putting

$$\mathbf{Q}_1 = \mathbf{Q}_2 = \ldots = \mathbf{Q}_i = 1.$$

We may represent this circumstance by the convenient notation

$$(D_m \mathbf{Q}_1^{k_1} \mathbf{Q}_2^{k_2} \ldots \mathbf{Q}_i^{k_i})_{Q=1} = F_q(m;\ 1^{k_1} 2^{k_2} \ldots i^{k_i}).$$

The number of partitions, of the unipartite number m, into k or fewer parts, is by the present investigation

$$\begin{aligned} &D_m F_k(\mathbf{Q})_{Q=1} \\ &= \Sigma \frac{(D_m \mathbf{Q}_1^{k_1} \mathbf{Q}_2^{k_2} \ldots \mathbf{Q}_i^{k_i})_{Q=1}}{1^{k_1} 2^{k_2} \ldots i^{k_i} .\ k_1!\, k_2! \ldots k_i!} \\ &= \Sigma \frac{F_q(m;\ 1^{k_1} 2^{k_2} \ldots i^{k_i})}{1^{k_1} 2^{k_2} \ldots i^{k_i} .\ k_1!\, k_2! \ldots k_i!} \\ &= \text{coefficient of } x^m \text{ in } \Sigma \frac{(1-x)^{-k_1}(1-x^2)^{-k_2} \ldots (1-x^i)^{-k_i}}{1^{k_1} 2^{k_2} \ldots i^{k_i} .\ k_1!\, k_2! \ldots k_i!}; \end{aligned}$$

the summation being for every partition

$$k_1 + 2k_2 + \ldots + ik_i$$

of the number k.

But we know, otherwise, that the number of partitions of m, into k or fewer parts, is given by the coefficients of x^m in

$$\frac{1}{(1-x)(1-x^2) \ldots (1-x^k)}.$$

Hence the identity

$$\Sigma \frac{(1-x)^{-k_1}(1-x^2)^{-k_2} \ldots (1-x^i)^{-k_i}}{1^{k_1} 2^{k_2} \ldots i^{k_i} k_1!\, k_2! \ldots k_i!} = \frac{1}{(1-x)(1-x^2) \ldots (1-x^k)},$$

which being a known result supplies an interesting verification of our work. The present investigation in any case supplies one proof of it.

Art. 9. There is now no difficulty in proceeding to the result

$$\begin{aligned} &D_{m_1} D_{m_2} \ldots D_{m_s} .\ \mathbf{Q}_1^{k_1} \mathbf{Q}_2^{k_2} \ldots \mathbf{Q}_i^{k_i} \\ &= F_q(m_1;\ 1^{k_1} 2^{k_2} \ldots i^{k_i}) .\ F_q(m_2;\ 1^{k_1} 2^{k_2} \ldots i^{k_i}) \ldots\ldots F_q(m_s;\ 1^{k_1} 2^{k_2} \ldots i^{k_i}) .\ \mathbf{Q}_1^{k_1} \mathbf{Q}_2^{k_2} \ldots \mathbf{Q}_i^{k_i}; \end{aligned}$$

and the number of partitions of the multipartite number

$$m_1 m_2 \ldots m_s,$$

into k or fewer parts, is

$$\mathrm{D}_{m_1}\mathrm{D}_{m_2}\ldots\mathrm{D}_{m_s}\mathrm{F}_k(\mathrm{Q})_{\mathrm{Q}=1}$$

$$= \Sigma \frac{(\mathrm{D}_{m_1}\mathrm{D}_{m_2}\ldots\mathrm{D}_{m_s}\mathrm{Q}_1^{k_1}\mathrm{Q}_2^{k_2}\ldots\mathrm{Q}_i^{k_i})_{\mathrm{Q}=1}}{1^{k_1}2^{k_2}\ldots i^{k_i}\,.\,k_1!\,k_2!\ldots k_i!}$$

$$= \Sigma \frac{\mathrm{F}_q(m_1;\ 1^{k_1}2^{k_2}\ldots i^{k_i})\,.\,\mathrm{F}_q(m_2;\ 1^{k_1}2^{k_2}\ldots i^{k_i})\ldots\ldots\mathrm{F}_q(m_s;\ 1^{k_1}2^{k_2}\ldots i^{k_i})}{1^{k_1}2^{k_2}\ldots i^{k_i}\,.\,k_1!\,k_2!\ldots k_i!}$$

This is the general solution of the problem of enumeration in the absence of any restriction upon the magnitudes of the constituents of the multipartite parts.*

Art. 10. It will be convenient at this point to give a few results derived from the function

$$(1-x)^{-k_1}(1-x^2)^{-k_2}\ldots(1-x^i)^{-k_i}$$

which will be useful in the sequel.

$$\mathrm{F}_q(m;\ 1^k) = \binom{m+k-1}{k-1},$$

$$\mathrm{F}_q(2m;\ 2^k) = \binom{m+k-1}{k-1},$$

$$\mathrm{F}_q(im;\ i^k) = \binom{m+k-1}{k-1},$$

$$\mathrm{F}_q(2m;\ 12) = m+1, \qquad \mathrm{F}_q(2m+1;\ 12) = m+1,$$

$$\mathrm{F}_q(2m;\ 1^2 2) = (m+1)^2, \qquad \mathrm{F}_q(2m+1;\ 1^2 2) = (m+1)(m+2),$$

$$\mathrm{F}_q(2m;\ 12^2) = \binom{m+2}{2}, \qquad \mathrm{F}_q(2m+1;\ 12^2) = \binom{m+2}{2},$$

* With regard to the algebraical identity met with above, the reader may compare 'SYLVESTER'S Mathematical Papers,' vol. III., p. 598, where it is shown that for the roots of the equation

$$z^q - \frac{1}{1-c}z^{q-1} + \frac{c}{(1-c)(1-c^2)}z^{q-2} - \frac{c^3}{(1-c)(1-c^2)(1-c^3)}z^{q-3} + \ldots = 0,$$

the general term being

$$(-)^n \frac{c^{\binom{n}{2}}}{(1-c)(1-c^2)\ldots(1-c^n)}z^{q-n},$$

the homogeneous product-sum of order n is

$$\frac{1}{(1-c)(1-c^2)\ldots(1-c^n)};$$

and the sum of the nth powers of the roots is

$$\frac{1}{1-c^n}.$$

The expression of the homogeneous product-sum of order k, in terms of the sums of the powers, by the formula quoted early in this paper, gives the identity in question.

O 2

$$F_q(3m;\ 13) = F_q(3m+1;\ 13) = F_q(3m+2;\ 13) = m+1,$$

$$F_q(3m;\ 1^23) = \tfrac{1}{2}(m+1)(3m+2), \qquad F_q(3m+1;\ 1^23) = \tfrac{1}{2}(m+1)(3m+4),$$

$$F_q(3m+2;\ 1^23) = \tfrac{1}{2}(m+1)(3m+6),$$

$$F_q(2m;\ 1^{k_1}2^{k_2})$$

$$= \binom{m+k_2-1}{k_2-1} + \binom{k_1+1}{k_1-1}\binom{m+k_2-2}{k_2-1} + \binom{k_1+3}{k_1-1}\binom{m+k_2-3}{k_2-1} + \ldots + \binom{2m+k_1-1}{k_1-1},$$

$$F_q(2m+1;\ 1^{k_1}2^{k_2})$$

$$= \binom{k_1}{k_1-1}\binom{m+k_2-1}{k_2-1} + \binom{k_1+2}{k_1-1}\binom{m+k_2-2}{k_2-1} + \ldots + \binom{2m+k_1}{k_1-1},$$

$$F_q(3m;\ 1^{k_1}3^{k_3})$$

$$= \binom{m+k_3-1}{k_3-1} + \binom{k_1+2}{k_1-1}\binom{m+k_3-2}{k_3-1} + \binom{k_1+5}{k_1-1}\binom{m+k_3-3}{k_3-1} + \ldots + \binom{3m+k_1-1}{k_1-1},$$

$$F_q(3m+1;\ 1^{k_1}3^{k_3})$$

$$= \binom{k_1}{k_1-1}\binom{m+k_3-1}{k_3-1} + \binom{k_1+3}{k_1-1}\binom{m+k_3-2}{k_3-1} + \ldots + \binom{3m+k_1}{k_1-1},$$

$$F_q(3m+2;\ 1^{k_1}3^{k_3})$$

$$= \binom{k_1+1}{k_1-1}\binom{m+k_3-1}{k_3-1} + \binom{k_1+4}{k_1-1}\binom{m+k_3-2}{k_3-1} + \ldots + \binom{3m+k_1+1}{k_1-1}.$$

The use of the Hammond operator D_m is convenient but not essential to this investigation. It is convenient from the algebraic point of view, and also because it brings into prominence properties of the operator which are in themselves important. The coefficient of $\alpha_1^{m_1}\alpha_2^{m_2}\ldots\alpha_s^{m_s}$ in the product $Q_1^{k_1}Q_2^{k_2}\ldots Q_i^{k_i}$ is readily obtained when we remember that

$$Q_i = \frac{1}{(1-\alpha^i)(1-\beta^i)(1-\gamma^i)\ldots\ldots}$$

and the various modifications are readily made for the allied functions $A_i, B_i, \ldots U_i$.

The Partitions of Multipartite Numbers into Two Parts.

Art. 11. The generating function which enumerates the partitions into two or fewer parts is

$$\tfrac{1}{2}(Q_1^2+Q_2);$$

and since, from the principles just stated,

$$D_{2m}Q_1^2 = (2m+1)\,Q_1^2, \qquad D_{2m+1}Q_1^2 = (2m+2)\,Q_1^2,$$

$$D_{2m}Q_2 = Q_2, \qquad D_{2m+1}Q_2 = 0;$$

we find

$$D_{2m}\tfrac{1}{2}(Q_1^2+Q_2) = \tfrac{1}{2}(2m+1)Q_1^2+\tfrac{1}{2}Q_2,$$

$$D_{2m+1}\tfrac{1}{2}(Q_1^2+Q_2) = \tfrac{1}{2}(2m+2)Q_1^2;$$

and, by reason of the important properties possessed by the Q products in their relations with the Hammond operators, we can at once proceed to the results

$$D_{2m}^s = \tfrac{1}{2}(2m+1)^s Q_1^2+\tfrac{1}{2}Q_2,$$

$$D_{2m+1}^s = \tfrac{1}{2}(2m+2)^s Q_1^2.$$

Thence we derive, by putting $Q_1 = Q_2 = 1$, the coefficients of the symmetric functions

$$(2m^s), \qquad (2m+1^s)$$

(the exponent s meaning the numbers $2m$, $2m+1$ respectively s times repeated) in the development of the function

$$\tfrac{1}{2}(Q_1^2+Q_2).$$

Thence we obtain the numbers

$$\tfrac{1}{2}(2m+1)^s+\tfrac{1}{2}, \qquad \tfrac{1}{2}(2m+2)^s$$

which, respectively, enumerate the ways of partitioning the *multipartite numbers*

$$(2m, 2m \ldots \text{repeated } s \text{ times}), \qquad (2m+1, 2m+1 \ldots \text{repeated } s \text{ times})$$

into two or fewer parts.

When the enumeration is concerned with exactly two parts we have clearly to subtract unity in each case. In fact the generating function is

$$\tfrac{1}{2}(Q_1^2+Q_2)-Q_1;$$

and

$$D_{2m}^s Q_1 = D_{2m+1}^s Q_1 = Q_1,$$

showing that unity must be subtracted.

The numbers then become

$$\tfrac{1}{2}(2m+1)^s-\tfrac{1}{2}, \qquad \tfrac{1}{2}(2m+2)^s-1.$$

These numbers also enumerate the ways of exhibiting the composite integers

$$(p_1 p_2 \ldots p_s)^{2m}, \qquad (p_1 p_2 \ldots p_s)^{2m+1},$$

as the product of two factors.

To obtain a general formula for the multipartite number

$$(m_1 m_2 \ldots m_s)$$

we write

$$D_m Q_1^2 = F_q(m;\ 1^2) . Q_1^2,$$

$$D_m Q_2 = F_q(m;\ 2) . Q_2,$$

then

$$D_{m_1} \tfrac{1}{2}(Q_1^2 + Q_2) = \tfrac{1}{2}\{F_q(m_1;\ 1^2) . Q_1^2 + F_q(m_1;\ 2) . Q_2\},$$

$$D_{m_1} D_{m_2} \tfrac{1}{2}(Q_1^2 + Q_2) = \tfrac{1}{2}\{F_q(m_1;\ 1^2)\, F_q(m_2;\ 1^2) . Q_1^2 + F_q(m_1;\ 2)\, F_q(m_2;\ 2) . Q_2\},$$

$$D_{m_1} D_{m_2} \ldots D_{m_s} \tfrac{1}{2}(Q_1^2 + Q_2) = \tfrac{1}{2}\prod_1^s F_q(m_i;\ 1^2) . Q_1^2 + \tfrac{1}{2}\prod_1^s F_q(m_i;\ 2) . Q_2,$$

leading us to the number

$$\tfrac{1}{2}\prod_1^s F_q(m_i;\ 1^2) + \tfrac{1}{2}\prod_1^s F_q(m_i;\ 2),$$

as the enumerator of the partitions, into two or fewer parts, of the multipartite number

$$(m_1 m_2 \ldots m_s).$$

The reader will observe that the algebraic expressions of

$$F_q(m_i;\ 1^2) \text{ and } F_q(m_i;\ 2)$$

will depend upon the parity of m_i.

The notation has been adopted so as to save a multiplicity of formulæ in certain cases.

This of course solves the question of the factorization into two or fewer factors of the composite integer

$$p_1^{m_1} p_2^{m_2} \ldots p_s^{m_s}.$$

We have, therefore, solved completely the question of enumerating the bipartitions of multipartite numbers.

What has been done as a question in the theory of distribution may be stated as follows. We are given an assemblage of any numerical specification and two boxes which cannot be distinguished from one another. We have found the number of ways of distributing the objects between the boxes. The similar question when the boxes are distinguished from one another is simpler and connected with the compositions of multipartite numbers.

Art. 12. At this point it may be appropriate to give a statement in regard to the nature of the solution given in this investigation.

The enumeration of the partitions of a unipartite number m_1, into k or fewer parts, is formed as a linear function of certain numbers $a_1, b_1, c_1 \ldots$; the linear function being

$$\lambda a_1 + \mu b_1 + \nu c_1 + \ldots$$

where the numbers λ, μ, ν, depend only upon k.

Associated with another unipartite number m_2 we have the linear function

$$\lambda a_2 + \mu b_2 + \nu c_2 + \dots .$$

It has then been shown that the number of partitions, into k or fewer parts, of the multipartite number $m_1 m_2$ is

$$\lambda a_1 a_2 + \mu b_1 b_2 + \nu c_1 c_2 + \dots ,$$

and in general the number of partitions, into k or fewer parts, of the multipartite number

$$m_1 m_2 \dots m_s$$

is

$$\lambda a_1 a_2 \dots a_s + \mu b_1 b_2 \dots b_s + \nu c_1 c_2 \dots c_s + \dots .$$

The multipartite solution is thus essentially derived from the solutions which appertain to the separate unipartite numbers whose conjunction defines the multipartite number. The numbers $\lambda, \mu, \nu, \dots$ are those well known in connexion with the expression of the homogeneous product sum h_k in terms of the sums of the powers $s_1, s_2, s_3, \dots s_k$, the whole question is therefore reduced to finding the numbers

$$a, b, c, \dots$$

appertaining to the unipartite number m.

This, as has been shown, depends upon finding the coefficient of x^m in a function

$$(1-x)^{-k_1}(1-x^2)^{-k_2} \dots (1-x^i)^{-k_i}$$

where

$$k_1 + k_2 + \dots + k_i = k.$$

The possibility of the solution rests upon the remarkable circumstances that when the operator D_m is performed upon the operand

$$Q_1^{k_1} Q_2^{k_2} \dots Q_i^{k_i}$$

its effect is to merely multiply it by an integer.

The Partitions of Multipartite Numbers into Three Parts.

Art. 13. I will, in future, merely deal with the partitions into k or fewer parts, since the result for exactly k parts is at once derived by subtracting the result for $k-1$ or fewer parts.

The operand is

$$\tfrac{1}{6}(Q_1^3 + 3Q_1Q_2 + 2Q_3),$$

and since the result depends upon the divisibility of m by both 2 and 3 it will be necessary to consider the operations of

$$D_{6m},\ D_{6m+1},\ D_{6m+2},\ D_{6m+3},\ D_{6m+4},\ D_{6m+5}.$$

The investigation is therefore in six parts.

(i.) Since

$$D_{6m}Q_1^3 = \binom{6m+2}{2} Q_1^3,$$

$$D_{6m}Q_1Q_2 = (3m+1) Q_1Q_2,$$

$$D_{6m}Q_3 = Q_3,$$

we find

$$D_{6m}\tfrac{1}{6}(Q_1^3+3Q_1Q_2+2Q_3) = \tfrac{1}{6}\left\{\binom{6m+2}{2} Q_1^3+3(3m+1) Q_1Q_2+2Q_3\right\};$$

and immediately

$$D_{6m_1}D_{6m_2}\dots D_{6m_s}\tfrac{1}{6}(Q_1^3+3Q_1Q_2+2Q_3) = \tfrac{1}{6}\left\{\binom{6m_1+2}{2}\binom{6m_2+2}{2}\dots\binom{6m_s+2}{2} Q_1^3 \right.$$
$$+3(3m_1+1)(3m_2+1)\dots(3m_s+1) Q_1Q_2$$
$$\left. +2Q_3\right\};$$

and in particular

$$D_{6m}^s\tfrac{1}{6}(Q_1^3+3Q_1Q_2+2Q_3) = \tfrac{1}{6}\left\{\binom{6m+2}{2}^s Q_1^3+3(3m+1)^s Q_1Q_2+2Q_3\right\};$$

results which establish that the partitions of the multipartite number

$$6m_1 6m_2 \dots 6m_s,$$

into three or fewer parts, are enumerated by

$$\tfrac{1}{6}\left\{\binom{6m_1+2}{2}\binom{6m_2+2}{2}\dots\binom{6m_s+2}{2}+3(3m_1+1)(3m_2+1)\dots(3m_s+1)+2\right\}.$$

(ii.) Since

$$D_{6m+1}Q_1^3 = \binom{6m+3}{2} Q_1^3,$$

$$D_{6m+1}Q_1Q_2 = (3m+1) Q_1Q_2,$$

$$D_{6m+1}Q_3 = 0,$$

we find

$$D_{6m+1}\tfrac{1}{6}(Q_1^3+3Q_1Q_2+2Q_3) = \tfrac{1}{6}\left\{\binom{6m+3}{2} Q_1^3+3(3m+1) Q_1Q_2\right\};$$

and immediately

$$D_{6m_1+1}D_{6m_2+1}\dots D_{6m_s+1}\tfrac{1}{6}(Q_1^3+3Q_1Q_2+2Q_3) = \tfrac{1}{6}\left\{\binom{6m_1+3}{2}\binom{6m_2+3}{2}\dots\binom{6m_s+3}{2} Q_1^3\right.$$
$$\left.+3(3m_1+1)(3m_2+1)\dots(3m_s+1) Q_1Q_2\right\};$$

establishing that the partitions of the multipartite number

$$6m_1+1\ 6m_2+1 \dots 6m_s+1,$$

into three or fewer parts are enumerated by

$$\tfrac{1}{6}\left\{\binom{6m_1+3}{2}\binom{6m_2+3}{2}\dots\binom{6m_s+3}{2}+3(3m_1+1)(3m_2+1)\dots(3m_s+1)\right\}.$$

(iii.) Since

$$D_{6m+2}Q_1^3=\binom{6m+4}{2}Q_1^3,$$

$$D_{6m+2}Q_1Q_2=(3m+2)Q_1Q_2,$$

$$D_{6m+2}Q_3=0,$$

we, as above, derive, for the partitions of the multipartite number

$$6m_1+2\ 6m_2+2\ \dots\ 6m_s+2,$$

the enumerating number

$$\tfrac{1}{6}\left\{\binom{6m_1+4}{2}\binom{6m_2+4}{2}\dots\binom{6m_s+4}{2}+3(3m_1+2)(3m_2+2)\dots(3m_s+2)\right\}.$$

(iv.) Since

$$D_{6m+3}Q_1^3=\binom{6m+5}{2}Q_1^3,$$

$$D_{6m+3}Q_1Q_2=(3m+2)Q_1Q_2,$$

$$D_{6m+3}Q_3=Q_3,$$

we obtain, for the multipartite number

$$6m_1+3\ 6m_2+3\ \dots\ 6m_s+3,$$

the enumerating number

$$\tfrac{1}{6}\left\{\binom{6m_1+5}{2}\binom{6m_2+5}{2}\dots\binom{6m_s+5}{2}+3(3m_1+2)(3m_2+2)\dots(3m_s+2)+2\right\}.$$

(v.) Also

$$D_{6m+4}Q_1^3=\binom{6m+6}{2}Q_1^3,$$

$$D_{6m+4}Q_1Q_2=(3m+3)Q_1Q_2,$$

$$D_{6m+4}Q_3=0,$$

and we obtain, for the multipartite number

$$6m_1+4\ 6m_2+4\ \dots\ 6m_s+4,$$

the enumerating number

$$\tfrac{1}{6}\left\{\binom{6m_1+6}{2}\binom{6m_2+6}{2}\cdots\binom{6m_s+6}{2}+3(3m_1+3)(3m_2+3)\ldots(3m_s+3)\right\}.$$

(vi.) **Lastly, since**

$$D_{6m+5}Q_1^3 = \binom{6m+7}{2}Q_1^3,$$

$$D_{6m+5}Q_1Q_2 = (3m+3).Q_1Q_2,$$

$$D_{6m+5}Q_3 = 0,$$

we obtain, for the multipartite number

$$6m_1+5\ 6m_2+5\ \ldots\ 6m_s+5,$$

the enumerating number

$$\tfrac{1}{6}\left\{\binom{6m_1+7}{2}\binom{6m_2+7}{2}\cdots\binom{6m_s+7}{2}+3(3m_1+3)(3m_2+3)\ldots(3m_s+3)\right\}.$$

Finally, in the notation employed for the bipartite case, for the multipartite number

$$m_1m_2\ldots m_s$$

we have the enumerating number

$$\tfrac{1}{6}\left\{\prod_1^s F_q(m_i;\ 1^3)+3\prod_1^s F_q(m_i;\ 1,2)+2\prod_1^s F_q(m_i;\ 3)\right\}.$$

Art. 14. I collect these results :—

Multipartite Numbers.	Number of Partitions into three or fewer parts.
$6m\ 6m\ 6m$ repeated s times	$\frac{1}{3!}\left\{\binom{6m+2}{2}^s+3(3m+1)^s+2\right\}$
$6m+1\ 6m+1\ 6m+1$ repeated s times	$\frac{1}{3!}\left\{\binom{6m+3}{2}^s+3(3m+1)^s\right\}$
$6m+2\ 6m+2\ 6m+2$,, ,,	$\frac{1}{3!}\left\{\binom{6m+4}{2}^s+3(3m+2)^s\right\}$
$6m+3\ 6m+3\ 6m+3$,, ,,	$\frac{1}{3!}\left\{\binom{6m+5}{2}^s+3(3m+2)^s+2\right\}$
$6m+4\ 6m+4\ 6m+4$,, ,,	$\frac{1}{3!}\left\{\binom{6m+6}{2}^s+3(3m+3)^s\right\}$
$6m+5\ 6m+5\ 6m+5$,, ,,	$\frac{1}{3!}\left\{\binom{6m+7}{2}^s+3(3m+3)^s\right\}.$

Multipartite Numbers.	Number of Partitions into exactly three parts.
$6m\ 6m\ 6m$ repeated s times	$\frac{1}{3!}\left\{\binom{6m+2}{2}^s+3(3m+1)^s-3(6m+1)^s-1\right\}$
$6m+1\ 6m+1\ 6m+1$ repeated s times	$\frac{1}{3!}\left\{\binom{6m+3}{2}^s+3(3m+1)^s-3(6m+2)^s\right\}$
$6m+2\ 6m+2\ 6m+2$,, ,,	$\frac{1}{3!}\left\{\binom{6m+4}{2}^s+3(3m+2)^s-3(6m+3)^s-3\right\}$
$6m+3\ 6m+3\ 6m+3$,, ,,	$\frac{1}{3!}\left\{\binom{6m+5}{2}^s+3(3m+2)^s-3(6m+4)^s+2\right\}$
$6m+4\ 6m+4\ 6m+4$,, ,,	$\frac{1}{3!}\left\{\binom{6m+6}{2}^s+3(3m+3)^s-3(6m+5)^s-3\right\}$
$6m+5\ 6m+5\ 6m+5$,, ,,	$\frac{1}{3!}\left\{\binom{6m+7}{2}^s+3(3m+3)^s-3(6m+6)^s\right\}$.

As a verification, connected with unipartite partitions, we put $s = 1$ in these last six formulæ, and reach the six numbers

$$3m^2,\quad 3m^2+m,\quad 3m^2+2m,\quad 3m^2+3m+1,\quad 3m^2+4m+1,\quad 3m^2+5m+2,$$

and since these may be exhibited in the forms

$$\frac{(6m)^2}{12},\quad \frac{(6m+1)^2}{12}-\tfrac{1}{12},\quad \frac{(6m+2)^2}{12}-\tfrac{1}{3},\quad \frac{(6m+3)^2}{12}+\tfrac{1}{4},\quad \frac{(6m+4)^2}{12}-\tfrac{1}{3},\quad \frac{(6m+5)^2}{12}-\tfrac{1}{12},$$

we verify the well-known theorem which states that the number of tripartitions of n is the nearest integer to $\frac{n^2}{12}$.

The Partitions of Multipartite Numbers into Four Parts.

Art. 15. The operand is

$$\tfrac{1}{24}(Q_1^4+6Q_1^2Q_2+3Q_2^2+8Q_1Q_3+6Q_4)$$

since the result depends upon the divisibility of m by the numbers 2, 3 and 4, and 12 is the least common multiple of those numbers, it will be necessary to take the operator suffix to the modulus 12, and the investigation is, therefore, in twelve parts.

P 2

We have

$$D_m Q_1^4 = \binom{m+3}{3} Q_1^4,$$

$$D_{2m} Q_1^2 Q_2 = (m+1)^2 Q_1^2 Q_2,$$

$$D_{2m+1} Q_1^2 Q_2 = (m+1)(m+2) Q_1^2 Q_2,$$

$$D_{2m} Q_2^2 = (m+1) Q_2^2,$$

$$D_{2m+1} Q_2^2 = 0,$$

$$D_{3m} Q_1 Q_3 = (m+1) Q_1 Q_3,$$

$$D_{3m+1} Q_1 Q_3 = (m+1) Q_1 Q_3,$$

$$D_{3m+2} Q_1 Q_3 = (m+1) Q_1 Q_3,$$

$$D_{4m} Q_4 = Q_4,$$

$$D_{4m+1} Q_4 = D_{4m+2} Q_4 = D_{4m+3} Q_4 = 0.$$

Utilising these results and taking as operators $D^s_{12m}, D^s_{12m+1}, \ldots D^s_{12m+11}$ in succession we find for partitions into four or fewer parts :—

Multipartite Numbers.	Number of Partitions into four or fewer parts.
$12m$ $12m$ $12m$ repeated s times	$\frac{1}{4!}\left\{\binom{12m+3}{3}^s + 6(6m+1)^{2s} + 3(6m+1)^s + 8(4m+1)^s + \right.$
$12m+1$ $12m+1$ $12m+1$ repeated s times	$\frac{1}{4!}\left\{\binom{12m+4}{3}^s + 12\binom{6m+2}{2}^s + 8(4m+1)^s\right.$
$12m+2$ $12m+2$ $12m+2$,, ,,	$\frac{1}{4!}\left\{\binom{12m+5}{3}^s + 6(6m+2)^{2s} + 3(6m+2)^s + 8(4m+1)^s\right.$
$12m+3$ $12m+3$ $12m+3$,, ,,	$\frac{1}{4!}\left\{\binom{12m+6}{3}^s + 12\binom{6m+3}{2}^s + 8(4m+2)^s\right.$
$12m+4$ $12m+4$ $12m+4$,, ,,	$\frac{1}{4!}\left\{\binom{12m+7}{3}^s + 6(6m+3)^{2s} + 3(6m+3)^s + 8(4m+2)^s + \right.$
$12m+5$ $12m+5$ $12m+5$,, ,,	$\frac{1}{4!}\left\{\binom{12m+8}{3}^s + 12\binom{6m+4}{2}^s + 8(4m+2)^s\right.$
$12m+6$ $12m+6$ $12m+6$,, ,,	$\frac{1}{4!}\left\{\binom{12m+9}{3}^s + 6(6m+4)^{2s} + 3(6m+4)^s + 8(4m+3)^s\right.$
$12m+7$ $12m+7$ $12m+7$,, ,,	$\frac{1}{4!}\left\{\binom{12m+10}{3}^s + 12\binom{6m+5}{2}^s + 8(4m+3)^s\right.$
$12m+8$ $12m+8$ $12m+8$,, ,,	$\frac{1}{4!}\left\{\binom{12m+11}{3}^s + 6(6m+5)^{2s} + 3(6m+5)^s + 8(4m+3)^s + \right.$
$12m+9$ $12m+9$ $12m+9$,, ,,	$\frac{1}{4!}\left\{\binom{12m+12}{3}^s + 12\binom{6m+6}{2}^s + 8(4m+4)^s\right.$
$12m+10$ $12m+10$ $12m+10$,, ,,	$\frac{1}{4!}\left\{\binom{12m+13}{3}^s + 6(6m+6)^{2s} + 3(6m+6)^s + 8(4m+4)^s\right.$
$12m+11$ $12m+11$ $12m+11$,, ,,	$\frac{1}{4!}\left\{\binom{12m+14}{3}^s + 12\binom{6m+7}{2}^s + 8(4m+4)^s\right.$

For the unipartite case we put $s = 1$ and find, reading by rows, the numbers

m^3.	m^2,	m.	1.
12	15	6	1
12	18	8	1
12	21	12	2
12	24	15	3
12	27	20	5
12	30	24	6
12	33	30	9
12	36	35	11
12	39	42	15
12	42	48	18
12	45	56	23
12	48	63	27

which admit of easy verification.

In the notation of this paper, for the multipartite number

$$m_1 m_2 \ldots m_s,$$

we have the enumerating number

$$\tfrac{1}{24}\left\{\prod_1^s F_q(m_i;\ 1^4)+6\prod_1^s F_q(m_i;\ 1^2 2)+3\prod_1^s F_q(m_i;\ 2^2)+8\prod_1^s F_q(m_i;\ 13)+6\prod_1^s F_q(m_i;\ 4)\right\}.$$

Section II.

Art. 16. The multipartite partitions which have been under consideration above have involved multipartite parts, and the integers which are constituents of those parts have been quite unrestricted in magnitude. We have now to consider the enumeration when these magnitudes are subject to various restrictions.

The first restriction to come before us is that in which no integer constituent of a multipartite partition is to exceed unity.

We form a fraction $\mathbf{A}_1$ from $\mathbf{Q}_1$ by striking out from the latter every partition which involves a part greater than unity.

Thus

$$\mathbf{A}_1 = 1+(1)+(1^2)+(1^3)+\ldots\ldots \text{ ad inf.}$$

We now form $\mathbf{A}_2$, $\mathbf{A}_3$, ... $\mathbf{A}_i$, ..., from $\mathbf{A}_1$, by doubling, trebling, ... multiplying by i, ... all the bracket numbers of $\mathbf{A}_1$, in the same way as we formed $\mathbf{Q}_2$, $\mathbf{Q}_3$, ... $\mathbf{Q}_i$, ... from $\mathbf{Q}_1$.

Thus

$$\mathbf{A}_2 = 1+(2)+(2^2)+(2^3)+\ldots \text{ ad inf.}$$
$$\mathbf{A}_3 = 1+(3)+(3^2)+(3^3)+\ldots \quad ,, \quad ,,$$
$$\cdot \quad \cdot \quad \cdot \quad \cdot \quad \cdot \quad \cdot \quad \cdot \quad \cdot \quad \cdot \quad ,, \quad ,,$$
$$\mathbf{A}_i = 1+(i)+(i^2)+(i^3)+\ldots \quad ,, \quad ,,$$
$$\cdot \quad \cdot \quad \cdot \quad \cdot \quad \cdot \quad \cdot \quad \cdot \quad \cdot \quad \cdot \quad ,, \quad ,,$$

We proceed in this manner because we desire the development of the generating symmetric function

$$\frac{1}{(1-a)(1-\alpha a)(1-\beta a)\ldots(1-\alpha\beta a)(1-\alpha\gamma a)\ldots(1-\alpha\beta\gamma a)\ldots(1-\alpha\beta\gamma\delta a)\ldots}$$

there being a denominator factor for every $\alpha, \beta, \gamma, \ldots$ product in which no letter is repeated. The expansion of this fraction involves the whole of the homogeneous product-sums of such $\alpha, \beta, \gamma, \ldots$ products; and we form these product-sums through the medium of the sums of the powers of the products which are, in fact, $\mathbf{A}_1, \mathbf{A}_2, \ldots \mathbf{A}_i, \ldots$. The development is

$$\begin{aligned} &1+a\mathbf{A}_1 \\ &+\frac{a^2}{2\,!}(\mathbf{A}_1^2+\mathbf{A}_2) \\ &+\frac{a^3}{3\,!}(\mathbf{A}_1^3+3\mathbf{A}_1\mathbf{A}_2+2\mathbf{A}_3) \\ &+\ldots\ldots \\ &+a^k\,\mathrm{F}_k(\mathbf{A}) \\ &+\ldots \end{aligned}$$

where

$$\mathrm{F}_k(\mathbf{A}) = \Sigma\frac{\mathbf{A}_1^{k_1}\mathbf{A}_2^{k_2}\ldots\mathbf{A}_i^{k_i}}{1^{k_1}2^{k_2}\ldots i^{k_i}.\,k_1!\,k_2!\,k_i!},$$

precisely similar to the $\mathbf{Q}$ development with $\mathbf{A}$ written for $\mathbf{Q}$.

It may at this point be worth stating that the two developments may be written

$$\exp(a\mathbf{Q}_1+\tfrac{1}{2}a^2\mathbf{Q}_2+\tfrac{1}{3}a^3\mathbf{Q}_3+\ldots), \qquad \exp(a\mathbf{A}_1+\tfrac{1}{2}a^2\mathbf{A}_2+\tfrac{1}{3}a^3\mathbf{A}_3+\ldots)$$

respectively.

We now examine the effect of the Hammond operators upon this infinite set of $\mathbf{A}$ functions. It is clear from the well-known fundamental property of the operators that

$$\mathrm{D}_i\mathbf{A}_i = \mathbf{A}_i,$$

and

$$\mathrm{D}_m\mathbf{A}_i = 0 \quad \text{when} \quad m \neq i,$$

results of great simplicity.

When D_m operates upon any product

$$A_1^{k_1}A_2^{k_2}\dots A_i^{k_i}$$

the demonstration proceeds as with the **Q** function *ante*.

Writing out the **A** product at length and underneath it any composition of m into $k_1+k_2+\dots+k_i$ parts, zero counting as a part, we note that if the composition operator is to have the effect of multiplying the product by unity and not by zero, every part under the first k_1 factors of the operand must be zero or unity; every part under the next k_2 factors must be zero or 2; every part under the next k_3 factors must be zero or 3; and so on, until finally every part under the last k_i factors must be zero or i.

The number of compositions of m which possess these properties is equal to the coefficients of x^m in the developments of

$$(1+x)^{k_1}(1+x^2)^{k_2}(1+x^3)^{k_3}\dots(1+x^i)^{k_i},$$

which may be written

$$\left(\frac{1-x^2}{1-x}\right)^{k_1}\left(\frac{1-x^4}{1-x^2}\right)^{k_2}\left(\frac{1-x^6}{1-x^3}\right)^{k_3}\dots\left(\frac{1-x^{2i}}{1-x^i}\right)^{k_i},$$

or, in CAYLEY's notation,

$$\frac{(2)^{k_1}(4)^{k_2}(6)^{k_3}\dots(2i)^{k_i}}{(1)^{k_1}(2)^{k_2}(3)^{k_3}\dots(i)^{k_i}}.$$

Let this coefficient be denoted by

$$F_a(m;\,1^{k_1}2^{k_2}\dots i^{k_i})$$

so that

$$D_m A_1^{k_1}A_2^{k_2}\dots A_i^{k_i} = F_a(m;\,1^{k_1}2^{k_2}\dots i^{k_i})\,A_1^{k_1}A_2^{k_2}\dots A_i^{k_i}.$$

Looking to the symmetric function expressions of $A_1, A_2 \dots A_i \dots$, it will be noted that the only portion of the product

$$A_1^{k_1}A_2^{k_2}\dots A_i^{k_i},$$

that is free from the elements $\alpha, \beta, \gamma \dots$, is unity.

Hence the portion of

$$D_m A_1^{k_1}A_2^{k_2}\dots A_i^{k_i},$$

that is free from the elements, is

$$F_a(m;\,1^{k_1}2^{k_2}\dots i^{k_i});$$

and we may write, as before,

$$(D_m A_1^{k_1}A_2^{k_2}\dots A_i^{k_i})_{A=1} = F_a(m;\,1^{k_1}2^{k_2}\dots i^{k_i}).$$

The number of partitions of the unipartite number m into k or fewer parts, restricted not to exceed unity, is therefore

$$D_m F_k(A)_{A=1} = \Sigma \frac{(D_m A_1^{k_1} A_2^{k_2} \dots A_i^{k_i})_{A=1}}{1^{k_1}2^{k_2}\dots i^{k_i}.\,k_1!\,k_2!\dots k_i!},$$

$$= \Sigma \frac{F_a(m;\,1^{k_1}2^{k_2}\dots i^{k_i})}{1^{k_1}2^{k_2}\dots i^{k_i}.\,k_1!\,k_2!\dots k_i!},$$

$$= \text{coefficients of } x^m \text{ in}$$

$$\Sigma \frac{\left(\frac{1-x^2}{1-x}\right)^{k_1}\left(\frac{1-x^4}{1-x^2}\right)^{k_2}\dots\left(\frac{1-x^{2i}}{1-x^i}\right)^{k_i}}{1^{k_1}2^{k_2}\dots i^{k_i}.\,k_1!\,k_2!\dots k_i!},$$

the summation being for every partition

$$k_1+2k_2+\dots+ik_i$$

of the number k.

Now, obviously, the number we seek is also the coefficient of x^m in $1+x+x^2+\dots+x^k$. Hence the formula

$$\Sigma \frac{\left(\frac{1-x^2}{1-x}\right)^{k_1}\left(\frac{1-x^4}{1-x^2}\right)^{k_2}\dots\left(\frac{1-x^{2i}}{1-x^i}\right)^{k_i}}{1^{k_1}2^{k_2}\dots i^{k_i}.\,k_1!\,k_2!\dots k_i!} = \frac{1-x^{k+1}}{1-x},$$

when k is 3 the identity is

$$\tfrac{1}{6}\{(1+x)^3+3(1+x)(1+x^2)+2(1+x^3)\} = 1+x+x^2+x^3.$$

We have now the result

$$D_{m_1}D_{m_2}\dots D_{m_s}.\,A_1^{k_1}A_2^{k_2}\dots A_i^{k_i}$$
$$= F_a(m_1;\,1^{k_1}2^{k_2}\dots i^{k_i}).\,F_a(m_2;\,1^{k_1}2^{k_2}\dots i^{k_i})\dots F(m_s;\,1^{k_1}2^{k_2}\dots i^{k_i}).\,A_1^{k_1}A_2^{k_2}\dots A_i^{k_i};$$

and the number of partitions of the multipartite number

$$m_1m_2\dots m_s$$

into k or fewer parts, no integer constituent of the multipartite parts exceeding unity, is

$$\Sigma \frac{F_a(m_1;\,1^{k_1}2^{k_2}\dots i^{k_i}).\,F_a(m_2;\,1^{k_1}2^{k_2}\dots i^{k_i})\dots F_a(m_s;\,1^{k_1}2^{k_2}\dots i^{k_i})}{1^{k_1}2^{k_2}\dots i^{k_i}.\,k_1!\,|k_2!\dots k_i!};$$

the general solution of the problem.

Some examples are now given.

For the partitions into two, or fewer parts, it is only necessary to consider the cases $m = 1$ and $m = 2$, since there are no partitions of the nature examined when $m > 2$.

$$D_1^{s_1}\tfrac{1}{2}(A_1^2+A_2) = \tfrac{1}{2}.\,2^{s_1}A_1^2,$$
$$D_2^{s_2}\tfrac{1}{2}(A_1^2+A_2) = \tfrac{1}{2}(A_1^2+A_2),$$

so that

$$D_1 \tfrac{1}{2}(A_1^2+A_2) = \tfrac{1}{2} \cdot 2A_1^2,$$

$$D_2 \tfrac{1}{2}(A_1^2+A_2) = \tfrac{1}{2}(A_1^2+A_2),$$

and

$$D_1^{s_1} D_2^{s_2} \tfrac{1}{2}(A_1^2+A_2) = \tfrac{1}{2} \cdot 2^{s_1} A_1^2,$$

results which show

(i.) that the multipartite number

$$111 \ldots s_1 \text{ times repeated,}$$

has 2^{s_1-1} partitions into two or fewer parts;

(ii.) that the number

$$222 \ldots s_2 \text{ times repeated,}$$

has one partition into two or fewer parts;

(iii.) that the multipartite number

$$222 \ldots s_2 \text{ times, } 111 \ldots s_1 \text{ times,}$$

has 2^{s_1-1} partitions into two or fewer parts.

Ex. gr. the multipartite number 2111 has the four partitions

(1111 1000), (1011 1100), (1101 1010), (1110 1001).

For the partitions, into three or fewer parts, we have

$$D_3 \tfrac{1}{6}(A_1^3+3A_1A_2+2A_3) = \tfrac{1}{6}(A_1^3+3A_1A_2+2A_3),$$

$$D_2 \tfrac{1}{6}(A_1^3+3A_1A_2+2A_3) = \tfrac{1}{6}(3A_1^3+3A_1A_2),$$

$$D_1 \tfrac{1}{6}(A_1^3+3A_1A_2+2A_3) = \tfrac{1}{6}(3A_1^3+3A_1A_2),$$

to which we may add for symmetry

$$D_0 \tfrac{1}{6}(A_1^3+3A_1A_2+2A_3) = \tfrac{1}{6}(A_1^3+3A_1A_2+2A_3);$$

we gather that

$$D_3^{s_3} \tfrac{1}{6}(A_1^3+3A_1A_2+2A_3) = \tfrac{1}{6}(A_1^3+3A_1A_2+2A_3),$$

$$D_2^{s_2} \tfrac{1}{6}(A_1^3+3A_1A_2+2A_3) = \tfrac{1}{6}(3^{s_2}A_1^3+3A_1A_2),$$

$$D_1^{s_1} \tfrac{1}{6}(A_1^3+2A_1A_2+2A_3) = \tfrac{1}{6}(3^{s_1}A_1^3+3A_1A_2),$$

$$D_3^{s_3} D_2^{s_2} D_1^{s_1} \tfrac{1}{6}(A_1^3+3A_1A_2+2A_3) = \tfrac{1}{6}(3^{s_1+s_2}A_1^3+3A_1A_2);$$

and it follows that the multipartite number

$$333 \ldots s_3 \text{ times, } 222 \ldots s_2 \text{ times, } 111 \ldots s_1 \text{ times,}$$

has $\tfrac{1}{6}(3^{s_1+s_2}+3)$ partitions, into three or fewer parts, of the nature we are considering.

Ex. gr. The multipartite number 3221 has the five partitions

$$(1111\ \ 1110\ \ 1000),\qquad (1111\ \ 1100\ \ 1010),$$
$$(1110\ \ 1110\ \ 1001),\qquad (1110\ \ 1011\ \ 1100),\qquad (1110\ \ 1101\ \ 1010).$$

Art. 17. Again, to pass to a different restriction, if no integer constituent of a multipartite part is to exceed 2, we strike out from the **Q** functions all partitions which involve integers greater than 2 and arrive at an infinite set of **B** functions which can be dealt with in a similar way. Thus

$$\mathbf{B}_1 = 1+(1)+(2)+(1^2)+(21)+(1^3)+(2^2)+(21^2)+(1^4)+\ldots \quad \text{ad inf.}$$
$$\mathbf{B}_2 = 1+(2)+(4)+(2^2)+(42)+(2^3)+(4^2)+(42^2)+(2^4)+\ldots \quad ,, \quad ,,$$
$$\mathbf{B}_i = 1+(i)+(2i)+(i^2)+(2i,\, i)+(i^3)+(2i,\, 2i)+(2i,\, i,\, i)+(i^4)+\ldots \quad ,, \quad ,,$$

In regard to the Hammond operators

$$\mathrm{D}_i\mathbf{B}_i = \mathrm{D}_{2i}\mathbf{B}_i = \mathbf{B}_i\,;$$

while every other operator causes $\mathbf{B}_i$ to vanish.

To find the effect of D_m upon the product

$$\mathbf{B}_1^{k_1}\mathbf{B}_2^{k_2}\ldots\mathbf{B}_i^{k_i}$$

we observe that D_m operates through the compositions of m into exactly $k_1+k_2+\ldots+k_i$ parts, zero counting as a part. In order that a particular composition operator shall not cause the product to vanish, the k_s factors of $\mathbf{B}_s^{k_s}$ must only be operated upon by $\mathrm{D}_0\,(\equiv 1)$, D_s and D_{2s}. Hence the number of compositions which multiply the product by unity and not by zero is given by the coefficient of x^m in the development of

$$(1+x+x^2)^{k_1}(1+x^2+x^4)^{k_2}\ldots(1+x^i+x^{2i})^{k_i},$$

which is

$$\left(\frac{1-x^3}{1-x}\right)^{k_1}\left(\frac{1-x^6}{1-x^2}\right)^{k_2}\ldots\left(\frac{1-x^{3i}}{1-x^i}\right)^{k_i}.$$

This establishes that the effect of D_m upon the product is to multiply it by this coefficient.

The generating function is

$$\begin{aligned}
&1+a\mathbf{B}_1\\
&\quad+\frac{a^2}{2!}(\mathbf{B}_1^2+\mathbf{B}_2)\\
&\quad+\frac{a^3}{3!}(\mathbf{B}_1^3+3\mathbf{B}_1\mathbf{B}_2+2\mathbf{B}_3)\\
&\quad+\ldots\\
&\quad+a^k\mathrm{F}_k(\mathbf{B})\\
&\quad+\ldots
\end{aligned}$$

where

$$F_k(B) = \Sigma \frac{B_1^{k_1} B_2^{k_2} \dots B_i^{k_i}}{1^{k_1} 2^{k_2} \dots i^{k_i} . k_1! \, k_2! \dots k_i!}.$$

Thence we find

$$(D_{m_1} D_{m_2} \dots D_{m_s} F_k(B))_{B=1}$$

$$= \Sigma \frac{F_b(m_1;\ 1^{k_1} 2^{k_2} \dots i^{k_i}) . F_b(m_2;\ 1^{k_1} 2^{k_2} \dots i^{k_i}) \dots F(m_s;\ 1^{k_1} 2^{k_2} \dots i^{k_i})}{1^{k_1} 2^{k_2} \dots i^{k_i} . k_1! \, k_2! \dots k_i!};$$

the solution of the problem of enumeration in respect of the multipartite number $m_1 m_2 \dots m_s$.

If, in the function $F_k(B)$, we substitute

$$\frac{1-x^{3s}}{1-x^s} \text{ for } B_s,$$

we obtain

$$\frac{(1-x^{k+1})(1-x^{k+2})}{(1-x)(1-x^2)},$$

because this function enumerates unipartite partitions whose parts are limited in number by k and in magnitude by 2.

As an example consider partitions into three parts. We have the symmetrical results

$$D_6 \tfrac{1}{6}(B_1^3 + 3B_1B_2 + 2B_3) = \tfrac{1}{6}(B_1^3 + 3B_1B_2 + 2B_3),$$

$$D_5 \tfrac{1}{6}(B_1^3 + 3B_1B_2 + 2B_3) = \tfrac{1}{6}(3B_1^3 + 3B_1B_2),$$

$$D_4 \tfrac{1}{6}(B_1^3 + 3B_1B_2 + 2B_3) = \tfrac{1}{6}(6B_1^3 + 3 . 2B_1B_2),$$

$$D_3 \tfrac{1}{6}(B_1^3 + 3B_1B_2 + 2B_3) = \tfrac{1}{6}(7B_1^3 + 3B_1B_2 + 2B_3),$$

and D_2, D_1, D_0, yield the same results as D_4, D_5, D_6, respectively.

The number m, not exceeding 6, D_m and D_{6-m} produce upon the operand the same result. This symmetry naturally follows from the known property of the function

$$\frac{(1-x^{k+1})(1-x^{k+2})}{(1-x)(1-x^2)}$$

which on expansion is, as regards coefficients, centrally symmetrical.

We now at once deduce that

$$D_6^s F_3(B) = D_0^s F_3(B) = \tfrac{1}{6}(B_1^3 + 3B_1B_2 + 2B_3)$$

$$D_5^s F_3(B) = D_1^s F_3(B) = \tfrac{1}{6}(3^s B_1^3 + 3B_1B_2)$$

$$D_4^s F_3(B) = D_2^s F_3(B) = \tfrac{1}{6}(6^s B_1^3 + 3 . 2^s B_1B_2)$$

$$D_3^s F_3(B) = \tfrac{1}{6}(7^s B_1^s + 3B_1B_2 + 2B_3).$$

Q 2

We deduce that the multipartite number

$$666 \ldots s \text{ times}$$

has only one partition of the nature we consider, and this of course is quite obvious.

That the multipartite numbers

$$555 \ldots s \text{ times}, \qquad 111 \ldots s \text{ times}$$

have each of them

$$\tfrac{1}{6}(3^s+3) \text{ partitions};$$

the multipartite numbers

$$444 \ldots s \text{ times}, \qquad 222 \ldots s \text{ times},$$

have each of them

$$\tfrac{1}{6}(6^s+3 \,.\, 2^s) \text{ partitions};$$

and the number

$$333 \ldots s \text{ times}$$

$$\tfrac{1}{6}(7^s+5) \text{ partitions}.$$

Also the multipartite number

$$333 \ldots s \text{ times}, \; 222 \ldots t \text{ times}$$

has

$$\tfrac{1}{6}(7^s \,.\, 6^t+3 \,.\, 2^s)$$

partitions, and various other results.

The symmetry shown above in the case of multipartite numbers is of general application in the subject and is very remarkable. I do not see any other *a priori* proof of it at the moment of writing.

Art. 18. In general, we consider the case in which no constituent of the multipartite parts is to exceed the integer j. We strike out from the functions **Q** all partitions which involve numbers exceeding j and reach the infinite series of functions

$$\mathbf{J}_1, \mathbf{J}_2, \ldots \mathbf{J}_i \ldots .$$

These functions are operated upon in the manner

$$\mathbf{D}_i\mathbf{J}_i = \mathbf{D}_{2i}\mathbf{J}_i = \mathbf{D}_{3i}\mathbf{J}_i = \ldots = \mathbf{D}_{ji}\mathbf{J}_i = \mathbf{J}_i \cdot$$

while every other Hammond operator causes $\mathbf{J}_i$ to vanish. By the same reasoning as was used in the special cases we find that

$$\mathbf{D}_m\mathbf{J}_1^{k_1}\mathbf{J}_2^{k_2} \ldots \mathbf{J}_i^{k_i} = \mathrm{F}_j(m\,;\; 1^{k_1}2^{k_2} \ldots i^{k_i}) \,.\, \mathbf{J}_1^{k_1}\mathbf{J}_2^{k_2} \ldots \mathbf{J}_i^{k_i},$$

where

$$\mathrm{F}_j(m\,;\; 1^{k_1}2^{k_2} \ldots i^{k_i})$$

is equal to the coefficient of x^m in the expansion of the function

$$(1+x+\ldots+x^j)^{k_1}(1+x^2+\ldots+x^{2j})^{k_2} \ldots (1+x^i+\ldots+x^{ij})^{k_i},$$

which, written in CAYLEY'S notation, is

$$\left\{\frac{(j+1)}{(1)}\right\}^{k_1}\left\{\frac{(2j+2)}{(2)}\right\}^{k_2}\cdots\left\{\frac{(ij+i)}{(i)}\right\}^{k_i}.$$

We have

$$F_k(J) = \Sigma\frac{J_1^{k_1}J_2^{k_2}\ldots J_i^{k_i}}{1^{k_1}2^{k_2}\ldots i^{k_i}.\,k_1!\,k_2!\ldots k_i!};$$

and if, in $F_k(J)$, we write

$$\frac{(sj+s)}{(s)}\text{ for }J_s$$

we must reach the expression

$$\frac{(k+1)(k+2)\ldots(k+j)}{(1)(2)\ldots(j)}\text{ otherwise written }\frac{(j+1)(j+2)\ldots(j+k)}{(1)(2)\ldots(k)}.$$

For the rest the enumeration, connected with this restriction, proceeds *pari passu* with the special cases involving the values of j, ∞, 1, 2, which have been already examined.

Art. 19. Coming now to the last stage of the investigation we have for consideration the partitions of multipartite numbers in which the constituents of the multipartite parts *must* be drawn from the particular set of numbers

$$u_1,\ u_2,\ \ldots\ u_s.$$

The partitions involved in the first function U_1 of the infinite series

$$U_1,\ U_2,\ U_3,\ \ldots\ U_k,\ \ldots$$

must involve the numbers u_1, u_2, ... u_s, and no others.

Thus

$$U_1 = 1+(u_1)+(u_1^2)+(u_2)+(u_1^3)+(u_1u_2)+(u_3)+\ldots\text{ ad inf.},$$

$$U_2 = 1+(2u_1)+(2u_12u_1)+(2u_2)+(2u_12u_12u_1)+(2u_12u_2)+(2u_3)+\ldots\text{ ad inf.},$$

. . . .

$$U_i = 1+(iu_1)+(iu_1iu_1)+(iu_2)+(iu_1iu_1iu_1)+(iu_1iu_2)+(iu_3)+\ldots\text{ ad inf.}$$

Clearly

$$D_{u_1}U_1 = D_{u_2}U_1 = \ldots = D_{u_s}U_1 = U_1;$$

and any other Hammond operator causes U_1 to vanish.

Also

$$D_{iu_1}U_i = D_{iu_2}U_i = \ldots = D_{iu_s}U_i = U_i;$$

and any other Hammond operator causes U_i to vanish.

We have presented to us the generating function

$$
\begin{aligned}
&1+a\mathrm{U}_1\\
&+\frac{a^2}{2!}(\mathrm{U}_1^2+\mathrm{U}_2)\\
&+\frac{a^3}{3!}(\mathrm{U}_1^3+3\mathrm{U}_1\mathrm{U}_2+2\mathrm{U}_3)\\
&+\dots\\
&+a^k\mathrm{F}^k(\mathrm{U})\\
&+\dots
\end{aligned}
$$

where

$$\mathrm{F}_k(\mathrm{U}) = \Sigma \frac{\mathrm{U}_1^{k_1}\mathrm{U}_2^{k_2}\dots\mathrm{U}_i^{k_i}}{1^{k_1}.\,2^{k_2}\dots i^{k_i}.\,k_1!\,k_2!\dots k_i!}.$$

The effect of D_m upon the product

$$\mathrm{U}_1^{k_1}\mathrm{U}_2^{k_2}\dots\mathrm{U}_i^{k_i}$$

is to multiply it by the coefficient of x^m in the function

$$(1+x^{u_1}+x^{u_2}+\dots+x^{u_s})^{k_1}(1+x^{2u_1}+x^{2u_2}+\dots+x^{2u_s})^{k_2}\dots(1+x^{iu_1}+x^{iu_2}+\dots+x^{iu_s})^{k_i},$$

which I write in the abbreviated notation

$$\mathrm{X}_u^{k_1}\mathrm{X}_{2u}^{k_2}\dots\mathrm{X}_{iu}^{k_i}.$$

If in the expression of $\mathrm{F}_k(\mathrm{U})$ we write X_{su} for U_s we must reach the expression which enumerates the partitions of unipartite numbers into k or fewer parts, such parts being drawn from the series $u_1, u_2, u_3 \dots u_s$. That is to say we must arrive at the coefficient of a^k in the expansion of

$$\frac{1}{(1-a)(1-ax^{u_1})(1-ax^{u_2})\dots(1-ax^{u_s})}.$$

Hence this coefficient has the expression

$$\Sigma \frac{\mathrm{X}_u^{k_1}\mathrm{X}_{2u}^{k_2}\dots\mathrm{X}_{iu}^{k_i}}{1^{k_1}.\,2^{k_2}\dots i^{k_i}.\,k_1!\,k_2!\dots k_i!}.$$

To enumerate the partitions we have

$$\mathrm{D}_{m_1}\mathrm{F}_k(\mathrm{U}) = \Sigma \frac{\mathrm{F}_u(m_1;\ 1^{k_1}2^{k_2}\dots i^{k_i})}{1^{k_1}.\,2^{k_2}\dots i^{k_i}.\,k_1!\,k_2!\dots k_i!}\mathrm{U}_1^{k_1}\mathrm{U}_2^{k_2}\dots\mathrm{U}_i^{k_i},$$

where $F_u(m_1;\ 1^{k_1}2^{k_2}\ldots i^{k_i})$ is equal to the coefficient of x^{m_1} in

$$X_u^{k_1}X_{2u}^{k_2}\ldots X_{iu}^{k_i}.$$

$$D_{m_1}D_{m_2}\ldots D_{m_s}F_k(U)$$

$$=\Sigma\frac{F_u(m_1;\ 1^{k_1}2^{k_2}\ldots i^{k_i}).F_u(m_2;\ 1^{k_1}2^{k_2}\ldots i^{k_i})\ldots F(m_s;\ 1^{k_1}2^{k_2}\ldots i^{k_i})}{1^{k_1}.2^{k_2}\ldots i^{k_i}.k_1!\,k_2!\ldots k_i!}.U_1^{k_1}U_2^{k_2}\ldots U_i^{k_i};$$

and thence the number of partitions of the multipartite number

$$m_1m_2\ldots m_s,$$

into k or fewer parts, such parts being drawn exclusively from the series

$$u_1, u_2, \ldots u_s,$$

is

$$\Sigma\frac{F_u(m_1;\ 1^{k_1}2^{k_2}\ldots i^{k_i}).F_u(m_2;\ 1^{k_1}2^{k_2}\ldots i^{k_i})\ldots F_u(m_s;\ 1^{k_1}2^{k_2}\ldots i^{k_i})}{1^{k_1}.2^{k_2}\ldots i^{k_i}.k_1!\,k_2!\ldots k_i!}$$

the summation being in regard to the partitions of k.

As an example, I will consider partitions of multipartite numbers where the numbers which are constituents of the multipartite parts are limited to be either 3, 5, or 7. For the partitions into three or fewer parts we have the function

$$\tfrac{1}{6}(U_1^3+3U_1U_2+2U_3),$$

and we have to find the coefficients of x^m in the three functions

$$(1+x^3+x^5+x^7)^3,$$

$$(1+x^3+x^5+x^7)(1+x^6+x^{10}+x^{14}),$$

$$(1+x^9+x^{15}+x^{21}).$$

Thence, as particular cases,

$$D_{12}U_1^3=6U_1^3;\quad D_{11}U_1^3=3U_1^3;\quad D_{10}U_1^3=9U_1^3;$$

$$D_{12}U_1U_2=0;\quad D_{11}U_1U_2=U_1U_2;\quad D_{10}U_1U_2=0;$$

$$D_{12}U_3=D_{11}U_3=D_{10}U_3=0.$$

Thence

$$D_{12}^{\sigma_1}\tfrac{1}{6}(U_1^3+3U_1U_2+2U_3)=\tfrac{1}{6}.6^{\sigma_1}U_1^3,$$

$$D_{11}^{\sigma_2}\tfrac{1}{6}(U_1^3+3U_1U_2+2U_3)=\tfrac{1}{6}(6^{\sigma_2}U_1^3+3U_1U_2),$$

$$D_{10}^{\sigma_3}\tfrac{1}{6}(U_1^3+3U_1U_2+2U_3)=\tfrac{1}{6}.9^{\sigma_3}U_1^3,$$

showing that the multipartite numbers

$$12^{\sigma_1},\qquad 11^{\sigma_2},\qquad 10^{\sigma_3},$$

have 6^{σ_1-1}, $\tfrac{1}{6}(3^{\sigma_2}+3)$, $\tfrac{1}{6}.9^{\sigma_3}$ partitions respectively into three or fewer parts.

Also

$$\mathbf{D}_{12}^{\sigma_1}\mathbf{D}_{11}^{\sigma_2}\mathbf{D}_{10}^{\sigma_3}\tfrac{1}{6}\,(\mathbf{U}_1{}^3+3\mathbf{U}_1\mathbf{U}_2+2\mathbf{U}_3) = \tfrac{1}{6}\,(6^{\sigma_1}\,.\,3^{\sigma_2}\,.\,9^{\sigma_3}\mathbf{U}_1{}^3+3\mathbf{U}_1\mathbf{U}_2),$$

where the term $3\mathbf{U}_1\mathbf{U}_2$ only appears if σ_1 and σ_2 are *both* zero.

Hence in general, if σ_1 and σ_2 are not both zero, the multipartite number

$$12^{\sigma_1}11^{\sigma_2}10^{\sigma_3}$$

possesses $6^{\sigma_1-1}\,.\,3^{\sigma_2}\,.\,9^{\sigma_3}$ partitions of the nature considered. In particular the multipartite number

$$12\ 12\ 11, \quad \text{for } \sigma_1 = 2,\ \sigma_2 = 1,\ \sigma_3 = 0,$$

possesses the 18 partitions

(775 553 003),	(755 573 003),	(573 705 053),
(775 053 503),	(575 753 003),	(753 073 505),
(773 555 003),	(755 073 503),	(573 703 055),
(773 553 005),	(575 703 053),	(555 073 703),
(773 505 053),	(753 573 005),	(553 705 073),
(773 055 503),	(753 075 503),	(553 075 703).

Art. 20. In considering the partitions of the multipartite number $m_1m_2\ldots m_s$, the partitions of the unipartite constituents of the number have been regarded as being subject to the same conditions and restrictions. This, however, is not necessary except in the case of the number of parts which has been denoted by k.

We may for m_1 choose any of the restrictions that have been denoted by the symbols $\mathbf{A, B},\ldots \mathbf{J}\ldots \mathbf{Q, U}$. For m_2 similarly and so on. For instance, suppose the numbers m_1, m_2, m_3 are subject to the restrictions denoted by $\mathbf{B}$, $\mathbf{Q}$, and $\mathbf{U}$; that is to say, the partitions of m_1 are such that no part exceeds 2; the partitions of m_2 are unrestricted; the partitions of m_3 are such that the parts are drawn exclusively from specified integers $u_1, u_2, \ldots u_s$.

For the partitions of the multipartite number $m_1m_2\ldots m_s$, subject to this combination of restrictions, we

(i.) Take

$$\mathbf{D}_{m_1}\sum'\frac{\mathbf{B}_1{}^{k_1}\mathbf{B}_2{}^{k_2}\ldots \mathbf{B}_i{}^{k_i}}{1^{k_1}2^{k_2}\ldots i^{k_i}\,.\,k_1!\ k_2!\ldots k_i!},$$

with the result

$$\sum \mathbf{F}_b\,(m_1;\ 1^{k_1}2^{k_2}\ldots i^{k_i})\,\frac{\mathbf{B}_1{}^{k_1}\mathbf{B}_2{}^{k_2}\ldots \mathbf{B}_i{}^{k_i}}{1^{k_1}2^{k_2}\ldots i^{k_i}\,.\,k_1!\ k_2!\ldots k_i!};$$

(ii.) We change the symbol **B** into the symbol **Q** and find

$$D_{m_2} \Sigma F_b(m_1;\ 1^{k_1}2^{k_2}\dots i^{k_i}) \frac{Q_1^{k_1}Q_2^{k_2}\dots Q_i^{k_i}}{1^{k_1}2^{k_2}\dots i^{k_i}.\,k_1!\,k_2!\dots k_i!}$$

$$= \Sigma F_b(m_1;\ 1^{k_1}2^{k_2}\dots i^{k_i}).\,F_q(m_2;\ 1^{k_1}2^{k_2}\dots i^{k_i}) \frac{Q_1^{k_1}Q_2^{k_2}\dots Q_i^{k_i}}{1^{k_1}2^{k_2}\dots i^{k_i}.\,k_1!\,k_2!\dots k_i!};$$

(iii.) Lastly we change the symbol **Q** into the symbol **U** and find that

$$D_{m_3} \Sigma F_b(m_1;\ 1^{k_1}2^{k_2}\dots i^{k_i}).\,F_q(m_2;\ 1^{k_1}2^{k_2}\dots i^{k_i}) \frac{U_1^{k_1}U_2^{k_2}\dots U_i^{k_i}}{1^{k_1}2^{k_2}\dots i^{k_i}.\,k_1!\,k_2!\dots k_i!}$$

$$= \Sigma F_b(m_1;\ 1^{k_1}2^{k_2}\dots i^{k_i}).\,F_q(m_2;\ 1^{k_1}2^{k_2}\dots i^{k_i}).\,F_u(m_3;\ 1^{k_1}2^{k_2}\dots i^{k_i}) \frac{U_1^{k_1}U_2^{k_2}\dots U_i^{k_i}}{1^{k_1}2^{k_2}\dots i^{k_i}.\,k_1!\,k_2!\dots k_i!},$$

and herein putting $U_1 = U_2 = \dots = U_i = 1$, we reach the conclusion that the multipartite number $m_1m_2\dots m_s$ has

$$\Sigma \frac{F_b(m_1;\ 1^{k_1}2^{k_2}\dots i^{k_i}).\,F_q(m_2;\ 1^{k_1}2^{k_2}\dots i^{k_i}).\,F_u(m_3;\ 1^{k_1}2^{k_2}\dots i^{k_i})}{1^{k_1}2^{k_2}\dots i^{k_i}.\,k_1!\,k_2!\dots k_i!}$$

partitions of the nature we are considering.

DIRICHLET SERIES AND THE THEORY OF PARTITIONS

By Major P. A. MacMahon.

[Received July 30th, 1923.—Read June 14th, 1923.]

1. Consider the infinite product

$$\prod_p \frac{1}{1-b^p p^{-s}}$$

where there is a factor for every prime number p. b and s are numerical magnitudes which are not further specified for the moment.

On development we have

$$\prod_p \frac{1}{1-b^p p^{-s}} = \sum_n b^{\pi_1 p_1+\pi_2 p_2+\pi_3 p_3+\dots} n^{-s}$$

where the integer n has the prime number expression

$$n = p_1^{\pi_1} p_2^{\pi_2} p_3^{\pi_3} \dots.$$

It is convenient to term the magnitude

$$\pi_1 p_1+\pi_2 p_2+\pi_3 p_3+\dots,$$

the "potency" of the integer n, and to write

$$pt(n) = \Sigma \pi p.$$

The magnitude $\pi_1+\pi_2+\pi_3+\dots$

will be termed the "multiplicity" of the integer n, and we write

$$mt(n) = \Sigma \pi.$$

The succession of integers $\pi_1 \pi_2 \pi_3 \ldots$, *when arranged in descending order of magnitude*, will be held to specify a multipartite number

$$\overline{\pi_1 \pi_2 \pi_3 \ldots}.$$

In the Theory of Partitions of numbers*

$$(p_1^{\pi_1} p_2^{\pi_2} p_3^{\pi_3} \ldots)$$

usually denotes a partition of the number

$$\Sigma \pi p$$

into π_s parts equal to p_s; $s = 1, 2, 3, \ldots$.

1.1. Hence every integer n may be regarded as specifying a partition of its potency $pt(n)$ into parts which are primes, the number of parts being equal to its multiplicity, viz. $mt(n)$. There is here a one-to-one correspondence.

Conversely every partition of an integer ν into prime parts, $\Sigma\pi$ in number, connotes an integer n of potency ν and multiplicity $\Sigma\pi$, which is associated with a definite multipartite number $\overline{\pi_1 \pi_2 \pi_3 \ldots}$. If the integer ν be given, every solution in integers of the equation

$$\Sigma \pi p = \nu$$

connotes an integer which has the potency ν.

Further, if $\Sigma\pi = \mu$ be given, every solution in integers of the simultaneous equations

$$\Sigma \pi p = \nu$$

$$\Sigma \pi = \mu$$

connotes an integer which has the potency ν and the multiplicity μ. Again, if the integers $\pi_1, \pi_2, \pi_3, \ldots$ be fixed, the indeterminate equation

$$\pi_1 p_1 + \pi_2 p_2 + \pi_3 p_3 + \ldots = \nu,$$

connotes integers which have the potency ν, the multiplicity μ, and are associated with the multipartite number $\overline{\pi_1 \pi_2 \pi_3 \ldots}$.

* In the Theory of the Partitions of Numbers what is termed "the partition notation" was initiated by Meyer Hirsch more than a century ago and has been developed by Cayley and others since that date.

1.2. The annexed table will make the correspondence clear.

Potency ν	Partition of potency	Integer n	Multiplicity $\Sigma\pi$	Multipartite number $\pi_1\pi_2\pi_3\ldots$
2	2	2	1	1
3	3	3	1	1
4	2^2	4	2	2
5	23	6	2	$\overline{11}$
,,	5	5	1	1
6	2^3	8	3	3
,,	3^2	9	2	2
7	$2^2 3$	12	3	$\overline{21}$
,,	25	10	2	$\overline{11}$
,,	7	7	1	1
8	2^4	16	4	4
,,	23^2	18	3	$\overline{21}$
,,	35	15	2	$\overline{11}$
9	$2^3 3$	24	4	$\overline{31}$
,,	3^3	27	3	3
,,	$2^2 5$	20	3	$\overline{21}$
,,	27	14	2	$\overline{11}$
10	2^5	32	5	5
,,	$2^2 3^2$	36	4	$\overline{22}$
,,	235	30	3	$\overline{111}$
,,	5^2	25	2	2
,,	37	21	2	$\overline{11}$
11	$2^4 3$	48	5	$\overline{41}$
,,	23^3	54	4	$\overline{31}$
,,	$2^3 5$	40	4	$\overline{31}$
,,	$3^2 5$	45	3	$\overline{21}$
,,	$2^2 7$	28	3	$\overline{21}$
,,	11	11	1	1

1.21. It will be noted, from the central column of numbers, that the integers 8, 9 have the same potency, viz. 6, but that their multiplicities are different. Also that the integers 27, 20 have the same potency, viz. 9, and the same multiplicity, viz. 3, but that they are associated with different multipartite numbers. Also that the integers 54, 40 have the same potency, viz. 11, the same multiplicity, viz. 4, and are associated with the same multipartite number, viz. $\overline{31}$.

1.22. In this case we observe that the multipartite number arises from the two partitions 23^3, 2^35 of the potency, and we have agreed above to regard both multipartites $\overline{13}$, $\overline{31}$ as included in the single multipartite $\overline{31}$. If we do not so agree, then a further classification takes place in regard to each permutation of the constituents $\pi_1, \pi_2, \pi_3, \ldots$ of the multipartite number $\overline{\pi_1\pi_2\pi_3\ldots}$. Thus the next following integers 45, 28 are not only associated with the same potency, viz. 11, and the same multiplicity, viz. 3, but also, through the partitions 3^25, 2^27, in a stricter sense, with the same multipartite, viz. $\overline{21}$.

It will be gathered from the above that we may group all positive integers according to potency, multiplicity, and multipartite number (in the two ways specified) or according to compatible combinations of these characteristics.

1.23. Goldbach's celebrated conjecture "Every even number is the sum of two primes" amounts, from the present point of view, to showing that every even potency > 4 is compatible with association with a multipartite number $\overline{11}$.

1.24. The potency of an integer n is $\nless n$.

If n is the power of a prime and we put

$$p_1^{\pi_1} = \pi_1 p_1 \quad \text{or} \quad p_1^{\pi_1 - 1} = \pi_1$$

we must have $(p_1, \pi_1) = (p_1, 1)$ or $(2, 2)$.

This shows that the integer n has a potency n

(i) when n is a prime,

(ii) when $n = 4$.

It may be recalled that in other parts of the Theory of Numbers, *e.g.* in the theory of primitive roots of numbers, the number 4 is exceptionally associated with prime numbers.

If we supposed integers to be grouped according to potency, we expand

$$\Pi \frac{1}{1-b^p} = \sum_\nu N_\nu b^\nu$$

and observe that N_ν integers are in the group of potency ν.

If we take any partition of ν, the potency, into primes and multiply the parts together, we obtain an integer whose potency is ν.

It is obvious that ν is a lower limit of the magnitude of the numbers which have a potency ν.

1.25. It is easy to find an upper limit because if ν be any real numerical quantity and it be separated into any number of parts, the continued product of such parts has an upper limit

$$e^{\nu/e}.$$

This fact is readily established by the ordinary process of the differential calculus.

When $\nu = e$, the upper limit is e. In general the upper limit is $>$ or $< e$ according as $\nu >$ or $< e$, an interesting property of the Napierian base. When ν is an integer, we note that 3 is the nearest integer to e. If $[\nu/3]$ denote the greatest integer in $\nu/3$, we find that the highest integer in the group is

$$3^{[\nu/3]} \text{ if } \nu \text{ be of form } 0 \pmod 3,$$

$$2^2.3^{[\nu/3]-1} \quad ,, \quad ,, \quad 1 \quad ,,$$

$$2.3^{[\nu/3]} \quad ,, \quad ,, \quad 2 \quad ,,$$

When ν is a prime, the number ν is invariably the smallest number in the group.

The lower limits can be raised when ν is not a prime.

If p be the smallest prime such that $\nu - p$ is also prime,

$$p(\nu - p)$$

is a lower limit.*

1.3. A group of integers, which has the potency ν, involves a prime number ν if ν be prime. The $N_\nu - 1$ other integers of the group are composite.

If ν be composite the N_ν integers in the group are all composite.

If the integer n be $\qquad p_1^{\pi_1} p_2^{\pi_2} p_3^{\pi_3} \ldots,$

and π_s involves a_s primes, in its expression in terms of prime factors, it is clear that p_s may be interchanged with any one of the a_s primes without altering the potency.

---

* In this connexion the inequality

$$\frac{pt(n)}{mt(n)} \nless n^{1/[mt(n)]},$$

derived from a well-known formula, may be noted.

In such a case the above expression of n may connote as many as

$$(a_1+1)(a_2+1)(a_3+1)\dots$$

integers which have the same potency.

Thus, *e.g.* the integer $\qquad 5^{2^2 . 3}$

has the same potency as each of the integers

$$2^{2 . 3 . 5}, \quad 3^{2^2 . 5}$$

and so forth.

2. In the next place I propose, apart from the question of convergence, to discuss briefly a Dirichlet series

$$f(s) = \sum_1^\infty a_n n^{-s}$$

from the point of view of the partition of numbers.

The summation is in respect of all positive integers.

I suppose that the integer n, expressed in terms of primes, is

$$p_1^{\pi} p_2^{\pi_2} p_3^{\pi_3} \dots .$$

I take a_n, in the series, to be the number which enumerates the partitions of the multipartite number

$$\overline{\pi_1 \pi_2 \pi_3 \dots .}$$

2.1. Then I say that

$$\prod_{n=2}^\infty \frac{1}{1-n^{-s}} = \sum_1^\infty a_n n^{-s}$$

that is to say that the infinite product on the left is the generating function which enumerates the partitions of all multipartite numbers.

It is in fact obvious, from the constructon of the infinite product, that a_n enumerates the ways in which the integer n can be expressed as a *product of positive integers.*

Moreover, since $\qquad n = p_1^{\pi_1} p_2^{\pi_2} p_3^{\pi_3} \dots,$

we find one such expession for each partition of the multipartite number

$$\overline{\pi_1 \pi_2 \pi_3 \dots .}$$

2.11. *E.g.* $\qquad n = 2^2 . 3 . 5 = 60.$

The multipartite number is $\overline{211}$.

Partitions of $\overline{211}$				*Partition of* $2^2.3.5$	*Expression as a Product of Positive Integers*
$\overline{211}$				$2^2.3.5$	60
$\overline{210}$	$\overline{001}$			$2^2.3\times5$	12.5
$\overline{201}$	$\overline{010}$			$2^2.5\times3$	20.3
$\overline{111}$	$\overline{100}$			$2.3.5\times2$	30.2
$\overline{200}$	$\overline{011}$			$2^2\times3.5$	4.15
$\overline{110}$	$\overline{101}$			2.3×2.5	6.10
$\overline{200}$	$\overline{010}$	$\overline{001}$		$2^2\times3\times5$	4.3.5
$\overline{110}$	$\overline{100}$	$\overline{001}$		$2.3\times2\times5$	6.2.5
$\overline{101}$	$\overline{100}$	$\overline{010}$		$2.5\times2\times3$	10.2.3
$\overline{011}$	$\overline{100}$	$\overline{100}$		$3.5\times2\times2$	15.2.2
$\overline{100}$	$\overline{100}$	$\overline{010}$	$\overline{001}$	$2\times2\times3\times5$	2.2.3.5

So that $a_{60}=11$, equal to the number of partitions of the multipartite $\overline{211}$.

On the right-hand side of the identity we regard the series of positive integers raised to the power $-s$ as arranged, it may be, into an infinite number of groups; for one group corresponds to every multipartite number; and each group contains an infinite number of members. For this purpose we may regard the different permutations of the constituents of the multipartite number

$$\overline{\pi_1\pi_2\pi_3\ldots}$$

as denoting the same or different multipartites.

2.2. How many integers of a given group

$$\overline{\pi_1\pi_2\pi_3\ldots}$$

are there which do not exceed a given magnitude?

This and a large number of other questions present themselves. The group which is associated with the unipartite number which is merely the unipartite number

unity,

includes the whole of the prime numbers exclusively.

The generating function

$$\prod_n \frac{1}{1-n^{-s}}$$

may be split up into an infinite number of factors of which the first now written down is the Riemann Zeta Function

$$\prod_n \frac{1}{1-n^{-s}} = \prod_p \frac{1}{1-p^{-s}}$$

$$\times \prod_p \prod_{p'} \frac{1}{1-(pp')^{-s}}$$

$$\times \prod_p \prod_{p'} \prod_{p''} \frac{1}{1-pp'p'')^{-s}}$$

$$\times \ldots,\ \textit{ad inf.}$$

The functions herein appearing seem to be worthy of study by the powerful analytical theory of numbers that has been successfully applied to the Zeta Function.

2.21. The function $\prod_{n=1}^{\infty} \frac{1}{1-(2n+1)^{-s}}$

may be separately studied. We are concerned with uneven numbers exclusively.

2.22. Similarly the infinite product

$$\Pi(1+n^{-s})$$

is a generating function which enumerates the partitions of the multipartite number

$$\overline{\pi_1 \pi_2 \pi_3 \ldots}$$

into unrepeated parts.

Further, the partitions into a given number of parts may be studied by means of the functions

$$\Pi(1-an^{-s})^{-1},$$

$$\Pi(1+an^{-s}).$$

The Enumeration of the Partitions of Multipartite Numbers. By Major P. A. MacMahon, Sc.D., St John's College.

[*Received* 1 September, *read* 26 October 1925.]

Introduction.

This paper is a study of a new method of enumeration of the partitions of multipartite numbers.

Incidentally an algebraic function, which is derived from the repetitional exponents of partitions of unipartite numbers, presents itself. The generating function which enumerates the partitions of unipartite numbers is expressible in terms of these functions and finds in such expression its fullest connection with the divisors of numbers. There are also similarly derived functions connected directly with bipartite, tripartite, etc. numbers. It has not been necessary to study these for the purposes of this paper.

An arithmetical function $\sigma_0^{(f)}(n)$ which enumerates what I have called the factor-divisor of n has appeared. It is the multipartite generalization of the well-known arithmetical function $\sigma_0(n)$ which enumerates the divisions of the unipartite number n. It gives rise to an allied arithmetical function $j(n)$ which possesses properties analogous to those which appertain to the ϕ-function of Euler. A short table which enumerates the partitions of some bipartite and tripartite numbers is appended.

§ 1. *The Dirichlet Series Generating Function.*

1·1. This subject has been dealt with by the author* and two methods explained for carrying out the evaluation. The matter has acquired a fresh interest from the discovery of the new Generating Function, viz. the ordinary Dirichlet series

$$\eta(s) = \prod_2^\infty (1 - n^{-s})^{-1}.$$

1·2. The development of this may be written

$$\Sigma p_{\pi_1 \pi_2 \ldots \pi_k} n^{-s},$$

where the prime number form of n is

$$p_1^{\pi_1} p_2^{\pi_2} \ldots p_k^{\pi_k};$$

and $p_{\overline{\pi_1 \pi_2 \ldots \pi_k}}$ denotes the number of partitions of the multipartite number $(\overline{\pi_1 \pi_2 \ldots \pi_k})$; or alternatively

$$\Sigma fc(n) n^{-s},$$

* *Combinatory Analysis*, vol. I, pp. 264 *et seq.*

where $fc\,(n)$ denotes the number of ways of factorizing the integer n. Thus for $n = 12 = 2^2.3$, p_{21} enumerates the four partitions of the bipartite number $(\overline{21})$, these being

$$(\overline{21}),\quad (\overline{20}, \overline{01}),\quad (\overline{11}, \overline{10}),\quad (\overline{10}, \overline{10}, \overline{01}),$$

and $fc\,(12)$ enumerates the four factorizations

$$12,\quad 2 \times 6,\quad 3 \times 4,\quad 3 \times 2^2.$$

If we expand the function $\eta\,(s)$ to the power l we find

$$1{\cdot}3.\quad \{\eta\,(s)\}^l = \Sigma\Sigma \binom{\nu_1 + l - 1}{\nu_1}\binom{\nu_2 + l - 1}{\nu_2}\cdots \binom{\nu_k + l - 1}{\nu_k}(n_1^{\nu_1} n_2^{\nu_2} \ldots n_k^{\nu_k})^{-s} \;\ldots\ldots(\text{A}),$$

where $n_1^{\nu_1} n_2^{\nu_2} \ldots n_k^{\nu_k}$ is a factorization of n, and the summation is for all such factorizations of n and for all integer values of $n > 1$.

1·31. To every integer n there appertains the number

$$\Sigma \binom{\nu_1 + l - 1}{\nu_1}\binom{\nu_2 + l - 1}{\nu_2}\cdots\binom{\nu_k + l - 1}{\nu_k} \;\ldots\ldots(\text{A})',$$

where the summation is for every factorization of n and involves $fc\,(n)$ terms. Before proceeding to the arithmetical interpretation of this numerical quantity reference is made to Bouniakowski*, who interpreted the powers of Riemann's Zeta Function

$$\sum_1^\infty n^{-s},$$

and to E. B. Elliott† who also threw light upon the powers of $\zeta\,(s)$. We write for comparison

$$\{\zeta\,(s)\}^l = \sum_n \binom{\pi_1 + l - 1}{\pi_1}\binom{\pi_2 + l - 1}{\pi_2}\cdots\binom{\pi_k + l - 1}{\pi_k}(p_1^{\pi_1} p_2^{\pi_2} \cdots p_k^{\pi_k})^{-s} \;\ldots\ldots(\text{B}).$$

1·32. When $l = 2$, the numerical coefficient in (B) enumerates the divisors of n. In (A) the coefficient enumerates those divisors of n which appertain to the factorization $n_1^{\nu_1} n_2^{\nu_2} \ldots n_k^{\nu_k}$ of n. In other words the divisors enumerated are drawn from the powers and products of powers of $n_1, n_2 \ldots n_k$ that present themselves in the factorization; unity is of course included. The sum in (A)′ enumerates the divisors appropriate to each factorization and adds the results. We write it $\sigma_0^{(f)}\,(n)$.

* *Mém. Ac. Sc. St Pétersbourg* (7), 4, 1862, No. 2, 35 pp.
† *Proc. Lond. Math. Soc.* 34, 1901, pp. 3–15.

We may write

$$\{\zeta(s)\}^2 = \Sigma\sigma_0(n)\, n^{-s}; \quad \{\eta(s)\}^2 = \Sigma\sigma_0^{(f)}(n) . n^{-s}.$$

When $n = 12$, the factorizations are

$$12, \quad 2\times 6, \quad 3\times 4, \quad 2^2\times 3,$$

and give $\sigma_0^{(f)}(12) = 2+4+4+6 = 16.$

1·33. When $l = 3$, we know that (Bouniakowski and Elliott, *l.c.*)

$$\{\zeta(s)\}^3 = \sum_n \sigma_{00}(n)\, n^{-s},$$

where $\sigma_{00}(n) = \Sigma\sigma_0(d)$, d a divisor of n.

We may write $\{\eta(s)\}^3 = \sum_n \sigma_{00}^{(f)}(n) . n^{-s},$

wherein $\sigma_{00}^{(f)}(n) = \Sigma\sigma_0^{(f)}(d)$, d a divisor of n.

1·34. And generally from the definitions

$$\left.\begin{aligned}\sigma_{00\ldots k\text{ times}}(n) &= \Sigma\sigma_{00\ldots k-1\text{ times}}(d)\\ \sigma^{(f)}_{00\ldots k\text{ times}}(n) &= \Sigma\sigma^{(f)}_{00\ldots k-1\text{ times}}(d)\end{aligned}\right\} d \text{ a divisor of } n,$$

we may write $\{\zeta(s)\}^l = \Sigma\sigma_{00\ldots l-1\text{ times}}(n) . n^{-s},$

$$\{\eta(s)\}^l = \Sigma\sigma^{(f)}_{00\ldots l-1\text{ times}}(n) . n^{-s}.$$

Ex. gr. When $n = 12$, $\sigma_0^{(f)}(12)$ enumerates the divisors which occur in the factorizations

$$12, \quad 2.6, \quad 3.4, \quad 2^2.3,$$

viz.

$$1, 12, \quad 1, 2, 6, 2\times 6, \quad 1, 3, 4, 3\times 4, \quad 1, 2, 3, 2^2, 2\times 3, 2^2\times 3,$$

altogether 16 in number. To obtain $\sigma_{00}^{(f)}(12)$ we enumerate the factor-divisors of these 16 numbers; we find

$$1+2; \quad 1+2+2+4; \quad 1+2+2+4; \quad 1+2+2+3+4+6,$$

giving $3+9+9+18 = 39$

in correspondence with

$$\binom{3}{2} + \binom{3}{2}\binom{3}{2} + \binom{3}{2}\binom{3}{2} + \binom{3}{2}\binom{4}{2} = 39$$

as required by the theory.

1·4 It is clear that by carrying out the multiplication to produce $\{\eta(s)\}^2$ we connect $\sigma_0^{(f)}(n)$ and $fc(n)$ in the manner

$$\sigma_0^{(f)}(n) = \Sigma fc(d) fc\left(\frac{n}{d}\right),$$

where d is a divisor of m.

§ 2. *The Bipartite Theory.*

2·1. Euler's function for the enumeration of the partitions of ordinary or unipartite numbers is

$$F(x) = 1/(1-x)(1-x^2)(1-x^3)\ldots \text{ ad inf.} = \Sigma p_n x^n$$

which has the advantage of being in close contact with elliptic functions.

That for bipartite numbers $\overline{n_1 n_2}$ is

$$1/(1-x)(1-y).(1-x^2)(1-xy)(1-y^2).(1-x^3)(1-x^2y)$$
$$\times (1-xy^2)(1-y^3). \ldots = \Sigma p_{n_1 n_2} x^{n_1} y^{n_2},$$

which can be expressed in the form

2·11. $F(x)\, F(y)\, F(xy)\, F(x^2y)\, F(xy^2) \ldots F(x^a y^b) \ldots,$

wherein a, b are relatively prime integers.

2·12. If $n_2 \nless 1$ we are only concerned with y to the power unity; the generating function becomes

$$1/1-x.1-x^2.1-x^3. \ldots 1-y.1-xy.1-x^2y. \ldots,$$

and since $1/1-y.1-xy.1-x^2y. \ldots = 1 + \frac{y}{1-x} + \ldots$

the coefficient of y is

$$1/(1-x)^2(1-x^2)(1-x^3)\ldots \text{ad inf.} = p_{01} + p_{11}x + p_{21}x^2 + p_{21}x^3 + \ldots$$

and thence $\qquad p_{n1} = \sum_0^n p_n$, where $p_0 = 1$.

2·13. If $n_2 \nless 2$, the generating function is, effectively,

$$1/(1-x)(1-x^2)(1-x^3)\ldots(1-y)(1-xy)(1-x^2y)\ldots$$
$$(1-y^2)(1-xy^2)(1-x^2y^2)\ldots$$

which is

$$F(x)\left\{1 + \frac{y}{1-x} + \frac{y^2}{(1-x)(1-x^2)} + \ldots\right\}\left\{1 + \frac{y^2}{1-x} + \ldots\right\},$$

wherein the coefficient of y^2 is

$$F(x)\left\{\frac{1}{1-x} + \frac{1}{(1-x)(1-x^2)}\right\} = p_{02} + p_{12}x + p_{22}x^2 + \ldots + p_{n2}x^n + \ldots$$

which leads easily to the formula

2·14. $p_{n2} = 2p_n + 2p_{n-1} + 3p_{n-2} + 3p_{n-3} + 4p_{n-4} + 4p_{n-5} + \ldots$

to $n+1$ terms.

We observe that the factor of $F(x)$, above, consists of two terms. Moreover, the unipartite number 2 has two partitions 2, 11 which we may write so as to exhibit the frequency with which the parts 1, 2 occur in the partitions in the form

$$2^1, \quad 1^2$$

and we will term these exponents "frequency exponents."

We next note that if we associate with the frequency exponents 1, 2 the algebraic fractions

$$\frac{1}{1-x}, \quad \frac{1}{(1-x)(1-x^2)} \text{ respectively,}$$

there is a simple correspondence between the generating function and the partitions of the unipartite number 2.

2·2. If $n_2 \leqslant 3$, we proceed in a similar manner. For brevity we write after Cayley

$$(1-x^k) = (\mathbf{k}),$$

and find in the first place

$$F(x)\left\{1+\frac{y}{(\mathbf{1})}+\frac{y^2}{(\mathbf{1})(\mathbf{2})}+\frac{y^3}{(\mathbf{1})(\mathbf{2})(\mathbf{3})}\right\}\left\{1+\frac{y^2}{(\mathbf{1})}\right\}\left\{1+\frac{y^3}{(\mathbf{1})}\right\},$$

whence taking out the coefficient of y^3

$$F(x)\left\{\frac{1}{(\mathbf{1})}+\frac{1}{(\mathbf{1})^2}+\frac{1}{(\mathbf{1})(\mathbf{2})(\mathbf{3})}\right\}$$
$$= p_{03} + p_{13}x + p_{23}x^2 + p_{33}x^3 + \ldots + p_{n3}x^n + \ldots .$$

We now remark that, in the repetitional notation, the partitions of 3 are

$$3^1, \quad 2^1 1^1, \quad 1^3,$$

and that we can derive the left-hand side of the identity just obtained by associating the algebraic fractions

$$\frac{1}{(\mathbf{1})}, \quad \frac{1}{(\mathbf{1})(\mathbf{2})}, \quad \frac{1}{(\mathbf{1})(\mathbf{2})(\mathbf{3})}$$

with the repetitional exponents 1, 2, 3 respectively.

It is also clear that this correspondence persists and that we may in general write down the co-factor of $F(x)$ by associating $1/(\mathbf{1})(\mathbf{2}) \ldots (\mathbf{k})$ with the repetitional exponents k which occur in the partitions of the unipartite number n_2.

Thus the reader will verify the formula

$$F(x)\left\{\frac{1}{(\mathbf{1})}+\frac{1}{(\mathbf{1})^2}+\frac{1}{(\mathbf{1})(\mathbf{2})}+\frac{1}{(\mathbf{1})^2(\mathbf{2})}+\frac{1}{(\mathbf{1})(\mathbf{2})(\mathbf{3})(\mathbf{4})}\right\}$$
$$= p_{04} + p_{14}x + p_{24}x^2 + \ldots + p_{n4}x^n + \ldots$$

as being derived from the frequencies of the partitions

$$4^1, \quad 3^1 1^1, \quad 2^2, \quad 2^1 1^2, \quad 1^4.$$

We have thus the cardinal theorem.

2·3. "If a partition of the unipartite number n_2 be

$$1^{c_1}2^{c_2}3^{c_3}\ldots s^{c_s},$$

$$F(x)\,\Sigma\frac{1}{(\mathbf{1})(\mathbf{2})\ldots(\mathbf{c_1})(\mathbf{1})(\mathbf{2})\ldots(\mathbf{c_2})(\mathbf{1})(\mathbf{2})\ldots(\mathbf{c_3})\ldots\ldots(\mathbf{1})(\mathbf{2})\ldots(\mathbf{c_s})}$$
$$=p_{0n_2}+p_{1n_2}x+p_{2n_2}x^2+\ldots+p_{n_1n_2}x^{n_1}+\ldots,$$

where the summation is for every partition

$$1^{c_1}2^{c_2}\ldots s^{c_s} \text{ of } n_2,$$

and $$F(x)=1/(\mathbf{1})(\mathbf{2})(\mathbf{3})\ldots \text{ ad inf.''}$$

§ 3. *Study of the Frequencies of Partitions.*

3·1. I study the partition

$$n_1^{c_1}n_2^{c_2}\ldots n_s^{c_s} \text{ of the number } n,$$

$$\Sigma c_t n_t=n,\quad n_1<n_2\ldots<n_s,\quad c_t>0,\quad t=1,2,\ldots s$$

with particular reference to the frequency integers $c_1, c_2, \ldots c_s$.

I write

$$\frac{1}{\prod\limits_1^\infty(1-x^n)}=\prod_1^\infty\left(1+\frac{x^n}{1-x^n}\right)$$
$$=1+X_1+X_2+\ldots+X_3+\ldots,$$

where $$X_s=\Sigma\frac{x^{c_1+2c_2+\ldots+sc_s}}{(1-x^{c_s})(1-x^{c_{s-1}+c_s})\ldots(1-x^{c_1+c_2+\ldots+c_s})}$$
$$=\Sigma\frac{x^{c_s}}{1-x^{c_s}}\cdot\frac{x^{c_{s-1}+c_s}}{1-x^{c_{s-1}+c_s}}\cdots\frac{x^{c_1+c_2+\ldots+c_s}}{1-x^{c_1+c_2+\ldots+c_s}},$$

where the summation is for every ordered set of s integers each >0, between which there may be any number of equalities.

It is obvious that the integer n possesses a number of partitions which have the set of frequencies $c_1, c_2, \ldots c_s$ and that these are enumerated by the coefficients of x^n in the development of X_s.

3·2. We may write *in extenso*, taking $1-x^s$ to be denoted by $(\mathbf{s})$,

$$\frac{1}{\prod\limits_1^\infty(1-x^n)}=1+\frac{x}{(\mathbf{1})}+\frac{x^2}{(\mathbf{2})}+\frac{x^3}{(\mathbf{3})}+\frac{x^4}{(\mathbf{4})}+\frac{x^5}{(\mathbf{5})}+\frac{x^6}{(\mathbf{6})}+$$
$$+\frac{x^3}{(\mathbf{1})(\mathbf{2})}+\frac{x^4}{(\mathbf{1})(\mathbf{3})}+\frac{x^5}{(\mathbf{1})(\mathbf{4})}+\frac{x^6}{(\mathbf{1})(\mathbf{5})}+\ldots$$
$$+\frac{x^5}{(\mathbf{2})(\mathbf{3})}+\frac{x^6}{(\mathbf{2})(\mathbf{4})}+\ldots$$
$$+\frac{x^6}{(\mathbf{1})(\mathbf{2})(\mathbf{3})}+\ldots$$

$$=1+\sum_1^\infty\Sigma\frac{x^n}{(\mathbf{n_1})(\mathbf{n_2})(\mathbf{n_3})\ldots},$$

where $(n_1 n_2 n_3 \ldots)$ is any partition of n into unequal parts and in the first place the summation is for all such partitions.

Ex. gr. We write down at once the next set of terms

$$\frac{x^7}{1-x^7} + \frac{x^7}{(1-x)(1-x^6)} + \frac{x^7}{(1-x^2)(1-x^5)} + \frac{x^7}{(1-x^3)(1-x^4)} + \frac{x^7}{(1-x)(1-x^2)(1-x^4)}.$$

We note at once that

$$X_1 = \Sigma\sigma_0(x)\, x^n,$$

where $\sigma_0(n)$ enumerates the divisors of n.

3·3. We have before us what we may regard as a generalization of the notion of a divisor of an integer and we may expect to obtain expressions for X_2, X_3, ... which involve the arithmetical functions

$$\sigma_0(n), \quad \sigma_1(n), \quad \sigma_2(n), \ldots,$$

where $\sigma_k(n)$ denote the sum of the kth powers of the divisors of n. To reach such expressions I consider the symmetric functions of the elements

$$\frac{x}{1-x}, \quad \frac{x^2}{1-x^2}, \quad \frac{x^3}{1-x^3}, \ldots$$

which occur in the expression of X_1.

I write $\left(\frac{x}{1-x}, \quad \frac{x^2}{1-x^2}, \quad \frac{x^3}{1-x^3}, \ldots\right) = (\alpha, \beta, \gamma, \ldots)$,

or alternatively $\qquad (\alpha_1, \alpha_2, \alpha_3, \ldots)$,

$$(\Sigma\alpha, \Sigma\alpha\beta, \Sigma\alpha\beta\gamma, \ldots) = (b, c, d, \ldots),$$

or alternatively $(a_1, a_2, a_3, \ldots)$ in the usual notation of symmetric functions of a single system of quantities.

Also $\qquad \Sigma\alpha^k = s_k.$

It is clear that we have

$$(X_1, X_2, X_3, \ldots) = (\Sigma\alpha, \Sigma\alpha\beta, \Sigma\alpha\beta\gamma, \ldots)$$

and that we obtain expressions for $\Sigma\alpha\beta$, $\Sigma\alpha\beta\gamma$, ... by first of all calculating expressions for the sums of powers s_2, s_3, s_4, ... and then making use of the formulae which express $\Sigma\alpha\beta$, $\Sigma\alpha\beta\gamma$, ... in terms of s_1, s_2, s_3, ...

We have already before us

$$s_1 = X_1 = \Sigma\sigma_0(n)\, x^n,$$

and we next see that

$$s_2 = \left(\frac{x}{1-x}\right)^2 + \left(\frac{x^2}{1-x^2}\right)^2 + \left(\frac{x^3}{1-x^3}\right)^2 + \ldots,$$

and since $$\left(\frac{x^t}{1-x^t}\right)^2 = \sum_{k=1}^{\infty} kx^{kt},$$

we see that $$s_2 = \sum_1^{\infty}{}_n \{\sigma_1(n) - \sigma_0(n)\} x^n.$$

3·31. Similarly, in succession,

$$s_3 = \Sigma \frac{1}{2!}\{\sigma_2(n) - 3\sigma_1(n) + 2\sigma_0(n)\} x^n,$$

$$s_4 = \Sigma \frac{1}{3!}\{\sigma_3(n) - 6\sigma_2(n) + 11\sigma_1(n) - 6\sigma_0(n)\} x^n,$$

....

$$s_k = \Sigma \frac{1}{(k-1)!}\left\{\sigma_{k-1}(n) - \binom{k}{2}\sigma_{k-2}(n) + \ldots + (-)^{k-1}(k-1)!\,\sigma_0(n)\right\} x^n,$$

the coefficients being those in the development of

$$(x-1)(x-2)\ldots(x-k+1).$$

In fact we may write symbolically

$$s_2 = \Sigma(\sigma - 1)(n)\, x^n,$$

$$s_3 = \Sigma \frac{1}{2!}(\sigma-1)(\sigma-2)(n)\, x^n$$

$$\vdots$$

$$s_k = \Sigma \frac{1}{(k-1)!}(\sigma-1)(\sigma-2)\ldots(\sigma-k+1)(n).x^n,$$

where after multiplication of the symbol σ, σ^k is to be replaced by σ_k.

Thus in general $s_k = \Sigma \binom{\sigma-1}{k-1}(n).x^n$ symbolically.

We now find

$$2!\,X_2 = 2!\,a_2 = s_1^2 - s_2$$

$$= \{\Sigma\sigma_0(n)\, x^n\}^2 - \Sigma\{\sigma_1(n) - \sigma_0(n)\} x^n$$

$$= \Sigma_n \left\{\sum_{m=1}^{n-1} \sigma_0(m)\,\sigma_0(n-m) - \sigma_1(n) + \sigma_0(n)\right\} x^n,$$

etc.

3·32. The inverse formulae for the expression of $\Sigma\sigma_k(n)\, x^n$ in terms of $s_1, s_2, s_3, \ldots$ are

$$\Sigma\sigma_0(n)\, x^n = s_1,$$

$$\Sigma\sigma_1(n)\, x^n = s_1 + s_2,$$

$$\Sigma\sigma_2(n)\, x^n = s_1 + 3s_2 + 2s_3,$$

$$\Sigma\sigma_3(n)\, x^n = s_1 + 7s_2 + 12s_3 + 6s_4,$$

....

where the table of coefficients

1			
1	1		
1	3	2	
1	7	12	6
....			

is such that if $u_{x,y}$ is the number in the xth column and yth row

$$u_{x,y} = xu_{x,y-1} + (x-1)\,u_{x-1,y-1}.$$

If we write $$X_k = \Sigma\tau^{(k)}(n)\,x^n,$$

we have the new arithmetical function

$$\tau^{(k)}(n).$$

3·321. We have obtained the expressions

$$\tau^{(1)}(n) = \sigma_0(n),$$

$$\tau^{(2)}(n) = \frac{1}{2!}\left\{\sum_{m=1}^{n-1}\sigma_0(m)\,\sigma_0(n-m) - \sigma_1(n) + \sigma_0(n)\right\},$$

and we have the means for the calculation of the expression of $\tau^{(k)}(n)$.

It must involve the arithmetical function $\sigma_t(m)$ where t takes the values $0, 1, 2, \ldots k-1$.

When we restrict the numbers $c_1, c_2, \ldots$ to be unity we enumerate partitions into unrepeated parts and the result gives

$$(1+x)(1+x^2)(1+x^3)\ldots = 1 + \frac{x}{(1)} + \frac{x^3}{(1)(2)} + \frac{x^6}{(1)(2)(3)} + \ldots$$

as is well known.

3·4. I now write

$$\Sigma\frac{x^{c_1+2c_2+\ldots+sc_s}}{(1-x^{c_s})(1-x^{c_{s-1}+c_s})\ldots(1-x^{c_1+c_2+\ldots+c_s})} = \phi_1(c_1, c_2, \ldots c_s),$$

where the summation is for all permutations of $c_1, c_2, \ldots c_s$.

3·41. I note the identities

$$\phi_1(c_1, c_2) = \phi_1(c_1+c_2)\{\phi_1(c_1)+\phi_1(c_2)\} = \phi_1(c_1)\,\phi_1(c_2) - \phi_1(c_1+c_2),$$

so that we have the additive theorem

3·42. $$\phi_1(c_1+c_2) = \frac{\phi_1(c_1)\,\phi_1(c_2)}{1+\phi_1(c_1)+\phi_1(c_2)}.$$

3·43. Also

$$\phi_1(c_1, c_2, c_3) = \phi_1(c_1+c_2+c_3)\{\phi_1(c_2, c_3) + \phi_1(c_3, c_1) + \phi_1(c_1, c_2)\},$$

and generally

3·44. $$\phi_1(c_1, c_2, \ldots c_s) = \phi_1(c_1+c_2+\ldots+c_s)\,\Sigma\phi_1(c_2, c_3, \ldots c_s),$$

where the summation is for the s terms obtained by the circular substitution.

3·45. Also

$$\phi_1(c_1 + c_2 + \dots + c_s)$$
$$= \frac{\phi_1(c_1)\,\phi_1(c_2)\dots\phi_1(c_s)}{\{1+\phi_1(c_1)\}\{1+\phi_1(c_2)\}\dots\{1+\phi_1(c_s)\} - \phi_1(c_1)\,\phi_1(c_2)\dots\phi_1(c_s)},$$

3·46. $\phi_1(c_1, c_2, c_3) = \phi_1(c_1 + c_2 + c_3)\{\phi_1(c_2)\,\phi_1(c_3) + \phi_1(c_3)\,\phi_1(c_1)$
$+ \phi_1(c_1)\,\phi_1(c_2) - \phi_1(c_2 + c_3) - \phi_1(c_3 + c_1) - \phi_1(c_1 + c_2)\}$.

§ 4. *Tripartite, etc. Numbers.*

4·1. The generating function has for denominator

$$(1-x)(1-y)(1-z)$$
$$\times (1-x^2)(1-xy)(1-xz)(1-y^2)(1-yz)(1-z^2)$$
$$\times \;\dots\dots\dots\dots$$

For the tripartite $\overline{n11}$ this denominator is effectively composed of the four infinite products

$$(1-x)(1-x^2)\dots; \quad (1-y)(1-xy)(1-x^2y)\dots;$$
$$(1-z)(1-xz)(1-x^2z)\dots; \quad (1-yz)(1-xyz)(1-x^2yz)\dots;$$

and the effective expansion of the algebraic fraction is thus

$$F(x)\left\{1+\frac{y}{(\mathbf{1})}\right\}\left\{1+\frac{z}{(\mathbf{1})}\right\}\left\{1+\frac{yz}{(\mathbf{1})}\right\},$$

where $(\mathbf{1})$ stands for $(1-x)$ (after Cayley).

Herein the coefficient of yz is

$$F(x)\left\{\frac{1}{(\mathbf{1})}+\frac{1}{(\mathbf{1})^2}\right\} = \sum_0^\infty p_{n11}x^n.$$

We remark that the multipartite number $\overline{11}$ has the two partitions which, with the frequency exponents inserted, are

$$11^1, \quad 10^1 01^1,$$

and we obtain a correspondence between these and the co-factor of $F(x)$ by associating $1/(\mathbf{1})$ with the exponent 1.

4·2. The resulting formula is

$$p_{n11} = (n+2) + (n+1)\,p_1 + np_2 + (n-1)\,p_3 + \dots \text{ to } n+1 \text{ terms.}$$

Similarly we may calculate the series for $p_{nn_2n_3}$, where n_2, n_3 are any assigned integers.

As an example consider the series for p_{n22}.

The partitions of the bipartite $\overline{22}$ are

$$\overline{22}^1;\quad \overline{21}^1\overline{01}^1;\quad \overline{12}^1\overline{10}^1;\quad \overline{20}^1\overline{02}^1;\quad \overline{11}^2;$$
$$\overline{20}^1\overline{01}^2;\quad \overline{02}^1\overline{10}^2;\quad \overline{11}^1\overline{10}^1\overline{01}^1;\quad \overline{10}^2\overline{01}^2;$$

and we are led to the formula

$$F(x)\left\{\frac{1}{(\mathbf{1})}+3\frac{1}{(\mathbf{1})^2}+\frac{1}{(\mathbf{1})(\mathbf{2})}+2\frac{1}{(\mathbf{1})^2(\mathbf{2})}+\frac{1}{(\mathbf{1})^3}+\frac{1}{(\mathbf{1})^2(\mathbf{2})^2}\right\}$$
$$=p_{022}+p_{122}x+\ldots+p_{n22}x^n+\ldots$$

for the calculation of p_{n22}.

4·3. For multipartite numbers in general we can now enumerate the Cardinal Theorem:

"Let $\overline{n_2 n_3 \ldots n_s}$ be any multipartite number and

$$\overline{a_1 a_2 \ldots a_s}^{\pi_1}\,\overline{b_1 b_2 \ldots b_s}^{\pi_2}\ldots$$

one of its partitions. Then

$$\frac{1}{(1-x)(1-x^2)(1-x^3)\ldots}$$
$$\times\Sigma\left\{\frac{1}{(1-x)(1-x^2)\ldots(1-x^{\pi_1})}\cdot\frac{1}{(1-x)(1-x^2)\ldots(1-x^{\pi_2})}\cdot\ldots\right\}$$
$$=\sum_{n_1=0}^{\infty}p_{n_1 n_2\ldots n_s}x^{n_1},$$

where the summation on the left-hand side is in regard to every partition of the multipartite number

$$\overline{n_2 n_3 \ldots n_s}."$$

§ 5. *The Factor-divisor Function* $\sigma_0^{(f)}(n)$.

5·1. I refer to § 1 and write

$$J(x)=j(1)\frac{x}{1-x}+j(2)\frac{x^2}{1-x^2}+\ldots=\Sigma j(n)\frac{x^n}{1-x^n}.$$

5·11. Then $\qquad J(x)=\Sigma\sigma_0^{(f)}(n)\,x^n$

if $\qquad \sigma_0^{(f)}(n)=\Sigma j(d)$, d a divisor of n.*

* This result is not difficult to arrive at, but we may if we please derive it from

$$\Sigma j(n)\frac{x^n}{1-x^n}=\Sigma\sigma_0^{(f)}(n)\,x^n$$

by referring to Laguerre, *Bull. Soc. Math. France*, 1, 1872–3, pp. 77–81, who showed that if

$$\sum_1^\infty f(n)\frac{x^n}{1-x^n}=\sum_1^\infty F(n)\,x^n,$$

then $\qquad F(n)=\Sigma f(d)$, d a divisor of n.

I refer to Dedekind* who proved that if

$$F(n) = \Sigma f(d),$$

where d ranges over the divisors of n,

$$f(n) = F(n) - \Sigma F\left(\frac{n}{p_1}\right) + \Sigma F\left(\frac{n}{p_1 p_2}\right) \ldots,$$

where $p_1, p_2, \ldots$ are the distinct prime factors of n.

5·2. Applying this theorem we get

$$j(n) = \sigma_0^{(f)}(n) - \Sigma\sigma_0^{(f)}\left(\frac{n}{p_1}\right) + \Sigma\sigma_0^{(f)}\left(\frac{n}{p_1 p_2}\right) - \ldots.$$

5·3. Thence

$$J(x) = \Sigma\sigma_0^{(f)}(x)\frac{x^n}{1-x^n} - \Sigma\sigma_0^{(f)}\left(\frac{n}{p_1}\right)\frac{x^n}{1-x^n} + \Sigma\sigma_0^{(f)}\left(\frac{n}{p_1 p_2}\right)\frac{x^n}{1-x^n} - \ldots,$$

wherein after the first term the summations are double, i.e. in regard (i) to the distinct prime factors of n, (ii) in regard to n.

5·4. Now by inversion we obtain

$$\sum_1^\infty \sigma_0^{(f)}(n)\frac{x^n}{1-x^n} = \sum_1^\infty J(x^n).$$

The arithmetical function $j(n)$ has been expressed in terms of the functions $\sigma_0^{(f)}(n)$. It will be remarked that it is $j(n)$ and not $\sigma_0^{(f)}(n)$ that possesses properties analogous to those possessed by allied arithmetical functions.

5·5. Thus if n_1, n_2 be relatively prime integers

$$j(n_1)\,j(n_2) = j(n_1 n_2),$$

whilst the result

$$\Sigma j\left(\frac{n}{d}\right) = \sigma_0^f(n)$$

seems to be the analogue of Gauss' property of the ϕ-functions.

* *Journ. für Math.* 54, 1857, pp. 21, 25.

Enumeration of Bipartite Partitions.

11	**21**	**31**	**41**	**51**	**61**	**71**	**81**	**91**	**101**
2	4	7	12	19	30	45	67	97	139
	22	**32**	**42**	**52**	**62**	**72**	**82**	**92**	**102**
	9	16	29	47	77	118	181	267	392
		33	**43**	**53**	**63**	**73**	**83**	**93**	**103**
		31	57	97	162	257	401	608	907
			44	**54**	**64**	**74**	**84**	**94**	**104**
			109	189	323	522	831	1279	1941
				55	**65**	**75**	**85**	**95**	**105**
				339*	589	975	1576	2472	3737
					66	**76**	**86**	**96**	**106**
					1043	1752	2876	4571	7128
						77	**87**	**97**	**107**
						2998	4987	8043	12693
							88	**98**	**108**
							8406	13748	21938

Enumeration of Tripartite Partitions.

111	**211**	**311**	**411**	**511**	**611**	**711**	**811**	**911**	**1011**
5	11	21	28	64	105	165	254	381	562
	221	**321**	**421**	**521**	**621**	**721**	**821**	**921**	
	26	52	98	171	289	467	737	1131	
		331	**431**	**531**	**631**	**731**	**831**		
		109	212	382	662	1097	1768		

222	**322**	**422**	**522**	**622**
66	137	269	484	843
	332	**432**	**532**	
	300	606	1129	

* This number is given, erroneously, as 336 in *Combinatory Analysis*, vol. I, p. 269. Read 'The multipartite **55** has 339 partitions.'

EULER'S ϕ-FUNCTION AND ITS CONNEXION WITH MULTIPARTITE NUMBERS

By P. A. MACMAHON.

[Received 8 August, 1925.—Read 12 November, 1925.]

Introduction.

This short paper is regarded by the author as a small contribution to the theory of multipartite numbers on classical lines. The theory of numbers has involved the ϕ-function to no small extent since the time of Euler. The prominent manner in which the function and its extensions enter the multipartite theory seems to suggest that other parts of the classical theory may prove to be similarly involved as soon as the suitable point of view has been reached.

1. *The generating function which enumerates multipartite numbers.*

In the case of unipartite numbers the function is

$$\prod_{1}^{\infty}{}_a(1-x^a)^{-1},$$

where the exponent a ranges over all the unipartite numbers. We write

$$F(x) = \prod_{1}^{\infty}{}_a(1-x^a)^{-1}.$$

For bipartite numbers we write

$$F(x, y) = \prod_{0}^{\infty}{}_a \prod_{0}^{\infty}{}_b (1-x^a y^b)^{-1},$$

wherein the exponents a, b are such that they range over every bipartite number $\overline{ab}$, but are *not* simultaneously zero.

If a, b be relatively prime unipartite numbers, each greater than or equal to 1,

$$F(x^a y^b)$$

is a factor of $F(x, y)$, and we write

$$F(x, y) = F(x)\,F(y)\,\Pi F(x^a y^b).$$

Thus, for a few factors,

$$F(x, y) = F(x)\,F(y)\,F(xy)\,F(x^2y)\,F(xy^2)\,F(x^3y)\,F(xy^3)\ldots$$

Similarly in the case of tripartite numbers

$$F(x, y, z) = F(x)\,F(y)\,F(z)\,\Pi F(x^a y^b)\,\Pi F(x^a z^c)\,\Pi F(y^b z^c)\,\Pi F(x^a y^b z^c),$$

where, in the four infinite products,

$F(x^a y^b)$,	a, b	are relatively prime and each is equal to or greater than 1,
$F(x^a z^c)$,	a, c	,, ,, ,, ,,
$F(y^b z^c)$,	b, c	,, ,, ,, ,,
$F(x^a y^b z^c)$,	a, b, c	,, ,, ,, ,,

and so on for quadripartite, etc., numbers.

A question that arises in dealing with the generating function in these forms, which involve m letters $x_1, x_2, \ldots, x_m$ is as to the enumeration of the relatively prime sets of numbers

$$a_1,\ a_2,\ \ldots,\ a_m,$$

which are such that $a_1+a_2+\ldots+a_m = n$,

where n is a given integer, and $a_s \geqslant 1$; $s = 1, 2, \ldots, m$. In other words, we require the enumeration of the multipartite numbers

$$\overline{a_1\,a_2\ldots a_m},$$

where the constituents $a_1, a_2, \ldots, a_m$ are relatively prime, the sum

$$a_1+a_2+\ldots+a_m = n,$$

a given integer, and $a_s \geqslant 1$ $(s = 1, 2, \ldots, m)$.

The constituents $a_1, a_2, \ldots$ may be alike or different, and we must consider them in all permutations, since a different multipartite number is in correspondence with each permutation.

In the case of bipartite numbers we find that the enumeration is given by the ϕ-function $\phi(n)$ of Euler.

For consider the particular case $n = 12$; the bipartite numbers which have relatively prime constituents are

$$\overline{11\ 1}$$
$$\overline{75}$$
$$\overline{57}$$
$$\overline{1\ 11}$$

If a, b be relatively prime and $a+b = n$, the relative primeness of a and $n-a$ involves the relative primeness of a and n and also of $n-a$ and n. It follows that the numbers in the first column of the bipartites, as above written, are relatively prime to n ($= 12$) and that the enumeration of the bipartites is of necessity given by $\phi(12)$. Thence it is obvious that the enumeration is, in general, given by $\phi(n)$.

We have here, in fact, a new definition of $\phi(n)$. That of Euler is "$\phi(n)$ denotes the number of integers not exceeding n which are relatively prime to n".

In this definition, and throughout this paper, two or more unipartite numbers are said to be relatively prime if they contain no common factor greater than 1. The new definition that is suggested is "$\phi(n)$ denotes the number of bipartite numbers

$$\overline{a,\ n-a},$$

which are such that a, $n-a$ are relative primes and both greater than or equal to 1". The merit of the new definition is that the idea may be extended to multipartite numbers in general.

In view of the many generalizations of the ϕ-function* that have been given it is somewhat difficult to decide upon an appropriate notation. The $\phi(n)$ of Euler I write $\mu_1(n)$.

2. *First generalization of the ϕ-function.*

Consider the application of the second definition of § 1 to tripartite numbers.

"$\mu_2(n)$ denotes the number of tripartite numbers

$$\overline{a,\ b,\ n-a-b},$$

which are such that a, b, and $n-a-b$ are relative primes and each greater than or equal to 1".

* L. E. Dickson, *History of the Theory of Numbers*, 1 (1919), 140 *et seq.*

Take the particular case, $n = 6$, and write down the tripartites in numerical order :

41	411
32	321
31	312
23	231
21	213
14	141
13	132
12	123
11	114

and to the left of them the bipartites which arise from deletion of the third constituent of each tripartite.

Since a, b, $n-a-b$ contain no common divisor greater than 1, we see that a, b contain no divisor greater than 1, which is also a divisor of n. Thence the bipartites which have been reached by deletion from the tripartites are such that the highest common divisor of the two constituents is prime to n. Moreover, we obtain thus a complete set of bipartites thus defined; for suppose a bipartite $\overline{\alpha\beta}$ were not included we could derive a tripartite $\overline{\alpha,\ \beta,\ n-\alpha-\beta}$ which is not included in the tripartite set, and this is contrary to the definition which gives rise to the tripartites.

Hence we have also the first definition of $\mu_2(n)$.

"$\mu_2(n)$ denotes the number of bipartite numbers

$$\overline{ab}$$

such that $a+b < n$ and the constituents a, b are such that their highest common divisor is a relative prime to n and each of them greater than or equal to 1".

In general the two definitions of $\mu_k(n)$ are as follows :—

(i) "$\mu_k(n)$ denotes the number of k-partite numbers

$$\overline{a_1\, a_2 \ldots a_k},$$

such that $a_1+a_2+ \ldots +a_k < n$, and the constituents $a_1, a_2, \ldots, a_k$ such that their highest common divisor is a relative prime to n and each of them greater than or equal to 1".

(ii) "$\mu_k(n)$ denotes the number of $k+1$-partite numbers,

$$\overline{a_1, a_2 \ldots a_k.\ n-\Sigma a},$$

which are such that $a_1, a_2, \dots a_k, n-\Sigma a$ are relatively prime positive integers, each of which is greater than or equal to 1".

Before discussing the properties of $\mu_k(n)$ it should be remarked that this new function has arisen in a natural manner from a discussion of multipartite numbers. Of other generalizations due to Schemmel*, C. Jordan†, W. E. Story‡ a few words may be in place, as they are concerned with *sets* of integers. Schemmel considered the $\Phi(m)$ sets of n consecutive integers each less than m and relatively prime to m.

C. Jordan, in connexion with the study of linear congruence groups, defined $J_k(n)$ as denoting the number of different sets of k (equal or distinct) positive integers less than or equal to n whose highest common divisor is prime to n. These sets involve one whose content is $nk-1$. The theory has given rise to much work by other mathematicians, but has no connexion with the present paper.

W. E. Story's generalization is concerned with "sets of k integers $\leqslant n$, not all divisible by any factor of n, such that we do not distinguish between two sets differing only by permutation of their numbers". This research also, it is clear, has no connexion with that before us.

3. *The properties of* $\mu_k(N)$.

The first property is $\mu_1(N) = \phi(N)$.

The second definition of $\mu_k(N)$ involves $k+1$-partite numbers of content N. These, from another point of view, are partitions of the unipartite number N, the order in which the parts are written being essential. If we discard the condition of relative primeness of parts the whole number of such *ordered* partitions (called by me in another place "compositions") is well known to be

$$(3.1) \qquad \binom{N-1}{k}$$

If the distinct primes which are contained in N be $p_1, p_2, p_3, \dots$ a well known principle thence establishes that

$$(3.2) \qquad \mu_k(N) = \binom{N-1}{k} - \Sigma\binom{(N/p_1)-1}{k} + \Sigma\binom{(N/p_1p_2)-1}{k} - \dots$$

* *Journal für Math.*, 70 (1869), 191–2.

† *Traité des substitutions*, Paris (1870), 95–97.

‡ *Johns Hopkins Univ. Circulars*, 1 (1881), 132, *ib.* 51.

which, for $k=1$, since the sum of the coefficients in

$$(1-x)^s \quad (s \text{ a positive integer}),$$

is zero, reduces to

$$\phi(N) = N - \Sigma\frac{N}{p_1} + \Sigma\frac{N}{p_1p_2} - \dots,$$

as it should.

It is very easy to show, from the known properties of binominal coefficients, that

$$(3.3) \qquad \mu_k(N)+\mu_{k-1}(N) = \binom{N}{k} - \Sigma\binom{N/p_1}{k} + \Sigma\binom{N/p_1p_2}{k} - \dots,$$

and, in general,

$$(3.33) \quad \mu_k(N)+\binom{s}{1}\mu_{k-1}(N)+\binom{s}{2}\mu_{k-2}(N)-\dots$$

$$= \binom{N+s-1}{k} - \Sigma\binom{(N/p_1)+s-1}{k} + \Sigma\binom{(N/p_1p_2)-1}{k} - \dots.$$

I write, for brevity, the expression found for $\mu_k(N)$ in the form

$$\mu_k(N) = \left[\binom{N-1}{k}\right].$$

Then, considering $\mu_k(N/d)$ for every divisor d of N,

$$\mu_k(N)+\Sigma\mu_k\left(\frac{N}{p_1}\right)+\Sigma\mu_k\left(\frac{N}{p_1^2}\right)+\dots$$

$$+\Sigma\mu_k\left(\frac{N}{p_1p_2}\right)+\dots$$

$$= \left[\binom{N-1}{k}\right] + \Sigma\left[\binom{(N/p_1)-1}{k}\right] + \Sigma\left[\binom{(N/p_1^2)-1}{k}\right]+\dots$$

$$+\Sigma\left[\binom{(N/p_1p_2)-1}{k}\right]+\dots.$$

If we develop the expressions in square brackets, the right-hand side of this identity vanishes, with the exception of

$$\binom{N-1}{k}.$$

So that

$$\Sigma\mu_k\left(\frac{N}{d}\right) = \binom{N-1}{k},$$

where d is any divisor of N except N itself, or

$$(3.4) \qquad \Sigma\mu_k(d) = \binom{N-1}{k} \quad \text{[see also } post \text{ the method of (5.3)],}$$

where d is any divisor of N except unity.

This is the analogue of the formula of Gauss.

It will be recalled that Euler* obtained the result $\phi(m)\ \phi(n) = \phi(mn)$, where m and n are relatively prime. In seeking for some analogue in the case of the new functions I have been led to some interesting expressions for $\mu_k(n)$, which involve $\mu_1(n)$ as an *algebraical* factor. From the expression for $\mu_2(N)$ above

$$\mu_2(p_1^{\pi_1}) = \tfrac{1}{2}p_1^{\pi_1-1}(p_1-1)\{p_1^{\pi_1-1}(p_1+1)-3\},$$

or writing $p_1^{\pi_1} = N$,

$$\mu_2(N) = \tfrac{1}{2}\mu_1(N)\left\{\frac{N}{p_1}\sigma_1(p_1)-3\right\},$$

where $\sigma_1(m)$ = sum of the divisors of m; and in general, if $N = p_1^{\pi_1}p_1^{\pi_2}\ldots$ and $m = p_1p_2\ldots$, we are led to

$$\mu_2(N) = \tfrac{1}{2}\mu_1(N)\left\{\frac{N}{m}\sigma_1(m)-3\right\}.$$

Similarly we find

$$\mu_3(N) = \frac{1}{3!}\mu_1(N)\left\{\left(\frac{N}{m}\right)^2\sigma_1(m^2)-6\left(\frac{N}{m}\right)\sigma_1(m)+11\right\},$$

where 1, —6, 11 are the first three coefficients which occur in the development of $(x-1)(x-2)(x-3)$.

It is easy to establish the final result

$$\mu_k(N) = \frac{1}{k!}\mu_1(N)\left\{\left(\frac{N}{m}\right)^{k-1}\sigma_1(m^{k-1})-a_2\left(\frac{N}{m}\right)^{k-2}\sigma_1(m^{k-2})+\ldots+(-)^{k+1}a_k\right\},$$

where $N = p_1^{\pi_1}p_2^{\pi_2}\ldots$, $m = p_1p_2\ldots$, and

$$1,\ -a_2,\ \ldots,\ (-)^{k+1}a_k$$

are the k coefficients in the development of $(x-1)(x-2)\ldots(x-k)$.

---

* If M be the highest common divisor of $N_1, N_2, \ldots, N_s$ it is easy to show that

$$\phi(N_1)\phi(N_2)\ldots\phi(N_s) = \phi\{(M)\}^{s-1}\phi\left(\frac{N_1N_2\ldots N_s}{M^{s-1}}\right).$$

4. *Further extensions of* $\phi(m)$.

Sections 1, 2, and 3 have dealt with extensions in regard to a unipartite number N as put in evidence by the notation $\mu_k(N)$. I now consider

$$\phi(\overline{N_1 N_2 \ldots N_s}), \tag{4.1}$$

where the argument is a multipartite number and the ϕ function is defined as enumerating the multipartite numbers which are $< \overline{N_1 N_2 \ldots N_s}$ and prime relatively to it.

The multipartite number $\overline{N'_1 N'_2 \ldots N'_s}$ is defined as being less than $\overline{N_1 N_2 \ldots N_s}$ if

$$N'_r \leqslant N_r \quad (r = 1, 2, \ldots, s)$$

and

$$\sum_1^s N'_r < \sum_1^s N_r.$$

If M be the highest common divisor of $N_1, N_2, \ldots, N_s$,

and M' „ „ „ $N'_1, N'_2, \ldots, N'_s$,

the two multipartite numbers are defined as being relatively prime if M, M' have no common divisor > 1. In all cases *zero constituents are excluded* from the multipartites.

Taking first the bipartite theory, and discarding the condition of relative primeness, it will be noticed that the whole number of bipartites which are $\leqslant$ than $\overline{N_1 N_2}$ is $N_1 N_2$.

Let M, the highest common divisor of N_1, N_2, be $p_1^{\pi_1} p_2^{\pi_2} \ldots$.

Of the bipartites $\leqslant \overline{N_1 N_2}$, there are $(N_1 N_2)/p_1^2$ which contain the divisor p_1.

Thence an appeal to the well known principle of Möbius gives

$$\phi(\overline{N_1 N_2}) = N_1 N_2 - \Sigma \frac{N_1 N_2}{p_1^2} + \Sigma \frac{N_1 N_2}{p_1^2 p_2^2} - \ldots$$

$$= N_1 N_2 \left(1 - \frac{1}{p_1^2}\right)\left(1 - \frac{1}{p_2^2}\right) \ldots.$$

By the same argument it is evident that

$$\phi(\overline{N_1 N_2 \ldots N_3}) = N_1 N_2 \ldots N_s \left(1 - \frac{1}{p_1^s}\right)\left(1 - \frac{1}{p_2^s}\right) \ldots, \tag{4.2}$$

where $p_1^{\pi_1} p_1^{\pi_2} \ldots$ is the highest common divisor of the unipartites $N_1, N_2, \ldots, N_s$.

Let $\overline{L_1 L_2 \dots L_s}$ be another multipartite such that the highest common divisor of $L_1, L_2, \dots L_s$ contains none of the primes $p_1, p_2, \dots$, and put

$$L_r N_r = P_r \quad (r = 1, 2, \dots, s).$$

Then

$$\phi(\overline{L_1 L_2 \dots L_s}) \,.\, \phi(\overline{N_1 N_2 \dots N_s}) = \phi(\overline{P_1 P_2 \dots P_s}) \tag{4.3}$$

is an obvious corollary.

Ex. gr. $\overline{N_1 N_2} = \overline{82}$; $\overline{L_1 L_2} = \overline{63}$,

$$\phi(\overline{82}) = 8\,.\,2\,.\,(1-\tfrac{1}{4}) = 12,$$

$$\phi(\overline{63}) = 6\,.\,3\,.\,(1-\tfrac{1}{9}) = 16,$$

$$\phi(\overline{48\ 6}) = 48\,.\,6(1-\tfrac{1}{4})(1-\tfrac{1}{9}) = 192 = 12\,.\,16.$$

If d denote any common divisor of $N_1, N_2, \dots N_s$,

$$\Sigma\phi\left(\overline{\frac{N_1}{d}\,.\,\frac{N_2}{d}\dots\frac{N_s}{d}}\right) = N_1 N_2 \dots N_s.$$

Ex. gr.

$$\begin{aligned} \phi(\overline{24\ 12}) &= 192, \\ \phi(\overline{12\ 6}) &= 48 \\ \phi(\overline{84}) &= 24 \\ \phi(\overline{63}) &= .16 \\ \phi(\overline{42}) &= 6 \\ \phi(\overline{21}) &= 2 \end{aligned}$$

$$\Sigma\left(\overline{\frac{24}{d}\,.\,\frac{12}{d}}\right) = \overline{288} = 24 \times 12.$$

To establish this the first remark is that the result shows that, if M be the highest common divisor of $N_1, N_2, \dots N_s$ and $N_r = MN'_r$, $r = 1, 2, \dots, s$.

$$\Sigma\phi\left(\overline{\frac{N_1}{d}\,\frac{N_2}{d}\dots\frac{N_s}{d}}\right) = \phi(\overline{N'_1 N'_2 \dots N'_s})\,\Sigma\phi\left(\overline{\frac{M}{d}\,.\,\frac{M}{d}\dots s \text{ factors}}\right),$$

because N'_r and M/d are relatively prime for $r = 1, 2, \dots, s$. Now $\phi(\overline{N'_1 N'_2 \dots N'_s}) = N'_1 N'_2 \dots N'_s$ and $\phi(\overline{M/d\,.\,M/d\dots})$ bears the same relation to M^s and d^s that $\phi(M/d)$ does to M and d; so that, since

$$\Sigma\phi\left(\frac{M}{d}\right) = M,$$

we have
$$\Sigma\phi\left(\overline{\frac{M}{d}\,\frac{M}{d}\,\ldots\, s \text{ factors})}\right) = M^s.$$

Thence

$$(4\,.\,4)\qquad \Sigma\phi\left(\overline{\frac{N_1}{d}\,\frac{N_2}{d}\,\ldots\,\frac{N_s}{d}}\right) = N_1'N_2'\ldots N_s'M^s = N_1N_2\ldots N_s,$$

as was required to be established.

(See also *post* the method of 5 . 3.)

The function $\phi(\overline{N_1N_2\ldots N_s})$ cannot be extended on lines parallel to $\mu_k(N)$ because, as it has been defined, it introduces zero constituents into the multipartites when attempts are made to extend it. We require a new definition in order to proceed in the direction indicated.

4 . 5. I define the multipartite $\overline{N_1'N_2'\ldots N_s'}$ to be s-fold less than $\overline{N_1N_2\ldots N_s}$ if

$$N_r' < N_r \quad (r = 1, 2, \ldots, s).$$

and I take $\mu_1(\overline{N_1N_2\ldots N_s})$ to enumerate the multipartites which are s-fold $<\overline{N_1N_2\ldots N_s}$ and relatively prime to it, two multipartites being relatively prime when the highest common divisors associated with them are relatively prime.

If relative primeness be ignored the enumeration of the s-fold lesser multipartites is given by

$$(4\,.\,51)\qquad (N_1-1)(N_2-1)\ldots(N_s-1).$$

For the present take bipartites, so that $s = 2$.

Of these lesser bipartites the number which involve the factor p_1 in each constituent, $M = p_1^{\pi_1}p_2^{\pi_2}\ldots$ being the highest common divisor of N_1, N_2, is

$$\left(\frac{N_1}{p_1}-1\right)\left(\frac{N_2}{p_1}-1\right).$$

Thence we are led to the relation

$$(4\,.\,52)\qquad \mu_1(\overline{N_1N_2}) = (N_1-1)(N_2-1) - \Sigma\left(\frac{N_1}{p_1}-1\right)\left(\frac{N_2}{p_1}-1\right)$$

$$+\Sigma\left(\frac{N_1}{p_1p_2}-1\right)\left(\frac{N_2}{p_1p_2}-1\right)-\ldots$$

$$= N_1N_2\left(1-\frac{1}{p_1^2}\right)\left(1-\frac{1}{p_2^2}\right)-(N_1+N_2)\left(1-\frac{1}{p_1}\right)\left(1-\frac{1}{p_2}\right)$$

$$= \phi(\overline{N_1N_2})-\phi(N_1)-\phi(N_2).$$

Similarly we find that

$$\mu_1(\overline{N_1N_2N_3}) = \phi(\overline{N_1N_2N_3}) - \phi(\overline{N_1N_2}) + \phi(N_1)$$
$$- \phi(\overline{N_1N_3}) + \phi(N_2)$$
$$- \phi(\overline{N_2N_3}) + \phi(N_3),$$

which we may write

$$\mu_1(\overline{N_1N_2N_3}) = \phi(\overline{N_1N_2N_3}) - \Sigma\phi(\overline{N_1N_2}) + \Sigma\phi(N_1),$$

and in general

$$(4.53) \quad \mu_1(\overline{N_1N_2\ldots N_s})$$
$$= \phi(\overline{N_1N_2\ldots N_s}) - \Sigma\phi(\overline{N_1N_2\ldots N_{s-1}}) + \ldots + (-)^{s+1}\Sigma\phi(N_1).$$

From this series we readily deduce the series

$$(4.54) \quad \phi(\overline{N_1N_2\ldots N_s}) = \mu_1(\overline{N_1N_2\ldots N_s}) + \Sigma\mu_1(\overline{N_1N_2\ldots N_{s-1}}) + \ldots + \Sigma\mu_1(N_1).$$

I now return to the bipartite case and take as an example

$$N_1 = 8, \quad N_2 = 4, \quad M = 4.$$

The number of 2-fold lesser and relatively prime bipartites is

$$\mu_1(\overline{84}) = \phi(84) - \phi(8) - \phi(4)$$
$$= 8.4.(1-\tfrac{1}{4}) - 8(1-\tfrac{1}{2}) - 4(1-\tfrac{1}{2}) = 24 - 6 = 18.$$

4.55. These are set forth in the adjoined left-hand double column and the subtraction of these from $\overline{84}$ in the right-hand double column. If a bipartite to the left be $\overline{ab}$, that to the right is $\overline{N_1-a,\ N_2-b}$. If M contain a factor δ and such a factor be common to a and b it must be also common to N_1-a and N_2-b. Hence the bipartites $\overline{ab}$, $\overline{N_1-a,\ N_2-b}$ are relative primes and are a partition of $\overline{N_1N_2}$ into two parts which are relatively prime.

73	11
72	12
71	13
63	21
61	23
53	31
52	32
51	33
43	41
41	43
33	51
32	52
31	53
23	61
21	63
13	71
12	72
11	73

Hence we have before us a new definition of $\mu_1(\overline{N_1N_2})$ which as in the unipartite case leads to the following two definitions of $\mu_k(\overline{N_1N_2})$.

First definition.

4.6. "$\mu_k(\overline{N_1N_2})$ enumerates sets of k bipartite numbers, free from zero constituents,

$$\overline{N_1^{(1)}N^{(1)}}\ \overline{N_1^{(2)}N_1^{(2)}}\ \ldots\ \overline{N_1^{(k)}N_2^{(k)}},$$

the sum of which is 2-fold less than $\overline{N_1 N_2}$, and which contain no common divisor which is also a divisor of $\overline{N_1 N_2}$; and where account is taken of the permutations in which the bipartites, which make up a given combination, can be written".

Second definition.

4.61. "$\mu_k(\overline{N_1 N_2})$ enumerates partitions of $\overline{N_1 N_2}$ into $k+1$ parts which are relatively prime to one another. Account is taken of the different permutations of the bipartite parts of these partitions and zero constituents of the bipartite parts are not permitted".

In general for the s-partite numbers we have

First definition.

4.7. "$\mu_k(\overline{N_1 N_2 \dots N_s})$ enumerates sets of k s-partite numbers, free from zero constituents,

$$\overline{N_1^{(1)} N_2^{(1)} \dots N_s^{(1)}}\ \overline{N_1^{(2)} N_2^{(2)} \dots N_s^{(2)}} \dots \overline{N_1^{(s)} N_2^{(s)} \dots N_s^{(s)}},$$

the sum of which is s-fold less than $\overline{N_1 N_2 \dots N_s}$; and which contain no common divisor which is also a divisor of $\overline{N_1 N_2 \dots N_s}$; and where account is taken of the permutations in which the s-partites, which make a given combination, can be written".

Second definition.

4.71. "$\mu_k(\overline{N_1 N_2 \dots N_s})$ enumerates partitions of $\overline{N_1 N_2 \dots N_s}$ into $k+1$ parts which are relatively prime to one another. Account is taken of the different permutations of the s-partite parts of these partitions and zero constituents of the s-partite parts are not permitted".

5. *Properties of* $\mu_k(\overline{N_1 N_2 \dots N_s})$.

In order to evaluate this arithmetical function we have to enumerate the forms which are specified in one of its two definitions. It is fortunate that the second definition provides a ready means for accomplishing this. It can be established that, if we ignore the condition of relative primeness, the number of *ordered* partitions into $k+1$ parts, which are s-fold less than $\overline{N_1 N_2 \dots N_s}$ and have no zero constituents, of the multipartite number $\overline{N_1 N_2 \dots N_s}$ is

$$\binom{N_1-1}{k}\binom{N_2-1}{k}\dots\binom{N_s-1}{k}.$$

For consider the $k+1$ parts of such a partition

$$A_1B_1C_1\ldots S_1,\ A_2B_2C_2\ldots S_2,\ \ldots,\ A_{k+1}B_{k+1}C_{k+1}\ldots S_{k+1}.$$

The constituents $A_1\ A_2\ldots A_{k+1}$ constitute an ordered partition of the unipartite number N_1 into $k+1$ non-zero parts. It is one of the commonplaces of combinatory analysis, as in 3.1 above, that the enumeration of these is given by

$$\binom{N_1-1}{k}$$

We have similar enumerations for the constituents $B_1B_2\ldots B_{k+1}$, $C_1C_2\ldots C_{k+1}$, etc., ..., $S_1S_2\ldots S_{k+1}$, and combining them we proceed to the desired enumeration

$$(5.1)\qquad \binom{N_1-1}{k}\binom{N_2-1}{k}\cdots\binom{N_s-1}{k}.$$

If M, the highest common divisor of $N_1, N_2, \ldots, N_s$, be $p_1^{\pi_1}p_2^{\pi_2}\ldots$, we proceed, as usual, to the formula

$$(5.2)\quad \mu_k(\overline{N_1N_2\ldots N_s}) = \binom{N_1-1}{k}\binom{N_2-1}{k}\cdots\binom{N_s-1}{k}$$
$$-\Sigma\binom{(N_1/p_1)-1}{k}\binom{(N_2/p_1)-1}{k}\cdots\binom{(N_s/p_1)-1}{k}$$
$$+\Sigma\binom{(N_1/p_1p_2)-1}{k}\binom{(N_2/p_1p_2)-1}{k}\cdots\binom{(N_s/p_1p_2)-1}{k}$$
$$-\ldots.$$

5.3. It can be shown that

$$\Sigma\mu_k\left(\frac{\overline{N_1N_2\ldots N_s}}{d\ d\ \ldots d}\right) = \binom{N_1-1}{k}\binom{N_2-1}{k}\cdots\binom{N_s-1}{k},$$

where d ranges over every divisor of

$$M = p_1^{\pi_1}p_2^{\pi_2}\ldots p_m^{\pi_m}$$

where M is the highest common divisor of $N_1, N_2, \ldots, N_s$.

For write the formula for $\mu_k(\overline{N_1N_2\ldots N_s})$ in the manner

$$\mu_k(\overline{N_1N_2\ldots N_s}) = A_0-\Sigma A_{p_1}+\Sigma A_{p_1p_2}-\Sigma A_{p_1p_2p_3}+\ldots+(-)^s\Sigma A_{p_1p_2\ldots p_s}$$

and suppose that, in succession, we substitute for N_r $(r = 1, 2, 3, \ldots)$ N_r/d, where d runs through every divisor of M.

When d is the particular divisor

$$d' = p_1^{\phi_1} p_2^{\phi_2} \dots p_i^{\phi_i}$$

consider the occurrence of the term

$$\binom{(N_1/d')-1}{k}\binom{(N_2/d')-1}{k}\dots\binom{(N_s/d')-1}{k}$$

on the right-hand side of the identities. I say that it will be found,

in A_0, once with the positive sign when N_r/d' is substituted for N_r;

in ΣA_{p_1}, i times with the negative sign when $\frac{p_1}{d'} N_r$ or $\frac{p_2}{d'} N_r$ or ... are substituted for N_r;

in $\Sigma A_{p_1 p_2}$, $\binom{i}{2}$ times with the positive sign when $\frac{p_1 p_2}{d'} N_r$, $\frac{p_1 p_3}{d'} N_r$, ... are substituted for N_r;

... ...

and finally

in A_i, $\binom{i}{i}$ times with the sign $(-)^i$ when $\frac{p_1 p_2 \dots p_i}{d'} N_r$ is substituted for N_r.

The term in question will thus occur on the right-hand sides in the case of $1+\binom{i}{1}+\binom{i}{2}+\dots+\binom{i}{i}$ divisors of M; *i.e.* in the case of 2^i divisors of M; and since, on proceeding to the sum

$$\Sigma \mu_k \left(\overline{\frac{N_1 N_2 \dots N_s}{d\, d \dots d}}\right),$$

the coefficient of the term is

$$1-\binom{i}{1}+\binom{i}{2}-\dots+(-)^i\binom{i}{i}=0,$$

it is clear that the term in question will not occur at all in such sum.

Thus it is manifest that every term in the sum except

$$\binom{N_1-1}{k}\binom{N_2-1}{k}\dots\binom{N_s-1}{k}$$

will vanish, and we are brought to the result

$$\Sigma\mu_k\binom{N_1 N_2 \ldots N_s}{d\, d \ldots d} = \binom{N_1-1}{k}\binom{N_2-1}{k}\cdots\binom{N_s-1}{k}. \qquad \text{Q.E.D.}$$

The table of values of $\mu_k(N)$.

Since $$\mu_k(N) = \binom{N-1}{k} - \Sigma\binom{(N/p_1)-1}{k} + \ldots$$

$p_1, p_2, \ldots$ being the prime divisors of N, we see that

$$\mu_k(N) = \binom{N-1}{k} \text{ when } N \text{ is prime for all values of } k.$$

Also, if p_1 be the least prime divisor of N,

$$\mu_k(N) = \binom{N-1}{k} \quad \text{when} \quad k > \frac{N}{p_1} - 1.$$

These values, being simple binomial coefficients, are not tabulated.

The value of $\mu_k(N)$ when less than or equal to $N/p_1 - 1$ is found at the intersection of the horizontal through N and the vertical through μ_k.

Thus $\mu_4(25)$ is tabulated because $4 = \frac{25}{5} - 1$, k in this case not exceeding $\frac{25}{5} - 1$.

The results of § 3 have been freely used to verify the numbers entered.

Values of $\mu_k(N)$, when $k \leqslant (N/P_1) - 1$, P_1 being the least prime divisor of N.

N	$\mu_1(N)$	μ_2	μ_3	μ_4	μ_5	μ_6	μ_7	μ_8	μ_9	μ_{10}	μ_{11}
4	2										
6	2	9									
8	4	18	34								
9	6	27									
10	4	30	80	125							
12	4	42	154	325	461						
14	6	63	266	700	1281	1715					
15	8	84	360	1000							
16	8	84	420	1330	2982	4998	6434				
18	6	99	614	2305	6131	12348	19440	24039			
20	8	132	884	3750	11502	27048	50352	75573	92377		
21	12	174	1120	4830	15498	38759					
22	10	165	1210	5775	20097	54054	116160	203445	293920	352715	
24	8	180	1572	8490	33166	100478	244826	490149	817135	1144055	1352077
25	20	270	2020	10625							

Chapter 9
Partitions

9.1 Introduction and Commentary

Surprisingly, MacMahon's work in ordinary (i.e., linear) partitions is not extensive. His most important contributions in this area perhaps lie outside his papers. He computed the first 200 values of $p(n)$, the number of partitions of n; this table, which appears in an extremely important paper by G. H. Hardy and S. Ramanujan (1918), turned out to be crucial in the development of the theory of partitions (see section 9.3 and 9.4). In volume 2 of *Combinatory Analysis*, MacMahon devoted all of section VII, chapter 3 to two important identities that he mistakenly believed to be unproven and which have since become known as the Rogers-Ramanujan identities (see section 9.5).

After a brief taste of the work on generating functions, asymptotics, and partition congruences, we shall present an account of a recently discovered partition theorem (Andrews, 1967b) that is related both to the Rogers-Ramanujan identities and to a theorem of MacMahon from section VII, chapter 4 of *Combinatory Analysis*. For an extensive treatment of the theory of partitions the reader is referred to Andrews (1976a).

9.2 Generating Functions

The study of generating functions for partitions dates back to Euler (1748), and many fascinating results concerning partitions and partition generating functions are due to him. Euler observed that if $p(n)$ denotes the number of partitions of n, then

$$(2.1)\qquad 1+\sum_{n=1}^{\infty} p(n)\, q^n = \sum_{k_1\geqq 0,\, k_2\geqq 0,\, k_3\geqq 0,\,\ldots} q^{k_1\cdot 1+k_2\cdot 2+k_3\cdot 3+\ldots}$$

$$= \frac{1}{(1-q)(1-q^2)(1-q^3)(1-q^4)\ldots}.$$

He also discovered and proved the following important identity which has come to be known as Euler's Pentagonal Number Theorem:

$$(2.2)\qquad \prod_{n=1}^{\infty}(1-q^n) = \sum_{n=-\infty}^{\infty}(-1)^n q^{n(3n-1)/2}.$$

The material in this chapter corresponds to section VII of *Combinatory Analysis*.

Combining (2.1) and (2.2), we may easily deduce the following recurrence for the partition function $p(n)$:

$$(2.3)\quad p(n) - p(n-1) - p(n-2) + p(n-5) + p(n-7) - \ldots + (-1)^r p(n - r(3r-1)/2) + (-1)^r p(n - r(3r+1)/2) + \ldots = 0, \qquad n > 0.$$

This formula for $p(n)$ was utilized by MacMahon in a very important computation that will be described in Section 9.3.

Table 9.1 illustrates some of the generating functions related to the simpler partition functions. Each formula is established in precisely the manner that (2.1) is. Generating functions play an important role throughout the study of partitions as will become obvious in the next three sections.

Table 9.1

Partition Function	Partitions Enumerated	Generating Function
$p(n)$	all ordinary partitions of n	$\prod_{j=1}^{\infty}(1-q^j)^{-1}$
$D(n)$	all ordinary partitions of n with distinct parts	$\prod_{j=1}^{\infty}(1+q^j)$
$q(n)$	all ordinary partitions of n with odd parts	$\prod_{j=1}^{\infty}(1-q^{2j-1})^{-1}$

9.3 Partition Asymptotics

MacMahon played an important role in one of the most exciting chapters in the entire theory of partitions. G. H. Hardy and S. Ramanujan (1918) had obtained in a joint work (described by J. E. Littlewood, 1953, page 90, as "... a singularly happy collaboration of two men, of quite unlike gifts, in which each contributed the best, most characteristic, and most fortunate work that was in him.") an asymptotic formula for $p(n)$ through an extremely subtle use of the fact that $f(x) = 1 + \sum_{n=1}^{\infty} p(n) x^n$ is substantially the reciprocal of the modular form $\eta(\tau)$:

$$\begin{aligned}\eta(\tau) &= \exp(\pi i\tau/12)\prod_{n=1}^{\infty}(1 - \exp(2\pi i n\tau))\\ &= \exp(\pi i\tau/12)/f(\exp(2\pi i\tau)).\end{aligned}$$

Hardy and Ramanujan were at first able to show that

$$p(n) = L_1(n)\phi_1(n) + L_2(n)\phi_2(n) + \ldots + L_Q(n)\phi_Q(n) + R(n),$$

where

$$L_k(n) = \sum_{\substack{p=1\\(p,k)=1}}^{k} w_{p,k}\exp(-2np\pi i/k),$$

$$\phi_k(n) = \frac{k^{1/2}}{2\pi\sqrt{2}}\frac{d}{dn}(\lambda_n^{-1}\exp(K\lambda_n/k)),$$

$$K = \pi\sqrt{\tfrac{2}{3}}, \qquad \lambda_n = \sqrt{n - 1/24},$$

$w_{p,k}$ is a certain 24th root of unity, and

$$R(n) = o(\exp(K\sqrt{n}/Q)).$$

We now quote Hardy (1940, page 119) (reprinted with the permission of the Cambridge University Press):

At this point we might have stopped had it not been for Major MacMahon's love of calculation. MacMahon was a practised and enthusiastic computer, and made us a table of $p(n)$ up to $n = 200$. In particular he found that

$$p(200) = 3972999029388,$$

and we naturally took this value as a test for our asymptotic formula. We expected a good result, with an error of perhaps one or two figures, but we had never dared to hope for such a result as we found. Actually 8 terms of our formula gave $p(200)$ with an error of 0.004. We were inevitably led to ask whether the formula could not be used to calculate $p(n)$ *exactly* for any large n.

It is plain that if this is possible, it will be necessary to use a "large" number of terms of the series, that is to say to make Q a function of n. Our final result was as follows. There are constants α, M such that

(3.1) $$p(n) = \sum_{q<\alpha n^{1/2}} P_q(n) + R(n)$$

where

$$|R(n)| < Mn^{-1/4}, \qquad [P_q(n) = L_q(n)\,\phi_q(n)]$$

and, since $p(n)$ is an integer, (3.1) will give its value exactly for sufficiently large n. The formula is one of the rare formulae which are both asymptotic and exact; it tells us all we want to know about the order and approximate form of $p(n)$, and it appears also to be adapted for exact calculation. It was in fact from this formula that D. H. Lehmer first calculated the value of $p(721)$.

As we mentioned earlier, MacMahon produced his table using formula (2.3); a further unexpected by-product of this computation is described in the next section. The work of Hardy and Ramanujan (1918) proved to be the breakthrough that led to widespread activity in the asymptotic theory of partitions. Perhaps the next most important result was due to H. Rademacher who showed that

$$p(n) = \sum_{k=1}^{\infty} L_k(n)\,\psi_k(n),$$

where

$$\psi_k(n) = \frac{k^{1/2}}{\pi\sqrt{2}}\,\frac{d}{dn}\left(\lambda_n^{-1}\sinh(K\lambda_n/k)\right).$$

Mention should also be made of A. E. Ingham's (1941) important paper in which very general asymptotic formulas are derived. Rademacher (1973) presents a recent account of partition asymptotics.

9.4 Congruence Properties of Partition Functions

The table constructed by MacMahon for the first 200 values of $p(n)$ was carefully studied by S. Ramanujan (see Paper 25, page 210, of Ramanujan, 1927), and he presented the following brief account of his observations:

A recent paper by Mr Hardy and myself contains a table, calculated by Major MacMahon, of the values of $p(n)$, the number of unrestricted partitions of n, for all values of n from 1 to 200. On studying the numbers in this table I observed a number of curious congruence properties, apparently satisfied by $p(n)$. Thus

$$\begin{array}{lllll}
(1)\ p(4), & p(9), & p(14), & p(19), & \ldots \equiv 0 \pmod{5},\\
(2)\ p(5), & p(12), & p(19), & p(26), & \ldots \equiv 0 \pmod{7},\\
(3)\ p(6), & p(17), & p(28), & p(39), & \ldots \equiv 0 \pmod{11},
\end{array}$$

$$
\begin{array}{lllll}
(4)\ p(24), & p(49), & p(74), & p(99), & \ldots \equiv 0 \pmod{25}, \\
(5)\ p(19), & p(54), & p(89), & p(124), & \ldots \equiv 0 \pmod{35}, \\
(6)\ p(47), & p(96), & p(145), & p(194), & \ldots \equiv 0 \pmod{49}, \\
(7)\ p(39), & p(94), & p(149), & \ldots & \equiv 0 \pmod{55}, \\
(8)\ p(61), & p(138), & \ldots & & \equiv 0 \pmod{77}, \\
(9)\ p(116), & \ldots & & & \equiv 0 \pmod{121}, \\
(10)\ p(99), & \ldots & & & \equiv 0 \pmod{125}.
\end{array}
$$

From these data I conjectured the truth of the following theorem:
If $\delta = 5^a 7^b 11^c$ *and* $24\lambda \equiv 1 \pmod{\delta}$, *then*

$$p(\lambda), \quad p(\lambda + \delta), \quad p(\lambda + 2\delta), \quad \ldots \equiv 0 \pmod{\delta}.$$

This theorem is supported by all the available evidence; but I have not yet been able to find a general proof.

A survey of subsequent work on Ramanujan's conjecture is given by Hardy (1940, pages 87–90); much of the following material is derived from this source.

S. Chowla noted from extended tables for $p(n)$ due to H. Gupta that

$$p(243) \not\equiv 0 \pmod{7^3},$$

but

$$24 \cdot 243 \equiv 1 \pmod{7^3}.$$

Thus some modification of Ramanujan's conjecture must be made:

If $\quad 24\lambda \equiv 1 \pmod{5^a 7^b 11^c}$,

then $\quad p(\lambda) \equiv 0 \pmod{5^a 7^{[(b+2)/2]} 11^c}$.

A number of special cases of the corrected conjecture were proved by Darling, Ramanujan, Mordell, and others. In 1938 G. N. Watson was able to prove the conjecture relative to all powers of 5 and 7 (i.e., $c = 0$). The powers of 11 seemed especially unyielding. The case $c = 1, 2$ was treated by Ramanujan, while $c = 3$ was obtained by J. Lehner in 1950. It was only in 1967 that A. O. L. Atkin managed to prove the complete conjecture for all $5^a 7^b 11^c$.

MacMahon did treat the question of when $p(n)$ is even (see [97] and [118]). This problem is extremely difficult as may be well appreciated from the paper by Parkin and Shanks (1967), and MacMahon made a valuable contribution to the computational aspects of the problem. Parkin and Shanks (1967) named the absolute constant m, written in binary as

$$m = 1.10111110000111011101\ldots,$$

for MacMahon (the kth bit to the right of the binary point is 0 or 1 according as $p(k)$ is even or odd). In decimal notation, $m = 1.74264258\ldots$.

D. B. Lahiri (1949) and A. O. L. Atkin (1969) have found other congruences not covered by Ramanujan's conjecture. For example Atkin (1969) proved that

$$p(206839n + 2623) \equiv 0 \pmod{17}.$$

Finally we remark that congruence properties for partition functions other than $p(n)$ have been found by O. Rødseth (1970), H. Gupta (1971, 1972), D. B. Lahiri (1969, 1970a, 1970b), G. E. Andrews (1971c), G. Dirdal (1975a, 1975b, 1976), and M. D. Hirschhorn and J. H. Loxton (1975).

9.5 Identities for Partition Functions

Although MacMahon did not contribute extensively to this aspect of partition theory, he did recognize the importance of it, and most of section VII in Combinatory Analysis, volume 2, is devoted to material of this nature. In chapter 3 of that section, MacMahon states the Rogers-Ramanujan identities:

(5.1)

$$1 + \sum_{n=1}^{\infty} \frac{q^{n^2}}{(1-q)(1-q^2)\ldots(1-q^n)} = \prod_{n=0}^{\infty} \frac{1}{(1-q^{5n+1})(1-q^{5n+4})},$$

(5.2)

$$1 + \sum_{n=1}^{\infty} \frac{q^{n^2+n}}{(1-q)(1-q^2)\ldots(1-q^n)} = \prod_{n=0}^{\infty} \frac{1}{(1-q^{5n+2})(1-q^{5n+3})}.$$

He deduces directly from these analytic identities, the following partition identities:

Theorem 1. The partitions of n into parts that involve neither repetitions nor consecutive integers are equinumerous with the partitions of n into parts congruent to 1 or 4 modulo 5.

Theorem 2. The partitions of n into parts that involve neither ones, nor repetitions, nor consecutive integers are equinumerous with the partitions of n into parts congruent to 2 or 3 modulo 5.

MacMahon states these results as unproved conjectures of Ramanujan. Surprisingly L. J. Rogers (1894) had proved (5.1) and (5.2) 20 years earlier, and they lay forgotten until S. Ramanujan empirically rediscovered them. G. H. Hardy (1940) presents a complete account of the history of the Rogers-Ramanujan identities prior to 1940.

A significant effort during the last 50 years has been devoted to the development of a coherent theory that would place the Rogers-Ramanujan identities in a general setting. Reasonable surveys of the accomplishments to date may be found in Alder (1969), Cheema (1969), Gupta (1970b), and Andrews (1970, 1972a, 1974a, 1974b, 1976a).

In section VII, chapter 4 of *Combinatory Analysis*, MacMahon proves (although he does not explicitly state) a partition identity very much like Theorem 2 above:

Theorem 3. The partitions of n into parts that involve neither ones nor consecutive integers are equinumerous with the partitions of n into parts congruent to 0, 2, 3, or 4 modulo 6.

To give a sample of recent work on partition identities, we shall prove a theorem (Theorem 4) which in some vague sense is intermediate between Theorem 2 and Theorem 3. Actually, we shall prove a general theorem (Theorem 5) about a family of partition identities; Theorem 4 will be a special case of Theorem 5.

Theorem 4. The partitions of n into parts that involve neither ones, nor consecutive integers, nor repeated repetitions (i.e., no integer appears 3 or more times as a part) are equinumerous with the partitions of n into parts congruent to 2, 3, or 4 modulo 6.

Theorem 5. Let $0 \leqq a < k$ be integers. Let $B_{k,a}(N)$ denote the number of partitions of N such that, for any $i > 0$ and $j \geqq 0$, if i appears $2j - 1$ or $2j$ times as a part, then $i + 1$ appears at most $2k - 2j$ times as a part and, furthermore, 1 appears at most $2a$ times as a part. Let $A_{k,a}(N)$ denote the number of partitions of N into parts not congruent to 0, $\pm(2a + 1)$ (modulo $4k + 2$). Then for all N,

$$A_{k,a}(N) = B_{k,a}(N).$$

Remark. Theorem 5 reduces to Theorem 4 in the special case $k = 1$, $a = 0$.

Proof. We introduce an auxiliary function studied by both L. J. Rogers (see Rogers and Ramanujan, 1919) and A. Selberg (1936):

$$C_{k,a}(x;q) = \sum_{n=0}^{\infty} (-1)^n x^{kn} q^{(2k+1)n(n+1)/2-an}(1 - x^a q^{(2n+1)a}) \times \frac{(1 - xq)(1 - xq^2)\ldots(1 - xq^n)}{(1 - q)(1 - q^2)\ldots(1 - q^n)}$$

We see by inspection of the above series that

$$C_{k,0}(x;q) \equiv 0, \tag{5.3}$$

$$C_{k,-a}(x;q) = -x^{-a}q^{-a}C_{k,a}(x;q). \tag{5.4}$$

Furthermore,

$$\begin{aligned}
(5.5)\quad & C_{k,a}(x;q) - C_{k,a-1}(x;q) \\
&= \sum_{n=0}^{\infty} (-1)^n x^{kn} q^{(2k+1)n(n+1)/2} \\
&\quad \times (q^{-an} - x^a q^{(n+1)a} - q^{(-a+1)n} + x^{a-1} q^{(n+1)(a-1)}) \\
&\quad \times \frac{(1 - xq)(1 - xq^2)\ldots(1 - xq^n)}{(1 - q)(1 - q^2)\ldots(1 - q^n)} \\
&= \sum_{n=0}^{\infty} (-1)^n x^{kn} q^{(2k+1)n(n+1)/2} \\
&\quad \times (q^{-an}(1 - q^n) + x^{a-1} q^{(n+1)(a-1)}(1 - xq^{n+1})) \\
&\quad \times \frac{(1 - xq)(1 - xq^2)\ldots(1 - xq^n)}{(1 - q)(1 - q^2)\ldots(1 - q^n)} \\
&= \sum_{n=1}^{\infty} (-1)^n x^{kn} q^{(2k+1)n(n+1)/2-an} \\
&\quad \times \frac{(1 - xq)(1 - xq^2)\ldots(1 - xq^n)}{(1 - q)(1 - q^2)\ldots(1 - q^{n-1})} \\
&\quad + (xq)^{a-1} \sum_{n=0}^{\infty} (-1)^n x^{kn} q^{(2k+1)n(n+1)/2+n(a-1)} \\
&\qquad \times \frac{(1 - xq)(1 - xq^2)\ldots(1 - xq^{n+1})}{(1 - q)(1 - q^2)\ldots(1 - q^n)}
\end{aligned}$$

$$= -x^k q^{2k+1-a} \sum_{n=0}^{\infty} (-1)^n x^{kn} q^{(2k+1)n(n+1)/2+(2k+1-a)n}$$
$$\times \frac{(1-xq)(1-xq^2)\dots(1-xq^{n+1})}{(1-q)(1-q^2)\dots(1-q^n)}$$
$$+ (xq)^{a-1} \sum_{n=0}^{\infty} (-1)^n x^{kn} q^{(2k+1)n(n+1)/2+n(a-1)}$$
$$\times \frac{(1-xq)(1-xq^2)\dots(1-xq^{n+1})}{(1-q)(1-q^2)\dots(1-q^n)}$$
$$= (xq)^{a-1} \sum_{n=0}^{\infty} (-1)^n (xq)^{kn} q^{(2k+1)n(n+1)/2-(k-a+1)n}$$
$$\times (1 - (xq)^{k-a+1} q^{(2n+1)(k-a+1)})$$
$$\times \frac{(1-xq)(1-xq^2)\dots(1-xq^{n+1})}{(1-q)(1-q^2)\dots(1-q^n)}$$
$$= (xq)^{a-1}(1-xq)\,C_{k,k-a+1}(xq;q).$$

Let us now define

$$R_{k,a}(x;q) = C_{k,a+1/2}(x^2;q^2)\prod_{j=1}^{\infty}(1-xq^j)^{-1}. \tag{5.6}$$

We see immediately that for $0 < a \leqq k$, we have, by (5.5),

$$R_{k,a}(x;q) - R_{k,a-1}(x;q) = (xq)^{2a-1}(1+xq)\,R_{k,k-a}(xq;q); \tag{5.7}$$

while (5.4) and (5.5) together imply that

$$R_{k,0}(x;q) = R_{k,k}(xq;q). \tag{5.8}$$

We consider a double series expansion of $R_{k,a}(x;q)$ valid for $|x| < q^{-1}$, $|q| < 1$:

$$R_{k,a}(x;q) = \sum_{M,\,N=-\infty}^{\infty} r_{k,a}(M,N)\,x^M q^N.$$

From (5.6), (5.7), and (5.8) we may now deduce equivalent identities for the $r_{k,a}(M,N)$:

$$r_{k,a}(M, N) = \begin{cases} 1 & \text{if } M = N = 0, \\ 0 & \text{if either } M \leqq 0 \text{ or } N \leqq 0 \text{ but } (M, N) \neq (0,0), \end{cases} \tag{5.9}$$

$$r_{k,0}(M, N) = r_{k,k}(M, N - M), \tag{5.10}$$

$$r_{k,a}(M, N) - r_{k,a-1}(M, N) = r_{k,k-a}(M - 2a + 1, N - M) + r_{k,k-a}(M - 2a, N - M), \quad 0 < a \leqq k. \tag{5.11}$$

It is easy to prove by mathematical induction that equations (5.9), (5.10), and (5.11) uniquely determine the $r_{k,a}(M, N)$.

Let $b_{k,a}(M, N)$ denote the number of those partitions enumerated by $B_{k,a}(N)$ that have exactly M parts. Our next object is to show that the $b_{k,a}(M, N)$ also satisfy (5.9), (5.10), and (5.11).

We verify (5.9) by noting that the only possible partition of a nonpositive number is the empty partition of zero, and the only partition with a nonpositive number of parts is also the empty partition of zero.

Let us now verify (5.10). We examine the partitions enumerated by $b_{k,0}(M, N)$: Since 1 does not appear as a part, each summand is larger than 1 in each partition considered. We now transform the set of partitions under consideration by subtracting 1 from each summand of each partition. This transformation reduces the number being partitioned from N to $N - M$; however, the number of parts is still M in each instance. Since 1 originally appeared 0 times, we see that 2 originally appeared at most $2k$ times, and therefore after the transformation, 1 appears at most $2k$ times. All the other conditions on the partitions under consideration are unaltered; hence, since our transformation is clearly reversible, we see that we have a bijection between the partitions enumerated by $b_{k,0}(M, N)$, and those enumerated by $b_{k,k}(M, N - M)$. Consequently, (5.10) is established.

Finally we treat (5.11). We first note that $b_{k,a}(M, N) - b_{k,a-1}(M, N)$ enumerates those partitions enumerated by $b_{k,a}(M, N)$ in which 1 appears either $2a - 1$ or $2a$ times. Case (i): 1 appears $2a - 1$ times. In this case, 2 appears at most $2k - 2a$ times. We transform the partitions under consideration by deleting all the $2a - 1$ ones and subtracting 1 from each of the remaining parts. The same reasoning used in the preceding paragraph shows that this transformation establishes a bijection between the partitions initially considered in the present case, and the partitions enumerated by $b_{k,k-a}(M - 2a + 1, N - M)$. Case (ii): 1 appears $2a$ times. In this case, 2 also appears at most $2k - 2a$ times. We now transform our partitions by deleting all the $2a$ ones and subtracting 1 from each of the remaining parts.

This transformation establishes a bijection between the partitions under consideration and those enumerated by $b_{k,k-a}(M - 2a, N - M)$.

Combining the results of Case (i) and Case (ii), we see that (5.11) is established. Therefore, by the remark following (5.11), we obtain

$$r_{k,a}(M, N) = b_{k,a}(M, N). \tag{5.12}$$

Hence for $0 \leqq a < k$,

$$\begin{aligned}\sum_{N=0}^{\infty} A_{k,a}(N) q^N &= \prod_{\substack{n=1 \\ n \not\equiv 0,\ \pm(2a+1) \pmod{4k+2}}}^{\infty} (1 - q^n)^{-1} \\ &= R_{k,a}(1; q) \\ &= \sum_{N=-\infty}^{\infty} \sum_{M=-\infty}^{\infty} b_{k,a}(M, N) q^N \\ &= \sum_{N=0}^{\infty} B_{k,a}(N) q^N,\end{aligned} \tag{5.13}$$

where the second equation follows from Jacobi's identity (see G. H. Hardy and E. M. Wright, 1960, page 283, or Andrews, 1965). Comparing coefficients in the extremes of (5.13), we see that $A_{k,a}(N) = B_{k,a}(N)$ for all N, and this is Theorem 5.

9.6 References

M. I. Aissen (1970a) Some identities involving partitions, *Ann. New York Acad. Sci.*, *175*, 7–22.

M. I. Aissen (1970b) Patterns, partitions, and polynomials, Trans. *New York Acad. Sci.*, *32*, 535–544.

H. L. Alder (1969) Partition identities—from Euler to the present, *American Math. Monthly*, *76*, 733–746.

G. E. Andrews (1965) A simple proof of Jacobi's triple product identity, *Proc. A.M.S.*, *16*, 333–334.

G. E. Andrews (1967a) A generalization of a partition theorem of MacMahon, *J. Combinatorial Th.*, *2*, 100–101.

G. E. Andrews (1967b) Some new partition theorems, *J. Combinatorial Th.*, *2*, 431–436.

G. E. Andrews (1969a) Some new partition theorems II, *J. Combinatorial Th.*, *7*, 262–263.

G. E. Andrews (1969b) A generalization of the classical partition theorems, *Trans. A.M.S.*, *145*, 205–221.

G. E. Andrews (1970) A polynomial identity which implies the Rogers-Ramanujan identities, *Scripta Math.*, *23*, 297–305.

G. E. Andrews (1971a) Number Theory, W. B. Saunders, Philadelphia.

G. E. Andrews (1971b) On the foundations of combinatorial theory V, Eulerian differential operators, *Studies in Appl. Math.*, *50*, 345–375.

G. E. Andrews (1971c) Congruence properties of the *m*-ary partition function, *J. Number Th.*, *3*, 104–110.

G. E. Andrews (1972a) Partition identities, *Advances in Math.*, *9*, 10–51.

G. E. Andrews (1972b) Sieves for theorems of Euler, Rogers, and Ramanujan, in Arithmetic Functions, Lecture Notes in Math. No. 251, Springer, New York.

G. E. Andrews (1972c) Sieves in the theory of partitions, *American J. Math.*, *94*, 1214, 1230.

G. E. Andrews (1974a) On the general Rogers-Ramanujan theorem, *Memoirs of the A.M.S.*, No. 152.

G. E. Andrews (1974b) A general theory of identities of the Rogers-Ramanujan type, *Bull. A.M.S.*, *80*, 1033–1052.

G. E. Andrews (1974c) An analytic generalization of the Rogers-Ramanujan identities for odd moduli, *Proc. Nat. Acad. Sci.*, *71*, 4082–4085.

G. E. Andrews (1974d) Applications of basic hypergeometric functions, *S.I.A.M. Review*, *16*, 441–484.

G. E. Andrews (1975a) Partially ordered sets and the Rogers-Ramanujan identities, *Aequationes Math.*, *12*, 94–107.

G. E. Andrews (1975b) On Rogers-Ramanujan type identities related to the modulus 11, *Proc. London Math. Soc.*, *30*, 330–346.

G. E. Andrews (1975c) On the Alder polynomials and a new generalization of the Rogers-Ramanujan identities, *Trans. A.M.S.*, *294*, 40–64.

G. E. Andrews (1976a) The Theory of Partitions, Encyclopedia of Mathematics and Its Applications, vol. 2, Addison-Wesley, Reading.

G. E. Andrews (1976b) On identities implying the Rogers-Ramanujan identities, *Houston J. Math.*, *2*, 289–298.

J. Arkin (1970) Researches on partitions, *Duke Math. J.*, *38*, 403–409.

J. Arkin (1971) Researches on some classical problems, Proc. Washington State Univ. Conf. on Number Theory, 130–147.

A. O. L. Atkin (1967) Proof of a conjecture of Ramanujan, *Glasgow Math. J.*, *8*, 14–32.

A. O. L. Atkin (1969) Congruence Hecke operators, *Proc. Symp. in Pure Math.*, *12*, 33–40.

A. O. L. Atkin and B. J. Birch (1971) Computers in Number Theory, Academic Press, London.

F. C. Auluck (1942) An asymptotic formula for $p_k(n)$, *J. Indian Math. Soc.(N.S.)*, *6*, 113–114.

L. Carlitz (1953) A note on partitions in $GF[q, x]$, *Proc. A.M.S.*, *4*, 464–469.

L. Carlitz (1965a) Generating functions and partition problems, *Proc. Symp. in Pure Math.*, *8*, 144–169.

L. Carlitz (1965b) Weighted two-line arrays, *Duke Math. J.*, *32*, 721–740.

L. Carlitz (1969) Particije, *Math. Biblioteka*, *39*, 5–14.

L. Carlitz and J. Riordan (1965) Enumeration of certain two-line arrays, *Duke Math. J.*, *32*, 529–540.

J. A. Carpenter and V. R. R. Uppuluri (1974) A problem of restricted partitions, *Naval Res. Logist. Quart.*, *21*, 201–205.

M. S. Cheema (1964) Vector partitions and combinatorial identities, *Math. Comp.*, *18*, 414–420.

M. S. Cheema (1969) Duality in the theory of partitions, *Res. Bull.(N.S.) Panjab University*, *20*, 201–206.

M. S. Cheema (1971) Computers in the theory of partitions, from Computers in Number Theory, edited by A. O. L. Atkin and B. J. Birch, Academic Press, London.

M. S. Cheema (1973) On more restricted partitions, *Rocky Mountain J. Math.*, *3*, 31–34.

M. P. Chen and C. S. Yu (1974) A combinatorial proof of a result of Andrews, *Tamkang J. Math.*, *5*, 261–264.

R. F. Churchhouse (1969) Congruence properties of the binary partition function, *Proc. Cambridge Phil. Soc.*, *66*, 371–376.

W. G. Connor (1975) Partition theorems related to some identities of Rogers and Watson, Trans. A.M.S., *214*, 95–111.

G. Dirdal (1975a) On restricted *m*-ary partitions, *Math. Scand.*, *37*, 51–60.

G. Dirdal (1975b) Congruences for *m*-ary partitions, *Math. Scand.*, *37*, 76–82.

G. Dirdal (1976) Congruence properties for a class of arithmetical functions, *Math. Scand.*, *38*, 247–261.

P. Erdös (1942) On an elementary proof of some asymptotic formulas in the theory of partitions, *Ann. Math.(2)*, *43*, 437–450.

L. Euler (1748) Introductio in analysin infinitorum, vol. 1, Marcum-Michaelem Bousquet, Lausannae, chapter 16.

D. C. Fielder (1964a) Partition enumeration by means of simpler partitions, *Fibonacci Quart.*, *2*, 115–118.

D. C. Fielder (1964b) Properties of certain partitions of numbers and applications to counting of switching tree distributions, presented at I.C.M.I., Tokyo.

D. C. Fielder (1964c) On Shannon's almost uniform distribution, *I.E.E.E. Trans. on Electronic Computers*, *EC-13*, No. 1, 53–54.

D. C. Fielder (1966) Enumeration of partitions subject to limitations on size of members, *Fibonacci Quart.*, *5*, 319–324.

D. C. Fielder (1967) Certain Lucas-like sequences and their generation by partitions of numbers, *Fibonacci Quart.*, *5*, 319–324.

D. C. Fielder (1968a) Partitions of numbers for finding certain graphs, *I.E.E.E. Trans. on Education*, *E-11*, No. 2, 134–136.

D. C. Fielder (1968b) Generation of Stirling numbers by means of special partitions of numbers, *Fibonacci Quart.*, *6*, 1–10.

N. J. Fine (1948) Some new results on partitions, *Proc. Nat. Acad. Sci.*, *34*, 616–618.

N. J. Fine (1956) On a system of modular functions connected with the Ramanujan identities, *Tohuku Math. J.(2)*, *8*, 149–164.

N. J. Fine (1975) Some Basic Hypergeometric Series and Applications, unpublished monograph.

H. Göllnitz (1967) Partitionen mit Differenzenbedingungen, *J. reine und angew. Math.*, *225*, 154–190.

B. Gordon (1961) A combinatorial generalization of the Rogers-Ramanujan identities, *American J. Math.*, *83*, 393–399.

B. Gordon (1965) Some continued fractions of the Rogers-Ramanujan type, *Duke Math. J.*, *31*, 741–748.

E. Grosswald (1962) Results, new and old, in the theory of partitions (Spanish), *Rev. Un. Argentina*, *20*, 40–57.

H. Gupta (1936) Minimum partitions into specified parts, *American J. Math.*, *58*, 573–576.

H. Gupta (1964) On the coefficients of the powers of Dedekind's modular form, *J. London Math.* Soc., *39*, 433–440.

H. Gupta (1969) Highly restricted partitions, *J. Res. Nat. Bureau Standards(B)*, *Math. Sci.*, *73B*, 329–350.

H. Gupta (1970a) Products of parts in partitions into primes, *Res. Bull.(N.S.) Panjab Univ.*, *21*, 251–253.

H. Gupta (1970b) Partitions—a survey, *J. Res. Nat. Bureau Standards(B)*, *Math. Sci.*, *74B*, 1–29.

H. Gupta (1971a) Proof of the Churchhouse conjecture concerning binary partitions, *Proc. Cambridge Phil. Soc.*, *70*, 53–56.

H. Gupta (1971b) On partitions of n into k summands, *Proc. Edinburgh Math. Soc.*, *17*, 337–339.

H. Gupta (1971c) On Sylvester's theorem in partitions, *Indian J. Pure and Appl. Math.*, *2*, 740–748.

H. Gupta (1971d) Partial fractions in partition theory, *Res. Bull.(N.S.) Panjab Univ.*, *22*, 23–25.

H. Gupta (1972) On m-ary partitions, *Proc. Cambridge Phil. Soc.*, *71*, 343–345.

H. Gupta (1974) A partition theorem of Subbarao, *Canadian Math. Bull.* *17*, 121–123.

H. Gupta (1975) A technique in partitions, *Univ. Beograd. Publ. Elektrotehn. Fak. (Ser. Mat. Fiz.)*, No. 498–541, 73–76.

H. Gupta (1976a) A direct proof of the Churchhouse conjecture concerning binary partitions, *Indian J. Math.*, *18*, 1–6.

H. Gupta (1976b) Combinatorial proof of a theorem on partitions into an even or odd number of parts, *J. Combinatorial Th.*, *A-21*, 100–103.

H. Gupta (1976c) Partitions embedded in a rectangle, *Utilitas Math.*, *10*, 229–240.

R. K. Guy (1958) Two theorems on partitions, *Math. Gaz.*, *42*, 84–86.

M. Hall (1958) Survey of combinatorial analysis, in Some Aspects of Analysis and Probability IV, J. Wiley, New York.

G. H. Hardy (1940) Ramanujan, Cambridge University Press, Cambridge (Reprinted: Chelsea, New York).

G. H. Hardy and S. Ramanujan (1918) Asymptotic formulae in combinatory analysis, *Proc. London Math. Soc.(2)*, *17*, 75–115.

G. H. Hardy and E. M. Wright (1960) An Introduction to the Theory of Numbers, 4th ed., Oxford University Press, Oxford.

D. R. Hickerson (1973) Identities relating the number of partitions into an even and odd number of parts, *J. Combinatorial Th.(A)*, *A-15*, 351–353.

D. R. Hickerson (1974) A partition identity of the Euler type, *American Math. Monthly*, *81*, 627–629.

M. D. Hirschhorn and J. H. Loxton (1975) Congruence properties of the binary partition function, *Math. Proc. Cambridge Phil. Soc.*, *78*, 437–442.

A. E. Ingham (1941) A Tauberian theorem for partitions, *Ann. Math.*, *42*, 1075–1090.

M. I. Knopp (1970) Modular Functions in Analytic Number Theory, Markham, Chicago.

D. E. Knuth (1971) Subspaces, subsets, and partitions, *J. Combinatorial Th.(A)*, *A-10*, 178–180.

D. Knutson (1972) A lemma on partitions, *American Math. Monthly*, *79*, 1111–1112.

D. B. Lahiri (1948) Some non-Ramanujan congruence properties of the partition function, *Proc. Nat. Inst. Sci. India*, *14*, 337–338.

D. B. Lahiri (1949) Further non-Ramanujan congruence properties of the partition function, *Science and Culture*, *14*, 336–337.

D. B. Lahiri (1969) Some restricted partition functions: congruences modulo 7, *Trans. A.M.S.*, *140*, 475–484.

D. B. Lahiri (1970a) Some restricted partition functions: congruences modulo 2, *Trans. A.M.S.*, *147*, 271–278.

D. B. Lahiri (1970b) Some restricted partition functions: congruences modulo 3, *J. Australian Math. Soc.*, *11*, 82–90.

J. Lehner (1950) Proof of Ramanujan's partition congruence for the modulus 11^3, *Proc. A.M.S.*, *1*, 172–181.

N. Metropolis and P. R. Stein (1970) An elementary solution to a problem in restricted partitions, *J. Combinatorial Th.*, *9*, 365–376.

E. Moore (1974a) Generalized Euler-type partition identities, *J. Combinatorial Th.(A)*, *A-17*, 78–83.

E. Moore (1974b) Partitions with parts appearing a specified number of times, *Proc. A.M.S.*, *46*, 205–210.

E. Netto (1901) Lehrbuch der Kombinatorik, Reprinted: Chelsea, New York.

A. Oppenheim (1926a) On an arithmetic function, *J. London Math. Soc.*, *1*, 205–211.

A. Oppenheim (1926b) On an arithmetic function II, *J. London Math. Soc.*, *2*, 123–130.

T. R. Parkin and D. Shanks (1967) On the distribution of parity in the partition function, *Math. Comp.*, *21*, 466–480.

J. K. Percus (1971) Combinatorial Methods, Springer, New York.

G. Polya (1969) On the number of certain lattice polygons, *J. Combinatorial Th.*, *6*, 102–105.

H. Rademacher (1973) Topics in Analytic Number Theory, Die Grundlehren der mathematischen Wissenschaften, Band 169, Springer, Berlin.

S. Ramanujan (1927) Collected Papers of S. Ramanujan, Cambridge University Press, Cambridge, Reprinted: Chelsea, New York.

S. Ramanujan and L. J. Rogers (1919) Proof of certain identities in combinatory analysis, *Proc. Cambridge Phil. Soc.*, *19*, 211–216.

L. B. Richmond (1972) On a conjecture of Andrews, *Utilitas Math.*, *2*, 3–8.

J. Riordan (1958) An Introduction to Combinatorial Analysis, J. Wiley, New York.

J. Riordan (1968) Combinatorial Identities, J. Wiley, New York.

M. M. Robertson (1976) Partitions with congruence conditions, *Proc. A.M.S.*, *57*, 45–49.

O. Rødseth (1970) Some arithmetical properties of *m*-ary partitions, *Proc. Cambridge Phil. Soc.*, *68*, 447–453.

L. J. Rogers (1894) Second memoir on the expansion of certain infinite products, *Proc. London Math. Soc.*, *25*, 318–343.

G.-C. Rota and J. Goldman (1970) On the foundations of combinatorial theory: IV Finite vector spaces and Eulerian generating functions, *Studies in Appl. Math.*, *49*, 239–258.

K. F. Roth and G. Szekeres (1954) Some asymptotic formulae in the theory of partitions, *Quart. J. Math.(2)*, *5*, 241–259.

L. von Schrutka (1916) Zur Systematik der additiven Zahlentheorie, *J. reine und angew. Math.*, *146*, 245–254.

L. von Schrutka (1917) Zur additiven Zahlentheorie, *Sitz. der Kaiserl. Akad. der Wiss. in Wien Abt. IIa*, *125*, 1081–1163.

W. Schwarz (1968) Schwache asymptotische Eigenschaften von Partitionen, *J. reine und angew. Math.*, *232*, 1–16.

W. Schwarz (1969) Asymptotische Formeln für Partitionen, *J. reine und angew. Math.*, *234*, 172–178.

A. Selberg (1936) Über einige arithmetische Identitäten, Avhandlinger Norske Akademie, No. 8.

L. J. Slater (1966) Generalized Hypergeometric Functions, Cambridge University Press, Cambridge.

R. P. Stanley (1972) Ordered structures and partitions, *Memoirs of the A.M.S.*, No. 119.

K. B. Stolarsky (1968) Higher partition functions and their relation to finitely generated nilpotent groups, Ph.D. Thesis, University of Wisconsin, Madison.

K. B. Stolarsky (1969) A vanishing sum associated with Jacobi's triple product identity, *J. Combinatorial Th.*, *6*, 392–398.

K. B. Stolarsky (1971) Generalization of the Bellavitis partition identities, *J. Number Theory*, *3*, 240–246.

M. V. Subbarao (1971a) Combinatorial proofs of some identities, Proc. Washington State Univ. Conf. on Number Theory.

M. V. Subbarao (1971b) On a partition theorem of MacMahon-Andrews, *Proc. A.M.S.*, *27*, 449–450.

M. V. Subbarao (1971c) Partition theorems for Euler pairs, *Proc. A.M.S.*, *28*, 330–336.

G. Szekeres (1951) An asymptotic formula in the theory of partitions, *Quart. J. Math.(2)*, *2*, 85–108.

G. Szekeres (1953) Some asymptotic formulae in the theory of partitions (II), *Quart. J. Math.(2)*, *4*, 96–111.

G. N. Watson (1938) Ramanujans Vermutung über Zerfällungsan-zahlen, *J. reine und angew. Math.*, *179*, 97–128.

9.7 Summaries of the Papers

[51] Memoir on the theory of the partition of numbers—Part I, *Phil. Trans.*, *187* (1897), 619–673.

This is the first of MacMahon's seven memoirs, with the above title, on partitions. He begins with a study of separations of partitions: a separation of a partition is a set of partitions which, when taken together, form the original partition; for example, $4 + 1$, $3 + 1$, $2 + 2$ form a separation of $4 + 3 + 2 + 2 + 1 + 1$; in MacMahon's notation, we would say that $(14)\,(13)\,(2^2)$

form a separation of $(1^2\,2^2\,34)$. The principal result in section 1 of this paper relates separations and multipartite numbers:

Theorem. The number of separations of the partition $(p_1^{\pi_1} p_2^{\pi_2} p_3^{\pi_3} \ldots)$ is identical with the number of partitions of the multipartite number $(\overline{\pi_1 \pi_2 \pi_3 \ldots})$.

MacMahon now moves to the theory of graphical representations of partitions. Here he relates compositions of the bipartite number $(\overline{pq})$ to partitions with at most q parts and largest part at most p. The relationship is best described by Figure 9.1 (with $q = 6, p = 7$): the dots form the Ferrers

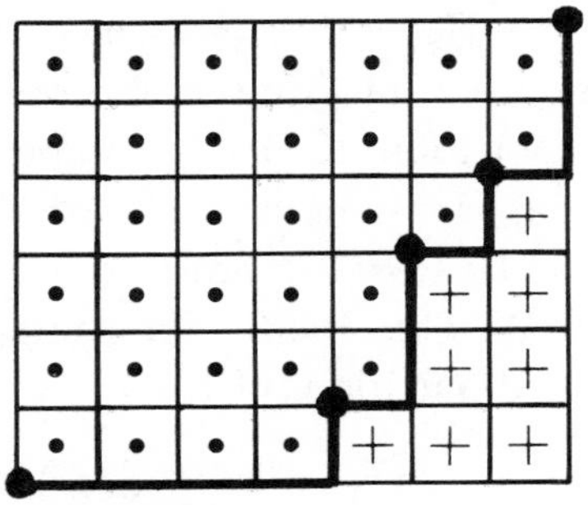

graph of the partition $(45^2 67^2)$, the crosses form the Ferrers graph of $(12^2 3)$, and the darkened path forms a composition of $(\overline{76})$, namely,

$$(\overline{41}) + (\overline{12}) + (\overline{11}) + (\overline{12}).$$

MacMahon studies the implications of this relationship in detail in section 2; he is able to deduce various explicit formulas for different types of partitions. The simplest such observation is that there are $\binom{p+q}{q}$ partitions with at most q parts and each part at most p. From these studies, MacMahon deduces various identities of the basic hypergeometric or q-series type. For example,

$$(1 - x)^{-1} \prod_{j=0}^{p} (1 - ax^j)^{-1} = (1 - ax^p)^{p+1} \sum_{j=0}^{p} U_j(x),$$

where

$$U_j(x) = \left\{ \prod_{h=0}^{j-1} (1 - ax^h)^{-1} \right\} ax^{j-1} \sum_{k=0}^{j-1} (1 - ax^j)^{j-1-k} (1 - ax^{j-1})^k.$$

In section 3 the results of the previous section are subjected to the natural generalization that relates partitions of m-partite numbers to compositions of $(m + 1)$-partite numbers. A number of explicit formulas are again deduced.

The final section of this paper is historically important and quite interesting. From the work given in the previous section, MacMahon illustrates how he is naturally led from the two-dimensional Ferrers graphs of one-dimensional partitions to $(m + 1)$-dimensional graphs of m-dimensional partitions. Herewith he presents numerous conjectures for the generating functions of various types of higher-dimensional partitions. For example, he conjectures that the number of two-dimensional partitions of n (plane partitions) is the coefficient of x^n in $\Pi_{j \geqq 1} (1 - x^j)^{-j}$, a conjecture for which he later sketched the proof in [78]. All of his conjectures concerning generating functions for two-dimensional partitions turned out to be true. Unfortunately, his conjectures on n-dimensional partitions for $n \geqq 3$ are all false (see the references for chapter 12—in particular, Atkin, Bratley, MacDonald, and McKay, 1967, Knuth, 1970a, and Wright, 1968). For example, the suggestion that the number of n-dimensional partitions of $\mathcal{N}$ is the coefficient of $x^{\mathcal{N}}$ in

$$(1 - x)^{-1} \prod_{m=2}^{\infty} (1 - x^m)^{-\binom{m+n-2}{n-1}},$$

turns out to be true for $\mathcal{N} \leqq 5$ and false for $\mathcal{N} = 6$, whenever $n \geqq 3$.

MacMahon undertakes the establishment of one of the simplest of his formulas, namely, the number of plane partitions of n with at most m columns and with each part $\leqq 2$. He shows that this number is the coefficient of x^n in

$$\frac{1}{(1 - x)(1 - x^2)^2(1 - x^3)^2 \ldots (1 - x^m)^2(1 - x^{m+1})},$$

a result first obtained by A. R. Forsyth (Some algebraical theorems connected with the theory of partitions, *Proc. London Math. Soc.*, *27* (1895), 18–35). MacMahon's technique for solving this problem forms the basis for his work in the second memoir in this series, [54].

[94] On partitions into unequal and into uneven parts, *Quart. J. Math.*, *49* (1923), 40–45.

Let $\mathbf{N}$ denote the set of all positive integers; let $\mathbf{O}$ denote the set of all odd positive integers. Let $\mathbf{P}$ be a set of primes, and let $\mathbf{N_P}$ (resp. $\mathbf{O_P}$) denote those elements of $\mathbf{N}$ (resp. $\mathbf{O}$) that are not divisible by any element of $\mathbf{P}$.

Theorem. The partitions of an integer n into distinct parts taken from

$\mathbf{N_P}$ are equinumerous with the partitions of n into parts taken from $\mathbf{O_P}$.

The proof follows the lines of Euler's theorem wherein $\mathbf{P}$ is the empty set, and MacMahon points out that these theorems may be translated into divisor function identities by means of logarithmic differentiation of the related generating functions. This topic has been greatly generalized in recent years (see Andrews, 1972a, section 3).

[97] Note on the parity of the number which enumerates the partitions of a number, *Proc. Cambridge Phil. Soc.*, *20* (1921), 281–283.

From

$$\prod_{m=1}^{\infty}(1-q^m)^{-1}=\sum_{n=0}^{\infty}p(n)q^n,$$

where $p(n)$ is the number of partitions of n, and

$$\begin{aligned}\prod_{m=1}^{\infty}(1-q^m)^{-1}&\equiv\prod_{m=1}^{\infty}(1+q^m)^{-1} &&\pmod 2\\ &=\prod_{m=1}^{\infty}(1-q^{2m-1}) &&\text{(Euler's formula)}\\ &\equiv\prod_{m=1}^{\infty}(1+q^{2m-1}) &&\pmod 2,\end{aligned}$$

we see that

$$\begin{aligned}\sum_{n=0}^{\infty}p(n)q^n&=\prod_{m=1}^{\infty}(1-q^m)^{-1}\\ &\equiv\prod_{m=1}^{\infty}(1+q^{4m-1})(1+q^{4m-3})(1-q^{4m})\\ &\qquad(1-q^{4m})^{-1} \pmod 2\\ &=\sum_{j=0}^{\infty}q^{j(j+1)/2}\prod_{m=1}^{\infty}(1-q^{4m})^{-1} &&\text{(by Jacobi's identity)}\\ &=\sum_{j=0}^{\infty}q^{j(j+1)/2}\sum_{n=0}^{\infty}p(n)q^{4n}.\end{aligned}$$

Comparison of the coefficients of q^{4n}, q^{4n+1}, q^{4n+2}, and q^{4n+3} in the above congruence yields rather efficient recurrences for determining the parity of $p(n)$. In fact, MacMahon is able to deduce "in about five minutes work" that $p(1000)$ is odd, using a table of $p(n)$ for $n \leqq 200$ and the above described recurrences.

[110] The theory of modular partitions, *Proc. Cambridge Phil. Soc., 21* (1923), 197–204.

In this paper MacMahon generalizes the Ferrers graph of a partition to what he calls a modular representation of the partition. This is done for the modulus m by writing each summand $s = km + n$ $(1 \leqq n \leqq m)$ of a partition as a row of $k + 1$ integers in which the first k entries are m and the last entry is n. For example, if we consider the partition $7 + 5 + 3$ of 15, the modular representations are those given in Table 9.2.

Table 9.2

Mod 1	Mod 2	Mod 3	Mod 4	Mod 5	Mod 6	Mod $n \geqq 7$
1111111	2221	331	43	52	61	7
11111	221	32	41	5	5	5
111	21	3	3	3	3	3

MacMahon uses standard techniques generalized from those used on Ferrers graphs to obtain various generating functions and partition function identities.

[118] The parity of $p(n)$, the number of partitions of n, when $n \leqq 1000$, *J. London Math. Soc., 1* (1926), 225–226.

By means of the techniques described in [97], MacMahon determines the parity of $p(n)$ for $n \leqq 1000$. He lists the 470 values of n such that $n \leqq 1000$ and $p(n)$ is even.

[119] The elliptic products of Jacobi and the theory of linear congruences, *Proc. Cambridge Phil. Soc., 23* (1927), 337–355.

This paper presents a dissection of the Jacobi triple product identity and a detailed analysis of the resulting identities. For example, let

$$(a, b) \equiv \prod_{m=1}^{\infty}(1 - q^{2ma-a-b})(1 - q^{2ma-a+b})(1 - q^{2ma})$$

$$= \sum_{m=-\infty}^{\infty} (-1)^m q^{am^2+bm},$$

and let $(a, b)_f$ denote the function (a, b) with q replaced by q^f. Replacing m by $\mu m + \varepsilon$, $-\infty < m < \infty, 0 \leqq \varepsilon < \mu$ in the above series, MacMahon easily derives the identity

$$\begin{aligned}(a, b) = &-q^{a-b}(\mu a, 2a - b)_\mu + q^{4a-2b}(\mu a, 4a - b)_\mu \\ &- \ldots + (-1)^{(\mu-1)/2} q^{(\mu-1)^2 a/4 - (\mu-1)b/2}(\mu a, \mu a - a - b)_\mu \\ &+ (\mu a, b)_\mu - q^{a+b}(\mu a, 2a + b)_\mu + q^{4a+2b}(\mu a, 4a + b)_\mu \\ &- \ldots + (-1)^{(\mu-1)/2} q^{(\mu-1)^2 a/4 + (\mu-1)b/2}(\mu a, \mu a - a + b),\end{aligned}$$

where μ is an odd prime.

A number of results are deduced from this and similar formulas, and two tables are constructed that exhibit interesting properties of these results.

Reprints of the Papers

XVI. *Memoir on the Theory of the Partition of Numbers.*—Part I.

By Major P. A. MacMahon, *R.A., F.R.S.*

Received December 31, 1895,—Read January 16, 1896.

§ 1.

Art. 1. I have under consideration multipartite numbers as defined in a former paper.*

I recall that the multipartite number

$$\overline{\alpha\beta\gamma\ldots},$$

may be regarded as specifying $\alpha + \beta + \gamma + \ldots$ things, α of one sort, β of a second, γ of a third, and so forth. If the things be of m different sorts the number is said to be multipartite of order m or briefly an m-partite number. It is convenient to call $\alpha, \beta, \gamma, \ldots$ the first, second, third ... figures of the multipartite number. If such a number be divided into parts each part is regarded as being m-partite; if the order in which the parts are written from left to right is essential we obtain a composition of the multipartite number; whereas if the parts themselves are alone specified, and not the order of arrangement, we have a partition of the multipartite number. This, and much more, is explained in the paper quoted, which is concerned only with the compositions of multipartite numbers.

Art. 2. The far more difficult subject of partitions is taken up in the present paper. The compositions admitted of easy treatment by a graphical process. An m-partite reticulation or lattice is taken to be the graph of an m-partite number, and on this graph every composition can be satisfactorily depicted.

A suitable graphical representation of the partitions appears to be difficult of attainment. As the Memoir proceeds, the extent to which the difficulties have been overcome will appear. There are several bonds of connection between partitions and compositions; in general, these do not exist between m-partite partitions and m-partite compositions, but arise from a general survey of the partitions and compositions of multipartite numbers of all orders.

These bonds are of considerable service in the gradual evolution of a theory of partitions. Two bonds have already been made known (*loc. cit.*). They both have

* "Memoir on the Theory of the Composition of Numbers," 'Phil. Trans.,' R.S. of London, vol. 184 (1893), A, pp. 835–901.

reference to the perfect partitions of unipartite numbers.* Firstly, there is a one-to-one correspondence between the compositions of the multipartite

$$\overline{\alpha\beta\gamma\ldots},$$

and the perfect partitions of the unipartite

$$a^\alpha b^\beta c^\gamma \ldots - 1;$$

$a, b, c, \ldots$ being any different primes.

Secondly, there is a one-to-one correspondence between the compositions of the unipartite number m and the perfect partitions, comprising m parts, of the whole assemblage of unipartite numbers.

These bonds are interesting but, for present purposes, trivial. We require some correspondence concerning partitions which are not subject to the restriction of being *perfect*.

Art. 3. I first proceed to explain an important link connecting unipartite with multipartite partitions arising from the notion of the separation of the partition of a number (whether unipartite or multipartite) into separates; a notion which leads to a theory of the separations of a partition of a number* which was partially set forth in the series of Memoirs referred to in the foot-note. The theory of separations arises in a perfectly natural manner in the evolution of the theory of symmetrical algebra, and is, I venture to think, of considerable algebraical importance. Up to the present time the theory has been worked out as it was required for algebraical purposes; the various definitions and theorems are scattered about several Memoirs in a manner which is inconvenient for reference, and it therefore will be proper, while explaining the connection with compositions, to bring the salient features of the theory together under the eye of the reader. It suffices for the most part to deal with unipartite numbers.

The Theory of Separations.

Definitions.

Art. 4. A number is partitioned into parts by writing down a set of positive numbers (it is convenient, but not necessary to assume positive parts, and occasionally to regard zero as a possible part) which, when added together, reproduce the original number.

The constituent numbers, termed parts, are written in descending numerical order from left to right and are usually enclosed in a bracket ( ). This succession of

* See also "The Perfect Partitions of Numbers and the Compositions of Multipartite Numbers." The Author, 'Messenger of Mathematics,' New Series, No. 235, November, 1890.

numbers is termed a partition of the original number; and this number, *quâ* partitions, is termed by SYLVESTER the partible number.

A partition of a number is separated into separates by writing down a set of partitions, each in its own brackets, such that when all the parts of the partitions are assembled in a single bracket and arranged in order, the partition which is separated is reproduced. The constituent partitions, which are the separates, are written down from left to right in descending numerical order as regards the weights of the partitions.

N.B. The partition (pqr) is said to have a weight $p+q+r$.

The partition separated may be termed the separable partition.

Taking as separable partition

$$(p_1p_2p_3p_4p_5),$$

two separations are

$$(p_1p_2)\,(p_3p_4)\,(p_5),$$
$$(p_1p_2p_3)\,(p_4p_5),$$

and there are many others.

If the successive weights of the separates be

$$w_1,\ w_2,\ w_3,\ \ldots,$$

the separation is said to have a specification

$$(w_1,\ w_2,\ w_3,\ \ldots);$$

the specification being denoted by a partition of the weight w of the separable partition.

The *degree* of a separation is the sum of the highest parts of the several separates.

If the separation be

$$(p_1\ldots)^{j_1}\,(p_s\ldots)^{j_2}\,(p_t\ldots)^{j_3}\ldots$$

the *multiplicity* of the separation is defined by the succession of indices

$$j_1,\ j_2,\ j_3,\ \ldots$$

The characteristics of a separation are

(i.) The weight.
(ii.) The separable partition.
(iii.) The specification.
(iv.) The degree.
(v.) The number of separates.
(vi.) The multiplicity.

Art. 5. The separations of a given partition may be grouped in a manner which is independent of their specifications.

Consider the separable partition

$$(p_1^3 p_2^2),$$

which is itself to be regarded as one amongst its own separations.

Viewed thus, it has *quâ* partition a multiplicity (32).

Write down any one of its separations, say

$$(p_1^2)(p_1 p_2)(p_2).$$

This separation may be regarded as being compounded of the two separations

$$(p_1^2)(p_1) \text{ of } (p_1^3)$$

and

$$(p_2)^2 \text{ of } (p_2^2).$$

Three other separations enjoy the same property, viz.,

$$\begin{array}{l} (p_1^2)(p_1)(p_2)^2, \\ (p_1^2 p_2)(p_1)(p_2), \\ (p_1^2 p_2)(p_1 p_2); \end{array}$$

for, on suppressing p_2 in each, we are left with

$$(p_1^2)(p_1);$$

and on suppressing p_1 there remains

$$(p_2)^2.$$

These four separations

$$\left.\begin{array}{l} (p_1^2)(p_1 p_2)(p_2) \\ (p_1^2)(p_1)(p_2)^2 \\ (p_1^2 p_2)(p_1)(p_2) \\ (p_1^2 p_2)(p_1 p_2) \end{array}\right\} \text{Set } \{(21), (1^2)\},$$

form a set which is defined by two partitions, one appertaining to each of the two numbers which define the multiplicity of the separable partition.

Thus the first number 3 of the multiplicity occurs in each separation of the group in the partition (21), and the second number 2 in the partition (1^2).

There are as many sets of separations as there are combinations of a partition of 3 with a partition of 2.

In the present instance there are six sets, viz.,

$$\begin{gathered} S\{(3),\ (2)\}, \\ S\{(3),\ (1^2)\}, \\ S\{(21),\ (2)\}, \\ S\{(21),\ (1^2)\}, \\ S\{(1^3),\ (2)\}, \\ S\{(1^3),\ (1^2)\}. \end{gathered}$$

In general, if

$$(p_1^{\pi_1}\, p_2^{\pi_2}\, p_3^{\pi_3} \ldots)$$

be the separable partition, the multiplicity is

$$(\pi_1\ \pi_2\ \pi_3 \ldots);$$

and if the unipartite π_s possess ρ_s partitions, the separations can be arranged in

$$\rho_1\ \rho_2\ \rho_3 \ldots$$

sets.

I observe that the notion of sets of separations enters in a fundamental manner into the theory of symmetric functions.

Art. 6. One of the first problems encountered in the arithmetical theory is the enumeration of the separations of a given partition. I shall prove that the number of separations of the partition

$$(p_1^{\pi_1} p_2^{\pi_2} p_3^{\pi_3} \ldots)$$

is identical with the number of partitions of the multipartite number

$$(\overline{\pi_1 \pi_2 \pi_3 \ldots}),$$

formed from the multiplicity of the separable partition.*

If we separate $(p_1^{\pi_1} p_2^{\pi_2} p_3^{\pi_3} \ldots)$ so that one separate of the separation is

$$(q_1^{\chi_1} q_2^{\chi_2} q_3^{\chi_3} \ldots),$$

it is clear that we can partition the multipartite $(\overline{\pi_1 \pi_2 \pi_3 \ldots})$ in such wise that one part of the partition is

$$(\overline{\chi_1 \chi_2 \chi_3 \ldots}).$$

* Note that the rule ———— distinguishes a multipartite number from a partition of a unipartite number.

It is also manifest that to each separate of the separation corresponds a part of the partition, and that we obtain a partition

$$(\overline{\chi_1\chi_2\chi_3\ldots} \quad \ldots \quad \ldots)$$

of the multipartite

$$(\overline{\pi_1\pi_2\pi_3\ldots}),$$

in correspondence with each separation

$$(q_1^{\chi_1}q_2^{\chi_2}q_3^{\chi_3}\ldots)\ (\ldots)\ (\ldots)\ldots$$

of the partition

$$(p_1^{\pi_1}p_2^{\pi_2}p_3^{\pi_3}\ldots).$$

There is, therefore, identity of enumeration. Also, the enumeration of the separations into k separates is identical with that of the partitions into k parts.

Ex. gr., the subjoined correspondence :—

Separations.	Partitions.
(p^2q^2)	$(\overline{22})$,
$(p^2)\ (q^2)$,	$(\overline{20}\ \overline{02})$,
$(p^2q)\ (q)$,	$(\overline{21}\ \overline{01})$,
$(p)\ (pq^2)$,	$(\overline{10}\ \overline{12})$,
$(pq)^2$,	$(\overline{11}^2)$,
$(p^2)\ (q)^2$,	$(\overline{20}\ \overline{01}^2)$,
$(p)^2\ (q^2)$,	$(\overline{10}^2\ \overline{02})$,
$(pq)\ (p)\ (q)$,	$(\overline{11}\ \overline{10}\ \overline{01})$,
$(p)^2\ (q)^2$,	$(\overline{10}^2\ \overline{01}^2)$.

Art. 7. It is important to take note of the fact that the subject of the separations of partitions of unipartite numbers necessitates the consideration of the partitions of multipartite numbers.

The partitions of a multipartite number are divisible into sets in the same manner as the separations of a unipartite number. In the case of the number

$$(\pi_1\pi_2\pi_3\ldots)$$

there are $\rho_1\rho_2\rho_3\ldots$ sets, where ρ_s is the number of partitions of the unipartite π_s.

The same succession of numbers may be employed to denote either a multipartite number or the partition of a unipartite number, so that we naturally find great similarity between the theories of unipartite partitions and of multipartite numbers. We see above that the separations of partitions of unipartites are in co-relation with the partitions of multipartites.

Art. 8. In a natural manner the separations of a partition of a multipartite present themselves for consideration. In correspondence we find what may be termed a double separation of a partition of a unipartite.

Ex. gr. Consider the partition $(\overline{20}\ \overline{01}^2)$ of the multipartite $(\overline{22})$, which is co-related to the separation $(p^2)\,(q)^2$ of the partition (p^2q^2) of the unipartite $2p + 2q$.

Of this partition we find a separation

$$(\overline{20}\ \overline{01})\,(\overline{01})$$

in correspondence with a *double* separation

$$\{(p^2)\,(q),\,(q)\}$$

of the partition (p^2q^2).

Hence the enumeration of the separations of a multipartite partition is identical with that of the double separations of a unipartite partition.

In general n-tuple separations of unipartite numbers correspond one-to-one with $n - 1$-tuple separations of multipartite numbers.

For the present I leave the subject of separations, merely remarking that the theory was made the basis of all the memoirs on symmetric functions to which reference has been given, and that algebraically considered they are of extraordinary interest.

§ 2. The Graphical Representation of Partitions.

Art. 9. In an important contribution to the theory of unipartite partitions Sylvester* adopted a graphical method which threw great light on the subject, and was fruitful in algebraical results.

The method consisted in arranging rows of nodes, each row corresponding to a part of the partition and containing as many nodes as the number expressing the magnitude of the part.

Ex. gr., the partition (32^21) has the graph

• • •
• •
• •
•

* "A Constructive Theory of Partitions," by J. J. Sylvester, with insertions by Dr. F. Franklin, 'American Journal of Mathematics,' vol. 5.

This method cannot be simply extended to the case of multipartite partitions, though some progress can be made in this direction as will be shown.

Art. 10. SYLVESTER'S graphs naturally present themselves in the graphical representation of the compositions of bipartite numbers.

As the graph of a bipartite $(\overline{pq})$ we take $p+1$ lines parallel and at equal distances apart and cut them by $q+1$ other lines at equal distances apart and at right angles to the former; (N.B. The right angle is not essential,) thus forming a reticulation or lattice.

The figure represents the reticulation of the bipartite $(\overline{76})$.

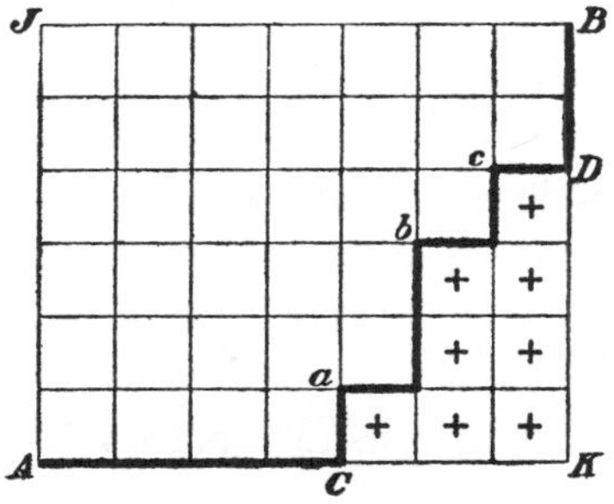

I recall that A, B are the initial and final points of the graph, and that the remaining intersections are termed the "points" of the graph.

The lines of the graph have either the direction AK (called the α direction) or the direction AJ (called the β direction).

Each line is made up of segments, and we speak of α-segments and of β-segments, indicating that the lines, on which lie the segments, are in the α and β directions.

If the bipartite $(\overline{pq})$ have a composition

$$(\overline{p_1q_1}\ \overline{p_2q_2}\ \overline{p_3q_3}\ldots),$$

the composition is delineated upon the graph, as follows:—

Starting from the point A we pass over p_1 α-segments and then over q_1 β-segments, and place a node at the point arrived at. Starting again from this node, we pass over p_2 α-segments and q_2 β-segments, and place a second node at the point then reached; we proceed similarly with the other parts of the composition until finally the point B is reached. At this point it is not necessary to place a node. In this manner the composition containing θ parts is represented by $\theta-1$ nodes placed at $\theta-1$ different "points" of the graph.

The segments, passed over in tracing the composition, form a line of route through the reticulation. In general many compositions have the same line of route. Along every line of route there are $p+q-1$ "points," which *may* be nodes. A certain number of these points *must* be nodes. These occur at all points where there is a change from the β to the α direction. They are termed "essential nodes."

In the line of route traced in the figure the points a, b, c are essential nodes.

Along a line of route there is a composition which is depicted by the essential nodes alone. This is termed the "principal composition along the line of route."

In the figure the principal composition is

$$(\overline{41}\ \overline{12}\ \overline{11}\ \overline{12}).$$

I have shown (*loc. cit.*) that the number of lines of route which possess s essential nodes is

$$\binom{p}{s}\binom{q}{s};$$

and that the total number of lines of route is

$$\sum_{s=0}^{s=q}\binom{p}{s}\binom{q}{s} = \binom{p+q}{p} \qquad (p \geq q).$$

We remark that the line of route divides the reticulation into two portions, an upper portion AJBDC, and a lower portion CKD.

Placing a Sylvester-node in each square of the lower portion, we recognise at once SYLVESTER'S regularised graph of the partition

$$(32^21)$$

of the unipartite number 8.

Similarly, from the upper portion, we obtain SYLVESTER'S regularised graph of the partition

$$(7^265^24),$$

of the unipartite number 34 (thirty-four).

Whatever be the line of route, we simultaneously exhibit two of SYLVESTER'S regularised graphs, one of a partition of the unipartite N, and one of a partition of the unipartite $pq - \mathrm{N}$.

The two partitions may be termed complementary in respect of the unipartite number pq.

Art. 11. This interesting bond between the partitions of unipartites and the compositions of bipartites, I propose to submit to a detailed examination. The partitions, with which we are concerned, are limited in magnitude of part to p, and in number of parts to q.

Moreover (as in the bipartite), we may suppose the numbers p, q, to be interchanged. This would simply amount to rotating the reticulation through a right angle.

4 L 2

The line of route partitions the reticulation into two parts, each of which may be regarded as a partition of a unipartite. In fact, a line of route graphically represents a pair of partitions.

These partitions can be equally depicted by the essential nodes only that occur along the line of route. It will be more convenient, for some purposes, to take the line of route, and not merely the essential nodes, to be the graphical representation. Attention for the present will be limited to the partition lying to the North-West of the line of route (*i.e.*, towards the point J). A line of route involves bends $\lrcorner$, termed "left-bends," and bends $\ulcorner$, termed "right-bends." Essential nodes occur at the angular points of the latter. The North-West partition has as many *different* parts as there are left-bends on the line of route. The number of lines of route which have s left-bends is equal to the number which have s right-bends. This may be seen by rotating the reticulation through two right angles.

This number is $\binom{p}{s}\binom{q}{s}$ (*loc. cit.*, Art. 22).

Hence :—

"The number of partitions of all numbers into s different parts limited in magnitude to p and in number to q is

$$\binom{p}{s}\binom{q}{s}."$$

Art. 12. The number of different partitions is equal to the number of lines of route, and this is

$$\sum_s \binom{p}{s}\binom{q}{s} = \binom{p+q}{p} \qquad (p \nless q).$$

Hence :—

"The number of partitions, of all numbers, into parts limited in magnitude to p, and in number to q is

$$\binom{p+q}{p}."$$

Art. 13. A line of route, with s left-bends, has either $s-1$, s or $s+1$ right-bends. If it commences by tracing an α-segment and ends by tracing a β-segment, the number is $s-1$. If it commences by tracing an α-segment and ends by tracing an α-segment, the number is s. If a β-segment begins and a β-segment ends the line of route, the number is s, and if a β-segment begins and an α-segment ends, the number is $s+1$.

If an α-segment begins the line, the North-West partition has exactly q parts. If an α-segment ends, the highest part is less than p. If a β-segment begins, the number of parts is less than q, and if a β-segment ends, the highest part is equal to p.

Inspection of the Memoir cited shows that the enumeration of the lines of route possessing s left-bends and $s-1$, s, $s+1$ right-bends respectively are given by

$$\binom{p-1}{s-1}\binom{q-1}{s-1},$$

$$\binom{p-1}{s}\binom{q-1}{s-1}+\binom{p-1}{s-1}\binom{q-1}{s},$$

$$\binom{p-1}{s}\binom{q-1}{s};$$

hence :—

"The number of partitions of all numbers which have exactly q parts, a highest part equal to p, and s different parts is

$$\binom{p-1}{s-1}\binom{q-1}{s-1}."$$

"The number of partitions of all numbers which have exactly q parts, a highest part less than p, and s different parts, *or* which have less than q parts, a highest part equal to p, and s different parts is

$$\binom{p-1}{s}\binom{q-1}{s-1}+\binom{p-1}{s-1}\binom{q-1}{s}."$$

"The number of partitions of all numbers which have less than q parts, a highest part less than p, and s different parts is

$$\binom{p-1}{s}\binom{q-1}{s}."$$

Ex. gr. If $p = q = 3$, $s = 2$ the partitions enumerated by these three theorems are

(31^2), (3^21), (32^2), (3^22)

(31), (21^2), (32), (2^21)

(21)

respectively.

Art. 14. It is clear that all identical relations between binomial coefficients yield results in this theory of partitions. *Ex. gr.*, such relations as

$$\binom{p+q}{p}-\binom{p+q-1}{p-1}=\binom{p+q-1}{p}$$

$$\binom{p-1}{s}\binom{q-1}{s}-\binom{p-2}{s}\binom{q-1}{s}=\binom{p-2}{s-1}\binom{q-1}{s}$$

admit of immediate interpretation.

Art. 15. A line of route has in general right and left bends. The whole number of bends may be even or uneven. To determine the number of lines, with a given number of bends, we must separate the two cases. If the bends be $2k$ in number, k must be right and k left and the enumeration gives

$$\binom{p-1}{k}\binom{q-1}{k-1}+\binom{p-1}{k-1}\binom{q-1}{k}.$$

If the bends be $2k+1$ in number we may have k right-bends and $k+1$ left-bends or $k+1$ right and k left. The enumeration gives

$$2\binom{p-1}{k}\binom{q-1}{k}.$$

Hence the number of pairs of complementary partitions which have each k different parts is

$$\binom{p-1}{k}\binom{q-1}{k-1}+\binom{p-1}{k-1}\binom{q-1}{k};$$

and the number of pairs in which the partitions have k and $k+1$ different parts respectively, is

$$2\binom{p-1}{k}\binom{q-1}{k}.$$

Adding the number of lines of route with $2k-1$ bends to the number with $2k$ bends we obtain the number

$$\frac{p+q}{k}\binom{p-1}{k-1}\binom{q-1}{k-1}.$$

Hence the identity,

$$(p+q)\sum_k \frac{1}{k}\binom{p-1}{k-1}\binom{q-1}{k-1}=\binom{p+q}{p};$$

or

$$\binom{p-1}{0}\binom{q-1}{0}+\tfrac{1}{2}\binom{p-1}{1}\binom{q-1}{1}+\tfrac{1}{3}\binom{p-1}{2}\binom{q-1}{2}+\ldots=\frac{(p+q-1)!}{p!\,q!};$$

which may be established independently.

Art. 16. Consider the lines of route which pass through a particular point P of the graph, say the point distant from A, a α-segments and b β-segments. In the reticulation AP we may draw any line of route, and any line also in the reticulation PB. In AB we can thus obtain

$$\binom{a+b}{a}\binom{p+q-a-b}{p-a}$$

lines of route passing through P.

In the associated North-West partitions, the highest part $\not> a$ and the number of parts $\not> q-b$.

Put $q-b=c$, and we obtain the theorem :—

"Of the partitions whose parts are limited in magnitude to p and in number to q, there are,

$$\binom{q+a-c}{a}\binom{p-a+c}{c},$$

such that the highest part $\nless a$ and number of parts $\nless c$."

In particular, if

$$a + b = p,$$

or

$$a - c = p - q,$$

the number is

$$\binom{p}{p-a}\binom{q}{p-a};$$

which enumerates the lines of route with $p - a$ left-bends. There is thus a one-to-one correspondence between the lines of route passing through the point for which $(a, b) = (a, p - a)$, and the lines of route with $p - a$ left-bends, and hence also between the partitions under consideration and those which involve parts of $p - a$ different kinds.

Art. 17. A line of route is a graphical representation of a principal composition of the bipartite. Such a composition being

$$(\overline{p_1q_1}\ \overline{p_2q_2}\ \overline{p_3q_3} \ldots \overline{p_sq_s}),$$

the numbers

$$q_1, p_2, q_2, p_3, q_3 \ldots p_s$$

are all superior to zero. A positive number q_1 of the bipart $\overline{p_1q_1}$ is adjacent to a positive number p_2 of the bipart $\overline{p_2q_2}$, and we may assert of the composition that all its contacts are positive-positive (see Art. *loc. cit.*).

Each line of route represents a composition with positive-positive contacts, and there is a one-to-one correspondence. Hence:

"There is a one-to-one correspondence between the compositions, with positive-positive contacts, of the bipartite $\overline{pq}$ and the partitions of all unipartite numbers into parts limited in magnitude by p and in number by q."

Further, there are as many such compositions with $s + 1$ parts as there are such partitions which involve s *different* parts.

Art. 18. Every line of route through the reticulation of $\overline{pq}$ may be represented by a permutation of the letters in $\alpha^p\beta^q$. We have merely to write down the α and β segments as they occur along the line of route to obtain such a permutation.

The composition

$$(\overline{p_1q_1}\ \overline{p_2q_2}\ \overline{p_3q_3} \ldots \overline{p_{s+1}\ q_{s+1}}),$$

gives the permutation

$$\alpha^{p_1}\beta^{q_1}\alpha^{p_2}\beta^{q_2}\alpha^{p_3}\beta^{q_3}\ldots\alpha^{p_{s+1}}\beta^{q_{s+1}}.$$

From what has gone before it will be seen that every permutation of $\alpha^p\beta^q$ corresponds to a partition of a unipartite number into parts limited in magnitude to p and in number to q. Every theorem in permutations of two *different* letters will thus yield a theorem in partitions of unipartite numbers.

The North-West partition associated with the above-written permutation is easily seen to be (writing the parts in ascending order as regards magnitude, viz.: in SYLVESTER's regularised orders reversed)

$$\left(p_1^{q_1}\,\overline{p_1+p_2}^{q_2}\,\overline{p_1+p_2+p_3}^{q_3}\ldots\overline{p_1+p_2+\ldots+p_{s+1}}^{q_{s+1}}\right).$$

This is a partition of the unipartite

$$q_1p_1 + q_2(p_1+p_2) + q_3(p_1+p_2+p_3) + \ldots + q_{s+1}(p_1+p_2+\ldots+p_{s+1})$$

into $q_1+q_2+q_3+\ldots+q_{s+1}$ parts, the highest part being $p_1+p_2+p_3+\ldots+p_{s+1}$.

If p_1 be zero there will be less than q parts. If q_{s+1} be zero the highest part will be less than p. On the other hand if p_1 be not zero there are exactly q parts, and if q_{s+1} be not zero the highest part is p.

Art. 19. Observe that we have a fourfold correspondence, viz., between

(1.) The lines of route in the reticulation of the bipartite $\overline{pq}$.

(2.) The compositions, with positive-positive contacts, of the bipartite number $\overline{pq}$.

(3.) The permutations of the letters forming the product $\alpha^p\beta^q$.

(4.) The partitions of all unipartite numbers into parts limited in magnitude to p and in number to q.

And also, in particular, between—

(1.) The lines of route with s right-bends or with s left-bends.

(2.) The compositions into $s+1$ parts.

(3.) The permutations with s, $\alpha\beta$ or with s, $\beta\alpha$ contacts.

(4.) The partitions involving s *different* parts.

Art. 20. The generating function for the number of lines of route through the reticulation which possess s left-bends is

$$1+\binom{p}{1}\binom{q}{1}\mu+\binom{p}{2}\binom{q}{2}\mu^2+\ldots+\binom{p}{s}\binom{q}{s}\mu^s+\ldots+\binom{p}{q}\binom{q}{p}\mu^q$$

which is the coefficient of $\alpha^p\beta^q$ in the product

$$(\alpha+\mu\beta)^p\,(\alpha+\beta)^q;$$

(see *loc. cit.*, Art. 24),

and this is the coefficient of $\alpha^p\beta^q$ in the development of the fraction

$$\frac{1}{1-\alpha-\beta+(1-\mu)\,\alpha\beta};$$

(see "Memoir on a Certain Class of Generating Functions in the Theory of Numbers," 'Phil. Trans.,' Roy. Soc. of London, vol. 185 (1894), A, pp. 111–160), and this fraction may be written

$$\sum_0 \frac{\alpha^s\beta^s}{(1-\alpha)^{s+1}(1-\beta)^{s+1}}\mu^s.$$

Hence

$$\frac{\alpha^s\beta^s}{(1-\alpha)^{s+1}(1-\beta)^{s+1}}$$

is the generating function for the lines of route in all bipartite reticulations which possess s left-bends *or* s right-bends, and also for the other entities in correspondence therewith. In particular it enumerates all unipartite partitions into s *different* parts limited, in any desired manner, in regard to number and magnitude.

Art. 21. Various theorems in algebra are derivable from the foregoing theorems.

The generating function for the partitions of all unipartite numbers into parts limited in magnitude to p and in number to q is

$$\frac{1}{1-a\,.\,1-x\,.\,1-ax\,.\,1-ax^2\,.\,1-ax^3\,\ldots\,.\,1-ax^p};$$

the enumeration being given by the coefficients of a^qx^{pq} in the ascending expansion.

The G.F. is redundant as we are only concerned with that portion, of the expanded form, which proceeds by powers of ax^p.

The foregoing theory enables us to isolate this portion, inasmuch as we know it to have the expression

$$1+\binom{p+1}{1}ax^p+\binom{p+2}{2}a^2x^{2p}+\binom{p+3}{3}a^3x^{3p}+\ldots.+\binom{p+q}{p}a^qx^{qp}+\ldots.$$

which may be written

$$(1-ax^p)^{-p-1}.$$

As a verification of the simplest cases we find the identities

$$\frac{1}{1-a.1-x.1-ax}=\frac{1}{(1-ax)^2}\left(1+\frac{a}{1-a}+\frac{x}{1-x}\right)$$

$$\frac{1}{1-a.1-x.1-ax.1-ax^2}=\frac{1}{(1-ax^2)^3}\left\{1+\frac{a}{1-a}+\frac{x}{1-x}+\frac{ax}{1-a}+\frac{ax\,(1-ax^2)}{1-a\,.\,1-ax}\right\},$$

a simple inspection of which demonstrates the validity of the theorem in these cases

MDCCCXCVI.—A. 4 M

To obtain a general formula write

$$\frac{(1-ax^p)^{p+1}}{1-a\,.\,1-x\,.\,1-ax\,.\,1-ax^2\ldots 1-ax^p.}$$
$$=\frac{(1-ax^{p-1})^p}{1-a\,.\,1-x\,.\,1-ax\,.\,1-ax^2\ldots 1-ax^{p-1}}+U_p(x);$$

then

$$U_p(x)=\frac{(1-ax^p)^{p+1}-(1-ax^p)(1-ax^{p-1})^p}{1-a\,.\,1-x\,.\,1-ax\,.\,.\,1-ax^2\ldots 1-ax^p}$$
$$=\frac{ax^{p-1}\{(1-ax^p)^{p-1}+(1-ax^p)^{p-2}(1-ax^{p-1})+\ldots+(1-ax^{p-1})^{p-1}\}}{1-a\,.\,1-ax\,.\,1-ax^2\ldots 1-ax^{p-1}}$$

and we have in succession

$$U_0(x)=\frac{1}{1-x},$$

$$U_1(x)=\frac{a}{1-a},$$

$$U_2(x)=\frac{ax}{1-a}+\frac{ax(1-ax^2)}{1-a.1-ax},$$

$$U_3(x)=\frac{ax^2(1-ax^2)}{1-a.1-ax}+\frac{ax^2(1-ax^3)}{1-a.1-ax}+\frac{ax^2(1-ax^3)^2}{1-a.1-ax.1-ax^2},$$

$$U_4(x)=\frac{ax^3(1-ax^3)^2}{1-a.1-ax.1-ax^2}+\frac{ax^3(1-ax^3)(1-ax^4)}{1-a.1-ax.1-ax^2}$$
$$+\frac{ax^3(1-ax^4)^2}{1-a.1-ax.1-ax^2}+\frac{ax^3(1-ax^4)^3}{1-a.1-ax.1-ax^2.1-ax^3};$$

and, in general,

$$U_p(x)=\frac{ax^{p-1}(1-ax^{p-1})^{p-2}}{1-a.1-ax\ldots 1-ax^{p-2}}+\frac{ax^{p-1}(1-ax^{p-1})^{p-3}(1-ax^p)}{1-a.1-ax\ldots 1-ax^{p-2}}$$
$$+\frac{ax^{p-1}(1-ax^{p-1})^{p-4}(1-ax^p)^2}{1-a.1-ax\ldots 1-ax^{p-2}}+\ldots$$
$$+\frac{ax^{p-1}(1-ax^p)^{p-2}}{1-a.1-ax\ldots 1-ax^{p-2}}+\frac{ax^{p-1}(1-ax^p)^{p-1}}{1-a.1-ax\ldots 1-ax^{p-2}.1-ax^{p-1}}.$$

Hence,

$$\frac{1}{1-x.1-a.1-ax.1-ax^2\ldots 1-ax^p}=\frac{1}{(1-ax^p)^{p+1}}\sum_0^p U_p(x);$$

a valuable expansion.

Simple inspection of this formula shows that $(1-ax^p)^{-p-1}$ represents that portion of the G.F. which is a function of ax^p only.

Art. 22. Again, the partitions of all unipartite numbers into s *different* parts, limited in magnitude to p and in number to q, are enumerated by the coefficient of $a^q b^s x^{pq}$ in the development of the product

$$\frac{1}{1-x}\cdot\frac{1}{1-a}\left(1+\frac{abx}{1-ax}\right)\left(1+\frac{abx^2}{1-ax^2}\right)\cdots\left(1+\frac{abx^p}{1-ax^p}\right);$$

or by the coefficient of $a^q x^{pq}$ in

$$\frac{a^s}{1-x\,.\,1-a}\Sigma\frac{x^{k_1+k_2+k_3+\ldots+k_s}}{(1-ax^{k_1})(1-ax^{k_2})(1-ax^{k_3})\ldots(1-ax^{k_s})},$$

where $k_1, k_2, k_3, \ldots k_s$ are any s different numbers drawn from the natural series

$$1, 2, 3, \;..\, p;$$

and the summation is in respect of all such selections. This is the coefficient of

$$a^{q-s}\,x^{pq-\binom{s+1}{2}}$$

in

$$\frac{1}{1-x\,.\,1-a}\Sigma\frac{x^{k_1+k_2+k_3+\ldots+k_s-\binom{s+1}{2}}}{(1-ax^{k_1})(1-ax^{k_2})(1-ax^{k_3})\ldots(1-ax^{k_s})}.$$

Taking the former of these two expressions; inasmuch as we know from the reticulation theory that the coefficient of $a^q x^{pq}$ is $\binom{p}{s}\binom{q}{s}$, we find that the effective portion of the generating function is

$$\binom{p}{s}\binom{s}{s}(ax^p)^s+\binom{p}{s}\binom{s+1}{s}(ax^p)^{s+1}+\ldots\ ad\ inf.,$$

which is

$$\binom{p}{s}\frac{(ax^p)^s}{(1-ax^p)^{s+1}};$$

viz., we have succeeded in isolating that portion of the generating function which proceeds by powers of ax^p only.

The latter expression of the generating function also, is seen to have an effective portion

$$\binom{p}{s}\frac{x^{ps-\binom{s+1}{2}}}{(1-ax^p)^{s+1}}.$$

The isolations thus effected would, I believe, be difficult to accomplish algebraically.

4 M 2

Art. 23. Again, regarding p and q as constant and s as variable, we know that the coefficient of $(ax^p)^q$ in the product

$$\frac{1}{1-x\,.\,1-a}\left(1+\frac{abx}{1-ax}\right)\left(1+\frac{abx^2}{1-ax^2}\right).\;.\left(1+\frac{abx^p}{1-ax^p}\right),$$

is

$$\binom{p}{0}\binom{q}{0}+\binom{p}{1}\binom{q}{1}b+\binom{p}{2}\binom{q}{2}b^2+\ldots+\binom{p}{q}\binom{q}{q}b^q;$$

calling this expression B_q, the effective portion of the generating function is written

$$1+B_1ax^p+B_2(ax^p)^2+B_3(ax^p)^3+\ldots \text{ ad inf.}$$

Art. 24. I recall now the generating function which enumerates the partitions of all unipartite numbers into parts limited in magnitude by p and in number by q, viz. :—

$$\frac{1}{(1-x)(1-a)(1-ax)(1-ax^2)\ldots(1-ax^p)}$$

The coefficient of a^q in this development is well known to be

$$\frac{(1-x^{q+1})(1-x^{q+2})\ldots(1-x^{q+p})}{(1-x)^2(1-x^2)(1-x^3)\ldots(1-x^p)}.$$

Hence the coefficient of a^qx^{pq} in the former is the same as the coefficient of x^{pq} in the latter. This we know to have the value $\binom{p+q}{q}$.

Art. 25. Numerous theorems of *isolation*, in the senses in which the word is employed in this paper, may be obtained from the reticulation theorems. I proceed to give some of those which present features of interest.

A previous result was to the effect that the number of partitions of all numbers which have exactly q parts, a highest part equal to p and s *different* parts is enumerated by

$$\binom{p-1}{s-1}\binom{q-1}{s-1}.$$

Hence, without specification of s, the number is

$$\sum_s\binom{p-1}{s}\binom{q-1}{s}=\binom{p+q-2}{p-1}.$$

This enumeration is also given by the coefficient of $a^{q-1}x^{p(q-1)}$ in the function

$$\frac{1}{(1-x)(1-ax)\ldots(1-ax^p)}$$

Hence we can isolate that portion of the generating function which contains only powers of ax^p. It is

$$1 + \binom{p}{1} ax^p + \binom{p+1}{2} (ax^p)^2 + \binom{p+2}{3} (ax^p)^3 + \ldots\ldots ;$$

or

$$\frac{1}{(1-ax^p)^p}.$$

This fact leads as before to an expansion theorem.

Putting

$$\frac{(1-ax^p)^p}{1-x \,.\, 1-ax \ldots . 1-ax^p} = \frac{(1-ax^{p-1})^{p-1}}{1-x \,.\, 1-ax \ldots . 1-ax^{p-1}} + \mathrm{V}_p(x),$$

then of course

$$\frac{1}{1-x \,.\, 1-ax \ldots . 1-ax^p} = \frac{1}{(1-ax^p)^p} \sum_{p=1}^{p=p} \mathrm{V}_p(x),$$

and we find in succession

$$\mathrm{V}_1(x) = \frac{1}{1-x},$$

$$\mathrm{V}_2(x) = \frac{ax}{1-ax},$$

$$\mathrm{V}_3(x) = \frac{ax^2}{1-ax} + \frac{ax^2(1-ax^3)}{1-ax \,.\, 1-ax^2}$$

$$\mathrm{V}_4(x) = \frac{ax^3(1-ax^3)}{1-ax \,.\, 1-ax^2} + \frac{ax^3(1-ax^4)}{1-ax \,.\, 1-ax^2} + \frac{ax^3(1-ax^4)^2}{1-ax \,.\, 1-ax^2 \,.\, 1-ax^3},$$

and in general

$$\mathrm{V}_p(x) = \frac{ax^{p-1}(1-ax^{p-1})^{p-3}}{1-ax \,.\, 1-ax^2 \ldots\ldots 1-ax^{p-2}} + \frac{ax^{p-1}(1-ax^{p-1})^{p-4}(1-ax^p)}{1-ax \,.\, 1-ax^2 \ldots . 1-ax^{p-2}}$$

$$+ \ldots\ldots + \frac{ax^{p-1}(1-ax^p)^{p-3}}{1-ax \,.\, 1-ax^2 \ldots . 1-ax^{p-2}} + \frac{ax^{p-1}(1-ax^p)^{p-2}}{1-ax \,.\, 1-ax^2 \ldots . 1-ax^{p-2} \,.\, 1-ax^{p-1}}.$$

The simplest cases, omitting the trivial one corresponding to $p = 1$, are

$$\frac{1}{1-x \,.\, 1-ax \,.\, 1-ax^2} = \frac{1}{(1-ax^2)^2}\left\{\frac{1}{1-x} + \frac{ax}{1-ax}\right\};$$

$$\frac{1}{1-x \,.\, 1-ax \,.\, 1-ax^2 \,.\, 1-ax^3} = \frac{1}{(1-ax^3)^3}\left\{\frac{1}{1-x} + \frac{ax}{1-ax} + \frac{ax^2}{1-ax} + \frac{ax^2(1-ax^3)}{1-ax \,.\, 1-ax^2}\right\};$$

$$\frac{1}{1-x.1-ax.1-ax^2.1-ax^3.1-ax^4}=\frac{1}{(1-ax^4)^4}\left\{\frac{1}{1-x}+\frac{ax}{1-ax}+\frac{ax^2}{1-ax}+\frac{ax^2(1-ax^3)}{1-ax.1-ax^2}\right.$$
$$\left.+\frac{ax^3(1-ax^3)}{1-ax.1-ax^2}+\frac{ax^3(1-ax^4)}{1-ax.1-ax^2}+\frac{ax^3(1-ax^4)^2}{1-ax.1-ax^2.1-ax^3}\right\}.$$

The fractions, in the brackets { }, may be united in batches, but I prefer to leave them as written, as the law of development is shown the better. Also the fraction $\frac{1}{1-ax^p}$ may be cancelled if desired. I have not done so, in order to keep in touch with the arithmetic.

Inspection of these expansions establishes the isolation therein independently.

Art. 26. In the function

$$\frac{1}{1-x.1-ax.1-ax^2....1-ax^p},$$

the coefficient of a^{q-1} is, as is well known,

$$\frac{1-x^q.1-x^{q+1}....1-x^{q+p-2}}{1-x.1-x^2....1-x^{p-1}}x^{q-1}.$$

Hence the coefficient of $a^{q-1}x^{p(q-1)}$ in the former is equal to the coefficient of $x^{p(q-1)}$ in the latter; *i.e.*, to the coefficient of $x^{(p-1)(q-1)}$ in

$$\frac{1-x^q.1-x^{q+1}....1-x^{q+p-2}}{(1-x)^2\,1-x^2....1-x^{p-1}},$$

which we know to be $\binom{p+q-2}{p-1}$, verifying our result.

For a given value of s the partitions are enumerated by the coefficient of $a^qb^sx^{pq}$ in the product

$$\frac{1}{1-x}\left(1+\frac{abx}{1-ax}\right)\left(1+\frac{abx^2}{1-ax^2}\right)\ldots\left(\frac{abx^p}{1-ax^p}\right);$$

or, the same thing, by the coefficient of $(ax^p)^{q-1}$ in the function

$$\frac{a^{s-1}}{1-x.1-ax^p}\sum_{\Xi}\frac{x^{k_1+k_2+\ldots+k_{s-1}}}{(1-ax^{k_1})(1-ax^{k_2})\ldots(1-ax^{k_{s-1}})},$$

wherein $k_1, k_2, \ldots k_{s-1}$ denote any selection of $s-1$ different integers drawn from the natural series 1, 2, 3, . . . $p-1$.

This coefficient, from previous work, has the value

$$\binom{p-1}{s-1}\binom{q-1}{s-1}$$

hence the portion of this function which consists only of powers of ax^p is

$$\binom{p-1}{s-1}\left\{(ax^p)^{s-1}+\binom{s}{1}(ax^p)^s+\binom{s+1}{2}(ax^p)^{s+1}+\ldots\right\}$$

or

$$\binom{p-1}{s-1}\frac{(ax^p)^{s+1}}{(1-ax^p)^s}.$$

Hence also from the function

$$\frac{a^{s-1}}{1-x}\Sigma\frac{x^{k_1+k_2+\ldots+k_{s-1}}}{(1-ax^{k_1})(1-ax^{k_2})\ldots(1-ax^{k_{s-1}})}$$

we can isolate the portion

$$\binom{p-1}{s-1}\left(\frac{ax^p}{1-ax^p}\right)^{s-1}.$$

Ex. gr. for $p=3$, $s=2$ we can verify that

$$2\frac{ax^3}{1-ax^3}$$

can be isolated from

$$\frac{ax}{1-x\,.\,1-ax}+\frac{ax^2}{1-x\,.\,1-ax^2}.$$

Art. 27. Before generalising the foregoing it will be proper to give another correspondence between the compositions and partitions of unipartite numbers which leads readily to theorems concerning the generating functions of partitions when the parts are *unrepeated.*

Writing down any composition of the unipartite p, viz.:

$$(p_1p_2p_3\,\ldots\,,p_s),$$

we can at once construct a regularised partition, viz.:—

$$(p_1,\,p_1+p_2,\,p_1+p_2+p_3,\,\ldots\,,p)$$

of the number $sp_1+(s-1)p_2+(s-2)p_3+\ldots+p_s$.

The correspondence is between the compositions of p into s parts and the partitions of all unipartite numbers into s *unequal* parts limited in magnitude to p and possessing a part p.

The numbers whose partitions appear are the natural series extending from

$$p+\binom{s}{2},\text{ to }sp-\binom{s}{2}.$$

For the enumeration we must take the coefficient of $a^s x^{sp-\binom{s}{2}}$ in the development of the generating function

$$\frac{ax^p}{1-x}(1+ax)(1+ax^2)(1+ax^3)\ldots.(1+ax^{p-1});$$

or the coefficient of $(ax^{p-\frac{1}{2}s})^{s-1}$ in

$$\frac{1}{1-x}(1+ax)(1+ax^2)(1+ax^3)\ldots(1+ax^{p-1}).$$

But the number of compositions of p into s parts is

$$\binom{p-1}{s-1},$$

and thence we see that a term in the development of

$$\frac{1}{1-x}(1+ax)(1+ax^2)(1+ax^3)\ldots(1+ax^{p-1}),$$

is

$$\binom{p-1}{s-1}a^{s-1}x^{(p-\frac{1}{2}s)(s-1)};$$

and, giving s successive values, we can isolate, from this product, a portion

$$1+\binom{p-1}{1}ax^{p-1}+\binom{p-1}{2}a^2x^{2p-3}+\binom{p-1}{3}a^3x^{3p-6}+\ldots+a^{p-1}x^{\binom{p}{2}};$$

or symbolically $(1+ax^p)^{p-1}$, where after expansion $x^{\mu p}$ is to be replaced by $x^{\mu p-\frac{1}{2}\mu(\mu+1)}$.

Art. 28. By leaving s unspecified we can readily reach a theorem concerning the product,

$$\frac{1}{1-x}(1+x)(1+x^2)(1+x^3)\ldots(1+x^{p-1}).$$

It is easy to show that the coefficient of $x^{\binom{p}{2}}$, in the development, is 2^{p-1}. We may say that the number of partitions of $\binom{p}{2}$ and all lower numbers into unrepeated parts not exceeding $p-1$ in magnitude is 2^{p-1}. For $p=5$ these partitions are:

4321				
432	43	31	4	
431	42	21	3	, 16 in number.
421	41		2	
321	32		1	
			0	

Art. 29. It is obvious that, by the same process, we can obtain a correspondence

between the compositions and partitions of multipartite numbers. In the bipartite case we pass from any composition

$$(\overline{p_1q_1}\ \overline{p_2q_2}\ \overline{p_3q_3}\cdot\ \ \cdot\overline{p_sq_s})$$

to the regularised partition

$$(\overline{p_1q_1}\ \overline{p_1+p_2,\ q_1+q_2},\ \overline{p_1+p_2+p_3,\ q_1+q_2+q_3}\ldots\overline{pq})$$

of a certain bipartite number.

The correspondence is between the compositions of $\overline{pq}$ into s parts and the partitions of all bipartite numbers into s *unrepeated* biparts, the parts of the biparts being limited in magnitude to p and q respectively, and the highest bipart being $\overline{pq}$.

Or, we may strike out the highest bipart $\overline{pq}$, and then the partition is into $s-1$ unrepeated biparts, the parts of the biparts being limited as before. The partitions are subject to the further restriction that they are regularised in the sense that the unipartite partitions of p and q, that appear in the bipartition, are separately regularised.

Art. 30. Instead of insisting upon this two-fold regularisation, we may, starting from the composition

$$(\overline{p_1q_1},\ \overline{p_2q_2},\ \overline{p_3q_3},\ldots\overline{p_sq_s}),$$

proceed to the singly regularised partition

$$(\overline{p_1q_1}\ \overline{p_1+p_2,\ q_2}\ \overline{p_1+p_2+p_3,\ q_3}\ldots\overline{pq_s}).$$

There are, in fact, various ways of forming connecting links between compositions and partitions of multipartite numbers whatever the order of multiplicity. These methods may be pursued at pleasure so as to obtain results of more or less interest.

§ 3.

Art. 31. The correspondence set forth between unipartite partitions and bipartite compositions naturally suggests the possibility of a similar correspondence between bipartite partitions and tripartite compositions, and generally between m-partite partitions and $m+1$-partite compositions.

For the graph of the tripartite number $\overline{pqr}$, we take $r+1$ similar graphs of the bipartite $\overline{pq}$, and place them similarly with corresponding lines parallel, and like points lying on straight lines; the graph is completed by drawing these straight lines, which are in a new direction, say the γ direction.

There are three directions through each point of the graph (see *loc. cit.*, Art. 31). There are $\binom{p+q+r}{p,\ q}$* lines of route along which the tripartite compositions are

* This notation explains $\frac{(p+q+r)!}{p!\ q!\ r!}$ and so in similar cases.

depicted, one line of route for each permutation of the symbols in the product $\alpha^p\beta^q\gamma^r$. A study of these permutations shows the connection with a certain class of bipartite partitions.

Consider a permutation

$$\alpha^{p_1}\beta^{q_1}\gamma^{r_1}\,\alpha^{p_2}\beta^{q_2}\gamma^{r_2}\,\alpha^{p_3}\beta^{q_3}\gamma^{r_3}\ldots\alpha^{p_s}\beta^{q_s}\gamma^{r_s},$$

which is *not* the most general permutation, but such that, in regard to any section

$$\alpha^{p_k}\beta^{q_k}\gamma^{r_k}$$

of the permutation

(1.) r_k must be superior to zero except when $k = s$.
(2.) p_k, q_k may be either, but not both, zero, except when $k = 1$.

The permutation has $\gamma\alpha$ and $\gamma\beta$ contacts, but no $\beta\alpha$ contact.

In the reticulation corresponding thereto, we have lines of route with $\gamma\alpha$ and $\gamma\beta$ bends but *not* with $\beta\alpha$ bends. All the lines of route with $\beta\alpha$ bends are excluded from consideration. From the permutation we can form a bipartite partition.

$$(\overline{p_1q_1}^{\,r_1}\ \overline{p_1+p_2, q_1+q_2}^{\,r_2}\ \overline{p_1+p_2+p_3, q_1+q_2+q_3}^{\,r_3}\ldots\overline{p_1+p_2+\ldots+p_s, q_1+q_2+\ldots+q_s}^{\,r_s}),$$

which is regularised in the sense that the partitions of the unipartites p, q, that appear are each separately regularised.

The two parts of the bipartite number thus partitioned are

$$r_1p_1 + r_2(p_1+p_2) + r_3(p_1+p_2+p_3) + \ldots + r_s(p_1+p_2+\ldots+p_s),$$
$$r_1q_1 + r_2(q_1+q_2) + r_3(q_1+q_2+q_3) + \ldots + r_s(q_1+q_2+\ldots+q_s).$$

The associated principal composition is

$$(\overline{p_1q_1r_1}\ \overline{p_2q_2r_2}\ \overline{p_3q_3r_3}\ldots\overline{p_sq_sr_s}).$$

As before, consider the contacts r_1p_2, r_2p_3, &c. ... Looking at the whole of the principal compositions, observe that a $\gamma\alpha$ contact in the permutation yields a contact r_kp_{k+1} in the composition in which r_k and p_{k+1} are both superior to zero, say a positive-positive contact. A $\gamma\beta$ contact yields a positive-zero contact and a $\beta\alpha$ contact a zero-positive contact. Hence the present correspondence is only concerned with compositions which possess positive-positive and positive-zero contacts, and not with those which involve contacts of other natures. The bipartite partitions are those of all bipartite numbers into biparts whose parts are limited to p and q respectively in magnitude and whose biparts are limited to r in number.

Art. 32. We have then a one-to-one correspondence. Each bipartite partition of the nature considered is represented graphically by a line of route in a tripartite reticulation. If we please we may regard a pair of bipartite partitions as represented by a line of route, for from the permutation we are also led to the complementary partition,

$$(\overline{p - p_1, q - q_1}^{\,r_1} \quad \overline{p - p_1 - p_2, q - q_1 - q_2}^{\,r_2} \dots),$$

in which p_1, q_1 *may* be both zero.

Art. 33. It has been shown that the number of lines of route which possess

$$\begin{aligned} &s_{21}\ \beta\alpha \text{ bends,} \\ &s_{32}\ \gamma\beta \quad ,, \quad , \\ &s_{31}\ \gamma\alpha \quad ,, \quad , \end{aligned}$$

is

$$\binom{s_{21} + s_{31}}{s_{21}} \binom{p}{s_{21} + s_{31}} \binom{q}{s_{32}} \binom{q + s_{31}}{s_{21} + s_{31}} \binom{r}{s_{32} + s_{31}};*$$

and that this number is the coefficient of

$$\lambda_{21}{}^{s_{21}} \lambda_{32}{}^{s_{32}} \lambda_{31}{}^{s_{31}} \alpha^p \beta^q \gamma^r$$

in the development of

$$(\alpha + \lambda_{21} \beta + \lambda_{31} \gamma)^p (\alpha + \beta + \lambda_{32})^q (\alpha + \beta + \gamma)^r.$$

Here $s_{21} = 0$, and the number in question becomes

$$\binom{p}{s_{31}} \binom{q}{s_{32}} \binom{q + s_{31}}{s_{31}} \binom{r}{s_{32} + s_{31}},$$

whilst the generating function becomes

$$(\alpha + \lambda_{31} \gamma)^p (\alpha + \beta + \lambda_{32} \gamma)^q (\alpha + \beta + \gamma)^r.$$

In this the coefficient of

$$\lambda_{32}{}^{s_{32}} \lambda_{31}{}^{s_{31}} \alpha^p \beta^q \gamma^r$$

is equal to the coefficient of the same term in the expansion of the fraction

$$\frac{1}{1 - \alpha - \beta - \gamma + \alpha\beta + (1 - \lambda_{32}) \beta\gamma + (1 - \lambda_{31}) \alpha\gamma - (1 - \lambda_{32}) \alpha\beta\gamma},$$

which is

$$\frac{1}{(1 - \alpha)(1 - \beta)(1 - \gamma) - \lambda_{31} \alpha\gamma - \lambda_{32} \beta\gamma (1 - \alpha)},$$

and the verification is readily carried out.

* 'Phil. Trans.,' vol. 184 (*loc. cit.*), Arts. 34, *et seq.*

Art. 34. The number of lines of route which possess exactly s_{31} $\gamma\alpha$ bends and no $\beta\alpha$ bends is

$$\binom{p}{s_{31}}\binom{q+s_{31}}{s_{31}}\sum_{s_{32}}\binom{q}{s_{32}}\binom{r}{s_{32}+s_{31}}$$

or

$$\binom{p}{s_{31}}\binom{q+s_{31}}{s_{31}}\binom{q+r}{q+s_{31}} \text{ or } \binom{p}{s_{31}}\binom{r}{s_{31}}\binom{q+r}{r}.$$

Also, by putting $\lambda_{31} = \lambda_{32} = \lambda$ in the generating function, we find that the number of lines of route which have s $\gamma\beta$ and $\gamma\alpha$ bends but no $\beta\alpha$ bend is given by the coefficient of $\alpha^p\beta^q\gamma^r$ in

$$\frac{(\alpha\gamma+\beta\gamma-\alpha\beta\gamma)^s}{(1-\alpha)^{s+1}(1-\beta)^{s+1}(1-\gamma)^{s+1}}.$$

Art. 35. It will be convenient to give the complete correspondence in the case of some simple tripartite number, say, $\overline{222}$.

S_{31}	S_{32}	Permutation.	Composition.	Partition.	Number partitioned.
0	0	$\alpha^2\beta^2\gamma^2$	$(\overline{222})$	$(\overline{22}^2)$	$\overline{44}$
1	0	$\alpha\beta^2\gamma\alpha\gamma$	$(\overline{121}\ \overline{101})$	$(\overline{12}\ \overline{22})$	$\overline{34}$
1	0	$\alpha\beta^2\gamma^2\alpha$	$(\overline{122}\ \overline{100})$	$(\overline{12}^2)$	$\overline{24}$
1	0	$\beta^2\gamma\alpha^2\gamma$	$(\overline{021}\ \overline{201})$	$(\overline{02}\ \overline{22})$	$\overline{24}$
1	0	$\alpha\beta\gamma\alpha\beta\gamma$	$(\overline{111}\ \overline{111})$	$(\overline{11}\ \overline{22})$	$\overline{33}$
1	0	$\alpha\gamma\alpha\beta^2\gamma$	$(\overline{101}\ \overline{121})$	$(\overline{10}\ \overline{22})$	$\overline{32}$
1	0	$\beta\gamma\alpha^2\beta\gamma$	$(\overline{011}\ \overline{211})$	$(\overline{01}\ \overline{22})$	$\overline{23}$
1	0	$\beta^2\gamma^2\alpha^2$	$(\overline{022}\ \overline{200})$	$(\overline{02}^2)$	$\overline{04}$
1	0	$\alpha\beta\gamma^2\alpha\beta$	$(\overline{112}\ \overline{110})$	$(\overline{11}^2)$	$\overline{22}$
1	0	$\gamma\alpha^2\beta^2\gamma$	$(\overline{001}\ \overline{221})$	$(\overline{00}\ \overline{22})$	$\overline{22}$
1	0	$\alpha\gamma^2\alpha\beta^2$	$(\overline{102}\ \overline{120})$	$(\overline{10}^2)$	$\overline{20}$
1	0	$\beta\gamma^2\alpha^2\beta$	$(\overline{012}\ \overline{210})$	$(\overline{01}^2)$	$\overline{02}$
1	0	$\gamma^2\alpha^2\beta^2$	$(\overline{002}\ \overline{220})$	$(\overline{00}^2)$	$\overline{00}$
0	1	$\alpha^2\beta\gamma\beta\gamma$	$(\overline{211}\ \overline{011})$	$(\overline{21}\ \overline{22})$	$\overline{43}$
0	1	$\alpha^2\beta\gamma^2\beta$	$(\overline{212}\ \overline{010})$	$(\overline{21}^2)$	$\overline{42}$
0	1	$\alpha^2\gamma\beta^2\gamma$	$(\overline{201}\ \overline{021})$	$(\overline{20}\ \overline{22})$	$\overline{42}$
0	1	$\alpha^2\gamma^2\beta^2$	$(\overline{202}\ \overline{020})$	$(\overline{20}^2)$	$\overline{40}$
1	1	$\alpha\beta\gamma\alpha\gamma\beta$	$(\overline{111}\ \overline{101}\ \overline{010})$	$(\overline{11}\ \overline{21})$	$\overline{32}$
1	1	$\alpha\beta\gamma\beta\gamma\alpha$	$(\overline{111}\ \overline{011}\ \overline{100})$	$(\overline{11}\ \overline{12})$	$\overline{23}$

S_{31}	S_{32}	Permutation.	Composition.	Partition.	Number partitioned.
1	1	$\alpha\gamma\alpha\beta\gamma\beta$	$(\overline{101}\ \overline{111}\ \overline{010})$	$(\overline{10}\ \overline{21})$	$\overline{31}$
1	1	$\beta\gamma\alpha^2\gamma\beta$	$(\overline{011}\ \overline{201}\ \overline{010})$	$(\overline{01}\ \overline{21})$	$\overline{22}$
1	1	$\alpha\gamma\beta^2\gamma\alpha$	$(\overline{101}\ \overline{021}\ \overline{100})$	$(\overline{10}\ \overline{12})$	$\overline{22}$
1	1	$\alpha\gamma\alpha\gamma\beta^2$	$(\overline{101}\ \overline{101}\ \overline{020})$	$(\overline{10}\ \overline{20})$	$\overline{30}$
1	1	$\beta\gamma\beta\gamma\alpha^2$	$(\overline{011}\ \overline{011}\ \overline{200})$	$(\overline{01}\ \overline{02})$	$\overline{03}$
1	1	$\gamma\alpha^2\beta\gamma\beta$	$(\overline{001}\ \overline{211}\ \overline{010})$	$(\overline{00}\ \overline{21})$	$\overline{21}$
1	1	$\alpha\gamma\beta\gamma\alpha\beta$	$(\overline{101}\ \overline{011}\ \overline{110})$	$(\overline{10}\ \overline{11})$	$\overline{21}$
1	1	$\gamma\alpha^2\gamma\beta^2$	$(\overline{001}\ \overline{201}\ \overline{020})$	$(\overline{00}\ \overline{20})$	$\overline{20}$
1	1	$\gamma\beta^2\gamma\alpha^2$	$(\overline{001}\ \overline{021}\ \overline{200})$	$(\overline{00}\ \overline{02})$	$\overline{02}$
1	1	$\gamma\beta\gamma\alpha^2\beta$	$(\overline{001}\ \overline{011}\ \overline{210})$	$(\overline{00}\ \overline{01})$	$\overline{01}$
2	0	$\beta^2\gamma\alpha\gamma\alpha$	$(\overline{021}\ \overline{101}\ \overline{100})$	$(\overline{02}\ \overline{12})$	$\overline{14}$
2	0	$\beta\gamma\alpha\beta\gamma\alpha$	$(\overline{011}\ \overline{111}\ \overline{100})$	$(\overline{01}\ \overline{12})$	$\overline{13}$
2	0	$\gamma\alpha\beta^2\gamma\alpha$	$(\overline{001}\ \overline{121}\ \overline{100})$	$(\overline{00}\ \overline{12})$	$\overline{12}$
2	0	$\beta\gamma\alpha\gamma\alpha\beta$	$(\overline{011}\ \overline{101}\ \overline{110})$	$(\overline{01}\ \overline{11})$	$\overline{12}$
2	0	$\gamma\alpha\beta\gamma\alpha\beta$	$(\overline{001}\ \overline{111}\ \overline{110})$	$(\overline{00}\ \overline{11})$	$\overline{11}$
2	0	$\gamma\alpha\gamma\alpha\beta^2$	$(\overline{001}\ \overline{101}\ \overline{120})$	$(\overline{00}\ \overline{10})$	$\overline{10}$
0	2	$\alpha^2\gamma\beta\gamma\beta$	$(\overline{201}\ \overline{011}\ \overline{010})$	$(\overline{20}\ \overline{21})$	$\overline{41}$

There are 36 partitions.

The first two columns show the nature of the permutation in regard to $\gamma\alpha$ and $\gamma\beta$ contacts and the nature of the composition in regard to positive-positive and positive-zero contacts. The partitions are into two parts, zero not excluded, and have regard to bipartite numbers extending from $\overline{44}$ to $\overline{00}$. They are doubly regularised by ascending magnitude, and the figures of the parts do not exceed 2, 2 the first two figures of the tripartite.

If we write down the partitions of 4 into two parts, zeros not excluded, limited not to exceed 2 in magnitude, viz. :—

22, 12, 02, 11, 01, 00,

the ascending order of part magnitude being adhered to, we can obtain one of the 36 partitions by combining any one of these partitions with itself or any other of the 6.

Thus the fourth of the above partitions is obtained by combining the unipartite partitions

02, 22,

and from any two unipartite partitions

ab, cd,

we proceed to the bipartite partition

$$(\overline{ac}\quad \overline{bd}).$$

The number is thus shown to be $6 \times 6 = 36$.

Art. 36. In general, when the tripartite is $\overline{pqr}$, the partitions are into r parts, zeros not excluded, the first and second figures of the biparts being limited to p and q respectively.

The bipartite numbers partitioned extend from

$$\overline{p \times r, q \times r} \quad \text{to} \quad \overline{00}.$$

The partitions are doubly regularised and may be enumerated by observing that we have to combine every partition of $p \times r$ and lower unipartite numbers into r parts, zeros not excluded, and no part exceeding p in magnitude, with every partition of $q \times r$ and lower numbers into r parts, zeros not excluded, and no part exceeding q in magnitude.

Hence (see *ante*, Art. 12) the number of partitions is

$$\binom{p+r}{r}\binom{q+r}{r}.$$

This expression also enumerates (1) the compositions which have only positive-positive and positive-zero contacts; (2) the lines of route in the tripartite reticulation which are without $\beta\alpha$ bends; (3) the permutations of $\alpha^p\beta^q\gamma^r$ which are without $\beta\alpha$ contacts.

Art. 37. The truth of the theorem may be seen also as follows :—Suppose a solid reticulation and take the directions α, β, γ as axes of x, y, and z meeting at the origin of the lines of route. The face of the solid in the plane xz is a *bipartite* reticulation in which $\binom{p+r}{r}$ lines of route may be drawn; similarly $\binom{q+r}{r}$ lines of route may be drawn in the bipartite reticulation which lies in the plane yz. One of the former lines of route is an orthogonal projection of a tripartite line of route on the plane xz; one of the latter is an orthogonal projection on the plane yz; any one of the former may be associated with any one of the latter, and such a pair uniquely determines a tripartite line of route which does not possess $\beta\alpha$ bends. This may be clearly seen by considering the permutation

$$\alpha^{p_1}\beta^{q_1}\gamma^{r_1}\ \alpha^{p_2}\beta^{q_2}\gamma^{r_2}\ldots;$$

suppression alternately of the letters β and α yields two permutations, viz. :—

$$\alpha^{p_1}\gamma^{r_1}\alpha^{p_2}\gamma^{r_2}\dots$$
$$\beta^{q_1}\gamma^{r_1}\beta^{q_2}\gamma^{r_2}\dots$$

which express the bipartite lines of route which are the projections on the planes xz, yz respectively. Since the tripartite permutation involves no $\beta\alpha$ contacts, we see that these two permutations uniquely determine the permutation

$$\alpha^{p_1}\beta^{q_1}\gamma^{r_1}\alpha^{p_2}\beta^{q_2}\gamma^{r_2}\dots$$

Hence the number of lines of route in question is

$$\binom{p+r}{r}\binom{q+r}{r}.$$

Art. 38. Hence also the interesting summation formula

$$\sum_{s_{31}}\sum_{s_{32}}\binom{p}{s_{31}}\binom{q}{s_{32}}\binom{q+s_{31}}{s_{31}}\binom{r}{s_{31}+s_{32}}=\binom{p+r}{r}\binom{q+r}{r}.$$

Observe that the expression further enumerates the lines of route with r, $\beta\alpha$ bends in the reticulation of the bipartite $\overline{p+r,\, q+r}$.

A generating function which enumerates these partitions is

$$\frac{1}{1-x\,.\,1-a\,.\,1-ax\dots 1-ax^p\,1-y\,.\,1-b\,.\,1-by\dots 1-by^q},$$

in which the coefficient of $(abx^py^q)^r$ must be sought.

The compositions that appear are the principal ones along lines of route which have no $\beta\alpha$ bends. We may strike out the last part of the composition whenever its last figure is zero, and then the compositions are not of the single tripartite $\overline{222}$, but of the 9 tripartites extending from $\overline{222}$ to $\overline{002}$, the last figure being 2, and the first two figures not exceeding 2, 2 respectively. The compositions are into 2, or fewer parts. Generally the compositions appear of the $(p+1)(q+1)$ tripartites extending from $\overline{pqr}$ to $\overline{00r}$, the last figure being r, and the first two figures not exceeding p, q, respectively. The compositions are into r, or fewer parts, no part having the last figure zero.

The partitions present themselves in complementary pairs. To every partition $(\overline{ab}\;\overline{cd}\dots)$ corresponds another $(\overline{p-a,\, q-b},\ \overline{p-c,\, q-d}\dots)$ the numbers partitioned being respectively $\overline{a+c+\dots,\, b+d+\dots}$ and $\overline{rp-a-c-\dots,\, rq-b-d-\dots}$ *Ex. gr.*, the complementary partitions $(\overline{02}\;\overline{22})$, $(\overline{20}\;\overline{00})$ of the bipartites $\overline{24}$, $\overline{20}$ Certain partitions are self-complementary. The number partitioned is then $\overline{\tfrac{1}{2}rp,\, \tfrac{1}{2}rq}$.

Art 39. We may enumerate the partitions which, excluding zero, involve k different parts. Let s_{23}, s_{13}, represent the number of $\beta\gamma$ and $\alpha\gamma$ contracts in a permutation. If $s_{23} + s_{13} = k$, the corresponding partition possesses k different parts other than zero. The lines of route are such as have no $\beta\alpha$ bends, s_{13} $\alpha\gamma$ bends and s_{23} $\beta\gamma$ bends. Reversing the permutation we have a similar number of lines of route which have no $\alpha\beta$ bends, s_{13} $\gamma\alpha$ bends, and s_{23} $\gamma\alpha$ bends. Now interchange α and β and replace the reticulation of the tripartite $\overline{pqr}$ by that of $\overline{qpr}$. In this new reticulation we have the same number of lines of route which have no $\beta\alpha$ bends, s_{13} $\gamma\beta$ bends, and s_{23} $\gamma\alpha$ bends. This number has been shown to be

$$\binom{q}{s_{23}}\binom{p+s_{23}}{s_{23}}\binom{p}{s_{13}}\binom{r}{s_{13}+s_{23}}.$$

Art. 40. Hence the bipartite partitions possessing k different parts other than zero are enumerated by

$$\binom{r}{k}\sum_{s_{23}}\binom{q}{s_{23}}\binom{p+s_{23}}{s_{23}}\binom{p}{k-s_{23}}.$$

Theorem.—Having under consideration the doubly-regularised partitions of all bipartite numbers into r parts, zero parts included, such that the figures of the parts are limited in magnitude to p and q respectively, the number of partitions which possess exactly k different parts, other than zero, is

$$\binom{r}{k}\sum_{s_{23}}\binom{q}{s_{23}}\binom{p+s_{23}}{s_{23}}\binom{p}{k-s_{23}},$$

s_{23} assuming all compatible values.

This result may be verified in the case of the tripartite $\overline{222}$ from the table given above. As an additional verification, consider the tripartite $\overline{123}$. For $k = 2$, we have

Permutations.	Compositions.	Partitions.
$\gamma\alpha\beta\gamma\beta\gamma$	$(\overline{001}\ \overline{111}\ \overline{011})$	$(\overline{00}\ \overline{11}\ \overline{12})$
$\gamma\alpha\gamma\beta^2\gamma$	$(\overline{001}\ \overline{101}\ \overline{021})$	$(\overline{00}\ \overline{10}\ \overline{12})$
$\gamma\alpha\gamma\beta\gamma\beta$	$(\overline{001}\ \overline{101}\ 011)$	$(\overline{00}\ \overline{10}\ \overline{11})$
$\alpha\beta\gamma^2\beta\gamma$	$(\overline{112}\ \overline{011})$	$(\overline{11}^2\ \overline{12})$
$\alpha\beta\gamma\beta\gamma^2$	$(\overline{111}\ \overline{012})$	$(\overline{11}\ \overline{12}^2)$
$\alpha\gamma^2\beta^2\gamma$	$(\overline{102}\ \overline{021})$	$(\overline{10}^2\ \overline{12})$
$\alpha\gamma^2\beta\gamma\beta$	$(\overline{102}\ \overline{011})$	$(\overline{10}^2\ \overline{11})$
$\alpha\gamma\beta\gamma^2\beta$	$(\overline{101}\ \overline{012})$	$(\overline{10}\ \overline{11}^2)$
$\alpha\gamma\beta^2\gamma^2$	$(\overline{101}\ \overline{022})$	$(\overline{10}\ \overline{12}^2)$

Permutations.	Compositions.	Partitions.
$\gamma\beta\gamma\beta\gamma\alpha$	$(\overline{001}\ \overline{011}\ \overline{011})$	$(\overline{00}\ \overline{01}\ \overline{02})$
$\gamma\beta\gamma\alpha\beta\gamma$	$(\overline{001}\ \overline{011}\ \overline{111})$	$(\overline{00}\ \overline{01}\ \overline{12})$
$\beta^2\gamma^2\alpha\gamma$	$(\overline{022}\ \overline{101})$	$(\overline{02}^2\ \overline{12})$
$\beta^2\gamma\alpha\gamma^2$	$(\overline{021}\ \overline{102})$	$(\overline{02}\ \overline{12}^2)$
$\beta\gamma^2\alpha\beta\gamma$	$(\overline{012}\ \overline{111})$	$(\overline{01}^2\ \overline{12})$
$\beta\gamma^2\alpha\gamma\beta$	$(\overline{012}\ \overline{101})$	$(\overline{01}^2\ \overline{11})$
$\beta\gamma^2\beta\gamma\alpha$	$(\overline{012}\ \overline{011})$	$(\overline{01}^2\ \overline{02})$
$\beta\gamma\alpha\beta\gamma^2$	$(\overline{011}\ \overline{112})$	$(\overline{01}\ \overline{12}^2)$
$\beta\gamma\alpha\gamma^2\beta$	$(\overline{011}\ \overline{102})$	$(\overline{01}\ \overline{11}^2)$
$\beta\gamma\beta\gamma^2\alpha$	$(\overline{011}\ \overline{012})$	$(\overline{01}\ \overline{02}^2)$
$\gamma\beta^2\gamma\alpha\gamma$	$(\overline{001}\ \overline{021}\ \overline{101})$	$(\overline{00}\ \overline{02}\ \overline{12})$
$\gamma\beta\gamma\alpha\gamma\beta$	$(\overline{001}\ \overline{011}\ \overline{101})$	$(\overline{00}\ \overline{01}\ \overline{11})$

21 partitions; while for the enumeration, giving s_{23} the values 0, 1, 2 in succession with $k = 2$, $p = 1$, $q = 2$, $r = 3$.

$$\binom{3}{2}\left\{\binom{2}{0}\binom{1+0}{0}\binom{1}{2}+\binom{2}{1}\binom{1+1}{1}\binom{1}{1}+\binom{2}{2}\binom{1+2}{2}\binom{1}{0}\right\}$$
$$= 3\,(1 \times 1 \times 0 + 2 \times 2 \times 1 + 1 \times 3 \times 1) = 3 \times 7 = 21.$$

The foregoing particular theory of the correspondence that exists between tripartite compositions and bipartite partitions is, for present purposes, sufficiently indicative of the general correspondence between $(m + 1)$-partite compositions and a certain regularised class of m-partite partitions.

§ 4. Constructive Theory.

Art. 41. Given a line of route in a bipartite reticulation it may be necessary to enumerate the lines of route which lie altogether on either side of it.

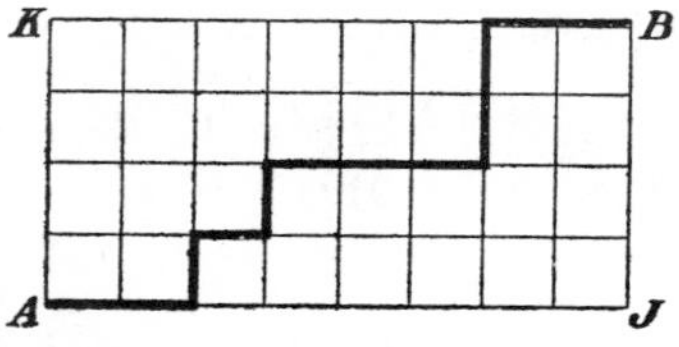

Thus in respect of the line of route delineated in the reticulation AB, lines of route exist which, throughout their entire course, are either coincident with it, or lie on the side of it towards J. Such lines of route may be termed inferior or subjacent to the given line of route. Similarly those lines of route which, everywhere, are either coincident with the given line, or on the side remote from J, may be termed superior or superjacent lines of route in respect of the given line. All lines are thus accounted for with the exception of those which cross the given line passing from the side towards J to the side remote from J, or *vice versâ*; these may be termed transverse lines in respect of the given line.

Art. 42. I am concerned, at present, with those lines which are subjacent to a given line, though it will be remarked that the superjacent and transverse lines also suggest questions of interest. A given line of route defines a bipartite principal composition

$$(\overline{p_1q_1}\ \overline{p_2q_2}\ \ldots),$$

and a unipartite south-easterly partition

$$(\overline{p-p_1}^{q_1}\ \overline{p-p_1-p_2}^{q_2}\ \ldots).$$

The bipartite compositions and the unipartite partitions, defined by the subjacent lines of route, are termed subjacent to the given composition and the given partition respectively.

We may draw a number of lines of route, each of which is subjacent to the given line and not transverse to any other of the number. We thus obtain what may be termed a subjacent succession of lines giving rise to a subjacent succession of unipartite partitions.

These regularised partitions may be

$$(a_1a_2a_3\ldots),\quad (b_1b_2b_3\ldots),\quad (c_1c_2c_3\ldots)\ \ldots.$$

and they are such that the partitions

$$(a_1b_1c_1\ldots),\ (a_2b_2c_2\ldots),\ (a_3b_3c_3\ldots).$$

are also regularised.

It is clear also that the subjacent succession of lines represents the multipartite partition

$$(\overline{a_1a_2a_3\ldots},\ \overline{b_1b_2b_3\ldots},\ \overline{c_1c_2c_3\ldots},\ \overline{\ldots\ldots})$$

of the multipartite numbers

$$(\overline{a_1+b_1+c_1+\ldots},\ \overline{a_2+b_2+c_2+\ldots},\ \overline{a_3+b_3+c_3+\ldots},\ \overline{\ldots}).$$

This partition may be termed "graphically regularised" by reason of its origination in a subjacent succession of lines in the bipartite graph. This species of regularisation is the natural extension to three dimensions of SYLVESTER's graphical method in two dimensions.

Art. 43. SYLVESTER represents the partition $(a_1 a_2 a_3 \ldots)$ of a unipartite number A by the graph

the lines containing $a_1, a_2, a_3 \ldots$ nodes successively.

The same graph also represents a multipartite number $(\overline{a_1 a_2 a_3 \ldots})$ whose content is A, viz.,

$$a_1 + a_2 + a_3 + \ldots = \mathrm{A}.$$

SYLVESTER's theory is, in fact, not only a theory of the partitions of a number A, but also a theory of the multipartite numbers whose content is A. For purpose of generalization I prefer to regard it from the latter point of view.

If we consider the graphically regularised partition

$$(\overline{a_1 a_2 a_3 \ldots}, \overline{b_1 b_2 b_3 \ldots}, \overline{c_1 c_2 c_3 \ldots}, \overline{\ldots\ldots})$$

of the multipartite number

$$(\overline{a_1 + b_1 + c_1 + \ldots}, \overline{a_2 + b_2 + c_2 + \ldots}, \overline{a_3 + b_3 + c_3 + \ldots}, \overline{\ldots\ldots})$$

and write down the Sylvester-graphs of the multipartite numbers which are the parts of the partition

it is clear that we may pile B upon A, and then C upon B, &c., and thus form a three-dimensional graph of the partition

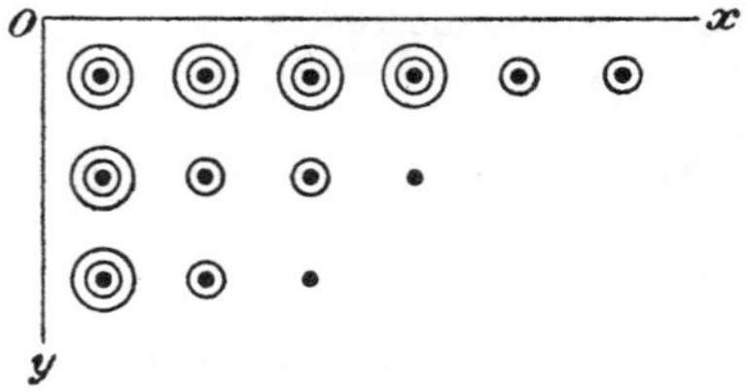

which is regularised in three-dimensions just as the Sylvester-graphs are regularised in two.

4 O 2

This representation is only possible when the subjacent succession of lines is insisted upon.

Art. 44. Every Sylvester-graph in two dimensions is representative of two unipartite partitions; it may, in fact, be read by lines or by columns, and when the two readings are identical the graph is said to be self conjugate.

In this enlarged theory every graph denotes 3! graphically-regularised multipartite partitions; of the same total content, but not, as a rule, appertaining to the same multipartite number.

Take coordinate axes as shown, the axis of z being perpendicular to the plane of the paper. We read as follows:—

Planes parallel to the plane xy and in direction Ox

$$(\overline{643}\ \overline{632}\ \overline{411}).$$

Planes parallel to plane xy and in direction Oy

$$(\overline{333211}\ \overline{332111}\ \overline{311100}).$$

Planes parallel to plane yz and in direction Oy

$$(\overline{333}\ \overline{331}\ \overline{321}\ \overline{211}\ \overline{110}\ \overline{110}).$$

Planes parallel to plane yz and in direction Oz

$$(\overline{333}\ \overline{322}\ \overline{321}\ \overline{310}\ \overline{200}\ \overline{200}).$$

Planes parallel to plane zx and in direction Oz

$$(\overline{333322}\ \overline{322100}\ \overline{321000}).$$

Planes parallel to plane zx and in direction Ox

$$(\overline{664}\ \overline{431}\ \overline{321}),$$

the multipartite numbers, of which these are partitions, being

$$\begin{array}{ll} (\overline{16, 8, 6}), & \\ (\overline{976422}), & \text{content} \\ (\overline{13, 11, 6}), & \quad 30 \\ (\overline{16, 8, 6}), & \\ (\overline{976422}), & \\ (\overline{13, 11, 6}). & \end{array}$$

The graph is therefore representative of three multipartite numbers and of two partitions of each.

Art. 45. A multipartite number has two characteristics. It may be r-partite, *i.e.*, it may consist of r figures, and its highest figure may be p. A multipartite partition has three characteristics. Each part may be r-partite; the highest figure may be p; the number of parts may be q. If the graph be formed of a multipartite partition with characteristics

$$r, p, q,$$

the five other readings yield partitions with characteristics :—

$$\begin{matrix} p, & r, & q \\ q, & r, & p \\ r, & q, & p \\ p, & q, & r \\ q, & p, & r. \end{matrix}$$

The six partitions correspond to the six permutations of the three symbols p, q, r.

The two partitions which are r-partite appertain to the same multipartite number; similarly for the pairs which are p-partite and q-partite respectively. Hence the three multipartite numbers involved correspond to the three pairs of permutations so formed that in any pair the commencing symbol of each permutation is the same.

Art. 46. The consideration of graphs formed with a *given number* of nodes now leads to the theorem: "The enumeration of the graphically regularised r-partite partitions, into q parts and having p for the highest figure, gives the same number for each of the six ways in which the numbers p, q, r may be permuted."

Also the theorem :—

"The enumeration of the graphically regularised partitions which are at most r-partite, into q or fewer parts, the highest figure not exceeding p, gives the same number for each of the six ways in which the numbers p, q, r may be permuted."

The first theorem is concerned with fixed values of p, q, and r; the second with restricted values of these numbers. It is also clear that we may fix one or two of the numbers and leave the remaining two or one restricted.

Observe that this six-fold conjugation obtains even though equalities exist between the numbers p, q, r; they must be regarded always as different numbers. Sometimes, as we shall see, the correspondence is less than six-fold, but this does not depend solely upon the assignment of the numbers p, q, r.

If we regard the multipartite number appertaining to a partition and not merely the total content, we find that the partitions occur in pairs.

Quâ a given multipartite number, a partition which has q parts and a highest figure p is in association with one which has p parts and a highest figure q.

Thus of the multipartite number $(\overline{13.11.6})$ we have the partitions

$$(\overline{333}\ \overline{331}\ \overline{321}\ \overline{211}\ \overline{110}\ \overline{110})$$

$$(\overline{664}\ \overline{431}\ \overline{321})$$

derived from the above written graph.

Art. 47. It is interesting to view the two-dimensional Sylvester-graphs from the three-dimensional standpoint.

Consider the graph

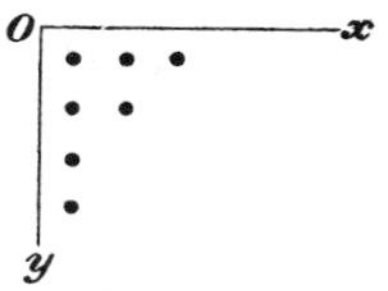

which, following SYLVESTER, denotes the unipartite partition (3211) of the unipartite number 7.

In this paper, the graph, read Sylvester-wise in the plane xy and in direction Ox, denotes the multipartite number $(\overline{3211})$ of content 7. SYLVESTER's conjugate reading, plane xy and direction Oy, gives the partition (421) but here denotes the multipartite number $(\overline{421})$. There are four other readings in this theory. The six readings are

Plane	xy	Direction	Ox	$(\overline{3211})$	$(p, q, r) = (3, 1, 4)$
,,	xy	,,	Oy	$(\overline{421})$	$(p, q, r) = (4, 1, 3)$
,,	yz	,,	Oy	(421)	$(p, q, r) = (4, 3, 1)$
,,	yz	,,	Oz	$(\overline{1111}\ \overline{1100}\ \overline{1000})$	$(p, q, r) = (1, 3, 4)$
,,	zx	,,	Oz	(3211)	$(p, q, r) = (3, 4, 1)$
,,	zx	,,	Ox	$(\overline{111}\ \overline{110}\ \overline{100}\ \overline{100})$	$(p, q, r) = (1, 4, 3)$.

The three multipartite numbers

$$(7),\ \overline{421},\ (\overline{3211})$$

appear each in two partitions.

In general we establish, in regard to Sylvester-graphs, the six-fold correspondence between

(1) r-partite partitions, containing 1 part and a highest figure p.

(2) p-partite partitions, containing 1 part and a highest figure r.

(3) unipartite partitions, containing p parts and a highest figure r.
(4) r-partite partitions, containing p parts and a highest figure 1.
(5) unipartite partitions, containing r parts and a highest figure p.
(6) p-partite partitions, containing r parts and a highest figure 1.

In this enunciation we may substitute for r or p, or for both, the phrases "not exceeding r," "not exceeding p."

Art. 48. For a given number of nodes, in the simplest cases, it will be suitable to view the graphs of the graphically regularised partitions.

Omitting the trivial case of a single node, we have

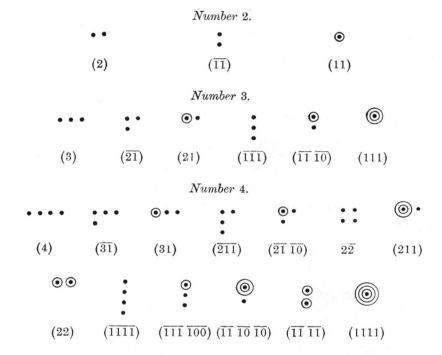

Number 2.

(2) $(\overline{11})$ (11)

Number 3.

(3) $(\overline{21})$ (21) $(\overline{111})$ $(\overline{11}\ \overline{10})$ (111)

Number 4.

(4) $(\overline{31})$ (31) $(\overline{211})$ $(\overline{21}\ \overline{10})$ $2\overline{2}$ (211)

(22) $(\overline{1111})$ $(\overline{111}\ \overline{100})$ $(\overline{11}\ \overline{10}\ \overline{10})$ $(\overline{11}\ \overline{11})$ (1111)

The table is continued in an obvious manner. The *essentially* distinct graphs are for the

Number 2.

Number 3.

Number 4.

and the whole of the partitions are obtainable by reading them in the various ways above explained.

Art. 49. It will be convenient to adopt in future another notation for the graphs; the number m will denote a vertical column of m nodes piled upon one another. The 13 graphs appertaining to the number 4 are written

1111	111 1	211	11 1 1	21 1	11 11	31
(4)	$(\overline{31})$	(31)	$(\overline{211})$	$(\overline{21}\ \overline{10})$	$(2\overline{2})$	(211)

22	1 1 1 1	2 1 1	3 1	2 2	4
(22)	$(\overline{1111})$	$(\overline{111}\ \overline{100})$	$(\overline{11}\ \overline{10}\ \overline{10})$	$(\overline{11}\ \overline{11})$	(1111).

The essentially distinct graphs with the partitions appertaining to them are

1111	111 1	21 1	11 11
(4) $(\overline{1111})$ (1111)	$(\overline{31})$ (31) $(\overline{211})$ (211) $(\overline{111}\ \overline{100})$ $(\overline{11}\ \overline{10}\ \overline{10})$	$(\overline{21}\ \overline{10})$	$(\overline{22})$ (22) $(\overline{11}\ \overline{11})$

Art. 50. Such graphs are either symmetrical, quasi-symmetrical, or unsymmetrical. The symmetrical graphs have three dimensional symmetry, and yield only one partition each.

The quasi-symmetrical have two-dimensional symmetry and yield three partitions each. The unsymmetrical yield each six partitions.

If F (x) be the enumerating generating function to a given content, we may write

$$F(x) = f_1(x) + 3f_2(x) + 6f_3(x),$$

* The interesting question arises as to the enumeration of the essentially distinct graphs of given content.

$f_1(x), f_2(x), f_3(x)$ being the generating functions for the essentially distinct graphs which are symmetrical, quasi-symmetrical, and unsymmetrical respectively.

Also we may write

$$F(x) = F_1(x) + F_2(x) + F_3(x),$$

where $F_1(x)$, $F_2(x)$, $F_3(x)$ are the generating functions of the partitions of the three natures.

The present theory is really the solidification of SYLVESTER's theory given in the 'American Journal of Mathematics' (*loc. cit.*). Already we have seen that the Sylvester-graphs are susceptible of a far wider interpretation than was at first anticipated. If we view these graphs from a two-dimensional standpoint, every graph is either symmetrical or unsymmetrical, the symmetrical class comprising all graphs which are self-conjugate. If, however, our standpoint be three-dimensional, there are no longer any symmetrical graphs. The two classes are the quasi-symmetrical and the unsymmetrical. A single exception to the above occurs where the graph is of unity. Moreover, the classes now do not comprise the same members. Certain graphs which were unsymmetrical from the first standpoint appear as quasi-symmetrical from the second.

Omitting the trivial symmetrical graph of unity every two-dimensional graph can be read either in three or six ways. The quasi-symmetrical class giving three readings, comprises the self-conjugate graphs and also those which consist of either a single line or a single column of nodes. The remaining graphs give six readings.

Ex. gr. The graph

$$11111$$

yields the three partitions (5), $(\overline{11111})$, (11111) being quasi-symmetrical from the three-dimensional standpoint although it is unsymmetrical in SYLVESTER's theory.

Also the self-conjugate graph

$$\begin{array}{l} 111 \\ 1 \\ 1 \end{array}$$

yields the three partitions $(\overline{311})$, (311), $(\overline{111}\ \overline{100}\ \overline{100})$.

Such a graph as

$$\begin{array}{l} 111 \\ 111 \end{array}$$

being unsymmetrical in both theories yields six partitions

$$(\overline{33}),\ (33),\ (\overline{222}),\ (222),\ (\overline{111}\ \overline{111}),\ (\overline{11}\ \overline{11}\ \overline{11}).$$

Art. 51. The enumeration of the three-dimensional graphs that can be formed with a given number of nodes, corresponding to the regularised partitions of all multipartite numbers of given content, is a weighty problem. I have verified to a high order that the generating function of the complete system is

$$(1-x)^{-1}(1-x^2)^{-2}(1-x^3)^{-3}(1-x^4)^{-4}\dots \textit{ad inf.},$$

and, so far as my investigations have proceeded, everything tends to confirm the truth of this conjecture.

I observe that, to negative signs *près*, the exponents are

$$1,\quad 2,\quad 3,\quad 4,\quad 5,\dots$$

viz., the figurate numbers of order 2.

The generating function which enumerates the two-dimensional graphs, is

$$(1-x)^{-1}(1-x^2)^{-1}(1-x^3)^{-1}(1-x^4)^{-1}\dots$$

where (notice) the exponents are

$$1,\quad 1,\quad 1,\quad 1,\quad 1,\dots$$

the figurate numbers of order 1.

Proceeding further back, we find that one-dimensional graphs are enumerated by

$$(1-x)^{-1}(1-x^2)^{0}(1-x^3)^{0}(1-x^4)^{0}\dots$$

the numbers

$$1,\quad 0,\quad 0,\quad 0,\quad 0,\dots$$

being the figurate numbers of order zero. Going forward again it is easy to verify up to a certain point that four-dimensional graphs (which it is quite easy to graphically realise in two dimensions) are enumerated by

$$(1-x)^{-1}(1-x^2)^{-3}(1-x^3)^{-6}\dots,$$

where the exponents involve the figurate numbers of order 3.

The law of enumeration appears, conjecturally, to involve the successive series of figurate numbers.

Art. 52. Before proceeding to establish certain results, it may be proper, as illustrating the method pursued in this difficult investigation, to give other results which, at first mere conjectures, are gradually having the mark of truth stamped upon them.

Consider graphs in which only the numbers 1 and 2 appear. These are two-layer partitions. The enumeration to a high order is given by the generating function

$$(2\,;\infty\,;\infty) = (1-x)^{-1}(1-x^2)^{-2}(1-x^3)^{-2}(1-x^4)^{-2}\dots$$

where the notation $(l\,;m\,;n)$ is employed to represent the generating function of partitions whose graphs are limited in height, breadth, and length by l, m, n respectively.

Similarly we shall find :—

$$(3\,;\infty\,;\infty) = (1-x)^{-1}(1-x^2)^{-2}[(1-x^3)(1-x^4)\ldots]^{-3},$$

$$(4\,;\infty\,;\infty) = (1-x)^{-1}(1-x^2)^{-2}(1-x^3)^{-3}[(1-x^4)(1-x^5)\ldots]^{-4},$$

$$(l\,;\infty\,;\infty) = (1-x)^{-1}(1-x^2)^{-2}\ldots(1-x^{l-1})^{-(l-1)}[(1-x^l)(1-x^{l+1})\ldots]^{-l},$$

$$(l\,;1\,;\infty) = (1-x)^{-1}(1-x^2)^{-1}\ldots(1-x^l)^{-1},$$

$$(l\,;2\,;\infty) = (1-x)^{-1}(1-x^2)^{-2}(1-x^3)^{-2}\ldots(1-x^l)^{-2}(1-x^{l+1})^{-1},$$

$$(l\,;3\,;\infty) = (1-x)^{-1}(1-x^2)^{-2}(1-x^3)^{-3}\ldots(1-x^l)^{-3}(1-x^{l+1})^{-2}(1-x^{l+2})^{-1},$$

$$(l\,;\mathrm{m}\,;\infty) = (1-x)^{-1}(1-x^2)^{-2}\ldots(1-x^{m-1})^{-(m-1)}\times[(1-x^m)\ldots(1-x^l)]^{-m},$$
$$\times(1-x^{l+1})^{-(m-1)}(1-x^{l+2})^{-(m-2)}\ldots(1-x^{l+m-1})^{-1},$$

if m be not greater than l;

with an equivalent form

$$(l\,;m\,;\infty) = (1-x)^{-1}(1-x^2)^{-2}\ldots(1-x^{l-1})^{-l-1}\times[(1-x^l)\ldots(1-x^m)]^{-l}$$
$$\times(1-x^{m+1})^{-(l-1)}(1-x^{m+2})^{-(l-2)}\ldots(1-x^{l+m-1})^{-1},$$

if m be greater than l;

and finally

$$(l\,;m\,;n) = \frac{1-x^{n+1}}{1-x}\cdot\frac{(1-x^{n+2})^2}{(1-x^2)^2}\cdots\frac{(1-x^{n+l-1})^{l-1}}{(1-x^{l-1})^{l-1}}$$
$$\times\left[\frac{1-x^{n+l}}{1-x^l}\cdot\frac{1-x^{n+l+1}}{1-x^{l+1}}\cdots\frac{1-x^{n+m}}{1-x^m}\right]^l$$
$$\times\frac{(1-x^{n+m+1})^{l-1}}{1-x^{m+1})^{l-1}}\cdot\frac{(1-x^{n+m+2})^{l-2}}{(1-x^{m+2})^{l-2}}\cdots\frac{1-x^{n+l+m-1}}{1-x^{l+m-1}},$$

a result which can be shown to be symmetrical in l, m and n, as ought, of course, to be the case.

This expression for $(l\,;\,m\,;\,n)$ can be exhibited in a more suggestive form, viz. :— Writing $1-x^s=(s)$

$$(l;\ m;\ n) = \frac{(n+1)(n+2)\ldots\ldots(l+m+n-1)}{(1)\quad(2)\quad\ldots\ldots\quad(l+m-1)}$$
$$\times\frac{(n+2)(n+3)\ldots\ldots(l+m+n-2)}{(2)\quad(3)\quad\ldots\ldots\quad(l+m-2)}\times\frac{(n+3)(n+4)\ldots\ldots(l+m+n-3)}{(3)\quad(4)\quad\ldots\ldots\quad(l+m-3)},$$

$\times$ . . . . to l factors or m factors, according as m or l is the greater.

4 P 2

Art. 53. In attempting to establish these results, it is easy to construct a generating function which contains implicitly the complete solution of the problems.

The problem itself may be enunciated in another manner which has points of great interest. A two-dimensional graph of SYLVESTER may be supposed formed by pushing

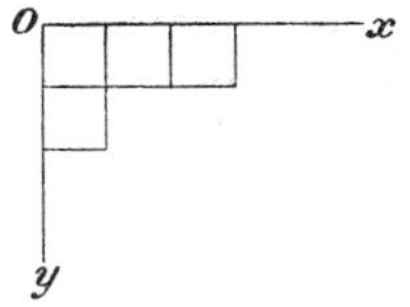

a number of cubes into a flat rectangular corner $y0x$ in such wise that the arrangement is immovable under the action of forces applied in the directions $x0$, $y0$.

It is clear that the number of such arrangements of n cubes is the number of two-dimensional graphs of n, or the number of partitions of the unipartite number n. Similarly, we may push a number of cubes into a three-dimensional rectangular corner, piling of cubes permissible, and such that the arrangement is immovable for forces applied in the three directions $x0$, $y0$, $z0$. The enumeration of these arrangements is the same as that in the problem under discussion.

Art. 54. First consider arrangements limited in the manner $(l\,;\ m\,;\ n) = (2\,;\ 1\,;\ \infty)$. We have such a graph as

$$\begin{matrix} 2 \\ 2 \\ 2 \\ 1 \\ 1, \end{matrix}$$

obtained by writing a column of nodes, and over it another column of nodes, not exceeding the former in number.

We may take, as the generating function,

$$\frac{1}{(1-ax)\,(1-x/a)},$$

in which we are only concerned with that portion of the expansion which is integral as regards a. The function is, in fact, redundant since it involves terms which are superfluous, and we obtain the reduced or condensed generating function by putting a equal to unity in the portion we retain.

Since

$$\frac{1}{(1-ax)\left(1-\dfrac{x}{a}\right)} = \frac{1}{1-x^2}\left\{\frac{1}{1-ax} + \frac{\dfrac{x}{a}}{1-\dfrac{x}{a}}\right\}$$

the reduced generating function is

$$\frac{1}{(1-x)(1-x^2)},$$

and this is obviously correct, because from the form of the graph we have merely to enumerate the ways of partitioning numbers with the parts 1 and 2.

Art. 55. Again if $(l\,;\ m\,;\ n) = (2\,;\ 2\,;\ \infty)$, we have graphs like

```
2 2
2 2
2 1
1 1
1
1
```

We are led to construct the function

$$\frac{1}{(1-ax)\left(1-\frac{x}{a}\right)(1-abx^2)\left(1-\frac{x^2}{ab}\right)},$$

in the expansion of which all terms involving negative powers of a and b have to be rejected. Isolating the integral portion and putting $a = b = 1$, we find the reduced generating function

$$\frac{1}{(1-x)(1-x^2)^2(1-x^3)}$$

a result which, unlike the previous one, is not obvious.

For the case $(l\,;\ m\,;\ n) = (2\,;\ 3\,;\ \infty)$ we introduce additional denominator factors

$$(1-abcx^3)\left(1-\frac{x^3}{abc}\right),$$

and with increasing labour of algebraical performance we arrive at the reduced generating function

$$\frac{1}{(1-x)(1-x^2)^2(1-x^3)^2(1-x^4)}.$$

Art. 56 In general for the case

$$(l\,;\ m\,;\ n) = (2\,;\ m\,;\ \infty)$$

the generating function is the reciprocal of the product of the $2m$ factors

$$\begin{array}{l} (1 - \alpha_1 x)\left(1 - \dfrac{x}{\alpha_1}\right) \\ (1 - \alpha_1\alpha_2 x^2)\left(1 - \dfrac{x^2}{\alpha_1\alpha_2}\right) \\ (1 - \alpha_1\alpha_2\alpha_3 x^3)\left(1 - \dfrac{x^3}{\alpha_1\alpha_2\alpha_3}\right) \\ \qquad \vdots \qquad\qquad\qquad \vdots \\ (1 - \alpha_1\alpha_2 \ldots \alpha_m x^m)\left(1 - \dfrac{x^m}{\alpha_1\alpha_2 \ldots \alpha_m}\right) \end{array}$$

We have to expand this fraction in ascending powers of x, reject all terms containing negative powers of $\alpha_1, \alpha_2, \ldots \alpha_m$, and then obtain the reduced generating function by putting $\alpha_1 = \alpha_2 = \ldots = \alpha_m = 1$ in the portion retained. This has proved a difficult algebraical problem. I am indebted to Professor FORSYTH for a beautiful solution which he will publish elsewhere.* He establishes that the reduced generating function is

$$\frac{1}{(1-x)\{(1-x^2)(1-x^3)\ldots(1-x^m)\}^2(1-x^{m+1})},$$

a result which agrees with the prediction. His method is that of selective summation. He forms the general term of the expansion

$$\left.\begin{array}{l} \alpha_1^{\,m_1-n_1+m_2-n_2+m_3-n_3+\ldots.} \\ \alpha_2^{\,m_2-n_2+m_3-n_3+\ldots.} \\ \alpha_3^{\,m_3-n_3+\ldots.} \\ \vdots \end{array}\right\} x^{m_1+n_1+2(m_2+n_2)+3(m_3+n_2)+\ldots.}$$

and performs the enumeration

$$\sum_{m_1}\sum_{n_1}\sum_{m_2}\sum_{n_2}\sum_{m_3}\sum_{n_3} \ldots . \; x^{m_1+n_1+2(m_2+n_2)+3(m_3+n)_3+\ldots.}$$

for all values of $m_1, n_1, m_2, n_2, m_3, n_3, \ldots \ldots$ which make each of the m expressions

$$\begin{array}{r} m_1 - n_1 + m_2 - n_2 + m_3 - n_3 + \ldots. \\ m_2 - n_2 + m_3 - n_3 + \ldots. \\ m_3 - n_2 + \ldots. \\ \ldots\ldots \end{array}$$

not less than zero.

The case $(l\,;\, m\,;\, n) = (2\,;\, m\,;\, \infty)$ is thus completely solved by a method which

* 'Proc. L.M.S.,' vol. 27.

appears to be of general application if the difficulties presented by the algebra can be surmounted.

Art. 57. At this point we may enquire into the meaning of the reduced generating function which has been so happily and ingeniously established. We may write it as the product of two fractions :—

$$\frac{1}{(1-x)(1-x^2)\ldots(1-x^m)} + \frac{1}{(1-x^2)(1-x^3)\ldots(1-x^{m+1})}$$

the indication being that every two-layer arrangement is derivable from a combination of two ordinary single-layer partitions whose parts are drawn from the two series of numbers,

$$1, 2, 3, \ldots m,$$
$$2, 3, \ldots m, m+1,$$

respectively. Otherwise we may say that a number N possesses as many two-layer partitions (2 ; m ; ∞) as there are modes of partitionment employing the parts

$$1_1, 2_1, 2_2, 3_1, 3_2, \ldots m_1, m_2, m+1_2.$$

Ex. gr. If $N = 4$ and $m = 3$, the graphs are 9 in number

111	11	11	1	211	21	2	22	2
1	11	1	1		1	1		2
		1	1			1		
			1					

and employing parts

$$\begin{matrix} 1_1 & 2_1 & 3_1 & \\ & 2_2 & 3_2 & 4_2 \end{matrix}$$

we can form 9 partitions, viz. :—

$$(1_1^4),\ (2_1 1_1^2),\ (2_2 1_1^2),\ (2_1^2),\ (2_1 2_2),\ (2_2^2),\ (3_1 1_1),\ (3_2 1_1),\ (4_2).*$$

Art. 58. The problem is therefore reduced to establishing a one-to-one correspondence, between the graphs and the partitions of the kind indicated, of general application. I will in part establish this correspondence, which is not very simple in character, later on. At present it is convenient to take a further survey of the general problems in order to obtain ideas concerning the difficulties that confront us.

I form a tableau of algebraic factors.

* The solution thus shows that the two-layer graphs may be exhibited as a one-layer graph by nodes of two colours, say black and red; nodes of different colours not appearing in any single line.

$$(1-p_1x)\left(1-\frac{p_2}{p_1}x\right)\left(1-\frac{p_3}{p_2}x\right)\ldots\left(1-\frac{p_{l-1}}{p_{l-2}}x\right)\left(1-\frac{x}{p_{l-1}}\right),$$

$$(1-p_1q_1x^2)\left(1-\frac{p_2q_2}{p_1q_1}x^2\right)\left(1-\frac{p_3q_3}{p_2q_2}x^2\right)\ldots\left(1-\frac{p_{l-1}q_{l-1}}{p_{l-2}q_{l-2}}x^2\right)\left(1-\frac{x^2}{p_{l-1}q_{l-1}}\right),$$

$$(1-p_1q_1r_1x^3)\left(1-\frac{p_2q_2r_2}{p_1q_1r_1}x^3\right)\left(1-\frac{p_3q_3r_3}{p_2q_2r_2}x^3\right)\ldots\left(1-\frac{p_{l-1}q_{l-1}r_{l-1}}{p_{l-2}q_{l-2}r_{l-2}}x^3\right)\left(1-\frac{x^3}{p_{l-1}q_{l-1}r_{l-1}}\right),$$

" " " .... " "

" " " .... " "

$$(1-p_1q_1\ldots x^m)\left(1-\frac{p_2q_2\ldots}{p_1q_1\ldots}x^m\right)\left(1-\frac{p_3q_3\ldots}{p_2q_2\ldots}x^m\right)\ldots\left(1-\frac{p_{l-1}q_{l-1}\ldots}{p_{l-2}q_{l-2}\ldots}x^m\right)\left(1-\frac{x^m}{p_{l-1}q_{l-1}\ldots}\right),$$

forming a rectangle of m rows and l columns, the letters $p, q, r, \ldots$, m in number, each occurring with $l-1$ different suffixes.

I say that forming a fraction with unit numerator, having the product of these factors for denominator, we obtain a generating function for the arrangements defined by $(l\,;\ m\,;\ \infty)$.

The number of layers is restricted to l (*i.e.*, l or less), and the breadth to m (*i.e.*, m or less), but the graphs are otherwise unrestricted. Reasoning of the same nature as that employed in the simple case of two layers, enables us readily to construct this function. The function is redundant, as we only require that portion of the expansion whose terms are altogether integral. In this portion we put the letters $p, q, r, \ldots$ all equal to unity, and thus arrive at the reduced generating function.

I recall that the predicted result is the reciprocal of

$$\begin{aligned}(1-x)(1-x^2)(1-x^3)\ldots(1-x^{m-2})(1-x^{m-1})(1-x^m)&\\ \times(1-x^2)(1-x^3)\ldots\ldots\ldots\ldots(1-x^{m-1})(1-x^m)(1-x^{m+1})&\\ \times(1-x^3)\ldots\ldots\ldots\ldots\ldots\ldots(1-x^m)(1-x^{m+1})(1-x^{m+2})&\\ \times\ldots&\\ \times(1-x^l)(1-x^{l+1})\ldots\ldots\ldots\ldots\ldots(1-x^{l+m-1}).&\end{aligned}$$

Professor FORSYTH has not yet succeeded in obtaining this result from his powerful method of selective summation. I hear from him that he has verified it in numerous particular cases, but that, so far, he has not been able to surmount the algebraic difficulties presented by the general case.

As regards the final result, the tableau of factors possesses row and column symmetry.

Simple rotation of the graphs through a right angle in the plane xy establishes this intuitively.

We get the same result from m rows and l columns as from l rows and m columns. Taking only the first row, we find that the fraction

$$\frac{1}{(1-p_1x)\left(1-\frac{p_2}{p_1}x\right)\ldots\left(1-\frac{p_{l-1}}{p_{l-2}}x\right)\left(1-\frac{x}{p_{l-1}}\right)}$$

leads to the same reduced generating function as the fraction

$$\frac{1}{(1-p_1x)(1-p_1q_1x^2)\ldots(1-p_1q_1\ldots x^l)},$$

and this is obviously

$$\frac{1}{(1-x)(1-x^2)\ldots(1-x^l)}.$$

Art. 59. If the result predicted be the true result we should be able to establish it by means of a one-to-one correspondence between the graphs and partitions of a certain kind. This presents difficulties to which I will advert in a moment.

Finally I construct the generating function for the case $(l\,;m\,;n)$, the graphs being restricted in all three dimensions.

The numerator is unity and the denominator the product of the factors exhibited in the subjoined tableau :—

$$\begin{array}{ccccc}
(1-gp_1x) & \left(1-\frac{p_2}{p_1}x\right) & \ldots & \left(1-\frac{p_{l-1}}{p_{l-2}}x\right) & \left(1-\frac{x}{p_{l-1}}\right) \\
(1-gp_1q_1x^2) & \left(1-\frac{p_2q_2}{p_1q_1}x^2\right) & \ldots & \left(1-\frac{p_{l-1}q_{l-1}}{p_{l-2}q_{l-2}}x^2\right) & \left(1-\frac{x^2}{p_{l-1}q_{l-1}}\right) \\
,, & ,, & \ldots & ,, & ,, \\
,, & ,, & \ldots & ,, & ,, \\
(1-gp_1q_1\ldots x^m) & \left(1-\frac{p_2q_2\ldots}{p_1q_1\ldots}x^m\right) & \ldots & \left(1-\frac{p_{l-1}q_{l-1}\ldots}{p_{l-2}q_{l-2}\ldots}x^m\right) & \left(1-\frac{x^m}{p_{l-1}q_{l-1}\ldots}\right)
\end{array}$$

in which the occurrence of the symbol g in the first column will be noticed.

We have as usual to neglect all terms in the expansion which involve negative powers of symbols and in addition we must now neglect all terms which involve g raised to a greater power than n.

This construction prevents the lower layer of the graph from having a greater extent than n in the direction Oy, and thus the whole graph is similarly restricted.

The reduced generating functions can be shown in simple instances to agree with the predicted results.

Ex. gr. take $(l\,;\ m\,;\ n) = (2\,;\ 2\,;\ 1)$; the fraction is

$$\frac{1}{(1-gp_1x)\left(1-\dfrac{x}{p_1}\right)(1-gp_1q_1x^2)\left(1-\dfrac{x^2}{p_1q_1}\right)}.$$

We have to retain the integral portion of

$$(1+p_1x+p_1q_1x^2)\left(1+\frac{x}{p_1}+\frac{x^2}{p_1q_1}\right);$$

selecting this and putting $p_1 = q_1 = 1$, we obtain

$$1+x+2x^2+x^3+x^4,$$

which is

$$\frac{(1-x^2)(1-x^3)^2(1-x^4)}{(1-x)(1-x^2)^2(1-x^3)}.$$

As in simpler cases I have not been able to overcome the algebraic difficulties, it is perhaps needless to say that in this most general case I cannot establish the form of the reduced generating function.

Art. 60. I return to consider various particular points of the problem. When the number of layers of nodes is restricted to two, we have seen that the generating function which enumerates the graphs that can be formed with a given number of nodes is

$$(1-x)^{-1}(1-x^2)^{-2}(1-x^3)^{-2}(1-x^4)^{-2}\ldots\ldots$$

In correspondence we have the regularised bipartitions (including uni-partitions) of multipartite numbers of given content.

Also if the breadth of the graph do not exceed m *or* the multipartite numbers be not more than m-partite the generating function is

$$(1-x)^{-1}\{(1-x^2)(1-x^3)\ldots(1-x^m)\}^{-2}(1-x^{m+1})^{-1}.$$

I propose to give another proof of these results based upon a certain mode of dissection of the graph.

In the notation that has been used, a graph may be written

$$\begin{matrix} 2^{\lambda} & 1^{\mu} \\ 2^{\lambda'} & 1^{\mu'} \\ 2^{\lambda''} & 1^{\mu''} \\ \vdots & \end{matrix}$$

where

$$\lambda + \mu, \quad \lambda' + \mu', \quad \lambda'' + \mu'', \ldots .$$

and also

$$\lambda, \quad \lambda' \quad \lambda'', \ldots .$$

are in descending order of magnitude.

A line of the graph has a certain weight $2\lambda + \mu$. Any number of lines may be identical and consequently of the same weight, but no two *different* lines may have the same weight in the same graph. Let us form a graph, beginning at the lowest line, taken to be of weight unity, and proceeding upwards through every superior weight. We will find that such a graph may have a variety of forms. Construct the subjoined scheme of graph lines.

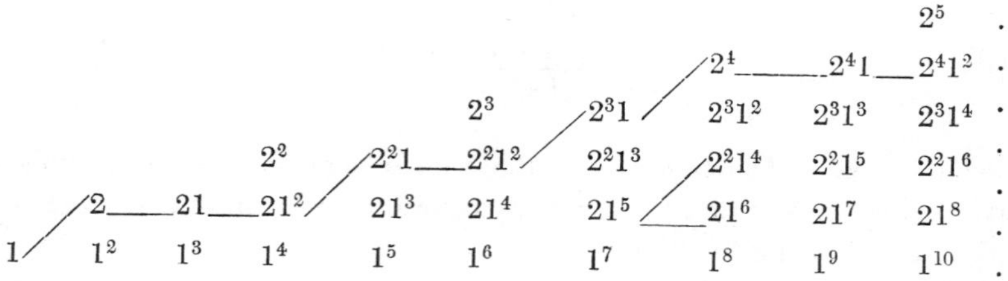

In each column every graph line has the same weight. In each line every graph line has the same number of twos. From any graph line, say $2^\lambda 1^\mu$ $(\mu > 0)$, of weight $2\lambda + \mu$, we can pass to a graph line of weight $2\lambda + \mu + 1$ in two ways; viz., by taking $2^\lambda 1^{\mu+1}$ by horizontal progression or $2^{\lambda+1}1^{\mu-1}$ by diagonal progression. From 2^λ we can only pass to $2^\lambda 1$ by horizontal progression. In accordance with these laws we can form a graph consisting of graph lines of all weights, from unity upwards, in a definite number of ways, depending upon the weight of the highest graph line. For example, we can select the graph whose successive lines are

$$1,\ 2,\ 21,\ 21^2,\ 2^21,\ 2^21^2,\ 2^31,\ 2^4,\ 2^41,\ 2^41^2 \ldots .$$

The progression from graph line to graph line is either horizontal or diagonal, which we can denote by A and B respectively. Then the graph may be denoted by

B A A B A B B A A.

The specification of the selected graph may be taken to be a collection of line graphs, each of which is reached by diagonal progression, and which proceeds by horizontal progression.

Thus, in the particular case before us, the specification is

$$2,\ 2^21,\ 2^4.$$

4 Q 2

To find the form of the graph between any two graph lines, say $2^2 1$, 2^4, we have merely to proceed forwards horizontally from $2^2 1$ and backwards diagonally from 2^4 till a junction is effected in the manner

$$2^2 1 \;\_\_\_\; 2^2 1^2 \;/\; 2^3 1 \;/\; 2^4$$

In the A, B notation the specification is given by the position of the BA contacts.

BAABABBAA.

Every graph that can be selected has its own specification. To enumerate the graphs we must enumerate the specifications.

We must discover the properties of the succession of graph lines which are able to constitute a specification.

If $2^\lambda 1^\mu$ be a graph line in a specification, λ must be greater than zero since the lower line of the scheme cannot be reached by diagonal succession. Also $2^\lambda 1^\mu$ may be followed by a graph line $2^{\lambda_1} 1^{\mu_1}$ so long as $\lambda_1 > \lambda$ and $\lambda_1 + \mu_1 > \lambda + \mu$; and may be preceded by a graph line $2^{\lambda_0} 1^{\mu_0}$ so long as $\lambda > \lambda_0$, $\lambda + \mu > \lambda_0 + \mu_0$.

These laws may be gathered by simple inspection of the scheme.

Hence we may form the graph lines of the specification into a graph of the specification, viz.:—

$$\begin{array}{c} 2^\lambda 1^\mu \\ 2^{\lambda_1} 1^{\mu_1} \\ 2^{\lambda_2 \mu_2} \\ \vdots \end{array}$$

which has the simple properties—

(i) $\lambda > \lambda_1 > \lambda_2 > \ldots$

(ii) $\lambda + \mu > \lambda_1 + \mu_1 > \lambda_2 + \mu_2 > \ldots$

(iii) The lowest graph line contains, at least, one 2.

Art. 61. Any selected graph has a certain specification graph. From the former we may suppose any graph lines, which do not also belong to the specification, to be absent, and thus obtain a number of graphs which all have the same specification, and may be considered to follow the same line of route through the scheme. Further, the graph lines, whether belonging to the specification or no, may occur any number of times, repeated without the graph of the specification being changed.

Finally, we may associate any partition of a unipartite number with any specification graph whatever, so as to form a two-layer graph. The partition of the unipartite must be interpreted upon that line of route through the scheme which is associated with the given specification graph.

Ex. gr. Suppose the specification graph to be

2 2 2 2
2 2 1
2

and the unipartite partition to be

(9 7 7 6 5 5 4 3 2 2).

Interpreted on the line of route concerned, which is that marked upon the scheme above, we obtain

2 2 2 2 1
S 2 2 2 2
2 2 2 1
2 2 2 1
2 2 1 1
2 2 1
2 2 1
S 2 2 1
2 1 1
2 1
2
2
S 2,

in which the specification graph lines, marked S, have been interpolated.

Hence, if F (x) be the generating function which enumerates specification graphs,

$$\frac{\mathrm{F}(x)}{(1-x)(1-x^2)(1-x^3)(1-x^4)\ldots \mathit{ad\ inf.}}$$

will be the generating function of all two-layer graphs—that is of forms specified by $(2\,;\ \infty\,;\ \infty)$.

We have next to determine the form of F (x).

Art. 62. A specification graph may contain no graph lines; this will be the case when the line of route through the scheme is the lowest horizontal line. There is only *one* such graph; generating function 1. If it contains one line, this line must be of the form $2^\lambda 1^\mu$ $(\lambda > 0)$, and the number of such graphs is given by the generating function

$$\frac{x^2}{(1-x)(1-x^2)}.$$

We may also take the following view of the matter. Let $k_1(x)$ be the generating function for the number of two-layer graphs in which the number of rows is limited to unity; that is, of the graphs specified by $(2; 1; \infty)$. The graph either does not or does contain a specification graph line. The former are enumerated by

$$\frac{1}{1-x};$$

the latter by $x^2 k_1(x)$.

Hence,

$$k_1(x) = \frac{1}{1-x} + x^2 k_1(x);$$

or

$$k_1(x) = \frac{1}{(1-x)(1-x^2)};$$

and the number of specification graphs containing one graph line is

$$x^2 k_1(x) \quad \text{or} \quad \frac{x^2}{(1-x)(1-x^2)}.$$

Next let $k_2(x)$ denote the number of two-layer graphs in which the number of rows is restricted to two. If it contain no specification graph line it must be of the form

$$\begin{matrix} 1^{\lambda} & \\ 1^{\mu} & \quad \lambda \geqq \mu. \end{matrix}$$

For these the generating function is

$$\frac{1}{(1-x)(1-x^2)}.$$

If it contain *one* specification graph line it must be of the form

$$\begin{matrix} 2^{\lambda+1} 1^{\mu} \\ 1^{\nu} \end{matrix} \quad (\lambda + \mu + 1 \geqq \nu)$$

and these are enumerated by

$$\frac{x^2 k_1(x)}{1-x}.$$

If it contain two specification graph lines its form must be

$$\begin{matrix} 2^{\lambda+2}\, 1^{\mu} \\ 2^{\lambda'+1}\, 1^{\mu'} \end{matrix}$$

where

$$\lambda \overline{\gtrless} \lambda' \text{ and } \lambda + \mu + 1 > \lambda' + \mu'.$$

These are enumerated by

$$x^6 k_2(x).$$

Hence

$$k_2(x) = \frac{1}{(1-x)(1-x^2)} + \frac{x^2 k_1(x)}{1-x} + x^6 k_2(x).$$

Therefore

$$k_2(x) = \frac{1}{(1-x)(1-x^2)^2(1-x^3)}$$

and the number of specification graphs containing two graph lines is

$$x^6 k_2(x) \quad \text{or} \quad \frac{x^6}{(1-x)(1-x^2)^2(1-x^3)}.$$

Similarly we shall find that $k_3(x)$ is composed of four parts corresponding to the occurrence of 0, 1, 2, or 3 specification graph lines. The first three are readily seen to be enumerated by generating functions

$$\frac{1}{(1-x)(1-x^2)(1-x^3)}, \quad \frac{x^2 k_1(x)}{(1-x)(1-x^2)}, \quad \frac{x^6 k_2(x)}{1-x};$$

When three specification graph lines occur, the form must be

$$\begin{matrix} 2^{\lambda+3}\, 1^{\mu} \\ 2^{\lambda'+2}\, 1^{\mu'} \\ 2^{\lambda''+1}\, 1^{\mu''} \end{matrix}$$

$$\lambda \overline{\gtrless} \lambda' \overline{\gtrless} \lambda'' \text{ and } \lambda + \mu + 2 > \lambda' + \mu' + 1 > \lambda'' + \mu'',$$

and the generating function $x^{12} k_3(x)$.

Hence

$$k_3(x) = \frac{1}{(1-x)(1-x^2)(1-x^3)} + \frac{x^2 k_1(x)}{(1-x)(1-x^2)} + \frac{x^6 k_2(x)}{1-x} + x^{12} k_3(x),$$

and we can show that

$$k_3(x) = \frac{1}{(1-x)(1-x^2)^2(1-x^3)^2(1-x^4)},$$

and the specification graphs containing three graph lines are given by

$$x^{12} k_3 (x) \quad \text{or} \quad \frac{x^{12}}{(1-x)(1-x^2)^2(1-x^3)^2(1-x^4)}.$$

In general we obtain the relation

$$k_m(x) = \frac{1}{(1-x)(1-x^2)\dots(1-x^m)} + \frac{x^2 k_1(x)}{(1-x)(1-x^2)\dots(1-x^{m-1})}$$

$$+\frac{x^6 k_2(x)}{(1-x)(1-x^2)\dots(1-x^{m-2})} + \dots + \frac{x^{s(s+1)} k_s(x)}{(1-x)(1-x^2)\dots(1-x^{m-s})} + \dots + x^{m(m+1)} k_m(x),$$

and also

$$k_\infty(x) = \frac{1 + x^2 k_1(x) + x^6 k_2(x) + \dots + x^{s(s+1)} k_s(x) + \dots}{(1-x)(1-x^2)(1-x^3)\dots},$$

where the numerator is the generating function for specification graphs of given content.

Art. 63. We can now establish that $k_s(x)$ is the expression

$$\frac{1}{(1-x)(1-x^2)^2(1-x^3)^2\dots(1-x^s)^2(1-x^{s+1})};$$

for assume the law true for values of s equal and inferior to $m-1$; substitute in the foregoing identity and writing $1 - x^s = (s)$,

$$(1)(2)\dots(m)(1-x^{m^2+m})k_m(x)$$

$$= 1 + \frac{x^2(m)}{(1)(2)} + \frac{x^6(m)(m-1)}{(1)(2)^2(3)} + \dots + \frac{x^{m^2-m}(m)(m-1)\dots(2)}{(1)(2)^2\dots(m-1)^2(m)}.$$

Recalling the well-known identity

$$\frac{1}{(1-ax)(1-ax^2)\dots(1-ax^m)} = 1 + \frac{(m)}{(1)}\cdot\frac{ax}{1-ax} + \frac{(m)(m-1)}{(1)(2)}\,\frac{a^2x^4}{1-ax\,.\,1-ax^2}$$

$$+\frac{(m)(m-1)(m-2)}{(1)(2)(3)}\,\frac{a^3x^9}{1-ax\,.\,1-ax^2\,.\,1-ax^3} + \dots$$

and putting therein $a = x$, we find

$$\frac{1}{(2)(3)\dots(m+1)} = 1 + \frac{x^2(m)}{(1)(2)} + \frac{x^6(m)(m-1)}{(1)(2)^2(3)} + \dots$$

$$+\frac{x^{m^2-m}(m)(m-1)\dots(2)}{(1)(2)^2(3)^2\dots(m-1^2)(m)} + \frac{x^{m^2+m}(m)(m-1)\dots(1)}{(1)(2)^2(3)^2\dots(m)^2(m+1)}.$$

Hence

$$(1)(2)\ldots(m)(1-x^{m^2+m})\,k_m(x)=\frac{1}{(2)(3)\ldots(m+1)}-\frac{x^{m^2+m}}{(2)(3)\ldots(m+1)}.$$

Therefore

$$k_m(x)=\frac{1}{(1)(2)^2(3)^2\ldots(m-1)^2(m)^2(m+1)}.$$

Hence, by induction, it has been established that $k_m(x)$ has this expression for all values of *m*.

Therefore the result

$$(2\,;\,m\,;\,\infty)=\frac{1}{(1-x)(1-x^2)^2(1-x^3)^2\ldots(1-x^m)^2(1-x^{m+1})}$$

agreeing with that obtained in a totally different manner by FORSYTH.

I hope to continue the theory, adumbrated in this paper, in a future communication to the Royal Society.

ON THE PARTITIONS INTO UNEQUAL AND INTO UNEVEN PARTS.

By Major P. A. MacMahon.

It is a well-known theorem of Euler that the partitions of a number into unequal parts are equi-numerous with the partitions into uneven parts.

This theorem arises from the identity

$$\prod_1^\infty {}_m (1+q^m) = \frac{1}{\prod_1^\infty {}_m (1-q^{2m-1})} \quad \ldots\ldots\ldots\ldots\ldots (1),$$

which is also of importance in the theory of elliptic products.

The object of the present communication is to further examine this connexion and to obtain theorems of great generality which include as particular cases theorems which at first sight seem to be remotely, if at all, related.

It will be found that the theorem in partitions whose analytical expression is

$$\prod_1^\infty {}_m (1+q^{2^{m-1}}) = \frac{1}{1-q} \quad \ldots\ldots\ldots\ldots\ldots (2),$$

and whose interpretation is that a number can be partitioned in only one way into parts which are unequal powers of 2 follows the same general law as that whose expression is (1) above.

From the identity

$$\prod_1^\infty {}_m (1+q^m) = \frac{1}{\prod_1^\infty {}_m (1-q^{2m-1})}$$

we derive, by putting q^{p_1} for q, where p_1 is any uneven prime,

$$\prod_1^\infty {}_m (1+q^{p_1 m}) = \frac{1}{\prod_1^\infty {}_m (1-q^{2p_1 m-p_1})},$$

and thence by division

$$\prod_1^\infty {}_m \frac{1+q^m}{1+q^{p_1 m}} = \prod_1^\infty {}_m \frac{1-q^{2p_1 m-p_1}}{1-q^{2m-1}} \quad \ldots\ldots\ldots\ldots (3),$$

the interpretation of which is that, if from the complete system of positive integers we delete all numbers which are multiples of p_1, and from the complete system of uneven integers all such numbers also, then the partitions of a number into *unrepeated* parts of the deleted first system are equi-numerous with the partitions into parts, repetitions allowable, of the deleted second system.

In the identity (3) we now put q^{p_2} for q, p_2 also being an uneven prime, and thence obtain by division

$$\prod_1^\infty {}_m \frac{(1+q^m)(1+q^{p_1p_2m})}{(1+q^{p_1m})(1+q^{p_2m})} = \prod_1^\infty {}_m \frac{(1-q^{2p_1m-p_1})(1-q^{2p_2m-p_2})}{(1-q^{2m-1})(1-q^{2p_1p_2m-p_1p_2})},$$

the interpretation of which is that from the two systems of numbers above defined we may delete any two primes and their multiples, and still the partitions drawn from the deleted systems will be equi-numerous. Similarly

$$\prod_1^\infty {}_m \frac{(1+q^m)(1+q^{p_1p_2m})(1+q^{p_1p_3m})(1+q^{p_2p_3m})}{(1+q^{p_1m})(1+q^{p_2m})(1+q^{p_3m})(1+q^{p_1p_2p_3m})}$$

$$= \prod_1^\infty {}_m \frac{(1-q^{2p_1m-p_1})(1-q^{2p_2m-p_2})(1-q^{2p_3m-p_3})(1-q^{2p_1p_2p_3m-p_1p_2p_3})}{(1-q^{2m-1})(1-q^{2p_1p_2m-p_1p_2})(1-q^{2p_1p_3m-p_1p_3})(1-q^{2p_2p_3m-p_2p_3})},$$

which is similarly interpreted with regard to three uneven primes p_1, p_2, p_3. These observations point at once to a general theorem, which may be enunciated as follows:

"Let there be

I. Numbers $a_1, a_2, a_3, \ldots$ the complete system of integers $1, 2, 3, \ldots, \infty$.

II. Numbers $b_1, b_2, b_3, \ldots$ the complete system of uneven integers $1, 3, 5, \ldots, \infty$.

III. Numbers $p_1, p_2, p_3, \ldots, p_n$ a system of n uneven primes > 1, where n may be any positive integer, zero included; and let there be deleted from systems I. and II. all numbers which are multiples of any of the numbers in system III.; then

$$\Pi\,(1+q^a) = \frac{1}{\Pi\,(1-q^b)}$$

a factor appearing in the products for each number in the deleted systems".

In this general theorem we now pay particular attention to what we may call the extreme cases.

(i) When $n=0$, the system III. does not exist and there are no deletions; we have therefore Euler's result

$$\prod_1^\infty {}_m (1+q^m) = \frac{1}{\prod_1^\infty {}_m (1-q^{2m-1})}.$$

(ii) When $n=\infty$, the system III. includes the whole of the uneven primes 3, 5, 7, 11, ...

The resulting deletions leave only powers of 2 in system I. and the single uneven number *unity* in system II. Hence

$$\prod_{1}^{\infty}{}_m (1+q^{2^{m-1}}) = \frac{1}{1-q}.$$

We may advantageously specify a few cases adjacent to the extremes.

Thus if the system III. contains the whole of the primes except 3,

$$\prod_{0}^{\infty}{}_{\alpha,\beta}(1+q^{2^\alpha.3^\beta}) = \frac{1}{\prod_{0}^{\infty}{}_\alpha (1-q^{3^\alpha})},$$

shewing that the partitions into unequal parts of the form $2^\alpha.3^\beta$ are equi-numerous with the partitions into parts which are powers of 3, repetitions allowed. In the case of the number 15 there are 9 such partitions of each kind, viz.

12, 3 12, 2, 1 9, 6 9, 4, 2 9, 3, 2, 1 8, 6, 1 8, 4, 3 8, 4, 2, 1

933, 93111, 9111111, 33333, 3333111, 333111111,

33111111111, 3111111111111, 111111111111111.

Similarly, if p_1 be any uneven prime

$$\prod_{0}^{\infty}{}_{\alpha,\beta}(1+q^{2^\alpha.p_1^\beta}) = \frac{1}{\prod_{1}^{\infty}{}_\alpha (1-q^{p_1^\alpha})},$$

shewing that the partitions into unequal parts of the form $2^\alpha.p_1^\beta$ are equi-numerous with the partitions into parts which are powers of p_1. If the system III. contains the whole of the primes except p_1, p_2,

$$\prod_{0}^{\infty}{}_{\alpha,\beta,\gamma}(1+q^{2^\alpha p_1^\beta p_2^\gamma}) = \frac{1}{\prod_{1}^{\infty}{}_{\alpha,\beta}(1-q^{p_1^\alpha.p_2^\beta})},$$

and generally

$$\prod_{0}^{\infty}{}_{\alpha_1\alpha_2\ldots\alpha_{n+1}}(1+q^{2^{\alpha_1}p_1^{\alpha_2}p_2^{\alpha_3}\ldots p_n^{\alpha_{n+1}}}) = \frac{1}{\prod_{1}^{\infty}{}_{\alpha_1\alpha_2\ldots\alpha_n}(1-q^{p_1^{\alpha_1}p_2^{\alpha_2}\ldots p_n^{\alpha_n}})}.$$

To pass to the other extreme, suppose the system III. to involve the single prime 3, we find that

$$(1+q)(1+q^2).(1+q^4)(1+q^5).(1+q^7)(1+q^8)\ldots$$
$$=\frac{1}{(1-q).(1-q^5)(1-q^7).(1-q^{11})(1-q^{13}).(1-q^{17})(1-q^{19})\ldots}.$$

We may similarly delete from the products in Euler's identity all factors which involve q with exponents which are multiples of any prime p_1 or of any system of primes $p_1, p_2, \ldots, p_n$.

We may have a further generalization by taking for the system III. any n uneven numbers $P_1, P_2, \ldots, P_n$, such that the components of every pair are relatively prime. In particular, the system may consist of any one uneven number or of two uneven numbers which have no common factor.

§ 2.

An application of the results of the preceding section to the Theory of the Divisors of Numbers is not far to seek.

If we take the identity

$$\prod_1^{\infty}{}_m (1+q^m) = \frac{1}{\prod_1^{\infty}{}_m (1-q^{2m-1})},$$

subject it to logarithmic differentiation and multiply throughout by q, we find

$$\sum_1^{\infty}{}_m \frac{mq^m}{1+q^m} = \sum_1^{\infty}{}_m \frac{(2m-1)q^{2m-1}}{1-q^{2m-1}},$$

or

$$\zeta'(n) = \Delta(n),$$

where, in Glaisher's notation,

$\zeta'(n)$ = sum of the divisors of n which have uneven conjugates

— " " " even "

$\Delta(n)$ = sum of the uneven divisors of n.

It is quite clear from the reasoning that has been employed above that we may enunciate the following theorem:

"Let there be

I. $a_1, a_2, a_3, \ldots$ the complete system of positive integers $1, 2, 3, \ldots, \infty$.

II. $b_1, b_2, b_3, \ldots$ the complete system of uneven positive integers $1, 3, 5, \ldots, \infty$.

III. $p_1, p_2, \ldots, p_\mu$ any set of μ uneven primes, unity excluded, where μ may be zero or any positive integer.

Let there be deleted from the systems I. and II. all numbers which are multiples of any of the primes in system III.

Let
$$\zeta'(n)_a$$
denote the arithmetical function $\zeta'(n)$ in respect of all divisors of n which appear in the deleted system I., and
$$\Delta(n)_b$$
the arithmetical function $\Delta(n)$ in respect of all divisors of n which appear in the deleted system II.

Then
$$\zeta'(n)_a = \Delta(n)_b\text{''}.$$

The extreme cases are

(i) $\mu = 0$; there are no deletions, and therefore
$$\zeta'(n) = \Delta(n),$$
the well-known theorem.

(ii) $\mu = \infty$ and the system III. involves every uneven prime > 1.

The system I. involves only powers of 2 and the system II. only unity.

Hence, if $\zeta'(n)_{2^a}$ denotes that the arithmetic function has reference only to those divisors of n which are powers of 2,
$$\zeta'(n)_{2^a} = 1.$$

This is indeed obvious because, if $n = 2^\beta n'$, where n' is uneven,
$$\zeta'(n)_{2^a} = 2^\beta - 1 - 2 - 2^2 - \ldots - 2^{\beta-1} = 1.$$

As another particular case to be noted, suppose that the system III. involves every uneven prime > 1 except the number 3; then the system I. comprises only numbers of the form
$$2^a . 3^\beta,$$
and the system II. only powers of 3.

Hence
$$\zeta'(n)_{2^a.3^\beta} = \Delta(n)_{3^a},$$
or, in words, the function $\zeta'(n)$ for divisors of the form $2^a . 3^\beta$ is equal to the function $\Delta(n)$ for divisors of the form 3^a.

Ex. gr. for $n = 540$, the divisors are

of the form	$2^a . 3^\beta$	1,	2,	3,	4,	6,	9,	12,	18,	27,	36,	54,	108,
,,	3^a	1		3			9			27			

$$\zeta'(n) = 4 + 12 + 36 + 108 - 1 - 2 - 3 - 6 - 9 - 18 - 27 - 54 = 40,$$
$$\Delta(n) = 1 + 3 + 9 + 27 = 40,$$
a verification.

Similarly $$\zeta'(n)_{2^\alpha . p^\beta} = \Delta(n)_{p^\beta},$$
if p be any uneven prime > 1.

$$\zeta'(n)_{2^\alpha p_1{}^\beta p_2{}^\gamma} = \Delta(n)_{p_1{}^\beta p_2{}^\gamma},$$

$$\zeta'(n)_{2^\alpha p_1{}^\beta p_2{}^\gamma p_3{}^\delta} = \Delta(n)_{p_1{}^\beta p_2{}^\gamma p_3{}^\delta},$$

&c.

$p_1, p_2, p_3, \ldots$ being any uneven primes > 1.

These appear to be the most interesting particular cases.

If we desire greater generality we may, as in section 1, substitute for the μ primes in system III. any μ uneven numbers such that no two of them possess a common factor.

Note on the parity of the number which enumerates the partitions of a number. By Major P. A. MacMahon.

[*Read* 25 October 1920.]

In a letter received by me from the late S. Ramanujan about a year ago he stated that it was his intention to calculate the number of the partitions of 1000 by the direct approximate formula which had been successfully used by him and G. H. Hardy in the calculation of the case $n = 200$. He enquired, at the same time, if I knew of any simple way of ascertaining whether the sought number is even or uneven as this information would be of importance to him. This note has arisen in consequence of this enquiry.

I shew, in particular, that in the case of the partitions of 1000 the parity can be found, in a few minutes, from certain congruence relations.

It is easy to derive, from the theory of the self-conjugate partitions of n, that

$$\frac{1}{\prod_1^\infty{}_m (1-q^m)} \equiv \prod_1^\infty{}_m (1+q^{2m-1}) \bmod 2.$$

Thence

$$\frac{1}{\prod_1^\infty{}_m (1-q^m)} \equiv \frac{\prod_1^\infty{}_m \{(1+q^{4m-1})(1+q^{4m-3})(1-q^{4m})\}}{\prod_1^\infty{}_m (1-q^{4m})} \bmod 2.$$

The numerator of the fraction on the right is one of Jacobi's elliptic products which has the series expression

$$1 + q + q^3 + q^6 + q^{10} + \ldots + q^{\frac12 s(s+1)} + \ldots.$$

Wherefore if p_n denote the number of partitions of n

$$\Sigma p_n q^n \equiv (1 + q + q^3 + q^6 + q^{10} + \ldots)\, \Sigma p_n q^{4n} \bmod 2.$$

Put $\qquad p_n \equiv a_n \bmod 2.$

Comparison of the coefficients of like powers of q yields the four relations

$$a_{4n} \equiv a_n + a_{n-7} + a_{n-9} + a_{n-30} + a_{n-34} + a_{n-69} + a_{n-75} + a_{n-124}$$
$$+ a_{n-132} + a_{n-195} + a_{n-205} + a_{n-282} + a_{n-294} + \ldots,$$
$$a_{4n+1} \equiv a_n + a_{n-5} + a_{n-11} + a_{n-26} + a_{n-38} + a_{n-63} + a_{n-81} + a_{n-116}$$
$$+ a_{n-140} + a_{n-185} + a_{n-215} + a_{n-270} + a_{n-306} + \ldots,$$
$$a_{4n+2} \equiv a_{n-1} + a_{n-2} + a_{n-16} + a_{n-19} + a_{n-47} + a_{n-52} + a_{n-94} + a_{n-101}$$
$$+ a_{n-157} + a_{n-166} + a_{n-236} + a_{n-247} + a_{n-331} + \ldots,$$
$$a_{4n+3} \equiv a_n + a_{n-3} + a_{n-13} + a_{n-22} + a_{n-42} + a_{n-57} + a_{n-87} + a_{n-108}$$
$$+ a_{n-148} + a_{n-175} + a_{n-225} + a_{n-258} + a_{n-318} + \ldots.$$

These relations may be written

$$\left.\begin{aligned} a_{4n} &= \sum_0^\infty {}_s\, a_{n-(8s^2\pm s)} \\ a_{4n+1} &\equiv \sum_0^\infty {}_s\, a_{n-(8s^2\pm 3s)} \\ a_{4n+2} &\equiv \sum_0^\infty {}_s\, a_{n-1-(8s^2\pm 7s)} \\ a_{4n+3} &\equiv \sum_0^\infty {}_s\, a_{n-(8s^2\pm 5s)} \end{aligned}\right\}.$$

In fact the four relations are connected with the four elliptic products

$$\prod_1^\infty {}_m (1-q^{16m-9})(1-q^{16m-7})(1-q^{16m}),$$

$$\prod_1^\infty {}_m (1-q^{16m-11})(1-q^{16m-5})(1-q^{16m}),$$

$$q\prod_1^\infty {}_m (1-q^{16m-15})(1-q^{16m-1})(1-q^{16m}),$$

$$\prod_1^\infty {}_m (1-q^{16m-13})(1-q^{16m-3})(1-q^{16m}).$$

The formulae which have been obtained soon involve high numbers and are therefore suitable for the calculation of parity of p_n when m is large.

As an example of the use of the formulae I append the calculation of a_{1000}, the parity of p_{1000}, making use of the enumeration of the partitions of n, as far as $n = 200$, calculated by me in connection with the valuable paper by G. H. Hardy and S. Ramanujan*.

We use the first of the four relations but we first of all require the parities of p_{250}, p_{243}, p_{241}, p_{220}, p_{216}.

From the third relation

$$\begin{aligned} a_{250} &\equiv a_{61} + a_{60} + a_{46} + a_{43} + a_{15} + a_{10} \\ &\equiv 1+1+0+1+0+0 \equiv 1. \end{aligned}$$

From the fourth relation

$$\begin{aligned} a_{243} &\equiv a_{60} + a_{57} + a_{47} + a_{38} + a_{18} + a_{3} \\ &\equiv 1+0+0+1+1+1 \\ &\equiv 0. \end{aligned}$$

From the second relation

$$\begin{aligned} a_{241} &\equiv a_{60} + a_{55} + a_{49} + a_{34} + a_{22} \\ &\equiv 1+0+1+0+0 \\ &\equiv 0. \end{aligned}$$

* *Proc. Lond. Math. Soc.* Vol. xv *et seq.*

From the first relation

$$a_{220} \equiv a_{55} + a_{48} + a_{46} + a_{25} + a_{21}$$
$$\equiv 0 + 1 + 0 + 0 + 0$$
$$\equiv 1.$$
$$a_{216} \equiv a_{54} + a_{47} + a_{45} + a_{24} + a_{20}$$
$$\equiv 1 + 0 + 0 + 1 + 1$$
$$\equiv 1.$$

Thence

$$a_{1000} \equiv a_{250} + a_{243} + a_{241} + a_{220} + a_{216} + a_{181} + a_{175} + a_{126} + a_{118} + a_{55} + a_{45}$$
$$\equiv 1 + 0 + 0 + 1 + 1 + 1 + 0 + 0 + 1 + 0 + 0$$
$$\equiv 1,$$

establishing, in about five minutes work, that p_{1000} is an uneven number.

The Theory of Modular Partitions. By Major P. A. MacMahon, Sc.D., F.R.S.

[*Received* 3 June, 1922.]

The Denotation of a Partition of a Number.

1. There are two methods of denoting partitions of numbers which have been of service in researches. The one simply denotes a part by a number which gives its magnitude and places the parts in descending order of magnitude—usually in a horizontal line. It is more convenient for my present purpose to suppose the parts placed underneath one another in a vertical line—say $\begin{smallmatrix}3\\2\end{smallmatrix}$ a partition of the number 5.

The other, after Ferrers, denotes a part by a succession of nodes, equal in number to the magnitude of the part, placed in a horizontal line and successive parts placed underneath one another in numerical order so that the left-hand nodes of the several parts are in a vertical line. Thus the Euler notation $\begin{smallmatrix}3\\2\end{smallmatrix}$ for a partition becomes

$$\begin{matrix}\bullet & \bullet & \bullet\\ \bullet & \bullet & \end{matrix}$$

in the Ferrers notation.

The connexion between these two modes was established in a previous paper* wherein it was shown that the Ferrers representation is fundamentally one which employs units instead of nodes and is most suitably denoted by, in the above special case,

$$\begin{matrix}1\,1\,1\\ 1\,1\end{matrix}$$

and is, like the Euler notation, numerical but in two dimensions of space and composed entirely of units. This appears when we base the theory on a deeper foundation than that furnished by Euler's intuitive method. There appears, moreover, to be no advantage in working with nodes rather than units. Every consideration and transformation of the Ferrers graphs is just as simple with units—but on the other hand, from the point of view of generalization, the unit representation possesses possibilities that are not shared by the graph of nodes. The Euler and Ferrers representations have both been of great service in the theory and it appears that the subject may be regarded from the point of view which is now taken up.

* *Phil. Trans. R. S. A.*, vol. cxcii, p. 356.

Let each part of a partition be expressed in the form

$$\nu \bmod \mu$$

so that a part $p = s\mu + \nu$.

We may denote the part by a succession of s numbers equal to μ followed by a number ν which may be any of the integers 0, 1, 2, ... $\mu - 1$; these integers being placed in a horizontal line and the zero being, where it occurs, omitted. All the parts of the partition may be similarly treated and the parts placed in successive rows with the left-hand integers of each in the same column. If the parts be placed in rows of descending order of magnitude, as regards numerical content, we obtain a representation of the partition to the modulus μ.

The partition 753221 of the number 20 would be represented to various moduli as follows:

Mod 1	2	3	4	5	6	7 to ∞
1111111	2221	331	43	52	61	7
11111	221	32	41	5	5	5
111	21	3	3	3	3	3
11	2	2	2	2	2	2
11	2	2	2	2	2	2
1	1	1	1	1	1	1

and it is evident that the Ferrers and Euler moduli are 1 and ∞ respectively.

It is convenient to call the number of integers which occur in the representation of a part to a given modulus its "*range*" with regard to the modulus.

The "range" of a partition, similarly, is a convenient phrase, such range being the same as the range of the highest part in the partition.

When the modulus is unity the representation by units has the valuable property that a partition to the same modulus is reached if it be read by columns instead of by rows. In other words the partition is conjugable and this fact has been applied by Sylvester and others to obtain interesting algebraic identities. This property is not enjoyed by the complete set of partitions to any modulus differing from unity.

When the modulus is ∞ indeed only one partition of n, viz. the partition, which is n itself, enjoys the property.

For every modulus we can uniquely select a set of partitions which enjoys the property. We call this the set of conjugable partitions to the modulus μ. If we write down the conjugable partitions of 6 to the modulus 2 we find

222	221	22	21	2
	1	2	2	2
			1	2

where observe that $\begin{smallmatrix}21\\21\end{smallmatrix}$ is not a member of the set because, read by columns, $\begin{smallmatrix}22\\11\end{smallmatrix}$ is not a partition expressed to the modulus 2.

In general, for the modulus 2, any partition which, in the Euler representation, involves an uneven part more than once cannot be comprised in the set. The enumerating generating function for the conjugable partitions of n to the modulus 2 is therefore

$$\frac{(1+q)(1+q^3)(1+q^5)\dots}{(1-q^2)(1-q^4)(1-q^6)\dots},$$

or in Cayley's notation $\qquad \dfrac{[1+q^{2m+1}]}{[1-q^{2m}]}$.

Since it may be also written

$$\frac{1}{(1-q)(1-q^3)(1-q^4)(1-q^5)(1-q^7)(1-q^8)(1-q^9)\dots},$$

we see that the partitions are equi-numerous with the system in which no parts occur of the form 2 mod 4.

We are already in a position to enunciate a general theorem in regard to any conjugable set of partitions to the modulus μ.

Theorem. "Of the conjugate set of partitions to the modulus μ of the number n there are as many partitions which have a range $\begin{pmatrix}\text{equal to}\\ \text{not exceeding}\end{pmatrix}$ k and a number of parts $\begin{pmatrix}\text{equal to}\\ \text{not exceeding}\end{pmatrix}$ i as there are partitions which have a range $\begin{pmatrix}\text{equal to}\\ \text{not exceeding}\end{pmatrix}$ i and a number of parts $\begin{pmatrix}\text{equal to}\\ \text{not exceeding}\end{pmatrix}$ k."

When the range is k and the greatest part has the value j,

$$k = \underset{\prec}{E}\frac{j}{\mu},$$

where $\underset{\prec}{E}x$ denotes the integer not less than x.

When the range is k, j must have one of the values

$$\mu k,\ \mu k-1,\ \mu k-2,\ \dots,\ \mu k-(\mu-1).$$

With the modulus μ, for a partition to appertain to the conjugable set, two adjacent parts must not terminate with numbers drawn from the series

$$1,\ 2,\ \dots,\ \mu-1,$$

or in other words the magnitudes of two adjacent parts must not be numbers drawn from the series

$$\mu r+1,\ \mu r+2,\ \dots,\ \mu r+\mu-1,$$

where r is any integer (zero included).

Hence it follows that the function which enumerates the number of partitions in the conjugable set is

$$\frac{[1 + q^{\mu m+1} + q^{\mu m+2} + \ldots + q^{\mu m+\mu-1}]}{[1 - q^{\mu m}]},$$

which may be thrown into the form

$$1 + \frac{q}{1-q} \cdot \frac{1-q^{\mu-1}}{1-q^{\mu}} + \frac{q^2}{(1-q)^2} \frac{q^{\mu}(1-q^{\mu-1})^2}{(1-q^{\mu})(1-q^{2\mu})} + \frac{q^3}{(1-q)^3} \cdot \frac{q^{3\mu}(1-q^{\mu-1})^3}{(1-q^{\mu})(1-q^{2\mu})(1-q^{3\mu})} + \ldots$$

with a denominator $\qquad [1 - q^{\mu m}]$,

and where the general term in the numerator is

$$\left(\frac{q}{1-q}\right)^s \frac{q^{\binom{s}{2}\mu}(1-q^{\mu-1})^s}{(1-q^{\mu})(1-q^{2\mu})\ldots(1-q^{s\mu})}.$$

Limitation of the Number of Parts.

2. I now proceed to consider the enumerating function when the number of parts in the partitions is limited by the number i.

For *modulus* 2 this is the coefficient of a^i in the expansion of

$$Q = \frac{(1+aq)(1+aq^3)(1+aq^5)\ldots}{(1-a)(1-aq^2)(1-aq^4)(1-aq^6)\ldots} = 1 + aQ_1 + a^2Q_2 + \ldots.$$

If in Q we write aq^2 for a and then multiply Q by

$$\frac{1+aq}{1-a},$$

the function Q is unaltered.

Thence we readily find that

$$Q_i = \frac{(1-q^2)(1-q^6)(1-q^{10})\ldots(1-q^{4i-2})}{(1-q)(1-q^2)(1-q^3)\ldots(1-q^{2i})},$$

the required enumerating function.

Also the function which enumerates the partitions into exactly i parts is

$$Q_i - Q_{i-1} = \frac{1+q}{1-q^{2i}} q^{2i-1} Q_{i-1}.$$

For *modulus* 3 we obtain similarly the enumerating function

$$\frac{(1+q+q^2)(1+q^4+q^5)\ldots(1+q^{3i-2}+q^{3i-1})}{(1-q^3)(1-q^6)\ldots(1-q^{3i})}$$

for the partitions into i or fewer parts; and for

Modulus μ

$$\frac{(1+q+\ldots+q^{\mu-1})(1+q^{\mu+1}+\ldots+q^{2\mu-1})\ldots(1+q^{i\mu-\mu+1}+\ldots+q^{i\mu-1})}{(1-q^{\mu})(1-q^{2\mu})\ldots(1-q^{i\mu})}.$$

Limitation of the Part Magnitude.

Modulus 2. If the part magnitude be $2j$ the function is

$$\frac{(1+q)(1+q^3)\dots(1+q^{2j-1})}{(1-q^2)(1-q^4)\dots(1-q^{2j})},$$

and if the part magnitude be $2j+1$ it is

$$\frac{(1+q)(1+q^3)\dots(1+q^{2j+1})}{(1-q^2)(1-q^4)\dots(1-q^{2j})}.$$

Modulus 3.

For part magnitude $=3j$ function is $\dfrac{(1+q+q^2)(1+q^4+q^5)\dots(1+q^{3j-2}+q^{3j-1})}{(1-q^3)(1-q^6)\dots(1-q^{3j})}$;

$3j+1$ „ $\dfrac{(1+q+q^2)(1+q^4+q^5)\dots(1+q^{3j+1}+0)}{(1-q^3)(1-q^6)\dots(1-q^{3j})}$;

$3j+2$ „ $\dfrac{(1+q+q^2)(1+q^4+q^5)\dots(1+q^{3j+1}+q^{3j+2})}{(1-q^3)(1-q^6)\dots(1-q^{3j})}$;

and in general for the

Modulus μ.

For part magnitude

μj function is $\dfrac{(1+q+\dots+q^{\mu-1})(1+q^{\mu+1}+\dots+q^{2\mu-1})\dots(1+q^{\mu j-\mu+1}+\dots+q^{\mu j-1})}{(1-q^{\mu})(1-q^{2\mu})\dots(1-q^{\mu j})}$;

$\mu j+1$ „ $\dfrac{(1+q+\dots+q^{\mu-1})\dots(1+q^{\mu j-\mu+1}+\dots+q^{\mu j-1})(1+q^{\mu j+1})}{(1-q^{\mu})(1-q^{2\mu})\dots(1-q^{\mu j})}$;

$\mu j+2$ „ $\dfrac{(1+q+\dots+q^{\mu-1})\dots(1+q^{\mu j-\mu+1}+\dots+q^{\mu j-1})(1+q^{\mu j+1}+q^{\mu j+2})}{(1-q^{\mu})(1-q^{2\mu})\dots(1-q^{\mu j})}$;

....

$mj+\mu-1$ „ $\dfrac{(1+q+\dots q^{\mu-1})(1+q^{\mu+1}+\dots+q^{2\mu-1})\dots(1+q^{\mu j+1}+\dots+q^{\mu j+\mu-1})}{(1-q^{\mu})(1-q^{2\mu})\dots(1-q^{\mu j})}$.

Limitation of the Range.

3. We have only to conjugate the partitions of a set to modulus μ to see that the function which enumerates partitions for a given limitation of range is identical with that which enumerates for the same limitation of the number of parts.

In fact when the range is i the part magnitude may be any one of

$$(i-1)\mu+1,\ (i-1)\mu+2,\ \dots,\ i\mu,$$

or may take μ values.

Thence we see that the part magnitude is limited not to exceed $i\mu$.

The expression obtained above, for the limitation of the part magnitude to μj, is, on writing i for j, identical with that which has been obtained above for the limitation of the number of parts so as not to exceed the integer i.

Limitation of the Range and Number of Parts.

4. When the range is limited to k and the number of parts to i we can now see from the preceding that the enumeration is given by the coefficient of a^i in

$$R_{i,k}=\frac{\{1+a(q+q^2+\ldots+q^{\mu-1})\}\{1+a(q^{\mu+1}+\ldots+q^{2\mu-1})\}\ldots\{1+a(q^{k\mu-\mu+1}+\ldots+q^{k\mu-1})\}}{(1-a)(1-aq^{\mu})(1-aq^{2\mu})\ldots(1-aq^{k\mu})}$$

and the fact of conjugation shows that this is also the coefficient of a^k in

$$R_{k,i}=\frac{\{1+a(q+q^2+\ldots+q^{\mu-1})\}\{1+a(q^{\mu+1}+\ldots+q^{2\mu-1})\}\ldots\{1+a(q^{i\mu-\mu+1}+\ldots+q^{i\mu-1})\}}{(1-a)(1-aq^{\mu})(1-aq^{2\mu})\ldots(1-aq^{i\mu})}.$$

It is easily verified for $k=1$, $i=2$ when the coefficient is found to be in each case

$$\frac{1-q^{2\mu+1}}{1-q}.$$

The Algebraic Genesis of the Modular Partitions.

5. I have in previous papers* considered the algebraic fraction

$$\frac{1}{(1-\lambda_1 X_1)\left(1-\frac{\lambda_2}{\lambda_1}X_2\right)\left(1-\frac{\lambda_3}{\lambda_2}X_3\right)\ldots\left(1-\frac{1}{\lambda_{i-1}}X_i\right)},$$

in which the general term is

$$\lambda_1^{\alpha_1-\alpha_2}\lambda_2^{\alpha_2-\alpha_3}\ldots\lambda_{i-1}^{\alpha_{i-1}-\alpha_i}X_1^{\alpha_1}X_2^{\alpha_2}\ldots X_i^{\alpha_i}.$$

If the numbers $\alpha_1, \alpha_2, \ldots, \alpha_i$ satisfy the conditions

$$\alpha_1 \geqslant \alpha_2 \geqslant \ldots \geqslant \alpha_i,$$

the term product $$X_1^{\alpha_1}X_2^{\alpha_2}\ldots X_i^{\alpha_i}$$

denotes by its exponent a partition

$$(\alpha_1\alpha_2\ldots\alpha_i)$$

of the number $\Sigma\alpha$.

In order to satisfy the system of Diophantine Inequalities we must, on expansion of the algebraic fraction, reject all negative powers of $\lambda_1, \lambda_2, \ldots, \lambda_i$ and we *may* afterwards put

$$\lambda_1=\lambda_2=\ldots=\lambda_i=1.$$

* *Phil. Trans. R. S. A.*, vol. CXCII, p. 356 *et seq.*

Denoting the performance of both of these operations by $\underset{\geqslant}{\Omega}$, we find

$$\underset{\geqslant}{\Omega}\frac{1}{(1-\lambda_1X_1)\left(1-\frac{\lambda_2}{\lambda_1}X_2\right)\left(1-\frac{\lambda_3}{\lambda_2}X_3\right)\ldots\left(1-\frac{\lambda_{i-1}}{\lambda_{i-2}}X_{i-1}\right)\left(1-\frac{1}{\lambda_{i-1}}X_i\right)}$$

$$=\frac{1}{(1-X_1)(1-X_1X_2)(1-X_1X_2X_3)\ldots(1-X_1X_2X_3\ldots X_i)},$$

and this a real generating function of all partitions to modulus 1 (of all numbers) whose range does not exceed i.

The generating function is real because an X product appears in correspondence with every partition. To obtain an enumerating function it is necessary to put

$$X_1 = X_2 = \ldots = X_i = x.$$

In fact the expansion exhibits the Ferrers representation.

In the next place consider the expression

$$\underset{\geqslant}{\Omega}\frac{1}{(1-\lambda_1X_1)\left(1-\frac{\lambda_2}{\lambda_1}X_1\right)\left(1-\frac{\lambda_3}{\lambda_2}X_2\right)\left(1-\frac{\lambda_4}{\lambda_3}X_2\right)\ldots\left(1-\frac{\lambda_{2i-1}}{\lambda_{2i-2}}X_i\right)\left(1-\frac{\lambda_{2i}}{\lambda_{2i-1}}X_i\right)},$$

where, for convenience only, the number of denominator factors is taken to be even. We thence reach the real generating function

$$\frac{1}{(1-X_1)(1-X_1^2)(1-X_1^2X_2)(1-X_1^2X_2^2)\ldots(1-X_1^2X_2^2\ldots X_{i-1}^2X_i)(1-X_1^2X_2^2\ldots X_i^2)}$$

of partitions to the modulus 2 and of range i.

The fundamental set of partitions

$$1,\ 2,\ 21,\ 22,\ \ldots\ 22\ldots1,\ 22\ldots2$$

is indicated.

The real generating function of the conjugate system is

$$\frac{(1+X_1)(1+X_1^2X_2)(1+X_1^2X_2^2X_3)\ldots(1+X_1^2X_2^2\ldots X_{i-1}^2X_2)}{(1-X_1^2)(1-X_1^2X_2^2)(1-X_1^2X_2^2X_3^2)\ldots(1-X_1^2X_2^2\ldots X_i^2)}.$$

Similarly for the modulus μ we take as generating function

$$\underset{\geqslant}{\Omega}\frac{1}{D},$$

where D is the product of

$$\mu \text{ factors } (1-\lambda_1X_1)\left(1-\frac{\lambda_2}{\lambda_1}X_1\right)\ldots\left(1-\frac{\lambda_\mu}{\lambda_{\mu-1}}X_1\right),$$

$$\mu \text{ factors } \left(1-\frac{\lambda_{\mu+1}}{\lambda_\mu}X_2\right)\left(1-\frac{\lambda_{\mu+2}}{\lambda_{\mu+1}}X_2\right)\ldots\left(1-\frac{\lambda_{2\mu}}{\lambda_{2\mu-1}}X_2\right),$$

....

μ factors $\left(1-\frac{\lambda_{(i-2)\mu+1}}{\lambda_{(i-2)\mu}}X_{i-1}\right)\left(1-\frac{\lambda_{(i-2)\mu+2}}{\lambda_{(i-2)\mu+1}}X_{i-1}\right)\ldots\left(1-\frac{\lambda_{(i-1)\mu}}{\lambda_{(i-1)\mu-1}}X_{i-1}\right)$,

μ factors $\left(1-\frac{\lambda_{(i-1)\mu+1}}{\lambda_{(i-1)\mu}}X_1\right)\left(1-\frac{\lambda_{(i-1)\mu+2}}{\lambda_{(i-1)\mu+1}}X_i\right)\ldots\left(1-\frac{\lambda_{i\mu}}{\lambda_{i\mu-1}}X_1\right)$,

where for convenience only the number of denominator factors is supposed to be of the form $\equiv 0 \bmod \mu$, and thence proceed to the real generating function

$$\frac{1}{(1-X_1)(1-X_1^2)\ldots(1-X_1^\mu)(1-X_1^\mu X_2)\ldots(1-X_1^\mu X_2^\mu)\ldots(1-X_1^\mu X_2^\mu\ldots X_i^\mu)}$$

and that of the conjugate system to the modulus μ and range i

$$\frac{N}{(1-X_1^\mu)(1-X_1^\mu X_2^\mu)(1-X_1^\mu X_2^\mu X_3^\mu)\ldots(1-X_1^\mu X_2^\mu\ldots X_i^\mu)},$$

where N is the product of factors

$$(1+X_1+X_1^2+\ldots+X_1^{\mu-1})$$
$$(1+X_1^\mu X_2+X_1^\mu X_2^2+\ldots+X_1^\mu X_2^{\mu-1})$$
$$(1+X_1^\mu X_2^\mu X_3+X_1^\mu X_2^\mu X_3^2+\ldots+X_1^\mu X_2^\mu X_3^{\mu-1})$$
$$\vdots$$
$$(1+X_1^\mu X_2^\mu\ldots X_{i-1}^\mu X_i+X_1^\mu X_2^\mu\ldots X_{i-1}^\mu X_i^2+\ldots+X_1^\mu X_2^\mu\ldots X_{i-1}^\mu X_i^{\mu-1}).$$

THE PARITY OF $p(n)$, THE NUMBER OF PARTITIONS OF n, WHEN $n \leqslant 1000$

P. A. MacMahon*.

I refer to a note† written by me in response to a letter from S. Ramanujan, who had inquired if I knew of a simple way of determining the parity of the number $p(n)$, the number of partitions of n. He stated, in particular, that he had the intention of calculating $p(1000)$ by the method explained in the paper‡ by G. H. Hardy and himself, and that

* Received 27 May, 1926; read 11 November, 1926.

† P. A. MacMahon, "Note on the parity of the number which enumerates the partitions of a number", *Proc. Camb. Phil. Soc.*, 20 (1921), 281-283.

‡ G. H. Hardy and S. Ramanujan, "Asymptotic formulae in combinatory analysis", *Proc. London Math. Soc.* (2), 17 (1918), 75-115.

information in regard to the parity of that number would be of assistance to him. I append a list of values of n for which $p(n)$ is an even number and $n \nless 1000$. This may be useful, because Ramanujan was led by a study of the values of $p(n)$, for $n \leqslant 200$, printed in *Proc. London Math. Soc.* (*l.c.*), to important congruence relations*. I find that in 470 cases $p(n)$ is an even number, viz., when n has the values :—

2, 8, 9, 10, 11, 15, 19, 21, 22, 25, 26, 27, 28, 30, 31, 34, 40, 42, 45, 46, 47, 50, 55, 57, 58, 59, 62, 64, 65, 66, 70, 74, 75, 78, 79, 80, 84, 86, 94, 96, 97, 98, 100.

101, 103, 106, 108, 109, 110, 112, 113, 116, 117, 120, 122, 124, 125, 126, 128, 129, 130, 131, 133, 135, 136, 137, 141, 142, 147, 149, 151, 153, 154, 158, 160, 163, 167, 170, 171, 174, 175, 176, 179, 180, 184, 187, 191, 197, 198, 200.

205, 206, 207, 213, 217, 224, 227, 228, 230, 231, 236, 241, 243, 245, 246, 247, 248, 253, 255, 256, 258, 260, 262, 264, 265, 267, 268, 271, 274, 278, 280, 282, 288, 290, 291, 292, 295, 297, 298, 300.

303, 305, 307, 310, 314, 315, 317, 318, 319, 322, 323, 327, 328, 329, 334, 336, 337, 339, 340, 341, 342, 344, 347, 348, 350, 351, 352, 353, 356, 358, 359, 364, 365, 370, 371, 372, 374, 375, 379, 380, 383, 384, 386, 387, 388, 391, 393, 397, 400.

404, 405, 408, 409, 410, 411, 412, 413, 414, 415, 416, 419, 422, 424, 427, 430, 431, 435, 439, 441, 444, 445, 447, 448, 449, 450, 452, 455, 456, 458, 465, 469, 471, 473, 476, 478, 481, 483, 485, 487, 492, 494, 496.

501, 503, 505, 506, 509, 510, 511, 512, 513, 515, 517, 518, 520, 521, 523, 524, 526, 527, 529, 530, 531, 532, 533, 538, 539, 541, 544, 545, 548, 549, 550, 551, 552, 553, 554, 556, 557, 561, 562, 563, 564, 566, 569, 570, 571, 577, 579, 580, 581, 588, 589, 590, 591, 596, 597, 599, 600.

602, 604, 605, 606, 607, 609, 615, 616, 618, 623, 628, 631, 632, 633, 637, 638, 639, 641, 643, 644, 645, 646, 648, 649, 651, 654, 657, 659, 662, 663, 669, 670, 671, 674, 675, 676, 677, 679, 680, 682, 684, 687, 688, 692, 694, 697, 698, 699.

701, 702, 703, 705, 706, 710, 711, 713, 715, 718, 721, 727, 732, 735, 738, 743, 744, 745, 748, 751, 752, 754, 758, 760, 761, 764, 765, 766, 767, 770, 771, 772, 773, 775, 776, 777, 780, 781, 782, 784, 785, 787, 789, 790, 793, 794, 795, 797, 798, 799.

803, 806, 810, 811, 813, 815, 819, 820, 822, 824, 827, 828, 832, 833, 836, 838, 839, 840, 843, 844, 845, 847, 848, 849, 855, 858, 859, 865, 867, 871, 873, 874, 877, 879, 880, 881, 885, 888, 889, 893, 895, 896, 899, 900.

904, 905, 908, 917, 919, 920, 921, 922, 923, 925, 927, 928, 929, 932, 934, 936, 938, 941, 942, 947, 948, 949, 950, 952, 953, 958, 959, 963, 964, 965, 966, 968, 969, 973, 974, 975, 976, 978, 980, 982, 983, 984, 985, 986, 989, 992, 994, 995, 999.

The corresponding question for $q(n)$, the number of partitions into odd parts (or into unequal parts), is simple. In fact,

$$1+\Sigma q(n)x^n = (1+x)(1+x^2)(1+x^3)\ldots \equiv (1-x)(1-x^2)(1-x^3)\ldots \pmod 2,$$

so that $q(n)$ is odd if n is of the form $\frac{1}{2}(3m^2 \pm m)$, and otherwise even.

* S. Ramanujan, "Some properties of $p(n)$, the number of partitions of n", *Proc. Camb. Phil. Soc.*, 19 (1920), 207–210.

The elliptic products of Jacobi and the theory of linear congruences. By Major P. A. MACMAHON, St John's College.

[*Received* 22 May, *read* 26 July 1926.]

In the application of Elliptic Functions to the Theory of Numbers the two formulae of Jacobi

$$\prod_1^\infty (1-q^{2ma-a-b})(1-q^{2ma-a+b})(1-q^{2ma}) = \sum_{-\infty}^{+\infty} (-)^m q^{am^2+bm},$$

$$\prod_1^\infty (1+q^{2ma-a-b})(1+q^{2ma-a+b})(1-q^{2ma}) = \sum_{-\infty}^{+\infty} q^{am^2+bm}$$

are of great importance.

I denote either side of the first identity by

$$(a, b)$$

and either side of the second by

$$[a, b].$$

For my present purpose I take the exponents of q which occur on the left-hand sides to be integers. The summations are in regard to m, and a, b must be simultaneously either both integers or both the halves of uneven integers. If a, b possess a common factor f so that $a = a'f$, $b = b'f$, (a, b) and $[a, b]$ can be derived from (a', b'), $[a', b']$ respectively by writing q^f for q. It is then convenient to write

$$(a, b) = (a', b')_f;\ [a, b] = [a', b']_f,$$

the subscript f implying that q^f is to be written for q.

With this understanding we may regard a, b as being relatively prime integers. Since moreover

$$(a, b) = -q^{a-b}(a, 2a-b) = -q^{a+b}(a, 2a+b) = (a, -b),$$

$$[a, b] = \quad q^{a-b}[a, 2a-b] = \quad q^{a+b}[a, 2a+b] = [a, -b].$$

we may always transform (a, b) so that $b \not> a$ and $[a, b]$ so that $b \not> a$.

In the first identity $\quad (a, a) = 0,$

In the second $\quad [a, a] = [\tfrac{1}{2}, \tfrac{1}{2}]_{2a} = 2[2, 1]_{2a}.$

Therefore in (a, b), for a given value of a, b may be any integer $< a$ and prime to a. It may therefore assume $\phi(a)$ values where ϕ is the ϕ function of Euler and $\phi(a)$ has been termed the totient of a by Sylvester and b one of the $\phi(a)$ totitives of a. If we have before us $\left(\frac{a}{2}, \frac{b}{2}\right)$, where a, b are uneven integers, we may similarly regard b as one of the totitives of a. The same remarks apply to $[a, b]$ and $\left[\frac{a}{2}, \frac{b}{2}\right]$ in the two cases except that here b may be equal to a and then we have

$$[a, a] = [1, 1]_a, \quad \left[\frac{a}{2}, \frac{a}{2}\right] = \left[\frac{1}{2}, \frac{1}{2}\right]_a,$$

so that in this case the difference from the forms (a, b), $\left(\frac{a}{2}, \frac{b}{2}\right)$ is that those are the exceptional cases $[1, 1]$, $[\frac{1}{2}, \frac{1}{2}]$.

In recent years G. H. Hardy, S. Ramanujan and H. B. C. Darling have been concerned with those enumerating generating functions which involve elliptic products. They have obtained valuable theorems in regard to that part of the function which is entirely expressed by powers of q whose exponents $\equiv \epsilon \bmod \mu$ for special values of μ and ϵ. As a contribution to this study I investigate some properties of elliptic products which depend upon the partitioning of the product into parts each of which is specified by a certain congruence which is satisfied by the q exponents. In this partition the modulus μ remains constant and the residue varies from part to part. The residue cannot be assigned at pleasure, because it must be of the form $a\epsilon^2 + b\epsilon$, where ϵ is an integer.

§ 1. *The elliptic product* (a, b).

We put $\mu m + \epsilon$ for m in the series given above and obtain

$$(-)^\epsilon q^{a\epsilon^2+b\epsilon} \sum_{-\infty}^{+\infty} (-)^{\mu m} q^{\mu^2 a m^2 + \mu(2a\epsilon+b)m}.$$

For the present I consider μ to be an uneven prime. We can then write the portion of (a, b), which has m of the form $\mu m + \epsilon$,

$$(-)^\epsilon q^{a\epsilon^2+b\epsilon} (\mu a, 2a\epsilon + b)_\mu.$$

If a, b be both integers, every exponent of q in this function

$$\equiv a\epsilon^2 + b\epsilon \bmod \mu.$$

Moreover if a, b be the halves of uneven integers

$$a = \alpha + \tfrac{1}{2}, \quad b = \beta + \tfrac{1}{2},$$

the power of q is in general

$$(\alpha + \tfrac{1}{2})\epsilon^2 + (\beta + \tfrac{1}{2})\epsilon + \mu^2(\alpha + \tfrac{1}{2})m^2 + \mu(2\alpha\epsilon + \epsilon + \beta + \tfrac{1}{2})m,$$

which is of the form $\qquad a\epsilon^2 + b\epsilon \bmod \mu$,

and $a\epsilon^2 + b\epsilon$ is an integer *provided* that

$$\binom{\mu m + 1}{2} \equiv 0 \bmod \mu;$$

and this is so because μ is uneven. Hence the transformed series is also

$$(-)^{\epsilon} q^{a\epsilon+b\epsilon}(\mu a,\ 2a\epsilon+b)_{\mu},$$

where a, b are the halves of uneven integers.

We now sum this expression, giving ϵ the values $0, 1, 2, \ldots \mu-1$, and apply the formula

$$(a,\ b)=-\ q^{a-b}(a,\ 2a-b),$$

so as to make the second element in the bracket less than the first element, obtaining the standard formula

$$\begin{aligned}(a,\ b)= \quad & -q^{a-b}(\mu a,\ 2a-b)_{\mu}+q^{4a-2b}(\mu a,\ 4a-b)_{\mu}-\ldots \\ & +(-)^{\frac{1}{2}(\mu-1)} q^{\left(\frac{\mu-1}{2}\right)^2 a-\frac{\mu-1}{2} b}(\mu a,\ \mu a-a-b)_{\mu}, \\ & +(\mu a,\ b)_{\mu}-q^{a+b}(\mu a,\ 2a+b)_{\mu}+q^{4a+2b}(\mu a,\ 4a+b)_{\mu}-\ldots \\ & +(-)^{\frac{1}{2}(\mu-1)} q^{\left(\frac{\mu-1}{2}\right)^2 a+\frac{\mu-1}{2} b}(\mu a,\ \mu a-a+b)_{\mu},\end{aligned}$$

a partition into μ parts.

This formula is valid when a, b are integers or the halves of integers and is so far a complete solution of the question.

The partition is usually effective with less than μ parts.

If μ be even, a, b integers, we follow the foregoing method and find that the part separated is

$$(-)^{\epsilon} q^{a\epsilon^2+b\epsilon}[\mu a,\ 2a\epsilon+b]_{\mu},$$

and it is to be observed that here the product (a, b) gives rise to products of type $[a, b]$.

Every exponent of q is clearly $\equiv a\epsilon^2+b\epsilon \bmod \mu$ and we obtain the crude partition

$$(a,\ b)=\sum_{0}^{\mu-1}(-)^{\epsilon} q^{a\epsilon^2+b\epsilon}[\mu a,\ 2a\epsilon+b]_{\mu},$$

which by reason of the relation

$$[a,\ b]=q^{a-b}[a,\ 2a-b]$$

is reducible at once to the standard formula.

The elliptic product $[a, b]$.

The separated part is

$$q^{a\epsilon^2+b\epsilon}[\mu a,\ 2a\epsilon+b]_{\mu}.$$

If μ be even or uneven and a, b integers *or* if μ be uneven and a, b the halves of uneven integers, the standard formula is

23—2

$$[a, b] = \quad + q^{a-b}[\mu a, 2a-b]_\mu + q^{4a-2b}[\mu a, 4a-b]_\mu + \dots$$
$$+ q^{\left(\frac{\mu-1}{2}\right)^2 a - \frac{\mu-1}{2} b}[\mu a, \mu a - a - b]_\mu,$$
$$+ [\mu a, b]_\mu + q^{a+b}[\mu a, 2a+b]_\mu + q^{4a+2b}[\mu a, 4a+b]_\mu + \dots$$
$$+ q^{\left(\frac{\mu-1}{2}\right)^2 a + \frac{\mu-1}{2} b}[\mu a, \mu a - a + b]_\mu.$$
$$(a, b) = \quad - q^{a-b}[\mu a, 2a-b]_\mu + q^{4a-2b}[\mu a, 4a-b]_\mu - \dots$$
$$(-)^{\frac{1}{2}(\mu-1)} q^{\left(\frac{\mu-1}{2}\right)^2 a - \frac{\mu-1}{2} b}[\mu a, \mu a - a - b]_\mu,$$
$$+ [\mu a, b]_\mu - q^{a+b}[\mu a, 2a+b]_\mu + q^{4a+2b}[\mu a, 4a+b]_\mu - \dots$$
$$(-)^{\frac{1}{2}(\mu-1)} q^{\left(\frac{\mu-1}{2}\right)^2 + \frac{\mu-1}{2} b}[\mu a, \mu a - a + b]_\mu.$$

When μ is even, and a, b the halves of integers, the formula is not so simple or interesting but is readily obtainable.

A leading property of these partitions, which will be dealt with later, is that, when a is prime to μ, in one and in only one of the μ terms involved in the partition is the second element, which is of form

$$2ha \pm b$$

divisible by μ. In this *special* term the subscript can be made equal to μ^2.

§ 2. *Important particular cases.*

The elliptic products which are of most importance from the present point of view are

$$(1, 0) = \prod_1^\infty (1 - q^{2m-1})^2 (1 - q^{2m}) = \sum_{-\infty}^{+\infty} (-)^m q^{m^2}$$
$$= \sqrt{\frac{2k'K}{\pi}} = \theta_{0,1}(0),$$

$$[1, 0] = \prod_1^\infty (1 + q^{2m-1})^2 (1 - q^{2m}) = \sum_{-\infty}^{+\infty} q^{m^2}$$
$$= \sqrt{\frac{2K}{\pi}} = \theta_{0,0}(0),$$

$$q^{\frac{1}{4}}[1, 1] = q^{\frac{1}{4}} \prod_1^\infty (1 + q^{2m-2})(1 - q^{4m}) = q^{\frac{1}{4}} \sum_{-\infty}^{+\infty} q^{m^2+m}$$
$$= \sqrt{\frac{2kK}{\pi}} = \theta_{1,0}(0),$$

$$(\tfrac{3}{2}, \tfrac{1}{2}) = \prod_1^\infty (1 - q^m) \qquad = \sum_{-}^{+\infty} (-)^m q^{\frac{3}{2}m^2 + \frac{1}{2}m}$$

(Euler's series),

$$(\tfrac{5}{2}, \tfrac{3}{2}) = \prod_{1}^{\infty} (1 - q^{5m-4})(1 - q^{5m-1})(1 - q^{5m}) = \sum_{-\infty}^{+\infty} (-)^m q^{\frac{5}{2}m^2 + \frac{3}{2}m},$$

$$(\tfrac{5}{2}, \tfrac{1}{2}) = \prod_{1}^{\infty} (1 - q^{5m-3})(1 - q^{5m-2})(1 - q^{5m}) = \sum_{-\infty}^{+\infty} (-)^m q^{\frac{5}{2}m^2 + \frac{1}{2}m}.$$

These last two arise in the Theory of Partitions by reason of properties that have been established in respect of the functions

$$\frac{(\frac{3}{2}, \frac{1}{2})_5}{(\frac{5}{2}, \frac{3}{2})}, \quad \frac{(\frac{3}{2}, \frac{1}{2})_5}{(\frac{5}{2}, \frac{1}{2})}$$

by L. Rogers* and S. Ramanujan†. They also arise in the elliptic function solutions of the general quintic equation as presented by H. B. C. Darling‡.

There is also the case of the Cube of Euler's series, the formula for which can be deduced from the fundamental formulae of Jacobi as a limiting case§.

The moduli of principal importance are prime numbers; in particular the early primes 2, 3, 5, 7, 11, ...

Postponing the case of the modulus 2, and generally of composite moduli, I proceed to the consideration of the simplest uneven prime moduli.

The modulus 3.

The standard formula is

$$\begin{aligned}(a, b) &= (3a, b)_3 - q^{a+b}(3a, 2a + b)_3 \\ &\quad - q^{a-b}(3a, 2a - b)_3, \\ [a, b] &= [3a, b]_3 + q^{a+b}[3a, 2a + b]_3 \\ &\quad + q^{a-b}[3a, 2a - b]_3.\end{aligned}$$

Thence $\qquad (1, 0) = (1, 0)_{3^2} - 2q(3, 2)_3.$

The special term establishes that

$$(1, 0) + 2q(3, 2)_3$$

is a transformation of the q series (1, 0), which is $\sqrt{\dfrac{2k'K}{\pi}}$, of order 3^2.

$$[1, 0] = [1, 0]_{3^2} + 2q[3, 2]_3,$$

$$[1, 1] = 2[3, 1]_3 + q^2[1, 1]_{3^2}$$

indicating transformation of order 3^2 of

$$[1, 0] \text{ which is } \sqrt{\frac{2K}{\pi}},$$

* On Two Theorems, *Proc. L.M.S.*, New Series, vol. XVI, p. 315 *et seq.*

† Some properties of $p(n)$, the number of partitions of n, *Camb. Phil. Soc. Proc.*, vol. XIX, p. 207 *et seq.*

‡ On The Trinomial Quintic, *Proc. L.M.S.*, New Series, vol. XXIII, p. 383 *et seq.*

§ Sylvester, *Collected Papers*, vol. IV, p. 60.

and $\qquad [1, 1]$ which is $q^{-\frac{1}{4}}\sqrt{\frac{2kK}{\pi}}$;

$$(\tfrac{3}{2}, \tfrac{1}{2}) = (\tfrac{9}{2}, \tfrac{1}{2})_3 - q(\tfrac{9}{2}, \tfrac{5}{2})_3 - q^2(\tfrac{9}{2}, \tfrac{7}{2})_3,$$

which, writing with Cayley $1 - q^s = (s)$, is

$$\begin{aligned}(1)(2)(3)\ldots = {} & (12)(39)(66)\ldots(15)(42)(69)\ldots(27)(54)(81)\ldots \\ & - q(6)(33)(60)\ldots(21)(48)(75)\ldots(27)(54)(81)\ldots \\ & - q^2(3)(30)(57)\ldots(24)(51)(78)\ldots(27)(54)(81)\ldots\end{aligned}$$

verifying a known result.

Since the leading element here is the half of an uneven number, viz. 3, which is not prime to the modulus 3, which we have before us, no special term arises.

The next two partitions are of particular interest and furnish the clue to an interesting theory

$$(\tfrac{5}{2}, \tfrac{3}{2}) = (\tfrac{5}{2}, \tfrac{1}{2})_{3^2} - q(\tfrac{15}{2}, \tfrac{7}{2})_3 - q\{q(\tfrac{15}{2}, \tfrac{13}{2})\}_3,$$

$$(\tfrac{5}{2}, \tfrac{1}{2}) = (\tfrac{15}{2}, \tfrac{1}{2})_3 - \{q(\tfrac{15}{2}, \tfrac{11}{2})\}_3 - q^2(\tfrac{5}{2}, \tfrac{3}{2})_{3^2}.$$

We find that the special term in $(\tfrac{5}{2}, \tfrac{3}{2})$ is $(\tfrac{5}{2}, \tfrac{1}{2})_{3^2}$,

" $\quad$ " $\quad$ " $\quad$ " $\quad (\tfrac{5}{2}, \tfrac{1}{2})$ is $-q^2(\tfrac{5}{2}, \tfrac{3}{2})_{3^2}$.

This fact is remarkable, but not surprising, because the work to which allusion has been made has shewn that the two functions are very intimately connected not only in the Theory of the Transformation of Elliptic Functions but also in the application of Elliptic Functions to the solution of the general algebraic quintic equation. The property reminds one of that possessed by the so-called 'amicable numbers.' In this parallel the functions (1, 0), [1, 0], [1, 1] considered above will be in correspondence with 'perfect numbers.'

We observe in the partitions of $(\tfrac{5}{2}, \tfrac{3}{2})$ and $(\tfrac{5}{2}, \tfrac{1}{2})$ that in the former case the special term is one part, the remaining two terms both belonging to the second part—whilst in the latter the special term is also one part and the remaining two terms constitute the second part.

Next we are led to consider the three products

$$(\tfrac{7}{2}, \tfrac{5}{2}),\ (\tfrac{7}{2}, \tfrac{3}{2}),\ (\tfrac{7}{2}, \tfrac{1}{2}),$$

a nd we find

$$(\tfrac{7}{2}, \tfrac{5}{2}) = (\tfrac{21}{2}, \tfrac{5}{2})_3 - \{q^2(\tfrac{21}{2}, \tfrac{19}{2})\}_3 - q(\tfrac{7}{2}, \tfrac{3}{2})_{3^2},$$

$$(\tfrac{7}{2}, \tfrac{3}{2}) = (\tfrac{7}{2}, \tfrac{1}{2})_{3^2} - q^2(\tfrac{21}{2}, \tfrac{11}{2})_3 - q^2\{q(\tfrac{21}{2}, \tfrac{17}{2})\}_3,$$

$$(\tfrac{7}{2}, \tfrac{1}{2}) = (\tfrac{21}{2}, \tfrac{1}{2})_3 - \{q(\tfrac{21}{2}, \tfrac{13}{2})\}_3 - q^4(\tfrac{7}{2}, \tfrac{5}{2})_{3^2}$$

exhibiting an amicable triad, for

$(\frac{7}{2}, \frac{5}{2})$ involves a definite *complete* part $-q\,(\frac{7}{2}, \frac{3}{2})_{3^2}$,

$(\frac{7}{2}, \frac{3}{2})$ " " " " $(\frac{7}{2}, \frac{1}{2})_{3^2}$,

$(\frac{7}{2}, \frac{1}{2})$ " " " " $-q^4(\frac{7}{2}, \frac{5}{2})_{3^2}$.

We have clearly before us a cyclic property.

So far we may say that it appears in the case of

$$(1, 0) \text{ as } (0),$$
$$[1, 0] \text{ as } (0),$$
$$[1, 1] \text{ as } (1),$$
$$\left.\begin{matrix}(\frac{5}{2}, \frac{3}{2})\\(\frac{5}{2}, \frac{1}{2})\end{matrix}\right\} \text{ as } (31),$$
$$\left.\begin{matrix}(\frac{7}{2}, \frac{5}{2})\\(\frac{7}{2}, \frac{3}{2})\\(\frac{7}{2}, \frac{1}{2})\end{matrix}\right\} \text{ as } (531),$$

the modulus being 3.

The modulus 5.

The standard partitions are

$$(a, b) = (5a, b)_5 - q^{a+b}(5a, 2a+b)_5 + q^{4a+2b}(5a, 4a+b)_5$$
$$- q^{a-b}(5a, 2a-b)_5 + q^{4a-2b}(5a, 4a-b)_5,$$
$$[a, b] = [5a, b]_5 + q^{a+b}[5a, 2a+b]_5 + q^{4a+2b}[5a, 4a+b]_5$$
$$+ q^{a-b}[5a, 2a-b]_5 + q^{4a-2b}[5a, 4a-b]_5.$$

We derive

$$(1, 0) = (1, 0)_{5^2} - 2q\,(5, 2)_5 + 2q^4\,(5, 4)_5 \text{ three parts,}$$
$$[1, 0] = [1, 0]_{5^2} + 2q\,[5, 2]_5 + 2q^4\,[5, 4]_5 \text{ three parts,}$$
$$[1, 1] = 2\,[5, 1]_5 + 2q^2\,[5, 3]_5 + q^6\,[1, 1]_{5^2},$$
$$(\tfrac{3}{2}, \tfrac{1}{2}) = (\tfrac{15}{2}, \tfrac{1}{2})_5 + \{q\,(\tfrac{15}{2}, \tfrac{11}{2})\}_5$$
$$-q\,(\tfrac{3}{2}, \tfrac{1}{2})_{5^2},$$
$$-q^2(\tfrac{15}{2}, \tfrac{7}{2})_5 + q^2\{q\,(\tfrac{15}{2}, \tfrac{13}{2})\}_5.$$

In the partition, last written, the parts are given in separate rows so as to bring out the better that the special term stands out. It will be noticed that 3, the uneven integer of $(\frac{3}{2}, \frac{1}{2})$, being prime to the modulus, the cyclic property presents itself.

$$(\tfrac{5}{2}, \tfrac{3}{2}) = (\tfrac{25}{2}, \tfrac{3}{2})_5 - q^2(\tfrac{25}{2}, \tfrac{7}{2})_5 - q^4(\tfrac{25}{2}, \tfrac{13}{2})_5 + q^7(\tfrac{25}{2}, \tfrac{17}{2})_5 + q^{13}(\tfrac{25}{2}, \tfrac{23}{2})_5,$$
$$(\tfrac{5}{2}, \tfrac{1}{2}) = (\tfrac{25}{2}, \tfrac{1}{2})_5 - q^2(\tfrac{25}{2}, \tfrac{9}{2})_5 - q^3(\tfrac{25}{2}, \tfrac{11}{2})_5 + q^9(\tfrac{25}{2}, \tfrac{19}{2})_5 + q^{11}(\tfrac{25}{2}, \tfrac{21}{2})_5.$$

There is no cyclic property in evidence, as we expected, because the leading element 5 is not prime to the modulus. Each partition exhibits the full number of parts, viz. 5.

Next we have the triad

$$\begin{aligned}
(\tfrac{7}{2}, \tfrac{5}{2}) = {} & (\tfrac{7}{2}, \tfrac{1}{2})_{5^2} \\
& - q\,(\tfrac{35}{2}, \tfrac{9}{2})_5 - q^6\,(\tfrac{35}{2}, \tfrac{19}{2})_5 \\
& + q^9\,(\tfrac{35}{2}, \tfrac{23}{2})_5 + q^{19}\,(\tfrac{35}{2}, \tfrac{33}{2})_5, \\
(\tfrac{7}{2}, \tfrac{3}{2}) = {} & (\tfrac{35}{2}, \tfrac{3}{2})_5 - q^5\,(\tfrac{35}{2}, \tfrac{17}{2})_5 \\
& + q^{11}\,(\tfrac{7}{2}, \tfrac{5}{2})_{5^2} \\
& - q^2\,(\tfrac{35}{2}, \tfrac{11}{2})_5 + q^{17}\,(\tfrac{35}{2}, \tfrac{31}{2})_5, \\
(\tfrac{7}{2}, \tfrac{1}{2}) = {} & (\tfrac{35}{2}, \tfrac{1}{2})_5 + q^{15}\,(\tfrac{35}{2}, \tfrac{29}{2})_5 \\
& - q^3\,(\tfrac{35}{2}, \tfrac{13}{2})_5 + q^{13}\,(\tfrac{35}{2}, \tfrac{27}{2})_5 \\
& - q^4\,(\tfrac{7}{2}, \tfrac{3}{2})_{5^2},
\end{aligned}$$

with the cyclic specification

$$(513) \qquad (\text{cf. mod} = 3).$$

The modulus 7.

The standard partitions are

$$\begin{aligned}
(a, b) = {} & (7a, b)_7 \\
& - q^{a+b}\,(7a, 2a+b)_7 + q^{4a+2b}\,(7a, 4a+b)_7 - q^{9a+3b}\,(7a, 6a+b)_7 \\
& - q^{a-b}\,(7a, 2a-b)_7 + q^{4a-2b}\,(7a, 4a-b)_7 - q^{9a-3b}\,(7a, 6a-b)_7, \\
[a, b] = {} & [7a, b]_7 \\
& + q^{a+b}\,[7a, 2a+b]_7 + q^{4a+2b}\,[7a, 4a+b]_7 + q^{9a+3b}\,[7a, 6a+b]_7 \\
& + q^{a-b}\,[7a, 2a-b]_7 + q^{4a-2b}\,[7a, 4a-b]_7 + q^{9a-3b}\,[7a, 6a-b]_7, \\
(1, 0) = {} & (1, 0)_{7^2} - 2q\,(7, 2)_7 + 2q^4\,(7, 4)_7 - 2q^6\,(7, 6)_7, \\
[1, 0] = {} & [1, 0]_{7^2} + 2q\,[7, 2]_7 + 2q^4\,[7, 4]_7 + 2q^6\,[7, 6]_7, \\
[1, 1] = {} & 2\,[7, 1]_7 + 2q^2\,[7, 3]_7 + 2q^6\,[7, 5]_7 + q^{12}\,[1, 1]_{7^2}, \\
(\tfrac{3}{2}, \tfrac{1}{2}) = {} & (\tfrac{21}{2}, \tfrac{1}{2})_7 + \{q\,(\tfrac{21}{2}, \tfrac{19}{2})\}_7 \\
& - q\,(\tfrac{21}{2}, \tfrac{5}{2})_7 - q\,\{q^2\,(\tfrac{21}{2}, \tfrac{19}{2})\}_7 \\
& - q^2\,(\tfrac{3}{2}, \tfrac{1}{2})_{7^2} \\
& + q^5\,(\tfrac{21}{2}, \tfrac{11}{2})_7 - q^5\,\{q\,(\tfrac{21}{2}, \tfrac{17}{2})\}_7,
\end{aligned}$$

having the cyclic specification (1).

$$
\begin{aligned}
(\tfrac{5}{2}, \tfrac{3}{2}) = \quad & (\tfrac{35}{2}, \tfrac{3}{2})_7 \quad + \{q\,(\tfrac{35}{2}, \tfrac{17}{2})\}_7 \\
& \qquad - q\,(\tfrac{5}{2}, \tfrac{1}{2})_{7^2} \\
& - q^4 (\tfrac{35}{2}, \tfrac{13}{2})_7 - q^4 \{q^2 (\tfrac{35}{2}, \tfrac{27}{2})\}_7 \\
& + q^6 (\tfrac{35}{2}, \tfrac{23}{3})_7 - q^6 \{q^3 (\tfrac{35}{2}, \tfrac{33}{2})\}_7, \\
(\tfrac{5}{2}, \tfrac{1}{2}) = \quad & (\tfrac{35}{2}, \tfrac{1}{2})_7 - \{q^3 (\tfrac{35}{2}, \tfrac{29}{2})\}_7, \\
& - q^2 (\tfrac{35}{2}, \tfrac{9}{2})_7 + q^2 \{q\,(\tfrac{35}{2}, \tfrac{19}{2})\}_7 \\
& - q^3 (\tfrac{35}{2}, \tfrac{11}{2})_7 - q^3 \{q^3 (\tfrac{35}{2}, \tfrac{31}{2})\}_7 \\
& + q^{11} (\tfrac{5}{2}, \tfrac{3}{2})_{7^2},
\end{aligned}
$$

a group of 2 having the cyclic specification (31).

The grouping of elliptic products is concerned with a product (a, b) or $[a, b]$ and a modulus μ such that μ and a are relatively prime and b one of the $\phi(a)$ totitives of a.

For the present I take μ to be a prime number so that the condition for a is that it is not a multiple of μ.

Before proceeding to its study as a branch of the Theory of Congruences in Higher Arithmetic it may be stated that the groups that will be considered or other groups have properties connected with the external powers of q which present themselves in the partitions.

The explanation of the Tables will be found in § 3.

The application to the partitions of elliptic products according to a given modulus may be instanced by reference to modulus 3 and leading element 14 of Table I.

The six totitives of the integer 14, viz. 13, 11, 9, 5, 3, 1, are given in two cycles of order three, viz.

$$(13, 5, 11) \quad (1, 9, 3),$$

this means that the elliptic product (14, 13), when partitioned according to mod 3, has a part to a factor *près* $(14, 5)_{5^2}$, that (14, 5) has a part $(14, 11)_{5^2}$, and that (14, 11) has a part $(14, 13)_{5^2}$. Also that (14, 1), (14, 9), (14, 3) have parts $(14, 9)_{5^2}$, $(14, 3)_{5^2}$, $(14, 1)_{5^2}$ respectively.

When the whole of the totitives present themselves in a single cycle, I term the leading element a an elliptic primitive of the modulus μ.

The Table II is derived from Table I, obviously, by restriction to uneven integers in both elements of the elliptic product. When the whole of the uneven totitives of the uneven leading element numerator a present themselves in a single cycle we may, provisionally, term the leading element an uneven primitive of the modulus.

TABLE I.

a, b integers; *b* a totitive of *a*.

Mod 3.

First element *a*	Cycles of $\phi(a)$ numbers
2	(1)
4	(3, 1)
5	(4, 2) (1, 3)
7	(6, 2, 4) (1, 5, 3)
8	(7, 3, 1, 5)
10	(9, 3, 1, 7)
11	(10, 4, 2, 6, 8) (1, 7, 9, 5, 3)
13	(11, 5, 7) (1, 9, 3) (12, 4, 10) (2, 8, 6)
14	(13, 5, 11) (1, 9, 3)
16	(15, 5, 9, 3, 1, 11, 7, 13)
17	(16, 6, 2, 12, 4, 10, 8, 14) (1, 11, 15, 5, 13, 7, 9, 3)
19	(18, 6, 2, 12, 4, 14, 8, 10, 16) (1, 13, 17, 7, 15, 5, 11, 9, 3)

Primitives to the modulus 3

2, 4, 8, 10, 16, ...

Mod 5.

First element *a*	Cycles of $\phi(a)$ numbers
2	(1)
3	(2) (1)
4	(3, 1)
6	(5, 1)
7	(6, 4, 2) (1, 3, 5)
8	(7, 5, 1, 3)
9	(8, 2, 4) (1, 7, 5)
11	(10, 2, 4, 8, 6) (1, 9, 7, 3, 5)
12	(11, 7) (1, 5)
13	(12, 8) (9, 7) (11, 3) (2, 10) (4, 6) (1, 5)
14	(13, 3, 5, 1, 11, 9)
16	(15, 3, 7, 5, 1, 13, 9, 11)
17	(16, 10, 2, 14, 4, 6, 8, 12) (1, 7, 15, 3, 13, 11, 9, 5)
18	(17, 11, 5, 1, 7, 13)
19	(18, 4, 16, 12, 10, 2, 8, 6, 14) (1, 15, 3, 7, 9, 17, 11, 13, 5)

Elliptic primitives to the modulus 5

2, 4, 6, 8, 14, 16, 18, ...

TABLE I (*cont.*)

Mod 7.

First element a	Cycles of $\phi(a)$ numbers
2	(1)
3	(2) (1)
4	(3) (1)
5	(4, 2) (1, 3)
6	(5, 1)
8	(7, 1) (5, 3)
9	(8, 4, 2) (1, 5, 7)
10	(9, 7, 1, 3)
11	(10, 8, 2, 6, 4) (1, 3, 9, 5, 7)
12	(11, 5) (1, 7)
13	(12, 2, 4, 8, 10, 6) (1, 11, 9, 5, 3, 7)
15	(14, 2, 4, 8) (1, 13, 11, 7)
16	(15, 7, 1, 9) (13, 11, 3, 5)
17	(16, 12, 8, 6, 4, 14, 2, 10) (1, 5, 9, 11, 13, 3, 15, 7)
18	(17, 13, 7, 1, 5, 11)
19	(18, 8, 12) (16, 14, 2) (10, 4, 6) (1, 11, 7) (3, 5, 17) (9, 15, 13)

Elliptic primitives to the modulus 7

2, 6, 10, 18, ...

Mod 11.

First element a	Cycles of $\phi(a)$ numbers
2	(1)
3	(2) (1)
4	(3, 1)
5	(4) (3) (2) (1)
6	(5) (1)
7	(6, 2, 4) (1, 5, 3)
8	(7, 5, 1, 3)
9	(8, 4, 2) (1, 5, 7)
10	(9, 1) (7, 3)
12	(11, 1) (7, 5)
13	(12, 6, 10, 8, 4, 2) (1, 7, 3, 5, 9, 11)
14	(13, 9, 11, 1, 5, 3)
15	(14, 4) (13, 7) (2, 8) (1, 11)
16	(15, 13, 7, 11, 1, 3, 9, 5)
17	(16, 14, 8, 10, 4, 12, 2, 6) (1, 3, 9, 7, 13, 5, 15, 11)
18	(17, 5, 7) (1, 13, 11)
19	(18, 12, 8) (16, 2, 14) (10, 6, 4) (1, 7, 11) (3, 17, 5) (9, 13, 15)

Elliptic primitives to the modulus 11

2, 4, 8, 14, 16, ...

TABLE I (*cont.*)

Mod 13.

First element a	Cycles of $\phi(a)$ numbers
2	(1)
3	(2) (1)
4	(3, 1)
5	(4, 2) (1, 3)
6	(5) (1)
7	(6) (5) (4) (3) (2) (1)
8	(7, 3, 1, 5)
9	(8, 2, 4) (1, 7, 5)
10	(9, 7, 1, 3)
11	(10, 6, 8, 4, 2) (1, 5, 3, 7, 9)
12	(11, 1) (7, 5)
14	(13, 1) (11, 3) (9, 5)
15	(14, 8, 4, 2) (1, 7, 11, 13)
16	(15, 11, 9, 13, 1, 5, 7, 3)
17	(16, 4) (14, 12) (10, 6) (8, 2) (9, 15) (7, 11) (3, 5) (1, 13)
18	(17, 7, 5) (1, 11, 13)
19	(18, 16, 10, 8, 14, 4, 12, 2, 6) (1, 3, 9, 11, 5, 15, 7, 17, 13)

Elliptic primitives to the modulus 13

2, 4, 8, 10, 16, ...

Mod 17.

First element a	Cycles of $\phi(a)$ numbers
2	(1)
3	(2) (1)
4	(3) (1)
5	(4, 2) (1, 3)
6	(5, 1)
7	(6, 2, 4) (1, 5, 3)
8	(7) (5) (3) (1)
9	(8) (7) (5) (4) (2) (1)
10	(9, 3, 1, 7)
11	(9, 7, 3, 5, 1) (2, 4, 8, 6, 10)
12	(11, 5) (7, 1)
13	(12, 10, 4) (8, 2, 6) (5, 11, 7) (1, 3, 9)
14	(13, 9, 11, 1, 5, 3)
15	(14, 8, 4, 2) (1, 7, 11, 13)
16	(15, 1) (13, 3) (11, 5) (9, 7)
18	(17, 1) (13, 5) (11, 7)
19	(18, 10, 14, 12, 6, 16, 8, 4, 2) (1, 9, 5, 7, 13, 3, 11, 15, 17)

Elliptic primitives to the modulus 17

2, 6, 10, 14, ...

TABLE I (*cont.*)

Mod 19.

First element a	Cycles of $\phi(a)$ numbers
2	(1)
3	(2) (1)
4	(3, 1)
5	(4) (3) (2) (1)
6	(5, 1)
7	(6, 4, 2) (1, 3, 5)
8	(7, 3, 1, 5)
9	(8) (7) (5) (4) (2) (1)
10	(9) (7) (3) (1)
11	(10, 4, 6, 2, 8) (1, 7, 5, 9, 3)
12	(11, 7) (1, 5)
13	(12, 2, 4, 8, 10, 6) (1, 11, 9, 5, 3, 7)
14	(13, 11, 5) (1, 3, 9)
15	(1, 11) (14, 4) (8, 2) (7, 13)
16	(15, 11, 9, 13, 1, 5, 7, 3)
17	(16, 8, 4, 2) (12, 6, 14, 10) (5, 11, 3, 7) (1, 9, 13, 15)
18	(17, 1) (13, 5) (11, 7)

Primitives to the modulus 19

2, 4, 6, 8, 16, ...

TABLE II.

a, *b* both uneven integers; *b* a totitive of *a*.

Mod 3.

First element $\frac{a}{2}$	Cycles of $\phi'(a)$ numbers
$\frac{5}{2}$	$(\frac{3}{2}, \frac{1}{2})$
$\frac{7}{2}$	$(\frac{5}{2}, \frac{3}{2}, \frac{1}{2})$
$\frac{11}{2}$	$(\frac{9}{2}, \frac{5}{2}, \frac{3}{2}, \frac{1}{2}, \frac{7}{2})$
$\frac{13}{2}$	$(\frac{11}{2}, \frac{5}{2}, \frac{7}{2})$ $(\frac{1}{2}, \frac{9}{2}, \frac{3}{2})$
$\frac{17}{2}$	$(\frac{1}{2}, \frac{11}{2}, \frac{15}{2}, \frac{5}{2}, \frac{13}{2}, \frac{7}{2}, \frac{9}{2}, \frac{3}{2})$
$\frac{19}{2}$	$(\frac{1}{2}, \frac{13}{2}, \frac{17}{2}, \frac{7}{2}, \frac{15}{2}, \frac{5}{2}, \frac{11}{2}, \frac{9}{2}, \frac{3}{2})$

Uneven elliptic primitives

5, 7, 11, 17, 19, ...

Mod 5.

First element $\frac{a}{2}$	Cycles of $\phi'(a)$ numbers
$\frac{3}{2}$	$(\frac{1}{2})$
$\frac{7}{2}$	$(\frac{1}{2}, \frac{3}{2}, \frac{5}{2})$
$\frac{9}{2}$	$(\frac{1}{2}, \frac{7}{2}, \frac{5}{2})$
$\frac{11}{2}$	$(\frac{1}{2}, \frac{9}{2}, \frac{7}{2}, \frac{3}{2}, \frac{5}{2})$
$\frac{13}{2}$	$(\frac{11}{2}, \frac{3}{2})$ $(\frac{9}{2}, \frac{7}{2})$ $(\frac{5}{2}, \frac{1}{2})$
$\frac{17}{2}$	$(\frac{15}{2}, \frac{3}{2}, \frac{13}{2}, \frac{11}{2}, \frac{9}{2}, \frac{5}{2}, \frac{1}{2}, \frac{7}{2})$
$\frac{19}{2}$	$(\frac{1}{2}, \frac{15}{2}, \frac{3}{2}, \frac{7}{2}, \frac{9}{2}, \frac{17}{2}, \frac{11}{2}, \frac{13}{2}, \frac{5}{2}$

Uneven elliptic primitives

3, 7, 9, 11, 17, 19, ...

TABLE II (*cont.*)

Mod 7.

First element $\frac{a}{2}$	Cycles of $\phi'(a)$ totitives
$\frac{3}{2}$	$(\frac{1}{2})$
$\frac{5}{2}$	$(\frac{3}{2}, \frac{1}{2})$
$\frac{9}{2}$	$(\frac{7}{2}, \frac{1}{2}, \frac{5}{2})$
$\frac{11}{2}$	$(\frac{9}{2}, \frac{5}{2}, \frac{7}{2}, \frac{1}{2}, \frac{3}{2})$
$\frac{13}{2}$	$(\frac{11}{2}, \frac{9}{2}, \frac{5}{2}, \frac{3}{2}, \frac{7}{2}, \frac{1}{2})$
$\frac{15}{2}$	$(\frac{13}{2}, \frac{11}{2}, \frac{7}{2}, \frac{1}{2})$
$\frac{17}{2}$	$(\frac{1}{2}, \frac{5}{2}, \frac{9}{2}, \frac{11}{2}, \frac{13}{2}, \frac{3}{2}, \frac{15}{2}, \frac{7}{2})$
$\frac{19}{2}$	$(\frac{1}{2}, \frac{11}{2}, \frac{7}{2})\ (\frac{3}{2}, \frac{5}{2}, \frac{17}{2})\ (\frac{9}{2}, \frac{15}{2}, \frac{13}{2})$

Uneven elliptic primitives

3, 5, 9, 11, 13, 15, 17, ...

Mod 11.

First element $\frac{a}{2}$	Cycles of $\phi'(a)$ totitives
$\frac{3}{2}$	$(\frac{1}{2})$
$\frac{5}{2}$	$(\frac{3}{2})\ (\frac{1}{2})$
$\frac{7}{2}$	$(\frac{5}{2}, \frac{3}{2}, \frac{1}{2})$
$\frac{9}{2}$	$(\frac{7}{2}, \frac{1}{2}, \frac{5}{2})$
$\frac{13}{2}$	$(\frac{11}{2}, \frac{1}{2}, \frac{7}{2}, \frac{3}{2}, \frac{5}{2}, \frac{9}{2})$
$\frac{15}{2}$	$(\frac{13}{2}, \frac{7}{2})\ (\frac{11}{2}, \frac{1}{2})$
$\frac{17}{2}$	$(\frac{1}{2}, \frac{3}{2}, \frac{9}{2}, \frac{7}{2}, \frac{13}{2}, \frac{5}{2}, \frac{15}{2}, \frac{11}{2})$
$\frac{19}{2}$	$(\frac{1}{2}, \frac{7}{2}, \frac{11}{2})\ (\frac{3}{2}, \frac{17}{2}, \frac{5}{2})\ (\frac{9}{2}, \frac{13}{2}, \frac{15}{2})$

Uneven elliptic primitives

3, 7, 9, 13, 17, ...

Mod 13.

First element $\frac{a}{2}$	Cycles of $\phi'(a)$ totitives
$\frac{3}{2}$	$(\frac{1}{2})$
$\frac{5}{2}$	$(\frac{3}{2}, \frac{1}{2})$
$\frac{7}{2}$	$(\frac{5}{2})\ (\frac{3}{2})\ (\frac{1}{2})$
$\frac{9}{2}$	$(\frac{7}{2}, \frac{5}{2}, \frac{1}{2})$
$\frac{11}{2}$	$(\frac{9}{2}, \frac{1}{2}, \frac{5}{2}, \frac{3}{2}, \frac{7}{2})$
$\frac{15}{2}$	$(\frac{13}{2}, \frac{1}{2}, \frac{7}{2}, \frac{11}{2})$
$\frac{17}{2}$	$(\frac{9}{2}, \frac{15}{2})\ (\frac{7}{2}, \frac{11}{2})\ (\frac{3}{2}, \frac{5}{2})\ (\frac{1}{2}, \frac{13}{2})$
$\frac{19}{2}$	$(\frac{1}{2}, \frac{3}{2}, \frac{9}{2}, \frac{11}{2}, \frac{5}{2}, \frac{15}{2}, \frac{7}{2}, \frac{17}{2}, \frac{13}{2})$

Uneven elliptic primitives

3, 5, 9, 15, 19, ...

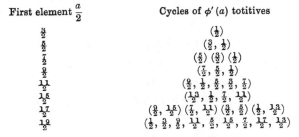

TABLE II (*cont.*)

Mod 17.

First element $\frac{a}{2}$	Cycles of $\phi'(a)$ totitives
$\frac{3}{2}$	$(\frac{1}{2})$
$\frac{5}{2}$	$(\frac{3}{2}, \frac{1}{2})$
$\frac{7}{2}$	$(\frac{5}{2}, \frac{3}{2}, \frac{1}{2})$
$\frac{9}{2}$	$(\frac{7}{2})\ (\frac{5}{2})$
$\frac{11}{2}$	$(\frac{9}{2}, \frac{7}{2}, \frac{3}{2}, \frac{5}{2}, \frac{1}{2})$
$\frac{13}{2}$	$(\frac{11}{2}, \frac{7}{2}, \frac{5}{2})\ (\frac{9}{2}, \frac{1}{2}, \frac{3}{2})$
$\frac{15}{2}$	$(\frac{13}{2}, \frac{11}{2})\ (\frac{7}{2}, \frac{1}{2})$
$\frac{19}{2}$	$(\frac{1}{2}, \frac{9}{2}, \frac{5}{2}, \frac{7}{2}, \frac{13}{2}, \frac{3}{2}, \frac{11}{2}, \frac{15}{2}, \frac{17}{2})$

Uneven elliptic primitives

3, 5, 7, 11, 19, ...

Mod 19.

First element $\frac{a}{2}$	Cycles of $\phi'(a)$ totitives
$\frac{3}{2}$	$(\frac{1}{2})$
$\frac{5}{2}$	$(\frac{3}{2})\ (\frac{1}{2})$
$\frac{7}{2}$	$(\frac{1}{2}, \frac{3}{2}, \frac{5}{2})$
$\frac{9}{2}$	$(\frac{7}{2})\ (\frac{5}{2})\ (\frac{1}{2})$
$\frac{11}{2}$	$(\frac{1}{2}, \frac{7}{2}, \frac{5}{2}, \frac{9}{2}, \frac{3}{2})$
$\frac{13}{2}$	$(\frac{1}{2}, \frac{11}{2}, \frac{9}{2}, \frac{5}{2}, \frac{3}{2}, \frac{7}{2})$
$\frac{15}{2}$	$(\frac{1}{2}, \frac{11}{2})\ (\frac{7}{2}, \frac{13}{2})$
$\frac{17}{2}$	$(\frac{5}{2}, \frac{11}{2}, \frac{3}{2}, \frac{7}{2})\ (\frac{1}{2}, \frac{9}{2}, \frac{13}{2}, \frac{15}{2})$

Uneven elliptic primitives

3, 7, 11, 13, ...

§ 3.

When we regard the annexed Tables certain properties appear to be in evidence which now come up for examination.

μ being an uneven prime, a any integer prime to μ and b any one of the $\phi(a)$ totitives of a, the integers b appear in the elliptic product discussion in two series each of which is in arithmetical progression, viz.

$$b,\ 2a+b,\ 4a+b,\ \ldots (\mu-1)a+b \quad \tfrac{1}{2}(\mu+1) \text{ terms} \qquad (A),$$

$$2a-b,\ 4a-b,\ \ldots (\mu-1)a-b \quad \tfrac{1}{2}(\mu-1) \text{ terms} \qquad (B).$$

Write
$$2ka+b \equiv \alpha_k \bmod \mu,$$
$$2ka-b \equiv \beta_k \bmod \mu,$$

so as to obtain the two series

$$\alpha_0,\ \alpha_1,\ \alpha_2,\ \ldots\ \alpha_{\frac{1}{2}(\mu-1)} \qquad (A'),$$
$$\beta_1,\ \beta_2,\ \ldots\ \beta_{\frac{1}{2}(\mu-1)} \qquad (B').$$

Since the common difference in each of the series (A), (B) is $2a$ and $2a$ is prime to μ, we know that the residues which appear in (A') are all different—so also the residues which appear in (B'). If $2a \equiv \delta \bmod \mu$, the residues in (A') and in (B') may be taken to have the common difference δ.

The series are thus

$$\alpha_0,\ \alpha_0+\delta,\ \alpha_0+2\delta,\ \ldots\ \alpha_0+\tfrac{1}{2}(\mu-1)\,\delta,$$
$$\beta_1,\quad \beta_1+\delta,\ \ldots\ \beta_1+\tfrac{1}{2}(\mu-3)\,\delta,$$

when reduced by the relations

$$\alpha_0+k\delta \equiv \alpha_k \bmod \mu,$$
$$\beta_1+(k-1)\,\delta \equiv \beta_k \bmod \mu.$$

If the series do not terminate with the terms $(\mu-1)\,a \pm b$ but are continued indefinitely, we write

$$b,\ 2a+b,\ 4a+b,\ \ldots\ (A_\infty),$$
$$2a-b,\ 4a-b,\ \ldots\ (B_\infty);$$
$$\alpha_0,\ \alpha_1,\ \alpha_2,\ \ldots\qquad (A'_\infty),$$
$$\beta_1,\ \beta_2,\ \ldots\qquad (B'_\infty).$$

By a known theorem the first μ terms in each of the series (A'_∞), (B'_∞) are the μ integers $0, 1, 2, \ldots \mu-1$ in some order. Thus for some values of i and j

$$\alpha_i = \beta_j.$$

This being so it is clear that the two series

$$\alpha_i,\ \alpha_{i+1},\ \ldots\ \alpha_{i+\mu-1},$$
$$\beta_j,\ \beta_{j+1},\ \ldots\ \beta_{j+\mu-1}$$

are identical.

Theorem I. **Either** the series (A') involves the residue zero and (B') does not **or** (B') involves zero and (A') does not.

For suppose k and l to be integers such that

$$2ka+b \equiv 0 \bmod \mu,\quad 2la-b \equiv 0 \bmod \mu,$$

giving $$2\,(k+l)\,a \equiv 0 \bmod \mu$$

and $$k+l \equiv 0 \bmod \mu.$$

If $k+l=\mu$, we can only have $k=\tfrac{1}{2}(\mu+1)$, $l=\tfrac{1}{2}(\mu-1)$, an impossibility because we cannot have simultaneously

$$(\mu-1)\,a+b \equiv 0 \bmod \mu,$$
$$(\mu-1)\,a-b \equiv 0 \bmod \mu.$$

Hence zeros cannot occur in each of the series (A'), (B'). But the first μ terms in each of the series (A'_∞), (B'_∞) involves a zero, so that we may have $k+l<2\mu$. Hence for some values of k, l, $k+l=\mu$. Hence one and one only of the series (A'), (B') involves zero.

In a similar manner it may be shewn that if the residue $+\epsilon$ occurs in the series (A') the residue $-\epsilon$ cannot occur in (B').

It follows that if b_1 be any one of the $\phi(a)$ totitives of a, we can uniquely select a term from the combined series (A), (B) which is congruent to zero for the modulus μ.

Dividing this term by μ we reach the number b_2 which is also a totitive of a; thence from b_2 we reach a totitive b_3 and so on until finally we reach a totitive b_{s+1} which is equal to b_1, and observe the cycle of totitives

$$(b_1, b_2, \ldots b_s) \text{ of order } s.$$

It thus appears that the whole of the totitives of a can be arranged in a set of cycles of various orders as exhibited in the table for given simple values of μ and a.

Theorem II. If in any cycle of totitives b' exceeds b, then in the same or some other cycle the totitive $a-b'$ succeeds the totitive $a-b$.

Let θ be a symbol such that $\theta^2=1$; θ may be therefore either $+1$ or -1. Since

$$2ia+\theta b=\mu b'$$

for some value of i, assume

$$2ja+\theta(a-b)=\mu(a-b')$$

for some value of j; by addition

$$2(i+j)a+\theta a=\mu a$$

or
$$2j=\mu-\theta-2i$$

and
$$j=\tfrac{1}{2}(\mu-\theta-2i)$$

uniquely. Hence the succession $a-b$, $a-b'$ follows from the succession b, b' and the theorem is proved.

In particular the succession b, b' implies the succession b', b, and there is a cycle of order 2, when

$$a=b+b',$$

$$2ia+\theta b=\mu(a-b),$$

$$(\mu-2i)a=(\mu+\theta)b,$$

giving
$$a=\mu+\theta,\ b=\mu-2i,$$

so that b is uneven and $a=\mu\pm1$.

Theorem III. A cycle of order 2

$$(b,\ a-b)$$

occurs when $a=\mu\pm 1$, b an uneven number.

Let $\theta_t^2=1$ so that we may put $\theta_t=+1$ or -1 at pleasure.

In the case of a cycle of order s we have s equations

$$\begin{aligned}
2i_1 a + \quad \theta_1 b_1 \quad &= \mu b_2,\\
2i_2 a + \quad \theta_2 b_2 \quad &= \mu b_3,\\
&\ldots\ldots\\
2i_{s-1} a + \theta_{s-1} b_{s-1} &= \mu b_s,\\
2i_s a + \quad \theta_s b_s \quad &= \mu b_1,
\end{aligned}$$

and we find:

Theorem IV. The integers in the cycle

$$(b_1,\ b_2,\ \ldots\ b_s)$$

are all of the same parity. As a result of this we have:

Theorem V. If the order of the cycle be the maximum $\phi(a)$, a must be an even number.

Theorem VI. An elliptic primitive to the uneven prime modulus μ must be an even number.

Theorem VII. The cycles in which the totitives of a appear for a given modulus are all of the same order.

From the s equations above written we can eliminate the $s-1$ quantities $b_2,\ b_3,\ \ldots\ b_s$ and obtain the relation

$$2\left(i_s\mu^{s-1}+\theta_s i_{s-1}\mu^{s-2}+\theta_s\theta_{s-1}i_{s-2}\mu^{s-3}+\ldots+\theta_s\theta_{s-1}\ldots\theta_2 i_1\right)a$$
$$=\left(\mu^s-\theta_1\theta_2\ldots\theta_s\right)b_1,$$

and by circular procession we can write down $s-1$ other relations of which the next is

$$2\left(i_1\mu^{s-1}+\theta_1 i_s\mu^{s-2}+\theta_1\theta_s i_{s-1}\mu^{s-3}+\ldots+\theta_1\theta_s\ldots\theta_3 i_2\right)a$$
$$=\left(\mu^s-\theta_1\theta_2\ldots\theta_s\right)b_2,$$

and the ultimate one is

$$2\left(i_{s-1}\mu^{s-1}+\theta_{s-1}i_{s-2}\mu^{s-2}+\theta_{s-1}\theta_{s-2}i_{s-3}\mu^{s-3}+\ldots+\theta_1\theta_2\ldots\theta_{s-1}i_s\right)a$$
$$=\left(\mu^s-\theta_1\theta_2\ldots\theta_s\right)b_s.$$

Any integer i_t may assume values

$$0,\ 1,\ 2,\ \ldots\ \tfrac{1}{2}(\mu-1),$$

and θ_t may be $+1$ or -1, but we may not put $i_t=0$, $\theta_t=-1$ simultaneously.

Adding the s equations

$$2\,(\mu^{s-1}\Sigma i_s + \mu^{s-2}\Sigma\theta_s i_{s-1} + \ldots + \Sigma\theta_s\theta_{s-1}\ldots\theta_2 i_1)\,a$$
$$= (\mu^s - \theta_1\theta_2\ldots\theta_s)\,\Sigma b.$$

The product $\theta_1\theta_2\ldots\theta_s = \pm 1$, and the bracketed factors on each side of the tth of the s equations must possess a common factor so as to reduce their ratio to $\frac{2a}{b_t}$. Moreover, in the last equation written, the ratio of the factors must reduce to $\frac{2a}{\Sigma b}$.

I hope to resume consideration of this subject on a future occasion.

Chapter 10
Partition Analysis

10.1 Introduction and Commentary

R. P. Stanley has totally dominated modern work in extending this area of MacMahon's research. In three long and important papers, Stanley (1972, 1973, 1974a) has provided deep insights into several combinatorial problems that MacMahon initiated in [54] and [67].

In the first paper, Stanley (1972) considers the enumeration of order reversing maps of finite ordered sets P into chains; this is modern terminology for MacMahon's basic idea in partition analysis. As Stanley points out, such a study includes in its purview such diverse combinatorial topics as (1) ordinary partitions, (2) compositions, (3) plane and multi-dimensional partitions with applications to Young tableaux, (4) Eulerian numbers, (5) tangent and secant numbers, (6) indices of permutations, (7) trees, and (8) stacks. The main tool used by Stanley is the generating function, and sufficient discoveries are made in this first memoir to suggest substantial developments (Stanley, 1973, 1974).

In the second paper of this series, Stanley (1973) proves the following two major conjectures on magic squares by greatly extending MacMahon's magic square analysis [67] and applying Hilbert's Syzygy Theorem at a crucial stage.

Anand-Dumir-Gupta Conjecture. The number $H_n(r)$ of $n \times n$ matrices of nonnegative integers, with each row and column having sum r, is a polynomial of degree $(n-1)^2$ in r such that $H_n(-1) = \ldots = H_n(-n+1) = 0$ and $H_n(r) = (-1)^{n-1} H_n(-n-r)$.

In *Combinatory Analysis*, volume 2, page 161, MacMahon actually proved that $H_3(r) = 3\binom{r+3}{4} + \binom{r+2}{2}$, which confirms the Anand-Dumir-Gupta Conjecture for $n = 3$.

Carlitz Conjecture (Carlitz, 1966). Let $S_n(r)$ denote the number of symmetric $n \times n$ matrices of nonnegative integers with each row and column having sum r. The $S_n(2r)$ and $S_n(2r+1)$ are polynomials in r.

The third paper (Stanley, 1974a) concludes with Stanley's "Monster Reciprocity Theorem" which provides general criteria for a certain duality between related enumeration problems. Actual application of these duality

---

The material in this chapter corresponds to section VIII of *Combinatory Analysis*.

results explains (and establishes) the second part of the Anand-Dumir-Gupta Conjecture, that $H_n(-1) = \ldots = H_n(-n+1) = 0$ and $H_n(r) = (-1)^{n-1} H_n(-n-r)$. This turns out to be an example of what Stanley (1974a) calls a combinatorial reciprocity theorem; an excellent introduction to such theorems is given in Stanley (1974b).

10.2 An Introduction to Stanley's Partition Analysis

We shall closely follow Stanley (1970), which will provide an introduction to some of his methods and discoveries, and shall occasionally draw attention to the related work in his major papers. We begin with some of the main ideas and theorems from Stanley (1972).

Let P denote a finite partially ordered set of cardinality p with largest totally ordered subset (i.e., chain) of cardinality $l+1$ (equivalently, of length l). Let $\mathbf{m} = \{1, 2, \ldots, m\}$.

Definition 1. A map $\sigma: P \to n$ is said to be *order-preserving* (resp. *strictly order preserving*) if $\sigma(X) \leqq \sigma(Y)$ whenever $X \leqq Y$ (resp. $\sigma(X) < \sigma(Y)$ whenever $X < Y$). We let $\Omega(n)$ (resp. $\overline{\Omega}(n)$) denote the number of order-preserving (resp. strict order perserving) maps $\sigma: P \to \mathbf{n}$.

Definition 2. We let e_n (resp. $\bar{e}_n$) denote the number of surjective order-preserving (resp. surjective strict order-preserving) maps $\sigma: P \to \mathbf{n}$.

For example, if $P = \mathbf{p}$, then $\Omega(n)$ is just the number of possible selections, with replacement, of p elements from a set of n elements; hence $\Omega(n) = \binom{n+p-1}{p}$ (Riordan, 1958, page 7). In this instance, $\overline{\Omega}(n)$ is the number of possible selections, without replacement, of p elements from a set of n elements; hence $\overline{\Omega}(n) = \binom{n}{p}$.

For any P a moment's reflection reveals that $e_p = \bar{e}_p$, and this number is precisely the number of ways of extending the partial order on P to a total (or linear) order.

Theorem 1. $\Omega(n)$ and $\overline{\Omega}(n)$ are polynomials in n of degree p and leading coefficient $e_p/p!$, namely,

$$\Omega(n) = \sum_{s=1}^{p} e_s \binom{n}{s}, \qquad \overline{\Omega}(n) = \sum_{s=1}^{p} \bar{e}_s \binom{n}{s}.$$

Proof. For each of the $\binom{n}{s}$ subsets S of n of cardinality s, there are exactly e_s (resp. $\bar{e}_s$) order-preserving (resp. strict order-preserving) maps of P onto

S; this establishes the formulas for $\Omega(n)$ and $\overline{\Omega}(n)$. The assertion about leading coefficients follows once we recall that $e_p = \bar{e}_p$.

The polynomial $\overline{\Omega}(n)$ is an ordered set analog of the chromatic polynomial of a graph; it counts the number of "colorings" of P with colors numbered $1, 2, \ldots, n$ such that no two comparable elements of P have the same color, and such that this coloring is compatible with the ordering of P. This is not an absolutely perfect analogy in that the chromatic polynomials have alternating coefficients; the smallest P for which this phenomenon does not occur in $\overline{\Omega}(n)$ has 5 elements.

Next we state Stanley's main lemma whose proof (omitted here) involves Stanley's theory of ω-separators. Let ω denote any surjective order-preserving map $P \to \mathbf{p}$; thus ω is an extension of $P = \{X_1, X_2, \ldots, X_p\}$ to a total order, where $\omega(X_i) = i$. List all permutations $i_1, i_2, \ldots, i_p$ of $1, 2, \ldots, p$ with the property that if $X < Y$ in P, then $\omega(X)$ appears before $\omega(Y)$ in $i_1, i_2, \ldots, i_p$. Place a "$\leqq$" sign between two consecutive terms i_j and i_{j+1} if $i_j \leqq i_{j+1}$; otherwise put in a "$<$" sign. Denote the array thus produced by $\mathscr{L}$. Denote by $\overline{\mathscr{L}}$ the array obtained from $\mathscr{L}$ by replacing all "$<$" by "$\leqq$" and vice versa. We shall say that a map $\sigma: P \to \mathbf{n}$ is *compatible* with a permutation $i_1, i_2, \ldots, i_p$ appearing in $\mathscr{L}$ (resp. $\overline{\mathscr{L}}$) if $\sigma(X_{i_1}) \leqq \sigma(X_{i_2}) \leqq \ldots \leqq \sigma(X_{i_p})$, and $\sigma(X_{i_j}) < \sigma(X_{i_{j+1}})$ whenever a "$<$" appears in $\mathscr{L}$ (resp. $\overline{\mathscr{L}}$) between i_j and i_{j+1}.

Example. Let P and ω be given diagrammatically by

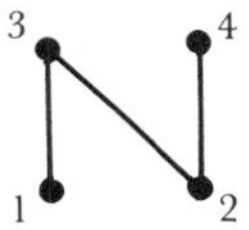

Then $\mathscr{L}$ and $\overline{\mathscr{L}}$ are given as follows.

$\mathscr{L}$	$\overline{\mathscr{L}}$
$1 \leqq 2 \leqq 3 \leqq 4$	$1 < 2 < 3 < 4$
$2 < 1 \leqq 3 \leqq 4$	$2 \leqq 1 < 3 < 4$
$1 \leqq 2 \leqq 4 < 3$	$1 < 2 < 4 \leqq 3$
$2 < 1 \leqq 4 < 3$	$2 \leqq 1 < 4 \leqq 3$
$2 \leqq 4 < 1 \leqq 3$	$2 < 4 \leqq 1 < 3$

Lemma (Stanley, 1972, p. 18). Every order-preserving (resp. strict order-preserving) map $\sigma: P \to \mathbf{n}$ is compatible with exactly one permutation in $\mathscr{L}$ (resp. $\overline{\mathscr{L}}$).

We may obtain from the Lemma new expressions for $\Omega(n)$ and $\overline{\Omega}(n)$ by summing the contributions from each permutation in $\mathscr{L}$ and $\overline{\mathscr{L}}$. If precisely s "<" signs occur in a given permutation, then this term contributes $\binom{n+p-1-s}{p}$ to $\Omega(n)$ (resp. $\overline{\Omega}(n)$). Hence we have the following:

Theorem 2. Let w_s (resp. $\overline{w}_s$) denote the number of permutations in $\mathscr{L}$ (resp. $\overline{\mathscr{L}}$) with exactly s "<" signs. Then

$$\Omega(n) = \sum_{s=0}^{p-1} w_s \binom{p+n-1-s}{p}, \qquad \overline{\Omega}(n) = \sum_{s=0}^{p-1} \overline{w}_s \binom{p+n-1-s}{p}.$$

The next three theorems are all of the type referred to by Stanley (1974a, 1974b) as combinatorial reciprocity theorems.

Theorem 3. $\overline{\Omega}(n) = (-1)^p \Omega(-n)$.

Proof. By the construction of $\mathscr{L}$ and $\overline{\mathscr{L}}$, we see that $\overline{w}_s = w_{p-1-s}$; substituting for $\overline{w}_s$ in Theorem 2 and comparing the resulting expression for $\overline{\Omega}(n)$ with that for $\Omega(n)$, we obtain the desired result.

The numbers w_s are natural generalizations of the Eulerian numbers (Riordan, 1958, pages 214–215). When P is a disjoint union of p points (i.e., no two distinct elements are comparable), then w_s is just the number of permutations of $1, 2, \ldots, p$ with exactly s decreases between consecutive terms. This is the combinatorial definition of the Eulerian numbers $A_{p,s+1}$. From Theorem 2 we may also obtain the following generating functions using the binomial series:

$$\sum_{n=0}^{\infty} \Omega(n) x^n = (1-x)^{-p-1} \sum_{s=0}^{p-1} w_s x^{s+1},$$

$$\sum_{n=0}^{\infty} \overline{\Omega}(n) x^n = (1-x)^{-p-1} \sum_{s=0}^{p-1} \overline{w}_s x^{s+1}.$$

From Theorem 3 we may deduce the nature of the integral zeros of $\Omega(n)$:

Corollary 1.

$$\Omega(-j) = 0, \qquad 0 \leqq j \leqq l,$$

$$(-1)^p \Omega(-l-n) \geqq \Omega(n) < 0, \qquad n < 0.$$

The circumstances in which equality holds in the final assertion of Corollary 1 are described in the following two theorems. The proofs (omitted here) rely on the construction of a strict order-preserving map $\tau : P \to n + l$ corresponding to a given order-preserving map $\sigma : P \to n$, and an analysis of when this correspondence is one-to-one (see Stanley, 1972, pages 73–76).

Theorem 4. $\Omega(-l - 1) = (-1)^p$ if and only if every element of P is contained in a chain of length l.

Theorem 5. The following three conditions are equivalent:

(i) $\Omega(-l - n) = (-1)^p \Omega(n)$ for some integer $n > 1$;
(ii) $\Omega(-l - n) = (-1)^p \Omega(n)$ for all n;
(iii) Every maximal chain of P has length l.

Corollary 2. If every maximal chain of P has length l, then

(i) $2e_{p-1} = (p + l - 1) e_p$;
(ii) $2\bar{e}_{p-1} = (p - l - 1) e_p$;
(iii) The coefficient of n^{p-1} in $\Omega(n)$ is $e_p/2(p - 1)!$.

Proof. From Theorem 5 (ii), we see that

$$\Omega(n) = \sum_{s=1}^{p} e_s \binom{n}{s} = (-1)^p \sum_{s=1}^{p} e_s \binom{-l - n}{s}.$$

Comparing coefficients of n^{p-1}, we obtain (i); (ii) is obtained in a similar manner from $\overline{\Omega}(n)$. Assertion (iii) immediately follows from (i).

From (i) and (ii) above, we see that if every maximal chain of P has length l, then either $p - l$ is odd or e_p is even; a direct combinatorial proof of this fact would be of interest. The previous observation motivates the following:

Conjecture. Let P be any finite partially ordered set. If the length of every maximal chain of P has the same parity as $|P| = p$, then e_p is even.

We conclude by mentioning an interesting result closely related to MacMahon's solution of the "generalized ballot problem" (see [73]). Suppose that P is the direct product of two chains $P = r \times s$; then it turns out that

$$\Omega(n) = \frac{\binom{r + n - 1}{r}\binom{r + n}{r}\cdots\binom{r + s + n - 2}{r}}{\binom{r}{r}\binom{r + 1}{r}\cdots\binom{r + s - 1}{r}}$$

(see Stanley, 1972, page 83 for details).

10.3 References

M. Abramson and W. O. J. Moser (1973) Arrays with fixed row and column sum, *Discr. Math.*, *6*, 1–14.

H. Anand, V. C. Dumir, and H. Gupta (1966) A combinatorial distribution problem, *Duke Math. J.*, *33*, 757–769.

L. Carlitz (1966) Enumeration of symmetric arrays, *Duke Math. J.*, *33*, 771–782.

L. Carlitz (1971) Enumeration of symmetric arrays II, *Duke Math. J.*, *38*, 717–731.

L. Carlitz (1972) Enumeration of 3 × 3 arrays, *Fibonacci Quart.*, *10*, 489–498.

C. J. Everett and P. R. Stein (1971) The asymptotic number of integer stochastic matrices, *Discr. Math.*, *1*, 55–72.

R. C. Grimson (1969) Enumeration of rectangular arrays, Ph.D. thesis, Duke University.

R. C. Grimson (1971) Some results on the enumeration of symmetric arrays, *Duke Math. J.*, *38*, 711–715.

R. C. Grimson (1974) The evaluation of certain arithmetic sums, *Fibonacci Quart.*, *12*, 373–380.

H. Gupta (1968) Enumeration of symmetric matrices, *Duke Math. J.*, *35*, 653–660.

H. Gupta and G. Baiknuth Nath (1973) Enumeration of stochastic cubes, *Indian J. Pure and Appl. Math.*, *4*, 545–567.

H. Gupta and S. Srinivasan (1971) The number of 3 by 3 magic matrices, *Res. Bull.(N.S.) Panjab Univ.*, *22*, 525–526.

M. Hall (1958) Survey of combinatorial analysis, in Some Aspects of Analysis and Probability IV, J. Wiley, New York.

D. Jackson and R. Entringer (1975/76) A preamble to an enumeration of certain generalized arrays associated with ballot problems, *Matrix Tensor Quart.*, *26*, 146–148.

J. Riordan (1958) An Introduction to Combinatorial Analysis, J. Wiley, New York.

J. Riordan and P. R. Stein (1972) Arrangements on chessboards, *J. Combinatorial Th.*, *12*, 72–80.

M. L. Stein and P. R. Stein (1970) Enumeration of stochastic matrices with integer elements, Los Alamos Scientific Lab. Publ. No. LA–4434.

R. P. Stanley (1970) A chromatic-like polynomial for ordered sets, from Combinatorial Mathematics and Its Applications, University of North Carolina at Chapel Hill, pp. 421–427.

R. P. Stanley (1972) Ordered structures and partitions, *Memoirs of the A.M.S.*, No. 119.

R. P. Stanley (1973) Linear homogeneous Diophantine equations and magic labelings of graphs, *Duke Math. J.*, *40*, 607–632.

R. P. Stanley (1974a) Combinatorial reciprocity theorems, *Advances in Math.*, *14*, 194–253.

R. P. Stanley (1974b) Combinatorial reciprocity theorems, from Combinatorics, Part 2: Graph Theory, Foundations, Partitions and Combinatorial Geometry, Math. Centre Tracts 56, Amsterdam, 107–118.

R. P. Stanley (1976) Some combinatorial aspects of the Schubert calculus, from Combinatoire et Representation du Groupe Symetrique, Strasbourg 1976, Lecture Notes in Math. No. 579, Springer-Verlag, Berlin-Heidelberg-New York.

[33] On play "à outrance," *Proc. London Math. Soc.*, *20* (1889), 195–198.

Suppose two players A and B have x and $n - x$ counters, respectively. They are to play a series of games in which the loser gives the winner a counter at the end of each game. The player who eventually obtains all the counters is declared the winner of the series.

MacMahon considers the probability of A winning the series, assuming that the chance of A winning a game when he possesses x counters is $\phi(x)/(\phi(x) + \phi(n - x))$. The probability that A wins the series is

$$\frac{1 + \sum_{j=1}^{x-1} \frac{\phi(n-1)\phi(n-2)\ldots\phi(n-j)}{\phi(1)\phi(2)\ldots\phi(j)}}{1 + \sum_{j=1}^{n-1} \frac{\phi(n-1)\phi(n-2)\ldots\phi(n-j)}{\phi(1)\phi(2)\ldots\phi(j)}}.$$

Certain special cases of this solution are investigated, and MacMahon concludes by pointing out that the problem, as considered by him, may be more applicable to practical situations than the classical case, when the probability of A winning a game is given as a constant p.

[54] Memoir on the theory of partitions of numbers—Part II, *Phil. Trans.* *192* (1899), 351–402.

This paper presents the fundamentals of what MacMahon calls "partition analysis," the subject comprising section VIII of his two volume work *Combinatory Analysis*, and contains sections 5–8 in the first three memoirs of this series.

He begins by observing that partitions and compositions may be viewed as two aspects of a general problem; for a partition of n is a representation

$$n = \alpha_1 + \alpha_2 + \ldots + \alpha_s, \text{ where } \alpha_1 \geqq \alpha_2 \geqq \ldots \geqq \alpha_s;$$

while a composition of n is a representation

$$n = \alpha_1 + \alpha_2 + \ldots + \alpha_s, \text{ where } \alpha_1 \lesseqqgtr \alpha_2 \lesseqqgtr \ldots \lesseqqgtr \alpha_s.$$

He proposes to analyze partitions and obtain generating functions, where the symbol between α_i and α_{i+1} is taken from among $\{>, =, <, \geqq, \leqq, \gtrless, \gtreqqless\}$.

More generally, he considers systems of linear inequalities, for the αs, of the form

$$\sum_{j=1}^{s} A_j^{(i)} \alpha_i \geqq 0, \qquad 1 \leqq i \leqq r.$$

He remarks that in every case the general solutions $(\alpha_1, \ldots, \alpha_n)$ are linear combinations of the fundamental solutions. The generating functions thus obtained have an elegant form that depends on the fundamental solutions and various syzygies (in analogy with the syzygies of invariant theory) that unite the fundamental solutions.

To make this theory amenable to computation, MacMahon introduces certain operators. For example, suppose that $F(X_1, X_2, \ldots, X_s, \alpha_1, \alpha_2, \ldots, \alpha_t)$ (a function of $s + t$ variables) is expressible in an $(s + t)$-fold Laurent series. Then

$$\underset{\geqq}{\Omega} F(X_1, X_2, \ldots, X_s, \alpha_1, \alpha_2, \ldots, \alpha_t)$$

is that portion of the series expansion of F in which each power of the αs is nonnegative, evaluated at $\alpha_1 = \alpha_2 = \ldots = \alpha_t = 1$. For example,

$$\begin{aligned} \underset{\geqq}{\Omega}(1 - ax)^{-1}(1 - y/a)^{-1} &= \underset{\geqq}{\Omega} \sum_{n=0}^{\infty} \sum_{m=0}^{\infty} a^{n-m} x^n y^m \\ &= \sum_{m=0}^{\infty} \sum_{n=m}^{\infty} x^n y^m \\ &= (1 - x)^{-1}(1 - xy)^{-1}. \end{aligned}$$

MacMahon spends the remainder of section 5 showing that the general types of partitions he is considering have generating functions easily expressible in terms of the Ω operators. He also develops a calculus of these operators in order to obtain explicit closed forms for many of the corresponding generating functions.

In section 6 he moves to two-dimensional partitions for which his new method is successful in many cases. Unfortunately it is inadequate, in its present form, for a proof that $\Pi_{m \geqq 1}(1 - x^m)^{-m}$ is the plane partition generating function. However, all the results he obtains suggest the truth of this conjecture (for which he sketches the proof in [78]).

In section 7 MacMahon considers three-dimensional partition problems which prove harder to handle. He determines, for example, the generating function for solid partitions in which the parts lie on the vertices of a cube, and the result is not nearly as simple as the two-dimensional result related to parts on the vertices of a square.

MacMahon concludes in section 8 by asking whether the conjectured generating functions for various restricted two-dimensional partitions are polynomials (such must be the case or his conjectures will be trivially false). He considers a number of special cases and asserts that he can prove these functions are always polynomials.

[58] Application of the partition analysis to the study of the properties of any system of consecutive integers, *Trans. Cambridge Phil Soc.*, *18* (1900), 12–34.

As is well-known, the product of any m consecutive integers is divisible by $m!$, i.e.,

$$\binom{n+m}{m} = \frac{n+1}{1} \cdot \frac{n+2}{2} \cdot \ldots \cdot \frac{n+m}{m}$$

is an integer. In this paper, MacMahon asks what conditions are necessary on $\alpha_1, \alpha_2, \ldots, \alpha_m$ in order that

$$\left(\frac{n+1}{1}\right)^{\alpha_1} \left(\frac{n+2}{2}\right)^{\alpha_2} \cdots \left(\frac{n+m}{m}\right)^{\alpha_m}$$

be an integer.

He points out that the problem arose initially in [54] where he treated $(1 - x^{n+s})/(1 - x^s)$ rather than $(n + s)/s$, and proposes to utilize the partition analysis developed there in order to treat this problem. In fact, only the cases $m = 2$, 3, 4 and 5 are treated, and he begins each one by presenting the various sets of inequalities that are necessary on the α_i. He then uses partition analysis on related crude forms of the generating function in order to obtain fundamental solutions and the syzygies among the solutions.

[62] The Diophantine inequality $\lambda x \geqq \mu y$, *Trans. Cambridge Phil. Soc.*, *19* (1904), 111–131.

As a result of his work in [54], MacMahon is motivated to study systems of Diophantine inequalities of the form

$$\lambda_1\alpha_1 + \lambda_2\alpha_2 + \ldots + \lambda_s\alpha_s \geqq \mu_1\beta_1 + \mu_2\beta_2 + \ldots + \mu_t\beta_t.$$

In this paper he treats the cases $s = 1$ (section 1) and $s = 2$ (section 2) and uses the partition analysis of [54] to show that in each case the solution set is a finitely generated module over the integers.

When $s = 1$ the principal result is the following: Every solution (α, β) of the Diophantine inequality $\lambda\alpha \geqq \mu\beta$ is a linear combination (over the integers) of (r_i, s_i), where s_i/r_i is any member of the ascending intermediate series of convergents to the continued fraction expansion of λ/μ. When $s = 2$ the result is more complex, but it does involve the continued fraction expansions of μ_1/λ_1 and μ_2/λ_2.

[65] Note on the Diophantine inequality $\lambda x \geqq \mu y$, *Quart. J. Math.*, *36* (1905), 80–93.

In paper [62] MacMahon proved that the positive integral solutions (x, y) of $\lambda x \geqq \mu y$ are all linear combinations of a finite set of fundamental solutions. Indeed, if the ascending intermediate series of convergents to the fraction λ/μ is $y_1/x_1, y_2/x_2, \ldots, y_\sigma/x_\sigma$, and $\lambda x_\alpha - \mu y_\alpha = z_\alpha$, then the general positive integral solution of $\lambda x = \mu y + z$ is

$$x = \sum_{i=1}^{\sigma} A_i x_i, \qquad y = \sum_{i=1}^{\sigma} A_i y_i, \qquad z = \sum_{i=1}^{\sigma} A_i z_i,$$

where $A_i \geqq 0$, A_i is integral.

In this note, MacMahon considers those cases in which the continued fraction for λ/μ possesses a reciprocal series of partial quotients, e.g.,

$$\frac{104}{77} = 1 + \frac{1}{2+}\,\frac{1}{1+}\,\frac{1}{5+}\,\frac{1}{1+}\,\frac{1}{2+}\,\frac{1}{1}.$$

In all such instances, he is able to show that either the x_i or y_i is connected with the z_i in a reciprocal or quasi-reciprocal manner. For example, when $\lambda = 104$, $\mu = 77$, x_i, y_i, and z_i are given by Table 10.1.

Table 10.1

i	x_i	y_i	z_i
1	1	0	104
2	1	1	27
3	3	4	4
4	20	27	1
5	77	104	0

[66] On a deficient multinomial expansion, *Proc. London Math. Soc.(2)*, *2* (1905), 478–485.

MacMahon investigates the following theorem of Cayley utilizing the techniques of his "partition analysis" that were established in paper [54].

Theorem. If $\{x_1 + x_2 + x_3 + \ldots\}^p$ denotes the expansion of $(x_1 + x_2 + x_3 + \ldots)^p$ in which only those terms $Cx_1^{a_1} x_2^{a_2} x_3^{a_3} \ldots$ are included that satisfy $\sum_{j \geqq 1} a_j \not> p - i + 1$ for all $i > 1$, then

$$X^n = \sum_{j=0}^{n} (-1)^j \binom{n}{j} \{x_1 + x_2 + \ldots + x_j\}^j (X + x_1 + x_2 + \ldots + x_{j+1})^{n-j}.$$

MacMahon first counts the number of terms $H_{p,q}$ $(q \geqq p)$ in $\{x_1 + x_2 + \ldots + x_p\}^q$; he proves in general that

$$H_{p,q} = \frac{q - p + 2}{q + p} \binom{q + p}{p - 1}.$$

Next he considers $C_{p,q}$, the sum of the coefficients in the expansion of $\{x_1 + x_2 + \ldots + x_p\}^q$; he is only able to prove that

$$C_{p,p} = (p + 1)^{p-1}, \qquad C_{p,q} = \sum_{j=p-1}^{q} \binom{q}{j} C_{p-1,j},$$

from which the $C_{p,q}$ can be calculated. He concludes (in article 4) by presenting a syzygetic theory of the distinct terms in the expansion of $\{x_1 + x_2 + \ldots + x_p\}^q$.

[67] Memoir on the theory of the partitions of numbers—Part III, *Phil. Trans.*, *205* (1906), 37–59.

This paper is devoted to a further study of magic squares (a study initiated in [52] and [57]), and contains sections 9 and 10 in the first three memoirs of this series.

Section 9 is devoted to using the "partition analysis" developed in [54] to treat magic squares. MacMahon shows that it is possible to create a syzygetic theory for such squares, gives complete details for 3 × 3 squares, and lists the 20 fundamental 4 × 4 squares.

In section 10 MacMahon recalls that in [57] he solved the "general magic square" problem in which row and column sums were all to be a given integer ω. His object now is to extend that work to the case where the two diagonals must sum to ω as well. Suppose we let h_n denote the sum of all monomial symmetric functions of degree n in the variables $\alpha_1, \alpha_2, \alpha_3, \ldots$.

Theorem. Let $h_\omega^{(s)}$ denote the image of h when α_s, α_{n-s+1} are replaced by $\lambda\alpha_s$, $\mu\alpha_{n-s+1}$, respectively. Then the coefficient of $(\lambda\mu\alpha_1\alpha_2 \ldots \alpha_n)^\omega$ in

$$h_\omega^{(1)} h_\omega^{(2)} \ldots h_\omega^{(n)}$$

is the number of general magic squares of order n corresponding to the sum ω.

A discussion of modifications of this theorem (e.g., variations of restrictions on the sth row) concludes the paper.

[68] The Diophantine equation $x^n - Ny^n = z$, *Proc. London Math. Soc. (2), 5* (1907), 45–58.

MacMahon extends his previous work on the Diophantine inequality $\lambda x \geqq \mu y$ to the case where μ/λ is irrational, and in particular he treats the case $\mu/\lambda = N^{1/m}$, where N is not a perfect mth power. The inequality is replaced by the Diophantine equation $x^n = Ny^n + z$, where only positive integral solutions are to be considered. MacMahon shows how the fundamental solutions for x and y depend in a linear manner on the continued fraction expansion of $N^{1/m}$, while z is obtained from a certain unimodular linear transformation of $x^n - Ny^n$. Thus the properties of z are obtainable from invariant theory applied to $x^n - Ny^n$. MacMahon treats in detail the cases $n = 2$, 3 and 4.

[73] Memoir on the theory of the partitions of numbers—Part IV. On the probability that the successful candidate at an election by ballot may never at any time have fewer votes than the one who is unsuccessful; on a generalization of this question; and on its connexion with other questions of partition, permutation, and combination, *Phil. Trans., 209* (1909). 153–175.

This paper is devoted to the solution of various "ballot problems."

Section 1 is devoted to the ballot problem described in the title. We assume that Pierre is the successful candidate with m votes and Paul is the unsuccessful candidate with n votes ($m > n$). MacMahon describes the problem in terms of the graphical representation of compositions, and then he utilizes his "partition analysis" (see [54]) to show that the total number of orders of the ballots in which Pierre is never behind is

$$\frac{m - n + 1}{m + 1}\binom{m + n}{m}.$$

Section 2 treats an extension of this problem to an n-candidate election:

Theorem. In an n-candidate election in which the ith candidate receives a_i votes with $a_1 \geqq a_2 \geqq a_3 \geqq \ldots \geqq a_n$, the number of orderings of the ballots in which the winner always has the most votes, the first runner-up always has the second most votes, etc., is

$$\frac{(a_1 + a_2 + \ldots + a_n)!}{a_1!a_2!\ldots a_{n-1}!}\prod_{t=1}^{n-1}\prod_{s=t+1}^{n}\left(1 - \frac{a_t}{a_s + s - t}\right).$$

Reprints of the Papers

[33] On play "à outrance"

[54] Memoir on the theory of the partitions of numbers—Part II

[58] Application of the partition analysis to the study of the properties of any system of consecutive integers

[62] The Diophantine inequality $\lambda x \geqq \mu y$

[65] Note on the Diophantine inequality $\lambda x \geqq \mu y$

[66] On a deficient multinomial expansion

[67] Memoir on the theory of the partitions of numbers—Part III

[68] The Diophantine equation $x^n - Ny^n = z$

[73] Memoir on the theory of the partitions of numbers—Part IV. On the probability that the successful candidate at an election ballot may never at any time have fewer votes than the one who is unsuccessful; on a generalization of this question; and on its connexion with other questions of partition, permutation, and combination

On Play "à outrance." *By* Major P. A. MacMahon, R.A.

[*Read March* 14*th*, 1889.]

Suppose two players A and B to have a and b counters respectively and to play a number of games, the winner receiving a counter from the loser until one of the players is deprived of all of his counters. We may consider the probability of all of the counters remaining finally with A, being given the probability of his winning any one game. When the probability of A winning each game is equal to a constant p, and no game can be drawn, the problem has been solved. (See Todhunter's *History of the Theory of Probability.*) I introduce here the consideration of the effect of supposing that the probability of a player winning a certain game depends in some manner upon the number of counters which he possesses at the commencement of that game.

In regard to the game which is commenced when the players A and B have respectively x and $n-x$ counters (where $a+b=n$), let the chances of winning be as $\phi(x)$ to $\phi(n-x)$, where $\phi(x)$ is a known function of x, which is finite for all integer values of the argument from 1 to n. Further, let u_x denote the chance of A winning the "partie" when he is in possession of x counters at the commencement of a game.

The chance of A winning the next game is then

$$\frac{\phi(x)}{\phi(x)+\phi(n-x)},$$

and the chance of his losing it is

$$\frac{\phi(n-x)}{\phi(x)+\phi(n-x)};$$

hence the difference equation

$$u_x = \frac{\phi(x)}{\phi(x)+\phi(n-x)}\,u_{x+1} + \frac{\phi(n-x)}{\phi(x)+\phi(n-x)}\,u_{x-1},$$

which may be written $\dfrac{\phi(x)}{\phi(n-x)} = \dfrac{u_x - u_{x-1}}{u_{x+1}-u_x}$,

and thence we obtain

$$\frac{\phi(n-1)}{\phi(1)}\cdot\frac{\phi(n-2)}{\phi(2)}\cdots\cdots\frac{\phi(n-s)}{\phi(s)} = \frac{u_{n-s}-u_{n-s-1}}{u_n-u_{n-1}}.$$

Put $$1+\frac{\phi(n-1)}{\phi(1)}+\frac{\phi(n-1)\,\phi(n-2)}{\phi(1)\,\phi(2)}+\ldots$$
$$+\frac{\phi(n-1)\,\phi(n-2)\ldots\phi(n-s)}{\phi(1)\,\phi(2)\ldots\phi(s)}=\Phi(s+1),$$

so that $$\frac{u_n-u_{n-s-1}}{u_n-u_{n-1}}=\Phi(s+1),$$

and thence the two results,

$$\frac{u_n-u_x}{u_n-u_{n-1}}=\Phi(n-x),$$

$$\frac{u_n-u_0}{u_n-u_{n-1}}=\Phi(n);$$

from these, by division, $\dfrac{u_n-u_x}{u_n-u_0}=\dfrac{\Phi(n-x)}{\Phi(n)}$.

The conditions of the problem are

$$u_n=1,\quad u_0=0;$$

hence $$1-u_x=\frac{\Phi(n-x)}{\Phi(n)},$$

and since, by definition,

$$\Phi(n-x)+\Phi(x)=\Phi(n),$$

we obtain $$u_x=\frac{\Phi(x)}{\Phi(n)};$$

this is the solution of the difference equation subject to the given conditions; that is to say, the probability u_x is the ratio of the sum of the first x terms to the sum of all of the terms of the series—

$$1+\frac{\phi(n-1)}{\phi(1)}+\frac{\phi(n-1)\,\phi(n-2)}{\phi(1)\,\phi(2)}+\ldots+\frac{\phi(n-1)}{\phi(1)}+1.$$

This reminds one of the binomial series,

$$(1+1)^{n-1},$$

with which it becomes identical when

$$\phi(x)=x.$$

When, in general, $$\phi(x)=x^m,$$

where m may be any number, positive, zero, or negative, commensurable or not, we are concerned with powers of binomial coefficients.

If we write
$$\frac{\phi(x)}{\phi(x)+\phi(n-x)}=f(x),$$
so that
$$f(x)+f(n-x)=1,$$
the difference equation becomes
$$u_x-u_{x-1}=f(x)\,(u_{x+1}-u_{x-1}),$$
and the solution may be written,
$$u_x=\frac{F(x)}{F(n)},$$
where
$$F(x)=1+\frac{f(n-1)}{f(1)}+\frac{f(n-1)\,f(n-2)}{f(1)\,f(2)}+\ldots$$
$$\ldots+\frac{f(n-1)\,f(n-2)\,\ldots\,f(n-x+1)}{f(1)\,f(2)\,\ldots\,f(x-1)}.$$

I now suppose that when A and B have respectively x and $n-x$ counters the chances of winning the next game are as $\phi(x)$ to $f(n-x)$, the most general case.

The difference equation becomes
$$u_x=\frac{\phi(x)}{\phi(x)+f(n-x)}\,u_{x+1}+\frac{f(n-x)}{\phi(x)+f(n-x)}\,u_{x-1},$$
or
$$\frac{\phi(x)}{f(n-x)}=\frac{u_x-u_{x-1}}{u_{x+1}-u_x},$$
putting
$$x=n-s,$$
$$\frac{\phi(n-s)}{f(s)}=\frac{u_{n-s}-u_{n-s-1}}{u_{n-s+1}-u_{n-s}},$$
thence
$$\frac{\phi(n-1)\,\phi(n-2)\,\ldots\,\phi(n-s)}{f(1)\,f(2)\,\ldots\,f(s)}=\frac{u_{n-s}-u_{n-s-1}}{u_n-u_{n-1}},$$
put
$$\left(\frac{\phi}{f}\right)_{s+1}=1+\frac{\phi(n-1)}{f(1)}+\frac{\phi(n-1)\,\phi(n-2)}{f(1)\,f(2)}+\ldots$$
$$\ldots+\frac{\phi(n-1)\,\ldots\,\phi(n-s)}{f(1)\,\ldots\,f(s)},$$
and then
$$\frac{u_n-u_{n-s-1}}{u_n-u_{n-1}}=\left(\frac{\phi}{f}\right)_{s+1},$$
which gives
$$\frac{u_n-u_x}{u_n-u_{n-1}}=\left(\frac{\phi}{f}\right)_{n-x},$$
and
$$\frac{u_n-u_0}{u_n-u_{n-1}}=\left(\frac{\phi}{f}\right)_n,$$

and remembering that $u_n = 1$, $u_0 = 0$, we obtain, by division,

$$1 - u_x = \frac{\left(\frac{\phi}{f}\right)_{n-x}}{\left(\frac{\phi}{f}\right)_n},$$

or
$$u_x = \frac{\left(\frac{\phi}{f}\right)_x}{\left(\frac{\phi}{f}\right)_n},$$

showing that to obtain u_x we have to divide the sum of the first x terms of the series,

$$\left(\frac{\phi}{f}\right)_n,$$

by the series itself.

Similarly, if v_x be the chance that B wins the partie when he has x counters,

$$v_x = \frac{\left(\frac{f}{\phi}\right)_x}{\left(\frac{f}{\phi}\right)_n},$$

and since
$$u_x + v_{n-x} = 1,$$
we obtain the interesting identity,

$$\frac{\left(\frac{\phi}{f}\right)_x}{\left(\frac{\phi}{f}\right)_n} + \frac{\left(\frac{f}{\phi}\right)_{n-x}}{\left(\frac{f}{\phi}\right)_n} = 1,$$

and the value of u_x may be presented in either of the forms shown by

$$u_x = \frac{\left(\frac{\phi}{f}\right)_x}{\left(\frac{\phi}{f}\right)_n} = \frac{\left(\frac{f}{\phi}\right)_n - \left(\frac{f}{\phi}\right)_{n-x}}{\left(\frac{f}{\phi}\right)_n}.$$

This result may be of service when it is necessary to take account of the courage of the two players. Thus putting

$$f(x) = px^{\alpha},$$
$$\phi(x) = qx^{-\beta},$$

where α and β are small quantities determined from experience, we would have the circumstance of A playing better when he is winning and of B playing better when he is losing, and doubtless a more correct estimate of the probability of either player winning the partie would thus be formed.

VIII. *Memoir on the Theory of the Partitions of Numbers.*—Part II.

By Major P. A. MACMAHON, *R.A., D.Sc., F.R.S.*

Received November 21,—Read November 24, 1898.

Introduction.

Art. 64. The subject of the partition of numbers, for its proper development, requires treatment in a new and more comprehensive manner. The subject-matter of the theory needs enlargement. This will be found to be a necessary consequence of the new method of regarding a partition that is here brought into prominence.

Let an integer n be broken up into any number of integers

$$\alpha_1,\ \alpha_2,\ \alpha_3,\ \ldots\ \alpha_s\,;$$

if we ascribe the conditions

$$\alpha_1 \geqq \alpha_2 \geqq \alpha_3 \geqq \ldots \geqq \alpha_s,$$

the succession

$$\alpha_1,\ \alpha_2,\ \alpha_3 \ldots \alpha_s$$

is what is known as a partition of n.

There are $s - 1$ conditions

$$\alpha_1 \geqq \alpha_2,\ \alpha_2 \geqq \alpha_3,\ \ldots\ \alpha_{s-1} \geqq \alpha_s,$$

to which we may add

$$\alpha_s \geqq 0$$

if the integers be all of them positive (or zero). For the present all the integers are restricted to be positive or zero by hypothesis, so that this last-written condition will not be further attended to.

If, on the other hand, the conditions be

$$\alpha_1 \gtreqless \alpha_2 \gtreqless \alpha_3 \ldots \gtreqless \alpha_s,$$

no order of magnitude is supposed to exist between the successive parts, and we obtain what has been termed a "composition" of the integer n.

Various other systems of partitions into s parts may be brought under view, because between two consecutive parts we may place either of the seven symbols

$$>,\ =,\ <,\ \geqq,\ \leqq,\ \gtrless,\ \gtreqless.$$

5.5.99

We thus obtain 7^{s-1} different sets of conditions that may be assigned; these are not all essentially different and in many cases they overlap.

Art. 65. For the moment I concentrate attention upon the symbol

$$\geqq,$$

and remark that the $s - 1$ conditions, which involve this symbol, set forth above, constitute one set of a large class of sets which involve the symbol. We may have the single condition

$$A_1^{(1)}\alpha_1 + A_2^{(1)}\alpha_2 + A_3^{(1)}\alpha_3 + \ldots + A_s^{(1)}\alpha_s \geqq 0,$$

wherein $A_1, A_2, A_3 \ldots A_s$ are integers $+$, zero or $-$, of which at least one must be positive, or we may have the set of conditions

$$\left.\begin{array}{l} A_1^{(1)}\alpha_1 + A_2^{(1)}\alpha_2 + A_3^{(1)}\alpha_3 + \ldots + A_s^{(1)}\alpha_s \geqq 0 \\ A_1^{(2)}\alpha_1 + A_2^{(2)}\alpha_2 + A_3^{(2)}\alpha_3 + \ldots + A_s^{(2)}\alpha_s \geqq 0 \\ A_1^{(3)}\alpha_1 + A_2^{(3)}\alpha_2 + A_3^{(3)}\alpha_3 + \ldots + A_s^{(3)}\alpha_s \geqq 0 \\ \cdot \quad \cdot \quad \cdot \quad \cdot \quad \cdot \quad \cdot \quad \cdot \quad \cdot \quad \cdot \quad \cdot \quad \cdot \quad \cdot \quad \cdot \\ A_1^{(r)}\alpha_1 + A_2^{(r)}\alpha_2 + A_3^{(r)}\alpha_3 + \ldots + A_s^{(r)}\alpha_s \geqq 0 \end{array}\right\}$$

as the definition of the partitions considered. If the symbol be $=$ instead of $\geqq$ the solution of the equations falls into the province of linear Diophantine analysis. The problem before us may be regarded as being one of linear partition analysis. There is much in common between the two theories; the problems may be treated by somewhat similar methods.

The partition analysis of degree higher than the first, like the Diophantine, is of a more recondite nature, and is left for the present out of consideration.

I treat the partition conditions by the method of generating functions. I seek the summation

$$\Sigma X_1^{\alpha_1} X_2^{\alpha_2} X_3^{\alpha_3} \ldots X_s^{\alpha_s}$$

for every set of values (integers)

$$\alpha_1, \alpha_2, \alpha_3, \ldots \alpha_s$$

which satisfy the assigned conditions.

It appears that there are, in every case, a finite number of ground or fundamental solutions of the conditions, viz.:—

$$\begin{array}{l} \alpha_1^{(1)}, \alpha_2^{(1)}, \alpha_3^{(1)} \ldots \alpha_s^{(1)} \\ \alpha_1^{(2)}, \alpha_2^{(2)}, \alpha_3^{(2)} \ldots \alpha_s^{(2)} \\ \cdot \quad \cdot \quad \cdot \quad \cdot \quad \cdot \quad \cdot \quad \cdot \\ \alpha_1^{(m)}, \alpha_2^{(m)}, \alpha_3^{(m)} \ldots \alpha_s^{(m)} \end{array}$$

such that every solution

$$\alpha_1, \alpha_2, \alpha_3, \ldots \alpha_s$$

is such that

$$\alpha_1 = \lambda_1 \alpha_1^{(1)} + \lambda_2 \alpha_1^{(2)} \ldots + \lambda_m \alpha_1^{(m)}$$
$$\alpha_2 = \lambda_1 \alpha_2^{(1)} + \lambda_2 \alpha_2^{(2)} \ldots + \lambda_m \alpha_2^{(m)}$$
$$\alpha_3 = \lambda_1 \alpha_3^{(1)} + \lambda_2 \alpha_3^{(2)} \ldots + \lambda_m \alpha_3^{(m)}$$
$$\cdots \cdots \cdots$$
$$\alpha_s = \lambda_1 \alpha_s^{(1)} + \lambda_2 \alpha_s^{(2)} \ldots + \lambda_m \alpha_s^{(m)}$$

$\lambda_1, \lambda_2, \ldots \lambda_m$ being positive integers.

This arises from the fact that every term

$$X_1^{\alpha_1} X_2^{\alpha_2} X_3^{\alpha_3} \ldots X_s^{\alpha_s}$$

of the summation is found to be expressible as a product

$$\{X_1^{\alpha_1^{(1)}} X_2^{\alpha_2^{(1)}} X_3^{\alpha_3^{(1)}} \ldots X_s^{\alpha_s^{(1)}}\}^{\lambda_1}$$
$$\times \{X_1^{\alpha_1^{(2)}} X_2^{\alpha_2^{(2)}} X_3^{\alpha_3^{(2)}} \ldots X_s^{\alpha_s^{(2)}}\}^{\lambda_2}$$
$$\times \cdots \cdots \cdots$$
$$\times \{X_1^{\alpha_1^{(m)}} X_2^{\alpha_2^{(m)}} X_3^{\alpha_3^{(m)}} \ldots X_s^{\alpha_s^{(m)}}\}^{\lambda_m}$$

Denoting this product by

$$P_1^{\lambda_1} P_2^{\lambda_2} \ldots P_m^{\lambda_m}$$

the generating function assumes the form

$$\frac{1 - (Q_1^{(1)} + Q_1^{(2)} + Q_1^{(3)} + \ldots) + (Q_2^{(1)} + Q_2^{(2)} + Q_2^{(3)} + \ldots) - (Q_3^{(1)} + \ldots) + \ldots}{(1 - P_1)(1 - P_2) \ldots (1 - P_m)}$$

wherein the denominator indicates the ground solutions and the numerator the simple and compound syzygies which unite them.

The terms

$Q_1^{(1)}, Q_1^{(2)}, Q_1^{(3)} \ldots$ denote first syzygies
$Q_2^{(1)}, Q_2^{(2)}, Q_2^{(3)} \ldots$ „ second „
$Q_3^{(1)}, Q_3^{(2)}, Q_3^{(3)} \ldots$ „ third „
. . . .

The reader will note the striking analogy with the generating functions of the theory of invariants.

Similar results are obtained as solutions of linear Diophantine equations.

The generating functions under view are *real* in the sense of CAYLEY and SYLVESTER. Enumerating generating functions of various kinds are obtained by assigning equalities between the suffixed capitals

$$X_1, X_2, \ldots X_s.$$

Putting, *e.g.*,

$$X_1 = X_2 = \ldots = X_s = x,$$

we obtain the function which enumerates by the coefficient of x^n, in the ascending expansion, the numbers of solutions for which

$$\alpha_1 + \alpha_2 + \ldots + \alpha_s = n.$$

It will be gathered that the note of the following investigation is the importation of the idea that the solution of any system of equations of the form

$$A_1\alpha_1 + A_2\alpha_2 + A_3\alpha_3 + \ldots + A_s\alpha_s \geqq 0$$

(all the quantities involved being integers) is a problem of partition analysis, and that the theory proceeds *pari passu* with that of the linear Diophantine analysis.

Section 5.

Art. 66. I propose to lead up to the general theory of partition analysis by considering certain simple particular cases in full detail.

Suppose we have a function

$$F(x, a)$$

which can be expanded in ascending powers of x. Such expansion being either finite or infinite, the coefficients of the various powers of x are functions of a which in general involve both positive and negative powers of a. We may reject all terms containing negative powers of a and subsequently put a equal to unity. We thus arrive at a function of x only, which may be represented after CAYLEY (modified by the association with the symbol $\geqq$) by

$$\underset{\geqq}{\Omega} F(x, a),$$

the symbol $\geqq$ denoting that the terms retained are those in which the power of a is $\geqq 0$.

Similarly we may indicate by the operation

$$\underset{=}{\Omega}$$

that the only terms retained are those in which a occurs to the power zero and the meaning of the operations

$$\underset{>}{\Omega}, \quad \underset{<}{\Omega}, \quad \underset{\geqq}{\Omega}, \quad \underset{\gtrless}{\Omega}$$

will be understood without further explanation. To generalise the notion we may consider

$$\underset{\geqq}{\Omega}\, F(X_1, X_2, \ldots X_n, a_1, a_2, \ldots a_t)$$

to mean that the function is to be expanded in ascending powers of $X_1, X_2, \ldots X_n$, the terms involving any negative powers of $a_1, a_2, \ldots a_t$ are to be rejected, and that subsequently we are to put

$$a_1 = a_2 = \ldots = a_t = 1.$$

In this case the operation Ω has reference to each of the letters $a_1, a_2, \ldots a_t$ and a term involving any negative power of either of these quantities is rejected.

If the quantities $a_1, a_2, \ldots a_t$ be not all subjected to the same operation we may denote the whole operation by

$$\overset{a_1}{\underset{\geqq}{\Omega}}\ \overset{a_2}{\underset{>}{\Omega}}\ \overset{a_3}{\underset{\gtrless}{\Omega}} \ldots \overset{a_t}{\underset{=}{\Omega}}\, F(X_1, X_2, \ldots X_n, a_1, a_2, a_3 \ldots a_t)$$

wherein $\overset{a_r}{\underset{\sigma_r}{\Omega}}$ operates upon a_r according to the law of the symbol σ_r.

The operation, *quâ* a single quantity and the symbol $\geqq$, have been studied by CAYLEY.* *Quâ* more than one quantity it has presented itself in a memoir on partitions by the present author.†

These Ω functions are of moment in all questions of partition and linear Diophantine analysis.

Art. 67. I will construct Ω functions to serve as generators of well-known solutions and enumerations in the theory of unipartite partition.

Problem I. To determine the number of partitions of w into i or fewer parts.

Graphically considered we have i rows of nodes

$$\begin{array}{l} a_1 \ \cdot\ \cdot\ \cdot\ \cdot\ \cdot\ \cdot\ \cdot \\ a_2 \ \cdot\ \cdot\ \cdot\ \cdot \\ a_3 \ \cdot\ \cdot\ \cdot \\ \ \vdots \\ a_i \ \cdot\ \cdot\ \cdot \end{array}$$

* "On an Algebraical Operation," 'Collected Papers,' vol. 9, p. 537.

† "Memoir on the Theory of the Partitions of Numbers," Part I., 'Phil. Trans.,' A, vol. 187, pp. 619–673, 1896.

$\alpha_1, \alpha_2, \ldots$ denoting the numbers of nodes in the first, second, &c., rows,

$$\alpha_1 \geqq \alpha_2$$
$$\alpha_2 \geqq \alpha_3$$
$$\vdots$$
$$\alpha_{i-1} \geqq \alpha_i$$

To find

$$\Sigma X_1^{\alpha_1} X_2^{\alpha_2} \ldots X_i^{\alpha_i}$$

for all sets of integers satisfying the conditions take

$$\underset{\geqq}{\Omega} \frac{1}{1 - a_1 X_1 \,.\, 1 - \frac{a_2}{a_1} X_2 \,.\, 1 - \frac{a_3}{a_2} X_3 \ldots 1 - \frac{a_i}{a_{i-1}} X_i}$$

where observe that the factors $\frac{1}{1 - a_1 X_1}, \frac{1}{1-(a_2/a_1)X_2}, \ldots$ generate the successive rows of nodes and that the method of placing the letters $a_1, a_2, \ldots$ ensures the satisfaction of the first, second, &c., conditions.

Continued application of the simple theorem

$$\underset{\geqq}{\Omega} \frac{1}{1 - ax \,.\, 1 - \frac{1}{a} y} = \frac{1}{1 - x \,.\, 1 - xy},$$

applied in respect of the quantities $a_1, a_2, \ldots$ in succession, reduces the Ω function to the form

$$\frac{1}{1 - X_1 \,.\, 1 - X_1 X_2 \,.\, 1 - X_1 X_2 X_3 \ldots 1 - X_1 X_2 X_3 \ldots X_i}$$

the *real* generating function.

The ground solutions or fundamental partitions are, as shown by the denominator factors,

$$(\alpha_1, \alpha_2, \alpha_3, \ldots \alpha_i)$$

$$= \begin{cases} (1, & 0, & 0, & \ldots 0) \\ (1, & 1, & 0, & \ldots 0) \\ (1, & 1, & 1, & \ldots 0) \\ \cdot & \cdot & \cdot & \cdot\;\cdot\;\cdot \\ (1, & 1, & 1, & \ldots 1) \end{cases}$$

and, as might have been anticipated, the graphical representation is in evidence.

Art. 68. By choosing to sum the expression

$$\Sigma X_1^{\alpha_1} X_2^{\alpha_2} \ldots X_i^{\alpha_i},$$

every solution of the given conditions has been generated. The same result might have been achieved by other summations such as

$$\Sigma X_1^{\lambda_1\alpha_1} X_2^{\mu_2\alpha_2} \ldots X_i^{\eta_i\alpha_i},$$

$\lambda_1, \lambda_2, \ldots \lambda_i$ being given positive integers, or as

$$\Sigma X_1^{\alpha_1-\alpha_2} X_2^{\alpha_2-\alpha_3} \ldots X_{i-1}^{\alpha_{i-1}-\alpha_i} X_i^{\alpha_i}.$$

We, in fact, may take as indices of $X_1, X_2, \ldots X_i$ any given linear functions of $\alpha_1, \alpha_2, \ldots \alpha_i$, and form the corresponding generating function.

For the two cases specified, the Ω functions are

$$\underset{\geq}{\Omega} \frac{1}{1 - a_1 X_1^{\lambda_1} . 1 - \frac{a_2}{a_1} X_2^{\mu_2} \ldots 1 - \frac{1}{a_{i-1}} X_i^{\eta_i}},$$

$$\underset{\geq}{\Omega} \frac{1}{1 - a_1 X_1 . 1 - \frac{a_2}{a_1}\frac{X_2}{X_1} . 1 - \frac{a_3}{a_2}\frac{X_3}{X_2} \ldots 1 - \frac{1}{a_{i-1}}\frac{X_i}{X_{i-1}}},$$

and the reduced functions

$$\frac{1}{1 - X_1^{\lambda_1} . 1 - X_1^{\lambda_1} X_2^{\mu_2} \ldots 1 - X_1^{\lambda_1} X_2^{\mu_2} \ldots X_i^{\eta_i}},$$

$$\frac{1}{1 - X_1 . 1 - X_2 . 1 - X_3 \ldots 1 - X_i}$$

respectively.

Generally for the sum

$$\Sigma X_1^{\lambda_1\alpha_1 + \mu_1\alpha_2 + \ldots} X_2^{\lambda_2\alpha_1 + \mu_2\alpha_2 + \ldots} \ldots X_i^{\lambda_i\alpha_1 + \mu_i\alpha_2 + \ldots + \eta_i\alpha_i}$$

the two functions are

$$\underset{\geq}{\Omega} \frac{1}{1 - a_1 X_1^{\lambda_1} X_2^{\lambda_2} \ldots X_i^{\lambda_i} . 1 - \frac{a_2}{a_1} X_1^{\mu_1} X_2^{\mu_2} \ldots X_i^{\mu_i} \ldots 1 - \frac{1}{a_{i-1}} X_1^{\eta_1} X_2^{\eta_2} \ldots X_i^{\eta_i}}$$

and

$$\frac{1}{1 - X_1^{\lambda_1} X_2^{\lambda_2} \ldots X_i^{\lambda_i} . 1 - X_1^{\lambda_1+\mu_1} X_2^{\lambda_2+\mu_2} \ldots X_i^{\lambda_i+\mu_i} \ldots 1 - X_1^{\lambda_1+\ldots+\eta_1} X_2^{\lambda_2+\ldots+\eta_2} \ldots X_i^{\lambda_i+\ldots+\eta_i}}.$$

Art. 69. In any of these instances we have i quantities at disposal, viz.:

$$X_1, X_2, \ldots X_i,$$

in order to derive enumerating generating functions corresponding to certain problems. In the last-written general case, the quantities $\lambda, \mu, \ldots \eta$ being given integers, put as a particular case,

$$X_1 = X_2 = \ldots = X_i = x.$$

The reduced function is

$$\frac{1}{1 - x^{\Sigma\lambda} . 1 - x^{\Sigma\lambda + \Sigma\mu} . - \ldots 1 - x^{\Sigma\lambda + \Sigma\mu + \ldots + \Sigma\eta}},$$

and herein the coefficients of x^w, in the expansion, give the number of partitions

$$\alpha_1, \alpha_2, \alpha_3, \ldots \alpha_i$$

of all numbers which satisfy the equation

$$\Sigma\lambda . \alpha_1 + \Sigma\mu . \alpha_2 + \ldots + \Sigma\eta . \alpha_i = w,$$

$\alpha_1, \alpha_2, \ldots \alpha_i$ being in descending order.

For the three particular cases considered above this equation takes the forms

$$\alpha_1 + \alpha_2 + \ldots + \alpha_i = w,$$
$$\lambda_1\alpha_1 + \mu_2\alpha_2 + \ldots + \eta_i\alpha_i = w,$$
$$\alpha_1 = w,$$

connected with the reduced generators,

$$\frac{1}{1 - x . 1 - x^2 . 1 - x^3 \ldots 1 - x^i},$$

$$\frac{1}{1 - x^{\lambda_1} . 1 - x^{\lambda_1 + \mu_2} \ldots 1 - x^{\lambda_1 + \mu_2 + \ldots + \eta_i}},$$

$$\frac{1}{(1 - x)^i},$$

respectively.

Further, we may separate $X_1, X_2, \ldots X_i$ in any manner into k sets and put those which are in the first set equal to x_1, those in the second equal to x_2, and so on, and so reach an enumerating function involving k quantities, $x_1, x_2, x_3, \ldots x_k$.

Ex. gr. Put

$$X_1 = X_3 = X_5 = \ldots = x_1,$$
$$X_2 = X_4 = X_6 = \ldots = x_2,$$

and suppose i even. We obtain

$$\frac{1}{1 - x_1 . 1 - x_1x_2 . 1 - x_1^2x_2 . 1 - x_1^2x_2^2 \ldots 1 - x_1^{\frac{1}{2}i}x_2^{\frac{1}{2}i}},$$

to enumerate by the coefficient of $x_1^{w_1}x_2^{w_2}$ those partitions of $w_1 + w_2$ for which

$$\alpha_1 + \alpha_3 + \alpha_5 + \ldots = w_1$$
$$\alpha_2 + \alpha_4 + \alpha_6 + \ldots = w_2.$$

This enumerating function, since it involves x_1 and x_2, is one connected also with the partitions of bipartite numbers. In general when k sets are taken, we have a theorem of k-partite partitions. When $k = i$, we have at once a *real* generating function for unipartites and an enumerating function for i-partites, for, from the latter point of view, the number unity which appears as the coefficient of $X_1^{\alpha_1}X_2^{\alpha_2}\ldots X_i^{\alpha_i}$ shows that the multipartite number

$$\overline{\alpha_1\alpha_2\ldots\alpha_i}$$

can be partitioned in one way only into the parts

$$\begin{array}{l} \overline{1\ 0\ .\ .\ .\ .\ .\ .} \\ \overline{1\ 1\ 0\ .\ .\ .\ .\ .} \\ \overline{1\ 1\ 1\ .\ .\ .\ .\ .} \\ .\ .\ .\ .\ .\ .\ .\ . \\ \overline{1\ 1\ 1\ .\ .\ .\ .\ 1} \end{array}$$

there being i figures in each part.

Art. 70. We may now enquire into the partitions of all numbers

$$\alpha_1, \quad \alpha_2, \quad \alpha_3, \ldots \alpha_i,$$

subject to the given conditional relations and also to the linear equations

$$\lambda_1\alpha_1 + \mu_2\alpha_2 + \ldots + \eta_i\alpha_i = w$$
$$\lambda'_1\alpha_1 + \mu'_2\alpha_2 + \ldots + \eta'_i\alpha_i = w'$$
$$\ldots\ldots\ldots\ldots$$
$$\lambda_1^{(s)}\alpha_1 + \mu_2^{(s)}\alpha_2 + \ldots + \eta_i^{(s)}\alpha_i = w^{(s)}.$$

To illustrate the method, it suffices to take $s = 2$, and then we have to perform the summation

$$\Sigma X_1^{\lambda_1\alpha_1}X_2^{\mu_2\alpha_2}\ldots X_i^{\eta_i\alpha_i}Y_1^{\lambda'_1\alpha_1}Y_2^{\mu'_2\alpha_2}\ldots Y_i^{\eta'_i\alpha_i}.$$

The Ω function reduced is

$$\frac{1}{1 - X_1^{\lambda_1}Y_1^{\lambda'_1}\,.\,1 - X_1^{\lambda_1}X_2^{\mu_2}Y_1^{\lambda'_1}Y_2^{\mu'_2}\ldots 1 - X_1^{\lambda_1}X_2^{\mu_2}\ldots X_i^{\eta_i}Y_1^{\lambda'_1}Y_2^{\mu'_2}\ldots Y_i^{\eta'_i}},$$

wherein putting

$$X_1 = X_2 = \ldots = X_i = x,$$
$$Y_1 = Y_2 = \ldots = Y_i = y,$$

we obtain the enumerating function

$$\frac{1}{1 - x^{\lambda_1}y^{\lambda'_1} \,.\, 1 - x^{\lambda_1+\mu_2}y^{\lambda'_1+\mu'_2} \ldots 1 - x^{\lambda_1+\mu_2+\ldots+\eta_i}y^{\lambda'_1+\mu'_2+\ldots+\eta'_i}},$$

in which we seek the coefficient of $x^w y^{w'}$.

Art. 71. *Ex. gr.* Consider the particular case

$$\alpha_1 + \alpha_2 + \ldots + \alpha_i = w,$$
$$\alpha_1 + 2\alpha_2 + \ldots + i\alpha_i = w',$$

$\alpha_1, \alpha_2, \ldots \alpha_i$ being, as usual, subject to the conditional relations.

The enumerating function is

$$\frac{1}{1 - xy \,.\, 1 - x^2y^3 \,.\, 1 - x^3y^6 \ldots 1 - x^iy^{\frac{1}{2}i(i+1)}},$$

and it is obvious also that the partitions of the bipartite $\overline{ww'}$ which satisfy the conditions may be composed by the biparts

$$\overline{11},\ \overline{23},\ \overline{36}, \ldots \overline{i, \tfrac{1}{2}i(i+1)}.$$

The corresponding graphical representation is not by superposition of lines of nodes, but by angles of nodes, of the natures

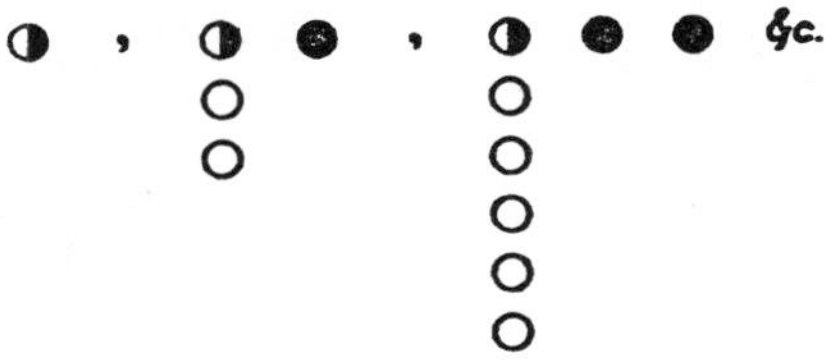

Art. 72. It is convenient, at this place, to give some elementary theorems concerning the Ω function which will be useful in what follows.

$$\underset{\geqq}{\Omega} \frac{1}{1 - ax \,.\, 1 - \frac{1}{a}y} = \frac{1}{1 - x \,.\, 1 - xy},$$

$$\underset{>}{\Omega} \frac{1}{1 - ax \,.\, 1 - ay \,.\, 1 - \frac{1}{a}z} = \frac{1 - xyz}{1 - x \,.\, 1 - y \,.\, 1 - xz \,.\, 1 - yz},$$

$$\underset{\geq}{\Omega}\frac{1}{1-ax.1-\frac{1}{a}y.1-\frac{1}{a}z}=\frac{1}{1-x.1-xy.1-xz},$$

$$\underset{\geq}{\Omega}\frac{1}{1-a^2x.1-\frac{1}{a}y}=\frac{1+xy}{1-x.1-xy^2},$$

$$\underset{\geq}{\Omega}\frac{1}{1-ax.1-\frac{1}{a^2}y}=\frac{1}{1-x.1-x^2y},$$

$$\underset{\geq}{\Omega}\frac{1}{1-a^3x.1-\frac{1}{a}y}=\frac{1+xy+xy^2}{1-x.1-xy^3},$$

$$\underset{\geq}{\Omega}\frac{1}{1-ax.1-\frac{1}{a^3}y}=\frac{1}{1-x.1-x^3y},$$

$$\underset{\geq}{\Omega}\frac{1}{1-a^2x.1-ay.1-\frac{1}{a}z}=\frac{1+xz-xyz-xyz^2}{1-x.1-y.1-yz.1-xz^2},$$

$$\underset{\geq}{\Omega}\frac{1}{1-a^2x.1-\frac{1}{a}y.1-\frac{1}{a}z}=\frac{1+xy+xz+xyz}{1-x.1-xy^2.1-xz^2},$$

$$\underset{\geq}{\Omega}\frac{1}{1-ax.1-ay.1-az.1-\frac{1}{a}w}=\frac{1-xyw-xzw-yzw+xyzw+xyzw}{1-x.1-y.1-z.1-xw.1-yw.1-zw},$$

$$\underset{\geq}{\Omega}\frac{1}{1-ax.1-ay.1-\frac{1}{a}z.1-\frac{1}{a}}=\frac{1-xyz-xyw-xyzw+xy^2zw+x^2yzw}{1-x.1-y.1-xz.1-xw.1-yz.1-yw}.$$

Art. 73. I pass on to consider the partitions of numbers into parts limited not to exceed i in magnitude.

The Ω function is clearly

$$\underset{\geq}{\Omega}\frac{1-(a_1X_1)^{i+1}}{1-a_1X_1}\cdot\frac{1-\left(\frac{a_2}{a_1}X_2\right)^{i+1}}{1-\frac{a_2}{a_1}X_2}\cdot\frac{1-\left(\frac{a_3}{a_2}X_3\right)^{i+1}}{1-\frac{a_3}{a_2}X_3}\ \ldots\ ad\ inf.$$

In this form I have not succeeded in effecting the reduction, but if we put at once

$$X_1=X_2=X_3=\ldots=x,$$

the reduced form is

$$\frac{1}{1-x.1-x^2.1-x^3\ldots1-x^i}$$

If the parts be limited to i in number and to j in magnitude, we find

$$\underset{\geq}{\Omega}\frac{1-(a_1x)^{j+1}}{1-a_1x}\cdot\frac{1-\left(\frac{a_2}{a_1}x\right)^{j+1}}{1-\frac{a_2}{a_1}x}\cdots\frac{1-\left(\frac{1}{a_{i-1}}x\right)^{j+1}}{1-\frac{1}{a_{i-1}}x}=\frac{1-x^{j+1}\,.\,1-x^{j+2}\,.\,1-x^{j+3}\ldots1-x^{j+i}}{1-x\,.\,1-x^2\,.\,1-x^3\ldots1-x^i}$$

the well-known result.

Art. 74. It is to be remarked that the generating function in question may also be written

$$\underset{\geq}{\Omega}\frac{\frac{1}{1-g}}{1-a_1gx\,.\,1-\frac{a_2}{a_1}x\,.\,1-\frac{a_3}{a_2}x^2\ldots1-\frac{1}{a_{i-1}}x},$$

in which we have to seek the coefficient of g^j. This function reduces to

$$\frac{1}{1-g\,.\,1-gx\,.\,1-gx^2\,.\,1-gx^3\ldots1-gx^i}$$

the well-known form.

In general, when a generating function reduces to the product of factors

$$\frac{1}{1-x^s},$$

the part-magnitude being unrestricted, we obtain a product of factors

$$\frac{1}{1-gx^s}$$

for the restricted case, and this is frequently exhibitable, as regards the coefficients of g^j, as a product of factors

$$\frac{1-x^{j+s}}{1-x^s}.$$

The Ω function is not altered by the interchange of the letters i, j.

Art. 75. If the successive parts of the partition are limited in magnitude by

$$j_1, j_2, \ldots j_i,$$

numbers necessarily in descending order, the generating function is

$$\underset{\geq}{\Omega}\frac{1-(a_1x)^{j_1+1}}{1-a_1x}\cdot\frac{1-\left(\frac{a_2}{a_1}x\right)^{j_2+1}}{1-\frac{a_2}{a_1}x}\cdots\frac{1-\left(\frac{1}{a_{i-1}}x\right)^{j_i+1}}{1-\frac{1}{a_{i-1}}x}.$$

For $i = 2$, this may be shown to be equal to

$$\frac{(1 - x^{j_2+1})(1 - x^{j_1+2})}{(1 - x)(1 - x^2)} + x^{j_2+1} \frac{(1 - x^{j_2+1})(1 - x^{j_1-j_2})}{(1 - x)(1 - x^2)},$$

but for $i > 2$, the functions are obtained with increasing labour, and are of increasing complexity.

Many cases present themselves, similar to the one before us, where the Ω function is written down with facility, but no serviceable reduced function appears to exist. On the other hand, we meet with astonishing instances of compact reduced functions which involve valuable theorems.

Art. 76. From the reduced function we can frequently proceed to an $\underset{\geq}{\Omega}$ function, thus inverting the usual process. If, for example, we require an Ω equivalent to

$$\frac{1}{1 - x^{P_1} . 1 - x^{P_2} . 1 - x^{P_3} \ldots 1 - x^{P_i}},$$

a little consideration leads us to

$$\underset{\geq}{\Omega} \frac{1}{1 - a_1 x^{P_1} . 1 - \frac{a_2}{a_1} x^{P_2 - P_1} . 1 - \frac{a_3}{a_2} x^{P_3 - P_2} \ldots 1 - \frac{1}{a_{i-1}} x^{P_i - P_{i-1}}}.$$

This indicates that a unipartite partition into the parts $P_1, P_2, \ldots P_i$ may be represented by a two-dimensional partition of another kind which involves the parts

$$P_1,\ P_2 - P_1,\ P_3 - P_2, \ldots P_i - P_{i-1}.$$

Ex. gr., the numbers P_1, P_2, P_3 being in ascending order, the line partition

$$P_3 P_3 P_3 P_2 P_2 P_1$$

can be thrown into the plane partition

$$\begin{array}{cccccc} P_1 & P_1 & P_1 & P_1 & P_1 & P_1 \\ P_2 - P_1 & P_2 - P_1 & P_2 - P_1 & P_2 - P_1 & P_2 - P_1 & \\ P_3 - P_2 & P_3 - P_2 & P_3 - P_2. & & & \end{array}$$

of the nature of a regularised graph in the elements $P_1, P_2 - P_1, P_3 - P_2$, though these quantities are not necessarily in any specified order of magnitude. We obtain, in fact, a mixed numerical and graphical representation of a partition of a new kind. If

$$(P_1, P_2, P_3) = (1, 3, 4),$$

the partition 4 3 3 3 1 has the mixed graph

1 1 1 1 1
2 2 2 2
1

as well as its ordinary unit-graph.

In one case the mixed graph is composed entirely of units, and is, moreover, the graph conjugate to the unit graph.

This happens when

$$(P_1, P_2, P_3, \ldots) = (1, 2, 3, \ldots).$$

Thus, *quâ* these elements,

4 3 3 3 1

has the mixed (here the conjugate) graph

1 1 1 1 1
1 1 1 1
1 1 1 1 .
1

Art. 77. Observe that a partition may be such *quâ* the parts which actually appear in it, *or* it may be *quâ*, in addition, certain parts which might appear, but which happen to be absent. A mixed graph corresponds to each such supposition.

Ex. gr. :—

Partition.	*Quâ* elements.	Graph.
4 3	4, 3	3 3 1
4 3	4, 3, 1	1 1 2 2 1
4 3	4, 3, 2	2 2 1 1 1
4 3	4, 3, 2. 1	1 1 1 1 1 1 1

We thus arrive at a generalization of the notion of a conjugate partition, and are convinced that the proper representation of a Ferrers-graph is not by nodes or points, but by units.

When the mixed elements

$$P_1,\ P_2 - P_1,\ P_3 - P_2, \ldots$$

are in descending order of magnitude we have a correspondence between unipartite partitions and multipartite partitions of a certain class.

Art. 78. It is usual to consider the parts of a partition arranged in descending order. The Ω function enables us to assign any desired order of magnitude between the successive parts.

In the case of three parts we have already considered the system

$$\alpha_1 \geqq \alpha_2,\ \alpha_2 \geqq \alpha_3.$$

For the system

$$\alpha_1 \geqq \alpha_2,\ \alpha_3 \geqq \alpha_2,$$

we have the solution

$$\underset{\geqq}{\Omega} \frac{1}{1 - a_1 X_1 . 1 - \frac{1}{a_1 a_2} X_2 . 1 - a_2 X_3},$$

and thence the *real* reduced generator

$$\frac{1}{1 - X_1 . 1 - X_1 X_2 X_3 . 1 - X_3},$$

and the enumerating function

$$\frac{1 + x}{(1 - x)(1 - x^2)(1 - x^2)}.$$

On the other hand, for the system

$$\alpha_2 \geqq \alpha_1,\ \alpha_2 \geqq \alpha_3,$$

we construct

$$\underset{\geqq}{\Omega} \frac{1}{1 - \frac{X_1}{a_1} . 1 - a_1 a_2 X_2 . 1 - \frac{X_3}{a_2}},$$

leading to the real and enumerating functions

$$\frac{1 - X_1 X_2^2 X_3}{1 - X_2 . 1 - X_1 X_2 . 1 - X_2 X_3 . 1 - X_1 X_2 X_3},$$

$$\frac{1 + x^2}{(1 - x)(1 - x^2)(1 - x^3)};$$

of the former, the denominator shows the ground solutions, *id est*, fundamental partitions,

$$(\alpha_1, \alpha_2, \alpha_3) = (0, 1, 0);\ (110);\ (011);\ (111);$$

and the enumerator points to the syzygy

$$X_2 . X_1 X_2 X_3 - X_1 X_2 . X_2 X_3 = 0.$$

Art. 79. If the partition be into i parts, we can assign 2^{i-1} different orders depending upon the symbols $\geqq$, $\eqslantless$, and these can all be expressed by conditional relations affecting $\alpha_1, \alpha_2, \ldots \alpha_i$, involving the symbol $\geqq$ only. These are not all *essentially* different, as one order does or does not give rise to a different order by inversions of parts. Denoting $\geqq$, $\eqslantless$ by the letters d, a, we have for $i = 3$ the orders *dd, da, ad, aa*; the orders *dd, aa* are not essentially different, because interchange of a and d combined with inversion converts the one into the other; *da, ad* are essentially different, because this two-fold operation leaves each of these unchanged. Hence there are three orders to be considered, and the results have been obtained above.

For $i = 4$ we have the essentially different orders *ddd, dda, dad, add.* The first of these has been obtained; the other three are solved by the Ω functions:

$$\underset{\geqq}{\Omega} \frac{1}{1 - a_1X_1 . 1 - \frac{a_2}{a_1} X_2 . 1 - \frac{X}{a_2 a} . 1 - a_3X_4};$$

$$\underset{\geqq}{\Omega} \frac{1}{1 - a_1X_1 . 1 - \frac{X_2}{a_1a_2} . 1 - a_2a_3X_3 . 1 - \frac{X_4}{a_3}};$$

$$\underset{\geqq}{\Omega} \frac{1}{1 - \frac{X_1}{a_1} . 1 - a_1a_2X_2 . 1 - \frac{a_3}{a_2} X_3 . 1 - \frac{X_4}{a_3}};$$

which reduce to the three expressions:

$$\frac{1}{1 - X_1 . 1 - X_4 . 1 - X_1X_2 . 1 - X_1X_2X_3X_4}$$

$$\frac{1 - X_1X_2X_3^2X_4}{1 - X_1 . 1 - X_3 . 1 - X_3X_4 . 1 - X_1X_2X_3 . 1 - X_1X_2X_3X}$$

$$\frac{1 - X_1X_2^2X_3 - X_1X_2^2X_3X_4 - X_1X_2^2X_3^2X_4 + X_1X_2^3X_3^2X_4 + X_1^2X_2^3X_3^2X}{1 - X_2 . 1 - X_1X_2 . 1 - X_2X_3 . 1 - X_1X_2X_3 . 1 - X_2X_3X_4 . 1 - X_1X_2X_3X};$$

and to the three enumerating functions:

$$\frac{1 + x + x^2}{1 - x . 1 - x^2 . 1 - x^3 . 1 - x^4};$$

$$\frac{1 + x + x^2 + x^3 + x^4}{1 - x . 1 - x^2 . 1 - x^3 . 1 - x^4};$$

$$\frac{1 + x^2 + x^3}{1 - x . 1 - x^2 . 1 - x^3 . 1 - x^4}.$$

The last real generating function that has been written down gives the solution of the system of conditions

$$\alpha_2 \geqq \alpha_1, \quad \alpha_2 \geqq \alpha_3, \quad \alpha_3 \geqq \alpha_4;$$

the ground solutions are

$$(\alpha_1, \alpha_2, \alpha_3, \alpha_4) = (0, 1, 0, 0), (1, 1, 0, 0), (0, 1, 1, 0), (1, 1, 1, 0), (0, 1, 1, 1), (1, 1, 1, 1);$$

the three simple syzygies are given by

$$\begin{aligned} X_2.X_1X_2X_3 - X_1X_2.X_2X_3 &= S_1 = 0, \\ X_2.X_1X_2X_3X_4 - X_1X_2.X_2X_3X_4 &= S_2 = 0, \\ X_2X_3.X_1X_2X_3X_4 - X_1X_2X_3.X_2X_3X_4 &= S_3 = 0, \end{aligned}$$

and the two compound syzygies by

$$\begin{aligned} X_2X_3X_4.S_1 - X_2X_3.S_2 &= 0, \\ X_1X_2X_3.S_2 - X_1X_2.S_3 &= 0. \end{aligned}$$

Art. 80. In general, when the number of parts is i, we have k_i orders which are altered by interchange of d and a, combined with inversion, and l_i which are unaltered where

$$2k_i + l_i = 2^{i-1}.$$

Hence the number of essentially different orders is

$$k_i + l_i = 2^{i-2} + \tfrac{1}{2}l_i$$

To determine l_i observe that an order

$$d^{\lambda_1}a^{\mu_1}\, d^{\lambda_2}a^{\mu_2} \ldots d^{\lambda_{s-1}}a^{\mu_{s-1}}\, d^{\lambda_s}a^{\mu_s}$$

will be unaltered by the operations spoken of when

$$\lambda_1 - \mu_s = \mu_1 - \lambda_s = \lambda_2 - \mu_{s-1} = \mu_2 - \lambda_{s-1} = \ldots = 0;$$

so that $i - 1$ must be even and there will be two such unaltered orders for each partition of $i - 1$ into even parts.

Hence the generating function for $k_i + l_i$ is

$$\frac{x^2}{1-2x} + \frac{x}{(1-x^2)(1-x^4)(1-x^6)\ldots \textit{ad. inf.}},$$

giving for

$$i = 2, 3, 4, 5, 6, 7, \ldots$$

the numbers

$$1, 3, 4, 10, 16, 35, \ldots$$

Section 6.

Art. 81. The theory, so far, has been concerned with partitions upon a line. The parts were supposed

$$\underset{\bullet}{\alpha_1}\text{---}\underset{\bullet}{\alpha_2}\text{---}\underset{\bullet}{\alpha_3}\text{---}\underset{\bullet}{\alpha_4}\text{---}\underset{\bullet}{\alpha_5}\ \ldots\ \underset{\bullet}{\alpha_{i-1}}\text{---}\underset{\bullet}{\alpha_i}$$

to be placed at the points upon a line with one of the symbols $\geqq$, $\leqq$ placed between every pair of consecutive points.

When the symbol was invariably $\geqq$ the enumerating function found was

$$\frac{(j+1)}{(1)}\cdot\frac{(j+2)}{(2)}\cdot\frac{(j+3)}{(3)}\cdots\frac{(j+i)}{(i)}$$

wherein (s) denotes $1 - x^s$. If we place these factors at the successive points of the line we obtain a diagrammatic exhibition of the generating function, viz. :—

$$\underset{\bullet}{\frac{(j+1)}{(1)}}\text{------}\underset{\bullet}{\frac{(j+2)}{(2)}}\text{------}\underset{\bullet}{\frac{(j+3)}{(3)}}\text{------}\underset{\bullet}{\frac{(j+4)}{(4)}}\ \ldots\ \underset{\bullet}{\frac{(j+i-1)}{(i-1)}}\text{------}\underset{\bullet}{\frac{(j+i)}{(i)}}$$

a simple fact that the following investigation shows to be fundamental in idea.

Art. 82. I pass on to consider partitions into parts placed at the points of a two-dimensional lattice.

For clearness take the elementary case of four parts placed at the points of a square.

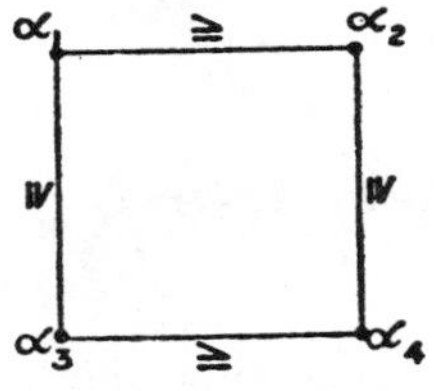

with symbols $\geqq$ placed as shown. We have to solve the conditional relations

$$\alpha_1 \geqq \alpha_2, \qquad \alpha_2 \geqq \alpha_4$$
$$\alpha_1 \geqq \alpha_3, \qquad \alpha_3 \geqq \alpha_4.$$

The four parts are subject to two descending orders. For the sum

$$\Sigma\, X_1^{\alpha_1} X_2^{\alpha_2} X_3^{\alpha_3} X_4^{\alpha_4}$$

we have the Ω function

$$\underset{\geq}{\Omega} \frac{1}{1 - abX_1 \,.\, 1 - \frac{d}{a}X_2 \;.\; 1 - \frac{c}{b}X_3 \,.\, 1 - \frac{1}{cd}X_4}$$

which reduces to

$$\frac{1 - X_1^2 X_2 X_3}{1 - X_1 \,.\, 1 - X_1X_2 \,.\, 1 - X_1X_3 \,.\, 1 - X_1X_2X_3 \,.\, 1 - X_1X_2X_3X_4},$$

establishing the ground solutions

$$(\alpha_1, \alpha_2, \alpha_3, \alpha_4) = (1, 0, 0, 0)\,;\quad (1, 1, 0, 0);\quad (1, 0, 1, 0)\,;\quad (1, 1, 1, 0)\,;\quad (1, 1, 1, 1).$$

connected by the syzygy indicated by

$$X_1 \,.\, X_1X_2X_3 - X_1X_2 \,.\, X_1X_3 = 0,$$

and leading to the enumerating function

$$\frac{1}{(1 - x)(1 - x^2)^2(1 - x^3)}.$$

Art. 83. If the parts be restricted not to exceed j in magnitude, we may take as Ω function

$$\underset{\geq}{\Omega} \frac{1 - (abX_1)^{j+1}}{1 - abX_1 \,.\, 1 - \frac{d}{a}X_2 \;.\; 1 - \frac{c}{b}X_3 \,.\, 1 - \frac{1}{cd}X_4},$$

and herein putting $X_1 = X_2 = X_3 = X_4 = x$, and reducing, we get

$$\frac{1 - x^{j+1}}{1 - x} \,.\, \left(\frac{1 - x^{j+2}}{1 - x^2}\right)^2 \,.\, \frac{1 - x^{j+3}}{1 - x^3},$$

and we notice that we may represent this diagrammatically on the points of the original lattice, viz. :—

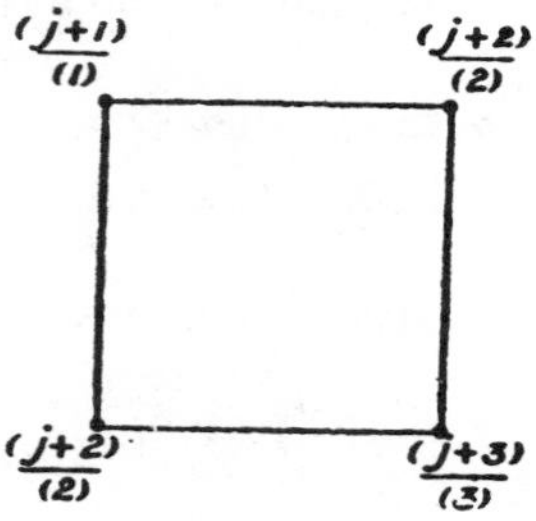

Art. 84. We next have to observe the identity

$$\underset{\geq}{\Omega}\frac{1}{1-abX_1\,.\,1-\frac{d}{a}X_2\,.\,1-\frac{c}{b}X_3\,.\,1-\frac{1}{cd}X_4}=\underset{\geq}{\Omega}\frac{1}{1-aX_1\,.\,1-abX_1X_2\,.\,1-\frac{1}{a}X_3\,.\,1-\frac{1}{ab}X_3X_4},$$

and to note that the dexter leads to the enumerating function

$$\underset{\geq}{\Omega}\frac{1}{1-ax\,.\,1-abx^2\,.\,1-\frac{1}{a}x\,.\,1-\frac{1}{ab}x^2},$$

corresponding to the problem of two superposable layers of units, each of two rows;

$$\begin{matrix}1\,1\,1\,1\,1\,1 & \quad & 1\,1\,1\,1\\ 1\,1\,1\,1 & & 1\,1\end{matrix},$$

in the case indicated superposition yields

$$\begin{matrix}2\,2\,2\,2\,1\,1\\ 2\,2\,1\,1\end{matrix};$$

the first row contains a combined number of two's and units ≥ the combined numbers in the second row, and further, the number of two's in first row, ≥ the number of two's in second row. In the Ω function these conditions are secured by the auxiliaries a, b, respectively, and it is established that the problem of partition at the points of the elementary (*i.e.*, simple square) lattice is identical with that of two superposable unit-graphs, each of at most two rows.

In fact, the graph

$$\begin{array}{|l}2\;2\;2\;2\;1\;1\;.\,.\,.\,.\,.\,.\,.\,.\;x\\ 2\;2\;1\;1\\ \\ \\ \\ \end{array}$$
y

the axis of z being perpendicular to the plane of the paper, is immediately convertible to the lattice form by projection, with summation of units, upon the plane $y\,z$. The numbers at the points of the square lattice would be 6, 4, 4, 2 respectively.

Art. 85. Observe too that the partition is also one upon another kind of lattice in which the part-magnitude is limited not to exceed 2.

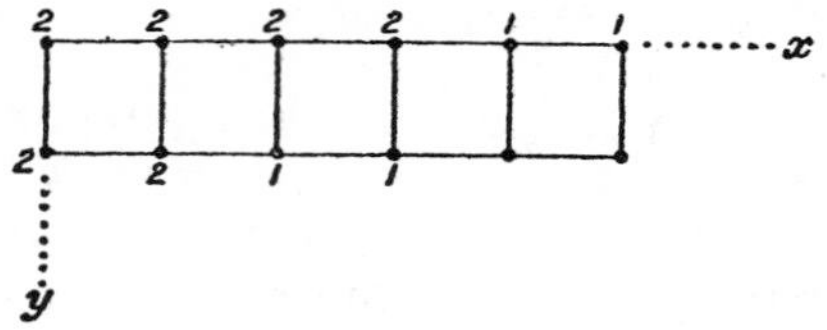

Here, starting from the origin, we may proceed to the opposite point of the lattice along any line of route which proceeds in the positive direction along either axis, and the condition is that along each line of route (here there are six) the numbers must be in descending order and limited in magnitude to 2.

Art. 86. We have, therefore, solved the system of conditions :

$$\begin{array}{ccccccccc} \alpha_1 & \geq & \alpha_2 & \geq & \alpha_3 & \geq & \ldots\ldots & \geq & \alpha_j \\ \text{IV} & & \text{IV} & & \text{IV} & & & & \text{IV} \\ \beta_1 & \geq & \beta_2 & \geq & \beta_3 & \geq & \ldots\ldots & \geq & \beta_j \end{array}$$

$$2 \geq \alpha_1 \geq 0,$$

which is seen to possess the same solution as the system

$$\begin{array}{ccc} \alpha_1 & \geq & \alpha_2 \\ \text{IV} & & \text{IV} \\ \alpha_3 & \geq & \alpha_4 \end{array}$$

$$j \geq \alpha_1 \geq 0;$$

and we remark the diagrammatic representation

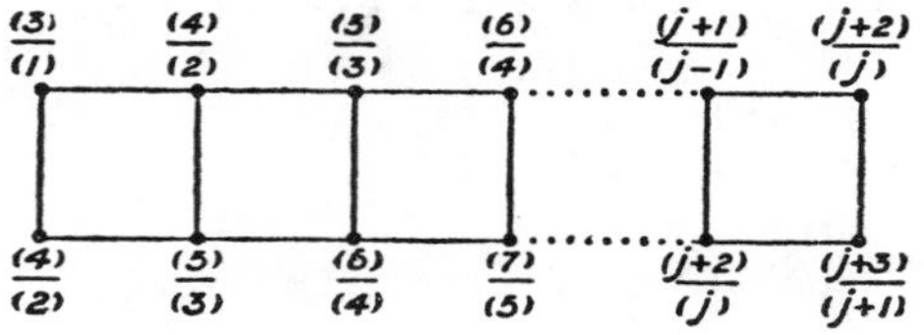

the product of all the factors being

$$\frac{(j+1)(j+2)^2(j+3)}{(1)\quad(2)^2\quad(3)}.$$

3 B 2

Art. 87. I return to the enumerating function

$$\frac{1}{(1-x)(1-x^2)^2(1-x^3)},$$

to note that it may be exhibited as

$$\underset{\geq}{\Omega} \frac{1}{1-ax \, . \, 1-\frac{b}{a}x \, . \, 1-\frac{c}{b} \, . \, 1-\frac{1}{c}x};$$

the interpretation of which is that the coefficient of x^w in the development gives the number of instances in which

$$\alpha_1 + \alpha_2 + \alpha_4 = w_1,$$

$\alpha_1, \alpha_2, \alpha_3, \alpha_4$ being integers satisfying the conditions

$$\alpha_1 \geq \alpha_2 \geq \alpha_3 \geq \alpha_4.$$

We arrive at the form in question if for these conditions we construct

$$\Sigma X_1^{\alpha_1} X_2^{\alpha_2} X_4^{\alpha_4}$$

and then put $X_1 = X_2 = X_4 = x$.
The graphical representation is of the form

1 1 1 1 1 1 1 1 . . .
1 1 1 1 1 . . .
0 0 0 0 . . .
1 1 . . .

the numbers of figures in the rows being in descending order and the third row of figures zeros.

Art. 88. As another instance of the elementary lattice take the system

$$\alpha_1 \geq \alpha_2, \qquad \alpha_1 \geq \alpha_3$$
$$\alpha_4 \geq \alpha_2, \qquad \alpha_4 \geq \alpha_3,$$

leading to

$$\underset{\geq}{\Omega} \frac{1}{1-abX_1 \, . \, 1-\frac{1}{ad}X_2 \, . \, 1-\frac{1}{bc}X_3 \, . \, 1-cdX_4},$$

reducing to

$$\frac{1-X_1^2X_2X_3X_4^2}{1-X_1 \, . \, 1-X_1X_2X_4 \, . \, 1-X_1X_3X_4 \, . \, 1-X_1X_2X_3X_4 \, . \, 1-X_4},$$

establishing the fundamental solutions

$(\alpha_1, \alpha_2, \alpha_3, \alpha_4) = (1, 0, 0, 0)$; $(1, 1, 0, 1)$; $(1, 0, 1, 1)$; $(1, 1, 1, 1,)$; $(0, 0, 0, 1)$;

connected by the syzygy indicated by

$$X_1 . X_1X_2X_3X_4 . X_4 - X_1X_2X_4 . X_1X_3X_4 = 0.$$

Art. 89. A more general generating function connected with the elementary lattice and descending orders is

$$\underset{\geq}{\Omega} \frac{\begin{array}{c} 1 - (abX_1)^{j_1+1} . 1 - \left(\dfrac{d}{a} X_2\right)^{j_2+1} \\ 1 - \left(\dfrac{c}{b} X_3\right)^{j_3+1} . 1 - \left(\dfrac{1}{cd} X_4\right)^{j_4+1} \end{array}}{\begin{array}{c} 1 - abX_1 . 1 - \dfrac{d}{a} X_2 \\ 1 - \dfrac{c}{b} X_3 . 1 - \dfrac{1}{cd} X_4 \end{array}},$$

where now $\alpha_1, \alpha_2, \alpha_3, \alpha_4$ are restricted not to exceed j_1, j_2, j_3, j_4 respectively, and of course

$$\begin{array}{ccc} j_1 & \geq & j_2 \\ \text{IV} & & \text{IV} \\ j_3 & \geq & j_4 \end{array}$$

are conditions.

It should be remarked that we examine the case of bipartite partitions with regular graphs by putting $X_2 = X_1$, $X_4 = X_3$.

Part-magnitude being unlimited, the reduced function is

$$\frac{1 - X_1^3X_2}{1 - X_1 . 1 - X_1^2 . 1 - X_1X_2 . 1 - X_1^2X_2 . 1 - X_1^2X_2^2},$$

and is *real*.

Art. 90. Leaving the particular case, I pass on to consider the general theory of partitions at the points of a lattice in two dimensions. It can be shown immediately that it is coincident with the theory of those partitions of all multipartite numbers which can be represented by regular graphs in three dimensions. For consider the superposition of any number of unit graphs, adding into single numbers the units in the same vertical line. We obtain a scheme of numbers

$$\begin{array}{lllllll} a_{11} & a_{12} & a_{13} & a_{14} & . & . & . \quad x \\ a_{21} & a_{22} & a_{23} & . & . & . & \\ a_{31} & a_{32} & . & . & . & & \\ a_{41} & & & & & & \\ . & & & & & & \\ . & & & & & & \\ . & & & & & & \\ y. & & & & & & \end{array}$$

in which all the rows and all the columns taken in the positive directions along the axes of x and y are in descending order. We may consider these numbers to be placed at the points of a lattice of which the sides involve m and l points along the sides parallel to the axes of x and y respectively; m will then be a limit to the number of units in any row of a unit graph, and l will be the limit to the number of rows.

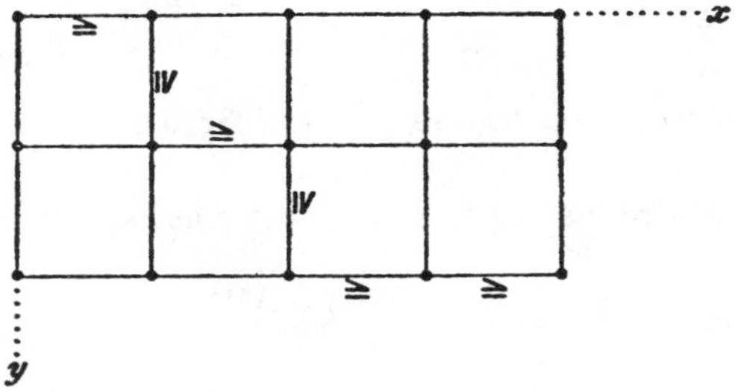

There is a descending order along each line of route from the origin to the opposite corner of the lattice, and there are altogether

$$\binom{l+m-2}{l-1} \text{ such lines of route.}$$

Art. 91. The theory of the regular partitions of multipartite numbers is thus reduced to a lattice partition into $l\,m$ parts *in plano*. The conditional relations may be written

$$\begin{array}{ccccccccc}
\alpha_{11} & \geq & \alpha_{12} & \geq & \alpha_{13} & \cdots & \alpha_{1.m-1} & \geq & \alpha_{1m} \\
\text{IV} & & \text{IV} & & \text{IV} & & \text{IV} & & \text{IV} \\
\alpha_{21} & \geq & \alpha_{22} & \geq & \alpha_{23} & \cdots & \alpha_{2.m-1} & \geq & \alpha_{2.m} \\
\cdot & & \cdot & & \cdot & \cdots & \cdot & & \cdot \\
\alpha_{l-1.1} & \geq & \alpha_{l-1.2} & \geq & \alpha_{l-1.3} & \cdots & \alpha_{l-1.m-1} & \geq & \alpha_{l-1.m} \\
\text{IV} & & \text{IV} & & \text{IV} & & \text{IV} & & \text{IV} \\
\alpha_{l.1} & \geq & \alpha_{l.2} & \geq & \alpha_{l.3} & \cdots & \alpha_{l.m-1} & \geq & \alpha_{l.m}
\end{array}$$

and for the sum

$$\Sigma \prod_{s=1}^{s=l} \prod_{t=1}^{t=m} X_{st}^{\alpha_{st}}$$

we at once write down the Ω generating function, viz. :—

$$\underset{\geqq}{\Omega} \frac{1}{\begin{array}{llll} 1 - a_1\alpha_1 X_{11} & .\,1 - \frac{\alpha_2}{\alpha_1}\beta_1 X_{12} & .\,1 - \frac{\alpha_3}{\alpha_2}\gamma_1 X_{13} & \ldots \text{ to } m \text{ factors} \\ 1 - b_1 \frac{\alpha_2}{\alpha_1} X_{21} & .\,1 - \frac{b_2}{b_1}\frac{\beta_2}{\beta_1} X_{22} & .\,1 - \frac{b_3}{b_2}\frac{\gamma_2}{\gamma_1} X_{23} & \ldots \text{ to } m \text{ factors} \\ 1 - c_1 \frac{\alpha_3}{\alpha_2} X_{31} & .\,1 - \frac{c_2}{c_1}\frac{\beta_3}{\beta_2} X_{32} & .\,1 - \frac{c_3}{c_2}\frac{\gamma_3}{\gamma_2} X_{33} & \ldots \text{ to } m \text{ factors} \\ \vdots & \vdots & \vdots & \&c. \\ \text{to } l \text{ factors} & \text{to } l \text{ factors} & \text{to } l \text{ factors} & \end{array}}$$

If the part-magnitude be limited to n, we must place as numerator in the function

$$\begin{array}{lll} 1 - \left(a_1\alpha_1 X_{11}\right)^{n+1} & .\,1 - \left(\frac{\alpha_2}{\alpha_1}\beta_1 X_{12}\right)^{n+1} & \ldots \text{ to } m \text{ factors} \\ 1 - \left(b_1 \frac{\alpha_2}{\alpha_1} X_{21}\right)^{n+1} & .\,1 - \left(\frac{b_2}{b_1}\frac{\beta_2}{\beta_1} X_{22}\right)^{n+1} & \ldots \text{ to } m \text{ factors} \\ \vdots & \vdots & \&c. \\ \text{to } l \text{ factors} & \text{to } l \text{ factors} & \end{array}$$

and if we please we may reject all the numerator factors except

$$1 - (a_1\alpha_1 X_{11})^{n+1}.$$

Art. 92. The existence of the three-dimensional graph shows that this function remains unaltered, when X_{st} is put equal to x, for every substitution impressed upon the numbers

$$l, m, n,$$

but there is a still more refined theorem of reciprocity connected with a more general generating function.

Suppose that the number of layers which involve 1, 2, 3, &c. rows be restricted to

$$l_1, l_2, l_3, \ldots;$$

that the successive layers are restricted to involve at most

$$m_1, m_2, m_3, \ldots \text{ rows};$$

and that the successive rows of the layers are restricted to contain at most

$$n_1, n_2, n_3, \ldots \text{ units.}$$

We have then the comprehensive Ω function :—

$$\mathop{\Omega}_{\geq}\frac{\begin{array}{llll} 1-(a_1\alpha_1X_{11})^{n_1+1} & .\,1-\left(\frac{a_2}{a_1}\beta_1X_{12}\right)^{n_2+1} & .\,1-\left(\frac{a_3}{a_2}\gamma_1X_{13}\right)^{n_3+1} & \text{to } m_1 \text{ factors} \\ 1-\left(b_1\frac{\alpha_2}{\alpha_1}X_{21}\right)^{n_1+1} & .\,1-\left(\frac{b_2}{b_1}\frac{\beta_2}{\beta_1}X_{22}\right)^{n_2+1} & .\,1-\left(\frac{b_3}{b_2}\frac{\gamma_2}{\gamma_1}X_{23}\right)^{n_3+1} & \text{to } m_2 \text{ factors} \\ 1-\left(c_1\frac{\alpha_3}{\alpha_2}X_{31}\right)^{n_1+1} & .\,1-\left(\frac{c_2}{c_1}\frac{\beta_3}{\beta_2}X_{32}\right)^{n_2+1} & .\,1-\left(\frac{c_3}{c_2}\frac{\gamma_3}{\gamma_2}X_{33}\right)^{n_3+1} & \text{to } m_3 \text{ factors} \\ \vdots & \vdots & \vdots & \&\text{c.} \\ \text{to } l_1 \text{ factors} & \text{to } l_2 \text{ factors} & \text{to } l_3 \text{ factors} & \end{array}}{\begin{array}{llll} 1-a_1\alpha_1X_{11} & .\,1-\frac{a_2}{a_1}\beta_1X_{12} & .\,1-\frac{a_3}{a_2}\gamma_1X_{13} & \ldots \text{ to } m_1 \text{ factors} \\ 1-b_1\frac{\alpha_2}{\alpha_1}X_{21} & .\,1-\frac{b_2}{b_1}\frac{\beta_2}{\beta_1}X_{22} & .\,1-\frac{b_3}{b_2}\frac{\gamma_2}{\gamma_1}X_{23} & \ldots \text{ to } m_2 \text{ factors} \\ 1-c_1\frac{\alpha_3}{\alpha_2}X_{31} & .\,1-\frac{c_2}{c_1}\frac{\beta_3}{\beta_2}X_{32} & .\,1-\frac{c_3}{c_2}\frac{\gamma_3}{\gamma_2}X_{33} & \ldots \text{ to } m_3 \text{ factors} \\ \vdots & \vdots & \vdots & \&\text{c.} \\ \text{to } l_1 \text{ factors} & \text{to } l_2 \text{ factors} & \text{to } l_3 \text{ factors} & \end{array}},$$

wherein, naturally, each of the series

$$l_1,\ l_2,\ l_3,\ \ldots$$
$$m_1,\ m_2,\ m_3,\ \ldots$$
$$n_1,\ n_2,\ n_3,\ \ldots$$

is in descending order, and the theorem of reciprocity involved in the fact of the existence of the graph consists in the circumstance that the function remains unaltered, when X_{st} is put equal to x, for any substitution impressed upon the unsuffixed symbols l, m, n.

In the corresponding lattice the conditions are :—

(i.) The first, second, &c., rows do not contain more than n_1, n_2, &c. numbers respectively ;

(ii.) The first, second, &c., rows do not contain higher numbers than l_1, l_2, &c. . . . ;

(iii.) No number so great as s occurs below row m_s for all values of s; $m_1, m_2, \ldots m_s \ldots$ being of course in descending order of magnitude.

Art. 93. The reduction of this Ω function presents great difficulties, and I propose to restrict consideration to the case

$$l_1 = l_2 = l_3 = \ldots = l$$
$$m_1 = m_2 = m_3 = \ldots = m$$
$$n_1 = n_2 = n_3 = \ldots = n.$$

To adapt the function to enumerate the partitions into at most m parts of l-partite numbers, such partitions being such as possess regular graphs *in solido*, put

$$X_{11} = X_{12} = X_{13} = \ldots = X_{1m} = x_1$$
$$X_{21} = X_{22} = X_{23} = \ldots = X_{2m} = x_2$$
$$\cdot \quad \cdot \quad \cdot \quad \cdot \quad \cdot \quad \cdot \quad \cdot \quad \cdot \quad \cdot \quad \cdot \quad \cdot \quad \cdot \quad \cdot$$
$$X_{l_1} = X_{l_2} = X_{l_3} = \ldots = X_{l_m} = x_l,$$

and the resulting function enumerates by the coefficients of

$$x_1^{p_1} x_2^{p_2} \ldots x_l^{p_l},$$

the number of partitions of the l-partite

$$\overline{(p_1 p_2 \ldots p_l)}$$

into at most m parts.

Art. 94. Further putting

$$x_1 = x_2 = x_3 = \ldots = x_l = x,$$

the coefficients of x gives the number of graphs *in solido or* unipartite partitions upon a two-dimensional lattice, limited, as indicated above, by the numbers l, m, n.

This function appears to be reducible to the product of factors shown in the tableau below :—

$$\frac{(n+1)}{(1)} \cdot \frac{(n+2)}{(2)} \cdot \frac{(n+3)}{(3)} \cdots \frac{(n+m)}{(m)};$$

$$\frac{(n+2)}{(2)} \cdot \frac{(n+3)}{(3)} \cdot \frac{(n+4)}{(4)} \cdots \frac{(n+m+1)}{(m+1)};$$

$$\frac{(n+3)}{(3)} \cdot \frac{(n+4)}{(4)} \cdot \frac{(n+5)}{(5)} \cdots \frac{(n+m+2)}{(m+2)};$$

$$\vdots$$

$$\frac{(n+l)}{(l)} \cdot \frac{(n+l+1)}{(l+1)} \cdot \frac{(n+l+2)}{(l+2)} \cdots \frac{(n+m+l-1)}{(l+m-1)}.$$

This result, verified in a multitude of particular cases, awaits demonstration. For $l = 2$ it has been proved independently by Professor FORSYTH and by the present author. The diagrammatic exhibition of the result at the points of the lattice is clear, and since the product is an invariant for any substitution impressed upon the letters l, m, n, it appears that such exhibition is six-fold. Taking a lattice whose sides contain l m points respectively, so that l m points in all are involved, we mark a corner point, regarding it as an origin of rectangular axes *one*, and proceed to the opposite corner, along any line of route, such that progression along any branch or section of the lattice is in the positive direction, marking the successive points reached *two*, *three*, &c.

For every point, marked s, we have a factor,

$$\frac{(n+s)}{(s)},$$

and express the generating function as a product of $l\,m$ such factors. If n be ∞, each factor is of the form

$$\frac{1}{(s)},$$

and if the number s appears σ times on the lattice, we have a factor $(s)^{-\sigma}$, and the complete result may be written

$$\frac{1}{(s_1)^{\sigma_1}(s_2)^{\sigma_2}(s_3)^{\sigma_3}\dots}.$$

Art. 95. Hence the enumeration is identical with that of the partitions of a unipartite number into an unlimited number of parts of $\sigma_1 + \sigma_2 + \sigma_3 + \dots$ different kinds, viz. :—

σ_1 of the numerical value s_1 but differently coloured.
σ_2 " " s_2 " "
σ_3 " " s_3 " "
. . . .

The number of distinct lines of route in a lattice of $l\,m$ points is

$$\binom{l+m-2}{l-1},$$

so that, in general, on the lattice we have partitions of a number into $l\,m$ parts subject to $\binom{l+m-2}{l-1}$ descending orders.

Such a partition is transformable ($l \geq m$) into one composed of the parts

1	of	1	colour
2	"	2	"
⋮		⋮	⋮
m	"	m	"
⋮		⋮	
l	"	m	"
$l+1$	"	$m-1$	"
⋮		⋮	
$l+m-1$	"	1	"

a theorem of reciprocity analogous to and including the well-known theorem connected with the partitions of a number on a line. There is also a lattice theory connected with unipartite partitions on a line, for the unit-graph of such a partition is nothing more than a number of units and zeros placed at the points of a two-dimensional lattice, such numbers being subject to the $\binom{l+m-2}{l-1}$ descending orders.

Art. 96. The fact is that the theory of the two-dimensional lattice, the part-magnitude being restricted to unity, is co-extensive with the whole theory of partitions upon a line. Hence for such partitions we may represent the generating function, diagrammatically, in two ways upon a lattice as well as in two ways upon a line.

The two representations upon a line are

$$\underset{\bullet\!\!-\!\!-\!\!-}{\frac{(l+1)}{(1)}}\quad \underset{\bullet\!\!-\!\!-\!\!-}{\frac{(l+2)}{(2)}}\quad \underset{\bullet\!\!-\!\!-\!\!-}{\frac{(l+3)}{(3)}}\quad \underset{\bullet}{\frac{(l+4)}{(4)}}\quad \cdots\quad \underset{\bullet\!\!-\!\!-\!\!-}{\frac{(l+m-1)}{(m-1)}}\quad \underset{\bullet}{\frac{(l+m)}{(m)}}.$$

$$\underset{\bullet\!\!-\!\!-\!\!-}{\frac{(m+1)}{(1)}}\quad \underset{\bullet\!\!-\!\!-\!\!-}{\frac{(m+2)}{(2)}}\quad \underset{\bullet\!\!-\!\!-\!\!-}{\frac{(m+3)}{(3)}}\quad \underset{\bullet}{\frac{(m+4)}{(4)}}\quad \cdots\quad \underset{\bullet\!\!-\!\!-\!\!-}{\frac{(l+m-1)}{(l-1)}}\quad \underset{\bullet}{\frac{(l+m)}{(l)}}.$$

Upon a lattice we have

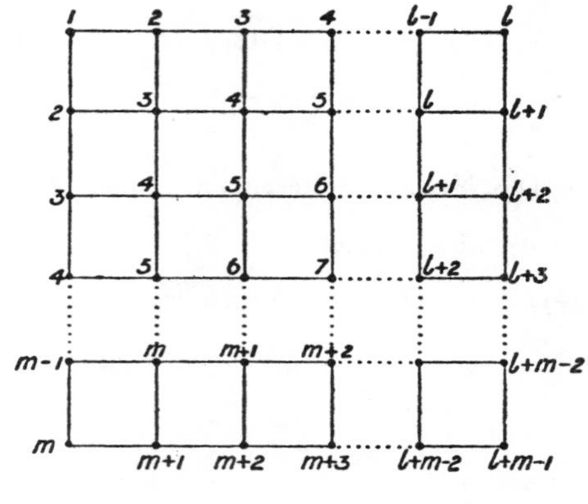

and at the point marked s we place the factor

$$\frac{(s+1)}{(s)}.$$

The second lattice is obtained by interchange of l and m.

The product thus obtained is

$$\prod_{s=1}^{s=l+m-1}\left\{\frac{(s+1)}{(s)}\right\}^{b_s-b_{s-l}-b_{s-m}}$$

3 c 2

b_s denoting the sth figurate number of the second order, and $b_s - b_{s-l} - b_{s-m}$ is easily shown to be equal to the number of points of the lattice marked s. We have to show that this is equal to

$$\prod_{s=1}^{s=m} \frac{(l+s)}{(s)}.$$

Taking $l \geqq m$, observe that $(l+s)$ occurs in the former to the power

$$\begin{aligned} &b_{l+s-1} - b_{s-1} - b_{l+s-1-m} \\ &- b_{l+s} + b_s + b_{l+s-m} \end{aligned}$$

which

$$\begin{aligned} &= 1 \text{ if } l + s > m \\ &= 0 \text{ if } l + s \leqq m; \end{aligned}$$

whilst (s) occurs to the power

$$\begin{aligned} &b_{s-1} - b_{s-1-l} - b_{s-1-m} \\ &- b_s + b_{s-l} + b_{s-m} \end{aligned}$$

which

$$\begin{aligned} &= 1 \quad \text{if} \quad s > m \\ &= 0 \quad \text{if} \quad s > l \text{ and } \leqq m \\ &= -1 \quad \text{if} \quad s \leqq l; \end{aligned}$$

the product is, therefore,

$$\frac{\{(l+1)(l+2)\dots(m)\}\,0\,\{(m+1)(m+2)\dots(l+m)\}}{\{(1)(2)\dots(l)\}\,\{(l+1)(l+2)\dots(m)\}\,0} = \prod_{s=1}^{s=m} \frac{(l+s)}{(s)}.$$

Art. 97. When $l = m = n = \infty$ the generating function is

$$\frac{1}{(1-x)(1-x^2)^2(1-x^3)^3(1-x^4)^4\dots},$$

which may be written

$$\underset{\geqq}{\Omega} \frac{1}{(1-a_1x)\left(1-\frac{a_2}{a_1}x \,.\, 1-\frac{a_3}{a_2}\right)\left(1-\frac{a_4}{a_3}x \,.\, 1-\frac{a_5}{a_4} \,.\, 1-\frac{a_6}{a_5}\right)(\dots)\dots}$$

from which is deduced a graphical representation in two dimensions involving units and zeros.

The graph is regular, and the successive rows involve the numbers

$$1\,;\ 1,\,0\,;\ 1,\,0,\,0\,;\ 1,\,0,\,0,\,0\,;\ \dots$$

respectively. In the general case there is a similar representation, proper restrictions being placed upon the numbers of figures in the rows.

Section 7.

Art. 98. It might have been conjectured that the lattice *in solido* would have afforded results of equal interest, but this on investigation does not appear to be the case. The simplest of such lattices is that in which the points are the summits of a cube and the branches the edges of the cube.

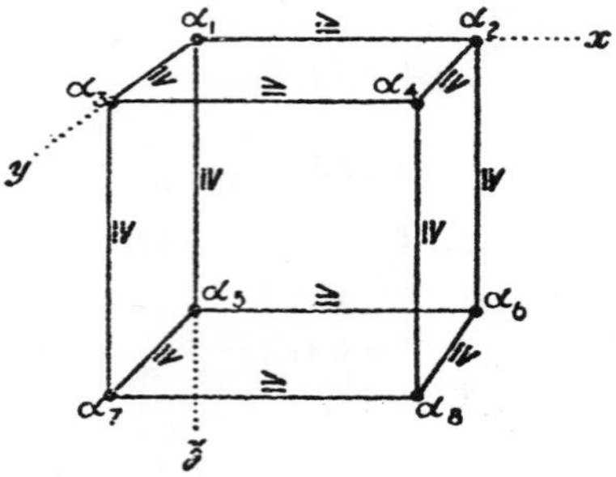

$a_1\ a_2\ a_3\ a_4\ a_5\ a_6\ a_7\ a_8$ is a partition of a number into eight parts, satisfying the conditional relations indicated by the symbols $\geqq$ as shown. The descending order is in the positive direction parallel to each axis. The Ω function

$$\underset{\geqq}{\Omega} \frac{1}{1 - a_1a_2a_3X_1 \, . \, 1 - \frac{a_4a_5}{a_1}X_2 \, . \, 1 - \frac{a_6a_7}{a_2}X_3 \, . \, 1 - \frac{a_8a_9}{a_3}X_4 \atop 1 - \frac{a_{10}}{a_4a_6}X_5 \, . \, 1 - \frac{a_{11}}{a_5a_8}X_6 \, . \, 1 - \frac{a_{12}}{a_7a_9}X_7 \, . \, 1 - \frac{1}{a_{10}a_{11}a_{12}}X_8}$$

is difficult to deal with, and the result which I have obtained too complicated to be worth preserving. I therefore put at once

$$X_1 = X_2 = X_3 = X_4 = X_5 = X_6 = X_7 = X_8 = x,$$

and seek the sum $\Sigma x^{a_1+a_2+a_3+a_4+a_5+a_6+a_7+a_8}$. I divide the calculation into eighteen parts as follows :—

Conditions.	Result.
$a_6 \geqq a_7 \geqq a_4$ $a_5 \geqq a_2, \quad a_5 \geqq a_3$	$\dfrac{1 + x^3 + x^4}{(1)(2)(3)(4)(5)(6)(7)(8)}$
$a_6 \geqq a_7 \geqq a_4$ $a_2 \geqq a_3, \quad a_2 > a_5$	$\dfrac{x^2 + x^3 + x^6}{(1)(2)(3)(4)(5)(6)(7)(8)}$

Conditions.	Result.
$\alpha_6 \geqq \alpha_7 \geqq \alpha_4$ $\alpha_3 > \alpha_2, \quad \alpha_3 > \alpha_5$	$\dfrac{x^2 + x^5}{(1)(2)(3)(4)(5)(6)(7)(8)}$
$\alpha_6 \geqq \alpha_4, \quad \alpha_4 > \alpha_7$ $\alpha_5 \geqq \alpha_2, \quad \alpha_5 \geqq \alpha_3$	$\dfrac{x^6 + x^9 + x^{10}}{(1)(2)(3)(4)(5)(6)(7)(8)}$
$\alpha_6 \geqq \alpha_4, \quad \alpha_4 > \alpha_7$ $\alpha_2 \geqq \alpha_3, \quad \alpha_2 > \alpha_5$	$\dfrac{x^8 + x^9 + x^{12}}{(1)(2)(3)(4)(5)(6)(7)(8)}$
$\alpha_6 \geqq \alpha_4, \quad \alpha_4 > \alpha_7$ $\alpha_3 > \alpha_2, \quad \alpha_3 > \alpha_5$	$\dfrac{x^8 + x^{11}}{(1)(2)(3)(4)(5)(6)(7)(8)}$
$\alpha_4 > \alpha_6, \quad \alpha_6 \geqq \alpha_7$ $\alpha_5 \geqq \alpha_2, \quad \alpha_5 \geqq \alpha_3$	$\dfrac{x^5 + x^8}{(1)(2)(3)(4)(5)(6)(7)(8)}$
$\alpha_4 > \alpha_6, \quad \alpha_6 \geqq \alpha_7$ $\alpha_2 \geqq \alpha_3, \quad \alpha_2 > \alpha_5$	$\dfrac{x^4 + x^7 + x^8}{(1)(2)(3)(4)(5)(6)(7)(8)}$
$\alpha_4 > \alpha_6, \quad \alpha_6 \geqq \alpha_7$ $\alpha_3 > \alpha_2, \quad \alpha_3 > \alpha_5$	$\dfrac{x^6 + x^7 + x^{10}}{(1)(2)(3)(4)(5)(6)(7)(8)}$
$\alpha_4 > \alpha_7, \quad \alpha_7 > \alpha_6$ $\alpha_5 \geqq \alpha_2, \quad \alpha_5 \geqq \alpha_3$	$\dfrac{x^{11} + x^{14}}{(1)(2)(3)(4)(5)(6)(7)(8)}$
$\alpha_4 > \alpha_7, \quad \alpha_7 > \alpha_6$ $\alpha_2 \geqq \alpha_3, \quad \alpha_2 > \alpha_5$	$\dfrac{x^{10} + x^{13} + x^{14}}{(1)(2)(3)(4)(5)(6)(7)(8)}$
$\alpha_4 > \alpha_7, \quad \alpha_7 > \alpha_6$ $\alpha_3 > \alpha_2, \quad \alpha_3 > \alpha_5$	$\dfrac{x^{12} + x^{13} + x^{16}}{(1)(2)(3)(4)(5)(6)(7)(8)}$
$\alpha_7 \geqq \alpha_4 > \alpha_6$ $\alpha_5 \geqq \alpha_2, \quad \alpha_5 \geqq \alpha_3$	$\dfrac{x^6 + x^9 + x^{10}}{(1)(2)(3)(4)(5)(6)(7)(8)}$
$\alpha_7 \geqq \alpha_4 > \alpha_6$ $\alpha_2 \geqq \alpha_3, \quad \alpha_2 > \alpha_5$	$\dfrac{x^8 + x^9}{(1)(2)(3)(4)(5)(6)(7)(8)}$

Conditions.	Result.
$\alpha_7 \geqq \alpha_4 > \alpha_6$ $\alpha_3 > \alpha_2, \quad \alpha_3 > \alpha_5$	$\dfrac{x^8 + x^{11} + x^{12}}{(1)(2)(3)(4)(5)(6)(7)(8)}$
$\alpha_7 > \alpha_6 \geqq \alpha_4$ $\alpha_5 \geqq \alpha_2, \quad \alpha_5 \geqq \alpha_3$	$\dfrac{x^4 + x^5 + x^8}{(1)(2)(3)(4)(5)(6)(7)(8)}$
$\alpha_7 > \alpha_6 \geqq \alpha_4$ $\alpha_2 \geqq \alpha_3, \quad \alpha_2 > \alpha_5$	$\dfrac{x^7 + x^8}{(1)(2)(3)(4)(5)(6)(7)(8)}$
$\alpha_7 > \alpha_6 \geqq \alpha_4$ $\alpha_3 > \alpha_2, \quad \alpha_3 > \alpha_5$	$\dfrac{x^6 + x^7 + x^{10}}{(1)(2)(3)(4)(5)(6)(7)(8)}$

and by addition the resulting generating function* is

$$\frac{1 + 2x^2 + 2x^3 + 3x^4 + 3x^5 + 5x^6 + 4x^7 + 8x^8 + 4x^9 + 5x^{10} + 3x^{11} + 3x^{12} + 2x^{13} + 2x^{14} + x^{16}}{(1)(2)(3)(4)(5)(6)(7)(8)}.$$

Art. 99. By analogy with the lattice *in plano* one might have conjectured that the result would have been

$$\frac{1}{(1)(2)^3(3)^3(4)};$$

but this is not so, although the two functions do coincide as far as the coefficient of x^5 inclusive. In fact, the two expansions yield respectively

$$1 + x + 4x^2 + 7x^3 + 14x^4 + 23x^5 + 41x^6 + 63x^7 + \ldots,$$
$$1 + x + 4x^2 + 7x^3 + 14x^4 + 23x^5 + 42x^6 + 63x^7 + \ldots,$$

the succeeding coefficients becoming widely divergent. This at first seemed surprising, but observe that analogy might also lead us to expect that, if the part-magnitude be limited to i, the result would be

$$\frac{(i+1)(i+2)^3(i+3)^3(i+4)}{(1)(2)^3(3)^3(4)};$$

but this does not happen to be expressible in a finite integral form for all values of i, a fact which necessitates the immediate rejection of the conjecture. The expression in question is only finite and integral when i is of the form $3p$ or $3p + 1$. We have,

* Mr. A. B. KEMPE, Treas. R.S., has verified this conclusion by a different and most ingenious method of summation, which also readily yields the result for any desired restriction on the part-magnitude.

further, the fact that the expression does give the enumeration when $i = 1$, for then the generating function is easily ascertainable to be

$$1 + x + 3x^2 + 3x^3 + 4x^4 + 3x^5 + 3x^6 + x^7 + x^8,$$

which may be exhibited in the forms

$$\frac{(4)^2(5)}{(1)(2)^2} \equiv \frac{(3)(4)^2(5)}{(1)(2)^2(3)} \equiv \frac{(2)(3)^3(4)^3(5)}{(1)(2)^3(3)^3(4)}.$$

Art. 100. The second of these forms immediately arrests the attention, for, *in plano*, it denotes the number of partitions on a lattice of four points (in fact, a square), the part-magnitude being limited not to exceed 2. The reason of this is as follows :—

Taking the cube with any distribution of units at the summits, we may project the summits upon the plane of $y\,z$, adding up the units on the cube edges at right

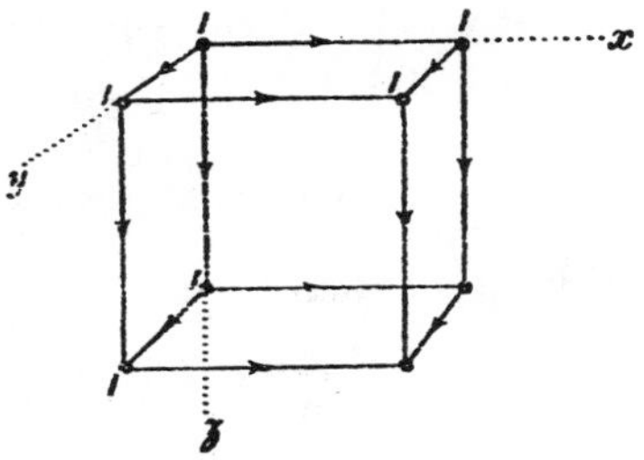

angles to that plane, and thus obtain a distribution, on the points of the cube face in that plane, of numbers limited in magnitude to 2.

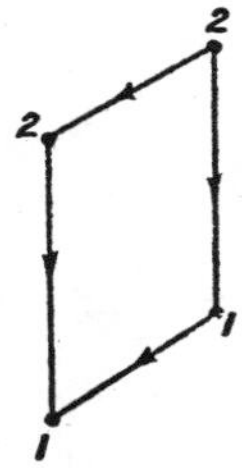

This projection establishes the theorem, which may now be generalized. Conceive a lattice *in solido* having l, m, n points along the axes of x, y, z respectively, and a distribution of units at the points of the lattice which form an unbroken succession along each line of route through the lattice from the origin to the opposite corner, a line of route always proceeding parallel to the axes in a positive sense. Now project and sum units on the plane of $y\,z$.

The result is a partition of the number at the points of a lattice *in plano* whose sides contain m and n points respectively, the part-magnitude being limited not to exceed l. The descending order in this lattice is clearly from the origin to the opposite corner in the plane $y\,z$ along each of its lines of route.

The enumerating generating function is

$$\begin{array}{ccccccc} & \frac{(l+1)}{(1)} \cdot & \frac{(l+2)}{(2)} & \cdot & \frac{(l+3)}{(3)} & \cdots & \frac{(l+m)}{(m)} \\ \times & \frac{(l+2)}{(2)} \cdot & \frac{(l+3)}{(3)} & \cdot & \frac{(l+4)}{(4)} & \cdots & \frac{(l+m+1)}{(m+1)} \\ \times & \frac{(l+3)}{(3)} \cdot & \frac{(l+4)}{(4)} & \cdot & \frac{(l+5)}{(5)} & \cdots & \frac{(l+m+2)}{(m+2)} \\ & \vdots & \vdots & & \vdots & & \vdots \\ \times & \frac{(l+n)}{(n)} \cdot & \frac{(l+n+1)}{(n+1)} & \cdot & \frac{(l+n+2)}{(n+2)} & \cdots & \frac{(l+m+n)}{(m+n)} . \end{array}$$

Each factor may be supposed at a point of the corresponding lattice; if any point is the sth along a line of route the factor is

$$\frac{(l+s)}{(s)} .$$

The number of points at which we place

$$\frac{(l+s)}{(s)}$$

is equal to the coefficient of x^s in the expansion of

$$x(1+x+x^2+\ldots+x^{m-1})(1+x+x^2+\ldots+x^{n-1})$$

that is of

$$\frac{x}{(1+x)^2}(1-x^m)(1-x^n).$$

If m, n be in ascending order and b_s denote the sth figurate number of the second order, this coefficient is

$$b_s - b_{(s-m)} - b_{(s-n)}$$

the term $+\,b_{s-m-n}$ being omitted because s is at most $m+n-1$.

Hence the generating function may be written

$$\prod_{s=1}^{s=m+n-1}\left\{\frac{(l+s)}{(s)}\right\}^{b_s - b_{s-m} - b_{s-n}} .$$

Art. 101. It is now important to show the connexion between this result and the original lattice *in solido*.

I say that this generating function may be exhibited by factors placed at the points of the lattice *in solido*. These factors are of form

$$\frac{(s+1)}{(s)},$$

and such a factor must be placed at every point which is the sth occurring along a line of route in the cubic reticulation.

I take l, m, n in ascending order, and remark that the number of points possessing this property is the coefficient of x^s in the product

$$x(1+x+x^2+\ldots+x^{l-1})(1+x+x^2+\ldots+x^{m-1})(1+x+x^2+\ldots+x^{n-1}),$$

which is

$$\frac{x}{(1-x)^3}(1-x^l)(1-x^m)(1-x^n),$$

and that, if c_s denote the sth of the third order of figurate numbers, this coefficient is

$$c_s - c_{s-l} - c_{s-m} - c_{s-n} + c_{s-l-m} + c_{s-l-n} + c_{s-m-n},$$

the term $-c_{s-l-m-n}$ being omitted, because s is at most $l+m+n-2$.
I propose, therefore, to prove the identity

$$\prod_{s=1}^{s=m+n-1}\left\{\frac{(l+s)}{(s)}\right\}^{b_s-b_{s-m}-b_{s-n}} = \prod_{s=1}^{s=l+m+n-2}\left\{\frac{(s+1)}{(s)}\right\}^{c_s-c_{s-l}-c_{s-m}-c_{s-n}+c_{s-l-m}+c_{s-l-n}+c_{s-m-n}}$$

The factor $(l+s)$ occurs to the power

$$-b_{l+s}+b_s-b_{s-m}-b_{s-n}+b_{l+s-m}+b_{l+s-n}$$

on the sinister side, and to the power

$$\begin{aligned}&-(c_{l+s}-c_{l+s-1})+(c_s-c_{s-1})-(c_{s-m}-c_{s-m-1})\\&-(c_{s-n}-c_{s-n-1})+(c_{l+s-m}-c_{l+s-m-1})+(c_{l+s-n}-c_{l+s-n-1})\end{aligned}$$

on the dexter. But

$$c_k - c_{k-1} = b_k = k.$$

Hence, under all circumstances, the two powers must be equal.

Again the factor (s) occurs to the power,

$$-b_s+b_{s-l}+b_{s-m}+b_{s-n}-b_{s-l-m}-b_{s-l-n}$$

on the sinister side, and to the power

$$-(c_s - c_{s-1}) + (c_{s-l} + c_{s-l-1}) + (c_{s-m} - c_{s-m-1})$$
$$+ (c_{s-n} - c_{s-n-1}) - (c_{s-l-m} - c_{s-l-m-1}) - (c_{s-l-n} - c_{s-l-n-1})$$

on the dexter, and again the two powers are equal.

Hence the identity under consideration is established, and this carries with it the proof of the diagrammatic representation of the generating function on the points of the solid reticulation.

Art. 102. I resume the general theory of the partitions on the summits of a cube. When the parts are unrestricted in magnitude the generating function has been found. A process similar to that employed leads to the theorem that when the parts are restricted not to exceed t in magnitude the generating function is the quotient of

$$\begin{aligned} 1 &+ a(2x^2 + 2x^3 + 3x^4 + 2x^5 + 2x^6) \\ &+ a^2(x^5 + 3x^6 + 4x^7 + 8x^8 + 4x^9 + 3x^{10} + x^{11}) \\ &+ a^3(2x^{10} + 2x^{11} + 3x^{12} + 2x^{13} + 2x^{14}) \\ &+ a^4 . x^{16} \end{aligned}$$

by

$$(1-a)(1-ax)(1-ax^2)(1-ax^3)(1-ax^4)(1-ax^5)(1-ax^6)(1-ax^7)(1-ax^8),$$

the required number being given by the coefficient of $a^t x^w$. Denoting the numerator by $1 + aP(x) + a^2Q(x) + a^3R(x) + a^4 . x^{16}$, the whole coefficient of a^t is

$$\frac{(9)(10)\dots(t+8)}{(1)(2)\dots(t)} + P(x)\frac{(9)(10)\dots(t+7)}{(1)(2)\dots(t-1)} + Q(x)\frac{(9)(10)\dots(t+6)}{(1)(2)\dots(t-2)}$$
$$+ R(x)\frac{(9)(10)\dots(t+5)}{(1)(2)\dots(t-3)} + x^{16}.\frac{(9)(10)\dots(t+4)}{(1)(2)\dots(t-4)}.$$

Denoting this generating function by $F_t(x)$, I find

$$P(x) = F_1(x) - \frac{(9)}{(1)},$$

$$Q(x) = F_2(x) - \frac{(9)}{(1)}F_1(x) + x\frac{(8)(9)}{(1)(2)},$$

$$R(x) = F_3(x) - \frac{(9)}{(1)}F_2(x) + x\frac{(8)(9)}{(1)(2)}F_1(x) - x^3\frac{(7)(8)(9)}{(1)(2)(3)},$$

$$x_{16} = F_4(x) - \frac{(9)}{(1)}F_3(x) + x\frac{(8)(9)}{(1)(2)}F_2(x) - x^3\frac{(7)(8)(9)}{(1)(2)(3)}F_1(x) + x^6\frac{(6)(7)(8)(9)}{(1)(2)(3)(4)},$$

whence

$$F_5(x) = \frac{(9)}{(1)}F_4(x) - x\frac{(8)(9)}{(1)(2)}F_3(x) + x^3\frac{(7)(8)(9)}{(1)(2)(3)}F_2(x)$$
$$- x^6\frac{(6)(7)(8)(9)}{(1)(2)(3)(4)}F_1(x) + x^{10}.\frac{(5)(6)(7)(8)(9)}{(1)(2)(3)(4)(5)},$$

and in general

$$F_t(x) = \frac{(9)(10)\ldots(t+4)}{(1)(2)\ldots(t-4)} F_4(x) - x\frac{(8)(9)\ldots(t+4)}{(1)(2)\ldots(t-3)}\cdot\frac{(t-4)}{(1)} F_3(x)$$

$$+ x^3\frac{(7)(8)\ldots(t+4)}{(1)(2)\ldots(t-2)}\cdot\frac{(t-4)(t-3)}{(1)(2)}\cdot F_2(x)$$

$$- x^6\frac{(6)(7)\ldots(t+4)}{(1)(2)\ldots(t-1)}\cdot\frac{(t-4)(t-3)(t-2)}{(1)(2)(3)} F_1(x)$$

$$+ x^{10}\frac{(5)(6)\ldots(t+4)}{(1)(2)\ldots(t)}\cdot\frac{(t-4)(t-3)(t-2)(t-1)}{(1)(2)(3)(4)}.$$

Art. 103. This appears to be the most symmetrical form in which the generating function can be exhibited, and it may be assumed that the like function for the solid reticulation in general will be of complicated nature. The argument that has been given shows that the theory of the n-dimensional lattice (easily realizable *in plano*), the part-magnitude being limited so as not to exceed unity, is co-extensive with the whole theory of partitions on the lattice of $n - 1$ dimensions.

Section 8.

Art. 104. The enumerating generating functions that are met with at the outset in the theory of the partitions of numbers are such as are formed by factors of the forms

$$\frac{1 - x^{n+s}}{1 - x^s},$$

written for brevity $\frac{(n+s)}{(s)}$. All those which appear in connection with regular graphs in two and three dimensions are so expressible, and the mere fact of such expression proves beyond question that the numerator of the generating function is exactly divisible by the denominator; in other words, it proves that the function can be put into a finite integral form. It is quite natural therefore to seek the general expression of functions of this form, which possesses this property of competency to generate a finite number of terms. Moreover, it is conceivable that such a determination will indicate the paths of future research in these matters: will be in fact a sign-post at the cross-ways. This is the reason why I undertook the investigation; but, as frequently happens in similar cases, the problem proves *à posteriori* to be *per se* of great interest and to involve in itself a notable theorem in partitions.

Art. 105. I consider the function

$$\frac{(n+1)^{a_1}(n+2)^{a_2}(n+3)^{a_3}\ldots(n+s)^{a_s}}{(1)^{a_1}\quad(2)^{a_2}\quad(3)^{a_3}\quad\ldots\quad(s)^{a_s}},$$

which I also write

$$X_1^{a_1}X_2^{a_2}X_3^{a_3}\ldots X_s^{a_s},$$

and investigate the sum

$$\Sigma X_1^{\alpha_1} X_2^{\alpha_2} X_3^{\alpha_3} \ldots X_s^{\alpha_s}$$

for all values of $\alpha_1, \alpha_2, \alpha_3, \ldots \alpha_s$, which render the expression under the sign of summation expressible in a finite integral form *for all values of the integer n.*

Art. 106. Let ξ_t be that factor of $1 - x^t$ which, when equated to zero, yields all the primitive roots of the equation

$$1 - x^t = 0.$$

Then $1 - x^t = \xi_1 \xi_{d_1} \xi_{d_2} \ldots \xi_t$ where $1, d_1, d_2, \ldots t$ are all the divisors of t. We must find the circumstances under which every expression ξ_t will occur at least as often in the numerator as in the denominator. We need not attend to ξ_1, since it occurs with equal frequency in numerator and denominator. In regard to ξ_2, we have equal frequency if $n + 1$ be uneven, but if $n + 1$ be even we must have

$$\alpha_1 + \alpha_3 + \alpha_5 + \ldots \geq \alpha_2 + \alpha_4 + \alpha_6 + \ldots$$

For ξ_3 if $n + 1 \equiv 0 \bmod 3$,

$$\alpha_1 + \alpha_4 + \alpha_7 + \ldots \geq \alpha_3 + \alpha_6 + \alpha_9 + \ldots,$$

and if $n + 1 \equiv 1 \bmod 3$,

$$\alpha_2 + \alpha_5 + \alpha_8 + \ldots \geq \alpha_3 + \alpha_6 + \alpha_9 + \ldots,$$

while the case of $n + 1 \equiv 2 \bmod 3$ need not be attended to.

Proceeding in this manner we find the following conditions :—

$$\alpha_1 + \alpha_3 + \alpha_5 + \ldots \geq \alpha_2 + \alpha_4 + \alpha_6 + \ldots$$

$$\begin{cases} \alpha_1 + \alpha_4 + \alpha_7 + \ldots \geq \alpha_3 + \alpha_6 + \alpha_9 + \ldots \\ \alpha_2 + \alpha_5 + \alpha_8 + \ldots \geq \alpha_3 + \alpha_6 + \alpha_9 + \ldots \end{cases}$$

$$\begin{cases} \alpha_1 + \alpha_5 + \alpha_9 + \ldots \geq \alpha_4 + \alpha_8 + \alpha_{12} + \ldots \\ \alpha_2 + \alpha_6 + \alpha_{10} + \ldots \geq \alpha_4 + \alpha_8 + \alpha_{12} + \ldots \\ \alpha_3 + \alpha_7 + \alpha_{11} + \ldots \geq \alpha_4 + \alpha_8 + \alpha_{12} + \ldots \end{cases}$$

$$\cdots\cdots\cdots\cdots\cdots\cdots$$

$$\cdots\cdots\cdots\cdots\cdots\cdots$$

$$\begin{cases} \alpha_1 + \alpha_s + \ldots\ldots \geq \alpha_{s-1} \\ \alpha_2 \qquad \geq \alpha_{s-1} \\ \vdots \qquad \vdots \\ \alpha_{s-2} \qquad \geq \alpha_{s-1} \end{cases}$$

$$\begin{cases} \alpha_1 \geq \alpha_s \\ \alpha_2 \geq \alpha_s \\ \vdots \qquad \vdots \\ \alpha_{s-1} \geq \alpha_s \end{cases}$$

$\frac{1}{2} s(s - 1)$ in number.

The next step is to construct an Ω function which shall express these conditions and lead practically to the desired summation.

Art. 107. First take $s = 2$; there is but one condition

$$\alpha_1 \geq \alpha_2,$$

and the function is

$$\underset{\geq}{\Omega} \frac{1}{1 - a_1 X_1 . 1 - \frac{1}{a_1} X_2} = \frac{1}{1 - X_1 . 1 - X_1 X_2},$$

and every term in the ascending expansion of this function is of the required form, and no other forms exist. The general term being

$$X_1^{\alpha_1 - \alpha_2} (X_1 X_2)^{\alpha_2} \qquad \alpha_1 \geq \alpha_2,$$

we may call X_1 and $X_1 X_2$ the ground forms from which all other forms are derived.

Art. 108. Next take $s = 3$. The conditions are

$$\left.\begin{array}{rl} \alpha_1 + \alpha_3 & \geq \alpha_2 \\ \alpha_1 & \geq \alpha_3 \\ \alpha_2 & \geq \alpha_3 \end{array}\right\},$$

leading to the summation formula

$$\underset{\geq}{\Omega} \frac{1}{1 - a_1 a_2 X_1 . 1 - \frac{a_3}{a_1} X_2 . 1 - \frac{a_1}{a_2 a_3} X},$$

the auxiliaries a_1, a_2, a_3 determining the first, second and third conditions respectively. The function is equal to

$$\underset{\geq}{\Omega} \frac{1}{1 - a_1 a_2 X_1 . 1 - \frac{1}{a_1} X_2 . 1 - \frac{1}{a_2} X_2 X_3}$$

$$= \underset{\geq}{\Omega} \frac{1}{1 - a_1 X_1 . 1 - \frac{1}{a_1} X_2 . 1 - a_1 X_1 X_2 X_3}$$

$$= \underset{\geq}{\Omega} \frac{1}{1 - a_1 X_1} \left\{ \frac{1}{1 - a_1 X_1 X_2 X_3 . 1 - X_1 X_2^2 X_3} + \frac{\frac{X_2}{a_1}}{1 - \frac{X_2}{a_1} . 1 - X_1 X_2^2 X_3} \right\}$$

$$= \frac{1}{1 - X_1 . 1 - X_1 X_2 X_3 . 1 - X_1 X_2^2 X_3} + \frac{X_1 X_2}{1 - X_1 . 1 - X_1 X_2 . 1 - X_1 X_2^2 X_3}$$

$$= \frac{1 - X_1^2 X_2^2 X_3}{1 - X_1 . 1 - X_1 X_2 . 1 - X_1 X_2 X_3 . 1 - X_1 X_2^2 X_3},$$

representing the complete solution.

The denominator factors yield the ground forms

$$X_1X_2X_3, \qquad X_1X_2^2X_3$$

in addition to those previously met with, whilst the numerator factor indicates the ground form syzygy

$$X_1 . X_1X_2^2X_3 - X_1X_2 . X_1X_2X_3 = 0.$$

Observe that

$$X_1X_2X_3 = \frac{1 - x^{n+1} . 1 - x^{n+2} . 1 - x^{n+3}}{1 - x . 1 - x^2 . 1 - x^3}$$

$$X_1X_2^2X_3 = \frac{(1 - x^{n+1})(1 - x^{n+2})^2(1 - x^{n+3})}{(1 - x)(1 - x^2)^2(1 - x^3)}$$

are those with which we are familiar in the theories of simple and compound partition respectively.

Art. 109. I pass on to the case $s = 4$; the conditions are

$$\begin{aligned} \alpha_1 + \alpha_3 &\geq \alpha_2 + \alpha_4 \\ \alpha_1 + \alpha_4 &\geq \alpha_3 \\ \alpha_2 &\geq \alpha_3 \\ \alpha_1 &\geq \alpha_4 \\ \alpha_2 &\geq \alpha_4 \\ \alpha_3 &\geq \alpha_4 \end{aligned}$$

We neglect the fifth of these as being implied by the remainder and from the function

$$\underset{\geq}{\Omega} \frac{1}{1 - a_1a_2a_4X_1 . 1 - \frac{a_3}{a_1}X_2 . 1 - \frac{a_1a_5}{a_2a_3}X_3 . 1 - \frac{a_2}{a_1a_4a_5}X_4}$$

which, when reduced, is

$$\frac{1}{1 - X_1 . 1 - X_1X_2 . 1 - X_1X_2X_3X_4 . 1 - X_1X_2^2X_3^2X_4}$$
$$+ \frac{X_1X_2^2X_3}{1 - X_1 . 1 - X_1X_2 . 1 - X_1X_2^2X_3 . 1 - X_1X_2^2X_3^2X_4}$$
$$+ \frac{X_1X_2X_3}{1 - X_1 . 1 - X_1X_2X_3 . 1 - X_1X_2^2X_3 . 1 - X_1X_2^2X_3^2X_4}$$

showing that the new ground forms are $X_1X_2X_3X_4$ and $X_1X_2^2X_3^2X_4$, both of which have presented themselves before.

The result may be written

$$\frac{1 - X_1^2X_2^2X_3 - X_1^2X_2^2X_3^2X_4 - X_1^2X_2^3X_3^2X_4^1 + X_1^3X_2^3X_3^2X_4 + X_1^3X_2^4X_3^3X_4}{1 - X_1 . 1 - X_1X_2 . 1 - X_1X_2X_3 . 1 - X_1X_2^2X_3 . 1 - X_1X_2X_3X_4 . 1 - X_1X_2^2X_3^2X_4},$$

and the numerator now indicates the existence of first and second syzygies between the ground forms.

We have the first syzygies

$$(A) = X_1X_2 \, . \, X_1X_2X_3 - X_1 \, . \, X_1X_2^2X_3 = 0,$$
$$(B_1) = X_1X_2X_3 \, . \, X_1X_2X_3X_4 - X_1 \, . \, X_1X_2^2X_3^2X_4 = 0,$$
$$(B_2) = X_1X_2^2X_3 \, . \, X_1X_2X_3X_4 - X_1X_2 \, . \, X_1X_2^2X_3^2X_4 = 0,$$

and the second syzygies

$$X_1 (B_2) - X_1X_2 (B_1) = 0,$$
$$X_1X_2X_3 (B_2) - X_1X_2^2X_3 (B_1) = 0.$$

Art. 110. For $s = 5$, the generating function is

$$\underset{\geqq}{\Omega} \frac{1}{1 - a_1a_2a_3a_4X_1 \, . \, 1 - \frac{b_2b_3b_4}{a_1}X_2 \, . \, 1 - \frac{a_1c_3c_4}{a_2b_2}X_3 \, . \, 1 - \frac{a_2d_1}{a_1a_3b_3c} X_4 \, . \, 1 - \frac{a_1b_2a_3}{a_4b_4c_4d_4}X_5}$$

and there is no difficulty in continuing the series. The obtaining, however, of the reduced forms soon becomes laborious.

Art. 111. There is another method of investigation. Guided by the results obtained let us restrict consideration to the forms

$$X_1^{\alpha_1}X_2^{\alpha_2} \dots X_s^{\alpha_s}$$

which are such that

$$\alpha_m = \alpha_{s+1-m} \, .^*$$

This is of great importance, because we are thus able, for any given order, to generate the functions of that order alone.

Put $X_mX_{s+1-m} = Y_m$ and seek $\Sigma Y_1^{\alpha_1}Y_2^{\alpha_2} \dots$

Art. 112. For $s = 2$, the generating function is simply

$$\frac{1}{1 - Y_1} = \frac{1}{1 - X_1X_2}.$$

Art. 113. For $s = 3$, the conditions

$$2\alpha_1 \geqq \alpha \geqq \alpha_1$$

lead to

$$\underset{\geqq}{\Omega} \frac{1}{1 - \frac{a^2}{b} Y_1 \, . \, 1 - \frac{b}{a} Y_2},$$

the letters a; b determining the first and second conditions respectively

* The validity of this assumption will be considered later.

This is on reduction

$$\frac{1}{1 - Y_1Y_2 \,.\, 1 - Y_1Y_2^2} = \frac{1}{1 - X_1X_2X_3 \,.\, 1 - X_1X_2^2X_3}$$

a real generating function.

Art. 114. For $s = 4$, the conditions are the same, viz.:—

$$2\alpha_1 \geq \alpha_2 \geq \alpha_1$$

and the Ω function, where now

$$Y_1 = X_1X_4, \quad Y_2 = X_2X_3,$$

is

$$\underset{\geq}{\Omega} \frac{1}{1 - \frac{a^2}{b} Y_1 \,.\, 1 - \frac{b}{a} Y_2} = \frac{1}{1 - Y_1Y_2 \,.\, 1 - Y_1Y_2^2}$$

$$= \frac{1}{1 - X_1X_2X_3X_4 \,.\, 1 - X_1X_2^2X_3^2X_4}$$

yielding the ground forms already found by the first method.

Art. 115. For $s = 5$, the conditions are

$$\alpha_1 + \alpha_2 \geq \alpha_3 \geq \alpha_2,$$
$$2\alpha_1 \geq \alpha_2 \geq \alpha_1;$$

leading to

$$\underset{\geq}{\Omega} \frac{1}{1 - \frac{ab^2}{d} Y_1 \,.\, 1 - \frac{ad}{bc} Y_2 \,.\, 1 - \frac{c}{a} Y},$$

where

$$Y_1 = X_1X_5, \quad Y_2 = X_2X_4, \quad Y_3 = X_3,$$

and this is

$$\underset{\geq}{\Omega} \frac{1}{1 - \frac{ab^2}{d} Y_1 \,.\, 1 - \frac{d}{b} Y_2Y_3 \,.\, 1 - \frac{1}{a} Y}$$

$$= \underset{\geq}{\Omega} \frac{1}{1 - abY_1Y_2Y_3 \,.\, 1 - \frac{1}{b} Y_2Y_3 \,.\, 1 - \frac{1}{a} Y_3}$$

$$= \underset{\geq}{\Omega} \frac{1}{1 - bY_1Y_2Y_3 \,.\, 1 - \frac{1}{b} Y_2Y_3 \,.\, 1 - bY_1Y_2Y_3^2}$$

$$= \frac{1 - Y_1^2Y_2^3Y_3^4}{1 - Y_1Y_2Y_3 \,.\, 1 - Y_1Y_2Y_3^2 \,.\, 1 - Y_1Y_2^2Y_3^2 \,.\, 1 - Y_1Y_2^2Y_3^3}$$

$$= \frac{1 - X_1X_2X_3X_4X_5}{1 - X_1X_2X_3X_4X_5 \,.\, 1 - X_1X_2X_3^2X_4X_5 \,.\, 1 - X_1X_2^2X_3^2X_4^2X_5 \,.\, 1 - X_1X_2^2X_3^3X_4^2X_5},$$

establishing the ground forms

$$X_1X_2X_3X_4X_5, \quad X_1X_2X_3^2X_4X_5$$
$$X_1X_2^2X_3^2X_4^2X_5, \quad X_1X_2^2X_3^3X_4^2X_5$$

connected by the simple syzygy

$$(X_1X_2X_3X_4X_5)(X_1X_2^2X_3^3X_4^2X_5) - (X_1X_2X_3^2X_4X_5)\, X_1X_2^2X_3^2X_4^2X_5) = 0.$$

Art. 116. I stop to remark that one of these ground forms, viz.:—

$$X_1X_2X_3^2X_4X_5$$

is new, not having so far presented itself in a partition theorem. It is one of an infinite system which merits, and will receive, separate consideration later on. The one before us is associated with partitions at the points of the dislocated lattice.

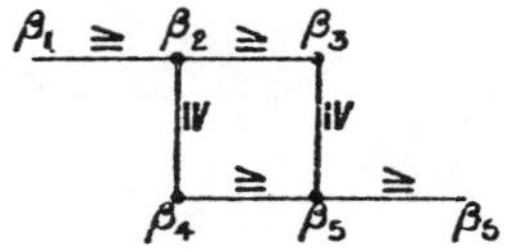

Art. 117 For $s = 6$, the conditions are:

$$2\alpha_2 \geq \alpha_1 + \alpha_3$$
$$2\alpha_1 \geq \alpha_2$$
$$\alpha_3 \geq \alpha_2,$$

leading to

$$\underset{\geq}{\Omega} \frac{1}{1 - \frac{b^2}{a} Y_1 \cdot 1 - \frac{a^2}{bc} Y_2 \cdot 1 - \frac{c}{a} Y_3},$$

where

$$Y_1 = X_1X_6, \quad Y_2 = X_2X_5, \quad Y_3 = X_3X_4.$$

This is

$$\underset{\geq}{\Omega} \frac{1}{1 - \frac{b^2}{a} Y_1 \cdot 1 - \frac{a}{b} Y_2Y_3 \cdot 1 - \frac{1}{a} Y}$$

$$= \underset{\geq}{\Omega} \frac{1}{1 - \frac{1}{a} Y_3 \cdot 1 - \frac{a}{b} Y_2Y_3 \cdot 1 - bY_1Y_2Y_3}$$

$$= \frac{1}{1 - Y_1Y_2Y_3 \cdot 1 - Y_1Y_2^2Y_3^2 \cdot 1 - Y_1Y_2^2Y_3^3},$$

establishing the ground forms:

$$X_1X_2X_3X_4X_5X_6$$
$$X_1X_2^2X_3^2X_4^2X_5^2X_6$$
$$X_1X_2^2X_3^3X_4^3X_5^2X_6,$$

unconnected by any syzygy.

Art. 118. For $s = 7$, the independent conditions are:

$$2\alpha_1 + 2\alpha_3 \geqq 2\alpha_2 + \alpha_4$$
$$2\alpha_2 \geqq \alpha_4$$
$$\alpha_1 + \alpha_2 \geqq \alpha_3$$
$$\alpha_4 \geqq \alpha_3$$
$$2\alpha_1 \geqq \alpha_2$$
$$\alpha_3 \geqq \alpha_2$$
$$\alpha_2 \geqq \alpha_1,$$

and these lead to

$$\underset{\geqq}{\Omega} \frac{}{1 - \frac{a^2ce^2}{g} Y_1 . 1 - \frac{b^2cg}{a^2ef} Y_2 . 1 - \frac{a^2f}{cd} Y_3 . 1 - \frac{d}{ab} Y_4}$$

and eliminating d, f, g in succession

$$= \underset{\geqq}{\Omega} \frac{1}{1 - \frac{a^2ce^2}{g} Y_1 . 1 - \frac{b^2cg}{a^2ef} Y_2 . 1 - \frac{af}{bc} Y_3Y_4 . 1 - \frac{1}{ab} Y_4}$$

$$= \underset{\geqq}{\Omega} \frac{1}{1 - \frac{a^2ce^2}{g} Y_1 . 1 - \frac{bg}{ae} Y_2Y_3Y_4 . 1 - \frac{a}{bc} Y_3Y_4 . 1 - \frac{1}{ab} Y_4},$$

$$= \underset{\geqq}{\Omega} \frac{1}{1 - abceY_1Y_2Y_3Y_4 . 1 - \frac{b}{ae} Y_2Y_3Y_4 . 1 - \frac{a}{bc} Y_3Y_4 . 1 - \frac{1}{ab} Y_4}.$$

and eliminating e

$$= \underset{\geqq}{\Omega} \frac{1}{1 - abcY_1Y_2Y_3Y_4 . 1 - b^2cY_1Y_2^2Y_3^2Y_4^2 . 1 - \frac{a}{bc} Y_3Y_4 . 1 - \frac{1}{ab} Y}$$

and eliminating c

$$= \underset{\geqq}{\Omega} \frac{1 - a^2b^2Y_1^2Y_2^3Y_3^4Y_4^4}{\begin{array}{c}1 - abY_1Y_2Y_3Y_4 . 1 - a^2Y_1Y_2Y_3^2Y_4^2 . 1 - b^2Y_1Y_2^2Y_3^2Y_4^2, \\ 1 - ab . Y_1Y_2^2Y_3^3Y_4^3 . 1 - \frac{1}{ab} Y_4 .\end{array}}$$

And on further reduction it is finally

$$\frac{1 - Y_1^2Y_2^3Y_3^4Y_4^4 . 1 - Y_1^2Y_2^3Y_3^4Y^5}{\begin{array}{c}1 - Y_1Y_2Y_3Y_4 . 1 - Y_1Y_2Y_3Y_4^2 . 1 - Y_1Y_2Y_3^2Y_4^2, \\ 1 - Y_1Y_2^2Y_3^2Y_4^2 . 1 - Y_1Y_2^2Y_3^3Y_4^3 . 1 - Y_1Y_2^2Y_3^3Y_4^4\end{array}}$$

establishing the ground forms

3 E 2

$$Y_1Y_2Y_3Y_4 \equiv X_1X_2X_3X_4X_5X_6X_7$$
$$Y_1Y_2Y_3Y_4^2 \equiv X_1X_2X_3X_4^2X_5X_6X_7$$
$$Y_1Y_2Y_3^2Y_4^2 \equiv X_1X_2X_3^2X_4^2X_5^2X_6X_7$$
$$Y_1Y_2^2Y_3^2Y_4^2 \equiv X_1X_2^2X_3^2X_4^2X_5^2X_6^2X_7$$
$$Y_1Y_2^2Y_3^3Y_4^3 \equiv X_1X_2^2X_3^3X_4^3X_5^3X_6^2X_7$$
$$Y_1Y_2^2Y_3^3Y_4^4 \equiv X_1X_2^2X_3^3X_4^4X_5^3X_6^2X_7$$

connected by the simple syzygies

$$(Y_1Y_2Y_3Y_4)\,(Y_1Y_2^2Y_3^3Y_4^3) - (Y_1Y_2Y_3^2Y_4^2)\,(Y_1Y_2^2Y_3^2Y_4^2) = 0,$$
$$(Y_1Y_2Y_3Y_4)\,(Y_1Y_2^2Y_3^3Y_4^4) - (Y_1Y_2Y_3Y_4^2)\,(Y_1Y_2^2Y_3^3Y_4^3) = 0,$$

and, denoting these respectively by A and B, the numerator term $+\ Y_1^4Y_2^6Y_3^8Y_4^9$ indicates the second, or compound, syzygy :—

$$(Y_1Y_2Y_3Y_4)\,(Y_1Y_2^2Y_3^3Y_4^4)\,(A) - (Y_1Y_2Y_3^2Y_4^2)\,(Y_1Y_2^2Y_3^2Y_4^2)\,(B) = 0.$$

Art. 119. I remark that the forms

$$Y_1Y_2Y_3Y_4^2, \qquad Y_1Y_2Y_3^2Y_4^2$$

are new to partition theory.

Art. 120. For $s = 8$, the reduced conditions are

$$\alpha_2 + \alpha_3 \geq \alpha_1 + \alpha_4$$
$$\alpha_1 + \alpha_2 \geq \alpha_3$$
$$\alpha_4 \geq \alpha_3$$
$$2\alpha_1 \geq \alpha_2$$
$$\alpha_3 \geq \alpha_2$$

leading to

$$\underset{\geq}{\Omega}\ \frac{1}{1-\frac{bd^2}{a}Y_1\,.\,1-\frac{ab}{dc}Y_2\,.\,1-\frac{ac}{bc}Y_3\,.\,1-\frac{c}{a}Y_4}$$

$$=\underset{\geq}{\Omega}\ \frac{1}{1-\frac{bd^2}{a}Y_1\,.\,1-\frac{a}{d}Y_2Y_3Y_4\,.\,1-\frac{1}{b}Y_3Y_4\,.\,1-\frac{1}{a}Y_4}$$

$$=\underset{\geq}{\Omega}\ \frac{1}{1-bdY_1Y_2Y_3Y_4\,.\,1-\frac{1}{d}Y_2Y_3Y_4\,.\,1-\frac{1}{b}Y_3Y_4\,.\,1-\frac{1}{d}Y_2Y_3Y_4^2}$$

$$=\underset{\geq}{\Omega}\ \frac{1}{1-bY_1Y_2Y_3Y_4\,.\,1-bY_1Y_2^2Y_3^2Y_4^2\,.\,1-\frac{1}{b}Y_3Y_4\,.\,1-bY_1Y_2^2Y_3^2Y_4^3}$$

$$=\frac{1-Y_1^2Y_2^3Y_3^4Y_4^4-Y_1^2Y_2^3Y_3^4Y_4^5-Y_1^2Y_2^4Y_3^5Y_4^6+Y_1^3Y_2^5Y_3^6Y_4^7+Y_1^3Y_2^5Y_3^7Y_4^8}{\begin{matrix}1-Y_1Y_2Y_3Y_4\,.\,1-Y_1Y_2Y_3^2Y_4^2\,.\,1-Y_1Y_2^2Y_3^2Y_4^2\\ 1-Y_1Y_2^2Y_3^2Y_4^3\,.\,1-Y_1Y_2^2Y_3^3Y_4^3\,.\,1-Y_1Y_2^2Y_3^3Y_4^4\end{matrix}}$$

indicating the ground forms

$$Y_1Y_2Y_3Y_4 \equiv X_1X_2X_3X_4X_5X_6X_7X_8$$
$$Y_1Y_2Y_3^2Y_4^2 \equiv X_1X_2X_3^2X_4^2X_5^2X_6^2X_7X_8$$
$$Y_1Y_2^2Y_3^2Y_4^2 \equiv X_1X_2^2X_3^2X_4^2X_5^2X_6^2X_7^2X_8$$
$$Y_1Y_2^2Y_3^2Y_4^3 \equiv X_1X_2^2X_3^2X_4^3X_5^3X_6^2X_7^2X_8$$
$$Y_1Y_2^2Y_3^3Y_4^3 \equiv X_1X_2^2X_3^3X_4^3X_5^3X_6^3X_7^2X_8$$
$$Y_1Y_2^2Y_3^3Y_4^4 \equiv X_1X_2^2X_3^3X_4^4X_5^4X_6^3X_7^2X_8$$

Art. 121. So far it appears that all products which can be placed in the form of a rectangle

$$\begin{array}{llll} X_1 X_2 & X_3 & \ldots & X_l \\ X_2 X_3 & X_4 & \ldots & X_{l+1} \\ \vdots\ \ \vdots & \vdots & & \vdots \\ X_m X_{m+1} & X_{m+2} & \ldots & X_{l+m-1} \end{array}$$

are ground forms for all values of l and m.

I have established this independently, and thus proved that the conjectured result for the general lattice *in plano* is, at any rate, finite and integral, as it should be.

It is desirable to obtain information concerning the ground forms which are not within the rectangular tableau.

The forms

$$X_1^{\alpha_1}X_2^{\alpha_2}\ldots X_s^{\alpha_s},$$

which appear in the tableau, may be eliminated from consideration, with the exception of the form

$$X_1X_2\ldots X_s,$$

by ascribing additional conditions such as

$$\alpha_1 = \alpha_2,$$

which are not true in the tableau.

The condition of this tableau is that if $\alpha_p = \alpha_{p+1}$, no index α_{p+q} is greater than α_p; after a repetition of index, no rise in index takes place. In the Y form, therefore, we may assign the conditions

$$\alpha_p = \alpha_{p+1} < \alpha_{p+2}$$

for any value of p, as one excluding the whole of the forms appertaining to the tableau.

We may impress the conditions

$$\alpha_1 = \alpha_2 < \alpha_3$$
$$\alpha_2 = \alpha_3 < \alpha_4$$
$$\alpha\ = \alpha_4 < \alpha_5$$
$$\ldots$$

in succession, and we may combine any number of such conditions as are independent.

Art. 122. I postpone further investigation into this interesting theory, and will now give a formal proof that the product tableau is, in fact, finite and integral. The product in question for $m \geqq l$ is

$$X_1 X_2^2 \ldots X_{l-1}^{l-1} (X_l X_{l+1} \ldots X_m)^l X_{m+1}^{l-1} X_{m+2}^{l-2} \ldots X_{m+l-2}^2 X_{m+l-1}$$

so that

$$\begin{array}{lll} \alpha_s = s & \text{for} & l \geqq s \\ \alpha_s = l & \text{for} & s > l \text{ and } < m+1 \\ \alpha_{m+s} = l - s & \text{for} & l - 1 \geqq s \end{array}$$

All the conditions may be resumed in the single formula

$$\alpha_s + \alpha_{2s+t} + \alpha_{3s+2t} + \ldots \geqq \alpha_{s+t} + \alpha_{2s+2t} + \alpha_{3s+3t} + \ldots$$

s and t being any integers.

Let the greatest integer in $\frac{l+t-1}{s+t}$ be denoted by $I_1 \frac{l+t-1}{s+t}$ or by I_1 simply for brevity. Similarly let I_2 refer to $\frac{m+t}{s+t}$, I_3 to $\frac{l+m+t-1}{s+t}$, J_1 to $\frac{l-1}{s+t}$, J_2 to $\frac{m}{s+t}$, and J_3 to $\frac{l+m-1}{s+t}$. We derive

$$\begin{array}{lll} I_1 = J_1 & \text{or} & J_1 + 1 \\ I_2 = J_2 & \text{or} & J_2 + 1 \\ I_3 = J_3 & \text{or} & J_3 + 1 \\ I_1 + I_2 = I_3 & \text{or} & I_3 + 1 \\ J_1 + J_2 = J_3 & \text{or} & J_3 - 1 \end{array}$$

and we have ten possible cases to consider, viz. :—

Case 1.

$$\begin{array}{ll} I_1 = J_1 & I_1 + I_2 = I_3 \\ I_2 = J_2 & J_1 + J_2 = J_3 \\ I_3 = J_3 & \end{array}$$

Case 2.

$$\begin{array}{ll} I_1 = J_1 + 1 & I_1 + I_2 = I_3 \\ I_2 = J_2 & J_1 + J_2 = J_3 - 1 \\ I_3 = J_3 & \end{array}$$

Case 3.

$$\begin{array}{ll} I_1 = J_1 + 1 & I_1 + I_2 = I_3 + 1 \\ I_2 = J_2 & J_1 + J_2 = J_3 \\ I_3 = J_3 & \end{array}$$

Case 4.

$I_1 = J_1$ $\qquad$ $I_1 + I_2 = I_3$

$I_2 = J_2 + 1$ $\qquad$ $J_1 + J_2 = J_3 - 1$

$I_3 = J_3$

Case 5.

$I_1 = J_1$ $\qquad$ $I_1 + I_2 = I_3 + 1$

$I_2 = J_2 + 1$ $\qquad$ $J_1 + J_2 = J_3$

$I_3 = J_3$

Case 6.

$I_1 = J_1 + 1$ $\qquad$ $I_1 + I_2 = I_3 + 1$

$I_2 = J_2 + 1$ $\qquad$ $J_1 + J_2 = J_3 - 1$

$I_3 = J_3$

Case 7.

$I_1 = J_1 + 1$ $\qquad$ $I_1 + I_2 = I_3$

$I_2 = J_2$ $\qquad$ $J_1 + J_2 = J_3$

$I_3 = J_3 + 1$

Case 8.

$I_1 = J_1$ $\qquad$ $I_1 + I_2 = I_3$

$I_2 = J_2 + 1$ $\qquad$ $J_1 + J_2 = J_3$

$I_3 = J_3 + 1$

Case 9.

$I_1 = J_1 + 1$ $\qquad$ $I_1 + I_2 = I_3$

$I_2 = J_2 + 1$ $\qquad$ $J_1 + J_2 = J_3 - 1$

$I_3 = J_3 + 1$

Case 10.

$I_1 = J_1 + 1$ $\qquad$ $I_1 + I_2 = I_3 + 1$

$I_2 = J_2 + 1$ $\qquad$ $J_1 + J_2 = J_3$

$I_3 = J_3 + 1$

For the series

$$\alpha_s + \alpha_{2s+t} + \alpha_{3s+2t} + \ldots,$$

we have, as far as α_{l-1}, I_1 terms; as far as α_m, I_2 terms; and, as far as α_{l+m-1}, I_3 terms.

Hence the summation gives :—

$$\begin{aligned} &\tfrac{1}{2} I_1 \{2s + (I_1 - 1)(s + t)\} + l(I_2 - I_1) \\ &\quad + \tfrac{1}{2}(I_3 - I_2)\{2l + 2m + 2t - 2(s + t)(I_2 + 1) - (s + t)(I_3 - I_2 - 1)\} \\ &= \tfrac{1}{2}(s + t)(I_1^2 + I_2^2 - I_3^2) + (\tfrac{1}{2}s - \tfrac{1}{2}t - \tfrac{1}{2}l) I_1 \\ &\quad + (\tfrac{1}{2}s - \tfrac{1}{2}t - m) I_2 + (l + m - \tfrac{1}{2}s + \tfrac{1}{2}t) I_3. \end{aligned}$$

Summing similarly the series

$$a_{s+t} + a_{2s+2t} + a_{3s+3t} \dots$$

we find

$$\tfrac{1}{2}(s+t)(J_1^2 + J_2^2 - J_3^2) + (\tfrac{1}{2}s + \tfrac{1}{2}t - l)J_1$$
$$+ (\tfrac{1}{2}s + \tfrac{1}{2}t - m)J_2 + (l + m - \tfrac{1}{2}s - \tfrac{1}{2}t)J_3,$$

and we have in each of the ten cases to establish the relation

$$\tfrac{1}{2}(s+t)(I_1^2 + I_2^2 - I_3^2) + (\tfrac{1}{2}s - \tfrac{1}{2}t - l)I_1$$
$$+ (\tfrac{1}{2}s - \tfrac{1}{2}t - m)I_2 + (l + m - \tfrac{1}{2}s + \tfrac{1}{2}t)I_3,$$
$$\geq \tfrac{1}{2}(s+t)(J_1^2 + J_2^2 - J_3^2) + (\tfrac{1}{2}s + \tfrac{1}{2}t - l)J_1$$
$$+ (\tfrac{1}{2}s + \tfrac{1}{2}t - m)J_2 + (l + m - \tfrac{1}{2}s - \tfrac{1}{2}t)J_3$$

for all values of s and t.

For Case 1 it reduces to

$$I_1 + I_2 \geq I_3,$$

which is true, for here $I_1 + I_2 = I_3$.

For Case 2, making use of $J_1 + J_2 = J_3 - 1$, the reduction is to

$$J_1 \geq \frac{l - s - t}{s + t},$$

and J_1 being the greatest integer in $\frac{l-1}{s+t}$, and moreover $s + t$ being at least unity, the relation is obviously satisfied.

For Case 3, making use of $J_1 + J_2 = J_3$, we find

$$J_1 \geq \frac{l - s}{s + t},$$

and this is satisfied as $s \geq 1$.

For Case 4, reducing by $J_1 + J_2 = J_3 - 1$, we find

$$J_2 \geq \frac{m}{s + t} - 1$$

obviously true from the definition of J_2.

For Case 5, reducing by $J_1 + J_2 = J_3$, we find

$$J_2 \geq \frac{m - s}{s + t}$$

obviously satisfied.

For Case 6, reducing by $J_1 + J_2 = J_3 - 1$, we find

$$J_3 \geq \frac{l + m - s}{s + t}$$

clearly satisfied.

For Case 7, reducing by $J_1 + J_2 = J_3$, we find

$$J_2 \geqq \frac{m}{s+t},$$

which is right.

For Case 8, reducing by $J_1 + J_2 = J_3$, we find

$$J_1 \geqq \frac{l}{s+t},$$

which is satisfied.

For Case 9, reducing by $J_1 + J_2 = J_3 - 1$, the ratio is one of equality.

For Case 10, reducing by $J_1 + J_2 = J_3$, we find

$$s + t \geqq 0,$$

which is right.

Hence the relation is universally satisfied, and we have proved that the expression

$$X_1 X_2^2 \ldots X_{l-1}^{l-1} (X_l X_{l+1} \ldots X_m)^l X_{m+1}^{l-1} X_{m+2}^{l-2} \ldots X_{l+m-1}$$

is in every case finite and integral.

Art. 123. In Part 3 of this Memoir I hope to treat of other systems of algebraical and arithmetical functions which fall within the domain of partition analysis and the theory of the linear composition of integers; also to take up the general theories of partition analysis and linear Diophantine analysis, with possible extensions to higher degrees.

II. *Application of the Partition Analysis to the study of the properties of any system of Consecutive Integers.* By Major P. A. MacMahon, R.A., D.Sc., F.R.S., Hon. Mem. C.P.S.

[Received 15 *May*, 1899.]

INTRODUCTION.

The object of this paper is to solve a problem, concerning any arbitrarily selected set of consecutive integers, by the application of a new method of Partition analysis. I will first explain the problem, and afterwards the analysis that will be used.

In the binomial and multinomial expansions, the exponent being a positive integer, every coefficient is an integer. This fact depends analytically upon the circumstance that the product of any m consecutive integers is divisible by factorial m; we have

$$\left(\frac{n+1}{1}\right)\left(\frac{n+2}{2}\right)\left(\frac{n+3}{3}\right)\ldots\left(\frac{n+m}{m}\right),$$

an integer for all values of n.

The present question is the determination of all values of $\alpha_1, \alpha_2, \alpha_3, \ldots \alpha_m$ for which the expression

$$\left(\frac{n+1}{1}\right)^{\alpha_1}\left(\frac{n+2}{2}\right)^{\alpha_2}\left(\frac{n+3}{3}\right)^{\alpha_3}\ldots\left(\frac{n+m}{m}\right)^{\alpha_m}$$

is an integer for all values of n; in particular the discovery of the finite number of ground or fundamental products of this form, from which all the forms may be generated by multiplication.

There is a parallel theory connected with the algebraic product

$$\left(\frac{1-x^{n+1}}{1-x}\right)^{\alpha_1}\left(\frac{1-x^{n+2}}{1-x^2}\right)^{\alpha_2}\left(\frac{1-x^{n+3}}{1-x^3}\right)^{\alpha_3}\ldots\left(\frac{1-x^{n+m}}{1-x^m}\right)^{\alpha_m},$$

where $\alpha_1, \alpha_2, \alpha_3, \ldots \alpha_m$ have to be assigned so that the product is finite and integral for all values of n. This has been discussed by me in the 'Memoir on the Theory of the Partitions of Numbers, Part II.' *Phil. Trans. R. S.* 1899. It will be observed that the algebraic product merges into the arithmetical product for the particular case $x = 1$, so that all algebraic products which are finite and integral produce in this manner arithmetical products which are integers. This, however, is as much as can be said, for otherwise the theories proceed on widely divergent lines; as might be expected the arithmetical products form a more extended group than the algebraical.

Denote, for brevity,

$$\frac{1-x^{n+s}}{1-x^s} \text{ and } \frac{n+s}{s}$$

by X_s and N_s respectively.

The principal X theorem, that has been obtained *loc. cit.*, is to the effect that constructing any X rectangle

$$\begin{matrix} X_1 & X_2 & X_3 & X_4 & \dots & X_l \\ X_2 & X_3 & X_4 & X_5 & \dots & X_{l+1} \\ X_3 & X_4 & X_5 & X_6 & \dots & X_{l+2} \\ X_4 & X_5 & X_6 & X_7 & \dots & X_{l+3} \\ \vdots & \vdots & \vdots & \vdots & & \vdots \\ X_m & X_{m+1} & X_{m+2} & X_{m+3} & \dots & X_{l+m+1}, \end{matrix}$$

l and m having any values, with the law that any X has a suffix one greater than the X above it or to the left of it, the product

$$X_1 X_2^2 X_3^3 \dots X_{l+m-1},$$

obtained by multiplying all the X's together, is finite and integral for all values of the integer n. There are other forms as well, e.g. the product

$$X_1 X_2 X_3^2 X_4 X_5,$$

which are not expressible in the rectangular lattice form, the theory of which is not yet complete.

We see therefore that the product of N's contained in the rectangle

$$\begin{matrix} N_1 & N_2 & N_3 & \dots & N_l \\ N_2 & N_3 & N_4 & \dots & N_{l+1} \\ N_3 & N_4 & N_5 & \dots & N_{l+2} \\ \vdots & \vdots & \vdots & & \vdots \\ N_m & N_{m+1} & N_{m+2} & \dots & N_{l+m-1} \end{matrix}$$

is an integer for all values of n.

It will appear moreover that no product exists which is free from N_1, so that all these products, being irreducible, are fundamental solutions of the problem.

The method of partition analysis is concerned with the solution of one or more relations of the type

$$\lambda_1\alpha_1 + \lambda_2\alpha_2 + \lambda_3\alpha_3 + \dots + \lambda_s\alpha_s$$

$$\geqslant \mu_1\beta_1 + \mu_2\beta_2 + \mu_3\beta_3 + \dots + \mu_t\beta_t,$$

the coefficients λ and μ being given positive integers, and it is required to find the general values of $\alpha_1, \alpha_2 \dots \beta_1, \beta_2, \dots$, being positive integers, which satisfy the one or more relations.

This is accomplished by constructing the sum

$$\Sigma x_1^{\alpha_1} x_2^{\alpha_2} x_3^{\alpha_3} \ldots x_s^{\alpha_s} y_1^{\beta_1} y_2^{\beta_2} y_3^{\beta_3} \ldots y_t^{\beta_t}$$

for all sets of values of $\alpha_1, \alpha_2 \ldots \beta_1, \beta_2, \ldots$ which satisfy the relation. The expression obtained is found to indicate the ground solutions of the relations and the syzygies that connect them.

The sum is expressible in the crude form

$$\underset{\geqslant}{\Omega} \frac{1}{1 - m^{\lambda_1} x_1 . 1 - m^{\lambda_2} x_2 \ldots 1 - m^{\lambda_s} x_s . 1 - m^{-\mu_1} y_1 . 1 - m^{-\mu_2} y_2 \ldots 1 - m^{-\mu_t} y_t}$$

where the symbol of operation

$$\underset{\geqslant}{\Omega}$$

is connected with the auxiliary symbol m in the following manner:—

The fraction is to be expanded in ascending powers of $x_1, x_2, \ldots y_1, y_2, \ldots$; all terms containing negative powers of m are to be then deleted; subsequently, in the remaining terms, m is to be put equal to unity.

Slight reflection will shew that the conditional relation will be satisfied in all products which survive this operation, and that if we can perform the operation so as to retain the fractional form we shall arrive at a reduced generating function which will establish the ground solutions and the syzygies which connect them.

As a simple example of reduction which is of great service in what follows take

$$\alpha_1 \geqslant \beta_1;$$

this leads to

$$\underset{\geqslant}{\Omega} \frac{1}{1 - m x_1 . 1 - \frac{1}{m} y_1},$$

and observing that

$$\frac{1}{1 - m x_1 . 1 - \frac{1}{m} y_1} = \frac{1}{1 - m x_1 . 1 - x_1 y_1} + \frac{\frac{1}{m} y_1}{1 - \frac{1}{m} y_1 . 1 - x_1 y_1},$$

we find

$$\underset{\geqslant}{\Omega} \frac{1}{1 - m x_1 . 1 - \frac{1}{m} y_1} = \frac{1}{1 - x_1 . 1 - x_1 y_1};$$

also

$$\underset{\geqslant}{\Omega} \frac{\frac{1}{m^s}}{1 - m x_1 . 1 - \frac{1}{m} y_1} = \frac{x_1^s}{1 - x_1 . 1 - x_1 y_1};$$

which is the solution of

$$\alpha_1 \geqslant \beta_1 + s;$$

so that the solution of

$$\alpha_1 > \beta_1$$

is given by

$$\underset{\geqslant}{\Omega} \frac{\frac{1}{m}}{1 - mx_1 \,.\, 1 - \frac{1}{m} y_1} = \frac{x_1}{1 - x_1 \,.\, 1 - x_1 y_1}.$$

Again, if

$$\alpha_1 \geqslant \beta_1 + \beta_2,$$

we have

$$\underset{\geqslant}{\Omega} \frac{1}{1 - mx_1 \,.\, 1 - \frac{1}{m} y_1 \,.\, 1 - \frac{1}{m} y_2}$$

$$= \underset{\geqslant}{\Omega} \frac{1}{1 - mx_1 \,.\, 1 - \frac{1}{m} y_1 \,.\, 1 - x_1 y_2},$$

$$= \frac{1}{1 - x_1 \,.\, 1 - x_1 y_1 \,.\, 1 - x_1 y_2},$$

the solution.

Also the solution of

$$\alpha_1 > \beta_1 + \beta_2$$

is

$$\frac{x_1}{1 - x_1 \,.\, 1 - x_1 y_1 \,.\, 1 - x_1 y_2}.$$

Lastly,

$$\alpha_1 + \alpha_2 \geqslant \beta_1$$

gives, by repeated application of the above simple theorem,

$$\frac{1 - x_1 x_2 y_1}{1 - x_1 \,.\, 1 - x_2 \,.\, 1 - x_1 y_1 \,.\, 1 - x_2 y_1}$$

In general the subsequent work merely involves processes easily derivable from these cases.

Particular theorems will be given as they become necessary, and for the general theory, which is here not needed, the reader is referred to Part III. of the Memoir on Partitions which may appear shortly in *Phil. Trans. R. S.*

To come to the object of the paper I commence with

ORDER 2.

$$\left(\frac{n+1}{1}\right)^{\alpha_1} \left(\frac{n+2}{2}\right)^{\alpha_2} = N_1^{\alpha_1} N_2^{\alpha_2};$$

this product is an integer when n is even, but when n is uneven we must have

$$\alpha_1 \geqslant \alpha_2;$$

and

$$\Sigma N_1^{\alpha_1} N_2^{\alpha_2} = \underset{\geqslant}{\Omega} \frac{1}{1 - aN \,.\, 1 - \frac{1}{a} N_2},$$

$$= \frac{1}{1 - N_1 \,.\, 1 - N_1 N_2};$$

shewing that the ground products are $N_1,\ N_1N_2$,

or $$(\alpha_1\alpha_2) = (1,\ 0);\ (1,\ 1).$$

ORDER 3.

$$\left(\frac{n+1}{1}\right)^{\alpha_1}\left(\frac{n+2}{2}\right)^{\alpha_2}\left(\frac{n+3}{3}\right)^{\alpha_3} = N_1^{\alpha_1}N_2^{\alpha_2}N_3^{\alpha_3}.$$

When n is of form	condition is	
$4m+1$,	$\alpha_1 + 2\alpha_3 \geqslant \alpha_2$	(a),
$4m+3$,	$2\alpha_1 + \alpha_3 \geqslant \alpha_2$	,
$3m+1$,	$\alpha_2 \geqslant \alpha_3$	(b),
$3m+2$,	$\alpha_1 \geqslant \alpha_2$	(c).

We may omit the second of these as being implied by the first and fourth and introducing the auxiliaries $a,\ b,\ c$ in the relations marked (a), (b), (c) respectively we write down the Ω function

$$\mathop{\Omega}_{\geqslant} \frac{1}{1-acN_1\,.\,1-\dfrac{b}{a}N_2\,.\,1-\dfrac{a^2}{bc}N_3},$$

as the expression of the sum $\Sigma N_1^{\alpha_1}N_2^{\alpha_2}N_3^{\alpha_3}$.

It must be observed that the operating symbol $\mathop{\Omega}_{\geqslant}$, has reference to each of the three auxiliaries $a,\ b,\ c$.

These must be dealt with in the most convenient order, so that unnecessary labour may be avoided; this order is not always obvious without some preliminary experiments. In the present instance it is clearly advisable to commence with b because a occurs to the second power, and operation upon c will introduce a^3. It should be remarked that operation upon one letter may cause two letters to vanish; this would indicate that the relations associated with these letters are not independent members of the system of relations. It does not follow conversely that if the relations are not all independent two letters must vanish as the result of operation upon some one letter. This does follow for a certain order of operation upon the letters, but not for all orders.

Eliminating b we obtain

$$\mathop{\Omega}_{\geqslant} \frac{1}{1-acN_1\,.\,1-\dfrac{1}{a}N_2\,.\,1-\dfrac{a}{c}N_2N_3}.$$

Observe that this expression would have presented itself if for the two relations

$$\alpha_1 + \alpha_3 \geqslant \alpha_2,$$

$$\alpha_1 \geqslant \alpha_3,$$

we had constructed the sum $$\Sigma N_1^{\alpha_1}N_2^{\alpha_2}(N_2N_3)^{\alpha_3}.$$

The fact is that we can reduce the three relations (*a*), (*b*), (*c*) to two by writing $\alpha_2+\alpha_3$ for α_2, a tranformation that the relation (*b*) permits, and then we have to write N_2N_3 for N_3 in the sum

$$\Sigma N_1^{\alpha_1}N_2^{\alpha_2}N_3^{\alpha_3}.$$

We next eliminate *c*, obtaining

$$\underset{\geqslant}{\Omega}\frac{1}{1-aN_1\,.\,1-\frac{1}{a}N_2\,.\,1-a^2N_1N_2N_3},$$

an expression that would have presented itself if we had been summing

$$\Sigma N_1^{\alpha_1}N_2^{\alpha_2}(N_1N_2N_3)^{\alpha_3}$$

for the single relation $$\alpha_1+2\alpha_3\geqslant\alpha_2,$$

obtained from the relation

$$\alpha_1+\alpha_3\geqslant\alpha_2,$$

by writing $\alpha_1+\alpha_3$ for α_1, a transformation permitted by the relation

$$\alpha_1\geqslant\alpha_3.$$

The process employed is therefore equivalent to a gradual reduction in the number of the conditional relations associated with a proper transformation of the product to be summed.

To eliminate *a* we require the subsidiary theorem

$$\underset{\geqslant}{\Omega}\frac{1}{1-a^2x\,.\,1-ay\,.\,1-\frac{1}{a}z}=\frac{1+xy-xyz-xyz^2}{1-x\,.\,1-y\,.\,1-yz\,.\,1-xz^2};$$

and thence we derive

$$\frac{1+N_1N_2^2N_3-N_1^2N_2^2N_3-N_1^2N_2^3N_3}{1-N_1\,.\,1-N_1N_2\,.\,1-N_1N_2N_3\,.\,1-N_1N_2^3N_3}$$

$$=\frac{1-N_1^2N_2^2N_3-N_1^2N_2^3N_3-N_1^2N_2^4N_3^2+N_1^3N_2^4N_3^2+N_1^3N_2^5N_3^2}{1-N_1\,.\,1-N_1N_2\,.\,1-N_1N_2N_3\,.\,1-N_1N_2^2N_3\,.\,1-N_1N_2^3N_3}$$

In this result the denominator indicates the ground products, and the numerator the simple and compound syzygies which connect them.

It is manifest that the ground products are

$$N_1,\ N_1N_2,\ N_1N_2N_3,\ N_1N_2^2N_3,\ N_1N_2^3N_3$$

connected by the simple syzygies

$$(A)=(N_1)(N_1N_2^2N_3)-(N_1N_2)(N_1N_2N_3)=0,$$

$$(B)=(N_1)(N_1N_2^3N_3)-(N_1N_2)(N_1N_2^2N_3)=0,$$

$$(C)=(N_1N_2N_3)(N_1N_2^3N_3)-(N_1N_2^2N_3)^2=0;$$

and the compound syzygies

$$(N_1)(C) - (N_1N_2N_3)(B) = 0,$$
$$(N_1N_2^2N_3)(B) - (N_1N_2^3N_3)(A) = 0;$$

indicated by the numerator terms:

$$-N_1^2N_2^2N_3, \quad -N_1^2N_2^3N_3, \quad -N_1^2N_2^4N_3^2, \quad +N_1^3N_2^4N_3^2, \quad +N_1^3N_2^5N_3^2$$

respectively.

The generating function takes also the suggestive form:—

$$\frac{1 - N_1^2N_2^3N_3}{1-N_1 \,.\, 1-N_1N_2 \,.\, 1-N_1N_2N_3 \,.\, 1-N_1N_2^3N_3}$$
$$+\frac{N_1N_2^2N_3}{1-N_1N_2 \,.\, 1-N_1N_2N_3 \,.\, 1-N_1N_2^3N_3}.$$

By proceeding in this manner we not only obtain the new ground products appertaining to the order but also those of lower orders previously obtained. It would be desirable to exclude the latter, and in the case before us we see *à posteriori* that this could have been secured by impressing the additional condition

$$\alpha_1 = \alpha_3;$$

but no method, similar to this, seems to be available for an order higher than 3, as no equation invariably connects the indices of the ground products.

ORDER 4.

$$\binom{n+1}{1}^{\alpha_1}\binom{n+2}{2}^{\alpha_2}\binom{n+3}{3}^{\alpha_3}\binom{n+4}{4}^{\alpha_4} = N_1^{\alpha_1}N_2^{\alpha_2}N_3^{\alpha_3}N_4^{\alpha_4}.$$

When n is of the form	condition is
$4m+1$,	$\alpha_1 + 2\alpha_3 \geqslant \alpha_2 + 2\alpha_4$,
$4m+2$,	$\alpha_2 \geqslant \alpha_4$,
$4m+3$,	$2\alpha_1 + \alpha_3 \geqslant \alpha_2 + 2\alpha_4$,
$3m+1$,	$\alpha_2 \geqslant \alpha_3$,
$3m+2$,	$\alpha_1 + \alpha_4 \geqslant \alpha_3$.

The Ω function which can be at once written down is somewhat troublesome to deal with, so that I find it appropriate to divide the generating function into two parts according as $\alpha_1 \geqslant \alpha_3$, $\alpha_3 > \alpha_1$.

Case 1. $\alpha_1 \geqslant \alpha_3$.

The conditions reduce to

$$\alpha_1 \geqslant \alpha_3 \qquad (a),$$
$$\alpha_1 + 2\alpha_3 \geqslant \alpha_2 + 2\alpha_4 \qquad (b),$$
$$\alpha_2 \geqslant \alpha_4 \qquad (c),$$
$$\alpha_2 \geqslant \alpha_3 \qquad (d),$$

and it is convenient to add the implied condition

$$\alpha_1 \geqslant \alpha_4 \qquad (e).$$

We obtain, for $\Sigma N_1^{\alpha_1} N_2^{\alpha_2} N_3^{\alpha_3} N_4^{\alpha_4}$,

$$\underset{\geqslant}{\Omega} \frac{1}{1-abeN_1 \,.\, 1-\frac{cd}{b}N_2 \,.\, 1-\frac{b^2}{ad}N_3 \,.\, 1-\frac{1}{b^2ce}N_4};$$

and, eliminating d and e, this is

$$\underset{\geqslant}{\Omega} \frac{1}{1-abN_1 \,.\, 1-\frac{c}{b}N_2 \,.\, 1-\frac{bc}{a}N_2N_3 \,.\, 1-\frac{a}{bc}N_1N_4},$$

which, eliminating c, is

$$\underset{\geqslant}{\Omega} \frac{1-\frac{1}{b}N_1N_2^2N_3N_4}{1-abN_1 \,.\, 1-\frac{1}{b}N_2 \,.\, 1-\frac{b}{a}N_2N_3 \,.\, 1-\frac{a}{b^2}N_1N_2N_4 \,.\, 1-N_1N_2N_3N_4};$$

and, eliminating a, this becomes

$$\underset{\geqslant}{\Omega} \frac{1-N_1^2N_2^2N_3N_4}{1-bN_1 \,.\, 1-\frac{1}{b}N_2 \,.\, 1-\frac{1}{b^2}N_1N_2N_4 \,.\, 1-b^2N_1N_2N_3 \,.\, 1-N_1N_2N_3N_4},$$

the term $1-\frac{1}{b}N_1N_2^2N_3N_4$ disappearing.

This is equal to

$$\frac{1-N_1^2N_2^2N_3N_4\Omega}{1-N_1N_2N_3N_4 \geqslant}\left\{\frac{1}{1-bN_1 \,.\, 1-N_1N_2}+\frac{\frac{1}{b}N_2}{1-\frac{1}{b}N_2 \,.\, 1-N_1N_2}\right\}$$

$$\times\left\{\frac{1}{1-b^2N_1N_2N_3 \,.\, 1-N_1^2N_2^2N_3N_4}+\frac{\frac{1}{b^2}N_1N_2N_4}{1-\frac{1}{b^2}N_1N_2N_4 \,.\, N_1^2N_2^2N_3N_4}\right\}$$

$$=\frac{1}{1-N_1 \,.\, 1-N_1N_2 \,.\, 1-N_1N_2N_3 \,.\, 1-N_1N_2N_3N_4}.$$

$$+\underset{\geqslant}{\Omega} \frac{\frac{1}{b}N_2}{1-b^2N_1N_2N_3 \,.\, 1-\frac{1}{b}N_2} \cdot \frac{1}{1-N_1N_2 \,.\, 1-N_1N_2N_3N_4}$$

$$+\underset{\geqslant}{\Omega} \frac{\frac{1}{b^2}N_1N_2N_4}{1-bN_1 \,.\, 1-\frac{1}{b^2}N_1N_2N_4} \cdot \frac{1}{1-N_1N_2 \,.\, 1-N_1N_2N_3N_4}$$

3—2

$$=\frac{1}{1-N_1\,.\,1-N_1N_2\,.\,1-N_1N_2N_3\,.\,1-N_1N_2N_3N_4}$$

$$+\frac{N_1N_2^2N_3+N_1N_2^3N_3}{1-N_1N_2\,.\,1-N_1N_2N_3\,.\,1-N_1N_2N_3N_4\,.\,1-N_1N_2^3N_3}$$

$$+\frac{N_1^3N_2N_4}{1-N_1\,.\,1-N_1N_2\,.\,1-N_1N_2N_3N_4\,.\,1-N_1^3N_2N_4}.$$

Case 2. $\alpha_3>\alpha_1$.

The conditions become

$$\alpha_3>\alpha_1 \qquad (a),$$

$$2\alpha_1+\alpha_3\geqslant\alpha_2+2\alpha_4 \qquad (b),$$

$$\alpha_2\geqslant\alpha_4 \qquad (c),$$

$$\alpha_2\geqslant\alpha_3 \qquad (d),$$

$$\alpha_1+\alpha_4\geqslant\alpha_3 \qquad (e);$$

to which it is convenient to add the implied conditions

$$\alpha_1\geqslant\alpha_4 \qquad (f),$$

$$\alpha_3\geqslant\alpha_4 \qquad (g);$$

the Ω function is

$$\underset{\geqslant}{\Omega}\frac{\frac{1}{a}}{1-\frac{b^2ef}{a}N_1\,.\,1-\frac{cd}{b}N_2\,.\,1-\frac{abg}{de}N_3\,.\,1-\frac{e}{b^2cfg}N_4}$$

$$=\underset{\geqslant}{\Omega}\frac{\frac{1}{a}.}{1-\frac{b^2e}{a}N_1\,.\,1-\frac{cd}{b}N_2\,.\,1-\frac{ab}{de}N_3\,.\,1-\frac{be}{cd}N_1N_3N_4}$$

$$=\underset{\geqslant}{\Omega}\frac{\frac{b}{de}N_3}{1-\frac{b^3}{d}N_1N_3\,.\,1-\frac{d}{b}N_2\,.\,1-\frac{b}{de}N_3\,.\,1-eN_1N_2N_3N_4}$$

$$=\frac{1}{1-N_1N_2N_3N_4}\,.\,\underset{\geqslant}{\Omega}\frac{\frac{b}{d}N_1N_2N_3^2N_4}{1-\frac{b^3}{d}N_1N_3\,.\,1-\frac{d}{b}N_2\,.\,1-\frac{b}{d}N_1N_2N_3^2N_4}$$

$$=\frac{N_1N_2^2N_3^2N_4}{1-N_1N_2N_3N_4}\underset{\geqslant}{\Omega}\frac{1}{1-b^2N_1N_2N_3\,.\,1-\frac{1}{b}N_2\,.\,1-N_1N_2^2N_3^2N_4}$$

$$=\frac{N_1N_2^2N_3^2N_4(1+N_1N_2^2N_3)}{1-N_1N_2N_3\,.\,1-N_1N_2^3N_3\,.\,1-N_1N_2N_3N_4\,.\,1-N_1N_2^2N_3^2N_4}.$$

Hence the complete sum

$$\Sigma N_1^{\alpha_1} N_2^{\alpha_2} N_3^{\alpha_3} N_4^{\alpha_4},$$

is

$$\frac{1}{1 - N_1 \,.\, 1 - N_1 N_2 \,.\, 1 - N_1 N_2 N_3 \,.\, 1 - N_1 N_2 N_3 N_4}$$

$$+ \frac{N_1 N_2^2 N_3 + N_1 N_2^3 N_3}{1 - N_1 N_2 \,.\, 1 - N_1 N_2 N_3 \,.\, 1 - N_1 N_2 N_3 N_4 \,.\, 1 - N_1 N_2^3 N_3}$$

$$+ \frac{N_1^3 N_2 N_4}{1 - N_1 \,.\, 1 - N_1 N_2 \,.\, 1 - N_1 N_2 N_3 N_4 \,.\, 1 - N_1^3 N_2 N_4}$$

$$+ \frac{N_1 N_2^2 N_3^2 N_4 (1 + N_1 N_2^2 N_3)}{1 - N_1 N_2 N_3 \,.\, 1 - N_1 N_2^3 N_3 \,.\, 1 - N_1 N_2 N_3 N_4 \,.\, 1 - N_1 N_2^2 N_3^2 N_4},$$

and we have three ground products of order 4, viz.:—

$$N_1 N_2 N_3 N_4,$$
$$N_1^3 N_2 N_4,$$
$$N_1 N_2^2 N_3^2 N_4,$$

and every product of order 4 can be compounded of these and of ground products of lower orders.

I pause to observe that the form $N_1^3 N_2 N_4$ is one of a kind that always presents itself for an even order. The system is

$$N_1^{\alpha_1} N_2^{\alpha_2} N_4^{\alpha_4} N_6^{\alpha_6} \ldots N_{2s}^{\alpha_{2s}},$$

and may be separately examined. For the order 6 the ground products

$$N_1^2 N_2 N_6,\ N_1^4 N_2 N_4 N_6,\ N_1^6 N_2 N_4^2 N_6,\ N_1^8 N_2 N_4^3 N_6,$$

and for the order 8

$$N_1^7 N_2 N_4 N_6 N_8,\ N_1^8 N_2 N_4 N_6^2 N_8,\ N_1^8 N_2^3 N_4 N_8,$$
$$N_1^{10} N_2 N_4^2 N_6^2 N_8,\ N_1^{22} N_2^2 N_4^3 N_6^5 N_8^3,$$

are easily obtained.

ORDER 5.

We now come to a very complicated system of forms, which includes no fewer than 13 ground products of order 5. These I find to be

$$N_1 N_2 N_3 N_4 N_5,\ N_1 N_2 N_3^2 N_4 N_5,\ N_1 N_2^2 N_3 N_4 N_5,$$
$$N_1 N_2^2 N_3^2 N_4 N_5,\ N_1 N_2^2 N_3^2 N_4^2 N_5,\ N_1 N_2^2 N_3^3 N_4^2 N_5,$$
$$N_1 N_2^3 N_3^2 N_4 N_5,\ N_1 N_2^3 N_3^3 N_4^2 N_5,\ N_1 N_2^4 N_3^2 N_4 N_5,$$
$$N_1^2 N_2^5 N_3^7 N_4^5 N_5^2,\ N_1^3 N_2^2 N_3 N_4^2 N_5,\ N_1^3 N_2^3 N_3^2 N_4^3 N_5^2,$$
$$N_1^3 N_2^4 N_3^3 N_4^4 F_5^3.$$

The complete generating function can be obtained without difficulty, but, on account of its great length, I restrict my endeavours to the establishing of the 13 ground products. I find it necessary to adopt abridged notations, and in future, where it is convenient, I denote

$$N_1^{\alpha_1}N_2^{\alpha_2}N_3^{\alpha_3}N_4^{\alpha_4}N_5^{\alpha_5} \text{ by } (\alpha_1\alpha_2\alpha_3\alpha_4\alpha_5).$$

Further, if a portion of the generating function presents itself, which involves merely ground products already obtained in the previous work, I enclose it in brackets [ ] and thenceforward omit it. For example, I write

$$A = [B] + C \equiv C = [D] + E \equiv E;$$

and so on.

For
$$\left(\frac{n+1}{1}\right)^{\alpha_1}\left(\frac{n+2}{2}\right)^{\alpha_2}\left(\frac{n+3}{3}\right)^{\alpha_3}\left(\frac{n+4}{4}\right)^{\alpha_4}\left(\frac{n+5}{5}\right)^{\alpha_5}$$
$$= N_1^{\alpha_1}N_2^{\alpha_2}N_3^{\alpha_3}N_4^{\alpha_4}N_5^{\alpha_5} = (\alpha_1\alpha_2\alpha_3\alpha_4\alpha_5).$$

When n is of form	condition is
$4p + 1$,	$\alpha_1 + 2\alpha_3 + \alpha_5 \geqslant \alpha_2 + 2\alpha_4$,
$4p + 2$,	$\alpha_2 \geqslant \alpha_4$,
$4p + 3$,	$2\alpha_1 + \alpha_3 + 2\alpha_5 \geqslant \alpha_2 + 2\alpha_4$,
$3p + 1$,	$\alpha_2 + \alpha_5 \geqslant \alpha_3$,
$3p + 2$,	$\alpha_1 + \alpha_4 \geqslant \alpha_3$,
$5p + 1$,	$\alpha_4 \geqslant \alpha_5$,
$5p + 2$,	$\alpha_3 \geqslant \alpha_5$,
$5p + 3$,	$\alpha_2 \geqslant \alpha_5$ omit,
$5p + 4$,	$\alpha_1 \geqslant \alpha_5$;

the eighth of these conditions may be omitted as being implied by the second and sixth. I separate the generating function into six portions corresponding to

Case 1. $\alpha_1 \geqslant \alpha_2,\ \alpha_2 \geqslant \alpha_3$;

Case 2. $\alpha_1 \geqslant \alpha_2,\ \alpha_3 > \alpha_2$;

Case 3. $\alpha_2 > \alpha_1,\ \alpha_2 \geqslant \alpha_3,\ \alpha_1 + \alpha_5 \geqslant \alpha_3$;

Case 4. $\alpha_2 > \alpha_1,\ \alpha_2 \geqslant \alpha_3,\ \alpha_3 > \alpha_1 + \alpha_5$;

Case 5. $\alpha_2 > \alpha_1,\ \alpha_3 > \alpha_2,\ \alpha_1 + \alpha_5 \geqslant \alpha_3$;

Case 6. $\alpha_2 > \alpha_1,\ \alpha_3 > \alpha_2,\ \alpha_3 > \alpha_1 + \alpha_5$.

For Case 1. The conditions become

$$\alpha_1 + 2\alpha_3 + \alpha_5 \geqslant \alpha_2 + 2\alpha_4 \quad \text{.......(a)},$$

$$\begin{cases}\alpha_1 \geqslant \alpha_2 & \text{.......(b)},\\ \alpha_2 \geqslant \alpha_3 & \text{.......(c)},\end{cases}$$

$$\alpha_4 \geqslant \alpha_5 \quad \text{.......(d)},$$

$$\alpha_3 \geqslant \alpha_5 \quad \text{.......(e)},$$

$$\alpha_2 \geqslant \alpha_4 \quad \text{.......(f)},$$

for which the generating function is

$$\underset{\geqslant}{\Omega} \frac{1}{1 - abN_1 \,.\, 1 - \frac{cf}{ab}N_2 \,.\, 1 - \frac{a^2e}{c}N_3 \,.\, 1 - \frac{d}{a^2f}N_4 \,.\, 1 - \frac{a}{de}N_5};$$

which, eliminating b and c, is

$$\underset{\geqslant}{\Omega} \frac{1}{1 - aN_1 \,.\, 1 - fN_1N_2 \,.\, 1 - a^2efN_1N_2N_3 \,.\, 1 - \frac{d}{a^2f}N_4 \,.\, 1 - \frac{a}{de}N_5};$$

and eliminating d, e, and f,

$$\underset{\geqslant}{\Omega} \frac{1-(2211)}{1 - N_1N_2 \,.\, 1 - a^2N_1N_2N_3 \,.\, 1 - \frac{1}{a^2}N_1N_2N_4 \,.\, 1 - aN_1 \,.\, 1 - a(11111) \,.\, 1 - (1111)}.$$

Now $\underset{\geqslant}{\Omega} \dfrac{1}{1 - a^2x \,.\, 1 - ay \,.\, 1 - az \,.\, 1 - \frac{1}{a^2}w}$

$$= \frac{1}{1 - xw} \underset{\geqslant}{\Omega} \left\{ \frac{1}{1 - a^2x \,.\, 1 - ay \,.\, 1 - az} + \frac{\frac{w}{a^2}}{1 - ay \,.\, 1 - az \,.\, 1 - \frac{w}{a^2}} \right\}$$

$$= \frac{1}{1 - x \,.\, 1 - y \,.\, 1 - z \,.\, 1 - xw} + \frac{z^2w + yzw}{1 - xw \,.\, 1 - z \,.\, 1 - z^2w}$$

$$+ \underset{\geqslant}{\Omega} \frac{wy^2}{1 - ay \,.\, 1 - az \,.\, 1 - \frac{w}{a^2} \,.\, 1 - xw}$$

$$= \frac{1}{1 - x \,.\, 1 - y \,.\, 1 - z \,.\, 1 - xw} + \frac{z^2w + yzw}{1 - xw \,.\, 1 - z \,.\, 1 - z^2w}$$

$$+ \frac{y^2w}{1 - y \,.\, 1 - z \,.\, 1 - z^2w \,.\, 1 - xw} + \frac{y^2w\,(y^2w + yzw)}{1 - y \,.\, 1 - y^2w \,.\, 1 - z^2w \,.\, 1 - xw}.$$

Hence, putting $x = N_1N_2N_3$, $y = N_1$, $z = (11111)$, $w = N_1N_2N_4$, we have

$$xw = (2211),$$
$$y^2w = N_1^3N_2N_4,$$
$$z^2w = (33232),$$
$$yzw = (32121);$$

and we arrive at the three ground products

$$(11111),$$
$$(32121),$$
$$(33232),$$

which, as far as this case is concerned, are irreducible.

Case 2. $\alpha_1 \geqslant \alpha_2,\ \alpha_3 > \alpha_2.$

The system of conditions reduces to

$$\alpha_1 \geqslant \alpha_2 \quad \text{.........} (a),$$
$$\alpha_3 > \alpha_2 \quad \text{.........} (b),$$
$$\alpha_2 \geqslant \alpha_4 \quad \text{.........} (c),$$
$$\alpha_2 + \alpha_5 \geqslant \alpha_3 \quad \text{.........} (d),$$
$$\alpha_4 \geqslant \alpha_5 \quad \text{.........} (e),$$

and
$$\Sigma N_1^{\alpha_1}N_2^{\alpha_2}N_3^{\alpha_3}N_4^{\alpha_4}N_5^{\alpha_5}$$

$$= \underset{\geqslant}{\Omega} \frac{\frac{1}{b}}{1 - aN_1 . 1 - \frac{cd}{ab}N_2 . 1 - \frac{b}{d}N_3 . 1 - \frac{e}{c}N_4 . 1 - \frac{d}{e}N_5}$$

$$= \underset{\geqslant}{\Omega} \frac{\frac{1}{d}N_3}{1 - aN_1 . 1 - \frac{c}{a}N_2N_3 . 1 - \frac{1}{d}N_3 . 1 - \frac{1}{c}N_4 . 1 - \frac{d}{c}N_4N_5}$$

$$= \underset{\geqslant}{\Omega} \frac{\frac{1}{c}N_3N_4N_5}{1 - N_1 . 1 - cN_1N_2N_3 . 1 - \frac{1}{c}N_3N_4N_5 . 1 - \frac{1}{c}N_4 . 1 - \frac{1}{c}N_4N_5}$$

$$= \frac{(11211)}{1 - N_1 . 1 - N_1N_2N_3 . 1 - N_1N_2N_3N_4 . 1 - (11111) . 1 - (11211)},$$

yielding the new ground product

$$(11211).$$

Case 3. $\alpha_2 > \alpha_1, \ \alpha_2 \geqslant \alpha_3, \ \alpha_1 + \alpha_5 \geqslant \alpha_3.$

The reduced conditions are

$$\alpha_2 > \alpha_1 \quad \ldots\ldots(a),$$

$$\alpha_2 \geqslant \alpha_3 \quad \ldots\ldots(b),$$

$$\alpha_1 + \alpha_5 \geqslant \alpha_3 \quad \ldots\ldots(c),$$

$$\alpha_1 + 2\alpha_3 + \alpha_5 \geqslant \alpha_2 + 2\alpha_4 \quad \ldots\ldots(d),$$

$$\alpha_1 + \alpha_4 \geqslant \alpha_3 \quad \ldots\ldots(e),$$

$$\alpha_2 \geqslant \alpha_4 \quad \ldots\ldots(f),$$

$$\alpha_4 \geqslant \alpha_5 \quad \ldots\ldots(g),$$

$$\alpha_3 \geqslant \alpha_5 \quad \ldots\ldots(h),$$

$$\alpha_1 \geqslant \alpha_5 \quad \ldots\ldots(i),$$

of which the generator is

$$\underset{\geqslant}{\Omega} \frac{\frac{1}{a}}{1 - \frac{cdei}{a} N_1 \,.\, 1 - \frac{abf}{d} N_2 \,.\, 1 - \frac{d^2 h}{bce} N_3 \,.\, 1 - \frac{eg}{d^2 f} N_4 \,.\, 1 - \frac{cd}{ghi} N_5}$$

$$= \underset{\geqslant}{\Omega} \frac{\frac{bf}{d} N_2}{1 - bcefN_1N_2 \,.\, 1 - \frac{bf}{d} N_2 \,.\, 1 - \frac{d^2}{bce} N_3 \,.\, 1 - \frac{e}{d^2 f} N_4 \,.\, 1 - ced\, N_1N_2N_3N_4N_5};$$

the result of eliminating a, g, h and i.

It might be thought advisable at this stage to eliminate b or f, but experiment shews an advantage in proceeding with d.

Consider

$$\underset{\geqslant}{\Omega} \frac{\frac{p}{d}}{1 - d^2 x \,.\, 1 - \frac{y}{d^2} \,.\, 1 - dz \,.\, 1 - \frac{w}{d}}$$

$$= \underset{\geqslant}{\Omega} \frac{\frac{p}{d}}{1 - xy \,.\, 1 - zw} \left(\frac{1}{1 - d^2 x} + \frac{\frac{y}{d^2}}{1 - \frac{y}{d^2}} \right) \left(\frac{1}{1 - dz} + \frac{\frac{w}{d}}{1 - \frac{w}{d}} \right)$$

$$= \underset{\geqslant}{\Omega} \frac{\frac{p}{d}}{1 - xy \,.\, 1 - zw \,.\, 1 - d^2 x \,.\, 1 - dz} + \underset{\geqslant}{\Omega} \frac{\frac{1}{d^2} pw}{1 - xy \,.\, 1 - zw \,.\, 1 - d^2 x \,.\, 1 - \frac{w}{d}}$$

$$+ \underset{\geqslant}{\Omega} \frac{\frac{1}{d^3} py}{1 - xy \,.\, 1 - zw \,.\, 1 - dz \,.\, 1 - \frac{y}{d^2}}$$

$$= \frac{px}{1-x\,.\,1-xy\,.\,1-zw} + \frac{pz}{1-x\,.\,1-z\,.\,1-xy\,.\,1-zw}$$

$$+ \frac{pxw\,(1+xw)}{1-x\,.\,1-xy\,.\,1-zw\,.\,1-xw^2} + \frac{pyz^3}{1-z\,.\,1-xy\,.\,1-zw\,.\,1-yz^2}.$$

Hence the generator is :—

$$\left(\text{putting } p = bfN_2,\ x = \frac{1}{bce}N_3,\ y = \frac{e}{f}N_4,\ z = ced\,(11111)\right)$$

$$\underset{\geqslant}{\Omega}\left\{\frac{\frac{f}{ce}N_2N_3}{1-\frac{1}{bce}N_3\,.\,1-\frac{1}{bcf}N_3N_4\,.\,1-bcef\,(12111)\,.\,1-bcefN_1N_2}\right.$$

$$+ \frac{bcef\,(12111)}{1-\frac{1}{bce}N_3\,.\,1-ce\,(11111)\,.\,1-\frac{1}{bcf}N_3N_4\,.\,1-bcef\,(12111)\,.\,1-bcefN_1N_2}$$

$$+ \frac{bc^3\,(34343)}{1-c\,(11111)\,.\,1-\frac{1}{bcf}N_3N_4\,.\,1-bcf\,(12111)\,.\,1-\frac{c^2}{f}\,(22232)\,.\,1-bcfN_1N_2}$$

$$\left.+ \frac{\frac{bf^2}{ce}N_2^2N_3\left(1+\frac{f}{ce}N_2N_3\right)}{1-\frac{1}{bce}N_3\,.\,1-\frac{1}{bcf}N_3N_4\,.\,1-bcef\,(12111)\,.\,1-\frac{bf^2}{ce}N_2^2N_3\,.\,1-bcefN_1N_2}\right\}$$

$= A + B + C + D$, suppose.

$$\text{Now } A = \underset{\geqslant}{\Omega}\frac{\frac{1}{ce}N_2N_3}{1-bceN_1N_2\,.\,1-bce\,(12111)\,.\,1-e\,(12221)\,.\,1-\frac{1}{bce}N_3}$$

$$+ \underset{\geqslant}{\Omega}\frac{\frac{1}{bc^2e}N_2N_3^2N_4}{1-bceN_1N_2\,.\,1-e\,(1111)\,.\,1-e\,(12221)\,.\,1-\frac{1}{bce}N_3}$$

$$= \underset{\geqslant}{\Omega}\frac{\frac{1}{ce}N_2N_3\,\{1-N_1N_2^2N_3\,(11111)\}}{1-ceN_1N_2\,.\,1-ce\,(12111)\,.\,1-e\,(12221)\,.\,1-N_1N_2N_3\,.\,1-(12211)}$$

$$+ \underset{\geqslant}{\Omega}\frac{be\,(2321)}{1-beN_1N_2\,.\,1-e\,(1111)\,.\,1-e\,(12221)\,.\,1-N_1N_2N_3}$$

$$= \frac{(13211)\,\{1-(121)\,(11111)\}}{1-(111)\,.\,1-(12111)\,.\,1-(12211)\,.\,1-(12221)}$$

$$+ \frac{(121)\,\{1-(121\}\,(11111)\}}{1-(11)\,.\,1-(111)\,.\,1-(12111)\,.\,1-(12211)\,.\,1-(12221)}$$

$$+ \frac{(121)\,(1111)}{1-(11)\,.\,1-(1111)\,.\,1-(111)\,.\,1-(12221)};$$

a result which indicates the new ground forms

$$(12111),$$
$$(12211),$$
$$(12221),$$
$$(13211).$$

B is easily shewn to have the expression

$$\frac{(12111)}{1-(11).1-(111).1-(1111).1-(11111).1-(12111)}$$
$$+\frac{(12211)}{1-(111).1-(1111).1-(11111).1-(12111).1-(12211)}.$$

C, by elimination of b and c (in one operation), becomes

$$\frac{(34343)}{1-(1111).1-(11111)}\underset{\geqslant}{\Omega}\frac{1}{1-f(12111).1-f(11).1-\frac{1}{f}(22232)}$$

$$+\frac{(34453)}{1-(1111).1-(11111).1-(12221)}\underset{\geqslant}{\Omega}\frac{\frac{1}{f}}{1-f(12111).1-\frac{1}{f}(22232)}$$

$$=\frac{(34343)\{1-(45343)\}}{1-(11).1-(1111).1-(11111).1-(12111).1-(34343).1-(33232)}$$
$$+\frac{(46564)}{1-(1111).1-(11111).1-(12221).1-(12111).1-(34343)};$$

wherein observe that $(45343)=(11)(34343),$

$$(46564)=(12221)(34343);$$

so that (34343) is the only new ground product that emerges.

Separating the numerator terms of D it can be written D_1+D_2.

For D_1 we require the result

$$\underset{\geqslant}{\Omega}\frac{\frac{w}{e}}{1-ex.1-ey.1-\frac{z}{e}.1-\frac{w}{e}}$$

$$=\frac{yw}{1-y.1-xz.1-yw}+\frac{xw}{1-x.1-y.1-xz.1-yw}+\frac{x^2w^2}{1-x.1-xz.1-xw.1-yw}$$
$$+\frac{y^2zw}{1-y.1-xz.1-yz.1-yw};$$

4—2

which $\left(\text{putting } x = bcf(12111),\ y = bcf(11),\ z = \frac{1}{bc}N_3,\ w = \frac{bf^2}{c}N_2^2N_3\right)$ brings it to

$$\underset{\geqslant}{\Omega}\frac{b^2f^3(131)}{1-\frac{1}{bcf}N_3N_4\,.\,1-f(12211)\,.\,1-b^2f^3(131)\,.\,1-bcf(11)}$$

$$+\underset{\geqslant}{\Omega}\frac{b^2f^3(14211)}{1-\frac{1}{bcf}N_3N_4\,.\,1-f(12211)\,.\,1-b^2f^3(131)\,.\,1-bcf(11)\,.\,1-bcf(12111)}$$

$$+\underset{\geqslant}{\Omega}\frac{b^4f^6(28422)}{1-\frac{1}{bcf}N_3N_4\,.\,1-f(12211)\,.\,1-b^2f^3(131)\,.\,1-bcf(12111)\,.\,1-b^2f^3(14211)}$$

$$+\underset{\geqslant}{\Omega}\frac{b^2f^4(121)^2}{1-\frac{1}{bcf}N_3N_4\,.\,1-f(12211)\,.\,1-b^2f^3(131)\,.\,1-bcf(11)\,.\,1-f(111)}$$

$$=\frac{(131)}{1-(11)\,.\,1-(1111)\,.\,1-(131)\,.\,1-(12211)}$$

$$+\frac{(14211)\{1-(11)(12221)\}}{1-11\,.\,1-(1111)\,.\,1-(131)\,.\,1-(12111)\,.\,1-(12211)\,.\,1-(12221)}$$

$$+\frac{(14211)^2}{1-(131)\,.\,1-(12111)\,.\,1-(12211)\,.\,1-(12221)\,.\,1-(14211)}$$

$$+\frac{(121)^2}{1-(11)\,.\,1-(111)\,.\,1-(1111)\,.\,1-(131)\,.\,1-(12211)};$$

yielding the single new ground product

(14211).

For D_2 we require the result

$$\underset{\geqslant}{\Omega}\frac{\frac{p}{e^2}}{1-ex\,.\,1-ey\,.\,1-\frac{z}{e}\,.\,1-\frac{w}{e}}$$

$$=\frac{p}{1-xz\,.\,1-yw}\left\{\frac{y^2}{1-y}+\frac{xy}{1-y}+\frac{x^2}{1-x\,.\,1-y}+\frac{x^3w}{1-x\,.\,1-xw}+\frac{y^3z}{1-y\,.\,1-yz}\right\},$$

and putting $p = \frac{bf^3}{c^2}N_2^3N_3^2,\ x = bcf(12111),\ y = bcf(11),$

$$z = \frac{1}{bc}N_3, \qquad w = \frac{bf^2}{c}N_2^2N_3,$$

and observing that we may put $b=f=1$ and that moreover

$$xz=(12111),\quad yw=(131),$$
$$py^2=(121)(131),\quad pxy=(121)(14211),$$
$$px^2=(13211)(14211),\quad px^3w=(13211)(14211)^2,$$
$$py^3z=(121)^3,\quad xw=(14211),\quad yz=(111),$$

while operation upon the remaining letter c produces no new form, it is clear that no new form arises.

Case 4. $\alpha_2>\alpha_1,\ \alpha_2\geqslant\alpha_3,\ \alpha_3>\alpha_1+\alpha_5$.

The reduced conditions are

$$\alpha_2>\alpha_1 \quad\ldots\ldots(a),$$
$$\alpha_2\geqslant\alpha_3 \quad\ldots\ldots(b),$$
$$\alpha_3>\alpha_1+\alpha_5 \quad\ldots\ldots(c),$$
$$2\alpha_1+\alpha_3+2\alpha_5\geqslant\alpha_2+2\alpha_4 \quad\ldots\ldots(d),$$
$$\alpha_1+\alpha_4\geqslant\alpha_3 \quad\ldots\ldots(e),$$
$$\alpha_2\geqslant\alpha_4 \quad\ldots\ldots(f),$$
$$\alpha_4\geqslant\alpha_5 \quad\ldots\ldots(g),$$
$$\alpha_1\geqslant\alpha_5 \quad\ldots\ldots(h);$$

leading to $\Sigma N_1^{\alpha_1}N_2^{\alpha_2}N_3^{\alpha_3}N_4^{\alpha_4}N_5^{\alpha_5}$

$$=\underset{\geqslant}{\Omega}\frac{\frac{1}{ac}}{1-\frac{d^2eh}{ac}N_1\,.\,1-\frac{abf}{d}N_2\,.\,1-\frac{cd}{be}N_3\,.\,1-\frac{eg}{d^2f}N_4\,.\,\frac{d^2}{cgh}N_5};$$

and this by elimination of a, b, c, e, g and h becomes

$$\underset{\geqslant}{\Omega}\frac{\frac{1}{d^2}N_2N_3N_4}{1-d^2f(111)\,.\,1-d^2f(12211)\,.\,1-\frac{f}{d}N_2\,.\,1-\frac{1}{d^2f}N_4\,.\,1-\frac{1}{d^2}N_2N_3N_4};$$

and since

$$\underset{\geqslant}{\Omega}\frac{1}{1-fx\,.\,1-fy\,.\,1-fz\,.\,1-\frac{1}{f}w}$$

$$=\frac{1}{1-x\,.\,1-y\,.\,1-z\,.\,1-zw}+\frac{yw}{1-y\,.\,1-yw\,.\,1-zw}$$

$$+\frac{xw}{1-x\,.\,1-y\,.\,1-yw\,.\,1-zw}+\left[\frac{x^2w^2}{1-x\,.\,1-xw\,.\,1-yw\,.\,1-zw}\right]$$

this becomes:—

$$\left(\text{putting } x=d^2(111),\ y=d^2(12211),\ z=\frac{1}{d}N_2,\ w=\frac{1}{d^2}N_4\right)$$

$$\underset{\geqslant}{\Omega}\frac{\frac{1}{d^2}N_2N_3N_4}{1-\frac{1}{d^2}N_2N_3N_4}\left\{\frac{1}{1-d^2(111).1-d^2(12211).1-\frac{1}{d}N_2.1-\frac{1}{d^3}N_2N_4}\right.$$

$$+\frac{(12221)}{1-d^2(12211).1-\frac{1}{d^3}N_2N_4.1-(12221)}$$

$$\left.+\frac{(1111)}{1-d^2(111).1-d^2(12211).1-\frac{1}{d^3}N_2N_4.1-(12221)}\right\};$$

the fourth fraction being omitted as obviously contributing nothing new.

Now writing $x=(111)$, $y=(12211)$, $z=N_2$, $w=N_2N_3N_4$,

$$p=N_2N_4,$$

$$\underset{\geqslant}{\Omega}\frac{\frac{w}{d^2}}{1-d^2y.1-\frac{w}{d^2}.1-\frac{p}{d^3}}=\frac{yw(1+y^2p)}{1-y.1-yw.1-y^3p^2};$$

$$\underset{\geqslant}{\Omega}\frac{\frac{w}{d^2}}{1-d^2x.1-d^2y.1-\frac{w}{d^2}.1-\frac{p}{d^3}}=\frac{yw(1+y^2p)}{1-y.1-xw.1-yw.1-y^3p^2}$$

$$+\frac{x^3(1+y+y^2)(1+p)}{1-x.1-x^3p^2.1-y^3p^2}\cdot\frac{xw}{1-xw}$$

$$+\frac{y^3(1+p)}{1-x.1-y.1-y^3p^2}\cdot\frac{xw}{1-xw}$$

$$+\frac{K}{1-x^3p^2.1-y^3p^2}\cdot\frac{xw}{1-xw},$$

where

$$K=1+x+y+x^2+xy+y^2+x^2y+xy^2+x^2y^2$$
$$+(x^2+xy+y^2+x^2y+xy^2+x^2y^2)p$$
$$+(x^2y+xy^2+x^2y^2)p^2;$$

$$\underset{\geqslant}{\Omega}\frac{\frac{w}{d^2}}{1-d^2x.1-d^2y.1-\frac{z}{d}.1-\frac{w}{d^2}.1-\frac{p}{d^3}}$$

$$=\frac{yw(1+y^2p)(1+yz)}{1-y.1-xw.1-yw.1-y^3p^2.1-yz^2}+\frac{y^3zwp}{1-xw.1-yw.1-y^3p^2.1-yz^2}$$

$$+\frac{xw(1+xz)(1+y^2p)}{1-x.1-y.1-xw.1-xz^2.1-y^3p^2}+\frac{xw(yz+y^2zp)}{1-y.1-xw.1-y^3p^2}$$

$$+\frac{xw(xyp+xyzp+x^3p^2+x^2yp^2+x^4zp^2+x^3y^2p^3+x^2y^2zp^2+x^3yz^2p^2+x^3y^2zp^3)}{1-x.1-xw.1-xz^2.1-x^3p^2.1-y^3p^2}.$$

In verifying these laborious calculations the relations

$$\frac{1}{1-d^2y.1-\frac{z}{d}}=\frac{1+dyz}{1-yz^2.1-d^2y}+\frac{\frac{z}{d}}{1-yz^2.1-\frac{z}{d}},$$

$$\frac{1}{1-d^2y.1-\frac{p}{d^3}}=\frac{1+dy^2p}{1-y^3p^2.1-d^2y}+\frac{\frac{p}{d^3}+\frac{y^2p^2}{d^2}+\frac{yp}{d}}{1-y^3p^2.1-\frac{p}{d^3}},$$

will be found useful.

On examining these results we find that

$$yw=(13321)$$

is a new ground form, and that every other term is expressible by means of it and of ground forms previously reached.

Case 5. $\alpha_2>\alpha_1,\ \alpha_3>\alpha_2,\ \alpha_1+\alpha_5\geqslant\alpha_3$.

The reduced conditions are

$$\begin{array}{rl} \alpha_2>\alpha_1 & (a),\\ \alpha_3>\alpha_2 & (b),\\ \alpha_1+\alpha_5\geqslant\alpha_3 & (c),\\ \alpha_2\geqslant\alpha_4 & (d),\\ \alpha_4\geqslant\alpha_5 & (e),\\ \alpha_1\geqslant\alpha_5 & (f); \end{array}$$

from which

$$\underset{\geqslant}{\Omega}\frac{\frac{1}{ab}}{1-\frac{cf}{a}N_1.1-\frac{ab}{d}N_2.1-\frac{b}{c}N_3.1-\frac{e}{d}N_4.1-\frac{c}{ef}N_5},$$

$$=\underset{\geqslant}{\Omega}\frac{\frac{d}{c^2}N_2N_3^2}{1-c(11111).1-\frac{d}{c}N_2N_3.1-\frac{1}{c}N_3.1-d(111).1-\frac{1}{d}N_4},$$

$$=\underset{\geqslant}{\Omega}\frac{d(23422)}{1-d(12211).1-d(111).1-\frac{1}{d}N_4.1-(11111).1-(11211)},$$

$$= \frac{(12211)(11211)}{1-(11111).1-(11211)} \left\{ \frac{(12211)}{1-(12211).1-(111).1-(1111)} + \frac{(12221)}{1-(12211).1-(12221).1-(1111)} \right\},$$

so that new forms do not arise.

Case 6. $\alpha_2 > \alpha_1$, $\alpha_3 > \alpha_2$, $\alpha_3 > \alpha_1 + \alpha_5$.

The reduced conditions are

$$\alpha_3 > \alpha_2 \qquad (a),$$

$$\alpha_3 > \alpha_1 + \alpha_5 \qquad (b),$$

$$2\alpha_1 + \alpha_3 + 2\alpha_5 \geqslant \alpha_2 + 2\alpha_4 \qquad (c),$$

$$\alpha_2 \geqslant \alpha_4 \qquad (d),$$

$$\alpha_2 + \alpha_5 \geqslant \alpha_3 \qquad (e),$$

$$\alpha_1 + \alpha_4 \geqslant \alpha_3 \qquad (f),$$

$$\alpha_1 \geqslant \alpha_5 \qquad (g);$$

leading to

$$\underset{\geqslant}{\Omega} \frac{\frac{1}{ab}}{1-\frac{c^2fg}{b}N_1 . 1-\frac{de}{ac}N_2 . 1-\frac{abc}{ef}N_3 . 1-\frac{f}{c^2d}N_4 . 1-\frac{c^2e}{bg}N_5},$$

which is readily thrown into the form

$$\underset{\geqslant}{\Omega} \frac{\frac{c^5}{b^2}N_1N_3N_5\left(1-\frac{c^6}{b^2}N_1^2N_2N_3N_5\right)}{1-c^2(111) . 1-\frac{b}{c^2}N_2N_3N_4 . 1-\frac{c^2}{b}N_1 . 1-\frac{c^4}{b}N_1N_2N_3N_5 . 1-\frac{c^5}{b}N_1N_3N_5 . 1-\frac{c^4}{b^2}N_1N_5},$$

and eliminating b this is

$$\underset{\geqslant}{\Omega} \frac{c(12321)-c^3(11)(12321)^2}{1-c^2(111) . 1-\frac{1}{c^2}N_2N_3N_4 . 1-(1111) . 1-c^2(12211) . 1-c^3(11211) . 1-(12221)}$$

$$= \underset{\geqslant}{\Omega} \frac{c(12321)}{1-(1111) . 1-(12221) . 1-c^3(11211) . 1-c^2(111) . 1-\frac{1}{c^2}N_2N_3N_4}$$

$$+ \underset{\geqslant}{\Omega} \frac{c^3(12211)(12321)}{1-(12221) . 1-c^3(11211) . 1-c^2(12211) . 1-c^2(111) . 1-\frac{1}{c^2}N_2N_3N_4}.$$

Now

$$\underset{\geqslant}{\Omega} \frac{c}{1-c^3x . 1-c^2y . 1-\frac{1}{c^2}z}$$

$$= \frac{1}{1-x . 1-y . 1-yz} + \frac{xz+xz^2+x^2z^3}{1-x . 1-yz . 1-x^2z^3}$$

and

$$\underset{\geqslant}{\Omega} \frac{c^3}{1-c^3x \,.\, 1-c^2y \,.\, 1-c^2w \,.\, 1-\frac{1}{c^2}z}$$

$$= \frac{1}{1-x \,.\, 1-y \,.\, 1-w} + \frac{z}{1-x \,.\, 1-y \,.\, 1-w \,.\, 1-wz}$$

$$+ \frac{xz^2}{1-x \,.\, 1-y \,.\, 1-wz \,.\, 1-yz} + \frac{yz^2}{1-y \,.\, 1-wz \,.\, 1-yz}$$

$$+ \frac{xz^3+x^2z^4+x^3z^5}{1-x \,.\, 1-wz \,.\, 1-yz \,.\, 1-x^2z^3}.$$

Putting now

$$x=(11211),\quad y=(111),\quad w=(12211),\quad z=(0111),$$

we can examine the generating function.

It is clear that

$$xz=(12321)$$

is a ground product.

Also
$$x^2z^3=(12321)(13431)=(25752)$$

is a ground product, (13431) not being a solution of the conditions.

Further

$$(12211)(12321)z=(12321)(13321),$$
$$wz=(13321),$$
$$(12211)(12321)yz^2=(121)(25752),$$
$$(12211)(12321)xz^3=(13321)(25752);$$

so that there are no more ground products.

We have therefore in Case 6 obtained the new fundamental forms :—

(12321),

(25752).

The investigation that has been given does not establish that the 13 forms obtained are ground products *quâ* the whole of the six cases, but it does prove that all the ground products are included amongst these 13. But it is clear that all forms in which $\alpha_1 = 1$ are necessarily ground products. This accounts for 9 of the 13 and it is easy by actual experiment to convince oneself that the remaining 4, viz. :—

(25752),

(32121),

(33232),

(34343),

are, in fact, irreducible.

Hence the 13 ground products of order 5 are established.

Finally, to resume the foregoing, it has been shewn, in respect of the arithmetical function

$$\left(\frac{n+1}{1}\right)^{a_1} \left(\frac{n+2}{2}\right)^{a_2} \left(\frac{n+3}{3}\right)^{a_3} \left(\frac{n+4}{4}\right)^{a_4} \left(\frac{n+5}{5}\right)^{a_5} \equiv (\alpha_1, \ \alpha_2, \ \alpha_3, \ \alpha_4, \ \alpha_5),$$

n being any integer whatever, that all integral forms are expressible as products of

$$
\begin{array}{ll}
\{(1) & \text{order 1,} \\
\{(11) & \text{order 2,} \\
\left\{\begin{array}{l}(111) \\ (121) \\ (131)\end{array}\right. & \text{order 3,} \\
\left\{\begin{array}{l}(1111) \\ (3101) \\ (1221)\end{array}\right. & \text{order 4,} \\
\left\{\begin{array}{l}(11111) \\ (11211) \\ (12111) \\ (12211) \\ (12221) \\ (12321) \\ (13211) \\ (13321) \\ (14211) \\ (25752) \\ (32121) \\ (33232) \\ (34343)\end{array}\right. & \text{order 5.}
\end{array}
$$

V. *The Diophantine Inequality* $\lambda x \geqslant \mu y$. By Major P. A. MacMahon, R.A., D.Sc., F.R.S., Hon. Mem. Camb. Phil. Soc.

[*Received* 7 April 1900.]

§ 1. I have shown elsewhere* that the theory of Partitions may be discussed by means of one or more inequalities in integers of the type

$$\lambda_1\alpha_1 + \lambda_2\alpha_2 + \ldots + \lambda_s\alpha_s \geqslant \mu_1\beta_1 + \mu_2\beta_2 + \ldots + \mu_t\beta_t,$$

where all the quantities denote positive integers, and $\lambda_1, \lambda_2, \ldots \lambda_s, \mu_1, \mu_2, \ldots \mu_t$ are given.

Hilbert has shown† that in respect of any such system of Diophantine *equations*, say of type

$$\lambda_1\alpha_1 + \lambda_2\alpha_2 + \ldots + \lambda_s\alpha_s = \mu_1\beta_1 + \mu_2\beta_2 + \ldots + \mu_t\beta_t,$$

there exists a finite number of solutions

$$\begin{array}{ll} \alpha_1', \alpha_2', \ldots \alpha_s'; & \beta_1', \beta_2', \ldots \beta_t', \\ \alpha_1'', \alpha_2'', \ldots \alpha_s''; & \beta_1'', \beta_2'', \ldots \beta_t'', \\ \ldots\ldots\ldots & \ldots\ldots\ldots \\ \alpha_1^{(k)}, \alpha_2^{(k)}, \ldots \alpha_s^{(k)}; & \beta_1^{(k)}, \beta_2^{(k)}, \ldots \beta_t^{(k)}, \end{array}$$

such that the most general solution may be written

$$\begin{aligned} \alpha_1 &= A_1\alpha_1' + A_2\alpha_1'' + \ldots + A_k\alpha_1^{(k)}, \\ \alpha_2 &= A_1\alpha_2' + A_2\alpha_2'' + \ldots + A_k\alpha_2^{(k)}, \\ &\ldots\ldots\ldots \\ \alpha_s &= A_1\alpha_s' + A_2\alpha_s'' + \ldots + A_k\alpha_s^{(k)}, \\ \beta_1 &= A_1\beta_1' + A_2\beta_1'' + \ldots + A_k\beta_1^{(k)}, \\ \beta_2 &= A_1\beta_2' + A_2\beta_2'' + \ldots + A_k\beta_2^{(k)}, \\ &\ldots\ldots\ldots \\ \beta_t &= A_1\beta_t' + A_2\beta_t'' + \ldots + A_k\beta_t^{(k)}, \end{aligned}$$

where $A_1, A_2, \ldots A_k$ have positive integral, including zero, values which may be assigned at pleasure.

* *Phil. Trans. R. S.*, Vol. 192, Series A, pp. 351—401.
† *Math. Ann.* t. xxx.

The like theorem obtains also for a system of Diophantine inequalities; this is intuitive directly the observation is made that the inequality

$$\lambda_1\alpha_1 + \lambda_2\alpha_2 + \ldots + \lambda_s\alpha_s \geqslant \mu_1\beta_1 + \mu_2\beta_2 + \ldots + \mu_t\beta_t$$

is equivalent to the equation

$$\lambda_1\alpha_1 + \lambda_2\alpha_2 + \ldots + \lambda_s\alpha_s = \mu_1\beta_1 + \mu_2\beta_2 + \ldots + \mu_t\beta_t + \beta_{t+1};$$

for this shows that the theory of Diophantine inequalities is, in reality, a special case of the theory of Diophantine equations. The inequality

$$\lambda\alpha \geqslant \mu\beta$$

is thus equivalent to the equation

$$\lambda\alpha = \mu\beta + \beta;$$

and I propose to determine the fundamental solutions of this equation as a contribution not only to Partition Analysis but also to the Theory of Hilbert. The quantities λ, μ are supposed to be given relatively prime positive integers and I seek the solutions from which all others can be obtained by addition. To reach the end in view I seek the sum $\Sigma x^\alpha y^\beta$, the summation being for all solutions of the given inequality $\lambda\alpha \geqslant \mu\beta$; and thence it is easy to form the sum

$$\Sigma x^\alpha y^\beta z^\gamma,$$

appertaining to the associated Diophantine equation. The summation can be carried out in a variety of ways and the result exhibited in a variety of forms; but it is not every form of result which establishes, in an irrefragable manner, the whole of the fundamental solutions. The desirable method of procedure is not the most obvious one and the desired result is not the most compact obtainable though it is one of great elegance.

The method of summation which first presents itself to the mind is first to sum α from $\overset{\geqslant}{E}\frac{\mu}{\lambda}\beta$ to ∞, where the symbol $\overset{\geqslant}{E}$ denotes the smallest integer $\geqslant$ than the quantity which follows it, and then to sum β from 0 to ∞.

The result easily obtainable is

$$\frac{1 + x^{\overset{\geqslant}{E}\frac{\mu}{\lambda}}y + x^{\overset{\geqslant}{E}\frac{2\mu}{\lambda}}y^2 + \ldots + x^{\overset{\geqslant}{E}\frac{\lambda-1}{\lambda}\mu}y^{\lambda-1}}{1 - x \,.\, 1 - x^\mu y^\lambda},$$

which certainly proves that the ground solutions are included in the exponents of the x, y products in numerator and denominator but does not indicate which of these are the ground solutions. The true method leads to a sum of algebraic fractions and not to a single fraction, as will appear.

We require a preliminary lemma which may be thus stated:—

Lemma. The relation

$$\lambda\alpha \geqslant \mu\beta$$

may be made to depend upon a similar relation in which λ is unchanged and $\lambda > \mu$.

We have

$$\Sigma x^{\alpha} y^{\beta} = \underset{\geqslant}{\Omega} \frac{1}{1 - a^{\lambda}x \,.\, 1 - \dfrac{y}{a^{\mu}}},$$

where $\underset{\geqslant}{\Omega}$ is an operator which expresses that when the fraction is expanded in ascending powers of x and y we are to reject all terms which involve negative powers of a and subsequently to put a equal to unity. Now if p be the greatest integer in $\frac{\mu}{\lambda}$, the given relation implies the relation $\alpha \geqslant p\beta$, and we may write

$$\Sigma x^{\alpha} y^{\beta} = \underset{\geqslant}{\Omega} \frac{1}{1 - a^{\lambda}bx \,.\, 1 - \dfrac{y}{a^{\mu}b^{p}}},$$

where $\underset{\geqslant}{\Omega}$ operates upon both a and b. Eliminating b we find

$$\Sigma x^{\alpha} y^{\beta} = \underset{\geqslant}{\Omega} \frac{1}{1 - a^{\lambda}x \,.\, 1 - \dfrac{x^{p}y}{a^{\mu-\lambda p}}},$$

which denotes also the sum

$$\Sigma x^{\alpha} (x^{p}y)^{\beta} = \Sigma x^{\alpha+p\beta} y^{\beta},$$

where α and β are connected by the relation

$$\lambda\alpha \geqslant (\mu - \lambda p)\,\beta.$$

We have therefore reduced the sum

$$\Sigma x^{\alpha} y^{\beta} \text{ for the relation } \lambda\alpha \geqslant \mu\beta$$

to the sum

$$\Sigma x^{\alpha+p\beta} y^{\beta} \text{ for the relation } \lambda\alpha \geqslant (\mu - \lambda p)\,\beta;$$

wherein, from the definition of p, $\lambda > \mu - \lambda p$.

This proves the lemma.

In the next place I say that there is a second lemma which involves a further reduction.

Lemma. The relation

$$\lambda\alpha \geqslant \mu\beta,$$

wherein $\lambda > \mu$, may be made to depend upon a similar relation in which μ is unchanged and $\lambda < \mu$.

This is really the key to the solution of the problem before us, as will appear.

Observe that the relation

$$\lambda\alpha \geqslant \mu\beta, \text{ where } \lambda > \mu,$$

may be broken up into the two simultaneous sets of relations:—

$$\begin{array}{ll} \lambda\alpha \geqslant \mu\beta, & \text{and } \lambda\alpha \geqslant \mu\beta, \\ \beta \geqslant \alpha; & \qquad \alpha > \beta; \end{array}$$

and that the second of these simultaneous sets,

$$\lambda\alpha \geqslant \mu\beta,$$
$$\alpha > \beta,$$

may be replaced by the single relation

$$\alpha > \beta,$$

since, λ being $> \mu$, the relation

$$\lambda\alpha \geqslant \mu\beta$$

is implied thereby.

We may therefore separate a portion of the generating function corresponding to

$$\alpha > \beta,$$

viz.:—

$$\frac{x}{1-x\,.\,1-xy};$$

and consider the remaining portion, corresponding to the first simultaneous set, viz.:—

$$\lambda\alpha \geqslant \mu\beta,$$
$$\beta \geqslant \alpha.$$

The crude generator is

$$\underset{\geqslant}{\Omega} \frac{1}{1-\dfrac{a^{\lambda}}{b}x\,.\,1-\dfrac{by}{a^{\mu}}}$$

$$= \underset{\geqslant}{\Omega} \frac{1}{1-a^{\lambda-\mu}xy\,.\,1-\dfrac{y}{a^{\mu}}},$$

which gives the sum

$$\Sigma\,(xy)^{\alpha}\,y^{\beta},$$

for the relation

$$(\lambda-\mu)\,\alpha \geqslant \mu\beta.$$

If $\lambda-\mu<\mu$ we have done what was required, but if $\lambda-\mu>\mu$ we may repeat the process; and just as we have found

$$\underset{\geqslant}{\Omega} \frac{1}{1-a^{\lambda}x\,.\,1-\dfrac{y}{a^{\mu}}} = \frac{x}{1-x\,.\,1-xy} + \underset{\geqslant}{\Omega} \frac{1}{1-a^{\lambda-\mu}xy\,.\,1-\dfrac{y}{a^{\mu}}},$$

we shall find

$$\underset{\geqslant}{\Omega} \frac{1}{1-a^{\lambda-\mu}xy \,.\, 1-\dfrac{y}{a^{\mu}}} = \frac{xy}{1-xy \,.\, 1-xy^2} + \underset{\geqslant}{\Omega} \frac{1}{1-a^{\lambda-2\mu}xy^2 \,.\, 1-\dfrac{y}{a^{\mu}}},$$

and, if q be the greatest integer in $\dfrac{\lambda}{\mu}$, we obtain finally

$$\underset{\geqslant}{\Omega} \frac{1}{1-a^{\lambda}x \,.\, 1-\dfrac{y}{a^{\mu}}}$$

$$= \frac{x}{1-x \,.\, 1-xy} + \frac{xy}{1-xy \,.\, 1-xy^2} + \dots \frac{xy^{q-1}}{1-xy^{q-1} \,.\, 1-xy^q}$$

$$+ \underset{\geqslant}{\Omega} \frac{1}{1-a^{\lambda-\mu q}xy^q \,.\, 1-\dfrac{y}{a^{\mu}}},$$

where the last written portion of the generator represents the sum $\Sigma (xy^q)^{\alpha} y^{\beta}$ for the relation

$$(\lambda - \mu q)\, \alpha \geqslant \mu\beta.$$

Hence

Lemma. The relation $\lambda\alpha \geqslant \mu\beta$, where $\lambda > \mu$, may be made to depend upon a similar relation, in which $\lambda < \mu$, as shewn by the identity

$$\underset{\geqslant}{\Omega} \frac{1}{1-a^{\lambda}x \,.\, 1-\dfrac{y}{a^{\mu}}}$$

$$= \underset{\geqslant}{\Omega} \frac{1}{1-a^{\lambda-\mu q}\, xy^q \,.\, 1-\dfrac{y}{a^{\mu}}} + \sum_{1}^{q} \frac{xy^{q-1}}{1-xy^{q-1} \,.\, 1-xy^q};$$

wherein q is the greatest integer in $\dfrac{\lambda}{\mu}$.

These two lemmas evidently yield a process of reduction which can be pushed to the last extent.

In what follows λ will be considered $> \mu$. If $\mu > \lambda$ it will be merely necessary to write $\mu - \lambda p$ for μ and $x^p y$ for y where p is the greatest integer in $\dfrac{\mu}{\lambda}$.

It will be surmised that the reduction depends upon the convergents of the continued fraction $\dfrac{\lambda}{\mu}$.

Let $$\frac{\lambda}{\mu} = a_1 + \frac{1}{a_2 +}\, \frac{1}{a_3 +}\, \frac{1}{a_4 +} \dots \frac{1}{+ a_n},$$

where n is uneven; this is always possible because a_n may, if convenient, be written

$$a_n - 1 + \frac{1}{1}.$$

Take as the first two convergents

$$\frac{0}{1} = \frac{p_{-1}}{q_{-1}}, \quad \frac{1}{0} = \frac{p_0}{q_0},$$

and write down the convergents of uneven order

$$\frac{p_{-1}}{q_{-1}}, \quad \frac{p_1}{q_1}, \quad \frac{p_3}{q_3}, \quad \ldots\ldots \frac{p_n}{q_n},$$

and form the ascending series of intermediate convergents, viz.:—

$$\frac{p_{-1}}{q_{-1}}, \quad \frac{p_{-1}+p_0}{q_{-1}+q_0}, \quad \frac{p_{-1}+2p_0}{q_{-1}+2q_0}, \quad \ldots\ldots \frac{p_{-1}+(a_1-1)p_0}{q_{-1}+(a_1-1)q_0},$$

$$\frac{p_1}{q_1}, \quad \frac{p_1+p_2}{q_1+q_2}, \quad \frac{p_1+2p_2}{q_1+2q_2}, \quad \ldots\ldots \frac{p_1+(a_3-1)p_2}{q_1+(a_3-1)q_2},$$

....

$$\frac{p_{n-2}}{q_{n-2}}, \quad \frac{p_{n-2}+p_{n-1}}{q_{n-2}+q_{n-1}}, \quad \frac{p_{n-2}+2p_{n-1}}{q_{n-2}+2q_{n-1}}, \quad \ldots\ldots \frac{p_{n-2}+(a_n-1)p_{n-1}}{q_{n-2}+(a_n-1)q_{n-1}}, \quad \frac{p_n}{q_n};$$

where of course $\frac{p_n}{q_n} = \frac{\lambda}{\mu}$.

Applying now the second lemma to the crude generator

$$\underset{\geqslant}{\Omega} \frac{1}{1 - a^{\lambda}x \,.\, 1 - \frac{y}{a^{\mu}}},$$

we obtain

$$\underset{\geqslant}{\Omega} \frac{1}{1 - a^{\lambda-\mu a_1} xy^{a_1} \,.\, 1 - \frac{y}{a^{\mu}}} + \sum_1^{a_1} \frac{xy^{a_1-1}}{1 - xy^{a_1-1} \,.\, 1 - xy^{a_1}},$$

where

$$\sum_1^{a_1} \frac{xy^{a_1-1}}{1 - xy^{a_1-1} \,.\, 1 - xy^{a_1}}$$

$$= \frac{x^1y^0}{1 - x^1y^0 \,.\, 1 - x^{1+0}y^{0+1}} + \frac{x^{1+0}y^{0+1}}{1 - x^{1+0}y^{0+1} \,.\, 1 - x^{1+2\cdot 0}y^{0+2\cdot 1}}$$

$$+ \ldots + \frac{x^{1+(a_1-1)\,0}\,y^{0+(a_1-1)\,1}}{1 - x^{1+(a_1-1)\,0}\,y^{0+(a_1-1)\,1} \,.\, 1 - x^{1+a_1 0}\,y^{0+a_1\cdot 1}}$$

$$= \frac{x^{q_{-1}}\,y^{p_{-1}}}{1 - x^{q_{-1}}\,y^{p_{-1}} \,.\, 1 - x^{q_{-1}+q_0}\,y^{p_{-1}+p_0}} + \frac{x^{q_{-1}+q_0}\,y^{p_{-1}+p_0}}{1 - x^{q_{-1}+q_0}\,y^{p_{-1}+p_0} \,.\, 1 - x^{q_{-1}+2q_0}\,y^{p_{-1}+2p_0}}$$

$$+ \ldots + \frac{x^{q_{-1}+(a_1-1)\,q_0}\,y^{p_{-1}+(a_1-1)\,p_0}}{1 - x^{q_{-1}+(a_1-1)\,q_0}\,y^{p_{-1}+(a_1-1)\,p_0} \,.\, 1 - x^{q_1}\,y^{p_1}},$$

wherein the exponents are derived from the intermediate convergents to $\frac{p_{-1}}{q_{-1}}, \frac{p_1}{q_1}$ inclusive of these principal convergents.

The remaining part, viz. :—

$$\underset{\geqslant}{\Omega} \frac{1}{1 - a^{\lambda - \mu a_1} xy^{a_1} \cdot 1 - \frac{y}{a^{\mu}}},$$

must now be subjected to the process of the first lemma and becomes

$$\underset{\geqslant}{\Omega} \frac{1}{1 - a^{\lambda - \mu a_1} bxy^{a_1} \cdot 1 - \frac{y}{a^{\mu} b^{a_2}}} = \underset{\geqslant}{\Omega} \frac{1}{1 - a^{\lambda - \mu a_1} x^{q_1} y^{p_1} \cdot 1 - \frac{x^{q_2} y^{p_2}}{a^{\mu p_2 - \lambda q_2}}};$$

and now the operation of the second lemma yields

$$\underset{\geqslant}{\Omega} \frac{1}{1 - a^{\lambda - \mu a_1 - a_3(\mu p_2 - \lambda q_2)} x^{q_1 + a_3 q_2} y^{p_1 + a_3 p_2} \cdot 1 - \frac{x^{q_2} y^{p_2}}{a^{\mu p_2 - \lambda q_2}}}$$

$$+ \sum_1^{a_3} \frac{x^{q_1} y^{p_1} (x^{q_2} y^{p_2})^{a_3 - 1}}{1 - x^{q_1} y^{p_1} (x^{q_2} y^{p_2})^{a_3 - 1} \cdot 1 - x^{q_1} y^{p_1} (x^{q_2} y^{p_2})^{a_3}}$$

$$= \underset{\geqslant}{\Omega} \frac{1}{1 - a^{\lambda q_3 - \mu p_3} x^{q_3} y^{p_3} \cdot 1 - \frac{x^{q_2} y^{p_2}}{a^{\mu p_2 - \lambda q_2}}}$$

$$+ \frac{x^{q_1} y^{p_1}}{1 - x^{q_1} y^{p_1} \cdot 1 - x^{q_1 + q_2} y^{p_1 + p_2}} + \frac{x^{q_1 + q_2} y^{p_1 + p_2}}{1 - x^{q_1 + q_2} y^{p_1 + p_2} \cdot 1 - x^{q_1 + 2q_2} y^{p_1 + 2p_2}}$$

$$+ \ldots + \frac{x^{q_1 + (a_3 - 1) q_2} y^{p_1 + (a_3 - 1) p_2}}{1 - x^{q_1 + (a_3 - 1) q_2} y^{p_1 + (a_3 - 1) p_2} \cdot 1 - x^{q_3} y^{p_3}};$$

and we have before us a new portion of the generating function corresponding to the intermediate convergents to the principal convergents $\frac{p_1}{q_1}, \frac{p_3}{q_3}$, the latter both included.

The portion remaining,

$$\underset{\geqslant}{\Omega} \frac{1}{1 - a^{\lambda q_3 - \mu p_3} x^{q_3} y^{p_3} \cdot 1 - \frac{x^{q_2} y^{p_2}}{a^{\mu p_2 - \lambda q_2}}},$$

is now subject to Lemma 1 and the Lemmas operate alternately until the final result is reached. The last remaining portion must be

$$\underset{\geqslant}{\Omega} \frac{1}{1 - a^{\lambda q_n - \mu p_n} x^{q_n} y^{p_n} \cdot 1 - \frac{x^{q_{n-1}} y^{p_{n-1}}}{a^{\mu p_{n-1} - \lambda q_{n-1}}}}$$

$$= \frac{1}{1 - x^{q_n} y^{p_n}} = 1 + \frac{x^{q_n} y^{p_n}}{1 - x^{q_n} y^{p_n}}.$$

Hence the complete generating function may be written

$$1+\frac{x}{1-x\,.\,1-xy}+\frac{xy}{1-xy\,.\,1-xy^2}+\ldots+\frac{x^{q_1}y^{p_1-1}}{1-x^{q_1}y^{p_1-1}\,.\,1-x^{q_1}y^{p_1}}$$

$$+\frac{x^{q_1}y^{p_1}}{1-x^{q_1}y^{p_1}\,.\,1-x^{q_1+q_2}y^{p_1+p_2}}+\frac{x^{q_1+q_2}y^{p_1+p_2}}{1-x^{q_1+q_2}y^{p_1+p_2}\,.\,1-x^{q_1+2q_2}y^{p_1+2p_2}}+\ldots$$

$$+\frac{x^{q_1+(a_3-1)q_2}y^{p_1+(a_3-1)p_2}}{1-x^{q_1+(a_3-1)q_2}y^{p_1+(a_3-1)p_2}\,.\,1-x^{q_3}y^{p_3}}$$

$$+\ldots\ldots$$

$$+\frac{x^{q_{n-2}}y^{p_{n-2}}}{1-x^{q_{n-2}}y^{p_{n-2}}\,.\,1-x^{q_{n-2}+q_{n-1}}y^{p_{n-2}+p_{n-1}}}+\ldots+\frac{x^{q_{n-2}+(a_n-1)q_{n-1}}y^{p_{n-2}+(a_n-1)p_{n-1}}}{1-x^{q_{n-2}+(a_n-1)q_{n-1}}y^{p_{n-2}+(a_n-1)p_{n-1}}\,.\,1-x^{q_n}y^{p_n}}$$

$$+\frac{x^{q_n}y^{p_n}}{1-x^{q_n}y^{p_n}},$$

and, from well-known properties of the convergents, every xy product occurring in these fractions gives a ground solution of the relation.

Hence the ground solutions are

$$\alpha=r,\quad \beta=s,$$

where $\frac{s}{r}$ is in its lowest terms and is any member of the ascending intermediate series of convergents to $\frac{\lambda}{\mu}$.

As an example, take

$$779\alpha \geqslant 207\beta,$$

where

$$\frac{779}{207}=3+\frac{1}{1+}\frac{1}{3+}\frac{1}{4+}\frac{1}{2+}\frac{1}{4+}\frac{1}{1}.$$

The principal convergents are

$$\frac{0}{1},\ \frac{1}{0},\ \frac{3}{1},\ \frac{4}{1},\ \frac{15}{4},\ \frac{64}{17},\ \frac{143}{38},\ \frac{636}{169},\ \frac{779}{207},$$

and the ascending intermediate series

$$\frac{0}{1},\ \frac{1}{1},\ \frac{2}{1},\ \frac{3}{1},\ \frac{7}{2},\ \frac{11}{3},\ \frac{15}{4},\ \frac{79}{21},\ \frac{143}{38},\ \frac{779}{207},$$

where, if the principal convergents are $2m+3$ in number, the number in the intermediate series is

$$m+2+a_1+a_3+a_5+\ldots+a_{2m+1}-m-1=1+a_1+a_3+a_5+\ldots+a_{2m+1}.$$

Hence $\Sigma x^\alpha y^\beta$ is

$$1+\frac{x}{1-x\,.\,1-xy}+\frac{xy}{1-xy\,.\,1-xy^2}+\frac{xy^2}{1-xy^2\,.\,1-xy^3}$$

$$+\frac{xy^3}{1-xy^3\,.\,1-x^2y^7}+\frac{x^2y^7}{1-x^2y^7\,.\,1-x^3y^{11}}+\frac{x^3y^{11}}{1-x^3y^{11}\,.\,1-x^4y^{15}}$$

$$+\frac{x^4y^{15}}{1-x^4y^{15}\,.\,1-x^{21}y^{79}}+\frac{x^{21}y^{79}}{1-x^{21}y^{79}\,.\,1-x^{38}y^{143}}$$

$$+\frac{x^{38}y^{143}}{1-x^{38}y^{143}\,.\,1-x^{207}y^{779}}+\frac{x^{207}y^{779}}{1-x^{207}y^{779}},$$

and the ground solutions of the relation are

α	β
1	0
1	1
1	2
1	3
2	7
3	11
4	15
21	79
38	143
207	779

As another example, take

$$77\alpha \geqslant 104\beta.$$

Lemma 1 shews that we have to sum

$$\Sigma x^{\alpha}(xy)^{\beta},$$

for the relation

$$77\alpha \geqslant 27\beta;$$

and

$$\frac{77}{27} = 2 + \frac{1}{1+}\frac{1}{5+}\frac{1}{1+}\frac{1}{3}.$$

The principal convergents are

$$\frac{0}{1}, \frac{1}{0}, \frac{2}{1}, \frac{3}{1}, \frac{17}{6}, \frac{20}{7}, \frac{77}{27};$$

and the intermediate series

$$\frac{0}{1}, \frac{1}{1}, \frac{2}{1}, \frac{5}{2}, \frac{8}{3}, \frac{11}{4}, \frac{14}{5}, \frac{17}{6}, \frac{37}{13}, \frac{57}{20}, \frac{77}{27}.$$

Hence the sum $\Sigma x^{\alpha}(xy)^{\beta}$ is

$$1 + \frac{x}{1-x\,.\,1-x^2y} + \frac{x^2y}{1-x^2y\,.\,1-x^3y^2}$$

$$+ \frac{x^3y^2}{1-x^3y^2\,.\,1-x^7y^5} + \frac{x^7y^5}{1-x^7y^5\,.\,1-x^{11}y^8} + \frac{x^{11}y^8}{1-x^{11}y^8\,.\,1-x^{15}y^{11}}$$

$$+ \frac{x^{15}y^{11}}{1-x^{15}y^{11}\,.\,1-x^{19}y^{14}} + \frac{x^{19}y^{14}}{1-x^{19}y^{14}\,.\,1-x^{23}y^{17}} + \frac{x^{23}y^{17}}{1-x^{23}y^{17}\,.\,1-x^{50}y^{37}}$$

$$+ \frac{x^{50}y^{37}}{1-x^{50}y^{37}\,.\,1-x^{77}y^{57}} + \frac{x^{77}y^{57}}{1-x^{77}y^{57}\,.\,1-x^{104}y^{77}} + \frac{x^{104}y^{77}}{1-x^{104}y^{77}}.$$

Hence the ground solutions are

α	β
1	0
2	1
3	2
7	5
11	8
15	11
19	14
23	17
50	37
77	57
104	77

The operation of Lemma 1 is however not necessary, for, since

$$\frac{77}{104} = 0 + \frac{1}{1+}\frac{1}{2+}\frac{1}{1+}\frac{1}{5+}\frac{1}{1+}\frac{1}{3},$$

we can form the series of intermediate convergents

$$\frac{0}{1}, \frac{1}{2}, \frac{2}{3}, \frac{5}{7}, \frac{8}{11}, \frac{11}{15}, \frac{14}{19}, \frac{17}{23}, \frac{37}{50}, \frac{57}{77}, \frac{77}{104},$$

which lead directly to the generating function and to the fundamental solution.

The generating function shews that every solution is of the form

$$\alpha = A_r\alpha^{(r)} + A_{r+1}\alpha^{(r+1)},$$

$$\beta = A_r\beta^{(r)} + A_{r+1}\beta^{(r+1)},$$

where $\frac{\beta^{(r)}}{\alpha^{(r)}}, \frac{\beta^{(r+1)}}{\alpha^{(r+1)}}$ are consecutive members of the ascending intermediate series of convergents to $\frac{\lambda}{\mu}$. The corresponding generating function, associated with the Diophantine equation

$$\lambda\alpha = \mu\beta + \gamma,$$

is obtained by multiplying each product $x^\alpha y^\beta$, in the generating function, by

$$z^{\lambda\alpha - \mu\beta}.$$

Ex. gr. for the equation

$$77\alpha = 104\beta + \gamma,$$

we find $\Sigma x^\alpha y^\beta z^\gamma$

$$= 1 + \frac{xz^{77}}{1 - xz^{77} \,.\, 1 - x^2yz^{50}} + \frac{x^2yz^{50}}{1 - x^2yz^{50} \,.\, 1 - x^3y^2z^{23}}$$

$$+ \frac{x^3y^2z^{23}}{1 - x^3y^2z^{23} \,.\, 1 - x^7y^5z^{19}} + \frac{x^7y^5z^{19}}{1 - x^7y^5z^{19} \,.\, 1 - x^{11}y^8z^{15}} + \frac{x^{11}y^8z^{15}}{1 - x^{11}y^8z^{15} \,.\, 1 - x^{15}y^{11}z^{11}}$$

$$+ \frac{x^{15}y^{11}z^{11}}{1 - x^{15}y^{11}z^{11} \,.\, 1 - x^{19}y^{14}z^7} + \frac{x^{19}y^{14}z^7}{1 - x^{19}y^{14}z^7 \,.\, 1 - x^{23}y^{17}z^3} + \frac{x^{23}y^{17}z^3}{1 - x^{23}y^{17}z^3 \,.\, 1 - x^{50}y^{37}z^2}$$

$$+ \frac{x^{50}y^{37}z^2}{1 - x^{50}y^{37}z^2 \,.\, 1 - x^{77}y^{57}z} + \frac{x^{77}y^{57}z}{1 - x^{77}y^{57}z \,.\, 1 - x^{104}y^{77}} + \frac{x^{104}y^{77}}{1 - x^{104}y^{77}},$$

establishing the ground solutions

α	β	γ
1	0	77
2	1	50
3	2	23
7	5	19
11	8	15
15	11	11
19	14	7
23	17	3
50	37	2
77	57	1
104	77	0

There is a connexion between the values of α and γ peculiar to this case which will be made the subject of investigation elsewhere.

It appears, from the form of the generating function, that every solution (α, β, γ) of the equation must be derivable from, at most, two of the fundamental solutions, say

$$(\alpha^{(r)}, \beta^{(r)}, \gamma^{(r)}), \quad (\alpha^{(r+1)}, \beta^{(r+1)}, \gamma^{(r+1)}),$$

viz.:—we must have

$$\alpha = A_r\alpha^{(r)} + A_{r+1}\alpha^{(r+1)},$$
$$\beta = A_r\beta^{(r)} + A_{r+1}\beta^{(r+1)},$$
$$\gamma = A_r\gamma^{(r)} + A_{r+1}\gamma^{(r+1)},$$

the two solutions being consecutive. Eliminating A_r and A_{r+1} from the three equations we obtain

$$\begin{vmatrix} \alpha & \beta & \gamma \\ \alpha^{(r)} & \beta^{(r)} & \gamma^{(r)} \\ \alpha^{(r+1)} & \beta^{(r+1)} & \gamma^{(r+1)} \end{vmatrix} = 0,$$

a linear relation between α, β, γ which must be identical with $\lambda\alpha - \mu\beta - \gamma = 0$; hence

$$\lambda = \begin{vmatrix} \gamma^{(r)} & \gamma^{(r+1)} \\ \beta^{(r)} & \beta^{(r+1)} \end{vmatrix}, \quad \mu = \begin{vmatrix} \gamma^{(r)} & \gamma^{(r+1)} \\ \alpha^{(r)} & \alpha^{(r+1)} \end{vmatrix},$$

$$-1 = \begin{vmatrix} \beta^{(r)} & \beta^{(r+1)} \\ \alpha^{(r)} & \alpha^{(r+1)} \end{vmatrix},$$

the latter relation verifying an elementary property of consecutive intermediate convergents. We have the identity

$$\lambda\alpha - \mu\beta - \gamma = A_r\{\lambda\alpha^{(r)} - \mu\beta^{(r)} - \gamma^{(r)}\} + A_{r+1}\{\lambda\alpha^{(r+1)} - \mu\beta^{(r+1)} - \gamma^{(r+1)}\},$$

so that in a sense the linear function

$$\lambda\alpha - \mu\beta - \gamma$$

is reducible *quâ* the fundamental linear functions

$$\lambda\alpha^{(r)} - \mu\beta^{(r)} - \gamma^{(r)}, \quad \lambda\alpha^{(r+1)} - \mu\beta^{(r+1)} - \gamma^{(r+1)}.$$

Ex. gr. consider

$$77\alpha - 104\beta = 815.$$

In the generating function we find a term

$$(x^2yz^{50})^{14}(x^3y^2z^{23})^5 = x^{43}y^{24}z^{815},$$

yielding the simplest solution, and since

$$77 \,.\, 43 - 104 \,.\, 24 - 815 = 14(77 \,.\, 2 - 104 \,.\, 1 - 50) + 5(77 \,.\, 3 - 104 \,.\, 2 - 23),$$

we find that the solution before us of

$$77\alpha - 104\beta - 815 = 0$$

is a linear function of the simplest solutions of the equations

$$77\alpha_1 - 104\beta_1 - 50 = 0,$$

$$77\alpha_2 - 104\beta_2 - 23 = 0;$$

viz.:—we have

$$\alpha = 14\alpha_1 + 5\alpha_2,$$

$$\beta = 14\beta_1 + 5\beta_2,$$

$$815 = 14 \,.\, 50 + 5 \,.\, 23.$$

The general solutions of

$$77\alpha_1 - 104\beta_1 - 50 = 0,$$

$$77\alpha_2 - 104\beta_2 - 23 = 0,$$

are

$$\alpha_1 = 2 + 104\theta_1, \qquad \beta_1 = 1 + 77\theta_1,$$

$$\alpha_2 = 3 + 104\theta_2, \qquad \beta_2 = 2 + 77\theta_2,$$

respectively, and these lead to the solutions

$$\alpha = 43 + 104\,(14\theta_1 + 5\theta_2),$$

$$\beta = 24 + \ \ 77\,(14\theta_1 + 5\theta_2),$$

of the equation

$$77\alpha - 104\beta = 815\,;$$

the general solution of this last equation being

$$\alpha = 43 + 104\theta,$$

$$\beta = 24 + \ \ 77\theta\,;$$

consequently the solutions derived are those for which θ is of the form

$$14\theta_1 + 5\theta_2.$$

§ 2. I next consider the simultaneous Diophantine inequalities

$$\lambda_1\alpha \geqslant \mu_1\beta,$$

$$\mu_2\beta \geqslant \lambda_2\alpha,$$

equivalent to the simultaneous Diophantine equations

$$\lambda_1\alpha = \mu_1\beta + \gamma,$$

$$\mu_2\beta = \lambda_2\alpha + \delta.$$

I observe that, if the given relations can be simultaneously satisfied by other than zero values of the arguments α_1, β_1, we must have

$$\begin{vmatrix} \lambda_1 & \lambda_2 \\ \mu_1 & \mu_2 \end{vmatrix} \geqslant 0.$$

It is convenient to reduce the relations to others in which $\lambda_1 > \mu_1$ and $\mu_2 > \lambda_2$, an operation which is always possible, as I establish in the following investigation. I assume λ_1, μ_1 to be relatively prime and also λ_2, μ_2, and I assume further that $\mu_1 > \lambda_1$.

The crude expression for

$$\Sigma x^\alpha y^\beta,$$

the sum being for every solution of the simultaneous relations, is

$$\underset{\geqslant}{\Omega} \frac{1}{1 - \dfrac{a^{\lambda_1}}{b^{\lambda_2}}x \,.\, 1 - \dfrac{b^{\mu_2}}{a^{\mu_1}}y}.$$

If the fraction $\dfrac{\mu_1}{\lambda_1}$ be developed, in the form of a continued fraction, the first step is to write

$$\frac{\mu_1}{\lambda_1} = a_1 + \frac{c_2}{c_1},$$

16—2

where $c_1 = \lambda_1$ and $c_2 < c_1$. Hence

$$\alpha \geqslant a_1\beta,$$

and the crude expression may be written

$$\underset{\geqslant}{\Omega} \frac{1}{1 - \frac{a^{\lambda_1}d}{b^{\lambda_2}} x \,.\, 1 - \frac{b^{\mu_2}}{a^{\mu_1}d^{a_1}}};$$

Ω operating upon a, b and d. Eliminating d this is

$$\underset{\geqslant}{\Omega} \frac{1}{1 - \frac{a^{\lambda_1}}{b^{\lambda_2}} x \,.\, 1 - \frac{b^{\mu_2 - a_1\lambda_2}}{a^{\mu_1 - a_1\lambda_1}} x^{a_1}y}.$$

In this expression $\mu_2 - a_1\lambda_2$ cannot be zero unless $\lambda_2 = 1$. Suppose then

$$\lambda_2 = 1, \qquad \mu_2 - a_1\lambda_2 = 0 = \mu_2 - a_1.$$

The expression becomes

$$\underset{\geqslant}{\Omega} \frac{1}{1 - \frac{x^{a_1}y}{a^{\mu_1 - a_1\lambda_1}}},$$

which is unity, except when $\mu_1 - a_1\lambda_1 = 0$, when we obtain the generating function

$$\frac{1}{1 - x^{a_1}y},$$

corresponding to the Diophantine equality

$$\alpha = a_1\beta.$$

Moreover, if $\mu_2 - a_1\lambda_2$ be negative, the whole expression reduces to unity so that there exists only the singular solution $\alpha = \beta = 0$.

Assume therefore

$$\mu_2 - a_1\lambda_2 > 0,$$

and putting $\lambda_2 = c_1'$, $\mu_2 - a_1\lambda_2 = c_2'$, write the expression

$$\underset{\geqslant}{\Omega} \frac{1}{1 - \frac{a^{c_1}}{b^{c_1'}} x \,.\, 1 - \frac{b^{c_2'}}{a^{c_2}} y}.$$

If now $c_2' > c_1'$, we have effected the reduction, because $c_1 > c_2$, and we have the two relations

$$c_1\alpha \geqslant c_2\beta,$$

$$c_2'\beta \geqslant c_1'\alpha,$$

connected with the sum

$$\Sigma x^{\alpha + a_1\beta} y^{\beta}.$$

If $c_2' = c_1'$, or $\mu_2 = (a_1 + 1)\lambda_2$, we must assume $\lambda_2 = 1$, because λ_2, μ_2 are relatively prime, and $\mu_2 > \lambda_2$. The expression then becomes

$$\underset{\geqslant}{\Omega} \frac{1}{1 - a^{(a_1+1)\lambda_1 - \mu_1} x^{a_1+1} y \,.\, 1 - \dfrac{x^{a_1} y}{a^{\mu_1 - a_1\lambda_1}}},$$

which denotes

$$\Sigma x^{(a_1+1)\alpha + a_1\beta} y^{\alpha+\beta},$$

for the single relation

$$\{(a_1 + 1)\lambda_1 - \mu_1\} \alpha \geqslant (\mu_1 - a_1\lambda_1)\beta.$$

Putting this aside, as a case already dealt with in § 1, we are left with

$$\underset{\geqslant}{\Omega} \frac{1}{1 - \dfrac{a^{c_1}}{b^{c_1'}} x \,.\, 1 - \dfrac{b^{c_2'}}{a^{c_2}} y},$$

in which $c_2' < c_1'$, and $c_1 > c_2$; and the expression must be reduced further.

I observe that $\dfrac{\mu_1}{\lambda_1}$ and $\dfrac{\mu_2}{\lambda_2}$ have the same first convergents, and that

$$\frac{\mu_2}{\lambda_2} = a_1 + \frac{c_2'}{c_1'}.$$

Write

$$\frac{\mu_2}{\lambda_2} = a_1 + \frac{1}{a_2 +} \frac{c_3'}{c_2'},$$

where $a_1 + \dfrac{1}{a_2}$ is the second convergent to $\dfrac{\mu_2}{\lambda_2}$, and $c_3' < c_2'$. Since

$$\frac{c_1'}{c_2'} = a_2 + \frac{c_3'}{c_2'},$$

we may take for the expression

$$\underset{\geqslant}{\Omega} \frac{1}{1 - \dfrac{a^{c_1}}{b^{c_1'} d^{a_2}} x \,.\, 1 - \dfrac{b^{c_2'} d}{a^{c_2}} x^{a_1} y},$$

from which, eliminating d, we obtain

$$\underset{\geqslant}{\Omega} \frac{1}{1 - \dfrac{a^{c_1 - a_2 c_2}}{b^{c_1' - a_2 c_2'}} x^{a_1 a_2 + 1} y^{a_2} \,.\, 1 - \dfrac{b^{c_2'}}{a^{c_2}} x^{a_1} y}.$$

Since $\dfrac{\mu_1}{\lambda_1} = a_1 + \dfrac{c_2}{c_1}$, where $c_2 < c_1$, we may write

$$\frac{c_1}{c_2} = a_2 + \frac{c_3}{c_2},$$

whenever $c_1 - a_2c_2$ is positive, and then

$$\frac{\mu_1}{\lambda_1} = a_1 + \frac{1}{a_2} \, \frac{c_3}{+ c_2},$$

where c_3 is not necessarily $< c_2$.

The expression becomes

$$\underset{\geqslant}{\Omega} \frac{1}{1 - \frac{a^{c_3}}{b^{c_3'}} x^{a_1a_2+1} y^{a_2} \,.\, 1 - \frac{b^{c_2'}}{a^{c_2}} x^{a_1} y}.$$

If c_3 be zero, c_3' must be zero also, and the generator is simply

$$\frac{1}{1 - x^{a_1a_2+1} y^{a_2}},$$

corresponding to

$$a_2\alpha = (a_1a_2 + 1)\beta.$$

If c_3 be not zero, we must have

$$c_3 >, = \text{ or } < c_2.$$

If $c_3 > c_2$ the reduction is complete because $c_2' > c_3'$ also. In this case $a_1 + \frac{1}{a_2}$ is the second convergent to $\frac{\mu_2}{\lambda_2}$ but is *not* the second convergent to $\frac{\mu_1}{\lambda_1}$.

If $c_3 = c_2$ we can at once eliminate a from the expression and obtain an expression that has been already dealt with.

If $c_3 < c_2$, $a_1 + \frac{1}{a_2}$ is the second convergent to both $\frac{\mu_1}{\lambda_1}$ and $\frac{\mu_2}{\lambda_2}$ and a further process of reduction is necessary.

To sum up the results so far reached:—

(i) if

$$\lambda_1 > \mu_1, \qquad \mu_2 > \lambda_2,$$
$$c_1 > \mu_1,$$

no reduction takes place;

(ii) if

$$\mu_2 - a_1\lambda_2 > \lambda_2, \qquad \lambda_1 > \mu_1 - a_1\lambda_1,$$
$$c_2' > c_1', \qquad c_1 > c_2,$$

the final reduced form is

$$\underset{\geqslant}{\Omega} \frac{1}{1 - \frac{a^{c_1}}{b^{c_1'}} x \,.\, 1 - \frac{b^{c_2'}}{a^{c_2}} x^{p_1} y^{q_1}},$$

wherein $\frac{\mu_1}{\lambda_1} = a_1 + \frac{c_2}{c_1}$, $p_1 = a_1$, $q_1 = 1$, and $a_1 = \frac{p_1}{q_1}$ is the first convergent to $\frac{\mu_1}{\lambda_1}$; also $\frac{\mu_2}{\lambda_2} = a_1 + \frac{c_2'}{c_1'}$. and a_1 is *not* the first convergent to $\frac{\mu_2}{\lambda_2}$;

(iii) if a_1 is the first convergent to both $\frac{\mu_1}{\lambda_1}$ and $\frac{\mu_2}{\lambda_2}$; $c_3 > c_2$, $c_2' > c_3'$; the final form is

$$\underset{\geqslant}{\Omega} \frac{1}{1 - \frac{a^{c_3}}{b^{c_3'}} x^{p_2} y^{q_2} \,.\, 1 - \frac{b^{c_2'}}{a^{c_2}} x^{p_1} y^{q_1}},$$

wherein
$$\frac{\mu_1}{\lambda_1} = a_1 + \frac{1}{a_2 +} \frac{c_3}{c_2},$$

$$\frac{\mu_2}{\lambda_2} = a_1 + \frac{1}{a_2 +} \frac{c_3'}{c_2'},$$

and $a_1 + \frac{1}{a_2} = \frac{p_2}{q_2}$ is not the second convergent to $\frac{\mu_1}{\lambda_1}$ but is the second convergent to $\frac{\mu_2}{\lambda_2}$.

If $a_1 + \frac{1}{a_2} = \frac{p_2}{q_2}$ is the second convergent to both $\frac{\mu_1}{\lambda_1}$ and $\frac{\mu_2}{\lambda_2}$ a further reduction becomes necessary.

Write
$$\frac{\mu_1}{\lambda_1} = a_1 + \frac{1}{a_2 +} \frac{1}{a_3 +} \frac{c_4}{c_3},$$

$$\frac{\mu_2}{\lambda_2} = a_1 + \frac{1}{a_2 +} \frac{1}{a_3 +} \frac{c_4'}{c_3'},$$

wherein $c_3 > c_4$, $c_4' > c_3'$, so that $a_1 + \frac{1}{a_2 +} \frac{1}{a_3} = \frac{p_3}{q_3}$ is the third convergent to $\frac{\mu_1}{\lambda_1}$ but is not the third convergent to $\frac{\mu_2}{\lambda_2}$.

The final reduced form is then

$$\underset{\geqslant}{\Omega} \frac{1}{1 - \frac{a^{c_3}}{b^{c_3'}} x^{p_2} y^{q_2} \,.\, 1 - \frac{b^{c_4'}}{a^{c_4}} x^{p_3} y^{q_3}},$$

which denotes the sum

$$\Sigma \left(x^{p_2} y^{q_2}\right)^{\alpha} \left(x^{p_3} y^{q_3}\right)^{\beta}$$
$$= \Sigma x^{p_2\alpha + p_3\beta} y^{q_2\alpha + q_3\beta},$$

for the relations
$$c_3\alpha \geqslant c_4\beta, \quad c_4'\beta \geqslant c_3'\alpha.$$

We proceed in this manner so as to establish the following general results.

CASE 1. If

$$\frac{\mu_1}{\lambda_1} = a_1 + \frac{1}{a_2 +}\frac{1}{a_3 +} \dots \frac{1}{+ a_n +}\frac{c_{n+1}}{c_n},$$

$$\frac{\mu_2}{\lambda_2} = a_1 + \frac{1}{a_2 +}\frac{1}{a_3 +} \dots \frac{1}{+ a_n +}\frac{c'_{n+1}}{c'_n},$$

where $c'_{n+1} > c'_n$, but $c_n > c_{n+1}$, so that $\frac{p_n}{q_n}$ is the nth convergent to $\frac{\mu_1}{\lambda_1}$ but *not* the nth convergent to $\frac{\mu_2}{\lambda_2}$, the final reduced form is

$$\underset{\geqslant}{\Omega} \frac{1}{1 - \frac{a^{c_n}}{b^{c'_n}} x^{p_{n-1}} y^{q_{n-1}} \,.\, 1 - \frac{b^{c'_{n+1}}}{a^{c_{n+1}}} x^{p_n} y^{q_n}},$$

which represents the sum

$$\Sigma \, (x^{p_{n-1}} y^{q_{n-1}})^{\alpha} \, (x^{p_n} y^{q_n})^{\beta}$$

$$= \Sigma \, x^{p_{n-1}\alpha + p_n\beta} \, y^{q_{n-1}\alpha + q_n\beta},$$

for the relations

$$c_n\alpha \geqslant c_{n+1}\beta,$$

$$c'_{n+1}\beta \geqslant c'_n\alpha,$$

wherein $c_n > c_{n+1}$, $c'_{n+1} > c'_n$.

CASE 2. If

$$\frac{\mu_1}{\lambda_1} = a_1 + \frac{1}{a_2 +}\frac{1}{a_3 +} \dots \frac{1}{+ a_{n+1} +}\frac{c_{n+2}}{c_{n+1}},$$

$$\frac{\mu_2}{\lambda_2} = a_1 + \frac{1}{a_2 +}\frac{1}{a_3 +} \dots \frac{1}{+ a_{n+1} +}\frac{c'_{n+2}}{c'_{n+1}},$$

where $c_{n+2} > c_{n+1}$, $c'_{n+1} > c'_{n+2}$, so that $\frac{p_{n+1}}{q_{n+1}}$ is the $n+1$th convergent to $\frac{\mu_2}{\lambda_2}$ but is *not* the $n+1$th convergent to $\frac{\mu_1}{\lambda_1}$, the final reduced form is

$$\underset{\geqslant}{\Omega} \frac{1}{1 - \frac{a^{c_{n+2}}}{b^{c'_{n+2}}} x^{p_{n+1}} y^{q_{n+1}} \,.\, 1 - \frac{b^{c'_{n+1}}}{a^{c_{n+1}}} x^{p_n} y^{q_n}},$$

which represents the sum

$$\Sigma \, (x^{p_{n+1}} y^{q_{n+1}})^{\alpha} \, (x^{p_n} y^{q_n})^{\beta}$$

$$= \Sigma x^{p_{n+1}\alpha + p_n\beta} \, y^{q_{n+1}\alpha + q_n\beta},$$

for the relations

$$c_{n+2}\alpha \geqslant c_{n+1}\beta,$$

$$c'_{n+1}\beta \geqslant c'_{n+2}\alpha,$$

wherein $c_{n+2} > c_{n+1}$, $c'_{n+1} > c'_{n+2}$.

This is the complete solution of the problem before us. As an example, suppose we have to find $\Sigma x^\alpha y^\beta$ for all values of α and β which satisfy the Diophantine inequalities

$$64\alpha \geqslant 275\beta,$$

$$142\beta \geqslant 33\alpha.$$

The first process is the reduction as above to a standard form. We develope the fractions $\frac{275}{64}$ and $\frac{142}{33}$ to the point where they fail to have the same convergents. Thus

$$\frac{275}{64} = 4 + \frac{1}{3+}\frac{1}{2+}\frac{5}{7},$$

$$\frac{142}{33} = 4 + \frac{1}{3+}\frac{1}{2+}\frac{4}{3},$$

and we observe that $\frac{4}{1}$ and $\frac{13}{3}$ are the first and second convergents to both fractions, but that $\frac{30}{7}$ is the third convergent to $\frac{275}{64}$ but *not* to $\frac{142}{33}$. This is Case 1 above and shews that the problem is reduced to finding the sum

$$\Sigma (x^{13}y^3)^\alpha (x^{30}y^7)^\beta$$

$$= \Sigma x^{13\alpha+30\beta} y^{3\alpha+7\beta},$$

for the relations

$$7\alpha \geqslant 5\beta,$$

$$4\beta \geqslant 3\alpha,$$

which are of the standard form.

We are now able to give a complete solution of the problem of finding the fundamental solutions of the simultaneous Diophantine inequalities

$$\lambda_1\alpha \geqslant \mu_1\beta,$$

$$\mu_2\beta \geqslant \lambda_2\alpha.$$

If these be of standard form, viz. $\lambda_1 > \mu_1$, $\mu_2 > \lambda_2$, consider the three systems

I.	II.	III.
$\lambda_1\alpha \geqslant \mu_1\beta$,	$\lambda_1\alpha \geqslant \mu_1\beta$,	$\lambda_1\alpha \geqslant \mu_1\beta$,
$\mu_2\beta \geqslant \lambda_2\alpha$,	$\mu_2\beta \geqslant \lambda_2\alpha$,	$\mu_2\beta \geqslant \lambda_2\alpha$,
$\beta \geqslant \alpha$,	$\alpha \geqslant \beta$,	$\alpha = \beta$.

If $F_1(x, y)$, $F_2(x, y)$, $F_3(x, y)$ be the corresponding generating functions it is clear that the generating function that we seek is

$$F_1(x, y) + F_2(x, y) - F_3(x, y).$$

Moreover we may reject as superfluous the second relation of I., the first relation of II. and the first and second relations of III., reducing them to

I.	II.	III.
$\lambda_1\alpha \geqslant \mu_1\beta$,	$\mu_2\beta \geqslant \lambda_2\alpha$,	
$\beta \geqslant \alpha$,	$\alpha \geqslant \beta$,	$\alpha = \beta$.

Hence $F_3(x, y) = \dfrac{1}{1-xy}$ and we have

$$F_1(x, y) + F_2(x, y) - F_3(x, y)$$

$$= \underset{\geqslant}{\Omega} \frac{1}{1 - \dfrac{a^{\lambda_1}}{b}x \,.\, 1 - \dfrac{b}{a^{\mu_1}}y} + \underset{\geqslant}{\Omega} \frac{1}{1 - \dfrac{b}{a^{\lambda_2}}x \,.\, 1 - \dfrac{a^{\mu_2}}{b}y} - \frac{1}{1-xy}$$

$$= \underset{\geqslant}{\Omega} \frac{1}{1 - a^{\lambda_1-\mu_1}xy \,.\, 1 - \dfrac{y}{a^{\mu_1}}} + \underset{\geqslant}{\Omega} \frac{1}{1 - a^{\mu_2-\lambda_2}xy \,.\, 1 - \dfrac{x}{a^{\lambda_2}}} - \frac{1}{1-xy}.$$

We now apply the theorem of § 1 to the two $\underset{\geqslant}{\Omega}$ expressions forming the ascending intermediate series of convergents to $\dfrac{\lambda_1 - \mu_1}{\mu_1}$ and to $\dfrac{\mu_2 - \lambda_2}{\lambda_2}$. We can thus determine the complete generating function and by inspection ascertain the fundamental solutions. As an example take the relations

$$7\alpha \geqslant 5\beta,$$

$$4\beta \geqslant 3\alpha\,;$$

we find

$$\underset{\geqslant}{\Omega} \frac{1}{1 - a^2xy \,.\, 1 - \dfrac{y}{a^5}} + \underset{\geqslant}{\Omega} \frac{1}{1 - axy \,.\, 1 - \dfrac{x}{a^3}} - \frac{1}{1-xy}.$$

The ascending series to $\frac{2}{5}$ is $\quad \frac{0}{1}, \frac{1}{3}, \frac{2}{5}$;

and to $\frac{1}{3}$ is $\quad \frac{0}{1}, \frac{1}{3}$.

Hence $\Sigma x^\alpha y^\beta$ is

$$1 + \frac{xy}{1 - xy \,.\, 1 - x^3y^4} + \frac{x^3y^4}{1 - x^3y^4 \,.\, 1 - x^5y^7} + \frac{x^5y^7}{1 - x^5y^7}$$

$$+ 1 + \frac{xy}{1 - xy \,.\, 1 - x^4y^3} + \frac{x^4y^3}{1 - x^4y^3} - \frac{1}{1-xy}\,;$$

and, without further reducing the expression, it is plain that the ground products are

$$xy, \quad x^3y^4, \quad x^4y^3, \quad x^5y^7\,;$$

giving the fundamental solutions

α	β
1	1
3	4
4	3
5	7

If we now wish to solve

$$64\alpha \geqslant 275\beta,$$
$$142\beta \geqslant 33\alpha,$$

which we have already reduced to the question of summing $\Sigma(x^{13}y^3)^\alpha(x^{30}y^7)^\beta$ for the relations

$$7\alpha \geqslant 5\beta,$$
$$4\beta \geqslant 3\alpha,$$

we have merely to write $x^{13}y^3$, $x^{30}y^7$ for x and y in the result obtained; we thus reach the ground products

$$(x^{13}y^3)\,(x^{30}y^7) = x^{43}y^{10},$$
$$(x^{13}y^3)^3\,(x^{30}y^7)^4 = x^{159}y^{37},$$
$$(x^{13}y^3)^4\,(x^{30}y^7)^3 = x^{142}y^{33},$$
$$(x^{13}y^3)^5\,(x^{30}y^7)^7 = x^{275}y^{64},$$

and the fundamental solutions

α	β
43	10
159	37
142	33
275	64

In general the number of fundamental solutions of two simultaneous Diophantine inequalities is the same as the number appertaining to the pair of standard inequalities to which the given inequalities are reducible.

NOTE ON THE DIOPHANTINE INEQAULITY $\lambda x \geq \mu y$.

By Major P. A. MacMahon, D.Sc., F.R.S.

In the paper "The Diophantine Inequality $\lambda x \geq \mu y$" *Camb. Phil. Trans.* Vol. XIX., Part 1, 1901. I shewed that the ground solutions depend upon the ascending intermediate series of convergents to the fraction $\frac{\lambda}{\mu}$, and I thence derived the ground solutions of the Diophantine Equality $\lambda x = \mu y + z$.

If the ascending intermediate series be

$$\frac{y_1}{x_1}, \frac{y_2}{x_2}, \ldots, \frac{y_\sigma}{x_\sigma}$$

and
$$\lambda x_a - \mu y_a = z_a,$$

the general Hilbertian solution may be written

$$x = A_1 x_1 + A_2 x_2 + \ldots + A_\sigma x_\sigma,$$
$$y = A_1 y_1 + A_2 y_2 + \ldots + A_\sigma y_\sigma,$$
$$z = A_1 z_1 + A_2 z_2 + \ldots + A_\sigma z_\sigma,$$

where the coefficients A are positive integers which may be assumed at pleasure.

The object of the present communication is to shew that when the expression of $\frac{\lambda}{\mu}$ as a continued fraction involves a reciprocal series of partial quotients the series of numbers

$$x_1, x_2, \ldots, x_\sigma,$$

or the series

$$y_1, y_2, \ldots, y_\sigma,$$

is connected with the series

$$z_1, z_2, \ldots, z_\sigma,$$

in a reciprocal or quasi-reciprocal manner.

The investigation involves four distinct cases.

Case I.

$\lambda > \mu$.

The partial quotients $2i-1$ *in number.*

Let the partial quotients be in order

$$a_1, a_2, \ldots, a_{i-1}, a_i, a_{i-1}, \ldots, a_2, a_1$$

and the successive principal convergents

$$\frac{0}{1}, \frac{1}{0}, \frac{p_1}{q_1}, \frac{p_2}{q_2}, \ldots, \frac{p_{i-1}}{q_{i-1}}, \frac{p_i}{q_i}, \ldots, \frac{p_{2i-2}}{q_{2i-2}}, \frac{p_{2i-1}}{q_{2i-1}}.$$

The ascending intermediate convergents are then

$$\frac{0}{1}, \frac{1}{1}, \frac{2}{1}, \ldots, \frac{a_1}{1} \left(=\frac{p_1}{q_1}\right),$$

$$\frac{a_1+p_2}{1+q_2}, \frac{a_1+2p_2}{1+2q_2}, \ldots \frac{a_1+a_3p_2}{1+a_3q_2} \left(=\frac{p_3}{q_3}\right),$$

$$\frac{a_1+a_3p_2+p_4}{1+a_3q_2+q_4}, \frac{a_1+a_3p_2+2p_4}{1+a_3q_2+2q_4}, \ldots, \frac{a_1+a_3p_2+a_5p_4}{1+a_3q_2+a_5q_4} \left(=\frac{p_5}{q_5}\right),$$

....

$$\frac{a_1+a_3p_2+a_5p_4+\ldots+a_3p_{2i-4}+a_1p_{2i-2}}{1+a_3q_2+a_5q_4+\ldots+a_3q_{2i-4}+a_1q_{2i-2}} \left(=\frac{p_{2i-1}}{q_{2i-1}}=\frac{\lambda}{\mu}\right).$$

If this series be denoted by

$$\frac{P_1}{Q_1}, \frac{P_2}{Q_2}, \ldots, \frac{P_\sigma}{Q_\sigma},$$

I shall shew that

$$\lambda Q_\alpha - \mu P_\alpha = P_{\sigma-\alpha+1}$$

for all values of α.

We must establish a property of the principal convergents which we call

Lemma I.

$$p_{2i-n}q_{2i-1} - p_{2i-1}q_{2i-n} = (-)^n p_{n-2}.$$

This is obviously true when $n=1$ or $n=2$ for

$$p_{2i-2}q_{2i-1} - p_{2i-1}q_{2i-n} = 1 = p_0.$$

Assuming the truth for values $n-1$ and n we will shew that it is true when $n+1$ is written for n.

We have

$$p_{2i-n-1}q_{2i-1} - p_{2i-1}q_{2i-n-1}$$
$$= (p_{2i-n+1} - a_{2i-n+1}p_{2i-n})\, q_{2i-1} - p_{2i-1}\,(q_{2i-n+1} - a_{2i-n+1}q_{2i-n})$$
$$= (-)^{n-1}p_{n-3} + a_{n-1}\,(p_{2i-1}q_{2i-n} - p_{2i-n}q_{2i-1})$$
$$= (-)^{n-1}p_{n-3} + a^{n-1}\,(-)^{n-1}p_{n-2}$$
$$= (-)^{n+1}p_{n-1}.$$

Hence the theorem is established.

Take now the intermediate convergent

$$\frac{P_{\sigma-s_1}}{Q_{\sigma-s_1}} = \frac{p_{2i-1} - s_1 p_{2i-2}}{q_{2i-1} - s_1 q_{2i-2}},$$

where s_1 is one of the numbers 0, 1, 2, ..., a_1,

$$\lambda Q_{\sigma-s_1} - \mu P_{\sigma-s_1} = \lambda q_{2i-1} - \mu p_{2i-1} + s_1\,(p_{2i-2}q_{2i-1} - p_{2i-1}q_{2i-2})$$
$$= s_1$$
$$= P_{s_1+1},$$

Also take

$$\frac{P_{\sigma-a_1-s_3}}{Q_{\sigma-a_1-s_3}} = \frac{p_{2i-1} - a_1 p_{2i-2} - s_3 p_{2i-4}}{q_{2i-1} - a_1 q_{2i-2} - s_3 q_{2i-4}},$$

where s_3 is one of the numbers 0, 1, 2, ..., a_3.

We have

$$\lambda Q_{\sigma-a_1-s_3} - \mu P_{\sigma-a_1-s_3} = P_{a_1+1} + s_3\,(p_{2i-4}q_{2i-1} - p_{2i-1}q_{2i-4})$$
$$= p_1 + s_3 p_2$$
$$= P_{a_1+s_3+1},$$

and obviously we have in general

$$\lambda Q_{\sigma-a_1-a_3-\ldots-a_{2n+1}-s_{2n+3}} - \mu P_{\sigma-a_1-a_3-\ldots-a_{2n+1}-s_{2n+3}}$$
$$= P_{a_1+a_3+\ldots+a_{2n+1}+s_{2n+3}+1},$$

where s_{2n+3} is one of the numbers 0, 1, 2, ..., a_{2n+3}.

Hence in general

$$\lambda Q_a - \mu P_a = P_{\sigma-a+1}.$$

We have now the

Theorem. If $\lambda > \mu$ and the fraction $\frac{\lambda}{\mu}$, when expressed as a continued fraction, involves a reciprocal series of partial quotients uneven in number, the ground solutions of the Diophantine equality

$$\lambda x = \mu y + z$$

are

x	y	z
x_1	y_1	y_σ
x_2	y_2	$y_{\sigma-1}$
x_3	y_3	$y_{\sigma-2}$
$\vdots$	$\vdots$	$\vdots$
$x_{\sigma-2}$	$y_{\sigma-2}$	y_3
$x_{\sigma-1}$	$y_{\sigma-1}$	y_2
x_σ	y_σ	y_1

where $\frac{y_1}{x_1}, \frac{y_2}{x_2}, \ldots, \frac{y_\sigma}{x_\sigma}$ is the ascending intermediate series of convergents to $\frac{\lambda}{\mu}$.

The Diophantine equality

$$\lambda x = \mu y + z$$

has, in this case, a reciprocal property, and the complete solution may be written

$$\left.\begin{aligned} x &= A_1x_1 + A_2x_2 \quad +\ldots+ A_{\sigma-1}x_{\sigma-1} + A_\sigma x_\sigma \\ y &= A_1y_1 + A_2y_2 \quad +\ldots+ A_{\sigma-1}y_{\sigma-1} + A_\sigma y_\sigma \\ z &= A_1y_\sigma + A_2y_{\sigma-1} +\ldots+ A_{\sigma-1}y_2 \quad + A_\sigma y_1 \end{aligned}\right\}.$$

As an example take

$$104x = 77y + z,$$

where

$$\frac{104}{77} = 1 + \frac{1}{2} + \frac{1}{1} + \frac{1}{5} + \frac{1}{1} + \frac{1}{2} + \frac{1}{1}.$$

The principal convergents are

$$\frac{0}{1}, \frac{1}{0}, \frac{1}{1}, \frac{3}{2}, \frac{4}{3}, \frac{23}{17}, \frac{27}{20}, \frac{77}{57}, \frac{104}{77},$$

G 2

and the ascending intermediate convergents

$$\frac{0}{1}, \frac{1}{1}, \frac{4}{3}, \frac{27}{20}, \frac{104}{77},$$

yielding the ground solutions

x	y	z
1	0	104
1	1	27
3	4	4
20	27	1
77	104	0

which possess the reciprocal property.

Case II.

$\lambda < \mu$.

The reciprocal partial quotients $2i-1$ *in number.*

The intermediate series of convergents is

$$\frac{0}{1}, \frac{1}{1+a_1}, \frac{2}{1+2a_1}, \ldots, \frac{a_2p_1}{1+a_2q_1}\left(=\frac{p_2}{q_2}\right),$$

$$\frac{a_2p_1+p_3}{1+a_2q_1+q_3}, \frac{a_2p_1+2p_3}{1+a_2q_1+2q_3}, \ldots, \frac{a_2p_1+a_4p_3}{1+a_2q_1+a_4q_3}\left(=\frac{p_4}{q_4}\right),$$

....

$$\frac{a_2p_1+a_4p_3+a_6p_5+\ldots+a_4p_{2i-5}+a_2p_{2i-3}}{1+a_2q_1+a_4q_3+a_6q_5+\ldots+a_4q_{2i-5}+a_2q_{2i-3}}\left(=\frac{p_{2i-2}}{q_{2i-2}}\right),$$

$$\frac{p_{2i-1}}{q_{2i-1}}\left(=\frac{\lambda}{\mu}\right).$$

If this series be denoted by

$$\frac{P'_1}{Q'_1}, \frac{P'_2}{Q'_2}, \ldots, \frac{P'_{\sigma'}}{Q'_{\sigma'}},$$

I shall shew that

$$\lambda Q_\alpha' - \mu P_\alpha' = Q'_{\sigma'-\alpha}$$

for the values of α, 1, 2, 3, ..., $\sigma'-1$.

We must establish a property of the principal convergents which we shall call

LEMMA II.

$$p_{2i-n}q_{2i-1} - p_{2i-1}q_{2i-n} = (-)^{n+1}q_{n-2}.$$

This is obviously true when $n=2$, and it is easy to verify when $n=3$.

Now

$$p_{2i-n-1}q_{2i-1} - p_{2i-1}q_{2i-n-1}$$

$$= (p_{2i-n+1} - a_{2i-n+1}p_{2i-n})\, q_{2i-1} - p_{2i-1}\,(q_{2i-n+1} - a_{2i-n+1}q_{2i-n})$$

$$= (-)^n q_{n-3} - a_{n-1}\,(-)^{n-1}q_{n-2}$$

$$= (-)^{n+2}q_{n-1}.$$

Hence by induction the Lemma is established.

Tako now the intermediate convergent

$$\frac{P'_{\sigma'-s_2-1}}{Q'_{\sigma'-s_2-1}} = \frac{p_{2i-2} - s_2 p_{2i-3}}{q_{2i-2} - s_2 q_{2i-3}},$$

where s_2 is one of the numbers 0, 1, 2, ..., a_2.

We have

$$\lambda Q'_{\sigma'-s_2-1} - \mu P'_{\sigma'-s_2-1} = 1 + s_2 q_1 = Q'_{s_2+1}.$$

Also take

$$\frac{P'_{\sigma'-a_2-s_4-1}}{Q'_{\sigma'-a_2-s_4-1}} = \frac{p_{2i-2} - a_2 p_{2i-3} - s_4 p_{2i-5}}{q_{2i-2} - a_2 q_{2i-3} - s_4 q_{2i-5}},$$

where s_4 is one of the numbers 0, 1, 2, ..., a_4.

We have

$$\lambda Q'_{\sigma'-a_2-s_4-1} - \mu P'_{\sigma'-a_2-s_4-1} = Q'_{a_2+1} + s_4\,(p_{2i-5}q_{2i-1} - p_{2i-1}q_{2i-5})$$

$$= Q'_{a_2+1} + s_4 q_3$$

$$= q_2 + s_4 q_3$$

$$= Q'_{a_2+s_4+1},$$

and it is now obvious that

$$\lambda Q'_{a} - \mu P'_{a} = Q'_{\sigma'-a}$$

for all values of α, excluding σ'.

We have now the

Theorem. If $\lambda < \mu$ and the fraction $\frac{\lambda}{\mu}$, when expressed as a continued fraction, involves a reciprocal series of partial quotients uneven in number, the ground solutions of the Diophantine equality

$$\lambda x = \mu y + z$$

are

x	y	z
x_1	y_1	$x_{\sigma-1}$
x_2	y_2	$x_{\sigma-2}$
x_3	y_3	$x_{\sigma-3}$
$\vdots$	$\vdots$	$\vdots$
$x_{\sigma-2}$	$y_{\sigma-2}$	x_2
$x_{\sigma-1}$	$y_{\sigma-1}$	x_1
x_σ	y_σ	0

where $\frac{y_1}{x_1}, \frac{y_2}{x_2}, \ldots, \frac{y_\sigma}{x_\sigma}$ is the ascending intermediate series of convergents to $\frac{\lambda}{\mu}$.

The Diophantine equality possesses a reciprocal property and the complete solution may be written:—

$$\left.\begin{aligned} x &= A_1x_1 + A_2x_2 + \ldots + A_{\sigma-1}x_{\sigma-1} + A_\sigma x_\sigma \\ y &= A_1y_1 + A_2y_2 + \ldots + A_{\sigma-1}y_{\sigma-1} + A_\sigma y_\sigma \\ z &= A_1x_{\sigma-1} + A_2x_{\sigma-2} + \ldots + A_{\sigma-1}x_1 \end{aligned}\right\}.$$

As an example take

$$77x = 104y + z,$$

where

$$\frac{77}{104} = \frac{1}{1+}\frac{1}{2+}\frac{1}{1+}\frac{1}{5+}\frac{1}{1+}\frac{1}{2+}\frac{1}{1}.$$

The principal convergents are

$$\frac{0}{1}, \frac{1}{1}, \frac{2}{3}, \frac{3}{4}, \frac{17}{23}, \frac{20}{27}, \frac{57}{77}, \frac{77}{104},$$

and the ascending intermediate convergents

$$\frac{0}{1},\ \frac{1}{2},\ \frac{2}{3},\ \frac{5}{7},\ \frac{8}{11},\ \frac{11}{15},\ \frac{14}{19},\ \frac{17}{23},\ \frac{37}{50},\ \frac{57}{77},\ \frac{77}{104},$$

yielding the ground solutions

x	y	z
1	0	77
2	1	50
3	2	23
7	5	19
11	8	15
15	11	11
19	14	7
23	17	3
50	37	2
77	57	1
104	77	0

which possess the reciprocal property of Case II.

CASE III.

$\lambda > \mu.$

The reciprocal partial quotients $2i$ *in number.*

Let the partial quotients be

$$a_1,\ a_2,\ \ldots,\ a_i,\ a_i\ \ldots,\ a_2,\ a_1,$$

The ascending intermediate convergents are

$$\frac{0}{1},\ \frac{1}{1},\ \frac{2}{1},\ \ldots,\ \frac{a_1}{1}\left(=\frac{p_1}{q_1}\right),$$

$$\frac{a_1+p_2}{1+q_2},\ \frac{a_1+2p_2}{1+2q_2},\ \ldots,\ \frac{a_1+a_3p_2}{1+a_3q_2}\left(=\frac{p_3}{q_3}\right),$$

....

$$\frac{a_1+a_3p_2+a_5p_4+\ldots+a_4p_{2i-4}+a_2p_{2i-2}}{1+a_3q_2+a_5q_4+\ldots+a_4q_{2i-4}+a_2q_{2i-2}}\left(=\frac{p_{2i-1}}{q_{2i-1}}\right),$$

$$\frac{p_{2i}}{q_{2i}}\left(=\frac{\lambda}{\mu}\right).$$

In this case we require also the descending intermediate convergents. These are

$$\frac{1}{0},\quad \frac{1+a_1}{1},\quad \frac{1+2a_1}{2},\ \dots,\ \frac{1+a_2a_1}{a_2}\left(=\frac{p_2}{q_2}\right),$$

$$\frac{1+a_2p_1+p_3}{a_2q_1+q_3},\quad \frac{1+a_2p_1+2p_3}{a_2q_1+2q_3},\ \dots,\ \frac{1+a_2p_1+a_4p_3}{a_2q_1+a_4q_3}\left(=\frac{p_4}{q_4}\right),$$

....

$$\frac{1+a_2p_1+a_4p_3+\dots+a_3p_{2i-3}+a_1p_{2i-1}}{a_2q_1+a_4q_3+\dots+a_3q_{2i-3}+a_1q_{2i-1}}\left(=\frac{p_{2i}}{q_{2i}}\right).$$

From the fact that the partial quotients are reciprocal it follows that the ascending and descending series $\left(\text{with the addition of } \frac{0}{1}\right)$ contain the same number of convergents.

Let then the ascending set be

$$\frac{P_1}{Q_1},\quad \frac{P_2}{Q_2},\ \dots,\ \frac{P_\sigma}{Q_\sigma},$$

and the descending set

$$\frac{R_2}{S_2},\quad \frac{R_3}{S_3},\ \dots,\ \frac{R_\sigma}{S_\sigma}.$$

Also let

$$\frac{R_1}{S_1}=\frac{P_1}{Q_1}=\frac{0}{1},$$

There is a property of the convergents as in the previous cases, viz:

Lemma III.

$$p_{2i-2n}q_{2i}-p_{2i}q_{2i-2n}=(-)^n p_{n-1}$$

which is easily established by induction.

Now

$$\frac{P_{\sigma-s_2-1}}{Q_{\sigma-s_2-1}}=\frac{p_{2i-1}-s_2p_{2i-2}}{q_{2i-1}-s_2q_{2i-2}},$$

where s_2 is one of the numbers 0, 1, 2, ..., a_2, therefore

$$\lambda Q_{\sigma-s_2-1} - \mu P_{\sigma-s_2-1} = 1 + s_2(p_{2i-2}q_{2i} - p_{2i}q_{2i-2})$$
$$= 1 + s_2 p_i$$
$$= R_{s_2+1},$$
$$\lambda Q_{\sigma-a_2-s_4-1} - \mu P_{\sigma-a_2-s_4-1} = R_{a_2+s_4+2},$$

where s_4 is one of the numbers 0, 1, 2, ..., a_4, and generally

$$\lambda Q_\alpha - \mu P_\alpha = R_{\sigma-\alpha+1}$$

for all values of α from 1 to $\sigma - 1$.

We have now the

Theorem. If $\lambda > \mu$, and $\frac{\lambda}{\mu}$, when expressed as a continued fraction, involves reciprocal partial quotients even in number, the ground solutions of the Diophantine equality

$$\lambda x = \mu y + z$$

are

x	y	z
x_1	y_1	$y'_{\sigma-1}$
x_2	y_2	$y'_{\sigma-2}$
$\vdots$	$\vdots$	$\vdots$
$x_{\sigma-1}$	$y_{\sigma-1}$	y'_1
x_σ	y_σ	0

where $\frac{y_1}{x_1}, \frac{y_2}{x_2}, \ldots, \frac{y_\sigma}{x_\sigma}$ is the ascending set,

and $\frac{y'_1}{x'_1}, \frac{y'_2}{x'_2}, \ldots, \frac{y'_\sigma}{x'_\sigma}$ is the descending set

of intermediate convergents to the fraction $\frac{\lambda}{\mu}$.

The Diophantine equality has in this case the quasi-reciprocal property indicated above, and the complete solution may be written

$$x = A_1 x_1 + A_2 x_2 + \ldots + A_{\sigma-1} x_{\sigma-1} + A_\sigma x_\sigma,$$
$$y = A_1 y_1 + A_2 y_2 + \ldots + A_{\sigma-1} y_{\sigma-1} + A_\sigma y_\sigma,$$
$$z = A_1 y'_{\sigma-1} + A_2 y'_{\sigma-2} + \ldots + A_{\sigma-1} y'_1.$$

As an example take

$$305x = 233y + z,$$

where the partial quotients are

$$1, 3, 4, 4, 3, 1.$$

The principal convergents are

$$\frac{0}{1}, \frac{1}{0}, \frac{1}{1}, \frac{4}{3}, \frac{17}{13}, \frac{72}{55}, \frac{233}{178}, \frac{305}{233}.$$

The intermediate ascending series

$$\frac{0}{1}, \frac{1}{1}, \frac{5}{4}, \frac{9}{7}, \frac{13}{10}, \frac{17}{13}, \frac{89}{68}, \frac{161}{123}, \frac{233}{178}, \frac{305}{233}.$$

The intermediate descending series

$$\frac{0}{1} \Bigg| \frac{1}{0}, \frac{2}{1}, \frac{3}{2}, \frac{4}{3}, \frac{21}{16}, \frac{38}{29}, \frac{55}{42}, \frac{72}{55}, \frac{305}{233},$$

yielding the ground solutions

x	y	z
1	0	305
1	1	72
4	5	55
7	9	38
10	13	21
13	17	4
68	89	3
123	161	2
178	233	1
233	305	0

which possess the quesi-reciprocal property.

CASE IV.

$$\lambda < \mu.$$

The reciprocal partial quotients 2i in number.

The ascending intermediate convergents are

$$\frac{0}{i}, \quad \frac{1}{1+a_1}, \quad \frac{2}{1+2a_1}, \quad \cdots, \quad \frac{a_2}{1+a_2 q_1} \left(=\frac{p_2}{q_2}\right),$$

$$\frac{a_2+p_3}{1+a_2 q_1 + q_3}, \quad \frac{a_2+2p_3}{1+a_2 q_1+2q_3}, \quad \cdots, \quad \frac{a_2+a_4 p_3}{1+a_2 q_1 + a_4 q_3} \left(=\frac{p_4}{q_4}\right),$$

$$\cdots\cdots\cdots\cdots\cdots\cdots\cdots\cdots\cdots\cdots\cdots\cdots$$

$$\frac{a_2+a_4 p_3+a_6 p_5+\ldots+a_3 p_{2i-3}+a_1 p_{2i-1}}{1+a_2 q_1+a_4 q_3 +a_6 q_5 +\ldots+ a_3 q_{2i-3}+a_1 q_{2i-1}} \left(=\frac{p_{2i}}{q_{2i}}\right).$$

The descending intermediate convergents are

$$\frac{1}{0}, \quad \frac{1}{1}, \quad \frac{1}{2}, \quad \cdots, \quad \frac{1}{a_1} \left(=\frac{p_1}{q_1}\right),$$

$$\frac{1+p_2}{a_1+q_2}, \quad \frac{1+2p_2}{a_1+2q_2}, \quad \cdots, \quad \frac{1+a_3 p_2}{a_1+a_3 q_2} \left(=\frac{p_3}{q_3}\right),$$

$$\cdots\cdots\cdots\cdots\cdots\cdots\cdots\cdots\cdots\cdots\cdots$$

$$\frac{1+a_3 p_2+a_5 p_4+\ldots+a_4 p_{2i-4}+a_2 p_{2i-2}}{a_1+a_3 q_2+a_5 q_4 +\ldots+ a_4 q_{2i-4}+a_2 q_{2i-2}} \left(=\frac{p_{2i-1}}{q_{2i-1}}\right),$$

$$\frac{p_{2i}}{q_{2i}} \left(=\frac{\lambda}{\mu}\right),$$

the number in the descending set exceeding by unity the number in the ascending set.

Let the ascending set be

$$\frac{P_1}{Q_1}, \quad \frac{P_2}{Q_2}, \quad \cdots, \quad \frac{P_\sigma}{Q_\sigma},$$

and the descending set

$$\frac{R_1}{S_1}, \quad \frac{R_2}{S_2}, \quad \cdots, \quad \frac{R_\sigma}{Q_\sigma}, \quad \frac{R_{\sigma+1}}{Q_{\sigma-1}}, \quad \text{where } \frac{R_{\sigma-1}}{S_{\sigma+1}} = \frac{P_\sigma}{Q_\sigma} = \frac{\lambda}{\mu}.$$

The property of the convergents is in this case

LEMMA IV.

$$p_{2i-n}q_{2i} - p_{2i}q_{2i-n} = (-)^{n+1}q_{n-1},$$

and as before it is easy to show that

$$\lambda Q_a - \mu P_a = S_{\sigma-a+1}$$

for all values of α from 1 to σ.

We have now the

Theorem. If $\lambda < \mu$, and $\frac{\lambda}{\mu}$, when expressed as a continued fraction, involves reciprocal partial quotients even in number, the ground solutions of the Diophantine equality

$$\lambda x = \mu y + z$$

are

x	y	z
x_1	y_1	x'_σ
x_2	y_2	$x'_{\sigma-1}$
$\vdots$	$\vdots$	$\vdots$
$x_{\sigma-1}$	$y_{\sigma-1}$	x'_2
x_σ	y_σ	x'_1

where $\frac{y_1}{x_1}, \frac{y_2}{x_2}, \ldots, \frac{y_\sigma}{x_\sigma}$ is the ascending set

and $\frac{y'_1}{x'_1}, \frac{y'_2}{x'_2}, \ldots, \frac{y'_\sigma}{x'_\sigma}$ is the descending set

of intermediate convergents to the fraction $\frac{\lambda}{\mu}$.

We have thus a quasi-reciprocal property, and the general solution is of the forms

$$x = (A_\theta + 1)\,x_\theta \quad + A_{\theta+1}x_{\theta+1},$$
$$y = (A_\theta + 1)\,y_\theta \quad + A_{\theta+1}y_{\theta+1},$$
$$z = (A_\theta + 1)\,x'_{\sigma-\theta+1} + A_{\theta+1}x'_{\sigma-\theta}.$$

As an example take

$$233x = 305y + z.$$

The ascending and descending series of convergents are

$$\frac{0}{1}, \frac{1}{2}, \frac{2}{3}, \frac{3}{4}, \frac{16}{21}, \frac{29}{33}, \frac{41}{55}, \frac{55}{72}, \frac{233}{305},$$

$$\frac{1}{0}, \frac{1}{1}, \frac{4}{5}, \frac{7}{9}, \frac{10}{13}, \frac{13}{17}, \frac{68}{89}, \frac{123}{161}, \frac{178}{233}, \frac{233}{305},$$

yielding the ground solutions

x	y	z
1	0	233
2	1	161
3	2	89
4	3	17
21	16	13
38	29	9
55	42	5
72	55	1
305	233	0

which possess the quasi-reciprocal property.

ON CERTAIN SERIES OF DISCONTINUOUS FUNCTIONS CONNECTED WITH THE MODULAR FUNCTIONS.

By G. H. Hardy, Trinity College, Cambridge.

§ 1. THE formulæ of the linear transformation theory of the elliptic and modular functions in many cases assume very curious and interesting forms when the parameter τ is made to tend in an appropriate manner to a real rational limiting value, or in other words when

$$q = e^{\pi i \tau}$$

ON A DEFICIENT MULTINOMIAL EXPANSION

By Major P. A. MACMAHON, R.A., Sc.D., F.R.S.

[Received November 23rd, 1904. — Read December 8th, 1904.]

ART. 1. In *Crelle*, t. I., p. 367, there is given a generalization of the binomial theorem which was restated by Cayley in 1851* in a better form. The statement is as follows:—

"If $\{x_1+x_2+x_3+\ldots\}^p$

denote the expansion of $(x_1+x_2+x_3+\ldots)^p$,

retaining those terms only $N x_1^{a_1} x_2^{a_2} x_3^{a_3} \ldots$

in which

$$\begin{aligned} a_2+a_3+a_4+\ldots &\ngtr p-1, \\ a_3+a_4+\ldots &\ngtr p-2, \\ \ldots \quad \ldots \quad &\ldots, \end{aligned}$$

then

$$\begin{aligned} X^n = \;& 1 \times && (X+x_1)^n \\ & -\binom{n}{1} \times \{x_1\}^1 && (X+x_1+x_2)^{n-1} \\ & +\binom{n}{2} \times \{x_1+x_2\}^2 && (X+x_1+x_2+x_3)^{n-2} \\ & -\binom{n}{3} \times \{x_1+x_2+x_3\}^3 && (X+x_1+x_2+x_3+x_4)^{n-3} \\ & +\ldots \text{"} \end{aligned}$$

The formula is curious, and, as remarked by Cayley, of some interest by reason of the introduction of the deficient multinomial expansion

$$\{x_1+x_2+x_3+\ldots\}^p.$$

I propose to investigate properties of the expansion by the application of a method which has been found to be valuable in similar questions.

ART. 2. We may first enquire into the number of distinct terms in the development of

$$\{x_1+x_2+\ldots+x_p\}^q \quad (q \geqslant p).$$

---

* Cayley, *Collected Papers*, Vol. II., p. 102.

A term

$$N x_1^{a_1} x_2^{a_2} \dots x_p^{a_p}$$

arises when

$$\begin{aligned} a_2+a_3+\dots+a_p &\leqslant q-1, \\ a_3+\dots+a_p &\leqslant q-2, \\ \dots \quad \dots \quad &\dots \\ a_p &\leqslant q-p+1, \end{aligned}$$

a number of Diophantine inequalities which may be replaced by the set

$$\begin{aligned} a_1 &\geqslant 1, \\ a_1+a_2 &\geqslant 2, \\ \dots \quad &\dots \\ a_1+a_2+\dots+a_p &\geqslant p, \end{aligned}$$

and, of course,

$$a_1+a_2+\dots+a_p = q.$$

To form a generating function for the number in question consider the case $p = 3$ and $q \geqslant 3$,

$$\begin{aligned} a_1 &\geqslant 1, \\ a_1+a_2 &\geqslant 2, \\ a_1+a_2+a_3 &= q, \end{aligned}$$

and form the expression

$$\underset{\geqslant}{\Omega} \frac{\dfrac{1}{a_1 a_2^2 a_3^3}}{(1-a_1 a_2 a_3 x_1)(1-a_2 a_3 x_2)(1-a_3 x_3)},$$

where $\underset{\geqslant}{\Omega}$ signifies that, in the ascending expansion, all terms involving negative powers of the quantities a are to be rejected, and those quantities then put equal to unity.

From the construction it is clear that x_1 must occur at least once, for otherwise every term would be rejected by the operation Ω; similarly the exponents of x_1 and x_2 must together amount to at least two, and those of x_1, x_2, and x_3 to at least three. The development will contain every product

$$x_1^{a_1} x_2^{a_2} x_3^{a_3}$$

which satisfies the given conditions, and the expansion is therefore a representative generating function.

We pass to the enumerating generating function by putting

$$x_1 = x_2 = x_3 = x,$$

and thence, eliminating a_1, a_2, a_3 in succession, we find

$$\Omega \frac{\dfrac{1}{a_1 a_2^2 a_3^3}}{(1-a_1 a_2 a_3 x)(1-a_2 a_3 x)(1-a_3 x)}$$

$$= \Omega \frac{\dfrac{x}{a_2 a_3^2}}{(1-a_2 a_3 x)^2 (1-a_3 x)}$$

$$= \Omega \left\{ \frac{\dfrac{x^2}{a_3}}{(1-a_3 x)^3} + \frac{\dfrac{x^2}{a_3}}{(1-a_3 x)^2} \right\}$$

$$= \Omega \frac{x^2}{(1-x)^3} + \frac{x^2}{(1-x)^2} - 2x^2$$

$$= \frac{5x^3 - 6x^4 + 2x^5}{(1-x)^3}$$

$$= \Sigma \tfrac{1}{2} (q+2)(q-1)\, x^q \quad (q \geqslant 3).$$

It is thus established that the number of terms in

$$\{x_1 + x_2 + x_3\}^q$$

is

$$\tfrac{1}{2}(q+2)(q-1).$$

Passing to the general case, we have

$$\Omega \frac{\dfrac{1}{a_1 a_2^2 \dots a_p^p}}{(1-a_1 \dots a_p x)(1-a_2 \dots a_p x) \dots (1-a_{p-1} a_p x)(1-a_p x)}.$$

The reduction will be made by employing the formula

$$\binom{s}{1} + \binom{s+1}{2} x + \binom{s+2}{3} x^2 + \dots = \frac{1}{1-x} + \frac{1}{(1-x)^2} + \dots + \frac{1}{(1-x)^s}.$$

The expression is

$$\Omega \frac{\dfrac{x}{a_2 a_3^2 \dots a_p^{p-1}}}{(1-a_2 \dots a_p x)^2 (1-a_3 \dots a_p x) \dots (1-a_{p-1} a_p x)(1-a_p x)}$$

$$= \Omega \frac{\dfrac{x^2}{a_3 a_4^2 \dots a_p^{p-2}} (2 + 3a_3 \dots a_p x + 4a_3^2 \dots a_p^2 x^2 + \dots)}{(1-a_3 \dots a_p x)(1-a_4 \dots a_p x) \dots (1-a_p x)},$$

and by the help of the above formula this is

$$\Omega \frac{\dfrac{x^2}{a_3 a_4^2 \ldots a_p^{p-2}}}{(1-a_3 \ldots a_p x)^2 (1-a_4 \ldots a_p x) \ldots (1-a_p x)}$$

$$+\Omega \frac{\dfrac{x^2}{a_3 a_4^2 \ldots a_p^{p-2}}}{(1-a_3 \ldots a_p x)^3 (1-a_4 \ldots a_p) x \ldots (1-a_p x)},$$

viz., after eliminating a_2 we have two fractions, and it may be readily seen that after elimination of a_s we will have a sum of s fractions.

Assume the result

$$\Omega \frac{\dfrac{(2s-4)!}{(s-2)!\,(s-1)!} \dfrac{x^{s-1}}{a_s a_{s+1}^2 \ldots a_p^{p-s+1}}}{(1-a_s \ldots a_p x)^2 (1-a_{s+1} \ldots a_p x) \ldots (1-a_p x)}$$

$$+\Omega \frac{\dfrac{(2s-5)!}{(s-3)!\,(s-1)!}\, 2\, \dfrac{x^{s-1}}{a_3 a_{s+1}^2 \ldots a_p^{p-s+1}}}{(1-a_s \ldots a_p x)^3 (1-a_{s+1} \ldots a_p x) \ldots (1-a_p x)}$$

$$+\Omega \frac{\dfrac{(2s-6)!}{(s-4)!\,(s-1)!}\, 3\, \dfrac{x^{s-1}}{a_s a_{s+1}^2 \ldots a_p^{p-s+1}}}{(1-a_s \ldots a_p x)^4 (1-a_{s+1} \ldots a_p x) \ldots (1-a_p x)}$$

$$+\ldots$$

$$+\Omega \frac{\dfrac{x^{s-1}}{a_s a_{s+1}^2 \ldots a_p^{p-s+1}}}{(1-a_s \ldots a_p x)^s (1-a_{s+1} \ldots a_p x) \ldots (1-a_p x)}$$

to accrue after elimination of a_{s-1}.

It consists of $s-1$ fractions and agrees with the result obtained when $s=2$.

It will be shown that the result when a_s is eliminated may be obtained from the above by writing $s+1$ for s.

The series above may be written

$$\sum_{m=2}^{m=s} \Omega \frac{\dfrac{(2s-m-2)!}{(s-m)!\,(s-1)!} (m-1) \dfrac{x^{s-1}}{a_s a_{s+1}^2 \ldots a_p^{p-s+1}}}{(1-a_s \ldots a_p x)^m (1-a_{s+1} \ldots a_p x) \ldots (1-a_p x)}.$$

This general term becomes, on elimination of a_s,

$$\Omega \frac{\dfrac{(2s-m-2)!}{(s-m)!\,(s-1)!} (m-1) \left\{ \binom{m}{1} + \binom{m+1}{2} a_{s+1} \ldots a_p + \ldots \right\} \dfrac{x^s}{a_{s+1} a_{s+2}^2 \ldots a_p^{p-s}}}{(1-a_{s+1} \ldots a_p x)(1-a_{s+2} \ldots a_p x) \ldots (1-a_p x)},$$

and this by the above quoted formula is

$$\sum_{t=2}^{t=m+1} \Omega \frac{\dfrac{(2s-m-2)!}{(s-m)!\,(s-1)!}(m-1)\dfrac{x^s}{a_{s+1}a_{s+2}^2\ldots a_p^{p-s}}}{(1-a_{s+1}\ldots a_p x)^t\,(1-a_{s+2}\ldots a_p x)\ldots(1-a_p x)}.$$

Summing this expression from $m=2$ to $m=s$, we find for the coefficient of the fraction whose denominator involves

$$(1-a_{s+1}\ldots a_p x)^n$$

the sum

$$\frac{1}{s-1}\left\{(n-2)\binom{2s-n-1}{s-2}+(n-1)\binom{2s-n-2}{s-2}+\ldots \text{ to } s-n+2 \text{ terms}\right\},$$

which by elementary algebra is

$$\frac{(2s-n)!}{(s-n+1)!\,s!}(n-1),$$

which is what the coefficient

$$\frac{(2s-n-2)!}{(s-n)!\,(s-1)!}(n-1),$$

which occurs in the assumed formula, becomes on writing $s+1$ for s. Hence the result is established.

Putting therein $s=p$, we obtain

$$\sum_{t=0}^{t=p-2} \Omega \frac{\dfrac{(2p-t-4)!}{(p-t-2)!\,(p-1)!}(t+1)\dfrac{x^{p-1}}{a_p}}{(1-a_p x)^{t+2}}$$

$$=\sum_{t=0}^{t=p-2}\frac{(2p-t-4)!}{(p-t-2)!\,(p-1)!}(t+1)\left\{\frac{1}{1-x}+\frac{1}{(1-x)^2}+\ldots+\frac{1}{(1-x)^{t+2}}\right\}x^p$$

$$=\frac{(2p-2)!}{(p-1)!\,p!}\frac{x^p}{1-x}+2\frac{(2p-3)!}{(p-2)!\,p!}\frac{x^p}{(1-x)^2}+\ldots+\frac{x^p}{(1-x)^p},$$

after some reductions.

Herein the coefficient of x^q is

$$\frac{(2p-2)!}{(p-1)!\,p!}+2(q-p+1)\frac{(2p-3)!}{(p-2)!\,p!}+3\binom{q-p+2}{2}\frac{(2p-4)!}{(p-3)!\,p!}+\ldots$$

to p terms.

This sum by elementary algebra is

$$\frac{q-p+2}{p-1}\binom{q+p-1}{q+1}=(q-p+2)\frac{(q+p-1)!}{(q+1)!\,(p-1)!};$$

which is therefore the number of distinct terms in

$$\{x_1+x_2+\ldots+x_p\}^q \quad \text{for} \quad q \geqslant p.$$

Denoting the number in question by

$$H_{p,q}, \quad q \geqslant p,$$

it is easy to prove the formula

$$H_{p,q} = H_{p-1,p-1}+H_{p-1,p}+\ldots+H_{p-1,q}$$

by considering the coefficient of b^q in

$$\Omega \frac{\dfrac{1}{x_{p-1}^{p-1} x_{p-2}^{p-2} \ldots x_2^2 x_1}}{(1-bx_{p-1})(1-bx_{p-1}x_{p-2}) \ldots (1-bx_{p-1}x_{p-2} \ldots x_2 x_1)},$$

and then eliminating x_{p-1}.

ART. 3. Consider next the sum of the coefficients in

$$\{x_1+x_2+\ldots+x_p\}^q.$$

If $q = p$, we find from Cayley's formula, by putting

$$X = x_1 = x_2 = \ldots = x_p = 1,$$

$$1 = 2^n - \binom{n}{1} C_{1,1} 3^{n-1} + \binom{n}{2} C_{2,2} 4^{n-2} - \ldots + (-)^n \binom{n}{n} C_{n,n},$$

where $C_{p,p}$ is the sum of the coefficients in

$$\{x_1+x_2+\ldots+x_p\}^p.$$

From this relation it is easy to show that

$$C_{p,p} = (p+1)^{p-1}.$$

In general, consider the expression

$$\Omega \frac{(1+x_{p-1}+x_{p-1}x_{p-2}+\ldots+x_{p-1}x_{p-2}\ldots x_2 x_1)^q}{x_{p-1}^{p-1} x_{p-2}^{p-2} \ldots x^2 x_1},$$

wherein Ω as usual operates by rejecting all terms in the development which involve negative powers of the quantities x, and subsequently replaces each of these quantities by unity.

The general term under the operator Ω is

$$\frac{q!}{a_1!\, a_2! \ldots a_{p-1}!} \frac{x_1^{a_1} x_2^{a_1+a_2} \ldots x_{p-1}^{a_1+a_2+\ldots+a_{p-1}}}{x_1 x_2^2 \ldots x_{p-1}^{p-1}},$$

and hence the conditions that the term may be integral are precisely those that define the deficient multinomial expansion.

Hence, $q \geqslant p$, the expression above written is equal to the sum of the coefficients in

$$\{x_1+x_2+\ldots+x_p\}^q,$$

which we will denote by $C_{p,q}$.

Write the expression

$$C_{p,q} = \Omega \frac{X_{p-1}^q}{X'_{p-1}} = \Omega \frac{(1+x_{p-1}X_{p-2})^q}{x_{p-1}^{p-1}X'_{p-2}}$$

$$= \binom{q}{p-1}\Omega \frac{X_{p-2}^{p-1}}{X'_{p-2}} + \binom{q}{p}\Omega \frac{X_{p-2}^{p}}{X'_{p-2}} + \ldots + \binom{q}{q}\Omega \frac{X_{p-2}^{q}}{X'_{p-2}}.$$

Hence the difference equation

$$C_{p,q} = \binom{q}{p-1} C_{p-1,p-1} + \binom{q}{p} C_{p-1,p} + \ldots + \binom{q}{q} C_{p-1,q},$$

from which the quantities $C_{p,q}$

can be calculated.

ART. 4. There is a syzygetic theory of the distinct terms in

$$\{x_1+x_2+\ldots+x_p\}^q, \quad q \geqslant p,$$

of which I give an illustration for $p=3$. The sum of the terms

$$x_1^{\alpha_1}x_2^{\alpha_2}\ldots x_p^{\alpha_p}$$

has been shown, for $p=3$, to be

$$\Omega \frac{\dfrac{1}{a_1a_2^2a_3^3}}{(1-a_1a_2a_3x_1)(1-a_2a_3x_2)(1-a_3x_3)}$$

which without difficulty may be given the expression

$$\frac{x_1x_2x_3+x_1x_2^2+x_1^2x_2+x_1^2x_3+x_1^3-(x_1x_2^2x_3+2x_1^2x_2x_3+x_1^2x_2^2+x_1^3x_2+x_1^3x_3) + x_1^2x_2^2x_3+x_1^3x_2x_3}{(1-x_1)(1-x_2)(1-x_3)},$$

which shows that every term $x_1^{\alpha_1}x_2^{\alpha_2}x_3^{\alpha_3}$

contains one of the products

$$x_1x_2x_3, \quad x_1x_2^2, \quad x_1^2x_2, \quad x_1^2x_3, \quad x_1^3,$$

which may be called ground products. The other products are formed by multiplying these into any power products of x_1, x_2, x_3, but the numerator indicates by the terms

$$x_1 x_2^2 x_3, \quad 2x_1^2 x_2 x_3, \quad x_1^2 x_2^2, \quad x_1^3 x_2, \quad x_1^3 x_3$$

six syzygies of the first order, viz.,

$$A = x_3(x_1 x_2^2) - x_2(x_1 x_2 x_3) = 0,$$
$$B = x_3(x_1^2 x_2) - x_1(x_1 x_2 x_3) = 0,$$
$$C = x_2(x_1^2 x_3) - x_1(x_1 x_2 x_3) = 0,$$
$$D = x_1(x_1 x_2^2) - x_2(x_1^2 x_2) = 0,$$
$$E = x_2(x_1^3) - x_1(x_1^2 x_2) = 0,$$
$$F = x_1(x_1^2 x_3) - x_3(x_1^3) = 0,$$

and by the terms $\quad x_1^2 x_2^2 x_3, \quad x_1^3 x_2 x_3$

two syzygies of the second order, viz.,

$$x_1 A - x_2 C = x_1 C - x_3 E = 0.$$

II. *Memoir on the Theory of the Partitions of Numbers.—Part* III.

By P. A. MacMahon, *Major R.A., Sc.D., F.R.S.*

Received November 21,—Read December 8, 1904.

Since Part II. of the Memoir appeared in November, 1898, the following papers by the author, bearing upon the Partition of Numbers, have been published :—

"Partitions of Numbers whose Graphs possess Symmetry," 'Cambridge Phil. Trans.,' vol. XVII., Part II. ;

"Application of the Partition Analysis to the Study of the Properties of any System of Consecutive Integers," 'Cambridge Phil. Trans.,' vol. XVIII. ;

"The Diophantine Inequality $\lambda x \geqq \mu y$," 'Cambridge Phil. Trans.,' vol. XIX. ;

"Combinatorial Analysis. The Foundations of a New Theory," 'Phil. Trans. Roy. Soc. London,' A, vol. 194, 1900.

In the present Part III. I consider problems of "Arithmetic of Position." In particular, I define a "general magic square" composed of integers and show that for a given order of square it is possible to construct a syzygetic theory. Such a theory is worked out in detail for the order 3 as an illustration. I further discuss the problem of the enumeration of the squares of given order associated with a given sum. I show that there is no difficulty in constructing a generating function for such squares even when the construction is specified in detail, and I obtain an analytical expression for the number when the sum, associated with rows, columns and diagonals, is unity or two.

§ 9.

Art. 124. A "general magic square" I take to consist of n^2 integers arranged in a square in such wise that the rows, columns and diagonals contain partitions of the same number, zero and repetitions of the same integer being permissible among the integers.

An ordinary magic square I define to be a general magic square in which the n^2 integers are restricted to be the first n^2 integers of the natural succession.

We may regard general magic squares as numerical magnitudes. To add two such magnitudes we add together the numbers in corresponding positions to form a magnitude which is obviously also a general magic square. We can, therefore, form a linear function of magnitudes of the same order, n, the coefficients being positive integers, and such linear functions will denote a general magic square.

The magnitudes, of the same given order, can be taken as the elements of a linear algebra, and since arithmetical addition can be made to depend upon algebraical multiplication, the properties of the magnitudes can be investigated by means of a non-linear algebra.

Art. 125. The properties of a general magic square can be exhibited by means of homogeneous linear Diophantine equations, and it thence immediately follows that there must be a syzygetic theory of such formations. There exists a finite number of ground forms, corresponding to the ground solutions of the equations, and the method of investigation determines these and the syzygies which connect them.

Generally speaking, there is a syzygetic theory associated with every system of linear homogeneous Diophantine equalities or inequalities, and it is because invariant theories depend upon such systems that they are connected with syzygetic theories.

Art. 126. The method of investigation about to be given applies not only to magic squares of different kinds but to all arrangements of integers, which are defined by homogeneous linear Diophantine equalities or inequalities, whose properties persist after addition of corresponding numbers.

For example, the partitions of all numbers into n, or fewer parts, are defined by the linear homogeneous Diophantine inequalities

$$\alpha_1 \geq \alpha_2 \geq \alpha_3 \ldots \geq \alpha_n,$$

and if another solution be

$$\beta_1 \geq \beta_2 \geq \beta_3 \ldots \geq \beta_n,$$

we have

$$\alpha_1+\beta_1 \geq \alpha_2+\beta_2 \geq \alpha_3+\beta_3 \ldots \geq \alpha_n+\beta_n,$$

and since the property persists after addition, a syzygetic theory results.

This is one of the simplest cases that could be adduced and is at the same time the true basis of the Theory of Partitions.

Many instances of configurations of integers *in plano* or *in solido* will occur to the mind as having been subjects of contemplation by mathematicians and others from the earliest times. These when defined by properties which persist after addition of corresponding parts fall under the present theory.

Art. 127. There is no general magic square of the order 2 except the trivial case $\begin{vmatrix} a & a \\ a & a \end{vmatrix}$, but we may consider squares of order 2 in which the row and column properties, but not the diagonal properties, are in evidence.

Let such a square be

$$\begin{vmatrix} \alpha_1 & \alpha_2 \\ \alpha_3 & \alpha_4 \end{vmatrix},$$

which must clearly have the form

$$\begin{vmatrix} \alpha_1 & \alpha_2 \\ \alpha_2 & \alpha_1 \end{vmatrix},$$

and we may associate with it the Diophantine equation

$$\alpha_1 + \alpha_2 = \alpha_5$$

and regard α_1, α_2 and α_5 as the unknowns.

The syzygetic theory is obtained by forming the sum

$$\Sigma X_1^{\alpha_1} X_2^{\alpha_2} X_5^{\alpha_5}$$

for all solutions of the equation, and the result is

$$\underset{=}{\Omega} \frac{1}{1-aX_1 \,.\, 1-aX_2 \,.\, 1-\frac{1}{a}X_5},$$

where a is an auxiliary quantity and the meaning of the prefixed symbol $\underset{=}{\Omega}$ is that after expansion of the algebraic fraction in ascending powers of X_1, X_2, X_5 we are to retain those terms only which are free from a.

The expression clearly has the value

$$\frac{1}{1-X_1X_5 \,.\, 1-X_2X_5}.$$

The denominator factors denote the ground solutions

α_1	α_2	α_5
1	0	1
0	1	1

and the absence of numerator terms shows that there are no syzygies.

Thus the fundamental squares are

$$\begin{vmatrix} 1 & 0 \\ 0 & 1 \end{vmatrix} \qquad \begin{vmatrix} 0 & 1 \\ 1 & 0 \end{vmatrix},$$

and this is otherwise evident. The case is trivial and is introduced only for the orderly presentation of the subject.

Art. 128. Passing on to the general magic squares of order 3 we have the square

$$\begin{vmatrix} \alpha_1 & \alpha_2 & \alpha_3 \\ \alpha_4 & \alpha_5 & \alpha_6 \\ \alpha_7 & \alpha_8 & \alpha_9 \end{vmatrix}$$

defined by the eight Diophantine equations

$$\alpha_1+\alpha_2+\alpha_3 = \alpha_{10}, \qquad \alpha_2+\alpha_5+\alpha_8 = \alpha_{10},$$
$$\alpha_4+\alpha_5+\alpha_6 = \alpha_{10}, \qquad \alpha_3+\alpha_6+\alpha_9 = \alpha_{10},$$
$$\alpha_7+\alpha_8+\alpha_9 = \alpha_{10}, \qquad \alpha_1+\alpha_5+\alpha_9 = \alpha_{10},$$
$$\alpha_1+\alpha_4+\alpha_7 = \alpha_{10}, \qquad \alpha_3+\alpha_5+\alpha_7 = \alpha_{10}.$$

We require all values of the quantities α which satisfy these equations.

To form the sum

$$\Sigma X_1^{\alpha_1}X_2^{\alpha_2}X_3^{\alpha_3}X_4^{\alpha_4}X_5^{\alpha_5}X_6^{\alpha_6}X_7^{\alpha_7}X_8^{\alpha_8}X_9^{\alpha_9}X_{10}^{\alpha_{10}},$$

for all solutions, introduce the auxiliary quantities

$$a,\ b,\ c,\ d,\ e,\ f,\ g,\ h$$

in association with the successive Diophantine equations. The sum in question may be written

$$\underset{=}{\Omega}\frac{1}{\left\{\begin{array}{l}(1-adgX_1)(1-aeX_2)(1-afhX_3)(1-bdX_4)(1-beghX_5)\\ (1-bfX_6)(1-cdhX_7)(1-ceX_8)(1-cfgX_9)\left(1-\dfrac{X_{10}}{abcdefgh}\right)\end{array}\right\}},$$

where after expansion we retain that portion only which is free from the auxiliaries.

Remarking that

$$\underset{=}{\Omega}\frac{1}{(1-aP_1)(1-aP_2)(1-aP_3)\left(1-\dfrac{P_4}{a}\right)} = \frac{1}{(1-P_1P_4)(1-P_2P_4)(1-P_3P_4)},$$

we eliminate the auxiliar a and obtain

$$\underset{=}{\Omega}\frac{1}{\left\{\begin{array}{l}\left(1-\dfrac{X_1X_{10}}{bcefh}\right)\left(1-\dfrac{X_2X_{10}}{bcdfgh}\right)\left(1-\dfrac{X_3X_{10}}{bcdeg}\right)(1-bdX_4)\\ (1-beghX_5)(1-bfX_6)(1-cdhX_7)(1-ceX_8)(1-cfgX_9)\end{array}\right\}}.$$

Put now $bd = A$, $be = B$, $bf = C$, $cd = D$, and we obtain

$$\underset{=}{\Omega}\frac{1}{\left\{\begin{array}{l}\left(1-\dfrac{AX_1X_{10}}{BCDh}\right)\left(1-\dfrac{X_2X_{10}}{CDgh}\right)\left(1-\dfrac{X_3X_{10}}{BDg}\right)(1-AX_4)\\ (1-BghX_5)(1-CX_6)(1-DhX_7)\left(1-\dfrac{BDX_8}{A}\right)\left(1-\dfrac{CDgX_9}{A}\right)\end{array}\right\}},$$

an artifice which reduces the number of auxiliars to be eliminated by unity.

Remarking that

$$\underset{=}{\Omega}\frac{1}{(1-a\mathrm{P}_1)(1-a\mathrm{P}_2)\left(1-\frac{1}{a}\mathrm{P}_3\right)\left(1-\frac{1}{a}\mathrm{P}_4\right)}$$

$$=\frac{1-\mathrm{P}_1\mathrm{P}_2\mathrm{P}_3\mathrm{P}_4}{(1-\mathrm{P}_1\mathrm{P}_3)(1-\mathrm{P}_1\mathrm{P}_4)(1-\mathrm{P}_2\mathrm{P}_3)(1-\mathrm{P}_2\mathrm{P}_4)}$$

$$=\frac{1}{(1-\mathrm{P}_1\mathrm{P}_4)(1-\mathrm{P}_2\mathrm{P}_3)(1-\mathrm{P}_2\mathrm{P}_4)}+\frac{\mathrm{P}_1\mathrm{P}_3}{(1-\mathrm{P}_1\mathrm{P}_3)(1-\mathrm{P}_1\mathrm{P}_4)(1-\mathrm{P}_2\mathrm{P}_3)}.$$

We eliminate A and find

$$\underset{=}{\Omega}\frac{1}{\left\{\begin{array}{l}\left(1-\frac{g}{\mathrm{B}h}\mathrm{X}_1\mathrm{X}_9\mathrm{X}_{10}\right)(1-\mathrm{BDX}_4\mathrm{X}_8)(1-\mathrm{CD}g\mathrm{X}_4\mathrm{X}_9)\left(1-\frac{1}{\mathrm{CD}gh}\mathrm{X}_2\mathrm{X}_{10}\right)\\ \left(1-\frac{1}{\mathrm{BD}g}\mathrm{X}_3\mathrm{X}_{10}\right)(1-\mathrm{B}gh\mathrm{X}_5)(1-\mathrm{CX}_6)(1-\mathrm{D}h\mathrm{X}_7)\end{array}\right\}}$$

$$+\underset{=}{\Omega}\frac{\frac{1}{\mathrm{C}h}\mathrm{X}_1\mathrm{X}_8\mathrm{X}_{10}}{\left\{\begin{array}{l}\left(1-\frac{1}{\mathrm{C}h}\mathrm{X}_1\mathrm{X}_8\mathrm{X}_{10}\right)\left(1-\frac{g}{\mathrm{B}h}\mathrm{X}_1\mathrm{X}_9\mathrm{X}_{10}\right)(1-\mathrm{BDX}_4\mathrm{X}_8)\left(1-\frac{1}{\mathrm{CD}gh}\mathrm{X}_2\mathrm{X}_{10}\right)\\ \left(1-\frac{1}{\mathrm{BD}g}\mathrm{X}_3\mathrm{X}_{10}\right)(1-\mathrm{B}gh\mathrm{X}_5)(1-\mathrm{CX}_6)(1-\mathrm{D}h\mathrm{X}_7)\end{array}\right\}}.$$

Eliminating B from the first fraction and C from the second, we have

$$\underset{=}{\Omega}\frac{1}{\left\{\begin{array}{l}(1-\mathrm{CD}g\mathrm{X}_4\mathrm{X}_9)(1-\mathrm{D}h\mathrm{X}_7)\left(1-\frac{1}{\mathrm{CD}gh}\mathrm{X}_2\mathrm{X}_{10}\right)\left(1-\frac{h}{\mathrm{D}}\mathrm{X}_3\mathrm{X}_5\mathrm{X}_{10}\right)\\ (1-\mathrm{CX}_6)(1-g^2\mathrm{X}_1\mathrm{X}_5\mathrm{X}_9\mathrm{X}_{10})\left(1-\frac{1}{g}\mathrm{X}_3\mathrm{X}_4\mathrm{X}_8\mathrm{X}_{10}\right)\end{array}\right\}}$$

$$+\underset{=}{\Omega}\frac{\frac{g\mathrm{D}}{h}\mathrm{X}_1\mathrm{X}_4\mathrm{X}_8\mathrm{X}_9\mathrm{X}_{10}}{\left\{\begin{array}{l}\left(1-\frac{g\mathrm{D}}{h}\mathrm{X}_1\mathrm{X}_4\mathrm{X}_8\mathrm{X}_9\mathrm{X}_{10}\right)(1-\mathrm{CD}g\mathrm{X}_4\mathrm{X}_9)(1-\mathrm{D}h\mathrm{X}_7)\left(1-\frac{1}{\mathrm{CD}gh}\mathrm{X}_2\mathrm{X}_{10}\right)\\ (1-\mathrm{CX}_6)(1-g^2\mathrm{X}_1\mathrm{X}_5\mathrm{X}_9\mathrm{X}_{10})\left(1-\frac{1}{g}\mathrm{X}_3\mathrm{X}_4\mathrm{X}_8\mathrm{X}_{10}\right)\end{array}\right\}}$$

$$+\underset{=}{\Omega}\frac{\frac{1}{h}\mathrm{X}_1\mathrm{X}_6\mathrm{X}_8\mathrm{X}_{10}}{\left\{\begin{array}{l}(1-\mathrm{BDX}_4\mathrm{X}_8)(1-\mathrm{B}gh\mathrm{X}_5)\left(1-\frac{1}{\mathrm{D}gh}\mathrm{X}_2\mathrm{X}_6\mathrm{X}_{10}\right)\left(1-\frac{g}{\mathrm{B}h}\mathrm{X}_1\mathrm{X}_9\mathrm{X}_{10}\right)\\ \left(1-\frac{1}{\mathrm{BD}g}\mathrm{X}_3\mathrm{X}_{10}\right)(1-\mathrm{D}h\mathrm{X}_7)\left(1-\frac{1}{h}\mathrm{X}_1\mathrm{X}_6\mathrm{X}_8\mathrm{X}_{10}\right)\end{array}\right\}}.$$

From the first fraction eliminate C, from the second C, and from the third B, obtaining

$$\underset{=}{\Omega}\frac{1}{\left\{\begin{array}{l}(1-\mathrm{D}h\mathrm{X}_7)\left(1-\frac{1}{\mathrm{D}gh}\mathrm{X}_2\mathrm{X}_6\mathrm{X}_{10}\right)\left(1-\frac{h}{\mathrm{D}}\mathrm{X}_3\mathrm{X}_5\mathrm{X}_{10}\right)\\(1-g^2\mathrm{X}_1\mathrm{X}_5\mathrm{X}_9\mathrm{X}_{10})\left(1-\frac{1}{g}\mathrm{X}_3\mathrm{X}_4\mathrm{X}_8\mathrm{X}_{10}\right)\left(1-\frac{1}{h}\mathrm{X}_2\mathrm{X}_4\mathrm{X}_9\mathrm{X}_{10}\right)\end{array}\right\}}$$

$$+\underset{=}{\Omega}\frac{\frac{g\mathrm{D}}{h}\mathrm{X}_1\mathrm{X}_4\mathrm{X}_8\mathrm{X}_9\mathrm{X}_{10}}{\left\{\begin{array}{l}(1-\mathrm{D}h\mathrm{X}_7)\left(1-\frac{g\mathrm{D}}{h}\mathrm{X}_1\mathrm{X}_4\mathrm{X}_8\mathrm{X}_9\mathrm{X}_{10}\right)\left(1-\frac{1}{\mathrm{D}gh}\mathrm{X}_2\mathrm{X}_6\mathrm{X}_{10}\right)\\(1-g^2\mathrm{X}_1\mathrm{X}_5\mathrm{X}_9\mathrm{X}_{10})\left(1-\frac{1}{g}\mathrm{X}_3\mathrm{X}_4\mathrm{X}_8\mathrm{X}_{10}\right)\left(1-\frac{1}{h}\mathrm{X}_2\mathrm{X}_4\mathrm{X}_9\mathrm{X}_{10}\right)\end{array}\right\}}$$

$$+\underset{=}{\Omega}\frac{\frac{1}{h}\mathrm{X}_1\mathrm{X}_6\mathrm{X}_8\mathrm{X}_{10}\,(1-g\mathrm{X}_1\mathrm{X}_3\mathrm{X}_4\mathrm{X}_5\mathrm{X}_8\mathrm{X}_9\mathrm{X}_{10}{}^2)}{\left\{\begin{array}{l}(1-\mathrm{D}h\mathrm{X}_7)\left(1-\frac{g\mathrm{D}}{h}\mathrm{X}_1\mathrm{X}_4\mathrm{X}_8\mathrm{X}_9\mathrm{X}_{10}\right)\left(1-\frac{1}{\mathrm{D}gh}\mathrm{X}_2\mathrm{X}_6\mathrm{X}_{10}\right)\\\left(1-\frac{h}{\mathrm{D}}\mathrm{X}_3\mathrm{X}_5\mathrm{X}_{10}\right)(1-g^2\mathrm{X}_1\mathrm{X}_5\mathrm{X}_9\mathrm{X}_{10})\left(1-\frac{1}{g}\mathrm{X}_3\mathrm{X}_4\mathrm{X}_8\mathrm{X}_{10}\right)\left(1-\frac{1}{h}\mathrm{X}_1\mathrm{X}_6\mathrm{X}_8\mathrm{X}_{10}\right)\end{array}\right\}}.$$

Eliminating D from each of the three fractions, we obtain

$$\underset{=}{\Omega}\frac{1}{\left\{\begin{array}{l}(1-g^2\mathrm{X}_1\mathrm{X}_5\mathrm{X}_9\mathrm{X}_{10})\left(1-\frac{1}{g}\mathrm{X}_2\mathrm{X}_6\mathrm{X}_7\mathrm{X}_{10}\right)\left(1-\frac{1}{g}\mathrm{X}_3\mathrm{X}_4\mathrm{X}_8\mathrm{X}_{10}\right)\\(1-h^2\mathrm{X}_3\mathrm{X}_5\mathrm{X}_7\mathrm{X}_{10})\left(1-\frac{1}{h}\mathrm{X}_2\mathrm{X}_4\mathrm{X}_9\mathrm{X}_{10}\right)\end{array}\right\}}$$

$$+\underset{=}{\Omega}\frac{\frac{1}{h^2}\mathrm{X}_1\mathrm{X}_2\mathrm{X}_4\mathrm{X}_6\mathrm{X}_8\mathrm{X}_9\mathrm{X}_{10}{}^2}{\left\{\begin{array}{l}(1-g^2\mathrm{X}_1\mathrm{X}_5\mathrm{X}_9\mathrm{X}_{10})\left(1-\frac{1}{g}\mathrm{X}_2\mathrm{X}_6\mathrm{X}_7\mathrm{X}_{10}\right)\left(1-\frac{1}{g}\mathrm{X}_3\mathrm{X}_4\mathrm{X}_8\mathrm{X}_{10}\right)\\\left(1-\frac{1}{h}\mathrm{X}_2\mathrm{X}_4\mathrm{X}_9\mathrm{X}_{10}\right)\left(1-\frac{1}{h^2}\mathrm{X}_1\mathrm{X}_2\mathrm{X}_4\mathrm{X}_6\mathrm{X}_8\mathrm{X}_9\mathrm{X}_{10}{}^2\right)\end{array}\right\}}$$

$$+\underset{=}{\Omega}\frac{\frac{1}{h}\mathrm{X}_1\mathrm{X}_6\mathrm{X}_8\mathrm{X}_{10}\,(1-\mathrm{X}_1\mathrm{X}_2\mathrm{X}_3\mathrm{X}_4\mathrm{X}_5\mathrm{X}_6\mathrm{X}_7\mathrm{X}_8\mathrm{X}_9\mathrm{X}_{10}{}^3)}{\left\{\begin{array}{l}(1-g^2\mathrm{X}_1\mathrm{X}_5\mathrm{X}_9\mathrm{X}_{10})\left(1-\frac{1}{g}\mathrm{X}_2\mathrm{X}_6\mathrm{X}_7\mathrm{X}_{10}\right)\left(1-\frac{1}{g}\mathrm{X}_3\mathrm{X}_4\mathrm{X}_8\mathrm{X}_{10}\right)\\(1-h^2\mathrm{X}_3\mathrm{X}_5\mathrm{X}_7\mathrm{X}_{10})\left(1-\frac{1}{h}\mathrm{X}_1\mathrm{X}_6\mathrm{X}_8\mathrm{X}_{10}\right)\left(1-\frac{1}{h^2}\mathrm{X}_1\mathrm{X}_2\mathrm{X}_4\mathrm{X}_6\mathrm{X}_8\mathrm{X}_9\mathrm{X}_{10}{}^2\right)\end{array}\right\}}.$$

Art. 129. Before proceeding to eliminate g and h, observe that if we now put $g = h = 1$, we obtain the generating function for the solutions of the first six of the Diophantine equations corresponding to the squares which possess row and column but not diagonal properties.

Putting $g = h = 1$, the generating function reduces to

$$\frac{1 - X_1X_2X_3X_4X_5X_6X_7X_8X_9X_{10}{}^3}{\left\{\begin{matrix}(1-X_1X_5X_9X_{10})(1-X_1X_6X_8X_{10})(1-X_2X_4X_9X_{10}) \\ (1-X_2X_6X_7X_{10})(1-X_3X_4X_8X_{10})(1-X_3X_5X_7X_{10})\end{matrix}\right\}},$$

indicating ground forms

$$\begin{matrix} X_1X_5X_9X_{10}, & X_2X_6X_7X_{10}, \\ X_1X_6X_8X_{10}, & X_3X_4X_8X_{10}, \\ X_2X_4X_9X_{10}, & X_3X_5X_7X_{10} \end{matrix}$$

connected by the ground syzygy

$$X_1X_5X_9X_{10} \,.\, X_2X_6X_7X_{10} \,.\, X_3X_4X_8X_{10} = X_1X_6X_8X_{10} \,.\, X_2X_4X_9X_{10} \,.\, X_3X_5X_7X_{10},$$

corresponding to the fundamental squares

$$\begin{vmatrix} 1 & 0 & 0 \\ 0 & 1 & 0 \\ 0 & 0 & 1 \end{vmatrix} \qquad \begin{vmatrix} 0 & 1 & 0 \\ 0 & 0 & 1 \\ 1 & 0 & 0 \end{vmatrix} \qquad \begin{vmatrix} 0 & 0 & 1 \\ 1 & 0 & 0 \\ 0 & 1 & 0 \end{vmatrix}$$

$$\begin{vmatrix} 1 & 0 & 0 \\ 0 & 0 & 1 \\ 0 & 1 & 0 \end{vmatrix} \qquad \begin{vmatrix} 0 & 1 & 0 \\ 1 & 0 & 0 \\ 0 & 0 & 1 \end{vmatrix} \qquad \begin{vmatrix} 0 & 0 & 1 \\ 0 & 1 & 0 \\ 1 & 0 & 0 \end{vmatrix}$$

connected by the fundamental syzygy

$$\begin{vmatrix} 1 & 0 & 0 \\ 0 & 1 & 0 \\ 0 & 0 & 1 \end{vmatrix} + \begin{vmatrix} 0 & 1 & 0 \\ 0 & 0 & 1 \\ 1 & 0 & 0 \end{vmatrix} + \begin{vmatrix} 0 & 0 & 1 \\ 1 & 0 & 0 \\ 0 & 1 & 0 \end{vmatrix}$$

$$= \begin{vmatrix} 1 & 0 & 0 \\ 0 & 0 & 1 \\ 0 & 1 & 0 \end{vmatrix} + \begin{vmatrix} 0 & 1 & 0 \\ 0 & 0 & 1 \\ 1 & 0 & 0 \end{vmatrix} + \begin{vmatrix} 0 & 0 & 1 \\ 0 & 1 & 0 \\ 1 & 0 & 0 \end{vmatrix},$$

each side being equal to

$$\begin{vmatrix} 1 & 1 & 1 \\ 1 & 1 & 1 \\ 1 & 1 & 1 \end{vmatrix}.$$

This is the complete syzygetic theory of these particular squares of order 3.

Art. 130. Resuming the discussion, we proceed to eliminate g and h and remark that the second fraction may be omitted as contributing no term free from h. Eliminating g from the first and g from the third, we have

$$\underset{=}{\Omega}\frac{1+X_1X_2X_3X_4X_5X_6X_7X_8X_9X_{10}^3}{\left\{\begin{array}{l}(1-X_1X_3^2X_4^2X_5X_8^2X_9X_{10}^3)(1-X_1X_2^2X_5X_6^2X_7^2X_9X_{10}^3)\\ (1-h^2X_3X_5X_7X_{10})\left(1-\frac{1}{h}X_2X_4X_9X_{10}\right)\end{array}\right\}}$$

$$+\underset{=}{\Omega}\frac{\frac{1}{h}X_1X_6X_8X_{10}(1-X_1^2X_2^2X_3^2X_4^2X_5^2X_6^2X_7^2X_8^2X_9^2X_{10}^6)}{\left\{\begin{array}{l}(1-X_1X_3^2X_4^2X_5X_8^2X_9X_{10}^3)(1-X_1X_2^2X_5X_6^2X_7^2X_9X_{10}^3)\\ (1-h^2X_3X_5X_7X_{10})\left(1-\frac{1}{h}X_1X_6X_8X_{10}\right)\left(1-\frac{1}{h^2}X_1X_2X_4X_6X_8X_9X_{10}^2\right)\end{array}\right\}}.$$

Art. 131. If the diagonal property associated with g is alone to be satisfied in addition to the row and column properties we may put $h=1$. Observe that the second of the three fractions cannot now be omitted. Simplifying we obtain

$$\frac{1-X_1^2X_2^2X_3^2X_4^2X_5^2X_6^2X_7^2X_8^2X_9^2X_{10}^6}{\left\{\begin{array}{l}(1-X_1X_6X_8X_{10})(1-X_2X_4X_9X_{10})(1-X_3X_5X_7X_{10})\\ (1-X_1X_2^2X_5X_6^2X_7^2X_9X_{10}^3)(1-X_1X_3^2X_4^2X_5X_8^2X_9X_{10}^3)\end{array}\right\}}.$$

Establishing the five ground products

$$X_1X_6X_8X_{10},$$
$$X_2X_4X_9X_{10},$$
$$X_3X_5X_7X_{10},$$
$$X_1X_2^2X_5X_6^2X_7^2X_9X_{10}^3,$$
$$X_1X_3^2X_4^2X_5X_8^2X_9X_{10}^3$$

connected by the ground syzygy

$$(X_1X_6X_8X_{10})^2(X_2X_4X_9X_{10})^2(X_3X_5X_7X_{10})^2$$
$$=(X_1X_2^2X_5X_6^2X_7^2X_9X_{10}^3)(X_1X_3^2X_4^2X_5X_8^2X_9X_{10}^3)$$

corresponding to the fundamental squares

$$\begin{matrix}1&0&0\\0&0&1\\0&1&0\end{matrix}\qquad\begin{matrix}0&1&0\\1&0&0\\0&0&1\end{matrix}\qquad\begin{matrix}0&0&1\\0&1&0\\1&0&0\end{matrix}$$

$$\begin{matrix}1&2&0\\0&1&2\\2&0&1\end{matrix}\qquad\begin{matrix}1&0&2\\2&1&0\\0&2&1\end{matrix}$$

connected by the fundamental syzygy

$$2\begin{matrix}1&0&0\\0&0&1\\0&1&0\end{matrix}+2\begin{matrix}0&1&0\\1&0&0\\0&0&1\end{matrix}+2\begin{matrix}0&0&1\\0&1&0\\1&0&0\end{matrix}$$

$$=\begin{matrix}1&2&0\\0&1&2\\2&0&1\end{matrix}+\begin{matrix}1&0&2\\2&1&0\\0&2&1\end{matrix},$$

involving the complete theory of the squares in which the property of one chosen diagonal is excluded.

Art. 132. Resuming and finally eliminating h, we obtain

$$\frac{1+X_1X_2X_3X_4X_5X_6X_7X_8X_9X_{10}^3}{(1-X_1X_2^2X_5X_6^2X_7^2X_9X_{10}^3)(1-X_1X_3^2X_4^2X_5X_8^2X_9X_{10}^3)(1-X_2^2X_3X_4^2X_5X_7X_9^2X_{10}^3)}$$

$$+\frac{X_1^2X_3X_5X_6^2X_7X_8^2X_{10}^3(1-X_1^2X_2^2X_3^2X_4^2X_5^2X_6^2X_7^2X_8^2X_9^2X_{10}^6)}{\left\{\begin{matrix}(1-X_1X_2^2X_5X_6^2X_7^2X_9X_{10}^3)(1-X_1X_3^2X_4^2X_5X_8^2X_9X_{10}^3)\\(1-X_1^2X_3X_5X_6^2X_7X_8^2X_{10}^3)(1-X_1X_2X_3X_4X_5X_6X_7X_8X_9X_{10}^3)\end{matrix}\right\}},$$

which may be written

$$\frac{(1-X_1^2X_2^2X_3^2X_4^2X_5^2X_6^2X_7^2X_8^2X_9^2X_{10}^6)^2}{\left\{\begin{matrix}(1-X_1X_2^2X_5X_6^2X_7^2X_9X_{10}^3)(1-X_1X_3^2X_4^2X_5X_8^2X_9X_{10}^3)(1-X_1^2X_3X_5X_6^2X_7X_8^2X_{10}^3)\\(1-X_2^2X_3X_4^2X_5X_7X_9^2X_{10}^3)(1-X_1X_2X_3X_4X_5X_6X_7X_8X_9X_{10}^3)\end{matrix}\right\}},$$

indicating the ground products

$$X_1X_2^2X_5X_6^2X_7^2X_9X_{10}^3,$$
$$X_1X_3^2X_4^2X_5X_8^2X_9X_{10}^3,$$
$$X_1^2X_3X_5X_6^2X_7X_8^2X_{10}^3,$$
$$X_2^2X_3X_4^2X_5X_7X_9^2X_{10}^3,$$
$$X_1X_2X_3X_4X_5X_6X_7X_8X_9X_{10}^3$$

connected by the fundamental syzygies

$$(X_1X_2^2X_5X_6^2X_7^2X_9X_{10}^3)(X_1X_3^2X_4^2X_5X_8^2X_9X_{10}^3)$$
$$=(X_1^2X_3X_5X_6^2X_7X_8^2X_{10}^3)(X_2^2X_3X_4^2X_5X_7X_9^2X_{10}^3)$$
$$=(X_1X_2X_3X_4X_5X_6X_7X_8X_9X_{10}^3)^2$$

corresponding to the fundamental general magic squares

$$\begin{matrix}1&2&0\\0&1&2\\2&0&1\end{matrix}\qquad\begin{matrix}1&0&2\\2&1&0\\0&2&1\end{matrix}\qquad\begin{matrix}2&0&1\\0&1&2\\1&2&0\end{matrix}\qquad\begin{matrix}0&2&1\\2&1&0\\1&0&2\end{matrix}$$

$$\begin{matrix}1&1&1\\1&1&1\\1&1&1\end{matrix}$$

connected by the fundamental syzygies

$$\begin{matrix}1&2&0\\0&1&2\\2&0&1\end{matrix}+\begin{matrix}1&0&2\\2&1&0\\0&2&1\end{matrix}=2\begin{matrix}1&1&1\\1&1&1\\1&1&1\end{matrix}=\begin{matrix}2&0&1\\0&1&2\\1&2&0\end{matrix}+\begin{matrix}0&2&1\\2&1&0.\\1&0&2\end{matrix}$$

Art. 133. If the sum of each row, column, and diagonal be $3n$, the number of general magic squares of order 3 that can be constructed is, from the generating function, the coefficient of x^{3n} in the expansion of

$$(1-x^6)^2(1-x^3)^{-5},$$

and this is found to be

$$n^2+(n+1)^2.$$

Art. 134. The ordinary magic squares, the component integers being 0, 1, 2, 3, 4, 5, 6, 7, 8, are eight in number and are easily found to be

$$\begin{matrix}1&2&0\\0&1&2\\2&0&1\end{matrix}+3\begin{matrix}2&0&1\\0&1&2\\1&2&0\end{matrix}=\begin{matrix}7&2&3\\0&4&8,\\5&6&1\end{matrix}$$

and seven others obtained from

$$3\begin{matrix}1&2&0\\0&1&2\\2&0&1\end{matrix}+\begin{matrix}2&0&1\\0&1&2,\\1&2&0\end{matrix}\quad\begin{matrix}1&2&0\\0&1&2\\2&0&1\end{matrix}+3\begin{matrix}0&2&1\\2&1&0,\\1&0&2\end{matrix}\quad3\begin{matrix}1&2&0\\0&1&2\\2&0&1\end{matrix}+\begin{matrix}0&2&1\\2&1&0,\\1&0&2\end{matrix}$$

$$\begin{matrix}1&0&2\\2&1&0\\0&2&1\end{matrix}+3\begin{matrix}2&0&1\\0&1&2,\\1&2&0\end{matrix}\quad3\begin{matrix}1&0&2\\2&1&0\\0&2&1\end{matrix}+\begin{matrix}2&0&1\\0&1&2,\\1&2&0\end{matrix}\quad\begin{matrix}1&0&2\\2&1&0\\0&2&1\end{matrix}+3\begin{matrix}0&2&1\\2&1&0,\\1&0&2\end{matrix}$$

$$3\begin{matrix}0&2&1\\2&1&0\\1&0&2\end{matrix}+\begin{matrix}0&2&1\\2&1&0.\\1&0&2\end{matrix}$$

Art. 135. There is no theoretical difficulty in proceeding to investigate the squares of higher orders, but even in the case of order 4 there is practical difficulty in handling the $\underset{=}{\Omega}$ generating function. There are 20 fundamental squares, viz. :—

$$\begin{matrix}1&0&0&0\\0&0&1&0\\0&0&0&1\\0&1&0&0\end{matrix}\qquad\begin{matrix}1&0&0&0\\0&0&0&1\\0&1&0&0\\0&0&1&0\end{matrix}\qquad\begin{matrix}0&1&0&0\\0&0&1&0\\1&0&0&0\\0&0&0&1\end{matrix}\qquad\begin{matrix}0&1&0&0\\0&0&0&1\\0&0&1&0\\1&0&0&0\end{matrix}$$

$$\begin{matrix}0&0&1&0\\1&0&0&0\\0&1&0&0\\0&0&0&1\end{matrix}\qquad\begin{matrix}0&0&1&0\\0&1&0&0\\0&0&0&1\\1&0&0&0\end{matrix}\qquad\begin{matrix}0&0&0&1\\1&0&0&0\\0&0&1&0\\0&1&0&0\end{matrix}\qquad\begin{matrix}0&0&0&1\\0&1&0&0\\1&0&0&0\\0&0&1&0\end{matrix}$$

$$\begin{matrix}1&1&0&0\\1&0&1&0\\0&1&0&1\\0&0&1&1\end{matrix}\qquad\begin{matrix}1&0&1&0\\0&0&1&1\\1&1&0&0\\0&1&0&1\end{matrix}\qquad\begin{matrix}0&1&0&1\\1&1&0&0\\0&0&1&1\\1&0&1&0\end{matrix}\qquad\begin{matrix}0&0&1&1\\0&1&0&1\\1&0&1&0\\1&1&0&0\end{matrix}$$

$$\begin{matrix}0&2&0&0\\1&0&1&0\\0&0&1&1\\1&0&0&1\end{matrix}\qquad\begin{matrix}0&0&2&0\\0&1&0&1\\1&1&0&0\\1&0&0&1\end{matrix}\qquad\begin{matrix}1&0&1&0\\0&0&0&2\\0&1&1&0\\1&1&0&0\end{matrix}\qquad\begin{matrix}1&1&0&0\\0&1&1&0\\0&0&0&2\\1&0&1&0\end{matrix}$$

$$\begin{matrix}1&0&0&1\\1&1&0&0\\0&1&0&1\\0&0&2&0\end{matrix}\qquad\begin{matrix}1&0&0&1\\0&0&1&1\\1&0&1&0\\0&2&0&0\end{matrix}\qquad\begin{matrix}0&0&1&1\\0&1&1&0\\2&0&0&0\\0&1&0&1\end{matrix}\qquad\begin{matrix}0&1&0&1\\2&0&0&0\\0&1&1&0\\0&0&1&1\end{matrix}$$

§ 10.

Art. 136. The direct enumeration of general magic squares of given order and sum of row.

Let h_w denote the sum of all the homogeneous products w together of the magnitudes

$$\alpha_1,\ \alpha_2,\ldots\alpha_{n-1},\ \alpha_n.$$

If h_w be raised to the power n and developed, the coefficient of

$$(\alpha_1\alpha_2\alpha_3\ldots\alpha_{n-1}\alpha_n)^w$$

is the number of squares that can be formed of order n, so that the sum of each row and column is w, but in which there is no diagonal property in evidence.*

* "Combinatorial Analysis—The Foundations of a New Theory," 'Phil. Trans.,' A, vol. 194, 1900, p. 369 *et seq.*

In fact, if

$$(x-\alpha_1)(x-\alpha_2)\dots(x-\alpha_n) = x^n - p_1 x^{n-1} + \dots + (-)^n p_n$$

and

$$w!\, D_w = (\partial_{p_1} + p_1 \partial_{p_2} + \dots + p_{n-1}\partial_{p_n})^w,$$

an operator of order w obtained by raising the linear operator to the power w symbolically as in TAYLOR's theorem, then the number in question is concisely expressed by the formula

$$D_w{}^n h_w{}^n,$$

a particular case of a general formula given by the author (*loc. cit.*).

Art. 137. To introduce the diagonal properties, proceed as follows :—

Let $h_w{}^{(s)}$ denote what h_w becomes when $\lambda\alpha_s$, $\mu\alpha_{n-s+1}$ are written for α_s, α_{n-s+1} respectively, and form the product $h_w{}^{(1)} h_w{}^{(2)} \dots h_w{}^{(n)}$.

I say that the coefficient of

$$(\lambda\mu\alpha_1\alpha_2\dots\alpha_n)^w$$

in the development of this product is the number of general magic squares of order n corresponding to the sum w.

To see how this is take $n = 4$, $w = 1$, and form a product

$$\begin{aligned} &(\lambda\alpha_1+\alpha_2+\alpha_3+\mu\alpha_4) \\ \times &(\alpha_1+\lambda\alpha_2+\mu\alpha_3+\alpha_4) \\ \times &(\alpha_1+\mu\alpha_2+\lambda\alpha_3+\alpha_4) \\ \times &(\mu\alpha_1+\alpha_2+\alpha_3+\lambda\alpha_4), \end{aligned}$$

and observe that, in picking out the terms

$$\lambda\mu\alpha_1\alpha_2\alpha_3\alpha_4,$$

one factor must be taken every time from each row, column and diagonal of the matrix.

Similarly, if $n = 2$, we form the product

$$\begin{aligned} &\{\lambda^2\alpha_1^2+\lambda\mu\alpha_1\alpha_4+\mu^2\alpha_4^2+(\lambda\alpha_1+\mu\alpha_4)(\alpha_2+\alpha_3)+\alpha_2^2+\alpha_2\alpha_3+\alpha_3^2\} \\ \times &\{\lambda^2\alpha_2^2+\lambda\mu\alpha_2\alpha_3+\mu^2\alpha_3^2+(\lambda\alpha_2+\mu\alpha_3)(\alpha_1+\alpha_4)+\alpha_1^2+\alpha_1\alpha_4+\alpha_4^2\} \\ \times &\{\lambda^2\alpha_3^2+\lambda\mu\alpha_2\alpha_3+\mu^2\alpha_2^2+(\lambda\alpha_3+\mu\alpha_2)(\alpha_1+\alpha_4)+\alpha_1^2+\alpha_1\alpha_4+\alpha_4^2\} \\ \times &\{\lambda^2\alpha_4^2+\lambda\mu\alpha_1\alpha_4+\mu^2\alpha_1^2+(\lambda\alpha_4+\mu\alpha_1)(\alpha_2+\alpha_3)+\alpha_2^2+\alpha_2\alpha_3+\alpha_3^2\}. \end{aligned}$$

In forming the term involving

$$\lambda^2\mu^2\alpha_1^2\alpha_2^2\alpha_3^2\alpha_4^2$$

regard the successive products as corresponding to the successive rows of the square, the suffix of the α as denoting the column, and λ, μ as corresponding to the diagonals.

Thus picking out the factors

$$\lambda^2\alpha_1^2, \quad \mu\alpha_3\alpha_4, \quad \mu\alpha_2\alpha_4, \quad \alpha_2\alpha_3,$$

we obtain the corresponding square

$$\begin{matrix} 2 & 0 & 0 & 0 \\ 0 & 0 & 1 & 1 \\ 0 & 1 & 0 & 1 \\ 0 & 1 & 1 & 0 \end{matrix}.$$

These examples are sufficient to establish the validity of the theorem.

Art. 138. If we wish to make any restriction in regard to the numbers that appear in the sth row, we have merely to strike out certain terms from the function

$$h_w^{(s)}.$$

E.g., if no number is to exceed t, we have merely to strike out all terms involving exponents which exceed t.

If the rows are to be drawn from certain specified partitions of w, we have merely to strike out from the functions

$$h_w^{(1)}, h_w^{(2)} \ldots h_w^{(n)}$$

all terms whose exponents do not involve these partitions.

We have thus unlimited scope for particularising and specially defining the squares to be enumerated.

Let us now consider the enumeration of the fundamental squares of order n, such that the sum of each row, column and diagonal is unity. Observe that if the diagonal properties are not essential the number is obviously $n!$

Art. 139. It is convenient to consider a more general problem and then to deduce what we require at the moment as a particular case. I propose to determine the number of squares of given order which have one unit in each row and in each column, and specified numbers of units in the two diagonals.

Consider an even order $2n$, and form the product

$$\begin{aligned} &(\lambda\alpha_1+\alpha_2+\alpha_3+\ldots+\alpha_{2n-2}+\alpha_{2n-1}+\mu\alpha_{2n}) \\ \times&(\alpha_1+\lambda\alpha_2+\alpha_3+\ldots+\alpha_{2n-2}+\mu\alpha_{2n-1}+\alpha_{2n}) \\ \times&(\alpha_1+\alpha_2+\lambda\alpha_3+\ldots+\mu\alpha_{2n-2}+\alpha_{2n-1}+\alpha_{2n}) \\ &\cdot \quad \cdot \quad \cdot \quad \cdot \quad \cdots \quad \cdot \quad \cdot \quad \cdot \quad \cdot \quad \cdot \quad \cdot \\ &\cdot \quad \cdot \quad \cdot \quad \cdot \quad \cdots \quad \cdot \quad \cdot \quad \cdot \quad \cdot \quad \cdot \quad \cdot \\ \times&(\alpha_1+\alpha_2+\mu\alpha_3+\ldots+\lambda\alpha_{2n-2}+\alpha_{2n-1}+\alpha_{2n}) \\ \times&(\alpha_1+\mu\alpha_2+\alpha_3+\ldots+\alpha_{2n-2}+\lambda\alpha_{2n-1}+\alpha_{2n}) \\ \times&(\mu\alpha_1+\alpha_2+\alpha_3+\ldots+\alpha_{2n-2}+\alpha_{2n-1}+\lambda\alpha_{2n}). \end{aligned}$$

We require the complete coefficient of

$$\alpha_1\alpha_2\alpha_3\ldots\alpha_{2n-2}\alpha_{2n-1}\alpha_{2n},$$

when the multiplication has been performed.

Writing
$$\Sigma\alpha = s,$$

the product is, taking the tth and $2n+1-t^{\text{th}}$ factors together,

$$\begin{aligned}&\{s+(\lambda-1)\,\alpha_1+(\mu-1)\,\alpha_{2n}\}\,\{s+(\mu-1)\,\alpha_1+(\lambda-1)\,\alpha_{2n}\}\\ \times&\{s+(\lambda-1)\,\alpha_2+(\mu-1)\,\alpha_{2n-1}\}\,\{s+(\mu-1)\,\alpha_2+(\lambda-1)\,\alpha_{2n-1}\}\\ \times&\{s+(\lambda-1)\,\alpha_3+(\mu-1)\,\alpha_{2n-2}\}\,\{s+(\mu-1)\,\alpha_3+(\lambda-1)\,\alpha_{2n-2}\}\\ &\cdots\cdots\cdots\cdots\cdots\cdots\\ \times&\{s+(\lambda-1)\,\alpha_n+(\mu-1)\,\alpha_{n+1}\}\,\{s+(\mu-1)\,\alpha_n+(\lambda-1)\,\alpha_{n+1}\}.\end{aligned}$$

Observing that we only require terms which involve the quantities α with unit exponents, the product of the first two complementary factors is *effectively*

$$s^2+(\lambda+\mu-2)\,(\alpha_1+\alpha_{2n})\,s+\{(\lambda-1)^2+(\mu-1)^2\}\alpha_1\alpha_{2n},$$

and the complete product

$$\prod_{t=1}^{t=n}\left[s^2+(\lambda+\mu-2)\,(\alpha_t+\alpha_{2n+1-t})\,s+\{(\lambda-1)^2+(\mu-1)^2\}\alpha_t\alpha_{2n+1-t}\right]$$

has, on development, the form

$$s^{2n}+A_1s^{2n-1}+A_2s^{2n-2}+\ldots+A_{2n},$$

where A_m is a linear function of products of the quantities α, each term of which contains m different factors α, each with the exponent unity.

Since, moreover, s^m gives rise to the term

$$m!\,\Sigma\alpha_1\alpha_2\ldots\alpha_m,$$

it follows that the coefficient of $\Sigma\alpha_1\alpha_2\ldots\alpha_{2n}$ in the product is obtained by putting each quantity α equal to unity and $s^m = m!$.

Hence, if $s^m = m!$ symbolically, the symbolic expression of the coefficient is

$$\{s^2+2\,(\lambda+\mu-2)\,s+(\lambda-1)^2+(\mu-1)^2\}^n,$$

or

$$\{s^2-4s+2+2\,(\lambda+\mu)\,(s-1)+\lambda^2+\mu^2\}^n,$$

or writing

$$s^2-4s+2 = \sigma_2,\quad s-1 = \sigma_1$$

$$\{\sigma_2+2\,(\lambda+\mu)\,\sigma_1+\lambda^2+\mu^2\}^n.$$

This is the complete solution of the problem for an even order $2n$.

For an uneven order $2n+1$, it is now evident that the symbolical expression of the coefficient of

$$\alpha_1\alpha_2\ldots\alpha_{2n+1}$$

is

$$\{\sigma_2+2\,(\lambda+\mu)\,\sigma_1+\lambda^2+\mu^2\}^n\,(\sigma_1+\lambda\mu),$$

the complete solution in respect of the uneven order $2n+1$.

Art. 140. To find the number of ground "general magic squares" corresponding to the sum unity, we have merely to pick out the coefficient of $\lambda\mu$; we thus find

$$\text{even order } 2n \text{ number is } 8\binom{n}{2}\sigma_2^{n-2}\sigma_1^2,$$

$$\text{uneven order } 2n+1 \text{ number is } 8\binom{n}{2}\sigma_2^{n-2}\sigma_1^3+\sigma_2^n,$$

wherein it must be remembered that the σ products are to be expanded in powers of s and then s^m put equal to m!

In the general results the coefficient of

$$\lambda^l\mu^m$$

gives the number of squares in which the row and column sums are unity and the dexter and sinister diagonals' sums are l, m respectively.

I give the following table of values of simple σ products :—

σ_1			
σ_1^2	σ_2		
σ_1^3	$\sigma_1\sigma_2$		
σ_1^4	$\sigma_1^2\sigma_2$	σ_2^2	
σ_1^5	$\sigma_1^3\sigma_2$	$\sigma_1\sigma_2^2$	
σ_1^6	$\sigma_1^4\sigma_2$	$\sigma_1^2\sigma_2^2$	σ_2^3
σ_1^7	$\sigma_1^5\sigma_2$	$\sigma_1^3\sigma_2^2$	$\sigma_1\sigma_2^3$

=

0			
1	0		
2	0		
9	4	4	
44	24	16	
265	168	116	80
1854	1280	920	672

The numbers $\sigma_1^p=(s-1)^p$ denote the number of permutations of p letters in which each letter is displaced and constitute a well-known series.

The remaining numbers are readily calculated from these by the formula

$$\sigma_1^p\sigma_2^q=\sigma_1^{p+2}\sigma_2^{q-1}-2\sigma_1^{p+1}\sigma_2^{q-1}-\sigma_1^p\sigma_2^{q-1}.$$

Art. 141. Another solution of the same problem yielding a more detailed result is now given.

For the even order $2n$ I directly determine the coefficient of

$$\lambda^l\mu^m\alpha_1\alpha_2\ldots\alpha_{2n}$$

in the product above set forth.

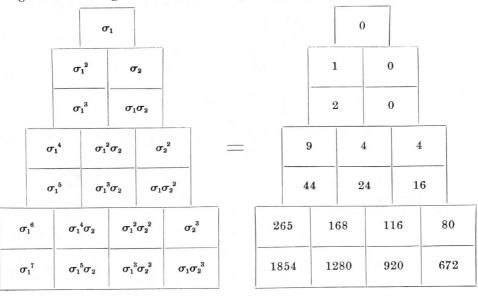

We have to pick out l λ's and m μ's and to find the associated factors, $2n-l-m$ in number, which are linear functions of the quantities α.

In any such selection of l λ's and m μ's there will be i pairs of λ's symmetrical about the sinister diagonals and j pairs of μ's symmetrical about the dexter diagonals, and the associated factors will depend upon the numerical values of i and j.

Consider then in the first place the number of ways of selecting l λ's in such wise that i pairs are symmetrical about the sinister diagonals.

This number is readily found to be

$$\binom{n}{i}\binom{n-i}{l-2i}2^{l-2i}.$$

With these l λ's we cannot associate any μ which is either in the same row or in the same column as one of the selected λ's.

Each of the i symmetrical *pairs* of λ's in this way accounts for 2 μ's, and each of the $l-2i$ remaining λ's accounts for 2 μ's.

Thus we must select m μ's out of $2n-2i-2(l-2i)$ μ's, *i.e.*, m μ's out of $2n-2l+2i$ μ's.

We may select these so as to involve i pairs symmetrical about the dexter diagonals in

$$\binom{n-l+i}{j}\binom{n-l+i-j}{m-2j}2^{m-2j} \text{ ways.}$$

This number is obtained by writing in the first formula $n-l+i$, j and m for

n, i and l respectively,

and observe that we may do this because the selection of a symmetrical pair of λ's or of one of the remaining λ's results in the rejection of a pair of μ's which is symmetrical about the dexter diagonals.

Consequently the $2n-2l+2i$ possible places for the m μ's are also symmetrically arranged about the dexter diagonal. Hence the formula is valid.

We have established at this point that we may pick out l λ's involving i symmetrical pairs and m μ's involving j symmetrical pairs in

$$\binom{n}{i}\binom{n-i}{l-2i}2^{l-2i}\cdot\binom{n-l+i}{j}\binom{n-l+i-j}{m-2j}2^{m-2j} \text{ ways.}$$

We must now determine the nature of the $2n-l-m$ associated factors, linear functions, of the quantities α.

In the matrix of the product delete the rows and columns which contain selected λ's and μ's. We thus delete $l+m$ rows and $l+m$ columns.

Consider the $2n-l-m$ remaining rows. There remain in these rows at most

$$2n-l-m$$

elements α, because $l+m$ columns have been deleted, but some of these elements must be rejected if they involve λ or μ as coefficients, because by hypothesis we are only concerned with l λ's and m μ's, and these have already been accounted for.

Observe now that the columns which contain a symmetrical selected pair of λ's only contain μ's which are in the same rows as these λ's, and therefore the deletion of these columns cannot delete μ's appertaining to any rows except those occupied by the selected pair of λ's. Observe further that the column which contains an unsymmetrical λ, say in the pth row, contains a μ in the $2n-p+1^{\text{th}}$ row, and that therefore the disappearance of a μ in the $2n-p+1^{\text{th}}$ row follows from the deletion of a column containing an unsymmetrical λ in the pth row.

Hence of the $2n-l-m$ rows in question

$$l+m-2i-2j \text{ rows contain } 2n-l-m-1, \ \alpha \text{ elements},$$

and thence

$$2n-2l-2m+2i+2j \text{ rows contain } 2n-l-m-2, \ \alpha \text{ elements}.$$

Accordingly if s is the sum of all the α elements except those which appear as coefficients of the selected λ's and μ's the co-factor of

$$\binom{n}{i}\binom{n-i}{l-2i}2^{l-2i}\cdot\binom{n-l+i}{j}\binom{n-l+i-j}{m-2j}2^{m-2j}$$

contains $l+m-2i-2j$ factors of type

$$(s-\alpha_u),$$

and $2n-2l-2m+2i+2j$ factors of type

$$(s-\alpha_v-\alpha_w),$$

or of $n-l-m+i+j$ squared factors of type

$$(s-\alpha_v-\alpha_w)^2,$$

since these factors occur in equal pairs.

Hence the co-factor is

$$\Pi\,(s-\alpha_u)\,\Pi\,(s-\alpha_v-\alpha_w)^2,$$

wherein the quantities α_u, $l+m-2i-2j$ in number, which appear in the first product, and the quantities α_v, α_w, $2n-2l-2m+2i+2j$ in number, which appear in the second product, are all different.

Also $(s-\alpha_v-\alpha_w)^2$ is effectively equal to

$$s^2-2(\alpha_v+\alpha_w)s+2\alpha_v\alpha_w$$

since squares of the α's may be rejected.

Hence, by the reasoning employed in the first solution we may put the quantities α equal to unity, regard s^p as equal to $p!$ symbolically, and say that the coefficient of

$$\alpha_1\alpha_2\ldots\alpha_{2n}$$

in the product

$$\Pi\,(s-\alpha_u)\,\Pi\,\{s^2-2(\alpha_v+\alpha_w)s+2\alpha_v\alpha_w\}$$

has the symbolical expression

$$(s-1)^{l+m-2i-2j}(s^2-4s+2)^{n-l-m+i+j},$$

or, putting

$$s-1=\sigma_1,\quad s^2-4s+2=\sigma_2,$$

we obtain

$$\binom{n}{i}\binom{n-i}{l-2i}2^{l-2i}\cdot\binom{n-l+i}{j}\binom{n-l+i-j}{m-2j}2^{m-2j}\sigma_1^{l+m-2i-2j}\sigma_2^{n-l-m+i+j}$$

for the number of squares such that—

(1) Sum associated with rows and columns is unity;
(2) There are l units involving i symmetrical pairs in the dexter diagonal;
(3) There are m units involving j symmetrical pairs in the sinister diagonal.

Giving i and j all possible values we find that the complete coefficient of

$$\alpha_1\alpha_2\ldots\alpha_{2n}$$

in the product, which we have already ascertained to have the expression

$$\{\sigma_2+2(\lambda+\mu)\sigma_1+\lambda^2+\mu^2\}^n,$$

may be also expressed in the form

$$\begin{aligned}\sum_l\sum_m\Bigg[&\binom{n}{0}\binom{n}{l}2^l\cdot\binom{n-l}{0}\binom{n-l}{m}2^m\sigma_1^{l+m}\sigma_2^{n-l-m}\\
&+\binom{n}{1}\binom{n-1}{l-2}2^{l-2}\cdot\binom{n-l+1}{0}\binom{n-l+1}{m}2^m\sigma_1^{l+m-2}\sigma_2^{n-l-m+1}\\
&+\binom{n}{0}\binom{n}{l}2^l\cdot\binom{n-l}{1}\binom{n-l-1}{m-2}2^{m-2}\sigma_1^{l+m-2}\sigma_2^{n-l-m+1}\\
&+\binom{n}{2}\binom{n-2}{l-4}2^{l-4}\cdot\binom{n-l+2}{0}\binom{n-l+2}{m}2^m\sigma_1^{l+m-4}\sigma_2^{n-l-m+2}\\
&+\binom{n}{1}\binom{n-1}{l-2}2^{l-2}\cdot\binom{n-l+1}{1}\binom{n-l}{m-2}2^{m-2}\sigma_1^{l+m-4}\sigma_2^{n-l-m+2}\\
&+\binom{n}{0}\binom{n}{l}2^l\cdot\binom{n-l}{2}\binom{n-l-2}{m-4}2^{m-4}\sigma_1^{l+m-4}\sigma_2^{n-l-m+2}\\
&+\binom{n}{3}\binom{n-3}{l-6}2^{l-6}\cdot\binom{n-l+3}{0}\binom{n-l+3}{m}2^m\sigma_1^{l+m-6}\sigma_2^{n-l-m+3}\\
&+\binom{n}{2}\binom{n-2}{l-4}2^{l-4}\cdot\binom{n-l+2}{1}\binom{n-l+1}{m-2}2^{m-2}\sigma_1^{l+m-6}\sigma_2^{n-l-m+3}\\
&+\binom{n}{1}\binom{n-1}{l-2}2^{l-2}\cdot\binom{n-l+1}{2}\binom{n-l-1}{m-4}2^{m-4}\sigma_1^{l+m-6}\sigma_2^{n-l-m+3}\\
&+\binom{n}{0}\binom{n}{l}2^l\cdot\binom{n-l}{3}\binom{n-l-3}{m-6}2^{m-6}\sigma_1^{l+m-6}\sigma_2^{n-l-m+3}\\
&+\ldots\Bigg]\lambda^l\mu^m.\end{aligned}$$

Further simplification of this series cannot be effected because each term of the sum must be considered on its merits and does or does not add to the numerical result as may appear.

Art. 142. Writing the result for even order $2n$

$$\{\sigma_2+2(\lambda+\mu)\sigma_1+\lambda^2+\mu^2\}^n$$
$$=\Sigma\Sigma F(n, l, m)\lambda^l\mu^m,$$

it appears that the result for uneven order $2n+1$ may be written

$$\{\sigma_2+2(\lambda+\mu)\sigma_1+\lambda^2+\mu^2\}^n(\sigma_1+\lambda\mu)$$
$$=\Sigma\Sigma\{\sigma_1 F(n, l, m)+F(n, l-1, m-1)\}\lambda^l\mu^m.$$

For the squares of simple orders we have the results—

ORDER 2.

m = \ l =	0	1	2
0	0	0	1
1	0	0	0
2	0	1	0

ORDER 3.

m = \ l =	0	1	2	3
0	0	2	0	0
1	2	0	0	1
2	0	0	0	0
3	0	1	0	0

ORDER 4.

m = \ l =	0	1	2	3	4
0	4	0	4	0	1
1	0	8	0	0	0
2	4	0	2	0	0
3	0	0	0	0	0
4	1	0	0	0	0

ORDER 5.

m = \ l =	0	1	2	3	4	5
0	16	16	8	4	0	0
1	16	20	4	4	0	1
2	8	4	8	0	0	0
3	4	4	0	2	0	0
4	0	0	0	0	0	0
5	0	1	0	0	0	0

Order 6.

m \ l =	0	1	2	3	4	5	6
0	80	96	60	16	12	0	1
1	96	96	48	24	0	0	0
2	60	48	24	0	3	0	0
3	16	24	0	0	0	0	0
4	12	0	3	0	0	0	0
5	0	0	0	0	0	0	0
6	1	0	0	0	0	0	0

Art. 143. I now proceed to consider the enumeration of the squares of even order $2n$, such that every row and column contains two units, and the dexter and sinister diagonals l and m units respectively.

I form the product

$$a_2^{(1)} a_2^{(2)} \ldots a_2^{(2n)},$$

where $a_2^{(s)}$ is the sum two together of the quantities

$$\alpha_1, \alpha_2, \ldots \lambda\alpha_s, \ldots \mu\alpha_{2n+1-s} \ldots \alpha_{2n-1}, \alpha_{2n},$$

and I seek the coefficient, a function of λ and μ, of

$$(\alpha_1 \alpha_2 \ldots \alpha_{2n})^2$$

in the product.

The coefficient of $\lambda^l \mu^m$ in the sought function of λ and μ is the required number.

Let p_1, p_2 be the sum and the sum two together of the quantities

$$\alpha_1, \alpha_2, \ldots \alpha_{2n},$$

then

$$a_2^{(1)} = p_2 + \{(\lambda-1)\alpha_1 + (\mu-1)\alpha_{2n}\}(p_1 - \alpha_1 - \alpha_{2n}) + (\lambda\mu - 1)\alpha_1\alpha_{2n},$$

$$a_2^{(2n)} = p_2 + \{(\mu-1)\alpha_1 + (\lambda-1)\alpha_{2n}\}(p_1 - \alpha_1 - \alpha_{2n}) + (\lambda\mu - 1)\alpha_1\alpha_{2n},$$

whence

$$\begin{aligned} a_2^{(1)} a_2^{(2n)} = {} & p_2^2 + (\lambda+\mu-2)(\alpha_1+\alpha_{2n}) p_2 p_1 - (\lambda+\mu-2)(\alpha_1+\alpha_{2n})^2 p_2 \\ & + 2(\lambda\mu-1)\alpha_1\alpha_{2n} p_2 + (\lambda\mu-1)^2 \alpha_1^2 \alpha_{2n}^2 \\ & + (\lambda-1)(\mu-1)(\alpha_1^2+\alpha_{2n}^2) + \{(\lambda-1)^2+(\mu-1)^2\}\alpha_1\alpha_{2n} p_1^2 \\ & - 2(\lambda^2+\lambda\mu+\mu^2-3\lambda-3\mu+3)\{\alpha_1\alpha_{2n}(\alpha_1+\alpha_{2n})p_1 - \alpha_1^2\alpha_{2n}^2\} \\ & + (\lambda\mu-1)(\lambda+\mu-2)\alpha_1\alpha_{2n}(\alpha_1+\alpha_{2n})p_1 - 2(\lambda\mu-1)(\lambda+\mu-2)\alpha_1^2\alpha_{2n}^2 \\ & + \text{terms involving powers of } \alpha_1, \alpha_{2n} \text{ above the second.} \end{aligned}$$

The product of $a_2^{(1)}, a_2^{(2n)}$ is thus, after re-arrangement, effectively equivalent to

$$\begin{aligned} &p_2^2+(\lambda+\mu-2)(\alpha_1+\alpha_{2n})p_2p_1-(\lambda+\mu-2)(\alpha_1^2+\alpha_{2n}^2)p_2 \\ &+2(\lambda-1)(\mu-1)\alpha_1\alpha_{2n}p_2+(\lambda-1)(\mu-1)(\alpha_1^2+\alpha_{2n}^2)p_1^2 \\ &+\{(\lambda-1)^2+(\mu-1)^2\}\alpha_1\alpha_{2n}p_1^2 \\ &+\{(\lambda+\mu)(\lambda\mu-\lambda-\mu)-(\lambda+\mu-1)(\lambda+\mu-4)\}\alpha_1\alpha_{2n}(\alpha_1+\alpha_{2n})p_1 \\ &+\{(\lambda\mu-\lambda-\mu)^2+(\lambda+\mu-1)(\lambda+\mu-3)\}\alpha_1^2\alpha_{2n}^2. \end{aligned}$$

Regarded apart from p_2, p_1 this expression is a function of α_1, α_{2n}; the product

$$a_2^{(2)}a_2^{(2n-1)}$$

is a function of α_2, α_{2n-1}, and generally the product

$$a_2^{(s)}a_2^{(2n+1-s)}$$

is a function of α_s, α_{2n+1-s}, and all of these products are of similar form in regard to p_2, p_1, λ, μ.

Remembering that we desire the coefficients of

$$(\alpha_1\alpha_2\ldots\alpha_{2n})^2$$

in the product

$$a_2^{(1)}a_2^{(2)}\ldots a_2^{(2n)},$$

we must distinguish between p_2 where it occurs as a multiplier of $\alpha_1^2+\alpha_{2n}^2$ and where it occurs as a multiplier of $\alpha_1\alpha_{2n}$, and make a similar distinction in respect of p_1^2.

Put then

$$\alpha_1\alpha_{2n}p_2 = \alpha_1\alpha_{2n}\Pi_2$$

$$(\alpha_1^2+\alpha_{2n}^2)p_1^2 = (\alpha_1^2+\alpha_{2n}^2)\Pi_1^2.$$

Putting further the quantities α equal to unity and regarding a product

$$p_2^a p_1^b \pi_2^c \pi_1^{2d}$$

as a symbol for the coefficient of symmetric function

$$(2^{a+d}1^{b+2c})$$

in the development of symmetric function

$$(1^2)^{a+c}(1)^{b+2d},$$

I say that

$$\begin{aligned} [&p_2^2+2(\lambda+\mu-2)p_2p_1-2(\lambda+\mu-2)p_2+2(\lambda-1)(\mu-1)(\pi_2+\pi_1^2) \\ &\qquad+\{(\lambda-1)^2+(\mu-1)^2\}p_1^2 \\ &+2\{(\lambda+\mu)(\lambda\mu-\lambda-\mu)-(\lambda+\mu-1)(\lambda+\mu-4)\}p_1 \\ &+(\lambda\mu-\lambda-\mu)^2+(\lambda+\mu-1)(\lambda+\mu-3)]^n \end{aligned}$$

is the symbolic expression of the required coefficient of

$$(\alpha_1\alpha_2\ldots\alpha_{2n})^2.$$

This may be written

$$\{\sigma_4+2(\lambda+\mu)\sigma_3+(\lambda^2+\mu^2)\sigma_2+2\lambda\mu\sigma'_2+2\lambda\mu(\lambda+\mu)\sigma_1+\lambda^2\mu^2\}^n,$$

where

$$\sigma_4 = p_2^2-4p_2p_1+4(p_2+\pi_2)+2(p_1^2+\pi_1^2)-8p_1+3,$$

$$\sigma_3 = p_2p_1-(p_2+\pi_2)-(p_1^2+\pi_1^2)+5p_1-2,$$

$$\sigma_2 = p_1^2-4p_1+2,$$

$$\sigma'_2 = \pi_2+\pi_1^2-4p_1+2,$$

$$\sigma_1 = p_1-1.$$

For the uneven order $2n+1$ it is easy to show that the coefficient is symbolically

$$\{\sigma_4+2(\lambda+\mu)\sigma_3+(\lambda^2+\mu^2)\sigma_2+2\lambda\mu\sigma'_2+2\lambda\mu(\lambda+\mu)\sigma_1+\lambda^2\mu^2\}^n\times(p_2-\sigma_1+\sigma_1\lambda\mu).$$

It is easy to calculate the values of

$$p_2^a p_1^b \pi_2^c \pi_1^{2d}$$

for small values of a, b, c, d.

Some results are, omitting the obvious result $p_1^b = b\,!$,

a.	b.	c.	d.	Value.
1				0
		1		1
			1	1
	1	1		3
	1		1	3
1		1		2
1			1	2
	2	1		12
1	2			5
		2		6
			2	6
		1	1	5

enabling the verification of the results

$$\sigma_4 = \sigma_3 = \sigma_2 = \sigma'_2 = \sigma_1 = 0,$$
$$\sigma_2^2 = 4, \quad \sigma_3\sigma_1 = 0, \quad \sigma'^2_2 = 2.$$

Hence for the even order 2 the whole coefficient is $\lambda^2\mu^2$, corresponding to the only possible square

1	1
1	1

and I find for the uneven order 3

$$\lambda^2+\mu^2+2\lambda^2\mu^2(\lambda+\mu).$$

Art. 144. To find in general the number of squares which have two units in each diagonal we find the coefficient of $\lambda^2\mu^2$ and obtain for even order $2n$.

$$\binom{n}{1}\sigma_4^{n-1}+\binom{n}{2}\sigma_4^{n-2}(2\sigma_2^2+4\sigma'^2_2+16\sigma_3\sigma_1)$$
$$+\binom{n}{3}\sigma_4^{n-3}(24\sigma_3^2\sigma_2+48\sigma_3^2\sigma'_2)+\binom{n}{4}\sigma_4^{n-4}96\sigma_3^4;$$

putting $n = 2$ we find for the order 4

$$2\sigma_4+2\sigma_2^2+4\sigma'^2_2+16\sigma_3\sigma_1 = 16,$$

and the verification of this number is easy.

For the uneven order $2n+1$ we obtain the number

$$\binom{n}{1}\sigma_4^{n-1}\,.\,2\sigma'_2\sigma_1+\binom{n}{2}\sigma_4^{n-2}8\sigma_3^2\sigma_1$$
$$+(p_2-\sigma_1)\left\{\binom{n}{1}\sigma_4^{n-1}+2\binom{n}{2}\sigma_4^{n-2}(\sigma_2^2+2\sigma'^2_2+8\sigma_3\sigma_1)\right.$$
$$\left.+\binom{n}{3}\sigma_4^{n-3}24(\sigma_3^2\sigma_2+2\sigma_3^2\sigma'_2)+\binom{n}{4}\sigma_4^{n-4}\,.\,96\sigma_3^4\right\}.$$

The general value of

$$p_2^a p_1^b \pi_2^c \pi_1^{2d}$$

may be obtained by means of the calculus of finite differences.

There is no theoretical difficulty in finding symbolical expressions for the enumeration of general magic squares associated with higher numbers, but the method does not lead to the determination of general magic squares. These must be regarded as arising from the generating function method of § 9.

I 2

THE DIOPHANTINE EQUATION $x^n - Ny^n = z$

By Major P. A. MacMahon, F.R.S.

[Received December 7th, 1906.—Read December 13th, 1906.]

1. The Diophantine equality

$$\lambda x = \mu y + z$$

is equivalent to the Diophantine inequality

$$\lambda x \geqslant \mu y.$$

It has been shewn* that the arithmetically independent solutions of this inequality are obtained by forming the descending intermediate series of convergents to the continued fraction

$$\frac{\mu}{\lambda}.$$

If this finite series be

$$\frac{1}{0}, \frac{a_0}{\beta_0}, \frac{a_1}{\beta_1}, \ldots, \frac{a_t}{\beta_t}, \frac{a_{t+1}}{\beta_{t+1}}, \ldots, \frac{a_n}{\beta_n},$$

the general solution is

$$x = (A+1)\,a_t + Ba_{t+1},$$

$$y = (A+1)\beta_t + B\beta_{t+1},$$

where A, B are arbitrary positive (including zero) integers.

The fundamental solutions are given by

$$x = a_t,$$

$$y = \beta_t,$$

and it was shewn long ago by Sylvester that, if

$$\frac{p_0}{q_0}, \frac{p_1}{q_1}, \ldots, \frac{p_n}{q_n}$$

---

* "The Diophantine Inequality $\lambda x \geqslant \mu y$," *Camb. Phil. Trans.*, Vol. XIX., Part I., 1900.

be the principal convergents to $\frac{\mu}{\lambda}$,

and $\frac{p'_0}{q'_0}, \frac{p'_1}{q'_1}, \ldots, \frac{p'_n}{q'_n}$

the principal convergents to the fraction whose partial quotients are the same as those of

$$\frac{\mu}{\lambda},$$

but taken in reverse order,

$$p_n q_s - p_s q_n \quad = (-)^{s+1} p_{n-s-1},$$
$$p_{n-1} q_s - p_s q_{n-1} = (-)^{s+1} q'_{n-s-1}.$$

These important results involve the complete solution of the problem of finding the arithmetically independent solutions of the Diophantine equality

$$\lambda x = \mu y + z;$$

for, taking the descending intermediate series of convergents to

$$\frac{p_n}{q_n} \text{ or } \frac{\mu}{\lambda},$$

$$p_n \beta_t - a_t q_n = p_n (q_{2m} + s q_{2m+1}) - (p_{2m} + s p_{2m+1}) q_n$$
$$= -p'_{n-2m-1} + s p'_{n-2m-2};$$

hence $\lambda a_t - \mu \beta_t = p'_{n-2m-3} + (a_{2m+1} - s) p'_{n-2m-2},$

where a_{2m+2} is the partial quotient which is the maximum value of s.

Now $p'_{n-2m-3} + (a_{2m+1} - s) p'_{n-2m-2}$

is the numerator of one of the descending intermediate series of convergents to the fraction which is the reverse of

$$\frac{\mu}{\lambda}.$$

Writing this series $\frac{a'_0}{\beta'_0}, \frac{a'_1}{\beta'_1}, \ldots, \frac{a'_{n'}}{\beta'_{n'}},$

$$\lambda a_t - \mu \beta_t = a'_{n'-t-1}$$

where, when $t = n'$, we take a'_{-1} to be zero.

We have therefore the following theorem :—

"The fundamental arithmetically independent solutions of the Diophantine equality

$$\lambda x = \mu y + z$$

are formed by constructing

"(i.) the descending intermediate series of convergents to the fraction $\frac{\mu}{\lambda}$, viz.,

$$\frac{a_0}{\beta_0}, \frac{a_1}{\beta_1}, \ldots, \frac{a_{n'}}{\beta_{n'}}.$$

"(ii.) The descending intermediate series to the fraction which is the reverse of $\frac{\mu}{\lambda}$, viz.,

$$\frac{a_0'}{\beta_0'}, \frac{a_1'}{\beta_1'}, \ldots, \frac{a_{n'}'}{\beta_{n'}'}.$$

"The solutions are then $n'+1$ in number, viz.,

$$x = a_t, \qquad y = \beta_t, \qquad z = a'_{n'-t-1},$$

where a'_{-1} is to be taken equal to zero."

As an example, take $31x = 222y + z$,

where $$\frac{222}{31} = 7 + \frac{1}{6+}\frac{1}{5}, \qquad \frac{222}{43} = 5 + \frac{1}{6+}\frac{1}{7}:$$

the descending series to $\frac{222}{31}$

is $$\frac{1}{0}, \frac{8}{1}, \frac{15}{2}, \frac{22}{3}, \frac{29}{4}, \frac{36}{5}, \frac{43}{6}, \frac{222}{31}.$$

and to $\frac{222}{43}$

$$\frac{1}{0}, \frac{6}{1}, \frac{11}{2}. \frac{16}{3}, \frac{21}{4}, \frac{26}{5}, \frac{31}{6}, \frac{222}{43},$$

and we derive the fundamental solutions

x	y	z
1	0	31
8	1	26
15	2	21
22	3	16
29	4	11
36	5	6
43	6	1
222	31	0

and every other solution is a linear function of these eight.

2. It is a corollary from the foregoing theorem that the fundamental (that is arithmetically independent) solutions of the Diophantine inequality

$$x \geqslant N^{1/m} y,$$

N not being a perfect m-th power, are obtained by forming the infinite descending intermediate series of convergents to

$$N^{1/m}.$$

If this series be $\frac{1}{0}, \frac{\alpha_1}{\beta_1}, \ldots, \frac{\alpha_t}{\beta_t}, \frac{\alpha_{t+1}}{\beta_{t+1}}, \ldots$, ad inf.,

the general solution is
$$x = (A+1)\,\alpha_t + B\alpha_{t+1},$$
$$y = (A+1)\beta_t + B\beta_{t+1}$$

Hence this is also the solution of the Diophantine inequality

$$x^m \geqslant Ny^m;$$

and thence we say, in regard to the Diophantine equality

$$x^m = Ny^m + z,$$

that the fundamental values of x and y are as above, and it remains to determine the properties of z.

My first aim is to connect z with the continued fraction which denotes

$$N^{1/m}.$$

Putting, in a usual notation,

$$N^{1/m} = x_1 = a_1 + \frac{1}{x_2},$$
$$x_2 = a_2 + \frac{1}{x_3},$$
$$\ldots \quad \ldots \quad \ldots,$$
$$x_k = a_k + \frac{1}{x_{k+1}},$$

let the principal convergents be

$$\frac{p_{-1}}{q_{-1}}, \frac{p_0}{q_0}, \frac{p_1}{q_1}, \ldots$$

or
$$\frac{0}{1}, \frac{1}{0}, \frac{a_1}{1}, \ldots;$$

we have
$$x_{n+2} = \frac{q_n N^{1/m} - p_n}{p_{n+1} - q_{n+1} N^{1/m}}$$

$$= \frac{(-)^m (p_n p_{n+1}^{m-1} - Nq_n q_{n+1}^{m-1}) + p_{n+1}^{m-2} N^{1/m} + p_{n+1}^{m-3} q_{n+1} N^{2/m} + \ldots + q_{n+1}^{m-2} N^{(m-1)/m}}{(-)^{n+1}(p_{n+1}^m - Nq_{n+1}^m)}.$$

The expressions $$(-)^n (p_n p_{n+1}^{m-1} - Nq_n q_{n+1}^{m-1}) = L_{n+1},$$
$$(-)^{n+1} (p_{n+1}^m - Nq_{n+1}^m) = M_{n+1}$$

are essentially positive; L_{n+1} has been termed a "rational dividend," and M_{n+1} a "divisor."

Substituting this value of x_{n+2} in the relation

$$x_{n+1} x_{n+2} = a_{n+1} x_{n+2} + 1$$

and comparing coefficients, we find, putting

$$(-)^n (p_n^{m-1} p_{n+1} - Nq_n^{m-1} q_{n+1}) = K_{n+1},$$

the two fundamental relations

$$L_n L_{n+1} + (-)^{n+1} N (p_{n+1}^{m-1} q_n^{m-1} - p_n^{m-1} q_{n+1}^{m-1}) = a_{n+1} L_{n+1} M_n + M_n M_{n+1}, \quad (1)$$

$$L_n + K_{n+1} = a_{n+1} M_n, \quad (2)$$

where, comparing the coefficients of $N^{s/m}$, the first relation corresponds to $s = 0$, and the second relation is given by all values of s from 1 to $m-1$, and no other relation is obtainable.

Now, the intermediate convergent of the descending series appertaining to

$$N^{1/m}$$

is, in general, $$\frac{a_t}{\beta_t},$$

where $$a_t = p_{2n} + sp_{2n+1},$$
$$\beta_t = q_{2n} + sq_{2n+1};$$

so that the general expression of z is

$$\begin{aligned} z = & \quad [(A+1)(p_{2n} + sp_{2n+1}) + B\{p_{2n} + (s+1)p_{2n+1}\}]^m \\ & - N[(A+1)(q_{2n} + sq_{2n+1}) + B\{q_{2n} + (s+1)q_{2n+1}\}]^m \\ = & \quad [(A+B+1)p_{2n} + \{s(A+B+1)+B\} p_{2n+1}]^m \\ & - N[(A+B+1)q_{2n} + \{s(A+B+1)+B\} q_{2n+1}]^m; \end{aligned}$$

or, writing $$A+B+1 = u,$$

$$s(A+B+1)+B = u_s,$$

we have

$$z = (p_{2n}^m - Nq_{2n}^m,\ p_{2n}^{m-1} p_{2n+1} - Nq_{2n}^{m-1} q_{2n+1},\ \ldots,\ p_{2n+1}^m - Nq_{2n+1}^m \between u, u_s)^m.$$

There are $m+1$ coefficients of which

the first $p_{2n}^m - Nq_{2n}^m = M_{2n}$,

the second $p_{2n}^{m-1} p_{2n+1} - Nq_{2n}^{m-1} q_{2n+1} = K_{2n+1} = a_{2n+1} M_{2n} - L_{2n}$,

the m-th $p_{2n} p_{2n+1}^{m-1} - Nq_{2n} q_{2n+1}^{m-1} = L_{2n+1}$,

the $m+1$-th $p_{2n+1}^m - Nq_{2n+1}^m = -M_{2n+1}$.

We have to express the remaining coefficients from the 3rd to the $(m-1)$-th inclusive in terms of the numbers L, M and the partial quotients a. Postponing the general consideration of this matter, I observe that the expression for z, viz.,

$$(up_{2n}+u_s p_{2n+1})^m - N(uq_{2n}+u_s q_{2n+1})^m,$$

is a linear transformation of $x^m - Ny^m$

to new variables u, u_s;

the modulus of transformation being

$$\begin{vmatrix} p_{2n} & p_{2n+1} \\ q_{2n} & q_{2n+1} \end{vmatrix} = 1.$$

Hence we may employ known results in the theory of invariants of binary forms in the investigation.

It is convenient to consider in particular and in detail the cases of $m = 2, 3, 4, \ldots$ in order to lead up to the theory of the general case.

3. *The case* $m = 2$.— $x^2 = Ny^2 + z$.

The general value of z is

$$(M_{2n}, L_{2n+1}, -M_{2n+1} \between u, u_s)^2,$$

where

$$u = A+B+1,$$

$$u_s = su+B,$$

A, B being arbitrary positive integers, and s a positive integer not exceeding a_{2n+2}.

Since here $K_{2n+1} = L_{2n+1}$, we have the relations

$$L_{2n} L_{2n+1} + N = a_{2n+1} L_{2n+1} M_{2n} + M_{2n} M_{2n+1},$$

$$L_{2n} + L_{2n+1} = a_{2n+1} M_{2n},$$

and thence $$M_{2n}M_{2n+1} + L_{2n+1}^2 = N,$$

where $M_{2n}M_{2n+1}$ is an essentially positive number. This relation at once establishes the periodicity of the numbers M, the numbers L, and the numbers a. Also, for given values of A, B, it is clear that z has a finite number of numerical values.

The relation $$M_{2n}M_{2n+1} + L_{2n+1}^2 = N$$

is obtained at once by equating the invariants of

$$x^2 - Ny^2$$

and $$(M_{2n},\ L_{2n+1},\ -M_{2n+1} \between u,\ u_s).$$

It leads to a relation between three values of z, if such values depend upon the same three numbers

$$M_{2m},\quad L_{2m+1},\quad L_{2m+1};$$

for write $$z_{A,B,s} = (M_{2m},\ L_{2m+1},\ -M_{2m+1} \between u_1,\ u_2)^2 = a_u^2 = \beta_u^2 = \ldots,$$
$$z_{C,D,t} = (M_{2m},\ L_{2m+1},\ -M_{2m+1} \between v_1,\ v_2)^2 = a_v^2 = \beta_v^2 = \ldots,$$
$$z_{E,F,r} = (M_{2m},\ L_{2m+1},\ -M_{2m+1} \between w_1,\ w_2)^2 = a_w^2 = \beta_w^2 = \ldots.$$

We may clearly eliminate

$$M_{2m},\quad L_{2m+1},\quad -M_{2m+1}$$

between these three equations and the invariant relation

$$M_{2n}M_{2n+1} + L_{2n+1}^2 = N;$$

for, squaring the symbolic identity

$$a_u\beta_v - a_v\beta_u = (a\beta)(uv),$$

we obtain $$a_u^2\beta_v^2 + a_v^2\beta_u^2 - (a\beta)^2(uv)^2 = 2a_u a_v \beta_u \beta_v$$

and two similar identities which lead to the relations

$$z_{C,D,t}\, z_{E,F,r} + N(vw)^2 = \{(M_{2m},\ L_{2m+1},\ -M_{2m+1} \between v_1,\ v_2 \between w_1,\ w_2)\}^2,$$
$$z_{E,F,r}\, z_{A,B,s} + N(wu)^2 = \{(M_{2m},\ L_{2m+1},\ -M_{2m+1} \between w_1,\ w_2 \between u_1,\ u_2)\}^2,$$
$$z_{A,B,s}\, z_{C,D,t} + N(uv)^2 = \{(M_{2m},\ L_{2m+1},\ -M_{2m+1} \between u_1,\ u_2 \between v_1,\ v_2)\}^2,$$

E 2

and, since $(\alpha\beta)^2 = -2N$,

$$a_v a_w = (M_{2m}, L_{2m+1}, -M_{2m+1} \between v_1, v_2 \between w_1, w_2).$$

Moreover $$(vw)\, a_u + (wu)\, a_v + (uv)\, a_w = 0;$$

whence $$(vw)^2 a_u^2 = (wu)^2 a_v^2 + (uv)^2 a_w^2 + 2\,(wu)(uv)\, a_v a_w,$$

or

$$(vw)^2 z_{A,B,s} = (wu)^2 z_{C,D,t} + (uv)^2 z_{E,F,r} + 2\,(wu)(uv)\sqrt{\{z_{C,D,t} z_{E,F,r} + N\,(vw)^2\}}$$

and rationalising,

$$(vw)^4 z_{A,B,s}^2 + (wu)^4 z_{C,D,t}^2 + (uv)^4 z_{E,F,r}^2 - 2\,(wu)^2 (uv)^2 z_{C,D,t} z_{E,F,r}$$
$$-2\,(uv)^2 (vw)^2 z_{E,F,r} z_{A,B,s} - 2\,(vw)^2 (wu)^2 z_{A,B,s} z_{C,D,t}$$
$$= 4\,(uv)^2 (vw)^2 (wu)^2 N.$$

This relation can also be obtained as follows:—

The four quadratics $$a_u^2,\ a_v^2,\ a_w^2,\ a_x^2$$

are connected by the relation

$$a_x^2 = \frac{(xv)(xw)}{(uv)(uw)}\, a_u^2 + \frac{(xw)(xu)}{(vw)(vu)}\, a_v^2 + \frac{(xu)(xv)}{(wu)(wv)}\, a_w^2,$$

since the right-hand side is a homogeneous quadratic function of x_1, x_2, and has the values

$$a_u^2,\ a_v^2,\ a_w^2$$

for $x = u, v, w$ respectively; if we compare the coefficients of

$$x_1^2,\ x_1 x_2,\ x_2^2$$

on the two sides of the identity, we obtain the expressions of

$$M_{2m},\ L_{2m+1},\ M_{2m+1}$$

as linear functions of $$a_u^2,\ a_v^2,\ a_w^2.$$

Moreover, equating the discriminants of the two quadratic functions of x_1, x_2, we obtain the sought quadratic relation between

$$a_u^2,\ a_v^2,\ a_w^2.$$

The expressions $$(vw),\ (wu),\ (uv)$$

involve arbitrary positive integers

$$A,\ B,\ C,\ D,\ E,\ F,$$

which can be determined so that

$$(vw)^2 = (wu)^2 = (uv)^2 = 1;$$

and then

$$z^2_{A,B,s} + z^2_{C,D,t} + z^2_{E,F,r} - 2z_{C,D,t}\,z_{E,F,r} - 2z_{E,F,r}\,z_{A,B,s} - 2z_{A,B,s}\,z_{C,D,t} = 4N,$$

a representation of $4N$ as a ternary quadratic form.

Also

$$z_{C,D,t}\,z_{E,F,r} + N,$$
$$z_{E,F,r}\,z_{A,B,s} + N,$$
$$z_{A,B,s}\,z_{C,D,t} + N$$

are, each of them, perfect squares.

Now, it is clear that (vw), (wu), (uv) cannot all be of the same sign.

If we choose the constants so that

$$(vw) = -(wu) = -(uv) = +1,$$

$$z_{A,B,s} = z_{C,D,t} + z_{E,F,r} + 2\sqrt{(z_{C,D,t}\,z_{E,F,r} + N)},$$
$$z_{C,D,t} = z_{E,F,r} + z_{A,B,s} - 2\sqrt{(z_{E,F,r}\,z_{A,B,s} + N)},$$
$$z_{E,F,r} = z_{A,B,s} + z_{C,D,t} - 2\sqrt{(z_{A,B,s}\,z_{C,D,t} + N)}.$$

In every case two of the signs attached to the radical will be negative and one positive.

Mr. G. B. Mathews has pointed out to me that, if

$$a = (vw)^2 a_u^2, \qquad b = (wu)^2 a_v^2, \qquad c = (uv)^2 a_w^2,$$

$$f = (wu)(uv)\,a_v a_w, \quad g = (uv)(vw)\,a_w a_u, \quad h = (vw)(wu)\,a_u a_v,$$

the ternary quadratic form

$$(a, b, c, f, g, h \between x, y, z)^2 = \{(vw)\,a_u x + (wu)\,a_v y + (uv)\,a_w z\}^2,$$
$$bc - f^2 = ca - g^2 = ab - h^2 = -(vw)^2 (wu)^2 (uv)^2 N,$$
$$gh - af = hf - bg = fg - ch = -(vw)^2 (wu)^2 (uv)^2 N,$$
$$a - b = b - g = c - h,$$
$$a - b - c - 2f = b - c - a - 2g = c - a - b - 2h = 0,$$

$$\begin{vmatrix} a & h & g \\ h & b & f \\ g & f & c \end{vmatrix} = 0,$$

and $$af+bg+ch = +2N(vw)^2(wu)^2(uv)^2,$$
$$gh+hf+fg = -\ N(vw)^2(wu)^2(uv)^2.$$

4. *The case* $m = 3$.— $x^3 = Ny^3+z$.

The general value of z is

$$(M_{2n},\ a_{2n+1}M_{2n}-L_{2n},\ L^{2n+1},\ -M_{2n+1})(u,\ u_s)^3;$$

and we have the relation

$$L_{2n}L_{2n+1}+N(p_{2n+1}q_{2n}+p_{2n}q_{2n+1}) = a_{2n+1}L_{2n+1}M_{2n}+M_{2n}M_{2n+1}.$$

Since $$(p_{2n+1}q_{2n}+p_{2n}q_{2n+1})^2 = 1+4p_{2n}q_{2n}p_{2n+1}q_{2n+1}$$

and $$(p_{2n+1}q_{2n}+p_{2n}q_{2n+1})-(p_{2n}q_{2n-1}+p_{2n-1}q_{2n}) = 2a_{2n+1}p_{2n}q_{2n},$$

there is no difficulty in eliminating p and q and obtaining a relation between the quantities L, M, a, and N.

The following is a better method of procedure.

Equating the invariants of x^3-Ny^3

and $$(M_{2n},\ a_{2n+1}M_{2n}-L_{2n},\ L_{2n+1},\ -M_{2n+1})(u,\ u_s)^3,$$

we obtain at once

$$M_{2n}^2M_{2n+1}^2+6M_{2n}M_{2n+1}L_{2n+1}(a_{2n+1}M_{2n}-L_{2n})-4(a_{2n+1}M_{2n}-L_{2n})^3M_{2n+1}$$
$$+4M_{2n}L_{2n+1}^3-3(a_{2n+1}M_{2n}-L_{2n})^2L_{2n+1}^2 = N^2.$$

This invariant relation between

$$M_{2n},\ M_{2n+1},\ L_{2n},\ L_{2n+1},\ a_{2n+1},\ \text{and}\ N$$

is the fundamental property of the continued fraction to

$$N^{\frac{1}{3}}.$$

It may be written

$$M_{2n}^2M_{2n+1}^2-6M_{2n}M_{2n+1}L_{2n}L_{2n+1}+4M_{2n+1}L_{2n}^3+4M_{2n}L_{2n+1}^3-3L_{2n}^2L^{2n+1}$$
$$+6(M_{2n}M_{2n+1}L_{2n+1}-2M_{2n+1}L_{2n}^2+L_{2n}L_{2n+1}^2)M_{2n}a_{2n+1}$$
$$+3(4M_{2n+1}L_{2n}-L_{2n+1}^2)M_{2n}^2a_{2n+1}^2$$
$$-4M_{2n+1}M_{2n}^3a_{2n+1}^3$$
$$= N^2.$$

From the Hessian of z, we obtain

$$[M_{2n} L_{2n+1} - (a_{2n+1} M_{2n} - L_{2n})^2,\ \tfrac{1}{2}\{-M_{2n} M_{2n+1} - L_{2n+1}(a_{2n+1} M_{2n} - L_{2n})\},$$
$$-M_{2n+1}(a_{2n+1} M_{2n} - L_{2n}) - L_{2n+1}^2](u+u_s)^2$$
$$= -Nxy = -N(up_{2n} + u_s p_{2n+1})(uq_{2n} + u_s q_{2n+1}),$$

yielding by comparison of coefficients

$$M_{2n} L_{2n+1} - (a_{2n+1} M_{2n} - L_{2n})^2 = -N p_{2n} q_{2n},$$
$$-M_{2n} M_{2n+1} - L_{2n+1}(a_{2n+1} M_{2n} - L_{2n}) = -N(p_{2n} q_{2n+1} + p_{2n+1} q_{2n}.$$
$$-M_{2n+1}(a_{2n+1} M_{2n} - L_{2n}) - L_{2n+1}^2 = -N p_{2n+1} q_{2n+1}.$$

Of these the second relation has been previously found as a fundamental equation derivable from the continued fraction. Similarly from the cubic covariant

$$-M_{2n}^2 M_{2n+1} - 3M_{2n}(a_{2n+1} M_{2n} - L_{2n}) L_{2n+1} + 2(a_{2n+1} M_{2n} - L_{2n})^3$$
$$= -N(p_{2n}^3 + N q_{2n}^3),$$

and three other relations.

By the solution of a cubic equation it is obvious that z can be expressed in terms of

$$M_{2n},\ M_{2n+1},\ L_{2n},\ L_{2n+1},\ N.$$

Let a value of z be

$$(M_{2n},\ K_{2n+1},\ L_{2n+1},\ -M_{2n+1} \between u_1,\ u_2)^3 = \alpha_u^3,$$

where

$$K_{2n+1} = a_{2n+1} L_{2n} - M_{2n},$$

and let other values of z, depending upon the same consecutive pair of principal convergents

$$\frac{p_{2n}}{q_{2n}},\ \frac{p_{2n+1}}{q_{2n+1}},$$

be

$$\alpha_v^3,\ \alpha_w^3,\ \alpha_x^3,\ \alpha_y^3;$$

then we have the obvious identity

$$\alpha_u^3 = \frac{(yv)(yw)(yx)}{(uv)(uw)(ux)}\alpha_u^3 + \frac{(yw)(yx)(yu)}{(vw)(vx)(vu)}\alpha_v^3 + \frac{(yx)(yu)(yv)}{(wx)(wu)(wv)}\alpha_w^3 + \frac{(yu)(yv)(yw)}{(xu)(xv)(xw)}\alpha_x^3,$$

which is symmetrical in u, v, w, x, y, in that an identity remains whatever substitution be impressed upon these letters. Each side of this identity is a homogeneous cubic function of

$$y_1,\ y_2.$$

and, equating the discriminants, we obtain a homogeneous quartic function of

$$a_u^3, \quad a_v^3, \quad a_w^3, \quad a_x^3$$

equal to a constant. This constant is a function of

$$M_{2n}, \quad M_{2n+1}, \quad L_{2n}, \quad L_{2n+1}, \quad a_{2n+1},$$

homogeneous and of degree 4, which, as shewn above, is equal to

$$N^2.$$

Hence, by giving special values to

$$u_1, \; u_2, \quad v_1, \; v_2, \quad w_1, \; w_2, \quad x_1, \; x_2,$$

we get a representation of N^2 by means of a quaternary quartic form.

5. *The case* $m = 4$.— $\quad x^4 = Ny^4 + z$.

We require an expression for

$$p_{2n}^2 p_{2n+1}^2 - N q_{2n}^2 q_{2n+1}^2 = X_{2n+1}.$$

Since $$p_{2n+2} = a_{2n+2} p_{2n+1} + p_{2n},$$

we have

$$M_{2n+2} = M_{2n} + 4(a_{2n+1} M_{2n} - L_{2n}) a_{2n+2} + 6 X_{2n+1} a_{2n+2}^2 + 4 L_{2n+1} a_{2n+2}^3 - M_{2n+1} a_{2n+2}^4,$$

$$-L_{2n+2} = a_{2n+1} M_{2n} - L_{2n} + 3 X_{2n+1} a_{2n+2} + 3 L_{2n+1} a_{2n+2}^2 - M_{2n+1} a_{2n+2}^3,$$

$$X_{2n+2} = X_{2n+1} + 2 L_{2n+1} a_{2n+2} - M_{2n+1} a_{2n+2}^2.$$

The second of these relations enables the required expression.

From the first two relations, eliminating X_{2n+1},

$$M_{2n+2} + 2 L_{2n+2} a_{2n+2}$$

$$= M_{2n} + 2(a_{2n+1} M_{2n} - L_{2n}) a_{2n+2} - 2 L_{2n+1} a_{2n+2}^3 + M_{2n+1} a_{2n+2}^4.$$

We now investigate the relations obtained from invariant theory. The expressions for z are

$$(u p_{2n} + u_s p_{2n+1})^4 - N(u q_{2n} + u_s q_{2n+1})^4$$

and $$(M_{2n}, \; a_{2n+1} M_{2n} - L_{2n}, \; X_{2n+1}, \; L_{2n+1}, \; -M^{2n+1} \between u, u_s)^4$$

Equating the two invariants of the two forms

$$-M_{2n}M_{2n+1}-4(a_{2n+1}M_{2n}-L_{2n})L_{2n+1}+3X_{2n+1}^2=-N,$$

$$-M_{2n}X_{2n+1}M_{2n+1}+2(a_{2n+1}M_{2n}-L_{2n})X_{2n+1}L_{2n+1}-M_{2n}L_{2n+1}^2$$
$$+(a_{2n+1}M_{2n}-L_{2n})^2M_{2n+1}-X_{2n+1}^3=0,$$

whence

$$\{-M_{2n}M_{2n+1}-4(a_{2n+1}M_{2n}-L_{2n})L_{2n+1}\}X_{2n+1}-3M_{2n}M_{2n+1}X_{2n+1}$$
$$+6(a_{2n+1}M_{2n}-L_{2n})L_{2n+1}X_{2n+1}-3M_{2n}L_{2n+1}^2$$
$$+3(a_{2n+1}M_{2n}-L_{2n})^2M_{2n+1}+NX_{2n+1}=0;$$

therefore

$$X_{2n+1}=\frac{3M_{2n}L_{2n+1}^2-3(a_{2n-1}M_{2n}-L_{2n})^2M_{2n+1}}{-M_{2n}M_{2n+1}-4(a_{2n+1}M_{2n}-L_{2n})L_{2n+1}-3M_{2n}M_{2n+1}+6(a_{2n+1}M_{2n}-L_{2n})L_{2n+1}+N}$$

$$=3\,\frac{(a_{2n+1}M_{2n}-L_{2n})^2M_{2n+1}-M_{2n}L_{2n+1}^2}{4M_{2n}M_{2n+1}-2(a_{2n+1}M_{2n}-L_{2n})L_{2n+1}-N},$$

the required expression of X_{2n+1}.

Thence, from the invariant relations,

$$M_{2n}M_{2n+1}+4(a_{2n+1}M_{2n}-L_{2n})L_{2n+1}-N$$
$$=27\left\{\frac{(a_{2n+1}M_{2n}-L_{2n})^2M_{2n+1}-M_{2n}L_{2n+1}^2}{4M_{2n}M_{2n+1}-2(a_{2n+1}M_{2n}-L_{2n})L_{2n+1}-N}\right\}^2,$$

the invariant relation between

$$M_{2n},\ M_{2n+1},\ L_{2n},\ L_{2n+1},\ a_{2n+1}\ \text{and}\ N,$$

which is the fundamental property of the continued fraction to

$$N^{\frac{1}{4}}.$$

Let the value of z

$$(u_1p_{2n}+u_2p_{2n+1})^4-N(u_1q_{2n}+u_2q_{2n+1})^4$$

be denoted by a_u^4;

then we have the identity

$$a_z^4=\frac{(zv)(zw)(zx)(zy)}{(uv)(uw)(ux)(uy)}a_u^4+\frac{(zw)(zx)(zy)(zu)}{(vw)(vx)(vy)(vu)}a_v^4+\frac{(zx)(zy)(zu)(zv)}{(wx)(wy)(wu)(wv)}a_w^4$$
$$+\frac{(zy)(zu)(zv)(zw)}{(xy)(xu)(xv)(xw)}a_x^4+\frac{(zu)(zv)(zw)(zx)}{(yu)(yv)(yw)(yx)}a_y^4;$$

and, equating the invariants of degrees 2 and 3, we find

(i.) a homogeneous quadratic function of

$$a_u^4, \quad a_v^4, \quad a_w^4, \quad a_x^4, \quad a_y^4$$

equal to N,

(ii.) a homogeneous cubic function of the same values of z equal to zero.

Results in integers are obtained by giving special values to

$$u_1, u_2, \quad v_1, v_2, \quad w_1, w_2, \quad x_1, x_2, \quad y_1, y_2$$

consistent with the definition of these quantities.

We obtain representation of N

by means of a quinary quadratic form.

In general the special properties of the continued fraction to

$$N^{1/m}$$

are obtainable from the invariant theory of the binary m-ic.

VII. *Memoir on the Theory of the Partitions of Numbers.—Part IV. On the Probability that the Successful Candidate at an Election by Ballot may never at any time have Fewer Votes than the One who is Unsuccessful; on a Generalization of this Question; and on its Connexion with other Questions of Partition, Permutation, and Combination.*

By Major P. A. MacMahon, *F.R.S.*

Received July 15,—Read November 19, 1908.

Section 1.

1. Consider a lattice in two dimensions, taking, for instance, one in which AB, BC are 7 and 5 segments in length respectively. It may be utilised for the study of permutations, combinations, and partitions in various ways, also for the study of certain questions in the theory of probabilities.

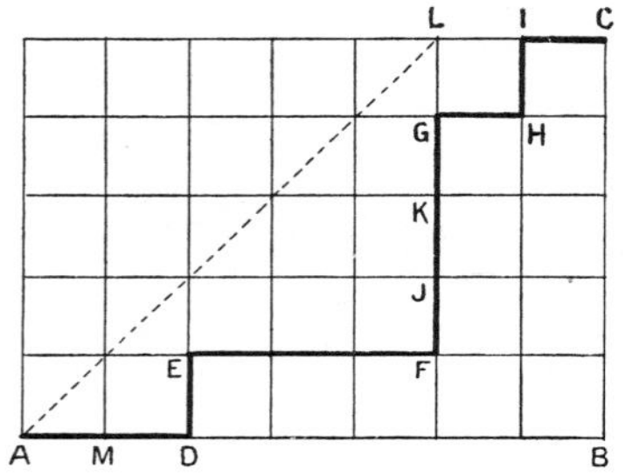

2.* A "line of route" through the lattice from A to C may be traced by moving over horizontal segments (α segments) in the direction AB, and over vertical segments (β segments) in the direction BC in any order. Thus one line of route is ADEFGHIC. The number of such lines of route is

$$\binom{12}{7},$$

or, in general, if AB, BC contain m, n segments respectively,

$$\binom{m+n}{m}.$$

* See "Memoir on the Theory of the Compositions of Numbers," 'Phil. Trans.,' A, 1893.

VOL. CCIX.—A 447. X 27.11.08

3. The line of route above depicted denotes a "principal composition" of the bipartite number $(\overline{75})$, viz.,

$$(\overline{21},\ \overline{33},\ \overline{11},\ \overline{10}),$$

and, in general, some principal composition of the bipartite number $(\overline{mn})$.

4. It also denotes the permutation

$$\alpha^2\beta\alpha^3\beta^3\alpha\beta\alpha$$

of the letters in the product $\alpha^7\beta^5$; and, in general, some permutation of the letters in the product $\alpha^m\beta^n$.

5. The line of route divides the lattice into two portions, each of which denotes the Sylvester-Ferrers graph of a partition of a unipartite number.

Consider, for example, the portion of the lattice bounded by ADEFGHICB. We obtain the graph of a partition in two ways :—

(i) By placing a unit (or node) in the centre of each square contained in the bounded area; thus

```
    1
   11
   11
   11
11111
```

denotes the partition 54111, or its conjugate 52221 of the number 12.

In general, we thus obtain a partition of some numbers into n, or fewer, parts, the part magnitude being limited not to exceed m, and its conjugate a partition of some numbers into m, or fewer, parts, the part magnitude being limited not to exceed n.

Similarly the remaining portion of the lattice denotes some partition and its conjugate.

(ii) By placing a unit (or node) at the centre of each segment to the right hand of the points A, E, J, K, G, I respectively; thus

```
      1
     11
     11
     11
  11111
1111111
```

denotes the partition 6522211, or its conjugate 752221 of the number 19.

In general, we thus obtain a partition of some numbers into m parts, the part magnitude being limited not to exceed $n+1$, and its conjugate, a partition having m for the highest part, the number of parts being limited not to exceed $n+1$.

The remaining portion of the lattice may be similarly interpreted.

6. The line of route also denotes the zig-zag graph of a composition of a unipartite number.

For placing nodes at all points passed over by the line of route we obtain

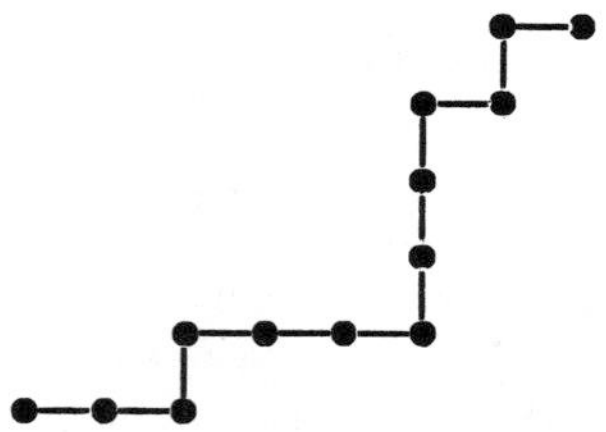

the graph of the composition 341122, and also of three other compositions

221143
12411211
11211421

of the number 13. (See 'Phil. Trans.,' Series A, vol. 207, pp. 65–134.)

In general, we thus obtain four compositions of the unipartite number $m+n+1$.

It will thus be noted that these four compositions of a unipartite number define two pairs of partitions of unipartite numbers, and clearly every theorem in partitions can be made to give a corresponding theorem of compositions.

This manifold interpretation of the line of route through a lattice must be borne in mind throughout the following investigation.

7. My object now is to show how certain questions of probability can be treated by means of the lattice.

Bertrand and Desiré André* have discussed a question which they have stated in the following terms :—

"Pierre et Paul sont soumis à un scrutin de ballottage ; l'urne contient m bulletins favorables à Pierre, n favorables à Paul ; m est plus grand que n, Pierre sera élu. Quelle est la probabilité pour que, pendant le dépouillement du scrutin, les bulletins sortent dans un ordre tel que Pierre ne cesse pas un seul instant d'avoir l'avantage ?"

The probability is found by an ingenious method to be

$$\frac{m-n}{m+n}.$$

8. I discuss the question by drawing in the lattice the line AL,† making an angle of 45° with the line AB. The problem of Bertrand and André is seen to be

* 'Calcul des Probabilités,' par J. Bertrand, Paris, 1888. J. Bertrand et D. André, 'Comptes Rendus de l'Académie des Sciences,' tome cv., p. 369 et 436, Paris, 1887.

† "Théories des Nombres," tome 1, par Edouard Lucas, 'Le Scrutin de ballottage' (pp. 83, 84, 164).

x 2

identical with that of enumerating the lines of route which neither cross nor touch the line AL, for each such line of route gives a permutation of the letters in $\alpha^m\beta^n$ which is required by the conditions.

I prefer in the first instance to alter the conditions of the problem so as to determine the probability that Pierre never at any instant has fewer votes than Paul. The lines of route to be enumerated are then those which do not cross, but which may touch the line AL.

Owing to the different interpretations that may be given to the line of route many courses of procedure are open. I select that one which in the special graphs gives the partition

$$752221$$

and, in general, a partition having m for the highest part and a number of parts not exceeding $n+1$.

Regarding zero as an admissible part, let the parts of such a partition be (in descending order)

$$\alpha_1, \alpha_2, \ldots \alpha_{n+1}.$$

These parts are subject to the conditions

$$\begin{array}{ll}
\alpha_1 \geq \alpha_2+1 & (a_1), \\
\alpha_2 \geq \alpha_3 & (a_2), \\
\alpha_1 \geq \alpha_3+2 & (a_3), \\
\alpha_3 \geq \alpha_4 & (a_4), \\
\alpha_1 \geq \alpha_4+3 & (a_5), \\
\vdots & \\
\alpha_{n'-1} \geq \alpha_{n'} & (a_{2n'-4}), \\
\alpha_1 \geq \alpha_{n'}+n' \quad 1 & (a_{2n'-3}),
\end{array}$$

where $n' = n+1$.

We can perform the summation

$$\Sigma x_1^{\alpha_1} x_2^{\alpha_2} \ldots x_{n'}^{\alpha_{n'}}$$

for all numbers satisfying the above $2n'-3$ conditions.

9. Suppose $n = 3$, the sum in question is

$$\underset{\geq}{\Omega} \frac{a_1^{-1} a_3^{-2}}{1-a_1 a_3 x_1 \,.\, 1-\dfrac{a_2}{a_1} x_2 \,.\, 1-\dfrac{1}{a_2 a_3} x_3},*$$

the meaning of the symbol $\underset{\geq}{\Omega}$ being that, after expansion in ascending powers of x_1, x_2, x_3, all terms involving negative powers of a_1, a_2, a_3 are to be rejected, and that in the surviving terms a_1, a_2, a_3 are, each of them, to be put equal to unity.

* See "Memoir on the Theory of the Partitions of Numbers.—Part II.," 'Phil. Trans.,' A, vol. 192, 1899.

The quantity a_2 is at once eliminated and we obtain

$$\underset{\geqq}{\Omega}\frac{a_1^{-1}a_3^{-2}}{1-a_1a_3x_1\,.\,1-\dfrac{1}{a_1}x_2\,.\,1-\dfrac{1}{a_1a_3}x_2x_3};$$

a_3 is now easily eliminated and we obtain

$$\underset{\geqq}{\Omega}\frac{a_1x_1^2}{1-a_1x_1\,.\,1-\dfrac{1}{a_1}x_2\,.\,1-x_1x_2x_3}.$$

To eliminate a_1 we require the easily established theorem

$$\underset{\geqq}{\Omega}\frac{c^s}{1-cx\,.\,1-\dfrac{1}{c}y}=\frac{1+y+\ldots+y^s-xy\,(1+y+\ldots+y^{s-1})}{1-x\,.\,1-xy}.$$

Thence we reach the final result

$$\frac{x_1^2+x_1^2x_2-x_1^3x_2}{1-x_1\,.\,1-x_1x_2\,.\,1-x_1x_2x_3},$$

which shows in the clearest possible manner how the partitions are constructed. The denominator factors indicate that we may write down any partition composed of three parts; the numerator terms x_1^2, $x_1^2x_2$ show that we may add either 2 to the first part or simultaneously 2 to the first and 1 to the second part. We in that manner obtain every partition satisfying the conditions, but the numerator term $-x_1^3x_2$ shows that certain partitions are in this manner obtained twice over.

10. As we are only concerned with the magnitude of the highest part and not at all with the weight of the partition, we may for the present purpose put $x_1 = x$, $x_2 = x_3 = 1$, and consider the result

$$\frac{2x^2-x^3}{(1-x)^3}$$

as the one to generalise.

11. I write down the expression

$$\underset{\geqq}{\Omega}\frac{a_1^{-1}a_3^{-2}a_5^{-3}\ldots a_{2n'-5}^{-(n'-2)}a_{2n'-3}^{-(n'-1)}}{1-a_1a_3a_5\ldots a_{2n-3}x\,.\,1-\dfrac{a_2}{a_1}\,.\,1-\dfrac{a_4}{a_2a_3}\,.\,1-\dfrac{a_6}{a_4a_5}\,.\,\ldots\,1-\dfrac{a_{2n'-4}}{a_{2n'-6}a_{2n'-5}}\,.\,1-\dfrac{1}{a_{2n'-4}a_{2n'-3}}}$$

as the crude expression for the sum.

We can immediately eliminate all the auxiliaries a which have an even suffix and reach the expression

$$\underset{\geqq}{\Omega}\frac{a_1^{-1}a_3^{-2}a_5^{-3}\ldots a_{2n'-5}^{-(n'-2)}a_{2n'-3}^{-(n'-1)}}{1-a_1a_3a_5\ldots a_{2n-3}x\,.\,1-\dfrac{1}{a_1}\,.\,1-\dfrac{1}{a_1a_3}\,.\,1-\dfrac{1}{a_1a_3a_5}\,.\,\ldots 1-\dfrac{1}{a_1a_3\ldots a_{2n'-3}}}.$$

We now require the auxiliary theorem

$$\underset{\geqq}{\Omega} \frac{c^{-s}}{1-cx_1 \,.\, 1-\frac{1}{c}x_2 \,.\, 1-\frac{1}{c}x_3 \ldots 1-\frac{1}{c}x_p}$$

$$= \frac{x_1^{\,s}}{1-x_1 \,.\, 1-x_1x_2 \,.\, 1-x_1x_3 \,, \ldots 1-x_1x_p};$$

so that, eliminating α_1, we reach

$$\underset{\geqq}{\Omega} \frac{\alpha_3^{-1}\alpha_5^{-2}\ldots\alpha_{2n'-5}^{-(n'-3)}\alpha_{2n'-3}^{-(n'-2)}x}{(1-\alpha_3\alpha_5\ldots\alpha_{2n'-3}x)^2(1-\alpha_5\ldots\alpha_{2n'-3}x)(1-\alpha_7\ldots\alpha_{2n'-3}x)\ldots(1-\alpha_{2n'-3}x)(1-x)};$$

and, eliminating α_3,

$$\underset{\geqq}{\Omega} \frac{2\alpha_5^{-1}\alpha_7^{-2}\ldots\alpha_{2n'-5}^{-(n'-4)}\alpha_{2n'-3}^{-(n'-3)}x^2-\alpha_7^{-1}\ldots\alpha_{2n'-5}^{-(n'-5)}\alpha_{2n'-3}^{-(n'-4)}x^3}{(1-\alpha_5\alpha_7\ldots\alpha_{2n'-3}x)^3(1-\alpha_7\ldots\alpha_{2n'-3}x)\ldots(1-\alpha_{2n'-3}x)(1-x)}.$$

Note that for $n' = 3$ or $n = 2$ this becomes the before obtained expression

$$\frac{2x^2-x^3}{(1-x)^3}.$$

To eliminate α_5, we have to substitute for

$$\frac{\alpha_5^{-1}}{(1-\alpha_5\alpha_7\ldots\alpha_{2n'-3}x)^3}$$

the expression

$$\frac{\alpha_5^{-1}}{(1-\alpha_5\alpha_7\ldots\alpha_{2n'-3}x)^3}-1,$$

and then put $\alpha_5 = 1$, thus getting

$$2\alpha_7^{-2}\alpha_9^{-3}\ldots\alpha_{2n'-5}^{-(n'-4)}\alpha_{2n'-3}^{-(n'-3)}(3\alpha_7\ldots\alpha_{2n'-3}x-3\alpha_7^{2}\ldots\alpha^{2}_{2n'-3}x^2+\alpha_7^{3}\ldots\alpha^{3}_{2n'-3}x^3)\,x^2$$

as the expression to be substituted for

$$2\alpha_5^{-1}\alpha_7^{-2}\ldots\alpha_{2n'-3}^{-(n'-3)}x^2$$

in the numerator, which thus becomes

$$\underset{\geqq}{\Omega} \frac{5\alpha_7^{-1}\alpha_9^{-2}\ldots\alpha_{2n'-3}^{-(n'-4)}x^3-6\alpha_9^{-1}\alpha_{11}^{-2}\ldots\alpha_{2n'-3}^{-(n'-5)}x^4+2\alpha_7\alpha_{11}^{-1}\alpha_{13}^{-2}\ldots\alpha_{2n'-3}^{-(n'-6)}x^5}{(1-\alpha_7\alpha_9\ldots\alpha_{2n'-3}x)^4(1-\alpha_9\ldots\alpha_{2n'-3}x)\ldots(1-\alpha_{2n'-3}x)(1-x)}.$$

This, for $n' = 4$ or $n = 3$, is

$$\frac{5x^3-6x^4+2x^5}{(1-x)^4},$$

which may be written

$$\frac{2x^2(3x-3x^2+x^3)-x^3}{(1-x)^4},$$

in a form showing its mode of derivation from

$$\frac{2x^2-x^3}{(1-x)^3}.$$

We now see that, when $n' = p+1$, we get a form which may be written

$$\frac{u_{p1}x^p - u_{p2}x^{p+1} + \ldots (-)^{p-1}u_{p,p}x^{2p-1}}{(1-x)^{p+1}};$$

and then, when $n' = p+2$, the form is

$$\frac{u_{p+1,1}x^{p+1} - u_{p+1,2}x^{p+2} + \ldots (-)^{p+1}u_{p+1,p+1}x^{2p+1}}{(1-x)^{p+2}},$$

where the numerator of the function last written is

$$u_{p1}x^p\left\{\binom{p+1}{1}x - \binom{p+1}{2}x^2 + \ldots (-)^p x^{p+1}\right\} - u_{p2}x^{p+1} + u_{p3}x^{p+2} - \ldots (-)^p u_{pp}x^{2p-1}.$$

Hence

$$u_{p+1,1} = \binom{p+1}{1}u_{p1} - u_{p2},$$

$$u_{p+1,2} = \binom{p+1}{2}u_{p1} - u_{p3},$$

$$u_{p+1,3} = \binom{p+1}{3}u_{p1} - u_{p4},$$

$$\cdot \quad \cdot \quad \cdot \quad \cdot \quad \cdot \quad \cdot \quad \cdot \quad \cdot \quad \cdot$$

$$u_{p+1,p-1} = \binom{p+1}{p-1}u_{p1} - u_{pp},$$

$$u_{p+1,p} = \binom{p+1}{p}u_{p1},$$

$$u_{p+1,p+1} = \binom{p+1}{p+1}u_{p1}.$$

These relations are satisfied by

$$u_{pq} = \frac{1}{p}\binom{p+q-2}{q-1}\binom{2p}{p-q},$$

so that the result of the summation is

$$\frac{\frac{1}{n}\binom{n-1}{0}\binom{2n}{n-1}x^n - \frac{1}{n}\binom{n}{1}\binom{2n}{n-2}x^{n+1} + \frac{1}{n}\binom{n+1}{2}\binom{2n}{n-3}x^{n+2} - \ldots (-)^n\frac{1}{n}\binom{2n-2}{n-1}\binom{2n}{0}x^{2n-1}}{(1-x)^{n+1}};$$

and there is no difficulty in showing that this in fact is equal to

$$\sum_{s=n}^{s=\infty}\left\{\binom{n+s-1}{n} - \binom{n+s-1}{n-2}\right\}x.$$

12. Hence the number of partitions having a highest part m and $n+1$ parts, zero being included as a part, subject to the given conditions as regards magnitude is

$$\binom{n+m-1}{n}-\binom{n+m-1}{n-2},$$

which may be also written

$$\frac{m-n+1}{m+1}\binom{m+n}{m}.$$

This, therefore, is the number of lines of route which do not cross the line AL. Hence the probability that Pierre is never in a minority is

$$\frac{m-n+1}{m+1}.$$

13. From this probability, which call F (m, n), is immediately derivable the probability discussed by Bertrand and André, which call P (m, n).

For in the lattice $\frac{m}{m+n}$ is the probability that the line ot route passes through the point M, and thence we find

$$\text{P}(m, n) = \frac{m}{m+n}\text{F}(m-1, n) = \frac{m-n}{m+n}.$$

Other probability questions may be discussed in a similar manner, with the advantage that light is at the same time thrown upon several other problems of partitions, compositions, and combinations of unipartite and bipartite numbers. In the above investigation we have had before us partitions of unipartite numbers which have a given number of parts, a given highest part and parts which in addition satisfy certain inequalities.

14. If we had had before us the parallel theory of the compositions of unipartite numbers there would have been the composition

$$\beta_1\beta_2\beta_3\ldots\beta_n$$

in correspondence with the partition

$$\alpha_1\alpha_2\alpha_3\ldots\alpha_n,$$

the weight $\Sigma\beta$ and the number of parts n would have been given and the parts $\beta_1, \beta_2,\ldots$ would have been subject to the inequalities

$$\begin{gathered}\beta_1 \geqq 2,\\ \beta_1+\beta_2 \geqq 4,\\ \beta_1+\beta_2+\beta_3 \geqq 6,\\ \vdots\\ \beta_1+\beta_2+\ldots+\beta_{n-1} \geqq 2n-2.\end{gathered}$$

In the former case the partitions of highest part m and n parts (zero not excluded) are enumerated by

$$\frac{m-n+2}{m+1}\binom{m+n-1}{m}.$$

In the latter the compositions of the number w, having n parts (zero excluded because $\beta_n = \alpha_n + 1$), are enumerated by

$$\frac{w-2n+2}{w-n+1}\binom{w-1}{n-1}.$$

Ex. gr. for $w = 6$, $n = 3$, we have the five compositions

411
321
312
231
222

which satisfy the given inequalities.

In this Section I have shown the connexion between a well-known question in probabilities and various other combinatorial questions in preparation for the generalization to which I now proceed in Section 2.

SECTION 2.

15. In the Second Memoir on the Partitions of Numbers I broached the subject of the two-dimensional partitions of numbers. I start with any Sylvester-Ferrers graph of an ordinary one-dimensional partition—say

● ● ● ● ● ●
● ● ● ●
● ● ● ●
● ●
●

and I consider the parts of the partition to be placed at the nodes in suchwise that the numbers in the rows, read from West to East, and also in the columns read from North to South, are in descending order of magnitude. Thus

433222
3222
2111
21
2

is a two-dimensional partition of the number 35.

The Memoir referred to contained some striking results in the theory, but the general result as conjectured and verified in numerous instances remained unproved.

The present paper is mainly concerned with the partitions into different parts placed at the nodes of any graph, and with the associated question in probabilities, a generalization of that of Section 1.

Taking any graph of n nodes and any n different integers, the inquiry is as to the number of ways of placing the numbers at the nodes so that the descending orders in the rows and columns, as above defined, are in evidence.

Consider in detail a simple case—that of six different numbers at the nodes of the graph

We find the 16 arrangements

654	654	653	653	652	652	651	651
32	31	42	41	43	41	43	42
1	2	1	2	1	3	2	3
631	632	641	642	641	643	642	643
52	51	52	51	53	51	53	52
4	4	3	3	2	2	1	1

the second row of eight arrangements being the conjugates of those in the first row because the graph is self-conjugate.

16. The problem is immediately transformable into one concerned with the conditioned permutations of the six numbers in a line.

Take the form

$$\begin{array}{l} 654\longrightarrow \\ 32\ \longrightarrow \\ 1\,\big\downarrow \\ \downarrow\downarrow \end{array}$$

and suppose the six numbers to be written down in a line so that four descending orders

$$\begin{array}{l} 654\longrightarrow \\ 32\ \longrightarrow \\ 631\longrightarrow \\ 52\ \longrightarrow \end{array}$$

are in evidence; I say that there is a one-to-one correspondence between such permutations and the two-dimensional partitions under investigation.

To see how this is, take any one of the 16 arrangements

$$\begin{array}{l} 632 \\ 51 \\ 4 \end{array}$$

and taking each number in succession, in order from the highest to the lowest, write a letter α, β, or γ, according as the number is in the first, second, or third row. Thus beginning with 6 we write down α, then for 5 β, for 4 γ, for 3 α, for 2 α, and, lastly, for 1 β, thus obtaining

$$\alpha,\ \beta,\ \gamma,\ \alpha,\ \alpha,\ \beta.$$

Now underneath the α's write 6, 5, 4 in order, under the β's, 3, 2 in order, and under γ 1, in accordance with the rows of the arrangement

$$\begin{matrix}654\\32\\1\end{matrix}$$

We thus obtain

$$\begin{matrix}\alpha & \beta & \gamma & \alpha & \alpha & \beta\\ 6 & 3 & 1 & 5 & 4 & 2.\end{matrix}$$

I say that

$$631542$$

is a permutation subject to the given conditions as defined by the descending orders in the arrangement

$$\begin{matrix}654\\32\\1\end{matrix}$$

The 16 permutations corresponding to the 16 graph arrangements are

$\alpha\alpha\alpha\beta\beta\gamma$	$\alpha\alpha\alpha\beta\gamma\beta$	$\alpha\alpha\beta\alpha\beta\gamma$	$\alpha\alpha\beta\alpha\gamma\beta$	$\alpha\alpha\beta\beta\alpha\gamma$	$\alpha\alpha\beta\gamma\alpha\beta$
654321	654312	653421	653412	653241	653142
$\alpha\alpha\beta\beta\gamma\alpha$	$\alpha\alpha\beta\gamma\beta\alpha$	$\alpha\beta\gamma\alpha\beta\alpha$	$\alpha\beta\gamma\alpha\alpha\beta$	$\alpha\beta\alpha\gamma\beta\alpha$	$\alpha\beta\alpha\gamma\alpha\beta$
653214	653124	631524	631542	635124	635142
	$\alpha\beta\alpha\beta\gamma\alpha$	$\alpha\beta\alpha\alpha\gamma\beta$	$\alpha\beta\alpha\beta\alpha\gamma$	$\alpha\beta\alpha\alpha\beta\gamma$	
	635214	635412	635241	635421.	

To show that there are no other permutations it is sufficient to prove that one can pass back from a permutation to a graph in a unique manner.

Thus take the permutation

$$635124\ ;$$

write α's under 6, 5, 4 ; β's under 3, 2 ; and γ under 1 :—

$$\begin{matrix}635124\\ \alpha\beta\alpha\gamma\beta\alpha\ ;\end{matrix}$$

the succession

$$\alpha\beta\alpha\gamma\beta\alpha$$

Y 2

indicates that 6 is in the first row corresponding to α,

5	,,	second	,,	,,	,,	β,
4	,,	first	,,	,,	,,	α,
3	,,	third	,,	,,	,,	γ,
2	,,	second	,,	,,	,,	β,
1	,,	first	,,	,,	,,	α.

Hence the arrangement

$$\begin{matrix} 641 \\ 52 \\ 3 \end{matrix}$$

The transformation is quite general; thus from

$$\begin{matrix} 12 & 9 & 7 & 4 \\ 11 & 8 & 6 & 2 \\ 10 & 5 & 3 & 1 \end{matrix}$$

we pass to

$$\begin{matrix} \alpha & \beta & \gamma & \alpha & \beta & \alpha & \beta & \gamma & \alpha & \gamma & \beta & \gamma \\ 12 & 8 & 4 & 11 & 7 & 10 & 6 & 3 & 9 & 2 & 5 & 1, \end{matrix}$$

a permutation in which the descending orders indicated by

$$\begin{matrix} 12 & 11 & 10 & 9 \\ 8 & 7 & 6 & 5 \\ 4 & 3 & 2 & 1 \end{matrix}$$

are in evidence.

17. The first question is the enumeration of the partitions where the graph and the set of unequal numbers are given.

Let the graph contain $a, b, c, \ldots$ nodes in the successive rows, and for a given set of $a+b+c+\ldots$ unequal numbers, let

$$(abc\ldots;)$$

denote the number of partitions.

Observe that above we found

$$(321;) = 16.$$

First we note that

$$(a;) = 1.$$

*Next take a graph of two rows. In any such graph

● ● ● ● ● ● ● ● ●
● ● ●

* Compare "Problème des deux files de soldats," 'Théories des Nombres,' tome 1, p. 86, par EDOUARD LUCAS.

if ϵ be the smallest number involved, the arrangements are of two types, viz.,

● ● ● ● ● ● ● ϵ
● ● ●

or

● ● ● ● ● ● ● ●
● ● ϵ

except when the rows contain the same number of nodes; then there is the one type

● ● ● ● ● ●
● ● ● ● ● ϵ

Hence, a moment's consideration establishes the relations

$$(ab\,;) = (a-1,\ b\,;)+(a,\ b-1\,;)$$

when $a > b$, and

$$(aa\,;) = (a,\ a-1\,;).$$

Treating these as difference equations it is easy to obtain the result

$$(ab\,;) = \frac{(a+b)\,!}{(a+1)\,!\;b\,!}(a-b+1) = \binom{a+b}{a}\frac{a-b+1}{a+1}\,;$$

$$(aa\,;) = \frac{(2a)\,!}{(a+1)\,!\;a\,!} = \binom{2a}{a}\frac{1}{a+1}.$$

18. This case of two rows is worth a special examination before proceeding to a greater number of rows. First consider the generating function of the numbers $(aa;)$:

$$u_x = \Sigma\,(aa;)\,x^a = 1+x+2x^2+5x^3+14x^4+\ldots$$

If we expand

$$(1-4x)^{\frac{1}{2}}$$

we find that the general term after the first is

$$-\frac{(2r)\,!}{(r+1)\,!\;r\,!}\,2x^{r+1},$$

and thence

$$u_x = \frac{1}{2x}\{1-(1-4x)^{\frac{1}{2}}\}$$

and

$$xu_x^2-u_x+1 = 0,$$

exhibiting a remarkable property of u_x.

Reverting to the difference equation

$$\begin{aligned}(aa;) &= (a,\ a-1\,;)\\ &= (a,\ a-2\,;)+(a-1,\ a-1\,;)\\ &= (a,\ a-3\,;)+2\,(a-1,\ a-2\,;)\\ &= (a,\ a-4\,;)+3\,(a-1,\ a-3\,;)+2\,(a-2,\ a-2),\end{aligned}$$

and observing that this last result may be written

$$(aa\,;) = (40\,;)(a,\ a-4\,;)+(31\,;)(a-1,\ a-3\,;)+(22\,;)(a-2,\ a-2\,;),$$

it is natural to suspect the law

$$(aa\,;) = \Sigma\,(st\,;)(a-t,\ a-s\,;),$$

where

$$s+t = \text{constant},$$

and it is easy to establish it.

For consider the graph

● ● ● ● ● ● ● ●x

● ● ● ● ● ●x ●x ●x

The four lowest numbers at the nodes are

(i) The last four of the second row,

(ii) The last of the first row and the last three of the second,

(iii) The last two of both rows.

Taking case (ii), the nodes marked x, the numbers may be

$$\begin{matrix} 2 & & 3 & & 4 \\ 431 & \text{or} & 421 & \text{or} & 321, \end{matrix}$$

and these arrangements are enumerated by (31;), we see, by subtracting each number from the number 5. Hence, in particular,

$$(88\,;) = (40\,;)(84\,;)+(31\,;)(75\,;)+(22\,;)(66\,;),$$

and, in general,

$$(aa\,;) = \Sigma\,(st\,;)(a-t,\ a-s\,;) \quad \text{where} \quad s+t = \text{constant}.$$

19. *Representation of* $(aa\,;)$ *as a Sum of Squares.*

Putting $s+t = a$, we find

$$(aa\,;) = (a0\,;)^2+(a-1,\ 1\,;)^2+(a-2,\ 2\,;)^2+\ldots,$$

the last term being

$$(\tfrac{1}{2}a,\ \tfrac{1}{2}a\,;)^2 \quad \text{or} \quad \{\tfrac{1}{2}(a+1),\ \tfrac{1}{2}(a-1)\}^2,$$

ascending as a is even or uneven.

Hence the identity

$$\frac{(2a)\,!}{(a+1)\,!\,a\,!} = 1^2+(a-1)^2+\{\tfrac{1}{2}a\,(a-3)\}^2+\left\{\frac{1}{3\,!}\,a\,(a-1)\,(a-5)\right\}^2+\ldots,$$

the last term of the series being the square of

$$\frac{a\,!}{\{\frac{1}{2}(a+2)\}\,!\,(\frac{1}{2}a)\,!} \quad \text{or of} \quad \frac{2\,.\,a!}{\{\frac{1}{2}(a+3)\}\,!\,\{\frac{1}{2}(a-1)\}\,!},$$

ascending as a is even or uneven.

The permutations enumerated by $(aa\,;)$ are those of $2a$ numbers

$$\begin{matrix} m_1 m_2 \ldots m_a, \\ n_1 n_2 \ldots n_a, \end{matrix}$$

all different and subject to $a+2$ descending orders corresponding to the a columns and the 2 rows.

20. Consider now other permutations such that whilst the row numbers are in descending order, exactly s of the a column pairs are not in descending order.

Let $(aa\,;\,s)$ be the number of such permutations. I propose to show that

$$(aa\,;\,s) = (aa\,;\,0) = (aa\,;)$$

for all values of s, from $s = 0$ to $s = a$.

Ex. gr., the permutations enumerated by

$$(22\,;\,0) \quad \text{are} \quad \begin{matrix} 43 & 42 \\ 21 & 31, \end{matrix}$$

$$(22\,;\,1) \quad \text{are} \quad \begin{matrix} 41 & 32 \\ 32 & 41, \end{matrix}$$

$$(22\,;\,2) \quad \text{are} \quad \begin{matrix} 31 & 21 \\ 42 & 43. \end{matrix}$$

To establish this theorem, since

$$u_x = 1+(11;)\,x+(22\,;)\,x^2+(33\,;)\,x^3+\ldots$$

and

$$xu_x^{\,2} = u_x-1,$$

we find

$$(aa\,;) = (a-1,\ a-1;)+(a-2,\ a-2\,;)\,(11;)+\ldots+(11;)\,(a-2,\ a-2\,;)+(a-1,\ a-1;).$$

The right-hand side of this identity is equal to

$$(aa\,;\,1),$$

for it consists of a terms, of which the first enumerates the permutations in which the pair m_1n_1 is out of order, the second those in which the pair m_2n_2 is out of order, and so on. Hence

$$(aa\,;) = (aa\,;\,1).$$

Consider in general the arrangement

$$\alpha_1\alpha_2\beta_3\beta_4\beta_5\alpha_6\beta_7\ldots,$$

$$\alpha'_1\alpha'_2\beta'_3\beta'_4\beta'_5\alpha'_6\beta'_7\ldots,$$

where the α pairs are in order and the β pairs out of order.

For this particular arrangement the enumeration is given by

$$(22\,;)\,(33\,;)\,(11;)\,(11;)\,\ldots.$$

Put

$$u_x = 1+\mathrm{X},$$

$$u_y = 1+\mathrm{Y}.$$

The generating function to be considered is

$$1+\mathrm{X}+\mathrm{Y}+\mathrm{XY}+\mathrm{YX}+\mathrm{XYX}+\mathrm{YXY}+\mathrm{XYXY}+\mathrm{YXYX}+\ldots.$$

The X, Y in a product occurring *alternately* in all possible ways.

This function is

$$\frac{(1+\mathrm{X})\,(1+\mathrm{Y})}{1-\mathrm{XY}}$$

or

$$\frac{u_xu_y}{u_x+u_y-u_xu_y}$$

or

$$\frac{xu_x-yu_y}{x-y},$$

since

$$xu_x^2-u_x+1 = yu_y^2-u_y+1 = 0.$$

Now

$$\frac{xu_x-yu_y}{x-y} = 1+(11;)\,(x+y)+(22\,;)\,(x^2+xy+y^2)+(33\,;)\,(x^3+x^2y+xy^2+y^3)+\ldots,$$

and the coefficient of $x^{a-s}y^s$ in the function is none other than

$$(aa\,;\,s).$$

Hence

$$(aa\,;\,s) = (aa\,;),$$

a remarkable theorem.

21. I will now obtain the generating function for the numbers $(ab\,;)$. Since

$$(ab\,;) = \binom{a+b}{a}\frac{a+1-b}{a+1},$$

$$\sum_b\sum_a (ab\,;)\,x^ay^b = \Sigma\Sigma\binom{a+b}{a}x^ay^b - \sum_0^\infty\sum_1^\infty\binom{a+b}{a}x^{a-1}y^{b+1}.$$

As yet we have assigned no meaning to $(ab\,;)$ when $a < b$, but retaining such terms and adding to the term

$$\Sigma\Sigma\binom{a+b}{a}x^{a-1}y^{b+1}$$

terms given by placing a equal to zero, we obtain the suggestive redundant generating function

$$\frac{1-\frac{y}{x}}{1-x-y},$$

in the expansion of which we require only those terms involving

$$x^a y^b$$

in which $a \geqq b$.

To eliminate the terms containing x^{-1} we have merely to add

$$\frac{\frac{y}{x}}{1-y},$$

and then we obtain

$$\frac{1-2y}{1-y,\ 1-x-y}.$$

We might also seek to remove those terms in $x^a y^b$ for which $a < b$, but for my present purpose the redundant form is quite as convenient and infinitely more suggestive.*

22. The result

$$(ab\,;) = \binom{a+b}{a}\frac{a-b+1}{a+1}$$

leads to the observation that the number obtained is precisely that obtained in Section 1 for the number of arrangements of the letters in

$$\alpha^a\beta^b$$

such that drawing a line between any two letters the number of α's to the left of the line $\geqq$ to the number of β's to the left of the line. Also that

$$(ab\,;) \div \binom{a+b}{a}$$

is the solution of the probability question for $a+b$ electors.

The one-to-one correspondence is easily established, for suppose

86531
742

* I have found that the reduced generating function is $\frac{1}{1-x-y} - \frac{1-\sqrt{(1-4xy)}}{2x(1-x-y)}$.

is an arrangement enumerated by (53 ;), I take the numbers 8, 7, 6, 5, 4, 3, 2, 1 in descending order and write down α when the number is in the first row and β when it is in the second, thus

$$\begin{matrix} 8 & 7 & 6 & 5 & 4 & 3 & 2 & 1 \\ \alpha & \beta & \alpha & \alpha & \beta & \alpha & \beta & \alpha, \end{matrix}$$

and we now have an arrangement of the letters in

$$\alpha^5\beta^3$$

such that, proceeding from left to right and stopping at any point, the number of α's met with is at least as large as the number of β's.

The result and correspondence are at once generalisable, for consider the arrangement

$$\begin{matrix} 9765 \\ 832 \\ 4 \\ 1 \end{matrix} \quad ;$$

we are led to the permutation of $\alpha^4\beta^3\gamma\delta$, viz.,

$$\alpha\beta\alpha\alpha\alpha\gamma\beta\beta\delta,$$

which is such that, in passing from left to right, we have at any instant

(i) passed over at least as many α's as β's,
(ii) ,, ,, ,, β's ,, γ's,
(iii) ,, ,, ,, γ's ,, δ's.

Hence, if in an election four candidates have a, b, c, d (these numbers being in descending order of magnitude) supporters respectively, and at any instant they have respectively A, B, C, D votes, the probability that always

$$\mathrm{A} \geq \mathrm{B} \geq \mathrm{C} \geq \mathrm{D}$$

is

$$(abcd\,;) \div \frac{(a+b+c+d)!}{a!\,b!\,c!\,d!},$$

and, in general, if the votes polled at any instant show invariably the final order of the candidates, we have a state of affairs of which the probability is

$$(abcde\ldots\,;) \div \frac{(a+b+c+d+e+\ldots)!}{a!\,b!\,c!\,d!\,e!\ldots}.$$

23. From the result

$$(ab\,;) = \frac{(a+b)!}{(a+1)!\,b!}(a-b+1)$$

we find the analytical results

$$(a-1,\ a\,;) = 0,$$

$$(b-1,\ a+1\,;) = -\,(ab\,;),$$

and the latter of these is not so far interpretable.

Passing to three rows

● ● ● ● ● ● ●
● ● ● ●
● ●

containing a, b, and c nodes respectively, it is seen that the smallest of the $a+b+c$ different numbers must be situated at the right-hand nodes of some row, unless such row contains as many nodes as the row beneath.

Hence the difference equation

$$(abc\,;) = (a-1,\, b,\, c\,;) + (a,\, b-1,\, c\,;) + (a,\, b,\, c-1\,;),$$

provided that $(abc\,;) = 0$ when either

$$a-b+1 = 0 \quad \text{or} \quad b-c+1 = 0.$$

I find such a solution of the difference equation to be

$$(abc\,;) = \frac{(a+b+c)!}{(a+2)!\,(b+1)!\,c!}\,(a-b+1)\,(b-c+1)\,(a-c+2).$$

This is only interpretable when

$$a \geq b \geq c,$$

but *analytically*

$$+(abc\,;) = +(b-1,\, c-1,\, a+2\,;) = +(c-2,\, a+1,\, b+1\,;)$$
$$= -(a,\, c-1,\, b+1\,;) = -(b-1,\, a+1,\, c\,;) = -(c-2,\, b,\, a+2\,;),$$

relations which are useful for the manipulation of the functions.

The sum

$$\Sigma\Sigma\Sigma\,(abc\,;)\,x^a y^b z^c$$

with the inclusion of redundant terms I find to be

$$\frac{1-\dfrac{y}{x}\,.\,1-\dfrac{z}{y}\,.\,1-\dfrac{z}{x}}{1-x-y-z},$$

an expression which is remarkably suggestive.

z 2

To establish the expression it is merely necessary to verify that the coefficient of $x^a y^b z^c$, viz. :—

$$\frac{(a+b+c)!}{c!\,b!\,c!}-\frac{(a+b+c)!}{(a+1)!\,(b-1)!\,c!}-\frac{(a+b+c)!}{a!\,(b+1)!\,(c-1)!}$$
$$+\frac{(a+b+c)!}{(a+2)!\,(b-1)!\,(c-1)!}+\frac{(a+b+c)!}{(a+1)!\,(b+1)!\,(c-2)!}-\frac{(a+b+c)!}{(a+2)!\,b!\,(c-2)!}$$

reduces to the value of $(abc;)$ above given.

This is easy because it appears at once that $a-b+1$, $b-c+1$, and $a-c+2$ are factors.

It is easy now to conjecture the form of the general result.

The truth of the generating function for $(abc;)$ is, perhaps, best seen by writing

$$(abc;)=\binom{a+b+c}{a,\,b}\left(1-\frac{b}{a+1}\right)\left(1-\frac{c}{b+1}\right)\left(1-\frac{c}{a+2}\right),$$

where $\binom{a+b+c}{a,\,b}$ stands for $\frac{(a+b+c)!}{a!\,b!\,c!}$.

24. I will now establish the result

$$(a_1, a_2, a_3 \ldots a_n;)$$
$$=\binom{a_1+a_2+a_3+\ldots+a_n}{a_1,\,a_2,\,\ldots\,a_{n-1}}$$
$$\times\left(1-\frac{a_2}{a_1+1}\right)\left(1-\frac{a_3}{a_2+1}\right)\left(1-\frac{a_4}{a_3+1}\right)\ldots\left(1-\frac{a_n}{a_{n-1}+1}\right)$$
$$\times\left(1-\frac{a_3}{a_1+2}\right)\left(1-\frac{a_4}{a_2+2}\right)\ldots\left(1-\frac{a_n}{a_{n-2}+2}\right)$$
$$\times\left(1-\frac{a_4}{a_1+3}\right)\ldots\left(1-\frac{a_n}{a_{n-3}+3}\right)$$
$$\cdots\cdots\cdots\cdots$$
$$\times\left(1-\frac{a_{n-1}}{a_1+n-2}\right)\left(1-\frac{a_n}{a_2+n-2}\right)$$
$$\times\left(1-\frac{a_n}{a_1+n-1}\right)$$

by showing that this expression satisfies the difference equation

$$(a_1, a_2, a_3, \ldots a_n;)=(a_1-1, a_2, a_3\ldots a_n;)+(a_1, a_2-1, a_3, \ldots a_n;)+\ldots+(a_1, a_2, a_3, \ldots a_n-1;).$$

Writing the co-factors of the multinomial coefficients in

$$(a_1, a_2, a_3, \ldots a_n;) \quad\text{and}\quad (a_1, a_2, \ldots a_s-1, \ldots a_n;)$$

as

$$\mathrm{C} \quad\text{and}\quad \mathrm{C}_s \text{ respectively,}$$

we find

$$(a_1+a_2+a_3+\ldots+a_n)\,C$$
$$= a_1C_1+a_2C_2+a_3C_3+\ldots+a_nC_n.$$

To prove this relation I will show that any factor

$$1-\frac{a_s}{a_t+s-t}(s>t) \text{ of } C$$

is also a factor of

$$a_1C_1+a_2C_2+\ldots+a_nC_n.$$

First observe that $1-\dfrac{a_s}{a_t+s-t}$ is a factor of C_m unless m is equal to s or t. Therefore consider merely

$$a_tC_t+a_sC_s.$$

The factors of C which involve either a_t or a_s or both a_s and a_t are

$$\left(1-\frac{a_s}{a_t+s-t}\right)\prod_{m=1}^{m=t-1}\left(1-\frac{a_s}{a_m+s-m}\right)\prod_{m=t+1}^{m=s-1}\left(1-\frac{a_s}{a_m+s-m}\right)\prod_{p=s+1}^{p=n}\left(1-\frac{a_p}{a_s+p-s}\right)$$
$$\prod_{l=1}^{l=t-1}\left(1-\frac{a_t}{a_l+t-l}\right)\prod_{q=s+1}^{q=n}\left(1-\frac{a_q}{a_t+q-t}\right)\prod_{q=t+1}^{q=s-1}\left(1-\frac{a_q}{a_t+q-t}\right).$$

Hence, disregarding a common factor, C_t involves the factors

$$\frac{a_t-a_s+s-t-1}{a_t+s-t-1}\prod_{m=1}^{m=t-1}\frac{a_m-a_s+s-m}{a_m+s-m}\prod_{m=t+1}^{m=s-1}\frac{a_m-a_s+s-m}{a_m+s-m}\prod_{p=s+1}^{p=n}\frac{a_s-a_p-s+p}{a_s-s+p}$$
$$\prod_{l=1}^{l=t-1}\frac{a_l-a_t+t-l+1}{a_l+t-l}\prod_{q=s+1}^{q=n}\frac{a_t-a_q+q-t-1}{a_t+q-t-1}\prod_{q=t+1}^{q=s-1}\frac{a_t-a_q+q-t-1}{a_t+q-t-1},$$

and C_s involves the factors

$$\frac{a_t-a_s+s-t+1}{a_t+s-t}\prod_{m=1}^{m=t-1}\frac{a_m-a_s+s-m+1}{a_m+s-m}\prod_{m=t+1}^{m=s-1}\frac{a_m-a_s+s-m+1}{a_m+s-m}\prod_{p=s+1}^{p=n}\frac{a_s-a_p+p-s-1}{a_s+p-s-1}$$
$$\prod_{l=1}^{l=t-1}\frac{a_l-a_t+t-l}{a_l+t-l}\prod_{q=s+1}^{q=n}\frac{a_t-a_q-t+q}{a_t-t+q}\prod_{q=t+1}^{q=s-1}\frac{a_t-a_q-t+q}{a_t-t+q}.$$

Discarding common factors from these expressions we find that C_t involves the factors

$$\frac{a_t-a_s+s-t-1}{a_t+s-t-1}(a_{t-1}-a_s+s-t+1)(a_{s-1}-a_s+1)\frac{a_s-a_n-s+n}{a_s-s+n}$$
$$\times(a_1-a_t+t)\frac{a_t-a_{s+1}+s-t}{a_t+s-t}\cdot\frac{a_t-a_{t+1}}{a_t},$$

and that C_s involves the factors

$$\frac{\alpha_t-\alpha_s+s-t+1}{\alpha_t+s-t}(\alpha_1-\alpha_s+s)(\alpha_{t+1}-\alpha_s+s-t)\frac{\alpha_s-\alpha_{s+1}}{\alpha_s}$$

$$\times(\alpha_{t-1}-\alpha_t+1)\frac{\alpha_t-\alpha_n-t+n}{\alpha_t-t+n}\cdot\frac{\alpha_t-\alpha_{s-1}+s-t-1}{\alpha_t+s-t-1}.$$

Hence $\alpha_t C_t+\alpha_s C_s$ involves a factor which by elimination of α_t by means of the relation

$$1-\frac{\alpha_s}{\alpha_t+s-t}=0$$

may be written

$$\frac{(-1)(\alpha_{t-1}-\alpha_s+s-t+1)(\alpha_{s-1}-\alpha_s+1)(\alpha_s-\alpha_n-s+n)(\alpha_1-\alpha_s+s)(\alpha_s-\alpha_{s+1})(\alpha_s-\alpha_{t+1}-s+t)}{(\alpha_s-1)\,\alpha_s\,(\alpha_s-s+n)}$$

$$+\frac{(+1)(\alpha_1-\alpha_s+s)(\alpha_{t+1}-\alpha_s+s-t)(\alpha_s-\alpha_{s+1})(\alpha_{t-1}-\alpha_s+s-t+1)(\alpha_s-\alpha_n-s+n)(\alpha_s-\alpha_{s-1}-1)}{\alpha_s\,(\alpha_s-s+\nu)\,(\alpha_s-1)},$$

and this is at once seen to be zero. Hence

$$\sum_{m=1}^{m=n}\alpha_m C_m$$

contains

$$1-\frac{\alpha_s}{\alpha_t+s-t}\quad(s>t)$$

as a factor. It therefore contains C as a factor.

It also contains another factor which is linear, and considerations of symmetry show it to be

$$\alpha_1+\alpha_2+\alpha_3+\ldots\alpha_n.$$

Hence the expression assumed for

$$(\alpha_1,\ \alpha_2,\ \alpha_3,\ \ldots\ \alpha_n\,;)$$

satisfies the difference equation, and is evidently the expression corresponding to the problem in hand.

25. We now observe that a redundant generating function, viz., an expression for the sum

$$\Sigma\,(\alpha_1,\ \alpha_2,\ \alpha_3,\ \ldots\ \alpha_n\,;)\,x_1^{\alpha_1}x_2^{\alpha_2}x_3^{\alpha_3}\ldots x_n^{\alpha_n},$$

is

$$\frac{\prod_{s=t+1}^{s=n}\prod_{t=1}^{t=n-1}\left(1-\frac{x_s}{x_t}\right)}{1-(x_1+x_2+x_3+\ldots+x_n)}.$$

The question in probabilities, here, I believe, solved for the first time, may be stated as follows :—

If n candidates at an election have

$$a_1,\ a_2,\ a_3,\ \dots\ a_n$$

voters in their favour respectively, where

$$a_1 \geqq a_2 \geqq a_3 \dots a_{n-1} \geqq a_n,$$

and if any instant A_1, A_2, ... A_{n-1}, A_n voters have recorded their votes in favour of the several candidates respectively, the probability that *always*

$$A_1 \geqq A_2 \geqq A_3 \dots A_{n-1} \geqq A_n$$

is

$$\prod_{s=t+1}^{s=n} \prod_{t=1}^{t=n-1} \left(1 - \frac{a_s}{a_t + s - t}\right).$$

The interesting point is the connection of the problem with the theory of the two-dimensional partitions of any set of Σa unequal numbers.

The graph solution proves at once the remarkable theorem that if $(b_1 b_2 b_3 \dots)$ be the partition conjugate to $(a_1 a_2 a_3 \dots)$,

$$(a_1 a_2 a_3 \dots ;) = (b_1 b_2 b_3 \dots ;),$$

a fact which it would be difficult to establish by pure algebra. In this case the corresponding probabilities are in the inverse ratio of

$$\frac{(a_1 + a_2 + a_3 + \dots)\,!}{a_1!\ a_2!\ a_3! \dots} \quad \text{to} \quad \frac{(b_1 + b_2 + b_3 + \dots)\,!}{b_1!\ b_2!\ b_3! \dots}.$$

Chapter 11
Plane Partitions (Part 1)

11.1 Introduction and Commentary

Significant portions of the first six memoirs in MacMahon's series on the partition of numbers relate to plane partitions. Plane partitions of n are two-dimensional arrays of integers whose sum is n and wherein numbers are nonincreasing along rows and columns. Thus the six plane partitions of 3 are

$$3, \quad 21, \quad \begin{matrix}2,\\1\end{matrix} \quad 111, \quad \begin{matrix}11,\\1\end{matrix} \quad \begin{matrix}1.\\1\\1\end{matrix}$$

MacMahon's crowning achievement in this area lies in papers [77] and [78] in which he describes (among many other interesting results) the infinite product form of the generating function for $p_{\infty,\infty}(n)$, the number of plane partitions of n:

$$\sum_{n \geqq 0} p_{\infty,\infty}(n)\, q^n = \prod_{n=1}^{\infty} (1 - q^n)^{-n}. \tag{1.1}$$

This result, probably the most famous of all his achievements, and related problems have been the subject of considerable research during the last 50 years. T. W. Chaundy (1931, 1932, 1933) was the first to extend MacMahon's work, and although there are some significant errors in Chaundy's work (see Gordon, 1962), he greatly simplified and advanced MacMahon's considerations.

In a two-part survey paper, R. P. Stanley (1971) presents a comprehensive history of plane partitions up to 1972, and describes the intricate interrelationships between plane partitions, symmetric functions, and the representation theory of the symmetric group. Stanley (1973), and Bender and Knuth (1972) have treated more complicated generating functions than (1.1), and we shall touch on some of their accomplishments in chapter 12. Plane partitions are discussed at length in Andrews (1976), Chapter 11.

11.2 The Generating Function for Plane Partitions

As a sample of recent work related to (1.1), we have chosen to present L.

The material in this chapter corresponds to section IX of *Combinatory Analysis.*

Carlitz's (1967b) proof of (1.1). While MacMahon's original work led to many important new discoveries in the theory of permutations (see [82] and [86]), his approach did require a lengthy study of lattice permutations before plane partitions could be treated. Carlitz's elegant proof requires only some elementary knowledge of determinants and Gaussian polynomials. (A. Lascoux has pointed out that the type of problem considered here has had applications in algebraic geometry. In particular, the case $q = 1$ of equation (2.9) below appears in a slightly altered form in Hodge and Pedoe, 1952, page 384, equation (16).)

We begin with the definition of the Gaussian polynomials:

$$\begin{bmatrix} n \\ m \end{bmatrix} = \begin{cases} (q)_n/(q)_m(q)_{n-m} & \text{if } 0 \leqq m \leqq n, \\ 0 & \text{otherwise,} \end{cases} \tag{2.1}$$

where

$$(q)_n = (1 - q)(1 - q^2) \dots (1 - q^n). \tag{2.2}$$

It is not immediately obvious that these apparent rational functions of q are polynomials; however, the following simple algebraic identities easily lead to a proof of this fact by mathematical induction.

$$\begin{bmatrix} n \\ m \end{bmatrix} = \begin{bmatrix} n-1 \\ m-1 \end{bmatrix} + q^m \begin{bmatrix} n-1 \\ m \end{bmatrix}; \tag{2.3}$$

$$\begin{bmatrix} n \\ m \end{bmatrix} = \begin{bmatrix} n-1 \\ m \end{bmatrix} + q^{n-m} \begin{bmatrix} n-1 \\ m-1 \end{bmatrix}. \tag{2.4}$$

Iteration of identity (2.4) yields the following summation:

$$\sum_{j=k}^{n} q^{j-k} \begin{bmatrix} j \\ k \end{bmatrix} = \begin{bmatrix} n+1 \\ k+1 \end{bmatrix}. \tag{2.5}$$

We now define $\pi_r(n_1, n_2, n_3, \dots, n_k; q)$ as the generating function for plane partitions with at most r columns, at most k rows, and n_i the first entry in the ith row. Immediately we see that

$$\pi_1(n_1, n_2, \dots, n_k; q) = q^{n_1+n_2+\dots+n_k}, \tag{2.6}$$

and

$$(2.7)\qquad \pi_{r+1}(n_1, n_2, \ldots, n_k; q)$$

$$= q^{n_1+n_2+\ldots+nk} \sum_{m_k=0}^{n_k} \sum_{m_{k-1}=m_k}^{n_{k-1}} \cdots \sum_{m_1=m_2}^{n_1} \pi_r(m_1, \ldots, m_k; q).$$

Using (2.6) and (2.7), let us compute a few of these generating functions for small values of r and k:

$$\pi_2(n, m; q) = q^{n+m} \sum_{m_1=0}^{m} \sum_{n_1=m_1}^{n} q^{n_1+m_1}$$

$$= q^{n+m} \sum_{m_1=0}^{m} q^{m_1} \frac{q^{m_1} - q^{n+1}}{1 - q}$$

$$= q^{n+m} \left\{ \frac{1 - q^{2m+2}}{1 - q^2} - q^{n+1} \frac{1 - q^{m+1}}{1 - q} \right\} (1 - q)^{-1}$$

$$= q^{n+m} \left\{ \frac{1 - q^{n+1}}{1 - q} \cdot \frac{1 - q^{m+1}}{1 - q} - q \frac{(1 - q^{m+1})(1 - q^m)}{(1 - q)(1 - q^2)} \right\}$$

$$= q^{n+m} \left\{ \begin{bmatrix} n+1 \\ 1 \end{bmatrix} \begin{bmatrix} m+1 \\ 1 \end{bmatrix} - q \begin{bmatrix} m+1 \\ 2 \end{bmatrix} \right\}.$$

We may use this formula, in conjunction with (2.5), to obtain $\pi_3(n, m; q)$:

$$\pi_3(n, m; q) = q^{n+m} \sum_{m_1=0}^{m} \sum_{n_1=m_1}^{n} q^{n_1+m_1} \left\{ \begin{bmatrix} n_1+1 \\ 1 \end{bmatrix} \begin{bmatrix} m_1+1 \\ 1 \end{bmatrix} - q \begin{bmatrix} m_1+1 \\ 2 \end{bmatrix} \right\}$$

$$= q^{n+m} \sum_{m_1=0}^{m} q^{m_1} \left\{ \left(\begin{bmatrix} n+2 \\ 2 \end{bmatrix} - \begin{bmatrix} m_1+1 \\ 1 \end{bmatrix} \right) \begin{bmatrix} m_1+1 \\ 1 \end{bmatrix} \right.$$

$$\left. - \left(\begin{bmatrix} n+2 \\ 1 \end{bmatrix} - \begin{bmatrix} m_1+1 \\ 1 \end{bmatrix} \right) \begin{bmatrix} m_1+1 \\ 2 \end{bmatrix} \right\} \qquad \text{(by (2.5))}$$

$$= q^{n+m} \sum_{m_1=0}^{m} q^{m_1} \left\{ \begin{bmatrix} n+2 \\ 2 \end{bmatrix} \begin{bmatrix} m_1+1 \\ 1 \end{bmatrix} - \begin{bmatrix} n+2 \\ 1 \end{bmatrix} \begin{bmatrix} m_1+1 \\ 2 \end{bmatrix} \right\}$$

$$= q^{n+m} \left(\begin{bmatrix} n+2 \\ 2 \end{bmatrix} \begin{bmatrix} m+2 \\ 2 \end{bmatrix} - q \begin{bmatrix} n+2 \\ 1 \end{bmatrix} \begin{bmatrix} m+2 \\ 3 \end{bmatrix} \right).$$

It is now a simple matter to establish by mathematical induction that

$$(2.8)\qquad \pi_r(n, m; q) = q^{n+m}\left(\begin{bmatrix} n+r-1 \\ r-1 \end{bmatrix}\begin{bmatrix} m+r-1 \\ r-1 \end{bmatrix} - q\begin{bmatrix} n+r-1 \\ r-2 \end{bmatrix}\begin{bmatrix} m+r-1 \\ r \end{bmatrix}\right)$$

$$= q^{n+m}\begin{vmatrix} \begin{bmatrix} n+r-1 \\ r-1 \end{bmatrix} & q\begin{bmatrix} m+r-1 \\ r \end{bmatrix} \\ \begin{bmatrix} n+r-1 \\ r-2 \end{bmatrix} & \begin{bmatrix} m+r-1 \\ r-1 \end{bmatrix} \end{vmatrix}.$$

The above procedure can be applied to three-row plane partitions, then four-row plane partitions, etc. In each case, the result that one obtains is a special case of the following identity:

$$(2.9)\qquad \pi_r(n_1, n_2, \ldots, n_r) = q^{n_1+n_2+\ldots+n_k}\det(A_{ij}),$$

where

$$A_{ij} = q^{(i-j)(i-j-1)/2}\begin{bmatrix} n_j + r - 1 \\ r - i + j - 1 \end{bmatrix}, \qquad 1 \leqq i,j \leqq k.$$

We may prove (2.9) by mathematical induction on r. We note that (2.9) reduces to (2.6) when $r = 1$, and to (2.8) when $r = 2$. Let us assume that (2.9) is valid for all integers up to and including r; then by (2.7),

$$(2.10)\qquad q^{-n_1-\ldots-n_k}\pi_{r+1}(n_1, \ldots, n_k; q)$$

$$= \sum_{m_k=0}^{n_k}\sum_{m_{k-1}=m_k}^{n_{k-1}}\cdots\sum_{m_1=m_2}^{n_1} q^{m_1+m_2+\ldots+m_k}\det\left(q^{(i-j)(i-j-1)/2}\begin{bmatrix} m_j + r - 1 \\ r - i + j - 1 \end{bmatrix}\right).$$

We now perform the summation with respect to m_1, using (2.5); this leaves a $(k-1)$-fold sum. Furthermore, the determinant in the expression is unaltered, except for the first column which is replaced by

$$q^{(i-1)(i-4)/2}\left(\begin{bmatrix} n_1 + r \\ r - i + 1 \end{bmatrix} - \begin{bmatrix} m_2 + r - 1 \\ r - i + 1 \end{bmatrix}\right), \qquad 1 \leqq i \leqq k.$$

Since the second column of our determinant is

$$q^{(i-2)(i-3)/2}\begin{bmatrix} m_2 + r - 1 \\ r - i + 1 \end{bmatrix}, \qquad 1 \leqq i \leqq k,$$

we may multiply this column by q^{-1} and add it to the first column; the resulting first column is

$$q^{(i-1)(i-4)/2}\begin{bmatrix} n_1 + r \\ r - i + 1 \end{bmatrix}, \qquad 1 \leqq i \leqq k.$$

We next sum with respect to m_2, then m_3, etc.; in general, when we sum with respect to m_j, we obtain a new jth column; we multiply the $(j + 1)$st column by q^{-j} and add it to our new jth column before we sum with respect to m_{j+1}. From this procedure, we obtain as the jth column

$$q^{(i-j)(i-j-3)/2}\begin{bmatrix} n_j + r \\ r - i + j \end{bmatrix}, \qquad 1 \leqq i \leqq k.$$

Therefore, after all the summations are performed, we see that (2.10) is reduced to

(2.11)
$$q^{-n_1-n_2-\dots-n_k}\pi_{r+1}(n_1, n_2, \dots, n_k; q) = \det\left(q^{(i-j)(i-j-3)/2}\begin{bmatrix} n_j + r \\ r - i + j \end{bmatrix}\right).$$

Now multiply the ith row of (2.11) by q^{i-1} and divide the jth column by q^{j-1}; this does not alter the value of the determinant, but the exponent of q becomes

$$(i - j)(i - j - 3)/2 + (i - 1) - (j - 1) = (i - j)(i - j - 1)/2.$$

Hence

(2.12)
$$q^{-n_1-n_2-\dots-n_k}\pi_{r+1}(n_1, n_2, \dots, n_k; q) = \det\left(q^{(i-j)(i-j-1)/2}\begin{bmatrix} n_j + r \\ r - i + j \end{bmatrix}\right),$$

which is (2.9) with r replaced by $r+1$. Hence (2.9) is established.

We now define

$$\pi_{k,r}(n;q) = \sum_{n_k \leqq \cdots \leqq n_1 \leqq n} \pi_r(n_1, \ldots, n_k; q); \tag{2.13}$$

it follows immediately from (2.7) that

$$q^{kn}\pi_{k,r}(n;q) = \pi_{r+1}(n, \ldots, n; q), \tag{2.14}$$

and therefore

$$\pi_{k,r}(n;q) = \det\left(q^{(i-j)(i-j-1)/2}\begin{bmatrix} n+r \\ r-i+j \end{bmatrix}\right), \tag{2.15}$$

a polynomial in q. Consequently, if $p_{k,r}(m,n)$ denotes the number of plane partitions of m with at most k rows, at most r columns, and largest part at most n, then by (2.13),

$$\pi_{k,r}(n;q) = \sum_{m=0}^{\infty} p_{k,r}(m,n)\, q^m. \tag{2.16}$$

From (2.9) we see directly that

$$\pi_{1,r}(n;q) = \begin{bmatrix} n+\mathrm{r} \\ r \end{bmatrix},$$

a well-known result concerning the Gaussian polynomials (see MacMahon's *Combinatory Analysis*, volume 2, section 7). If we evaluate the determinant in (2.9) with $k = 2$, we find that

$$\pi_{2,r}(n;q) = \frac{1-q}{1-q^{n+r+1}}\begin{bmatrix} n+r+1 \\ r \end{bmatrix}\begin{bmatrix} n+r+1 \\ r+1 \end{bmatrix}.$$

Similar evaluations may be performed for other small values of k, and the results are all special cases of the following general formula:

$$\pi_{k,r}(n;q) = \prod_{j=0}^{k-2}\left\{\frac{(1-q^{j+1})^{k-1-j}}{(1-q^{r+1+j})^{k-1-j}}\begin{bmatrix} n+r+j \\ r \end{bmatrix}\right\} \tag{2.17}$$

$$= \frac{(q)_1(q)_2\cdots(q)_{k-1}}{(q)_r(q)_{r+1}\cdots(q)_{r+k-1}} \cdot \frac{(q)_{n+r}(q)_{n+r+1}\cdots(q)_{n+r+k-1}}{(q)_n(q)_{n+1}\cdots(q)_{n+k-1}}.$$

Our next goal is to prove (2.17); from this result we shall immediately obtain several interesting results on the generating functions of plane partitions, including (1.1).

Let $D(k, r, n)$ denote the right-hand side of (2.15). To establish (2.17) we must show that $D(k, r, n)$ is equal to the right-hand side of that equation. Let us define

$$W(k, r) = \det\left(q^{ri + i(i-1)/2}\begin{bmatrix} j \\ i \end{bmatrix}\right), \qquad 0 \leqq i, j \leqq k - 1.$$

Therefore

$$D(k, r, n)\, W(k, r) = \det(c_{ij}), \qquad 0 \leqq i, j \leqq k - 1,$$

where

$$\begin{aligned} c_{ij} &= \sum_{s=0}^{j} q^{s(s-1)/2}\begin{bmatrix} j \\ s \end{bmatrix} q^{rs + (i-s)(i-s-1)/2}\begin{bmatrix} n + r \\ n + i - s \end{bmatrix} \\ &= q^{i(i-1)/2}\begin{bmatrix} n + r + j \\ n + i \end{bmatrix}; \end{aligned}$$

the final summation here is done by using the q-analog of the Vandermonde summation (a combinatorial proof of one of the forms of the q-analog of the Vandermonde summation was given by MacMahon in [83]). Hence

$$\begin{aligned} D(k, r, n)\, W(k, r) &= \det\left(q^{i(i-1)/2}\begin{bmatrix} n + r + j \\ n + i \end{bmatrix}\right), \qquad 0 \leqq i, j \leqq k - 1, \\ &= \frac{(q)_{n+r}(q)_{n+r+1} \cdots (q)_{n+r+k-1}}{(q)_n (q)_{n+1} \cdots (q)_{n+k-1}} \det\left(\frac{q^{i(i-1)/2}}{(q)_{r-i+j}}\right). \end{aligned}$$

Therefore

$$D(k, r, n) = \frac{(q)_{n+r}(q)_{n+r+1} \cdots (q)_{n+r+k-1}}{(q)_n (q)_{n+1} \cdots (q)_{n+k-1}} C(k, r),$$

where $C(k, r)$ does not depend on n. Setting $n = 0$ in the above equation, we see that

$$C(k,r) = \frac{(q)_1 (q)_2 \cdots (q)_{k-1}}{(q)_r (q)_{r+1} \cdots (q)_{r+k-1}},$$

and so (2.17) is established.

If we now assume that $|q| < 1$, then we may let one or more of the parameters tend to infinity in (2.17). Hence

$$\sum_{m=0}^{\infty} p_{k,\infty}(m) q^m = \prod_{j=1}^{\infty} (1 - q^j)^{-min(k,j)}, \tag{2.18}$$

$$\sum_{m=0}^{\infty} p_{\infty,\infty}(m) q^m = \prod_{j=1}^{\infty} (1 - q^j)^{-j}; \tag{2.19}$$

the second equation is MacMahon's famous result (1.1).

11.3 References

G. E. Andrews (1971) On a conjecture of Guinand for the plane partition function, *Proc. Edinburgh Math. Soc.(2)*, *17*, 275–276.

G. E. Andrews (1976) The Theory of Partitions, Encyclopedia of Mathematics and Its Applications, vol. 2, Addison-Wesley, Reading.

F. C. Auluck and C. B. Haselgrove (1952) On Ingham's Tauberian for partitions, *Proc. Cambridge Phil. Soc.*, *48*, 566–570.

E. A. Bender and D. Knuth (1972) Enumeration of plane partitions, *J. Combinatorial Th.*, *13*, 40–54.

C. Berge (1971) Principles of Combinatorics, Academic Press, New York.

L. Carlitz (1956) The expansion of certain products, *Proc. A.M.S.*, *7*, 558–564.

L. Carlitz (1963a) Some generating functions, *Duke Math. J.*, *30*, 191–202.

L. Carlitz (1963b) A problem in partitions, *Duke Math. J.*, *30*, 203–214.

L. Carlitz (1965a) Generating functions and partition problems, *Proc. Symp. in Pure Math.*, *8*, 144–169.

L. Carlitz (1965b) Weighted two-line arrays, *Duke Math. J.*, *32*, 721–740.

L. Carlitz (1967a) Some determinants of q-binomial coefficients, *J. reine und angew. Math.*, *226*, 216–220.

L. Carlitz (1967b) Rectangular arrays and plane partitions, *Acta Arith.*, *13*, 29–47.

L. Carlitz (1973) Enumeration of two-line arrays, *Fibonacci Quart.*, *11*, 113–130.

L. Carlitz (1976) A special class of triangular arrays, *Collect. Math.*, *27*, 23–58.

L. Carlitz and J. Riordan (1965) Enumeration of certain two-line arrays, *Duke Math. J.*, *32*, 529–540.

L. Carlitz and D. P. Roselle (1972) Triangular arrays subject to MacMahon's conditions, *Fibonacci Quart.*, *10*, 591–597, 658.

L. Carlitz and R. Scoville (1975) A generating function for triangular partitions. Collection of articles dedicated to Derrick Henry Lehmer on the occasion of his seventieth birthday, *Math. Comp.*, *29*, 67–77.

L. Carlitz and R. P. Stanley (1975) Branchings and partitions, Proc. A.M.S., *53*, 246–249.

T. W. Chaundy (1931) Partition-generating functions, *Quart. J. Math.*, *2*, 234–240.

T. W. Chaundy (1932) The unrestricted plane partition, *Quart. J. Math.*, *3*, 76–80.

T. W. Chaundy (1933) Plane partitions, *Proc. London Math. Soc.(2)*, *35*, 14–22.

M. S. Cheema and W. E. Conway (1972) Numerical investigation of certain asymptotic results in the theory of partitions, *Math. of Comp.*, *26*, 999–1005.

M. S. Cheema and B. Gordon (1964) Some remarks on two- and three-line partitions, *Duke Math. J.*, *31*, 267–274.

M. S. Cheema and C. T. Haskell (1967) Multirestricted and rowed partitions, *Duke Math. J.*, *34*, 443–452.

L. Comtet (1974) Advanced Combinatorics, Reidel, Dordrecht.

B. Gordon (1962) Two new representations of the partition function, *Proc. A.M.S.*, *13*, 869–873.

B. Gordon and L. Houten (1968a) Notes on plane partitions I, *J. Combinatorial Th.*, *4*, 72–80.

B. Gordon and L. Houten (1968b) Notes on plane partitions II, *J. Combinatorial Th.*, *4*, 81–99.

B. Gordon and L. Houten (1969) Notes on plane partitions III, *Duke Math. J.*, *36*, 801–824.

B. Gordon (1971a) Notes on plane partitions IV: multirowed partitions and strict decrease along columns, *Proc. Symp. in Pure Math.*, *19*, 91–100.

B. Gordon (1971b) Notes on plane partitions V, *J. Combinatorial Th.*, *11*, 157–168.

H. Gupta (1970) Partitions—a survey, *J. Res. Nat. Bureau Standards(B)*, *Math. Sci.*, *74B*, 1–29.

M. J. Hodel (1975a) Enumerations of weighted rectangular arrays, *Ann. Mat. Pura Appl.(4)*, *106*, 329–351.

M. J. Hodel (1975b) Enumeration of weighted *p*-line arrays, *Pacific J. Math.*, *60*, 141–167.

W. V. D. Hodge and D. Pedoe (1952) Methods of Algebraic Geometry, vol. 2, Cambridge University Press, Cambridge.

A. Lascoux (1974) Polynômes symétriques et coefficients d'intersection de cycles de Schubert, *C. R. Acad. Sc. Paris*, *279*, 201–204.

A. Lascoux (1974) Tableaux de Young et fonctions de Schur-Littlewood, Séminaire Delange-Pisot-Poitou, No. 4, 1–7.

J. Southam (1970) A combinatorial proof of the generating function of three-line partitions, Ph.D. thesis, Oregon State University.

R. P. Stanley (1971) Theory and application of plane partitions I and II, *Studies in Appl. Math.*, *50*, 167–188, 259–279.

R. P. Stanley (1972) Ordered structures and partitions, Memoirs of the A.M.S., No. 119.

R. P. Stanley (1973) The conjugate trace and trace of plane partitions, *J. Combinatorial Th. (A)*, *14*, 53–65.

E. M. Wright (1931) Asymptotic partition formulae I: Plane partitions, *Quart. J. Math.(2)*, *2*, 177–189.

[77] Memoir on the theory of the partitions of numbers—Part V. Partitions in two-dimensional space, *Phil. Trans.*, *211* (1912), 75–110.

This paper and [78] are devoted to the derivation of formulas for the generating functions of various types of plane partitions. The most elegant result sought (the proof is sketched in [78]) is the identity

$$1 + \sum_{m=1}^{\infty} p_{\infty,\infty}(m)\, q^m = \prod_{n=1}^{\infty} (1 - q^n)^{-n},$$

where $p_{\infty,\infty}(m)$ is the number of plane partitions of m.

In this paper MacMahon lays the groundwork for the identity

$$\mathrm{GF}(l, m, n) = \prod_{n=1}^{n} |\mathrm{LM}|_i,$$

where

$$|\mathrm{LM}|_i = \frac{(1 - x^{l+i})(1 - x^{l+i+1}) \ldots (1 - x^{l+m+i-1})}{(1 - x^i)(1 - x^{i+1}) \ldots (1 - x^{m+i-1})},$$

and where $\mathrm{GF}(l, m, n)$ is the generating function for plane partitions in which each part is $\leqq n$, there are $\leqq m$ rows, and there are $\leqq l$ columns. This result is of course just equation (2.17) from section 2 of this chapter. His use of the theory of permutations for the study of plane partitions leads to an extensive independent study of permutations in [82].

MacMahon begins with the discussion of partitions defined relative to an incomplete lattice (i.e., Young tableau), namely, he considers $\mathrm{GF}(l; a, b, c, \ldots)$, which denotes the generating function for plane partitions with nonnegative parts, in which each part is $\leqq l$, the first row has a parts, the second b parts, the third c parts, etc.

We now define a lattice permutation of an assemblage of letters (i.e., multiset), such as $\alpha^3\beta^2\gamma$: If a dividing line be made between any two adjacent letters of the permutation, the succession of letters to the left of the line is like the whole permutation, such that α occurs at least as often as β, β at least as often as γ, etc.; in other words, the numbers that specify the occurrences of α, β, γ, etc. are in descending order of magnitude. Thus $\alpha\beta\gamma\alpha\beta\alpha$ is a lattice permutation of $\alpha^3\beta^2\gamma$; however $\alpha\gamma\alpha\beta\alpha\beta$ is not.

We then separate each lattice permutation into a minimal number of groups of letters such that in each group the letters are in alphabetical order. Thus the grouping of $\alpha\beta\gamma\alpha\beta\gamma$ is $\alpha\beta\gamma|\alpha\beta|\alpha$. With each permutation we associate a set of numbers $p_1, p_2, p_3, \ldots$, where p_s denotes the number of letters to the left of the sth dividing line. Corresponding to the assemblage $\alpha^a\beta^b\gamma^c\ldots$, we define the lattice function

$$L(\infty; a, b, c, \ldots) = \sum x^{p_1+p_2+p_3+\cdots},$$

where the sum is over all lattice permutations of the given assemblage.

Interrelationships between the above and various plane partition generating functions and lattice functions form the bulk of this paper. For example, the identity

$$\mathrm{GF}(\infty; a, b, c, \ldots) = \frac{\mathrm{L}(\infty; a, b, c, \ldots)}{(1-x)(1-x^2)\ldots(1-x^{a+b+c+\cdots})}$$

is of great importance in this work. The paper concludes with a postscript which provides an introduction to [82].

[105] The connexion between the sum of the squares of the divisors and the number of partitions of a given number, *Messenger of Math.*, *52* (1922), 113–116.

Let $p(n)$ denote the number of (one-dimensional) partitions of n, and $B(n)$ denote the number of plane partitions of n. From

$$\sum_{n=0}^{\infty} p(n)q^n = \prod_{n=1}^{\infty}(1-q^n)^{-1},$$

Euler was able to deduce (using the logarithmic derivative) that

$$np(n) = \sum_{j=0}^{n-1} \sigma(n-j)p(j),$$

where $\sigma(j)$ is the sum of the divisors of j. From

$$\sum_{n=0}^{\infty} B(n)q^n = \prod_{i=1}^{\infty}(1-q^n)^{-n},$$

MacMahon deduces (using the logarithmic derivative) that

$$nB(n) = \sum_{j=0}^{n-1} \sigma_2(n - j)\, B(j),$$

where $\sigma_2(j)$ is the sum of the squares of the divisors of j. This is an efficient recurrence for computing a table of values of $B(n)$. MacMahon then discusses other partition functions where similar recurrences may be determined.

Reprints of the Papers

III. *Memoir on the Theory of the Partitions of Numbers.—Part V. Partitions in Two-dimensional Space.*

By Major P. A. MacMahon, *R.A., D.Sc., F.R.S.*

Received December 31, 1910,—Read January 26, 1911.

Introduction.

In previous papers* I have broached the question of the two-dimensional partitions of numbers—or, say, the partitions in a plane—without, however, having succeeded in establishing certain conjectured formulas of enumeration. The parts of such partitions are placed at the nodes of a complete, or of an incomplete, lattice in two dimensions, in such wise that descending order of magnitude is in evidence in each horizontal row of nodes and in each vertical column. No decided advance was made in regard to the complete lattice, and the question of the incomplete lattice is considered for the first time in the present paper.

I return to the subject because I am now able to throw a considerable amount of fresh light upon the problem, and have succeeded in overcoming most of the difficulties which surround it. In fact, I am now able to show how the generating functions may be constructed in respect of any lattice, complete or incomplete, in forms which are free from redundant terms. I have not succeeded, so far, in giving a general algebraic expression to the functions, but, in the case of the complete lattice, I have shown that an assumption as to form, consistent with all results that have been arrived at in particular cases, leads at once to the expression that has been for so long the conjectured result. For the complete lattice of two rows, and for the incomplete lattice of two rows, the results have been obtained without any assumption in regard to form, and must be regarded as rigidly established.

Before proceeding to explain the new method of research which enables this paper to make a notable advance, I must hasten to correct an error which I had not detected at the time a former paper was written.

It will be remembered that partitions in a plane are such that there is a graphical

* "Memoir on the Theory of the Partitions of Numbers," 'Phil. Trans. Roy. Soc.,' A, 1896, vol. 187, pp. 619–673; 1899, vol. 192, pp. 351–401; 1905, vol. 205, pp. 37–59.

VOL. CCXI.—A 473. L 2 27.4.11

representation by nodes upon a three-dimensional lattice, just as for partitions on a line there is a graphical representation by nodes upon a two-dimensional lattice. It is convenient to replace these nodes by units, and to regard partitions on a line as being in one-to-one correspondence with partitions in a plane when the part magnitude of such is restricted to be not greater than unity; thus, instead of saying with FERRERS that

· · ·
· ·
·

is a graphical representation of the line partition 321, I regard the plane partition of units

111
11
1

as being in one-to-one correspondence with the line partition.

Just so the plane partition

331
22
1

is graphically represented by piles of nodes perpendicular to the plane of the paper, say

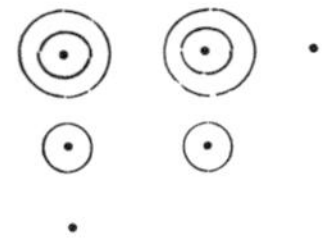

or we may replace the nodes by units, and say that it is in one-to-one correspondence with a space partition, the part magnitude being restricted to unity. The plane partition arises by projection of the space partition upon one of the co-ordinate planes, just as the line partition arises by projection of the plane partition, with which it is in correspondence, upon one of the co-ordinate axes.

Every two-dimensional graph of nodes may be interpreted either by rows or by columns, and every plane partition of units may be projected in two ways. The graphs *in solido* admit of one, *two*, three, or six readings.

In previous papers I omitted to notice that a three-dimensional graph may admit of two readings. The omission came to my notice when I was trying to verify that the number of partitions of w *in plano* the numbers of rows and columns, and also the part magnitude, being unrestricted is given by the coefficient of x^w in the ascending expansion of the algebraic fraction

$$\frac{1}{(1-x)(1-x^2)^2(1-x^3)^3(1-x^4)^4 \ldots \textit{ad inf.}}.$$

I counted, as far as weight 16, the numbers of the partitions by separately counting those whose graphs possess one, three, and six readings. At weight 13 a discrepancy appeared, because of that weight there are only two graphs which have one reading, and, on the assumption that the remaining graphs could be read in either three or six ways, it was clear that the number of the partitions must be $\equiv 2 \bmod 3$; but the coefficients of x^{13} in the *supposed* generating function was found to be 2485, which is $\equiv 1 \bmod 3$. It thus became clear either that the reasoning from the graphs was wrong, or that the generating function was at fault. The discrepancy was cleared up by the discovery that at weight 13 graphs with two readings present themselves for the first time. The simplest of these is

331
211 of weight 13.
2

The property possessed by these partitions is that the successive rows are the conjugates of the successive columns without being identical with them; that is to say, that the successive rows are not to be self-conjugate partitions. Thus, 331, 211, 2 are conjugates of 322, 31, 11 respectively. The reading of the corresponding three-dimensional graph in the six modes gives either

331		322
211	or	31
2		11.

The separate enumeration of these forms is a matter for future enquiry.

Art. 1. Turning now to the substance of this communication, I shall introduce a new plan of procedure which is applicable when the places for the parts of the partitions are given by the nodes of two-dimensional lattices, which may be complete or incomplete. In every case I suppose the part magnitude to be not greater than l, and when the lattice is complete, I suppose it to have m rows and n columns. The generating function which gives by the coefficients of x^w the number of the partitions of w of the nature considered will be denoted by GF (l, m, n).

In the excellent notation of Cayley and Sylvester I shall denote the algebraic expression $1-x^s$ by $(\mathbf{s})$, employing Clarendon type for the letter s, and thus $1-x$ by $(\mathbf{1})$ and $1-x^{l+1}$ by $(\mathbf{l+1})$, using always the Clarendon type in order to differentiate such notation from that in which between the brackets the ordinary Roman type is employed; the latter will, in general, denote integers s, 1, $l+1$, as the case may be. The notation is perfect for the purpose in hand, because it merely exhibits and concentrates attention upon the exponent s, which is the essential part of the expression, and the only part that in many cases it is necessary to handle algebraically. Further, in several instances, identities involving such expressions in

Clarendon type can be transformed into identities involving Roman type by simply changing the type in the bracketed factors; this ensues because the fractions $(\mathbf{s})\div(\mathbf{t})$ in Clarendon becomes equal to $(s)\div(t)$ in Roman in the limit when x is equal to unity.

Art. 2. The graphical representation by a three-dimensional lattice shows that the generating function GF (l, m, n) is unaltered by any permutation of the letters l, m, n. The subjoined notation is designed to show clearly the six alternative expressions of the generating function, arising from this circumstance, which it is a principal object of this paper to establish.

Thus, write

$$|\mathrm{LM}|_s = \frac{(\mathbf{l+s})\,(\mathbf{l+s+1})\ldots(\mathbf{l+m+s-1})}{(\mathbf{s})\,(\mathbf{s+1})\ldots(\mathbf{m+s-1})},$$

$$|\mathrm{ML}|_s = \frac{(\mathbf{m+s})\,(\mathbf{m+s+1})\ldots(\mathbf{m+l+s-1})}{(\mathbf{s})\,(\mathbf{s+1})\ldots(\mathbf{l+s-1})},$$

$$|\mathrm{MN}|_s = \frac{(\mathbf{m+s})\,(\mathbf{m+s+1})\ldots(\mathbf{m+n+s-1})}{(\mathbf{s})\,(\mathbf{s+1})\ldots(\mathbf{n+s-1})},$$

$$|\mathrm{NM}|_s = \frac{(\mathbf{n+s})\,(\mathbf{n+s+1})\ldots(\mathbf{n+m+s-1})}{(\mathbf{s})\,(\mathbf{s+1})\ldots(\mathbf{m+s-1})},$$

$$|\mathrm{NL}|_s = \frac{(\mathbf{n+s})\,(\mathbf{n+s+1})\ldots(\mathbf{n+l+s-1})}{(\mathbf{s})\,(\mathbf{s+1})\ldots(\mathbf{l+s-1})},$$

$$|\mathrm{LN}|_s = \frac{(\mathbf{l+s})\,(\mathbf{l+s+1})\ldots(\mathbf{l+n+s-1})}{(\mathbf{s})\,(\mathbf{s+1})\ldots(\mathbf{n+s-1})}.$$

It is to be shown that

$$\begin{aligned}
\mathrm{GF}(l, m, n) &= |\mathrm{LM}|_1\ |\mathrm{LM}|_2 \ldots |\mathrm{LM}|_n, \\
&= |\mathrm{ML}|_1\ |\mathrm{ML}|_2 \ldots |\mathrm{ML}|_n, \\
&= |\mathrm{MN}|_1\ |\mathrm{MN}|_2 \ldots |\mathrm{MN}|_l, \\
&= |\mathrm{NM}|_1\ |\mathrm{NM}|_2 \ldots |\mathrm{NM}|_l, \\
&= |\mathrm{NL}|_1\ |\mathrm{NL}|_2 \ldots |\mathrm{NL}|_m, \\
&= |\mathrm{LN}|_1\ |\mathrm{LN}|_2 \ldots |\mathrm{LN}|_m.
\end{aligned}$$

Art. 3. Every known particular case agrees with these formulæ, but only two general results have been established prior to this paper. One is the well-known case of partitions on a line, viz. :—

$$\mathrm{GF}(l, 1, n) = \frac{(\mathbf{l+1})\,(\mathbf{l+2})\ldots(\mathbf{l+n})}{(\mathbf{1})\,(\mathbf{2})\ldots(\mathbf{n})} = \frac{(\mathbf{n+1})\,(\mathbf{n+2})\ldots(\mathbf{n+l})}{(\mathbf{1})\,(\mathbf{2})\ldots(\mathbf{l})},$$

$$= |\mathrm{LN}|_1 = |\mathrm{NL}|_1,$$

and the other is that given in Part II. of this Memoir, and also by FORSYTH,

$$\mathrm{GF}(\infty, 2, n) = \frac{1}{(\mathbf{1})\{(\mathbf{2})(\mathbf{3})\ldots(\mathbf{n})\}^2(\mathbf{n}+\mathbf{1})} = |\mathrm{LN}|_1\,|\mathrm{LN}|_2, \text{ when } l = \infty.$$

This generating function may be regarded as enumerating partitions—

(i) At the nodes of a lattice of 2 rows and n columns (or of n rows and 2 columns)

$$\begin{array}{cccccc} \cdot & \cdot & \cdot & \cdot & \cdot & \cdot \\ \cdot & \cdot & \cdot & \cdot & \cdot & \cdot \end{array}$$

the part magnitude being unrestricted;

(ii) At the nodes of a lattice which has the number of $\begin{smallmatrix}\text{rows}\\ \text{columns}\end{smallmatrix}$ unrestricted, the number of $\begin{smallmatrix}\text{columns}\\ \text{rows}\end{smallmatrix}$ equal to n and the part magnitude restricted to be $\ngtr 2$.

The found result shows that the number of partitions of w is equal to the number of ways of composing w with

one kind of unity,
two kinds of twos,
two kinds of threes,
. . . .
two kinds of n's,
one kind of $n+1$,

but all attempts to establish a one-to-one correspondence have failed. Had this proved to have been feasible it *might* have been extended to prove the similar results for GF (∞, m, n) where $m > 2$.

Art. 4. The linear Diophantine Analysis, which was applied to the same question in an earlier part of this Memoir, having also failed to establish general results, recourse has been had to a plan suggested by Part IV. of the Memoir,* and a considerable advance has been made. In that paper I considered the number of different ways in which k *different* numbers can be placed at the nodes of a lattice, complete or incomplete, the number of nodes being k, and the numbers being placed in such wise that descending order of magnitude is in evidence in each row from West to East and in each column from North to South.

In the paper quoted I showed that if the rows involve $\alpha_1, \alpha_2, \ldots, \alpha_m$ nodes respectively, where, of course, $\alpha_1 \geqq \alpha_2 \geqq \ldots \geqq \alpha_m$, the number of ways of arranging the $\Sigma\alpha$ different numbers at the nodes is

$$\frac{(\Sigma\alpha)!}{(\alpha_1+m-1)!\,(\alpha_2+m-2)!\,\ldots\,(\alpha_{m-1}+1)!\,\alpha_m!}\prod_{s,}(\alpha_s-\alpha_t-s+t),$$

* "Memoir on the Theory of the Partitions of Numbers," 'Phil. Trans.,' A, 1908, vol. 209, pp. 153–175.

where $s < t$ and the product Π has reference to every pair of numbers a_s, a_t drawn from the succession $a_1, a_2, \ldots, a_m$.

This result will be found to furnish an important key to the solution of the problems before us.

It is possible, by the method employed, to consider the generating functions for partitions at the nodes of an incomplete lattice, and I shall use GF $(l; a, b, c, \ldots)$ to denote that which has reference to a lattice whose successive rows involve $a, b, c, \ldots$ nodes, respectively, the part magnitude being restricted by the number l. In this notation GF (l, m, n) may alternately be written GF $(l; n^m)$ or GF $(l; m^n)$, wherein n^m will denote m rows each of n nodes.

I derive from every lattice, complete or incomplete, a lattice-function of x, and this function depends, like the generating function, not only upon the specification of the lattice, but also upon the number l which limits the part magnitude. I denote this function by L (l, m, n) or by L $(l; a, b, c, \ldots)$, according as the lattice is complete or incomplete. In cases where no confusion can arise, I simply write L for brevity.

Art. 5. I will now explain the formation of the functions

$$\mathrm{L}(\infty, m, n) \quad \text{and} \quad \mathrm{L}(\infty; a, b, c, \ldots);$$

and then establish the fundamental propositions

$$\mathrm{GF}(\infty, m, n) = \frac{\mathrm{L}(\infty, m, n)}{(\mathbf{1})(\mathbf{2})\ldots(\mathbf{mn})},$$

$$\mathrm{GF}(\infty; a, b, c, \ldots) = \frac{\mathrm{L}(\infty; a, b, c, \ldots)}{(\mathbf{1})(\mathbf{2})\ldots(\boldsymbol{\Sigma}\mathbf{a})}.$$

In the next place I will explain the formation of the functions

$$\mathrm{L}(l, m, n) \quad \text{and} \quad \mathrm{L}(l; a, b, c, \ldots),$$

and establish the fundamental propositions

$$\mathrm{GF}(l, m, n) = \frac{\mathrm{L}(l, m, n)}{(\mathbf{1})(\mathbf{2})\ldots(\mathbf{mn})},$$

$$\mathrm{GF}(l; a, b, c, \ldots) = \frac{\mathrm{L}(l; a, b, c, \ldots)}{(\mathbf{1})(\mathbf{2})\ldots(\boldsymbol{\Sigma}\mathbf{a})}.$$

Art. 6. Consider an incomplete lattice having 3, 2, 1 nodes in the rows respectively, and let any six different integers (say the first six) be placed in any manner at the nodes in such wise that descending order of magnitude is in evidence in each row and in each column; such an arrangement may be

631
52
4

Let the Greek letters α, β, γ be associated with the first, second, and third rows, respectively, and consider each number in the lattice in succession in descending order of magnitude. Thus, beginning with 6: since it is in the first row I commence a succession of Greek letters with α; passing to 5, since it is in the second row, I follow with β; then 4 gives γ, since it is in the third row; then 3 gives α; 2, β; and finally 1 gives α.

$$\alpha\beta\gamma\alpha\beta\alpha.$$

In this way I obtain a permutation of the letters in $\alpha^3\beta^2\gamma$, where the exponents 3, 2, 1 enumerate the nodes in the successive rows of the lattice. This permutation possesses the property :—

> "If a dividing line be made between any two adjacent letters of the permutation, the succession of letters to the left of the dividing line is like the whole permutation, such that α occurs at least as often as β, β at least as often as γ; in other words, the numbers which specify the occurrences of α, β, γ, are in descending order of magnitude."

In fact, if the process of forming the Greek letter succession (or permutation) be arrested at any point, the lattice numbers that have been dealt with occupy a set of nodes which also constitute a lattice, complete or incomplete.

It follows, of course, that the first letter of the permutation must be α. The lattice arrangement of numbers is recoverable from the permutation, for it is merely necessary to write the numbers in descending order underneath the letters when we see that the successive lattice rows are indicated by the letters α, β, γ, respectively,

$$\begin{array}{cccccc}\alpha&\beta&\gamma&\alpha&\beta&\alpha\\ 6&5&4&3&2&1\end{array}.$$

The process is thus unique, and there will be as many different Greek letter permutations having the properties above specified as of arrangements of unequal numbers at the nodes of the lattice having the specified descending orders.

Every Greek letter permutation can be separated into groups, each of which contains letters in alphabetical order; in the case before us this is accomplished by two dividing lines

$$\alpha\beta\gamma\,|\,\alpha\beta\,|\,\alpha,$$

each of which separates a letter from one which follows it, but is prior to it in alphabetical order.

I associate a power of x with each permutation by taking for the exponent a sum of numbers $p_1+p_2+p_3+\ldots$, where p_s denotes that the sth dividing line has p_s letters to the left of it. Thus in the above instance $p_1 = 3$, $p_2 = 5$, and the associated power of x is $x^{3+5} = x^8$.

Every one of the

$$\frac{(\Sigma a)!}{(a_1+m-1)!\,(a_2+m-2)!\,\ldots\,(a_{m-1}+1)!\,a_m!}\,\Pi_{s,t}\,(a_s-a_t-s+t)$$

arrangements of the different integers at the nodes of the lattice will thus have a power of x associated with it, and taking the sum of them all I obtain the lattice function

$$L(\infty\,;\,a_1, a_2, a_3, \ldots) = \Sigma x^{p_1+p_2+p_3+\ldots}.$$

Art. 7. I will set out at length the formation of $L(\infty\,;\,3, 2, 1)$.

654	654	653	653	652	652	651	651
32	31	42	41	43	41	43	42
1	2	1	2	1	3	2	3
$\alpha\alpha\alpha\beta\beta\gamma$,	$\alpha\alpha\alpha\beta\gamma\mid\beta$,	$\alpha\alpha\beta\mid\alpha\beta\gamma$,	$\alpha\alpha\beta\mid\alpha\gamma\mid\beta$,	$\alpha\alpha\beta\beta\mid\alpha\gamma$,	$\alpha\alpha\beta\gamma\mid\alpha\beta$,	$\alpha\alpha\beta\beta\gamma\mid\alpha$,	$\alpha\alpha\beta\gamma\mid\beta\mid\alpha$
x^0	x^5	x^3	x^8	x^4	x^4	x^5	x^9

631	632	641	642	641	643	642	643
52	51	52	51	53	51	53	52
4	4	3	3	2	2	1	1
$\alpha\beta\gamma\mid\alpha\beta\mid\alpha$,	$\alpha\beta\gamma\mid\alpha\alpha\beta$,	$\alpha\beta\mid\alpha\gamma\mid\beta\mid\alpha$,	$\alpha\beta\mid\alpha\gamma\mid\alpha\beta$,	$\alpha\beta\mid\alpha\beta\gamma\mid\alpha$,	$\alpha\beta\mid\alpha\alpha\gamma\mid\beta$,	$\alpha\beta\mid\alpha\beta\mid\alpha\gamma$,	$\alpha\beta\mid\alpha\alpha\beta\gamma$
x^8	x^3	x^{11}	x^6	x^7	x^7	x^6	x^2

so that

$$L(\infty\,;\,3, 2, 1) = 1+x^2+2x^3+2x^4+2x^5+2x^6+2x^7+2x^8+x^9+x^{11}.$$

It is obvious that this process can be carried out in respect of any lattice, complete or incomplete, and that the number of different Greek letters involved will be equal to the number of rows.

Art. 8. To show the connexion between such a lattice function and the corresponding generating of partitions at the nodes of the lattice I proceed as follows :—

We have to establish the relation

$$GF(\infty\,;\,a_1, a_2, a_3, \ldots) = \frac{L(\infty\,;\,a_1, a_2, a_3, \ldots)}{(1)\,(2)\ldots(\Sigma a)}.$$

As the simplest possible case (with a trivial exception) consider the complete lattice of 2 rows and 2 columns

· ·

· ·

and any numbers, equal or unequal, to be placed at the four nodes in such wise that there is descending order of magnitude in both rows and in both columns; say that the numbers are

$$\begin{matrix} p & q \\ r & s \end{matrix}$$

subject to the conditions $p \geqq q \geqq s$, $p \geqq r \geqq s$.

It is clear that we must either have

$$\text{(i) } p \geqq q \geqq r \geqq s \quad \text{or} \quad \text{(ii) } p \geqq r > q \geqq s;$$

and that these two systems do not overlap.

If (i) obtains we may perform the summation $\Sigma x^{p+q+r+s}$ by writing $r = s+\text{A}$, $q = s+\text{A}+\text{B}$, $p = s+\text{A}+\text{B}+\text{C}$, where A, B, C are arbitrary positive integers, zero included; the sum is thus

$$\Sigma x^{\text{C}+2\text{B}+3\text{A}+4s};$$

and since C, B, A, s may each of them assume all values ranging from zero to infinity, the sum is clearly

$$\frac{1}{(1)(2)(3)(4)};$$

if, on the other hand, the parts of the partition have such values that (ii) obtains, we may write

$$q = s+\text{A}, \quad r = s+\text{A}+\text{B}+1, \quad p = s+\text{A}+\text{B}+\text{C}+1,$$

and we have the sum

$$\Sigma x^{\text{C}+2\text{B}+3\text{A}+4s+2},$$

which is equal to

$$\frac{x^2}{(1)(2)(3)(4)}.$$

By addition we have

$$\text{GF}(\infty; 2, 2) = \frac{1+x^2}{(1)(2)(3)(4)} = \frac{1}{(1)(2)^2(3)};$$

and it will be noted that $1+x^2 = \text{L}(\infty; 2, 2)$, derived, as above, from the lattice arrangements

43	42
21	31
$\alpha\alpha\beta\beta$	$\alpha\beta \mid \alpha\beta$
x^0	x^2

The fact is that the alternatives (i) and (ii) exist *because* there are two lattice arrangements of *unequal* numbers, and the signs of equality and inequality are arranged in (i) and (ii) so that the required sum may be separated into two non-overlapping systems in correspondence with the lattice arrangements. The fact that

M 2

$r > q$ in (ii), q being a letter prior to r in alphabetical order, is in direct correspondence with the Greek letter permutation $\alpha\beta|\alpha\beta$, in which a β precedes an α which is prior to it in alphabetical order. Moreover, r occurring in the second place in the condition (ii), $p \geqq r > q \geqq s$ clearly contributes the integer 2 to the exponent $C+2B+3A+4s+2$. Thus the numerator finally determined is necessarily the lattice function $L(\infty; 2, 2)$ found by the specified rules.

Art. 9. Next take a case which is not quite so simple

$$\begin{matrix} p & q & r \\ s & t & u, \end{matrix}$$

where $p \geqq q \geqq r$, $s \geqq t \geqq u$, $p \geqq s$, $q \geqq t$, $r \geqq u$; the associated lattice arrangements and the Greek letter permutations are

654	643	653	652	642
321	521	421	431	531
$\alpha\alpha\alpha\beta\beta\beta$	$\alpha\beta\|\alpha\alpha\beta\beta$	$\alpha\alpha\beta\|\alpha\beta\beta$	$\alpha\alpha\beta\beta\|\alpha\beta$	$\alpha\beta\|\alpha\beta\|\alpha\beta$
x^0	x^2	x^3	x^4	x^6

yielding

$$L(\infty; 3, 3) = 1+x^2+x^3+x^4+x^6 = \frac{(5)(6)}{(2)(3)}.$$

We have five non-overlapping systems

$$\begin{array}{ll} \text{(i) } p \geqq q \geqq r \geqq s \geqq t \geqq u, & \alpha\alpha\alpha\beta\beta\beta, \\ \text{(ii) } p \geqq s > q \geqq r \geqq t \geqq u, & \alpha\beta|\alpha\alpha\beta\beta, \\ \text{(iii) } p \geqq q \geqq s > r \geqq t \geqq u, & \alpha\alpha\beta|\alpha\beta\beta, \\ \text{(iv) } p \geqq q \geqq s \geqq t > r \geqq u, & \alpha\alpha\beta\beta|\alpha\beta, \\ \text{(v) } p \geqq s > q \geqq t > r \geqq u, & \alpha\beta|\alpha\beta|\alpha\beta, \end{array}$$

wherein the positions occupied by the symbol $>$ are to be compared with the positions of the dividing lines in the corresponding Greek letter permutations. It is clear that the summations derived from the systems (i), (ii), (iii), (iv), (v) give powers of x in the numerator of the generating function exactly corresponding to those which enter into the lattice function by the rules given. Hence

$$GF(\infty; 3, 3) = \frac{L(\infty; 3, 3)}{(1)(2)\dots(6)} = \frac{(5)(6)}{(2)(3)} \cdot \frac{1}{(1)(2)\dots(6)},$$

$$= \frac{1}{(1)(2)^2(3)^2(4)}.$$

This short demonstration suffices to establish the general relations

$$\mathrm{GF}(\infty\,;\,a_1, a_2, a_3, \ldots) = \frac{\mathrm{L}(\infty\,;\,a_1, a_2, a_3, \ldots)}{(\mathbf{1})(\mathbf{2})\ldots(\mathbf{\Sigma a})}.$$

$$\mathrm{GF}(\infty\,;\,m, n) = \frac{\mathrm{L}(\infty\,;\,m, n)}{(\mathbf{1})(\mathbf{2})\ldots(\mathbf{mn})}.$$

Art. 10. Remarkable properties of the lattice functions will present themselves as the investigation proceeds. A few observations may be usefully made at this point. In every case the zero power of x presents itself in correspondence with that permutation of the Greek letters which is in alphabetical order.

In the case of partitions on a line the lattice is a single row of nodes; the Greek letter succession is composed entirely of the letter α and the lattice function is unity.

A most useful property arises simply from the definition of the function, viz., putting x equal to unity we find that the sum of the coefficient is

$$\frac{(\Sigma a)!}{(a_1+m-1)!\,(a_2+m-2)!\ldots(a_{m-1}+1)!\,a_m!}\prod_{s,t}(a_s-a_t-s+t),$$

a verification of constant service.

I seek a representation of the lattice function that shall be a constant reminder of this enumerating function, and with this object in view I write the latter in the form

$$\frac{(1\,.\,2\,.\ldots\,\Sigma a)}{(m\,.\,m+1\,.\ldots\,a_1+m-1)\,(m-1\,.\,m\,.\ldots\,a_2+m-2)\ldots\{2\,.\,3\ldots(a_{m-1}+1)\}\,(1\,.\,2\,.\ldots\,a_m)}\cdot\frac{\prod_{s,t}(a_s-a_t+t-s)}{\prod_{s,t}(t-s)},$$

and I then write

$$\mathrm{L}(\infty\,;\,a_1, a_2, a_3, \ldots, a_m)$$

$$= \frac{(\mathbf{1})(\mathbf{2})\ldots(\mathbf{\Sigma a})}{(\mathbf{m})(\mathbf{m+1})\ldots(\mathbf{a_1+m-1})\,.\,(\mathbf{m-1})(\mathbf{m})\ldots(\mathbf{a_2+m-2})\ldots(\mathbf{2})(\mathbf{3})\ldots(\mathbf{a_{m-1}+1})\,.\,(\mathbf{1})(\mathbf{2})\ldots(\mathbf{a_m})}\,\mathrm{IL}(\infty\,;\,a_1, a_2, \ldots, a_m)$$

where the algebraic fraction on the dexter, which I term the outer lattice function, is of *fixed form*, and the remaining algebraic factor $\mathrm{IL}(\infty\,;\,a_1, a_2, \ldots, a_m)$, which I term the inner lattice function, has to be determined.

The outer function reduces to the corresponding part of the arithmetical function when x is put equal to unity; under the same circumstances the inner function reduces to the sum of its own coefficients, viz., to

$$\prod_{s,t}(a_s-a_t+t-s) \div \prod_{s,t}(t-s).$$

There is a convenience in thus postulating the expression of an outer lattice function, because in every known result in regard to complete lattices the inner function turns out to be simply unity; a principal object of this investigation is to establish that for the complete lattice the inner function is invariably unity. This is consistent with the result conjectured in Art. 2.

In regard to incomplete lattices the inner function is unity in special cases. The determination of its form for the general incomplete lattice is apparently a very difficult matter, which is reserved for future consideration. Its actual form for the lattice of two unequal rows will be determined presently.

Art. 11. There is also a vitally essential representation of the lattice function as a sum of sub-lattice functions, which forms a natural bridge from the function GF $(\infty;\alpha_1,\alpha_2,\alpha_3,\ldots)$ to the general function GF $(l;\alpha_1,\alpha_2,\alpha_3,\ldots)$. When the lattice function was formed from the permutations of the Greek letters, every permutation had s dividing lines where s ranged from zero up to a maximum value μ, which has not yet been determined. That portion of the lattice function which is derived from those permutations which involve precisely s dividing lines I name the sub-lattice function of order s and write it

$$L_s(\infty;\alpha_1\alpha_2\alpha_3\ldots),\text{ or } L_s(\infty;m,n),\text{ or simply } L_s,$$

if no confusion arises from the abbreviation.

Thus
$$L=\sum_0^\mu L_s.$$

In the elementary examples already dealt with

$$\begin{aligned}
L(\infty,2,2)&=L_0(\infty;2,2)+L_1(\infty;2,2),\\
&=\quad 1\quad+\quad x^2\quad,\\[1ex]
L(\infty,2,3)&=L_0(\infty;2,3)+L_1(\infty;2,3)+L_2(\infty;2,3),\\
&=\quad 1\quad+x^2+x^3+x^4+\quad x^6\quad,\\[1ex]
L(\infty;3,2,1)&=L_0\quad+L_1\quad+L_2\quad+L_3,\\
&=1\ +x^2+2x^3+2x^4+2x^5\ +2x^6+2x^7+2x^8+x^9\ +x^{11}.
\end{aligned}$$

It will be observed that L_0 is invariably unity.

Art. 12. In terms of these sub-lattice functions I now define the new and more general lattice function $L(l;\alpha_1,\alpha_2,\alpha_3,\ldots)$, in which l replaces ∞. I write

$$\begin{aligned}
L(l;\alpha_1,\alpha_2,\alpha_3,\ldots)&=(\mathfrak{l}+1)(\mathfrak{l}+2)\ldots(\mathfrak{l}+\Sigma\mathbf{a})\,L_0(\infty;\alpha_1,\alpha_2,\alpha_3,\ldots)\\
&+(\mathfrak{l})(\mathfrak{l}+1)\ldots(\mathfrak{l}+\Sigma\mathbf{a}-1)\,L_1(\infty;\alpha_1,\alpha_2,\alpha_3,\ldots)\\
&+\ldots\\
&+(\mathfrak{l}-\mu+1)(\mathfrak{l}-\mu+2)\ldots(\mathfrak{l}+\Sigma\mathbf{a}-\mu)\,L_\mu(\infty;\alpha_1,\alpha_2,\alpha_3,\ldots);
\end{aligned}$$

and also a general sub-lattice function

$$L_s(l;\alpha_1,\alpha_2,\alpha_3,\ldots)=(\mathfrak{l}-\mathbf{s}+1)(\mathfrak{l}-\mathbf{s}+2)\ldots(\mathfrak{l}+\Sigma\mathbf{a}-\mathbf{s})\,L_s(\infty;\alpha_1,\alpha_2,\alpha_3,\ldots).$$

Art. 13. The next step is to establish the fundamental relations

$$\mathrm{GF}\,(l\,;\,a_1, a_2, a_3, \ldots) = \frac{\mathrm{L}\,(l\,;\,a_1, a_2, a_3, \ldots)}{(1)\,(2)\ldots(\Sigma \mathbf{a})}\,;$$

$$\mathrm{GF}\,(l, m, n) = \frac{\mathrm{L}\,(l, m, n)}{(1)\,(2)\ldots(\mathbf{mn})}.$$

We have to take account of the circumstance that the part magnitude is now restricted not to exceed l. Take again the case, previously considered, of two rows and three columns. I recall the five distinct parts of the summation

$$\begin{array}{lll}
\text{(i)} & p \geqq q \geqq r \geqq s \geqq t \geqq u & \text{giving} \quad \Sigma x^{E+2D+3C+4B+5A+6u}\,, \\
\text{(ii)} & p \geqq s > q \geqq r \geqq t \geqq u & \quad ,, \quad \Sigma x^{E+2D+3C+4B+5A+6u+2}\,, \\
\text{(iii)} & p \geqq q \geqq s > r \geqq t \geqq u & \quad ,, \quad \Sigma x^{E+2D+3C+4B+5A+6u+3}\,, \\
\text{(iv)} & p \geqq q \geqq s \geqq t > r \geqq u & \quad ,, \quad \Sigma x^{E+2D+3C+4B+5A+6u+4}\,, \\
\text{(v)} & p \geqq s > q \geqq t > r \geqq u & \quad ,, \quad \Sigma x^{E+2D+3C+4B+5A+6u+6}\,.
\end{array}$$

For the condition (i) we put

$$t = u+\mathrm{A}, \qquad r = u+\mathrm{A}+\mathrm{B}, \qquad q = u+\mathrm{A}+\mathrm{B}+\mathrm{C}, \qquad s = u+\mathrm{A}+\mathrm{B}+\mathrm{C}+\mathrm{D},$$

$$p = u+\mathrm{A}+\mathrm{B}+\mathrm{C}+\mathrm{D}+\mathrm{E},$$

from which it is clear that

$$u+\mathrm{A}+\mathrm{B}+\mathrm{C}+\mathrm{D}+\mathrm{E}$$

cannot exceed l in magnitude; hence the sum

$$\Sigma x^{E+2D+3C+4B+5A+6u}$$

is the generating function of partitions on a line into l parts not exceeding 6 in magnitude, and is therefore

$$\frac{(\mathfrak{l}+1)\,(\mathfrak{l}+2)\ldots(\mathfrak{l}+6)}{(1)\,(2)\ldots(6)}.$$

Similarly in each of the cases (ii), (iii), (iv), belonging to $\mathrm{L}_1\,(\infty, 3, 2)$, we put $p = u+\mathrm{A}+\mathrm{B}+\mathrm{C}+\mathrm{D}+\mathrm{E}+1$, and it is clear that $u+\mathrm{A}+\mathrm{B}+\mathrm{C}+\mathrm{D}+\mathrm{E}$ cannot exceed $l-1$ in magnitude; the corresponding portion of the generating function is therefore

$$\frac{(\mathfrak{l})\,(\mathfrak{l}+1)\ldots(\mathfrak{l}+5)}{(1)\,(2)\ldots(6)}\,\mathrm{L}_1\,(\infty, 3, 2).$$

Finally, since in (v) we put

$$p = u+\mathrm{A}+\mathrm{B}+\mathrm{C}+\mathrm{D}+\mathrm{E}+2,$$

we obtain a part of the generating function

$$\frac{(l-1)(l)\ldots(l+4)}{(1)(2)\ldots(6)} L_2(\infty, 3, 2).$$

Thence

$$\text{GF}(l, 3, 2) = \frac{(l+1)\ldots(l+6)\,L_0(\infty, 3, 2) + (l)\ldots(l+5)\,L_1(\infty, 3, 2) + (l-1)\ldots(l+4)\,L_2(\infty, 3, 2)}{(1)(2)\ldots(6)},$$

$$= \frac{L(l, 3, 2)}{(1)(2)\ldots(6)}.$$

In general, when the lattice has $\Sigma\alpha$ nodes, we have a set of inequalities belonging to $L_s(\infty; \alpha_1, \alpha_2, \alpha_3, \ldots)$ which give rise to the generating function

$$\frac{(l-s+1)(l-s+2)\ldots(l+\Sigma a-s)}{(1)(2)\ldots(\Sigma a)} L_s(\infty; \alpha_1, \alpha_2, \alpha_3, \ldots);$$

and thus the above-given fundamental relations are established.

Art. 14. The generating functions for two-dimensional partitions GF (l, m, n) has been found in terms of lattice functions in the form

$$\frac{(l+1)\ldots(l+mn)\,L_0 + (l)\ldots(l+mn-1)\,L_1 + \ldots\ldots + (l-\mu+1)\ldots(l-\mu+mn)\,L_\mu}{(1)\ldots(mn)}.$$

If we subtract these partitions from those enumerated by GF (∞, m, n), we are left with those partitions which contain one part at least equal to or greater than $l+1$. I shall show how to determine directly the generating function for these in terms of lattice functions. To lead up to the proof, I will give an inductive proof of the theorem—

$$\text{GF}(l, 1, n) = \frac{(l+1)\ldots(l+n)}{(1)\ldots(n)}.$$

Taking the parts at n nodes in one row

$$\cdot \quad \cdot \quad \cdot \quad \cdot \quad \cdot \quad \cdots$$

the partitions which have a highest part equal to $l+1$ will be obtained by placing the part $l+1$ to the left of each of the partitions enumerated by GF $(l+1, 1, n-1)$.

Hence the whole of the partitions which have one part at least equal or greater than $l+1$ are enumerated by

$$\sum_l^\infty x^{l+1}\,\text{GF}(l+1, 1, n-1)$$

and

$$\text{GF}(l, 1, n) = \text{GF}(\infty, 1, n) - \sum_l^\infty x^{l+1}\,\text{GF}(l+1, 1, n-1).$$

Assume the truth of the theorem in the case of

$$\mathrm{GF}\,(l+1,\,1,\,n-1)$$

for all values of l; then

$$\mathrm{GF}\,(l,\,1,\,n) = \frac{1}{(1)\ldots(\mathrm{n})} - \sum_{l}^{\infty} x^{l+1}\frac{(\mathrm{l}+2)\ldots(\mathrm{l}+\mathrm{n})}{(1)\ldots(\mathrm{n}-1)}.$$

Putting $x^l = \theta$,

$$x^{l+1}\frac{(\mathrm{l}+2)\ldots(\mathrm{l}+\mathrm{n})}{(1)\ldots(\mathrm{n}-1)}$$

$$= \frac{\theta x}{(1)\ldots(\mathrm{n}-1)}\left\{1-\theta x^2\frac{(\mathrm{n}-1)}{(1)}+\theta^2x^5\frac{(\mathrm{n}-1)(\mathrm{n}-2)}{(1)(2)}-\ldots+(-)^{n-1}\theta^{n-1}x^{\frac{1}{2}(n-1)(n+2)}\right\};$$

and, therefore,

$$\sum_{l}^{\infty} x^{l+1}\frac{(\mathrm{l}+2)\ldots(\mathrm{l}+\mathrm{n})}{(1)\ldots(\mathrm{n}-1)} = \frac{\theta x}{(1)^2(2)\ldots(\mathrm{n}-1)} - \frac{\theta^2x^3\dfrac{(\mathrm{n}-1)}{(1)}}{(1)(2)^2(3)\ldots(\mathrm{n}-1)}$$

$$+\frac{\theta^3x^6\dfrac{(\mathrm{n}-1)(\mathrm{n}-2)}{(1)(2)}}{(1)(2)(3)^2(4)\ldots(\mathrm{n}-1)} - \ldots + (-)^{n-1}\theta^n\frac{x^{\frac{1}{2}n(n+1)}}{(1)\ldots(\mathrm{n})};$$

and thence

$$\frac{1}{(1)\ldots(\mathrm{n})} - \Sigma x^{l+1}\frac{(\mathrm{l}+2)\ldots(\mathrm{l}+\mathrm{n})}{(1)\ldots(\mathrm{n}-1)}$$

$$= \frac{1}{(1)\ldots(\mathrm{n})}\left\{1-\theta x\frac{(\mathrm{n})}{(1)}+\theta^2x^3\frac{(\mathrm{n})(\mathrm{n}-1)}{(1)(2)}-\ldots+(-)^n\theta^nx^{\frac{1}{2}n(n+1)}\right\},$$

$$= \frac{(\mathrm{l}+1)\ldots(\mathrm{l}+\mathrm{n})}{(1)\ldots(\mathrm{n})}.$$

Hence

$$\mathrm{GF}\,(l,\,1,\,n) = \frac{(\mathrm{l}+1)\ldots(\mathrm{l}+\mathrm{n})}{(1)\ldots(\mathrm{n})}$$

by induction.

To generalize this method, I take a lattice which is complete but for the node at the left-hand top corner

    · · · · ·
  · · · · · ·
  · · · · · ·
  · · · · · ·

and first determine the generating function for partitions such that the descending order of part magnitude is in evidence in each row and in each column. I take the number of rows to be m, and the number of columns n. A slight consideration shows that if L_s be the sub-lattice function of order s for the complete lattice, that of the deficient lattice now under consideration is $x^{-s}L_s$; hence, if the part magnitude be unrestricted, the generating function is

$$\frac{1+x^{-1}L_1+x^{-2}L_2+\ldots+x^{-\mu}L_\mu}{(1)\ldots(mn-1)};$$

and if the part magnitude be restricted not to exceed l,

$$\frac{(l+1)\ldots(l+mn-1)+x^{-1}(l)\ldots(l+mn-2)\,L_1+\ldots+x^{-\mu}(l-\mu+1)\ldots(l-\mu+mn-1)\,L_\mu}{(1)\ldots(mn-1)}.$$

A simple example, that may be at once verified, is found by taking $m = n = 2$ and the defective lattice

$$\begin{array}{cc} & \cdot \\ \cdot & \cdot \end{array}$$

Here $L_1 = L_\mu = x^2$ and the generating functions are

$$\frac{1+x}{(1)(2)(3)} = \frac{1}{(1)^2(3)},$$

$$\frac{(l+1)(l+2)(l+3)+x(l)(l+1)(l+2)}{(1)(2)(3)};$$

putting $l = 1$ we obtain $1+2x+x^2+x^3$, verified by

$$\begin{array}{ccccccccccc} & \cdot & & & 1 & & & \cdot & & & 1 & & & 1 \\ \cdot & \cdot & & \cdot & \cdot & & 1 & \cdot & & 1 & \cdot & & 1 & 1 \\ & 1 & & & x & & & x & & & x^2 & & & x^3. \end{array}$$

Now consider the partitions at the nodes of the complete lattice such that one part at least is equal to $l+1$ and no part exceeds $l+1$. We obtain all such by placing the part $l+1$ at the node situated at the left-hand top corner and connecting with it all of the partitions at the nodes of the incomplete lattice, which are such that the part magnitude is restricted not to exceed $l+1$ in magnitude.

We thus derive a generating function

$$x^{l+1}\frac{(l+2)\ldots(l+mn)+x^{-1}(l+1)\ldots(l+mn-1)\,L_1+\ldots\ldots+x^{-\mu}(l-\mu+2)\ldots(l-\mu+mn)}{(1)\ldots(mn-1)},$$

and thence the generating function, which enumerates all partitions at the nodes of the complete lattice, which are such that each has one or more parts at least as great as $l+1$, is

$$\Sigma x^{l+1}\frac{(l+2)\ldots(l+mn)+x^{-1}(l+1)\ldots(l+mn-1)\,L_1+\ldots\ldots+x^{-\mu}(l-\mu+2)\ldots(l-\mu+mn)}{(1)\ldots(mn-1)},$$

and it is easy to verify that this expression added to the expression already found for GF (l, m, n) is, in fact, equal to

$$\frac{L_0+L_1+\ldots+L_\mu}{(1)\ldots(mn)},$$

that is, to GF (∞, m, n).

Art. 15. This main proposition involves the whole theory of the partitions at the nodes of an incomplete lattice; it gives the true generating function without redundant terms, and this only needs examination and, where possible, simplification. Such simplification is apparently always possible when the lattice is complete. Moreover, there is the task of exhibiting $L(\infty; a_1, a_2, a_3, \ldots)$ as a product of outer and inner lattice functions and of finding the algebraic expression of $L_s(\infty; a_1, a_2, a_3, \ldots)$. There is an important and quite general property of the lattice function which must now be explained. If a lattice be read by columns instead of by rows its specification changes from a partition to the conjugate partition, and it is a trivial remark that the generating function of partitions at the nodes is not altered. In fact, if the rows possess $a_1, a_2, \ldots, a_m$ nodes and the columns $b_1, b_2, \ldots, b_n$ nodes

$$\mathrm{GF}(l; a_1, a_2, \ldots, a_m) = \mathrm{GF}(l; b_1, b_2, \ldots, b_n).$$

Moreover, since the generating function is the quotient of the lattice function by an algebraic function which depends merely upon the number of nodes, it is clear that

$$L(\infty; a_1, a_2, \ldots, a_m) = L(\infty; b_1, b_2, \ldots, b_n),$$

$$L(l; a_1, a_2, \ldots, a_m) = L(l; b_1, b_2, \ldots, b_n),$$

$(a_1, a_2, \ldots, a_m)$ and $(b_1, b_2, \ldots, b_n)$ being conjugate partitions.

From the last written relation we find

$$(\mathfrak{l}+1)\ldots(\mathfrak{l}+\Sigma\mathfrak{a})+(\mathfrak{l})\ldots(\mathfrak{l}+\Sigma\mathfrak{a}-1)\,L_1(\infty; a_1, a_2, \ldots, a_m)$$
$$+(\mathfrak{l}-1)\ldots(\mathfrak{l}+\Sigma\mathfrak{a}-2)\,L_2(\infty; a_1, a_2, \ldots, a_m)+\ldots$$
$$= (\mathfrak{l}+1)\ldots(\mathfrak{l}+\Sigma\mathfrak{a})+(\mathfrak{l})\ldots(\mathfrak{l}+\Sigma\mathfrak{a}-1)\,L_1(\infty; b_1, b_2, \ldots, b_n)$$
$$+(\mathfrak{l}-1)\ldots(\mathfrak{l}+\Sigma\mathfrak{a}-2)\,L_2(\infty; b_1, b_2, \ldots, b_n)+\ldots$$

Putting herein $l = 1, 2, \ldots$ in succession, we establish that

$$L_s(\infty; a_1, a_2, \ldots, a_m) = L_s(\infty; b_1, b_2, \ldots, b_n),$$

and thence

$$L_s(l; a_1, a_2, \ldots, a_m) = L_s(l; b_1, b_2, \ldots, b_n),$$

proving that the sub-lattice functions also do not change in passing from a lattice to the conjugate lattice.

N 2

As a rule; with some exceptions, the inner lattice function changes in passing from a lattice to its conjugate.

Thus it will be found that

$$\mathrm{IL}(\infty\,;\,22221)=\frac{(5)}{(2)}\,\mathrm{IL}(\infty\,;\,54),$$

(22221) and (54) being conjugate partitions.

Exceptionally, if it be proved that the inner lattice function of a complete lattice is unity, the function obviously does not change on passing to the conjugate lattice.

Another exception *appears* to be

$$\mathrm{IL}(\infty\,;\,m1^n)=\mathrm{IL}(\infty\,;\,n+1\,.\,1^{n-1}),$$

$m1^n$ and $n+1\,.\,1^{m-1}$ being conjugate partitions and there may be others.

Art. 16. My next object is to obtain the lattice function for $l=\infty$ which appertains to a lattice of two unequal rows and to find the form of the inner lattice function.

The first step is to establish the relation

$$\mathrm{L}(\infty\,;\,ab)=\mathrm{L}(\infty\,;\,a,b-1)+x^{a+b-1}\mathrm{L}(\infty\,;\,a-1,b-1)$$
$$+x^{a+b-2}\mathrm{L}(\infty\,;\,a-2,b-1)+\ldots+x^{2b}\mathrm{L}(\infty\,;\,b,b-1).$$

Consider the Greek letter succession $\alpha^a\beta^b$, where $a\geqq b$.

The whole of the permutations derived from the lattice terminate in one of the following ways

$$\beta\,;\ \beta\,|\,\alpha\,;\ \beta\,|\,\alpha^2\,;\ \ldots\ \beta\,|\,\alpha^{a-b},$$

since α cannot occur more than $a-b$ times at the end of the permutation by reason of the fundamental property of a permutation. Permutations which terminate in the manner $\beta\,|\,\alpha^s$ where $s>0$ clearly give rise to a factor x^{a+b-s} in the associated powers of x; the other factor will be due to all of the permutations of the succession $\alpha^{a-s}\beta^b$ *which terminate with* β; that is to say, the other factor will be

$$\mathrm{L}(\infty\,;\,a-s,b-1).$$

Hence

$$\mathrm{L}(\infty\,;\,ab)=\mathrm{L}(\infty\,;\,a,b-1)+\sum_1^{a-b}x^{a+b-s}\mathrm{L}(\infty\,;\,a-s,b-1),$$

as was to be shown.

Now assume the truth of the relation

$$\mathrm{L}(\infty\,;\,as)=\frac{(1)(2)\ldots(a+s)}{(2)(3)\ldots(a+1)\,.\,(1)(2)\ldots(s)}\cdot\frac{x^{s+1}(a-s)+(1)}{(1)},$$

when $s = b-1$, for all values of a. Then

$$\mathrm{L}(\infty\,;\,ab) = \frac{(1)(2)\ldots(a+b-1)}{(2)(3)\ldots(a+1)\,.\,(1)(2)\ldots(b-1)}\cdot\frac{x^b(a-b+1)+(1)}{(1)}$$
$$+x^{a+b-1}\frac{(1)(2)\ldots(a+b-2)}{(2)(3)\ldots(a)\,.\,(1)(2)\ldots(b-1)}\cdot\frac{x^b(a-b)+(1)}{(1)}$$
$$+\ \ldots$$
$$+x^{2b}\frac{(1)(2)\ldots(2b-1)}{(2)(3)\ldots(b+1)\,.\,(1)(2)\ldots(b-1)}\cdot\frac{x^b(1)+(1)}{(1)}.$$

The right-hand side has $a-b+1$ terms; assume that the sum of the last p terms may be written

$$x^{2b}\frac{(1)(2)\ldots(2b+p-1)}{(1)(2)\ldots(b+p)\,.\,(1)(2)\ldots(b)}\cdot(p),$$

an assumption which is obviously justified when $p = 1$; then the sum of the last $p+1$ terms ($p \not> a-b$) is

$$x^{2b}\frac{(1)(2)\ldots(2b+p-1)(p)}{(1)(2)\ldots(b+p)\,.\,(1)(2)\ldots(b)}+x^{2b+p}\frac{(1)(2)\ldots(2b+p-1)\{x^b(p+1)+(1)\}}{(1)(2)\ldots(b+p-1)\,.\,(1)(2)\ldots(b-1)};$$

and this on simplification proves to be

$$x^{2b}\frac{(1)(2)\ldots(2b+p)}{(1)(2)\ldots(b+p+1)\,.\,(1)(2)\ldots(b)}(p+1),$$

which is a justification of the assumption. Hence the right-hand side of the expression of $\mathrm{L}(\infty\,;\,ab)$ is, leaving out the first term,

$$x^{2b}\frac{(1)(2)\ldots(a+b-1)}{(1)(2)\ldots(a)\,.\,(1)(2)\ldots(b)}(a-b);$$

leading to

$$\mathrm{L}(\infty\,;\,ab) = \frac{(1)(2)\ldots(a+b-1)}{(2)(3)\ldots(a+1)\,.\,(1)(2)\ldots(b-1)}\cdot\frac{x^b(a-b+1)+(1)}{(1)}$$
$$+x^{2b}\frac{(1)(2)\ldots(a+b-1)}{(1)(2)\ldots(a)\,.\,(1)(2)\ldots(b)}(a-b)$$
$$= \frac{(1)(2)\ldots(a+b)}{(2)(3)\ldots(a+1)\,.\,(1)(2)\ldots(b)}\cdot\frac{x^{b+1}(a-b)+(1)}{(1)}.$$

This result, being true when $b = 0$, is thus established universally. The outer function is of the required form, and the inner function is

$$\mathrm{IL}(\infty\,;\,ab) = \frac{x^{b+1}(a-b)+(1)}{(1)} = 1+x^{b+1}\frac{(a-b)}{(1)}.$$

This leads to the new result

$$\mathrm{GF}\,(\,\infty;\,ab) = \frac{x^{b+1}\,(\mathbf{a-b})+(\mathbf{1})}{(\mathbf{1})\,(\mathbf{2})\,\ldots\,(\mathbf{a+1})\,.\,(\mathbf{1})\,(\mathbf{2})\,\ldots\,(\mathbf{b})},$$

and as a particular case,

$$\mathrm{GF}\,(\,\infty;\,nn) = \mathrm{GF}\,(\,\infty,\,2,\,n) = \frac{1}{(\mathbf{1})\,\{(\mathbf{2})\,(\mathbf{3})\,\ldots\,(\mathbf{n})\}^2\,(\mathbf{n+1})},$$

a result already known.

Art. 17. The determination of $\mathrm{L}\,(\,\infty;\,abc)$ presents great difficulties, so that the investigation proceeds in the path of least resistance. When the lattice is complete, the Greek letter succession is conveniently taken to be

$$\alpha_1^{\,n}\alpha_2^{\,n}\,\ldots\,\alpha_m^{\,n}.$$

It is clear that each permutation, that arises from the lattice, must terminate with α_m; hence this latter may be always deleted, and we find

$$\mathrm{L}\,(\,\infty;\,n^{m-1},\,n-1) = \mathrm{L}\,(\,\infty;\,n^m)$$

and the sub-lattice functions are also equal, but the inner lattice functions differ; thus it will be found that

$$\mathrm{IL}\,(\,\infty;\,nn) = 1 \qquad \text{but} \qquad \mathrm{IL}\,(\,\infty;\,n,\,n-1) = 1+x^n.$$

The Sub-lattice Functions.

Art. 18. It is necessary to inquire as to the highest order of sub-lattice function that presents itself. For a lattice of m rows and n columns I form the rectangular scheme

$$\begin{array}{ccccc} \alpha_1 & \alpha_1 & \alpha_1 & \ldots & \alpha_1 \\ \alpha_2 & \alpha_2 & \alpha_2 & \ldots & \alpha_2 \\ \nearrow\;\alpha_3 & \alpha_3 & \alpha_3 & \ldots & \alpha_3 \\ \cdot & \cdot & \cdot & \ldots & \cdot \\ \alpha_m & \alpha_m & \alpha_m & \ldots & \alpha_m \end{array}$$

where there are n columns.

Reading this parallel to the arrow (inclined at 45 degrees), commencing with the left-hand top corner, I obtain the permutation

$$\alpha_1,\,\alpha_2\,|\,\alpha_1,\,\alpha_3\,|\,\alpha_2\,|\,\alpha_1,\,\alpha_4\,|\,\alpha_3\,|\,\alpha_2\,|\,\alpha_1\,\ldots\,\alpha_m\,|\,\alpha_{m-1}\,|\,\alpha_{m-2},\,\alpha_m\,|\,\alpha_{m-1},\,\alpha_m.$$

This is the permutation which involves the maximum number of dividing lines and corresponds to the sub-lattice function of highest order; the permutation is unique, yielding a single power of x, which is the sub-lattice function in question. The dividing lines may be counted.

Since n^m and m^n are conjugate partitions, we may take $n \geqq m$ without loss of generality. The number of dividing lines is

$$1+2+3+\ldots+m-2+(n-m+1)(m-1)+(m-2)+\ldots+3+2+1,$$
$$=(n-1)(m-1).$$

Hence we have sub-lattice functions of all orders from zero to $(n-1)(m-1)$. It will be observed that the permutation, above written, possesses symmetry in that it is unchanged by writing α_{m-s+1} for α_s and inverting the order.

The same method is applicable to the determination of the maximum number of dividing lines appertaining to permutations derived from an incomplete lattice. Thus, if the letters be $\alpha_1^3\alpha_2^2\alpha_3$,

$$\nearrow \quad \begin{array}{ccc} \alpha_1 & \alpha_1 & \alpha_2 \\ \alpha_2 & \alpha_2 & \\ \alpha_3 & & \end{array}$$

the reading parallel to the arrow gives

$$\alpha_1,\ \alpha_2 \,|\, \alpha_1,\ \alpha_3 \,|\, \alpha_2 \,|\, \alpha_1.$$

The highest order of sub-lattice functions when the letters are

$$\alpha_1^{n_1}\alpha_2^{n_2}\ldots\alpha_m^{n}$$

will be found to have higher and lower limits $\Sigma n-n_1$ and $\Sigma n-n_1-m+1$ respectively, the actual value depending upon the magnitudes of $n_1, n_2, \ldots, n_m$. The lower limit is the actual value when the lattice is complete.

Art. 19. The next point is the determination of the expression of $L_{(n-1)(m-1)}(\infty; n^m)$, or of $L_{(n-1)(m-1)}(\infty, m, n)$ as it may be also written.

The dividing lines occur in groups—

(i) In $m-2$ groups, containing $1, 2, \ldots, m-2$ lines respectively;

(ii) In $n-m+1$ groups, each containing $m-1$ lines;

(iii) In $m-2$ groups, containing $m-2, m-1, \ldots, 2, 1$ lines respectively.

Let the exponent of x sought be $\pi_1+\pi_2+\pi_3$; π_1, π_2, π_3, corresponding to (i), (ii), and (iii), respectively.

$$\begin{aligned}\pi_1 &= 2+(4+5)+(7+8+9)+\ldots+\{\tfrac{1}{2}(m^2-3m+4)+\ldots+\tfrac{1}{2}(m^2-m-2)\},\\ &= \tfrac{1}{2}(1\,.\,2^2+2\,.\,3^2+3\,.\,4^2+\ldots \text{ to } m-2 \text{ terms}),\\ &= \tfrac{1}{8}(m-2)^2(m-1)^2+\tfrac{1}{6}(m-2)(m-1)(2m-3)+\tfrac{1}{4}(m-2)(m-1),\\ &= \tfrac{1}{24}(m)(m-1)(m-2)(3m-1).\end{aligned}$$

$$\begin{aligned}\pi_2 &= \tfrac{1}{2}(m-1)m^2+\tfrac{1}{2}(m-1)m(m+2)+\tfrac{1}{2}(m-1)m(m+4)+\ldots \text{ to } n-m+1 \text{ terms},\\ &= \tfrac{1}{2}(m-1)mn(n-m+1).\end{aligned}$$

$$\begin{aligned}\pi_3 &= (mn-2)+(mn-4+mn-5)+(mn-7+mn-8+mn-9)+\ldots \text{ to } m-2 \text{ terms},\\ &= \tfrac{1}{2}mn(m-1)(m-2)-\tfrac{1}{24}m(m-1)(m-2)(3m-1).\end{aligned}$$

Whence

$$\pi_1+\pi_2+\pi_3 = \tfrac{1}{2} n(n-1)\, m(m-1);$$

and

$$L_{(n-1)(m-1)}(\infty, m, n) = x^{\frac{1}{2}n(n-1)m(m-1)}.$$

If, in the succession

$$\alpha_1, \alpha_2 \mid \alpha_1, \alpha_3 \mid \alpha_2 \mid \alpha_1, \ldots, \alpha_m \mid \alpha_{m-1}, \alpha_m,$$

we fix upon any dividing line and arrange the letters to the right of it in alphabetical order, thus obliterating the lines to the right of the one fixed upon, we obtain a permutation involving (suppose) s lines which yields x to the lowest power that occurs in the sub-lattice function of order s. When the lattice is complete we may, in any derived permutation, write α_{m-s+1} for α_s and invert the order, and we thus obtain another permutation belonging to the same sub-lattice function as the former. For a succession $\alpha_p \mid \alpha_q$ $p > q$ in the former becomes by the stated operations $\alpha_{m-q+1} \mid \alpha_{m-p+1}$, where $m-q+1 > m-p+1$ in the latter; and if α_p is the kth letter from the left of the former permutation, α_{m-q+1} is the $mn-k^{\text{th}}$ letter from the left of the latter. Hence, if the power of x given by the former permutation be

$$x^{p_1+p_2+\ldots+p_s},$$

that given by the latter is

$$x^{mns-p_1-p_2-\ldots-p_s}.$$

Thus, for every term x^p in L_s, there is a corresponding term x^{mns-p}.

Hence we may say that L_s is centrically symmetrical both as regards the powers of x and the coefficients.

If e be the lowest power of x in L_s, determined as above, the highest power of x will be $mns-e$.

Ex. gr.,

$$\begin{aligned} L_0(\infty, 3, 3) &= 1, \\ L_1(\infty, 3, 3) &= x^2+2x^3+2x^4+2x^5+2x^6+x^7, \\ L_2 &= 2(x^6+x^7+2x^8+2x^9+2x^{10}+x^{11}+x^{12}), \\ L_3 &= x^{11}+2x^{12}+2x^{13}+2x^{14}+2x^{15}+x^{16}, \\ L_4 &= x^{18}; \end{aligned}$$

and it will be noted that

in L_1, x^k and x^{9-k}; in L_2, x^k and x^{18-k}; in L_3, x^k and x^{27-k};

occur in pairs, whilst the theorem is clearly verified in L_0 and in L_4.

The result of writing $\dfrac{1}{x}$ for x in L_s is the acquisition of the factor x^{-mns} by L_s.

Art. 20. Let e_s and f_s be the least and greatest exponents of x that occur in the expression of $L_s(\infty, m, n)$. Consider again the permutation

$$\alpha_1, \alpha_2 \mid \alpha_1, \alpha_3 \mid \alpha_2 \mid \alpha_1, \ldots, \alpha_m \mid \alpha_{m-1}, \alpha_m;$$

e_s is the exponent of x due to the dividing lines when only the first s lines from the left are retained, the letters to the right of the sth line being arranged in alphabetical order. If $\mu = (m-1)(n-1)$ we know that $e_\mu = f_\mu = \frac{1}{2}mn\mu$. What is the relation between e_s and $e_{\mu-s}$? To obtain $e_{\mu-s}$ we must clearly obliterate the last s lines on the right and arrange the affected letters in alphabetical order. Since the number of letters is mn, if for e_s we retain s lines which give

$$e_s = p_1+p_2+\ldots+p_s,$$

we must for $e_{\mu-s}$ reject s lines of power values

$$mn-p_1,\ mn-p_2,\ \ldots,\ mn-p_s.$$

Hence

$$e_{\mu-s} = e_\mu - smn + e_s = e_s + \tfrac{1}{2}mn(\mu-2s);$$

and from the symmetry of the permutation we find also

$$f_{\mu-s} = f_s + \tfrac{1}{2}mn(\mu-2s);$$

so that

$$x^{e_{\mu-s}} + x^{f_{\mu-s}} = x^{1/2 mn(\mu-2s)}\left(x^{e_s}+x^{f_s}\right);$$

an interesting result which foreshadows the theorem

$$L_{\mu-s} = x^{1/2 mn(\mu-2s)}\, L_s.$$

The circumstance that the lattice function, when the lattice is complete, involves x to the power $\frac{1}{2}mn(m-1)(n-1)$, which is the greatest exponent of x that occurs in the outer function, is consistent with the inner function being simply unity.

Art. 21. I will now investigate an expression for $L_1(\infty, m, n)$.

Suppose that a certain power of x arises therein from the conjunction α_v/α_u, where $m \geqq v > u$, and let the Greek letter succession be

$$\text{A})\, \alpha_v \,|\, \alpha_u\, (\text{B},$$

where in the space A there is any suitable succession of letters in ascending order (of subscripts) to v, and in the space B any suitable succession such that the subscripts are in ascending order from u.

The least power of x is obtained when in the space A there is the succession

$$\alpha_1{}^n \alpha_2{}^n \ldots \alpha^n{}_{u-1} \alpha_u \alpha_{u+1} \ldots \alpha_{v-1}.$$

This gives the term $x^{v+(n-1)u-n+1}$.

The greatest power arises when in the space A is

$$\alpha_1{}^n \alpha_2{}^n \ldots \alpha^n{}_{u-1} \alpha_u{}^{n-1} \alpha_{u+1}^{n-1} \ldots \alpha_{v-1}^{n-1} \alpha_v{}^{n-2};$$

and this gives the term $x^{(n-1)v+u-1}$.

It must be remembered, in assigning these successions to the space A, that only one dividing line is to be in the whole permutation, and that the latter must possess the fundamental property which is the attribute of all such.

Writing the above succession

$$\alpha_1{}^n\alpha_2{}^n\ldots\alpha^n{}_{u-1}\alpha_u\alpha_{u+1}\ldots\alpha_{v-1}\alpha_v\,|\,\alpha_u\ldots,$$

$$\alpha_1{}^n\alpha_2{}^n\ldots\alpha^n{}_{u-1}\alpha_u{}^{n-1}\alpha_{u+1}^{n-1}\ldots\alpha_{v-1}^{n-1}\alpha_v{}^{n-1}\,|\,\alpha_u\ldots,$$

we have to determine all of the successions ranging from

$$\alpha_u\alpha_{u+1}\ldots\alpha_{v-1}\alpha_v \quad \text{to} \quad \alpha_u{}^{n-1}\alpha_{u+1}^{n-1}\ldots\alpha_{v-1}^{n-1}\alpha_v{}^{n-1}$$

that may be placed between $\alpha_1{}^n\alpha_2{}^n\ldots\alpha^n{}_{u-1}$ and the dividing line in order to form an L_1 succession involving $\alpha_v\,|\,\alpha_u$.

Since u, $u+1$, ..., v is a succession of $v-u+1$ numbers in ascending order we are clearly concerned with the whole of the partitions at the points of a one-row lattice of $v-u+1$ nodes which are such that the part magnitude lies between $n-2$ and zero.

If $g_1g_2\ldots g_{v-u+1}$ be one such partition

$$\Sigma x^{\Sigma g} = \frac{(\mathbf{n-1})\,(\mathbf{n})\ldots(\mathbf{n+v-u-1})}{(\mathbf{1})\,(\mathbf{2})\ldots(\mathbf{v-u+1})}.$$

Denote by $L_{1,vu}$ that portion of L_1 which is associated with the conjunction $\alpha_v\,|\,\alpha_u$; then

$$L_{1,vu} = x^{v+(n-1)u-n+1}\frac{(\mathbf{n-1})\,(\mathbf{n})\ldots(\mathbf{n+v-u-1})}{(\mathbf{1})\,(\mathbf{2})\ldots(\mathbf{v-u+1})};$$

leading to

$$L_1(\infty, m, n) = \sum_v\sum_u x^{v+(n-1)u-n+1}\frac{(\mathbf{n-1})\,(\mathbf{n})\ldots(\mathbf{n+v-u-1})}{(\mathbf{1})\,(\mathbf{2})\ldots(\mathbf{v-u+1})}.$$

Put herein $v = u+j$, then, for a constant value of j,

$$L_1(\infty, m, n)_j = \{x^{j+1}+x^{j+n+1}+\ldots+x^{nm-(n-1)(j+1)}\}\frac{(\mathbf{n-1})\,(\mathbf{n})\ldots(\mathbf{n+j-1})}{(\mathbf{1})\,(\mathbf{2})\ldots(\mathbf{j+1})}$$

$$= x^{j+1}\frac{(\mathbf{nm-nj})}{(\mathbf{n})}\cdot\frac{(\mathbf{n-1})\,(\mathbf{n})\ldots(\mathbf{n+j-1})}{(\mathbf{1})\,(\mathbf{2})\ldots(\mathbf{j+1})};$$

and, consequently, giving j all values from 1 to $m-1$,

$$\begin{aligned} L_1(\infty, m, n) = {} & x^2\frac{(\mathbf{nm-n})}{(\mathbf{n})}\cdot\frac{(\mathbf{n-1})\,(\mathbf{n})}{(\mathbf{1})\,(\mathbf{2})} \\ & +x^3\frac{(\mathbf{nm-2n})}{(\mathbf{n})}\cdot\frac{(\mathbf{n-1})\,(\mathbf{n})\,(\mathbf{n+1})}{(\mathbf{1})\,(\mathbf{2})\,(\mathbf{3})} \\ & +\ldots \\ & +x^m\frac{(\mathbf{n})}{(\mathbf{n})}\cdot\frac{(\mathbf{n-1})\,(\mathbf{n})\ldots(\mathbf{n+m-2})}{(\mathbf{1})\,(\mathbf{2})\ldots(\mathbf{m})}. \end{aligned}$$

Art. 22. It is not, I think, immediately obvious that this series can be effectively summed. That this is, in fact, the case appears when the problem is looked at from another point of view.

I shall show that

$$L_1(\infty, m, n) = \frac{(\mathbf{n+1})(\mathbf{n+2})\ldots(\mathbf{n+m})}{(\mathbf{1})(\mathbf{2})\ldots(\mathbf{m})} - \frac{(\mathbf{mn+1})}{(\mathbf{1})};$$

for suppose that the permutation

$$\alpha_1^{k_1}\alpha_2^{k_2}\ldots\alpha_m^{k_m} \mid \alpha_1^{n-k_1}\alpha_2^{n-k_2}\ldots\alpha_m^{n-k_m}$$

gives rise to the term $x^{\Sigma k}$, where $(k_1k_2\ldots k_m)$ is a one-row partition of Σk. The part to the left of the dividing line may be as small as $\alpha_1\alpha_2$, and the part to the right as small as $\alpha_{m-1}\alpha_m$; hence Σk has values ranging from 2 to $mn-2$; the partition $(k_1k_2\ldots k_m)$ has parts limited in magnitude to n and in number to m; and if there were no deductions from the total of partitions the value of $L_1(\infty, m, n)$ would be

$$\frac{(\mathbf{n+1})(\mathbf{n+2})\ldots(\mathbf{n+m})}{(\mathbf{1})(\mathbf{2})\ldots(\mathbf{m})};$$

but there must be deductions, because every partition of the form $n^i j$ must be absent; for the corresponding succession of letters is

$$\alpha_1^n\alpha_2^n\ldots\alpha_i^n\alpha_{i+1}^j;$$

and a dividing line cannot be placed after this succession because every letter prior to α_{i+1} has *already* appeared to the left of α_{i+1}^j. Of these omitted partitions there is one, and only one, of each weight, viz. :—

For

$$\begin{array}{ll} i = 0 & 1,\ \alpha_1,\ \alpha_1^2, \ldots \alpha_1^n, \\ i = 1 & \alpha_1^n\alpha_2,\ \alpha_1^n\alpha_2^2, \ldots \alpha_1^n\alpha_2^n, \\ i = 2 & \alpha_1^n\alpha_2^n\alpha_3,\ \alpha_1^n\alpha_2^n\alpha_3^2, \ldots \alpha_1^n\alpha_2^n\alpha_3^n, \\ i = m-1 & \alpha_1^n\alpha_2^n\ldots\alpha_{m-1}^n\alpha_m, \ldots \alpha_1^n\alpha_2^n\ldots\alpha_m^n. \end{array}$$

Hence we must subtract

$$\frac{(\mathbf{mn+1})}{(\mathbf{1})};$$

and

$$L_1(\infty, m, n) = \frac{(\mathbf{n+1})(\mathbf{n+2})\ldots(\mathbf{n+m})}{(\mathbf{1})(\mathbf{2})\ldots(\mathbf{m})} - \frac{(\mathbf{mn+1})}{(\mathbf{1})}.$$

Since $L_0 = 1$, we have also

$$L_0(\infty, m, n) + L_1(\infty, m, n) = \frac{(\mathbf{n+1})(\mathbf{n+2})\ldots(\mathbf{n+m})}{(\mathbf{1})(\mathbf{2})\ldots(\mathbf{m})} - x\frac{(\mathbf{mn})}{(\mathbf{1})}.$$

o 2

The Fundamental Relation.

Art. 23. We must now examine the fundamental relation

$$\mathrm{GF}(l, m, n) = \frac{(\mathfrak{l}+1)(\mathfrak{l}+2)\ldots(\mathfrak{l}+\mathfrak{mn})\,L_0 + (\mathfrak{l})(\mathfrak{l}+1)\ldots(\mathfrak{l}+\mathfrak{mn}-1)\,L_1 + \ldots + (\mathfrak{l}-\mu+1)(\mathfrak{l}-\mu+2)\ldots(\mathfrak{l}-\mu+\mathfrak{mn})\,L_\mu}{(1)(2)\ldots(\mathfrak{mn})}$$

$$= \frac{l_0L_0+l_1L_1+\ \ldots\ +l_\mu L_\mu}{(1)(2)\ldots(\mathfrak{mn})}$$

for brevity, where

$$l_s = (\mathfrak{l}-\mathfrak{s}+1)(\mathfrak{l}-\mathfrak{s}+2)\ldots(\mathfrak{l}-\mathfrak{s}+\mathfrak{mn})$$

and

$$\mu = (m-1)(n-1).$$

When a partition enumerated by GF (l, m, n) is represented graphically by nodes in three dimensions, we see that the nodes form a portion of a parallelopiped of nodes, the sides having l, m, n nodes respectively; the unoccupied nodes graphically represent another partition of the number $lmn-w$ if the former partition be of the number w. Hence, if GF (l, m, n) be F (x), we have (writing Co as short for coefficient), Co x^w in F (x) equal to Co x^{lmn-w} in F (x) or equal to Co x^w in $x^{lmn}\mathrm{F}\left(\frac{1}{x}\right)$. From which it appears that $\mathrm{F}\left(\frac{1}{x}\right) = x^{-lmn}\mathrm{F}(x)$, and this property may be directly verified in the fundamental relation by means of the formulæ

$$(-\mathfrak{s}) = -x^{-s}(\mathfrak{s})\,;\quad L_s\left(\frac{1}{x}\right) = x^{-smn}L_s(x).$$

From another point of view we may suppose the nodes of the lattice of m rows and n columns to be all occupied by parts, zero being taken as a part, and then if we diminish each part by l, we obtain a partition of the negative integer $-(lmn-w)$ into negative parts $0, -1, -2, \ldots -l$; the effect upon the generating function F (x) is alternatively to substitute $\frac{1}{x}$ for x or to divide it by x^{lmn}. It will be noted that in this respect $L_s(\infty, m, n)$ possesses the same property as GF (s, m, n).

The numerator function $l_0L_0+l_1L_1+\ \ldots\ +l_\mu L_\mu$ has the factor

$$(\mathfrak{l}+1)(\mathfrak{l}+2)\ldots(\mathfrak{l}+\mathfrak{m}+\mathfrak{n}-1)$$

which stands as a determined factor of the generating function.

Writing

$$l_s = (\mathfrak{l}+1)(\mathfrak{l}+2)\ldots(\mathfrak{l}+\mathfrak{m}+\mathfrak{n}-1)\,l_s',$$

$$l_s' = (\mathfrak{l}-\mathfrak{s}+1)(\mathfrak{l}-\mathfrak{s}+2)\ldots(\mathfrak{l})(\mathfrak{l}+\mathfrak{m}+\mathfrak{n})(\mathfrak{l}+\mathfrak{m}+\mathfrak{n}+1)\ldots(\mathfrak{l}-\mathfrak{s}+\mathfrak{mn})\,;$$

l_s' involving $(m-1)(n-1)$ or μ bracket factors.

I observe at this point that the substitution of $-l-m-n$ for l converts the factor $(\mathfrak{l}+1)(\mathfrak{l}+2)\ldots(\mathfrak{l}+\mathfrak{m}+\mathfrak{n}-1)$ into

$$(-)^{m+n-1}x^{-(m+n-1)(l+\frac{1}{2}m+\frac{1}{2}n)}(\mathfrak{l}+1)(\mathfrak{l}+2)\ldots(\mathfrak{l}+\mathfrak{m}+\mathfrak{n}-1),$$

so that it is unchanged except as to sign and a power of x *près*.

Art. 24. Some particular cases of the fundamental relation are instructive. Thus

$$\mathrm{GF}(l, 2, 2) = \frac{(\mathfrak{l}+1)\ldots(\mathfrak{l}+4)+(\mathfrak{l})\ldots(\mathfrak{l}+3)\,x^2}{(1)\ldots(4)},$$

$$= \frac{(\mathfrak{l}+1)(\mathfrak{l}+2)^2(\mathfrak{l}+3)}{(1)(2)^2(3)} = |\mathrm{LM}|_1|\mathrm{LM}|_2 \quad \text{when} \quad m = 2.$$

Observe that

$$(\mathfrak{l}+1)\ldots(\mathfrak{l}+4)+(\mathfrak{l})\ldots(\mathfrak{l}+3)\,x^2 = (\mathfrak{l}+1)(\mathfrak{l}+2)(\mathfrak{l}+3)\{(\mathfrak{l}+4)+(\mathfrak{l})\,x^2\},$$

$$= (\mathfrak{l}+1)(\mathfrak{l}+2)^2(\mathfrak{l}+3)\,\mathrm{L}(\infty, 2, 2),$$

showing that $\mathrm{L}(\infty, 2, 2)$ is a factor of the numerator.

It appears that in general $\mathrm{L}(\infty, m, n)$ is a factor of the numerator. Thus

$$\mathrm{GF}(l, 3, 3) = \frac{(\mathfrak{l}+1)(\mathfrak{l}+2)^2(\mathfrak{l}+3)^3(\mathfrak{l}+4)^2(\mathfrak{l}+5)\,\mathrm{L}(\infty, 3, 3)}{(1)(2)\ldots(9)},$$

and since

$$\mathrm{L}(\infty, 3, 3) = \frac{(6)(7)(8)(9)}{(2)(3)^2(4)}$$

$$\mathrm{GF}(l, 3, 3) = \frac{(\mathfrak{l}+1)(\mathfrak{l}+2)^2(\mathfrak{l}+3)^3(\mathfrak{l}+4)^2(\mathfrak{l}+5)}{(1)(2)^2(3)^3(4)^2(5)}$$

$$= |\mathrm{LM}|_1|\mathrm{LM}|_2|\mathrm{LM}|_3 \quad \text{when} \quad m = 3.$$

Art. 25. I have arrived at the expression for $\mathrm{GF}(l, 2, n)$ in the following manner. We have

$$\mathrm{GF}(l, 2, n) = \sum_{s=0}^{s=n-1} \frac{(\mathfrak{l}-\mathfrak{s}+1)\ldots(\mathfrak{l}-\mathfrak{s}+2\mathfrak{n})}{(1)\ldots(2\mathfrak{n})}\,\mathrm{L}_s(\infty, 2, n),$$

and I determine $\mathrm{L}_s(\infty, 2, n)$ from the Greek letter succession; for suppose $s = 3$ and a succession to be

$$\alpha^{p_1+1}\beta^{q_1+1}\,|\,\alpha^{p_2+1}\beta^{q_2+1}\,|\,\alpha^{p_3+1}\beta^{q_3+1}\,|\,\alpha^{p_4+1}\beta^{q_4+1},$$

where p_1, p_2, p_3, p_4; q_1, q_2, q_3, q_4 may have all integer (including zero) values subject to the conditions

$$\Sigma p = \Sigma q = n-4;$$

$$p_1 \geq q_1;\ p_1+p_2 \geq q_1+q_2;\ p_1+p_2+p_3 \geq q_1+q_2+q_3.$$

A proper permutation with three dividing lines is thus secured, and we have to perform the summation

$$\Sigma x^{3(p_1+q_1)+2(p_2+q_2)+p_3+q_3+12},$$

which is the expression of $\mathrm{L}_3(\infty, 2, n)$.

But I have shown in Part II. of this Memoir that

$$\Sigma x^{3(p_1+q_1)+2(p_2+q_2)+p_3+q_3}$$

subject to the stated conditions is GF $(n-4, 2, 3)$.

In general I thus establish that

$$L_s(\infty, 2, n) = x^{s(s+1)}\text{GF}(n-s-1, 2, s);$$

so that I am able to write

$$\text{GF}(l, 2, n) = \sum_{s=0}^{s=n-1} x^{s(s+1)} \frac{(\mathfrak{l}-\mathfrak{s}+1)\dots(\mathfrak{l}-\mathfrak{s}+2\mathfrak{n})}{(1)\dots(2\mathfrak{n})}\text{GF}(n-s-1, 2, s);$$

the expression of GF $(l, 2, n)$ is thus made to depend upon the expression of GF $(l', 2, n')$, where l', n' have all values such that

$$l'+n' = n-1.$$

Now assume

$$\text{GF}(l', 2, n') = \frac{(\mathfrak{l}'+1)\{(\mathfrak{l}'+2)\dots(\mathfrak{l}'+\mathfrak{n}')\}^2(\mathfrak{l}'+\mathfrak{n}'+1)}{(1)\{(2)\dots(\mathfrak{n}')\}^2(\mathfrak{n}'+1)},$$

so that

$$\text{GF}(l, 2, n) = \sum_{0}^{n-1} x^{s(s+1)} \frac{(\mathfrak{l}-\mathfrak{s}+1)\dots(\mathfrak{l}-\mathfrak{s}+2\mathfrak{n})}{(1)\dots(2\mathfrak{n})}\cdot\frac{(\mathfrak{n}-\mathfrak{s})\{(\mathfrak{n}-\mathfrak{s}+1)\dots(\mathfrak{n}-1)\}^2(\mathfrak{n})}{(1)\{(2)\dots(\mathfrak{s})\}^2(\mathfrak{s}+1)},$$

or

$$\begin{aligned}\frac{(1)\dots(2\mathfrak{n})}{(\mathfrak{l}+1)\dots(\mathfrak{l}+\mathfrak{n}+1)}\text{GF}(l, 2, n) &= (\mathfrak{l}+\mathfrak{n}+2)\dots(\mathfrak{l}+2\mathfrak{n})\\ &+(\mathfrak{l})(\mathfrak{l}+\mathfrak{n}+2)\dots(\mathfrak{l}+2\mathfrak{n}-1)\,x^2\frac{(\mathfrak{n}-1)(\mathfrak{n})}{(1)(2)}\\ &+(\mathfrak{l}-1)(\mathfrak{l})(\mathfrak{l}+\mathfrak{n}+2)\dots(\mathfrak{l}+2\mathfrak{n}-2)\,x^6\frac{(\mathfrak{n}-2)(\mathfrak{n}-1)^2(\mathfrak{n})}{(1)(2)^2(3)}\\ &+\dots\\ &+(\mathfrak{l}-\mathfrak{n}+2)\dots(\mathfrak{l})\,x^{(n-1)n}.\end{aligned}$$

It is easy to show that $(\mathfrak{l}+2)$, $(\mathfrak{l}+3)$, ... $(\mathfrak{l}+\mathfrak{n})$ are each of them factors of the right-hand side; the remaining factor is free from l, and is, in fact, the lattice function $L(\infty, 2, n)$.

Giving l to the special value zero, we readily find

$$L(\infty, 2, n) = \frac{(\mathfrak{n}+2)\dots(2\mathfrak{n})}{(2)\dots(\mathfrak{n})},$$

and thence

$$\text{GF}(l, 2, n) = \frac{(\mathfrak{l}+1)\{(\mathfrak{l}+2)\dots(\mathfrak{l}+\mathfrak{n})\}^2(\mathfrak{l}+\mathfrak{n}+1)}{(1)\{(2)\dots(\mathfrak{n})\}^2(\mathfrak{n}+1)},$$

as was to be shown.

This proof rests upon the assumption that the law can be shown to hold for GF $(n-s-1, 2, s)$ for all values of s from 0 to $n-1$, and for all values of n. Now suppose the law to hold for GF $(l, 2, \nu)$ for all values of ν inferior to n; then it obviously holds for GF $(n-s-1, 2, s)$ for all values of s inferior to n, and thence, as has just been proved, it holds for GF $(l, 2, n)$; but the law does hold when $\nu = 0$ or 1, and thence by induction the law holds in general. This method of proof seems to be of application only when $m = 2$, for then only can the function $L_s(\infty, 2, n)$ be identified with a form GF $(l', 2, n')$ where the sum of $l'+n'$ is less than n.

Art. 26. I turn again to the relation

$$\mathrm{GF}(l, m, n) = \frac{l_0 L_0 + l_1 L_1 + \dots + l_\mu L_\mu}{(1)(2)\dots(\mathrm{mn})}$$

in order to establish relations between the functions GF (l, m, n) and the sub-lattice functions $L_s(\infty, m, n)$. The relation, as it stands, exhibits GF (l, m, n) as a linear function of the sub-lattice functions, but giving l the special values $0, 1, 2, \dots$ in succession, we obtain

$$\mathrm{GF}(0, m, n) = L_0(\infty, m, n) = 1,$$

$$\mathrm{GF}(1, m, n) = \frac{(\mathrm{mn}+1)}{(1)} + L_1(\infty, m, n),$$

$$\mathrm{GF}(2, m, n) = \frac{(\mathrm{mn}+1)(\mathrm{mn}+2)}{(1)(2)} + \frac{(\mathrm{mn}+1)}{(1)} L_1 + L_2,$$

$$\dots\dots\dots\dots\dots\dots$$

$$\mathrm{GF}(\mu, m, n) = \frac{(\mathrm{mn}+1)\dots(\mathrm{mn}+\mu)}{(1)\dots(\mu)} + \frac{(\mathrm{mn}+1)\dots(\mathrm{mn}+\mu-1)}{(1)\dots(\mu-1)} L_1 + \dots\dots + L_\mu,$$

and thence

$$L_0 = 1,$$

$$L_1 = \mathrm{GF}(1, m, n) - \frac{(\mathrm{mn}+1)}{(1)},$$

$$L_2 = \mathrm{GF}(2, m, n) - \frac{(\mathrm{mn}+1)}{(1)} \mathrm{GF}(1, m, n) + x\frac{(\mathrm{mn})(\mathrm{mn}+1)}{(1)(2)},$$

$$\dots\dots\dots\dots\dots\dots$$

$$L_\mu = \mathrm{GF}(\mu, m, n) - \frac{(\mathrm{mn}+1)}{(1)} \mathrm{GF}(\mu-1, m, n) + x\frac{(\mathrm{mn})(\mathrm{mn}+1)}{(1)(2)} \mathrm{GF}(\mu-2, m, n),$$

$$- \dots\dots + (-)^k x^{1/2(k-1)k} \frac{(\mathrm{mn}-\mathrm{k}+2)\dots(\mathrm{mn}+1)}{(1)\dots(\mathrm{k})} \mathrm{GF}(\mu-k, m, n)$$

$$+ \dots$$

$$+ (-)^\mu x^{1/2(\mu-1)\mu} \frac{(\mathrm{mn}-\mu+2)\dots(\mathrm{mn}+1)}{(1)\dots(\mu)}.$$

Since $L_s = 0$, when $s > \mu$, the series may be continued,

$$0 = \mathrm{GF}(\mu+1, m, n) - \frac{(\mathrm{mn}+1)}{(1)}\mathrm{GF}(\mu, m, n) + \ldots + (-)^{\mu+1}x^{\frac{1}{2}\mu(\mu+1)}\frac{(\mathrm{mn}-\mu+1)\ldots(\mathrm{mn}+1)}{(1)\ldots(\mu+1)};$$

and for values of l ranging from $\mu+1$ to mn,

$$0 = \mathrm{GF}(l, m, n) - \frac{(\mathrm{mn}+1)}{(1)}\mathrm{GF}(l-1, m, n) + \ldots + (-)^{l}x^{\frac{1}{2}(l-1)l}\frac{(\mathrm{mn}-\mathrm{l}+1)\ldots(\mathrm{mn}+1)}{(1)\ldots(\mathrm{l}+1)},$$

the series having $l+1$ terms; but, when $l > mn$,

$$0 = \mathrm{GF}(l, m, n) - \frac{(\mathrm{mn}+1)}{(1)}\mathrm{GF}(l-1, m, n) + \ldots + (-)^{mn+1}x^{\frac{1}{2}mn(mn+1)}\mathrm{GF}(l-mn-1, m, n),$$

the series having $mn+2$ terms.

Art. 27. We have thus a number of difference equations satisfied by the functions $\mathrm{GF}(l, m, n)$, and we can now show that if

$$\mathrm{GF}(l, m, n) = |\mathrm{LM}|_1 |\mathrm{LM}|_2 \ldots |\mathrm{LM}|_n = \mathrm{J}(l, m, n)$$

for all values of l not exceeding mn, the law is true universally.

For $\mathrm{J}(l, m, n)$ is of the form

$$\mathrm{P}_0 - \mathrm{P}_1 x^l + \mathrm{P}_2 x^{2l} - \ldots (-)^{mn}\mathrm{P}_{mn}x^{lmn},$$

where the coefficients P are functions of x independent of l.

Then

$$\sum_0^\infty \mathrm{J}(l, m, n)\,\theta^l = \frac{\mathrm{P}_0}{1-\theta} - \frac{\mathrm{P}_1}{1-\theta x} + \frac{\mathrm{P}_2}{1-\theta x^2} - \ldots\ldots (-)^{mn}\frac{\mathrm{P}_{mn}}{1-\theta x^{mn}};$$

from which it appears that

$$(1-\theta)(1-\theta x)\ldots(1-\theta x^{mn})\sum_0^\infty \mathrm{J}(l, m, n)\,\theta^l$$

is of degree mn in θ, at most, and hence, when $l > mn$, since

$$(1-\theta)(1-\theta x)\ldots(1-\theta x^{mn}) = 1 - \theta\frac{(\mathrm{mn}+1)}{(1)} + \theta^2 x\frac{(\mathrm{mn})(\mathrm{mn}+1)}{(1)(2)} - \ldots\ldots$$

we have

$$\mathrm{J}(l, m, n) - \frac{(\mathrm{mn}+1)}{(1)}\mathrm{J}(l-1, m, n) + \ldots + (-)^{mn+1}x^{\frac{1}{2}mn(mn+1)}\mathrm{J}(l-mn-1, m, n) = 0;$$

but it has been shown that $l > mn$

$$\mathrm{GF}(l, m, n) - \frac{(\mathrm{mn}+1)}{(1)}\mathrm{GF}(l-1, m, n) + \ldots\ldots + (-)^{mn+1}x^{\frac{1}{2}mn(mn+1)}\mathrm{GJ}(l-mn-1, m, n) = 0.$$

Assume that $\mathrm{GF}(l, m, n) = \mathrm{J}(l, m, n)$ when l does not exceed mn; then putting $l = mn+1$ in our equations, we find that

$$\mathrm{GF}(mn+1, m, n) = \mathrm{J}(mn+1, m, n);$$

and thence by induction

$$\mathrm{GF}(l, m, n) = \mathrm{J}(l, m, n) \quad \text{when} \quad l > mn.$$

Art. 28. I now write the fundamental relation

$$\mathrm{GF}(l, m, n) = \frac{l_0\mathrm{L}_0 + l_1\mathrm{L}_1 + \ldots + l_\mu\mathrm{L}_\mu}{(1)(2)\ldots(\mathbf{mn})},$$

in the form

$$\mathrm{GF}(l, m, n) = \frac{\mathrm{E}_l}{\mathrm{E}_0},$$

where

$$\mathrm{E}_l = l_0\mathrm{L}_0 + l_1\mathrm{L}_1 + \ldots + l_\mu\mathrm{L}_\mu,$$

and assume that $\mathrm{GF}(l, m, n)$ is a product of powers of factors of the two types $1-x^{l+s}$, $1-x^s$, or $(\mathbf{l}+\mathbf{s})$, $(\mathbf{s})$, where the powers may be positive or negative integers.

I thus write

$$\frac{\mathrm{E}_l}{\mathrm{E}_0} = \Pi_1(\mathbf{l}+\mathbf{s}) \,.\, \Pi_2(\mathbf{s}),$$

leading to

$$\frac{\mathrm{E}_{l+1}}{\mathrm{E}_l} = \frac{\Pi_1(\mathbf{l}+\mathbf{s}+\mathbf{1})}{\Pi_1(\mathbf{l}+\mathbf{s})}.$$

Now

$$\frac{\mathrm{E}_1}{\mathrm{E}_0} = \frac{(\mathbf{mn}+\mathbf{1})}{(\mathbf{1})} + \mathrm{L}_1 = \frac{(\mathbf{n}+\mathbf{1})(\mathbf{n}+\mathbf{2})\ldots(\mathbf{n}+\mathbf{m})}{(\mathbf{1})(\mathbf{2})\ldots(\mathbf{m})},$$

therefore

$$\frac{\Pi_1(\mathbf{s}+\mathbf{1})}{\Pi_1(\mathbf{s})} = \frac{(\mathbf{n}+\mathbf{1})(\mathbf{n}+\mathbf{2})\ldots(\mathbf{n}+\mathbf{m})}{(\mathbf{1})(\mathbf{2})\ldots(\mathbf{m})}$$

and

$$\frac{\mathrm{E}_{l+1}}{\mathrm{E}_l} = \frac{(\mathbf{l}+\mathbf{n}+\mathbf{1})(\mathbf{l}+\mathbf{n}+\mathbf{2})\ldots(\mathbf{l}+\mathbf{n}+\mathbf{m})}{(\mathbf{l}+\mathbf{1})(\mathbf{l}+\mathbf{2})\ldots(\mathbf{l}+\mathbf{m})} = |\mathrm{NM}|_{l+1};$$

therefore

$$\frac{\mathrm{E}_{l+2}}{\mathrm{E}_l} = |\mathrm{NM}|_{l+1}|\mathrm{NM}|_{l+2},$$

and

$$\frac{\mathrm{E}_{l+l'}}{\mathrm{E}_l} = |\mathrm{NM}|_{l+1}|\mathrm{NM}|_{l+2}\ldots|\mathrm{NM}|_{l+l'}.$$

Putting herein $l = 0$, $l' = l$,

$$\frac{\mathrm{E}_l}{\mathrm{E}_0} = |\mathrm{NM}|_1|\mathrm{NM}|_2\ldots|\mathrm{NM}|_l.$$

Hence

$$\mathrm{GF}(l, m, n) = |\mathrm{NM}|_1|\mathrm{NM}|_2\ldots|\mathrm{NM}|_l,$$

and on the assumption as to form the main theorem is established.

Art. 29. We may now obtain properties of the sub-lattice functions—

For, in the relation

$$\frac{E_l}{E_0} = |NM|_1 |NM|_2 \dots |NM|_l,$$

put $l = -l-m-n$ and observe that $(-\mathbf{s}) = -x^{-s}(\mathbf{s})$.

We find

$$E_{-l-m-n} = \frac{(-)^{mn}}{x^{lmn+\frac{1}{2}mn(mn+1)}}\{(\mathbf{l}+\mathbf{1})\dots(\mathbf{l}+\mathbf{mn})L_\mu + x^{mn}(\mathbf{l})\dots(\mathbf{l}+\mathbf{mn}-\mathbf{1})L_{\mu-1}$$
$$+\dots+x^{\mu mn}(\mathbf{l}-\mu+\mathbf{1})\dots(\mathbf{l}+\mathbf{m}+\mathbf{n}-\mathbf{1})L_0\};$$

and $|NM|_1 |NM|_2 \dots |NM|_l$ is unaltered to a factor $\frac{(-)^{mn}}{x^{lmn+\frac{1}{2}mn(m+n)}}$ *près.*

Hence the identity

$$(\mathbf{l}+\mathbf{1})\dots(\mathbf{l}+\mathbf{mn})L_\mu + x^{mn}(\mathbf{l})\dots(\mathbf{l}+\mathbf{mn}-\mathbf{1})L_{\mu-1}+\dots+x^{\mu mn}(\mathbf{l}-\mu+\mathbf{1})\dots(\mathbf{l}-\mu+\mathbf{mn})L_0$$
$$= x^{\frac{1}{2}\mu mn}\{(\mathbf{l}+\mathbf{1})\dots(\mathbf{l}+\mathbf{mn})L_0 + (\mathbf{l})\dots(\mathbf{l}+\mathbf{mn}-\mathbf{1})L_1+\dots+(\mathbf{l}-\mu+\mathbf{1})\dots(\mathbf{l}-\mu+\mathbf{mn})L_\mu\}.$$

Herein putting

$l = 0$ we find $L_\mu = x^{\frac{1}{2}\mu mn} L_0$,

$l = 1$ „ $L_{\mu-1} = x^{\frac{1}{2}(\mu-2)mn} L_1$,

$l = 2$ „ $L_{\mu-2} = x^{\frac{1}{2}(\mu-n)mn} L_2$,

. . . .

$l = t$ „ $L_{\mu-t} = x^{\frac{1}{2}(\mu-2t)mn} L_t$,

exhibiting an elegant property of the functions.

From the relation

$$\frac{E_l}{E_0} = |NM|_1 |NM|_2 \dots |NM|_l$$

we find, giving l the values 0, 1, 2, ... in succession,

$$L_0 = 1,$$

$$L_1 = |NM|_1 - \frac{(\mathbf{mn}+\mathbf{1})}{(\mathbf{1})},$$

$$L_2 = |NM|_1 |NM|_2 - \frac{(\mathbf{mn}+\mathbf{1})}{(\mathbf{1})}|NM|_1 + x\frac{(\mathbf{mn})(\mathbf{mn}+\mathbf{1})}{(\mathbf{1})(\mathbf{2})},$$

. . . .

$$L_\mu = |NM|_1 |NM|_2 \dots |NM|_\mu - \frac{(\mathbf{mn}+\mathbf{1})}{(\mathbf{1})}|NM|_1 |NM|_2 \dots |NM|_{\mu-1}$$
$$+\dots+(-)^\mu x^{\frac{1}{2}(\mu-1)\mu}\frac{(\mathbf{mn}-\mu+\mathbf{2})\dots(\mathbf{mn}+\mathbf{1})}{(\mathbf{1})\dots(\mu)}$$

The relations between the functions L_s and $L_{\mu-s}$ yield remarkable algebraical identities.

So far I have established a new method for the discussion of these questions of the arrangements of numbers, and have made some progress with the simplification of the fundamental expression arrived at for the generating function. I have further shown the great probability of the outer lattice function being the whole lattice function whenever the lattice is complete. Before this can be rigidly established, I believe that a further study of the theory of the incomplete lattice will be necessary. From many particular incomplete lattices that have already been worked out this investigation promises well, and I hope in due course to lay the results before the Society.

Postscript.

There is an analogous theory which is concerned with the totality of the permutations of $\alpha_1^{p_1}\alpha_2^{p_2}\ldots\alpha_n^{p_n}$. We thus obtain permutation functions which possess elegant properties. The functions also arise from the theory of partitions.

Suppose that we desire the number of two-dimensional partitions of a number such that the nodes of the lattice descending order of part magnitude is in evidence in each row but *not necessarily* in each column. It is immediately evident that the generating function of such at the nodes of a lattice which contains $p_1, p_2, \ldots, p_n$ nodes in the successive rows is

$$\frac{1}{(1)\ldots(\mathrm{p}_1)\,(1)\ldots(\mathrm{p}_2)\ldots\ldots(1)\ldots(\mathrm{p}_n)},$$

whether the numbers $p_1, p_2, \ldots, p_n$ be in descending order of magnitude or not. This fact enables us to determine the lattice function and the sub-lattice functions derivable from the whole of the permutations of the letters $\alpha_1^{p_1}\alpha_2^{p_2}\ldots\alpha_n^{p_n}$ when we may suppose the exponents $p_1, p_2, \ldots, p_n$ to be in descending order of magnitude and establishes also that *these functions are invariant for any permutation of* $p_1, p_2, \ldots, p_n$ *in the product* $\alpha_1^{p_1}\alpha_2^{p_2}\ldots\alpha_n^{p_n}$.

We may proceed in exactly the same manner as when the restricted permutations were under view. Taking the lattice corresponding to $\alpha_1^3\alpha_2^2\alpha_3$ and arranging 6 different numbers in any way so that descending order is in evidence in the rows

$$\begin{array}{lll} 3 & 2 & 1 \\ 6 & 5 & \\ 4 & & \end{array} \qquad \alpha_2\alpha_2\alpha_3 \mid \alpha_1\alpha_1\alpha_1$$

we have the arrangement figured and the corresponding Greek-letter succession, yielding a portion

$$\frac{x^3}{(1)\ldots(6)}$$

of the generating function.

For the whole of the permutations derived as above from the lattice which are, in fact, the whole of the permutations of $\alpha_1^3\alpha_2^2\alpha_3$ we derive a permutation function

$$\text{PF}(\infty; 321),$$

such that the generating function sought is

$$\frac{\text{PF}(\infty; 321)}{(1)\ldots(6)},$$

and this we know otherwise to have the value

$$\frac{1}{(1)(2)(3).(1)(2).(1)}.$$

Hence

$$\text{PF}(\infty; 321) = \frac{(1)(2)(3)(4)(5)(6)}{(1)(2)(3).(1)(2).(1)},$$

and, in general,

$$\text{PF}(\infty; p_1p_2\ldots p_n) = \frac{(1)\ldots(\Sigma p)}{(1)\ldots(p_1).(1)\ldots(p_2)\ldots\ldots(1)\ldots(p_n)},$$

an expression which is to be compared with the number which enumerates the permutations of the letters in $\alpha_1^{p_1}\alpha_2^{p_2}\ldots\alpha_n^{p_n}$. The former becomes equal to the latter when $x = 1$.

When the part magnitude is limited by the number l, the enumerating generating function of the partition is

$$\frac{(l+1)\ldots(l+p_1).(l+1)\ldots(l+p_2)\ldots\ldots(l+1)\ldots(l+p_n)}{(1)\ldots(p_1).(1)\ldots(p_2)\ldots\ldots(1)\ldots(p_n)},$$

but, from previous work, if $\text{PF}_s(\infty; p_1p_2\ldots p_n)$ is the sub-permutation function derived from the permutations possessing s dividing lines, this generating function is also

$$\frac{(l+1)\ldots(l+\Sigma p)\,\text{PF}_0+(l)\ldots(l+\Sigma p-1)\,\text{PF}_1+\ldots\ldots+(l-\nu+1)\ldots(l-\nu+\Sigma p)\,\text{PF}_\nu}{(1)\ldots(\Sigma p)},$$

where $\nu = \Sigma p - p_1$ (see 'Phil. Trans. Roy. Soc.,' A, vol. 207, p. 119).

Equating the two expressions for the generating function, and giving l the values 0, 1, 2, ... in succession, we find the relations

$$1 = \text{PF}_0,$$

$$\frac{(p_1+1)(p_2+1)\ldots(p_n+1)}{(1)^n} = \frac{(\Sigma p+1)}{(1)} + \text{PF}_1,$$

$$\frac{(p_1+1)(p_1+2).(p_2+1)(p_2+2)\ldots(p_n+1)(p_n+2)}{(1)^n(2)^n} = \frac{(\Sigma p+1)(\Sigma p+2)}{(1)(2)} + \frac{(\Sigma p+1)}{(1)}\text{PF}_1+\text{PF}_2,$$

$$\&\text{c.} = \&\text{c.},$$

from which the general expression for PF_s is readily obtainable.

Putting $p_1 = p_2 = \ldots = p_n = p$ for a complete lattice, we find

$$\mathrm{PF}_0 = 1,$$

$$\mathrm{PF}_1 = \left\{\frac{(\mathrm{p}+1)}{(1)}\right\}^{\mathrm{n}} - \frac{(\mathrm{np}+1)}{(1)},$$

$$\mathrm{PF}_2 = \left\{\frac{(\mathrm{p}+1)(\mathrm{p}+2)}{(1)(2)}\right\}^{\mathrm{n}} - \frac{(\mathrm{np}+1)}{(1)}\left\{\frac{(\mathrm{p}+1)}{(1)}\right\}^{\mathrm{n}} + x\frac{(\mathrm{np})(\mathrm{np}+1)}{(1)(2)},$$

$$\mathrm{PF}_s = \left\{\frac{(\mathrm{p}+1)\ldots(\mathrm{p}+\mathrm{s})}{(1)\ldots(\mathrm{s})}\right\}^{\mathrm{n}} - \frac{(\mathrm{np}+1)}{(1)}\left\{\frac{(\mathrm{p}+1)\ldots(\mathrm{p}+\mathrm{s}-1)}{(1)\ldots(\mathrm{s}-1)}\right\}^{\mathrm{n}}$$
$$+\ldots+(-)^s x^{\frac{1}{2}(s-1)(s)}\frac{(\mathrm{np}-\mathrm{s}+2)\ldots(\mathrm{np}+1)}{(1)\ldots(\mathrm{s})}.$$

A simplification, when $n = 2$, is interesting; for then

$$\mathrm{PF}_s(\infty; pp)_x = x^{s^2}\left\{\frac{(\mathrm{p})\ldots(\mathrm{p}-\mathrm{s}+1)}{(1)\ldots(\mathrm{s})}\right\}^2.$$

In fact, more generally it will be found that

$$\mathrm{PF}_s(\infty; pq) = x^{s^2}\frac{(\mathrm{p})\ldots(\mathrm{p}-\mathrm{s}+1).(\mathrm{q})\ldots(\mathrm{q}-\mathrm{s}+1)}{\{(1)\ldots(\mathrm{s})\}^2}.$$

An interesting verification is supplied by a result in a previous paper.* It was therein shown that the number of permutations of the letters composing the product

$$\alpha^p\beta^q,$$

which have s dividing lines, is the coefficient of $\lambda^s\alpha^p\beta^q$ in the expansion of the product

$$(\alpha+\lambda\beta)^p(\alpha+\beta)^q.$$

From this expression I derive a function of x, viz.,

$$(\alpha+\lambda\beta x)(\alpha+\lambda\beta x^2)\ldots(\alpha+\lambda\beta x^p).(\beta+\alpha)(\beta+\alpha x)\ldots(\beta+\alpha x^{q-1}),$$

and therein the coefficient of $\lambda^s\alpha^p\beta^q$ is readily shown to be

$$x^{s^2}\frac{(\mathrm{p})\ldots(\mathrm{p}-\mathrm{s}+1).(\mathrm{q})\ldots(\mathrm{q}-\mathrm{s}+1)}{\{(1)\ldots(\mathrm{s})\}^2},$$

as already obtained.

When the lattice is complete the functions PF_s possess elegant properties, just as when the permutations are restricted.

* "Memoir on the Theory of the Compositions of Numbers," 'Phil. Trans.,' A, 1893, Art. 24.

For, in the identity

$$\left\{\frac{(\mathfrak{l}+1)(\mathfrak{l}+2)\ldots(\mathfrak{l}+\mathfrak{p})}{(1)(2)\ldots(\mathfrak{p})}\right\}^{\mathfrak{n}}$$

$$=\frac{(\mathfrak{l}+1)\ldots(\mathfrak{l}+\mathfrak{np})\,\mathrm{PF}_0+(\mathfrak{l})\ldots(\mathfrak{l}+\mathfrak{np}-1)\,\mathrm{PF}_1+\ldots+(\mathfrak{l}-\mathfrak{np}+\mathfrak{p}+1)\ldots(\mathfrak{l}+\mathfrak{p})\,\mathrm{PF}_{(n-1)p}}{(1)(2)\ldots(\mathfrak{np})},$$

substitute $-l-p-1$ for l and we find that the left-hand side is merely multiplied by $x^{-lnp-\frac{1}{2}np(p+1)}$, whilst on the right hand the coefficient of PF_s is multiplied by $x^{-lnp+\frac{1}{2}np(np-2p-2s-1)}$. An identity thence arises, and putting therein $l_1 = 0, 1, 2, \ldots$ in succession, we find the relations

$$\mathrm{PF}_{np-p} = x^{\frac{1}{2}n(n-1)p^2}\,\mathrm{PF}_0,$$

$$\mathrm{PF}_{np-p-1} = x^{\frac{1}{2}n(n-1)p^2-np}\,\mathrm{PF}_1,$$

$$\cdot\quad\cdot\quad\cdot\quad\cdot\quad\cdot\quad\cdot\quad\cdot\quad\cdot$$

$$\mathrm{PF}_{np-p-s} = x^{\frac{1}{2}n(n-1)p^2-snp}\,\mathrm{PF}_s,$$

giving very noteworthy algebraical identities.

THE CONNEXION BETWEEN THE SUM OF THE SQUARES OF THE DIVISORS AND THE NUMBER OF THE PARTITIONS OF A GIVEN NUMBER.

By *Major P. A. MacMahon.*

IN various papers* I have considered the partitions of a number as determined by a succession of integers in *descending order of magnitude.*

I have, from this point of view, dealt with an array of numbers ordered in such wise that a descending order of magnitude is in evidence in each column and in each row of the array and have defined such an array as a two-dimensional partition.

The enumerating generating function I found to be

$$F(q) = \frac{1}{(1-q)(1-q^2)^2(1-q^3)^3 \dots (1-q^s)^s \dots},$$

when the partitions are unrestricted both in regard to number and magnitude.

Thus the 13 partitions of 4 are

4	31	3	22	2	211	21	2	1111	111	11	11	1
		1		2		1	1		1	11	1	1
							1				1	1
												1

and the expansion of the above fraction gives a term $13q^4$.

We find that

$$q\frac{d}{dq}\log F(q) = \frac{q}{1-q} + \frac{2^2 q^2}{1-q^2} + \frac{3^2 q^3}{1-q^3} + \dots = \Sigma\sigma_2(n)\,q^n,$$

where $\sigma_2(n)$ denotes the sum of the squares of the divisors of n.

Writing

$$F(q) = 1 + B(1)\,q + B(2)\,q^2 + \dots + B(n)\,q^n + \dots,$$

and operating with $q\dfrac{d}{dq}\log$, we find by comparison

$$nB(n) = \sigma_2(n) + B(1)\,\sigma_2(n-1) + B(2)\,\sigma_2(n-2) + \dots$$
$$+ B(s)\,\sigma_2(n-s) + \dots .$$

* *Combinatory Analysis* (Camb. Univ. Press, 1915–16).

The denominator of $F(q)$ may be written

$$\begin{array}{r}(1-q)(1-q^2)(1-q^3)(1-q^4)\ldots\\ (1-q^2)(1-q^3)(1-q^4)\ldots\\ (1-q^3)(1-q^4)\ldots\\ (1-q^4)\ldots\\ \ldots\end{array}$$

and I have shown (*loc. cit.*) that if s rows only are retained the function enumerates the partitions when the array is restricted to have at most s rows. Thus, when s is 1, we have the case of ordinary linear partitions for which $\sigma_1(n)$, the sum of the divisors of n, is in evidence, and we have seen above that, when s is ∞, $\sigma_2(n)$ presents itself. It is interesting to enquire concerning the intermediate stages when s has some value between unity and infinity. For s rows,

$$F_s(q)=\frac{1}{(1-q)(1-q^2)^2(1-q^3)^3\ldots(1-q^s)^s\{(1-q^{s+1})(1-q^{s+2})\ldots\}^s}$$

and

$$q\frac{d}{dq}\log F_s(q)$$

$$=\frac{q}{1-q}+\frac{2^2q^2}{1-q^2}+\frac{3^2q^3}{1-q^3}+\ldots$$

$$+\frac{s^2q^3}{1-q^s}+\frac{s(s+1)q^{s+1}}{1-q^{s+1}}+\frac{s(s+2)}{1-q^{s+2}}+\ldots \textit{ ad inf.}$$

As regards the coefficient herein of q^n, if d be a divisor $\leq s$, let it be squared, but if it be $>s$, let it be multiplied by s. The coefficient of q^n is then

$$\sum_{\leq s} d^2+s\sum_{>s} d\,;$$

and we may write this

$$\underset{\leq s}{\sigma_2}(n)+s\underset{>s}{\sigma_1}(n).$$

Hence $$q\frac{d}{dq}\log F_s(q)=\sum_1^\infty n\,\{\underset{\leq s}{\sigma_2}(n)+s\underset{>s}{\sigma_1}(n)\}\,q^n\,;$$

and it will be noted that the arithmetical function becomes $\sigma_1(n)$, $\sigma_2(n)$ for $s=1$ and ∞ respectively.

Writing

$$F_s(q)=1+B_s(1)\,q+B_s(2)\,q^2+\ldots+B_s(n)\,q^n+\ldots,$$

we find, as before,

$$nB_s(n) = \{\underset{\overline{<}s}{\sigma_2}(n) + s\underset{>s}{\sigma_1}(n)\} + B_s(1)\,\{\underset{\overline{<}s}{\sigma_2}(n-1) + s\underset{>s}{\sigma_1}(n-1)\}$$
$$+ B_s(2)\,\{\underset{\overline{<}s}{\sigma_2}(n-2) + s\underset{>s}{\sigma_1}(n-2)\} + \ldots.$$

The value of $\underset{\overline{<}s}{\sigma_2}(n) + s\underset{>s}{\sigma_1}(n)$ is given in the annexed Table so far as $s = 10$, $n = 10$, and in the second Table the value of $B_s(n)$.

I.

$s.n=1$		2	3	4	5	6	7	8	9	10	...
1	1	3	4	7	6	12	8	15	13	18	...
2		5	7	13	11	23	15	29	25	35	...
3			10	17	16	32	22	41	37	50	...
4				21	21	38	29	53	46	65	...
5					26	44	36	61	55	80	...
6						53	43	69	64	90	...
7							50	77	73	100	...
8								85	82	110	...
9									91	120	...
10										130	...
											...

II.

$n=1$		2	3	4	5	6	7	8	9	10	...
1	1	2	3	5	7	11	15	22	30	42	...
2		3	5	10	16	29	45	75	115	181	...
3			6	12	21	40	67	117	193	319	...
4				13	23	45	78	141	239	409	...
5					24	47	83	152	263	457	...
6						48	85	157	274	481	...
7							86	159	279	492	...
8								160	281	497	...
9									282	499	...
10										500	...
											...

In Table I. the values of $\underset{\overline{<}s}{\sigma_2}(n) + s\underset{>s}{\sigma_1}(n)$, for a fixed value of s and successive values of n, are obtained by reading down the slanting side of the Table as far as the sth row and then proceeding along that row to the right.

Similarly in Table II. the values of $B_s(n)$, for a fixed value of s and successive values of n, are obtained by reading down the slanting side as far as the sth row and then proceeding along that row.

It may be added that if

$$f(q) = \{(1-q)(1-q^2)(1-q^3)\ldots\}^{-1},$$

and, after Euler, if ϕ_m denote the number of primitive m^{th} roots of unity,

$$F(q) = \prod_{m_1}^{\infty} \{f(q^m)\}^{\phi_m},$$

this establishes a connection with Elliptic Functions.

ELECTROMAGNETISM AND DYNAMICS.

By *Dr. H. Bateman.*

DIFFERENT pictures of physical phenomena may be obtained by adopting different conventions with regard to the types of discontinuity that are to be regarded as admissible in the mathematical specification of physical quantities. A picture of considerable interest is based on the idea that, when all types of energy and momentum are taken into consideration, these physical quantities are distributed throughout space in such a manner that we can speak of densities of energy and momentum that are continuous functions of the rectangular coordinates (x, y, z), used to specify the position of a point, and of the time t. A different picture is obtained if the densities of energy and momentum are allowed to change suddenly in value as the point (x, y, z) crosses the boundary of an electron or some other entity such as a hypothetical light quantum of limited size.

It is doubtful whether the first picture is adequate for a complete description of all the physical phenomena with which man is acquainted, but in any case it is well worth while to give it a fair trial. The type of analysis associated with this picture of phenomena will be called *continuous analysis.* The main principles of the analysis are already familiar, as they play an important part in Maxwell's electromagnetic theory, the theory of electrons and the theory of relativity in both the restricted and general forms.*

* In all these theories there seems, however, to be some type of discontinuity at the boundary of a particle of matter when the density of electricity is not zero at the boundary.

Chapter 12
Plane Partitions (Part 2) and Solid Partitions

12.1 Introduction and Commentary

In chapter 11, we discussed MacMahon's study of plane partitions and presented L. Carlitz's elegant proof of MacMahon's main theorem. Beyond this work a number of results are known for restricted plane partitions. B. Gordon (1962, 1971a, 1971b), and B. Gordon and L. Houten (1968a, 1968b, 1969) have presented an extensive account of restricted plane partitions for which concise infinite product representations of the generating functions can be found. Also L. Carlitz (in work not described in chapter 11) has found a number of theorems on the generating functions for many integer arrays (see Carlitz, 1963a, 1963b, 1965).

One of the most important works in this area concerns what has become known as the Knuth correspondence (Knuth, 1970b) and its application to restricted plane partitions (Knuth and Bender, 1972). In section 12.2 we shall present an introduction to their work.

There is an extensive interaction between the theory of plane partitions and the representation theory of the symmetric group (see Littlewood, 1950; Robinson, 1961; Rutherford, 1948; and Stanley, 1971). At the end of section 12.2 we shall briefly describe this relationship.

Young tableaux are arrays closely connected with plane partitions; they are defined as follows: Form a Ferrers graph of some partition using boxes instead of nodes; then insert positive integers in each box so that there is strict increase along rows and columns. The tableau is called a standard tableau if for some n, the integers appearing in the tableau are the first n positive integers. For example, Figure 12.1

Figure 12.1

1	3	5	7
2	4	6	9
8	10	11	
12			

The material in this chapter corresponds to section X of *Combinatory Analysis.*

is a standard tableau associated with the partition (134^2). There is an extensive literature on Young Tableaux (see the papers—and books—by S. Blaha, D. Knuth, G. Kreweras, D. Littlewood, G.-C. Rota, D. Rutherford, C. Schensted, M. Schützenberger, and A. Young listed in section 12.3). A highly readable account of some of the basic properties of Young Tableaux is given by C. Berge (1970, pages 59–71).

The subject of solid partitions has certainly not received the attention that has been given plane partitions. This is undoubtedly for the very simple reason that very few elegant results concerning the generating functions appear to be true (see D. Knuth, 1970a; Andrews, 1976a). MacMahon ([51], page 658) originally conjectured that the generating function for three-dimensional partitions is

$$\prod_{n=1}^{\infty} (1 - q^n)^{-n(n+1)/2},$$

a result proved false by A. O. L. Atkin et al. (1967) and later, in a different manner, by E. M. Wright (1968). Beyond these papers H. Gupta (1972), L. Houten (1968), D. Mitchell (1972), and E. M. Wright (1966) have thrown some light on solid partitions. D. Mitchell was able to prove that if $r(n)$ denotes the number of solid partitions of n, where for each j the number of appearances of j in any row is neither less than the number of appearances of j in the next row of the same layer, nor less than the number of appearances of j in the same row of the next layer, then

$$\sum_{n \geqq 0} r(n) q^n = \prod_{n=1}^{\infty} (1 - q^n)^{-\sigma(n)},$$

where $\sigma(n)$ denotes the sum of the divisors of n.

Finally we mention the very important work of R. P. Stanley. Most of the work in section 12.2 follows the survey of Stanley (1971) very closely. Stanley (1973) has also obtained some elegant interpretations for the two-variable generating function

$$\prod_{n=1}^{\infty} (1 - aq^n)^{-n} = \sum_{m,\, n \geqq 0} p_2(m, n) a^m q^n.$$

He has shown that $p_2(m, n)$ denotes the number of plane partitions of n in which exactly m of the parts are at least as large as the number of the row in which they appear; furthermore, he has shown that $p_2(m, n)$ is the number of plane partitions of n wherein the sum of the parts appearing on the main diagonal is m. For example, the terms in the above expansion that involve q^4 are:

$$4aq^4 + 6a^2q^4 + 2a^3q^4 + a^4q^4.$$

The first of Stanley's results on $p_2(m, n)$ divides the thirteen plane partitions of 4 into the classes shown in Table 12.1.

Table 12.1

$p_2(1,4)$ counts	$p_2(2,4)$ counts	$p_2(3,4)$ counts	$p_2(4,4)$ counts
4	3 1	2 1 1	1 1 1 1
2	2 2	1 1 1	
1		1	
1	2		
	2		
3			
1	2 1		
	1		
1			
1	1 1		
1	1 1		
1			
	1 1		
	1		
	1		

The second of Stanley's results on $p_2(m, n)$ divides the thirteen plane partitions of 4 into the classes shown in Table 12.2.

12.2 An Introduction to the Knuth-Bender Theorems on Plane Partitions

In section 1.2, we studied several bases for A_n, the space of all homogeneous symmetric functions of degree n over $\mathbf{Q}$, the rational numbers. We now introduce the Schur functions e_λ (our notation follows that of section 1.2):

$$e_\lambda = \sum_\pi M(\pi),$$

where $\lambda = (\lambda_1, \lambda_2, \ldots)$ is a partition of n, $\sum_\pi$ is over all plane partitions π

Table 12.2

$p_2(1,4)$ counts	$p_2(2,4)$ counts	$p_2(3,4)$ counts	$p_2(4,4)$ counts
1 1 1 1	2 2	3 1	4
1 1 1 1	2 2	3 1	
1 1 1 1	2 1 1		
	2 1 1		
1 1 1 1	2 1 1		
	1 1 1 1		

with λ_i parts in the ith row and strict decrease along columns, and $M(\pi) = x_1^{a_1} x_2^{a_2} \ldots$, where a_i is the number of times i appears in π.

It is certainly not obvious that e_λ is a symmetric function; however, we shall be able to deduce this from the following theorem which presents the Knuth correspondence (Knuth, 1970b) and which generalizes a correspondence due to Schensted (1961).

Theorem 1. There exists a one-to-one correspondence, $K: A \to (\pi, \sigma)$, between matrices $A = (a_{ij})$ of nonnegative integers ($i, j \geqq 1$) with only finitely many non-zero entries, and ordered pairs of plane partitions (π, σ) with strict decrease along columns and the same shape. In this correspondence, i occurs in σ exactly $\sum_j a_{ij}$ times and j occurs in π exactly $\sum_i a_{ij}$ times.

Remark. For this and the other theorems in this section we shall follow closely the development of R. Stanley (1971).

Proof. We shall not present a complete proof, but rather we shall content ourselves with an example from which the reader can construct a proof in full generality. Suppose

$$A = \begin{pmatrix} 1 & 1 & 0 & 2 \\ 2 & 0 & 0 & 1 \\ 1 & 1 & 0 & 1 \\ 1 & 0 & 1 & 2 \end{pmatrix}.$$

From A we construct a two-line array wherein the top row is nonincreasing and the column $\binom{i}{j}$ occurs a_{ij} times. Thus the array here is

$$\begin{matrix} 4 & 4 & 4 & 4 & 3 & 3 & 3 & 2 & 2 & 2 & 1 & 1 & 1 & 1 \\ 4 & 4 & 3 & 1 & 4 & 2 & 1 & 4 & 1 & 1 & 4 & 4 & 2 & 1. \end{matrix}$$

Using this array, we shall form the plane partition π from the bottom row, while σ will be formed from the top row. To form π, we begin from the left on the bottom row, and we successively insert the entries in the top row of the evolving plane partition as far to the right as possible. If there is already a smaller number appearing where the insertion is to take place, then the smaller number is "bumped" down to the second row where the same insertion procedure is again followed, etc. The plane partition σ is formed by inserting each corresponding upper entry so that the evolving plane partition σ always maintains the same shape as the evolving π. The following presents the formation of π and σ step by step.

π	σ
4	4
4 4	4 4
4 4 3	4 4 4
4 4 3 1	4 4 4 4
4 4 4 1 3	4 4 4 4 3
4 4 4 2 3 1	4 4 4 4 3 3
4 4 4 2 1 3 1	4 4 4 4 3 3 3
4 4 4 4 1 3 2 1	4 4 4 4 3 3 3 2
4 4 4 4 1 1 3 2 1	4 4 4 4 3 2 3 3 2

4 4 4 4 1 1 1 3 2 1	4 4 4 4 3 2 2 3 3 2
4 4 4 4 4 1 1 3 2 1 1	4 4 4 4 3 2 2 3 3 1 2
4 4 4 4 4 4 1 3 2 1 1 1	4 4 4 4 3 2 2 3 3 1 1 2
4 4 4 4 4 4 2 3 2 1 1 1 1	4 4 4 4 3 2 2 3 3 1 1 1 2
4 4 4 4 4 4 2 1 3 2 1 1 1 1	4 4 4 4 3 2 2 1 3 3 1 1 1 2

The last pair of plane partitions represent π and σ. Most of the assertions of the theorem are clear from this construction. The fact that the mapping K is one-to-one is, however, not obvious, and we refer the reader to Knuth (1970b) for details.

Theorem 2. The e_λ are all symmetric functions.

Proof. Here we follow Bender and Knuth (1972). Let $(\mu_1, \mu_2, \ldots)$ be a sequence of nonnegative integers, only finitely many of which are nonzero. Clearly

$$h_\mu = h_{\mu_1} h_{\mu_2} \cdots = \sum x_1^{\nu_1} x_2^{\nu_2} \ldots, \tag{2.1}$$

where the summation is over all matrices $A = (a_{ij})$ of nonnegative integers with column sums $\sum_i a_{ij} = \mu_j$, with row sums $\sum_i a_{ij} = \nu_j$, and where μ is the partition with parts μ_i. Comparing (2.1) with Theorem 1, we see that

$$h_\mu = \sum x_1^{\nu_1} x_2^{\nu_2} \ldots,$$

where the sum is now over all ordered pairs (π, σ) of plane partitions, with strict decrease along columns, and of the same shape such that π contains μ_i parts equal to i, and σ contains ν_i parts equal to i.

We now define $K_{\lambda\mu}$ to be the number of plane partitions with strict decrease along columns, with shape λ, and with μ_i parts equal to i. Hence

$$h_\mu = \sum_\lambda K_{\lambda\mu} e_\lambda .$$

Since the h_μ are linearly independent (see the Corollary of Theorem 6 in section 1.2), we see that the $p(n) \times p(n)$ matrix $(K_{\lambda\mu})$ must be invertible. Hence the e_λ may be expressed as linear combinations of the h_μ, and so the e_λ are symmetric functions.

We refer the reader to Knuth (1970b) for the proof of the next theorem which provides further information concerning the Knuth correspondence.

Theorem 3. There exists a one-to-one correspondence between symmetric matrices $A = (a_{ij})$ of nonnegative integers $(i, j \geqq 1)$ with finitely many nonzero entries, and plane partitions π with strict decrease along columns. In this correspondence, i occurs exactly $\sum_j a_{ij}$ times.

Theorem 4. $\displaystyle\sum_\lambda e_\lambda = \prod_i (1 - x_i)^{-1} \prod_{i<j} (1 - x_i x_j)^{-1}.$

Proof. By Theorem 3,

$$\sum_\lambda e_\lambda = \sum_A \prod_{i,j=1} x_i^{a_{ij}},$$

where the sum is over all symmetric matrices $A = (a_{ij})$ of nonnegative integers with finitely many nonzero entries. Hence

$$\begin{aligned}
\sum_\lambda e_\lambda = \sum_A \prod x_i^{a_{ij}} &= \sum_A \left(\prod_{i=j} x_i^{a_{ij}}\right)\left(\prod_{i<j} x_i^{a_{ij}} x_j^{a_{ji}}\right) \\
&= \sum_A \left(\prod_i x_i^{a_{ii}}\right)\left(\prod_{i<j} (x_i x_j)^{a_{ij}}\right) \\
&= \left(\prod_i \sum_{a_{ii}=0}^{\infty} x_i^{a_{ii}}\right)\left(\prod_{i<j} \sum_{a_{ij}=0}^{\infty} (x_i x_j)^{a_{ij}}\right) \\
&= \prod_i (1 - x_i)^{-1} \prod_{i<j} (1 - x_i x_j)^{-1}.
\end{aligned}$$

We conclude with what is perhaps the most striking result on plane partitions that Bender and Knuth (1972) obtained.

Theorem 5. Let S be any subset of the positive integers. Let $B_S(n)$ denote the number of plane partitions with strict decrease along columns and with all parts in S. Then

$$\sum_{n \geqq 0} B_S(n) q^n = \prod_{i \in S} (1 - q^i)^{-1} \prod_{\substack{i, j \in S \\ i < j}} (1 - q^{i+j})^{-1}.$$

Proof. From the definition of the e_λ, we see that this result follows immediately from Theorem 4 by the substitutions $x_i = q^i$ if $i \in S$, $x_i = 0$ otherwise.

It is clear from the proof of Theorem 2 that $\{e_\lambda\}$ is also a basis for A_n. Recalling the power sum symmetric functions $\{s_\lambda\}$ from section 1.2, and noting from Theorem 6 of that chapter that they also form a basis for A_n, we see that there exist χ_λ^μ such that

$$s_\lambda = \sum_{\mu \vdash n} \chi_\lambda^\mu e_\mu.$$

The following theorem of Frobenius shows the relationship between our work on plane partitions and the representation theory of the symmetric group S_n. The reader is referred to Littlewood (1950, section 5.2) for details.

Theorem 6. The matrix (χ_λ^μ) is the character table of the symmetric group S_n. Specifically, χ_λ^μ is the character χ^μ corresponding to the partition μ, evaluated at the conjugacy class of S_n corresponding to the partition λ.

12.3 References

G. E. Andrews (1971) On a conjecture of Guinand for the plane partition function, *Proc. Edinburgh Math. Soc.(2)*, *17*, 275–276.

G. E. Andrews (1976a) The Theory of Partitions, Encyclopedia of Mathematics and Its Applications, vol. 2, Addison-Wesley.

G. E. Andrews (1976b) Implications of the MacMahon conjecture, from Combinatoire et Representation du Groupe Symetrique, Strasbourg 1976, Lecture Notes in Mathematics No. 579, Springer-Verlag, Berlin-Heidelberg-New York.

G. E. Andrews (1977a) MacMahon's conjecture on symmetric plane partitions, *Proc. Nat. Acad. Sci. U.S.A.*, *74*, 426–429.

G. E. Andrews (1977b) Plane partitions II: the equivalence of the Bender-Knuth and MacMahon conjectures, *Pacific J. Math.*, *72*, 283–291.

G. E. Andrews (1978) Plane partitions I: the MacMahon conjecture, *Advances in Math.*

A. O. L. Atkin, P. Bratley, I. G. Macdonald, and J. K. S. McKay (1967) Some computations for *m*-dimensional partitions, *Proc. Cambridge Math. Soc.*, *63*, 1097–1100.

F. C. Auluck and C. B. Haselgrove (1952) On Ingham's Tauberian theorem for partitions, *Proc. Cambridge Phil. Soc.*, *48*, 566–570.

E. A. Bender and D. Knuth (1972) Enumeration of plane partitions, *J. Cambinatorial Th.*, *A-13*, 40–54.

J. H. Bennett (1956) Partitions in more than one dimension, *J. Royal Stat. Soc.(B)*, *18*, 104–112.

J. H. Bennett (1957) The enumeration of genotype-phenotype correspondences, *Heredity*, *11*, 403–409.

J. H. Bennett (1967) A general class of enumerations arising in genetics, *Biometrics*, *23*, 517–537.

C. Berge (1971) Principles of Combinatorics, Academic Press, New York.

S. Blaha (1969) Character analysis of U(N) and SU(N), *J. Math. Physics*, *10*, 2156–2168.

W. H. Burge (1974) Four correspondences between graphs and generalized Young tableaux, J. *Combinatorial Th.*, *A-17*, 12–30.

L. Carlitz (1956) The expansion of certain products, *Proc. A.M.S.* *7*, 558–564.

L. Carlitz (1963a) Some generating functions, *Duke Math. J.*, *30*, 191–202.

L. Carlitz (1963b) A problem in partitions, *Duke Math. J.*, *30*, 203–214.

L. Carlitz (1965) Generating functions and partition problems, *Proc. Symp. in Pure Math.*, *8*, 144–169.

P. Cartier (1971) La serie géneratric exponentielle applications probabilistes et algébriques, version préliminaire, Institut de Recherche Mathématique Avancée, Strasbourg.

L. Comtet (1974) Advanced Combinatorics, Reidel, Dordrecht.

P. Doubilet (1972) on foundations of combinatorial theory VII: Symmetric functions through the theory of distribution and occupancy, *Studies in Appl. Math.*, *51*, 377–396.

P. Doubilet (1973) An inversion formula involving partitions, Bull A.M.S., *79*, 177–179.

P. Doubilet, G.-C. Rota and J. Stein (1974) On the foundations of combinatorial theory IX: Combinatorial methods in invariant theory, *Studies in Appl. Math.*, *53*, 185–216.

D. Foata (1976) Une propriété du vidage-remplissage des tableaux de Young, from Combinatoire et Representation du Groupe Symetrique, Strasbourg 1976, Lecture Notes in Math. No. 579, Springer-Verlag, Berlin-Heidelberg-New York.

M. L. Fredman (1975) On computing the length of longest increasing subsequences, *Discr. Math.*, *11*, 29–35.

B. Gordon (1962) Two new representations of the partition function, *Proc. A.M.S.* *13*, 869–873.

B. Gordon (1971a) Notes on plane partitions IV: Multirowed partitions with strict decrease along columns, *Proc. Symp. in Pure Math.*, *19*, 91–100.

B. Gordon (1971b) Notes on plane partitions V, *J. Combinatorial Th.*, *B-11*, 157–168.

B. Gordon and L. Houten (1968a) Notes on plane partitions I, *J. Combinatorial Th.*, *4*, 72–80.

B. Gordon and L. Houten (1968b) Notes on plane partitions II, *J. Combinatorial Th.*, *4*, 81–99.

B. Gordon and L. Houten (1969) Notes on plane partitions III, *Duke Math. J.*, *36*, 801–824.

R. M. Grassl and A. P. Hillman (1976) Reverse plane partitions and tableau hook numbers, *J. Combinatorial Th.*, *A-21*, 216–221.

C. Greene (1974) An extension of Schensted's theorem, *Advances in Math.*, *14*, 254–265.

C. Greene (1976a) Some order theoretic properties of the Robinson-Schensted correspondence, from Combinatoire et Representation du Groupe Symetrique, Strasbourg 1976, Lecture Notes in Math. No. 579, Springer-Verlag, Berlin-Heidelberg-New York.

C. Greene (1976b) Some partitions associated with a partially ordered set, *J. Combinatorial Th.*, *A-20*, 69–79.

J. A. Green (1955) The characters of the finite general linear groups, *Trans. A.M.S.* *80*, 402–447.

H. Gupta (1970) Partitions—a survey, *J. Res. Nat. Bureau Standards(B)*, *Math. Sci.*, *74B*, 1–29.

H. Gupta (1972) Restricted solid partitions, *J. Combinatorial Th.*, *A-13*, 140–144.

P. Hall (1957) The algebra of partitions, Proc. 4th Canadian Math. Congress, 147–159.

M. Henle (1972) Dissection of generating functions, *Studies in Appl. Math.*, *51*, 397–410.

M. Henle (1975) Binomial enumeration on dissects, Trans. A.M.S., *202*, 1–39.

W. V. D. Hodge and D. Pedoe (1952) Methods of Algebraic Geometry, vol. 2, Cambridge University Press, Cambridge.

L. Houten (1968) A note on solid partitions, *Acta Arith.*, *15*, 71–76.

A. E. Ingham (1941) A Tauberian theorem for partitions, *Ann. Math.*, *42*, 1075–1090.

D. E. Knuth (1970a) A note on solid partitions, *Math. Comp. 24*, 955–961.

D. E. Knuth (1970b) Permutations, matrices, and generalized Young tableaux, *Pacific J. Math.*, *34*, 709–727.

D. E. Knuth (1970c) The Art of Computer Programming, vol. 3: Sorting and Searching, Addison-Wesley, Reading.

D. Knutson (1973) λ-Rings and the Representation Theory of the Symmetric Group, Lecture Notes in Math., No. 308, Springer, New York.

G. Kreweras (1965) Sur une classe de problèmes de dènombrement liés au treillis des partitions des entiers, Cahiers du B.U.R.O., No. 6, Paris.

G. Kreweras (1967) Traitment simultané du "Problème de Young" et "Problème Simon Newcomb," Cahiers du B.U.R.O., No. 10, Paris.

G. Kreweras (1969) Dénombrement systematiques de relations binaires externes, *Math. et Sc. Humaines*, *7*, No. 26, 5–15.

G. Kreweras (1972) Classification des permutations suivant certaines propriétés ordinales de leur représentation plane, Actes du colloque sur les permutations, Juillet 1972, Gauthier-Villars, 97–115.

A. Lascoux (1974a) Polynômes symétriques et coefficients d'intersection de cycles de Schubert, *C. R. Acad. Sc. Paris*, *279*, 201–204.

A. Lascoux (1974b) Tableaux de Young et fonctions de Schur-Littlewood, Séminaire Delange-Pisot-Poitou, No. 4, 1–7.

A. Lascoux (1976) Calcul de Schur et extensions grassmanniennes des λ-anneaux, from Combinatoire et Représentation du Groupe Symétrique, Strasbourg 1976, Lecture Notes in Math. No. 579, Springer-Verlag, Berlin-Heidelberg-New York.

D. E. Littlewood (1950) The theory of Group Characters and Matrix Representations of Groups, 2nd ed., Oxford University Press, Oxford.

D. E. Littlewood (1958) A University Algebra: An Introduction to Classic and Modern Algebra, Reprinted: Dover, New York.

D. E. Littlewood and A. R. Richardson (1934) Group characters and algebra, *Phil. Trans.*, *A-233*, 99–141.

D. Mitchell (1972) Generating functions for various sets of solid partitions, Ph. D. Thesis, Pennsylvania State University, University Park.

A. O. Morris (1963a) The characters of the group $GL(n, q)$, *Math. Zeit.*, *81*, 112–123.

A. O. Morris (1963b) The multiplication of Hall functions, *Proc. London Math. Soc.(3)*, *13*, 733–742.

A. O. Morris (1964) A note on symmetric functions, *American Math. Monthly*, *71*, 50–53.

A. O. Morris (1965) On an algebra of symmetric functions, *Quart. J. Math.(2)*, *16*, 53–64.

A. O. Morris (1967) A note on lemmas of Green and Kondo, *Proc. Cambridge Phil. Soc.*, *63*, 83–85.

A. O. Morris (1971) Generalizations of the Cauchy and Schur identities, *J. Combinatorial Th.*, *11*, 163–169.

G. deB. Robinson (1961) Representation Theory of the Symmetric Group, University of Toronto Press, Toronto.

G.-C. Rota (1971) Combinatorial theory and invariant theory, Notes by L. Guibas, Bowdoin College, Brunswick, Maine.

D. E. Rutherford (1948) Substitutional Analysis, Reprinted: Hafner, New York.

C. Schensted (1961) Longest increasing and decreasing subsequences, *Canadian J. Math.*, *13*, 179–191.

M. P. Schützenberger (1963) Quelques remarques sur une construction de Schensted, *Math. Scand.*, *12*, 117–128.

R. P. Stanley (1971) Theory and application of plane partitions (I) and (II), *Studies in Appl. Math.*, *50*, 167–188, 259–279.

R. P. Stanley (1972) Ordered structures and partitions, *Memoirs of the A.M.S.*, No. 119.

R. P. Stanley (1973) The conjugate trace and trace of a plane partition, *J. Combinatorial Th.*, *A-14*, 53–65.

R. P. Stanley (1976) Some combinatorial aspects of the Schubert calculus, from Combinatoire et Représentation du Groupe Symétrique, Strasbourg 1976, Lecture Notes in Math. No. 579, Springer-Verlag, Berlin-Heidelberg-New York.

M.-P. Schützenberger (1971) Sur un théorème de G. de B. Robinson, *C.R. Acad. Sc. Paris*, *272*, 420–421.

M.-P. Schützenberger (1976) La correspondance de Robinson, from Combinatoire et Représentation du Groupe Symétrique, Strasbourg 1976, Lecture Notes in Math. No. 579, Springer-Verlag, Berlin-Heidelberg-New York.

G. P. Thomas (1974) Baxter algebras and Schur functions, Ph.D. Thesis, Univ. of Wales, Swansea, 1974.

G. P. Thomas (1976a) Further results on Baxter sequences and generalized Schur functions, from Combinatoire et Représentation du Groupe Symétrique, Strasbourg 1976, Lecture Notes in Math. No. 579, Springer-Verlag, Berlin-Heidelberg-New York.

G. P. Thomas (1976b) A combinatorial interpretation of the wreath product of Schur functions, *Canadian J. Math.*, *28*, 879–884.

G. P. Thomas (1977) Frames, Young tableaux, and Baxter sequences, *Advances in Math.* (to appear).

G. Viennot (1976) Une forme géometrique de la correspondance de Robinson-Schensted, from Combinatoire et Représentation du Groupe Symétrique, Strasbourg 1976, Lecture Notes in Math. No. 579, Springer-Verlag, Berlin-Heidelberg-New York.

E. M. Wright (1931) Asymptotic partition formulae I: Plane partitions, *Quart. J. Math.(2)*, 177–189.

E. M. Wright (1966) The generating function for solid partitions, *Proc. Royal Soc. Edinburgh*, *A*-67, 185–195.

E. M. Wright (1968) Rotatable partitions, *J. London Math. Soc.*, *43*, 501–505.

A. Young (1977) The collected papers of Alfred Young, University of Toronto Press, Toronto.

[56] Partitions of numbers whose graphs possess symmetry, *Trans. Cambridge Phil. Soc.*, *17* (1899), 149–170.

In this paper, MacMahon considers a three-dimensional Ferrers graph representation of plane partitions (see [51] for definitions). He restricts his consideration to those partitions whose graphs are first *xy*-symmetric (i.e., invariant under $(x, y, z) \to (y, x, z)$), and then he considers *xyz*-symmetry (i.e., invariance under any permutation of the coordinates).

Perhaps the most striking aspect of this paper is the following conjecture: The generating function of *xy*-symmetric graphs which have at most i nodes along each of the axes, x, y, and at most j nodes along the axis of z, is conjectured to be

$$\frac{\prod_{1 \leqq s \leqq i/2} \prod_{k=0}^{2i-4s} (2j + 4s + 2k)}{\prod_{1 \leqq s \leqq i/2} \prod_{k=0}^{2i-4s} (4s + 2k)} \cdot \frac{(j + 1)(j + 3) \ldots (j + 2i - 1)}{(1)(3) \ldots (2i - 1)},$$

where $(s) = 1 - x^s$. We refer to Knuth and Bender (1972, page 50) for a discussion of this and similar conjectures. Solutions to these conjectures appear in Andrews (1978a), (1978b).

Results on *xyz*-symmetry do not provide any such interesting questions. It is shown, for example, that

$$1 + \sum_{k \geqq 2} \frac{x^{(k-1)^3 - (k-2)^3}}{(1 - x^3)(1 - x^9) \ldots (1 - x^{6k-9})}$$
$$= 1 + x + \sum_{k \geqq 2} x^{3k-2} (1 + x^3)(1 + x^9) \ldots (1 + x^{6k-9})$$

through the use of two different enumerations of the partitions whose graphs possess *xyz*-symmetry with every node in one of the coordinate planes.

[78] Memoir on the theory of the partitions of numbers—Part VI. Partitions in two-dimensional space, to which is added an adumbration of the theory of partitions in three-dimensional space, Phil. Trans., *211* (1912), 345–373.

This paper concludes the work begun in [77], which introduces the terminology and functions studied here. MacMahon's basic approach is to use functional equations for appropriate generating functions in order to prove that they have a desired form. In this way he is able to sketch the proof of the main conjectures in [77].

He concludes by indicating the lines of research for treating three-dimensional partitions. He states, "We have evidently, potentially, the complete solution of the problem of three-dimensional partition, and it remains to work it out and bring it to the same completeness as has been secured in this part for the problem of two dimensions. This will form the subject of Part VII of this Memoir."

Unfortunately, all of MacMahon's conjectures on higher dimensional partitions have proved false in later years (see the discussion of [51], section 9.6), and indeed Part VII of this series of memoirs [87] treats enumeration of the partitions of multipartite numbers and not three-dimensional partitions.

Reprints of the Papers

[56] Partitions of numbers whose graphs possess symmetry

[78] Memoir on the theory of the partitions of numbers—Part VI. Partitions in two-dimensional space, to which is added an adumbration of the theory of the partitions in three-dimensional space

V. *Partitions of Numbers whose Graphs possess Symmetry.*

By Major P. A. MacMahon, R.A., D.Sc., F.R.S., Hon. Mem. C.P.S.

[*Received* and *read* 28 November, 1898.]

It will be remembered that in *Phil. Trans. R. S. of London,* Vol. 187, 1896 A. pp. 619—673, I undertook the extension to three dimensions of Sylvester's constructive theory of Partitions. In Sylvester's two-dimensional theory every partition of a unipartite number can be associated with a regular two-dimensional graph. In the present theory only a limited number of the partitions of multipartite numbers can be represented by regular graphs in three dimensions. But whereas Sylvester was only concerned with unipartite numbers, the three-dimensional theory has to do with multipartite numbers of unrestricted multiplicity. Though the partitions of such are not all involved the field is infinitely greater, and all which come within the purview of the regular graph are brought harmoniously together. If in this new theory we restrict ourselves to two dimensions but view the graphs from a three-dimensional standpoint, we obtain in general six interpretations of the graphs instead of two and multipartite numbers are brought under consideration as well as those which are unipartite. The enumeration of the three-dimensional graphs of given weight (number of nodes), the numbers of nodes along the axes being restricted not to exceed l, m, n respectively, was conjectured in Part I. but only established for some particular values of l, m, n.

For $m \geqq l$ it may be written

$$\frac{1-x^{n+1}}{1-x} \cdot \left(\frac{1-x^{n+2}}{1-x^2}\right)^2 \cdots \left(\frac{1-x^{n+l-1}}{1-x^{l-1}}\right)^{l-1}$$
$$\times \left\{\frac{1-x^{n+l}}{1-x^l} \cdot \frac{1-x^{n+l+1}}{1-x^{l+1}} \cdots \frac{1-x^{n+m}}{1-x^m}\right\}^l$$
$$\times \left(\frac{1-x^{n+m+1}}{1-x^{m+1}}\right)^{l-1} \cdot \left(\frac{1-x^{n+m+2}}{1-x^{m+2}}\right)^{l-2} \cdots \frac{1-x^{n+l+m-1}}{1-x^{l+m-1}}.$$

The symmetry of this expression and its real nature are best shewn by a symbolic crystalline form.

Observing that it is composed of factors of the form $(1-x^s)^t$, where t may be positive or negative, put

$$1-x^s = \exp.(-u^s) \text{ in the case of every factor,}$$

and it will be found, after a few simplifications, to take the form

$$\exp. \frac{u}{(1-u)^2}(1-u^l)(1-u^m)(1-u^n).$$

In the two-dimensional theory the generating function

$$\frac{(1-x^{l+1})(1-x^{l+2})\dots(1-x^{l+m})}{(1-x)(1-x^2)\dots(1-x^m)}$$

has the symbolic crystalline form

$$\exp. \frac{u}{1-u}(1-u^l)(1-u^m),$$

whilst in one-dimensional theory

$$\frac{1-x^{l+1}}{1-x}$$

obviously leads to

$$\exp. u(1-u^l).$$

Hence we seem to have before us a system in κ dimensions associated with the crystalline form

$$\exp. \frac{u}{(1-u)^{\kappa-1}}(1-u^{l_1})(1-u^{l_2})\dots(1-u^{l_\kappa}).$$

In general a graph by rotations about the axes of x, y, and z

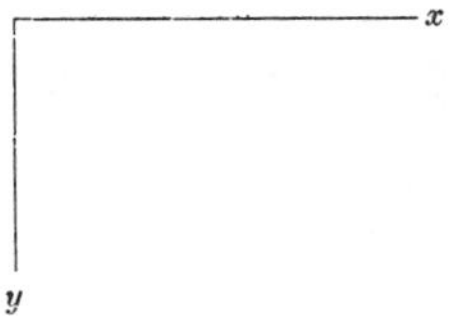

may assume six forms.

When these forms are identical the graph is said to be symmetrical or to possess xyz-symmetry.

Such ex. gr. is

When the six forms reduce to three the graph is said to be quasi-symmetrical. If it be such that each layer of nodes is symmetrical in two dimensions or, the same thing, is a Sylvester self-conjugate graph, it is said to possess xy-symmetry. Ex. gr.

Similarly the graph

possesses yz-symmetry, and by rotation about the y axis, or that of z, may be converted into one possessing xy or zx-symmetry.

It is proposed to investigate generating functions for the enumeration of graphs possessing xy and xyz-symmetry, the former naturally including the latter.

Algebraic theorems will be evolved in the course of the work by the method initiated by Sylvester.

xy-SYMMETRICAL GRAPHS.

The self-conjugate Sylvester graphs which have i nodes along each axis can be formed by fitting into an angle of $2i-1$ nodes any number of angles of nodes, any angle containing an uneven number, less than $2i-1$, of nodes and no two angles possessing the same number of nodes. Ex. gr. for $i=7$ we have the angles

which by selection of the 1st, 3rd and 4th of the angles may be formed up into the graph

Hence, as Sylvester shewed, the generating function of such graphs is immediately seen to be

$$x^{2i-1}(1+x)(1+x^3)\dots(1+x^{2i-3}).$$

Each layer of the three-dimensional graph has this form, and if there be two layers at most we may construct a generating function

$$\Omega a_1a_2\dots a_i x^{2i-1}(1+a_1x)(1+a_1a_2x^3)\dots(1+a_1a_2\dots a_{i-1}x^{2i-3})$$

$$\times\left\{\left(1+\frac{x}{a_1}\right)\left(1+\frac{x^3}{a_1a_2}\right)\left(1+\frac{x^5}{a_1a_2a_3}\right)\dots \text{ ad inf.}\right\}$$

where Ω is a symbol of Cayley's which means that after multiplication all terms containing negative powers of a_1, a_2, $a_3\dots a_i$ are to be struck out and then each of these letters put equal to unity.

The first line of the expression following Ω is derived from

$$x^{2i-1}(1+x)(1+x^3)\dots(1+x^{2i-3})$$

by placing as coefficient to each x^{2s-1} the product $a_1a_2\dots a_s$.

The angles of the first or lower layer correspond to the powers of x in the first line, those of the second layer to the powers in the second line, and the operation of Ω is such as to prevent any combinations of the former and the latter which give rise to an irregular graph.

Summing this function from $i=1$ to $i=i$ and supposing its value unity when $i=0$ (a convention that is made only for convenience; no form exists for $i=0$) we obtain

$$\Omega\,(1+a_1x)(1+a_1a_2x^3)(1+a_1a_2a_3x^5)\ldots(1+a_1a_2\ldots a_ix^{2i-1})$$
$$\times\left\{\left(1+\frac{x}{a_1}\right)\left(1+\frac{x^3}{a_1a_2}\right)\left(1+\frac{x^5}{a_1a_2a_3}\right)\ldots\text{ ad inf.}\right\}$$

as the generating function which enumerates xy-symmetrical graphs of at most two layers, the number of nodes along an x or y axis being limited not to exceed i.

Further if i be infinite this becomes:—

$$\Omega\,(1+a_1x)(1+a_1a_2x^3)(1+a_1a_2a_3x^5)\ldots\ldots\text{ ad inf.}$$
$$\times\left(1+\frac{x}{a_1}\right)\left(1+\frac{x^3}{a_1a_2}\right)\left(1+\frac{x^5}{a_1a_2a_3}\right)\ldots\ldots\text{ ad inf.}$$

It is moreover clear that the generating function of xy-symmetrical graphs which have at most i nodes along each of the axes x, y and at most j nodes along the axis of z (i.e. which involve at most j layers) is:—

$$\Omega\,(1+a_1x)(1+a_1a_2x^3)(1+a_1a_2a_3x^5)\ldots\ldots(1+a_1a_2\ldots a_ix^{2i-1})$$
$$\times\left(1+\frac{b_1}{a_1}x\right)\left(1+\frac{b_1b_2}{a_1a_2}x^3\right)\left(1+\frac{b_1b_2b_3}{a_1a_2a_3}x^5\right)\ldots\ldots\text{ ad inf.}$$
$$\times\left(1+\frac{c_1}{b_1}x\right)\left(1+\frac{c_1c_2}{b_1b_2}x^3\right)\left(1+\frac{c_1c_2c_3}{b_1b_2b_3}x^5\right)\ldots\ldots\text{ ad inf.}$$
$$\times\left(1+\frac{d_1}{c_1}x\right)\left(1+\frac{d_1d_2}{c_1c_2}x^3\right)\left(1+\frac{d_1d_2d_3}{c_1c_2c_3}x^5\right)\ldots\ldots\text{ ad inf.}$$
$$\cdots\cdots\cdots\cdots\cdots\cdots\cdots\cdots\cdots\cdots$$

j rows,

Ω operating in regard to all the symbols, a, b, c, d, &c. ...

If the graphs be unrestricted, as regards i, we put $i=\infty$; and, if they be totally unrestricted, we regard the tableau, upon which Ω operates, as possessing an unlimited number of rows and columns.

The generating function is crude. One, which only involves x, is ultimately to be desired. It should be possible, by algebraic processes, to perform the operation Ω and thus to pick out the terms of the product which constitute the reduced generating function. This appears to be a matter of considerable difficulty, and in order to determine the probable form of the reduced function I have examined many particular cases and attempted its construction. My conclusion is that, writing (s) to denote $1-x^s$, the reduced function is, in all probability, an algebraic fraction of which the numerator is

$$\begin{array}{l} (j+1)(j+3)(j+5) \quad \ldots\ldots (j+2i-1) \\ \times (2j+4)(2j+6)(2j+8) \quad \ldots\ldots (2j+4i-4) \\ \times (2j+8)(2j+10)(2j+12) \quad \ldots\ldots (2j+4i-8) \\ \times \ldots\ldots\ldots\ldots\ldots\ldots\ldots\ldots\ldots\ldots \\ \times (2j+4s)(2j+4s+2)(2j+4s+4) \ldots\ldots (2j+4i-4s) \\ \times \ldots\ldots\ldots\ldots\ldots\ldots\ldots\ldots\ldots\ldots \end{array}$$

wherein, if i be even, there are $\frac{1}{2}i$ rows the last of which is

$$(2j+2i);$$

and, if i be uneven, there are $\frac{1}{2}(i-1)$ rows the last of which is

$$(2j+2i-2)(2j+2i)(2j+2i+2);$$

and the denominator is obtained from the numerator by putting $j=0$, viz. :—it is

$$\begin{array}{l} (1)(3)(5) \quad \ldots\ldots (2i-1) \\ \times (4)(6)(8) \quad \ldots\ldots (4i-4) \\ \times (8)(10)(12) \quad \ldots\ldots (4i-8) \\ \times \ldots\ldots\ldots\ldots\ldots\ldots\ldots\ldots \\ \times (4s)(4s+2)(4s+4) \ldots\ldots (4i-4s) \\ \times \ldots\ldots\ldots\ldots\ldots\ldots\ldots\ldots \end{array}$$

the last row being $(2i)$ or $(2i-2)(2i)(2i+2)$ according as i is even or uneven.

The proof of this formula, the truth of which seems unquestionable, is much to be desired.

When the number of layers of nodes is unrestricted we put $j=\infty$ and the numerator reduces to unity. When moreover both i and j are unrestricted in magnitude the reduced function becomes

$$\frac{1}{(1)(3)(5)(7)\ldots(4)(6)(8)^2(10)^2(12)^3(14)^3(16)^4(18)^4\ldots\ldots},$$

or as it may be also written

$$\frac{(1+x)(1+x^3)(1+x^5)(1+x^7)\ldots\ldots}{(2)(4)(6)^2(8)^2(10)^3(12)^3(14)^4(16)^4\ldots\ldots},$$

wherein the numerator denotes the generating function of Sylvester's unrestricted self-conjugate graphs in two dimensions.

Some particular cases are interesting.

By putting $j=1$ we should obtain Sylvester's result in two dimensions.

We find

$$\frac{(2)(6)(10)(14)\dots(4i-2)}{(1)(3)(5)(7)\dots(2i-1)},$$

which may be written

$$(1+x)(1+x^3)(1+x^5)(1+x^7)\dots(1+x^{2i-1})$$

and is right.

When $j=2$, we find

$$\frac{(2i+1)}{(1)}\cdot\frac{(2i+4)(2i+6)\dots(4i-2)(4i)}{(4)\quad(6)\quad\dots(2i-2)(2i)},$$

or

$$\frac{(2i+2)(2i+4)\dots(4i-2)(4i)}{(2)\quad(4)\quad\dots(2i-2)(2i)}+x\frac{(2i+4)(2i+6)\dots(4i-2)(4i)}{(2)(4)(6)\quad\dots(2i-2)}.$$

For an even weight $2w$ we must take the coefficients of x^w in

$$\frac{(i+1)(i+2)\dots(2i-1)(2i)}{(1)\quad(2)\quad\dots\quad(i-1)\ (i)},$$

and this is the generating function of two-dimensional graphs of weight w, not more than i nodes being allowed along either the x or y axis. Hence a correspondence between the at-most-two-layer xy-symmetrical graphs of weight $2w$ restricted as to the x and y axes by the number i and the graphs in two dimensions of weight w restricted as to the axes by the number i.

Ex. gr. for $w=4$, $i=3$ the correspondence is

1 1 1	1 1	1 1	2 2 1	2 2	1 1 1
1	1 1	1	2	2 2	1 1 1
		1	1		1 1

For an uneven weight $2w+1$ we take the coefficients of x^w in

$$\frac{(i+2)(i+3)\dots(2i-1)(2i)}{(1)(2)(3)\quad\dots\quad(i-1)},$$

and this is the generating function of two-dimensional graphs of weight w, not more than $i+1$ nodes being allowed along the x axis nor more than $i-1$ along the y axis.

The correspondence established is that between the at-most-two-layer xy-symmetrical graphs of weight $2w+1$ restricted as to each of the x and y axes by the number i and the graphs of two dimensions of weight w restricted as to the x axis by the number $i+1$ and as to the y axis by the number $i-1$.

Ex. gr. for $w=5$, $i=4$ we have the five to five correspondence

1 1 1 1 1	1 1 1 1	1 1 1	1 1 1	1 1	1 1 1 1	2 1 1 1	2 2 1 1	2 2 1	2 2 2
	1	1 1	1	1 1	1 1 1	1 1 1	2 1	2 1 1	2 1
			1	1	1 1 1	1 1	1	1 1	2
					1	1	1		

where i is indefinite the generating function becomes

$$\frac{1+x}{(1-x^2)(1-x^4)(1-x^6)\ldots\ldots \text{ ad inf.}}.$$

This curious result shews that the number of at-most-two-layer xy-symmetrical graphs of weight w is equal to the whole number of partitions of $\frac{1}{2}w$ or of $\frac{1}{2}(w-1)$ according as w is even or uneven.

There is another solution of the problem that has been under consideration.

Instead of constructing a generating function from successive layers of nodes parallel to the plane of xy, we may build one up by first considering all the exterior angles of nodes; then those which become exterior when the former are removed; and so on. Thus if any graph were

$$\begin{array}{lllll} 4 & 3 & 2 & 2 & 1 \\ 3 & 2 & 2 & 1 & \\ 2 & 2 & 1 & 1 & \\ 2 & 1 & 1 & & \\ 1 & & & & \end{array}$$

we first take

$$\begin{array}{lllll} 4 & 3 & 2 & 2 & 1 \\ 3 & & & & \\ 2 & & & & \\ 2 & & & & \\ 1 & & & & \end{array}$$

as constructed by the superposition of

$$\begin{array}{lllll} 1 & 1 & 1 & 1 & 1 \\ 1 & & & & \\ 1 & & & & \\ 1 & & & & \\ 1 & & & & \end{array}\qquad \begin{array}{llll} 1 & 1 & 1 & 1 \\ 1 & & & \\ 1 & & & \\ 1 & & & \end{array}\qquad \begin{array}{ll} 1 & 1 \\ 1 & \end{array}\qquad 1;$$

then

$$\begin{array}{lll} 2 & 2 & 1 \\ 2 & & \\ 1 & & \end{array}\quad \text{made up of}\quad \begin{array}{lll} 1 & 1 & 1 \\ 1 & & \\ 1 & & \end{array}\quad \begin{array}{ll} 1 & 1 \\ 1 & \end{array}\quad ;$$

then

$$\begin{array}{ll} 1 & 1 \\ 1 & \end{array}$$

We are then led to the crude generating function

$$\Omega \frac{1}{(1-m)(1-mx)(1-max^3)(1-mabx^5)(1-mabcx^7)\ldots(1-mabc\ldots x^{2i-1})}$$

$$\left(1-\frac{x}{a}\right)\left(1-\frac{a'}{ab}x^3\right)\left(1-\frac{a'b'}{abc}x^5\right)\ldots \text{ ad inf.},$$

$$\left(1-\frac{x}{a'}\right)\left(1-\frac{a''}{a'b'}x^3\right)\left(1-\frac{a''b''}{a'b'c'}x^5\right)\ldots \text{ ad inf.}$$

$$\vdots$$

ad inf.

in the ascending expansion of which we must take the coefficient of m^j, Ω operating in regard to the letters

$$\begin{matrix} a\ , & b\ , & c\ ,\ldots \\ a'\ , & b'\ , & c'\ ,\ldots \\ a'', & b'', & c'',\ldots \\ \ldots & \ldots & \ldots \end{matrix}$$

We have, therefore, the identity

$$\begin{aligned} &\Omega(1+a_1x)(1+a_1a_2x^3)(1+a_1a_2a_3x^5)\ldots(1+a_1a_2\ldots a_ix^{2i-1}) \\ &\quad\times\left(1+\frac{b_1}{a_1}x\right)\left(1+\frac{b_1b_2}{a_1a_2}x^3\right)\left(1+\frac{b_1b_2b_3}{a_1a_2a_3}x^5\right)\ldots \text{ ad inf.} \\ &\quad\times\left(1+\frac{c_1}{b_1}x\right)\left(1+\frac{c_1c_2}{b_1b_2}x^3\right)\left(1+\frac{c_1c_2c_3}{b_1b_2b_3}x^5\right)\ldots \text{ ad inf.} \\ &\quad\ldots\ldots\ldots\ldots\ldots\ldots\ldots\ldots\ldots\ldots\ldots\ldots \\ &\qquad\qquad j \text{ rows} \end{aligned}$$

$$\begin{aligned} = \text{Co } m^j\,\Omega\,&\frac{1}{(1-m)(1-mx)(1-max^3)(1-mabx^5)(1-mabcx^7)\ldots(1-mabc\ldots x^{2i-1})} \\ &\left(1-\frac{x}{a}\right)\left(1-\frac{a'}{ab}x^3\right)\left(1-\frac{a'b'}{abc}x^5\right)\ldots \text{ ad inf.} \\ &\left(1-\frac{x}{a'}\right)\left(1-\frac{a''}{a'b'}x^3\right)\left(1-\frac{a''b''}{a'b'c'}x^5\right)\ldots \text{ ad inf.} \\ &\qquad\vdots \\ &\qquad\text{ad inf.} \end{aligned}$$

and, when j is unrestricted,

$$\begin{aligned} &\Omega(1+a_1x)(1+a_1a_2x^3)(1+a_1a_2a_3x^5)\ldots(1+a_1a_2\ldots a_ix^{2i-1}) \\ &\quad\times\left(1+\frac{b_1}{a_1}x\right)\left(1+\frac{b_1b_2}{a_1a_2}x^3\right)\left(1+\frac{b_1b_2b_3}{a_1a_2a_3}x^5\right)\ldots \text{ ad inf.} \\ &\quad\times\left(1+\frac{c_1}{b_1}x\right)\left(1+\frac{c_1c_2}{b_1b_2}x^3\right)\left(1+\frac{c_1c_2c_3}{b_1b_2b_3}x^5\right)\ldots \text{ ad inf.} \\ &\qquad\vdots \\ &\qquad\text{ad inf.} \end{aligned}$$

$$\begin{aligned} = \Omega\,&\frac{1}{(1-x)(1-ax^3)(1-abx^5)(1-abcx^7)\ldots(1-abc\ldots x^{2i-1})} \\ &\left(1-\frac{x}{a}\right)\left(1-\frac{a'}{ab}x^3\right)\left(1-\frac{a'b'}{abc}x^5\right)\ldots \text{ ad inf.} \\ &\left(1-\frac{x}{a'}\right)\left(1-\frac{a''}{a'b'}x^3\right)\left(1-\frac{a''b''}{a'b'c'}x^5\right)\ldots \text{ ad inf.} \\ &\qquad\vdots \\ &\qquad\text{ad inf.,} \end{aligned}$$

a remarkable result, which it would be difficult to establish algebraically.

As it is necessary in the sequel we will now determine the generating function which enumerates the xy-symmetrical graphs, limited as above, but subject to a new restriction, viz. each layer of nodes is to be formed by, at most, s plane angles of nodes.

The enumeration, it is easy to see, is given by the coefficients of $m^s x^w$ in the development of

$$\frac{1}{1-m}\,\Omega\,(1+ma_1x)(1+ma_1a_2x^3)(1+ma_1a_2a_3x^5)\dots(1+ma_1a_2\dots a_ix^{2i-1})$$

$$\times\left(1+\frac{b_1}{a_1}x\right)\left(1+\frac{b_1b_2}{a_1a_2}x^3\right)\left(1+\frac{b_1b_2b_3}{a_1a_2a_3}x^5\right)\dots \text{ ad inf.}$$

$$\times\left(1+\frac{c_1}{b_1}x\right)\left(1+\frac{c_1c_2}{b_1b_2}x^3\right)\left(1+\frac{c_1c_2c_3}{b_1b_2b_3}x^5\right)\dots \text{ ad inf.}$$

$$\vdots$$

j rows,

and also by the coefficients of $m^j x^w$ in the development of

$$\frac{1}{1-m}\,\Omega\,\frac{1}{(1-mx)(1-max^3)(1-mabx^5)(1-mabcx^7)\dots(1-mabc\dots x^{2i-1})}$$

$$\left(1-\frac{x}{a}\right)\left(1-\frac{a'}{ab}x^3\right)\left(1-\frac{a'b'}{abc}x^5\right)\dots \text{ ad inf.}$$

$$\left(1-\frac{x}{a'}\right)\left(1-\frac{a''}{a'b'}x^3\right)\left(1-\frac{a''b''}{a'b'c'}x^5\right)\dots \text{ ad inf.}$$

$$\vdots$$

s rows.

Let the coefficients of m^s in the former of these generating functions be denoted by $F_{j,s}(x)$, and denoting the generating functions by A and B respectively, we have:—

$$A = 1 + mF_{j,1}(x) + m^2F_{j,2}(x) + \dots + m^sF_{j,s}(x) + \dots,$$

$$B = 1 + mF_{1,s}(x) + m^2F_{2,s}(x) + \dots + m^jF_{j,s}(x) + \dots.$$

Moreover for $j=\infty$, we have

$$A = 1 + mF_{\infty,1}(x) + m^2F_{\infty,2}(x) + m^3F_{\infty,3}(x) + \dots,$$

and for $s=\infty$,

$$B = 1 + mF_{1,\infty}(x) + m^2F_{2,\infty}(x) + m^3F_{3,\infty}(x) + \dots.$$

THE xyz-SYMMETRICAL GRAPHS.

Just as Sylvester dissected the xy-symmetrical graph in two dimensions into plane angles we may dissect the xyz-symmetrical graphs in three dimensions into solid angles. Each solid angle is in the shape of a symmetrical fragment of half of a hollow cube.

In each of the planes xy, yz, zx we find the same symmetrical two-dimensional graph. If this graph has i columns or rows the number of nodes which lie on one or other of the three axes is $1+3(i-1)$ or $3i-2$. In the plane of xy we can place plane angles of nodes so as to form a symmetrical graph in two dimensions. If w be the weight of the solid angle we have $w-3i+2$ nodes to dispose symmetrically in the three planes and this can be done in a number of ways which is given by the coefficients of

$$x^{\frac{1}{3}(w-3i+2)}$$

in
$$(1+x)(1+x^3)(1+x^5)\dots(1+x^{2i-3}),$$

that is, by the coefficients of x^w in

$$Q_i = x^{3i-2}(1+x^3)(1+x^9)(1+x^{15})\dots(1+x^{6i-9}),$$

which is therefore the generating function of the solid angles in question which have exactly i nodes along each axis. Observe that $i-1$ factors follow x^{3i-2}, and that, when it is convenient, we suppose the expression to have the value unity when $i=0$.

Hence the solid angles which possess i or fewer nodes along the axes are enumerated by

$$\begin{aligned}
&1\\
&+x\\
&+x^4(1+x^3)\\
&+x^7(1+x^3)(1+x^9)\\
&+\dots\dots\\
&+x^{3i-2}(1+x^3)(1+x^9)(1+x^{15})\dots(1+x^{6i-9}).
\end{aligned}$$

Fitting solid angle graphs together when possible produces xyz-symmetrical graphs.

When $i=2$, $Q_2 = x^4(1+x^3)$, the two solid angles being

⊙ •   ⊙ ⊙
•   ⊙ •

of contents 4 and 7 respectively.

We cannot fit a solid angle into the first of these, for there is no node upon which it can rest. In the case of the second we can fit in the solid angle for which $i=1$, $Q_1 = x$ represented by a single node •, and thus form the symmetrical graph

⊙ ⊙
⊙ ⊙ of content 8.

Synthetically we form the generating function

$$\Omega x^4\left(1+ax^3\right)\left(1+\frac{1}{a}x\right) = x^4 + x^7 + x^8$$

of all symmetric graphs having $i=2$.

Observe that the construction of the factors, following the operator Ω, permits the association of

⊙ ⊙
⊙ •   and •

and does not permit that of

⊙ •
•   and •

Restricting ourselves to two solid angles when

$$i = 3, \quad Q_3 = x^7(1+x^3)(1+x^9)$$

we are similarly led to the construction of the generating function

$$\Omega x^7(1+ax^3)(1+abx^9) \times \left\{1 + \frac{x}{a} + \frac{x^4}{ab}\left(1+\frac{x^3}{a}\right)\right\},$$

whence after expansion and operation we find

$$(x^7 + x^{10} + x^{16} + x^{19}) + (x^{11} + x^{17} + x^{20}) + (x^{20} + x^{23}) + x^{26},$$

and the correspondence is

x^7	$x^7 . ax^3$	$x^7 \times abx^9$	$x^7 . ax^3 . abx^9$
3 1 1	3 2 1	3 3 2	3 3 3
1	2 1	3 1 1	3 1 1
1	1	2 1	3 1 1

$x^7 . ax^3 . \frac{x}{a}$	$x^7 . abx^9 . \frac{x}{a}$	$x^7 . ax^3 . abx^9 . \frac{x}{a}$
3 2 1	3 3 2	3 3 3
2 2	3 2 1	3 2 1
1	2 1	3 1 1

$x^7 . abx^9 . \frac{x^4}{ab}$	$x^7 . ax^3 . abx^9 . \frac{x^4}{ab}$
3 3 2	3 3 3
3 3 2	3 3 2
2 2	3 2 1

$$x^7 . ax^3 . abx^9 . \frac{x^4}{ab} . \frac{x^3}{a}.$$

3 3 3
3 3 3
3 3 2

In the form which arises from the product $x^7 . abx^9 . \frac{x}{a}$, the largest solid angle is given by $x^7 . abx^9$; that is, x^7 gives the axial portion $\begin{matrix} 3 & 1 & 1 \\ 1 & & \\ 1 & & \end{matrix}$, x^9 yields $\begin{matrix} 1 & 1 \\ 1 & \end{matrix}$ in each of the three planes, so that the resulting angle is $\begin{matrix} 3 & 3 & 2 \\ 3 & 1 & 1 \\ 2 & 1 & \end{matrix}$; the next largest solid angle is given by $\frac{x}{a}$ and this fits into the larger.

Again from $x^7 . ax^3 . abx^9 . \frac{x^4}{ab} . \frac{x^3}{a}$ we get first $\begin{matrix}3&1&1\\1&&\\1&&\end{matrix}$ from x^7, and then $\begin{matrix}1&1\\1&\end{matrix}$ and 1, in each plane, from x^9 and x^3, yielding $\begin{matrix}3&3&3\\3&1&1\\3&1&1\end{matrix}$ the outer solid angle; and $\frac{x^4}{ab} . \frac{x^3}{a}$ gives a solid angle, composed of $\begin{matrix}2&1\\1&\end{matrix}$ and 1 fitting in each plane, viz.:— $\begin{matrix}2&2\\2&1\end{matrix}$, and this fits into the larger solid angle yielding $\begin{matrix}3&3&3\\3&3&3\\3&3&2\end{matrix}$.

It will be clear now that the generating function for symmetrical graphs having i nodes along each axis and formed of at most two solid angles is

$$\Omega x^{3i-2}(1 + a_1x^3)(1 + a_1a_2x^9) \dots (1 + a_1a_2 \dots a_{i-1}x^{6i-9})$$

$$\times \left\{1 + \frac{x}{a_1} + \frac{x^4}{a_1a_2}\left(1 + \frac{x^3}{a_1}\right) + \frac{x^7}{a_1a_2a_3}\left(1 + \frac{x^3}{a_1}\right)\left(1 + \frac{x^9}{a_1a_2}\right) + \dots \text{ad inf.}\right\},$$

the general term in the series to infinity being

$$\frac{x^{3s-5}}{a_1a_2 \dots a_{s-1}}\left(1 + \frac{x^3}{a_1}\right)\left(1 + \frac{x^9}{a_1a_2}\right) \dots \left(1 + \frac{x^{6s-15}}{a_1a_2 \dots a_{s-2}}\right).$$

Summing this function, for values of i, it is found that the generating function, for the graphs composed of at most two solid angles and having at most i nodes along each axis, is

$$\Omega\,\{1 + x + x^4(1 + a_1x^3) + x^7(1 + a_1x^3)(1 + a_1a_2x^9) + \dots$$
$$+ x^{3i-2}(1 + a_1x^3)(1 + a_1a_2x^9) \dots (1 + a_1a_2 \dots a_{i-1}x^{6i-9})\}$$
$$\times \left\{1 + \frac{x}{a_1} + \frac{x^4}{a_1a_2}\left(1 + \frac{x^3}{a_1}\right) + \frac{x^7}{a_1a_2a_3}\left(1 + \frac{x^3}{a_1}\right)\left(1 + \frac{x^9}{a_1a_2}\right) + \dots \text{ad inf.}\right\}.$$

If the graphs are to be composed of at most two solid angles but to be otherwise unrestricted we obtain

$$\Omega\,\{1 + x + x^4(1 + a_1x^3) + x^7(1 + a_1x^3)(1 + a_1a_2x^9) + \dots \text{ad inf.}\}$$
$$\times \left\{1 + \frac{x}{a_1} + \frac{x^4}{a_1a_2}\left(1 + \frac{x^3}{a_1}\right) + \frac{x^7}{a_1a_2a_3}\left(1 + \frac{x^3}{a_1}\right)\left(1 + \frac{x^9}{a_1a_2}\right) + \dots \text{ad inf.}\right\}.$$

It is now easy to pass to the general case in which the composition is to be from at most s solid angles. The generating function is

$$\Omega\,\{1 + x + x^4(1 + a_1x^3) + x^7(1 + a_1x^3)(1 + a_1a_2x^9) + \dots$$
$$+ x^{3i-2}(1 + a_1x^3)(1 + a_1a_2x^9)\dots(1 + a_1a_2\dots a_{i-1}x^{6i-9})\}$$
$$\times \left\{1 + \frac{x}{a_1} + \frac{x^4}{a_1a_2}\left(1 + \frac{b_1}{a_1}x^3\right) + \frac{x^7}{a_1a_2a_3}\left(1 + \frac{b_1}{a_1}x^3\right)\left(1 + \frac{b_1b_2}{a_1a_2}x^9\right) + \dots \text{ ad inf.}\right\}$$
$$\times \left\{1 + \frac{x}{b_1} + \frac{x^4}{b_1b_2}\left(1 + \frac{c_1}{b_1}x^3\right) + \frac{x^7}{b_1b_2b_3}\left(1 + \frac{c_1}{b_1}x^3\right)\left(1 + \frac{c_1c_2}{b_1b_2}x^9\right) + \dots \text{ ad inf.}\right\}$$
$$\times \left\{1 + \frac{x}{c_1} + \frac{x^4}{c_1c_2}\left(1 + \frac{d_1}{c_1}x^3\right) + \frac{x^7}{c_1c_2c_3}\left(1 + \frac{d_1}{c_1}x^3\right)\left(1 + \frac{d_1d_2}{c_1c_2}x^9\right) + \dots \text{ ad inf.}\right\}$$
$$\times \dots\dots$$

s rows.

When the first row is also continued to infinity, and the number of rows is infinite, we have the crude form of generating function for xyz-symmetrical graphs quite unrestricted.

When $s = 1$ and $i = \infty$ it may be easily proved that the generating function may be written

$$1 + \frac{x}{1-x^3} + \frac{x^7}{(1-x^3)(1-x^9)} + \frac{x^{19}}{(1-x^3)(1-x^9)(1-x^{15})} + \dots$$
$$+ \frac{x^{(k-1)^3-(k-2)^3}}{(1-x^3)(1-x^9)(1-x^{15})\dots(1-x^{6k-9})} + \dots$$

There is another mode of enumeration of xyz-symmetrical graphs which it is important to consider.

Durfee has shewn how to dissect a symmetrical graph in two dimensions into a square of nodes and two appendages lateral and subjacent.

Ex. gr. the graph

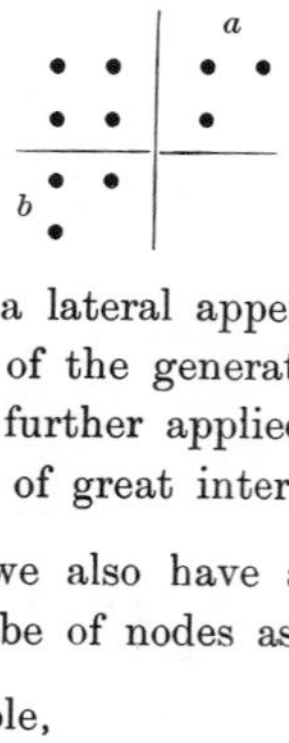

where this is a square of four nodes, a lateral appendage a and one which is subjacent b. This dissection leads to the expression of the generating function in the form of an infinite series of algebraic fractions. Sylvester further applied the same dissection to unsymmetrical graphs and derived algebraic identities of great interest.

In the case of three dimensions we also have a dissection of the same nature. This is not based upon the isolation of a cube of nodes as might at first appear.

If we take such a cube, for example,

2 2
2 2

we may, it is true, attach appropriate lateral, subjacent, and superjacent graphs and thus obtain an xyz-symmetrical graph; but a slight consideration shews that a large number of symmetrical graphs escape enumeration by this process. Ex. gr. the graph

$$\begin{array}{cc|c} 3 & 3 & 2 \\ 3 & 2 & 1 \\ \hline 2 & 1 & \end{array}$$

is based upon the cube in question, whereas the graph

$$\begin{array}{cc|c} 3 & 3 & 3 \\ 3 & 2 & 1 \\ \hline 3 & 1 & 1 \end{array}$$

is not based upon that or any other cube, yet it is without doubt symmetrical.

In the former of the two graphs observe that the appendages are

$$\begin{array}{rl} \text{lateral} & \begin{array}{c} 2 \\ 1 \end{array} \\ \\ \text{subjacent} & 2\ 1 \\ \\ \text{superjacent} & \begin{array}{cc} 1 & 1 \\ 1 & \end{array} \end{array}$$

The fact is that symmetrical graphs are based also upon graphs other than those which are perfect cubes.

The whole series is formed as follows:—

We have, first, those based upon the cube 1, viz.

the base is 1.

Secondly, we have those based upon graphs such that there is a square of four nodes in each of the three planes of reference. These are of two kinds, viz.:—

$$\begin{array}{cc} 2 & 2 \\ 2 & 1 \end{array} \qquad \begin{array}{cc} 2 & 2 \\ 2 & 2 \end{array}$$

where the nodes of the former are in the shape of the half of a hollow cube and the latter is obtained from the former by combining with it the cube 1.

Thirdly, we have four bases derived from the graph which has the shape of a half-hollow-cube of side 3; viz.:—

$$\begin{array}{ccc} 3 & 3 & 3 \\ 3 & 1 & 1 \\ 3 & 1 & 1 \end{array} \qquad \begin{array}{ccc} 3 & 3 & 3 \\ 3 & 2 & 1 \\ 3 & 1 & 1 \end{array} \qquad \begin{array}{ccc} 3 & 3 & 3 \\ 3 & 3 & 3 \\ 3 & 3 & 2 \end{array} \qquad \begin{array}{ccc} 3 & 3 & 3 \\ 3 & 3 & 3 \\ 3 & 3 & 3 \end{array}$$

where observe that the three latter bases are derived from the former by combination with the three bases previously constructed, viz.:—

$$1 \qquad \begin{array}{cc} 2 & 2 \\ 2 & 1 \end{array} \qquad \begin{array}{cc} 2 & 2 \\ 2 & 2 \end{array}.$$

Similarly, of the fourth order, we have eight bases, viz.:—

4 4 4 4	4 4 4 4	4 4 4 4	4 4 4 4
4 1 1 1	4 2 1 1	4 3 3 1	4 3 3 1
4 1 1 1	4 1 1 1	4 3 2 1	4 3 3 1
4 1 1 1	4 1 1 1	4 1 1 1	4 1 1 1

4 4 4 4	4 4 4 4	4 4 4 4	4 4 4 4
4 4 4 4	4 4 4 4	4 4 4 4	4 4 4 4
4 4 2 2	4 4 3 2	4 4 4 4	4 4 4 4
4 4 2 2	4 4 2 2	4 4 4 3	4 4 4 4

the seven latter being derived from the former by combination with the seven forms previously constructed.

The way in which the bases are built up is now plain and we see that, of order n, we can construct 2^{n-1} bases of which $2^{n-1}-1$ are derived by combining with the half-hollow-square of order n, all the bases of lower orders in number,

$$1 + 2 + 2^2 + \dots + 2^{n-2} = 2^{n-1} - 1.$$

As one illustration take the graph

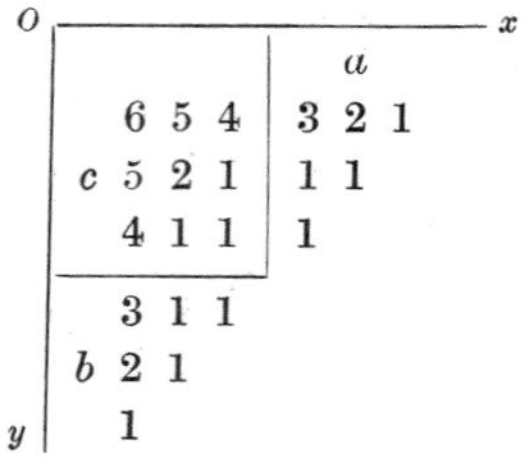

the z axis being perpendicular to the plane of the paper.

The graph is built upon the base

3	3	3
3	2	1
3	1	1;

a the lateral appendage being

3	2	1
1	1	
1;		

b the subjacent appendage being

3	1	1
2	1	
1;		

c the superjacent appendage being

3	2	1
2		
1.		

From the lateral appendage we derive in succession the subjacent and superjacent appendages.

The rule is to face the origin and give the lateral graph right-handed rotations through 90° about the axes of z and y in succession. We thus derive the subjacent graph, and a repetition of the process upon the latter then gives the superjacent graph.

Thus starting with the lateral

$$\begin{array}{lll} 3 & 2 & 1 \\ 1 & 1 & \\ 1 & & \end{array} \quad ,$$

the two rotations give in succession

$$\begin{array}{rrr} 1 & 1 & 3 \\ & 1 & 2 \\ & & 1 \end{array} \quad \text{and} \quad \begin{array}{lll} 3 & 1 & 1 \\ 2 & 1 & \\ 1 & & \end{array} \quad ,$$

the latter being the subjacent, and operating similarly on the latter we obtain in succession

$$\begin{array}{rrr} 1 & 2 & 3 \\ & 1 & 1 \\ & & 1 \end{array} \quad \text{and} \quad \begin{array}{lll} 3 & 2 & 1 \\ 2 & & \\ 1 & & \end{array} \quad ,$$

the last written graph being the superjacent.

As another example, if the lateral be

$$\begin{array}{lll} 2 & 2 & 1 \\ 1 & 1 & \end{array} \quad ,$$

we obtain by operation

$$\begin{array}{rr} 1 & 2 \\ 1 & 2 \\ & 1 \end{array} \quad \text{and} \quad \begin{array}{ll} 2 & 1 \\ 2 & 1 \\ 1 & \end{array} \quad ,$$

giving $\begin{array}{ll} 2 & 1 \\ 2 & 1 \\ 1 & \end{array}$ the subjacent; operating upon this

$$\begin{array}{rrr} 1 & 2 & 2 \\ & 1 & 1 \end{array} \quad \text{and} \quad \begin{array}{ll} 3 & 2 \\ 2 & \end{array}$$

giving $\begin{array}{ll} 3 & 2 \\ 2 & \end{array}$ the superjacent.

Compare the graph

$$\begin{array}{ll|lll} 5 & 4 & 2 & 2 & 1 \\ 4 & 1 & 1 & 1 & \\ \hline 2 & 1 & & & \\ 2 & 1 & & & \\ 1 & & & & \end{array}$$

upon the base

$$\begin{array}{ll} 2 & 2 \\ 2 & 1 \end{array} .$$

We have arrived at the point of shewing the construction of the bases and we have seen how to construct the graph, being given the base and the lateral; the base and the lateral completely determine the graph, and if they be of contents w_1, w_2 respectively the complete graph is of content $w_1 + 3w_2$. For a given base we have now to determine the possible forms of lateral appendage preparatory to attempting their enumeration.

Every line of numbers parallel to the axis of y in a symmetrical graph is of necessity a self-conjugate partition of a number, for otherwise more than one interpretation of the graph would be obtainable. Ex. gr. in the graph

5 4 2 2 1
4 1 1 1
2 1
2 1
1

5 4 2 2 1 is a self-conjugate partition of the number 14,
2 1 „ „ „ „ „ 3,

the corresponding symmetrical two-dimensional graphs being

• • • • •   • •
• • • •   •
• •
• •
•

Hence this self-conjugate property appertains also to the lateral appendage, the lines of numbers being taken parallel to the axis of y, *not* parallel to the axis of x. The reverse would naturally be the case if we were considering the subjacent appendage. This property imposes a limitation upon the possible forms of lateral appendage.

Let w_1 be the content of the base, i_1 its order i.e. the number of nodes along an axis; also let w_2 and i_2 refer to the lateral.

Then for the complete symmetrical graph we have content $w_1 + 3w_2$ and order $i_1 + i_2$, or say, w, i referring to the complete graph,

$$w = w_1 + 3w_2, \quad i = i_1 + i_2.$$

For the base 1

$w_1 = 1$, $i_1 = 1$ the lateral must have the form

1 1 1 1 ...,

and the generating function for such laterals whose order does not exceed i_2 is

$$\frac{1 - x^{i_2+1}}{1 - x};$$

therefore if $F(x)$ denote the generating function of the associated symmetrical graphs

$$^{*}\mathrm{Co}\, x^{w} F(x) = \mathrm{Co}\, x^{1+3w_2} F(x) = \mathrm{Co}\, x^{w_2} \frac{1 - x^{i_2+1}}{1 - x},$$

$$\therefore\ F(x) = x \frac{1 - x^{3i_2+3}}{1 - x^3},$$

the generating function of symmetrical graphs on the base 1.

Since $i_2 = i - 1$, we may write this

$$F(x) = x \frac{1 - x^{3i}}{1 - x^3}.$$

For the base $\begin{matrix} 2 & 2 \\ 2 & 1 \end{matrix}$ the lateral may involve $\begin{matrix} 2 \\ 1 \end{matrix}$ and 1 but not $\begin{matrix} 1 \\ 1 \end{matrix}$; hence its form must be

$$\begin{matrix} 2 & 2 & 2 \ldots 1 & 1 & 1 \ldots \\ 1 & 1 & 1 & & & \end{matrix}$$

If i_2 be unrestricted the lateral generating function is

$$\frac{1}{1 - x \,.\, 1 - x^3};$$

otherwise we have to seek the coefficient of $m^{i_2} x^{w_2}$ in

$$\frac{1}{1 - m \,.\, 1 - mx \,.\, 1 - mx^3};$$

and since $\qquad w_1 = 7,\ i_1 = 2,\ i_2 = i - 2,$

we obtain the generating function of symmetrical graphs

$$\frac{x^7}{1 - m \,.\, 1 - mx^3 \,.\, 1 - mx^9},$$

in which we seek the coefficient of $m^{i-2} x^w$.

Similarly for the base $\begin{matrix} 2 & 2 \\ 2 & 2 \end{matrix}$ since the lateral must be of the form $\begin{matrix} 2 & 2 \\ 2 & 2 \end{matrix} \ldots \begin{matrix} 2 & 2 \\ 1 & 1 \end{matrix} \ldots \begin{matrix} 1 & 1 \ldots \\ & \end{matrix}$ we are led to the generating function

$$\frac{x^8}{1 - m \,.\, 1 - mx^3 \,.\, 1 - mx^9 \,.\, 1 - mx^{12}},$$

in which we seek the coefficient of $m^{i-2} x^w$.

If the base is at most of order 2 we may say that the enumeration of symmetrical graphs of content w and having at most i nodes along an axis is given by the coefficient of $m^0 x^w$ in

$$1 + \frac{m^{1-i} x}{1 - m \,.\, 1 - mx^3} + \frac{m^{2-i} x^7}{1 - m \,.\, 1 - mx^3 \,.\, 1 - mx^9} + \frac{m^{2-i} x^8}{1 - m \,.\, 1 - mx^3 \,.\, 1 - mx^9 \,.\, 1 - mx^{12}}.$$

* $\mathrm{Co}\, x^w F(x)$ denotes the coefficient of x^w in the expansion of $F(x)$.

If i_2 and therefore i be unrestricted this expression naturally becomes

$$1+\frac{x}{1-x^3}+\frac{x^7}{1-x^3 \,.\, 1-x^9}+\frac{x^8}{1-x^3 \,.\, 1-x^9 \,.\, 1-x^{12}}.$$

The question now arises as to the direct formation of the fractions appertaining to the bases of order i_1.

The form of the lateral appendage depends, as we have seen, upon the self-conjugate unipartite partition represented by the right-hand column or boundary of the base. We will call this partition the base-lateral. So far of the first four orders we have met with certain base-laterals, viz.:—

Order $i_1 =$	Base-lateral			
1	1			
2	2	2		
	1	2		
	3	3	3	
3	1	3	3	
	1	2	3	
	4	4	4	4
4	1	4	4	4
	1	2	4	4
	1	2	3	4

Of order i_1 there are i_1 different base-laterals; for consider the formation of the base of order n from those of inferior orders. Combination of the half-hollow-cube form of order i_1 with the bases of orders less than $i_1 - 1$ can only result in base-laterals identical with that of the half-hollow-cube base; and assuming that base-laterals of order $i_1 - 1$ are $i_1 - 1$ in number, it is plain that the combination referred to can only produce $i_1 - 1$ additional base-laterals. Hence, on the assumption made, the whole number of base-laterals of order i_1 is $1 + i_1 - 1 = i_1$. By induction the theorem is established.

The i_1 base-laterals of order i_1 are (writing them for convenience horizontally instead of vertically)

$$i_1 1^{i_1-1}, \quad i_1^2 2^{i_1-2}, \quad i_1^3 3^{i_1-3}, \;\ldots\ldots\; i_1^{i_1}.$$

We must discover the generating function of bases having a given base-lateral $i_1^s s^{i_1-s}$.

The base-lateral in question is associated with 2^{i_1-s-1} different bases if $s < n$, while $i_1^{i_1}$ is associated with but a single base. Taking $s = 1$, the simplest base, having $i_1 1^{i_1-1}$ for base-lateral, is the half-hollow-graph of content $i_1^3 - (i_1 - 1)^3$. The remaining bases with this base-lateral are obtained by combination with the bases of the first $i_1 - 2$ orders.

22—2

Denote by $u_s - 1$ the generating function of the bases of the first s orders; then

$$u_{s+1} = \{1 + x^{(s+1)^3 - s^3}\} u_s,$$

or

$$u_{i_1-2} = \{1 + x^{(i_1-2)^3 - (i_1-3)^3}\} u_{i_1-3},$$

whence

$$u_{i_1-2} = (1+x)(1+x^{2^3-1^3})(1+x^{3^3-2^3}) \ldots \{1 + x^{(i_1-2)^3-(i_1-3)^3}\}.$$

Therefore the generating function of bases, having the base-lateral $i_1 1^{i_1-1}$, is

$$x^{i_1^3-(i_1-1)^3}[(1+x)(1+x^{2^3-1^3})(1+x^{3^3-2^3}) \ldots \{1 + x^{(i_1-2)^3-(i_1-3)^3}\}].$$

Next consider the bases having the base-lateral

$$i_1^2 2^{i_1-2}.$$

The simplest base of this nature is derived by combining the half-hollow-cube form of content $i_1^3 - (i_1 - 1)^3$ with the similar form of content $(i_1 - 1)^3 - (i_1 - 2)^3$ and thus it has the content $i_1^3 - (i_1 - 2)^3$. With this we can again combine every base of the first $i_1 - 3$ orders without altering the base-lateral, which remains $i_1^2 2^{i_1-2}$. Hence the bases are enumerated by the generating function

$$x^{i_1^3-(i_1-2)^3}[(1+x)(1+x^{2^3-1^3})(1+x^{3^3-2^3}) \ldots \{1 + x^{(i_1-3)^3-(i_1-4)^3}\}].$$

In general the simplest base with base-lateral $i_1^s s^{i_1-s}$ is obtained by combining the half-hollow-cube forms of orders i_1, $i_1 - 1$, $i_1 - 2, \ldots i_1 - s + 1$, and thus has the content

$$i_1^3 - (i_1 - s)^3.$$

By reasoning before employed we arrive at the fact that the bases with base-lateral $i_1^s s^{i_1-s}$ are enumerated by

$$x^{i_1^3-(i_1-s)^3}[(1+x)(1+x^{2^3-1^3})(1+x^{3^3-2^3}) \ldots \{1 + x^{(i_1-s-1)^3-(i_1-s-2)^3}\}].$$

In this expression, between the brackets [ ] there are $i_1 - s - 1$ factors; if $s = i_1 - 1$ or i_1 we take merely $x^{i_1^3-1^3}$ and $x^{i_1^3}$ respectively.

The next question is the ascertainment of the generating function which enumerates the lateral appendages that can be associated with the base-lateral $i_1^s s^{i_1-s}$. When $s = 1$ this is easy because the lateral must be composed of columns

$$\begin{array}{cccc} 1 & 2 & 3 \ldots & i_1 \\ & 1 & 1 & 1 \\ & & 1 & 1 \\ & & & 1 \\ & & & \vdots \end{array}$$

not more than i_2 being taken.

The generating function is

$$\frac{1}{1-m \,.\, 1-mx \,.\, 1-mx^3 \,.\, 1-mx^5 \ldots 1-mx^{2i_1-1}},$$

where we seek the coefficient of $m^{i_2}x^{w_2}$ or of $m^{i-i_1}x^{w_2}$ since $i = i_1 + i_2$.

Finally for the symmetrical graphs constructed on bases whose base-laterals are $i_1 1^{i_1-1}$ we have the generating function

$$\frac{x^{i_1^3-(i_1-1)^3}\left[(1+x)(1+x^{2^3-1^3})(1+x^{3^3-2^3})\ldots\{1+x^{(i_1-2)^3-(i_1-3)^3}\}\right]}{(1-m)(1-mx^3)(1-mx^9)\ldots(1-mx^{6i_1-3})},$$

in the expansion of which we must seek the coefficient of $m^{i-i_1}x^w$.

When the base-lateral is $i_1^s s^{i_1-s}$ the matter is by no means so simple. The lateral appendage is composed of columns each of which is a self-conjugate partition of a number, and the possible forms of the columns are further limited by the form of the base-lateral. To explain take $i_1=5$, $s=3$ so that the base-lateral (written horizontally) is

5 5 5 3 3.

This has a graph

• • • • •
• • • • •
• • • • •
• • •
• • •

formed of three plane angles. Any column of the lateral must have a graph which can be superposed; the condition for this is obviously that it must be composed of not more than three plane angles, the largest angle containing not more than 9 nodes. So with base-lateral $i_1^s s^{i_1-s}$ a lateral column must have a graph composed of not more than s plane angles, the largest angle containing not more than $2i_1-1$ nodes.

The complete lateral appendage constitutes a multipartite partition whose graph is symmetrical in two dimensions. We have therefore to enumerate the graphs of this nature, each layer being composed of at most s plane angles and no angle containing more than $2i_1-1$ nodes and the number of layers not exceeding $i-i_1$ or i_2.

The crude form of this generating function was found earlier in the paper to be

$$\Omega\,\frac{1}{(1-m)(1-mx)(1-max^3)(1-mabx^5)\ldots(1-mabc\ldots x^{2i_1-1})},$$

$$\left(1-\frac{x}{a}\right)\left(1-\frac{a'}{ab}x^3\right)\left(1-\frac{a'b'}{abc}x^5\right)\ldots\text{ ad inf.},$$

$$\left(1-\frac{x}{a'}\right)\left(1-\frac{a''}{a'b'}x^3\right)\left(1-\frac{a''b''}{a'b'c'}x^5\right)\ldots\text{ ad inf.},$$

....

s rows,

in which we take the coefficient of $m^{i-i_1}x^{w_2}$.

We can now assert that the symmetrical graphs which appertain to the base-lateral $i_1{}^s s^{i_1-s}$, and have at most i nodes along an axis, are enumerated by the coefficient of $m^{i-i_1}x^w$ in

$$\Omega \frac{x^{i_1{}^3-(i_1-s)^3}\left[(1+x)(1+x^{2^3-1^3})(1+x^{3^3-2^3})\ldots\{1+x^{(i_1-s-1)^3-(i_1-s-2)^3}\}\right]}{(1-m)(1-mx^3)(1-max^9)(1-mabx^{15})\ldots(1-mabc\ldots x^{6i_1-3})},$$

$$\left(1-\frac{x^3}{a}\right)\left(1-\frac{a'}{ab}x^9\right)\left(1-\frac{a'b'}{abc}x^{15}\right)\ldots \text{ad inf.},$$

$$\left(1-\frac{x^3}{a'}\right)\left(1-\frac{a''}{a'b'}x^9\right)\left(1-\frac{a''b''}{a'b'c'}x^{15}\right)\ldots \text{ad inf.},$$

....

s rows.

Denoting this expression by $S(m, x)$, we see that

$$\sum_{i_1=1}^{i_1=i}\sum_{s=1}^{s=i_1} S(m, x) \,.\, m^{i_1-i}$$

enumerates, by the coefficient of m^0x^w, the whole of the symmetrical graphs subject to the single restriction that more than i nodes are not to occur along an axis.

IX. *Memoir on the Theory of the Partitions of Numbers.—Part VI. Partitions in Two-dimensional Space, to which is added an Adumbration of the Theory of the Partitions in Three-dimensional Space.*

By Major P. A. MacMahon, *R.A., D.Sc., LL.D., F.R.S.*

Received June 13,—Read June 29, 1911.

Introduction.

I resume the subject of Part V.* of this Memoir by inquiring further into the generating function of the partitions of a number when the parts are placed at the nodes of an incomplete lattice, viz., of a lattice which is regular but made up of unequal rows. Such a lattice is the graph of the line partition of a number. In Part V. I arrived at the expression of the generating function in respect of a two-row lattice when the past magnitude is unrestricted. This was given in Art. 16 in the form

$$\text{GF}(\infty; a, b) = \frac{(1)+x^{b+1}(\mathbf{a}-\mathbf{b})}{(\mathbf{1})(\mathbf{2})\ldots(\mathbf{a}+\mathbf{1}).(\mathbf{1})(\mathbf{2})\ldots(\mathbf{b})}.$$

I remind the reader that the determination of the generating function, when the part magnitude is unrestricted, depends upon the determination of the associated lattice function (see Art. 5, *loc. cit.*). This function is assumed to be the product of an expression of known form and of another function which I termed the inner lattice function (see Art. 10, *loc. cit.*), and it is on the form of this function that the interest of the investigation in large measure depends. All that is known about it *à priori* is its numerical value when x is put equal to unity (Art. 10, *loc. cit.*). The lattice function was also exhibited as a sum of sub-lattice functions, and it was shown that the generating function, when the part magnitude is restricted, may be expressed as a linear function of them. These sub-lattice functions are intrinsically interesting, but it will be shown in what follows that they are not of vital importance to the investigation. In fact, the difficulty of constructing them has been turned by the

* 'Phil. Trans.,' A, vol. 211, 1911.

formation and solution of certain functional equations which lead in the first place to the required generating functions, and in the second place to an exhibition of the forms of the sub-lattice functions. To previous definitions I here add the definition of the inner lattice function when there is a restriction upon the part magnitude, and it will be shown that the generating, lattice, and inner lattice functions satisfy certain functional equations both when there is not and when there is a restriction upon the part magnitude.

There are two methods of investigation available. We may commence with a study of the Greek-letter successions (Art. 6, *et seq.*, *loc. cit.*) from which the lattice functions are derived, and having obtained the functional equations which they satisfy, proceed thence to those satisfied by the generating and inner lattice functions; or we may reverse the process, and, by a prior determination of the equations appertaining to the generating functions, arrive at those satisfied by the lattice and inner lattice functions.

Both methods have been of service.

The results, herein achieved, are complete so far as the lattice of unequal rows and the particular question under consideration are concerned. They are elegant and algebraically interesting. In proof of this, it will suffice to say that the generating function is unaltered when the lattice is changed into its conjugate. The subject thus swarms with algebraical relations which are established intuitively.

Other results are obtained of a more general and extensive character which mark out the path of further investigation.

Art. 1. I recall that for the lattice of two unequal rows, containing a, b nodes respectively, the established results are

$$\text{Inner lattice function} = \mathrm{IL}\,(\infty\,;\,a,b) = 1+x^{b+1}\frac{(\mathfrak{a}-\mathfrak{b})}{(1)};$$

$$\text{Lattice function} = \mathrm{L}\,(\infty\,;\,a,b) = \frac{(1)\,(2)\ldots(\mathfrak{a}+\mathfrak{b})}{(2)\,(3)\ldots(\mathfrak{a}+1)\,.\,(1)\,(2)\ldots(\mathfrak{b})}\left\{1+x^{b+1}\frac{(\mathfrak{a}-\mathfrak{b})}{(1)}\right\};$$

$$\text{Generating function} = \mathrm{GF}\,(\infty\,;\,a,b) = \frac{1+x^{b+1}\dfrac{(\mathfrak{a}-\mathfrak{b})}{(1)}}{(2)\,(3)\ldots(\mathfrak{a}+1)\,.\,(1)\,(2)\ldots(\mathfrak{b})}.$$

We have yet to determine $\mathrm{IL}\,(l\,;\,a,b)$, $\mathrm{L}\,(l\,;\,a,b)$, $\mathrm{GF}\,(l\,;\,a,b)$, where an inner lattice function, for a restricted part magnitude, is defined by the relation

$$\mathrm{L}\,(l\,;\,a_1, a_2, \ldots, a_n)$$

$$= (1)\,(2)\ldots(\Sigma\mathfrak{a})\frac{(\mathfrak{l}+\mathfrak{n})\ldots(\mathfrak{l}+\mathfrak{a}_1+\mathfrak{n}-1)\,.\,(\mathfrak{l}+\mathfrak{n}-1)\ldots\ldots(\mathfrak{l}+\mathfrak{a}_2+\mathfrak{n}-2)\ldots\ldots(\mathfrak{l}+1)\ldots(\mathfrak{l}+\mathfrak{a}_n)}{(\mathfrak{n})\ldots(\mathfrak{a}_1+\mathfrak{n}-1)\,.\,(\mathfrak{n}-1)\ldots(\mathfrak{a}_2+\mathfrak{n}-2)\ldots\ldots(1)\ldots(\mathfrak{a}_n)}\,\mathrm{IL}\,(l\,;\,a_1, a_2, \ldots, a_n).$$

The Functional Equations.

Art. 2. It is convenient to begin by establishing the functional equations satisfied by the generating functions.

Suppose the lattice to have three unequal rows of a, b, c nodes respectively, and let the part magnitude be restricted by the number l,

Subject to the parts being in descending order of magnitude in each row and in each column, in every partition each node is either occupied by zero (that is, is unoccupied) or by a number greater than zero and not greater than l. In certain partitions every node is occupied; such partitions may be constructed by

(i) Placing a unit at each node,
(ii) Superposing every partition enumerated by GF $(l-1\,;\ a, b, c)$.

Hence these special, full-based partitions are clearly enumerated by

$$x^{a+b+c}\,\mathrm{GF}\,(l-1\,;\ a, b, c).$$

Similarly those partitions which are full-based upon a *contained* lattice specified by the line partition $(a'b'c')$ are enumerated by

$$x^{a'+b'+c'}\mathrm{GF}\,(l-1\,;\ a', b', c')\,;$$

and we are led to the relation

$$\mathrm{GF}\,(l\,;\ a, b, c) = \Sigma x^{a'+b'+c'}\mathrm{GF}\,(l-1\,;\ a', b', c'),$$

where the summation is in regard to every lattice, specified by $(a'b'c')$, which is contained in the lattice specified by (abc).

Art. 3. If from the partitions enumerated by GF $(l\,;\ a, b, c)$ we subtract those enumerated by $x^{a+b+c}\mathrm{GF}\,(l-1\,;\ a, b, c)$, we have remaining, in the case of three unequal rows, partitions which *include* those enumerated by each of the three generating functions

$$\mathrm{GF}\,(l\,;\ a-1, b, c),\quad \mathrm{GF}\,(l\,;\ a, b-1, c),\quad \mathrm{GF}\,(l\,;\ a, b, c-1)\,;$$

and which, by a well-known principle of the combinatory analysis, are enumerated by

$$\begin{aligned}&\mathrm{GF}\,(l\,;\ a-1, b, c)+\mathrm{GF}\,(l\,;\ a, b-1, c)+\mathrm{GF}\,(l\,;\ a, b, c-1)\\&\quad-\mathrm{GF}\,(l\,;\ a-1, b-1, c)-\mathrm{GF}\,(l\,;\ a-1, b, c-1)-\mathrm{GF}\,(l\,;\ a, b-1, c-1)\\&\quad+\mathrm{GF}\,(l\,;\ a-1, b-1, c-1).\end{aligned}$$

2 Y 2

Hence the functional equation

$$\begin{aligned}\mathrm{GF}\,(l\,;\,a,\,b,\,c)-x^{a+b+c}\mathrm{GF}\,(l-1\,;\,a,\,b,\,c)\\ &= \mathrm{GF}\,(l\,;\,a-1,\,b,\,c)+\mathrm{GF}\,(l\,;\,a,\,b-1,\,c)+\mathrm{GF}\,(l\,;\,a,\,b,\,c-1)\\ &-\mathrm{GF}\,(l\,;\,a-1,\,b-1,\,c)-\mathrm{GF}\,(l;\,a-1,\,b,\,c-1)-\mathrm{GF}\,(l\,;\,a,\,b-1,\,c-1)\\ &+\mathrm{GF}\,(l\,;\,a-1,\,b-1,\,c-1).\end{aligned}$$

In the general case of n unequal rows we have the theorem for $\mathrm{GF}\,(l\,;\,a_1,\,a_2,\,\ldots,\,a_n)$; for if p_s be a symbol such that

$$p_s\mathrm{GF}\,(l\,;\,a_1,\,a_2,\,\ldots,\,a_n) = \mathrm{GF}\,(l\,;\,a_1,\,a_2,\,\ldots,\,a_s-1,\,\ldots,\,a_n),$$

it is readily seen that

$$(1-p_1)\,(1-p_2)\ldots(1-p_n)\,\mathrm{GF}\,(l\,;\,a_1,\,a_2,\,\ldots,\,a_n) = x^{\Sigma a}\mathrm{GF}\,(l-1\,;\,a_1,\,a_2,\,\ldots,\,a_n).$$

This equation is, at first sight, only true when there are no equalities between the numbers $a_1,\,a_2,\,\ldots,\,a_n$; but in the sequel, when an algebraic expression of $\mathrm{GF}\,(l;\,a_1,\,a_2,\,\ldots,\,a_n)$ has been found, it will be seen to be true universally as an algebraical identity.

Art. 4. However, the formula *may* be modified, in the direction of simplification, when the rows are not *all* unequal.

For a given lattice we require to know how many nodes may be *singly* detached and yet leave a *contained* lattice. Thus in the three-row lattice illustrated above it is clear, the rows presenting no equalities, that we may detach singly either of the nodes lettered A, B, C; but in the case now given

$$\begin{array}{lll}\cdot\;\cdot\;\cdot\;\cdot\;\cdot\;\cdot\;\cdot\;\cdot\;\overset{\mathrm{A}}{\cdot} & & a \text{ nodes}\\ \cdot\;\cdot\;\cdot\;\cdot\;\cdot & & b \text{ nodes}\\ \cdot\;\cdot\;\cdot\;\cdot\;\overset{\mathrm{C}}{\cdot} & & b \text{ nodes,}\end{array}$$

it is seen that we can detach either A or C only, so that the resulting functional equation is

$$\begin{aligned}\mathrm{GF}\,(l\,;\,a,\,b,\,b)-x^{a+2b}\mathrm{GF}\,(l-1\,;\,a,\,b,\,b)\\ &= \mathrm{GF}\,(l;\,a-1,\,b,\,b)+\mathrm{GF}\,(l\,;\,a,\,b,\,b-1)-\mathrm{GF}\,(l\,;\,a-1,\,b,\,b-1),\end{aligned}$$

which is to be compared with the equation appertaining to a lattice of two unequal rows

$$\mathrm{GF}\,(l\,;a,\,b)-x^{a+b}\mathrm{GF}\,(l-1\,;a,\,b) = \mathrm{GF}\,(l\,;a-1,\,b)+\mathrm{GF}\,(l\,;a,\,b-1)-\mathrm{GF}\,(l\,;a-1,\,b-1).$$

Similarly we derive the equations

$$\begin{aligned}\mathrm{GF}\,(l\,;\,a,\,a,\,b)-x^{2a+b}\mathrm{GF}\,(l-1\,;\,a,\,a,\,b)\\ &= \mathrm{GF}\,(l\,;\,a,\,a-1,\,b)+\mathrm{GF}\,(l;\,a,\,a,\,b-1)-\mathrm{GF}\,(l\,;\,a,\,a-1,\,b-1),\end{aligned}$$

$$\mathrm{GF}\,(l\,;\,a,\,a,\,a)-x^{3a}\mathrm{GF}\,(l-1\,;\,a,\,a,\,a) = \mathrm{GF}\,(l\,;\,a,\,a,\,a-1)\,;$$

and also

$$\mathrm{GF}(l\,;\,a^n)-x^{na}\mathrm{GF}(l-1\,;\,a^n)=\mathrm{GF}(l\,;\,a^{n-1},a-1).$$

The formation of the relation in any particular case presents no difficulty. When the lattice has k singly detachable nodes, the right-hand side of the relation involves 2^k-1 terms.

Art. 5. When the part magnitude is unrestricted, or $l=\infty$, the equations become

$$\begin{aligned}&(\mathbf{a}+\mathbf{b}+\mathbf{c})\,\mathrm{GF}(\infty\,;\,a,b,c)\\&\quad=\mathrm{GF}(\infty\,;\,a-1,b,c)+\mathrm{GF}(\infty\,;\,a,b-1,c)+\mathrm{GF}(\infty\,;\,a,b,c-1)\\&\quad-\mathrm{GF}(\infty\,;\,a-1,b-1,c)-\mathrm{GF}(\infty\,;\,a-1,b,c-1)-\mathrm{GF}(\infty\,;\,a,b-1,c-1)\\&\quad+\mathrm{GF}(\infty\,;\,a-1,b-1,c-1)\end{aligned}$$

$$(\Sigma\mathbf{a})\,\mathrm{GF}(\infty\,;\,a_1,a_2,\ldots,a_n)=\{1-(1-p_1)(1-p_2)\ldots(1-p_n)\}\,\mathrm{GF}(\infty\,;\,a_1,a_2,\ldots,a_n)$$

and the modified forms are easily written down.

Art. 6. The next step is to deduce the corresponding relations between lattice functions. From the relation

$$\mathrm{GF}(\infty\,;\,a_1,a_2,\ldots,a_n)=\frac{\mathrm{L}(\infty\,;\,a_1,a_2,\ldots,a_n)}{(\mathbf{1})(\mathbf{2})\ldots(\Sigma\mathbf{a})},$$

we find

$$\mathrm{L}(\infty\,;\,a,b,c)=\Sigma x^{a'+b'+c'}\frac{(\mathbf{1})(\mathbf{2})\ldots(\mathbf{a}+\mathbf{b}+\mathbf{c})}{(\mathbf{1})(\mathbf{2})\ldots(\mathbf{a}'+\mathbf{b}'+\mathbf{c}')}\,\mathrm{L}(\infty\,;\,a',b',c');$$

$$\mathrm{L}(\infty;a,a)=\mathrm{L}(\infty;a,a-1);$$

$$\mathrm{L}(\infty;a,b)=\mathrm{L}(\infty;a-1,b)+\mathrm{L}(\infty;a,b-1)-(\mathbf{a}+\mathbf{b}-\mathbf{1})\,\mathrm{L}(\infty;a-1,b-1);$$

$$\mathrm{L}(\infty;a,a,a)=\mathrm{L}(\infty;a,a,a-1);$$

$$\mathrm{L}(\infty;a,a,b)=\mathrm{L}(\infty;a,a-1,b)+\mathrm{L}(\infty;a,a,b-1)-(\mathbf{2a}+\mathbf{b}-\mathbf{1})\,\mathrm{L}(\infty;a,a-1,b-1);$$

$$\mathrm{L}(\infty;a,b,b)=\mathrm{L}(\infty;a-1,b,b)+\mathrm{L}(\infty;a,b,b-1)-(\mathbf{a}+\mathbf{2b}-\mathbf{1})\,\mathrm{L}(\infty;a-1,b,b-1);$$

$$\begin{aligned}\mathrm{L}(\infty;a,b,c)&=\mathrm{L}(\infty;a-1,b,c)+\mathrm{L}(\infty;a,b-1,c)+\mathrm{L}(\infty;a,b,c-1)\\&\quad-(\mathbf{a}+\mathbf{b}+\mathbf{c}-\mathbf{1})\,\{\mathrm{L}(\infty;a-1,b-1,c)+\mathrm{L}(\infty;a-1,b,c-1)\\&\qquad+\mathrm{L}(\infty;a,b-1,c-1)\}\\&\quad+(\mathbf{a}+\mathbf{b}+\mathbf{c}-\mathbf{2})(\mathbf{a}+\mathbf{b}+\mathbf{c}-\mathbf{1})\,\mathrm{L}(\infty;a-1,b-1,c-1).\end{aligned}$$

In the case of n unequal rows, if we write symbolically,

$$p_s\,\mathrm{L}(\infty;a_1,a_2,\ldots,a_n)=\mathrm{L}(\infty;a_1,a_2,\ldots,a_s-1,\ldots,a_n),$$

$$(\Sigma\mathbf{a})(\Sigma\mathbf{a}-\mathbf{1})\ldots(\Sigma\mathbf{a}-\mathbf{m}+\mathbf{1})=\mathrm{X}^m;$$

then

$$(1-\mathrm{X})\,\mathrm{L}(\infty\,;\,a_1,a_2,\ldots,a_n)=(1-p_1\mathrm{X})(1-p_2\mathrm{X})\ldots\ldots(1-p_n\mathrm{X})\,\mathrm{L}(\infty\,;\,a_1,a_2,\ldots,a_n).$$

Art. 7. Continuing, for the present, to regard the part magnitude as unrestricted, we now proceed to the equations satisfied by the inner lattice functions.

Guided by the relation

$$L(\infty; a_1, a_2, \ldots, a_n) = \frac{(1)\ldots(\Sigma\mathbf{a})}{(\mathbf{n})\ldots(\mathbf{a}_1+\mathbf{n}-1).(\mathbf{n}-1)\ldots(\mathbf{a}_2+\mathbf{n}-2)\ldots \ldots(2)\ldots(\mathbf{a}_{n-1}+1).(1)\ldots(\mathbf{a}_n)} IL(\infty; a_1, a_2, \ldots, a_n),$$

we find

$$(2\mathbf{a})\, IL(\infty; a, a) = (\mathbf{a})\, IL(\infty; a, a-1);$$

$$(\mathbf{a}+\mathbf{b})\, IL(\infty; a, b) = (\mathbf{a}+1)\, IL(\infty; a-1, b) + (\mathbf{b})\, IL(\infty; a, b-1) - (\mathbf{a}+1)(\mathbf{b})\, IL(\infty; a-1, b-1);$$

$$(3\mathbf{a})\, IL(\infty; a, a, a) = (\mathbf{a})\, IL(\infty; a, a, a-1);$$

$$(2\mathbf{a}+\mathbf{b})\, IL(\infty; a, a, b) = (\mathbf{a}+1)\, IL(\infty; a, a-1, b) + (\mathbf{b})\, IL(\infty; a, a, b-1) - (\mathbf{a}+1)(\mathbf{b})\, IL(\infty; a, a-1, b-1);$$

$$(\mathbf{a}+2\mathbf{b})\, IL(\infty; a, b, b) = (\mathbf{a}+2)\, IL(\infty; a-1, b, b) + (\mathbf{b})\, IL(\infty; a, b, b-1) - (\mathbf{a}+2)(\mathbf{b})\, IL(\infty; a-1, b, b-1);$$

$$\begin{aligned}(\mathbf{a}+\mathbf{b}+\mathbf{c})\, IL(\infty; c, b, c) = {} & (\mathbf{a}+2)\, IL(\infty; a-1, b, c) + (\mathbf{b}+1)\, IL(\infty; a, b-1, c) + (\mathbf{c})\, IL(\infty; a, b, c-1) \\ & - (\mathbf{a}+2)(\mathbf{b}+1)\, IL(\infty; a-1, b-1, c) - (\mathbf{a}+2)(\mathbf{c})\, IL(\infty; a-1, b, c-1) \\ & - (\mathbf{b}+1)(\mathbf{c})\, IL(\infty; a, b-1, c-1) \\ & + (\mathbf{a}+2)(\mathbf{b}+1)(\mathbf{c})\, IL(\infty; a-1, b-1, c-1),\end{aligned}$$

and, in general, for n unequal rows, if we write symbolically,

$$q_s\, IL(\infty; a_1, a_2, \ldots, a_n) = (\mathbf{a}_s+\mathbf{n}-\mathbf{s})\, IL(\infty; a_1, a_2, \ldots, a_s-1, \ldots, a_n),$$

$$x^{\Sigma a}\, IL(\infty; a_1, a_2, \ldots, a_n) = (1-q_1)(1-q_2)\ldots(1-q_n)\, IL(\infty; a_1, a_2, \ldots, a_n).$$

To these may be added

$$IL(\infty; a, b, c) = \Sigma x^{a'+b'+c'} \frac{(3)\ldots(\mathbf{a}+2).(2)\ldots(\mathbf{b}+1).(1)\ldots(\mathbf{c})}{(3)\ldots(\mathbf{a}'+2).(2)\ldots(\mathbf{b}'+1).(1)\ldots(\mathbf{c}')} IL(\infty; a', b', c'),$$

which can be readily generalized.

Art. 8. I proceed at once to find an expression for the inner lattice function.

It appears to be right to seek an expression of the function which shall show at once that the sum of the coefficients therein is that which it is otherwise known to be. Thus the result

$$IL(\infty; ab) = 1 + x^{b+1}\frac{(\mathbf{a}-\mathbf{b})}{(1)}$$

shows at once that the sum of the coefficients is $a-b+1$.

Since the sum of the coefficients in $\mathrm{IL}(\infty; a, b, c)$ is

$$(a-b+1)(a-c+2)(b-c+1)$$

it was at first conjectured that the expression might be

$$\left\{1+x^{b+1}\frac{(\mathbf{a}-\mathbf{b})}{(1)}\right\}\left\{1+x^{c+2}\frac{(\mathbf{a}-\mathbf{c})}{(2)}\right\}\left\{1+x^{c+1}\frac{(\mathbf{b}-\mathbf{c})}{(1)}\right\},$$

but this neither satisfies the functional equation nor verifies in simple particular cases.

If, in the formula

$$\mathrm{GF}(\infty; a, b, c) = \frac{\mathrm{IL}(\infty; a, b, c)}{(3)\ldots(\mathbf{a}+2).(2)\ldots(\mathbf{b}+1).(1)\ldots(\mathbf{c})},$$

we put $\mathrm{IL}(\infty; a, b, c)$ equal to the expression above and then put $c = 0$, we obtain the known result for $\mathrm{GF}(\infty; a, b)$, as may be readily seen by putting the expression in the form

$$\left\{1+x^{b+1}\frac{(\mathbf{a}-\mathbf{b})}{(1)}\right\}\left\{\frac{(\mathbf{a}+2)-x^2(\mathbf{c})}{(2)}\right\}\left\{\frac{(\mathbf{b}+1)-x(\mathbf{c})}{(1)}\right\}.$$

We are therefore justified in putting $\mathrm{IL}(\infty; a, b, c)$ equal to the expression with an added term which contains the factor $(\mathbf{c})$.

Write, therefore,

$$\mathrm{IL}(\infty; a, b, c) = \left\{1+x^{b+1}\frac{(\mathbf{a}-\mathbf{b})}{(1)}\right\}\left\{1+x^{c+2}\frac{(\mathbf{a}-\mathbf{c})}{(2)}\right\}\left\{1+x^{c+1}\frac{(\mathbf{b}-\mathbf{c})}{(1)}\right\}+(\mathbf{c})\,\mathrm{F}(\infty; a, b, c).$$

By working out several particular cases I was led to the conjecture

$$\mathrm{F}(\infty; a, b, c) = \frac{1}{(2)}\{x^{b+3}(\mathbf{a}-\mathbf{b})-x^{c+2}(\mathbf{b}-\mathbf{c})\};$$

and I then found that the expression

$$\left\{1+x^{b+1}\frac{(\mathbf{a}-\mathbf{b})}{(1)}\right\}\left\{1+x^{c+2}\frac{(\mathbf{a}-\mathbf{c})}{(2)}\right\}\left\{1+x^{c+1}\frac{(\mathbf{b}-\mathbf{c})}{(1)}\right\}+\frac{(\mathbf{c})}{(2)}\{x^{b+3}(\mathbf{a}-\mathbf{b})-x^{c+2}(\mathbf{b}-\mathbf{c})\}$$

does, as a fact, satisfy the functional equation.

Art. 9. Having thus, beyond doubt, established the forms of $\mathrm{IL}(\infty; a, b)$ and $\mathrm{IL}(\infty; a, b, c)$, I proceed to a study of the functional equations.

In the equation

$$(\mathbf{a}+\mathbf{b})\,\mathrm{IL}(\infty; a, b)$$
$$= (\mathbf{a}+1)\,\mathrm{IL}(\infty; a-1, b)+(\mathbf{b})\,\mathrm{IL}(\infty; a, b-1)-(\mathbf{a}+1)(\mathbf{b})\,\mathrm{IL}(\infty; a-1, b-1),$$

put

$$x^{-a}\{\mathrm{IL}(\infty; a, b)-(a+1)\,\mathrm{IL}(\infty; a-1, b)\} = \mathrm{V}_1(\infty; a, b);$$

then

$$(\mathbf{a}+\mathbf{b})\, V_1(\infty; a, b)$$
$$= (\mathbf{a}+\mathbf{1})\, V_1(\infty; a-1, b) + (\mathbf{b})\, V_1(\infty; a, b-1) - (\mathbf{a}+\mathbf{1})(\mathbf{b})\, V_1(\infty; a-1, b-1),$$

which is of the same form as the original equation.

Hence, if $\text{IL}(\infty; a, b)$ be a solution of the equation,

$$x^{-a}\{\text{IL}(\infty; a, b) - (\mathbf{a}+\mathbf{1})\,\text{IL}(\infty; a-1, b)\}$$

is also a solution.

I write

$$x^{-a}\{\text{IL}(\infty; a, b) - (\mathbf{a}+\mathbf{1})\,\text{IL}(\infty; a-1, b)\} = O_a \text{IL}(\infty; a, b),$$

exhibiting the new solution as the result of the performance of a certain operation upon the original one.

Again put, in the original equation,

$$x^{-b}\{\text{IL}(\infty; a, b) - (\mathbf{b})\,\text{IL}(\infty; a, b-1)\} = V_2(\infty; a, b) = O_b \text{IL}(\infty; a, b),$$

and we find

$$(\mathbf{a}+\mathbf{b})\, V_2(\infty; a, b)$$
$$= (\mathbf{a}+\mathbf{1})\, V_2(\infty; a-1, b) + (\mathbf{b})\, V_2(\infty; a, b-1) - (\mathbf{a}+\mathbf{1})(\mathbf{b})\, V_2(\infty; a-1, b-1);$$

so that another solution is

$$V_2(\infty; a, b) = O_b \text{IL}(\infty; a, b).$$

I write further

$$O_{ab} \text{IL}(\infty; a, b) = x^{-a-b}\{\text{IL}(\infty; a, b) - (\mathbf{a}+\mathbf{1})\,\text{IL}(\infty; a-1, b)$$
$$- (\mathbf{b})\,\text{IL}(\infty; a, b-1) + (\mathbf{a}+\mathbf{1})(\mathbf{b})\,\text{IL}(\infty; a-1, b-1)\};$$

so that, from the functional equation itself,

$$O_{ab} \text{IL}(\infty; a, b) = \text{IL}(\infty; a, b);$$

$$O_{ab} = 1;$$

and it is easy to verify that

$$O_a O_b = O_{ab}.$$

Art. 10. Since we know one solution of the equation

$$1 + x^{b+1}\frac{(\mathbf{a}-\mathbf{b})}{(\mathbf{1})}, \text{ or more conveniently } (\mathbf{1}) + x^{b+1}(\mathbf{a}-\mathbf{b}),$$

or better still $(\mathbf{a}+\mathbf{1}) - x(\mathbf{b})$, we may at once apply the operators O_a, O_b. Operating $s-1$ times successively with O_a, I find

$$O_a^{s-1}\{(\mathbf{a}+\mathbf{1}) - x(\mathbf{b})\} = (\mathbf{a}+\mathbf{1}) - x^s(\mathbf{b});$$

and

$$O_b\{(\mathbf{a}+\mathbf{1}) - x(\mathbf{b})\} = (\mathbf{a}+\mathbf{1}) - (\mathbf{b}),$$

$$O_b^{s+1}\{(\mathbf{a}+\mathbf{1}) - x(\mathbf{b})\} = (\mathbf{a}+\mathbf{1}) - x^{-s}(\mathbf{b}).$$

We may therefore take $(\mathbf{a+1})$ and $(\mathbf{b})$ as the two fundamental solutions, and clearly we may always multiply a solution by any function of x which does not involve a or b.

The final expression of IL $(\infty; a, b)$ which I adopt is

$$\mathrm{IL}(\infty; a, b) = \begin{vmatrix} (\mathbf{a+1}), & (\mathbf{b}) \\ x\;, & 1 \end{vmatrix} \div \begin{vmatrix} 1, & 1 \\ x, & 1 \end{vmatrix};$$

and we will find that, expressed thus as the quotient of two determinants, it is generalizable. I might now, knowing *à posteriori* the expression for IL $(\infty; abc)$, proceed in a simpler manner than what follows; but I think it better to put before the reader the actual course that the investigation took.

Art. 11. In the functional equation satisfied by IL $(\infty; a, b, c)$, which may be written

$$\mathrm{IL}(\infty; a, b, c) - (\mathbf{b+1})\,\mathrm{IL}(\infty; a, b-1, c) - (\mathbf{c})\,\mathrm{IL}(\infty; a, b, c-1) + (\mathbf{b+1})(\mathbf{c})\,\mathrm{IL}(\infty; a, b-1, c-1)$$
$$= x^{a+b+c}\,\mathrm{IL}(\infty; a, b, c) + (\mathbf{a+2})\,\{\mathrm{IL}(\infty; a-1, b, c) - (\mathbf{b+1})\,\mathrm{IL}(\infty; a-1, b-1, c)$$
$$-(\mathbf{c})\,\mathrm{IL}(\infty; a-1, b, c-1) + (\mathbf{b+1})(\mathbf{c})\,\mathrm{IL}(\infty; a-1, b-1, c-1)\},$$

I write

$$\mathrm{V}_1(\infty; a, b, c) = x^{-b-c}\,\{\mathrm{IL}(\infty; a, b, c) - (\mathbf{b+1})\,\mathrm{IL}(\infty; a, b-1, c)$$
$$-(\mathbf{c})\,\mathrm{IL}(\infty; a, b, c-1) + (\mathbf{b+1})(\mathbf{c})\,\mathrm{IL}(\infty; a, b-1, c-1\};$$

and thence derive the relation

$$(\mathbf{a+b+c})\,\mathrm{V}_1(\infty; a, b, c)$$
$$= (\mathbf{a+2})\,\mathrm{V}_1(\infty; a-1, b, c) + (\mathbf{b+1})\,\mathrm{V}_1(\infty; a, b-1, c) + (\mathbf{c})\,\mathrm{V}_1(\infty; a, b, c-1)$$
$$-(\mathbf{a+2})(\mathbf{b+1})\,\mathrm{V}_1(\infty; a-1, b-1, c) - (\mathbf{a+2})(\mathbf{c})\,\mathrm{V}_1(\infty; a-1, b, c-1)$$
$$-(\mathbf{b+1})(\mathbf{c})\,\mathrm{V}_1(\infty; a, b-1, c-1) + (\mathbf{a+2})(\mathbf{b+1})(\mathbf{c})\,\mathrm{V}_1(\infty; a-1, b-1, c-1).$$

Comparing this with the functional equation it is clear that $\mathrm{V}_1(\infty; a, b, c)$, as defined, is a solution.

Proceeding similarly we find six solutions which I exhibit as operations performed upon IL $(\infty; a, b, c)$ as follows:—

$$x^{-a}\{\mathrm{IL}(\infty; a, b, c) - (\mathbf{a+2})\,\mathrm{IL}(\infty; a-1, b, c)\} = \mathrm{O}_a\mathrm{IL}(\infty; a, b, c);$$

$$x^{-b}\{\mathrm{IL}(\infty; a, b, c) - (\mathbf{b+1})\,\mathrm{IL}(\infty; a, b-1, c)\} = \mathrm{O}_b\mathrm{IL}(\infty; a, b, c);$$

$$x^{-c}\{\mathrm{IL}(\infty; a, b, c) - \;(\mathbf{c})\;\; \mathrm{IL}(\infty; a, b, c-1)\} = \mathrm{O}_c\mathrm{IL}(\infty; a, b, c);$$

$$x^{-b-c}\{\mathrm{IL}(\infty; a, b, c)-(\mathbf{b+1})\,\mathrm{IL}(\infty; a, b-1, c)-(\mathbf{c})\,\mathrm{IL}(\infty; a, b, c-1)$$
$$+(\mathbf{b+1})(\mathbf{c})\,\mathrm{IL}(\infty; a, b-1, c-1)\} = \mathrm{O}_{bc}\mathrm{IL}(\infty; a, b, c);$$

$$x^{-c-a}\{\mathrm{IL}(\infty; a, b, c)-(\mathbf{c})\,\mathrm{IL}(\infty; a, b, c-1)-(\mathbf{a+2})\,\mathrm{IL}(\infty; a-1, b, c)$$
$$+(\mathbf{c})(\mathbf{a+2})\,\mathrm{IL}(\infty; a-1, b, c-1)\} = \mathrm{O}_{ca}\mathrm{IL}(\infty; a, b, c);$$

$$x^{-a-b}\{\mathrm{IL}(\infty; a, b, c)-(\mathbf{a+2})\,\mathrm{IL}(\infty; a-1, b, c)-(\mathbf{b+1})\,\mathrm{IL}(\infty; a, b-1, c)$$
$$+(\mathbf{a+2})(\mathbf{b+1})\,\mathrm{IL}(\infty; a-1, b-1, c)\} = \mathrm{O}_{ab}\mathrm{IL}(\infty; a, b, c).$$

I further write

$$\mathrm{O}_{abc}\mathrm{IL}(\infty; a, b, c)$$
$$= x^{-a-b-c}\{\mathrm{IL}(\infty; a, b, c)-(\mathbf{a+2})\,\mathrm{IL}(\infty; a-1, b, c)-(\mathbf{b+1})\,\mathrm{IL}(\infty; a, b-1, c)$$
$$-(\mathbf{c})\,\mathrm{IL}(\infty; a, b, c-1)+(\mathbf{a+2})(\mathbf{b+1})\,\mathrm{IL}(\infty; a-1, b-1, c)$$
$$+(\mathbf{a+2})(\mathbf{c})\,\mathrm{IL}(\infty; a-1, b, c-1)+(\mathbf{b+1})(\mathbf{c})\,\mathrm{IL}(\infty; a, b-1, c-1)$$
$$-(\mathbf{a+2})(\mathbf{b+1})(\mathbf{c})\,\mathrm{IL}(\infty; a-1, b-1, c-1)\};$$

and it is easy to establish the operator relations

$$\mathrm{O}_a\mathrm{O}_b\mathrm{O}_c = 1,$$

$$\mathrm{O}_a\mathrm{O}_b = \mathrm{O}_{ab},\quad \mathrm{O}_a\mathrm{O}_c = \mathrm{O}_{ac},\quad \mathrm{O}_b\mathrm{O}_c = \mathrm{O}_{bc},$$

$$\mathrm{O}_a\mathrm{O}_{bc} = \mathrm{O}_b\mathrm{O}_{ac} = \mathrm{O}_c\mathrm{O}_{ab} = \mathrm{O}_{abc} = 1.$$

Art. 12. I now operate with these operators upon the known solution of the functional equation. To clear it of fractions I multiply throughout by $(1)^2(2)$. Operating m times in succession with O_c I obtain the result

$$(\mathbf{1})-x^{a+1}(\mathbf{2})(\mathbf{b+1})-x^{a+2}(\mathbf{a+1})+x^{b+2}(\mathbf{b})+x^{a+2b+4}(\mathbf{a-b})$$
$$-x^{m+1}\{(\mathbf{2})+x^{b+2}(1-x)(\mathbf{a-b-1})-x^{2b+3}(\mathbf{2a-2b})\}(\mathbf{c})$$
$$+x^{-2m+3}\{(\mathbf{1})+x^{b+2}(\mathbf{a-b})\}(\mathbf{c-1})(\mathbf{c}).$$

Whence I conclude that

$$\mathrm{P}_1 = (\mathbf{1})-x^{a+1}(\mathbf{2})(\mathbf{b+1})-x^{a+2}(\mathbf{a+1})+x^{b+2}(\mathbf{b})+x^{a+2b+4}(\mathbf{a-b}),$$

$$\mathrm{P}_2 = \{(\mathbf{2})+x^{b+2}(1+x)(\mathbf{a-b-1})-x^{2b+3}(\mathbf{2a-2b})\}(\mathbf{c}),$$

$$\mathrm{P}_3 = \{(\mathbf{1})+x^{b+2}(\mathbf{a-b})\}(\mathbf{c-1})(\mathbf{c}),$$

are solutions of the functional equation.

I find that

$$\mathrm{O}_c\mathrm{P}_1 = \mathrm{P}_1,\quad \mathrm{O}_c\mathrm{P}_2 = x^{-1}\mathrm{P}_2,\quad \mathrm{O}_c\mathrm{P}_3 = x^{-2}\mathrm{P}_3;$$

but new solutions are obtained by operating upon P_1, P_2, and P_3 with O_a and O_b.

Operating with O_b, m times successively, upon P_3 I obtain

$$-x(\mathbf{c}-\mathbf{1})(\mathbf{c})\{(\mathbf{b}+\mathbf{1})-x^{m-1}(\mathbf{a}+\mathbf{2})\},$$

and I draw the inference that

$$(\mathbf{c}-\mathbf{1})(\mathbf{c})(\mathbf{b}+\mathbf{1}) \quad \text{and} \quad (\mathbf{c}-\mathbf{1})(\mathbf{c})(\mathbf{a}+\mathbf{2})$$

are solutions.

Further, operating with O_b, m times successively, upon P_2, I obtain

$$(\mathbf{a}+\mathbf{1})(\mathbf{a}+\mathbf{2})(\mathbf{c})x^m-(\mathbf{b})(\mathbf{b}+\mathbf{1})(\mathbf{c})x^{-m+2};$$

and it thence appears that

$$(\mathbf{a}+\mathbf{1})(\mathbf{a}+\mathbf{2})(\mathbf{c}) \quad \text{and} \quad (\mathbf{b})(\mathbf{b}+\mathbf{1})(\mathbf{c})$$

are solutions.

Again, operating with O_b, m times successively, upon P_1, I obtain

$$(\mathbf{a}+\mathbf{1})(\mathbf{a}+\mathbf{2})(\mathbf{b}+\mathbf{1})-(\mathbf{b})(\mathbf{b}+\mathbf{1})(\mathbf{a}+\mathbf{2})x^{-m+1};$$

and the conclusion is that

$$(\mathbf{a}+\mathbf{1})(\mathbf{a}+\mathbf{2})(\mathbf{b}+\mathbf{1}) \quad \text{and} \quad (\mathbf{b})(\mathbf{b}+\mathbf{1})(\mathbf{a}+\mathbf{2})$$

are solutions.

No other fundamental solutions are obtainable by operating with O_a, O_b, and O_c upon P_1, P_2, and P_3, and clearly we have no need to consider the other operators because of the relations between them.

We have thus six fundamental solutions

$$(\mathbf{a}+\mathbf{1})(\mathbf{a}+\mathbf{2})(\mathbf{b}+\mathbf{1}), \quad (\mathbf{a}+\mathbf{1})(\mathbf{a}+\mathbf{2})(\mathbf{c}), \quad (\mathbf{b})(\mathbf{b}+\mathbf{1})(\mathbf{c}),$$

$$(\mathbf{b})(\mathbf{b}+\mathbf{1})(\mathbf{a}+\mathbf{2}), \quad (\mathbf{c}-\mathbf{1})(\mathbf{c})(\mathbf{a}+\mathbf{2}), \quad (\mathbf{c}-\mathbf{1})(\mathbf{c})(\mathbf{b}+\mathbf{1}).$$

Art. 13. The known solution of the functional equation from which these solutions have been derived can now be expressed in terms of these. Since it has been found that

$$O_c^m(\mathbf{1})^2(\mathbf{2})\,\mathrm{IL}(\infty; a, b, c) = P_1-x^{m+1}P_2+x^{-2m+3}P_3,$$

we have

$$(\mathbf{1})^2(\mathbf{2})\,\mathrm{IL}(\infty; a, b, c) = P_1-xP_2+x^3P_3;$$

and, putting $m=0$ in results obtained above, it appears that

$$P_1 = (\mathbf{a}+\mathbf{1})(\mathbf{a}+\mathbf{2})(\mathbf{b}+\mathbf{1})-x(\mathbf{b})(\mathbf{b}+\mathbf{1})(\mathbf{a}+\mathbf{2}),$$

$$P_2 = (\mathbf{a}+\mathbf{1})(\mathbf{a}+\mathbf{2})(\mathbf{c})-x^2(\mathbf{b})(\mathbf{b}+\mathbf{1})(\mathbf{c}),$$

$$P_3 = -x(\mathbf{c}-\mathbf{1})(\mathbf{c})(\mathbf{b}+\mathbf{1})+(\mathbf{c}-\mathbf{1})(\mathbf{c})(\mathbf{a}+\mathbf{2}).$$

whence

$$(1)^2(2)\,\mathrm{IL}(\infty; a, b, c) = (\mathbf{a}+1)(\mathbf{a}+2)(\mathbf{b}+1) - x(\mathbf{b})(\mathbf{b}+1)(\mathbf{a}+2) - x(\mathbf{a}+1)(\mathbf{a}+2)(\mathbf{c})$$
$$+x^3(\mathbf{b})(\mathbf{b}+1)(\mathbf{c}) + x^3(\mathbf{c}-1)(\mathbf{c})(\mathbf{a}+2) - x^4(\mathbf{c}-1)(\mathbf{c})(\mathbf{b}+1),$$

or

$$\mathrm{IL}(\infty; a, b, c) = \begin{vmatrix} (\mathbf{a}+1)(\mathbf{a}+2), & (\mathbf{b})(\mathbf{b}+1), & x(\mathbf{c}-1)(\mathbf{c}) \\ x(\mathbf{a}+2), & (\mathbf{b}+1), & (\mathbf{c}) \\ x^3, & x, & 1 \end{vmatrix} \div \begin{vmatrix} 1, & 1, & x \\ x, & 1, & 1 \\ x^3, & x, & 1 \end{vmatrix},$$

a satisfactory representation of the inner lattice function.

Art. 14. Passing now to the consideration of the inner lattice function of the order 4, viz., $\mathrm{IL}(\infty; a, b, c, d)$, and guided by the above results, I put

$$\begin{array}{lll} A_1 = (\mathbf{a}+3), & A_2 = (\mathbf{a}+2)(\mathbf{a}+3), & A_3 = (\mathbf{a}+1)(\mathbf{a}+2)(\mathbf{a}+3), \\ B_1 = (\mathbf{b}+2), & B_2 = (\mathbf{b}+1)(\mathbf{b}+2), & B_3 = (\mathbf{b})(\mathbf{b}+1)(\mathbf{b}+2), \\ C_1 = (\mathbf{c}+1), & C_2 = (\mathbf{c})(\mathbf{c}+1), & C_3 = (\mathbf{c}-1)(\mathbf{c})(\mathbf{c}+1), \\ D_1 = (\mathbf{d}), & D_2 = (\mathbf{d}-1)(\mathbf{d}), & D_3 = (\mathbf{d}-1)(\mathbf{d}-1)(\mathbf{d}), \end{array}$$

and I consider the twenty-four products

$$\begin{array}{llll} A_3B_2C_1, & A_3B_2D_1, & A_3D_2C_1, & D_3B_2C_1, \\ A_3C_2B_1, & A_3D_2B_1, & A_3C_2D_1, & D_3C_2B_1, \\ B_3A_2C_1, & B_3A_2D_1, & D_3A_2C_1, & B_3D_2C_1, \\ B_3C_2A_1, & B_3D_2A_1, & D_3C_2A_1, & B_3C_2D_1, \\ C_3A_2B_1, & D_3A_2B_1, & C_3A_2D_1, & C_3D_2B_1, \\ C_3B_2A_1, & D_3B_2A_1, & C_3D_2A_1, & C_3B_2D_1, \end{array}$$

which, to suffices 3, 2, 1 in descending order, involve every permutation of the letters ABCD, three at a time.

Art. 15. I shall show that each of these products is a solution of the functional equation

$$x^{\Sigma a}\,\mathrm{IL}(\infty; a, b, c, d) = (1-q_1)(1-q_2)(1-q_3)(1-q_4)\,\mathrm{IL}(\infty; a, b, c, d);$$

for, looking at the definition of the symbol q, it is clear that

$$q_1A_3B_2C_1 = (\mathbf{a})A_3B_2C_1,\ q_2A_3B_2C_1 = (\mathbf{b})A_3B_2C_1,\ q_3A_3B_2C_1 = (\mathbf{c})A_3B_2C_1,\ q_4A_3B_2C_1 = (\mathbf{d})A_3B_2C_1,$$
$$q_1q_2A_3B_2C_1 = (\mathbf{a})(\mathbf{b})A_3B_2C_1,\ q_1q_2q_3A_3B_2C_1 = (\mathbf{a})(\mathbf{b})(\mathbf{c})A_3B_2C_1,\ \&c.$$

Hence

$$(1-q_1)(1-q_2)(1-q_3)(1-q_4)\,A_3B_2C_1$$
$$= \{1-(\mathbf{a})\}\{1-(\mathbf{b})\}\{1-(\mathbf{c})\}\{1-(\mathbf{d})\}\,A_3B_2C_1 = x^{\Sigma a}A_3B_2C_1,$$

establishing that $A_3B_2C_1$ is a solution.

In general, put

$$A_0 = B_0 = C_0 = D_0 = 1,$$

and consider the product $A_\alpha B_\beta C_\gamma D_\delta$ where A, B, C, D are in fixed alphabetical order and α, β, γ, δ is some permutation of 3, 2, 1, 0.

We find

$$q_1 A_\alpha B_\beta C_\gamma D_\delta = (\mathbf{a}-\alpha+3)\, A_\alpha B_\beta C_\gamma D_\delta,$$

and the effect of the symbols q_2, q_3, q_4 is to multiply by $(\mathbf{b}-\beta+2)$, $(\mathbf{c}-\gamma+1)$, $(\mathbf{d}-\delta)$ respectively. Hence

$$(1-q_1)(1-q_2)(1-q_3)(1-q_4)\, A_\alpha B_\beta C_\gamma D_\delta$$

$$= \{1-(\mathbf{a}-\alpha+3)\}\,\{1-(\mathbf{b}-\beta+2)\}\,\{1-(\mathbf{c}-\gamma+1)\}\,\{1-(\mathbf{d}-\delta)\}\, A_\alpha B_\beta C_\gamma D_\delta,$$

$$= x^{a-\alpha+3+b-\beta+2+c-\gamma+1+d-\delta}\, A_\alpha B_\beta C_\gamma D_\delta,$$

$$= x^{a+b+c+d}\, A_\alpha B_\beta C_\gamma D_\delta,$$

since

$$\alpha+\beta+\gamma+\delta = 1+2+3.$$

It is thus established that each of the products in question is a solution of the functional equation.

Art. 16. Hence the determinant, which is a linear function of these products, viz. :—

$$\begin{vmatrix} (\mathbf{a}+1)(\mathbf{a}+2)(\mathbf{a}+3), & (\mathbf{b})(\mathbf{b}+1)(\mathbf{b}+2), & x(\mathbf{c}-1)(\mathbf{c})(\mathbf{c}+1), & x^3(\mathbf{d}-2)(\mathbf{d}-1)(\mathbf{d}) \\ x(\mathbf{a}+2)(\mathbf{a}+3), & (\mathbf{b}+1)(\mathbf{b}+2), & (\mathbf{c})(\mathbf{c}+1), & x(\mathbf{d}-1)(\mathbf{d}) \\ x^3(\mathbf{a}+3), & x(\mathbf{b}+2), & (\mathbf{c}+1), & (\mathbf{d}) \\ x^6, & x^3, & x, & 1 \end{vmatrix};$$

and I shall show that this determinant, divided by the determinant

$$\begin{vmatrix} 1, & 1, & x, & x^3 \\ x, & 1, & 1, & x \\ x^3, & x, & 1, & 1 \\ x^6, & x^3, & x, & 1 \end{vmatrix},$$

is the actual expression of IL (∞ ; a, b, c, d).

Art. 17. I first take the test of the sum of the coefficients which we know otherwise to be

$$\tfrac{1}{12}(a-b+1)(a-c+2)(a-d+3)(b-c+1)(b-d+2)(c-d+1).$$

The denominator determinant has the value $(\mathbf{1})^3\,(\mathbf{2})^2\,(\mathbf{3})$; dividing numerator and denominator by $(1)^6$, and then putting x equal to unity, we find

$$\frac{1}{12}\begin{vmatrix} (a+1)(a+2)(a+3), & b(b+1)(b+2), & (c-1)c(c+1), & (d-2)(d-1)(d) \\ (a+2)(a+3), & (b+1)(b+2), & c(c+1), & (d-1)d \\ a+3, & b+2, & c+1, & d \\ 1, & 1, & 1, & 1 \end{vmatrix},$$

and herein putting $a-b+1$, $a-c+2$, $a-d+3$, $b-c+1$, $b-d+2$, $c-d+1$ separately equal to zero, we in each case find two columns becoming identical and the determinant vanishing. Hence the sum of the coefficients has the proper numerical value.

Art. 18. As a second test I will show that the quotient of determinants becomes unity on putting $a = b = c = d$.

The numerator determinant becomes

$$\begin{vmatrix} (\mathbf{a}+\mathbf{1})(\mathbf{a}+\mathbf{2})(\mathbf{a}+\mathbf{3}), & (\mathbf{a})(\mathbf{a}+\mathbf{1})(\mathbf{a}+\mathbf{2}), & x(\mathbf{a}-\mathbf{1})(\mathbf{a})(\mathbf{a}+\mathbf{1}), & x^3(\mathbf{a}-\mathbf{2})(\mathbf{a}-\mathbf{1})(\mathbf{a}) \\ x(\mathbf{a}+\mathbf{2})(\mathbf{a}+\mathbf{3}), & (\mathbf{a}+\mathbf{1})(\mathbf{a}+\mathbf{2}), & (\mathbf{a})(\mathbf{a}+\mathbf{1}), & x(\mathbf{a}-\mathbf{1})(\mathbf{a}) \\ x^3(\mathbf{a}+\mathbf{3}), & x(\mathbf{a}+\mathbf{2}), & (\mathbf{a}+\mathbf{1}), & (\mathbf{a}) \\ x^6, & x^3, & x, & 1 \end{vmatrix}.$$

Transform this by taking

For New First Row—

$$1^{\text{st}}\text{ Row}+x^a(1+x+x^2)\times 2^{\text{nd}}\text{ Row}+x^{2a+1}(1+x+x^2)\times 3^{\text{rd}}\text{ Row}+x^{3a+3}\times 4^{\text{th}}\text{ Row},$$

For New Second Row—

$$2^{\text{nd}}\text{ Row}+x^a(1+x)\times 3^{\text{rd}}\text{ Row}+x^{2a+1}\times 4^{\text{th}}\text{ Row},$$

For New Third Row—

$$3^{\text{rd}}\text{ Row}+x^a\times 4^{\text{th}}\text{ Row},$$

and it becomes

$$\begin{vmatrix} 1, & 1, & x, & x^3 \\ x, & 1, & 1, & x \\ x^3, & x, & 1, & 1 \\ x^6, & x^3, & x, & 1 \end{vmatrix},$$

and thus the quotient of determinants is unity.

This verifies numerous particular cases.

Art. 19. A third test is to show that the quotient of determinants has the value

$$1+x^a+x^{2a}+x^{3a}$$

when $b = c = a$, $d = a-1$.

The proof is too long to find a place here.

All of the processes employed above are obviously valid when applied to the functional equation of order n and lead to the expression of $\mathrm{IL}(\infty; a_1, a_2, \ldots, a_n)$ as a quotient of determinants.

Art. 20. Before proceeding further I collect together the chief results obtained above.

$$\mathrm{IL}(\infty; a, b) = \begin{vmatrix} (\mathbf{a}+1), & (\mathbf{b}) \\ x, & 1 \end{vmatrix} \div \begin{vmatrix} 1, & 1 \\ x, & 1 \end{vmatrix};$$

$$\mathrm{L}(\infty; a, b) = \frac{(1)(2)\ldots(\mathbf{a}+\mathbf{b}) \begin{vmatrix} (\mathbf{a}+1), & (\mathbf{b}) \\ x, & 1 \end{vmatrix} \div \begin{vmatrix} 1, & 1 \\ x, & 1 \end{vmatrix}}{(2)(3)\ldots(a+1).(1)(2)\ldots(b)};$$

$$\mathrm{GF}(\infty; a, b) = \frac{\begin{vmatrix} (\mathbf{a}+1), & \mathbf{b} \\ x, & 1 \end{vmatrix} \div \begin{vmatrix} 1, & 1 \\ x, & 1 \end{vmatrix}}{(2)(3)\ldots(\mathbf{a}+1).(1)(2)\ldots(\mathbf{b})};$$

$$\mathrm{IL}(\infty; a, b, c) = \begin{vmatrix} (\mathbf{a}+1)(\mathbf{a}+2), & (\mathbf{b})(\mathbf{b}+1), & x(\mathbf{c}-1)(\mathbf{c}) \\ x(\mathbf{a}+2), & (\mathbf{b}+1), & (\mathbf{c}) \\ x^3, & x, & 1 \end{vmatrix} \div \begin{vmatrix} 1, & 1, & 1 \\ x, & 1, & 1 \\ x^3, & x, & 1 \end{vmatrix};$$

$$\mathrm{L}(\infty; a, b, c) = \frac{\begin{vmatrix} (\mathbf{a}+1)(\mathbf{a}+2), & (\mathbf{b})(\mathbf{b}+1), & x(\mathbf{c}-1)(\mathbf{c}) \\ x(\mathbf{a}+2), & (\mathbf{b}+1), & (\mathbf{c}) \\ x^3, & x, & 1 \end{vmatrix} \div \begin{vmatrix} 1, & 1, & 1 \\ x, & 1, & 1 \\ x^3, & x, & 1 \end{vmatrix}}{(3)(4)\ldots(\mathbf{a}+2).(2)(3)\ldots(\mathbf{b}+1).(1)(2)\ldots(\mathbf{c})}(1)(2)\ldots(\mathbf{a}+\mathbf{b}+\mathbf{c});$$

and, not putting the denominator determinant in evidence,

$$\mathrm{GF}(\infty; a, b, c) = \frac{\begin{vmatrix} (\mathbf{a}+1)(\mathbf{a}+2), & (\mathbf{b})(\mathbf{b}+1), & x(\mathbf{c}-1)(\mathbf{c}) \\ x(\mathbf{a}+2), & (\mathbf{b}+1), & (\mathbf{c}) \\ x^3, & x, & 1 \end{vmatrix}}{(1)\ldots(\mathbf{a}+2).(1)\ldots(\mathbf{b}+1).(1)\ldots(\mathbf{c})};$$

$$\mathrm{GF}(\infty; a, b, c, d) = \frac{\begin{vmatrix} (\mathbf{a}+1)(\mathbf{a}+2)(\mathbf{a}+3), & (\mathbf{b})(\mathbf{b}+1)(\mathbf{b}+2), & x(\mathbf{c}-1)(\mathbf{c})(\mathbf{c}+1), & x^3(\mathbf{d}-2)(\mathbf{d}-1)(\mathbf{d}) \\ x(\mathbf{a}+2)(\mathbf{a}+3), & (\mathbf{b}+1)(\mathbf{b}+2), & (\mathbf{c})(\mathbf{c}+1), & x(\mathbf{d}-1)(\mathbf{d}) \\ x^3(\mathbf{a}+3), & x(\mathbf{b}+2), & (\mathbf{c}+1), & (\mathbf{d}) \\ x^6, & x^3, & x, & 1 \end{vmatrix}}{(1)\ldots(\mathbf{a}+3).(1)\ldots(\mathbf{b}+2).(1)\ldots(\mathbf{c}+1).(1)\ldots(\mathbf{d})}$$

In general, the determinant numerator of GF (∞ ; $a_1, a_2, \ldots, a_n$) involves x explicitly as exhibited in the determinant

$$\begin{vmatrix} 1, & 1, & x, & x^3, & \ldots, & x^{\binom{s-1}{2}} \\ x, & 1, & 1, & x, & \ldots, & x^{\binom{s-2}{2}} \\ x^3, & x, & 1, & 1, & \ldots, & x^{\binom{s-3}{2}} \\ x^6, & x^3, & x, & 1, & \ldots, & x^{\binom{s-4}{2}} \\ \vdots & \vdots & \vdots & \vdots & & \vdots \\ x^{\binom{r}{2}}, & x^{\binom{r-1}{2}}, & x^{\binom{r-2}{2}}, & x^{\binom{r-3}{2}}, & \ldots, & 1 \end{vmatrix}$$

the exponents of x being figurate numbers of order 3,

$$\mathrm{GF}(\infty\,;\,a^n) = \frac{1}{(\mathbf{n})\ldots(\mathbf{a+n-2})\,.\,(\mathbf{n-1})\ldots(\mathbf{a+n-3})\ldots\ldots(\mathbf{2})\ldots(\mathbf{a+1})\,.\,(\mathbf{1})\ldots(\mathbf{a})}.$$

The Restriction on the Part Magnitude.

Art. 21. I pass on to consider the case in which the part magnitude is restricted by the integer l.

I take as my point of departure the functional equation

$$(1-p_1)(1-p_2)\ldots(1-p_n)\,\mathrm{GF}(l\,;\,a_1, a_2, \ldots, a_n) = x^{\Sigma a}\,\mathrm{GF}(l-1\,;\,a_1, a_2, \ldots, a_n),$$

and, by means of the relation

$$\mathrm{GF}(l\,;\,a_1, a_2, \ldots, a_n) = \frac{\mathrm{L}(l\,;\,a_1, a_2, \ldots, a_n)}{(\mathbf{1})(\mathbf{2})\ldots(\mathbf{\Sigma a})},$$

convert it into a functional equation for the lattice function.

For the orders 2 and 3 we have

$$\mathrm{L}(l\,;\,a, b) - x^{a+b}\,\mathrm{L}(l-1\,;\,a, b)$$
$$= (\mathbf{a+b})\{\mathrm{L}(l\,;\,a-1, b) + \mathrm{L}(l\,;\,a, b-1)\} - (\mathbf{a+b-1})(\mathbf{a+b})\,\mathrm{L}(l\,;\,a-1, b-1)\,;$$

$$\mathrm{L}(l\,;\,a, b, c) - x^{a+b+c}\,\mathrm{L}(l-1\,;\,a, b, c)$$
$$= (\mathbf{a+b+c})\{\mathrm{L}(l\,;\,a-1, b, c) + \mathrm{L}(l\,;\,a, b-1, c) + \mathrm{L}(l\,;\,a, b, c-1)\}$$
$$-(\mathbf{a+b+c-1})(\mathbf{a+b+c})\{\mathrm{L}(l; a-1, b-1, c) + \mathrm{L}(l; a-1, b, c-1) + \mathrm{L}(l; a, b-1, c-1)\}$$
$$+(\mathbf{a+b+c-2})(\mathbf{a+b+c-1})(\mathbf{a+b+c})\,\mathrm{L}(l\,;\,a-1, b-1, c-1)\,;$$

and in general

$$(1-p_1 X)(1-p_2 X)\ldots(1-p_n X)\,L(l;\,a_1, a_2, \ldots, a_n) = x^{\Sigma a}\,L(l-1;\,a_1, a_2, \ldots, a_n);$$

wherein p is the symbol of Art. 3, and symbolically

$$X^m = (\Sigma\mathrm{a})(\Sigma\mathrm{a}-1)\ldots(\Sigma\mathrm{a}-\mathrm{m}+1).$$

Art. 22. Also, from the relation

$$L(l;\,a_1, a_2, \ldots, a_n)$$

$$= (1)(2)\ldots(\Sigma\mathrm{a})\,\frac{(\mathrm{l}+\mathrm{n})\ldots(\mathrm{l}+\mathrm{a}_1+\mathrm{n}-1).(\mathrm{l}+\mathrm{n}-1)\ldots\ldots(\mathrm{l}+\mathrm{a}_2+\mathrm{n}-2)\ldots\ldots(\mathrm{l}+1)\ldots(\mathrm{l}+\mathrm{a}_\mathrm{n})}{(\mathrm{n})\ldots(\mathrm{a}_1+\mathrm{n}-1).(\mathrm{n}-1)\ldots(\mathrm{a}_2+\mathrm{n}-2)\ldots\ldots(1)\ldots(\mathrm{a}_\mathrm{n})}\,\mathrm{IL}(l;\,a_1, a_2, \ldots, a_n),$$

for the orders 2 and 3 we have

$$(\mathrm{l}+\mathrm{a}+1)(\mathrm{l}+\mathrm{b})\,\mathrm{IL}(l;a,b)-(\mathrm{a}+1)(\mathrm{l}+\mathrm{b})\,\mathrm{IL}(l;a-1,b)-(\mathrm{l}+\mathrm{a}+1)(\mathrm{b})\,\mathrm{IL}(l;a,b-1)$$
$$+(\mathrm{a}+1)(\mathrm{b})\,\mathrm{IL}(l;\,a-1,b-1) = x^{a+b}(\mathrm{l})(\mathrm{l}+1)\,\mathrm{IL}(l-1;\,a,b);$$

$$(\mathrm{l}+\mathrm{a}+2)(\mathrm{l}+\mathrm{b}+1)(\mathrm{l}+\mathrm{c})\,\mathrm{IL}(l;\,a,b,c)$$
$$-(\mathrm{a}+2)(\mathrm{l}+\mathrm{b}+1)(\mathrm{l}+\mathrm{c})\,\mathrm{IL}(l;\,a-1,b,c)-(\mathrm{l}+\mathrm{a}+2)(\mathrm{b}+1)(\mathrm{l}+\mathrm{c})\,\mathrm{IL}(l;\,a,b-1,c)$$
$$-(\mathrm{l}+\mathrm{a}+2)(\mathrm{l}+\mathrm{b}+1)(\mathrm{c})\,\mathrm{IL}(l;\,a,b,c-1)$$

$$+(\mathrm{a}+2)(\mathrm{b}+1)(\mathrm{l}+\mathrm{c})\,\mathrm{IL}(l;\,a-1,b-1,c)+(\mathrm{a}+2)(\mathrm{l}+\mathrm{b}+1)(\mathrm{c})\,\mathrm{IL}(l;\,a-1,b,c-1)$$
$$+(\mathrm{l}+\mathrm{a}+2)(\mathrm{b}+1)(\mathrm{c})\,\mathrm{IL}(l;\,a,b-1,c-1)-(\mathrm{a}+2)(\mathrm{b}+1)(\mathrm{c})\,\mathrm{IL}(l;\,a-1,b-1,c-1)$$
$$= x^{a+b+c}(\mathrm{l})(\mathrm{l}+1)(\mathrm{l}+2)\,\mathrm{IL}(l-1;\,a,b,c).$$

While in general, if r_s be a symbol such that

$$r_s\,\mathrm{IL}(l;\,a_1, a_2, \ldots, a_n) = \frac{(\mathrm{a}_\mathrm{s}+\mathrm{n}-\mathrm{s})}{(\mathrm{l}+\mathrm{a}_\mathrm{s}+\mathrm{n}-\mathrm{s})}\,\mathrm{IL}(l;\,a_1, a_2, \ldots, a_s-1, \ldots, a_n),$$

$$(\mathrm{l}+\mathrm{a}_1+\mathrm{n}-1)(\mathrm{l}+\mathrm{a}_2+\mathrm{n}-2)\ldots(\mathrm{l}+\mathrm{a}_\mathrm{n})(1-r_1)(1-r_2)\ldots(1-r_n)\,\mathrm{IL}(l;\,a_1, a_2, \ldots, a_n)$$
$$= x^{\Sigma a}(\mathrm{l})(\mathrm{l}+1)\ldots(\mathrm{l}+\mathrm{n}-1)\,\mathrm{IL}(l-1;\,a_1, a_2, \ldots, a_n).$$

Art. 23. I propose to obtain solutions of these functional equations. In order to ascertain the form of the required solutions it was necessary to examine several particular cases appertaining to the order 2; the result was the conjecture that

$$\mathrm{GF}(l;\,a,b) = \frac{(\mathrm{l}+2)(\mathrm{l}+3)\ldots(\mathrm{l}+\mathrm{a}+1).(\mathrm{l}+1)(\mathrm{l}+2)\ldots(\mathrm{l}+\mathrm{b})}{(2)(3)\ldots(\mathrm{a}+1).(1)(2)\ldots(\mathrm{b})}\left\{1+x^{b+1}\frac{(\mathrm{a}-\mathrm{b})(\mathrm{l})}{(1)(\mathrm{l}+\mathrm{a}+1)}\right\}.$$

This expression was found to satisfy the functional equation, so that certainly

$$\mathrm{IL}(l\,;\,a,b) = 1+x^{b+1}\frac{(\mathfrak{a}-\mathfrak{b})(\mathfrak{l})}{(1)(\mathfrak{l}+\mathfrak{a}+1)};$$

and then, observing that we may write

$$\mathrm{IL}(l\,;\,a,b) = \frac{1}{(1)(\mathfrak{l}+\mathfrak{a}+1)}\begin{vmatrix}(\mathfrak{a}+1), & (\mathfrak{b})\\ x(\mathfrak{l}), & (\mathfrak{l}+1)\end{vmatrix},$$

and remembering the nature of solution when $l=\infty$, it became clear that we should seek solutions for the order 2 of the forms

$$\frac{(\mathfrak{a}+1)}{(\mathfrak{l}+\mathfrak{a}+1)}\mathrm{F}_l,\quad \frac{(\mathfrak{b})}{(\mathfrak{l}+\mathfrak{a}+1)}\mathrm{F}_l\,;$$

where F_l is a function of l to be determined in each case.

I therefore substitute $\frac{(\mathfrak{a}+1)}{(\mathfrak{l}+\mathfrak{a}+1)}\mathrm{F}_l$ for $\mathrm{IL}(l\,;\,a,b)$ in the functional equation and arrive at the relation

$$(\mathfrak{l})\,\mathrm{F}_l = (\mathfrak{l}+1)\,\mathrm{F}_{l-1}\,;$$

from which I deduce

$$\mathrm{F}_l = (\mathfrak{l}+1),$$

yielding for me the fundamental solution

$$\frac{(\mathfrak{a}+1)(\mathfrak{l}+1)}{(\mathfrak{l}+\mathfrak{a}+1)}.$$

Similarly I find that another fundamental solution is

$$\frac{(\mathfrak{b})(\mathfrak{l})}{(\mathfrak{l}+\mathfrak{a}+1)};$$

and, in terms of these two solutions, I find

$$\mathrm{IL}(l\,;\,a,b) = \frac{1}{(1)}\left\{\frac{(\mathfrak{a}+1)(\mathfrak{l}+1)}{(\mathfrak{l}+\mathfrak{a}+1)} - x\frac{(\mathfrak{b})(\mathfrak{l})}{(\mathfrak{l}+\mathfrak{a}+1)}\right\}.$$

Art. 24. This simple exposition for the second order clearly points out the path of investigation for the third order. For, guided by the six fundamental solutions when $l=\infty$, it is natural to seek for solutions of the functional equation of the six types

$$\frac{(\mathfrak{a}+1)(\mathfrak{a}+2)(\mathfrak{b}+1)\mathrm{F}_l}{(\mathfrak{l}+\mathfrak{a}+1)(\mathfrak{l}+\mathfrak{a}+2)(\mathfrak{l}+\mathfrak{b}+1)};\qquad \frac{(\mathfrak{a}+1)(\mathfrak{a}+2)(\mathfrak{c})\mathrm{F}_l}{(\mathfrak{l}+\mathfrak{a}+1)(\mathfrak{l}+\mathfrak{a}+2)(\mathfrak{l}+\mathfrak{b}+1)};$$

$$\frac{(\mathfrak{b})(\mathfrak{b}+1)(\mathfrak{a}+2)\mathrm{F}_l}{(\mathfrak{l}+\mathfrak{a}+1)(\mathfrak{l}+\mathfrak{a}+2)(\mathfrak{l}+\mathfrak{b}+1)};\qquad \frac{(\mathfrak{c}-1)(\mathfrak{c})(\mathfrak{a}+2)\mathrm{F}_l}{(\mathfrak{l}+\mathfrak{a}+1)(\mathfrak{l}+\mathfrak{a}+2)(\mathfrak{l}+\mathfrak{b}+1)};$$

$$\frac{(\mathfrak{b})(\mathfrak{b}+1)(\mathfrak{c})\mathrm{F}_l}{(\mathfrak{l}+\mathfrak{a}+1)(\mathfrak{l}+\mathfrak{a}+2)(\mathfrak{l}+\mathfrak{b}+1)};\qquad \frac{(\mathfrak{c}-1)(\mathfrak{c})(\mathfrak{b}+1)\mathrm{F}_l}{(\mathfrak{l}+\mathfrak{a}+1)(\mathfrak{l}+\mathfrak{a}+2)(\mathfrak{l}+\mathfrak{b}+1)};$$

where F_l is a function of l to be determined in each case.

Substituting the first of these for IL$(l; a, b, c)$ in the functional equation of order 3 I find

$$\mathrm{F}_l = \frac{(l+1)(l+2)}{(l)^2}\mathrm{F}_{l-1};$$

but the solution of the equation

$$\mathrm{F}_l = \phi_l \mathrm{F}_{l-1}$$

is clearly

$$\mathrm{F}_l = \phi_l \phi_{l-1} \phi_{l-2} \ldots;$$

so that, in the present case,

$$\mathrm{F}_l = \frac{(l+1)(l+2)}{(l)^2}\cdot\frac{(l)(l+1)}{(l-1)^2}\cdot\frac{(l-1)(l)}{(l-2)^2}\ldots = (l+1)^2(l+2);$$

so that I obtain a fundamental solution

$$\frac{(a+1)(a+2)(b+1)(l+1)^2(l+2)}{(l+a+1)(l+a+2)(l+b+1)}.$$

Art. 25. Similarly I arrive at five other fundamental solutions,

$$\frac{(a+1)(a+2)(c)(l)(l+1)(l+2)}{(l+a+1)(l+a+2)(l+b+1)},$$

$$\frac{(b)(b+1)(a+2)(l)(l+1)(l+2)}{(l+a+1)(l+a+2)(l+b+1)},$$

$$\frac{(c-1)(c)(a+2)(l)^2(l+1)}{(l+a+1)(l+a+2)(l+b+1)},$$

$$\frac{(b)(b+1)(c)(l-1)(l)(l+2)}{(l+a+1)(l+a+2)(l+b+1)},$$

$$\frac{(c-1)(c)(b+1)(l-1)(l)(l+1)}{(l+a+1)(l+a+2)(l+b+1)};$$

and I next seek, guided by previous work, to construct the function IL$(l; a, b, c)$ by a linear function of these six solutions.

It is natural to write

$$(a+1)(a+2)(b+1)(l+1)^2(l+2) - x(b)(b+1)(a+2)(l)(l+1)(l+2)$$
$$-x(a+1)(a+2)(c)(l)(l+1)(l+2) + x^3(c-1)(c)(a+2)(l)^2(l+1)$$
$$+x^3(b)(b+1)(c)(l-1)(l)(l+2) - x^4(c-1)(c)(b+1)(l-1)(l)(l+1),$$

with a denominator

$$(1)^2(2)(l+a+1)(l+a+2)(l+b+1),$$

3 A 2

which, in determinant form, is

$$\begin{vmatrix} (\mathfrak{a}+1)(\mathfrak{a}+2), & (\mathfrak{b})(\mathfrak{b}+1), & x(\mathfrak{c}-1)(\mathfrak{c}) \\ x(\mathfrak{a}+2)(\mathfrak{l}), & (\mathfrak{b}+1)(\mathfrak{l}+1), & (\mathfrak{c})(\mathfrak{l}+2) \\ x^3(\mathfrak{l}-1)(\mathfrak{l}), & x(\mathfrak{l})(\mathfrak{l}+1), & (\mathfrak{l}+1)(\mathfrak{l}+2) \end{vmatrix}$$

divided by

$$(1)^2(2)(\mathfrak{l}+\mathfrak{a}+1)(\mathfrak{l}+\mathfrak{a}+2)(\mathfrak{l}+\mathfrak{b}+1);$$

and it will be shown that it is, in fact, the expression of IL $(l;\, a, b, c)$.

Art. 26. In a general manner we may take the following view—

Recalling a previous notation for the order 3

$$A_2 = (\mathfrak{a}+1)(\mathfrak{a}+2),\quad A_1 = (\mathfrak{a}+2),\quad B_2 = (\mathfrak{b})(\mathfrak{b}+1),\quad B_1 = (\mathfrak{b}+1),\quad C_2 = (\mathfrak{c}-1)(\mathfrak{c}),$$
$$C_1 = (\mathfrak{c}),\quad A_0 = B_0 = C_0 = 1;$$

and taking a product

$$A_pB_qC_r,$$

where p, q, r denotes some permutation of the numbers 2, 1, 0, suppose

$$\frac{A_pB_qC_rF_l}{(\mathfrak{l}+\mathfrak{a}+1)(\mathfrak{l}+\mathfrak{a}+2)(\mathfrak{l}+\mathfrak{b}+1)}$$

substituted for IL $(l;\, a, b, c)$ is the functional equation. The result on reduction is

$$F_l = \frac{(\mathfrak{l})(\mathfrak{l}+1)(\mathfrak{l}+2)}{(\mathfrak{l}-2+\mathfrak{p})(\mathfrak{l}-1+\mathfrak{q})(\mathfrak{l}+\mathfrak{r})}F_{l-1};$$

and now writing

$$\frac{(\mathfrak{l})}{(\mathfrak{l}-2+\mathfrak{p})}\cdot\frac{(\mathfrak{l}-1)}{(\mathfrak{l}-3+\mathfrak{p})}\cdot\frac{(\mathfrak{l}-2)}{(\mathfrak{l}-4+\mathfrak{p})}\cdot\ldots\ \textit{ad inf.} = \phi_{ap},$$

$$\frac{(\mathfrak{l}+1)}{(\mathfrak{l}-1+\mathfrak{q})}\cdot\frac{(\mathfrak{l})}{(\mathfrak{l}-2+\mathfrak{q})}\cdot\frac{(\mathfrak{l}-1)}{(\mathfrak{l}-3+\mathfrak{q})}\cdot\ldots\ \textit{ad inf.} = \phi_{bq},$$

$$\frac{(\mathfrak{l}+2)}{(\mathfrak{l}+\mathfrak{r})}\cdot\frac{(\mathfrak{l}+1)}{(\mathfrak{l}-1+\mathfrak{r})}\cdot\frac{(\mathfrak{l})}{(\mathfrak{l}-2+\mathfrak{r})}\cdot\ldots\ \textit{ad inf.} = \phi_{cr};$$

$$F_l = \phi_{ap}\phi_{bq}\phi_{cr};$$

and the inner lattice function IL $(l;\, a, b, c)$ is, to a divisor

$$(1)^2(2)(\mathfrak{l}+\mathfrak{a}+1)(\mathfrak{l}+\mathfrak{a}+2)(\mathfrak{l}+\mathfrak{b}+1)$$

près, equal to the determinant

$$\begin{vmatrix} A_2\phi_{a2}, & B_2\phi_{b2}, & xC_2\phi_{c2} \\ xA_1\phi_{a1}, & B_1\phi_{b1}, & C_1\phi_{c1} \\ x^3A_0\phi_{a0}, & xB_0\phi_{b0}, & C_0\phi_{c0} \end{vmatrix},$$

and this determinant is clearly

$$\begin{vmatrix} A_2 & , & B_2 & , & xC_2 \\ xA_1(\mathfrak{l}) & , & B_1(\mathfrak{l}+1) & , & C_1(\mathfrak{l}+2) \\ x^3A_0(\mathfrak{l}-1)(\mathfrak{l}), & & xB_0(\mathfrak{l})(\mathfrak{l}+1), & & C_0(\mathfrak{l}+1)(\mathfrak{l}+2) \end{vmatrix}.$$

Art. 27. This is evidently a perfectly general process and suffices to establish that a solution of the functional equation of order n is a determinant of order n of which the constituent in the sth row and t^{th} column is

$$x^{\binom{s-t+1}{2}}(\mathfrak{a}_t+\mathfrak{s}-\mathfrak{t}+1)\ldots(\mathfrak{a}_t+\mathfrak{n}-\mathfrak{t}).(\mathfrak{l}-\mathfrak{s}+\mathfrak{t}+1)\ldots(\mathfrak{l}+\mathfrak{t}-1);$$

and when this determinant is divided by

$$(1)^{n-1}(2)^{n-2}\ldots(\mathfrak{n}+1)(\mathfrak{l}+\mathfrak{a}_1+1)\ldots(\mathfrak{l}+\mathfrak{a}_1+\mathfrak{n}-1).(\mathfrak{l}+\mathfrak{a}_2+1)\ldots(\mathfrak{l}+\mathfrak{a}_2+\mathfrak{n}-2)\ldots(\mathfrak{l}+\mathfrak{a}_{n-1}+1),$$

we have, as will be proved, the expression of

$$\text{IL}(l\,;\,a_1, a_2, \ldots, a_n).$$

Art. 28. To establish this we may apply a series of tests.

Thus, take the expression of IL $(l\,;\,a, b, c, d)$

$$\begin{vmatrix} (\mathfrak{a}+1)(\mathfrak{a}+2)(\mathfrak{a}+3), & (\mathfrak{b})(\mathfrak{b}+1)(\mathfrak{b}+2), & x(\mathfrak{c}-1)(\mathfrak{c})(\mathfrak{c}+1), & x^3(\mathfrak{d}-2)(\mathfrak{d}-1)(\mathfrak{d}) \\ x(\mathfrak{l})(\mathfrak{a}+2)(\mathfrak{a}+3), & (\mathfrak{l}+1)(\mathfrak{b}+1)(\mathfrak{b}+2), & (\mathfrak{l}+2)(\mathfrak{c})(\mathfrak{c}+1), & x(\mathfrak{l}+3)(\mathfrak{d}-1)(\mathfrak{d}) \\ x^3(\mathfrak{l}-1)(\mathfrak{l})(\mathfrak{a}+3), & x(\mathfrak{l})(\mathfrak{l}+1)(\mathfrak{b}+2), & (\mathfrak{l}+1)(\mathfrak{l}+2)(\mathfrak{c}+1), & (\mathfrak{l}+2)(\mathfrak{l}+3)(\mathfrak{d}) \\ x^6(\mathfrak{l}-2)(\mathfrak{l}-1)(\mathfrak{l}), & x^3(\mathfrak{l}-1)(\mathfrak{l})(\mathfrak{l}+1), & x(\mathfrak{l})(\mathfrak{l}+1)(\mathfrak{l}+2), & (\mathfrak{l}+1)(\mathfrak{l}+2)(\mathfrak{l}+3) \end{vmatrix}$$

divided by

$$(1)^3(2)^2(3)(\mathfrak{l}+\mathfrak{a}+1)(\mathfrak{l}+\mathfrak{a}+2)(\mathfrak{l}+\mathfrak{a}+3)(\mathfrak{l}+\mathfrak{b}+1)(\mathfrak{l}+\mathfrak{b}+2)(\mathfrak{l}+\mathfrak{c}+1).$$

It clearly reduces to IL $(\infty\,;\,a, b, c, d)$ when l is put equal to ∞.

Moreover, it can be shown that when $d = c = b = a$, the determinant involves l and a always in the combination $l+a$; for consider the determinant in question when l is put equal to $-a$.

It is

$$\begin{vmatrix} (\mathrm{a}+1)(\mathrm{a}+2)(\mathrm{a}+3), & (\mathrm{a})(\mathrm{a}+1)(\mathrm{a}+2), & x(\mathrm{a}-1)(\mathrm{a})(\mathrm{a}+1), & x^3(\mathrm{a}-2)(\mathrm{a}-1)(\mathrm{a}) \\ x(-\mathrm{a})(\mathrm{a}+2)(\mathrm{a}+3), & (-\mathrm{a}+1)(\mathrm{a}+1)(\mathrm{a}+2), & (-\mathrm{a}+2)(\mathrm{a})(\mathrm{a}+1), & x(-\mathrm{a}+3)(\mathrm{a}-1)(\mathrm{a}) \\ x^3(-\mathrm{a}-1)(-\mathrm{a})(\mathrm{a}+3), & x(-\mathrm{a})(-\mathrm{a}+1)(\mathrm{a}+2), & (-\mathrm{a}+1)(-\mathrm{a}+2)(\mathrm{a}+1), & (-\mathrm{a}+2)(-\mathrm{a}+3)(\mathrm{a}) \\ x^6(-\mathrm{a}-2)(-\mathrm{a}-1)(-\mathrm{a}), & x^3(-\mathrm{a}-1)(-\mathrm{a})(-\mathrm{a}+1), & x(-\mathrm{a})(-\mathrm{a}+1)(-\mathrm{a}+2), & (-\mathrm{a}+1)(-\mathrm{a}+2)(-\mathrm{a}+3) \end{vmatrix},$$

and since $(-\mathrm{s}) = -x^{-s}(\mathrm{s})$, it may be written

$$x^{10-6a}\begin{vmatrix} (\mathrm{a}+1)(\mathrm{a}+2)(\mathrm{a}+3), & (\mathrm{a})(\mathrm{a}+1)(\mathrm{a}+2), & (\mathrm{a}-1)(\mathrm{a})(\mathrm{a}+1), & (\mathrm{a}-2)(\mathrm{a}-1)(\mathrm{a}) \\ (\mathrm{a})(\mathrm{a}+2)(\mathrm{a}+3), & (\mathrm{a}-1)(\mathrm{a}+1)(\mathrm{a}+2), & (\mathrm{a}-2)(\mathrm{a})(\mathrm{a}+1), & (\mathrm{a}-3)(\mathrm{a}-1)(\mathrm{a}) \\ (\mathrm{a})(\mathrm{a}+1)(\mathrm{a}+3), & (\mathrm{a}-1)(\mathrm{a})(\mathrm{a}+2), & (\mathrm{a}-2)(\mathrm{a}-1)(\mathrm{a}+1), & (\mathrm{a}-3)(\mathrm{a}-2)(\mathrm{a}) \\ (\mathrm{a})(\mathrm{a}+1)(\mathrm{a}+2), & (\mathrm{a}-1)(\mathrm{a})(\mathrm{a}+1), & (\mathrm{a}-2)(\mathrm{a}-1)(\mathrm{a}), & (\mathrm{a}-3)(\mathrm{a}-2)(\mathrm{a}-1) \end{vmatrix};$$

and now, putting $-\alpha$ for α, we find after a few transformations by multiplication and division of columns and rows that it is unchanged; this suffices to show that the expression is not a function of α at all since it is clearly not of the form $k+\phi(\alpha)+\phi(-\alpha)$. Hence the determinant under examination is a function of $l+\alpha$. Put therein $l = p-\alpha$ and then put $\alpha = 0$; we find

$$\begin{vmatrix} (1)(2)(3), & 0, & 0, & 0 \\ x(\mathrm{p})(2)(3), & (\mathrm{p}+1)(1)(2), & 0, & 0 \\ x^3(\mathrm{p}-1)(\mathrm{p})(3), & x(\mathrm{p})(\mathrm{p}+1)(2), & (\mathrm{p}+1)(\mathrm{p}+2)(1), & 0 \\ x^6(\mathrm{p}-2)(\mathrm{p}-1)(\mathrm{p}), & x^3(\mathrm{p}-1)(\mathrm{p})(\mathrm{p}+1), & x(\mathrm{p})(\mathrm{p}+1)(\mathrm{p}+2), & (\mathrm{p}+1)(\mathrm{p}+2)(\mathrm{p}+3) \end{vmatrix},$$

which has the value

$$(1)^3(2)^2(3)(\mathrm{l}+\mathrm{a}+1)^3(\mathrm{l}+\mathrm{a}+2)^2(\mathrm{l}+\mathrm{a}+3),$$

and thus

$$\mathrm{IL}(l;\alpha^4) = 1.$$

This is a verification because in all particular cases of this nature that have been examined the inner lattice function is unity.

Art. 29. We may now resume the results.

$$\mathrm{GF}\,(l\,;\,ab)$$

$$=\frac{(1)\ldots(\mathfrak{l}+\mathfrak{a})}{(1)\ldots(\mathfrak{l}+1)\,.\,(1)\ldots(\mathfrak{a}+1)}\cdot\frac{(1)\ldots(\mathfrak{l}+\mathfrak{b})}{(1)\ldots(\mathfrak{l})\,.\,(1)\ldots(\mathfrak{b})}\cdot\begin{vmatrix}(\mathfrak{a}+1), & (\mathfrak{b})\\ x(\mathfrak{l}), & (\mathfrak{l}+1)\end{vmatrix};$$

$$\mathrm{GF}\,(l\,;\,a^2)$$

$$=\frac{(\mathfrak{l}+2)\ldots(\mathfrak{l}+\mathfrak{a}+1)}{(2)\ldots(\mathfrak{a}+1)}\cdot\frac{(\mathfrak{l}+1)\ldots(\mathfrak{l}+\mathfrak{a})}{(1)\ldots(\mathfrak{a})};$$

$$\mathrm{GF}\,(l\,;\,abc)$$

$$=\frac{(1)\ldots(\mathfrak{l}+\mathfrak{a})}{(1)\ldots(\mathfrak{l}+2).(1)\ldots(\mathfrak{a}+2)}\cdot\frac{(1)\ldots(\mathfrak{l}+\mathfrak{b})}{(1)\ldots(\mathfrak{l}+1).(1)\ldots(\mathfrak{b}+1)}\cdot\frac{(1)\ldots(\mathfrak{l}+\mathfrak{c})}{(1)\ldots(\mathfrak{l}).(1)\ldots(\mathfrak{c})}\cdot\begin{vmatrix}(\mathfrak{a}+1)(\mathfrak{a}+2), & (\mathfrak{b})(\mathfrak{b}+1), & x(\mathfrak{c}-1)(\mathfrak{c})\\ x(\mathfrak{l})(\mathfrak{a}+2), & (\mathfrak{l}+1)(\mathfrak{b}+1), & (\mathfrak{l}+2)(\mathfrak{c})\\ x^3(\mathfrak{l}-1)(\mathfrak{l}), & x(\mathfrak{l})(\mathfrak{l}+1), & (\mathfrak{l}+1)(\mathfrak{l}+2)\end{vmatrix};$$

$$\mathrm{GF}\,(l\,;\,a^3)$$

$$=\frac{(\mathfrak{l}+3)\ldots(\mathfrak{l}+\mathfrak{a}+2)}{(3)\ldots(\mathfrak{a}+2)}\cdot\frac{(\mathfrak{l}+2)\ldots(\mathfrak{l}+\mathfrak{a}+1)}{(2)\ldots(\mathfrak{a}+1)}\cdot\frac{(\mathfrak{l}+1)\ldots(\mathfrak{l}+\mathfrak{a})}{(1)\ldots(\mathfrak{a})};$$

$$\mathrm{GF}\,(l\,;\,a,b,c,d)$$

$$=\frac{(1)\ldots(\mathfrak{l}+\mathfrak{a})}{(1)\ldots(\mathfrak{l}+3)\,.\,(1)\ldots(\mathfrak{a}+3)}\cdot\frac{(1)\ldots(\mathfrak{l}+\mathfrak{b})}{(1)\ldots(\mathfrak{l}+2)\,.\,(1)\ldots(\mathfrak{b}+2)}\cdot\frac{(1)\ldots(\mathfrak{l}+\mathfrak{c})}{(1)\ldots(\mathfrak{l}+1)\,.\,(1)\ldots(\mathfrak{c}+1)}\cdot\frac{(1)\ldots(\mathfrak{l}+\mathfrak{d})}{(1)\ldots(\mathfrak{l})\,.\,(1)\ldots(\mathfrak{d})}$$

$$\times\begin{vmatrix}(\mathfrak{a}+1)(\mathfrak{a}+2)(\mathfrak{a}+3), & (\mathfrak{b})(\mathfrak{b}+1)(\mathfrak{b}+2), & x(\mathfrak{c}-1)(\mathfrak{c})(\mathfrak{c}+1), & x^3(\mathfrak{d}-2)(\mathfrak{d}-1)(\mathfrak{d})\\ x(\mathfrak{l})(\mathfrak{a}+2)(\mathfrak{a}+3), & (\mathfrak{l}+1)(\mathfrak{b}+1)(\mathfrak{b}+2), & (\mathfrak{l}+2)(\mathfrak{c})(\mathfrak{c}+1), & x(\mathfrak{l}+3)(\mathfrak{d}-1)(\mathfrak{d})\\ x^3(\mathfrak{l}-1)(\mathfrak{l})(\mathfrak{a}+3), & x(\mathfrak{l})(\mathfrak{l}+1)(\mathfrak{b}+2), & (\mathfrak{l}+1)(\mathfrak{l}+2)(\mathfrak{c}+1), & (\mathfrak{l}+2)(\mathfrak{l}+3)(\mathfrak{d})\\ x^6(\mathfrak{l}-2)(\mathfrak{l}-1)(\mathfrak{l}), & x^3(\mathfrak{l}-1)(\mathfrak{l})(\mathfrak{l}+1), & x(\mathfrak{l})(\mathfrak{l}+1)(\mathfrak{l}+2), & (\mathfrak{l}+1)(\mathfrak{l}+2)(\mathfrak{l}+3)\end{vmatrix};$$

and so forth, the law of formation being quite clear.

It is a remarkable fact that this elegant result appears to be valid whatsoever the equalities may be that present themselves between the integers $a, b, c, d, \ldots$. The discussion of this and the interpretation of $\text{GF}(l\,;\,a, b, c, d, \ldots)$ when $a, b, c, d, \ldots$, are not in descending order must be deferred to another occasion.

I add the general result for equal rows which has now been established

$$\text{GF}(l\,;\,a^n)$$
$$= \frac{(\mathbf{l+n})\ldots(\mathbf{l+a+n-1})}{(\mathbf{n})\ldots(\mathbf{a+n-1})}\cdot\frac{(\mathbf{l+n-1})\ldots(\mathbf{l+a+n-2})}{(\mathbf{n-1})\ldots(\mathbf{a+n-2})}\cdots\cdots\frac{(\mathbf{l+1})\ldots(\mathbf{l+a})}{(\mathbf{1})\ldots(\mathbf{a})}.$$

Art. 30. I recall that, in Part V., the formula was given

$$\text{GF}(l\,;\,a, b, c, d, \ldots) = \text{L}_0\text{GF}(l\,;\,\Sigma a)+\text{L}_1\text{GF}(l-1\,;\,\Sigma a)+\ldots+\text{L}_\mu\text{GF}(l-\mu\,;\,\Sigma a),$$

wherein L_s is the sub-lattice function of order s derived from the lattice whose specification is $(abcd\ldots)$. So far it has not been feasible to directly determine an expression for L_s, but as we now know the expression for $\text{GF}(l\,;\,a, b, c, d, \ldots)$ it is possible to find expressions for $\text{L}_1, \text{L}_2, \text{L}_3, \ldots$, by giving l the values $1, 2, 3, \ldots$, in succession in the resulting identity. One interesting result was, however, directly determined, viz. :—

$$\text{L}_s(\infty\,;\,a, b)$$
$$= x^{s(s+1)}\frac{(\mathbf{a-s+1})\ldots(\mathbf{a-1})}{(\mathbf{1})\ldots(\mathbf{s})}\cdot\frac{(\mathbf{b-s+1})\ldots(\mathbf{b})}{(\mathbf{1})\ldots(\mathbf{s+1})}\{(\mathbf{1})(\mathbf{a-s})+x^{b-s+1}(\mathbf{a-b})(\mathbf{s})\}\,;$$

but I do not give the proof of it at present, as the subject of the sub-lattice functions has not yet been worked out, and they are, in fact, no longer necessary for this part of the general investigation.

Art. 31. Valuable information, concerning line or one-dimensional partitions, is furnished by putting $l = 1$ in the general formula.

The partitions that are then enumerated are those in which every part is unity, there being not more than a_s units in the sth row; if thence we proceed to line partitions by adding the units that appear in each row we clearly get a system of line partitions such that the sth part is limited in magnitude by the integer a_s; or the system comprises all partitions contained in or subordinate to $(a_1, a_2, \ldots, a_n)$, viz., such that the first part $\not> a_1$, the second $\not> a_2, \ldots$, the $n^{\text{th}} \not> a_n$.

Denoting the generating function of these by

$$\text{GF}(a_1, a_2, \ldots, a_n\,;\,n),$$

and, denoting $\dfrac{(\mathbf{1})(\mathbf{2})\ldots(\mathbf{p})}{(\mathbf{1})(\mathbf{2})\ldots(\mathbf{q})\,.\,(\mathbf{1})(\mathbf{2})\ldots(\mathbf{p-q})}$ by X_{pq} or $\text{X}_{p,p-q}$, I find

$$\text{GF}(a, b\,;\,2) = \frac{(\mathbf{b+1})}{(\mathbf{1})(\mathbf{2})}\cdot\begin{vmatrix}(\mathbf{a+1}), & (\mathbf{b}) \\ x\;\;, & \text{X}_{21}\end{vmatrix}\,;$$

$$\mathrm{GF}(a, b, c\,;\,3) = \frac{(\mathbf{c}+1)}{(1)(2)(3)} \begin{vmatrix} (\mathbf{a}+1), & (\mathbf{b})(\mathbf{b}+1), & x(\mathbf{c}-1)(\mathbf{c}) \\ x\ , & X_{21}(\mathbf{b}+1), & X_{31}(\mathbf{c}) \\ 0\ , & x\ , & X_{32} \end{vmatrix};$$

$$\mathrm{GF}(a, b, c, d\,;\,4)$$

$$= \frac{(\mathbf{d}+1)}{(1)\ldots(4)} \begin{vmatrix} (\mathbf{a}+1), & (\mathbf{b})(\mathbf{b}+1), & x(\mathbf{c}-1)(\mathbf{c})(\mathbf{c}+1), & x^3(\mathbf{d}-2)(\mathbf{d}-1)(\mathbf{d}) \\ x\ , & X_{21}(\mathbf{b}+1), & X_{31}(\mathbf{c})(\mathbf{c}+1)\ , & xX_{41}(\mathbf{d}-1)(\mathbf{d}) \\ 0\ , & x\ , & X_{32}(\mathbf{c}+1)\ , & X_{42}(\mathbf{d}) \\ 0\ , & 0\ , & x\ , & X_{43} \end{vmatrix};$$

$$\mathrm{GF}(a, b, c, d, e\,;\,5)$$

$$= \frac{(\mathbf{e}+1)}{(1)\ldots(5)} \begin{vmatrix} (\mathbf{a}+1), & (\mathbf{b})(\mathbf{b}+1), & x(\mathbf{c}-1)(\mathbf{c})(\mathbf{c}+1), & x^3(\mathbf{d}-2)(\mathbf{d}-1)(\mathbf{d})(\mathbf{d}+1), & x^6(\mathbf{e}-3)(\mathbf{e}-2)(\mathbf{e}-1)(\mathbf{e} \\ x\ , & X_{21}(\mathbf{b}+1), & X_{31}(\mathbf{c})(\mathbf{c}+1)\ , & xX_{41}(\mathbf{d}-1)(\mathbf{d})(\mathbf{d}+1)\ , & x^3X_{51}(\mathbf{e}-2)(\mathbf{e}-1)(\mathbf{e}) \\ 0\ , & x\ , & X_{32}(\mathbf{c}+1),\ , & X_{42}(\mathbf{d})(\mathbf{d}+1)\ , & xX_{52}(\mathbf{e}-1)(\mathbf{e}) \\ 0\ , & 0\ , & x\ , & X_{43}(\mathbf{d}+1)\ , & X_{53}(\mathbf{e}) \\ 0\ , & 0\ , & 0\ , & x\ , & X_{54} \end{vmatrix}$$

and the law is evident.

Art. 32. In general, supposing the lattice to be in the plane of xy, that of the paper and the axis of z perpendicular to the plane of the paper, if we project the partition on to the plane of yz, we obtain a partition at the nodes of a lattice of l rows in which the part magnitude in the sth columns is limited by the number a_s.

The general formula for $\mathrm{GF}(l\,;\,a_1, a_2, \ldots, a_n)$ is remarkable from the fact that

$$\mathrm{GF}(l\,;\,a_1, a_2, \ldots, a_n) = \mathrm{GF}(l\,;\,b_1, b_2, \ldots, b_m),$$

where $(a_1, a_2, \ldots, a_n)$, $(b_1, b_2, \ldots, b_m)$ are any two conjugate line partitions.

Adumbration of the Three-dimensional Theory.

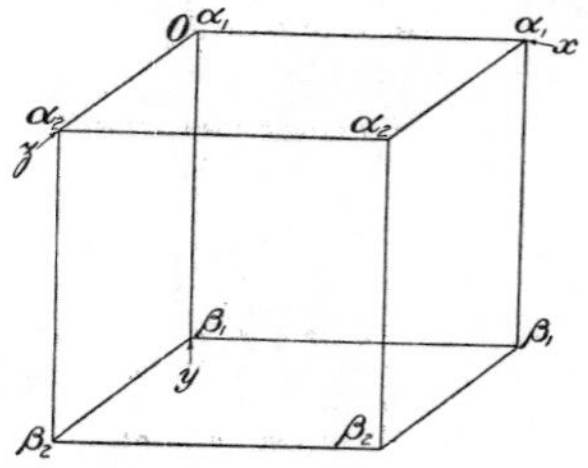

Art. 33. I conclude this Part by pointing out a path of future investigation into the Theory of Partitions in space of three dimensions.

I consider a complete or incomplete lattice in three dimensions, the lines of the lattice being in the direction of three rectangular axes of x, y, z respectively. Just as an incomplete lattice in two dimensions is defined by a one-dimensional partition whose successive parts specify the successive rows of the lattice, so an incomplete lattice in three dimensions is defined by a two-dimensional partition whose successive rows specify the successive layers of the lattice.

I shall suppose these layers to be in or parallel to the plane of xy which is the plane of the paper, and the axis of z to be perpendicular to the plane of the paper. Descending order of magnitude of parts placed at the points of the lattice is to be in evidence in the three directions Ox, Oy, Oz.

Art. 34. Consider the simplest case of a complete lattice, the points forming the summits of a cube. The two-dimensional lattices $\alpha_1\alpha_1\beta_1\beta_1$, $\alpha_2\alpha_2\beta_2\beta_2$, in and parallel to the plane of the paper are superposed to form the three-dimensional lattice.

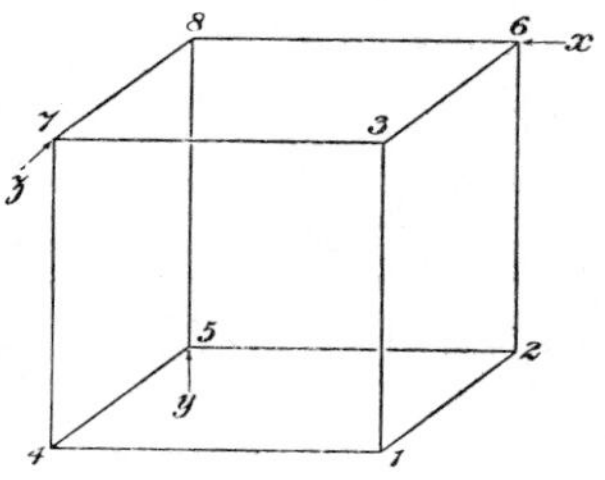

Suppose that the first 8 integers are placed at the points of the lattice so that descending order of magnitude is in evidence in the directions Ox, Oy, Oz, *e.g.*, one of 48 such arrangements is as shown.

I associate with the first and second rows of the first layer the letters α_1, β_1 respectively, and with the first and second rows of the second layer the letters α_2, β_2 respectively, and then from the illustrated arrangement of the first 8 numbers I derive a Greek-letter succession in the following manner :—I take the numbers in descending order of magnitude and write down the Greek letter with which the *position* of each number is associated : thus the arrangement above gives

$$\begin{matrix} 8 & 7 & 6 & 5 & 4 & 3 & 2 & 1 \\ \alpha_1 & \alpha_2 & \alpha_1 & \beta_1 & \beta_2 & \alpha_2 & \beta_1 & \beta_2. \end{matrix}$$

Art. 35. In this Greek-letter succession we have to *note*

(i.) A β which is succeeded by an α,

(ii.) An α which is succeeded by an α with a smaller suffix,

(iii.) A β which is succeeded by a β with a smaller suffix.

If a letter which is thus noted is the sth letter in the permutation I associate with the permutation the power $x^{\Sigma s}$, and taking the sum of these powers in respect of the whole of the permutations associated with and derived from the lattice I obtain the lattice function

$$\Sigma x^{\Sigma s} ;$$

and, following the reasoning of Part V., Art. 6, I derive the generating function for partitions at the points of the lattice, the part magnitude being unrestricted, viz.,

$$\text{GF}(\infty ; 22 ; 22) = \frac{\Sigma x^{\Sigma s}}{(1)(2)\ldots(8)} = \frac{\text{L}(\infty ; 22 ; 22)}{(1)(2)\ldots(8)}.$$

Similarly, from the Greek-letter successions which involve *t noted* letters, I derive the sub-lattice function of order t, and thence, by previous reasoning, arrive at the generating function when the part magnitude is restricted by the integer l,

$$\mathrm{GF}(l\,;\,22\,;\,22) = \frac{\Sigma \mathrm{L}_t(\infty\,;\,22\,;\,22)\,(\mathrm{l}-\mathrm{t}+1)(\mathrm{l}-\mathrm{t}+2)\ldots(\mathrm{l}-\mathrm{t}+8)}{(1)(2)\ldots(8)}.$$

Art. 36. The method is generally applicable to any incomplete lattice in three dimensions. I work out in detail the case in which the points of the lattice form the 8 summits of a cube, in order to show that the result obtained, in Part II., Section 7, in quite a different manner, is verified. That result, with modified notation, was

Generating Function

$$= \frac{(\mathrm{l}+1)\ldots(\mathrm{l}+8)}{(1)\ldots(8)} + \mathrm{P}(x)\frac{(\mathrm{l})\ldots(\mathrm{l}+7)}{(1)\ldots(8)} + \mathrm{Q}(x)\frac{(\mathrm{l}-1)\ldots(\mathrm{l}+6)}{(1)\ldots(8)}$$
$$+ \mathrm{R}(x)\frac{(\mathrm{l}-2)\ldots(\mathrm{l}+5)}{(1)\ldots(8)} + x^{16}\frac{(\mathrm{l}-3)\ldots(\mathrm{l}+4)}{(1)\ldots(8)};$$

where

$$\mathrm{P}(x) = 2x^2+2x^3+3x^4+2x^5+2x^6,$$

$$\mathrm{Q}(x) = x^5+3x^6+4x^7+8x^8+4x^9+3x^{10}+x^{11},$$

$$\mathrm{R}(x) = 2x^{10}+2x^{11}+3x^{12}+2x^{13}+2x^{14}.$$

I shall now show that

$$1,\quad \mathrm{P}(x),\quad \mathrm{Q}(x),\quad \mathrm{R}(x),\quad x^{16}$$

are, in fact, the sub-lattice functions of orders 0 to 4 which appertain to the lattice formed by the summits of a cube.

I write down the 48 permutations of the Greek letters and over each the arrangement of the first 8 integers from which it is derived, the lower layer of numbers being placed to the left :—

3 B 2

87 65 , 43 21 $\alpha_1\alpha_1\alpha_2\alpha_2\beta_1\beta_1\beta_2\beta_2$,	85 64 , 73 21 $\alpha_1\beta_1 \mid \alpha_2 \mid \alpha_1\alpha_2\beta_1\beta_2\beta_2$,	85 73 , 64 21 $\alpha_1\alpha_2\beta_1 \mid \alpha_1\beta_1 \mid \alpha_2\beta_2\beta_2$,	85 63 , 74 21 $\alpha_1\beta_1 \mid \alpha_2 \mid \alpha_1\beta_1 \mid \alpha_2\beta_2\beta_2$,
86 75 , 43 21 $\alpha_1\alpha_2 \mid \alpha_1\alpha_2\beta_1\beta_1\beta_2\beta_2$,	86 74 , 53 21 $\alpha_1\alpha_2 \mid \alpha_1\beta_1 \mid \alpha_2\beta_1\beta_2\beta_2$,	85 73 , 62 41 $\alpha_1\alpha_2\beta_1 \mid \alpha_1\beta_2 \mid \alpha_2\beta_1\beta_2$,	85 63 , 72 41 $\alpha_1\beta_1 \mid \alpha_2 \mid \alpha_1\beta_2 \mid \alpha_2\beta_1\beta_2$,
86 54 , 73 21 $\alpha_1\beta_1 \mid \alpha_1\alpha_2\alpha_2\beta_1\beta_2\beta_2$,	86 43 , 75 21 $\alpha_1\beta_1 \mid \alpha_1\beta_1 \mid \alpha_2\alpha_2\beta_2\beta_2$,	86 54 , 72 31 $\alpha_1\beta_1 \mid \alpha_1\alpha_2\alpha_2\beta_2 \mid \beta_1\beta_2$,	85 64 , 72 31 $\alpha_1\beta_1 \mid \alpha_2 \mid \alpha_1\alpha_2\beta_2 \mid \beta_1\beta_2$,
87 54 , 63 21 $\alpha_1\alpha_1\beta_1 \mid \alpha_2\alpha_2\beta_1\beta_2\beta_2$,	84 63 , 72 51 $\alpha_1\beta_1 \mid \alpha_2\beta_2 \mid \alpha_1\alpha_2\beta_1\beta_2$,	86 52 , 74 31 $\alpha_1\beta_1 \mid \alpha_1\alpha_2\beta_1\beta_2 \mid \alpha_2\beta_2$,	85 62 , 74 31 $\alpha_1\beta_1 \mid \alpha_2 \mid \alpha_1\beta_1\beta_2 \mid \alpha_2\beta_2$,
85 74 , 63 21 $\alpha_1\alpha_2\beta_1 \mid \alpha_1\alpha_2\beta_1\beta_2\beta_2$,	86 73 , 54 21 $\alpha_1\alpha_2 \mid \alpha_1\beta_1\beta_1 \mid \alpha_2\beta_2\beta_2$,	87 54 , 62 31 $\alpha_1\alpha_1\beta_1 \mid \alpha_2\alpha_2\beta_2 \mid \beta_1\beta_2$,	86 74 , 52 31 $\alpha_1\alpha_2 \mid \alpha_1\beta_1 \mid \alpha_2\beta_2 \mid \beta_1\beta_2$,
87 64 , 53 21 $\alpha_1\alpha_1\alpha_2\beta_1 \mid \alpha_2\beta_1\beta_2\beta_2$,	86 73 , 52 41 $\alpha_1\alpha_2 \mid \alpha_1\beta_1\beta_2 \mid \alpha_2\beta_1\beta_2$,	87 52 , 64 31 $\alpha_1\alpha_1\beta_1 \mid \alpha_2\beta_1\beta_2 \mid \alpha_2\beta_2$,	86 42 , 75 31 $\alpha_1\beta_1 \mid \alpha_1\beta_1 \mid \alpha_2\beta_2 \mid \alpha_2\beta_2$,
87 43 , 65 21 $\alpha_1\alpha_1\beta_1\beta_1 \mid \alpha_2\alpha_2\beta_2\beta_2$,	86 53 , 74 21 $\alpha_1\beta_1 \mid \alpha_1\alpha_2\beta_1 \mid \alpha_2\beta_2\beta_2$,	85 74 , 62 31 $\alpha_1\alpha_2\beta_1 \mid \alpha_1\alpha_2\beta_2 \mid \beta_1\beta_2$,	84 62 ; 73 51 $\alpha_1\beta_1 \mid \alpha_2\beta_2 \mid \alpha_1\beta_1 \mid \alpha_2\beta_2$,
84 73 , 62 51 $\alpha_1\alpha_2\beta_1\beta_2 \mid \alpha_1\alpha_2\beta_1\beta_2$,	86 53 , 72 41 $\alpha_1\beta_1 \mid \alpha_1\alpha_2\beta_2 \mid \alpha_2\beta_1\beta_2$,	85 72 , 64 31 $\alpha_1\alpha_2\beta_1 \mid \alpha_1\beta_1\beta_2 \mid \alpha_2\beta_2$,	86 72 , 53 41 $\alpha_1\alpha_2 \mid \alpha_1\beta_1\beta_2 \mid \beta_1 \mid \alpha_2\beta_2$,
87 63 , 54 21 $\alpha_1\alpha_1\alpha_2\beta_1\beta_1 \mid \alpha_2\beta_2\beta_2$,	87 53 , 64 21 $\alpha_1\alpha_1\beta_1 \mid \alpha_2\beta_1 \mid \alpha_2\beta_2\beta_2$,	87 64 , 52 31 $\alpha_1\alpha_1\alpha_2\beta_1 \mid \alpha_2\beta_2 \mid \beta_1\beta_2$,	86 52 , 73 41 $\alpha_1\beta_1 \mid \alpha_1\alpha_2\beta_2 \mid \beta_1 \mid \alpha_2\beta_2$,
87 63 , 52 41 $\alpha_1\alpha_1\alpha_2\beta_1\beta_2 \mid \alpha_2\beta_1\beta_2$,	87 53 , 62 41 $\alpha_1\alpha_1\beta_1 \mid \alpha_2\beta_2 \mid \alpha_2\beta_1\beta_2$,	87 42 , 65 31 $\alpha_1\alpha_1\beta_1\beta_1 \mid \alpha_2\beta_2 \mid \alpha_2\beta_2$,	87 52 , 63 41 $\alpha_1\alpha_1\beta_1 \mid \alpha_2\beta_2 \mid \beta_1 \mid \alpha_2\beta_2$,
87 65 , 42 31 $\alpha_1\alpha_1\alpha_2\alpha_2\beta_1\beta_2 \mid \beta_1\beta_2$,	86 75 , 42 31 $\alpha_1\alpha_2 \mid \alpha_1\alpha_2\beta_1\beta_2 \mid \beta_1\beta_2$,	84 72 , 63 51 $\alpha_1\alpha_2\beta_1\beta_2 \mid \alpha_1\beta_1 \mid \alpha_2\beta_2$,	85 72 , 63 41 $\alpha_1\alpha_2\beta_1 \mid \alpha_1\beta_2 \mid \beta_1 \mid \alpha_2\beta_2$,
87 62 , 54 31 $\alpha_1\alpha_1\alpha_2\beta_1\beta_1\beta_2 \mid \alpha_2\beta_2$,	86 72 , 54 31 $\alpha_1\alpha_2 \mid \alpha_1\beta_1\beta_1\beta_2 \mid \alpha_2\beta_2$,	87 62 , 53 41 $\alpha_1\alpha_1\alpha_2\beta_1\beta_2 \mid \beta_1 \mid \alpha_2\beta_2$,	85 62 , 73 41 $\alpha_1\beta_1 \mid \alpha_2 \mid \alpha_1\beta_2 \mid \beta_1 \mid \alpha_2\beta_2$.

A dividing line has been placed after each letter that has to be noted. Thence, by the rules given,

$$L_0(\infty ; 22 ; 22) = 1,$$
$$L_1(\infty ; 22 ; 22) = 2x^2+2x^3+3x^4+2x^5+2x^6,$$
$$L_2(\infty ; 22 ; 22) = x^5+3x^6+4x^7+8x^8+4x^9+3x^{10}+x^{11},$$
$$L_3(\infty ; 22 ; 22) = 2x^{10}+2x^{11}+3x^{12}+2x^{13}+2x^{14},$$
$$L_4(\infty ; 22 ; 22) = x^{16},$$

supplying a complete verification of the work in Part II.

We have, therefore,

$$\mathrm{GF}(l ; 22 ; 22)$$
$$= L_0\frac{(l+1)\ldots(l+8)}{(1)\ldots(8)}+L_1\frac{(l)\ldots(l+7)}{(1)\ldots(8)}+L_2\frac{(l-1)\ldots(l+6)}{(1)\ldots(8)}$$
$$+L_3\frac{(l-2)\ldots(l+5)}{(1)\ldots(8)}+L_4\frac{(l-3)\ldots(l+4)}{(1)\ldots(8)}.$$

We have evidently, potentially, the complete solution of the problem of three-dimensional partition, and it remains to work it out and bring it to the same completeness as has been secured in this Part for the problem in two dimensions.

This will form the subject of Part VII. of this Memoir.

Contents of Volume II